Analysis of Pesticide in Tea
茶叶农药多残留检测方法学研究

Analysis of Pesticide in Tea
Chromatography–Mass Spectrometry
Methodology
茶叶农药多残留检测方法学研究

Main Researchers
主要研究者 / 著者

Guo-Fang Pang
庞国芳

Chinese Academy of Inspection and Quarantine, Beijing, China

中国检验检疫科学研究院，北京，中国

Chun-Lin Fan
范春林

Chinese Academy of Inspection and Quarantine, Beijing, China

中国检验检疫科学研究院，北京，中国

Qiao-Ying Chang
常巧英

Chinese Academy of Inspection and Quarantine, Beijing, China

中国检验检疫科学研究院，北京，中国

Hui-Qin Wu
吴惠勤

Guangdong Institute of Analysis, Guangzhou, China

广东省测试分析研究所，广州，中国

Xiao-Lan Huang
黄晓兰

Guangdong Institute of Analysis, Guangzhou, China

广东省测试分析研究所，广州，中国

Fang Yang
杨方

Technology Center of Fuzhou Customs, Fuzhou, China

福州海关技术中心，福州，中国

Yan-Zhong Cao
曹彦忠

Qinhuangdao Customs, Qinhuangdao, China

秦皇岛海关，秦皇岛，中国

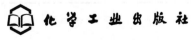 化学工业出版社

· 北京 ·

 ELSEVIER

本书详细介绍了一系列测定茶叶中农药残留的快速高通量分析方法，具有高精度、高可靠性、高灵敏度的特点，适用广泛。全书共7章：茶叶农药残留分析基础研究和检测方法建立；不同样品制备技术提取净化效能对比研究；茶叶水化对农药多残留方法效率的影响；茶叶农药残留测定基质效应及其补偿作用研究；方法耐用性系统评价，误差原因分析，关键控制点建立；茶叶陈化样品和污染样品农药降解规律研究；经11个国家和地区的30个实验室国际协同研究，建立茶叶中653种农药化学污染物高通量分析方法AOAC标准。

本书可作为科研单位、高等院校、质检机构等各类专业技术人员从事食品安全、环境保护、农业科技及农药开发等技术研究与应用的参考书，也可作为大学教学参考书。

图书在版编目（CIP）数据

茶叶农药多残留检测方法学研究=Analysis of Pesticide in Tea：Chromatography-Mass Spectrometry Methodology：英文 / 庞国芳著. —北京：化学工业出版社，2019.9

ISBN 978-7-122-34411-3

Ⅰ.①茶⋯　Ⅱ.①庞⋯　Ⅲ.① 茶叶-农药残留量分析-英文　Ⅳ.① S481

中国版本图书馆CIP数据核字（2019）第082249号

责任编辑：成荣霞　　　　　　　　　　　　　　　　　封面设计：关　飞
责任校对：边　涛

出版发行：化学工业出版社（北京市东城区青年湖南街13号　邮政编码100011）
印　　装：三河市航远印刷有限公司
889mm×1194mm　1/16　印张56½　字数1713千字　2019年10月北京第1版第1次印刷

购书咨询：010-64518888　　　　　　　　　　　售后服务：010-64518899
网　　址：http://www.cip.com.cn
凡购买本书，如有缺损质量问题，本社销售中心负责调换。

定　　价：888.00元　　　　　　　　　　　　　　　版权所有　违者必究

Contents

3 Study on the Influences of Tea Hydration for the Method Efficiency and Uncertainty Evaluation of the Determination of Pesticide Multiresidues in Tea Using Three Sample Preparation Methods/GC–MS/MS

4 Matrix Effect for Determination of Pesticide Residues in Tea

Preface

High-throughput detection of pesticides and chemical pollutants remains an important issue in the field of international food safety, and it has become a hot research topic for food safety researchers all over the world. The research team led by academician Guo-Fang Pang has been engaged in this field for more than 30 years and they have attained considerable achievements attracting worldwide attention. As early as more than 10 years ago, the high-throughput testing technology for the analysis of more than 1000 kinds of pesticide residues in agricultural products such as fruit, vegetables, grains, herbs, edible fungus, tea, animal tissue, aquatic products, dairy products, and honey using GC–MS and LC–MS/MS, as well as more than 20 China national standards, had been established by the team. Meanwhile, the standard system for the detection of more than 1000 kinds of pesticide residues was also constructed by the team and the system was widely used in China. Most of these studies are at the international advanced level, thus making China's research level in the field of trace analysis of multipesticide residues ranks the world forefront and promotes the technological progress of this field. At the same time, these standardized testing systems, which have been in line with international standards, have improved the quality of China's agricultural products and promoted the development of international trade in agricultural products. Therefore, research professor Guo-Fang Pang won AOAC international highest scientific Honor Award, Harvey W. Wiley Award, in 2014.

This monograph is another masterpiece of Guo-Fang Pang research team and their scientific research on the theory and application practice of pesticide residues detection technology in recent 10 years, and it is a true portrayal of their persistence, temper forward, and painstaking efforts.

The main aspects are as follows: (1) comparative study of extraction efficiencies of the three sample preparation techniques and the influences of tea hydration for the method efficiency of multipesticide residues; (2) the evaluation of the cleanup efficiency of SPE cartridge and the development process of Cleanert TPT tea purification column; (3) study on matrix effects of different tea varieties from different producing areas and the compensation of it; (4) uncertainty evaluation of the determination of multipesticide residues in tea; (5) the evaluation of the ruggedness of the method, error analysis, and the key control points of the method; and (6) study on the degradation of pesticide residues and its application in predicting the residue concentrations of target pesticides.

The book centers on a series of studies on the detection of pesticide residues in tea, and it is the most thorough and comprehensive systematic study of the high-throughput detection of pesticide residues in tea. The book not only introduces the high-throughput detection method for rapid simultaneous determination of multiclasses and multikinds of pesticide residues and chemical pollutants in tea using a variety of cutting-edge chromatography–mass spectrometry techniques established by the authors, but also shows the readers the concept of innovative technology and meticulous and precise thinking during the development of the method.

The research series of this book have important academic value, and the methods established have the characteristics of advanced technology, wide range of products, fast, reliable, and practical. Related research papers have been published in well-known international journals and they were widely recognized by people all around the world. In addition, the presentations made at international meetings were very well received by many international peers, which helped China to exert international influence in the arenas of international residual analysis and trace analytical chemistry and has induced the technological progress of the said area.

More importantly, the methods established in this study were selected as priority research projects of AOAC international in 2010, and the international collaborative research work, participated by a total of 30 laboratories from 11 countries and regions, was organized and implemented by academician Guo-Fang Pang as AOAC international supervisor, and it was successfully completed in 2013. In 2014, this method was adopted by AOAC international and approved as the official method of AOAC for the detection of pesticide residues in tea, and it was awarded the 2015 AOAC international excellent method. The method can provide technical support for monitoring the quality of tea, and thus promotes the healthy development of international trade of tea products.

We believe this book will be a great aid for scientific researchers engaged in related research, agricultural products inspection technicians, and college teachers and scholars.

Wei Fusheng
October 10, 2017

Introduction

In China, tea has a long history. It is a unique Chinese health drink which is popular with people all over the world. It is also an important cash crop and a traditional bulk export agricultural product in China. As tea plants like to grow in hot and humid environment, so their growth process is very susceptible to diseases and insect pests and weeds, and chemical pesticide spraying is the most effective solution so far. However, chemical pesticides leave undesirable residues in tea in accompany with the increment of yield and these residues will affect human health through the food chain. In addition, the developed countries and regions, such as the European Union, the United States, and Japan, are important importers of Chinese tea. The increasingly strict pesticide residue limit standards (MRLs) have become an important bottleneck in China's tea export. Therefore, from the protection of human health, to eliminate trade barriers, and to safeguard the perspective of international trade of tea, the thorough study of the detection technique of multipesticide residues in tea and the establishment of a high-sensitivity, accuracy, selectivity, and throughput detection method have extremely important practical significance.

This book is the systematic summarization of the research of our team on the detection technology theory and application of the multipesticide residues in tea products in last 10 years. The book describes in detail the high-throughput analytical techniques for 653 multiclasses and multikinds of pesticide residues and chemical pollutants in tea, while the mainstream and advanced detection means such as gas chromatography–mass spectrometry (GC–MS), gas chromatography–tandem mass spectrometry (GC-MS/MS), liquid chromatography–tandem mass spectrometry (LC–MS/MS), gas chromatography quadrupole tandem time-of-flight mass spectrometry (GC-Q-TOF), and liquid chromatography quadrupole tandem time-of-flight mass spectrometry (LC-Q-TOF) are concerned.

Compared with other plant-derived agricultural products such as vegetables and fruits, the composition of tea is extremely complex for it is rich in physiologically active compounds such as polyphenols, alkaloids, pigments, aromatic substances, amino acids, vitamins, and so on, and the composition varies greatly with processing procedure, so it is considerable difficult for the detection of multiclasses and multikinds of trace pesticide residues in tea.

Considering multiclasses and multikinds of pesticides may exist in different tea matrix, deep research on the comparison of the efficiency of various sample preparation technology and sample purification technology is introduced in this book, and the Cleanert TPT tea purification column, which has their own independent intellectual property rights, was developed. The simultaneous extraction and purification pretreatment method was established, while the matrix effect and its compensation of 28 different areas and species tea was deeply studied and the solution, using protecting agent for the compensation of matrix effect, was put forward.

In view of people's drinking habits, the comparison of two extraction methods, hydration method and nonhydration method, on the extraction efficiency, purification effect of residual pesticides in tea, and the applicability of different pesticides was conducted in-depth. At the same time, each component of the uncertainty is discussed in detail, and the measurement uncertainty is regarded as evaluation criteria of the method.

In addition, in order to further evaluate the repeatability, reproducibility, and durability of the method, the authors also carried out the evaluation of the ruggedness of the method for up to 3 months. At the same time, concrete analysis has been made to error types from different analytical stages and their causes, while elaborated discussions have been conducted on the key control points in different stages of the method to ensure the accurate and reliable of the method.

Furthermore, the dynamics of pesticide degradation of aged oolong tea samples and contaminated green tea samples at different field test and room temperature storage conditions using the high-throughput detection method established were carried by the author's work team. The stability of the contaminated tea samples within 3 months was also investigated, so as the degradation rule of pesticides in tea can be obtained. So, the prediction of the content of pesticide residues using degradation dynamic equation can be accomplished and the results are basically consistent with the measured values. This has a very important significance in offering guidance to tea growers to the applications of pesticide and to the import and export trade of tea.

Based on the aforementioned series of important research, the international collaborative research, participated by a total of 30 laboratories from 11 countries and regions, was organized and implemented by our team as the Study Director of AOAC international and the purpose of the research is the evaluation of the reproducibility of the analysis method for the 653 kinds of pesticide residues in tea and investigation whether the method can meet the requirements of AOAC international official method. This is overall comprehensive collaborative research of AOAC international on the detection technical standards of pesticide residue in tea, meanwhile the detailed report of the study is also included in this book.

In short, a series of important studies about the detection technology of multipesticide residues in tea are involved in this book and they are the deepest and most comprehensive summary of high-throughput detection of multipesticide residues in tea till now. The method, high-throughput detection of 653 multiclasses and multikinds of pesticide residues and chemical pollutants using GC–MS, GC–MS/MS, and LC–MS/MS, was selected as AOAC international priority research projects in 2010, and the method passed the AOAC international collaborative study and was approved as AOAC international official method for the detection of pesticide residues in tea (AOAC Official Method 2014.09) in 2014. Nevertheless, due to the limitations of the level, there may be unavoidable errors. We would kindly ask the users of this publication to provide feedback to the authors so that subsequent editions may be improved upon.

Guo-Fang Pang

October 10, 2017

1 Fundamental Research: Analytical Methods for Multiresidues in Tea

Chapter 1.1

Simultaneous Determination of 653 Pesticide Residues in Teas by Solid Phase Extraction (SPE) with Gas Chromatography– Mass Spectrometry (GC–MS) and Liquid Chromatography– Tandem Mass Spectrometry (LC–MS/MS)

1.1.1 INTRODUCTION

As a healthy beverage, tea is very popular in the general public worldwide and is consumed in huge amounts. Recently, pesticides have been widely used in tea cultivation to prevent various pests and diseases. Because of abuse, misuse, and unreasonable use, tea has inevitably come into contact with pesticides [1]. With an increasing public concern for food safety, many countries have established strict limitations of pesticide residue on agricultural products [2]. Pesticide residue in teas has received more and more attention [3]. EU has created stricter standards for the maximum residue limits (MRLs) of pesticides [4]. According to the standard, MRLs of dibromoethane, diazinon, dripping acid, fluvalinate, and dichlorvos have been reduced significantly, and some botanical pesticides such as azadirachtin, rotenone, and pyrethrins have been put on the limit list for the first time. In recent years, the requirements for determining pesticide in tea have increased continuously in Japan and the EU [5]. The number of limited pesticides in Japan was increased from 34 in 1986 to 51 in 1993 and then 276 in 2006, and in the EU the increases were from 6 in 1988 to 134 in 2004 and then 210 in 2006. Similarly, Canada, Germany, and other countries have also established stricter MRLs for pesticides in food [6]. Therefore, a more efficient method is urgently required for analysis of multiresidues of pesticide, not only for protecting consumer health, but also for promoting foreign trade. In order to have a better overall control of pesticide residues in teas, many pesticides should be taken into consideration for determination. The traditional single-residue detection method, however, cannot meet the requirements of more and more strict food safety regulations, especially on the high efficiency and the high throughput. In recent years, many works on the development of multiresidues analysis were reported. Jiang et al. [7] adopted gas chromatograph equipped with flame photometric detector (GC-FPD) for determination of 7 organophosphorus pesticide residues in tea. Shen et al. [8] using ethyl acetate extraction and SPE cartridges cleanup, by gas chromatography negative chemical ionization mass spectrometry (GC-NCI/MS) for determination of 11 pyrethroid pesticides residues in teas. Yuan et al. [9] developed a method of microwave assisted extraction-solid phase microextraction-gas chromatography (MAE-SPME-GC) for determination of 12 organochlorine and pyrethroid pesticides in teas. Yue et al. [10] adopted high-pressure thin-layer chromatography (HPTLC) for the determination of 9 pesticides in teas, and the results showed that the recoveries of the pesticides were 90.7%–105.5%, and relative standard deviations were 7.3%–13.5%. Cho et al. [11] proposed a method for the determination of 14 pesticides in green tea, using pressurized liquid extraction (PLE). Analysis was performed by GC with an electron capture detector (GC-ECD), and the pesticide identity of the positive samples was confirmed by GC–MS in a selected ion-monitoring (SIM) mode. Hu et al. [12] applied the techniques of accelerated solvent extraction (ASE), gel permeation chromatography (GPC), SPE, and GC-NCI-MS (SIM) for determination of 23 kinds of organochlorine and pyrethroid pesticides in tea. Zeng et al. [13] used hexane-dichloromethane (1:1 V/V) extraction, alumina neutral-florisil SPE cleanup, and GC-ECD to determine 18 organochlorine and 9 pyrethroid pesticide residues in tea. Jin et al. [14] developed a method composed of n-hexane-acetone (2:1) extraction, florisil SPE cleaned-up, and GC-ECD for determination of 25 organochlorine pesticides in tea. Hu et al. [15] reported a combined method of matrix solid-phase dispersion extraction (MSPD) and GC-MS-SIM for determination of 19 pesticide residues in tea. Schurek et al. [16] have developed a method for the determination of 36 pesticides in teas with gas chromatography/time-of-flight mass spectrometry (GC–TOFMS), and the limits of quantification ranged from 1 to 28 μg/kg. Ochiai et al. [17] reported a method of stir bar sorptive extraction (SBSE) coupled to thermal desorption (TD) and retention time locked (RTL) GC–MS for determination of 85 pesticides, including organochlorine pesticides, carbamates, organophosphorus pesticides, and pyrethroids in vegetables, fruits, and green tea. Lou et al. [18] established a method for 92 pesticides inspection in teas. The samples were extracted with acetonitrile followed by SPE cleanup, and gas chromatography-flame photometric detection (GC-FPD) and gas chromatography-electron capture detection (GC-ECD). The recoveries of those pesticides ranged from 80.3% to 117.1% by a fortification test with the relative standard deviations being from 1.5% to 9.8%. The limits of detection were 0.0025–0.10 mg/kg. With the development of detection technology, some methods have been already established for simultaneous determination of hundreds of pesticides. Huang et al. [19] reported a method based on the gas chromatography–mass spectrometry for the determination of 102 pesticides residues in teas. Yang [20] developed a method for determination of 118 pesticides using GC–MS. The method involved extraction with ethyl acetate-hexane, cleanup using GPC and solid phase extraction (SPE). At the low, medium, and high three fortification levels of 0.05–2.5 mg/kg, the average recoveries of 118 pesticides ranged from 61% to 121% and relative standard deviations (RSD) were in the range of 0.6%–9.2% for all analytes. The limits of detection for the method were 0.00030–0.36 mg/kg, depending on sensitivity of individual pesticide. Hirahara et al. [21] adopted GC, GC-MS, and LC–MS-MS for determination of 140 pesticides residues in 12 crops, including tea. Mol et al. [22] applied dispersive SPE, GC–MS, and LC–MS-MS for the determination of 341 pesticide residues in fruits, vegetables, spices, milk powder, honey, batter, flour, corn syrup, tea, fruit juice.

The tea matrices are complex, belonging to a material that is hard to analyze in microanalytical chemistry. The several hundred pesticide residues with different molecule structures and different polarities are extracted and cleaned up simultaneously from different teas, which is a tough topic in residual analysis. On the basis of our previous study of the simultaneous determination of 446 pesticide residues in fruits and vegetables [23], 405 in grains [24], 660 in animal tissues [25], and 450 in honey [26], a comparative study in our laboratory was conducted on the extraction efficiency of six solvents used either singly or combined against more than 600 pesticides in teas. At the same time, a detailed comparative study was carried out on the influence of multiple factors such as cleanup materials (six), including single filling materials and combined materials, also filling quantities and filling sequences, on cleanup efficiency, and eventually on a three-component Cleanert TPT cartridge which has a unique cleanup effect on 635 pesticides in teas. A rapid high-throughput analytical method has been established for the determination of 635 pesticides in teas using both GC–MS and LC–MS/MS, of which 490 pesticides are suitable for GC–MS and 448 are suitable for LC–MS/MS.

1.1.2 EXPERIMENTAL

1.1.2.1 Reagents and Materials

1. Solvents. Acetonitrile, dichlormethane, isooctane and methanol (LC grade), purchased from Dikma Co. (Beijing, China).
2. Pesticides standard and internal standard. Purity, ≥95% (LGC Promochem, Wesel, Germany).

3. Stock standard solutions. Weigh 5–10 mg of individual pesticide standards (accurate to 0.1 mg) into a 10 mL volumetric flask. Dissolve and dilute to volume with toluene, toluene-acetone combination, or cyclohexane, depending on each individual compound solubility. Stock standard solutions should be stored in the dark below 4°C.

4. Mixed standard solution. Depending on properties and retention time of each pesticide, all the 490 pesticides for GC/MS analysis are divided into six groups, A–F. All the 448 pesticides for LC–MS/MS analysis are divided into seven groups, A–G. The concentration of mixed standard solutions depended on the sensitivity of each compound for the instrument used for analysis. Mixed standard solutions should be stored in the dark below 4°C.

1.1.2.2 Apparatus

1. GC-MS system. Model 6890N gas chromatograph connected to a Model 5973N MSD and equipped with a Model 7683 autosampler (Agilent Technologies, Wilmington, DE). The column used was a DB-1701 capillary column (30 m×0.25 mm×0.25 μm, J&W Scientific, Folsom, CA).
2. LC-MS-MS system. An Agilent Series 1200 HPLC system connected to a 6410 QQQ Trap tandem quadrupole mass spectrometer equipped with ESI (Agilent Technologies, Santa Clara, CA, USA). The column used was a Zorbax SB-C18, 2.1×100 mm, 3.5 μm (Agilent Technologies, Santa Clara, CA, USA).
3. SPE. Cleanert-TPT (Agela, Tianjin, China).
4. Homogenizer. T-25B (Janke & Kunkel, Staufen, Germany).
5. Rotary evaporator. Buchi EL131 (Flawil, Switzerland).
6. Centrifuge. Z 320 (B. HermLe AG, Gosheim, Germany).
7. Nitrogen evaporator. EVAP 112 (Organomation Associates, Inc., New Berlin, MA).

1.1.2.3 Extraction

Weigh 5 g of sample (accurate to 0.01 g) into an 80 mL centrifuge tube. Add 15 mL of acetontrile, homogenize at 15,000 rpm for 1 min, and centrifuge for 5 min at 4200 rpm/min. Transfer the supernatants to a glass funnel. Then the sample was reextracted in the same way and the supernatants combined, and then were carefully evaporated just to 1 mL at 40°C with a vacuum evaporator in a water bath for the following cleanup procedure.

1.1.2.4 Cleanup

Before sample application, anhydrous sodium sulfate (ca. 2 cm) was placed on the top of the Cleanet-TPT cartridge, which connected a pear-shaped flask. The Cleanet-TPT was conditioned with 10 mL of acetontrile-toluene (3:1). When the conditioning solution reached the top of the sodium sulphate, the concentrated extract was added to the Cleanet-TPT, the flask was rinsed with 3 mL of acetonitrile-toleune twice, and the washings were also applied to the Cleanet-TPT. A 50 mL of reservoir was attached to the Cleanet-TPT and the pesticides were eluted with 25 mL of acetontrile-toluene (3:1). The eluted portion was concentrated to ca 0.5 mL by rotary evaporation at 40°C.

For determination of pesticides by GC-MS, add 5 mL of hexane for solvent exchange on a rotary evaporator in a water bath of 40°C and repeat it twice and make up the final solution volume to 1 mL. Add 40 μL of internal standard solution and mix thoroughly for the following determination. For determination of pesticides by LC–MS/MS, the residue was evaporated to dryness with nitrogen gas, and then it was dissolved in 1.0 mL of acetonitrile-water (3:2) and mixed thoroughly. Then pass the 0.20 μm filter membrane for determination.

1.1.2.5 Determination

1. GC-MS operating conditions. The oven temperature was programmed as follows: 40°C held for 1 min to 130°C at 30°C/min and to 250°C at 5°C/min and to 300°C at 10°C/min, held for 5 min; carrier gas, helium, purity ≥99.999%; flow rate,1.2 mL/min; injection port temperature, 290°C; inject volume, 1 μL; injection mode, splitless, purge on after 1.5 min; ionization voltage, 70 eV; ion source temperature, 230°C; GC-MS interface temperature, 280°C; selected ion monitoring (SIM) mode, selected one quantitative ion and 2–3 qualitative ions for each compound; heptachlor epoxide was used as an internal standard for quantitation.
2. LC-MS-MS operating conditions: injection volumn, 10 μL; flow rate: 400 μL/min; column temperature, 40°C; the nebulizer gas was nitrogen, nebulizer gas pressure: 0.28 Mpa; the ionization voltage, 4 kV; desolvation temperature, 350°C; drying gas flow rate: 10 L/min. For groups A, B, C, D, E, and F, the mass spectrometer was operated in the positive electrospray ionization (ESI) mode, and the mobile phases of LC were 0.1% formic acid (A) and acetonitrile (B). For group

G, the mass spectrometer was operated in the negative ESI mode, and the mobile phases were 5 mmol/L ammonium acetate (A) and acetonitrile (B). The gradient profile was 1% to 30% of B, from 0 to 3 min in linear conditions; 30% to 40% of B from 3 to 6 min (linear); 40% of B from 6 to 9 min (isocratic); 40% to 60% of B from 9 to 15 min (linear); 60% to 99% of B from 15 to 19 min (linear); 99% of B from 19 to 23 min (isocratic).

1.1.3 RESULTS AND DISCUSSION

1.1.3.1 Optimization of Gas Chromatography–Mass Spectrometry Conditions and Selection of Pesticide Varieties Suitable for Analysis

The DB-1701 column was often selected for GC–MS multiresidue pesticide analysis [24,25], which could be used for more than 400 pesticide residues detection [27]. In order to analyze more pesticides, a three-ramps oven temperature program was developed for the baseline separation.

Screening for 827 commonly used pesticides was carried out in the following 3 steps to find out the suitable pesticides varieties for GC–MS analysis. (1) The individual stock standards of total 827 pesticides were prepared. The compounds were analyzed in the full-scan mode under optimized GC–MS condition before the mass spectrum and the retention time of each compound were obtained. The experimental results showed that 245 compounds do not have signals in the mass spectrum, and 12 compounds with poor thermal stability are decomposed in the GC–MS system, while 10 compounds were mixtures, but we could not find which peak was the target ion's in the scanned mass spectrum. All the pesticides mentioned earlier were eliminated from the objective pesticides list. (2) According to their retention times, the remaining 560 compounds were divided into 6 groups: A, B, C, D, E, and F (ca. 90 compounds in each group), and the mixed standard solutions were prepared, the concentration of each compound on which depended their sensitivities in the instrument. Tested by GC–MS, it was found that 21 compounds did not have signals or had low signal intensities that were eliminated. (3) The remaining 539 kinds of compounds were spiked, respectively, to the green tea, black tea, Oolong tea, and Pu'er tea, and the samples were extracted and cleaned up and determined by the developed GC–MS method. Forty-nine compounds with low recoveries were eliminated. By the aforementioned step-by-step selection, the pesticides that have proved to be unsuitable for GC–MS are listed in Table 1.1.1. Finally, the remaining 490 pesticides were chosen for determination by GC–MS. GC–MS chromatograms of 490 pesticide standards in tea blank matrix extracts for groups A, B, C, D, E, and F are shown in Figs. 1.1.1–1.1.6. A quantitative and two qualitative ions were selected for each compound, and three qualitative ions were chosen for the banned pesticides, such as HCH, DDT, nitrogen, aldrin, and chlordimeform. Retention times, quantitative ions, qualitative ions, and relative abundance of 490 pesticides are shown in Table 1.1.2.

1.1.3.2 Optimization of Liquid Chromatography–Tandem Mass Spectrometry Conditions and Selection of Pesticide Varieties Suitable for Analysis

In the positive and negative ionization mode, the parent ions of pesticides were selected by Q1 scan. And then in the product ion scan mode, fragment ions were scanned and the ion transitions of pesticides for quantitation and qualification acquired. The optimal sensitivities were obtained by optimizing the fragmentor voltage and collision-induced dissociation (CID).

Screening of 673 commonly used pesticides was carried out in the following four steps to find out the suitable pesticides varieties for LC–MS/MS analysis. First, the individual stock standard solutions of total 673 compounds were prepared. The target compounds were directly introduced into the ion source, scanned in positive and negative ionization mode. The findings were: (1) 110 compounds did not find the precursor ions. (2) 18 compounds did not find the product ions during the product ion scan regarding the 620 compounds that had found the precursor ions. (3) Mixed standard solutions were prepared using the remaining 602 compounds at concentration of ca 1 μg/mL each, with each group for 20 compounds. It was found that 33 compounds had extremely low sensitivity or did not have peaks in the mixed standard solutions. All the pesticeds mentioned earlier were eliminated from the objective pesticides list. (4) The remaining 569 compounds were divided into 7 groups: A, B, C, D, E, F, and G, and were spiked to the green tea, black tea, Oolong tea, and Pu'er tea matrix, respectively. It was determined that 64 compounds had no signals in the LC–MS/MS system. Table 1.1.3 summarizes the 255 compounds that were not suitable for LC–MS/MS analysis. As for the remaining 448 compounds suitable for LC–MS/MS analysis, their retention times, quantitative ions, qualitative ions, and relative abundance are shown in Table 1.1.4.

TABLE 1.1.1 The 337 Pesticides No Suitable for the GC–MS

No.	Name	No.	Name	No.	Name
(1) The 245 compounds do not have signals in the mass spectrum					
1	1,2-Dichloro ethane	83	Dicamba	165	Methiocarb sulfoxide
2	1-Naphthy1 acetic acid	84	Dichlone	166	Methoxyfenozide
3	1,2-Dibromo-3-chloropropane	85	Dichlorprop	167	Methyl isothiocyanate
4	1,2-Dichloropropane	86	Diclocymet	168	Metosulam
5	1,3-Dichloropropene(cis+trans)	87	Diclomezine	169	Metsulfuron-methyl
6	2-(2,4,5-Trichloro-phenoxy)propionic acid	88	Dienochlor	170	Milbemectin A3
7	2,6-Difluorobenzoic acid	89	Difenzoquat-methyl sulfate(Difenzoquat)	171	Milbemectin A4
8	6-Chloro-4-hydroxy-3-phenyl-pyridazin	90	Diflufenzopyr-sodium	172	Monuron
9	Abamectin	91	Dimehypo	173	Naptalam
10	Acequinocyl	92	Dimethirimol	174	Neburon
11	Acifluorfen	93	Dinocap technical mixture of isomers	175	Nitenpyram
12	Acrylamide	94	Dinotefuran	176	Novaluron
13	Alanycarb	95	Diquat dibromide hydrate(Diquat)	177	Oryzalin
14	Aldicarb	96	Dithianon	178	Oxabetrinil
15	Aldicarb sulfone(aldoxycarb)	97	Diuron	179	Oxamyl-oxime
16	Aldicarb sulfoxide	98	DMST	180	Paraquat Dichloride
17	Aldimorph	99	DNOC	181	Phenmedipham
18	Aldoxycarb	100	Dodine	182	Phorate sulfoxide
19	Alloxydim-sodium	101	Emamectin-benzoate	183	Phoxim
20	Amidithion	102	Ethephon	184	Phthalic acid di-(2-ethylhexyl) ester
21	Aminopyralid	103	Ethidimuron	185	Phthalic acid dibutyl ester
22	Amitrole	104	Ethiofencarb-sulfone	186	Phthalic acid dicyclohexyl ester
23	Amobam	105	Ethiofencarb-sulfoxid	187	Picloram
24	Anilazine	106	Ethiprole	188	Pirimicarb-desmethyl-formamido
25	Asulam	107	Ethirimol	189	Pirmicarb-desmethyl
26	Azocyclotin	108	Ethylene thiourea	190	Primisulfuron-methyl
27	Benazolin	109	Etobenzanid	191	Probenazole
28	Bensultap	110	Famoxadone	192	Prohexadione-calcium
29	Benzofenap	111	Fenaminosulf	193	Propaquizafop
30	Benzoximate	112	Fenazaflor	194	Propineb
31	Benzyladenine (6-benzylamino-purine)	113	Fenbutatin oxide	195	Propoxycarbazone-sodium
32	Bromide	114	Fenoprop (Silvex,2,4,5-TP)	196	Propylene oxide
33	Bromochloromethane	115	Fenthion oxon	197	Prosulfuron
34	Bromoxynil	116	Fentin acetate	198	Pymetrozin
35	Brompyrazon	117	Fentin-chloride	199	Pyrazolynate (Pyrazolate)
36	Butocarboxim	118	Fentrazamide	200	Pyrazosulfuron-ethyl
37	Butocarboxim-sulfoxide	119	Flazasulfuron	201	Pyrazoxyfen
38	Butoxycarboxim	120	Florasulam	202	Pyridalyl
39	Butoxycarboxim-sulfoxid	121	Fluazuron	203	Pyridate

(Continued)

TABLE 1.1.1 The 337 Pesticides No Suitable for the GC–MS (*Cont.*)

No.	Name	No.	Name	No.	Name
40	Buturon	122	Flucarbazone-sodium	204	Pyrithiobac sodium
41	Camphechlor	123	Flumethrin	205	Pyrithlobac sodium
42	Carbendazim	124	Flumetsulam	206	Quinclorac
43	Carbetamide	125	Fluoroimide	207	Rimsulfuron
44	Carbofuran-3-hydroxy	126	Flupropanate	208	Rotenone
45	Carbon disulphide	127	Fluroxypyr	209	Spinosad
46	Carbonyl sulfide	128	Flusulfamide	210	Sulfanitran
47	Carpropamid	129	Fluthiacet-Methyl	211	Sulfentrazone
48	Cartap hydrochloride	130	Fomesafen	212	TCA-sodium
49	Chloridazon	131	Forchlorfenuron	213	Tebufenozide
50	Chlormequat	132	Formetanate hydrochloride	214	Temephos
51	Chlormequat chloride	133	Furathiocarb	215	TEPP
52	Chlorobenzuron	134	Gibberellic acid	216	Tepraloxydim
53	Chloropicrin	135	Glyphosate	217	Terbucarb
54	Chloroxuron	136	Haloxyfop-ehyoxyethyl (或Haloxyfop)	218	Tert-butyl-4-hydroxyanisole
55	Chlorsulfuron	137	Hydramethylnon	219	Tert-Butylamine
56	Cinidon-ethyl	138	Hymexazol	220	Thiacloprid
57	Cinosulfuron	139	Imazamox	221	Thidiazuron
58	Clodinafop acid	140	Imazapic	222	Thifensulfuron-methyl
59	Clodinafop free acid	141	Imazapyr	223	Thiodicarb
60	Clophen A 30	142	Imazethapyr	224	Thiofanox
61	Clophen A 40	143	Imibenconazole	225	Thiofanox sulfone
62	Clophen A 50	144	Imidacloprid	226	Thiofanox-sulfoxide
63	Clophen A 60	145	Iminoctadine triacetate	227	Thiophanate-methyl
64	Cloprop	146	Indoxacarb	228	Thiophanat-ethyl
65	Clopyralld	147	Iodosulfuron-methyl	229	Tiamulin-fumerate
66	Cloransulam-methyl	148	Iodosulfuron-methyl sodium	230	Triasulfuron
67	Clothianidin	149	Ioxynil	231	Triazoxide
68	Coumatetralyl	150	Isoproturon	232	Trichlamide
69	Cumyluron	151	Isouron	233	Trichlorfon
70	Cyazofamid	152	Isoxaflutole	234	Trichloro acetic acid sodium salt
71	Cyclanilide	153	Kadethrin	235	Trichlorphon
72	Cycloprothrin	154	Kelevan	236	Triclopyr
73	Cyclosulfamuron	155	Leptophos oxon	237	Triflumuron
74	Cyhexatin	156	Loxynil	238	Triflusulfuron-methyl
75	Cymoxanil	157	Lufenuron	239	Triforine
76	Cyromazine	158	Maleic hydrazide	240	Trimethylsulfonium iodide
77	Daimuron	159	MCPA	241	Triticonazole
78	Dalapon	160	MCPB	242	Vamidothion
79	Dalapon acid	161	Mepiquat chloride	243	Vamidothion sulfone
80	Daminozide	162	Mercapto dimethur sulfone	244	Warfarin
81	Demeton-s-methyl sulfoxide	163	Mesotrion	245	Ziram
82	Diafenthiuron	164	Methazole		

TABLE 1.1.1 The 337 Pesticides No Suitable for the GC–MS (*Cont.*)

No.	Name	No.	Name	No.	Name
(2) 12 compounds with poor thermal stability which then broke down and the decomposition peak appeared when it got into the GC–MS system					
1	2,4-DB	5	Chlorimuronethyl	9	Ethoxysulfuron
2	4-Bromo-3.5-dimethylphenyln-methylcarbamate	6	Chlorphoxim	10	Promecarb
3	Barban	7	Chlortoluron	11	Pyrethrins
4	Bensulfuron-methyl	8	Clofentezine	12	Toxaphene
(3) 10 compounds were mixtures but we could not find which peak was the target ion's in the scanned mass spectrum					
1	Aroclor 1221	5	Aroclor 1254	8	Aroclor 1268
2	Aroclor 1232	6	Aroclor 1260	9	PCB-1232
3	Aroclor 1242	7	Aroclor 1262	10	PCB-1254
4	Aroclor 1248				
1	2,4-DB	8	Aroclor 1260	15	Clofentezine
2	4-Bromo-3.5-dimethylphenyln-methylcarbamate	9	Aroclor 1262	16	Ethoxysulfuron
3	Aroclor 1221	10	Aroclor 1268	17	PCB-1232
4	Aroclor 1232	11	Barban	18	PCB-1254
5	Aroclor 1242	12	Bensulfuron-methyl	19	Promecarb
6	Aroclor 1248	13	Chlorphoxim	20	Pyrethrins
7	Aroclor 1254	14	Chlortoluron	21	Toxaphene
(4) 21 compounds did not have signals or with low signal intensities in the mixed standard solution					
1	1-Naphthyl acetamide	8	Pentachlorophenol	15	Terbufos sulfone
2	4-Aminopyridine	9	Dinoseb	16	Chromafenozide
3	Benfuracarb	10	Dinoseb acetate	17	Dinoterb
4	Bromoxynil octanoate	11	Chlorothalonil	18	Fluometuron
5	Thiram	12	Crotoxyphos	19	Bensulide
6	Chlordecone (kepone)	13	Uniconazole	20	Fosthiazate
7	Propylene thiourea	14	Paraoxon-methyl	21	Isoxaben
(5) 49 compounds with low recoveries in tea matrix					
1	Parathion	18	Dinobuton	34	Tralkoxydim
2	Fenothiocarb	19	Ditalimfos	35	Fluridone
3	Folpet	20	XMC	36	Triflumizole
4	Phosmet	21	Acenaphthene	37	Acephate
5	Oxycarboxin	22	Naled	38	Endothal
6	Terbumeton	23	Fenfuram	39	Desmedipham
7	Chlorbromuron	24	Benfuresate	40	Pyrifenox -2
8	Endosulfan-2	25	Dithiopyr	41	Dimethametryn
9	Captafol	26	Phosphamidon-2	42	Endrin aldehyde
10	Simetryn	27	Thiamethoxam	43	Halosulfuran-methyl
11	Dimethipin	28	Captan	44	Tricyclazole
12	Fenoxycarb	29	TCMTB	45	Cythioate
13	Acrinathrin	30	Clethodim	46	Acetamiprid
14	Cycloxydim	31	Chrysene	47	Tralomethrin-1
15	Tri-iso-butyl phosphate	32	Famphur	48	Tralomethrin-2
16	Musk ketone	33	Picolinafen	49	Azoxystrobin
17	Methfuroxam				

(Continued)

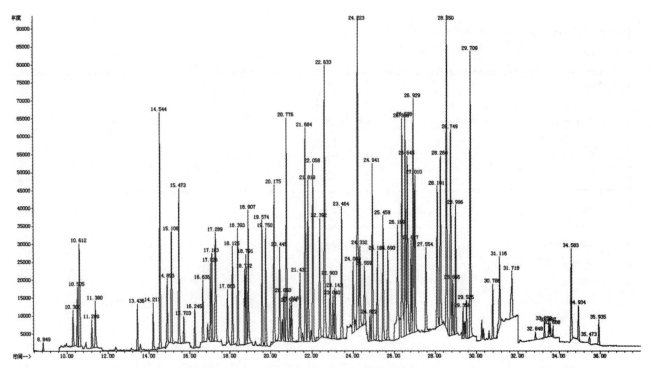

FIGURE 1.1.1 GC–MS total ion chromatogram of 90 pesticides in tea samples (A group).

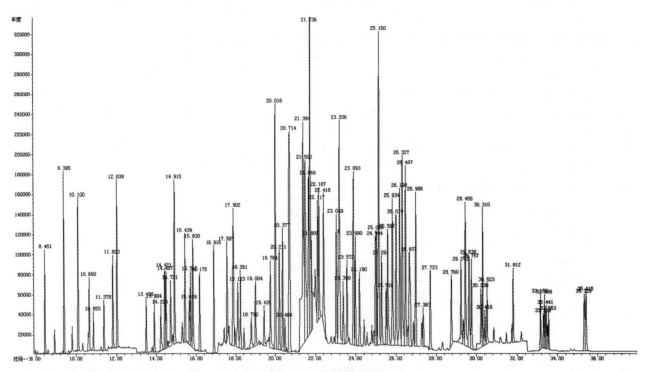

FIGURE 1.1.2 GC–MS total ion chromatogram of 97 pesticides in tea samples (B group).

1.1.3.3 Optimization of Sample Extraction Conditions

For simultaneous determination of hundreds of pesticides in tea, extraction and cleanup techniques call for more strict requirements inevitably due to different molecule structures and different polarities of pesticides. In addition, tea contains a great many pigments, theophylline, and caffeine, which are removed before instrumental analysis. Therefore, choosing a suitable extraction and cleanup condition becomes prominent, among other things. In order to optimize the extraction

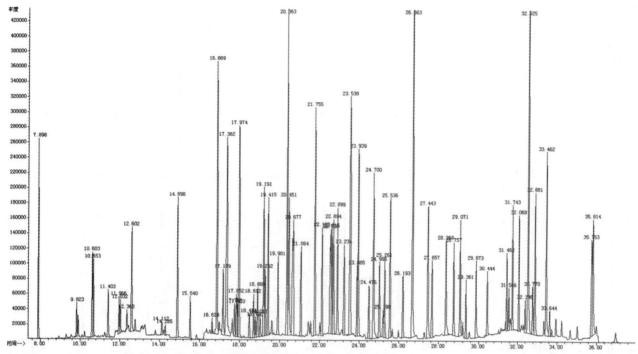

FIGURE 1.1.3   GC–MS total ion chromatogram of 79 pesticides in tea samples (C group).

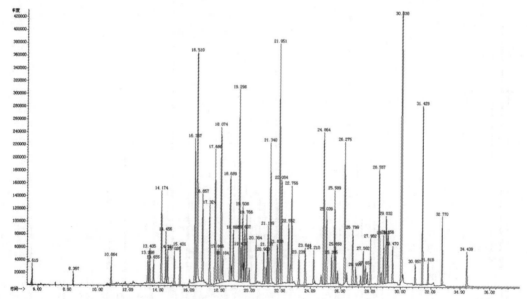

FIGURE 1.1.4   GC–MS total ion chromatogram of 92 pesticides in tea samples (D group).

procedure, the extraction efficiency of several solvents frequently used in multiresidue analysis including ethyl acetate (EtAc), acetone, acetonitrile (MeCN) [28], and some other solvents, such as methanol, N-hexane, and dichloromethane were investigated. The results indicated that acetone gave good extraction efficiency; however, the complex coextracting compounds make the subsequent cleanup procedure more difficult. And the solvent of acetoacetate, dichlormethane, and N-hexane had low extraction efficiency because of their poor penetrative ability to the tea tissue. The acetonitrile, by contrast, was preferred as providing both a high-extracting recovery and a relatively low levels of matrix interference. For these reasons, acetonitrile was adopted as the extracting solvent in this method. In addition, the effect of extraction times on extraction efficiency was investigated, which is described in Table 1.1.5. The results showed that better recoveries and lower solvent consumption can be obtained from the twice extraction.

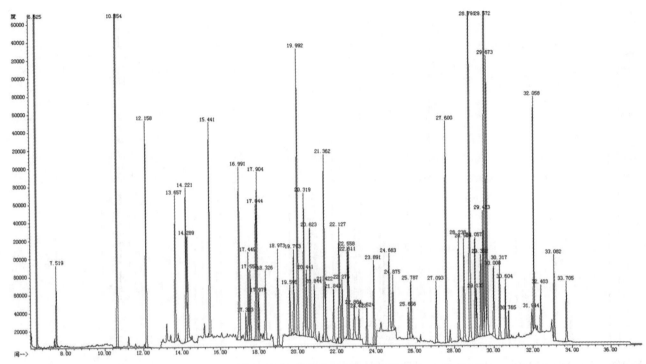

FIGURE 1.1.5 GC–MS total ion chromatogram of 79 pesticides in tea samples (E group).

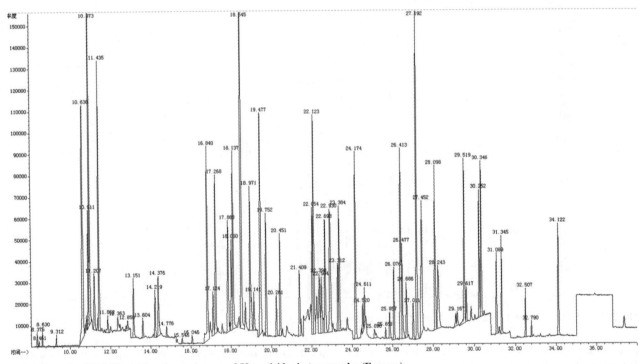

FIGURE 1.1.6 GC–MS total ion chromatogram of 53 pesticides in tea samples (F group).

1.1.3.4 Optimization of Sample Cleanup Conditions

The SPE technique is widely used in areas of analysis of pesticide residues because of its simple operation and high efficiency and can be applied to different matrices by changing its cartridge filling materials [29]. The tea matrices are complex, containing a great amount of chlorophyll, renieratene, tea polyphenol, and other uncertain interfering matters. The crucial point of this cleanup technique is how to effectively separate more than 600 trace pesticide residues from a great

TABLE 1.1.2 The Groups, Retention Times, Quantifying Ions and Qualifying Ions of the 490 Pesticides to be Analyzed in Teas and Internal Standard Compounds by GC–MS

No.	Compounds	Retention time (min)	Quantifying ion	Qualifying ion 1	Qualifying ion 2	Qualifying ion 3
ISTD	Heptachlor-epoxide	22.10	353(100)	355(79)	351(52)	
A group						
1	Allidochlor	8.78	138(100)	158(10)	173(15)	
2	Dichlormid	9.74	172(100)	166(41)	124(79)	
3	Etridiazol	10.42	211(100)	183(73)	140(19)	
4	Chlormephos	10.53	121(100)	234(70)	154(70)	
5	Propham	11.36	179(100)	137(66)	120(51)	
6	Cycloate	13.56	154(100)	186(5)	215(12)	
7	Diphenylamine	14.55	169(100)	168(58)	167(29)	
8	Chlordimeform	14.93	196(100)	198(30)	195(18)	183(23)
9	Ethalfluralin	15.00	276(100)	316(81)	292(42)	
10	Phorate	15.46	260(100)	121(160)	231(56)	153(3)
11	Thiometon	16.20	88(100)	125(55)	246(9)	
12	Quintozene	16.75	295(100)	237(159)	249(114)	
13	Atrazine-desethyl	16.76	172(100)	187(32)	145(17)	
14	Clomazone	17.00	204(100)	138(4)	205(13)	
15	Diazinon	17.14	304(100)	179(192)	137(172)	
16	Fonofos	17.31	246(100)	137(141)	174(15)	202(6)
17	Etrimfos	17.92	292(100)	181(40)	277(31)	
18	Propetamphos	17.97	138(100)	194(49)	236(30)	
19	Secbumeton	18.36	196(100)	210(38)	225(39)	
20	Pronamide	18.72	173(100)	175(62)	255(22)	
21	Dichlofenthion	18.80	279(100)	223(78)	251(38)	
22	Mexacarbate	18.83	165(100)	150(66)	222(27)	
23	Dimethoate	19.25	125(100)	143(16)	229(11)	
24	Dinitramine	19.35	305(100)	307(38)	261(29)	
25	Aldrin	19.67	263(100)	265(65)	293(40)	329(8)
26	Ronnel	19.80	285(100)	287(67)	125(32)	
27	Prometryne	20.13	241(100)	184(78)	226(60)	
28	Cyprazine	20.18	212(100)	227(58)	170(29)	
29	Vinclozolin	20.29	285(100)	212(109)	198(96)	
30	*Beta*-HCH	20.31	219(100)	217(78)	181(94)	254(12)
31	Metalaxyl	20.67	206(100)	249(53)	234(38)	
32	Methyl-parathion	20.82	263(100)	233(66)	246(8)	200(6)
33	Chlorpyrifos (-ethyl)	20.96	314(100)	258(57)	286(42)	
34	*Delta*-HCH	21.16	219(100)	217(80)	181(99)	254(10)
35	Anthraquinone	21.49	208(100)	180(84)	152(69)	
36	Fenthion	21.53	278(100)	169(16)	153(9)	
37	Malathion	21.54	173(100)	158(36)	143(15)	
38	Paraoxon-ethyl	21.57	275(100)	220(60)	247(58)	263(11)
39	Fenitrothion	21.62	277(100)	260(52)	247(60)	
40	Triadimefon	22.22	208(100)	210(50)	181(74)	

(Continued)

TABLE 1.1.2 The Groups, Retention Times, Quantifying Ions and Qualifying Ions of the 490 Pesticides to be Analyzed in Teas and Internal Standard Compounds by GC–MS (*Cont.*)

No.	Compounds	Retention time (min)	Quantifying ion	Qualifying ion 1	Qualifying ion 2	Qualifying ion 3
41	Linuron	22.44	61(100)	248(30)	160(12)	
42	Pendimethalin	22.59	252(100)	220(22)	162(12)	
43	Chlorbenside	22.96	268(100)	270(41)	143(11)	
43	Chlorbenside	22.96	268(100)	270(41)	143(11)	
44	Bromophos-ethyl	23.06	359(100)	303(77)	357(74)	
45	Quinalphos	23.10	146(100)	298(28)	157(66)	
46	*Trans*-chlordane	23.29	373(100)	375(96)	377(51)	
47	Phenthoate	23.30	274(100)	246(24)	320(5)	
48	Metazachlor	23.32	209(100)	133(120)	211(32)	
49	Prothiophos	24.04	309(100)	267(88)	162(55)	
50	Chlorfurenol	24.15	215(100)	152(40)	274(11)	
51	Procymidone	24.36	283(100)	285(70)	255(15)	
52	Dieldrin	24.43	263(100)	277(82)	380(30)	345(35)
53	Methidathion	24.49	145(100)	157(2)	302(4)	
54	Napropamide	24.84	271(100)	128(111)	171(34)	
55	Cyanazine	24.94	225(100)	240(56)	198(61)	
56	Oxadiazone	25.06	175(100)	258(62)	302(37)	
57	Fenamiphos	25.29	303(100)	154(56)	288(31)	217(22)
58	Tetrasul	25.85	252(100)	324(64)	254(68)	
59	Bupirimate	26.00	273(100)	316(41)	208(83)	
60	Flutolanil	26.23	173(100)	145(25)	323(14)	
61	Carboxin	26.25	235(100)	143(168)	87(52)	
62	*p,p'*-DDD	26.59	235(100)	237(64)	199(12)	165(46)
63	Ethion	26.69	231(100)	384(13)	199(9)	
64	Etaconazole-1	26.81	245(100)	173(85)	247(65)	
65	Sulprofos	26.87	322(100)	156(62)	280(11)	
66	Etaconazole-2	26.89	245(100)	173(85)	247(65)	
67	Myclobutanil	27.19	179(100)	288(14)	150(45)	
68	Fensulfothion	27.94	292(100)	308(22)	293(73)	
69	Diclofop-methyl	28.08	253(100)	281(50)	342(82)	
70	Propiconazole-1	28.15	259(100)	173(97)	261(65)	
71	Propiconazole-2	28.15	259(100)	173(97)	261(65)	
72	Bifenthrin	28.57	181(100)	166(25)	165(23)	
73	Mirex	28.72	272(100)	237(49)	274(80)	
74	Carbosulfan	28.80	160(100)	118(95)	323(30)	
75	Nuarimol	28.90	314(100)	235(155)	203(108)	
76	Benodanil	29.14	231(100)	323(38)	203(22)	
77	Methoxychlor	29.38	227(100)	228(16)	212(4)	
78	Oxadixyl	29.50	163(100)	233(18)	278(11)	
79	Tebuconazole	29.51	250(100)	163(55)	252(36)	
80	Tetramethirn	29.59	164(100)	135(3)	232(1)	
81	Norflurazon	29.99	303(100)	145(101)	102(47)	
82	Pyridaphenthion	30.17	340(100)	199(48)	188(51)	

TABLE 1.1.2 The Groups, Retention Times, Quantifying Ions and Qualifying Ions of the 490 Pesticides to be Analyzed in Teas and Internal Standard Compounds by GC–MS (*Cont.*)

No.	Compounds	Retention time (min)	Quantifying ion	Qualifying ion 1	Qualifying ion 2	Qualifying ion 3
83	Tetradifon	30.70	227(100)	356(70)	159(196)	
84	*Cis*-permethrin	31.42	183(100)	184(15)	255(2)	
85	Pyrazophos	31.60	221(100)	232(35)	373(19)	
86	*Trans*-permethrin	31.68	183(100)	184(15)	255(2)	
87	Cypermethrin	33.19	181(100)	152(23)	180(16)	
88	Fenvalerate-1	34.45	167(100)	225(53)	419(37)	181(41)
89	Fenvalerate-2	34.79	167(101)	225(54)	419(38)	181(42)
90	Deltamethrin	35.77	181(100)	172(25)	174(25)	
B group						
91	EPTC	8.54	128(100)	189(30)	132(32)	
92	Butylate	9.49	156(100)	146(115)	217(27)	
93	Dichlobenil	9.75	171(100)	173(68)	136(15)	
94	Pebulate	10.18	128(100)	161(21)	203(20)	
95	Nitrapyrin	10.89	194(100)	196(97)	198(23)	
96	Mevinphos	11.23	127(100)	192(39)	164(29)	
97	Chloroneb	11.85	191(100)	193(67)	206(66)	
98	Tecnazene	13.54	261(100)	203(135)	215(113)	
99	Heptanophos	13.78	124(100)	215(17)	250(14)	
100	Ethoprophos	14.40	158(100)	200(40)	242(23)	168(15)
101	Hexachlorobenzene	14.69	284(100)	286(81)	282(51)	
102	Propachlor	14.73	120(100)	176(45)	211(11)	
103	*Cis*-diallate	14.75	234(100)	236(37)	128(38)	
104	Trifluralin	15.23	306(100)	264(72)	335(7)	
105	*Trans*-diallate	15.29	234(100)	236(37)	128(38)	
106	Chlorpropham	15.49	213(100)	171(59)	153(24)	
107	Sulfotep	15.55	322(100)	202(43)	238(27)	266(24)
108	Sulfallate	15.75	188(100)	116(7)	148(4)	
109	*Alpha*-HCH	16.06	219(100)	183(98)	221(47)	254(6)
110	Terbufos	16.83	231(100)	153(25)	288(10)	186(13)
111	Profluralin	17.36	318(100)	304(47)	347(13)	
112	Dioxathion	17.51	270(100)	197(43)	169(19)	
113	Propazine	17.67	214(100)	229(67)	172(51)	
114	Chlorbufam	17.85	223(100)	153(53)	164(64)	
115	Dicloran	17.89	206(100)	176(128)	160(52)	
116	Terbuthylazine	18.07	214(100)	229(33)	173(35)	
117	Monolinuron	18.15	61(100)	126(45)	214(51)	
118	Cyanophos	18.73	243(100)	180(8)	148(3)	
119	Flufenoxuron	18.83	305(100)	126(67)	307(32)	
120	Chlorpyrifos-methyl	19.38	286(100)	288(70)	197(5)	
121	Desmetryn	19.64	213(100)	198(60)	171(30)	
122	Dimethachlor	19.80	134(100)	197(47)	210(16)	
123	Alachlor	20.03	188(100)	237(35)	269(15)	
124	Pirimiphos-methyl	20.30	290(100)	276(86)	305(74)	

(Continued)

TABLE 1.1.2 The Groups, Retention Times, Quantifying Ions and Qualifying Ions of the 490 Pesticides to be Analyzed in Teas and Internal Standard Compounds by GC–MS (*Cont.*)

No.	Compounds	Retention time (min)	Quantifying ion	Qualifying ion 1	Qualifying ion 2	Qualifying ion 3
125	Terbutryn	20.61	226(100)	241(64)	185(73)	
126	Aspon	20.62	211(100)	253(52)	378(14)	
127	Thiobencarb	20.63	100(100)	257(25)	259(9)	
128	Dicofol	21.33	139(100)	141(72)	250(23)	251(4)
129	Metolachlor	21.34	238(100)	162(159)	240(33)	
130	Pirimiphos-ethyl	21.59	333(100)	318(93)	304(69)	
131	Oxy-chlordane	21.63	387(100)	237(50)	185(68)	
132	Dichlofluanid	21.68	224(100)	226(74)	167(120)	
133	Methoprene	21.71	73(100)	191(29)	153(29)	
134	Bromofos	21.75	331(100)	329(75)	213(7)	
135	Ethofumesate	21.84	207(100)	161(54)	286(27)	
136	Isopropalin	22.10	280(100)	238(40)	222(4)	
137	Propanil	22.68	161(100)	217(21)	163(62)	
138	Crufomate	22.93	256(100)	182(154)	276(58)	
139	Isofenphos	22.99	213(100)	255(44)	185(45)	
140	Endosulfan -1	23.10	241(100)	265(66)	339(46)	
141	Chlorfenvinphos	23.19	323(100)	267(139)	269(92)	
142	Tolylfluanide	23.45	238(100)	240(71)	137(210)	
143	Cis-chlordane	23.55	373(100)	375(96)	377(51)	
144	Butachlor	23.82	176(100)	160(75)	188(46)	
145	Chlozolinate	23.83	259(100)	188(83)	331(91)	
146	*P,p*′-DDE	23.92	318(100)	316(80)	246(139)	248(70)
147	Iodofenphos	24.33	377(100)	379(37)	250(6)	
148	Tetrachlorvinphos	24.36	329(100)	331(96)	333(31)	
149	Profenofos	24.65	339(100)	374(39)	297(37)	
150	Buprofezin	24.87	105(100)	172(54)	305(24)	
151	Hexaconazole	24.92	214(100)	231(62)	256(26)	
152	*o,p*′-DDD	24.94	235(100)	237(65)	165(39)	199(14)
153	Chlorfenson	25.05	302(100)	175(282)	177(103)	
154	Fluorochloridone	25.14	311(100)	313(64)	187(85)	
155	Endrin	25.15	263(100)	317(30)	345(26)	
156	Paclobutrazol	25.21	236(100)	238(37)	167(39)	
157	*o,p*′-DDT	25.56	235(100)	237(63)	165(37)	199(14)
158	Methoprotryne	25.63	256(100)	213(24)	271(17)	
159	Chloropropylate	25.85	251(100)	253(64)	141(18)	
160	Flamprop-methyl	25.90	105(100)	77(26)	276(11)	
161	Nitrofen	26.12	283(100)	253(90)	202(48)	139(15)
162	Oxyfluorfen	26.13	252(100)	361(35)	300(35)	
163	Chlorthiophos	26.52	325(100)	360(52)	297(54)	
164	Flamprop-isopropyl	26.70	105(100)	276(19)	363(3)	
165	Carbofenothion	27.19	157(100)	342(49)	199(28)	
166	*p,p*′-DDT	27.22	235(100)	237(65)	246(7)	165(34)

TABLE 1.1.2 The Groups, Retention Times, Quantifying Ions and Qualifying Ions of the 490 Pesticides to be Analyzed in Teas and Internal Standard Compounds by GC–MS (*Cont.*)

No.	Compounds	Retention time (min)	Quantifying ion	Qualifying ion 1	Qualifying ion 2	Qualifying ion 3
167	Benalaxyl	27.54	148(100)	206(32)	325(8)	
168	Edifenphos	27.94	173(100)	310(76)	201(37)	
169	Triazophos	28.23	161(100)	172(47)	257(38)	
170	Cyanofenphos	28.43	157(100)	169(56)	303(20)	
171	Chlorbenside sulfone	28.88	127(100)	99(14)	89(33)	
172	Endosulfan-sulfate	29.05	387(100)	272(165)	389(64)	
173	Bromopropylate	29.30	341(100)	183(34)	339(49)	
174	Benzoylprop-ethyl	29.40	292(100)	365(36)	260(37)	
175	Fenpropathrin	29.56	265(100)	181(237)	349(25)	
176	EPN	30.06	157(100)	169(53)	323(14)	
177	Hexazinone	30.14	171(100)	252(3)	128(12)	
178	Leptophos	30.19	377(100)	375(73)	379(28)	
179	Bifenox	30.81	341(100)	189(30)	310(27)	
180	Phosalone	31.22	182(100)	367(30)	154(20)	
181	Azinphos-methyl	31.41	160(100)	132(71)	77(58)	
182	Fenarimol	31.65	139(100)	219(70)	330(42)	
183	Azinphos-ethyl	32.01	160(100)	132(103)	77(51)	
184	Cyfluthrin	32.94	206(100)	199(63)	226(72)	
185	Prochloraz	33.07	180(100)	308(59)	266(18)	
186	Coumaphos	33.22	362(100)	226(56)	364(39)	
187	Fluvalinate	34.94	250(100)	252(38)	181(18)	
C group						
188	Dichlorvos	7.80	109(100)	185(34)	220(7)	
189	Biphenyl	9.00	154(100)	153(40)	152(27)	
190	Propamocarb	9.40	58(100)	129(6)	188(5)	
191	Vernolate	9.82	128(100)	146(17)	203(9)	
192	3,5-Dichloroaniline	11.20	161(100)	163(62)	126(10)	
193	Methacrifos	11.86	125(100)	208(74)	240(44)	
194	Molinate	11.92	126(100)	187(24)	158(2)	
195	2-phenylphenol	12.47	170(100)	169(72)	141(31)	
196	*cis*-1,2,3,6-Tetrahydroph-thalimide	13.39	151(100)	123(16)	122(16)	
197	Fenobucarb	14.60	121(100)	150(32)	107(8)	
198	Benfluralin	15.23	292(100)	264(20)	276(13)	
199	Hexaflumuron	16.20	176(100)	279(28)	277(43)	
200	Prometon	16.66	210(100)	225(91)	168(67)	
201	Triallate	17.12	268(100)	270(73)	143(19)	
202	Pyrimethanil	17.28	198(100)	199(45)	200(5)	
203	*Gamma*-HCH	17.48	183(100)	219(93)	254(13)	221(40)
204	Disulfoton	17.61	88(100)	274(15)	186(18)	
205	Atrizine	17.64	200(100)	215(62)	173(29)	
206	Iprobenfos	18.44	204(100)	246(18)	288(17)	
207	Heptachlor	18.49	272(100)	237(40)	337(27)	

(Continued)

TABLE 1.1.2 The Groups, Retention Times, Quantifying Ions and Qualifying Ions of the 490 Pesticides to be Analyzed in Teas and Internal Standard Compounds by GC–MS (*Cont.*)

No.	Compounds	Retention time (min)	Quantifying ion	Qualifying ion 1	Qualifying ion 2	Qualifying ion 3
208	Isazofos	18.54	161(100)	257(53)	285(39)	313(15)
209	Plifenate	18.87	217(100)	175(96)	242(91)	
210	Fluchloralin	18.89	306(100)	326(87)	264(54)	
211	Transfluthrin	19.04	163(100)	165(23)	335(7)	
212	Fenpropimorph	19.22	128(100)	303(5)	129(9)	
213	Tolclofos-methyl	19.69	265(100)	267(36)	250(10)	
214	Propisochlor	19.89	162(100)	223(200)	146(17)	
215	Metobromuron	20.07	61(100)	258(11)	170(16)	
216	Ametryn	20.11	227(100)	212(53)	185(17)	
217	Metribuzin	20.33	198(100)	199(21)	144(12)	
218	Dipropetryn	20.82	255(100)	240(42)	222(20)	
219	Formothion	21.42	170(100)	224(97)	257(63)	
220	Diethofencarb	21.43	267(100)	225(98)	151(31)	
221	Dimepiperate	22.28	119(100)	145(30)	263(8)	
222	Bioallethrin-1	22.29	123(100)	136(24)	107(29)	
223	Bioallethrin-2	22.34	123(100)	136(24)	107(29)	
224	Fenson	22.54	141(100)	268(53)	77(104)	
225	o,p'-DDT	22.64	246(100)	318(34)	176(26)	248(70)
226	Diphenamid	22.87	167(100)	239(30)	165(43)	
227	Penconazole	23.17	248(100)	250(33)	161(50)	
228	Tetraconazole	23.35	336(100)	338(33)	171(10)	
229	Mecarbam	23.46	131(100)	296(22)	329(40)	
230	Propaphos	23.92	304(100)	220(108)	262(34)	
231	Flumetralin	24.10	143(100)	157(25)	404(10)	
232	Triadimenol-1	24.22	112(100)	168(81)	130(15)	
233	Triadimenol-2	24.94	112(100)	168(71)	130(10)	
234	Pretilachlor	24.67	162(100)	238(26)	262(8)	
235	Kresoxim-methyl	25.04	116(100)	206(25)	131(66)	
236	Fluazifop-butyl	25.21	282(100)	383(44)	254(49)	
237	Chlorfluazuron	25.27	321(100)	323(71)	356(8)	
238	Chlorobenzilate	25.90	251(100)	253(65)	152(5)	
239	Flusilazole	26.19	233(100)	206(33)	315(9)	
240	Fluorodifen	26.59	190(100)	328(35)	162(34)	
241	Diniconazole	27.03	268(100)	270(65)	232(13)	
242	Piperonyl butoxide	27.46	176(100)	177(33)	149(14)	
243	Dimefuron	27.82	140(100)	105(75)	267(36)	
244	Propargite	27.87	135(100)	350(7)	173(16)	
245	Mepronil	27.91	119(100)	269(26)	120(9)	
246	Diflufenican	28.45	266(100)	394(25)	267(14)	
247	Fludioxonil	28.93	248(100)	127(24)	154(21)	
248	Fenazaquin	28.97	145(100)	160(46)	117(10)	
249	Phenothrin	29.08	123(100)	183(74)	350(6)	

TABLE 1.1.2 The Groups, Retention Times, Quantifying Ions and Qualifying Ions of the 490 Pesticides to be Analyzed in Teas and Internal Standard Compounds by GC–MS (*Cont.*)

No.	Compounds	Retention time (min)	Quantifying ion	Qualifying ion 1	Qualifying ion 2	Qualifying ion 3
250	Amitraz	30.00	293(100)	162(138)	132(168)	
251	Anilofos	30.68	226(100)	184(52)	334(10)	
252	Lambda-cyhalothrin	31.11	181(100)	197(100)	141(20)	
253	Mefenacet	31.29	192(100)	120(35)	136(29)	
254	Permethrin	31.57	183(100)	184(14)	255(1)	
255	Pyridaben	31.86	147(100)	117(11)	364(7)	
256	Fluoroglycofen-ethyl	32.01	447(100)	428(20)	449(35)	
257	Bitertanol	32.25	170(100)	112(8)	141(6)	
258	Etofenprox	32.75	163(100)	376(4)	183(6)	
259	Alpha-cypermethrin	33.35	163(100)	181(84)	165(63)	
260	Flucythrinate-1	33.58	199(100)	157(90)	451(22)	
261	Flucythrinate-2	33.85	199(101)	157(91)	451(23)	
262	Esfenvalerate	34.65	419(100)	225(158)	181(189)	
263	Difenconazole-1	35.40	323(100)	325(66)	265(83)	
264	Difenonazole-2	35.49	323(100)	325(69)	265(70)	
265	Flumioxazin	35.50	354(100)	287(24)	259(15)	
266	Flumiclorac-pentyl	36.34	423(100)	308(51)	318(29)	
D group						
267	Dimefox	5.62	110(100)	154(75)	153(17)	
268	Disulfoton-sulfoxide	8.41	212(100)	153(61)	184(20)	
269	Pentachlorobenzene	11.11	250(100)	252(64)	215(24)	
270	Crimidine	13.13	142(100)	156(90)	171(84)	
271	BDMC-1	13.25	200(100)	202(104)	201(13)	
272	Chlorfenprop-methyl	13.57	165(100)	196(87)	197(49)	
273	Thionazin	14.04	143(100)	192(39)	220(14)	
274	2,3,5,6-tetrachloroaniline	14.22	231(100)	229(76)	158(25)	
275	*Tri-N*-butyl phosphate	14.33	155(100)	211(61)	167(8)	
276	2,3,4,5-Tetrachloroanisole	14.66	246(100)	203(70)	231(51)	
277	Pentachloroanisole	15.19	280(100)	265(100)	237(85)	
278	Tebutam	15.30	190(100)	106(38)	142(24)	
279	Methabenzthiazuron	16.34	164(100)	136(81)	108(27)	
280	Desisopropyl-atrazine	16.69	173(100)	158(84)	145(73)	
281	Simetone	16.69	197(100)	196(40)	182(38)	
282	Atratone	16.70	196(100)	211(68)	197(105)	
283	Tefluthrin	17.24	177(100)	197(26)	161(5)	
284	Bromocylen	17.43	359(100)	357(99)	394(14)	
285	Trietazine	17.53	200(100)	229(51)	214(45)	
286	2,6-Dichlorobenzamide	17.93	173(100)	189(36)	175(62	
287	Cycluron	17.95	89(100)	198(36)	114(9)	
288	*de*-PCB 28	18.15	256(100)	186(53)	258(97)	
289	*de*-PCB 31	18.19	256(100)	186(53)	258(97)	
290	Desethyl-sebuthylazine	18.32	172(100)	174(32)	186(11)	

(*Continued*)

TABLE 1.1.2 The Groups, Retention Times, Quantifying Ions and Qualifying Ions of the 490 Pesticides to be Analyzed in Teas and Internal Standard Compounds by GC–MS (*Cont.*)

No.	Compounds	Retention time (min)	Quantifying ion	Qualifying ion 1	Qualifying ion 2	Qualifying ion 3
291	2,3,4,5-Tetrachloroaniline	18.55	231(100)	229(76)	233(48)	
292	Musk ambrette	18.62	253(100)	268(35)	223(18)	
293	Musk xylene	18.66	282(100)	297(10)	128(20)	
294	Pentachloroaniline	18.91	265(100)	263(63)	230(8)	
295	Aziprotryne	19.11	199(100)	184(83)	157(31)	
296	Isocarbamid	19.24	142(100)	185(2)	143(6)	
297	Sebutylazine	19.26	200(100)	214(14)	229(13)	
298	Musk moskene	19.46	263(100)	278(12)	264(15)	
299	*de*-PCB 52	19.48	292(100)	220(88)	255(32)	
300	Prosulfocarb	19.51	251(100)	252(14)	162(10)	
301	Dimethenamid	19.55	154(100)	230(43)	203(21)	
302	Bdmc-2	19.74	200(100)	202(101)	201(12)	
303	Monalide	20.02	197(100)	199(31)	239(45)	
304	Musk tibeten	20.40	251(100)	266(25)	252(14)	
305	Isobenzan	20.55	311(100)	375(31)	412(7)	
306	Octachlorostyrene	20.60	380(100)	343(94)	308(120)	
307	Isodrin	21.01	193(100)	263(46)	195(83)	
308	Isomethiozin	21.06	225(100)	198(86)	184(13)	
309	Dacthal	21.25	301(100)	332(31)	221(16)	
310	4,4-Dichlorobenzophe-none	21.29	250(100)	252(62)	215(26)	
311	Nitrothal-isopropyl	21.69	236(100)	254(54)	212(74)	
312	Rabenzazole	21.73	212(100)	170(26)	195(19)	
313	Cyprodinil	21.94	224(100)	225(62)	210(9)	
314	Isofenphos oxon	22.04	229(100)	201(2)	314(12)	
315	Fuberidazole	22.10	184(100)	155(21)	129(12)	
316	Dicapthon	22.44	262(100)	263(10)	216(10)	
317	Mcpa-butoxyethyl ester	22.61	300(100)	200(71)	182(41)	
318	*de*-PCB 101	22.62	326(100)	254(66)	291(18)	
319	Isocarbophos	22.87	136(100)	230(26)	289(22)	
320	Phorate sulfone	23.15	199(100)	171(30)	215(11)	
321	Chlorfenethol	23.29	251(100)	253(66)	266(12)	
322	Trans-nonachlor	23.62	409(100)	407(89)	411(63)	
323	DEF	24.08	202(100)	226(51)	258(55)	
324	Flurochloridone	24.31	311(100)	187(74)	313(66)	
325	Bromfenvinfos	24.62	267(100)	323(56)	295(18)	
326	Perthane	24.81	223(100)	224(20)	178(9)	
327	*de*-PCB 118	25.08	326(100)	254(38)	184(16)	
328	Mephosfolan	25.29	196(100)	227(49)	168(60)	
329	4,4-Dibromobenzophe-none	25.30	340(100)	259(30)	185(179)	
330	Flutriafol	25.31	219(100)	164(96)	201(7)	
331	*de*-PCB 153	25.64	360(100)	290(62)	218(24)	

TABLE 1.1.2 The Groups, Retention Times, Quantifying Ions and Qualifying Ions of the 490 Pesticides to be Analyzed in Teas and Internal Standard Compounds by GC–MS (*Cont.*)

No.	Compounds	Retention time (min)	Quantifying ion	Qualifying ion 1	Qualifying ion 2	Qualifying ion 3
332	Diclobutrazole	25.95	270(100)	272(68)	159(42)	
333	Disulfoton sulfone	26.16	213(100)	229(4)	185(11)	
334	Hexythiazox	26.48	227(100)	156(158)	184(93)	
335	de-PCB 138	26.84	360(100)	290(68)	218(26)	
336	Cyproconazole	27.23	222(100)	224(35)	223(11)	
337	Resmethrin-1	27.26	171(100)	143(83)	338(7)	
338	Resmethrin-2	27.43	171(100)	143(80)	338(7)	
339	Phthalic acid,benzyl butyl ester	27.56	206(100)	312(4)	230(1)	
340	Clodinafop-propargyl	27.74	349(100)	238(96)	266(83)	
341	Fenthion sulfoxide	28.06	278(100)	279(290)	294(145)	
342	Fluotrimazole	28.39	311(100)	379((60)	233(36)	
343	Fluroxypr-1-methylheptyl ester	28.45	366(100)	254(67)	237(60)	
344	Fenthion sulfone	28.55	310(100)	136(25)	231(10)	
345	Metamitron	28.63	202(100)	174(52)	186(12)	
346	Triphenyl phosphate	28.65	326(100)	233(16)	215(20)	
347	de-PCB 180	29.05	394(100)	324(70)	359(20)	
348	Tebufenpyrad	29.06	318(100)	333(78)	276(44)	
349	Cloquintocet-mexyl	29.32	192(100)	194(32)	220(4)	
350	Lenacil	29.70	153(100)	136(6)	234(2)	
351	Bromuconazole-1	29.90	173(100)	175(65)	214(15)	
352	Bromuconazole-2	30.72	173(100)	175(67)	214(14)	
353	Nitralin	30.92	316(100)	274(58)	300(15)	
354	Fenamiphos sulfoxide	31.03	304(100)	319(29)	196(22)	
355	Fenamiphos sulfone	31.34	320(100)	292(57)	335(7)	
356	Fenpiclonil	32.37	236(100)	238(66)	174(36)	
357	Fluquinconazole	32.62	340(100)	342(37)	341(20)	
358	Fenbuconazole	34.02	129(100)	198(51)	125(31)	
E group						
359	Propoxur-1	6.58	110(100)	152(16)	111(9)	
360	Isoprocarb -1	7.56	121(100)	136(34)	103(20)	
361	Terbucarb-1	10.89	205(100)	220(51)	206(16)	
362	Dibutyl succinate	12.20	101(100)	157(19)	175(5)	
363	Chlorethoxyfos	13.43	153(100)	125(67)	301(19)	
364	Isoprocarb -2	13.69	121(100)	136(34)	103(20)	
365	Tebuthiuron	14.25	156(100)	171(30)	157(9)	
366	Pencycuron	14.30	125(100)	180(65)	209(20)	
367	Demeton-s-methyl	15.19	109(100)	142(43)	230(5)	
368	Propoxur-2	15.48	110(100)	152(19)	111(8)	
369	Phenanthrene	16.97	188(100)	160(9)	189(16)	
370	Fenpyroximate	17.49	213(100)	142(21)	198(9)	
371	Tebupirimfos	17.61	318(100)	261(107)	234(100)	

(*Continued*)

TABLE 1.1.2 The Groups, Retention Times, Quantifying Ions and Qualifying Ions of the 490 Pesticides to be Analyzed in Teas and Internal Standard Compounds by GC–MS (*Cont.*)

No.	Compounds	Retention time (min)	Quantifying ion	Qualifying ion 1	Qualifying ion 2	Qualifying ion 3
372	Prohydrojasmon	17.80	153(100)	184(41)	254(7)	
373	Fenpropidin	17.85	98(100)	273(5)	145(5)	
374	Dichloran	18.10	176(100)	206(87)	124(101)	
375	Pyroquilon	18.28	173(100)	130(69)	144(38)	
376	Propyzamide	19.01	173(100)	255(23)	240(9)	
377	Pirimicarb	19.08	166(100)	238(23)	138(8)	
378	Benoxacor	19.62	120(100)	259(38)	176(19)	
379	Phosphamidon -1	19.66	264(100)	138(62)	227(25)	
380	Acetochlor	19.84	146(100)	162(59)	223(59)	
381	Tridiphane	19.90	173(100)	187(90)	219(46)	
382	Esprocarb	20.01	222(100)	265(10)	162(61)	
383	Terbucarb-2	20.06	205(100)	220(52)	206(16)	
384	Acibenzolar-s-methyl	20.42	182(100)	135(64)	153(34)	
385	Mefenoxam	20.91	206(100)	249(46)	279(11)	
386	Malaoxon	21.17	127(100)	268(11)	195(15)	
387	Chlorthal-dimethyl	21.39	301(100)	332(27)	221(17)	
388	Simeconazole	21.41	121(100)	278(14)	211(34)	
389	Terbacil	21.50	161(100)	160(70)	117(39)	
390	Thiazopyr	21.91	327(100)	363(73)	381(34)	
391	Dimethylvinphos	22.21	295(100)	297(56)	109(74)	
392	Zoxamide	22.30	187(100)	242(68)	299(9)	
393	Allethrin	22.60	123(100)	107(24)	136(20)	
394	Quinoclamine	22.89	207(100)	172(259)	144(64)	
395	Flufenacet	23.09	151(100)	211(61)	363(6)	
396	Fenoxanil	23.58	140(100)	189(14)	301(6)	
397	Furalaxyl	23.97	242(100)	301(24)	152(40)	
398	Bromacil	24.73	205(100)	207(46)	231(5)	
399	Picoxystrobin	24.97	335(100)	303(43)	367(9)	
400	Butamifos	25.41	286(100)	200(57)	232(37)	
401	Imazamethabenz-methyl	25.50	144(100)	187(117)	256(95)	
402	Methiocarb sulfone	25.56	200(100)	185(40)	137(16)	
403	Metominostrobin	25.61	191(100)	238(56)	196(75)	
404	Imazalil	25.72	215(100)	173(66)	296(5)	
405	Isoprothiolane	25.87	290(100)	231(82)	204(88)	
406	Cyflufenamid	26.02	91(100)	412(11)	294(11)	
407	Isoxathion	26.51	313(100)	105(341)	177(208)	
408	Quinoxyphen	27.14	237(100)	272(37)	307(29)	
409	Trifloxystrobin	27.71	116(100)	131(40)	222(30)	
410	Imibenconazole-des-benzyl	27.86	235(100)	270(35)	272(35)	
411	Imiprothrin-1	28.31	123(100)	151(55)	107(54)	
412	Fipronil	28.34	367(100)	369(69)	351(15)	

TABLE 1.1.2 The Groups, Retention Times, Quantifying Ions and Qualifying Ions of the 490 Pesticides to be Analyzed in Teas and Internal Standard Compounds by GC–MS (*Cont.*)

No.	Compounds	Retention time (min)	Quantifying ion	Qualifying ion 1	Qualifying ion 2	Qualifying ion 3
413	Imiprothrin-2	28.50	123(100)	151(21)	107(17)	
414	Epoxiconazole -1	28.58	192(100)	183(24)	138(35)	
415	Pyributicarb	28.87	165(100)	181(23)	108(64)	
416	Pyraflufen ethyl	28.91	412(100)	349(41)	339(34)	
417	Thenylchlor	29.12	127(100)	288(25)	141(17)	
418	Mefenpyr-diethyl	29.55	227(100)	299(131)	372(18)	
419	Etoxazole	29.64	300(100)	330(69)	359(65)	
420	Epoxiconazole-2	29.73	192(100)	183(13)	138(30)	
421	Pyriproxyfen	30.06	136(100)	226(8)	185(10)	
422	Iprodione	30.24	187(100)	244(65)	246(42)	
423	Ofurace	30.36	160(100)	232(83)	204(35)	
424	Piperophos	30.42	320(100)	140(123)	122(114)	
425	Clomeprop	30.48	290(100)	288(279)	148(206)	
426	Fenamidone	30.66	268(100)	238(111)	206(32)	
427	Pyraclostrobin	31.98	132(100)	325(14)	283(21)	
428	Lactofen	32.06	442(100)	461(25)	346(12)	
429	Pyraclofos	32.18	360(100)	194(79)	362(38)	
430	Dialifos	32.27	186(100)	357(143)	210(397)	
431	Spirodiclofen	32.50	312(100)	259(48)	277(28)	
432	Flurtamone	32.78	333(100)	199(63)	247(25)	
433	Pyriftalid	32.94	318(100)	274(71)	303(44)	
434	Silafluofen	33.18	287(100)	286(274)	258(289)	
435	Pyrimidifen	33.63	184(100)	186(32)	185(10)	
436	Butafenacil	33.85	331(100)	333(34)	180(35)	
437	Cafenstrole	34.36	100(100)	188(69)	119(25)	
F group						
438	Tribenuron-methyl	9.34	154(100)	124(45)	110(18)	
439	Ethiofencarb	11.00	107(100)	168(34)	77(26)	
440	Dioxacarb	11.10	121(100)	166(44)	165(36)	
441	Dimethyl phthalate	11.54	163(100)	194(7)	133(5)	
442	4-chlorophenoxy acetic acid	11.84	200(100)	141(93)	111(61)	
443	Phthalimide	13.21	147(100)	104(61)	103(35)	
444	Diethyltoluamide	14.00	119(100)	190(32)	191(31)	
445	2,4-T	14.35	199(100)	234(63)	175(61)	
446	Carbaryl	14.42	144(100)	115(100)	116(43)	
447	Cadusafos	15.14	159(100)	213(14)	270(13)	
448	Demetom-s	16.88	88(100)	170(15)	143(11)	
449	Spiroxamine -1	17.26	100(100)	126(7)	198(5)	
450	Dicrotophos	17.31	127(100)	237(11)	109(8)	
451	3,4,5-trimethacarb	17.70	136(100)	193(32)	121(31)	
452	2,4,5-T	17.75	233(100)	268(49)	209(36)	

(Continued)

TABLE 1.1.2 The Groups, Retention Times, Quantifying Ions and Qualifying Ions of the 490 Pesticides to be Analyzed in Teas and Internal Standard Compounds by GC–MS (*Cont.*)

No.	Compounds	Retention time (min)	Quantifying ion	Qualifying ion 1	Qualifying ion 2	Qualifying ion 3
453	3-Phenylphenol	18.11	170(100)	141(23)	115(17)	
454	Furmecyclox	18.22	123(100)	251(6)	94(10)	
455	Spiroxamine-2	18.23	100(100)	126(5)	198(5)	
456	Dmsa	18.45	200(100)	92(123)	121(8)	
457	Sobutylazine	18.63	172(100)	174(32)	186(11)	
458	Cinmethylin	18.96	105(100)	169(16)	154(14)	
459	Monocrotophos	19.18	127(100)	192(2)	223(4)	164(20)
460	*S421*(octachlorodipropyl ether)-1	19.31	130(100)	132(96)	211(8)	
461	*S421*(octachlorodipropyl ether)-2	19.57	130(100)	132(97)	211(7)	
462	Dodemorph	19.62	154(100)	281(12)	238(10)	
463	Fenchlorphos	19.84	285(100)	287(69)	270(6)	
464	Difenoxuron	20.85	241(100)	226(21)	242(15)	
465	Butralin	22.18	266(100)	224(16)	295(9)	
466	Pyrifenox-1	23.46	262(100)	294(18)	227(15)	
467	Thiabendazole	24.97	201(100)	174(87)	175(9)	
468	Iprovalicarb-1	26.13	119(100)	134(126)	158(62)	
469	Azaconazole	26.50	217(100)	173(59)	219(64)	
470	Iprovalicarb-2	26.54	134(100)	119(75)	158(48)	
471	Diofenolan -1	26.76	186(100)	300(60)	225(24)	
472	Diofenolan -2	27.09	186(100)	300(60)	225(29)	
473	Aclonifen	27.24	264(100)	212(65)	194(57)	
474	Chlorfenapyr	27.47	247(100)	328(54)	408(51)	
475	Bioresmethrin	27.55	123(100)	171(54)	143(31)	
476	Isoxadifen-ethyl	27.90	204(100)	222(76)	294(44)	
477	Carfentrazone-ethyl	28.09	312(100)	330(52)	290(53)	
478	Fenhexamid	28.86	97(100)	177(33)	301(13)	
479	Spiromesifen	29.56	272(100)	254(27)	370(14)	
480	Fluazinam	30.04	387(100)	417(44)	371(29)	
481	Bifenazate	30.38	300(100)	258(99)	199(100)	
482	Endrin ketone	30.40	317(100)	250(28)	281(35)	
483	Norflurazon-desmethyl	30.80	145(100)	289(76)	88(35)	
484	*Gamma*-cyhaloterin-1	31.10	181(100)	197(84)	141(28)	
485	Metoconazole	31.12	125(100)	319(14)	250(17)	
486	Cyhalofop-butyl	31.40	256(100)	357(74)	229(79)	
487	*Gamma*-cyhalothrin-2	31.40	181(100)	197(77)	141(20)	
488	Halfenprox	32.81	263(100)	237(5)	476(5)	
489	Boscalid	34.16	342(100)	140(229)	112(71)	
490	Dimethomorph	37.40	301(100)	387(32)	165(28)	

TABLE 1.1.3 255 Pesticides No Suitable for the LC–MS/MS

No.	Name	No.	Name	No.	Name
110 compounds did not find the precursor ions					
1	1,2-Dibromo ethane	38	Deltamethrin	75	Methiocarb Sulfone
2	2,3,5,6-Tetrachloroaniline	39	DE-PCB 101	76	Methoxychlor
3	3,5-Dichloroaniline	40	DE-PCB 118	77	Mirex
4	Acequinocyl	41	DE-PCB 153	78	Nitrapyrin
5	Amitraz	42	DE-PCB 180	79	Pentachloroanisole
6	Amitrole	43	DE-PCB 52	80	Pentachlorobenzene
7	Amobam	44	Dichlobenil	81	Perthane
8	Anilazine	45	Dichlone	82	Phthalic acid di-(2-ethylhexyl) ester
9	Aramite	46	Dienochlor	83	Primisulfuron-methyl
10	Azocyclotin	47	Dimethipin	84	Profluralin
11	Bediocarb	48	Dinoseb acetate	85	Prohexadione-calcium
12	Bifenox	49	Dioxathion	86	Prosulfuron
13	Biphenyl	50	DMSA	87	Quinmerac
14	Bromocylen	51	Endrin	88	Quintozene
15	Bromopropylate	52	Erbon	89	Rimsulfuron
16	Bromoxynil octanoate	53	Ethalfluralin	90	TCMTB
17	Camphechlor	54	Fenazaflor	91	Tecnazene
18	Captan PESTANAL	55	Fenbutatin oxide	92	Tefluthrin
19	Carbofenothion	56	Fenchlorphos oxon	93	Tetradifon
20	Chinomethionat (Quinomethionate)	57	Fenpiclonil	94	Tetrasul
21	Chlorbenside	58	Fenson	95	Thiocyclam hydrogenoxalate
22	Chlordane	59	Fentin acetate	96	Thiram
23	Chlordecone (Kepone)	60	Flumetralin	97	Tralomethrin
24	Chlorfenprop-methyl	61	Fluquinconazole	98	Trans-Chlodane
25	Chlorfenson	62	Fluroxypr-1-methylheptyl ester	99	Tribenuron-methyl
26	Chlorobenzilate	63	Formetanate hydrochloride	100	Tridiphane
27	Chloroneb	64	Gamma-HCH	101	Triflumizole
28	Chlorothalonil	65	Glyphosate	102	Trans-nonachlor
29	Chlozolinate	66	alpha-HCH	103	1,2-dichloropropane
30	Chrysene (D12.98%)	67	Heptachlor-epoxide	104	Bromochloromethane
31	Cis-Chlordanc	68	Hexaconazole	105	Methazole
32	Cycloprothrin	69	Imiprothrin	106	Endosulfan
33	Cyhexatin	70	Iodofenphos	107	Halfenprox
34	Cypermethin	71	Iprodione	108	Isobenzan
35	Dacthal	72	Isodrin	109	Propylene oxide
36	P'p-DDE	73	Leptophos	110	Phthalide
37	P'p-DDT	74	MCPA-butoxyethyl ester		
18 compounds did not find the product ions					
1	Abamectin	7	DE-PCB 28	13	Nitrofen
2	Acenaphthene(D10.99%)	8	DE-PCB 31	14	Nitrothal-isopropyl

(Continued)

TABLE 1.1.3 255 Pesticides No Suitable for the LC–MS/MS (*Cont.*)

No.	Name	No.	Name	No.	Name
3	Barban	9	Flumethrin	15	Paraquat dichloride
4	BENFURESATE	10	Formothion	16	Transfluthrin
5	Binapacryl	11	Methyl isothiocyanate	17	Diclomezine
6	Chlorthal-dimethyl	12	Musk ambrette	18	1,3-Dichloropropene (cis+trans) obenzan

33 compounds with extremely low sensitivity or did not have peak in the mixed standard solutions

No.	Name	No.	Name	No.	Name
1	Lambda-Cyhalothrin	12	Cyfluthrin	23	Hexachlorobenzene
2	1,2-Dibromo-3-chloropropane	13	Captafol	24	4,4-Dibromobenzophenone
3	Chlorfenapyr	14	Guazatine triacetate	25	Parathion
4	Phenanthrene	15	Chloropicrin	26	Cyhalofop-butyl
5	Propineb	16	Chlorethoxyfos	27	Chlorfenvinphos
6	Triflusulfuron-methyl	17	DE-PCB 138	28	Prohydrojasmon
7	Bromobutide	18	Diquat dibromide hydrate	29	DMSA
8	Butamifos	19	Dinocap technical mixture of isomers	30	Biphenyl
9	Dichlormid	20	Bifenthrin	31	Pentachlorobenzene
10	Isoxathion	21	Tiamulin-fumerate	32	Chlorothalonil
11	Cinmethylin	22	2,3,4,5-tetrachloroanisole	33	Pentachloroanisole

64 compounds had no peaks in the tea matrix

No.	Name	No.	Name	No.	Name
1	Desmedipham	23	Mecoprop	44	Heptachlor
2	Phthalic acid,dicyclobexyl ester	24	MCPB	45	Plifenate
3	Bensultap	25	Fenaminosulf	46	Ioxynil
4	Chlorimuron ethyl	26	Dichlorprop	47	Metsulfuron-methyl
5	Fentin-chloride	27	Bentazone	48	Sulfentrazone
6	Alachlor	28	Forchlorfenuron	49	Propoxycarbzone-sodium
7	Cinosulfuron	29	Fluroxypyr	50	Flazasulfuron
8	Pyrazosulfuron-ethyl	30	Fenoprop	51	Flusulfamide
9	Dimepiperate	31	Cyclanilide	52	Cyclosulfamuron
10	Florasulam	32	Bromoxynil	53	Triforine
11	Thidiazuron PESTANAL	33	Pentachlorophenol	54	Halosulfuron-methyl
12	Iminoctadine triacetate	34	Isocarbophos	55	Fomesafen
13	Pyriminobac-methyl(Z)	35	Alloxydim-sodium	56	Tecloftalam
14	Triphenyl phosphate	36	Pyrithlobac sodium	57	Fluazuron
15	Dieldrin	37	Dinobuton	58	Iodosulfuron-methyl sodium
16	Etoxazole	38	Fluorodifen	59	Thifluzamide
17	Ethephon	39	Dimehypo	60	Iodosulfuron-methyl
18	Flupropanate	40	Fempxaprop-ethyl	61	Octachlorostyrene
19	2,6-difluorobenzoic acid	41	Diflufenzopyr-sodium	62	Azaconazole
20	Trichloroacetic acid sodium salt	42	Sulfanitran	63	Triclopyr
21	Tert-butyl-4-hydroxyanisole	43	Gibberellic acid	64	Naled
22	Aminopyralid				

TABLE 1.1.4 The Retention Times, Quantifying Ion, Qualifying Ions, CID, and Collision Energy for the Determination of 448 Pesticides by LC–MS/MS

No.	Compounds	Retention time (min)	Quantifying ion	Qualifying ions	CID (V)	Collision energy (V)
A group						
1	Propham	8.80	180.1/138.0	180.1/138.0; 180.1/120.0	80	5; 15
2	Isoprocarb	8.38	194.1/95.0	194.1/95.0; 194.1/137.1	80	20; 5
3	3,4,5-Trimethacarb	8.38	194.2/137.2	194.2/137.2; 194.2/122.2	80	5; 20
4	Cycluron	7.73	199.4/72.0	199.4/72.0; 199.4/89.0	120	25; 15
5	Carbaryl	7.45	202.1/145.1	202.1/145.1; 202.1/127.1	80	10; 5
6	Propachlor	8.75	212.1/170.1	212.1/170.1; 212.1/94.1	100	10; 30
7	Rabenzazole	7.54	213.2/172.0	213.2/172; 213.2/118.0	120	25; 25
8	Simetryn	5.32	214.2/124.1	214.2/124.1; 214.2/96.1	120	20; 25
9	Monolinuron	7.82	215.1/126.0	215.1/126.0; 215.1/148.1	100	15; 10
10	Mevinphos	5.17	225.0/127.0	225.0/127.0; 225.0/193.0	80	15; 1
11	Aziprotryne	10.40	226.1/156.1	226.1/156.1; 226.1/198.1	100	10; 10
12	Secbumeton	5.56	226.2/170.1	226.2/170.1; 226.2/142.1	120	20; 25
13	Cyprodinil	9.24	226.0/93.0	226.0/93.0; 226.0/108.0	120	40; 30
14	Buturon	9.38	237.1/84.1	237.1/84.1; 237.1/126.1	120	30; 15
15	Carbetamide	5.80	237.1/192.1	237.1/192.1; 237.1/118.1	80	5; 10
16	Pirimicarb	4.20	239.2/72.0	239.2/72.0; 239.2/182.2	120	20; 15
17	Clomazone	9.36	240.1/125.0	240.1/125.0; 240.1/89.1	100	20; 50
18	Cyanazine	6.38	241.1/214.1	241.1/214.1; 241.1/174.0	120	15; 15
19	Prometryne	7.66	242.2/158.1	242.2/158.1; 242.2/200.2	120	20; 20
20	Paraoxon methyl	6.20	248.0/202.1	248.0/202.1; 248.0/90.0	120	20; 30
21	4,4-Dichlorobenzophenone	12.00	251.1/111.1	251.1/111.1; 251.1/139.0	100	35; 20
22	Thiacloprid	5.65	253.1/126.1	253.1/126.1; 253.1/186.1	120	20; 10
23	Imidacloprid	4.73	256.1/209.1	256.1/209.1; 256.1/175.1	80	10; 10
24	Ethidimuron	4.62	265.1/208.1	265.1/208.1; 265.1/162.1	80	10; 25
25	Isomethiozin	14.20	269.1/200.0	269.1/200.0; 269.1/172.1	120	15; 25
26	Diallate	17.40	270.0/86.0	270.0/86.0; 270.0/109.0	100	15; 35
27	Acetochlor	13.70	270.2/224.0	270.2/224; 270.2/148.2	80	5; 20
28	Nitenpyram	3.87	271.1/224.1	271.1/224.1; 271.1/237.1	100	15; 15
29	Methoprotryne	6.47	272.2/198.2	272.2/198.2; 272.2/170.1	140	25; 30
30	Dimethenamid	10.50	276.1/244.1	276.1/244.1; 276.1/168.1	120	10; 15
31	Terbucarb	16.50	278.2/166.1	278.2/166.1; 278.2/109.0	80	15; 30

(Continued)

TABLE 1.1.4 The Retention Times, Quantifying Ion, Qualifying Ions, CID, and Collision Energy for the Determination of 448 Pesticides by LC–MS/MS (Cont.)

No.	Compounds	Retention time (min)	Quantifying ion	Qualifying ions	CID (V)	Collision energy (V)
32	Penconazole	13.70	284.1/70.0	284.1/70.0; 284.1/159.0	120	15; 20
33	Myclobutanil	12.10	289.1/125.0	289.1/125.0; 289.1/70.0	120	20; 15
34	Imazethapyr	5.60	290.2/177.1	290.2/177.1; 290.2/245.2	120	25; 20
35	Paclobutrazol	10.32	294.2/70.0	294.2/70.0; 294.2/125.0	100	15; 25
36	Fenthion sulfoxide	7.31	295.1/109.0	295.1/109.0; 295.1/280.0	140	35; 20
37	Triadimenol	10.15	296.1/70.0	296.1/70.0; 296.1/99.1	80	10; 10
38	Butralin	18.60	296.1/240.1	296.1/240.1; 296.1/222.1	100	10; 20
39	Spiroxamine	9.90	298.2/144.2	298.2/144.2; 298.2/100.1	120	20; 35
40	Tolclofos methyl	16.60	301.2/269	301.2/269.0; 301.2/125.2	120	15; 20
41	Methidathion	10.69	303.0/145.1	303.0/145.1; 303.0/85.0	80	5; 10
42	Allethrin	18.10	303.2/135.1	303.2/135.1; 303.2/123.2	60	10; 20
43	Diazinon	15.95	305.0/169.1	305.0/169.1; 305.0/153.2	160	20; 20
44	Edifenphos	3.00	311.1/283.0	311.1/283.0; 311.1/109.0	100	10; 35
45	Pretilachlor	17.15	312.1/252.1	312.1/252.1; 312.1/176.2	100	15; 30
46	Flusilazole	13.60	316.1/247.1	316.1/247.1; 316.1/165.1	120	15; 20
47	Iprovalicarb	12.00	321.1/119.0	321.1/119.0; 321.1/203.2	100	25; 5
48	Benodanil	9.80	324.1/203.0	324.1/203; 324.1/231.0	120	25; 40
49	Flutolanil	14.00	324.2/262.1	324.2/262.1; 324.2/282.1	120	20; 10
50	Famphur	10.30	326.0/217.0	326.0/217; 326.0/281.0	100	20; 10
51	Benalyxyl	15.19	326.2/148.1	326.2/148.1; 326.2/294.0	120	1; 5
52	Diclobutrazole	12.20	328.0/159.0	328.0/159.0; 328.0/70.0	120	35; 30
53	Etaconazole	11.75	328.1/159.1	328.1/159.1; 328.1/205.1	80	25; 20
54	Fenarimol	12.20	331.0/268.1	331.0/268.1; 331.0/81.0	120	25; 30
55	Tetramethrin	17.85	332.2/164.1	332.2/164.1; 332.2/135.1	100	15; 15
56	Dichlofluanid	15.16	333.0/123.0	333.0/123.0; 333.0/224.0	80	20; 10
57	Cloquintocet mexyl	17.36	336.1/238.1	336.1/238.1; 336.1/192.1	120	15; 20
58	Bitertanol	13.90	338.2/70.0	338.2/70.0; 338.2/269.2	60	5; 1
59	Chlorprifos methyl	16.72	322.0/125.0	322.0/125.0; 322.0/290.0	80	15; 15
60	Azinphos ethyl	14.00	346.0/233	346.0/233.0; 346.0/261.1	120	10; 5
61	Clodinafop propargyl	16.09	350.1/266.1	350.1/266.1; 350.1/238.1	120	15; 20
62	Triflumuron	15.59	359.0/156.1	359.0/156.1; 359.0/139.0	120	15; 30
63	Isoxaflutole	12.00	360.0/251.1	360.0/251.1; 360.0/220.1	120	10; 45

64	Anilofos	17.35	367.9/145.2	367.9/145.2; 367.9/205.0	120	20; 5
65	Quizalofop-ethyl	17.40	373.0/299.1	373.0/299.1; 373.0/91.0	140	15; 30
66	Haloxyfop-methyl	17.11	376.0/316.0	376.0/316.0; 376.0/288.0	120	15; 20
67	Fluazifop butyl	18.24	384.1/282.1	384.1/282.1; 384.1/328.1	120	20; 15
68	Bromophos-ethyl	19.15	393.0/337.0	393.0/337.0; 393.0/162.1	100	20; 30
69	Bensulide	16.18	398.0/158.1	398.0/158.1; 398.0/314.0	80	20; 5
70	Bromfenvinfos	15.22	402.9/170.0	402.9/170.0; 402.9/127.0	100	35; 20
71	Azoxystrobin	12.50	404.0/372.0	404.0/372.0; 404.0/344.1	120	10; 15
72	Pyrazophos	16.20	374.0/222.0	374.0/222.0; 374.0/194.0	120	20; 30
73	Flufenoxuron	18.30	489.0/158.1	489.0/158.1; 489.0/141.1	80	10; 15
74	Indoxacarb	17.43	528.0/150.0	528.0/150.0; 528.0/218.0	120	20; 20
B group						
75	Ethylene thiourea	0.74	103.0/60.0	103.0/60.0; 103.0/86.0	100	35; 10
76	Daminozide	0.74	161.1/143.1	161.1/143.1; 161.1/102.2	80	15; 15
77	Dazomet	3.80	163.1/120.0	163.1/120.0; 163.1/77.0	80	10; 35
78	Nicotine	0.74	163.2/130.1	163.2/130.1; 163.2/117.1	100	25; 30
79	Fenuron	4.50	165.1/72.0	165.1/72.0; 165.1/120.0	120	15; 15
80	Crimidine	4.47	172.1/107.1	172.1/107.1; 172.1/136.2	120	30; 25
81	Molinate	11.30	188.1/126.1	188.1/126.1; 188.1/83.0	120	10; 15
82	Carbendazim	3.30	192.1/160.1	192.1/160.1; 192.1/132.1	80	15; 20
83	6-Chloro-4-hydroxy-3-phenyl-pyridazin	12.86	207.1/77.0	207.1/77; 207.1/104.0	120	25; 35
84	Propoxur	6.79	210.1/111.0	210.1/111.0; 210.1/168.1	80	10; 5
85	Isouron	6.11	212.2/167.1	212.2/167.1; 212.2/72.0	120	15; 25
86	Chlorotoluron	7.23	213.1/72.0	213.1/72.0; 213.1/140.1	80	25; 25
87	Thiofanox	1.00	241.0/184.0	241.0/184.0; 241/57.1	120	15; 5
88	Chlorbufam	11.67	224.1/172.1	224.1/172.1; 224.1/154.1	120	5; 15
89	Bendiocarb	6.87	224.1/109.0	224.1/109; 224.1/167.1	80	5; 10
90	Propazine	9.37	229.9/146.1	229.9/146.1; 229.9/188.1	120	20; 15
91	Terbuthylazine	10.15	230.1/174.1	230.1/174.1; 230.1/132.0	120	15; 20
92	Diuron	7.82	233.1/72.0	233.1/72.0; 233.1/160.1	120	20; 20
93	Chlormephos	13.70	235.0/125.0	235.0/125.0; 235.0/75.0	100	10; 10
94	Carboxin	7.67	236.1/143.1	236.1/143.1; 236.1/87.0	120	15; 20
95	Clothianidin	4.40	250.2/169.1	250.2/169.1; 250.2/132.0	80	10; 15

(Continued)

TABLE 1.1.4 The Retention Times, Quantifying Ion, Qualifying Ions, CID, and Collision Energy for the Determination of 448 Pesticides by LC–MS/MS (*Cont.*)

No.	Compounds	Retention time (min)	Quantifying ion	Qualifying ions	CID (V)	Collision energy (V)
96	Pronamide	11.81	256.1/190.1	256.1/190.1; 256.1/173.0	80	10; 20
97	Dimethachloro	8.96	256.1/224.2	256.1/224.2; 256.1/148.2	120	10; 20
98	Methobromuron	8.25	259.0/170.1	259.0/170.1; 259/148.0	80	15; 15
99	Phorate	16.55	261.0/75.0	261.0/75.0; 261/199.0	80	10; 5
100	Aclonifen	14.70	265.1/248.0	265.1/248.0; 265.1/193.0	120	15; 15
101	Mephosfolan	5.97	270.1/140.1	270.1/140.1; 270.1/168.1	100	25; 15
102	Imibenzonazole-des-benzyl	5.96	271.0/174.0	271.0/174.0; 271.0/70.0	120	25; 25
103	Neburon	14.17	275.1/57.0	275.1/57; 275.1/88.1	120	20; 15
104	Mefenoxam	7.92	280.1/192.1	280.1/192.1; 280.1/220.0	100	15; 10
105	Prothoate	4.78	286.1/227.1	286.1/227.1; 286.1/199.0	100	5; 15
106	Ethofume sate	12.86	287/121.0	287.0/121.0; 287.0/161.0	80	10; 20
107	Iprobenfos	13.50	289.1/91.0	289.1/91.0; 289.1/205.1	80	25; 5
108	TEPP	5.64	291.1/179.0	291.1/179.0; 291.1/199.0	100	20; 35
109	Cyproconazole	10.59	292.1/70.0	292.1/70.0; 292.1/125	120	15; 15
110	Thiamethoxam	4.05	292.1/211.2	292.1/211.2; 292.1/181.1	80	10; 20
111	Crufomate	11.56	292.1/236.0	292.1/236.0; 292.1/108.1	120	20; 30
112	Etrimfos	6.16	293.1/125.0	293.1/125.0293.1/265.1	80	20; 15
113	Coumatetralyl	4.68	293.2/107.0	293.2/107; 293.2/175.1	140	35; 25
114	Cythioate	6.59	298/217.1	298.0/217.1; 298.0/125.0	100	15; 25
115	Phosphamidon	5.77	300.1/174.1	300.1/174.1; 300.1/127.0	120	10; 20
116	Phenmedipham	10.69	301.1/168.1	301.1/168.1; 301.1/136	80	5; 20
117	Bifenazate	13.28	301.2/198.1	301.2/198.1; 301.2/170.1	60	5; 20
118	Fenhexamid	12.33	302.0/97.1	302.0/97.1; 302.0/55.0	80	30; 25
119	Flutriafol	7.55	302.1/70.0	302.1/70; 302.1/123.0	120	15; 20
120	Furalaxyl	10.77	302.2/242.2	302.2/242.2; 302.2/270.2	100	15; 5
121	Bioallethrin	18.00	303.1/135.1	303.1/135.1; 303.1/107.0	80	10; 20
122	Cyanofenphos	16.44	304.0/157.0	304.0/157.0; 304.0/276.0	100	20; 10
123	Pirimiphos methyl	15.50	306.2/164.0	306.2/164.0; 306.2/108.1	120	20; 30
124	Buprofezin	13.34	306.2/201.0	306.2/201.0; 306.2/116.1	120	15; 10
125	Disulfoton sulfone	9.79	307.0/97.0	307.0/97.0; 307.0/125.0	100	30; 10
126	Fenazaquin	18.80	307.2/57.1	307.2/57.1; 307.2/161.2	120	20; 15
127	Triazophos	13.80	314.1/162.1	314.1/162.1; 314.1/286	120	20; 10
128	DEF	19.21	315.1/169.0	315.1/169.0; 315.1/113	100	10; 20

129	Pyrifalid	12.00	319.0/139.1	319.0/139.1; 319/179	140	35; 35
130	Metconazole	13.77	320.2/70.0	320.2/70.0; 320.2/125.0	140	35; 55
131	Pyriproxyfen	18.00	322.1/96.0	322.1/96.0; 322.1/227.1	120	15; 10
132	Isoxaben	13.21	333.1/165.0	333.1/165.0; 333.1/150.1	120	15; 50
133	Flurtamone	11.25	334.1/247.1	334.1/247.1; 334.1/303.0	120	30; 20
134	Trifluralin	12.86	336.0/138.9	336.0/138.9; 336.0/103.0	120	20; 45
135	Flamprop methyl	13.20	336.1/105.1	336.1/105.1; 336.1/304.0	80	20; 5
136	Bioresmethrin	19.39	339.2/171.1	339.2/171.1; 339.2/143.1	100	15; 25
137	Propiconazole	14.29	342.1/159.1	342.1/159.1; 342.1/69.0	120	20; 20
138	Chlorpyrifos	18.29	350.0/198.0	350.0/198.0; 350.0/79.0	100	20; 35
139	Fluchloralin	17.68	356.0/186.0	356.0/314.1; 356.0/63.0	80	15; 30
140	Chlorsulfuron	6.96	358.0/141.1	358.0/141.1; 358.0/167.0	120	15; 15
141	Flamprop isopropyl	16.00	364.1/105.1	364.1/105.1; 364.1/304.1	80	20; 5
142	Tetrachlorvinphos	13.70	365.0/127.0	365.0/127.0; 365.0/239.0	120	15; 15
143	Propargite	18.77	368.1/231.0	368.1/231; 368.1/175.1	100	5; 15
144	Bromuconazole	12.70	376.0/159.0	376.0/159.0; 376.0/70.0	80	20; 20
145	Picolinafen	17.74	377.0/238.0	377.0/238.0; 377.0/359.0	120	20; 20
146	Fluthiacet methyl	14.80	404.0/215.0	404.0/215.0; 404.0/274.0	180	50; 10
147	Trifloxystrobin	17.44	409.3/186.1	409.3/186.1; 409.3/206.2	120	15; 10
148	Hexaflumuron	16.90	461.0/141.1	461.1/141.1; 461.0/158.1	120	35; 35
149	Novaluron	17.39	493.0/158.0	493.0/158.0; 493.0/141.1	80	15; 55
150	Flurazuron	18.10	506.0/158.1	506.0/158.1; 506.0/141.1	120	15; 50
C group						
151	Maleic hydrazide	0.73	113.1/67.1	113.1/67.1; 113.1/85.0	100	20; 20
152	Methamidophos	0.74	142.1/94.0	142.1/94.0; 142.1/125.0	80	15; 10
153	EPTC	14.00	190.2/86.0	190.2/86.0; 190.2/128.1	100	10; 10
154	Diethyltoluamide	7.70	192.2/119.0	192.2/119.0; 192.2/91.0	100	15; 30
155	Monuron	5.94	199.0/72.0	199.0/72.0; 199.0/126.0	120	15; 15
156	Pyrimethanil	6.70	200.2/107.0	200.2/107.0; 200.2/183.1	120	25; 25
157	Fenfuram	7.48	202.1/109.0	202.1/109.0; 202.1/83.0	120	20; 20
158	Quinoclamine	6.09	208.1/105.0	208.1/105.0; 208.1/154.1	120	30; 20
159	Fenobucarb	9.92	208.2/95.0	208.2/95.0; 208.2/152.1	80	10; 5
160	Propanil	9.09	218.0/162.1	218.0/162.1; 218.0/127.0	120	15; 20
161	Carbofuran	6.81	222.3/165.1	222.3/165.1; 222.3/123.1	120	5; 20

(Continued)

TABLE 1.1.4 The Retention Times, Quantifying Ion, Qualifying Ions, CID, and Collision Energy for the Determination of 448 Pesticides by LC–MS/MS (*Cont.*)

No.	Compounds	Retention time (min)	Quantifying ion	Qualifying ions	CID (V)	Collision energy (V)
162	Acetamiprid	4.86	223.2/126.0	223.2/126.0; 223.2/56.0	120	15; 15
163	Mepanipyrim	12.23	224.2/77.0	224.2/77.0; 224.2/106.0	120	30; 25
164	Prometon	5.40	226.2/142.0	226.2/142.0; 226.2/184.1	120	20; 20
165	Methiocarb	4.51	226.2/121.1	226.2/121.1; 226.2/169.1	80	10; 5
166	Metoxuron	5.59	229.1/72.0	229.1/72.0; 229.1/156.1	120	20; 20
167	Dimethoate	4.88	230.0/199.0	230.0/199.0; 230.0/171.0	80	5; 10
168	Fluometuron	7.27	233.1/72.0	233.1/72.0; 233.1/160.0	120	20; 20
169	Dicrotophos	3.97	238.1/112.1	238.1/112.1; 238.1/193.0	80	10; 5
170	Monalide	14.50	240.1/85.1	240.1/85.1; 240.1/57.0	120	15; 35
171	Diphenamid	9.00	240.1/134.1	240.1/134.1; 240.1/167.1	120	20; 25
172	Ethoprophos	11.98	243.1/173.0	243.1/173.0; 243.1/215.0	120	10; 10
173	Fonofos	16.10	247.1/109.0	247.1/109.0; 247.1/137.1	80	15; 5
174	Etridiazol	17.20	247.1/183.1	247.1/183.1; 247.1/132.0	120	15; 15
175	Hexazinone	5.66	253.2/171.1	253.2/171.1; 253.2/71.0	120	15; 20
176	Dimethametryn	8.79	256.2/186.1	256.2/186.1; 256.2/96.1	140	20; 35
177	Trichlorphon	4.21	257.0/221.0	257.0/221.0; 257.0/109.0	120	10; 20
178	Demeton(o+s)	8.59	259.1/89.0	259.1/89.0; 259.1/61.0	60	10; 35
179	Benoxacor	10.83	260.0/149.2	260.0/149.2; 260.0/134.1	120	15; 20
180	Bromacil	5.78	261.0/205.0	261.0/205.0; 261.0/188.0	80	10; 20
181	Phorate sulfoxide	7.34	277.0/143.0	277.0/143.0; 277.0/199.0	100	15; 5
182	Brompyrazon	4.69	266.0/92.0	266.0/92.0; 266.0/104.0	120	30; 30
183	Oxycarboxin	5.38	268.0/175.0	268.0/175.0; 268.0/147.1	100	10; 20
184	Mepronil	13.15	270.2/119.1	270.2/119.1; 270.2/228.2	100	30; 15
185	Disulfoton	16.80	275.0/89.0	275.0/89.0; 275/61.0	80	5; 20
186	Fenthion	15.54	279.0/169.1	279.0/169.1; 279.0/247.0	120	15; 10
187	Metalaxyl	7.75	280.1/192.2	280.1/192.2; 280.1/220.2	120	15; 20
188	Ofurace	7.65	282.1/160.2	282.1/160.2; 282.1/254.2	120	20.1
189	Fosthiazate	4.38	284.1/228.1	284.1/228.1; 284.1/104.0	80	5; 20
190	Imazamethabenz-methyl	5.33	289.1/229.0	289.1/229.0; 289.1/86.0	120	15; 25
191	Disulfoton-sulfoxide	7.38	291.0/185.0	291.0/185.0; 291.0/157.0	80	10; 20
192	Isoprothiolane	13.17	291.1/189.1	291.1/189.1; 291.1/231.1	80	20; 5
193	Imazalil	6.86	297.0/159.0	297.0/159.0; 297.0/255.0	120	20; 20
194	Phoxim	16.80	299.0/77.0	299.0/77.0; 299.0/129.0	80	20; 10
195	Quinalphos	14.80	299.1/147.1	299.1/147.1; 299.1/163.1	120	20; 20

196	Fenoxycarb	18.10	362.1/288.0	362.1/288.0; 362.1/244.0	120	20; 20
197	Pyrimitate	14.00	306.1/170.2	306.1/170.2; 306.1/154.2	120	20; 20
198	Fensulfothin	8.55	309.0/157.1	309.0/157.1; 309.0/253.0	120	25; 15
199	Fluorochloridone	13.80	312.1/292.1	312.1/292.1; 312.1/89.0	100	25; 25
200	Butachlor	18.00	312.2/238.1	312.2/238.1; 312.2/162.0	80	10; 20
201	Kresoxim-methyl	15.20	314.1/267	314.1/267.0; 314.1/206.0	80	5; 5
202	Triticonazole	10.55	318.2/70.0	318.2/70.0; 318.2/125.1	120	15; 35
203	Fenamiphos sulfoxide	5.87	320.1/171.1	320.1/171.1; 320.1/292.1	140	25; 15
204	Thenylchlor	14.00	324.1/127.0	324.1/127.0; 324.1/59.0	80	10; 45
205	Fenoxanil	18.81	329.1/302.0	329.1/302.0; 329.1/189.1	80	5; 30
206	Fluridone	10.30	330.1/309.1	330.1/309.1; 330.1/259.2	160	40; 55
207	Epoxiconazole	18.81	330.1/141.1	330.1/141.1; 330.1/121.1	120	20; 20
208	Chlorphoxim	17.15	333.0/125.0	333.0/125.0; 333.0/163.1	80	5; 5
209	Fenamiphos sulfone	6.63	336.1/188.2	336.1/188.2; 336.1/266.2	120	30; 20
210	Fenbuconazole	13.40	337.1/70.0	337.1/70.0; 337.1/125.0	120	20; 20
211	Isofenphos	17.25	346.1/217.0	346.1/217.0; 346.1/245.0	80	20; 10
212	Phenothrin	19.70	351.1/183.2	351.1/183.2; 351.1/237.0	100	15; 5
213	Piperophos	17.00	354.1/171.0	354.1/171.0; 354.1/143.0	100	20; 30
214	Piperonyl butoxide	17.75	356.2/177.1	356.2/177.1; 356.2/119.0	100	10; 35
215	Oxyflurofen	18.00	362.0/316.1	362.0/316.1; 362/237.1	120	10; 25
216	Flufenacet	14.00	364.0/194.0	364.0/194.0; 364.0/152.0	80	5; 10
217	Phosalone	16.79	368.1/182.0	368.1/182.0; 368.1/322.0	80	10; 5
218	Methoxyfenozide	13.41	313.0/149.0	313.0/149.0; 313.0/91.0	100	10; 35
219	Aspon	19.22	379.1/115.0	379.1/115.0; 379.1/210.0	80	30; 15
220	Ethion	18.46	385.0/199.1	385.0/199.1; 385.0/171.0	80	5; 15
221	Diafenthiuron	18.90	385.0/329.2	385.0/329.2; 385.0/278.2	140	15; 35
222	Dithiopyr	17.81	402.0/354.0	402.0/354.0; 402.0/272.0	120	20; 30
223	Spirodiclofen	19.28	411.1/71.0	411.1/71.0; 411.1/313.1	100	10; 5
224	Fenpyroximate	18.66	422.2/366.2	422.2/366.2; 422.2/135.0	120	10; 35
225	Flumiclorac-pentyl	18.00	441.1/308.0	441.1/308.0; 441.1/354.0	100	25; 10
226	Temephos	18.30	467.0/125.0	467.0/125.0; 467.0/155.0	100	30; 30
227	Butafenacil	15.00	492.0/180.0	492.0/180.0; 492.0/331.0	120	35; 25
228	Spinosad	14.30	732.4/142.2	732.4/142.2; 732.4/98.1	180	30; 75

D group

(Continued)

TABLE 1.1.4 The Retention Times, Quantifying Ion, Qualifying Ions, CID, and Collision Energy for the Determination of 448 Pesticides by LC–MS/MS (Cont.)

No.	Compounds	Retention time (min)	Quantifying ion	Qualifying ions	CID (V)	Collision energy (V)
229	Mepiquat chloride	0.71	114.1/98.1	114.1/98.1; 114.1/58.0	140	30; 30
230	Allidochlor	5.78	174.1/98.1	174.1/98.1; 174.1/81.0	100	10; 15
231	Tricyclazole	5.06	190.1/136.1	190.1/136.1; 190.1/163.1	120	30; 25
232	Metamitron	4.18	203.1/175.1	203.1/175.1; 203.1/104.0	120	15; 20
233	Isoproturon	7.44	207.2/72.0	207.2/72.0; 207.2/165.1	120	15; 15
234	Atratone	4.46	212.2/170.2	212.2/170.2; 212.2/100.1	120	15; 30
235	Oesmetryn	4.92	214.1/172.1	214.1/172.1; 214.1/82.1	120	15; 25
236	Metribuzin	7.16	215.1/187.2	215.1/187.2; 215.1/131.1	120	15; 20
237	Dmst	7.06	215.3/106.1	215.3/106.1; 215.3/151.2	80	10; 5
238	Cycloate	15.95	216.2/83.0	216.2/83.0; 216.2/154.1	120	15; 10
239	Atrazine	7.20	216.0/174.2	216.0/174.2; 216.0/132.0	120	15; 20
240	Butylate	17.20	218.1/57.0	218.1/57.0; 218.1/156.2	80	10; 5
241	Pymetrozin	0.73	218.1/105.1	218.1/105.1; 218.1/78.0	100	20; 40
242	Chloridazon	4.35	222.1/104.0	222.1/104.0; 222.1/92.0	120	25; 35
243	Sulfallate	15.25	224.1/116.1	224.1/116.1; 224.1/88.2	100	10; 20
244	Ethiofencarb	4.48	227.0/107.0	227.0/107.0; 227/164.0	80	5; 5
245	Terbumeton	5.25	226.2/170.1	226.2/170.1; 226.2/114	120	15; 20
246	Cyprazine	7.15	228.2/186.1	228.2/186.1; 228.2/108.1	120	15; 25
247	Ametryn	5.85	228.2/186.0	228.2/186.0; 228.2/68.0	120	20; 35
248	Tebuthiuron	5.30	229.2/172.2	229.2/172.2; 229.2/116.0	120	15; 20
249	Trietazine	12.00	230.1/202.0	230.1/202.0; 230.1/132.1	160	20; 20
250	Sebutylazine	8.65	230.1/174.1	230.1/174.1; 230.1/104.0	12	15; 30
251	Dibutyl succinate	14.80	231.1/101.0	231.1/101; 231.1/157.1	60	1; 10
252	Tebutam	13.04	234.2/91.1	234.2/91.1; 234.2/192.2	120	20; 15
253	Thiofanox-sulfoxide	4.08	235.1/104.0	235.1/104.0; 235.1/57.0	60	5; 20
254	Cartap hydrochloride	5.90	238.0/73.0	238.0/73.0; 238.0/150	100	30; 10
255	Methacrifos	10.03	241.0/209.0	241.0/209.0; 241.0/125.0	60	5; 20
256	Thionazin	8.84	249.1/97.0	249.1/97.0; 249.1/193.0	80	30; 10
257	Linuron	9.84	249.0/160.1	249.0/160.1; 249/182.1	100	15; 15
258	Heptanophos	7.85	251.0/127.0	251.0/127.0; 251.0/109.0	80	10; 30
259	Prosulfocarb	17.10	252.1/91.0	252.1/91.0; 252.1/128.1	120	15; 10
260	Dipropetryn	8.58	256.1/144.1	256.1/144.1; 256.1/214.0	140	30; 20

261	Thiobencarb	15.80	258.1/125.0	258.1/125.0; 258.1/89.0	80	20; 55
262	Tri-iso-butyl phosphate	15.45	267.1/99.0	267.1/99.0; 267.1/155.1	80	20; 5
263	Tri-n-butyl phosphate	15.45	267.2/99.0	267.2/99.0; 267.2/155.1	80	5; 15
264	Diethofencarb	10.40	268.1/226.2	268.1/226.2; 268.1/152.1	80	5; 20
265	Cadusafos	15.27	271.1/159.1	271.1/159.1; 271.1/131	80	10; 20
266	Metazachlor	8.36	278.1/134.1	278.1/134.1; 278.1/210.1	80	20; 5
267	Propetamphos	13.60	282.1/138	282.1/138.0; 282.1/156.1	80	15; 10
268	Terbufos	13.70	289.0/57.0	289.0/57.0; 289.0/103.1	80	20; 5
269	Simeconazole	11.00	294.2/70.1	294.2/70.1; 294.2/135.1	120	15; 15
270	Triadimefon	11.88	294.2/69.0	294.2/69.0; 294.2/197.1	100	20; 15
271	Phorate sulfone	9.34	293.0/171.0	293.0/171.0; 293/143.1	60	5; 15
272	Tridemorph	14.00	298.3/130.1	298.3/130.1; 298.3/57.1	160	25; 35
273	Mefenacet	11.60	299.1/148.1	299.1/148.1; 299.1/120.1	100	15; 25
274	Fenamiphos	8.97	304.0/216.9	304.0/216.9; 304.0/202.0	100	20; 35
275	Fenpropimorph	9.10	304.0/147.2	304.0/147.2; 304.0/130.0	120	30; 30
276	Tebuconazole	12.44	308.2/70.0	308.2/70.0; 308.2/125.0	100	25; 25
277	Isopropalin	19.05	310.2/225.7	310.2/225.7; 310.2/207.7	120	15; 20
278	Nuarimol	9.20	315.1/252.1	315.1/252.1; 315.1/81.0	120	25; 30
279	Bupirimate	9.52	317.2/166.0	317.2/166; 317.2/272.0	120	25; 20
280	Azinphos-methyl	10.45	318.1/125.0	318.1/125; 318.1/160.0	80	15; 10
281	Tebupirimfos	18.15	319.1/277.1	319.1/277.1; 319.1/153.2	120	10; 30
282	Phenthoate	15.57	321.1/247.0	321.1/247; 321.1/163.1	80	5; 10
283	Sulfotep	16.35	323.0/171.1	323.0/171.1; 323.0/143.0	120	10; 20
284	Sulprofos	18.40	323.0/219.1	323.0/219.1; .0/247.0	120	15; 10
285	Epn	17.10	324.0/296.0	324.0/296.0; 324.0/157.1	120	10; 20
286	Diniconazole	13.67	326.1/70.0	326.1/70.0; 326.1/159.0	120	25; 30
287	Sethoxydim	5.36	328.2/282.2	328.2/282.2; 328.2/178.1	100	10; 15
288	Pencycuron	16.33	329.2/125.0	329.2/125.0; 329.2/218.1	120	20; 15
289	Mecarbam	14.46	330.0/227.0	330.0/227.0; 330.0/199.0	80	5; 10
290	Tralkoxydim	18.09	330.2/284.2	330.2/284.2; 330.2/138.1	100	10; 20
291	Malathion	13.20	331.0/127.1	331.0/127.1; 331.0/99.0	80	5; 10
292	Pyributicarb	18.26	331.1/181.1	331.1/181.1; 331.1/108.0	120	10; 20
293	Pyridaphenthion	12.32	341.1/189.2	341.1/189.2; 341.1/205.2	120	20; 20
294	Pirimiphos-ethyl	17.75	334.2/198.2	334.2/198.2; 334.2/182.2	120	20; 25

(Continued)

TABLE 1.1.4 The Retention Times, Quantifying Ion, Qualifying Ions, CID, and Collision Energy for the Determination of 448 Pesticides by LC–MS/MS (*Cont.*)

No.	Compounds	Retention time (min)	Quantifying ion	Qualifying ions	CID (V)	Collision energy (V)
295	Thiodicarb	6.55	355.1/88.0	355.1/88.0; 355.1/163.0	80	15; 5
296	Pyraclofos	15.34	361.1/257.0	361.1/257.0; 361.1/138.0	120	25; 35
297	Picoxystrobin	15.40	368.1/145.0	368.1/145.0; 368.1/205.0	80	20; 5
298	Tetraconazole	12.54	372.0/159.0	372.0/159.0; 372.0/70.0	120	35; 35
299	Mefenpyr-diethyl	16.80	373.0/327.0	373.0/327.0; 373.0/160.0	80	15; 35
300	Profenefos	16.74	373.0/302.9	373.0/302.9; 373.0/345.0	120	15; 10
301	Pyraclostrobin	16.04	388.0/163.0	388.0/163.0; 388.0/194.0	120	20; 10
302	Dimethomorph	16.04	388.1/165.1	388.1/165.1; 388.1/301.1	120	25; 20
303	Kadethrin	17.95	397.1/171.1	397.1/171.1; 397.1/128.0	100	15; 55
304	Thiazopyr	16.15	397.1/377.0	397.1/377; 397.1/335.1	140	20; 30
305	Chlorfluazuron	18.53	540.0/383.0	540.0/383.0; 540/158.2	120	15; 15
E group						
306	4-Aminopyridine	0.72	95.1/52.1	95.1/52.1; 95.1/78.1	120	25; 5
307	Methomyl	3.76	163.2/88.1	163.2/88.1; 163.2/106.1	80	5; 10
308	Pyroquilon	5.87	174.1/117.1	174.1/117.1; 174.1/132.2	140	35; 25
309	Fuberidazole	3.66	185.2/157.2	185.2/157.2; 185.2/92.1	120	20; 25
310	Isocarbamid	4.35	186.2/87.1	186.2/87.1; 186.2/130.1	80	20; 5
311	Butocarboxim	5.30	213.0/75.1	213.0/75.1; 213.0/156.1	100	15; 5
312	Chlordimeform	4.13	197.2/117.1	197.2/117.1; 197.2/89.1	120	25; 50
313	Cymoxanil	4.95	199.1/111.1	199.1/111.1; 199.1/128.1	80	20; 15
314	Chlorthiamid	5.80	206.0/189.0	206.0/189.0; 206.0/119.0	80	15; 50
315	Aminocarb	0.75	209.3/137.1	209.3/137.1; 209.3/152.1	100	20; 10
316	Omethoate	0.75	214.1/125.0	214.1/125.0; 214.1/183.0	80	20; 5
317	Ethoxyquin	7.19	218.2/174.2	218.2/174.2; 218.2/160.1	120	30; 35
318	Aldicarb sulfone	3.50	223.1/76.0	223.1/76.0; 223.1/148.0	80	5; 5
319	Dioxacarb	4.70	224.1/123.1	224.1/123.1; 224.1/167.1	80	15; 5
320	Demeton-s-methyl	6.25	253.0/89.0	253.0/89.0; 253.0/61.0	80	10; 35
321	Cyanohos	6.89	244.2/180.0	244.2/180.0; 244.2/125.0	120	20; 15
322	Thiometon	7.16	247.1/171.0	247.1/171.0; 247.1/89.1	100	10; 10
323	Folpet	12.82	260.0/130.0	260.0/130.0; 260.0/102.3	100	10; 40
324	Demeton-s-methyl sulfone	3.96	263.1/169.1	263.1/169.1; 263.1/125.0	80	15; 20
325	Fenpropidin	8.96	274.0/147.1	274.0/147.1; 274.0/86.1	160	25; 25
326	Amidithion	14.25	274.1/97.0	274.1/97.0; 274.1/122.0	140	20; 15

No.	Compound					
327	Imazapic	4.80	276.2/163.2	276.2/163.2; 276.2/216.2; 276.2/86.1	120	20; 20; 25
328	Paraoxon-ethyl	8.00	276.2/220.1	276.2/220.1; 276.2/94.1	100	10; 40
329	Aldimorph	14.10	284.4/57.2	284.4/57.2; 284.4/98.1	160	30; 30
330	Vinclozolin	14.66	286.1/242	286.1/242; 286.1/145.1	100	5; 45
331	Uniconazole	11.69	292.1/70.1	292.1/70.1; 292.1/125.1	120	30; 30
332	Pyrifenox	7.42	295.0/93.1	295.0/93.1; 295.0/163.0	120	15; 15
333	Chlorthion	14.45	298.0/125.0	298.0/125.0; 298.0/109.0	100	15; 20
334	Dicapthon	14.47	298.0/125.0	298.0/125.0; 298.0/266.1	80	10; 10
335	Clofentezine	16.18	303.0/138.0	303.0/138.0; 303.0/156.0	100	25; 25
336	Norflurazon	8.08	304.0/284.0	304.0/284.0; 304.0/160.1	140	25; 35
337	Triallate	18.52	304.0/143.0	304.0/143.0; 304.0/86.1	120	25; 15
338	Quinoxyphen	17.05	308.0/197.0	308.0/197.0; 308.0/272.0	180	35; 35
339	Fenthion sulfone	8.71	311.1/125.0	311.1/125.0; 311.1/109.0	140	15; 20
340	Flurochloridone	13.34	312.2/292.2	312.2/292.2; 312.2/53.1	140	25; 30
341	Phthalic acid, benzyl butyl ester	17.34	313.2/91.1	313.2/91.1; 313.2/149.0; 313.2/205.1	80	10; 10; 5
342	Isazofos	13.67	314.1/162.1	314.1/162.1; 314.1/120.0	100	10; 35
343	Dichlofenthion	18.15	315.0/259.0	315.0/259.0; 315.0/287.0	100	10; 5
344	Vamidothion sulfone	2.45	178.0/87.0	178.0/87.0; 178.0/60.0	100	15; 10
345	Terbufos sulfone	12.57	321.2/171.1	321.2/171.1; 321.2/143.0	80	5; 15
346	Dinitramine	15.80	323.1/305.0	323.1/305.0; 323.1/247.0	120	10; 15
347	Cyazofamid	5.10	325.2/261.3	325.2/261.3; 325.2/108.0	80	5; 15
348	Trichloronat	18.98	333.1/304.9	333.1/304.9; 333.1/161.8	100	10; 45
349	Resmethrin-2	12.35	339.2/171.1	339.2/171.1; 339.2/143.1	80	10; 25
350	Boscalid	12.20	343.2/307.2	343.2/307.2; 343.2/271.0	140	20; 35
351	Nitralin	15.15	346.1/304.1	346.1/304.1; 346.1/262.1	100	10; 20
352	Fenpropathrin	19.00	350.2/125.2	350.2/125.2; 350.2/97	120	5; 20
353	Hexythiazox	18.23	353.1/168.1	353.1/168.1; 353.1/228.1	120	20; 10
354	Benzoximate	17.00	386.1/197.0	386.1/197; 386.1/199.2	140	30; 30
355	Benzoylprop-ethyl	16.00	366.1/105.0	366.1/105.0; 366.1/77.0	80	15; 35
356	Pyrimidifen	13.69	378.2/184.1	378.2/184.1; 378.2/150.2	140	15; 40
357	Furathiocarb	17.85	383.3/195.1	383.3/195.1; 383.3/252.1; 383.3/167	100	10; 5; 25
358	Trans-permethin	21.00	391.3/149.1	391.3/149.1; 391.3/167.1	100	10; 10

(Continued)

TABLE 1.1.4 The Retention Times, Quantifying Ion, Qualifying Ions, CID, and Collision Energy for the Determination of 448 Pesticides by LC–MS/MS (*Cont.*)

No.	Compounds	Retention time (min)	Quantifying ion	Qualifying ions	CID (V)	Collision energy (V)
359	Etofenprox	19.73	394.0/177.0	394.0/177.0; 394/359.0	100	15; 5
360	Pyrazoxyfen	14.30	403.2/91.1	403.2/91.1; 403.2/105.1; 403.2/139.1	140	25; 20; 20
361	Flubenzimine	14.48	417.0/397.0	417.0/397; 417.0/167.1	100	10; 25
362	Zeta cypermethrin	20.45	433.3/416.2	433.3/416.2; 433.3/191.2	100	5; 10
363	Haloxyfop-2-ethoxyethyl	17.65	434.1/316.0	434.1/316.0; 434.1/288.0; 434.1/91.2	120	15; 20; 45
364	Esfenvalerate	8.28	437.2/206.9	437.2/206.9; 437.2/154.2	80	35; 20
365	Fluoroglycofen-ethyl	17.70	344.0/300.0	344.0/300.0; 344.0/233.0	120	15; 20
366	Tau-fluvalinate	19.58	503.2/181.2	503.2/181.2; 503.2/208.1	80	25; 15
F group						
367	Acrylamide	0.73	72.0/55.0	72.0/55.0; 72.0/27.0	100	10; 10
368	Tert-butylamine	0.65	74.1/46.0	74.1/46.0; 74.1/56.8	120	5; 5
369	Hymexazol	2.65	100.1/54.1	100.1/54.1; 100.1/44.2; 100.1/28	100	10; 15; 15
370	Phthalimide	0.74	148.0/130.1	148.0/130.1; 148.0/102.0	100	10; 25
371	Dimefox	3.88	155.1/110.1	155.1/110.1; 155.1/135.0	120	20; 10
372	Metolcarb	6.50	166.2/109.0	166.2/109.0; 166.2/97.1	80	15; 50
373	Diphenylamin	13.06	170.2/93.1	170.2/93.1; 170.2/152	120	30; 30
374	1-naphthy acetamide	5.30	186.2/141.1	186.2/141.1; 186.2/115.1	100	15; 45
375	Atrazine-desethyl	4.43	188.2/146.1	188.2/146.1; 188.2/104.1	120	10; 20
376	2,6-Dichlorobenzamide	3.85	190.1/173.0	190.1/173.0; 190.1/145.0	100	20; 30
377	Aldicarb	5.42	213.0/89.0	213/89; 213.0/116.0	100	30; 10
378	Dimethyl phthalate	3.50	217.0/86.0	217.0/86.0; 217.0/156.0	100	15; 20
379	Chlordimeform hydrochloride	4.00	197.2/117.1	197.2/117.1; 197.2/89.1	120	25; 50
380	Simeton	3.94	198.2/100.1	198.2/100.1; 198.2/128.2	120	25; 20
381	Dinotefuran	3.06	203.3/129.2	203.3/129.2; 203.3/87.1	80	5; 10
382	Pebulate	16.05	204.2/72.1	204.2/72.1; 204.2/128.0	100	10; 10
383	Acibenzolar-s-methyl	10.00	211.1/91.0	211.1/91.0; 211.1/136.0	120	20; 30
384	Dioxabenzofos	10.15	217.0/77.1	217.0/77.1; 217.0/107.1	100	40; 30
385	Oxamyl	3.46	241.0/72.0	241.0/72.0; 242.0/121.0	120	15; 10
386	Methabenzthiazuron	6.80	222.2/165.1	222.2/165.1; 222.2/149.9	100	15; 35
387	Butoxycarboxim	3.30	223.2/63.0	223.2/63; 223.2/106.1	80	10; 5

388	Mexacarbate	233.2/151.2	233.2/151.2; 233.2/166.2	100	15; 10
389	Demeton-s-methyl sulfoxide	247.1/109.0	247.1/109.0; 247.1/169.1	80	20; 10
390	Thiofanox sulfone	251.1/57.2	251.1/57.2; 251.1/76.1	80	5; 5
391	Phosfolan	256.2/140.0	256.2/140.0; 256.2/228.0	100	25; 10
392	Demeton-s	259.1/89.1	259.1/89.1; 259.1/61.0	60	10; 35
393	Fenthion oxon	263.2/230.0	263.2/230.0; 263.2/216.0	100	10; 20
394	Napropamide	272.2/171.1	272.2/171.1; 272.2/129.2	120	15; 15
395	Fenitrothion	278.1/125.0	278.1/125.0; 278.1/246.0	140	15; 15
396	Phthalic acid, dibutyl ester	279.2/149.0	279.2/149.0; 279.2/121.1	80	10; 45
397	Metolachlor	284.1/252.2	284.1/252.2; 284.1/176.2	120	10; 15
398	Procymidone	284.0/256.0	284.0/256.0; 284.0/145.0	140	10; 45
399	Vamidothion	288.2/146.1	288.2/146.1; 288.2/118.1	80	10; 20
400	Chloroxuron	291.2/72.1	291.2/72.1; 291.2/218.1	120	20; 30
401	Triamiphos	295.2/135.1	295.2/135.1; 295.2/92.0	100	25; 35
402	Prallethrin	301.0/105.0	301.0/105.0; 301/169.0	80	5; 20
403	Cumyluron	303.3/185.1	303.3/185.1; 303.3/125.0	100	5; 45
404	Imazamox	304.2/260.0	304.2/260.0; 304.2/186.0	100	5; 40
405	Warfarin	309.2/163.1	309.2/163.1; 309.2/251.2	100	20; 15
406	Phosmet	318.0/160.1	318.0/160.1; 318.0/133.0	80	10; 35
407	Ronnel	320.9/125.0	320.9/125.0; 320.9/288.8	120	10; 10
408	Pyrethrin	329.2/161.1	329.2/161.1; 329.2/133.1	100	5; 15
409	Phthalic acid, biscyclohexyl ester	331.3/149.1	331.3/149.1; 331.3/167.1; 331.3/249	80	10; 5; 5
410	Carpropamid	334.2/196.1	334.2/196.1; 334.2/139.1	120	10; 15
411	Tebufenpyrad	334.3/147	334.3/147; 334.3/117.1	160	25; 40
412	Chlorthiophos	361.0/305.0	361.0/305.0; 361/225	100	10; 15
413	Dialifos	394.0/208	394.0/208; 394.0/187	100	5; 20
414	Cinidon-ethyl	394.2/348.1	394.2/348.1; 394.2/107.1	120	15; 45
415	Rotenone	395.3/213.2	395.3/213.2; 395.3/192.2	160	20; 20
416	Imibenconazole	411.0/125.1	411.0/125.1; 411.0/171.1; 411/342	120	25; 15; 10
417	Propaquiafop	444.2/100.1	444.2/100.1; 444.2/299.1	140	15; 25
418	Lactofen	479.1/344.0	479.1/344.0; 479.1/223	120	15; 35
419	Benzofenap	431.0/105.0	431.0/105.0; 431.0/119.0	140	30; 20

(Continued)

TABLE 1.1.4 The Retention Times, Quantifying Ion, Qualifying Ions, CID, and Collision Energy for the Determination of 448 Pesticides by LC–MS/MS (*Cont.*)

No.	Compounds	Retention time (min)	Quantifying ion	Qualifying ions	CID (V)	Collision energy (V)
420	Dinoseb acetate	0.75	283.1/89.2	283.1/89.2; 283.1/133.1; 283.1/177.2	120	10; 10; 10
421	Propisochlor	15.00	284.0/224.0	284.0/224.0; 284.0/212.0	80	5; 15
422	Silafluofen	20.80	412.0/91.0	412.0/91.0; 412/72.1	100	40; 30
423	Etobenzanid	15.65	340.0/149.0	340.0/149.0; 340.0/121.1	120	20; 30
424	Fentrazamide	16.00	372.1/219.0	372.1/219.0; 372.1/83.2	200	5; 35
425	Pentachloroaniline	14.30	285.0/99.1	285.0/99.1; 285.0/127.0	100	15; 5
426	Carbosulfan	19.50	381.2/118.1	381.2/118.1381.2/160.2	100	10; 10
427	Cyphenothrin	19.40	376.2/151.2	376.2/151.2; 376.2/123.2	100	5; 15
428	Dimefuron	10.30	339.1/167.0	339.1/167.0; 339.1/72.1	140	20; 30
429	Malaoxon	13.80	331.0/99.0	331.0/99.0; 331.0/127.0	120	20; 5
430	Chlorbenside sulfone	9.86	299.0/235.0	299.0/235.0; 299.0/125.0	100	5; 25
431	Dodine	7.46	228.2/57.3	228.2/57.3; 228.2/60.1	160	25; 20
G group						
432	Dalapon	0.60	140.8/58.8	140.8/58.8; 140.8/62.9	100	10; 15
433	2-phenylphenol	9.78	169.0/115.0	169.0/115.0; 169.0/93.0	140	35; 20
434	3-phenylphenol	9.78	169.0/115.0	169.0/115.0; 169.0/141.1	140	35; 35
435	Dicloran	8.82	205.1/169.3	205.1/169.3; 205.1/123.2	120	15; 30
436	Chlorpropham	12.55	212.0/152.0	212.0/152.0; 212.0/57.0	80	5; 20
437	Terbacil	5.94	215.1/159.0	215.1/159.0; 215.1/73.0	120	10; 40
438	2,4-D	4.28	218.9/161.0	218.9/161.0; 218.9/125.0	80	5; 20
439	Fludioxonil	11.10	247.0/180.0	247.0/180; 247.0/126.0	140	10; 10
440	Chlorfenethol	11.81	265.0/96.7	265.0/96.7; 265.0/152.7	120	15; 5
441	Naptalam	4.30	290.0/246.0	290.0/246.0; 290.0/168.3	100	10; 30
442	Chlorobenzuron	14.05	306.9/154.0	306.9/154; 306.9/125.9	100	5; 20
443	Chloramphenicolum	5.07	321.0/152.0	321.0/152.0; 321.0/257.0	100	15; 10
444	Famoxadone	16.52	373.0/282.0	373.0/282.0; 373.0/328.9	120	20; 15
445	Diflufenican	17.30	393.1/329.1	393.1/329.1; 393.1/272.0	100	10; 10
446	Ethiprole	10.74	394.9/331.0	394.9/331.0; 394.9/250.0	100	5; 25
447	Fluazinam	17.25	462.9/415.9	462.9/415.9; 462.9/398.0	120	20; 15
448	Kelevan	19.50	628.1/169.0	628.1/169.0; 628.1/422.6	120	24; 22

TABLE 1.1.5 Effect of Extraction Times on the Recoveries of 50 Pesticides

No	Pesticides	First Ave%	Second Ave%	Third Ave%
1	Phorate	74.82	6.94	0.00
2	Alpha-hch	77.85	3.66	0.00
3	Vinclozolin	83.97	0.00	0.00
4	Lamda-cyhalothrin	83.98	1.94	0.00
5	Amitraz	85.17	2.59	0.00
6	Gamma-hch	85.48	0.00	0.00
7	p,p'-DDT	85.54	4.17	0.00
8	Cis-permethrin	85.98	9.37	0.00
9	Captan	87.67	0.00	0.00
10	Pirimiphos-methyl	88.20	3.34	0.00
11	Pyrimethanil	88.72	8.21	0.00
12	Ametryn	89.25	8.22	1.01
13	o,p'-DDT	89.47	6.44	2.59
14	Diazinon	89.78	10.94	0.00
15	Endrin	90.05	0.00	0.00
16	Alachlor	91.20	7.65	0.00
17	Dimethenamid	91.99	9.90	0.00
18	Metolachlor	92.16	8.54	0.00
19	Chlorpyrifos-methyl	92.31	2.71	0.00
20	Prothiophos	92.38	0.00	0.00
21	Prometryne	92.39	8.64	0.00
22	Flucythrinate	92.54	0.00	0.00
23	Chlorpyrifos	92.64	0.00	0.00
24	Fluazifop-butyl	92.69	8.09	0.00
25	Dieldrin	92.79	0.00	0.00
26	Triadimefon	93.54	12.26	0.00
27	Acetochlor	93.59	8.37	0.00
28	Metalaxyl	93.79	7.46	0.00
29	Atrazine	93.81	7.61	0.00
30	Fenthion	93.83	6.03	0.00
31	Parathion	93.84	2.20	0.00
32	Iprobenfos	94.29	12.18	0.00
33	Beta-HCH	94.59	0.00	0.00
34	o,p'-DDE	94.61	10.48	0.00
35	EPN	94.71	4.71	0.00
36	Propargite	94.90	24.91	0.00
37	p,p'-DDE	96.10	10.45	0.00
38	Bifenthrin	96.74	9.01	0.00
39	Flusilazole	97.08	7.14	0.00
40	Quinalphos	97.54	0.00	0.00
41	Pyridaben	98.25	6.81	0.00
42	Chlorfenapyr	98.81	0.00	0.00
43	Diethofencarb	100.01	5.54	0.00

(Continued)

TABLE 1.1.5 Effect of Extraction Times on the Recoveries of 50 Pesticides (*Cont.*)

No	Pesticides	First Ave%	Second Ave%	Third Ave%
44	Difenconazole	100.97	5.73	0.00
45	Fenpropathrin	101.33	8.05	0.00
46	p,p'-DDD	102.77	9.16	0.00
47	o,p'-DDD	102.81	10.79	4.44
48	Malathion	104.41	2.21	0.00
49	Oxadixyl	105.84	7.54	0.00
50	Dicofol	108.67	10.81	8.12

deal of these coextraction interfering matters. Therefore, developing an exclusive SPE for cleaning up tea samples turns out to be a top priority for the moment. The core of the SPE technique has been studied first things first—the effect of different cartridge fillings on cleanup.

As the main interfering matter in tea, pigments are the first class that should be removed, and the depigmenting filling materials include Florisil (FS), aluminium oxide, PestiCarb (PC), and amid silica (NH_2) [18,30]. A comparative study was conducted on six filling materials such as FS, PC, neutral aluminum oxide (Al-N), PC/NH_2, PC/FS, PC/Al-N/FS. The results indicated that PC/NH_2 SPE cartridge turned out to be the best for depigmenting, which could possibly be traced to the fact that PC is of homogeneous graphited regular polyhedron that is capable of absorbing pigments and reducing the background interferences to a very large extent. However, NH_2 has polarity solid phase and weak anion exchanger, which may retain chemical compounds through polar stationary phase and weak anion exchanger and is applicable to removing foreign matters, such as volatile organic acids, tea polyphenols. Therefore, the PC/NH_2 was selected as the materials packed in the SPE tube for removed pigments in the sample extraction.

After PC/NH_2 was decided to be the main filling material, Cleanert series SPE cartridges have been selected so as to further study the influences of their filling quantities and the filling sequences on the cleanup effect. The test protocol was designed in such a way that PC was placed on top layer, NH_2 on bottom layer: Cleanert PC/NH_2 (1 g), Cleanert PC/NH_2 (1.6 g), Cleanert PC/NH_2 (2 g); PC on the bottom layer and NH_2 on top layer: Cleanert NH_2/PC (1 g), Cleanert NH_2/PC (2 g). Test results are shown in Table 1.1.6. The comprehensive statistical analysis sorted on the basis of Table 1.1.6 was tabulated in Table 1.1.7, the test data of which demonstrated that the increase of the filling quantities can obviously improve the depigmenting effect. Where the recoveries of 82% compounds fell within 80%–110% when using Cleanert PC/NH_2 (1 g) for cleanup and the recoveries of 91% compounds ranged 80%–110% when using Cleanert PC/NH_2 (2 g) for cleanup. Therefore, the pigments can be basically removed at 2.0 g filling quantity and the recoveries were also satisfactory. The further increase of the filling quantity was not considered due to the cost of the method. Consequently, the influence of the filling sequences on the cleanup was investigated, and the test found that NH_2 proved to have better cleanup effect on the top layer instead of on the bottom. Where recoveries of 65% pesticides concentrated in the range of 90%–110% when using Cleanert NH_2/PC (2 g) for cleanup, recoveries of only 32% pesticides concentrated in the range of 90%–110% when using Cleanert PC/NH_2 (2 g) for cleanup, which might possibly be traced to the fact that NH_2 on the top layer can first absorb part of foreign matters and pigments and relieved the burden of PC on the lower layer, making it possible to have greater capacity to absorb big quantities of pigments in tea.

Besides pigments, there are other materials that interfere, such as tea polyphenols and caffeine, that need to be removed. The commonly used filling materials include ion exchange sorbent, C_{18}, amide polystyrene. Representative propanesulfonic acid (PRS), C18, and acetylated polystyrene as the third filling material have been selected to clean up other interfering matters in tea. The test protocol was designed in such a way that Cleanert NH_2/PRS/PC (1 g), Cleanert NH2/PRS/PC (2 g), Cleanert NH_2/C18/PC (1 g), Cleanert NH_2/C18/PC (2 g), and Cleanert TPT were used, among which the filling materials for Cleanert TPT were filled with A (PC), B (polyamine silica), and C (Amide polystyrene) on separate layers per certain proportions. Test results are shown in Table 1.1.8. The statistical analysis sorted on the basis of Table 1.1.8 was tabulated in Table 1.1.9, which demonstrated C18 had relatively low recoveries, being found unfit for the method. Where the recoveries of 86% compounds were less than 60% when using Cleanert NH_2/C18/PC (1 g) for cleanup, and ion exchange filling materials had better recoveries than that of C18 fillers, the recoveries mainly concentrated 60%–80%, while Cleanert TPT cartridge obtained the best cleanup effect, with recoveries of 94% compounds 60%–110%, of which the recoveries of 54% compounds fell 80%–110%. The recoveries proved to be ideal. The earlier test results illustrated that PC mainly

TABLE 1.1.6 The Influence of Two Components of SPE With Different Filling Materials, Different Filling Quantities and Different Filling Sequences on the Recoveries of 79 Pesticides

Tno	Pesticides	CleanertPC/NH2(2 g)	CleanertPC/NH2(1.6 g)	CleanertPC/NH2(1 g)	Cleanert NH2/PC (2 g)	Cleanert NH2/PC (1 g)	Cleanert PSA/PC (2 g)	Cleanert PSA/PC (1 g)
1	Alachlor	88.6	89.3	86.4	95.7	80.2	82.0	84.3
2	Alpha-HCH	108.4	80.3	63.9	118.9	71.7	115.7	184.8
3	Azinphos-ethyl	93.7	96.6	70.7	94.1	81.5	-	-
4	Azinphos-methyl	95.7	89.4	80.0	94.9	93.8	-	82.2
5	Benalaxyl	93.0	92.1	85.2	96.0	91.3	85.9	91.9
6	Benzoylprop-ethyl	85.5	90.2	86.5	98.4	82.5	84.5	78.9
7	Bifenox	108.6	100.6	80.6	84.5	88.9	102.9	119.1
8	Bromofos	85.9	89.2	88.1	96.4	85.2	82.1	83.2
9	Bromopropylate	84.5	88.7	86.5	101.7	85.1	76.0	81.6
10	Butachlor	88.0	90.9	87.2	95.2	86.5	80.5	84.4
11	Butylate	93.0	78.6	54.5	75.6	109.2	103.8	53.6
12	Carbofenothion	84.7	92.5	88.9	90.3	89.4	79.6	84.8
13	Chlorbenside sulfone	95.2	91.9	91.2	99.2	99.8	91.2	0
14	Chlorfenson	93.4	92.3	87.9	61.1	89.3	81.3	79.6
15	Chlorfenvinphos	105.9	99.6	81.7	109.3	83.3	85.1	73.5
16	Chloroneb	83.1	85	81.5	75.8	75.0	95.1	66.7
17	Chlorprifos-methyl	84.1	87.3	86.1	93.8	84.3	83.7	85
18	Chlorpropham	87.6	87.6	87.3	91.9	83.4	86.3	87.7
19	Chlorpropylate	85.2	91.9	83.3	96.8	85.8	80.0	84.1
20	Chlorthiophos	94.4	97.7	84.8	79.4	84.3	92.3	92.9
21	Chlozolinate	63.0	66.5	36.4	60.3	58.6	61.1	38.7
22	Cis-chlordane	87.3	91.8	87.1	94.8	84.7	81.4	83.6
23	Cis-diallate	77.1	89.8	84.2	89.6	81.2	82.9	76.1
24	Coumaphos	90.4	87.7	86.4	101.7	93.7	81.3	98.2
25	Crufomate	87.9	89.7	77.4	94.6	89.1	80.2	93.2
26	Cyanofenphos	97.7	98.1	93.4	100.4	270.7	99.4	93.4
27	Cyfluthrin	86.3	100.4	87.7	99.3	73.8	81.7	89.9
28	Desmetryn	87.0	16.6	76.5	71.7	75.3	77.3	76.1
29	Dichlobenil	75.7	60.2	77	65.2	71.2	74.3	78.2
30	Dichlofluanid	87.7	88.1	73	79.7	94.6	85.2	74.6

(Continued)

TABLE 1.1.6 The Influence of Two Components of SPE With Different Materials, Different Filling Quantities and Different Filling Sequences on the Recoveries of 79 Pesticides (Cont.)

TNo	Pesticides	CleanertPC/NH2(2 g)	CleanertPC/NH2(1.6 g)	CleanertPC/NH2(1 g)	Cleanert NH2/PC (2 g)	Cleanert NH2/PC (1 g)	Cleanert PSA/PC (2 g)	Cleanert PSA/PC (1 g)
31	Dicofol	88.0	98.0	96.2	112.7	123.6	129.4	77.2
32	Dimethachloro	86.2	89.3	87.3	99.3	84.1	82.5	83.4
33	Edifenphos	80.8	83.0	79.4	115.9	88.6	83.8	93.9
34	Endosulfan-1	95.5	105.2	89.5	93.5	97.1	100.7	76.8
35	Endosulfan-2	96.7	108.4	97.9	89.6	97.1	109.7	73.6
36	Endosulfan-sulfate	102.7	98.1	90.2	97.4	104.9	108.6	80.4
37	Endrin	90.1	86.7	87.3	96.2	88.6	85.0	90.9
38	EPN	87.8	86.2	87.8	98.8	83.8	80.3	94.3
39	EPTC	69.1	57.7	74.1	67.4	65.8	73.7	40.9
40	Ethofumesate	89.7	91.0	83.2	94.3	89.0	102.1	84.9
41	Ethoprophos	86.6	82.1	85.7	90.9	84.6	86.6	82
42	Fenarimol	92.8	84.3	86.1	82.8	86.8	79.6	78.9
43	Fenpropathrin	103.3	88.3	99.6	111.1	111.8	97.0	80.2
44	Flamprop-methyl	86.9	88.0	81.2	100.9	89.7	82.4	79.8
45	Flufenoxuron	112.3	101.3	106.5	119.1	92.9	71.4	77
46	Fluvalinate	92.0	101.2	102.5	83.4	77.0	75.8	78.4
47	Heptanophos	82.8	81.0	82.5	92.3	90.2	88.9	85.4
48	Hexachlorobenzene	77.8	81.4	80	76.7	73.3	84.9	44.6
49	Iodofenphos	86.9	86.6	84.8	98.5	85.7	81.8	93.5
50	Isofenphos	87.9	90.9	84.6	96.6	81.5	86.0	96.4
51	Isopropalin	86.1	87.4	91.4	92.8	88.2	82.6	90.6
52	Leptophos	93.4	88.6	84.8	97.7	92.2	82.0	86.1
53	Methoprene	83.7	91.3	85.2	95.1	80.8	79.2	83.1
54	Metolachlor	102	92.5	88.2	103.1	75.0	86.6	81.1
55	Mevinphos	82.4	82.2	84.5	86	69.3	87.2	87.3
56	Nitrapyrin	70.9	63.9	71.9	72.9	66.1	86.2	64.1
57	Nitrofen	85.3	87.7	85.8	96.0	87.3	80.6	94.6
58	o,p'-DDD	89.2	89.6	90.3	101.2	90.3	83.8	80.9
59	o,p'-DDT	94.9	93.8	87.7	85.2	97.6	87.8	94.1
60	Oxyflurofen	86.2	89.7	85.2	100	84.9	79.2	98.9

61	p,p'-DDE	87.3	94.7	87.2	95.2	90.0	85.5	82.5
62	p,p'-DDT	86.9	84.3	79.5	79.6	86.5	80.2	96.5
63	Pebulate	89.1	67.8	83.0	104.2	82.8	91.0	50.1
64	Phosalone	93.6	94.2	96.6	98.2	89.2	84.0	80.5
65	Pirimiphos-ethyl	84.6	16.5	87.4	98.2	84.5	81.1	83.9
66	Profenofos	84.5	88.9	83.7	98.7	85.5	82.6	93.2
67	Profluralin	84.9	89.8	86.3	93.3	82.1	83.1	88.4
68	Propachlor	84.3	88.9	83.3	90.7	89.1	81.8	67.4
69	Propanil	88.6	91.5	82.7	98.3	97.4	81.2	68.9
70	Sulfallate	83.2	85.4	83.1	80	79.6	91.5	78.7
71	Sulfotep	85.7	87.9	87.1	90.1	81.5	85.4	78.5
72	Tecnazene	81.3	83.1	83.1	77.5	77.6	88.9	70.9
73	Terbufos	102.9	85.4	85	84.1	85.1	93.6	101.1
74	Terbuthylazine	93.5	53.7	92.8	96.7	98.8	98.0	83.3
75	Tetrachlorvinphos	86.4	87.5	84.4	109.6	86.3	84.6	102
76	Thiobencarb	88.1	90.7	85.5	96.1	82.8	80.8	80.6
77	Trans-diallate	89.6	86.0	83	84.3	83.7	91.8	76.5
78	Triazophos	88.3	80.8	112.8	107	130	105.1	-
79	Trifluralin	84.8	80.7	84.9	91.8	81.3	95.4	83.2

TABLE 1.1.7 A Comprehensive Statistical Analysis of the Influence of Two Components of SPE With Different Filling Materials, Different Filling Quantities and Different Filling Sequences on Cleanup Efficiencies (Sorted on the Basis of Table 1.1.6)

Rec. range	CleanertPC/ NH2(2 g)	CleanertPC/ NH_2 (1.6 g)	CleanertPC/ NH2(1 g)	Cleanert NH_2/PC (2 g)	Cleanert NH_2/PC (1 g)	Cleanert PSA/PC (2 g)	Cleanert PSA/PC (1 g)
<60	0	4	2	0	1	2	9
60–80	6	5	11	13	14	12	21
80–90	47	41	53	10	45	45	29
90–110	25	29	12	51	16	18	18
>110	1	0	1	5	4	2	2

TABLE 1.1.8 The Influence of Three Components of SPE With Different Filling Materials, Different Filling Quantities and Different Filling Sequences on the Recoveries of 79 Pesticides

No	Pesticides	Cleanert TPT	Cleanert NH2/ PRS/PC (2 g)	Cleanert NH2/ PRS/PC (1 g)	CleanertNH2/ C18/PC (2 g)	Cleanert NH2/ C18/PC (1 g)
1	Alachlor	81.6	63.8	74.8	60.7	55.5
2	α-HCH	79.9	71.4	104.1	56.5	50.1
3	Azinphos-Ethyl	80.1	–	–	–	–
4	Azinphos-Methyl	78.8	–	–	–	–
5	Benalaxyl	80.2	73.5	72.5	65.4	47.5
6	Benzoylprop-Ethyl	81.6	69.3	75.0	58.7	52.8
7	Bifenox	85.9	70.3	65.3	76.9	59.2
8	Bromofos	78.0	67.0	73.5	59.7	52.7
9	Bromopropylate	81.1	72.7	71.0	62.5	48.9
10	Butachlor	80.5	70.7	75.2	62.7	52.5
11	Butylate	63.9	75.9	77.9	61.6	67.1
12	Carbofenothion	81.5	66.5	69.1	60.4	52.4
13	Chlorbenside Sulfone	81.9	122.2	80.5	67.5	72.1
14	Chlorfenson	80.7	68.6	73.4	66.2	47.2
15	Chlorfenvinphos	78.4	92.4	114.4	66.4	44.9
16	Chloroneb	69.4	127.5	77.6	65.5	79.8
17	Chlorprifos-Methyl	79.4	70.6	74.1	61.9	52.2
18	Chlorpropham	82.9	69.3	71.9	62.4	53.2
19	Chlorpropylate	80.1	76.1	73.4	62.9	51.2
20	Chlorthiophos	81.9	74.8	104.4	67.7	47.1
21	Chlozolinate	60.1	42.7	61.6	53.9	39.7
22	Cis-Chlordane	79.3	67.7	75.9	63.7	51.8
23	Cis-Diallate	77.1	69.8	75.7	59.2	49.3
24	Coumaphos	78.1	70.5	71.7	53.2	56.9
25	Crufomate	79.6	67.5	66.4	60.4	51.8
26	Cyanofenphos	81.5	76.1	73.1	63.3	60.5
27	Cyfluthrin	64.1	72.3	67.0	60.3	37.0
28	Desmetryn	78.5	13.7	73.0	62.0	13.7
29	Dichlobenil	63.1	45.2	57.9	56.6	42.1

TABLE 1.1.8 The Influence of Three Components of SPE With Different Filling Materials, Different Filling Quantities and Different Filling Sequences on the Recoveries of 79 Pesticides (*Cont.*)

No	Pesticides	Cleanert TPT	Cleanert NH2/ PRS/PC (2 g)	Cleanert NH2/ PRS/PC (1 g)	CleanertNH2/ C18/PC (2 g)	Cleanert NH2/ C18/PC (1 g)
30	Dichlofluanid	34.2	84.4	76.0	75.8	63.6
31	Dicofol	113.4	115.6	79.6	66.6	65.0
32	Dimethachloro	80.1	69.1	71.5	63.3	51.2
33	Edifenphos	84.8	67.4	64.0	65.4	51.0
34	Endosulfan-1	86.1	85.9	82.9	87.2	62.7
35	Endosulfan-2	0.0	89.2	89.4	104.6	63.7
36	Endosulfan-Sulfate	97.6	94.8	87.5	89.9	64.7
37	Endrin	78.8	69.1	76.0	61.2	50.6
38	EPN	86.2	73.8	72.3	58.0	90.6
39	EPTC	48.8	41.5	52.4	53.0	44.9
40	Ethofumesate	79.0	64.9	71.4	60.6	55.8
41	Ethoprophos	79.2	62.1	65.1	60.8	50.9
42	Fenarimol	79.3	51.8	76.2	59.2	49.4
43	Fenpropathrin	92.0	82.0	95.5	71.2	58.3
44	Flamprop-Methyl	80.3	71.8	73.6	62.7	52.7
45	Flufenoxuron	84.2	62.5	75.6	52.7	48.6
46	Fluvalinate	84.3	63.2	69.3	55.1	49.4
47	Heptanophos	82.0	71.0	74.0	63.9	53.5
48	Hexachlorobenzene	34.5	63.0	68.4	57.8	46.1
49	Iodofenphos	81.1	70.8	72.6	58.7	55.7
50	Isofenphos	80.2	70.9	82.0	60.4	66.6
51	Isopropalin	82.5	69.3	70.5	56.4	56.1
52	Leptophos	82.8	72.8	69.4	58.8	57.4
53	Methoprene	78.5	66.2	73.4	63.0	49.1
54	Metolachlor	81.0	72.6	74.4	62.5	52.3
55	Mevinphos	68.4	75.5	73.5	62.1	47.5
56	Nitrapyrin	60.3	44.5	67.3	63.1	52.0
57	Nitrofen	84.2	67.0	71.5	53.1	56.0
58	o,p′-DDD	83.6	67.4	73.3	64.8	49.2
59	o,p′-DDT	80.5	75.7	73.9	56.0	56.3
60	Oxyflurofen	82.9	67.1	69.8	55.3	54.9
61	p,p′-DDE	86.8	72.5	75.6	64.1	52.2
62	p,p′-DDT	81.7	66.9	70.2	50.4	53.3
63	Pebulate	61.6	–	–	–	–
64	Phosalone	88.5	133.1	84.1	67.9	80.3
65	Pirimiphos-Ethyl	80.5	9.8	73.3	62.3	8.6
66	Profenofos	75.0	68.1	70.3	63.1	51.0
67	Profluralin	81.6	66.7	71.6	58.4	52.0
68	Propachlor	79.3	75.6	66.7	56.2	46.9
69	Propanil	80.3	70.8	70.3	95.4	50.0
70	Sulfallate	75.7	66.3	70.7	61.3	51.0
71	Sulfotep	73.3	67.5	71.2	61.6	51.4

(*Continued*)

TABLE 1.1.8 The Influence of Three Components of SPE With Different Filling Materials, Different Filling Quantities and Different Filling Sequences on the Recoveries of 79 Pesticides (*Cont.*)

No	Pesticides	Cleanert TPT	Cleanert NH2/ PRS/PC (2 g)	Cleanert NH2/ PRS/PC (1 g)	CleanertNH2/ C18/PC (2 g)	Cleanert NH2/ C18/PC (1 g)
72	Tecnazene	71.5	63.9	69.5	59.7	49.5
73	Terbufos	78.2	66.0	71.4	70.3	50.1
74	Terbuthylazine	80.6	49.1	82.6	0.0	33.1
75	Tetrachlorvinphos	81.7	70.2	70.5	66.4	54.0
76	Thiobencarb	81.7	65.9	67.7	66.9	45.2
77	Trans-Diallate	78.9	67.0	73.0	57.5	50.5
78	Triazophos	105.0	87.0	79.5	61.1	53.8
79	Trifluralin	79.9	66.7	72.0	59.9	51.0

TABLE 1.1.9 A Comprehensive Statistical Analysis of the Influence of Three Components of SPE With Different Filling Materials, Different Filling Quantities and Different Filling Sequences on Cleanup Efficiencies (Sorted on the Basis of Table 1.1.8)

Rec range	Cleanert TPT	Cleanert NH2/ PRS/PC (2 g)	Cleanert NH2/ PRS/PC(1 g)	Cleanert NH2/ C18/PC(2 g)	Cleanert NH2/ C18/PC(1 g)
<60%	4	12	5	29	67
60%–80%	31	57	63	46	10
80%–110%	43	6	10	4	2
>110%	1	4	1	0	0

functions as removing the pigments in tea instead of absorbing the target compounds; polyamine silica removes foreign matters like volatile organic acids, tea polyphenol. Amide polystyrene mainly serves as removing foreign matters except for the pigments and theophylline. The combination of these three filling materials has realized the highly efficient cleanup of tea samples, while Cleanert TPT may serve as the exclusive cleanup cartridge for tea samples.

1.1.3.5 Evaluation of Method Efficiency—LOD, LOQ, Recovery and Relative Standard Deviations of Both Gas Chromatography–Mass Spectrometry and Liquid Chromatography–Tandem Mass Spectrometry

The LOD and LOQ of the GC-MS and LC–MS/MS methods were obtained with control samples fortified with 490 pesticides for GC–MS and 448 pesticides for LC–MS/MS at different concentrations. The results are listed in Tables 1.1.10–1.1.12. A good linear relationship was shown in the linear regression analysis for the analysis. The linear correlation coefficient for 96% pesticides determined by GC–MS method is $r \geq 0.980$; the linear correlation coefficient for 90% pesticides determined by LC–MS-MS method is $r \geq 0.980$. The fortified concentration at signal/noise ratio ≥ 5 for each pesticide is fixed as LOD of the method while the fortified concentration at signal/noise ratio ≥ 10 is fixed as LOQ of the method. LOD of the GC–MS method is 1.0–500 μg/kg, LOQ 2.0–1000 μg/kg, LOD of the LC–MS/MS method 0.03–4820 μg/kg and LOQ 0.06–9640 μg/kg. LOD comparison data of both GC–MS and LC–MS/MS are shown in Table 1.1.13. It can be seen from Table 1.1.13 that there are 482 pesticides with LOD ≤ 100 μg/kg for GC–MS method, accounting for 98% of the pesticides tested and there are 417 pesticides for LC–MS/MS, accounting for 93% of the pesticides analyzed; there are 264 pesticides with LOD ≤ 10 μg/kg for GC–MS method, accounting for 54% of the pesticides tested, 325 pesticides for LC–MS/MS method, making up 73% of the pesticides analyzed. There are 270 pesticides that can be analyzed by both GC–MS and LC–MS/MS. There are 264 pesticides with LOD ≤ 100 μg/kg for GC–MS method, accounting for 98% of the pesticides tested, 247 for LC–MS/MS, making up 91% of the pesticides analyzed; there are, however, 133 pesticides with LOD ≤ 10 μg/kg for GC–MS method, accounting for 49% of the pesticides tested, 200 pesticides for LC–MS/MS

TABLE 1.1.10 The LOD, LOQ and Linearity of 270 Pesticides Applicable for Both LC–MS/MS and GC–MS

No.	Pesticides	LC–MS/MS			GC–MS		
		LOD µg/kg	LOQ µg/kg	Correlation coefficient(r)	LOQ µg/kg	LOD µg/kg	Correlation coefficient(r)
1	2,4-D	5.93	11.86	0.9913	200.00	100.00	0.9930
2	2,6-Dichlorobenzamide	2.25	4.50	0.9914	20.00	10.00	0.9983
3	2-Phenylphenol	84.94	169.88	0.9965	25.00	12.50	0.9946
4	3-Phenylphenol	2.00	4.00	0.9950	60.00	30.00	0.9990
5	4,4-Dichlorobenzophenone	6.80	13.60	0.9940	10.00	5.00	0.9959
6	Acetochlor	23.70	47.40	0.9966	50.00	25.00	1.0000
7	Acibenzolar-s-methyl	1.54	3.08	0.9998	50.00	25.00	
8	Aclonifen	12.10	24.20	0.9988	500.00	250.00	0.9940
9	Allethrin	30.20	60.40	0.9941	40.00	20.00	0.9980
10	Allidochlor	20.52	41.04	0.9951	20.00	10.00	0.9986
11	Ametryn	0.48	0.96	0.9997	30.00	15.00	0.9989
12	Anilofos	0.36	0.71	0.7724	20.00	10.00	0.9954
13	Aspon	0.87	1.73	0.9908	20.00	10.00	0.9994
14	Atratone	0.09	0.18	0.9999	25.00	12.50	0.9988
15	Atrazine-desethyl	0.31	0.62	0.9979	10.00	5.00	0.9956
16	Azinphos ethyl	54.46	108.93	0.9925	50.00	25.00	0.9930
17	Azinphos-methyl	552.17	1104.33	0.9969	150.00	75.00	0.9971
18	Aziprotryne	0.69	1.38	0.9965	80.00	40.00	0.9979
19	Benalyxyl	0.62	1.24	0.9997	10.00	5.00	0.9972
20	Benodanil	1.74	3.48	0.9998	30.00	15.00	0.9899
21	Benoxacor	3.45	6.90	0.9997	50.00	25.00	0.9990
22	Benzoylprop-ethyl	154.00	308.00	0.9927	30.00	15.00	0.9983
23	Bifenazate	11.40	22.80	0.9947	80.00	40.00	0.9990
24	Bioresmethrin	3.71	7.42	0.9950	20.00	10.00	0.9990
25	Bitertanol	16.70	33.40	0.9948	30.00	15.00	0.9931
26	Boscalid	2.38	4.76	0.9990	40.00	20.00	
27	Bromacil	11.80	23.60	0.9989	50.00	25.00	0.9990
28	Bromfenvinfos	1.51	3.02	0.9963	10.00	5.00	0.9937
29	Bromophos-ethyl	283.85	567.69	0.9941	10.00	5.00	0.9973
30	Bupirimate	0.35	0.70	0.9993	10.00	5.00	0.9954

(Continued)

TABLE 1.1.10 The LOD, LOQ and Linearity of 270 Pesticides Applicable for Both LC–MS/MS and GC–MS (*Cont.*)

No.	Pesticides	LC–MS/MS			GC–MS		
		LOD µg/kg	LOQ µg/kg	Correlation coefficient(*r*)	LOQ µg/kg	LOD µg/kg	Correlation coefficient(*r*)
31	Buprofezin	0.44	0.88	0.9994	20.00	10.00	0.9987
32	Butachlor	10.03	20.07	0.9937	20.00	10.00	0.9970
33	Butafenacil	4.75	9.50	0.9976	10.00	5.00	0.9860
34	Butralin	0.95	1.90	1.0000	40.00	20.00	0.9940
35	Butylate	151.00	302.00	0.9927	30.00	15.00	0.9994
36	Cadusafos	0.58	1.15	0.9971	40.00	20.00	0.9990
37	Carbaryl	5.16	10.32	0.9937	30.00	15.00	0.9990
38	Carbosulfan	0.80	0.40		30.00	15.00	0.9987
39	Carboxin	0.28	0.56	0.9919	30.00	15.00	0.9930
40	Chlorbenside sulfone	0.80	0.40		20.00	10.00	0.9966
41	Chlorbufam	91.50	183.00	0.9941	50.00	25.00	0.9921
42	Chlordimeform	0.67	1.33	0.9992	10.00	5.00	0.9966
43	Chlorfenethol	82.15	164.30	0.9236	10.00	5.00	0.9975
44	Chlorfluazuron	4.34	8.68	0.9958	10.00	5.00	0.9980
45	Chlormephos	224.00	448.00	0.9989	20.00	10.00	0.9928
46	Chlorpropham	7.88	15.77	0.9956	10.00	5.00	0.9972
47	Chlorthiophos	15.90	31.80	0.9956	30.00	15.00	0.9978
48	Clodinafop propargyl	1.22	2.44	0.9949	20.00	10.00	0.9820
49	Clomazone	0.21	0.42	0.9961	10.00	5.00	0.9976
50	Cloquintocet mexyl	0.94	1.88	0.9986	100.00	50.00	0.9915
51	Crimidine	0.78	1.56	0.9994	10.00	5.00	0.9987
52	Crufomate	0.26	0.52	0.9995	60.00	30.00	0.9914
53	Cyanazine	0.08	0.16	0.9998	30.00	15.00	0.9952
54	Cyanofenphos	10.40	20.80	0.9941	10.00	5.00	0.9967
55	Cyanohos	0.00	0.00	0.2539	20.00	10.00	0.9980
56	Cycloate	2.22	4.44	0.9971	10.00	5.00	0.9986
57	Cycluron	0.10	0.21	0.9994	30.00	15.00	0.9982
58	Cyphenothrin	8.40	16.80	0.9981	30.00	15.00	0.9190

59	Cyprazine	0.03	0.06	0.9997	10.00	5.00	0.9978
60	Cyproconazole	0.37	0.73	0.9995	25.00	12.50	0.9983
61	Cyprodinil	0.37	0.74	0.9997	10.00	5.00	0.9977
62	DEF	0.81	1.61	0.9994	20.00	10.00	0.9971
63	Demeton-s	40.00	80.00	0.0015	40.00	20.00	0.9700
64	Demeton-s-methyl	2.65	5.30	0.9903	40.00	20.00	0.9990
65	Dialifos	78.50	157.00	0.9900	800.00	400.00	
66	Diazinon	0.36	0.71	1.0000	10.00	5.00	0.9980
67	Dibutyl succinate	111.20	222.40	0.9933	20.00	10.00	1.0000
68	Dicapthon	0.12	0.24	0.3815	50.00	25.00	0.9946
69	Dichlofenthion	14.98	29.95	0.9967	10.00	5.00	0.9986
70	Dichlofluanid	1.30	2.60	0.7612	60.00	30.00	0.9977
71	Diclobutrazole	0.23	0.47	0.9996	40.00	20.00	0.9957
72	Dicloran	24.28	48.56	0.9990	20.00	10.00	0.9963
73	Dicrotophos	0.57	1.14	0.9985	80.00	40.00	0.9990
74	Diethofencarb	1.00	2.00	0.9995	60.00	30.00	0.9977
75	Diethyltoluamide	0.28	0.55	0.9998	8.00	4.00	0.9970
76	Diflufenican	14.14	28.27	0.9977	10.00	5.00	0.9950
77	Dimefox	34.10	68.20	0.9963	30.00	15.00	0.9992
78	Dimefuron	2.00	4.00	0.9984	10.00	5.00	0.9885
79	Dimethachloro	0.95	1.90	0.9950	30.00	15.00	0.9989
80	Dimethenamid	2.15	4.30	0.9980	10.00	5.00	0.9990
81	Dimethoate	3.80	7.60	0.9868	40.00	20.00	
82	Dimethomorph	0.18	0.35	0.8778	20.00	10.00	0.9980
83	Dimethyl phthalate	6.60	13.20		40.00	20.00	0.9330
84	Diniconazole	0.67	1.34	0.9997	10.00	5.00	0.9936
85	Dinitramine	0.90	1.79	0.9991	40.00	20.00	0.9956
86	Dioxacarb	1.68	3.36	0.9939	80.00	40.00	
87	Diphenamid	0.07	0.14	0.9999	10.00	5.00	0.9989
88	Dipropetryn	0.14	0.27	1.0000	10.00	5.00	0.9976
89	Disulfoton	234.85	469.70	0.9934	10.00	5.00	0.9983
90	Disulfoton sulfone	1.23	2.46	0.9978	50.00	25.00	0.9966
91	Disulfoton-sulfoxide	1.42	2.84	0.4886	20.00	10.00	0.9993
92	Edifenphos	0.38	0.75	0.9951	20.00	10.00	0.9912

(Continued)

TABLE 1.1.10 The LOD, LOQ and Linearity of 270 Pesticides Applicable for Both LC–MS/MS and GC–MS (Cont.)

No.	Pesticides	LC–MS/MS			GC–MS		
		LOD µg/kg	LOQ µg/kg	Correlation coefficient(r)	LOQ µg/kg	LOD µg/kg	Correlation coefficient(r)
93	EPN	16.50	33.00	0.9947	40.00	20.00	0.9913
94	EPTC	18.67	37.34	0.9990	30.00	15.00	0.9993
95	Esfenvalerate	208.00	416.00	0.2962	40.00	20.00	0.9983
96	Ethiofencarb	2.46	4.92	0.9981	100.00	50.00	0.9870
97	Ethion	1.48	2.96	0.9839	20.00	10.00	0.9944
98	Ethofume sate	186.00	372.00	0.9969	20.00	10.00	0.9992
99	Ethoprophos	1.38	2.76	0.9992	30.00	15.00	0.9982
100	Etofenprox	1140.14	2280.28	0.9833	25.00	12.50	0.9961
101	Etridiazol	50.21	100.42	0.9958	30.00	15.00	0.9945
102	Etrimfos	9.38	18.76	0.9995	10.00	5.00	0.9976
103	Fenamiphos	0.10	0.21	0.9785	30.00	15.00	0.9864
104	Fenamiphos sulfone	0.22	0.45	0.9921	40.00	20.00	0.9954
105	Fenamiphos sulfoxide	0.37	0.74	0.9970	100.00	50.00	0.9561
106	Fenarimol	0.30	0.61	0.9997	20.00	10.00	0.9993
107	Fenazaquin	0.16	0.32	0.9909	25.00	12.50	0.9950
108	Fenbuconazole	0.82	1.65	0.9998	50.00	25.00	0.9896
109	Fenhexamid	0.47	0.95	0.9694	500.00	250.00	0.9940
110	Fenitrothion	13.40	26.80	0.9998	20.00	10.00	0.9957
111	Fenobucarb	2.95	5.90	0.9984	30.00	15.00	0.9982
112	Fenoxanil	19.70	39.40	0.9942	20.00	10.00	0.9960
113	Fenpropathrin	122.50	245.00	0.9926	20.00	10.00	0.9966
114	Fenpropidin	0.09	0.18	0.9996	50.00	25.00	0.9970
115	Fenpropimorph	0.09	0.18	1.0000	20.00	10.00	0.9989
116	Fenpyroximate	0.68	1.36	0.9991	10.00	5.00	0.9960
117	Fensulfothin	1.00	2.00	0.9980	50.00	25.00	0.9900
118	Fenthion	26.00	52.00	0.9952	10.00	5.00	0.9973
119	Fenthion sulfone	8.73	17.46	0.9934	40.00	20.00	0.9839
120	Fenthion sulfoxide	0.16	0.31	0.9997	100.00	50.00	0.9863
121	Flamprop isopropyl	0.22	0.43	0.9999	10.00	5.00	0.9974
122	Fluazifop butyl	0.13	0.26	1.0000	10.00	5.00	0.9961

123	Fluchloralin	244.00	488.00	0.9922	40.00	20.00	0.9952
124	Fludioxonil	31.08	62.16	0.9972	10.00	5.00	0.9963
125	Flufenacet	2.65	5.30	0.9996	200.00	100.00	0.9990
126	Flufenoxuron	1.58	3.17	0.9978	30.00	15.00	0.9929
127	Flumiclorac-pentyl	5.30	10.61	0.9976	20.00	10.00	0.9903
128	Fluorochloridone	6.89	13.78	0.9965	20.00	10.00	0.9994
129	Fluoroglycofen-ethyl	2.50	5.00	0.9986	120.00	60.00	0.9826
130	Flurochloridone	0.65	1.29	0.9916	20.00	10.00	0.9979
131	Flurtamone	0.22	0.44	0.9996	50.00	25.00	0.9330
132	Flusilazole	0.29	0.58	0.9999	10.00	5.00	0.9983
133	Flutolanil	0.57	1.15	0.9989	10.00	5.00	0.9951
134	Flutriafol	4.29	8.58	0.9983	20.00	10.00	0.9985
135	Fonofos	3.73	7.46	0.9976	10.00	5.00	0.9975
136	Fuberidazole	0.95	1.89	0.9907	50.00	25.00	
137	Furalaxyl	0.39	0.77	0.9936	20.00	10.00	1.0000
138	Heptanophos	2.92	5.84	0.9971	30.00	15.00	0.9979
139	Hexaflumuron	12.60	25.20	0.9975	60.00	30.00	
140	Hexazinone	0.06	0.12	0.9992	30.00	15.00	0.9970
141	Hexythiazox	11.80	23.60	0.9950	80.00	40.00	0.9981
142	Imazalil	1.00	2.00	0.9994	40.00	20.00	0.9980
143	Imazamethabenz-methyl	0.08	0.16	0.9986	30.00	15.00	0.9960
144	Imibenzonazole-des-benzyl	3.11	6.22	0.9993	40.00	20.00	
145	Iprobenfos	4.14	8.28	0.9947	30.00	15.00	0.9895
146	Isazofos	0.09	0.18	0.9988	20.00	10.00	0.9994
147	Isocarbamid	0.85	1.70	0.9960	50.00	25.00	0.9965
148	Isofenphos	109.34	218.67	0.9659	20.00	10.00	0.9982
149	Isomethiozin	0.53	1.07	0.9988	20.00	10.00	0.9970
150	Isopropalin	15.00	30.00	0.9906	20.00	10.00	
151	Isoprothiolane	0.92	1.85	0.9967	20.00	10.00	0.9990
152	Kresoxim-methyl	50.29	100.58	0.9850	20.00	10.00	0.9976
153	Lactofen	31.00	62.00	0.9905	80.00	40.00	0.9930
154	Linuron	5.82	11.63	0.9999	40.00	20.00	0.9730
155	Malaoxon	2.34	4.69	0.2845	50.00	25.00	0.9880

(Continued)

TABLE 1.1.10 The LOD, LOQ and Linearity of 270 Pesticides Applicable for Both LC–MS/MS and GC–MS (Cont.)

No.	Pesticides	LC–MS/MS			GC–MS		
		LOD μg/kg	LOQ μg/kg	Correlation coefficient(r)	LOQ μg/kg	LOD μg/kg	Correlation coefficient(r)
156	Malathion	2.82	5.64	0.9935	40.00	20.00	0.9974
157	Mecarbam	9.80	19.60	0.9934	40.00	20.00	0.9986
158	Mefenacet	1.10	2.21	0.9998	30.00	15.00	0.9928
159	Mefenoxam	0.77	1.54	0.9955	20.00	10.00	
160	Mefenpyr-diethyl	6.28	12.56	0.9918	30.00	15.00	1.0000
161	Mephosfolan	1.16	2.32	0.9981	20.00	10.00	0.9890
162	Mepronil	0.19	0.38	0.9981	25.00	12.50	0.9941
163	Metalaxyl	0.25	0.50	0.9986	30.00	15.00	0.9982
164	Metamitron	3.18	6.36	0.9993	100.00	50.00	0.9994
165	Metazachlor	0.49	0.98	0.9994	30.00	15.00	0.9984
166	Methabenzthiazuron	0.04	0.07	0.9999	100.00	50.00	0.9930
167	Methacrifos	1211.85	2423.70	0.9936	25.00	12.50	0.9994
168	Methidathion	5.33	10.66	0.9947	50.00	25.00	0.9948
169	Methoprotryne	0.12	0.24	0.9999	40.00	20.00	0.9975
170	Metolachlor	0.20	0.39	0.9987	10.00	5.00	0.9977
171	Metribuzin	0.27	0.54	0.9652	30.00	15.00	0.9986
172	Mevinphos	0.78	1.57	0.9995	20.00	10.00	0.9957
173	Mexacarbate	0.47	0.94	0.9997	30.00	15.00	0.9924
174	Molinate	1.05	2.10	0.9994	10.00	5.00	0.9996
175	Monalide	0.60	1.20	0.9992	20.00	10.00	0.9988
176	Monolinuron	1.78	3.56	0.9994	40.00	20.00	0.9888
177	Myclobutanil	0.50	1.00	0.9991	10.00	5.00	0.9958
178	Napropamide	0.64	1.27	0.9935	30.00	15.00	0.9963
179	Nitralin	17.20	34.40	0.9975	100.00	50.00	0.9907
180	Norflurazon	0.13	0.26	0.9997	10.00	5.00	0.9879
181	Nuarimol	0.50	1.00	0.9998	20.00	10.00	0.9962
182	Ofurace	0.50	1.00	0.9933	30.00	15.00	0.9990
183	Oxamyl	274.03	548.06	0.9832	0.00	0.00	
184	Oxyflurofen	29.27	58.55	0.9883	40.00	20.00	0.9945
185	Paclobutrazol	0.29	0.57	0.9999	30.00	15.00	0.9941

186	Paraoxon-ethyl	0.24	0.47	0.9986	320.00	160.00	0.9892
187	Pebulate	1.70	3.40	0.9991	30.00	15.00	0.9994
188	Penconazole	1.00	2.00	0.9996	30.00	15.00	0.9992
189	Pencycuron	0.14	0.27	0.9995	40.00	20.00	0.9980
190	Pentachloroaniline	1.87	3.74	0.9035	10.00	5.00	0.9990
191	Phenothrin	169.60	339.20	0.9957	10.00	5.00	0.9950
192	Phenthoate	46.18	92.35	0.9907	20.00	10.00	0.9970
193	Phorate	157.00	314.00	0.9987	10.00	5.00	0.9969
194	Phorate sulfone	21.00	42.00	0.9955	10.00	5.00	0.9946
195	Phosalone	24.02	48.04	0.9946	20.00	10.00	0.9955
196	Phosphamidon	1.94	3.88	0.9986	20.00	10.00	0.9770
197	Phthalic acid,benzyl butyl ester	316.00	632.00	0.9949	10.00	5.00	0.9973
198	Phthalimide	21.50	43.00	0.9988	50.00	25.00	0.9980
199	Picoxystrobin	4.22	8.44	0.9936	20.00	10.00	0.9990
200	Piperonyl butoxide	0.57	1.13	0.9952	30.00	15.00	0.9919
201	Piperophos	4.62	9.24	0.9952	30.00	15.00	0.9990
202	Pirimicarb	0.08	0.15	0.9998	20.00	10.00	1.0000
203	Pirimiphos methyl	0.10	0.20	0.9999	20.00	10.00	0.9980
204	Pirimiphos-ethyl	0.03	0.06	0.9982	10.00	5.00	0.9977
205	Pretilachlor	0.17	0.33	0.9985	75.00	37.50	0.9980
206	Procymidone	43.30	86.60	0.9863	10.00	5.00	0.9979
207	Profenefos	1.01	2.02	0.9990	60.00	30.00	0.9969
208	Prometon	0.07	0.13	0.9996	30.00	15.00	0.9988
209	Prometryne	0.08	0.16	0.9992	10.00	5.00	0.9961
210	Pronamide	7.69	15.38	0.9978	10.00	5.00	0.9905
211	Propachlor	0.14	0.27	0.9998	30.00	15.00	0.9998
212	Propanil	10.80	21.59	0.9994	20.00	10.00	0.9930
213	Propargite	34.30	68.60	0.9909	40.00	20.00	0.9995
214	Propazine	0.16	0.32	0.9998	10.00	5.00	0.9988
215	Propetamphos	27.00	54.00	0.9997	10.00	5.00	0.9968
216	Propham	55.00	110.00	0.9974	10.00	5.00	0.9975
217	Propisochlor	0.40	0.80	0.9955	10.00	5.00	0.9970
218	Prosulfocarb	0.18	0.37	0.9997	10.00	5.00	0.9988

(Continued)

TABLE 1.1.10 The LOD, LOQ and Linearity of 270 Pesticides Applicable for Both LC–MS/MS and GC–MS (Cont.)

No.	Pesticides	LC–MS/MS			GC–MS		
		LOD µg/kg	LOQ µg/kg	Correlation coefficient(r)	LOQ µg/kg	LOD µg/kg	Correlation coefficient(r)
219	Pyraclofos	0.50	1.00	0.9996	80.00	40.00	0.9620
220	Pyraclostrobin	0.25	0.51	0.9993	300.00	150.00	0.9910
221	Pyrazophos	0.81	1.62	0.9991	20.00	10.00	0.9904
222	Pyributicarb	0.17	0.34	0.9996	50.00	25.00	0.9990
223	Pyridaphenthion	0.44	0.87	0.9998	10.00	5.00	0.9841
224	Pyrifenox	0.13	0.27	0.9953	80.00	40.00	0.9990
225	Pyriftalid	0.31	0.62	0.9997	25.00	12.50	0.9980
226	Pyrimethanil	0.34	0.68	1.0000	10.00	5.00	0.9980
227	Pyrimidifen	7.00	14.00	0.8789	50.00	25.00	0.9720
228	Pyriproxyfen	0.22	0.43	0.9904	10.00	5.00	0.9960
229	Pyroquilon	1.74	3.48	0.9922	25.00	12.50	1.0000
230	Quinalphos	1.00	2.00	0.9976	10.00	5.00	0.9959
231	Quinoclamine	3.96	7.92	0.9997	40.00	20.00	0.9980
232	Quinoxyphen	76.70	153.40	0.9871	10.00	5.00	0.9990
233	Rabenzazole	0.67	1.33	0.9999	10.00	5.00	0.9931
234	Resmethrin-2	0.15	0.30	0.9984	50.00	25.00	0.9953
235	Ronnel	6.57	13.13	0.9962	20.00	10.00	0.9978
236	Sebutylazine	0.16	0.31	0.9998	10.00	5.00	0.9989
237	Secbumeton	0.04	0.07	0.9998	10.00	5.00	0.9952
238	Silafluofen,	304.00	608.00	0.1262	1800.00	900.00	0.9868
239	Simeconazole	1.47	2.94	0.9998	20.00	10.00	1.0000
240	Simeton	0.55	1.10	0.9954	20.00	10.00	0.9984
241	Spirodiclofen	4.95	9.91	0.9952	200.00	100.00	0.9950
242	Sulfallate	103.60	207.20	0.9962	20.00	10.00	0.9958
243	Sulfotep	1.30	2.60	0.9945	10.00	5.00	0.9991
244	Sulprofos	2.92	5.84	0.9976	20.00	10.00	0.9959
245	Tebuconazole	1.12	2.23	0.9998	75.00	37.50	0.9941
246	Tebufenpyrad	0.13	0.25	0.9930	10.00	5.00	0.9965
247	Tebupirimfos	0.06	0.13	0.9946	80.00	40.00	0.9980

248	Tebutam	0.07	0.14	0.9995	20.00	10.00	0.9990
249	Tebuthiuron	0.11	0.22	0.9999	20.00	10.00	0.9960
250	Terbacil	0.44	0.88	0.9990	20.00	10.00	0.9977
251	Terbufos	1120.00	2240.00	0.8747	20.00	10.00	0.9886
252	Terbuthylazine	0.23	0.47	0.9999	25.00	12.50	0.9975
253	Tetrachlorvinphos	1.11	2.22	0.9998	30.00	15.00	0.9992
254	Tetraconazole	0.86	1.72	0.9996	30.00	15.00	0.9928
255	Tetramethrin	0.91	1.82	0.9947	25.00	12.50	1.0000
256	Thenylchlor	12.07	24.14	0.9918	20.00	10.00	0.9990
257	Thiazopyr	0.98	1.96	0.9981	20.00	10.00	0.9987
258	Thiobencarb	1.65	3.30	0.9971	20.00	10.00	0.9967
259	Thiometon	289.00	578.00	0.9975	10.00	5.00	0.9969
260	Thionazin	11.34	22.68	0.9990	10.00	5.00	0.9991
261	Tolclofos methyl	33.28	66.56	0.9947	10.00	5.00	0.9911
262	Trans-permethin	2.40	4.80	0.9773	25.00	12.50	0.9976
263	Triadimefon	3.94	7.88	0.9998	20.00	10.00	0.9993
264	Triallate	23.10	46.20	0.9960	30.00	15.00	0.9944
265	Triazophos	0.34	0.68	0.9998	30.00	15.00	0.9987
266	Trietazine	0.30	0.60	0.9998	10.00	5.00	0.9990
267	Trifloxystrobin	1.00	2.00	0.9991	40.00	20.00	0.9930
268	Trifluralin	167.40	334.80	0.9958	20.00	10.00	0.9972
269	Tri-n-butyl phosphate	0.19	0.37	0.9983	20.00	10.00	0.9973
270	Vinclozolin	1.27	2.54	0.9985	10.00	5.00	

TABLE 1.1.11 The LOD, LOQ and Linearity of 178 Pesticides Applicable for LC–MS/MS

No	Pesticides	LOD μg/kg	LOQ μg/kg	Corelation coefficient (r)
1	1-Naphthy acetamide	0.41	0.81	0.9989
2	3,4,5-Trimethacarb	0.17	0.34	0.9988
3	4-Aminopyridine	0.43	0.87	1.0000
4	6-Chloro-4-hydroxy-3-phenyl-pyrid-azin	0.83	1.65	0.9484
5	Acetamiprid	0.72	1.44	0.9969
6	Acrylamide	8.90	17.80	0.9938
7	Aldicarb	130.50	261.00	0.9792
8	Aldicarb sulfone	10.68	21.36	0.9906
9	Aldimorph	1.58	3.16	0.9902
10	Amidithion	95.20	190.40	0.9963
11	Aminocarb	8.21	16.42	0.9901
12	Atrazine	0.18	0.36	0.9987
13	Azoxystrobin	0.23	0.45	0.9999
14	Bendiocarb	1.59	3.18	0.9801
15	Bensulide	17.10	34.20	0.9903
16	Benzofenap	0.04	0.08	0.9998
17	Benzoximate	9.83	19.66	0.9861
18	Bioallethrin	99.00	198.00	0.9652
19	Brompyrazon	1.80	3.60	0.9936
20	Bromuconazole	1.57	3.14	0.9936
21	Butocarboxim	0.79	1.57	0.9962
22	Butoxycarboxim	13.30	26.60	0.9980
23	Buturon	4.48	8.96	0.9913
24	Carbendazim	0.23	0.47	0.9950
25	Carbetamide	1.82	3.64	0.9975
26	Carbofuran	6.53	13.06	0.9949
27	Carpropamid	2.60	5.20	0.9979
28	Cartap	1040.00	2080.00	0.7874
29	Chloramphenicolum	1.94	3.88	0.9955
30	Chlordimeform hydrochloride	1.32	2.64	0.9923
31	Chloridazon	1.16	2.33	0.9989
32	Chlorobenzuron	10.20	20.40	0.9996
33	Chlorotoluron	0.31	0.62	0.9988
34	Chloroxuron	0.22	0.44	0.9927
35	Chlorphoxim	38.79	77.57	0.9916
36	Chlorprifos methyl	8.00	16.00	0.9949
37	Chlorpyrifos	26.90	53.80	0.9792
38	Chlorsulfuron	1.37	2.74	0.9966
39	Chlorthiamid	4.41	8.82	0.6681
40	Chlorthion	66.80	133.60	0.0710
41	Cinidon-ethyl	7.29	14.58	0.9940
42	Clofentezine	0.38	0.76	0.9938
43	Clothianidin	31.50	63.00	0.9932

TABLE 1.1.11 The LOD, LOQ and Linearity of 178 Pesticides Applicable for LC–MS/MS (*Cont.*)

No	Pesticides	LOD µg/kg	LOQ µg/kg	Corelation coefficient (*r*)
44	Coumatetralyl	0.68	1.35	0.2086
45	Cumyluron	0.66	1.32	0.9999
46	Cyazofamid	2.25	4.50	0.9578
47	Cymoxanil	27.80	55.60	0.9982
48	Cythioate	40.00	80.00	0.9970
49	Dalapon	115.37	230.74	0.9926
50	Daminozide	1.30	2.60	0.9906
51	Dazomet	63.50	127.00	0.9834
52	Demeton(o+s)	3.39	6.77	0.9991
53	Demeton-s-methyl sulfone	9.88	19.76	0.9897
54	Demeton-s-methyl sulfoxide	1.96	3.92	0.9992
55	Diafenthiuron	0.14	0.28	0.9983
56	Diallate	44.60	89.20	0.9952
57	Dimethametryn	0.06	0.11	0.9998
58	Dinoseb acetate	20.64	41.28	0.9988
59	Dinotefuran	5.09	10.18	0.9940
60	Dioxabenzofos	6.92	13.84	0.0984
61	Diphenylamin	0.21	0.41	0.9991
62	Dithiopyr	5.20	10.40	0.9917
63	Diuron	0.78	1.56	0.9919
64	DMST	20.00	40.00	0.9920
65	Dodine	4.00	8.00	0.9762
66	Epoxiconazole	2.03	4.06	0.9925
67	Etaconazole	0.89	1.78	0.9996
68	Ethidimuron	0.75	1.50	0.9952
69	Ethiprole	19.93	39.85	1.0000
70	Ethoxyquin	1.76	3.52	0.9999
71	Ethylene thiourea	26.10	52.20	0.9904
72	Etobenzanid	0.40	0.80	0.9992
73	Famoxadone	22.64	45.29	0.9973
74	Famphur	1.80	3.60	0.9996
75	Fenfuram	0.39	0.78	0.9982
76	Fenoxycarb	9.14	18.27	0.8951
77	Fenthion oxon	0.59	1.19	0.9992
78	Fentrazamide	6.20	12.40	0.9943
79	Fenuron	0.52	1.03	0.9989
80	Flamprop methyl	10.10	20.20	0.9919
81	Fluazinam	35.30	70.60	0.9916
82	Fluazuron	0.01	0.02	0.9977
83	Flubenzimine	3.89	7.78	0.9955
84	Fluometuron	0.46	0.92	0.9922
85	Fluridone	0.09	0.18	0.9995
86	Fluthiacet methyl	2.65	5.30	0.9991
87	Folpet	69.30	138.60	0.9971

(Continued)

TABLE 1.1.11 The LOD, LOQ and Linearity of 178 Pesticides Applicable for LC–MS/MS (*Cont.*)

No	Pesticides	LOD µg/kg	LOQ µg/kg	Corelation coefficient (*r*)
88	Fosthiazate	0.28	0.57	0.7209
89	Furathiocarb	0.96	1.92	0.9948
90	Haloxyfop-2-ethoxyethyl	1.25	2.50	0.9971
91	Haloxyfop-methyl	1.32	2.64	0.9998
92	Hymexazol	112.07	224.14	0.9993
93	Imazamox	0.90	1.80	0.3015
94	Imazapic	2.95	5.90	0.9979
95	Imazethapyr	0.56	1.13	0.9966
96	Imibenconazole	5.13	10.26	0.9931
97	Imidacloprid	11.00	22.00	0.9957
98	Indoxacarb	3.77	7.54	0.9996
99	Iprovalicarb	1.16	2.32	1.0000
100	Isoprocarb	1.15	2.30	0.9988
101	Isoproturon	0.07	0.14	0.9999
102	Isouron	0.20	0.41	0.9983
103	Isoxaben	0.09	0.19	0.9969
104	Isoxaflutole	1.95	3.90	0.9975
105	Kadethrin	1.66	3.33	0.9974
106	Kelevan	4821.41	9642.82	0.9928
107	Maleic hydrazide	40.00	80.00	0.9869
108	Mepanipyrim	0.16	0.32	0.9999
109	Mepiquat chloride	0.45	0.90	0.9956
110	Metconazole	0.66	1.32	0.9996
111	Methamidophos	2.47	4.93	0.9853
112	Methiocarb	20.60	41.20	0.0003
113	Methobromuron	8.42	16.84	0.9938
114	Methomyl	4.78	9.56	0.9906
115	Methoxyfenozide	1.85	3.70	0.9979
116	Metolcarb	12.70	25.40	0.9926
117	Metoxuron	0.32	0.64	0.9964
118	Monuron	17.37	34.74	0.9902
119	Naptalam	0.97	1.95	0.9946
120	Neburon	3.55	7.10	0.9966
121	Nicotine	1.10	2.20	0.9980
122	Nitenpyram	8.56	17.12	0.9949
123	Novaluron	4.02	8.04	0.9904
124	Oesmetryn	0.09	0.17	0.9994
125	Omethoate	4.83	9.65	0.9947
126	Oxycarboxin	0.45	0.90	0.9905
127	Paraoxon methyl	0.38	0.76	1.0000
128	Phenmedipham	2.24	4.48	0.6849
129	Phorate sulfoxide	184.14	368.28	0.9991
130	Phosfolan	0.24	0.49	0.9994
131	Phosmet	8.86	17.72	0.9933

TABLE 1.1.11 The LOD, LOQ and Linearity of 178 Pesticides Applicable for LC–MS/MS (*Cont.*)

No	Pesticides	LOD µg/kg	LOQ µg/kg	Corelation coefficient (r)
132	Phoxim	41.40	82.80	0.9935
133	Phthalic acid, biscyclohexyl ester	0.34	0.68	0.9996
134	Phthalic acid, dibutyl ester	19.80	39.60	0.9950
135	Picolinafen	0.36	0.73	0.9901
136	Prallethrin	0.05	0.10	0.9999
137	Propaquiafop	0.62	1.24	1.0000
138	Propiconazole	0.88	1.76	1.0000
139	Propoxur	12.20	24.40	0.9925
140	Prothoate	1.23	2.46	0.0248
141	Pymetrozin	17.14	34.28	0.5514
142	Pyrazoxyfen	0.16	0.33	0.9997
143	Pyrethrin	17.90	35.80	0.9995
144	Pyrimitate	0.09	0.17	0.9999
145	Quizalofop-ethyl	0.34	0.68	1.0000
146	Rotenone	1.16	2.32	1.0000
147	Sethoxydim	44.80	89.60	0.6459
148	Simetryn	0.07	0.14	0.9997
149	Spinosad	0.28	0.57	1.0000
150	Spiroxamine	0.03	0.05	0.9997
151	Tau-fluvalinate	115.00	230.00	0.9922
152	Temephos	0.61	1.22	0.9996
153	TEPP	5.20	10.40	0.9999
154	Terbufos sulfone	44.30	88.60	0.9957
155	Terbumeton	0.05	0.10	0.9997
156	Terrbucarb	1.05	2.10	0.9996
157	Tert-butylamine	19.48	38.95	0.9966
158	Thiacloprid	0.19	0.37	0.9999
159	Thiamethoxam	16.50	33.00	0.9959
160	Thiodicarb	19.68	39.37	
161	Thiofanox	78.50	157.00	0.9958
162	Thiofanox sulfone	12.04	24.08	0.9935
163	Thiofanox-sulfoxide	4.15	8.29	0.9970
164	Tralkoxydim	0.16	0.32	0.9980
165	Triadimenol	5.28	10.55	0.9997
166	Triamiphos	0.01	0.01	0.8937
167	Trichloronat	33.40	66.80	0.9995
168	Trichlorphon	0.56	1.12	0.0416
169	Tricyclazole	0.62	1.25	0.5466
170	Tridemorph	1.30	2.60	0.9982
171	Triflumuron	1.96	3.92	0.9999
172	Tri-iso-butyl phosphate	1.79	3.58	0.9983
173	Triticonazole	1.51	3.02	0.9996
174	Uniconazole	1.20	2.40	0.9994
175	Vamidothion	2.28	4.56	0.9935

(*Continued*)

TABLE 1.1.11 The LOD, LOQ and Linearity of 178 Pesticides Applicable for LC–MS/MS (*Cont.*)

No	Pesticides	LOD μg/kg	LOQ μg/kg	Corelation coefficient (*r*)
176	Vamidothion sulfone	238.00	476.00	0.9780
177	Warfarin	1.34	2.68	0.9935
178	Zeta cypermethrin	0.34	0.68	0.0084

TABLE 1.1.12 The LOD, LOQ and Linearity of 220 Pesticides Applicable for GC–MS

No.	Pesticides	LOD μg/kg	LOQ μg/kg	Corelation coefficient (*r*)
1	2,3,4,5-Tetrachloroaniline	10.00	20.00	0.9989
2	2,3,4,5-Tetrachloroanisole	500.00	1000.00	0.9997
3	2,3,5,6-Tetrachloroaniline	5.00	10.00	0.9994
4	2,4,5-T	100.00	200.00	0.9960
5	3,5-Dichloroaniline	5.00	10.00	0.9905
6	3.4.5-Trimethacarb	40.00	80.00	
7	4,4-Dibromobenzophenone	5.00	10.00	0.9945
8	4-Chlorophenoxy acetic acid	6.30	12.50	0.9940
9	Alachlor	15.00	30.00	0.9990
10	Aldrin	10.00	20.00	0.9990
11	Alpha-cypermethrin	25.00	50.00	0.9974
12	α-HCH	5.00	10.00	0.9994
13	Amitraz	15.00	30.00	0.9993
14	Anthraquinone	12.50	25.00	0.9862
15	Atrizine	5.00	10.00	0.9991
16	Azaconazole	20.00	40.00	
17	BDMC-1	10.00	20.00	0.9978
18	BDMC-2	25.00	50.00	0.9912
19	Benfluralin	5.00	10.00	0.9888
20	β-HCH	5.00	10.00	0.9992
21	Bifenox	10.00	20.00	0.9826
22	Bifenthrin	5.00	10.00	0.9936
23	Bioallethrin-1	50.00	100.00	0.9941
24	Bioallethrin-2	50.00	100.00	0.9983
25	Biphenyl	5.00	10.00	0.9998
26	Bromocylen	5.00	10.00	0.9990
27	Bromofos	10.00	20.00	0.9987
28	Bromopropylate	10.00	20.00	0.9963
29	Bromuconazole-1	10.00	20.00	0.9916
30	Bromuconazole-2	10.00	20.00	0.9994
31	Butamifos	5.00	10.00	0.9710
32	Cafenstrole	20.00	40.00	0.9810
33	Carbofenothion	10.00	20.00	0.9955
34	Carfentrazone-ethyl	10.00	20.00	0.9990
35	Chlorbenside	10.00	20.00	0.9946
36	Chlorethoxyfos	10.00	20.00	1.0000

TABLE 1.1.12 The LOD, LOQ and Linearity of 178 Pesticides Applicable for LC–MS/MS (*Cont.*)

No.	Pesticides	LOD µg/kg	LOQ µg/kg	Corelation coefficient (*r*)
37	Chlorfenapyr	100.00	200.00	0.9990
38	Chlorfenprop-methyl	5.00	10.00	0.9989
39	Chlorfenson	10.00	20.00	0.9980
40	Chlorfenvinphos	15.00	30.00	0.9976
41	Chlorfurenol	15.00	30.00	0.9960
42	Chlorobenzilate	15.00	30.00	0.9977
43	Chloroneb	5.00	10.00	0.9990
44	Chloropropylate	5.00	10.00	0.9970
45	Chlorpropham	10.00	20.00	0.9971
46	Chlorpyifos(ethyl)	5.00	10.00	0.9985
47	Chlorthal-dimethyl	10.00	20.00	0.9980
48	Chlozolinate	10.00	20.00	0.9994
49	Cinmethylin	25.00	50.00	
50	Cis-1,2,3,6tetrahydrophthalimide	5.00	10.00	0.9985
51	Cis-chlordane	10.00	20.00	0.9994
52	Cis-diallate	10.00	20.00	0.9998
53	Cis-permethrin	5.00	10.00	0.9935
54	Clomeprop	5.00	10.00	0.9970
55	Coumaphos	30.00	60.00	0.9955
56	Cyflufenamid	80.00	160.00	
57	Cyfluthrin	120.00	240.00	0.9966
58	Cyhalofop-butyl	10.00	20.00	
59	Dacthal	5.00	10.00	0.9995
60	δ-HCH	10.00	20.00	0.9986
61	Deltamethrin	75.00	150.00	0.9921
62	De-PCB 101	5.00	10.00	0.9994
63	De-PCB 118	5.00	10.00	0.9985
64	De-PCB138	12.50	25.00	0.9991
65	De-PCB153	5.00	10.00	0.9992
66	De-PCB180	5.00	10.00	0.9989
67	De-PCB28	5.00	10.00	0.9997
68	De-PCB31	5.00	10.00	0.9997
69	De-PCB52	5.00	10.00	0.9997
70	Desethyl-sebuthylazine	10.00	20.00	0.9985
71	Desisopropyl-atrazine	40.00	80.00	0.9975
72	Desmetryn	5.00	10.00	0.9964
73	Dichlobenil	1.00	2.00	0.9998
74	Dichloran	10.00	20.00	0.9990
75	Dichlormid	10.00	20.00	0.9981
76	Dichlorofop-methyl	5.00	10.00	0.9958
77	Dichlorvos	30.00	60.00	0.9990
78	Dicofol	10.00	20.00	0.9980
79	Dieldrin	10.00	20.00	0.9994
80	Difenonazole-1	30.00	60.00	0.9860

(Continued)

TABLE 1.1.12 The LOD, LOQ and Linearity of 178 Pesticides Applicable for LC–MS/MS (*Cont.*)

No.	Pesticides	LOD µg/kg	LOQ µg/kg	Corelation coefficient (*r*)
81	Difenonazole-2	30.00	60.00	0.9860
82	Difenoxuron	40.00	80.00	0.9980
83	Dimepiperate	10.00	20.00	1.0000
84	Dimethylvinphos	25.00	50.00	0.9980
85	Diofenolan -1	10.00	20.00	0.9970
86	Diofenolan -2	10.00	20.00	0.9990
87	Dioxathion	50.00	100.00	0.9993
88	Diphenylamin	5.00	10.00	0.9975
89	DMSA	40.00	80.00	0.9910
90	Dodemorph	15.00	30.00	0.9990
91	Endosulfan-1	30.00	60.00	0.9991
92	Endosulfan-sulfate	15.00	30.00	0.9985
93	Endrin	60.00	120.00	0.9982
94	Endrin ketone	80.00	160.00	0.9980
95	Epoxiconazole -1	100.00	200.00	0.9790
96	Epoxiconazole-2	100.00	200.00	0.9990
97	Esprocarb	20.00	40.00	0.9970
98	Etaconazole-1	15.00	30.00	0.9948
99	Etaconazole-2	15.00	30.00	0.9976
100	Ethalfluralin	20.00	40.00	0.9926
101	Etoxazole	30.00	60.00	
102	Fenamidone	12.50	25.00	0.9980
103	Fenchlorphos	20.00	40.00	0.9990
104	Fenpiclonil	20.00	40.00	0.9981
105	Fenson	5.00	10.00	0.9993
106	Fenvalerate-1	20.00	40.00	0.9928
107	Fenvalerate-2	20.00	40.00	0.9928
108	Fipronil	100.00	200.00	0.9960
109	Flamprop-methyl	5.00	10.00	0.9983
110	Fluazinam	100.00	200.00	0.9950
111	Flucythrinate-1	10.00	20.00	
112	Flucythrinate-2	10.00	20.00	
113	Flumetralin	10.00	20.00	0.9873
114	Flumioxazin	10.00	20.00	
115	Fluorodifen	15.00	30.00	
116	Fluotrimazole	5.00	10.00	0.9981
117	Fluquinconazole	5.00	10.00	0.9991
118	Fluroxypr-1-methylheptyl ester	5.00	10.00	0.9974
119	Fluvalinate	60.00	120.00	0.9954
120	Formothion	25.00	50.00	0.9993
121	Furmecyclox	15.00	30.00	0.9890
122	Gamma-cyhaloterin-1	4.00	8.00	
123	Gamma-cyhalothrin-2	4.00	8.00	0.9780
124	γ-HCH	10.00	20.00	0.9998

TABLE 1.1.12 The LOD, LOQ and Linearity of 178 Pesticides Applicable for LC–MS/MS (*Cont.*)

No.	Pesticides	LOD µg/kg	LOQ µg/kg	Corelation coefficient (*r*)
125	Halfenprox	25.00	50.00	
126	Heptachlor	15.00	30.00	0.9990
127	Hexachlorobenzene	5.00	10.00	0.9996
128	Hexaconazole	30.00	60.00	0.9989
129	Imiprothrin-1	10.00	20.00	0.9850
130	Imiprothrin-2	10.00	20.00	0.9900
131	Iodofenphos	10.00	20.00	0.9940
132	Iprodione	20.00	40.00	0.9980
133	Iprovalicarb-1	20.00	40.00	0.9990
134	Iprovalicarb-2	20.00	40.00	0.9990
135	Isobenzan	5.00	10.00	0.9995
136	Isocarbophos	10.00	20.00	
137	Isodrin	5.00	10.00	0.9982
138	Isofenphos oxon	10.00	20.00	
139	Isoprocarb -1	10.00	20.00	0.9670
140	Isoprocarb -2	10.00	20.00	0.9670
141	Isoxadifen-ethyl	10.00	20.00	0.9970
142	Isoxathion	100.00	200.00	
143	Lambda-cyhalothrin	5.00	10.00	0.9982
144	Lenacil	5.00	10.00	0.9967
145	Leptophos	10.00	20.00	
146	Mcpa-butoxyethyl ester	5.00	10.00	0.9966
147	Methiocarb sulfone	160.00	320.00	0.9930
148	Methobromuron	30.00	60.00	0.9986
149	Methoprotryne	15.00	30.00	0.9973
150	Methoxychlor	5.00	10.00	0.9921
151	Methyl-parathion	20.00	40.00	0.9892
152	Metoconazole	20.00	40.00	0.9980
153	Metominostrobin	20.00	40.00	0.9980
154	Mirex	5.00	10.00	0.9974
155	Monocrotophos	100.00	200.00	0.9941
156	Musk ambrette	5.00	10.00	
157	Musk moskene	5.00	10.00	0.9970
158	Musk tibeten	5.00	10.00	0.9970
159	Musk xylene	5.00	10.00	
160	Nitrapyrin	15.00	30.00	0.9957
161	Nitrofen	30.00	60.00	0.9905
162	Nitrothal-isopropyl	10.00	20.00	0.9911
163	Norflurazon-desmethyl	50.00	100.00	0.9880
164	o,p′-DDD	5.00	10.00	
165	o,p′-DDE	12.50	25.00	0.9998
166	o,p′-DDT	10.00	20.00	0.9978
167	Octachlorostyrene	5.00	10.00	0.9997
168	Oxadiazone	5.00	10.00	0.9980

(Continued)

TABLE 1.1.12 The LOD, LOQ and Linearity of 178 Pesticides Applicable for LC–MS/MS (*Cont.*)

No.	Pesticides	LOD µg/kg	LOQ µg/kg	Corelation coefficient (*r*)
169	Oxychlordane	12.50	25.00	0.9997
170	p,p′-DDD	5.00	10.00	0.9930
171	p,p′-DDE	5.00	10.00	0.9994
172	p,p′-DDT	10.00	20.00	0.9971
173	Pendimethalin	20.00	40.00	0.9960
174	Pentachloroanisole	5.00	10.00	0.9996
175	Pentachlorobenzene	5.00	10.00	0.9998
176	Permethrin	10.00	20.00	0.9967
177	Perthane	12.50	25.00	0.9974
178	Phenanthrene	12.50	25.00	0.9990
179	Plifenate	5.00	10.00	0.9992
180	Prochloraz	60.00	120.00	0.9919
181	Profluralin	20.00	40.00	0.9955
182	Prohydrojasmon	10.00	20.00	
183	Propamocarb	37.50	75.00	0.9912
184	Propaphos	10.00	20.00	0.9940
185	Propiconazole-1	15.00	30.00	0.9942
186	Propiconazole-2	15.00	30.00	0.9942
187	Propoxur-1	100.00	200.00	0.9570
188	Propoxur-2	50.00	100.00	0.9890
189	Propyzamide	5.00	10.00	0.9990
190	Prothiophos	5.00	10.00	0.9957
191	Pyraflufen ethyl	10.00	20.00	0.9990
192	Pyridaben	5.00	10.00	0.9958
193	Quintozene	10.00	20.00	0.9958
194	Resmethrin-1	25.00	50.00	0.9823
195	S 421(octachlorodipropyl ether)-1	100.00	200.00	0.9980
196	S 421(octachlorodipropyl ether)-2	100.00	200.00	0.9990
197	Sobutylazine	10.00	20.00	0.9780
198	Spiromesifen	50.00	100.00	0.9490
199	Spiroxamine -1	10.00	20.00	0.9950
200	Spiroxamine -2	10.00	20.00	0.9980
201	Tecnazene	10.00	20.00	0.9985
202	Tefluthrin	5.00	10.00	0.9980
203	Terbucarb-1	10.00	20.00	1.0000
204	Terbucarb-2	10.00	20.00	1.0000
205	Terbutryn	10.00	20.00	0.9980
206	Tetradifon	5.00	10.00	0.9966
207	Tetrasul	5.00	10.00	0.9958
208	Thiabendazole	100.00	200.00	0.9990
209	Tolyfluanide	15.00	30.00	0.9981

TABLE 1.1.12 The LOD, LOQ and Linearity of 178 Pesticides Applicable for LC–MS/MS (*Cont.*)

No.	Pesticides	LOD µg/kg	LOQ µg/kg	Corelation coefficient (*r*)
210	Trans-chlodane	5.00	10.00	0.9989
211	Trans-diallate	10.00	20.00	0.9994
212	Transfluthrin	5.00	10.00	0.9991
213	Trans-nonachlor	5.00	10.00	0.9990
214	Triadimenol-1	15.00	30.00	0.9972
215	Triadimenol-2	10.00	20.00	0.9972
216	Tribenuron-methyl	5.00	10.00	0.9760
217	Tridiphane	10.00	20.00	
218	Triphenyl phosphate	5.00	10.00	0.9990
219	Vernolate	10.00	20.00	0.9997
220	Zoxamide	10.00	20.00	0.9990

TABLE 1.1.13 LODs Comparison of Both GC–MS and LC–MS-MS Method (Sorted on Basis of Tables 1.1.10–1.1.12)

LOD(µg/kg)	GC–MS(490)		LC–MS-MS(448)		GC–MS(270)		LC–MS-MS(270)	
	Varieties	%	Varieties	%	Varieties	%	Varieties	%
≤10	264	53.9	325	75.5	133	49.3	200	74.1
11–50	195	39.8	75	16.7	126	46.7	38	14.1
51–100	23	4.7	17	3.8	5	1.9	9	3.3
101–500	7	1.4	26	5.8	5	1.9	19	7.0
501–1000	1	0.2	1	0.2	1	0.4	1	0.4
>1000	0	0	4	0.9	0	0	3	1.1

method, making up 74% of the pesticides analyzed. It is evident that the LC–MS/MS method is more sensitive than the GC–MS method. Yet there are 59 pesticides that have more sensitivity with GC–MS than LC–MS/MS, which shows that these two methods are reciprocally complementary, especially the 270 pesticides that can be analyzed by both these two methods, one of which is the confirmatory method for the other. Thus, the accuracy and the reliability of the test results have been improved.

Fortification recoveries and precision tests are conducted with four kinds of samples like black tea, green tea, Pu Er tea, and oolong tea without containing pesticides and related chemicals. After the samples are fortified with pesticides and related chemical standard solutions, they are placed for 30 min so as to make sure the pesticides and the related chemicals are fully absorbed before going through extraction, cleanup, and determination per the method, with test results tabulated in Tables 1.1.14–1.1.16. Based on these three tables and the low fortification level is taken for an instance, the average recoveries and RSD statistical analytical results are shown in Table 1.1.17. It can be seen from Table 1.1.17 that at low fortification levels there are 424 pesticides with recoveries 60%–120% among the 451 pesticides tested by GC–MS, accounting for 94% of the pesticides and related chemicals, of which pesticides with RSD < 20% account for 77% of the pesticides and related chemicals tested. There are 91% of 439 pesticides and related chemicals tested by LC–MS/MS with average recoveries 60%–120%, 76% pesticides and related chemicals have RSD < 20%. The 270 pesticides that can be determined by both GC–MS and LC–MS/MS with average recoveries 60%–120% account for 96% and 94%, respectively, pesticides with RSD < 20% account for 75% and 79%, respectively, which explains that LC–MS/MS and GC–MS is of very good reproducibility and repeatability for determination of tea matrix pesticide residues.

TABLE 1.1.14 Recovery and Precision Date for 270 Pesticides Residues in Teas Applicable for Both LC–MS/MS and GC–MS (n = 4)

No	Pesticides	LC–MS/MS						GC–MS					
		Levels (μg/kg)	Ave. %	RSD %	Levels μg/kg	Ave.	RSD %	Levels (μg/kg)	Ave. %	RSD%	Levels (μg/kg)	Ave%	RSD%
1	2,4-D	11.9	25.6	41.3	47.4	19.7	42.4	200.0	–	–	2000	93.3	15.5
2	2,6-Dichlorobenzamide	1.5	98.8	3.0	6.0	99.8	11.5	20.0	90.0	4.2	200	74.5	24.3
3	2-Phenylphenol	169.9	46.9	10.5	679.5	47.5	3.6	10.0	78.6	4.4	100	75.2	10.6
4	3-Phenylphenol	4.0	46.9	10.5	16.0	47.5	3.6	60.0	68.2	11.5	600	68.9	40.8
5	4,4-Dichlorobenzophenone	13.6	95.7	10.0	54.4	84.0	29.8	10.0	92.6	6.0	100	95.3	5.6
6	Acetochlor	47.4	96.0	1.6	189.6	92.0	9.2	20.0	71.8	28.5	200	92.8	11.8
7	Acibenzolar-S-methyl	1.0	75.0	25.3	4.1	96.6	13.7	20.0	72.0	25.0	200	73.5	4.2
8	Aclonifen	24.2	92.2	15.4	96.8	93.9	7.1	200.0	69.8	6.8	2000	83.2	15.9
9	Allethrin	60.4	91.6	5.8	241.6	88.9	5.0	40.0	75.9	29.4	400	90.9	11.9
10	Allidochlor	41.0	106.6	8.8	164.2	98.5	11.9	20.0	70.0	18.5	200	74.0	25.4
11	Ametryn	1.0	91.8	9.2	3.8	99.2	7.7	30.0	80.6	10.9	300	71.0	10.2
12	Anilofos	0.7	85.8	21.5	2.9	66.4	18.5	20.0	64.6	9.3	200	83.0	14.4
13	Aspon	1.7	93.3	9.5	6.9	118.1	25.5	20.0	84.5	7.5	–	–	–
14	Atratone	0.2	95.5	15.9	0.7	104.9	10.8	10.0	81.7	5.2	100	91.8	12.6
15	Atrazine-Desethyl	0.2	–	–	0.8	92.2	7.8	10.0	–	–	100	62.2	28.9
16	Azinphos ethyl	108.9	89.0	18.9	435.7	91.7	10.7	20.0	–	–	200	72.7	10.5
17	Azinphos-methyl	1104.3	93.7	14.5	4417.3	95.8	12.6	60.0	–	–	600	71.6	8.9
18	Aziprotryne	1.4	91.6	14.4	5.5	95.1	9.9	80.0	84.0	0.8	800	87.5	13.7
19	Benalyxyl	1.2	95.8	18.3	5.0	94.0	10.4	10.0	74.4	7.1	100	80.8	12.2
20	Benodanil	3.5	97.5	18.5	13.9	96.1	13.0	30.0	69.2	13.2	300	66.1	21.0
21	Benoxacor	6.9	89.6	15.6	27.6	87.3	11.6	20.0	71.4	22.9	200	85.3	10.2
22	Benzoylprop-ethyl	10.3	100.6	9.9	41.1	98.6	13.4	30.0	73.1	6.5	300	79.1	13.9
23	Bifenazate	22.8	58.1	96.8	91.2	51.9	94.6	80.0	78.9	23.0	800	77.2	29.3
24	Bioresmethrin	7.4	94.6	19.8	29.7	71.9	16.6	20.0	53.8	22.0	200	64.7	11.6
25	Bitertanol	33.4	95.3	15.3	133.6	94.5	7.0	30.0	83.2	7.6	300	76.0	6.6
26	Boscalid	0.2	94.6	16.5	0.6	92.4	14.8	200.0	75.6	4.5	2000	84.2	11.3
27	Bromacil	23.6	78.8	18.5	94.4	81.3	18.0	20.0	70.1	18.3	200	72.0	24.5
28	Bromfenvinfos	3.0	98.0	16.8	12.1	94.1	12.3	10.0	94.0	9.9	100	98.7	11.9
29	Bromophos-ethyl	567.7	95.7	7.6	2270.8	92.6	14.6	10.0	67.8	21.7	100	87.3	4.0
30	Bupirimate	0.7	79.0	13.0	2.8	90.7	10.2	10.0	69.2	9.6	100	83.3	7.7

No.	Name											
31	Buprofezin	0.9	95.1	8.5	3.5	85.5	11.6	20.0	81.6	13.4	200	85.2	9.0
32	Butachlor	20.1	90.8	7.5	80.3	97.6	5.8	20.0	76.7	9.2	200	79.9	11.1
33	Butafenacil	9.5	79.9	15.5	38.0	87.9	12.9	10.0	76.3	13.9	100	90.2	10.2
34	Butralin	1.9	98.5	5.8	7.6	91.3	6.2	40.0	77.1	16.2	400	89.8	12.8
35	Butylate	302.0	273.8	80.1	1208.0	109.6	8.2	30.0	94.6	29.6	300	59.8	19.3
36	Cadusafos	1.2	89.4	20.5	4.6	70.2	22.3	40.0	75.4	3.8	–	–	–
37	Carbaryl	10.3	90.3	19.1	41.3	93.2	16.3	30.0	79.7	23.9	300	79.1	29.1
38	Carbosulfan	0.4	83.7	11.6	1.6	97.0	10.3	30.0	85.8	10.2	–	–	–
39	Carboxin	0.6	59.4	93.2	2.2	33.8	72.4	30.0	35.8	47.8	300	41.7	65.8
40	Chlorbenside sulfone	0.4	83.0	19.7	2.0	96.8	21.8	20.0	77.6	19.2	200	77.2	14.3
41	Chlorbufam	183.0	97.9	12.8	732.0	97.9	3.6	20.0	–	–	200	72.5	11.4
42	Chlordimeform	0.0	108.3	10.5	0.2	85.1	23.8	10.0	58.3	4.2	100	76.0	8.7
43	Chlorfenethol	164.3	69.3	26.6	657.2	68.0	8.4	10.0	74.7	7.9	–	–	–
44	Chlorfluazuron	8.7	67.0	17.4	34.7	86.8	7.6	30.0	44.7	22.8	300	55.1	20.6
45	Chlormephos	19540.0	104.5	14.2	78160.0	82.3	10.1	20.0	42.3	29.3	200	70.9	10.6
46	Chlorpropham	15.8	49.3	7.3	63.1	51.2	6.9	20.0	77.4	12.0	200	76.1	12.8
47	Chlorthiophos	10.6	58.6	74.2	42.4	107.6	17.4	30.0	76.0	5.5	300	78.0	13.0
48	Clodinafop propargyl	2.4	101.5	10.6	9.8	77.4	18.6	20.0	84.8	8.5	200	100.2	16.8
49	Clomazone	0.4	87.3	15.6	1.7	90.1	5.8	10.0	66.5	8.4	100	85.3	4.3
50	Cloquintocet mexyl	1.9	90.4	16.4	7.5	98.5	7.5	10.0	69.7	10.8	100	90.5	19.1
51	Crimidine	1.6	106.5	6.6	6.2	95.7	10.0	10.0	78.2	21.6	100	80.5	16.2
52	Crufomate	0.5	89.7	20.5	2.1	96.8	5.4	60.0	72.7	13.5	600	76.9	2.3
53	Cyanazine	0.2	86.9	18.2	0.7	103.3	18.8	30.0	–	–	300	74.2	25.5
54	Cyanofenphos	20.8	97.3	5.1	83.2	95.8	3.1	10.0	81.3	8.8	100	79.0	14.6
55	Cyanohos	0.4	106.1	28.4	1.6	111.7	18.5	20.0	76.4	10.5	–	–	–
56	Cycloate	4.4	100.7	16.6	17.8	102.9	6.9	10.0	61.3	3.7	100	80.5	4.6
57	Cycluron	0.2	97.5	15.3	0.8	93.4	6.4	30.0	89.0	9.2	300	94.8	12.4
58	Cyphenothrin	5.6	98.2	15.1	22.4	89.7	19.8	30.0	90.6	6.9	300	83.7	16.8
59	Cyprazine	0.0	91.9	9.3	0.2	99.2	7.7	10.0	75.0	11.9	–	–	–
60	Cyproconazole	0.7	101.6	5.9	2.9	97.2	3.0	10.0	105.0	28.2	100	88.8	10.3
61	Cyprodinil	0.7	97.2	24.7	3.0	94.2	10.3	10.0	86.2	12.3	100	88.9	11.2
62	DEF	1.6	103.0	8.1	6.5	101.4	4.8	20.0	86.2	9.4	200	92.9	14.5
63	Demeton-S	0.4	–	–	1.6	91.2	16.8	40.0	65.4	11.0	400	87.5	29.4

(Continued)

TABLE 1.1.14 Recovery and Precision Date for 270 Pesticides Residues in Teas Applicable for Both LC–MS/MS and GC–MS (*n* = 4) (*Cont.*)

No	Pesticides	LC–MS/MS Levels (µg/kg)	Ave. %	RSD %	Levels µg/kg	Ave.	RSD %	GC–MS Levels (µg/kg)	Ave. %	RSD%	Levels (µg/kg)	Ave%	RSD%
64	Demeton-s-methyl	0.2	92.1	11.8	0.7	87.5	9.9	40.0	64.0	25.4	400	73.0	22.4
65	Dialifos	52.3	69.2	10.2	209.3	76.5	28.9	320.0	63.2	27.2	3200	86.4	21.6
66	Diazinon	0.7	94.3	13.4	2.9	92.3	2.8	10.0	70.1	9.6	100	87.1	4.6
67	Dibutyl succinate	222.4	102.1	7.2	889.6	98.1	7.7	20.0	70.5	13.2	200	84.0	10.6
68	dicapthon	0.0	71.0	18.3	0.0	90.0	21.6	50.0	87.7	15.3	500	96.2	18.4
69	Dichlofenthion	1.0	100.4	13.3	4.0	92.5	18.0	10.0	66.9	9.4	100	82.0	16.7
70	Dichlofluanid	2.6	112.5	15.3	10.4	100.5	3.9	60.0	–	–	600	28.4	20.9
71	Diclobutrazole	0.5	90.8	23.5	1.9	94.0	9.8	40.0	83.7	3.6	400	89.0	13.1
72	Dicloran	48.6	53.6	16.1	194.2	69.1	18.2	20.0	74.3	25.8	200	75.2	9.0
73	Dicrotophos	1.1	94.0	28.9	4.6	79.7	10.2	80.0	67.7	27.6	800	63.0	28.9
74	Diethofencarb	2.0	88.9	10.5	8.0	94.1	7.8	60.0	77.8	18.6	600	77.6	8.0
75	Diethyltoluamide	0.6	102.2	8.8	2.2	95.1	18.9	10.0	87.4	6.6	–	–	–
76	Diflufenican	28.3	46.3	4.9	113.1	47.3	3.3	10.0	78.9	11.7	100	79.7	12.0
77	Dimefox	22.7	79.5	4.1	90.9	108.6	8.3	30.0	22.4	30.2	300	28.7	59.7
78	Dimefuron	1.3	106.3	29.0	5.3	89.5	6.5	40.0	–	–	400	62.2	28.5
79	Dimethachloro	1.9	106.3	6.1	7.6	98.3	4.1	30.0	74.1	8.7	300	77.3	14.1
80	Dimethenamid	4.3	87.4	11.6	17.2	90.0	7.7	10.0	83.2	5.7	100	89.1	13.8
81	Dimethoate	7.6	83.0	16.6	30.4	92.3	13.3	40.0	–	–	400	60.4	27.6
82	Dimethomorph	0.4	76.5	22.7	1.4	63.1	21.6	20.0	67.2	9.2	200	93.0	27.3
83	Dimethyl phthalate	4.4	79.2	12.2	17.6	89.8	14.2	40.0	82.1	14.3	400	80.8	23.2
84	Diniconazole	1.3	84.8	2.6	5.4	87.6	11.1	30.0	82.3	12.6	300	79.9	15.9
85	Dinitramine	0.1	80.3	10.2	0.2	87.9	3.4	40.0	64.4	5.0	400	86.8	4.6
86	Dioxacarb	0.1	75.2	4.6	0.4	76.5	13.5	80.0	77.2	13.5	800	78.6	26.0
87	Diphenamid	0.1	–	–	0.6	99.6	7.5	10.0	83.9	9.9	100	83.5	10.7
88	Dipropetryn	0.3	94.9	10.5	1.1	93.5	9.1	10.0	83.6	11.2	100	80.4	8.4
89	Disulfoton	469.7	112.0	17.4	1878.8	89.2	21.3	10.0	65.3	10.6	100	61.2	15.3
90	Disulfoton sulfone	2.5	102.9	14.1	9.8	95.6	3.4	20.0	–	–	200	90.5	29.5
91	Disulfoton-sulfoxide	2.8	81.8	14.5	11.4	112.4	13.6	20.0	83.7	11.5	200	92.1	11.7
92	Edifenphos	0.8	81.4	21.7	3.0	94.3	20.7	20.0	–	–	200	76.1	8.8

93	EPN	33.0	87.1	11.3	132.0	100.5	8.9	40.0	72.8	21.3	400	77.9	9.4
94	EPTC	37.3	100.3	7.7	149.4	91.6	7.4	30.0	44.9	23.1	300	46.5	17.8
95	Esfenvalerate	13.9	–	–	55.5	97.5	115.8	40.0	103.7	12.2	400	79.6	13.7
96	Ethiofencarb	4.9	86.4	15.6	19.7	85.1	21.2	100.0	56.9	18.8	1000	65.1	21.9
97	Ethion	3.0	103.0	21.3	11.8	97.6	10.3	20.0	74.7	17.0	200	84.1	3.6
98	Ethofume sate	372.0	100.7	3.4	1488.0	97.2	4.5	20.0	76.6	12.6	200	77.1	15.5
99	Ethoprophos	2.8	104.8	9.9	11.1	86.1	16.0	30.0	69.1	21.5	300	75.7	14.4
100	Etofenprox	76.0	105.9	22.1	304.0	91.7	22.5	10.0	85.3	11.0	100	63.8	28.7
101	Etridiazol	100.4	90.4	13.1	401.7	94.4	6.7	30.0	–	–	300	60.5	14.7
102	Etrimfos	18.8	48.4	72.5	75.0	83.4	14.7	10.0	81.5	13.4	100	86.4	9.8
103	Fenamiphos	0.2	106.4	22.2	0.8	95.3	29.2	30.0	64.7	25.1	300	74.4	9.2
104	Fenamiphos sulfone	0.4	78.6	22.1	1.8	83.7	20.1	40.0	92.1	6.8	400	82.5	26.0
105	Fenamiphos sulfoxide	0.7	77.3	23.6	3.0	90.7	14.8	40.0	77.2	34.0	400	70.2	25.2
106	Fenarimol	0.6	69.6	25.8	2.4	99.0	7.8	20.0	80.5	12.9	200	77.0	12.4
107	Fenazaquin	0.3	101.4	5.3	1.3	98.3	3.0	10.0	81.4	7.9	100	76.6	9.7
108	Fenbuconazole	1.6	82.4	20.7	6.6	89.8	12.5	20.0	74.8	21.6	200	92.3	10.8
109	Fenhexamid	0.9	76.0	14.4	3.8	97.2	9.0	500.0	82.9	10.9	–	–	–
110	Fenitrothion	8.9	59.3	78.6	35.7	98.5	5.0	20.0	70.7	18.4	200	82.4	7.8
111	Fenobucarb	5.9	95.4	14.0	23.6	92.8	8.4	20.0	90.6	20.9	200	80.0	10.4
112	Fenoxanil	39.4	84.3	10.5	157.6	91.2	8.9	20.0	70.5	15.0	200	87.1	10.1
113	Fenpropathrin	10.5	95.6	8.7	41.8	87.8	27.3	20.0	86.2	17.8	200	80.0	10.6
114	Fenpropidin	0.0	63.1	26.6	0.0	70.7	27.1	20.0	71.4	11.4	200	60.9	7.3
115	Fenpropimorph	0.2	72.0	8.8	0.7	90.8	5.7	10.0	82.4	13.2	100	80.1	11.6
116	Fenpyroximate	1.4	84.8	14.6	5.4	95.7	5.4	80.0	71.1	19.8	800	89.7	2.4
117	Fensulfothin	2.0	64.6	27.1	8.0	81.6	8.5	20.0	65.5	2.9	–	–	–
118	Fenthion	52.0	79.5	9.6	208.0	100.2	10.3	10.0	103.5	7.2	100	76.2	24.5
119	fenthion sulfone	0.6	95.4	11.8	2.3	88.9	17.3	40.0	–	–	400	96.1	15.3
120	Fenthion sulfoxide	0.3	97.4	7.5	1.3	93.5	7.4	40.0	63.2	21.1	400	84.6	19.7
121	Flamprop isopropyl	0.4	81.8	22.9	1.7	100.8	27.3	10.0	71.7	6.5	100	79.3	15.0
122	Fluazifop butyl	0.3	94.4	11.7	1.1	92.6	3.9	10.0	83.5	5.3	100	78.7	7.0
123	Fluchloralin	488.0	98.0	9.0	1952.0	91.7	13.6	40.0	81.6	14.6	400	80.1	8.6
124	Fludioxonil	62.2	53.2	15.1	248.6	72.9	17.0	10.0	68.2	29.0	100	27.9	63.4
125	Flufenacet	5.3	83.2	10.8	21.2	88.0	5.4	80.0	76.4	8.5	800	75.5	13.0

(Continued)

TABLE 1.1.14 Recovery and Precision Date for 270 Pesticides Residues in Teas Applicable for Both LC–MS/MS and GC–MS (n = 4) (Cont.)

No	Pesticides	LC–MS/MS Levels (μg/kg)	Ave. %	RSD %	Levels μg/kg	Ave.	RSD %	GC–MS Levels (μg/kg)	Ave. %	RSD%	Levels (μg/kg)	Ave%	RSD%
126	Flufenoxuron	3.2	96.5	3.3	12.7	89.3	3.2	30.0	76.8	11.1	300	78.1	16.3
127	Flumiclorac-pentyl	10.6	87.8	25.6	42.4	81.3	18.6	20.0	77.1	17.3	200	74.6	10.4
128	Fluorochloridone	13.8	84.9	6.7	55.1	92.8	12.8	20.0	86.7	24.4	200	75.8	6.9
129	fluoroglycofen-ethyl	0.2	82.0	3.2	0.7	81.4	10.0	120.0	89.2	28.6	1200	90.1	13.6
130	flurochloridone	0.0	93.5	8.4	0.2	99.6	17.2	20.0	101.9	27.8	200	94.7	8.8
131	Flurtamone	0.4	87.6	29.3	1.8	97.0	10.7	20.0	72.7	3.0	200	64.2	27.6
132	Flusilazole	0.6	93.4	21.2	2.3	93.0	7.6	30.0	85.3	12.8	300	76.7	5.6
133	Flutolanil	1.1	100.5	8.4	4.6	94.8	7.1	10.0	69.3	5.1	100	82.5	8.9
134	Flutriafol	8.6	100.7	5.2	34.3	100.6	3.0	20.0	86.3	2.0	200	88.1	15.7
135	Fonofos	7.5	102.9	16.2	29.8	92.9	6.1	10.0	68.2	5.2	100	85.2	4.2
136	Fuberidazole	0.1	94.3	27.1	0.3	91.5	11.2	50.0	80.5	26.9	500	61.7	49.3
137	Furalaxyl	0.8	101.8	9.2	3.1	99.3	4.2	20.0	74.1	15.9	200	89.9	9.2
138	Heptanophos	5.8	92.3	10.9	23.4	84.7	8.8	30.0	70.5	24.0	300	73.6	10.4
139	Hexaflumuron	25.2	93.5	6.0	100.8	92.9	4.1	60.0	79.3	7.0	600	75.8	26.2
140	Hexazinone	0.1	77.8	17.6	0.5	88.8	15.5	30.0	67.7	26.3	300	70.2	10.5
141	Hexythiazox	0.8	94.5	12.6	3.1	88.2	22.8	80.0	99.7	26.4	800	83.4	15.2
142	Imazalil	2.0	89.1	14.2	8.0	86.8	17.5	40.0	69.6	15.3	400	73.6	24.6
143	Imazamethabenz-methyl	0.2	76.0	15.2	0.7	82.4	18.9	30.0	68.1	8.7	–	–	–
144	Imibenzonazole-des-benzyl	6.2	96.5	12.1	24.9	97.6	5.7	40.0	63.4	4.3	400	55.9	38.6
145	Iprobenfos	8.3	105.2	6.8	33.1	99.0	2.6	30.0	87.9	13.9	300	83.1	9.4
146	Isazofos	0.0	105.2	16.3	0.0	99.2	13.4	20.0	87.5	8.4	200	79.7	17.0
147	Isocarbamid	0.1	104.0	4.8	0.2	90.7	13.8	50.0	84.0	16.0	500	84.1	21.4
148	Isofenphos	218.7	86.2	14.3	874.7	84.1	26.4	20.0	76.4	5.0	200	79.6	11.7
149	Isomethiozin	1.1	81.5	15.4	4.3	78.0	9.2	20.0	74.7	18.9	200	75.7	13.3
150	Isopropalin	30.0	90.2	7.0	120.0	101.1	19.1	20.0	72.8	10.2	200	77.1	11.3
151	Isoprothiolane	1.8	90.2	4.6	7.4	98.4	9.4	20.0	78.1	21.3	200	90.3	10.3
152	Kresoxim-methyl	100.6	83.9	10.8	402.3	91.6	8.2	10.0	84.3	9.4	100	84.0	10.0
153	Lactofen	20.7	63.3	77.1	82.7	118.8	26.7	80.0	93.7	13.3	–	–	–
154	Linuron	11.6	83.8	6.0	46.5	83.9	12.4	60.0	–	–	640	69.9	23.3
155	Malaoxon	1.6	93.4	8.3	6.3	75.1	12.6	160.0	72.8	4.6	1600	84.7	15.6

156	Malathion	5.6	95.4	16.1	22.6	86.5	8.7	40.0	75.8	25.9	400	88.5	4.9
157	Mecarbam	19.6	93.5	9.0	78.4	85.4	3.9	40.0	87.9	19.5	400	73.1	10.1
158	Mefenacet	2.2	92.2	15.8	8.8	94.6	4.3	30.0	66.2	39.0	300	71.7	11.5
159	Mefenoxam	1.5	104.1	6.5	6.2	100.1	5.4	20.0	95.1	29.9	200	78.2	25.0
160	Mefenpyr-diethyl	12.6	93.6	14.1	50.2	95.8	7.6	30.0	71.1	20.5	300	83.4	8.6
161	Mephosfolan	2.3	101.8	11.2	9.3	98.8	4.7	20.0	83.3	23.0	200	88.5	25.1
162	Mepronil	0.4	73.2	9.1	1.5	87.5	10.1	10.0	86.4	11.2	100	79.0	10.1
163	Metalaxyl	0.5	78.9	15.0	2.0	99.3	11.1	10.0	77.9	17.5	100	80.4	10.6
164	Metamitron	6.4	86.4	10.1	25.4	98.7	9.7	100.0	–	–	1000	60.8	58.0
165	Metazachlor	1.0	88.8	14.3	3.9	94.7	5.6	30.0	76.2	16.6	300	89.2	4.9
166	Methabenzthiazuron	0.0	63.8	71.1	0.1	87.3	15.7	100.0	78.1	6.3	1000	90.3	11.7
167	Methacrifos	2423.7	101.6	9.8	9694.8	92.0	8.3	10.0	71.3	13.1	100	75.2	11.7
168	Methidathion	10.7	85.9	16.6	42.6	89.9	6.4	20.0	64.3	4.1	200	84.3	19.9
169	Methoprotryne	0.2	75.4	8.5	1.0	84.9	4.9	30.0	71.2	8.8	300	77.4	12.4
170	Metolachlor	0.1	93.9	7.2	0.5	97.3	5.9	10.0	87.5	7.3	100	78.4	16.1
171	Metribuzin	0.5	86.9	7.7	2.2	94.2	13.1	30.0	82.7	11.3	300	74.8	10.0
172	Mevinphos	1.6	81.5	20.4	6.3	85.3	5.2	20.0	88.7	18.6	200	66.5	6.7
173	Mexacarbate	0.3	94.7	23.3	1.3	78.0	13.7	30.0	65.4	23.5	300	68.0	29.8
174	Molinate	2.1	99.2	11.5	8.4	83.4	27.6	10.0	74.5	2.4	100	72.1	11.1
175	Monalide	1.2	82.1	15.8	4.8	90.0	16.0	20.0	76.5	21.2	200	73.4	22.5
176	Monolinuron	3.6	89.0	7.8	14.2	89.3	7.1	40.0	87.6	26.2	400	70.2	6.9
177	Myclobutanil	1.0	86.9	26.6	4.0	94.1	9.9	10.0	68.3	9.9	100	85.8	5.9
178	Napropamide	0.4	61.5	76.9	1.7	98.3	6.7	30.0	65.2	7.7	300	89.5	5.0
179	Nitralin	1.1	89.2	10.6	4.6	89.8	24.2	100.0	68.4	29.4	1000	96.0	13.3
180	Norflurazon	0.0	102.5	7.8	0.0	94.5	9.0	10.0	66.7	21.3	100	71.5	8.3
181	Nuarimol	1.0	93.4	10.4	4.0	105.6	20.6	20.0	72.5	8.4	200	82.3	9.2
182	Ofurace	1.0	75.1	21.0	4.0	89.5	13.9	30.0	73.1	17.5	–	–	–
183	Oxamyl	182.7	93.4	7.8	730.7	97.6	11.2	10.0	85.9	28.5	100	77.2	14.8
184	Oxyflurofen	58.5	84.6	9.8	234.2	101.3	13.2	40.0	70.5	12.3	400	77.3	11.6
185	Paclobutrazol	0.6	101.3	18.9	2.3	97.5	9.0	30.0	73.3	13.9	300	78.7	10.8
186	Paraoxon-ethyl	0.0	97.8	26.1	0.1	85.4	18.1	320.0	73.6	5.0	–	–	–
187	Pebulate	1.1	97.9	12.5	4.5	111.0	13.6	30.0	61.5	16.7	300	61.0	5.0
188	Penconazole	2.0	92.0	18.7	8.0	96.6	16.4	30.0	81.4	11.0	300	78.3	8.7

(Continued)

TABLE 1.1.14 Recovery and Precision Date for 270 Pesticides Residues in Teas Applicable for Both LC-MS/MS and GC-MS (n = 4) (Cont.)

No	Pesticides	LC-MS/MS Levels (µg/kg)	Ave. %	RSD %	Levels µg/kg	Ave.	RSD %	GC-MS Levels (µg/kg)	Ave. %	RSD%	Levels (µg/kg)	Ave%	RSD%
189	Pencycuron	0.3	97.6	9.8	1.1	97.7	7.2	40.0	73.2	10.5	400	87.3	11.6
190	pentachloroaniline	1.2	86.0	8.2	5.0	106.2	10.6	10.0	67.7	9.0	100	79.8	16.8
191	Phenothrin	339.2	92.2	11.1	1356.8	111.5	18.9	10.0	84.6	11.4	100	82.7	8.2
192	Phenthoate	92.4	91.4	13.4	369.4	93.3	4.2	20.0	74.2	17.4	200	91.0	15.4
193	Phorate	314.0	100.5	6.3	1256.0	87.5	7.3	10.0	60.1	23.8	100	76.9	6.8
194	Phorate sulfone	42.0	91.0	9.7	168.0	91.8	2.6	10.0	101.6	7.3	100	97.1	10.5
195	Phosalone	48.0	85.2	7.2	192.2	93.1	4.4	20.0	74.9	1.5	200	78.9	8.3
196	Phosphamidon	3.9	98.4	9.9	15.5	98.2	4.0	20.0	70.9	34.2	200	89.4	8.5
197	Phthalic acid,benzyl butyl ester	21.1	106.7	22.4	84.3	107.1	10.7	10.0	82.2	3.0	100	91.1	11.8
198	Phthalimide	14.3	94.1	15.1	57.3	83.9	10.9	80.0	55.2	27.1	800	34.7	70.7
199	Picoxystrobin	8.4	93.1	17.2	33.8	95.3	6.5	20.0	77.9	12.4	200	89.6	7.2
200	Piperonyl butoxide	1.1	89.7	10.0	4.5	102.8	6.4	10.0	85.7	12.7	100	81.4	7.3
201	Piperophos	9.2	84.8	8.5	37.0	95.7	4.6	30.0	62.5	6.4	300	77.7	13.2
202	Pirimicarb	0.2	95.6	27.1	0.6	85.8	12.1	20.0	72.9	11.1	200	88.8	13.1
203	Pirimiphos methyl	0.2	98.4	6.4	0.8	99.8	4.4	20.0	72.3	8.4	200	76.9	15.0
204	Pirimiphos-ethyl	0.0	89.6	21.6	0.2	103.3	8.6	10.0	74.6	7.7	100	73.9	18.4
205	Pretilachlor	0.3	85.6	16.8	1.3	76.0	26.1	20.0	74.3	12.2	200	80.8	30.2
206	Procymidone	28.9	89.8	10.5	115.5	96.8	6.0	10.0	74.8	13.5	100	82.0	11.9
207	Profenefos	2.0	77.4	17.1	8.1	80.7	11.0	60.0	72.2	17.1	600	67.0	9.7
208	Prometon	0.1	81.9	7.9	0.5	84.3	9.9	30.0	83.7	9.7	300	81.1	12.9
209	Prometryne	0.2	97.4	13.4	0.6	89.5	7.2	10.0	67.6	15.9	100	80.3	9.4
210	Pronamide	15.4	105.1	3.9	61.5	93.9	12.0	10.0	68.6	10.8	100	71.5	30.4
211	Propachlor	0.3	84.6	21.9	1.1	91.8	2.0	30.0	98.8	38.7	300	76.1	10.6
212	Propanil	21.6	79.0	15.3	86.4	77.4	4.7	20.0	61.2	20.1	200	68.2	13.5
213	Propargite	68.6	98.9	10.6	274.4	87.0	20.3	20.0	–	–	200	77.7	9.8
214	Propazine	0.3	93.6	16.0	1.3	96.4	11.9	10.0	73.7	7.6	100	79.0	16.1
215	Propetamphos	54.0	90.0	2.7	216.0	94.3	10.7	10.0	70.2	8.6	100	85.2	4.6
216	Propham	110.0	87.1	12.3	440.0	86.8	6.8	10.0	71.6	21.7	100	83.0	25.3
217	Propisochlor	0.3	89.4	11.3	1.1	98.0	11.4	10.0	77.2	10.0	–	–	–
218	Prosulfocarb	0.4	90.5	17.9	1.5	103.2	7.9	10.0	80.8	5.8	100	89.4	12.7

219	Pyraclofos	1.0	86.0	15.5	4.0	96.8	7.7	80.0	75.2	10.4	–	–	–
220	Pyraclostrobin	0.5	91.1	26.5	2.0	97.2	7.7	120.0	–	–	1200	88.3	2.0
221	Pyrazophos	1.6	91.1	28.8	6.5	85.3	13.3	20.0	73.7	28.0	200	82.3	17.4
222	Pyributicarb	0.3	98.8	11.0	1.4	103.4	10.8	20.0	68.9	8.3	200	88.2	7.1
223	Pyridaphenthion	0.9	98.5	13.1	3.5	95.2	4.9	10.0	75.6	25.1	100	82.8	17.3
224	pyrifenox	0.0	101.7	20.5	0.0	94.3	13.1	80.0	78.5	20.9	800	89.8	10.8
225	Pyrifralid	0.6	96.3	6.4	2.5	86.9	20.3	10.0	74.3	13.3	100	86.9	19.9
226	Pyrimethanil	0.7	82.5	29.0	2.7	90.0	13.3	10.0	82.3	12.2	100	78.6	9.6
227	Pyrimidifen	0.5	53.1	89.1	1.9	80.5	87.9	20.0	43.6	62.9	–	–	–
228	Pyriproxyfen	0.4	103.0	6.3	1.7	97.9	1.9	10.0	74.4	21.1	100	88.1	11.4
229	Pyroquilon	0.1	97.7	7.4	0.5	89.8	11.9	10.0	80.8	21.4	100	85.2	8.0
230	Quinalphos	2.0	86.5	14.7	8.0	92.4	9.0	10.0	71.6	15.8	100	87.7	7.7
231	Quinoclamine	7.9	73.7	29.0	31.7	91.6	13.4	40.0	–	–	400	36.6	17.4
232	Quinoxyphen	5.1	109.0	6.4	20.5	98.2	16.6	10.0	73.5	13.4	100	81.8	4.9
233	Rabenzazole	1.3	68.1	22.3	5.3	73.3	11.4	10.0	79.8	16.5	100	87.7	13.3
234	Resmethrin-2	0.0	93.1	14.8	0.0	96.9	14.5	20.0	76.6	29.1	200	60.5	6.6
235	Ronnel	4.4	91.0	9.3	17.5	100.8	10.3	20.0	70.7	13.1	200	83.8	4.8
236	Sebutylazine	0.3	95.9	14.6	1.3	90.9	21.5	10.0	83.6	4.0	100	89.2	12.9
237	Secbumeton	0.1	77.6	25.2	0.3	87.2	4.4	10.0	68.5	8.7	100	86.7	5.9
238	Silafluofen,	202.7	89.2	1.9	810.7	89.9	13.2	10.0	65.9	6.9	100	90.9	15.6
239	Simeconazole	2.9	92.9	18.0	11.8	98.6	5.8	20.0	79.6	16.1	200	88.2	11.9
240	Simeton	0.4	59.4	77.0	1.5	94.4	7.1	20.0	81.6	4.7	200	88.9	14.0
241	Spirodiclofen	9.9	66.9	13.1	39.6	58.4	28.2	80.0	48.7	18.8	800	61.8	26.3
242	Sulfallate	207.2	102.2	15.0	828.8	100.8	12.6	20.0	68.4	12.3	200	70.9	9.1
243	Sulfotep	2.6	98.6	12.0	10.4	97.7	3.2	10.0	71.2	10.1	100	74.8	12.7
244	Sulprofos	5.8	83.4	16.1	23.4	88.2	12.4	20.0	66.4	12.4	200	80.0	5.9
245	Tebuconazole	2.2	92.3	13.9	8.9	91.4	6.5	30.0	66.8	9.2	300	79.6	7.9
246	Tebufenpyrad	0.4	89.8	10.5	1.6	99.1	7.9	10.0	85.5	6.5	100	89.2	12.8
247	Tebupirimfos	0.1	96.8	6.5	0.5	95.3	11.2	20.0	70.6	10.5	200	88.6	8.5
248	Tebutam	0.1	83.4	7.0	0.5	94.6	10.5	20.0	77.8	4.7	200	88.5	12.6
249	Tebuthiuron	0.2	100.6	19.8	0.9	98.3	12.2	40.0	76.3	28.6	400	75.4	9.0
250	Terbacil	0.9	53.2	11.9	3.5	49.9	5.3	20.0	–	–	200	76.2	28.9
251	Terbufos	2240.0	97.4	8.6	8960.0	70.2	15.9	20.0	68.6	9.8	200	74.6	10.3

(Continued)

TABLE 1.1.14 Recovery and Precision Date for 270 Pesticides Residues in Teas Applicable for Both LC–MS/MS and GC–MS (n = 4) (Cont.)

No	Pesticides	LC–MS/MS						GC–MS					
		Levels (µg/kg)	Ave. %	RSD %	Levels µg/kg	Ave.	RSD %	Levels (µg/kg)	Ave. %	RSD%	Levels (µg/kg)	Ave%	RSD%
252	Terbuthylazine	0.5	107.8	3.6	1.9	95.6	4.1	10.0	81.2	11.2	100	80.2	10.2
253	Tetrachlorvinphos	2.2	101.1	5.7	8.9	96.3	4.2	30.0	81.5	28.4	300	76.0	5.9
254	Tetraconazole	1.7	91.1	8.9	6.9	90.5	14.9	30.0	82.0	9.8	300	77.7	7.9
255	Tetramethrin	1.8	86.0	10.5	7.3	95.0	10.0	10.0	70.6	16.6	100	87.2	9.9
256	Thenylchlor	24.1	83.0	13.1	96.6	94.5	8.4	20.0	73.9	13.6	200	95.3	11.8
257	Thiazopyr	2.0	99.3	9.9	7.8	96.6	8.6	20.0	71.5	13.7	200	90.9	7.9
258	Thiobencarb	3.3	92.5	13.0	13.2	96.8	13.1	20.0	73.8	12.3	200	78.1	15.2
259	Thiometon	19.3	104.2	29.7	77.1	83.8	10.8	10.0	60.1	10.9	100	67.5	15.9
260	Thionazin	22.7	96.6	9.3	90.7	89.2	23.0	10.0	71.4	24.5	100	79.6	5.5
261	Tolclofos methyl	66.6	90.3	5.9	266.2	83.2	3.8	10.0	85.2	13.0	100	80.3	11.1
262	trans-Permethin	0.2	95.4	20.4	0.6	84.5	19.0	10.0	72.7	6.8	100	85.4	8.1
263	Triadimefon	7.9	96.2	17.4	31.5	98.7	7.8	20.0	67.3	9.2	200	71.4	23.2
264	Triallate	1.5	106.1	11.2	6.2	97.8	23.7	20.0	80.7	6.7	200	81.7	12.5
265	Triazophos	0.7	93.8	15.5	2.7	99.0	5.3	30.0	–	–	300	84.2	19.4
266	Trietazine	0.6	97.9	9.6	2.4	98.9	2.6	10.0	80.5	4.9	100	85.4	13.5
267	Trifloxystrobin	2.0	106.6	9.3	8.0	96.6	5.6	40.0	81.7	16.4	400	88.9	12.1
268	Trifluralin	1240.0	96.2	11.4	4960.0	98.5	8.1	20.0	71.9	12.3	200	75.5	9.6
269	Tri-n-butyl phosphate	0.4	102.6	5.3	1.5	98.3	3.0	20.0	79.4	5.8	200	88.7	13.8
270	Vinclozolin	0.1	31.1	108.4	0.3	60.7	70.1	10.0	66.4	9.1	100	75.2	8.6

TABLE 1.1.15 Recovery and Precision Date for 178 Pesticides Residues in Teas Applicable for LC–MS/MS (n = 4)

No.	Pesticides	Levels µg/kg	Ave %	RSD %	levels µg/kg	Ave %	RSD %
1	1-Naphthy acetamide	0.3	93.7	7.2	1.1	99.8	13.1
2	3,4,5-Trimethacarb	0.3	106.1	22.6	1.4	90.0	8.0
3	4-Aminopyridine	0.0	103.3	10.9	0.1	97.4	7.4
4	6-Chloro-4-hydroxy-3-phenyl-pyridazin	1.7	99.3	4.9	6.6	103.0	8.2
5	Acetamiprid	1.4	87.8	12.5	5.8	97.7	8.9
6	Acrylamide	5.9	–	–	23.7	86.0	26.5
7	Aldicarb	87.0	94.8	6.2	348.0	101.1	17.9
8	Aldicarb sulfone	0.7	83.5	27.7	2.9	84.8	17.1
9	Aldimorph	0.4	107.0	6.5	1.6	100.7	13.8
10	Amidithion	21.9	103.1	17.6	87.7	79.4	19.0
11	Aminocarb	0.6	88.7	17.2	2.2	92.4	25.3
12	Atrazine	0.4	102.8	19.8	1.4	107.6	18.0
13	Azoxystrobin	0.5	105.4	6.5	1.8	93.1	10.1
14	Bendiocarb	3.2	87.3	24.0	12.7	90.1	11.9
15	Bensulide	34.2	95.7	13.2	136.8	96.0	7.2
16	Benzofenap	0.0	–	–	0.1	96.4	75.7
17	Benzoximate	0.7	63.3	15.9	2.6	75.5	4.9
18	Bioallethrin	198.0	99.4	6.2	792.0	94.5	5.9
19	Brompyrazon	3.6	77.0	26.8	14.4	93.3	19.0
20	Bromuconazole	3.1	102.3	6.0	12.6	99.1	3.7
21	Butocarboxim	0.1	98.8	12.4	0.2	99.6	11.1
22	Butoxycarboxim	8.9	–	–	35.5	93.2	12.9
23	Buturon	9.0	88.0	22.4	35.8	96.5	12.0
24	Carbendazim	0.5	101.4	4.6	1.9	105.7	13.9
25	Carbetamide	3.6	91.4	22.4	14.6	94.7	15.2
26	Carbofuran	13.1	86.4	14.3	52.2	84.9	16.1
27	carpropamid	1.7	98.6	7.8	6.9	88.9	22.9
28	Cartap hydrochloride	2080.0	99.6	5.7	8320.0	85.7	25.1
29	Chloramphenicolum	3.9	57.7	22.9	15.5	58.6	19.3
30	Chlordimeform hydrochloride	0.9	–	–	3.5	84.7	28.2
31	Chloridazon	2.3	72.8	8.3	9.3	94.6	6.8

(Continued)

TABLE 1.1.15 Recovery and Precision Date for 178 Pesticides Residues in Teas Applicable for LC–MS/MS (n = 4) (Cont.)

No.	Pesticides	Levels µg/kg	Ave %	RSD %	levels µg/kg	Ave %	RSD %
32	Chlorobenzuron	20.4	55.8	7.0	81.6	56.2	15.8
33	Chlorotoluron	0.6	103.2	8.7	2.5	95.1	2.3
34	Chloroxuron	0.2	96.9	21.3	0.6	72.5	20.5
35	Chlorphoxim	77.6	85.9	6.5	310.3	96.9	11.8
36	Chlorprifos methyl	16.0	91.3	5.6	64.0	85.5	4.6
37	Chlorpyrifos	53.8	101.4	6.5	215.2	93.2	9.2
38	Chlorsulfuron	2.7	11.2	46.5	11.0	4.4	4.6
39	Chlorthiamid	0.3	45.9	78.1	1.2	95.9	83.6
40	Chlorthion	0.4	94.6	11.2	1.6	98.4	12.4
41	Cinidon-ethyl	4.9	–	–	19.4	73.4	9.1
42	Cis and trans diallate	89.2	94.3	6.6	356.8	93.3	6.9
43	Clofentezine	0.0	96.1	10.7	0.1	85.7	19.2
44	Clothianidin	63.0	103.3	4.1	252.0	99.3	3.9
45	Coumatetralyl	1.4	95.5	21.0	5.4	78.1	20.6
46	Cumyluron	0.4	59.2	77.3	1.8	97.3	5.4
47	Cyazofamid	0.2	50.2	88.8	0.7	90.1	138.3
48	Cymoxanil	1.9	101.8	8.2	7.4	99.2	6.2
49	Cythioate	80.0	104.2	3.3	320.0	95.8	2.7
50	Dalapon	230.7	46.8	27.6	923.0	97.7	25.0
51	Daminozide	2.6	84.9	21.0	10.4	86.9	6.7
52	Dazomet	127.0	93.9	10.5	508.0	93.1	29.1
53	Demeton(O+S)	6.8	69.5	20.9	27.1	71.6	17.3
54	Demeton-s-methyl sulfone	0.7	97.7	6.8	2.6	85.5	10.9
55	Demeton-S-methyl sulfoxide	1.3	85.8	5.4	5.2	87.0	13.3
56	Diafenthiuron	0.3	89.3	12.7	1.1	114.3	15.4
57	Dimethametryn	0.1	93.1	5.9	0.4	84.9	4.0
58	Dinoseb acetate	13.8	101.4	18.2	55.0	84.6	22.3
59	Dinotefuran	3.4	48.2	80.3	13.6	65.1	60.0
60	Dioxabenzofos	4.6	81.6	30.0	18.5	93.5	6.2
61	Diphenylamin	0.1	94.4	7.3	0.6	74.7	12.6
62	Dithiopyr	10.4	88.6	5.8	41.6	102.9	19.8

No.	Compound						
63	Diuron	1.6	102.7	8.8	6.2	102.7	3.3
64	DMST	40.0	89.2	17.3	160.0	89.0	6.7
65	Dodine	2.7	72.7	69.8	10.7	104.5	16.7
66	Epoxiconazole	4.1	85.5	18.4	16.2	85.9	17.0
67	Etaconazole	1.8	90.8	23.5	7.1	94.0	9.8
68	Ethidimuron	1.5	82.1	24.7	6.0	103.0	17.9
69	Ethiprole	39.9	57.3	10.6	159.4	56.3	13.4
70	Ethoxyquin	0.1	79.6	9.6	0.5	29.6	41.2
71	Ethylene thiourea	52.2	81.2	11.5	208.8	83.2	7.5
72	Etobenzanid	0.3	82.5	9.1	1.1	76.6	21.1
73	Famoxadone	45.3	47.7	6.3	181.2	47.8	2.9
74	Famphur	3.6	90.6	11.4	14.4	85.8	7.7
75	Fenfuram	0.8	84.1	16.1	3.1	79.2	16.2
76	Fenoxycarb	18.3	73.1	29.9	73.1	89.4	24.0
77	Fenthion oxon	0.4	75.7	20.7	1.6	94.1	14.1
78	Fentrazamide	4.1	97.2	13.9	16.5	92.8	14.5
79	Fenuron	1.0	104.7	4.8	4.1	99.5	1.0
80	Flamprop methyl	20.2	104.9	7.7	80.8	98.1	2.5
81	Fluazinam	70.6	47.4	7.0	282.4	49.2	1.7
82	Flubenzimine	0.3	79.4	23.1	1.0	21.6	109.3
83	Fluometuron	0.9	95.9	19.1	3.7	92.6	8.8
84	Flurazuron	26.8	97.5	8.1	107.2	92.1	9.2
85	Fluridone	0.2	84.1	24.8	0.7	93.7	11.6
86	Fluthiacet methyl	5.3	85.0	6.6	21.2	83.9	16.7
87	Folpet	4.6	97.6	9.2	18.5	96.2	5.5
88	Fosthiazate	0.4	100.1	6.5	1.6	69.6	13.8
89	Furathiocarb	0.1	100.6	8.5	0.3	119.6	9.9
90	Haloxyfop-2-ethoxyethyl	0.1	101.5	14.5	0.3	92.1	20.3
91	Haloxyfop-methyl	2.6	94.2	8.5	10.6	93.4	4.5
92	Hymexazol	74.7	71.7	24.1	298.9	117.1	13.0
93	Imazamox	0.4	66.9	18.2	1.6	86.3	14.6
94	Imazapic	0.2	23.8	173.9	0.8	1.5	28.8
95	Imazethapyr	1.1	2.5	46.0	4.5	1.0	63.9
96	Imibenconazole	3.4	84.2	20.5	13.7	92.6	7.8

(Continued)

TABLE 1.1.15 Recovery and Precision Date for 178 Pesticides Residues in Teas Applicable for LC–MS/MS (n = 4) (Cont.)

No.	Pesticides	Levels µg/kg	Ave %	RSD %	levels µg/kg	Ave %	RSD %
97	Imidacloprid	22.0	90.8	10.5	88.0	104.8	14.5
98	Indoxacarb	7.5	98.6	8.3	30.2	95.2	10.7
99	Iprovalicarb	2.3	93.5	19.4	9.3	90.3	11.7
100	Isoprocarb	2.3	97.8	6.1	9.2	86.5	9.4
101	Isoproturon	0.1	102.0	25.3	0.5	96.5	25.5
102	Isouron	0.4	99.0	6.7	1.6	101.9	5.3
103	Isoxaben	0.2	99.4	9.7	0.7	99.2	3.9
104	Isoxaflutole	3.9	72.7	23.1	15.6	62.9	27.3
105	Kadethrin	3.3	49.1	75.6	13.3	89.1	18.1
106	Kelevan	9642.8	47.7	10.0	38571.3	49.2	15.6
107	Maleic hydrazide	80.0	96.4	21.9	320.0	88.7	11.4
108	Mepanipyrim	0.3	90.5	8.8	1.3	92.7	18.2
109	Mepiquat chloride	0.9	101.8	14.3	3.6	103.3	4.5
110	Metconazole	1.3	101.6	5.2	5.3	100.2	5.6
111	Methamidophos	4.9	65.1	10.8	19.7	76.7	11.5
112	Methiocarb	41.2	91.3	14.3	164.8	99.5	12.3
113	Methobromuron	16.8	99.6	9.8	67.4	96.6	4.3
114	Methomyl	0.3	107.3	13.9	1.3	90.0	15.7
115	Methoxyfenozide	3.7	81.5	18.1	14.8	92.9	13.7
116	Metolcarb	8.5	94.5	17.8	33.9	79.5	11.9
117	Metoxuron	0.6	86.7	15.0	2.6	91.9	9.7
118	Monuron	34.7	85.6	8.7	138.9	87.5	14.9
119	Naptalam	2.0	4.4	200.0	7.8	75.8	200.0
120	Neburon	7.1	100.6	11.6	28.4	94.6	12.2
121	Nicotine	2.2	85.2	19.2	8.8	79.9	26.0
122	Nitenpyram	17.1	91.3	6.7	68.5	72.0	28.1
123	Novaluron	8.0	100.1	9.3	32.2	95.9	6.8
124	Oesmetryn	0.2	83.9	18.6	0.7	91.6	10.9
125	Omethoate	0.3	78.4	12.4	1.3	88.2	17.4
126	Oxycarboxin	0.9	30.9	72.6	3.6	16.8	15.7
127	Paraoxon methyl	0.8	83.1	18.3	3.1	78.4	9.0

No.	Name						
128	Phenmedipham	17.9	72.2	17.9	15.1	84.7	4.5
129	Phorate sulfoxide	3.5	78.7	1473.1	23.5	104.4	368.3
130	Phosfolan	10.4	102.3	0.7	4.5	92.5	0.2
131	Phosmet	10.4	69.6	23.6	22.5	93.9	5.9
132	Phoxim	10.7	97.2	331.2	9.6	87.3	82.8
133	Phthalic acid, biscyclohexyl ester	20.0	113.4	0.9	5.2	97.1	0.2
134	Phthalic acid, dibutyl ester	9.9	96.3	52.8	6.7	98.0	13.2
135	Picolinafen	10.9	94.1	2.9	14.2	94.9	0.7
136	Prallethrin	6.2	99.1	78.4	23.2	101.9	19.6
137	Propaquizafop	22.0	88.3	1.7	13.5	92.3	0.4
138	Propiconazole	6.2	93.3	7.0	10.5	111.5	1.8
139	Propoxur	2.9	96.1	97.6	8.3	104.5	24.4
140	Prothoate	18.2	101.1	9.8	17.3	74.7	2.5
141	Pymetrozin	8.8	74.8	137.1	8.1	87.1	34.3
142	Pyrazoxyfen	29.0	104.1	0.0	7.9	103.2	0.0
143	Pyrethrin	19.5	114.2	47.7	27.1	79.8	11.9
144	Pyrimitate	8.7	93.3	0.7	14.2	95.6	0.2
145	Quizalofop-ethyl	10.4	93.7	2.7	8.3	92.8	0.7
146	Rotenone	10.5	88.8	3.1	77.9	57.4	0.8
147	Sethoxydim	20.7	88.3	1.6	11.7	79.0	0.4
148	Simetryn	8.0	85.7	0.5	23.2	77.1	0.1
149	Spinosad	25.2	72.3	2.3	8.0	97.4	0.6
150	Spiroxamine	21.0	78.1	0.2	22.6	74.5	0.1
151	Tau-Fluvalinate	13.5	100.6	30.7	8.3	97.6	7.7
152	Temephos	19.9	94.7	4.9	11.2	86.2	1.2
153	TEPP	17.4	11.9	41.6	35.4	5.8	10.4
154	Terbufos sulfone	11.9	96.5	11.8	5.1	97.3	3.0
155	Terbumeton	6.4	98.3	0.4	17.5	88.4	0.1
156	Terrbucarb	8.2	90.6	8.4	18.2	97.4	2.1
157	Tert-Butylamine	22.1	97.5	1.6	15.9	93.4	0.4
158	Thiacloprid	12.8	90.3	1.5	26.3	78.4	0.4
159	Thiamethoxam	5.2	99.9	132.0	7.3	105.3	33.0
160	Thiodicarb	12.4	104.1	157.5	27.2	103.0	39.4
161	Thiofanox	14.8	78.7	628.0	6.6	91.9	157.0

(Continued)

TABLE 1.1.15 Recovery and Precision Date for 178 Pesticides Residues in Teas Applicable for LC–MS/MS ($n = 4$) (Cont.)

No.	Pesticides	Levels µg/kg	Ave %	RSD %	levels µg/kg	Ave %	RSD %
162	Thiofanox sulfone	8.0	143.6	105.5	32.1	79.1	8.6
163	Thiofanox-sulfoxide	8.3	82.3	19.9	33.2	97.7	21.1
164	Tralkoxydim	0.3	86.8	15.3	1.3	86.7	5.5
165	Triadimenol	10.6	92.2	18.6	42.2	95.3	10.8
166	Triamiphos	0.0	93.2	15.7	0.0	86.3	16.8
167	Trichloronat	2.2	95.2	13.5	8.9	88.5	21.3
168	Trichlorphon	1.1	87.3	16.3	4.5	82.3	15.3
169	Tricyclazole	0.4	110.2	12.8	1.6	98.1	12.4
170	Tridemorph	2.6	78.3	23.9	10.4	72.9	9.0
171	Triflumuron	3.9	97.1	8.3	15.7	91.4	3.7
172	Tri-iso-butyl phosphate	0.4	100.5	8.7	1.6	98.0	3.4
173	Triticonazole	3.0	77.9	15.7	12.1	88.2	11.6
174	Uniconazole	0.1	97.2	9.1	0.3	91.3	15.3
175	Vamidothion	1.5	48.8	72.8	6.1	93.1	17.1
176	Vamidothion sulfone	15.9	39.0	103.6	63.5	71.8	120.5
177	Warfarin	0.9	100.7	12.5	3.6	95.0	7.6
178	Zeta cypermethrin	0.0	98.3	9.2	0.1	99.1	25.7

TABLE 1.1.16 Recovery and Precision Date for 220 Pesticides Residues in Teas Applicable for GC–MS (n = 4)

No.	Pesticides	Levels µg/kg	Ave.%	RSD%	Levels µg/kg	Ave.%	RSD%
1	2,3,4,5-Tetrachloroaniline	20	71.6	8.5	200	82.0	15.5
2	2,3,4,5-Tetrachloroanisole	10	74.2	5.8	100	82.1	15.9
3	2,3,5,6-Tetrachloroaniline	10	74.0	7.6	100	81.9	15.8
4	2,4,5-T	200	–	–	2000	97.2	21.8
5	3,4,5-Trimethacarb	80	71.7	16.0	–	–	–
6	3,5-Dichloroaniline	10	60.2	11.0	100	57.7	24.2
7	4,4-Dibromobenzophenone	10	84.9	8.4	100	87.3	15.0
8	4-Chlorophenoxy acetic acid	100	–	–	1000	75.8	21.9
9	Alachlor	30	98.2	98.2	300	78.2	13.2
10	Aldrin	20	74.8	8.0	–	–	–
11	Alpha-cypermethrin	20	97.4	23.4	200	84.1	17.0
12	α-HCH	10	106.3	8.8	100	75.5	12.0
13	Amitraz	30	45.5	27.2	300	44.7	10.7
14	Anthraquinone	10	–	–	–	–	–
15	Atrizine	10	83.7	6.9	100	78.9	9.3
16	Azaconazole	40	–	–	400	86.2	12.3
17	BDMC-1	20	83.7	4.1	200	85.6	19.2
18	BDMC-2	20	77.0	12.8	200	96.3	21.8
19	Enfluralin	10	80.7	11.3	100	77.2	10.0
20	Beta-hch	10	69.3	18.8	100	83.5	13.9
21	Bifenox	20	75.0	16.5	200	76.8	11.7
22	Bifenthrin	10	88.5	12.6	100	107.8	22.6
23	Bioallethrin-1	40	81.2	9.1	400	80.0	12.1
24	Bioallethrin-2	40	87.0	18.4	400	85.3	8.1
25	Biphenyl	10	79.0	5.6	0		
26	Bromocylen	10	77.0	5.2	100	84.7	15.5
27	Bromofos	20	76.0	16.4	200	74.0	9.2
28	Bromopropylate	20	70.4	8.4	200	77.5	13.9
29	Bromuconazole-1	20	93.5	5.1	200	95.8	13.3
30	Bromuconazole-2	20	89.2	9.5	200	94.4	13.1
31	Butamifos	10	76.6	13.5	–	–	–

(Continued)

TABLE 1.1.16 Recovery and Precision Date for 220 Pesticides Residues in Teas Applicable for GC–MS ($n = 4$) (Cont.)

No.	Pesticides	Levels μg/kg	Ave.%	RSD%	Levels μg/kg	Ave.%	RSD%
32	Cafenstrole	40	61.8	41.6	–	–	–
33	Carbofenothion	20	72.4	11.2	200	74.1	16.3
34	Carfentrazone-ethyl	20	80.5	15.7	200	93.1	15.2
35	Chlorbenside	20	67.3	4.8	200	78.9	8.4
36	Chlorethoxyfos	20	–	–	200	83.0	9.9
37	Chlorfenapyr	80	–	–	800	77.2	7.8
38	Chlorfenprop-methyl	10	81.1	14.5	100	85.6	14.1
39	Chlorfenson	20	77.9	11.4	200	77.2	13.7
40	Chlorfenvinphos	30	77.8	19.9	300	77.4	9.6
41	Chlorfurenol	30	70.0	17.2	300	85.7	9.4
42	Chlorobenzilate	10	84.7	10.4	100	83.6	15.7
43	Chloroneb	10	–	–	100	71.1	9.9
44	Chlorprifos-methyl	10	74.7	13.1	100	74.8	11.6
45	Chlorpropylate	10	74.7	8.3	100	77.7	15.3
46	Chlorpyifos(ethyl)	30	78.8	19.3	300	94.1	11.7
47	Chlorthal-dimethyl	20	72.1	10.8	200	87.5	6.8
48	Chlozolinate	20	45.6	21.9	200	51.5	21.4
49	Cinmethylin	20	–	–	200	93.5	11.3
50	Cis-1,2,3,6-tetrahydrophthalimide	30	–	–	300	45.8	33.2
51	Cis-chlordane	20	74.0	9.1	200	77.1	15.0
52	Cis-diallate	20	72.2	11.6	200	75.9	11.2
53	Cis-permethrin	–	74.6	12.6	–	87.7	10.3
54	Clomeprop	10	78.6	12.0	–	–	–
55	Coumaphos	60	70.3	9.3	600	67.7	15.1
56	Cyflufenamid	160	78.2	6.3	–	–	–
57	Cyfluthrin	120	75.1	6.2	1200	71.8	14.2
58	Cyhalofop-butyl	20	79.2	25.1	200	92.6	16.9
59	Dacthal	10	82.7	6.3	100	88.1	12.6
60	δ-HCH	20	87.6	15.2	200	73.6	14.8
61	Deltamethrin	60	66.9	9.7	600	73.8	28.6
62	de-PCB101	10	80.6	4.8	100	86.6	12.9
63	de-PCB 118	10	79.6	3.9	100	82.2	15.7

64	de-PCB 138	10	73.4	17.1	100	86.9	12.6
65	de-PCB 153	10	75.4	8.8	–	–	–
66	de-PCB180	10	80.6	7.7	100	86.3	13.7
67	de-PCB 28	10	81.2	7.6	100	86.1	13.2
68	de-PCB 31	10	78.0	6.1	100	83.1	17.7
69	de-PCB 52	10	76.3	7.2	–	–	–
70	Desethyl-sebuthylazine	20	77.0	7.6	200	80.1	28.8
71	Desisopropyl-atrazine	80	74.4	29.1	–	–	–
72	Desmetryn	10	72.1	11.6	100	77.0	12.9
73	Dichlobenil	0	–	–	20	52.0	15.9
74	Dichloran	20	72.5	15.6	–	–	–
75	Dichlormid	20	67.1	4.9	–	–	–
76	Dichlorofop-methyl	10	71.7	12.6	100	84.1	3.7
77	Dichlorvos	60	34.2	21.4	600	38.6	36.3
78	Dicofol	20	83.7	16.7	200	82.4	25.3
79	Dieldrin	20	56.3	27.7	200	90.2	11.1
80	Difenconazole-1	60	86.9	22.4	600	80.5	8.8
81	Difenconazole-2	60	82.7	14.3	600	72.9	6.1
82	Difenoxuron	80	73.2	2.6	800	75.4	39.2
83	Dimepiperate	20	88.0	15.2	200	77.5	5.5
84	Dimethylvinphos	–	–	–	200	86.3	3.3
85	Diofenolan-1	20	59.9	50.8	200	91.2	11.6
86	Diofenolan-2	20	72.5	11.6	200	90.9	12.9
87	Dioxathion	40	–	–	400	95.4	24.7
88	Diphenylamin	10	72.0	15.5	100	70.9	17.8
89	DMSA	80	–	–	800	70.4	9.5
90	Dodemorph	30	67.0	10.1	300	71.3	6.7
91	Endosulfan-1	60	81.4	9.3	600	82.0	12.0
92	Endosulfan-sulfate	30	75.9	14.4	300	84.6	11.5
93	Endrin	120	84.7	28.6	1200	75.2	10.8
94	Endrin ketone	160	82.2	20.2	1600	85.7	8.8
95	Epoxiconazole-1	80	70.5	14.0	800	90.7	11.2
96	Epoxiconazole-2	80	86.5	11.0	800	89.0	10.1
97	Esprocarb	20	68.7	6.6	200	83.0	11.1
98	Etaconazole-1	30	72.1	19.5	300	98.1	5.6

(Continued)

TABLE 1.1.16 Recovery and Precision Date for 220 Pesticides Residues in Teas Applicable for GC–MS ($n = 4$) (Cont.)

No.	Pesticides	Levels μg/kg	Ave.%	RSD%	Levels μg/kg	Ave.%	RSD%
99	Etaconazole-2	30	69.4	6.9	300	87.3	6.2
100	Ethalfluralin	40	62.0	11.6	400	86.0	4.1
101	Etoxazole	60	71.0	4.6	600	89.5	11.6
102	Fenamidone	10	77.6	19.5	-	-	-
103	Fenchlorphos	40	72.5	7.0	400	88.7	18.3
104	Fenpiclonil	40	85.0	34.4	400	25.5	28.0
105	Fenson	10	105.4	31.0	100	79.7	8.9
106	Fenvalerate-1	40	86.0	15.2	400	72.6	24.1
107	Fenvalerate-2	40	83.7	14.8	400	86.5	19.6
108	Fipronil	80	74.7	15.0	800	80.4	15.2
109	Flamprop-methyl	10	67.6	23.7	100	78.0	15.4
110	Fluazinam	80	-	-	800	59.4	6.0
111	Flucythrinate-1	20	87.7	14.5	200	81.5	7.7
112	Flucythrinate-2	20	83.8	3.2	200	81.8	13.1
113	Flumetralin	20	89.2	14.0	200	85.1	9.6
114	Flumioxazin	20	78.2	21.2	200	74.0	11.1
115	Fluorodifen	10	-	-	100	82.1	11.1
116	Fluotrimazole	10	83.3	1.3	100	94.3	8.7
117	Fluquinconazole	10	76.5	7.9	100	89.2	12.4
118	Fluroxypr-1-methylheptyl ester	10	88.2	3.6	100	92.3	10.9
119	Fluvalinate	120	72.3	13.8	1200	76.2	11.2
120	Formothion	20	88.3	6.8	200	78.6	9.6
121	Furmecyclox	30	46.7	50.5	300	32.2	29.5
122	Gamma-cyhaloterin-1	10	76.7	26.5	80	90.1	14.7
123	Gamma-cyhalothrin-2	10	115.8	21.9	80	95.5	12.1
124	γ-HCH	20	90.7	13.0	200	80.0	11.8
125	Halfenprox	20	72.3	10.3	200	93.1	18.3
126	Heptachlor	30	84.0	10.4	300	81.8	13.6
127	Hexachlorobenzene	10	41.9	48.0	100	31.4	57.4
128	Hexaconazole	60	73.0	7.5	600	78.5	12.8
129	Imiprothrin-1	20	77.6	8.5	-	-	-
130	Imiprothrin-2	20	80.5	9.6	-	-	-

131	Iodofenphos	20	75.9	20.0	200	76.5	7.5
132	Iprodione	40	75.1	12.5	–	–	–
133	Iprovalicarb-1	40	76.9	3.5	400	89.2	14.0
134	Iprovalicarb-2	40	72.9	6.3	400	89.6	13.9
135	Isobenzan	10	76.0	7.4	–	–	–
136	Isocarbophos	20	90.6	7.9	200	100.9	15.5
137	Isodrin	10	83.2	2.4	100	86.2	14.4
138	Isofenphos oxon	20	101.9	3.8	180	91.1	7.1
139	Isoprocarb-1	20	85.8	14.6	200	92.1	9.9
140	Isoprocarb-2	20	78.9	21.6	200	87.0	16.3
141	Isoxadifen-ethyl	20	73.1	14.9	–	–	–
142	Isoxathion	80	82.3	15.3	–	–	–
143	Lambda-cyhalothrin	10	98.1	10.7	100	77.1	13.3
144	Lenacil	100	84.3	3.2	1000	82.7	18.7
145	Leptophos	20	73.4	17.6	200	73.9	8.4
146	Mcpa-butoxyethyl ester	10	83.3	7.0	100	89.8	13.0
147	Methiocarb sulfone	320	75.9	14.2	–	–	–
148	Methoprene	40	72.9	8.2	400	79.5	17.2
149	Methoxychlor	10	72.5	20.5	100	77.1	14.5
150	Methyl-parathion	40	74.5	25.2	400	79.2	15.6
151	Metobromuron	60	78.1	32.5	–	–	–
152	Metoconazole	40	75.0	4.6	400	88.3	15.2
153	Metominostrobin-1	40	76.0	8.8	–	–	–
154	Mirex	10	84.2	27.7	100	97.4	21.7
155	Monocrotophos	400	60.2	28.0	–	–	–
156	Musk ambrette	10	–	–	100	84.3	14.8
157	Musk moskene	10	74.5	6.3	–	–	–
158	Musk tibeten	10	73.0	13.8	–	–	–
159	Musk xylene	10	75.8	29.1	100	77.7	10.3
160	Nitrapyrin	30	89.6	41.3	300	55.6	12.2
161	Nitrofen	60	67.0	7.4	600	75.3	11.8
162	Nitrothal-isopropyl	20	86.4	12.5	200	94.8	9.3
163	Norflurazon-desmethyl	40	84.8	12.5	400	92.0	10.6
164	o,p'-DDD	10	67.7	3.9	100	82.5	19.7
165	o,p'-DDE	10	84.3	5.6	100	82.6	15.8

(Continued)

TABLE 1.1.16 Recovery and Precision Date for 220 Pesticides Residues in Teas Applicable for GC–MS (n = 4) (Cont.)

No.	Pesticides	Levels μg/kg	Ave.%	RSD%	Levels μg/kg	Ave.%	RSD%
166	o,p'-DDT	20	108.8	56.6	200	72.2	10.1
167	Octachlorostyrene	10	74.0	12.0	–	–	–
168	Oxadiazone	10	70.2	2.7	100	92.3	5.6
169	Oxy-chlordane	10	78.3	12.0	100	62.4	12.4
170	p,p'-DDD	10	74.2	15.3	100	76.3	10.9
171	p,p'-DDE	10	77.4	11.2	100	79.3	16.5
172	p,p'-DDT	20	–	–	200	72.6	11.4
173	Pendimethalin	40	66.1	7.3	400	85.4	3.5
174	Pentachloroanisole	10	70.3	4.0	100	83.2	12.9
175	Pentachlorobenzene	10	58.9	8.2	100	64.2	25.6
176	Permethrin	20	84.8	7.6	200	81.0	8.8
177	Perthane	10	83.8	5.9	100	90.6	12.9
178	Phenanthrene	10	69.1	16.8	100	84.3	7.5
179	Plifenate	20	83.2	12.8	200	84.5	14.3
180	Prochloraz	60	72.7	10.0	600	70.8	11.2
181	Profluralin	40	72.5	11.9	400	76.6	10.8
182	Prohydrojamon	40	74.8	5.4	–	–	–
183	Propamocarb	30	54.9	21.5	300	22.0	49.3
184	Propaphos	20	74.2	9.2	–	–	–
185	Propiconazole-1	30	71.0	10.4	300	85.1	6.4
186	Propiconazole-2	30	70.7	9.9	300	85.0	10.3
187	Propoxur-1	80	77.2	6.9	800	87.1	8.3
188	Propoxur-2	80	69.7	29.6	800	88.9	19.3
189	Propyzamide	20	78.4	16.3	200	88.6	6.3
190	Prothiophos	10	69.9	8.8	100	87.8	4.1
191	Pyraflufen ethyl	20	–	–	200	–	–
192	Pyridaben	10	91.7	8.2	100	79.5	8.2
193	Quintozene	20	65.6	10.4	200	84.5	6.3
194	Resmethrin-1	20	96.2	8.7	200	63.5	5.4
195	S 421(octachlorodipropyl ether)-1	200	73.8	19.1	2000	85.2	17.0
196	S 421(octachlorodipropyl ether)-2	200	79.4	25.2	2000	86.1	16.3
197	Sobutylazine	20	74.5	29.5	200	66.8	16.8

198	Spiromesifen	100	57.9	16.4	1000	57.6	19.4
199	Spiroxamine-1	20	52.4	17.4	–	–	–
200	Spiroxamine-2	20	54.9	13.1	–	–	–
201	Tecnazene	20	69.0	18.0	200	67.1	8.2
202	Tefluthrin	10	81.5	12.2	100	90.3	10.9
203	Terbucarb-1	20	82.1	28.4	200	93.3	10.8
204	Terbucarb-2	20	–	–	200	94.0	11.0
205	Terbutryn	20	67.4	28.9	200	77.0	14.4
206	Tetradifon	10	75.6	17.0	100	79.4	6.4
207	Tetrasul	10	67.6	6.7	100	83.6	7.1
208	Thiabendazole	200	41.9	68.4	2000	28.0	83.9
209	Tolylfluanide	30	66.1	7.5	300	40.8	19.1
210	Trans-chlodane	10	77.8	7.2	–	–	–
211	Trans-diallate	20	78.0	17.7	200	77.5	11.2
212	Transfluthrin	10	81.3	11.2	100	81.8	15.8
213	Trans-nonachlor	10	82.6	3.6	100	91.2	11.8
214	Triadimenol-1	30	83.6	11.3	300	76.3	6.7
215	Triadimenol-2	30	88.1	2.7	300	84.9	3.6
216	Tribenuron-methyl	10	60.4	7.6	100	48.7	56.7
217	Tridiphane	40	70.0	4.4	400	84.2	10.4
218	Triphenyl phosphate	10	87.0	2.1	100	90.9	12.1
219	Vernolate	10	64.7	7.5	100	65.6	7.5
220	Zoxamide	20	76.6	7.1	200	87.4	10.2

TABLE 1.1.17 A Comparative Statistical Analysis of Recoveries and RSD for Both GC–MS and LC–MS-MS Methods (Sorted on Basis of Tables 1.1.14–1.1.16)

Item		GC–MS (490)		LC–MS-MS (448)		GC–MS (270)		LC–MS-MS (270)	
		Varieties	%	Varieties	%	Varieties	%	Varieties	%
Rec. range (%)	60–120	424	94.0	400	91.1	239	95.6	249	93.6
	<60	27	6.0	37	8.4	11	4.4	16	6.0
	>120	0	0	2	0	0	0	1	0
	Not added	39	/	9	/	20	/	4	/
RSD range (%)	<20	348	77.2	333	75.9	188	75.2	211	79.3
	20–30	86	19.0	77	17.5	56	22.4	42	15.8
	>30	17	3.8	29	6.6	6	2.4	13	4.9
	Not added	39	/	9	/	20	/	4	/

REFERENCES

[1] Wan, H., Xia, H., Chen, Z., 1991. Food Addit. Contam. 8, 497–500.

[2] Weixuan, L., 2002. The Compilation of Residue Limits Standards for Pesticides and Veterinary Drugs in Foodstuffs in the World—Export. Dalian Maritime University Press, Dalian, China.

[3] Weixuan, L., 2004. Chin. J.Teas 26 (2), 4–7.

[4] EC149/2008: http://eur lex.europa.eu/LexUriServ/LexUriServ.do?uri=CELEX:32008R0149:EN:NOT.

[5] Zhao, H.X., Xiao, G.Y., Peng, G.L., Luo, Z.Y., 2006. Food Sci. 27 (12), 894–896.

[6] Wong, J.W., Webster, M.G., Halverson, C.A., Hengel, M.J., Ngim, K.K., Ebeler, S.E., 2003. J. Agric. Food Chem. 51, 1148–1161.

[7] Jiang, Y.X., Ye, L., Tang, M.R., 2007. Fenxi Shiyan. 26 (1), 97–101.

[8] Shen, C.Y., Shen, W.J., Jiang, Y., Zhao, Z.Y., Chen, H.L., Wu, B., Xu, J.Z., 2006. Fenxi Huaxue 34, 36–40, Supp.

[9] Yuan, N., Yu, B.B., Zhang, M.S., Zeng, J.B., Chen, X., 2006. Sepu 24 (6), 636–640.

[10] Yue, Y., Zhang, R., Fan, W., Tang, F., 2008. J. AOAC Int. 5, 1210–1217.

[11] Cho, S.K., Abd El Aty, A.M., Choi, J.H., Jeong, Y.M., Shin, H.C., Chang, B.J., Lee, C., Shim, J.H., 2008. J. Sep. Sci. 10, 1750–1760.

[12] Hu, B.Z., Shen, G.J., Shao, T.F., Xie, L.P., 2009. Fenxi Shiyan. 28 (1), 80–83.

[13] Zeng, X.X., Wan, Y.Q., Xie, M., 2008. Y. Fenxi Ceshi Xuebao 24 (6), 636–640.

[14] Jin, B.H., Chen, P.J., Xie, L.Q., Zhao, Q.H., Wu, W.D., Lan, F., Lin, L., 2007. Fenxi Ceshi Xuebao 26 (1), 104–106.

[15] Hu, Y.Y., Zheng, P., He, Y.Z., Sheng, G.P., 2005. J. Chromatogr. A 1098 (12), 188–193.

[16] Schurek, J., Portolés, T., Hajslova, J., Riddellova, K., Hernández, F., 2008. Anal. Chim. Acta 2, 163–172.

[17] Ochiai, N., Sasamoto, K., Kanda, H., Yamagami, T., David, F., Tienpont, B., Sandra, P., 2005. J. Sep. Sci. 28 (9–10), 1083–1092.

[18] Zhenyun, L., Zongmao, C., Fengjian, L., Fubing, T., Guangmin, L., 2008. Chin. J. Chromatogr. 26 (5), 568–576.

[19] Huang, Z., Li, Y., Chen, B., Yao, S., 2007. J. Chromatogr. B 853, 154–162.

[20] Yang, X., Xu, D.C., Qiu, J.W., Zhang, H., Zhang, Y.C., Dong, A.J., Ma, Y., Wang, J., 2009. Chem. Papers 1, 39–46.

[21] Hirahara, Y., Kimura, M., Inoue, T., Uchikawa, S., Otani, S., Hirose, H., Suzuki, S., Uchida, Y., 2006. J. Food Hyg. Soc. Jpn. 47 (5), 225–231.

[22] Mol, H.G.J., Rooseboom, A., Dam, R., van Roding, M., Arondeus, K., Sunarto, S., 2007. Anal. Bioanal. Chem. 389 (6), 1715–1754.

[23] Pang, G.F., Fan, C.L., Liu, Y.M., Cao, Y.Z., Zhang, J.J., Li, X.M., Li, Z.Y., Wu, Y.P., Guo, T.T., 2006. J. AOAC Int. 89 (3), 740–771.

[24] Pang, G.F., Liu, Y.M., Fan, C.L., Li, Z.Y., Wu, Y.P., Guo, T.T., 2006. Anal. Bioanal. Chem. 384, 1366–1408.

[25] Pang, G.F., Cao, Y.Z., Zhang, J.J., Fan, C.L., Liu, Y.M., Li, X.M., Jia, G.Q., Li, Z.Y., Shi, Y.Q., Wu, Y.P., Guo, T.T., 2006. J. Chromatogr. A 1125, 1–30.

[26] Pang, G.F., Fan, C.L., Liu, Y.M., Cao, Y.Z., Zhang, J.J., Fu, B.L., Li, X.M., Li, Z.Y., Wu, Y.P., 2006. Food Addit. Contam. 23 (8), 777–810.

[27] Wang, L.B., Li, Cao, Peng, C.F., Li, X.Q., Xu, C.L., 2008. J. Chromatogr. Sci. 46, 424–429.

[28] Maštovská, K., Lehotay, S.J., 2004. J. Chromatogr. A 1040, 259–272.

[29] Gülbakan, B., Uzun, C., Celikbıçak, Ö., Güven, O., Salih, B., 2008. React. Funct. Polym. 68, 580–593.

[30] Peng, C.F., Kuang, H., Li, X.Q., Xu, C.L., 2007. Chem. Papers 61, 1–5.

Chapter 1.2

Research on the Evaluation of the Effectiveness of the Nontargeting, High-Throughput Method for the Detection of 494 Pesticides Residue Using GC-Q-TOF/MS Technique in Tea

Chapter Outline

1.2.1 INTRODUCTION

As a high-resolution mass spectrometry method, GC-Q-TOF/MS has had, in recent years, rapid growth with the following features: (1) accurate-mass full-spectrum data, the full-spectrum data acquisition was conducted in scan mode, the mass deviation was less than 5 ppm; (2) narrow window-XICs, the adoption of narrow window 20 ppm extraction ion chromatogram can reduce the interference of background and equal mass ion, thus increased signal-to-noise ratio, compared with traditional quadruple mass analyzer, this method can provide reliable basis for the identification of compound; (3) the MassHunter series software was an effective method of analysis because it can be applied in the identification of known or unknown substances, so the software can be adopted in the analysis of compound using GC-Q-TOF/MS. With the continuous development of the software, the matching analysis using PCDL database can enhance the reliability of pesticide detection and confirmatory analysis, meanwhile it can save analysis time. The confirmatory analysis can be conducted without a standard substance using this method, and the method was an important approach for the analysis of target compound and has important significance in the field of residue analysis.

Considering the main characteristics of GC-Q-TOF/MS, this method is suited for the analysis of multiresidue and unknown substances, and rapid identification of target compounds and nontarget compounds can be realized by the application of the software. The method was proved to be of a low detection limit, and the resolution was up to 10,000 (FWHM), so it was an effective method for detection and identification of compounds. The application of GC-Q-TOF/MS can promote the pesticide residue detection technology to enter a new era of accurate identification without standard substances.

Tea is an important industrial crop as well as a traditional large export product. In recent years, the main tea-importing countries, such as Japan and the European Union, have been expanding the range of pesticide residues and enhancing the testing standard continuously, and they have issued very demanding maximum residue limits (MRLs) for these pesticides in tea. So it is of great practical value to establish an accurate method for the determination of multipesticide residues in tea.

The tea matrix is relatively complex because it contains a large number of pigments, alkaloids, and phenolic compounds. As a consequence, sample pretreatment is one of the most important steps in the analysis of pesticides, which have various physical and chemical properties, and it is an important factor to ensure the reliability, accuracy, and reproducibility of results. Sample pretreatment techniques include extraction, separation, purification, and concentration of samples. The aim is to extract the target compounds as well as remove the impurities that coexist with the target compound, as far as possible, thus reducing with interference of test results and avoiding the contamination of testing instruments.

The analysis of 494 pesticide residues in four kinds of tea matrix was conducted by the establishment of a PCDL database. Target pesticides can be detected and confirmed within 1 h using this method, so it is a rapid and effective method for pesticide residue screening. The characteristic ion formula, exact mass, retention time, mass spectrum, and ion abundance of the 494 pesticides are included in the PCDL database, and the database covers most commercially available pesticides suited for gas chromatography analysis. The flow chart of the pesticide residue detection process can be seen in Fig. 1.2.1.

1.2.2 EXPERIMENTAL

1.2.2.1 Reagents, Standard, and Materials

1. Solvents. Acetonitrile, toluene, hexane, acetone (LC grade), purchased from Honeywell, Co. (USA). Acetonitrile–toluene (3:1 V/V);
2. Pesticides standard. Purity, $\geq$95% (Dr. Ehrenstorfer, Augsburg, Germany);
3. Solid phase extraction column. Cleanert-TPT (Agela, China);
4. Stock standard solutions. Weigh 10 mg (accurate to 0.01 mg) of individual pesticide standards (accurate to 0.1 mg) into a 10 mL volumetric flask. Dissolve and dilute to volume with toluene, toluene–acetone combination, cyclohexane, depending on the solubility of each compound. Stock standard solutions should be stored in the dark below 4°C;
5. Mixed standard solution. Depending on properties and retention time of each pesticide, all the 494 pesticides for GC-Q-TOF/MS analysis are divided into 8 groups, A–H, and each group contains about 60 kinds of pesticide. Except for the

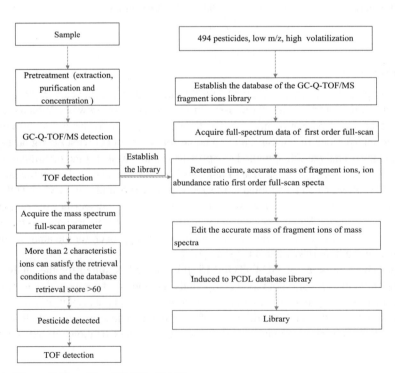

FIGURE 1.2.1 Flow chart of the pesticide residue detection process.

concentration of some pesticides dependent on the sensitivity for analysis, the concentration of mixed standard solutions was 10 mg/L. Mixed standard solutions should be stored in the dark below 4°C.

The tea samples were purchased at local supermarkets.

1.2.2.2   Equipment

Agilent 7200GC-Q-TOF/MS (Agilent Technologies, USA), transferpettor (Eppendorf, Germany); SR-2DS horizontal oscillator (TAITEC, Japan), rotary evaporator (Anhui Zhongke Zhongjia Science Instrument Co. Ltd., China), N-EVAP112 Nitrogen evaporator (OA-SYS, USA), 8893 ultrasonic cleaner (Cole-Pamer, USA), TL-602L balance (Mettler, Germany), 50 mL plastic centrifuge tube.

1.2.2.3   Condition of Gas Chromatography and Mass Spectrometry

A VF-1701 ms capillary column (30 m × 0.25 mm (i.d.) × 0.25 μm, 0.25 μm of film thickness, Agilent Technologies, USA) was used; temperature programming started at 40°C, held for 1.0 min, and then increased to 130°C at a rate of 10°C/min, increased to 250°C at a rate of 5°C/min, increased to 300°C at a rate of 10°C/min, maintained at 300°C for 5 min, helium gas as carrier gas, injection temperature: 290°C, flow rate: 1.2 mL/min, volume injected, 1.0 μL; split ratio, 10:1, injection mode: spitless.

Ionization model, EI, electron energy, 70 eV; ion source temperature, 230°C, translating line temperature, solvent delay: 6 min; the mass scan ranged from 50 to 600 m/z, acquisition rate, 2 spectra/s, heptachlor epoxide was selected for the calibration of retention time. Agilent MassHunter workstation was selected.

1.2.2.4   Pretreatment Methods for the Screening Analysis of Tea Samples

Weigh 2 g testing sample (accurate to 0.01 g) into 80 mL centrifugal tube, add 15 mL acetonitrile, extract homogeneously at 15,000 rpm 1 min, centrifuge 5 min at 4200 rpm and transfer the supernatants into 150 mL pear-shaped flask. Repeat extracting the dregs with 15 mL acetonitrile once, centrifuge, consolidate the supernatants over two times, and place them in a 40°C water bath and rotary evaporate (40 rpm) to about 2 mL before standby for cleanup.

Add about 2 cm high anhydrous sodium sulfate into Cleanert-TPT SPE cartridge, precleanse with 5 mL acetonitrile–toluene (3:1), and discard the effluents. Connect the pear-shaped flask at the lower end and place it into the fixing bracket. Transfer the aforementioned concentrated sample solutions into Cleanert-TPT cartridge, cleanse the bottle with 2 mL acetonitrile–toluene (3:1) and repeat cleansing two times, transfer the cleaning solutions into the cartridge, while attaching a 50 mL solution storage device onto the cartridge, then cleanse with 25 mL acetonitrile–toluene, collect the aforementioned effluents into pear-shaped flask, and place in a 40°C water bath and rotary evaporate (40 rpm) to about 0.5 mL. Add 40 μL heptachlor epoxide internal standard solution, and nitrogen blow to dryness and dilute immediately with hexane to 1 mL and mix uniformly to pass filtering membrane before being used for GC-Q-TOF/MS detection.

1.2.3   ESTABLISHMENT OF DATABASE

The 494 kinds of pesticide were prepared to standard solution and then determined using TOF in full-scan mode under the aforementioned gas chromatography condition, thus the retention time, accurate mass of fragment ions, formula, isotopic distribution, and abundance can be obtained. More than five fragment ions, which have relative high m/z and sensibility, were selected as target monitoring ions, and the formula, accurate mass, isotopic distribution, and abundance of each target monitoring ions of the compound, as well as retention time, were introduced to the PCDL software: thus, the generation of the database in the format of *.cdb. The TOF/MS accurate mass database of the 494 kinds of pesticide can be realized using the aforementioned method; it can be adopted for the qualitative analysis during the screening process; and the confirmation can be conducted by the retention time, accurate mass, and ion abundance ratio.

1.2.4   EVALUATION OF THE EFFICIENCY OF THE METHOD

The evaluation of the efficiency of the GC-Q-TOF/MS screening method was carried out by the selection of four kinds of tea; black tea (completely fermented), green tea (nonfermented), oolong (partially fermented), and puer (ripe tea, completely fermented). For GC-Q-TOF/MS, the investigation of spiked recovery and reproducibility was conducted at concentrations of 10, 50, and 200 µg/kg. In addition, the spiked matrix experiment was conducted at concentrations of 5, 20, and 100 µg/kg to acquire the screening limit of the aforementioned six concentration levels.

1.2.4.1 Qualitative Analysis

The 494 kinds of pesticides were included in the GC-Q-TOF/MS database, and the verification was carried out by the appraisal of the qualitative screening of the spiked 494 kinds of pesticides to the four groups of tea matrix.

The qualitative analysis was conducted by the retrieval of the GC-Q-TOF/MS first order accurate mass full-scan mass spectrum with library, so the retrieval parameters should be optimized. The optimized retrieval parameters were: the window of retention time: 0.15 min, the window of extract accurate mass: 20 ppm, signal to noise (S/N) = 3. The positive results can be achieved when at least three ions (base peak ion, as quantitative ion, as well as two qualitative ions) of the five ions, which have relative high abundance, were detected. The criteria for the selection of the two qualitative ions were: (in terms of its respective abundance, the five ions were recorded as ion 1, ion 2, ion 3, ion 4, and ion 5, and X is one of the ions) ion

TABLE 1.2.1 Analysis of False Positive Pesticides

No.	Compound	CAS	Methods of treatment	Methods of detection	Concern	Remarks
1	2-Phenylphenol	90-43-7	Considerable attention	GC-Q-TOF	Whether the matrix is ubiquity, if so, delete it	It is a fungicide used for fruits and vegetables. It is no longer a permitted food additive in the European Union, but is still allowed as a post harvest treatment (in only 4 EU countries). It is also used for disinfection of seed boxes. It is a general surface disinfectant, used in households, hospitals, nursing homes, farms, laundries, barber shops, and food processing plants. It can be used on fibers and other materials. It is used to disinfect hospital and veterinary equipment. Other uses are in rubber industry and as a laboratory reagent. It is also used in the manufacture of other fungicides, dye stuffs, resins and rubber chemicals
2	Allethrin	584-79-2	Considerable attention	GC-Q-TOF	Whether the matrix is ubiquity, if so, delete it	It is used almost exclusively in homes and gardens for control of flies and mosquitoes
3	Tert-butyl-4-hydroxyanisole	25013-16-5	Considerable attention	GC-Q-TOF	Whether the matrix is ubiquity, if so, delete it	It is primarily used as an antioxidant food additive, and it is also used as an antioxidant in products such as cosmetics, pharmaceuticals, rubber, electrical transformer oil and embalming fluid
4	Diethyltoluamide	134-62-3	Considerable attention	GC-Q-TOF and LC-Q-TOF		
5	Dimethyl phthalate	131-11-3	Considerable attention	GC-Q-TOF		Dimethyl phthalate is a broad-spectrum, highly effective insect repellent, it is colorless to pale yellow transparent liquid, is an effective insect repellent ingredient in toilet water, it has a good repellent effect on flies, lice, ants, mosquitoes, cockroaches, midges, gadfly, flat fleas, sand fleas, sand midges, sandflies, cicadas; its repellent effect lasts for a long time, and it can be used in different climatic conditions. Under the conditions of using, it's chemical stable, displaying both high thermal stability and high resistance to sweat. It has good compatibility with common cosmetic and pharmaceutical agents, can be made of solutions, emulsions, pastes, coating agents, gel, aerosol, mosquito coils, micro-capsules and other special repellent agents, and also can be added to other products or materials (such as toilet water), so that product displays the repellent effect as well. Compared with standard mosquito repellent agent and mosquito repellent gel, it is less toxic, and less irritating, with longer repellent time and other notable features, is a repellent gel replacement

TABLE 1.2.1 Analysis of False Positive Pesticides (*Cont.*)

No.	Compound	CAS	Methods of treatment	Methods of detection	Concern	Remarks
6	Phenanthrene	1985/1/8	Considerable attention	GC-Q-TOF	Whether the matrix is ubiquity, if so, delete it	Phenanthrene can be used in the manufacture of pesticides and dyes, but also be used as the stabilizer of the high efficiency and low toxicity pesticides and smokeless powder explosives
7	Phthalic acid, bis-2-ethylhexyl ester	117-81-7	Considerable attention	GC-Q-TOF		Phthalate metabolite, responsible for inducing apoptosis in germ and Sertoli cells by disrupting junction complexes
8	Tributyl phosphate	126-73-8	Considerable attention	GC-Q-TOF and LC-Q-TOF		Tributyl phosphate is a solvent and plasticizer for cellulose esters such as nitrocellulose and cellulose acetate. The major uses of tributyl phosphate in industry are as a component of aircraft hydraulic fluid, brake fluid, and as a solvent for extraction and purification of rare earthmetals from their ores. It finds use as antifoaming agent in detergent solutions, and in various emulsions, paints, and adhesives. It is also found as a defoamer in ethylene glycol-borax antifreeze solutions. It is used also in mercerizing liquids, where it improves their wetting properties. It is also used as a heat exchange medium. Tributyl phosphate is used in some consumer products such as herbicides and water thinned paints and tinting bases
9	Triphenyl phosphate	115-86-6	Considerable attention	GC-Q-TOF		Triphenyl phosphate has been used widely as a flame retardant and plasticizer
10	Phthalic acid, Bis-butyl ester	84-74-2	Considerable attention	GC-Q-TOF and LC-Q-TOF		Dibutyl phthalate (DBP) is a commonly used plasticizer. It is also used as an additive to adhesives or printing inks. It is soluble in various organic solvents, e.g. in alcohol, ether and benzene. DBP is also used as an ectoparasiticide. DBP is also a putative endocrine disruptor
11	Metolcarb	1129-41-5	Considerable attention	GC-Q-TOF		The rention time of metolcarb was short, at that time the instrument was not stable, so excess background ions will result in false positive result

2 and ion 3 were usually selected as qualitative ions, but when the abundance ratio of ion X and X + 1 was relatively small (<15%), meanwhile the ion X was not detected, ion X + 1 can be selected as qualitative ion.

In this experiment, some kinds of false positive pesticides were detected in solvent blank and matrix blank. To avoid the phenomenon of false positive, further research was conducted and the results are listed in the next sections (Table 1.2.1).

1.2.4.2   Screening Limit

1.2.4.2.1   Pesticides Can Be Screened

Considering that pesticides differ in chemical and physical properties and matrix effects, the screening limit of the same pesticide may be different in the matrix. The results are listed in Appendix Table 1.2.1.

The number and percentage of pesticides that can be screened using GC-Q-TOF/MS are listed in Table 1.2.2. It can be seen that there is variance in quantity in different matrices. The number of pesticides that can be screened in the four kinds of matrix range from 451 to 468, accounting for 91.3%–94.7% of the 494 spiked pesticides. Among the four kinds of matrix, the number and percentage of pesticides that could be screened in the matrix of black tea, 468 kinds and 94.7%, was the largest amount, while in the matrix of oolong, 451 kinds and 91.3%, was the least. The results demonstrated that the applicability of the nontargeted screened method in the four kinds of matrix was excellent. The difference of the matrix in chemical and physical properties results in the inhibition or enhancement of the ion in different compounds, resulting in the difference in the number of pesticides are detected. The number of pesticides cannot be screened in the four kinds of matrix were ranged from 26 to 43. The results are listed in Appendix Table 1.2.1.

The number and percentage of pesticides that can be screened in the four kinds of matrix according to the screening limit are listed in Table 1.2.3. It can be seen that the number and percentage of pesticides that can be screened was the largest,

APPENDIX TABLE 1.2.1 Results of Nontarget Screening for 494 Pesticides in Two Kinds of Tea by GC-Q-TOF/MS

序号	农药名称	CAS 号	红茶 筛查限 (μg/kg)	10 μg/kg AVE	RSD, % (n=5)	50 μg/kg AVE	RSD, % (n=5)	200 μg/kg AVE	RSD, % (n=5)	绿茶 筛查限 (μg/kg)	10 μg/kg AVE	RSD, % (n=5)	50 μg/kg AVE	RSD, % (n=5)	200 μg/kg AVE	RSD, % (n=5)
1	1-Naphthylacetamide	86-86-2	5	88.6	4.2	50.9	5.3	111.4	41.2	10	98.2	10.4	109.0	11.4	102.2	9.3
2	1-Naphthylaceticacid	86-87-3	10	42.0	13.7	48.7	18.4	59.7	8.1	20	NA	NA	64.2	9.3	44.5	22.1
3	2,3,5,6-Tetrachloroaniline	3481-20-7	5	86.3	3.6	93.4	3.9	91.8	4.1	5	89.0	4.9	96.9	6.6	91.6	2.0
4	2,4-DB	94-82-6	20	NA	NA	362.2	6.6	95.9	4.3	50	NA	NA	NA	NA	93.0	4.4
5	2,4'-DDD	53-19-0	5	101.7	2.2	93.5	2.9	93.7	3.7	5	91.3	4.7	91.8	4.7	90.9	2.1
6	2,4'-DDE	3424-82-6	5	91.8	1.9	92.6	3.4	96.2	1.7	5	87.7	4.1	93.3	2.7	89.9	2.2
7	2,4'-DDT	789-02-6	5	97.8	6.8	93.1	2.0	97.1	1.6	5	99.7	5.1	97.5	5.4	98.0	1.3
8	2,6-Dichlorobenzamide	2008-58-4	5	101.9	60.2	NA	NA	155.8	80.7	5	88.0	14.2	91.5	15.3	86.1	10.3
9	2-Phenylphenol	90-43-7	5	94.6	5.8	92.1	5.5	94.7	3.3	5	95.3	11.5	90.6	7.1	93.5	2.3
10	3,4,5-Trimethacarb	2686-99-9	20	NA	NA	85.2	5.1	99.9	6.6	5	132.3	4.0	118.7	6.5	121.4	2.1
11	3,5-Dichloroaniline	626-43-7	20	NA	NA	78.8	7.3	86.9	19.5	20	NA	NA	74.3	18.5	84.1	19.4
12	3-Chloro-4-methylaniline	95-74-9	5	44.9	38.2	38.8	10.3	58.1	26.2	5	75.4	5.1	71.8	18.9	101.9	2.7
13	3-Phenylphenol	580-51-8	5	110.8	15.2	90.7	6.2	93.7	17.1	5	95.3	18.2	84.8	10.6	102.3	1.9
14	4,4'-DDD	72-54-8	5	93.5	2.9	93.0	4.1	99.5	1.7	5	92.7	6.3	97.4	3.8	99.6	2.3
15	4,4'-Dibromobenzophenone	3988-03-2	5	105.6	2.2	99.1	4.1	94.5	3.0	5	93.6	8.2	92.9	4.8	93.5	2.4
16	4-Bromo-3,5-Dimethylphenyl-N-Methylcarbamate	672-99-1	5	92.0	6.8	90.2	4.2	97.9	1.6	5	79.2	8.4	90.2	7.8	89.3	9.2
17	4-Chloronitrobenzene	100-00-5	5	72.7	13.1	160.1	26.9	83.4	14.1	5	141.8	5.9	127.6	8.4	72.0	11.3
18	4-Chlorophenoxy acetic acid	122-88-3	5	91.3	6.2	90.7	10.3	98.2	4.8	5	95.6	7.9	98.9	5.1	96.7	5.4
19	Acenaphthene	83-32-9	5	44.7	18.9	112.2	7.8	112.8	16.9	5	101.2	13.5	132.3	18.7	81.6	11.2
20	Acetochlor	34256-82-1	10	98.8	7.4	92.5	5.0	93.9	1.9	5	98.2	2.9	96.1	8.1	107.3	1.5
21	Acibenzolar-S-methyl	135158-54-2	5	73.2	9.6	102.2	7.9	87.4	10.3	5	75.7	2.6	67.4	4.8	83.7	11.5
22	Aclonifen	74070-46-5	5	109.1	5.2	93.5	5.1	100.4	9.2	5	103.3	9.0	106.3	9.7	121.9	2.4
23	Acrinathrin	101007-06-1	20	NA	NA	31.4	8.7	42.7	10.8	20	NA	NA	29.3	16.0	70.8	10.6
24	Akton	1757-18-2	5	89.6	4.9	92.6	4.0	97.7	1.4	5	90.6	5.4	95.5	2.7	99.9	2.3
25	Alachlor	15972-60-8	5	93.8	4.9	87.7	10.6	93.8	2.9	5	93.3	9.2	87.4	5.1	94.9	2.2
26	Alanycarb	83130-01-2	100	NA	NA	NA	NA	76.5	13.8	NA	NA	NA	NA	NA	NA	NA
27	Aldicarb-S-methyl	1646-88-4	200	NA	NA	NA	NA	NA	NA	50	NA	NA	100.1	39.7	95.5	5.4
28	Aldimorph	91315-15-0	20	NA	NA	79.9	3.8	79.8	8.6	10	97.8	4.8	92.1	3.7	93.2	17.2
29	Aldrin	309-00-2	5	94.7	6.9	89.4	3.5	92.4	3.0	5	91.0	3.8	89.3	3.8	91.7	1.7
30	Allethrin	584-79-2	NA	NA	NA	NA	NA	NA	NA	NA	NA	NA	NA	NA	NA	NA
31	Allidochlor	93-71-0	5	93.4	8.2	91.3	4.4	91.7	11.5	5	89.0	19.8	60.1	9.2	65.9	11.8

#	Compound	CAS														
32	Alpha-Cypermethrin	67375-30-8	50	NA	NA	NA	NA	94.4	8.5	200	NA	NA	NA	NA	NA	NA
33	Alpha-Endosulfan	959-98-8	5	104.7	2.7	89.7	6.1	92.7	4.1	5	83.9	4.0	93.8	3.9	89.3	3.1
34	Alpha-HCH	319-84-6	5	92.7	3.8	90.4	5.1	93.0	2.0	5	95.9	5.6	91.2	5.8	102.3	1.6
35	Ametryn	834-12-8	5	82.9	5.0	94.7	1.6	89.7	5.5	5	91.0	5.1	81.4	1.5	90.9	19.3
36	Amidosulfuron	120923-37-7	10	86.1	6.9	85.8	12.1	94.9	8.0	10	90.3	7.5	91.9	12.7	102.5	1.8
37	Aminocarb	2032-59-9	5	63.1	9.1	101.2	7.9	88.6	5.1	20	NA	NA	89.2	4.1	97.0	4.3
38	Amitraz	33089-61-1	20	NA	NA	NA	NA	95.0	9.6	20	NA	NA	79.6	9.4	98.0	4.8
39	Ancymidol	12771-68-5	5	96.5	4.1	94.1	3.0	92.1	7.7	5	75.7	3.1	67.5	5.6	86.0	16.6
40	Anilofos	64249-01-0	5	100.7	14.0	88.7	5.7	104.4	8.4	5	104.9	19.8	116.9	7.2	129.2	3.0
41	AnthraceneD10	1719-06-8	5	89.2	4.6	87.3	5.0	84.7	4.0	5	84.5	4.4	84.7	4.8	85.1	2.0
42	Aramite	140-57-8	100	NA	NA	NA	NA	98.5	2.2	50	NA	NA	100.0	6.9	109.3	2.3
43	Atraton	1610-17-9	5	87.5	16.3	94.2	6.1	93.8	2.2	5	77.3	11.7	92.6	8.7	107.3	1.6
44	Atrazine	1912-24-9	5	87.3	5.2	91.6	5.1	91.7	1.7	5	90.9	7.1	90.5	6.1	100.0	1.4
45	Atrazine-desethyl	6190-65-4	5	85.9	8.1	101.1	3.9	86.3	8.5	5	70.3	6.5	64.7	6.3	83.5	14.9
46	Atrazine-desisopropyl	1007-28-9	5	111.7	10.3	109.3	5.5	92.4	16.5	5	63.9	18.8	95.0	14.5	110.7	12.0
47	Azaconazole	60207-31-0	5	90.4	10.5	89.7	6.9	93.6	3.6	5	86.7	5.0	96.4	8.5	107.4	3.9
48	Azinphos-ethyl	2642-71-9	10	104.9	5.7	89.6	3.9	96.2	9.5	20	NA	NA	87.6	5.8	89.4	18.3
49	Aziprotryne	4658-28-0	10	91.4	5.4	89.6	3.1	99.4	5.2	5	88.4	7.5	85.7	3.3	94.8	18.8
50	Azoxystrobin	131860-33-8	50	NA	NA	111.4	6.8	76.2	14.3	50	NA	NA	NA	NA	NA	NA
51	Beflubutamid	113614-08-7	5	111.4	4.7	94.9	4.2	94.5	1.4	10	110.8	8.7	99.4	4.4	107.4	1.5
52	Benalaxyl	71626-11-4	5	91.4	2.0	90.6	1.7	94.6	8.2	5	81.9	4.7	80.4	2.9	88.5	15.9
53	Bendiocarb	22781-23-3	5	89.0	4.3	88.2	3.2	92.7	6.7	5	79.2	7.3	73.3	4.6	96.6	18.3
54	Benfluralin	1861-40-1	5	95.8	4.8	90.7	5.6	92.9	3.8	5	104.4	7.0	88.3	8.2	90.0	3.1
55	Benfuracarb	82560-54-1	200	NA	NA	NA	NA	103.3	7.1	50	NA	NA	112.8	13.8	143.6	2.6
56	Benfuresate	68505-69-1	10	137.7	14.3	89.5	8.4	94.4	2.9	50	NA	NA	NA	NA	93.4	2.7
57	Benodanil	15310-01-7	20	NA	NA	87.5	7.4	93.5	8.2	5	207.9	11.5	117.0	7.9	112.7	5.5
58	Benoxacor	98730-04-2	5	94.5	7.2	90.4	1.6	94.2	7.8	5	87.8	7.5	80.8	3.8	85.2	19.1
59	Benzoximate	29104-30-1	5	84.7	6.3	87.8	2.7	97.4	7.8	5	96.8	7.9	96.9	3.2	98.8	14.7
60	Benzoylprop-Ethyl	22212-55-1	5	97.4	7.8	92.9	3.4	95.9	8.2	5	158.2	3.8	96.8	3.0	101.6	13.2
61	Beta-Endosulfan	33213-65-9	5	100.0	6.9	85.9	7.6	91.8	4.6	5	89.1	7.9	89.0	7.3	89.0	3.1
62	Beta-HCH	319-85-7	5	87.8	7.9	85.3	3.7	100.1	4.7	5	111.0	2.1	103.1	2.8	118.3	12.7
63	Bifenazate	149877-41-8	50	NA	NA	120.0	6.1	98.0	12.3	50	NA	NA	162.1	15.5	78.3	16.9
64	Bifenox	42576-02-3	50	NA	NA	99.7	12.1	81.8	8.0	50	NA	NA	NA	NA	92.6	13.0
65	Bifenthrin	82657-04-3	5	86.3	3.0	97.0	2.3	95.1	1.3	5	102.7	4.0	92.3	3.2	93.7	2.7
66	Bioresmethrin	28434-01-7	20	NA	NA	76.9	4.8	74.7	11.2	50	NA	NA	89.2	4.7	108.5	13.5
67	Biphenyl	92-52-4	NA	NA	NA	NA	NA	NA	NA	5	NA	NA	86.1	17.3	96.5	9.9
68	Bitertanol	55179-31-2	20	NA	NA	97.5	14.5	88.4	7.4	20	NA	NA	103.2	9.0	106.6	8.4
69	Boscalid	188425-85-6	5	88.6	4.1	87.6	4.1	98.6	3.8	5	81.1	9.9	96.5	6.6	80.6	8.8
70	Bromfenvinfos	33399-00-7	5	96.6	3.5	93.6	2.7	98.8	8.5	5	96.4	3.2	83.1	6.3	87.3	19.0

(Continued)

APPENDIX TABLE 1.2.1 Results of Nontarget Screening for 494 Pesticides in Two Kinds of Tea by GC-Q-TOF/MS (Cont.)

序号	农药名称	CAS 号	红茶 筛查限 (μg/kg)	10 μg/kg AVE	RSD, % (n = 5)	50 μg/kg AVE	RSD, % (n = 5)	200 μg/kg AVE	RSD, % (n = 5)	绿茶 筛查限 (μg/kg)	10 μg/kg AVE	RSD, % (n = 5)	50 μg/kg AVE	RSD, % (n = 5)	200 μg/kg AVE	RSD, % (n = 5)
71	Bromfenvinfos-Methyl	13104-21-7	5	106.4	7.3	94.7	4.2	103.7	1.8	5	80.4	10.4	101.2	9.8	84.1	7.5
72	Bromobutide	74712-19-9	10	85.3	5.3	90.2	2.3	95.4	6.9	5	79.8	3.1	78.9	2.6	88.0	16.0
73	Bromocyclen	1715-40-8	5	89.3	4.1	90.7	3.4	92.0	3.3	5	92.3	2.5	87.8	5.0	89.9	2.1
74	Bromophos-Ethyl	4824-78-6	5	93.0	3.4	91.6	2.4	96.9	5.7	5	117.7	1.1	114.6	3.4	116.1	10.3
75	Bromophos-Methyl	2104-96-3	5	104.2	4.3	96.2	4.1	93.7	3.6	5	102.3	4.9	89.3	6.3	93.4	2.9
76	Bromopropylate	18181-80-1	20	NA	NA	105.1	3.3	92.7	5.3	20	NA	NA	95.1	6.3	90.0	5.3
77	Bromoxyniloctanoate	1689-99-2	20	NA	NA	110.8	14.5	102.3	12.6	50	NA	NA	NA	NA	89.4	18.1
78	Bromuconazole	116255-48-2	5	92.5	7.2	90.1	4.0	92.6	10.9	5	78.5	2.6	72.0	6.5	84.1	12.3
79	Bupirimate	41483-43-6	5	88.6	3.6	89.3	1.9	94.8	8.1	5	91.2	3.0	86.0	2.4	94.0	14.8
80	Buprofezin	69327-76-0	50	NA	NA	97.5	6.0	93.9	2.5	20	NA	NA	100.0	5.7	107.2	2.2
81	Butachlor	23184-66-9	5	92.6	4.3	93.1	1.7	96.5	7.1	5	100.0	5.8	94.7	2.6	101.2	13.4
82	Butafenacil	134605-64-4	5	100.2	14.8	90.1	6.4	99.1	9.5	5	105.9	5.8	108.2	7.0	116.4	4.2
83	Butamifos	36335-67-8	5	128.4	11.0	95.2	4.1	104.8	6.4	5	106.9	5.8	102.4	6.8	117.6	3.1
84	Butralin	33629-47-9	5	103.1	8.4	95.8	4.2	95.3	3.2	5	101.3	3.9	97.1	5.4	108.5	1.7
85	Cadusafos	95465-99-9	5	95.1	6.8	92.3	4.5	92.3	1.8	5	95.4	6.9	96.0	5.7	103.4	1.4
86	Cafenstrole	125306-83-4	20	NA	NA	98.4	7.4	94.5	5.6	100	NA	NA	NA	NA	72.1	11.3
87	Captafol	2425-06-1	NA	NA	NA	NA	NA	NA	NA	NA	NA	NA	NA	NA	NA	NA
88	Captan	133-06-2	200	NA	NA	NA	NA	91.7	10.5	NA	NA	NA	NA	NA	NA	NA
89	Carbaryl	63-25-2	5	82.6	6.5	92.0	3.5	93.9	2.4	5	83.4	14.3	105.1	7.4	99.6	2.8
90	Carbofuran	1563-66-2	5	104.3	6.0	93.9	15.2	87.1	4.4	5	94.1	20.0	80.7	12.2	94.9	6.5
91	Carbofuran-3-hydroxy	16655-82-6	20	NA	NA	42.9	34.2	101.5	7.4	5	78.3	19.9	83.7	18.1	105.8	3.9
92	Carbophenothion	786-19-6	5	97.2	6.3	94.8	4.5	102.4	2.5	5	117.7	8.6	113.3	3.7	100.3	3.2
93	Carbosulfan	55285-14-8	NA	NA	NA	NA	NA	NA	NA	NA	NA	NA	NA	NA	NA	NA
94	Carboxin	5234-68-4	100	NA	NA	NA	NA	94.5	66.9	5	NA	NA	16.8	82.2	37.5	43.6
95	Chlorbenside	103-17-3	5	101.9	2.8	101.2	3.4	95.6	3.3	5	98.0	6.0	91.7	4.0	93.0	2.4
96	Chlorbensidesulfone	7082-99-7	5	100.4	2.2	92.6	7.6	94.3	3.4	5	88.5	10.6	90.3	6.9	95.5	2.5
97	Chlorbromuron	13360-45-7	NA	NA	NA	NA	NA	NA	NA	NA	NA	NA	NA	NA	NA	NA
98	Chlorbufam	1967-16-4	5	85.7	10.7	87.4	2.7	88.9	7.1	5	82.6	12.3	82.6	7.3	88.7	2.4
99	Chlordane	57-74-9	5	87.4	4.4	92.8	6.4	98.3	1.9	5	88.7	7.3	88.6	3.1	88.9	2.3
100	Chlordecone	143-50-0	NA	NA	NA	NA	NA	NA	NA	NA	NA	NA	NA	NA	NA	NA
101	Chlordimeform	6164-98-3	100	NA	NA	NA	NA	80.5	7.5	100	NA	NA	NA	NA	78.2	22.2
102	Chlorethoxyfos	54593-83-8	5	74.9	6.3	98.3	4.4	90.3	6.9	5	93.3	7.5	89.6	13.8	85.3	3.9
103	Chlorfenapyr	122453-73-0	10	80.0	4.5	91.2	9.0	96.8	2.5	50	NA	NA	110.1	16.2	88.5	10.0

No.	Compound	CAS No.														
104	Chlorfenethol	80-06-8	5	102.3	4.3	91.8	5.1	93.5	3.4	5	92.1	7.5	92.7	5.1	93.6	2.5
105	Chlorfenprop-methyl	14437-17-3	5	84.4	4.1	97.4	4.6	92.6	3.7	5	91.1	8.5	91.0	10.7	89.6	3.2
106	Chlorfenson	80-33-1	5	100.6	3.2	95.2	3.8	95.8	3.0	5	93.2	6.7	93.3	4.3	93.4	1.9
107	Chlorfenvinphos	470-90-6	5	98.4	9.1	81.0	8.8	92.5	3.6	5	90.1	9.8	85.4	7.7	94.2	1.4
108	Chlorfluazuron	71422-67-8	10	44.1	28.0	30.6	38.3	47.3	23.5	5	86.0	5.7	102.2	18.1	82.0	6.3
109	Chlorflurenol-methyl	2536-31-4	5	91.0	5.8	88.4	2.4	98.3	1.9	5	81.8	8.4	98.1	7.2	82.7	7.8
110	Chloridazon	1698-60-8	200	NA	NA	NA	NA	100.4	5.8	NA	NA	NA	NA	NA	NA	NA
111	Chlorobenzilate	510-15-6	20	NA	NA	100.3	3.6	90.1	5.6	20	87.7	7.9	93.8	8.3	89.7	5.6
112	Chloroneb	2675-77-6	5	74.0	7.8	97.6	4.7	91.8	7.4	5	NA	NA	97.0	14.8	86.5	5.0
113	Chloropropylate	5836-10-2	20	NA	NA	99.7	3.4	91.2	5.3	20	NA	NA	94.3	7.9	90.4	6.1
114	Chlorothalonil	1897-45-6	5	64.4	25.7	65.6	12.9	87.5	12.9	5	149.0	9.6	102.3	12.9	113.7	1.1
115	Chlorotoluron	15545-48-9	20	NA	NA	90.0	18.7	73.8	9.2	50	NA	NA	NA	NA	98.2	17.3
116	Chlorpropham	101-21-3	5	85.5	5.1	84.6	3.2	90.3	8.3	5	78.3	13.6	94.0	5.6	89.8	2.9
117	Chlorpyrifos	2921-88-2	5	91.6	4.0	93.5	5.4	93.1	2.7	5	107.3	6.1	91.9	6.2	94.5	2.0
118	Chlorpyrifos-methyl	5598-13-0	5	102.8	4.1	96.4	5.7	93.8	2.7	5	98.3	9.3	89.0	5.5	95.2	2.4
119	Chlorsulfuron	64902-72-3	5	86.7	14.9	83.3	5.1	92.9	14.4	5	112.1	4.9	93.5	12.9	97.6	5.1
120	Chlorthal-dimethyl	1861-32-1	5	97.6	1.4	88.3	6.2	92.9	3.0	5	90.3	6.8	89.4	3.4	91.5	2.1
121	Chlorthiamid	1918-13-4	20	NA	NA	93.8	7.5	85.5	5.0	20	NA	NA	97.1	5.1	89.3	6.0
122	Chlorthion	500-28-7	5	118.3	6.2	97.2	6.2	92.8	5.4	5	110.5	16.2	88.9	11.1	95.7	5.6
123	Chlorthiophos	60238-56-4	5	89.1	1.8	91.9	2.1	96.0	6.3	5	113.3	4.1	109.4	3.3	118.2	11.9
124	Chlozolinate	84332-86-5	5	118.6	1.5	102.1	6.9	114.7	3.5	5	91.5	7.3	100.8	4.0	97.6	2.1
125	Cinidon-Ethyl	142891-20-1	50	NA	NA	98.7	6.5	88.3	8.8	50	NA	NA	NA	NA	79.0	9.7
126	cis-1,2,3,6-Tetrahydrophthalimide	1469-48-3	20	96.3	2.6	76.4	19.2	65.5	8.1	20	101.6	4.2	79.6	11.6	93.5	10.6
127	cis-Permethrin	61949-76-6	5	86.9	7.7	91.4	3.4	100.0	2.4	5	84.5	8.5	97.4	2.8	94.5	2.2
128	Clodinafop	114420-56-3	5	107.5	11.1	94.3	2.6	96.2	5.6	5	117.6	5.9	88.8	3.7	97.7	17.5
129	Clodinafop-propargyl	105512-06-9	5	97.7	4.8	92.8	5.6	104.1	3.7	10	89.1	6.1	118.4	6.8	87.2	7.4
130	Clomazone	81777-89-1	5	93.4	5.6	92.3	5.2	94.7	1.1	5	96.0	17.1	93.4	5.8	105.4	1.4
131	Clopyralid	1702-17-6	5	88.2	8.5	80.3	16.3	96.1	4.9	5	98.8	12.2	97.7	3.9	92.1	6.6
132	Coumaphos	56-72-4	5	104.9	16.8	91.3	5.9	95.0	9.3	5	88.6	12.2	77.5	4.8	92.1	12.8
133	Crufomate	299-86-5	10	104.0	4.8	89.3	4.9	90.2	2.0	5	95.0	12.2	100.4	5.7	112.7	1.8
134	Cyanofenphos	13067-93-1	5	107.7	6.6	93.3	7.2	93.6	2.7	5	93.1	8.7	90.3	6.5	94.9	3.2
135	Cyanophos	2636-26-2	5	NA	NA	94.4	9.0	95.2	2.9	5	94.4	11.1	89.4	6.0	94.8	2.9
136	Cycloate	1134-23-2	20	86.1	2.7	86.9	7.7	91.1	5.9	10	86.9	5.6	90.2	7.3	99.6	2.1
137	Cycloprothrin	63935-38-6	5	NA	NA	88.7	4.4	97.1	2.2	5	86.9	6.8	95.7	4.3	87.0	2.8
138	Cyflufenamid	180409-60-3	20	NA	NA	92.5	3.0	95.7	7.0	10	NA	NA	96.1	4.7	92.1	2.8
139	Cyfluthrin	68359-37-5	200	NA	NA	NA	NA	114.9	15.0	20	NA	NA	NA	NA	104.1	15.2
140	Cypermethrin	52315-07-8	200	NA	NA	NA	NA	92.4	7.4	NA	NA	NA	NA	NA	NA	NA
141	Cyphenothrin	39515-40-7	50	NA	NA	NA	NA	100.5	7.3	200	NA	NA	NA	NA	NA	NA
142	Cyprazine	22936-86-3	5	97.3	6.6	96.2	8.9	95.0	3.0	5	98.2	9.6	84.0	11.7	100.7	14.4

(Continued)

APPENDIX TABLE 1.2.1 Results of Nontarget Screening for 494 Pesticides in Two Kinds of Tea by GC-Q-TOF/MS (Cont.)

序号	农药名称	CAS 号	红茶 筛查限 (μg/kg)	10 μg/kg AVE	10 μg/kg RSD, % (n=5)	50 μg/kg AVE	50 μg/kg RSD, % (n=5)	200 μg/kg AVE	200 μg/kg RSD, % (n=5)	绿茶 筛查限 (μg/kg)	10 μg/kg AVE	10 μg/kg RSD, % (n=5)	50 μg/kg AVE	50 μg/kg RSD, % (n=5)	200 μg/kg AVE	200 μg/kg RSD, % (n=5)
143	Cyproconazole	94361-06-5	5	104.3	12.7	86.7	16.6	87.5	9.6	5	98.8	7.4	100.9	7.2	90.9	16.8
144	Cyprodinil	121552-61-2	5	81.6	3.1	90.6	4.9	89.8	1.5	5	85.3	5.6	83.5	12.3	101.8	1.4
145	Cyprofuram	69581-33-5	5	89.6	2.5	92.0	2.2	94.7	5.7	5	71.8	3.3	68.7	3.3	80.3	19.8
146	DDT	50-29-3	5	104.0	2.2	97.4	2.0	85.3	3.8	5	92.8	7.7	82.9	6.3	89.0	4.9
147	Delta-HCH	319-86-8	5	94.5	4.2	90.0	4.1	94.1	1.9	5	91.0	5.8	86.8	7.1	101.2	2.0
148	Deltamethrin	52918-63-5	200	NA	NA	NA	NA	NA	NA	NA	NA	NA	NA	NA	NA	NA
149	Desethylterbuthylazine	30125-63-4	5	84.5	6.5	82.3	6.0	91.8	4.1	5	72.9	8.5	95.0	7.9	77.6	16.0
150	Desmetryn	1014-69-3	5	91.9	2.1	91.3	1.4	95.9	6.2	5	81.0	3.3	77.7	2.9	86.9	17.1
151	Dialifos	10311-84-9	100	NA	NA	NA	NA	94.4	3.6	NA	NA	NA	NA	NA	NA	NA
152	Diallate	2303-16-4	5	91.1	4.1	89.8	1.2	94.6	6.5	5	104.9	3.7	97.9	4.4	104.5	13.0
153	Dibutylsuccinate	141-03-7	5	93.7	3.8	89.4	7.6	89.3	6.6	5	97.7	5.6	91.4	6.0	99.3	1.2
154	Dicapthon	2463-84-5	5	100.7	2.0	90.1	8.0	91.5	5.4	5	103.5	10.7	86.5	11.2	95.9	4.4
155	Dichlobenil	1194-65-6	5	34.5	23.9	107.2	7.3	93.3	31.7	5	76.6	13.2	112.8	18.1	81.7	18.8
156	Dichlofenthion	97-17-6	5	102.6	4.9	93.7	5.5	92.5	0.5	5	101.0	3.9	94.2	4.5	104.5	0.7
157	Dichlofluanid	1085-98-9	20	NA	NA	92.8	13.0	78.2	17.3	20	NA	NA	20.2	53.7	46.7	45.4
158	Dichlormid	37764-25-3	5	53.4	15.3	104.2	11.3	89.1	18.3	5	85.8	18.6	112.6	13.5	75.3	15.6
159	Dichlorprop	120-36-5	10	88.6	5.2	87.0	7.9	100.3	4.1	10	82.9	15.2	95.6	4.1	91.9	4.8
160	Dichlorvos	62-73-7	5	38.6	18.0	115.2	9.7	104.8	18.3	5	89.4	18.6	114.1	16.9	75.7	17.7
161	Diclocymet	139920-32-4	20	NA	NA	101.9	8.2	94.0	13.3	20	NA	NA	87.4	8.0	103.0	14.5
162	Diclofop-methyl	51338-27-3	5	104.3	2.3	97.3	4.2	93.1	2.9	5	90.4	4.9	90.5	5.7	92.8	2.7
163	Dicloran	99-30-9	5	89.7	9.0	94.0	2.9	95.3	2.2	5	83.0	9.4	104.4	6.1	83.2	7.0
164	Dicofol	115-32-2	5	105.7	10.0	96.2	5.7	110.1	4.0	5	93.5	5.0	91.1	4.1	94.9	3.4
165	Dieldrin	60-57-1	5	95.0	2.2	91.0	4.7	94.9	3.4	50	NA	NA	90.3	5.8	92.9	2.6
166	Diethatyl-Ethyl	38727-55-8	5	92.5	7.1	92.9	5.0	92.1	2.0	5	91.3	5.9	95.9	5.5	103.0	2.0
167	Diethofencarb	87130-20-9	5	95.5	4.5	73.3	8.2	90.6	10.9	20	NA	NA	92.6	3.7	84.4	3.0
168	Diethyltoluamide	134-62-3	20	NA	NA	91.6	4.1	90.9	0.9	20	NA	NA	74.9	9.2	93.8	4.9
169	Difenoxuron	14214-32-5	5	91.3	12.1	112.9	8.3	93.1	9.7	5	87.6	18.6	56.4	12.8	78.1	3.7
170	Diflufenican	83164-33-4	5	94.0	5.6	93.7	4.3	90.5	1.3	5	97.1	5.2	95.1	4.7	105.4	1.7
171	Diflufenzopyr	109293-97-2	5	83.5	7.8	87.6	6.3	90.0	4.9	5	84.3	12.3	86.9	7.2	89.3	4.3
172	Dimethachlor	50563-36-5	5	94.7	9.7	93.9	4.9	96.0	2.4	5	95.0	5.9	93.7	5.8	108.3	2.0
173	Dimethametryn	22936-75-0	20	NA	NA	93.4	6.0	93.7	4.4	20	NA	NA	89.4	5.7	91.7	5.4
174	Dimethenamid	87674-68-8	5	95.1	2.4	89.3	1.7	96.0	6.1	5	75.0	4.1	70.6	3.3	79.0	18.7
175	Dimethipin	55290-64-7	20	NA	NA	NA	NA	87.8	9.8	20	NA	NA	83.8	10.6	95.1	2.5

176	Dimethoate	60-51-5	5	118.0	7.6	81.5	3.0	88.0	8.7	5	100.1	12.8	105.9	9.0	108.5	3.6
177	Dimethylphthalate	131-11-3	5	71.1	5.1	100.4	13.5	92.4	7.3	5	98.4	7.8	100.9	13.1	85.4	7.0
178	Dimetilan	644-64-4	5	111.4	8.6	93.4	4.3	91.2	1.9	5	93.3	7.0	98.9	6.5	104.6	2.4
179	Diniconazole	83657-24-3	5	114.7	9.9	89.6	5.4	99.9	5.7	10	103.7	6.7	102.5	7.1	113.0	3.3
180	Dinitramine	29091-05-2	5	113.2	8.7	93.8	4.6	103.2	4.8	20	NA	NA	101.0	6.6	114.7	2.9
181	Dinobuton	973-21-7	100	NA	NA	NA	NA	92.2	11.2	50	NA	NA	76.0	17.7	79.9	13.6
182	Dinoterb	1420-07-1	NA	NA	NA	NA	NA	NA	NA	NA	NA	NA	NA	NA	NA	NA
183	Diofenolan	63837-33-2	5	99.7	3.8	96.1	3.0	94.7	2.6	5	94.1	8.8	91.4	5.9	94.9	2.6
184	Dioxabenzofos	3811-49-2	5	100.4	3.8	95.4	7.5	94.0	2.5	5	93.2	10.1	89.6	7.3	93.1	3.0
185	Dioxacarb	6988-21-2	NA	NA	NA	NA	NA	NA	NA	NA	NA	NA	NA	NA	NA	NA
186	Dioxathion	78-34-2	5	95.4	4.4	92.6	6.3	94.9	4.0	5	98.2	9.2	95.9	6.7	96.8	3.2
187	Diphenamid	957-51-7	5	89.7	2.9	90.0	2.0	95.8	6.9	5	71.7	1.1	66.7	2.1	76.7	19.0
188	Diphenylamine	122-39-4	5	97.7	3.5	93.3	4.1	94.1	2.4	5	103.5	4.2	90.6	5.5	90.5	2.7
189	Dipropetryn	4147-51-7	5	97.0	6.3	93.9	3.7	93.0	2.6	5	93.0	4.5	92.9	5.1	106.4	2.1
190	Disulfoton	298-04-4	5	97.2	15.3	115.8	9.7	135.8	3.3	5	136.8	7.4	190.0	3.8	102.4	3.2
191	Disulfotonsulfone	2497-06-5	5	99.8	5.1	91.4	2.9	53.0	66.8	5	95.0	5.6	89.0	3.7	93.6	11.4
192	Disulfotonsulfoxide	2497-07-6	5	102.1	10.8	91.8	5.3	96.2	2.3	5	90.5	8.5	93.6	7.4	104.3	2.5
193	Ditalimfos	5131-24-8	5	103.7	18.3	91.3	2.8	88.4	4.4	5	100.0	7.0	101.9	5.8	103.2	4.2
194	Dithiopyr	97886-45-8	5	87.1	5.2	95.0	2.9	87.8	0.7	5	90.4	5.6	90.6	6.1	99.2	1.6
195	Dodemorph	1593-77-7	5	62.9	4.4	62.7	7.6	59.3	11.6	5	70.1	1.6	58.6	9.9	71.3	6.4
196	Edifenphos	17109-49-8	5	108.4	5.5	96.8	3.7	103.6	9.4	20	NA	NA	70.4	12.2	85.5	18.4
197	Endosulfan-sulfate	1031-07-8	5	86.0	4.2	91.9	2.2	100.7	1.4	5	86.7	8.6	95.0	8.1	97.9	4.5
198	Endrin	72-20-8	5	86.0	6.6	91.1	3.3	98.1	2.0	5	77.8	4.5	94.9	4.7	102.4	2.5
199	Endrin-aldehyde	7421-93-4	NA	NA	NA	NA	NA	NA	NA	200	NA	NA	NA	NA	NA	NA
200	Endrin-ketone	53494-70-5	5	90.1	4.4	87.1	3.4	99.0	1.5	5	83.0	8.6	94.5	6.6	101.4	3.7
201	EPN	2104-64-5	5	101.4	6.7	96.4	7.6	88.4	5.2	5	100.4	11.0	85.6	11.2	96.0	6.4
202	EPTC	759-94-4	5	89.9	9.7	60.1	12.0	81.7	17.1	5	85.8	16.0	87.7	15.0	75.8	17.6
203	Esprocarb	85785-20-2	5	46.5	5.0	76.4	14.8	96.4	1.5	5	136.3	15.9	96.3	5.9	103.2	1.1
204	Ethalfluralin	55283-68-6	5	93.7	7.0	91.4	5.2	91.5	5.8	5	104.9	5.6	91.8	9.2	91.8	3.5
205	Ethion	563-12-2	5	99.9	5.0	91.4	2.9	79.1	16.3	5	95.0	5.6	89.0	3.7	93.7	11.3
206	Ethofumesate	26225-79-6	5	99.7	3.4	88.3	8.9	94.2	2.9	5	83.5	9.2	94.2	4.6	93.1	1.7
207	Ethoprophos	13194-48-4	5	87.5	4.4	87.7	2.1	94.9	6.8	5	79.7	5.8	75.0	2.0	81.6	18.3
208	Etofenprox	80844-07-1	20	NA	NA	101.0	3.6	89.4	3.4	5	106.4	16.2	90.0	4.8	91.2	4.2
209	Etridiazole	2593-15-9	5	115.8	19.5	74.5	6.4	86.7	15.6	5	105.0	12.0	106.8	14.2	81.6	14.1
210	Famphur	52-85-7	5	107.1	4.6	90.4	11.2	90.9	5.5	5	96.9	14.0	88.7	8.6	96.6	5.1
211	Fenamidone	161326-34-7	5	90.9	1.3	89.4	2.7	94.7	7.0	5	77.7	11.6	79.5	4.0	94.8	17.6
212	Fenamiphos	22224-92-6	20	NA	NA	102.1	7.1	98.3	12.3	20	NA	NA	93.6	13.3	81.1	6.5
213	Fenarimol	60168-88-9	10	107.9	15.6	92.0	5.4	91.7	4.9	5	88.3	8.0	98.6	9.9	112.2	3.3

(Continued)

APPENDIX TABLE 1.2.1 Results of Nontarget Screening for 494 Pesticides in Two Kinds of Tea by GC-Q-TOF/MS (Cont.)

序号	农药名称	CAS 号	筛查限 (μg/kg)	红茶						筛查限 (μg/kg)	绿茶					
				10 μg/kg		50 μg/kg		200 μg/kg			10 μg/kg		50 μg/kg		200 μg/kg	
				AVE	RSD, % (n = 5)	AVE	RSD, % (n = 5)	AVE	RSD, % (n = 5)		AVE	RSD, % (n = 5)	AVE	RSD, % (n = 5)	AVE	RSD, % (n = 5)
214	Fenazaflor	14255-88-0	5	2.6	59.3	43.2	49.4	4.0	80.0	5	54.4	19.3	38.6	10.0	71.7	13.4
215	Fenazaquin	120928-09-8	5	89.5	5.3	87.1	1.1	94.7	6.6	5	105.2	3.0	98.6	2.2	106.2	13.5
216	Fenchlorphos	299-84-3	5	94.8	4.5	92.1	3.2	98.9	2.1	5	93.2	5.4	100.5	3.8	88.5	3.3
217	Fenchlorphos-Oxon	3983-45-7	5	104.6	8.3	81.4	6.2	96.0	9.4	5	110.3	6.8	116.8	9.4	133.0	4.2
218	Fenfuram	24691-80-3	100	NA	NA	NA	NA	24.6	121.2	NA	NA	NA	NA	NA	NA	NA
219	Fenitrothion	122-14-5	5	104.7	8.0	94.9	4.2	101.3	3.2	5	109.2	9.1	106.0	5.2	88.9	5.1
220	Fenobucarb	3766-81-2	5	131.0	4.1	91.6	3.3	99.3	5.4	5	NA	NA	88.3	11.9	100.7	15.8
221	Fenoprop	93-72-1	20	NA	NA	116.1	18.4	94.3	4.5	20	NA	NA	52.9	13.9	72.0	10.3
222	Fenothiocarb	62850-32-2	5	99.3	4.1	94.3	4.5	94.5	2.3	5	99.8	3.4	95.8	4.1	107.6	1.4
223	Fenoxaprop-Ethyl	66441-23-4	5	93.1	4.7	92.8	3.2	97.1	8.1	5	106.4	4.7	95.4	3.4	100.5	15.5
224	Fenoxycarb	72490-01-8	20	NA	NA	91.6	11.6	86.0	5.5	20	NA	NA	NA	NA	71.2	3.0
225	Fenpiclonil	74738-17-3	NA	NA	NA	NA	NA	NA	NA	50	NA	NA	60.6	7.6	131.3	36.1
226	Fenpropathrin	64257-84-7	20	NA	NA	97.2	1.9	93.4	4.1	20	NA	NA	99.8	7.3	89.9	5.6
227	Fenpropidin	67306-00-7	20	NA	NA	58.3	16.2	42.3	13.6	20	NA	NA	45.3	11.7	60.9	7.9
228	Fenpropimorph	67564-91-4	5	76.2	8.4	91.9	4.4	80.0	3.0	5	88.6	6.6	90.4	6.8	98.4	1.9
229	Fenson	80-38-6	5	113.6	1.5	100.4	5.3	95.5	2.8	5	101.6	5.1	91.4	5.1	94.3	2.3
230	Fensulfothion	115-90-2	5	93.0	3.4	86.4	3.0	94.6	2.9	5	77.9	8.5	92.3	5.9	90.3	4.1
231	Fensulfothion-oxon	6552-21-2	200	NA	NA	NA	NA	101.9	19.7	10	119.3	12.3	116.2	19.7	138.3	4.8
232	Fensulfothion-sulfone	14255-72-2	5	113.5	8.8	93.1	4.3	90.2	4.2	5	107.6	6.2	94.1	8.1	104.0	1.2
233	Fenthion	55-38-9	5	92.7	7.0	106.3	4.2	113.4	1.6	5	106.3	8.6	119.8	1.6	102.5	1.6
234	Fenvalerate	51630-58-1	20	NA	NA	93.1	9.2	100.6	3.6	50	NA	NA	102.1	7.4	91.0	6.2
235	Fipronil	120068-37-3	5	89.3	10.4	86.9	8.5	93.8	10.8	5	108.6	12.4	113.0	9.7	94.5	3.6
236	Flamprop-isopropyl	52756-22-6	5	86.3	11.6	95.8	4.5	92.1	2.7	5	87.6	5.9	92.8	6.4	108.7	1.5
237	Flamprop-methyl	52756-25-9	5	83.2	12.8	95.2	4.9	88.0	1.6	5	84.9	6.4	91.1	6.6	103.0	2.1
238	Fluazinam	79622-59-6	20	NA	NA	10.1	73.6	11.4	115.2	100	NA	NA	NA	NA	NA	NA
239	Flubenzimine	37893-02-0	NA	NA	NA	NA	NA	NA	NA	NA	NA	NA	NA	NA	NA	NA
240	Fluchloralin	33245-39-5	5	105.4	4.9	91.8	3.2	100.7	2.8	5	93.5	8.9	99.7	5.2	98.6	4.1
241	Flucythrinate	70124-77-5	10	82.7	18.9	92.9	4.0	101.2	1.1	20	NA	NA	106.7	6.6	104.3	2.2
242	Flufenacet	142459-58-3	5	101.8	12.6	84.7	4.5	97.6	6.1	5	97.0	6.7	107.4	8.2	125.2	2.0
243	Flumetralin	62924-70-3	5	103.2	4.5	89.7	5.1	91.9	5.9	5	100.8	8.2	87.4	9.5	88.1	4.5
244	Flumioxazin	103361-09-7	50	NA	NA	82.9	10.0	99.4	8.5	20	NA	NA	118.8	12.6	97.6	9.4
245	Fluopyram	658066-35-4	5	91.3	3.2	91.8	2.1	95.6	7.4	5	86.9	6.3	82.3	2.4	92.3	16.5
246	Fluorodifen	15457-05-3	5	107.7	10.4	91.9	3.2	100.4	3.4	5	103.8	6.3	109.8	7.5	98.1	7.3

#	Name	CAS														
247	Fluoroglycofen-ethyl	77501-90-7	20	NA	NA	91.9	3.7	90.0	2.5	20	NA	NA	107.8	3.3	111.9	4.6
248	Fluotrimazole	31251-03-3	5	91.0	9.0	87.8	2.3	99.8	1.4	5	92.6	8.8	95.8	5.5	99.4	4.6
249	Flurochloridone	61213-25-0	5	97.3	5.2	91.7	5.0	98.7	2.5	5	88.5	3.9	99.6	3.8	89.6	4.3
250	fluroxypyr-mepthyl	81406-37-3	5	95.6	3.9	92.9	3.3	99.0	1.4	5	99.1	7.0	97.4	3.4	99.6	3.2
251	Flurprimidol	56425-91-3	5	93.4	6.8	94.8	4.0	92.4	1.9	5	89.5	6.4	93.7	6.2	105.5	2.0
252	Flusilazole	85509-19-9	5	101.1	5.8	79.9	18.5	91.9	6.3	5	86.6	9.8	89.7	8.8	92.2	3.2
253	Flutolanil	66332-96-5	5	91.2	7.6	91.4	3.8	89.4	2.2	5	91.8	5.1	94.2	5.7	103.7	2.0
254	Flutriafol	76674-21-0	5	95.4	7.3	90.5	3.6	91.8	8.7	5	72.2	5.2	70.5	4.8	79.0	19.7
255	Fluxapyroxad	907204-31-3	5	96.7	5.5	81.5	11.6	89.9	6.5	5	85.3	10.7	90.0	7.9	89.0	3.5
256	Folpet	133-07-3	200	NA	NA	NA	NA	109.2	16.6	NA	NA	NA	NA	NA	NA	NA
257	Fonofos	944-22-9	5	93.6	2.5	89.9	2.1	95.5	6.3	5	103.4	3.9	98.1	2.8	101.0	13.6
258	Formothion	2540-82-1	5	89.7	5.4	76.7	12.7	81.9	19.4	5	77.6	17.2	100.3	9.8	77.9	20.0
259	Fuberidazole	3878-19-1	100	NA	NA	NA	NA	52.4	28.4	20	NA	NA	58.0	24.6	71.2	10.1
260	Furalaxyl	57646-30-7	5	88.9	2.8	89.9	1.7	96.5	6.0	5	75.6	6.8	71.3	2.5	79.8	18.0
261	Furathiocarb	65907-30-4	10	102.1	3.2	94.7	3.4	95.5	6.6	50	NA	NA	102.4	16.6	80.3	9.3
262	Furmecyclox	60568-05-0	5	60.8	13.2	50.2	13.4	51.4	7.7	5	73.7	7.4	80.6	11.3	56.2	2.2
263	Gamma-Cyhalothrin	76703-62-3	5	94.2	16.8	96.4	5.6	106.9	3.6	5	116.4	6.5	104.1	2.5	99.6	1.3
264	Haloxyfop	69806-34-4	NA	NA	NA	NA	NA	NA	NA	NA	NA	NA	NA	NA	NA	NA
265	Haloxyfop-methyl	69806-40-2	5	90.5	1.5	90.4	1.9	96.7	5.4	5	102.1	3.5	99.7	3.0	108.2	13.1
266	Heptachlor	76-44-8	5	95.9	6.9	91.7	2.0	96.2	2.4	5	93.5	7.5	100.7	5.6	86.3	3.2
267	Heptachlor-exo-epoxide	1024-57-3	5	91.3	4.7	91.8	4.7	90.9	2.1	5	101.7	2.2	93.5	2.9	93.7	3.7
268	Heptenophos	23560-59-0	5	95.1	18.3	92.4	3.8	94.8	8.4	10	94.9	15.8	70.0	5.2	75.9	14.3
269	Hexaconazole	79983-71-4	50	NA	NA	92.9	5.3	95.6	5.8	50	NA	NA	86.8	15.9	102.7	4.1
270	Hexaflumuron	86479-06-3	NA	NA	NA	NA	NA	NA	NA	NA	NA	NA	NA	NA	NA	NA
271	Hexazinone	51235-04-2	10	93.7	7.2	88.1	4.1	87.1	4.7	10	74.0	11.4	90.8	10.1	109.3	2.7
272	Imazamethabenz-methyl	81405-85-8	20	NA	NA	78.2	19.4	82.6	8.1	20	NA	NA	82.5	6.6	82.3	6.2
273	Indanofan	133220-30-1	NA	NA	NA	NA	NA	NA	NA	NA	NA	NA	NA	NA	NA	NA
274	Indoxacarb	144171-61-9	200	NA	NA	NA	NA	87.9	2.2	200	NA	NA	NA	NA	70.6	7.0
275	Iodofenphos	18181-70-9	5	104.2	5.7	95.6	4.5	93.4	4.3	5	97.1	7.1	87.8	6.4	95.5	4.1
276	Iprobenfos	26087-47-8	20	NA	NA	90.6	2.9	95.8	7.9	5	119.2	4.1	77.3	3.5	82.5	18.0
277	Iprodione	36734-19-7	NA	NA	NA	NA	NA	NA	NA	5	57.5	73.8	166.9	57.0	22.1	52.7
278	Iprovalicarb	140923-17-7	50	NA	NA	78.9	5.7	94.4	4.6	50	NA	NA	NA	NA	87.1	10.1
279	Isazofos	42509-80-8	5	93.7	1.4	91.2	2.3	97.4	6.7	5	88.4	4.2	83.8	2.3	92.8	15.5
280	Isocarbamid	30979-48-7	5	96.8	14.1	80.2	4.3	90.8	12.9	10	94.0	8.6	96.7	9.1	98.8	2.0
281	isocarbophos	24353-61-5	5	94.1	6.7	87.5	2.6	87.4	3.0	5	87.6	7.5	86.4	7.2	105.3	1.2
282	Isofenphos-Methyl	99675-03-3	10	111.0	13.9	98.5	4.3	88.8	4.0	5	96.0	13.0	88.4	7.5	97.8	0.7
283	Isofenphos-oxon	31120-85-1	5	102.2	12.1	91.3	4.2	95.9	6.9	5	106.7	9.0	113.9	8.0	119.6	4.1
284	Isoprocarb	2631-40-5	5	87.5	2.6	90.9	1.3	94.7	7.6	5	76.0	7.5	70.4	0.1	84.8	14.5

(Continued)

APPENDIX TABLE 1.2.1 Results of Nontarget Screening for 494 Pesticides in Two Kinds of Tea by GC-Q-TOF/MS (Cont.)

序号	农药名称	CAS 号	红茶 筛查限 (µg/kg)	10 µg/kg AVE	10 µg/kg RSD, % (n = 5)	50 µg/kg AVE	50 µg/kg RSD, % (n = 5)	200 µg/kg AVE	200 µg/kg RSD, % (n = 5)	绿茶 筛查限 (µg/kg)	10 µg/kg AVE	10 µg/kg RSD, % (n = 5)	50 µg/kg AVE	50 µg/kg RSD, % (n = 5)	200 µg/kg AVE	200 µg/kg RSD, % (n = 5)
285	Isopropalin	33820-53-0	5	92.3	3.0	90.5	2.1	94.9	7.7	5	119.5	0.4	117.1	4.0	117.3	10.7
286	Isoprothiolane	50512-35-1	5	94.5	5.0	94.9	5.0	90.8	0.9	5	92.5	5.5	93.8	5.2	102.2	1.4
287	Isoproturon	34123-59-6	NA	NA	NA	NA	NA	NA	NA	NA	NA	NA	NA	NA	NA	NA
288	Isoxadifen-ethyl	163520-33-0	5	58.2	11.7	91.5	2.8	94.5	5.2	50	NA	NA	100.3	7.8	110.6	12.3
289	Isoxaflutole	141112-29-0	10	76.7	14.2	82.9	4.9	75.1	11.2	5	113.7	7.5	97.4	6.1	98.7	17.8
290	Isoxathion	18854-01-8	5	86.3	11.6	95.8	4.5	92.1	2.7	5	87.6	5.9	92.8	6.4	108.7	1.5
291	Kinoprene	42588-37-4	20	NA	NA	93.8	6.6	90.4	6.1	5	NA	NA	134.4	5.2	118.7	2.4
292	Kresoxim-methyl	143390-89-0	5	83.1	2.0	90.3	2.0	96.7	6.3	10	98.8	5.1	89.2	1.4	95.6	14.1
293	Lactofen	77501-63-4	20	NA	NA	90.9	3.6	91.5	11.1	20	NA	NA	105.3	7.6	98.4	19.9
294	Lambda-Cyhalothrin	91465-08-6	NA	NA	NA	NA	NA	NA	NA	NA	NA	NA	NA	NA	NA	NA
295	Lindane	58-89-9	5	96.3	5.4	90.8	2.2	98.1	2.0	5	88.2	6.0	95.3	3.1	86.6	2.8
296	Malathion	121-75-5	20	NA	NA	94.0	6.4	98.5	2.1	20	NA	NA	97.7	6.5	110.5	1.9
297	McpaButoxyethylEster	19480-43-4	5	89.8	5.0	95.6	5.8	102.4	3.2	5	93.7	13.6	96.6	4.8	90.7	2.7
298	Mecoprop	7085-19-0	5	86.6	3.9	84.3	13.1	95.8	3.5	5	81.9	12.9	97.0	4.1	94.7	4.1
299	Mefenacet	73250-68-7	50	NA	NA	89.2	5.6	91.9	7.3	20	NA	NA	109.3	9.9	117.0	3.7
300	Mefenpyr-diethyl	135590-91-9	5	88.7	3.3	91.2	3.1	95.0	7.1	5	83.6	4.6	81.6	2.5	91.0	15.4
301	Mepanipyrim	110235-47-7	5	106.0	5.7	90.4	5.8	94.1	3.3	5	95.0	5.6	99.7	7.1	105.0	2.4
302	Mepronil	55814-41-0	5	101.0	5.8	91.5	4.6	92.4	2.9	5	97.1	8.5	101.1	5.2	106.4	2.1
303	Metalaxyl	57837-19-1	5	101.0	2.6	79.5	13.8	92.3	4.9	10	71.2	6.8	94.5	9.0	86.4	2.6
304	Metamitron	41394-05-2	100	NA	NA	NA	NA	NA	NA	200	NA	NA	NA	NA	NA	NA
305	Metazachlor	67129-08-2	5	85.1	11.2	93.1	5.0	89.2	2.4	5	103.6	2.2	97.0	5.7	102.8	2.6
306	Metconazole	125116-23-6	50	NA	NA	97.5	11.4	94.8	3.1	50	NA	NA	NA	NA	89.9	7.0
307	Methabenzthiazuron	18691-97-9	5	92.6	4.6	89.0	2.5	93.4	7.1	5	76.5	4.4	71.1	2.9	85.7	15.5
308	Methacrifos	62610-77-9	5	70.3	8.2	93.8	7.5	91.0	6.9	5	92.3	9.0	89.8	18.7	86.5	6.0
309	Methamidophos	10265-92-6	20	NA	NA	197.0	34.4	76.0	17.1	5	79.5	19.7	102.6	19.0	103.5	6.1
310	Methfuroxam	28730-17-8	5	16.8	35.2	12.3	12.9	11.4	26.2	5	56.1	10.3	59.8	9.5	50.8	10.6
311	Methidathion	950-37-8	10	102.8	8.8	94.3	3.5	102.3	3.0	10	77.9	NA	107.4	6.7	100.0	4.1
312	Methoprene	40596-69-8	5	95.6	11.1	91.0	3.1	100.3	2.3	5	81.7	17.8	113.3	9.7	90.3	3.0
313	Methoprotryne	841-06-5	5	92.8	4.9	91.9	2.1	92.3	8.2	5	84.5	5.9	77.2	3.1	86.0	17.3
314	Methothrin	34388-29-9	5	91.1	3.3	93.6	3.4	96.0	2.8	20	NA	NA	NA	NA	93.1	2.6
315	Methoxychlor	72-43-5	5	107.0	3.8	96.1	3.4	85.4	4.6	5	101.2	6.9	82.9	7.9	88.0	5.9
316	Metolachlor	51218-45-2	5	90.0	2.7	88.6	1.9	95.1	8.0	5	81.0	4.4	76.2	2.4	83.7	16.6
317	Metolcarb	1129-41-5	5	93.4	9.8	104.5	5.4	105.0	5.6	NA	NA	NA	NA	NA	NA	NA

318	Metribuzin	21087-64-9	10	102.9	5.7	84.7	14.2	92.7	5.5	5	98.4	15.8	91.5	10.3	92.8	3.6
319	Mevinphos	7786-34-7	5	96.1	7.0	94.5	12.8	93.8	5.7	5	90.2	15.9	87.3	13.7	94.1	4.1
320	Mexacarbate	315-18-4	10	53.9	35.2	94.8	6.1	94.6	6.7	10	71.7	16.3	81.9	10.6	90.1	18.9
321	Mgk264	113-48-4	20	NA	NA	91.0	7.5	94.6	3.3	10	97.5	8.4	89.2	2.3	91.3	3.1
322	Mirex	2385-85-5	5	98.8	1.2	91.0	3.1	90.3	4.3	5	90.8	3.1	89.7	4.4	88.8	2.3
323	Molinate	2212-67-1	5	82.6	6.4	89.3	1.6	93.3	7.7	5	83.4	1.8	78.1	5.4	80.5	15.6
324	Monalide	7287-36-7	5	97.4	3.5	91.4	3.8	98.4	1.2	5	77.6	8.9	96.5	3.0	99.3	2.8
325	Monuron	150-68-5	NA	NA	NA	NA	NA	NA	NA	NA	NA	NA	NA	NA	NA	NA
326	MuskAmbrette	83-66-9	5	94.9	7.6	94.0	2.7	100.2	2.4	5	88.9	7.5	105.5	5.7	86.9	3.7
327	Muskketone	81-14-1	5	97.7	6.5	92.4	2.6	99.5	1.3	5	98.1	6.8	101.8	3.5	100.2	3.2
328	Myclobutanil	88671-89-0	5	99.4	7.3	96.4	2.4	95.0	7.4	5	75.6	10.5	76.0	4.7	87.0	16.9
329	Naled	300-76-5	100	NA	NA	NA	NA	71.6	14.2	50	NA	NA	77.5	2.7	68.5	32.2
330	Napropamide	15299-99-7	10	73.8	18.4	94.8	3.6	96.0	7.8	10	82.8	16.1	61.2	6.7	78.5	18.7
331	Nitralin	4726-14-1	20	NA	NA	107.9	14.5	96.6	4.5	20	NA	NA	96.5	9.9	95.6	9.3
332	Nitrapyrin	1929-82-4	5	104.8	26.0	76.3	5.6	89.5	11.3	5	94.1	20.0	106.6	12.2	85.3	10.6
333	Nitrofen	1836-75-5	5	101.8	8.1	97.3	4.6	102.7	3.7	5	110.6	12.0	113.2	5.7	99.4	5.9
334	Nitrothal-Isopropyl	10552-74-6	5	95.0	7.9	87.2	8.5	92.4	4.8	5	93.1	3.5	98.5	8.8	84.2	3.5
335	Norflurazon	27314-13-2	5	98.3	7.2	93.1	5.5	91.1	9.1	5	79.0	11.4	69.6	4.3	87.1	13.3
336	Nuarimol	63284-71-9	20	NA	NA	92.1	3.0	90.3	2.5	20	NA	NA	77.7	6.1	72.7	2.9
337	Octachlorostyrene	29082-74-4	5	100.3	1.9	91.9	3.1	91.9	3.3	5	92.5	3.5	91.1	3.1	92.3	2.2
338	Octhilinone	26530-20-1	5	94.6	6.5	93.3	3.1	91.5	7.2	5	43.8	8.7	39.5	13.6	38.3	23.7
339	Ofurace	58810-48-3	5	82.5	13.7	87.7	2.9	81.1	6.7	5	87.0	16.8	101.5	7.2	105.6	3.1
340	Orbencarb	34622-58-7	5	90.8	2.8	90.4	1.9	97.1	5.5	5	107.0	5.2	96.8	2.7	106.8	15.0
341	Oxabetrinil	74782-23-3	5	93.1	2.3	94.3	3.1	101.0	2.8	5	43.4	64.2	46.5	116.1	99.6	3.0
342	Oxadiazon	19666-30-9	5	91.3	3.1	90.9	4.0	98.3	1.0	5	84.3	4.7	93.4	2.6	100.7	2.8
343	Oxadixyl	77732-09-3	5	90.9	11.0	86.9	4.1	75.1	6.0	5	82.3	8.7	90.0	10.3	99.1	3.5
344	Oxycarboxin	5259-88-1	NA	NA	NA	NA	NA	NA	NA	NA	NA	NA	NA	NA	NA	NA
345	Paclobutrazol	76738-62-0	5	91.9	1.9	90.6	3.6	90.5	10.4	5	85.2	3.2	78.0	4.5	89.3	17.4
346	Paraoxon-Methyl	950-35-6	NA	NA	NA	NA	NA	NA	NA	100	NA	NA	NA	NA	185.9	80.6
347	Parathion	56-38-2	5	95.1	6.7	93.7	2.0	98.7	2.4	5	88.9	9.3	102.7	4.8	88.7	4.4
348	Parathion-Methyl	298-00-0	5	86.2	11.0	95.7	1.9	102.4	4.3	5	83.5	11.5	108.8	3.9	100.4	4.5
349	Pebulate	1114-71-2	5	82.6	7.5	93.1	3.0	90.3	10.4	10	105.1	3.9	94.3	8.5	92.6	11.0
350	Penconazole	66246-88-6	5	90.6	4.5	90.8	1.9	95.2	7.3	5	83.8	6.1	82.7	2.4	89.0	15.4
351	Pendimethalin	40487-42-1	5	96.0	5.7	91.4	3.0	98.5	2.8	10	95.6	11.5	101.1	4.1	101.0	3.6
352	Pentachloroaniline	527-20-8	5	45.6	12.5	45.8	9.7	71.5	3.2	5	67.1	6.1	78.2	8.5	73.0	2.5
353	Pentachloroanisole	1825-21-4	5	85.3	7.5	84.4	7.8	92.9	1.9	5	88.2	5.2	96.7	4.3	85.9	3.0
354	Pentachlorobenzene	608-93-5	5	88.1	19.6	71.9	8.7	83.1	9.5	5	81.2	18.5	95.6	8.4	77.7	7.6
355	Pentachlorocyanobenzene	20925-85-3	5	78.3	5.3	71.0	3.8	86.5	8.9	5	75.1	7.1	93.6	4.8	92.9	3.2

(Continued)

APPENDIX TABLE 1.2.1 Results of Nontarget Screening for 494 Pesticides in Two Kinds of Tea by GC-Q-TOF/MS (Cont.)

序号	农药名称	CAS 号	红茶 筛查限 (µg/kg)	10 µg/kg AVE	10 µg/kg RSD, % (n = 5)	50 µg/kg AVE	50 µg/kg RSD, % (n = 5)	200 µg/kg AVE	200 µg/kg RSD, % (n = 5)	绿茶 筛查限 (µg/kg)	10 µg/kg AVE	10 µg/kg RSD, % (n = 5)	50 µg/kg AVE	50 µg/kg RSD, % (n = 5)	200 µg/kg AVE	200 µg/kg RSD, % (n = 5)
356	Pentanochlor	2307-68-8	5	87.4	3.3	92.0	1.4	97.8	5.9	5	95.4	4.6	89.8	1.8	99.0	15.0
357	Permethrin	52645-53-1	10	96.3	2.6	91.4	3.4	100.0	2.4	10	101.6	4.2	97.4	2.8	94.5	2.2
358	Perthane	72-56-0	5	93.0	2.3	91.9	3.6	99.2	2.2	5	94.0	5.0	95.6	3.4	90.2	2.7
359	Phenanthrene	85-01-8	5	98.4	4.3	90.9	5.2	94.2	1.8	5	112.7	3.3	98.0	8.3	108.6	2.1
360	Phenthoate	2597-03-7	5	97.3	4.3	95.5	2.3	102.4	8.2	5	111.0	4.4	99.2	3.5	103.3	14.2
361	Phorate-Sulfone	2588-04-7	5	118.6	8.9	94.1	2.9	97.3	3.3	5	96.4	6.9	100.4	5.5	111.3	2.1
362	Phorate-Sulfoxide	2588-03-6	50	NA	NA	119.2	11.3	91.7	6.2	NA	NA	NA	102.7	8.5	111.7	6.4
363	Phosalone	2310-17-0	5	116.5	11.3	89.1	5.8	102.9	8.1	5	111.4	5.7	109.7	7.1	119.5	2.1
364	Phosfolan	947-02-4	5	108.1	11.8	81.0	4.7	91.9	9.0	5	98.8	11.7	111.1	10.8	119.0	1.8
365	Phosmet	732-11-6	200	NA	NA	NA	NA	NA	NA	200	NA	NA	NA	NA	NA	NA
366	Phosphamidon	13171-21-6	5	94.4	4.6	96.2	4.0	97.4	9.6	5	70.4	3.5	60.4	4.3	77.1	19.4
367	Phthalic Acid,Benzyl Butyl Ester	85-68-7	5	95.6	4.8	94.4	4.6	92.3	1.1	5	96.6	4.7	95.2	4.7	104.9	1.4
368	Phthalic Acid,bis-2-ethylhexyl ester	117-81-7	NA	NA	NA	NA	NA	NA	NA	NA	NA	NA	NA	NA	NA	NA
369	Phthalic Acid,Bis-Butyl Ester	84-74-2	NA	NA	NA	NA	NA	NA	NA	NA	NA	NA	NA	NA	NA	NA
370	Phthalic Acid,Bis-Cyclohexyl Ester	84-61-7	20	NA	NA	93.9	5.5	90.0	1.0	20	NA	NA	96.2	5.6	104.6	0.8
371	Phthalimide	85-41-6	5	77.2	24.8	71.2	15.4	72.0	11.1	20	NA	NA	73.6	14.8	97.8	10.6
372	Picolinafen	137641-05-5	5	92.3	4.3	92.6	2.3	96.1	6.7	5	114.2	2.5	109.0	3.6	114.1	12.9
373	Picoxystrobin	117428-22-5	5	97.1	4.4	89.6	1.4	96.7	5.9	5	91.0	7.4	90.8	2.5	98.5	13.9
374	Piperonyl Butoxide	51-03-6	20	NA	NA	94.0	3.3	92.4	3.1	20	NA	NA	92.0	7.2	99.2	3.3
375	Piperophos	24151-93-7	5	71.8	20.4	90.0	2.5	95.8	8.1	100	NA	NA	NA	NA	76.5	17.0
376	Pirimicarb	23103-98-2	5	91.1	7.8	92.8	4.6	94.1	1.7	5	87.0	6.2	91.9	6.1	105.4	1.7
377	Pirimiphos-Ethyl	23505-41-1	5	97.9	15.9	90.8	2.7	95.3	6.8	20	NA	NA	102.5	2.5	112.1	10.9
378	Pirimiphos-Methyl	29232-93-7	5	108.6	5.2	95.5	4.7	95.5	1.8	5	101.6	4.8	96.1	6.1	105.3	1.7
379	Pirimiphos-methyl-N-desethyl	67018-59-1	10	112.5	2.4	94.9	3.4	88.6	3.6	10	104.2	8.4	91.0	7.3	99.9	0.6
380	Plifenate	21757-82-4	5	97.5	7.2	88.9	5.3	98.2	1.8	5	95.8	4.3	98.2	3.2	97.6	2.9
381	Prallethrin	23031-36-9	20	NA	NA	52.5	1.6	183.2	9.3	20	NA	NA	70.3	3.2	95.7	13.0
382	Pretilachlor	51218-49-6	5	98.5	2.8	92.1	1.9	97.1	7.7	5	89.8	7.9	80.4	2.8	88.6	15.7
383	Probenazole	27605-76-1	20	NA	NA	96.0	6.4	80.8	11.0	20	NA	NA	94.2	19.7	82.5	11.6
384	Procyazine	32889-48-8	5	81.3	8.1	70.8	7.0	85.5	10.4	10	77.6	7.1	97.5	14.0	81.4	7.0
385	Procymidone	32809-16-8	5	95.9	11.7	90.8	2.8	98.1	1.7	5	78.8	11.7	97.2	4.1	98.4	3.1
386	Profenofos	41198-08-7	5	101.3	7.3	87.1	5.4	95.7	5.2	5	103.0	6.0	105.2	6.3	116.8	2.4
387	Profluralin	26399-36-0	5	100.6	9.1	94.7	2.9	101.6	3.2	5	93.3	7.6	106.6	5.4	87.1	3.7
388	Promecarb	2631-37-0	50	NA	NA	108.3	4.4	85.4	6.4	20	NA	NA	86.2	5.9	94.9	4.3

No.	Compound	CAS														
389	Prometon	1610-18-0	5	88.4	6.9	91.6	3.7	91.2	1.5	5	96.3	6.1	93.3	6.8	101.3	1.5
390	Prometryn	7287-19-6	5	101.3	4.8	87.6	9.5	93.1	3.8	5	87.4	10.5	90.4	6.0	91.5	2.0
391	Propachlor	1918-16-7	10	95.0	10.5	93.4	5.8	89.8	1.7	20	NA	NA	93.6	6.9	101.1	1.8
392	Propamocarb	24579-73-5	NA	NA	NA	NA	NA	NA	NA	NA	NA	NA	NA	NA	NA	NA
393	Propanil	709-98-8	5	100.6	2.7	97.8	1.5	96.3	6.4	5	91.7	4.8	84.9	3.9	93.5	19.3
394	Propaphos	7292-16-2	5	114.7	8.4	98.5	2.5	94.4	8.3	5	72.4	14.1	71.7	11.3	83.4	18.2
395	Propargite	2312-35-8	100	NA	NA	NA	NA	44.9	13.1	50	NA	NA	118.5	5.4	106.4	4.1
396	Propazine	139-40-2	5	93.0	2.8	91.2	1.2	96.1	6.4	5	84.8	7.4	82.0	2.4	92.3	15.7
397	Propetamphos	31218-83-4	5	99.2	3.3	89.4	1.5	96.4	7.0	5	97.3	2.9	93.0	3.8	100.5	15.1
398	Propham	122-42-9	20	NA	NA	87.7	6.8	93.4	4.8	50	NA	NA	90.4	10.5	95.3	4.5
399	Propisochlor	86763-47-5	5	91.3	0.8	89.5	1.9	96.2	6.0	5	89.6	2.9	83.9	3.5	92.1	15.8
400	PropyleneThiourea	2122-19-2	NA	NA	NA	NA	NA	NA	NA	NA	NA	NA	NA	NA	NA	NA
401	Propyzamide	23950-58-5	5	105.1	4.6	89.4	9.5	95.2	3.6	5	86.8	10.6	89.4	6.0	93.9	2.0
402	Prosulfocarb	52888-80-9	50	NA	NA	94.6	5.5	92.3	1.2	5	91.6	10.4	92.1	4.5	105.5	1.5
403	Prothiofos	34643-46-4	5	86.7	2.0	93.7	3.5	99.5	3.5	5	101.1	5.2	101.9	2.8	90.9	3.1
404	Pyracarbolid	24691-76-7	5	90.0	5.0	89.2	4.9	97.0	2.3	5	79.2	10.1	95.6	4.4	98.6	5.2
405	Pyraclostrobin	175013-18-0	100	NA	NA	NA	NA	98.5	6.3	5	NA	NA	107.1	NA	56.6	84.1
406	Pyrazophos	13457-18-6	5	97.4	8.9	84.6	3.0	95.8	4.8	20	NA	NA	NA	4.9	80.9	6.5
407	Pyrethrins	8003-34-7	NA	NA	NA	NA	NA	NA	NA	NA	NA	NA	NA	NA	NA	NA
408	Pyributicarb	88678-67-5	5	104.4	11.5	95.8	4.5	96.7	3.0	5	103.5	7.0	98.9	5.2	110.0	1.4
409	Pyridaben	96489-71-3	20	NA	NA	93.4	6.8	95.1	6.5	50	NA	NA	102.8	6.3	112.2	1.8
410	Pyridalyl	179101-81-6	20	NA	NA	29.7	20.1	27.3	43.5	20	NA	NA	54.4	11.5	66.4	7.0
411	Pyridaphenthion	119-12-0	5	114.8	17.7	88.2	4.8	99.3	7.0	5	110.4	10.2	110.8	7.7	115.1	3.0
412	Pyrifenox	88283-41-4	50	NA	NA	91.8	3.3	89.5	5.7	20	NA	NA	93.5	12.0	101.9	2.7
413	Pyriftalid	135186-78-6	20	NA	NA	90.0	5.4	86.1	2.8	20	NA	NA	79.3	9.0	94.7	7.7
414	Pyrimethanil	53112-28-0	5	86.7	3.1	88.9	1.8	94.3	5.9	5	82.5	4.8	78.7	2.3	86.4	17.5
415	Pyriproxyfen	95737-68-1	5	97.4	9.6	95.6	5.1	94.5	3.2	5	103.8	5.2	102.9	4.8	112.5	2.2
416	Pyroquilon	57369-32-1	5	95.5	3.4	90.8	2.0	95.5	6.4	5	70.0	2.6	60.6	2.9	78.0	19.2
417	Quinalphos	13593-03-8	5	110.3	5.8	92.7	5.4	92.4	1.4	5	100.1	6.6	96.3	5.3	104.2	1.8
418	Quinoclamine	2797-51-5	20	NA	NA	72.5	9.6	84.0	12.5	20	NA	NA	60.0	19.6	52.4	32.4
419	Quinoxyfen	124495-18-7	5	93.4	3.0	89.6	2.6	97.8	2.0	5	86.6	6.6	95.3	3.8	86.9	4.1
420	Quintozene	82-68-8	5	92.6	7.2	88.5	1.8	98.9	2.3	5	93.4	5.9	102.4	4.5	85.2	4.4
421	Quizalofop-Ethyl	76578-14-8	10	98.2	3.3	89.6	6.9	91.3	7.5	5	110.7	4.2	105.4	7.0	113.6	2.2
422	Rabenzazole	40341-04-6	5	58.8	26.5	74.1	12.9	89.7	5.8	5	72.0	5.2	73.5	8.3	87.0	17.9
423	S421	127-90-2	10	96.0	15.7	95.7	4.6	49.3	4.6	10	91.7	10.8	98.8	7.5	44.2	7.8
424	Sebuthylazine	7286-69-3	5	95.6	5.4	91.8	4.6	90.9	0.9	5	89.4	6.2	92.7	6.1	100.7	1.8
425	Sebuthylazine-desethyl	37019-18-4	5	100.7	15.4	80.3	4.3	97.9	11.6	5	88.7	9.2	92.0	9.4	105.1	4.0

(Continued)

APPENDIX TABLE 1.2.1 Results of Nontarget Screening for 494 Pesticides in Two Kinds of Tea by GC-Q-TOF/MS (Cont.)

序号	农药名称	CAS 号	红茶						绿茶					
			筛查限 (μg/kg)	10 μg/kg AVE	RSD, % (n = 5)	50 μg/kg AVE	RSD, % (n = 5)	200 μg/kg AVE	RSD, % (n = 5)	筛查限 (μg/kg)	10 μg/kg AVE	RSD, % (n = 5)	50 μg/kg AVE	RSD, % (n = 5)	200 μg/kg AVE	RSD, % (n = 5)
426	Secbumeton	26259-45-0	5	94.1	3.2	90.3	2.7	94.9	8.0	5	80.7	7.0	77.2	3.3	83.8	17.0
427	Silafluofen	105024-66-6	5	92.6	3.7	91.9	3.9	98.3	2.8	5	101.8	4.8	96.8	3.4	91.8	2.6
428	Simazine	122-34-9	5	105.8	9.3	90.8	4.5	94.1	2.1	5	90.4	10.0	94.4	6.8	103.7	1.8
429	Simeconazole	149508-90-7	10	119.0	11.8	93.7	6.5	92.8	1.8	10	86.8	12.5	97.3	6.6	101.4	3.2
430	Simeton	673-04-1	5	98.1	5.2	91.4	2.1	94.4	8.3	5	82.2	11.3	72.9	2.1	87.3	13.3
431	Simetryn	1014-70-6	5	93.4	4.6	90.2	3.6	98.2	1.1	5	74.6	10.1	91.4	6.7	95.2	5.8
432	Spirodiclofen	148477-71-8	50	NA	NA	98.4	30.2	78.5	12.9	50	NA	NA	104.3	13.4	140.8	3.8
433	Spiromesifen	283594-90-1	5	96.7	6.9	85.1	4.1	99.3	2.4	5	87.7	5.9	96.1	5.1	86.9	5.3
434	Spiroxamine	118134-30-8	5	64.2	7.1	63.5	3.3	64.7	6.3	20	NA	NA	59.8	5.5	70.9	10.2
435	Sulfallate	95-06-7	5	92.7	4.4	90.2	1.8	96.1	7.2	5	101.4	2.6	96.4	4.2	95.4	14.3
436	Sulfotep	3689-24-5	5	94.6	4.8	86.6	4.2	98.8	2.1	5	92.9	6.9	100.4	4.5	85.8	3.6
437	Sulprofos	35400-43-2	5	92.4	11.9	107.7	5.0	118.4	3.5	5	132.8	6.6	145.4	3.4	103.4	3.1
438	Tau-Fluvalinate	102851-06-9	5	85.7	5.1	93.3	2.7	92.2	3.7	5	92.3	3.7	89.5	6.0	90.0	3.0
439	TCMTB	21564-17-0	20	NA	NA	93.0	6.2	88.7	6.3	50	NA	NA	107.7	6.5	119.7	9.1
440	Tebuconazole	107534-96-3	10	93.9	10.3	94.1	4.3	92.0	9.6	5	90.7	7.6	93.8	6.6	92.2	17.5
441	Tebufenpyrad	119168-77-3	5	93.4	6.9	94.2	5.5	93.6	2.2	5	99.6	4.8	94.4	5.7	109.4	1.7
442	Tebupirimfos	96182-53-5	5	96.4	4.9	90.6	2.7	97.8	2.7	5	98.2	7.0	101.0	4.0	87.4	2.9
443	Tebutam	35256-85-0	5	82.8	5.7	105.2	7.1	95.2	7.2	10	91.7	19.5	89.0	4.9	90.3	18.1
444	Tebuthiuron	34014-18-1	10	88.4	3.0	80.4	14.3	85.8	13.5	5	73.8	6.5	67.3	5.5	81.5	13.2
445	Tecnazene	117-18-0	5	70.8	6.1	96.7	6.0	90.6	7.9	5	94.3	10.4	91.0	14.6	87.5	3.8
446	Teflubenzuron	83121-18-0	5	101.0	14.1	85.9	2.6	98.9	2.5	5	90.0	6.6	103.3	11.9	79.6	18.6
447	Tefluthrin	79538-32-2	5	90.5	2.2	92.2	2.1	98.1	0.5	5	87.1	4.4	95.8	3.5	98.8	2.0
448	Tepraloxydim	149979-41-9	50	NA	NA	88.6	3.4	82.6	10.7	20	NA	NA	71.5	4.3	89.2	12.1
449	Terbucarb	1918-11-2	5	90.2	2.0	90.7	1.6	95.4	7.6	10	97.0	4.0	93.0	2.5	101.7	13.0
450	Terbufos	13071-79-9	5	95.4	5.3	97.1	6.4	93.4	4.6	5	99.2	6.9	88.8	5.6	90.7	2.8
451	Terbufos-Sulfone	56070-16-7	10	166.9	26.7	93.8	5.1	100.3	7.4	20	NA	NA	71.0	8.5	88.2	17.1
452	Terbumeton	33693-04-8	5	93.7	1.6	89.7	3.3	96.7	6.6	5	80.3	5.7	74.9	3.2	84.3	17.0
453	Terbuthylazine	5915-41-3	5	95.7	5.1	93.2	5.0	92.9	1.5	5	91.6	5.4	92.8	5.8	104.1	1.4
454	Terbutryn	886-50-0	5	95.0	6.0	91.4	1.6	99.1	1.4	5	92.8	11.5	97.2	4.1	98.3	3.9
455	Tert-butyl-4-Hydroxyanisole	25013-16-5	5	85.3	3.4	98.3	3.0	88.8	5.3	10	78.6	15.5	109.7	15.8	79.3	15.9
456	Tetrachlorvinphos	22248-79-9	5	101.0	5.4	83.0	5.6	94.0	8.0	5	105.1	9.7	118.3	8.4	129.0	3.5
457	Tetraconazole	112281-77-3	5	94.1	8.4	92.6	5.7	89.0	1.9	5	86.2	5.6	90.8	6.4	98.6	3.0
458	Tetradifon	116-29-0	5	87.5	1.5	94.3	3.2	98.5	2.3	5	86.0	6.4	96.7	4.1	88.7	3.4

No.	Compound	CAS														
459	Tetramethrin	7696-12-0	5	100.1	6.0	92.5	5.3	95.6	4.2	10	97.0	7.1	101.0	5.6	111.6	1.9
460	Tetrasul	2227-13-6	5	99.6	2.3	96.4	1.7	93.0	4.0	5	96.8	3.4	90.7	3.9	93.0	2.2
461	Thenylchlor	96491-05-3	5	92.8	4.1	92.3	2.1	97.2	6.9	5	81.0	9.5	72.9	4.1	80.3	19.5
462	Thiabendazole	148-79-8	NA	NA	NA	NA	NA	NA	NA	200	NA	NA	NA	NA	NA	NA
463	Thiazopyr	117718-60-2	5	84.9	4.7	90.1	2.3	96.8	6.4	5	94.1	3.6	95.4	2.9	102.1	13.4
464	Thiobencarb	28249-77-6	5	99.8	3.2	95.4	4.3	88.9	0.8	5	95.4	4.0	93.2	5.3	102.7	0.9
465	Thiocyclam	31895-21-3	200	NA	NA	NA	NA	107.5	7.7	50	NA	NA	117.4	17.3	88.3	8.5
466	Thiofanox	39196-18-4	10	150.0	28.3	107.0	2.9	87.9	16.5	20	NA	NA	78.6	17.3	98.9	18.6
467	Thionazin	297-97-2	100	NA	NA	NA	NA	88.1	4.5	5	156.1	11.5	99.5	2.8	104.6	2.2
468	Tiocarbazil	36756-79-3	50	NA	NA	50.9	9.2	54.7	6.4	50	NA	NA	51.8	3.6	61.6	10.8
469	Tolclofos-Methyl	57018-04-9	5	98.6	3.4	93.3	4.5	91.7	0.6	5	96.6	5.0	94.4	5.2	102.2	1.1
470	Tolfenpyrad	129558-76-5	20	NA	NA	89.6	9.6	88.8	10.6	20	NA	NA	112.0	13.3	111.1	3.2
471	Tolylfluanid	731-27-1	5	81.3	9.3	79.1	11.9	103.0	10.9	20	NA	NA	96.7	10.4	79.3	15.3
472	Tralkoxydim	87820-88-0	50	NA	NA	85.7	5.9	92.8	9.5	100	NA	NA	NA	NA	92.4	15.9
473	Trans-Chlordane	5103-74-2	5	97.5	2.4	87.7	3.8	92.9	3.4	5	86.2	3.4	91.2	3.8	88.2	2.0
474	Transfluthrin	118712-89-3	5	101.0	2.6	93.4	2.9	93.8	3.3	5	91.9	3.5	91.6	3.5	93.0	1.8
475	Trans-Nonachlor	39765-80-5	5	91.4	2.1	91.9	2.5	96.9	1.7	5	84.1	6.0	90.6	3.2	89.1	2.6
476	Trans-Permethrin	61949-77-7	10	101.0	3.9	99.7	2.5	90.8	3.6	20	NA	NA	NA	NA	94.5	3.3
477	Triadimefon	43121-43-3	20	NA	NA	119.2	2.5	102.5	6.0	NA	NA	NA	NA	NA	NA	NA
478	Triadimenol	55219-65-3	10	100.5	15.7	92.7	6.7	94.8	3.1	10	91.7	13.2	97.4	6.6	103.1	3.2
479	Triallate	2303-17-5	5	89.7	5.3	92.4	11.8	96.0	6.5	10	165.0	20.4	108.7	3.1	119.0	13.0
480	Triapenthenol	76608-88-3	10	110.8	9.3	94.4	5.5	91.8	1.5	5	94.5	6.0	96.3	5.8	105.3	2.9
481	Triazophos	24017-47-8	5	104.8	3.5	92.7	3.8	95.4	8.6	5	108.9	12.1	97.0	4.7	96.5	18.4
482	Tribufos	78-48-8	5	86.7	2.8	91.6	2.5	97.1	6.3	5	114.8	3.8	109.2	3.9	117.2	13.2
483	TributylPhosphate	126-73-8	5	106.0	12.2	94.5	4.0	96.1	2.7	5	98.3	6.6	96.0	5.8	108.1	1.6
484	Triclopyr	55335-06-3	100	NA	NA	NA	NA	114.2	8.2	50	NA	NA	48.7	17.9	52.7	6.6
485	Tricyclazole	41814-78-2	10	69.3	72.4	99.3	8.6	76.4	18.3	100	NA	NA	NA	NA	76.8	8.9
486	Tridiphane	58138-08-2	5	111.6	7.8	99.0	4.3	92.6	4.6	5	115.6	12.2	86.7	8.6	90.0	3.9
487	Trietazine	1912-26-1	5	91.7	5.1	93.5	5.2	91.9	1.0	5	87.9	5.4	91.3	5.9	101.5	1.1
488	Trifenmorph	1420-06-0	5	94.9	7.5	99.4	4.6	90.7	6.9	5	100.1	6.7	92.2	4.5	113.8	3.1
489	Trifloxystrobin	141517-21-7	20	NA	NA	89.5	3.2	94.2	8.3	50	NA	NA	98.8	4.3	106.8	12.5
490	Trifluralin	1582-09-8	5	93.9	5.4	91.3	5.8	92.3	5.7	5	103.9	5.1	92.4	9.0	91.0	3.9
491	Triphenylphosphate	115-86-6	5	94.5	2.9	90.3	2.7	97.1	6.8	5	96.2	4.9	89.1	2.8	94.5	16.0
492	Uniconazole	83657-22-1	5	97.7	6.8	92.1	3.8	94.0	9.4	5	89.5	2.8	81.6	5.3	89.6	17.3
493	Vernolate	1929-77-7	5	94.0	17.7	74.8	6.5	87.8	10.8	5	84.4	11.8	95.0	7.7	76.4	8.4
494	Vinclozolin	50471-44-8	5	88.7	4.5	91.6	4.2	98.0	1.9	10	84.7	4.8	94.4	3.5	85.8	4.7

NA: 响应低或回收率不合格或 RSD 不合格.

TABLE 1.2.2 The Number and Percentage of Pesticides Can Be Screened in the Four Kinds of Matrix

	Black tea	Green tea	Puer	Oolong
Number of pesticides in the library	494	494	494	494
Number of pesticides can be screened	468	463	465	451
Number of pesticides not not spiked	13	13	13	13
Number of pesticides not detected	26	31	29	43
Percentage of pesticides can be screened	94.7%	93.7%	94.1%	91.3%

TABLE 1.2.3 The Number and Percentage of Pesticides Can Be Screened in the Four Kinds of Matrix According to the Screening Limit

Screening concentration		Matrix			
		Black tea	Green tea	Puer	Oolong
5 µg/kg	Number of pesticides	331	322	337	339
	Percentage of pesticides	67.0%	65.2%	68.2%	68.6%
10 µg/kg	Number of pesticides	35	33	21	22
	Percentage of pesticides	7.1%	6.7%	4.3%	4.5%
20 µg/kg	Number of pesticides	56	58	68	51
	Percentage of pesticides	11.3%	11.7%	13.8%	10.3%
50 µg/kg	Number of pesticides	20	34	21	23
	Percentage of pesticides	4.1%	6.9%	4.3%	4.7%
100 µg/kg	Number of pesticides	14	8	10	10
	Percentage of pesticides	2.8%	1.6%	2.0%	2.0%
200 µg/kg	Number of pesticides	12	8	8	6
	Percentage of pesticides	2.4%	1.6%	1.6%	1.2%

322–339 and 65.2%–68.6%, at the screening concentration of 5 µg/kg within the concentration ranged from 5 to 200 µg/kg in the four kinds of matrix. The degree of interference to certain pesticides due to the variance in the chemical and physical properties of different matrices will lead to a difference in the screening limit of the same pesticide in different matrices, and the difference is significant. The number and percentage of pesticides that can be screened in the four kinds of matrix, black tea, green tea, oolong, and puer, were 366, 355, 358, and 361 kinds, and 74.1%, 71.9%, 72.5%, and 73.1%, respectively, at the screening concentration of 10 µg/kg of the 494 kinds of pesticides. It can be seen that the number and percentage of pesticides that can be screened in the matrix of black tea was the largest amount, 366 kinds and 74.1%, while in the matrix of green tea, 355 kinds and 71.9%, was the least at the screening concentration of 10 µg/kg. The earlier results illustrate the universal applicability of GC-Q-TOF/MS in the four kinds of matrix and they are listed in Appendix Table 1.2.1.

The number and percentage of pesticides that can be screened according to the kind of matrix are listed in Table 1.2.4. It can be seen that 478 of the 494 kinds of pesticides can be detected in at least one matrix, and the number and percentage of pesticides that can be screened in all four kinds of matrix were 436 kinds and 88.3%, while in three or more kinds 462 kinds and 93.6%. The results demonstrate that most of the pesticides can be screened in different tea matrices and the universal applicability of the method.

1.2.4.2.2 Pesticides Cannot Be Screened

The number and percentage of pesticides that cannot be screened, of the 494 kinds of pesticides spiked in the four kinds of matrix, were 16 and 3.2%. (The compounds are listed in Table 1.2.5.) The 16 kinds of pesticides were not detected when spiked in the four kinds of matrix at the concentration of 200 µg/kg; this is because the sensitivity of these 16 kinds of pesticides. Since the enhancement of the concentration spiked was required, or they are not suited for the analysis using GC method, further evaluation was not conducted. The results are listed in Appendix Table 1.2.1.

TABLE 1.2.4 The Number and Percentage of Pesticides Can Be Screened According to the Kinds of Matrix

Kinds of matrix can be screened	Pesticides	
	Number	Percentage (%)
4	436	88.3
3	26	5.3
2	9	1.8
1	7	1.4
0	16	3.2

TABLE 1.2.5 The Pesticides Cannot Be Screened Using GC-Q-TOF/MS

No.	Compound	CAS	No.	Compound	CAS
1	Captafol	2425-06-1	9	Indanofan	133220-30-1
2	Carbosulfan	55285-14-8	10	Isoproturon	34123-59-6
3	Chlorbromuron	13360-45-7	11	Lambda-Cyhalo-thrin	91465-08-6
4	Chlordecone	143-50-0	12	Monuron	150-68-5
5	Dinoterb	1420-07-1	13	Oxycarboxin	5259-88-1
6	Dioxacarb	6988-21-2	14	Propamocarb	24579-73-5
7	Flubenzimine	37893-02-0	15	Propylene Thiourea	2122-19-2
8	Hexaflumuron	86479-06-3	16	Pyrethrins	8003-34-7

1.2.4.3 Recovery (Rec. 60%–120%&RSD < 20%)

1.2.4.3.1 Analysis of Recovery at Three Spiked Level Using GC-Q-TOF/MS

According to European Union guidance document (SANCO/10684/2009), the recovery (60%–120%) and the relative standard deviation (RSD) <20% serves as "recovery and RSD double standard." A total of 494 kinds of pesticides were included in the GC-Q-TOF/MS database, and the verification was conducted by the spiking of the 494 kinds of pesticides in the four kinds of tea matrix.

The number and percentage of pesticides can be detected in the four kinds of matrix at different spiked concentration levels were listed in Table 1.2.6. The 494 kinds of pesticide were spiked to the four kinds of matrix at concentrations of 10, 50, and 200 μg/kg. The number of pesticides that can meet the requirement of the "recovery and RSD double standard" at the concentration of 10 μg/kg in the matrix of black tea, green tea, oolong, and puer were 341, 336, 334, and 324, while the percentages were 69.0%, 68.0%, 67.6%, and 65.6%. Among these the number and percentage of pesticides which can meet the requirement of the "recovery and RSD double standard" in the matrix of black tea were the largest amount, 341 kinds and 69.0%, while the matrix of oolong were the least, 324 kinds and 65.6%. The number and percentage of pesticides that meet the requirement will increase with the increased concentration. The number of pesticides that can meet the requirement at the concentration of 200 μg/kg in the matrix of black tea, green tea, oolong, and puer were 441, 430, 422, and 411 kinds, while the percentage was 89.3%, 87.0%, 85.4%, and 83.2%, among these the number and percentage of pesticides that can meet the requirement in the matrix of black tea were the largest number, 441 kinds and 89.3%, while in the matrix of oolong they were the smallest, 411 kinds and 83.2%. Compared with high concentration, poor recovery was obtained at low concentration, and the phenomenon, matrix-induced response enhancement, or inhibition, was more obvious, meanwhile the parallelism was not excellent, so as the recovery of some pesticides cannot meet the requirement of "recovery and RSD double standard." If the recovery of all the 494 kinds of pesticides were evaluated by the internationally accepted uniform standard 10 μg/kg, the percentage of pesticides that can meet the requirement of the "recovery standard" in the matrix of the four kinds of matrix can exceed 65%. The results demonstrated the adequate accuracy of the method, and they are listed in Appendix Table 1.2.1.

The number and percentage of pesticides that can meet the requirement of the "recovery and RSD double standard" in the four kinds of matrix are listed in Table 1.2.7. The number of pesticides that can be recovered exceeds 440, while the

TABLE 1.2.6 The Number and Percentage of Pesticides Can Be Detected in the Four Kinds of Matrix at Different Spiked Concentration

Spiked concentration		Matrix			
		Black tea	Green tea	Puer	Oolong
10 μg/kg	Number of pesticides	341	336	334	324
	Percentage of pesticides	69.0%	68.0%	67.6%	65.6%
50 μg/kg	Number of pesticides	419	412	403	377
	Percentage of pesticides	84.8%	83.4%	81.6%	76.3%
200 μg/kg	Number of pesticides	441	430	422	411
	Percentage of pesticides	89.3%	87.0%	85.4%	83.2%

TABLE 1.2.7 The Number and Percentage of Pesticides With Qualified Recovery

	Black tea	Green tea	Puer	Oolong
Total pesticides	494	494	494	494
Number of pesticides with qualified recovery	448	446	442	431
Percentage of pesticides with qualified recovery	90.7%	90.3%	89.5%	87.3%
Number of pesticides with unqualified recovery	46	48	52	63
Number of pesticides which can meet the requirement of the "recovery and RSD double standard" at three concentration levels	334	324	310	299
Percentage of pesticides which can meet the requirement of the "recovery and RSD double standard" at three concentration levels	67.6%	65.6%	62.8%	60.5%

"Qualified recovery" means at least one concentration level can meet the requirement of the "recovery and RSD double standard" of three concentration levels.

percentage was more than 89.0%. The results showed the high accuracy of the method, and the method can be applied in the accurate quantitative of more than 89.0% kinds of pesticides. In addition, the number and percentage of pesticides that can meet the requirement of the "recovery and RSD double standard" in the four kinds of matrix at the spiked concentration of 10, 50, and 200 μg/kg are also listed in Table 1.2.7. Among these the number and percentage of pesticides that can meet the requirement of the "recovery and RSD double standard" in the matrix of black tea were 334 kinds and 67.6%; while in the matrix of oolong there were 299 kinds and 60.5%; in the matrix of green tea and oolong, the percentages were 65.6% and 62.8%, respectively. The results illustrated that good recovery can be obtained for the majority of pesticides in most of the matrix and have proven the universal applicability and accuracy of this method.

There was a great difference in the number of pesticides with unqualified recovery of the aforementioned matrices. This was due to the obvious differences in the composition of matrix, and will result in different degrees of matrix inhibition or enhancement effects for different pesticide. Even the ion of individual pesticides cannot be attained owing to the interfere of some special matrices. Among the four kinds of matrix, the number of pesticides with unqualified recovery in the matrix of oolong were the greatest number, 63 kinds, while in the matrix of oolong they were the least, 46 kinds. The results are listed in Appendix Table 1.2.1.

1.2.4.3.2 The Comprehensive Analysis of the Recovery of the Four Kinds of Matrix

The number and percentage of pesticides with qualified recovery according to the kind of matrix are listed in Table 1.2.8. It can be seen from Table 1.2.8 that of the 494 pesticides, 470 kinds of pesticides can meet the requirement of the "recovery and RSD double standard" in at least one kind of matrix, the number and percentage of pesticides with qualified recovery of all the four kinds of matrix was 396 and 80.2%, while the number and percentage of pesticides with qualified recovery in at least two (including two) kinds of matrix was 457 and 92.5%. The results demonstrated that good recovery can be obtained for the majority pesticides in most of matrix and proved the universal applicability of this method (Table 1.2.9).

The number and percentage of pesticides with unqualified recovery in all the four kinds of matrix was 24 and 4.9%. The 24 kinds of pesticides were not detected when spiked in the four kinds of matrix at the concentration of 200 μg/kg. This is

TABLE 1.2.8 The Number and Percentage of Pesticides With Qualified Recovery According to the Kind of Matrix

Number of matrices	Pesticide	
	Number	Percentage (%)
4	396	80.2
3	48	9.7
2	13	2.6
1	13	2.6
0	24	4.9

"Qualified recovery" means at least one concentration level can meet the requirement of the "recovery and RSD double standard" of three concentration levels.

TABLE 1.2.9 The Pesticides Which Cannot Meet the Requirement of the "Recovery and RSD Double Standard" in all the Four Kind of Matrix Using GC-Q-TOF/MS at Three Spiked Concentration Levels

No.	Compound	CAS	No.	Compound	CAS
1	1-Naphthylacetic acid	86-87-3	13	Chlordecone	143-50-0
2	Carboxin	5234-68-4	14	Dinoterb	1420-07-1
3	Endrin-aldehyde	7421-93-4	15	Dioxacarb	6988-21-2
4	Fenpropidin	67306-00-7	16	Flubenzimine	37893-02-0
5	Metamitron	41394-05-2	17	Hexaflumuron	86479-06-3
6	Fenfuram	24691-80-3	18	Indanofan	133220-30-1
7	Thiabendazole	148-79-8	19	Isoproturon	34123-59-6
8	Haloxyfop	69806-34-4	20	Lambda-cyhalothrin	91465-08-6
9	Captafol	2425-06-1	21	Monuron	150-68-5
10	Carbosulfan	55285-14-8	22	Propamocarb	24579-73-5
11	Iprodione	36734-19-7	23	Propylene thiourea	2122-19-2
12	Chlorbromuron	13360-45-7	24	Pyrethrins	8003-34-7

because the sensitivity of these 24 kinds of pesticides was too low; it required the enhancement of the concentration spiked, or else they are not suited for analysis using GC method. The results are listed in Appendix Table 1.2.1.

1.2.4.4 Comprehensive Analysis (Comparative Analysis the Influence of Matrix on the Screening Limit and Recovery of Pesticides)

The screening limit and recovery of the 494 kinds of pesticides in the four kinds of matrix were different from each other, and it can be seen from Table 1.2.10 that the number and percentage of pesticides with qualified recovery of all the four kinds of matrix was 396 and 80.2%. Among the pesticides that can be screened in all the four kinds of matrix, the number of pesticides with unqualified recovery in at least one kind of matrix was 38. This was because part of the pesticide was absorbed by the filter or because of the matrix enhancement effect and the matrix weakening effect during the detection process, thus unqualified recovery. (Rec. 60%–120%&RSD < 20%) was acquired. The recovery of the two kinds of peptides, metamitron and fluazinam, were unqualified in all the four kinds of matrix for the low sensitivity and poor stability of the pesticides in GC analysis. Among the pesticides that cannot be screened in all four kinds of matrix, the number and percentage of pesticides that cannot acquire qualified recovery were 36 and 7.3%, while the number of pesticides that can acquire unqualified recovery in all four kinds of matrix was 16. This is because the sensitivity of these 16 kinds of pesticides was too low, and required the enhancement of the concentration spiked, or they are not suited for the analysis using the GC method. The results are listed in Appendix Table 1.2.1.

TABLE 1.2.10 Comprehensive Comparative Analysis of Screening Limit and Recovery in the Four Kinds of Tea Matrix

	Number	Percentage (%)	Remarks
Number of pesticides spiked	494		
Can be screened and can acquire qualified recovery in all the four kinds of matrix	396	80.2	
Can not be screened and can not acquire qualified recovery in all the four kinds of matrix	16	3.2	
Can be screened in all the four kinds of matrix and acquire qualified recovery in part of matrix	38	7.7	
Can be screened and can not acquire qualified recovery in all the four kinds of matrix	2	0.4	
Can be screened and can acquire qualified recovery in part of matrix	36	7.3	
Can be screened in part of matrix and can not acquire qualified recovery in all the four kinds of matrix	6	1.2	

1.2.5 CONCLUSION

A nontargeting, high-throughput method for the screening of 499 pesticides using GC-Q-TOF/MS technique in tea samples was developed, and the pretreatment can be done at just once. The method is based on the establishment of the GC-Q-TOF/MS accurate mass library of 499 pesticides, and qualitative analysis was conducted by the retrieval of the GC-Q-TOF/MS accurate mass spectrum with library. The number of pesticides that can meet the requirement of the "recovery and RSD double standard" at the concentration of 10 μg/kg exceeded 324 (percentage 65.6%), and the results demonstrate the universal applicability and accuracy of this technique. At the same time, the technique achieves the substitution of the electronic identification standard for the traditional quality method: using real pesticides as a reference substance, and realizing the great-leap-forward development from traditional targeted detection to nontargeting screening. The determination of pesticide residues can be conducted automated, digitally, and informationalized, and the technique has great advantages over traditional methods. Above all, the technique has great capability in the determination of pesticide residues, and it will be effective tool for the detection of pesticide residues, playing a great role in the supervision of food and agriculture product pesticide residue safety.

Chapter 1.3

A Study of Efficiency Evaluation for Nontarget and High-Throughput Screening of 556 Pesticides Residues by LC-Q-TOF/MS

1.3.1 INTRODUCTION

On account of the high resolution, accurate-mass detection, and high sensitivity under full-scan models, liquid chromatography quadrupole time-of-flight mass spectrometry (LC-Q-TOF/MS) has developed rapidly in recent years, confirming compounds from complex matrices. Due to the high scan rate, there is no limitation for compound numbers, making it possible to screen large quantities of pesticides at the same time. In addition, full-scan technology brought nontarget screening and data tracing to reality. This method has been widely applied to pesticides screening and verification in soil, water, and food.

Software such as MassHunter from Agilent, MarkerLynx from Waters, and Swathtm from AB sceix is an effective tool for compound structural identification in LC-Q-TOF/MS. With the development of this software, the application of personal compound database and library (PCDL) for matching analysis has not only improved the reliability of pesticide detection and confirmation but also reduced the time spent. The application of LC-Q-TOF/MS has promoted pesticide residue detection into a new age, with no reference standards for verification, and will play an important role in residual analyzing.

Tea, as an important economic crop, is a commodity export in China. However, in recent years, the importing nations, such as Japan and the European Union, have extended the detection range and enacted a more strict maximum residue limit standard. Thus, establishing an accurate analyzing method to evaluate pesticide residue in tea has an important practical application.

Due to the complex matrices in tea, which have plenty of pigment, alkaloids, and phenols, an efficient pretreatment, including extraction, cleanup, and concentration, aims to extract the target compounds as much as possible and wipe out most of impurities, reducing interference and contamination. Pretreatment is an important procedure to guarantee the reliability, accuracy, and reproducibility of pesticide screening methods.

In this chapter, we set up an accurate mass database and an MS/MS library for conclusive confirmation of pesticide residuals in four kinds of tea. Five hundred and fifty-six pesticides in agricultural products were screened and verified by LC-Q-TOF/MS—this detecting coupled with one form of pretreatment. The screening process is shown in Fig. 1.3.1.

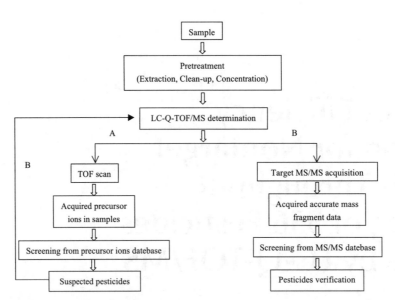

FIGURE 1.3.1 **Screening process for pesticides residual by LC-Q-TOF/MS.**

1.3.2 EXPERIMENTAL

1.3.2.1 Reagents and Materials

Pesticides standards (≥95%) were purchased from Dr. Ehrenstorfer (Augsburg, Germany). All HPLC grade reagents, including acetonitrile (ACN), toluene, and methanol (MeOH) were obtained from Fisher Scientific (New Jersey, USA). Formic acid was obtained from Duksan Pure Chemicals (Ansan, Korea). Milli-Q water was prepared with the Millipore water purification system (MA, USA). Na_2SO_4 and NH_4OAc were purchased from Beijing Chemical Reagent Factory (Beijing, China). The triple phase of tea (TPT) SPE column was purchased from Agela Technologies Inc. (DE, USA). Individual stock standard solutions of the pesticides were prepared at 1000 mg/L in MeOH, ACN, or Actone, and stored in darkness at 4°C.

1.3.2.2 HPLC Conditions

An Agilent 1290 series HPLC system (Agilent Technologies, Santa Clara, CA, USA) coupled with an ODS-C18column (ZORBAX SB-C18 column, 2.1 mm × 100 mm, 3.5 μm; Agilent Technologies, Santa Clara, CA, USA) was applied for chromatographic separations. The mobile phase consisted of 5 mmol/L ammonium acetate and 0.1% formic acid in water (A) and ACN (B). The gradient was as follows: 0–3 min, 1%–30% B; 3–6 min, 30%–40% B; 6–9 min, 40% B; 9–15 min, 40%–60% B; 15–19 min, 60%–90% B; 19–23 min, 90% B and 23–23.01 min, and 90%–100% B, followed by a reequilibration time for 4 min. Chromatography employed an injection volume of 10 μL, a flow rate of 0.4 mL/min, and a column temperature of 30°C.

1.3.2.3 Mass Spectrometry Conditions

The HPLC system was coupled to an Agilent 6530 LC-Q-TOF/MS, equipped with a dual-spray Agilent Jet Stream electrospray ionization source (Agilent 1290 series, Agilent technologies, Santa Clara, CA, USA). Pesticides were analyzed in positive ion mode. The capillary voltage, cone voltage, and fragmentation voltage were 4000, 60, and 140 V, respectively. Drying gas temperature and sheath gas temperature were both 325°C. Drying gas flow, sheath gas flow, and nebulizer gas pressure were 10, 11 L/min, and 40 psi, respectively. MS spectra were acquired across the range 50–1600 m/z. The TOF mass spectrometer was calibrated routinely by an internal reference solution that contains the accurate masses at m/z 121.0509 and 922.0098.

Data were evaluated using Agilent MassHunter Workstation Software (version B.05.00). Accurate mass databases were stored in Microsoft Excel (2010) by CSV format. A fragment mass library was established by Agilent MassHunter PCDL Manager (B.04.00).

1.3.2.4 Sample Preparation

A 2 g portion of sample was accurately weighed (precision of 0.01 g), homogenized in 15 mL ACN less than 15,000 rpm for 1 min, and then centrifuged at 4200 rpm for 5 min. The supernatant was decanted into a 150 mL pear-shaped flask and the residue was extracted one more time using 15 mL ACN. The combined extracts were concentrated to 2 mL by a rotary evaporator (40°C, 40 rpm) and then purified by a TPTSPE column. The column, with added 2 cm anhydroussodium sulfate, was washed with 5 mL ACN/toluene (3:1, v/v). The concentrated sample was loaded and washed by 2 mL ACN/toluene (3:1, v/v) 3 times. Then, the column was eluted by 25 mL ACN/toluene (3:1, v/v). Effluent was collected by an 80 mL pear-shaped flask and concentrated (40°C, 80 rpm) to 0.5 mL. The sample was dried by nitrogen, resuspended in 1 mL ACN/H2O (3:2, v/v), and filtered through a 0.2 μm filter for LC-Q-TOF/MS analyzing.

1.3.3 CREATION OF THE PESTICIDE DATABASE

Accurate mass of the 556 pesticides was acquired in a TOF model by standard solutions in 1000 μg/L. Positive identification of pesticide was reported if the compound was detected by the Find by Molecular Formula data-mining algorithm with a mass error below 5 ppm and with a score exceeding 90. Compound name, molecular feature, accurate mass, and retention time were recorded in Microsoft Excel for screening.

For each pesticide, fragment data was obtained in target MS/MS mode by inputting the accurate mass, retention time, and collision energies (5–60 V). Compound fragment data were imported into the PCDL by CEF file. Four fragment spectras for each pesticide with abundant information under different collision energies were recorded in PCDL for verification.

1.3.4 METHOD EVALUATION

Four kinds of tea, including black, green, oolong, and puer, were chosen for method evaluation. Recoveries and precisions were investigated at spiked concentrations of 10, 50, and 200 μg/kg. Screening limits for pesticides were investigated at six spiked concentrations (5, 10, 20, 50, 100, and 200 μg/kg).

1.3.4.1 Qualitative Analysis

For pesticide verification, the LC-Q-TOF/MS system was first operated in TOF mode and a search from precursor ion databases was conducted. A targeted MS/MS method with a list of suspected pesticides was used and the obtained spectra were compared to the MS/MS library for verification. It was determined that the same precursor ions for isomers cannot be distinguished by a TOF scan. However, there are big differences in their fragment mass spectra. The following: prometryn and terbutryn and dimethirimol and ehirimol were taken as examples for analysis.

1. Prometryn and terbutryn have the same precursor ions and similar retention time (Table 1.3.1) so they cannot be distinguished in a TOF scan. However, due to the different positions of the methyl group, their fragmentation modes were different, and prometryn and terbutryn could be identified by MS/MS library according to their fragment mass spectra. When retrieving accurate mass from a precursor ion database for a sample spiked with terbutryn, the scores for prometryn and terbutryn were 90.24 and 80.31, respectively. A second injection under targeted MS/MS scan was analyzed and the fragment mass spectra were compared to the MS/MS library. Scores for prometryn and terbutryn were 92.05 and 5.9, respectively (Fig. 1.3.2). Thus, we could identify terbutryn because of the difference of fragment mass spectra between prometryn and terbutryn.

2. In some cases, isomers had the same fragments, such as dimethirimol and ethirimol. There was a dimethyl amino in dimethirimol and an ethylamino in ethirimol (Table 1.3.2). The same fragments were obtained in an MS/MS scan. However, we could distinguish these two pesticides from the abundance of their fragments. When comparing the spectrum from a tea sample with an accurate mass database and MS/MS library, the scores of precursor ions of dimethirimol and ethirimol were 85.22 and 84.39, and the scores of their fragments were 73.27 and 96.2, respectively. Both of them had the same fragments (Table 1.3.2), but the abundance of their fragments was different, which could be seen from their mirror representation (Fig. 1.3.3). It was obvious that the suspect pesticide was ethirimol.

1.3.4.2 Screening Limits

1.3.4.2.1 Pesticides Can Be Screened

Five hundred and fifty-six pesticides were involved in this evaluation. According to their different physicochemical properties and matrix effects, the screening limits were different in various kinds of tea (Appendix Table 1.3L.1).

TABLE 1.3.1 Formula, Retention Time, Precursor Ion and Structure for Two Sets of Isomer

Compound	Formula	Retention time/min	Precursor ion (m/z)	Structure
Prometryn	$C_{10}H_{19}N_5S$	9.25	241.1361	
Terbutryn	$C_{10}H_{19}N_5S$	9.53	241.1361	
Dimethirimol	$C_{11}H_{19}N_3O$	3.72	209.1528	
Ethirimol	$C_{11}H_{19}N_3O$	3.72	209.1528	

Due to matrix effects, enhancing/suppressing the ionization efficiency of pesticides, different matrices generated various effective screening rates (Table 1.3.3); 509–531 pesticides (91.6%–95.5% of the target pesticides) were screened from the target four kinds of tea, of which the highest effective screening rate (95.5%, 531 pesticides) appeared in oolong. Black tea and green tea had the same, lowest effective screening rate (91.6%, 509 pesticides). These results demonstrate that LC-Q-TOF/MS is an efficient method to screen pesticides from tea.

Statistical results of screening limits are presented in Table 1.3.4, with 443–491 pesticides accounting for 79.3%–88.3% of the target 556 pesticides, obtaining the screening limits of 5 μg/kg. On account of matrix effects, pesticides have different screening limits in various kinds of tea (Table 1.3.4). Confirmatory analysis was found to be successful for 470, 476, 483, and 508 pesticides, accounted for 84.6%, 85.6%, 86.9%, and 91.4%, at spiked levels less than or equal to 10 μg/kg, in black tea, green tea, puer tea, and oolong tea, respectively. Details of screening limits are presented in Appendix Table 1.3L.1. The majority of pesticides can be confirmed under low spiked levels, implying the suitability of the developed method for pesticide screening of different kinds of tea.

Five hundred and thirty-four pesticides can be confirmed in at least one kind of tea, and 498 or 517 pesticides were confirmed in the four or at least three kinds of tea. These contributed 89.6% or 93.0% to the 556 pesticides, respectively (Table 1.3.5). The developed method has been successfully applied in verifying the most targeted pesticides in different kinds of tea, demonstrated the universal applicability of this method for pesticide screening.

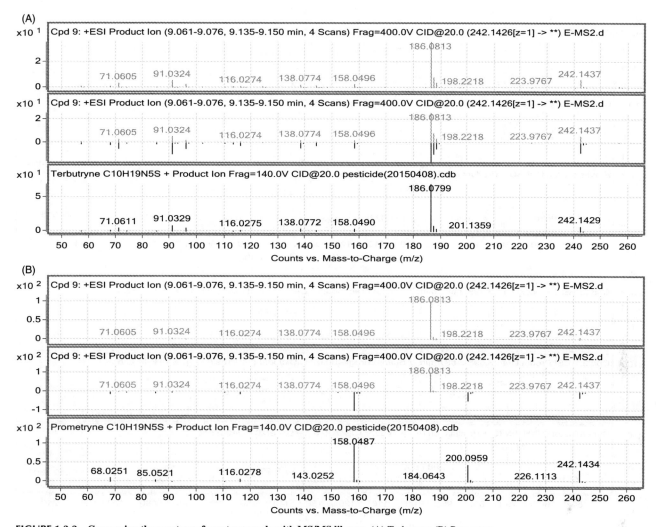

FIGURE 1.3.2 Comparing the spectrum from tea sample with MS/MS library. (A) Terbutryn; (B) Prometryn.

TABLE 1.3.2 Fragment Ions and Their Relative Abundance for Dimethirimol and Ethirimol

Fragment ions (relative abundance/%)	
Dimethirimol	140.1070 (26.5); 210.1600 (21.9); 98.0601 (38.6); 70.0655 (16.2)
Ethirimol	140.1070 (74.6); 210.1600 (18.9); 98.0601 (100); 70.0655 (37.8)

1.3.4.2.2 Pesticides Cannot Be Screened

Among the target 556 pesticides, 22 pesticides cannot be screened in any of the four kinds of tea, which were undetected at spiked levels of 200 µg/kg, indicating the unsuitability for screening by LC-Q-TOF/MS. Therefore, these pesticides were not considered further in the evaluation part of this study (Table 1.3.6).

1.3.4.3 Accuracy and Precision

1.3.4.3.1 Recoveries and Relative Standard Deviations from Different Kinds of Tea

According to EU documents SANCO/10684/2009, recoveries in the range of 60%–120% and relative standard deviations (RSDs) less than 20% were conformed to Recovery-RSD standard; 556 pesticides were spiked into four kinds of tea for verification. Details are recorded in Appendix Table 1.3L.1.

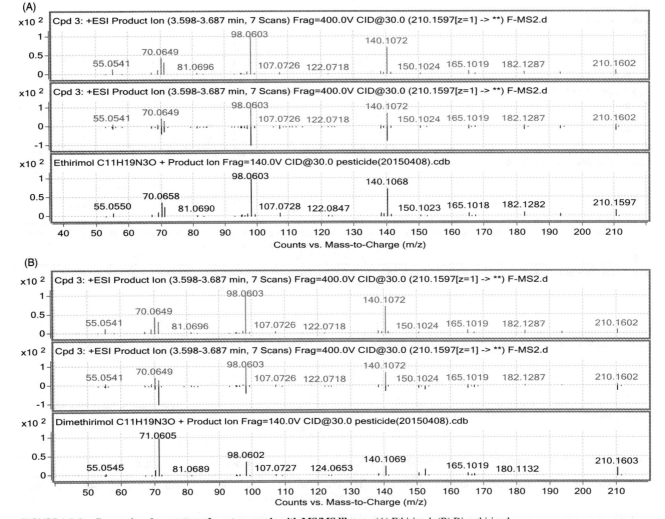

FIGURE 1.3.3 **Comparing the spectrum from tea sample with MS/MS library.** (A) Ethirimol; (B) Dimethirimol.

TABLE 1.3.3 Amount of the Detected and Nondetected Pesticides

	Black tea	Green tea	Puer tea	Oolong tea
Target pesticides	556	556	556	556
Detected pesticides	509	509	525	531
Nonspiked pesticides	9	9	9	9
Nondetected pesticides	47	47	31	25
Proportion of the detected pesticides	91.6%	91.6%	94.4%	95.5%

The accuracies and precisions of 556 pesticides in four kinds of tea were assessed by spiking a standard mixture at three concentrations (10, 50, and 200 μg/kg). Statistical results are presented in Table 1.3.7. The proportion of pesticide, which obtained acceptable recoveries and precisions under the spiked level of 10 μg/kg, were 70.1%, 70.3%, 65.8%, and 60.4% in black tea, green tea, puer tea, and oolong tea, respectively. In addition, more than 70% (71.6%–77.0%) pesticide obtained acceptable recoveries and precisions that met the Recovery-RSD standard at the spiked level of 50 μg/kg, in all four kinds of matrix. In general, a high proportion of pesticides obtained acceptable recoveries and precisions under the three spiked

TABLE 1.3.4 Proportions of the Detected Pesticides Under Different Spiked Levels

µg/kg		Matrices			
		Black tea	Green tea	Puer tea	Oolong tea
5	Amount	443	441	461	491
	Proportions/%	79.7%	79.3%	82.9%	88.3%
10	Amount	27	35	22	17
	Proportions/%	4.9%	6.3%	4.0%	3.1%
20	Amount	23	22	19	10
	Proportions/%	4.1%	4.0%	3.4%	1.8%
50	Amount	16	7	18	6
	Proportions/%	2.9%	1.3%	3.2%	1.1%
	Amount	0	4	5	3
	Proportions/%	0.0%	0.7%	0.9%	0.5%
200	Amount	0	0	0	4
	Proportions/%	0.0%	0.0%	0.0%	0.7%

TABLE 1.3.5 Amount of the Detected Pesticides From Different Species Number

Number of species	Pesticides	
	Amount	Proportion/%
4 matrices	498	89.6
3 matrices	19	3.4
2 matrices	8	1.4
1 matrix	9	1.6
Unable to screen	22	4.0

TABLE 1.3.6 Pesticides Which Were Unable to Screen

No.	Name	CAS number	No.	Name	CAS number
1	Allethrin	584-79-2	12	Thiram	137-26-8
2	Carbophenothion	786-19-6	13	Aclonifen	74070-46-5
3	Dazomet	533-74-4	14	Bromophos-ethyl	4824-78-6
4	Metamitron-desamino	36993-94-9	15	Clomeprop	84496-56-0
5	Pirimicarb-desmethyl-formamido	27218-04-8	16	Dibutyl succinate	141-03-7
6	Cartap	22042-59-7	17	Isoxaflutole	141112-29-0
7	Chlorimuron-ethyl	90982-32-4	18	Orthosulfamuron	213464-77-8
8	Bensultap	17606-31-4	19	Terbufos-oxon-sulfone	56070-15-6
9	Ethoxysulfuron	126801-58-9	20	Tribenuron-methyl	101200-48-0
10	Lactofen	77501-63-4	21	Phosmet	732-11-6
11	Metolcarb	1129-41-5	22	Primisulfuron-methyl	86209-51-0

TABLE 1.3.7 Amount of Pesticides Which Met the Recovery-RSD Standard Under Different Spiked Levels

Added (μg/kg)		Black tea	Green tea	Puer tea	Oolong tea
10	Amount	390	391	366	336
	Proportion /%	70.1%	70.3%	65.8%	60.4%
50	Amount	418	428	406	398
	Proportion /%	75.2%	77.0%	73.0%	71.6%
200	Amount	429	437	431	424
	Proportion /%	77.2%	78.6%	77.5%	76.3%

TABLE 1.3.8 Amount of Pesticides Which Met the Recovery-RSD Standard Under All the Spiked Levels

	Black tea	Green tea	Puer tea	Oolong tea
Target pesticides	572	572	572	572
Amount of recovered pesticides	432	440	434	428
Proportion of recovered pesticides/%	77.7	79.1	78.1	77.0
Unspiked pesticides	16	16	16	16
Nonrecoverable pesticides	124	116	122	128
Amount of pesticides which met the recovery-RSD standard under all the spiked levels	383	390	360	331
Proportion of pesticides which met the recovery-RSD standard under all the spiked levels/%	68.9	70.1	64.8	59.5

Note: Recovered pesticides were the pesticides which met the Recovery-RSD standard under at least one kind of tea.

levels in all four kinds of tea. According to the international recognized uniform standard (10 μg/kg), more than 60% pesticides had acceptable recoveries. These results demonstrate the sufficient accuracy of this method.

Table 1.3.8 lists the amount of pesticides that conform to the Recovery-RSD standard. More than 420 pesticides account for more than 77.0%, obtaining gratifying recoveries and RSDs in four kinds of tea, which reflects the high accuracy of the screening method. Furthermore, the proportion of pesticides that conform to the Recovery-RSD standard under all three spiked levels were 70.1%, 59.5%, 68.9%, and 64.8% in green tea, oolong tea, blank tea, and puer tea, respectively.

Comparing the number of pesticides that met the Recovery-RSD standard between different spiked levels shows that fewer pesticides met the standard at the level of 10 μg/kg than 200 μg/kg, due to poor recoveries in pretreatment and serious matrix effect under a lower spiked level.

1.3.4.3.2 Comprehensive Analysis of Recoveries and RSDs From All Four Kinds of Tea

Table 1.3.9 lists the amount of pesticides that met the Recovery-RSD standard from different kinds of tea; 394 pesticides, which contributed 70.9% to the 556 pesticides, have acceptable recoveries and RSDs in all four kinds of tea. A majority of the pesticides could be recovered from different matrices, which demonstrates the universal applicability of this method. Ninety-seven pesticides have unacceptable recoveries or RSDs in all four kinds of tea, part of which have no response under all spiked levels, and a higher concentration was needed for further verification. Among the 97 pesticides, because of the influence of the pretreatment, 59 pesticides could be detected with poor recoveries (Table 1.3.10).

TABLE 1.3.9 Amount of Pesticides Which Met the Recovery-RSD Standard From Different Species Number

Number of species	Amount	Proportion/%
4 matrices	394	70.9
3 matrices	39	7.0
2 matrices	15	2.7
1 matrix	11	2.0
Nonrecoverable	97	17.5

TABLE 1.3.10 Pesticides Which Were Detectable and Nonrecoverable

No.	Name	CAS number	No.	Name	CAS number
1	6-Benzylaminopurine	1214-39-7	31	Ethametsulfuron-methyl	97780-06-8
2	Carbendazim	10605-21-7	32	Flamprop	58667-63-3
3	Chlorsulfuron	64902-72-3	33	Fluazifop	69335-91-7
4	Cinosulfuron	94593-91-6	34	Foramsulfuron	173159-57-4
5	Florasulam	145701-23-1	35	Imazamox	114311-32-9
6	Forchlorfenuron	68157-60-8	36	Imazapic	104098-48-8
7	Imazapyr	81334-34-1	37	Metosulam	139528-85-1
8	Imazethapyr	81335-77-5	38	Metsulfuron-methyl	74223-64-6
9	Mesosulfuron-methyl	208465-21-8	39	Propoxycarbazone	181274-15-7
10	Pyridafol	40020-01-7	40	Pyrazosulfuron-ethyl	93697-74-6
11	Tepp	107-49-3	41	Quinclorac	84087-01-4
12	Triazoxide	72459-58-6	42	Quinmerac	90717-03-6
13	Asulam	3337-71-1	43	Quizalofop	76578-12-6
14	Cloransulam-methyl	147150-35-4	44	Thiabendazole	148-79-8
15	Diafenthiuron	80060-09-9	45	Thidiazuron	51707-55-2
16	Flazasulfuron	104040-78-0	46	Thifensulfuron-methyl	79277-27-3
17	Flumetsulam	98967-40-9	47	Triasulfuron	82097-50-5
18	Halosulfuron-methyl	100784-20-1	48	Flucarbazone	181274-17-9
19	Haloxyfop	69806-34-4	49	Hydramethylnon	67485-29-4
20	Imazaquin	81335-37-7	50	Nicosulfuron	111991-09-4
21	Inabenfide	82211-24-3	51	Penoxsulam	219714-96-2
22	Iodosulfuron-methyl	144550-36-7	52	Pyrasulfotole	365400-11-9
23	Propamocarb	24579-73-5	53	Pyraflufen	129630-17-7
24	Pymetrozine	123312-89-0	54	Saflufenacil	372137-35-4
25	Albendazole	54965-21-8	55	Sulcotrione	99105-77-8
26	Bensulfuron-methyl	83055-99-6	56	Tembotrione	335104-84-2
27	Cyclosulfamuron	136849-15-5	57	Thiabendazole-5-hydroxy	948-71-0
28	Cyromazine	66215-27-8	58	Thiencarbazone-methyl	317815-83-1
29	Diclosulam	145701-21-9	59	Validamycin	37248-47-8
30	Emamectin	155569-91-8			

TABLE 1.3.11 Comprehensive Analysis of Screening Limits and Recoveries

	Amount	Proportion/%
Target pesticides	556	–
Pesticides were detectable and recoverable from all the matrices	394	70.9
Pesticides were undetectable and nonrecoverable from all the matrices	22	4.0
Pesticides were detectable from all the matrices and recoverable from part of the matrices	45	8.1
Pesticides were detectable from all the matrices and recoverable from none kind of matrix	59	10.6
Pesticides were detectable and recoverable from part of the matrices	20	3.6
Pesticides were detectable from part of the matrices and recoverable from none kind of matrix	16	2.9

1.3.4.4 Comprehensive Analysis of Screening Limits and Recoveries

According to the earlier result, 394 pesticides, accounting for 70.9% of the 556 pesticides, could be screened and recovered from all four kinds of tea (Table 1.3.11). Forty-five pesticides could be screened from all matrices but recovered from only one to three kinds of tea, which was due to the loss during pretreatment or matrix effects. Meanwhile, 59 pesticides were screened from all kinds of matrices but recovered from none of the four kinds of tea. There were 20 pesticides, accounting for 3.6% of the spiked pesticides, which were screened and recovered from part of the four matrices. Also, 16 pesticides with poor sensitivity and serious matrix effect could be screened from part of the four matrices but recovered from none of the four kinds of tea. The remaining 22 pesticides could be screened and recovered from none of the kinds of tea because of poor sensitivity, which illustrated that these pesticides were unsuitable for LC–MS detecting.

1.3.5 CONCLUSIONS

In this study, a new method with one pretreatment coupled with LC-Q-TOF/MS detection was developed for nontarget and high-throughput screening of 556 pesticides residues in four kinds of tea. Pesticides were identified by matching the data acquired from tea samples with the created accurate mass LC-Q-TOF/MS database and MS/MS library. There were 336 pesticides, accounting for 60.4% of the total target pesticides, meeting the Recovery-RSD standard at 10 μg/kg, which demonstrates the applicability and accuracy of this method. In addition, there was no need for Reference: standards in this study; MS/MS fragment information was typically matched to the database to enable conclusive confirmation of chemical identity. What's more, full-scan technology has brought nontarget screening into its own, and pesticide residue detection has achieved rapid progress in becoming automatic, digitized, and informationalized. In conclusion, the application of LC-Q-TOF/MS is an effective tool for rapid screening of pesticide residues in tea and will play an important role in food safety supervision for pesticide residue.

APPENDIX TABLE 1.3L.1 Results of Nontarget Screening for 556 Pesticides in Four Kinds of Tea by LC-Q-TOF/MS

	Name	CAS number	Puer tea							Oolong tea						
			Screening limits (µg/kg)	10 µg/kg AVE	RSD (%) (n = 5)	50 µg/kg AVE	RSD (%) (n = 5)	200 µg/kg AVE	RSD (%) (n = 5)	Screening limits (µg/kg)	10 µg/kg AVE	RSD (%) (n = 5)	50 µg/kg AVE	RSD (%) (n = 5)	200 µg/kg AVE	RSD (%) (n = 5)
1	1,3-Diphenyl Urea	102-07-8	5	15.3	35.3	23.0	61.4	65.0	49.4	5	33.9	26.3	37.8	14.3	57.6	7.2
2	3,4,5-Trimethacarb	2686-99-9	50	NA	NA	86.3	7.7	87.5	2.2	5	78.2	11.9	71.0	5.6	73.3	4.9
3	6-Benzylaminopurine	1214-39-7	5	22.9	5.0	7.0	6.4	2.2	4.9	5	1.8	61.1	0.9	17.6	1.0	10.0
4	Acetochlor	34256-82-1	5	81.5	7.6	82.6	5.3	86.9	7.1	5	72.8	7.0	71.8	5.7	70.1	7.3
5	Allethrin	584-79-2	NA	NA	NA	NA	NA	NA	NA	NA	NA	NA	NA	NA	NA	NA
6	Amidosulfuron	120923-37-7	20	NA	NA	NA	NA	NA	NA	5	176.0	26.0	33.8	24.0	37.6	34.4
7	Anilofos	64249-01-0	5	80.7	8.2	85.8	10.2	88.1	2.5	5	76.0	9.4	71.1	6.5	79.9	3.0
8	Atraton	1610-17-9	5	72.2	2.4	77.4	7.8	88.1	3.6	5	57.9	20.9	63.8	6.8	75.9	2.7
9	Atrazine-desisopropyl	1007-28-9	5	62.8	7.8	64.8	1.6	65.3	10.0	5	50.7	15.3	48.1	6.3	53.6	3.8
10	Azaconazole	60207-31-0	5	67.3	2.7	71.2	3.7	111.2	7.1	5	61.0	16.5	57.9	8.4	67.1	3.6
11	Azamethiphos	35575-96-3	5	68.2	1.2	72.4	3.8	80.0	2.9	5	58.2	18.7	59.5	5.5	71.1	2.4
12	Azinphos-ethyl	2642-71-9	5	83.4	2.3	82.7	14.1	88.0	14.4	5	93.1	10.4	91.6	19.1	96.3	16.5
13	Aziprotryne	4658-28-0	5	80.2	6.5	81.6	10.3	92.4	3.6	5	74.7	6.2	70.3	6.3	75.3	10.0
14	Azoxystrobin	131860-33-8	5	65.6	8.9	72.4	10.8	80.2	3.7	5	77.4	9.3	70.3	4.0	75.8	1.4
15	Bendiocarb	22781-23-3	5	76.5	7.4	74.0	10.8	84.2	6.3	5	71.8	8.1	71.5	7.5	70.6	3.4
16	Benodanil	15310-01-7	5	76.8	9.6	87.6	15.4	87.4	2.9	5	70.6	6.8	76.0	7.5	78.7	5.0
17	Bensulide	741-58-2	5	193.9	16.9	95.4	11.6	99.2	12.1	5	89.8	14.3	98.6	16.9	95.4	20.0
18	Benzoximate	29104-30-1	5	89.7	11.0	70.1	9.4	82.4	14.2	5	81.3	15.4	91.5	10.3	86.3	13.0
19	Bifenazate	149877-41-8	5	120.0	22.9	31.6	26.5	91.0	9.7	5	187.4	21.8	81.9	3.6	91.7	11.9
20	Bioresmethrin	28434-01-7	50	NA	NA	27.3	100.6	73.4	7.3	100	NA	NA	NA	NA	84.6	38.3
21	Bitertanol	55179-31-2	5	80.7	17.6	81.2	13.3	88.0	3.1	5	71.7	14.7	74.3	5.9	77.2	3.1
22	Bromfenvinfos	33399-00-7	5	84.4	4.3	82.7	7.6	88.3	2.3	5	73.2	10.2	73.1	4.0	77.2	4.5
23	Brompyrazon	3042-84-0	5	54.2	17.4	64.7	3.2	70.6	1.5	5	59.2	9.4	53.7	4.7	60.0	6.9
24	Bromuconazole	116255-48-2	5	70.3	6.9	71.6	4.1	84.2	4.5	5	63.6	22.7	59.9	8.7	70.2	2.2
25	Buprofezin	69327-76-0	5	82.0	3.6	89.3	5.7	89.8	5.7	5	92.3	5.9	81.4	9.8	81.3	9.8
26	Butamifos	36335-67-8	5	89.5	5.1	102.5	19.8	90.7	3.2	5	76.0	20.0	74.0	7.3	81.2	4.6
27	Butocarboxim-Sulfoxide	34681-24-8	5	82.6	15.7	70.7	12.5	71.2	7.1	5	50.0	12.3	56.7	10.4	57.8	2.6
28	Butralin	33629-47-9	5	89.0	7.2	85.5	8.4	96.0	1.9	5	94.0	19.0	80.2	9.4	78.2	5.5
29	Cadusafos	95465-99-9	5	82.7	7.3	81.9	5.9	90.5	3.3	5	80.9	7.7	75.9	6.4	85.8	2.8
30	Carbaryl	63-25-2	20	NA	NA	138.6	14.8	97.7	2.7	200	NA	NA	NA	NA	84.7	5.3
31	Carbendazim	10605-21-7	5	1.4	45.2	0.1	37.3	0.6	188.7	5	1.7	128.2	0.6	116.8	0.5	10.1

(Continued)

APPENDIX TABLE 1.3L.1 Results of Nontarget Screening for 556 Pesticides in Four Kinds of Tea by LC-Q-TOF/MS (cont.)

	Name	CAS number	Puer tea Screening limits (µg/kg)	10 µg/kg AVE	10 µg/kg RSD (%) (n = 5)	50 µg/kg AVE	50 µg/kg RSD (%) (n = 5)	200 µg/kg AVE	200 µg/kg RSD (%) (n = 5)	Oolong tea Screening limits (µg/kg)	10 µg/kg AVE	10 µg/kg RSD (%) (n = 5)	50 µg/kg AVE	50 µg/kg RSD (%) (n = 5)	200 µg/kg AVE	200 µg/kg RSD (%) (n = 5)
32	Carbetamide	16118-49-3	5	76.0	13.6	78.7	6.6	77.2	1.8	5	43.1	124.5	70.1	2.7	70.4	1.6
33	Carbophenothion	786-19-6	NA	NA	NA	NA	NA	NA	NA	NA	NA	NA	NA	NA	NA	NA
34	Carboxin	5234-68-4	5	43.0	58.3	67.6	24.3	59.8	15.0	5	42.7	24.5	76.1	1.6	99.5	8.5
35	Carfentrazone-ethyl	128639-02-1	5	217.1	68.7	82.9	14.3	102.9	5.8	20	NA	NA	83.8	9.3	93.9	6.8
36	chlorotoluron	15545-48-9	5	74.0	1.9	86.1	8.3	83.0	2.2	5	72.4	5.2	75.9	5.6	78.4	2.4
37	Chlorpyrifos	2921-88-2	50	NA	NA	81.1	5.8	91.3	4.0	5	110.1	14.4	96.8	5.7	95.7	4.7
38	Chlorsulfuron	64902-72-3	20	NA	NA	26.7	20.8	5.5	34.5	50	NA	NA	15.4	60.0	8.7	16.6
39	Cinmethylin	87818-31-3	100	NA	NA	NA	NA	75.3	5.6	50	NA	NA	NA	NA	76.0	7.8
40	Cinosulfuron	94593-91-6	5	1.4	7.9	0.2	20.3	0.1	0.0	5	0.3	38.7	0.1	0.0	NA	NA
41	Clomazone	81777-89-1	5	82.2	8.2	92.0	12.7	84.4	2.1	5	77.0	10.6	71.2	6.0	74.8	5.7
42	Cloquintocet-mexyl	99607-70-2	5	84.6	0.2	88.1	4.4	90.4	3.4	5	75.4	10.5	72.5	7.6	91.1	9.9
43	Clothianidin	210880-92-5	5	54.7	4.7	76.2	14.4	73.4	4.2	5	45.7	10.0	49.2	5.2	51.6	9.0
44	Crufomate	299-86-5	5	78.3	3.0	85.5	5.5	94.0	3.9	5	78.9	12.7	70.5	6.8	81.6	2.9
45	Cyanazine	21725-46-2	5	70.1	0.5	78.6	7.3	76.3	3.1	5	60.5	19.4	59.7	7.0	70.3	3.0
46	Cycluron	2163-69-1	5	76.4	1.7	79.8	2.3	88.8	2.3	5	73.6	8.9	72.5	7.3	82.0	4.0
47	Cyproconazole	94361-06-5	5	72.5	3.7	77.5	5.9	87.6	4.8	5	73.7	10.8	72.4	7.6	72.3	2.9
48	Cyprodinil	121552-61-2	5	59.8	13.3	70.6	11.1	81.5	1.2	5	29.8	34.5	46.5	11.9	72.1	1.2
49	Daminozide	1596-84-5	10	NA	NA	NA	NA	NA	NA	5	176.0	24.4	300.3	13.2	52.1	18.3
50	Dazomet	533-74-4	NA	NA	NA	NA	NA	NA	NA	NA	NA	NA	NA	NA	NA	NA
51	Demeton-S-Sulfoxide	2496-92-6	5	63.6	5.6	71.2	2.7	86.9	1.9	5	62.5	23.3	65.7	3.9	81.5	1.6
52	Diallate	2303-16-4	5	83.3	12.8	85.4	6.3	90.4	4.7	5	103.4	4.1	72.6	4.1	83.1	7.3
53	Diazinon	333-41-5	5	82.9	3.5	82.6	4.5	88.9	2.5	5	79.5	7.9	76.4	5.9	89.3	4.6
54	Diclobutrazol	75736-33-3	5	75.6	5.2	70.4	1.3	80.9	4.6	5	73.1	10.5	72.1	7.1	74.9	3.7
55	Dimethachlor	50563-36-5	5	68.9	2.5	77.6	4.9	76.6	13.9	5	71.3	11.2	71.0	2.3	70.7	3.8
56	Dimethenamid	87674-68-8	5	74.8	7.0	87.6	15.4	84.3	3.4	5	72.5	14.5	71.0	6.2	70.8	3.5
57	Disulfoton sulfone	2497-06-5	5	76.2	6.0	88.8	11.5	96.0	5.2	5	70.1	16.2	76.4	6.2	74.8	6.0
58	Diuron	330-54-1	5	78.5	3.0	96.8	12.6	94.7	1.4	5	73.8	8.2	71.4	5.3	83.4	3.6
59	Edifenphos	17109-49-8	5	78.2	7.9	82.0	7.5	89.2	2.4	5	72.3	10.0	70.5	4.7	78.2	2.0
60	Esprocarb	85785-20-2	5	82.2	13.2	99.6	15.0	89.5	2.9	5	89.6	12.8	83.7	10.0	84.0	4.0
61	Etaconazole	60207-93-4	5	73.3	2.6	72.7	4.7	85.0	4.0	5	62.0	24.0	66.7	9.1	75.0	6.1
62	Ethidimuron	30043-49-3	5	55.7	8.2	67.9	2.6	74.5	2.8	5	49.3	11.2	50.8	7.3	70.4	0.6
63	Ethoprophos	13194-48-4	5	83.1	4.6	87.9	6.9	94.3	1.8	5	75.3	12.0	70.2	5.4	78.2	6.4
64	Etrimfos	38260-54-7	5	82.7	7.3	81.9	5.9	90.5	3.3	5	81.0	7.7	75.9	6.4	85.6	3.0

#	Name	CAS														
65	Famphur	52-85-7	5	73.0	8.5	74.5	12.9	95.8	3.0	5	75.0	7.6	70.2	5.2	79.7	4.3
66	Fenarimol	60168-88-9	5	73.7	7.0	74.0	16.0	77.4	9.7	5	49.2	26.4	84.4	15.0	71.1	10.5
67	Fenazaquin	120928-09-8	5	86.0	4.2	85.4	4.9	92.0	5.3	5	85.4	16.3	70.0	13.3	84.7	3.4
68	Fenhexamid	126833-17-8	5	58.5	11.6	76.0	6.7	73.0	12.3	5	45.5	27.9	57.6	9.5	57.9	6.2
69	Fenthion-sulfoxide	3761-41-9	5	71.7	2.8	82.6	9.3	86.4	2.2	5	71.5	10.3	78.9	4.6	80.7	1.9
70	Fenuron	101-42-8	5	70.2	3.6	72.9	6.3	83.6	5.4	5	60.5	14.0	62.3	4.3	75.4	5.8
71	Flamprop-isopropyl	52756-22-6	5	92.7	6.9	92.6	13.4	93.5	4.1	5	71.5	16.0	75.0	7.0	81.1	5.1
72	Flamprop-methyl	52756-25-9	5	76.6	3.5	91.5	16.8	90.3	4.0	5	71.2	10.6	72.9	6.7	75.6	6.3
73	Florasulam	145701-23-1	5	NA	NA	NA	NA	NA	NA	5	8.6	52.7	0.4	44.3	0.2	85.7
74	Fluazifop-butyl	69806-50-4	5	88.5	3.6	102.3	15.3	91.2	2.6	5	92.0	7.3	84.0	6.7	90.7	5.1
75	Flufenoxuron	101463-69-8	NA	NA	NA	NA	NA	NA	NA	5	NA	NA	NA	NA	118.4	12.9
76	Flurtamone	96525-23-4	5	70.4	6.0	77.7	8.1	86.8	3.7	5	75.1	12.6	70.6	2.8	77.9	4.1
77	Fluthiacet-methyl	117337-19-6	5	78.8	11.0	83.9	9.6	92.2	3.1	5	72.4	9.6	71.4	3.0	75.7	5.0
78	Flutolanil	66332-96-5	5	76.6	11.9	85.6	12.1	87.2	1.8	5	71.1	7.6	70.2	2.6	76.5	4.5
79	Flutriafol	76674-21-0	5	67.1	3.8	59.6	3.2	85.9	4.2	5	62.4	14.2	63.0	7.7	77.7	4.9
80	Forchlorfenuron	68157-60-8	5	NA	NA	NA	NA	NA	NA	5	0.1	0.0	0.1	0.0	0.1	0.0
81	Furalaxyl	57646-30-7	5	66.0	5.5	73.3	8.9	85.4	4.9	5	73.1	14.6	74.6	5.4	77.2	4.3
82	Haloxyfop-methyl	69806-40-2	5	87.6	4.9	105.0	13.1	91.8	2.2	5	87.3	7.9	79.6	7.6	88.1	2.5
83	Imazapyr	81334-34-1	5	NA	NA	NA	NA	NA	NA	5	20.3	88.0	2.1	58.4	0.5	115.7
84	Imazethapyr	81335-77-5	5	NA	NA	NA	NA	NA	NA	5	0.9	59.5	0.2	37.3	0.2	0.0
85	Imazosulfuron	122548-33-8	50	NA	NA	NA	NA	NA	NA	100	NA	NA	NA	NA	NA	NA
86	Imidacloprid	138261-41-3	5	53.0	14.5	70.1	4.3	85.9	4.1	5	62.6	21.0	50.2	7.9	61.7	9.6
87	Indoxacarb	144171-61-9	5	101.3	9.3	80.7	10.9	83.2	14.6	5	93.8	15.5	112.5	14.1	95.3	12.4
88	Iprobenfos	26087-47-8	5	76.9	17.3	80.0	18.0	91.2	3.2	5	76.9	12.1	72.6	5.4	79.7	4.3
89	Iprovalicarb	140923-17-7	5	73.1	5.8	81.4	7.6	84.5	3.2	5	74.7	13.4	72.9	3.0	73.9	2.2
90	Isomethiozin	57052-04-7	5	79.2	8.0	84.0	8.8	86.5	1.8	5	71.4	6.1	70.8	3.6	79.9	2.3
91	Isoprocarb	2631-40-5	5	87.6	11.3	92.0	6.3	83.6	8.1	20	NA	NA	71.3	2.8	72.9	4.5
92	Isouron	55861-78-4	5	71.3	4.1	74.5	8.9	84.0	3.2	5	61.5	19.7	61.6	9.4	79.6	1.5
93	Isoxaben	82558-50-7	5	73.6	1.1	75.2	19.3	84.7	3.0	5	70.6	8.9	71.0	4.2	80.6	1.4
94	Isoxadifen-ethyl	163520-33-0	20	NA	NA	92.8	18.7	78.2	8.5	50	NA	NA	NA	NA	74.7	12.6
95	Karbutilate	4849-32-5	5	71.4	1.5	73.2	4.5	77.7	2.7	5	72.3	9.4	73.1	4.9	73.7	1.8
96	Mephosfolan	950-10-7	5	71.3	2.0	80.5	6.4	88.8	2.8	5	62.6	22.5	64.6	6.9	77.5	1.1
97	Mesosulfuron-methyl	208465-21-8	5	63.8	1.2	70.7	4.8	NA	NA	5	4.3	45.7	0.7	36.5	0.5	17.4
98	Metalaxyl-M	70630-17-0	5	NA	NA	NA	NA	81.2	4.2	NA	NA	NA	NA	NA	NA	NA
99	Metamitron-desamino	36993-94-9	NA	NA	NA	NA	NA	NA	NA	NA	NA	NA	NA	NA	NA	NA
100	Methiocarb-sulfone	2179-25-1	20	NA	NA	79.0	13.0	73.9	6.6	10	NA	NA	50.5	4.2	50.6	3.1
101	Methomyl	16752-77-5	5	77.2	18.4	73.8	13.1	100.1	12.9	20	NA	NA	89.9	14.4	94.0	14.9

(Continued)

APPENDIX TABLE 1.3L.1 Results of Nontarget Screening for 556 Pesticides in Four Kinds of Tea by LC-Q-TOF/MS (cont.)

			Puer tea							Oolong tea						
	Name	CAS number	Screening limits (µg/kg)	10 µg/kg AVE	RSD (%) (n = 5)	50 µg/kg AVE	RSD (%) (n = 5)	200 µg/kg AVE	RSD (%) (n = 5)	Screening limits (µg/kg)	10 µg/kg AVE	RSD (%) (n = 5)	50 µg/kg AVE	RSD (%) (n = 5)	200 µg/kg AVE	RSD (%) (n = 5)
102	Methoprotryne	841-06-5	5	76.1	1.6	82.4	5.0	87.0	1.6	5	71.5	8.3	72.7	3.3	84.1	1.1
103	Neburon	555-37-3	5	82.9	12.2	100.4	16.4	95.0	3.6	5	81.2	7.2	75.3	5.7	83.4	7.5
104	Nitenpyram	120738-89-8	5	42.5	10.0	41.2	3.5	64.4	6.8	5	43.2	32.2	34.9	16.4	55.3	7.6
105	Paclobutrazol	76738-62-0	5	73.6	4.3	74.2	7.0	83.8	4.1	5	71.7	7.5	70.5	8.9	70.4	2.8
106	Paraoxon-Ethyl	311-45-5	5	76.3	7.3	104.3	10.3	87.9	2.4	5	61.6	13.6	62.4	7.8	70.3	5.5
107	Penconazole	66246-88-6	5	70.4	18.5	73.6	17.2	83.0	2.5	5	61.6	20.8	63.0	9.5	70.3	2.7
108	Phenmedipham	13684-63-4	5	75.7	6.3	82.4	8.7	88.2	2.2	5	74.6	11.2	71.5	2.5	77.3	3.0
109	Phosphamidon	13171-21-6	5	68.3	1.8	73.7	3.1	85.5	3.5	5	71.8	10.7	71.0	5.3	78.3	1.2
110	Phoxim	14816-18-3	50	NA	NA	82.1	17.8	91.1	3.8	50	NA	NA	50.7	31.2	77.8	4.4
111	Picolinafen	137641-05-5	5	109.8	43.1	100.6	14.0	88.7	11.3	5	48.2	60.3	78.5	7.7	82.7	7.8
112	Pirimicarb	23103-98-2	5	73.1	2.7	79.4	3.4	88.0	2.0	5	73.0	4.7	72.4	6.1	81.5	1.4
113	Pirimicarb-Desmethyl-Formamido	27218-04-8	NA	NA	NA	NA	NA	NA	NA	NA	NA	NA	NA	NA	NA	NA
114	Pirimiphos-Methyl	29232-93-7	5	87.5	5.8	90.4	6.4	93.5	3.1	5	95.2	17.9	75.3	13.6	83.5	13.9
115	Pretilachlor	51218-49-6	5	84.0	5.6	91.3	10.4	87.2	3.3	5	79.2	10.8	71.8	6.9	80.3	2.1
116	Profenofos	41198-08-7	5	86.9	9.6	91.1	6.6	94.7	4.1	5	81.6	12.0	73.9	7.5	86.2	4.9
117	Prometryn	7287-19-6	5	81.3	1.2	85.3	3.4	87.6	2.3	5	75.0	8.7	70.5	6.0	80.3	1.7
118	Propachlor	1918-16-7	5	76.3	4.5	81.4	10.2	82.1	1.9	5	61.0	23.1	60.0	8.6	67.0	4.2
119	Propaphos	7292-16-2	5	76.6	6.7	80.6	8.3	87.0	2.9	5	77.4	10.3	73.8	5.1	81.7	3.0
120	Propargite	2312-35-8	5	141.4	39.7	82.8	7.8	87.9	11.6	5	97.7	15.0	75.4	12.7	81.2	6.5
121	Propazine	139-40-2	5	80.0	5.6	81.5	6.2	87.6	3.7	5	79.5	8.7	70.9	5.0	77.3	5.5
122	Propiconazol	#N/A	5	73.8	8.6	78.7	7.6	87.2	4.3	5	79.6	10.9	73.2	6.6	77.6	4.7
123	Propisochlor	86763-47-5	5	83.1	4.5	84.2	6.3	88.7	3.5	5	76.8	9.5	70.8	5.4	74.2	5.3
124	Propoxur	114-26-1	5	89.8	7.8	76.4	9.5	82.5	2.2	10	NA	NA	72.9	1.5	77.3	3.6
125	Propyzamide	23950-58-5	5	75.4	19.3	84.4	7.4	88.0	3.1	5	80.8	19.4	71.2	5.1	77.2	6.0
126	Pyraclostrobin	175013-18-0	5	88.4	4.7	86.8	7.4	88.7	2.6	5	82.8	12.2	78.5	5.7	87.2	5.4
127	Pyridafol	40020-01-7	5	NA	NA	NA	NA	NA	NA	5	0.2	50.0	0.1	NA	0.1	NA
128	Quizalofop-Ethyl	76578-14-8	5	85.4	12.5	95.1	14.0	94.8	2.9	5	73.1	18.7	74.5	6.4	81.9	5.5
129	Rabenzazole	40341-04-6	5	62.4	21.7	71.8	11.2	78.0	7.4	5	56.1	16.7	50.4	20.2	71.8	5.1
130	Secbumeton	26259-45-0	5	75.5	2.8	81.8	5.5	88.1	2.9	5	71.2	5.0	70.3	5.6	81.7	2.2
131	Simetryn	1014-70-6	5	67.6	9.3	75.1	2.8	83.3	1.9	5	46.0	22.4	57.6	8.0	78.7	1.9
132	Spiroxamine	118134-30-8	5	40.4	13.0	47.2	10.1	56.4	12.5	5	40.6	25.5	42.2	9.5	53.0	5.2
133	Sulfentrazone	122836-35-5	5	54.5	10.1	57.9	4.8	71.5	9.1	5	50.3	21.2	52.8	14.6	58.5	11.7
134	Tebufenozide	112410-23-8	5	82.5	3.7	72.0	1.1	71.6	7.4	5	81.5	11.6	75.7	11.4	85.0	7.5

No.	Name	CAS														
135	Tepp	107-49-3	5	33.7	56.1	35.9	37.3	53.2	31.9	20	NA	NA	299.4	74.5	26.0	55.8
136	Terbucarb	1918-11-2	5	81.6	14.7	79.0	9.3	84.3	5.3	5	85.3	8.3	76.1	13.0	73.3	19.6
137	Tetrachlorvinphos	22248-79-9	5	73.8	16.8	74.8	16.9	89.7	3.2	5	71.1	7.6	71.8	4.2	73.4	6.4
138	Thiacloprid	111988-49-9	5	66.1	3.3	72.1	6.3	75.3	4.4	5	49.1	10.3	51.1	8.8	57.3	2.4
139	Thiamethoxam	153719-23-4	5	56.6	4.8	56.4	1.0	72.6	5.7	5	58.7	22.9	48.1	8.5	56.9	6.9
140	Thiophanate-Ethyl	23564-06-9	5	7.2	69.6	4.5	96.2	23.5	60.3	5	18.1	46.3	22.3	71.0	21.8	79.6
141	Tolfenpyrad	129558-76-5	5	100.5	7.1	102.5	14.3	88.7	8.4	5	105.6	16.2	108.4	4.1	104.8	5.9
142	Triadimenol	55219-65-3	5	73.3	6.2	73.9	6.2	82.6	4.6	5	67.7	35.7	63.2	8.9	70.7	2.9
143	Triazophos	24017-47-8	5	74.9	16.6	79.4	14.7	91.5	3.8	5	73.5	7.0	71.5	5.9	78.6	2.6
144	Triazoxide	72459-58-6	5	NA	NA	NA	NA	NA	NA	5	0.2	NA	0.1	NA	NA	NA
145	Tribufos	78-48-8	5	85.2	3.5	95.6	13.9	85.5	7.0	5	71.8	4.8	70.7	7.9	84.0	3.2
146	Trichlorfon	52-68-6	5	51.2	5.1	71.8	10.9	71.8	2.4	5	37.1	15.9	40.1	9.7	45.7	9.6
147	Trietazine	1912-26-1	5	81.0	6.1	84.7	7.1	86.0	1.5	5	77.5	7.2	71.7	5.3	81.6	1.6
148	Trifloxystrobin	141517-21-7	5	89.2	4.2	96.6	11.8	87.3	18.9	5	77.7	7.2	77.5	6.3	87.6	3.5
149	Triflumuron	64628-44-0	20	NA	NA	85.5	11.8	88.4	2.9	5	76.9	16.0	73.6	8.8	82.7	4.0
150	Vamidothion	2275-23-2	5	62.8	7.3	71.0	1.9	75.5	17.7	5	49.2	19.6	56.1	7.0	64.1	2.7
151	Vamidothion Sulfoxide	20300-00-9	5	60.0	2.8	62.0	7.2	71.5	3.4	5	55.2	27.2	54.6	6.4	64.2	4.2
152	Zoxamide	156052-68-5	5	86.6	6.8	93.2	19.7	94.9	5.5	5	81.6	12.0	78.2	5.9	87.7	8.2
153	Acetamiprid	135410-20-7	5	70.2	5.2	70.4	3.3	70.1	3.4	5	66.4	12.4	67.0	3.8	60.5	9.6
154	Ametryn	834-12-8	5	81.6	12.2	75.1	1.3	84.6	1.4	5	65.3	8.9	81.4	1.3	87.3	2.2
155	Aspon	3244-90-4	5	84.3	4.5	84.6	3.1	89.3	5.2	5	75.0	10.4	85.0	9.6	83.1	6.7
156	Asulam	3337-71-1	5	0.6	35.9	0.3	60.1	NA	NA	5	NA	NA	NA	NA	NA	NA
157	Azinphos-methyl	86-50-0	10	95.0	11.4	90.1	5.1	99.6	8.5	10	NA	NA	NA	NA	88.1	13.6
158	Benoxacor	98730-04-2	50	NA	NA	89.6	9.9	80.7	4.5	20	NA	NA	NA	NA	70.6	3.6
159	Benzoylprop	22212-56-2	20	NA	NA	NA	NA	NA	NA	20	NA	NA	NA	NA	NA	NA
160	Bromacil	314-40-9	20	NA	NA	57.6	12.8	70.2	3.7	10	NA	NA	74.6	4.2	71.3	2.4
161	Bupirimate	41483-43-6	5	75.8	1.2	78.2	1.1	86.5	2.1	5	71.6	3.1	81.0	1.4	87.2	1.9
162	Butachlor	23184-66-9	5	85.0	2.1	81.4	1.7	86.9	2.9	5	78.7	12.2	71.1	7.9	77.0	7.0
163	Butafenacil	134605-64-4	5	78.7	2.7	79.0	1.7	91.1	1.3	5	73.4	11.0	81.4	8.5	85.3	3.2
164	Carbofuran	1563-66-2	5	70.6	2.9	73.8	1.8	74.2	1.4	5	71.9	3.5	70.6	2.5	71.1	2.0
165	Carbofuran-3-Hydroxy	16655-82-6	5	62.0	6.5	61.8	8.0	65.4	3.9	5	54.6	19.1	71.5	7.7	70.0	2.6
166	Cartap	15263-53-3	NA	NA	NA	NA	NA	NA	NA	NA	NA	NA	NA	NA	NA	NA
167	Chlorfenvinphos	470-90-6	5	83.4	2.7	81.2	1.5	87.4	1.8	5	71.7	9.5	76.3	7.1	75.0	2.4
168	Chloridazon	1698-60-8	5	63.1	8.8	68.7	6.1	64.0	6.3	5	45.7	17.2	70.4	8.0	70.8	5.2
169	Chlorimuron-ethyl	90982-32-4	NA	NA	NA	NA	NA	NA	NA	NA	NA	NA	NA	NA	NA	NA
170	Chlorphonium	7695-87-6	5	54.5	5.9	58.7	3.0	70.7	1.7	5	17.3	48.6	43.0	36.5	80.7	1.8

(Continued)

APPENDIX TABLE 1.3L.1 Results of Nontarget Screening for 556 Pesticides in Four Kinds of Tea by LC-Q-TOF/MS (cont.)

	Name	CAS number	Puer tea Screening limits (µg/kg)	Puer 10 µg/kg AVE	RSD (%) (n=5)	Puer 50 µg/kg AVE	RSD (%) (n=5)	Puer 200 µg/kg AVE	RSD (%) (n=5)	Oolong tea Screening limits (µg/kg)	Oolong 10 µg/kg AVE	RSD (%) (n=5)	Oolong 50 µg/kg AVE	RSD (%) (n=5)	Oolong 200 µg/kg AVE	RSD (%) (n=5)
171	Chlorphoxim	14816-20-7	50	NA	NA	82.9	12.0	86.6	8.1	5	70.5	5.6	72.0	12.3	71.9	7.7
172	Clethodim	99129-21-2	5	42.9	12.3	46.0	12.6	51.5	10.4	5	35.2	24.0	36.9	17.9	54.1	12.8
173	Clodinafop-propargyl	105512-06-9	5	84.5	1.3	86.8	1.1	90.9	1.2	5	63.0	11.4	73.0	4.1	77.3	3.2
174	Cloransulam-Methyl	147150-35-4	5	NA	NA	NA	NA	NA	NA	5	NA	NA	NA	NA	NA	NA
175	Cycloate	1134-23-2	5	78.9	5.2	81.9	1.4	84.6	14.2	5	77.9	15.6	79.8	7.1	78.7	4.8
176	Cyprazine	22936-86-3	5	71.9	4.1	74.8	1.3	83.6	5.0	5	70.2	4.2	82.9	4.2	78.5	2.8
177	Desmedipham	13684-56-5	5	74.3	3.4	81.4	1.7	83.0	4.6	5	71.1	9.7	81.0	4.6	76.7	4.3
178	Diafenthiuron	80060-09-9	50	NA	NA	NA	NA	NA	NA	50	NA	NA	NA	NA	NA	NA
179	Dicrotophos	141-66-2	5	70.3	5.2	70.4	5.8	74.1	2.7	5	58.7	18.4	81.1	0.9	81.9	5.2
180	Diethofencarb	87130-20-9	5	92.8	6.6	87.6	1.8	106.2	14.9	5	71.3	5.5	75.3	5.1	70.8	2.2
181	Diethyltoluamide	134-62-3	5	76.5	10.6	76.4	3.9	83.7	3.6	5	87.8	20.6	84.8	7.0	89.8	2.9
182	Difenoxuron	14214-32-5	5	72.9	1.8	75.9	1.8	84.2	2.0	5	72.3	7.5	79.8	2.9	82.1	1.4
183	Dimethametryn	22936-75-0	5	77.1	1.1	80.0	0.9	89.0	3.5	5	71.4	7.3	89.0	3.0	91.3	1.5
184	Dimethoate	60-51-5	5	64.9	3.8	70.6	7.6	72.0	3.5	5	58.6	12.4	77.4	5.9	72.1	5.2
185	Dimethomorph	110488-70-5	5	70.8	4.3	70.3	1.5	73.4	2.7	5	71.9	11.5	76.2	5.9	71.5	3.1
186	Diniconazole	83657-24-3	5	72.6	2.7	74.7	1.1	81.0	3.2	5	72.7	7.4	83.1	5.1	76.9	3.0
187	Diphenamid	957-51-7	5	67.1	1.2	71.9	1.4	83.7	1.4	5	70.7	4.4	75.7	1.2	92.4	2.1
188	Dipropetryn	4147-51-7	5	92.3	10.7	79.0	8.7	90.5	1.8	5	71.4	7.2	85.3	1.7	93.0	2.0
189	Disulfoton sulfoxide	2497-07-6	5	70.6	0.6	75.5	2.0	93.1	3.6	5	72.2	5.7	81.6	2.7	81.6	2.6
190	Dithiopyr	97886-45-8	20	NA	NA	79.1	3.0	88.8	2.8	10	89.8	7.3	72.4	11.2	70.5	3.2
191	Dodemorph	1593-77-7	5	72.7	2.3	74.3	1.0	81.3	3.5	5	65.6	8.2	77.0	2.6	86.9	3.6
192	Drazoxolon	5707-69-7	10	51.9	15.8	58.4	7.6	71.5	5.2	5	51.5	15.0	52.4	11.9	58.4	7.1
193	Epoxiconazole	133855-98-8	5	73.9	2.0	73.7	1.9	81.7	2.6	5	65.7	11.4	82.0	2.8	74.7	3.4
194	Ethiofencarb	29973-13-5	5	55.9	6.4	56.9	9.5	64.3	3.2	5	42.6	22.7	57.4	11.9	60.5	10.5
195	Ethiofencarb-sulfone	53380-23-7	5	58.5	3.5	70.6	1.6	70.1	2.4	5	53.9	15.8	70.7	5.1	70.9	4.4
196	Ethion	563-12-2	10	77.8	13.8	84.7	4.3	99.4	3.2	5	83.2	14.8	75.4	14.6	75.8	12.1
197	Ethoxyquin	91-53-2	50	NA	NA	NA	NA	NA	NA	50	NA	NA	NA	NA	NA	NA
198	Fenamiphos	22224-92-6	5	72.1	4.5	71.1	5.3	83.7	3.4	5	64.0	9.7	74.2	5.4	84.9	3.4
199	Fenamiphos-sulfone	31972-44-8	5	63.1	2.2	67.5	1.4	77.6	2.4	5	73.4	6.2	76.9	1.3	88.2	3.3
200	Fenamiphos-sulfoxide	31972-43-7	5	70.5	2.3	74.3	2.3	84.4	3.8	5	70.7	15.7	83.2	4.9	86.7	1.5
201	Fenbuconazole	114369-43-6	5	70.8	2.2	71.4	1.3	79.8	2.0	5	74.7	2.9	75.0	4.0	75.7	4.2
202	Fenfuram	24691-80-3	5	53.4	3.8	55.7	3.4	58.7	4.3	5	66.2	7.2	91.7	5.3	92.2	3.9
203	Fenobucarb	3766-81-2	5	72.3	7.4	76.0	2.4	77.4	3.8	5	62.5	17.9	66.9	4.7	67.6	4.5

No.	Name	CAS														
204	Fenothiocarb	62850-32-2	5	83.8	1.2	85.3	0.9	91.7	1.1	5	73.9	10.1	77.5	3.4	72.3	2.2
205	Fenoxanil	115852-48-7	5	80.0	2.5	82.4	1.8	91.1	1.0	5	71.9	10.1	73.4	1.7	71.7	1.9
206	Fenoxaprop-Ethyl	66441-23-4	5	78.4	9.1	79.5	8.5	89.1	4.0	5	73.5	7.6	85.0	4.5	84.6	3.5
207	Fenoxycarb	72490-01-8	5	85.5	7.3	83.0	2.1	91.1	2.0	5	63.2	10.2	77.1	2.6	77.7	4.4
208	Fenpropimorph	67564-91-4	5	72.7	2.3	74.3	1.0	81.3	3.5	5	65.6	8.2	77.0	2.6	86.9	3.6
209	fenpyroximate	134098-61-6	5	79.0	5.7	82.5	2.5	92.6	1.6	5	72.0	7.1	92.6	7.2	84.8	3.9
210	Fensulfothion	115-90-2	5	76.1	3.9	78.9	3.0	90.4	2.5	5	73.7	11.3	77.0	4.7	83.4	2.3
211	Flazasulfuron	104040-78-0	5	NA	NA	NA	NA	NA	NA	5	NA	NA	NA	NA	NA	NA
212	Flufenacet	142459-58-3	5	79.1	1.8	81.8	1.9	89.4	2.4	5	72.8	0.6	75.9	4.9	78.5	2.8
213	Flumequine	42835-25-6	5	39.7	4.8	42.2	5.1	47.9	6.5	5	18.6	22.2	21.2	17.8	22.4	5.9
214	Flumetsulam	98967-40-9	5	0.8	107.9	2.0	84.4	0.7	116.9	5	NA	NA	NA	NA	NA	NA
215	Flumiclorac-Pentyl	87546-18-7	5	86.9	3.6	83.0	3.1	83.8	1.5	5	75.3	10.9	71.6	6.8	70.7	7.1
216	Fluometuron	2164-17-2	5	71.5	0.9	75.3	1.1	87.5	3.4	5	73.0	3.9	78.9	1.9	88.5	2.2
217	Fluridone	59756-60-4	5	65.5	2.5	70.6	1.9	84.3	1.0	5	64.3	12.1	78.1	1.4	91.4	3.4
218	Fonofos	944-22-9	20	NA	NA	NA	NA	93.0	4.0	10	NA	NA	68.9	3.1	71.5	3.3
219	Fosthiazate	98886-44-3	5	66.9	1.8	72.5	2.8	82.9	4.7	5	72.0	6.8	86.3	4.0	80.7	3.9
220	Halosulfuron-methyl	100784-20-1	10	NA	NA	NA	NA	NA	NA	5	NA	NA	NA	NA	NA	NA
221	Haloxyfop	69806-34-4	10	NA	NA	NA	NA	NA	NA	10	NA	NA	NA	NA	4.3	43.1
222	Heptenophos	23560-59-0	5	70.5	3.3	72.5	1.6	76.0	6.1	5	10.0	99.3	13.8	48.8	73.0	2.4
223	Hexazinone	51235-04-2	5	61.8	1.5	66.9	1.4	75.9	1.4	5	58.2	12.8	72.5	0.4	79.2	2.7
224	Imazalil	35554-44-0	5	27.3	111.5	33.4	13.0	37.7	54.3	5	62.4	13.6	84.8	7.9	76.1	5.0
225	Imazamethabenz-methyl	81405-85-8	5	74.0	3.4	70.6	2.3	80.9	2.8	5	71.6	3.8	86.9	2.8	90.6	7.5
226	Imazaquin	81335-37-7	5	1.6	73.2	0.2	34.2	0.2	0.0	5	NA	NA	NA	NA	NA	NA
227	Inabenfide	82211-24-3	5	NA	NA	1.9	124.0	4.2	94.1	5	16.6	28.8	24.0	54.7	14.2	9.1
228	Iodosulfuron-methyl	144550-06-1	5	NA	NA	NA	NA	NA	NA	5	NA	NA	NA	NA	NA	NA
229	Isofenphos	25311-71-1	5	103.6	9.8	80.3	5.4	91.7	1.3	5	75.9	16.7	58.2	20.3	72.6	3.1
230	Isofenphos-oxon	31120-85-1	5	71.7	3.7	89.3	7.8	78.8	5.3	5	73.8	8.7	82.7	5.2	80.5	3.2
231	Isopropalin	33820-53-0	5	80.6	2.4	85.5	2.9	86.0	2.3	5	97.8	11.6	91.8	15.7	81.8	9.0
232	Isoprothiolane	50512-35-1	5	71.5	3.7	78.0	2.6	78.4	2.1	5	71.0	2.3	79.0	4.1	83.8	4.1
233	Isoproturon	34123-59-6	5	72.4	2.1	75.7	0.9	85.2	2.8	5	71.5	11.2	84.5	4.0	88.3	1.9
234	Kresoxim-methyl	143390-89-0	5	76.3	3.3	81.3	0.9	90.1	4.6	5	70.4	13.6	72.6	7.1	72.7	2.5
235	Linuron	330-55-2	5	80.6	3.9	80.6	0.7	88.1	2.5	5	78.3	6.8	80.9	5.2	74.1	3.8
236	Malathion	121-75-5	5	75.1	2.2	82.5	1.1	89.4	1.4	5	75.2	9.1	73.6	4.2	70.7	2.0
237	Mecarbam	2595-54-2	5	86.7	11.2	80.9	5.7	90.8	9.8	5	71.8	8.2	75.7	5.7	72.6	2.0
238	Mefenacet	73250-68-7	5	74.3	2.4	76.3	1.8	83.9	2.0	5	70.7	11.1	79.3	4.9	79.6	2.4
239	Mefenpyr-diethyl	135590-91-9	5	87.0	2.8	83.6	2.5	94.2	2.7	5	71.0	19.1	83.7	10.4	82.9	2.5
240	Mepanipyrim	110235-47-7	5	80.0	3.1	80.4	1.8	88.6	3.9	5	70.9	2.9	81.7	4.6	84.3	1.8

(Continued)

APPENDIX TABLE 1.3L.1 Results of Nontarget Screening for 556 Pesticides in Four Kinds of Tea by LC-Q-TOF/MS (cont.)

	Name	CAS number	Puer tea Screening limits (μg/kg)	Puer 10 μg/kg AVE	Puer 10 μg/kg RSD (%) (n = 5)	Puer 50 μg/kg AVE	Puer 50 μg/kg RSD (%) (n = 5)	Puer 200 μg/kg AVE	Puer 200 μg/kg RSD (%) (n = 5)	Oolong tea Screening limits (μg/kg)	Oolong 10 μg/kg AVE	Oolong 10 μg/kg RSD (%) (n = 5)	Oolong 50 μg/kg AVE	Oolong 50 μg/kg RSD (%) (n = 5)	Oolong 200 μg/kg AVE	Oolong 200 μg/kg RSD (%) (n = 5)
241	Mepronil	55814-41-0	5	81.3	2.4	82.3	1.8	90.0	2.4	5	70.2	9.8	79.8	1.2	81.3	2.3
242	Metamitron	41394-05-2	5	64.2	2.1	59.5	1.7	70.2	3.7	5	44.4	17.1	75.5	8.4	77.8	18.1
243	Metazachlor	67129-08-2	5	70.6	2.7	70.2	0.5	73.5	3.5	5	71.5	4.2	74.2	3.8	70.1	4.2
244	Metconazole	125116-23-6	5	79.7	3.3	76.1	0.8	86.0	1.7	5	71.7	9.5	84.9	2.6	80.6	3.0
245	Methamidophos	10265-92-6	5	60.1	13.2	54.0	1.8	58.3	3.8	5	48.6	14.5	48.2	9.5	50.3	18.7
246	Methiocarb	2032-65-7	5	72.7	8.6	78.9	3.0	86.7	2.0	5	78.1	7.2	70.1	4.3	72.7	2.5
247	Metribuzin	21087-64-9	5	60.1	5.1	63.0	3.8	63.2	2.9	5	52.8	12.2	67.1	2.9	66.7	2.2
248	Mevinphos	7786-34-7	5	76.1	4.2	70.6	4.3	71.0	1.9	5	70.6	7.0	73.9	3.6	76.1	3.0
249	Monocrotophos	6923-22-4	5	60.8	1.7	59.9	9.6	65.2	3.2	5	55.2	19.0	82.6	2.5	78.7	7.3
250	Monuron	150-68-5	5	66.6	3.3	69.2	2.0	75.6	1.8	5	60.2	12.5	73.0	1.1	71.9	2.4
251	Myclobutanil	88671-89-0	10	70.3	2.5	70.2	1.4	77.5	2.7	5	71.2	3.5	73.5	1.8	70.8	2.4
252	Nuarimol	63284-71-9	5	64.8	2.3	68.3	1.8	79.1	2.6	5	89.6	19.4	117.0	14.7	97.6	3.6
253	Ofurace	58810-48-3	5	57.8	4.1	63.6	2.0	75.9	2.4	5	59.8	14.3	73.2	2.1	84.0	2.1
254	Oxadixyl	77732-09-3	5	45.0	13.1	54.3	3.8	58.7	2.4	5	57.0	13.1	74.2	5.2	70.2	2.7
255	Oxycarboxin	5259-88-1	5	65.1	5.6	60.2	1.2	66.6	3.0	5	54.8	15.8	62.6	2.6	56.2	4.0
256	Phenthoate	2597-03-7	50	NA	NA	79.2	11.7	98.2	7.3	10	78.5	7.1	72.1	8.6	72.6	9.0
257	Phorate-Sulfone	2588-04-7	5	61.3	7.1	73.0	3.7	70.6	3.5	10	56.2	14.1	53.7	11.8	57.6	10.1
258	Phosalone	2310-17-0	100	NA	NA	NA	NA	86.8	14.6	200	NA	NA	NA	NA	103.9	15.1
259	Phthalic Acid, Bis-Butyl Ester	84-74-2	5	104.2	10.2	101.8	13.8	105.3	8.1	5	73.9	11.8	78.1	14.4	81.5	12.4
260	Picloram	1918-02-1	100	NA	NA	NA	NA	NA	NA	200	NA	NA	NA	NA	NA	NA
261	Piperonyl Butoxide	51-03-6	5	84.0	0.9	84.7	0.5	93.1	1.3	5	77.5	6.7	88.1	4.4	85.4	2.0
262	Piperophos	24151-93-7	5	81.2	1.0	81.7	1.0	90.8	1.2	5	74.3	8.4	86.9	2.6	89.1	1.4
263	Pirimicarb-desmethyl	30614-22-3	5	65.6	4.4	64.4	9.2	75.4	2.4	5	57.0	16.7	80.3	2.6	75.5	3.5
264	Pirimiphos-Ethyl	23505-41-1	5	82.6	3.0	85.3	1.1	95.4	0.8	5	78.2	7.8	91.9	2.3	96.7	1.8
265	Prochloraz	67747-09-5	5	75.8	6.1	72.3	1.4	81.2	3.4	5	73.3	5.9	90.6	6.7	71.6	4.3
266	Prometon	1610-18-0	5	74.6	1.6	73.6	1.2	82.3	2.6	5	62.2	12.1	78.7	1.4	84.3	2.8
267	Propamocarb	24579-73-5	5	0.1	0.0	0.1	NA	NA	NA	5	NA	NA	NA	NA	NA	NA
268	Propanil	709-98-8	5	74.6	6.9	80.1	2.0	95.3	3.5	5	73.0	9.3	76.6	2.7	90.3	2.8
269	Prosulfocarb	52888-80-9	5	82.8	3.8	85.0	1.8	89.9	1.5	5	75.3	7.7	90.0	4.7	86.2	3.5
270	Pymetrozine	123312-89-0	5	2.0	90.6	0.6	103.8	0.4	59.5	5	NA	NA	NA	NA	NA	NA
271	Pyraclofos	89784-60-1	5	87.6	3.7	85.9	2.7	94.9	2.7	5	76.0	7.8	81.4	2.7	80.3	2.9
272	Pyraflufen-Ethyl	129630-19-9	5	85.8	3.8	83.3	1.8	92.8	1.9	5	70.9	7.7	78.8	2.7	83.1	2.6

No.	Name	CAS														
273	Pyributicarb	88678-67-5	5	88.4	3.1	84.5	2.2	96.8	1.7	5	77.8	9.5	86.4	1.9	92.0	2.7
274	Pyridaphenthion	119-12-0	5	76.6	4.4	82.2	4.1	85.7	7.6	5	74.1	6.6	75.6	3.7	73.4	2.3
275	Pyridate	55512-33-9	5	84.3	4.5	83.4	2.5	106.4	5.2	5	73.5	10.4	85.0	9.5	83.1	6.7
276	Pyrimethanil	53112-28-0	5	70.4	4.7	74.5	2.5	82.8	1.6	5	70.0	4.0	73.6	4.5	79.5	1.8
277	Quinalphos	13593-03-8	5	88.7	2.5	83.3	2.0	91.4	1.8	5	76.4	8.1	83.7	4.9	83.0	3.0
278	Quinoclamine	2797-51-5	5	62.2	8.3	58.1	5.3	58.7	3.8	5	54.4	9.8	53.7	2.0	50.7	2.1
279	Sebuthylazine	7286-69-3	5	77.7	1.7	78.4	1.5	85.2	1.7	5	70.3	10.9	76.8	4.7	84.6	3.8
280	Simazine	122-34-9	5	51.8	6.1	55.2	1.5	58.3	2.7	5	53.5	11.3	67.3	2.9	63.9	3.2
281	Simeconazole	149508-90-7	5	84.7	1.1	72.3	1.1	76.3	2.5	5	71.9	9.3	82.4	5.6	74.0	3.2
282	Simeton	673-04-1	5	73.6	10.5	69.1	1.8	79.6	2.1	5	55.5	14.0	71.4	13.0	78.1	4.7
283	Spinosad	168316-95-8	5	6.1	99.3	6.9	49.9	10.3	60.4	5	59.3	11.3	59.8	43.7	37.0	8.4
284	Spirodiclofen	148477-71-8	5	80.7	10.9	80.6	7.6	83.3	2.8	5	83.1	9.3	85.0	9.4	75.6	6.2
285	Sulfotep	3689-24-5	5	84.6	4.1	80.6	2.6	84.5	4.5	5	76.0	8.8	83.0	11.0	75.5	14.0
286	Sulprofos	35400-43-2	20	NA	NA	76.2	3.8	89.3	5.7	10	NA	NA	71.7	3.9	74.7	10.3
287	Tebuconazole	107534-96-3	5	77.1	3.6	79.1	3.0	82.0	6.4	5	77.3	8.6	85.6	5.7	76.6	2.6
288	Tebupirimfos	96182-53-5	5	79.6	2.6	80.8	0.7	86.3	1.6	5	80.0	7.0	83.2	2.9	85.9	2.4
289	Tebutam	35256-85-0	5	72.1	3.3	74.6	1.6	80.5	2.1	5	71.0	9.7	77.1	5.5	83.3	2.2
290	Tebuthiuron	34014-18-1	5	63.1	4.4	64.9	2.6	70.2	1.8	5	59.1	12.5	70.7	2.3	73.7	2.1
291	Temephos	3383-96-8	5	90.5	3.9	89.4	1.4	95.1	2.2	5	72.5	5.3	71.6	6.2	71.2	4.1
292	Terbumeton	33693-04-8	5	75.8	1.5	77.9	1.0	90.8	2.4	5	61.8	11.3	81.2	0.8	92.8	1.9
293	Tetraconazole	112281-77-3	5	72.2	2.6	75.6	2.4	76.8	4.5	5	70.4	2.6	76.3	1.3	71.2	3.2
294	Thiazopyr	117718-60-2	5	85.5	2.4	81.7	3.1	91.3	2.3	5	71.6	12.1	80.3	7.8	81.7	2.0
295	Thiobencarb	28249-77-6	5	82.4	1.9	84.3	1.1	99.2	0.7	5	74.0	11.0	77.9	2.2	76.1	1.9
296	Thiodicarb	59669-26-0	5	58.7	5.4	62.0	3.9	53.9	4.5	5	48.9	16.0	66.1	3.7	61.8	9.1
297	Thiofanox-Sulfoxide	39184-27-5	5	77.2	13.9	71.6	11.7	70.7	7.8	5	79.2	9.4	86.1	14.5	102.7	18.2
298	Thionazin	297-97-2	5	72.1	5.1	75.0	1.1	85.8	10.6	5	70.2	10.5	73.6	4.4	75.0	3.6
299	Tralkoxydim	87820-88-0	5	77.0	3.9	71.5	1.9	75.7	3.2	5	75.7	4.5	66.4	5.5	64.4	4.5
300	Triadimefon	43121-43-3	5	79.1	2.9	77.4	1.6	85.4	2.3	5	72.6	10.5	81.9	4.1	73.4	2.6
301	Tributyl Phosphate	126-73-8	5	85.2	3.8	84.7	2.4	92.9	1.4	5	88.3	6.3	90.6	5.7	88.2	2.0
302	Tricyclazole	41814-78-2	5	54.5	5.2	53.2	3.6	61.5	4.5	5	48.0	16.7	61.3	4.5	70.9	4.7
303	Tridemorph	81412-43-3	5	80.4	6.1	75.9	6.2	80.3	4.4	5	70.6	0.8	72.4	3.3	71.4	3.6
304	Triticonazole	131983-72-7	5	73.8	3.1	73.9	1.7	80.4	2.2	5	71.0	9.8	80.8	4.2	76.5	4.6
305	1-Naphthyl Acetamide	86-86-2	5	73.6	6.9	81.0	1.9	78.7	2.1	5	118.1	5.0	86.2	3.5	78.5	4.0
306	2,6-Dichlorobenzamide	2008-58-4	50	NA	NA	70.9	8.0	72.2	5.7	5	NA	NA	78.5	6.4	71.0	4.1
307	Albendazole	54965-21-8	5	4.1	37.0	2.4	41.3	3.3	68.1	5	3.9	34.7	2.3	41.5	3.6	70.6
308	Aldicarb	116-06-3	20	NA	NA	89.2	9.2	72.9	1.8	5	51.0	8.7	70.6	5.7	74.3	10.7

(Continued)

APPENDIX TABLE 1.3L.1 Results of Nontarget Screening for 556 Pesticides in Four Kinds of Tea by LC-Q-TOF/MS (cont.)

#	Name	CAS number	Puer tea Screening limits (µg/kg)	10 µg/kg AVE	10 µg/kg RSD (%) (n = 5)	50 µg/kg AVE	50 µg/kg RSD (%) (n = 5)	200 µg/kg AVE	200 µg/kg RSD (%) (n = 5)	Oolong tea Screening limits (µg/kg)	10 µg/kg AVE	10 µg/kg RSD (%) (n = 5)	50 µg/kg AVE	50 µg/kg RSD (%) (n = 5)	200 µg/kg AVE	200 µg/kg RSD (%) (n = 5)
309	Aldicarb-sulfone	1646-88-4	5	101.1	6.1	72.7	11.9	73.3	15.6	5	74.5	8.6	71.8	3.2	72.1	7.3
310	Aldimorph	91315-15-0	5	71.8	6.8	70.0	1.4	78.9	5.5	5	66.0	4.1	64.1	2.9	63.5	4.1
311	Aminocarb	2032-59-9	5	113.2	8.6	94.2	2.4	98.9	4.3	5	56.0	2.0	80.8	2.6	85.7	2.0
312	Aminopyralid	150114-71-9	NA	NA	NA	NA	NA	NA	NA	5	NA	NA	2.5	47.5	0.4	31.5
313	Ancymidol	12771-68-5	5	60.0	4.7	83.6	13.1	86.1	6.1	5	82.8	6.6	74.9	12.8	80.5	6.6
314	Atrazine	1912-24-9	5	95.5	7.0	104.6	3.6	104.5	2.7	5	112.6	1.9	97.5	2.2	90.8	3.9
315	Atrazine-desethyl	6190-65-4	5	60.1	5.6	76.6	4.4	80.0	2.7	5	160.5	5.7	87.1	3.8	78.7	1.9
316	Benalaxyl	71626-11-4	5	102.1	7.5	105.7	1.1	107.0	0.8	5	88.2	2.1	94.2	1.4	95.0	1.8
317	Bensulfuron-methyl	83055-99-6	5	12.3	15.3	2.9	76.7	1.2	33.5	5	9.9	15.4	2.1	9.5	0.7	48.4
318	Bensultap	17606-31-4	NA	NA	NA	NA	NA	NA	NA	NA	NA	NA	NA	NA	NA	NA
319	Benzoylprop-Ethyl	22212-55-1	5	81.9	19.2	114.2	2.1	109.8	2.6	5	90.5	19.1	95.7	2.2	95.0	2.6
320	Bromobutide	74712-19-9	5	91.4	14.9	108.4	5.0	93.8	6.1	5	85.1	4.2	94.4	3.6	84.3	3.2
321	Butocarboxim	34681-10-2	5	85.9	14.1	86.7	11.9	71.2	9.4	5	38.1	14.9	77.1	14.4	75.5	17.0
322	Butoxycarboxim	34681-23-7	5	101.9	12.4	89.5	6.6	83.8	18.4	5	73.1	7.6	115.2	9.0	70.7	3.7
323	Cafenstrole	125306-83-4	5	89.3	10.1	111.6	4.8	110.3	7.9	5	92.4	11.8	88.5	3.9	99.3	7.0
324	Carpropamid	104030-54-8	5	92.2	12.8	114.7	3.1	106.8	3.0	5	91.7	2.4	97.1	2.5	92.2	3.0
325	Chlordimeform	6164-98-3	5	46.1	19.7	150.9	22.1	137.0	13.1	10	NA	NA	NA	NA	NA	NA
326	Chloroxuron	1982-47-4	5	107.2	3.4	113.1	2.3	109.4	3.2	5	93.4	2.2	93.3	2.3	92.5	4.0
327	Chlorthiophos	60238-56-4	NA	NA	NA	NA	NA	NA	NA	10	NA	NA	NA	NA	NA	NA
328	Chromafenozide	143807-66-3	5	109.9	10.3	110.1	2.4	106.1	4.6	5	85.8	0.7	89.1	1.2	89.7	4.1
329	Clodinafop	114420-56-3	10	NA	NA	NA	NA	NA	NA	5	97.4	5.0	82.8	5.9	116.9	3.9
330	Clofentezine	74115-24-5	10	142.0	31.9	115.7	5.5	112.3	2.3	5	116.0	7.1	106.1	7.8	97.4	3.4
331	Cumyluron	99485-76-4	5	102.9	3.5	106.3	2.9	100.7	4.6	5	89.7	1.8	84.9	2.5	82.7	4.1
332	Cyclosulfamuron	136849-15-5	50	NA	NA	NA	NA	NA	NA	5	39.4	66.8	1.8	55.0	1.0	42.4
333	Cyflufenamid	180409-60-3	5	98.5	11.5	112.0	3.0	113.0	1.7	5	89.8	2.4	92.9	3.4	96.9	3.3
334	Cyprofuram	69581-33-5	5	77.0	4.0	89.9	4.4	86.2	4.7	5	76.4	3.6	78.1	3.3	75.4	4.6
335	Cyromazine	66215-27-8	5	8.5	19.3	1.4	10.4	0.5	16.3	5	8.6	18.4	1.4	6.1	0.5	27.2
336	Demeton-S-Methyl	919-86-8	20	NA	NA	NA	NA	NA	NA	10	NA	NA	NA	NA	NA	NA
337	Demeton-S-methyl-sulfone	17040-19-6	5	93.6	10.1	87.6	3.9	90.7	1.4	5	112.1	8.4	74.9	3.5	81.8	1.4
338	Desmetryn	1014-69-3	5	94.4	5.1	93.5	3.6	94.0	0.8	5	115.7	3.5	82.6	0.4	87.5	1.9
339	Dialifos	10311-84-9	5	73.1	5.3	105.2	4.5	103.5	2.8	5	96.6	6.3	96.2	5.2	88.8	3.3
340	Dichlofenthion	97-17-6	NA	NA	NA	NA	NA	NA	NA	5	NA	NA	262.9	24.9	112.9	11.0

No.	Compound	CAS No.	Level	Rec	RSD	Rec	RSD	Rec	RSD	Level	Rec	RSD	Rec	RSD	Rec	RSD
341	Diclosulam	145701-21-9	5	NA	NA	NA	NA	NA	NA	5	NA	NA	NA	NA	NA	NA
342	Diethatyl-Ethyl	38727-55-8	5	89.0	7.6	111.5	4.1	99.7	5.1	5	101.0	2.8	96.4	2.3	85.3	3.7
343	Difenoconazole	119446-68-3	5	794.1	19.1	129.7	3.3	113.6	2.2	5	111.1	7.9	107.1	2.5	95.5	2.8
344	Dimefuron	34205-21-5	5	113.8	3.2	108.1	1.4	109.1	5.0	5	91.6	2.1	95.5	1.2	95.4	4.3
345	Dimepiperate	61432-55-1	50	NA	NA	100.7	9.9	118.8	7.9	10	NA	NA	84.0	5.2	87.9	3.6
346	Dimethirimol	5221-53-4	5	42.0	4.1	42.0	4.4	50.8	1.8	5	51.5	5.5	42.1	4.5	58.7	2.3
347	Dinotefuran	165252-70-0	5	96.7	5.1	87.4	3.1	85.0	6.7	5	111.1	4.2	85.1	3.1	77.4	7.6
348	Ditalimfos	5131-24-8	5	96.7	7.7	97.8	10.7	109.3	2.4	5	82.3	4.8	81.1	10.0	88.8	3.4
349	Emamectin	119791-41-2	5	11.5	78.3	15.7	34.5	19.0	44.5	5	11.7	98.0	13.8	32.5	17.4	44.3
350	Ethametsulfuron-Methyl	97780-06-8	5	NA	NA	NA	NA	NA	NA	5	3.8	85.5	0.6	73.5	0.3	29.9
351	Ethiofencarb-sulfoxide	53380-22-6	5	87.8	12.1	81.1	4.6	82.0	2.0	5	96.0	8.0	72.1	4.4	73.8	2.2
352	Ethiprole	181587-01-9	10	97.8	8.1	107.4	4.5	106.9	3.3	10	6.1	13.3	106.0	5.2	88.2	4.5
353	Ethirimol	23947-60-6	5	42.0	4.0	41.9	4.5	49.2	1.9	5	51.5	5.6	42.2	4.7	59.6	3.9
354	Ethoxysulfuron	126801-58-9	NA	NA	NA	NA	NA	NA	NA	NA	NA	NA	NA	NA	NA	NA
355	Etobenzanid	79540-50-4	5	72.4	16.1	75.3	7.9	74.7	3.7	5	71.2	6.2	70.1	3.6	71.3	0.4
356	Etoxazole	153233-91-1	5	85.9	6.8	119.2	5.7	111.0	3.7	5	104.5	2.9	94.8	4.7	101.3	0.9
357	Fenamidone	161326-34-7	5	88.8	8.4	106.6	4.2	110.4	0.7	5	110.0	2.3	107.5	4.3	91.8	1.5
358	Fenpropidin	67306-00-7	5	60.5	14.3	59.9	13.4	61.4	16.4	5	52.2	13.4	50.8	7.2	70.9	2.4
359	Fenthion	55-38-9	10	98.6	8.9	113.6	3.8	104.9	4.6	5	101.7	15.5	107.5	1.8	105.4	3.8
360	Fenthion-sulfone	3761-42-0	5	111.2	8.2	95.6	4.7	94.2	2.1	5	106.9	4.6	105.2	2.6	86.4	2.0
361	Flamprop	58667-63-3	10	NA	NA	NA	NA	NA	NA	10	19.6	45.4	6.4	31.2	0.4	98.7
362	Fluazifop	69335-91-7	5	NA	NA	NA	NA	NA	NA	5	21.5	114.0	1.5	60.9	0.2	66.7
363	Flucycloxuron	94050-52-9	5	95.6	6.8	97.9	9.2	99.5	4.4	5	87.9	8.2	92.7	9.2	84.1	4.6
364	Fluoroglycofen-ethyl	77501-90-7	NA	NA	NA	NA	NA	NA	NA	20	NA	NA	NA	NA	114.4	15.0
365	Fluquinconazole	136426-54-5	5	104.1	6.2	102.2	2.0	98.6	4.2	5	88.6	7.6	94.3	3.8	81.6	4.2
366	Flurochloridone	61213-25-0	50	NA	NA	92.0	10.0	116.0	6.6	5	87.8	15.1	111.0	18.3	102.3	6.1
367	Flusilazole	85509-19-9	5	91.2	7.8	103.4	3.7	105.0	1.4	5	81.5	1.8	91.6	4.7	93.9	1.3
368	Foramsulfuron	173159-57-4	10	NA	NA	NA	NA	NA	NA	5	NA	NA	23.6	27.5	3.5	31.4
369	Fuberidazole	3878-19-1	5	41.6	23.6	51.3	7.4	70.4	2.1	5	39.0	30.7	36.2	12.6	76.8	10.7
370	Furathiocarb	65907-30-4	5	27.1	15.6	119.4	10.6	116.2	2.7	5	90.4	4.3	85.7	10.8	100.3	2.2
371	Furmecyclox	60568-05-0	5	92.1	9.0	92.1	6.0	85.9	5.2	5	85.3	3.4	74.5	4.2	74.6	4.9
372	Haloxyfop-2-ethoxyethyl	87237-48-7	5	75.3	13.4	109.9	5.5	113.6	4.4	5	104.5	3.4	111.7	3.4	106.5	2.8
373	Hexaconazole	79983-71-4	5	91.4	6.3	108.3	3.9	104.0	1.8	5	87.1	3.7	94.0	3.7	88.5	1.7
374	Hexythiazox	78587-05-0	5	83.1	13.6	111.0	6.6	112.4	4.0	5	108.5	5.2	102.3	6.8	96.6	3.5
375	Imazamox	114311-32-9	5	3.7	102.4	0.5	45.1	0.4	11.8	5	15.1	49.3	0.7	130.8	0.6	139.9
376	Imazapic	104098-48-8	5	41.5	63.5	3.9	85.7	0.4	11.8	5	13.6	37.1	1.4	12.6	0.5	0.0
377	Imibenconazole	86598-92-7	5	71.7	5.8	108.3	6.2	103.1	2.8	5	93.9	5.3	90.7	6.8	84.9	2.6

(Continued)

APPENDIX TABLE 1.3L.1 Results of Nontarget Screening for 556 Pesticides in Four Kinds of Tea by LC-Q-TOF/MS (cont.)

	Name	CAS number	Puer tea Screening limits (µg/kg)	Puer 10 µg/kg AVE	Puer 10 µg/kg RSD (%) (n=5)	Puer 50 µg/kg AVE	Puer 50 µg/kg RSD (%) (n=5)	Puer 200 µg/kg AVE	Puer 200 µg/kg RSD (%) (n=5)	Oolong tea Screening limits (µg/kg)	Oolong 10 µg/kg AVE	Oolong 10 µg/kg RSD (%) (n=5)	Oolong 50 µg/kg AVE	Oolong 50 µg/kg RSD (%) (n=5)	Oolong 200 µg/kg AVE	Oolong 200 µg/kg RSD (%) (n=5)
378	Isazofos	42509-80-8	5	94.5	11.0	112.4	3.4	114.3	1.2	5	100.1	6.4	98.4	5.8	96.3	1.6
379	Isocarbamid	30979-48-7	5	72.6	2.7	76.6	4.2	80.0	5.0	5	191.9	13.7	87.1	7.3	82.0	5.0
380	isocarbophos	24353-61-5	5	73.9	9.4	101.0	3.6	112.2	7.2	5	89.8	5.2	84.4	6.0	119.5	4.0
381	Isoxathion	18854-01-8	5	102.2	18.5	117.9	3.1	119.2	3.2	5	106.4	4.6	107.3	3.1	106.3	2.5
382	Kadethrin	58769-20-3	5	22.7	22.3	76.4	15.4	78.0	5.1	5	70.6	3.7	74.7	7.6	76.6	3.3
383	Lactofen	77501-63-4	NA	NA	NA	NA	NA	NA	NA	NA	NA	NA	NA	NA	NA	NA
384	Malaoxon	1634-78-2	5	96.3	5.2	95.0	1.4	94.5	4.8	5	96.3	2.9	86.0	1.5	83.7	4.2
385	Mepiquat	15302-91-7	5	42.5	39.0	37.1	26.0	296.8	9.2	5	36.6	37.3	32.0	22.1	41.6	8.3
386	Metalaxyl	57837-19-1	5	100.1	6.0	98.0	2.8	97.3	4.2	5	80.4	2.4	78.2	2.3	85.3	4.2
387	Methabenzthiazuron	18691-97-9	5	89.1	6.2	93.3	1.4	92.6	4.1	5	90.1	2.5	81.5	1.0	82.0	4.1
388	Methiocarb-sulfoxide	2635-10-1	5	85.3	9.9	70.9	5.8	61.8	7.0	5	95.8	10.4	70.5	5.6	70.5	2.8
389	Methoxyfenozide	161050-58-4	5	82.3	13.3	116.6	4.4	111.1	2.1	5	99.1	13.4	95.8	4.7	93.1	3.5
390	Metobromuron	3060-89-7	5	90.7	9.2	98.5	1.7	96.5	2.8	5	117.9	5.4	106.4	2.8	86.8	3.5
391	Metolachlor	51218-45-2	5	96.5	5.4	107.5	1.3	104.3	4.1	5	94.1	2.2	94.7	3.2	91.8	3.9
392	Metolcarb	1129-41-5	NA	NA	NA	NA	NA	NA	NA	5	NA	NA	NA	NA	NA	NA
393	Metosulam	139528-85-1	5	NA	NA	NA	NA	NA	NA	NA	NA	NA	NA	NA	NA	NA
394	Metoxuron	19937-59-8	5	92.3	5.6	90.0	2.8	90.2	1.4	5	119.1	3.2	76.3	2.4	80.9	1.4
395	Metsulfuron-Methyl	74223-64-6	5	NA	NA	NA	NA	NA	NA	5	5.5	11.6	1.9	45.0	1.8	36.1
396	Mexacarbate	315-18-4	5	90.7	13.6	81.1	4.6	79.9	8.5	5	92.7	4.5	118.3	8.4	74.4	4.3
397	Napropamide	15299-99-7	5	100.8	5.7	105.7	3.0	103.7	3.4	5	92.3	1.5	90.5	1.6	91.2	3.0
398	Naptalam	132-66-1	NA	NA	NA	NA	NA	NA	NA	NA	NA	NA	NA	NA	NA	NA
399	Norflurazon	27314-13-2	5	80.9	4.6	99.1	3.7	100.5	1.2	5	99.4	2.5	89.5	1.9	89.3	2.0
400	Octhilinone	26530-20-1	5	75.2	4.2	83.7	4.7	73.8	5.8	5	80.0	3.2	75.3	4.1	70.5	5.5
401	Omethoate	1113-02-6	5	115.3	6.1	83.2	4.1	81.1	2.1	5	119.2	2.3	79.8	3.2	70.6	2.8
402	Orbencarb	34622-58-7	5	109.5	4.8	118.3	1.5	119.9	0.9	5	99.0	3.8	98.4	2.4	96.7	2.9
403	Oxamyl	23135-22-0	5	99.1	10.7	72.2	3.8	77.0	18.5	5	27.0	10.3	43.0	9.4	76.7	40.9
404	Oxamyl-Oxime	30558-43-1	5	77.0	6.6	89.3	10.9	93.2	10.2	5	130.0	5.6	91.7	6.0	80.2	12.9
405	oxydemeton-methyl	301-12-2	5	157.3	15.9	77.2	12.9	78.3	6.3	5	89.0	8.6	74.4	9.3	71.9	7.2
406	Pebulate	1114-71-2	5	105.3	11.7	102.1	12.0	108.8	5.4	5	113.8	5.7	103.9	11.7	91.2	6.2
407	Pencycuron	66063-05-6	5	99.0	13.2	109.9	2.2	111.2	2.6	5	100.3	5.9	102.1	2.3	96.4	2.0
408	Pentanochlor	2307-68-8	5	99.8	11.2	114.3	2.4	109.6	5.0	5	109.4	3.4	108.3	1.3	94.6	2.8
409	Phorate-Sulfoxide	2588-03-6	5	98.5	4.5	98.6	2.2	99.3	3.7	5	80.6	2.8	85.5	2.8	87.3	4.0

#	Compound	CAS														
410	Phosfolan	947-02-4	5	77.5	4.8	86.5	2.6	87.4	4.7	5	104.0	2.6	81.3	2.1	81.6	4.3
411	Phosmet	732-11-6	NA	NA	NA	NA	NA	NA	NA	NA	NA	NA	NA	NA	NA	NA
412	Phthalic Acid, Benzyl Butyl Ester	85-68-7	5	20.7	42.8	102.2	12.4	93.9	9.7	5	105.4	4.9	94.7	10.7	92.5	3.8
413	Phthalic Acid, Bis-Cyclohexyl Ester	84-61-7	5	45.0	44.3	97.9	3.9	90.5	8.8	5	107.1	9.2	78.1	17.8	91.9	3.1
414	Picoxystrobin	117428-22-5	5	107.3	15.3	117.1	2.8	114.4	1.0	5	82.3	1.6	88.8	1.5	95.8	2.3
415	Primisulfuron-Methyl	86209-51-0	NA	NA	NA	NA	NA	NA	NA	NA	NA	NA	NA	NA	NA	NA
416	Propaquizafop	111479-05-1	5	74.4	8.9	98.4	3.6	98.6	3.2	5	97.4	10.5	98.9	2.8	90.6	2.7
417	Propoxycarbazone	145026-81-9	5	NA	NA	NA	NA	NA	NA	5	13.5	77.3	1.6	113.4	0.7	NA
418	Pyrazolynate	58011-68-0	5	103.0	10.2	99.4	5.2	108.9	4.7	5	89.9	8.1	87.0	6.1	93.8	4.8
419	Pyrazophos	13457-18-6	5	102.6	16.4	119.9	2.1	108.9	5.4	5	100.9	18.8	99.0	2.3	90.1	4.8
420	Pyrazosulfuron-Ethyl	93697-74-6	5	NA	NA	NA	NA	NA	NA	5	0.3	NA	NA	NA	NA	NA
421	Pyrazoxyfen	71561-11-0	5	98.2	6.3	103.0	2.8	106.1	2.1	5	111.2	4.2	102.9	1.8	91.9	2.2
422	Pyridaben	96489-71-3	5	734.4	6.4	111.7	11.8	107.6	3.3	5	96.6	4.2	103.1	10.7	97.7	2.5
423	Pyridalyl	179101-81-6	5	6.2	60.0	94.1	9.8	99.7	11.6	5	153.4	9.5	118.2	7.1	75.3	8.9
424	Pyrifenox	88283-41-4	5	61.9	4.9	99.8	12.9	89.4	8.6	5	73.4	14.0	107.6	4.7	87.0	3.6
425	Pyrimidifen	105779-78-0	5	71.8	9.7	90.8	6.0	92.6	3.8	5	71.8	8.1	73.6	4.9	83.0	3.6
426	Pyriproxyfen	95737-68-1	5	75.9	5.3	111.2	10.9	109.0	4.0	5	112.0	2.1	106.3	5.5	100.9	2.3
427	Pyroquilon	57369-32-1	5	73.8	2.4	81.1	1.6	82.0	1.1	5	115.2	3.9	74.7	3.4	77.7	2.7
428	Quinclorac	84087-01-4	5	NA	NA	NA	NA	NA	NA	5	4.1	33.8	0.7	50.5	0.7	48.1
429	Quinmerac	90717-03-6	5	NA	NA	NA	NA	NA	NA	5	NA	NA	NA	NA	NA	NA
430	Quinoxyfen	124495-18-7	5	73.6	12.3	101.3	4.2	96.5	4.1	5	98.5	7.9	98.3	4.3	87.8	4.4
431	Quizalofop	76578-12-6	5	NA	NA	NA	NA	NA	NA	5	26.5	97.8	14.9	69.0	1.6	55.5
432	Rimsulfuron	122931-48-0	NA	NA	NA	NA	NA	NA	NA	20	NA	NA	NA	NA	3.4	52.5
433	Rotenone	83-79-4	5	95.2	4.4	112.9	3.4	114.2	6.2	5	113.6	2.6	96.2	2.2	94.5	5.3
434	Sebuthylazine-desethyl	37019-18-4	5	88.7	4.4	87.5	3.8	83.3	5.7	5	140.3	4.7	84.4	2.3	71.3	5.3
435	Tebufenpyrad	119168-77-3	5	94.5	11.0	115.0	4.0	108.9	2.6	5	92.3	7.7	87.4	3.6	90.8	3.1
436	Tepraloxydim	149979-41-9	5	79.7	13.2	91.4	3.8	98.9	4.1	5	96.2	2.7	89.2	5.7	81.3	6.5
437	Terbuthylazine	5915-41-3	5	88.5	11.9	110.9	2.0	104.6	3.7	5	107.7	3.0	105.1	2.3	91.3	3.5
438	Terbutryn	886-50-0	5	103.2	6.3	109.9	1.8	110.6	1.3	5	90.8	1.4	88.9	1.8	96.9	2.0
439	Tetramethrin	7696-12-0	5	12.3	35.4	115.5	12.2	99.7	6.6	5	113.3	15.2	100.8	13.1	99.8	4.7
440	Thenylchlor	96491-05-3	5	70.5	7.8	112.9	5.2	107.3	13.5	5	73.6	3.3	82.0	3.8	96.5	10.3
441	Thiabendazole	148-79-8	5	4.6	90.0	6.9	29.0	17.0	24.0	5	5.8	96.2	6.8	32.2	18.4	22.3
442	Thiazafluron	25366-23-8	5	100.6	8.8	87.3	1.9	82.7	2.9	5	139.0	8.1	79.3	3.2	74.6	2.1
443	Thidiazuron	51707-55-2	5	11.2	45.4	3.5	12.0	1.0	15.1	5	16.9	7.7	3.1	4.8	0.8	13.6
444	Thifensulfuron-Methyl	79277-27-3	5	0.3	75.0	0.1	40.0	0.1	NA	5	0.6	21.9	0.1	39.1	NA	NA

(Continued)

APPENDIX TABLE 1.3L.1 Results of Nontarget Screening for 556 Pesticides in Four Kinds of Tea by LC-Q-TOF/MS (cont.)

	Name	CAS number	Puer tea Screening limits (µg/kg)	10 µg/kg AVE	RSD (%) (n = 5)	50 µg/kg AVE	RSD (%) (n = 5)	200 µg/kg AVE	RSD (%) (n = 5)	Oolong tea Screening limits (µg/kg)	10 µg/kg AVE	RSD (%) (n = 5)	50 µg/kg AVE	RSD (%) (n = 5)	200 µg/kg AVE	RSD (%) (n = 5)
445	Thiofanox	39196-18-4	5	79.0	2.7	89.7	10.0	77.4	16.4	5	67.7	96.9	51.0	16.8	78.8	25.0
446	Thiofanox-Sulfone	39184-59-3	5	93.7	6.8	104.3	5.1	105.6	9.4	5	32.3	7.2	87.1	13.2	115.1	43.8
447	Thiophanate-Methyl	23564-05-8	10	0.7	46.9	21.6	54.9	31.1	56.3	5	4.1	62.8	14.7	62.5	29.1	76.2
448	Thiram	137-26-8	NA	NA	NA	NA	NA	NA	NA	NA	NA	NA	NA	NA	NA	NA
449	Tiocarbazil	36756-79-3	5	49.6	30.9	106.3	11.7	101.1	5.0	5	92.7	2.6	95.2	7.6	94.1	1.7
450	Triapenthenol	76608-88-3	5	86.3	7.0	108.6	4.3	105.8	3.1	5	103.7	2.4	103.5	3.7	89.1	3.5
451	Triasulfuron	82097-50-5	5	NA	NA	NA	NA	NA	NA	5	2.8	NA	0.2	34.6	NA	NA
452	Triflumizole	99387-89-0	5	90.8	6.7	109.0	9.5	109.3	0.5	5	78.9	3.7	82.7	10.9	86.3	2.9
453	Triflusulfuron-Methyl	126535-15-7	NA	NA	NA	NA	NA	NA	NA	20	NA	NA	92.7	6.3	90.4	16.7
454	Trinexapac-Ethyl	95266-40-3	5	69.6	30.1	59.9	14.1	61.1	5.0	5	50.0	16.1	55.2	11.9	56.3	7.7
455	Triphenyl phosphate	115-86-6	5	114.4	6.4	114.4	1.9	110.7	4.0	5	104.2	3.2	98.9	2.1	94.0	3.7
456	Uniconazole	83657-22-1	5	91.4	7.0	108.2	3.4	102.3	2.5	5	99.3	4.8	99.6	1.0	83.8	2.5
457	Abamectin	71751-41-2	5	NA	NA	71.6	4.6	79.9	3.0	5	74.3	9.4	84.2	5.8	103.8	6.3
458	Acephate	30560-19-1	20	NA	NA	17.3	34.1	53.2	8.1	5	50.1	14.2	50.5	12.8	52.2	3.0
459	Acetamiprid-N-Desmethyl	190604-92-3	5	62.4	9.7	75.1	2.3	70.1	3.4	5	NA	NA	62.5	8.1	71.6	6.4
460	Aclonifen	74070-46-5	NA	NA	NA	NA	NA	NA	NA	NA	NA	NA	NA	NA	NA	NA
461	Aldicarb-sulfoxide	1646-87-3	5	242.3	29.9	175.4	29.7	200.1	42.8	5	103.6	14.0	72.1	7.4	70.8	4.6
462	Allidochlor	93-71-0	5	32.3	13.6	41.4	23.1	71.8	12.6	5	119.2	2.1	70.4	3.0	70.0	3.6
463	Ametoctradin	865318-97-4	5	81.0	4.1	75.4	6.3	87.1	1.2	5	71.2	8.3	86.3	4.2	80.5	2.6
464	Amicarbazone	129909-90-6	5	75.7	8.2	79.5	5.3	74.2	6.9	5	53.4	9.3	70.1	4.4	64.3	2.8
465	Beflubutamid	113614-08-7	5	91.0	11.5	73.7	8.7	83.7	2.8	5	88.1	7.2	96.9	6.5	78.4	2.6
466	Benthiavalicarb-Isopropyl	177406-68-7	5	87.6	4.9	74.3	10.5	83.1	2.1	5	91.7	3.9	88.8	3.1	74.1	2.4
467	Benzofenap	82692-44-2	5	85.3	6.9	76.7	8.3	85.7	2.4	5	85.8	9.4	86.4	2.2	79.7	1.3
468	Boscalid	188425-85-6	5	84.8	6.6	75.2	4.3	76.5	3.3	5	73.9	12.6	86.1	3.2	84.6	6.1
469	Bromophos-Ethyl	4824-78-6	NA	NA	NA	NA	NA	NA	NA	NA	NA	NA	NA	NA	NA	NA
470	Butylate	2008-41-5	5	91.5	17.8	74.5	5.3	82.6	8.2	5	105.9	12.4	95.1	8.7	80.6	6.2
471	Chlorantraniliprole	500008-45-7	5	77.5	7.1	70.5	2.9	71.3	2.3	5	81.6	2.0	80.5	4.2	70.0	0.4
472	Chlorpyrifos-methyl	5598-13-0	NA	NA	NA	NA	NA	NA	NA	100	NA	NA	NA	NA	71.3	15.1
473	Clomeprop	84496-56-0	NA	NA	NA	NA	NA	NA	NA	NA	NA	NA	NA	NA	NA	NA
474	Coumaphos	56-72-4	5	97.6	4.6	75.1	9.8	77.7	5.0	5	82.1	16.1	92.2	7.9	74.6	2.0
475	Crotoxyphos	7700-17-6	5	92.6	5.6	73.7	11.4	90.4	2.6	5	81.3	4.4	84.8	5.2	74.0	2.8
476	Cyazofamid	120116-88-3	10	83.8	11.4	79.5	5.5	83.4	4.4	5	83.4	11.9	104.4	6.5	78.4	3.8
477	Dibutyl succinate	141-03-7	NA	NA	NA	NA	NA	NA	NA	NA	NA	NA	NA	NA	NA	NA

478	Diflubenzuron	35367-38-5	5	102.4	5.6	80.3	5.2	87.8	3.5	5	77.7	14.9	91.2	7.0	73.1	2.6
479	Dimefox	115-26-4	5	37.1	46.7	25.1	61.3	35.7	58.6	5	148.9	46.4	58.5	19.1	70.8	6.8
480	Dimethenamid-P	163515-14-8	5	82.1	7.6	82.7	9.0	79.8	3.0	5	78.9	4.7	83.0	5.5	75.3	3.0
481	Dimethylvinphos (Z)	67628-93-7	5	87.8	5.3	90.5	11.2	86.2	3.2	5	81.8	2.7	88.8	3.7	79.9	2.9
482	Dimetilan	644-64-4	5	72.0	9.1	70.1	4.2	78.6	6.6	5	223.0	3.6	75.0	4.4	70.2	2.7
483	Dimoxystrobin	149961-52-4	5	83.1	5.0	72.3	9.2	119.7	1.0	5	82.1	3.5	85.9	4.2	82.7	1.8
484	Dinitramine	29091-05-2	10	82.7	11.2	74.6	9.7	89.3	3.4	5	92.8	7.7	101.9	3.9	80.8	3.5
485	Fensulfothion-oxon	6552-21-2	5	75.4	6.3	70.6	6.2	78.1	2.9	5	117.6	2.7	72.5	3.6	75.9	2.2
486	Fensulfothion-sulfone	14255-72-2	5	94.3	6.4	84.7	10.7	88.9	2.3	5	81.0	10.5	90.9	4.6	82.1	3.9
487	Fenthion-oxon	6552-12-1	5	79.8	3.7	77.9	10.5	87.3	2.2	5	98.7	3.7	82.6	6.2	75.9	3.7
488	Fenthion-oxon-sulfone	14086-35-2	5	70.2	5.1	70.2	6.8	72.7	3.1	5	131.5	3.5	61.1	5.3	65.5	3.0
489	Fenthion-oxon-sulfoxide	6552-13-2	5	75.5	9.5	71.6	5.6	73.2	2.0	5	148.9	2.2	79.3	6.6	79.9	3.3
490	Fentrazamide	158237-07-1	5	109.6	4.9	70.5	7.9	77.8	6.2	5	78.4	3.4	105.7	17.3	82.1	9.7
491	Fluazifop-P-Butyl	79241-46-6	5	94.9	2.5	75.2	11.3	96.1	2.5	5	94.1	6.8	113.5	9.9	90.2	5.3
492	Flubendiamide	272451-65-7	5	88.7	4.3	75.6	3.6	88.4	2.0	5	96.4	9.1	100.3	10.2	86.1	3.6
493	Flucarbazone	145026-88-6	5	NA	NA	NA	NA	NA	NA	5	NA	NA	NA	NA	NA	NA
494	Flufenpy-Ethyl	188489-07-8	5	96.0	4.9	90.7	11.5	83.6	3.4	10	2.9	5.6	86.7	5.6	78.5	1.4
495	Flumorph	211867-47-9	5	75.6	7.5	71.2	2.9	72.8	2.6	5	73.3	4.9	81.2	4.8	72.8	3.4
496	Fluopicolide	239110-15-7	5	95.7	6.1	81.8	8.7	86.8	3.0	5	85.8	3.4	90.5	4.2	77.8	2.4
497	Fluopyram	658066-35-4	5	92.1	5.5	78.8	9.4	87.2	2.5	5	89.7	3.3	90.9	3.9	80.2	2.2
498	Fluoxastrobin	361377-29-9	5	77.3	5.0	71.5	7.8	77.3	1.6	5	82.3	3.8	82.6	4.1	78.7	1.8
499	Flurprimidol	56425-91-3	5	79.2	6.1	82.4	7.9	77.2	2.3	5	100.8	5.3	89.4	5.0	74.6	3.4
500	Fluxapyroxad	907204-31-3	5	81.0	6.8	72.6	9.7	72.8	1.8	5	96.1	1.2	91.6	4.3	73.3	1.9
501	Halofenozide	112226-61-6	5	93.3	6.8	73.5	7.2	86.6	1.9	5	110.5	6.7	105.2	15.0	100.7	6.2
502	Hydramethylnon	67485-29-4	5	NA	NA	NA	NA	NA	NA	5	0.8	61.2	0.1	NA	NA	NA
503	Imidacloprid-Urea	120868-66-8	5	66.8	7.1	66.0	12.1	65.3	2.1	5	177.6	3.5	67.3	5.7	61.5	1.2
504	Ipconazole	125225-28-7	5	97.1	5.9	74.7	8.2	85.2	4.0	5	73.4	10.2	79.9	3.9	76.1	2.7
505	Isoxaflutole	141112-29-0	NA	NA	NA	NA	NA	NA	NA	NA	NA	NA	NA	NA	NA	NA
506	Mandipropamid	374726-62-2	5	86.9	7.7	71.5	7.8	74.9	2.7	5	76.3	3.4	84.1	1.8	72.0	2.0
507	Methidathion	950-37-8	5	86.7	11.7	113.7	4.9	91.5	4.1	5	81.4	23.8	92.4	12.1	95.4	1.8
508	Metominostrobin-(E)	133408-50-1	5	80.6	4.8	70.4	9.4	75.0	0.9	5	85.2	3.5	81.3	2.2	75.5	1.9
509	Metrafenone	220899-03-6	5	90.6	4.8	77.2	9.9	87.0	2.3	5	99.4	5.1	95.6	7.5	80.6	3.1
510	Molinate	2212-67-1	10	90.7	15.6	70.4	8.0	77.0	10.1	5	92.5	17.4	101.1	4.9	72.0	3.6
511	Monolinuron	1746-81-2	5	80.2	3.3	72.1	7.9	75.5	2.0	5	109.5	5.2	99.6	4.5	76.0	3.1
512	Naproanilide	52570-16-8	5	95.5	3.0	79.6	6.6	93.0	3.2	5	105.3	8.0	106.0	2.8	86.7	2.7
513	Nicosulfuron	111991-09-4	5	418.2	44.0	33.3	47.4	4.7	43.5	5	NA	NA	23.8	20.8	5.6	20.2
514	Nitralin	4726-14-1	10	82.2	16.6	73.9	9.4	85.6	3.7	5	94.1	16.1	89.4	9.2	76.9	3.8

(Continued)

APPENDIX TABLE 1.3L.1 Results of Nontarget Screening for 556 Pesticides in Four Kinds of Tea by LC-Q-TOF/MS (cont.)

	Name	CAS number	Puer tea							Oolong tea						
			Screening limits (µg/kg)	10 µg/kg AVE	10 µg/kg RSD (%) (n=5)	50 µg/kg AVE	50 µg/kg RSD (%) (n=5)	200 µg/kg AVE	200 µg/kg RSD (%) (n=5)	Screening limits (µg/kg)	10 µg/kg AVE	10 µg/kg RSD (%) (n=5)	50 µg/kg AVE	50 µg/kg RSD (%) (n=5)	200 µg/kg AVE	200 µg/kg RSD (%) (n=5)
515	Orthosulfamuron	213464-77-8	NA	NA	NA	NA	NA	NA	NA	NA	NA	NA	NA	NA	NA	NA
516	Oxaziclomefone	153197-14-9	5	88.3	2.7	73.0	3.5	93.2	2.8	5	70.5	7.5	91.2	16.0	84.8	8.7
517	Oxyfluorfen	42874-03-3	50	NA	NA	NA	NA	NA	NA	NA	NA	NA	NA	NA	NA	NA
518	Pendimethalin	40487-42-1	10	86.3	12.7	81.4	2.9	104.2	2.3	5	86.8	11.5	109.2	13.6	96.6	5.9
519	Penoxsulam	219714-96-2	5	1.1	60.6	2.4	15.0	0.3	94.3	5	NA	NA	NA	NA	NA	NA
520	Phorate	298-02-2	100	NA	NA	NA	NA	88.6	6.1	20	NA	NA	98.0	11.1	88.9	3.8
521	Picaridin	119515-38-7	5	79.8	7.3	71.9	10.2	77.7	4.2	5	88.0	3.2	81.3	2.7	72.1	2.6
522	Pinoxaden	243973-20-8	5	58.2	3.5	73.9	6.9	70.1	3.5	5	58.3	5.2	62.2	8.0	71.4	8.0
523	Pirimiphos-methyl-N-desethyl	67018-59-1	5	79.8	6.0	75.9	5.4	87.8	3.4	5	87.6	6.5	83.0	7.2	78.9	1.0
524	Prallethrin	23031-36-9	50	NA	NA	NA	NA	105.4	8.3	200	NA	NA	NA	NA	NA	NA
525	Promecarb	2631-37-0	5	84.8	7.8	84.7	9.1	85.1	4.3	5	86.9	8.1	85.3	5.8	79.7	2.1
526	Propetamphos	31218-83-4	50	NA	NA	79.4	9.9	90.3	3.8	5	NA	NA	101.7	5.0	85.1	2.6
527	Proquinazid	189278-12-4	5	43.0	17.0	52.4	26.7	45.1	7.2	5	78.4	8.1	97.3	8.1	73.9	11.6
528	Prothioconazole	178928-70-6	20	NA	NA	54.6	48.2	11.0	19.4	10	NA	NA	NA	NA	NA	NA
529	Pyraflufen	129630-17-7	10	NA	NA	NA	NA	NA	NA	5	NA	NA	NA	NA	NA	NA
530	Pyrasulfotole	365400-11-9	5	1.4	55.5	1.1	90.9	0.1	0.0	5	14.7	32.2	1.5	20.5	0.3	34.4
531	Pyriftalid	135186-78-6	5	86.5	5.8	75.2	2.8	85.5	2.6	5	76.1	3.6	80.3	4.1	77.2	2.3
532	Pyriminobac-Methyl (Z)	147411-70-9	5	77.0	4.9	71.6	9.6	74.8	2.1	5	80.4	3.0	81.6	4.4	77.6	2.1
533	Quizalofop-P-Ethyl	100646-51-3	5	96.7	4.2	77.3	6.4	91.9	2.5	5	107.5	5.7	99.9	6.1	82.5	2.8
534	Resmethrin	10453-86-8	20	NA	NA	70.4	42.3	86.3	13.6	10	NA	NA	NA	NA	108.7	13.5
535	RH 5849	112225-87-3	20	NA	NA	64.6	25.0	62.6	5.5	10	NA	NA	82.0	2.0	70.9	3.8
536	Saflufenacil	372137-35-4	5	NA	NA	NA	NA	NA	NA	5	0.9	96.1	0.4	59.5	0.9	96.0
537	Siduron	1982-49-6	5	89.1	7.2	74.1	11.0	86.5	2.1	5	93.4	1.5	88.9	4.9	70.7	3.5
538	S-Metolachlor	87392-12-9	5	86.8	4.4	79.8	5.6	79.1	12.7	5	101.9	5.5	94.3	2.5	74.7	2.5
539	Spinetoram	187166-40-1	5	70.8	5.5	72.5	5.9	70.0	1.8	5	77.3	2.7	83.5	3.8	72.1	4.0
540	Spirotetramat	203313-25-1	5	77.3	7.0	72.6	7.5	70.5	3.2	5	64.3	8.4	65.8	8.7	50.3	12.3
541	Sulcotrione	99105-77-8	5	NA	NA	NA	NA	NA	NA	5	1.6	21.7	0.2	69.3	0.2	22.2
542	Sulfallate	95-06-7	50	NA	NA	61.9	21.4	90.0	4.1	5	NA	NA	103.2	9.0	86.0	6.5
543	Sulfoxaflor	946578-00-3	5	63.7	7.0	70.7	11.7	70.0	3.5	5	221.7	4.3	58.2	7.1	74.0	2.9
544	Tembotrione	335104-84-2	5	NA	NA	NA	NA	NA	NA	5	NA	NA	NA	NA	0.1	0.0
545	Terbufos	13071-79-9	20	NA	NA	81.5	4.6	95.9	3.2	5	104.1	9.1	96.2	7.5	84.1	2.8
546	Terbufos-Oxon-Sulfone	56070-15-6	NA	NA	NA	NA	NA	NA	NA	NA	NA	NA	NA	NA	NA	NA
547	Terbufos-Sulfone	56070-16-7	10	93.1	12.7	76.9	7.8	84.8	5.0	5	82.0	5.5	84.1	4.6	74.4	2.6

548	Thiabendazole-5-Hydroxy	948-71-0	5	1.5	112.9	4.6	150.6	0.3	95.2	5	0.3	54.5	0.3	84.9	NA	NA
549	Thiencarbazone-Methyl	317815-83-1	5	NA	NA	NA	NA	NA	NA	5	0.1	0.0	NA	NA	NA	NA
550	Thiocyclam	31895-21-3	20	NA	NA	83.4	35.7	34.8	16.1	5	43.3	51.6	45.0	6.0	39.1	5.6
551	Tolclofos-Methyl	57018-04-9	5	NA	NA	39.2	75.8	28.6	41.3	5	77.3	0.9	88.9	13.7	91.4	15.4
552	Triallate	2303-17-5	10	83.8	9.1	74.1	14.2	96.2	2.2	5	115.6	5.6	109.6	7.3	89.9	9.8
553	Tribenuron-Methyl	101200-48-0	NA	NA	NA	NA	NA	NA	NA	NA	NA	NA	NA	NA	NA	NA
554	Validamycin	37248-47-8	100	NA	NA	NA	NA	15.6	49.6	5	192.0	44.4	14.4	56.2	2.0	73.0
555	Valifenalate	283159-90-0	5	84.0	6.2	74.4	10.1	79.7	2.1	5	76.1	1.6	85.2	1.6	71.5	2.6
556	Vamidothion Sulfone	70898-34-9	5	72.6	7.1	72.6	2.4	70.6	7.1	5	113.3	1.3	70.1	2.3	70.5	4.8

NA: No response or recoveries/RSDs not met the recovery-RSD standard.

Chapter 1.4

A Study of Efficiency Evaluation for Nontarget Screening of 1050 Pesticide Residues by GC–Q–TOF/MS and LC–Q–TOF/MS

1.4.1 INTRODUCTION

With the broad mass range, high scan rate, high sensitivity, high resolution, and accurate mass detection, time-of-flight mass spectrum has become a reliable method for qualitative analysis of trace compounds in complex matrices. Time-of-flight mass spectrum provides an accurate and efficient confirmatory analyzing method when coupled with chromatographic technology. It plays an increasingly important role in pesticide multiresidue high-throughput and rapid screening, unknown compound screening, compound separating, and analyzing from a complex matrix. Recently, gas chromatography quadrupole time-of-flight mass spectrometry (GC–Q–TOF/MS) and liquid chromatography quadrupole time-of-flight mass spectrometry (LC–Q–TOF/MS) have been the main hyphenated techniques within Q–TOF. From earlier chapters, it is obvious that on account of the performance and principles of the equipment, there is a big difference in pesticide screening between GC–Q–TOF/MS and LC–Q–TOF/MS, and that is the primary subject for discussion in this chapter. Based on accurate mass databases and MS/MS libraries, including 494 pesticides in GC–Q–TOF/MS and 556 pesticides in LC–Q–TOF/MS, a nontarget and high throughput method by GC–Q–TOF/MS combined with LC–Q–TOF/MS has been successfully applied to the screening of 765 pesticide residues, which is more than 30% higher than GC–Q–TOF/MS or LC–Q–TOF/MS running alone.

Screening limits, recoveries, and RSDs (relative standard deviations) of the hyphenated techniques were assessed by spiking a standard mixture at six concentrations. The results showed that 235–270 and 352–402 pesticides were suitable for analyzing by GC–Q–TOF/MS and LC–Q–TOF/MS, respectively. Meanwhile, these two screening methods had the same performance for 272 pesticides. The developed hyphenated method was successfully applied to the determination of 1050 pesticides in tea.

1.4.2 EXPERIMENTAL

1.4.2.1 Reagents and Materials

Pesticides standards (≥95%) were purchased from Dr. Ehrenstorfer (Augsburg, Germany). All HPLC grade reagents, including acetonitrile (ACN), toluene, and methanol (MeOH) were obtained from Fisher Scientific (NJ, USA). Formic acid was obtained from Duksan Pure Chemicals (Ansan, Korea). Milli-Q water was prepared with the Millipore water purification system (MA, USA). Na_2SO_4 and NH_4OAc were purchased from Beijing Chemical Reagent Factory (Beijing, China). Triple phases of tea (TPT) solid phase extraction (SPE) cartridge was purchased from Agela Technologies Inc. (Tianjin, China).

1.4.2.2 Standard Solution Preparation

Individual stock standard solutions of the pesticides were prepared at 1000 mg/L in MeOH, toluene or acetone, depending on the physicochemical properties of the substance or requirement of equipment. Mixed working solutions of the pesticides were prepared by dilution of the individual stock standard solutions to 10 mg/L in MeOH. Pesticides were divided into 16 groups, named as A1–H1 for LC–Q–TOF/MS analyzing and A2–H2 for GC–Q–TOF/MS analyzing. All the standard solutions were stored in darkness at 4°C.

1.4.2.3 Sample Collection and Pretreatment

All tea samples were obtained from supermarkets in Beijing, smashed in blenders, and sieved for analyzing.

A 2 g portion of the sample was accurately weighed (precision of 0.01 g), homogenized in 15 mL ACN under 15,000 rpm for 1 min, and then centrifuged at 4200 rpm for 5 min. The supernatant was decanted into a 150-mL pear-shaped flask and the residue was extracted one more time using 15 mL ACN. The combined extracts were concentrated to 2 mL by a rotary evaporator (40, 40 rpm) and then purified by a TPTSPE cartridge. The column added 2 cm of hydrous sodium sulfate, and was washed with 5 mL ACN/toluene (3:1, v/v). A concentrated sample was loaded and washed by 2 mL ACN/toluene (3:1, v/v) for three times. Then, the column was eluted by 25 mL ACN/toluene (3:1, v/v). The effluent was collected by an 80-mL pear-shaped flask and concentrated (40, 80 rpm) to 0.5 mL. For GC–Q–TOF/MS analyzing, the concentrated eluate was spiked with 40 μL heptachlor epoxide solution as an internal standard, and evaporated to dryness by nitrogen under 35. The residue was resuspended in 1 mL hexane and filtered through 0.2 μm filter for GC–Q–TOF/MS analyzing. Before injection into the LC–Q–TOF/MS system, the concentrated eluate was dried under nitrogen at 35 and resuspended in 1 mL ACN/H$_2$O (3:2, v/v) and filtered through 0.2 μm filter for LC–Q–TOF/MS analyzing.

1.4.2.4 Instrumental Analysis

1.4.2.4.1 LC–Q–TOF/MS Conditions

Pesticides were analyzed on an Agilent 1290 infinity UHPL C system (Santa Clara, CA, USA) for separation and an Agilent 6550 Quadrupole Time-of-Flight LC-MS system with electrospray Jet Stream technology for detection. Chromatographic separation utilized a reversed-phase column (ZORBAX SB-C18 column, 2.1 mm × 100 mm, 3.5 μm; Agilent Technologies, Santa Clara, CA, USA). The mobile phase consisted of 5 mM ammonium acetate and 0.1% formic acid in water (A) and acetonitrile (B). The gradient was: 0–3 min, 1%–30% B; 3–6 min, 30%–40% B; 6–9 min, 40% B; 9–15 min, 40%–60% B; 15–19 min, 60%–90% B; 19–23 min, 90% B; and 23–23.01 min, 90%–1% B, followed by a reequilibration time of 4 min. Chromatography employed an injection volume of 10 μL, a flow rate of 0.4 mL/min, and a column temperature of 30°C. Over a 27-min run, MS spectra were acquired across the range 50–1600 *m/z* under a positive ion mode. The capillary voltage, cone voltage, and fragmentation voltage were 4000 V, 60 V, and 140 V, respectively. Drying gas temperature and sheath gas temperature were both 325°C. Drying gas flow, sheath gas flow, and nebulizer gas pressure were 10 L/min, 11 L/min, and 40 psi, respectively. The TOF mass spectrometer was calibrated routinely by an internal reference solution, which contained the accurate masses at *m/z* 121.0509 and 922.0098.

Data was acquired and analyzed by Agilent MassHunter Workstation Software (version B.07.00). Accurate mass databases were stored in Microsoft Excel (2010) by CSV format. Fragment mass databases were established by Agilent Mass-Hunter PCDL Manager (B.07.00).

1.4.2.4.2 GC–Q–TOF/MS Conditions

The extracted samples were determined with an Agilent 7890A GC system coupled to an Agilent 7200 high-resolution accurate mass Q–TOF system. Full-scan electron impact ionization data were acquired under the following conditions: solvent delay: 6 min, electron energy: 70 eV, transfer line temperature: 280°C, ion source temperature: 230°C, mass scan range: 50–600 *m/z*, scanrate:2 spectrum/s. Accurately, a 1 μL sample was injected into a GC–MS system with no split mode. The injector temperature was maintained at 280 °C. Chromatographic separation utilized a VF-1701 MS column (30 m × 0.25 mm, 0.25 μm). The oven temperature was set at 40°C, held for 1 min, and raised to 150°C at a rate of 30°C/min, increased from 150°C to 250°C at 5°C/min, and finally ramped at 10°C/min to 300°C and then held for 5 min. Helium was used as carrier gas at a constant flow rate of 1.2 mL/min.

Data was acquired and analyzed by Agilent MassHunter Workstation Software (version B.07.00).

1.4.2.5 Creation of the Pesticides Database

1.4.2.5.1 LC–Q–TOF/MS Database

An accurate mass of the 556 pesticides was acquired in TOF models by standard solutions in 1 mg/L. Positive identifications of pesticides were reported if the compound was detected by the Find by Molecular Formula data-mining algorithm with a mass error below 5 ppm and with a score that exceeded 90. Compound names, molecular features, accurate mass, and retention time were recorded in Microsoft Excel for pesticide confirmation.

For each pesticide, fragment data were obtained in target MS/MS mode by inputting accurate mass, retention time, and collision energies (5–80 V). Compound fragment data were imported into the PCDL by a CEF file. Eight fragment spectras for each pesticide with abundant information under different collision energies were recorded in PCDL for verification.

1.4.2.5.2 GC–Q–TOF/MS Database

For each pesticide, I μL individual standard solution under 1 mg/L was injected into the GC–Q–TOF/MS system, operating in full-spectrum acquisition mode. The retention time was recorded and compound information, including name, formula, accurate mass, and fragment date, were obtained from NIST library by a search library function. A fragment ion library, including 703 pesticides, was set up by importing this information into PCDL manager software.

1.4.2.6 Method Evaluation

Four kinds of tea were chosen for method evaluation by investigating the sensitivity, accuracy, and precision. Screening limits for 494 pesticides in GC–Q–TOF/MS and 556 pesticides in LC–Q–TOF/MS were assessed by spiking a standard mixture at six concentrations (5, 10, 20, 50, 100, and 200 μg/kg). Meanwhile, three spiked levels, including 10, 50, and 200 μg/kg, were analyzed for recoveries and precisions evaluation for GC–Q–TOF/MS, which was 10, 50, and 100 μg/kg for LC–Q–TOF/MS.

1.4.3 RESULTS AND DISCUSSION

1.4.3.1 Screening Limits Analysis

1.4.3.1.1 Screening Limits From Different Kinds of Tea

Four hundred and seventy eight pesticides were detected by GC–Q–TOF/MS, accounting for 96.8% of the larger group of 494 pesticides (Tables 1.4.1 and 1.4.2). Varying amounts of pesticides, ranging from 451 to 468, were detected in different kinds of tea, with the highest effective screening rate (94.7%, 468 pesticides) appeared in black tea and the lowest (91.3%, 451 pesticides) appeared in oolong tea (Table 1.4.1). There were 534 pesticides detected by LC–Q–TOF/MS, accounting for 96.0% of the larger group of 556 pesticides (Table 1.4.2), and 509–531 pesticides detected in other kinds of tea. Unlike the findings with GC–Q–TOF/MS, the highest effective screening rate (95.5%, 531 pesticides) was determined in oolong tea by LC–Q–TOF/MS, while black and green tea had the lowest effective screening rate (91.6%, 509 pesticides). Only 16 and 22 pesticides (3.2%) were undetected in samples by GC–Q–TOF/MS and LC–Q–TOF/MS, respectively, which demonstrates the applicability of these two nontarget screening methods.

Table 1.4.1 lists the statistics of screening limits for both GC–Q–TOF/MS and LC–Q–TOF/MS. At the spiked level of 5 μg/kg, 65.2%–68.6% of the target pesticides were detected by GC–Q–TOF/MS, which was lower than that of LC–Q–TOF/MS (79.3%–88.3%). Under the spiked level of 10 μg/kg, another 4.3%–7.1% and 3.1%–6.3% of pesticides were detected by GC–Q–TOF/MS and LC–Q–TOF/MS, respectively. These results reflect the sufficient sensitivity of these two screening methods. Under the uniform standard (10 μg/kg), the highest and lowest detection rates were 74.1% and 71.9%, presented in black tea and green tea, respectively, by GC–Q–TOF/MS, and they were 91.4% and 84.5%, appearing in oolong tea and black tea, respectively, by LC–Q–TOF/MS. These results reflect that more than 70% pesticides had screening limits less than or equal to 10 μg/kg in both GC–Q–TOF/MS and LC–Q–TOF/MS. With the combination of these two methods, more than 83% of pesticides were screened out at 10 μg/kg, which demonstrates the high sensitivity and high efficiency of this hyphenated method, and the developed method met the requirement of the strict maximum residue limit (MRL) standard.

Statistic results of the amount of detected pesticides are presented in Table 1.4.2. For GC–Q–TOF/MS, 478 pesticides were screened out from at least one kind of tea, and 436 or 462 pesticides, accounting for 88.3% or 93.5%, were screened out from all four or at least three kinds of tea, respectively. For LC–Q–TOF/MS, 534 pesticides could be confirmed in at least one kind of tea, and 498 or 517 pesticides, accounting for 89.6% or 93.0% of the 556 pesticides, were confirmed in all four or at least three kinds of tea, respectively. The majority of the target pesticides could be screened out from all four kinds of tea. Furthermore, 91.0% and 94.1% of the target 765 pesticides could be screened out from all four or at least three kinds of tea, respectively, with the combination of GC–Q–TOF/MS and LC–Q–TOF/MS, which further illustrates the applicability of the developed method and the complementarity of these two technologies.

1.4.3.1.2 Comparison of Screening Limits for Pesticides by GC–Q–TOF/MS and LC–Q–TOF/MS

When comparing the screening limits of these two nontarget screening methods, it was obvious that most of the pesticides were detected at 5 μg/kg. However, LC–Q–TOF/MS had a higher screening ability than GC–Q–TOF/MS, which was

TABLE 1.4.1 Comprehensive Analysis of Screening Limits by GC-Q-TOF/MS, LC-Q-TOF/MS and the Hyphenated method

		5 µg/kg		10 µg/kg		20 µg/kg		50 µg/kg		100 µg/kg		200 µg/kg		5 µg/kg and 10 µg/kg		Detectable Pesticides		Undetectable Pesticides		Target Pesticides
		Amount	Proportion (%)	Amount	Proportion (%)	Amount	Proportion (%)	Amount	Proportion (%)	Amount	Proportion (%)	Amount	Proportion (%)	Amount	Proportion (%)	Amount	Proportion (%)	Amount	Proportion (%)	Amount
Black tea	GC	331	67.0	35	7.1	56	11.3	20	4.0	14	2.8	12	2.4	366	74.1	468	94.7	26	5.3	494
	LC	443	79.7	27	4.9	23	4.1	16	2.9	0	0.0	0	0.0	470	84.5	509	91.6	47	8.4	556
	GC+LC	615	80.4	33	4.3	43	5.6	12	1.6	7	0.9	8	1.0	648	84.7	718	93.9	47	6.1	765
Green tea	GC	322	65.2	33	6.7	58	11.7	34	6.9	8	1.6	8	1.6	355	71.9	463	93.7	31	6.3	494
	LC	441	79.3	35	6.3	22	4.0	7	1.3	4	0.7	0	0.0	476	85.6	509	91.6	47	8.4	556
	GC+LC	610	79.7	31	4.1	42	5.5	25	3.3	5	0.7	5	0.7	641	83.8	718	93.9	47	6.1	765
Puer tea	GC	337	68.2	21	4.3	68	13.8	21	4.3	10	2.0	8	1.6	358	72.5	465	94.1	29	5.9	494
	LC	461	82.9	22	4.0	19	3.4	18	3.2	5	0.9	0	0.0	483	86.9	525	94.4	31	5.6	556
	GC+LC	631	82.5	26	3.4	38	5.0	19	2.5	7	0.9	5	0.7	657	85.9	726	94.9	39	5.1	765
Oolong tea	GC	339	68.6	22	4.5	51	10.3	23	4.7	10	2.0	6	1.2	361	73.1	451	91.3	43	8.7	494
	LC	491	88.3	17	3.1	10	1.8	6	1.1	3	0.5	4	0.7	508	91.4	531	95.5	25	4.5	556
	GC+LC	647	84.6	24	3.1	34	4.4	11	1.4	5	0.7	4	0.5	671	87.7	725	94.8	40	5.2	765

TABLE 1.4.2 Amount of the Detectable Pesticides by GC–Q–TOF/MS, LC–Q–TOF/MS and the Hyphenated Method From Different Species Number

	GC–Q–TOF/MS		LC–Q–TOF/MS		Hyphenated Method	
	Amount	Proportion (%)	Amount	Proportion (%)	Amount	Proportion (%)
4 Matrices	436	88.3	498	89.6	696	91.0
3 Matrices	26	5.3	19	3.4	24	3.1
2 Matrices	9	1.8	8	1.4	11	1.4
1 Matrix	7	1.4	9	1.6	9	1.2
Sum of the detectable pesticides	478	96.8	534	96.0	740	96.7
Unable to screen	16	3.2	22	4.0	25	3.3
Sum of target pesticides	494	—	556	—	765	—

TABLE 1.4.3 Comparatabtion of the Screening Limits for Pesticides by GC–Q–TOF/MS and LC–Q–TOF/MS

			Black Tea	Green Tea	Puer Tea	Oolong Tea
Suitable for GC	Amount	235	233	230	212	248
	Proportion (%)	30.7	30.5	30.1	27.7	33.8
Suitable for LC	Amount	326	330	334	336	402
	Proportion (%)	42.6	43.1	43.7	43.9	54.8
Suitable for GC and LC	Amount	157	155	162	177	47
	Proportion (%)	20.5	20.3	21.2	23.1	6.4
Unable to screen	Amount	47	47	39	40	36
	Proportion (%)	6.1	6.1	5.1	5.2	4.9

Note: Suitable for GC: Pesticides had the lower screening limits for GC–Q–TOF/MS, Suitable for GC: Pesticides had the lower screening limits for LC–Q–TOF/MS, Suitable for GC and LC: Pesticides had the same screening limits for GC–Q–TOF/MS and LC–Q–TOF/MS.

probably due to the large injection volume and soft ionization in LC–Q–TOF/MS. Table 1.4.3 lists the compared results of the screening limits for these two screening methods. On the whole, the amount of pesticides that were detected with LC–Q–TOF/MS was 12% higher than GC–Q–TOF/MS.

1.4.3.2   Accuracy and Precision

1.4.3.2.1   Recoveries and RSDs From Different Kinds of Tea

According to EU documents SANCO/10684/2009, recoveries in the range of 60%–120% and RSDs less than 20% were conformed to recovery-RSD standards. Results of recoveries and RSDs are presented in Table 1.4.4. The proportion of pesticides that obtained acceptable recoveries and precisions under the spiked level of 10 µg/kg were in the range of 65.6%–69.0% and 60.4%–70.3% for GC–Q–TOF/MS and LC–Q–TOF/MS, respectively, and they were 76.3%–84.8% and 71.6%–77.0% at the spiked level of 50 µg/kg, and 83.2%–89.3% and 76.4%–7.6% at the spiked level of 200 µg/kg. These results demonstrated the satisfactory accuracy of the developed method. According to the internationally recognized uniform standard (10 µg/kg), 65.6%–69.0% and 60.4%–70.3% of the target pesticides in GC–Q–TOF/MS and LC–Q–TOF/MS, respectively, obtained acceptable recoveries and precisions, which reflected sufficient sensitivity and accuracy. There were 511–556 pesticides that obtained the satisfactory recoveries and precisions when combining GC–Q–TOF/MS with LC–Q–TOF/MS, far more than the singe-screening method, which had further illustrated the complementarity in these two technologies. For GC–Q–TOF/MS, 470 pesticides could be recovered from at least one kind of tea, and 396 and 444 pesticides were recovered from all four or at least three kinds of tea, accounting for 80.2% and 89.9% of the 494 pesticides, respectively (Table 1.4.5). For LC–Q–TOF/MS, there were 459 pesticides obtained the satisfactory recoveries and precisions in at least one kind of tea, and it was 394 and 433 pesticides in all four or at least three kinds of tea, accounting

TABLE 1.4.4 Recoveries and RSDs for Pesticides by GC–Q–TOF/MS, LC–Q–TOF/MS and the Hyphenated Method

| | | GC–Q–TOF/MS | | | | | | | | LC–Q–TOF/MS | | | | | | | | | Hyphenated Method | |
| | | Rec. 60%–120% | | RSD ≤ 20% (n = 5) | | Rec. 60%–120% and RSD ≤ 20% | | | | | Rec. 60%–120% | | RSD ≤ 20% (n = 5) | | Rec. 60%–120% and RSD ≤ 20% | | | | Rec. 60%–120% and RSD ≤ 20% (10 µg/kg) | |
	Spiked level (µg/kg)	Amount	Proportion (%)	Amount	Proportion (%)	Amount	Proportion (%)	AVE Rec.%	AVE RSD%	Spiked level (µg/kg)	Amount	Proportion (%)	Amount	Proportion (%)	Amount	Proportion (%)	AVE Rec.%	AVE RSD%	Amount	Proportion (%)
Black tea	10	347	70.2	351	71.1	341	69.0	94.6	6.7	10	391	70.3	408	73.4	390	70.1	85.6	6.5	556	72.7
	50	420	85.0	430	87.0	419	84.8	91.7	5.4	50	419	75.4	437	78.6	418	75.2	91.4	4.9		
	200	444	89.9	451	91.3	441	89.3	93.8	5.8	200	430	77.3	440	79.1	429	77.2	97.0	5.2		
Green tea	10	338	68.4	348	70.4	337	68.2	92.2	7.8	10	396	71.2	406	73.0	391	70.3	92.3	7.0	554	72.4
	50	413	83.6	430	87.0	412	83.4	92.9	6.9	50	429	77.2	435	78.2	428	77.0	95.6	9.2		
	200	433	87.7	443	89.7	430	87.0	94.6	7.1	200	438	78.8	450	80.9	437	78.6	95.9	6.7		
Puer tea	10	336	68.0	347	70.2	334	67.6	83.9	5.8	10	370	66.5	406	73.0	366	65.8	81.7	6.8	539	70.5
	50	411	83.2	426	86.2	403	81.6	83.9	6.6	50	411	73.9	437	78.6	406	73.0	83.9	6.2		
	200	424	85.8	449	90.9	422	85.4	88.4	7.0	200	432	77.7	457	82.2	431	77.5	87.5	4.5		
Oolong tea	10	332	67.2	342	69.2	324	65.6	85.7	6.7	10	347	62.4	400	71.9	336	60.4	82.5	8.6	511	66.8
	50	380	76.9	413	83.6	377	76.3	81.1	5.6	50	398	71.6	447	80.4	398	71.6	82.4	6.0		
	200	413	83.6	436	88.3	411	83.2	82.0	6.0	200	429	77.2	464	83.5	425	76.4	81.4	4.7		

TABLE 1.4.5 Amounts of Pesticides Which Met the Recovery-RSD Standard from Different Species Numbers

Number of species	GC–Q–TOF/MS		LC–Q–TOF/MS		Hyphenated Method	
	Amount	Proportion (%)	Amount	Proportion (%)	Amount	Proportion (%)
4 Matrices	396	80.2	394	70.9	585	76.5
3 Matrices	48	9.7	39	7.0	48	6.3
2 Matrices	13	2.6	15	2.7	17	2.2
1 Matrix	13	2.6	11	2.0	16	2.1
Nonrecoverable	24	4.9	97	17.4	99	12.9
Target pesticides	494	—	556	—	765	—

for 70.9% or 77.9% of the 556 pesticides, respectively. If we combined GC–Q–TOF/MS with LC–Q–TOF/MS, the proportions of pesticides that confirmed to the recovery-RSD standard were 76.5% and 82.8% in all four and at least three kinds of tea, respectively, and the detected capacity of pesticides residual was greatly improved.

1.4.3.2.2 Comparison of the Recoveries for GC–Q–TOF/MS and LC–Q–TOF/MS

For GC–Q–TOF/MS, 396 pesticides, accounting for 80.2% of the 494 pesticides, could be screened and recovered from all four kinds of tea (Table 1.4.6). Thirty-eight pesticides were screened from all matrices but could be recovered from only one to three kinds of tea. At the same time, 36 pesticides were screened and recovered from part of the four matrices. For LC–Q–TOF/MS, 394 pesticides accounting for 70.9% of the 556 pesticides were screened and recovered from all four kinds of tea. Forty-five pesticides were screened from all matrices but recovered from only one to three kinds of tea. Meanwhile, 20 pesticides were screened and recovered from part of the four matrices.

TABLE 1.4.6 Comprehensive Analysis of Screening Limits and Recoveries for Pesticides by GC–Q–TOF/MS and LC–Q–TOF/MS

	GC–Q–TOF/MS		LC–Q–TOF/MS	
	Amount	Proportion (%)	Amount	Proportion (%)
Target pesticides	494	—	556	—
Pesticides were detectable and recoverable from all the matrices	396	80.2	394	70.9
Pesticides were undetectable and nonrecoverable from all the matrices	16	3.2	22	4.0
Pesticides were detectable from all the matrices and recoverable from part of the matrices	38	7.7	45	8.1
Pesticides were detectable from all the matrices and recoverable from none kind of matrix	2	0.4	59	10.6
Pesticides were detectable and recoverable from part of the matrices	36	7.3	20	3.6
Pesticides were detectable from part of the matrices and recoverable from none kind of matrix	6	1.2	16	2.9

According to the internationally recognized uniform standards for pesticide residual, the amount of pesticides with recoveries that met the recovery-RSD standard at 10 μg/kg by the combination of GC–Q–TOF/MS and LC–Q–TOF/MS are presented in Table 1.4.4. The hyphenated screening method obtained more pesticides (511–556 pesticides) that met the recovery-RSD standard than GC–Q–TOF/MS (324–341 pesticides) or LC–Q–TOF/MS (336–391 pesticides) that demonstrated the higher efficiency and sensitivity. In conclusion, the hyphenated screening method not only guaranteed the high level of screening ability but also met the requirement of accurate quantification at 10 μg/kg.

1.4.3.3 Comparison of the Common-Detected Pesticides

1.4.3.3.1 Screening Limits for Common-Detected Pesticides

Among the target 765 pesticides, 272 pesticides were screened from at least one kind of tea by both GC–Q–TOF/MS and LC–Q–TOF/MS, which were named for common-detected pesticides. Due to the matrix effects and different physicochemical properties for pesticides, there were 259 pesticides, accounting for 88.2% of the 272 common-detected pesticides, detected from at least three kinds of tea by both GC–Q–TOF/MS and LC–Q–TOF/MS, and 240 common-detected pesticides were detected from all four kinds of tea. These results demonstrated the adaptability of common-detected pesticides for different matrices. Nonetheless, common-detected pesticides from different matrices were slightly different, and details are in Appendix Table C-1.

Table 1.4.7 lists the amount of common-detected pesticides in different matrices by GC–Q–TOF/MS, LC–Q–TOF/MS, and the combination of these two methods. The results show that 83.1%–93.8% and 65.8%–71.3% of the target 272 common-detected pesticides were detected at the spiked level of 5 μg/kg by LC–Q–TOF/MS and GC–Q–TOF/MS, respectively, indicating the higher sensitivity of LC–Q–TOF/MS as compared to GC–Q–TOF/MS, which is in accordance with the screening limit results seen earlier. In addition, 254–264 pesticides, accounting for 93.4%–97.1% of the target common-detected pesticides, were detected by the hyphenated method from all four kinds of tea, reflecting similar common-detected pesticides in different matrices.

Comparing the screening limits of the common-detected pesticides (Table 1.4.8) between GC–Q–TOF/MS and LC–Q–TOF/MS, 6.6%–13.2% and 28.3%–30.2% of the 272 pesticides had lower screening limits by GC–Q–TOF/MS and LC–Q–TOF/MS, respectively. The remaining 57.0%–65.1% common-detected pesticides had the common limits by these two screening methods. These results reflected the consistency of these two methods for the screening of the common-detected pesticides and the great improvement of the detected capacity when combined with GC–Q–TOF/MS and LC–Q–TOF/MS.

1.4.3.3.2 Recoveries of Common-Detected Pesticides

When we evaluated the recoveries of the 272 common-detected pesticides, there were four pesticides that could only be recovered by LC–Q–TOF/MS and nine pesticides could only be recovered by GC–Q–TOF/MS. The rest, 259 pesticides, could be recovered from at least three kinds of tea, and 240 pesticides accounted for 88.2% of the common-detected pesticides recovered from all four kinds of tea by both GC–Q–TOF/MS and LC–Q–TOF/MS.

Table 1.4.9 lists the statistics of the recoveries and relative standard deviations (RSDs) for the 272 common-detected pesticides, in which the recoveries were in the range of 60%–120% or RSDs were no more than 20%. Pesticides with recoveries in the range of 60%–120% and RSDs with no more than 20% (recovery-RSD standard) were also listed. There were 182–197 pesticides, accounting for 66.9%–72.4%, and 210–236 pesticides, accounting for 77.2%–86.8%, which conformed to recovery-RSD standard at spiked levels of 10 μg/kg by GC–Q–TOF/MS and LC–Q–TOF/MS, respectively, while there were 209–245 pesticides, accounting for 76.8%–90.1%, and 241–249 pesticides, accounting for 88.6%–91.5%, at the spiked level of 50 μg/kg by GC–Q–TOF/MS and LC–Q–TOF/MS, respectively. Under the spiked level of 200 μg/kg, 236–254 pesticides, accounting for 86.8%–93.4%, and 250–257 pesticides, accounting for 91.9%–94.5%, obtained recoveries that met the recovery-RSD standard by GC–Q–TOF/MS and LC–Q–TOF/MS, respectively. In general, more than 66% of pesticides met the recovery-RSD standard, implying the high sensitivity and accuracy.

1.4.3.3.3 Common-Detected Pesticides That Met the Recovery-RSD Standard at 10 μg/kg

According to the internationally recognized uniform standard for pesticides residuals, there were 182–197 and 210–236 pesticides that contributed 66.9%–72.4% and 77.2%–86.8% to the common-detected pesticides, met the recovery-RSD standard at 10 μg/kg when screened by GC–Q–TOF/MS and LC–Q–TOF/MS, respectively (Table 1.4.9). Due to the different matrix effects and different screening methods, it was more difficult to meet the recovery-RSD standard for both GC–Q–TOF/MS and LC–Q–TOF/MS. However, there were still more than 50% (52.3%–61.4%) of the common-detected pesticides that met the recovery-RSD standard. Table 1.4.10 lists the evaluation results of 115 common-detected pesticides in detail. All these pesticides obtained a good reproducibility, accuracy, and reliability with the average recoveries more than 79.1% and average RSDs less than 8.0%. Thus, we selected these 115 pesticides as the internal quality control standard to verify the screening results and to improve the accuracy of the hyphenated screening method.

TABLE 1.4.7 Comprehensive Analysis of Screening Limits for Common-Detected Pesticides by GC-Q-TOF/MS, LC–Q–TOF/MS and Their Hyphenated Method

		5 µg/kg		10 µg/kg		20 µg/kg		50 µg/kg		100 µg/kg		200 µg/kg		Undetectable Pesticides		Hyphenated Method	
		Amount	Proportion (%)	Amount	Proportion (%)	Amount	Proportion (%)	Amount	Proportion (%)	Amount	Proportion (%)	Amount	Proportion (%)	Amount	Proportion (%)	Amount	Proportion (%)
Black tea	GC	184	67.6	24	8.8	32	11.8	15	5.5	9	3.3	5	1.8	3	1.1	259	95.2
	LC	235	86.4	9	3.3	8	2.9	10	3.7	0	0.0	0	0.0	10	3.7		
Green tea	GC	179	65.8	24	8.8	36	13.2	17	6.3	6	2.2	3	1.1	7	2.6	254	93.4
	LC	226	83.1	20	7.4	10	3.7	3	1.1	2	0.7	0	0.0	11	4.0		
Puer tea	GC	191	70.2	11	4.0	43	15.8	12	4.4	8	2.9	3	1.1	4	1.5	264	97.1
	LC	237	87.1	11	4.0	9	3.3	10	3.7	1	0.4	0	0.0	4	1.5		
Oolong tea	GC	194	71.3	10	3.7	27	9.9	16	5.9	7	2.6	3	1.1	15	5.5	257	94.5
	LC	255	93.8	7	2.6	3	1.1	2	0.7	2	0.7	3	1.1	0	0.0		

TABLE 1.4.8 Comparisons of the Screening Limits for Common-Detected Pesticides by GC–Q–TOF/MS and LC–Q–TOF/MS

		Black Tea	Green Tea	Puer Tea	Oolong Tea
Suitable for GC	Amount	36	35	33	18
	Proportion (%)	13.2	12.9	12.1	6.6
Suitable for LC	Amount	79	82	77	77
	Proportion (%)	29.0	30.1	28.3	28.3
Suitable for GC and LC	Amount	157	155	162	177
	Proportion (%)	57.7	57.0	59.6	65.1

Note: Suitable for GC: Pesticides had the lower screening limits for GC–Q–TOF/MS, Suitable for LC: Pesticides had the lower screening limits for LC–Q–TOF/MS, Suitable for GC and LC: Pesticides had the same screening limits for GC–Q–TOF/MS and LC–Q–TOF/MS.

1.4.4 CONCLUSION

In this study, a new method with a one-forming pretreatment coupled with GC–Q–TOF/MS and LC–Q–TOF/MS detection was developed for the nontarget and high-resolution screening of 1050 pesticide residues in tea. Pesticides were identified by comparing the data acquired from tea samples with the accurate mass GC–Q–TOF/MS database for 494 pesticides and LC–Q–TOF/MS database for 556 pesticides. When combined with these two screening methods, the hyphenated method achieved simultaneous determination of 765 pesticides, which was 30% higher than GC–Q–TOF/MS and LC–Q–TOF/MS. Meanwhile, 511 pesticides obtained acceptable accuracy and precision, which met the recovery-RSD standard at the spiked level of 10 μg/kg by the hyphenated method, and that was far more than the single-screening technology. At the same time, no reference standards were needed in this study and MS/MS fragment information was typically matched to the database to enable conclusive confirmation of chemical identity. What's more, full-scan technology brought forward the nontarget screening, and pesticide residue detection achieved rapid progress in automation, digitization, and informationalization. In conclusion, the application of LC–Q–TOF/MS is an effective tool for rapid screening of pesticide residues in tea and is playing an important role in food safety supervision for pesticide residue.

TABLE 1.4.9 Recoveries and RSDs for Common-Detected Pesticides by GC-Q-TOF/MS, LC–Q–TOF/MS and Their Hyphenated Method

| | | GC-Q-TOF/MS | | | | | | | | LC–Q–TOF/MS | | | | | | | | | Hyphenated Method | |
| | | Rec. 60–120% | | RSD ≤ 20% (n = 5) | | Rec. 60%–120% and RSD ≤ 20% | | | | | Rec. 60%–120% | | RSD ≤ 20% (n = 5) | | Rec. 60%–120% and RSD ≤ 20% | | | | Rec. 60%–120% and RSD ≤ 20% (10 µg/kg) | |
	Spiked level (µg/kg)	Amount	Proportion (%)	Amount	Proportion (%)	Amount	Proportion (%)	AVE Rec.%	AVE RSD%	Spiked level (µg/kg)	Amount	Proportion (%)	Amount	Proportion (%)	Amount	Proportion (%)	AVE Rec.%	AVE RSD%	Amount	Proportion (%)
Black tea	10	200	73.5	201	73.9	197	72.4	95.0	6.8	10	236	86.8	240	88.2	236	86.8	85.0	6.1	175	61.4
	50	246	90.4	250	91.9	245	90.1	91.7	5.0	50	248	91.2	253	93.0	248	91.2	90.6	4.2		
	200	256	94.1	260	95.6	254	93.4	93.5	6.0	200	250	91.9	256	94.1	250	91.9	97.4	5.1		
Green tea	10	194	71.3	199	73.2	193	71.0	91.7	7.4	10	235	86.4	236	86.8	233	85.7	92.7	6.4	173	60.7
	50	241	88.6	249	91.5	240	88.2	91.6	6.5	50	249	91.5	249	91.5	249	91.5	96.4	9.2		
	200	252	92.6	258	94.9	251	92.3	96.5	8.4	200	255	93.8	256	94.1	255	93.8	95.7	6.6		
Puer tea	10	194	71.3	199	73.2	193	71.0	80.3	5.6	10	224	82.4	233	85.7	221	81.3	81.2	6.7	161	56.5
	50	234	86.0	246	90.4	230	84.6	81.6	6.7	50	244	89.7	253	93.0	241	88.6	85.4	5.9		
	200	238	87.5	257	94.5	237	87.1	85.2	7.9	200	253	93.0	262	96.3	253	93.0	89.1	4.5		
Oolong tea	10	188	69.1	190	69.9	182	66.9	83.3	7.0	10	217	79.8	234	86.0	210	77.2	82.1	8.8	149	52.3
	50	210	77.2	238	87.5	209	76.8	76.4	5.9	50	243	89.3	256	94.1	243	89.3	82.1	5.8		
	200	237	87.1	251	92.3	236	86.8	76.1	4.9	200	259	95.2	264	97.1	257	94.5	82.2	4.5		

TABLE 1.4.10 Recoveries and RSDs for 115 Common-Detected Pesticides

Name	Black Tea GC Rec.%	GC RSD% (n=5)	LC Rec.%	LC RSD% (n=5)	Green Tea GC Rec.%	GC RSD% (n=5)	LC Rec.%	LC RSD% (n=5)	Puer Tea GC Rec.%	GC RSD% (n=5)	LC Rec.%	LC RSD% (n=5)	Oolong Tea GC Rec.%	GC RSD% (n=5)	LC Rec.%	LC RSD% (n=5)
1 1-Naphthyl acetamide	88.6	4.2	70.9	7.8	98.2	10.4	85.2	10.0	79.9	8.8	73.6	6.9	84.7	7.3	118.1	5.0
2 Acetochlor	98.8	7.4	93.2	6.7	98.2	2.9	110.6	2.9	77.5	1.4	81.5	7.6	70.8	2.1	72.8	7.0
3 Ametryn	82.9	5.0	82.4	9.6	91.0	5.1	83.9	4.2	80.9	7.4	81.6	12.2	68.6	6.3	65.3	8.9
4 Atrazine	87.3	5.2	79.9	2.3	90.9	7.1	95.7	4.0	76.1	8.0	95.5	7.0	85.9	13.0	112.6	1.9
5 Azaconazole	90.4	10.5	81.7	3.2	86.7	5.0	91.7	1.6	72.6	8.7	67.3	2.7	72.6	6.7	61.0	16.5
6 Aziprotryne	91.4	5.4	91.0	4.0	88.4	7.5	105.0	1.7	82.9	12.8	80.2	6.5	91.1	13.7	74.7	6.2
7 Bendiocarb	89.0	4.4	82.3	6.0	79.2	7.4	103.5	2.8	78.1	8.3	76.5	7.4	81.8	13.4	71.8	8.1
8 Benzoximate	84.7	6.3	71.2	1.6	96.8	8.0	85.3	7.0	84.4	8.3	89.7	11.0	86.7	9.4	81.3	15.4
9 Boscalid	88.6	4.1	96.0	15.8	81.1	9.9	72.2	12.0	76.5	4.5	84.8	6.6	113.8	5.4	73.9	12.6
10 Bromfenvinfos	96.6	3.5	84.2	2.5	96.4	3.2	106.8	4.4	78.8	5.3	84.4	4.3	83.0	6.1	73.2	10.3
11 Bromobutide	85.3	5.3	76.8	3.7	79.8	3.1	95.1	4.2	93.6	5.3	91.4	14.9	79.9	5.1	85.1	4.2
12 Bupirimate	88.6	3.6	82.1	7.7	91.2	3.0	81.2	3.1	77.3	2.3	75.8	1.2	85.4	8.1	71.6	3.1
13 Butachlor	92.6	4.3	98.1	8.5	100.0	5.8	86.6	3.2	86.4	1.9	85.0	2.1	91.9	8.2	78.7	12.2
14 Butafenacil	100.2	14.8	81.0	9.8	105.9	5.8	89.5	3.3	80.0	4.0	78.7	2.7	77.1	8.0	73.4	11.0
15 Cadusafos	95.1	6.8	88.7	3.8	95.4	6.9	103.1	3.1	82.7	4.0	82.7	7.3	85.4	5.0	80.9	7.7
16 Chlorfenvinphos	98.4	9.1	87.0	7.0	90.1	9.9	79.2	4.7	95.1	14.6	83.4	2.7	100.3	13.3	71.7	9.5
17 Clomazone	97.7	4.8	81.5	8.1	89.1	6.1	93.7	3.7	79.6	3.9	82.2	8.2	79.8	6.1	77.0	10.7
18 Coumaphos	88.2	8.5	103.7	1.5	98.8	12.2	88.5	13.2	79.7	12.8	97.6	4.6	84.1	7.6	82.1	16.1
19 Crufomate	104.9	16.8	83.5	3.5	88.6	12.2	119.9	2.4	76.2	19.6	78.3	3.0	73.7	8.1	78.9	12.7
20 Cyprazine	97.3	6.6	82.7	7.8	98.2	9.6	88.5	10.5	70.4	3.5	71.9	4.1	89.8	11.4	70.2	4.2
21 Cyproconazole	104.3	12.7	98.2	3.7	98.8	7.4	108.0	3.7	73.6	2.9	72.5	3.7	78.3	6.1	73.7	10.8
22 Cyprofuram	89.6	2.5	80.9	2.8	71.8	3.3	98.9	1.0	70.1	5.2	77.0	4.0	76.7	6.2	76.4	3.6
23 Desmetryn	91.9	2.1	84.1	2.9	81.0	3.3	95.3	2.1	71.1	11.3	94.4	5.1	69.7	6.8	115.7	3.5
24 Diallate	91.1	4.1	61.8	9.1	104.9	3.7	97.2	19.3	80.9	4.1	83.3	12.8	87.2	4.2	103.4	4.1
25 Diethatyl-ethyl	92.5	7.1	82.2	2.1	91.3	5.9	102.6	1.6	74.8	2.9	89.0	7.6	77.1	6.5	101.0	2.8

26	Dimethachlor	94.7	9.7	81.6	3.9	95.0	5.9	103.4	2.4	72.0	2.9	68.9	2.6	79.4	13.6	71.3	11.2
27	Dimethenamid	95.1	2.4	75.0	3.9	75.0	4.1	108.6	3.5	76.7	3.9	74.8	7.0	82.0	5.0	72.5	14.5
28	Diniconazole	114.7	9.9	81.6	8.5	103.7	6.7	77.8	3.3	74.5	7.3	72.6	2.7	75.3	6.9	72.7	7.4
29	Diphenamid	89.7	2.9	83.2	7.8	71.7	1.1	85.3	5.4	77.7	2.5	67.1	1.2	78.6	4.4	70.7	4.4
30	Dipropetryn	97.0	6.3	71.6	6.3	93.0	4.5	90.7	2.9	80.2	4.6	92.3	10.7	79.3	5.0	71.4	7.2
31	Disulfoton sulfone	99.8	5.1	84.2	4.7	95.0	5.6	112.7	3.3	82.8	3.5	76.2	6.0	94.9	5.4	70.1	16.2
32	Disulfoton sulfoxide	102.1	10.9	86.4	6.8	90.5	8.5	89.2	5.1	76.8	4.7	70.6	0.6	84.8	4.1	72.2	5.7
33	Ditalimfos	103.7	18.3	87.2	3.5	100.0	7.0	97.0	14.1	65.0	6.2	96.7	7.7	79.4	19.3	82.3	4.8
34	Ethion	99.9	5.0	85.0	9.4	95.0	5.6	116.0	8.9	82.8	3.5	77.8	13.8	94.9	5.4	83.2	14.8
35	Ethoprophos	87.5	4.4	85.5	4.9	79.7	5.8	111.1	1.8	81.9	2.6	83.1	4.6	88.2	4.6	75.3	12.0
36	Famphur	107.1	4.6	88.9	4.3	96.9	14.0	112.6	1.8	89.3	3.8	73.0	8.5	81.3	9.8	75.0	7.6
37	Fenamidone	90.9	1.3	84.2	5.4	77.7	11.6	107.8	1.2	74.4	3.5	88.8	8.5	71.7	12.2	110.0	2.3
38	Fenazaquin	89.5	5.3	82.9	2.9	105.2	3.0	99.9	2.9	79.3	1.9	86.0	4.2	95.1	6.0	85.4	16.3
39	Fenoxaprop-ethyl	93.1	4.7	94.3	7.4	106.4	4.7	96.3	7.4	82.2	1.7	78.4	9.1	87.4	7.5	73.5	7.6
40	Fenpropimorph	76.2	8.4	71.3	7.0	88.6	6.6	81.0	2.6	72.4	4.2	72.7	2.3	70.7	5.1	65.6	8.2
41	Fensulfothion	93.0	3.4	90.7	7.9	77.9	8.5	86.5	5.9	73.5	6.7	76.1	3.9	90.3	7.9	73.7	11.3
42	Fensulfothion-sulfone	113.5	8.8	93.7	8.9	107.6	6.2	93.9	7.5	70.3	5.7	94.3	6.4	75.0	7.1	81.0	10.5
43	Fenthion	92.7	7.0	74.9	11.8	106.3	8.6	106.8	12.4	86.8	4.3	98.6	8.9	74.0	3.1	101.7	15.5
44	Flamprop-isopropyl	86.3	11.6	84.6	1.9	87.6	5.9	106.5	7.1	72.8	3.7	92.7	6.9	77.4	7.8	71.5	16.0
45	Flamprop-methyl	83.2	12.8	76.6	3.1	84.9	6.4	108.5	1.3	73.2	4.4	76.6	3.5	72.3	6.7	71.2	10.6
46	Flufenacet	101.8	12.6	82.1	9.0	97.0	6.7	82.1	4.9	80.2	8.7	79.1	1.8	80.5	7.0	72.8	0.6
47	Fluopyram	91.3	3.2	111.6	2.0	86.9	6.3	94.1	10.0	81.6	2.8	92.1	5.5	84.2	6.0	89.7	3.3
48	Flusilazole	101.1	5.8	83.5	2.8	86.6	9.8	95.0	3.1	81.9	5.9	91.2	7.8	79.0	2.6	81.5	1.8
49	Flutolanil	91.2	7.6	89.3	3.2	91.8	5.1	105.1	1.6	78.0	1.6	76.6	11.9	80.0	4.4	71.1	7.6
50	Flutriafol	95.4	7.3	82.9	5.3	72.2	5.2	94.8	2.6	70.0	2.4	67.1	3.9	70.6	13.0	62.4	14.2

(Continued)

TABLE 1.4.10 Recoveries and RSDs for 115 Common-Detected Pesticides (cont.)

Name		Black Tea GC Rec.%	RSD% (n = 5)	LC Rec.%	RSD% (n = 5)	Green Tea GC Rec.%	RSD% (n = 5)	LC Rec.%	RSD% (n = 5)	Puer Tea GC Rec.%	RSD% (n = 5)	LC Rec.%	RSD% (n = 5)	Oolong Tea GC Rec.%	RSD% (n = 5)	LC Rec.%	RSD% (n = 5)
51	Fluxapyroxad	96.7	5.5	105.8	7.1	85.3	10.7	90.4	8.5	80.4	4.7	81.0	6.8	87.8	4.3	96.1	1.2
52	Furalaxyl	88.9	2.8	72.0	5.1	75.6	6.8	111.6	2.6	69.6	1.5	66.0	5.5	81.7	3.8	73.1	14.6
53	Haloxyfop-methyl	90.5	1.5	78.2	1.6	102.1	3.5	101.9	3.8	83.9	3.6	87.6	4.9	94.5	4.4	87.3	7.9
54	Heptenophos	95.1	18.3	85.5	7.1	94.9	15.8	77.0	13.8	83.0	15.5	70.5	3.3	91.2	6.2	70.6	9.1
55	Isazofos	93.7	1.4	78.8	2.1	88.4	4.2	100.6	3.3	80.9	1.8	94.5	11.0	90.8	7.2	100.1	6.4
56	isocarbophos	94.1	6.7	70.7	5.1	87.6	7.5	87.4	7.5	71.2	5.8	73.9	9.4	77.4	5.7	89.8	5.2
57	Isofenphos-oxon	102.2	12.1	83.8	8.8	106.7	9.0	85.2	2.1	78.9	4.3	71.7	3.7	81.5	5.2	73.8	8.7
58	Isopropalin	92.3	3.0	101.9	10.0	119.5	0.4	83.1	17.8	86.8	4.5	80.6	2.4	93.5	4.8	97.8	11.6
59	Isoprothiolane	94.5	5.0	76.1	4.4	92.5	5.5	81.5	3.9	72.0	2.8	71.5	3.8	80.0	5.4	71.0	2.3
60	Kresoxim-methyl	83.1	2.0	81.4	10.7	98.8	5.1	69.7	18.3	83.8	3.8	76.3	3.3	85.5	6.2	70.4	13.6
61	Mefenpyr-diethyl	88.7	3.3	90.2	9.6	83.6	4.6	103.9	2.9	79.0	3.1	87.0	2.8	85.0	3.6	71.0	19.1
62	Mepanipyrim	106.0	5.7	82.5	10.0	95.0	5.6	75.3	2.6	80.9	3.2	80.0	3.1	75.9	5.0	70.9	2.9
63	Mepronil	101.0	5.8	92.6	7.6	97.1	8.5	86.2	5.2	74.2	6.5	81.3	2.4	74.4	2.5	70.2	9.8
64	Metalaxyl	101.0	2.6	87.4	1.7	71.2	6.8	104.7	1.0	74.5	4.3	100.1	6.0	79.4	13.3	80.4	2.4
65	Metazachlor	85.1	11.2	75.5	9.8	103.6	2.2	75.6	4.4	79.2	3.9	70.6	2.7	88.6	9.4	71.5	4.2
66	Methabenz-thiazuron	92.6	4.6	81.3	2.8	76.5	4.4	93.4	3.5	76.6	8.1	89.1	6.2	80.5	4.1	90.1	2.5
67	Methopro-tryne	92.8	4.9	79.3	2.8	84.5	5.9	97.9	2.2	77.5	6.3	76.1	1.6	76.5	5.7	71.5	8.4
68	Mevinphos	96.1	7.0	82.5	9.5	90.2	15.9	81.0	12.7	87.9	7.2	76.1	4.2	87.7	7.2	70.6	7.0
69	Myclobutanil	99.4	7.3	81.8	9.0	75.6	10.5	83.4	8.1	96.2	8.0	70.3	2.5	70.1	5.2	71.2	3.5
70	Norflurazon	98.3	7.2	80.9	1.9	79.0	11.4	99.3	2.5	78.8	7.0	80.9	4.6	76.0	7.4	99.4	2.5
71	Orbencarb	90.8	2.8	81.4	2.4	107.0	5.2	99.8	2.7	86.0	1.6	109.5	4.8	80.6	6.6	99.0	3.8
72	Paclobutrazol	91.9	1.9	105.3	4.2	85.2	3.2	106.4	2.5	75.5	4.3	73.6	4.3	74.6	8.5	71.7	7.5
73	Pentanochlor	87.4	3.3	85.2	2.5	95.4	4.6	109.9	4.7	85.3	3.4	99.8	11.2	89.4	6.4	109.4	3.4
74	Phosfolan	108.1	11.8	85.3	2.6	98.8	11.7	103.8	1.8	84.6	6.6	77.5	4.8	82.0	6.9	104.0	2.6
75	Phosphamidon	94.4	4.6	80.8	2.4	70.4	3.5	101.1	2.6	75.5	8.6	68.3	1.8	81.4	8.2	71.8	10.7

76	Picoxystrobin	97.1	4.4	84.9	2.8	91.0	7.4	101.8	1.3	87.2	3.9	107.3	15.3	80.1	11.3	82.3	1.6
77	Pirimicarb	91.1	7.8	86.7	2.2	87.0	6.2	93.2	3.7	73.9	3.4	73.1	2.7	84.3	7.3	73.0	4.7
78	Pirimiphos-methyl	108.6	5.2	97.6	2.5	101.6	4.8	90.7	8.0	74.2	4.4	87.5	5.8	84.9	8.4	95.2	17.9
79	Pirimiphos-methyl-N-desethyl	112.5	2.4	82.7	7.0	104.2	8.4	90.3	7.1	73.8	9.4	79.8	6.0	81.3	5.4	87.6	6.5
80	Pretilachlor	98.5	2.8	84.1	3.3	89.8	7.9	103.8	2.9	81.7	4.4	84.0	5.6	88.4	7.1	79.2	10.8
81	Profenofos	101.3	7.3	98.1	1.7	103.0	6.1	95.7	3.3	86.1	5.9	86.9	9.6	82.8	2.7	81.6	12.0
82	Prometon	88.4	6.9	72.0	6.1	96.3	6.1	83.5	5.5	72.1	3.9	74.6	1.6	96.4	9.5	62.2	12.1
83	Prometryn	101.3	4.8	89.3	1.8	87.4	10.5	102.4	1.9	88.4	2.1	81.3	1.2	85.1	2.9	75.0	8.7
84	Propanil	100.6	2.7	86.0	8.0	91.7	4.8	84.5	6.8	80.6	4.0	74.6	6.9	88.5	4.6	73.0	9.3
85	Propaphos	114.7	8.4	88.8	6.0	72.4	14.1	113.7	4.7	82.1	2.8	76.6	6.7	80.9	5.6	77.4	10.3
86	Propazine	93.0	2.8	86.9	3.8	84.8	7.4	102.7	2.5	79.4	3.6	80.0	5.6	81.6	3.4	79.5	8.7
87	Propisochlor	91.3	0.8	96.2	6.0	89.6	2.9	109.4	4.1	78.9	2.9	83.1	4.5	85.2	6.2	76.8	9.5
88	Propyzamide	105.1	4.6	91.5	5.3	86.8	10.6	102.5	11.3	92.2	1.2	75.4	19.3	86.1	3.9	80.8	19.4
89	Pyributicarb	104.4	11.5	95.7	10.2	103.5	7.0	91.2	4.3	76.0	4.3	88.4	3.1	87.8	5.9	77.8	9.5
90	Pyridaphenthion	114.8	17.7	89.1	11.4	110.4	10.2	86.5	7.1	83.0	6.2	76.6	4.4	83.1	7.4	74.1	6.6
91	Pyrimethanil	86.7	3.1	83.8	7.5	82.5	4.8	81.3	8.5	82.7	2.7	70.4	4.7	65.6	6.4	70.0	4.0
92	Pyriproxyfen	97.4	9.6	84.2	0.5	103.8	5.2	104.7	4.7	79.3	4.4	75.9	5.3	90.8	9.4	112.0	2.1
93	Pyroquilon	95.5	3.4	80.8	2.7	70.0	2.6	91.0	3.8	75.7	5.2	73.8	2.5	78.0	3.0	115.2	3.9
94	Quinalphos	110.3	5.8	93.9	8.2	100.1	6.6	85.9	4.1	75.4	4.9	88.7	2.5	81.0	5.9	76.4	8.1
95	Quinoxyfen	93.4	3.0	84.4	2.1	86.6	6.6	114.9	9.2	85.0	3.5	73.6	12.3	78.0	1.9	98.5	8.0
96	Sebuthylazine	95.6	5.4	85.8	7.6	89.4	6.2	83.5	5.6	74.6	3.9	77.7	1.7	80.7	5.3	70.3	10.9
97	Secbumeton	94.1	3.2	78.2	2.2	80.7	7.0	93.7	4.1	80.5	2.3	75.5	2.8	73.0	3.4	71.2	5.0
98	Simeconazole	119.0	11.9	80.1	9.0	86.8	12.5	76.6	3.6	70.4	9.6	84.7	1.1	70.0	4.5	71.9	9.3
99	Tebuconazole	93.9	10.3	85.8	12.6	90.7	7.6	80.8	2.6	83.2	10.6	77.1	3.6	82.3	6.1	77.3	8.6
100	Tebufenpyrad	93.4	6.9	80.3	2.0	99.6	4.8	100.2	4.9	77.3	4.6	94.5	11.0	83.3	3.9	92.3	7.7
101	Tebupirimfos	96.4	4.9	97.8	7.3	98.2	7.0	90.3	3.2	81.8	3.2	79.6	2.6	80.7	1.6	80.0	7.0

(Continued)

TABLE 1.4.10 Recoveries and RSDs for 115 Common-Detected Pesticides (cont.)

		Black Tea				Green Tea				Puer Tea				Oolong Tea		
		GC		LC		GC		LC		GC		LC		GC		LC	
Name		Rec.%	RSD% (n = 5)	Rec.%	RSD% (n = 5)	Rec.%	RSD% (n = 5)	Rec.%	RSD% (n = 5)	Rec.%	RSD% (n = 5)	Rec.%	RSD% (n = 5)	Rec.%	RSD% (n = 5)	Rec.%	RSD% (n = 5)
102	Terbucarb	90.2	2.0	89.3	3.1	97.0	4.0	83.1	10.2	80.2	2.3	81.6	14.7	88.2	3.9	85.3	8.3
103	Terbumeton	93.7	1.6	80.1	7.4	80.3	5.7	83.1	7.6	81.3	4.8	75.8	1.5	76.0	2.3	61.8	11.3
104	Terbuthyla-zine	95.7	5.1	85.8	3.5	91.6	5.4	95.6	3.6	73.7	3.3	88.5	11.9	81.6	4.5	107.7	3.0
105	Terbutryn	95.0	6.0	86.4	1.9	92.8	11.5	103.1	1.9	85.6	4.5	103.2	6.3	70.2	4.2	90.8	1.4
106	Tetrachlorvin-phos	101.0	5.4	93.2	1.9	105.1	9.7	102.0	2.8	83.5	3.2	73.8	16.8	80.8	5.1	71.1	7.6
107	Tetraconazole	94.1	8.4	78.7	9.3	86.2	5.6	75.7	3.2	71.1	6.0	72.2	2.6	76.2	7.1	70.4	2.6
108	Thenylchlor	92.8	4.1	73.9	5.8	81.0	9.5	82.1	4.7	77.0	3.9	70.5	7.8	76.3	6.6	73.6	3.3
109	Thiazopyr	84.9	4.7	84.4	9.2	94.1	3.6	91.2	4.5	79.5	2.2	85.5	2.4	86.7	4.7	71.6	12.1
110	Thiobencarb	99.8	3.2	94.3	5.8	95.4	4.0	89.3	11.0	78.5	2.8	82.4	1.9	84.8	4.5	74.0	11.1
111	Triapenthenol	110.8	9.3	86.0	1.6	94.5	6.1	103.0	2.7	72.4	4.2	86.3	7.0	76.6	5.0	103.7	2.4
112	Triazophos	104.8	3.5	92.4	3.2	108.9	12.2	113.9	11.2	88.5	5.1	74.9	16.6	79.0	4.4	73.5	7.0
113	Tribufos	86.7	2.8	97.7	10.6	114.8	3.8	77.0	1.1	83.1	6.5	85.2	3.5	84.0	6.8	71.8	4.8
114	Trietazine	91.7	5.1	95.7	3.9	87.9	5.4	100.9	2.2	73.7	3.1	81.0	6.1	80.3	7.2	77.5	7.3
115	Uniconazole	97.7	6.8	83.8	3.3	89.5	2.9	104.1	5.0	77.5	4.2	91.4	7.0	78.7	6.3	99.3	4.8
	Average (n = 115)	95.7	6.2	85.3	5.5	91.6	6.7	94.8	5.2	79.1	5.0	81.3	5.9	82.0	6.5	80.6	8.0

APPENDIX TABLE C-1 Evaluation Results of the Common-Detected Pesticides by GC-Q-TOF/MS Combined with LC–Q-TOF/MS

Black Tea

No.	Pesticides	CAS Number	GC-Q-TOF/MS SDL (µg/kg)	10 µg/kg AVE	RSD% (n=5)	50 µg/kg AVE	RSD% (n=5)	200 µg/kg AVE	RSD% (n=5)	LC-QTOF/MS SDL (µg/kg)	10 µg/kg AVE	RSD% (n=5)	50 µg/kg AVE	RSD% (n=5)	200 µg/kg AVE	RSD% (n=5)
1	1-Naphthyl acetamide	86-86-2	5	88.6	4.2	NA	5.3	111.4	NA	5	70.9	7.8	101	5.6	96.6	4.5
20	2,6-Dichlorobenzamide	2008-58-4	5	101.9	NA	NA	NA	NA	NA	50	NA	NA	87.7	13.4	87.5	13.3
3	3,4,5-Trimethacarb	2686-99-9	20	NA	NA	85.2	5.2	99.9	6.6	5	105.8	15.2	105.5	5.1	89.4	2.2
4	Acetochlor	34256-82-1	10	98.8	7.4	92.5	5	93.9	1.9	5	93.2	6.7	86	5	85.3	4.8
5	Aldicarb-sulfone	1646-88-4	200	NA	NA	NA	NA	NA	NA	5	78.2	4.4	74	11.4	90.8	7.9
6	Aldimorph	91315-15-0	20	NA	NA	79.9	3.8	79.8	8.6	5	76	6.9	85.3	4	71.1	16.4
7	Allidochlor	93-71-0	5	93.4	8.2	91.3	4.4	91.7	11.5	5	72.6	6.3	79.1	6.1	88.3	3.8
8	Ametryn	834-12-8	5	82.9	5	94.7	1.6	89.7	5.5	5	82.4	9.6	84.1	2.2	94	6.4
9	Amidosulfuron	120923-37-7	10	86.1	6.9	85.8	12.1	94.9	8	NA	NA	NA	NA	NA	NA	NA
10	Aminocarb	2032-59-9	5	63.1	9.1	101.2	7.9	88.6	5.1	5	82.3	2.5	99.8	3	98.5	5.2
11	Ancymidol	12771-68-5	5	96.5	4.1	94.1	3	92.1	7.7	5	86.2	2.6	102.9	1.4	97.7	4.6
12	Anilofos	64249-01-0	5	100.7	14	88.7	5.7	104.4	8.5	5	80.8	1.5	83.8	1.3	92.4	1.6
13	Atraton	1610-17-9	5	87.5	16.3	94.2	6.1	93.8	2.2	5	79.5	3.2	82.1	6.5	85.6	1.3
14	Atrazine	1912-24-9	5	87.3	5.2	91.6	5.1	91.7	1.7	5	79.9	2.3	103.2	2.1	103	4.2
15	Atrazine-desethyl	6190-65-4	5	85.9	8.2	101.1	3.9	86.3	8.5	5	87.6	5	100.3	8.7	119.5	7.6
16	Atrazine-desisopropyl	1007-28-9	5	111.7	10.3	109.3	5.5	92.4	16.5	5	75.4	7	87.4	6.5	70.1	1.7
17	Azaconazole	60207-31-0	5	90.4	10.5	89.7	6.9	93.6	3.6	5	81.7	3.2	87.9	4.2	79	1.4

(Continued)

APPENDIX TABLE C-1 Evaluation Results of the Common-Detected Pesticides by GC–Q–TOF/MS Combined with LC–Q–TOF/MS (cont.)

| No. | Pesticides | CAS Number | Black Tea GC–Q–TOF/MS | | | | | | | LC–QTOF/MS | | | | | | |
			SDL (μg/kg)	10 μg/kg AVE	RSD% (n = 5)	50 μg/kg AVE	RSD% (n = 5)	200 μg/kg AVE	RSD% (n = 5)	SDL (μg/kg)	10 μg/kg AVE	RSD% (n = 5)	50 μg/kg AVE	RSD% (n = 5)	200 μg/kg AVE	RSD% (n = 5)
18	Azinphos-ethyl	2642-71-9	10	104.9	5.7	89.6	3.9	96.2	9.5	5	83.3	4.2	78.8	10.3	80.6	10.9
19	Aziprotryne	4658-28-0	10	91.4	5.4	89.6	3.1	99.4	5.2	5	91	4	87.6	4	90.2	4.5
20	Azoxystrobin	131860-33-8	50	NA	NA	111.4	6.8	76.2	14.3	5	76.4	3.4	71.4	2.8	82.8	2
21	Beflubutamid	113614-08-7	5	111.4	4.7	94.9	4.2	94.5	1.4	5	87.6	7.6	95.7	4	98.1	11.6
22	Benalaxyl	71626-11-4	5	91.4	2	90.6	1.7	94.6	8.2	5	84.3	1.8	109.7	1.7	104	5.6
23	Bendiocarb	22781-23-3	5	89	4.4	88.2	3.2	92.7	6.7	5	82.3	6	80.7	2	82.6	1.3
24	Benodanil	15310-01-7	20	NA	NA	87.5	7.4	93.5	8.2	5	90.2	3.7	89.5	2.3	86.8	0.8
25	Benoxacor	98730-04-2	5	94.5	7.2	90.4	1.6	94.2	7.8	20	NA	NA	82.1	17.1	100.4	4.2
26	Benzoximate	29104-30-1	5	84.7	6.3	87.8	2.7	97.4	7.8	5	71.2	1.6	70.7	0.5	89	2.1
27	Benzoylprop-ethyl	22212-55-1	5	97.4	7.8	92.9	3.4	95.9	8.3	5	75.9	3.3	118.2	1.4	107.5	4.6
28	Bifenazate	149877-41-8	50	NA	NA	120	6.1	98	12.3	5	81.3	11.1	78.7	16	84.2	7.8
29	Bioresmethrin	28434-01-7	20	NA	NA	76.9	4.8	74.7	11.2	50	NA	NA	NA	NA	NA	NA
30	Bitertanol	55179-31-2	20	NA	NA	97.5	14.5	88.4	7.4	5	83.8	5.2	100.7	3	93.8	2.4
31	Boscalid	188425-85-6	5	88.6	4.1	87.6	4.2	98.6	3.9	5	96	15.8	85.6	18.9	106.8	4.9
32	Bromfenvinfos	33399-00-7	5	96.6	3.5	93.6	2.7	98.8	8.5	5	84.2	2.5	86.4	0.7	93.4	1.6
33	Bromobutide	74712-19-9	10	85.3	5.3	90.2	2.3	95.4	6.9	5	76.8	3.7	106.4	3.2	104.2	3.3
34	Bromuconazole	116255-48-2	5	92.5	7.2	90.1	4	92.6	10.9	5	84.7	3.2	87	3.9	91.5	2.1

	Name	CAS														
35	Bupirimate	41483-43-6	5	88.6	3.6	89.3	1.9	94.8	8.1	5	82.1	7.7	80	2.6	95.1	7.1
36	Buprofezin	69327-76-0	50	NA	NA	97.5	6	93.9	2.5	5	92	1.8	82.3	1.5	94.7	1.3
37	Butachlor	23184-66-9	5	92.6	4.3	93.1	1.7	96.5	7.1	5	98.1	8.5	85.6	4.1	98.5	9
38	Butafenacil	134605-64-4	5	100.2	14.8	90.1	6.4	99.1	9.5	5	81	9.8	76.9	2.4	111.1	3.9
39	Butamifos	36335-67-8	5	NA	11	95.2	4.1	104.8	6.4	5	95.3	5.7	71.3	11.6	87.5	2.4
40	Butralin	33629-47-9	5	103.1	8.4	95.8	4.2	95.3	3.2	5	85.2	6.6	75.2	9.4	70.4	9
41	Cadusafos	95465-99-9	5	95.1	6.8	92.3	4.5	92.3	1.8	5	88.7	3.8	78.1	3.6	89.2	1.7
42	Cafenstrole	125306-83-4	20	NA	NA	98.4	7.4	94.5	5.6	5	76.4	3.5	113.9	1.2	108.7	4.2
43	Carbaryl	63-25-2	5	82.6	6.5	92	3.5	93.9	2.4	50	NA	NA	NA	NA	111.1	5.5
44	Carbofuran	1563-66-2	5	104.3	6	93.9	15.2	87.1	4.4	5	77.6	7.7	79.9	3.6	94	5.5
45	Carbofuran-3-hydroxy	16655-82-6	20	NA	NA	NA	NA	101.5	7.4	5	72	9.4	72.2	7.8	80.1	12.2
46	Carboxin	5234-68-4	100	NA	NA	NA	NA	94.5	NA	NA	NA	NA	NA	NA	NA	NA
47	Chlordimeform	6164-98-3	100	NA	NA	NA	NA	80.5	7.5	NA	NA	NA	NA	NA	NA	NA
48	Chlorfenvinphos	470-90-6	5	98.4	9.1	81	8.8	92.5	3.7	5	87	7	83.1	1.9	111.8	4.4
49	Chloridazon	1698-60-8	200	NA	NA	NA	NA	100.4	5.8	5	72.4	10.1	75.7	4.9	75.6	10.5
50	Chlorotoluron	15545-48-9	20	NA	NA	90	18.7	73.8	9.2	5	87.9	1.7	107.7	2.4	93.9	1.7
51	Chlorpyrifos	2921-88-2	5	91.6	4	93.5	5.4	93.1	2.7	10	93.9	8.7	76.8	1.7	82.6	2.6
52	Chlorpyrifos-methyl	5598-13-0	5	102.8	4.1	96.4	5.7	93.8	2.7	NA	NA	NA	NA	NA	NA	NA

(Continued)

APPENDIX TABLE C-1 Evaluation Results of the Common-Detected Pesticides by GC–Q–TOF/MS Combined with LC–Q–TOF/MS (cont.)

Black Tea

| No. | Pesticides | CAS Number | GC–Q–TOF/MS | | | | | | | LC–QTOF/MS | | | | | | |
			SDL (μg/kg)	10 μg/kg AVE	RSD% (n=5)	50 μg/kg AVE	RSD% (n=5)	200 μg/kg AVE	RSD% (n=5)	SDL (μg/kg)	10 μg/kg AVE	RSD% (n=5)	50 μg/kg AVE	RSD% (n=5)	200 μg/kg AVE	RSD% (n=5)
53	Chlorsulfuron	64902-72-3	5	86.7	14.9	83.3	5.1	92.9	14.4	20	NA	NA	NA	NA	NA	12.7
54	Chlorthiophos	60238-56-4	5	89.1	1.8	91.9	2.1	96	6.3	NA	NA	NA	NA	NA	NA	NA
55	Clodinafop	114420-56-3	5	86.9	7.7	94.3	2.6	96.2	5.6	20	NA	NA	NA	NA	NA	NA
56	Clodinafop-propargyl	105512-06-9	5	107.5	11.1	92.8	5.6	104.1	3.7	5	89.6	8.3	85.1	2.1	104.3	5
57	Clomazone	81777-89-1	5	97.7	4.8	92.3	5.2	94.7	1.1	5	81.5	8.1	91.4	2.1	84.1	2
58	Coumaphos	56-72-4	5	88.2	8.5	91.3	6	95	9.3	5	103.7	1.5	107.6	4.7	104.9	3.3
59	Crufomate	299-86-5	10	104.9	16.8	89.3	4.9	90.2	2	5	83.5	3.5	86.5	2.8	102.9	2.6
60	Cycloate	1134-23-2	20	NA	NA	86.9	7.7	91.1	5.9	5	84.7	9.7	119.6	14.5	114.3	5.9
61	Cyflufenamid	180409-60-3	20	NA	NA	92.5	3	95.7	7	5	78.6	1.8	106.4	2.4	107.2	4.1
62	Cyprazine	22936-86-3	5	97.3	6.6	96.2	8.9	95	3	5	82.7	7.8	83.8	2.5	88.6	9.6
63	Cyproconazole	94361-06-5	5	104.3	12.7	86.7	16.6	87.5	9.6	5	98.2	3.7	100.2	4	103.3	1.6
64	Cyprodinil	121552-61-2	5	81.6	3.1	90.6	4.9	89.8	1.5	5	90.3	2.5	73.8	1.4	85.8	1.4
65	Cyprofuram	69581-33-5	5	89.6	2.5	92	2.2	94.7	5.7	5	80.9	2.8	109.5	2.4	102	4
66	Desmetryn	1014-69-3	5	91.9	2.1	91.3	1.4	95.9	6.2	5	84.1	2.9	103.3	1.8	96.5	4.1
67	Dialifos	10311-84-9	100	NA	NA	NA	NA	94.4	3.6	10	78.1	6.3	106.2	2.7	104.4	4.6
68	Diallate	2303-16-4	5	91.1	4.1	89.8	1.2	94.6	6.5	10	61.8	9.1	71.4	6.2	92.7	2.2
69	Dichlofenthion	97-17-6	5	102.6	4.9	93.7	5.5	92.5	0.5	NA	NA	NA	NA	NA	NA	NA
70	Diethatyl-ethyl	38727-55-8	5	92.5	7.1	92.9	5	92.1	2	5	82.2	2.1	109.3	1.8	105.9	4.5

No.	Name	CAS														
71	Diethofencarb	87130-20-9	5	95.5	4.5	73.3	8.2	90.6	10.9	5	95.6	5.7	88.9	4.4	97.6	5.2
72	Diethyltoluamide	134-62-3	20	NA	NA	91.6	4.1	90.9	0.9	5	96.5	12.1	97	2.3	107.7	2.3
73	Difenoxuron	14214-32-5	5	91.3	12.1	112.9	8.3	93.1	9.7	5	86.8	7.1	84.6	2.7	101.2	5.3
74	Dimethachlor	50563-36-5	5	94.7	9.7	93.9	4.9	96	2.4	5	81.6	3.9	81.5	2.2	87.5	2
75	Dimethametryn	22936-75-0	20	NA	NA	93.4	6	93.7	4.4	5	86.2	7.4	82.2	4.2	102.3	3
76	Dimethenamid	87674-68-8	5	95.1	2.4	89.3	1.7	96	6.1	5	75	3.9	78.8	1.8	94.4	1.8
77	Dimethoate	60-51-5	5	118	7.6	81.5	3	88	8.8	5	84.4	10	73.5	9.3	84.7	9.3
78	Dimetilan	644-64-4	5	111.4	8.6	93.4	4.3	91.2	1.9	5	90.5	19.2	96.8	13.2	103.2	7.2
79	Diniconazole	83657-24-3	5	114.7	9.9	89.6	5.4	99.9	5.7	5	81.6	8.5	79.4	1.2	97.2	7.8
80	Dinitramine	29091-05-2	5	113.2	8.7	93.8	4.6	103.2	4.8	10	95.1	13.8	95.5	6.5	101.5	3.3
81	Diphenamid	957-51-7	5	89.7	2.9	90	2	95.8	6.9	5	83.2	7.8	81.5	2.9	103.6	3
82	Dipropetryn	4147-51-7	5	97	6.3	93.9	3.8	93	2.6	5	71.6	6.3	85.4	2.2	98.5	6.1
83	Disulfoton sulfone	2497-06-5	5	99.8	5.1	91.4	2.9	NA	NA	5	84.2	4.7	84.6	2	96.2	1.6
84	Disulfoton sulfoxide	2497-07-6	5	102.1	10.9	91.8	5.3	96.2	2.3	5	86.4	6.8	84.4	3.3	107.6	2.8
85	Ditalimfos	5131-24-8	5	103.7	18.3	91.3	2.8	88.4	4.4	5	87.2	3.5	101.8	0.7	104.8	6
86	Dithiopyr	97886-45-8	5	87.1	5.2	95	2.9	87.8	0.7	5	89	3.8	87.9	3.8	106.6	4.6
87	Dodemorph	1593-77-7	5	62.9	4.4	62.7	7.6	NA	11.6	5	71.3	7	70	0.8	97.8	9.8
88	Edifenphos	17109-49-8	5	108.4	5.5	96.8	3.7	103.6	9.4	5	87.3	2.5	90.2	1.8	94.5	1.3
89	Esprocarb	85785-20-2	5	NA	5	76.4	14.8	96.4	1.5	5	85.6	2.7	85.8	1.3	92.8	1.6
90	Ethion	563-12-2	5	99.9	5	91.4	2.9	79.1	16.3	5	85	9.4	97.5	7.5	98.4	8.7

(Continued)

APPENDIX TABLE C-1 Evaluation Results of the Common-Detected Pesticides by GC–Q–TOF/MS Combined with LC–Q–TOF/MS (*cont.*)

No.	Pesticides	CAS Number	Black Tea GC–Q–TOF/MS SDL (μg/kg)	10 μg/kg AVE	10 μg/kg RSD% (n=5)	50 μg/kg AVE	50 μg/kg RSD% (n=5)	200 μg/kg AVE	200 μg/kg RSD% (n=5)	LC–QTOF/MS SDL (μg/kg)	10 μg/kg AVE	10 μg/kg RSD% (n=5)	50 μg/kg AVE	50 μg/kg RSD% (n=5)	200 μg/kg AVE	200 μg/kg RSD% (n=5)
91	Ethoprophos	13194-48-4	5	87.5	4.4	87.7	2.1	94.9	6.8	5	85.5	4.9	90.5	3.6	98.8	1.1
92	Famphur	52-85-7	5	107.1	4.6	90.4	11.2	90.9	5.5	5	88.9	4.3	94	2.2	117.9	3.6
93	Fenamidone	161326-34-7	5	90.9	1.3	89.4	2.7	94.7	7	5	84.2	5.4	108	1.4	103.7	5.7
94	Fenamiphos	22224-92-6	20	NA	NA	102.1	7.1	98.3	12.3	5	94.5	5.3	79.5	6.4	116.3	6.1
95	Fenarimol	60168-88-9	10	107.9	15.6	92	5.4	91.7	4.9	5	NA	4.5	NA	2.5	70.2	4.7
96	Fenazaquin	120928-09-8	5	89.5	5.3	87.1	1.1	94.7	6.6	5	82.9	2.9	74.3	1.5	84.9	2.1
97	Fenfuram	24691-80-3	100	NA	NA	NA	NA	NA	NA	5	89.7	6.4	86.6	5.6	77.7	2.8
98	Fenobucarb	3766-81-2	5	NA	4.1	91.6	3.3	99.3	5.4	5	94.3	10	88.9	2.9	107	7.4
99	Fenothiocarb	62850-32-2	5	99.3	4.1	94.3	4.5	94.5	2.3	5	94.9	7.3	92.4	1.3	96.4	8
100	Fenoxaprop-ethyl	66441-23-4	5	93.1	4.7	92.8	3.2	97.1	8.1	5	94.3	7.4	90.1	1.8	108.8	3.5
101	Fenoxycarb	72490-01-8	20	NA	NA	91.6	11.6	86	5.5	5	89.1	11	91.3	1.8	100	3.8
102	Fenpropidin	67306-00-7	20	NA	NA	NA	16.2	NA	13.6	5	78.8	6.4	79.4	2.5	70.2	15.3
103	Fenpropimorph	67564-91-4	5	76.2	8.4	91.9	4.4	80	3	5	71.3	7	71.4	2.9	97.8	9.8
104	Fensulfothion	115-90-2	5	93	3.4	86.4	3	94.6	2.9	5	90.7	7.9	85.3	3.9	119.6	4.6
105	Fensulfothion-oxon	6552-21-2	200	NA	NA	NA	NA	101.9	19.7	5	79.1	10.2	97.1	3.7	101.8	3.2
106	Fensulfothion-sulfone	14255-72-2	5	113.5	8.8	93.1	4.3	90.2	4.2	10	93.7	8.9	111	2.7	105.6	1.6
107	Fenthion	55-38-9	5	92.7	7	106.3	4.2	113.4	1.7	10	74.9	11.8	116.7	12.7	119.9	3.8
108	Flamprop-isopropyl	52756-22-6	5	86.3	11.6	95.8	4.5	92.1	2.7	5	84.6	1.9	100.8	2	100.7	2.9
109	Flamprop-methyl	52756-25-9	5	83.2	12.8	95.2	4.9	88	1.6	5	76.6	3.1	88.4	2.3	94.2	2.8

No.	Name	CAS															
110	Flufenacet	142459-58-3	5	101.8	12.6	84.7	4.5	97.6	6.1	5	82.1	9	83.7	2.3	105.9	3.9
111	Fluopyram	658066-35-4	5	91.3	3.2	91.8	2.1	95.6	7.4	5	111.6	2	113	4	110.4	1.6
112	Fluorogly-cofen-ethyl	77501-90-7	20	NA	NA	91.9	3.7	90	2.5	NA	NA	NA	NA	NA	NA	NA
113	Flurochlori-done	61213-25-0	5	97.3	5.2	91.7	5.1	98.7	2.5	50	NA	NA	99.8	10.3	88.6	8.1
114	Flurprimidol	56425-91-3	5	93.4	6.8	94.8	4	92.4	1.9	5	92.9	7.4	99	10.6	106.1	3.7
115	Flusilazole	85509-19-9	5	101.1	5.8	79.9	18.6	91.9	6.3	5	83.5	2.8	110	1.8	101.3	3.8
116	Flutolanil	66332-96-5	5	91.2	7.6	91.4	3.8	89.4	2.2	5	89.3	3.2	81.6	1.1	86.5	2.3
117	Flutriafol	76674-21-0	5	95.4	7.3	90.5	3.6	91.8	8.7	5	82.9	5.3	88	3.1	92.6	3.3
118	Fluxapyroxad	907204-31-3	5	96.7	5.5	81.5	11.6	89.9	6.5	5	105.8	7.1	112.6	5.4	109.2	2.9
119	Fonofos	944-22-9	5	93.6	2.5	89.9	2.1	95.5	6.3	20	NA	NA	93.2	5.5	111.7	3.5
120	Fuberidazole	3878-19-1	100	NA	NA	NA	NA	NA	NA	5	106.2	13.1	70.5	6	92.8	16.3
121	Furalaxyl	57646-30-7	5	88.9	2.8	89.9	1.7	96.5	6	5	72	5.1	70.3	2.7	92.2	1.8
122	Furathiocarb	65907-30-4	10	102.1	3.2	94.7	3.4	95.5	6.6	5	77.9	2	114.2	2.5	114.1	2.7
123	Furmecyclox	60568-05-0	5	60.8	13.2	NA	13.4	NA	7.7	5	76.9	9.8	71.1	6.4	70.6	3.4
124	Haloxyfop	69806-34-4	NA	NA	NA	NA	NA	NA	NA	5	NA	NA	NA	NA	NA	NA
125	Haloxyfop-methyl	69806-40-2	5	90.5	1.5	90.4	1.9	96.7	5.4	5	78.2	1.6	86.1	2.1	91.8	1.6
126	Heptenophos	23560-59-0	5	95.1	18.3	92.4	3.8	94.8	8.4	5	85.5	7.1	93	6.1	100.3	5
127	Hexacon-azole	79983-71-4	50	NA	NA	92.9	5.3	95.6	5.9	5	82.2	2.1	103.1	0.7	100.6	4.2

(Continued)

APPENDIX TABLE C-1 Evaluation Results of the Common-Detected Pesticides by GC-Q-TOF/MS Combined with LC-Q-TOF/MS (cont.)

			Black Tea													
			GC–Q-TOF/MS							LC–QTOF/MS							
			SDL (µg/kg)	10 µg/kg		50 µg/kg		200 µg/kg		SDL (µg/kg)	10 µg/kg		50 µg/kg		200 µg/kg	
No.	Pesticides	CAS Number		AVE	RSD% (n = 5)	AVE	RSD% (n = 5)	AVE	RSD% (n = 5)		AVE	RSD% (n = 5)	AVE	RSD% (n = 5)	AVE	RSD% (n = 5)
128	Hexazinone	51235-04-2	10	93.7	7.2	88.1	4.1	87.1	4.7	5	73.9	7.2	73.4	3	95.7	3.2
129	Imazamethabenz-methyl	81405-85-8	20	NA	NA	78.2	19.4	82.6	8.1	5	70.1	12.9	73.9	3	99.6	2.9
130	Indoxacarb	144171-61-9	200	NA	NA	NA	NA	87.9	2.2	5	74.3	3.4	76.6	1.4	87.8	2.9
131	Iprobenfos	26087-47-8	20	NA	NA	90.6	2.9	95.8	8	5	85.3	2	81.1	2.3	91.7	2.7
132	Iprovalicarb	140923-17-7	50	NA	NA	78.9	5.7	94.4	4.6	5	76.6	3.1	76.8	3.5	90.7	1.8
133	Isazofos	42509-80-8	5	93.7	1.4	91.2	2.3	97.4	6.7	5	78.8	2.1	108.3	2.3	106.4	3.8
134	Isocarbamid	30979-48-7	5	96.8	14.1	80.2	4.3	90.8	12.9	5	66.3	14.7	98.1	6.5	111.5	10.9
135	isocarbophos	24353-61-5	5	94.1	6.7	87.5	2.6	87.4	3	5	70.7	5.1	116.7	2	NA	3.6
136	Isofenphos-oxon	31120-85-1	5	102.2	12.1	91.3	4.2	95.9	7	5	83.8	8.8	81.3	3.7	108.7	4.2
137	Isoprocarb	2631-40-5	5	87.5	2.6	90.9	1.3	94.7	7.6	5	NA	NA	NA	NA	NA	3.9
138	Isopropalin	33820-53-0	5	92.3	3	90.5	2.1	94.9	7.7	5	101.9	10	95.2	7.9	98.6	11.2
139	Isoprothiolane	50512-35-1	5	94.5	5	94.9	5	90.8	0.9	5	76.1	4.4	79.8	2.4	106	6.4
140	Isoxadifen-ethyl	163520-33-0	5	NA	11.7	91.5	2.8	94.5	5.2	NA	NA	NA	NA	NA	NA	NA
141	Isoxathion	18854-01-8	5	86.3	11.6	95.8	4.5	92.1	2.7	5	81.9	1.5	109.7	4	113.9	4.1
142	Kresoxim-methyl	143390-89-0	5	83.1	2	90.3	2	96.7	6.3	5	81.4	10.7	83.8	1.2	116.1	3.8
143	Malathion	121-75-5	20	NA	NA	94	6.4	98.5	2.1	5	89.7	7.7	88.6	1.7	106.3	6.6
144	Mefenacet	73250-68-7	50	NA	NA	89.2	5.7	91.9	7.3	5	84.1	8.4	82.6	3.2	104	7.4
145	Mefenpyr-diethyl	135590-91-9	5	88.7	3.3	91.2	3.1	95	7.1	5	90.2	9.6	82.5	3.5	114.6	7

No.	Name	CAS														
146	Mepanipyrim	110235-47-7	5	106	5.7	90.4	5.8	94.1	3.3	5	82.5	10	83.6	2.1	99.4	4.1
147	Mepronil	55814-41-0	5	101	5.8	91.5	4.6	92.4	2.9	5	92.6	7.6	89.9	2	102.1	7.5
148	Metalaxyl	57837-19-1	5	101	2.6	79.5	13.8	92.3	4.9	5	87.4	1.7	107.2	1.4	104.9	5.6
149	Metamitron	41394-05-2	100	NA	NA	NA	NA	NA	NA	5	70.9	10.9	71.1	6.6	71.1	5.6
150	Metazachlor	67129-08-2	5	85.1	11.2	93.1	5	89.2	2.4	5	75.5	9.8	75.9	3.2	100.1	8.3
151	Metconazole	125116-23-6	50	NA	NA	97.5	11.4	94.8	3.1	5	84.2	10.4	81.9	1.9	108.7	4.6
152	Methabenz-thiazuron	18691-97-9	5	92.6	4.6	89	2.5	93.4	7.1	5	81.3	2.8	107.1	1.5	106.6	4.5
153	Methamido-phos	10265-92-6	20	NA	NA	NA	NA	76	17.1	5	89.1	8.1	75.6	16.8	70.3	7.5
154	Methidathion	950-37-8	10	102.8	8.8	94.3	3.5	102.3	3	50	NA	NA	116.8	6.2	105.6	2.1
155	Methopro-tryne	841-06-5	5	92.8	4.9	91.9	2.2	92.3	8.2	5	79.3	2.8	85.3	3.2	87.5	2.5
156	Metolachlor	51218-45-2	5	90	2.7	88.6	1.9	95.1	8	5	82.6	1.7	109.8	2.3	105.1	4.9
157	Metribuzin	21087-64-9	10	102.9	5.7	84.7	14.2	92.7	5.5	5	72.1	7.9	71.5	2.6	73.2	13
158	Mevinphos	7786-34-7	5	96.1	7	94.5	12.8	93.8	5.7	5	82.5	9.5	78.2	0.9	88.5	5.1
159	Mexacarbate	315-18-4	10	NA	NA	94.8	6.1	94.6	6.7	5	76.1	8.9	112.4	7.2	117.1	5.3
160	Molinate	2212-67-1	5	82.6	6.4	89.3	1.6	93.3	7.7	50	NA	NA	114.4	4.6	112.2	9
161	Myclobutanil	88671-89-0	5	99.4	7.3	96.4	2.4	95	7.4	5	81.8	9	77.7	3.8	101.5	7.7
162	Napropamide	15299-99-7	10	73.8	18.4	94.8	3.6	96	7.8	5	85.4	2.2	106	2.5	105.2	2.9
163	Nitralin	4726-14-1	20	NA	NA	107.9	14.5	96.6	4.5	50	NA	NA	94.2	12.9	105.4	13.5

(Continued)

APPENDIX TABLE C-1 Evaluation Results of the Common-Detected Pesticides by GC-Q-TOF/MS Combined with LC-Q-TOF/MS (cont.)

			Black Tea												
			GC-Q-TOF/MS						LC-QTOF/MS						
			SDL (μg/kg)	10 μg/kg		50 μg/kg		200 μg/kg		SDL (μg/kg)	10 μg/kg		50 μg/kg		200 μg/kg	
No.	Pesticides	CAS Number		AVE	RSD% (n = 5)	AVE	RSD% (n = 5)	AVE	RSD% (n = 5)		AVE	RSD% (n = 5)	AVE	RSD% (n = 5)	AVE	RSD% (n = 5)
164	Norflurazon	27314-13-2	5	98.3	7.2	93.1	5.5	91.1	9.1	5	80.9	1.9	106.8	0.7	102.9	3.9
165	Nuarimol	63284-71-9	20	NA	NA	92.1	3	90.3	2.5	5	75	8.3	75	2.5	101.6	6.2
166	Octhilinone	26530-20-1	5	94.6	6.5	93.3	3.1	91.5	7.2	5	83.8	1.4	104.3	2.8	104.7	2.7
167	Ofurace	58810-48-3	5	82.5	13.7	87.7	2.9	81.1	6.7	5	75.6	8.3	76.3	2.4	103.4	3.3
168	Orbencarb	34622-58-7	5	90.8	2.8	90.4	1.9	97.1	5.6	5	81.4	2.4	108.6	0.6	107.2	2.9
169	Oxadixyl	77732-09-3	5	90.9	11	86.9	4.1	75.1	6	5	70.1	6	72.5	3.1	90.7	9.6
170	Paclobutrazol	76738-62-0	5	91.9	1.9	90.6	3.6	90.5	10.4	5	105.3	4.2	96	4.9	95	2.9
171	Pebulate	1114-71-2	5	82.6	7.5	93.1	3	90.3	10.4	20	NA	NA	NA	9	NA	15.1
172	Penconazole	66246-88-6	5	90.6	4.5	90.8	1.9	95.2	7.3	5	95.9	2.2	79.2	2.5	86.9	1
173	Pendimeth-alin	40487-42-1	5	96	5.7	91.4	3	98.5	2.8	10	95.7	4	110.9	6.8	115.7	3
174	Pentanochlor	2307-68-8	5	87.4	3.3	92	1.4	97.8	5.9	5	85.2	2.5	97.4	6.5	99.3	6.3
175	Phenthoate	2597-03-7	5	97.3	4.3	95.5	2.3	102.4	8.2	20	NA	NA	83.4	9	96.7	13.4
176	Phorate-sulfone	2588-04-7	5	118.6	8.9	94.1	2.9	97.3	3.3	5	83.8	13	80.5	1.8	84.2	13.9
177	Phorate-sulfoxide	2588-03-6	50	NA	NA	119.2	11.3	91.7	6.2	5	87.5	2.1	108.4	2.4	111	3.7
178	Phosalone	2310-17-0	5	116.5	11.3	89.1	5.8	102.9	8.1	20	NA	NA	71.3	19.2	NA	15
179	Phosfolan	947-02-4	5	108.1	11.8	81	4.7	91.9	9	5	85.3	2.6	103.4	2.1	104.4	4.6
180	Phosphami-don	13171-21-6	5	94.4	4.6	96.2	4	97.4	9.6	5	80.8	2.4	83.8	1.7	90.4	1.8

No.	Name	CAS														
181	Phthalic acid, Benzyl butyl ester	85-68-7	5	95.6	4.8	94.4	4.6	92.3	1.1	5	64	5.7	NA	2.1	118.7	5.5
182	Phthalic acid, Bis-butyl ester	84-74-2	NA	NA	NA	NA	NA	NA	NA	5	117	11.4	114.1	16	114.7	5.5
183	Phthalic acid, Bis-cyclohexyl ester	84-61-7	20	NA	NA	93.9	5.5	90	1	5	NA	12.5	101.2	4.5	108.5	6.1
184	Picolinafen	137641-05-5	5	92.3	4.3	92.6	2.3	96.1	6.7	5	89.4	2.5	81.8	2.1	81.7	5.4
185	Picoxystrobin	117428-22-5	5	97.1	4.4	89.6	1.4	96.7	5.9	5	84.9	2.8	103.2	2.5	103.9	3.6
186	Piperonyl butoxide	51-03-6	20	NA	NA	94	3.3	92.4	3.1	5	94.4	6.9	89.4	2.5	108.8	3.6
187	Piperophos	24151-93-7	5	71.8	NA	90	2.5	95.8	8.1	5	89.3	7.4	85.6	2.7	108.1	3.1
188	Pirimicarb	23103-98-2	5	91.1	7.8	92.8	4.6	94.1	1.7	5	86.7	2.2	86	1.2	84.6	0.9
189	Pirimiphos-ethyl	23505-41-1	5	97.9	15.9	90.8	2.7	95.3	6.8	5	95.5	7.6	91.6	2.5	99.5	5.1
190	Pirimiphos-methyl	29232-93-7	5	108.6	5.2	95.5	4.7	95.5	1.8	5	97.6	2.5	79.5	1.6	90.5	1.5
191	Pirimiphos-methyl-N-desethyl	67018-59-1	10	112.5	2.4	94.9	3.4	88.6	3.6	5	82.7	7	103.2	3.6	105.1	2
192	Prallethrin	23031-36-9	20	NA	NA	NA	1.6	NA	9.3	NA	NA	NA	NA	NA	NA	NA
193	Pretilachlor	51218-49-6	5	98.5	2.8	92.1	1.9	97.1	7.7	5	84.1	3.3	78.9	2	91.5	2.6
194	Profenofos	41198-08-7	5	101.3	7.3	87.1	5.4	95.7	5.2	5	98.1	1.7	85.1	1.7	90.7	3
195	Promecarb	2631-37-0	50	NA	NA	108.3	4.4	85.4	6.4	5	91.9	5.4	107.5	6	108.4	3.2
196	Prometon	1610-18-0	5	88.4	6.9	91.6	3.7	91.2	1.5	5	72	6.1	77.8	2.4	88.8	6.7

(Continued)

APPENDIX TABLE C-1 Evaluation Results of the Common-Detected Pesticides by GC–Q-TOF/MS Combined with LC–Q-TOF/MS (cont.)

No.	Pesticides	CAS Number	Black Tea GC–Q-TOF/MS SDL (µg/kg)	10 µg/kg AVE	RSD% (n = 5)	50 µg/kg AVE	RSD% (n = 5)	200 µg/kg AVE	RSD% (n = 5)	LC-QTOF/MS SDL (µg/kg)	10 µg/kg AVE	RSD% (n = 5)	50 µg/kg AVE	RSD% (n = 5)	200 µg/kg AVE	RSD% (n = 5)
197	Prometryn	7287-19-6	5	101.3	4.8	87.6	9.5	93.1	3.8	5	89.3	1.8	87.7	1.9	91.1	2.2
198	Propachlor	1918-16-7	10	95	10.5	93.4	5.8	89.8	1.7	5	87.8	3.4	92.5	2.4	81.8	1.6
199	Propanil	709-98-8	5	100.6	2.7	97.8	1.5	96.3	6.4	5	86	8	89.7	2.9	100.5	3.8
200	Propaphos	7292-16-2	5	114.7	8.4	98.5	2.5	94.4	8.3	5	88.8	6	81.8	5.2	96.1	2
201	Propargite	2312-35-8	100	NA	NA	NA	NA	NA	13.1	5	92.2	4.9	71.8	4.5	77.7	4.8
202	Propazine	139-40-2	5	93	2.8	91.2	1.2	96.1	6.4	5	86.9	3.8	91.3	2.2	87	1.6
203	Propetamphos	31218-83-4	5	99.2	3.3	89.4	1.5	96.4	7.1	50	NA	NA	111.9	6.5	110.6	2.4
204	Propisochlor	86763-47-5	5	91.3	0.8	89.5	1.9	96.2	6	5	96.2	6	84.6	2.2	89.5	2
205	Propyzamide	23950-58-5	5	105.1	4.6	89.4	9.5	95.2	3.6	5	91.5	5.3	79.2	2.4	80.3	3.2
206	Prosulfocarb	52888-80-9	50	NA	NA	94.6	5.5	92.3	1.2	5	97.3	6	93	2.6	98.4	8.5
207	Pyraclostrobin	175013-18-0	100	NA	NA	NA	NA	98.5	6.3	5	85.9	1.9	78.5	2.4	89.2	1.7
208	Pyrazophos	13457-18-6	5	97.4	8.9	84.6	3	95.8	4.8	5	79.4	3.1	113.3	3.1	107.1	5.4
209	Pyributicarb	88678-67-5	5	104.4	11.5	95.8	4.5	96.7	3.1	5	95.7	10.2	89.6	3.2	96.5	5.8
210	Pyridaben	96489-71-3	20	NA	NA	93.4	6.8	95.1	6.5	5	79.1	3.2	103.7	2.9	107.4	3.5
211	Pyridalyl	179101-81-6	20	NA	NA	NA	NA	NA	NA	5	NA	19.9	NA	NA	NA	NA
212	Pyridaphenthion	119-12-0	5	114.8	17.7	88.2	4.9	99.3	7	5	89.1	11.4	92.2	5.4	112.1	9.9
213	Pyrifenox	88283-41-4	50	NA	NA	91.8	3.3	89.5	5.7	5	83.3	3	105.9	1.6	101	5
214	Pyriftalid	135186-78-6	20	NA	NA	90	5.4	86.1	2.8	5	95.1	7.3	119.6	2.6	109.6	1.2

No.	Name	CAS															
215	Pyrimethanil	53112-28-0	5	86.7	3.1	88.9	1.8	94.3	5.9	5	83.8	7.5	83.3	2.6	94.2	4.1
216	Pyriproxyfen	95737-68-1	5	97.4	9.6	95.6	5.1	94.5	3.2	5	84.2	0.5	109.3	1.4	113.4	2.8
217	Pyroquilon	57369-32-1	5	95.5	3.4	90.8	2	95.5	6.4	5	80.8	2.7	99.5	2.4	93.3	6.1
218	Quinalphos	13593-03-8	5	110.3	5.8	92.7	5.4	92.4	1.4	5	93.9	8.2	89.3	2	102.6	8.2
219	Quinoclamine	2797-51-5	20	NA	NA	72.5	9.6	84	12.5	5	63.1	9.5	61.4	3.2	79.9	5.1
220	Quinoxyfen	124495-18-7	5	93.4	3	89.6	2.6	97.8	2	5	84.4	2.1	111.8	6.2	107	3.1
221	Quizalofop-ethyl	76578-14-8	10	98.2	3.3	89.6	6.9	91.3	7.5	5	85.2	2.2	82.9	1.9	85.4	2.1
222	Rabenzazole	40341-04-6	5	NA	NA	74.1	12.9	89.7	5.8	5	72.3	10.8	96	20	71.6	9.8
223	Sebuthylazine	7286-69-3	5	95.6	5.4	91.8	4.6	90.9	0.9	5	85.8	7.6	86	2.3	98.7	8.1
224	Sebuthylazine-desethyl	37019-18-4	5	100.7	15.4	80.3	4.3	97.9	11.6	5	78.5	1.8	96.9	1.2	93.8	4
225	Secbumeton	26259-45-0	5	94.1	3.2	90.3	2.7	94.9	8	5	78.2	2.2	81.8	3.7	91.8	4.2
226	Simazine	122-34-9	5	105.8	9.3	90.8	4.5	94.1	2.1	5	72.3	1.5	70.4	3.1	91.3	8.9
227	Simeconazole	149508-90-7	10	119	11.9	93.7	6.5	92.8	1.8	5	80.1	9	77.7	1.6	95.3	8.7
228	Simeton	673-04-1	5	98.1	5.2	91.4	2.1	94.4	8.3	5	79.5	10.7	78.6	3.6	90.7	3.5
229	Simetryn	1014-70-6	5	93.4	4.6	90.2	3.6	98.2	1.1	5	83.5	1.8	86.7	2.8	80.3	1
230	Spirodiclofen	148477-71-8	50	NA	NA	98.4	NA	78.5	12.9	5	79.7	6.4	80.5	6.8	100.9	4.9
231	Spiroxamine	118134-30-8	5	64.2	7.1	63.5	3.3	64.7	6.3	5	NA	6.2	NA	4.6	60	6.7
232	Sulfallate	95-06-7	5	92.7	4.4	90.2	1.8	96.1	7.2	50	NA	NA	119.3	6	106.3	7.5

(Continued)

APPENDIX TABLE C-1 Evaluation Results of the Common-Detected Pesticides by GC–Q-TOF/MS Combined with LC-Q-TOF/MS (cont.)

No.	Pesticides	CAS Number	Black Tea GC–Q-TOF/MS SDL (μg/kg)	10 μg/kg AVE	RSD% (n = 5)	50 μg/kg AVE	RSD% (n = 5)	200 μg/kg AVE	RSD% (n = 5)	LC–QTOF/MS SDL (μg/kg)	10 μg/kg AVE	RSD% (n = 5)	50 μg/kg AVE	RSD% (n = 5)	200 μg/kg AVE	RSD% (n = 5)
233	Sulfotep	3689-24-5	5	94.6	4.9	86.6	4.2	98.8	2.1	5	117.6	8.3	96.9	8.1	104.5	7.3
234	Sulprofos	35400-43-2	5	92.4	11.9	107.7	5	118.4	3.5	10	80.2	13	92.4	14.4	101.6	10.8
235	Tebuconazole	107534-96-3	10	93.9	10.3	94.1	4.3	92	9.6	5	85.8	12.6	85.5	4	103.6	9.3
236	Tebufenpyrad	119168-77-3	5	93.4	6.9	94.2	5.5	93.6	2.2	5	80.3	2	107.3	1	109.5	2.7
237	Tebupirimfos	96182-53-5	5	96.4	4.9	90.6	2.7	97.8	2.8	5	97.8	7.3	92.2	4.7	98.4	8.5
238	Tebutam	35256-85-0	5	82.8	5.7	105.2	7.1	95.2	7.3	5	81.9	7.6	87	5.1	99.3	7
239	Tebuthiuron	34014-18-1	10	88.4	3	80.4	14.3	85.8	13.5	5	73.4	7.8	74.2	4.1	78.1	9.3
240	Tepraloxydim	149979-41-9	50	NA	NA	88.6	3.5	82.6	10.7	5	94	16.7	82.7	17.6	98.7	15.7
241	Terbucarb	1918-11-2	5	90.2	2	90.7	1.6	95.4	7.6	5	89.3	3.1	71.1	1.6	85.2	2.3
242	Terbufos	13071-79-9	5	95.4	5.3	97.1	6.4	93.4	4.6	NA	NA	NA	NA	NA	NA	NA
243	Terbufos-sulfone	56070-16-7	10	NA	NA	93.8	5.1	100.3	7.4	5	113.7	6.4	110.8	8.5	113.5	2.4
244	Terbumeton	33693-04-8	5	93.7	1.6	89.7	3.3	96.7	6.6	5	80.1	7.4	79.6	2.8	98.7	2.5
245	Terbuthylazine	5915-41-3	5	95.7	5.1	93.2	5	92.9	1.5	5	85.8	3.5	101.4	1.1	102.9	3.7
246	Terbutryn	886-50-0	5	95	6	91.4	1.6	99.1	1.4	5	86.4	1.9	107.7	1.7	106.5	4.3
247	Tetrachlorvinphos	22248-79-9	5	101	5.4	83	5.6	94	8	5	93.2	1.9	81.7	1.2	97.1	2.4
248	Tetraconazole	112281-77-3	5	94.1	8.4	92.6	5.7	89	1.9	5	78.7	9.3	78.1	2	92.5	9.3
249	Tetramethrin	7696-12-0	5	100.1	6	92.5	5.3	95.6	4.3	5	71.7	3.1	118.9	4.8	116.1	1.3
250	Thenylchlor	96491-05-3	5	92.8	4.1	92.3	2.1	97.2	6.9	5	73.9	5.8	101.9	1.6	101.8	3.8

#	Name	CAS														
251	Thiabendazole	148-79-8	NA	NA	NA	NA	NA	NA	NA	5	NA	NA	NA	NA	NA	NA
252	Thiazopyr	117718-60-2	5	84.9	4.7	90.1	2.3	96.8	6.4	5	84.4	9.2	82	3	114.8	7.7
253	Thiobencarb	28249-77-6	5	99.8	3.2	95.4	4.3	88.9	0.8	5	94.3	5.8	93.5	2.6	102.6	8
254	Thiocyclam	31895-21-3	200	NA	NA	NA	NA	107.5	7.7	20	NA	NA	NA	10.9	NA	14.9
255	Thiofanox	39196-18-4	10	NA	NA	107	2.9	87.9	16.5	50	NA	NA	NA	NA	NA	NA
256	Thionazin	297-97-2	100	NA	NA	NA	NA	88.1	4.5	5	89	12.5	98.4	8.8	107.8	5.9
257	Tiocarbazil	36756-79-3	50	NA	NA	NA	9.2	NA	6.4	5	80.9	3.7	107.3	4.7	110.3	3
258	Tolclofos-methyl	57018-04-9	5	98.6	3.4	93.3	4.5	91.7	0.6	10	NA	NA	103.1	15.4	98.1	20
259	Tolfenpyrad	129558-76-5	20	NA	NA	89.6	9.6	88.8	10.6	5	87.1	3.2	81.7	2	72	7.5
260	Tralkoxydim	87820-88-0	50	NA	NA	85.7	5.9	92.8	9.5	5	87.5	4.6	71.7	8.3	97.4	7.9
261	Triadimefon	43121-43-3	20	NA	NA	119.2	2.6	102.5	6	5	89	9.7	85.1	3	107.4	7.2
262	Triadimenol	55219-65-3	10	100.5	15.7	92.7	6.7	94.8	3.1	5	103.4	4.1	90.8	5.5	89.9	2.1
263	Triallate	2303-17-5	5	89.7	5.3	92.4	11.8	96	6.5	5	107.8	8	112.3	9	116.9	4.6
264	Triapenthenol	76608-88-3	10	110.8	9.3	94.4	5.5	91.8	1.5	5	86	1.6	106.4	1.9	104.1	4.1
265	Triazophos	24017-47-8	5	104.8	3.5	92.7	3.8	95.4	8.6	5	92.4	3.2	87.2	6.5	90.3	1.7
266	Tributos	78-48-8	5	86.7	2.8	91.6	2.5	97.1	6.3	5	97.7	10.6	84.6	3.7	70.7	2.6
267	Tributyl phosphate	126-73-8	5	106	12.2	94.5	4	96.1	2.7	5	95.1	10.8	94.6	4.1	103.3	7
268	Tricyclazole	41814-78-2	10	69.3	NA	99.3	8.6	76.4	18.3	5	62.4	10.5	64.5	3	80.2	6.3

(Continued)

APPENDIX TABLE C-1 Evaluation Results of the Common-Detected Pesticides by GC-Q-TOF/MS Combined with LC-Q-TOF/MS (cont.)

Black Tea

No.	Pesticides	CAS Number	GC-Q-TOF/MS SDL (µg/kg)	10 µg/kg AVE	RSD% (n=5)	50 µg/kg AVE	RSD% (n=5)	200 µg/kg AVE	RSD% (n=5)	LC-QTOF/MS SDL (µg/kg)	10 µg/kg AVE	RSD% (n=5)	50 µg/kg AVE	RSD% (n=5)	200 µg/kg AVE	RSD% (n=5)
269	Trietazine	1912-26-1	5	91.7	5.1	93.5	5.2	91.9	1	5	95.7	3.9	85.6	1.9	86.1	1.2
270	Trifloxystrobin	141517-21-7	20	NA	NA	89.5	3.2	94.2	8.3	5	86.4	1.7	83.9	2.1	90.7	2.1
271	Triphenyl phosphate	115-86-6	5	94.5	2.9	90.3	2.7	97.1	6.8	5	81	1.2	109.6	1.8	107.6	3.2
272	Uniconazole	83657-22-1	5	97.7	6.8	92.1	3.8	94	9.4	5	83.8	3.3	102.8	2.2	99.8	4.5

Black Tea

No.	Pesticides	CAS Number	GC-Q-TOF/MS SDL (µg/kg)	10 µg/kg AVE	RSD% (n=5)	50 µg/kg AVE	RSD% (n=5)	200 µg/kg AVE	RSD% (n=5)	LC-QTOF/MS SDL (µg/kg)	10 µg/kg AVE	RSD% (n=5)	50 µg/kg AVE	RSD% (n=5)	200 µg/kg AVE	RSD% (n=5)
1	1-Naphthylacetamide	86-86-2	10	98.2	10.4	109	11.4	102.2	9.4	5	85.2	10	99.2	6.4	96.1	4.5
2	2,6-Dichlorobenzamide	2008-58-4	5	88	14.2	91.5	15.3	86.1	10.3	20	NA	NA	75.3	3.3	72.9	6.5
3	3,4,5-Trimethacarb	2686-99-9	5	NA	4	118.7	6.5	NA	2.1	10	88.2	15.3	82.4	17.3	90.5	18.2
4	Acetochlor	34256-82-1	5	98.2	2.9	96.1	8.1	107.3	1.5	5	110.6	2.9	101	16.8	98.4	12
5	Aldicarbsulfone	1646-88-4	50	NA	NA	100.1	NA	95.5	5.4	5	81.5	14.2	90.2	13	82.7	6.3
6	Aldimorph	91315-15-0	10	97.8	4.8	92.1	3.7	93.2	17.3	5	79	4	88.5	5.6	85.9	17.7
7	Allidochlor	93-71-0	5	89	19.8	60.1	9.2	65.9	11.9	5	NA	NA	74	5.5	89.8	3
8	Ametryn	834-12-8	5	91	5.1	81.4	1.5	90.9	19.3	5	83.9	4.2	87.4	11.8	95.8	2.4
9	Amidosulfuron	120923-37-7	10	90.3	7.5	91.9	12.7	102.5	1.8	NA	NA	NA	NA	NA	NA	NA
10	Aminocarb	2032-59-9	20	NA	NA	89.2	4.1	97	4.3	5	102.5	3.3	108.5	4.3	100.6	1.7
11	Ancymidol	12771-68-5	5	75.7	3.1	67.5	5.6	86	16.6	5	96.3	5.4	102.4	2.9	94.5	2.8

12	Anilofos	64249-01-0	5	104.9	19.9	116.9	7.2	NA	3.1	5	104	3.4	97.5	2.1	92.7	17.4
13	Atraton	1610-17-9	5	77.3	11.7	92.6	8.7	107.3	1.6	5	96.2	2.8	92.2	2.5	86.8	18.5
14	Atrazine	1912-24-9	5	90.9	7.1	90.5	6.1	100	1.4	5	95.7	4	105.6	3.1	96.2	2.9
15	Atrazine-desethyl	6190-65-4	5	70.3	6.5	64.7	6.3	83.5	14.9	5	90.8	9.3	105	6.7	103	2.9
16	Atrazine-desisopropyl	1007-28-9	5	63.9	18.8	95	14.5	110.7	12	5	72.1	3.4	70.3	8	70	4.9
17	Azaconazole	60207-31-0	5	86.7	5	96.4	8.5	107.4	3.9	5	91.7	1.6	82.8	19	82.8	13.6
18	Azinphos-ethyl	2642-71-9	20	NA	NA	87.6	5.8	89.4	18.3	5	88.8	10	98.1	12.9	99.3	17.6
19	Aziprotryne	4658-28-0	5	88.4	7.5	85.7	3.3	94.8	18.8	5	105	1.7	91.5	17.8	90.3	16.9
20	Azoxystrobin	131860-33-8	50	NA	NA	NA	NA	NA	NA	5	110.1	2.4	99.7	12.3	98.8	14.7
21	Beflubutamid	113614-08-7	10	110.8	8.8	99.4	4.4	107.4	1.5	5	85.2	9.7	92.9	8.3	99	6.9
22	Benalaxyl	71626-11-4	5	81.9	4.7	80.4	2.9	88.5	15.9	5	108.2	1.4	112	3.1	104.4	1.6
23	Bendiocarb	22781-23-3	10	79.2	7.4	73.3	4.6	96.6	18.3	5	103.5	2.8	90.5	14.9	93.9	11.8
24	Benodanil	15310-01-7	5	NA	11.5	117	7.9	112.7	5.5	5	106.8	2	89.8	19.2	91.6	15.8
25	Benoxacor	98730-04-2	5	87.8	7.5	80.8	3.8	85.2	19.1	NA	NA	NA	NA	NA	NA	NA
26	Benzoximate	29104-30-1	5	96.8	8	96.9	3.2	98.8	14.7	5	85.3	7	86.5	5.4	80.5	9.3
27	Benzoylprop-ethyl	22212-55-1	5	NA	3.8	96.8	3	101.6	13.2	5	100.6	2	108.9	4	92	1.8
28	Bifenazate	149877-41-8	50	NA	NA	NA	15.5	78.3	16.9	20	NA	NA	NA	NA	NA	15.2

(Continued)

APPENDIX TABLE C-1 Evaluation Results of the Common-Detected Pesticides by GC–Q-TOF/MS Combined with LC–Q-TOF/MS (cont.)

			Black Tea																
			GC–QTOF/MS							LC–QTOF/MS									
				10 μg/kg		50 μg/kg		200 μg/kg			10 μg/kg		50 μg/kg		200 μg/kg				
	Pesticides	CAS Number	SDL (μg/kg)	AVE	RSD% (n=5)	AVE	RSD% (n=5)	AVE	RSD% (n=5)	SDL (μg/kg)	AVE	RSD% (n=5)	AVE	RSD% (n=5)	AVE	RSD% (n=5)			
29	Bioresmethrin	28434-01-7	50	NA	NA	89.2	4.7	108.5	13.5	NA	NA	NA	NA	NA	NA	NA
30	Bitertanol	55179-31-2	20	NA	NA	103.2	9	106.6	8.4	5	114.9	3.1	112.3	10.9	108.8	13.3
31	Boscalid	188425-85-6	5	81.1	9.9	96.5	6.6	80.6	8.8	5	72.2	12	88.6	10.3	80.6	9.7
32	Bromfenvinfos	33399-00-7	5	96.4	3.2	83.1	6.3	87.3	19	5	106.8	4.4	99.8	2.2	95.6	17.9
33	Bromobutide	74712-19-9	5	79.8	3.1	78.9	2.6	88	16	5	95.1	4.2	107	4.4	99.1	1.9
34	Bromuconazole	116255-48-2	5	78.5	2.6	72	6.5	84.1	12.3	5	108.5	2.1	94.2	2.3	86.5	16.8
35	Bupirimate	41483-43-6	5	91.2	3	86	2.4	94	14.8	5	81.2	3.1	93.7	13.1	99.4	2.6
36	Buprofezin	69327-76-0	20	NA	NA	100	5.7	107.2	2.2	5	86.2	3.4	92.8	14.3	92.5	11.7
37	Butachlor	23184-66-9	5	100	5.8	94.7	2.6	101.2	13.4	5	86.6	3.2	95.1	4.7	99.4	4.4
38	Butafenacil	134605-64-4	5	105.9	5.8	108.2	7	116.4	4.2	5	89.5	3.3	101.8	13.1	104.8	4.7
39	Butamifos	36335-67-8	5	106.9	5.8	102.4	6.8	117.6	3.1	5	119.3	3.7	95.8	5.8	88.1	11.5
40	Butralin	33629-47-9	5	101.3	3.9	97.1	5.4	108.5	1.7	10	NA	NA	90.9	8.7	90.9	11.4
41	Cadusafos	95465-99-9	5	95.4	6.9	96	5.7	103.4	1.4	5	103.1	3.1	89.9	13.1	88.8	12.5
42	Cafenstrole	125306-83-4	100	NA	NA	NA	NA	72.1	11.4	5	99.8	4.6	111.8	3.2	108	1.9
43	Carbaryl	63-25-2	5	83.4	14.3	105.1	7.4	99.6	2.8	50	NA	NA	NA	NA	84.7	12.3
44	Carbofuran	1563-66-2	5	94.1	NA	80.7	12.2	94.9	6.5	5	70.3	6.8	86.7	9.5	97.9	2.3
45	Carbofuran-3-hydroxy	16655-82-6	5	78.3	19.9	83.7	18.1	105.8	3.9	5	75.9	16.3	81.3	9.4	83.9	5.1

No.	Name	CAS														
46	Carboxin	5234-68-4	5	NA	NA	NA	NA	NA	NA	5	NA	NA	NA	NA	73.1	15.2
47	Chlordimeform	6164-98-3	100	NA	NA	NA	NA	78.2	NA	NA	NA	NA	NA	NA	NA	NA
48	Chlorfenvinphos	470-90-6	5	90.1	9.9	85.4	7.7	94.2	1.4	5	79.2	4.7	96.5	10.6	103.5	4.5
49	Chloridazon	1698-60-8	NA	NA	NA	NA	NA	NA	NA	5	75.4	NA	73.2	8.3	81.3	4.2
50	Chlorotoluron	15545-48-9	50	NA	NA	NA	NA	98.2	17.3	5	93.5	4.7	83.1	17.9	85.5	15.9
51	Chlorpyrifos	2921-88-2	5	107.3	6.1	91.9	6.2	94.5	2	20	NA	NA	93.5	17.4	92.2	5
52	Chlorpyrifos-methyl	5598-13-0	5	98.3	9.3	89	5.5	95.2	2.4	NA	NA	NA	NA	NA	NA	NA
53	Chlorsulfuron	64902-72-3	5	112.1	4.9	93.5	12.9	97.6	5.1	50	NA	NA	NA	NA	NA	6
54	Chlorthiophos	60238-56-4	5	113.3	4.1	109.4	3.3	118.2	11.9	NA	NA	NA	NA	NA	NA	NA
55	Clodinafop	114420-56-3	5	84.5	8.5	88.8	3.7	97.7	17.5	5	91.2	7.5	113.2	3.1	119.2	0.7
56	Clodinafop-propargyl	105512-06-9	10	117.6	5.9	118.4	6.8	87.2	7.4	5	87.7	7.4	98.3	10.9	86.5	7.3
57	Clomazone	81777-89-1	5	89.1	6.1	93.4	5.8	105.4	1.4	5	93.7	3.7	88.1	12.8	88.2	11.3
58	Coumaphos	56-72-4	5	98.8	12.2	77.5	4.8	92.1	12.8	5	88.5	13.2	98.5	10.2	93.2	16.9
59	Crufomate	299-86-5	5	88.6	12.2	100.4	5.7	112.7	1.8	5	119.9	2.4	100.7	19.9	104.7	18.2
60	Cycloate	1134-23-2	10	94.4	5.6	90.2	7.3	99.6	2.1	5	87.1	9.2	98.5	2.4	97.8	5.2
61	Cyflufenamid	180409-60-3	20	NA	NA	96.1	4.7	104.1	15.2	5	97.1	2.1	105.8	3.2	103.9	2.5
62	Cyprazine	22936-86-3	5	98.2	9.6	84	11.7	100.7	14.5	5	88.5	10.5	82.5	12.1	95	2.2
63	Cyproconazole	94361-06-5	5	98.8	7.4	100.9	7.2	90.9	16.8	5	108	3.7	93.7	12.1	99.7	10.4

(Continued)

APPENDIX TABLE C-1 Evaluation Results of the Common-Detected Pesticides by GC-Q-TOF/MS Combined with LC-Q-TOF/MS (cont.)

	Pesticides	CAS Number	Black Tea GC–QTOF/MS							LC–QTOF/MS						
			SDL (µg/kg)	10 µg/kg AVE	RSD% (n = 5)	50 µg/kg AVE	RSD% (n = 5)	200 µg/kg AVE	RSD% (n = 5)	SDL (µg/kg)	10 µg/kg AVE	RSD% (n = 5)	50 µg/kg AVE	RSD% (n = 5)	200 µg/kg AVE	RSD% (n = 5)
64	Cyprodinil	121552-61-2	5	85.3	5.7	83.5	12.3	101.8	1.4	5	85.5	7	81	13.9	84.9	13.7
65	Cyprofuram	69581-33-5	5	71.8	3.3	68.7	3.3	80.3	19.8	5	98.9	1	106.9	3.5	104.6	3.7
66	Desmetryn	1014-69-3	5	81	3.3	77.7	2.9	86.9	17.1	5	95.3	2.1	103.7	3	94.3	1.3
67	Dialifos	10311-84-9	NA	NA	NA	NA	NA	NA	NA	5	95.9	7.2	101.4	3.2	98	2.2
68	Diallate	2303-16-4	5	104.9	3.7	97.9	4.4	104.5	13	5	97.2	19.3	80.4	7.9	89.1	16.8
69	Dichlofenthion	97-17-6	5	101	3.9	94.2	4.5	104.5	0.7	NA	NA	NA	NA	NA	NA	NA
70	Diethatyl-ethyl	38727-55-8	5	91.3	5.9	95.9	5.5	103	2	5	102.6	1.6	110.9	3	105.9	1.9
71	Diethofencarb	87130-20-9	20	NA	NA	92.6	3.7	84.4	3	10	84.9	8.5	91.9	13.5	93.7	5.6
72	Diethyltoluamide	134-62-3	20	NA	NA	74.9	9.2	93.8	4.9	5	105.8	12.9	106.2	9.6	101.8	1.8
73	Difenoxuron	14214-32-5	5	87.6	18.6	NA	12.9	78.1	3.7	5	83.5	4.8	91.2	12.2	101.4	1.6
74	Dimethachlor	50563-36-5	5	95	5.9	93.7	5.8	108.3	2	5	103.4	2.4	92.7	14	91.2	12.6
75	Dimethametryn	22936-75-0	20	NA	NA	89.4	5.7	91.7	5.4	5	88.2	5.6	92.3	15	109.6	11.1
76	Dimethenamid	87674-68-8	5	75	4.1	70.6	3.3	79	18.8	5	108.6	3.5	95	15.1	100.7	14.2
77	Dimethoate	60-51-5	5	100.1	12.8	105.9	9	108.5	3.6	5	81.5	7.9	92.7	10.1	80	7
78	Dimetilan	644-64-4	5	93.3	7	98.9	6.5	104.6	2.4	5	91.8	11.3	89.1	6.5	99.4	6.7
79	Diniconazole	83657-24-3	10	103.7	6.7	102.5	7.1	113	3.3	5	77.8	3.3	90.6	9.9	97.7	2.7
80	Dinitramine	29091-05-2	20	NA	NA	101	6.6	114.7	2.9	10	90.3	8.8	98.7	8.4	113.4	4.3
81	Diphenamid	957-51-7	5	71.7	1.1	66.7	2.1	76.7	19	5	85.3	5.4	92.8	13.6	106.5	2.7

No	Name	CAS														
82	Dipropetryn	4147-51-7	5	93	4.5	92.9	5.1	106.4	2.1	5	90.7	2.9	92.1	12.9	99	4.6
83	Disulfoton sulfone	2497-06-5	5	95	5.6	89	3.7	93.6	11.4	5	112.7	3.3	100.4	13.2	101.4	13.1
84	Disulfoton sulfoxide	2497-07-6	5	90.5	8.5	93.6	7.4	104.3	2.5	5	89.2	5.1	89.5	12.3	101.9	1
85	Ditalimfos	5131-24-8	5	100	7	101.9	5.8	103.2	4.2	5	97	14.1	110.3	3.8	108.6	1.6
86	Dithiopyr	97886-45-8	5	90.4	5.6	90.6	6.1	99.2	1.6	10	84.4	8.2	99.1	1.8	93.7	6.5
87	Dodemorph	1593-77-7	5	70.1	1.6	NA	9.9	71.3	6.4	5	81	2.6	95.2	14.3	106.9	1.1
88	Edifenphos	17109-49-8	20	NA	NA	70.4	12.2	85.5	18.4	5	109.9	1.9	102.9	1.6	98.4	17.1
89	Esprocarb	85785-20-2	5	NA	15.9	96.3	5.9	103.2	1.1	5	107.2	6.1	92.6	11.2	92.1	11.6
90	Ethion	563-12-2	5	95	5.6	89	3.7	93.7	11.3	10	116	8.9	91.5	7.2	93.8	2.9
91	Ethoprophos	13194-48-4	5	79.7	5.8	75	2	81.6	18.3	5	111.1	1.8	92.9	17.8	94.6	14.9
92	Famphur	52-85-7	5	96.9	14	88.7	8.6	96.6	5.1	5	112.6	1.8	102.4	4.9	111.4	10.3
93	Fenamidone	161326-34-7	5	77.7	11.6	79.5	4	94.8	17.6	5	107.8	1.2	114.2	2.5	110.3	1.1
94	Fenamiphos	22224-92-6	20	NA	NA	93.6	13.3	81.1	6.5	5	82.4	9.9	99.5	13.6	113.6	1.9
95	Fenarimol	60168-88-9	5	88.3	8	98.6	9.9	112.2	3.4	5	108.4	3.2	83.8	19.6	79.1	9.3
96	Fenazaquin	120928-09-8	5	105.2	3	98.6	2.2	106.2	13.5	5	99.9	2.9	119	18.5	84.3	10.7
97	Fenfuram	24691-80-3	NA	NA	NA	NA	NA	NA	NA	5	85.2	10.5	81.7	9.1	84.7	1.9
98	Fenobucarb	3766-81-2	50	NA	NA	88.3	11.9	100.7	15.8	5	95.5	14.5	100.8	13.6	97.5	5.9
99	Fenothiocarb	62850-32-2	5	99.8	3.4	95.8	4.1	107.6	1.4	5	87.4	7.4	101	10.3	98.6	4

(Continued)

APPENDIX TABLE C-1 Evaluation Results of the Common-Detected Pesticides by GC–Q-TOF/MS Combined with LC–Q-TOF/MS (cont.)

Black Tea

	Pesticides	CAS Number	GC-QTOF/MS							LC-QTOF/MS						
			SDL (µg/kg)	10 µg/kg AVE	RSD% (n=5)	50 µg/kg AVE	RSD% (n=5)	200 µg/kg AVE	RSD% (n=5)	SDL (µg/kg)	10 µg/kg AVE	RSD% (n=5)	50 µg/kg AVE	RSD% (n=5)	200 µg/kg AVE	RSD% (n=5)
100	Fenoxaprop-ethyl	66441-23-4	5	106.4	4.7	95.4	3.4	100.5	15.5	5	96.3	7.4	98.2	8.5	101	4.7
101	Fenoxycarb	72490-01-8	100	NA	NA	NA	NA	71.2	3	10	85.2	15.8	102	10.8	99.1	2.2
102	Fenpropidin	67306-00-7	20	NA	NA	NA	11.7	60.9	7.9	5	76.9	3.5	87.5	2.8	79	5.9
103	Fenpropimorph	67564-91-4	5	88.6	6.6	90.4	6.8	98.4	1.9	5	81	2.6	95.2	14.3	108.6	1.1
104	Fensulfothion	115-90-2	5	77.9	8.5	92.3	5.9	90.3	4.2	5	86.5	5.9	98.5	11.1	113	2.1
105	Fensulfothi-on-oxon	6552-21-2	10	119.3	12.3	116.2	19.7	NA	4.8	5	95.2	7.1	92.3	3.1	105.3	2.5
106	Fensulfothi-on-sulfone	14255-72-2	5	107.6	6.2	94.1	8.1	104	1.2	5	93.9	7.5	99.9	3.6	117.7	4.2
107	Fenthion	55-38-9	5	106.3	8.6	119.8	1.6	102.5	1.6	5	106.8	12.4	115.9	4	88.6	3.1
108	Flamprop-isopropyl	52756-22-6	5	87.6	5.9	92.8	6.4	108.7	1.5	5	106.5	7.1	101.4	3.2	95.6	18.1
109	Flamprop-methyl	52756-25-9	5	84.9	6.4	91.1	6.7	103	2.1	5	108.5	1.3	103.3	2.4	94	18.8
110	Flufenacet	142459-58-3	5	97	6.7	107.4	8.2	NA	2	10	82.1	4.9	97.2	9.4	100.3	4.1
111	Fluopyram	658066-35-4	5	86.9	6.3	82.3	2.5	92.3	16.5	5	94.1	10	103	6.5	115.2	3.3
112	Fluorogly-cofen-ethyl	77501-90-7	20	NA	NA	107.8	3.3	111.9	4.6	NA	NA	NA	NA	NA	NA	NA
113	Flurochlori-done	61213-25-0	5	88.5	3.9	99.6	3.8	89.6	4.3	10	89.3	16.1	106	7.9	77.1	13.9
114	Flurprimidol	56425-91-3	5	89.5	6.4	93.7	6.2	105.5	2	5	90.8	10.7	97.3	8.4	100.8	7.4
115	Flusilazole	85509-19-9	5	86.6	9.8	89.7	8.8	92.2	3.2	5	95	3.1	108.3	3	102.4	1.1
116	Flutolanil	66332-96-5	5	91.8	5.1	94.2	5.7	103.7	2	5	105.1	1.6	92.6	18.1	92.9	13.7
117	Flutriafol	76674-21-0	5	72.2	5.2	70.5	4.8	79	19.7	5	94.8	2.6	87.1	14.4	87.4	11.5

#		CAS													
118	Fluxapyroxad	907204-31-3	5	85.3	10.7	90	7.9	89	3.5	5	90.4	8.5	107.2	6	114.8	4.4
119	Fonofos	944-22-9	5	103.4	3.9	98.1	2.8	101	13.6	10	NA	NA	98	5.2	103.1	6.3
120	Fuberidazole	3878-19-1	20	NA	NA	NA	NA	71.2	10.1	5	NA	11.8	78.1	16.5	82.1	4.1
121	Furalaxyl	57646-30-7	5	75.6	6.8	71.3	2.5	79.8	18	5	111.6	2.6	99.8	9.8	105.1	9.5
122	Furathiocarb	65907-30-4	50	NA	NA	102.4	16.6	80.3	9.3	5	102.8	10.1	110.8	3.1	98	2.9
123	Furmecyclox	60568-05-0	5	73.7	7.4	80.6	11.3	NA	2.2	5	64.9	12.1	77	5.4	74.6	3.3
124	Haloxyfop	69806-34-4	NA	NA	NA	NA	NA	NA	NA	20	NA	NA	NA	NA	NA	NA
125	Haloxyfop-methyl	69806-40-2	5	102.1	3.5	99.7	3	108.2	13.1	5	101.9	3.8	81.3	9.2	95.3	11.2
126	Heptenophos	23560-59-0	10	94.9	15.8	70	5.2	75.9	14.3	5	77	13.8	93.8	9.3	101.4	1.8
127	Hexaconazole	79983-71-4	50	NA	NA	86.8	15.9	102.7	4.1	5	95.5	4.1	113.6	5.8	104.9	4
128	Hexazinone	51235-04-2	10	74	11.4	90.8	10.1	109.3	2.7	5	81.3	5.2	85.2	12.2	100.5	0.8
129	Imazametha-benz-methyl	81405-85-8	20	NA	NA	82.5	6.6	82.3	6.2	5	83.1	5.8	98.8	12.1	101.3	2
130	Indoxacarb	144171-61-9	200	NA	NA	NA	NA	70.6	7	5	91.9	8.9	91.3	6.2	90.1	9.1
131	Iprobenfos	26087-47-8	5	119.2	4.1	77.3	3.5	82.5	18	5	112.7	1.6	92.7	17.2	96.4	15
132	Iprovalicarb	140923-17-7	50	NA	NA	NA	NA	87.1	10.1	5	112	2.8	96.8	11.3	102.1	10.4
133	Isazofos	42509-80-8	5	88.4	4.2	83.8	2.3	92.8	15.5	5	100.6	3.3	108.9	3.5	103.5	0.8
134	Isocarbamid	30979-48-7	10	94	8.6	96.7	9.1	98.8	2	5	105.6	8.2	107.7	10.1	85.9	5.9

(Continued)

APPENDIX TABLE C-1 Evaluation Results of the Common-Detected Pesticides by GC-Q-TOF/MS Combined with LC-Q-TOF/MS (*cont.*)

	Pesticides	CAS Number	Black Tea GC–QTOF/MS SDL (µg/kg)	10 µg/kg AVE	RSD% (n = 5)	50 µg/kg AVE	RSD% (n = 5)	200 µg/kg AVE	RSD% (n = 5)	LC–QTOF/MS SDL (µg/kg)	10 µg/kg AVE	RSD% (n = 5)	50 µg/kg AVE	RSD% (n = 5)	200 µg/kg AVE	RSD% (n = 5)
135	isocarbophos	24353-61-5	5	87.6	7.5	86.4	7.2	105.3	1.2	5	87.4	7.5	111.2	3.3	117.8	1.6
136	Isofenphos-oxon	31120-85-1	5	106.7	9	113.9	8	119.6	4.1	5	85.2	2.1	97.4	14.8	106.9	2.2
137	Isoprocarb	2631-40-5	5	76	7.5	70.4	0.1	84.8	14.5	10	95.6	12	79	16.4	98.2	5.5
138	Isopropalin	33820-53-0	5	119.5	0.4	117.1	4	117.3	10.7	5	83.1	17.8	110.1	16.1	96	8.6
139	Isoprothiolane	50512-35-1	5	92.5	5.5	93.8	5.2	102.2	1.4	5	81.5	3.9	85	13.8	98.3	6.4
140	Isoxadifen-ethyl	163520-33-0	50	NA	NA	100.3	7.8	110.6	12.3	NA	NA	NA	NA	NA	NA	NA
141	Isoxathion	18854-01-8	5	87.6	5.9	92.8	6.4	108.7	1.5	5	99.4	6.1	110.6	3.6	98.1	0.8
142	Kresoxim-methyl	143390-89-0	10	98.8	5.1	89.2	1.4	95.6	14.1	5	69.7	18.3	97.7	8.1	99.3	5.5
143	Malathion	121-75-5	20	NA	NA	97.7	6.5	110.5	1.9	5	78.9	9.5	101.5	11.5	88.8	2.8
144	Mefenacet	73250-68-7	20	NA	NA	109.3	9.9	117	3.7	5	85.7	3.8	96.1	12.7	103.1	3.9
145	Mefenpyr-diethyl	135590-91-9	5	83.6	4.6	81.6	2.5	91	15.4	5	103.9	2.9	106	10.1	109.3	4.1
146	Mepanipyrim	110235-47-7	5	95	5.6	99.7	7.1	105	2.4	5	75.3	2.6	91.2	12.3	99.6	1.5
147	Mepronil	55814-41-0	5	97.1	8.5	101.1	5.2	106.4	2.1	5	86.2	5.2	103.4	10.3	95.7	4
148	Metalaxyl	57837-19-1	10	71.2	6.8	94.5	9.1	86.4	2.6	5	104.7	1	111.9	3.4	104.8	2.5
149	Metamitron	41394-05-2	200	NA	NA	NA	NA	NA	NA	5	71.8	12.9	77.1	12.9	78.7	7.8
150	Metazachlor	67129-08-2	5	103.6	2.2	97	5.7	102.8	2.6	5	75.6	4.4	88.4	14.7	102.2	2
151	Metconazole	125116-23-6	50	NA	NA	NA	NA	89.9	7	5	86.7	5.9	97.3	8.9	103.4	2.3

152	Methabenz-thiazuron	18691-97-9	5	76.5	4.4	71.1	2.9	85.7	15.5	5	93.4	3.5	102.7	2.5	98.3	3.1
153	Methamido-phos	10265-92-6	5	79.5	19.7	102.6	19	103.5	6.1	5	60.4	15.4	71.1	17.5	86.7	2.4
154	Methidathion	950-37-8	10	77.9	NA	107.4	6.7	100	4.1	10	98.2	19.2	101.2	2.4	112.4	5.9
155	Methopro-tryne	841-06-5	5	84.5	5.9	77.2	3.1	86	17.3	5	97.9	2.2	89.2	12.7	90	10.9
156	Metolachlor	51218-45-2	5	81	4.4	76.2	2.4	83.7	16.6	5	106	1.6	112.3	3.5	103.8	2.6
157	Metribuzin	21087-64-9	5	98.4	15.8	91.5	10.3	92.8	3.6	5	65.7	9.7	73.8	10.5	87.2	3.2
158	Mevinphos	7786-34-7	5	90.2	15.9	87.3	13.7	94.1	4.1	5	81	12.7	82.6	8.1	90.2	2.1
159	Mexacarbate	315-18-4	10	71.7	16.3	81.9	10.6	90.1	18.9	5	84.6	9.3	93.3	6.6	91.5	4.7
160	Molinate	2212-67-1	5	83.4	1.8	78.1	5.4	80.5	15.6	10	NA	NA	110.5	6.4	110.5	4.7
161	Myclobutanil	88671-89-0	5	75.6	10.5	76	4.7	87	16.9	5	83.4	8.1	93.3	12.3	92.5	3.1
162	Napropamide	15299-99-7	10	82.8	16.1	61.2	6.7	78.5	18.7	5	105	1.3	112.1	3.2	103.5	1.5
163	Nitralin	4726-14-1	20	NA	NA	96.5	9.9	95.6	9.3	20	NA	NA	115.3	17.4	101.3	6.5
164	Norflurazon	27314-13-2	5	79	11.4	69.6	4.3	87.1	13.3	5	99.3	2.5	107.3	3.4	103	1.7
165	Nuarimol	63284-71-9	20	NA	NA	77.7	6.1	72.7	3	5	75.7	8.8	89.6	11.3	101.4	2.4
166	Octhilinone	26530-20-1	5	NA	8.7	NA	13.6	NA	NA	5	87.3	4.7	105.7	4.3	87.5	5.2
167	Ofurace	58810-48-3	5	87	16.8	101.5	7.2	105.6	3.1	5	73.3	5.4	84	10.5	102.1	2.8
168	Orbencarb	34622-58-7	5	107	5.2	96.8	2.7	106.8	15	5	99.8	2.7	109.4	3.1	94.7	1.8
169	Oxadixyl	77732-09-3	5	82.3	8.7	90	10.3	99.1	3.5	5	73.3	9.6	81.7	12.7	96.7	1.4

(Continued)

APPENDIX TABLE C-1 Evaluation Results of the Common-Detected Pesticides by GC–Q-TOF/MS Combined with LC–Q-TOF/MS (cont.)

			Black Tea GC–QTOF/MS							LC–QTOF/MS						
			SDL (µg/kg)	10 µg/kg AVE	RSD% (n = 5)	50 µg/kg AVE	RSD% (n = 5)	200 µg/kg AVE	RSD% (n = 5)	SDL (µg/kg)	10 µg/kg AVE	RSD% (n = 5)	50 µg/kg AVE	RSD% (n = 5)	200 µg/kg AVE	RSD% (n = 5)
	Pesticides	CAS Number														
170	Paclobutrazol	76738-62-0	5	85.2	3.2	78	4.5	89.3	17.4	5	106.4	2.5	109.6	6.2	98.4	16.2
171	Pebulate	1114-71-2	10	105.1	3.9	94.3	8.5	92.6	11.1	5	83.6	16.8	96.6	4.9	93	4.5
172	Penconazole	66246-88-6	5	83.8	6.1	82.7	2.4	89	15.4	5	105.1	1.5	104	3.4	91.9	18.5
173	Pendimethalin	40487-42-1	10	95.6	11.5	101.1	4.1	101	3.6	10	102	17.7	91	8	98.2	7.8
174	Pentanochlor	2307-68-8	5	95.4	4.6	89.8	1.8	99	15	5	109.9	4.7	109.5	2.5	107.7	6.7
175	Phenthoate	2597-03-7	5	111	4.4	99.2	3.6	103.3	14.2	10	89.8	14.7	94	14.7	71.8	17
176	Phorate-sulfone	2588-04-7	5	96.4	7	100.4	5.5	111.3	2.1	5	61.5	NA	87.2	10.3	89.8	5.4
177	Phorate-sulfoxide	2588-03-6	20	NA	NA	102.7	8.5	111.7	6.4	5	102.6	3	110.3	3.4	110	2.5
178	Phosalone	2310-17-0	5	111.4	5.7	109.7	7.1	119.5	2.1	20	NA	NA	NA	NA	NA	NA
179	Phosfolan	947-02-4	5	98.8	11.7	111.1	10.8	119	1.8	5	103.8	1.8	109.9	2.9	101.8	1.7
180	Phosphamidon	13171-21-6	5	70.4	3.5	60.4	4.3	77.1	19.4	5	101.1	2.6	88.5	14.7	93.2	11
181	Phthalic acid, Benzyl butyl ester	85-68-7	5	96.6	4.7	95.2	4.7	104.9	1.4	5	78.7	19.3	93.5	2.1	78.4	3.2
182	Phthalic acid, Bis-butyl ester	84-74-2	NA	NA	NA	NA	NA	NA	NA	5	103.9	6.1	99.2	4.8	103.7	3.9
183	Phthalic acid, Bis-cyclohexyl ester	84-61-7	20	NA	NA	96.2	5.6	104.6	0.8	NA	NA	NA	NA	NA	NA	NA
184	Picolinafen	137641-05-5	5	114.2	2.5	109	3.6	114.1	12.9	5	71.7	6.3	100.1	3.2	91.5	7.8
185	Picoxystrobin	117428-22-5	5	91	7.4	90.8	2.5	98.5	13.9	5	101.8	1.3	109	2.4	101.4	1.5
186	Piperonyl butoxide	51-03-6	20	NA	NA	92	7.2	99.2	3.3	5	107.3	5.5	96.4	9.6	105	2.1

187	Piperophos	24151-93-7	100	NA	NA	NA	NA	76.5	17	5	119.5	3.5	95.5	11.8	100.9	4.7
188	Pirimicarb	23103-98-2	5	87	6.2	91.9	6.1	105.4	1.7	5	93.2	3.7	87.4	8.9	89.3	4.7
189	Pirimiphos-ethyl	23505-41-1	20	NA	NA	102.5	2.5	112.1	10.9	5	94.3	3.8	94.9	5.1	102.8	1.7
190	Pirimiphos-methyl	29232-93-7	5	101.6	4.8	96.1	6.1	105.3	1.7	5	90.7	8	118.8	4.1	80.2	17.9
191	Pirimiphos-methyl-N-desethyl	67018-59-1	5	104.2	8.4	91	7.3	99.9	0.6	5	90.3	7.1	92.6	7.9	108.7	2.5
192	Prallethrin	23031-36-9	20	NA	NA	70.3	3.2	95.7	13	50	NA	NA	NA	NA	96	15.9
193	Pretilachlor	51218-49-6	5	89.8	7.9	80.4	2.8	88.6	15.7	5	103.8	2.9	99.6	12.8	96.5	11.8
194	Profenofos	41198-08-7	5	103	6.1	105.2	6.3	116.8	2.4	5	95.7	3.3	102	11.8	85.5	6.5
195	Promecarb	2631-37-0	20	NA	NA	86.2	5.9	94.9	4.3	5	98.9	8.2	103.9	6.2	112.4	2.8
196	Prometon	1610-18-0	5	96.3	6.1	93.3	6.8	101.3	1.6	5	83.5	5.5	87.8	15	100.8	1.9
197	Prometryn	7287-19-6	5	87.4	10.5	90.4	6	91.5	2	5	102.4	1.9	94.1	12.3	91.5	8.1
198	Propachlor	1918-16-7	20	NA	NA	93.6	6.9	101.1	1.8	5	100	2.3	85.8	12.9	84.7	10.7
199	Propanil	709-98-8	5	91.7	4.8	84.9	3.9	93.5	19.4	5	84.5	6.8	93.9	6.9	102.4	2.4
200	Propaphos	7292-16-2	5	72.4	14.1	71.7	11.3	83.4	18.2	5	113.7	4.7	105.9	1.5	99.9	17.8
201	Propargite	2312-35-8	50	NA	NA	118.5	5.4	106.4	4.1	20	NA	NA	102.4	13.2	85.6	12.2
202	Propazine	139-40-2	5	84.8	7.4	82	2.4	92.3	15.7	5	102.7	2.5	89.8	19	89.1	15.6
203	Propetamphos	31218-83-4	5	97.3	2.9	93	3.8	100.5	15.1	10	76	11.9	88.6	14.4	102.4	4.1

(Continued)

APPENDIX TABLE C-1 Evaluation Results of the Common-Detected Pesticides by GC–Q-TOF/MS Combined with LC–Q-TOF/MS (cont.)

	Pesticides	CAS Number	Black Tea GC–QTOF/MS SDL (µg/kg)	10 µg/kg AVE	RSD% (n=5)	50 µg/kg AVE	RSD% (n=5)	200 µg/kg AVE	RSD% (n=5)	LC–QTOF/MS SDL (µg/kg)	10 µg/kg AVE	RSD% (n=5)	50 µg/kg AVE	RSD% (n=5)	200 µg/kg AVE	RSD% (n=5)
204	Propisochlor	86763-47-5	5	89.6	2.9	83.9	3.5	92.1	15.8	5	109.4	4.1	95.4	15.3	90.4	14.4
205	Propyzamide	23950-58-5	5	86.8	10.6	89.4	6	93.9	2	5	102.5	11.3	94.1	9.1	88	5.9
206	Prosulfocarb	52888-80-9	5	91.6	10.4	92.1	4.5	105.5	1.5	5	90.4	4.4	97.4	3.6	97.3	4.7
207	Pyraclos-trobin	175013-18-0	NA	NA	NA	NA	NA	NA	NA	5	99.5	6.2	94.3	13.3	92	11.9
208	Pyrazophos	13457-18-6	20	NA	NA	107.1	4.9	80.9	6.5	5	107	2.5	109.5	4.1	111.8	2.4
209	Pyributicarb	88678-67-5	5	103.5	7	98.9	5.2	110	1.4	5	91.2	4.3	95.6	6.5	101.1	0.9
210	Pyridaben	96489-71-3	50	NA	NA	102.8	6.3	112.2	1.8	5	NA	NA	NA	NA	95.4	4.2
211	Pyridalyl	179101-81-6	20	NA	NA	NA	11.5	66.4	7	5	73.5	13	79.8	11	70.8	2.3
212	Pyridaphen-thion	119-12-0	5	110.4	10.2	110.8	7.7	115.1	3	5	86.5	7.1	97.3	11.3	106.1	3
213	Pyrifenox	88283-41-4	20	NA	NA	93.5	12	101.9	2.7	5	99.9	4.3	102.1	1.8	92.5	2.1
214	Pyriftalid	135186-78-6	20	NA	NA	79.3	9	94.7	7.7	5	96.8	7.6	104.3	3.4	116.7	3.6
215	Pyrimethanil	53112-28-0	5	82.5	4.8	78.7	2.3	86.4	17.5	5	81.3	8.5	89.5	11.4	100.3	2
216	Pyriproxyfen	95737-68-1	5	103.8	5.2	102.9	4.8	112.5	2.2	5	104.7	4.7	112.3	2.6	106.7	2
217	Pyroquilon	57369-32-1	5	70	2.6	60.6	2.9	78	19.2	5	91	3.8	105.1	3.9	88.7	2.1
218	Quinalphos	13593-03-8	5	100.1	6.6	96.3	5.3	104.2	1.8	5	85.9	4.1	98.8	10	88.7	5.2
219	Quinocla-mine	2797-51-5	20	NA	NA	60	19.6	NA	NA	10	NA	NA	66.8	9.1	80	4.2
220	Quinoxyfen	124495-18-7	5	86.6	6.6	95.3	3.8	86.9	4.1	5	114.9	9.2	117.8	2	103.6	3.1

No.	Name	CAS														
221	Quizalofop-ethyl	76578-14-8	5	110.7	4.2	105.4	7	113.6	2.2	5	101.1	2.1	92.7	11.5	89	9.9
222	Rabenzazole	40341-04-6	5	72	5.2	73.5	8.3	87	17.9	5	78.7	16.2	74.8	18.3	71.9	12.1
223	Sebuthylazine	7286-69-3	5	89.4	6.2	92.7	6.1	100.7	1.8	5	83.5	5.6	93.1	11.1	98.2	2.3
224	Sebuthyla-zine-desethyl	37019-18-4	5	88.7	9.2	92	9.4	105.1	4	5	96.5	6.4	102.2	3	86.1	2.8
225	Secbumeton	26259-45-0	5	80.7	7	77.2	3.3	83.8	17	5	93.7	4.1	87.6	14.9	86.1	12.5
226	Simazine	122-34-9	5	90.4	10	94.4	6.8	103.7	1.8	5	73.7	2.8	84.8	13.3	96.7	0.7
227	Simeconazole	149508-90-7	5	86.8	12.5	97.3	6.6	101.4	3.2	5	76.6	3.6	89.7	12.2	99.1	1.8
228	Simeton	673-04-1	5	82.2	11.3	72.9	2.1	87.3	13.3	5	78.5	10.6	89.9	12.3	92.5	3
229	Simetryn	1014-70-6	5	74.6	10.1	91.4	6.7	95.2	5.8	5	90.4	3	81.2	16.7	82.1	14.1
230	Spirodiclofen	148477-71-8	50	NA	NA	104.3	13.4	NA	3.8	100	NA	NA	NA	NA	NA	NA
231	Spiroxamine	118134-30-8	20	NA	NA	NA	5.5	70.9	10.2	5	84.4	10.3	88.3	20	80.3	8.2
232	Sulfallate	95-06-7	5	101.4	2.6	96.4	4.2	95.4	14.3	20	NA	NA	98.7	9	117	4.7
233	Sulfotep	3689-24-5	5	92.9	6.9	100.4	4.5	85.8	3.6	5	NA	8.5	94.9	15	85.7	12
234	Sulprofos	35400-43-2	5	NA	6.6	NA	3.4	103.4	3.1	10	NA	18.5	83.3	9.7	86.5	1.9
235	Tebuconazole	107534-96-3	5	90.7	7.6	93.8	6.6	92.2	17.5	5	80.8	2.6	95.7	13.1	99.5	2.1
236	Tebufenpyrad	119168-77-3	5	99.6	4.8	94.4	5.7	109.4	1.7	5	100.2	4.9	108.1	4	85.3	3.4
237	Tebupirimfos	96182-53-5	5	98.2	7	101	4	87.4	2.9	5	90.3	3.2	94.4	5.2	98.4	4.8
238	Tebutam	35256-85-0	10	91.7	19.5	89	4.9	90.3	18.1	5	77.4	2.9	91.6	11.7	101.1	1.8

(Continued)

APPENDIX TABLE C-1 Evaluation Results of the Common-Detected Pesticides by GC-Q-TOF/MS Combined with LC-Q-TOF/MS (*cont.*)

		Black Tea													
		GC-QTOF/MS						LC-QTOF/MS							
		SDL (µg/kg)	10 µg/kg		50 µg/kg		200 µg/kg		SDL (µg/kg)	10 µg/kg		50 µg/kg		200 µg/kg	
	Pesticides	CAS Number		AVE	RSD% (n = 5)	AVE	RSD% (n = 5)	AVE	RSD% (n = 5)		AVE	RSD% (n = 5)	AVE	RSD% (n = 5)	AVE	RSD% (n = 5)
239	Tebuthiuron	34014-18-1	5	73.8	6.5	67.3	5.5	81.5	13.2	5	74	6.2	79.8	12.4	90.8	1.6
240	Tepraloxydim	149979-41-9	20	NA	NA	71.5	4.3	89.2	12.1	5	92.1	7.4	117.9	3.5	95.4	4.7
241	Terbucarb	1918-11-2	10	97	4	93	2.5	101.7	13	5	83.1	10.2	113.6	11.2	82.5	13.9
242	Terbufos	13071-79-9	5	99.2	6.9	88.8	5.6	90.7	2.8	NA	NA	NA	NA	NA	NA	NA
243	Terbufos-sulfone	56070-16-7	20	NA	NA	71	8.5	88.2	17.1	5	96.1	8.9	107.1	6.9	118.3	4.6
244	Terbumeton	33693-04-8	5	80.3	5.7	74.9	3.2	84.3	17	5	83.1	7.6	87.8	13.6	102.6	1.2
245	Terbuthylazine	5915-41-3	5	91.6	5.4	92.8	5.8	104.1	1.4	5	95.6	3.6	108.3	2.7	103.7	2.3
246	Terbutryn	886-50-0	5	92.8	11.5	97.2	4.1	98.3	3.9	5	103.1	1.9	112.5	3.4	102.1	0.7
247	Tetrachlorvinphos	22248-79-9	5	105.1	9.7	118.3	8.5	NA	3.5	10	102	2.8	87.8	19.6	86.2	15.7
248	Tetraconazole	112281-77-3	5	86.2	5.6	90.8	6.4	98.6	3	5	75.7	3.2	88.8	12.7	94.3	3.1
249	Tetramethrin	7696-12-0	10	97	7.1	101	5.6	111.6	1.9	5	95.7	15	108.6	3.2	95.1	5.2
250	Thenylchlor	96491-05-3	5	81	9.5	72.9	4.1	80.3	19.5	5	82.1	4.7	94.7	4.3	85.8	3.9
251	Thiabendazole	148-79-8	200	NA	NA	NA	NA	NA	NA	5	NA	NA	NA	NA	NA	NA
252	Thiazopyr	117718-60-2	5	94.1	3.6	95.4	3	102.1	13.4	5	91.2	4.5	104.1	9.7	101.2	3.8
253	Thiobencarb	28249-77-6	5	95.4	4	93.2	5.3	102.7	1	5	89.3	11	97.8	4.9	97.2	6.6
254	Thiocyclam	31895-21-3	50	NA	NA	117.4	17.4	88.3	8.5	100	NA	NA	NA	NA	70.6	6
255	Thiofanox	39196-18-4	20	NA	NA	78.6	17.3	98.9	18.6	20	NA	NA	96.7	8.8	91	3.5

#	Name	CAS															
256	Thionazin	297-97-2	5	NA	11.5	99.5	2.8	104.6	2.2	5	77.9	5.9	95.7	11	78.7	2.7
257	Tiocarbazil	36756-79-3	50	NA	NA	NA	3.6	61.6	10.8	5	102.1	15.7	111.1	4.8	71.1	6.2
258	Tolclofos-methyl	57018-04-9	5	96.6	5	94.4	5.2	102.2	1.1	20	NA	NA	NA	NA	NA	NA
259	Tolfenpyrad	129558-76-5	20	NA	NA	112	13.4	111.1	3.2	5	72.1	10.3	80.9	7.6	89.3	7.9
260	Tralkoxydim	87820-88-0	100	NA	NA	NA	NA	92.4	16	5	88.8	4.3	85.1	7.4	93.9	4
261	Triadimefon	43121-43-3	NA	NA	NA	NA	NA	NA	NA	5	95	6.3	98.9	11.8	106.1	2.7
262	Triadimenol	55219-65-3	10	91.7	13.2	97.4	6.7	103.1	3.2	5	109.7	2.3	93.6	14.9	94.6	15.1
263	Triallate	2303-17-5	10	NA	NA	108.7	3.1	119	13	5	86.3	10.2	94.7	6.8	106.7	8.6
264	Triapenthenol	76608-88-3	5	94.5	6.1	96.3	5.8	105.3	2.9	5	103	2.7	113.4	3.1	106.1	1.7
265	Triazophos	24017-47-8	5	108.9	12.2	97	4.7	96.5	18.4	5	113.9	11.2	102	19.1	95.5	13.2
266	Tributos	78-48-8	5	114.8	3.8	109.2	3.9	117.2	13.3	10	77	1.1	101.8	2.2	81.5	16.3
267	Tributyl phosphate	126-73-8	5	98.3	6.6	96	5.8	108.1	1.6	5	84.8	4.6	98.8	11	103.5	2.9
268	Tricyclazole	41814-78-2	100	NA	NA	NA	NA	76.8	8.9	5	64.5	4.4	72.4	13.2	90.7	1.7
269	Trietazine	1912-26-1	5	87.9	5.4	91.3	5.9	101.5	1.1	5	100.9	2.2	85.8	19.6	88.5	16.1
270	Trifloxystrobin	141517-21-7	50	NA	NA	98.8	4.3	106.8	12.5	5	101.5	3.9	87.2	11.5	87.2	10.1
271	Triphenyl phosphate	115-86-6	5	96.2	4.9	89.1	2.9	94.5	16	5	105.4	5.6	110.5	2.8	105.2	1.7
272	Uniconazole	83657-22-1	5	89.5	2.9	81.6	5.3	89.6	17.3	5	104.1	5	106.7	10.7	104.4	1.7

(Continued)

APPENDIX TABLE C-1 Evaluation Results of the Common-Detected Pesticides by GC–Q-TOF/MS Combined with LC–Q-TOF/MS (cont.)

			Puer Tea													
			GC–QTOF/MS						LC–QTOF/MS							
			SDL (µg/kg)	10 µg/kg		50 µg/kg		200 µg/kg		SDL (µg/kg)	10 µg/kg		50 µg/kg		200 µg/kg	
	Pesticides	CAS Number		AVE	RSD% (n = 5)	AVE	RSD% (n = 5)	AVE	RSD% (n = 5)		AVE	RSD% (n = 5)	AVE	RSD% (n = 5)	AVE	RSD% (n = 5)
1	1-Naphthyl acetamide	86-86-2	5	79.9	8.8	NA	5.2	NA	NA	5	73.6	6.9	81	1.9	78.7	2.1
2	2,6-Dichlorobenzamide	2008-58-4	5	75.3	6.6	73.5	11.3	NA	NA	50	NA	NA	70.9	8	72.2	5.7
3	3,4,5-Trimethacarb	2686-99-9	50	NA	NA	80.4	12.9	81.8	19.3	50	NA	NA	86.3	7.7	87.5	2.2
4	Acetochlor	34256-82-1	5	77.5	1.4	70	10.4	75.6	11.3	5	81.5	7.6	82.6	5.4	86.9	7.1
5	Aldicarb-sulfone	1646-88-4	5	70.4	8.8	NA	19.9	NA	16.5	5	101.1	6.1	72.7	11.9	73.3	15.6
6	Aldimorph	91315-15-0	10	77.5	3.7	83.5	2.5	89	1.7	5	71.8	6.8	70	1.4	78.9	5.5
7	Allidochlor	93-71-0	20	NA	NA	117.8	9	114.7	15.7	5	NA	13.6	NA	NA	71.8	12.6
8	Ametryn	834-12-8	5	80.9	7.4	84.8	1.9	86.9	2.7	5	81.6	12.2	75.1	1.4	84.6	1.4
9	Amidosulfuron	120923-37-7	10	96.4	11.1	NA	NA	72.5	19.7	20	NA	NA	NA	NA	NA	NA
10	Aminocarb	2032-59-9	5	70.3	1.3	61.7	9.5	72.1	8.8	5	113.2	8.6	94.2	2.4	98.9	4.3
11	Ancymidol	12771-68-5	10	77.5	5.7	77.5	8.3	NA	5.4	5	NA	4.7	83.6	13.1	86.1	6.1
12	Anilofos	64249-01-0	5	70.9	11.5	72.7	9.9	77.4	15.5	5	80.7	8.2	85.8	10.3	88.1	2.5
13	Atraton	1610-17-9	5	81.5	6.1	70.5	4.5	71.1	12.4	5	72.2	2.4	77.4	7.8	88.1	3.6
14	Atrazine	1912-24-9	5	76.1	8	75.1	10.2	77.5	11.6	5	95.5	7	104.6	3.6	104.5	2.7
15	Atrazine-desethyl	6190-65-4	20	NA	NA	86.2	14.4	93	17.7	5	60.1	5.6	76.6	4.4	80	2.7
16	Atrazine-desisopropyl	1007-28-9	20	NA	NA	83.5	13.6	98.5	9.3	5	62.8	7.8	64.8	1.6	65.3	10
17	Azaconazole	60207-31-0	5	72.6	8.7	NA	13.1	70.2	14.6	5	67.3	2.7	71.2	3.7	111.2	7.1
18	Azinphos-ethyl	2642-71-9	20	NA	NA	83.3	6.5	87.6	11.4	5	83.4	2.3	82.7	14.1	88	14.4

No.	Name	CAS														
19	Aziprotryne	4658-28-0	5	82.9	12.8	92	2.9	93.2	6.7	5	80.2	6.5	81.6	10.3	92.4	3.6
20	Azoxystrobin	131860-33-8	20	NA	NA	84.3	10.6	70.3	8.3	5	65.6	8.9	72.4	10.8	80.2	3.7
21	Beflubutamid	113614-08-7	5	78	7.7	82.2	8.8	82.5	12.3	5	91	11.5	73.7	8.7	83.7	2.8
22	Benalaxyl	71626-11-4	5	77.7	4.1	84.6	3.8	89.8	4	5	102.1	7.5	105.7	1.1	107	0.8
23	Bendiocarb	22781-23-3	5	78.1	8.3	85.8	5.1	87.8	6.3	5	76.5	7.4	74	10.8	84.2	6.3
24	Benodanil	15310-01-7	5	81.7	4.6	NA	NA	70.7	12.4	5	76.8	9.6	87.6	15.4	87.4	2.9
25	Benoxacor	98730-04-2	5	79.9	3.5	86.1	2.6	91.7	5.5	50	NA	NA	89.6	9.9	80.7	4.5
26	Benzoximate	29104-30-1	5	84.4	8.3	85.2	3.7	94.3	4.3	5	89.7	11	70.1	9.5	82.4	14.2
27	Benzoylprop-ethyl	22212-55-1	5	112.3	12.4	93.8	4.5	90.9	3.9	5	81.9	19.3	114.2	2.1	109.8	2.6
28	Bifenazate	149877-41-8	50	NA	NA	90.2	3.3	101.7	4.7	5	120	NA	NA	NA	91	9.8
29	Bioresmethrin	28434-01-7	20	NA	NA	NA	19.6	NA	8.2	50	NA	NA	NA	NA	73.4	7.3
30	Bitertanol	55179-31-2	50	NA	NA	NA	NA	71.2	3.5	5	80.7	17.6	81.2	13.3	88	3.1
31	Boscalid	188425-85-6	5	76.5	4.5	75.5	5.6	87.3	5.8	5	84.8	6.6	75.2	4.3	76.5	3.3
32	Bromfenvinfos	33399-00-7	5	78.8	5.3	86.4	3.3	92.6	5.5	5	84.4	4.3	82.7	7.6	88.3	2.3
33	Bromobutide	74712-19-9	5	93.6	5.3	89.2	2.6	91.1	2.9	5	91.4	14.9	108.4	5	93.8	6.1
34	Bromuconazole	116255-48-2	5	73.5	6.2	75.6	17.1	NA	6.5	5	70.3	6.9	71.6	4.1	84.2	4.5
35	Bupirimate	41483-43-6	5	77.3	2.3	86.6	2.4	90.4	3.5	5	75.8	1.2	78.2	1.1	86.5	2.1

(Continued)

APPENDIX TABLE C-1 Evaluation Results of the Common-Detected Pesticides by GC-Q-TOF/MS Combined with LC-Q-TOF/MS (cont.)

	Pesticides	CAS Number	Puer Tea GC–QTOF/MS SDL (µg/kg)	10 µg/kg AVE	10 µg/kg RSD% (n = 5)	50 µg/kg AVE	50 µg/kg RSD% (n = 5)	200 µg/kg AVE	200 µg/kg RSD% (n = 5)	LC–QTOF/MS SDL (µg/kg)	10 µg/kg AVE	10 µg/kg RSD% (n = 5)	50 µg/kg AVE	50 µg/kg RSD% (n = 5)	200 µg/kg AVE	200 µg/kg RSD% (n = 5)
36	Buprofezin	69327-76-0	10	81.9	9.7	83.7	7.7	81.1	8.5	5	82	3.6	89.3	5.7	89.8	5.7
37	Butachlor	23184-66-9	5	86.4	1.9	86.6	2.1	90.8	3	5	85	2.1	81.4	1.7	86.9	2.9
38	Butafenacil	134605-64-4	5	80	4	70.3	2.7	72.3	14.1	5	78.7	2.7	79	1.7	91.1	1.3
39	Butamifos	36335-67-8	5	82.6	3.6	75	10.8	76.2	13.2	5	89.5	5.1	102.5	19.8	90.7	3.2
40	Butralin	33629-47-9	5	77.6	8.1	82.4	11.6	79.8	9.8	5	89	7.3	85.5	8.4	96	1.9
41	Cadusafos	95465-99-9	5	82.7	4	78.6	8.1	82.7	10.4	5	82.7	7.3	81.9	5.9	90.5	3.3
42	Cafenstrole	125306-83-4	50	NA	NA	66.7	NA	73.1	NA	5	89.3	10.1	111.6	4.8	110.3	7.9
43	Carbaryl	63-25-2	5	80.3	7.3	71.4	3.3	72	12.4	20	NA	NA	NA	14.8	97.7	2.7
44	Carbofuran	1563-66-2	50	NA	NA	76.4	7.1	88.3	9.4	5	70.6	2.9	73.8	1.8	74.2	1.5
45	Carbofuran-3-hydroxy	16655-82-6	5	88.9	17.8	71.1	11.8	86	8	5	62	6.5	61.8	8	65.4	3.9
46	Carboxin	5234-68-4	20	NA	NA	NA	12.7	NA	5.6	5	NA	NA	67.6	NA	NA	15
47	Chlordimeform	6164-98-3	100	NA	NA	NA	NA	95.4	4.7	5	NA	19.7	NA	NA	NA	13.1
48	Chlorfenvinphos	470-90-6	5	95.1	14.6	83.4	8	85.2	2.8	5	83.4	2.7	81.2	1.5	87.4	1.8
49	Chloridazon	1698-60-8	NA	NA	NA	NA	NA	NA	NA	5	63.1	8.8	68.7	6.1	64	6.3
50	Chlorotoluron	15545-48-9	20	NA	NA	99.4	6.9	77.7	10.6	5	74	1.9	86.1	8.3	83	2.2
51	Chlorpyrifos	2921-88-2	5	89.8	2.7	90.3	4	97.2	2.1	50	NA	NA	81.1	5.8	91.3	4
52	Chlorpyrifos-methyl	5598-13-0	5	93.6	2.2	88.3	3.8	92.9	3.1	NA	NA	NA	NA	NA	NA	NA
53	Chlorsulfuron	64902-72-3	5	88.4	8.8	NA	11.7	NA	8	20	NA	NA	NA	NA	NA	NA

No.	Compound	CAS														
54	Chlorthiophos	60238-56-4	5	86.3	1.9	85.3	1.3	91.5	3.3	NA	NA	NA	NA	NA	NA	NA	NA
55	Clodinafop	114420-56-3	5	84.6	8.9	90.3	3.6	93.5	4.6	10	NA	NA	NA	NA	NA	NA	NA
56	Clodinafop-propargyl	105512-06-9	5	82.2	7.8	89.8	8.1	100.4	5	84.5	5	1.3	86.8	1.2	90.9	1.2
57	Clomazone	81777-89-1	5	79.6	3.9	73.1	10.5	80.5	11.6	82.2	5	8.2	92	12.8	84.4	2.1
58	Coumaphos	56-72-4	5	79.7	12.8	81.6	9.5	86.7	12.1	97.6	5	4.6	75.1	9.8	77.7	5
59	Crufomate	299-86-5	5	76.2	19.6	71.9	10.6	74.6	11.6	78.3	5	3	85.5	5.5	94	3.9
60	Cycloate	1134-23-2	50	NA	NA	77.1	7.3	84.1	10.4	78.9	5	5.2	81.9	1.4	84.6	14.2
61	Cyflufenamid	180409-60-3	10	89	9.4	83	1.5	87	4.4	98.5	5	11.5	112	3	113	1.7
62	Cyprazine	22936-86-3	5	70.4	3.5	98.2	3.9	75.1	6	71.9	5	4.1	74.8	1.3	83.6	5.1
63	Cyproconazole	94361-06-5	5	73.6	2.9	62.6	12.1	70.8	10.7	72.5	5	3.7	77.5	5.9	87.6	4.8
64	Cyprodinil	121552-61-2	5	73.1	2.3	79.6	5.6	76.3	11	NA	5	13.3	70.6	11.1	81.5	1.2
65	Cyprofuram	69581-33-5	5	70.1	5.2	81.8	4.8	83.9	3.5	77	5	4	89.9	4.4	86.2	4.7
66	Desmetryn	1014-69-3	5	71.1	11.3	87.7	2	89.8	4	94.4	5	5.1	93.5	3.6	94	0.8
67	Dialifos	10311-84-9	100	NA	NA	NA	NA	90.6	13	73.1	5	5.3	105.2	4.5	103.5	2.8
68	Diallate	2303-16-4	5	80.9	4.1	90.2	2.5	95.6	3.9	83.3	5	12.8	85.4	6.3	90.4	4.7
69	Dichlofenthion	97-17-6	5	78.2	3.3	83.1	8.8	84	8.8	NA	NA	NA	NA	NA	NA	NA	NA
70	Diethatyl-ethyl	38727-55-8	5	74.8	2.9	70.6	9.9	72.5	10.7	89	5	7.6	111.5	4.1	99.7	5.1
71	Diethofencarb	87130-20-9	20	NA	NA	78.5	10.6	82.3	5.4	92.8	5	6.6	87.6	1.8	106.2	14.9

(Continued)

APPENDIX TABLE C-1 Evaluation Results of the Common-Detected Pesticides by GC-Q-TOF/MS Combined with LC-Q-TOF/MS (cont.)

			Puer Tea													
			GC-QTOF/MS							LC-QTOF/MS						
			SDL (µg/kg)	10 µg/kg		50 µg/kg		200 µg/kg		SDL (µg/kg)	10 µg/kg		50 µg/kg		200 µg/kg	
	Pesticides	CAS Number		AVE	RSD% (n = 5)	AVE	RSD% (n = 5)	AVE	RSD% (n = 5)		AVE	RSD% (n = 5)	AVE	RSD% (n = 5)	AVE	RSD% (n = 5)
72	Diethyltoluamide	134-62-3	20	NA	NA	74.6	3.4	82.9	5.6	5	76.5	10.6	76.4	3.9	83.7	3.6
73	Difenoxuron	14214-32-5	5	71.2	19.1	83.4	NA	76.1	5.6	5	72.9	1.8	75.9	1.8	84.2	2
74	Dimethachlor	50563-36-5	5	72	2.9	71.2	8.5	71.5	12.1	5	68.9	2.6	77.6	4.9	76.6	13.9
75	Dimethametryn	22936-75-0	20	NA	NA	72.8	3.1	78.7	10.5	5	77.1	1.1	80	0.9	89	3.5
76	Dimethenamid	87674-68-8	5	76.7	3.9	85.5	2.2	87.4	4.1	5	74.8	7	87.6	15.4	84.3	3.4
77	Dimethoate	60-51-5	5	NA	11.8	NA	9.1	NA	13.9	5	64.9	3.8	70.6	7.6	72	3.5
78	Dimetilan	644-64-4	5	80.2	1.8	60.1	13.1	71.2	12.3	5	72	9.1	70.1	4.2	78.6	6.6
79	Diniconazole	83657-24-3	5	74.5	7.3	71	12	73.9	13.6	5	72.6	2.7	74.7	1.1	81	3.2
80	Dinitramine	29091-05-2	5	77.4	7	77.6	10.8	79.7	12.6	10	82.7	11.2	74.6	9.7	89.3	3.4
81	Diphenamid	957-51-7	5	77.7	2.5	85.3	4.1	89.3	2.8	5	67.1	1.2	71.9	1.4	83.7	1.4
82	Dipropetryn	4147-51-7	5	80.2	4.6	79.1	11.2	79.8	8.9	5	92.3	10.7	79	8.7	90.5	1.8
83	Disulfoton sulfone	2497-06-5	5	82.8	3.5	86.7	2.1	91.3	5.4	5	76.2	6	88.8	11.5	96	5.2
84	Disulfoton sulfoxide	2497-07-6	5	76.8	4.7	71.6	2.5	72.2	13.1	5	70.6	0.6	75.5	2	93.1	3.6
85	Ditalimfos	5131-24-8	5	65	6.2	NA	11.1	75.5	3.1	5	96.7	7.7	97.8	10.7	109.3	2.4
86	Dithiopyr	97886-45-8	5	74.9	1.6	82.9	10.7	81.3	7.1	20	NA	NA	79.1	3	88.8	2.8
87	Dodemorph	1593-77-7	5	NA	2.6	60.3	NA	79.9	3.3	5	72.7	2.3	74.3	1	81.3	3.5
88	Edifenphos	17109-49-8	20	NA	NA	79.2	14.7	108	16.5	5	78.2	7.9	82	7.6	89.2	2.4
89	Esprocarb	85785-20-2	5	NA	12.1	88.2	13	89.1	10.6	5	82.2	13.2	99.6	15	89.5	2.9
90	Ethion	563-12-2	5	82.8	3.5	86.7	2.1	91.3	5.4	10	77.8	13.8	84.7	4.3	99.4	3.2

#	Name	CAS													
91	Ethoprophos	13194-48-4	5	81.9	2.6	87.9	1.6	93.2	5.5	5	83.1	4.6	87.9	6.9	94.3	1.8
92	Famphur	52-85-7	5	89.3	3.8	82.3	6.2	89.8	7.5	5	73	8.5	74.5	12.9	95.8	3
93	Fenamidone	161326-34-7	5	74.4	3.5	77.3	4.2	89.7	4.2	5	88.8	8.5	106.6	4.2	110.4	0.7
94	Fenamiphos	22224-92-6	20	NA	NA	75.1	6.7	82.6	5.2	5	72.1	4.5	71.1	5.3	83.7	3.4
95	Fenarimol	60168-88-9	5	74.4	6.2	70.7	3.2	72.2	12.4	5	73.7	7	74	16	77.4	9.7
96	Fenazaquin	120928-09-8	5	79.3	1.9	84.8	2.8	89	4.2	5	86	4.2	85.4	4.9	92	5.3
97	Fenfuram	24691-80-3	200	NA	NA	NA	NA	NA	7.6	5	NA	3.8	NA	3.4	NA	4.4
98	Fenobucarb	3766-81-2	5	77.5	8.4	90.6	6.8	85.7	5.2	5	72.3	7.4	76	2.4	77.4	3.8
99	Fenothiocarb	62850-32-2	20	NA	NA	83.3	9.7	86.3	10.1	5	83.8	1.2	85.3	0.9	91.7	1.1
100	Fenoxaprop-ethyl	66441-23-4	10	82.2	1.7	84.2	4.4	90	8.4	5	78.4	9.1	79.5	8.5	89.1	4
101	Fenoxycarb	72490-01-8	20	NA	NA	83.6	14.5	79.5	15.2	5	85.5	7.3	83	2.1	91.1	2
102	Fenpropidin	67306-00-7	50	NA	NA	NA	8.3	NA	12.5	5	60.5	14.3	NA	13.4	61.4	16.4
103	Fenpropimorph	67564-91-4	5	72.4	4.2	71.1	12	70.8	9.7	5	72.7	2.3	74.3	1	81.3	3.5
104	Fensulfothion	115-90-2	5	73.5	6.7	86.2	3.7	99.4	2.2	5	76.1	3.9	78.9	3	90.4	2.5
105	Fensulfothion-oxon	6552-21-2	20	NA	NA	NA	8.7	NA	18.2	5	75.4	6.4	70.6	6.2	78.1	2.9
106	Fensulfothion-sulfone	14255-72-2	5	70.3	5.7	70.2	11.1	73.6	11.8	5	94.3	6.4	84.7	10.7	88.9	2.3
107	Fenthion	55-38-9	5	86.8	4.3	104.2	4.1	100.9	4.1	10	98.6	8.9	113.6	3.8	104.9	4.6
108	Flamprop-isopropyl	52756-22-6	5	72.8	3.7	74.5	10.2	75.8	9.7	5	92.7	6.9	92.6	13.4	93.5	4.1

(Continued)

APPENDIX TABLE C-1 Evaluation Results of the Common-Detected Pesticides by GC-Q-TOF/MS Combined with LC-Q-TOF/MS (cont.)

			Puer Tea													
			GC–QTOF/MS							LC–QTOF/MS						
			SDL (µg/kg)	10 µg/kg		50 µg/kg		200 µg/kg		SDL (µg/kg)	10 µg/kg		50 µg/kg		200 µg/kg	
	Pesticides	CAS Number		AVE	RSD% (n=5)	AVE	RSD% (n=5)	AVE	RSD% (n=5)		AVE	RSD% (n=5)	AVE	RSD% (n=5)	AVE	RSD% (n=5)
109	Flamprop-methyl	52756-25-9	5	73.2	4.4	71.2	3.4	71.7	9.7	5	76.6	3.5	91.5	16.8	90.3	4
110	Flufenacet	142459-58-3	5	80.2	8.7	75.1	11	77.4	15.2	5	79.1	1.8	81.8	1.9	89.4	2.4
111	Fluopyram	658066-35-4	5	81.6	2.8	85.7	2.6	90.5	2.6	5	92.1	5.5	78.8	9.4	87.2	2.5
112	Fluorogly-cofen-ethyl	77501-90-7	20	NA	NA	85.9	2.5	89.4	2.4	NA	NA	NA	NA	NA	NA	NA
113	Flurochlori-done	61213-25-0	5	87.8	5.4	90.8	4.4	97.2	3.9	50	NA	NA	92	10	116	6.6
114	Flurprimidol	56425-91-3	50	NA	NA	71.7	10.6	75.2	10	5	79.2	6.1	82.4	8	77.2	2.3
115	Flusilazole	85509-19-9	5	81.9	5.9	79.9	4.5	84.3	2.7	5	91.2	7.8	103.4	3.7	105	1.4
116	Flutolanil	66332-96-5	5	78	1.6	73.4	9.5	77.7	10.5	5	76.6	11.9	85.6	12.1	87.2	1.8
117	Flutriafol	76674-21-0	5	70	2.4	76.8	5.7	NA	5.2	5	67.1	3.9	NA	3.2	85.9	4.2
118	Fluxapyroxad	907204-31-3	5	80.4	4.7	75.9	4.2	85.3	4.9	5	81	6.8	72.6	9.7	72.8	1.8
119	Fonofos	944-22-9	5	83.3	2.6	88.3	1.2	93.4	4.1	20	NA	NA	77.2	1.9	93	4
120	Fuberidazole	3878-19-1	50	NA	NA	NA	NA	NA	NA	5	NA	NA	NA	7.4	70.4	2.1
121	Furalaxyl	57646-30-7	5	69.6	1.5	81.6	4.6	81.8	4.1	5	66	5.5	73.3	8.9	85.4	5
122	Furathiocarb	65907-30-4	20	NA	NA	89.9	4.6	90.9	7.4	5	NA	15.6	119.4	10.6	116.2	2.7
123	Furmecyclox	60568-05-0	5	NA	12.1	NA	19.1	NA	11.6	5	92.1	9	92.1	6	85.9	5.2
124	Haloxyfop	69806-34-4	NA	NA	NA	NA	NA	NA	NA	10	NA	NA	NA	NA	NA	NA
125	Haloxyfop-methyl	69806-40-2	5	83.9	3.6	86.6	1.6	89.7	1.4	5	87.6	4.9	105	13.1	91.8	2.2
126	Heptenophos	23560-59-0	5	83	15.5	89.8	1.2	97.3	5.8	5	70.5	3.3	72.5	1.6	76	6.1

No.	Name	CAS														
127	Hexaconazole	79983-71-4	20	NA	NA	85.2	6.2	71.3	10.3	5	91.4	6.3	108.3	3.9	104	1.8
128	Hexazinone	51235-04-2	20	NA	NA	NA	12.8	71.8	9.6	5	61.8	1.6	66.9	1.4	75.9	1.4
129	Imazametha-benz-methyl	81405-85-8	20	NA	NA	66.2	9.1	NA	11.3	5	74	3.4	70.6	2.3	80.9	2.8
130	Indoxacarb	144171-61-9	200	NA	NA	NA	NA	71.3	12.5	5	101.3	9.3	80.7	10.9	83.2	14.7
131	Iprobenfos	26087-47-8	5	97.9	6.3	90.9	3.3	91.9	6.4	5	76.9	17.3	80	18	91.2	3.2
132	Iprovalicarb	140923-17-7	50	NA	NA	NA	NA	71.5	7.2	5	73.1	5.8	81.4	7.6	84.5	3.2
133	Isazofos	42509-80-8	5	80.9	1.8	86.7	1.9	92.1	4	5	94.5	11	112.4	3.4	114.3	1.2
134	Isocarbamid	30979-48-7	20	NA	NA	60.7	7.5	63.8	3.1	5	72.6	2.7	76.6	4.2	80	5
135	isocarbophos	24353-61-5	5	71.2	5.8	72.4	4.4	74.7	11.6	5	73.9	9.4	101	3.6	112.2	7.2
136	Isofenphos-oxon	31120-85-1	5	78.9	4.3	70.8	3.7	70.8	15.1	5	71.7	3.7	89.3	7.8	78.8	5.3
137	Isoprocarb	2631-40-5	5	78.4	3.8	85.7	1.6	93.5	6.2	5	87.6	11.3	92	6.3	83.6	8.1
138	Isopropalin	33820-53-0	5	86.8	4.5	85.7	2.1	86.1	3.3	5	80.6	2.4	85.5	2.9	86	2.4
139	Isoprothio-lane	50512-35-1	5	72	2.8	74.5	10	77.8	10.2	5	71.5	3.8	78	2.6	78.4	2.1
140	Isoxadifen-ethyl	163520-33-0	20	NA	NA	88.8	3.5	88.5	2.8	20	NA	NA	92.8	18.7	78.2	8.5
141	Isoxathion	18854-01-8	5	72.8	3.7	74.5	10.2	75.8	9.7	5	102.2	18.5	117.9	3.1	119.2	3.2
142	Kresoxim-methyl	143390-89-0	5	83.8	3.8	86.1	1.7	88.4	4.2	5	76.3	3.3	81.3	0.9	90.1	4.6
143	Malathion	121-75-5	5	NA	19.9	NA	16.2	76.8	11.5	5	75.1	2.2	82.5	1.1	89.4	1.4

(Continued)

APPENDIX TABLE C-1 Evaluation Results of the Common-Detected Pesticides by GC–Q-TOF/MS Combined with LC–Q-TOF/MS (cont.)

Puer Tea

	Pesticides	CAS Number	GC–QTOF/MS SDL (µg/kg)	10 µg/kg AVE	RSD% (n = 5)	50 µg/kg AVE	RSD% (n = 5)	200 µg/kg AVE	RSD% (n = 5)	LC–QTOF/MS SDL (µg/kg)	10 µg/kg AVE	RSD% (n = 5)	50 µg/kg AVE	RSD% (n = 5)	200 µg/kg AVE	RSD% (n = 5)
144	Mefenacet	73250-68-7	20	NA	NA	70.7	4.4	74.8	14.1	5	74.3	2.4	76.3	1.9	83.9	2
145	Mefenpyr-diethyl	135590-91-9	5	79	3.1	83.6	2	87.3	4.2	5	87	2.8	83.6	2.5	94.2	2.7
146	Mepanipyrim	110235-47-7	5	80.9	3.2	73	12.2	77.2	12.2	5	80	3.1	80.4	1.8	88.6	3.9
147	Mepronil	55814-41-0	5	74.2	6.5	77.6	10.3	80.2	11	5	81.3	2.4	82.3	1.8	90	2.4
148	Metalaxyl	57837-19-1	5	74.5	4.3	78.3	6.7	77.1	3.1	5	100.1	6	98	2.8	97.3	4.2
149	Metamitron	41394-05-2	100	NA	NA	NA	NA	NA	19.4	5	64.2	2.1	NA	1.7	70.2	3.7
150	Metazachlor	67129-08-2	5	79.2	3.9	61.7	12.4	71.8	15.1	5	70.6	2.7	70.2	0.5	73.5	3.5
151	Metconazole	125116-23-6	20	NA	NA	72.9	15.2	83	6.4	5	79.7	3.4	76.1	0.8	86	1.7
152	Methabenz-thiazuron	18691-97-9	5	76.6	8.1	88.4	1.8	92.3	8	5	89.1	6.2	93.3	1.4	92.6	4.1
153	Methamido-phos	10265-92-6	5	84.3	NA	84.4	NA	101.5	10.9	5	60.1	13.2	NA	1.8	NA	3.8
154	Methidathion	950-37-8	5	84.8	7.5	84.6	5.7	102.1	4.1	5	86.7	11.7	113.7	4.9	91.5	4.1
155	Methopro-tryne	841-06-5	5	77.5	6.3	85.2	1.2	90.8	3.7	5	76.1	1.6	82.4	5	87	1.6
156	Metolachlor	51218-45-2	5	110	12	93.4	2.1	92.9	3.1	5	96.5	5.4	107.5	1.3	104.3	4.1
157	Metribuzin	21087-64-9	20	NA	NA	78.8	3.6	85.1	5	5	60.1	5.1	63	3.8	63.2	2.9
158	Mevinphos	7786-34-7	5	87.9	7.2	77.1	2.8	83.3	7.3	5	76.1	4.2	70.6	4.4	71	1.9
159	Mexacarbate	315-18-4	5	117.2	6	80.4	6.5	84.7	9.4	5	90.7	13.6	81.1	4.6	79.9	8.5
160	Molinate	2212-67-1	5	76.3	3.9	99.8	4.4	104.4	8.4	10	90.7	15.6	70.4	8	77	10.1
161	Myclobutanil	88671-89-0	5	96.2	8	NA	NA	NA	3.9	10	70.3	2.5	70.2	1.4	77.5	2.8

No.	Name	CAS													
162	Napropamide	15299-99-7	5	NA	NA	78.8	2.8	92.3	3.6	5	100.8	5.7	105.7	3	103.7	3.4
163	Nitralin	4726-14-1	20	NA	NA	75.9	5.6	74.8	11.4	10	82.2	16.6	73.9	9.4	85.6	3.7
164	Norflurazon	27314-13-2	5	78.8	7	73.6	10.1	89.2	8	5	80.9	4.6	99.1	3.7	100.5	1.2
165	Nuarimol	63284-71-9	20	NA	NA	71	3.3	81.1	2.3	5	64.8	2.3	68.3	1.8	79.1	2.6
166	Octhilinone	26530-20-1	5	80.2	6.9	87.5	2	91.9	6.3	5	75.2	4.2	83.7	4.7	73.8	5.8
167	Ofurace	58810-48-3	5	70.1	4.9	NA	7.8	NA	13.5	5	NA	4.1	63.6	2	75.9	2.4
168	Orbencarb	34622-58-7	5	86	1.6	88	1	92.9	2.5	5	109.5	4.8	118.3	1.5	119.9	0.9
169	Oxadixyl	77732-09-3	5	71.2	8.7	NA	11.5	NA	12.7	5	NA	13.1	NA	3.8	NA	2.4
170	Paclobutrazol	76738-62-0	5	75.5	4.3	73.8	14.6	NA	3.6	5	73.6	4.3	74.2	7	83.8	4.1
171	Pebulate	1114-71-2	5	76.7	6.6	107.5	5.6	114.2	10.4	5	105.3	11.7	102.1	12	108.8	5.4
172	Penconazole	66246-88-6	5	79.7	3.1	80.4	7.2	115.6	2.3	5	70.4	18.5	73.6	17.2	83	2.5
173	Pendimethalin	40487-42-1	100	NA	NA	NA	NA	95.9	3.7	10	86.3	12.7	81.4	2.9	104.2	2.3
174	Pentanochlor	2307-68-8	10	85.3	3.4	83.4	5	90.6	2.8	5	99.8	11.2	114.3	2.4	109.6	5
175	Phenthoate	2597-03-7	5	78.9	4.9	77.3	6.4	80.6	9.8	50	NA	NA	79.2	11.7	98.2	7.3
176	Phorate-sulfone	2588-04-7	5	82	4.8	71.5	2.6	74.5	14.2	5	61.3	7.1	73	3.7	70.6	3.5
177	Phorate-sulfoxide	2588-03-6	20	NA	NA	77.8	3.7	62.3	12.5	5	98.5	4.5	98.6	2.2	99.3	3.7
178	Phosalone	2310-17-0	5	78.7	2.4	77.3	8.4	81.7	13.7	100	NA	NA	NA	NA	86.8	14.6
179	Phosfolan	947-02-4	5	84.6	6.6	NA	13.1	70.8	13.8	5	77.5	4.8	86.5	2.6	87.4	4.7

(Continued)

APPENDIX TABLE C-1 Evaluation Results of the Common-Detected Pesticides by GC–Q-TOF/MS Combined with LC–Q-TOF/MS (cont.)

	Pesticides	CAS Number	Puer Tea													
			GC–QTOF/MS							LC–QTOF/MS						
			SDL (µg/kg)	10 µg/kg		50 µg/kg		200 µg/kg		SDL (µg/kg)	10 µg/kg		50 µg/kg		200 µg/kg	
				AVE	RSD% (n = 5)	AVE	RSD% (n = 5)	AVE	RSD% (n = 5)		AVE	RSD% (n = 5)	AVE	RSD% (n = 5)	AVE	RSD% (n = 5)
180	Phosphamidon	13171-21-6	5	75.5	8.6	85.5	3.2	90.3	7.3	5	68.3	1.8	73.7	3.1	85.5	3.5
181	Phthalic acid, Benzyl butyl ester	85-68-7	5	77.9	3.5	84.2	9.6	85	9.2	5	NA	NA	102.2	12.4	93.9	9.7
182	Phthalic acid, Bis-butyl ester	84-74-2	5	106.4	4.5	110.8	6.6	104.9	6.7	5	104.2	10.2	101.8	13.8	105.3	8.1
183	Phthalic acid, Bis-cyclohexyl ester	84-61-7	20	NA	NA	85.1	9.4	86.2	8.6	5	NA	NA	97.9	4	90.5	8.8
184	Picolinafen	137641-05-5	5	82.3	2.4	85.6	2.3	91.1	4.1	5	109.8	NA	100.6	14	88.7	11.3
185	Picoxystrobin	117428-22-5	5	87.2	3.9	87.4	2.8	87.2	3.1	5	107.3	15.3	117.1	2.8	114.4	1
186	Piperonyl butoxide	51-03-6	20	NA	NA	85.2	2.4	93.2	4	5	84	0.9	84.7	0.5	93.1	1.3
187	Piperophos	24151-93-7	20	NA	NA	92.8	4.8	89.9	7.2	5	81.2	1	81.7	1	90.8	1.2
188	Pirimicarb	23103-98-2	5	73.9	3.4	70.5	2.1	72	13	5	73.1	2.7	79.4	3.4	88	2
189	Pirimiphos-ethyl	23505-41-1	5	NA	NA	85	2.8	89.8	3.3	5	82.6	3	85.3	1.1	95.4	0.8
190	Pirimiphos-methyl	29232-93-7	5	74.2	4.4	79.9	9.1	82.2	9.9	5	87.5	5.8	90.4	6.4	93.5	3.1
191	Pirimiphos-methyl-N-desethyl	67018-59-1	5	73.8	9.4	71.6	10	73.5	10.2	5	79.8	6	75.9	5.4	87.8	3.4
192	Prallethrin	23031-36-9	10	71.5	11	85.7	5.9	88.1	5	50	NA	NA	NA	NA	105.4	8.3
193	Pretilachlor	51218-49-6	5	81.7	4.4	86.2	3.7	92.4	3.3	5	84	5.6	91.3	10.4	87.2	3.3
194	Profenofos	41198-08-7	5	86.1	5.9	81.1	9.7	83.3	12.5	5	86.9	9.6	91.1	6.6	94.7	4.1

195	Promecarb	2631-37-0	5	76.8	5.3	75.7	12.1	81.5	4.2	5	84.8	7.8	84.7	9.1	85.1	4.3
196	Prometon	1610-18-0	5	72.1	3.9	71.9	10.8	74.1	11.5	5	74.6	1.6	73.6	1.2	82.3	2.6
197	Prometryn	7287-19-6	5	88.4	2.1	86.1	3.9	92.1	4	5	81.3	1.2	85.3	3.4	87.6	2.3
198	Propachlor	1918-16-7	20	NA	NA	71.8	9	71.5	13	5	76.3	4.5	81.4	10.2	82.1	1.9
199	Propanil	709-98-8	5	80.6	4	84.8	2.3	94.1	5.7	5	74.6	6.9	80.1	2	95.3	3.5
200	Propaphos	7292-16-2	5	82.1	2.8	96.8	3.4	90.1	4.7	5	76.6	6.7	80.6	8.3	87	2.9
201	Propargite	2312-35-8	20	NA	NA	72.9	4.3	NA	12.4	5	NA	NA	82.8	7.8	87.9	11.6
202	Propazine	139-40-2	5	79.4	3.6	87.3	1.5	90.9	2.4	5	80	5.6	81.5	6.2	87.6	3.7
203	Propetamphos	31218-83-4	5	76.4	0.8	85.4	1.8	89.4	3.4	50	NA	NA	79.4	9.9	90.3	3.8
204	Propisochlor	86763-47-5	5	78.9	2.9	86.6	1.5	89	3.8	5	83.1	4.5	84.2	6.3	88.7	3.5
205	Propyzamide	23950-58-5	5	92.2	1.2	88.3	3.7	90.5	2.5	5	75.4	19.3	84.4	7.4	88	3.1
206	Prosulfocarb	52888-80-9	5	77	10.9	84.7	5.9	86.1	7.7	5	82.8	3.8	85	1.8	89.9	1.5
207	Pyraclostrobin	175013-18-0	NA	NA	NA	NA	NA	70.5	4.8	5	88.4	4.7	86.8	7.4	88.7	2.7
208	Pyrazophos	13457-18-6	50	NA	NA	83.8	7.7	88.6	6.5	5	102.6	16.4	119.9	2.1	108.9	5.4
209	Pyributicarb	88678-67-5	5	76	4.3	83.5	10	82.5	9.2	5	88.4	3.1	84.5	2.2	96.8	1.7
210	Pyridaben	96489-71-3	10	89.6	7.6	80.5	11.2	87.3	9.3	5	NA	6.4	111.7	11.8	107.6	3.3
211	Pyridalyl	179101-81-6	20	NA	NA	NA	NA	NA	NA	5	NA	NA	94.1	9.8	99.7	11.6
212	Pyridaphenthion	119-12-0	5	83	6.2	71.3	4.5	73.9	15.1	5	76.6	4.4	82.2	4.2	85.7	7.6

(Continued)

APPENDIX TABLE C-1 Evaluation Results of the Common-Detected Pesticides by GC–Q-TOF/MS Combined with LC–Q-TOF/MS (cont.)

	Pesticides	CAS Number	Puer Tea GC–QTOF/MS SDL (μg/kg)	10 μg/kg AVE	RSD% (n = 5)	50 μg/kg AVE	RSD% (n = 5)	200 μg/kg AVE	RSD% (n = 5)	LC–QTOF/MS SDL (μg/kg)	10 μg/kg AVE	RSD% (n = 5)	50 μg/kg AVE	RSD% (n = 5)	200 μg/kg AVE	RSD% (n = 5)
213	Pyrifenox	88283-41-4	5	73.7	3.8	70.7	15.6	70.5	14.4	5	61.9	4.9	99.8	12.9	89.4	8.6
214	Pyriftalid	135186-78-6	20	NA	NA	72.3	5.2	78.8	4.7	5	86.5	5.8	75.2	2.8	85.5	2.6
215	Pyrimethanil	53112-28-0	5	82.7	2.7	87.3	1.4	89.9	5.2	5	70.4	4.7	74.5	2.5	82.8	1.6
216	Pyriproxyfen	95737-68-1	5	79.3	4.4	86	9.3	87.5	9.2	5	75.9	5.3	111.2	10.9	109	4
217	Pyroquilon	57369-32-1	5	75.7	5.2	88	2.4	89.1	3.6	5	73.8	2.5	81.1	1.6	82	1.1
218	Quinalphos	13593-03-8	5	75.4	4.9	77.8	10.3	80.5	9.9	5	88.7	2.5	83.3	2	91.4	1.8
219	Quinoclamine	2797-51-5	20	NA	NA	87.3	8	71.6	7.6	5	62.2	8.3	NA	5.3	NA	3.8
220	Quinoxyfen	124495-18-7	5	85	3.5	91.2	4	95.2	2.9	5	73.6	12.3	101.3	4.2	96.5	4.2
221	Quizalofop-ethyl	76578-14-8	100	NA	NA	NA	NA	84.8	9.7	5	85.4	12.5	95.1	14	94.8	2.9
222	Rabenzazole	40341-04-6	5	71.9	12.5	70.3	4.8	NA	NA	5	62.4	NA	71.8	11.2	78	7.4
223	Sebuthylazine	7286-69-3	5	74.6	3.9	75.6	10.3	78.7	10.2	5	77.7	1.7	78.4	1.5	85.2	1.7
224	Sebuthylazine-desethyl	37019-18-4	5	86.4	2.7	60.4	10.3	61	11.1	5	88.7	4.4	87.5	3.8	83.3	5.7
225	Secbumeton	26259-45-0	5	80.5	2.3	84.2	2.2	93.6	3.9	5	75.5	2.8	81.8	5.5	88.1	2.9
226	Simazine	122-34-9	5	78.8	3.3	72.5	10.1	74.8	12.6	5	NA	6.1	NA	1.5	NA	2.7
227	Simeconazole	149508-90-7	10	70.4	9.6	72	3.6	70.1	12.1	5	84.7	1.1	72.3	1.1	76.3	2.5
228	Simeton	673-04-1	5	81.3	7.2	83.7	5.7	92.6	11.6	5	73.6	10.5	69.1	1.8	79.6	2.1
229	Simetryn	1014-70-6	5	79.2	5.1	84.8	4.8	97.2	3	5	67.6	9.3	75.1	2.8	83.3	1.9
230	Spirodiclofen	148477-71-8	NA	NA	NA	NA	NA	NA	NA	5	80.7	10.9	80.6	7.6	83.3	2.8

No.	Compound	CAS No.													
231	Spiroxamine	118134-30-8	5	NA	3.4	60.7	8.1	NA	NA	5	NA	13	NA	10.1	NA	12.5
232	Sulfallate	95-06-7	5	82.7	1	90.5	3.5	94.8	5.8	50	NA	NA	61.9	NA	90	4.1
233	Sulfotep	3689-24-5	5	78.5	6	89.1	6.9	94.2	3.7	5	84.6	4.1	80.6	2.6	84.5	4.5
234	Sulprofos	35400-43-2	5	98.3	4.9	115.8	4.4	105.1	4.5	20	NA	NA	76.2	3.8	89.3	5.7
235	Tebuconazole	107534-96-3	5	83.2	10.6	83.5	4.9	NA	NA	5	77.1	3.6	79.1	3	82	6.5
236	Tebufenpyrad	119168-77-3	5	77.3	4.6	84.2	10.4	84.8	9.2	5	94.5	11	115	4	108.9	2.6
237	Tebupirimfos	96182-53-5	5	81.8	3.2	90.4	4.1	95.7	2.9	5	79.6	2.6	80.8	0.8	86.3	1.6
238	Tebutam	35256-85-0	5	87.3	4.9	95.2	2.8	92.6	3.5	5	72.1	3.3	74.6	1.6	80.5	2.1
239	Tebuthiuron	34014-18-1	20	NA	NA	81.6	4	93.8	4.2	5	63.1	4.4	64.9	2.6	70.2	1.8
240	Tepraloxydim	149979-41-9	20	NA	NA	82.2	7.8	116.8	6.5	5	79.7	13.2	91.4	3.8	98.9	4.1
241	Terbucarb	1918-11-2	5	80.2	2.3	85.5	2.5	90.8	2.6	5	81.6	14.7	79	9.3	84.3	5.3
242	Terbufos	13071-79-9	5	101.9	3.3	95.5	4.1	104.9	5.4	20	NA	NA	81.5	4.6	95.9	3.2
243	Terbufos-sulfone	56070-16-7	5	82.8	9.7	86.3	2.2	89.3	5.4	10	93.1	12.7	76.9	7.8	84.8	5
244	Terbumeton	33693-04-8	5	81.3	4.8	85.1	2.4	90	4	5	75.8	1.5	77.9	1	90.8	2.4
245	Terbuthylazine	5915-41-3	5	73.7	3.3	75.3	10.2	77	9.5	5	88.5	11.9	110.9	2	104.6	3.7
246	Terbutryn	886-50-0	5	85.6	4.5	90.4	3.6	98.8	2.4	5	103.2	6.3	109.9	1.8	110.6	1.3
247	Tetrachlorvinphos	22248-79-9	5	83.5	3.2	71.4	9.6	75.4	15.3	5	73.8	16.8	74.8	16.9	89.7	3.2
248	Tetraconazole	112281-77-3	5	71.1	6	72.8	2.8	71.1	10.9	5	72.2	2.6	75.6	2.4	76.8	4.5

(Continued)

APPENDIX TABLE C-1 Evaluation Results of the Common-Detected Pesticides by GC-Q-TOF/MS Combined with LC-Q-TOF/MS (cont.)

	Pesticides	CAS Number	Puer Tea GC–QTOF/MS SDL (µg/kg)	10 µg/kg AVE	RSD% (n = 5)	50 µg/kg AVE	RSD% (n = 5)	200 µg/kg AVE	RSD% (n = 5)	LC–QTOF/MS SDL (µg/kg)	10 µg/kg AVE	RSD% (n = 5)	50 µg/kg AVE	RSD% (n = 5)	200 µg/kg AVE	RSD% (n = 5)
249	Tetramethrin	7696-12-0	5	79.1	3.6	79.5	10.2	83.3	10.8	5	NA	NA	115.5	12.2	99.7	6.6
250	Thenylchlor	96491-05-3	5	77	3.9	84.5	2.9	87	4.7	5	70.5	7.8	112.9	5.2	107.3	13.5
251	Thiabenda-zole	148-79-8	200	NA	NA	NA	NA	NA	NA	5	NA	NA	NA	NA	NA	NA
252	Thiazopyr	117718-60-2	5	79.5	2.2	85.8	3.2	88.5	1.6	5	85.5	2.4	81.7	3.1	91.3	2.3
253	Thiobencarb	28249-77-6	5	78.5	2.8	85.1	9	86	8.5	5	82.4	1.9	84.3	1.1	99.2	0.7
254	Thiocyclam	31895-21-3	100	NA	NA	NA	NA	NA	NA	20	NA	NA	83.4	NA	NA	16.1
255	Thiofanox	39196-18-4	20	NA	NA	115.3	8.3	96.9	14.9	5	79	2.7	89.7	10.1	77.4	16.4
256	Thionazin	297-97-2	100	NA	NA	NA	NA	NA	NA	5	72.1	5.1	75	1.1	85.8	10.6
257	Tiocarbazil	36756-79-3	20	NA	NA	96.9	9.7	90.3	3	5	NA	NA	106.3	11.7	101.1	5
258	Tolclofos-methyl	57018-04-9	5	77.2	2.8	82.4	9.4	83.1	9.1	5	NA	NA	NA	NA	NA	NA
259	Tolfenpyrad	129558-76-5	20	NA	NA	83.5	9.3	82.4	11.3	5	100.5	7.2	102.5	14.3	88.7	8.4
260	Tralkoxydim	87820-88-0	50	NA	NA	78.7	4.1	91.7	5.8	5	77	3.9	71.5	1.9	75.7	3.2
261	Triadimefon	43121-43-3	100	NA	NA	NA	NA	96.2	5.3	5	79.1	2.9	77.4	1.6	85.4	2.3
262	Triadimenol	55219-65-3	10	63	14.4	74.4	11.4	71.6	13.4	5	73.3	6.2	73.9	6.2	82.6	4.6
263	Triallate	2303-17-5	5	88.2	6.1	93.6	3.1	93.1	1.8	10	83.8	9.1	74.1	14.2	96.2	2.2
264	Triapenthenol	76608-88-3	5	72.4	4.2	70.3	2	70.1	12	5	86.3	7	108.6	4.3	105.8	3.1
265	Triazophos	24017-47-8	5	88.5	5.1	86.9	2.7	94.4	7.2	5	74.9	16.6	79.4	14.7	91.5	3.8
266	Tribufos	78-48-8	5	83.1	6.5	86.8	2.7	90.8	3.2	5	85.2	3.5	95.6	13.9	85.5	7

No.	Pesticides	CAS Number	GC SDL (μg/kg)	10 μg/kg AVE	RSD% (n=5)	50 μg/kg AVE	RSD% (n=5)	200 μg/kg AVE	RSD% (n=5)	LC SDL (μg/kg)	10 μg/kg AVE	RSD% (n=5)	50 μg/kg AVE	RSD% (n=5)	200 μg/kg AVE	RSD% (n=5)
267	Tributyl phosphate	126-73-8	5	78.7	4.9	81.1	9.4	85.1	10.5	5	85.2	3.8	84.7	2.4	92.9	1.4
268	Tricyclazole	41814-78-2	5	83	4.8	72.3	13.4	NA	NA	5	NA	5.2	NA	3.6	61.5	4.5
269	Trietazine	1912-26-1	5	73.7	3.1	80.5	10.5	81.9	9.6	5	81	6.1	84.7	7.2	86	1.5
270	Trifloxystrobin	141517-21-7	5	87.7	8.8	87.8	3.5	92	4.3	5	89.2	4.2	96.6	11.8	87.3	18.9
271	Triphenyl phosphate	115-86-6	20	NA	NA	85.9	1.4	92.1	2.6	5	114.4	6.4	114.4	1.9	110.7	4
272	Uniconazole	83657-22-1	5	77.5	4.2	72.2	15.2	NA	4.7	5	91.4	7	108.2	3.4	102.3	2.5

Oolong Tea

No.	Pesticides	CAS Number	GC–QTOF/MS SDL (μg/kg)	10 μg/kg AVE	RSD% (n=5)	50 μg/kg AVE	RSD% (n=5)	200 μg/kg AVE	RSD% (n=5)	LC–QTOF/MS SDL (μg/kg)	10 μg/kg AVE	RSD% (n=5)	50 μg/kg AVE	RSD% (n=5)	200 μg/kg AVE	RSD% (n=5)
1	1-Naphthyl acetamide	86-86-2	5	84.7	7.3	NA	4.9	64.2	7.1	5	118.1	5	86.2	3.5	78.5	4
2	2,6-Dichlorobenzamide	2008-58-4	5	82.9	9.2	NA	12.1	72.5	7.6	5	NA	NA	78.5	6.4	71	4.1
3	3,4,5-Trimethacarb	2686-99-9	50	NA	NA	87	2.8	77.2	3.2	5	78.2	11.9	71	5.6	73.3	4.9
4	Acetochlor	34256-82-1	5	70.8	2.1	65.6	3.1	72.8	1.5	5	72.8	7	71.8	5.8	70.1	7.3
5	Aldicarb-sulfone	1646-88-4	5	94.2	16.5	71.8	8.7	72.1	8	5	74.5	8.6	71.8	3.2	72.1	7.3
6	Aldimorph	91315-15-0	NA	NA	NA	NA	NA	NA	NA	5	66	4.1	64.1	2.9	63.5	4.1
7	Allidochlor	93-71-0	50	NA	NA	76.3	12.6	81.7	15.1	5	119.2	2.1	70.4	3	70	3.6
8	Ametryn	834-12-8	5	68.6	6.3	67.6	5.2	66.4	3.5	5	65.3	8.9	81.4	1.3	87.3	2.2
9	Amidosulfuron	120923-37-7	5	74.8	7.4	NA	NA	70	4.7	5	NA	NA	NA	NA	NA	NA

(Continued)

APPENDIX TABLE C-1 Evaluation Results of the Common-Detected Pesticides by GC-Q-TOF/MS Combined with LC-Q-TOF/MS (cont.)

	Pesticides	CAS Number	GC–QTOF/MS Oolong Tea SDL (µg/kg)	10 µg/kg AVE	10 µg/kg RSD% (n = 5)	50 µg/kg AVE	50 µg/kg RSD% (n = 5)	200 µg/kg AVE	200 µg/kg RSD% (n = 5)	LC–QTOF/MS SDL (µg/kg)	10 µg/kg AVE	10 µg/kg RSD% (n = 5)	50 µg/kg AVE	50 µg/kg RSD% (n = 5)	200 µg/kg AVE	200 µg/kg RSD% (n = 5)
10	Aminocarb	2032-59-9	5	NA	NA	NA	5.9	60.1	8.5	5	NA	2	80.8	2.6	85.7	2
11	Ancymidol	12771-68-5	200	NA	NA	NA	NA	71.2	3.1	5	82.8	6.6	74.9	12.8	80.5	6.6
12	Anilofos	64249-01-0	20	NA	NA	83.3	8.5	76.5	2.1	5	76	9.4	71.1	6.5	79.9	3
13	Atraton	1610-17-9	5	82.2	10.6	NA	6.4	66.4	3.3	5	NA	NA	63.8	6.8	75.9	2.7
14	Atrazine	1912-24-9	5	85.9	13	67.8	3.4	73.7	3.8	5	112.6	1.9	97.5	2.2	90.8	3.9
15	Atrazine-desethyl	6190-65-4	5	74.1	5.8	63.4	12.2	70.8	2.2	5	NA	5.7	87.1	3.8	78.7	1.9
16	Atrazine-desisopropyl	1007-28-9	5	94.7	9.7	67.4	7	71.6	5.1	5	NA	15.3	NA	6.3	NA	3.9
17	Azaconazole	60207-31-0	5	72.6	6.7	NA	10.5	62.7	6.4	5	61	16.5	NA	8.5	67.1	3.6
18	Azinphos-ethyl	2642-71-9	100	NA	NA	NA	NA	79.2	0.6	5	93.1	10.4	91.6	19.1	96.3	16.5
19	Aziprotryne	4658-28-0	10	91.1	13.7	78.3	4.3	80.1	4.4	5	74.7	6.2	70.3	6.3	75.3	10
20	Azoxystrobin	131860-33-8	20	NA	NA	78.7	8.6	NA	7	5	77.4	9.3	70.3	4	75.8	1.4
21	Beflubutamid	113614-08-7	5	NA	6.5	NA	4.3	NA	2.9	5	88.1	7.2	96.9	6.5	78.4	2.6
22	Benalaxyl	71626-11-4	50	NA	NA	73.7	8.4	77.3	8.2	5	88.2	2.1	94.2	1.4	95	1.8
23	Bendiocarb	22781-23-3	5	81.8	13.4	66.6	8.4	65.6	5.1	5	71.8	8.1	71.5	7.5	70.6	3.4
24	Benodanil	15310-01-7	10	86.6	5.5	89.7	5.9	65.5	4	5	70.6	6.8	76	7.5	78.7	5
25	Benoxacor	98730-04-2	10	86.9	7.8	82.3	3.3	79.2	1.8	20	NA	NA	NA	NA	70.6	3.6
26	Benzoximate	29104-30-1	5	86.7	9.4	71.2	8.5	78.1	3.2	5	81.3	15.4	91.5	10.3	86.3	13

No	Name	CAS														
27	Benzoylprop-Ethyl	22212-55-1	5	92	4.9	60.7	13.8	78.1	3.7	5	90.5	19.1	95.7	2.2	95	2.6
28	Bifenazate	149877-41-8	50	NA	NA	92.4	17.8	78.1	11.5	5	NA	NA	81.9	3.6	91.7	11.9
29	Bioresmethrin	28434-01-7	NA	NA	NA	NA	NA	NA	NA	100	NA	NA	NA	NA	84.6	NA
30	Bitertanol	55179-31-2	20	NA	NA	82.1	5.3	65.1	5.5	5	71.7	14.7	74.3	5.9	77.2	3.1
31	Boscalid	188425-85-6	5	113.8	5.4	103.6	3	113.9	5.2	5	73.9	12.6	86.1	3.2	84.6	6.1
32	Bromfenvinfos	33399-00-7	5	83	6.1	72.2	5.9	73.9	1.8	5	73.2	10.3	73.1	4	77.2	4.5
33	Bromobutide	74712-19-9	5	79.9	5.1	70.3	4.3	75.8	3.2	5	85.1	4.2	94.4	3.6	84.3	3.2
34	Bromuconazole	116255-48-2	5	70.6	9.4	NA	19.5	63.2	6.7	5	63.6	NA	NA	8.7	70.2	2.2
35	Bupirimate	41483-43-6	5	85.4	8.1	72.2	8.7	74.6	2.5	5	71.6	3.1	81	1.4	87.2	1.9
36	Buprofezin	69327-76-0	5	NA	NA	87.2	4.5	88.3	3.2	5	92.3	5.9	81.4	9.8	81.3	9.8
37	Butachlor	23184-66-9	5	91.9	8.2	80.2	4.3	80.3	2.1	5	78.7	12.2	71.1	7.9	77	7
38	Butafenacil	134605-64-4	5	77.1	8	70.4	6.1	64.8	0.8	5	73.4	11	81.4	8.5	85.3	3.2
39	Butamifos	36335-67-8	5	91.5	6.1	75.6	4.7	81.6	3.3	5	76	20	74	7.4	81.2	4.6
40	Butralin	33629-47-9	5	104.3	8.4	77.8	3.2	84.5	6.9	5	94	19	80.2	9.4	78.2	5.6
41	Cadusafos	95465-99-9	5	85.4	5	72.2	3.1	77.7	1.6	5	80.9	7.7	75.9	6.4	85.8	2.9
42	Cafenstrole	125306-83-4	20	NA	NA	60.2	9.8	70.8	5.4	5	92.4	11.8	88.5	3.9	99.3	7
43	Carbaryl	63-25-2	5	71.4	7.2	NA	6.7	73.9	5.7	200	NA	NA	NA	NA	84.7	5.3

(Continued)

APPENDIX TABLE C-1 Evaluation Results of the Common-Detected Pesticides by GC-Q-TOF/MS Combined with LC-Q-TOF/MS (cont.)

			Oolong Tea														
			GC-QTOF/MS							LC-QTOF/MS								
	Pesticides	CAS Number	SDL (µg/kg)	10 µg/kg		50 µg/kg		200 µg/kg		SDL (µg/kg)	10 µg/kg		50 µg/kg		200 µg/kg	
				AVE	RSD% (n=5)	AVE	RSD% (n=5)	AVE	RSD% (n=5)		AVE	RSD% (n=5)	AVE	RSD% (n=5)	AVE	RSD% (n=5)
44	Carbofuran	1563-66-2	5	84.2	6.4	65.4	4.9	71.2	16.1	5	71.9	3.5	70.6	2.5	71.1	2
45	Carbofuran-3-Hydroxy	16655-82-6	5	108.8	13.5	73.5	14.7	68.4	6	5	NA	19.1	71.5	7.7	70	2.6
46	Carboxin	5234-68-4	NA	NA	NA	NA	NA	NA	NA	5	NA	NA	76.1	1.6	99.5	8.5
47	Chlordimeform	6164-98-3	10	NA	NA	NA	NA	NA	19.3	10	NA	NA	NA	NA	NA	NA
48	Chlorfenvinphos	470-90-6	5	100.3	13.3	74.9	5.4	81.7	9.2	5	71.7	9.5	76.3	7.1	75	2.4
49	Chloridazon	1698-60-8	NA	NA	NA	NA	NA	NA	NA	5	NA	17.2	70.4	8	70.8	5.2
50	Chlorotoluron	15545-48-9	20	NA	NA	79	13	80.6	10.7	5	72.4	5.2	75.9	5.6	78.4	2.4
51	Chlorpyrifos	2921-88-2	5	109.1	8.1	94.4	5.1	86.3	19.6	5	110.1	14.4	96.8	5.7	95.7	4.7
52	Chlorpyrifos-methyl	5598-13-0	5	96.3	4.8	83.8	2	86.4	7.4	100	NA	NA	NA	NA	71.3	15.1
53	Chlorsulfuron	64902-72-3	5	118.3	11.1	70.1	3.2	70.9	6.6	50	NA	NA	NA	NA	NA	16.6
54	Chlorthiophos	60238-56-4	5	100.2	4.4	82.7	4.7	87.5	2.6	10	NA	NA	NA	NA	NA	NA
55	Clodinafop	114420-56-3	5	68.1	NA	82.3	6	80.7	1.9	5	97.4	5.1	82.8	5.9	116.9	3.9
56	Clodinafop-propargyl	105512-06-9	5	NA	6.5	79.1	6.9	75	7.2	5	63	11.4	73	4.1	77.3	3.2
57	Clomazone	81777-89-1	5	79.8	6.1	70.3	0.9	73	1.8	5	77	10.7	71.2	6	74.8	5.7
58	Coumaphos	56-72-4	5	84.1	7.6	73.6	3.5	71.9	1.3	5	82.1	16.1	92.2	7.9	74.6	2.1
59	Crufomate	299-86-5	5	73.7	8.1	72.1	2	74.7	2.4	5	78.9	12.7	70.5	6.8	81.6	2.9
60	Cycloate	1134-23-2	5	85.8	6.1	71.3	3.5	81.9	2.6	5	77.9	15.6	79.8	7.1	78.7	4.8
61	Cyflufenamid	180409-60-3	NA	NA	NA	NA	NA	NA	NA	5	89.8	2.4	92.9	3.4	96.9	3.3

62	Cyprazine	22936-86-3	10	89.8	11.4	90.9	17.9	113.1	10.1	5	70.2	4.2	82.9	4.2	78.5	2.8
63	Cyproconazole	94361-06-5	5	78.3	6.1	NA	9.4	67	5.4	5	73.7	10.8	72.4	7.6	72.3	2.9
64	Cyprodinil	121552-61-2	5	NA	NA	NA	11.9	60.1	4.1	5	NA	NA	NA	11.9	72.1	1.2
65	Cyprofuram	69581-33-5	5	76.7	6.2	71.4	7	66.9	2.7	5	76.4	3.6	78.1	3.3	75.4	4.6
66	Desmetryn	1014-69-3	5	69.7	6.8	70.9	5.4	67.3	3.7	5	115.7	3.5	82.6	0.4	87.5	1.9
67	Dialifos	10311-84-9	NA	NA	NA	NA	NA	97.1	11.8	5	96.6	6.4	96.2	5.2	88.8	3.3
68	Diallate	2303-16-4	5	87.2	4.2	75.8	3.2	82.4	2.5	5	103.4	4.1	72.6	4.1	83.1	7.3
69	Dichlofenthion	97-17-6	5	86.1	3.2	77.4	1	85.2	2.9	5	NA	NA	NA	NA	112.9	11
70	Diethatyl-Ethyl	38727-55-8	5	77.1	6.5	67.3	3.2	73.1	4.5	5	101	2.8	96.4	2.4	85.3	3.7
71	Diethofencarb	87130-20-9	NA	NA	NA	NA	NA	NA	NA	5	71.3	5.5	75.3	5.1	70.8	2.2
72	Diethyltoluamide	134-62-3	20	NA	NA	76.9	1.9	73.1	6	5	87.8	NA	84.8	7	89.8	2.9
73	Difenoxuron	14214-32-5	10	70.6	NA	77.9	16.9	71.9	8.8	5	72.3	7.5	79.8	2.9	82.1	1.4
74	Dimethachlor	50563-36-5	5	79.4	13.6	63.1	4.3	65.9	2	5	71.3	11.2	71	2.3	70.7	3.8
75	Dimethametryn	22936-75-0	20	NA	NA	75.9	4.8	62.1	3.3	5	71.4	7.3	89	3	91.3	1.5
76	Dimethenamid	87674-68-8	5	82	5	73.2	3.7	74.9	2.1	5	72.5	14.5	71	6.2	70.8	3.5
77	Dimethoate	60-51-5	5	77.9	7.9	70.2	3.2	71.7	5.8	5	NA	12.4	77.4	5.9	72.1	5.2
78	Dimetilan	644-64-4	5	74.6	5.8	66	3.5	67.9	4.5	5	NA	3.6	75	4.4	70.2	2.7
79	Diniconazole	83657-24-3	5	75.3	6.9	70.6	4.4	70.3	3.3	5	72.7	7.4	83.1	5.1	76.9	3.1

(Continued)

APPENDIX TABLE C-1 Evaluation Results of the Common-Detected Pesticides by GC-Q-TOF/MS Combined with LC-Q-TOF/MS (cont.)

	Pesticides	CAS Number	Oolong Tea GC-QTOF/MS							LC-QTOF/MS						
			SDL (μg/kg)	10 μg/kg AVE	RSD% (n=5)	50 μg/kg AVE	RSD% (n=5)	200 μg/kg AVE	RSD% (n=5)	SDL (μg/kg)	10 μg/kg AVE	RSD% (n=5)	50 μg/kg AVE	RSD% (n=5)	200 μg/kg AVE	RSD% (n=5)
80	Dinitramine	29091-05-2	5	92.1	11.7	72.4	2.2	82.4	5.8	5	92.8	7.7	101.9	3.9	80.8	3.5
81	Diphenamid	957-51-7	5	78.6	4.4	70	3.8	70	2.2	5	70.7	4.4	75.7	1.2	92.4	2.1
82	Dipropetryn	4147-51-7	5	79.3	5	71.2	2	76.7	1.7	5	71.4	7.2	85.3	1.7	93	2
83	Disulfoton sulfone	2497-06-5	5	94.9	5.4	92.1	5.8	89.8	2.2	5	70.1	16.2	76.4	6.2	74.8	6
84	Disulfoton sulfoxide	2497-07-6	5	84.8	4.1	64.1	1.9	75.4	5.6	5	72.2	5.7	81.6	2.7	81.6	2.7
85	Ditalimfos	5131-24-8	5	79.4	19.3	NA	NA	NA	NA	5	82.3	4.8	81.1	10.1	88.8	3.4
86	Dithiopyr	97886-45-8	5	80.9	6.1	78.8	2.3	80	3.4	10	89.8	7.3	72.4	11.2	70.5	3.2
87	Dodemorph	1593-77-7	NA	NA	NA	NA	NA	NA	NA	5	65.6	8.2	77	2.6	86.9	3.6
88	Edifenphos	17109-49-8	5	86.8	6.7	78	5.8	79.3	2	5	72.3	10	70.5	4.7	78.2	2
89	Esprocarb	85785-20-2	5	119.8	12.2	81.4	3.6	81.9	3.1	5	89.6	12.8	83.7	10	84	4
90	Ethion	563-12-2	5	94.9	5.4	92	5.8	89.8	2.2	5	83.2	14.8	75.4	14.6	75.8	12.1
91	Ethoprophos	13194-48-4	5	88.2	4.6	81.6	2.5	81.5	2.4	5	75.3	12	70.2	5.4	78.2	6.4
92	Famphur	52-85-7	5	81.3	9.8	76.5	3.8	72.3	5.3	5	75	7.6	70.2	5.2	79.7	4.3
93	Fenamidone	161326-34-7	5	71.7	12.2	67.2	12.1	70.1	4.3	5	110	2.3	107.5	4.3	91.8	1.5
94	Fenamiphos	22224-92-6	20	NA	NA	89.1	10.8	72.7	5.9	5	64	9.7	74.2	5.4	84.9	3.4
95	Fenarimol	60168-88-9	5	NA	8.8	60.8	9.6	64.9	11.5	5	NA	NA	84.4	15	71.1	10.5
96	Fenazaquin	120928-09-8	5	95.1	6	76.5	2	83.1	2.8	5	85.4	16.3	70	13.3	84.7	3.4
97	Fenfuram	24691-80-3	200	NA	NA	NA	NA	NA	NA	5	66.2	7.2	91.7	5.3	92.2	3.9

#	Name	CAS															
98	Fenobucarb	3766-81-2	100	NA	NA	NA	NA	75	6.2	5	62.5	17.9	66.9	4.7	67.6	4.5
99	Fenothiocarb	62850-32-2	5	87.8	2.3	81.2	3.6	80.1	1.2	5	73.9	10.1	77.5	3.4	72.3	2.2
100	Fenoxaprop-ethyl	66441-23-4	5	87.4	7.5	78.5	3.6	81.6	2.4	5	73.5	7.6	85	4.5	84.6	3.5
101	Fenoxycarb	72490-01-8	NA	NA	NA	NA	NA	NA	17.9	5	63.2	10.2	77.1	2.6	77.7	4.4
102	Fenpropidin	67306-00-7	50	NA	NA	NA	11.1	NA	10.8	5	NA	13.4	NA	7.2	70.9	2.4
103	Fenpropimorph	67564-91-4	5	70.7	5.1	NA	5.6	63.3	5.6	5	65.6	8.2	77	2.6	86.9	3.6
104	Fensulfothion	115-90-2	5	90.3	7.9	108.8	6.1	NA	6	5	73.7	11.3	77	4.7	83.4	2.3
105	Fensulfothion-oxon	6552-21-2	10	102.2	8.8	95.6	7.2	70.3	11.2	5	117.6	2.7	72.5	3.6	75.9	2.2
106	Fensulfothion-sulfone	14255-72-2	5	75	7.1	79.7	2.8	72.9	7.1	5	81	10.5	90.9	4.6	82.1	3.9
107	Fenthion	55-38-9	5	74	3.1	87.1	5.3	89.6	4.6	5	101.7	15.5	107.5	1.8	105.4	3.8
108	Flamprop-isopropyl	52756-22-6	5	77.4	7.8	70.3	2.6	71.8	1.6	5	71.5	16	75	7	81.1	5.1
109	Flamprop-methyl	52756-25-9	5	72.3	6.7	66.4	3	70.1	4.1	5	71.2	10.6	72.9	6.8	75.6	6.3
110	Flufenacet	142459-58-3	5	80.5	7	86.2	5.8	76.9	2.3	5	72.8	0.6	75.9	4.9	78.5	2.8
111	Fluopyram	658066-35-4	5	84.2	6	71.1	8	75.1	1.9	5	89.7	3.3	90.9	3.9	80.2	2.2
112	Fluroglycofen-ethyl	77501-90-7	20	NA	NA	82.2	7	69.5	4.2	20	NA	NA	NA	NA	114.4	15
113	Flurochloridone	61213-25-0	5	70.6	6.1	81.2	6.2	84.7	6.6	5	87.8	15.1	111	18.3	102.3	6.1
114	Flurprimidol	56425-91-3	5	71.8	3.4	60.9	5.4	72.8	1.4	5	100.8	5.3	89.4	5	74.6	3.4
115	Flusilazole	85509-19-9	5	79	2.6	72.7	3	77.9	7.1	5	81.5	1.8	91.6	4.7	93.9	1.3

(Continued)

APPENDIX TABLE C-1 Evaluation Results of the Common-Detected Pesticides by GC-Q-TOF/MS Combined with LC-Q-TOF/MS (cont.)

	Pesticides	CAS Number	Oolong Tea GC–QTOF/MS SDL (μg/kg)	10 μg/kg AVE	10 μg/kg RSD% (n = 5)	50 μg/kg AVE	50 μg/kg RSD% (n = 5)	200 μg/kg AVE	200 μg/kg RSD% (n = 5)	LC–QTOF/MS SDL (μg/kg)	10 μg/kg AVE	10 μg/kg RSD% (n = 5)	50 μg/kg AVE	50 μg/kg RSD% (n = 5)	200 μg/kg AVE	200 μg/kg RSD% (n = 5)
116	Flutolanil	66332-96-5	5	80	4.4	71.1	2.7	73.3	3.6	5	71.1	7.6	70.2	2.6	76.5	4.5
117	Flutriafol	76674-21-0	5	70.6	13	66.3	11.7	66.4	10.2	5	62.4	14.2	63	7.7	77.7	4.9
118	Fluxapyroxad	907204-31-3	5	87.8	4.3	73	1.5	71.5	3.1	5	96.1	1.2	91.6	4.3	73.3	1.9
119	Fonofos	944-22-9	5	89.3	5.6	79.1	3.5	83.6	2.5	10	NA	NA	68.9	3.1	71.5	3.3
120	Fuberidazole	3878-19-1	5	NA	4.2	NA	2.5	NA	2.8	5	NA	NA	NA	12.7	76.8	10.7
121	Furalaxyl	57646-30-7	5	81.7	3.8	71.4	3.7	72.3	2.5	5	73.1	14.6	74.6	5.4	77.2	4.3
122	Furathiocarb	65907-30-4	100	NA	NA	NA	NA	74.9	16.9	5	90.4	4.3	85.7	10.8	100.3	2.2
123	Furmecyclox	60568-05-0	5	71.8	9.1	NA	4.3	60.8	13.4	5	85.3	3.4	74.5	4.2	74.6	4.9
124	Haloxyfop	69806-34-4	20	NA	NA	NA	NA	NA	NA	5	NA	NA	NA	NA	NA	NA
125	Haloxyfop-methyl	69806-40-2	5	94.5	4.4	79.5	1.8	82	1.8	5	87.3	7.9	79.6	7.6	88.1	2.5
126	Heptenophos	23560-59-0	5	91.2	6.2	75.2	7.6	75	3.3	5	70.6	9.1	70.4	1.2	73	2.4
127	Hexaconazole	79983-71-4	NA	NA	NA	NA	NA	NA	NA	5	87.1	3.7	94	3.7	88.5	1.7
128	Hexazinone	51235-04-2	200	NA	NA	NA	NA	NA	NA	5	NA	12.8	72.5	0.4	79.2	2.7
129	Imazamethabenz-methyl	81405-85-8	20	NA	NA	NA	9.5	NA	NA	5	71.6	3.8	86.9	2.8	90.6	7.5
130	Indoxacarb	144171-61-9	20	NA	NA	97.3	11.1	77.3	5.3	5	93.8	15.5	112.5	14.1	95.3	12.4
131	Iprobenfos	26087-47-8	20	NA	NA	NA	NA	71.6	10.9	5	76.9	12.1	72.6	5.4	79.7	4.3
132	Iprovalicarb	140923-17-7	50	NA	NA	NA	NA	73.2	3.1	5	74.7	13.4	72.9	3	73.9	2.2
133	Isazofos	42509-80-8	5	90.8	7.2	78.6	4.7	80.9	2.9	5	100.1	6.4	98.4	5.8	96.3	1.6

No.	Compound	CAS														
134	Isocarbamid	30979-48-7	5	80.4	3.4	65.6	3.5	71.6	6.3	5	NA	13.7	87.1	7.3	82	5
135	isocarbophos	24353-61-5	5	77.4	5.7	67.2	2.9	75	3.3	5	89.8	5.2	84.4	6	119.5	4
136	Isofenphos-oxon	31120-85-1	5	81.5	5.2	74.5	3	71.9	3.8	5	73.8	8.7	82.7	5.2	80.5	3.2
137	Isoprocarb	2631-40-5	5	92.7	8.9	76	4.7	79.4	2.9	20	NA	NA	71.3	2.8	72.9	4.5
138	Isopropalin	33820-53-0	5	93.5	4.8	87.3	2.3	88.1	2.7	5	97.8	11.6	91.8	15.7	81.8	9.1
139	Isoprothiolane	50512-35-1	5	80	5.4	70.6	2.9	76.2	3.6	5	71	2.3	79	4.1	83.8	4.1
140	Isoxadifen-ethyl	163520-33-0	20	NA	NA	63.7	10.8	72.1	5.9	50	NA	NA	NA	NA	74.7	12.6
141	Isoxathion	18854-01-8	20	NA	NA	85	9.3	NA	6.1	5	106.4	4.6	107.3	3.1	106.3	2.5
142	Kresoxim-methyl	143390-89-0	5	85.5	6.2	83.4	5.7	81.1	1.6	5	70.4	13.6	72.6	7.1	72.7	2.5
143	Malathion	121-75-5	5	72.6	7.7	75.3	1.7	79.1	1.8	5	75.2	9.1	73.6	4.2	70.7	2
144	Mefenacet	73250-68-7	5	75	8.4	74.9	6.7	66.2	5.3	5	70.7	11.1	79.3	4.9	79.6	2.4
145	Mefenpyr-diethyl	135590-91-9	5	85	3.6	79.5	3	79.3	2.1	5	71	19.1	83.7	10.4	82.9	2.5
146	Mepanipyrim	110235-47-7	5	75.9	5	66.8	4.7	70.3	1.5	5	70.9	2.9	81.7	4.6	84.3	1.8
147	Mepronil	55814-41-0	5	74.4	2.5	72.3	3.7	74	3.4	5	70.2	9.8	79.8	1.2	81.3	2.3
148	Metalaxyl	57837-19-1	5	79.4	13.3	70.3	2.8	66.9	4.3	5	80.4	2.4	78.2	2.3	85.3	4.2
149	Metamitron	41394-05-2	20	NA	NA	69.9	NA	NA	12.6	5	NA	17.1	75.5	8.4	77.8	18.1
150	Metazachlor	67129-08-2	5	88.6	9.4	65.7	3.4	70.1	4.9	5	71.5	4.2	74.2	3.8	70.1	4.2
151	Metconazole	125116-23-6	20	NA	NA	82.8	10.5	72.5	17.1	5	71.7	9.5	84.9	2.6	80.6	3

(Continued)

APPENDIX TABLE C-1 Evaluation Results of the Common-Detected Pesticides by GC-Q-TOF/MS Combined with LC-Q-TOF/MS (cont.)

	Pesticides	CAS Number	Oolong Tea GC-QTOF/MS SDL (µg/kg)	10 µg/kg AVE	10 µg/kg RSD% (n = 5)	50 µg/kg AVE	50 µg/kg RSD% (n = 5)	200 µg/kg AVE	200 µg/kg RSD% (n = 5)	LC-QTOF/MS SDL (µg/kg)	10 µg/kg AVE	10 µg/kg RSD% (n = 5)	50 µg/kg AVE	50 µg/kg RSD% (n = 5)	200 µg/kg AVE	200 µg/kg RSD% (n = 5)
152	Methabenz-thiazuron	18691-97-9	5	80.5	4.1	74.9	7.1	73.5	2.1	5	90.1	2.5	81.5	1	82	4.1
153	Methamidophos	10265-92-6	5	62.3	11.5	NA	NA	61.1	9.8	5	NA	14.5	NA	9.5	NA	18.7
154	Methidathion	950-37-8	50	NA	NA	75.4	8.1	97.5	6.5	5	81.4	NA	92.4	12.1	95.4	1.8
155	Methoprotryne	841-06-5	5	76.5	5.7	66.3	7.4	67.7	3.2	5	71.5	8.4	72.7	3.3	84.1	1.1
156	Metolachlor	51218-45-2	5	NA	NA	75.3	3.1	75.4	2.5	5	94.1	2.2	94.7	3.2	91.8	3.9
157	Metribuzin	21087-64-9	5	81.5	7.8	73.1	2.3	72	3.1	5	NA	12.2	67.1	2.9	66.7	2.2
158	Mevinphos	7786-34-7	5	87.7	7.2	80.6	4.8	81.2	15.2	5	70.6	7	73.9	3.6	76.1	3
159	Mexacarbate	315-18-4	20	NA	NA	63.2	5.7	71.4	12.8	5	92.7	4.5	118.3	8.4	74.4	4.3
160	Molinate	2212-67-1	5	83.1	6.1	76.2	6.5	77.7	6.3	5	92.5	17.4	101.1	4.9	72	3.6
161	Myclobutanil	88671-89-0	5	70.1	5.2	72.3	10.8	71.9	8.4	5	71.2	3.5	73.5	1.8	70.8	2.4
162	Napropamide	15299-99-7	5	62.1	19.4	70	1.7	71.8	2.7	5	92.3	1.5	90.5	1.6	91.2	3
163	Nitralin	4726-14-1	50	NA	NA	113.5	8.4	62.4	4.9	5	94.1	16.1	89.4	9.2	76.9	3.8
164	Norflurazon	27314-13-2	5	76	7.4	70.4	6	68.2	3.5	5	99.4	2.5	89.5	1.9	89.3	2
165	Nuarimol	63284-71-9	20	NA	NA	71.3	12.2	72.5	12	5	89.6	19.4	117	14.7	97.6	3.6
166	Octhilinone	26530-20-1	5	71.8	5.9	NA	3.3	NA	7	5	80	3.2	75.3	4.1	70.5	5.5
167	Ofurace	58810-48-3	10	NA	18.9	65	5.5	64.6	7	5	NA	14.3	73.2	2.2	84	2.1
168	Orbencarb	34622-58-7	5	80.6	6.6	72.2	1.4	74.2	2.3	5	99	3.8	98.4	2.4	96.7	2.9
169	Oxadixyl	77732-09-3	5	89.2	13.4	NA	4.5	63.4	6.6	5	NA	13.1	74.2	5.2	70.2	2.7

170	Paclobutrazol	76738-62-0	5	74.6	8.5	NA	17.2	65.7	5.8	5	71.7	7.5	70.5	8.9	70.4	2.8
171	Pebulate	1114-71-2	5	85.3	6.1	75.9	7.2	79.7	7.9	5	113.8	5.7	103.9	11.7	91.2	6.2
172	Penconazole	66246-88-6	5	80.8	7.8	71.5	7.2	70.9	4.1	5	61.6	NA	63	9.5	70.3	2.8
173	Pendimeth-alin	40487-42-1	10	77.9	4.6	86.5	3.7	98.6	5.5	5	86.8	11.5	109.2	13.6	96.6	5.9
174	Pentanochlor	2307-68-8	5	89.4	6.4	81.6	2.4	83.2	2.1	5	109.4	3.4	108.3	1.3	94.6	2.8
175	Phenthoate	2597-03-7	5	79.2	NA	79.4	4.5	77.6	2.7	10	78.5	7.1	72.1	8.7	72.6	9
176	Phorate-sulfone	2588-04-7	5	81.4	4.9	70.4	2.8	72.3	2.9	10	NA	14.1	NA	11.8	NA	10.1
177	Phorate-sulfoxide	2588-03-6	20	NA	NA	83.5	9.2	72.8	4.5	5	80.6	2.8	85.5	2.8	87.3	4
178	Phosalone	2310-17-0	5	83.8	5.2	91.2	4.3	77.3	1.6	200	NA	NA	NA	NA	103.9	15.1
179	Phosfolan	947-02-4	5	82	6.9	67	3.4	70.7	1.7	5	104	2.6	81.3	2.1	81.6	4.3
180	Phosphami-don	13171-21-6	5	81.4	8.2	77.2	5	74.9	2.6	5	71.8	10.7	71	5.3	78.3	1.2
181	Phthalic acid, Benzyl butyl ester	85-68-7	5	82	5	78.1	2	78.6	2.5	5	105.4	4.9	94.7	10.7	92.5	3.8
182	Phthalic acid, Bis-butyl ester	84-74-2	50	NA	NA	86.2	2.4	NA	4.7	5	73.9	11.8	78.1	14.4	81.5	12.4
183	Phthalic acid, Bis-cyclohex-yl ester	84-61-7	50	NA	NA	77	4	78.7	3.3	5	107.1	9.3	78.1	17.8	91.9	3.1
184	Picolinafen	137641-05-5	5	88.5	4.9	80.2	2.4	83.5	2.5	5	NA	NA	78.5	7.7	82.7	7.8
185	Picoxystrobin	117428-22-5	5	80.1	11.3	82.1	6.6	79	1.5	5	82.3	1.6	88.8	1.5	95.8	2.3
186	Piperonyl butoxide	51-03-6	20	NA	NA	76.1	2.6	83.9	4.1	5	77.5	6.7	88.1	4.4	85.4	2.1

(Continued)

APPENDIX TABLE C-1 Evaluation Results of the Common-Detected Pesticides by GC-Q-TOF/MS Combined with LC-Q-TOF/MS (cont.)

	Pesticides	CAS Number	Oolong Tea GC-QTOF/MS							LC-QTOF/MS						
			SDL (µg/kg)	10 µg/kg AVE	RSD% (n = 5)	50 µg/kg AVE	RSD% (n = 5)	200 µg/kg AVE	RSD% (n = 5)	SDL (µg/kg)	10 µg/kg AVE	RSD% (n = 5)	50 µg/kg AVE	RSD% (n = 5)	200 µg/kg AVE	RSD% (n = 5)
187	Piperophos	24151-93-7	5	NA	NA	NA	NA	83.7	6.1	5	74.3	8.4	86.9	2.6	89.1	1.4
188	Pirinicarb	23103-98-2	5	84.3	7.3	62.5	3.3	70.6	2.3	5	73	4.7	72.4	6.1	81.5	1.4
189	Pirimiphos-ethyl	23505-41-1	5	108.1	NA	88.7	2.6	91.4	2.1	5	78.2	7.8	91.9	2.3	96.7	1.9
190	Pirimiphos-methyl	29232-93-7	5	84.9	8.4	76.9	2.8	82.3	1.6	5	95.2	17.9	75.3	13.6	83.5	13.9
191	Pirimiphos-methyl-N-desethyl	67018-59-1	5	81.3	5.4	78.9	2.3	NA	NA	5	87.6	6.5	83	7.2	78.9	1
192	Prallethrin	23031-36-9	5	98.3	17.7	83	10.5	82.9	2.9	200	NA	NA	NA	NA	NA	NA
193	Pretilachlor	51218-49-6	5	88.4	7.1	77	6.9	77.2	2	5	79.2	10.8	71.8	6.9	80.3	2.1
194	Profenofos	41198-08-7	5	82.8	2.7	85.6	4.5	80.6	3.7	5	81.6	12	73.9	7.6	86.2	4.9
195	Promecarb	2631-37-0	50	NA	NA	NA	NA	71	3.6	5	86.9	8.1	85.3	5.8	79.7	2.1
196	Prometon	1610-18-0	5	96.4	9.5	64.9	3.6	71.7	3.8	5	62.2	12.1	78.7	1.4	84.3	2.8
197	Prometryn	7287-19-6	5	85.1	2.9	77.2	3.4	74.6	3.7	5	75	8.7	70.5	6	80.3	1.7
198	Propachlor	1918-16-7	100	NA	NA	NA	NA	71.8	5.3	5	61	NA	60	8.6	67	4.2
199	Propanil	709-98-8	5	88.5	4.6	84.5	4.9	84	3.2	5	73	9.3	76.6	2.7	90.3	2.8
200	Propaphos	7292-16-2	5	80.9	5.6	71	8.6	74.2	7.6	5	77.4	10.3	73.8	5.1	81.7	3
201	Propargite	2312-35-8	NA	NA	NA	NA	NA	80.1	2.9	5	97.7	15	75.4	12.7	81.2	6.5
202	Propazine	139-40-2	5	81.6	3.4	77.9	3.9	75.8	1.7	5	79.5	8.7	70.9	5	77.3	5.5
203	Propetamphos	31218-83-4	5	89.5	5.6	75.2	3.8	77	2.7	5	NA	NA	101.7	5	85.1	2.6

No.	Name	CAS														
204	Propisochlor	86763-47-5	5	85.2	6.2	82	2.7	79.3	2.1	5	76.8	9.5	70.8	5.4	74.2	5.3
205	Propyzamide	23950-58-5	5	86.1	3.9	79.1	2.8	79.7	9.5	5	80.8	19.4	71.2	5.1	77.2	6
206	Prosulfocarb	52888-80-9	5	NA	18.5	76.6	15.3	78.2	3.4	5	75.3	7.7	90	4.7	86.2	3.5
207	Pyraclostrobin	175013-18-0	NA	NA	NA	NA	NA	NA	NA	5	82.8	12.3	78.5	5.7	87.2	5.4
208	Pyrazophos	13457-18-6	50	NA	NA	117	6.1	103.7	6.6	5	100.9	18.8	99	2.3	90.1	4.8
209	Pyributicarb	88678-67-5	5	87.8	5.9	77.8	4.2	77.8	1.1	5	77.8	9.5	86.4	1.9	92	2.7
210	Pyridaben	96489-71-3	50	NA	NA	119	11	101.3	11.3	5	96.6	4.2	103.1	10.7	97.7	2.5
211	Pyridalyl	179101-81-6	5	93	10.5	74.7	8.4	83.4	8.7	5	NA	9.5	118.2	7.1	75.3	8.9
212	Pyridaphenthion	119-12-0	5	83.1	7.4	73.2	4.8	70.7	3.2	5	74.1	6.6	75.6	3.7	73.4	2.3
213	Pyrifenox	88283-41-4	5	74.1	15.5	NA	11.6	83.8	NA	5	73.4	14	107.6	4.7	87	3.6
214	Pyriftalid	135186-78-6	20	NA	NA	82.9	4.9	95.1	5.3	5	76.1	3.6	80.3	4.1	77.2	2.3
215	Pyrimethanil	53112-28-0	5	65.6	6.4	65.9	6	71.5	2.6	5	70	4	73.6	4.5	79.5	1.8
216	Pyriproxyfen	95737-68-1	5	90.8	9.4	85.2	2.9	79.6	1.6	5	112	2.1	106.3	5.5	100.9	2.3
217	Pyroquilon	57369-32-1	5	78	3	67.4	5.5	66.1	2.8	5	115.2	3.9	74.7	3.4	77.7	2.7
218	Quinalphos	13593-03-8	5	81	5.9	77.1	3	79.5	3.2	5	76.4	8.1	83.7	4.9	83	3
219	Quinoclamine	2797-51-5	20	NA	NA	NA	10.7	NA	10.2	5	NA	9.8	NA	2	NA	2.1
220	Quinoxyfen	124495-18-7	5	78	1.9	84	4.1	92.1	6.2	5	98.5	8	98.3	4.3	87.8	4.4

(Continued)

APPENDIX TABLE C-1 Evaluation Results of the Common-Detected Pesticides by GC-Q-TOF/MS Combined with LC-Q-TOF/MS (cont.)

Oolong Tea

	Pesticides	CAS Number	GC–QTOF/MS SDL (µg/kg)	GC 10 µg/kg AVE	GC 10 µg/kg RSD% (n=5)	GC 50 µg/kg AVE	GC 50 µg/kg RSD% (n=5)	GC 200 µg/kg AVE	GC 200 µg/kg RSD% (n=5)	LC–QTOF/MS SDL (µg/kg)	LC 10 µg/kg AVE	LC 10 µg/kg RSD% (n=5)	LC 50 µg/kg AVE	LC 50 µg/kg RSD% (n=5)	LC 200 µg/kg AVE	LC 200 µg/kg RSD% (n=5)
221	Quizalofop-ethyl	76578-14-8	100	NA	NA	NA	NA	72.3	4.3	5	73.1	18.7	74.5	6.4	81.9	5.5
222	Rabenzazole	40341-04-6	5	NA	NA	NA	NA	NA	NA	5	NA	16.7	NA	NA	71.8	5.1
223	Sebuthyla-zine	7286-69-3	5	80.7	5.3	68.5	2.9	71.5	3.3	5	70.3	10.9	76.8	4.7	84.6	3.8
224	Sebuthyla-zine-desethyl	37019-18-4	5	77.6	5.9	60.2	3.4	65.8	5	5	NA	4.7	84.4	2.3	71.3	5.3
225	Secbumeton	26259-45-0	5	73	3.4	65	7.9	66.3	2.6	5	71.2	5	70.3	5.6	81.7	2.2
226	Simazine	122-34-9	5	80.9	7.7	71.6	2.7	73	2.4	5	NA	11.3	67.3	2.9	63.9	3.2
227	Simecon-azole	149508-90-7	5	70	4.5	62.8	7.4	71	4.9	5	71.9	9.3	82.4	5.6	74	3.2
228	Simeton	673-04-1	5	NA	NA	NA	18.9	NA	8.8	5	NA	14	71.4	13	78.1	4.7
229	Simetryn	1014-70-6	5	NA	13.3	NA	18.4	74.3	8.4	5	NA	NA	NA	8	78.7	1.9
230	Spirodiclofen	148477-71-8	20	NA	NA	83.4	10	73.6	5.5	5	83.1	9.3	85	9.4	75.6	6.2
231	Spiroxamine	118134-30-8	5	NA	12.3	NA	19.5	NA	7.5	5	NA	NA	NA	9.5	NA	5.2
232	Sulfallate	95-06-7	5	86.8	5.6	74.8	6.2	82.4	4.9	5	NA	NA	103.2	9	86	6.5
233	Sulfotep	3689-24-5	5	81.9	3.8	81.4	3.9	97.7	6.5	5	76	8.8	83	11	75.5	14
234	Sulprofos	35400-43-2	5	68.7	3.3	89.2	4.4	81.1	5	10	NA	NA	71.7	3.9	74.7	10.3
235	Tebucon-azole	107534-96-3	5	82.3	6.1	73.6	8.7	73.3	7.6	5	77.3	8.6	85.6	5.7	76.6	2.6
236	Tebufenpyrad	119168-77-3	5	83.3	3.9	77.4	2.8	76.3	1.3	5	92.3	7.7	87.4	3.6	90.8	3.1
237	Tebupirimfos	96182-53-5	5	80.7	1.6	78.8	3.8	90.4	5.8	5	80	7	83.2	2.9	85.9	2.4
238	Tebutam	35256-85-0	5	87.6	NA	81.3	8.4	77.5	3.3	5	71	9.7	77.1	5.5	83.3	2.2
239	Tebuthiuron	34014-18-1	5	76.5	6.2	66	4.3	65.6	2.7	5	NA	12.5	70.7	2.3	73.7	2.1

No.	Compound	CAS																
240	Tepraloxydim	149979-41-9	100	NA	NA	NA	NA	NA	5.9	5	96.2	2.7	89.2	5.7	81.3	6.5
241	Terbucarb	1918-11-2	5	88.2	3.9	73.2	2.5	76.7	2	5	85.3	8.3	76.1	13	73.3	19.6
242	Terbufos	13071-79-9	5	89.1	3.5	82.9	1.8	84.9	4	5	104.1	9.1	96.2	7.5	84.1	2.8
243	Terbufos-sulfone	56070-16-7	5	95.9	NA	73.1	10.3	71.2	6.5	5	82	5.5	84.1	4.7	74.4	2.6
244	Terbumeton	33693-04-8	5	76	2.3	71.5	4.6	70.8	2.7	5	61.8	11.3	81.2	0.9	92.8	1.9
245	Terbuthyla-zine	5915-41-3	5	81.6	4.5	68.1	2	72.4	1.4	5	107.7	3	105.1	2.3	91.3	3.5
246	Terbutryn	886-50-0	5	70.2	4.2	81.9	4.9	75.5	5.4	5	90.8	1.4	88.9	1.8	96.9	2
247	Tetrachlorvin-phos	22248-79-9	5	80.8	5.1	87.2	3.4	76.3	1.7	5	71.1	7.6	71.8	4.2	73.4	6.4
248	Tetracon-azole	112281-77-3	5	76.2	7.1	61	7.7	70.2	4.1	5	70.4	2.6	76.3	1.3	71.2	3.2
249	Tetramethrin	7696-12-0	5	85.9	7.1	79.8	3.1	73.5	1.5	5	113.3	15.2	100.8	13.1	99.8	4.7
250	Thenylchlor	96491-05-3	5	76.3	6.6	75	6.2	73.8	2	5	73.6	3.3	82	3.8	96.5	10.3
251	Thiabenda-zole	148-79-8	100	NA	NA	NA	NA	NA	NA	5	NA	NA	NA	NA	NA	NA
252	Thiazopyr	117718-60-2	5	86.7	4.7	73.1	2.7	79.2	2.4	5	71.6	12.1	80.3	7.8	81.7	2
253	Thiobencarb	28249-77-6	5	84.8	4.5	80.3	2.5	78.7	3.4	5	74	11.1	77.9	2.2	76.1	1.9
254	Thiocyclam	31895-21-3	NA	NA	NA	NA	NA	NA	NA	5	NA	NA	NA	6	NA	5.6
255	Thiofanox	39196-18-4	20	NA	NA	NA	12.8	74	7	5	67.7	NA	NA	16.9	78.8	NA
256	Thionazin	297-97-2	50	NA	NA	NA	0.8	77	6.7	5	70.2	10.5	73.6	4.4	75	3.6
257	Tiocarbazil	36756-79-3	50	NA	NA	95.8	4	84.5	6.7	5	92.7	2.6	95.2	7.6	94.1	1.7

(Continued)

APPENDIX TABLE C-1 Evaluation Results of the Common-Detected Pesticides by GC–Q-TOF/MS Combined with LC–Q-TOF/MS (cont.)

			Oolong Tea														
			GC–QTOF/MS							LC–QTOF/MS							
			SDL (µg/kg)	10 µg/kg		50 µg/kg		200 µg/kg		SDL (µg/kg)	10 µg/kg		50 µg/kg		200 µg/kg	
	Pesticides	CAS Number		AVE	RSD% (n = 5)	AVE	RSD% (n = 5)	AVE	RSD% (n = 5)		AVE	RSD% (n = 5)	AVE	RSD% (n = 5)	AVE	RSD% (n = 5)
258	Tolclofos-methyl	57018-04-9	5	84.7	4.8	75.2	2.7	79	2.6	5	77.3	0.9	88.9	13.7	91.4	15.4
259	Tolfenpyrad	129558-76-5	50	NA	NA	NA	4.3	81	2.1	5	105.6	16.2	108.4	4.1	104.8	5.9
260	Tralkoxydim	87820-88-0	NA	NA	NA	NA	NA	NA	NA	5	75.7	4.5	66.4	5.5	64.4	4.5
261	Triadimefon	43121-43-3	10	93.7	5.8	NA	4.1	99.6	3.9	5	72.6	10.5	81.9	4.2	73.4	2.6
262	Triadimenol	55219-65-3	20	NA	NA	79.1	6.3	71.1	5.6	5	67.7	NA	63.2	8.9	70.7	2.9
263	Triallate	2303-17-5	5	86.9	4.1	73.3	2.5	75.3	2.7	5	115.6	5.6	109.6	7.3	89.9	9.8
264	Triapenthenol	76608-88-3	5	76.6	5	66.5	2.4	70.3	3.8	5	103.7	2.4	103.5	3.7	89.1	3.5
265	Triazophos	24017-47-8	5	79	4.4	72.9	3.1	75.6	2.5	5	73.5	7	71.5	5.9	78.6	2.6
266	Tribufos	78-48-8	5	84	6.8	72.2	7.1	79	2.5	5	71.8	4.8	70.7	7.9	84	3.2
267	Tributyl phosphate	126-73-8	5	86.5	7.1	70.7	2.3	77	1.6	5	88.3	6.3	90.6	5.7	88.2	2
268	Tricyclazole	41814-78-2	NA	NA	NA	NA	NA	NA	NA	5	NA	16.7	61.3	4.5	70.9	4.7
269	Trietazine	1912-26-1	5	80.3	7.2	71.1	2.4	76.5	3.2	5	77.5	7.3	71.7	5.3	81.6	1.6
270	Trifloxystrobin	141517-21-7	5	92.3	9.2	81.2	3.2	81.9	2.4	5	77.7	7.2	77.5	6.3	87.6	3.5
271	Triphenyl phosphate	115-86-6	20	NA	NA	70.4	5.4	71.3	2.7	5	104.2	3.2	98.9	2.1	94	3.7
272	Uniconazole	83657-22-1	5	78.7	6.3	71	9.5	73.1	6.5	5	99.3	4.8	99.6	1	83.8	2.5

NA, No response or recoveries/RSDs not met the Recovery-RSD standard.

2 Comparative Study of Extraction and Cleanup Efficiencies of Residue Pesticides in Tea

Chapter 2.1

Review of Sample Preparation Techniques for Residue Pesticides in Tea

Chapter Outline

2.1.1 REVIEW OF SOLID PHASE EXTRACTION TECHNIQUE

The solid phase extraction (SPE) technique has already seen rapid development since it was commercialized in 1978 [1]. There have been more than 50 companies manufacturing SPE products worldwide since 1999 [2], with products applied to different analytical fields requiring cleanup, one of which is the pesticide residual analysis.

As far as SPE inorganic fillers are concerned, Park et al. [3] developed a method for determining 18 kinds of insecticides and germicides in ginseng. The method adopts acetonitrile for extraction, Florisil cartridge for cleanup, and ECD for detection, with recoveries 72.3%•117.2% at 0.01 and 14.9 mg/kg spiked levels as well as RSD < 5%. Yahya et al. [4] used petroleum ether-ethyl acetate (80:20, v/v) for extraction and Florisil cartridge for cleanup to determine 11 pesticide residues in honey, with recoveries 86%•105% and RSD <10% at three spiked levels of 10, 30, 50 α/4g/kg. Baugros et al. [5] utilized acetonitrile-isopropanol (1:1) for extraction and silica gel cartridge for cleanup to make an LC•MS/MS determination of 12 pesticide residues in sludge, with recoveries falling 67%•127% and RSD < 13%.

As far as the SPE-bound silane filler is concerned, Albero et al. [6], for example, developed a method for determining 50 pesticide residues in fruit juices and adopted a C18 cartridge for cleanup, with recoveries higher than 91% and RSD lower than 9% at 0.02•0.1 α/4g/mL spiked levels. Chen et al. [7] churned out a method for easy and rapid determination of 21 pesticide residues in fish. The sample is extracted with acetonitrile and cleaned up by NH_2 cartridge and determined by GC•MS, with recoveries falling 81.3%•113.7% and RSD ≤ 13.5% at the spiked levels of 0.05, 0.02 and 0.1 mg/kg.

In terms of polymer fillers, Gervais et al. [8] adopted an Oasis HLB cartridge for cleanup and acetonitrile-methylene dichloride (1:1) for elution, making UPLC-MS/MS determination of 34 pesticide residues in water with average recoveries 82%•109%. Hernandez et al. [9] used acidified methanol-water-mixed solutions for extraction and OASIS HLB for cleanup, making an LC-MS/MS determination of 43 pesticide residues in tomatoes, lemons, raisins, and avocados with good recoveries for the method.

Fourth, with regard to SPE-mixed type fillers, which is also the most widely used SPE, Yague et al. [10] employed acetone for extraction and C18 and a neutral oxidized aluminum cartridge for cleanup, making a GC-ECD determination of 25 pesticide residues in milk cheeses, with average recoveries 74%•102%. Kitagawa et al. [11] used ethyl acetate for extraction, GCB/PSA cartridges for cleanup, detected by GC•MS. Recovery tests of 222 pesticides in five kinds of processed foods of dumplings, curry, French fries, fried chicken, and fried fish were performed at two different fortification levels of 0.02 and 0.1 α/4g/g, and 100 pesticides showed acceptable recovery (70%•120%) with RSD ≤ 20%. Okihashi et al. [12] adopted acetonitrile for extraction and GCB/PSA cartridges for cleanup, making a GC•MS/MS determination of 260 pesticides in fruits, vegetables, rice, and so on, with recoveries falling 70%•120% and RSD ≤ 20 for the majority of pesticides at the two spiked levels of 0.02•0.1 α/4g/g.

Lou et al. [14] established the GC method for determination of 92 pesticides in tea. The tea samples were extracted with acetonitrile only one time, after which the organophosphorus pesticides were cleaned up via Envi-Carb SPE cartridge and acetonitrile-toluene (volume ratio 3:1) as eluates, using GC-FPD for determination; organochlorines and pyrethroids

Analysis of Pesticide in Tea. http://dx.doi.org/10.1016/B978-0-12-812727-8.00005-5

pesticides pass through Envi-Carb and NH$_2$ SPE in tandem, acetonitrile-toluene (volume ratio 3:1) as eluates and submitted for GC-ECD detection, with the test results that the average recoveries for 92 pesticides fall within 80.3%•117.1% and RSD 1.5%•9.8%. Limits of the method detection are 0.0025•1.0 mg/kg. Huang et al. [14] used acetonitrile for extraction and GCB-NH2 cartridge for cleanup, making LC•MS/MS (MRM) determination of 103 pesticides in tea, with recoveries 65%•114% at three spiked levels. Fillion et al. [15] employed acetonitrile salt out for extraction against 251 pesticides in fruits and vegetables, such as apples, bananas, cabbages, and C18 in combination with Carb/NH$_2$ composite columns for cleanup, with limit of detection ≤ 0.04 mg/kg for 80% pesticides.

Wong et al. [16] diluted grape wine samples directly, adopted NH$_2$ cartridge combined with HLB for cleanup, and made GC•MS determination of 153 pesticides in grape wine, with results that were 116 pesticides in red wines and 124 in white wines with recoveries > 70% at the spiked level of 0.01 mg/L, while at the spiked level of 0.1 mg/L there were 123 in red wines and 128 in white wines. Wong et al. [17] made an analysis of 168 pesticides in Ginseng dried powders, used acetonitrile and acetone mixed solvents (acetone:cyclohexane:ethyl acetate = 2:1:1) as extraction solutions, with C8 as dispersive agent, as well as graphite carbon-PSA in tandem for cleanup, and made a respective GC•MS and GC•MS/MS determination, with average recoveries for acetonitrile extraction of GC•MS (87 ± 10)%, (88 ± 8)%, and (86 ± 10)% at three spiked levels of 25, 100, and 500 α/4g/kg, average recoveries for acetone mixed solutions (88 ± 13)%, (88 ± 12)%, and (88 ± 14)%, while average recoveries for acetonitrile extraction of GC-MS/MS were (83 ± 19)%, (90 ± 13)%, and (89 ± 11)%, average recoveries for acetone mixed solutions (98 ± 20)%, (91 ± 13)%, and (88 ± 14)%.

In addition, Schenck et al. [18] compared the cleanup effects of five different SPE cartridges of GCB, C18, SAX, NH2, and PSA on fruits and vegetable matrices and found that NH$_2$ and PSA were capable of effectively removing the interfering matters in matrices while GCB removed the pigments in matrixes effectively, yet had no apparent results on fat acids. Amvrazi et al. [19] compared, respectively, neutral alumina cartridge, florisil, C18, and Envi-Carb in the method of GC determination of 35 pesticides in olive oil. Tests found that Envi-Carb obtained the best results. The earlier described literatures found that the complexities of the subject test sample constituents and the wide polarities of simultaneous determination of several hundreds of pesticides decided that a single filler was not able to satisfy the requirement for sample cleanup, while the main orientation at present should be the study of SPE cleanup technique of combined cartridges of multiple kinds of fillers.

In recent years, the author's team has studied the residual analytical techniques of an accumulating total of more than 1000 pesticides and environment contaminants in agricultural products for human consumption, focusing on sample cleanup, as well; for example, Envi-Carb and Sep-Pak-NH$_2$ combined with SPE cleanup [20,21] were adopted for analytical standards of multigroups of pesticide residues in honeys, fruit juices, fruit wines. and edible fungi; Envi-18, Envi-Carb, and Sep-Pak-NH$_2$ combined with SPE cleanup were used for analytical standards of multigroups of pesticide residues in fruits, vegetables, and grains [22,23]; Envi-18 SPE was adopted for milk samples [24].

2.1.2 REVIEW OF QUECHERS METHOD

In 2003, Lehotay et al. [25] first proposed the QuEchERS method and adopted GC/MS to analyze 22 residue pesticides in vegetables and fruits such as tomatoes, pumpkins, apples, and strawberries, with recoveries 85%•105% and RSD < 5%. In 2005, Lehotay et al. [26] used his method to make a LC/MS/MS determination of 229 residue pesticides in lettuces and oranges, with recoveries 70%•120% and RSD < 20% at the spiked level of 10•100 ng/g. In 2007, Lehotay et al. [27] used the same method to analyze 16 residue pesticides in olives, with recoveries 70%•130% and RSD < 20%. In the same year, Lehotay [28] organized 13 laboratories from seven countries to conduct an intercollaborative study on the QuEChERS analytical method for multiresidues in matrices of fruits and vegetables and developed the first QuEChERS AOAC official method.

The announcement of the QuEChERS method caught the attention of our scientific counterparts, and similar studies have also been reported, successively. Wong et al. [29] used QuEChERS method to make a determination of 191 pesticides in oranges, peaches, spinaches, and ginsengs, extracted with 1% acetic acid•acetonitrile and vibrated for 1 min by Geno Grinder medical osilator, cleaned up by PSA, and analyzed by LC/MS/MS. More than 79% of pesticides had recoveries 80%•120%, and more than 94% had limits of detection 0.5•5 ppb. Wang et al. [30] used QuEChERS method to make a LC-TOF/MS determination of 142 pesticide residues in apples, bananas, peaches, fruit juices, peas, corn, pumpkins, and carrots, with recoveries 80%•110% at three spike levels of 10, 50, and 80 α/4g/kg, and at medium level spike precision <20%. The author [31] further used the method to analyze 138 pesticide residues in baby foods made of vegetables. Kmellar et al. [32] used QuEChERS for extraction and cleanup, and made an LC/MS/MS determination of 160 pesticide residues in tomatoes, pears, and oranges. Under the spiked levels of 10•100 α/4g/kg and recoveries 70%•120%, pesticide ratios for tomatoes, pears, and oranges are respectively 97%, 98%, and 97%. Nguyen et al. [33] used the method to make a GC/MS determination of 107 pesticide residues in cabbages and turnips, with recoveries 80%•115% and RSD < 15%.

Recently, Lehotay et al. [34] has improved his original QuEChERS method, and used GC-TOF/MS to determine 180 pesticide residues in grains and cereals. The method adopted water:acetonitrile (1:1), which was oscilliated for 1 h before extraction with cleanup by PSA and C18 two fillers, with recoveries 70%•120% and RSD < 20% for the majority of medium polarity pesticides. Walorczyk [35] used the improved QuEChERS method to extract with 10 mL water and 15 mL acetonitrile osilliated for 5 min, as well as cleanup by PSA and C18 two fillers for GC-MS/MS determination of 144 pesticide residues in wheat and fodders, with average recoveries 70%•120% and RSD < 20% at 0.01 mg/kg spiked level. Nguyen et al. [36] adopted the improved QuEChERS method by using 10 mL water and 10 mL 0.5% acetic acid in acetonitrile as extracting solvents and PSA and graphited carbon for cleanup for making GC/MS determination of 203 pesticide residues in rice, with average recoveries 75%•115% and RSD 2%•15%. Przybylski et al. [37] also made modifications to the QuEChERS method in combination with IT/MS for a determination of 236 residue pesticides in baby foods, with recoveries 70%•121% and RSD 2%•15% at three spiked levels. The linear correlation coefficient ≥0.9814.

REFERENCES

[1] Telepchak, J.M., August, F.T., Chaney, G., 2004. Totowa. Humana Press, New Jersey, United States.

[2] Hennion, M.C., 1999. J. Chromatogr. A 856, 3–54.

[3] Park, Y.S., Abd El-Aty, A.M., Choi, J.H., Cho, S.K., Shin, D.H., Shim, J.H., 2007. Biomed. Chromatogr. 21 (1), 29–39.

[4] Yahya, R.T., Mohammad, F.Z., Thaer, A.B., 2006. Biomed. Chromatogr. 558 (1•2), 62–68.

[5] Baugros, J.B., Cren-Olivèc), C., Giroud, B., Gauvrit, J.Y., Lantèc)ri, P., Grenier-Loustalot, M.F., 2009. J. Chromatogr. A 1216 (25), 4941–4949.

[6] Albero, B., Sánchez-Brunete, C., Tadeo, J.L., 2005. Talanta 66 (4), 917–924.

[7] Chen, S.B., Yu, X.J., He, X.Y., Xie, D.H., Fan, Y.M., Peng, J.F., 2009. Food Chem. 113 (4), 1297–1300.

[8] Gervais, G., Brosillon, S., Laplanche, A., Helen, C., 2008. J. Chromatogr. A 1202 (2), 163–172.

[9] Hernández, F., Pozo, O.J., Sancho, J.V., Bijlsma, L., Barreda, M., Pitarch, E., 2006. J. Chromatogr. A 1109 (2), 242–252.

[10] Yagñ/4e, C., Herrera, A., Ariño, A., Lázaro, R., Bayarri, S., Conchello, P., 2002. J. AOAC Int. 85 (5), 1181–1186.

[11] Kitagawa, Y., Okihashi, M., Takatori, S., Okamoto, Y., Fukui, N., Murata, H., Sumimoto, T., Obana, H., 2009. Shokuhin Eiseigaku Zasshi 50 (5), 198–207.

[12] Okihashi, M., Takatori, S., Kitagawa, Y., Tanaka, Y., 2007. J. AOAC Int. 90 (4), 1165–1179.

[13] Lou, Z.Y., Chen, Z.M., Luo, F.J., Tang, F.B., Liu, G.M., 2008. Chin. J. Chromatogr. 5, 568–576.

[14] Huang, Z., Zhang, Y., Wang, L., Ding, L., Wang, M., Yan, H., Li, Y., Zhu, S., 2009. J. Sep. Sci. 32 (9), 1294–1301.

[15] Fillion, J., Sauvèc), F., Selwyn, J., 2000. J. AOAC Int. 83 (3), 698–713.

[16] Wong, J.W., Webster, M.G., Halverson, C.A., Hengel, M.J., Ngim, K.K., Ebeler, S.E., 2003. J. Agric. Food Chem. 51 (5), 1148–1161.

[17] Wong, J.W., Zhang, K., Tech, K., Hayward, D.G., Krynitsky, A.J., Cassias, I., Schenck, F.J., Banerjee, K., Dasgupta, S., Brown, D., 2010. J. Agric. Food Chem. 58 (10), 5884–5896.

[18] Schenck, F.J., Lehotay, S.J., Vega, V., 2002. J. High Resolut. Chromatogr. 25 (14), 883–890.

[19] Amvrazi, E.G., Albanis, T.A., 2006. J. Agric. Food Chem. 54 (26), 9642–9651.

[20] Standardization Administration of P.R. China, 2008. GB/T 19426-2006. Method for the Determination of 497 Pesticides and Related Chemicals Residues in Honey, Fruit Juice and Wine by Gas Chromatography-Mass Spectrometry Method. Standards Press, Beijing, China.

[21] Standardization Administration of P.R. China, 2008. GB/T 23216-2008. Determination of 503 Pesticides and Related Chemicals Residues in Mushrooms•GC-MS Method. Standards Press, Beijing, China.

[22] Standardization Administration of P.R. China, 2008. GB/T 19648-2006. Method for Determination of 500 Pesticides and Related Chemicals Residues in Fruits and Vegetables by Gas Chromatography-Mass Spectrometry Method. Standards Press, Beijing, China.

[23] Standardization Administration of P.R. China, 2008. GB/T 19649-2006. Method for Determination of 475 Pesticides and Related Chemicals Residues in Grains by Gas Chromatography-Mass Spectrometry Method. Standards Press, Beijing, China.

[24] Standardization Administration of P.R. China, 2008. GB/T 23210-2008. Determination of 511 Pesticides and Related Chemicals Residues in Milk and Milk Powder•GC-MS Method. Standards Press, Beijing, China.

[25] Anastassiades, M., Lehotay, S.J., Stajnbaher, D., Schenck, F., 2003. J. AOAC Int. 86 (2), 412–431.

[26] Lehotay, S.J., de Kok, A., Hiemstra, M., Van Bodegraven, P.J., 2005. AOAC Int. 88 (2), 595–614.

[27] Cunha, S.C., Lehotay, S.J., Mastovska, K., Fernandes, J.O., Beatriz, M., Oliveira, P.P., 2007. J. Sep. Sci. 30 (4), 620–632.

[28] Lehotay, S.J., 2007. J. AOAC Int. 90 (2), 485–520.

[29] Fintschenko, Y., Krynitsky, A.J., Wong, J.W., 2010. J. Agric. Food Chem. 58, 5897–5903.

[30] Wang, J., Leung, D., 2009. J. AOAC Int. 92 (1), 279–301.

[31] Wang, J., Leung, D., 2009. J. Agric. Food Chem. 57 (6), 2162–2173.

[32] Kmellár, B., Fodor, P., Pareja, L., Ferrer, C., Martínez-Uroz, M.A., Valverde, A., Fernandez-Alba, A.R., 2008. J. Chromatogr. A 1215 (1•2), 37–50.

[33] Nguyen, T.D., Yu, J.E., Lee, D.M., 2008. J. Food Chem. 110, 207–213.

[34] Koesukwiwat, U., Lehotay, S.J., Mastovska, K., Dorweiler, K.J., Leepipatpiboon, N., 2010. J. Agric. Food Chem. 58, 5950–5958.

[35] Walorczyk, S., 2008. J. Chromatogr. A 1208, 202–214.

[36] Nguyen, T.D., Han, E.M., Seo, M.S., Kim, S.R., Yun, M.Y., Lee, D.M., Lee, G.H., 2008. Anal. Chim. Acta 619 (1), 67–74.

[37] Przybylski, C., Segard, C., 2009. J. Sep. Sci. 32 (11), 1858–1867.

Chapter 2.2

Comparative Study of Extraction Efficiencies of the Three Sample Preparation Techniques

2.2.1 INTRODUCTION

The disparities of different pretreatment approaches have been reported in some related literatures. In 2008, Schenck et al. [1] compared solid phase extraction (SPE) with QuEChERS to find that the cleanup effect of SPE is better than that of the original QuEChERS. Lee et al. [2] used QuEChERS to analyze 49 pesticide residues in tobacco and compared the three methods of liquid–liquid extraction, pressurized liquid extration with SPE and QuEChERS. The improved QuEChERS method adopted water:acetonitrile (1:1) by extracting at 2 times oscillation with 1 min for each and cleanup by PSA fillers. Tests found that QuEChERS was superior to liquid–liquid extraction, pressurized liquid extration with SPE in recoveries and RSD. Hercegova et al. [3] compared the four pretreatment methods of QuEChERS, improved QuEChERS, SPE, and MSPD through their applications in the determination of 20 pesticide residues in baby foods. The results through comparison showed that QuEChERS < improved QuEChERS < MSPD < SPE in terms of testing time, chemicals consumptions, and so on; SPE did the best in cleanup; QuEChERS showed no apparent differences with SPE in terms of recoveries.

    The author's team has, over the past 10 years, focused on the study on the high-throughput residual analysis techniques of agricultural products for human consumption and established 20 China National Standards based respectively on GC/MS and LC/MS/MS analytical techniques for determination of the accumulated total of more than 1000 pesticides and environmental pollutants. These occurred in fruits and vegetables [4], grains and cereals [5], animal tissues [6], acquatic products [7], honey, fruit, and vegetable juices, fruit wines [8], edible fugi [9], tea [10], milk powders [11], and Chinese medicinal herbs [12]. For the purpose of designing the best AOAC intercollaborative study protocol for multiresidues in teas, literature retrieval has been carried out, but so far no analytical methods for multiresidues in tea have yet been found in any association with the QuEChERS method. Therefore, a total of three methods, such as China national standards (Method-1), the original QuEChERS (Method-2), and a tea hydrolic sample preparation (Method-3), have been chosen to see their effects on sample prepartions. Tests found that when a recovery test was conducted the spiked sample of the day on the tolerances of recoveries and precision for Methods 1 and 2 were permissible within the range of the between-lab, while Method-1 is a little better than Method-2 in cleanup effects. However, tests of the 201 pesticides incurred samples found that the analytical results were much different from each other for the two methods, be it green tea, or oolong tea, or be it GC/MS or GC/MS/MS. The content of the pesticides determined by Method-1 was generally higher than that of Method-2, 30%–50% higher on average. Moreover, the number of pesticides of the content higher than that of Method-2 determined by Method-1 (ratio > 1) accounts for 95.5%–98%, with the ratio range distribution mostly falling 1.10–1.70. This means that Method-1 was capable of extracting more pesticides from 165-day pesticide-incurred samples than Method-2. The cause can be attributed to that the high-speed homogeneous extraction efficacy is slightly stronger than the osillating extraction efficacy, which led to the fact that Method-1 had better extraction efficiency than Method-2. From this point, it can be deduced that Method-1, if adopted as the AOAC collaborative extraction method, would make the test results more accurate. Therefore, the comparative study of the extracting efficencies of the three methods suggested that Method-1 will still be adopted as the sample preparation technique for AOAC intercollaborative study protocol.

2.2.2 EXPERIMENTAL

2.2.2.1 Reagents and Materials

1. Solvents. Acetonitrile, toluene, hexane, acetone (LC grade), purchased from Dikma Co. (Beijing, China).
2. $MgSO_4$, NaCl, and CH_3COONa. Reagent grade anhydrous $MgSO_4$ in powder form and ACS grade NaCl. The $MgSO_4$ were baked for 4 h at 650°C in a muffle furnace.
3. Organic acids. Glacial acetic acid (HAc).
4. Pesticides standard and internal standard. Purity, ≥95% (LGC Promochem, Wesel, Germany).
5. Stock standard solutions. Weigh 5–10 mg of individual pesticide standards (accurate to 0.1 mg) into a 10 mL volumetric flask. Dissolve and dilute to volume with toluene, toluene-acetone combination, or cyclohexane, depending on each individual compound solubility. Stock standard solutions should be stored in the dark below 4°C.
6. Mixed standard solution. Depending on properties and retention time of each pesticide, all the 201 pesticides for GC/MS and GC/MS/MS analysis are divided into 3 groups, A–C. The concentration of mixed standard solutions was dependent on the sensitivity of each compound for the instrument used for analysis. Mixed standard solutions should be stored in the dark below 4°C.

2.2.2.2 Apparatus

1. GC/MS system. Model 6890N gas chromatograph connected to a Model 5973N MSD and equipped with a Model 7683 autosampler (Agilent Technologies, Wilmington, DE, USA). The column used was a DB-1701 capillary column (30 m × 0.25 mm × 0.25 μm; J&W Scientific, Folsom, CA, USA).
2. GC/MS/MS system. Model 7890 gas chromatograph connected to a Model 7000A and equipped with a Model 7693 autosampler (Agilent Technologies, Wilmington, DE, USA). The column used was a DB-1701 capillary column (30 m × 0.25 mm × 0.25 μm; J&W Scientific, Folsom, CA, USA).
3. SPE. Cleanert-TPT (Agela, Tianjin, China).
4. PSA. Filler (Agela, Tianjin, China).
5. Graphite Carbon.Filler (Agela, Tianjin, China).
6. Homogenizer. T-25B (Janke & Kunkel, Staufen, Germany).
7. Rotary evaporator. Buchi EL131 (Flawil, Switzerland).
8. Centrifuge. Z 320 (B. HermLe AG, Gosheim, Germany).
9. Nitrogen evaporator. EVAP 112 (Organomation Associates, Inc., New Berlin, MA, USA).

2.2.2.3 Experimental Method

Method-1 [10]: weigh 5 g testing sample (accurate to 0.01 g) into 80 mL centrifugal tube, add 15 mL 1% acetic acid–acetonitrile, extract homogeneously at 15,000 r/m 1 min, centrifuge 5 min at 4200 r/m, and transfer the supernatants into 100 mL pear-shaped flask. Repeat extracting the dregs with 15 mL with 1% acetic acid–acetonitrile once, centrifuge, consolidate the supernatants over 2 times, and place them in 45 °C water bath and rotary evaporate to about 1 mL before standby for cleanup.

Add about 2 cm high anhydrous sodium sulfate into Cleanert-TPT SPE cartridge, precleanse with 10 mL acetonitrile-toluene (3 + 1), and discard the effluents. Connect the pear-shaped flask at the lower end and place it into the fixing bracket. Transfer the aforementioned concentrated sample solutions into a Cleanert-TPT cartridge, cleanse the bottle with 2 mL acetonitrile-toluene (3 + 1), and repeat the cleansing 2 times, transfer the cleaning solutions into the cartridge, while attaching a 50 mL solution storage device onto the cartridge, then cleanse with 25 mL acetonitrile-toluene, collect the aforementioned effluents into a pear-shaped flask, and place in a 40 °C water bath and rotary evaporate to about 0.5 mL. Add 5 mL hexane for solvent exchange, repeat 2 times, dilute to about 1 mL, add 40 µL internal standard solutions, and mix uniformly to pass filtering membrane before being used for GC/MS detection.

Method-2 [13]: weigh 10 g testing sample (accurate to 0.01 g) into an 80 mL centrifuge tube with a cover, add 40 mL 1% acetic acid–acetonitrile solutions and vigorously vortex in a vortex mixer for 2 min. Add 1.5 g anhydrous sodium acetate into a centrifuge tube with a cover, shake another 1 min, then add 2 g anhydrous magnesium sulfate into the centrifuge tube, shake 2 min, centrifuge at 4200 r/min for 5 min, take 20 mL supernatants and transfer into another centrifuge tube containing 0.30 g anhydrous magnesium sulfate, 0.13 g PSA powder and 0.13 g graphite carbon, and shake 2 min, centrifuge at 4200 r/min for 5 min. Transfer the supernatants into a 100 mL pear-shaped flask and rotate in a 40 °C water bath and concentrate to about 0.5 mL. Add 5 mL hexane for solvent exchange, repeat 2 times, dilute 1 mL, add 40 µL internal standard solutions, mix uniformly to pass filtering membrane before being used for GC/MS determination.

Method-3 [14]: weigh 5 g testing sample (accurate to 0.01 g) into an 80 mL centrifuge tub, add 20 mL water to soak 1 h before adding 20 mL acetonitrile and 10 g sodium chloride, extract homogeneously at 15,000 r/min for 1 min, centrifuge at 4200 r/min for 5 min, take acetonitrile layer to pass the barrel funnel containing anhydrous sodium sulfate, collect into a 100 mL pear-shaped flask, repeat extracting the dregs one time with 20 mL with acetonitrile before passing centrifuge filtering, consolidate the extracting solutions more than 2 times and place the extracting solutions in a 45 °C water bath and rotate in a rotary evaporator to dryness to about 1 mL. Cleanup for Cleanert-TPT is similar to those in Method-1.

2.2.3 PREPARING PESTICIDES INCURRED TEA SAMPLES AND DECIDING THE PRECIPITATED CONTENT OF THE TARGET PESTICIDE

1. Take green tea and oolong tea that are found to be free from target pesticides after testing and pass them respectively through 10 mesh and 16 mesh sieves after initial blending via blender. Take 10–16 mesh 500 g oolong tea and green tea each, and spread them uniformly over the bottom of a stainless steel vessel with 40 cm diameter and wait for spraying.
2. Accurately transfer a certain amount of pesticide-mixed standard solutions into the full-glass sprayer and spray the tea leaves. Spray while stirring the tea leaves with a glass rod so as to make a uniform spray. Wait until the fluid in the sprayer is finished, then put 3 × 5 mL toluene into the sprayer, and shake a couple of times and mix homogeneously before spraying onto the tea leaves once again with spraying and stirring going on at the same time.
3. After the completion of spraying, continue to stir the tea leaves for half an hour to wait for the volatile solvents in the tea leaves have thoroughly volatized, after which, place the sprayed tea leaves into a 4 L brown bottle, and vibrate on an oscillator overnight and avoid exposure to light for storage at room temperature.
4. Spread the incurred tea samples onto the bottom of a flat-bottomed vessel, draw a cross and weigh a total of 5 portions of incurred tea samples located the symmetrical four points of the cross and the central part. Submit them for GC/MS and GC/MS/MS determination, and calculate the average value of the pesticide content of the incurred samples and RSD. When the ratio of RSD of the pesticide analyzed being within 15% reached 90%, it could be judged that tea samples have been sprayed and mixed homogeneously, and the average values of the content obtained more than 5 times shall serve as the fixed value concentration of the pesticides in the incurred samples.

Determination conditions are the same as reported by Pang et al. [10,15]. Retention times, limit of detection (LOD), limit of quantity (LOQ), quantitative ions, qualitative ions, and the abundance ratios of 201 pesticides for GC/MS are shown in Table 2.2.1, retention times, LOD, LOQ, monitoring ions and collision energy for GC/MS/MS are shown in Table 2.2.2, SIM acquisition of three group ions for GC/MS are shown in Table 2.2.3, MRM acquisition of three group ions for GC/MS/MS are shown in Table 2.2.4.

2.2.4 OPTIMIZATION OF EXPERIMENTAL CONDITIONS FOR METHOD-3

1. Deciding the hydrate conditions of teas: weigh 6 portions of tea, with each portion 5 g and sequential numbers being 1#, 2#, 3#, 4#, 5#, and 6#. Add respectively 5, 10, 15, 20, 25, and 30 mL water, soak for 1 h until flowing water is hardly visible on the surface. Teas 1# and 2# are not well soaked, which demonstrates that teas are not thoroughly hydrated.

TABLE 2.2.1 Retention Times, LOD, LOQ, Quantifying Ion, Qualifying Ions, the Abundance Ratios for 201 Pesticides by GC/MS

No.	Pesticides	Retention time (min)	Quantifying ion	Qualifying ion 1	Qualifying ion 2	LOQ (μg/kg)	LOD (μg/kg)
ISTD	Heptachlor-epoxide	22.15	353(100)	355(79)	351(52)		
A group							
1	2,3,4,5-Tetrachloroaniline	18.72	231(100)	229(76)	233(48)	20.0	10.0
2	2,3,5,6-Tetrachloroaniline	14.32	231(100)	229(76)	158(25)	10.0	5.0
3	4,4-Dibromobenzophe-none	25.49	340(100)	259(30)	185(179)	10.0	5.0
4	4,4-Dichlorobenzophe-none	21.46	250(100)	252(62)	215(26)	10.0	5.0
5	Acetochlor	19.80	146(100)	162(59)	223(59)	50.0	25.0
6	Alachlor	20.21	188(100)	237(35)	269(15)	30.0	15.0
7	Atratone	16.93	196(100)	211(68)	197(105)	25.0	12.5
8	Benodanil	28.95	231(100)	323(38)	203(22)	30.0	15.0
9	Benoxacor	19.62	120(100)	259(38)	176(19)	50.0	25.0
10	Bromophos-ethyl	23.11	359(100)	303(77)	357(74)	10.0	5.0
11	Butralin	22.18	266(100)	224(16)	295(9)	40.0	20.0
12	Chlorfenapyr	27.46	247(100)	328(54)	408(51)	200.0	100.0
13	Clomazone	17.12	204(100)	138(4)	205(13)	10.0	5.0
14	Cycloate	13.53	154(100)	186(5)	215(12)	10.0	5.0
15	Cycluron	18.24	89(100)	198(36)	114(9)	30.0	15.0
16	Cyhalofop-butyl	31.38	256(100)	357(74)	229(79)	20.0	10.0
17	Cyprodinil	22.03	224(100)	225(62)	210(9)	10.0	5.0
18	Dacthal	21.39	301(100)	332(31)	221(16)	10.0	5.0
19	de-PCB 101	22.63	326(100)	254(66)	291(18)	10.0	5.0
20	de-PCB 118	25.12	326(100)	254(38)	184(16)	10.0	5.0
21	de-PCB 138	26.86	360(100)	290(68)	218(26)	25.0	12.5
22	de-PCB 180	29.07	394(100)	324(70)	359(20)	10.0	5.0
23	de-PCB 28	18.21	256(100)	186(53)	258(97)	10.0	5.0
24	de-PCB 31	18.21	256(100)	186(53)	258(97)	10.0	5.0
25	Dichlorofop-methyl	28.16	253(100)	281(50)	342(82)	10.0	5.0
26	Dimethenamid	19.81	154(100)	230(43)	203(21)	10.0	5.0
27	Diofenolan-1	26.75	186(100)	300(60)	225(24)	20.0	10.0
28	Diofenolan-2	27.08	186(100)	300(60)	225(29)	20.0	10.0
29	Fenbuconazole	34.44	129(100)	198(51)	125(31)	50.0	25.0
30	Fenpyroximate	17.49	213(100)	142(21)	198(9)	10.0	5.0
31	Fluotrimazole	28.54	311(100)	379((60)	233(36)	10.0	5.0
32	Fluroxypr-1-methylheptyl ester	28.70	366(100)	254(67)	237(60)	10.0	5.0
33	Iprovalicarb-1	26.11	119(100)	134(126)	158(62)	40.0	20.0
34	Iprovalicarb-2	26.51	134(100)	119(75)	158(48)	40.0	20.0
35	Isodrin	20.99	193(100)	263(46)	195(83)	10.0	5.0
36	Isoprocarb-1	7.59	121(100)	136(34)	103(20)	20.0	10.0
37	Isoprocarb-2	13.71	121(100)	136(34)	103(20)	20.0	10.0

No.	Pesticides	Retention time (min)	Quantifying ion	Qualifying ion 1	Qualifying ion 2	LOQ (µg/kg)	LOD (µg/kg)
38	Lenacil	30.05	153(100)	136(6)	234(2)	10.0	5.0
39	Metalaxyl	20.84	206(100)	249(53)	234(38)	30.0	15.0
40	Metazachlor	23.54	209(100)	133(120)	211(32)	30.0	15.0
41	Methabenzthiazuron	16.65	164(100)	136(81)	108(27)	100.0	50.0
42	Mirex	28.83	272(100)	237(49)	274(80)	10.0	5.0
43	Monalide	20.43	197(100)	199(31)	239(45)	20.0	10.0
44	Paraoxon-ethyl	21.94	275(100)	220(60)	247(58)	320.0	160.0
45	Pebulate	10.16	128(100)	161(21)	203(20)	30.0	15.0
46	Pentachloroaniline	19.03	265(100)	263(63)	230(8)	10.0	5.0
47	Pentachloroanisole	15.07	280(100)	265(100)	237(85)	10.0	5.0
48	Pentachlorobenzene	11.02	250(100)	252(64)	215(24)	10.0	5.0
49	Perthane	24.89	223(100)	224(20)	178(9)	25.0	12.5
50	Phenanthrene	17.01	188(100)	160(9)	189(16)	25.0	12.5
51	Pirimicarb	19.05	166(100)	238(23)	138(8)	20.0	10.0
52	Procymidone	24.63	283(100)	285(70)	255(15)	10.0	5.0
53	Prometrye	20.25	241(100)	184(78)	226(60)	10.0	5.0
54	Propham	11.48	179(100)	137(66)	120(51)	10.0	5.0
55	Prosulfocarb	19.63	251(100)	252(14)	162(10)	10.0	5.0
56	Secbumeton	18.47	196(100)	210(38)	225(39)	10.0	5.0
57	Silafluofen	33.04	287(100)	286(274)	258(289)	1800.0	900.0
58	Tebupirimfos	17.56	318(100)	261(107)	234(100)	80.0	40.0
59	Tebutam	15.47	190(100)	106(38)	142(24)	20.0	10.0
60	Tebuthiuron	14.25	156(100)	171(30)	157(9)	20.0	10.0
61	Tefluthrin	17.35	177(100)	197(26)	161(5)	10.0	5.0
62	Thenylchlor	29.03	127(100)	288(25)	141(17)	20.0	10.0
63	Thionazin	14.26	143(100)	192(39)	220(14)	10.0	5.0
64	Trichloronat	21.20	297(100)	269(86)	196(16)	10.0	5.0
65	Trifluralin	15.44	306(100)	264(72)	335(7)	20.0	10.0
B group							
66	2,4'-DDT	25.53	235(100)	237(63)	165(37)	20.0	10.0
67	4,4'-DDE	23.92	318(100)	316(80)	246(139)	10.0	5.0
68	Benalaxyl	27.69	148(100)	206(32)	325(8)	10.0	5.0
69	Benzoylprop-ethyl	29.59	292(100)	365(36)	260(37)	30.0	15.0
70	Bromofos	21.83	331(100)	329(75)	213(7)	20.0	10.0
71	Bromopropylate	29.43	341(100)	183(34)	339(49)	20.0	10.0
72	Buprofenzin	24.97	105(100)	172(54)	305(24)	20.0	10.0
73	Butachlor	23.97	176(100)	160(75)	188(46)	20.0	10.0
74	Butylate	9.45	156(100)	146(115)	217(27)	30.0	15.0
75	Carbofenothion	27.36	157(100)	342(49)	199(28)	20.0	10.0
76	Chlorfenson	25.32	302(100)	175(282)	177(103)	20.0	10.0
77	Chlorfenvinphos	23.37	323(100)	267(139)	269(92)	30.0	15.0
78	Chlormephos	10.59	121(100)	234(70)	154(70)	20.0	10.0
79	Chloroneb	11.90	191(100)	193(67)	206(66)	10.0	5.0
80	Chloropropylate	26.00	251(100)	253(64)	141(18)	10.0	5.0

(Continued)

TABLE 2.2.1 Retention Times, LOD, LOQ, Quantifying Ion, Qualifying Ions, the Abundance Ratios for 201 Pesticides by GC/MS (*cont.*)

No.	Pesticides	Retention time (min)	Quantifying ion	Qualifying ion 1	Qualifying ion 2	LOQ (µg/kg)	LOD (µg/kg)
81	Chlorpropham	15.72	213(100)	171(59)	153(24)	20.0	10.0
82	Chlorpyifos(ethyl)	20.96	314(100)	258(57)	286(42)	10.0	5.0
83	Chlorthiophos	26.65	325(100)	360(52)	297(54)	30.0	15.0
84	cis-chlordane	23.59	373(100)	375(96)	377(51)	20.0	10.0
85	cis-diallate	14.77	234(100)	236(37)	128(38)	20.0	10.0
86	Cyanofenphos	28.54	157(100)	169(56)	303(20)	10.0	5.0
87	Desmetryn	19.79	213(100)	198(60)	171(30)	10.0	5.0
88	Dichlobenil	9.90	171(100)	173(68)	136(15)	2.0	1.0
89	Dicloran	18.22	206(100)	176(128)	160(52)	20.0	10.0
90	Dicofol	21.45	139(100)	141(72)	250(23)	20.0	10.0
91	Dimethachlor	20.03	134(100)	197(47)	210(16)	30.0	15.0
92	Dioxacarb	11.02	121(100)	166(44)	165(36)	80.0	40.0
93	Endrin	25.15	263(100)	317(30)	345(26)	120.0	60.0
94	Epoxiconazole-2	29.66	192(100)	183(13)	138(30)	200.0	100.0
95	EPTC	8.51	128(100)	189(30)	132(32)	30.0	15.0
96	Ethofumesate	22.17	207(100)	161(54)	286(27)	20.0	10.0
97	Ethoprophos	14.52	158(100)	200(40)	242(23)	30.0	15.0
98	Etrimfos	17.95	292(100)	181(40)	277(31)	10.0	5.0
99	Fenamidone	30.57	268(100)	238(111)	206(32)	25.0	12.5
100	Fenarimol	31.80	139(100)	219(70)	330(42)	20.0	10.0
101	Flamprop-isopropyl	26.93	105(100)	276(19)	363(3)	10.0	5.0
102	Flamprop-methyl	26.15	105(100)	77(26)	276(11)	10.0	5.0
103	Fonofos	17.40	246(100)	137(141)	174(15)	10.0	5.0
104	Hexachlorobenzene	14.45	284(100)	286(81)	282(51)	10.0	5.0
105	Hexazinone	30.49	171(100)	252(3)	128(12)	30.0	15.0
106	Iodofenphos	24.43	377(100)	379(37)	250(6)	20.0	10.0
107	Isofenphos	23.17	213(100)	255(44)	185(45)	20.0	10.0
108	Isopropalin	22.39	280(100)	238(40)	222(4)	20.0	10.0
109	Methoprene	21.70	73(100)	191(29)	153(29)	40.0	20.0
110	Methoprotryne	25.82	256(100)	213(24)	271(17)	30.0	15.0
111	Methoxychlor	29.40	227(100)	228(16)	212(4)	10.0	5.0
112	Methyl-parathion	21.12	263(100)	233(66)	246(8)	40.0	20.0
113	Metolachlor	21.50	238(100)	162(159)	240(33)	10.0	5.0
114	Nitrapyrin	10.94	194(100)	196(97)	198(23)	30.0	15.0
115	Oxyfluorfen	26.44	252(100)	361(35)	300(35)	40.0	20.0
116	Pendimethalin	22.71	252(100)	220(22)	162(12)	40.0	20.0
117	Picoxystrobin	24.83	335(100)	303(43)	367(9)	20.0	10.0
118	Piperophos	30.27	320(100)	140(123)	122(114)	30.0	15.0
119	Pirimiphos-ethyl	21.65	333(100)	318(93)	304(69)	10.0	5.0
120	Pirimiphos-methyl	20.47	290(100)	276(86)	305(74)	20.0	10.0
121	Profenofos	24.80	339(100)	374(39)	297(37)	60.0	30.0
122	Profluralin	17.57	318(100)	304(47)	347(13)	40.0	20.0

No.	Pesticides	Retention time (min)	Quantifying ion	Qualifying ion 1	Qualifying ion 2	LOQ (µg/ kg)	LOD (µg/kg)
123	Propachlor	14.96	120(100)	176(45)	211(11)	30.0	15.0
124	Propiconazole	28.15	259(100)	173(97)	261(65)	30.0	15.0
125	Propyzamide	18.99	173(100)	255(23)	240(9)	10.0	5.0
126	Ronnel	19.85	285(100)	287(67)	125(32)	20.0	10.0
127	Sulfotep	15.73	322(100)	202(43)	238(27)	10.0	5.0
128	Tebufenpyrad	29.06	318(100)	333(78)	276(44)	10.0	5.0
129	Terbutryn	20.61	226(100)	241(64)	185(73)	20.0	10.0
130	Thiobencarb	20.73	100(100)	257(25)	259(9)	20.0	10.0
131	Tralkoxydim	32.03	283(100)	226(7)	268(8)	120.0	60.0
132	Trans-chlodane	23.32	373(100)	375(96)	377(51)	10.0	5.0
133	Trans-diallate	15.35	234(100)	236(37)	128(38)	20.0	10.0
134	Trifloxystrobin	27.53	116(100)	131(40)	222(30)	40.0	20.0
135	Zoxamide	22.30	187(100)	242(68)	299(9)	20.0	10.0
C group							
136	2,4'-DDE	22.73	246(100)	318(34)	176(26)	25.0	12.5
137	Ametryn	20.39	227(100)	212(53)	185(17)	30.0	15.0
138	Bifenthrin	28.56	181(100)	166(25)	165(23)	10.0	5.0
139	Bitertanol	32.48	170(100)	112(8)	141(6)	30.0	15.0
140	Boscalid	34.18	342(100)	140(229)	112(71)	40.0	20.0
141	Butafenacil	33.61	331(100)	333(34)	180(35)	10.0	5.0
142	Carbaryl	14.57	144(100)	115(100)	116(43)	30.0	15.0
143	Chlorobenzilate	26.10	251(100)	253(65)	152(5)	30.0	15.0
144	Chlorthal-dimethyl	21.39	301(100)	332(27)	221(17)	20.0	10.0
145	Dibutyl succinate	12.21	101(100)	157(19)	175(5)	20.0	10.0
146	Diethofencarb	21.76	267(100)	225(98)	151(31)	60.0	30.0
147	Diflufenican	28.73	266(100)	394(25)	267(14)	10.0	5.0
148	Dimepiperate	22.54	119(100)	145(30)	263(8)	20.0	10.0
149	Dimethametryn	22.77	212(100)	255(9)	213(2)	20.0	10.0
150	Dimethomorph	37.45	301(100)	387(32)	165(28)	20.0	10.0
151	Dimethylphthalate	11.56	163(100)	194(7)	133(5)	40.0	20.0
152	Diniconazole	27.43	268(100)	270(65)	232(13)	10.0	5.0
153	Diphenamid	23.24	167(100)	239(30)	165(43)	10.0	5.0
154	Dipropetryn	21.07	255(100)	240(42)	222(20)	10.0	5.0
155	Ethalfluralin	15.17	276(100)	316(81)	292(42)	40.0	20.0
156	Etofenprox	32.83	163(100)	376(4)	183(6)	25.0	12.5
157	Etridiazol	10.38	211(100)	183(73)	140(19)	30.0	15.0
158	Fenazaquin	29.07	145(100)	160(46)	117(10)	25.0	12.5
159	Fenchlorphos	19.86	285(100)	287(69)	270(6)	40.0	20.0
160	Fenoxanil	23.53	140(100)	189(14)	301(6)	20.0	10.0
161	Fenpropidin	17.86	98(100)	273(5)	145(5)	50.0	25.0
162	Fenson	22.94	141(100)	268(53)	77(104)	10.0	5.0
163	Flufenacet	22.99	151(100)	211(61)	363(6)	200.0	100.0
164	Furalaxyl	23.88	242(100)	301(24)	152(40)	20.0	10.0
165	Heptachlor	18.54	272(100)	237(40)	337(27)	30.0	15.0

(Continued)

TABLE 2.2.1 Retention Times, LOD, LOQ, Quantifying Ion, Qualifying Ions, the Abundance Ratios for 201 Pesticides by GC/MS (*cont.*)

No.	Pesticides	Retention time (min)	Quantifying ion	Qualifying ion 1	Qualifying ion 2	LOQ (μg/kg)	LOD (μg/kg)
166	Iprobenfos	18.71	204(100)	246(18)	288(17)	30.0	15.0
167	Isazofos	18.89	161(100)	257(53)	285(39)	20.0	10.0
168	Isoprothiolane	25.78	290(100)	231(82)	204(88)	20.0	10.0
169	Kresoxim-methyl	25.23	116(100)	206(25)	131(66)	20.0	10.0
170	Mefenacet	31.55	192(100)	120(35)	136(29)	30.0	15.0
171	Mepronil	28.34	119(100)	269(26)	120(9)	25.0	12.5
172	Metribuzin	20.72	198(100)	199(21)	144(12)	30.0	15.0
173	Molinate	12.06	126(100)	187(24)	158(2)	10.0	5.0
174	Napropamide	25.01	271(100)	128(111)	171(34)	30.0	15.0
175	Nuarimol	29.07	314(100)	235(155)	203(108)	20.0	10.0
176	Permethrin	31.70	183(100)	184(14)	255(1)	20.0	10.0
177	Phenothrion	29.32	123(100)	183(74)	350(6)	10.0	5.0
178	Piperonyl butoxide	27.62	176(100)	177(33)	149(14)	30.0	15.0
179	Pretilachlor	24.96	162(100)	238(26)	262(8)	75.0	37.5
180	Prometon	16.92	210(100)	225(91)	168(67)	30.0	15.0
181	Pronamide	19.00	173(100)	175(62)	255(22)	10.0	5.0
182	Propetamphos	18.20	138(100)	194(49)	236(30)	10.0	5.0
183	Propoxur-1	6.59	110(100)	152(16)	111(9)	200.0	100.0
184	Propoxur-2	15.50	110(100)	152(19)	111(8)	100.0	50.0
185	Prothiophos	24.07	309(100)	267(88)	162(55)	10.0	5.0
186	Pyridaben	32.04	147(100)	117(11)	364(7)	10.0	5.0
187	Pyridaphenthion	30.20	340(100)	199(48)	188(51)	10.0	5.0
188	Pyrimethanil	17.45	198(100)	199(45)	200(5)	10.0	5.0
189	Pyriproxyfen	29.99	136(100)	226(8)	185(10)	10.0	5.0
190	Quinalphos	23.21	146(100)	298(28)	157(66)	10.0	5.0
191	Quinoxyphen	27.14	237(100)	272(37)	307(29)	10.0	5.0
192	Telodrin	20.55	311(100)	375(35)	103(134)	50.0	25.0
193	Tetrasul	25.79	252(100)	324(64)	254(68)	10.0	5.0
194	Thiazopyr	21.80	327(100)	363(73)	381(34)	20.0	10.0
195	Tolclofos-methyl	19.94	265(100)	267(36)	250(10)	10.0	5.0
196	Transfluthrin	19.27	163(100)	165(23)	335(7)	10.0	5.0
197	Triadimefon	22.46	208(100)	210(50)	181(74)	20.0	10.0
198	Triadimenol	24.68	112(100)	168(81)	130(15)	30.0	15.0
199	Triallate	17.23	268(100)	270(73)	143(19)	30.0	15.0
200	Tribenuron-methyl	9.43	154(100)	124(45)	110(18)	10.0	5.0
201	Vinclozolin	20.53	285(100)	212(109)	198(96)	10.0	5.0

Add 5 g sodium chloride into 3#, 4#, 5#, and 6# sequentially, with 15 mL acetonitrile homogenous extraction for 1 min, respectively, centrifuge to find that 3# has only acetonitrile layer, while 4#, 5#, and 6# have both a water layer and an acetonitrile layer, with the acetonitrile layer volumes being, respectively: 8.2, 7.5, and 4.5 mL. It suggests that when 5 g tea is added into 20 mL (4#) water tea leaves can be fully hydrate, with little loss of acetonitrile during liquid–liquid partitioning, so 5 g of tea added into 20 mL water is decided to be suitable.

TABLE 2.2.2 Retention Times, LOD, LOQ, Monitoring Ions and Collision Energy for 201 Pesticides by GC/MS/MS

6.996 mm	Pesticides	Retention time (min)	Quantifying for precursor, product ion	Qualifying for precursor, product ion	Collision energy, V	LOQ (µg/kg)	LOD (µg/kg)
ISTD	Heptachlor-epoxide	22.15	353/282	353/282;353/263	17;17		
A group							
1	2,3,4,5-Tetrachloroaniline	18.74	231/160	231/160;231/158	15;20	7.5	3.8
2	2,3,5,6-Tetrachloroaniline	14.34	231/160	231/160;231/158	25;25	5.0	2.5
3	4,4-Dibromobenzophenone	25.55	340/185	340/185;340/183	15;15	25.0	12.5
4	4,4-Dichlorobenzophenone	21.51	250/215	250/215;250/139	5;10	10.0	5.0
5	Acetochlor	19.75	146/131	146/131;146/118	10;10	10.0	5.0
6	Alachlor	20.16	237/160	237/160;237/146	8;20	5.0	2.5
7	Atratone	16.87	211/196	211/196;211/169	10;5	10.0	5.0
8	Benodanil	28.87	323/231	323/231;323/196	10;5	25.0	12.5
9	Benoxacor	19.51	259/176	259/176;259/120	10;25	20.0	10.0
10	Bromophos-ethyl	23.16	359/331	359/331;359/303	10;10	12.5	6.3
11	Butralin	22.16	266/190	266/190;266/174	10;10	30.0	15.0
12	Chlorfenapyr	27.37	408/363	408/363;408/59	5;15	175.0	87.5
13	Clomazone	17.06	204/107	204/107;204/78	25;25	2.5	1.3
14	Cycloate	13.58	154/83	154/83;154/72	10;10	1.3	0.6
15	Cycluron	18.15	198/89	198/89;198/72	5;15	25.0	12.5
16	Cyhalofop-butyl	31.36	357/256	357/256;357/229	10;15	70.0	35.0
17	Cyprodinil	22.09	224/222	224/222;224/208	15;15	10.0	5.0
18	Dacthal	21.39	301/273	301/273;301/223	15;25	3.0	1.5
19	de-PCB101	22.74	326/256	326/256;326/254	25;25	2.0	1.0
20	de-PCB118	25.26	326/256	326/256;326/254	25;25	5.0	2.5
21	de-PCB138	26.98	360/325	360/325;360/290	15;15	15.0	7.5
22	de-PCB180	29.2	394/359	394/359;394/324	15;25	15.0	7.5
23	de-PCB28	18.29	256.01/186	256.01/186;256.01/151	25;25	0.4	0.2
24	de-PCB31	18.29	258/186	258/186;256/186	15;25	0.4	0.2
25	Dichlorofop-methyl	28.22	342/255	342/255;342/184	15;25	50.0	25.0
26	Dimethenamid	19.73	230/154	230/154;230/111	8;25	2.5	1.3
27	Diofenolan-1	26.84	186/158	186/158;186/109	5;15	25.0	12.5
28	Diofenolan-2	27.16	186/158	186/158;186/109	5;15	25.0	12.5
29	Fenbuconazole	34.33	198/129	198/129;198/102	15;25	5.0	2.5

(Continued)

TABLE 2.2.2 Retention Times, LOD, LOQ, Monitoring Ions and Collision Energy for 201 Pesticides by GC/MS/MS (cont.)

6.996 mm	Pesticides	Retention time (min)	Quantifying for precursor, product ion	Qualifying for precursor, product ion	Collision energy, V	LOQ (µg/kg)	LOD (µg/kg)
30	Fenpyroximate	17.38	213/212	213/212;213/77	10;25	25.0	12.5
31	Fluotrimazole	28.54	379/276	379/276;379/262	10;15	30.0	15.0
32	Fluroxypr-1-methylheptylester	28.72	366/209	366/209;366/181	15;15	250.0	125.0
33	Iprovalicarb-1	26.46	134/93	134/93;134/91	15;15	140.0	70.0
34	Iprovalicarb-2	26.46	134/93	134/93;134/91	15;15	140.0	70.0
35	Isodrin	21.07	193/157	193/157;193/123	15;25	12.5	6.3
36	Isoprocarb-1	7.6	121/103	121/103;121/77	10;15	1.0	0.5
37	Isoprocarb-2	13.58	121/103	121/103;121/77	10;15	2.5	1.3
38	Lenacil	29.95	153/136	153/136;153/110	15;15	27.5	13.8
39	Metalaxyl	20.73	206/132	206/132;206/105	15;15	10.0	5.0
40	Metazachlor	23.43	209/133	209/133;209/132	5;15	10.0	5.0
41	Methabenzthiazuron	16.53	164/136	164/136;164/108	10;25	15.0	7.5
42	Mirex	29.02	272/237	272/237;272/235	10;10	2.5	1.3
43	Monalide	20.34	239/197	239/197;239/85	5;15	10.0	5.0
44	Paraoxon-ethyl	21.77	275/149	275/149;275/99	5;10	375.0	187.5
45	Pebulate	10.23	161/128	161/128;128/72	5;7	10.0	5.0
46	Pentachloroaniline	19.03	263/192	263/192;263/156	15;25	10.0	5.0
47	Pentachloroanisole	15.26	280/265	280/265;280/237	10;15	10.0	5.0
48	Pentachlorobenzene	11.17	250/215	250/215;250/177	15;25	10.0	5.0
49	Perthane	24.98	223/193	223/193;223/179	25;25	2.5	1.3
50	Phenanthrene	17.16	189/185	189/185;189/161	25;25	7.5	3.8
51	Pirimicarb	18.98	238/166	238/166;238/96	15;25	5.0	2.5
52	Procymidone	24.55	283/255	283/255;283/96	10;10	2.5	1.3
53	Prometrye	20.23	241/199	241/199;241/184	5;5	5.0	2.5
54	Propham	11.41	179/137	179/137;179/93	10;10	5.0	2.5
55	Prosulfocarb	19.66	251/128	251/128;251/86	5;10	10.0	5.0
56	Secbumeton	18.43	225/169	225/169;225/154	5;15	5.0	2.5
57	Silafluofen	33.14	287/259	287/259;287/179	5;25	20.0	10.0
58	Tebupirimfos	17.61	318/276	318/276;318/152	5;10	12.5	6.3
59	Tebutam	15.44	190/106	190/106;190/57	10;10	2.5	1.3

60	Tebuthiuron	14.12	156/89	156/89;156/74	10.0	5.0
61	Tefluthrin	17.4	177/127	177/127;177/101	1.0	0.5
62	Thenylchlor	28.97	288/174	288/174;288/141	22.5	11.3
63	Thionazin	14.18	143/79	143/79;143/52	7.5	3.8
64	Trichloronat	21.25	297/269	297/269;297/223	5.0	2.5
65	Trifluralin	15.41	306/264	306/264;306/206	6.0	3.0
B group						
66	2,4'-DDT	25.62	235/199	235/199;235/165	1.5	0.8
67	4,4'-DDE	24.01	318/248	318/248;318/246	5.0	2.5
68	Benalaxyl	27.66	148/105	148/105;148/79	2.5	1.3
69	Benzoylprop-ethyl	29.55	292/105	292/105;292/77	5.0	2.5
70	Bromofos	21.84	331/316	331/316;331/286	75.0	37.5
71	Bromopropylate	29.46	341/185	341/185;341/183	10.0	5.0
72	Buprofenzin	24.99	172/116	172/116;105/77	1.3	0.6
73	Butachlor	24.01	176/150	176/150;176/126	10.0	5.0
74	Butylate	9.52	146/90	146/90;146/57	0.8	0.4
75	Carbofenothion	27.39	342/199	342/199;157/121	125.0	62.5
76	Chlorfenson	25.27	302/175	302/175;302/111	12.5	6.3
77	Chlorfenviphos	23.31	323/267	323/267;323/159	37.5	18.8
78	Chlormephos	10.55	234/154	234/154;234/121	5.0	2.5
79	Chlorneb	11.92	191/141	191/141;191/113	5.0	2.5
80	Chloropropylate	26.03	251/139	251/139;251/111	5.0	2.5
81	Chlorpropham	15.68	213/171	213/171;213/127	40.0	20.0
82	Chlorpyifos(ethyl)	21.01	314/286	314/286;314/258	10.0	5.0
83	Chlorthiophos	26.66	360/325	360/325;360/297	50.0	25.0
84	Cis-chlordane	23.62	373/301	373/301;373/266	20.0	10.0
85	Cis-diallate	14.81	234/192	234/192;234/150	5.0	2.5
86	Cyanofenphos	28.66	157/110	157/110;157/77	2.5	1.3
87	Desmetryn	19.74	213/198	213/198;213/171	5.0	2.5
88	Dichlobenil	9.88	171/136	171/136;171/100	0.2	0.1
89	Dicloran	18.12	206/176	206/176;206/124	25.0	12.5
90	Dicofol	21.51	250/215	250/215;250/139	2.0	1.0
91	Dimethachlor	19.94	197/148	197/148;197/120	25.0	12.5
92	Dioxaccarb	11.01	166/165	166/165;166/121	40.0	20.0
93	Endrin	25.18	263/193	263/193;263/191	50.0	25.0

(Continued)

TABLE 2.2.2 Retention Times, LOD, LOQ, Monitoring Ions and Collision Energy for 201 Pesticides by GC/MS/MS (cont.)

6.996 mm	Pesticides	Retention time (min)	Quantifying for precursor, product ion	Qualifying for precursor, product ion	Collision energy, V	LOQ (µg/kg)	LOD (µg/kg)
94	Epoxiconazole-2	29.61	192/138	192/138;192/111	10;25	1.3	0.6
95	EPTC	8.56	132/90	132/90;132/62	10;15	8.0	4.0
96	Ethofumesate	22.02	207/161	207/161;207/137	5;15	2.5	1.3
97	Ethoprophos	14.46	158/114	158/114;158/97	7;12	5.0	2.5
98	Etrimfos	17.94	292/181	292/181;292/153	5;25	12.5	6.3
99	Fenamidone	32.23	268/180	268/180;268/77	15;25	12.5	6.3
100	Fenarimol	31.79	330/251	330/251;330/139	5;5	10.0	5.0
101	Flamprop-isopropyl	26.88	276/105	276/105;276/77	15;25	2.5	1.3
102	Flamprop-methyl	26.88	276/105	276/105;276/77	10;25	1.3	0.6
103	Fonofos	17.36	246/137	246/137;246/109	5;15	2.5	1.3
104	Hexachloroenzene	14.7	284/249	284/249;284/214	18;25	10.0	5.0
105	Hexazinone	30.34	171/85	171/85;171/71	15;15	1.0	0.5
106	Iodofenphos	24.44	379/364	379/364;379/334	15;25	250.0	125.0
107	Isofenphos	23.13	255/213	255/213;255/121	5;25	10.0	5.0
108	Isopropalin	22.38	280/238	280/238;280/180	10;10	3.0	1.5
109	Methoprene	21.86	153/111	153/111;153/83	5;15	20.0	10.0
110	Methoprotryne	25.8	256/212	256/212;256/170	10;15	7.5	3.8
111	Methoxychlor	29.44	227/212	227/212;227/169	15;15	10.0	5.0
112	Methyl-parathion	20.96	263/246	263/246;263/109	5;12	25.0	12.5
113	Metolachlor	21.45	238/162	238/162;238/133	15;25	0.8	0.4
114	Nitrapyrin	10.91	194/158	194/158;194/133	15;15	25.0	12.5
115	Oxyfluofen	26.38	361/317	361/317;361/300	5;10	62.5	31.3
116	Pendimethalin	22.68	252/162	252/162;252/161	10;25	10.0	5.0
117	Picoxystrobin	24.75	335/303	335/303;335/173	10;10	12.5	6.3
118	Piperophos	30.22	321/123	321/123;321/122	5;15	500.0	250.0
119	Pirimiphos-ethyl	21.67	333/180	333/180;333/168	15;25	12.5	6.3
120	Pirimiphos-methyl	20.36	290/233	290/233;290/125	5;15	12.5	6.3
121	Profenofos	24.78	374/339	374/339;374/337	5;10	1000.0	500.0
122	Profluralin	17.55	318/199	318/199;318/55	10;10	25.0	12.5
123	Propachlor	14.83	176/120	176/120;176/77	10;25	30.0	15.0
124	Propiconazole	28.3	259/173	259/173;259/69	15;10	7.5	3.8

125	Propyzamide	18.91	173/145	173/145;173/109	15;25	1.3	0.6
126	Ronnel	19.84	285/270	285/270;285/240	15;25	7.5	3.8
127	Sulfotep	15.63	322/294	322/294;322/202	5;10	7.5	3.8
128	Tebufenpyrad	29.26	333/276	333/276;333/171	5;15	10.0	5.0
129	Terbutryn	20.72	226/96	226/96;226/68	15;25	7.5	3.8
130	Thiobencarb	20.75	257/100	257/100;257/72	5;25	5.0	2.5
131	Tralkoxydim	31.98	283/227	283/227;283/137	10;15	112.5	56.3
132	Trans-chlordane	23.62	375/303	375/303;375/266	10;15	25.0	12.5
133	Trans-diallate	15.37	234/192	234/192;234/150	10;15	2.5	1.3
134	Trifloxystrobin	27.54	222/190	222/190;222/162	5;10	35.0	17.5
135	Zoxamide	22.4	242/214	242/214;242/187	10;15	15.0	7.5
C group							
136	2,4'-DDE	22.79	318/248	318/248;318/246	15;15	7.5	3.8
137	Ametryn	20.32	227/170	227/170;227/58	10;25	5.0	2.5
138	Bifenthrin	28.67	181/166	181/166;181/165	15;25	0.8	0.4
139	Bitertanol	32.47	170/115	170/115;170/141	25;25	5.0	2.5
140	Boscalid	34.12	342/112	342/112;342/140	25;10	40.0	20.0
141	Butafenacil	33.55	331/152	331/152;331/180	25;12	5.0	2.5
142	Carbaryl	14.34	144/116	144/116;144/115	10;15	75.0	37.5
143	Chlorobenzilate	26.18	251/139	251/139;251/111	15;25	25.0	12.5
144	Chlorthal-dimethyl	21.38	301/273	301/273;301/223	15;25	5.0	2.5
145	Dibutylsuccinate	12.2	101/100	101/100;101/73	25;10	0.4	0.2
146	Diethofencarb	21.68	225/168	225/168;225/96	10;25	50.0	25.0
147	Diflufenican	28.73	266/246	266/246;266/218	10;25	25.0	12.5
148	Dimepiperate	22.54	119/91	119/91;119/65	10;25	0.8	0.4
149	Dimethametryn	22.74	212/122	212/122;212/94	10;15	10.0	5.0
150	Dimethomorph	37.36	301/139	301/139;301/165	15;10	10.0	5.0
151	Dimethylphthalate	11.42	163/133	163/133;163/77	10;15	1.3	0.6
152	Diniconazole	27.34	268/232	268/232;268/136	10;25	80.0	40.0
153	Diphenamid	23.1	167/165	167/165;167/152	15;15	1.5	0.8
154	Dipropetryn	21.05	255/222	255/222;255/138	10;25	5.0	2.5
155	Ethalfluralin	15.13	316/276	316/276;316/202	10;25	40.0	20.0
156	Etofenprox	32.93	163/107	163/107;163/135	15;10	1.5	0.8

(Continued)

TABLE 2.2.2 Retention Times, LOD, LOQ, Monitoring Ions and Collision Energy for 201 Pesticides by GC/MS/MS (cont.)

6.996 mm	Pesticides	Retention time (min)	Quantifying for precursor, product ion	Qualifying for precursor, product ion	Collision energy, V	LOQ (µg/kg)	LOD (µg/kg)
157	Etridiazol	10.39	211/183	211/183;211/140	10;15	10.0	5.0
158	Fenazaquin	29.17	145/117	145/117;145/91	10;25	1.3	0.6
159	Fenchlorphos	19.83	287/272	287/272;287/242	15;25	20.0	10.0
160	Fenoxanil	23.51	140/85	140/85;140/71	10;25	25.0	12.5
161	Fenpropidin	18.06	98/70	98/70;98/69	10;15	2.5	1.3
162	Fenson	22.79	268/141	268/141;268/77	5;25	10.0	5.0
163	Flufenacet	22.87	211/123	211/123;211/96	10;15	160.0	80.0
164	Furalaxyl	23.77	301/224	301/224;242/95	15;10	2.5	1.3
165	Heptachlor	18.6	272/237	272/237;272/235	10;10	12.5	6.3
166	Iprobenfos	18.64	204/122	204/122;204/91	15;5	2.5	1.3
167	Isazofos	18.77	257/162	257/162;257/119	10;25	25.0	12.5
168	Isoprothiolane	25.65	290/204	290/204;290/118	5;15	12.5	6.3
169	Kresoxim-methyl	25.21	131/130	131/130;131/89	10;25	10.0	5.0
170	Mefenacet	31.49	192/109	192/109;192/136	25;15	37.5	18.8
171	Mepronil	28.24	119/91	119/91;119/65	10;25	1.0	0.5
172	Metribuzin	20.56	198/110	198/110;198/82	15;15	10.0	5.0
173	Molinate	12.03	126/83	126/83;126/55	5;10	0.8	0.4
174	Napropamide	24.94	271/128	271/128;271/72	5;10	5.0	2.5
175	Nuarimol	28.9	314/139	314/139;314/111	5;25	12.5	6.3
176	Permethrin	31.5	183/153	183/153;183/168	15;15	7.5	3.8
177	Phenothrin	29.42	123/81	123/81;123/79	10;12	2.5	1.3
178	Piperonylbutoxide	27.71	176/103	176/103;176/131	15;15	2.5	1.3
179	Pretilachlor	24.95	162/147	162/147;162/132	10;15	5.0	2.5
180	Prometon	16.84	225/183	225/183;225/168	5;10	5.0	2.5
181	Pronamide	18.89	173/145	173/145;173/109	15;25	1.3	0.6
182	Propetamphos	18.08	194/166	194/166;194/94	10;25	7.5	3.8
183	Propoxur-1	6.61	110/64	110/64;110/63	15;25	0.2	0.1
184	Propoxur-2	15.29	110/64	110/64;110/63	15;25	1.3	0.6
185	Prothiophos	24.12	309/239	309/239;309/221	15;25	10.0	5.0
186	Pyridaben	32.09	147/117	147/117;147/132	25;15	2.5	1.3
187	Pyridaphenthion	30.29	340/109	340/109;340/199	15;5	10.0	5.0

188	Pyrimethanil	17.42	200/199	200/199;183/102	10;30	7.5	3.8
189	Pyriproxyfen	30.06	136/96	136/96;136/78	15;25	5.0	2.5
190	Quinalphos	23.17	157/129	157/129;157/102	15;25	10.0	5.0
191	Quinoxuphen	27.18	237/208	237/208;237/182	25;25	100.0	50.0
192	Telodrin	20.61	311/241	311/241;311/240	25;25	35.0	17.5
193	Tetrasul	25.89	324/254	324/254;324/252	15;15	5.0	2.5
194	Thiazopyr	21.72	363/300	363/300;363/272	15;25	35.0	17.5
195	Tolclofos-methyl	19.87	267/252	267/252;267/93	15;25	12.5	6.3
196	Transfluthrin	19.28	163/143	163/143;163/91	15;15	2.5	1.3
197	Triadimefon	22.37	210/183	210/183;210/129	5;10	12.5	6.3
198	Triadimenol	24.56	168/70	168/70;128/100	10;15	10.0	5.0
199	Triallate	17.26	270/228	270/228;270/186	10;15	7.5	3.8
200	Tribenuron-methyl	9.35	154/124	154/124;124/83	5;10	60.0	30.0
201	Vinclozolin	20.43	285/212	285/212;285/178	10;10	15.0	7.5

TABLE 2.2.3 SIM Acquisition of 201 Pesticides for Three Group Ions by GC/MS

No.	Time (min)	Ions (amu)	Dwell time (ms)
A group			
1	7.00	103, 121, 128, 136, 161, 203	80
2	10.60	120, 137, 179, 215, 250, 252	80
3	12.41	103, 121, 136, 151, 154, 156, 171, 186, 215	55
4	14.00	143, 156, 151, 158, 171, 192, 220, 229, 231	50
5	14.70	106, 142, 190, 237, 264, 265, 280, 306, 335	50
6	16.10	108, 136, 138, 142, 160, 161, 164, 177, 188, 189, 196, 197, 198, 204, 205, 211, 213, 234, 261, 318	23
7	17.90	89, 114, 138, 166, 186, 196, 198, 210, 225, 229, 231, 233, 238, 256, 258	33
8	18.90	138, 166, 230, 238, 263, 265	80
9	19.25	120, 146, 154, 162, 176, 203, 223, 230, 251, 252, 259	33
10	20.01	184, 188, 197, 199, 206, 226, 234, 237, 239, 241, 249, 269	41
11	20.61	193, 195, 196, 206, 215, 221, 234, 249, 250, 252, 263, 269, 297, 301, 332	33
12	21.81	210, 220, 224, 225, 247, 254, 263, 266, 275, 291, 295, 326, 351, 353, 355	33
13	22.88	133, 209, 211, 303, 357, 359,	80
14	23.90	178, 184, 223, 224, 254, 255, 283, 285, 326	40
15	25.31	119, 134, 158, 185, 259, 340	50
16	25.96	119, 134, 158	140
17	26.33	119, 134, 158, 186, 218, 225, 247, 290, 300, 328, 360, 408	40
18	27.70	203, 231, 233, 237, 253, 254, 272, 274, 281, 311, 323, 342, 366, 379	33
19	28.80	127, 141, 203, 231, 237, 272, 274, 288, 323, 324, 359, 394	40
20	29.60	136, 153, 229, 234, 256, 356, 357	80
21	32.30	125, 129, 198, 258, 286, 287	80
B group			
1	8.00	128, 132, 136, 146, 156, 171, 173, 189, 217	50
2	10.30	121, 154, 165, 166, 194, 196, 198, 234	60
3	11.50	191, 193, 206	150
4	14.00	120, 128, 158, 168, 176, 200, 203, 211, 215, 234, 236, 242, 261, 282, 284, 286	38
5	15.15	128, 153, 171, 202, 213, 234, 236, 238, 266, 322	50
6	17.00	137, 160, 174, 176, 181, 202, 206, 246, 277, 292, 304, 318, 347	35
7	18.60	173, 240, 255	150
8	19.50	125, 134, 171, 198, 210, 213, 276, 285, 287, 290, 305	40
9	20.30	100, 185, 200, 226, 233, 241, 246, 257, 258, 259, 263, 276, 286, 290, 305, 314	30
10	21.27	73, 139, 141, 153, 161, 162, 191, 207, 213, 238, 240, 250, 251, 286, 304, 318, 329, 331, 333	25
11	22.00	161, 162, 185, 187, 207, 213, 220, 222, 238, 242, 252, 255, 280, 286, 299, 351, 353, 355	27
12	23.00	185, 213, 255, 267, 269, 323, 373, 375, 377	50
13	23.75	160, 176, 188, 246, 248, 316, 318	70
14	24.23	250, 303, 335, 345, 367, 377, 379	83
15	24.61	105, 172, 175, 177, 263, 297, 302, 303, 305, 317, 335, 339, 345, 367, 374	30
16	25.30	77, 105, 141, 165, 175, 177, 199, 213, 251, 235, 237, 253, 256, 276, 271, 302	31
17	25.96	77, 105, 141, 251, 252, 253, 276, 297, 300, 325, 360, 361, 363	30

(Continued)

TABLE 2.2.3 SIM Acquisition of 201 Pesticides for Three Group Ions by GC/MS (*cont.*)

No.	Time (min)	Ions (amu)	Dwell time (ms)
A group			
18	27.18	116, 131, 148, 157, 161, 173, 199, 206, 222, 259, 325, 342	41
19	28.00	157, 169, 173, 259, 261, 303	80
20	29.00	122, 138, 140, 183, 192, 212, 227, 228, 260, 276, 292, 318, 320, 333, 339, 341, 365	29
21	30.10	122, 128, 140, 171, 206, 238, 252, 268, 320	50
22	31.50	139, 219, 226, 268, 283, 330	80
C group			
1	6.30	110, 111, 124, 152, 154	90
2	10.00	133, 140, 163, 183, 194, 211	80
3	11.90	101, 126, 157, 158, 175, 187	80
4	14.00	110, 111, 115, 116, 144, 152, 276, 292, 316	41
5	16.50	98, 138, 143, 145, 168, 196, 198, 199, 200, 210, 225, 236, 268, 270, 273	27
6	18.00	98, 138, 145, 161, 173, 175, 194, 196, 204, 236, 237, 246, 255, 257, 272, 273, 285, 288, 337	20
7	19.14	163, 165, 185, 212, 227, 250, 265, 267, 270, 285, 287, 335	27
8	20.35	103, 144, 185, 198, 199, 212, 221, 222, 227, 240, 255, 285, 301, 311, 332, 375	31
9	21.28	151, 221, 225, 267, 301, 327, 332, 363, 381	41
10	22.00	181, 208, 210, 351, 353, 355	83
11	22.30	77, 119, 141, 145, 146, 151, 157, 165, 167, 176, 181, 208, 210, 211, 212, 213, 239, 246, 255, 263, 268, 298, 318, 363	20
12	23.10	112, 130, 140, 146, 152, 157, 162, 165, 167, 168, 189, 239, 242, 267, 298, 301, 309	29
13	24.50	112, 116, 128, 130, 131, 162, 168, 171, 204, 206, 231, 238, 262, 271, 290	33
14	25.53	152, 204, 231, 251, 252, 253, 254, 290, 324	50
15	26.95	119, 120, 149, 176, 177, 232, 237, 268, 269, 270, 272, 307	33
16	28.10	119, 120, 165, 166, 181, 203, 235, 266, 267, 269, 314, 394	41
17	28.90	117, 123, 145, 160, 183, 203, 235, 314, 350	50
18	29.80	136, 185, 188, 199, 226, 340	55
19	31.10	117, 120, 136, 147, 183, 184, 192, 255, 364	41
20	32.29	112, 141, 163, 170, 183, 274, 303, 318, 376	50
21	33.30	112, 140, 180, 331, 333, 342	80
22	37.00	165, 301, 387	150

2. Deciding the use amount of the NaCl: weigh 4 portions of tea, with each 5 g. Add 20 mL water respectively, soak 1 h, add 5, 10, 15, and 20 g NaCl, sequentially, add again with 15 mL acetonitrile, homogenize and centrifuge to find that the volumes of acetonitrile layers are 8.2, 9.0, 8.5, and 8.0 mL. It is evident that acetonitrile lost the least during liquid–liquid partitioning when 10 g NaCl is added, so 10 g NaCl is decided to be the use amount.

3. Deciding the use amount of acetonitrile: weigh 3 portions of teas, with each 5 g. Add 20 mL water and 10 g NaCl, respectively, add again 15, 20, and 25 mL acetonitrile, respectively, homogenize and centrifuge to find that the volumes of the acetonitrile are 9, 14, and 19 mL, respectively. Tests demonstrated that under such conditions there is a loss of 6.0 mL acetonitirle when different volumes of acetonitrile are added; 15 mL acetonitrile, however, suffers the greatest loss. Therefore, 20 and 25 mL acetonitriles have been chosen for further extraction experiments to check on the spiked recoveries of 201 pesticides.

TABLE 2.2.4 MRM Acquisition of 201 Pesticides for Three Group Ions by GC/MS/MS

No.	Time (min)	Ions (amu)	Dwell time (ms)
A group			
1	4.00	121/103, 121/77; 161/128, 128/72; 250/215, 250/177; 179/137, 179/93	25
2	12.50	154/83, 154/72; 121/103, 121/77; 156/89, 156/74; 143/79, 143/52; 231/160, 231/158	19
3	14.80	280/265, 280/237; 306/264, 306/206; 190/106, 190/57; 164/136, 164/108; 211/196, 211/169; 204/107, 204/78; 189/185, 189/161; 213/212, 213/77; 177/127, 177/101; 318/276, 318/152	9
4	18.00	198/89, 198/72; 256/186, 256/151; 258/186, 256/186; 225/169, 225/154	19
5	18.70	231/160, 231/158; 238/166, 238/96; 263/192, 263/156; 259/176, 259/120; 251/128, 251/86; 230/154, 230/111; 146/131, 146/118	12
6	20.10	237/160, 237/146; 241/199, 241/184; 239/197, 239/85; 206/132, 206/105; 193/157, 193/123; 297/269, 297/223; 301/273, 301/223; 250/215, 250/139	16
7	21.60	275/149, 275/99; 224/222, 224/208; 266/190, 266/174; 326/256, 326/254; 359/331, 359/303; 209/133, 209/132	10
8	24.00	283/255, 283/96; 223/193, 223/179; 326/256, 326/254; 340/185, 340/183	21
9	26.00	134/93, 134/91; 134/93, 134/91; 186/158, 186/109; 360/325, 360/290	25
10	27.10	186/158, 186/109; 408/363, 408/59; 342/255, 342/184	25
11	28.40	379/276, 379/262; 366/209, 366/181; 323/231, 323/196; 288/174, 288/141; 272/237, 272/235; 394/359, 394/324	19
12	29.80	153/136, 153/110; 357/256, 357/229; 287/259, 287/179; 198/129, 198/102	16
B group			
1	5.00	132/90, 132/62; 146/90, 146/57; 171/136, 171/100; 234/154, 234/121; 194/158, 194/133; 166/165, 166/121; 191/141, 191/113; 234/154, 234/121	8
2	13.00	158/114, 158/97; 284/249, 284/214; 234/192, 234/150; 176/120, 176/77	24
3	15.10	234/192, 234/150; 322/294, 322/202; 213/171, 213/127	25
4	16.50	246/137, 246/109; 318/199, 318/55; 292/181, 292/153; 206/176, 206/124	25
5	18.50	173/145, 173/109; 213/198, 213/171; 285/270, 285/240; 197/148, 197/120; 290/233, 290/125; 226/96, 226/68; 257/100, 257/72; 263/246, 263/109; 314/286, 314/258	10
6	21.20	238/162, 238/133; 250/215, 250/139; 333/180, 333/168; 331/316, 331/286; 153/111, 153/83; 207/161, 207/137; 280/238, 280/180; 242/214, 242/187; 252/162, 252/161	9
7	22.80	255/213, 255/121; 323/267, 323/159; 373/301, 373/266; 375/303, 375/266; 318/248, 318/246; 176/150, 176/126	13
8	24.20	379/364, 379/334; 335/303, 335/173; 374/339, 374/337; 172/116, 105/77; 263/193, 263/191; 302/175, 302/111	18
9	25.40	235/199, 235/165; 256/212, 256/170; 251/139, 251/111; 361/317, 361/300; 360/325, 360/297;p 276/105, 276/77; 276/105, 276/77	16
10	27.10	342/199, 157/121; 222/190, 222/162; 148/105, 148/79; 259/173, 259/69; 157/110, 157/77; 333/276, 333/171; 227/212, 227/169; 341/185, 341/183; 292/105, 292/77; 192/138, 192/111	9
11	30.00	321/123, 321/122; 171/85, 171/71; 330/251, 330/139; 283/227, 283/137; 268/180, 268/77	19
C group			
1	6.00	110/64, 110/63; 154/124, 124/83; 211/183, 211/140; 163/133, 163/77; 126/83, 126/55; 101/100, 101/73	16
2	12.50	144/116, 144/115; 316/276, 316/202; 110/64, 110/63	34
3	16.00	225/183, 225/168; 270/228, 270/186; 200/199, 183/102	32

(Continued)

TABLE 2.2.4 MRM Acquisition of 201 Pesticides for Three Group Ions by GC/MS/MS (*cont.*)

No.	Time (min)	Ions (amu)	Dwell time (ms)
4	17.70	98/70, 98/69; 194/166, 194/94; 272/237, 272/235; 204/122, 204/91; 257/162, 257/119; 173/145, 173/109; 163/143, 163/91	16
5	19.60	287/272, 287/242; 267/252, 267/93; 227/170, 227/58; 285/212, 285/178; 198/110, 198/82; 311/241, 311/240	16
6	20.80	255/222, 255/138; 301/273, 301/223; 225/168, 225/96; 363/300, 363/272	17
7	22.00	210/183, 210/129; 119/91, 119/65; 212/122, 212/94; 318/248, 318/246; 268/141, 268/77; 211/123, 211/96; 167/165, 167/152; 157/129, 157/102; 140/85, 140/71; 301/224, 242/95; 309/239, 309/221	7.5
8	24.30	168/70, 128/100; 271/128, 271/72; 162/147, 162/132	20
9	25.20	290/204, 290/118; 324/254, 324/252; 251/139, 251/111; 131/130, 131/89	20
10	26.90	237/208, 237/182; 268/232, 268/136; 176/103, 176/131	34
11	27.90	119/91, 119/65; 181/166, 181/165; 266/246, 266/218; 314/139, 314/111	20
12	29.00	145/117, 145/91; 123/81, 123/79	25
13	29.90	136/96, 136/78; 340/109, 340/199; 192/109, 192/136; 183/153, 183/168; 147/117, 147/132	18
14	32.25	170/115, 170/141; 163/107, 163/135; 331/152, 331/180; 342/112, 342/140; 301/139, 301/165	17

2.2.5 OPTIMIZATION OF THE EVAPORATION TEMPERATURES AND DEGREES OF THE SAMPLE SOLUTIONS FOR THREE METHODS

2.2.5.1 Optimization Selection of Temperatures of Rotary Evaporation

Transfer three groups of mixed standard solutions, A, B, and C (201 pesticides included), 50 μL each into sample bottles, add 40 μL internal standards, nitrogen blow to dryness and dilute immediately with hexane to 1 mL before being used for standard solutions for quantification. Transfer three groups of A, B, and C mixed standard solutions 50 μL each into a pear-shaped flask, add again 30 mL acetonitrile: toluene (3:1) solutions, rotary evaporate respectively at 40, 60, and 80 °C, or conduct solvent exchange with hexane, finally add 40 μL internal standards, mix uniformly to pass filtering membrane for determination. See Table 2.2.5 for analytical results.

It can be seen from Table 2.2.5 that the average recoveries for 201 pesticides tend to drop with the increase of the temperature of the rotary evaporation, while the number of pesticides with recoveries less than 80% tend to increase. This phenomenon indicates that with the increase of temperatures certain temperature-sensitive pesticides have decomposed in varying degrees. Based on this discovery, this method has chosen the temperature of 40 °C as the optimum rotary evaporation temperature that is capable of both evaporating and removing acetonitrile and toluene solvents and preventing decomposition of certain pesticides at the same time.

2.2.5.2 Optimum Selection of the Degree of Rotary Evaporation

The recoveries of 201 pesticides under three circumstances of solvents being evaporated to near dryness, just dryness and complete dryness while the rotary evaporation temperature is set at 40 °C. See Table 2.2.6. It can be seen from the data in Table 2.2.6 that under the three circumstances of solvents being evaporated to near dryness, just dryness and full dryness according to the degree of rotary evaporations average recoveries for 201 pesticides tend to drop. At the same time, the number of pesticides with recoveries less than 80% tends to increase rapidly. This phenomenon demonstrated sample solutions, if evaporated to dryness, will make recoveries of certain drugs drop markedly. Based on this fact, the method requires that during the stage of sample concentration sample solutions shall be ensured from being evaporated to dryness, which may be left to about 0.3–0.5 mL, while the pear-shaped flask is removed, after which the remaining heat resorts to natural volatility so as to reduce the loss of part of the drugs during the stage of rotary evaporation.

TABLE 2.2.5 Influences of the Temperatures of Rotary Evaporation and Solvent Exchanges on Recoveries

Instrument	GC/MS/MS			GC/MS		
Temperatures/°C	40	60	80	40	60	80
Average recoveries for 201 pesticides/%	92.3	84.9	82.4	99.6	87.2	83.4
Number of pesticides (recoveries <80%)	27	53	80	23	45	58

TABLE 2.2.6 Influences of the Degrees of Three Rotary Evaporations on Recoveries

Instrument	GC/MS/MS			GC/MS		
Degree	Near dryness (0.3–0.5 mL)	Just dryness	Complete dryness	Near dryness (0.3–0.5 mL)	Just dryness	Complete dryness
Average recoveries for 201 pesticides/%	89.4	77.7	75.0	89.0	80.6	76.7
Number of pesticides (recoveries <80%)	30	84	107	32	77	96

2.2.6 STAGE I: EVALUATION ON THE EXTRACTION EFFICIENCIES OF THE MULTIRESIDUE PESTICIDES IN GREEN TEA SAMPLES BY METHOD-1, METHOD-2, AND METHOD-3

Add 201 pesticide standard solutions to green tea samples at the spiked levels of 10–600 μg/kg, leave them still for 30 min and prepare 3 spiked samples for each method in order to carry out parallel tests, with GC/MS and GC/MS/MS test results tabulated in Table 2.2.7.

2.2.6.1 Test Data Analysis

The statistical analysis of average recoveries by GC/MS/MS and GC/MS determination in Table 2.2.7 is tabulated in Table 2.2.8, from which it can be seen that Method-1 accounts for 3.7% for pesticides with average recoveries less than 70%, with Method-1 7.2%, while Method-3 makes up more than 90% for pesticides extracted under two conditions; Method-1 accounts for 92.0% for pesticides with a good average recovery range of 70%–110%, with Method-2 91.8%, while Method-3 makes up no more than 4% for pesticides extracted under two conditions; it is evident that no big differences exist for the recoveries between Method-1 and Method-2, while Method-1 turns out slightly better, and Method-3 obtains relatively low recoveries with the majority of average recoveries less than 60% no matter whether 20 or 25 mL acetonitrile is adopted.

The statistical analysis of RSD by GC/MS/MS and GC/MS determination in Table 2.2.7 is tabulated in Table 2.2.8, from it can be seen that the number of pesticides with RSD < 15% accounts for 98.2% and 92.8%, respectively, for Method-1 and Method-2 determinations, among which the number of pesticides with RSD < 10% accounts for 95.0% and 95.5%, respectively, with hardly any differences between the two methods, while Method-3 tend to get markedly low RSD values in comparison with Method-1 and Method-2 no matter whether 20 or 25 mL acetonitrile is adopted.

2.2.6.2 Analysis of Experimental Phenomenon

It is found in the practical experiment that the testing solutions finally obtained by Method-1 are basically the lightest and transparent. With Method-2, transfer the extracted sample solutions into the centrifuge tube containing 0.30 g anhydrous magnesium sulfate and 0.13 g PSA powders, oscillate and centrifuge to find that the solutions present a dark green color; add again 0.13 g graphite carbon, oscillate 2 min, centrifuge at 4200 r/min for 5 min to find that the solutions turn pale

TABLE 2.7 Test Results of Average Recoveries and Coefficients of Variations of 201 Pesticides in Green Tea Extracted per the Three Sample Preparation Methods (n = 3)

No.	Pesticides	Spiked levels, μg/kg	GC/MS Method-1 AVE (%)	RSD (%)	Method-2 AVE (%)	RSD (%)	Method-3 (20 mL acetonitrile) AVE (%)	RSD (%)	Method-3 (25 mL acetonitrile) AVE (%)	RSD (%)	GC/MS/MS Method-1 AVE (%)	RSD (%)	Method-2 AVE (%)	RSD (%)	Method-3 (20 mL acetonitrile) AVE (%)	RSD (%)	Method-3 (25 mL acetonitrile) AVE (%)	RSD (%)
1	2,3,4,5-Tetrachloroaniline	10.0	74.3	4.7	82.6	6.0	50.0	8.1	48.1	8.0	81.4	6.2	88.8	7.9	41.5	6.4	36.4	25.5
2	2,3,5,6-Tetrachloroaniline	50.0	82.5	2.6	88.8	3.3	50.4	5.1	51.9	6.4	83.5	3.5	80.4	8.0	47.2	8.6	49.2	0.6
3	2,4'-DDE	50.0	71.2	10.0	87.5	2.1	70.1	0.0	65.5	8.8	74.2	3.4	87.3	1.8	65.1	0.3	62.6	11.3
4	2,4'-DDT	50.0	91.2	4.3	35.2	4.2	40.6	4.0	40.3	1.8	77.3	2.0	32.8	6.9	26.2	9.7	19.4	1.5
5	4,4'-DDE	100.0	84.3	2.9	87.3	3.2	60.7	5.7	61.7	9.3	88.1	3.5	85.4	3.6	60.5	8.9	61.6	12.9
6	4,4-Dibromobenzophenone	150.0	100.8	7.7	101.8	6.9	64.4	9.3	74.5	8.0	91.0	2.2	93.9	4.0	59.0	5.3	56.7	3.2
7	4,4-Dichlorobenzophenone	50.0	87.6	2.7	92.1	5.2	55.4	4.1	53.8	5.0	87.9	4.3	89.9	4.3	54.4	5.3	52.1	5.3
8	Acetochlor	150.0	81.4	2.4	86.9	4.8	43.2	7.7	43.4	7.5	80.4	2.1	86.4	4.1	45.6	7.1	44.0	4.6
9	Alachlor	100.0	82.7	0.5	90.9	4.1	44.5	9.1	44.7	21.8	79.0	2.6	88.5	5.0	45.3	5.1	42.4	5.7
10	Ametryn	50.0	81.8	4.0	87.8	4.5	29.2	6.0	28.8	23.5	83.8	6.4	84.8	2.2	29.9	8.9	28.8	23.0
11	Atratone	200.0	88.5	4.6	94.4	4.0	15.7	7.4	16.2	31.5	87.5	2.4	89.1	4.8	15.8	10.9	13.9	43.8
12	Benalaxyl	400.0	102.7	5.0	86.5	5.5	45.3	0.5	35.5	8.2	93.2	3.8	83.6	4.3	34.2	2.5	30.3	10.9
13	Benodanil	50.0	85.7	7.5	101.7	8.0	28.2	20.9	23.9	43.4	80.0	6.9	95.3	6.9	21.0	25.7	16.0	44.4
14	Benoxacor	50.0	76.6	11.8	56.2	11.5	37.3	4.5	32.6	13.4	67.0	8.3	51.3	7.7	29.5	1.8	24.9	22.3
15	Benzoylprop-ethyl	150.0	91.0	1.7	90.9	3.8	40.8	1.6	38.0	7.0	91.1	2.5	86.6	3.7	40.5	2.1	36.5	7.1
16	Bifenthrin	100.0	91.3	2.8	92.0	2.8	72.7	4.9	74.3	5.1	88.5	5.2	84.4	2.3	69.1	6.8	65.8	10.6
17	Bitertanol	50.0	85.8	11.6	86.3	9.1	14.2	11.2	13.0	50.1	86.5	0.6	88.8	4.4	13.7	11.7	11.2	46.4
18	Boscalid	50.0	120.4	6.7	85.0	7.5	21.5	12.5	17.8	42.1	115.9	0.9	81.6	4.4	18.9	11.6	15.7	38.5
19	Bromofos	50.0	78.5	3.2	83.5	5.6	61.4	2.5	62.0	4.7	84.4	9.6	82.8	3.8	59.4	8.5	58.2	6.5
20	Bromophos-ethyl	50.0	84.9	1.2	93.5	5.9	67.0	3.6	62.0	4.2	86.6	3.9	91.4	5.9	66.8	5.1	61.7	7.9
21	Bromopropylate	50.0	99.9	2.1	90.6	5.0	54.4	3.7	58.1	9.0	93.4	3.4	86.2	4.5	57.0	6.6	55.4	9.7
22	Buprofenzin	50.0	87.7	3.8	91.7	1.7	48.3	2.7	54.6	4.3	87.0	3.6	84.8	8.6	49.5	3.3	47.4	8.3
23	Butachlor	50.0	69.2	9.3	87.7	4.6	56.6	1.8	55.3	0.3	84.0	2.5	83.2	3.5	62.0	6.5	62.6	8.1

(Continued)

TABLE 2.2.7 Test Results of Average Recoveries and Coefficients of Variations of 201 Pesticides in Green Tea Extracted per the Three Sample Preparation Methods (n = 3) (cont.)

No.	Pesticides	Spiked levels, µg/kg	GC/MS Method-1		Method-2		Method-3 (20 mL acetonitrile)		Method-3 (25 mL acetonitrile)		GC/MS/MS Method-1		Method-2		Method-3 (20 mL acetonitrile)		Method-3 (25 mL acetonitrile)	
			AVE (%)	RSD (%)	AVE (%)	RSD (%)	AVE (%)	RSD (%)	AVE (%)	RSD (%)	AVE (%)	RSD (%)	AVE (%)	RSD (%)	AVE (%)	RSD (%)	AVE (%)	RSD (%)
24	Butafenacil	50.0	125.4	8.5	87.2	5.0	36.2	14.4	25.2	40.4	123.9	2.1	82.2	3.1	31.6	14.1	23.1	34.0
25	Butralin	50.0	92.9	2.9	82.1	9.3	74.7	7.7	78.0	7.0	97.5	8.8	78.1	5.4	58.3	8.6	51.3	16.3
26	Butylate	50.0	88.1	4.1	97.8	8.4	116.3	3.0	126.3	3.6	77.1	3.3	62.6	2.7	48.2	2.1	49.1	4.6
27	Carbaryl	100.0	85.3	2.4	92.6	2.7	25.7	18.6	22.7	32.7	75.8	10.3	108.4	3.6	26.0	24.4	21.9	30.5
28	Carbofenothion	100.0	78.0	0.4	82.1	8.0	57.7	4.8	58.7	6.1	83.1	1.6	85.3	6.0	64.1	14.0	60.6	6.3
29	Chlorfenapyr	100.0	81.3	1.8	93.9	4.3	60.8	4.5	54.3	3.4	84.5	5.1	92.1	4.0	58.9	6.6	52.4	3.0
30	Chlorfenson	400.0	98.2	4.0	85.0	7.3	63.3	2.9	66.7	1.5	107.9	1.4	81.4	3.2	55.1	4.7	55.2	4.4
31	Chlorfenviphos	50.0	77.4	6.0	80.6	5.4	36.3	2.6	31.8	10.3	73.9	3.2	76.0	3.9	34.3	0.6	29.8	10.5
32	Chlormephos	50.0	77.6	3.1	75.6	11.0	52.7	4.4	59.5	11.4	77.4	2.1	67.1	3.2	52.1	0.3	52.9	3.1
33	Chlorobenzilate	200.0	98.2	3.9	85.2	4.6	43.2	47.3	58.9	2.1	95.5	4.5	84.1	4.2	56.7	5.3	54.6	3.6
34	Chlorneb	200.0	82.1	2.5	84.0	2.2	58.2	4.4	66.0	6.2	81.5	4.9	79.0	3.1	52.0	0.3	51.7	1.0
35	Chlorpropylate	50.0	99.0	1.6	88.2	4.4	41.7	45.0	41.2	49.9	95.5	4.5	84.1	4.2	56.7	5.3	54.6	3.6
36	Chlorpropham	100.0	88.7	0.3	87.7	3.1	56.6	0.1	51.2	6.9	85.0	3.0	87.1	2.4	50.4	0.9	46.4	2.6
37	Chlorpyifos(ethyl)	100.0	84.4	2.0	86.3	3.7	61.7	3.7	63.0	6.4	84.2	3.0	86.5	2.6	62.4	7.3	60.7	11.6
38	Chlorthal-dimethyl	240.0	82.8	3.2	89.6	3.1	56.5	1.7	55.7	2.3	84.7	3.4	85.9	2.6	57.5	5.7	55.0	1.4
39	Chlorthiophos	150.0	89.0	4.4	87.9	3.5	62.5	5.5	64.9	3.8	86.3	2.6	85.3	2.4	63.3	10.4	61.2	9.5
40	Cis-chlordane	150.0	89.2	1.6	84.8	4.1	58.3	5.4	58.2	8.3	91.6	5.6	88.5	2.9	60.1	7.2	60.4	14.3
41	Cis-diallate	500.0	91.3	3.1	86.6	5.8	56.7	1.5	58.9	10.3	86.6	4.4	84.4	2.0	54.7	3.0	55.5	8.1
42	Clomazone	50.0	93.1	1.6	89.0	5.3	34.6	8.4	32.8	28.8	100.4	0.7	85.8	3.9	35.2	5.8	31.6	19.3
43	Cyanofenphos	100.0	105.5	4.9	84.2	2.3	54.0	4.7	48.6	9.2	102.6	3.2	81.1	4.3	51.4	8.3	46.8	2.4
44	Cycloate	200.0	84.3	3.1	90.4	2.0	55.0	1.0	57.5	0.7	83.6	2.9	84.5	3.3	49.6	4.2	49.9	1.0
45	Cycluron	150.0	74.5	3.0	93.6	1.3	55.4	17.2	48.3	32.6	73.0	6.3	103.2	2.2	6.2	11.4	7.1	22.8
46	Cyhalofop-butyl	50.0	100.4	5.3	94.0	9.1	57.6	2.8	55.7	3.8	108.9	2.6	90.0	4.6	55.4	6.1	46.4	0.8
47	Cyprodinil	50.0	92.6	2.4	85.5	6.6	39.2	11.6	40.1	0.5	87.3	3.5	78.6	4.7	34.6	6.8	34.2	6.9
48	Dacthal	50.0	82.8	3.2	89.7	3.1	56.5	1.7	55.7	2.3	84.7	3.4	85.9	2.6	57.5	5.7	55.0	1.4
49	de-PCB101	50.0	84.4	3.5	94.1	2.9	62.6	0.6	58.9	3.5	88.1	4.6	89.4	2.8	63.6	5.4	60.0	9.9

50	de-PCB118	50.0	87.5	2.7	91.1	3.4	60.4	0.3	60.0	10.0	86.8	4.1	85.6	3.2	61.3	3.4	57.0	10.3
51	de-PCB138	100.0	85.7	2.6	92.1	4.8	60.5	0.4	57.6	8.2	87.2	2.7	85.3	4.8	62.0	3.5	58.1	11.6
52	de-PCB180	50.0	84.7	2.3	91.7	3.7	58.3	2.2	56.0	11.3	87.1	2.8	86.5	4.8	58.0	0.6	54.6	16.9
53	de-PCB28	50.0	85.1	2.3	90.8	2.3	64.3	0.8	62.7	2.0	86.5	2.4	87.4	3.6	62.7	2.5	60.4	7.0
54	de-PCB31	50.0	83.8	1.7	88.4	2.9	60.6	2.6	59.6	5.3	86.5	2.4	87.4	3.6	62.7	2.5	60.4	7.0
55	Desmetryn	50.0	85.2	2.8	88.0	3.7	22.2	4.2	19.8	29.0	86.1	4.9	87.2	2.4	22.5	0.4	18.7	19.6
56	Dibutylsuccinate	50.0	75.7	4.9	82.8	2.0	60.0	3.5	63.1	5.6	79.1	2.7	78.9	2.6	56.7	8.0	58.7	1.8
57	Dichlobenil	50.0	75.1	1.5	60.2	6.5	44.8	2.5	47.0	2.8	77.2	4.6	57.3	6.7	44.8	2.5	43.5	7.2
58	Dichlorofop-methyl	100.0	126.1	4.4	93.6	4.2	61.8	4.0	58.7	3.4	115.0	3.7	90.5	4.0	61.9	5.1	57.4	6.2
59	Dicloran	100.0	77.2	4.0	82.0	16.5	52.1	22.6	38.8	21.0	61.0	6.5	69.2	15.4	39.2	7.3	35.7	17.7
60	Dicofol	200.0	94.2	2.9	95.2	3.8	60.0	3.9	60.3	1.9	87.9	4.3	89.9	4.3	54.4	5.3	52.1	5.3
61	Diethofencarb	50.0	87.7	3.0	88.3	4.1	36.9	3.8	36.4	19.1	85.0	3.2	86.2	3.4	35.8	13.1	31.5	18.7
62	Diflufenican	100.0	90.4	1.1	83.1	4.1	57.6	6.6	51.7	3.2	89.7	4.3	81.1	4.8	52.1	7.5	48.6	1.6
63	Dimepiperate	50.0	87.9	13.2	91.7	3.0	62.2	22.8	56.3	7.1	85.2	5.5	86.1	2.7	54.0	3.2	53.4	0.1
64	Dimethachlor	50.0	72.3	4.0	81.1	5.8	31.1	2.1	27.4	25.4	72.9	0.8	80.8	3.6	30.2	1.6	25.6	24.1
65	Dimethametryn	100.0	83.6	4.0	87.1	2.8	36.0	0.2	34.8	11.0	89.1	5.1	87.3	4.9	35.9	8.4	35.8	12.8
66	Dimethenamid	100.0	76.4	2.4	89.4	5.0	37.4	9.5	34.9	20.0	80.0	1.6	88.1	5.4	37.0	8.4	33.7	18.7
67	Dimethomorph	50.0	80.0	6.0	90.6	4.6	3.9	11.3	3.7	67.8	80.6	5.5	89.3	4.4	3.7	10.6	3.1	65.5
68	Dimethylphthalate	50.0	79.8	1.1	81.5	6.5	44.9	13.6	43.8	24.2	81.7	2.0	77.2	5.1	44.8	17.0	42.7	21.6
69	Diniconazole	150.0	97.3	3.5	93.4	2.9	22.9	7.1	22.0	30.3	93.4	4.5	92.1	2.0	25.3	9.2	23.2	21.7
70	Diofenolan-1	100.0	93.3	2.0	98.0	3.8	61.8	4.1	60.2	0.9	90.0	3.9	93.3	4.1	57.0	6.4	55.7	0.8
71	Diofenolan-2	100.0	93.7	3.3	103.6	6.6	55.2	7.6	50.5	1.1	90.0	3.9	94.6	8.8	55.8	1.4	53.5	2.2
72	Dioxacarb	100.0	81.3	11.3	89.1	7.1	17.0	14.2	25.4	51.5	79.6	9.8	98.4	1.5	13.3	29.3	12.1	15.2
73	Diphenamid	100.0	87.6	10.9	93.1	3.6	38.0	10.6	40.8	7.2	84.3	3.0	85.6	3.2	25.6	6.4	23.4	32.9
74	Dipropetryn	150.0	77.2	21.1	90.8	2.5	45.1	3.4	43.3	4.8	88.1	3.7	86.4	2.9	41.6	7.0	41.1	2.9
75	Endrin	100.0	83.6	3.0	79.7	4.5	54.5	2.1	52.7	1.3	85.7	3.2	78.6	3.1	49.6	4.6	47.9	7.8
76	Epoxiconazole-2	100.0	91.8	2.0	80.6	4.3	13.6	3.7	11.0	55.1	92.5	1.9	80.6	4.6	11.0	1.0	8.3	49.1
77	EPTC	150.0	63.3	1.5	59.7	4.6	43.6	5.6	46.3	1.5	73.2	2.5	50.9	4.8	45.3	0.8	45.1	1.6
78	Ethalfluralin	100.0	89.1	5.2	80.9	5.1	76.0	6.6	78.9	4.7	93.4	12.4	83.2	2.0	67.8	9.9	68.9	9.0
79	Ethofumesate	50.0	52.9	10.7	118.0	6.0	0.0	0.0	0.0	0.0	86.6	2.2	82.8	4.6	41.5	1.0	35.9	8.5
80	Ethoprophos	100.0	82.7	1.1	86.3	4.6	37.9	2.3	36.3	13.5	82.4	1.8	85.8	4.0	39.0	3.2	36.5	14.4
81	Etofenprox	50.0	95.4	3.9	102.6	13.8	80.0	7.0	83.2	3.5	100.8	4.6	87.7	1.0	92.6	11.5	94.7	8.8

(Continued)

TABLE 2.2.7 Test Results of Average Recoveries and Coefficients of Variations of 201 Pesticides in Green Tea Extracted per the Three Sample Preparation Methods (n = 3) (cont.)

No.	Pesticides	Spiked levels, µg/kg	GC/MS Method-1 AVE (%)	RSD (%)	Method-2 AVE (%)	RSD (%)	Method-3 (20 mL acetonitrile) AVE (%)	RSD (%)	Method-3 (25 mL acetonitrile) AVE (%)	RSD (%)	GC/MS/MS Method-1 AVE (%)	RSD (%)	Method-2 AVE (%)	RSD (%)	Method-3 (20 mL acetonitrile) AVE (%)	RSD (%)	Method-3 (25 mL acetonitrile) AVE (%)	RSD (%)
82	Etridiazol	50.0	70.9	11.9	42.6	7.1	44.4	5.0	50.2	0.6	63.4	3.7	37.9	6.9	33.9	3.9	30.5	10.0
83	Etrimfos	150.0	79.3	2.2	86.2	4.3	57.5	4.4	55.1	8.5	84.7	3.4	83.9	2.8	53.0	2.9	51.7	2.5
84	Fenamidone	100.0	86.9	1.2	90.9	5.3	33.6	5.7	32.2	15.7	0.0	0.0	0.0	0.0	0.0	0.0	0.0	0.0
85	Fenarimol	100.0	96.2	2.2	88.9	4.4	26.4	0.7	21.6	6.9	91.3	3.5	92.3	4.7	19.4	1.2	16.1	32.1
86	Fenazaquin	50.0	90.0	18.4	75.4	7.0	48.8	1.5	47.6	4.5	90.1	5.6	81.3	3.8	48.5	2.9	46.9	6.1
87	Fenbuconazole	50.0	83.8	6.1	97.4	4.9	13.9	21.4	13.9	40.8	84.3	8.8	95.9	6.8	10.0	18.7	7.5	52.4
88	Fenchlorphos	10.0	75.1	1.1	82.5	4.3	60.5	2.6	59.3	0.2	75.7	5.0	80.5	2.7	60.6	4.4	59.4	6.7
89	Fenoxanil	100.0	81.8	3.3	87.0	3.6	39.0	4.3	36.4	18.4	81.1	4.5	86.1	6.4	45.9	7.9	47.3	17.0
90	Fenpropidin	100.0	79.9	5.4	77.6	3.8	2.8	8.2	1.6	141.4	83.3	4.7	72.6	4.9	2.7	4.5	2.5	27.8
91	Fenpyroximate	150.0	66.8	5.5	89.0	6.5	44.1	0.2	38.4	9.4	66.4	5.3	76.1	4.8	45.6	3.8	41.6	3.9
92	Fenson	400.0	89.5	5.8	83.7	2.5	58.6	3.5	57.1	3.7	96.2	2.1	80.5	6.5	53.4	12.6	53.1	7.6
93	Flamprop-isopropyl	600.0	96.9	7.6	91.1	7.5	49.6	19.4	42.3	7.0	89.4	2.3	87.2	4.3	43.1	2.2	39.4	1.8
94	Flamprop-methyl	400.0	90.0	2.4	88.5	5.2	34.5	1.4	31.2	16.1	89.4	2.3	87.2	4.3	43.1	2.2	39.4	1.8
95	Flufenacet	150.0	82.9	9.1	71.5	11.9	36.3	5.4	32.5	23.1	77.9	3.6	73.8	4.1	32.6	7.5	27.3	22.4
96	Fluotrimazole	100.0	96.0	2.6	101.4	2.5	47.8	5.3	38.0	6.8	100.5	1.2	108.7	0.8	50.0	9.7	42.7	11.1
97	Fluroxypr-1-methylheptylester	150.0	105.7	6.4	95.9	4.3	65.3	4.5	61.7	3.2	93.5	5.7	96.1	9.1	59.9	13.0	54.5	10.3
98	Fonofos	50.0	84.8	2.2	89.1	5.6	54.5	0.6	54.2	2.1	86.7	4.6	87.9	4.5	54.6	0.3	54.2	5.5
99	Furalaxyl	50.0	86.5	1.7	85.9	2.4	24.8	9.1	21.8	33.8	84.5	1.7	85.8	3.1	24.7	12.6	21.5	31.4
100	Heptachlor	100.0	71.5	1.8	61.3	3.2	57.2	0.5	54.4	0.0	72.7	3.6	64.4	4.0	50.1	2.4	44.5	2.7
101	Hexachloroenzene	50.0	73.8	5.8	65.2	5.5	51.1	0.7	56.7	11.6	74.7	6.6	64.4	5.4	49.6	1.0	53.3	12.1
102	Hexazinone	50.0	84.6	8.1	91.8	5.6	6.2	13.3	5.1	61.7	84.9	7.6	87.3	5.9	3.1	5.5	2.3	69.9
103	Iodofenphos	50.0	83.0	9.2	81.3	6.4	60.3	4.8	62.5	5.4	77.1	0.8	77.7	2.0	58.7	8.4	57.3	3.2
104	Iprobenfos	50.0	83.2	2.8	90.2	3.1	34.5	6.2	34.4	15.5	82.4	4.3	91.5	2.5	34.5	9.5	33.6	13.9
105	Iprovalicarb-1	150.0	91.8	5.1	84.2	4.9	25.8	21.6	24.5	21.2	0.0	0.0	0.0	0.0	0.0	0.0	0.0	0.0

106	Iprovalicarb-2	100.0	88.2	2.2	88.7	6.4	18.8	14.4	16.3	33.8	0.0	0.0	0.0	0.0	0.0	0.0	0.0
107	Isazofos	100.0	80.4	2.3	87.9	2.4	52.0	8.1	50.2	8.5	85.8	1.3	83.1	4.8	51.4	9.6	47.3	1.7
108	Isodrin	100.0	85.1	2.0	91.9	2.1	64.3	4.7	59.7	0.8	87.1	4.2	85.1	3.5	62.7	0.6	58.3	9.1
109	Isofenphos	200.0	84.5	2.0	89.1	3.6	53.7	3.2	52.6	4.7	85.7	3.5	88.5	3.6	55.1	5.0	52.3	6.2
110	Isoprocarb-1	150.0	85.5	2.0	99.3	3.4	37.6	9.7	34.0	24.5	85.9	1.5	94.4	4.8	35.5	9.4	33.4	25.1
111	Isoprocarb-2	50.0	125.8	13.6	81.3	10.8	45.8	16.3	42.6	16.8	99.0	10.1	79.3	2.0	41.4	8.3	36.4	10.6
112	Isopropalin	200.0	84.2	3.4	88.1	5.3	65.7	8.6	64.4	16.5	83.1	5.7	83.0	3.2	59.2	12.1	54.6	12.1
113	Isoprothiolane	50.0	87.9	3.9	90.1	4.6	46.4	7.5	43.8	12.4	91.1	4.7	85.6	2.4	47.9	9.9	45.4	8.6
114	Kresoxim-methyl	150.0	94.8	2.9	87.3	6.8	48.4	7.9	46.6	5.8	107.2	4.6	81.6	3.4	47.9	5.2	46.3	2.3
115	Lenacil	200.0	106.5	2.4	90.4	7.8	97.1	4.7	102.9	8.1	140.1	5.5	89.7	4.7	7.3	14.4	6.2	54.8
116	Mefenacet	200.0	113.3	0.6	88.8	1.3	115.3	4.5	158.4	17.2	81.4	2.5	75.2	4.3	27.8	7.7	25.6	33.7
117	Mepronil	100.0	90.6	0.9	90.3	3.8	39.6	10.3	31.4	28.7	86.8	1.4	86.9	4.5	35.2	15.0	32.7	11.2
118	Metalaxyl	150.0	85.1	3.2	90.0	7.3	19.5	11.0	13.6	32.8	88.1	1.7	90.0	4.1	20.0	12.8	16.1	41.9
119	Metazachlor	100.0	81.3	6.3	89.0	6.8	22.0	8.8	22.4	50.5	71.4	2.2	79.5	3.8	19.7	14.3	15.7	37.3
120	Methabenzthiazuron	50.0	85.1	1.2	89.3	4.8	15.8	22.7	14.7	41.0	87.5	0.5	87.6	4.6	16.5	14.7	13.9	42.9
121	Methoprene	300.0	73.7	1.3	85.3	4.2	62.4	5.1	60.1	9.2	90.2	3.5	87.7	5.7	61.5	9.5	61.9	9.0
122	Methoprotryne	200.0	89.6	2.7	90.5	3.8	20.8	4.3	18.4	31.7	92.9	3.4	85.9	2.7	20.2	3.0	17.2	34.6
123	Methoxychlor	150.0	88.8	1.0	27.6	12.5	35.5	5.6	30.8	5.8	78.6	2.7	32.6	8.2	23.4	6.7	16.5	0.1
124	Methyl-parathion	150.0	85.1	2.7	76.7	11.0	52.0	4.6	49.1	11.2	74.7	4.2	74.9	5.3	44.7	5.7	37.5	10.5
125	Metolachlor	100.0	82.3	1.7	88.5	4.3	44.0	3.3	43.0	8.3	78.9	1.0	84.4	4.5	40.6	3.4	38.1	9.4
126	Metribuzin	100.0	82.1	4.5	86.3	1.8	25.1	19.1	20.1	41.3	87.3	7.5	86.3	2.9	23.0	17.1	19.1	39.9
127	Mirex	50.0	80.1	6.0	61.3	7.3	58.2	0.1	53.3	6.2	82.2	1.6	62.3	1.4	51.3	0.5	45.6	11.4
128	Molinate	50.0	81.0	3.5	75.8	5.5	59.0	9.1	59.5	3.6	78.1	5.5	74.3	2.0	45.9	7.4	48.2	6.8
129	Monalide	100.0	98.6	2.8	83.8	1.2	76.8	9.5	66.6	19.4	87.8	2.7	92.0	4.7	45.4	9.8	43.1	10.8
130	Napropamide	100.0	85.0	3.7	88.0	2.6	24.9	8.0	23.8	31.7	88.1	4.5	84.6	2.9	24.4	11.2	22.3	25.4
131	Nitrapyrin	400.0	68.2	2.8	40.3	11.0	45.2	7.6	44.3	3.4	58.1	5.5	34.0	5.6	28.5	12.8	23.1	8.7
132	Nuarimol	50.0	89.7	2.7	82.9	1.7	16.0	9.0	16.6	32.7	89.8	5.9	86.9	3.8	16.3	9.5	13.6	46.8
133	Oxyfluofen	100.0	76.8	1.5	82.0	9.9	66.6	12.1	66.2	8.7	75.3	5.2	79.1	10.9	57.5	17.8	52.7	12.0
134	Paraoxon-ethyl	200.0	65.2	4.4	92.6	1.3	95.6	4.3	92.4	0.3	82.3	5.6	71.4	8.2	16.8	30.3	12.0	15.6
135	Pebulate	100.0	99.6	5.8	83.2	3.4	55.3	4.2	56.0	2.2	82.5	1.8	73.5	1.6	50.2	1.9	51.9	7.6
136	Pendimethalin	50.0	78.2	1.9	79.7	8.5	60.8	7.5	63.1	9.2	79.4	4.1	76.4	7.1	54.3	7.5	52.7	9.5
137	Pentachloroaniline	150.0	80.1	1.3	71.2	4.7	46.9	4.3	48.2	3.9	82.0	5.0	74.6	3.4	47.5	10.4	47.9	2.1

(Continued)

TABLE 2.2.7 Test Results of Average Recoveries and Coefficients of Variations of 201 Pesticides in Green Tea Extracted per the Three Sample Preparation Methods (n = 3) (cont.)

No.	Pesticides	Spiked levels, μg/kg	GC/MS Method-1		Method-2		Method-3 (20 mL acetonitrile)		Method-3 (25 mL acetonitrile)		GC/MS/MS Method-1		Method-2		Method-3 (20 mL acetonitrile)		Method-3 (25 mL acetonitrile)	
			AVE (%)	RSD (%)	AVE (%)	RSD (%)	AVE (%)	RSD (%)	AVE (%)	RSD (%)	AVE (%)	RSD (%)	AVE (%)	RSD (%)	AVE (%)	RSD (%)	AVE (%)	RSD (%)
138	Pentachloroanisole	50.0	86.0	2.1	89.6	2.7	58.4	3.8	57.5	4.0	85.3	2.6	84.3	3.3	56.2	4.5	56.5	4.1
139	Pentachlorobenzene	150.0	76.5	2.7	70.9	1.8	51.2	2.6	52.1	3.0	78.4	3.5	65.8	2.8	50.8	4.8	52.2	8.3
140	Permethrin	200.0	94.6	4.5	89.9	3.6	69.6	4.3	68.8	5.1	87.6	9.4	81.7	5.0	67.8	5.9	66.0	5.2
141	Perthane	50.0	95.1	1.2	96.5	3.1	71.4	2.9	64.9	12.2	93.8	3.1	89.1	4.3	65.8	3.0	59.3	6.0
142	Phenanthrene	150.0	84.5	1.9	80.2	4.3	47.8	7.8	47.9	7.3	89.4	2.5	76.8	4.8	50.1	15.0	48.6	3.7
143	Phenothrin	50.0	88.7	5.9	98.3	0.9	71.1	5.8	71.8	0.3	85.1	6.8	79.3	16.5	69.4	7.1	62.3	11.5
144	Picoxystrobin	100.0	87.9	1.3	91.4	4.3	46.9	3.5	42.4	3.2	89.1	2.5	91.2	3.4	46.2	3.3	40.8	0.5
145	Piperonylbutoxide	100.0	91.3	7.6	84.0	3.5	55.6	5.7	56.4	8.1	88.8	5.5	91.8	3.5	54.7	4.6	53.8	2.5
146	Piperophos	300.0	84.9	1.6	90.2	5.9	43.6	2.1	40.4	6.0	86.4	1.5	85.7	5.7	42.6	6.9	42.0	8.1
147	Pirimicarb	50.0	90.7	4.2	94.4	5.8	14.9	2.3	12.7	55.1	88.5	1.1	90.5	4.2	13.9	4.8	11.2	46.8
148	Pirimiphos-ethyl	100.0	85.7	2.3	89.2	3.2	56.9	5.0	56.7	7.8	86.9	5.0	85.6	3.3	54.8	5.5	55.5	13.6
149	Pirimiphos-methyl	50.0	81.2	0.6	87.6	4.0	52.6	0.7	52.4	5.2	84.6	5.3	90.9	4.2	53.9	3.5	52.4	7.7
150	Pretilachlor	100.0	65.8	22.0	80.6	3.5	48.3	2.7	47.1	11.4	71.8	3.1	79.0	1.8	47.4	4.8	44.7	3.8
151	Procymidone	200.0	87.3	2.6	92.3	5.3	42.7	9.0	41.0	15.9	87.5	3.3	93.0	3.9	44.2	12.7	41.3	8.2
152	Profenofos	150.0	107.9	10.4	67.2	16.9	50.5	0.3	50.9	0.3	102.1	4.4	66.0	5.4	44.7	2.3	42.2	1.5
153	Profluralin	50.0	89.8	3.5	79.6	7.3	70.8	11.1	75.7	11.3	97.4	9.1	80.5	3.1	64.0	15.6	59.6	7.2
154	Prometon	50.0	82.7	4.5	88.0	2.5	20.5	6.4	19.4	33.9	85.1	4.9	85.2	2.6	20.9	9.5	19.3	31.7
155	Prometrye	200.0	87.6	2.8	97.8	3.6	37.0	6.9	36.8	11.3	92.7	3.4	89.6	3.3	36.1	7.0	34.4	15.1
156	Pronamide	50.0	84.4	7.0	90.9	7.1	39.8	3.4	37.7	20.6	85.4	2.7	86.0	3.1	38.4	3.9	34.9	13.0
157	Propachlor	150.0	65.6	3.8	67.4	8.4	28.8	4.9	25.6	19.5	68.6	1.8	75.4	5.2	34.0	1.8	29.4	19.3
158	Propetamphos	50.0	95.0	1.3	84.1	6.8	51.4	7.8	50.0	5.8	90.7	4.8	81.4	3.2	52.0	5.8	49.2	3.2
159	Propham	200.0	87.9	8.4	92.0	5.3	51.2	12.1	50.2	13.5	84.1	2.1	92.1	4.3	48.9	7.5	46.8	13.7
160	Propiconazole	100.0	87.1	1.8	89.3	2.8	25.6	36.1	22.3	63.9	93.9	5.4	90.1	1.3	19.3	2.0	15.2	33.3
161	Propoxur-1	100.0	78.7	1.9	89.5	2.7	31.0	13.5	26.5	31.5	78.7	2.5	85.4	3.6	30.0	16.1	25.1	31.7
162	Propoxur-2	50.0	81.1	8.1	83.8	7.7	27.1	12.2	24.9	35.6	128.6	7.5	71.1	3.1	24.3	15.3	18.7	20.9

163	Propyzamide	400.0	84.5	1.6	88.7	3.8	39.3	1.2	38.7	6.2	85.4	2.7	86.0	3.1	38.4	3.9	34.9	13.0
164	Prosulfocarb	100.0	88.6	2.6	93.6	3.7	58.0	7.5	54.5	0.5	86.0	2.1	90.6	3.0	54.2	3.1	52.3	4.7
165	Prothiophos	150.0	90.3	2.7	95.5	3.9	71.1	0.9	70.8	4.1	88.7	5.8	85.3	2.5	69.2	4.9	66.0	8.4
166	Pyridaben	150.0	104.5	4.7	81.8	6.7	68.2	5.8	67.8	3.3	104.9	5.0	81.4	4.8	58.5	5.1	53.9	2.5
167	Pyridaphenthion	100.0	87.1	5.2	89.4	7.7	31.0	10.8	26.3	44.5	82.2	0.9	86.2	2.8	27.4	14.5	21.1	33.0
168	Pyrimethanil	100.0	80.6	2.8	80.0	2.0	30.6	7.4	31.0	20.7	86.0	3.8	76.5	6.8	28.7	6.6	27.8	13.6
169	Pyriproxyfen	50.0	89.9	5.3	90.5	2.7	63.4	3.2	61.5	2.3	121.0	2.2	85.5	0.9	114.8	6.4	155.6	23.0
170	Quinalphos	150.0	85.0	2.1	84.9	2.4	56.3	0.6	58.6	9.0	84.5	5.7	79.9	3.1	49.8	6.2	47.4	3.6
171	Quinoxuphen	50.0	88.9	4.5	82.3	5.1	36.7	0.0	39.8	15.4	87.4	3.4	78.0	4.2	34.1	1.1	33.3	16.9
172	Ronnel	150.0	75.6	0.8	82.5	4.3	60.5	2.6	59.9	1.6	75.9	6.0	82.4	3.5	60.9	3.8	58.4	4.1
173	Secbumeton	50.0	93.8	2.5	97.5	6.5	20.7	11.4	19.6	30.9	84.9	3.4	92.4	2.7	17.4	7.9	16.1	29.9
174	Silafluofen	150.0	86.8	6.2	87.1	5.8	101.8	20.8	69.7	1.5	90.7	2.6	90.2	4.6	69.3	0.3	62.6	11.7
175	Sulfotep	100.0	81.8	1.6	87.6	3.8	56.7	2.2	56.2	4.0	80.7	4.6	91.4	4.3	55.0	0.9	55.9	7.5
176	Tebufenpyrad	100.0	89.5	2.4	93.2	4.2	49.6	2.4	48.1	0.5	91.6	3.6	90.7	3.7	50.2	3.7	46.8	2.8
177	Tebupirimfos	50.0	87.2	2.7	95.5	3.8	61.0	2.8	58.6	5.2	87.6	6.1	96.5	6.8	61.8	6.6	59.5	5.8
178	Tebutam	50.0	85.6	2.9	93.9	2.8	39.2	10.5	36.7	14.6	89.1	2.3	91.8	3.3	38.3	9.0	36.9	13.9
179	Tebuthiuron	100.0	84.8	4.3	99.3	5.0	7.1	30.6	5.4	71.7	85.7	3.7	102.0	3.5	6.5	17.8	5.2	59.1
180	Tefluthrin	150.0	92.1	3.8	98.8	1.9	76.3	10.9	71.2	11.5	88.8	1.9	91.7	5.9	66.6	6.4	61.6	10.1
181	Telodrin	50.0	98.9	3.0	82.3	2.9	63.7	0.3	64.0	3.7	105.6	6.5	83.2	2.2	62.4	4.4	59.0	5.3
182	Terbutryn	50.0	85.8	2.9	89.3	3.8	35.4	4.4	35.9	3.7	86.3	3.1	86.4	0.5	35.1	6.1	32.1	7.8
183	Tetrasul	400.0	82.1	3.5	81.9	3.7	59.9	0.4	59.4	5.6	83.5	4.8	76.4	3.2	61.7	1.2	61.9	9.9
184	Thenylchlor	50.0	67.3	8.8	74.2	9.1	31.7	9.3	27.8	18.7	71.4	0.6	74.9	3.9	32.7	11.0	28.7	14.7
185	Thiazopyr	50.0	86.4	2.6	91.4	3.5	49.8	6.4	50.1	11.0	85.9	8.0	93.1	4.9	54.3	4.8	50.2	3.1
186	Thiobencarb	50.0	86.9	2.4	92.4	6.5	50.1	0.1	51.2	0.6	87.2	3.5	87.9	3.5	51.5	3.2	50.4	0.9
187	Thionazin	50.0	84.9	7.3	89.7	8.7	48.8	15.2	45.6	15.5	84.3	1.9	87.6	4.8	43.5	12.8	40.3	13.7
188	Tolclofos-methyl	50.0	76.6	2.1	85.2	2.6	59.3	1.3	57.0	7.2	74.4	6.6	88.1	3.5	59.3	6.5	54.2	3.5
189	Tralkoxydim	50.0	81.0	4.3	78.4	2.5	42.4	2.7	42.7	1.2	80.2	3.0	73.5	3.9	40.6	0.3	40.6	12.1
190	Trans-chlordane	50.0	85.9	2.9	84.5	3.4	59.3	4.4	60.3	5.6	93.8	7.6	83.2	3.6	60.6	4.8	59.7	13.3
191	Trans-diallate	50.0	92.0	2.3	88.6	4.1	58.1	11.0	56.2	0.9	89.9	4.3	83.7	3.2	55.4	2.9	54.9	8.1
192	Transfluthrin	200.0	84.7	4.4	87.0	2.8	68.1	3.0	66.4	4.2	84.2	5.5	80.8	1.9	67.1	6.7	64.2	10.4
193	Triadimefon	50.0	79.7	2.4	90.6	2.1	29.0	8.4	27.0	24.5	84.9	1.7	87.1	2.4	30.5	14.2	26.3	23.0
194	Triadimenol	100.0	92.1	2.4	91.3	4.2	16.6	10.3	16.0	52.6	87.0	2.1	94.5	2.0	16.9	12.0	14.9	41.0

(Continued)

TABLE 2.2.7 Test Results of Average Recoveries and Coefficients of Variations of 201 Pesticides in Green Tea Extracted per the Three Sample Preparation Methods (n = 3) (cont.)

No.	Pesticides	Spiked levels, μg/kg	GC/MS															GC/MS/MS															
			Method-1		Method-2		Method-3 (20 mL acetonitrile)		Method-3 (25 mL acetonitrile)		Method-1		Method-2		Method-3 (20 mL acetonitrile)		Method-3 (25 mL acetonitrile)	
			AVE (%)	RSD (%)	AVE (%)	RSD (%)	AVE (%)	RSD (%)	AVE (%)	RSD (%)	AVE (%)	RSD (%)	AVE (%)	RSD (%)	AVE (%)	RSD (%)	AVE (%)	RSD (%)
195	Triallate	50.0	90.5	3.7	88.0	3.3	66.9	0.8	64.8	4.7	93.7	7.2	85.9	1.2	62.8	2.6	62.8	7.6
196	Tribenuron-methyl	50.0	140.5	17.6	68.1	18.3	0.0	0.0	0.0	0.0	126.9	8.6	51.4	22.0	6.9	17.1	5.7	65.9
197	Trichloronat	100.0	89.6	1.9	91.3	3.5	64.1	3.9	61.8	6.7	90.9	2.8	89.7	4.6	65.6	3.2	62.9	10.7
198	Trifloxystrobin	150.0	113.5	2.4	91.8	4.2	53.9	6.6	52.3	4.2	107.3	3.0	87.2	3.0	53.5	7.0	49.1	4.4
199	Trifluralin	100.0	92.1	2.9	89.0	6.3	72.3	7.5	72.1	6.9	93.9	10.3	86.5	5.5	64.8	9.2	59.3	9.1
200	Vinclozolin	50.0	91.2	4.2	88.1	3.4	52.2	5.5	52.9	7.2	96.5	4.2	86.8	1.7	51.7	8.3	49.3	9.7
201	Zoxamide	50.0	80.1	2.0	86.7	3.7	33.4	4.4	26.7	18.0	81.6	4.0	84.7	2.6	36.4	3.3	30.4	29.3

TABLE 2.2.8 The Statistical Analysis of Average Recoveries and RSD Values for 201 Pesticides in Green Tea Using the Three Methods

	GC/MS/MS								GC/MS							
	Method-1		Method-2		Method-3 20mL		Method-3 25 mL		Method-1		Method-2		Method-3 20 mL		Method-3 25mL	
AVE Range	Number	Percentage (%)	Number	Percentage (%)	Number	Percentage (%)	Number	Percentage (%)	Number	Percentage (%)	Number	Percentage (%)	Number	Percentage (%)	Number	Percentage (%)
<60%	1	0.5	8	4.0	161	80.1	173	86.1	1	0.5	6	3.0	147	73.1	155	77.1
60%-70%	5	2.5	8	4.0	35	17.4	23	11.4	8	4.0	7	3.5	35	17.4	30	14.9
70%-80%	31	15.4	32	15.9	0	0.0	0	0.0	29	14.4	13	6.5	11	5.5	9	4.5
80%-90%	110	54.7	112	55.7	0	0.0	0	0.0	106	52.7	104	51.7	1	0.5	1	0.5
90%-100%	33	16.4	34	16.9	1	0.5	1	0.5	42	20.9	65	32.3	2	1.0	1	0.5
100%-110%	11	5.5	4	2.0	0	0.0	0	0.0	8	4.0	5	2.5	1	0.5	0	0.0
110%-120%	2	1.0	0	0.0	1	0.5	0	0.0	2	1.0	1	0.5	2	1.0	0	0.0
>120%	5	2.5	0	0.0	0	0.0	1	0.5	5	2.5	0	0.0	0	0.0	2	1.0
N.D	3	1.5	3	1.5	3	1.5	3	1.5	0	0.0	0	0.0	2	1.0	2	1.0
Total	201	100.0	201	100.0	201	100.0	201	100.0	201	100.0	201	100.0	201	100.0	201	100.0
RSD range																
<10%	194	96.5	194	96.5	156	77.6	103	51.2	188	93.5	190	94.5	157	78.1	117	58.2
10%-15%	4	2.0	1	0.5	27	13.4	37	18.4	9	4.5	8	4.0	23	11.4	18	9.0
15%-20%	0	0.0	2	1.0	11	5.5	13	6.5	2	1.0	3	1.5	6	3.0	15	7.5
>20%	0	0.0	1	0.5	4	2.0	45	22.4	2	1.0	0	0.0	11	5.5	48	23.9
N.D	3	1.5	3	1.5	3	1.5	3	1.5	0	0.0	0	0.0	4	2.0	3	1.5
Total	201	100.0	201	100.0	201	100.0	201	100.0	201	100.0	201	100.0	201	100.0	201	100.0

in color, presenting a light green color. It is evident that the pigments have been effectively removed by adding graphite carbon. The tea leaves are processed per Method-3, and tests find that sample solutions are of relatively dark color: orange after cleanup. After concentrations with rotary evaporations, lots of solid matters condensed on the walls of a pear-shaped flask. The hexane is added for solvent exchange, a part of liquids condensed, which is attributed to the separating-out from extracting a lot of water-soluble foreign matters. See Figs. 2.2.1 and Figs. 2.2.2 for the comparison charts of matrix colors of green tea sample solutions after processing with the three methods.

It can be seen in Fig. 2.2.1 that with Method-1 the color of sample solutions processed is the lightest; with Method-2, the color of sample solutions remains medium; with Method-3 the color is the darkest, posing the worst contaminations against the instrument as well. Fig. 2.2.2 shows that during the interval of pesticide determination (7.2–35 min) the matrix baselines of Method-1 and Method-2 are superior to those of Method-3. Matrix influences of Method-3 are the biggest, with the highest baselines; next is Method-2. Method-1 have the best cleanup effect.

2.2.7 STAGE II: EVALUATION ON THE EXTRACTION EFFICIENCIES OF THE MULTIRESIDUE PESTICIDES IN GREEN TEA AND OOLONG TEA SAMPLES BY METHOD-1 AND METHOD-2

Six months after Stage I comparative experiments, the comparative tests were conducted on the green tea and oolong tea samples spiked with 201 pesticides (10–600 µg/kg) per Method-1 and Method-2, with five parallel tests carried out for each tea, calculating the average recoveries and RSD over 5 times analysis. See Table 2.2.9 for test results.

The statistical analysis of average recoveries by GC/MS/MS and GC/MS determination in Table 2.2.9 is tabulated in Table 2.2.10, from which it can be seen that Method-1 accounts for 1.3% for pesticides in green tea with average recoveries less than 70%, with Method-2 3.0%, while Method-1 accounts for 3.7 for pesticides in oolong tea with an average

FIGURE 2.2.1 Comparison of colors of Green tea sample solutions prepared with the three methods.

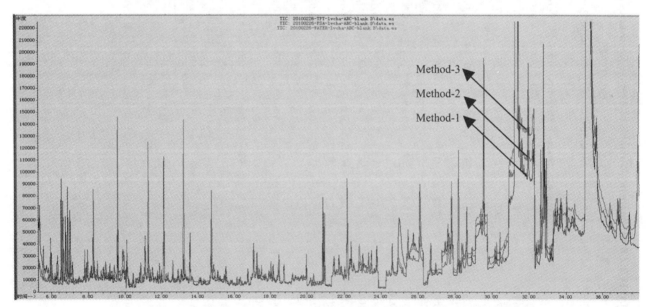

FIGURE 2.2.2 Comparison of the matrix baselines of Green tea blank sample solutions prepared with the three methods.

TABLE 2.2.9 Recoveries and Precision Test Results for 201 Pesticides in Green Tea and oolong Tea Extracted per Method-1 and Method-2 (n = 5)

No.	Pesticides	Spiked level, µg/kg	GC/MS								GC/MS/MS							
			Green tea				Oolong tea				Green tea				Oolong tea			
			Method-1		Method-2		Method-1		Method-2		Method-1		Method-2		Method-1		Method-2	
			AVE (%)	RSD (%)	AVE (%)	RSD (%)	AVE (%)	RSD (%)	AVE (%)	RSD (%)	AVE (%)	RSD (%)	AVE (%)	RSD (%)	AVE (%)	RSD (%)	AVE (%)	RSD (%)
1	2,3,4,5-Tetrachloroaniline	10.0	75.7	4.6	87.3	2.6	77.3	3.1	84.2	10.9	75.6	4.3	83.9	4.0	77.0	4.4	80.3	11.6
2	2,3,5,6-Tetrachloroaniline	50.0	78.9	2.1	86.6	4.7	81.8	2.6	85.0	11.4								
3	4,4-Dibromobenzophenone	50.0	83.6	1.9	93.5	2.8	84.8	3.3	101.8	10.7	81.9	2.6	88.3	3.2	82.0	3.7	93.6	11.1
4	4,4-Dichlorobenzophenone	50.0	84.3	1.6	93.6	3.0	100.3	4.2	151.3	12.9	81.1	3.2	87.9	3.4	101.9	4.3	145.1	12.9
5	Acetochlor	100.0	81.0	3.2	84.7	4.5	80.0	1.78	90.7	4.5	79.6	4.2	85.5	6.5	78.7	2.4	91.2	4.9
6	Alachlor	150.0	90.3	4.5	93.5	3.2	83.7	5.4	82.3	4.6	85.2	4.0	96.5	6.2	80.8	4.4	91.3	4.1
7	Atratone	50.0	67.5	12.4	105.0	14.5	92.5	17.9	101.3	12.0	82.8	2.8	89.1	3.3	79.6	4.6	84.0	12.5
8	Benodanil	150.0	84.9	9.5	103.0	39.8	100.3	8.8	118.1	12.6	116.7	22.5	61.0	51.3	83.3	9.0	88.7	4.7
9	Benoxacor	100.0	77.0	8.6	80.5	7.8	75.9	3.87	93.1	4.5	77.2	6.0	87.1	6.2	79.0	2.1	89.4	5.1
10	Bromophos-ethyl	50.0	81.7	5.8	92.3	6.8	86.9	4.5	85.2	2.4	91.9	4.3	84.5	11.1	77.5	5.2	87.0	4.1
11	Butralin	200.0	73.0	3.0	93.8	1.9	64.7	4.6	96.9	10.2	76.5	2.1	93.0	2.7	64.3	3.4	92.0	11.2
12	Chlorfenapyr	400.0	71.5	2.6	89.1	5.6					79.7	1.8	100.6	2.5	78.9	5.0	87.0	6.8
13	Clomazone	50.0	94.2	7.2	84.6	8.0	86.5	4.0	95.9	3.5	83.4	31.0	65.7	38.2	68.2	28.1	96.8	18.4
14	Cycloate	50.0	93.2	1.8	100.5	7.3	88.9	2.4	86.5	3.3	84.9	6.2	77.7	46.5	115.1	83.1	56.5	11.6
15	Cycluron	150.0	71.4	11.4	89.2	4.7	67.3	19.6	104.3	11.2	71.2	8.8	86.2	4.5	67.0	15.7	97.4	9.7
16	Cyhalofop-butyl	100.0	92.0	2.9	102.4	1.8	75.2	3.2	94.4	8.5	83.5	3.3	98.5	1.1	75.0	3.0	89.7	7.4
17	Cyprodinil	50.0	80.6	2.9	66.2	5.4	74.4	9.2	49.9	15.5	79.5	3.9	65.0	6.0	78.5	7.3	44.7	16.5
18	Dacthal	50.0	82.7	1.8	92.3	3.2	81.6	1.7	106.9	10.6	80.3	2.3	89.6	3.8	82.0	1.3	103.0	10.5
19	de-PCB 101	50.0	82.3	1.7	90.7	3.1	80.7	0.9	102.3	10.5	80.9	1.7	87.5	4.4	80.7	2.0	99.9	11.8
20	de-PCB 118	50.0	81.0	2.1	89.2	3.1	80.2	1.6	99.0	10.9	78.8	2.5	86.5	3.6	80.7	1.7	95.3	10.7
21	de-PCB 138	50.0	77.6	9.7	84.1	5.7	80.0	7.6	100.4	11.7	79.8	1.0	88.4	2.6	80.0	1.5	98.7	9.9
22	de-PCB 180	50.0	80.7	2.1	89.1	3.0	79.4	2.0	98.9	10.9	77.9	3.1	86.7	3.8	78.8	1.8	96.0	11.2
23	de-PCB 28	50.0	80.3	1.9	89.6	3.6	80.4	1.2	100.4	11.1	79.4	2.1	88.2	3.8	81.1	1.5	97.5	10.7
24	de-PCB 31	50.0	81.1	1.6	89.6	3.2	79.4	1.0	100.4	10.8	79.4	2.1	88.2	3.8	81.1	1.5	97.5	10.7

(Continued)

TABLE 2.2.9 Recoveries and Precision Test Results for 201 Pesticides in Green Tea and oolong Tea Extracted per Method-1 and Method-2 (n = 5) (cont.)

No.	Pesticides	Spiked level, μg/kg	GC/MS Green tea Method-1 AVE (%)	RSD (%)	GC/MS Green tea Method-2 AVE (%)	RSD (%)	GC/MS Oolong tea Method-1 AVE (%)	RSD (%)	GC/MS Oolong tea Method-2 AVE (%)	RSD (%)	GC/MS/MS Green tea Method-1 AVE (%)	RSD (%)	GC/MS/MS Green tea Method-2 AVE (%)	RSD (%)	GC/MS/MS Oolong tea Method-1 AVE (%)	RSD (%)	GC/MS/MS Oolong tea Method-2 AVE (%)	RSD (%)
25	Dichlorofop-methyl	50.0	110.1	11.4	88.8	5.2	88.0	5.1	98.4	5.2								
26	Dimethenamid	50.0	82.2	2.7	92.6	3.5	81.8	1.8	105.8	10.6	80.9	2.0	89.4	3.7	81.8	2.2	102.1	10.4
27	Diofenolan-1	100.0	91.5	2.3	101.5	1.1	76.8	2.7	91.6	7.1	81.5	3.2	100.2	0.7	78.6	3.6	89.9	8.3
28	Diofenolan-2	100.0	91.4	2.1	101.0	1.2	76.8	2.6	91.3	7.1	81.5	3.2	98.9	0.5	79.0	3.4	89.8	7.2
29	Fenbuconazole	100.0	79.9	4.2	98.3	2.7	73.3	7.1	95.0	11.6	79.8	2.2	90.2	2.8	75.8	8.5	89.8	10.7
30	Fenpyroximate	400.0	86.0	3.5	83.7	6.9	76.7	8.53	67.2	5.9	74.7	21.0	92.1	8.1	75.6	15.8	75.3	4.0
31	Fluotrimazole	50.0	90.7	2.5	99.3	3.0	90.8	1.6	105.5	11.0	83.5	2.8	92.2	3.5	85.1	2.5	98.8	12.7
32	Fluroxypr-1-methyl heptyl ester	50.0	83.3	1.9	95.2	2.6	83.4	3.3	97.8	11.3	80.3	2.4	85.5	5.8	88.2	4.7	93.4	11.2
33	Iprovalicarb-1	200.0	94.0	2.7	99.4	3.4	72.0	6.9	88.9	7.5	80.4	2.1	101.1	1.0	77.0	4.1	90.6	7.4
34	Iprovalicarb-2	200.0	93.0	3.0	102.8	2.8	73.1	1.8	92.4	7.5	80.4	2.1	101.1	1.0	77.0	4.1	90.6	7.4
35	Isodrin	50.0	81.7	2.3	95.2	6.4	81.6	1.7	107.3	12.8	82.2	2.6	86.5	3.9	82.0	1.5	97.8	10.5
36	Isoprocarb-1	100.0	80.4	8.0	106.9	6.9	76.6	6.69	100.7	3.6	77.4	8.3	118.5	7.6	110.2	6.5	112.1	6.9
37	Isoprocarb-2	100.0	91.3	13.5	60.9	11.3	89.8	16.44	84.0	7.4	83.2	16.4	53.4	15.2	80.4	36.0	86.4	6.3
38	Lenacil	240.0	N.D	N.D	N.D	N.D	N.D	N.D	N.D	N.D	N.D	N.D	N.D	N.D	N.D	N.D	N.D	N.D
39	Metalaxyl	150.0	81.6	3.7	79.6	19.0	82.5	3.4	114.1	3.7	76.0	11.4	78.6	12.8	81.5	5.9	101.9	4.0
40	Metazachlor	150.0	84.1	7.1	79.7	12.2	84.3	6.6	106.2	3.9	78.4	20.5	75.6	10.8	75.5	10.0	86.2	7.0
41	Methabenzthia-zuron	500.0	82.9	2.1	87.4	3.0	82.7	3.3	81.9	12.5	80.1	3.1	82.7	3.2	82.2	2.9	77.8	11.5
42	Mirex	50.0	95.2	8.4	96.6	11.0	85.0	5.3	90.1	4.6	82.5	3.2	86.1	20.2	87.5	2.2	111.2	4.9
43	Monalide	100.0	87.3	4.1	96.1	4.7	99.1	9.1	121.1	11.4	81.8	3.4	93.0	3.0	82.7	2.3	106.1	11.0
44	Paraoxon-ethyl	200.0	110.6	16.2	104.5	1.0	71.6	8.5	115.2	5.2	78.3	8.6	86.8	12.1	84.7	4.5	86.3	6.7
45	Pebulate	150.0	81.9	6.3	89.5	3.7	85.4	7.2	80.5	14.8	79.6	5.2	82.0	4.0	79.2	6.8	86.5	15.1
46	Pentachloroaniline	50.0	80.4	4.9	79.4	3.2	82.7	2.3	66.1	12.1	79.6	5.5	75.9	3.7	82.5	3.0	63.5	12.4
47	Pentachloroanisole	50.0	79.6	1.7	86.5	4.3	82.5	2.7	82.8	11.2	78.3	2.5	86.7	4.9	85.0	2.7	81.0	11.5

		50.0	77.3	2.6	79.7	12.0	85.9	8.4	48.7	19.1	75.3	3.5	76.8	13.6	85.8	9.0	47.1	18.0
48	Pentachlorobenzene	50.0	77.3	2.6	79.7	12.0	85.9	8.4	48.7	19.1	75.3	3.5	76.8	13.6	85.8	9.0	47.1	18.0
49	Perthane	50.0	81.9	2.0	93.1	3.1	81.6	2.0	106.5	10.6	82.3	3.2	90.1	4.0	83.1	3.0	102.2	9.9
50	Phenanthrene	50.0	78.4	4.0	74.6	4.3	76.4	2.84	65.6	4.4	73.5	6.0	74.5	6.6	73.0	4.7	65.2	5.3
51	Pirimicarb	100.0	88.6	1.8	100.3	1.1	77.0	2.7	86.3	6.1	80.7	3.0	99.3	1.3	79.4	3.7	85.5	6.8
52	Procymidone	50.0	83.2	2.9	89.5	8.3	88.7	4.3	96.6	4.1	85.4	2.5	90.1	6.0	84.7	3.2	87.8	5.4
53	Prometrye	50.0	81.1	2.1	95.3	4.0	85.7	5.0	90.4	3.4	81.0	2.3	83.7	14.3	85.9	1.4	100.2	3.0
54	Propham	50.0	85.4	3.8	87.2	6.1	93.1	2.9	100.0	5.0	88.4	2.9	85.8	11.7	86.2	3.8	93.0	5.2
55	Prosulfocarb	50.0	83.1	1.4	94.2	3.0	83.1	2.7	105.8	10.6	84.1	4.0	91.5	4.5	83.0	2.4	100.5	11.8
56	Secbumeton	50.0	84.8	2.0	89.0	8.0	87.8	3.4	88.7	3.1	83.8	3.6	94.9	6.4	86.5	2.5	84.6	4.3
57	Silafluofen	50.0	104.2	11.4	95.0	5.0	103.6	3.80	96.3	5.2	81.2	6.0	89.0	6.5	77.7	2.9	83.3	4.7
58	Tebupirimfos	100.0	80.7	3.3	88.9	4.6	78.4	3.28	88.3	4.7	81.2	4.1	90.1	5.6	78.3	3.8	87.5	5.3
59	Tebutam	100.0	80.6	1.6	91.7	3.6	81.0	1.0	104.4	10.4	79.8	2.4	90.4	4.1	81.3	1.3	99.9	10.9
60	Tebuthiuron	200.0	80.2	4.4	94.5	3.6	81.8	5.01	92.0	5.0	77.5	5.5	91.2	4.8	79.4	7.8	87.2	5.2
61	Tefluthrin	50.0	83.4	1.5	93.3	3.1	84.4	0.9	107.1	10.3	81.4	3.5	90.6	3.8	82.4	2.3	103.0	10.8
62	Thenylchlor	100.0	83.0	3.8	78.6	7.4	75.1	6.39	93.2	3.3	78.5	2.9	82.0	6.7	79.6	8.2	90.7	6.2
63	Thionazin	50.0	73.8	6.5	93.4	10.2	88.5	7.4	119.4	10.5	79.8	1.6	91.6	6.1	82.4	2.5	93.3	9.8
64	Trichloronat	50.0	82.2	2.1	88.9	3.1	81.4	1.7	99.1	11.0	81.3	2.9	87.5	3.8	82.2	2.2	95.1	10.8
65	Trifluralin	100.0	85.8	10.3	94.9	2.5	90.8	4.5	83.0	6.8	89.0	8.6	97.4	4.1	90.7	5.0	88.6	5.6
66	o,p'-DDT	100.0	80.7	6.8	81.7	7.1	78.0	4.5	81.5	7.8	85.0	6.8	86.8	5.1	87.8	2.7	89.9	6.2
67	p,p'-DDE	50.0	67.8	16.4	87.3	10.4	65.6	4.0	87.5	3.6	71.3	17.5	91.4	7.3	65.2	5.0	90.4	3.0
68	Benalaxyl	50.0	84.8	5.3	92.3	3.2	85.5	3.1	85.8	2.9	92.4	3.9	89.0	6.6	86.4	4.8	85.4	2.0
69	Benzoylprop-ethyl	150.0	91.3	7.0	98.3	3.4	90.7	2.7	91.9	2.0	89.5	4.2	89.9	4.7	84.9	3.5	85.5	2.5
70	Bromofos	100.0	73.8	9.8	91.5	3.8	84.0	4.2	80.0	7.3	82.2	5.6	90.5	1.8	82.2	4.5	79.9	5.0
71	Bromopropylate	100.0	94.2	10.1	87.3	6.8	87.9	6.2	94.5	7.1	93.0	4.9	89.7	3.9	87.9	2.9	84.5	4.0
72	Bupirimate	100.0	81.4	2.6	87.7	11.8	88.1	4.1	92.9	3.2	66.8	4.7	66.1	35.1	101.5	8.5	93.5	5.6
73	Butachlor	100.0	81.3	5.6	94.7	3.6	82.8	3.2	89.9	4.6	69.0	11.7	85.2	9.7	65.5	5.8	85.7	2.8
74	Butylate	150.0	93.7	23.3	91.2	6.0	110.5	21.3	83.6	21.6	76.2	5.9	79.6	4.9	81.2	8.9	78.1	16.2
75	Carbofenothion	100.0	82.4	4.0	91.1	3.9	83.7	2.1	84.3	5.1	85.2	4.1	83.6	6.8	88.5	4.0	77.1	6.3
76	Chlorfenson	100.0	100.8	11.9	98.3	7.7	104.3	5.5	101.9	5.7	99.4	11.3	91.3	3.5	93.1	2.2	91.5	3.7
77	Chlorfenvinphos	150.0	84.1	9.7	92.6	3.6	89.4	2.3	91.1	5.4	90.4	9.5	92.8	3.8	87.3	2.1	87.4	7.7
78	Chlormephos	100.0	86.1	4.8	96.5	14.1	90.3	5.6	83.6	12.8	80.9	5.7	94.1	8.1	84.3	2.9	87.3	6.5
79	Chloroneb	50.0	87.6	7.7	110.5	74.4	92.4	13.3	84.1	7.4	82.8	3.7	86.2	4.0	83.3	4.4	86.5	7.4

(Continued)

TABLE 2.2.9 Recoveries and Precision Test Results for 201 Pesticides in Green Tea and oolong Tea Extracted per Method-1 and Method-2 (n = 5) (cont.)

No.	Pesticides	Spiked level, µg/kg	GC/MS Green tea Method-1 AVE (%)	RSD (%)	Method-2 AVE (%)	RSD (%)	GC/MS Oolong tea Method-1 AVE (%)	RSD (%)	Method-2 AVE (%)	RSD (%)	GC/MS/MS Green tea Method-1 AVE (%)	RSD (%)	Method-2 AVE (%)	RSD (%)	GC/MS/MS Oolong tea Method-1 AVE (%)	RSD (%)	Method-2 AVE (%)	RSD (%)
80	Chlorpropham	100.0	85.2	5.7	84.0	4.0	89.6	2.6	88.9	3.5	87.9	4.2	91.2	5.7	85.7	2.2	86.4	2.9
81	Chlorpropylate	50.0	87.9	5.4	92.5	7.6	90.9	7.0	85.4	8.2	91.1	4.0	89.2	4.3	86.4	3.4	86.4	3.8
82	Chlorpyifos(ethyl)	50.0	87.6	5.9	91.1	6.6	86.5	4.1	88.8	4.0								
83	Chlorthiophos	150.0	83.8	4.4	91.0	3.6	85.1	2.6	73.4	9.2	83.1	3.1	89.6	5.3	88.3	5.3	73.1	7.6
84	Cis-chlordane	100.0	83.2	5.5	88.2	6.3	84.6	2.5	89.8	3.4	87.5	2.4	90.6	4.5	82.9	4.5	87.0	2.5
85	Cis-diallate	100.0	89.4	7.2	96.3	5.7	92.1	2.9	88.9	7.6	87.6	4.2	88.1	3.9	82.7	3.8	83.7	4.3
86	Cyanofenphos	50.0	98.0	8.6	93.6	2.9	88.3	3.8	85.3	3.5	99.3	9.8	91.5	3.9	90.9	3.5	84.7	2.3
87	Desmetryn	50.0	80.9	6.0	93.5	4.0	81.0	4.1	80.9	3.4	83.4	5.3	88.3	3.2	82.6	5.4	84.4	4.9
88	Dichlobenil	10.0	76.2	8.8	93.3	6.4	88.1	8.0	80.4	26.1	79.1	9.9	91.5	5.7	93.1	9.2	82.3	23.6
89	Dicloran	100.0	86.6	10.6	95.9	5.4	98.2	5.4	88.6	8.6	87.4	8.6	88.7	6.2	90.8	4.0	86.3	8.9
90	Dicofol	100.0	91.7	4.1	93.4	4.8	117.1	9.1	99.4	3.8	87.0	3.2	85.8	5.7	120.1	9.9	99.8	2.6
91	Dimethachloro	150.0	80.2	5.5	92.7	4.0	84.6	3.2	89.0	3.7	85.2	3.5	92.0	4.5	81.4	4.4	87.6	2.6
92	Dioxacarb	400.0	82.4	12.0	111.4	3.3	110.4	16.0	85.0	11.3	72.7	13.3	109.0	2.1	112.8	17.2	86.8	10.5
93	Endrin	600.0	82.0	3.7	93.4	3.5	80.0	4.3	88.1	2.6	85.2	2.3	84.1	3.8	82.1	4.6	87.1	3.7
94	Epoxiconazole-2	400.0	81.9	3.2	84.9	4.4	79.6	2.35	88.5	4.7	77.8	4.1	87.6	4.7	79.5	3.1	88.2	4.5
95	EPTC	150.0	70.2	11.1	77.9	8.6	87.6	9.9	78.5	24.0	74.9	6.3	75.8	3.1	82.3	8.8	79.5	11.2
96	Ethofumesate	100.0	73.5	3.6	86.1	3.7	76.6	2.4	89.2	3.4	88.8	2.9	91.0	4.6	82.5	4.4	89.7	1.8
97	Ethoprophos	150.0	89.2	5.1	95.7	3.0	84.5	3.0	88.4	6.3	84.5	2.5	88.1	4.6	82.0	2.9	86.2	4.5
98	Etrimfos	50.0	85.1	4.6	91.0	6.2	87.6	2.9	90.9	3.4	93.2	8.1	82.8	11.3	88.6	2.0	77.9	9.3
99	Fenamidone	50.0	86.5	3.7	91.2	4.0	84.1	4.82	92.7	5.3	78.7	3.8	92.4	5.5	81.4	5.0	88.5	5.8
100	Fenarimol	100.0	87.4	5.8	91.7	4.9	81.3	5.0	89.1	1.7	88.1	3.1	86.6	3.3	85.3	5.5	83.0	3.9
101	Flamprop-iso-propyl	50.0	78.3	4.3	94.1	5.0			88.7	3.4	90.8	3.0	86.6	5.0	85.3	3.5	86.4	2.6
102	Flamprop-methyl	50.0	87.4	5.9	95.1	4.1	88.7	3.2	86.6	2.4	90.8	3.0	86.6	5.0	85.3	3.5	86.4	2.6
103	Fonofos	50.0	82.4	2.5	92.9	6.5	89.4	3.2	90.7	3.4	45.4	16.5	65.0	42.1	69.3	23.9	84.1	18.8
104	Heptanophos	50.0	80.5	9.7	93.9	3.1	85.4	2.7	85.9	6.6	86.3	9.5	93.6	2.7	88.1	3.7	90.4	6.7
105	Hexazinone	150.0	89.1	4.9	95.9	6.7	75.7	23.0	90.8	10.9	82.9	5.2	83.8	8.9	67.3	20.4	84.0	10.6
106	Iodofenphos	100.0	79.0	9.2	89.1	2.6	87.0	4.4	75.2	8.3	79.6	9.8	82.5	9.3	83.1	6.1	69.9	12.2

No.	Compound	Conc.																
107	Isofenphos	100.0	87.4	4.9	94.5	4.1	86.8	6.2	87.4	5.2	85.8	2.8	90.4	3.5	85.5	3.2	87.0	2.5
108	Isopropalin	100.0	86.0	5.7	96.8	3.6	87.1	3.0	84.6	5.1	88.1	3.9	86.3	6.0	83.6	5.3	82.6	2.1
109	Methoprene	200.0	89.0	10.2	93.0	4.3	91.8	16.1	80.0	7.6	87.7	3.2	87.2	5.2	82.6	4.0	80.9	3.9
110	Methoprotryne	150.0	85.6	6.2	93.9	3.6	84.0	3.2	78.8	3.9	88.4	3.9	85.4	5.8	86.8	3.8	76.1	4.7
111	Methoxychlor	50.0	94.5	7.7	89.1	9.8	82.0	4.8	85.8	4.1	74.7	14.3	78.5	15.4	81.0	5.9	89.7	5.5
112	Methyl-parathion	200.0	98.1	13.1	84.2	12.4	81.7	7.4	101.8	3.7	131.7	36.2	76.5	19.8	76.6	9.6	66.6	13.8
113	Metolachlor	50.0	84.6	3.7	93.1	3.6	84.7	3.5	89.0	3.5	85.6	2.8	87.7	4.3	83.0	3.3	86.0	2.7
114	Nitrapyrin	150.0	86.6	5.4	86.3	2.7	98.6	17.3	79.2	22.2	83.0	13.8	82.1	6.8	91.4	4.6	78.2	24.1
115	Oxyflurofen	200.0	88.2	8.2	89.9	4.7	91.5	3.8	81.3	8.5	93.2	11.5	88.2	4.1	95.4	6.7	80.2	6.8
116	Pendimethalin	200.0	92.6	7.6	93.0	10.1	85.0	3.5	85.7	4.5	116.4	6.1	110.8	6.3	121.6	9.9	138.7	15.6
117	Picoxystrobin	100.0	85.8	3.2	90.2	4.2	79.9	2.89	89.9	4.8	79.3	4.4	90.8	5.2	80.5	4.0	88.2	4.1
118	Piperophos	150.0	81.5	3.5	89.4	4.7	80.2	1.83	94.3	3.2	79.0	4.6	90.5	6.1	74.2	2.6	90.6	5.9
119	Pirimiphos-ethyl	100.0	87.0	4.0	97.8	4.3	84.1	3.0	86.7	4.5	87.6	2.7	91.9	4.9	84.9	3.9	81.2	3.8
120	Pirimiphos-methyl	50.0	87.1	5.4	85.5	5.2	85.7	8.7	82.3	2.9	85.6	4.7	93.2	5.4	84.0	3.5	85.0	4.7
121	Profenofos	300.0	115.0	29.2	97.6	8.9	121.9	3.2	91.8	8.4	133.1	28.3	85.7	9.8	134.4	8.7	76.0	15.1
122	Profluralin	200.0	88.8	11.7	96.1	2.8	93.8	5.1	85.7	6.8	98.4	10.5	90.0	6.8	97.2	4.7	77.7	6.8
123	Propachlor	150.0	79.1	3.5	92.7	3.2	89.4	5.3	84.7	5.5	80.5	5.0	95.2	3.4	83.6	4.7	91.3	4.6
124	Propiconazole	150.0	89.3	4.7	81.4	11.2	87.1	4.4	97.3	2.4	86.6	5.1	92.8	6.5	85.1	1.9	85.0	6.8
125	Propyzamide	100.0	81.2	3.5	86.6	4.3	80.8	3.50	89.2	4.5	78.7	4.5	88.3	5.1	79.8	5.5	89.0	4.3
126	Ronnel	100.0	85.3	8.0	89.3	7.5	83.1	6.0	89.2	3.7	81.1	5.5	92.0	8.9	85.5	4.9	88.4	6.0
127	Sulfotep	50.0	88.7	3.5	101.8	3.0	90.6	3.2	87.0	4.4	83.6	4.2	90.2	4.0	82.0	4.2	85.1	3.6
128	Tebufenpyrad	50.0	83.5	1.4	96.1	2.5	82.4	3.4	103.0	10.8	82.4	2.1	89.1	3.8	87.4	4.3	97.3	10.5
129	Terbutryn	100.0	81.6	5.6	93.7	5.0	82.5	3.0	84.8	3.1	84.2	4.0	86.6	4.6	85.7	2.5	85.3	1.7
130	Thiobencarb	100.0	84.7	4.4	93.4	3.4	85.9	2.6	87.5	3.7	88.3	2.8	88.9	4.0	85.6	4.9	87.5	3.1
131	Tralkoxydim	400.0	85.8	5.8	88.5	4.5	82.2	8.70	84.0	4.5	81.6	5.1	84.7	5.7	80.6	7.4	83.6	6.2
132	Trans-chlodane	50.0	85.3	2.7	95.6	6.5	87.3	3.3	91.7	4.1								
133	Trans-diallate	100.0	82.3	4.9	94.3	4.0	85.4	3.5	86.5	5.2	87.6	4.2	88.1	3.9	82.7	3.8	83.7	4.3
134	Trifloxystrobin	200.0	83.0	8.4	85.2	4.2	82.8	0.95	89.6	5.1	76.0	7.6	89.5	4.4	78.7	1.5	87.6	5.9
135	Zoxamide	100.0	77.3	5.7	88.7	5.2	81.8	7.48	89.3	3.9	76.3	7.1	92.8	5.9	82.8	8.8	88.2	4.9
136	o,p-DDE	50.0	80.5	2.9	88.1	5.8	81.4	2.0	80.8	7.3	80.2	2.4	88.6	6.6	81.0	2.7	81.0	8.0
137	Ametryn	150.0	81.0	4.4	84.5	5.0	74.5	1.7	76.7	6.6	94.8	2.8	86.1	6.5	82.2	2.1	83.4	5.3
138	Bifenthrin	50.0	84.4	2.1	95.3	5.8	104.7	6.2	103.2	3.5	78.1	6.0	95.1	5.2	82.1	4.2	84.9	4.7

(Continued)

TABLE 2.2.9 Recoveries and Precision Test Results for 201 Pesticides in Green Tea and oolong Tea Extracted per Method-1 and Method-2 (n = 5) (cont.)

No.	Pesticides	Spiked level, μg/kg	GC/MS								GC/MS/MS							
			Green tea				Oolong tea				Green tea				Oolong tea			
			Method-1		Method-2		Method-1		Method-2		Method-1		Method-2		Method-1		Method-2	
			AVE (%)	RSD (%)	AVE (%)	RSD (%)	AVE (%)	RSD (%)	AVE (%)	RSD (%)	AVE (%)	RSD (%)	AVE (%)	RSD (%)	AVE (%)	RSD (%)	AVE (%)	RSD (%)
139	Bitertanol	150.0	81.3	2.0	85.4	5.8	84.0	5.9	80.3	8.6	85.3	3.8	81.2	5.3	84.9	4.1	83.9	9.6
140	Boscalid	200.0	84.1	2.5	100.5	2.3	77.1	3.7	90.3	6.7	78.8	4.0	97.8	1.5	76.7	3.3	85.4	5.4
141	Butafenacil	50.0	83.2	12.9	85.2	3.7	85.8	1.67	92.1	4.6	77.4	12.6	88.9	5.2	78.2	1.7	88.5	5.1
142	Carbaryl	150.0	92.1	3.1	103.0	0.8	88.6	5.1	90.5	5.9	82.2	4.4	101.2	1.7	89.9	8.1	91.5	5.0
143	Chlorobenzilate	50.0	81.2	6.5	89.0	4.6	88.0	4.6	80.1	7.2	85.1	3.2	86.2	6.2	83.1	2.1	82.9	7.3
144	Chlorthal-di-methyl	100.0	81.4	3.3	86.1	4.4	79.6	2.17	89.3	4.6	78.0	3.3	88.7	5.1	79.0	3.8	89.2	4.4
145	Dibutyl succinate	100.0	77.0	6.5	89.7	3.8	73.3	3.85	81.8	7.7	74.0	5.8	91.1	3.8	72.6	3.0	82.8	8.2
146	Diethofencarb	300.0	82.6	3.4	88.3	6.3	85.8	1.9	78.5	6.9	89.3	1.9	84.0	6.4	83.5	2.6	79.1	8.0
147	Diflufenican	50.0	83.4	3.8	85.2	6.3	84.6	3.0	50.6	7.8	87.0	3.5	80.3	6.1	82.6	2.9	52.1	9.0
148	Dimepiperate	100.0	80.9	4.0	89.3	4.4	77.2	4.2	87.7	12.1	83.6	1.7	85.6	5.4	82.6	2.1	82.1	8.0
149	Dimethametryn	50.0	88.1	1.9	99.6	1.4	75.1	2.2	87.1	6.3	81.0	2.3	99.9	1.2	78.4	4.0	86.9	6.1
150	Dimethomorph	100.0	88.9	6.2	109.4	1.9	94.8	8.2	89.3	5.4	75.2	9.1	104.7	1.6	88.3	10.8	88.4	6.4
151	Dimethyl phthalate	200.0	98.8	2.1	101.7	3.4	75.5	4.7	84.6	10.6	80.2	2.9	103.4	3.3	77.5	5.4	87.2	11.1
152	Diniconazole	150.0	80.5	5.2	84.9	5.9	80.6	2.8	66.0	6.9	87.7	2.9	79.7	8.5	81.1	2.7	73.6	8.0
153	Diphenamid	50.0	82.7	3.4	89.1	5.8	83.8	2.9	83.6	6.5								
154	Dipropetryn	50.0	83.2	4.5	86.7	5.6	73.1	6.7	77.0	7.0	85.8	3.5	82.1	7.1	82.3	2.0	80.1	8.3
155	Ethalfluralin	200.0	97.0	7.8	91.3	9.8	86.8	3.7	88.0	4.8	85.3	7.3	90.9	7.1	87.2	6.6	77.9	3.9
156	Etofenprox	50.0	87.3	9.6	93.8	5.7	87.6	5.5	72.0	6.7	92.1	4.5	92.1	5.5	82.6	3.4	75.9	6.6
157	Etridiazol	150.0	89.9	10.2	92.6	18.9	84.9	9.7	74.6	13.1	82.7	6.9	89.8	6.5	84.6	3.2	85.8	2.5
158	Fenazaquin	50.0	83.2	10.5	81.5	6.4	76.9	1.7	52.0	5.0	83.1	2.6	78.0	6.9	83.0	2.5	53.0	5.6
159	Fenchlorphos	200.0	80.9	2.5	96.5	1.8	75.0	2.4	85.8	6.4	79.5	3.3	96.3	1.9	75.9	3.3	84.8	7.7
160	Fenoxanil	100.0	84.4	13.2	105.5	6.0	82.8	8.21	82.9	3.6	77.0	19.2	130.8	14.2	70.8	28.5	75.7	4.6
161	Fenpropidin	100.0	78.0	5.2	81.1	3.8	76.6	2.96	85.2	4.8	75.2	6.4	83.4	4.6	75.2	5.7	84.9	4.5
162	Fenson	50.0	84.4	1.8	87.9	4.0	85.4	7.1	82.8	7.1	81.9	3.8	91.5	5.1	81.4	2.5	81.7	8.0
163	Flufenacet	400.0	87.6	14.0	113.9	4.4	51.8	14.6	113.2	15.9	77.3	14.2	89.3	4.9	62.5	6.4	94.4	6.8
164	Furalaxyl	100.0	84.1	3.5	90.4	4.2	79.8	3.41	91.2	4.6	78.9	4.0	88.7	5.0	78.7	4.4	89.6	4.5
165	Heptachlor	150.0	77.9	1.3	83.2	3.6	76.9	2.8	85.2	6.8	80.4	0.4	83.1	5.1	78.2	1.7	80.5	7.3

No.	Compound																
166	Iprobenfos	150.0	85.7	4.2	88.4	4.7	84.5	1.2	81.4	7.1	94.2	2.4	82.1	7.6	85.0	1.5	85.4	8.7
167	Isazofos	100.0	82.2	3.0	93.6	8.0	87.6	10.4	89.6	2.4	84.7	2.4	87.9	6.6	83.0	3.7	82.5	8.3
168	Isoprothiolane	100.0	82.8	3.1	87.3	4.3	80.5	2.91	89.9	4.5	79.2	4.0	90.0	5.2	80.1	3.8	90.5	5.1
169	Kresoxim-methyl	50.0	78.8	11.0	76.9	5.3	86.0	3.2	80.7	6.6	82.3	8.7	87.8	4.6	77.6	1.1	78.5	9.5
170	Mefenacet	150.0	113.1	8.7	94.1	6.2	77.0	3.2	75.8	9.3	106.2	4.0	78.6	4.8	79.4	2.2	79.4	8.0
171	Mepronil	50.0	80.3	3.7	130.5	35.2	85.2	2.0	76.4	7.7	91.7	3.1	85.8	6.9	85.2	3.3	83.3	9.1
172	Metribuzin	150.0	77.0	4.5	84.4	4.9	86.6	4.1	82.7	9.3	82.5	2.9	81.7	6.3	83.5	3.7	83.0	9.8
173	Molinate	50.0	77.2	5.4	71.0	5.8	75.7	3.4	111.7	7.8	75.5	2.8	65.9	5.8	72.1	4.3	111.6	8.3
174	Napropamide	150.0	82.8	2.4	86.6	14.9	87.8	3.0	92.7	2.7	88.0	8.2	86.9	30.2	87.2	7.9	103.0	23.0
175	Nuarimol	100.0	90.3	5.5	81.4	19.8	88.3	3.3	108.9	2.5	87.0	6.0	91.8	4.5	81.4	2.9	86.8	4.7
176	Permethrin	100.0	87.8	4.1	88.0	6.3	85.1	2.3	76.9	6.4	78.6	3.4	86.0	6.3	81.7	2.4	77.0	7.8
177	Phenothrin	50.0					85.6	1.4	76.0	6.5	70.7	14.0	90.7	4.8	95.2	22.6	77.6	7.6
178	Piperonyl bu-toxide	50.0					84.8	4.4	74.5	5.7	88.1	2.7	85.9	6.8	85.1	2.5	76.5	6.8
179	Pretilachlor	100.0	76.9	2.9	87.2	5.3	81.0	2.2	81.5	7.4	78.0	2.3	86.4	6.0	78.9	1.7	79.9	7.0
180	Prometon	150.0	81.4	3.8	87.3	5.8	81.3	0.9	79.8	7.6	84.0	1.7	86.4	6.7	82.1	1.0	79.3	8.4
181	Pronamide	50.0	85.5	2.5	85.6	13.4	87.8	3.7	99.0	2.4	95.9	10.6	86.0	7.0	84.9	5.0	89.4	4.6
182	Propetamphos	50.0	94.9	6.9	85.5	9.3	88.8	3.7	95.7	2.7	84.6	3.9	89.3	12.2	107.7	3.4	96.1	5.6
183	Propoxur-1	400.0	80.1	4.5	91.3	4.1	81.2	3.66	97.3	3.5	78.7	4.8	92.5	5.6	83.1	8.9	98.2	3.3
184	Propoxur-2	50.0	84.7	13.7	67.9	11.9	116.6	8.43	77.9	9.4	70.1	10.0	63.6	10.4	64.2	29.8	78.9	9.0
185	Prothiophos	50.0	86.5	3.8	93.1	7.9	84.8	5.1	87.8	4.2	90.4	15.0	81.9	10.6	83.7	4.7	64.5	8.2
186	Pyrimethanil	50.0	82.3	4.0	75.1	5.2	81.6	1.0	44.7	5.7	83.9	3.4	71.3	7.6	79.4	2.4	44.5	4.9
187	Pyridaben	50.0	87.4	7.3	88.8	4.2	82.9	1.4	84.6	5.0	88.1	8.3	83.8	5.8	83.1	2.9	80.7	7.9
188	Pyridaphenthion	50.0	88.6	8.0	88.1	9.1	89.8	7.7	111.9	4.8	79.8	6.9	87.1	6.2	83.5	7.5	79.5	4.9
189	Pyriproxyfen	50.0	83.6	3.4	88.1	4.2	81.4	3.91	88.7	4.7	78.1	4.1	91.2	5.3	77.9	2.6	85.9	7.0
190	Quinalphos	50.0	83.1	6.9	88.1	5.8	88.0	4.1	85.1	4.1	98.0	23.6	68.0	39.3	73.7	24.0	78.4	6.4
191	Quinoxyphen	50.0	82.3	3.6	78.3	4.5	79.8	2.98	66.5	3.6	78.3	5.4	79.7	7.5	79.0	4.8	65.1	7.8
192	Telodrin	200.0	86.5	3.2	99.8	1.5	74.6	3.1	87.7	6.2	81.3	2.8	97.0	1.6	75.7	3.8	85.2	6.6
193	Tetrasul	50.0	82.2	2.6	89.8	8.8	86.1	3.8	79.3	3.3	112.2	6.7	105.1	9.1	113.8	5.8	134.1	7.6
194	Thiazopyr	100.0	83.5	3.3	87.5	4.0	80.0	2.75	88.7	4.5	80.8	5.6	93.7	7.0	82.2	6.8	88.3	5.3
195	Tolclofos-methyl	50.0	79.2	2.7	86.7	4.8	82.6	1.9	80.1	6.8	83.8	3.4	85.5	6.0	81.2	3.7	83.4	6.4
196	Transfluthrin	50.0	79.3	3.7	87.0	5.1	80.4	1.6	80.2	7.4	84.1	1.1	85.7	5.7	82.2	2.8	81.5	5.8
197	Tribenuron-methyl	100.0	110.1	6.0	116.3	3.0	51.2	16.0	89.6	14.0	75.7	8.7	107.1	3.9	64.2	11.6	84.6	8.5

(Continued)

TABLE 2.2.9 Recoveries and Precision Test Results for 201 Pesticides in Green Tea and oolong Tea Extracted per Method-1 and Method-2 (n = 5) (cont.)

No.	Pesticides	Spiked level, µg/kg	GC/MS														GC/MS/MS											
			Green tea				Oolong tea								Green tea				Oolong tea						
			Method-1		Method-2		Method-1		Method-2						Method-1		Method-2		Method-1		Method-2	
			AVE (%)	RSD (%)	AVE (%)	RSD (%)	AVE (%)	RSD (%)	AVE (%)	RSD (%)					AVE (%)	RSD (%)	AVE (%)	RSD (%)	AVE (%)	RSD (%)	AVE (%)	RSD (%)
198	Triadimefon	150.0	87.4	6.1	86.0	13.5	90.1	7.8	102.3	4.9					84.6	5.5	92.7	8.4	86.2	3.3	92.7	6.1
199	Triadimenol	100.0	81.7	5.1	82.9	5.8	82.5	3.2	80.3	8.5					84.6	3.5	83.2	5.7	82.8	2.4	85.7	9.5
200	Triallate	50.0	79.6	4.7	85.2	5.0	80.6	2.2	80.0	6.4					80.3	1.5	85.3	4.6	80.8	1.5	83.0	6.9
201	Vinclozolin	50.0	87.3	4.0	89.5	7.1	87.7	3.7	94.0	3.1												

TABLE 2.2.10 Statistical Analysis of Average Recoveries and RSD for 201 Pesticides in Green Tea and Oolong Tea by Two Sample Preparation Methods

No.	Range	GC/MS								GC/MS/MS							
		Green tea				Oolong tea				Green tea				Oolong tea			
		Method-1		Method-2		Method-1		Method-2		Method-1		Method-2		Method-1		Method-2	
		Number	Percentage (%)	Number	Percentage (%)	Number	Percentage (%)	Number	Percentage (%)	Number	Percentage (%)	Number	Percentage (%)	Number	Percentage (%)	Number	Percentage (%)
Ave																	
1	<60%	0	0	0	0	2	1.0	5	2.5	1	0.5	1	0.5	0	0	6	3.0
2	60%–70%	2	1.0	3	1.5	3	1.5	5	2.5	2	1.0	8	4.0	10	5.0	6	3.0
3	70%–80%	31	15.4	11	5.5	44	21.9	21	10.4	63	31.3	15	7.5	49	24.4	27	13.4
4	80%–90%	134	66.7	82	40.8	121	60.2	93	46.3	103	51.2	100	49.8	116	57.7	104	51.7
5	90%–100%	24	11.9	81	40.3	18	9.0	43	21.4	19	9.5	57	28.4	9	4.5	36	17.9
6	100%–110%	2	1.0	16	8.0	5	2.5	23	11.4	1	0.5	10	5.0	3	1.5	9	4.5
7	110%–120%	5	2.5	4	2.0	4	2.0	7	3.5	3	1.5	2	1.0	4	2.0	3	1.5
8	>120%	0	0	1	0.5	1	0.5	2	1.0	2	1.0	1	0.5	3	1.5	3	1.5
	Total 70%–110%	191	95.0	190	94.5	188	93.5	180	89.6	186	92.5	182	90.5	177	88.1	176	87.6
RSD																	
9	<10%	173	86.1	176	87.6	186	92.5	153	76.1	170	84.6	171	85.1	180	89.6	151	75.1
10	10%–15%	21	10.4	16	8.0	3	1.5	39	19.4	13	6.5	12	6.0	2	1.0	32	15.9
11	15%–20%	2	1.0	3	1.5	7	3.5	3	1.5	4	2.0	3	1.5	3	1.5	8	4.0
12	>20%	2	1.0	3	1.5	2	1.0	4	2.0	7	3.5	8	4.0	9	4.5	3	1.5
13	N.D	3	1.5	3	1.5	3	1.5	2	1.0	7	3.5	7	3.5	7	3.5	7	3.5

recovery <70%, Method-2, 5.5%; Method-1 for green tea makes up 93.8% for pesticide within a good recovery range of 70%–110%, Method-2, 92.5%, while Method-1 for oolong tea accounts for 90.8%, Method-2, 88.6%; it is evident that no big differences exist for the recoveries between Method-1 and Method-2, yet Method-1 turns out slightly better.

The statistical analysis of RSD values by GC/MS/MS and GC/MS determination in Table 2.2.9 is tabulated in Table 2.2.10, from which it can be seen that regarding green tea samples the number of pesticides with RSD < 15% with Method-1 and Method-2 accounts for 93.8% and 93.3%, respectively, among which the number of pesticides with RSD < 10% accounts for 85.4% and 86.3%, respectively, with hardly any differences between the two methods; regarding oolong tea samples, the number of pesticides with RSD values <15% by Method-1 and Method-2 accounts for 92.3% and 93.3%, respectively, among which the number of pesticides with RSD < 10% accounts for 91.0% and 75%, respectively, with no marked differences between the two methods, except that the number of pesticides with RSD < 10% is slightly greater by Method-1.

2.2.8 STAGE III: EVALUATION ON THE EXTRACTION EFFICIENCIES OF GREEN TEA AND OOLONG TEA SAMPLE THE INCURRED 201 PESTICIDES AFTER 165 DAYS BY METHOD-1 AND METHOD-2

In order to further evaluate the extraction efficiencies of Method-1 and Method-2, the comparative experiment conducted on the incurred green tea and oolong tea samples. The incurred green tea and oolong tea samples used here were derived from an earlier section, the preparing pesticides incurred tea samples, which were kept at room temperature for 165 days from this experiment. Sample preparations were carried out per Method-1 and Method-2, respectively, each paralleling 5 samples by both GC/MS and GC/MS/MS determination to obtain a total 8040 data (2 methods × 2 kinds of tea samples × 5 parallel samples × 201 pesticides × 2 instrumental detections) and to calculate the average values and RSD of the content obtained more than 5 times. Analysis results are shown in Table 2.2.11.The statistical analysis of RSD and the average value ratios from the content derived from Method-1 and Method-2 in Table 2.2.11 are tabulated in Tables 2.2.12 and Tables 2.2.13.

2.2.8.1 Test Data Analysis

It can be seen from the test data in Tables 2.2.11 and Tables 2.2.12 that the content values of the pesticides determined by Method-1 are generally higher than those by Method-2 no matter whether it is green tea or oolong tea and regardless of GC/MS or GC/MS/MS, 30%–50% higher on the average. Moreover, the number of pesticides of the content higher than that of Method-2 determined by Method-1 (ratio > 1) accounts for 95.5%–98%, with the ratio range distribution mostly falling 1.10–1.70, which suggests that Method-1 is capable of extracting more residual pesticides from the precipitated 165-day incurred samples. The reason can be traced to the fact that Method-1 adopts high-speed homogeneous extraction, which has higher efficiency than the oscillating extraction by Method-2, leading to the higher extraction efficiency for Method-1 than Method-2. At this point, it can be said that the analytical results through sample preparation by Method-1 are slightly more accurate.

It can be seen from the data of RSD ≤ 10% in Table 2.2.13 that the precision of Method-1 is slightly better than that of Method-2 for green tea samples; the precision of Method-2 is slightly better for oolong tea. Viewed from the data of RSD ≤ 15% in Table 2.2.13, the ratios made by RSD ≤ 15% in the analytical results from both the methods are greater than 95%, proving that both methods have gained good reproducibility and are able to meet the requirements for residue analysis.

2.2.8.2 Test Phenomenon Analysis

The sample solution colors processed by Method-1 and Method-2 are shown in Fig. 2.2.3, from which it can be seen that Method-1 has a lighter color than Method-2 no matter whether it is green tea or oolong tea, and the green tea and oolong tea by Method-1 are of identical color—almost colorless, while Method-2 hasn't reached the ideal results in removing the pigments in green tea, which will produce contamination against the instrumental separation system over long-term injection of great consignments of samples.

TABLE 2.2.11 Comparative Test Results of the Content of 201 Pesticides in the Incurred Green Tea and Oolong Tea Samples by Method-1 and Method-2 (n = 5)

No.	Pesticides	GC/MS										GC/MS/MS									
		Green tea					Oolong tea					Green tea					Oolong tea				
		Method-1		Method-2		Ratio[a]	Method-1		Method-2		ratio[a]	Method-1		Method-2		Ratio[a]	Method-1		Method-2		Ratio[a]
		Con-tents	RSD (%)	Con-tents	RSD (%)		RSD (%)	Con-tents	RSD (%)	Con-tents		Con-tents	RSD (%)	Con-tents	RSD (%)		RSD (%)	Con-tents	RSD (%)	Con-tents	
1	2,3,4,5-TEtrachloroaniline	0.31	3.4	0.25	5.9	1.21	0.32	6.3	0.19	5.3	1.67	0.19	10.6	0.17	5.2	1.12	0.27	17.3	0.17	8.8	1.56
2	2,3,5,6-Tetrachloroaniline	23.5	4.3	18.6	7.2	1.27	15.3	6.1	14.6	3.9	1.05	23.4	4.3	18.2	9.6	1.28	25.7	3.6	18.2	5.3	1.41
3	4,4-Dibromobenzophenone	33.9	5.9	26.9	10.4	1.26	33.8	8.6	16.9	5.6	2.01	30.8	3.2	24.2	8.3	1.27	31.1	6.6	19.5	3.9	1.59
4	4,4-Dichlorobenzophenone	32.8	5.1	26.0	9.7	1.26	48.5	10.3	34.1	4.4	1.42	31.9	3.4	25.4	7.2	1.26	49.8	10.0	36.1	4.8	1.38
5	Acetochlor	66.8	6.4	59.3	6.4	1.13	72.3	7.2	38.3	5.2	1.89	75.6	3.5	59.0	5.4	1.28	74.6	6.4	48.5	3.9	1.54
6	Alachlor	73.1	5.7	63.7	5.8	1.15	76.6	4.5	47.0	5.8	1.63	72.7	4.2	62.0	6.4	1.17	71.8	5.4	48.7	2.6	1.48
7	Atratone	36.8	6.6	32.4	9.0	1.14	36.0	7.6	23.0	3.6	1.56	43.6	2.9	35.3	8.1	1.23	42.8	4.1	30.7	4.4	1.39
8	Benodanil	95.0	4.7	86.5	12.9	1.10	71.7	44.3	68.8	5.3	1.04	84.8	5.3	78.0	12.2	1.09	65.2	67.3	69.3	3.6	0.94
9	Benoxacor	55.0	13.3	46.3	4.1	1.19	49.8	7.6	39.6	4.7	1.26	52.9	4.1	41.2	4.9	1.28	45.6	7.1	38.7	6.3	1.18
10	Bromophos-ethyl	72.8	7.2	57.2	7.9	1.27	72.9	8.4	41.2	3.5	1.77	71.7	4.2	56.7	7.9	1.26	73.7	4.6	42.4	3.7	1.74
11	Butralin	83.0	12.0	59.5	7.0	1.39	78.9	7.2	52.3	4.6	1.51	72.6	6.7	51.9	6.3	1.40	70.5	9.7	47.0	5.6	1.50
12	Chlorfenapyr	224.6	4.5	192.3	8.5	1.17	243.2	5.1	158.4	3.7	1.54	231.7	3.6	198.5	7.7	1.17	251.2	2.8	173.9	6.0	1.44
13	Clomazone	29.5	7.8	24.7	5.5	1.20	26.0	0.9	18.9	2.7	1.37	28.9	4.1	23.9	4.9	1.21	27.3	6.5	19.4	4.3	1.41
14	Cycloate	44.1	10.9	34.6	9.8	1.27	40.8	2.2	27.5	3.2	1.48	43.4	6.8	35.0	9.3	1.24	42.1	9.4	29.7	2.9	1.42
15	Cycluron	96.8	3.5	86.5	5.7	1.12	94.3	9.3	53.3	3.2	1.77	63.3	11.5	68.3	4.5	0.93	80.9	14.4	54.7	3.0	1.48
16	Cyhalofop-butyl	65.3	7.2	50.7	9.9	1.29	43.1	1.7	29.8	3.8	1.44	59.6	5.1	49.5	8.5	1.20	45.9	3.3	31.6	4.1	1.45
17	Cyprodinil	24.6	9.7	16.8	13.8	1.46	24.3	7.9	10.2	6.9	2.39	23.1	4.2	16.9	12.2	1.37	24.3	3.1	9.3	6.1	2.61
18	Dacthal	25.1	6.8	18.4	7.7	1.36	24.2	9.8	15.2	3.2	1.59	25.2	4.4	18.2	7.4	1.39	24.4	5.5	16.5	3.1	1.48
19	de-PCB101	24.6	5.9	20.0	8.7	1.23	25.9	8.4	14.7	4.9	1.76	24.0	3.8	18.9	7.3	1.27	24.7	6.3	15.4	3.2	1.61
20	de-PCB118	26.3	5.3	21.4	7.9	1.23	24.6	8.9	15.5	1.8	1.58	24.9	3.8	20.5	7.1	1.22	26.3	4.8	16.3	3.5	1.61
21	de-PCB138	24.5	6.8	19.6	9.5	1.25	24.8	8.3	14.6	4.3	1.70	24.1	5.7	18.8	8.6	1.29	24.8	5.8	15.1	2.0	1.64
22	de-PCB180	23.7	5.9	18.8	11.9	1.26	22.6	1.4	13.7	3.9	1.65	22.4	4.6	18.3	9.5	1.23	25.3	3.3	14.0	4.1	1.81
23	de-PCB28	22.3	8.2	20.0	7.0	1.11	27.7	9.5	15.9	4.9	1.74	27.1	3.5	22.9	7.7	1.18	28.3	5.0	18.6	2.7	1.52
24	de-PCB31	28.7	5.0	24.3	8.4	1.18	35.9	8.6	20.4	1.9	1.76	27.1	3.5	22.9	7.7	1.18	28.3	5.0	18.6	2.7	1.52

(Continued)

TABLE 2.2.11 Comparative Test Results of the Content of 201 Pesticides in the Incurred Green Tea and Oolong Tea Samples by Method-1 and Method-2 (n = 5) (cont.)

No.	Pesticides	GC/MS Green tea M-1 Contents	M-1 RSD(%)	M-2 Contents	M-2 RSD(%)	Ratio[a]	GC/MS Oolong tea M-1 Contents	M-1 RSD(%)	M-2 Contents	M-2 RSD(%)	Ratio[a]	GC/MS/MS Green tea M-1 Contents	M-1 RSD(%)	M-2 Contents	M-2 RSD(%)	Ratio[a]	GC/MS/MS Oolong tea M-1 Contents	M-1 RSD(%)	M-2 Contents	M-2 RSD(%)	Ratio[a]
25	Dichlorofop-methyl	57.3	8.9	41.9	9.4	1.37	1.9	42.8	3.5	29.8	1.44	54.1	7.2	40.6	5.5	1.33	6.9	48.3	5.7	31.8	1.52
26	Dimethenamid	23.6	4.5	20.3	6.7	1.17	7.4	25.3	2.8	15.9	1.60	23.4	3.0	20.0	7.3	1.17	4.6	25.1	3.3	16.1	1.56
27	Diofenolan-1	84.9	6.8	74.4	9.7	1.14	3.5	79.9	3.1	53.0	1.51	82.2	3.8	71.1	7.7	1.16	7.6	86.2	2.4	57.8	1.49
28	Diofenolan-2	87.5	7.4	77.3	9.5	1.13	3.5	79.7	7.7	58.2	1.37	82.2	3.8	73.1	8.2	1.12	7.6	86.1	4.3	57.0	1.51
29	Fenbuconazole	48.7	7.0	55.0	13.7	0.89	13.9	53.7	5.8	40.0	1.34	47.2	8.5	51.3	12.6	0.92	10.5	55.0	5.6	43.1	1.28
30	Fenpyroximate	174.6	9.0	197.5	15.3	0.88	10.8	220.0	8.1	118.4	1.86	184.2	8.8	191.2	12.5	0.96	6.7	202.1	8.7	124.7	1.62
31	Fluotrimazole	48.1	5.4	35.4	11.4	1.36	7.9	43.9	4.8	27.2	1.61	41.4	6.2	34.6	9.4	1.19	7.1	70.6	5.2	29.0	2.44
32	Fluroxypr-1-methylheptylester	9.3	6.7	7.2	9.9	1.28	9.4	8.9	3.9	5.5	1.61	9.0	7.7	7.0	7.6	1.28	8.8	8.6	10.3	5.4	1.58
33	Iprovalicarb-1	144.7	5.8	124.0	8.9	1.17	9.1	145.4	4.2	109.0	1.33	140.0	5.1	122.3	9.7	1.15	5.5	155.9	1.7	110.8	1.41
34	Iprovalicarb-2	138.6	6.9	122.6	10.1	1.13	1.7	135.8	4.8	112.0	1.21	140.0	5.1	120.1	8.5	1.17	4.5	147.2	5.0	110.1	1.34
35	Isodrin	22.3	6.1	17.9	10.8	1.24	9.4	23.3	4.6	13.3	1.75	22.9	6.1	17.1	8.5	1.34	5.0	22.4	3.5	12.8	1.75
36	Isoprocarb-1	83.1	4.8	76.5	10.9	1.09	9.6	91.0	5.2	53.6	1.70	77.8	3.3	71.7	9.2	1.08	1.3	102.6	1.1	104.4	0.98
37	Isoprocarb-2	102.0	13.5	73.6	4.4	1.39	16.7	69.0	6.4	63.5	1.09	128.8	3.8	100.8	8.3	1.28	14.3	61.4	7.2	62.7	0.98
38	Lenacil	33.4	12.3	25.0	9.9	1.33	9.0	23.3	12.7	23.8	0.97	51.5	14.5	234.2	3.5	0.22	6.2	28.3	6.3	28.1	1.01
39	Metalaxyl	46.0	5.1	39.8	8.9	1.15	7.6	47.3	4.3	31.9	1.48	42.9	3.7	37.6	8.7	1.14	3.4	45.9	3.2	32.0	1.43
40	Metazachlor	73.0	4.9	71.9	6.3	1.02	4.3	77.4	6.4	50.3	1.54	70.6	2.4	64.8	5.1	1.09	2.0	75.5	4.7	57.1	1.32
41	Methabenzthiazuron	476.7	6.5	380.7	9.8	1.25	8.9	454.7	4.7	253.7	1.25	465.9	4.4	376.7	9.4	1.24	3.6	473.6	4.2	269.0	1.76
42	Mirex	18.9	6.5	12.8	6.4	1.48	7.9	18.9	4.4	11.0	1.48	18.7	4.5	12.6	5.5	1.48	5.3	18.6	5.4	13.2	1.41
43	Monalide	79.7	11.5	87.1	9.9	0.92	7.5	67.6	5.1	47.7	1.42	65.5	6.1	55.6	8.9	1.18	4.8	65.8	6.4	42.4	1.55
44	Paraoxon-ethyl	211.4	4.9	192.8	3.6	1.10	2.9	141.4	6.1	169.2	0.84										
45	Pebulate	72.3	11.4	50.1	9.3	1.44	8.8	55.7	6.5	45.0	1.24	64.2	11.5	49.6	11.4	1.30	4.8	54.9	5.9	51.0	1.08
46	Pentachloroaniline	25.5	5.5	18.8	11.0	1.36	7.2	18.6	5.1	10.8	1.72	23.8	3.7	17.7	10.2	1.34	7.5	25.7	4.4	12.5	2.05
47	Pentachloroanisole	48.4	7.3	39.2	9.5	1.23	3.9	47.2	3.7	29.9	1.58	48.1	5.3	38.0	10.1	1.26	6.5	51.4	3.4	30.6	1.68

No.	Compound																				
48	Pentachlorobenzene	19.8	11.7	14.9	12.7	1.33	18.1	6.3	12.4	7.2	1.46	19.8	9.7	14.7	12.5	1.35	19.9	5.1	15.9	5.4	1.25
49	Perthane	23.9	6.4	18.9	8.8	1.27	24.0	7.9	14.4	3.3	1.67	23.3	5.7	18.4	8.4	1.27	24.3	4.5	14.8	3.1	1.64
50	Phenanthrene	29.5	5.4	21.7	10.2	1.36	30.5	9.7	15.7	3.7	1.94	29.9	4.5	21.6	10.2	1.39	29.5	4.2	17.8	8.0	1.66
51	Pirimicarb	48.3	7.5	40.5	8.4	1.19	48.9	9.5	31.5	3.1	1.55	47.6	5.6	40.9	9.6	1.16	50.4	5.7	32.9	3.6	1.53
52	Procymidone	29.9	7.1	25.8	8.9	1.16	30.9	9.0	20.1	3.3	1.54	29.4	4.6	25.0	6.7	1.17	30.6	4.2	20.2	2.8	1.51
53	Prometryne	36.9	9.7	31.4	8.1	1.18	36.2	4.8	23.2	5.2	1.56	35.5	5.8	29.9	8.1	1.19	37.5	4.9	23.7	4.2	1.58
54	Propham	62.1	6.3	54.8	7.9	1.13	59.0	3.3	50.0	5.0	1.18	61.3	4.9	54.0	8.1	1.13	64.6	6.0	48.4	3.7	1.34
55	Prosulfocarb	71.8	4.8	63.1	7.9	1.14	78.0	8.8	46.1	2.1	1.69	76.8	4.1	61.4	7.8	1.25	74.1	6.0	49.5	2.8	1.50
56	Secbumeton	45.3	6.3	39.3	9.5	1.15	47.0	5.7	28.5	4.3	1.65	45.1	4.9	37.8	10.0	1.20	44.8	5.1	30.5	4.6	1.47
57	Silafluofen	30.4	8.0	23.7	16.4	1.28	23.7	4.0	14.5	4.2	1.63	24.9	4.7	21.3	13.6	1.17	27.3	6.2	15.7	7.4	1.74
58	Tebupirimfos	54.9	8.0	42.0	9.3	1.31	52.0	2.4	30.5	3.0	1.71	55.3	7.1	41.3	10.7	1.34	55.5	6.2	31.5	4.5	1.76
59	Tebutam	47.5	6.9	39.8	7.9	1.19	48.7	9.9	29.4	3.3	1.66	47.7	3.0	38.1	7.5	1.25	49.0	6.1	31.3	3.6	1.56
60	Tebuthiuron	110.2	5.5	95.6	11.9	1.15	111.5	7.3	74.4	4.0	1.50	109.2	3.2	93.3	11.3	1.17	118.5	4.6	80.4	4.3	1.47
61	Tefluthrin	22.4	4.2	16.6	9.2	1.34	23.2	7.4	12.7	4.1	1.83	20.8	5.2	15.4	8.0	1.35	21.5	6.5	12.6	3.2	1.71
62	Thenylchlor	60.4	2.4	52.9	7.8	1.14	46.7	1.8	39.7	10.2	1.18	52.2	4.9	49.2	5.7	1.06	52.3	1.9	44.0	3.7	1.19
63	Thionazin	69.9	6.3	58.5	8.8	1.19	64.7	5.1	51.0	3.2	1.27	70.1	5.0	59.5	7.0	1.18	69.0	5.7	49.9	4.3	1.38
64	Trichloronat	47.6	6.9	37.5	8.1	1.27	45.7	1.1	27.6	3.5	1.65	47.6	5.6	38.0	7.4	1.25	49.7	5.6	29.2	3.6	1.70
65	Trifluralin	45.7	8.0	32.1	8.0	1.42	45.7	3.6	26.8	3.1	1.70	43.7	5.7	28.3	10.7	1.54	41.4	4.0	25.4	4.6	1.63
66	2,4'-DDT	48.4	11.8	33.0	6.0	1.47	42.3	14.8	27.6	8.3	1.54	1.8	10.9	0.9	13.6	1.91	1.4	6.2	1.0	9.5	1.45
67	4,4'-DDE	28.8	4.5	21.6	8.0	1.33	27.5	4.3	17.0	3.0	1.62	30.1	4.1	22.3	9.5	1.35	27.0	3.7	17.3	3.3	1.56
68	Benalaxyl	26.0	9.1	20.1	9.0	1.29	22.5	1.3	14.8	2.4	1.52	26.2	9.2	20.2	10.0	1.30	23.2	4.3	15.5	5.2	1.49
69	Benzoylprop-ethyl	83.1	7.7	65.6	9.5	1.27	75.2	3.2	47.4	3.2	1.59	84.1	8.8	66.1	11.8	1.27	75.9	6.9	50.6	4.3	1.50
70	Bromofos	49.0	9.5	39.1	3.2	1.25	43.0	2.4	27.4	3.3	1.57	52.5	9.3	41.7	7.2	1.26	41.8	7.1	28.6	4.9	1.46
71	Bromopropylate	58.9	11.4	43.4	10.7	1.36	52.2	10.8	36.7	8.1	1.42	60.8	9.5	43.8	11.2	1.39	53.2	6.1	33.6	3.9	1.58
72	Buprofenzin	51.7	5.2	35.8	11.8	1.44	58.9	15.2	31.4	7.4	1.88	52.9	6.2	39.1	11.4	1.35	48.3	7.7	31.1	4.8	1.56
73	Butachlor	46.7	12.3	35.8	6.1	1.30	45.1	13.4	26.3	5.0	1.72	27.9	8.8	19.4	15.9	1.44	53.4	2.8	34.5	4.8	1.55
74	Butylate	86.2	12.0	67.6	17.9	1.27	65.7	40.0	61.4	12.2	1.07	59.6	8.4	37.7	14.3	1.58	40.2	9.2	38.6	10.7	1.04
75	Carbofenothion	21.5	12.6	17.1	7.4	1.26	18.7	2.3	12.3	3.6	1.52	19.9	9.9	15.6	4.6	1.28	17.4	17.5	11.1	5.8	1.57
76	Chlorfenson	79.0	6.0	52.3	4.7	1.51	53.1	10.2	40.9	2.8	1.30	79.6	6.8	54.4	4.9	1.46	55.7	8.0	40.7	4.6	1.37
77	Chlorfenviphos	78.2	10.9	54.9	6.8	1.42	72.0	7.8	47.6	6.1	1.51	81.9	8.7	61.1	7.8	1.34	63.6	8.1	46.1	3.0	1.38
78	Chlormephos	54.2	7.6	39.1	11.7	1.39	40.6	13.4	35.2	11.3	1.15	54.5	9.8	39.3	14.3	1.39	38.3	7.0	38.1	7.2	1.01

(Continued)

TABLE 2.2.11 Comparative Test Results of the Content of 201 Pesticides in the Incurred Green Tea and Oolong Tea Samples by Method-1 and Method-2 (n = 5) (cont.)

No.	Pesticides	GC/MS												GC/MS/MS										
		Green tea						Oolong tea						Green tea						Oolong tea				
		Method-1		Method-2		Ratio[a]	Method-1		Method-2		ratio[a]		Method-1		Method-2		Ra-tio[a]	Method-1		Method-2		Ratio[a]		
		Con-tents	RSD (%)	Con-tents	RSD (%)		RSD (%)	Con-tents	RSD (%)	Con-tents			Con-tents	RSD (%)	Con-tents	RSD (%)		RSD (%)	Con-tents	RSD (%)	Con-tents			
79	Chlorneb	31.8	4.1	26.3	8.2	1.21	28.7	5.8	23.2	4.7	1.24		31.2	8.6	26.1	10.3	1.19	25.4	3.0	22.1	3.9	1.15		
80	Chlorpropham	125.0	9.8	78.2	8.5	1.60	49.6	2.3	80.0	3.5	0.62		69.9	8.4	57.6	7.8	1.21	61.4	7.3	43.2	4.7	1.42		
81	Chlorpropylate	33.4	8.5	28.0	6.7	1.19	29.4	1.8	21.7	3.9	1.36		63.0	7.9	39.8	9.4	1.58	53.7	4.3	15.7	3.6	3.43		
82	Chlorpyrifos(ethyl)	77.2	10.1	57.1	6.1	1.35	64.9	1.0	39.2	3.1	1.65		77.5	8.5	59.8	9.1	1.30	68.0	9.2	41.7	5.1	1.63		
83	Chlorthiophos	54.7	10.2	40.9	7.2	1.34	45.1	0.9	26.2	4.0	1.72		54.4	8.0	42.4	9.4	1.28	44.7	9.6	27.8	4.8	1.60		
84	Cis-chlordane	54.1	7.1	38.0	7.5	1.42	47.3	2.0	27.7	3.4	1.71		53.0	7.1	38.8	8.9	1.36	50.4	7.5	29.3	2.4	1.72		
85	Cis-diallate	44.7	9.4	30.8	8.3	1.45	38.7	3.3	22.9	3.4	1.69		45.0	8.7	33.6	8.7	1.34	39.7	9.1	25.6	4.9	1.55		
86	Cyanofenphos	60.4	9.8	45.0	5.7	1.34	47.9	1.6	32.8	2.5	1.46		58.5	10.2	46.4	8.4	1.26	46.0	0.3	33.4	3.6	1.38		
87	Desmetryn	35.2	6.6	29.0	7.1	1.21	32.6	9.9	19.9	3.4	1.64		34.9	5.6	28.0	10.0	1.25	30.8	6.8	19.9	4.8	1.55		
88	Dichlobenil	5.9	9.1	4.6	11.5	1.28	4.3	13.3	4.3	13.9	0.98		6.1	8.2	4.4	11.8	1.37	4.2	16.2	4.8	12.7	0.87		
89	Dicloran	65.8	6.4	42.4	9.4	1.55	50.1	6.7	34.5	4.5	1.46		61.0	5.5	43.4	7.4	1.40	50.4	4.3	34.4	4.5	1.46		
90	Dicofol	61.5	4.9	52.7	9.7	1.17	104.8	9.9	74.7	4.0	1.40		61.6	3.9	53.5	11.3	1.15	95.4	7.9	72.5	4.2	1.32		
91	Dimethachlor	73.8	8.6	59.9	5.2	1.23	67.8	3.3	44.8	3.8	1.51		73.2	6.5	61.4	7.7	1.19	66.8	6.2	45.6	2.8	1.46		
92	Dioxaccarb	182.0	8.1	198.8	9.2	0.92	195.5	13.9	152.6	4.9	1.28		191.8	7.4	203.5	14.6	0.94	207.6	10.5	158.3	3.9	1.31		
93	Endrin	402.0	11.0	293.6	9.0	1.37	336.0	7.6	205.3	6.5	1.64		399.1	7.5	284.1	11.6	1.40	343.5	9.0	224.7	3.0	1.53		
94	Epoxiconazole-2	181.1	7.2	148.5	9.8	1.22	157.4	4.0	111.2	6.1	1.42		178.2	8.0	150.9	13.1	1.18	159.6	7.6	117.5	4.1	1.36		
95	EPTC	63.4	9.5	44.5	16.7	1.43	39.6	14.9	45.0	12.0	0.88		65.9	3.4	40.6	16.0	1.62	41.1	15.3	49.9	7.6	0.82		
96	Ethofumesate	67.0	9.6	54.2	7.0	1.24	58.7	4.1	38.6	2.5	1.52		67.2	10.1	54.9	9.0	1.22	57.8	2.9	39.9	3.1	1.45		
97	Ethoprophos	73.6	9.3	57.7	7.4	1.27	64.3	2.6	42.6	3.1	1.51		74.6	9.3	57.8	9.0	1.29	64.5	9.9	43.5	4.0	1.48		
98	Etrimfos	41.5	10.3	32.4	6.2	1.28	35.6	2.8	22.6	2.4	1.58		43.4	9.8	33.3	9.0	1.30	35.6	1.6	23.5	3.9	1.51		
99	Fenamidone	27.1	6.6	23.3	11.3	1.17	26.6	9.1	17.9	5.1	1.48		27.0	6.2	23.2	10.8	1.16	26.1	3.9	19.0	6.6	1.38		
100	Fenarimol	44.0	6.7	39.4	13.3	1.11	41.4	6.4	31.3	4.2	1.32		45.3	5.8	37.8	12.1	1.20	42.8	5.0	29.1	4.3	1.47		
101	Flamprop-iso-propyl	28.1	6.7	25.1	10.6	1.12	27.8	11.8	17.4	7.7	1.60		28.2	6.7	22.1	11.7	1.28	24.9	7.4	16.3	4.2	1.53		
102	Flamprop-methyl	29.7	8.4	23.0	10.2	1.29	25.7	3.2	16.6	3.8	1.55		28.2	6.7	22.1	11.7	1.28	24.9	7.4	16.3	4.2	1.53		
103	Fonofos	33.8	10.3	27.3	6.8	1.24	29.6	2.8	19.1	3.2	1.55		35.0	10.0	28.0	8.0	1.25	30.2	11.0	20.0	3.4	1.51		
104	Hexachloroenzene	18.8	14.8	13.7	12.4	1.38	12.1	8.7	8.4	4.8	1.44		19.3	14.6	13.6	14.2	1.41	12.4	7.2	9.0	5.0	1.38		

105	Hexazinone	80.5	11.9	69.0	14.2	1.17	69.9	23.2	35.1	19.5	1.99	67.6	11.6	81.5	13.3	0.83	71.8	12.4	60.4	5.8	1.19
106	Iodofenphos	77.5	11.3	59.4	3.4	1.31	69.7	3.9	41.7	4.4	1.67	77.8	9.3	62.5	4.6	1.24	61.6	7.0	43.6	2.8	1.41
107	Isofenphos	70.7	7.6	52.7	8.5	1.34	59.1	1.7	37.5	2.5	1.58	69.3	9.1	53.2	9.8	1.30	58.5	2.5	38.3	2.9	1.53
108	Isopropalin	48.4	7.4	33.2	9.4	1.46	43.5	7.4	24.2	4.7	1.80	44.7	8.5	30.2	11.4	1.48	40.9	5.6	22.4	4.0	1.82
109	Methoprene	80.4	13.0	57.3	10.5	1.40	66.6	2.3	36.0	2.7	1.85	85.2	7.3	59.5	10.8	1.43	67.5	4.4	38.2	7.5	1.77
110	Methoprotryne	64.2	7.1	52.0	9.2	1.23	57.0	2.5	37.1	2.8	1.54	63.8	7.6	52.7	10.9	1.21	58.6	5.6	39.6	3.8	1.48
111	Methoxychlor	31.4	7.7	20.6	7.5	1.53	23.8	18.2	16.4	13.8	1.44	31.8	4.2	24.3	13.6	1.31	29.5	10.8	17.6	5.4	1.68
112	Methyl-parathion	90.9	13.0	67.9	2.8	1.34	76.4	6.8	50.5	6.5	1.51	92.0	10.4	66.1	7.4	1.39	69.0	4.9	52.1	5.1	1.32
113	Metolachlor	26.2	9.1	20.6	7.6	1.27	22.0	7.9	14.5	2.8	1.52	25.4	7.6	19.8	10.7	1.29	23.0	7.4	14.4	3.8	1.59
114	Nitrapyrin	83.2	8.9	57.5	7.3	1.45	53.7	29.2	51.0	8.3	1.05	93.6	9.2	61.6	10.3	1.52	57.3	9.3	62.7	9.6	0.91
115	Oxyfluofen	126.1	7.4	75.7	10.3	1.66	94.9	7.3	57.2	11.0	1.66	120.7	10.8	74.5	12.6	1.62	93.8	8.9	60.0	8.5	1.56
116	Pendimethalin	89.4	9.4	57.5	9.6	1.55	76.5	8.4	43.3	7.4	1.77	83.5	9.3	55.1	10.5	1.52	68.4	4.1	43.7	6.0	1.56
117	Picoxystrobin	57.6	7.6	43.6	9.3	1.32	49.9	1.7	31.4	2.9	1.59	58.2	7.4	44.7	9.8	1.30	51.6	6.4	32.3	4.2	1.59
118	Piperophos	76.9	11.1	61.0	9.4	1.26	69.3	0.9	44.0	2.3	1.57	80.6	8.6	63.4	8.9	1.27	71.4	9.7	47.2	6.2	1.51
119	Pirimiphos-ethyl	37.1	9.7	27.2	8.5	1.36	31.9	2.2	18.4	2.6	1.73	37.3	5.0	27.8	9.4	1.34	31.1	1.1	19.7	5.2	1.58
120	Pirimiphos-methyl	49.5	9.4	38.3	6.9	1.29	43.1	2.2	26.1	3.1	1.65	51.0	8.6	39.9	9.3	1.28	44.3	3.0	27.7	5.1	1.60
121	Profenofos	224.3	9.1	146.2	11.1	1.53	138.3	9.1	88.1	11.2	1.57	209.4	10.4	110.5	7.3	1.90	90.6	8.1	79.2	5.6	1.14
122	Profluralin	99.2	8.2	65.6	7.8	1.51	88.2	6.7	48.1	6.2	1.83	93.9	6.7	58.9	10.0	1.60	80.1	4.4	47.2	4.6	1.70
123	Propachlor	69.9	3.5	56.0	4.6	1.25	59.5	8.3	39.5	4.1	1.51	68.6	8.2	58.8	6.9	1.17	60.7	7.1	43.8	3.6	1.39
124	Propiconazole	90.5	7.7	69.8	11.2	1.30	76.3	2.8	51.5	3.8	1.48	86.6	5.2	72.1	11.7	1.20	79.4	8.1	60.5	20.4	1.31
125	Propyzamide	58.5	7.9	47.4	7.0	1.23	50.3	7.4	35.6	2.8	1.41	59.6	8.4	47.1	7.9	1.26	49.3	6.8	37.3	4.2	1.32
126	Ronnel	46.9	8.1	34.5	3.8	1.36	41.7	2.2	27.7	2.8	1.51	49.6	6.0	35.7	5.9	1.39	42.3	5.7	29.8	4.0	1.42
127	Sulfotep	21.2	9.8	16.4	7.8	1.30	17.9	2.8	11.5	3.1	1.55	21.4	8.5	16.5	8.8	1.30	18.4	2.9	12.0	4.4	1.53
128	Tebufenpyrad	73.7	8.2	58.8	11.0	1.25	66.0	2.8	40.9	3.4	1.61	74.7	7.6	60.1	11.8	1.24	64.5	2.5	42.8	3.2	1.51
129	Terbutryn	37.0	8.1	28.9	7.5	1.28	32.6	2.3	20.7	3.8	1.57	37.5	7.7	29.2	8.7	1.28	34.3	10.1	20.9	3.7	1.64
130	Thiobencarb	53.0	9.7	42.2	7.6	1.26	45.3	2.6	29.6	2.8	1.53	55.0	9.1	44.2	7.7	1.24	47.7	8.8	31.9	3.9	1.50
131	Tralkoxydim	120.4	4.9	94.6	11.2	1.27	105.1	7.7	75.0	5.5	1.40	128.2	4.7	95.4	12.5	1.34	114.1	13.0	86.4	3.4	1.32
132	Trans-chlordane	25.5	6.3	18.3	8.0	1.39	23.8	9.7	13.5	3.5	1.77	27.6	8.9	18.7	6.2	1.48	24.6	5.9	14.3	5.2	1.72
133	Trans-diallate	45.1	7.6	34.5	9.5	1.31	41.6	3.1	25.7	2.2	1.62	45.9	9.5	34.0	9.2	1.35	39.7	9.1	25.4	3.4	1.56
134	Trifloxystrobin	108.5	10.4	74.5	7.9	1.46	83.8	0.5	53.2	2.6	1.57	106.9	9.8	71.8	10.1	1.49	84.9	4.7	55.9	4.4	1.52
135	Zoxamide	44.9	3.9	43.7	9.0	1.03	51.0	11.8	32.4	6.9	1.57	50.3	6.3	46.3	10.7	1.09	50.7	5.7	36.3	4.3	1.40
136	2,4-DDE	23.1	6.4	16.0	8.5	1.45	23.2	7.0	14.5	4.1	1.60	28.6	5.1	16.3	9.7	1.76	27.1	3.1	17.6	6.1	1.54

(Continued)

TABLE 2.2.11 Comparative Test Results of the Content of 201 Pesticides in the Incurred Green Tea and Oolong Tea Samples by Method-1 and Method-2 (n = 5) (cont.)

No.	Pesticides	GC/MS												GC/MS/MS												
		Green tea					Oolong tea					Green tea					Oolong tea						
		Method-1		Method-2		Ratio[a]	Method-1		Method-2		ratio[a]	Method-1		Method-2		Ratio[a]	Method-1		Method-2		Ratio[a]
		Contents	RSD (%)	Contents	RSD (%)		RSD (%)	Contents	RSD (%)	Contents		Contents	RSD (%)	Contents	RSD (%)		RSD (%)	Contents	RSD (%)	Contents	
137	Ametryn	60.4	12.5	40.5	8.5	1.49	51.9	7.2	34.9	3.9	1.49	59.4	6.2	39.7	9.3	1.50	55.4	3.3	36.0	4.4	1.54
138	Bifenthrin	26.1	6.4	15.4	11.5	1.70	48.2	10.3	34.0	4.0	1.42	24.2	6.2	15.2	11.0	1.59	47.8	8.9	33.7	3.8	1.42
139	Bitertanol	74.4	12.4	51.2	11.8	1.45	62.1	7.0	51.7	5.9	1.20	70.5	2.4	54.6	12.5	1.29	73.2	5.3	54.4	5.9	1.35
140	Boscalid	150.2	6.6	92.3	9.7	1.63	106.4	8.3	92.6	7.0	1.15	137.0	12.0	84.8	11.2	1.62	108.2	5.5	89.1	5.2	1.21
141	Butafenacil	34.5	9.7	22.8	7.6	1.51	29.5	9.7	23.7	3.4	1.25	37.3	7.0	23.6	7.9	1.58	29.9	3.5	25.0	5.5	1.20
142	Carbaryl	86.2	2.9	59.2	7.2	1.46	81.5	11.0	60.7	4.1	1.34	88.6	4.5	62.4	7.1	1.42	85.8	12.1	58.9	2.3	1.46
143	Chlorobenzilate	67.2	12.4	39.4	9.0	1.70	52.3	4.6	39.1	3.9	1.34	30.0	9.1	38.8	8.2	0.77	54.2	4.6	36.1	4.4	1.50
144	Chlorthal-di-methyl	50.9	6.7	36.2	6.2	1.41	46.7	2.3	30.3	4.3	1.54	54.1	6.9	35.8	7.7	1.51	49.5	6.2	31.3	4.5	1.58
145	Dibutylsuccinate	44.5	6.7	27.0	10.6	1.65	24.8	4.8	17.9	4.2	1.39	37.3	6.0	23.9	9.0	1.56	22.5	2.8	18.1	3.0	1.24
146	Diethofencarb	194.8	10.0	137.6	6.6	1.42	174.6	1.3	132.6	5.0	1.32	198.9	7.8	136.6	8.8	1.46	180.6	6.0	123.3	3.6	1.46
147	Diflufenican	32.2	6.6	20.6	11.8	1.56	29.2	9.6	16.9	6.8	1.73	30.3	7.9	19.2	12.0	1.58	28.0	4.8	15.8	7.2	1.76
148	Dimepiperate	52.3	11.7	31.3	7.8	1.67	43.0	3.8	29.6	2.9	1.45	49.9	9.1	32.0	8.3	1.56	46.7	9.4	29.5	4.0	1.58
149	Dimethametryn	18.9	6.6	13.0	9.7	1.45	17.3	9.1	11.2	4.5	1.54	19.5	8.1	12.6	8.5	1.52	17.2	3.8	11.3	5.3	1.52
150	Dimethomorph	86.0	6.2	78.9	12.2	1.09	92.9	9.7	82.2	5.5	1.13	92.9	4.6	77.5	12.6	1.20	96.9	4.8	82.8	4.1	1.17
151	Dimethylphthalate	111.7	9.9	81.9	8.1	1.36	95.5	6.8	80.9	6.6	1.18	114.5	9.0	79.5	7.9	1.44	98.4	4.3	84.1	5.4	1.17
152	Diniconazole	69.1	10.8	41.8	13.5	1.65	59.1	7.7	39.1	5.7	1.51	65.6	5.7	43.8	12.9	1.50	61.9	4.2	40.0	5.3	1.55
153	Diphenamid	27.2	4.4	19.4	8.1	1.40	25.2	6.6	18.2	4.3	1.38	27.2	4.5	19.3	7.6	1.41	25.3	5.8	18.5	3.1	1.37
154	Dipropetryn	20.2	5.7	14.3	8.3	1.42	25.3	21.2	12.1	2.8	2.09	22.5	4.4	14.1	8.5	1.60	19.8	7.4	12.6	3.7	1.58
155	Ethalfluralin	96.8	7.7	55.4	7.7	1.75	86.2	3.8	57.2	2.8	1.51	91.6	6.4	48.6	9.7	1.88	85.9	3.6	52.8	3.8	1.63
156	Etofenprox	37.7	6.4	28.1	9.8	1.34	23.5	3.7	15.3	6.2	1.54	38.3	7.0	27.2	9.5	1.41	22.5	3.1	15.1	4.8	1.49
157	Etridiazol	82.2	11.4	45.4	10.8	1.81	54.3	10.8	62.1	7.9	0.87	76.5	8.6	38.7	14.5	1.98	50.9	10.9	58.2	9.4	0.87
158	Fenazaquin	26.3	7.5	17.4	10.9	1.51	22.6	5.1	12.7	3.9	1.79	21.0	7.0	13.5	12.3	1.55	19.6	7.0	11.6	4.5	1.69
159	Fenchlorphos	97.4	8.5	69.9	5.1	1.39	83.2	2.4	55.8	3.1	1.49	99.0	6.6	69.9	5.0	1.42	87.2	7.4	56.1	3.7	1.55
160	Fenoxanil	81.6	5.8	53.8	9.7	1.52	86.0	9.1	56.7	3.8	1.52	76.9	5.0	61.6	13.8	1.25	107.0	17.6	49.1	6.4	2.18
161	Fenpropidin	44.6	5.1	33.5	11.7	1.33	103.1	5.8	66.2	3.5	1.56	46.3	5.0	33.7	12.4	1.37	121.2	5.1	80.2	3.8	1.51
162	Fenson	33.1	11.9	19.1	5.6	1.73	24.2	6.6	19.8	6.1	1.22	34.3	9.4	21.1	5.5	1.62	27.6	8.9	20.1	5.2	1.38

163	Flufenacet	200.8	9.0	118.4	3.9	1.70	153.3	4.8	117.9	3.8	1.30	197.1	6.3	123.5	4.9	1.60	140.2	3.1	117.1	4.7	1.20
164	Furalaxyl	52.5	5.4	36.8	10.1	1.43	48.8	7.7	34.6	3.3	1.41	53.7	5.4	36.5	10.0	1.47	48.3	4.9	35.3	3.5	1.37
165	Heptachlor	65.7	11.5	38.1	8.4	1.73	54.1	5.2	39.2	2.8	1.38	65.5	12.2	36.5	9.3	1.79	58.9	6.7	37.4	4.1	1.57
166	Iprobenfos	71.7	8.3	45.5	8.4	1.58	63.0	1.0	41.8	3.4	1.51	70.3	7.6	42.9	9.0	1.64	66.4	7.0	42.0	4.3	1.58
167	Isazofos	46.5	6.1	30.7	7.5	1.51	39.2	1.4	28.0	3.4	1.40	51.9	3.6	34.8	8.7	1.49	46.9	5.2	31.3	4.6	1.50
168	Isoprothiolane	70.5	6.9	48.5	8.4	1.45	65.6	9.4	43.9	3.1	1.50	71.0	7.7	48.2	9.7	1.47	66.4	7.0	44.9	2.8	1.48
169	Kresoxim-methyl	17.9	8.1	10.2	8.5	1.76	15.2	8.1	9.4	5.1	1.61	17.9	7.2	10.6	7.4	1.68	14.8	5.2	9.8	3.7	1.51
170	Mefenacet	172.6	6.9	149.5	13.0	1.15	84.8	9.2	62.8	7.3	1.35	129.9	6.6	86.7	12.2	1.50	78.6	2.5	64.8	5.0	1.21
171	Mepronil	32.0	3.9	23.6	4.7	1.36	32.4	7.0	22.1	4.1	1.47	31.8	4.9	23.2	9.9	1.37	30.5	3.7	22.1	4.1	1.38
172	Metribuzin	34.8	4.4	23.9	6.3	1.46	35.6	2.4	27.2	4.4	1.31	36.1	7.5	24.0	7.1	1.51	36.3	5.8	26.4	4.3	1.37
173	Molinate	23.5	6.6	15.9	10.6	1.47	22.3	7.0	18.8	7.0	1.19	23.3	9.6	15.7	8.6	1.49	19.7	4.5	16.8	4.0	1.17
174	Napropamide	79.8	8.7	53.5	7.0	1.49	74.8	9.9	48.6	3.9	1.54	80.7	6.1	53.7	8.5	1.50	74.0	4.5	49.2	3.8	1.50
175	Nuarimol	62.7	5.0	45.3	10.0	1.38	62.0	8.0	44.2	4.3	1.40	62.7	5.8	45.2	9.6	1.39	60.9	2.6	45.9	5.4	1.33
176	Permethrin	56.0	6.1	38.2	11.1	1.47	47.2	3.5	32.1	4.0	1.47	51.6	4.1	35.8	10.6	1.44	47.2	5.7	31.3	5.7	1.51
177	Phenothrin	33.1	9.5	15.5	11.8	2.14	20.6	8.4	12.8	4.7	1.61	53.8	8.1	15.7	11.7	3.43	22.2	10.9	39.3	31.2	0.56
178	Piperonylbutoxide	25.3	8.3	17.4	10.2	1.45	21.5	2.7	14.2	3.6	1.52	25.3	8.7	17.0	10.2	1.49	22.7	8.9	14.5	4.6	1.56
179	Pretilachlor	47.7	4.0	33.6	8.3	1.42	44.8	7.3	29.9	3.5	1.50	47.5	4.8	32.7	5.6	1.46	42.4	4.4	30.0	4.4	1.41
180	Prometon	83.8	7.2	59.8	8.5	1.40	76.0	3.1	50.3	4.5	1.51	81.8	6.5	55.9	8.9	1.46	75.1	5.5	49.4	3.2	1.52
181	Pronamide	30.1	9.3	24.0	7.0	1.25	25.0	7.7	17.7	3.9	1.41	30.5	9.0	23.0	7.8	1.33	25.1	7.7	17.8	3.3	1.41
182	Propetamphos	53.7	6.3	34.7	6.3	1.55	48.5	6.5	32.7	3.0	1.48	53.9	7.8	35.2	5.2	1.53	47.0	5.0	31.7	3.7	1.48
183	Propoxur-1	211.5	3.8	166.7	7.5	1.27	209.3	7.7	144.6	5.3	1.45	212.8	3.4	160.7	8.2	1.32	211.6	6.0	153.8	4.4	1.38
184	Propoxur-2	33.5	7.8	19.0	5.5	1.76	21.6	18.0	20.3	3.5	1.06	32.6	9.6	17.6	11.3	1.85	22.5	2.3	18.1	5.0	1.24
185	Prothiophos	50.9	7.1	32.2	7.6	1.58	46.0	8.5	27.7	2.8	1.66	49.9	6.5	31.6	9.3	1.58	44.4	6.6	27.9	5.2	1.59
186	Pyridaben	29.9	20.4	15.7	13.8	1.91	26.0	16.2	16.5	11.3	1.58	59.1	7.9	49.5	13.2	1.19	29.0	5.6	16.8	4.8	1.73
187	Pyridaphenthion	79.4	10.5	54.1	9.1	1.47	75.4	5.3	53.9	4.8	1.40	77.0	5.8	55.4	8.9	1.39	73.6	1.4	56.0	4.7	1.32
188	Pyrimethanil	24.3	6.1	15.7	9.0	1.55	22.8	9.2	10.9	4.3	2.10	24.9	5.5	15.7	9.7	1.58	23.6	6.1	11.4	7.0	2.07
189	Pyriproxyfen	29.1	7.2	19.9	10.3	1.46	26.7	2.5	17.7	2.4	1.51	59.4	7.5	49.4	13.4	1.20	27.0	7.0	18.1	4.2	1.49
190	Quinalphos	66.1	7.3	34.7	8.2	1.91	49.5	5.6	32.6	3.7	1.52	53.3	8.7	36.5	6.7	1.46	47.7	8.2	31.5	3.9	1.52
191	Quinoxuphen	22.3	7.6	14.6	11.2	1.52	21.2	9.7	10.7	5.8	1.99	22.9	7.6	14.9	10.1	1.54	21.4	5.8	11.5	6.3	1.87
192	Telodrin	100.9	5.8	60.2	9.6	1.68	82.4	3.0	53.1	7.7	1.55	104.2	6.3	58.4	8.3	1.78	90.2	4.8	56.3	4.5	1.60
193	Tetrasul	27.0	6.1	18.0	9.6	1.50	24.0	9.1	14.8	3.9	1.62	26.1	6.6	17.1	6.9	1.53	26.7	4.2	15.9	4.3	1.68
194	Thiazopyr	48.0	6.2	30.9	8.3	1.55	47.6	8.0	29.6	4.0	1.61	48.9	7.3	33.9	8.8	1.44	47.8	7.4	30.6	8.1	1.56

(Continued)

TABLE 2.2.11 Comparative Test Results of the Content of 201 Pesticides in the Incurred Green Tea and Oolong Tea Samples by Method-1 and Method-2 (n = 5) (cont.)

No.	Pesticides	GC/MS										GC/MS/MS										
		Green tea					Oolong tea					Green tea					Oolong tea				
		Method-1		Method-2		Ratio[a]	Method-1		Method-2		ratio[a]	Method-1		Method-2		Ratio[a]	Method-1		Method-2		Ratio[a]
		Contents	RSD (%)	Contents	RSD (%)		RSD (%)	Contents	RSD (%)	Contents		Contents	RSD (%)	Contents	RSD (%)		RSD (%)	Contents	RSD (%)	Contents	
195	Tolclofos-methyl	27.0	5.8	18.8	5.6	1.43	25.2	9.0	16.5	3.4	1.53	27.0	7.5	18.4	4.7	1.46	25.7	9.0	16.8	2.1	1.53
196	Transfluthrin	24.7	6.4	15.8	8.1	1.56	23.9	8.9	14.2	5.0	1.69	25.6	8.4	16.3	8.2	1.56	23.5	6.1	15.0	4.7	1.57
197	Triadimefon	45.4	6.4	26.8	7.4	1.69	43.5	3.1	30.7	8.5	1.42	49.5	3.5	33.3	10.1	1.49	41.4	5.0	29.1	2.1	1.42
198	Triadimenol	69.7	5.2	51.8	11.4	1.35	70.2	6.7	50.2	4.1	1.40	70.6	5.0	52.1	11.7	1.36	67.0	4.3	49.0	2.4	1.37
199	Triallate	45.7	7.9	28.9	7.0	1.58	38.5	2.5	24.8	4.1	1.56	46.0	7.7	27.7	7.2	1.66	39.6	7.3	25.2	3.2	1.57
200	Tribenuron-methyl	95.3	11.2	56.9	11.3	1.67	49.4	4.3	60.6	5.6	0.82	117.5	8.0	58.9	11.4	1.99	53.4	7.6	57.1	6.8	0.94
201	Vinclozolin	30.1	5.1	21.2	6.5	1.42	28.0	9.3	18.9	3.6	1.48	31.5	8.5	20.1	10.2	1.57	28.1	5.0	18.3	2.4	1.54
	Average	65.1	7.8	49.6	8.8	1.36	58.0	7.2	39.9	4.8	1.51	63.4	6.8	48.9	9.3	1.36	58.1	6.6	40.5	4.8	1.48

aThe ratios of the content determined by Method-1 and Method-2.

TABLE 2.2.12 Summary of the Distribution Range of the Ratio Values of Content of 201 Residue Pesticides in Method-1 and Method-2

The range of ratio (Method 1/ Method 2)	GC/MS Green tea		GC/MS Oolong tea		GC/MS/MS Green tea		GC/MS/MS Oolong tea	
	Number	Percentage (%)	Number	Percentage (%)	Number	Percentage (%)	Number	Percentage (%)
<1.00	4	2.0	7	3.5	7	3.5	9	4.5
1.00–1.10	5	2.5	6	3.0	5	2.5	4	2.0
1.10–1.20	33	16.4	8	4.0	33	16.4	10	5.0
1.20–1.30	48	23.9	10	5.0	51	25.4	6	3.0
1.30–1.40	37	18.4	20	10.0	33	16.4	30	14.9
1.40–1.50	34	16.9	31	15.4	30	14.9	43	21.4
1.50–1.60	20	10.0	57	28.4	24	11.9	62	30.8
1.60–1.70	11	5.5	30	14.9	7	3.5	16	8.0
1.70–1.80	5	2.5	19	9.5	3	1.5	11	5.5
1.80–1.90	1	0.5	6	3.0	3	1.5	3	1.5
1.90–2.00	2	1.0	3	1.5	3	1.5	0	0.0
>2.00	1	0.5	4	2.0	1	0.5	6	3.0
N.D	0	0.0	0	0.0	1	0.5	1	0.5
Total >2.00	197	98.0	194	96.5	193	96.5	191	95.5
Total 1.10–1.70	183	91.0	156	77.6	178	88.6	167	83.1

TABLE 2.2.13 Precision Data of the Two Kinds of Incurred Tea Samples by Two Kinds of Determinations With Two Sample Preparation Methods (n = 5)

The range of Precision	GC/MS								GC/MS/MS							
	Green tea				Oolong tea				Green tea				Oolong tea			
	Method-1		Method-2		Method-1		Method-2		Method-1		Method-2		Method-1		Method-2	
	Number	Percentage (%)	Number	Percentage (%)	Number	Percentage (%)	Number	Percentage (%)	Number	Percentage (%)	Number	Percentage (%)	Number	Percentage (%)	Number	Percentage (%)
RSD ≤ 10%	164	81.6	148	73.6	175	87.1	190	94.5	186	92.5	129	64.2	182	90.5	195	97.0
RSD ≤ 15%	200	99.5	197	98.0	191	95.0	200	99.5	200	99.5	198	98.5	194	96.5	198	98.5
RSD > 15%	1	0.5	4	2.0	10	5.0	1	0.5	0	0	2	1.0	6	3.0	2	1.0
N.D	0	0	0	0	0	0	0	0	1	0.5	1	0.5	1	0.5	1	0.5

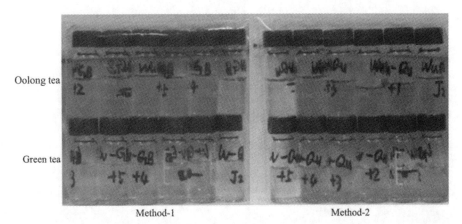

FIGURE 2.2.3 Comparison of sample solution colors from Green tea and oolong tea prepared with Method-1 and Method-2.

2.2.9   CONCLUSIONS

The statistical analysis of 20,904 original test data obtained by the three methods over the three stages shows that the recoveries and RSD values with Method-1 and Method-2 have all met the technical requirements of residual analysis, with recoveries a bit low for Method-3, which is deemed not applicable for the determination of multiresidue pesticides in tea. In terms of pigment removal, Method-1 is superior to Method-2, while in terms of easy operation, Method-2 outweighs Method-1. In the comparative tests of pesticide-incurred tea samples the extraction efficiency of Method-1 is evidently better than that of Method-2, which reflects that using high-speed homogenizing facility (Method-1) has better extraction efficiency than using vortex and oscillation (Method-2). Therefore, Method-1 has been chosen as the sample preparation technique on multiresidue pesticides analysis in tea for AOAC collaborative study.

REFERENCES

[1] Schenck, F.J., Brown, A.N., Podhorniak, L.V., 2008. J. AOAC Int. 91 (2), 422–438.

[2] Lee, J.M., Park, J.W., Jang, G.C., 2008. J. Chromatogr. A 1187, 25–33.

[3] Hercegová, A., Dömötörová, M., Kruzlicová, D., Matisová, E., 2006. J. Sep. Sci. 29, 1102–1109.

[4] Standardization Administration of P.R. China GB/T 19648-2006, Method for Determination of 500 Pesticides and Related Chemicals Residues in fruits and Vegetables by Gas Chromatography-Mass Spectrometry Method, Standards Press, Beijing, China, 2006.

[5] Standardization Administration of P.R. China GB/T 19649-2006, Method for Determination of 475 Pesticides and Related Chemicals Residues in Grains—GC–MS Method Standardization Administration of P.R. China, Standards Press Beijing, China, 2006.

[6] Standardization Administration of P.R. China GB/T 19650-2006, Method for Determination of 478 Pesticides and Related Chemicals Residues in Animal Muscles—GC-MS Method, Standards Press, Beijing, China, 2006.

[7] Standardization Administration of P.R. China GB/T 23207-2008, Determination of 485 Pesticides and Related Chemicals Residues in Fugu, Eel and Prawn—GC-MS Method, Standards Press, Beijing, China, 2008.

[8] Standardization Administration of P.R. China GB/T 19426-2006, Method for the Determination of 497 Pesticides and Related Chemicals Residues in Honey, Fruit Juice and Wine—GC-MS Method, Standards Press, Beijing, China, 2006.

[9] Standardization Administration of P.R. China GB/T 23216-2008, Determination of 503 Pesticides and Related Chemicals Residues in Mushrooms—GC-MS Method, Standards Press, Beijing, China, 2008.

[10] Standardization Administration of P.R. China GB/T 23204-2008, Determination of 519 Pesticides and Related Chemicals Residues in Tea—GC-MS Method, Standards Press, Beijing, China, 2008.

[11] Standardization Administration of P.R. China GB/T 23210-2008, Determination of 511 Pesticides and Related Chemicals Residues in Milk and Milk Powder—GC-MS Method, Standards Press, Beijing, China, 2008.

[12] Standardization Administration of P.R. China GB/T 23200-2008, Determination of 488 Pesticides and Related Chemicals Residues in Mulberry Twig, Honeysuckle, Barbary Wolfberry Fruit and Lotus Leaf—GC-MS Method, Standards Press, Beijing, China, 2008.

[13] Lehotay, S.J., 2007. J. AOAC Int. 90 (2), 485–520.

[14] Nguyen, T.D., Han, E.M., Seo, M.S., Kim, S.R., Yun, M.Y., Lee, D.M., Lee, G.H., 2008. Anal. Chim. Acta 619 (1), 67–74.

[15] Pang, G.F., Fan, C.L., Zhang, F., Li, Y., Chang, Q.Y., Cao, Y.Z., Wang, Q.J., Hu, X.Y., Liang, P., 2011. J. AOAC Int. 94 (4), 1253–1296.

CONCLUSIONS

[illegible]

REFERENCES

[illegible]

Chapter 2.3

The Evaluation of the Cleanup Efficiency of SPE Cartridge Newly Developed for Multiresidues in Tea

Chapter Outline

2.3.1 INTRODUCTION

The tea matrix is one of the samples that is hard to analyze for pesticide residue owing to its complexity, because it contains polyphenols, alkaloids, proteins, amino acids, vitamins, pectins, organic acids, lipopolysaccharide, saccharides, pigments, and so on. A further complication is that more than 500 pesticides have varying chemical constitutions and different kinds of chemical properties and polarities. It's easy to imagine how difficult it is to simultaneously manage extraction, cleanup, and determination. A three-ingredient Cleanert TPT cartridge has been developed for this purpose, which consists of graphite carbon, multiamine-based silicone, and amide polystyrene. To evaluate its cleanup efficiency on a comprehensive scale, three stages of comparative study were arranged. In Stage I, through the review of influences of 12 tandem Solid Phase Extraction (SPE) cartridges on the cleanup efficiencies of 84 representative pesticides in green tea and Oolong tea, the best combined cleanup cartridge, 4# Envi-Carb+PSA, was selected. In Stage II, a comparative study was conducted on the extraction efficiencies of the spiked 201 pesticides in different teas using Cleanert TPT and 4# Envi-Carb+PSA.

Test results show that Cleanert TPT is on a par with Envi-Carb+PSA only in terms of recoveries; concerning RSD, the former is 10% better than the latter. In Stage III, comparative tests were continued using the aforementioned two SPE cartridges on the cleanup efficiencies of two Youden pair samples of incurred 201 pesticides green tea and Oolong tea, once every 5 days with 19 times stability tests over a succession of 3 months. Just in terms of content values of the 201 pesticides detected from the incurred green tea and Oolong tea samples, the experiment identified 187 pesticides (93.0%) with average content obtained from the 4# Envi-Carb+PSA cleanup greater than those from Cleanert TPT for green tea samples. For Oolong tea samples there were 179 pesticides (89.1%) with average content obtained from Cleanert TPT cleanup greater than those from Envi-Carb+PSA. The statistical analysis also proves that above 93% pesticides have a deviation less than 15% for their analytical results from the two cartridges. In terms of RSD data, Cleanert TPT is 10% better than Envi-Carb+PSA, which totally agrees with the test results in Stage II. It proves once more that Cleanert TPT is superior to Envi-Carb+PSA in the reproducibility of cleanup effects. Based on the evaluations of the cleanup results from the three stages, both Cleanert TPT and Envi-Carb+PSA are capable of fulfilling the technical requirements for cleanup of more than 500 pesticide residues in tea, but as far as cleanup efficiencies are concerned, priority is given to Cleanert-TPT over Envi-Carb+PSA.

2.3.2 REAGENTS AND MATERIALS

1. Solvents. Acetonitrile, toluene, hexane, acetone (LC grade), purchased from Dikma Co. (Beijing, China).
2. MgSO$_4$, NaCl. Reagent grade anhydrous MgSO$_4$ in powder form and ACS grade NaCl. The MgSO$_4$ were baked for 4 h at 650°C in a muffle furnace.
3. Glacial acetic acid (AR grade) was purchased from Beijing Chemical Works.
4. Pesticides standard and internal standard. Purity, ≥95% (LGC Promochem, Wesel, Germany).
5. Stock standard solutions. Weigh 5–10 mg of individual pesticide standards (accurate to 0.1 mg) into a 10 mL volumetric flask. Dissolve and dilute to volume with toluene, toluene-acetone combination, cyclohexane, depending on individual compound solubility. Stock standard solutions should be stored in the dark below 4°C.
6. Mixed standard solution. Depending on properties and retention time of each pesticide, all the 201 pesticides for GC/MS and GC/MS/MS analysis are divided into 3 groups, A–C. The concentration of mixed standard solutions was dependent on the sensitivity of each compound for the instrument used for analysis. Mixed standard solutions should be stored in the dark below 4°C.

2.3.3 APPARATUS

1. GC/MS system. Model 6890N gas chromatograph connected to a Model 5973N MSD and equipped with a Model 7683 autosampler (Agilent Technologies, Wilmington, DE). The column used was a DB-1701 capillary column (30 m × 0.25 mm × 0.25 μm, J&W Scientific, Folsom, CA, USA).
2. GC/MS/MS system. Model 7890 gas chromatograph connected to a Model 7000A and equipped with a Model 7693 autosampler (Agilent Technologies, Wilmington, DE). The column used was a DB-1701 capillary column (30 m × 0.25 mm × 0.25 μm, J&W Scientific, Folsom, CA, USA).
3. SPE cartridge. Cleanert-TPT (10 mL, 2000 mg, Agela, Tianjin, China); Envi-Carb (5 mL, 500 mg, Supelco, USA); Envi-Carb (5 mL, 1000 mg, Supelco, USA); PSA (5 mL, 500 mg, Varian, USA); PSA (5 mL, 1000 mg, Varian, USA); NH$_2$ (5 mL, 500 mg, Supelco, USA); NH$_2$ (5 mL, 1000 mg, Supelco, USA); C18 (10 mL, 1000 mg, Supelco, USA); C18 (10 mL, 2000 mg, Supelco, USA).
4. Homogenizer. T-25B (Janke & Kunkel, Staufen, Germany).
5. Rotary evaporator. Buchi EL131 (Flawil, Switzerland).
6. Centrifuge. Z 320 (B. HermLe AG, Gosheim, Germany).
7. Nitrogen evaporator. EVAP 112 (Organomation Associates Inc., New Berlin, MA, USA).

2.3.4 EXPERIMENTAL

Weigh 5 g test samples (accurate to 0.01 g) into 80 centrifuge tube, add 5 g NaCl and 15 mL 1% acetic acetonitrile homogenize and extract at 15,000 r/min for 1 min, centrifuge 4200 r/min for 5 min and transfer the supernatants into 100 mL pear-shaped flask. Repeat extracting the dregs with 15 mL 1% acetic acetonitrile one time, centrifuge, consolidate the extractions over two times and rotary evaporate in water bath at 40°C to about 1 mL before waiting for cleanup.

Add about 2 cm high sodium sulfates anhydrous into the top of SPE cartridge, prewash with 10 mL acetonitrile-toulene (3:1) and discard the effluents. Connect the pear-shaped flask at the lower end and place it onto the fixing bracket. Transfer the earlier-described sample concentrates into SPE cartridge, rinse the sample solution bottle with 2 mL acetonitrile-toulene, repeat this step 3 times, transfer the rinsing liquids into the cartridge, attach a 50 mL storage device onto the cartridge, then elute with 25 mL acetonitrile-toulene, collect the aforementioned effluents into the pear-shaped flask and rotary evaporate in water bath at 40°C to about 0.5 mL. Add 5 mL hexane for solvent exchange on a rotary evaporator in a water bath of 40°C and repeat twice, dilute to about 1 mL, add 40 μL internal standards (heptachlor-epoxide), and mix uniformly before passing the 0.2 μm filtering membrane and being used for GC–MS determination.

2.3.5 STAGE I: THE COMPARATIVE TEST OF THE CLEANUP EFFICIENCIES OF 12 SPE COMBINED CLEANUP CARTRIDGES AGAINST THE SPIKED 84 PESTICIDES IN TEA

To effectively clean up more than 500 residual pesticides in tea, different cartridges of C18, GCB, PSA, and NH$_2$ SPE were purchased in markets and combined into 12 kinds of SPE cartridges according to different sequences (top layer, medium layer, and lower layer). See Table 2.3.1.

TABLE 2.3.1 Combined Solid Phase Extraction (SPE) Cartridges

SPE cartridge No.	Filling volumes and sequences
1#	PSA (1 g/5 mL) + Envi-Carb (1 g/5 mL)
2#	PSA (0.5 g/5 mL) + Envi-Carb (0.5 g/5 mL)
3#	Envi-Carb (1 g/5 mL) + PSA (1 g/5 mL)
4#	Envi-Carb (0.5 g/5 mL) + PSA (0.5 g/5 mL)
5#	C18 (2 g/10 mL) + NH_2 (0.5 g/5 mL) + Envi-Carb (0.5 g/5 mL)
6#	C18 (1 g/10 mL) + NH_2 (0.5 g/5 mL) + Envi-Carb (0.5 g/5 mL)
7#	Envi-Carb (0.5 g/5 mL) + NH_2 (0.5 g/5 mL)
8#	Envi-Carb (1 g/5 mL) + NH_2 (1 g/5 mL)
9#	NH_2 (1 g/5 mL) + Envi-Carb (1 g/5 mL)
10#	NH_2 (0.5 g/5 mL) + Envi-Carb (0.5 g/5 mL)
11#	C18 (1 g/10 mL) + Envi-Carb (0.5 g/5 mL) +NH_2 (0.5 g/5 mL)
12#	C18 (2 g/10 mL) +Envi-Carb (0.5 g/5 mL) +NH_2 (0.5 g/5 mL)

Because a total of more than 500 compounds have different chemical constituents, different chemical properties, and a wide range of polarities, in order to evaluate scientifically the cleanup effects of these 12 SPE cartridges on multiresidue pesticide in tea matrixes, a total of 84 representative compounds of different types were selected. These compounds represent different polarity ranges from weak (deltamethrin) to strong polarity (allidochlor), including organochlorines (p,p'-DDD, beta-HCH, and *trans*-chlordane), organophosphorous (chlormephos, and thiometon and fonofos), and pyrethroids (deltamethrin, tetramethirn, and fenvalerate). Green tea and Oolong tea samples were used for a spiked recovery test at the spiked concentration of 10–800 µg/kg, repeated 3 times for each sample, cleaned up, respectively, with Cleanert TPT and Envi-Carb+PSA, making a GC–MS determination with average recoveries and RSD data as shown in Table 2.3.2. Statistical calculation was conducted on the average recoveries and RSD values in Table 2.3.2 per different ranges, with their results tabulated in Table 2.3.3. Pesticide varieties with a good range of recoveries 70%–110% and RSD less than 20% and their percentage of the total number of tested pesticides are also listed in Table 2.3.3 and expressed in a bar diagram (see Fig. 2.3.1).

Table 2.3.3 and Fig. 2.3.1 show that acceptable cleanup results have been obtained with 10 cartridges except for 1# and 2#, which is capable of satisfying the technical requirement for residual analysis. For green tea samples there were 69 pesticides, being the largest number of pesticides having good analytical results with 4# cartridge, and accounting for 82.1% of the total number. For Oolong tea samples, there were 66 pesticides, the largest number of pesticides having good analytical results with 9# and 10# cartridges, accounting for 78.6% of the total. There were 65 with 4# cartridge, ranking the second, accounting for 77.4% of the total number. Based on the aforementioned test results, be it green tea or Oolong tea, the recoveries and RSD values obtained with 4# cartridges are better in comparison with the rest of SPE cartridges. Therefore, 4# SPE cartridge and the Cleanert TPT recommended earlier were chosen for a deeper comparative test on cleanup efficiencies.

2.3.6 STAGE II: THE COMPARATIVE TEST ON CLEANUP EFFICIENCIES OF SPIKED 201 PESTICIDES IN TEA WITH 4# ENVI-CARB+PSA TANDEM CARTRIDGE AND CLEANERT TPT

Green tea and Oolong tea samples were used for a recovery test on 201 pesticides at the spiked concentration of 10–600 µg/kg, and cleaned up respectively with 4# Envi-Carb+PSA and Cleanert TPT, making GC–MS and GC–MS/MS determination by running six parallel tests for each with recoveries and RSD data as shown in Table 2.3.4. Statistical calculation is conducted on the average recoveries and RSD values in Table 2.3.4 for different ranges, with their results tabulated in Table 2.3.5. Pesticide varieties with recoveries 70%–110% and RSD less than 20% are also listed in Table 2.3.5 and expressed in bar diagram (see Fig. 2.3.2).

Table 2.3.5 shows that Cleanert TPT is on a par with 4# Envi-Carb+PSA for cleanup results in terms of recoveries, no matter whether it is green tea or Oolong tea, or GC–MS or GC–MS/MS determination, results of which for both are within the acceptable range and capable of fulfilling the technical requirement for residual analysis; concerning RSD data, the

TABLE 2.3.2 Average Recoveries and RSD Data for 84 Pesticides Spiked in Green Tea and Oolong Tea With 12 Solid Phase Extraction Cartridge Cleanup ($n = 3$)

No.	Pesticides	Spiked concentration (μg/kg)	Sample	1# AVE (%)	1# RSD (%)	2# AVE (%)	2# RSD (%)	3# AVE (%)	3# RSD (%)	4# AVE (%)	4# RSD (%)	5# AVE (%)	5# RSD (%)	6# AVE (%)	6# RSD (%)	7# AVE (%)	7# RSD (%)	8# AVE (%)	8# RSD (%)	9# AVE (%)	9# RSD (%)	10# AVE (%)	10# RSD (%)	11# AVE (%)	11# RSD (%)	12# AVE (%)	12# RSD (%)
1	Allidochlor	50	Green tea	88.8	0.8	86.1	4.1	85.4	1.4	94.5	8.6	95.6	5.0	88.7	4.7	85.5	11.6	87.4	8.3	84.2	13.1	96.5	3.7	87.6	4.2	79.8	8.8
			Oolong tea	97.1	8.3	82.9	2.0	88.5	2.4	92.9	1.8	96.8	11.5	98.6	0.9	91.5	0.5	94.9	4.3	106.7	3.1	82.1	6.0	96.4	1.5	87.7	6.5
2	Dichlormid	50	Green tea	76.6	11.8	83.3	43.0	73.0	6.3	103.1	5.7	74.7	5.4	84.1	6.8	85.3	10.9	79.1	5.9	85.1	5.5	88.7	1.6	74.1	5.8	168.7	105.6
			Oolong tea	80.2	6.1	73.0	5.6	81.0	6.5	89.6	6.0	90.0	9.4	89.6	2.5	83.9	3.6	94.9	11.3	101.5	6.5	75.8	12.1	N.D	N.D	79.9	4.6
3	Etridiazol	75	Green tea	71.9	4.1	72.9	2.8	75.0	3.8	88.0	5.8	85.8	10.8	75.2	5.4	71.5	10.5	75.5	6.4	91.0	11.2	86.3	7.3	71.1	2.6	78.7	10.5
			Oolong tea	77.2	9.2	83.3	3.2	80.9	3.2	79.1	9.7	89.6	7.7	86.9	2.8	83.8	2.8	83.4	4.1	92.1	5.4	78.8	13.8	88.7	8.6	69.8	6.8
4	Chlormephos	50	Green tea	83.5	7.9	85.8	3.5	80.4	6.4	95.0	6.8	91.6	4.2	78.9	4.0	81.5	8.6	82.6	11.8	89.2	2.4	87.7	1.2	79.5	8.3	78.5	1.6
			Oolong tea	95.2	14.9	81.3	5.2	95.3	6.3	91.5	1.0	96.5	6.8	91.9	3.2	87.8	3.2	82.9	9.3	100.6	4.6	87.6	14.7	90.6	9.2	88.5	5.4
5	Propham	25	Green tea	91.2	27.7	66.2	8.5	83.3	10.4	94.9	7.9	N.D	N.D	81.8	6.1	79.3	15.7	87.9	3.7	82.5	6.8	100.0	3.8	82.8	3.8	81.9	13.0
			Oolong tea	91.0	14.7	90.8	1.9	89.4	1.9	91.8	5.6	N.D	N.D	N.D	N.D	85.1	14.6	86.9	14.5	99.5	2.7	86.2	11.9	87.4	17.1	95.4	7.9
6	Cycloate	25	Green tea	102.5	16.3	71.7	3.4	76.9	1.6	90.8	5.0	78.5	13.0	80.4	5.8	81.8	10.3	83.0	9.7	91.8	4.6	96.5	3.8	80.0	4.5	96.7	2.6
			Oolong tea	95.6	6.1	81.3	3.1	83.8	3.1	86.6	5.3	73.4	14.7	77.0	3.7	78.9	3.7	86.0	7.1	103.1	0.7	85.0	9.0	84.8	1.7	71.2	7.2
7	Diphenylamin	25	Green tea	95.0	7.7	98.4	7.2	95.6	7.2	105.5	13.9	92.1	10.4	89.3	8.2	94.9	3.7	94.4	6.4	94.5	20.5	79.9	13.9	108.4	12.1	96.4	11.1
			Oolong tea	75.3	28.8	97.8	1.6	86.3	3.5	83.3	3.5	77.4	18.9	125.3	11.3	77.7	3.8	82.5	3.8	102.9	9.6	88.2	6.0	120.7	17.5	156.6	41.5
8	Ethalfluralin	100	Green tea	71.7	4.7	75.4	3.1	74.0	2.9	72.8	5.7	97.7	6.3	88.0	8.1	73.4	7.4	81.8	3.3	91.0	2.8	93.8	6.3	78.1	4.0	79.4	7.1
			Oolong tea	75.8	11.6	72.1	2.2	91.2	3.9	83.4	3.9	88.0	7.4	77.3	7.7	82.2	2.2	82.2	4.7	96.7	7.0	84.6	11.1	90.6	6.8	80.4	8.2
9	Thiometon	25	Green tea	63.1	23.4	76.8	1.2	62.7	7.0	81.7	3.3	N.D	N.D	N.D	N.D	64.6	14.6	60.0	14.6	76.5	9.1	79.5	4.4	72.6	5.7	76.5	4.0
			Oolong tea	N.D	N.D	65.8	39.2	62.2	23.8	61.8	16.1	N.D	N.D	N.D	N.D	N.D	N.D	N.D	N.D	93.8	4.4	80.6	6.8	21.0	173.2	64.8	18.1
10	Quintozene	50	Green tea	68.3	0.8	66.8	4.8	76.8	2.0	83.5	13.7	91.7	7.3	85.4	13.9	80.8	19.6	81.3	5.4	102.7	11.9	76.0	13.0	72.3	6.8	87.8	10.2
			Oolong tea	84.3	12.6	73.4	5.9	80.4	9.6	75.9	0.1	96.2	13.5	78.0	2.1	85.0	1.3	83.6	5.5	88.4	8.2	85.2	3.7	88.3	11.5	82.8	9.0
11	Atrazine-desethyl	25	Green tea	76.8	19.4	89.8	4.6	79.7	9.3	94.7	6.2	91.4	9.7	83.5	11.4	75.5	4.5	83.4	11.1	85.7	18.5	96.4	6.1	83.8	10.2	71.7	2.8
			Oolong tea	106.0	4.1	118.3	6.0	111.6	6.0	107.3	3.4	95.1	28.6	77.2	6.3	93.4	5.2	90.7	5.3	109.4	11.4	81.6	6.5	112.7	3.7	93.3	8.0
12	Clomazone	25	Green tea	80.1	2.3	73.7	3.0	80.2	3.0	91.8	2.8	89.0	3.5	79.1	3.5	78.7	9.9	82.4	5.3	89.2	4.8	90.8	1.7	79.0	3.4	80.6	4.0
			Oolong tea	76.4	2.2	91.7	10.9	85.3	4.3	84.7	1.2	90.0	5.0	83.4	9.0	92.2	16.6	81.2	9.9	103.1	3.4	87.5	6.2	84.6	3.1	80.2	5.4
13	Diazinon	25	Green tea	81.5	3.6	76.4	2.7	79.4	2.7	89.8	3.6	N.D	N.D	83.4	3.0	79.7	3.0	82.8	6.3	N.D	N.D	94.8	3.6	N.D	N.D	81.0	4.3
			Oolong tea	85.6	2.7	76.7	0.5	56.0	80.1	84.8	3.2	N.D	N.D	N.D	N.D	N.D	N.D	80.6	6.3	N.D	N.D	N.D	N.D	56.3	86.6	N.D	N.D
14	Fonofos	25	Green tea	79.0	5.5	80.9	1.1	78.3	2.1	90.5	4.4	88.6	3.0	82.8	6.7	81.1	6.2	80.6	8.8	91.7	2.8	102.3	4.0	81.1	3.8	81.1	3.2
			Oolong tea	59.0	10.0	76.7	0.5	82.2	3.3	85.0	0.9	89.4	0.9	78.9	7.9	79.0	5.0	79.6	5.1	96.4	2.8	84.8	6.6	84.6	7.5	79.7	5.9
15	Etrimfos	25	Green tea	81.1	1.5	80.4	1.7	80.1	3.1	89.7	3.1	92.7	3.8	85.2	3.8	79.7	4.0	98.5	12.5	93.0	3.6	89.9	3.6	81.8	1.7	81.9	3.5
			Oolong tea	77.4	3.1	79.5	5.2	82.5	5.2	83.9	4.7	91.6	11.4	81.3	11.4	87.9	5.6	82.0	4.0	97.0	1.7	69.4	1.7	83.5	1.3	78.9	4.9
16	Propetamphos	25	Green tea	76.8	1.0	74.5	1.0	85.3	2.7	91.6	5.2	89.6	5.2	89.0	9.1	75.0	2.6	82.1	6.7	90.7	6.2	85.1	5.5	83.8	2.6	80.7	5.0
			Oolong tea	71.1	18.2	96.6	8.3	96.1	0.9	96.9	2.7	92.2	9.1	76.2	9.1	87.9	1.3	76.3	18.7	103.5	2.5	87.7	4.9	78.9	1.4	83.6	4.4
17	Secbumeton	25	Green tea	74.3	3.5	77.6	1.4	88.5	4.4	92.6	4.4	99.5	9.7	93.8	8.8	77.8	8.5	82.3	8.5	90.8	2.7	93.6	5.0	81.1	2.8	75.7	4.2
			Oolong tea	76.5	3.7	85.4	6.4	85.1	6.4	84.5	2.7	99.9	8.8	100.9	5.7	83.0	1.1	79.3	7.6	104.5	2.8	82.6	2.9	85.3	2.1	81.3	5.4
18	Dichlofenthion	25	Green tea	99.0	12.4	107.1	6.1	94.1	6.1	112.0	5.1	92.9	5.7	N.D	N.D	97.7	12.5	103.0	12.5	101.5	6.7	102.4	6.7	93.9	4.0	98.2	3.1
			Oolong tea	102.2	12.4	101.4	3.9	98.3	3.9	104.6	6.1	N.D	N.D	95.2	3.4	94.5	7.4	85.3	10.4	N.D	N.D	N.D	N.D	101.1	1.0	N.D	N.D
19	Pronamide	25	Green tea	82.4	5.3	85.3	5.2	83.1	5.2	95.6	5.2	99.5	3.4	82.3	10.5	83.3	10.9	86.4	10.9	83.2	14.0	92.1	5.6	70.7	5.6	82.0	2.2
			Oolong tea	80.6	10.4	87.1	8.4	77.7	8.4	87.6	1.4	89.7	10.5	92.0	3.0	86.5	5.0	87.8	5.0	103.8	5.6	84.5	2.2	87.9	2.9	85.1	3.7
20	Mexacarbate	75	Green tea	89.9	35.6	97.6	12.7	88.2	12.7	107.1	8.0	90.0	21.9	92.0	21.9	80.8	27.3	83.3	27.3	109.4	11.5	73.1	11.5	74.9	16.4	99.6	6.5
			Oolong tea	81.7	11.0	128.1	8.7	80.9	8.7	69.9	1.7	78.2	14.0	71.8	14.0	87.5	4.2	97.1	4.2	103.9	1.5	75.3	1.5	90.6	16.4	84.3	6.5

(Continued)

No.	Pesticide	Conc.	Sample																								
21	Dimethoate	100	Green tea	69.6	9.6	89.1	12.9	78.1	7.8	95.8	3.6	79.2	9.0	119.2	3.7	86.2	11.1	80.8	9.3	64.6	52.1	108.1	20.0	103.8	9.0	74.2	12.8
			Oolong tea	50.4	3.8	73.9	19.8	87.8	15.7	70.1	6.1	95.3	30.2	79.5	18.0	N.D	N.D	N.D	N.D	102.0	12.4	78.3	5.1	92.8	2.9	89.8	10.4
22	Dinitramine	100	Green tea	62.1	7.5	71.5	4.8	69.0	1.6	51.3	7.1	96.2	5.8	84.0	6.8	71.4	4.3	83.3	2.1	87.4	3.5	90.8	5.8	77.8	5.6	78.5	10.2
			Oolong tea	64.4	6.5	53.8	5.2	78.5	5.3	81.6	1.7	79.4	12.6	76.7	5.7	79.5	2.8	73.8	2.0	98.2	3.2	82.0	4.9	84.3	5.8	75.5	8.0
23	Ronnel	50	Green tea	79.5	3.1	77.7	1.0	79.5	1.9	87.3	5.3	89.5	2.8	81.2	5.1	80.4	11.0	83.5	5.3	95.4	4.0	88.5	3.1	79.5	2.7	82.4	5.8
			Oolong tea	77.6	3.1	73.6	3.3	82.2	3.0	88.9	4.0	88.6	12.1	75.2	3.1	84.8	5.6	84.1	2.2	93.6	6.2	77.0	3.2	87.0	1.0	79.6	6.5
24	Prometrye	25	Green tea	81.2	7.9	88.5	6.2	81.0	3.5	80.4	5.9	89.9	1.0	83.9	8.3	83.0	7.9	86.0	8.1	94.8	6.0	85.0	4.4	85.0	6.4	82.5	2.7
			Oolong tea	63.9	4.0	84.5	10.6	82.6	2.8	89.6	1.5	77.4	17.7	75.6	6.9	80.8	4.7	82.1	N.D	93.6	N.D	83.7	4.6	76.9	2.5	81.4	10.0
25	Chlorothalonil	25	Green tea	N.D	N.D	N.D	N.D	N.D	N.D	N.D	N.D	N.D	N.D	N.D	N.D	N.D	N.D	N.D	N.D	N.D	N.D	N.D	N.D	N.D	N.D	N.D	N.D
			Oolong tea	N.D	N.D	N.D	N.D	N.D	N.D	N.D	N.D	47.8	33.6	25.2	173.2	N.D	N.D	N.D	N.D	N.D	N.D	N.D	N.D	N.D	N.D	N.D	N.D
26	Cyprazine	25	Green tea	85.4	4.5	66.4	11.5	83.3	3.2	88.5	9.2	90.6	4.9	77.7	7.3	N.D	N.D	83.2	4.0	99.0	10.8	99.9	5.2	92.2	5.6	69.4	15.2
			Oolong tea	N.D	N.D	N.D	N.D	N.D	N.D	31.4	74.8	88.9	12.0	79.8	3.4	N.D	N.D	N.D	N.D	N.D	N.D	N.D	N.D	N.D	N.D	N.D	N.D
27	Vinclozolin	25	Green tea	77.5	0.3	68.9	4.2	81.3	5.2	94.4	2.9	86.8	3.6	75.2	3.2	84.0	4.1	88.4	4.8	104.3	3.9	89.1	4.4	77.9	6.7	84.4	3.7
			Oolong tea	75.5	7.9	82.6	4.1	73.3	0.7	74.6	5.9	93.4	16.9	78.5	6.1	86.9	10.5	77.8	6.3	91.4	6.0	78.7	9.8	92.5	3.1	84.2	5.9
28	Beta-HCH	25	Green tea	85.1	5.2	71.6	7.1	82.6	2.1	91.2	7.6	84.4	6.6	84.7	3.5	81.9	10.6	82.8	6.4	93.6	6.5	83.2	6.2	78.4	9.5	90.5	5.2
			Oolong tea	71.5	5.8	84.2	2.8	79.8	4.0	77.0	3.3	93.8	14.9	79.8	1.6	81.2	8.8	79.8	2.3	93.4	3.7	78.2	1.9	87.5	12.8	78.8	5.4
29	Metalaxyl	75	Green tea	78.1	5.7	79.2	9.1	77.8	4.5	86.0	3.7	97.8	3.3	89.7	7.3	72.6	9.0	85.4	3.5	104.0	5.6	94.9	7.4	84.9	12.1	85.0	5.9
			Oolong tea	82.5	11.1	68.4	3.9	77.5	2.7	76.1	7.9	88.1	15.3	81.8	1.1	100.1	14.0	83.2	5.2	76.3	13.9	84.0	5.1	98.4	3.9	81.8	6.0
30	Chlorpyrifos(ethyl)	25	Green tea	92.2	4.0	87.7	1.4	82.1	4.9	101.7	5.0	102.5	4.8	88.5	5.7	86.6	8.2	88.7	7.0	104.3	6.4	98.0	1.5	92.2	1.9	84.1	7.6
			Oolong tea	80.6	2.3	80.6	3.7	91.9	3.8	88.2	3.7	90.5	10.1	78.5	3.2	86.5	4.8	82.7	1.0	101.2	1.3	83.4	6.5	85.5	6.2	81.0	5.9
31	Methyl-parathion	100	Green tea	65.6	14.8	76.1	2.2	84.6	4.9	89.4	3.7	103.6	4.7	85.9	7.7	77.2	6.9	88.4	4.7	97.7	5.9	42.2	3.7	79.9	13.1	81.3	11.0
			Oolong tea	76.8	12.4	77.4	1.9	79.8	3.4	78.3	1.8	84.7	11.8	77.9	6.4	93.5	9.0	99.8	9.2	87.1	7.3	82.8	6.2	117.3	2.8	80.4	11.0
32	Anthraquinone	25	Green tea	N.D	N.D	109.3	10.4	94.2	14.4	104.8	6.0	90.4	18.3	86.8	12.1	91.1	8.5	80.5	12.9	95.2	2.7	N.D	N.D	90.4	1.7	89.8	13.4
			Oolong tea	82.7	33.0	71.9	2.1	64.8	1.0	100.7	9.2	86.1	18.7	77.1	12.5	87.1	11.1	45.4	41.9	N.D	N.D	93.4	1.5	60.1	N.D	56.6	10.2
33	Delta-HCH	50	Green tea	126.7	30.2	103.1	6.1	64.1	1.0	104.4	21.4	84.5	14.6	121.1	10.7	N.D	N.D	N.D	N.D	76.4	29.5	N.D	N.D	N.D	6.8	N.D	N.D
			Oolong tea	75.5	44.9	69.4	13.7	82.5	7.2	81.7	16.8	87.3	10.7	80.2	10.8	109.2	18.5	90.9	17.3	97.2	6.0	65.6	6.9	76.4	N.D	93.7	3.8
34	Fenthion	25	Green tea	N.D	N.D	N.D	N.D	N.D	N.D	N.D	N.D	N.D	N.D	N.D	N.D	N.D	N.D	N.D	N.D	N.D	N.D	N.D	N.D	N.D	173.2	N.D	N.D
			Oolong tea	N.D	N.D	N.D	N.D	N.D	N.D	N.D	N.D	N.D	N.D	N.D	N.D	N.D	N.D	N.D	N.D	N.D	N.D	N.D	N.D	26.6	173.2	81.5	12.2
35	Malathion	100	Green tea	86.4	12.6	79.1	11.5	74.7	7.9	108.9	11.0	95.7	13.5	71.2	14.7	72.0	9.0	80.5	3.4	97.7	12.7	97.5	7.4	80.0	4.3	50.5	87.1
			Oolong tea	73.3	8.4	91.2	22.4	86.5	4.1	81.5	8.2	90.7	10.1	78.4	7.4	89.7	8.4	90.6	7.3	101.0	2.6	83.1	2.4	90.8	5.9	87.7	6.3
36	Fenitrothion	50	Green tea	71.9	16.3	65.6	9.0	78.1	9.3	97.5	2.2	95.3	3.9	81.8	19.4	75.1	9.2	80.6	11.3	88.3	19.7	64.9	9.2	74.6	9.1	80.7	6.5
			Oolong tea	89.6	19.5	98.5	4.3	89.4	10.3	85.9	2.2	91.1	10.2	82.9	5.3	80.7	16.6	68.9	18.1	93.1	5.0	88.1	7.4	77.6	16.8	83.9	15.4
37	Paraoxon-ethyl	800	Green tea	71.5	4.9	77.5	4.9	79.4	6.7	85.4	6.6	99.1	5.0	80.7	7.5	77.5	3.8	84.6	4.3	81.3	5.8	90.8	14.1	86.1	11.7	81.6	7.1
			Oolong tea	65.6	11.5	81.4	6.3	72.6	5.9	73.8	13.3	88.3	9.9	85.2	1.4	92.3	2.7	84.8	17.3	90.2	3.7	83.0	7.3	82.3	14.0	72.9	14.6
38	Triadimefon	50	Green tea	95.2	24.9	85.5	6.0	89.7	4.0	124.0	14.6	92.4	2.8	96.3	1.5	80.0	13.2	85.5	8.0	86.1	16.1	89.0	19.7	81.0	4.1	72.2	12.5
			Oolong tea	95.4	35.3	78.2	3.2	83.5	2.3	85.0	1.0	88.3	12.1	81.1	2.4	83.5	4.1	71.5	2.6	84.7	4.1	84.7	4.3	63.3	7.6	81.4	7.6
39	Pendimethalin	100	Green tea	68.2	12.1	73.3	3.9	77.5	2.4	85.7	5.0	100.3	5.2	80.3	5.9	75.4	4.3	86.4	2.2	93.2	4.9	93.7	9.2	80.2	59.9	508.3	6.8
			Oolong tea	79.5	10.6	79.2	5.3	84.2	7.1	82.5	0.6	85.9	12.9	73.4	6.7	83.4	1.2	83.7	2.7	91.6	2.5	83.8	6.2	83.3	8.4	78.1	11.8
40	Linuron	100	Green tea	N.D	N.D	N.D	N.D	N.D	N.D	N.D	N.D	70.0	2.4	N.D	N.D	N.D	N.D	N.D	N.D	N.D	N.D	N.D	N.D	N.D	6.2	76.3	7.9
			Oolong tea	N.D	N.D	N.D	N.D	N.D	N.D	N.D	N.D	68.2	20.0	66.3	17.1	N.D	N.D	N.D	N.D	N.D	N.D	N.D	N.D	N.D	N.D	N.D	N.D
41	Chlorbenside	50	Green tea	75.3	9.9	73.5	6.3	79.8	4.2	91.2	4.9	51.9	11.0	69.0	2.8	77.7	11.5	68.3	7.1	90.3	3.3	86.9	8.1	74.3	7.6	78.6	4.3
			Oolong tea	N.D	N.D	72.4	4.4	77.4	8.1	83.0	0.8	68.4	14.7	59.2	12.5	58.3	17.0	60.8	17.7	94.0	1.8	83.0	7.3	79.3	0.5	75.4	7.7
42	Bromophos-ethyl	25	Green tea	80.6	4.9	81.5	11.9	78.6	1.8	94.8	9.3	93.6	4.8	83.7	4.1	71.3	8.9	83.6	5.9	93.3	3.4	82.2	3.5	74.7	9.7	86.2	12.4
			Oolong tea	80.1	8.6	76.8	3.0	81.9	2.7	84.3	1.8	90.2	8.9	79.7	1.9	74.5	4.2	74.5	1.0	94.5	5.1	86.1	5.6	83.0	9.2	78.4	5.2
43	Quinalphos	25	Green tea	72.0	3.7	84.2	1.9	80.5	6.8	95.1	4.3	94.0	4.8	82.8	12.8	75.7	9.8	96.3	10.6	102.5	15.5	85.2	18.4	78.7	5.5	89.6	10.3
			Oolong tea	79.7	6.4	100.8	4.5	97.4	2.7	87.3	3.8	88.8	10.0	79.2	9.0	83.8	5.2	86.1	4.3	101.9	4.0	89.4	3.5	83.1	8.7	75.1	13.7

TABLE 2.3.2 Average Recoveries and RSD Data for 84 Pesticides Spiked in Green Tea and Oolong Tea With 12 Solid Phase Extraction Cartridge Cleanup (n = 3) (cont.)

No.	Pesticides	Spiked concentration (μg/kg)	Sample	1# AVE (%)	1# RSD (%)	2# AVE (%)	2# RSD (%)	3# AVE (%)	3# RSD (%)	4# AVE (%)	4# RSD (%)	5# AVE (%)	5# RSD (%)	6# AVE (%)	6# RSD (%)	7# AVE (%)	7# RSD (%)	8# AVE (%)	8# RSD (%)	9# AVE (%)	9# RSD (%)	10# AVE (%)	10# RSD (%)	11# AVE (%)	11# RSD (%)	12# AVE (%)	12# RSD (%)
44	Trans-chlodane	25	Green tea	80.8	3.6	76.0	0.7	79.4	1.7	88.7	4.0	89.6	3.2	81.3	6.7	80.1	9.6	82.5	6.6	93.8	3.5	92.0	3.7	80.6	3.1	81.3	3.8
			Oolong tea	78.8	2.8	77.2	3.0	82.7	2.4	84.7	2.0	88.2	9.1	78.4	3.5	82.7	3.5	81.4	0.9	93.6	4.4	85.0	6.0	84.2	1.0	78.5	5.6
45	Metazachlor	75	Green tea	82.8	3.0	83.4	3.0	84.4	3.5	83.9	7.4	90.2	2.5	82.9	5.3	86.3	11.4	87.4	3.8	87.1	16.5	90.7	6.1	80.8	3.5	79.5	7.9
			Oolong tea	78.2	8.6	71.3	8.6	84.2	4.2	87.9	6.0	88.4	12.9	81.5	1.6	84.8	7.5	85.5	5.9	103.2	4.5	82.9	2.6	86.6	1.9	81.6	7.3
46	Prothiophos	25	Green tea	77.9	7.3	79.4	5.3	80.3	0.9	92.1	4.3	95.1	5.6	82.8	5.7	82.1	11.4	74.6	4.9	96.4	3.9	91.2	3.2	N.D	N.D	77.6	4.1
			Oolong tea	81.7	2.7	80.2	6.0	84.8	3.0	88.8	6.5	96.7	14.4	70.8	4.5	79.0	5.4	83.7	0.4	93.4	6.3	89.5	6.7	84.3	4.3	80.3	7.7
47	Folpet	100	Green tea	N.D	N.D	N.D	N.D	N.D	N.D	N.D	N.D	N.D	N.D	N.D	N.D	N.D	N.D	N.D	N.D	N.D	N.D	N.D	N.D	N.D	N.D	N.D	N.D
			Oolong tea	N.D	N.D	50.7	18.2	N.D	N.D	N.D	N.D	54.5	23.6	51.1	17.6	N.D	N.D	N.D	N.D	42.2	8.9	41.8	11.1	74.5	89.7	N.D	N.D
48	Chlorfurenol	75	Green tea	65.5	20.3	69.9	4.2	79.4	2.7	92.9	10.3	90.6	6.0	82.2	7.9	77.7	8.9	80.6	N.D	75.5	20.5	85.9	4.8	78.9	3.7	79.9	6.4
			Oolong tea	67.9	12.3	76.1	2.3	82.6	3.8	82.5	3.5	86.1	17.8	82.7	0.3	86.0	3.0	85.5	3.8	106.8	7.3	80.6	1.8	86.8	3.8	78.6	6.5
49	Procymidone	25	Green tea	90.8	7.0	82.1	2.5	80.1	2.2	88.2	2.2	93.0	3.4	85.1	6.1	86.3	7.7	88.3	2.6	88.4	9.6	93.0	5.5	84.0	8.2	77.9	3.9
			Oolong tea	83.8	8.6	76.2	2.7	112.0	0.8	86.4	2.3	87.9	8.0	80.8	4.8	86.4	2.6	85.8	1.1	98.6	2.7	87.0	6.5	86.5	2.1	79.1	5.4
50	Methidathion	50	Green tea	N.D	N.D	N.D	N.D	84.0	2.5	N.D	N.D	90.0	N.D	N.D	N.D	N.D	N.D	N.D	N.D	N.D	N.D	N.D	N.D	N.D	N.D	N.D	N.D
			Oolong tea	50.2	75.0	N.D	N.D	N.D	N.D	N.D	N.D	N.D	N.D	73.1	7.1	N.D	N.D	N.D	N.D	N.D	N.D	N.D	N.D	N.D	N.D	N.D	N.D
51	Cyanazine	75	Green tea	55.2	36.9	73.5	10.8	84.5	14.9	91.3	6.3	86.7	16.7	66.9	2.0	69.4	9.9	54.4	13.7	90.7	16.6	76.2	2.2	71.8	9.2	85.3	17.3
			Oolong tea	77.0	5.6	89.9	5.6	82.0	8.6	87.2	13.5	85.6	6.8	76.1	14.4	N.D	N.D	96.6	9.9	106.6	10.0	70.9	7.9	98.0	7.9	78.8	10.1
52	Napropamide	75	Green tea	80.5	6.0	76.6	1.3	78.6	2.3	90.1	4.2	89.1	3.1	80.4	6.0	79.8	8.9	79.9	3.0	88.3	8.5	94.5	4.0	81.5	3.4	82.0	3.1
			Oolong tea	57.6	8.7	66.0	7.0	81.6	3.7	70.4	1.4	88.0	9.3	77.5	4.6	73.1	16.3	76.7	4.6	102.0	4.6	78.9	12.0	85.3	2.5	78.9	5.3
53	Oxadiazone	25	Green tea	122.8	19.1	76.2	N.D	71.5	1.2	95.9	2.7	90.9	4.6	78.9	8.2	87.2	13.7	99.1	5.9	107.5	1.3	77.4	2.6	N.D	N.D	N.D	N.D
			Oolong tea	N.D	N.D	N.D	N.D	N.D	N.D	N.D	N.D	95.7	8.2	78.9	4.9	N.D	N.D	N.D	N.D	103.9	6.8	84.7	4.4	N.D	N.D	N.D	N.D
54	Fenamiphos	75	Green tea	72.3	5.3	78.0	4.0	72.3	5.3	88.5	5.1	11.4	31.4	45.1	10.6	67.0	8.0	69.9	8.0	84.5	4.5	84.6	2.5	77.1	3.2	71.7	8.5
			Oolong tea	N.D	N.D	76.2	N.D	65.3	4.0	78.9	2.6	28.3	30.0	19.4	30.0	N.D	N.D	18.7	N.D	96.0	7.5	69.4	4.7	70.4	0.5	68.5	7.4
55	Tetrasul	25	Green tea	86.0	17.6	80.5	11.6	67.7	3.0	79.0	6.3	96.1	7.0	63.4	15.9	81.5	6.4	69.7	7.3	76.7	3.3	87.0	4.2	76.4	5.3	81.1	3.2
			Oolong tea	74.9	5.1	78.0	3.0	84.1	2.3	83.7	5.8	88.4	5.9	78.2	9.1	83.2	7.3	76.8	1.2	94.9	4.0	90.5	5.2	76.2	2.2	76.2	5.0
56	Bupirimate	25	Green tea	81.8	4.7	76.2	2.3	78.8	4.2	85.5	1.9	88.6	3.2	73.5	2.9	78.0	9.7	82.6	5.6	88.2	6.3	93.0	4.3	80.8	2.6	81.2	2.7
			Oolong tea	62.0	7.8	74.7	4.2	83.8	1.3	87.8	1.3	84.5	2.6	73.3	10.5	70.2	5.6	77.0	N.D	103.4	5.6	82.9	6.3	87.3	4.0	80.2	5.0
57	Carboxin	600	Green tea	80.6	29.7	86.0	9.8	31.2	23.4	48.1	17.3	N.D	N.D	N.D	N.D	44.7	30.6	25.3	7.2	28.1	33.3	28.8	46.8	59.4	12.3	58.4	15.3
			Oolong tea	105.3	88.9	90.6	N.D	19.6	37.6	37.1	6.7	N.D	N.D	N.D	N.D	N.D	N.D	N.D	N.D	81.3	N.D	84.3	15.6	N.D	N.D	27.7	24.2
58	Flutolanil	25	Green tea	81.7	8.2	79.4	2.0	80.6	1.3	89.6	5.9	89.5	5.9	90.6	2.6	81.9	7.9	84.7	4.1	74.1	31.9	98.4	5.4	86.8	2.4	75.0	0.9
			Oolong tea	72.5	8.8	68.4	5.8	80.8	3.4	84.9	5.4	85.6	5.4	83.6	24.7	87.6	4.1	81.2	7.5	105.7	10.3	81.1	3.8	86.4	1.1	79.3	7.1
59	p,p'-DDD	25	Green tea	77.5	1.0	75.6	4.0	79.7	2.0	92.3	8.9	89.2	4.1	73.3	5.1	76.3	10.8	82.8	3.0	96.1	4.7	84.3	6.4	76.2	5.3	83.7	5.3
			Oolong tea	78.9	4.0	75.6	3.3	67.5	15.4	84.4	3.4	93.4	3.4	82.0	7.4	83.7	3.9	81.9	5.6	99.5	1.1	77.5	4.0	85.9	2.1	77.5	6.2
60	Ethion	50	Green tea	82.5	3.6	77.2	1.7	82.1	2.4	90.8	3.9	97.1	4.0	84.3	2.9	82.7	2.9	86.7	11.1	98.6	5.7	92.0	2.3	81.4	5.7	84.2	4.8
			Oolong tea	74.1	4.6	77.2	3.4	83.4	2.6	85.1	2.6	92.5	9.9	79.7	9.9	86.3	5.7	83.2	2.0	97.2	2.9	87.0	5.1	85.8	1.1	79.7	6.1
61	Sulprofos	50	Green tea	66.2	25.3	71.9	6.2	73.2	3.5	89.0	3.9	16.2	2.5	40.0	3.9	73.4	9.3	75.2	5.7	96.7	8.2	91.1	4.2	74.0	5.2	75.5	3.9
			Oolong tea	18.9	N.D	60.1	12.8	62.0	16.1	78.1	11.7	33.1	30.5	28.3	24.7	N.D	N.D	57.4	N.D	97.9	0.9	81.6	10.7	58.8	5.5	72.2	8.3
62	Etaconazole-1	75	Green tea	81.6	3.7	76.2	1.1	77.5	1.9	96.8	5.6	90.9	3.6	79.9	1.5	82.6	5.0	87.8	2.7	90.7	14.1	95.3	5.3	82.8	5.3	79.3	4.0
			Oolong tea	61.6	7.1	78.2	1.9	85.0	6.0	89.3	4.1	89.2	4.1	82.6	12.7	99.1	0.8	88.8	8.1	103.4	3.9	83.5	6.7	88.3	6.7	76.3	5.4
63	Etaconazole-2	75	Green tea	75.9	4.3	75.4	6.1	80.7	2.8	100.9	2.8	88.6	4.4	91.5	1.9	84.6	10.4	80.2	10.4	89.5	11.1	90.2	7.8	82.0	7.8	75.3	3.4
			Oolong tea	66.5	2.8	90.2	3.1	79.7	4.1	82.8	4.1	101.3	5.1	83.5	0.6	92.1	7.5	89.2	7.5	104.2	11.0	88.9	3.2	87.9	4.2	78.6	4.8

No.	Compound	Conc.	Tea	1	2	3	4	5	6	7	8	9	10	11	12	13	14	15	16	17	18	19	20	21	22	23	24
64	Myclobutanil	25	Green tea	89.6	13.4	67.4	22.5	86.2	2.2	99.8	5.7	92.9	0.7	88.1	6.7	82.8	8.7	86.1	8.1	76.1	26.7	101.3	2.0	91.4	3.5	79.0	8.4
			Oolong tea	79.2	22.9	80.9	4.4	85.2	4.4	88.8	3.9	109.5	8.3	101.6	2.2	87.6	3.5	84.8	8.8	100.1	16.9	94.9	2.8	90.2	2.0	82.1	6.3
65	Dichlorofop-methyl	25	Green tea	74.0	5.5	82.1	7.9	89.8	5.3	97.1	4.5	87.5	3.2	78.7	1.5	86.7	14.0	78.7	7.7	76.8	9.8	72.5	5.2	86.0	1.8	86.6	6.9
			Oolong tea	105.4	12.5	N.D	N.D	N.D	N.D	N.D	N.D	N.D	N.D	N.D	2.4	78.0	10.5	97.8	13.0	100.2	2.9	78.7	24.8	74.7	13.0	80.7	11.3
66	Propiconazole	75	Green tea	78.1	11.1	73.4	4.0	67.9	17.7	95.1	14.5	93.2	4.9	92.4	9.2	85.0	8.3	90.0	14.1	87.9	10.4	98.9	5.2	76.1	3.9	77.4	9.0
			Oolong tea	73.9	15.9	88.8	10.1	75.5	1.3	63.1	4.9	80.6	14.9	60.3	13.6	84.3	3.1	91.2	9.0	95.2	4.2	85.9	6.0	83.4	2.3	79.1	8.7
67	Fensulfothion	50	Green tea	95.7	6.4	83.9	0.8	86.0	6.1	105.7	9.9	58.4	11.0	77.3	12.1	91.6	15.6	84.1	8.3	81.9	37.7	107.7	14.5	89.6	3.5	77.7	14.3
			Oolong tea	65.3	31.3	N.D	N.D	N.D	N.D	96.2	5.2	N.D	N.D	77.8	9.8	N.D	N.D	N.D	N.D	107.6	11.1	67.1	3.4	N.D	N.D	N.D	N.D
68	Bifenthrin	25	Green tea	112.3	7.9	106.0	6.3	110.5	3.8	121.3	5.7	109.3	7.0	93.6	7.4	99.5	8.6	97.9	9.5	105.0	2.9	113.2	3.5	103.8	2.8	3.1	17.9
			Oolong tea	99.5	6.7	99.4	7.2	116.7	3.4	116.5	2.1	112.1	9.9	104.9	3.2	92.0	5.0	97.6	2.0	743.3	145.8	107.4	6.5	113.9	3.0	105.6	6.5
69	Mirex	25	Green tea	76.7	4.6	83.1	5.8	84.6	9.9	91.7	7.6	91.2	11.4	88.7	12.2	88.0	10.7	85.2	4.3	95.3	3.0	91.1	11.6	74.6	1.8	87.7	6.4
			Oolong tea	81.2	20.8	88.6	22.2	85.2	8.1	82.8	4.6	82.8	13.6	77.3	2.2	78.6	1.9	83.6	4.6	91.0	1.4	85.2	7.2	79.4	5.5	75.1	4.6
70	Benodanil	75	Green tea	59.8	20.0	69.5	5.0	75.7	3.3	84.5	4.7	94.4	8.0	123.4	3.1	70.7	2.6	92.5	4.8	55.1	46.1	103.0	14.2	86.6	15.0	77.5	6.7
			Oolong tea	67.6	27.1	79.7	3.2	84.8	3.2	79.3	7.8	83.7	32.2	69.1	5.8	91.2	8.9	87.8	17.2	102.9	14.8	88.0	2.6	83.6	4.5	78.2	15.5
71	Nuarimol	50	Green tea	79.2	7.3	74.6	0.5	78.5	2.7	91.4	4.7	88.2	3.3	90.5	6.9	79.0	6.9	82.4	5.3	78.7	24.1	92.7	4.7	83.3	3.1	48.0	3.7
			Oolong tea	64.7	11.0	73.7	0.9	79.6	2.2	80.9	1.3	82.6	18.4	40.9	0.5	83.5	7.1	79.5	4.0	104.6	5.8	78.3	4.8	85.3	3.2	75.0	6.7
72	Methoxychlor	200	Green tea	74.6	4.2	74.8	5.7	91.5	3.7	89.5	5.3	89.9	3.1	68.7	11.3	85.3	8.2	98.4	3.7	102.6	2.9	84.8	1.6	77.8	5.8	76.9	11.8
			Oolong tea	69.5	13.3	82.0	12.5	92.4	13.0	76.6	7.1	89.8	7.2	82.0	5.2	80.1	6.7	87.5	8.7	94.7	4.1	81.4	6.3	80.3	8.4	71.6	9.2
73	Oxadixyl	25	Green tea	92.2	8.8	105.0	8.3	81.2	8.7	77.8	3.9	74.4	9.4	N.D	N.D	N.D	N.D	N.D	N.D	N.D	N.D	N.D	N.D	90.8	16.1	70.4	9.5
			Oolong tea	N.D	N.D	N.D	N.D	N.D	N.D	N.D	N.D	N.D	N.D	N.D	N.D	N.D	N.D	N.D	N.D	N.D	N.D	N.D	N.D	N.D	N.D	N.D	N.D
74	Tetramethrin	50	Green tea	63.3	12.1	76.2	3.0	76.8	3.9	89.6	39.9	N.D	N.D	70.2	13.1	62.7	5.1	78.0	7.1	60.1	39.4	83.9	4.1	71.3	13.6	97.5	6.5
			Oolong tea	54.3	9.8	50.1	12.3	73.7	2.7	80.4	3.2	N.D	N.D	N.D	N.D	66.9	9.9	82.5	3.8	83.9	16.7	84.9	5.3	N.D	N.D	78.0	15.6
75	Tebuconazole	75	Green tea	80.5	4.2	84.0	19.5	66.2	1.5	77.6	19.1	91.2	2.7	86.9	5.6	81.9	10.4	76.5	11.8	78.8	3.2	93.4	4.1	80.1	2.4	78.5	2.9
			Oolong tea	45.7	10.3	75.8	1.0	80.6	3.4	85.3	5.7	89.1	24.9	78.2	5.1	80.5	5.2	77.8	1.1	110.0	16.7	70.1	2.2	83.7	2.9	75.6	3.6
76	Norflurazon	25	Green tea	77.2	8.4	80.4	4.9	76.1	1.8	87.7	5.9	82.8	4.5	N.D	N.D	75.7	5.8	63.9	2.5	54.4	56.5	104.6	20.7	93.1	7.6	66.7	4.7
			Oolong tea	N.D	N.D	N.D	N.D	N.D	N.D	N.D	N.D	N.D	N.D	N.D	N.D	N.D	N.D	N.D	N.D	N.D	N.D	N.D	N.D	N.D	N.D	N.D	N.D
77	Pyridaphenthion	25	Green tea	75.4	18.1	76.6	1.4	86.1	2.3	92.8	5.3	91.0	3.8	81.4	7.2	77.4	10.3	88.0	3.3	84.4	13.0	87.1	5.0	78.2	3.1	77.0	11.0
			Oolong tea	78.6	2.3	66.9	16.6	78.0	4.7	92.0	9.1	82.2	16.5	87.0	6.0	64.2	4.6	N.D	N.D	N.D	N.D	87.8	2.5	82.2	6.5	73.0	6.2
78	Phosmet	25	Green tea	53.4	68.2	52.6	31.4	96.5	11.1	88.7	6.9	96.2	9.6	84.5	15.5	85.9	10.8	103.1	4.2	87.4	11.4	86.3	11.4	N.D	N.D	50.5	43.1
			Oolong tea	53.3	79.8	N.D	N.D	N.D	N.D	N.D	N.D	N.D	N.D	N.D	N.D	N.D	N.D	N.D	N.D	N.D	N.D	N.D	N.D	N.D	N.D	N.D	N.D
79	Tetradifon	25	Green tea	15.7	14.1	78.6	27.2	67.2	14.3	109.3	4.6	90.2	7.0	N.D	N.D	N.D	N.D	N.D	N.D	N.D	N.D	N.D	N.D	N.D	N.D	72.2	33.2
			Oolong tea	N.D	N.D	N.D	N.D	N.D	N.D	N.D	N.D	N.D	N.D	N.D	N.D	N.D	N.D	N.D	N.D	N.D	N.D	N.D	N.D	N.D	N.D	N.D	N.D
80	Chloridazon	10	Green tea	72.0	27.9	390.5	139.3	N.D	2.4	87.3	2.4	N.D	N.D	82.6	8.8	116.4	8.8	8.4	47.6	91.8	9.4	53.9	3.2	16.1	1.7	73.7	5.9
			Oolong tea	113.3	65.0	84.7	5.1	85.1	0.5	91.6	1.5	101.9	50.1	N.D	N.D	86.7	6.6	88.4	11.4	83.6	36.1	94.8	17.8	88.8	5.4	80.7	7.6
81	Pyrazophos	50	Green tea	91.7	14.0	99.1	8.9	87.2	4.9	96.0	4.9	97.1	3.3	81.0	10.8	93.4	8.8	93.3	5.3	94.3	7.3	90.9	4.3	78.9	4.6	78.0	8.4
			Oolong tea	90.7	9.7	75.8	3.4	114.5	35.0	85.0	3.1	100.8	18.3	79.6	3.2	86.7	3.8	82.9	3.0	105.1	1.4	87.1	3.9	82.6	1.6	78.7	6.7
82	Cypermethrin	75	Green tea	N.D	N.D	N.D	N.D	17.7	24.7	N.D	N.D	N.D	N.D	N.D	N.D	N.D	N.D	N.D	8.6	90.8	8.6	N.D	N.D	N.D	N.D	98.1	9.2
			Oolong tea	N.D	N.D	N.D	N.D	N.D	N.D	N.D	N.D	108.7	5.6	97.5	5.8	90.8	8.6	N.D	N.D	N.D	N.D	N.D	N.D	N.D	N.D	N.D	N.D
83	Fenvalerate	100	Green tea	N.D	N.D	N.D	N.D	67.6	19.4	125.9	8.4	N.D	N.D	83.7	4.9	77.7	2.7	77.7	6.7	93.2	17.3	107.9	13.8	109.3	9.1	113.1	26.0
			Oolong tea	107.3	18.7	101.1	5.9	109.5	9.3	104.7	12.7	89.8	28.7	116.1	10.0	91.2	6.7	71.9	12.5	N.D	N.D	N.D	N.D	N.D	N.D	94.4	6.8
84	Deltamethrin	150	Green tea	N.D	N.D	N.D	N.D	47.7	25.7	N.D	N.D	N.D	N.D	N.D	N.D	71.9	12.5	N.D	N.D	N.D	N.D	N.D	N.D	N.D	N.D	236.2	33.6
			Oolong tea	N.D	N.D	N.D	N.D	N.D	N.D	N.D	N.D	82.6	17.3	68.2	6.6	N.D	N.D	N.D	N.D	N.D	N.D	N.D	N.D	N.D	N.D	67.0	11.6

TABLE 2.3.3 Average Recoveries and RSD Data Distribution for 84 Pesticides Spiked in Green Tea and Oolong Tea With 12 Solid Phase Extraction Cartridge Cleanup

Average Rec. and RSD			1#	2#	3#	4#	5#	6#	7#	8#	9#	10#	11#	12#
Green tea														
AVE	AVE < 70%		12	10	11	2	4	6	5	8	4	5	2	5
	70% ≤ AVE ≤ 110%		59	64	67	71	68	61	66	68	66	65	67	69
	AVE > 110%		4	1	1	4	0	3	1	0	0	1	0	3
	N.D.		9	9	5	7	12	14	12	8	14	13	15	7
RSD	RSD < 20%		63	70	76	75	70	70	68	74	60	67	68	72
	RSD ≥ 20%		12	5	3	2	2	0	4	2	10	4	1	5
	N.D		9	9	5	7	12	14	12	8	14	13	15	7
70% ≤ AVE ≤ 110% and RSD < 20%			52 (61.9%)	61 (72.6%)	67 (79.8%)	69 (82.1%)	67 (79.8%)	61 (72.6%)	65 (77.4%)	67 (79.8%)	61 (72.6%)	63 (75.0%)	66 (78.6%)	68 (81.0%)
Oolong tea														
AVE	AVE < 70%		19	10	7	5	6	10	3	5	1	5	6	8
	70% ≤ AVE ≤ 110%		50	58	59	67	64	60	58	60	68	67	60	61
	AVE > 110%		1	2	4	0	1	2	0	0	1	0	4	2
	N.D.		14	14	14	12	13	12	23	19	14	12	14	13
RSD	RSD < 20%		56	66	66	70	60	71	61	63	67	71	64	68
	RSD ≥ 20%		13	3	4	1	11	1	0	2	3	1	6	3
	N.D		14	15	14	13	13	12	23	19	14	12	14	13
70% ≤ AVE ≤ 110% and RSD < 20%			42 (50.0%)	55 (65.5%)	59 (70.2%)	65 (77.4%)	57 (67.9%)	60 (71.4%)	58 (69.0%)	60 (71.4%)	66 (78.6%)	66 (78.6%)	59 (70.2%)	61 (72.6%)

Numbers in brackets are the percentages of the total number of pesticides.

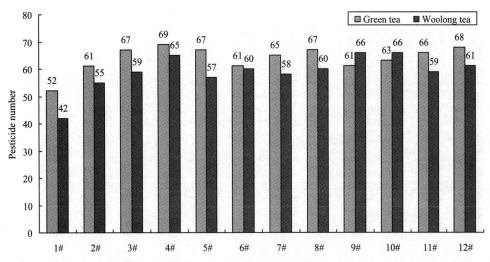

FIGURE 2.3.1   Comparison of recoveries and RSD for 84 pesticides in tea with 12 solid phase extraction (SPE) cartridges cleanup.

former is slightly better than the latter by 10%. In terms of meeting the two conditions of both recoveries 70%–110% and RSD < 20%, however, Table 2.3.5 and Fig. 2.3.2 demonstrate that for green tea samples GC–MS test results indicate that there are 193 pesticides and 183 pesticides, respectively, that can satisfy the conditions with Cleanert TPT and 4# Envi-Carb+PSA cartridges, accounting for 96.0% and 91.0% of total number, respectively. GC–MS/MS test results indicate that there are 191 pesticides and 184 pesticides, respectively, that can satisfy the conditions with Cleanert TPT and 4# Envi-Carb + PSA cartridge, accounting for 95.0% and 91.5% of the total number, respectively. For Oolong tea samples, GC–MS test results indicate there are 192 pesticides and 177 pesticides, respectively, that can meet the conditions with Cleanert TPT and 4# Envi-Carb + PSA cartridge, accounting for 95.5% and 88.1% of the total number, respectively. GC–MS/MS test results indicate that there are 195 and 184 pesticides, respectively, that can satisfy the conditions with Cleanert TPT and 4# Envi-Carb + PSA cartridge, accounting for 97.0% and 91.5% of the total number, respectively. In other words, based on the aforementioned analysis, whether it is green tea or Oolong tea or be it GC–MS or GC–MS/MS determination, the results from the clean up of 201 pesticides in tea with the Cleanert TPT cartridge are slightly better than those from Envi-Carb + PSA combined cartridge by 5%.

2.3.7   STAGE III: THE COMPARATIVE TEST ON THE CLEANUP EFFICIENCIES OF GREEN TEA AND OOLONG TEA YOUDEN PAIR SAMPLES INCURRED BY 201 PESTICIDES WITH CLEANERT TPT CARTRIDGE AND 4# ENVI-CARB+PSA COMBINED CARTRIDGE

The green and Oolong tea samples showing evidence of 201 pesticides used here were prepared in accordance with the following: samples were stored under laboratory conditions, taken for inspection once every 5 days, cleaned up respectively with 4# Envi-Carb+PSA and Cleanert TPT, and detected by GC–MS and GC–MS/MS, running a parallel analysis of three samples for each kind of sample at one time. Such circulative tests last 3 months, with a total of 19 determinations and 183,312 original data of pesticide content (2 kinds of tea × 2 concentrations × 2 kinds of SPE cartridges cleanup × 2 instruments × 19 determinations × 3 parallel samples × 201 pesticides) obtained. Here, the average content and RSD of three parallel samples in every instance was calculated for each pesticide, respectively obtaining 61104 data (see Annex 1 and Annex 2).

2.3.7.1   Comparison of the Determined Values of Target Pesticide Content

To comprehensively evaluate the influences of the two kinds of cartridges on the determined values of the target pesticide content, the average values of a total of four analytical results, corresponding to each cartridge from two determined values of two concentrations of each tea and two kinds of instruments (Annex 1) were calculated. Total average values were calculated for the average values obtained from 19 times stability tests over 3 months for each pesticide in Tables 2.3.6 and 2.3.7. Calculations were also made per the following Eq. (2.3.1) to obtain the variance ratio (R) of the determined results from the two corresponding cartridges. These are listed in Tables 2.3.6 and 2.3.7, as well. See Table 2.3.8 and Fig. 2.3.3 for statistical calculation of variance ratio distributions.

TABLE 2.3.4 Average Recoveries and RSD Data for 201 Pesticides in Green Tea and Oolong Tea With Two Solid Phase Extraction Cartridges Cleanup and GC–MS and GC–MS/MS Determinations (n = 6)

No.	Pesticides	Spiked concentration (µg/kg)	Cleanert TPT								Envi-Carb+PSA(0.5 g)							
			Green tea				Oolong tea				Green tea				Oolong tea			
			GC–MS		GC–MS/MS		GC–MS		GC–MS/MS		GC–MS		GC–MS/MS		GC–MS		GC–MS/MS	
			AVE (%)	RSD (%)	AVE (%)	RSD (%)	AVE (%)	RSD (%)	AVE (%)	RSD (%)	AVE (%)	RSD (%)	AVE (%)	RSD (%)	AVE (%)	RSD (%)	AVE (%)	RSD (%)
1	2,3,4,5-Tetrachloroani…	10	97.7	8.0	96.8	6.8	92.3	8.6	91.2	8.4	79.3	7.4	79.3	6.9	77.6	7.6	78.4	8.6
2	2,3,5,6-Tetrachloroani…	50	93.6	5.6	93.6	6.1	86.2	7.1	90.1	7.8	80.9	5.4	81.2	5.8	75.7	5.7	80.8	7.1
3	2,4'-DDE	100	100.3	8.3	97.0	8.9	95.4	4.4	92.1	6.0	78.5	9.6	75.2	10.7	82.4	7.5	80.9	8.1
4	2,4'-DDT	50	76.1	8.5	77.6	5.9	78.4	5.9	80.2	8.5	82.4	20.9	83.0	16.9	88.5	12.7	77.9	9.9
5	4,4'-DDE	50	91.1	5.7	87.3	5.1	96.3	2.5	92.3	3.0	77.2	7.3	76.1	6.0	86.9	5.9	86.6	6.7
6	4,4-Dibromobenzophenone	50	112.2	7.3	96.1	6.3	94.3	5.5	98.0	6.2	79.7	6.3	81.6	8.5	92.7	5.6	86.7	6.3
7	4,4-Dichlorobenzophenone	50	97.9	5.5	95.5	5.3	90.1	4.4	91.1	4.7	81.1	7.6	83.2	5.6	93.5	9.0	94.9	7.9
8	Acetochlor	100	91.0	8.2	99.8	3.2	80.1	2.9	88.4	5.6	86.4	5.3	83.0	4.3	80.5	7.8	85.2	6.8
9	Alachlor	150	99.3	11.9	96.8	4.6	82.5	6.3	86.5	4.7	82.2	10.1	80.7	5.0	82.3	8.0	78.6	5.8
10	Ametryn	150	38.8	28.5	96.3	9.1	89.9	20.0	88.1	3.3	73.0	24.7	81.8	4.7	84.2	32.2	75.2	8.0
11	Atratone	50	96.9	4.6	95.6	5.3	91.9	6.6	94.6	6.5	83.5	5.2	87.2	4.3	82.1	6.6	82.0	6.6
12	Benalaxyl	50	88.3	4.4	87.5	7.1	92.2	2.4	89.1	3.7	84.2	13.2	84.4	9.5	88.6	8.1	86.0	6.8
13	Benodanil	150	106.4	8.8	81.6	11.8	65.7	11.1	76.3	4.5	113.0	31.4	94.5	15.6	140.2	18.3	95.2	18.8
14	Benoxacor	100	91.3	16.1	89.9	7.5	68.0	8.9	75.4	5.9	93.0	10.1	89.3	12.5	95.8	12.9	97.5	10.9
15	Benzoylprop-ethyl	150	93.1	4.9	88.4	5.3	94.0	2.0	90.9	3.2	84.1	9.2	80.3	9.3	88.1	6.6	87.8	5.7
16	Bifenthrin	50	101.5	4.6	96.8	8.6	174.8	25.8	172.4	25.4	88.1	3.1	85.6	2.8	170.6	29.2	160.6	30.1
17	Biteranol	150	97.0	5.9	96.4	9.9	92.0	3.4	86.8	4.0	80.5	10.5	86.2	3.8	93.6	8.6	89.4	8.4
18	Boscalid	200	92.4	15.6	76.8	16.5	85.3	5.2	70.1	11.7	96.1	10.6	104.0	23.8	96.0	22.3	98.2	22.2
19	Bromofos	100	81.6	6.2	81.2	6.0	84.8	5.5	79.4	4.4	78.7	11.4	86.1	9.8	78.7	6.9	90.6	14.4
20	Bromophos-ethyl	50	91.5	5.5	92.5	5.5	84.6	6.1	86.1	5.6	86.0	5.7	85.4	5.6	80.6	8.0	79.6	7.7
21	Bromopropylate	100	76.1	11.1	82.2	7.8	83.7	5.5	85.3	6.7	94.6	28.4	85.1	21.1	103.8	18.9	89.0	17.4
22	Buprofenzin	100	95.9	5.3	91.5	5.5	97.5	2.7	92.7	4.5	81.6	8.0	77.2	5.9	91.5	7.6	83.0	6.8
23	Butachlor	100	89.2	4.8	91.9	5.5	96.9	6.7	92.4	3.8	75.1	7.7	77.0	6.9	67.0	5.1	85.0	8.6
24	Butafenacil	50	92.2	10.3	74.1	23.4	76.2	6.2	66.5	14.2	123.5	33.9	126.6	36.6	91.2	26.4	87.1	27.3
25	Butralin	200	87.2	11.6	79.4	11.1	66.4	7.1	66.3	5.3	102.4	14.4	110.0	30.4	107.4	22.8	86.0	22.5
26	Butylate	150	90.2	6.9	82.3	6.5	92.9	1.3	84.3	2.3	114.5	34.6	72.1	6.2	106.9	31.1	83.0	6.6
27	Carbaryl	150	101.0	6.5	97.8	10.4	97.3	4.3	100.5	8.0	84.4	6.4	84.8	5.9	78.6	6.7	82.9	6.6
28	Carbofenothion	100	85.3	6.6	87.6	5.8	88.8	3.1	92.2	6.0	79.8	10.4	81.5	11.4	88.9	6.6	84.1	7.7
29	Chlorfenapyr	400	97.2	5.5	95.4	4.3	80.1	8.4	83.6	8.5	84.0	5.3	85.2	5.6	128.0	20.5	127.6	19.6
30	Chlorfenson	100	72.9	11.5	72.4	12.5	77.9	6.0	73.3	8.4	100.3	24.5	98.7	32.1	95.6	21.9	82.0	16.3
31	Chlorfenvinphos	150	80.4	9.9	87.0	5.4	82.5	7.2	82.0	7.3	77.7	13.1	83.5	18.2	84.8	8.0	81.6	6.9
32	Chlormephos	100	83.3	6.3	83.2	5.8	86.9	1.8	85.6	2.2	80.6	5.8	74.6	7.2	84.5	5.9	77.4	8.9
33	Chlorobenzilate	50	79.9	15.9	86.5	6.1	123.2	4.4	90.1	3.4	100.6	30.4	75.0	6.7	98.1	25.2	83.7	7.5
34	Chloroneb	50	89.9	3.8	81.8	14.5	100.2	3.5	74.2	7.1	88.1	11.6	97.0	20.8	78.5	7.2	104.6	22.4

#	Compound	Spike																
35	Chloropropylate	50	86.1	12.9	85.5	6.8	81.7	8.8	90.7	3.8	87.0	22.5	78.1	12.6	103.0	13.7	79.6	8.2
36	Chlorpropham	100	83.6	8.8	85.0	6.6	90.7	2.3	72.6	7.0	81.7	17.0	81.4	14.5	92.4	8.2	84.9	10.9
37	Chlorpyrifos(ethyl)	50	82.2	6.0	85.0	6.0	84.9	3.8	85.5	3.8	82.7	14.3	82.5	13.9	82.7	8.6	79.9	7.5
38	Chlorthal-dimethyl	100	91.7	8.5	93.3	8.1	88.1	3.2	86.8	3.5	85.3	5.1	83.4	3.2	83.8	8.6	87.4	8.9
39	Chlorthiophos	150	87.1	5.0	88.3	6.0	91.6	2.6	90.8	4.1	79.7	10.4	79.8	9.2	85.3	6.0	89.6	10.3
40	cis-Chlordane	100	86.5	4.6	86.4	6.4	91.1	2.8	88.9	2.5	81.0	10.7	77.6	8.0	86.3	6.8	84.5	7.1
41	cis-Diallate	100	88.2	4.5	86.0	4.8	93.2	1.5	88.2	2.7	80.1	9.1	77.6	7.9	83.1	5.4	81.1	10.1
42	Clomazone	50	92.7	9.1	86.3	9.9	80.8	3.5	80.3	5.6	97.8	12.9	102.4	13.1	86.1	15.6	83.3	16.4
43	Cyanofenphos	50	102.9	5.4	80.0	9.9	91.7	4.8	77.3	7.8	100.9	8.6	97.0	26.1	115.9	6.7	88.1	14.8
44	Cycloate	50	94.2	4.8	92.3	5.1	87.8	6.9	91.4	7.6	83.1	6.0	83.0	6.4	77.0	5.1	80.7	7.0
45	Cycluron	150	101.6	5.2	99.9	7.0	95.0	6.6	92.8	6.9	85.2	6.4	82.2	9.2	77.3	4.8	66.9	6.1
46	Cyhalofop-butyl	100	96.7	9.1	87.8	8.1	85.2	5.1	83.9	6.4	97.9	9.6	101.3	13.8	85.0	12.7	83.7	15.0
47	Cyprodinil	50	96.8	4.7	93.5	4.9	98.6	10.3	88.6	7.7	84.4	5.8	84.1	5.8	91.0	12.9	80.4	2.3
48	Dacthal	50	93.4	7.9	93.4	6.2	85.0	3.5	89.0	3.4	86.6	6.2	84.8	4.1	83.8	8.5	82.5	7.6
49	DE-PCB101	50	93.5	5.4	92.9	4.9	90.9	6.8	89.8	7.6	81.4	6.0	83.8	5.3	80.1	6.9	79.4	5.9
50	DE-PCB118	50	95.9	7.9	90.7	4.9	87.9	6.3	89.9	7.5	80.9	4.6	81.7	6.2	80.0	7.8	77.4	5.9
51	DE-PCB138	50	92.9	5.0	92.2	5.4	90.6	6.8	90.1	6.4	81.0	5.6	80.9	5.7	79.3	6.5	77.4	4.8
52	DE-PCB180	50	96.1	14.2	90.3	5.0	88.8	6.6	88.7	6.8	78.9	5.6	80.9	7.6	80.7	6.4	78.6	7.6
53	DE-PCB28	50	92.9	4.8	92.5	4.8	91.1	6.4	91.1	7.3	81.7	5.8	82.8	5.7	79.9	6.1	79.9	6.4
54	DE-PCB31	50	92.9	4.7	92.5	4.8	91.8	6.4	91.1	7.3	81.6	6.0	82.8	5.7	80.7	6.7	79.9	6.4
55	Desmetryn	50	91.4	6.1	88.8	5.9	99.8	4.1	95.5	3.0	80.8	8.4	77.5	7.5	86.6	5.5	84.8	6.1
56	Dibutyl succinate	100	95.7	8.3	95.7	9.0	88.3	2.8	85.7	2.6	82.9	5.0	82.9	4.5	82.6	6.7	78.4	7.7
57	Dichlobenil	10	88.1	6.6	76.8	7.1	89.3	1.9	81.0	2.7	74.5	7.3	70.8	4.9	80.5	6.6	80.4	6.5
58	Dichlorofop-methyl	50	98.8	10.1	79.8	14.4	81.2	10.5	70.5	8.2	113.9	22.6	118.1	27.2	84.8	21.5	82.0	23.2
59	Dicloran	100	74.3	10.1	84.2	6.1	97.4	9.4	87.6	7.6	90.7	7.4	78.1	16.1	99.0	7.6	80.1	6.8
60	Dicofol	100	100.2	5.5	94.8	5.8	90.5	4.3	88.2	5.1	80.1	7.0	77.1	7.3	94.8	8.2	88.4	10.3
61	Diethofencarb	300	90.8	10.3	92.6	9.7	86.1	5.8	85.9	2.8	90.6	13.2	85.1	5.3	80.9	9.0	91.9	11.5
62	Diflufenican	50	92.4	14.0	88.7	10.9	85.0	2.9	83.2	4.0	89.2	15.6	85.8	6.6	93.7	14.4	101.1	15.5
63	Dimepiperate	100	94.6	9.3	93.4	9.0	88.1	2.5	87.2	2.6	88.3	4.4	81.9	4.5	87.9	8.6	84.9	8.7
64	Dimethachlor	150	69.9	7.5	90.2	4.2	85.9	4.3	86.6	5.2	74.6	11.2	76.0	7.1	84.0	6.0	84.0	4.6
65	Dimethametryn	50	97.9	8.2	95.8	10.2	88.5	3.6	87.5	3.0	82.9	5.4	82.4	3.1	79.8	6.2	78.5	6.9
66	Dimethenamid	50	88.2	6.8	96.4	5.0	82.2	7.1	87.3	5.4	76.9	5.4	81.5	5.0	73.1	7.9	79.8	6.1
67	Dimethomorph	100	93.9	11.5	94.3	9.7	92.4	2.9	90.5	3.1	77.1	4.9	82.1	5.4	84.1	5.0	79.7	5.7
68	Dimethylphthalate	200	95.6	8.0	94.1	8.5	89.4	2.4	85.0	1.4	84.0	4.5	83.1	4.4	83.0	5.8	79.6	7.3
69	Diniconazole	150	95.9	9.2	97.3	9.6	88.1	2.3	87.0	2.7	86.0	9.1	81.6	3.2	86.1	6.8	86.9	8.5
70	Diofenolan-1	100	97.4	5.4	96.6	4.3	93.6	6.6	94.7	7.8	84.1	5.1	86.0	5.3	82.2	6.4	82.7	6.8
71	Diofenolan-2	100	96.2	5.3	96.6	4.3	93.9	6.7	96.7	7.3	84.1	5.3	87.1	5.2	82.4	6.6	82.7	6.8
72	Dioxacarb	400	90.1	5.8	86.9	7.7	94.6	4.1	94.7	5.1	83.6	11.8	80.5	12.5	94.7	6.8	82.0	12.1
73	Diphenamid	50	90.5	8.6	98.9	8.4	94.3	17.6	89.2	2.5	78.9	7.0	81.2	4.0	92.9	14.7	77.6	6.7
74	Dipropetryn	50	98.2	8.6	94.6	10.0	90.4	3.4	87.8	2.4	83.2	4.4	82.6	2.9	77.3	5.6	83.1	8.8
75	Endrin	600	88.2	5.9	90.3	6.7	93.3	2.4	92.2	2.7	78.1	9.2	74.8	8.1	88.6	6.9	88.4	8.1
76	Epoxiconazole-2	400	91.3	5.3	90.0	5.7	94.7	1.8	92.8	3.3	82.0	10.0	78.1	10.0	87.7	5.3	86.3	5.5

(Continued)

TABLE 2.3.4 Average Recoveries and RSD Data for 201 Pesticides in Green Tea and Oolong Tea With Two Solid Phase Extraction Cartridges Cleanup and GC–MS and GC–MS/MS Determinations (n = 6) (cont.)

No.	Pesticides	Spiked concentration (μg/kg)	Cleanert TPT								Envi-Carb+PSA(0.5 g)							
			Green tea				Oolong tea				Green tea				Oolong tea			
			GC–MS		GC–MS/MS		GC–MS		GC–MS/MS		GC–MS		GC–MS/MS		GC–MS		GC–MS/MS	
			AVE (%)	RSD (%)	AVE (%)	RSD (%)	AVE (%)	RSD (%)	AVE (%)	RSD (%)	AVE (%)	RSD (%)	AVE (%)	RSD (%)	AVE (%)	RSD (%)	AVE (%)	RSD (%)
77	EPTC	150	83.9	7.7	79.3	6.0	83.2	2.5	78.7	3.0	76.1	5.5	71.5	4.2	80.4	8.6	81.0	6.7
78	Ethalfluralin	200	91.8	9.6	84.6	11.7	83.8	2.8	78.4	3.3	89.8	12.2	92.6	11.7	77.5	11.9	82.4	14.3
79	Ethfumesate	100	68.2	7.4	88.0	6.0	112.6	22.9	91.3	3.7	94.1	10.9	80.2	8.9	173.0	22.8	83.0	10.2
80	Ethoprophos	150	85.0	6.6	86.3	5.2	89.5	2.1	88.6	3.1	79.4	11.9	77.9	9.3	84.5	6.2	85.3	5.9
81	Etofenprox	50	111.2	5.4	108.8	7.5	88.9	2.9	87.0	2.8	97.4	6.6	95.5	10.0	85.3	9.7	83.5	9.9
82	Etridiazol	150	81.8	13.7	82.3	9.2	78.5	4.2	76.4	10.4	87.3	4.5	87.2	6.2	68.0	8.5	79.3	13.9
83	Etrimfos	50	85.7	5.6	87.1	4.4	88.6	3.4	87.7	2.5	80.4	10.0	79.7	8.8	81.7	5.4	83.6	5.7
84	Fenamidone	50	96.1	5.5	91.0	6.6	98.6	2.9	95.2	4.5	83.1	9.4	80.5	8.6	90.0	5.5	83.0	4.8
85	Fenaromol	100	127.4	4.9	90.9	6.9	99.5	3.3	97.6	3.9	86.1	13.4	76.4	9.0	91.2	7.5	80.1	9.2
86	Fenazaquin	50	100.0	7.4	97.5	9.4	92.2	4.5	86.8	2.8	82.9	6.3	82.2	3.9	86.0	8.1	79.8	7.7
87	Fenbuconazole	100	90.0	4.6	97.7	3.8	95.6	7.0	95.2	6.2	84.0	4.7	83.0	6.1	81.7	5.9	82.3	5.8
88	Fenchlorphos	200	89.8	6.1	90.1	8.1	84.2	4.8	83.3	2.8	84.3	3.1	84.2	6.2	73.3	5.2	79.5	7.6
89	Fenoxanil	100	100.0	8.2	108.4	13.5	95.5	4.1	91.0	6.9	86.1	9.7	84.2	6.2	80.9	5.4	71.0	7.9
90	Fenpropidin	100	93.2	10.3	95.6	11.5	88.0	4.9	87.3	4.4	79.2	4.5	79.3	3.9	69.1	12.2	81.4	6.2
91	Fenpyroximate	400	103.9	6.9	111.5	10.4	103.7	8.2	113.7	9.6	77.2	17.3	75.0	22.6	70.8	16.6	73.2	18.0
92	Fenson	50	87.0	12.6	80.7	13.7	79.8	4.8	78.2	3.9	79.7	16.1	98.9	15.7	100.5	21.0	86.6	17.5
93	Flamprop-isopropyl	50	92.5	5.4	90.3	5.9	96.9	3.0	93.6	2.8	81.0	8.3	77.8	7.7	89.7	6.2	87.8	5.9
94	Flamprop-methyl	50	92.4	9.5	90.3	5.9	92.8	3.5	93.6	2.8	90.6	12.5	82.6	12.4	76.3	11.0	87.8	5.9
95	Flufenacet	400	96.9	8.5	94.5	10.9	81.0	11.4	92.3	11.1	93.2	15.5	94.2	7.5	67.5	8.5	79.6	9.7
96	Fluotrimazole	50	96.3	5.0	96.4	4.1	92.2	6.6	95.8	7.1	84.8	5.2	84.9	5.9	82.5	6.1	81.3	6.7
97	Fluroxypr-1-methylhept...	50	95.6	5.3	85.7	6.4	84.5	4.7	86.0	7.5	93.0	10.6	105.6	14.5	87.2	11.7	87.8	10.9
98	Fonofos	50	88.8	5.8	87.7	4.9	93.8	2.5	91.2	2.7	78.5	8.6	75.9	7.1	85.6	5.8	82.6	6.2
99	Furalaxyl	100	99.1	8.2	98.5	9.0	90.9	2.5	88.8	2.3	84.0	5.4	83.7	3.6	82.6	5.7	82.1	6.6
100	Heptachlor	150	91.5	8.8	91.3	9.5	82.0	6.0	82.1	3.7	83.5	5.1	85.8	3.3	76.3	8.6	89.7	10.5
101	Hexachlorobenzene	50	81.2	7.6	83.4	8.2	85.1	6.9	84.3	6.7	72.7	7.8	71.8	5.3	83.6	7.5	82.5	7.2
102	Hexazinone	150	93.9	5.8	90.5	6.4	99.4	2.6	93.6	4.8	81.6	11.7	81.3	12.6	84.5	7.5	82.9	6.6
103	Iodofenphos	100	90.7	7.9	82.7	2.6	84.0	6.5	82.2	9.0	82.9	13.6	79.9	18.1	79.4	7.3	89.4	11.2
104	Iprobenfos	150	94.3	9.1	95.0	8.7	83.4	5.1	88.3	3.3	82.3	4.0	83.2	4.1	83.2	10.6	88.1	8.1
105	Iprovalicarb-1	200	96.3	10.3	91.0	5.1	84.1	5.0	89.1	5.8	94.7	14.6	86.2	7.3	106.4	13.6	94.5	10.4
106	Iprovalicarb-2	200	88.5	7.3	90.3	4.5	81.6	4.9	88.5	5.4	92.4	15.8	86.6	6.8	92.5	12.0	88.5	10.7
107	Isazofos	100	96.4	7.0	93.8	8.1	96.2	2.3	86.4	2.9	86.0	5.3	85.0	2.9	97.1	7.3	78.9	8.3
108	Isodrin	50	110.9	11.7	93.5	5.1	87.3	8.6	92.4	7.3	81.3	5.0	80.2	6.5	78.5	3.6	82.8	6.1
109	Isofenphos	100	84.0	3.1	88.1	5.6	94.9	3.2	92.7	2.1	79.8	14.5	79.2	9.6	89.8	7.1	83.0	6.9
110	Isopeopalin	100	85.6	6.5	97.7	4.9	92.4	2.3	89.5	3.4	80.7	11.4	76.4	10.4	86.8	8.7	79.3	7.0

111	Isoprocarb-1	100	101.0	4.7	110.7	10.4	98.0	7.1	104.1	6.1	81.6	7.6	81.9	8.5	82.2	7.2	126.4	9.4
112	Isoprocarb-2	100	101.9	4.7	83.7	7.4	75.4	7.0	73.6	6.5	114.7	15.4	99.5	8.3	100.6	8.8	98.5	8.4
113	Isoprothiolane	100	101.3	9.2	95.0	8.8	90.2	3.0	87.8	2.4	86.2	3.7	82.9	3.5	84.6	6.2	80.4	6.8
114	Kresoxim-methyl	50	106.5	11.5	121.4	14.7	93.6	4.4	90.8	5.0	86.1	7.0	69.4	29.0	85.9	6.5	88.0	15.5
115	Lenacil	240	95.4	7.0	123.2	16.3	80.1	2.3	72.2	5.5	99.4	31.6	100.3	11.9	111.8	23.5	103.6	27.0
116	Mefenacet	150	85.8	11.0	86.0	12.9	80.2	11.2	80.8	6.4	97.1	8.4	97.2	9.6	233.2	31.2	102.7	15.5
117	Mepronil	50	99.9	6.2	98.7	7.5	95.3	5.4	91.5	4.3	90.9	7.2	80.8	5.7	85.3	7.9	86.1	7.2
118	Metalaxyl	150	99.5	5.4	96.3	4.3	92.0	7.4	93.7	6.6	92.2	6.4	86.8	2.8	86.6	5.7	82.3	6.5
119	Metazachlor	150	89.7	8.0	95.6	4.5	76.2	7.2	79.1	4.5	84.1	6.0	77.7	5.4	87.2	6.8	82.3	5.3
120	Methabenzthiazuron	500	94.9	5.3	91.8	4.7	93.7	7.8	93.1	6.7	85.7	4.4	83.9	5.5	82.0	8.2	79.3	8.6
121	Methoprene	200	91.1	5.8	92.8	9.3	109.1	9.9	94.2	3.1	80.2	10.0	80.9	8.5	88.5	5.4	92.1	3.4
122	Methoprotryne	150	81.2	7.7	89.2	5.9	97.7	2.1	93.7	2.9	79.5	7.9	76.9	8.4	86.8	5.4	90.3	7.2
123	Methoxychlor	50	76.6	9.5	84.9	8.5	78.9	6.9	87.6	5.0	82.5	17.6	79.4	14.1	94.6	15.7	85.4	10.2
124	Methyl-parathion	200	83.4	8.3	85.7	3.5	90.4	5.0	87.3	9.0	81.4	9.9	80.4	15.1	83.4	10.0	79.3	8.9
125	Metolachlor	50	96.1	12.2	88.3	5.4	100.3	3.7	90.2	3.0	75.6	15.6	75.8	7.4	90.3	8.5	79.7	8.6
126	Metribuzin	150	90.9	11.6	85.4	14.3	82.3	3.7	77.7	4.7	89.3	15.5	96.2	13.8	83.9	15.5	81.0	13.9
127	Mirex	50	82.7	6.2	86.9	6.1	73.5	4.9	74.0	5.2	84.2	16.2	82.4	10.1	90.2	17.0	81.1	14.2
128	Molinate	50	100.5	6.6	94.2	8.8	86.0	3.2	84.6	2.1	87.0	8.8	80.1	4.6	81.5	6.5	78.5	7.7
129	Monalide	100	88.0	8.8	96.8	5.0	89.9	8.8	94.8	7.3	82.2	8.9	88.5	6.2	80.8	5.4	82.3	6.6
130	Napropamide	150	98.7	14.6	96.8	9.6	91.2	3.0	88.1	3.2	90.3	8.1	83.6	4.2	93.5	10.1	79.8	7.6
131	Nitrapyrin	150	88.1	5.4	79.4	4.7	84.0	7.7	77.6	9.4	80.6	7.6	71.7	11.5	85.9	7.1	82.1	9.3
132	Nuarimol	100	104.4	7.5	96.5	10.3	84.5	8.4	86.3	3.4	80.6	8.9	82.1	3.7	84.7	7.9	83.9	5.9
133	Oxyfluorfen	200	82.1	8.7	76.4	10.2	88.2	3.4	87.6	13.7	81.5	15.0	87.8	20.1	98.9	13.7	82.3	12.8
134	Paraoxon-ethyl	200	114.9	6.1	87.1	12.6	96.9	4.4	79.9	10.2	107.9	8.2	96.7	12.7	85.3	9.6	87.1	14.8
135	Pebulate	150	91.8	4.9	90.4	4.5	86.5	6.0	87.0	7.3	81.6	6.3	N.D	N.D	77.4	6.4	78.0	6.7
136	Pendimethalin	200	72.1	14.4	75.7	8.6	78.5	4.6	85.5	7.4	90.6	27.8	82.6	23.4	98.9	20.4	70.7	10.4
137	Pentachloroaniline	50	95.5	9.9	93.0	8.5	88.4	11.5	89.3	10.4	83.2	5.8	78.3	7.1	78.3	10.0	81.5	9.2
138	Pentachloroanisole	50	97.8	5.4	88.2	4.2	114.6	12.6	89.1	7.8	86.5	7.4	79.8	5.2	101.1	13.4	78.2	6.5
139	Pentachlorobenzene	50	87.7	6.2	87.5	6.7	84.1	6.8	84.3	7.4	75.6	6.1	76.5	6.8	74.7	6.7	74.3	6.3
140	Permethrin	130	95.9	7.6	93.9	9.5	89.9	2.3	85.9	2.9	92.5	4.3	77.8	4.1	85.9	7.3	79.3	8.0
141	Perthane	50	94.2	5.1	92.2	4.4	93.7	6.9	90.6	6.2	78.8	8.4	88.1	4.1	84.0	6.9	82.6	9.1
142	Phenanthrene	50	92.7	5.0	93.1	6.6	90.4	7.3	92.7	9.0	79.9	5.9	76.0	8.6	79.8	6.9	82.0	7.9
143	Phenothrin	50	102.1	11.1	106.2	35.4	95.3	3.1	88.4	7.2	87.1	6.4	99.3	30.3	85.0	7.0	104.1	20.3
144	Picoxystrobin	100	92.1	5.9	90.2	5.5	96.9	2.0	94.5	2.2	81.6	8.3	78.6	8.2	86.8	5.7	86.0	5.2
145	Piperonyl butoxide	50	99.9	8.0	98.3	9.0	97.9	4.1	89.4	2.9	81.7	5.3	82.1	4.2	86.5	6.4	82.9	7.5
146	Piperophos	150	95.9	6.2	85.7	6.3	86.5	4.2	89.8	7.1	81.0	14.1	86.3	12.6	89.1	14.0	84.9	9.2
147	Pirimicarb	100	94.5	4.3	92.4	4.7	92.5	7.4	93.1	7.0	86.5	5.5	86.8	4.4	82.8	12.8	84.6	8.3
148	Pirimiphoe-ethyl	100	90.7	5.6	90.0	6.9	94.7	1.9	93.3	2.7	79.5	7.7	76.3	7.8	86.5	5.7	85.9	6.2
149	Pirimiphos-methyl	50	88.5	4.9	86.7	5.5	92.5	1.7	91.4	2.8	78.8	8.8	79.9	7.9	84.1	5.3	77.4	6.2
150	Pretilachlor	100	96.9	6.1	98.6	12.7	87.7	5.8	87.5	5.3	84.4	6.2	85.1	4.3	74.9	6.2	80.0	7.3
151	Procymidone	50	98.2	4.8	95.0	4.8	93.6	6.4	95.4	7.3	84.4	5.6	86.9	5.3	83.2	6.0	81.5	6.9
152	Profenofos	300	74.0	10.0	77.3	11.0	80.4	13.8	83.3	12.6	82.5	18.3	75.7	24.9	76.2	12.2	93.7	26.4
153	Profluralin	200	90.6	4.6	77.0	7.3	86.0	3.9	85.8	4.6	101.3	12.2	88.6	19.2	85.4	10.6	83.8	12.7

(Continued)

TABLE 2.3.4 Average Recoveries and RSD Data for 201 Pesticides in Green Tea and Oolong Tea With Two Solid Phase Extraction Cartridges Cleanup and GC–MS and GC–MS/MS Determinations (n = 6) (cont.)

No.	Pesticides	Spiked concentration (µg/kg)	Cleanert TPT								Envi-Carb+PSA(0.5 g)							
			Green tea				Oolong tea				Green tea				Oolong tea			
			GC–MS		GC–MS/MS		GC–MS		GC–MS/MS		GC–MS		GC–MS/MS		GC–MS		GC–MS/MS	
			AVE (%)	RSD (%)	AVE (%)	RSD (%)	AVE (%)	RSD (%)	AVE (%)	RSD (%)	AVE (%)	RSD (%)	AVE (%)	RSD (%)	AVE (%)	RSD (%)	AVE (%)	RSD (%)
154	Prometon	150	99.4	7.4	95.1	9.1	89.2	3.7	88.4	3.1	84.4	5.4	83.1	4.7	81.4	6.3	77.4	7.6
155	Prometryne	50	96.1	4.7	93.2	4.9	91.1	4.9	93.6	7.6	85.4	6.9	84.4	5.7	81.2	4.9	82.3	6.9
156	Pronamide	50	88.6	5.8	89.3	6.8	93.3	2.1	91.5	1.5	73.7	31.1	80.5	5.0	84.8	6.2	86.7	6.7
157	Propachlor	150	85.9	4.3	89.5	4.5	93.4	3.8	84.8	5.8	75.5	7.7	71.8	6.9	91.1	5.5	90.8	7.5
158	Propetamphos	50	96.5	9.4	86.1	14.2	75.8	6.7	77.2	3.4	97.6	12.3	94.3	11.3	81.3	11.1	82.7	14.3
159	Propham	50	90.5	5.8	91.0	5.0	88.5	6.0	89.1	5.8	84.3	10.9	85.9	6.0	93.6	7.9	85.8	9.0
160	Propiconazole	150	94.6	6.0	92.7	4.3	97.9	3.8	93.9	2.6	76.0	8.4	78.5	7.9	81.8	6.0	88.2	5.3
161	Propoxur-1	400	99.7	7.9	97.4	8.9	95.6	3.5	89.1	2.9	82.5	6.3	81.7	6.0	85.3	5.9	80.7	6.0
162	Propoxur-2	400	106.6	8.8	72.2	27.4	137.5	20.0	57.1	11.7	106.6	10.5	167.1	48.9	109.1	22.7	82.2	35.5
163	Propyzamide	100	89.4	5.4	90.7	6.7	92.4	2.1	90.9	2.0	81.7	6.2	78.0	5.5	85.7	5.8	81.0	5.7
164	Prosulfocarb	50	104.9	7.4	94.4	5.3	92.1	6.7	94.2	7.5	83.5	6.3	83.6	6.3	80.6	7.7	82.2	6.7
165	Prothiophos	50	95.3	8.1	91.0	11.7	85.8	3.2	85.1	2.5	85.5	4.8	90.7	4.2	79.4	8.1	78.3	8.4
166	Pyridaben	50	93.5	8.8	113.1	6.3	79.0	3.9	93.4	15.1	84.1	19.9	95.1	8.5	106.0	25.5	161.6	20.1
167	Pyridaphenthion	50	88.2	10.2	89.5	12.5	88.3	8.1	83.9	4.9	88.4	8.6	90.7	9.4	79.7	11.0	99.2	11.9
168	Pyrimethanil	50	96.8	8.1	98.3	10.1	90.9	3.4	87.6	3.0	84.0	6.8	82.0	4.1	81.3	6.6	81.6	7.7
169	Pyriproxyfen	50	99.7	7.4	112.2	5.3	84.6	4.4	94.5	15.4	79.5	6.7	96.5	7.5	86.5	5.6	163.5	21.2
170	Quinalphos	50	87.8	9.7	91.1	10.1	83.6	4.2	83.3	1.8	87.1	5.3	92.4	5.1	82.8	8.3	81.4	7.9
171	Quinoxyphen	50	104.4	10.5	95.3	9.6	82.0	4.6	85.6	3.2	83.7	4.7	80.0	4.5	79.1	7.1	82.2	7.9
172	Ronnel	100	88.2	8.1	90.4	8.7	82.4	3.6	79.7	3.9	81.0	6.2	80.5	6.2	77.9	6.5	76.5	6.5
173	Secbumeton	50	95.7	4.8	95.9	4.8	92.0	6.4	93.8	7.4	82.7	5.0	84.0	6.0	82.7	7.0	83.9	7.4
174	Silafluofen	50	101.8	8.7	96.0	5.7	90.2	7.5	95.9	7.6	84.3	9.2	87.1	8.8	85.0	9.7	86.7	6.7
175	Sulfotep	50	88.5	6.2	88.8	5.5	93.3	1.8	90.5	2.6	86.9	12.0	78.2	7.8	85.5	5.5	94.5	11.6
176	Tebufenpyrad	50	95.3	5.1	91.9	4.6	95.5	3.7	92.5	3.8	81.3	5.5	78.8	5.9	85.9	6.0	83.5	6.6
177	Tebupirimfos	100	90.0	4.8	92.6	4.8	90.1	6.7	91.6	7.6	54.8	13.4	85.9	5.7	78.6	6.7	81.8	6.0
178	Tebutam	100	95.4	4.6	94.8	4.8	91.3	6.8	95.2	6.9	83.4	5.4	87.0	5.6	81.0	6.2	83.5	6.7
179	Tebuthiuron	200	98.9	4.9	94.7	4.3	93.0	5.7	98.8	7.4	82.2	5.5	83.8	6.4	76.9	5.5	82.8	6.9
180	Tefluthrin	50	96.3	3.9	92.9	4.9	95.0	7.5	92.9	7.5	86.3	5.3	85.0	5.0	86.8	6.7	81.4	7.4
181	Telodrin	200	87.5	12.8	84.7	15.2	74.8	4.8	74.3	5.5	94.3	11.2	98.1	11.8	77.6	15.2	78.2	12.8
182	Terbutryn	100	91.1	6.2	89.7	5.6	95.8	2.3	92.8	3.4	78.5	8.8	76.4	7.9	86.6	5.8	88.2	5.8
183	Tetrasul	50	94.2	9.2	94.3	9.2	88.6	3.2	86.5	3.0	78.2	5.7	76.9	3.9	79.8	6.4	79.3	7.9
184	Thenylchlor	100	89.6	10.6	101.4	4.2	75.4	8.9	78.4	5.0	82.4	7.1	75.6	5.7	90.3	10.6	80.6	5.6
185	Thiazopyr	100	98.4	8.1	89.2	9.1	91.1	2.5	93.0	5.1	85.8	4.5	87.5	5.1	82.2	6.4	80.7	8.7
186	Thiobencarb	100	91.4	6.2	88.8	5.5	95.8	2.4	93.9	2.9	79.0	7.1	76.6	6.5	87.8	5.6	87.5	5.5

187	Thionazin	50	93.2	4.8	93.4	4.7	89.9	4.3	89.7	6.2	87.7	4.0	83.9	4.9	81.7	7.1	81.4	7.0
188	Tolclofos-methyl	50	94.5	7.4	95.2	7.9	85.5	2.8	85.8	2.5	84.8	3.7	83.1	5.0	75.8	6.5	81.2	8.6
189	Tralkoxydim	400	86.3	7.1	264.0	23.2	96.2	1.7	95.2		83.8	27.2	99.5	4.3	84.3	13.6	6281.2	16.7
190	trans-Chlordane	50	88.4	6.1	91.5	5.7	93.3	2.7		3.5	80.5	9.6	74.3	10.4	87.9	6.6	86.1	5.1
191	trans-Diallate	100	89.2	6.4	86.1	4.8	92.5	2.2	88.2	2.7	78.7	8.6	77.3	7.1	85.7	5.8	36.5	9.7
192	Transfluthrin	50	96.1	6.9	93.9	9.3	90.1	3.4	87.8	2.6	85.2	4.3	81.7	4.1	82.6	5.5	79.8	7.7
193	Triadimefon	100	95.0	8.4	96.7	9.2	91.2	4.5	90.2	3.6	82.5	4.5	84.0	3.8	85.1	6.4	79.2	7.0
194	Triadimenol	150	99.8	7.6	97.6	9.1	82.8	6.3	89.0	2.4	84.9	5.5	83.0	3.3	85.5	6.2	81.6	7.3
195	Triallate	100	90.3	10.1	89.9	12.1	81.7	2.1	80.9	2.9	89.9	6.9	90.5	6.8	79.7	11.2	77.4	11.3
196	Tribenuron-methyl	50	86.0	13.9	77.1	13.2	80.6	4.8	81.3	10.2	107.3	23.4	118.9	31.3	93.8	21.2	103.6	19.0
197	Trichloronat	50	91.7	5.3	89.5	4.2	84.3	5.3	87.3	5.4	87.0	7.1	87.5	6.2	78.4	8.5	80.6	8.4
198	Trifloxystrobin	200	80.4	10.3	80.5	11.9	79.8	6.9	72.8	8.9	101.0	28.1	103.9	27.7	87.8	19.2	78.6	14.9
199	Trifluralin	100	88.7	7.2	88.6	7.4	80.2	5.1	80.8	6.5	89.6	12.9	103.2	13.4	82.8	12.4	78.8	11.8
200	Vinclozolin	50	88.1	9.1	89.9	11.9	86.6	3.8	84.3	3.7	83.1	5.1	96.0	8.4	84.1	9.7	85.1	11.3
201	Zoxamide	100	96.4	7.7	87.1	6.1	90.3	5.4	90.2	4.2	87.1	10.8	76.9	8.9	83.7	3.0	90.0	7.6

TABLE 2.3.5 Average Recoveries and RSD Data Distribution for 201 Pesticides in Green Tea and Oolong Tea With Two Solid Phase Extraction Cartridge Cleanup

Sample	Average Rec. and RSD		Cleanert TPT		Carb-Carb + PSA	
			GC–MS	GC–MS/MS	GC–MS	GC–MS/MS
Green tea	AVE	AVE < 70%	3	0	1	1
		70% ≤ AVE ≤ 110%	193	194	195	194
		AVE > 110%	5	7	5	5
		N.D.	0	0	0	1
	RSD	RSD < 20%	200	197	185	183
		RSD ≥ 20%	1	4	16	17
		N.D.	0	0	0	1
		70% ≤ AVE ≤ 110% and RSD < 20%	193 (96.0%)	191 (95.0%)	183 (91.0%)	184 (91.5%)
Oolong tea	AVE	AVE < 70%	3	3	4	2
		70% ≤ AVE ≤ 110%	193	195	190	193
		AVE > 110%	5	2	7	6
		N.D.	0	1	0	0
	RSD	RSD < 20%	197	199	183	189
		RSD ≥ 20%	4	1	18	12
		N.D.	0	1	0	0
		70% ≤ AVE ≤ 110% and RSD < 20%	192 (95.5%)	195 (97.0%)	177 (88.1%)	184 (91.5%)

Numbers in brackets are the percentages of the total number of pesticides.

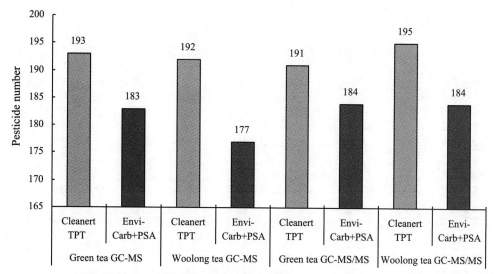

FIGURE 2.3.2 Comparison of the cleanup efficiencies of Cleanert TPT and Envi-Carb+PSA for 201 pesticides in Green tea and oolong tea.

$$R = \frac{|A - B|}{(A + B)/2} \times 100\%$$

(2.3.1)

where R is the variance ratios of the determined results from two SPE cartridges %; A is the total average value of determined results from Cleanert TPT cartridge; B is the total average value of determined results from Envi-Carb+PSA cartridge.

Table 2.3.8 and Figure 2.3.3 show that there is no marked differences for the pesticide content results between Cleanert TPT and 4# Envi-Carb-PSA, no matter whether it is green tea or oolong tea, of which more than 93% of pesticides have variance results less than 15% for the two SPE cartridges. At the same time, one interesting phenomenon has been discovered: there are 179 pesticides (89.1%) for green tea, having obtained the average content with Cleanert TPT greater than that with Envi-Carb+PSA cartridge. However, there are 187 pesticides (93%) for Oolong tea samples having obtained an average content with 4#Envi-Carb-PSA cleanup greater than that with Cleanert TPT. See the statistical analysis tabulated in Table 2.3.9 and Fig. 2.3.4. Such a phenomenon demonstrates that green tea is very different from Oolong tea in matrices due to varieties, which is also reflected in the cleanup efficiencies of two SPE cartridges. In spite of such a marked difference, they are still in conformity with the aforementioned over 93% pesticides with a variance less than 15% obtained for the analytical results from the two SPE cartridges.

2.3.7.2 Comparison of the Reproducibility of the Target Pesticide Determination

To comprehensively evaluate the influences of Clearnet TPT and Envi-Carb+PSA on the reproducibility of the determined values of the target pesticide content, statistical calculation was conducted on 61104 RSD data obtained from 19 times stability tests over 3 months (Annex 2) per distribution ranges. The results are listed in Table 2.3.10. RSD from 19 times stability tests also appears in Table 2.3.10, and RSD distribution ranges are listed in Table 2.3.11. The distribution ranges expressed in bar diagrams are shown in Figs. 2.3.5 and 2.3.6.

Tables 2.3.10 and 2.3.11 and Figs. 2.3.5 and 2.3.6 show that for green tea samples with RSD ≤ 10% pesticide number, GC–MS analytical results indicate there are 6716 and 5800, respectively, with Cleanert TPT and Envi-Carb+PSA, accounting for 87.9% and 75.9%, respectively. GC–MS/MS analytical results indicate there are respectively 6744 and 5634 with Cleanert TPT and Envi-Carb+PSA, accounting for 88.3% and 73.8%, respectively. For Oolong tea samples with RSD ≤ 10% pesticide number, GC–MS analytical results indicate there are respectively 6899 and 6188 with Cleanert TPT and Envi-Carb+PSA, accounting for 90.3% and 81.0%, respectively. GC–MS/MS analytical results indicate that there are respectively 6931 and 6266 with Cleanert TPT and Envi-Carb+PSA, accounting for 90.7% and 82.0%, respectively. This fully demonstrates that no matter whether it is green tea or Oolong tea or be it GC–MS/MS or GC–MS determination, there are more RSD ≤ 10% results obtained with Cleanert TPT than with Envi-Carb-PSA—by around 10%, which totally vtea

TABLE 2.3.6 The Determination Results for 201 Pesticides in Incurred Green Tea Youden Pair Samples Under 4 Conditions in 3 Months

Pesticide content determined With GC–MS and GC–MS/MS, two concentration of Youden pair samples (μg/kg)

No.	Pesticides	Solid Phase Extraction	November 9, 2009 (n = 5)	November 14, 2009 (n = 3)	November 19, 2009 (n = 3)	November 24, 2009 (n = 3)	November 29, 2009 (n = 3)	December 4, 2009 (n = 3)	December 9, 2009 (n = 3)	December 14, 2009 (n = 3)	December 19, 2009 (n = 3)	December 24, 2009 (n = 3)	December 14, 2009 (n = 3)	January 3, 2010 (n = 3)	January 8, 2010 (n = 3)	January 13, 2010 (n = 3)	January 18, 2010 (n = 3)	January 23, 2010 (n = 3)	January 28, 2010 (n = 3)	February 2, 2010 (n = 3)	February 7, 2010 (n = 3)	AVE (μg/kg)	Deviation rates (%)
1	2,3,4,5-Tetrachloroaniline	Cleanert TPT	7.5	5.5	5.4	4.6	4.4	4.8	4.4	4.1	4.6	4.0	4.0	4.1	3.5	4.1	3.3	2.6	3.6	3.1	3.9	4.3	11.6
		Envi-Carb + PSA	5.7	6.7	5.9	5.6	5.7	5.1	4.5	4.5	4.6	4.8	4.0	8.1	3.6	4.1	3.2	4.2	3.7	3.6	3.7	4.8	
2	2,3,5,6-Tetrachloroaniline	Cleanert TPT	42.9	36.5	38.1	39.1	30.3	35.4	33.3	31.6	36.2	32.9	33.7	33.9	34.1	33.4	33.1	25.5	30.9	28.0	36.7	34.0	13.5
		Envi-Carb + PSA	37.2	37.7	38.2	38.6	42.0	38.8	35.7	36.2	35.9	36.4	32.8	76.0	36.1	40.7	35.5	36.7	37.0	32.8	34.4	38.9	
3	4,4-Dibromobenzophenone	Cleanert TPT	50.2	41.8	46.0	42.8	37.6	44.3	45.1	44.5	54.4	36.5	49.4	50.5	55.5	45.3	47.6	58.6	44.6	39.7	66.7	47.4	11.0
		Envi-Carb + PSA	57.3	51.2	50.8	50.6	47.6	47.6	62.7	54.4	59.6	48.2	50.4	46.7	49.1	49.3	56.6	51.3	65.8	52.8	54.0	52.9	
4	4,4-Dichlorobenzophenone	Cleanert TPT	120.3	97.2	110.5	106.0	84.9	97.7	96.8	94.7	113.8	87.3	99.9	106.8	101.8	96.6	95.2	83.0	93.9	83.5	119.5	99.4	10.2
		Envi-Carb + PSA	114.8	112.8	113.4	115.3	112.1	105.2	119.4	111.2	122.3	99.8	102.3	75.3	98.5	117.0	114.2	121.8	121.1	110.3	106.3	110.2	
5	Acetochlor	Cleanert TPT	88.9	72.6	71.0	92.0	57.3	72.9	62.2	68.3	74.5	56.4	67.7	62.7	62.9	80.8	72.7	57.5	73.2	53.6	62.1	68.9	4.2
		Envi-Carb + PSA	74.2	82.8	72.4	70.6	80.8	74.2	71.3	65.3	79.5	80.0	67.9	76.0	58.6	72.5	70.3	69.3	77.4	61.5	61.3	71.9	
6	Alachlor	Cleanert TPT	150.2	104.8	104.1	128.7	83.3	108.3	92.1	90.5	106.0	87.0	100.8	83.3	94.0	87.4	90.7	87.9	79.6	73.9	95.3	97.3	4.8
		Envi-Carb + PSA	114.2	116.7	104.1	101.4	113.7	111.4	108.6	91.6	88.9	126.7	97.3	122.5	86.0	104.8	91.7	90.5	101.9	78.0	89.3	102.1	
7	Atratone	Cleanert TPT	56.6	44.7	44.9	45.8	32.7	57.3	35.1	33.4	40.3	38.8	43.6	34.7	37.2	34.4	34.4	27.5	31.2	27.2	38.0	38.8	5.1
		Envi-Carb + PSA	45.4	45.4	45.8	44.4	49.9	51.0	38.1	36.9	37.2	58.9	35.4	39.8	33.9	39.1	34.4	36.5	36.6	31.8	35.7	40.9	
8	Benodanil	Cleanert TPT	157.3	149.0	123.3	115.0	94.3	183.0	99.9	91.0	131.5	101.0	186.2	139.5	189.9	112.6	118.7	227.3	98.5	94.5	195.0	137.2	5.7
		Envi-Carb + PSA	130.3	142.9	145.7	86.3	145.8	149.6	153.7	118.0	150.7	138.2	137.7	127.5	115.5	103.1	109.1	96.4	181.5	105.1	125.8	129.6	
9	Benoxacor	Cleanert TPT	109.0	92.0	71.9	99.4	63.9	87.4	69.5	70.4	86.0	62.6	105.9	70.8	85.7	80.4	84.6	88.3	66.1	54.4	72.9	80.1	1.0
		Envi-Carb + PSA	88.3	94.5	86.7	60.1	86.4	104.2	77.7	74.1	65.7	77.8	82.8	92.5	79.4	68.9	72.8	63.4	86.2	69.2	74.7	79.2	
10	Bromophos-ethyl	Cleanert TPT	41.9	34.1	32.6	35.7	28.0	38.2	28.7	28.3	35.2	30.3	40.3	31.6	32.6	31.2	28.8	25.7	26.1	24.7	32.3	31.9	9.8
		Envi-Carb + PSA	36.2	37.9	36.5	32.5	39.7	38.8	33.4	32.2	33.4	58.5	34.3	36.9	30.5	34.8	31.8	30.8	32.7	27.5	30.3	35.2	
11	Butralin	Cleanert TPT	214.1	147.1	125.5	148.1	94.7	321.3	118.5	115.6	179.4	102.2	209.8	130.3	160.7	132.4	155.6	282.2	104.4	104.9	177.9	159.2	6.9
		Envi-Carb + PSA	158.2	170.1	159.7	94.8	157.8	196.2	169.1	149.6	133.1	146.1	195.4	188.0	136.5	109.5	124.5	114.1	181.4	110.2	128.2	148.6	

No.	Compound	Cartridge	1	2	3	4	5	6	7	8	9	10	11	12	13	14	15	16	17	18	19	20	21
12	Chlorfenapyr	Cleanert TPT	356.6	299.3	313.4	343.4	255.0	284.5	265.9	255.9	299.3	282.0	284.7	272.6	280.0	289.7	260.2	212.5	242.1	223.8	284.4	279.2	4.9
		Envi-Carb + PSA	333.4	329.4	310.9	322.2	326.3	319.3	300.1	277.3	284.4	286.3	264.4	278.9	271.0	306.1	271.9	282.2	288.9	248.9	271.2	293.3	
13	Clomazone	Cleanert TPT	54.1	38.7	34.3	36.3	30.3	49.0	30.0	30.3	36.2	33.2	45.8	35.6	34.7	36.0	35.5	43.5	30.0	27.1	36.3	36.7	5.1
		Envi-Carb + PSA	40.5	48.1	43.4	30.5	41.1	41.2	36.6	36.6	39.4	42.8	49.0	40.1	33.3	37.3	41.1	32.8	36.0	29.2	34.2	38.6	
14	Cycloate	Cleanert TPT	39.4	31.1	32.4	32.2	26.0	31.9	28.7	27.1	30.9	27.4	30.4	28.0	28.9	27.2	26.7	21.1	24.9	22.5	28.7	28.7	11.6
		Envi-Carb + PSA	33.4	33.1	33.4	32.8	37.2	33.9	31.1	30.5	31.7	36.1	27.8	41.4	28.2	34.9	30.0	30.9	30.3	27.4	28.2	32.2	
15	Cycluron	Cleanert TPT	130.0	103.9	118.1	125.7	101.4	148.8	95.6	92.0	108.7	99.4	101.6	100.5	110.6	95.6	91.7	69.7	83.5	86.9	104.0	103.6	2.7
		Envi-Carb + PSA	103.1	112.4	110.0	124.7	122.7	109.1	98.3	91.5	106.8	134.2	112.2	109.8	87.8	112.4	97.8	102.1	96.4	89.8	99.9	106.4	
16	Cyhalofop-butyl	Cleanert TPT	114.9	83.2	76.8	68.2	65.3	118.2	62.1	59.8	70.1	91.6	82.2	83.1	79.9	77.6	76.0	77.2	59.6	58.0	86.2	78.4	9.4
		Envi-Carb + PSA	81.0	101.5	89.5	66.6	85.5	82.4	65.9	71.7	116.4	62.4	91.7	156.7	86.8	97.9	87.5	72.8	83.5	63.1	73.8	86.2	
17	Cyprodinil	Cleanert TPT	46.1	36.5	37.5	37.4	27.6	36.2	29.7	29.1	34.8	30.0	36.8	31.1	31.9	30.4	28.7	24.1	26.6	23.3	31.1	32.0	8.0
		Envi-Carb + PSA	38.3	38.9	38.6	39.5	41.0	39.8	33.7	32.7	33.7	49.6	29.1	33.4	28.5	32.9	28.9	32.1	30.7	28.0	30.0	34.7	
18	Dacthal	Cleanert TPT	128.0	102.2	99.8	107.7	84.2	94.8	88.4	84.5	105.0	91.2	100.1	92.0	94.1	94.3	91.0	91.1	78.8	73.9	96.7	94.6	2.2
		Envi-Carb + PSA	105.1	107.9	106.8	100.7	114.0	109.3	101.6	95.0	96.4	91.5	93.7	69.1	89.8	101.1	92.9	93.2	99.1	80.8	90.2	96.7	
19	DE-PCB101	Cleanert TPT	40.4	32.0	33.0	33.4	26.6	29.6	28.6	26.9	31.1	29.0	30.1	30.2	30.2	29.5	27.8	22.2	26.3	24.2	30.1	29.5	7.0
		Envi-Carb + PSA	33.6	34.1	33.7	33.2	37.6	34.2	30.9	30.7	31.1	29.3	28.0	32.6	29.4	33.4	30.6	31.7	30.4	28.3	29.5	31.7	
20	DE-PCB118	Cleanert TPT	40.8	33.1	33.9	33.9	27.5	30.8	28.9	27.0	32.5	30.1	30.8	30.6	31.2	31.1	28.7	22.8	27.2	24.9	32.3	30.4	6.6
		Envi-Carb + PSA	34.5	34.9	34.6	34.3	37.4	35.1	31.4	31.5	32.3	30.2	28.6	32.7	29.9	35.4	31.3	32.9	31.1	29.0	30.2	32.5	
21	DE-PCB138	Cleanert TPT	40.0	32.8	34.3	31.9	26.4	29.9	28.3	26.8	31.6	28.9	29.9	30.0	29.7	31.5	27.5	21.8	26.3	24.1	30.6	29.6	5.9
		Envi-Carb + PSA	33.6	34.1	34.5	33.7	36.4	33.6	31.1	30.8	30.9	28.2	27.4	34.0	28.1	33.1	29.4	31.1	30.0	28.0	28.4	31.4	
22	DE-PCB180	Cleanert TPT	40.0	32.4	33.2	33.5	26.4	28.1	27.0	25.7	30.5	27.2	29.4	29.2	28.9	29.2	26.4	21.5	25.0	23.0	29.5	28.7	7.2
		Envi-Carb + PSA	33.9	34.4	34.1	32.6	35.9	33.1	32.4	29.6	30.1	27.4	27.3	33.3	27.6	33.6	29.1	28.6	29.7	26.2	28.0	30.9	
23	DE-PCB28	Cleanert TPT	41.2	33.9	36.2	38.6	29.0	33.6	31.2	29.5	34.5	32.0	32.7	32.4	32.1	31.5	30.5	24.0	29.4	26.6	33.2	32.2	7.8
		Envi-Carb + PSA	35.1	35.1	35.7	35.9	39.9	35.8	33.6	33.8	33.8	34.2	31.4	43.3	31.0	37.2	33.6	34.9	33.9	31.3	31.9	34.8	
24	DE-PCB31	Cleanert TPT	41.8	33.7	36.1	38.4	28.8	32.7	31.3	29.5	34.5	31.3	32.9	32.8	32.4	31.4	30.6	23.8	29.2	26.7	33.2	32.2	6.5
		Envi-Carb + PSA	35.0	35.0	35.9	35.9	39.9	36.1	31.2	33.9	33.6	34.8	31.3	35.9	30.7	37.4	33.7	35.1	33.6	31.1	31.9	34.3	
25	Dichlorofop-methyl	Cleanert TPT	65.7	42.8	31.8	29.1	33.7	56.6	26.4	27.8	33.6	33.7	50.5	37.3	34.9	34.0	34.3	56.5	25.0	25.6	37.9	37.8	5.5
		Envi-Carb + PSA	41.9	58.2	45.7	25.3	39.9	40.4	35.6	40.6	46.9	39.1	57.0	43.6	38.0	36.0	40.7	29.4	33.2	28.4	38.0	39.9	
26	Dimethenamid	Cleanert TPT	41.8	36.0	35.7	45.3	28.3	36.8	30.4	30.6	35.8	29.2	34.3	29.6	31.7	32.3	31.1	27.6	27.6	25.0	32.1	32.7	3.9
		Envi-Carb + PSA	37.8	39.1	35.6	34.7	38.7	38.0	34.6	30.1	31.0	42.6	31.3	35.2	28.9	34.0	31.3	31.3	34.5	26.8	30.6	34.0	
27	Diofenolan-1	Cleanert TPT	91.1	73.3	73.5	78.4	60.0	75.8	64.3	58.4	70.7	62.9	72.8	67.5	68.0	71.6	62.1	49.9	57.6	54.0	71.9	67.6	6.4
		Envi-Carb + PSA	77.9	77.2	78.5	76.3	81.2	79.4	67.0	70.5	70.1	84.0	59.3	73.3	64.4	76.0	68.3	70.4	69.4	61.2	64.9	72.1	
28	Diofenolan-2	Cleanert TPT	92.9	75.1	78.9	81.3	63.5	78.7	66.5	58.7	73.9	66.0	74.5	71.6	70.1	73.9	64.3	52.9	59.9	56.9	76.5	70.3	5.2

(Continued)

TABLE 2.3.6 The Determination Results for 201 Pesticides in Incurred Green Tea Youden Pair Samples Under 4 Conditions in 3 Months (cont.)

Pesticide content determined With GC–MS and GC–MS/MS, two concentration of Youden pair samples (μg/kg)

No.	Pesticides	Solid Phase Extraction	November 9, 2009 (n=5)	November 14, 2009 (n=3)	November 19, 2009 (n=3)	November 24, 2009 (n=3)	November 29, 2009 (n=3)	December 4, 2009 (n=3)	December 9, 2009 (n=3)	December 14, 2009 (n=3)	December 19, 2009 (n=3)	December 24, 2009 (n=3)	December 14, 2009 (n=3)	January 3, 2010 (n=3)	January 8, 2010 (n=3)	January 13, 2010 (n=3)	January 18, 2010 (n=3)	January 23, 2010 (n=3)	January 28, 2010 (n=3)	February 2, 2010 (n=3)	February 7, 2010 (n=3)	AVE (μg/kg)	Deviation rates (%)
		Envi-Carb + PSA	76.4	79.7	78.5	78.3	81.5	78.2	70.6	72.9	72.5	89.2	62.6	64.9	64.8	97.9	68.0	70.6	70.5	63.5	66.4	74.0	
29	Fenbuconazole	Cleanert TPT	94.0	78.9	88.4	89.3	67.9	79.7	65.3	62.7	75.9	60.4	83.1	69.8	76.3	66.8	58.1	48.8	58.2	54.6	72.6	71.1	5.4
		Envi-Carb + PSA	78.5	76.4	85.5	85.5	91.7	85.0	61.1	73.2	75.9	102.7	67.7	69.5	66.8	72.0	67.1	73.5	69.3	60.1	64.8	75.1	
30	Fenpyroximate	Cleanert TPT	280.1	238.5	397.1	383.5	236.3	207.4	297.4	243.2	263.0	229.0	191.4	220.4	207.9	252.4	203.5	96.2	228.5	228.2	229.2	243.9	10.2
		Envi-Carb + PSA	299.9	255.9	297.3	427.6	327.7	263.7	255.6	256.1	266.8	292.1	171.9	269.9	211.4	283.5	238.7	298.4	241.9	248.8	224.9	270.1	
31	Fluotrimazole	Cleanert TPT	46.5	37.8	39.1	30.6	28.3	39.9	29.1	26.7	31.7	27.4	33.3	27.9	26.7	30.4	27.2	21.3	25.4	22.8	29.3	30.6	7.4
		Envi-Carb + PSA	38.9	39.6	38.3	37.7	38.8	38.3	30.6	30.3	33.7	37.6	26.6	35.9	27.2	32.5	28.6	30.0	28.3	25.5	28.2	33.0	
32	Fluoxypr-1-methylheptyl ester	Cleanert TPT	52.1	39.8	40.2	35.9	31.6	42.3	31.4	31.5	36.0	33.0	42.6	36.2	37.1	34.8	33.8	40.7	28.4	27.9	40.5	36.6	4.7
		Envi-Carb + PSA	41.2	55.2	44.1	33.4	39.7	39.9	36.3	38.1	43.1	36.9	39.9	44.4	33.5	35.2	40.0	32.0	36.9	29.2	30.3	38.4	
33	Iprovalicarb-1	Cleanert TPT	201.0	168.9	174.4	171.2	141.0	201.3	151.3	150.3	160.4	139.8	215.1	145.3	156.6	174.8	156.1	235.3	122.3	107.9	171.2	165.5	0.0
		Envi-Carb + PSA	172.2	204.1	186.6	163.1	185.0	193.1	163.1	162.6	200.1	211.7	151.2	152.4	135.4	141.0	145.8	133.1	182.7	114.9	145.5	165.5	
34	Iprovalicarb-2	Cleanert TPT	211.1	175.1	165.4	172.0	134.1	213.0	135.8	136.2	169.6	133.4	228.2	154.3	159.9	157.4	155.5	238.1	126.3	108.5	178.8	165.9	0.8
		Envi-Carb + PSA	176.3	205.3	181.5	153.7	186.5	196.6	161.8	162.6	146.6	247.9	150.3	152.8	135.1	142.2	151.9	130.9	188.6	114.0	142.7	164.6	
35	Isodrin	Cleanert TPT	44.9	34.0	33.8	37.3	29.4	31.4	30.4	29.0	32.7	27.9	31.2	33.5	32.1	30.4	30.4	23.8	27.9	24.8	31.1	31.4	6.0
		Envi-Carb + PSA	39.0	36.2	34.4	35.5	39.3	35.7	30.0	29.5	34.1	29.1	29.7	43.3	30.0	36.5	32.3	32.7	30.6	27.9	27.1	33.3	
36	Isoprocarb-1	Cleanert TPT	83.0	70.7	81.9	81.6	61.8	69.8	67.3	62.1	72.6	62.6	65.5	68.6	69.4	59.8	61.3	45.7	61.6	55.7	71.6	67.0	8.6
		Envi-Carb + PSA	75.1	73.9	75.2	82.5	84.0	76.3	68.9	68.8	70.2	75.0	62.0	88.9	65.4	78.1	68.2	74.1	70.4	64.4	66.0	73.0	
37	Isoprocarb-2	Cleanert TPT	96.2	95.7	65.7	67.3	66.5	86.3	65.2	62.3	71.4	73.8	103.8	64.6	56.1	60.6	75.6	147.2	57.0	58.5	68.0	75.9	20.7
		Envi-Carb + PSA	87.8	131.4	104.6	68.3	93.8	92.3	73.6	96.4	87.3	76.6	118.2	105.5	95.8	94.5	99.7	100.6	72.4	92.8	81.8	93.3	
38	Lenacil	Cleanert TPT	293.8	225.4	190.4	197.3	172.1	197.4	149.6	178.9	228.7	188.0	339.5	260.8	227.2	199.1	228.4	581.2	126.0	106.3	305.2	231.4	0.7
		Envi-Carb + PSA	224.3	290.8	240.1	150.4	216.0	231.5	235.5	213.7	246.5	257.7	204.5	222.9	201.8	176.3	292.0	213.8	354.6	198.0	256.8	233.0	
39	Metalaxyl	Cleanert TPT	132.5	106.9	108.0	120.5	84.8	100.8	88.2	83.9	100.1	82.2	108.3	88.7	100.8	95.4	93.7	79.0	87.0	79.4	105.1	97.1	6.5
		Envi-Carb + PSA	110.4	113.6	115.2	109.5	118.5	112.2	90.7	92.0	94.5	131.0	93.1	111.6	90.6	105.7	95.9	100.7	100.5	89.1	95.6	103.7	
40	Metazachlor	Cleanert TPT	125.4	112.1	103.7	153.6	93.8	115.7	103.4	102.4	122.8	88.5	113.9	85.2	99.7	92.7	96.1	99.6	80.2	77.2	105.2	103.8	1.4
		Envi-Carb + PSA	116.4	135.0	101.3	102.3	115.0	121.5	97.4	97.9	107.9	139.0	107.1	96.9	89.3	99.9	94.0	84.9	118.2	83.4	90.8	105.2	

#	Compound	Sorbent	437.7	355.5	377.9	398.7	292.9	415.4	307.4	288.3	349.0	286.7	355.6	312.5	344.8	296.9	294.4	280.4	279.6	256.5	420.2	334.2	
41	Methabenzthiazuron	Cleanet TPT	392.9	391.3	407.0	376.7	424.3	400.9	326.1	347.1	332.8	476.8	292.5	360.4	317.4	362.3	323.2	329.6	359.3	295.0	324.3	360.0	7.4
		Envi-Carb + PSA																					
42	Mirex	Cleanet TPT	43.6	31.9	27.1	32.0	23.7	30.6	23.4	24.8	30.2	24.7	32.8	27.7	26.8	27.9	28.1	34.2	21.0	19.3	27.2	28.3	0.1
		Envi-Carb + PSA	34.0	34.0	31.0	23.9	33.1	31.7	25.3	25.1	24.5	22.2	31.7	37.9	27.8	28.6	25.7	22.1	31.7	19.9	26.2	28.2	
43	Monalide	Cleanet TPT	93.5	85.3	82.1	86.8	72.5	83.9	71.4	77.5	81.9	86.0	85.4	84.2	73.0	74.7	76.1	74.1	68.5	64.7	78.5	79.0	2.7
		Envi-Carb + PSA	79.5	85.6	74.6	82.7	97.6	88.3	69.2	84.6	86.1	75.6	77.7	85.1	72.9	88.0	81.6	83.3	77.1	78.7	74.2	81.2	
44	Paraoxon-ethyl	Cleanet TPT	199.5	130.9	141.6	167.7	166.7	249.6	135.5	151.8	191.0	146.5	189.7	148.0	196.6	165.0	155.2	169.4	125.9	142.0	104.7	165.1	1.5
		Envi-Carb + PSA	198.4	237.0	168.4	154.3	151.4	197.1	47.0	177.0	165.8	138.7	181.7	165.6	168.8	163.7	181.4	150.6	157.2	205.6	205.8	167.7	
45	Pebulate	Cleanet TPT	107.9	83.5	91.4	92.4	74.9	84.6	81.2	77.2	85.0	77.7	70.6	70.8	72.3	70.2	64.3	55.0	73.4	67.3	92.9	78.8	8.7
		Envi-Carb + PSA	99.1	95.2	95.0	94.5	107.9	95.5	85.4	88.2	83.3	64.2	81.3	73.9	63.2	63.7	91.5	88.3	95.0	84.4	83.6	86.0	
46	Pentachloroaniline	Cleanet TPT	66.8	35.7	37.9	37.7	28.8	34.7	29.0	30.4	35.0	32.0	31.9	33.1	32.8	30.9	29.8	24.8	28.6	25.6	34.7	33.7	4.2
		Envi-Carb + PSA	36.9	37.0	36.2	37.0	40.8	38.5	33.3	33.9	32.3	35.3	32.4	35.0	32.8	38.0	31.7	35.5	36.4	30.6	33.6	35.1	
47	Pentachloroanisole	Cleanet TPT	40.1	31.5	31.6	31.5	27.2	32.1	30.7	29.6	33.3	29.5	30.8	30.1	31.4	29.2	28.4	22.5	27.0	24.0	31.4	30.1	7.6
		Envi-Carb + PSA	39.1	40.3	33.6	35.7	37.8	34.0	34.1	33.4	33.8	32.2	29.7	33.0	31.0	35.2	30.4	30.8	31.0	28.5	29.6	32.5	
48	Pentachlorobenzene	Cleanet TPT	35.9	29.1	31.2	32.4	25.5	29.4	29.2	28.7	32.6	27.6	28.6	29.0	30.7	27.9	28.0	22.6	26.1	23.3	30.0	28.8	10.7
		Envi-Carb + PSA	33.0	31.9	32.2	31.7	37.2	33.2	32.7	32.8	32.0	29.2	29.2	35.9	31.2	35.5	30.4	31.4	32.8	28.3	29.1	32.1	
49	Perthane	Cleanet TPT	48.2	35.4	40.8	37.5	28.6	35.4	28.3	28.2	33.8	30.8	38.1	32.9	32.7	33.6	30.9	26.9	27.6	25.3	33.6	33.1	5.2
		Envi-Carb + PSA	39.1	40.3	37.7	35.7	39.5	38.0	35.1	32.9	34.2	38.6	33.3	35.1	29.3	36.7	32.7	32.8	32.1	28.6	30.7	34.9	
50	Phenanthrene	Cleanet TPT	42.6	36.7	36.2	41.9	30.9	36.2	34.4	31.2	36.6	33.5	35.6	34.7	35.9	32.4	33.5	25.7	32.0	28.0	36.7	34.5	7.8
		Envi-Carb + PSA	37.3	37.1	38.6	40.5	42.0	40.3	36.4	36.4	36.6	38.1	33.5	37.3	33.6	41.7	36.8	37.2	35.7	33.6	35.5	37.3	
51	Pirimicarb	Cleanet TPT	96.0	72.9	77.7	73.9	53.6	82.5	58.5	53.1	66.8	54.4	76.4	55.6	60.0	59.0	59.9	52.8	53.2	46.3	64.3	64.0	5.3
		Envi-Carb + PSA	77.1	75.1	76.6	69.2	78.7	75.6	62.6	62.8	64.7	106.6	58.6	60.3	55.2	66.2	60.9	59.9	61.9	53.2	57.7	67.5	
52	Procymidone	Cleanet TPT	46.2	37.1	38.9	40.8	31.4	34.6	33.2	30.7	36.4	34.0	33.1	36.2	37.6	38.1	31.9	25.9	30.9	28.4	35.8	34.8	6.7
		Envi-Carb + PSA	39.0	40.0	39.8	39.8	41.7	40.2	32.0	35.5	36.7	35.7	32.1	39.7	34.5	43.5	36.2	37.0	35.3	32.9	35.3	37.2	
53	Prometryne	Cleanet TPT	47.1	38.7	40.6	37.6	28.0	41.2	30.2	28.2	34.1	28.3	38.5	29.3	31.0	28.8	28.1	22.9	27.4	22.9	29.9	32.3	8.4
		Envi-Carb + PSA	39.9	44.4	41.4	38.8	41.7	42.3	32.8	32.6	31.9	54.7	29.0	32.2	26.9	33.4	29.2	31.5	30.2	27.7	29.2	35.1	
54	Propham	Cleanet TPT	45.3	39.9	41.0	37.8	33.1	44.1	35.1	34.4	40.4	36.4	42.0	37.4	40.7	35.0	38.0	45.4	33.1	29.6	44.6	38.6	4.6
		Envi-Carb + PSA	39.4	42.9	44.2	40.0	44.3	41.6	40.7	41.4	38.8	48.7	38.4	39.3	37.9	40.8	39.3	36.5	43.7	32.7	37.3	40.4	
55	Prosulfocarb	Cleanet TPT	43.3	34.3	34.7	38.6	27.4	36.2	30.4	28.2	33.5	27.6	35.0	29.7	32.0	30.9	29.2	22.5	27.5	24.9	31.7	31.5	8.8
		Envi-Carb + PSA	36.1	36.1	37.1	36.4	40.2	37.7	34.7	32.3	32.2	50.1	29.2	33.1	29.7	34.2	31.0	32.3	31.4	29.1	29.7	34.3	

(Continued)

TABLE 2.3.6 The Determination Results for 201 Pesticides in Incurred Green Tea Youden Pair Samples Under 4 Conditions in 3 Months (cont.)

Pesticide content determined With GC–MS and GC–MS/MS, two concentration of Youden pair samples (μg/kg)

No.	Pesticides	Solid Phase Extraction	November 9, 2009 (n=5)	November 14, 2009 (n=3)	November 19, 2009 (n=3)	November 24, 2009 (n=3)	November 29, 2009 (n=3)	December 4, 2009 (n=3)	December 9, 2009 (n=3)	December 14, 2009 (n=3)	December 19, 2009 (n=3)	December 24, 2009 (n=3)	December 14, 2009 (n=3)	January 3, 2010 (n=3)	January 8, 2010 (n=3)	January 13, 2010 (n=3)	January 18, 2010 (n=3)	January 23, 2010 (n=3)	January 28, 2010 (n=3)	February 2, 2010 (n=3)	February 7, 2010 (n=3)	AVE (μg/kg)	Deviation rates (%)
56	Secbumeton	Cleanert TPT	40.6	30.8	30.6	28.9	20.0	29.5	23.1	21.4	26.2	20.0	31.5	21.0	25.5	24.8	23.5	18.3	21.6	18.2	25.0	25.3	33.7
		Envi-Carb + PSA	40.9	42.2	43.4	37.9	40.9	45.5	34.2	33.5	33.2	48.2	28.6	30.2	27.4	34.6	30.4	32.9	31.2	28.5	30.8	35.5	
57	Silafluofen	Cleanert TPT	52.1	47.6	35.8	36.6	28.6	35.5	31.6	27.1	32.4	31.2	40.8	36.3	36.6	33.2	32.5	23.1	27.2	24.3	32.7	34.0	2.4
		Envi-Carb + PSA	40.8	52.1	37.9	35.6	39.1	35.1	32.3	32.2	33.0	37.3	27.0	34.6	30.3	34.6	37.1	31.5	31.0	28.2	31.0	34.8	
58	Tebupirimfos	Cleanert TPT	151.2	114.8	104.7	107.2	73.8	88.8	70.3	38.5	48.8	38.0	51.1	36.7	44.4	41.5	38.2	30.5	37.7	33.5	41.8	62.7	27.6
		Envi-Carb + PSA	120.9	131.0	133.8	106.6	149.2	127.3	62.2	59.0	58.1	100.9	55.7	64.6	52.2	66.6	58.7	58.4	59.6	54.4	54.1	82.8	
59	Tebutam	Cleanert TPT	86.5	67.7	71.6	72.4	54.8	66.6	58.8	56.2	66.1	57.6	65.4	59.6	61.6	57.7	58.8	45.7	55.7	48.1	61.7	61.7	7.9
		Envi-Carb + PSA	73.0	72.5	72.2	71.3	78.4	73.0	64.3	61.7	63.3	79.4	57.9	64.4	59.3	71.1	62.2	65.1	62.1	57.6	59.8	66.8	
60	Tebuthiuron	Cleanert TPT	191.6	144.5	169.4	148.5	115.1	181.1	124.0	113.4	137.1	105.7	130.2	118.0	140.8	115.3	117.9	93.2	114.2	101.4	145.1	131.9	6.5
		Envi-Carb + PSA	158.8	154.2	159.2	154.9	163.7	159.1	123.3	134.5	130.1	199.4	117.7	132.4	121.5	133.2	125.5	133.1	131.9	116.7	125.0	140.8	
61	Tefluthrin	Cleanert TPT	40.9	31.8	30.9	34.9	25.3	29.6	27.9	26.2	30.7	25.7	30.1	28.2	28.2	31.5	26.9	21.2	25.7	22.3	28.1	28.7	7.6
		Envi-Carb + PSA	33.7	33.9	33.6	32.1	36.6	33.6	30.0	30.0	30.1	35.8	27.7	33.4	26.7	31.6	28.7	29.5	28.6	25.8	27.4	31.0	
62	Thenylchlor	Cleanert TPT	79.9	80.3	68.2	112.4	62.7	74.9	64.4	62.5	75.8	59.7	70.6	59.4	66.7	72.4	65.5	69.8	54.2	59.0	72.5	70.0	0.8
		Envi-Carb + PSA	80.9	101.7	66.3	69.5	75.5	75.0	72.9	59.5	58.7	68.7	71.2	70.3	65.3	75.0	66.3	61.6	85.1	53.4	64.1	70.6	
63	Thionazin	Cleanert TPT	44.9	35.2	36.4	35.5	28.1	42.8	30.7	28.8	34.1	27.1	32.9	28.5	32.3	28.0	28.8	24.5	26.7	24.3	31.4	28.7	8.6
		Envi-Carb + PSA	36.3	37.6	37.8	35.1	41.4	36.7	36.6	32.4	30.9	46.8	31.4	35.8	31.4	34.8	32.2	30.7	31.8	27.1	28.6	34.5	
64	Trichloronat	Cleanert TPT	47.6	35.2	34.7	31.8	27.7	39.4	28.7	28.0	34.6	28.8	40.3	30.2	31.9	30.7	29.5	27.0	26.4	23.9	30.8	32.0	5.7
		Envi-Carb + PSA	36.6	39.0	39.3	30.9	40.5	39.0	32.4	32.6	33.9	49.3	35.6	17.9	29.7	34.1	32.9	30.1	31.3	27.8	30.1	33.8	
65	Trifluralin	Cleanert TPT	98.0	72.9	73.4	64.8	52.0	115.4	61.0	57.7	77.4	57.8	78.7	62.7	86.0	64.1	69.3	83.9	55.0	60.6	83.4	72.3	0.8
		Envi-Carb + PSA	72.9	75.4	76.4	60.8	79.0	83.2	69.3	65.0	67.0	78.1	79.9	72.3	68.5	38.7	71.8	61.3	118.3	52.3	72.7	71.7	
66	2,4-DDT	Cleanert TPT	92.2	75.8	65.1	111.9	62.2	59.3	62.4	61.3	61.5	54.8	72.6	51.4	57.8	57.9	62.0	55.2	43.7	49.3	60.6	64.1	2.5
		Envi-Carb + PSA	93.8	79.9	71.6	67.3	71.2	78.9	65.1	53.3	67.4	58.2	65.9	70.9	59.8	59.7	40.2	53.2	63.0	56.8	71.9	65.7	
67	4,4-DDE	Cleanert TPT	42.4	35.0	36.6	42.7	32.0	29.6	35.4	32.1	34.1	30.2	31.4	31.6	30.9	31.8	31.6	26.7	27.0	27.6	33.6	32.7	8.7
		Envi-Carb + PSA	46.8	38.9	36.7	35.7	37.7	37.4	58.9	32.7	32.1	32.4	31.8	34.3	32.0	32.5	30.7	34.6	29.3	31.4	32.5	35.7	
68	Benalaxyl	Cleanert TPT	47.2	39.1	40.0	49.0	34.4	31.6	36.2	31.9	34.5	30.1	35.4	31.9	30.2	29.7	30.0	27.6	26.2	29.8	34.1	34.2	6.6
		Envi-Carb + PSA	53.9	45.1	51.9	47.6	39.5	37.4	35.3	30.6	34.3	33.1	33.5	34.1	32.5	32.2	24.9	31.6	31.2	29.3	35.4	36.5	

No.	Compound	Cartridge																					
69	Benzoylprop-ethyl	Cleanert TPT	142.2	120.3	121.8	150.0	104.9	98.0	112.6	103.3	109.9	94.5	110.1	103.6	98.9	94.3	95.2	85.1	81.0	86.2	105.7	106.2	3.3
		Envi-Carb + PSA	153.6	135.6	117.9	113.4	120.9	119.2	106.8	99.0	107.0	103.3	102.1	112.7	103.9	104.4	86.1	100.7	96.4	95.1	106.3	109.7	
70	Bromofos	Cleanert TPT	82.9	72.8	69.2	100.7	67.9	64.7	73.7	69.0	69.1	60.4	68.0	55.3	60.7	59.3	63.8	58.0	50.7	54.8	61.2	66.4	6.7
		Envi-Carb + PSA	101.8	76.9	78.5	83.8	78.3	81.3	82.3	60.7	71.8	62.6	64.0	70.2	61.0	65.7	52.0	64.9	61.0	60.3	72.0	71.0	
71	Bromopropylate	Cleanert TPT	95.6	85.3	78.8	120.0	70.2	65.7	74.3	66.9	71.4	60.3	88.3	57.1	66.6	61.5	66.8	60.8	47.1	60.1	72.2	72.0	3.8
		Envi-Carb + PSA	101.7	104.0	79.7	72.3	80.4	82.3	69.2	63.0	86.0	75.5	72.8	76.6	73.0	64.7	42.6	51.5	83.0	67.4	75.7	74.8	
72	Buprofenzin	Cleanert TPT	91.5	74.3	74.7	89.6	66.4	61.6	72.6	66.2	67.7	61.3	67.1	52.5	57.0	68.6	63.9	51.9	59.8	63.9	65.4	67.2	2.1
		Envi-Carb + PSA	98.4	81.0	69.4	67.2	79.7	79.6	75.7	63.7	60.6	65.9	58.5	58.3	57.0	64.6	70.0	76.6	53.6	59.4	64.3	68.6	
73	Butachlor	Cleanert TPT	81.0	72.1	71.2	85.2	64.0	59.4	69.5	62.5	67.2	57.9	61.5	58.9	57.9	60.9	63.2	54.4	50.6	54.7	62.6	63.9	2.0
		Envi-Carb + PSA	85.0	74.5	70.3	69.9	73.3	66.8	68.6	61.1	62.7	62.1	61.0	62.1	61.8	61.2	54.8	63.3	57.9	58.8	64.4	65.2	
74	Butylate	Cleanert TPT	105.3	85.0	90.0	102.7	83.7	73.3	79.7	64.5	84.7	57.2	55.6	44.3	53.0	52.8	48.2	41.9	42.2	62.9	71.1	68.3	20.8
		Envi-Carb + PSA	207.6	147.7	87.4	74.0	97.5	102.7	93.2	71.5	58.6	79.2	111.9	87.7	56.9	63.8	42.6	51.2	49.6	76.3	73.3	84.2	
75	Carbotenothion	Cleanert TPT	84.9	74.7	68.3	88.9	62.4	54.1	64.5	63.5	64.5	50.7	62.0	58.8	49.3	49.1	47.1	38.4	36.8	37.1	42.5	57.8	10.0
		Envi-Carb + PSA	105.9	81.3	100.6	96.9	72.6	68.0	73.2	56.0	59.0	63.8	58.8	61.5	51.6	52.5	37.0	46.4	41.7	42.3	43.6	63.8	
76	Chlorfenson	Cleanert TPT	100.5	95.8	72.3	177.1	77.6	77.1	77.9	67.6	78.0	64.7	105.0	104.7	70.1	70.0	71.3	66.3	48.0	63.2	72.5	82.1	0.3
		Envi-Carb + PSA	107.1	99.5	83.8	83.7	82.8	89.2	87.3	65.1	93.2	77.9	84.1	85.9	75.2	71.2	47.0	55.4	93.8	75.4	97.8	81.9	
77	Chlorfenvinphos	Cleanert TPT	127.2	126.6	103.6	172.4	95.2	87.5	104.5	96.9	97.3	83.0	103.9	89.8	85.3	90.3	96.7	81.4	73.0	76.4	90.0	98.5	4.5
		Envi-Carb + PSA	146.3	124.8	106.6	108.0	110.8	142.7	107.9	78.8	110.3	90.1	100.4	113.1	79.7	89.3	68.1	91.3	99.7	82.1	106.8	103.0	
78	Chlormephos	Cleanert TPT	78.0	64.6	67.8	76.7	64.3	57.5	68.8	59.9	65.9	57.7	62.6	63.7	61.9	62.7	61.8	50.4	49.1	52.8	62.6	62.6	6.2
		Envi-Carb + PSA	89.8	70.6	62.7	61.9	75.4	69.8	70.6	65.2	62.5	69.8	59.5	74.4	65.1	66.6	51.1	62.6	60.5	61.1	65.8	66.6	
79	Chloroneb	Cleanert TPT	45.6	38.2	40.0	45.6	36.4	33.3	40.2	36.0	39.3	34.1	34.9	38.8	35.3	34.2	34.3	32.1	30.4	32.3	37.7	36.8	5.1
		Envi-Carb + PSA	48.0	41.2	41.4	39.8	42.7	37.6	39.3	38.2	36.1	37.4	36.2	37.4	39.9	38.3	33.8	38.6	34.5	35.9	38.7	38.7	
80	Chloropropylate	Cleanert TPT	138.2	40.3	38.3	57.1	29.2	33.2	35.6	36.4	34.4	30.3	40.2	34.2	32.0	32.9	34.0	30.1	26.2	27.8	36.2	40.4	9.0
		Envi-Carb + PSA	50.4	43.8	39.3	36.7	38.4	39.2	44.4	32.7	40.2	37.8	35.6	32.7	34.3	31.9	25.0	29.2	38.0	34.1	37.1	36.9	
81	Chlorpropham	Cleanert TPT	93.6	85.4	82.6	111.8	76.4	70.2	82.8	78.5	81.4	71.1	88.5	80.0	74.4	73.8	76.1	68.3	63.7	55.4	83.9	78.8	6.0
		Envi-Carb + PSA	105.6	96.8	96.4	89.2	89.4	88.5	92.2	75.0	85.7	84.4	78.2	77.4	82.5	75.3	61.6	72.4	81.0	77.6	81.1	83.7	
82	Chlorpyrifos(ethyl)	Cleanert TPT	43.4	36.3	34.1	48.5	32.0	29.5	34.4	32.7	33.6	27.8	36.4	30.6	30.2	29.2	31.1	26.7	24.2	25.3	31.0	32.5	5.9
		Envi-Carb + PSA	46.1	38.9	34.8	34.2	38.8	42.7	36.2	29.6	34.8	34.1	33.3	36.8	31.6	33.0	25.4	28.9	31.9	29.6	34.0	34.5	
83	Chlorthiophos	Cleanert TPT	125.8	108.4	103.2	129.8	89.4	79.2	95.1	85.1	86.0	68.9	82.2	58.5	77.0	76.0	76.7	63.5	62.2	57.7	64.3	83.6	8.8
		Envi-Carb + PSA	145.9	117.5	110.4	105.4	106.0	116.4	118.5	79.8	82.1	80.7	77.5	95.7	79.6	80.2	68.3	75.2	68.0	59.6	68.6	91.3	
84	Cis-chlordane	Cleanert TPT	88.0	72.2	71.6	92.9	65.4	61.4	68.9	63.6	66.5	60.0	68.0	59.0	62.9	57.6	63.0	53.3	50.7	54.2	63.1	65.4	6.2
		Envi-Carb + PSA	92.3	80.9	69.2	68.0	75.9	76.5	93.0	62.5	65.0	63.3	62.2	79.7	61.1	64.0	54.3	62.5	60.4	61.6	68.7	69.5	

(Continued)

TABLE 2.3.6 The Determination Results for 201 Pesticides in Incurred Green Tea Youden Pair Samples Under 4 Conditions in 3 Months (cont.)

Pesticide content determined With GC–MS and GC–MS/MS, two concentration of Youden pair samples (μg/kg)

No.	Pesticides	Solid Phase Extraction	November 9, 2009 (n=5)	November 14, 2009 (n=3)	November 19, 2009 (n=3)	November 24, 2009 (n=3)	November 29, 2009 (n=3)	December 4, 2009 (n=3)	December 9, 2009 (n=3)	December 14, 2009 (n=3)	December 19, 2009 (n=3)	December 24, 2009 (n=3)	December 14, 2009 (n=3)	January 3, 2010 (n=3)	January 8, 2010 (n=3)	January 13, 2010 (n=3)	January 18, 2010 (n=3)	January 23, 2010 (n=3)	January 28, 2010 (n=3)	February 2, 2010 (n=3)	February 7, 2010 (n=3)	AVE (μg/kg)	Deviation rates (%)
85	Cis-diallate	Cleanert TPT	84.7	68.7	71.1	86.8	62.6	59.9	69.6	62.5	68.1	57.7	65.2	52.9	61.4	57.7	61.5	51.6	52.3	45.3	64.0	63.3	7.5
		EnviCarb + PSA	93.2	76.8	71.0	68.9	77.2	73.9	70.9	61.6	62.9	65.6	64.4	67.0	71.0	65.4	56.2	62.2	60.5	61.4	66.7	68.2	
86	Cyanofenphos	Cleanert TPT	50.2	46.0	38.9	76.1	37.5	38.1	41.6	36.5	38.9	31.9	45.7	35.7	36.0	32.6	34.5	31.9	25.5	30.9	34.1	39.1	14.7
		EnviCarb + PSA	54.5	48.8	47.6	47.2	41.9	40.3	78.4	35.7	48.1	39.9	51.1	44.8	36.9	42.7	26.1	41.6	43.3	44.5	46.7	45.3	
87	Desmetryn	Cleanert TPT	50.0	40.4	39.9	44.8	33.0	30.0	34.7	32.3	32.6	29.2	35.9	26.1	29.1	27.3	29.3	25.2	25.2	18.7	29.5	32.3	7.5
		EnviCarb + PSA	52.4	44.1	36.7	34.7	39.3	41.2	35.8	31.8	32.2	38.0	32.4	39.7	32.2	29.6	26.5	32.1	26.5	26.8	28.4	34.8	
88	Dichlobenil	Cleanert TPT	8.2	6.6	6.9	7.5	6.6	6.1	7.1	6.1	6.9	6.0	6.3	6.3	6.4	5.9	6.0	5.1	5.0	5.8	6.6	6.4	10.0
		EnviCarb + PSA	9.0	7.1	7.0	6.7	8.0	6.9	6.9	7.2	6.7	7.3	6.4	11.1	6.6	6.2	5.0	6.2	5.9	6.5	7.6	7.1	
89	Dicloran	Cleanert TPT	116.8	96.7	78.1	132.6	88.3	85.4	81.6	88.2	104.5	67.0	108.1	77.4	74.6	68.1	71.5	67.0	62.6	44.5	80.3	83.9	5.1
		EnviCarb + PSA	123.8	100.5	113.2	118.6	91.9	99.5	78.2	73.1	113.7	90.8	90.3	86.9	72.6	78.7	56.2	68.1	73.8	66.8	80.1	88.3	
90	Dicofol	Cleanert TPT	143.8	146.5	159.8	187.6	138.5	126.1	158.2	149.4	152.2	127.2	135.6	155.5	130.3	127.1	138.8	118.5	124.1	113.0	151.3	141.2	7.2
		EnviCarb + PSA	164.5	170.8	210.1	212.0	156.8	150.6	112.2	153.2	150.0	140.3	133.1	102.8	149.2	148.3	145.9	163.4	131.2	142.8	147.7	151.8	
91	Dimethachlor	Cleanert TPT	119.3	109.0	108.2	127.1	98.1	91.7	108.1	103.7	102.4	90.3	99.3	80.1	93.0	88.5	98.4	83.1	79.3	79.1	95.3	97.6	2.5
		EnviCarb + PSA	150.6	109.4	85.8	89.8	116.9	115.6	115.0	87.0	100.4	90.0	98.9	104.3	92.1	94.3	80.4	97.4	89.0	85.7	99.3	100.1	
92	Dioxacarb	Cleanert TPT	338.9	281.4	323.4	381.2	275.5	240.7	308.7	256.2	279.2	252.9	277.6	289.9	239.7	264.5	235.2	214.1	244.4	243.2	277.9	275.0	0.7
		EnviCarb + PSA	395.0	314.2	273.7	279.7	314.6	283.1	303.1	266.9	290.5	280.4	249.6	287.4	273.0	248.1	207.5	241.7	233.2	251.2	270.3	277.0	
93	Endrin	Cleanert TPT	523.5	427.2	420.2	500.0	367.6	339.7	397.4	370.0	386.5	338.1	384.9	287.5	360.8	358.8	361.4	303.9	295.4	316.4	379.3	374.7	8.2
		EnviCarb + PSA	559.2	462.6	430.7	416.1	447.6	436.7	717.0	359.4	357.4	373.3	355.2	359.5	371.8	367.9	319.4	375.3	322.2	342.5	357.3	406.9	
94	Epoxiconazole-2	Cleanert TPT	393.8	332.2	327.0	371.1	273.9	236.0	290.9	264.0	274.0	218.0	308.4	245.8	243.6	234.2	240.9	209.0	197.1	199.7	243.5	268.6	4.1
		EnviCarb + PSA	408.1	352.4	335.0	323.3	311.5	299.9	277.6	256.5	258.6	277.0	253.8	290.0	258.4	250.2	217.3	251.3	223.4	223.1	246.9	279.7	
95	EPTC	Cleanert TPT	97.8	78.3	83.3	110.5	80.2	72.1	84.7	74.1	83.5	72.6	66.7	55.5	71.7	66.9	64.6	54.7	55.9	58.7	74.7	74.0	7.4
		EnviCarb + PSA	104.6	85.0	76.2	73.0	102.4	86.7	95.2	85.2	74.8	89.1	70.5	93.9	70.7	65.0	52.2	70.2	64.7	75.9	80.0	79.8	
96	Ethifumesate	Cleanert TPT	95.5	80.0	79.7	97.2	66.4	75.5	74.7	82.8	85.8	64.0	71.1	68.4	68.3	65.2	68.0	59.0	57.7	58.4	71.9	73.2	3.6
		EnviCarb + PSA	100.9	84.4	81.5	78.1	83.0	78.8	67.1	86.0	88.3	70.3	69.7	68.9	71.5	73.4	61.9	69.5	68.3	67.6	72.3	75.9	
97	Ethoprophos	Cleanert TPT	122.6	106.1	103.3	129.6	93.6	86.6	101.9	93.2	98.0	82.9	100.1	78.4	86.1	83.1	87.6	76.4	72.6	72.8	92.6	93.0	7.4
		EnviCarb + PSA	144.5	117.8	107.1	101.5	112.5	111.8	110.9	90.3	94.1	97.7	94.7	105.3	95.9	90.2	76.0	87.3	86.1	85.8	93.8	100.2	
98	Etrimfos	Cleanert TPT	43.9	35.8	35.5	43.3	32.2	30.1	35.7	33.5	33.7	29.1	35.0	27.7	29.4	30.6	31.8	26.6	27.5	17.4	31.0	32.1	9.2
		EnviCarb + PSA	47.4	39.1	36.3	35.0	38.6	42.9	41.7	30.6	32.4	35.9	33.7	37.2	34.8	33.5	27.8	31.3	29.1	29.5	31.9	35.2	

No.	Compound	Cartridge																					
99	Fenamidone	Cleanert TPT	47.5	42.8	44.1	50.7	38.3	33.9	40.5	37.0	39.6	32.6	39.3	33.6	35.6	34.6	34.9	29.5	29.1	29.2	35.9	37.3	3.7
		Envi-Carb + PSA	53.1	45.7	43.0	40.5	41.6	42.9	38.9	36.1	36.1	38.3	36.5	37.4	39.2	36.3	31.6	36.8	32.4	33.0	35.0	38.7	
100	Fenaromol	Cleanert TPT	97.4	84.6	86.1	100.8	74.2	64.5	74.2	70.4	70.4	62.0	70.4	66.7	64.6	67.0	65.9	56.9	53.9	52.0	66.0	70.9	1.9
		Envi-Carb + PSA	99.8	90.3	84.6	81.9	84.2	76.0	70.3	67.9	64.6	69.6	68.7	70.6	68.1	68.5	58.6	67.9	59.7	58.4	64.4	72.3	
101	Flamprop-iso-propyl	Cleanert TPT	46.4	39.0	39.1	44.7	33.7	31.6	36.5	33.2	35.8	30.5	35.5	31.6	31.7	33.2	32.4	28.0	27.7	28.0	34.0	34.3	3.8
		Envi-Carb + PSA	50.0	41.9	39.2	37.7	39.4	38.4	38.6	32.5	32.7	34.6	31.9	34.8	33.6	33.6	29.9	34.7	29.1	32.4	32.9	35.7	
102	Flamprop-methyl	Cleanert TPT	49.6	42.1	42.6	61.1	36.7	34.5	39.3	35.3	36.8	32.8	37.4	36.4	32.7	32.6	32.2	28.7	26.7	29.0	35.3	36.9	1.7
		Envi-Carb + PSA	50.3	47.8	43.6	41.6	42.3	40.1	34.8	34.9	39.1	35.1	35.5	35.7	34.2	34.8	28.1	33.3	33.9	32.0	36.2	37.5	
103	Fonofos	Cleanert TPT	46.5	36.6	37.3	42.3	33.0	30.2	36.2	33.9	35.0	29.6	35.6	33.2	31.2	37.1	33.6	27.0	26.8	34.0	32.4	34.3	3.7
		Envi-Carb + PSA	47.6	47.5	38.2	35.5	40.3	40.6	35.2	33.0	33.1	37.5	33.6	36.3	34.7	33.4	30.2	33.4	29.4	30.6	32.2	35.6	
104	Hexachlorobenzene	Cleanert TPT	35.9	37.4	31.7	38.8	28.0	27.0	32.2	28.3	31.3	25.8	28.7	28.2	27.6	25.4	27.7	23.4	22.2	19.9	30.4	28.6	10.6
		Envi-Carb + PSA	41.6	35.4	33.7	31.7	35.8	34.3	36.6	30.3	29.1	30.1	28.6	31.2	31.6	31.2	25.8	30.1	27.9	29.9	29.5	31.8	
105	Hexazinone	Cleanert TPT	123.1	113.8	130.8	155.8	111.4	98.7	123.1	103.3	109.6	101.0	104.8	103.8	104.7	102.1	98.7	91.3	106.1	98.1	110.3	110.0	2.4
		Envi-Carb + PSA	151.8	119.1	123.6	132.2	125.6	108.4	101.1	105.5	109.8	104.9	108.2	107.9	110.7	109.8	102.5	116.3	96.0	100.0	107.1	112.7	
106	Iodofenphos	Cleanert TPT	78.8	77.7	68.7	107.9	63.8	66.3	76.9	67.3	70.5	60.5	71.0	56.9	60.6	63.8	64.8	54.4	52.2	54.3	60.2	66.9	5.6
		Envi-Carb + PSA	99.8	78.1	73.1	79.4	79.6	79.5	77.2	58.7	72.1	63.0	66.9	75.8	58.5	69.0	50.0	64.8	63.9	59.4	74.3	70.7	
107	Isofenphos	Cleanert TPT	96.3	76.4	78.2	92.6	68.4	60.1	71.5	66.5	70.5	59.3	81.7	64.1	62.7	61.1	60.4	50.7	47.7	48.3	61.4	67.3	7.3
		Envi-Carb + PSA	97.7	91.1	80.5	73.8	83.3	86.6	69.9	64.3	78.4	85.1	71.2	75.1	65.8	68.7	49.6	56.0	58.7	55.4	60.4	72.3	
108	Isopeopalin	Cleanert TPT	98.0	76.4	72.0	93.2	63.5	56.6	69.1	64.9	69.1	57.1	81.0	71.1	62.9	61.4	66.3	55.6	52.5	52.3	67.8	67.9	4.0
		Envi-Carb + PSA	97.5	87.3	74.9	67.0	78.3	85.3	74.5	60.8	60.5	78.5	71.1	71.6	68.9	67.4	52.3	61.8	59.7	61.1	65.0	70.7	
109	Methoprene	Cleanert TPT	171.9	127.6	129.0	156.3	112.4	101.3	125.6	110.7	121.6	98.7	119.7	116.8	103.3	104.3	107.9	87.5	88.8	90.8	109.1	114.9	9.5
		Envi-Carb + PSA	177.2	161.7	131.1	166.1	135.2	129.4	144.6	114.3	113.1	131.0	108.7	148.0	108.6	112.1	100.2	113.4	96.3	102.8	108.0	126.4	
110	Methoprotryne	Cleanert TPT	145.6	119.8	128.0	138.7	105.0	90.0	109.3	101.8	101.5	81.4	102.1	86.3	91.1	92.8	91.9	78.3	74.7	73.3	92.6	100.2	6.0
		Envi-Carb + PSA	156.1	146.4	125.5	117.4	123.3	127.2	108.0	97.3	94.6	102.4	93.6	99.6	99.0	94.0	86.5	96.6	82.8	83.9	87.8	106.4	
111	Methoxychlor	Cleanert TPT	46.9	44.1	35.6	62.0	35.1	32.0	37.9	35.1	35.3	28.9	39.9	38.1	34.9	31.0	33.6	30.1	24.6	27.7	34.0	36.1	3.1
		Envi-Carb + PSA	44.7	56.7	40.5	36.3	38.9	39.4	37.9	31.7	37.5	37.5	36.5	36.3	38.0	31.5	25.4	31.3	37.2	31.5	39.7	37.3	
112	Methyl-parathion	Cleanert TPT	211.5	169.3	139.4	172.3	181.7	186.9	152.3	148.4	153.8	123.5	170.5	161.4	138.2	135.9	140.9	125.9	114.6	112.3	138.4	151.4	7.1
		Envi-Carb + PSA	219.2	175.5	156.5	199.7	207.0	261.8	140.8	129.8	180.0	159.3	153.6	171.6	130.9	137.6	101.3	133.6	139.0	130.7	161.3	162.6	
113	Metolachlor	Cleanert TPT	39.9	36.4	35.2	41.8	30.9	29.3	35.0	32.5	34.4	29.4	32.1	28.1	30.8	28.5	31.1	27.7	26.4	28.2	33.3	32.2	7.0
		Envi-Carb + PSA	48.5	38.0	34.0	34.0	35.9	35.1	41.7	31.6	33.5	31.9	33.7	40.0	32.0	32.7	28.6	32.7	28.6	30.2	32.5	34.5	
114	Nitrapyrin	Cleanert TPT	117.5	112.2	95.3	137.0	100.4	93.0	115.5	105.1	110.1	92.1	100.3	95.0	95.9	100.8	101.7	85.5	82.1	93.9	98.0	101.7	6.4
		Envi-Carb + PSA	137.9	102.9	108.0	107.4	128.1	117.4	118.0	104.9	123.0	107.1	104.9	111.1	93.6	96.3	77.5	99.2	103.0	95.5	119.1	108.4	

(Continued)

TABLE 2.3.6 The Determination Results for 201 Pesticides in Incurred Green Tea Youden Pair Samples Under 4 Conditions in 3 Months (cont.)

Pesticide content determined With GC–MS and GC–MS/MS, two concentration of Youden pair samples (µg/kg)

No.	Pesticides	Solid Phase Extraction	November 9, 2009 (n = 5)	November 14, 2009 (n = 3)	November 19, 2009 (n = 3)	November 24, 2009 (n = 3)	November 29, 2009 (n = 3)	December 4, 2009 (n = 3)	December 9, 2009 (n = 3)	December 14, 2009 (n = 3)	December 19, 2009 (n = 3)	December 24, 2009 (n = 3)	December 14, 2009 (n = 3)	January 3, 2010 (n = 3)	January 8, 2010 (n = 3)	January 13, 2010 (n = 3)	January 18, 2010 (n = 3)	January 23, 2010 (n = 3)	January 28, 2010 (n = 3)	February 2, 2010 (n = 3)	February 7, 2010 (n = 3)	AVE (µg/kg)	Deviation rates (%)
115	Oxyfluorfen	Cleanert TPT	217.2	174.1	153.0	268.6	139.1	113.0	142.9	141.6	154.5	112.8	205.7	117.0	137.7	132.6	140.1	125.6	107.3	101.3	159.5	149.7	4.4
		Envi-Carb + PSA	211.8	221.6	167.9	148.3	161.4	226.1	135.8	124.8	162.0	174.8	156.5	158.4	151.7	136.7	88.5	120.0	135.7	133.6	156.2	156.4	
116	Pendimethalin	Cleanert TPT	199.0	151.7	134.9	255.1	125.2	109.5	128.5	122.4	131.0	98.1	174.0	118.6	118.3	115.7	125.1	107.0	82.5	90.1	126.3	132.3	5.7
		Envi-Carb + PSA	188.3	201.1	147.5	131.8	151.2	179.2	136.8	109.3	151.7	152.2	139.9	149.0	140.3	116.4	77.5	99.8	136.9	113.3	137.3	140.0	
117	Picoxystrobin	Cleanert TPT	94.3	78.4	80.1	90.1	68.2	62.6	75.0	66.7	71.5	61.0	70.8	55.3	65.0	66.3	65.2	56.2	55.8	55.5	69.2	68.8	4.9
		Envi-Carb + PSA	100.4	84.5	80.1	75.3	80.3	82.4	64.6	67.1	68.5	70.8	67.8	66.7	72.3	69.7	59.8	69.4	60.5	64.1	68.6	72.3	
118	Piperophos	Cleanert TPT	129.1	121.7	113.3	167.4	100.8	87.2	109.4	106.0	106.2	82.4	121.4	110.5	94.5	97.7	96.5	85.3	74.9	76.5	103.2	104.4	3.0
		Envi-Carb + PSA	156.0	135.0	115.7	111.3	115.6	126.4	120.2	92.7	109.9	106.0	113.6	113.1	92.4	94.9	74.3	84.3	91.4	87.6	104.4	107.6	
119	Pirimiphoc-ethyl	Cleanert TPT	88.4	68.8	68.4	79.5	58.9	52.3	67.2	61.6	65.9	53.3	70.1	51.4	59.4	56.9	60.1	50.5	48.9	51.1	61.6	61.8	8.5
		Envi-Carb + PSA	94.0	77.4	70.0	64.3	72.3	76.2	66.3	61.5	59.7	72.5	64.8	74.6	66.1	63.4	57.8	63.7	55.5	58.1	60.0	67.3	
120	Pirimiphos-methyl	Cleanert TPT	44.8	35.5	35.8	43.6	32.4	29.9	35.6	32.6	33.9	28.7	38.6	35.9	29.6	30.7	31.1	25.6	25.2	24.0	31.1	32.9	5.4
		Envi-Carb + PSA	47.2	38.5	36.2	34.3	39.1	39.1	35.9	31.1	31.8	36.9	36.2	39.7	31.3	31.8	28.5	32.2	29.4	28.7	31.7	34.7	
121	Profenofos	Cleanert TPT	282.0	311.2	187.9	368.3	197.8	200.2	200.2	190.6	194.9	145.1	240.0	165.3	168.7	171.8	171.7	146.9	124.2	150.1	171.5	199.4	13.4
		Envi-Carb + PSA	320.6	425.8	209.3	253.5	247.9	247.8	237.7	143.5	256.4	214.1	222.3	215.1	148.5	184.4	128.8	163.0	253.5	210.6	250.1	228.1	
122	Profluralin	Cleanert TPT	254.9	188.1	179.0	408.0	175.0	171.2	193.0	125.4	129.8	107.4	157.3	125.7	117.4	115.1	120.6	104.0	88.6	102.7	125.1	157.3	2.9
		Envi-Carb + PSA	249.2	221.5	192.1	198.1	203.9	232.8	152.6	111.8	189.6	132.9	138.0	149.3	146.4	131.6	92.0	113.4	135.5	125.0	159.7	161.9	
123	Propachlor	Cleanert TPT	116.2	108.3	106.2	123.9	97.2	90.2	109.8	105.3	100.7	90.8	90.9	70.6	90.0	90.2	94.1	82.5	72.2	81.2	90.9	95.3	8.9
		Envi-Carb + PSA	148.5	108.8	108.8	110.4	115.0	117.3	117.0	88.5	98.7	86.2	96.7	140.0	95.5	94.8	83.1	97.1	88.0	88.1	97.1	104.2	
124	Propiconazole	Cleanert TPT	156.6	123.7	124.4	125.9	95.5	86.6	102.0	94.6	98.4	82.7	109.4	81.2	85.2	91.3	84.4	73.9	69.0	67.3	84.4	96.7	9.2
		Envi-Carb + PSA	158.7	134.5	124.5	109.1	112.4	111.2	97.6	90.5	95.8	104.2	91.5	190.8	89.3	91.4	82.9	93.0	74.7	78.5	82.3	105.9	
125	Propyzamide	Cleanert TPT	159.1	132.0	131.8	153.5	113.6	104.2	122.4	114.7	116.6	107.3	124.6	109.9	109.2	104.8	110.2	97.4	96.3	93.6	122.4	117.0	2.8
		Envi-Carb + PSA	160.6	140.5	131.5	127.9	133.5	133.8	106.6	111.4	118.5	127.6	117.0	100.3	111.3	113.3	99.4	116.8	106.3	110.1	120.5	120.4	
126	Ronnel	Cleanert TPT	233.0	211.4	204.6	283.9	195.4	190.2	218.5	202.9	198.7	184.5	199.9	178.5	181.3	173.2	183.6	160.3	147.9	155.0	179.8	193.8	4.2
		Envi-Carb + PSA	283.0	221.5	210.4	221.1	235.5	273.3	173.6	177.7	208.9	197.2	196.7	173.7	180.4	188.6	165.1	184.1	178.8	171.4	200.1	202.2	
127	Sulfotep	Cleanert TPT	45.2	35.8	36.3	40.8	31.8	29.7	34.6	33.1	34.0	29.6	35.0	25.7	31.3	28.6	30.6	26.2	25.2	27.8	30.8	32.2	9.3
		Envi-Carb + PSA	48.1	39.4	36.4	35.2	39.7	43.0	39.6	31.8	32.0	36.0	32.9	34.6	35.2	33.1	29.4	33.1	30.6	29.7	32.0	35.3	
128	Tebufenpyrad	Cleanert TPT	74.9	62.1	64.8	71.5	54.1	49.2	58.0	53.2	57.2	47.7	53.0	43.6	49.7	50.2	51.6	42.8	42.7	43.6	53.3	53.9	2.9
		Envi-Carb + PSA	81.3	60.2	62.4	58.1	62.2	63.2	49.3	52.1	52.4	53.9	52.0	45.2	55.4	54.4	48.3	55.3	47.1	49.3	51.1	55.4	

No.	Compound	Cartridge																				
129	Terbutryn	Cleanert TPT	98.7	79.7	78.2	87.9	64.9	58.5	68.6	64.7	64.9	57.6	68.6	53.8	56.8	59.9	60.3	49.2	53.2	33.9	58.1	64.1	6.4
		Envi-Carb + PSA	104.6	85.3	78.2	73.2	77.9	78.7	72.8	61.9	62.9	68.5	62.3	72.1	61.0	62.0	53.0	59.9	52.3	54.1	57.0	68.3	
130	Thiobencarb	Cleanert TPT	91.5	74.5	76.7	94.1	67.9	63.2	75.0	70.1	73.2	63.3	70.3	62.1	64.9	62.2	66.6	56.9	55.9	59.3	70.0	69.4	5.2
		Envi-Carb + PSA	96.4	80.3	76.6	74.1	80.5	81.7	70.5	69.9	69.4	72.8	67.3	69.9	73.6	70.7	64.2	74.2	62.9	66.4	66.3	73.0	
131	Tralkoxydim	Cleanert TPT	363.7	306.4	312.3	390.0	263.3	244.4	272.9	265.7	269.3	233.4	245.6	284.5	238.4	239.7	249.1	206.6	201.4	196.8	244.1	264.6	5.8
		Envi-Carb + PSA	465.0	359.3	316.8	250.4	303.2	295.8	283.4	242.5	290.2	342.8	231.2	295.0	264.4	225.4	214.5	241.1	230.1	256.8	220.5	280.5	
132	Trans-chlordane	Cleanert TPT	44.5	35.5	35.6	42.9	31.7	28.3	34.2	30.7	33.5	28.5	33.1	28.3	29.9	28.4	30.1	25.9	25.2	27.3	32.3	31.9	5.9
		Envi-Carb + PSA	46.4	39.5	36.9	35.9	37.3	38.6	34.6	30.7	30.6	31.6	30.6	38.0	30.9	33.4	27.3	31.9	28.4	29.1	30.7	33.8	
133	Trans-diallate	Cleanert TPT	84.7	68.0	68.5	81.8	63.3	59.1	68.8	61.9	66.1	58.7	73.5	61.5	68.1	54.6	66.3	61.8	62.0	42.7	72.3	65.5	11.0
		Envi-Carb + PSA	91.1	74.2	69.4	66.7	76.1	73.1	71.3	61.9	63.1	75.4	72.4	82.2	80.0	73.5	68.3	75.9	68.8	69.5	75.7	73.1	
134	Trifloxystrobin	Cleanert TPT	222.3	188.9	151.0	300.7	140.2	138.5	145.2	123.2	134.1	110.7	196.4	139.3	131.3	116.5	126.5	114.1	91.1	115.8	121.4	147.8	4.4
		Envi-Carb + PSA	225.2	194.9	165.3	165.1	160.5	155.0	166.6	117.3	174.4	157.7	164.0	106.7	149.1	144.3	91.9	108.3	177.2	139.8	170.0	154.4	
135	Zoxamide	Cleanert TPT	93.7	82.6	90.3	110.4	77.5	69.6	86.1	78.7	79.1	75.5	75.2	74.4	71.0	74.3	70.0	64.5	67.4	61.0	73.9	77.6	0.7
		Envi-Carb + PSA	106.6	85.4	89.6	91.9	86.5	82.6	68.4	72.4	80.6	72.3	75.0	78.6	73.6	74.8	61.5	72.1	67.2	67.9	78.1	78.2	
136	2,4'-DDE	Cleanert TPT	44.3	36.7	39.8	38.1	34.8	34.1	35.4	32.2	34.3	31.8	34.5	34.1	33.6	32.4	33.3	26.7	28.5	28.1	32.9	34.0	12.8
		Envi-Carb + PSA	51.7	36.9	41.9	40.7	40.9	43.5	37.2	43.0	36.9	35.6	33.0	32.9	34.6	40.3	37.0	40.9	35.3	34.7	36.9	38.6	
137	Ametryn	Cleanert TPT	142.2	110.5	131.5	118.4	61.3	89.2	85.8	86.6	109.2	91.4	97.1	88.5	84.9	102.8	91.6	78.1	75.4	69.2	79.6	94.4	8.6
		Envi-Carb + PSA	144.3	117.0	120.7	118.5	118.6	114.1	120.1	117.6	72.0	89.6	83.3	127.7	87.0	93.1	83.3	94.4	86.2	82.5	85.2	102.9	
138	Bifenthrin	Cleanert TPT	44.7	38.0	36.4	35.5	32.4	35.0	33.9	32.7	32.4	31.7	32.6	32.8	29.7	31.7	31.6	26.5	25.8	26.5	28.4	32.5	6.6
		Envi-Carb + PSA	45.7	37.4	37.7	36.7	38.7	36.4	32.3	38.0	33.5	32.3	29.6	44.1	31.4	30.8	30.4	33.1	31.5	30.1	30.7	34.8	
139	Bitertanol	Cleanert TPT	153.2	126.8	145.1	125.9	108.7	99.7	109.4	114.8	102.1	92.8	107.3	106.1	88.4	143.9	98.4	95.5	87.4	84.4	93.8	109.7	5.6
		Envi-Carb + PSA	157.9	133.2	130.6	121.5	128.8	109.6	101.8	128.2	95.3	101.3	95.7	198.0	102.5	100.4	95.4	109.9	98.2	96.0	100.3	116.0	
140	Boscalid	Cleanert TPT	224.7	195.3	183.8	166.9	130.5	190.7	158.3	199.2	156.7	184.0	159.6	173.8	117.6	222.4	144.7	164.0	120.9	123.3	136.8	165.9	4.3
		Envi-Carb + PSA	224.4	220.2	170.2	173.4	163.0	189.9	163.8	154.9	174.7	165.7	152.2	283.9	128.4	144.0	138.7	159.3	162.3	136.4	164.7	173.2	
141	Butafenacil	Cleanert TPT	64.8	64.9	66.5	40.8	33.7	51.8	37.9	48.6	39.8	48.6	40.0	41.2	29.6	60.3	37.4	42.1	31.2	32.6	34.2	44.5	6.0
		Envi-Carb + PSA	43.8	70.6	44.7	47.4	38.8	41.3	54.0	37.0	51.5	45.0	36.4	128.7	31.1	38.8	34.2	32.9	47.8	33.1	40.8	47.3	
142	Carbaryl	Cleanert TPT	97.6	92.8	87.7	139.7	146.6	124.1	101.4	114.5	94.4	124.3	93.2	87.3	105.7	118.9	104.2	86.9	90.9	96.2	100.0	105.6	15.0
		Envi-Carb + PSA	161.3	140.0	123.0	127.5	127.8	145.3	108.4	130.5	110.9	119.1	90.7	165.0	102.0	116.4	106.2	121.5	110.8	108.4	115.8	122.7	
143	Chlorobenzilate	Cleanert TPT	51.8	41.2	38.5	40.4	30.9	39.0	35.6	42.9	35.3	36.2	35.6	38.9	28.9	39.6	34.7	35.0	28.1	27.5	31.8	36.4	15.5
		Envi-Carb + PSA	80.6	62.0	39.4	38.0	39.3	44.5	50.5	34.9	34.3	36.3	32.3	59.5	28.6	34.1	41.9	41.3	39.0	29.8	35.6	42.5	
144	Chlorthal-dimethyl	Cleanert TPT	128.3	111.1	154.5	112.0	95.3	104.0	101.4	97.9	100.9	96.8	100.0	108.3	92.4	102.6	95.0	86.7	80.3	80.5	87.3	101.9	2.8
		Envi-Carb + PSA	141.1	108.4	111.3	108.4	115.8	117.1	106.3	115.5	98.6	97.4	91.0	92.5	93.7	104.2	98.6	103.6	98.1	92.3	96.6	104.8	

(Continued)

TABLE 2.3.6 The Determination Results for 201 Pesticides in Incurred Green Tea Youden Pair Samples Under 4 Conditions in 3 Months (cont.)

Pesticide content determined With GC-MS and GC-MS/MS, two concentration of Youden pair samples (μg/kg)

No.	Pesticides	Solid Phase Extraction	November 9, 2009 (n=5)	November 14, 2009 (n=3)	November 19, 2009 (n=3)	November 24, 2009 (n=3)	November 29, 2009 (n=3)	December 4, 2009 (n=3)	December 9, 2009 (n=3)	December 14, 2009 (n=3)	December 19, 2009 (n=3)	December 24, 2009 (n=3)	December 29, 2009 (n=3)	January 3, 2010 (n=3)	January 8, 2010 (n=3)	January 13, 2010 (n=3)	January 18, 2010 (n=3)	January 23, 2010 (n=3)	January 28, 2010 (n=3)	February 2, 2010 (n=3)	February 7, 2010 (n=3)	AVE (μg/kg)	Deviation rates (%)
145	Dibutylsuccinate	Cleanert TPT	78.9	61.9	66.1	65.8	56.5	61.0	60.4	54.9	57.1	52.4	55.3	57.3	51.7	57.1	50.3	43.2	43.1	45.2	45.5	56.0	10.5
		Envi-Carb + PSA	85.8	64.5	68.5	65.0	72.0	68.1	61.7	70.1	56.7	58.5	53.2	81.0	52.8	59.0	51.4	56.6	49.5	53.9	53.4	62.2	
146	Diethofencarb	Cleanert TPT	288.2	236.4	268.7	253.3	211.5	212.5	223.1	237.7	218.1	203.8	224.0	258.6	208.0	263.3	214.6	202.7	184.1	174.3	209.6	225.9	9.5
		Envi-Carb + PSA	406.7	292.2	255.3	237.9	252.4	278.6	244.4	249.3	206.6	214.6	209.2	305.3	207.9	225.6	238.1	250.0	229.1	195.4	222.9	248.5	
147	Diflufenican	Cleanert TPT	48.1	41.2	45.1	39.2	32.7	37.0	36.0	39.2	35.0	35.1	35.5	40.2	31.0	41.2	34.2	32.6	28.3	27.7	32.2	36.4	8.9
		Envi-Carb + PSA	69.0	53.0	40.4	38.4	40.8	45.5	42.5	38.8	31.9	34.5	33.9	41.2	31.6	32.2	37.9	39.5	36.0	32.2	36.5	39.8	
148	Dimepiperate	Cleanert TPT	87.5	72.3	70.8	73.4	59.8	61.2	65.6	63.7	63.8	55.7	65.4	76.5	59.5	68.4	60.5	57.1	53.3	51.7	57.0	64.4	8.1
		Envi-Carb + PSA	98.8	78.7	72.9	67.8	76.1	74.0	68.5	75.9	60.1	62.7	60.5	84.4	60.4	68.0	64.2	69.6	62.2	59.4	61.7	69.8	
149	Dimethametryn	Cleanert TPT	48.6	37.5	29.0	38.8	32.2	32.2	32.4	30.9	28.4	29.2	32.4	29.4	27.9	30.6	29.3	25.1	24.6	22.6	26.9	30.9	9.6
		Envi-Carb + PSA	49.7	39.9	39.9	37.3	39.3	37.8	33.1	38.0	29.9	31.1	27.9	43.4	27.7	31.5	27.2	30.4	28.5	27.2	27.5	34.1	
150	Dimethomorph	Cleanert TPT	193.6	146.1	206.5	83.9	78.5	76.2	79.0	74.1	74.6	64.5	76.3	76.3	66.9	111.2	71.4	66.6	62.3	60.8	71.0	91.6	14.3
		Envi-Carb + PSA	109.8	89.6	92.3	86.6	85.4	59.6	69.0	90.7	72.4	72.2	66.1	88.2	74.8	79.6	71.5	81.2	74.7	68.6	74.7	79.3	
151	Dimethylphthalate	Cleanert TPT	178.2	90.2	145.4	144.4	131.9	141.6	138.6	129.0	135.2	134.1	135.7	144.8	126.5	162.2	132.7	116.4	111.8	114.5	124.4	133.6	8.1
		Envi-Carb + PSA	188.3	143.0	154.1	151.0	160.1	154.6	135.3	163.5	137.5	138.5	129.5	142.5	128.7	146.6	130.8	145.8	130.0	135.9	135.4	144.8	
152	Diniconazole	Cleanert TPT	152.9	126.7	146.5	117.2	98.6	91.0	99.8	100.0	91.9	89.4	92.3	91.6	87.2	110.4	91.6	79.8	73.9	69.9	79.5	99.5	6.8
		Envi-Carb + PSA	168.0	128.1	121.0	114.9	119.6	122.3	103.5	114.3	84.5	96.1	87.4	138.1	89.7	92.2	87.9	94.6	90.7	82.0	89.2	106.5	
153	Diphenamid	Cleanert TPT	47.4	37.1	40.9	38.9	29.7	33.0	33.2	31.4	33.7	33.3	34.8	34.8	33.4	41.7	35.2	30.7	30.2	29.1	32.4	34.8	7.3
		Envi-Carb + PSA	48.4	38.7	41.7	40.6	41.5	43.0	38.0	41.1	26.2	33.3	31.4	41.2	33.7	39.7	33.1	36.6	34.4	33.6	34.6	37.4	
154	Dipropetryn	Cleanert TPT	49.9	38.8	43.8	38.4	33.3	32.9	34.0	32.3	31.6	29.2	31.7	32.7	30.6	32.7	30.2	25.6	26.0	23.4	26.9	32.8	6.9
		Envi-Carb + PSA	48.3	37.2	39.1	36.9	40.3	37.1	34.7	38.8	31.7	31.1	28.3	57.7	29.8	32.6	27.3	31.4	29.4	27.6	29.3	35.2	
155	Ethalfluralin	Cleanert TPT	188.1	141.2	143.5	134.3	112.3	138.3	126.4	130.8	126.2	119.6	130.6	130.7	105.7	138.5	123.1	115.2	103.2	96.3	113.3	127.2	13.3
		Envi-Carb + PSA	179.9	158.9	144.3	143.4	147.5	187.6	134.7	130.4	128.7	126.5	111.5	273.9	114.8	135.2	121.9	129.1	133.0	123.3	137.9	145.4	
156	Etofenprox	Cleanert TPT	55.7	43.5	61.9	41.5	42.4	41.6	43.5	39.3	38.6	38.4	37.7	37.8	37.9	46.1	40.4	34.6	36.4	38.9	39.7	41.9	9.1
		Envi-Carb + PSA	48.0	43.7	47.8	54.9	46.5	38.7	40.9	58.3	53.6	40.8	39.2	34.3	39.6	39.5	42.9	52.0	48.2	49.7	52.8	45.9	
157	Etridiazol	Cleanert TPT	112.1	90.2	92.5	98.9	79.6	98.5	87.1	85.4	86.0	87.8	90.5	92.1	77.5	105.1	77.8	70.2	65.1	71.7	77.2	86.6	12.3
		Envi-Carb + PSA	144.2	97.7	98.5	98.3	103.8	146.8	94.9	98.0	90.9	92.1	92.0	133.1	78.5	78.0	80.6	84.8	68.7	89.5	90.1	97.9	

#	Compound	Cartridge																					
158	Fenazaquin	Cleanert TPT	45.4	39.6	34.6	36.3	31.6	31.4	31.9	30.1	29.2	27.2	30.2	26.1	28.0	30.8	28.9	24.5	26.6	26.8	26.6	30.8	7.0
		Envi-Carb + PSA	44.9	35.1	36.0	35.0	37.8	33.9	31.4	36.3	28.5	28.7	26.0	37.2	29.8	32.8	28.2	30.4	31.9	32.0	32.5	33.1	
159	Fenchlorphos	Cleanert TPT	237.3	202.1	265.7	219.2	192.4	214.4	202.2	199.1	198.8	205.7	198.7	215.4	186.0	184.2	191.1	160.1	163.7	160.8	166.1	198.0	5.2
		Envi-Carb + PSA	266.6	222.7	223.1	232.3	232.3	229.1	201.1	221.9	197.0	205.4	178.9	225.9	186.9	200.7	183.5	187.1	189.6	180.9	197.2	208.5	
160	Fenoxanil	Cleanert TPT	95.8	75.8	86.9	76.6	76.8	65.6	70.3	65.1	68.4	52.8	68.0	74.6	72.3	80.0	65.5	60.8	59.8	58.5	65.6	70.5	9.9
		Envi-Carb + PSA	107.1	76.3	91.9	76.6	89.0	85.7	72.0	84.1	67.5	74.0	60.0	95.5	69.5	77.8	69.7	76.9	67.3	68.3	70.1	77.9	
161	Fenpropidin	Cleanert TPT	91.8	63.5	82.5	68.7	62.1	58.0	61.9	64.9	58.1	51.6	65.3	61.1	59.0	85.7	59.5	53.0	52.3	47.7	59.2	63.5	2.3
		Envi-Carb + PSA	97.2	65.4	71.5	69.6	72.3	65.4	47.7	74.8	66.2	56.9	60.0	77.6	54.4	65.8	54.8	62.4	57.8	54.1	59.4	64.9	
162	Fenson	Cleanert TPT	50.7	39.0	47.9	39.8	32.8	41.5	38.2	39.8	37.9	39.9	38.7	44.8	32.9	39.2	36.3	37.0	29.9	31.1	35.5	38.6	7.5
		Envi-Carb + PSA	59.7	49.2	41.1	41.3	41.9	46.2	45.2	40.6	39.8	39.6	36.9	46.7	35.0	39.2	37.5	40.0	38.1	33.9	37.7	41.6	
163	Flufenacet	Cleanert TPT	336.3	326.3	287.5	305.9	261.3	295.6	255.3	293.6	240.6	326.1	275.9	261.4	222.6	250.0	263.3	206.4	208.6	202.9	221.4	265.3	10.7
		Envi-Carb + PSA	354.5	332.5	296.2	327.2	303.1	342.7	289.0	278.1	261.9	273.5	248.6	462.1	235.3	281.4	268.9	261.9	310.9	234.2	247.3	295.2	
164	Furalaxyl	Cleanert TPT	90.6	95.9	87.5	74.0	67.0	66.9	66.9	63.4	63.8	61.3	65.0	73.0	60.9	75.4	62.6	55.2	54.7	52.0	57.9	68.1	1.5
		Envi-Carb + PSA	94.4	73.0	78.0	76.6	78.1	76.8	62.6	77.5	65.8	61.8	58.1	66.6	60.9	69.9	60.2	66.8	61.4	60.4	63.9	69.1	
165	Heptachlor	Cleanert TPT	117.6	94.0	91.8	103.7	82.7	93.3	89.7	89.8	91.8	85.0	92.3	96.0	80.4	89.7	86.2	75.2	71.1	69.3	77.5	88.3	7.6
		Envi-Carb + PSA	138.5	103.6	99.9	100.4	105.5	115.6	89.4	97.7	81.1	88.9	85.8	94.0	82.9	88.3	93.2	95.0	84.3	80.6	84.5	95.2	
166	Iprobenfos	Cleanert TPT	137.1	113.3	114.2	105.3	81.8	81.1	93.9	88.6	85.6	73.0	88.4	100.9	79.4	105.4	88.2	77.1	76.0	65.8	78.6	91.3	6.3
		Envi-Carb + PSA	154.2	112.8	105.5	95.0	113.7	108.2	90.3	105.7	77.4	87.2	82.9	110.0	82.4	94.4	86.3	93.2	84.0	79.2	84.2	97.2	
167	Isazofos	Cleanert TPT	91.5	69.6	82.1	82.0	63.2	65.9	67.5	69.7	70.2	65.1	71.2	67.5	67.1	66.9	64.1	57.0	56.8	53.0	59.9	67.9	6.8
		Envi-Carb + PSA	97.0	74.7	76.1	78.2	79.5	70.8	68.8	73.4	73.0	70.5	67.9	86.5	67.8	71.0	65.4	67.5	65.1	65.6	62.8	72.7	
168	Isoprothiolane	Cleanert TPT	91.5	68.0	60.4	74.4	64.6	66.9	66.7	63.3	63.3	62.1	65.5	58.2	60.7	74.6	62.0	52.9	51.5	51.1	56.1	63.9	9.8
		Envi-Carb + PSA	96.1	73.8	77.9	74.3	78.0	78.4	67.9	74.4	68.1	63.6	59.1	87.7	63.0	69.1	61.1	65.1	60.8	58.4	61.6	70.4	
169	Kresoxim-methyl	Cleanert TPT	77.1	61.8	70.4	38.0	42.5	35.0	39.7	42.2	36.3	32.5	35.0	33.2	29.6	37.9	32.5	30.5	28.6	31.4	32.4	40.3	9.7
		Envi-Carb + PSA	49.0	59.1	36.7	46.2	42.2	31.1	30.4	47.6	40.7	33.7	27.5	24.9	29.3	34.1	28.7	32.3	30.0	29.9	42.5	36.6	
170	Mefenacet	Cleanert TPT	126.2	98.4	115.2	133.4	94.4	115.4	133.0	130.8	110.2	123.7	120.9	141.1	79.5	151.9	144.7	139.9	131.6	132.9	85.3	121.5	2.1
		Envi-Carb + PSA	129.9	135.7	121.2	116.1	114.7	147.5	145.1	122.1	124.9	105.4	109.4	107.4	86.8	124.6	139.4	174.6	130.3	139.5	83.2	124.1	
171	Mepronil	Cleanert TPT	50.4	54.3	47.3	40.5	39.4	39.0	38.7	34.9	37.3	33.1	36.8	42.4	34.0	47.2	38.8	33.5	31.4	32.2	36.6	39.4	2.5
		Envi-Carb + PSA	47.9	45.9	45.2	40.0	44.2	33.8	33.6	43.8	38.6	40.4	35.6	46.5	37.6	38.8	37.0	41.0	38.5	38.0	39.9	40.3	
172	Metribuzin	Cleanert TPT	139.5	102.5	97.3	82.2	60.4	64.1	60.2	63.5	55.5	56.4	51.7	57.6	47.1	64.3	51.2	46.8	42.7	40.4	43.1	64.5	6.5
		Envi-Carb + PSA	143.4	110.2	81.8	76.9	71.8	72.3	68.1	64.0	58.4	57.9	46.2	106.7	48.4	58.7	46.5	51.6	52.9	43.0	50.7	68.9	
173	Molinate	Cleanert TPT	39.7	28.3	34.3	33.4	30.4	32.1	32.3	27.6	30.6	29.6	30.8	27.7	30.9	33.5	30.4	26.2	27.1	26.6	28.4	30.5	9.1
		Envi-Carb + PSA	42.4	32.6	35.5	34.7	37.6	36.4	32.2	36.3	31.2	30.9	29.1	33.8	29.6	37.5	28.9	33.4	30.4	31.3	31.2	33.4	
174	Napropamide	Cleanert TPT	136.9	110.9	119.6	111.5	99.3	101.8	101.6	96.8	98.2	93.4	101.6	106.1	85.4	108.0	97.5	84.5	78.0	78.2	90.5	100.0	6.2
		Envi-Carb + PSA	145.1	109.9	119.1	113.6	118.7	121.8	98.1	119.7	98.3	95.9	88.8	122.4	96.2	100.5	85.7	101.7	95.0	92.6	97.9	106.4	

(Continued)

TABLE 2.3.6 The Determination Results for 201 Pesticides in Incurred Green Tea Youden Pair Samples Under 4 Conditions in 3 Months (cont.)

Pesticide content determined With GC–MS and GC–MS/MS, two concentration of Youden pair samples (μg/kg)

No.	Pesticides	Solid Phase Extraction	November 9, 2009 (n=5)	November 14, 2009 (n=3)	November 24, 2009 (n=3)	November 29, 2009 (n=3)	December 4, 2009 (n=3)	December 9, 2009 (n=3)	December 14, 2009 (n=3)	December 19, 2009 (n=3)	December 24, 2009 (n=3)	December 29, 2009 (n=3)	January 3, 2010 (n=3)	January 8, 2010 (n=3)	January 13, 2010 (n=3)	January 18, 2010 (n=3)	January 23, 2010 (n=3)	January 28, 2010 (n=3)	February 2, 2010 (n=3)	February 7, 2010 (n=3)	AVE (μg/kg)	Deviation rates (%)
175	Nuarimol	Cleanert TPT	97.6	79.5	77.6	68.4	68.6	67.1	66.1	63.0	59.5	64.7	63.6	55.9	73.1	60.2	51.7	49.9	49.4	53.5	66.5	3.5
		Envi-Carb + PSA	96.1	65.2	82.4	79.7	73.3	61.7	76.7	63.6	66.0	56.9	80.1	60.1	72.1	55.7	62.6	59.9	57.0	57.6	68.8	
176	Permethrin	Cleanert TPT	92.9	75.8	75.4	65.9	73.1	70.7	68.5	69.6	68.3	70.4	69.7	61.0	71.7	65.1	56.9	56.1	55.2	60.0	69.1	6.3
		Envi-Carb + PSA	90.3	84.3	75.0	78.2	73.8	76.6	77.2	74.3	68.5	62.3	87.7	68.5	68.2	63.2	70.3	68.4	64.9	68.2	73.5	
177	Phenothrin	Cleanert TPT	49.3	35.2	33.5	32.0	33.4	32.7	30.5	30.4	30.3	37.1	33.8	39.7	33.5	32.7	26.3	25.5	27.4	26.4	33.2	5.1
		Envi-Carb + PSA	48.8	35.1	35.6	45.4	36.0	33.7	37.1	31.4	32.3	28.7	43.9	29.6	35.7	30.0	33.1	30.7	29.5	29.7	34.9	
178	Piperonylbutoxide	Cleanert TPT	46.6	36.4	39.7	32.6	33.1	34.7	32.5	31.6	29.4	31.2	29.8	29.6	31.2	30.1	26.5	26.1	24.4	28.0	32.4	8.8
		Envi-Carb + PSA	50.4	39.3	36.9	41.4	40.3	35.1	40.6	29.9	32.0	28.7	41.0	32.1	32.4	30.3	32.4	30.3	30.1	30.8	35.4	
179	Pretilachlor	Cleanert TPT	73.2	64.3	80.1	63.2	65.2	64.8	64.2	63.8	63.7	61.0	73.7	54.6	59.8	60.2	48.8	51.6	50.5	52.9	62.4	11.9
		Envi-Carb + PSA	100.0	66.9	74.4	73.2	79.6	58.5	69.3	58.9	58.2	52.0	164.4	56.9	60.5	55.2	59.1	58.7	57.7	61.2	70.3	
180	Prometon	Cleanert TPT	149.7	118.4	122.3	103.9	103.5	109.4	101.3	102.3	97.2	110.9	112.6	93.5	111.5	105.1	90.3	87.5	82.6	96.6	107.0	7.6
		Envi-Carb + PSA	155.2	121.7	123.6	129.4	121.7	108.3	124.4	102.6	100.1	93.7	168.7	96.0	113.7	95.8	108.3	102.7	98.5	103.3	115.5	
181	Pronamide	Cleanert TPT	163.5	136.2	148.0	118.3	125.5	120.1	118.5	119.6	115.3	123.5	130.2	104.6	140.2	115.8	104.6	102.6	98.7	115.2	123.9	3.0
		Envi-Carb + PSA	184.6	127.6	141.5	147.3	139.3	118.6	143.2	123.0	124.3	111.5	101.2	113.4	123.9	112.7	124.1	116.9	115.1	120.8	127.7	
182	Propetamphos	Cleanert TPT	55.7	43.6	40.0	29.0	36.8	35.3	36.3	33.4	33.1	35.4	36.6	30.8	40.7	33.6	30.9	28.0	27.8	30.2	35.7	9.3
		Envi-Carb + PSA	55.6	50.9	38.9	40.0	36.3	40.4	33.8	35.5	37.1	33.9	70.6	30.9	34.8	33.0	33.5	35.0	30.2	35.4	39.2	
183	Propoxur-1	Cleanert TPT	352.7	287.0	315.0	283.7	284.1	291.4	264.9	289.7	274.3	282.0	282.0	264.6	381.6	268.2	243.2	236.2	235.0	261.9	285.9	4.9
		Envi-Carb + PSA	382.8	283.6	322.5	337.6	325.5	268.0	352.4	284.8	281.0	262.6	269.9	272.2	316.6	277.9	308.1	279.9	271.5	278.0	300.3	
184	Propoxur-2	Cleanert TPT	436.4	438.5	311.4	253.1	377.7	252.9	387.5	310.1	322.8	341.7	342.3	271.0	402.4	349.8	303.9	290.2	322.9	319.8	334.5	6.1
		Envi-Carb + PSA	419.1	724.5	331.6	297.2	287.0	520.4	236.6	374.1	325.1	335.9	367.2	289.8	364.9	321.2	341.7	297.4	300.3	324.4	355.7	
185	Prothiophos	Cleanert TPT	42.6	34.5	35.3	29.7	32.2	32.6	31.4	30.8	28.3	31.7	29.9	29.2	32.6	30.4	26.8	25.5	26.6	27.5	31.3	7.5
		Envi-Carb + PSA	42.8	37.1	35.4	38.1	35.6	33.2	36.2	31.4	31.4	27.5	45.2	30.2	31.1	29.4	30.4	30.2	28.6	32.0	33.8	
186	Pyridaben	Cleanert TPT	65.5	51.6	50.0	39.3	46.4	44.1	47.6	41.4	47.7	40.3	36.5	38.8	56.7	49.2	47.0	39.8	39.2	44.9	46.4	12.9
		Envi-Carb + PSA	69.2	41.8	44.4	40.3	48.6	39.7	39.9	32.0	36.9	31.9	74.0	31.6	34.4	34.4	36.3	34.9	30.4	36.1	40.7	
187	Pyridaphenthion	Cleanert TPT	43.2	43.3	40.2	31.8	35.5	36.1	38.6	34.3	36.9	36.5	47.6	28.2	42.1	32.3	29.9	27.0	28.5	31.2	36.1	7.9
		Envi-Carb + PSA	45.2	42.2	38.1	41.1	48.8	34.1	38.7	33.5	32.5	32.0	80.6	32.2	35.5	32.0	35.7	33.2	33.3	34.8	39.1	

#	Compound	Cartridge																					
188	Pyrimethanil	Cleanert TPT	46.7	36.9	41.6	39.0	32.6	34.5	34.1	31.9	32.7	31.1	32.8	36.1	30.3	33.6	32.3	27.7	27.3	25.7	29.5	33.5	26.2
		Envi-Carb + PSA	52.3	39.5	46.5	48.1	44.0	42.8	40.9	37.4	48.5	39.8	40.9	44.9	39.3	39.7	42.8	47.0	44.1	44.2	45.7	43.6	
189	Pyriproxyfen	Cleanert TPT	56.6	42.1	51.3	43.9	43.4	41.1	45.1	34.6	39.8	43.8	38.0	40.9	37.9	51.3	46.0	39.5	28.2	39.5	44.4	42.5	4.4
		Envi-Carb + PSA	51.6	41.5	47.9	48.8	45.8	40.9	34.3	49.9	48.1	40.1	39.9	48.6	40.1	39.9	43.1	47.7	44.9	45.0	45.9	44.4	
190	Quinalphos	Cleanert TPT	40.5	37.8	38.6	38.8	31.5	36.8	34.8	35.9	35.5	34.4	34.1	40.0	34.3	36.3	33.1	30.0	27.6	27.9	29.7	34.6	7.3
		Envi-Carb + PSA	49.1	42.3	37.9	38.2	39.0	43.7	37.2	37.9	32.9	34.9	31.7	51.7	32.0	36.4	31.4	34.0	32.7	31.1	32.8	37.2	
191	Quinoxyphen	Cleanert TPT	44.6	35.3	40.8	36.6	32.1	33.6	32.9	31.6	32.2	29.3	32.3	35.3	28.1	33.5	31.4	27.9	26.8	24.7	27.7	32.5	6.9
		Envi-Carb + PSA	47.6	36.5	38.8	37.8	38.3	38.5	33.7	38.4	31.0	30.6	30.0	37.4	32.0	32.9	31.2	33.6	31.7	29.2	31.8	34.8	
192	Telodrin	Cleanert TPT	178.9	150.9	145.2	141.1	114.3	145.5	121.2	125.0	123.9	136.2	127.4	120.6	109.0	126.8	121.7	108.9	101.8	102.6	106.2	126.7	3.5
		Envi-Carb + PSA	154.8	161.1	129.2	144.9	142.7	131.9	139.5	127.6	139.2	134.4	107.8	154.4	106.2	131.5	105.5	107.4	130.1	113.0	131.7	131.2	
193	Tetrasul	Cleanert TPT	41.0	32.7	38.1	34.2	31.5	33.6	32.0	30.2	31.1	30.0	31.3	30.1	30.4	31.7	31.4	29.6	29.4	29.5	32.1	32.1	7.8
		Envi-Carb + PSA	43.0	34.3	37.2	35.6	38.2	37.7	32.6	38.0	31.5	31.8	28.9	33.9	32.3	33.7	33.0	35.5	34.1	33.4	34.8	34.7	
194	Thiazopyr	Cleanert TPT	98.0	70.8	80.3	67.4	66.3	64.9	65.7	64.7	66.2	65.1	63.2	71.1	62.8	74.0	63.9	55.3	53.5	52.6	59.0	66.6	4.9
		Envi-Carb + PSA	95.9	74.3	75.0	74.6	82.5	73.8	67.2	77.8	64.3	64.0	57.7	70.9	62.5	67.5	62.2	66.0	65.1	61.5	64.7	69.9	
195	Tolclofos-methyl	Cleanert TPT	40.9	35.6	38.6	38.2	33.0	35.0	35.2	34.3	34.2	32.8	35.4	35.4	32.9	34.4	33.1	28.8	28.3	28.7	29.7	33.9	7.6
		Envi-Carb + PSA	46.0	36.1	38.6	38.0	40.2	39.1	34.1	40.4	33.3	34.8	32.1	43.1	33.8	35.3	33.1	35.3	33.7	33.2	35.0	36.6	
196	Transfluthrin	Cleanert TPT	45.7	34.7	38.3	38.2	31.0	33.6	33.7	31.4	32.5	30.9	32.7	32.3	31.6	35.6	33.1	27.6	28.6	26.8	29.4	33.0	5.0
		Envi-Carb + PSA	43.7	37.8	36.0	37.3	38.4	34.3	33.9	37.7	32.3	33.7	28.7	40.1	31.4	35.1	30.0	34.2	32.4	31.1	32.2	34.7	
197	Triadimefon	Cleanert TPT	102.7	65.4	88.0	84.2	69.7	70.0	75.2	75.3	73.7	69.1	78.1	81.4	66.5	84.4	72.1	62.7	64.3	57.3	63.4	75.4	17.6
		Envi-Carb + PSA	122.4	107.7	96.6	92.0	98.3	97.2	85.4	97.2	80.6	82.2	77.4	108.1	76.4	94.5	78.3	82.0	78.1	75.0	81.2	90.0	
198	Triadimenol	Cleanert TPT	149.5	115.8	131.1	112.6	97.2	88.2	100.1	97.9	96.6	79.9	91.0	89.8	86.2	131.5	93.0	82.2	79.8	75.9	85.4	99.1	7.9
		Envi-Carb + PSA	149.7	105.4	102.2	95.4	103.4	101.4	79.5	90.7	75.5	72.1	71.2	140.7	77.8	81.5	77.3	86.2	78.3	74.4	77.6	91.6	
199	Triallate	Cleanert TPT	88.6	70.9	78.5	67.1	58.4	67.9	63.9	62.0	62.5	58.6	63.6	72.5	54.7	62.3	60.4	53.0	53.1	53.1	54.4	63.4	24.3
		Envi-Carb + PSA	61.1	54.4	51.1	50.5	53.9	48.8	49.5	49.8	47.2	47.8	40.9	75.2	41.8	49.0	42.0	43.0	47.8	44.2	45.9	49.7	
200	Tribenuron-methyl	Cleanert TPT	59.5	62.0	51.3	40.1	31.2	46.5	39.6	47.0	40.2	33.6	34.1	33.4	35.9	51.1	30.9	34.0	24.0	34.8	40.6	40.5	49.4
		Envi-Carb + PSA	70.1	115.4	66.9	77.2	61.7	78.1	123.2	74.4	71.9	69.3	38.3	7.5	64.6	57.3	37.4	53.0	63.5	67.6	77.1	67.1	
201	Vinclozolin	Cleanert TPT	47.3	41.2	42.5	40.8	33.1	39.3	36.8	36.7	35.9	37.9	36.5	38.8	33.4	38.2	35.8	31.9	33.7	30.7	32.0	37.0	2.6
		Envi-Carb + PSA	45.9	44.1	42.3	41.4	42.5	42.0	40.6	41.1	40.1	37.7	32.5	24.6	33.5	38.1	32.5	34.7	37.8	34.2	35.5	38.0	

TABLE 2.3.7 The Determination Results for 201 Pesticides in Incurred Oolong Tea Youden Pair Samples Under 4 Conditions in 3 Months

No.	Pesticides	Solid Phase Extraction	November 9, 2009 (n=5)	November 14, 2009 (n=3)	November 19, 2009 (n=3)	November 24, 2009 (n=3)	November 29, 2009 (n=3)	December 4, 2009 (n=3)	December 9, 2009 (n=3)	December 14, 2009 (n=3)	December 19, 2009 (n=3)	December 24, 2009 (n=3)	December 14, 2009 (n=3)	January 3, 2010 (n=3)	January 8, 2010 (n=3)	January 13, 2010 (n=3)	January 18, 2010 (n=3)	January 23, 2010 (n=3)	January 28, 2010 (n=3)	February 2, 2010 (n=3)	February 7, 2010 (n=3)	AVE (μg/kg)	Deviation rates (%)
1	2,3,4,5-Tetrachloroaniline	Cleanert TPT	6.8	6.1	5.1	5.3	5.5	4.2	4.2	4.6	4.9	4.6	4.5	4.8	4.3	4.3	4.8	4.2	3.2	3.9	4.0	4.7	1.9
		Envi-Carb + PSA	7.9	6.9	6.2	4.6	6.0	4.9	5.4	4.4	4.4	4.7	4.5	3.9	3.3	3.5	3.6	3.1	3.2	3.3	3.8	4.6	
2	2,3,5,6-Tetrachloroaniline	Cleanert TPT	37.3	37.5	32.3	34.1	35.5	29.8	29.5	36.7	35.6	35.5	35.6	35.2	34.9	36.7	31.4	33.5	30.6	32.0	32.1	34.0	3.3
		Envi-Carb + PSA	42.0	37.1	37.6	33.4	37.0	33.6	37.1	31.3	37.2	33.2	32.6	29.3	30.3	31.9	31.9	26.3	25.9	29.5	27.9	32.9	
3	4,4-Dibromobenzophenone	Cleanert TPT	52.5	51.0	42.8	44.2	46.5	37.9	54.2	64.0	47.5	40.4	62.0	39.5	47.3	46.5	50.1	53.8	49.5	42.0	41.6	48.1	7.2
		Envi-Carb + PSA	63.7	41.2	40.6	39.4	47.2	38.2	43.9	65.1	38.6	39.5	37.7	37.9	45.2	45.0	37.7	32.9	52.0	57.1	46.4	44.7	
4	4,4-Dichlorobenzophenone	Cleanert TPT	58.5	68.1	51.0	52.7	60.2	51.5	52.0	63.9	68.9	55.4	63.4	55.3	49.3	64.3	66.9	65.6	62.8	62.5	56.0	59.4	1.0
		Envi-Carb + PSA	69.0	51.0	60.5	55.4	56.7	54.4	57.6	67.7	60.7	56.9	56.1	53.4	50.5	62.0	61.8	46.3	55.3	68.4	73.5	58.8	
5	Acetochlor	Cleanert TPT	81.9	84.2	65.8	67.2	61.8	56.4	64.5	76.9	74.0	68.7	73.3	68.8	64.9	69.4	60.2	64.2	63.1	57.1	63.9	67.7	6.5
		Envi-Carb + PSA	74.9	76.0	81.5	67.3	77.4	64.3	68.9	67.1	70.1	63.3	63.5	53.7	57.5	55.6	59.9	43.8	48.0	56.3	55.8	63.4	
6	Alachlor	Cleanert TPT	117.1	122.4	95.6	100.0	88.7	82.3	90.4	113.4	105.2	102.9	106.7	102.0	96.7	102.8	86.4	97.0	82.8	89.6	88.8	98.5	6.5
		Envi-Carb + PSA	100.4	109.8	118.8	98.1	109.5	91.7	104.4	96.7	105.6	92.7	91.6	76.7	82.3	86.3	88.1	72.2	70.1	79.3	78.5	92.3	
7	Atratone	Cleanert TPT	46.7	43.2	38.5	37.4	36.3	32.7	32.7	40.3	37.7	39.0	42.1	38.5	36.3	37.3	34.3	34.9	32.5	35.3	31.3	37.2	2.0
		Envi-Carb + PSA	52.3	43.1	43.9	39.3	48.3	37.2	41.1	36.6	48.9	39.6	40.4	32.4	32.1	34.9	34.6	28.2	27.0	31.3	29.6	37.9	
8	Benodanil	Cleanert TPT	238.8	139.0	122.7	201.0	349.8	171.2	106.2	132.3	136.8	91.8	318.5	69.0	109.8	79.3	114.8	118.0	121.8	78.6	74.9	146.0	29.3
		Envi-Carb + PSA	108.5	130.7	134.1	73.5	183.8	78.7	78.1	196.4	46.4	68.1	77.5	93.8	114.1	118.6	98.9	71.0	104.8	168.2	121.0	108.7	
9	Benoxacor	Cleanert TPT	91.8	93.9	74.4	84.4	72.0	62.3	76.6	88.3	73.5	74.5	106.4	65.9	78.0	75.3	66.5	75.6	70.1	60.9	69.5	76.8	5.1
		Envi-Carb + PSA	70.4	74.7	85.9	70.2	80.3	75.0	75.4	92.6	63.2	66.9	66.4	67.8	73.1	70.2	68.1	53.6	73.0	82.7	77.1	73.0	
10	Bromophos-ethyl	Cleanert TPT	37.8	36.6	31.7	33.1	31.2	27.1	27.7	35.7	34.5	33.2	37.2	33.3	32.0	33.4	29.6	31.4	27.1	28.6	28.3	32.1	3.0
		Envi-Carb + PSA	36.9	36.5	36.1	31.7	36.1	30.6	34.2	31.6	35.8	31.7	32.1	28.3	27.8	29.6	29.6	23.9	24.6	27.3	27.2	31.1	
11	Butralin	Cleanert TPT	170.3	171.1	129.8	142.8	162.2	130.5	198.9	188.5	149.1	125.6	168.8	111.8	132.9	126.5	131.2	147.0	160.4	126.1	117.2	146.9	0.8
		Envi-Carb + PSA	129.7	126.8	130.9	120.9	232.6	142.7	116.6	200.6	140.8	120.6	129.3	122.3	139.7	198.6	97.3	89.5	171.8	224.6	134.1	145.8	

No.	Compound	Cartridge																					
12	Chlorfenapyr	Cleanert TPT	330.5	353.9	295.4	297.2	299.2	244.6	259.2	323.1	305.6	304.6	323.1	315.5	293.7	315.9	281.0	288.2	257.5	270.5	266.0	296.0	3.7
		Envi-Carb + PSA	346.3	330.0	346.0	304.4	318.5	279.5	316.1	276.8	323.1	290.9	274.0	250.0	254.9	265.0	279.1	226.7	226.7	258.9	251.7	285.2	
13	Clomazone	Cleanert TPT	40.7	40.1	34.4	35.7	36.5	29.5	31.6	37.8	36.9	33.2	45.3	32.9	32.7	34.5	31.1	33.6	32.2	31.4	28.7	34.7	4.8
		Envi-Carb + PSA	39.2	38.9	40.0	33.0	39.1	33.7	34.3	35.5	35.6	33.8	31.7	30.4	30.3	34.3	28.7	24.6	26.9	28.5	29.3	33.0	
14	Cycloate	Cleanert TPT	83.7	81.6	73.9	73.5	84.0	61.5	59.0	71.0	74.1	66.5	106.8	63.2	65.9	70.0	68.4	67.1	58.8	57.8	52.4	70.5	0.2
		Envi-Carb + PSA	97.1	84.1	80.1	68.1	83.0	64.4	70.6	75.6	80.5	66.2	63.2	58.0	66.0	81.6	57.4	46.8	60.7	77.0	61.4	70.6	
15	Cycluron	Cleanert TPT	33.5	32.3	27.1	26.9	25.7	22.9	25.0	30.7	31.2	30.6	31.6	29.1	28.6	29.5	26.4	26.3	23.7	26.6	25.5	28.1	1.9
		Envi-Carb + PSA	36.9	30.9	32.9	29.3	32.2	28.1	31.0	26.0	29.8	27.9	27.5	24.3	25.3	26.0	26.8	21.8	20.6	22.7	23.2	27.5	
16	Cyhalofop-butyl	Cleanert TPT	116.8	137.6	102.9	110.6	106.1	94.5	89.7	108.9	104.0	153.9	193.0	175.0	141.1	117.1	106.3	104.6	84.2	106.6	98.5	118.5	2.6
		Envi-Carb + PSA	147.3	140.6	128.9	117.0	136.8	95.4	119.7	94.8	107.4	154.7	154.7	124.7	127.9	103.5	108.5	81.4	77.2	91.4	82.4	115.5	
17	Cyprodinil	Cleanert TPT	38.6	38.5	32.1	33.7	32.4	27.9	28.1	37.1	36.3	36.2	34.2	36.1	33.2	35.6	33.0	32.1	28.7	31.3	28.8	33.4	0.9
		Envi-Carb + PSA	47.1	41.4	38.3	32.5	37.4	34.8	34.7	33.1	32.0	34.3	35.0	28.8	31.1	31.9	31.8	25.0	25.6	25.7	27.2	33.0	
18	Dacthal	Cleanert TPT	113.7	111.6	92.5	98.3	95.8	82.3	90.6	110.0	101.6	99.1	108.6	99.1	96.4	99.0	88.0	93.5	89.1	87.7	84.1	96.9	5.0
		Envi-Carb + PSA	113.2	107.2	107.1	95.5	101.7	93.9	101.5	95.9	98.5	93.8	90.7	83.6	86.9	85.1	89.9	73.2	73.0	78.8	82.5	92.2	
19	DE-PCB101	Cleanert TPT	33.9	35.8	29.7	31.8	30.3	26.5	26.7	32.7	33.7	32.9	32.8	33.1	31.3	33.7	28.2	30.2	27.0	29.5	29.1	31.0	2.8
		Envi-Carb + PSA	42.6	35.3	35.2	31.7	34.4	30.0	34.1	27.6	33.1	31.3	29.9	27.3	27.1	28.9	29.1	24.0	21.9	25.2	24.2	30.2	
20	DE-PCB118	Cleanert TPT	35.2	36.8	30.5	32.7	32.1	27.4	27.8	32.9	34.2	33.3	34.4	33.7	30.8	33.6	29.1	31.4	27.1	29.6	29.2	31.7	3.5
		Envi-Carb + PSA	44.4	36.0	34.7	32.7	34.5	30.8	34.6	27.3	32.9	31.4	29.7	26.9	27.0	29.1	30.0	23.8	23.3	25.9	25.8	30.6	
21	DE-PCB138	Cleanert TPT	35.6	35.6	30.1	31.7	31.3	27.3	27.0	32.8	33.8	31.3	32.8	32.7	30.2	32.2	28.8	29.6	26.9	29.1	27.4	30.9	2.8
		Envi-Carb + PSA	44.6	35.7	35.0	31.5	32.6	30.0	33.5	27.4	31.8	30.9	28.6	27.0	26.3	28.0	29.3	23.6	22.7	25.4	25.9	30.0	
22	DE-PCB180	Cleanert TPT	34.9	34.5	29.3	32.1	30.9	26.1	26.6	32.7	32.9	30.9	32.7	31.3	29.5	31.0	27.6	26.8	26.4	27.7	26.6	30.1	2.3
		Envi-Carb + PSA	44.7	34.5	34.4	31.0	32.7	28.6	31.8	29.2	31.5	29.9	28.2	26.0	26.7	27.0	29.0	22.8	22.2	25.2	24.0	29.4	
23	DE-PCB28	Cleanert TPT	34.1	35.5	30.8	33.0	32.7	28.2	28.4	34.5	35.0	34.2	35.9	33.3	33.0	34.6	29.9	31.8	27.7	30.9	30.2	32.3	1.5
		Envi-Carb + PSA	47.8	35.8	35.8	34.3	33.9	31.8	35.4	29.2	35.1	32.3	30.7	28.2	28.3	29.7	31.3	25.7	24.5	27.6	27.0	31.8	
24	DE-PCB31	Cleanert TPT	34.7	35.9	30.8	33.1	32.6	28.3	28.2	34.3	34.9	34.2	35.4	33.9	32.6	34.5	30.2	31.7	28.0	30.8	30.0	32.3	3.6
		Envi-Carb + PSA	36.9	35.8	35.7	34.3	33.6	31.7	35.3	29.4	35.1	32.2	30.4	28.3	28.1	29.7	31.1	25.7	24.4	27.4	27.1	31.2	
25	Dichlorofop-methyl	Cleanert TPT	40.2	44.2	38.1	37.3	43.8	28.1	31.0	36.2	41.2	31.9	60.2	27.6	31.0	32.0	30.0	32.2	29.5	27.6	23.8	35.0	2.8
		Envi-Carb + PSA	44.1	47.5	41.0	33.4	42.3	31.9	34.1	36.5	36.9	32.5	31.9	31.7	27.5	36.9	25.7	21.6	28.8	34.4	28.8	34.1	
26	Dimethenamid	Cleanert TPT	40.4	41.2	32.2	34.0	30.5	26.7	30.6	38.3	33.9	34.4	35.8	34.2	32.3	34.5	29.5	32.2	28.3	29.8	31.0	33.1	5.5
		Envi-Carb + PSA	34.5	37.3	39.9	33.2	36.6	31.4	34.9	32.8	36.9	32.1	30.5	27.0	28.7	28.7	29.4	23.5	24.2	27.3	27.0	31.4	
27	Diofenolan-1	Cleanert TPT	77.1	72.5	65.0	69.3	69.4	57.6	58.5	71.9	69.9	67.8	73.2	68.2	65.3	70.5	63.8	64.1	56.2	62.5	58.5	66.4	0.9
		Envi-Carb + PSA	91.9	75.4	75.1	69.8	77.3	62.4	71.8	61.3	73.4	65.7	64.1	56.1	56.7	65.5	61.6	48.3	53.3	62.8	56.9	65.8	

(Continued)

TABLE 2.3.7 The Determination Results for 201 Pesticides in Incurred Oolong Tea Youden Pair Samples Under 4 Conditions in 3 Months (cont.)

No.	Pesticides	Solid Phase Extraction	November 9, 2009 (n = 5)	November 14, 2009 (n = 3)	November 19, 2009 (n = 3)	November 24, 2009 (n = 3)	November 29, 2009 (n = 3)	December 4, 2009 (n = 3)	December 9, 2009 (n = 3)	December 14, 2009 (n = 3)	December 19, 2009 (n = 3)	December 24, 2009 (n = 3)	December 29, 2009 (n = 3)	January 3, 2010 (n = 3)	January 8, 2010 (n = 3)	January 13, 2010 (n = 3)	January 18, 2010 (n = 3)	January 23, 2010 (n = 3)	January 28, 2010 (n = 3)	February 2, 2010 (n = 3)	February 7, 2010 (n = 3)	AVE (μg/kg)	Deviation rates (%)
																						Pesticide content determined With GC–MS and GC–MS/MS, two concentration of Youden pair samples (μg/kg)		
28	Diofenolan-2	Cleanert TPT	77.0	76.5	66.3	70.4	71.9	58.4	58.4	72.9	70.1	70.5	75.7	70.9	66.2	70.9	65.9	66.6	57.5	61.6	59.0	67.7	8.8
		Envi-Carb + PSA	93.7	76.5	76.5	70.0	41.0	63.6	72.2	62.1	72.8	66.6	63.8	29.6	58.3	37.1	63.2	49.4	56.8	65.4	59.3	62.0	
29	Fenbuconazole	Cleanert TPT	86.9	79.5	73.7	75.7	71.4	65.3	59.1	74.6	68.1	67.6	77.2	73.4	66.1	68.3	62.5	59.9	53.4	57.1	52.8	68.0	0.8
		Envi-Carb + PSA	107.7	84.1	78.0	77.4	96.8	64.6	73.9	69.6	86.4	63.3	65.7	46.6	54.9	56.0	65.9	48.7	47.8	59.1	56.2	68.6	
30	Fenpyroximate	Cleanert TPT	283.0	305.7	207.0	229.1	235.7	219.0	199.0	255.2	234.3	303.3	201.6	337.5	242.8	292.5	216.7	204.7	166.4	244.5	244.8	243.3	2.8
		Envi-Carb + PSA	484.6	255.1	268.4	327.5	285.7	232.1	347.1	197.6	244.7	254.1	226.8	189.0	201.0	212.3	288.3	227.1	144.1	188.8	181.4	250.3	
31	Fluotrimazole	Cleanert TPT	39.2	35.8	33.4	31.9	30.1	27.4	27.9	32.8	31.0	30.8	32.9	32.0	30.6	31.1	27.5	27.4	24.5	27.5	25.1	30.5	3.4
		Envi-Carb + PSA	48.8	40.3	37.4	32.5	36.7	30.3	33.9	29.2	47.4	32.6	30.1	25.2	26.8	26.6	28.3	22.8	21.7	24.1	24.3	31.5	
32	Fluoxypr-1-methylheptyl ester	Cleanert TPT	39.7	42.8	38.3	38.3	38.1	29.6	31.7	36.3	38.2	35.4	47.0	33.4	33.9	35.0	35.9	31.4	31.2	31.3	26.3	35.5	0.4
		Envi-Carb + PSA	46.7	41.4	41.4	35.4	41.0	33.0	35.3	37.7	41.0	34.6	31.9	30.6	30.6	34.9	30.1	25.6	31.5	35.0	33.4	35.3	
33	Iprovalicarb-1	Cleanert TPT	208.4	152.7	168.7	167.3	156.1	155.0	193.4	186.0	155.8	146.9	161.4	155.8	151.2	155.3	153.7	156.6	155.0	140.2	125.9	160.3	4.6
		Envi-Carb + PSA	181.0	157.5	166.5	151.7	185.7	150.3	154.6	176.0	194.5	136.5	136.7	124.2	132.9	159.4	145.8	112.9	142.9	154.4	145.3	153.1	
34	Iprovalicarb-2	Cleanert TPT	206.1	158.6	148.1	159.2	149.8	132.7	224.2	203.8	158.3	145.3	171.4	144.6	152.6	149.7	142.5	147.8	155.3	136.3	128.8	158.7	4.6
		Envi-Carb + PSA	172.5	159.9	161.5	146.6	198.4	139.5	149.8	197.0	194.9	143.0	146.4	125.1	128.2	144.0	139.0	108.7	129.5	150.8	143.9	151.5	
35	Isodrin	Cleanert TPT	38.6	37.5	31.3	33.6	32.1	28.4	29.0	35.5	36.3	35.3	35.6	35.2	32.1	34.3	29.8	34.3	30.3	30.9	29.6	33.1	3.5
		Envi-Carb + PSA	43.3	37.6	38.1	34.2	34.4	32.2	35.7	30.2	34.1	32.8	29.8	30.7	30.1	29.1	33.8	25.9	24.2	26.0	25.8	32.0	
36	Isoprocarb-1	Cleanert TPT	83.7	88.9	78.7	80.8	90.4	86.9	78.3	89.0	100.2	92.7	96.7	99.9	102.7	105.8	103.2	102.8	81.3	87.0	89.2	91.5	2.1
		Envi-Carb + PSA	106.0	85.2	96.6	83.4	86.3	86.1	94.4	84.7	129.7	94.9	90.4	81.9	89.2	86.1	100.5	83.8	66.3	75.8	81.5	89.6	
37	Isoprocarb-2	Cleanert TPT	86.7	89.0	71.5	76.0	96.7	60.7	79.8	100.1	107.7	73.6	116.7	62.4	98.7	75.0	56.8	68.6	81.2	87.6	87.9	83.0	5.2
		Envi-Carb + PSA	74.5	78.1	103.9	81.5	93.6	75.9	76.9	68.8	77.2	73.1	93.9	89.7	84.5	86.6	60.6	54.3	65.7	73.9	83.5	78.8	
38	Lenacil	Cleanert TPT	324.3	202.0	231.5	229.4	228.8	199.0	407.4	262.8	220.8	158.1	314.1	150.8	192.7	154.9	172.0	225.7	237.0	145.3	187.5	223.4	12.1
		Envi-Carb + PSA	195.7	198.1	200.2	171.2	257.4	179.6	160.3	368.1	218.4	169.2	169.5	184.5	169.5	252.5	148.8	119.3	198.5	176.4	222.8	197.9	
39	Metalaxyl	Cleanert TPT	116.6	114.0	95.8	101.8	93.1	85.6	84.4	108.5	99.6	96.5	106.8	109.5	100.5	104.8	93.2	99.9	93.1	95.0	98.5	99.9	2.9
		Envi-Carb + PSA	127.8	109.9	113.7	98.8	112.9	95.2	105.7	93.7	111.7	95.2	91.4	80.7	91.0	91.1	99.8	79.4	75.5	85.3	84.9	97.0	

No.	Compound	Cartridge																					
40	Metazachlor	Cleanert TPT	125.3	141.4	100.1	105.7	95.0	87.0	131.2	163.2	107.3	113.3	124.8	102.7	99.3	102.1	88.4	100.5	92.2	80.7	99.5	108.4	9.4
		Envi-Carb + PSA	98.1	118.5	123.6	109.0	119.4	97.4	109.0	150.4	93.8	95.0	93.2	74.9	84.3	82.6	90.0	71.3	76.1	96.5	91.0	98.6	
41	Methabenzthia-zuron	Cleanert TPT	405.7	400.7	334.3	343.1	332.4	299.3	302.0	359.3	327.8	344.5	346.0	342.1	314.6	336.5	310.9	329.2	285.8	307.5	281.7	331.8	3.8
		Envi-Carb + PSA	481.8	399.8	379.4	341.5	409.4	337.1	360.8	357.4	405.0	328.7	323.1	282.6	304.1	321.2	304.4	259.9	295.7	355.1	303.5	344.8	
42	Mirex	Cleanert TPT	37.5	37.0	28.7	32.4	32.5	26.5	34.0	37.8	33.9	28.9	37.0	26.6	29.5	27.2	26.0	29.0	29.8	26.1	25.0	30.8	10.3
		Envi-Carb + PSA	34.3	32.4	33.3	28.5	32.6	28.8	29.2	32.5	24.5	27.7	26.9	26.1	25.5	28.3	24.9	20.6	22.4	25.3	24.1	27.8	
43	Monalide	Cleanert TPT	90.0	95.7	76.5	80.2	84.3	70.9	67.2	88.7	90.4	82.7	100.9	78.6	83.9	82.9	71.9	81.5	79.3	74.1	70.0	81.6	1.9
		Envi-Carb + PSA	98.4	91.1	92.9	79.8	84.9	79.3	83.4	76.6	77.4	83.4	79.5	75.9	71.3	77.9	72.5	101.0	57.1	63.8	73.6	80.0	
44	Paraoxon-ethyl	Cleanert TPT	179.5	214.2	190.4	161.8	141.3	154.9	180.9	181.2	104.4	161.8	206.4	165.5	154.8	162.1	156.9	98.4	93.7	81.8	96.9	151.9	3.2
		Envi-Carb + PSA	178.5	185.1	181.7	154.6	137.0	160.9	177.7	136.7	167.2	153.3	150.9	130.2	147.5	127.6	141.0	134.9	110.4	109.6	110.5	147.1	
45	Pebulate	Cleanert TPT	101.6	90.3	74.4	83.8	82.7	71.2	68.3	82.2	83.1	81.6	87.2	83.2	77.7	85.7	81.2	79.5	84.4	89.5	66.1	81.8	2.8
		Envi-Carb + PSA	105.0	82.6	86.0	82.8	89.7	80.7	85.4	65.0	86.2	75.5	74.2	66.1	68.1	84.8	85.5	69.9	71.3	76.3	75.2	79.5	
46	Pentachloroani-line	Cleanert TPT	34.3	32.8	28.8	30.1	31.0	27.8	28.5	37.0	33.4	33.8	32.8	31.3	30.0	31.2	28.9	29.0	26.4	28.1	27.2	30.6	2.2
		Envi-Carb + PSA	37.5	32.4	33.9	30.5	31.6	31.4	38.9	32.5	33.3	32.8	30.4	26.4	27.2	27.7	27.9	22.9	22.0	25.1	25.2	30.0	
47	Pentachloroan-isole	Cleanert TPT	32.6	29.9	25.7	28.7	29.6	26.0	26.0	32.5	29.5	29.7	32.5	29.2	29.1	29.0	27.1	28.1	24.6	26.9	25.9	28.5	5.1
		Envi-Carb + PSA	37.6	28.2	31.3	28.0	29.6	28.0	31.6	26.6	29.0	27.6	27.4	24.5	26.2	25.4	25.8	21.5	20.9	22.9	23.6	27.1	
48	Pentachloroben-zene	Cleanert TPT	39.2	38.7	32.1	34.3	34.8	29.0	31.0	37.4	36.0	34.5	38.5	33.6	32.6	35.7	31.3	32.1	30.2	31.3	29.2	33.8	4.8
		Envi-Carb + PSA	42.1	38.1	37.3	33.9	35.4	32.6	35.3	31.4	35.2	33.1	31.9	29.1	28.2	30.8	30.6	25.4	25.3	27.5	28.0	32.2	
49	Perthane	Cleanert TPT	111.0	38.3	33.0	34.5	36.8	29.8	29.3	34.9	34.5	34.3	36.8	33.1	33.9	33.5	30.2	32.5	29.2	30.0	31.3	37.2	12.8
		Envi-Carb + PSA	43.2	36.8	37.9	34.8	34.8	33.9	36.5	32.3	36.7	34.0	32.4	29.0	30.2	31.0	31.4	25.4	26.2	29.0	26.3	32.7	
50	Phenanthrene	Cleanert TPT	37.1	39.0	30.7	35.1	35.9	30.2	30.1	36.5	37.6	36.4	37.1	36.1	34.4	37.7	31.9	33.0	30.7	35.0	30.6	34.5	3.6
		Envi-Carb + PSA	41.3	36.7	38.8	35.4	37.1	33.7	36.9	31.0	38.7	35.2	33.1	29.1	30.6	32.5	31.4	26.8	25.8	29.8	28.3	33.3	
51	Pirimicarb	Cleanert TPT	87.0	72.6	63.0	66.5	59.7	53.1	56.2	67.4	63.1	62.0	69.1	62.9	59.4	63.9	57.0	58.9	53.7	57.1	52.2	62.4	0.8
		Envi-Carb + PSA	84.4	74.0	72.1	65.9	77.2	61.8	64.6	61.1	73.6	61.2	63.3	53.4	55.4	56.4	55.9	45.8	46.8	52.2	50.5	61.9	
52	Procymidone	Cleanert TPT	38.1	41.3	33.6	36.9	35.6	30.8	30.9	38.0	38.3	37.4	38.5	40.4	37.7	43.4	32.9	34.8	31.2	33.9	32.4	36.1	3.5
		Envi-Carb + PSA	47.4	40.1	40.8	36.3	38.3	34.6	39.2	32.4	37.8	35.5	33.3	30.0	33.4	37.5	32.7	27.7	26.0	29.5	30.0	34.9	
53	Prometrye	Cleanert TPT	41.8	42.7	33.2	34.4	31.8	28.6	28.6	35.0	33.0	32.5	33.2	33.5	30.6	33.6	28.5	29.8	27.2	29.6	27.1	32.4	0.1
		Envi-Carb + PSA	48.7	40.5	39.2	34.8	40.4	32.6	35.9	30.0	38.2	32.7	32.4	26.2	28.3	28.3	30.0	23.9	22.5	25.4	25.3	32.4	
54	Propham	Cleanert TPT	43.6	40.1	35.8	37.8	37.1	32.0	42.9	43.5	39.0	38.1	45.3	37.7	38.7	40.0	37.2	39.4	38.4	37.1	35.7	38.9	2.1
		Envi-Carb + PSA	39.9	37.8	41.2	35.6	42.0	36.5	38.5	40.5	44.3	36.5	35.3	32.1	34.4	36.2	34.3	29.5	42.3	48.2	39.1	38.1	

(Continued)

TABLE 2.3.7 The Determination Results for 201 Pesticides in Incurred Oolong Tea Youden Pair Samples Under 4 Conditions in 3 Months (cont.)

No.	Pesticides	Solid Phase Extraction	Pesticide content determined With GC–MS and GC–MS/MS, two concentration of Youden pair samples (μg/kg)																			AVE (μg/kg)	Deviation rates (%)
			November 9, 2009 (n = 5)	November 14, 2009 (n = 3)	November 19, 2009 (n = 3)	November 24, 2009 (n = 3)	November 29, 2009 (n = 3)	December 4, 2009 (n = 3)	December 9, 2009 (n = 3)	December 14, 2009 (n = 3)	December 19, 2009 (n = 3)	December 24, 2009 (n = 3)	December 14, 2009 (n = 3)	January 3, 2010 (n = 3)	January 8, 2010 (n = 3)	January 13, 2010 (n = 3)	January 18, 2010 (n = 3)	January 23, 2010 (n = 3)	January 28, 2010 (n = 3)	February 2, 2010 (n = 3)	February 7, 2010 (n = 3)		
55	Prosulfocarb	Cleanert TPT	36.6	35.5	30.8	32.5	30.4	26.8	26.5	33.4	32.1	33.0	32.3	33.2	31.0	33.9	29.3	30.1	27.1	28.8	27.8	31.1	0.9
		Envi-Carb + PSA	41.0	35.7	36.8	32.4	35.2	31.3	34.6	28.8	35.4	31.7	31.3	26.5	27.8	29.4	29.5	24.2	23.8	25.3	25.2	30.8	
56	Secbumeton	Cleanert TPT	43.5	37.5	34.6	33.9	29.5	28.4	28.8	33.4	32.2	33.8	36.8	31.9	31.2	33.9	31.0	32.0	29.1	32.3	29.4	32.8	3.0
		Envi-Carb + PSA	49.7	38.1	38.9	34.1	44.3	32.9	36.4	29.4	43.6	33.9	36.4	26.8	29.6	30.2	31.5	25.2	24.7	28.2	28.2	33.8	
57	Silafluofen	Cleanert TPT	40.4	34.0	33.0	30.0	35.2	29.3	25.3	34.5	36.4	33.0	37.1	16.2	32.7	33.9	29.4	29.7	27.0	29.4	27.0	31.2	4.1
		Envi-Carb + PSA	55.5	37.2	36.4	32.5	37.4	30.2	34.3	30.4	34.0	31.9	30.6	27.8	34.9	32.4	30.9	23.9	24.9	25.5	27.6	32.5	
58	Tebupirimfos	Cleanert TPT	150.3	125.7	111.2	115.5	87.9	71.8	75.4	63.8	59.4	57.0	65.2	61.5	53.7	62.5	53.2	54.7	47.7	53.9	50.0	74.8	2.0
		Envi-Carb + PSA	151.9	128.3	131.8	114.4	122.7	94.6	62.3	55.3	75.0	60.4	60.8	48.1	51.9	52.7	55.9	44.8	42.4	48.2	48.2	76.3	
59	Tebutam	Cleanert TPT	73.4	71.8	59.8	65.5	61.1	54.2	53.9	66.6	66.0	66.2	66.9	65.4	61.5	66.0	57.0	59.5	53.9	59.6	55.5	62.3	2.9
		Envi-Carb + PSA	80.8	71.1	72.9	64.1	68.2	61.9	67.2	55.6	68.6	62.5	60.8	53.6	56.0	56.8	58.0	48.0	44.4	49.6	49.7	60.5	
60	Tebuthiuron	Cleanert TPT	162.0	157.6	133.0	137.8	115.1	117.1	114.6	142.7	122.9	147.8	151.7	133.6	127.1	129.2	124.8	126.7	116.7	121.7	116.0	131.5	4.6
		Envi-Carb + PSA	195.3	163.1	150.4	135.4	175.1	130.4	143.7	126.1	210.2	128.4	133.6	108.3	115.7	128.9	130.0	102.9	102.5	119.4	115.4	137.6	
61	Tefluthrin	Cleanert TPT	35.1	33.7	29.2	31.5	29.4	25.3	25.6	33.0	32.1	31.7	31.3	32.2	28.6	30.2	27.1	27.0	25.2	27.8	25.9	29.6	1.8
		Envi-Carb + PSA	39.2	33.3	35.3	31.1	31.8	30.2	32.1	27.0	33.2	29.6	29.8	26.4	26.6	27.3	27.7	21.8	21.6	24.1	23.7	29.0	
62	Thenylchlor	Cleanert TPT	86.6	95.1	66.7	67.8	65.8	57.0	69.1	84.5	73.3	74.3	84.1	71.0	67.1	71.6	62.5	69.0	59.2	59.0	70.9	71.3	9.3
		Envi-Carb + PSA	69.4	79.4	83.4	72.1	74.8	63.2	69.1	81.0	58.4	61.9	61.3	52.8	58.8	58.7	59.0	50.4	53.2	63.0	63.8	64.9	
63	Thionazin	Cleanert TPT	39.1	37.3	33.4	35.0	30.6	29.0	29.4	35.3	32.7	32.6	36.0	33.0	32.3	33.5	30.7	32.6	28.9	30.8	30.6	32.8	3.4
		Envi-Carb + PSA	39.0	35.9	38.6	34.6	37.0	32.3	34.8	31.9	36.2	31.5	31.2	26.2	28.1	27.9	30.7	25.6	24.9	27.8	28.1	31.7	
64	Trichloronat	Cleanert TPT	39.6	35.2	33.0	34.0	31.8	27.9	28.0	34.3	33.3	31.3	37.0	32.3	31.0	32.9	28.7	30.5	27.4	29.0	27.4	31.8	1.7
		Envi-Carb + PSA	39.2	36.6	36.0	31.9	39.5	31.0	33.6	30.9	38.2	31.8	32.0	27.2	27.8	29.8	29.1	23.8	23.6	26.2	26.0	31.3	
65	Trifluralin	Cleanert TPT	77.3	83.7	67.8	70.1	76.1	61.1	67.1	76.1	71.0	72.7	90.9	67.2	70.5	73.6	70.0	77.4	78.3	70.0	59.1	72.6	2.8
		Envi-Carb + PSA	76.8	73.2	73.2	63.5	101.2	68.5	67.9	107.8	94.5	66.4	67.1	61.0	73.1	86.3	62.4	50.1	74.8	83.2	67.9	74.7	
66	2,4'-DDT	Cleanert TPT	75.5	42.7	36.3	49.8	37.0	30.1	33.1	47.7	29.7	32.1	34.9	30.9	36.4	33.6	30.2	34.1	32.7	28.2	27.0	37.0	45.2
		Envi-Carb + PSA	91.9	52.1	67.7	53.9	93.3	63.8	69.8	53.6	42.3	50.8	52.5	57.2	50.0	61.2	62.3	58.6	41.0	46.1	44.3	58.5	
67	4,4'-DDE	Cleanert TPT	39.7	36.8	37.7	42.7	35.6	36.2	33.1	41.2	36.8	34.7	35.0	36.0	35.7	34.9	32.4	34.2	31.3	34.2	30.6	35.7	7.2
		Envi-Carb + PSA	41.7	36.1	38.2	34.3	36.6	33.7	34.5	28.8	33.9	31.4	33.3	31.3	30.7	32.1	33.6	32.9	29.1	29.4	30.2	33.2	

#	Compound	Cartridge																					
68	Benalaxyl	Cleanert TPT	43.4	41.5	38.1	44.0	36.1	35.5	31.4	39.6	33.6	32.9	33.7	31.5	32.4	32.4	31.0	34.1	30.5	32.3	27.7	34.8	8.8
		Envi-Carb + PSA	44.7	35.5	38.1	33.6	38.8	33.6	33.2	27.6	30.8	30.1	30.6	28.3	26.0	29.1	33.1	31.4	26.5	27.6	27.7	31.9	
69	Benzoylprop-ethyl	Cleanert TPT	126.4	121.8	114.3	131.6	111.0	109.0	101.6	119.7	109.9	101.1	106.6	104.0	102.8	98.9	94.0	96.5	91.1	97.4	84.9	106.5	8.6
		Envi-Carb + PSA	136.7	104.6	114.9	107.4	115.7	98.4	100.0	87.0	96.6	92.9	95.6	90.3	84.5	94.1	91.1	91.7	82.8	87.5	83.9	97.7	
70	Bromofos	Cleanert TPT	84.5	84.4	75.3	79.4	70.0	63.9	59.6	94.8	66.1	61.1	65.6	62.0	69.9	66.5	58.8	61.9	61.9	59.8	54.8	68.4	11.9
		Envi-Carb + PSA	84.4	63.6	75.4	63.0	72.0	62.4	64.2	53.6	60.1	55.1	60.7	47.6	54.2	65.0	57.9	58.3	51.4	50.1	55.1	60.7	
71	Bromopropylate	Cleanert TPT	98.5	100.0	75.4	95.4	80.7	69.6	71.3	128.2	68.3	69.4	70.6	67.3	77.1	65.9	65.6	73.9	65.1	64.0	52.0	76.7	14.8
		Envi-Carb + PSA	99.9	64.0	73.5	68.8	83.7	66.6	66.3	59.2	65.5	63.7	62.1	66.7	57.7	61.2	64.6	64.2	55.8	57.4	55.9	66.2	
72	Buprofenzin	Cleanert TPT	87.5	81.0	80.2	90.7	77.0	79.7	65.3	84.2	74.5	72.9	73.2	81.3	71.2	73.1	65.0	59.7	66.2	68.1	64.4	74.5	6.9
		Envi-Carb + PSA	91.5	76.0	85.1	72.5	79.6	74.9	71.5	59.7	66.2	67.4	65.9	63.3	62.2	66.1	66.2	63.4	61.4	63.2	65.5	69.6	
73	Butachlor	Cleanert TPT	76.5	71.3	73.5	84.6	66.6	72.2	63.3	80.7	68.2	65.3	65.7	67.2	66.0	64.1	59.3	63.6	62.8	59.9	55.3	67.7	7.1
		Envi-Carb + PSA	84.1	68.6	74.1	67.5	70.3	67.7	64.6	53.5	68.9	58.1	63.7	58.1	52.7	60.1	62.4	60.3	53.3	57.3	53.4	63.1	
74	Butylate	Cleanert TPT	97.3	77.1	85.7	95.8	73.2	73.9	69.7	79.6	85.0	67.2	67.7	68.2	70.9	67.8	66.6	64.0	62.7	64.1	60.9	73.5	6.6
		Envi-Carb + PSA	80.9	81.5	91.5	71.8	80.5	77.2	69.2	64.3	66.2	67.3	66.8	62.7	66.4	63.0	67.9	59.6	54.5	60.5	56.3	68.9	
75	Carbofenothion	Cleanert TPT	93.3	80.8	67.7	83.3	71.1	65.0	59.0	110.6	54.4	56.3	57.0	52.5	57.9	48.8	48.4	45.3	47.5	42.4	37.2	62.0	18.3
		Envi-Carb + PSA	82.9	62.5	66.9	60.3	70.5	60.4	57.7	49.7	55.0	50.6	47.8	44.1	44.3	47.4	40.7	42.5	32.4	33.9	31.7	51.6	
76	Chlorfenson	Cleanert TPT	106.9	115.2	77.3	102.4	82.7	64.4	71.0	114.6	66.9	70.5	80.8	68.8	82.8	70.1	70.8	81.9	67.3	62.8	51.2	79.4	16.5
		Envi-Carb + PSA	90.4	58.2	70.8	61.7	94.7	70.0	69.4	63.5	69.5	66.0	65.8	68.7	62.4	64.7	66.4	71.1	56.2	58.3	50.4	67.3	
77	Chlorfenvinphos	Cleanert TPT	136.3	145.3	107.0	131.2	111.3	95.7	96.0	165.4	95.6	90.9	99.3	89.5	106.1	94.5	87.6	97.0	103.0	97.8	78.0	106.7	15.7
		Envi-Carb + PSA	147.4	89.0	116.6	100.1	114.1	96.9	96.9	80.8	88.9	81.9	85.9	72.5	77.2	89.5	86.3	88.2	70.8	74.6	75.0	91.2	
78	Chlormephos	Cleanert TPT	76.7	64.6	65.4	75.9	59.4	59.5	57.1	74.1	65.8	59.0	59.3	59.2	62.1	57.2	55.5	56.5	53.6	57.0	47.1	61.3	9.4
		Envi-Carb + PSA	75.1	55.3	66.2	56.3	62.5	60.4	57.5	52.4	55.7	55.0	55.4	52.4	54.0	53.9	55.2	52.2	46.0	48.0	47.1	55.8	
79	Chloroneb	Cleanert TPT	39.3	36.2	38.3	42.7	34.9	34.2	32.2	39.2	39.1	37.3	37.8	38.8	37.9	36.3	34.9	35.7	34.3	35.6	30.6	36.6	8.6
		Envi-Carb + PSA	38.7	33.3	36.3	32.7	37.0	33.2	33.4	29.5	32.4	33.7	34.1	33.5	33.6	35.2	35.3	33.7	30.7	31.1	30.8	33.6	
80	Chloropropylate	Cleanert TPT	95.4	93.3	78.8	95.4	78.4	71.9	69.2	115.9	74.8	73.5	74.3	72.5	75.2	71.9	70.7	77.6	69.7	72.4	61.6	78.6	11.2
		Envi-Carb + PSA	105.7	69.2	76.4	59.4	82.9	72.2	70.5	63.1	71.8	64.7	67.9	69.9	65.5	67.9	71.6	70.1	61.5	63.8	60.3	70.2	
81	Chlorpropham	Cleanert TPT	46.8	45.8	38.6	47.8	39.7	36.6	34.7	53.8	36.6	36.6	37.7	36.8	38.9	35.3	34.5	37.3	33.5	34.9	28.6	38.7	5.0
		Envi-Carb + PSA	47.0	34.3	37.7	35.1	61.6	35.2	34.7	30.6	33.2	32.7	33.6	33.7	30.7	33.4	49.9	33.8	29.4	43.0	29.1	36.8	
82	Chlorpyrifos(ethyl)	Cleanert TPT	42.6	37.8	34.0	42.3	33.9	31.7	29.5	45.8	31.7	30.9	32.1	31.4	33.3	31.2	29.5	31.4	28.8	30.2	25.7	33.3	10.7
		Envi-Carb + PSA	42.8	30.0	35.6	31.7	35.5	30.3	31.7	26.3	29.1	28.1	29.6	26.4	28.9	29.0	28.9	29.1	25.1	25.6	25.0	29.9	
83	Chlorthiophos	Cleanert TPT	128.2	112.1	101.6	116.2	95.8	90.0	81.9	120.5	82.7	78.1	78.1	81.7	85.8	80.6	70.9	69.2	69.2	61.4	53.7	87.2	10.3
		Envi-Carb + PSA	131.4	95.3	106.8	93.3	101.1	82.7	85.4	68.4	75.6	72.7	70.9	68.7	72.4	73.2	78.6	65.7	49.5	53.8	49.7	78.7	

(Continued)

TABLE 2.3.7 The Determination Results for 201 Pesticides in Incurred Oolong Tea Youden Pair Samples Under 4 Conditions in 3 Months (*cont.*)

Pesticide content determined With GC–MS and GC–MS/MS, two concentration of Youden pair samples (µg/kg)

No.	Pesticides	Solid Phase Extraction	November 9, 2009 (n=5)	November 14, 2009 (n=3)	November 19, 2009 (n=3)	November 24, 2009 (n=3)	November 29, 2009 (n=3)	December 4, 2009 (n=3)	December 9, 2009 (n=3)	December 14, 2009 (n=3)	December 19, 2009 (n=3)	December 24, 2009 (n=3)	December 29, 2009 (n=3)	January 3, 2010 (n=3)	January 8, 2010 (n=3)	January 13, 2010 (n=3)	January 18, 2010 (n=3)	January 23, 2010 (n=3)	January 28, 2010 (n=3)	February 2, 2010 (n=3)	February 7, 2010 (n=3)	AVE (µg/kg)	Deviation rates (%)
84	Cis-chlordane	Cleanert TPT	78.8	73.3	73.7	85.0	68.6	66.6	59.7	78.0	70.5	66.5	68.1	67.8	67.9	66.1	59.8	64.8	59.2	63.4	55.8	68.1	7.9
		Envi-Carb + PSA	81.2	65.5	74.6	65.7	72.8	63.2	65.9	57.6	63.3	60.3	63.0	60.1	56.0	60.2	60.2	60.7	53.7	56.7	54.6	62.9	
85	Cis-diallate	Cleanert TPT	72.7	68.2	68.0	78.6	61.9	62.1	55.5	71.7	66.1	60.7	63.1	62.7	61.3	61.5	58.0	59.6	53.7	59.8	50.5	62.9	8.0
		Envi-Carb + PSA	73.8	60.2	67.4	60.9	66.3	59.8	60.7	51.1	58.1	56.1	57.8	55.0	56.0	57.2	57.1	54.6	49.5	52.5	50.1	58.1	
86	Cyanofenphos	Cleanert TPT	51.1	49.4	40.5	49.1	42.1	39.6	36.9	56.7	37.8	37.8	41.2	35.7	39.9	36.5	34.5	39.9	32.9	32.1	29.1	40.2	15.9
		Envi-Carb + PSA	48.3	31.0	39.0	34.7	45.3	38.1	37.8	34.2	32.8	34.8	30.8	31.3	29.5	34.1	33.3	32.4	26.4	29.7	27.3	34.3	
87	Desmetryn	Cleanert TPT	43.6	38.5	37.2	44.1	36.6	34.4	30.3	38.8	33.5	32.3	31.6	31.3	32.0	30.6	28.4	28.4	26.4	28.1	24.9	33.2	8.1
		Envi-Carb + PSA	46.5	35.3	37.8	34.2	37.0	32.4	31.9	26.4	30.3	29.5	27.9	28.7	28.3	27.9	28.5	27.4	23.8	24.5	23.7	30.6	
88	Dichlobenil	Cleanert TPT	7.7	6.5	6.7	7.6	5.7	6.2	5.7	6.8	6.6	5.7	6.2	6.0	6.5	5.9	5.6	5.6	5.5	5.6	4.8	6.2	8.7
		Envi-Carb + PSA	7.5	5.3	6.6	5.6	6.3	6.2	5.9	5.1	6.0	5.6	5.6	5.3	5.4	5.5	5.6	5.4	4.7	5.1	4.7	5.7	
89	Dicloran	Cleanert TPT	100.5	96.0	84.4	112.8	98.7	76.3	75.3	112.2	80.5	84.0	79.0	70.0	89.1	70.6	78.1	77.1	90.5	65.2	61.0	84.3	15.2
		Envi-Carb + PSA	85.4	60.6	88.0	78.7	106.3	82.2	73.3	79.9	62.5	69.9	67.6	65.2	62.1	67.1	66.4	73.3	66.4	63.3	56.8	72.4	
90	Dicofol	Cleanert TPT	105.5	98.4	104.6	136.7	107.4	107.1	135.6	139.5	109.9	113.0	96.2	113.1	128.8	110.0	111.3	110.1	112.1	103.6	113.2	113.5	1.5
		Envi-Carb + PSA	116.2	110.7	132.9	112.4	128.7	114.7	112.0	90.9	110.5	96.3	99.3	112.7	106.3	103.7	117.8	109.8	125.9	100.1	123.0	111.8	
91	Dimethachlor	Cleanert TPT	128.4	120.9	108.2	123.5	103.4	102.6	96.4	144.2	101.5	96.8	100.3	99.9	100.7	95.7	87.3	92.5	96.6	97.9	86.3	104.4	11.6
		Envi-Carb + PSA	134.8	100.2	122.2	101.2	107.4	98.1	100.9	83.0	89.5	85.7	90.7	77.3	85.5	87.4	84.1	86.1	74.5	76.7	79.7	92.9	
92	Dioxacarb	Cleanert TPT	358.7	336.4	292.8	343.0	324.1	286.9	238.5	589.7	268.5	263.0	262.3	271.2	266.7	251.6	232.1	247.7	231.4	233.8	189.5	288.8	15.3
		Envi-Carb + PSA	347.0	288.3	281.3	274.1	276.9	256.4	246.4	229.3	250.0	229.5	255.1	223.6	233.1	231.5	228.0	229.2	195.8	216.6	213.8	247.7	
93	Endrin	Cleanert TPT	479.9	437.6	415.3	487.9	399.9	395.5	353.5	471.9	400.7	373.5	387.5	387.3	395.9	376.6	348.8	369.1	343.1	367.6	320.4	395.4	8.8
		Envi-Carb + PSA	481.0	389.7	424.6	380.2	415.5	377.7	382.5	319.4	361.2	341.7	356.2	346.2	335.0	346.7	349.6	346.4	302.2	317.1	309.3	362.2	
94	Epoxiconazole-2	Cleanert TPT	361.0	318.3	285.1	345.7	291.1	277.7	242.8	356.0	253.4	243.3	238.3	250.5	259.8	234.8	221.1	230.3	213.8	219.9	191.8	265.0	11.1
		Envi-Carb + PSA	376.1	268.2	283.0	272.9	296.2	242.7	251.2	207.4	241.3	219.7	212.9	207.5	203.7	217.0	220.3	211.7	184.6	202.1	187.0	237.1	
95	EPTC	Cleanert TPT	93.5	73.8	78.9	91.4	67.5	73.5	71.6	82.3	82.6	70.5	71.2	70.1	73.0	68.8	64.4	66.5	62.7	69.8	50.5	72.8	9.1
		Envi-Carb + PSA	93.8	59.2	83.5	66.2	76.4	77.2	68.9	63.3	68.6	69.3	66.2	62.4	62.7	65.0	63.7	60.5	49.9	55.2	49.8	66.4	
96	Ethofumesate	Cleanert TPT	89.5	81.8	78.9	85.0	65.1	71.8	88.6	86.0	76.6	76.9	78.8	73.4	69.8	70.3	63.6	68.4	84.7	67.1	58.8	75.5	11.6
		Envi-Carb + PSA	90.4	71.7	76.4	66.1	70.0	74.9	76.4	59.9	65.6	64.4	68.6	62.9	66.4	62.9	66.3	63.3	55.4	60.2	56.6	67.3	

#	Compound	Method																					
97	Ethoprophos	Cleanert TPT	121.6	111.1	101.2	119.7	96.6	92.3	84.0	124.4	93.6	89.3	91.0	88.8	89.6	87.8	82.5	84.7	79.3	87.2	75.0	94.7	10.5
		Envi-Carb + PSA	119.9	90.5	105.3	92.0	100.4	89.4	87.6	73.8	83.3	81.1	83.0	76.3	80.0	80.4	81.7	80.3	69.7	74.6	71.3	85.3	10.5
98	Etrimfos	Cleanert TPT	41.7	36.7	35.1	41.8	33.5	32.0	29.0	44.6	32.2	31.1	31.4	30.8	31.1	30.8	27.9	29.5	28.6	29.5	25.9	32.8	10.5
		Envi-Carb + PSA	42.1	31.9	35.9	32.7	34.0	30.2	30.8	25.7	28.8	27.9	28.8	25.9	28.0	28.3	28.1	27.8	24.3	25.3	24.6	29.5	
99	Fenamidone	Cleanert TPT	45.2	40.6	40.0	47.3	40.6	38.6	33.8	47.6	36.8	34.7	34.4	36.5	36.7	35.0	32.8	32.2	31.1	33.5	28.6	37.2	10.5
		Envi-Carb + PSA	48.2	36.3	38.4	36.6	39.8	34.1	34.2	29.0	33.5	31.9	30.9	30.4	29.6	31.5	32.7	30.3	28.0	31.5	28.9	33.5	
100	Fenaromol	Cleanert TPT	92.4	78.3	76.4	90.4	74.9	72.9	62.5	91.6	67.2	65.8	66.2	66.6	70.8	64.1	59.4	59.6	56.4	62.0	54.2	70.1	10.7
		Envi-Carb + PSA	97.1	70.9	71.3	69.4	73.2	65.0	63.4	58.0	63.3	57.5	56.9	53.7	57.3	58.8	59.2	55.2	52.4	60.5	53.0	62.9	
101	Flamprop-iso-propyl	Cleanert TPT	41.6	39.0	37.2	42.9	35.9	36.0	32.3	41.0	36.8	34.1	34.9	35.0	34.0	33.1	30.7	33.0	31.1	33.0	29.1	35.3	7.8
		Envi-Carb + PSA	44.2	35.0	38.0	34.5	37.3	33.0	34.0	28.6	32.4	31.6	31.8	31.4	29.7	30.8	32.5	31.5	27.0	29.1	28.2	32.7	
102	Flamprop-methyl	Cleanert TPT	41.1	39.0	40.5	44.9	39.8	38.3	32.5	40.4	36.0	33.6	34.2	32.5	36.0	34.1	31.9	33.5	30.8	33.8	28.8	35.9	7.2
		Envi-Carb + PSA	44.7	34.4	40.5	36.0	38.9	36.4	32.5	28.6	32.4	30.6	32.7	33.8	30.5	32.2	32.8	31.7	27.3	29.4	28.4	33.4	
103	Fonofos	Cleanert TPT	40.2	35.4	35.3	42.2	34.1	33.6	29.1	39.1	33.9	31.7	31.6	31.5	31.9	31.4	29.5	29.8	28.2	30.4	26.1	32.9	8.4
		Envi-Carb + PSA	39.6	32.0	35.8	32.6	35.7	31.2	31.6	26.5	30.4	28.8	29.9	28.3	29.3	28.7	29.6	28.2	25.2	26.2	25.2	30.2	
104	Hexachloroben-zene	Cleanert TPT	31.8	29.3	29.2	35.1	24.4	26.6	22.1	34.7	26.9	26.3	28.5	28.2	24.6	26.2	25.7	23.4	22.2	26.9	21.0	27.0	0.5
		Envi-Carb + PSA	33.2	26.7	30.6	28.3	29.5	28.1	28.4	24.1	27.4	26.5	27.7	25.9	26.6	25.4	27.3	24.0	22.9	25.0	22.9	26.9	
105	Hexazinone	Cleanert TPT	132.9	117.0	109.8	128.1	124.1	115.9	90.6	155.9	104.4	99.2	101.2	109.3	102.0	98.8	86.4	89.3	88.2	90.1	77.8	106.4	12.8
		Envi-Carb + PSA	139.4	107.9	107.9	105.1	105.4	93.3	91.1	84.9	94.8	85.7	92.3	79.2	82.0	91.6	85.3	83.4	75.9	85.3	87.2	93.6	
106	Iodofenphos	Cleanert TPT	88.7	83.4	74.0	80.1	72.7	64.9	63.5	99.5	70.4	63.7	75.1	63.1	73.2	65.6	61.4	60.6	64.4	61.1	53.3	70.5	14.7
		Envi-Carb + PSA	83.9	61.4	73.0	63.1	71.1	63.8	69.3	56.3	60.4	56.9	60.6	49.9	54.2	61.9	57.0	57.7	52.2	50.2	52.0	60.8	
107	Isofenphos	Cleanert TPT	86.6	82.3	72.5	94.1	78.7	71.9	65.4	94.7	71.1	70.3	70.9	67.5	68.7	59.7	56.0	58.9	53.9	57.6	50.0	70.0	10.6
		Envi-Carb + PSA	91.4	68.2	77.8	71.4	80.1	66.9	66.9	55.8	62.1	61.4	61.1	62.5	57.5	54.4	58.7	54.7	47.6	50.6	48.1	63.0	
108	Isopeopalin	Cleanert TPT	87.5	74.7	69.4	89.1	75.1	71.2	60.4	96.7	68.5	69.6	67.2	63.1	66.1	63.4	62.8	65.7	61.9	65.0	55.5	70.2	10.3
		Envi-Carb + PSA	89.3	64.4	73.1	68.9	75.4	66.3	61.5	51.7	61.2	61.6	59.0	64.0	60.9	60.8	64.7	60.8	51.8	55.5	51.3	63.3	
109	Methoprene	Cleanert TPT	149.6	129.8	129.6	148.3	205.6	116.1	103.2	144.6	119.6	117.8	108.0	110.8	105.1	105.2	96.0	102.8	91.4	105.0	89.1	119.9	3.6
		Envi-Carb + PSA	162.8	119.6	130.7	121.9	208.4	112.1	211.6	219.8	114.5	109.3	99.1	93.6	101.0	97.7	102.4	96.1	84.8	88.2	86.3	124.2	
110	Methoprotryne	Cleanert TPT	125.0	117.1	112.0	133.6	113.1	105.8	91.2	125.9	87.9	97.0	85.9	95.3	94.5	89.8	82.7	86.2	82.3	84.1	75.5	99.2	7.1
		Envi-Carb + PSA	142.9	107.6	119.2	108.7	117.4	96.6	95.3	78.1	92.6	86.0	79.7	79.3	80.9	82.1	84.0	82.4	72.7	76.8	73.3	92.4	
111	Methoxychlor	Cleanert TPT	43.5	45.3	36.3	45.9	37.2	32.3	34.5	60.0	33.3	34.2	36.2	33.6	36.7	33.5	31.3	33.4	31.1	28.7	27.4	36.6	13.2
		Envi-Carb + PSA	43.6	24.8	31.9	33.1	42.2	36.5	29.4	28.4	46.3	30.4	30.4	34.2	27.0	29.8	30.8	33.9	20.9	28.6	26.2	32.0	

(Continued)

TABLE 2.3.7 The Determination Results for 201 Pesticides in Incurred Oolong Tea Youden Pair Samples Under 4 Conditions in 3 Months (cont.)

No.	Pesticides	Solid Phase Extraction	November 9, 2009 (n = 5)	November 14, 2009 (n = 3)	November 19, 2009 (n = 3)	November 24, 2009 (n = 3)	November 29, 2009 (n = 3)	December 4, 2009 (n = 3)	December 9, 2009 (n = 3)	December 14, 2009 (n = 3)	December 19, 2009 (n = 3)	December 24, 2009 (n = 3)	December 14, 2009 (n = 3)	January 3, 2010 (n = 3)	January 8, 2010 (n = 3)	January 13, 2010 (n = 3)	January 18, 2010 (n = 3)	January 23, 2010 (n = 3)	January 28, 2010 (n = 3)	February 2, 2010 (n = 3)	February 7, 2010 (n = 3)	AVE (µg/kg)	Deviation rates (%)
			Pesticide content determined With GC–MS and GC–MS/MS, two concentration of Youden pair samples (µg/kg)																				
112	Methyl-parathion	Cleanert TPT	194.7	175.5	183.5	249.9	298.0	171.1	177.5	212.0	177.7	132.8	188.8	124.7	153.2	135.3	129.8	187.6	177.5	173.7	165.5	179.4	17.3
		Envi-Carb + PSA	185.9	120.3	180.2	196.7	258.4	181.0	166.9	163.3	118.3	113.1	168.2	101.3	161.1	127.5	129.3	123.9	104.4	109.2	155.7	150.8	
113	Metolachlor	Cleanert TPT	42.4	39.4	35.6	41.0	33.5	34.4	30.4	45.9	36.3	34.2	34.6	35.4	34.6	34.2	30.8	32.9	32.1	31.4	27.2	35.1	10.4
		Envi-Carb + PSA	45.1	32.4	37.5	33.1	35.3	32.5	33.2	28.5	30.4	30.5	31.5	28.2	30.4	30.5	31.2	30.8	25.8	27.2	26.4	31.6	
114	Nitrapyrin	Cleanert TPT	102.9	124.7	97.5	127.7	88.8	84.4	89.9	118.1	95.3	86.3	93.8	91.1	98.2	96.8	88.2	89.7	87.5	84.4	74.3	95.8	14.3
		Envi-Carb + PSA	116.1	66.4	99.4	83.4	104.8	97.0	101.3	82.6	68.1	75.7	80.0	69.7	74.7	92.5	82.7	81.8	70.2	65.5	65.4	83.0	
115	Oxyfluorfen	Cleanert TPT	200.4	176.8	138.4	230.9	198.2	147.8	134.6	358.1	131.9	146.2	143.7	122.6	152.9	128.7	135.3	155.7	146.8	129.3	123.5	163.2	22.1
		Envi-Carb + PSA	197.8	111.5	153.0	154.3	184.4	145.5	110.5	107.8	126.1	126.3	107.0	123.2	116.6	124.7	141.8	130.9	104.0	112.5	106.4	130.8	
116	Pendimethalin	Cleanert TPT	183.5	168.7	129.1	179.0	205.7	125.5	122.0	247.3	118.7	124.4	125.5	111.4	135.9	118.5	117.4	147.7	118.0	120.1	105.5	142.3	19.6
		Envi-Carb + PSA	177.3	114.6	139.4	128.7	159.4	127.5	108.4	98.5	114.2	107.1	100.8	114.1	106.6	108.6	121.4	113.1	91.1	99.0	92.2	116.9	
117	Picoxystrobin	Cleanert TPT	82.2	78.1	75.8	90.4	70.4	72.2	65.1	83.2	71.6	67.5	68.8	70.7	68.6	64.9	62.2	66.5	62.5	66.0	59.6	70.9	8.1
		Envi-Carb + PSA	86.9	72.8	75.9	68.6	74.4	67.5	68.0	57.6	65.9	61.2	63.6	63.1	60.1	60.4	63.1	63.1	54.5	58.2	55.9	65.3	
118	Piperophos	Cleanert TPT	149.6	138.2	108.3	142.2	123.8	103.9	102.0	234.7	94.1	98.9	100.5	92.7	105.9	93.9	88.3	92.4	92.3	87.5	78.1	112.0	17.8
		Envi-Carb + PSA	142.4	92.1	114.3	107.8	123.4	99.9	95.3	77.3	90.6	87.9	88.1	83.5	86.3	86.0	88.6	84.5	74.5	80.3	76.0	93.6	
119	Pirimiphoc-ethyl	Cleanert TPT	78.8	68.2	65.7	78.6	62.5	63.8	56.6	81.2	64.8	62.6	62.2	62.2	61.7	60.8	56.7	58.4	55.5	58.8	50.6	63.7	7.7
		Envi-Carb + PSA	80.6	63.2	67.2	62.4	66.6	61.4	62.3	51.1	56.6	56.8	57.2	55.9	56.0	56.6	59.9	55.6	48.5	51.7	50.4	58.9	
120	Pirimiphos-methyl	Cleanert TPT	41.0	37.3	34.9	43.4	34.3	33.1	29.7	42.8	32.8	32.3	34.3	30.8	32.3	30.9	29.4	29.5	28.4	30.0	27.1	33.4	9.1
		Envi-Carb + PSA	42.9	32.2	37.4	32.9	35.9	31.0	31.9	27.5	29.9	29.5	29.1	29.0	28.3	28.2	29.3	27.9	24.9	26.1	25.1	30.5	
121	Profenofos	Cleanert TPT	319.4	379.5	201.1	276.8	216.5	166.9	170.8	266.2	171.4	167.3	212.6	166.6	215.4	180.7	164.3	185.1	173.6	160.3	120.7	206.1	20.0
		Envi-Carb + PSA	272.6	125.2	190.5	167.5	268.2	189.3	180.9	143.3	155.4	156.3	170.5	130.0	139.1	178.1	153.7	169.6	135.6	147.7	129.1	168.6	
122	Profluralin	Cleanert TPT	207.0	199.3	175.6	239.0	352.5	193.5	118.2	243.6	128.1	128.6	130.1	125.0	138.0	125.8	127.1	148.9	116.1	122.0	103.2	164.3	18.8
		Envi-Carb + PSA	208.3	152.2	178.1	166.1	228.6	181.8	113.1	105.0	115.6	115.9	115.2	120.2	115.7	119.0	127.1	117.3	101.3	106.4	99.0	136.1	
123	Propachlor	Cleanert TPT	122.2	118.0	112.1	119.0	100.4	100.2	100.6	130.7	103.2	93.6	98.6	101.7	103.8	93.6	88.1	89.5	92.5	94.4	78.7	102.1	11.8
		Envi-Carb + PSA	124.1	99.4	117.2	97.8	104.3	96.8	102.1	83.6	84.6	84.9	89.2	75.7	86.4	87.4	84.8	85.1	75.1	73.4	71.9	90.7	
124	Propiconazole	Cleanert TPT	128.7	112.3	104.5	122.7	103.9	100.1	89.2	115.9	93.2	95.7	86.2	88.3	89.9	84.8	76.3	79.1	74.0	79.1	69.5	94.4	8.2
		Envi-Carb + PSA	146.1	100.9	108.6	97.0	101.7	87.5	92.4	77.0	86.7	87.7	76.0	77.0	76.1	77.5	77.2	76.0	68.4	70.1	67.6	86.9	

No.	Compound	Cartridge																					
125	Propyzamide	Cleanert TPT	128.4	124.5	123.5	153.7	118.5	114.6	100.6	137.8	120.8	110.8	114.3	111.4	115.4	114.5	103.5	105.5	102.8	106.3	98.3	116.1	10.1
		Envi-Carb + PSA	146.1	107.1	121.0	106.9	132.5	107.2	111.1	94.1	94.8	93.8	101.8	103.1	100.1	102.8	108.1	100.7	84.1	89.3	87.8	104.9	
126	Ronnel	Cleanert TPT	261.0	235.4	213.8	246.3	207.6	195.3	182.5	266.3	204.3	185.6	208.0	193.7	195.5	200.7	175.5	179.8	173.1	185.2	165.3	203.9	11.8
		Envi-Carb + PSA	247.9	191.1	224.7	191.1	215.9	185.6	197.7	163.1	180.3	170.3	182.5	152.7	166.5	178.7	169.6	168.7	147.4	154.9	154.7	181.2	
127	Sulfotep	Cleanert TPT	39.6	35.2	34.6	41.2	34.3	32.5	28.9	40.3	33.6	31.2	31.2	31.5	31.5	31.8	29.2	29.6	28.3	30.4	25.7	32.6	9.0
		Envi-Carb + PSA	40.4	31.7	35.6	33.3	34.4	30.4	30.6	27.0	29.6	28.7	29.6	26.9	28.3	29.5	29.5	26.9	24.1	25.3	25.2	29.8	
128	Tebutenpyrad	Cleanert TPT	69.2	62.2	59.2	69.0	58.6	57.0	50.3	68.0	56.6	52.9	54.2	53.9	54.0	51.8	48.2	48.2	47.3	51.5	44.0	55.6	8.4
		Envi-Carb + PSA	74.2	56.2	59.7	56.1	59.7	52.2	51.1	44.1	51.3	48.9	48.0	46.7	47.8	47.3	49.1	45.9	42.7	46.5	43.3	51.1	
129	Terbutryn	Cleanert TPT	84.1	74.6	74.1	87.1	68.5	66.9	59.0	76.2	65.0	62.7	61.0	61.5	60.5	58.6	55.4	56.0	55.3	56.5	51.6	65.0	6.9
		Envi-Carb + PSA	91.3	69.8	74.3	68.3	73.1	64.6	63.3	52.3	59.7	57.7	57.1	56.9	55.6	54.6	57.2	53.0	48.6	49.3	46.4	60.7	
130	Thiobencarb	Cleanert TPT	79.2	73.5	71.6	84.8	70.5	68.6	60.7	79.6	70.6	66.8	65.7	68.6	66.1	65.8	61.4	64.2	58.9	64.4	55.5	68.2	7.8
		Envi-Carb + PSA	83.4	67.6	72.8	66.2	72.4	65.9	64.8	56.1	62.0	60.8	61.0	60.2	60.7	60.6	61.3	61.2	53.6	55.2	53.4	63.1	
131	Tralkoxydim	Cleanert TPT	318.3	269.2	296.1	414.5	265.2	261.3	230.7	397.4	244.4	239.6	271.8	252.2	274.9	270.9	256.9	241.2	210.2	223.4	202.6	270.6	13.6
		Envi-Carb + PSA	339.0	248.7	307.3	235.7	343.1	243.2	270.9	196.8	166.1	187.6	210.5	231.6	211.4	236.1	286.3	222.7	183.6	199.8	167.6	236.2	
132	Trans-chlordane	Cleanert TPT	39.8	37.9	36.0	40.4	33.2	33.4	30.6	39.2	35.2	32.5	34.6	33.9	32.8	31.6	29.6	31.1	30.5	33.3	26.7	33.8	7.9
		Envi-Carb + PSA	40.4	33.2	36.4	31.8	36.5	32.2	32.8	27.9	32.2	29.8	30.9	29.4	29.0	29.6	31.1	29.3	26.6	27.9	26.9	31.3	
133	Trans-diallate	Cleanert TPT	73.2	67.6	66.7	77.0	61.8	62.9	57.4	74.8	68.9	71.0	72.0	73.4	70.7	67.8	62.1	69.6	61.3	72.3	61.9	68.0	4.9
		Envi-Carb + PSA	73.9	59.8	66.3	59.1	65.0	57.0	60.0	86.6	60.5	65.3	66.1	63.2	63.4	64.2	70.0	67.3	59.8	62.6	60.2	64.8	
134	Trifloxystrobin	Cleanert TPT	190.0	196.6	146.7	194.5	162.4	133.6	125.6	180.6	129.7	133.0	141.3	134.0	143.3	135.1	126.7	139.8	123.4	122.2	102.1	145.3	13.2
		Envi-Carb + PSA	193.7	120.2	140.9	133.9	173.7	127.2	132.0	114.6	130.4	121.7	120.5	124.0	112.2	122.0	120.2	122.0	101.9	108.1	98.8	127.3	
135	Zoxamide	Cleanert TPT	87.5	79.4	82.9	89.3	80.8	80.4	68.8	87.2	81.4	72.8	76.0	76.6	73.9	72.3	65.1	62.0	67.5	73.4	63.2	75.8	9.7
		Envi-Carb + PSA	97.8	80.4	87.6	78.7	78.0	66.7	70.1	61.1	65.6	64.1	66.1	57.9	62.5	63.5	64.9	58.8	58.5	61.3	63.5	68.8	
136	2,4'-DDE	Cleanert TPT	39.6	33.3	34.2	39.8	34.3	34.2	31.6	36.7	36.2	36.0	35.5	38.2	34.2	32.5	33.6	31.7	32.2	32.6	28.5	34.5	2.5
		Envi-Carb + PSA	40.3	38.3	40.9	36.6	34.8	32.4	33.8	29.9	33.7	31.7	35.2	31.7	32.3	34.8	34.3	28.5	26.1	34.9	28.6	33.6	
137	Ametryn	Cleanert TPT	121.2	98.3	103.1	126.4	97.8	100.1	66.7	83.7	97.6	123.9	87.9	97.6	94.5	84.6	87.5	84.7	83.6	83.6	72.2	94.5	2.8
		Envi-Carb + PSA	162.1	134.2	123.0	109.5	102.4	115.0	78.7	69.3	101.0	92.0	103.2	78.8	89.5	94.5	88.6	76.2	68.8	89.0	70.7	97.2	
138	Bifenthrin	Cleanert TPT	50.7	41.9	43.9	51.8	48.8	52.3	40.6	47.3	52.5	48.6	48.6	49.9	56.2	45.0	47.0	40.4	41.9	47.2	43.0	47.2	1.3
		Envi-Carb + PSA	47.7	46.8	57.9	49.6	48.7	46.0	44.2	70.8	48.4	49.8	48.7	38.8	46.5	43.4	47.6	38.5	39.2	52.7	43.9	47.8	
139	Bitertanol	Cleanert TPT	137.2	99.9	109.7	127.1	102.9	109.0	93.8	102.1	105.3	134.5	90.2	101.5	101.6	90.7	96.3	91.6	88.9	91.9	77.9	102.7	4.6
		Envi-Carb + PSA	134.5	116.5	122.6	117.0	104.2	96.6	101.7	74.1	100.6	106.3	98.0	78.8	96.2	91.0	96.6	79.0	74.4	97.1	78.7	98.1	
140	Boscalid	Cleanert TPT	239.0	141.6	155.3	191.2	177.4	177.0	129.9	155.8	135.5	178.2	151.6	129.1	144.8	115.5	164.5	157.6	129.0	139.4	132.0	155.0	3.3
		Envi-Carb + PSA	210.8	110.7	173.2	149.1	230.5	229.7	129.0	134.1	131.4	128.8	162.1	157.6	142.7	122.3	160.8	107.9	99.3	131.0	138.8	150.0	
141	Butafenacil	Cleanert TPT	50.2	37.6	39.1	47.0	44.2	40.4	31.8	38.2	35.0	43.4	36.8	33.9	37.0	30.6	39.1	38.9	33.9	37.2	34.6	38.4	2.7
		Envi-Carb + PSA	48.6	28.4	41.2	36.4	55.0	61.7	34.5	33.8	36.6	30.8	41.7	37.7	33.5	29.7	37.1	28.4	23.4	33.7	37.0	37.3	

(Continued)

TABLE 2.3.7 The Determination Results for 201 Pesticides in Incurred Oolong Tea Youden Pair Samples Under 4 Conditions in 3 Months (cont.)

No.	Pesticides	Solid Phase Extraction	November 9, 2009 (n = 5)	November 14, 2009 (n = 3)	November 19, 2009 (n = 3)	November 24, 2009 (n = 3)	November 29, 2009 (n = 3)	December 4, 2009 (n = 3)	December 9, 2009 (n = 3)	December 14, 2009 (n = 3)	December 19, 2009 (n = 3)	December 24, 2009 (n = 3)	December 29, 2009 (n = 3)	January 3, 2010 (n = 3)	January 8, 2010 (n = 3)	January 13, 2010 (n = 3)	January 18, 2010 (n = 3)	January 23, 2010 (n = 3)	January 28, 2010 (n = 3)	February 2, 2010 (n = 3)	February 7, 2010 (n = 3)	AVE (μg/kg)	Deviation rates (%)
								Pesticide content determined With GC–MS and GC–MS/MS, two concentration of Youden pair samples (μg/kg)															
142	Carbaryl	Cleanert TPT	112.5	137.1	89.7	103.6	115.3	107.1	95.4	108.6	109.7	94.8	110.0	126.4	109.2	118.9	100.6	96.4	98.9	100.5	97.9	107.0	0.9
		Envi-Carb + PSA	131.9	118.5	127.0	114.7	121.1	113.4	110.7	92.8	111.9	87.8	121.3	99.0	116.6	98.4	107.5	88.8	82.2	110.5	97.8	108.0	
143	Chlorobenzilate	Cleanert TPT	48.5	33.9	36.4	43.2	36.7	39.4	32.7	37.5	33.5	44.6	33.6	34.4	35.4	30.8	36.0	36.5	33.2	33.9	31.4	36.4	1.6
		Envi-Carb + PSA	50.0	33.2	43.3	37.1	43.0	46.1	32.2	31.4	34.4	33.9	37.9	34.2	33.5	34.3	38.0	27.8	25.4	33.1	31.5	35.8	
144	Chlorthal-dimethyl	Cleanert TPT	158.6	96.4	97.6	117.1	96.8	99.1	86.8	103.9	100.6	104.3	102.9	105.6	95.5	94.1	93.7	92.8	92.7	93.0	82.1	100.7	4.9
		Envi-Carb + PSA	112.6	93.5	120.8	103.9	103.6	99.2	96.4	85.9	99.7	92.3	102.3	86.3	93.3	97.2	99.3	82.3	73.8	99.4	80.3	95.9	
145	Dibutylsuccinate	Cleanert TPT	69.9	50.1	50.1	60.6	48.1	48.3	40.7	47.9	46.9	47.4	43.8	45.4	41.8	38.2	38.2	37.1	34.7	36.8	32.1	45.2	2.5
		Envi-Carb + PSA	67.4	55.9	64.9	54.4	50.3	47.4	46.7	37.5	44.6	43.1	44.0	33.3	39.0	40.2	39.0	32.8	29.5	37.7	28.7	44.0	
146	Diethofencarb	Cleanert TPT	301.4	208.2	219.7	254.7	211.3	233.3	202.3	216.2	211.8	306.8	231.5	251.6	245.2	200.9	216.6	218.0	215.0	220.0	191.4	229.3	4.7
		Envi-Carb + PSA	271.5	218.9	252.4	232.5	230.4	230.1	205.5	182.6	220.3	231.3	244.2	215.5	220.6	224.4	217.0	184.8	170.6	221.1	183.1	218.8	
147	Diflufenican	Cleanert TPT	47.7	32.4	34.2	42.1	36.2	35.9	31.9	37.1	34.2	42.5	35.3	35.6	37.2	32.7	34.4	34.7	31.1	33.1	31.4	35.8	2.7
		Envi-Carb + PSA	45.5	35.0	42.0	38.1	38.2	37.5	32.1	30.2	34.2	33.5	37.3	30.1	33.9	34.2	37.4	28.3	27.8	36.3	30.3	34.8	
148	Dimepiperate	Cleanert TPT	83.6	59.8	64.0	75.5	60.5	62.8	55.6	70.8	67.4	81.4	62.8	70.0	68.6	60.8	63.1	61.2	61.2	59.5	49.9	65.2	3.3
		Envi-Carb + PSA	74.2	66.5	77.9	69.2	63.9	64.8	60.3	57.3	65.3	65.4	67.5	60.4	61.3	63.2	66.9	54.0	49.6	62.9	47.8	63.1	
149	Dimethametryn	Cleanert TPT	41.0	31.9	33.6	38.7	32.0	32.8	29.6	32.2	32.7	40.2	28.4	31.4	31.2	28.3	28.9	27.3	28.7	27.1	23.6	31.6	1.8
		Envi-Carb + PSA	44.5	37.0	41.2	36.7	33.5	31.1	31.5	24.6	31.5	31.3	30.2	26.6	30.4	28.8	30.0	25.2	22.8	28.8	23.0	31.0	
150	Dimethomorph	Cleanert TPT	102.3	69.9	74.9	86.1	74.3	73.9	65.1	73.4	79.0	82.4	69.4	77.5	75.4	67.2	68.7	65.3	66.6	69.4	59.2	73.7	5.0
		Envi-Carb + PSA	87.3	72.4	85.1	82.1	74.3	65.7	75.9	58.1	71.4	68.2	69.4	60.3	68.3	66.4	73.0	61.8	58.4	74.7	58.1	70.0	
151	Dimethylphthalate	Cleanert TPT	160.2	122.1	127.5	155.3	124.8	128.3	109.3	124.8	127.5	130.4	126.9	132.0	124.6	117.1	117.6	114.5	116.9	122.4	104.4	125.6	0.8
		Envi-Carb + PSA	154.7	130.2	164.2	134.9	138.8	126.9	126.7	112.1	128.0	117.7	128.7	104.7	115.0	119.1	126.8	106.4	100.5	129.7	103.0	124.6	
152	Diniconazole	Cleanert TPT	137.2	97.6	110.6	125.3	92.4	100.3	89.2	98.7	100.4	135.4	91.2	89.7	96.0	88.8	87.9	85.0	84.6	80.2	75.5	98.2	3.8
		Envi-Carb + PSA	135.5	102.5	120.5	114.5	100.2	93.8	96.5	70.5	97.3	100.8	95.4	85.6	87.8	91.1	93.1	79.6	70.9	89.7	70.9	94.5	
153	Diphenamid	Cleanert TPT	43.0	35.2	34.0	41.4	35.4	35.8	30.7	33.8	36.5	37.8	34.4	39.0	35.9	34.0	32.7	33.4	33.9	34.0	29.5	35.3	3.3
		Envi-Carb + PSA	42.0	37.5	43.5	37.0	36.0	35.1	31.8	30.9	33.6	33.1	35.8	29.3	33.5	33.8	33.4	29.8	27.3	36.3	29.0	34.1	

No.	Name	Cartridge	C1	C2	C3	C4	C5	C6	C7	C8	C9	C10	C11	C12	C13	C14	C15	C16	C17	C18	C19	C20	C21
154	Dipropetryn	Cleanert TPT	44.6	33.5	33.8	40.1	32.6	33.4	31.0	34.3	37.7	38.3	28.7	37.7	33.8	34.7	28.7	29.0	28.2	27.3	24.8	33.3	0.9
		Envi-Carb + PSA	41.8	61.7	43.2	33.7	32.2	32.8	33.2	25.7	33.6	30.6	29.0	29.0	28.9	32.5	29.2	31.0	24.0	30.8	23.5	33.0	
155	Ethalfluralin	Cleanert TPT	161.7	118.7	127.4	163.3	132.1	134.2	105.9	128.4	126.9	149.2	129.0	130.0	132.4	114.3	139.6	139.2	121.8	126.4	112.6	131.2	2.3
		Envi-Carb + PSA	150.1	122.5	158.5	138.6	144.7	159.9	113.5	93.4	127.4	122.7	136.8	116.1	144.0	126.8	151.1	105.8	96.7	118.0	111.0	128.3	
156	Etofenprox	Cleanert TPT	41.4	32.9	34.1	39.2	35.0	34.8	28.7	33.6	35.6	35.5	33.7	36.1	33.9	33.5	32.2	30.5	30.1	28.1	26.9	33.5	7.2
		Envi-Carb + PSA	45.4	43.2	40.9	47.7	44.5	41.8	45.2	26.9	32.4	30.6	32.5	27.9	30.6	32.1	30.6	37.6	25.0	35.5	32.8	36.0	
157	Etridiazol	Cleanert TPT	102.9	77.4	81.8	107.6	83.4	86.8	76.7	84.2	73.9	88.0	89.1	74.5	81.3	70.6	79.9	76.1	77.5	81.6	73.5	82.5	1.9
		Envi-Carb + PSA	112.3	64.0	111.1	84.5	103.4	102.1	74.7	70.8	68.1	77.5	85.5	73.4	72.4	75.5	82.5	67.3	58.4	82.1	71.3	80.9	
158	Fenazaquin	Cleanert TPT	39.9	31.0	31.1	36.2	30.5	31.5	27.3	30.2	28.2	33.3	26.3	30.8	31.2	27.5	27.1	26.8	30.4	31.1	28.8	30.5	0.7
		Envi-Carb + PSA	41.5	35.1	38.9	34.1	30.8	29.0	29.7	23.7	29.7	27.7	30.0	21.2	30.3	28.5	28.1	26.6	25.8	34.2	30.0	30.3	
159	Fenchlorphos	Cleanert TPT	251.2	203.9	193.8	232.1	195.8	192.3	176.6	209.7	190.0	201.6	196.1	195.4	185.1	186.0	175.5	178.1	188.8	182.3	162.5	194.6	3.6
		Envi-Carb + PSA	243.2	211.1	252.8	212.0	203.1	197.7	188.9	166.9	193.6	169.2	189.6	159.2	170.9	181.1	175.6	158.5	140.4	192.4	159.4	187.7	
160	Fenoxanil	Cleanert TPT	89.3	71.6	69.9	78.3	71.0	68.0	61.4	72.7	80.9	78.5	68.1	70.0	72.1	62.2	65.6	66.2	67.8	64.8	60.5	70.5	1.2
		Envi-Carb + PSA	84.2	113.3	81.9	80.9	68.9	67.1	74.4	54.6	75.3	71.9	67.6	58.7	68.6	71.2	69.9	60.0	56.3	71.9	58.0	71.3	
161	Fenpropidin	Cleanert TPT	81.9	66.2	83.7	100.5	79.5	86.3	76.4	88.6	118.2	126.9	115.7	106.6	108.5	97.0	95.7	94.8	95.6	99.2	87.8	95.2	6.0
		Envi-Carb + PSA	79.7	82.8	95.0	87.3	82.6	75.9	85.6	76.1	47.0	107.5	116.8	107.0	94.7	95.7	106.2	87.2	83.1	110.5	83.8	89.7	
162	Fenson	Cleanert TPT	47.1	34.9	37.5	45.9	39.3	39.6	33.1	35.1	32.9	39.9	36.6	36.1	34.4	31.9	36.5	37.6	34.1	35.2	31.4	36.8	1.9
		Envi-Carb + PSA	42.2	35.4	43.7	37.2	41.7	48.0	33.6	34.2	35.7	33.1	38.8	36.6	34.6	34.3	36.9	28.3	26.1	34.7	30.7	36.1	
163	Flufenacet	Cleanert TPT	328.7	332.1	281.9	326.0	307.1	277.3	256.6	295.2	241.6	237.1	301.1	287.5	272.7	257.0	266.5	257.5	260.6	249.2	256.1	278.5	4.5
		Envi-Carb + PSA	319.3	299.5	341.1	277.9	336.4	339.5	254.7	251.0	266.3	184.9	274.6	236.8	241.7	241.2	262.9	234.3	174.0	263.9	257.3	266.2	
164	Furalaxyl	Cleanert TPT	83.3	65.5	66.4	77.2	65.1	65.8	59.8	67.4	68.1	73.5	64.8	71.4	66.8	63.0	60.1	61.3	62.8	61.9	53.7	66.2	2.5
		Envi-Carb + PSA	81.2	73.8	81.3	71.5	67.4	65.3	66.4	55.5	69.3	61.7	65.4	55.9	62.1	62.0	63.9	54.8	50.1	66.7	51.8	64.5	
165	Heptachlor	Cleanert TPT	108.4	87.1	90.2	108.2	85.5	89.8	81.5	99.7	89.9	99.8	92.4	91.9	85.4	83.6	89.1	88.4	86.8	88.2	75.4	90.6	2.2
		Envi-Carb + PSA	114.1	93.6	118.4	98.2	96.9	94.4	80.4	76.2	86.1	85.3	92.2	76.3	80.1	87.5	96.5	74.9	66.4	91.8	74.0	88.6	
166	Iprobenfos	Cleanert TPT	126.1	83.7	94.1	110.7	80.0	91.1	85.1	94.0	85.8	112.3	80.0	82.6	90.0	81.4	85.1	84.5	84.4	84.8	73.5	90.0	2.7
		Envi-Carb + PSA	120.9	92.9	117.0	102.1	88.6	91.8	81.9	64.8	87.5	91.7	87.8	68.3	83.3	86.5	91.9	75.1	68.6	92.1	71.2	87.6	
167	Isazofos	Cleanert TPT	80.6	60.2	66.7	81.7	63.1	65.1	58.8	75.3	71.2	82.1	67.4	69.6	72.5	63.3	55.4	62.9	64.8	60.7	54.8	67.2	1.0
		Envi-Carb + PSA	78.6	67.2	81.9	73.3	67.2	65.7	70.0	64.8	65.0	70.5	72.5	63.2	67.8	67.3	65.0	55.0	50.5	65.5	52.6	66.5	
168	Isoprothiolane	Cleanert TPT	80.5	65.5	66.6	79.8	65.7	63.6	57.8	67.3	68.0	74.4	63.4	69.5	66.9	57.0	59.8	60.1	58.3	58.4	51.1	64.9	2.0
		Envi-Carb + PSA	81.9	74.2	80.9	71.4	66.5	64.2	64.1	54.0	66.1	60.4	65.8	57.2	63.6	62.7	62.5	52.8	48.2	63.0	50.2	63.7	
169	Kresoxim-methyl	Cleanert TPT	44.0	32.9	34.8	35.5	32.9	42.2	33.0	40.1	35.7	43.0	34.0	35.6	32.7	32.1	31.7	33.0	31.7	31.4	29.1	35.0	7.4
		Envi-Carb + PSA	44.5	29.6	40.9	38.0	26.5	22.5	43.2	35.6	34.1	33.3	30.4	31.4	32.9	33.5	31.9	26.1	22.2	31.7	29.5	32.5	
170	Meienacet	Cleanert TPT	171.1	104.1	117.9	144.4	110.4	119.2	96.3	121.8	93.9	134.1	107.3	102.0	114.7	89.4	107.8	106.5	102.4	105.4	100.1	113.1	1.5
		Envi-Carb + PSA	129.2	92.3	126.0	113.2	131.9	132.5	94.8	219.6	93.9	101.6	113.8	93.9	101.0	92.3	117.3	83.5	76.4	103.7	100.4	111.4	

(Continued)

TABLE 2.3.7 The Determination Results for 201 Pesticides in Incurred Oolong Tea Youden Pair Samples Under 4 Conditions in 3 Months (cont.)

Pesticide content determined With GC–MS and GC–MS/MS, two concentration of Youden pair samples (µg/kg)

No.	Pesticides	Solid Phase Extraction	November 9, 2009 (n = 5)	November 14, 2009 (n = 3)	November 19, 2009 (n = 3)	November 24, 2009 (n = 3)	November 29, 2009 (n = 3)	December 4, 2009 (n = 3)	December 9, 2009 (n = 3)	December 14, 2009 (n = 3)	December 19, 2009 (n = 3)	December 24, 2009 (n = 3)	December 29, 2009 (n = 3)	January 3, 2010 (n = 3)	January 8, 2010 (n = 3)	January 13, 2010 (n = 3)	January 18, 2010 (n = 3)	January 23, 2010 (n = 3)	January 28, 2010 (n = 3)	February 2, 2010 (n = 3)	February 7, 2010 (n = 3)	AVE (µg/kg)	Deviation rates (%)
171	Mepronil	Cleanert TPT	42.8	35.9	37.8	41.3	38.2	39.0	32.8	38.0	36.2	38.4	34.6	40.8	39.8	36.1	34.3	34.3	35.7	34.7	32.2	37.0	0.5
		Envi-Carb + PSA	47.4	47.7	49.7	39.9	40.7	34.8	37.3	29.9	38.0	35.9	36.7	27.7	36.5	35.4	36.8	31.9	29.9	39.2	31.2	37.2	
172	Metribuzin	Cleanert TPT	127.0	87.0	71.1	84.1	65.4	73.8	58.0	63.8	54.8	68.8	59.6	53.9	57.9	53.4	55.9	52.5	49.0	49.7	45.2	64.8	1.5
		Envi-Carb + PSA	119.2	79.0	83.3	76.9	78.9	84.8	67.1	48.0	66.3	54.2	61.1	53.0	55.4	51.5	56.4	44.3	38.8	48.7	45.6	63.8	
173	Molinate	Cleanert TPT	35.3	27.8	25.7	35.6	28.4	29.3	25.5	29.1	30.4	32.0	31.4	32.8	31.4	30.1	29.6	29.7	28.2	30.0	25.4	29.9	2.5
		Envi-Carb + PSA	35.4	30.3	36.7	30.6	30.2	28.9	28.3	23.2	29.5	28.3	29.4	25.5	29.2	30.3	31.4	25.6	25.1	31.0	25.1	29.2	
174	Napropamide	Cleanert TPT	122.8	100.0	101.3	118.6	100.1	101.2	90.7	101.7	105.9	112.1	98.7	109.8	102.3	96.1	90.5	92.2	95.5	97.7	82.0	101.0	2.0
		Envi-Carb + PSA	126.0	111.0	123.7	110.0	103.7	97.6	100.0	83.9	102.5	96.5	102.2	91.6	92.9	98.0	99.1	86.0	75.7	100.8	79.8	99.0	
175	Nuarimol	Cleanert TPT	88.5	72.5	72.9	80.9	70.1	67.8	58.9	63.8	71.8	75.6	60.3	65.9	64.6	58.7	58.2	55.1	57.9	57.5	49.7	65.8	5.6
		Envi-Carb + PSA	89.9	73.1	78.6	73.8	67.4	62.2	67.2	51.4	66.0	59.0	60.1	49.7	59.7	58.4	56.3	51.4	46.1	63.0	49.6	62.3	
176	Permethrin	Cleanert TPT	83.8	65.9	67.9	77.7	72.0	72.5	60.6	69.1	71.5	75.6	70.2	69.6	69.5	63.1	66.2	62.5	62.6	62.2	55.8	68.3	3.3
		Envi-Carb + PSA	84.4	69.2	80.6	74.2	74.4	72.2	65.5	56.4	62.6	63.0	67.7	60.1	63.1	65.8	66.9	54.7	51.4	67.5	56.3	66.1	
177	Phenothrin	Cleanert TPT	40.2	31.6	32.3	39.1	32.7	32.5	27.5	39.1	43.4	40.1	45.9	36.6	57.7	27.3	39.5	37.6	40.4	32.7	24.6	36.9	7.9
		Envi-Carb + PSA	33.8	39.6	37.6	51.1	50.1	37.9	38.7	24.6	47.5	52.1	50.4	36.1	35.0	44.2	41.0	36.4	29.9	37.4	35.2	39.9	
178	Piperonylbutoxide	Cleanert TPT	40.7	31.6	35.6	40.6	33.6	33.8	28.9	32.8	34.2	41.7	26.6	31.1	32.0	28.5	28.4	28.5	27.6	28.5	24.0	32.0	4.2
		Envi-Carb + PSA	41.5	34.9	40.2	37.0	33.1	31.1	31.5	24.8	30.6	31.8	30.3	22.6	30.4	29.8	30.4	25.7	23.1	30.4	24.2	30.7	
179	Pretilachlor	Cleanert TPT	79.1	68.3	62.2	74.1	65.6	63.7	60.8	69.1	64.4	61.7	66.8	67.5	62.4	60.4	55.4	56.9	62.8	58.8	52.9	63.8	2.2
		Envi-Carb + PSA	90.3	76.1	85.8	68.0	68.1	65.4	61.3	56.1	62.6	54.2	60.8	53.9	54.4	58.7	57.1	51.3	45.9	66.9	49.5	62.4	
180	Prometon	Cleanert TPT	133.1	103.2	111.3	126.9	100.0	104.3	96.9	108.6	107.6	126.9	93.7	110.7	110.5	98.9	99.1	99.6	99.1	97.8	86.4	106.0	1.2
		Envi-Carb + PSA	136.7	124.0	135.5	114.4	105.8	105.3	104.2	83.5	104.0	103.7	107.0	95.4	97.7	106.3	105.0	89.0	82.2	105.4	84.5	104.7	
181	Pronamide	Cleanert TPT	158.9	119.6	124.0	137.2	126.0	114.5	101.0	116.8	118.3	139.2	117.8	115.0	113.9	105.5	116.3	103.6	103.0	104.3	93.5	117.3	3.1
		Envi-Carb + PSA	166.1	123.9	138.1	124.4	139.8	114.4	112.9	91.4	106.8	104.8	115.8	103.1	112.2	112.5	118.6	98.1	80.7	106.7	90.4	113.7	
182	Propetamphos	Cleanert TPT	46.1	32.4	35.2	43.1	35.1	36.2	31.0	37.5	39.6	47.5	34.6	29.5	34.7	31.3	33.5	33.5	31.0	32.2	28.0	35.3	5.2
		Envi-Carb + PSA	44.6	32.0	41.5	36.9	38.5	41.3	32.9	27.6	34.7	34.3	34.0	29.7	32.8	32.2	33.0	27.5	24.4	31.7	27.9	33.6	
183	Propoxur-1	Cleanert TPT	334.4	260.3	272.4	328.3	268.4	271.4	245.9	288.1	283.4	305.8	258.2	320.2	279.4	274.3	259.4	264.3	265.4	268.0	230.8	277.8	2.6
		Envi-Carb + PSA	307.8	304.8	342.7	294.1	279.0	255.8	286.9	238.3	285.4	265.8	282.0	218.9	267.2	273.0	273.1	235.4	222.7	287.4	221.7	270.6	

#	Compound	Cartridge																						
184	Propoxur-2	Cleanert TPT	303.9	334.1	298.2	309.9	431.3	409.7	329.9	355.5	226.4	241.7	324.0	250.7	359.4	338.3	376.3	309.5	343.0	363.3	276.0	325.3	14.2
		Envi-Carb + PSA	344.2	262.6	364.0	306.1	381.8	396.8	254.3	277.2	275.8	196.0	267.6	238.2	231.8	317.2	271.5	239.5	168.6	278.8	291.4	282.3	
185	Prothiophos	Cleanert TPT	40.3	31.0	32.0	37.7	30.6	31.7	28.0	32.7	31.7	36.8	29.7	33.0	30.9	30.5	30.0	29.6	30.3	30.7	26.5	31.8	2.8
		Envi-Carb + PSA	40.0	34.0	38.1	34.8	32.0	31.9	30.9	25.5	31.6	30.4	31.7	27.1	29.0	30.5	31.8	26.3	23.7	31.7	25.9	30.9	
186	Pyridaben	Cleanert TPT	56.3	38.0	42.4	50.0	47.7	54.4	40.8	44.7	47.4	49.6	41.9	44.6	47.3	38.3	48.2	45.0	43.1	43.8	39.2	45.4	10.1
		Envi-Carb + PSA	39.4	32.2	50.1	32.1	47.5	50.4	42.6	37.1	33.8	45.8	47.4	38.1	44.5	30.8	51.4	36.0	35.7	44.6	40.4	41.0	
187	Pyridaphenthion	Cleanert TPT	50.9	34.9	38.0	45.2	37.7	35.4	31.9	38.3	32.6	42.8	33.4	30.9	34.1	29.5	31.7	31.2	34.6	31.2	31.2	35.5	7.3
		Envi-Carb + PSA	45.2	33.1	44.2	37.8	39.4	38.5	29.8	28.1	32.5	32.2	34.0	26.5	28.1	30.1	33.4	27.2	24.1	33.5	30.1	33.0	
188	Pyrimethanil	Cleanert TPT	41.2	31.6	33.3	39.5	32.0	34.0	29.0	34.2	32.2	37.1	30.7	33.6	33.1	30.5	31.3	30.5	30.4	30.8	27.0	32.7	2.0
		Envi-Carb + PSA	40.5	36.1	41.3	35.9	33.5	32.6	31.7	25.6	34.2	32.1	33.5	26.5	30.1	33.3	32.2	27.2	25.4	32.1	26.1	32.1	
189	Pyriproxyfen	Cleanert TPT	42.9	34.6	34.5	41.3	35.0	34.4	31.4	34.7	35.2	37.6	31.7	35.4	34.1	31.1	30.3	31.0	30.4	31.0	27.2	33.9	3.8
		Envi-Carb + PSA	44.0	37.3	41.6	38.9	34.5	31.7	32.4	26.8	33.2	31.9	33.0	24.7	33.3	32.6	32.3	27.4	24.1	33.3	27.2	32.6	
190	Quinalphos	Cleanert TPT	50.0	33.7	35.3	42.7	34.4	35.5	30.7	36.9	35.7	43.6	36.6	34.8	34.8	33.1	30.0	29.9	31.9	31.2	29.2	35.3	6.5
		Envi-Carb + PSA	43.1	34.1	42.8	37.2	35.0	35.0	31.2	28.7	32.8	31.5	36.8	30.0	30.4	31.6	33.4	28.1	24.7	32.9	28.2	33.0	
191	Quinoxyphen	Cleanert TPT	39.0	30.6	34.9	36.5	31.2	33.1	27.7	33.4	33.3	34.2	31.1	34.4	31.6	30.5	29.3	28.5	27.4	28.1	26.7	31.7	0.1
		Envi-Carb + PSA	41.5	37.9	39.6	35.7	33.8	31.2	31.0	27.1	32.1	30.1	31.5	29.0	29.5	31.6	31.2	26.6	24.4	32.5	26.0	31.7	
192	Telodrin	Cleanert TPT	159.6	126.2	125.1	153.0	120.3	129.6	111.8	131.6	122.5	134.2	132.3	127.2	120.6	122.2	122.1	118.9	121.9	125.8	104.7	126.8	1.0
		Envi-Carb + PSA	150.0	128.3	156.6	131.7	143.2	144.9	122.0	112.2	128.3	110.8	137.8	119.9	118.6	122.5	128.2	105.9	92.1	123.7	108.7	125.5	
193	Tetrasul	Cleanert TPT	36.7	31.2	31.3	36.6	31.9	31.8	27.9	31.9	33.0	33.3	32.2	34.3	30.3	30.5	31.6	33.5	33.5	33.4	29.7	32.3	1.2
		Envi-Carb + PSA	40.0	37.0	38.0	34.0	31.9	29.7	31.1	26.3	31.6	29.9	31.4	28.3	29.7	31.5	34.5	30.6	27.2	35.1	29.2	31.9	
194	Thiazopyr	Cleanert TPT	83.3	66.6	67.8	80.4	65.3	67.5	60.3	65.0	70.2	76.7	70.5	73.5	65.6	63.3	61.9	62.0	62.2	64.0	56.1	67.5	2.3
		Envi-Carb + PSA	78.6	75.0	83.6	72.4	66.0	65.7	67.1	55.0	69.0	64.4	70.4	62.7	66.2	67.5	65.0	55.4	49.7	64.9	54.6	66.0	
195	Tolclofos-methyl	Cleanert TPT	42.0	32.4	34.1	39.4	33.4	33.3	31.3	36.0	33.6	37.5	34.4	35.7	35.0	33.6	32.4	31.8	33.2	33.2	29.0	34.3	3.8
		Envi-Carb + PSA	41.7	35.1	43.2	36.0	34.9	32.6	32.9	28.9	34.1	32.0	33.9	28.8	31.5	33.1	31.6	28.6	25.6	34.4	28.3	33.0	
196	Transfluthrin	Cleanert TPT	41.1	31.3	32.4	38.7	31.8	33.0	29.0	34.0	35.6	35.2	31.9	35.5	32.6	30.9	31.0	30.4	30.0	31.4	27.1	32.8	2.5
		Envi-Carb + PSA	38.7	33.8	41.0	35.6	33.0	32.2	32.6	27.5	33.5	30.8	33.4	26.3	31.8	32.1	32.7	28.0	25.2	32.4	26.7	32.0	
197	Triadimefon	Cleanert TPT	82.3	63.5	67.6	80.6	64.8	67.8	60.4	65.6	69.6	89.6	71.0	70.1	68.3	69.4	64.2	61.6	61.3	59.7	50.2	67.8	1.4
		Envi-Carb + PSA	83.1	71.9	81.1	73.4	67.6	67.5	66.3	53.8	66.5	66.1	73.0	68.9	66.6	66.1	71.4	58.6	55.1	62.6	50.1	66.8	
198	Triadimenol	Cleanert TPT	144.9	108.2	115.4	132.8	91.7	102.6	90.5	96.3	101.2	124.7	90.5	92.3	102.7	93.2	90.8	86.9	88.4	87.8	76.6	100.9	2.4
		Envi-Carb + PSA	136.1	114.2	135.6	124.0	102.5	97.3	102.4	73.8	101.2	89.9	98.6	89.5	88.5	92.1	94.5	78.9	73.6	100.2	78.5	98.5	

(Continued)

TABLE 2.3.7 The Determination Results for 201 Pesticides in Incurred Oolong Tea Youden Pair Samples Under 4 Conditions in 3 Months (*cont.*)

No.	Pesticides	Solid Phase Extraction	Pesticide content determined With GC–MS and GC–MS/MS, two concentration of Youden pair samples (μg/kg)																				AVE (μg/kg)	Deviation rates (%)
			November 9, 2009 (n = 5)	November 14, 2009 (n = 3)	November 19, 2009 (n = 3)	November 24, 2009 (n = 3)	November 29, 2009 (n = 3)	December 4, 2009 (n = 3)	December 9, 2009 (n = 3)	December 14, 2009 (n = 3)	December 19, 2009 (n = 3)	December 24, 2009 (n = 3)	December 14, 2009 (n = 3)	January 3, 2010 (n = 3)	January 8, 2010 (n = 3)	January 13, 2010 (n = 3)	January 18, 2010 (n = 3)	January 23, 2010 (n = 3)	January 28, 2010 (n = 3)	February 2, 2010 (n = 3)	February 7, 2010 (n = 3)		
199	Triallate	Cleanert TPT	73.4	57.7	58.9	72.5	59.5	60.1	51.3	61.5	59.6	65.2	61.3	61.9	60.6	58.2	57.8	58.3	56.5	58.7	48.5	60.1	2.1
		Envi-Carb + PSA	70.2	60.0	73.4	64.0	63.9	63.5	58.0	49.7	59.8	55.2	63.5	55.3	57.7	59.8	60.8	49.9	45.0	58.1	49.6	58.8	
200	Tribenuron-methyl	Cleanert TPT	51.2	31.2	47.6	56.7	45.3	59.5	34.6	41.3	36.4	52.1	36.2	35.9	41.3	28.6	40.3	42.7	35.0	40.9	39.3	41.9	6.0
		Envi-Carb + PSA	59.4	25.0	42.3	38.6	59.9	66.3	33.4	37.2	35.7	32.0	38.2	36.6	40.0	32.3	43.2	28.2	30.1	34.8	36.8	39.5	
201	Vinclozolin	Cleanert TPT	45.1	34.5	35.3	43.4	36.3	35.3	31.7	37.6	36.3	39.3	36.7	37.9	35.5	35.8	34.4	33.6	34.6	35.2	30.5	36.3	2.5
		Envi-Carb + PSA	40.5	37.3	43.2	37.8	39.0	39.0	34.8	31.6	36.6	32.2	37.5	33.7	32.3	34.2	36.3	31.5	27.5	36.1	30.8	35.4	

TABLE 2.3.8 Variance Distributions of the Total Average Value of Pesticide Content Determinated 19 Times for 201 Pesticide With Cleanert TPT and 4# Envi-Carb + PSA for Cleanup

Deviation rate between two Cartridge (*R*) (%)	Green tea (% in total)	Oolong tea (% in total)
<10	172 (85.6)	158 (78.6)
10–15	20 (10.0)	29 (14.4)
>15	9 (4.5)	14 (7.0)

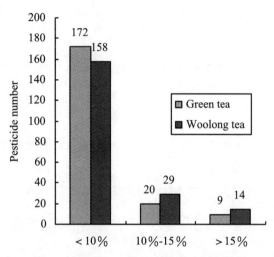

FIGURE 2.3.3 Average value variances of pesticide content for 201 pesticides in Green tea and oolong tea with Cleanert TPT and Envi-Carb+PSA cleanup.

TABLE 2.3.9 Comparison of Pesticide Content Derived from 201 Pesticides With Two Solid Phase Extraction Cartridges Cleanup

Comparison of deviation rates between two Cartridge	Green tea (% in total)	Oolong tea (% in total)
Cleanert TPT > ENVI-CARB + PSA	14 (7.0)	179 (89.1)
Cleanert TPT < ENVI-CARB + PSA	187 (93.0)	22 (10.9)

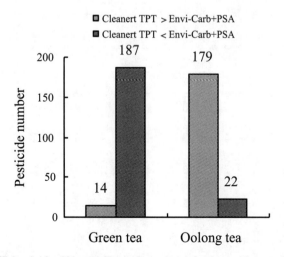

FIGURE 2.3.4 Variances of cleanup efficiencies for 201 pesticides in Green tea and oolong with two solid phase extraction cartridge cleanup.

TABLE 2.3.10 RSD Values for 201 Pesticides in Incurred Green Tea and Oolong Tea Youden Pair Samples With Cleanert TPT and Envi-Carb-PSA for Cleanup and GC–MS and GC–MS/MS for Determination (n = 3)

Date	Youden pair	Green tea Cleanert TPT GC–MS					GC–MS/MS					ENVI-CARB+PSA GC–MS					GC–MS/MS				
		<10	≤10–15	≤15–20	>20	No data	<10	≤10–15	≤15–20	>20	No data	<10	≤10–15	≤15–20	>20	No data	<10	≤10–15	≤15–20	>20	No data
9-11-2009	B	188	7	4	2	0	187	10	1	2	1	137	28	17	19	0	117	18	20	45	1
	A	180	18	2	1	0	183	11	4	2	1	119	49	18	15	0	125	30	23	22	1
14-11-2009	B	179	21	0	1	0	163	31	6	0	1	166	26	2	7	0	132	61	5	2	1
	A	188	10	2	1	0	188	8	4	0	1	171	18	3	9	0	170	16	5	9	1
19-11-2009	B	196	5	0	0	0	197	2	1	0	1	197	3	0	1	0	193	7	0	0	1
	A	175	24	1	1	0	197	3	0	0	1	161	20	10	10	0	164	22	5	9	1
24-11-2009	B	147	50	4	0	0	169	25	6	0	1	175	20	2	4	0	169	16	9	6	1
	A	195	6	0	0	0	173	19	2	6	1	182	16	2	1	0	189	10	1	0	1
29-11-2009	B	188	13	0	0	0	198	2	0	0	1	177	16	4	4	0	178	13	3	6	1
	A	172	29	0	0	0	119	63	15	3	1	175	23	2	1	0	186	10	3	1	1
4-12-2009	B	136	62	3	0	0	181	14	5	0	1	181	11	3	6	0	182	13	2	3	1
	A	186	15	0	0	0	191	8	1	0	1	158	12	14	17	0	152	16	13	19	1
9-12-2009	B	196	5	0	0	0	195	5	1	0	0	168	23	3	6	1	157	25	8	11	0
	A	170	22	7	2	0	164	27	5	5	0	124	27	15	34	1	167	24	5	5	0
14-12-2009	B	174	25	2	0	0	169	25	4	3	0	158	26	7	10	0	162	19	5	15	0
	A	163	34	3	1	0	159	31	6	5	0	162	21	7	11	0	160	14	14	13	0
19-12-2009	B	196	5	0	0	0	182	14	5	0	0	156	22	13	10	0	157	25	8	11	0
	A	175	24	1	1	0	164	23	5	5	0	160	24	7	10	0	167	24	5	5	0
24-12-2009	B	147	50	4	0	0	129	60	9	3	0	162	23	7	7	0	174	15	6	5	1
	A	195	6	0	0	0	195	4	2	0	0	189	11	1	0	0	191	7	1	5	1
29-12-2009	B	188	13	0	0	0	193	6	1	1	0	7	13	53	128	0	11	32	54	104	0
	A	172	29	0	0	0	179	18	4	0	0	159	35	5	2	0	136	48	12	5	0
3-1-2010	B	136	62	3	0	0	160	28	10	3	0	143	40	10	8	0	159	23	10	5	4
	A	186	15	0	0	0	197	3	0	1	0	149	24	10	18	0	155	20	9	13	4
8-1-2010	B	177	19	5	0	0	169	18	7	7	0	126	24	9	42	0	134	20	18	29	0
	A	170	31	0	0	0	141	50	6	4	0	194	4	2	1	0	167	23	10	1	0
13-1-2010	B	174	25	2	0	0	190	8	1	2	0	191	7	1	1	1	93	55	34	19	0
	A	168	28	5	0	0	171	20	8	2	0	158	23	10	9	1	103	52	30	16	0
18-1-2010	B	173	24	3	1	0	189	9	3	0	0	47	86	37	31	0	51	92	34	24	0
	A	186	12	2	1	0	198	2	1	0	0	177	20	1	3	0	179	15	1	6	0
23-1-2010	B	194	5	2	0	0	193	7	1	0	0	180	15	3	3	0	172	11	1	7	0
	A	192	9	0	0	0	188	11	1	1	0	138	36	11	16	0	142	40	12	7	0
28-1-2010	B	151	41	5	4	0	160	31	7	3	0	191	3	5	2	0	176	17	3	5	0
	A	185	11	5	0	0	183	13	3	2	0	71	116	10	4	0	115	70	8	8	0
2-2-2010	B	178	20	1	2	0	189	9	3	1	0	106	52	12	31	0	59	92	27	22	0
	A	152	38	2	9	0	163	25	7	6	0	136	29	13	23	0	173	18	6	3	0
7-2-2010	B	190	10	1	0	0	183	12	3	2	1	159	31	10	0	1	139	40	16	5	1
	A	198	3	0	0	0	195	4	0	0	1	190	8	2	0	1	178	15	3	4	1
		6716	826	69	27	0	6744	659	152	69	14	5800	985	343	504	6	5634	1068	439	471	26

Date	D/C																			
9-11-2009	D	186	10	3	2	0	178	18	1	2	2	177	17	4	3	0	170	21	4	5	1
	C	176	19	2	4	0	177	11	5	6	2	163	23	10	5	0	149	31	13	7	1
14-11-2009	D	193	7	1	0	0	184	10	3	2	2	189	7	0	5	0	191	6	1	2	1
	C	195	6	0	0	0	183	12	3	1	2	183	16	1	1	0	191	6	2	1	1
19-11-2009	D	195	5	1	0	0	194	4	1	1	1	192	9	0	0	0	190	9	13	0	1
	C	183	16	2	0	0	179	12	7	2	1	134	55	10	2	0	86	95	9	6	1
24-11-2009	D	196	4	1	3	0	195	3	1	1	1	181	10	7	3	0	173	14	8	4	1
	C	179	16	3	7	0	181	16	2	1	1	190	7	3	2	0	181	10	3	1	2
29-11-2009	D	171	16	7	0	0	186	7	2	5	1	175	16	8	12	0	186	5	11	5	2
	C	188	11	1	1	0	191	7	1	1	1	65	87	37	0	0	120	65	3	3	1
4-12-2009	D	196	4	1	10	0	192	5	2	1	1	194	6	1	4	0	193	4	3	0	0
	C	180	18	2	2	0	185	9	3	3	0	173	17	10	4	0	172	23	3	2	0
9-12-2009	D	136	51	4	6	0	143	45	10	3	0	177	18	2	16	0	190	10	0	0	0
	C	177	19	3	3	0	190	9	2	0	0	172	14	11	2	0	165	16	10	10	0
14-12-2009	D	142	43	10	1	0	187	7	3	4	0	136	28	21	13	0	157	10	17	17	0
	C	185	9	4	0	0	169	25	7	0	1	173	23	3	5	0	190	3	1	0	0
19-12-2009	D	182	16	2	0	1	179	17	0	4	1	165	15	16	9	1	198	91	0	0	0
	C	188	11	2	1	1	155	26	10	9	0	51	128	11	1	1	95	16	10	5	0
24-12-2009	D	175	24	1	3	0	194	6	0	1	0	143	38	9	0	0	177	10	2	6	0
	C	185	14	16	5	0	189	4	3	5	0	180	11	14	2	0	182	13	6	3	0
29-12-2009	D	134	48	13	1	0	181	11	7	2	0	185	14	10	2	0	184	11	3	1	0
	C	153	30	3	0	0	181	14	3	3	0	143	46	8	0	0	189	23	12	0	0
3-1-2010	D	168	28	1	2	0	190	7	3	1	0	169	22	1	5	0	163	8	6	2	1
	C	188	11	0	0	0	197	3	0	1	0	191	9	7	1	0	184	19	13	2	1
8-1-2010	D	191	8	0	0	0	167	25	6	3	0	176	13	7	0	0	158	15	3	11	0
	C	198	3	1	1	0	167	29	4	1	0	188	11	2	8	0	181	10	2	2	0
13-1-2010	D	197	3	0	0	0	193	5	2	1	0	190	9	8	0	0	186	70	14	2	1
	C	194	6	0	0	0	197	4	0	0	0	163	22	2	3	0	107	68	13	9	0
18-1-2010	D	182	19	3	1	0	176	22	1	2	0	193	6	23	18	0	116	117	17	4	0
	C	176	20	1	0	0	173	20	4	4	0	92	83	7	0	0	63	13	5	4	0
23-1-2010	D	183	16	0	2	0	181	12	6	1	1	160	16	7	4	0	171	10	1	12	0
	C	196	5	0	1	0	194	5	0	1	1	195	6	15	6	0	190	21	6	0	1
28-1-2010	D	190	11	1	0	0	184	12	3	1	1	77	113	10	5	0	173	12	8	0	1
	C	195	5	2	0	0	188	9	2	1	1	158	22	18	9	0	174	25	14	6	1
2-2-2010	D	188	7	0	4	0	181	9	4	6	1	163	23	2	3	0	149	15	11	12	1
	C	189	11	1	1	0	189	6	1	4	1	155	19	3	0	0	152	16	3	22	1
7-2-2010	D	193	7	4	0	0	182	13	3	2	1	186	10	0	0	0	177	5	0	4	1
	C	176	17	2	4	0	179	11	7	3	1	191	7	3	0	0	193	5	0	2	1
		6899	574	98	65	2	6931	470	122	89	26	6188	996	297	155	2	6266	926	250	172	24

TABLE 2.3.11 RSD Value Distribution for 201 Pesticides in Incurred Green Tea and Oolong Tea Youden Pair Samples With Two Solid Phase Extraction Cartridge Cleanup and GC–MS and GC–MS/MS Determination (n = 3)

| | TPT | | | | Envi-Carb+PSA | | | |
| | Green tea | | Oolong tea | | Green tea | | Oolong tea | |
RSD (%)	GC–MS (%)	GC–MS/MS (%)	GC–MS (%)	GC–MS/MS (%)	GC–MS (%)	GC–MS/MS (%)	GC–MS (%)	GC–MS/MS (%)
≤10	6716 (87.9)	6744 (88.3)	6899 (90.3)	6931 (90.7)	5800 (75.9)	5634 (73.8)	6188 (81.0)	6266 (82.0)
15–10	826 (10.8)	659 (8.6)	574 (7.5)	470 (6.2)	985 (12.9)	1068 (14.0)	996 (13.0)	926 (12.1)
20–15	69 (0.9)	152 (2.0)	98 (1.3)	122 (1.6)	343 (4.5)	439 (5.7)	297 (3.9)	250 (3.3)
>0	27 (0.4)	69 (0.9)	65 (0.9)	89 (1.2)	504 (6.6)	471 (6.2)	155 (2.0)	172 (2.3)
No data	0 (0)	14 (0.2)	2 (0.03)	26 (0.3)	6 (0.1)	26 (0.3)	2 (0.03)	24 (0.3)
Total	7638 (100)	7638 (100)	7638 (100)	7638 (100)	7638 (100)	7638 (100)	7638 (100)	7638 (100)

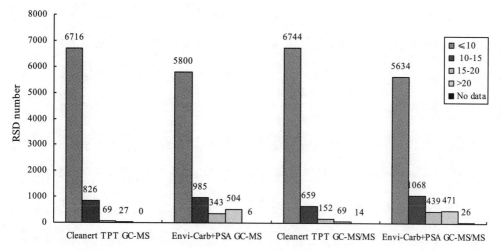

FIGURE 2.3.5   RSD distribution for 201 pesticides in incurred Green tea Youden pair samples with two solid phase extraction cartridge cleanup.

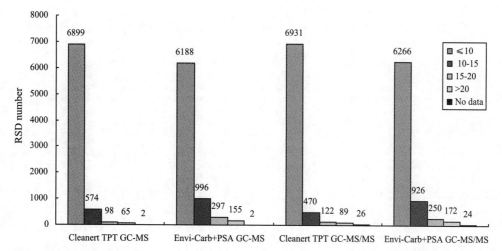

FIGURE 2.3.6   RSD distribution for 201 pesticides in incurred Oolong tea Youden pair samples with two solid phase extraction cartridge cleanup.

in Stage II. This once more proves that the Cleanert TPT newly studied and developed is superior to Envi-Carb-PSA in reproducibility of the cleanup results.

2.3.8 CONCLUSIONS

The comparative study of the cleanup efficiencies of the newly developed SPE Cleanert TPT in the three stages proves that it is as good as Envi-Carb+PSA in cleanup efficiencies and general use, and slightly better in reproducibility. Therefore, it is absolutely suitable for cleanup of more than 500 residual pesticides in tea, and will ensure a promising prospect for application in the future.

Annex 1: Content change for 201 pesticides in incurred tea Youden pair samples under 8 conditions with two SPE cartridge cleanup in three months (Nov.9, 2009-Feb.7, 2010)

No	Pesticides	SPE	Method	Sample	Youden pair	Nov.9 2009 (n=5)	Nov.14 2009 (n=3)	Nov.19 2009 (n=3)	Nov.24 2009 (n=3)	Nov.29 2009 (n=3)	Dec.4 2009 (n=3)	Dec.9 2009 (n=3)	Dec.14 2009 (n=3)	Dec.19 2009 (n=3)	Dec.24 2009 (n=3)	Jan.3 2010 (n=3)	Jan.8 2010 (n=3)	Jan.13 2010 (n=3)	Jan.18 2010 (n=3)	Jan.23 2010 (n=3)	Jan.28 2010 (n=3)	Feb.2 2010 (n=3)	Feb.7 2010 (n=3)	AVE μg/kg
1	2,3,4,5-tetrachloroaniline	Cleanert TPT	GC-MS	Green tea	A	10.0	5.7	4.6	3.4	5.1	4.2	4.7	4.5	5.1	4.3	4.3	3.6	4.0	3.2	2.7	4.1	3.4	4.2	4.5
					B	7.1	5.4	4.8	3.2	4.7	4.2	4.2	3.9	4.4	4.2	4.0	3.5	4.1	3.2	2.7	3.6	3.3	3.8	4.1
				Woolong tea	A*	7.3	6.6	5.7	6.1	5.8	4.6	4.8	4.4	5.4	5.0	5.2	5.1	4.8	6.5	4.9	3.9	5.0	4.7	5.3
					B	6.9	5.4	4.4	4.8	5.1	3.7	3.7	4.0	4.5	4.2	4.4	4.0	3.9	5.3	3.8	3.1	4.4	3.9	4.4
			GC-MS/MS	Green tea	A	6.6	5.5	6.5	6.2	3.9	5.6	4.6	4.1	4.5	4.1	4.0	3.7	3.9	3.5	2.5	3.4	2.9	4.0	4.3
					B	6.2	5.2	5.6	5.6	4.1	5.2	4.1	3.8	4.3	3.6	4.1	3.4	4.2	3.5	2.6	3.3	2.7	3.6	4.1
				Woolong tea	A	6.8	6.7	5.9	5.5	6.2	4.5	4.9	5.5	5.4	3.9	5.3	4.6	4.9	4.3	4.7	3.2	3.4	3.8	5.1
					B	6.3	5.8	4.6	4.8	4.8	3.9	3.5	4.2	4.3	3.9	4.1	3.5	3.7	3.2	3.4	2.5	2.8	3.6	4.0
		Envi-Carb+PSA	GC-MS	Green tea	A	5.6	7.7	7.5	6.1	5.9	5.6	4.3	4.8	4.7	4.8	4.7	3.9	4.2	3.8	3.4	4.3	4.3	4.1	5.0
					B	5.4	6.9	6.0	5.5	5.5	4.9	4.4	4.8	4.7	5.0	4.6	3.6	4.2	3.2	4.7	4.2	3.9	4.1	4.7
				Woolong tea	A	9.0	6.9	6.4	4.5	7.0	5.5	6.2	4.8	5.1	5.7	4.4	3.1	3.9	4.3	4.1	3.9	3.9	4.1	5.1
					B	7.3	5.9	5.5	3.9	5.6	4.3	5.2	4.0	3.8	4.4	3.6	3.3	4.0	3.0	2.8	3.2	2.9	3.7	4.2
			GC-MS/MS	Green tea	A*	5.9	6.7	5.2	6.0	5.8	5.0	4.8	4.5	4.6	4.8	11.9	3.5	4.0	3.4	4.3	3.2	3.3	2.9	4.9
					B*	5.8	5.4	5.0	4.9	5.6	4.9	4.4	4.1	4.5	3.6	11.3	3.4	4.1	2.6	3.7	3.2	3.0	3.9	4.6
				Woolong tea	A	8.5	8.1	7.4	5.0	6.2	5.6	5.4	5.0	5.1	5.0	4.3	3.9	3.9	4.2	3.7	3.1	3.8	3.8	5.1
					B	6.8	6.6	5.4	5.0	5.1	4.3	4.7	3.8	3.8	3.9	3.1	3.1	3.2	2.9	2.5	2.5	2.6	3.7	4.1
2	2,3,5,6-tetrachloroaniline	Cleanert TPT	GC-MS	Green tea	A	44.3	36.2	38.9	39.8	33.2	37.3	34.2	34.1	35.8	34.4	34.7	35.3	36.3	33.8	23.7	31.5	29.4	38.2	35.0
					B	41.3	34.8	34.8	37.3	29.7	35.4	31.3	31.2	35.7	32.8	32.4	33.4	33.5	30.3	23.2	30.5	27.5	33.4	32.6
				Woolong tea	A	39.8	40.9	35.0	37.8	37.6	31.6	31.8	38.8	39.5	38.5	39.2	38.5	38.2	35.8	36.5	32.8	34.9	37.1	37.0
					B	37.0	34.4	28.9	32.0	34.6	26.4	26.6	32.8	33.4	33.0	33.0	31.3	33.8	29.6	31.5	28.1	31.3	30.9	31.5
			GC-MS/MS	Green tea	A	43.3	38.7	41.9	42.2	30.0	34.9	35.4	31.4	36.8	32.6	35.3	34.9	31.9	35.2	28.1	31.1	27.8	39.6	35.1
					B	42.7	36.4	36.7	37.2	28.3	33.9	32.4	29.8	36.3	31.9	33.4	33.0	31.9	33.3	27.2	30.4	27.4	35.5	33.2
				Woolong tea	A	37.2	40.9	35.3	36.2	36.4	33.0	32.7	40.4	38.3	37.6	37.7	37.3	39.0	33.0	35.4	32.4	32.2	32.9	36.3
					B	35.2	33.7	29.7	30.6	33.5	28.0	26.7	35.0	31.2	32.8	30.9	32.6	35.8	27.2	30.7	29.1	29.6	27.7	31.1
		Envi-Carb+PSA	GC-MS	Green tea	A	41.1	41.0	37.6	41.0	42.4	39.7	35.5	37.2	37.8	34.6	38.6	37.8	42.7	37.0	40.1	36.7	36.7	35.0	38.3
					B	38.2	34.8	37.8	37.0	40.9	36.4	37.1	36.1	34.4	33.8	37.0	34.1	37.9	33.4	34.1	34.7	31.3	32.7	35.5
				Woolong tea	A	46.0	39.0	37.3	35.5	40.8	36.4	41.4	32.0	41.5	37.3	31.6	32.5	35.0	36.1	29.9	29.6	31.9	31.3	35.8
					B	35.7	33.1	33.5	31.4	33.3	30.8	35.2	28.7	32.9	30.1	27.5	28.5	29.4	28.8	24.8	25.2	26.4	28.0	30.2
			GC-MS/MS	Green tea	A*	35.7	42.0	38.9	40.5	42.5	41.1	36.2	37.5	36.7	39.7	116.2	36.7	43.5	38.4	39.1	41.0	33.7	35.0	42.5
					B*	33.5	33.1	38.5	35.7	42.0	38.0	33.9	34.1	34.5	37.6	112.4	35.8	38.8	33.4	33.7	35.4	29.7	34.9	39.3
				Woolong tea	A	48.2	41.2	46.3	35.6	40.5	36.6	37.7	34.8	41.5	36.8	30.7	31.5	33.3	35.7	27.5	25.6	27.4	35.5	35.7
					B	38.2	35.1	33.6	31.1	33.5	30.7	33.9	29.7	32.9	28.8	27.4	28.8	29.8	26.8	22.9	23.2	33.0	26.5	29.9
3	4,4-dibromobenzophenone	Cleanert TPT	GC-MS	Green tea	A	47.2	44.7	44.7	38.4	40.7	44.7	51.7	49.0	66.3	39.3	59.8	56.3	50.9	56.8	81.9	51.5	48.1	43.8	51.4
					B*	45.6	40.7	46.6	34.4	36.8	42.6	46.8	45.8	52.7	37.2	51.4	52.0	49.7	50.4	79.6	48.5	43.9	38.5	46.9
				Woolong tea	A	43.6	41.3	44.2	46.0	55.2	39.7	64.6	67.7	43.1	44.7	40.8	51.6	48.3	56.7	59.0	61.6	37.2	40.4	50.5
					B*	39.4	37.1	48.3	40.4	47.6	35.0	59.3	64.0	38.1	39.5	35.7	43.2	41.2	45.0	50.7	51.0	36.4	35.3	43.8
			GC-MS/MS	Green tea	A	52.6	41.0	42.0	52.9	38.6	47.6	42.9	42.1	51.6	36.0	45.4	60.2	40.5	42.6	36.8	39.3	33.8	105.3	47.5
					B	55.3	40.9	49.4	46.7	34.3	42.5	39.2	41.0	47.0	33.4	45.1	53.3	40.2	40.4	36.1	39.1	33.0	79.3	43.9
				Woolong tea	A	64.2	67.2	43.7	46.5	41.6	40.0	50.7	63.5	57.6	41.5	44.2	51.5	51.9	54.7	56.3	47.0	47.4	48.1	52.2
					B	63.1	58.5	43.6	39.3	41.5	37.0	42.4	60.7	51.1	36.0	37.2	42.8	44.4	43.9	49.0	38.2	47.1	42.6	45.6
		Envi-Carb+PSA	GC-MS	Green tea	A	55.3	46.8	37.1	53.4	49.1	51.1	62.4	57.7	67.3	47.9	58.9	51.9	44.4	63.4	57.2	92.0	63.2	51.0	56.9
					B	54.2	43.5	53.6	49.5	48.1	45.2	58.4	59.0	59.9	55.0	51.4	50.3	51.3	60.2	45.5	73.7	52.0	51.0	53.1
				Woolong tea	A*	56.0	44.3	42.9	46.0	58.1	43.5	52.4	93.6	42.5	45.0	44.2	58.3	44.9	44.5	38.6	49.6	36.8	41.4	49.8
					B*	47.9	38.9	42.5	40.4	47.5	35.4	40.6	76.9	34.7	36.6	36.9	50.6	65.6	35.4	33.0	37.8	31.3	38.7	41.3
			GC-MS/MS	Green tea	A	63.2	67.4	47.8	52.9	47.8	50.2	57.2	52.0	59.3	45.9	37.4	49.8	55.4	53.5	56.1	50.3	53.5	58.7	53.0
					B	56.6	47.2	47.1	46.7	45.3	43.7	72.8	48.8	51.7	44.0	39.0	44.6	45.5	49.2	46.2	47.0	42.4	55.9	48.7
				Woolong tea	A*	76.6	43.4	40.4	37.4	44.0	40.2	42.8	47.5	42.5	43.5	38.5	38.4	36.6	40.0	33.1	66.0	95.8	55.2	47.5
					B	74.3	38.3	36.8	34.0	39.2	33.8	39.7	42.5	34.7	33.1	32.1	33.4	31.3	30.9	27.0	54.7	64.5	50.2	40.3

Note: this table is printed rotated on the page. Column headers (individual pesticides) appear on the preceding page; the 20 data columns below are the continuation and are unlabelled here (shown as C1–C20). The data column C20 is the one adjacent to the replicate (A/B) labels.

No.	Compound	Sorbent	Instrument	Tea	Rep	C1	C2	C3	C4	C5	C6	C7	C8	C9	C10	C11	C12	C13	C14	C15	C16	C17	C18	C19	C20
4	4,4-dichlorobenzophenone	Cleanert TPT	GC-MS	Green tea	A	103.7	103.7	96.7	102.4	87.6	105.2	102.2	105.5	108.9	105.5	90.9	122.8	104.2	104.8	103.1	90.7	107.2	114.6	96.7	116.8
				Green tea	B	95.8	90.0	87.4	96.0	85.0	89.9	96.7	92.3	103.1	98.6	87.4	112.5	94.9	95.6	93.5	81.0	101.6	109.2	94.1	111.0
				Woolong tea	A	60.6	57.7	60.4	71.3	69.3	72.6	64.0	65.8	53.7	49.5	53.7	67.2	67.1	52.4	53.7	60.4	52.9	51.0	69.0	59.0
				Woolong tea	B	58.5	57.4	63.6	61.6	62.4	67.0	63.2	62.1	57.7	49.3	56.3	64.4	60.9	52.4	48.4	64.5	51.5	49.3	62.0	57.7
			GC-MS/MS	Green tea	A	103.2	156.8	75.6	89.2	79.8	96.7	95.8	107.2	110.4	106.5	87.5	115.7	92.2	97.4	101.1	88.2	112.4	119.5	100.5	128.3
				Green tea	B	95.2	127.7	74.4	88.0	79.7	89.0	91.8	94.5	104.7	96.5	83.5	104.3	87.6	89.4	93.1	79.8	102.8	98.8	97.5	125.2
				Woolong tea	A	59.7	54.3	60.7	61.1	67.4	65.5	65.5	66.5	53.0	47.2	54.2	74.6	66.7	51.6	54.3	54.9	53.8	52.4	74.0	57.5
				Woolong tea	B	58.7	54.4	65.4	57.1	63.4	62.6	64.4	59.0	56.9	51.1	57.3	69.5	60.6	51.6	49.6	61.0	52.5	51.3	67.2	59.7
		Envi-Carb+PSA	GC-MS	Green tea	A	118.6	102.4	133.5	138.5	143.0	123.1	130.6	102.9	116.9	107.0	96.2	135.3	116.0	126.1	109.2	111.9	121.7	115.0	112.0	112.7
				Green tea	B*	110.1	100.2	103.8	120.7	114.4	115.0	113.6	106.1	113.0	103.7	96.5	120.4	115.7	126.5	96.4	111.0	111.7	116.4	98.9	108.3
				Woolong tea	A*	60.3	67.6	55.8	53.5	49.5	69.4	70.1	58.2	55.7	53.8	58.5	61.7	73.2	62.0	55.6	67.5	60.1	55.2	48.5	70.2
				Woolong tea	B*	56.9	72.9	51.9	49.2	46.1	59.7	70.3	52.8	53.3	51.3	55.6	59.8	73.9	54.3	54.4	51.7	53.4	59.7	54.0	56.1
			GC-MS/MS	Green tea	A*	110.5	112.3	112.2	119.0	125.7	112.9	118.5	97.5	37.0	91.2	105.3	124.6	110.5	112.3	115.1	113.8	120.4	112.3	136.2	123.4
				Green tea	B*	101.4	110.1	91.7	106.2	104.1	105.7	105.1	102.7	34.4	92.0	101.3	109.1	102.6	112.7	100.0	111.8	107.5	110.1	104.2	114.7
				Woolong tea	A	60.9	73.8	90.4	60.3	46.2	64.6	51.1	59.8	51.5	49.9	58.8	61.7	62.4	58.3	54.8	59.5	57.4	66.8	47.4	82.8
				Woolong tea	B	57.1	79.6	75.5	58.2	43.4	53.4	56.4	53.5	53.0	47.1	54.6	59.8	61.1	55.8	52.7	48.1	50.9	60.0	54.1	66.8
5	acetochlor	Cleanert TPT	GC-MS	Green tea	A	72.8	59.2	52.1	83.6	60.1	82.3	89.2	74.8	61.7	62.4	56.4	85.4	73.2	63.9	72.3	58.5	69.3	83.9	76.6	92.2
				Green tea	B	68.4	53.5	51.2	73.3	58.5	74.2	104.2	65.3	58.9	61.2	54.1	73.7	66.8	57.2	67.1	56.7	75.4	71.0	73.2	88.9
				Woolong tea	A	71.9	65.1	56.8	70.3	70.1	65.6	75.7	77.2	76.4	72.7	72.6	78.9	73.1	67.1	57.9	63.0	62.3	73.1	92.7	86.6
				Woolong tea	B	60.4	57.9	52.7	60.1	59.8	52.9	63.4	56.3	62.9	57.8	62.4	66.6	60.8	50.8	49.4	57.8	61.3	58.9	72.7	83.0
			GC-MS/MS	Green tea	A	69.4	69.6	56.7	71.7	56.4	67.6	64.8	68.2	67.5	64.4	55.5	70.0	68.6	68.3	78.2	57.2	110.4	65.6	71.8	87.0
				Green tea	B	65.1	65.9	54.4	64.4	55.1	66.6	65.1	62.7	62.9	63.7	59.5	68.9	64.5	59.2	74.0	57.0	72.6	63.6	68.7	87.4
				Woolong tea	A	74.2	68.9	62.4	64.3	64.7	66.5	73.7	90.7	73.8	67.1	75.3	80.3	96.0	72.7	64.2	65.1	74.4	72.1	95.5	82.0
				Woolong tea	B	74.3	63.6	56.5	57.5	62.3	55.7	64.7	68.8	62.2	61.8	64.6	70.3	77.7	67.6	54.0	61.1	62.4	59.2	76.0	75.8
		Envi-Carb+PSA	GC-MS	Green tea	A	75.9	61.0	68.0	85.2	79.7	69.0	83.4	68.3	68.9	63.0	97.8	75.8	70.1	80.2	78.9	83.7	70.1	73.8	96.5	69.3
				Green tea	B	71.1	61.1	61.6	81.5	69.2	73.0	71.7	67.6	69.5	57.2	88.2	71.2	61.3	71.1	71.0	82.9	69.3	75.5	78.9	69.0
				Woolong tea	A*	69.3	56.2	54.8	52.2	50.0	71.5	60.4	71.6	59.5	61.8	71.8	79.3	76.2	68.7	67.7	91.8	75.4	78.1	83.0	86.2
				Woolong tea	B	59.1	52.1	46.5	46.2	44.4	53.7	53.0	57.1	52.6	55.7	59.0	60.8	70.4	68.8	58.4	69.4	62.3	64.6	70.7	77.4
			GC-MS/MS	Green tea	A	72.8	60.3	61.0	73.6	69.9	71.0	72.7	68.2	85.9	53.8	65.7	100.3	66.8	72.0	77.6	77.9	73.6	69.9	85.6	76.4
				Green tea	B	67.8	62.8	55.6	69.3	58.3	68.0	62.3	67.6	79.8	60.5	68.4	70.7	63.1	62.1	69.3	78.5	69.6	70.4	70.2	82.0
				Woolong tea	A*	68.2	56.2	67.1	48.0	42.1	65.1	57.4	65.4	53.7	57.5	67.7	79.3	65.0	76.0	71.1	81.9	68.0	112.9	80.0	80.9
				Woolong tea	B	57.1	58.7	56.8	45.7	38.7	49.3	51.7	59.7	49.1	54.9	54.8	60.8	57.0	62.2	60.0	66.3	63.4	70.2	70.3	55.1
6	alachlor	Cleanert TPT	GC-MS	Green tea	A	100.6	94.5	74.8	83.0	97.5	95.3	90.0	109.7	83.6	96.4	92.2	110.9	100.7	98.3	100.6	88.3	139.0	120.0	109.2	127.3
				Green tea	B	93.4	82.2	68.6	82.1	91.5	87.8	85.5	98.0	81.0	90.3	85.7	104.0	90.1	87.7	94.5	83.2	128.4	102.4	105.2	125.6
				Woolong tea	A	106.9	98.7	93.5	95.1	102.9	92.6	104.4	114.3	117.7	110.8	116.0	115.8	121.2	103.2	85.0	92.0	112.8	109.8	132.0	113.7
				Woolong tea	B	89.6	101.8	97.5	76.5	90.5	77.1	95.7	87.9	95.5	88.3	94.2	105.2	94.0	74.1	73.2	83.7	93.6	86.0	105.2	107.7
			GC-MS/MS	Green tea	A	101.8	109.4	77.6	79.7	83.1	91.4	86.3	102.3	87.3	96.6	83.0	105.3	88.2	97.7	123.2	81.5	149.5	94.8	103.3	193.5
				Green tea	B	93.3	95.1	74.7	73.7	79.4	88.2	87.7	93.1	81.3	92.6	87.1	104.0	83.2	84.6	115.0	80.3	97.7	99.4	101.4	154.2
				Woolong tea	A*	107.2	88.7	88.4	91.3	107.1	96.4	112.6	128.5	105.7	101.1	107.2	108.6	130.7	106.3	93.5	93.0	106.0	100.5	139.3	131.9
				Woolong tea	B	90.2	81.4	88.8	68.2	87.7	79.7	98.4	95.9	89.3	86.8	94.3	91.3	108.0	78.1	77.4	86.3	87.8	85.8	113.0	115.3
		Envi-Carb+PSA	GC-MS	Green tea	A*	106.6	86.9	73.5	117.8	98.8	88.0	112.5	100.1	93.3	95.3	164.2	97.0	106.7	124.2	118.1	120.5	98.4	108.1	132.9	88.2
				Green tea	B*	99.5	85.4	76.9	106.6	87.2	89.0	99.1	100.4	95.9	87.8	146.8	88.6	80.6	114.7	106.7	115.0	95.4	108.9	132.0	97.2
				Woolong tea	A	100.8	83.5	80.9	66.1	79.3	101.4	95.8	100.7	95.2	85.4	105.2	118.5	117.9	99.5	99.5	133.1	110.9	112.1	114.0	118.0
				Woolong tea	B	85.8	76.8	67.8	95.7	67.7	82.2	82.3	76.7	85.2	76.7	83.6	92.7	92.1	86.8	86.8	100.1	92.6	94.2	99.2	107.3
			GC-MS/MS	Green tea	A*	103.2	89.3	83.3	95.7	96.5	96.4	113.8	77.8	168.9	77.8	96.5	78.4	91.4	104.3	115.8	107.8	108.0	120.8	104.6	119.9
				Green tea	B	99.0	95.7	78.3	87.7	79.4	93.5	93.7	82.0	131.9	83.2	99.2	91.8	87.5	104.3	104.8	111.4	103.7	98.8	100.3	151.3
				Woolong tea	A	100.6	78.3	94.5	70.8	78.2	95.2	88.3	90.3	75.4	90.3	99.8	118.5	96.4	115.8	96.7	113.0	99.4	100.3	122.5	108.6
				Woolong tea	B	81.8	75.4	74.0	68.0	63.6	73.5	78.9	76.7	70.2	76.7	82.1	92.7	80.5	100.8	83.9	91.9	89.5	97.4	103.6	67.8

(Continued)

Annex 1: Content change for 201 pesticides in incurred tea Youden pair samples under 8 conditions with two SPE cartridge cleanup in three months (Nov.9, 2009-Feb.7, 2010) (cont.)

No	Pesticides	SPE	Method	Sample	Youden pair	Nov.9 2009 (n=5)	Nov.14 2009 (n=3)	Nov.19 2009 (n=3)	Nov.24 2009 (n=3)	Nov.29 2009 (n=3)	Dec.4 2009 (n=3)	Dec.9 2009 (n=3)	Dec.14 2009 (n=3)	Dec.19 2009 (n=3)	Dec.24 2009 (n=3)	Dec.14 2009 (n=3)	Jan.3 2010 (n=3)	Jan.8 2010 (n=3)	Jan.13 2010 (n=3)	Jan.18 2010 (n=3)	Jan.23 2010 (n=3)	Jan.28 2010 (n=3)	Feb.2 2010 (n=3)	Feb.7 2010 (n=3)	AVE μg/kg
7	atratone	Cleanert TPT	GC-MS	Green tea	A	60.2	42.8	43.0	43.8	33.5	33.7	32.8	33.7	40.7	41.0	48.0	31.7	35.8	31.9	30.6	22.9	29.6	25.3	36.0	36.7
				Woolong tea	B	54.2	41.2	36.8	39.5	29.7	32.9	30.6	30.5	38.7	35.8	38.9	28.8	32.5	30.4	27.1	23.3	28.8	25.2	31.0	33.5
			GC-MS/MS	Green tea	A*	47.7	42.2	37.0	38.0	35.6	31.0	32.9	38.4	40.9	38.1	45.9	38.8	37.2	34.1	35.5	34.7	33.0	32.7	30.5	37.1
				Woolong tea	B*	45.0	37.0	32.1	32.4	33.0	26.6	26.1	31.6	30.5	31.8	32.7	30.9	29.2	31.0	28.9	29.4	26.6	30.2	27.6	31.2
			GC-MS	Green tea	A*	56.8	48.3	53.9	52.9	35.2	89.6	40.1	35.8	41.2	38.1	46.8	41.0	41.6	36.9	41.4	31.8	33.3	28.7	45.6	44.2
				Woolong tea	B*	55.1	46.4	45.9	47.2	32.6	72.9	36.9	33.4	40.7	40.3	40.7	37.4	38.8	38.4	38.7	32.0	33.1	29.4	39.4	41.0
			GC-MS/MS	Green tea	A	48.2	50.2	45.5	41.7	39.6	39.3	40.6	49.8	43.2	47.3	51.8	45.8	42.2	45.4	40.7	40.4	38.9	39.9	36.1	43.5
				Woolong tea	B	45.8	43.5	39.2	37.3	37.1	33.8	31.1	41.5	36.3	38.9	38.0	38.3	36.4	38.9	31.9	35.0	31.5	38.4	30.9	37.0
		Envi-Carb +PSA	GC-MS	Green tea	A*	48.2	41.9	44.8	41.9	49.3	52.6	34.0	35.2	33.8	85.0	36.9	37.1	32.1	32.8	32.8	36.1	28.9	32.2	30.5	40.8
				Woolong tea	B*	41.6	39.7	43.0	37.3	45.6	44.3	34.5	33.4	33.0	59.8	35.0	33.9	28.6	32.5	29.0	28.5	28.1	28.1	31.0	36.2
			GC-MS/MS	Green tea	A*	54.2	42.6	41.1	40.4	65.0	36.6	40.0	39.4	55.1	43.1	48.8	29.2	31.6	37.2	35.6	28.2	28.9	31.4	29.5	39.9
				Woolong tea	B*	44.8	36.5	35.5	34.9	43.8	30.8	35.4	32.1	42.7	32.5	35.0	27.6	27.9	30.6	28.2	22.2	24.4	25.4	25.3	32.4
			GC-MS	Green tea	A	47.5	55.9	48.7	52.4	53.0	57.5	41.9	41.4	40.9	45.7	36.5	46.5	37.1	48.6	39.8	44.9	44.6	35.9	40.6	45.2
				Woolong tea	B	44.4	44.2	46.7	45.8	51.9	49.8	42.0	37.5	41.2	45.3	33.2	41.8	37.8	37.8	35.9	36.3	40.0	31.2	40.9	41.2
			GC-MS/MS	Green tea	A*	60.3	50.2	57.2	43.4	47.4	43.9	46.4	40.4	55.1	46.3	41.8	38.6	36.4	38.9	43.1	34.6	28.1	38.9	33.6	43.4
				Woolong tea	B*	50.1	43.0	42.0	38.6	37.1	37.5	42.7	34.4	42.7	36.3	36.0	34.2	32.5	33.0	31.4	28.1	26.7	29.7	29.9	36.1
8	benodanil	Cleanert TPT	GC-MS	Green tea	A	140.2	149.0	122.7	118.2	84.0	111.2	103.0	88.9	144.7	96.8	200.8	209.4	190.2	101.7	133.0	337.2	133.4	121.8	119.0	140.8
				Woolong tea	B*	154.1	140.0	107.5	102.4	83.8	105.2	110.0	85.4	137.2	94.0	173.9	169.0	166.4	122.2	128.5	322.6	109.7	98.8	82.9	131.2
			GC-MS/MS	Green tea	A*	238.9	120.6	113.2	226.8	352.0	252.0	110.6	80.8	158.8	73.6	399.7	79.6	97.3	91.5	127.2	120.4	155.8	77.9	77.9	155.1
				Woolong tea	B*	162.6	120.9	127.2	157.9	263.1	142.7	133.4	98.2	114.9	111.5	287.2	55.3	130.9	89.8	103.4	103.1	131.5	45.5	78.9	128.5
			GC-MS	Green tea	A*	161.7	158.5	133.1	126.1	109.6	290.4	93.6	91.4	108.5	113.0	201.2	93.3	221.4	111.4	111.8	125.3	92.3	80.8	354.2	146.6
				Woolong tea	B*	173.2	148.6	129.9	113.3	99.6	225.3	93.2	182.9	135.7	100.3	168.8	86.3	181.6	115.1	101.5	124.0	88.7	76.7	224.0	130.4
			GC-MS/MS	Green tea	A	336.1	154.9	131.0	241.1	415.8	181.1	111.8	184.8	155.1	68.8	313.4	81.3	88.3	65.3	128.1	130.3	106.9	118.6	72.4	162.3
				Woolong tea	B	217.6	159.8	119.6	178.1	368.5	109.0	76.2	116.3	118.4	113.4	273.6	59.9	122.8	70.9	100.3	118.2	92.9	72.3	70.3	138.2
		Envi-Carb +PSA	GC-MS	Green tea	A*	113.9	119.5	157.8	84.8	155.8	138.2	137.7	116.3	211.1	155.6	161.2	160.8	118.5	92.7	127.2	109.7	300.4	94.9	120.8	140.9
				Woolong tea	B*	146.7	150.3	155.2	73.5	141.8	158.9	155.0	103.4	128.4	154.2	122.9	139.6	114.8	96.1	103.9	88.3	233.1	111.8	124.9	131.7
			GC-MS/MS	Green tea	A*	114.6	141.4	120.2	70.1	310.3	76.4	87.5	388.2	67.3	56.1	81.9	107.3	166.2	161.1	116.0	93.6	91.2	99.4	99.2	128.9
				Woolong tea	B*	124.4	104.2	131.2	87.3	214.5	88.1	61.2	193.6	25.4	80.2	83.2	97.9	117.3	167.6	101.9	61.4	81.2	82.1	82.1	103.6
			GC-MS	Green tea	A	125.9	127.4	132.2	101.1	144.0	146.1	139.1	140.1	136.3	119.1	153.9	108.5	110.1	125.5	108.4	105.7	102.1	97.0	133.7	124.0
				Woolong tea	B	134.7	174.3	137.6	86.0	141.7	155.0	183.2	112.3	127.1	123.8	112.6	101.3	118.7	98.2	97.0	82.0	90.5	116.6	123.7	121.9
			GC-MS/MS	Green tea	A*	103.8	158.2	147.0	61.8	123.9	70.1	80.2	124.2	67.3	63.7	71.6	84.7	102.9	65.0	97.4	82.3	131.1	279.9	203.0	111.5
				Woolong tea	B*	91.2	119.1	137.9	74.7	86.6	80.1	83.4	79.7	25.4	72.7	73.2	85.3	69.9	80.7	80.1	46.4	115.7	211.7	116.7	91.1
9	benoxacor	Cleanert TPT	GC-MS	Green tea	A	112.4	104.0	90.2	114.6	65.2	73.0	80.9	73.1	91.4	63.6	108.0	87.4	82.1	91.4	93.3	90.0	80.5	60.9	51.1	84.9
				Woolong tea	B	110.4	96.2	77.3	113.9	66.6	66.9	73.1	72.6	83.6	65.7	103.6	80.1	77.8	101.6	94.1	87.6	83.1	53.1	50.2	82.0
			GC-MS/MS	Green tea	A*	87.4	106.3	88.4	99.5	71.0	68.2	84.7	95.8	87.9	83.3	121.9	76.8	88.7	77.1	62.1	69.9	75.0	59.4	71.0	82.9
				Woolong tea	B*	80.0	89.0	76.2	76.7	64.9	56.2	56.0	78.9	70.5	78.8	84.7	60.9	77.8	75.1	56.3	66.8	67.3	54.6	70.8	70.6
			GC-MS	Green tea	A*	109.8	85.6	55.2	98.2	61.7	104.4	65.7	70.4	78.9	58.0	108.9	61.3	94.8	61.0	76.9	86.3	49.2	51.9	102.3	77.9
				Woolong tea	B*	103.5	82.1	65.1	70.8	62.0	105.2	58.2	65.6	89.9	63.1	103.2	54.3	87.9	67.5	74.3	89.3	51.5	51.9	88.1	75.4
			GC-MS/MS	Green tea	A	107.1	94.4	69.3	90.3	80.7	71.2	93.7	95.2	71.6	67.3	127.8	72.6	73.1	78.5	84.0	89.6	77.2	71.1	71.9	83.5
				Woolong tea	B	92.7	86.1	63.6	71.3	71.5	53.5	72.0	83.1	64.0	68.6	91.2	53.4	72.5	70.5	63.6	76.1	60.8	58.5	64.5	70.4
		Envi-Carb +PSA	GC-MS	Green tea	A*	84.5	111.0	93.0	60.1	98.2	109.5	90.3	82.3	77.0	95.9	94.8	104.8	88.2	66.4	72.4	76.2	112.4	82.7	74.0	88.1
				Woolong tea	B*	92.4	100.5	97.2	53.7	89.0	110.8	83.9	74.3	70.5	92.5	79.1	100.9	84.9	70.8	75.9	66.8	99.8	78.8	75.8	84.1
			GC-MS/MS	Green tea	A*	69.9	79.8	90.8	85.5	111.1	88.1	72.4	128.3	73.4	71.1	76.1	84.8	92.4	78.1	79.6	66.0	59.1	59.7	66.2	80.6
				Woolong tea	B*	73.7	71.0	77.4	80.9	78.4	78.5	78.6	92.2	53.1	67.7	64.0	82.3	78.9	82.9	78.1	50.9	57.0	59.6	60.2	71.9
			GC-MS	Green tea	A	78.5	75.7	77.1	65.9	80.4	100.1	73.8	70.7	52.4	57.8	89.1	88.8	69.3	69.8	74.7	54.8	69.9	56.0	70.1	72.4
				Woolong tea	B	97.7	91.1	79.3	60.9	77.9	96.4	62.7	69.0	62.8	65.0	68.4	75.5	75.2	68.8	68.1	55.7	62.8	59.2	78.9	72.4
			GC-MS/MS	Green tea	A*	80.3	80.0	107.2	57.3	74.3	67.3	80.6	82.9	73.4	63.2	66.5	50.4	66.2	59.4	59.6	56.9	62.8	111.1	98.0	74.9
				Woolong tea	B*	57.7	68.1	68.1	57.3	57.4	66.1	70.0	66.8	53.1	65.7	58.8	53.7	54.9	60.2	55.2	40.5	87.4	100.5	84.0	64.5

(Continued)

No.	Compound	Cartridge	Method	Tea																					
10	bromophos-ethyl	Cleanert TPT	GC-MS	Green tea	A*	43.9	35.0	34.2	39.4	28.8	30.8	30.4	29.7	37.6	31.5	50.3	36.6	32.7	31.3	29.3	25.7	27.5	25.7	31.7	33.3
					B	42.3	33.9	30.5	35.5	27.0	29.5	27.8	27.7	36.0	28.9	43.6	33.3	30.5	29.9	26.8	25.3	26.6	24.5	28.0	30.9
				Woolong tea	A	38.0	37.8	32.9	34.9	33.0	29.0	31.7	37.1	39.1	37.0	45.2	37.0	34.9	34.5	32.1	33.5	31.2	30.3	30.2	34.7
					B	35.8	32.9	28.1	29.4	30.2	24.5	24.1	32.2	31.6	32.1	32.8	30.0	28.3	30.5	26.7	28.6	25.7	27.5	27.1	29.4
			GC-MS/MS	Green tea	A*	40.6	34.5	33.4	36.1	28.9	49.8	29.6	28.3	33.7	31.1	34.9	29.0	34.9	31.7	30.1	26.1	25.8	24.4	37.5	32.7
					B	40.8	33.2	32.3	32.0	27.2	42.7	26.9	27.5	33.5	29.6	32.4	29.0	32.4	32.1	28.9	25.5	24.4	24.2	31.9	30.8
				Woolong tea	A	41.5	41.1	35.9	36.2	32.2	29.6	31.0	39.5	36.7	33.8	40.2	36.6	36.1	36.7	32.8	34.1	28.5	29.2	29.6	34.8
					B	35.8	34.8	29.7	31.8	29.3	25.5	24.2	34.1	30.5	30.2	30.5	29.5	28.7	31.8	26.8	29.3	23.1	27.3	26.3	29.4
		Envi-Carb +PSA	GC-MS	Green tea	A*	34.8	41.6	38.8	33.0	42.7	41.3	35.5	33.8	34.9	90.7	37.7	38.8	30.9	35.8	32.1	34.1	35.9	29.5	29.4	38.5
					B*	34.2	35.4	37.6	30.6	39.5	37.0	31.8	30.3	32.7	75.8	35.1	35.4	28.4	33.2	31.4	28.4	32.0	27.2	29.1	35.0
				Woolong tea	A	42.3	38.2	35.8	34.3	47.3	32.6	36.3	37.7	40.1	36.2	38.5	30.2	30.2	33.8	34.7	26.7	26.2	27.5	27.3	34.5
					B	35.4	32.5	31.3	30.1	34.7	28.5	31.4	29.8	31.4	29.4	30.7	27.4	26.6	28.4	27.4	22.6	22.7	23.4	24.9	28.9
			GC-MS/MS	Green tea	A	37.8	40.4	35.2	33.9	38.7	40.7	35.0	33.5	32.8	33.8	33.6	37.6	31.9	37.8	33.1	32.9	33.3	27.6	31.6	34.8
					B	38.0	34.4	34.3	32.4	37.8	36.3	31.3	31.4	33.2	33.6	30.9	35.9	30.6	32.3	30.5	27.9	29.4	25.8	31.2	32.5
				Woolong tea	A	39.4	40.5	45.5	33.1	34.5	32.8	36.5	32.4	40.1	33.8	30.4	29.3	29.2	29.8	31.8	25.3	25.9	32.5	29.3	33.3
					B	30.4	34.9	31.6	29.3	27.7	28.4	26.5	26.6	31.4	27.6	28.6	26.2	25.2	26.4	24.6	21.0	23.8	25.7	27.5	27.9
11	butralin	Cleanert TPT	GC-MS	Green tea	A	219.9	154.2	139.7	145.7	92.4	125.2	134.9	115.1	193.5	109.3	231.3	165.0	197.3	140.4	181.5	389.7	112.7	151.8	107.2	163.5
					B	203.3	149.3	118.5	127.4	95.9	121.9	126.2	123.6	183.5	98.2	214.0	148.3	170.0	132.0	184.0	364.3	122.8	123.8	97.4	152.9
				Woolong tea	A*	164.3	132.6	140.1	170.5	235.8	145.5	199.9	187.2	161.0	138.3	192.8	135.8	151.4	108.2	158.0	171.2	228.8	132.1	123.2	161.9
					B	153.0	123.7	133.7	145.3	152.1	104.6	137.1	169.2	122.7	120.9	128.3	102.9	138.7	133.8	119.4	157.0	177.1	120.3	123.6	134.9
			GC-MS/MS	Green tea	A*	227.0	144.3	121.7	187.7	99.4	582.2	108.0	114.8	148.8	100.3	203.3	107.1	148.3	125.6	129.7	186.9	89.7	71.7	283.5	167.4
					B	206.2	140.4	122.1	131.6	91.1	455.8	104.7	108.8	192.0	101.1	190.4	100.9	127.3	131.7	127.3	187.8	92.4	72.4	223.7	153.0
				Woolong tea	A	186.1	226.5	128.4	132.6	136.4	154.5	267.6	212.5	172.8	130.9	212.2	114.6	123.1	138.8	136.0	136.3	134.2	138.9	109.9	157.5
					B	177.8	201.7	117.2	122.9	124.5	117.3	190.9	185.2	140.1	112.2	141.7	93.8	118.6	125.1	111.4	123.6	101.4	113.2	111.9	133.2
		Envi-Carb +PSA	GC-MS	Green tea	A*	137.2	162.1	192.2	85.3	189.5	222.2	244.9	180.6	146.2	193.3	221.0	206.3	158.7	88.0	130.1	139.1	292.6	149.7	111.3	171.3
					B	149.0	166.8	176.3	65.5	159.3	223.5	162.2	146.7	156.8	156.8	174.6	191.1	147.8	102.0	124.4	117.3	248.1	122.1	116.3	151.4
				Woolong tea	A*	148.8	130.4	136.3	131.7	487.8	169.4	103.5	310.0	155.8	133.9	165.4	145.8	196.7	313.3	80.1	117.6	114.0	114.6	114.3	172.1
					B*	137.9	113.5	108.3	120.5	210.0	153.6	100.7	192.4	125.8	119.2	120.6	141.0	149.2	282.4	109.7	60.6	112.0	95.7	100.8	134.4
			GC-MS/MS	Green tea	A	162.4	135.3	134.7	122.9	144.3	176.1	199.6	145.2	110.5	117.1	217.7	191.5	111.0	135.0	124.5	107.2	112.4	76.3	131.6	139.8
					B	184.0	216.0	135.7	105.2	138.1	163.0	69.8	126.0	119.1	117.1	168.2	163.3	128.6	113.3	119.0	92.6	101.8	92.6	149.0	131.7
				Woolong tea	A*	130.4	142.5	167.5	118.3	135.1	130.3	138.9	184.1	155.8	123.9	125.4	101.0	116.1	107.3	106.8	105.3	256.4	393.4	172.7	153.2
					B*	101.8	121.0	111.6	113.1	97.4	117.6	123.4	115.6	125.8	105.6	105.7	101.6	96.7	91.3	92.8	74.7	204.7	294.5	148.7	123.3
12	chlorfenapyr	Cleanert TPT	GC-MS	Green tea	A	352.3	304.4	328.8	363.7	264.7	297.4	278.9	276.4	297.0	284.1	296.6	302.4	276.7	286.4	268.4	207.2	256.5	225.4	295.0	287.5
					B	354.4	289.2	298.7	344.9	241.2	269.1	252.6	250.6	291.5	279.7	256.7	281.3	264.9	274.2	241.1	201.1	248.6	220.6	264.9	269.8
				Woolong tea	A*	316.5	377.1	314.0	328.9	297.0	253.7	285.1	301.0	328.5	298.2	357.2	339.5	321.4	327.5	297.0	315.4	297.1	228.8	293.9	316.2
					B*	303.7	318.6	270.8	276.2	294.5	236.3	238.9	257.1	280.9	298.0	280.7	290.7	271.9	291.4	254.6	269.1	254.6	270.6	262.6	276.8
			GC-MS/MS	Green tea	A	361.1	306.2	325.2	361.5	269.6	295.1	278.9	257.1	307.8	279.3	308.0	257.4	297.1	295.3	278.4	227.4	231.6	229.7	301.8	287.8
					B	358.5	297.5	301.0	303.6	244.4	276.5	253.4	239.5	300.7	284.7	277.4	249.4	281.4	302.8	252.8	214.2	231.7	219.6	275.8	271.8
				Woolong tea	A	377.4	386.8	313.4	315.4	306.9	258.3	276.1	345.4	330.4	300.3	365.9	337.9	318.5	345.2	309.1	307.5	265.1	268.6	267.6	315.6
					B	324.3	333.1	283.5	268.4	298.3	230.0	236.5	309.7	282.5	281.1	288.6	293.9	263.0	299.7	263.5	260.9	219.7	260.2	239.9	275.6
		Envi-Carb +PSA	GC-MS	Green tea	A	298.4	371.0	312.3	332.9	332.6	321.4	310.6	295.5	303.3	275.6	277.2	308.0	276.0	333.3	274.9	312.5	311.8	269.5	272.6	304.7
					B	308.4	307.9	313.7	310.8	327.9	297.5	300.0	260.6	278.3	274.6	262.5	290.4	250.3	289.6	257.9	264.0	274.9	236.1	272.5	283.6
				Woolong tea	A	394.8	348.7	323.5	329.1	350.6	297.6	336.6	294.8	357.1	317.1	297.0	272.7	274.9	292.8	327.1	250.0	248.7	262.2	260.1	307.1
					B	347.2	308.8	308.7	289.5	297.4	262.8	296.7	261.1	289.0	262.3	260.4	240.0	253.1	246.0	258.0	219.4	216.3	223.3	244.5	267.6
			GC-MS/MS	Green tea	A	352.4	349.6	310.9	339.8	319.2	347.0	292.5	284.0	282.7	290.9	266.3	265.6	293.3	325.0	288.1	302.8	304.0	258.5	278.6	302.7
					B	374.6	289.0	306.9	305.4	325.6	311.3	297.4	269.2	273.3	294.3	251.4	251.5	264.6	276.5	266.7	249.6	264.9	231.4	261.2	282.4
				Woolong tea	A	360.2	350.1	424.5	312.6	338.7	293.6	333.0	294.7	357.1	319.9	277.4	256.1	254.4	277.2	300.8	236.3	225.8	306.6	257.8	304.0
					B	283.1	312.3	327.3	286.3	287.5	264.0	298.3	257.7	289.0	264.4	261.3	231.3	237.4	244.1	230.5	201.2	216.0	243.5	244.4	262.1

Annex 1: Content change for 201 pesticides in incurred tea Youden pair samples under 8 conditions with two SPE cartridge cleanup in three months (Nov.9, 2009-Feb.7, 2010) (cont.)

No	Pesticides	SPE	Method	Sample	Youden pair	Nov.9, 2009 (n=5)	Nov.14, 2009 (n=3)	Nov.19, 2009 (n=3)	Nov.24, 2009 (n=3)	Nov.29, 2009 (n=3)	Dec.4, 2009 (n=3)	Dec.9, 2009 (n=3)	Dec.14, 2009 (n=3)	Dec.19, 2009 (n=3)	Dec.24, 2009 (n=3)	Dec.14, 2009 (n=3)	Jan.3, 2010 (n=3)	Jan.8, 2010 (n=3)	Jan.13, 2010 (n=3)	Jan.18, 2010 (n=3)	Jan.23, 2010 (n=3)	Jan.28, 2010 (n=3)	Feb.2, 2010 (n=3)	Feb.7, 2010 (n=3)	AVE, µg/kg.
13	clomazone	Cleanert TPT	GC-MS	Green tea	A	56.1	39.9	34.1	37.5	30.2	36.5	31.5	30.8	34.4	32.1	48.6	39.6	37.0	37.6	35.2	47.0	32.5	27.7	36.4	37.1
					B	52.3	37.7	31.9	33.9	30.6	34.0	28.7	29.8	38.6	31.9	44.3	35.0	33.2	41.0	38.4	45.3	29.6	26.9	31.0	35.5
				Woolong tea	A*	42.0	41.1	36.8	40.9	41.9	31.6	35.2	39.6	38.4	33.7	51.7	35.3	36.4	33.6	34.2	35.8	36.0	33.1	30.8	37.3
					B*	38.8	35.4	32.1	32.0	34.6	25.4	27.8	33.9	32.1	30.0	38.8	29.3	30.7	30.7	27.9	31.0	30.4	29.6	28.3	31.5
			GC-MS/MS	Green tea	A*	55.5	40.5	37.2	39.1	29.4	66.0	31.4	30.6	33.6	34.1	46.3	35.5	35.4	30.7	33.4	41.6	29.0	26.2	42.7	37.8
					B*	52.6	36.6	33.9	34.6	30.8	59.3	28.4	29.8	38.1	35.0	44.2	32.1	33.2	34.6	34.9	40.2	28.8	27.5	35.1	36.3
				Woolong tea	A*	43.8	45.7	37.3	38.3	36.8	34.3	34.2	42.0	42.9	36.5	52.2	37.6	34.3	38.5	33.9	35.1	33.1	33.5	29.9	37.9
					B	38.3	38.5	31.2	31.6	32.5	26.7	29.2	35.7	34.2	32.6	38.4	29.6	29.4	35.0	28.4	32.3	29.2	29.4	26.0	32.0
		Envi-Carb +PSA	GC-MS	Green tea	A	44.7	46.5	47.7	29.6	43.3	41.0	43.3	39.0	38.0	54.1	55.0	51.0	36.6	39.2	41.2	37.6	41.8	29.3	31.9	41.6
					B	37.8	45.5	42.9	26.3	39.7	41.4	31.3	37.3	35.7	43.5	45.1	45.2	33.7	36.8	41.0	29.9	34.9	31.5	32.2	37.5
				Woolong tea	A*	43.9	41.9	41.3	34.3	51.5	35.8	35.8	41.7	40.0	35.9	33.1	33.1	33.3	41.9	30.3	27.2	27.8	28.1	30.2	36.2
					B*	35.0	35.2	35.7	31.5	37.1	31.5	33.1	33.4	31.3	30.3	28.0	30.6	27.7	37.9	27.0	20.6	24.8	24.7	25.9	30.6
			GC-MS/MS	Green tea	A*	42.4	50.6	43.2	35.1	42.4	41.0	43.7	36.8	43.5	38.3	52.8	32.2	28.4	40.6	41.5	34.7	35.4	27.6	37.1	39.3
					B*	37.2	49.9	39.9	31.0	39.2	41.3	28.0	33.1	40.3	35.3	43.3	32.0	34.5	32.7	40.9	29.1	32.1	28.3	35.6	36.0
				Woolong tea	A*	42.9	43.1	48.5	34.5	37.6	36.1	35.4	36.9	40.0	37.6	35.0	29.5	33.2	30.0	31.9	28.7	29.1	33.2	32.1	35.5
					B	35.0	35.6	34.4	31.6	30.2	31.5	33.0	30.2	31.3	31.6	30.8	28.3	26.9	27.3	25.6	21.8	26.0	27.9	29.1	29.9
14	cycloate	Cleanert TPT	GC-MS	Green tea	A	41.6	31.6	33.0	34.0	27.8	30.7	29.8	29.8	30.8	29.4	31.8	27.9	31.0	29.0	27.4	19.2	25.2	23.0	31.9	29.7
					B	38.5	30.4	30.2	32.4	25.4	28.1	27.2	27.1	30.6	27.4	28.9	26.3	29.1	28.5	24.6	18.9	24.1	22.4	27.4	27.8
				Woolong tea	A*	90.7	80.0	71.6	90.3	102.2	73.6	58.3	70.1	80.4	66.5	138.9	64.9	69.9	68.7	78.3	74.3	68.1	64.0	57.5	77.3
					B*	84.3	73.9	64.7	63.4	78.2	53.0	58.0	67.6	68.0	65.7	101.9	59.1	64.5	68.4	68.4	64.7	59.7	59.0	54.6	67.2
			GC-MS/MS	Green tea	A*	39.2	31.8	35.1	30.0	26.0	36.2	30.3	26.1	31.1	26.9	31.9	29.7	28.4	25.0	28.0	23.2	25.6	22.4	29.3	29.3
					B*	38.3	30.7	31.3	32.3	24.6	32.8	27.6	25.4	30.9	25.8	29.0	28.1	27.2	26.1	26.6	23.0	24.6	22.2	26.1	28.0
				Woolong tea	A*	81.9	89.6	83.7	76.0	79.9	63.2	60.0	75.2	80.0	68.5	103.6	68.3	64.4	72.7	67.9	68.8	57.1	57.2	50.3	72.0
					B	78.0	82.7	75.4	64.2	75.6	56.0	59.5	70.9	67.9	65.1	82.6	60.3	64.9	70.1	59.1	60.8	50.5	51.0	47.3	65.4
		Envi-Carb +PSA	GC-MS	Green tea	A	36.7	36.1	33.6	34.1	37.7	35.5	31.2	31.5	39.0	41.0	30.3	35.1	30.7	36.6	32.4	35.3	30.0	31.4	28.5	34.0
					B	32.7	30.6	33.3	30.9	36.9	32.2	30.8	30.9	28.5	37.2	27.8	32.9	26.8	32.9	28.0	28.1	29.7	26.9	28.3	30.8
				Woolong tea	A*	98.4	86.1	76.8	69.3	110.2	67.7	81.7	97.0	89.5	69.3	66.8	60.9	78.5	112.9	62.7	55.6	53.3	54.0	58.1	76.3
					B*	87.3	77.7	76.2	67.2	86.1	61.3	64.5	79.4	71.4	60.9	58.3	57.9	68.1	103.9	57.4	44.2	48.6	53.0	53.5	67.2
			GC-MS/MS	Green tea	A*	33.2	36.5	33.3	35.0	37.3	35.7	31.9	30.8	30.3	33.6	27.3	52.4	27.4	36.0	31.2	32.4	31.7	27.5	28.6	33.3
					B*	31.0	29.3	33.2	31.0	37.0	32.1	30.5	28.6	29.3	32.6	25.8	45.3	28.0	34.0	28.4	27.7	29.8	23.7	27.6	30.8
				Woolong tea	A*	106.0	90.3	86.9	69.5	72.4	67.5	71.4	67.2	89.5	72.4	67.8	57.3	64.6	56.2	59.1	49.0	73.8	107.3	69.7	73.6
					B	96.6	82.4	80.5	66.4	63.2	61.3	64.7	58.9	71.4	62.2	59.9	56.0	52.7	53.3	50.4	38.6	66.9	93.6	64.4	65.4
15	cycluron	Cleanert TPT	GC-MS	Green tea	A	140.3	113.4	128.5	150.2	121.8	116.4	110.4	107.6	129.9	114.5	136.8	111.2	129.4	122.6	108.6	80.2	96.3	88.8	108.5	116.6
					B	143.7	113.2	122.2	144.1	111.8	112.5	99.9	96.2	112.6	101.4	123.8	112.2	126.8	117.0	96.6	73.0	87.9	84.2	96.8	109.3
				Woolong tea	A*	36.3	35.7	29.5	26.2	24.1	21.3	27.3	33.3	33.7	32.5	36.3	32.5	31.9	31.1	29.8	28.4	26.1	28.7	28.6	30.2
					B*	33.8	29.4	24.7	22.2	21.9	18.7	22.3	27.8	27.8	27.6	27.7	27.0	26.1	27.1	24.3	24.6	22.3	25.4	24.7	25.5
			GC-MS/MS	Green tea	A*	113.6	95.7	110.3	98.9	88.4	199.2	92.1	86.9	100.0	83.5	80.1	91.1	94.8	66.8	84.5	64.9	78.2	88.8	111.1	96.3
					B*	122.2	93.2	111.3	109.7	83.7	167.1	79.8	77.4	92.5	98.2	65.8	87.6	91.6	75.8	77.3	60.7	71.6	85.8	99.5	92.1
				Woolong tea	A*	33.0	34.9	29.2	32.3	29.7	28.4	28.2	33.1	34.3	33.8	35.5	31.0	29.0	32.2	28.5	27.7	24.8	27.6	26.5	30.5
					B	30.9	29.0	25.1	27.0	27.0	23.3	22.4	28.4	28.9	28.5	27.0	26.1	27.4	27.4	23.2	24.5	21.5	24.9	22.0	26.0
		Envi-Carb +PSA	GC-MS	Green tea	A	113.0	132.7	123.7	133.9	144.0	122.7	115.7	104.2	119.9	187.5	176.4	116.3	109.1	120.1	110.5	116.3	105.2	100.7	102.0	123.9
					B	107.9	106.1	125.6	117.8	147.1	109.9	110.6	96.8	111.3	164.3	109.5	113.1	93.1	116.7	101.5	97.9	94.5	89.2	100.3	111.2
				Woolong tea	A*	41.7	33.8	31.9	32.3	37.7	31.0	34.9	27.0	32.9	31.4	30.3	26.9	26.9	28.9	31.4	25.2	25.5	26.0	26.3	30.6
					B*	32.1	28.2	28.1	28.1	28.6	25.8	28.8	23.7	26.7	24.0	25.0	23.5	23.0	23.7	24.2	20.4	20.1	21.9	23.2	25.2
			GC-MS/MS	Green tea	A*	102.0	123.7	97.6	131.6	98.4	108.6	83.5	84.2	93.9	89.2	82.0	107.9	73.1	126.6	94.2	101.6	99.9	92.9	96.0	99.3
					B*	89.7	87.1	92.9	115.4	101.4	95.1	83.5	81.1	102.0	96.0	80.8	102.1	76.0	86.3	85.1	92.7	85.9	76.7	101.1	91.1
				Woolong tea	A*	42.3	33.5	43.3	30.6	34.6	30.2	32.1	28.9	32.9	31.6	29.5	24.5	26.2	27.6	29.6	22.8	19.4	23.6	22.7	29.8
					B	31.3	28.3	28.5	26.4	27.6	25.6	28.3	24.2	26.7	24.5	25.4	22.4	25.2	23.6	21.8	18.6	17.3	19.1	20.8	24.5

No.	Compound	Cartridge	Method	Tea	Row																				
16	cyhalofop-butyl	Cleanert TPT	GC-MS	Green tea	A	114.2	84.0	76.6	73.0	64.7	76.0	65.5	63.4	91.1	77.9	92.0	96.2	91.0	76.9	82.3	84.3	64.9	70.1	79.9	80.2
				Woolong tea	B	106.4	80.1	69.1	69.5	63.1	69.0	59.3	60.8	98.4	74.3	90.7	94.4	78.7	84.5	82.6	78.2	62.4	59.5	71.0	76.4
				Green tea	A*	119.4	161.8	120.7	125.5	118.5	101.3	102.9	118.8	105.7	276.1	399.2	323.3	254.0	120.5	125.2	120.2	83.6	120.9	112.9	159.4
				Woolong tea	B*	131.9	131.1	99.8	133.0	127.3	109.2	82.6	106.2	108.3	115.5	127.8	166.2	115.2	130.3	108.7	96.6	108.1	138.4	108.6	116.9
			GC-MS/MS	Green tea	A	121.7	86.8	87.3	57.1	68.6	179.5	65.2	58.1	41.4	69.9	75.6	75.7	78.2	69.0	69.6	73.9	56.4	50.5	107.5	78.5
				Woolong tea	B	117.2	81.8	74.0	73.1	65.0	148.2	58.3	57.0	49.6	144.1	70.6	66.0	71.7	80.0	69.7	72.3	54.6	51.7	86.4	78.5
				Green tea	A*	113.1	142.9	106.4	101.4	92.0	92.1	98.2	117.7	110.8	123.9	138.6	112.1	106.6	123.2	106.0	112.6	80.7	84.1	94.5	108.3
				Woolong tea	B*	102.8	114.5	84.8	82.5	86.5	75.4	75.0	93.1	91.1	100.2	106.3	98.4	88.5	94.5	85.2	88.8	70.0	83.1	78.1	89.4
		Envi-Carb+PSA	GC-MS	Green tea	A	94.2	91.7	99.4	67.3	86.8	79.2	84.3	76.0	99.8	88.3	91.0	109.4	97.0	117.5	105.0	89.6	108.5	62.8	65.4	90.2
				Woolong tea	B	79.7	100.8	89.1	59.0	83.3	81.1	70.8	73.6	92.7	83.8	86.2	104.5	96.2	97.9	106.7	68.2	97.8	66.4	69.9	83.6
				Green tea	A*	173.0	143.6	132.2	137.8	173.1	111.6	132.3	109.7	120.2	317.6	322.3	242.5	236.2	124.3	141.4	96.3	93.5	110.3	93.8	158.5
				Woolong tea	B*	151.5	129.8	127.3	122.4	134.8	97.7	120.5	99.0	94.6	110.7	132.9	103.2	102.2	111.3	111.3	82.8	85.2	90.4	87.9	110.7
			GC-MS/MS	Green tea	A	80.5	107.7	88.6	76.0	87.6	85.5	87.8	71.6	148.9	40.9	101.3	212.9	71.2	99.7	79.4	75.2	66.4	63.2	81.2	90.8
				Woolong tea	B	69.5	105.9	81.0	64.0	84.1	83.6	20.9	65.8	124.3	36.8	88.5	199.8	82.7	76.5	79.1	58.4	61.4	60.0	78.8	80.1
				Green tea	A*	151.4	157.0	147.3	114.4	126.2	93.3	119.6	91.3	120.2	113.6	83.4	86.6	84.3	93.8	114.3	73.5	66.3	95.4	75.0	105.6
				Woolong tea	B*	113.4	131.9	108.7	93.5	113.1	79.2	106.4	79.3	94.6	77.0	80.5	66.5	80.9	84.7	67.1	73.1	63.8	69.3	73.0	87.2
17	cyprodinil	Cleanert TPT	GC-MS	Green tea	A	48.8	37.5	40.5	39.3	30.5	31.3	30.1	31.3	35.2	30.7	43.0	30.5	35.4	29.8	29.8	23.7	27.4	24.6	32.8	33.3
				Woolong tea	B	45.1	35.9	34.1	34.6	26.5	30.5	27.3	29.6	35.6	27.4	38.3	30.4	32.1	28.5	26.8	23.3	26.6	22.7	28.1	30.7
				Green tea	A*	41.1	41.8	34.8	37.5	33.2	30.3	31.8	40.1	39.4	40.7	37.3	42.2	39.3	37.5	39.6	36.0	33.2	34.3	30.6	36.9
				Woolong tea	B*	39.1	37.7	29.5	31.5	30.8	24.5	25.2	37.5	33.0	33.0	29.1	34.1	33.5	35.5	32.1	31.6	26.2	31.1	29.3	31.8
			GC-MS/MS	Green tea	A	46.2	36.9	40.7	40.2	27.6	44.7	32.1	28.3	34.9	31.6	34.6	32.5	30.9	31.7	29.9	24.7	26.6	23.0	34.3	33.2
				Woolong tea	B	44.3	35.5	34.7	35.6	25.8	38.4	29.3	27.2	33.3	30.2	31.1	31.0	29.0	31.7	28.4	24.6	25.8	31.0	28.9	30.9
				Green tea	A*	37.6	39.8	34.3	35.3	33.8	30.6	29.5	39.9	39.4	39.1	40.5	36.9	31.9	37.4	32.6	32.7	30.0	31.0	28.6	34.8
				Woolong tea	B*	36.6	34.5	29.7	30.6	31.5	26.3	25.9	31.1	33.2	32.0	30.0	31.3	28.0	31.8	27.6	28.1	25.3	28.7	26.4	29.9
		Envi-Carb+PSA	GC-MS	Green tea	A	40.3	41.2	40.7	41.8	43.2	41.9	34.6	34.1	35.1	71.9	32.2	35.1	33.1	35.2	29.8	37.1	30.5	31.1	27.9	37.7
				Woolong tea	B	39.2	36.6	38.9	36.1	39.7	36.8	33.8	32.6	33.0	56.7	28.9	32.2	29.9	30.6	26.5	29.4	27.0	26.8	29.6	33.9
				Green tea	A*	48.8	47.3	36.5	33.1	42.6	41.2	36.6	40.8	35.3	41.0	43.5	31.4	35.8	37.8	37.4	30.9	31.2	27.7	30.1	37.3
				Woolong tea	B*	41.7	40.1	35.7	30.7	34.8	34.3	31.2	28.7	28.7	30.5	33.7	29.8	31.2	31.7	30.7	24.7	26.1	21.1	27.8	31.2
			GC-MS/MS	Green tea	A	37.8	42.8	38.0	42.6	41.3	43.3	33.6	33.2	33.9	35.2	28.3	34.4	24.5	34.2	31.2	33.8	34.8	29.1	31.3	34.9
				Woolong tea	B	35.8	35.2	36.8	37.4	39.9	37.3	32.7	30.8	32.9	34.7	26.9	32.1	26.7	31.6	27.8	28.3	30.6	24.8	31.2	32.3
				Green tea	A*	53.7	41.8	46.3	36.1	39.9	34.4	37.2	33.6	35.3	36.9	33.5	28.2	29.8	30.8	33.5	23.9	23.7	30.1	25.9	34.5
				Woolong tea	B*	44.1	36.4	34.8	30.3	32.1	29.2	33.9	29.2	28.7	28.7	29.4	25.8	27.6	27.3	25.6	20.6	21.4	23.9	24.9	29.2
18	dacthal	Cleanert TPT	GC-MS	Green tea	A	126.3	102.1	104.6	113.6	86.3	100.6	90.3	88.0	103.3	94.9	101.6	101.0	96.3	94.5	95.8	104.4	82.1	76.4	98.0	97.9
				Woolong tea	B	123.1	98.9	94.4	104.3	81.5	92.7	83.4	83.4	104.0	92.1	91.6	93.8	90.4	91.2	90.6	100.1	81.6	73.8	86.7	92.5
				Green tea	A*	107.7	119.9	101.7	106.8	99.9	87.8	105.8	116.2	109.8	105.9	123.4	110.9	106.3	102.2	96.3	101.9	107.7	93.4	90.9	105.0
				Woolong tea	B*	101.2	99.6	84.2	90.5	89.6	73.3	77.3	101.3	92.6	91.5	93.2	88.6	86.0	91.1	78.9	86.7	87.1	85.1	81.5	88.4
			GC-MS/MS	Green tea	A	131.5	105.6	103.1	113.4	86.4	96.4	93.5	84.9	106.5	90.0	108.5	88.2	98.9	95.7	90.7	80.0	77.7	73.3	110.3	96.6
				Woolong tea	B	130.9	102.2	96.9	99.4	82.3	89.7	86.4	81.8	106.2	87.8	98.8	85.2	91.0	95.9	86.7	79.8	73.9	72.1	91.7	91.5
				Green tea	A*	132.5	122.8	100.1	107.4	103.1	91.5	103.0	119.5	113.0	107.6	126.5	109.5	107.8	108.0	97.7	98.9	88.7	89.7	87.2	106.0
				Woolong tea	B*	113.6	104.0	84.2	88.7	90.7	76.8	76.2	102.9	90.9	91.5	91.5	87.3	85.6	94.7	79.1	86.6	72.7	82.8	76.8	88.2
		Envi-Carb+PSA	GC-MS	Green tea	A	101.7	119.0	107.1	102.5	115.6	111.4	104.9	100.5	96.9	82.6	99.0	107.9	102.0	102.0	99.0	99.3	109.6	81.3	87.8	101.2
				Woolong tea	B	100.2	99.8	106.3	94.5	110.9	103.4	96.0	89.2	92.8	84.5	90.8	102.5	87.5	96.8	91.7	85.3	97.7	80.5	87.4	94.6
				Green tea	A*	127.4	113.5	107.9	101.5	116.1	100.6	109.0	108.5	110.0	104.8	97.5	92.1	94.2	97.7	100.9	81.6	84.0	84.5	85.6	100.9
				Woolong tea	B*	103.6	96.2	94.3	90.5	90.9	86.5	94.2	90.6	87.0	84.7	83.0	78.0	80.9	82.3	81.1	66.5	71.5	70.0	76.1	84.6
			GC-MS/MS	Green tea	A*	112.4	113.1	107.2	107.5	115.9	115.9	102.4	98.5	97.4	100.3	97.4	34.2	91.3	108.1	93.9	100.8	97.8	84.8	92.2	98.5
				Woolong tea	B*	106.3	99.5	106.6	98.2	113.5	106.3	103.0	91.8	98.5	98.6	87.4	31.8	85.5	97.4	87.0	87.3	91.2	76.8	93.5	92.6
				Green tea	A*	120.1	118.4	129.8	101.7	110.9	102.0	107.6	101.2	110.0	103.5	96.4	87.6	94.1	85.8	100.6	80.0	70.8	89.8	87.8	99.9
				Woolong tea	B	101.8	100.6	96.4	88.3	89.0	86.4	95.2	83.3	87.0	82.2	85.9	76.6	78.4	74.7	77.0	64.5	65.8	70.8	80.5	83.4

(Continued)

Annex 1: Content change for 201 pesticides in incurred tea Youden pair samples under 8 conditions with two SPE cartridge cleanup in three months (Nov.9, 2009-Feb.7, 2010) (cont.)

No	Pesticides	SPE	Method	Sample	Youden pair	Nov.9, 2009 (n=5)	Nov.14, 2009 (n=3)	Nov.19, 2009 (n=3)	Nov.24, 2009 (n=3)	Nov.29, 2009 (n=3)	Dec.4, 2009 (n=3)	Dec.9, 2009 (n=3)	Dec.14, 2009 (n=3)	Dec.14, 2009 (n=3)	Dec.19, 2009 (n=3)	Dec.24, 2009 (n=3)	Dec.29, 2009 (n=3)	Jan.3, 2010 (n=3)	Jan.8, 2010 (n=3)	Jan.13, 2010 (n=3)	Jan.18, 2010 (n=3)	Jan.23, 2010 (n=3)	Jan.28, 2010 (n=3)	Feb.2, 2010 (n=3)	Feb.7, 2010 (n=3)	AVE µg/kg
19	DE-PCB101	Cleanert TPT	GC-MS	Green tea	A	42.0	31.7	34.0	35.9	28.1	32.4	29.8	28.2	30.8	30.3	31.5	30.8	31.5	32.0	30.2	29.3	20.4	27.0	24.9	32.6	30.7
					B	39.3	30.7	30.6	33.5	26.2	29.6	27.3	26.2	31.3	28.6	27.7	31.3	29.7	30.4	28.5	25.7	20.6	26.2	24.2	28.5	28.7
				Woolong tea	A	35.0	40.0	32.3	34.2	31.4	28.9	29.1	35.6	35.8	36.4	37.3	35.8	35.8	36.1	35.0	31.4	33.4	30.2	31.3	31.7	33.7
					B	33.1	33.1	26.3	29.0	28.9	24.7	23.8	30.1	30.7	31.0	29.4	30.7	29.7	27.8	31.2	26.4	28.8	25.6	28.6	27.4	28.7
			GC-MS/MS	Green tea	A	40.5	33.2	36.2	31.1	26.6	28.9	23.8	27.4	31.9	29.4	32.4	31.9	30.8	30.0	30.4	28.8	24.4	26.4	24.0	31.2	30.2
					B	39.8	32.3	31.2	33.2	25.4	27.3	27.3	25.7	30.5	27.7	28.9	30.5	28.7	28.5	29.1	27.5	23.5	25.4	23.9	28.2	28.6
				Woolong tea	A	34.5	38.5	32.9	34.4	31.8	28.4	30.1	35.1	37.4	34.3	37.2	37.4	37.4	34.0	36.5	29.7	31.3	27.6	30.3	30.5	33.3
					B	33.0	31.5	27.5	29.6	29.3	24.1	23.9	30.0	31.0	29.8	27.4	31.0	29.6	27.2	32.0	25.3	27.4	24.5	28.0	26.9	28.3
		Envi-Carb +PSA	GC-MS	Green tea	A	37.1	37.2	33.9	35.1	36.9	35.1	31.4	32.0	32.0	25.3	29.2	32.0	34.0	31.1	35.8	32.0	35.9	31.1	32.1	29.3	33.0
					B	33.3	31.3	33.5	32.4	35.9	33.2	30.7	30.5	30.3	26.9	27.3	30.3	32.4	29.3	31.6	29.5	29.8	28.2	27.7	28.8	30.7
				Woolong tea	A	44.3	38.4	35.1	34.3	39.1	32.5	37.2	28.0	37.7	35.5	32.8	37.7	29.5	28.8	32.9	33.7	27.3	25.2	28.2	26.4	33.0
					B	36.0	32.1	31.3	29.6	31.4	27.9	32.2	25.1	28.4	28.7	27.4	28.4	25.2	25.5	26.1	25.8	23.6	21.6	23.0	24.1	27.6
			GC-MS/MS	Green tea	A	33.6	37.5	34.3	34.0	39.2	35.9	32.2	31.1	31.6	32.8	28.2	31.6	33.1	30.3	35.3	32.2	33.2	31.8	28.5	30.0	32.9
					B	30.2	30.4	33.2	31.3	38.3	32.7	31.9	29.1	30.7	32.3	27.2	30.7	30.8	26.9	31.0	28.5	27.9	30.4	25.0	30.0	30.3
				Woolong tea	A	49.0	38.2	42.2	33.6	37.2	32.4	35.0	30.9	37.7	34.1	31.7	37.7	29.3	28.9	30.3	32.5	24.5	21.2	27.7	23.7	32.6
					B	41.2	32.6	31.9	29.1	29.9	27.4	32.0	26.5	28.4	26.8	27.9	28.4	25.1	25.1	26.5	24.5	20.4	19.4	22.0	22.6	27.3
20	DE-PCB118	Cleanert TPT	GC-MS	Green tea	A	42.9	32.7	35.0	36.4	29.4	33.2	30.4	28.4	31.9	31.8	31.8	31.9	32.2	32.5	28.6	28.6	21.6	28.3	26.5	34.4	31.4
					B	40.3	31.4	31.3	34.2	26.6	30.1	27.3	26.6	33.2	30.5	28.5	33.2	30.4	29.6	28.4	27.5	21.4	26.7	25.3	29.8	29.4
				Woolong tea	A	35.5	40.6	32.9	34.7	33.6	30.4	30.8	35.1	37.2	35.8	38.5	37.2	37.4	33.7	35.3	31.1	34.4	31.0	31.7	31.8	34.3
					B	33.8	33.9	27.4	29.6	30.5	25.4	25.2	29.9	31.9	32.1	30.8	31.9	30.8	26.7	31.4	27.1	29.7	26.0	29.7	28.5	29.5
			GC-MS/MS	Green tea	A	40.6	35.0	37.3	32.3	27.8	31.0	30.2	27.2	33.1	30.2	27.6	33.1	31.1	32.3	34.1	29.8	29.7	27.2	23.9	34.9	31.3
					B	39.4	33.5	32.0	32.7	26.1	28.7	27.8	25.7	31.8	28.0	27.6	31.8	28.8	30.5	33.2	28.8	24.3	26.8	23.8	30.2	29.5
				Woolong tea	A	36.7	39.3	33.0	35.6	33.1	29.0	30.2	35.2	37.1	35.3	39.1	37.1	36.9	35.1	35.7	31.6	32.5	27.9	29.3	29.9	33.8
					B	34.9	33.6	28.5	31.1	31.1	24.8	25.1	31.5	30.5	30.0	29.4	30.5	29.9	27.8	31.9	26.6	28.9	23.3	27.9	26.5	29.1
		Envi-Carb +PSA	GC-MS	Green tea	A	37.9	37.0	34.3	36.3	37.2	36.2	31.6	32.4	32.9	26.4	29.9	32.9	34.6	31.5	35.7	32.7	36.6	32.6	32.7	30.0	33.6
					B	34.7	31.8	34.0	33.4	36.1	33.8	31.1	31.1	31.5	27.6	29.1	31.5	33.1	28.0	33.4	29.7	31.5	29.5	27.8	28.8	31.4
				Woolong tea	A	46.9	38.9	33.8	35.8	38.2	33.1	37.6	27.0	36.7	35.4	33.0	36.7	28.2	28.9	33.1	34.6	28.2	26.6	27.8	27.7	33.2
					B	38.8	33.2	31.5	31.1	30.9	28.7	32.5	24.3	29.1	27.6	27.7	29.1	24.1	25.8	27.2	26.6	23.5	22.8	23.7	24.4	28.1
			GC-MS/MS	Green tea	A	34.2	38.7	35.0	35.3	38.8	37.1	32.7	32.3	33.7	33.7	28.2	33.7	33.1	31.7	38.4	32.5	35.0	32.8	30.0	31.3	33.9
					B	31.2	32.0	35.1	32.2	37.6	33.5	30.4	30.1	31.3	33.3	27.3	31.3	30.2	28.3	34.1	30.4	28.4	29.4	25.8	30.7	31.1
				Woolong tea	A	48.9	38.5	40.5	34.1	37.4	33.4	35.8	30.9	36.7	34.8	31.2	36.7	30.0	28.8	30.0	33.3	23.7	23.2	28.9	26.4	33.0
					B	42.8	33.4	33.1	30.0	31.6	28.2	32.6	27.0	29.1	27.6	26.8	29.1	25.4	24.7	25.9	25.5	19.9	20.8	23.2	24.7	28.0
21	DE-PCB138	Cleanert TPT	GC-MS	Green tea	A	42.1	37.1	33.8	35.7	36.0	35.1	31.4	31.0	31.5	24.5	28.5	31.5	33.8	30.8	30.0	28.5	20.4	26.9	24.8	32.2	30.5
					B	40.0	32.1	33.4	32.4	35.3	32.1	30.6	30.2	29.9	25.6	27.4	29.9	31.9	27.1	28.3	25.9	20.2	26.5	23.6	28.4	28.5
				Woolong tea	A	35.6	38.0	34.4	35.1	35.9	32.4	36.7	30.2	35.5	33.8	31.5	35.5	29.0	29.2	33.8	31.4	33.0	30.2	30.7	29.6	33.2
					B	33.9	33.0	32.6	30.7	28.7	28.0	31.9	26.6	28.2	27.6	27.2	28.2	25.0	25.5	30.1	26.3	28.0	24.8	28.5	26.7	28.7
			GC-MS/MS	Green tea	A	39.8	34.5	35.8	30.7	37.7	34.4	32.8	32.1	32.3	33.8	27.2	32.3	36.1	29.0	35.2	29.8	23.4	25.8	24.3	33.3	30.3
					B	38.2	33.4	32.6	26.8	36.8	32.8	29.5	29.7	29.7	27.6	27.2	29.7	34.4	26.4	31.1	28.6	23.2	26.0	23.6	28.6	29.0
				Woolong tea	A	37.3	37.0	33.0	34.9	30.4	29.9	29.1	31.2	31.7	32.5	31.5	31.7	36.6	29.0	35.2	30.7	30.7	26.0	29.1	28.6	33.0
					B	35.5	32.2	27.9	29.0	30.4	25.7	24.8	31.0	31.4	32.1	30.6	31.4	29.9	26.9	29.7	26.6	26.8	24.2	28.2	24.7	28.5
		Envi-Carb +PSA	GC-MS	Green tea	A	37.6	37.1	33.8	35.7	36.0	35.1	31.4	31.0	31.5	24.5	28.5	31.5	33.8	30.0	34.9	31.1	34.9	31.1	31.0	28.5	32.5
					B	34.6	32.1	33.4	32.4	35.3	32.1	30.6	30.2	29.9	25.6	27.4	29.9	31.9	27.1	31.1	28.3	28.4	27.8	26.1	28.2	30.1
				Woolong tea	A	47.0	38.0	34.4	35.1	35.9	32.4	36.7	30.2	35.5	33.8	31.5	35.5	29.0	29.2	31.0	34.3	26.5	26.3	28.2	28.0	32.6
					B	39.5	33.0	32.6	30.7	28.7	28.0	31.9	26.6	28.2	27.6	27.2	28.2	25.0	25.5	25.8	26.8	22.2	22.3	23.4	25.3	27.8
			GC-MS/MS	Green tea	A	32.0	36.9	35.8	34.8	37.7	34.4	32.8	32.1	32.3	33.8	27.2	32.3	36.1	29.0	35.2	29.8	33.8	32.0	29.5	28.8	32.7
					B	30.1	30.4	35.1	31.9	36.8	32.8	29.5	29.7	26.4	26.4	26.4	26.4	34.4	26.4	31.1	28.6	27.3	29.0	25.2	28.4	30.2
				Woolong tea	A	48.9	37.7	39.3	31.9	35.7	32.0	34.3	30.5	29.9	34.9	29.5	29.9	29.1	26.5	29.2	32.2	24.9	21.6	28.0	26.3	32.0
					B	42.8	34.0	33.8	28.5	29.9	27.7	31.1	26.8	28.2	27.4	26.4	28.2	24.9	24.0	26.1	24.0	20.9	20.4	21.9	24.2	27.5

No.	Compound	Cartridge	Method	Tea	A/B																				
22	DE-PCB180	Cleanert TPT	GC-MS	Green tea	A	40.0	32.3	34.1	35.2	27.5	29.9	28.6	26.4	30.4	28.1	30.5	31.2	30.8	28.2	27.9	21.7	25.6	24.1	31.1	29.7
					B	39.3	31.0	30.6	31.2	24.7	27.4	26.5	25.3	30.3	27.4	27.1	28.6	28.4	27.0	25.3	20.9	24.9	22.4	27.9	27.7
				Woolong tea	A	35.3	37.7	32.0	34.0	31.5	28.4	27.9	33.6	35.0	33.3	37.3	33.3	32.9	32.4	31.1	32.7	31.0	29.2	28.4	32.5
					B	33.5	32.3	27.5	29.0	30.2	24.8	24.8	30.4	29.9	29.4	29.5	28.7	27.5	28.8	26.1	27.6	25.0	26.9	25.6	28.3
			GC-MS/MS	Green tea	A	40.6	33.9	36.3	35.7	28.4	29.0	27.5	25.9	31.3	26.7	32.1	28.6	29.1	30.8	27.2	22.1	25.1	23.0	31.9	29.7
					B	40.1	32.5	31.8	31.7	25.1	26.2	25.4	25.3	30.1	26.4	28.1	28.5	27.3	30.6	25.2	21.5	24.3	22.4	27.1	27.9
				Woolong tea	A	36.2	35.5	31.2	35.7	31.8	27.2	28.6	34.1	36.6	32.4	35.9	34.0	30.7	33.7	28.7	29.6	27.0	28.5	27.6	31.8
					B	34.6	32.4	26.4	29.7	30.0	23.9	25.0	32.8	30.1	28.4	27.9	29.2	26.9	29.1	24.6	25.3	22.7	26.2	24.9	27.9
		Envi-Carb +PSA	GC-MS	Green tea	A	36.7	36.6	33.8	34.1	35.3	34.3	37.7	30.2	30.5	23.7	28.1	32.3	28.4	32.2	30.3	30.0	31.5	29.6	27.4	31.7
					B	34.9	32.6	33.8	30.5	34.6	31.1	29.5	29.7	28.9	24.9	26.6	30.5	26.5	30.9	27.3	23.9	27.6	25.1	27.6	29.3
				Woolong tea	A	46.8	35.7	33.5	33.8	37.4	27.2	35.6	30.7	34.8	32.9	31.0	28.9	29.7	30.5	33.7	26.4	26.3	28.3	25.4	32.3
					B	40.0	31.8	32.4	30.2	31.2	35.4	30.6	27.3	28.2	26.9	26.6	24.6	25.6	25.2	26.2	21.6	22.0	23.7	23.0	27.6
			GC-MS/MS	Green tea	A	33.7	37.4	34.6	34.0	37.1	31.6	32.2	30.5	31.4	31.1	27.5	36.1	28.8	38.5	30.6	33.2	31.7	22.9	28.2	32.6
					B	30.4	31.1	34.1	31.8	36.6	30.0	30.3	27.7	29.7	30.0	26.9	34.4	26.8	32.7	28.2	27.2	28.1	27.4	24.7	30.0
				Woolong tea	A	48.4	37.2	37.9	31.7	33.9	25.8	31.7	30.7	34.8	33.6	29.7	26.4	28.3	28.0	32.9	23.6	21.2	21.3	22.8	31.2
					B	43.8	33.3	33.7	28.1	28.4	37.2	29.5	28.2	28.2	26.0	25.7	24.1	23.2	24.4	23.3	19.5	19.3	27.1	35.5	26.8
23	DE-PCB28	Cleanert TPT	GC-MS	Green tea	A	41.9	33.9	38.6	41.4	32.6	33.3	32.6	30.9	33.8	33.8	33.3	33.6	33.2	33.3	30.8	22.7	30.7	26.0	30.9	33.5
					B	40.0	33.4	32.8	38.4	26.8	29.9	29.5	29.0	35.0	32.2	30.1	31.1	31.3	32.1	28.4	22.2	28.9	33.0	33.3	31.1
				Woolong tea	A	37.0	40.3	33.2	35.3	34.4	25.3	30.9	37.2	38.5	37.3	40.3	36.6	37.0	36.1	33.1	34.5	30.6	29.9	29.3	35.2
					B	34.8	33.6	27.5	30.6	31.5	32.8	25.4	31.6	32.8	31.8	32.0	30.8	30.1	32.3	27.8	30.0	25.8	26.7	35.7	30.2
			GC-MS/MS	Green tea	A	42.1	34.8	39.3	39.6	28.8	32.7	32.7	29.6	34.7	31.5	35.2	33.5	32.7	30.4	32.1	25.8	29.2	26.4	35.8	33.0
					B	40.8	33.6	34.1	35.1	27.6	30.9	30.1	28.6	34.4	30.4	26.9	31.5	31.1	30.1	30.6	25.3	28.6	31.6	30.8	31.2
				Woolong tea	A	34.0	37.1	33.9	35.7	34.0	31.3	31.8	36.9	37.1	36.2	40.2	35.8	33.7	37.6	32.2	33.3	28.7	31.6	31.4	34.3
					B	30.6	30.9	28.5	30.5	31.1	26.6	25.5	32.4	31.7	31.7	30.8	30.2	31.3	32.5	26.7	29.5	25.6	28.9	27.0	29.6
		Envi-Carb +PSA	GC-MS	Green tea	A*	39.7	37.5	34.7	37.4	39.9	36.7	32.9	34.8	35.6	31.5	33.4	36.3	34.1	38.9	35.0	38.4	34.4	35.0	31.7	35.7
					B	34.8	32.1	35.3	34.4	38.7	33.5	33.4	33.4	32.6	31.4	31.1	35.1	31.5	35.2	32.3	32.6	31.2	29.5	31.1	33.1
				Woolong tea	A	45.6	38.4	36.1	38.6	36.1	34.9	39.2	28.9	38.7	35.5	33.6	31.1	30.5	32.9	36.3	29.2	28.8	30.8	29.3	34.4
					B	36.0	32.3	36.2	34.2	29.8	29.3	33.2	27.4	31.4	29.1	28.5	26.9	26.8	27.9	28.6	24.3	24.8	25.7	26.3	29.2
			GC-MS/MS	Green tea	A	34.5	38.9	36.9	37.8	41.1	38.4	35.2	34.6	33.9	37.8	31.0	57.0	29.1	39.0	34.8	37.2	36.5	32.6	32.7	36.8
					B	31.5	31.9	36.1	34.1	39.8	34.4	33.0	32.4	32.9	36.3	30.2	44.8	29.2	35.5	32.4	31.6	33.4	28.1	32.1	33.7
				Woolong tea	A	59.6	39.2	43.1	34.4	38.3	34.2	36.7	32.2	38.7	35.7	32.1	28.6	29.7	31.1	34.3	26.8	23.1	29.8	27.1	34.5
					B	50.1	33.3	32.0	30.2	31.6	29.0	32.6	28.2	31.4	28.7	28.6	26.3	26.3	27.1	25.8	22.6	21.3	24.1	25.3	29.2
24	DE-PCB31	Cleanert TPT	GC-MS	Green tea	A	43.2	33.8	38.3	40.5	30.9	35.3	32.7	31.0	33.9	32.4	34.0	34.2	33.7	33.0	31.0	22.3	30.4	27.7	35.4	33.4
					B	41.1	32.7	32.9	38.2	27.9	31.7	29.8	28.8	34.9	31.0	30.5	32.1	32.2	36.0	28.6	21.8	28.5	25.9	30.9	31.1
				Woolong tea	A	36.3	39.2	33.3	35.5	34.1	30.1	30.4	36.7	37.9	36.9	39.1	37.8	36.1	31.9	28.1	34.2	31.2	32.9	29.7	35.0
					B	34.2	32.8	27.5	30.6	31.3	25.3	25.1	31.1	32.8	32.0	31.3	31.7	29.2	30.4	32.1	29.7	26.4	29.7	29.0	30.0
			GC-MS/MS	Green tea	A	42.1	34.8	39.3	39.6	28.8	32.8	32.7	29.6	34.7	31.5	35.2	33.5	32.7	32.1	30.6	25.8	29.2	26.7	35.7	33.0
					B	40.8	33.6	34.1	35.1	27.6	30.9	30.1	28.6	34.4	30.4	32.1	31.5	31.1	30.1	30.6	25.3	28.6	26.4	30.8	31.2
				Woolong tea	A	35.3	38.0	33.9	35.7	34.0	31.3	31.8	36.9	37.1	36.2	40.2	35.8	33.7	37.6	32.2	33.3	28.7	31.6	31.4	34.4
					B	32.8	33.6	28.5	30.5	31.1	26.6	25.5	32.4	31.7	31.7	30.8	30.2	31.3	32.5	26.7	29.5	25.6	28.9	27.0	29.8
		Envi-Carb +PSA	GC-MS	Green tea	A	39.4	37.6	35.1	37.3	39.9	37.3	28.7	35.0	35.4	32.8	31.1	37.6	33.4	39.3	35.1	38.8	33.7	34.8	27.0	29.8
					B	34.5	31.8	35.5	34.4	38.9	34.2	27.9	33.7	32.4	32.8	32.8	35.5	31.2	35.7	32.6	33.0	31.6	31.0	31.6	29.5
				Woolong tea	A	44.2	38.6	36.0	38.5	35.2	34.4	38.5	29.3	38.7	35.5	31.1	31.5	30.2	32.9	35.9	28.9	28.9	30.8	31.0	34.2
					B	35.1	32.3	32.1	34.2	29.1	29.2	33.3	27.7	31.4	28.7	32.8	26.7	26.3	27.7	28.3	24.3	24.3	25.0	29.5	29.0
			GC-MS/MS	Green tea	A	34.5	38.9	36.9	37.8	41.1	38.4	35.2	34.6	33.9	37.8	28.2	36.0	29.1	39.0	34.8	37.2	36.5	32.6	26.6	35.7
					B	31.5	31.9	36.1	34.1	39.8	34.4	33.0	32.4	32.9	36.3	31.0	34.7	29.2	35.5	32.4	31.6	33.4	28.1	32.7	33.1
				Woolong tea	A	37.0	39.2	42.8	34.4	38.3	34.2	36.7	32.2	38.7	35.7	30.2	28.6	29.7	31.1	34.3	26.8	23.1	29.8	27.1	33.3
					B	31.3	33.3	32.0	30.2	31.6	29.0	32.6	28.2	31.4	28.7	28.6	26.3	26.3	27.1	25.8	22.6	21.3	24.1	25.3	28.2

(Continued)

Annex 1: Content change for 201 pesticides in incurred tea Youden pair samples under 8 conditions with two SPE cartridge cleanup in three months (Nov.9, 2009–Feb.7, 2010) (cont.)

No	Pesticides	SPE	Method	Sample	Youden pair	Nov.9, 2009 (n=5)	Nov.14, 2009 (n=3)	Nov.19, 2009 (n=3)	Nov.24, 2009 (n=3)	Nov.29, 2009 (n=3)	Dec.4, 2009 (n=3)	Dec.9, 2009 (n=3)	Dec.14, 2009 (n=3)	Dec.19, 2009 (n=3)	Dec.24, 2009 (n=3)	Dec.14, 2009 (n=3)	Jan.3, 2010 (n=3)	Jan.8, 2010 (n=3)	Jan.13, 2010 (n=3)	Jan.18, 2010 (n=3)	Jan.23, 2010 (n=3)	Jan.28, 2010 (n=3)	Feb.2, 2010 (n=3)	Feb.7, 2010 (n=3)	AVE µg/kg.
25	dichlorofop-methyl	Cleanert TPT	GC-MS	Green tea	A	62.8	44.3	30.1	30.5	30.9	42.4	27.4	28.8	30.9	30.8	52.0	48.6	35.3	32.6	33.5	68.5	24.9	29.0	32.8	37.7
				Woolong tea	B*	58.7	41.4	28.7	28.0	34.4	36.8	25.4	26.9	38.9	29.2	48.5	41.0	31.9	32.0	36.4	63.5	25.0	27.4	28.2	35.9
				Green tea	A*	39.2	46.2	39.0	41.2	52.3	31.6	30.1	36.5	46.0	31.5	82.4	28.3	34.5	32.3	33.6	36.8	36.5	32.5	26.1	38.8
				Woolong tea	B*	35.6	38.9	34.3	31.0	41.5	22.9	32.3	31.9	36.2	33.0	53.7	23.7	29.9	27.8	24.5	27.8	26.9	25.1	22.4	31.5
			GC-MS/MS	Green tea	A	71.9	44.9	36.0	25.8	34.0	78.6	28.1	28.6	28.8	36.0	51.3	32.0	38.5	33.0	32.3	47.3	24.5	22.9	51.2	39.2
				Woolong tea	B	69.4	40.5	32.5	32.3	35.3	68.6	24.7	26.9	35.8	38.8	50.2	27.6	34.1	38.5	34.9	46.9	25.7	23.3	39.1	38.2
				Green tea	A*	45.3	48.5	42.3	44.0	44.1	32.6	31.0	41.6	45.1	32.3	62.0	33.0	32.4	37.3	35.1	36.8	31.3	30.4	26.0	38.5
				Woolong tea	B	40.6	43.0	36.9	33.0	37.2	25.2	30.5	34.9	37.7	30.7	42.6	25.3	27.2	30.6	26.9	27.3	23.4	22.4	20.8	31.4
		Envi-Carb +PSA	GC-MS	Green tea	A	48.8	46.4	56.9	25.5	40.9	38.7	39.9	42.2	45.4	46.8	60.6	55.5	38.5	40.1	39.8	33.1	47.1	30.0	33.1	42.4
				Woolong tea	B	39.4	59.2	42.4	21.9	37.6	42.9	28.3	42.2	42.1	41.4	48.8	47.6	35.0	34.7	38.4	25.2	33.4	25.4	29.6	38.1
				Green tea	A*	46.4	51.6	41.6	36.5	59.7	34.4	41.4	46.5	43.6	32.6	34.6	36.1	34.0	52.1	28.5	28.4	27.3	20.9	23.1	38.4
				Woolong tea	B*	40.9	42.4	39.2	33.0	44.8	31.8	30.2	37.7	30.1	27.4	27.4	30.5	24.2	44.8	23.8	17.6	21.7	26.5	23.1	31.1
			GC-MS/MS	Green tea	A	44.8	60.4	45.1	28.4	42.4	38.2	54.3	42.3	55.1	36.3	65.5	37.1	34.8	42.1	40.4	34.9	27.1	26.5	44.9	37.0
				Woolong tea	B	34.6	66.9	38.2	25.5	38.7	42.0	20.1	35.6	44.9	32.1	53.2	34.4	43.9	27.3	44.3	24.5	25.1	30.1	41.9	36.3
				Green tea	A*	44.3	52.1	41.4	33.0	28.7	32.2	34.1	34.0	43.6	37.3	36.3	31.9	29.6	26.6	21.8	25.0	36.6	51.8	35.1	30.5
				Woolong tea	B	44.6	44.0	41.9	31.0	30.2	34.6	30.5	27.9	30.1	32.7	29.4	28.1	22.1	24.2	33.5	15.6	29.7	39.6	27.4	34.5
26	dimethenamid	Cleanert TPT	GC-MS	Green tea	A	43.8	37.8	41.5	48.7	30.2	34.6	32.4	34.2	37.2	29.6	33.3	31.2	32.5	36.0	30.8	34.5	35.8	26.7	30.3	32.4
				Woolong tea	B	42.7	35.9	35.4	45.7	28.1	32.6	27.6	30.3	35.3	28.6	28.8	31.2	30.4	32.6	33.5	25.3	27.6	28.6	27.6	36.0
				Green tea	A*	41.4	37.4	29.0	37.8	31.5	23.5	25.1	40.8	37.9	38.5	28.8	38.6	25.9	27.7	26.8	21.8	23.9	23.6	25.0	30.0
				Woolong tea	B	38.8	35.7	32.4	31.9	29.2	41.8	31.7	32.4	31.1	31.6	34.6	30.4	31.3	31.8	28.4	33.0	33.1	28.1	29.2	33.0
			GC-MS/MS	Green tea	A	40.4	34.8	33.3	34.4	27.9	38.2	28.3	30.1	35.5	29.8	31.8	27.8	34.9	28.2	30.8	29.6	31.7	25.9	36.8	30.9
				Woolong tea	B	40.2	34.8	34.6	36.1	27.0	30.3	36.6	44.2	35.0	29.8	43.7	37.0	34.9	38.2	32.1	35.1	23.7	25.6	33.0	36.5
				Green tea	A*	43.5	45.1	29.3	30.1	29.6	25.5	26.5	35.9	36.7	36.4	32.7	30.6	30.8	32.8	25.8	29.6	39.6	31.0	28.0	30.1
				Woolong tea	B	37.7	35.7	36.0	35.0	40.8	40.2	34.4	30.3	37.9	38.5	28.2	30.4	31.3	36.0	28.4	34.3	31.9	32.2	29.2	33.6
		Envi-Carb +PSA	GC-MS	Green tea	A	31.5	46.2	41.4	33.7	31.8	41.8	31.7	30.1	31.1	31.6	34.6	30.2	31.3	31.8	30.8	27.0	26.0	25.9	29.2	34.7
				Woolong tea	B	33.4	36.6	35.7	34.4	28.1	32.6	25.1	27.7	35.5	28.8	34.6	27.8	29.6	24.2	31.1	26.2	29.6	24.1	25.0	29.1
				Green tea	A*	41.6	39.5	32.3	31.8	29.2	29.2	34.0	31.9	32.2	28.7	28.8	26.9	25.9	27.7	26.8	21.8	23.9	23.6	30.0	33.7
				Woolong tea	B*	35.0	33.6	34.6	36.0	37.6	40.2	34.0	31.2	28.1	33.0	32.4	36.9	26.3	35.6	32.7	33.0	33.1	28.1	30.0	32.6
			GC-MS/MS	Green tea	A	38.5	39.0	34.6	34.2	37.4	35.6	38.1	29.7	32.6	34.1	29.4	35.2	27.5	30.3	30.7	27.9	29.7	25.5	31.6	34.1
				Woolong tea	B	39.0	34.7	57.7	33.1	38.7	33.4	33.7	32.9	41.7	34.6	30.9	26.0	31.2	29.1	32.6	25.4	23.4	31.4	28.3	27.6
				Green tea	A*	47.7	41.1	32.0	36.1	38.7	28.8	33.7	27.0	32.2	27.6	27.2	23.6	27.2	25.5	24.7	21.4	21.9	25.5	27.2	34.1
				Woolong tea	B	37.9	35.1	29.6	33.1	31.8	33.4	38.1	35.9	32.6	34.1	31.9	26.0	27.5	25.5	28.4	21.4	23.4	27.2	28.0	27.6
27	diofenolan-1	Cleanert TPT	GC-MS	Green tea	A	93.7	72.6	71.8	79.5	64.0	71.3	67.8	62.8	71.6	62.8	78.3	73.7	70.9	67.3	61.6	47.6	59.1	56.8	72.3	68.7
				Woolong tea	B	88.5	73.1	65.0	74.0	55.9	65.2	62.2	58.4	70.0	60.5	70.6	67.2	66.9	65.4	57.8	46.7	57.4	52.9	63.2	64.3
				Green tea	A*	81.7	75.3	68.5	73.0	72.8	60.1	61.9	73.7	75.8	69.8	82.9	67.2	71.7	72.9	71.6	69.7	64.8	68.2	62.9	70.9
				Woolong tea	B	76.5	65.1	59.6	63.2	67.8	53.6	55.1	65.7	60.8	60.6	63.7	60.1	61.5	63.0	59.6	58.9	52.6	62.1	56.6	61.4
			GC-MS/MS	Green tea	A	93.5	75.1	84.5	87.2	62.1	87.9	66.4	65.7	72.5	65.5	75.9	66.3	70.4	77.0	65.7	53.0	57.0	53.2	82.1	71.2
				Woolong tea	B	88.7	72.5	72.7	72.9	58.0	78.7	60.8	55.6	68.8	62.7	66.4	63.0	64.0	76.7	63.3	52.5	56.9	53.0	70.1	66.2
				Green tea	A*	76.5	78.7	70.1	75.4	69.9	61.8	62.7	78.1	78.8	76.2	83.2	78.1	64.3	78.0	68.0	68.8	58.4	63.0	61.5	71.1
				Woolong tea	B	73.7	70.8	61.8	65.7	67.1	54.9	54.4	69.9	64.2	64.7	63.0	63.9	63.6	68.1	55.8	59.0	49.1	56.5	53.1	62.1
		Envi-Carb +PSA	GC-MS	Green tea	A	81.6	81.8	82.4	78.4	81.7	84.4	65.2	71.9	72.7	94.2	59.7	75.3	70.8	81.4	72.5	78.7	74.8	69.2	63.7	75.8
				Woolong tea	B	76.9	71.0	76.9	71.1	79.8	73.9	65.1	69.5	66.4	90.1	59.6	69.2	64.7	72.3	64.5	64.7	66.0	58.3	62.9	69.6
				Green tea	A*	93.6	79.9	72.0	74.6	94.4	67.6	79.6	63.7	80.2	72.8	72.8	61.4	64.9	82.4	72.1	55.1	58.1	60.0	60.7	71.9
				Woolong tea	B*	82.3	69.6	70.0	68.5	73.6	57.2	66.2	59.0	66.5	59.4	58.7	53.6	56.6	63.4	55.6	45.9	48.4	50.6	53.7	61.0
			GC-MS/MS	Green tea	A	79.1	84.8	78.4	82.8	82.5	84.6	70.6	73.2	72.9	76.6	59.2	75.8	62.2	81.0	71.2	74.9	72.0	62.5	66.0	74.2
				Woolong tea	B	73.9	71.2	76.0	72.9	80.9	74.7	67.1	67.5	68.3	75.0	58.5	72.8	59.8	67.3	65.2	63.3	64.9	54.6	67.1	68.6
				Green tea	A*	103.5	79.9	86.4	72.1	76.7	67.2	74.9	65.1	80.2	72.5	66.7	57.1	54.9	62.2	67.3	49.8	56.9	78.6	59.1	70.1
				Woolong tea	B	88.1	72.1	72.1	64.0	64.7	57.4	66.4	57.4	66.5	58.2	58.1	52.4	50.4	54.1	51.5	42.2	50.0	54.0	62.0	60.1

No.	Compound	Cartridge	Method	Tea																					
28	diofenolan-2	Cleanert TPT	GC-MS	Green tea	A	93.4	73.3	77.6	79.9	65.9	74.0	68.5	64.5	73.2	65.7	77.6	77.1	72.5	69.0	64.2	48.6	61.2	58.4	74.5	70.5
				Green tea	B	87.7	71.2	70.2	74.9	59.1	66.2	62.5	60.2	70.6	63.9	68.5	70.7	67.4	65.8	58.5	47.5	59.3	54.4	64.6	65.4
				Woolong tea	A	81.7	85.4	70.8	75.6	76.1	62.4	62.7	76.2	76.2	75.4	86.2	74.8	73.8	73.3	74.0	71.9	66.3	68.0	64.9	73.5
				Woolong tea	B	77.4	72.6	61.1	64.7	71.9	54.7	56.5	66.5	63.7	65.1	65.4	63.2	62.4	64.6	61.4	59.8	54.2	61.7	56.4	63.3
			GC-MS/MS	Green tea	A	95.4	78.2	88.2	90.4	66.2	90.1	68.9	54.3	75.7	67.8	79.2	70.2	71.6	78.6	67.3	55.4	58.9	56.9	88.8	73.8
				Green tea	B	95.0	77.9	79.6	80.0	62.8	84.3	66.1	55.5	76.2	66.8	72.6	68.4	69.1	82.0	67.1	60.2	60.3	58.0	78.1	71.6
				Woolong tea	A	75.9	79.0	70.3	75.4	71.7	61.8	60.1	79.1	77.5	76.8	86.2	79.1	65.1	77.3	69.3	73.8	58.8	62.3	61.5	71.6
				Woolong tea	B	72.9	69.1	62.9	65.7	68.0	54.8	54.3	69.8	62.9	64.8	64.9	66.6	63.6	68.5	59.1	60.9	50.7	54.5	53.1	62.5
		Envi-Carb+PSA	GC-MS	Green tea	A	82.6	82.7	79.1	82.1	82.4	78.7	69.9	73.9	75.0	102.6	65.2	76.8	71.3	80.8	73.4	79.4	77.0	70.1	65.1	77.3
				Green tea	B*	77.8	73.1	77.2	73.5	80.9	72.3	70.2	70.0	68.6	98.5	62.1	71.5	64.0	71.7	65.7	65.3	66.8	59.2	64.5	77.3
				Woolong tea	A	99.9	82.0	73.6	76.3	91.0	69.8	80.8	65.1	79.6	73.1	70.2	63.8	66.3	82.6	71.9	57.7	59.8	61.3	61.8	71.2
				Woolong tea	B	87.6	71.7	72.0	67.5	73.0	60.1	66.2	61.1	66.1	59.5	58.7	54.6	57.2	65.9	56.2	48.0	49.7	51.8	55.1	73.0
			GC-MS/MS	Green tea	A*	77.4	89.2	80.2	83.5	82.4	86.6	73.8	77.0	76.8	78.5	63.4	52.2	63.0	134.2	69.3	74.7	73.1	67.5	67.9	62.2
				Green tea	B	67.9	73.7	77.6	74.0	80.2	75.2	68.4	70.7	69.7	77.4	59.7	59.3	60.9	104.9	63.6	62.9	64.9	57.2	68.0	77.4
				Woolong tea	A	102.2	80.0	88.2	72.1	0.0	66.7	74.9	65.1	79.6	74.0	67.7	0.0	57.6	0.0	70.5	49.8	62.4	83.9	62.2	70.3
				Woolong tea	B	85.1	72.2	72.1	64.0	0.0	57.6	66.8	57.4	66.1	60.0	58.5	0.0	52.1	0.0	54.2	42.2	55.5	64.5	58.1	60.9
29	fenbuconazole	Cleanert TPT	GC-MS	Green tea	A	101.4	81.5	89.3	80.9	73.6	69.5	64.8	67.5	79.0	62.3	104.0	74.1	80.2	60.0	61.6	46.8	61.9	58.6	73.5	51.9
				Green tea	B	93.5	79.2	81.5	77.8	62.2	67.6	62.0	60.1	75.0	58.6	78.9	69.3	77.9	63.9	51.7	45.4	56.3	51.8	67.2	73.2
				Woolong tea	A	95.2	77.6	74.3	76.5	74.3	72.2	58.7	75.8	78.6	68.1	92.6	77.7	73.7	69.2	68.4	67.8	61.1	62.7	60.0	67.4
				Woolong tea	B	84.8	78.1	70.5	74.7	75.5	63.3	58.6	74.1	61.6	60.6	69.9	67.1	61.8	61.4	56.5	51.7	50.6	54.8	47.7	72.9
			GC-MS/MS	Green tea	A	89.3	77.5	96.8	116.8	74.6	97.4	68.3	64.4	78.4	63.3	81.3	69.6	73.6	63.2	63.6	51.3	60.4	55.9	79.8	64.4
				Green tea	B	91.9	77.3	86.1	81.8	61.4	84.2	66.0	58.9	71.3	57.4	68.1	66.3	73.5	80.1	55.6	51.8	54.2	52.2	69.8	75.0
				Woolong tea	A*	89.6	85.8	79.7	79.0	66.7	66.6	62.3	76.1	72.6	77.8	82.6	80.1	63.5	76.9	68.4	66.9	55.3	58.5	57.6	68.8
				Woolong tea	B*	78.0	76.3	70.1	72.5	69.1	59.2	56.9	72.3	59.7	64.1	63.7	68.8	65.4	65.6	56.6	53.3	46.4	52.4	45.8	71.9
		Envi-Carb+PSA	GC-MS	Green tea	A	81.3	76.1	87.9	89.4	92.3	88.5	56.6	77.4	82.7	138.9	76.6	78.8	72.4	76.0	70.8	87.4	76.0	68.3	65.0	63.0
				Green tea	B	83.1	75.9	84.0	78.5	95.2	81.3	56.7	72.4	70.8	109.0	71.5	75.6	58.8	76.0	65.6	67.5	64.8	57.2	64.4	81.2
				Woolong tea	A	111.6	94.7	76.6	81.2	136.8	68.8	87.7	82.7	93.3	80.0	80.3	46.7	61.7	70.1	87.5	55.7	52.8	65.2	58.4	73.8
				Woolong tea	B	96.1	82.8	84.6	80.3	96.4	57.3	69.4	67.5	79.6	50.2	66.7	41.5	55.2	64.1	59.3	47.5	44.8	50.2	53.4	78.2
			GC-MS/MS	Green tea	A*	73.5	80.1	85.9	94.2	87.5	88.7	54.8	75.7	79.9	85.5	61.6	63.7	67.4	54.3	68.2	74.8	73.1	53.4	65.4	65.1
				Green tea	B*	76.1	73.4	84.3	79.8	91.7	81.4	76.5	67.5	70.2	77.4	61.2	60.1	68.4	78.0	63.8	64.4	63.1	51.8	64.3	74.8
				Woolong tea	A	117.7	83.8	75.7	76.8	81.3	72.0	73.4	67.9	93.3	75.4	60.7	52.4	58.4	63.9	70.1	49.3	48.9	69.3	59.0	70.5
				Woolong tea	B	105.2	75.2	75.2	71.5	72.5	60.4	65.0	60.4	79.6	47.5	55.1	45.8	44.6	54.7	46.8	42.1	44.6	51.6	54.0	70.5
30	fenpyroximate	Cleanert TPT	GC-MS	Green tea	A	331.5	236.3	395.6	350.3	310.2	285.2	302.1	279.7	271.4	269.1	205.0	162.5	219.3	286.1	212.1	81.4	247.1	200.4	282.2	259.3
				Green tea	B	282.6	240.5	350.3	346.7	232.4	254.1	271.3	234.0	235.0	219.5	162.8	171.6	218.2	268.7	160.9	81.3	214.9	200.1	259.3	231.8
				Woolong tea	A	336.5	392.7	234.7	235.3	234.6	239.9	174.4	285.0	241.8	359.5	176.2	375.4	282.7	344.2	234.9	230.8	164.5	277.3	298.1	269.4
				Woolong tea	B	337.8	306.4	171.8	217.8	260.3	218.9	179.1	226.0	214.8	263.5	173.3	350.5	209.6	268.4	214.7	193.0	146.3	268.0	222.6	233.8
			GC-MS/MS	Green tea	A	256.0	233.3	472.2	466.2	229.0	148.0	316.2	240.6	321.1	218.0	223.6	279.5	193.0	244.0	244.8	114.8	239.1	267.6	189.4	257.7
				Green tea	B	250.5	244.0	370.3	371.0	173.6	142.5	299.9	218.6	224.5	209.5	174.2	267.8	201.3	210.7	196.0	107.3	213.0	244.5	186.0	226.6
				Woolong tea	A	231.7	298.0	219.3	233.1	211.0	213.1	182.0	278.6	260.0	333.1	241.7	320.4	258.9	304.1	222.9	214.9	190.0	214.7	265.0	247.0
				Woolong tea	B	226.0	225.5	202.2	230.1	237.0	204.0	260.5	231.3	219.8	257.0	215.2	303.6	220.1	253.2	194.2	180.0	164.7	217.9	193.7	223.0
		Envi-Carb+PSA	GC-MS	Green tea	A	287.9	293.1	275.8	499.7	301.0	308.7	217.3	288.4	296.1	317.9	158.4	220.2	237.4	325.6	245.5	324.6	195.3	301.6	236.5	280.6
				Green tea	B	304.4	222.3	290.6	432.6	323.5	213.9	268.8	242.5	254.8	299.5	197.1	229.0	216.8	271.9	227.5	278.9	186.4	220.6	231.5	258.6
				Woolong tea	A	572.6	319.9	259.5	389.0	213.4	279.9	397.7	160.2	250.6	320.5	258.5	214.2	181.1	185.2	389.2	221.7	197.7	277.1	226.7	279.7
				Woolong tea	B	440.3	273.0	255.6	304.1	200.6	201.1	314.7	186.9	238.8	216.3	209.8	150.7	180.7	142.3	221.5	258.4	159.8	216.3	223.1	231.3
			GC-MS/MS	Green tea	A	285.6	307.4	295.8	415.6	323.0	328.4	205.3	253.4	275.0	272.7	147.5	339.7	197.9	249.6	249.1	306.2	311.6	276.5	219.2	276.8
				Green tea	B	321.6	200.6	326.8	362.3	363.4	204.0	330.9	240.2	241.3	278.4	184.8	290.6	193.3	287.0	232.5	283.9	274.5	196.6	219.2	264.5
				Woolong tea	A	535.6	216.4	329.3	340.9	378.1	259.9	368.3	215.9	250.6	288.2	234.4	218.6	223.3	304.4	334.7	208.9	113.5	148.4	129.2	268.3
				Woolong tea	B	389.8	211.2	229.3	275.9	350.6	187.4	307.8	227.5	238.8	191.6	204.5	172.4	218.9	217.5	207.9	219.4	105.6	113.3	146.6	221.9

(Continued)

Annex 1: Content change for 201 pesticides in incurred tea Youden pair samples under 8 conditions with two SPE cartridge cleanup in three months (Nov.9, 2009-Feb.7, 2010) (cont.)

No	Pesticides	SPE	Method	Sample	Youden pair	Nov.9, 2009 (n=5)	Nov.14, 2009 (n=3)	Nov.19, 2009 (n=3)	Nov.24, 2009 (n=3)	Nov.29, 2009 (n=3)	Dec.4, 2009 (n=3)	Dec.9, 2009 (n=3)	Dec.14, 2009 (n=3)	Dec.19, 2009 (n=3)	Dec.24, 2009 (n=3)	Dec.14, 2009 (n=3)	Jan.3, 2010 (n=3)	Jan.8, 2010 (n=3)	Jan.13, 2010 (n=3)	Jan.18, 2010 (n=3)	Jan.23, 2010 (n=3)	Jan.28, 2010 (n=3)	Feb.2, 2010 (n=3)	Feb.7, 2010 (n=3)	AVE, µg/kg.
31	fluotrimazole	Cleanert TPT	GC-MS	Green tea	A	50.2	38.5	38.5	35.2	32.2	29.3	30.6	28.9	30.1	24.9	34.4	27.2	28.5	29.8	28.2	20.0	25.9	23.2	31.2	30.9
					B	45.4	36.3	34.8	31.9	27.8	26.6	27.6	26.4	30.7	24.1	31.7	24.4	25.7	28.5	25.3	19.7	24.8	22.0	26.8	28.4
				Woolong tea	A	43.5	36.2	32.1	34.8	33.0	30.0	29.1	32.8	36.9	31.1	33.5	34.4	34.1	32.3	30.7	31.3	28.5	29.9	28.7	32.8
					A*	40.5	34.2	30.1	31.0	32.1	26.1	26.3	29.6	29.4	29.1	28.2	29.5	28.3	28.2	25.9	22.5	23.2	26.8	25.0	28.9
			GC-MS/MS	Green tea	A*	45.8	39.0	46.0	21.2	28.9	60.5	26.3	26.2	34.0	31.3	36.2	29.8	27.1	29.8	28.6	22.9	25.8	23.0	32.1	32.5
					B	44.6	37.8	37.1	34.3	24.4	43.3	27.9	25.4	32.0	29.5	30.8	30.3	25.7	33.5	26.8	28.1	25.2	23.0	27.1	30.6
				Woolong tea	A*	37.1	37.8	37.4	31.9	26.5	28.5	27.9	36.3	30.5	33.6	39.4	35.3	30.5	33.9	28.4	24.0	24.9	27.2	25.1	31.7
					B	35.7	35.0	34.2	30.0	29.0	25.2	26.3	32.6	27.3	29.4	30.4	28.8	29.2	30.1	24.9	24.0	21.4	25.9	21.7	28.5
		Envi-Carb +PSA	GC-MS	Green tea	A	44.1	40.8	40.4	39.7	39.3	36.6	30.8	31.8	35.0	40.6	25.8	31.6	30.0	35.9	30.3	34.0	30.6	29.0	27.4	34.4
					B	40.1	38.5	36.3	34.9	40.3	32.0	31.3	30.1	35.0	36.1	27.9	28.6	26.7	31.2	27.7	27.5	26.6	24.4	27.6	31.7
				Woolong tea	A	50.4	43.7	35.5	36.4	50.8	33.1	39.1	33.4	53.7	39.4	35.0	26.9	29.1	28.3	33.4	25.9	25.6	27.4	26.5	35.5
					B*	43.2	38.2	33.8	32.9	39.5	28.3	31.7	28.1	41.2	36.1	25.7	24.6	24.8	24.3	24.9	21.8	21.3	22.6	23.9	29.5
			GC-MS/MS	Green tea	B	36.7	43.4	39.1	40.5	38.3	48.5	30.9	31.2	33.4	35.9	27.0	44.1	24.8	35.9	26.1	23.9	29.6	26.9	29.0	34.8
					B*	34.7	35.9	37.3	35.5	37.2	36.0	34.0	28.2	53.7	35.2	25.7	39.2	27.4	26.8	30.8	26.3	26.4	21.8	24.3	31.0
				Woolong tea	A	55.3	41.7	44.0	31.1	30.6	32.5	30.8	29.3	41.2	27.0	31.8	26.1	28.5	28.3	24.0	19.8	21.2	25.6	22.3	33.0
					B*	46.5	37.7	36.3	29.7	25.7	27.3	27.3	25.9	35.3	34.0	27.8	23.2	24.6	25.4	34.3	18.6	18.6	20.9	20.9	28.1
32	fluoxypr-1-methylheptyl ester	Cleanert TPT	GC-MS	Green tea	A	54.1	41.5	38.6	36.3	32.8	37.3	31.9	31.5	40.2	34.2	44.9	43.1	36.9	32.6	34.2	43.7	30.0	33.0	37.0	37.3
					B	49.5	39.2	34.4	33.8	31.6	34.7	29.1	30.4	41.1	37.8	40.6	39.5	33.9	32.2	38.0	42.3	29.1	27.2	31.7	35.2
				Woolong tea	A*	42.9	44.0	37.9	41.5	41.8	31.0	30.3	37.2	34.2	34.5	56.4	36.5	38.8	35.3	30.5	37.8	36.6	34.6	31.5	38.5
					B	40.3	38.1	35.9	33.3	38.2	26.7	31.3	35.7	32.5	31.9	42.1	31.0	34.5	32.9	33.4	31.5	29.7	31.0	28.0	33.7
			GC-MS/MS	Green tea	A	52.3	41.5	45.9	36.7	31.7	50.0	32.6	31.9	36.1	31.8	44.2	32.6	40.1	37.9	33.3	38.7	28.8	25.7	52.6	37.9
					B	52.6	37.3	36.8	36.8	30.1	47.1	32.0	32.1	41.1	36.1	40.7	29.6	37.6	36.4	39.8	38.0	25.7	25.5	40.6	36.1
				Woolong tea	A*	38.9	46.5	41.4	43.7	38.1	32.6	34.8	39.3	34.8	33.0	52.2	34.8	32.0	41.5	35.3	30.2	33.4	30.6	25.5	37.5
					B	50.4	42.6	42.6	34.9	34.2	28.0	30.2	33.0	30.2	40.7	37.4	31.4	30.2	30.2	40.2	25.1	25.1	29.1	20.2	32.2
		Envi-Carb +PSA	GC-MS	Green tea	A	44.6	46.2	47.9	35.7	42.6	40.5	40.0	39.5	41.2	37.5	47.2	48.3	35.9	39.6	40.7	36.8	46.4	30.8	32.5	40.8
					B	42.8	47.4	43.9	30.5	41.7	40.5	29.5	39.0	37.7	36.3	40.7	44.7	32.8	35.0	35.8	29.8	38.5	30.5	33.3	37.5
				Woolong tea	A*	39.3	41.6	39.5	37.0	47.2	33.9	42.2	44.9	45.4	30.4	35.6	33.4	35.4	43.7	29.4	30.7	31.1	30.3	32.7	38.1
					B*	47.9	36.9	38.5	34.3	38.5	29.6	32.9	38.8	36.5	35.5	30.4	29.6	29.3	37.7	41.1	23.6	26.5	27.0	28.9	32.8
			GC-MS/MS	Green tea	A*	43.9	65.3	43.6	36.5	36.6	41.1	44.6	39.7	49.8	34.1	39.8	44.3	31.5	37.8	38.1	35.4	31.1	26.8	28.6	39.6
					B	43.0	62.0	40.9	31.0	37.7	37.6	31.3	34.2	43.4	34.1	32.1	40.4	33.8	28.3	29.1	25.8	31.5	28.8	26.7	35.7
				Woolong tea	A*	39.8	45.9	46.1	35.7	43.5	36.1	35.2	36.3	45.4	37.9	32.8	30.7	32.7	30.5	26.1	27.9	36.7	44.8	40.0	37.8
					B*	44.6	41.2	41.6	34.6	34.6	32.4	30.8	30.5	36.5	33.8	29.0	28.8	24.9	27.8	20.3	20.3	31.5	37.9	32.0	32.6
33	iprovalicarb-1	Cleanert TPT	GC-MS	Green tea	A*	199.0	173.2	180.4	182.7	152.7	140.3	165.4	177.7	159.5	135.3	215.3	173.2	163.4	151.2	166.0	330.9	123.4	114.5	150.7	171.3
					B*	187.8	164.1	175.3	166.9	139.3	139.6	173.6	181.5	178.9	133.0	190.7	154.3	152.7	150.1	165.8	321.6	127.3	111.4	138.9	165.9
				Woolong tea	A	193.5	137.8	188.0	185.5	172.3	169.8	167.3	168.6	168.1	150.9	154.1	178.0	170.7	156.5	180.2	181.5	191.8	152.2	134.9	168.5
					B	203.6	145.7	184.3	170.0	162.8	161.5	165.8	177.1	127.7	136.3	144.0	161.0	156.4	149.5	154.6	153.5	160.9	136.6	132.0	157.0
			GC-MS/MS	Green tea	A*	211.1	173.7	174.8	177.8	144.9	284.1	137.2	117.8	142.9	145.8	230.3	128.8	161.2	191.3	145.0	144.1	122.0	103.5	213.2	165.8
					B	206.2	164.7	167.1	157.5	127.1	241.3	128.9	124.2	160.4	145.3	224.3	125.0	149.2	206.6	147.5	144.7	116.7	102.1	181.9	159.0
				Woolong tea	A*	235.4	161.6	158.8	167.0	147.1	154.1	226.3	194.3	169.9	159.6	199.6	154.2	123.1	166.4	152.8	158.2	147.9	144.8	116.4	165.1
					B*	201.3	165.4	143.9	146.6	142.2	134.6	214.4	204.2	157.5	140.7	148.1	130.2	114.8	148.7	127.0	133.0	119.4	127.3	120.5	150.5
		Envi-Carb +PSA	GC-MS	Green tea	A*	156.1	163.2	201.4	179.9	191.1	203.8	198.8	185.4	211.3	146.3	161.0	175.5	150.0	138.7	165.8	134.2	251.6	101.8	138.8	178.6
					B*	163.2	191.5	209.6	151.5	194.3	180.4	136.3	155.1	187.8	255.7	135.4	164.5	132.9	145.6	141.5	120.2	207.5	128.7	151.9	166.0
				Woolong tea	A	218.4	148.7	162.0	156.3	257.1	170.4	151.6	244.3	212.8	148.3	145.6	133.1	157.3	214.5	180.9	131.8	161.4	138.1	141.7	172.3
					B	200.6	146.1	164.7	163.4	190.7	158.1	168.9	188.3	176.2	123.3	127.8	119.7	141.7	173.1	144.5	106.0	135.5	126.0	129.3	151.8
			GC-MS/MS	Green tea	A*	174.9	184.6	169.5	173.0	177.6	206.2	185.6	165.4	204.7	170.6	170.8	140.4	131.0	141.2	143.6	145.6	146.7	114.7	138.4	162.3
					B	194.7	208.5	166.0	147.8	177.1	181.9	170.7	144.5	196.8	164.1	137.7	129.2	127.8	138.6	132.5	132.4	125.0	114.4	153.1	154.9
				Woolong tea	A*	161.4	174.1	177.8	148.1	162.4	146.0	156.8	152.2	212.8	150.6	143.9	115.2	130.9	130.2	142.3	119.3	145.2	194.1	164.8	154.1
					B*	143.5	160.9	161.6	139.0	132.6	126.8	141.0	119.4	176.2	123.7	129.7	128.9	101.5	119.6	115.4	94.5	129.5	159.5	145.5	134.1

#	Compound	Cartridge	Method	Tea																					
34	iprovalicarb-2	Cleanert TPT	GC-MS	Green tea	A*	216.2	178.6	163.0	185.3	135.9	133.4	137.5	143.9	197.4	130.5	284.2	185.4	169.8	145.7	166.8	338.1	129.6	116.1	147.2	173.9
					B*	208.8	179.1	141.9	161.9	121.8	133.3	131.9	136.7	170.4	117.6	243.2	161.3	155.1	150.0	163.7	325.8	130.5	107.6	133.0	161.8
				Woolong tea	A	195.2	156.0	147.0	167.2	159.5	133.4	221.4	203.4	168.8	144.6	192.0	157.3	159.7	142.7	151.4	158.0	188.6	140.8	136.5	164.4
					B*	192.5	151.5	142.7	155.9	148.8	119.9	211.1	206.7	126.5	131.5	141.6	129.7	149.3	139.1	127.4	133.0	148.7	128.8	126.7	148.0
			GC-MS/MS	Green tea	A*	212.3	177.4	182.0	181.1	147.9	318.3	141.6	133.9	153.1	140.7	202.3	140.8	163.8	159.5	148.6	145.3	126.2	105.5	234.6	169.2
					B	207.1	165.4	174.8	159.6	130.6	267.2	132.1	130.2	157.4	144.9	182.9	129.7	150.8	174.4	143.0	143.2	118.9	104.7	200.5	158.8
				Woolong tea	A	235.4	161.6	158.8	167.0	146.3	148.2	240.9	198.6	191.0	161.1	202.3	158.3	148.6	163.8	158.5	162.0	156.0	146.9	126.5	170.1
					B*	201.3	165.4	143.9	146.6	144.6	129.2	223.6	206.3	146.8	144.1	149.7	133.1	152.9	153.3	132.6	138.2	127.9	128.6	125.6	152.3
		Envi-Carb +PSA	GC-MS	Green tea	A*	158.2	216.8	197.9	154.0	201.4	203.2	176.8	180.3	142.7	395.9	158.3	171.6	157.3	140.9	169.9	128.2	256.4	95.1	135.4	181.1
					B	177.5	192.9	188.8	132.7	190.5	182.3	161.4	151.1	136.5	278.8	143.7	147.7	132.3	148.3	143.9	120.1	214.8	122.0	139.8	163.4
				Woolong tea	A	200.6	158.0	158.3	153.0	300.8	149.1	160.6	301.9	212.0	158.9	181.0	128.9	156.5	176.6	156.2	122.1	137.6	129.7	135.5	172.5
					B	184.6	146.4	148.2	146.2	197.7	132.3	140.2	214.6	177.8	130.8	135.6	130.8	132.6	149.4	130.9	97.4	119.0	112.9	121.7	144.7
			GC-MS/MS	Green tea	A*	174.4	201.4	171.1	177.3	177.9	212.1	158.7	170.0	150.5	161.2	166.5	149.4	124.1	145.5	153.1	144.3	151.3	119.1	141.8	160.5
					B	195.2	210.1	167.9	150.7	176.1	188.9	150.5	149.0	156.9	155.9	132.7	142.6	126.6	134.1	140.5	130.8	131.9	119.8	153.6	153.4
				Woolong tea	A*	161.4	174.1	177.8	148.1	162.4	148.0	157.9	152.2	212.0	154.7	141.6	123.0	114.4	130.2	140.3	119.9	137.8	199.6	168.3	153.9
					B*	143.5	160.9	161.6	139.0	132.6	128.7	140.3	119.4	177.8	127.7	127.5	117.6	109.3	119.6	128.8	95.3	123.8	161.0	150.2	135.0
35	isodrin	Cleanert TPT	GC-MS	Green tea	A	46.0	35.0	33.7	38.5	32.4	34.0	32.1	31.3	33.4	27.9	30.8	36.1	33.2	32.1	32.6	23.6	29.6	27.9	34.4	32.9
					B	43.9	34.2	31.2	35.1	29.8	32.4	28.9	28.7	32.4	27.5	29.0	34.5	31.4	30.7	30.6	23.0	29.8	24.2	30.0	30.9
				Woolong tea	A	40.7	41.4	34.0	37.8	34.3	30.3	34.8	40.7	35.4	39.5	36.8	39.0	36.5	37.1	33.9	35.6	32.9	32.2	32.2	36.2
					B	39.6	35.1	28.0	30.2	31.0	25.5	26.9	32.8	35.4	33.6	31.6	32.3	29.5	31.8	28.1	32.8	27.7	27.9	27.9	30.9
			GC-MS/MS	Green tea	A	46.1	34.3	36.9	40.3	27.9	30.5	31.4	28.0	33.1	29.3	34.9	33.0	33.6	29.8	30.7	24.7	27.7	23.9	32.5	32.0
					B	43.7	32.4	33.4	35.4	27.4	28.8	29.1	28.0	31.9	26.8	30.3	30.4	30.2	28.7	27.8	24.1	24.4	23.4	27.6	29.7
				Woolong tea	A*	38.0	40.5	34.0	36.2	32.8	31.5	30.7	36.5	39.4	36.9	42.5	39.1	33.8	36.4	32.4	34.3	31.7	33.6	31.5	35.4
					B*	36.1	33.2	29.0	30.0	30.2	26.1	23.7	32.0	32.9	31.2	31.4	30.5	28.6	31.7	24.9	34.7	28.7	30.0	26.6	30.1
		Envi-Carb +PSA	GC-MS	Green tea	A	44.6	40.7	33.7	35.8	40.9	37.4	30.5	32.0	32.9	31.2	32.8	30.0	32.0	37.1	33.8	35.8	30.5	28.7	25.4	33.6
					B	37.3	34.6	33.7	34.5	37.5	35.5	26.2	29.1	37.7	22.1	29.2	29.4	31.4	33.8	31.3	30.2	30.3	26.4	27.3	31.4
				Woolong tea	A	46.5	39.1	36.8	36.7	38.4	35.1	38.2	27.6	34.5	25.5	27.3	35.1	31.4	31.5	39.4	27.9	28.1	28.6	27.1	34.3
					B	43.7	32.7	33.7	30.8	29.8	30.7	35.4	31.6	37.6	34.6	29.3	30.6	28.5	26.7	31.5	24.1	24.9	25.1	23.7	29.9
			GC-MS/MS	Green tea	A*	37.4	38.2	35.0	36.5	39.3	36.5	31.9	31.3	32.7	34.6	29.3	57.6	28.4	39.5	33.7	34.8	31.9	29.5	28.2	35.1
					B*	36.6	31.2	35.3	35.1	39.5	33.2	31.3	30.2	31.7	34.2	27.8	56.3	28.1	35.4	30.6	30.0	29.7	26.8	27.4	33.2
				Woolong tea	A	46.2	42.9	48.0	35.2	37.9	33.8	36.3	32.9	37.6	38.7	33.3	30.1	32.7	30.7	37.4	27.6	23.5	28.0	26.7	34.7
					B	37.0	35.9	34.0	34.2	31.4	29.2	32.9	28.4	30.6	29.4	29.1	27.0	27.8	27.2	27.0	23.9	20.5	22.5	25.7	29.1
36	isoprocarb-1	Cleanert TPT	GC-MS	Green tea	A	80.2	70.1	82.0	82.9	65.8	74.0	70.3	69.9	73.8	65.9	65.6	69.9	75.3	66.3	62.4	42.2	61.3	57.8	78.0	69.1
					B	78.4	70.6	74.4	79.5	58.8	65.8	64.9	60.4	72.4	59.5	56.4	64.6	70.7	63.1	56.9	41.3	59.0	55.4	67.3	64.2
				Woolong tea	A	72.3	86.0	72.2	76.1	68.9	63.8	63.1	77.0	74.4	78.2	75.9	81.8	77.1	77.1	76.3	75.7	68.0	69.2	73.0	74.0
					B	69.9	70.0	59.2	66.3	64.1	54.5	51.8	63.4	63.3	63.1	59.2	68.1	62.2	66.9	62.5	64.2	56.1	64.0	59.9	62.6
			GC-MS/MS	Green tea	A	87.7	71.5	92.1	84.8	63.9	72.3	69.7	61.1	73.6	63.3	73.8	71.8	66.2	53.5	66.4	50.2	64.4	55.5	75.3	69.3
					B	85.7	70.8	79.1	79.3	58.8	67.1	64.3	57.1	70.6	62.0	66.2	68.1	65.4	56.2	59.5	49.2	61.7	53.9	65.8	65.3
				Woolong tea	A*	91.1	101.6	90.9	90.2	109.1	116.1	98.3	108.0	131.1	115.9	130.7	125.5	131.2	136.8	138.2	130.4	98.1	103.1	111.7	113.6
					B*	101.7	97.8	92.7	90.6	119.5	113.1	100.0	107.5	131.8	113.7	121.1	124.3	140.1	142.3	135.6	140.8	103.1	111.5	112.2	115.8
		Envi-Carb +PSA	GC-MS	Green tea	A	80.5	77.5	72.9	87.4	81.0	77.3	63.7	71.3	72.9	63.6	65.1	76.7	66.9	80.1	71.2	82.9	71.2	73.7	65.0	73.7
					B*	74.0	67.7	73.9	79.0	80.7	71.9	70.3	71.9	66.6	65.6	61.4	73.5	61.7	71.1	65.5	68.2	63.9	60.6	64.8	69.1
				Woolong tea	A*	100.9	82.6	73.8	77.3	81.2	74.4	84.4	60.0	130.7	76.2	67.9	64.6	66.0	67.5	73.3	60.6	60.8	66.2	62.8	75.3
					B*	77.6	67.3	64.8	67.4	63.4	60.6	69.1	54.7	128.7	57.0	58.1	53.1	56.3	54.6	57.3	50.8	50.6	51.4	55.2	63.0
			GC-MS/MS	Green tea	A*	75.5	83.7	77.6	87.2	87.4	80.4	68.4	68.1	72.6	88.1	61.7	96.6	66.5	87.4	71.5	78.8	76.2	67.3	67.8	77.0
					B*	70.4	66.8	76.5	76.4	86.8	75.8	73.4	63.9	68.5	82.7	59.9	68.9	66.3	73.9	64.5	66.3	70.1	56.2	66.3	72.3
				Woolong tea	A*	132.4	94.9	126.7	92.8	102.2	103.2	109.8	110.1	130.7	127.6	116.3	100.2	108.9	110.5	143.7	112.8	74.2	93.7	100.0	110.0
					B	112.9	95.9	121.2	96.0	98.5	106.4	114.5	114.0	128.7	118.9	119.1	109.8	125.7	111.7	127.7	111.1	79.6	91.9	108.0	110.1

(Continued)

Annex 1: Content change for 201 pesticides in incurred tea Youden pair samples under 8 conditions with two SPE cartridge cleanup in three months (Nov.9, 2009-Feb.7, 2010) (cont.)

No	Pesticides	SPE	Method	Sample	Youden pair	Nov.9, 2009 (n=5)	Nov.14, 2009 (n=3)	Nov.19, 2009 (n=3)	Nov.24, 2009 (n=3)	Nov.29, 2009 (n=3)	Dec.4, 2009 (n=3)	Dec.9, 2009 (n=3)	Dec.14, 2009 (n=3)	Dec.19, 2009 (n=3)	Dec.24, 2009 (n=3)	Dec.14, 2009 (n=3)	Jan.3, 2010 (n=3)	Jan.8, 2010 (n=3)	Jan.13, 2010 (n=3)	Jan.18, 2010 (n=3)	Jan.23, 2010 (n=3)	Jan.28, 2010 (n=3)	Feb.2, 2010 (n=3)	Feb.7, 2010 (n=3)	AVE, μg/kg.
37	isoprocarb-2	Cleanert TPT	GC-MS	Green tea	A*	69.4	105.0	67.7	66.7	69.5	79.4	69.1	60.0	64.3	72.8	99.0	68.1	52.1	63.7	78.3	172.4	67.2	48.4	63.4	75.6
				Green tea	B*	86.6	88.3	66.0	65.5	70.4	72.3	63.5	57.8	75.1	85.3	104.7	68.8	49.8	65.6	72.4	174.8	64.8	48.0	62.0	75.9
				Woolong tea	A	104.0	98.6	80.0	87.3	90.3	71.0	84.2	93.3	94.9	79.2	122.4	67.2	77.5	77.5	59.4	69.1	55.1	67.8	55.8	80.8
				Woolong tea	B	88.0	83.8	63.8	67.2	83.6	62.9	77.1	82.5	83.5	77.4	103.0	58.1	62.7	70.7	49.3	58.4	54.3	57.4	56.5	70.5
			GC-MS/MS	Green tea	A	118.0	102.1	66.2	70.9	58.2	96.4	69.2	67.8	66.9	59.9	110.7	61.6	62.6	48.2	68.5	125.8	48.6	68.1	75.8	76.1
				Green tea	B	110.7	87.4	62.8	66.2	67.8	97.0	59.1	63.5	79.1	77.2	100.8	59.7	60.1	65.0	83.2	115.8	47.3	69.5	70.7	75.9
				Woolong tea	A	81.7	94.9	74.0	85.5	98.3	59.3	83.3	106.7	123.8	64.6	123.8	67.3	116.6	76.0	65.6	74.6	103.2	106.5	116.6	90.7
				Woolong tea	B	73.0	78.6	68.2	64.0	114.5	49.3	74.6	118.1	128.5	73.2	117.4	57.1	138.1	75.9	53.1	72.2	112.1	118.7	122.8	90.0
		Envi-Carb +PSA	GC-MS	Green tea	A*	93.0	142.6	101.5	64.1	86.3	83.4	87.2	95.3	91.0	82.5	132.8	133.8	75.8	65.3	92.0	67.9	66.3	57.1	65.2	88.6
				Green tea	B*	89.7	122.1	101.7	57.2	90.9	81.3	58.1	87.0	91.6	70.1	103.2	125.2	69.3	67.5	78.7	60.7	66.3	77.5	64.0	82.2
				Woolong tea	A	74.5	83.4	95.3	83.2	108.9	78.0	75.9	75.5	91.8	73.1	80.4	79.8	62.7	74.2	79.6	65.4	48.6	52.1	58.7	75.8
				Woolong tea	B	64.4	73.0	88.6	71.1	91.9	72.3	78.6	63.3	62.7	65.3	72.1	76.8	66.0	76.9	64.2	54.6	54.6	53.1	58.5	68.5
			GC-MS/MS	Green tea	A	80.2	126.9	106.1	76.9	94.7	109.0	92.1	104.5	74.6	74.2	130.3	82.3	112.1	125.0	112.5	142.8	80.5	120.7	87.7	101.8
				Green tea	B	88.2	134.1	109.0	74.8	103.3	95.6	57.1	98.8	91.9	79.6	106.6	80.7	126.0	120.2	115.7	130.9	76.4	115.9	110.1	100.8
				Woolong tea	A	92.8	86.9	134.0	86.1	88.4	79.3	81.7	73.5	91.8	78.4	107.3	95.3	96.2	94.4	52.4	56.7	78.3	96.0	100.8	87.9
				Woolong tea	B	66.1	68.9	98.0	85.5	85.1	74.0	71.6	63.0	62.7	75.5	115.7	106.7	113.3	101.0	46.3	40.7	87.8	94.5	116.3	82.8
38	lenacil	Cleanert TPT	GC-MS	Green tea	A*	262.9	235.7	189.6	179.0	158.3	197.6	151.2	190.6	170.6	169.8	359.1	368.8	223.1	176.2	237.5	798.1	132.4	139.2	191.6	238.5
				Green tea	B*	258.0	215.7	172.8	166.4	159.5	183.0	150.5	186.3	295.8	174.9	337.1	310.7	194.5	187.3	270.7	795.3	152.7	134.7	157.3	237.0
				Woolong tea	A	230.3	177.1	228.4	261.0	266.3	198.8	456.6	246.1	207.6	142.8	376.4	162.1	187.1	141.0	201.3	194.8	349.5	174.4	122.5	227.6
				Woolong tea	B*	210.9	178.3	259.5	226.2	199.6	152.3	450.6	263.8	172.1	144.2	263.9	135.0	207.1	163.8	154.9	175.1	279.2	150.0	142.1	206.8
			GC-MS/MS	Green tea	A	327.1	233.6	206.4	257.2	185.0	208.6	148.0	163.4	185.9	208.8	340.9	195.0	263.3	208.7	188.2	344.2	95.9	66.7	405.5	225.5
				Green tea	B	326.9	216.4	193.0	186.8	185.4	200.3	148.6	175.4	262.5	209.9	321.0	168.9	227.8	224.5	217.2	387.2	123.1	84.7	465.3	224.5
				Woolong tea	A*	502.0	183.0	226.4	239.0	246.0	249.9	371.5	273.7	310.4	175.9	376.6	167.0	166.4	153.2	180.1	309.6	142.8	129.1	198.7	242.2
				Woolong tea	B*	354.0	269.6	211.7	191.3	203.1	195.0	351.1	267.5	193.1	169.5	239.4	139.0	210.1	161.7	155.1	223.4	176.5	127.5	286.7	217.1
		Cleanert TPT	GC-MS	Green tea	A*	207.5	235.4	262.7	151.9	222.7	217.5	289.2	210.7	202.9	328.8	305.2	286.0	225.2	147.7	241.1	166.2	427.5	118.4	144.0	231.1
				Green tea	B	223.6	270.1	250.0	117.0	202.7	238.8	220.3	212.1	209.4	291.0	195.8	249.6	189.3	166.4	184.5	134.3	323.0	155.3	152.4	209.8
				Woolong tea	A*	228.5	205.0	215.6	173.5	428.5	187.8	180.7	684.3	245.5	178.2	181.3	182.0	206.8	375.6	154.1	159.4	207.8	144.5	187.0	238.2
				Woolong tea	B*	219.7	176.7	209.1	177.5	276.1	175.6	143.1	381.4	191.3	149.6	156.1	184.4	158.9	355.6	160.7	102.6	178.7	134.6	159.1	194.2
			GC-MS/MS	Green tea	A	229.3	273.9	229.8	185.1	224.1	228.6	235.9	213.0	292.6	222.7	145.4	181.5	206.8	222.4	358.1	296.4	332.0	270.9	363.5	248.0
				Green tea	B	236.7	383.7	217.8	147.7	214.6	241.2	196.3	219.1	281.0	188.2	171.6	174.6	186.1	168.5	384.2	175.1	335.7	247.4	367.1	243.2
				Woolong tea	A*	159.0	217.1	178.3	169.5	187.4	186.2	163.3	242.3	245.5	186.3	180.3	175.5	183.4	141.6	141.9	132.8	198.4	219.3	267.9	188.2
				Woolong tea	B*	175.6	193.6	197.8	164.1	137.6	169.0	154.3	164.3	191.3	162.9	160.4	196.2	128.7	137.2	138.6	82.6	209.0	207.2	277.3	170.9
39	metalaxyl	Cleanert TPT	GC-MS	Green tea	A*	135.5	108.2	108.8	115.6	89.5	91.3	90.8	97.2	103.0	84.6	122.5	86.0	116.5	108.3	104.5	86.4	96.5	88.3	116.7	102.6
				Green tea	B*	123.9	103.8	98.9	112.9	80.2	81.2	85.1	80.0	103.0	70.7	103.5	82.5	104.9	107.0	93.9	85.3	93.3	85.9	101.6	94.2
				Woolong tea	A	123.8	116.4	102.2	97.2	99.7	91.5	92.0	118.0	109.3	100.3	118.5	129.6	119.1	109.1	114.4	121.9	117.6	109.9	119.1	111.9
				Woolong tea	B	118.1	103.7	92.6	154.4	83.5	78.8	71.3	96.9	79.7	77.3	88.9	107.3	97.7	99.5	94.1	101.4	94.0	103.0	100.3	94.0
			GC-MS/MS	Green tea	A	139.2	110.6	119.4	106.3	88.4	121.0	93.0	82.3	97.9	87.6	111.0	95.9	93.0	80.1	91.9	73.1	81.1	70.9	107.4	99.9
				Green tea	B	131.5	105.2	104.8	105.5	81.3	109.7	84.1	76.1	96.6	85.7	96.1	90.3	88.7	86.3	84.5	71.3	77.2	72.6	94.6	91.7
				Woolong tea	A	107.1	108.0	102.1	105.5	88.8	94.4	99.7	120.9	113.2	114.5	127.0	114.4	99.5	113.7	90.7	95.7	79.5	85.5	93.5	105.5
				Woolong tea	B	111.4	121.8	123.3	110.7	90.5	77.7	74.7	98.3	96.1	93.9	92.9	86.7	85.8	96.9	73.5	80.6	71.3	81.5	81.1	88.2
		Envi-Carb +PSA	GC-MS	Green tea	A*	110.2	102.8	116.0	99.0	121.2	116.1	89.7	91.2	91.6	175.7	102.3	128.3	107.9	122.0	106.9	116.0	116.5	98.9	100.8	113.3
				Green tea	B*	136.8	119.0	109.6	103.7	117.3	98.3	92.3	88.9	94.2	145.5	97.7	106.3	97.7	109.8	96.9	93.8	102.0	91.7	101.4	103.5
				Woolong tea	A	121.0	97.6	96.5	94.0	138.9	103.7	115.3	108.7	125.3	113.7	101.8	84.9	103.8	106.8	125.2	97.0	93.2	105.0	94.1	109.8
				Woolong tea	B	112.4	121.0	111.9	121.3	120.2	123.7	99.4	88.7	98.1	79.5	88.2	77.0	88.2	91.5	97.6	80.9	82.7	83.3	86.3	91.2
			GC-MS/MS	Green tea	A	107.4	103.5	109.4	107.0	115.5	110.9	86.9	97.2	95.7	102.5	89.3	110.2	78.4	103.5	95.1	106.1	95.7	88.5	88.2	102.5
				Green tea	B	142.6	120.1	145.6	103.8	116.8	103.3	93.9	90.7	96.7	100.4	83.2	101.5	78.2	87.4	84.5	86.7	87.8	77.3	91.7	95.5
				Woolong tea	A	110.7	102.9	103.0	93.5	94.5	86.0	100.4	80.5	98.1	109.2	97.8	86.1	91.1	86.6	102.7	76.2	66.0	87.6	82.1	102.5
				Woolong tea	B	107.0	102.9	102.9	73.8	94.5	86.0	100.4	80.5	84.3	78.4	73.8	74.9	81.0	79.6	73.8	63.6	60.2	65.2	77.4	84.6

No.	Compound	Cartridge	Method	Tea	Rep																				
40	metazachlor	Cleanert TPT	GC-MS	Green tea	A	120.9	120.5	126.0	177.0	99.0	117.9	122.2	117.9	132.8	89.8	129.8	94.6	95.4	101.9	105.4	108.4	85.8	76.3	85.5	110.9
					B*	125.8	109.9	115.4	171.6	96.4	106.6	108.5	106.6	127.2	86.4	114.2	85.5	94.3	100.8	92.7	107.2	89.3	78.0	81.2	105.2
				Woolong tea	A*	109.1	160.2	117.1	116.0	87.8	88.0	162.3	88.0	116.8	130.7	131.7	110.5	102.0	98.2	85.6	95.6	103.1	83.2	102.4	116.4
					B*	106.1	117.6	99.5	103.6	86.3	77.1	123.6	77.1	96.6	109.7	99.5	87.9	84.9	87.6	77.3	87.1	84.8	82.6	89.6	99.0
			GC-MS/MS	Green tea	A	127.0	111.8	76.4	163.4	90.7	120.8	97.9	120.8	116.5	82.4	110.8	83.4	104.1	77.0	97.2	92.1	76.6	78.2	132.3	101.9
					B	128.0	106.1	97.0	102.5	89.0	117.4	85.0	117.4	114.8	95.6	100.9	77.3	105.2	91.3	88.9	90.8	69.2	76.4	121.9	97.0
				Woolong tea	A*	154.4	159.2	98.9	113.9	105.8	102.2	137.7	102.2	118.3	112.8	152.7	116.2	115.8	120.3	103.2	121.8	106.1	81.9	110.0	119.7
					B*	131.8	128.7	84.9	89.5	100.1	80.6	101.4	80.6	97.5	100.1	115.5	96.3	94.4	102.2	87.6	97.7	74.8	75.2	96.1	98.6
		Envi-Carb +PSA	GC-MS	Green tea	A	79.7	165.7	103.1	98.4	121.7	125.4	95.0	125.4	139.5	198.9	112.7	93.3	93.4	93.2	89.0	92.4	153.0	91.0	93.2	113.6
					B*	100.2	125.0	109.0	98.1	117.7	123.3	106.9	123.3	132.6	190.9	113.7	92.8	87.1	92.1	91.0	81.6	133.5	93.5	93.5	109.2
				Woolong tea	A*	113.8	123.9	116.6	132.8	144.9	108.3	95.9	108.3	105.1	114.3	106.5	81.7	90.3	84.5	102.4	80.4	74.7	78.4	83.1	108.8
					B*	104.8	106.2	104.5	108.0	113.4	98.0	117.5	98.0	82.6	85.2	86.5	73.1	79.9	83.5	88.4	66.6	68.0	73.1	82.3	94.3
			GC-MS/MS	Green tea	A	117.3	123.7	95.3	106.8	108.6	123.8	97.9	123.8	66.6	77.9	108.7	107.7	84.1	122.8	102.4	85.7	101.9	73.6	80.9	98.7
					B	168.3	125.5	97.6	106.0	112.0	113.7	89.7	113.7	93.0	88.4	93.2	93.9	92.6	91.6	93.7	79.8	84.6	75.2	95.7	99.2
				Woolong tea	A*	106.2	130.4	168.0	103.1	121.1	96.3	120.2	96.3	105.1	100.1	93.2	73.7	89.7	79.9	95.5	75.2	80.7	126.9	102.2	104.3
					B*	67.8	113.6	105.5	92.2	98.3	87.0	102.5	87.0	82.6	80.4	86.5	71.0	77.4	82.4	73.6	63.1	80.9	107.8	96.5	87.2
41	methabenzthiazuron	Cleanert TPT	GC-MS	Green tea	A	478.9	358.1	379.5	390.2	308.8	354.6	322.8	354.6	354.8	274.1	398.0	306.7	365.1	314.5	311.8	299.0	284.5	279.6	373.9	340.9
					B	438.6	347.2	337.8	349.7	280.9	324.0	293.0	324.0	336.2	258.4	345.4	306.6	327.5	298.8	264.7	290.9	275.8	259.1	322.0	311.3
				Woolong tea	A*	457.0	421.3	345.1	379.4	374.9	329.0	263.0	329.0	366.2	366.2	382.0	368.2	357.4	345.7	368.4	361.5	324.2	323.2	303.4	362.4
					B*	417.3	370.4	296.0	315.9	317.2	277.4	325.0	277.4	275.8	313.3	269.4	304.5	281.1	307.7	288.8	296.1	261.8	305.5	281.0	303.3
			GC-MS/MS	Green tea	A*	417.4	365.7	428.9	492.3	301.4	529.5	289.0	529.5	358.5	307.0	363.1	342.9	361.3	275.6	312.7	269.0	281.2	239.7	550.5	358.2
					B*	415.7	351.1	365.3	362.6	280.6	453.7	344.5	453.7	346.4	307.4	315.9	319.9	325.2	298.6	288.5	262.9	276.8	247.7	434.3	326.6
				Woolong tea	A	379.3	433.3	377.1	366.5	328.0	315.1	266.9	315.1	369.3	382.4	427.9	391.8	328.5	376.0	324.1	354.8	303.8	310.9	278.5	357.0
					B	369.2	377.7	318.9	310.7	309.3	275.6	309.9	275.6	300.0	316.3	304.5	303.8	291.6	316.8	262.2	304.2	253.4	290.3	264.0	304.3
		Envi-Carb +PSA	GC-MS	Green tea	A*	441.6	405.1	431.5	396.6	451.0	436.2	354.4	436.2	339.5	621.5	292.1	332.6	342.3	383.6	354.9	365.7	430.0	333.8	299.1	385.6
					B*	410.0	370.0	429.0	347.4	440.3	367.5	354.4	367.5	329.1	524.4	300.8	317.5	301.0	349.8	305.2	300.0	358.7	281.2	304.3	354.5
				Woolong tea	A	528.9	445.1	380.9	369.0	547.9	371.9	416.3	371.9	458.8	372.2	363.1	296.5	329.6	397.3	349.3	294.4	281.6	308.5	292.0	381.8
					B	458.4	377.8	341.1	330.6	390.3	300.7	336.9	300.7	351.1	283.7	298.6	262.8	276.0	321.8	284.1	258.1	239.9	261.2	261.3	315.2
			GC-MS/MS	Green tea	A*	369.2	421.1	384.1	411.2	412.7	424.3	300.9	424.3	328.7	389.3	297.8	411.7	324.9	385.7	338.4	357.5	333.0	264.9	280.8	362.4
					B	350.8	369.0	383.5	351.8	393.1	375.4	339.2	375.4	333.8	372.2	279.3	379.9	301.6	330.3	294.3	295.4	315.6	270.1	353.0	337.5
				Woolong tea	A*	508.5	413.2	448.8	350.0	384.0	372.5	359.0	372.5	458.8	373.4	335.4	298.2	325.3	298.9	328.0	269.5	351.5	496.0	353.2	372.1
					B	431.3	363.1	346.8	316.2	315.4	303.4	331.0	303.4	351.1	285.6	295.1	272.8	285.6	266.7	256.1	217.6	309.7	354.8	307.5	309.9
42	mirex	Cleanert TPT	GC-MS	Green tea	A	41.9	31.3	30.2	32.6	23.3	32.1	25.1	32.1	27.8	26.0	32.4	33.1	27.8	28.1	30.7	42.6	20.3	20.2	24.8	29.2
					B	42.0	30.6	26.6	30.4	22.8	27.6	22.4	27.6	28.8	25.6	30.5	29.2	25.9	27.7	30.6	39.2	22.0	19.9	22.6	27.8
				Woolong tea	A*	35.8	38.0	32.4	36.3	34.9	27.2	38.1	27.2	34.7	30.9	42.0	29.7	30.9	25.1	27.5	31.1	36.0	26.9	26.5	32.9
					B*	33.1	31.2	28.4	29.4	29.8	22.3	30.8	22.3	29.1	28.3	31.8	23.9	26.5	25.8	23.3	28.3	28.8	25.4	24.7	28.1
			GC-MS/MS	Green tea	A	46.5	33.7	26.1	37.0	24.6	32.4	24.1	32.4	25.3	23.8	35.4	25.6	25.8	27.7	25.8	28.3	20.9	18.4	32.3	28.7
					B	44.2	31.8	25.7	28.0	24.3	30.1	22.0	30.1	30.2	23.4	32.7	22.9	27.6	28.1	25.3	26.9	20.9	18.5	29.1	27.3
				Woolong tea	A*	42.5	42.7	28.8	35.6	34.2	32.2	36.5	32.2	34.0	29.3	42.9	28.8	25.8	31.4	28.8	29.9	30.6	27.8	25.3	33.9
					B	38.5	36.0	25.3	28.3	31.3	24.4	30.5	24.4	40.6	27.4	31.2	24.0	34.0	26.5	24.3	26.8	23.8	24.3	23.4	28.4
		Envi-Carb +PSA	GC-MS	Green tea	A	30.2	36.6	32.0	23.6	34.3	32.5	24.9	32.5	31.2	20.3	34.6	35.8	26.7	24.9	26.0	23.0	42.4	19.1	26.0	28.8
					B	32.2	34.0	31.2	21.3	31.9	32.8	21.9	32.8	24.9	21.0	28.0	32.4	27.8	26.1	25.6	19.9	35.3	21.6	26.4	27.1
				Woolong tea	A*	39.3	33.9	32.8	30.3	40.9	30.9	27.6	30.9	25.4	29.0	28.6	28.4	26.0	33.0	26.2	24.3	23.7	23.3	24.3	30.3
					B	34.2	29.3	29.5	27.9	30.4	28.7	28.9	28.7	28.0	26.3	25.8	26.4	29.5	30.6	25.2	17.5	22.4	21.4	22.2	26.5
			GC-MS/MS	Green tea	A	35.9	30.6	30.4	26.3	33.5	30.7	33.3	30.7	21.0	23.2	34.5	45.1	25.2	35.2	26.1	24.7	25.7	19.8	25.3	29.3
					B	37.7	35.0	30.4	24.3	32.7	31.0	21.1	31.0	22.7	24.2	29.8	38.5	28.4	28.1	25.2	20.7	23.3	19.4	26.1	27.7
				Woolong tea	A*	35.1	35.8	40.1	29.2	33.1	28.9	32.1	28.9	24.8	28.9	27.6	25.1	29.2	25.6	26.0	22.9	22.3	30.1	27.3	29.2
					B	28.4	30.8	30.8	26.7	25.9	26.7	28.4	26.7	21.0	26.5	25.7	24.4	21.0	23.9	22.4	17.7	21.4	26.3	23.9	25.1

(Continued)

Annex 1: Content change for 201 pesticides in incurred tea Youden pair samples under 8 conditions with two SPE cartridge cleanup in three months (Nov.9, 2009-Feb.7, 2010) (cont.)

No	Pesticides	SPE	Method	Sample	Youden pair	Nov.9, 2009 (n=5)	Nov.14, 2009 (n=3)	Nov.19, 2009 (n=3)	Nov.24, 2009 (n=3)	Nov.29, 2009 (n=3)	Dec.4, 2009 (n=3)	Dec.9, 2009 (n=3)	Dec.14, 2009 (n=3)	Dec.19, 2009 (n=3)	Dec.24, 2009 (n=3)	Dec.29, 2009 (n=3)	Jan.3, 2010 (n=3)	Jan.8, 2010 (n=3)	Jan.13, 2010 (n=3)	Jan.18, 2010 (n=3)	Jan.23, 2010 (n=3)	Jan.28, 2010 (n=3)	Feb.2, 2010 (n=3)	Feb.7, 2010 (n=3)	AVE μg/kg
43	monalide	Cleanert TPT	GC-MS	Green tea	A	90.9	97.9	85.3	75.0	77.4	103.2	79.0	90.4	86.5	94.1	94.3	96.7	76.6	81.2	79.9	84.7	70.0	67.8	89.1	85.3
				Green tea	B	90.5	85.6	79.1	75.3	84.1	92.7	69.1	87.5	89.5	111.7	93.1	89.6	74.6	80.5	81.8	90.1	73.7	65.9	78.4	83.8
				Woolong tea	A*	105.2	118.6	89.4	97.3	98.5	86.3	69.3	101.8	105.1	81.5	127.6	84.4	97.2	87.1	82.5	90.7	93.3	82.6	78.1	93.5
				Woolong tea	B	99.4	103.1	76.0	72.6	93.0	69.2	73.3	91.5	97.2	91.9	110.8	78.0	93.8	85.9	75.2	91.8	89.6	78.0	71.9	86.4
			GC-MS/MS	Green tea	A	98.4	82.6	88.3	112.6	66.2	72.4	72.4	68.8	76.5	69.2	81.4	78.3	71.1	67.2	73.1	61.4	66.2	65.3	78.3	76.3
				Green tea	B	94.0	79.2	75.8	84.2	62.1	67.3	65.0	63.5	74.9	68.9	72.8	72.4	69.8	69.6	69.6	60.4	64.2	59.9	68.4	70.6
				Woolong tea	A	80.4	86.8	75.6	81.7	76.5	68.9	70.1	87.2	87.5	85.0	93.2	82.2	76.9	84.9	70.9	77.0	72.1	68.8	69.0	78.7
				Woolong tea	B	75.0	74.4	64.9	69.1	69.3	59.3	56.2	74.3	72.0	72.3	71.8	69.8	67.7	73.8	59.2	66.6	62.0	66.8	60.8	67.6
		Envi-Carb +PSA	GC-MS	Green tea	A*	83.3	96.4	63.2	82.9	105.7	82.4	70.5	104.1	107.2	71.9	100.2	117.3	79.4	95.5	93.8	96.1	82.2	86.1	73.7	89.0
				Green tea	B	80.9	80.6	62.8	78.1	106.9	97.9	57.7	90.8	78.9	71.4	78.5	98.3	84.6	99.7	86.5	82.6	76.1	83.3	74.4	82.6
				Woolong tea	A*	109.6	113.1	97.0	86.4	90.2	91.8	88.7	85.0	85.1	94.9	80.2	88.1	82.1	87.0	80.8	168.0	64.3	70.4	84.7	92.0
				Woolong tea	B*	90.6	88.8	100.6	81.0	94.7	87.3	80.8	81.9	69.7	91.0	100.9	88.6	72.3	95.7	75.8	121.2	63.6	63.0	82.7	85.3
			GC-MS/MS	Green tea	A	80.4	91.4	87.6	90.3	90.2	90.8	75.7	73.9	80.9	80.0	68.0	64.7	63.4	81.7	75.6	84.5	78.3	77.1	74.7	79.4
				Green tea	B	73.6	74.2	84.8	79.3	87.8	82.0	73.0	69.5	77.5	79.0	64.2	60.3	64.3	75.2	70.5	70.0	71.7	68.4	74.0	73.6
				Woolong tea	A	105.5	87.5	98.4	81.0	90.5	73.5	86.2	75.1	85.1	81.7	73.0	66.6	70.8	68.4	75.0	62.9	52.0	66.5	63.5	77.0
				Woolong tea	B	88.2	75.0	73.6	70.8	74.1	64.5	78.0	64.6	69.7	66.0	64.0	60.4	60.0	60.2	58.4	51.8	48.5	55.2	63.6	65.7
44	paraoxon-ethyl	Cleanert TPT	GC-MS	Green tea	A	193.7	210.2	179.8	184.1	198.2	205.4	152.7	202.7	212.5	154.9	179.8	211.8	198.4	189.2	206.5	192.7	197.4	226.4	201.9	194.6
				Green tea	B	190.4	205.9	136.0	184.8	192.8	197.4	129.7	194.8	201.2	164.2	182.6	210.1	206.6	190.9	206.0	197.4	202.2	212.8	216.7	193.3
				Woolong tea	A*	186.5	201.0	223.3	215.0	175.0	208.8	194.3	199.3	206.2	207.9	186.3	247.8	189.2	189.7	180.5	203.6	188.2	154.2	204.7	198.0
				Woolong tea	B	189.1	185.4	202.2	216.8	169.3	214.3	217.8	207.3	211.4	201.0	196.2	207.6	185.5	211.5	186.5	190.0	186.6	172.8	183.0	196.5
			GC-MS/MS	Green tea	A	206.8	178.8	85.2	170.0	133.8	321.2	136.4	105.6	164.5	105.0	200.0	92.1	216.1	122.0	101.5	143.8	57.8	65.2	0.0	137.4
				Green tea	B	207.1	168.5	115.2	131.7	142.1	274.5	123.2	104.2	185.9	156.4	196.3	77.8	165.1	158.0	106.9	143.7	46.2	63.5	0.0	135.1
				Woolong tea	A*	183.6	242.1	178.1	118.6	118.3	97.6	176.3	174.7	0.0	115.1	259.2	116.9	121.9	131.6	149.5	0.0	0.0	0.0	0.0	114.9
				Woolong tea	B	158.9	228.3	158.0	96.9	102.8	98.8	135.2	143.3	211.7	123.0	183.9	89.9	122.4	115.6	111.1	0.0	0.0	0.0	0.0	98.3
		Envi-Carb +PSA	GC-MS	Green tea	A	269.7	188.6	134.3	191.6	179.2	195.6	0.0	219.3	211.7	153.1	197.1	186.9	228.0	207.8	205.5	213.1	190.0	203.0	202.4	190.9
				Green tea	B	207.6	204.7	132.1	197.3	208.5	196.7	0.0	205.2	212.7	162.0	201.7	188.9	220.0	202.6	205.2	197.9	191.8	208.2	209.2	189.6
				Woolong tea	A*	207.7	204.2	196.7	218.4	190.3	191.3	211.6	171.8	196.6	209.6	190.9	178.6	196.2	172.9	206.1	203.9	225.5	221.5	216.1	200.5
				Woolong tea	B	204.7	215.9	210.4	199.7	212.5	187.3	204.6	179.7	137.8	191.9	193.7	176.4	187.4	180.6	181.9	184.8	216.1	217.0	226.0	195.2
			GC-MS/MS	Green tea	A*	153.9	215.3	148.0	115.4	100.3	217.4	98.3	157.2	116.2	115.2	189.6	146.4	114.6	131.4	169.5	85.2	145.9	172.8	183.0	142.3
				Green tea	B	162.5	219.5	159.3	112.8	117.5	178.6	89.6	126.2	122.6	124.7	138.5	140.3	112.7	113.1	145.6	106.0	100.8	65.2	0.0	133.5
				Woolong tea	A	184.0	174.6	195.7	90.0	76.0	141.9	161.5	116.9	196.6	114.2	114.2	86.9	108.2	66.6	89.8	90.1	0.0	0.0	0.0	105.6
				Woolong tea	B	117.6	145.7	123.8	110.3	69.1	123.0	133.3	118.5	137.8	97.3	104.7	88.8	98.2	60.8	86.1	60.8	0.0	0.0	0.0	87.1
45	pebulate	Cleanert TPT	GC-MS	Green tea	A	112.9	90.2	95.8	93.6	78.5	88.7	84.0	80.6	84.6	81.6	74.2	73.0	73.9	71.0	68.6	54.5	73.7	67.6	96.6	81.2
				Green tea	B	103.0	86.8	87.1	91.2	71.2	80.4	78.4	73.8	85.5	72.9	66.9	68.7	70.8	69.3	60.0	55.6	73.0	67.0	89.2	76.4
				Woolong tea	A	107.1	100.2	81.3	92.3	87.8	78.6	76.8	90.3	92.3	89.3	100.1	92.0	86.5	87.4	88.9	84.9	85.1	94.2	71.0	88.7
				Woolong tea	B	96.0	80.4	67.5	75.2	77.6	63.7	59.7	74.0	73.8	74.0	74.3	74.5	68.9	84.1	73.5	74.2	83.7	84.8	61.3	74.8
			GC-MS/MS	Green tea	A																				
				Green tea	B																				
				Woolong tea	A*																				
				Woolong tea	B*																				
		Envi-Carb +PSA	GC-MS	Green tea	A	103.0	104.1	94.3	99.1	107.8	98.5	87.3	89.6	87.1	98.8	86.7	77.0	67.6	66.7	98.6	99.2	98.7	85.7	82.1	91.1
				Green tea	B	95.2	86.3	95.7	89.8	108.1	92.6	83.5	86.9	79.5	93.7	76.0	70.7	58.9	60.8	84.4	77.3	91.3	83.1	85.1	84.1
				Woolong tea	A	121.4	90.5	89.6	88.7	102.0	89.0	94.3	71.1	97.2	86.2	81.1	72.6	79.4	92.6	98.5	72.2	78.0	82.0	83.0	87.9
				Woolong tea	B	88.6	74.6	82.4	76.9	77.4	72.4	76.6	59.0	75.2	64.9	67.3	59.6	56.9	77.1	72.5	67.5	64.5	70.7	67.3	71.1
			GC-MS/MS	Green tea	A										0.0										0.0
				Green tea	B																				
				Woolong tea	A*																				
				Woolong tea	B*																				

#	Compound	Cartridge	Method	Tea																				
46	pentachloroaniline	Cleanert TPT	GC-MS	Green tea	A	97.8	36.3	38.9	30.3	37.2	29.3	32.7	33.6	33.1	35.5	34.7	32.3	30.1	29.0	24.7	27.8	35.5	36.7
					B	86.3	35.2	36.8	27.0	34.2	27.1	30.0	32.1	30.9	30.5	31.1	30.8	28.9	27.7	22.8	25.8	32.2	33.7
				Woolong tea	A	35.2	36.8	33.2	32.4	31.2	32.8	45.0	39.5	38.6	33.7	33.5	33.4	31.8	28.8	31.8	31.1	29.5	34.0
					B	42.0	35.7	28.1	29.3	26.7	29.8	39.9	34.0	33.1	28.3	27.4	29.7	26.2	24.8	27.5	27.5	26.1	29.2
			GC-MS/MS	Green tea	A*	41.2	35.4	39.8	29.3	35.2	31.0	30.3	31.1	30.5	33.9	33.7	29.6	30.9	29.3	26.5	24.5	38.5	33.1
					B	36.2	34.6	35.1	28.7	32.0	28.5	28.5	31.3	36.0	32.6	31.6	30.9	29.1	28.3	25.3	24.4	32.5	31.9
				Woolong tea	A	33.1	29.6	32.1	31.8	29.2	23.1	32.7	28.9	27.7	28.3	27.1	28.9	25.8	27.1	38.4	25.1	24.6	27.5
					B	39.7	38.8	27.1	30.3	24.0	33.7	30.3	34.9	34.9	38.3	37.0	39.2	32.9	38.4	33.1	33.4	32.5	36.5
		Envi-Carb +PSA	GC-MS	Green tea	A	35.9	33.5	39.2	42.0	38.8	32.5	35.2	35.0	33.4	36.1	32.5	36.4	27.8	30.0	32.6	30.0	31.9	33.7
					B	42.3	34.6	35.2	39.7	35.9	46.2	31.7	40.0	34.6	27.8	29.5	30.8	32.3	25.9	21.8	27.7	27.2	33.4
				Woolong tea	A	32.0	29.3	33.9	35.0	30.7	45.6	35.3	31.3	28.7	24.2	25.6	26.0	25.6	21.8	26.0	23.2	24.2	28.8
					B	37.3	41.3	29.8	27.9	41.0	34.1	39.9	36.6	31.2	33.5	32.1	39.6	35.8	38.0	39.0	31.7	37.1	33.7
			GC-MS/MS	Green tea	A	34.6	34.5	36.1	42.3	38.1	33.0	35.9	35.6	36.6	32.0	29.5	36.9	30.2	32.6	35.8	27.4	33.0	36.6
					B	43.1	35.4	31.7	39.2	32.3	33.7	33.0	33.4	35.6	28.5	28.5	28.7	30.9	23.9	21.4	27.7	26.3	33.6
				Woolong tea	A	32.4	30.2	26.5	35.0	32.3	30.1	29.2	26.5	33.4	25.2	25.2	25.5	22.9	19.7	18.8	22.0	23.3	31.6
					B	43.9	32.0	34.9	28.6	26.9	33.5	25.5	30.8	26.5	31.1	32.0	31.0	29.0	22.0	27.6	24.7	32.9	26.1
47	pentachloroanisole	Cleanert TPT	GC-MS	Green tea	A	41.0	30.8	33.2	26.5	34.1	30.7	33.3	30.8	29.3	29.4	30.5	31.1	26.7	21.5	26.8	23.8	28.9	31.6
					B*	33.9	33.0	32.2	30.9	30.7	29.6	30.8	33.0	33.0	32.8	32.5	31.0	30.7	30.6	27.5	28.7	27.9	29.8
				Woolong tea	A	30.9	26.9	26.3	27.6	27.8	24.6	36.3	27.7	27.7	27.5	26.4	31.1	24.8	26.5	23.5	24.4	23.9	31.3
					B	37.8	32.4	26.0	27.5	22.7	30.5	30.3	30.1	32.7	31.0	32.6	27.4	29.9	23.4	27.2	23.9	34.2	26.3
			GC-MS/MS	Green tea	A	37.6	31.0	29.9	26.2	30.2	28.1	26.3	27.8	31.8	28.9	30.3	26.7	28.0	22.9	26.5	23.4	29.6	30.2
					B	33.8	32.6	27.4	31.7	29.4	27.3	33.9	31.4	31.0	31.3	31.8	31.0	28.8	29.5	25.4	29.0	28.0	30.7
				Woolong tea	A	31.9	27.0	23.8	28.1	24.0	22.3	29.6	26.7	26.4	25.1	25.6	26.4	24.0	25.8	22.1	25.4	23.9	26.0
					B	31.5	36.3	33.3	37.2	35.7	36.4	35.7	30.6	32.9	30.0	32.0	35.2	32.1	34.5	32.5	31.3	29.0	33.6
		Envi-Carb +PSA	GC-MS	Green tea	A	35.5	30.4	33.4	36.7	33.3	37.2	36.3	30.1	30.6	28.4	28.9	32.6	29.1	29.3	29.7	27.1	28.4	31.8
					B	40.6	30.3	31.6	32.6	31.1	36.1	29.4	31.3	29.1	26.4	28.8	29.0	30.5	25.1	24.7	25.4	25.3	29.9
				Woolong tea	A	29.3	25.6	30.7	25.7	25.9	30.7	25.7	24.1	31.3	22.5	24.7	24.0	23.3	19.7	20.4	20.7	22.5	24.8
					B*	33.5	36.5	27.0	38.8	34.8	32.0	32.2	35.0	24.1	38.5	33.2	39.4	31.3	32.1	32.1	29.7	30.5	33.6
			GC-MS/MS	Green tea	A	31.7	36.5	34.1	38.7	30.6	32.0	29.4	33.2	26.7	34.8	33.7	33.7	29.2	27.4	29.9	25.9	30.3	31.0
					B	46.6	30.0	30.6	33.3	33.0	30.9	27.4	31.3	28.5	26.5	26.5	26.5	29.1	23.2	20.3	25.4	24.3	29.6
				Woolong tea	A	33.8	31.0	29.0	26.9	24.9	31.5	23.7	23.6	26.1	22.5	24.7	22.1	20.5	18.0	18.3	20.3	22.2	24.2
					B	31.5	26.1	25.5	27.3	30.9	28.1	30.6	29.5	29.6	25.1	32.9	31.8	29.4	21.4	26.9	23.5	31.7	30.1
48	pentachlorobenzene	Cleanert TPT	GC-MS	Green tea	A	37.5	36.3	31.9	27.3	27.8	30.7	28.3	26.8	32.9	30.0	31.1	29.5	26.1	21.5	26.2	23.2	27.6	28.0
					B	34.8	30.4	34.2	24.8	30.8	35.8	38.6	36.8	30.6	28.4	36.2	35.6	34.3	35.4	34.7	32.2	30.9	36.2
				Woolong tea	A	37.4	41.8	31.6	36.4	26.4	26.8	34.1	32.6	42.5	37.5	30.1	32.6	34.3	35.4	28.1	30.3	28.7	31.1
					B	35.7	35.2	30.7	34.4	30.8	30.3	28.5	28.6	32.4	30.7	30.4	32.0	28.9	30.0	25.9	23.0	31.9	29.7
			GC-MS/MS	Green tea	A	36.4	29.8	36.7	25.5	26.7	30.3	27.5	25.7	30.3	30.1	30.4	25.4	27.4	23.8	25.9	23.4	28.8	27.7
					B	34.8	28.4	30.6	24.4	30.6	27.8	40.6	27.5	27.7	28.5	28.5	24.7	34.0	33.5	25.4	32.4	30.3	36.3
				Woolong tea	A	44.6	41.3	30.6	35.3	31.6	34.1	36.4	36.5	45.7	36.0	33.1	40.2	34.0	31.5	26.5	30.2	30.3	31.4
					B	39.0	36.6	31.5	33.3	27.1	27.2	32.9	32.3	33.3	30.4	31.2	35.0	27.8	26.5	33.4	32.5	27.0	33.5
		Envi-Carb +PSA	GC-MS	Green tea	A	36.1	34.7	33.2	36.6	33.4	34.9	34.4	26.6	31.4	34.6	33.1	37.0	33.7	35.0	30.7	27.2	29.1	30.8
					B	32.4	28.7	30.2	36.5	32.3	33.5	34.1	26.3	27.6	32.4	28.2	34.3	28.2	29.6	30.7	28.7	28.2	35.3
				Woolong tea	A	46.4	41.0	36.6	39.9	35.6	39.2	33.2	36.8	34.1	32.5	32.0	35.8	34.8	28.6	24.5	24.2	30.1	30.1
					B	40.3	34.9	32.6	31.8	30.9	32.2	29.1	30.1	29.5	28.6	27.8	29.1	28.0	23.8	34.8	28.7	27.2	27.2
			GC-MS/MS	Green tea	A	32.8	36.0	33.5	37.8	34.7	33.4	32.4	32.9	30.4	38.0	33.8	36.3	32.4	32.8	34.8	28.7	29.0	33.4
					B	30.9	28.3	29.9	37.8	32.3	28.8	30.1	25.7	29.7	38.5	29.6	34.5	27.4	28.3	32.3	24.9	28.8	30.7
				Woolong tea	A	43.2	40.4	34.9	38.6	34.4	36.9	34.4	35.7	34.4	28.3	28.9	30.7	33.2	27.2	24.9	31.3	28.4	34.0
					B	38.7	36.1	31.4	31.5	29.6	33.1	29.0	29.6	29.7	26.8	24.2	27.6	26.2	22.1	23.3	25.9	26.3	29.3

(Continued)

Annex 1: Content change for 201 pesticides in incurred tea Youden pair samples under 8 conditions with two SPE cartridge cleanup in three months (Nov.9, 2009-Feb.7, 2010) (cont.)

No	Pesticides	SPE	Method	Sample	Youden pair	Nov.9, 2009 (n=5)	Nov.14, 2009 (n=3)	Nov.19, 2009 (n=3)	Nov.24, 2009 (n=3)	Nov.29, 2009 (n=3)	Dec.4, 2009 (n=3)	Dec.9, 2009 (n=3)	Dec.14, 2009 (n=3)	Dec.19, 2009 (n=3)	Dec.24, 2009 (n=3)	Dec.29, 2009 (n=3)	Jan.3, 2010 (n=3)	Jan.8, 2010 (n=3)	Jan.13, 2010 (n=3)	Jan.18, 2010 (n=3)	Jan.23, 2010 (n=3)	Jan.28, 2010 (n=3)	Feb.2, 2010 (n=3)	Feb.7, 2010 (n=3)	AVE, μg/kg
49	perthane	Cleanert TPT	GC-MS	Green tea	A	53.1	35.4	35.9	37.3	30.7	35.8	27.5	30.1	33.4	32.3	40.3	36.7	36.4	32.1	32.4	26.8	28.5	26.9	34.6	34.0
					B	45.3	33.8	32.6	34.6	28.4	32.7	26.2	28.5	35.1	30.7	36.1	33.4	32.4	32.3	29.2	26.1	28.1	25.5	31.1	31.7
				Woolong tea	A	191.8	43.1	33.9	37.3	39.1	34.7	32.9	38.2	38.0	37.3	38.6	35.8	37.7	35.1	34.9	36.6	33.6	33.5	31.8	44.4
					B	180.0	37.1	29.1	32.8	33.0	26.5	26.2	31.5	33.4	32.1	29.7	31.6	31.3	33.1	28.7	32.1	28.2	28.7	29.5	38.7
			GC-MS/MS	Green tea	A	48.2	36.9	49.6	43.3	28.6	38.4	31.3	27.6	33.0	31.0	40.4	31.6	32.0	34.9	31.8	27.5	27.1	24.6	37.0	34.5
					B	46.3	35.6	45.0	34.9	26.7	34.8	28.4	26.6	33.6	29.2	35.7	29.8	30.0	35.0	30.3	27.0	26.8	24.4	31.9	32.2
				Woolong tea	A	36.9	39.0	37.9	37.7	39.8	31.9	32.1	37.9	36.4	36.4	44.9	35.5	35.9	34.9	31.6	33.1	28.4	30.0	33.1	35.5
					B	35.3	33.9	31.1	30.4	35.3	26.1	25.8	32.0	30.3	31.6	34.0	29.3	30.6	30.9	25.8	28.1	26.5	27.8	30.7	30.3
		Envi-Carb +PSA	GC-MS	Green tea	A	40.7	42.5	38.7	36.8	39.7	39.6	35.2	35.3	35.7	41.8	35.9	38.3	33.1	39.1	33.8	38.0	34.8	31.6	29.1	36.8
					B	47.6	36.9	37.6	32.9	38.0	37.1	34.4	30.8	33.2	41.0	32.9	35.0	29.8	33.9	31.4	30.1	30.9	28.4	29.1	33.7
				Woolong tea	A	37.9	38.6	38.3	39.1	39.7	36.4	39.5	35.9	40.6	37.2	34.6	31.5	33.2	34.9	35.0	28.6	28.4	31.4	29.1	35.8
					B	40.3	44.0	32.5	35.2	30.5	31.2	34.4	30.0	32.8	31.2	30.8	27.5	28.9	29.9	29.0	23.1	24.7	25.2	26.2	30.3
			GC-MS/MS	Green tea	A	37.6	38.0	37.9	39.0	41.0	39.9	36.6	34.4	34.3	35.8	33.9	34.7	26.4	39.5	34.1	34.6	33.3	28.8	32.3	33.1
					B	48.4	38.9	47.7	34.5	39.4	35.6	34.2	33.6	40.6	37.3	30.4	32.5	27.8	34.3	31.7	28.7	29.2	25.6	26.1	35.2
				Woolong tea	A	38.9	35.0	33.2	30.4	37.5	31.6	34.5	29.6	32.8	30.2	33.7	29.6	31.4	31.5	34.8	27.6	27.5	33.3	23.7	29.7
					B	43.2	36.0	38.9	42.8	31.4	39.3	34.9	32.7	35.8	34.4	30.5	27.2	27.3	27.6	26.9	22.4	24.4	26.0	38.5	35.3
50	phenanthrene	Cleanert TPT	GC-MS	Green tea	A	41.0	34.7	34.7	40.8	30.8	35.1	32.0	30.1	36.4	32.8	31.9	32.0	35.1	33.6	30.0	24.2	30.8	26.9	33.7	33.0
					B	39.1	42.2	34.5	37.0	38.3	31.9	31.6	39.1	39.0	38.4	40.3	39.3	38.4	38.1	36.4	37.1	32.1	35.1	34.0	36.9
				Woolong tea	A	36.7	34.9	28.5	31.8	34.2	27.0	26.7	32.6	33.4	33.1	31.9	33.3	31.8	33.9	29.7	26.7	27.9	31.8	30.6	31.7
					B	43.2	38.6	37.2	46.3	30.0	36.4	36.2	31.2	37.5	33.6	38.5	37.4	36.9	29.1	35.8	26.7	32.3	28.9	40.0	35.6
			GC-MS/MS	Green tea	A	43.0	37.3	33.9	37.6	28.9	34.2	34.6	30.9	36.8	33.1	42.7	35.0	34.6	41.2	33.7	27.4	32.8	27.6	34.7	34.0
					B	37.0	43.0	31.1	38.0	36.0	33.0	34.1	39.9	42.9	39.0	42.7	38.9	36.4	37.6	33.7	32.7	32.5	37.9	30.4	36.9
				Woolong tea	A	35.5	35.8	28.7	33.6	35.0	28.8	28.1	34.2	35.2	35.0	35.0	33.0	31.0	37.6	28.0	30.1	30.1	35.1	27.2	32.4
					B	41.2	39.5	37.4	40.3	42.2	40.2	35.5	38.1	38.4	34.3	43.5	38.2	38.4	43.6	37.4	41.3	36.4	35.4	34.3	38.3
		Envi-Carb +PSA	GC-MS	Green tea	A	36.8	34.1	37.6	37.0	41.1	37.4	35.5	35.6	35.1	34.2	43.1	36.6	34.7	39.2	34.2	35.3	33.1	30.6	33.4	35.5
					B	45.4	39.4	37.3	39.3	41.0	36.3	40.0	31.8	42.0	37.4	34.2	33.1	32.6	36.9	36.3	30.4	30.2	31.9	31.9	36.2
				Woolong tea	A	35.6	32.9	33.3	34.7	32.7	30.9	33.9	29.2	35.3	30.5	29.9	28.7	28.9	30.5	29.5	25.5	25.3	26.6	28.4	30.7
					B	36.6	41.9	39.8	44.3	42.7	43.9	37.9	38.1	37.4	42.4	34.8	38.1	29.6	45.3	39.1	38.7	38.0	37.4	28.4	39.1
			GC-MS/MS	Green tea	A	34.8	32.9	39.4	40.5	41.8	39.6	36.9	33.7	35.6	41.4	30.7	36.0	31.8	38.9	36.6	33.6	35.3	31.2	36.7	36.2
					B	46.7	39.5	43.7	35.3	40.1	37.1	38.7	34.3	42.0	40.3	36.3	27.5	30.8	32.5	34.4	27.3	24.7	34.3	27.4	35.7
				Woolong tea	A	37.4	34.9	35.7	32.2	34.6	30.7	34.9	28.8	35.3	32.4	32.2	27.0	30.1	29.9	25.2	24.2	22.8	26.5	25.4	30.5
51	pirimicarb	Cleanert TPT	GC-MS	Green tea	A	103.8	74.4	83.5	73.3	57.7	63.7	59.8	56.8	67.6	52.7	91.5	52.3	66.1	62.6	61.7	54.8	54.6	47.5	65.4	65.8
					B	95.9	74.2	63.8	66.0	53.6	60.6	58.5	53.8	69.3	48.6	76.9	49.3	60.2	62.2	57.2	53.7	52.3	46.1	57.5	61.1
				Woolong tea	A	99.1	75.4	67.3	73.7	64.6	55.1	63.6	69.0	72.7	64.3	79.2	69.8	69.2	65.6	63.4	65.2	63.2	60.3	56.8	68.3
					B	91.5	67.4	58.9	64.3	55.6	47.0	51.0	61.0	55.0	54.6	53.6	55.4	55.5	57.9	51.5	54.9	50.3	56.5	51.6	57.6
			GC-MS/MS	Green tea	A*	93.8	72.7	88.3	86.2	53.2	110.7	60.3	51.6	64.9	57.0	72.5	62.4	58.9	54.2	61.5	51.7	53.4	46.3	72.7	67.0
					B*	90.4	70.4	73.1	70.3	49.8	94.9	55.2	50.1	65.4	59.4	64.7	58.2	55.0	57.2	59.2	51.1	52.4	45.4	61.7	62.3
				Woolong tea	A	82.7	79.7	68.0	68.9	62.6	59.9	63.2	75.1	68.8	70.4	84.3	70.8	58.6	70.4	61.9	61.8	56.4	57.3	63.4	67.1
					B	74.6	67.9	57.9	59.0	56.1	50.4	47.0	64.4	55.9	58.8	59.2	55.7	54.2	61.6	51.4	53.8	44.9	54.2	46.9	56.5
		Envi-Carb +PSA	GC-MS	Green tea	A*	83.8	79.0	77.8	70.7	85.7	79.5	58.8	65.4	66.7	164.9	65.0	54.7	62.9	67.2	64.9	67.0	67.8	57.2	55.3	73.4
					B*	75.9	66.8	77.8	61.9	76.9	69.7	62.0	61.5	62.8	119.5	58.6	49.8	57.5	62.4	58.6	54.5	59.6	52.1	55.3	65.6
				Woolong tea	A	93.4	83.7	81.2	73.3	72.0	68.1	72.9	74.1	83.6	69.7	76.5	52.5	60.8	65.0	63.2	52.1	53.5	56.3	55.3	70.6
					B	79.4	66.2	77.4	66.7	77.3	57.4	55.3	57.5	63.7	53.0	56.5	51.7	51.8	53.5	50.9	42.2	44.5	44.6	48.4	56.6
			GC-MS/MS	Green tea	A	77.3	83.4	74.8	76.9	75.2	81.4	64.0	65.0	64.4	72.2	77.7	69.8	48.5	75.5	62.2	64.8	63.2	54.2	59.5	68.0
					B	71.6	71.4	72.6	67.3	75.2	71.9	65.8	59.3	65.0	69.8	53.2	66.9	51.8	59.9	58.1	53.4	56.9	49.4	59.8	63.1
				Woolong tea	A	89.4	79.7	86.7	65.1	70.3	65.8	68.3	62.0	83.6	68.7	65.3	57.2	58.5	57.5	62.5	49.4	47.2	60.9	52.2	65.8
					B	75.5	66.4	63.4	58.5	56.0	56.0	61.9	51.0	63.7	53.5	54.9	52.2	50.5	49.8	46.9	39.6	42.0	47.1	52.2	54.5

52	procymidone	Cleanert TPT	GC-MS	Green tea	A	47.9	36.8	40.4	41.7	33.5	37.6	34.7	33.8	36.5	35.6	31.0	38.6	39.5	42.3	32.4	24.7	32.2	29.3	38.8	36.2
					B	45.0	35.5	36.2	39.0	30.6	33.8	31.6	30.2	35.6	34.7	27.6	36.1	42.9	43.5	29.2	24.1	31.1	27.9	33.8	34.1
				Woolong tea	A	39.5	47.1	36.4	38.8	36.7	33.5	33.7	39.9	40.5	40.4	40.1	47.3	46.1	48.3	36.5	38.3	35.3	35.3	35.5	39.4
					B*	37.2	39.3	30.6	33.5	34.3	28.4	27.9	33.7	34.2	35.1	36.0	39.0	37.6	48.1	31.2	32.8	29.9	33.2	32.1	34.4
			GC-MS/MS	Green tea	A	46.9	39.2	42.8	44.1	31.9	34.6	34.7	30.2	37.1	34.0	39.1	35.7	35.1	32.6	33.5	27.6	30.5	28.3	38.0	35.6
					B	45.2	37.1	36.3	38.3	29.8	32.3	31.7	28.7	36.3	31.7	34.8	34.7	33.0	34.0	32.5	27.4	29.9	28.1	32.5	33.4
				Woolong tea	A	38.8	42.2	36.3	40.5	36.7	32.7	33.9	42.4	42.3	39.8	44.7	41.1	34.7	41.1	34.9	36.5	32.3	34.0	32.7	37.8
					B	36.8	36.5	31.2	34.9	34.8	28.6	28.1	35.9	36.0	34.2	33.3	34.3	32.5	36.0	29.1	31.7	27.5	32.9	29.4	32.8
		Envi-Carb +PSA	GC-MS	Green tea	A	41.8	43.0	40.7	41.7	40.9	41.3	28.9	36.4	37.8	32.8	33.7	40.0	37.6	47.8	41.1	41.4	36.3	37.3	34.0	38.7
					B	38.6	36.7	39.2	37.9	40.5	37.8	29.1	35.0	36.7	33.3	31.7	35.6	39.5	41.2	33.9	34.1	32.2	31.8	34.1	35.7
				Woolong tea	A*	50.9	43.6	39.4	39.3	41.8	37.4	43.0	32.0	41.8	39.3	34.9	31.2	40.4	46.2	38.9	30.6	30.4	33.3	32.5	38.3
					B*	41.9	36.6	36.7	34.7	34.1	32.3	37.6	29.6	33.8	31.4	31.3	28.2	33.9	42.0	27.8	26.3	26.4	27.4	29.5	32.7
			GC-MS/MS	Green tea	A	39.0	43.8	40.3	41.9	43.5	42.7	35.3	36.7	36.6	38.6	32.2	43.0	30.3	46.2	36.0	39.6	38.9	33.4	36.4	38.6
					B	36.6	36.3	39.0	37.8	42.1	39.0	34.8	33.8	35.5	37.9	30.8	40.2	30.5	39.1	33.7	32.9	33.9	29.1	36.8	35.8
				Woolong tea	A	53.1	43.1	49.2	37.7	42.4	37.1	39.7	36.4	41.8	39.6	35.3	31.8	32.3	33.1	36.3	29.2	24.2	31.7	30.7	37.1
					B	43.7	37.0	38.0	33.4	34.8	31.5	36.6	31.6	33.8	31.5	31.7	28.7	26.9	28.9	28.0	24.5	23.0	25.6	27.4	31.4
53	prometrye	Cleanert TPT	GC-MS	Green tea	A	53.6	41.1	45.4	39.8	30.4	32.7	31.7	31.2	35.2	27.8	47.9	26.5	33.7	29.8	27.2	21.6	27.7	23.1	31.7	33.6
					B	43.6	37.6	37.4	36.2	28.0	33.0	29.3	28.0	34.1	26.0	39.5	26.5	31.7	28.0	25.0	20.5	26.7	22.6	27.6	30.6
				Woolong tea	A	41.6	43.6	35.6	37.4	34.5	31.4	30.3	37.7	37.8	32.8	35.2	37.5	34.7	35.5	31.6	32.5	30.2	30.7	29.3	34.7
					B	40.2	39.0	30.6	33.5	31.5	26.1	24.5	31.1	27.8	28.3	25.4	30.4	27.6	30.7	26.9	27.3	24.2	28.4	26.3	29.4
			GC-MS/MS	Green tea	A*	46.3	38.6	43.1	38.7	27.8	54.2	31.1	27.4	33.9	29.8	35.3	33.3	30.2	28.0	31.4	25.1	27.6	23.0	32.5	33.5
					B*	45.1	37.5	36.6	35.7	26.0	45.1	28.5	26.3	33.1	29.7	31.3	30.6	28.2	29.4	28.8	24.6	27.5	22.9	28.0	31.3
				Woolong tea	A	42.7	46.7	35.4	35.4	31.3	30.6	33.1	38.4	35.8	37.0	42.3	36.1	31.4	36.6	30.4	32.2	30.5	29.8	28.0	34.9
					B	42.9	41.5	31.4	31.1	30.0	26.1	26.4	33.0	30.4	32.0	29.9	30.2	28.9	31.6	25.4	27.2	23.7	29.4	24.8	30.3
		Envi-Carb +PSA	GC-MS	Green tea	A*	46.1	44.0	44.6	40.2	44.8	46.1	33.6	34.3	33.0	86.7	31.2	31.9	30.8	36.5	31.5	34.9	30.0	30.2	28.1	38.9
					B*	39.0	38.5	41.7	35.5	41.5	38.9	32.5	32.5	30.7	61.9	29.5	27.8	28.1	32.3	26.7	27.8	26.9	26.1	28.4	34.0
				Woolong tea	A	52.6	45.3	39.3	38.8	54.7	34.9	39.6	33.9	43.2	36.9	38.5	26.0	29.6	32.4	35.7	28.1	25.7	28.4	27.7	36.4
					B	45.8	38.0	34.1	33.8	36.7	29.9	33.8	26.0	33.2	28.4	28.5	24.8	26.4	25.7	26.3	23.8	22.0	22.8	24.4	29.7
			GC-MS/MS	Green tea	A	38.4	46.1	40.2	42.6	41.2	45.0	33.4	33.0	32.3	35.4	28.3	35.7	23.9	35.1	31.4	34.9	33.4	29.1	29.8	35.2
					B	35.9	36.9	39.0	37.2	39.3	39.1	31.5	30.6	31.5	34.9	27.0	33.5	24.9	29.7	27.1	28.6	30.4	25.5	30.6	32.3
				Woolong tea	A	53.1	42.2	47.9	34.8	38.8	35.6	36.6	32.6	43.2	36.8	34.3	28.5	30.5	29.8	33.0	23.1	22.0	28.2	25.4	34.5
					B	43.0	36.5	35.4	32.0	31.5	29.8	33.7	27.7	33.2	28.8	28.4	25.5	26.8	25.2	25.0	20.7	20.4	22.0	23.7	28.9
54	propham	Cleanert TPT	GC-MS	Green tea	A*	44.3	41.2	40.7	34.7	34.4	38.6	37.2	37.3	40.3	38.5	44.0	40.0	43.1	38.1	40.0	58.6	34.0	30.9	40.4	39.8
					B*	42.4	39.3	37.0	32.0	32.5	35.6	33.7	35.2	41.3	37.3	40.3	37.5	40.9	37.9	38.4	54.9	33.5	29.9	36.5	37.7
				Woolong tea	A	44.2	41.5	37.7	42.3	40.4	33.1	45.0	46.4	42.2	39.5	52.4	42.1	42.8	40.8	42.9	42.2	45.9	40.7	42.6	42.3
					B	40.8	35.5	33.3	35.9	35.9	28.2	40.5	43.4	36.2	34.9	38.5	35.5	37.5	38.6	36.1	37.9	39.8	38.0	35.2	36.9
			GC-MS/MS	Green tea	A	47.4	40.7	44.6	44.7	33.6	53.8	36.0	32.9	39.6	35.1	43.3	37.5	40.5	31.1	37.3	34.3	32.6	28.7	54.9	39.4
					B	47.2	38.4	41.9	39.9	31.7	48.3	33.5	32.1	40.4	34.8	40.1	34.8	38.3	33.0	36.3	33.8	32.2	29.0	46.7	37.5
				Woolong tea	A	47.6	44.5	39.1	39.5	38.0	36.3	47.5	43.7	42.9	42.3	52.7	40.6	38.1	42.8	38.7	40.9	35.8	37.1	33.7	41.1
					B	41.7	38.7	32.9	33.5	33.9	30.6	38.6	40.3	34.9	35.7	37.6	32.7	36.6	37.8	31.3	36.5	32.2	32.7	31.1	35.2
		Envi-Carb +PSA	GC-MS	Green tea	A	38.1	44.6	47.0	41.5	45.3	43.8	41.4	46.0	41.0	55.0	44.4	45.3	41.4	36.8	43.2	38.4	52.0	32.6	35.2	42.8
					B	37.3	39.3	45.6	35.7	44.1	38.0	40.4	41.5	39.8	49.6	38.0	42.8	36.4	37.7	37.3	34.9	45.9	32.4	37.0	39.7
				Woolong tea	A*	46.3	39.8	41.4	38.6	51.8	41.3	43.8	51.6	48.7	40.8	39.5	34.7	37.9	43.1	38.9	34.3	51.9	55.1	45.0	43.4
					B*	37.1	33.8	36.6	34.7	40.4	33.4	35.9	41.1	39.8	33.2	34.0	31.7	33.6	37.5	33.1	28.5	50.9	55.1	40.6	37.4
			GC-MS/MS	Green tea	A	40.4	46.5	42.5	44.7	44.0	44.1	39.1	40.9	36.8	46.5	38.7	35.6	37.5	46.2	40.5	38.3	39.8	34.3	38.5	40.8
					B	41.9	41.2	41.7	38.0	43.7	40.5	41.8	37.2	37.8	43.6	32.4	33.4	36.1	42.7	36.2	34.2	37.0	31.5	38.6	38.4
				Woolong tea	A	42.6	42.0	50.9	36.5	41.9	38.9	38.9	38.2	48.7	40.0	36.2	31.7	34.6	33.7	36.7	30.4	35.0	46.2	37.1	39.0
					B	33.7	35.5	35.9	32.7	33.8	32.5	35.3	31.2	39.8	31.9	31.5	30.2	31.4	30.5	28.4	24.5	31.3	36.2	33.6	32.6

(Continued)

Annex 1: Content change for 201 pesticides in incurred tea Youden pair samples under 8 conditions with two SPE cartridge cleanup in three months (Nov.9, 2009-Feb.7, 2010) (cont.)

No	Pesticides	SPE	Method	Sample	Youden pair	Nov.9, 2009 (n=5)	Nov.14, 2009 (n=3)	Nov.19, 2009 (n=3)	Nov.24, 2009 (n=3)	Nov.29, 2009 (n=3)	Dec.4, 2009 (n=3)	Dec.9, 2009 (n=3)	Dec.14, 2009 (n=3)	Dec.19, 2009 (n=3)	Dec.24, 2009 (n=3)	Dec.14, 2009 (n=3)	Jan.3, 2010 (n=3)	Jan.8, 2010 (n=3)	Jan.13, 2010 (n=3)	Jan.18, 2010 (n=3)	Jan.23, 2010 (n=3)	Jan.28, 2010 (n=3)	Feb.2, 2010 (n=3)	Feb.7, 2010 (n=3)	AVE. μg/kg.
55	prosulfocarb	Cleanert TPT	GC-MS	Green tea	A	46.8	35.2	39.3	36.9	29.9	33.4	32.0	31.0	34.4	28.5	39.0	29.7	35.2	33.0	30.2	20.4	28.8	25.2	33.5	32.8
					B	42.1	34.4	32.9	34.9	27.3	31.5	29.2	28.2	33.9	26.1	33.7	27.4	31.3	29.7	26.5	20.3	27.7	24.1	29.6	30.0
				Woolong tea	A	39.7	39.0	32.4	35.6	31.9	29.5	28.2	35.5	36.4	35.7	35.9	37.0	37.1	35.6	33.3	33.1	30.7	32.9	30.8	34.2
					B	37.3	33.5	27.9	30.7	29.3	25.0	24.1	29.5	29.0	29.2	25.4	29.6	27.8	30.7	26.7	28.7	24.7	28.4	27.4	28.7
			GC-MS/MS	Green tea	A	42.8	34.4	36.4	47.3	27.1	42.8	31.6	27.1	33.1	28.4	35.4	31.7	31.7	31.0	31.0	25.2	27.7	25.5	34.4	32.9
					B	41.4	33.0	30.2	35.1	25.4	37.2	28.7	26.4	32.7	27.5	32.0	29.9	29.9	29.9	29.2	24.0	26.0	24.8	29.3	30.1
				Woolong tea	A	36.1	37.7	33.8	34.2	31.3	28.5	30.0	36.9	34.2	36.7	38.7	36.1	29.9	36.6	31.6	30.5	28.5	28.2	27.9	33.0
					B	33.5	31.6	29.1	29.5	28.9	24.4	23.9	31.7	29.0	30.5	29.1	30.1	29.0	32.5	25.6	27.9	24.3	25.7	25.0	28.5
		Envi-Carb +PSA	GC-MS	Green tea	A*	40.2	38.7	37.5	37.4	41.7	40.1	37.5	34.0	33.7	69.0	30.6	34.4	33.3	37.9	32.4	34.8	32.0	33.0	28.6	37.2
					B*	35.7	33.5	37.4	34.0	39.3	35.6	32.8	31.2	30.8	58.3	29.1	31.5	30.8	33.6	30.6	30.1	29.1	27.5	29.1	33.7
				Woolong tea	A	46.2	39.5	35.9	35.5	44.0	33.6	38.3	32.0	39.7	36.5	36.2	29.1	30.5	34.9	34.3	27.7	28.4	29.1	26.9	34.6
					B	36.1	32.4	30.6	31.1	32.3	28.7	32.4	25.9	31.0	27.9	27.4	25.1	25.7	26.3	26.3	22.5	22.9	23.8	24.5	28.1
			GC-MS/MS	Green tea	A	35.5	39.9	37.1	39.6	40.4	40.1	34.3	33.4	32.5	36.7	29.0	33.4	27.1	34.3	31.8	34.5	34.2	30.4	31.0	34.5
					B	32.9	32.4	36.4	34.5	39.6	34.8	34.1	30.7	31.9	36.4	28.2	33.2	27.6	30.9	29.1	29.8	30.3	25.5	30.0	32.0
				Woolong tea	A	45.7	38.5	47.9	33.3	35.5	34.2	35.8	30.9	39.7	34.9	33.3	27.0	29.0	30.5	33.0	25.3	22.8	27.2	25.4	33.2
					B	35.9	32.6	32.7	29.5	28.7	28.5	31.8	26.3	31.0	27.2	28.3	24.7	26.0	25.9	24.3	21.0	21.2	21.3	23.9	27.4
56	secbumeton	Cleanert TPT	GC-MS	Green tea	A	62.1	43.1	43.8	40.9	30.7	27.8	32.6	31.0	35.7	26.5	52.0	26.1	26.8	35.0	24.3	23.1	30.0	24.9	36.2	35.3
					B	54.9	41.0	36.1	37.5	24.9	27.9	29.8	31.0	35.8	23.6	40.6	25.2	33.2	33.2	33.4	23.1	28.7	24.2	31.7	32.1
				Woolong tea	A	50.7	37.5	34.7	35.3	31.5	29.5	31.4	33.4	38.3	34.4	43.7	34.9	34.3	34.9	35.0	35.8	33.7	34.5	32.3	35.6
					B	46.4	35.9	31.2	33.0	29.3	25.5	24.9	28.4	28.8	29.1	29.0	28.0	27.5	31.4	29.5	30.3	26.8	32.6	29.3	30.4
			GC-MS/MS	Green tea	A*	0.0	0.0	0.0	0.0	0.0	0.0	0.0	0.0	0.0	0.0	0.0	0.0	0.0	0.0	0.0	0.0	0.0	0.0	0.0	0.0
					B*	45.5	39.1	42.4	37.1	24.5	62.4	30.0	26.7	33.4	29.7	33.3	32.9	32.0	30.9	30.2	27.0	27.8	23.5	32.0	33.7
				Woolong tea	A	39.1	40.9	38.5	34.7	28.4	31.4	32.9	38.0	33.0	39.2	42.7	35.4	31.9	37.0	32.2	33.7	30.1	31.8	29.1	34.8
					B	38.0	35.7	34.0	32.5	28.8	27.1	26.0	33.6	28.8	32.4	32.0	29.3	31.0	32.2	27.0	28.2	25.6	30.1	27.0	30.5
		Envi-Carb +PSA	GC-MS	Green tea	A*	45.2	46.4	48.7	39.4	44.3	52.2	33.6	34.5	32.9	71.3	29.8	33.1	32.0	37.5	33.0	37.3	32.7	31.5	29.7	39.2
					B*	41.5	40.1	45.0	32.4	38.6	42.2	33.5	32.6	30.8	48.4	28.0	27.9	28.5	32.3	28.8	29.3	29.1	27.2	30.2	34.0
				Woolong tea	A	55.5	41.4	40.3	36.9	70.2	35.3	39.0	32.1	49.5	39.4	45.4	25.1	31.8	33.6	35.2	28.1	29.0	31.5	31.1	38.4
					B	46.2	35.4	33.8	35.9	42.9	30.1	34.1	26.7	37.7	29.0	34.5	25.7	27.9	26.3	28.5	23.0	24.6	25.9	27.3	31.3
			GC-MS/MS	Green tea	A	39.4	45.6	40.9	43.2	41.1	48.5	34.7	35.1	34.4	37.1	29.3	30.8	24.3	37.4	32.1	35.9	32.9	29.8	31.7	36.0
					B	37.4	36.8	39.2	36.7	39.5	39.2	35.0	31.8	35.0	36.2	27.3	29.2	24.8	31.3	27.9	29.2	30.1	25.7	31.7	32.8
				Woolong tea	A	53.1	40.0	46.1	33.1	35.0	36.0	37.8	31.8	49.5	37.4	35.4	29.5	31.2	32.9	35.9	27.0	23.6	31.3	27.9	35.5
					B	44.1	35.5	35.2	30.6	29.3	30.3	34.7	27.1	37.7	29.8	30.2	27.0	27.3	28.1	26.6	22.5	21.6	24.2	26.6	29.9
57	silafluofen	Cleanert TPT	GC-MS	Green tea	A	67.6	55.7	35.7	41.5	32.3	43.5	34.6	27.2	28.0	33.9	51.6	44.2	46.6	31.4	34.9	23.1	27.6	25.1	33.5	37.8
					B	50.4	62.7	31.0	40.6	26.2	37.4	29.2	24.9	31.9	32.3	44.6	38.9	38.8	29.1	33.8	21.1	28.8	24.5	31.6	34.6
				Woolong tea	A	43.6	33.1	35.0	30.7	37.3	31.9	24.7	34.1	39.6	35.1	42.4	0.0	36.0	32.3	34.4	31.2	30.9	29.6	31.5	32.4
					B	42.6	28.7	33.2	23.8	34.5	29.1	23.8	37.0	33.0	31.8	33.9	0.0	36.5	31.3	27.9	27.4	24.6	26.6	26.6	29.2
			GC-MS/MS	Green tea	A	46.1	36.5	41.9	28.8	29.5	32.1	32.8	27.9	34.9	29.6	35.8	32.7	31.4	35.7	31.4	24.0	27.0	23.6	35.2	32.5
					B	44.3	35.6	34.5	35.6	26.6	29.0	29.8	28.5	34.7	31.4	31.4	29.5	29.7	36.5	29.8	24.1	25.2	24.0	30.4	31.0
				Woolong tea	A	38.5	39.7	33.1	35.3	34.8	29.6	27.0	35.0	40.0	34.3	39.9	34.4	28.3	38.3	30.2	32.3	28.7	30.1	26.8	33.5
					B	36.7	34.4	30.5	30.4	34.1	26.6	25.9	31.8	33.1	30.8	32.1	30.2	30.2	33.7	25.4	28.0	23.7	26.5	23.2	29.8
		Envi-Carb +PSA	GC-MS	Green tea	A	49.5	64.9	37.7	34.9	38.7	33.0	31.8	32.6	33.5	41.8	30.1	35.8	36.0	34.4	48.8	34.7	30.7	31.3	30.3	37.4
					B	43.3	67.8	37.9	31.9	39.5	31.2	31.6	29.8	30.5	39.0	21.3	35.2	27.2	30.5	38.1	27.4	28.2	25.6	31.4	34.1
				Woolong tea	A*	75.2	40.7	32.8	30.5	43.9	33.9	38.2	33.8	37.5	35.7	32.8	30.0	51.4	41.2	36.3	27.5	29.9	25.5	29.8	37.2
					B	47.3	34.7	32.8	29.8	33.1	27.6	32.0	27.6	30.5	27.4	29.6	27.4	37.1	32.6	28.2	22.2	25.3	22.8	27.3	30.4
			GC-MS/MS	Green tea	A	35.9	40.9	38.2	40.6	39.0	40.1	34.0	34.7	35.6	34.7	28.5	34.5	28.2	34.8	32.0	34.9	34.1	30.5	31.2	35.1
					B	34.7	34.8	37.7	35.1	39.2	36.2	31.9	31.6	32.2	33.8	28.3	33.0	29.9	29.7	29.7	29.1	31.1	25.7	31.3	32.6
				Woolong tea	A	54.6	38.4	43.2	36.6	39.8	32.2	35.7	31.8	37.5	36.6	32.9	28.0	27.4	34.8	34.0	25.0	24.2	29.3	27.6	33.9
					B	44.8	35.0	35.1	33.2	33.0	27.1	31.2	28.6	30.5	28.0	27.1	25.8	23.6	25.7	25.1	20.8	20.4	24.5	25.7	28.7

No.	Compound	Cartridge	Method	Tea	Row																			
58	tebupirimfos	Cleanert TPT	GC-MS	Green tea	A	275.4	203.1	192.7	191.6	126.3	125.9	119.3	54.3	69.1	53.0	79.3	48.3	63.3	57.1	54.4	39.2	51.4	44.5	60.6	100.5
					B	248.2	194.6	168.0	169.9	121.6	122.2	110.2	50.8	65.1	48.3	66.3	46.1	58.1	54.1	49.0	38.4	49.2	42.9	53.0	92.4
				Woolong tea	A	237.9	193.0	168.2	182.6	132.3	103.6	111.7	67.5	69.9	62.1	76.1	66.1	62.8	63.5	59.2	60.3	53.9	58.5	54.7	99.1
					B	225.1	182.1	155.2	162.2	108.4	83.1	86.7	56.2	53.4	53.1	54.4	54.5	50.4	56.0	48.9	51.4	45.3	52.4	49.0	85.7
			GC-MS/MS	Green tea	A*	0.0	0.0	0.0	0.0	0.0	0.0	0.0	0.0	0.0	0.0	0.0	0.0	0.0	0.0	0.0	0.0	0.0	0.0	0.0	0.0
					B*	81.3	61.4	58.2	67.3	47.3	107.2	51.8	48.7	61.0	50.6	58.8	52.4	56.1	55.0	49.4	44.5	50.1	46.5	53.7	58.0
				Woolong tea	A	70.2	69.1	65.7	61.9	56.2	54.3	57.8	70.4	61.4	61.3	74.2	70.5	53.1	68.3	57.8	56.9	48.4	54.3	51.5	61.2
					B	67.9	58.7	55.8	55.5	54.7	46.1	45.6	61.0	52.9	51.7	55.9	54.9	48.5	62.4	46.9	50.0	43.2	50.4	44.9	53.0
		Envi-Carb+PSA	GC-MS	Green tea	A	184.3	195.8	204.2	154.2	229.5	187.2	56.3	61.2	62.4	157.1	58.3	61.2	58.2	75.4	59.6	65.8	59.4	58.1	52.7	107.4
					B	165.4	190.1	192.5	136.7	208.1	169.8	72.9	58.1	57.8	113.3	56.9	55.5	52.4	66.7	55.3	54.6	54.5	50.5	51.6	98.0
				Woolong tea	A	245.4	202.6	207.4	177.5	240.2	137.3	69.0	65.7	82.9	67.3	76.1	49.1	54.8	58.8	64.3	49.9	49.8	51.8	52.3	105.4
					B	206.3	173.3	175.0	164.3	134.2	122.5	55.8	53.1	57.8	51.7	55.1	45.9	47.7	48.6	50.2	41.5	41.6	42.8	45.3	85.4
			GC-MS/MS	Green tea	A	69.0	77.2	69.8	70.6	80.4	81.7	61.4	60.8	67.2	68.6	53.7	73.8	49.0	65.4	61.9	60.8	64.3	59.3	56.3	65.3
					B	65.1	61.0	68.7	64.8	78.8	70.5	58.3	55.7	56.5	64.6	53.8	67.8	49.2	58.9	57.9	52.5	60.3	49.7	55.9	60.5
				Woolong tea	A	89.4	74.7	88.2	61.6	63.5	64.7	65.9	54.2	82.9	67.9	60.6	50.0	55.1	55.4	62.4	47.5	41.3	55.0	51.6	62.7
					B	66.4	62.6	56.6	54.1	52.9	53.6	58.4	48.4	67.2	54.8	51.3	47.3	49.8	48.0	46.7	40.2	37.0	43.0	43.6	51.7
59	tebutam	Cleanert TPT	GC-MS	Green tea	A	90.7	68.5	70.9	73.8	58.1	66.0	60.5	62.5	65.5	59.8	69.6	57.9	65.8	64.4	60.9	41.7	57.7	48.9	66.5	63.7
					B	84.6	66.3	64.1	68.7	53.4	61.2	55.8	55.4	64.8	56.7	60.1	54.8	61.4	59.8	56.4	40.9	55.2	47.5	58.5	59.2
				Woolong tea	A	81.7	77.9	66.1	73.3	64.9	57.5	59.9	72.6	72.3	70.2	73.8	72.9	69.0	69.5	64.2	65.8	58.8	62.3	60.8	68.1
					B	76.9	65.4	54.3	60.7	59.0	47.5	47.7	59.1	59.3	59.5	57.3	59.3	54.5	60.6	52.7	56.6	49.0	56.3	53.5	57.3
			GC-MS/MS	Green tea	A*	87.3	69.1	81.9	77.6	55.2	72.8	62.2	54.4	68.2	57.5	69.7	64.8	60.9	52.7	60.6	50.3	55.9	48.3	65.2	63.9
					B*	83.3	66.9	69.5	69.4	52.4	66.4	56.6	52.3	66.1	56.5	62.3	61.0	58.2	53.8	57.2	50.1	53.9	47.7	56.7	60.0
				Woolong tea	A	69.5	79.5	64.3	69.1	63.2	61.6	60.5	73.2	71.5	74.0	78.9	71.9	63.8	72.5	60.7	61.3	57.9	63.1	58.1	67.1
					B	65.5	64.6	54.3	58.9	57.2	50.3	47.3	61.5	61.1	60.8	57.6	57.7	58.6	61.2	50.4	54.2	49.9	56.6	49.8	56.7
		Envi-Carb+PSA	GC-MS	Green tea	A	82.1	77.9	73.9	73.2	79.5	77.7	63.7	64.4	66.3	93.6	57.6	68.2	63.4	74.6	64.8	72.1	49.9	63.5	58.7	70.6
					B*	70.7	67.0	72.6	66.2	76.3	70.6	64.7	61.6	61.4	84.3	61.1	63.3	57.2	65.7	59.3	59.3	61.9	54.1	57.4	64.6
				Woolong tea	A	90.1	78.1	71.8	70.3	80.8	68.1	73.3	58.9	76.5	69.0	58.0	58.2	59.1	62.8	66.8	54.2	57.3	57.1	54.7	66.8
					B	68.8	63.3	61.9	61.1	60.1	57.1	63.2	51.3	60.7	54.6	65.7	50.3	51.5	52.5	52.4	45.4	53.5	46.3	48.9	55.2
			GC-MS/MS	Green tea	A*	71.8	80.1	71.7	77.2	80.0	76.5	65.2	63.2	63.1	71.6	58.0	65.7	58.2	77.9	64.2	70.0	44.9	60.7	61.5	68.6
					B	67.3	65.2	70.7	68.5	77.8	67.3	63.4	57.8	62.4	68.1	54.6	60.5	58.3	66.1	60.6	59.1	67.2	52.2	61.4	63.3
				Woolong tea	A*	94.1	77.3	96.1	66.7	73.1	66.5	69.7	61.2	76.5	70.9	67.0	55.4	60.3	60.3	64.2	50.8	62.2	52.4	48.9	65.9
					B	70.4	65.5	61.6	58.4	58.8	56.0	62.6	51.0	60.7	55.5	55.7	50.3	53.0	51.8	48.5	41.7	41.4	42.4	46.3	54.1
60	tebuthiuron	Cleanert TPT	GC-MS	Green tea	A	214.5	142.8	168.7	158.0	128.1	113.2	129.0	129.8	152.6	109.8	123.9	107.8	152.4	109.8	124.9	89.5	117.3	106.4	149.6	133.1
					B	196.4	133.0	139.9	126.2	114.7	114.0	119.1	112.2	130.9	98.5	126.4	105.6	138.3	122.5	107.0	88.5	111.1	101.5	132.3	122.0
				Woolong tea	A	191.9	153.1	129.0	154.4	123.2	131.0	127.5	155.3	140.5	153.6	187.3	146.1	144.9	114.7	141.1	141.7	131.9	129.8	124.1	143.2
					B	175.4	142.7	110.7	131.1	115.6	108.1	102.7	131.0	117.7	132.5	143.8	122.3	116.9	117.5	115.6	114.7	106.0	123.2	114.6	123.3
			GC-MS/MS	Green tea	A*	180.5	154.1	203.3	158.7	112.8	272.9	128.1	110.1	135.3	108.4	144.3	131.5	140.8	104.7	127.6	99.2	116.1	99.1	162.8	141.6
					B*	175.1	148.2	165.8	151.0	104.9	224.4	119.7	101.3	129.8	100.0	126.5	127.1	133.7	124.1	112.3	95.8	121.1	98.7	135.7	131.1
				Woolong tea	A	143.5	159.2	158.8	139.7	115.3	122.6	129.3	153.1	126.3	168.8	157.6	149.4	133.7	155.0	133.5	135.7	125.8	119.9	121.0	140.4
					B	137.2	155.2	133.4	125.9	106.2	106.9	99.1	131.5	107.2	136.4	118.0	116.7	113.1	129.4	108.8	114.6	103.0	114.0	104.4	119.0
		Envi-Carb+PSA	GC-MS	Green tea	A*	176.9	158.9	166.1	160.2	173.4	168.5	111.0	138.5	126.5	136.4	122.1	128.6	131.7	130.4	137.8	153.1	142.1	128.6	118.4	149.9
					B*	169.5	143.7	161.0	140.3	159.1	140.1	137.4	134.1	127.1	200.4	117.5	115.0	115.0	119.5	117.2	119.0	122.1	110.8	121.4	135.3
				Woolong tea	A*	215.3	174.6	149.9	149.9	258.1	136.7	161.5	147.0	235.3	153.7	154.3	108.5	125.8	157.7	158.2	115.0	110.6	126.2	117.8	155.6
					B*	179.2	146.2	136.1	130.1	177.6	117.4	136.6	119.2	185.1	105.3	128.4	97.8	103.8	133.4	124.0	98.1	97.6	96.7	107.5	127.4
			GC-MS/MS	Green tea	A	146.0	171.2	155.5	171.8	164.7	181.9	110.1	140.1	133.5	164.4	117.6	147.2	122.3	155.7	131.5	145.2	139.5	121.3	128.0	144.6
					B	142.8	143.0	154.4	147.5	157.8	145.9	134.7	125.1	133.3	157.0	113.4	138.6	116.8	127.2	155.7	115.3	124.0	106.2	132.0	133.2
				Woolong tea	A*	211.0	178.7	178.3	135.4	143.7	144.6	142.4	130.3	235.3	149.4	134.7	120.8	121.6	117.2	138.2	109.0	104.8	146.3	124.8	145.6
					B	175.7	152.7	137.2	126.2	121.0	123.1	134.2	107.8	185.1	105.4	116.9	106.0	111.6	107.4	99.7	89.4	97.0	108.3	111.5	121.9

(Continued)

Annex 1: Content change for 201 pesticides in incurred tea Youden pair samples under 8 conditions with two SPE cartridge cleanup in three months (Nov.9, 2009-Feb.7, 2010) (cont.)

No	Pesticides	SPE	Method	Sample	Youden pair	Nov.9 2009 (n=5)	Nov.14 2009 (n=3)	Nov.19 2009 (n=3)	Nov.24 2009 (n=3)	Nov.29 2009 (n=3)	Dec.4 2009 (n=3)	Dec.9 2009 (n=3)	Dec.14 2009 (n=3)	Dec.19 2009 (n=3)	Dec.24 2009 (n=3)	Dec.14 2009 (n=3)	Jan.3 2010 (n=3)	Jan.8 2010 (n=3)	Jan.13 2010 (n=3)	Jan.18 2010 (n=3)	Jan.23 2010 (n=3)	Jan.28 2010 (n=3)	Feb.2 2010 (n=3)	Feb.7 2010 (n=3)	AVE μg/kg
61	tefluthrin	Cleanert TPT	GC-MS	Green tea	A	43.7	32.3	33.1	34.3	26.8	29.8	28.9	28.5	29.9	26.8	31.2	28.7	30.3	34.9	27.8	19.7	29.4	22.8	30.9	30.0
				Green tea	B	39.5	31.2	29.5	31.9	24.7	27.8	26.8	26.5	30.8	25.4	27.9	27.4	28.5	36.8	26.2	19.8	25.9	22.0	26.8	28.2
				Woolong tea	A	37.3	37.3	32.1	33.6	31.0	26.4	28.3	36.6	34.4	34.1	35.2	36.1	32.8	30.9	30.8	30.0	28.3	29.6	28.3	32.3
				Woolong tea	B	35.5	31.6	27.0	29.6	28.7	22.5	23.5	30.2	29.4	28.9	27.6	30.8	27.1	27.3	25.9	26.3	23.7	27.4	25.2	27.8
			GC-MS/MS	Green tea	A	41.0	32.4	33.0	41.5	25.8	31.6	31.5	25.1	31.5	26.2	32.4	29.2	28.0	27.6	27.7	22.8	24.2	22.2	29.0	29.5
				Green tea	B	39.4	31.2	27.9	32.0	24.1	29.0	26.7	24.7	30.5	24.3	28.8	27.5	25.9	26.7	25.9	22.3	23.3	22.2	25.5	27.3
				Woolong tea	A	34.7	35.6	33.2	33.7	30.1	28.3	28.5	34.9	35.1	34.9	35.9	34.2	29.6	33.6	28.1	27.8	26.4	28.2	27.2	31.5
				Woolong tea	B	32.8	30.3	26.6	29.3	27.7	23.8	22.1	30.4	29.4	28.9	26.5	27.6	25.0	28.8	23.5	23.8	22.3	26.0	22.9	26.7
		Envi-Carb +PSA	GC-MS	Green tea	A	36.9	36.3	34.4	33.1	36.9	35.2	31.1	31.2	31.6	39.3	29.7	33.3	29.9	33.1	30.5	33.3	28.4	28.7	27.1	32.6
				Green tea	B	32.6	31.4	33.8	30.4	35.8	31.9	29.7	30.5	29.4	37.1	28.1	30.3	27.3	30.4	28.1	26.7	26.2	25.7	26.2	30.1
				Woolong tea	A	45.0	36.1	34.7	34.1	35.7	33.1	35.5	28.7	37.4	33.0	33.6	29.3	28.6	30.6	32.1	24.1	25.7	27.0	26.4	32.1
				Woolong tea	B	34.7	30.8	30.4	30.2	27.6	27.9	30.9	25.8	28.9	26.6	29.2	25.5	25.3	25.1	25.7	20.2	21.8	23.0	22.4	26.9
			GC-MS/MS	Green tea	A	34.2	37.5	33.5	33.9	37.2	35.3	30.9	30.3	28.9	34.2	27.0	36.3	25.1	32.9	29.3	31.9	31.0	26.0	27.8	31.8
				Green tea	B	30.9	30.5	32.6	30.8	36.3	31.8	28.5	27.9	29.1	32.8	26.0	33.8	24.6	29.9	27.1	26.0	28.9	22.7	28.7	29.4
				Woolong tea	A	43.9	36.2	46.1	32.4	35.4	32.8	32.7	28.9	37.4	33.1	30.6	26.4	28.2	29.1	30.1	23.7	20.3	25.3	24.2	31.4
				Woolong tea	B	33.1	30.1	30.1	27.9	28.5	27.0	29.3	24.5	28.9	25.7	25.9	24.4	24.4	24.5	22.9	19.4	18.5	21.0	21.7	25.7
62	thenylchlor	Cleanert TPT	GC-MS	Green tea	A*	73.6	90.2	85.2	125.8	65.2	72.5	70.6	62.5	81.2	60.4	70.4	71.9	62.9	80.7	67.8	78.5	58.7	66.0	63.9	74.1
				Green tea	B	79.6	79.5	73.2	126.5	62.4	66.5	62.8	58.5	72.3	62.3	67.1	69.3	61.3	82.9	72.5	78.2	61.2	59.0	62.3	71.4
				Woolong tea	A*	73.5	115.9	78.4	70.3	59.2	54.3	71.9	80.2	77.5	85.8	89.4	79.3	70.3	70.6	58.3	69.5	70.5	57.1	69.5	73.8
				Woolong tea	B	75.8	85.0	61.9	64.2	62.4	49.4	52.3	72.3	66.5	72.2	66.5	66.1	59.3	66.0	52.3	60.2	59.0	55.6	65.0	63.8
			GC-MS/MS	Green tea	A*	81.3	77.9	44.4	126.5	61.5	79.6	67.0	68.2	76.8	52.2	75.0	49.8	69.9	57.0	62.0	61.0	51.8	55.7	85.8	68.7
				Green tea	B	84.9	73.5	69.9	70.8	61.9	81.1	57.3	60.7	73.1	63.9	69.9	46.4	72.7	69.2	59.6	61.4	45.3	53.1	77.9	65.9
				Woolong tea	A	106.8	99.7	67.0	77.4	71.2	68.0	87.7	100.4	81.6	72.0	101.1	73.9	76.0	80.7	75.6	82.2	63.1	66.8	80.1	80.6
				Woolong tea	B	90.3	79.6	59.5	59.1	70.4	56.5	64.4	85.1	67.8	67.1	79.4	64.6	62.7	69.4	63.9	64.2	44.2	56.5	69.2	67.0
		Envi-Carb +PSA	GC-MS	Green tea	A*	54.9	130.8	65.3	65.7	78.3	73.2	81.7	72.2	66.1	83.6	69.7	67.3	65.2	71.6	59.9	66.2	113.0	58.0	67.7	74.2
				Green tea	B	69.8	95.8	73.6	68.8	78.5	71.0	73.6	46.0	61.0	87.0	69.4	68.3	64.2	66.3	60.8	59.4	100.4	56.3	69.3	70.5
				Woolong tea	A*	79.2	76.0	72.6	85.2	82.7	69.8	54.2	111.4	64.0	73.6	67.9	67.1	62.1	63.0	67.4	60.8	53.9	56.1	62.0	69.9
				Woolong tea	B	76.7	71.0	72.3	68.0	74.4	62.9	72.0	83.5	52.8	56.5	58.5	54.6	57.4	62.1	54.2	53.2	46.3	50.7	60.2	62.5
			GC-MS/MS	Green tea	A*	77.0	91.5	61.1	71.1	69.4	79.7	74.4	60.3	44.4	47.8	75.5	76.0	60.4	94.7	73.9	63.6	68.4	51.0	55.2	68.2
				Green tea	B*	121.9	88.8	65.0	72.7	75.9	76.1	61.8	59.4	63.3	56.5	70.1	69.7	71.3	67.4	70.8	57.3	58.4	48.3	64.3	69.4
				Woolong tea	A	76.3	89.1	74.4	69.9	76.9	62.1	82.0	72.7	64.0	64.0	60.6	46.8	62.5	54.4	64.9	47.4	57.4	77.2	68.5	69.0
				Woolong tea	B	45.3	81.6	74.3	65.5	65.3	57.9	68.2	56.6	52.8	53.3	58.4	42.5	53.3	55.3	49.5	40.4	55.3	68.0	64.3	58.3
63	thionazin	Cleanert TPT	GC-MS	Green tea	A*	49.6	36.3	37.7	37.4	27.4	30.6	31.8	31.3	34.8	27.1	34.6	27.7	34.2	29.2	29.9	24.4	27.6	26.0	31.7	32.3
				Green tea	B*	46.0	35.3	35.1	35.0	27.4	28.9	29.9	28.2	33.6	26.4	30.4	25.8	32.3	29.2	26.1	23.0	26.2	24.7	28.3	30.1
				Woolong tea	A	42.5	42.0	39.9	42.3	30.7	32.3	33.6	38.0	35.4	33.2	41.3	37.9	36.6	36.1	35.8	38.1	34.3	35.0	35.2	36.8
				Woolong tea	B	38.0	33.8	31.9	34.4	32.8	27.7	27.8	34.1	28.5	29.6	31.5	32.1	30.1	34.0	30.6	33.8	29.3	33.2	33.5	31.9
			GC-MS/MS	Green tea	A*	42.0	35.2	38.6	34.8	27.7	62.3	31.8	28.7	33.9	27.1	34.8	31.2	31.7	25.7	30.2	25.8	27.2	23.3	35.0	33.0
				Green tea	B*	42.1	33.9	34.4	34.8	25.9	49.6	29.2	27.0	34.1	27.8	31.9	29.4	30.8	28.0	28.9	25.0	25.9	23.4	30.8	31.2
				Woolong tea	A	40.7	40.6	33.4	34.1	30.8	30.5	31.7	37.2	36.4	37.0	40.9	33.8	32.2	33.9	31.2	31.5	28.4	29.0	30.8	33.8
				Woolong tea	B	35.4	33.0	28.2	29.1	28.1	25.3	24.7	31.8	30.5	30.7	30.4	28.2	30.3	30.1	25.4	26.9	23.8	25.9	29.3	28.5
		Envi-Carb +PSA	GC-MS	Green tea	A*	36.1	42.4	39.5	36.8	45.6	38.3	42.0	33.8	30.8	66.7	32.8	32.7	34.2	34.7	32.7	32.8	33.8	29.1	26.8	36.9
				Green tea	B*	34.7	35.3	39.2	33.6	41.9	34.8	36.6	31.0	29.0	49.9	30.1	31.5	30.8	32.2	31.0	28.7	30.0	25.8	27.0	33.3
				Woolong tea	A	48.1	40.2	42.0	42.9	48.7	36.2	39.0	37.1	40.2	36.1	36.4	27.5	30.5	30.9	37.9	31.0	29.0	31.8	32.0	36.7
				Woolong tea	B	37.8	32.4	33.7	34.3	34.3	31.3	33.4	31.9	32.1	29.0	30.2	25.7	27.1	27.1	30.3	26.5	25.1	27.8	29.7	30.5
			GC-MS/MS	Green tea	A	37.0	39.4	36.6	36.8	39.2	39.0	33.8	33.6	32.0	36.1	32.8	41.8	30.2	37.5	33.9	32.7	33.2	28.1	30.3	34.9
				Green tea	B	37.4	33.5	35.9	33.2	39.0	34.7	34.0	31.2	31.8	34.4	30.0	37.3	30.4	34.7	31.4	28.6	30.3	25.5	30.2	32.8
				Woolong tea	A	40.3	38.4	47.5	32.2	36.3	33.3	35.8	32.1	40.2	34.0	31.0	27.0	27.5	28.7	31.2	24.6	24.3	28.7	27.0	32.6
				Woolong tea	B	29.9	32.6	31.2	28.8	28.7	28.3	31.1	26.5	32.1	27.2	27.2	24.6	27.1	25.0	23.5	20.4	21.3	23.0	23.8	27.0

No.	Pesticide	Cartridge	Method	Tea		V1	V2	V3	V4	V5	V6	V7	V8	V9	V10	V11	V12	V13	V14	V15	V16	V17	V18	V19	V20
64	trichloronat	Cleanert TPT	GC-MS	Green tea	A	53.3	36.4	35.0	36.3	28.4	30.3	30.1	29.9	36.2	29.5	48.3	32.0	33.6	31.2	29.9	27.4	27.3	25.1	32.0	33.3
					B	49.9	35.9	31.3	32.2	26.7	29.4	27.9	27.9	36.2	27.3	42.0	29.3	31.1	30.2	28.2	26.8	26.4	24.2	28.1	31.1
				Woolong tea	A*	42.6	36.6	35.3	37.0	36.4	31.7	31.2	36.6	38.4	33.0	43.7	35.4	34.3	33.9	31.4	32.8	31.1	31.0	28.8	34.8
					B*	40.2	32.5	30.0	32.1	31.0	25.6	23.6	31.1	30.4	29.3	31.8	28.9	27.9	30.3	26.0	28.1	25.8	28.0	26.1	29.4
			GC-MS/MS	Green tea	A*	43.6	35.0	37.9	25.8	28.9	51.9	29.7	27.7	32.3	29.9	36.8	30.6	32.8	31.0	30.4	27.0	26.6	23.0	34.3	32.4
					B*	43.7	33.5	34.7	33.1	27.0	46.1	27.2	26.5	33.7	28.6	34.3	29.0	30.2	30.5	29.2	27.0	25.4	23.3	28.9	31.1
				Woolong tea	A	40.1	38.9	36.1	35.9	31.0	29.2	32.2	37.0	35.1	33.4	41.7	36.0	33.7	35.9	31.5	33.2	28.5	30.3	28.8	34.1
					B	35.6	32.9	30.4	31.0	28.9	25.1	25.1	32.7	29.4	29.5	30.7	28.7	28.1	31.4	25.8	27.7	24.4	26.8	25.6	28.9
		Envi-Carb +PSA	GC-MS	Green tea	A*	38.1	41.0	43.1	32.3	44.1	41.0	35.5	34.2	35.1	73.2	39.0	37.6	31.6	32.2	34.2	33.6	34.0	29.1	28.8	37.8
					B*	35.6	37.5	40.6	28.7	40.5	37.1	30.3	31.8	32.3	55.8	35.6	33.9	29.3	32.5	33.0	27.6	30.4	27.4	28.7	34.1
				Woolong tea	A	46.5	39.7	38.7	34.6	57.1	34.1	36.6	36.9	43.3	36.3	39.3	27.1	30.2	33.8	33.0	26.8	26.8	27.1	28.0	35.6
					B	37.4	33.1	30.9	29.9	37.2	28.8	32.0	29.2	33.2	29.1	29.7	25.8	26.0	29.1	26.9	21.8	22.8	23.2	24.8	29.0
			GC-MS/MS	Green tea	A	37.8	40.9	37.7	32.7	39.6	40.6	35.5	33.4	35.0	34.6	36.1	0.0	29.1	37.7	32.9	32.2	32.0	28.2	31.2	33.0
					B	34.9	36.5	35.9	30.1	37.7	37.3	28.3	31.1	33.3	33.8	31.6	0.0	28.9	34.2	31.4	27.2	28.8	26.5	31.5	30.5
				Woolong tea	A*	40.2	39.8	43.7	33.5	35.2	33.2	34.6	31.1	43.3	34.2	31.6	29.4	29.4	29.6	32.0	26.0	23.4	29.5	26.5	33.0
					B*	32.8	33.8	30.8	29.6	28.5	27.9	31.2	26.3	33.2	27.7	27.3	26.5	25.8	26.7	24.5	20.7	21.6	24.9	24.7	27.6
65	trifluralin	Cleanert TPT	GC-MS	Green tea	A	107.2	74.0	70.2	71.2	53.2	64.3	65.2	59.0	81.9	59.9	80.2	66.9	102.8	65.8	74.1	107.2	55.2	84.6	65.7	74.1
					B	96.2	70.0	61.5	63.3	52.4	61.0	60.5	56.8	76.5	56.7	70.1	60.6	86.9	64.0	68.4	95.6	55.0	66.9	57.7	67.4
				Woolong tea	A*	84.0	64.2	67.4	78.8	103.8	65.8	75.9	82.3	80.5	71.3	111.1	71.0	77.4	68.8	84.3	88.5	103.5	75.1	63.4	80.4
					B*	76.9	65.4	58.6	65.6	76.7	48.4	55.2	66.8	62.1	61.3	77.4	56.6	63.0	65.6	61.9	76.6	81.4	67.3	58.4	65.5
			GC-MS/MS	Green tea	A*	96.5	75.7	87.9	54.1	52.3	188.7	61.4	57.8	71.9	56.2	84.9	63.8	79.1	61.2	68.8	66.0	54.4	45.4	119.0	76.1
					B*	92.1	71.9	74.0	70.4	50.0	147.7	56.8	57.1	79.2	58.4	79.7	59.5	75.2	65.5	66.1	66.8	55.5	45.5	91.1	71.7
				Woolong tea	A	73.1	108.1	78.0	71.4	63.9	73.1	79.8	83.2	77.7	86.6	102.4	80.2	75.1	85.1	72.2	77.6	73.0	72.3	59.3	78.5
					B	75.1	87.0	67.2	64.8	59.9	56.9	57.3	72.2	63.8	71.6	72.6	61.1	66.3	74.8	61.5	66.8	55.2	65.3	55.2	66.0
		Envi-Carb +PSA	GC-MS	Green tea	A*	77.3	76.3	82.8	59.0	87.2	87.8	83.8	64.3	68.1	93.1	80.6	81.8	70.5	0.0	75.9	62.3	208.5	52.9	62.8	77.6
					B*	68.6	73.4	77.4	51.9	79.3	82.6	63.2	61.3	64.2	76.9	71.6	73.2	64.8	117.7	72.9	53.3	127.6	51.7	60.5	67.1
				Woolong tea	A	86.9	77.4	74.6	70.0	185.8	76.2	68.5	183.2	106.8	73.4	75.7	61.1	90.1	102.3	61.7	61.2	64.5	61.7	64.1	87.4
					B	69.1	62.7	60.6	61.7	95.3	65.7	62.5	118.6	82.2	58.2	58.6	56.4	72.6	84.9	60.3	39.3	56.5	52.4	54.7	67.9
			GC-MS/MS	Green tea	A	74.1	75.2	73.7	70.5	76.9	83.8	83.5	71.3	72.0	72.7	89.1	71.4	69.2	69.8	70.9	71.1	70.3	55.6	85.0	74.8
					B	71.5	76.7	71.6	61.7	72.7	78.6	46.9	63.1	63.9	69.8	78.3	62.7	69.5	67.8	67.5	58.5	66.6	48.9	82.5	67.4
				Woolong tea	A*	85.6	83.7	96.3	63.6	69.8	70.9	74.6	73.4	106.8	74.3	73.9	64.9	70.3	57.2	71.1	57.9	98.5	126.1	82.2	79.6
					B*	65.8	69.2	61.2	58.6	53.9	61.3	65.9	56.1	82.2	59.8	60.1	61.4	59.5	56.3	56.7	42.1	79.8	92.7	70.5	63.9
66	2,4'-DDT	Cleanert TPT	GC-MS	Green tea	A*	89.8	77.8	73.5	128.4	62.8	64.2	61.7	64.4	57.0	56.2	74.8	59.3	59.2	56.3	64.5	56.6	41.4	44.0	64.6	66.1
					B*	84.3	73.8	64.0	119.8	63.6	57.4	62.1	63.2	57.0	53.9	61.6	54.9	55.8	56.8	69.1	55.5	45.4	51.5	56.5	64.1
				Woolong tea	A	77.7	92.9	73.9	107.8	75.7	55.6	70.4	70.4	49.9	64.7	75.0	69.2	76.7	65.4	62.5	68.2	71.5	58.5	54.1	73.2
					B	73.2	78.1	71.3	91.5	72.1	54.9	62.0	62.0	61.9	63.9	64.7	54.3	68.8	68.9	58.2	68.4	59.5	54.2	54.0	66.7
			GC-MS/MS	Green tea	A	101.4	75.7	61.2	101.7	60.2	59.5	65.9	61.2	57.1	56.5	80.9	45.9	61.7	59.7	58.0	53.4	42.9	48.8	64.1	64.0
					B	93.3	76.0	61.8	97.7	62.3	56.1	60.0	56.5	62.9	52.5	73.3	45.4	54.5	58.9	56.3	55.1	44.9	52.9	57.0	62.0
				Woolong tea	A	0.0	0.0	0.0	0.0	0.0	0.0	0.0	0.0	0.0	0.0	0.0	0.0	0.0	0.0	0.0	0.0	0.0	0.0	0.0	0.0
					B	0.0	0.0	0.0	0.0	0.0	0.0	0.0	0.0	0.0	0.0	0.0	0.0	0.0	0.0	0.0	0.0	0.0	0.0	0.0	0.0
		Envi-Carb +PSA	GC-MS	Green tea	A*	72.2	70.0	71.7	69.7	72.7	79.9	58.8	56.3	69.5	59.3	62.7	73.5	58.1	62.9	36.6	58.0	69.3	53.2	69.5	64.4
					B*	74.8	77.3	70.3	62.8	69.5	81.9	54.6	50.4	69.1	60.6	55.8	71.0	58.3	62.0	37.0	49.1	62.3	55.6	72.0	62.9
				Woolong tea	A	84.0	54.9	66.9	54.8	101.0	66.8	54.4	58.2	49.9	48.4	54.4	64.9	53.7	65.4	62.6	60.1	44.7	44.0	47.4	59.8
					B	82.2	47.7	67.2	55.5	77.2	68.6	64.5	49.6	34.6	53.3	49.7	68.4	48.1	68.2	66.8	46.3	46.9	49.0	42.9	57.2
			GC-MS/MS	Green tea	A	104.6	94.0	72.7	67.8	74.7	77.0	80.8	54.1	67.3	52.1	78.1	72.8	65.6	59.6	44.3	57.2	62.0	57.7	71.4	69.1
					B	123.6	78.4	71.9	69.1	67.8	76.6	66.3	52.3	63.7	60.6	67.0	66.3	57.1	54.5	42.8	48.6	58.3	60.8	74.7	66.3
				Woolong tea	A*	109.4	55.8	73.9	54.2	111.0	57.5	92.7	59.2	49.9	45.7	52.8	44.8	52.9	55.5	61.3	65.3	34.9	45.8	45.6	61.5
					B*	91.8	50.0	62.9	51.3	84.0	62.3	67.5	47.3	34.6	55.7	53.1	50.5	45.2	55.8	58.4	62.8	37.4	45.5	41.5	55.7

(Continued)

Annex 1: Content change for 201 pesticides in incurred tea Youden pair samples under 8 conditions with two SPE cartridge cleanup in three months (Nov.9, 2009–Feb.7, 2010) (cont.)

No	Pesticides	SPE	Method	Sample	Youden pair	Nov.9, 2009 (n=5)	Nov.14, 2009 (n=3)	Nov.19, 2009 (n=3)	Nov.24, 2009 (n=3)	Nov.29, 2009 (n=3)	Dec.4, 2009 (n=3)	Dec.9, 2009 (n=3)	Dec.14, 2009 (n=3)	Dec.19, 2009 (n=3)	Dec.24, 2009 (n=3)	Dec.29, 2009 (n=3)	Jan.3, 2010 (n=3)	Jan.8, 2010 (n=3)	Jan.13, 2010 (n=3)	Jan.18, 2010 (n=3)	Jan.23, 2010 (n=3)	Jan.28, 2010 (n=3)	Feb.2, 2010 (n=3)	Feb.7, 2010 (n=3)	AVE, µg/kg
67	4,4'-DDE	Cleanert TPT	GC-MS	Green tea	A	42.6	35.1	37.8	42.9	33.8	30.8	37.0	33.4	35.5	31.0	32.7	31.1	32.1	30.9	33.0	27.6	28.3	27.3	35.8	33.6
				Green tea	B	40.5	34.4	34.2	40.2	31.3	28.9	33.9	31.4	34.3	29.6	29.4	29.2	30.4	30.0	30.1	27.4	28.1	26.5	31.6	31.7
				Woolong tea	A	40.8	38.7	39.2	44.0	34.4	37.7	34.6	41.7	37.8	36.4	38.1	38.3	38.2	35.2	34.6	34.5	33.7	34.5	31.5	37.0
				Woolong tea	B	40.1	34.9	35.0	40.7	35.3	34.9	31.7	38.7	34.4	33.4	32.1	33.8	33.9	33.7	31.4	31.9	30.0	34.7	30.5	34.3
			GC-MS/MS	Green tea	A	44.1	35.7	35.7	45.7	32.4	30.2	35.7	33.0	33.8	31.2	33.5	33.9	30.8	34.7	33.4	26.1	25.7	28.3	35.0	33.8
				Green tea	B	42.2	35.0	34.8	41.9	30.3	28.7	34.9	30.6	32.8	28.9	30.0	32.1	30.4	31.5	30.1	25.6	25.8	28.3	31.9	31.9
				Woolong tea	A	38.8	38.3	40.2	45.2	35.5	38.0	35.4	43.7	39.4	36.3	37.3	38.3	38.0	35.6	33.5	37.0	30.8	34.0	31.9	37.2
				Woolong tea	B	39.2	35.3	36.4	40.9	37.0	34.2	30.7	40.7	35.7	32.6	32.4	33.9	32.8	35.1	30.2	33.5	30.5	33.9	28.6	34.4
		Envi-Carb+PSA	GC-MS	Green tea	A*	49.9	43.0	37.0	37.1	38.9	39.2	103.3	33.0	32.3	31.4	30.9	33.3	33.3	36.3	32.4	38.8	30.9	32.7	31.7	39.2
				Green tea	B*	46.5	36.0	36.5	34.6	38.2	35.7	60.4	31.8	31.0	31.9	30.0	31.8	31.1	32.4	30.2	32.2	28.1	28.1	31.5	34.6
				Woolong tea	A	46.0	37.2	37.6	36.1	39.9	35.3	35.3	30.1	36.6	34.0	33.3	32.9	32.9	35.2	38.1	31.9	30.6	32.7	29.8	35.0
				Woolong tea	B	39.7	35.0	37.1	33.4	33.9	32.9	33.4	29.0	31.1	29.6	30.8	30.7	31.0	33.5	32.1	29.1	28.3	29.0	29.1	32.0
			GC-MS/MS	Green tea	A	47.1	43.0	43.0	43.0	37.2	39.7	36.5	33.5	33.5	33.1	33.7	34.9	33.7	31.5	31.7	36.1	29.5	34.0	33.7	35.7
				Green tea	B	43.7	33.6	36.3	33.6	36.5	35.2	35.5	32.4	31.6	33.1	32.3	31.9	30.0	30.0	28.7	31.1	28.8	31.0	33.2	33.3
				Woolong tea	A	42.7	37.2	37.2	34.7	38.9	34.5	35.2	29.5	36.6	33.1	33.9	29.5	29.8	30.3	35.9	37.0	28.7	30.0	30.9	34.3
				Woolong tea	B	38.4	34.9	36.5	33.2	33.7	32.2	34.0	26.7	31.1	28.9	35.1	32.0	28.8	29.2	28.2	33.5	28.7	26.1	31.1	31.6
68	benalaxyl	Cleanert TPT	GC-MS	Green tea	A	49.0	39.6	42.3	55.4	36.9	33.4	39.5	34.2	35.6	29.4	38.1	34.7	29.8	28.6	31.0	28.7	27.6	32.7	37.0	35.8
				Green tea	B	44.3	37.3	39.3	51.3	34.0	30.9	35.1	31.6	35.7	28.9	33.7	31.0	27.8	28.0	28.5	29.8	28.5	32.3	32.9	33.7
				Woolong tea	A	43.0	42.9	41.1	45.7	38.5	38.3	34.0	40.7	36.3	34.3	36.8	28.0	33.9	33.7	35.1	35.8	32.7	33.4	28.9	36.7
				Woolong tea	B	41.3	39.4	37.0	42.0	38.3	35.4	30.0	38.7	31.7	32.0	30.0	31.7	29.5	30.3	31.2	31.6	32.7	34.0	26.6	33.5
			GC-MS/MS	Green tea	A	48.7	39.8	41.9	46.0	33.9	32.0	35.4	32.2	33.2	31.2	35.8	31.9	34.6	32.4	31.4	26.3	24.1	26.8	35.8	34.5
				Green tea	B	46.9	39.5	36.5	43.2	32.7	30.0	35.0	29.8	33.7	31.0	33.9	32.6	28.6	30.0	29.1	25.5	24.6	27.2	30.8	32.7
				Woolong tea	A	45.9	43.9	39.4	46.9	33.7	36.4	32.2	41.3	35.6	33.6	37.0	34.0	35.2	35.3	30.6	36.4	31.9	31.6	28.1	36.3
				Woolong tea	B	43.2	39.7	35.0	41.4	33.9	31.8	29.3	37.9	30.8	31.9	31.1	30.7	30.8	30.5	27.3	32.6	27.9	30.3	27.4	32.8
		Envi-Carb+PSA	GC-MS	Green tea	A	59.9	49.3	42.5	42.5	41.1	42.0	34.3	33.1	36.1	33.3	34.7	38.0	31.2	32.7	25.3	35.9	32.9	29.3	33.8	37.3
				Green tea	B	48.4	43.2	41.2	37.5	40.3	39.7	35.0	29.7	34.2	32.7	30.0	34.7	28.1	29.8	23.0	27.4	31.3	29.0	35.2	34.2
				Woolong tea	A	46.8	37.9	39.2	35.7	44.1	36.0	35.4	30.1	33.1	33.0	32.2	31.0	28.5	31.3	37.4	29.1	28.9	28.7	29.3	34.1
				Woolong tea	B	42.0	34.6	39.2	34.2	36.5	33.3	33.5	26.5	28.5	28.3	29.2	28.0	26.1	30.3	37.0	27.5	26.1	28.5	29.4	31.5
			GC-MS/MS	Green tea	A	56.4	45.6	65.0	59.5	38.9	30.3	41.8	30.9	34.0	33.0	38.5	31.7	37.9	36.7	26.7	35.0	32.8	29.6	36.4	39.0
				Green tea	B	51.0	42.4	59.0	51.0	37.4	37.5	30.1	28.9	32.8	33.3	30.6	31.9	32.6	29.5	24.7	28.1	27.8	29.1	36.2	35.5
				Woolong tea	A	46.9	36.0	36.3	33.1	39.7	33.7	31.8	28.3	33.1	31.4	30.4	27.3	25.2	29.0	31.9	36.4	26.0	25.0	26.0	32.2
				Woolong tea	B	43.3	33.5	35.5	31.3	34.9	31.4	32.0	25.6	28.5	27.7	30.5	26.8	24.1	26.0	25.9	32.8	25.2	26.2	26.2	29.8
69	benzoylprop-ethyl	Cleanert TPT	GC-MS	Green tea	A	142.4	123.1	129.0	165.7	111.7	101.3	118.6	107.9	112.0	94.0	114.4	103.0	100.0	93.7	96.5	86.0	83.4	87.5	108.7	109.4
				Green tea	B	136.0	117.1	115.9	151.1	101.8	92.6	109.0	100.1	110.2	95.5	102.3	95.7	96.2	92.4	90.7	85.3	81.1	85.6	95.8	102.9
				Woolong tea	A	131.9	125.1	117.0	136.8	112.9	112.5	106.5	118.6	114.2	101.9	116.9	110.6	106.9	107.8	99.6	100.4	97.3	101.8	89.8	111.0
				Woolong tea	B	124.4	116.4	109.0	124.7	116.0	104.5	104.2	117.1	103.0	100.2	98.1	99.1	100.8	96.0	91.4	89.5	85.2	96.8	84.2	103.2
			GC-MS/MS	Green tea	A	148.2	124.2	128.8	145.8	106.5	104.0	111.7	106.4	109.8	94.9	116.7	110.6	104.9	95.3	100.0	85.2	79.8	85.3	115.7	109.2
				Green tea	B	142.4	116.9	113.5	137.4	99.6	94.3	111.2	98.7	107.6	93.7	106.9	105.0	94.3	95.9	93.7	83.9	79.7	86.4	102.6	109.3
				Woolong tea	A	126.8	125.5	110.5	139.1	106.6	114.5	99.2	124.2	116.9	103.8	114.3	108.7	107.5	97.8	98.2	104.9	95.5	97.6	85.0	109.8
				Woolong tea	B	122.7	120.2	110.7	125.7	108.4	104.5	96.6	119.0	105.4	98.5	97.0	97.7	96.1	94.0	86.6	91.3	95.5	93.5	80.4	109.8
		Envi-Carb+PSA	GC-MS	Green tea	A	161.4	147.1	125.2	126.0	121.4	124.8	108.1	101.9	110.3	103.3	104.6	113.8	104.1	111.3	87.1	111.8	105.3	100.1	102.1	114.2
				Green tea	B	153.2	133.4	120.6	112.4	121.3	118.7	102.7	96.1	101.0	98.8	97.4	105.6	96.6	96.4	83.7	90.7	92.4	89.4	101.8	106.0
				Woolong tea	A	143.8	105.0	111.6	111.9	125.4	104.3	107.1	90.6	102.4	101.4	97.4	95.4	92.6	105.0	100.8	91.9	87.8	94.3	86.5	102.9
				Woolong tea	B	131.7	99.8	115.4	109.2	107.2	96.6	96.4	87.0	90.8	87.8	89.8	89.9	82.3	99.9	84.9	78.6	79.6	84.9	81.5	94.4
			GC-MS/MS	Green tea	A	152.4	139.0	114.3	108.3	123.0	116.1	121.5	103.2	114.6	108.0	108.1	122.4	111.9	112.6	90.2	110.1	99.3	97.5	109.4	113.6
				Green tea	B	147.2	123.1	119.9	107.0	117.7	117.3	94.8	94.9	102.0	103.1	98.2	108.9	103.2	97.3	83.5	90.4	88.6	93.2	110.7	105.0
				Woolong tea	A	140.6	109.3	119.9	106.7	121.9	100.8	99.8	87.5	102.4	97.5	97.1	91.2	85.1	89.2	98.8	104.9	83.0	90.7	85.2	100.6
				Woolong tea	B	130.9	104.3	112.8	101.8	108.5	91.9	96.7	83.0	90.8	84.8	98.2	84.9	77.9	82.2	80.0	91.3	80.6	80.2	82.5	92.8

No.	Pesticide	Cartridge	Method	Tea																					
70	bromofos	Cleanert TPT	GC-MS	Green tea	A*	80.2	76.3	75.0	114.3	66.7	65.9	78.0	71.3	64.8	64.4	70.4	61.4	61.2	63.3	64.5	54.9	50.6	53.2	62.1	68.3
					B*	79.9	73.1	70.4	108.5	65.8	61.0	71.5	66.9	74.8	59.6	60.7	60.7	62.1	60.2	66.0	54.7	51.7	56.5	57.3	66.4
				Woolong tea	A*	79.6	90.4	82.0	88.5	73.6	65.7	68.4	88.0	64.2	66.5	72.2	66.2	76.0	66.6	60.3	62.8	66.0	59.8	58.8	71.3
					B	78.1	72.3	70.6	78.4	70.8	60.0	53.8	76.3	59.6	57.0	62.4	59.5	60.4	63.2	54.9	58.6	56.9	58.0	54.0	63.4
			GC-MS/MS	Green tea	A	85.6	70.9	63.9	92.1	70.3	67.4	77.7	71.0	66.2	59.7	79.7	53.7	55.9	60.6	63.7	60.3	48.0	53.9	65.3	66.6
					B	85.9	70.7	67.6	88.0	69.0	64.5	67.6	66.9	70.6	57.9	61.3	45.4	63.5	53.0	61.0	62.3	52.7	55.5	60.2	64.4
				Woolong tea	A*	94.2	92.9	79.3	82.2	68.6	70.0	64.5	122.6	80.8	58.8	67.9	64.6	79.5	71.7	62.0	68.2	68.4	62.5	52.7	74.3
					B	86.2	81.9	69.4	68.4	66.8	59.7	51.5	92.3	59.8	62.3	60.2	57.8	63.6	64.4	50.7	57.9	56.5	59.0	53.7	64.7
		Envi-Carb+PSA	GC-MS	Green tea	A	88.9	79.2	73.5	76.7	76.8	74.5	88.0	63.3	71.6	68.4	67.9	76.9	58.7	67.9	54.5	65.9	67.3	54.8	71.2	70.6
					B	85.8	69.8	72.4	73.1	78.1	70.6	82.7	52.9	69.3	66.1	60.1	77.0	55.9	63.7	62.1	56.9	61.9	55.7	71.4	67.3
				Woolong tea	A	77.4	67.5	74.8	69.9	72.8	64.1	53.7	58.3	64.9	61.6	63.2	57.3	54.6	66.6	53.6	56.2	52.5	52.3	51.6	62.2
					B	67.8	60.0	72.0	61.7	64.4	61.2	62.5	51.4	55.3	51.1	54.6	56.6	52.7	63.2	52.8	51.0	49.7	50.1	49.8	57.3
			GC-MS/MS	Green tea	A*	110.2	82.9	80.3	87.9	80.1	105.2	89.3	63.4	77.4	55.1	65.6	65.3	68.0	74.7	50.0	74.2	59.0	61.8	75.6	75.2
					B*	122.3	75.7	87.8	97.7	78.2	75.0	69.3	63.3	68.9	60.8	62.4	61.7	61.2	56.7	63.7	62.6	55.8	69.1	69.8	71.0
				Woolong tea	A	110.9	66.8	89.3	63.9	83.2	64.2	79.9	54.8	64.9	56.1	62.0	38.0	57.9	67.4	52.1	68.2	51.3	52.8	63.1	66.2
					B	81.4	60.3	65.4	56.6	67.7	60.2	60.8	49.9	55.3	51.4	63.1	38.5	51.6	62.9	65.9	57.9	51.9	45.4	55.9	57.3
71	bromopropylate	Cleanert TPT	GC-MS	Green tea	A	93.2	87.0	82.8	146.0	73.0	67.1	74.7	66.6	66.6	59.5	89.6	68.3	66.6	58.6	69.0	62.1	42.7	62.7	77.2	74.2
					B	92.2	83.1	74.5	132.1	69.4	63.1	73.0	66.0	81.3	58.3	82.1	61.8	62.9	59.9	69.9	62.3	45.3	66.3	69.4	72.2
				Woolong tea	A	87.4	86.2	73.9	100.7	94.0	72.1	72.4	96.3	74.8	72.2	76.2	70.1	78.1	61.8	63.5	79.1	69.1	70.6	55.4	76.9
					B	83.4	86.7	76.5	91.5	90.5	67.8	74.6	104.2	63.8	69.3	61.4	60.2	77.3	66.5	68.0	72.9	61.3	69.4	52.9	73.4
			GC-MS/MS	Green tea	A*	100.2	87.5	81.9	103.3	71.5	68.6	75.0	70.2	64.1	61.9	100.0	52.2	75.3	63.1	64.3	59.3	49.4	54.7	75.5	72.7
					B*	96.8	83.6	75.8	98.6	66.7	64.1	74.6	64.9	73.5	61.3	81.4	46.0	61.6	64.3	68.3	59.5	51.2	56.8	66.6	69.0
				Woolong tea	A	119.2	110.5	77.1	99.0	68.6	72.9	68.3	157.0	74.7	67.3	80.0	72.8	79.3	68.4	60.6	74.4	68.0	61.2	49.6	80.9
					B*	103.9	116.7	73.9	90.4	69.6	65.6	69.8	155.2	60.0	69.0	64.8	66.0	73.8	66.9	41.1	69.1	62.2	54.7	50.0	75.9
		Envi-Carb+PSA	GC-MS	Green tea	A	88.4	94.6	84.1	82.4	82.3	87.9	53.9	68.3	89.8	69.5	81.4	88.0	72.2	57.2	32.9	47.5	91.2	53.3	72.8	74.5
					B	99.8	94.1	80.3	63.0	79.9	85.7	73.9	56.2	95.1	75.8	60.0	79.1	61.5	66.9	68.8	42.4	81.8	75.4	78.6	72.8
				Woolong tea	A	90.9	62.5	73.9	66.7	95.5	68.2	68.0	88.8	70.6	67.4	64.4	66.7	64.8	61.8	67.4	64.1	59.9	58.5	60.8	68.5
					B	89.9	61.4	75.0	68.7	75.3	65.7	63.7	54.4	60.5	62.3	56.6	70.4	57.3	66.5	50.9	49.4	54.7	60.0	59.7	64.2
			GC-MS/MS	Green tea	A	102.6	104.3	78.0	75.8	82.0	74.6	88.0	66.8	81.9	73.6	86.2	70.3	84.4	70.3	45.7	62.8	84.0	67.4	71.5	77.6
					B	116.1	123.2	76.5	68.1	77.4	81.2	60.9	60.8	77.0	73.3	63.7	69.0	73.8	64.5	64.4	53.5	74.9	73.4	80.0	74.4
				Woolong tea	A	109.5	67.1	71.8	70.7	89.2	68.2	68.7	61.5	70.6	64.6	63.7	63.5	57.2	60.3	57.8	74.4	54.9	56.9	52.4	67.9
					B	109.3	65.1	73.0	69.2	74.9	64.5	64.9	52.2	60.5	60.6	63.7	66.2	51.7	56.2	66.7	69.1	53.8	54.1	50.9	64.1
72	buprofenzin	Cleanert TPT	GC-MS	Green tea	A	94.2	75.5	80.0	93.4	70.2	63.7	77.1	69.0	66.2	60.3	73.2	56.6	45.8	66.0	58.8	52.2	61.9	70.0	68.8	69.0
					B	87.7	71.3	70.5	85.2	63.5	60.8	72.3	63.9	65.9	53.8	64.3	50.2	45.5	62.3	66.7	50.6	62.1	68.0	61.5	64.1
				Woolong tea	A	90.4	86.6	83.8	92.8	84.5	86.7	70.8	89.1	83.0	74.4	78.9	85.0	81.7	78.0	60.7	57.7	68.9	64.5	63.6	78.5
					B	92.2	77.8	78.1	88.4	82.5	83.0	63.0	79.9	74.1	66.0	72.0	90.3	65.9	67.1	69.6	51.1	60.7	58.6	61.5	73.3
			GC-MS/MS	Green tea	A	94.3	76.7	78.5	94.7	67.8	63.0	70.4	69.1	68.3	65.1	63.0	54.0	70.2	71.9	60.4	53.1	69.6	58.5	69.6	69.4
					B	89.6	73.8	69.6	85.2	64.2	58.7	70.7	62.7	70.4	76.3	68.1	49.2	66.6	74.3	67.6	51.5	65.4	59.0	61.6	66.1
				Woolong tea	A	83.4	83.2	85.6	95.0	70.5	79.2	70.5	85.1	75.4	62.4	76.1	78.2	73.6	76.4	64.9	67.7	69.0	67.5	69.1	76.2
					B	84.1	76.2	73.5	86.6	70.7	70.1	69.8	82.9	65.7	63.5	65.9	71.7	63.8	70.8	74.4	62.3	65.4	72.0	63.3	70.0
		Envi-Carb+PSA	GC-MS	Green tea	A	106.1	91.8	73.3	73.3	79.4	90.4	77.5	65.2	57.0	61.5	57.9	66.8	54.5	76.5	70.9	82.8	53.9	62.7	65.2	72.2
					B	96.3	75.8	71.1	67.6	78.5	81.5	79.3	61.1	56.9	61.5	55.4	56.7	41.2	65.6	73.5	73.5	57.1	49.1	63.2	66.4
				Woolong tea	A	99.4	83.3	86.1	81.2	85.9	81.4	74.2	63.6	70.9	74.2	70.7	75.9	65.8	75.0	56.4	56.7	65.1	62.0	60.4	74.4
					B	84.2	77.7	87.4	73.3	79.7	73.9	71.7	58.0	61.6	65.3	65.5	64.7	75.7	68.9	71.7	66.9	62.3	66.9	75.4	69.8
			GC-MS/MS	Green tea	A	100.1	76.6	74.1	72.2	78.9	70.2	74.2	65.6	66.2	68.2	63.0	55.9	72.8	62.6	62.9	71.7	51.3	56.9	66.9	70.6
					B	91.3	70.0	58.7	55.5	81.8	76.2	71.8	63.0	62.5	70.4	57.6	53.9	59.6	53.6	69.7	71.7	51.9	58.7	66.0	65.1
				Woolong tea	A	97.7	73.8	92.0	70.9	80.7	76.7	71.4	61.6	70.9	70.9	63.4	58.7	53.6	63.1	69.7	67.7	56.7	72.3	65.2	70.1
					B	84.6	69.3	74.9	64.7	72.2	67.7	68.6	55.5	61.6	59.0	64.1	53.8	53.7	57.5	56.2	62.3	61.5	61.7	61.5	63.9

(Continued)

Annex 1: Content change for 201 pesticides in incurred tea Youden pair samples under 8 conditions with two SPE cartridge cleanup in three months (Nov.9, 2009-Feb.7, 2010) (cont.)

No	Pesticides	SPE	Method	Sample	Youden pair	Nov.9 2009 (n=5)	Nov.14 2009 (n=3)	Nov.19 2009 (n=3)	Nov.24 2009 (n=3)	Nov.29 2009 (n=3)	Dec.4 2009 (n=3)	Dec.9 2009 (n=3)	Dec.14 2009 (n=3)	Dec.19 2009 (n=3)	Dec.24 2009 (n=3)	Dec.14 2009 (n=3)	Jan.3 2010 (n=3)	Jan.8 2010 (n=3)	Jan.13 2010 (n=3)	Jan.18 2010 (n=3)	Jan.23 2010 (n=3)	Jan.28 2010 (n=3)	Feb.2 2010 (n=3)	Feb.7 2010 (n=3)	AVE μg/kg.
73	butachlor	Cleanert TPT	GC-MS	Green tea	A	76.5	73.0	73.5	83.7	64.6	62.2	72.7	64.4	64.9	56.8	63.3	54.1	53.8	58.8	63.3	54.4	48.7	48.8	63.3	63.2
					B	74.5	71.3	65.9	80.5	59.8	56.5	69.0	59.3	66.9	53.9	53.0	52.8	53.4	55.3	59.3	52.9	50.8	52.0	57.0	60.2
				Woolong tea	A	76.7	80.6	75.0	83.0	64.8	71.5	68.1	81.2	62.7	68.5	66.6	65.1	69.0	61.8	57.2	62.3	64.4	52.8	56.0	67.8
					B	74.9	66.6	64.5	75.8	66.7	68.0	57.7	82.2	57.7	56.0	59.2	57.6	55.4	56.8	52.8	56.4	52.2	54.0	51.6	61.4
			GC-MS/MS	Green tea	A	88.4	71.7	77.4	92.4	68.2	60.6	67.4	65.1	69.0	61.3	66.3	66.0	65.2	67.4	68.3	54.8	52.3	57.8	67.3	67.7
					B	84.6	72.3	68.0	84.2	63.5	58.2	69.0	61.3	68.0	59.7	63.6	62.8	59.3	62.0	62.0	55.5	50.4	60.1	62.8	64.6
				Woolong tea	A	77.4	72.9	82.7	94.2	67.8	78.9	69.0	83.4	78.8	71.6	74.2	77.3	73.5	70.8	68.9	69.7	68.5	69.4	54.8	73.9
					B	77.2	65.1	72.0	85.2	67.1	70.2	58.3	75.9	73.6	64.9	62.7	68.9	66.1	70.8	58.5	65.9	66.2	63.4	58.6	67.7
		Envi-Carb+PSA	GC-MS	Green tea	A	82.8	79.2	71.4	74.7	71.5	75.5	67.6	61.5	59.7	57.5	57.7	63.7	54.4	63.0	48.4	61.9	60.3	54.8	65.2	64.8
					B	77.5	70.1	72.8	72.3	70.7	70.7	62.7	52.7	58.5	56.1	56.0	62.7	53.1	55.7	49.4	53.8	55.0	52.3	63.0	61.3
				Woolong tea	A	96.4	69.4	71.5	70.5	72.4	68.6	57.3	51.2	75.1	58.2	60.4	54.9	51.0	59.1	64.5	55.6	51.3	56.8	49.5	62.8
					B	68.1	63.4	71.1	64.4	63.0	63.5	65.1	44.2	62.6	49.2	53.6	53.7	48.5	56.3	53.8	50.1	49.4	51.1	45.6	56.7
			GC-MS/MS	Green tea	A	93.2	82.4	68.8	66.5	77.1	50.1	72.6	67.7	67.1	66.8	68.3	62.9	75.7	64.3	63.5	73.5	58.6	65.8	67.0	69.1
					B	86.5	66.4	68.1	66.3	73.8	70.9	71.4	62.5	65.4	67.9	62.1	59.0	64.1	61.7	57.8	63.9	57.5	62.3	62.6	65.8
				Woolong tea	A	90.6	72.4	81.9	71.4	77.9	71.7	69.5	60.7	75.1	66.1	68.4	63.7	56.1	65.6	73.8	69.8	55.7	64.5	58.1	69.1
					B	81.3	69.1	71.8	63.7	68.0	66.9	66.6	57.7	62.6	58.9	72.3	59.9	55.1	59.4	57.4	65.8	56.7	56.9	60.3	63.7
74	butylate	Cleanert TPT	GC-MS	Green tea	A*	114.4	92.7	99.3	108.6	97.7	79.5	78.6	58.1	91.9	46.1	36.3	37.8	32.2	28.9	25.9	24.7	24.0	60.9	71.0	63.6
					B	102.5	88.7	92.0	109.0	85.1	73.0	73.7	58.2	88.9	42.6	34.5	35.3	32.2	32.1	24.5	24.2	24.7	71.5	64.8	60.8
				Woolong tea	A	110.4	86.6	88.0	107.3	73.7	81.6	77.3	83.8	93.3	68.3	73.1	75.9	74.3	74.6	80.6	64.1	67.0	64.9	79.3	79.8
					B	105.5	76.7	98.7	92.3	85.1	74.1	68.1	79.3	87.7	64.8	58.9	64.1	69.4	63.7	60.2	63.9	60.3	58.8	57.9	73.1
			GC-MS/MS	Green tea	A*	106.3	80.2	88.8	98.0	78.5	72.4	83.8	73.5	79.2	74.2	78.3	53.2	79.6	75.5	74.3	57.9	59.7	57.5	77.8	76.3
					B	98.1	78.4	60.0	95.0	73.6	68.1	82.5	68.2	78.8	65.9	73.2	50.9	70.4	74.6	68.3	60.6	60.4	61.8	70.8	72.6
				Woolong tea	A	84.3	79.7	84.0	102.1	70.8	77.9	73.2	82.9	87.0	72.7	77.2	73.7	76.1	75.8	68.9	67.9	66.5	71.1	58.0	76.6
					B	396.7	154.4	92.6	70.7	86.8	121.2	105.3	72.5	72.0	52.4	61.6	59.3	63.7	64.4	56.5	60.0	57.1	61.7	48.3	64.6
		Envi-Carb+PSA	GC-MS	Green tea	A	218.4	139.1	118.3	118.3	107.7	131.7	101.0	64.5	46.0	92.7	263.2	36.8	28.3	75.4	27.5	30.2	30.4	83.1	68.0	96.5
					B	80.8	94.1	66.4	93.5	94.6	85.0	79.6	62.2	43.4	92.7	30.9	30.6	23.1	24.8	18.8	25.7	29.3	75.7	66.7	75.5
				Woolong tea	A*	53.8	103.8	105.7	78.1	83.7	79.1	69.2	85.5	74.1	76.7	66.7	64.1	85.3	67.3	77.3	61.1	55.7	63.8	58.9	75.5
					B*	109.5	93.8	66.4	77.3	97.4	71.7	87.7	59.8	74.1	61.5	66.4	62.8	57.3	63.7	65.4	49.2	57.4	70.2	57.8	68.5
			GC-MS/MS	Green tea	A	105.9	71.5	72.4	66.1	98.2	86.4	78.7	82.0	74.0	85.8	83.9	154.8	96.8	80.5	68.1	79.2	72.0	75.5	79.8	85.5
					B	109.5	69.0	103.5	65.7	80.0	77.7	66.4	77.4	71.0	85.8	69.6	128.5	79.4	74.6	55.9	69.8	66.6	71.0	78.8	79.3
				Woolong tea	A*	79.3	58.9	70.5	69.2	63.9	67.1	61.8	61.3	74.1	74.1	68.1	65.3	65.9	65.7	74.9	67.9	55.4	59.2	56.8	71.8
					B*	78.2	72.3	71.5	62.8	64.3	56.7	59.9	50.7	58.2	57.1	65.8	58.5	57.2	55.4	54.2	60.0	49.4	49.0	51.8	59.6
75	carbofenothion	Cleanert TPT	GC-MS	Green tea	A	80.8	72.3	71.0	79.7	64.9	62.7	49.1	58.0	67.3	50.5	51.0	54.4	53.6	48.7	48.3	41.2	35.6	39.4	45.4	59.8
					B	77.1	76.3	79.7	71.0	64.5	53.0	73.8	63.9	62.2	50.9	64.4	56.2	51.7	52.5	38.0	41.0	38.0	35.7	40.2	57.4
				Woolong tea	A*	91.5	85.9	72.2	65.0	75.3	63.7	94.9	57.0	60.4	63.6	51.0	48.7	61.0	50.8	47.8	47.0	47.8	41.4	31.4	63.4
					B	89.2	76.3	72.6	62.9	62.5	51.5	60.8	65.9	49.3	51.4	48.9	47.5	50.7	50.8	44.2	40.3	38.4	39.7	42.4	56.4
			GC-MS/MS	Green tea	A*	115.5	88.3	70.2	87.0	64.9	72.3	59.2	146.0	112.9	58.6	68.8	53.9	64.6	48.8	50.9	57.4	55.2	50.0	32.1	58.4
					B*	99.5	83.7	66.6	79.7	64.5	62.7	49.1	112.9	94.9	50.9	51.0	49.6	55.4	46.4	48.3	36.6	48.5	38.6	47.7	55.5
				Woolong tea	A	82.6	85.9	75.6	71.0	75.3	78.1	73.8	63.9	64.3	67.2	64.4	61.9	54.4	54.3	43.5	49.9	44.8	38.6	41.8	68.6
					B	87.2	76.3	71.0	62.9	72.9	71.0	61.8	50.7	58.2	57.1	65.8	58.5	55.3	55.4	54.2	49.4	49.0	49.0	47.7	59.8
		Envi-Carb+PSA	GC-MS	Green tea	A	76.9	76.3	72.6	65.0	64.9	63.7	55.8	57.0	62.2	55.7	52.6	51.5	48.5	50.1	47.2	41.0	38.6	38.1	33.0	62.6
					B	77.0	62.8	62.0	61.6	61.6	58.0	55.3	47.4	47.9	60.4	42.3	51.0	44.7	46.9	40.5	35.2	33.4	33.0	40.2	58.5
				Woolong tea	A*	117.8	84.0	113.7	118.1	70.0	59.6	71.2	53.5	56.7	60.7	63.7	66.4	53.4	55.5	34.1	34.1	46.0	46.5	40.2	55.1
					B*	135.9	79.0	141.0	135.7	71.9	63.3	79.7	53.3	53.0	60.4	53.6	61.7	53.4	49.2	28.8	48.8	35.5	46.7	47.0	50.0
			GC-MS/MS	Green tea	A	92.5	62.9	62.9	59.2	81.8	62.0	66.2	51.9	60.8	52.8	46.9	39.2	42.6	48.2	38.6	57.4	30.8	32.6	34.3	66.3
					B	85.3	60.6	60.6	55.4	63.1	57.7	53.4	42.5	49.3	46.2	49.3	34.9	41.1	44.7	36.4	36.6	27.7	31.7	26.4	67.8
				Woolong tea	A	92.5	62.9	62.9	59.2	81.8	62.0	66.2	51.9	60.8	52.8	46.9	39.2	42.6	48.2	38.6	57.4	30.8	32.6	34.3	53.9
					B	85.3	60.6	60.6	55.4	63.1	57.7	53.4	42.5	49.3	46.2	49.3	34.9	41.1	44.7	36.4	36.6	27.7	31.7	26.4	47.6

No.	Compound	Cartridge	Method	Tea																					
76	chlorfenson	Cleanert TPT	GC-MS	Green tea	A*	94.0	100.5	73.8	244.3	76.0	79.1	76.7	69.7	73.0	61.5	97.5	74.6	72.5	64.1	68.1	69.0	45.8	59.5	74.0	82.8
					B*	97.1	91.0	70.1	215.5	78.1	73.6	75.4	66.2	89.1	64.1	94.0	66.2	68.7	63.0	82.1	68.0	47.4	62.3	68.9	81.1
				Woolong tea	A	88.6	104.1	76.5	112.8	109.4	71.3	74.9	92.1	72.0	68.3	91.0	74.3	86.4	66.2	76.8	83.5	71.4	69.4	49.3	80.9
					B	85.9	99.9	77.6	92.9	95.0	61.0	70.1	91.4	63.8	70.0	72.7	61.1	81.8	71.1	67.7	79.5	62.5	67.2	51.0	74.8
			GC-MS/MS	Green tea	A*	105.8	97.3	71.5	124.0	74.5	80.8	84.4	70.3	65.6	65.2	120.8	146.3	71.5	74.4	67.0	64.0	48.6	62.7	78.9	82.8
					B*	105.2	94.6	73.6	124.5	81.8	75.0	75.0	64.0	84.2	68.0	107.7	131.8	67.7	78.5	68.0	64.3	50.0	68.1	68.4	81.6
				Woolong tea	A*	144.2	129.4	80.0	110.1	65.9	69.0	70.3	153.1	73.5	68.6	90.2	75.4	83.5	71.8	74.7	87.0	70.3	61.2	50.7	85.7
					B*	109.1	127.3	75.0	93.9	60.5	56.4	68.9	121.7	58.3	75.1	69.2	64.4	79.3	71.3	64.0	77.8	65.1	53.3	54.0	76.0
		Envi-Carb +PSA	GC-MS	Green tea	A*	92.8	87.6	90.1	92.8	87.7	91.0	79.5	70.1	87.4	83.2	95.1	109.3	78.1	68.9	41.0	57.7	100.3	63.6	96.9	82.8
					B	98.4	95.7	79.2	77.4	82.1	100.1	77.5	61.0	89.4	76.6	64.4	99.0	74.5	73.1	40.6	48.4	89.9	73.5	98.8	78.9
				Woolong tea	A*	85.0	57.0	72.6	57.8	104.7	72.0	67.9	69.1	76.7	65.8	68.3	70.9	66.1	66.2	67.1	69.8	57.3	55.0	93.4	68.6
					B	82.7	53.6	70.8	62.2	82.3	74.0	63.4	59.4	62.3	65.0	61.3	77.0	57.4	71.1	73.1	49.7	56.2	56.9	48.6	64.6
			GC-MS/MS	Green tea	A*	115.6	96.0	90.8	87.9	86.1	69.4	124.0	66.8	101.6	77.6	104.6	68.9	73.0	79.6	53.5	64.0	96.2	76.8	94.0	85.6
					B	121.7	118.8	75.1	76.6	75.6	96.5	68.2	62.5	94.5	74.1	72.3	66.3	75.2	63.3	52.7	51.5	88.7	87.7	101.6	80.1
				Woolong tea	A	98.4	62.1	71.5	62.7	108.7	66.2	76.6	68.9	76.7	65.1	67.3	60.2	68.9	60.3	64.6	87.0	55.3	63.9	52.1	70.3
					B	95.6	60.3	68.2	63.9	83.3	68.0	69.7	56.5	62.3	68.0	66.4	66.5	57.1	61.2	60.9	77.8	56.2	57.4	47.3	65.6
77	chlorfenvinphos	Cleanert TPT	GC-MS	Green tea	A*	126.4	121.6	114.8	203.8	98.9	92.2	106.5	102.5	82.5	87.3	110.6	87.2	84.5	102.4	102.9	78.4	70.6	71.8	92.8	102.0
					B*	120.2	116.4	103.1	198.5	92.7	85.2	103.1	97.5	109.7	79.3	91.7	86.7	87.1	96.8	107.1	83.5	77.6	79.6	78.1	99.7
				Woolong tea	A	115.8	141.0	114.5	139.2	125.2	92.4	108.3	141.8	95.4	100.6	98.4	100.0	120.2	94.6	89.5	101.2	109.5	90.0	89.7	108.8
					B	115.0	118.6	105.8	127.3	120.2	88.0	86.6	127.3	87.7	83.8	75.6	95.6	92.8	93.9	87.4	95.4	90.7	88.1	81.9	97.3
			GC-MS/MS	Green tea	A	131.2	113.3	93.3	145.2	93.3	87.4	106.7	99.5	99.1	80.2	116.7	92.4	86.1	78.8	91.3	84.3	69.8	75.7	100.8	96.8
					B	131.3	114.9	103.1	142.0	96.0	85.0	101.8	87.9	98.1	85.2	96.5	92.9	83.8	83.0	85.4	85.3	74.1	78.5	88.5	95.4
				Woolong tea	A*	164.6	162.5	108.6	136.5	100.7	108.8	106.1	224.1	113.1	89.4	121.1	96.7	116.9	99.6	92.1	101.1	119.5	116.1	67.2	118.1
					B*	149.8	159.1	99.1	121.7	99.1	93.5	83.1	168.5	86.3	89.8	94.6	85.5	94.5	89.8	81.5	90.2	92.3	97.0	73.3	102.6
		Envi-Carb +PSA	GC-MS	Green tea	A	119.3	115.2	108.3	116.0	103.3	115.8	105.4	88.4	109.5	99.0	108.4	109.3	80.5	88.5	62.5	104.0	110.8	67.3	103.2	100.8
					B	121.1	121.8	103.3	99.3	103.8	108.9	104.0	69.0	107.8	95.8	82.0	109.9	71.5	87.2	65.3	84.0	103.5	71.4	111.4	95.8
				Woolong tea	A	117.6	87.9	114.0	109.6	120.6	106.0	66.2	86.9	99.2	94.1	91.1	85.5	77.6	92.3	95.4	86.0	75.9	74.7	74.7	92.4
					B	112.8	80.5	108.5	98.6	98.2	100.2	92.8	71.0	78.5	73.7	73.8	90.2	73.2	94.6	91.7	75.5	73.0	75.2	72.2	86.0
			GC-MS/MS	Green tea	A*	148.5	126.4	96.7	97.3	121.8	133.1	133.1	79.4	115.9	80.6	113.2	117.7	86.4	101.1	70.0	94.7	93.3	93.8	103.3	111.2
					B*	196.3	136.1	118.2	119.2	114.5	89.0	89.0	78.6	108.2	84.8	98.1	115.5	80.4	80.5	74.4	82.5	91.3	96.0	109.2	104.2
				Woolong tea	A	209.8	95.7	141.2	101.6	132.3	107.3	133.4	92.6	99.2	83.5	88.3	58.0	84.2	86.7	85.5	101.1	66.2	77.7	82.0	100.6
					B	149.5	92.1	102.8	90.6	105.3	89.1	95.2	72.9	78.5	76.1	90.3	56.4	73.7	84.3	72.5	90.2	68.1	70.7	70.7	85.8
78	chlormephos	Cleanert TPT	GC-MS	Green tea	A	79.8	64.6	72.9	76.6	66.4	59.9	70.9	63.2	68.3	60.0	62.6	58.4	64.8	64.3	67.2	51.9	49.3	51.2	64.9	64.1
					B	75.9	62.9	64.2	74.4	62.0	55.5	67.4	58.2	67.1	54.5	57.6	55.4	62.7	59.8	61.8	53.0	49.6	52.0	58.2	60.6
				Woolong tea	A	80.3	69.2	70.7	83.3	63.0	65.3	63.5	74.5	70.7	64.6	66.7	67.7	69.4	60.0	63.1	60.0	58.3	60.8	52.8	66.5
					B	75.4	61.4	62.0	70.5	59.2	56.1	52.8	71.1	59.9	55.4	51.8	56.6	60.0	56.4	55.3	55.5	49.5	54.1	46.3	58.4
			GC-MS/MS	Green tea	A*	80.4	66.2	64.0	79.5	66.8	59.5	68.5	62.2	64.7	61.6	70.0	72.2	60.6	65.0	61.1	47.2	48.2	52.3	66.8	64.4
					B	75.7	64.9	76.3	81.7	62.0	55.2	68.2	56.2	63.4	54.6	60.4	69.0	59.3	61.9	57.1	49.5	49.2	55.7	60.4	61.2
				Woolong tea	A	78.3	69.0	68.9	68.0	60.1	64.0	61.7	81.1	72.6	61.8	66.0	62.3	63.9	59.0	57.0	58.7	54.9	59.2	48.4	64.7
					B	72.6	58.7	59.9	73.8	55.1	52.4	50.3	69.6	59.8	54.0	52.7	50.3	55.2	53.4	46.3	51.7	51.7	54.0	40.9	65.6
		Envi-Carb +PSA	GC-MS	Green tea	A	92.4	79.8	70.2	68.1	74.3	75.3	74.1	69.5	64.7	68.1	64.1	66.0	69.2	69.8	55.2	68.1	61.9	62.6	64.4	69.7
					B	89.0	65.4	71.9	68.1	74.1	70.7	70.3	64.2	61.0	67.0	54.0	62.1	59.4	68.0	46.9	60.5	58.4	56.7	64.5	64.8
				Woolong tea	A	79.7	61.5	62.2	59.6	68.9	65.6	60.2	57.0	61.3	62.7	56.8	54.9	57.6	58.8	64.2	54.3	50.0	53.3	46.7	59.7
					B	64.0	51.2	59.9	56.2	55.4	57.4	56.3	50.5	50.1	50.6	51.9	51.0	51.6	54.8	52.2	44.3	43.5	46.5	44.7	52.2
			GC-MS/MS	Green tea	A	86.5	76.6	60.0	60.9	77.1	62.5	70.4	65.5	64.1	73.1	65.9	86.8	70.8	70.9	56.8	64.2	63.3	63.9	67.4	68.8
					B	91.1	60.6	48.7	45.1	76.0	70.6	67.6	61.6	60.4	71.1	54.1	82.7	61.1	57.6	45.4	57.8	58.6	61.3	66.9	63.1
				Woolong tea	A	91.3	57.7	85.7	57.0	69.7	61.9	59.6	54.6	61.3	59.2	55.7	53.9	56.5	53.7	58.5	58.7	47.5	49.7	49.5	60.1
					B	65.2	50.8	57.0	52.4	56.0	56.6	53.9	47.3	50.1	47.6	56.9	49.6	50.3	48.2	45.8	51.7	43.0	42.3	47.5	51.2

(Continued)

Annex 1: Content change for 201 pesticides in incurred tea Youden pair samples under 8 conditions with two SPE cartridge cleanup in three months (Nov.9, 2009-Feb.7, 2010) (cont.)

No	Pesticides	SPE	Method	Sample	Youden pair	Nov.9, 2009 (n=5)	Nov.14, 2009 (n=3)	Nov.19, 2009 (n=3)	Nov.24, 2009 (n=3)	Nov.29, 2009 (n=3)	Dec.4, 2009 (n=3)	Dec.9, 2009 (n=3)	Dec.14, 2009 (n=3)	Dec.19, 2009 (n=3)	Dec.24, 2009 (n=3)	Dec.29, 2009 (n=3)	Jan.3, 2010 (n=3)	Jan.8, 2010 (n=3)	Jan.13, 2010 (n=3)	Jan.18, 2010 (n=3)	Jan.23, 2010 (n=3)	Jan.28, 2010 (n=3)	Feb.2, 2010 (n=3)	Feb.7, 2010 (n=3)	AVE. µg/kg.
79	chloroneb	Cleanert TPT	GC-MS	Green tea	A	48.0	38.6	41.8	45.4	38.9	34.7	41.7	37.8	41.8	34.8	34.3	34.3	36.7	35.8	35.8	32.7	31.9	33.7	40.2	37.8
					B	45.4	37.2	38.7	43.8	35.4	32.3	38.9	34.2	39.7	32.7	32.7	32.5	34.4	34.3	33.3	32.4	31.1	32.4	35.4	35.6
				Woolong tea	A	42.3	38.9	40.6	44.3	36.7	37.2	34.6	42.5	43.0	40.4	43.5	45.2	43.6	40.3	38.9	38.9	35.4	32.7	32.7	39.8
					B	40.3	34.1	36.4	39.4	34.6	32.2	30.8	38.6	39.0	39.4	35.9	39.2	38.5	38.1	35.5	36.1	33.8	36.7	32.4	36.4
			GC-MS/MS	Green tea	A	45.8	39.6	41.9	48.0	36.4	33.9	40.5	37.5	38.1	35.7	37.0	45.5	34.9	34.3	35.5	31.6	29.6	31.4	39.7	37.7
					B	43.3	37.6	37.7	45.1	34.8	32.2	39.6	34.5	37.8	33.1	35.5	42.9	35.3	32.7	32.5	31.5	29.2	31.7	35.7	35.9
				Woolong tea	A	37.2	38.9	41.2	46.4	34.9	36.2	34.0	39.7	39.0	36.1	39.6	37.4	37.4	33.9	34.8	35.2	34.0	35.4	30.2	36.9
					B	37.4	33.0	35.1	40.6	33.4	31.1	29.4	36.0	35.4	33.1	32.2	33.3	32.1	33.1	30.6	32.8	33.8	32.3	26.9	33.2
		Envi-Carb+PSA	GC-MS	Green tea	A	52.0	47.3	41.4	41.2	43.1	43.3	40.6	40.0	38.2	35.1	36.6	37.9	40.2	41.4	35.9	41.9	35.9	38.0	38.0	40.4
					B	47.2	38.7	41.1	37.4	43.0	39.8	40.2	37.9	34.5	34.8	33.5	36.3	36.3	38.1	32.1	36.4	33.2	32.8	37.4	37.4
				Woolong tea	A	42.7	34.6	33.2	33.9	40.4	34.1	35.8	31.3	34.9	37.6	38.3	37.9	36.3	40.3	41.0	35.2	33.7	35.7	32.1	36.3
					B	33.5	31.4	32.9	32.3	33.8	31.3	32.5	27.7	30.0	32.3	34.6	34.6	34.7	39.2	34.6	31.6	30.6	31.7	31.2	32.7
			GC-MS/MS	Green tea	A	47.3	44.1	43.8	43.1	42.8	28.0	40.1	38.7	37.4	40.9	39.6	39.7	44.1	38.1	36.1	40.3	35.2	38.0	39.5	39.8
					B	45.6	34.7	35.4	37.4	41.9	39.3	36.2	36.3	34.4	38.9	35.2	35.8	39.2	35.7	31.2	35.6	33.6	34.7	39.6	37.1
				Woolong tea	A	43.9	35.6	45.4	33.0	39.2	34.9	32.9	30.8	34.9	34.9	32.1	32.0	33.1	31.4	36.1	35.2	30.0	30.4	29.7	34.5
					B	34.7	31.4	33.7	31.6	34.5	32.4	32.5	28.3	30.0	29.8	31.3	29.7	30.3	29.8	29.5	32.8	28.5	26.5	30.2	30.9
80	chloropropylate	Cleanert TPT	GC-MS	Green tea	A	229.5	42.6	35.7	73.1	23.3	35.3	35.7	48.0	32.2	29.7	43.0	35.2	32.1	31.5	33.9	31.5	25.9	26.1	38.7	46.5
					B	224.4	37.4	39.4	59.6	25.7	34.7	34.5	31.3	37.3	30.4	40.7	33.7	31.0	33.2	35.7	31.8	27.4	26.8	34.3	44.7
				Woolong tea	A	85.3	80.8	75.9	96.3	82.5	72.8	66.8	96.5	76.9	75.2	73.9	75.4	74.3	67.3	73.4	79.3	72.1	73.3	61.5	76.8
					B	83.4	82.2	75.5	87.2	81.2	67.5	70.0	92.5	66.7	68.1	62.8	62.7	70.0	68.4	67.2	73.0	62.1	71.1	61.0	72.3
			GC-MS/MS	Green tea	A	50.2	41.3	41.1	49.5	35.0	32.2	36.2	33.9	32.8	31.1	39.9	35.6	34.0	34.1	34.2	28.6	25.3	28.6	61.0	35.9
					B	48.4	39.9	37.0	46.3	32.7	30.6	36.1	32.3	35.3	30.2	37.4	32.4	31.0	33.0	32.0	28.4	26.0	29.8	33.8	34.3
				Woolong tea	A*	113.0	105.7	86.7	104.5	76.4	78.2	75.8	144.8	84.3	77.7	89.1	81.6	81.6	80.3	76.4	81.8	75.0	75.3	60.4	86.8
					B*	99.6	104.3	76.9	93.8	73.5	69.4	64.0	129.7	71.4	73.0	71.6	70.1	75.1	71.7	65.9	76.4	69.4	69.8	63.5	78.4
		Envi-Carb+PSA	GC-MS	Green tea	A	47.4	45.6	4.9	42.1	37.4	42.5	36.7	35.2	43.2	39.1	40.1	41.2	36.5	29.8	24.5	29.4	42.0	31.5	36.3	38.0
					B	49.8	41.5	40.8	33.5	37.0	40.8	36.4	31.2	43.7	38.1	32.4	36.0	29.2	34.5	20.8	26.0	37.4	39.7	38.1	36.2
				Woolong tea	A	130.0	69.1	68.8	41.9	88.9	71.3	69.8	65.5	76.5	65.2	68.3	68.8	63.4	66.6	73.6	66.4	61.0	64.1	58.9	70.4
					B	104.9	63.9	75.6	52.2	72.0	68.8	67.5	58.2	67.0	58.8	57.6	75.3	57.5	69.8	67.4	55.8	57.4	57.8	56.9	65.5
			GC-MS/MS	Green tea	A*	51.0	37.2	38.5	37.6	40.6	33.4	67.7	33.4	37.4	36.5	39.4	27.8	39.1	33.6	28.7	33.1	37.6	31.8	35.4	37.9
					B	53.3	50.8	36.2	33.5	38.6	40.1	36.8	31.1	36.7	37.5	30.7	25.8	32.5	29.7	25.7	28.3	35.0	33.3	38.7	35.5
				Woolong tea	A*	98.7	74.9	88.2	73.2	92.9	77.1	72.4	68.7	76.5	71.9	72.4	68.4	73.6	69.1	79.3	81.8	63.6	71.3	64.2	75.7
					B*	89.1	69.0	73.1	70.1	77.7	71.8	72.1	59.8	67.0	63.0	73.3	67.2	67.6	66.1	66.2	76.4	64.0	62.1	61.1	69.3
81	chlorpropham	Cleanert TPT	GC-MS	Green tea	A	92.2	85.5	86.7	126.8	79.7	72.1	85.2	82.4	82.1	71.8	87.0	75.3	77.2	72.8	78.4	69.6	67.7	40.6	85.1	79.9
					B	94.1	84.0	80.1	114.1	74.5	68.5	81.3	76.0	86.0	70.0	78.8	71.1	74.3	85.3	77.6	69.4	66.0	39.3	77.7	77.3
				Woolong tea	A	44.4	43.7	41.4	51.6	45.4	40.2	38.3	45.5	41.4	39.7	42.9	42.4	43.2	37.5	39.9	41.0	37.2	38.4	31.0	41.3
					B	43.1	41.4	33.6	46.9	42.4	36.6	33.6	43.1	36.5	36.5	35.2	35.5	38.5	36.9	35.1	37.9	32.9	38.4	31.3	37.9
			GC-MS/MS	Green tea	A	93.1	86.8	84.0	105.7	78.0	72.0	82.7	81.0	77.0	71.4	105.6	89.5	74.4	69.0	75.5	67.3	60.6	70.3	92.3	80.8
					B	94.8	85.4	79.6	100.7	73.5	68.1	82.0	74.4	80.5	71.3	82.5	83.9	71.7	68.2	72.8	66.8	60.7	71.3	80.5	77.3
				Woolong tea	A*	52.3	48.6	33.8	48.5	35.3	33.0	34.9	64.7	37.6	36.1	40.4	37.2	39.0	33.7	33.3	36.1	33.8	32.0	35.4	39.2
					B*	47.2	49.3	35.7	44.4	35.5	33.0	32.0	61.9	30.8	34.2	32.4	32.2	35.0	33.2	29.9	34.0	30.2	31.0	26.0	36.2
		Envi-Carb+PSA	GC-MS	Green tea	A	99.5	101.0	89.2	88.0	89.7	93.1	92.0	78.3	87.9	83.8	81.9	86.9	82.7	73.0	61.4	71.2	86.1	70.4	77.0	83.9
					B	102.6	89.9	87.1	75.0	88.3	87.6	103.6	70.4	89.6	81.7	69.2	83.5	72.3	78.1	54.5	65.4	79.0	78.1	80.9	80.9
				Woolong tea	A	43.1	36.9	40.6	36.5	45.9	38.4	37.2	35.6	35.9	37.0	36.1	35.4	35.4	37.5	39.3	35.1	32.8	34.7	31.7	37.1
					B	39.5	32.9	37.1	35.5	37.3	35.4	34.7	30.1	30.5	31.9	32.7	34.9	32.2	36.9	35.1	30.6	30.7	31.4	30.1	33.7
			GC-MS/MS	Green tea	A	106.3	100.7	99.9	95.7	91.7	83.4	84.0	77.9	84.6	85.7	90.5	71.1	94.0	77.3	69.7	79.5	82.8	79.5	80.3	86.0
					B	114.2	95.4	109.5	98.0	87.8	89.9	89.4	73.3	80.8	86.5	71.3	68.1	80.9	72.9	60.9	73.6	76.3	82.6	86.3	84.1
				Woolong tea	A*	54.3	34.5	37.8	34.7	92.2	34.5	33.7	30.6	35.9	32.6	32.4	32.3	29.1	30.9	68.0	36.1	27.1	56.2	27.5	40.0
					B*	51.0	33.2	35.3	33.8	70.9	32.6	33.4	25.9	30.5	29.3	33.2	32.2	26.0	28.4	57.2	33.3	27.2	49.6	27.0	36.3

#	Compound	Cartridge	Method	Tea																					
82	chlorpyrifos(ethyl)	Cleanert TPT	GC-MS	Green tea	A*	43.8	37.0	36.4	54.7	32.5	30.0	35.9	34.7	30.1	28.0	39.0	30.6	31.3	29.3	31.8	27.5	25.0	23.6	32.5	33.4
					B	44.4	36.1	33.3	49.9	30.7	28.6	33.8	32.7	35.8	25.9	33.8	28.8	30.1	29.5	31.0	27.2	24.7	24.1	29.2	32.1
				Woolong tea	A	41.5	37.9	35.3	43.8	36.9	33.4	32.7	44.1	32.7	33.3	34.4	33.1	36.0	31.3	31.1	32.1	30.8	30.5	27.4	34.6
					B	40.2	36.1	32.7	40.2	34.7	30.2	25.9	38.1	27.7	29.0	28.2	29.0	30.4	30.4	28.2	29.7	26.7	29.6	26.1	31.2
			GC-MS/MS	Green tea	A*	42.9	36.1	34.1	45.8	32.5	30.6	35.0	33.5	33.2	29.2	38.7	32.3	29.7	29.4	31.8	26.0	23.4	26.8	32.0	32.8
					B*	42.3	35.9	32.7	43.7	32.3	28.6	33.1	30.2	35.2	28.0	34.2	30.6	29.7	28.7	29.7	26.2	23.9	26.8	30.2	31.7
				Woolong tea	A	46.2	40.5	35.9	44.6	32.9	33.7	32.7	55.1	37.1	31.1	36.0	34.0	36.0	32.1	30.8	33.8	31.2	30.9	25.0	35.8
					B*	42.4	36.5	32.1	40.5	31.1	29.3	26.5	45.8	29.4	30.0	30.0	29.5	30.6	30.9	27.9	29.8	26.6	29.8	24.3	31.7
		Envi-Carb +PSA	GC-MS	Green tea	A	43.3	41.1	35.8	36.2	38.9	38.8	38.2	32.2	36.9	38.1	36.1	41.4	31.5	33.0	24.0	30.6	34.3	28.5	33.5	35.4
					B	42.5	36.9	35.0	32.2	37.4	36.3	35.9	28.0	35.6	35.1	31.1	37.0	29.6	31.8	24.2	26.2	31.4	29.2	33.4	33.1
				Woolong tea	A*	42.1	30.7	35.5	33.0	39.6	31.6	30.4	28.0	31.9	30.8	30.0	28.9	28.9	31.1	31.8	28.3	25.9	27.0	25.6	31.1
					B	37.4	27.8	32.2	30.9	31.3	29.6	30.9	24.0	26.3	26.3	25.7	28.2	26.7	30.0	28.6	24.5	24.2	24.9	24.2	28.1
			GC-MS/MS	Green tea	A	47.1	39.1	34.4	34.2	41.1	59.1	40.9	29.8	34.9	31.6	35.2	35.4	34.0	37.1	26.5	31.5	31.4	30.4	34.3	36.2
					B	51.4	38.4	33.9	34.1	37.9	36.9	30.0	28.4	31.9	31.4	30.8	33.2	31.4	30.2	26.8	27.3	30.5	30.4	34.8	33.2
				Woolong tea	A	51.5	31.9	43.8	32.3	39.3	30.9	34.6	28.4	31.9	29.0	31.1	24.7	31.6	28.3	30.0	33.8	25.5	26.6	25.4	32.1
					B	40.3	29.7	31.1	30.6	31.6	29.2	31.1	24.7	26.3	26.1	31.5	24.0	28.5	26.6	25.3	29.8	24.9	24.0	24.7	28.4
83	chlorthiophos	Cleanert TPT	GC-MS	Green tea	A	126.8	110.8	107.4	138.7	93.4	80.5	98.0	90.3	85.7	69.9	85.7	70.4	80.9	79.4	83.9	69.6	70.0	58.2	67.0	87.7
					B	126.7	105.1	98.5	128.3	85.4	76.4	91.1	84.7	91.8	65.7	73.8	67.9	85.6	81.4	87.7	72.4	72.4	59.1	60.0	84.9
				Woolong tea	A	124.4	112.4	105.2	120.4	100.5	94.5	90.4	114.2	86.8	83.7	80.7	94.2	101.5	86.3	83.6	80.5	77.6	62.9	59.0	92.6
					B	119.7	105.2	97.2	110.3	99.1	86.9	75.3	101.6	73.3	71.2	66.7	78.8	84.1	85.9	72.8	67.7	72.3	60.4	52.6	83.2
			GC-MS/MS	Green tea	A*	127.2	108.8	107.0	131.8	93.1	82.0	96.2	86.0	83.5	69.7	89.2	48.4	74.8	73.0	69.2	56.4	52.1	56.3	68.8	82.8
					B*	122.6	109.0	99.8	120.3	85.8	77.7	95.1	79.5	83.1	70.6	80.3	47.6	66.6	70.1	65.9	55.6	54.1	57.2	61.4	79.1
				Woolong tea	A	140.6	121.5	108.0	122.0	91.6	94.5	86.5	141.2	93.6	80.2	92.0	81.7	88.1	81.8	69.7	67.9	69.7	63.8	55.6	92.1
					B	128.0	109.2	95.9	112.2	91.8	84.1	75.3	124.9	77.3	77.2	73.2	72.0	69.6	68.5	57.6	60.6	57.0	58.4	47.4	81.1
		Envi-Carb +PSA	GC-MS	Green tea	A*	131.5	127.7	108.9	102.2	105.4	103.7	121.6	84.7	82.3	82.3	78.3	90.5	83.9	91.4	74.7	90.3	77.2	57.9	69.5	92.8
					B*	136.4	112.9	106.4	93.8	104.2	95.4	152.8	76.7	80.0	81.4	72.9	82.4	86.7	82.0	75.1	74.9	75.6	56.3	68.5	90.2
				Woolong tea	A	129.9	98.5	107.5	95.9	108.8	87.2	82.8	73.9	81.4	78.9	74.1	78.3	84.4	86.8	67.9	67.9	51.0	59.2	50.6	84.2
					B	121.2	92.7	105.9	89.2	89.3	80.5	79.4	66.2	69.9	65.9	62.2	71.7	74.2	85.9	65.5	66.5	46.4	51.3	46.8	76.3
			GC-MS/MS	Green tea	A*	153.3	122.8	117.9	117.2	111.1	168.7	104.2	80.5	86.1	77.8	83.8	108.1	77.9	80.4	63.0	73.9	61.2	63.1	67.1	95.7
					B*	162.4	106.6	108.4	108.4	103.4	97.8	95.3	77.3	79.8	81.3	75.1	101.8	70.0	67.0	60.4	61.8	58.1	60.9	69.2	86.6
				Woolong tea	A	149.9	96.7	117.4	97.7	111.7	84.0	96.5	72.2	81.4	78.0	75.2	64.4	68.5	64.0	71.2	67.9	50.7	56.3	51.2	81.8
					B	124.7	93.3	96.5	90.5	94.7	79.3	82.9	61.4	69.9	67.8	72.2	60.4	62.4	56.4	54.7	60.6	50.1	48.4	50.3	72.5
84	cis-chlordane	Cleanert TPT	GC-MS	Green tea	A	89.1	73.6	75.5	103.1	66.2	62.4	72.1	65.9	67.5	60.6	69.0	62.1	63.1	58.3	63.7	54.8	51.4	52.5	68.9	67.4
					B	86.2	71.2	68.3	95.1	63.2	58.4	66.7	62.2	69.6	58.4	62.6	58.0	59.8	56.4	59.6	54.8	51.5	52.2	61.1	64.0
				Woolong tea	A*	80.9	81.0	77.0	87.2	68.8	72.0	67.7	82.3	73.0	67.4	76.4	72.6	73.4	66.5	65.9	68.4	64.6	65.7	58.0	72.0
					B*	78.7	72.2	68.0	78.8	66.3	64.1	55.6	74.4	65.1	62.2	61.8	62.1	62.3	62.9	58.7	62.1	56.2	65.6	56.3	64.9
			GC-MS/MS	Green tea	A	89.6	71.3	75.3	88.4	65.1	63.6	68.5	64.4	63.5	62.1	72.8	61.6	66.9	60.2	65.4	50.9	50.7	53.7	65.7	66.3
					B	87.0	72.8	67.2	85.0	67.1	61.1	68.3	61.9	65.2	59.0	67.5	54.2	62.0	55.6	63.3	52.5	49.0	58.2	66.6	63.9
				Woolong tea	A	80.1	73.7	81.3	91.0	71.1	69.8	63.4	82.4	78.5	70.8	75.5	73.3	76.4	70.6	61.5	67.3	60.2	63.0	56.6	71.9
					B	75.6	66.5	68.8	83.1	68.5	60.7	52.2	73.0	65.3	65.5	58.7	63.1	59.6	64.3	53.0	61.6	56.0	59.3	52.8	63.6
		Envi-Carb +PSA	GC-MS	Green tea	A*	93.9	85.8	74.3	74.8	78.4	78.5	117.2	63.4	65.8	61.8	58.7	70.5	64.3	70.2	55.0	68.9	64.8	60.9	66.1	72.5
					B*	89.7	74.0	72.2	69.7	76.0	73.8	122.9	60.4	63.6	61.8	63.6	67.0	59.8	63.8	52.7	56.3	58.8	57.7	66.5	68.7
				Woolong tea	A	84.3	70.1	73.5	67.5	77.8	66.4	68.4	61.3	68.8	65.5	65.6	63.3	60.6	66.5	70.2	60.7	56.5	60.5	55.1	66.5
					B	73.2	62.7	69.4	62.6	65.1	61.7	65.2	54.5	65.2	56.6	60.1	58.9	57.4	62.9	60.4	53.0	52.3	53.0	52.5	60.0
			GC-MS/MS	Green tea	A*	94.9	88.2	69.1	69.4	77.3	79.3	74.1	64.0	67.1	65.2	66.7	96.8	60.6	65.4	56.6	69.0	61.2	66.3	71.9	71.7
					B*	90.9	75.8	61.3	58.3	71.8	74.6	57.6	62.3	63.5	64.5	59.7	84.3	59.6	56.7	53.0	56.0	57.0	61.4	57.2	65.2
				Woolong tea	A	91.3	68.3	87.6	68.3	80.9	63.3	66.5	62.8	68.8	63.4	61.8	61.1	55.4	59.3	61.4	67.3	52.8	61.9	57.3	66.3
					B	75.9	60.8	67.7	64.3	67.7	61.3	63.4	51.8	57.9	55.6	64.3	57.0	50.4	52.3	48.7	61.6	53.4	51.4	53.5	58.9

(Continued)

Annex 1: Content change for 201 pesticides in incurred tea Youden pair samples under 8 conditions with two SPE cartridge cleanup in three months (Nov.9, 2009-Feb.7, 2010) (cont.)

No	Pesticides	SPE	Method	Sample	Youden pair	Nov.9 2009 (n=5)	Nov.14 2009 (n=3)	Nov.19 2009 (n=3)	Nov.24 2009 (n=3)	Nov.29 2009 (n=3)	Dec.4 2009 (n=3)	Dec.9 2009 (n=3)	Dec.14 2009 (n=3)	Dec.19 2009 (n=3)	Dec.24 2009 (n=3)	Dec.14 2009 (n=3)	Jan.3 2010 (n=3)	Jan.8 2010 (n=3)	Jan.13 2010 (n=3)	Jan.18 2010 (n=3)	Jan.23 2010 (n=3)	Jan.26 2010 (n=3)	Feb.2 2010 (n=3)	Feb.7 2010 (n=3)	AVE μg/kg
85	cis-diallate	Cleanert TPT	GC-MS	Green tea	A	85.7	67.9	72.7	88.2	64.9	61.2	71.6	63.9	68.2	58.0	67.3	59.6	62.0	59.1	62.4	51.4	56.1	34.0	66.3	64.2
					B	82.5	66.5	67.1	82.4	61.5	57.8	65.7	60.9	68.7	55.4	61.3	55.6	58.9	55.1	57.4	51.4	53.7	33.5	58.9	60.7
				Woolong tea	A	77.2	71.8	71.1	81.0	65.7	67.1	61.0	75.5	69.7	63.9	70.1	67.9	67.5	62.4	63.5	63.4	56.5	63.1	52.4	66.9
					B	74.7	64.2	63.9	74.3	62.9	59.8	51.0	67.4	60.8	58.4	56.4	58.5	58.2	60.8	55.7	57.0	51.5	59.7	50.0	60.3
			GC-MS/MS	Green tea	A	86.2	71.0	76.1	90.8	63.6	62.2	71.2	64.4	67.8	60.6	68.7	48.5	65.8	59.8	65.1	52.6	50.0	56.0	69.4	65.8
					B	84.2	69.2	68.3	85.6	60.5	58.6	70.0	60.7	67.7	56.9	63.4	47.9	58.6	56.8	60.9	51.2	49.4	57.6	61.5	62.6
				Woolong tea	A	69.6	72.8	72.9	85.2	60.3	65.3	59.5	75.9	69.4	62.3	68.1	67.0	64.1	64.7	60.0	62.0	55.3	59.3	51.9	65.6
					B	69.4	64.2	64.0	74.1	58.7	56.3	50.4	67.8	64.5	58.4	57.8	57.3	55.3	58.1	52.7	55.9	51.8	57.1	47.4	59.0
		Envi-Carb+PSA	GC-MS	Green tea	A	100.0	82.6	71.7	71.4	77.6	74.6	72.4	63.5	64.7	66.6	65.1	73.1	62.9	68.8	58.0	67.1	63.9	62.7	65.4	70.1
					B	88.4	71.2	70.2	66.3	75.0	69.6	73.6	61.5	60.5	63.3	59.4	66.7	58.5	61.4	54.5	56.7	58.8	58.0	64.4	65.2
				Woolong tea	A	78.3	63.6	67.2	64.2	73.0	63.1	67.8	54.1	62.9	61.5	60.2	57.0	58.6	62.4	62.8	53.6	51.6	57.4	51.3	61.6
					B	68.7	56.3	62.2	59.5	59.6	56.6	59.7	49.6	53.3	52.6	53.9	53.3	54.1	60.4	53.6	46.8	48.3	50.7	47.4	55.1
			GC-MS/MS	Green tea	A	94.9	84.7	72.3	71.0	78.5	76.7	73.8	59.1	65.2	66.8	70.2	65.9	87.7	71.4	58.4	67.3	63.1	63.9	69.1	71.7
					B	89.3	68.7	69.7	66.9	76.8	74.6	63.8	59.1	61.2	65.9	63.0	62.3	74.9	59.8	54.0	57.7	56.2	60.9	68.0	65.9
				Woolong tea	A	81.1	63.9	79.2	61.9	71.6	61.9	58.9	53.2	62.9	59.2	58.1	56.1	58.6	54.8	61.3	62.0	50.3	54.2	51.3	61.1
					B	67.1	57.2	61.0	57.9	61.1	57.6	56.5	47.6	53.3	51.2	59.0	53.5	52.7	51.3	50.7	55.9	47.8	47.6	50.3	54.7
86	cyanofenphos	Cleanert TPT	GC-MS	Green tea	A*	48.9	47.8	41.7	102.7	38.5	39.2	44.9	39.4	41.7	32.8	46.9	34.5	37.8	32.8	34.0	31.8	24.7	29.6	36.7	41.4
					B*	50.6	45.0	38.4	88.3	38.5	37.4	43.2	38.1	44.0	32.1	42.3	30.8	34.9	33.1	35.8	32.5	26.1	32.3	32.7	39.8
				Woolong tea	A*	46.8	46.4	41.9	53.7	49.8	45.3	40.0	56.0	40.8	43.6	44.7	38.6	43.7	36.6	35.8	44.9	30.1	33.1	29.8	42.5
					B	44.9	45.5	42.5	47.4	48.2	43.3	38.9	54.1	38.4	40.3	39.5	34.1	39.5	38.5	34.4	40.8	30.7	32.7	30.7	40.2
			GC-MS/MS	Green tea	A	51.2	46.5	38.1	57.3	35.4	39.3	41.3	35.4	31.7	30.0	48.2	40.3	38.9	31.8	34.5	31.3	25.0	30.1	35.6	38.0
					B	50.3	44.6	37.3	56.3	37.4	36.4	37.2	33.2	38.1	32.6	45.3	37.1	32.3	32.8	33.8	31.9	26.0	31.6	31.7	37.2
				Woolong tea	A*	61.1	53.9	39.9	51.7	35.7	38.0	34.0	64.0	39.2	32.2	44.0	36.4	41.0	36.9	36.0	38.6	35.1	32.9	28.2	41.0
					B	51.8	51.7	37.7	43.5	34.7	31.9	34.9	52.6	32.6	35.1	36.5	33.9	35.6	34.1	31.8	35.2	31.1	29.7	27.7	37.0
		Envi-Carb+PSA	GC-MS	Green tea	A*	51.1	49.3	45.8	45.7	44.1	44.0	119.6	41.0	52.9	44.7	71.3	56.4	35.8	49.4	23.5	57.8	47.2	49.1	45.6	51.3
					B*	53.7	51.7	42.3	39.6	42.4	43.9	106.1	38.2	49.0	41.8	48.2	49.3	33.9	46.1	25.4	47.4	41.4	51.7	47.6	47.4
				Woolong tea	A	48.4	31.1	40.6	35.8	50.7	43.2	40.8	40.4	36.1	41.0	31.3	33.8	34.3	36.7	35.2	31.0	28.0	31.6	29.5	36.8
					B	46.1	30.1	42.3	37.2	42.2	43.0	37.9	35.1	29.5	36.5	25.5	36.0	29.9	38.5	34.7	24.7	25.8	30.9	28.5	32.2
			GC-MS/MS	Green tea	A	56.7	51.2	52.0	51.2	42.8	30.3	59.5	32.9	47.9	37.0	47.2	37.6	39.2	42.7	27.5	34.8	23.9	35.8	45.9	42.9
					B	56.4	42.9	50.3	52.3	33.3	43.1	28.3	30.5	42.4	35.9	37.8	35.8	38.8	32.4	27.8	26.5	24.3	41.1	47.9	39.5
				Woolong tea	A	50.0	40.3	36.1	33.0	47.5	34.4	38.8	30.5	36.1	31.1	33.1	27.3	29.5	30.6	33.9	38.6	25.4	29.3	27.0	33.9
					B	48.8	31.1	36.9	33.0	40.7	33.4	33.6	28.9	29.5	30.6	33.3	28.3	24.5	30.5	29.2	35.2	26.2	26.9	24.4	31.8
87	desmetryn	Cleanert TPT	GC-MS	Green tea	A	53.5	41.3	42.7	46.1	34.9	30.1	36.9	35.0	32.6	30.5	43.7	31.8	30.5	28.0	29.5	25.7	26.4	29.7	31.9	34.1
					B	51.5	40.2	37.3	41.3	31.4	29.0	33.9	30.8	33.9	28.2	36.5	29.8	28.6	26.9	27.9	26.0	26.1	16.6	28.3	31.8
				Woolong tea	A	49.6	38.7	38.3	45.6	38.4	37.2	33.1	41.3	36.9	36.7	35.0	33.5	36.9	32.6	33.2	30.0	29.1	28.6	26.7	35.9
					B	46.5	37.9	35.6	42.8	37.0	33.4	30.1	36.7	30.4	32.3	27.6	28.7	31.1	29.5	28.3	25.5	24.1	28.0	25.8	32.2
			GC-MS/MS	Green tea	A	48.5	40.8	42.4	47.9	34.2	31.7	33.9	33.1	32.0	29.1	32.5	21.7	29.9	26.1	30.6	24.7	23.9	20.0	30.4	32.3
					B	46.4	39.2	37.1	43.8	31.4	29.3	34.1	30.4	31.7	29.0	30.8	21.0	27.3	28.2	29.3	24.5	24.3	21.3	27.3	30.9
				Woolong tea	A	38.5	40.3	39.8	46.4	35.7	35.8	32.0	30.5	35.4	31.3	35.8	34.1	33.4	32.1	27.6	26.5	28.4	28.1	27.3	34.3
					B	39.6	37.1	35.1	41.6	35.1	31.2	26.2	36.4	31.3	28.9	28.0	29.0	26.7	28.3	27.6	27.2	23.9	27.7	22.7	30.6
		Envi-Carb+PSA	GC-MS	Green tea	A	57.7	49.8	39.8	39.4	40.8	44.3	37.3	33.0	34.1	42.3	35.6	41.0	31.0	34.0	28.2	27.2	28.2	28.1	28.0	37.5
					B	52.5	42.0	39.3	35.6	39.3	38.4	37.2	31.1	32.9	43.3	31.2	34.6	28.7	30.9	25.7	29.4	25.4	24.4	28.5	34.2
				Woolong tea	A	50.2	36.7	39.1	35.3	43.1	34.7	33.9	28.2	33.4	33.1	28.3	29.7	29.4	31.3	32.9	27.4	25.6	27.7	24.2	32.8
					B	45.9	33.7	35.9	33.1	32.9	30.7	31.4	25.1	27.2	27.7	24.4	28.8	27.8	27.8	26.5	24.1	22.8	22.7	22.4	29.0
			GC-MS/MS	Green tea	A	51.4	46.9	35.5	34.1	39.4	43.3	34.7	32.8	31.0	33.3	33.2	44.6	38.0	29.1	27.5	32.3	27.5	28.4	28.4	35.3
					B	47.9	37.6	32.3	29.7	37.6	38.8	34.0	30.4	30.8	33.1	29.5	38.6	31.1	24.6	24.6	27.2	25.0	26.3	28.9	32.0
				Woolong tea	A	47.3	36.8	41.4	34.9	39.5	33.8	30.8	27.6	33.4	31.3	29.3	29.6	29.1	27.8	30.3	31.0	23.8	26.3	24.4	32.0
					B	42.6	33.9	35.0	33.4	32.6	30.5	31.2	24.6	27.2	26.1	29.4	26.7	27.0	24.9	24.2	27.2	23.0	21.4	23.9	28.7

No.	Pesticide	Cartridge	Method	Tea	A/B																		
88	dichlobenil	Cleanert TPT	GC-MS	Green tea	A	8.5	6.6	7.3	7.2	7.1	6.2	7.6	6.5	7.1	6.3	6.7	7.0	6.4	6.0	5.0	5.2	5.9	6.8	6.6
				Green tea	B	7.8	6.4	6.7	6.9	6.5	5.7	7.0	5.8	6.6	5.7	6.3	6.6	6.2	5.5	5.0	5.2	5.8	6.1	6.2
				Woolong tea	A	8.3	6.7	7.2	8.3	5.9	6.9	6.3	6.3	7.0	6.0	7.0	6.9	7.0	6.4	5.8	5.7	6.2	5.3	6.6
				Woolong tea	B	7.6	5.9	6.2	6.9	5.5	6.0	5.3	6.1	5.9	5.4	5.9	6.0	6.2	5.4	5.3	5.1	5.6	4.7	5.8
			GC-MS/MS	Green tea	A	8.7	7.0	7.3	8.1	6.6	6.4	7.1	6.1	6.9	6.2	6.0	6.2	6.9	6.5	5.1	4.7	5.6	7.2	6.6
				Green tea	B	7.9	6.5	6.5	7.8	6.3	6.0	6.7	5.9	7.0	5.9	6.3	5.7	6.2	6.1	5.3	4.9	6.0	6.4	6.3
				Woolong tea	A	7.8	7.1	7.2	8.4	6.1	6.6	6.0	6.9	7.3	5.5	6.6	5.2	6.8	5.9	5.9	5.8	5.4	4.6	6.5
				Woolong tea	B	7.3	6.2	6.2	6.7	5.4	5.3	5.4	7.5	6.3	6.9	5.5	7.0	6.1	4.9	6.5	5.4	5.1	4.6	5.7
		Envi-Carb +PSA	GC-MS	Green tea	A	9.6	8.3	7.4	7.9	8.0	7.6	7.5	7.2	7.0	6.7	6.6	6.8	6.4	5.4	5.9	5.8	7.1	7.4	7.2
				Green tea	B	9.3	6.5	7.7	7.3	8.2	7.2	7.0	4.6	6.5	6.2	5.5	5.7	5.3	4.5	5.5	5.6	6.3	7.1	6.6
				Woolong tea	A	8.5	5.8	6.2	5.9	6.7	6.6	6.7	4.7	6.6	5.1	5.8	5.1	5.7	6.8	4.8	5.2	5.6	4.8	6.1
				Woolong tea	B	6.7	5.1	6.3	5.5	5.6	5.7	5.7	7.1	5.4	8.0	5.5	15.4	5.1	5.1	6.5	4.6	4.7	4.6	5.3
			GC-MS/MS	Green tea	A*	8.5	7.8	6.1	6.0	7.9	5.5	7.0	6.8	6.9	7.7	7.5	15.4	8.0	5.5	5.7	6.4	6.6	7.9	7.4
				Green tea	B*	8.5	5.9	6.8	5.8	8.0	7.4	6.0	6.0	6.4	6.1	5.8	5.3	6.9	4.7	5.9	5.9	6.1	7.9	7.1
				Woolong tea	A*	8.3	5.6	7.9	5.6	7.2	6.5	5.7	6.0	6.6	5.0	5.6	5.1	5.6	5.9	5.3	4.6	5.4	4.8	6.0
				Woolong tea	B*	6.5	4.8	6.1	5.5	5.9	6.0	5.6	5.0	5.4	5.0	5.0	5.1	5.1	4.7	4.5	4.3	4.5	4.5	5.2
89	dicloran	Cleanert TPT	GC-MS	Green tea	A	117.2	107.9	95.0	165.5	104.9	110.5	90.5	109.3	138.1	76.7	118.6	86.4	85.7	79.9	69.6	74.2	25.3	79.9	95.0
				Green tea	B	144.6	104.6	84.4	145.0	97.5	101.4	82.7	95.7	128.6	72.5	104.1	85.7	81.4	72.2	65.3	67.0	24.2	79.5	90.0
				Woolong tea	A	92.0	104.1	102.4	130.0	137.9	85.8	104.6	114.4	97.8	115.3	82.5	78.7	97.9	89.7	83.5	107.5	64.9	64.5	96.0
				Woolong tea	B	89.6	100.6	93.3	113.1	117.3	83.8	73.0	79.7	75.4	87.5	75.8	69.6	90.2	75.5	82.1	108.9	65.6	59.1	84.8
			GC-MS/MS	Green tea	A	103.1	87.0	59.3	111.2	74.9	64.5	77.7	78.1	74.8	57.0	117.7	68.5	67.4	66.3	63.9	51.8	62.9	88.1	75.5
				Green tea	B	102.2	87.4	73.8	108.7	75.8	65.0	75.4	69.9	76.5	61.8	92.0	69.0	63.7	67.5	69.1	57.5	65.7	73.5	74.9
				Woolong tea	A	120.7	96.1	72.0	106.6	69.5	74.1	69.0	147.7	84.2	63.5	91.0	71.6	88.4	80.0	71.6	77.9	71.3	60.4	83.5
				Woolong tea	B	99.7	83.2	70.1	101.5	70.0	61.3	54.7	107.1	64.5	69.5	66.6	60.2	80.0	67.0	71.3	67.7	59.1	60.1	72.8
		Envi-Carb +PSA	GC-MS	Green tea	A	127.0	113.5	107.5	133.7	100.8	109.3	75.0	85.3	155.2	106.2	115.1	118.6	66.4	55.4	76.5	88.3	58.6	78.0	98.0
				Green tea	B	126.3	111.8	107.4	104.8	100.6	98.9	77.6	70.5	130.2	101.0	82.7	66.4	66.1	59.0	70.1	85.3	63.7	78.3	91.0
				Woolong tea	A	82.0	67.1	104.7	88.2	135.6	99.3	74.3	110.9	65.7	91.7	84.0	90.6	62.9	69.1	79.2	78.8	65.4	57.9	83.3
				Woolong tea	B	85.1	62.8	102.1	90.9	106.0	92.3	70.6	86.6	59.2	69.4	65.2	87.7	60.5	76.2	71.0	79.7	64.7	55.0	76.8
			GC-MS/MS	Green tea	A*	111.7	87.8	109.1	115.7	84.9	96.3	85.6	68.1	88.5	73.2	90.7	64.4	81.1	55.8	63.8	58.0	74.8	80.7	81.9
				Green tea	B*	130.1	88.9	128.9	120.1	81.1	93.5	74.6	68.5	80.9	82.7	72.8	62.8	76.6	54.4	62.2	63.8	70.1	83.5	82.1
				Woolong tea	A*	97.7	56.6	82.6	68.5	103.0	68.9	85.0	66.9	65.7	60.2	61.3	38.8	67.1	61.5	71.6	52.0	65.4	59.0	68.0
				Woolong tea	B*	76.9	55.7	62.5	67.3	80.6	68.2	63.3	55.2	59.2	58.1	60.0	43.7	57.7	58.9	71.3	55.1	57.5	55.2	61.4
90	dicofol	Cleanert TPT	GC-MS	Green tea	A	104.5	145.4	168.7	209.6	149.0	134.5	170.1	160.5	166.9	129.8	135.2	128.4	132.8	154.9	123.0	132.3	114.3	155.0	144.9
				Green tea	B	113.3	144.1	151.1	185.5	132.8	124.2	155.0	146.4	147.6	125.2	120.5	121.5	126.2	134.2	121.3	125.8	110.1	130.4	133.8
				Woolong tea	A	107.4	101.6	105.3	135.0	107.9	114.6	136.2	134.7	112.2	97.4	97.4	117.3	138.7	122.7	120.8	123.8	107.1	123.8	117.4
				Woolong tea	B	106.5	91.9	103.8	132.6	117.0	103.5	135.2	120.1	104.6	93.3	93.3	112.7	122.5	108.1	103.5	103.1	105.1	105.1	110.7
			GC-MS/MS	Green tea	A	182.8	150.6	172.9	185.9	143.1	128.6	152.1	151.3	152.8	130.7	155.2	189.9	132.6	140.9	116.2	121.9	114.5	171.0	148.5
				Green tea	B	174.4	145.8	146.5	169.4	129.0	117.2	155.8	139.3	141.6	123.3	131.5	182.3	129.4	125.0	113.3	116.3	113.2	148.8	137.8
				Woolong tea	A	104.9	104.9	107.1	142.9	96.9	111.7	136.4	161.9	115.0	111.4	99.4	108.5	123.2	109.2	110.5	114.5	97.8	107.0	114.2
				Woolong tea	B	103.0	95.2	102.1	136.1	107.6	98.8	134.6	141.4	107.9	118.8	94.5	113.9	125.9	105.3	105.4	107.1	104.4	108.7	111.6
		Envi-Carb +PSA	GC-MS	Green tea	A	126.1	189.5	157.1	168.7	160.4	165.5	69.2	160.4	163.5	135.8	124.6	134.0	149.5	161.0	189.4	139.4	155.3	146.2	150.9
				Green tea	B	136.4	150.3	155.3	161.3	162.0	143.5	71.2	159.4	144.1	135.5	130.7	131.3	147.4	152.6	152.7	125.0	125.7	145.2	140.7
				Woolong tea	A	125.9	104.7	122.0	118.5	144.1	117.1	122.7	89.8	112.5	102.9	105.6	122.4	102.6	137.0	118.2	139.9	114.2	125.7	118.6
				Woolong tea	B	100.8	116.3	134.0	144.1	113.8	113.9	109.6	91.4	108.5	95.3	91.8	113.7	103.2	111.9	105.1	105.1	95.3	121.8	108.4
			GC-MS/MS	Green tea	A*	204.2	199.2	296.3	294.6	151.0	145.8	160.4	151.6	155.1	147.5	136.9	75.1	156.0	139.9	170.3	137.2	154.7	152.2	167.4
				Green tea	B*	191.3	144.1	231.7	223.2	153.8	147.6	148.0	141.3	137.1	142.2	140.2	70.6	143.9	130.5	141.3	123.4	135.5	147.3	148.2
				Woolong tea	A*	128.2	103.8	143.5	119.3	142.4	114.4	106.3	91.4	112.5	96.3	100.4	106.6	107.0	120.8	110.5	119.2	99.6	115.2	112.1
				Woolong tea	B*	109.7	118.0	132.0	106.4	114.4	113.5	109.4	91.1	108.5	90.8	99.4	107.9	102.3	101.3	105.4	122.4	91.2	129.5	108.1

(Continued)

Annex 1: Content change for 201 pesticides in incurred tea Youden pair samples under 8 conditions with two SPE cartridge cleanup in three months (Nov.9, 2009-Feb.7, 2010) (cont.)

No	Pesticides	SPE	Method	Sample	Youden pair	Nov.9, 2009 (n=5)	Nov.14, 2009 (n=3)	Nov.19, 2009 (n=3)	Nov.24, 2009 (n=3)	Nov.29, 2009 (n=3)	Dec.4, 2009 (n=3)	Dec.9, 2009 (n=3)	Dec.14, 2009 (n=3)	Dec.19, 2009 (n=3)	Dec.24, 2009 (n=3)	Jan.3, 2010 (n=3)	Jan.8, 2010 (n=3)	Jan.13, 2010 (n=3)	Jan.18, 2010 (n=3)	Jan.23, 2010 (n=3)	Jan.28, 2010 (n=3)	Feb.2, 2010 (n=3)	Feb.7, 2010 (n=3)	AVE, μg/kg
91	dimethachlor	Cleanert TPT	GC-MS	Green tea	A	117.0	111.5	120.3	119.8	103.2	97.2	108.0	113.3	93.7	93.8	91.9	91.6	95.8	102.8	85.7	78.0	69.7	99.2	99.9
					B	116.4	107.1	106.3	119.1	96.2	89.7	105.0	100.8	106.0	88.6	90.8	95.6	90.4	102.2	84.1	81.2	75.7	90.7	96.6
				Woolong tea	A	119.8	130.9	120.3	128.5	103.0	105.6	109.5	141.0	102.3	106.9	109.1	112.1	96.4	89.9	95.2	104.8	92.7	94.8	108.9
					B	117.3	105.7	102.9	119.5	104.0	98.4	81.7	117.4	93.8	89.5	94.6	90.3	92.7	84.0	88.5	84.9	92.0	87.1	96.6
			GC-MS/MS	Green tea	A	121.5	109.6	99.6	139.5	96.8	91.6	113.1	107.4	108.6	88.5	69.5	94.1	81.9	88.3	81.3	79.4	84.0	101.5	98.5
					B	122.4	107.7	106.7	129.8	95.3	88.2	106.5	93.5	101.3	90.5	68.4	90.7	68.4	90.4	81.4	78.5	86.9	89.8	95.4
				Woolong tea	A*	143.8	133.0	112.7	129.4	104.1	111.9	111.2	178.0	113.9	96.4	103.3	111.7	100.4	92.3	99.4	108.5	108.8	83.3	113.4
					B*	132.8	114.1	96.8	116.8	102.4	94.4	83.3	140.4	96.2	94.4	92.5	88.7	93.0	82.8	86.8	88.4	97.9	80.1	98.6
		Envi-Carb+PSA	GC-MS	Green tea	A	116.2	112.3	100.4	112.3	111.4	118.5	122.6	94.2	107.8	93.0	108.9	86.5	105.1	71.4	104.6	93.1	81.6	100.6	102.0
					B	119.0	99.1	103.2	108.7	114.5	108.8	102.3	72.2	105.2	94.4	107.1	83.9	91.9	77.1	92.6	89.6	83.1	100.8	97.2
				Woolong tea	A*	111.1	105.5	114.4	112.6	110.1	105.4	73.1	87.0	99.4	98.3	92.7	85.4	94.0	90.2	82.5	79.2	81.7	76.9	94.4
					B*	96.9	93.6	103.7	99.1	94.0	96.4	94.8	78.3	79.6	77.2	84.4	80.4	90.0	79.4	75.7	74.0	73.6	74.1	85.9
			GC-MS/MS	Green tea	A	161.8	116.7	58.2	57.9	122.7	126.5	110.5	90.6	96.2	82.6	103.5	107.2	101.8	87.9	101.5	89.3	88.2	96.0	100.5
					B	205.5	109.2	79.3	80.5	119.1	108.5	124.7	91.0	92.7	89.9	97.9	90.7	78.3	92.8	91.0	84.1	89.7	100.1	100.7
				Woolong tea	A*	203.7	105.5	165.0	102.5	122.9	99.3	137.6	91.2	99.4	89.4	68.5	90.7	83.9	73.9	99.4	70.7	81.5	84.7	104.2
					B*	127.5	96.2	103.8	90.5	102.4	91.2	98.1	75.8	79.6	77.9	63.4	85.5	81.6	73.9	86.8	74.2	70.2	83.3	87.0
92	dioxacarb	Cleanert TPT	GC-MS	Green tea	A	342.8	273.1	343.4	447.7	294.4	248.9	318.9	277.6	303.0	274.6	248.9	244.5	229.6	271.9	224.0	262.7	245.8	298.9	287.1
					B*	354.5	277.8	310.2	411.2	271.2	235.7	316.3	245.8	288.1	232.2	246.1	256.3	262.6	223.2	234.7	247.7	234.6	262.6	271.8
				Woolong tea	A*	359.0	308.9	300.2	351.1	364.9	302.6	250.3	476.6	292.6	281.7	302.3	290.3	258.2	258.2	267.3	249.9	242.9	213.0	298.3
					B*	364.3	296.4	290.0	312.9	342.9	288.6	222.9	520.2	265.2	250.6	253.5	254.6	241.7	238.2	230.5	221.1	256.6	201.7	278.3
			GC-MS/MS	Green tea	A	317.3	284.2	336.4	340.0	278.6	245.6	288.3	265.8	266.2	274.2	336.4	232.5	262.2	250.9	193.5	240.3	242.7	289.5	276.0
					B	340.8	290.4	303.5	325.9	257.7	232.7	311.4	235.6	259.7	230.7	328.0	225.5	303.6	194.8	204.2	236.9	249.6	260.5	265.0
				Woolong tea	A*	360.9	365.4	306.2	371.5	296.8	288.1	259.6	705.7	279.6	272.8	284.5	277.2	242.2	226.4	259.2	239.0	219.5	171.4	302.1
					B*	350.5	375.0	274.6	336.3	291.6	268.3	221.2	656.2	236.7	246.7	244.6	246.6	242.0	205.7	233.7	215.3	216.2	171.8	276.6
		Envi-Carb+PSA	GC-MS	Green tea	A	414.9	327.5	299.2	341.5	310.4	303.6	379.4	276.3	303.5	273.0	304.3	280.7	274.2	231.7	283.4	274.0	240.0	248.8	296.4
					B	414.5	302.2	326.3	292.8	326.9	309.6	272.7	249.8	299.9	280.2	286.5	245.9	266.0	210.0	226.5	238.3	237.4	267.1	284.4
				Woolong tea	A	346.0	298.6	270.6	270.7	293.7	267.9	250.8	224.1	258.4	195.9	251.6	231.6	246.6	262.4	218.0	194.8	248.4	210.6	259.1
					B	322.5	267.2	287.5	285.7	271.1	255.2	216.0	273.1	241.5	288.3	215.9	226.9	238.3	205.4	206.0	202.1	186.7	211.0	239.5
			GC-MS/MS	Green tea	A	367.9	323.4	244.8	277.8	305.2	209.4	238.0	254.8	279.0	286.8	287.6	305.7	259.9	208.7	247.5	221.8	262.1	272.1	269.5
					B	382.9	303.8	224.6	206.7	316.0	309.9	219.6	236.1	279.7	263.4	271.1	259.8	192.3	179.8	209.3	198.9	265.5	293.3	257.9
				Woolong tea	A*	360.1	306.4	270.6	268.5	281.7	258.1	242.5	207.1	258.4	178.7	223.1	243.8	215.0	248.1	259.1	184.1	245.8	214.6	252.4
					B*	359.4	281.2	256.5	271.6	261.3	244.3	985.1	240.8	241.5	273.4	203.8	230.1	226.1	196.3	233.7	202.3	185.7	185.7	239.7
93	endrin	Cleanert TPT	GC-MS	Green tea	A	534.0	426.7	441.1	512.2	388.5	348.6	413.8	385.8	393.3	347.6	359.8	369.1	338.3	380.8	314.0	314.0	324.9	402.5	387.6
					B	514.8	413.4	395.5	471.7	359.6	330.1	387.9	360.7	397.6	327.0	339.7	348.4	326.6	355.0	312.3	297.9	314.2	351.7	366.0
				Woolong tea	A	482.9	451.5	436.8	503.6	416.7	423.9	393.6	494.6	418.3	387.2	410.8	418.7	375.0	375.8	387.9	371.1	373.4	328.2	414.4
					B	469.5	416.5	392.2	463.2	411.4	381.1	316.9	452.5	364.6	347.9	343.3	376.0	361.1	343.8	358.3	317.7	382.8	315.5	376.8
			GC-MS/MS	Green tea	A	532.7	437.1	445.4	527.0	372.6	352.2	392.1	379.4	377.2	346.4	232.9	390.0	394.3	368.6	297.6	298.3	310.9	400.8	383.2
					B	512.5	431.5	398.9	489.2	349.5	328.0	395.7	354.3	378.0	331.5	217.6	335.7	376.2	341.1	291.7	291.9	315.6	362.4	361.8
				Woolong tea	A	493.6	461.8	438.7	512.1	387.5	417.0	387.7	499.9	439.7	396.9	427.6	431.8	398.4	359.5	379.0	364.8	358.9	326.9	416.4
					B	473.8	420.7	393.5	472.6	383.8	360.0	315.9	440.6	380.2	361.8	367.3	357.1	372.1	316.1	351.3	318.8	355.2	310.9	373.9
		Envi-Carb+PSA	GC-MS	Green tea	A*	536.2	508.5	432.0	422.5	449.9	470.4	1065.0	372.9	363.0	370.5	394.3	370.5	405.3	331.5	401.3	343.7	345.8	359.1	432.4
					B*	545.0	429.2	426.3	388.7	440.0	424.8	414.2	348.1	357.1	366.6	367.9	346.1	359.7	303.0	332.0	319.3	321.6	361.4	412.2
				Woolong tea	A	500.9	407.9	435.2	396.1	452.5	394.4	378.5	341.9	389.8	309.8	356.4	354.9	375.0	386.7	349.1	311.3	336.7	379.9	379.9
					B	445.1	373.5	402.9	372.1	459.4	361.6	363.5	302.5	332.6	371.9	335.6	332.7	354.1	331.8	306.1	287.1	309.3	292.1	342.5
			GC-MS/MS	Green tea	A	578.7	492.4	444.0	440.9	460.3	428.8	365.6	365.6	362.7	384.2	346.5	396.6	378.1	339.3	412.3	317.5	359.2	354.4	402.2
					B	576.9	420.1	420.6	412.2	440.1	422.9	412.4	351.2	346.6	367.4	329.4	374.0	328.7	303.7	355.7	308.4	343.5	354.2	380.8
				Woolong tea	A	521.6	407.0	471.3	392.9	456.1	392.8	414.2	336.1	389.8	316.0	354.2	343.6	344.3	373.4	379.0	307.7	333.2	319.9	382.7
					B	456.5	370.1	389.1	359.9	383.8	361.9	373.8	297.1	332.6	316.0	338.5	308.9	313.4	306.6	351.3	302.7	289.0	315.4	343.8

No.	Compound	Cleanup	Method	Tea	Rep																				
94	epoxiconazole-2	Cleanert TPT	GC-MS	Green tea	A	401.2	336.4	344.3	393.9	293.3	238.6	301.5	276.7	271.5	210.8	337.7	240.6	246.8	227.3	251.9	210.9	201.7	204.5	255.4	276.1
				Green tea	B	390.3	323.9	308.1	360.1	256.9	226.3	283.5	251.8	278.5	185.5	282.6	219.2	235.2	235.4	233.0	210.0	197.3	196.1	224.0	257.8
				Woolong tea	A	373.4	296.8	276.7	352.0	302.3	289.5	252.1	327.7	269.7	259.1	230.2	264.3	278.7	241.3	239.2	250.3	230.7	232.5	200.7	272.0
				Woolong tea	B	346.2	312.8	277.3	331.7	303.4	268.5	237.0	330.0	213.4	222.1	193.8	232.0	252.7	227.8	212.2	216.5	194.2	215.1	181.3	251.0
			GC-MS/MS	Green tea	A*	397.5	341.0	345.2	374.2	288.5	247.9	287.9	276.8	275.2	236.8	327.6	268.2	262.3	225.8	250.3	208.8	196.2	199.1	262.7	277.5
				Green tea	B*	386.2	327.4	310.5	356.0	257.0	231.1	290.7	250.8	270.8	238.9	285.6	255.1	230.0	248.3	228.6	206.4	192.9	199.0	232.0	263.0
				Woolong tea	A*	381.0	338.0	305.9	357.4	274.6	289.9	249.8	390.4	286.6	252.7	291.1	265.7	272.3	237.2	231.7	242.0	229.7	225.6	200.9	280.1
				Woolong tea	B*	343.5	325.7	280.4	341.9	284.2	262.8	232.3	375.7	243.9	239.2	237.9	239.9	235.4	232.8	200.8	212.5	200.5	206.3	184.1	256.8
		Envi-Carb +PSA	GC-MS	Green tea	A	392.9	377.2	336.3	331.7	312.5	330.8	295.7	267.1	263.7	304.3	249.7	294.9	259.5	269.2	226.3	277.5	244.5	231.7	240.9	289.8
				Green tea	B	423.2	360.4	316.9	292.4	316.7	298.0	290.2	244.5	239.3	269.1	235.4	251.7	240.4	241.5	213.3	231.4	214.4	204.9	243.4	269.8
				Woolong tea	A	393.3	268.2	279.3	275.6	329.5	254.7	269.3	220.8	258.8	248.2	214.9	202.5	229.9	241.3	251.6	210.3	194.6	217.6	188.8	250.0
				Woolong tea	B	385.0	251.7	287.9	275.7	270.9	232.1	231.8	189.6	223.8	188.4	176.2	198.5	207.3	228.6	205.3	182.3	180.2	191.2	180.5	225.6
			GC-MS/MS	Green tea	A*	400.2	352.4	342.4	343.8	308.9	260.7	255.2	269.4	280.7	271.5	281.3	316.2	250.1	265.7	223.6	270.2	232.1	236.1	253.0	286.5
				Green tea	B*	416.1	319.7	344.5	325.4	307.7	310.2	269.3	244.8	250.5	262.9	248.8	297.1	194.4	224.5	205.8	226.3	202.8	219.8	191.3	272.6
				Woolong tea	A*	366.3	283.2	276.6	272.6	311.5	250.7	263.9	219.4	258.8	245.5	229.6	222.4	183.0	202.2	234.6	242.0	181.8	215.0	187.6	245.4
				Woolong tea	B*	359.9	269.7	288.0	267.7	273.1	233.2	239.6	199.8	223.8	196.9	230.9	206.6	196.1	196.1	189.8	212.3	181.6	184.4	187.6	227.6
95	EPTC	Cleanert TPT	GC-MS	Green tea	A	101.6	76.6	86.5	123.0	86.9	75.8	87.5	81.9	89.2	75.7	64.8	61.3	60.7	60.3	59.1	47.9	50.2	55.2	75.7	74.7
				Green tea	B	92.3	76.7	80.9	124.9	79.6	70.9	85.5	72.6	87.0	68.2	62.1	57.7	58.4	59.9	52.7	50.4	51.8	58.2	69.5	71.5
				Woolong tea	A	102.3	83.2	87.3	104.7	75.3	85.9	80.3	92.6	95.5	78.6	80.8	83.9	82.6	71.1	75.1	73.4	70.0	76.9	52.8	81.7
				Woolong tea	B	89.2	70.6	78.3	85.3	66.9	70.4	70.1	81.0	77.0	70.4	63.2	65.2	70.0	62.6	61.2	63.1	59.3	66.3	44.2	69.2
			GC-MS/MS	Green tea	A*	103.4	80.5	85.9	97.2	79.3	74.3	84.1	74.2	78.1	78.2	67.0	51.9	89.2	73.3	75.7	60.2	60.7	56.5	81.4	76.4
				Green tea	B*	94.1	79.5	79.9	97.2	74.7	67.6	81.7	67.7	79.6	68.3	72.8	51.3	78.4	74.0	71.1	60.2	61.0	72.0	72.0	73.5
				Woolong tea	A*	95.5	78.0	81.3	100.1	69.2	77.8	74.8	84.8	89.0	71.4	78.8	71.6	74.1	78.7	67.4	70.4	63.2	73.8	58.4	76.8
				Woolong tea	B*	86.9	63.3	68.8	75.4	58.6	60.0	61.3	70.8	68.9	61.5	62.1	59.4	65.2	62.7	54.1	59.3	58.2	62.4	46.7	63.4
		Envi-Carb +PSA	GC-MS	Green tea	A	106.9	100.7	89.7	92.4	101.8	96.9	105.0	91.2	76.2	90.4	68.0	91.1	56.5	53.0	51.6	70.6	60.0	77.9	81.3	82.2
				Green tea	B	104.8	78.7	96.3	84.2	107.0	93.6	105.8	84.2	74.6	84.0	56.0	85.7	46.8	52.3	40.9	59.5	56.1	73.1	75.0	76.8
				Woolong tea	A	102.6	64.5	78.8	74.3	90.4	87.3	79.5	73.2	77.3	84.7	71.9	71.3	69.0	71.1	71.9	65.1	50.7	60.0	48.0	73.2
				Woolong tea	B	78.2	53.9	81.5	67.2	70.2	74.9	68.4	60.8	60.0	61.7	63.5	60.5	57.7	62.6	54.4	47.2	42.5	54.6	44.0	61.2
			GC-MS/MS	Green tea	A*	104.0	91.7	59.2	61.1	97.9	67.2	89.8	84.7	76.0	90.9	88.9	101.8	98.1	85.0	64.6	81.1	73.5	76.5	81.6	82.8
				Green tea	B*	102.6	69.0	59.6	54.3	102.8	89.3	80.4	80.8	77.3	91.2	68.9	96.9	81.5	69.7	51.6	69.6	69.2	76.2	82.0	77.3
				Woolong tea	A*	114.9	63.8	103.7	63.7	82.0	78.9	66.4	67.3	60.0	75.4	66.4	62.7	67.0	68.7	74.9	70.4	56.6	58.5	55.5	72.3
				Woolong tea	B*	79.6	54.5	70.1	59.7	63.1	67.8	61.3	51.8	103.0	55.6	62.9	54.9	57.3	57.6	53.5	59.3	49.8	47.7	51.7	58.8
96	ethiofumesate	Cleanert TPT	GC-MS	Green tea	A	99.5	82.3	86.2	102.6	63.1	86.2	79.4	94.6	97.0	65.7	72.5	68.6	68.3	66.6	68.9	61.2	59.5	56.9	76.6	77.2
				Green tea	B	96.0	78.0	77.8	93.1	59.3	86.6	71.4	101.0	85.0	64.4	66.7	65.1	68.3	69.6	66.1	60.1	58.8	55.8	68.2	74.1
				Woolong tea	A	93.2	86.6	86.9	86.2	60.6	77.5	113.0	102.8	78.5	93.4	96.0	78.4	78.1	74.8	67.3	73.3	103.7	69.7	61.2	83.5
				Woolong tea	B	88.8	76.9	75.6	80.1	71.8	70.3	113.4	97.6	71.3	81.9	74.7	71.5	68.5	69.5	63.2	68.6	110.7	69.1	60.2	77.9
			GC-MS/MS	Green tea	A*	94.6	80.9	81.4	100.4	67.8	67.2	74.6	70.8	72.0	63.2	74.3	71.7	69.5	63.5	71.2	57.7	56.1	60.0	75.0	72.4
				Green tea	B*	91.8	78.8	73.5	92.8	71.4	62.1	73.3	65.1	74.7	62.7	71.1	68.2	62.5	61.1	65.9	56.6	56.6	60.8	67.7	69.0
				Woolong tea	A*	90.3	85.4	81.0	91.3	69.0	74.8	65.3	76.5	68.3	68.2	79.7	76.8	71.1	71.0	65.9	69.1	62.7	65.9	57.3	73.6
				Woolong tea	B*	85.6	78.4	72.0	82.4	82.4	64.5	62.7	67.3	116.7	64.0	64.9	66.9	61.6	65.7	57.8	62.8	61.5	63.8	56.6	67.1
		Envi-Carb +PSA	GC-MS	Green tea	A	102.1	93.6	83.9	84.6	84.6	88.4	61.3	100.5	93.4	70.7	70.9	75.5	68.8	69.5	53.9	74.2	73.1	68.4	67.5	80.5
				Green tea	B	97.9	82.7	82.1	77.5	86.8	86.4	60.8	106.7	72.0	71.9	64.6	69.3	65.1	72.5	59.3	62.5	67.0	64.2	67.7	75.7
				Woolong tea	A	97.7	76.2	77.4	69.9	65.9	89.2	92.6	87.5	59.2	75.5	75.6	65.6	70.8	69.9	76.6	67.0	59.2	67.9	57.8	73.0
				Woolong tea	B	85.4	69.5	75.6	63.3	61.5	77.8	74.9	34.0	73.2	59.1	69.8	60.3	65.3	66.9	64.3	64.6	56.4	59.5	57.4	65.0
			GC-MS/MS	Green tea	A*	103.9	85.7	80.8	77.4	81.0	63.8	71.4	70.3	70.1	68.7	77.9	66.5	79.3	76.8	65.2	76.0	69.8	70.7	75.0	75.6
				Green tea	B*	99.4	75.6	79.1	72.8	77.8	76.5	74.2	66.6	72.0	69.8	65.3	64.3	72.8	65.1	58.9	65.4	63.4	67.0	78.9	71.6
				Woolong tea	A*	93.5	73.6	82.6	67.2	81.5	68.2	70.5	61.4	59.2	66.5	62.8	65.3	68.9	58.9	68.3	69.1	52.8	60.3	55.9	68.6
				Woolong tea	B*	85.0	67.4	70.1	63.9	71.0	64.4		56.8	56.5	56.5	66.1	60.5	60.7	55.8	55.8	62.8	53.2	53.1	55.4	62.5

(Continued)

Annex 1: Content change for 201 pesticides in incurred tea Youden pair samples under 8 conditions with two SPE cartridge cleanup in three months (Nov.9, 2009-Feb.7, 2010) (cont.)

No	Pesticides	SPE	Method	Sample	Youden pair	Nov.9, 2009 (n=5)	Nov.14, 2009 (n=3)	Nov.19, 2009 (n=3)	Nov.24, 2009 (n=3)	Nov.29, 2009 (n=3)	Dec.4, 2009 (n=3)	Dec.9, 2009 (n=3)	Dec.14, 2009 (n=3)	Dec.19, 2009 (n=3)	Dec.24, 2009 (n=3)	Dec.14, 2009 (n=3)	Jan.3, 2010 (n=3)	Jan.8, 2010 (n=3)	Jan.13, 2010 (n=3)	Jan.18, 2010 (n=3)	Jan.23, 2010 (n=3)	Jan.28, 2010 (n=3)	Feb.2, 2010 (n=3)	Feb.7, 2010 (n=3)	AVE, µg/kg.
97	ethoprophos	Cleanert TPT	GC-MS	Green tea	A	120.5	106.6	108.1	137.0	97.8	90.0	104.9	96.4	97.0	84.4	106.5	87.9	86.4	84.6	90.3	78.0	73.0	63.4	97.4	95.3
					B	125.4	104.7	99.3	127.5	90.8	85.2	99.3	89.1	102.8	79.9	94.9	83.0	82.2	87.5	85.7	76.6	71.7	65.3	88.2	91.5
				Woolong tea	A	121.8	112.4	106.3	125.2	104.3	98.9	93.5	115.6	99.6	96.9	99.7	98.0	94.0	85.9	88.2	87.5	83.6	88.9	80.9	99.0
					B	117.4	100.9	95.1	112.6	96.2	87.5	73.6	103.6	84.9	83.9	80.1	80.5	79.3	81.8	77.9	79.0	71.2	86.0	74.1	87.8
			GC-MS/MS	Green tea	A	123.4	107.4	105.3	130.8	95.1	88.1	102.6	97.3	95.6	83.2	105.4	74.2	91.8	79.2	89.8	75.2	73.1	80.3	96.9	94.5
					B	121.2	105.8	100.4	123.1	91.0	83.3	101.0	90.0	96.7	84.2	93.7	68.5	84.1	81.2	84.8	75.7	72.6	82.0	87.9	90.9
				Woolong tea	A*	128.3	122.9	108.1	127.1	95.9	98.7	92.8	148.6	103.6	92.0	101.8	95.9	99.6	98.1	87.9	89.5	85.8	90.4	75.0	102.1
					B*	118.9	108.3	95.0	114.1	89.8	84.0	75.9	129.9	86.4	84.5	82.6	81.0	85.6	85.5	76.2	82.7	76.5	83.3	70.0	90.0
		Envi-Carb +PSA	GC-MS	Green tea	A	143.7	126.2	106.7	105.9	111.9	115.2	126.0	96.1	97.5	100.2	100.1	106.3	90.0	92.7	73.3	90.2	87.4	83.8	91.9	102.4
					B	136.1	113.4	107.8	94.9	112.7	105.9	111.5	86.0	94.7	95.0	89.0	97.8	83.1	88.1	70.5	78.0	81.5	82.2	91.5	95.8
				Woolong tea	A	119.0	94.5	102.9	97.4	110.5	94.7	85.9	79.8	90.7	89.7	89.0	79.9	80.4	85.9	87.7	79.5	73.3	81.6	73.7	89.1
					B	102.2	83.8	91.8	88.1	87.4	85.2	83.8	67.9	76.0	75.0	73.6	76.2	73.9	81.0	75.7	69.6	64.4	71.9	67.6	78.7
			GC-MS/MS	Green tea	A	143.3	122.0	107.5	105.4	114.5	118.3	108.5	91.8	94.7	97.8	100.7	111.4	111.5	100.0	84.2	96.0	90.6	88.5	94.9	104.3
					B	155.0	109.7	106.3	99.8	110.8	107.7	97.7	87.4	89.4	98.1	88.9	105.6	98.9	79.9	76.2	84.9	84.7	88.5	96.9	98.2
				Woolong tea	A*	148.7	97.4	135.7	95.2	112.4	92.6	95.9	79.6	90.7	86.0	86.4	76.7	87.6	81.1	90.7	89.5	71.3	77.6	74.1	93.1
					B*	109.7	86.4	90.9	87.2	91.3	85.1	84.9	68.1	76.0	73.5	86.0	72.4	78.1	73.4	72.6	82.7	69.9	67.2	69.9	80.3
98	etrimfos	Cleanert TPT	GC-MS	Green tea	A*	45.6	36.3	38.1	43.7	33.5	30.1	37.4	36.0	30.8	30.1	40.3	31.1	29.8	31.5	32.7	32.1	31.8	8.4	32.1	33.0
					B*	46.5	35.3	34.4	40.4	31.1	28.8	35.1	33.0	36.1	28.0	33.6	29.5	28.8	32.9	32.7	26.6	27.8	8.4	28.9	31.5
				Woolong tea	A	43.0	36.3	36.4	42.6	35.6	33.8	32.7	42.1	33.5	34.6	34.5	31.1	34.9	31.7	30.6	30.4	30.6	30.2	27.3	34.3
					B	41.3	33.9	33.0	39.5	34.8	30.9	26.4	38.9	28.0	29.6	27.9	28.5	29.1	29.3	27.4	28.9	26.5	29.5	25.5	31.0
			GC-MS/MS	Green tea	A	42.1	36.4	35.9	46.0	33.1	31.5	35.4	33.4	33.8	29.5	35.2	25.2	30.3	28.8	31.7	26.3	25.6	26.1	33.9	32.6
					B	41.3	35.4	33.8	43.2	31.0	30.1	34.7	31.3	34.1	28.6	31.0	25.1	28.8	29.3	30.0	26.3	24.8	26.7	29.0	31.3
				Woolong tea	A*	42.7	40.6	38.2	44.6	32.2	33.8	31.4	52.4	36.0	30.2	34.6	33.7	32.6	32.5	28.7	30.6	30.0	29.7	26.8	34.8
					B*	39.9	36.1	32.9	40.7	31.6	29.3	25.4	45.0	31.2	29.9	28.6	30.1	27.8	29.6	24.8	28.0	27.4	28.7	24.0	31.1
		Envi-Carb +PSA	GC-MS	Green tea	A	44.9	43.4	36.9	36.9	38.9	39.9	46.2	33.0	34.6	40.5	34.8	41.9	32.1	36.1	28.0	32.6	31.6	29.2	31.9	36.5
					B	42.9	37.3	36.3	33.9	38.4	35.9	48.9	28.6	32.3	37.4	31.3	36.5	29.9	34.0	27.2	28.3	28.9	27.6	31.3	34.1
				Woolong tea	A	43.5	33.8	36.3	34.4	36.5	31.8	28.8	27.5	31.0	31.2	31.0	27.9	28.0	31.1	27.2	27.2	26.3	27.1	25.1	31.0
					B	36.9	30.1	32.5	31.3	29.7	29.2	29.6	24.3	26.6	26.0	26.3	27.5	26.3	29.3	26.8	25.2	23.9	24.4	23.6	27.9
			GC-MS/MS	Green tea	A	48.7	40.3	36.6	34.9	39.1	58.8	37.5	31.0	32.5	32.4	36.4	36.0	41.1	36.0	28.9	33.9	28.8	31.0	31.6	36.6
					B	53.0	35.2	35.6	34.3	37.8	37.0	34.3	29.7	30.3	33.2	32.2	34.3	36.1	27.8	27.3	30.2	27.2	30.2	32.8	33.6
				Woolong tea	A*	50.6	33.6	44.7	34.1	37.9	30.9	34.1	26.8	31.0	28.9	28.4	24.5	30.4	27.5	29.7	30.6	23.6	26.4	25.1	31.5
					B*	37.3	30.2	30.1	31.2	31.8	28.7	30.6	24.1	26.6	25.4	29.6	23.7	27.4	25.2	24.5	28.0	23.6	23.1	24.4	27.6
99	fenamidone	Cleanert TPT	GC-MS	Green tea	A	47.4	42.9	46.5	54.1	40.7	34.5	42.0	38.7	39.5	31.3	43.0	34.3	34.8	33.1	36.0	30.2	29.4	31.7	38.2	38.3
					B	46.4	42.7	41.3	50.7	37.1	34.1	40.0	35.2	39.3	30.5	36.9	32.5	33.3	33.6	32.6	30.3	28.8	27.9	33.9	36.2
				Woolong tea	A	46.1	39.7	40.2	47.4	41.2	40.2	32.7	43.0	38.6	37.3	36.1	36.7	38.4	34.6	34.8	35.1	33.1	34.6	30.4	37.9
					B	42.8	40.2	40.2	45.5	42.9	38.0	33.6	44.4	33.5	33.2	29.9	33.5	36.5	33.8	31.4	30.7	29.7	33.1	29.3	35.9
			GC-MS/MS	Green tea	A	49.0	44.0	46.8	50.5	39.5	34.9	39.6	38.6	40.3	34.2	41.0	34.2	40.2	36.0	37.4	28.8	29.7	29.1	38.6	38.5
					B	47.2	41.8	41.8	47.7	35.7	32.3	40.5	35.6	39.4	34.4	36.6	33.2	34.1	35.4	33.7	28.9	28.4	28.0	32.9	36.2
				Woolong tea	A*	47.4	41.9	40.6	49.7	37.8	39.7	34.1	51.1	40.5	34.1	38.6	38.8	38.2	36.9	33.9	33.5	32.1	33.6	28.1	38.4
					B*	44.3	40.7	39.0	46.6	40.7	36.4	34.7	51.9	34.9	34.2	33.0	36.8	33.6	34.8	30.9	29.5	29.5	32.9	26.5	33.6
		Envi-Carb +PSA	GC-MS	Green tea	A	54.6	49.9	43.8	44.3	41.6	46.5	39.2	37.2	34.9	38.2	37.0	39.2	36.8	39.0	32.9	39.8	34.8	34.4	34.3	40.2
					B	55.4	45.4	42.7	37.9	42.1	42.7	40.9	34.6	34.5	37.3	34.8	36.0	34.2	35.8	30.8	33.6	30.8	30.2	34.3	37.6
				Woolong tea	A	49.9	36.4	37.5	36.3	43.7	36.2	37.1	30.4	35.0	35.4	31.9	31.1	32.6	34.6	37.9	31.0	29.5	33.1	29.9	35.2
					B	47.1	34.9	39.4	37.4	36.5	32.7	31.6	28.3	32.1	28.7	28.0	30.8	30.1	33.9	30.8	27.1	27.3	29.3	28.6	32.3
			GC-MS/MS	Green tea	A	50.4	48.4	43.3	41.7	41.3	40.6	36.0	37.7	37.4	38.3	38.7	38.7	45.5	35.9	32.5	39.3	33.8	35.7	35.5	39.5
					B	51.9	43.0	42.4	38.3	41.5	41.9	39.4	34.8	34.1	37.8	35.7	35.6	40.2	34.3	30.1	34.5	30.3	31.7	35.7	37.5
				Woolong tea	A	48.3	37.1	37.1	36.3	41.9	34.9	34.8	29.6	35.0	34.8	31.1	31.0	28.5	29.1	35.0	33.5	27.9	33.5	28.7	34.1
					B	47.5	36.6	39.6	36.3	37.0	32.5	33.3	27.7	32.1	28.8	32.6	28.6	27.4	28.2	27.1	29.5	27.4	30.2	28.4	32.1

No.	Compound	Cartridge	Method	Tea	Rep	1	2	3	4	5	6	7	8	9	10	11	12	13	14	15	16	17	18	19	20
100	fenaromol	Cleanert TPT	GC-MS	Green tea	A	101.3	89.3	83.9	112.8	79.0	69.9	76.2	77.1	74.6	62.6	70.7	71.7	66.4	67.1	71.6	58.0	56.4	54.2	70.2	74.4
					B	96.4	82.8	81.1	105.8	70.8	66.7	69.7	67.0	70.2	58.2	62.0	57.3	65.1	74.5	67.8	61.7	56.9	53.8	61.6	70.0
				Woolong tea	A	96.1	78.4	77.9	94.4	77.5	78.3	63.4	84.4	71.3	67.4	68.6	70.1	75.6	66.8	66.9	65.1	61.1	71.2	61.6	73.5
					B	91.9	74.6	69.6	84.6	78.3	72.7	59.5	85.1	58.6	63.5	59.5	67.2	72.5	64.2	59.6	57.5	54.0	70.3	58.2	68.4
			GC-MS/MS	Green tea	A*	97.0	93.0	87.9	94.9	77.4	62.0	59.5	70.8	69.3	63.5	79.4	70.9	65.8	61.4	66.2	53.5	52.1	49.8	70.5	71.5
					B*	95.0	79.2	85.6	89.6	69.4	59.6	77.2	66.7	67.3	63.6	69.4	66.9	61.0	64.8	58.0	54.4	50.0	50.2	61.6	67.9
				Woolong tea	A	93.7	79.2	83.5	94.9	70.4	73.4	63.5	94.9	75.3	68.7	74.8	68.3	71.1	64.6	59.5	61.3	58.6	56.6	50.3	71.7
					B	88.0	73.4	82.0	87.9	73.3	67.3	63.7	101.8	63.7	65.5	61.9	60.6	64.0	61.0	51.6	54.5	51.7	50.0	46.8	66.8
		Envi-Carb +PSA	GC-MS	Green tea	A	97.8	86.0	93.1	93.1	87.4	80.6	77.6	72.0	62.5	69.2	64.8	79.8	65.9	81.6	64.0	80.5	64.8	59.7	65.3	76.1
					B	101.0	86.5	87.2	76.8	92.8	71.6	71.9	65.3	58.2	65.8	77.1	65.6	68.9	65.4	60.5	64.2	59.2	55.0	70.3	71.8
				Woolong tea	A	105.8	67.2	72.2	71.4	75.7	70.5	68.2	62.1	67.4	63.9	59.2	51.2	64.5	67.4	67.2	56.0	59.3	69.3	56.6	67.1
					B	95.1	66.7	66.7	67.0	67.0	62.7	61.3	61.1	59.2	48.8	50.7	51.8	62.5	63.5	57.1	49.0	53.2	63.0	54.1	61.2
			GC-MS/MS	Green tea	A	97.8	69.8	69.8	83.2	78.0	75.7	65.6	70.1	73.4	73.9	70.7	69.7	71.2	68.6	58.6	69.0	59.6	61.9	61.1	73.1
					B	102.4	84.5	84.3	74.4	78.5	76.0	66.2	64.0	64.4	69.6	62.3	67.3	66.6	58.6	51.2	58.0	55.2	57.1	60.8	68.3
				Woolong tea	A	95.5	81.5	73.6	69.5	80.7	64.8	63.2	56.7	67.4	64.7	58.9	57.0	55.0	54.1	63.5	61.3	48.8	57.8	52.1	64.1
					B	92.1	73.5	70.9	69.8	69.3	61.8	60.8	52.2	59.2	52.6	58.9	54.6	47.2	50.1	48.5	54.5	48.5	52.0	49.2	59.3
101	flamprop-isopropyl	Cleanert TPT	GC-MS	Green tea	A	45.8	40.7	38.7	45.6	35.3	32.4	38.2	35.4	37.4	30.5	36.1	32.4	32.5	32.0	34.6	28.9	28.2	29.2	36.3	35.3
					B	43.8	36.8	38.0	42.6	32.8	30.2	35.4	32.0	35.4	29.6	31.6	30.0	32.2	31.5	32.2	28.7	28.2	27.8	31.9	33.1
				Woolong tea	A	44.8	38.0	35.2	43.7	36.8	37.9	34.0	41.9	40.1	36.0	40.0	38.8	32.9	34.1	33.0	34.7	33.8	33.2	30.3	37.3
					B	42.9	35.2	41.9	40.4	36.8	35.8	31.1	39.9	34.1	33.3	30.9	32.1	35.9	31.9	30.1	31.4	28.4	32.7	28.5	33.9
			GC-MS/MS	Green tea	A	49.4	41.9	40.9	47.3	34.6	33.0	36.3	34.2	36.0	31.0	37.7	32.4	33.7	36.1	32.7	27.2	27.1	27.0	35.8	35.5
					B	46.8	36.8	39.3	43.5	32.2	30.8	36.3	31.3	34.4	30.8	36.6	31.4	29.6	33.4	30.1	27.4	27.2	27.9	32.2	33.6
				Woolong tea	A	39.4	39.6	40.8	45.7	34.9	37.0	33.8	42.3	39.2	34.8	37.5	36.7	35.3	34.1	31.6	34.4	32.7	33.6	29.4	36.5
					B	39.5	36.0	38.5	41.7	35.2	33.2	30.4	40.0	33.9	32.4	31.1	32.3	30.7	32.3	28.1	31.6	29.6	32.5	28.1	33.5
		Envi-Carb +PSA	GC-MS	Green tea	A	52.3	40.0	45.6	39.6	39.1	40.8	47.5	34.1	32.9	32.6	32.2	36.2	33.4	32.1	32.1	29.5	31.0	32.7	32.5	37.5
					B	49.9	38.7	39.7	36.2	38.9	37.0	35.6	31.9	31.7	32.9	31.0	35.9	30.8	37.3	29.2	31.5	27.7	28.4	32.7	34.2
				Woolong tea	A	46.3	38.4	35.9	35.9	40.2	35.2	35.7	29.6	34.6	35.9	33.5	34.0	31.7	32.7	36.7	31.7	28.9	32.4	28.3	34.7
					B	41.7	38.1	33.6	34.1	34.0	31.7	32.9	28.1	30.3	28.8	30.2	34.3	29.3	33.8	30.2	27.9	26.4	27.3	26.4	31.2
			GC-MS/MS	Green tea	A	50.4	39.5	45.1	38.7	40.5	38.7	36.4	32.9	33.8	36.8	33.4	35.5	37.1	36.3	30.5	37.4	30.2	34.2	32.9	36.9
					B	47.4	38.4	37.3	36.3	39.1	37.3	34.9	30.9	32.4	36.1	31.2	34.2	33.0	36.3	27.7	30.3	27.4	34.5	33.6	34.2
				Woolong tea	A	45.9	38.5	36.3	34.7	39.8	33.9	34.2	29.5	34.6	33.3	31.6	31.5	30.4	28.2	34.7	34.7	26.4	30.2	29.4	33.7
					B	43.0	34.3	34.3	33.2	35.0	31.3	33.1	27.2	30.3	28.6	32.0	29.1	27.6	27.3	28.3	31.5	26.1	26.4	28.6	31.0
102	flamprop-methyl	Cleanert TPT	GC-MS	Green tea	A	51.2	47.4	42.6	77.6	40.7	36.1	41.6	34.1	39.5	31.7	40.9	36.2	33.1	30.7	33.1	29.5	28.0	28.9	38.1	39.3
					B	49.8	42.4	37.9	71.5	37.5	35.6	42.7	35.3	39.5	35.8	39.7	35.9	34.0	31.6	35.9	29.4	27.4	28.5	33.8	38.2
				Woolong tea	A	44.2	43.5	42.4	45.8	44.7	42.5	35.7	42.1	37.5	33.8	35.7	34.0	34.9	36.7	35.0	35.9	33.9	32.4	30.8	38.1
					B	41.2	43.1	37.8	46.7	44.4	40.6	31.2	34.3	32.1	32.3	30.4	28.4	34.5	34.6	32.0	31.8	29.3	27.3	30.8	35.3
			GC-MS/MS	Green tea	A	49.7	41.7	42.6	49.3	35.0	34.5	36.8	34.4	34.3	31.9	36.2	37.8	34.0	33.9	32.6	27.9	25.9	28.4	37.3	36.0
					B	47.6	37.1	40.8	45.8	33.6	31.6	36.0	31.5	34.1	31.9	32.8	36.0	30.3	34.2	30.8	27.9	25.5	30.1	32.2	34.2
				Woolong tea	A	39.4	39.6	38.5	45.7	34.9	37.0	33.1	44.4	39.7	34.9	38.6	36.2	37.9	33.2	34.7	34.7	31.3	32.9	28.4	36.6
					B	39.5	36.0	34.3	41.7	35.2	33.2	30.0	40.7	34.7	33.3	32.0	31.4	32.2	31.8	28.0	31.5	28.6	32.9	27.0	33.6
		Envi-Carb +PSA	GC-MS	Green tea	A	52.0	45.2	36.3	46.3	42.8	43.0	34.2	36.1	45.4	36.1	40.3	41.8	33.9	39.1	28.5	38.4	37.9	32.8	34.5	40.4
					B	49.2	49.3	57.1	40.9	45.7	42.0	34.2	36.1	40.1	34.8	33.1	38.1	31.4	35.0	27.3	32.7	32.7	31.7	36.0	37.7
				Woolong tea	A	47.2	42.2	37.9	38.8	44.1	41.6	26.8	29.3	34.6	31.8	35.3	34.9	33.3	35.0	37.6	32.9	29.0	32.5	29.7	35.6
					B	42.6	44.2	29.2	37.4	36.8	38.8	35.8	28.5	30.3	29.1	32.0	39.6	30.6	36.7	32.6	28.0	27.7	28.7	27.9	35.4
			GC-MS/MS	Green tea	A	52.0	41.1	44.7	40.6	41.5	36.7	40.9	33.5	36.9	34.9	36.6	32.1	37.2	34.6	29.2	36.1	34.2	33.0	27.9	37.5
					B	48.0	38.6	40.9	38.5	39.3	38.5	30.0	31.3	34.0	34.5	32.2	30.7	34.2	30.3	27.6	28.7	30.9	30.7	37.4	34.5
				Woolong tea	A	45.9	38.9	36.3	34.7	39.8	33.9	34.2	29.5	34.6	33.0	31.6	31.5	30.4	30.4	33.8	34.4	26.4	30.2	27.6	33.5
					B	43.0	36.6	34.3	33.2	35.0	31.3	33.1	27.2	30.3	28.6	32.0	29.1	29.1	27.3	26.9	31.7	26.1	26.4	28.2	30.9

(Continued)

Annex 1: Content change for 201 pesticides in incurred tea Youden pair samples under 8 conditions with two SPE cartridge cleanup in three months (Nov.9, 2009-Feb.7, 2010) (cont.)

No	Pesticides	SPE	Method	Sample	Youden pair	Nov.9 2009 (n=5)	Nov.14 2009 (n=3)	Nov.19 2009 (n=3)	Nov.24 2009 (n=3)	Nov.29 2009 (n=3)	Dec.4 2009 (n=3)	Dec.9 2009 (n=3)	Dec.14 2009 (n=3)	Dec.19 2009 (n=3)	Dec.24 2009 (n=3)	Dec.14 2009 (n=3)	Jan.3 2010 (n=3)	Jan.8 2010 (n=3)	Jan.13 2010 (n=3)	Jan.18 2010 (n=3)	Jan.23 2010 (n=3)	Jan.28 2010 (n=3)	Feb.2 2010 (n=3)	Feb.7 2010 (n=3)	AVE. µg/kg.
103	fonofos	Cleanert TPT	GC-MS	Green tea	A	49.1	36.4	39.0	41.9	34.3	30.7	37.6	36.2	33.8	30.1	38.5	31.4	32.8	41.7	36.1	27.8	28.1	40.3	34.7	35.8
				Green tea	B*	48.7	36.0	35.4	38.3	31.6	29.1	35.3	33.3	36.1	27.5	34.1	29.3	30.9	47.1	36.5	27.5	27.9	39.1	30.7	34.5
				Woolong tea	A	42.5	35.6	36.3	42.1	36.2	36.0	31.7	40.2	35.8	34.3	34.0	33.0	35.2	31.8	31.7	31.4	29.6	31.5	27.4	34.5
				Woolong tea	B	40.9	34.0	33.2	40.2	34.4	32.3	26.9	36.6	29.7	29.8	28.0	29.2	30.7	30.0	28.1	29.0	26.3	30.3	25.9	31.3
			GC-MS/MS	Green tea	A	44.8	37.7	39.5	46.2	34.4	31.5	36.0	34.2	35.6	31.6	36.9	36.8	31.1	29.9	32.1	26.4	26.0	28.4	33.9	34.4
				Green tea	B	43.4	36.3	35.4	42.9	31.6	29.4	35.9	31.8	34.6	29.1	32.8	35.1	30.0	29.6	29.7	26.3	25.4	28.2	30.3	32.5
				Woolong tea	A	39.3	38.2	38.3	45.5	33.0	35.1	31.4	41.8	37.2	33.0	35.2	34.1	33.1	33.1	31.1	30.4	29.4	30.5	26.5	34.5
				Woolong tea	B	38.2	33.9	33.5	41.2	32.8	30.9	26.2	37.9	32.8	29.9	29.2	29.5	28.7	30.7	27.0	28.5	27.3	29.2	24.6	31.2
		Envi-Carb +PSA	GC-MS	Green tea	A	48.5	41.1	39.6	36.4	42.2	41.3	35.4	34.2	34.8	40.7	35.3	39.9	34.3	35.4	31.1	36.5	31.3	31.8	31.5	37.2
				Green tea	B	46.3	33.5	38.1	34.7	39.6	36.5	33.2	32.6	32.3	37.4	33.1	35.2	31.9	32.7	29.3	31.0	27.9	28.0	31.3	34.3
				Woolong tea	A	43.6	29.9	36.3	33.8	40.8	33.1	33.2	28.4	32.6	31.8	30.1	28.8	30.3	31.8	33.0	28.5	26.8	29.0	25.8	32.2
				Woolong tea	B	36.9	42.4	32.6	31.7	32.0	29.9	30.0	25.1	28.1	26.6	25.8	27.4	27.7	30.0	27.9	25.2	24.3	25.3	24.3	28.5
			GC-MS/MS	Green tea	A	48.6	35.4	37.6	36.3	40.3	47.0	36.9	33.5	33.9	36.4	34.7	35.2	38.4	36.2	31.7	35.1	30.2	32.8	33.1	26.8
				Green tea	B	47.1	34.1	37.5	34.7	38.9	37.7	35.3	31.6	31.2	35.6	31.3	34.8	34.4	29.4	28.6	30.9	28.3	30.0	32.8	34.0
				Woolong tea	A	42.3	31.7	42.3	33.4	31.5	32.2	32.3	27.4	32.6	30.5	31.2	29.2	31.0	27.9	32.0	30.4	25.0	26.9	25.5	31.8
				Woolong tea	B	35.4	30.5	32.1	31.4	31.4	31.4	30.8	24.9	28.1	26.5	32.6	27.5	28.2	25.0	25.7	28.5	24.8	23.5	25.2	28.5
104	hexachlorobenzene	Cleanert TPT	GC-MS	Green tea	A	36.8	31.1	32.9	40.8	28.9	27.6	33.6	28.9	31.6	25.8	29.7	26.9	28.6	26.1	27.9	24.5	22.0	18.5	31.8	29.2
				Green tea	B	36.0	30.6	30.5	38.7	27.2	25.8	31.6	28.3	32.9	25.1	27.1	25.2	27.0	24.7	26.6	23.7	23.0	20.8	28.4	28.1
				Woolong tea	A	30.2	27.3	29.6	37.5	26.2	28.4	24.2	34.7	24.2	28.8	27.1	30.0	26.1	26.1	28.0	26.2	22.8	27.7	22.4	28.3
				Woolong tea	B	32.6	31.6	27.2	32.7	23.8	25.5	20.2	32.0	25.9	24.7	25.5	26.3	23.5	26.2	24.2	22.3	22.0	26.5	20.2	25.7
			GC-MS/MS	Green tea	A	35.6	31.3	32.6	39.3	28.3	28.3	31.9	28.1	29.9	27.0	29.5	32.1	28.1	26.2	29.0	23.5	21.5	18.7	32.6	29.1
				Green tea	B	35.0	31.5	31.0	36.5	27.5	26.4	31.5	28.0	30.8	25.4	28.4	28.7	26.5	25.9	27.5	22.1	22.4	21.6	22.1	28.1
				Woolong tea	A	31.7	27.9	31.9	38.1	24.8	28.0	24.2	33.9	27.9	26.8	31.8	26.6	25.8	26.4	26.9	24.5	21.9	27.0	22.1	28.4
				Woolong tea	B	32.7	38.3	28.0	32.2	22.6	24.6	19.8	31.4	27.0	24.9	25.2	25.2	23.1	20.8	23.6	20.8	22.0	26.3	19.5	25.6
		Envi-Carb +PSA	GC-MS	Green tea	A	43.1	33.2	33.5	32.9	35.4	36.2	40.7	30.3	30.2	29.7	29.3	31.7	31.8	31.7	26.6	32.1	29.3	30.0	29.5	32.8
				Green tea	B	41.6	27.8	33.6	30.2	35.6	33.7	36.8	25.9	28.3	29.3	27.1	29.9	29.3	30.5	24.5	27.8	27.6	27.6	29.1	30.8
				Woolong tea	A	37.0	25.6	29.7	29.3	32.8	29.0	30.6	23.1	29.7	28.4	28.6	26.7	27.2	26.1	29.2	26.9	24.4	27.4	23.7	28.4
				Woolong tea	B	29.7	38.0	28.0	27.8	26.9	26.6	27.6	30.8	25.0	24.7	26.1	25.1	24.9	26.2	26.4	22.4	22.0	24.1	22.3	25.5
			GC-MS/MS	Green tea	A	41.8	32.2	33.8	33.3	36.8	34.4	34.1	28.7	29.8	30.9	30.3	32.1	34.3	33.1	27.7	32.2	29.1	32.0	29.8	32.9
				Green tea	B	40.0	28.1	33.8	30.2	35.5	33.1	34.7	25.2	28.0	30.5	27.9	31.1	30.8	29.6	24.5	28.4	25.7	30.1	29.6	30.8
				Woolong tea	A	36.2	25.4	36.5	29.1	31.9	29.5	28.9	22.3	29.7	28.5	28.2	26.4	28.4	25.2	29.0	24.8	22.9	26.1	23.2	28.3
				Woolong tea	B	29.9	27.5	28.2	27.2	26.5	27.4	26.6	23.5	25.0	24.5	27.8	25.3	25.8	24.2	24.6	21.9	22.4	22.4	22.3	25.2
105	hexazinone	Cleanert TPT	GC-MS	Green tea	A*	116.7	112.0	134.7	188.2	121.3	99.6	126.0	108.9	118.0	108.7	117.2	98.1	101.0	97.5	108.6	95.3	115.9	104.6	120.8	115.4
				Green tea	B*	118.0	114.8	127.4	172.8	106.0	91.6	123.8	94.7	105.8	87.6	93.5	93.7	104.5	104.2	89.9	94.2	100.7	95.7	100.6	106.3
				Woolong tea	A	136.9	115.3	109.3	133.5	133.0	123.7	91.2	128.5	110.9	107.6	108.6	112.4	109.1	107.0	91.7	99.9	93.4	91.6	75.4	109.4
				Woolong tea	B	129.9	116.5	103.8	133.6	130.3	113.7	90.4	126.4	94.9	92.4	90.2	105.3	99.0	96.4	87.0	84.7	82.9	91.6	74.6	101.8
			GC-MS/MS	Green tea	A	126.9	114.7	136.4	132.5	113.2	106.1	117.0	111.8	112.3	109.1	115.2	110.0	109.7	92.7	107.3	89.2	109.0	99.1	117.9	112.0
				Green tea	B	130.9	113.7	124.9	129.8	104.8	97.3	125.8	97.6	102.4	98.6	93.2	113.3	103.7	114.1	89.1	114.1	98.7	93.1	102.0	106.4
				Woolong tea	A*	137.6	120.7	121.0	136.2	116.0	118.5	92.2	193.1	111.8	102.0	111.2	112.8	108.7	102.2	88.5	92.8	92.7	89.1	87.2	112.3
				Woolong tea	B*	127.1	115.5	105.2	119.2	117.0	107.7	88.8	175.6	100.1	94.8	94.8	106.6	91.4	89.8	78.3	79.9	84.0	88.2	73.9	102.0
		Envi-Carb +PSA	GC-MS	Green tea	A	155.5	132.0	123.9	142.7	122.0	120.8	116.0	110.1	111.5	113.2	106.9	120.0	105.5	119.2	105.5	125.6	105.0	108.7	102.7	118.2
				Green tea	B	149.7	116.6	128.3	125.1	130.2	119.6	102.1	104.1	98.1	107.1	109.4	107.3	99.4	104.5	99.2	102.9	91.5	91.4	103.9	110.0
				Woolong tea	A	149.5	109.9	102.0	107.9	108.4	98.1	99.6	85.3	99.4	102.8	91.5	84.1	91.4	103.9	102.1	82.1	78.9	96.1	88.4	99.0
				Woolong tea	B	140.2	100.7	118.3	108.1	101.0	89.7	87.0	81.1	90.2	71.4	88.6	76.7	81.7	93.0	75.2	78.6	75.1	78.0	90.9	90.8
			GC-MS/MS	Green tea	A	150.5	123.1	121.0	134.1	122.4	118.4	97.8	107.0	125.2	101.1	107.1	104.7	120.4	126.9	104.3	128.9	100.2	108.0	111.6	114.2
				Green tea	B	151.5	104.6	121.3	127.0	127.7	95.0	88.3	100.9	104.6	98.3	109.3	99.4	117.4	88.4	100.8	107.8	87.3	95.2	110.3	108.3
				Woolong tea	A	134.2	114.9	96.2	101.7	107.3	95.5	84.5	87.4	99.4	98.7	90.4	82.8	81.2	84.6	95.5	93.0	73.8	94.2	82.8	94.6
				Woolong tea	B	133.8	105.9	115.3	102.6	105.0	90.1	93.2	85.7	90.2	69.9	98.6	73.0	73.5	84.9	68.5	80.0	75.6	72.8	86.6	89.7

No.	Compound	Cartridge	Method	Tea	Rep																				
106	iodofenphos	Cleanert TPT	GC-MS	Green tea	A*	75.7	73.8	72.9	126.2	65.5	67.7	78.8	73.1	61.9	64.4	71.4	63.0	60.6	65.9	66.9	55.1	50.9	49.2	62.7	68.7
				Woolong tea	B*	77.2	72.3	68.6	121.3	63.6	62.7	74.3	69.6	74.8	60.0	59.8	63.1	62.9	62.6	69.1	55.6	53.6	53.8	56.5	67.4
				tea	A*	79.7	89.5	82.3	82.8	79.1	64.1	65.6	93.3	67.9	68.5	75.9	69.2	80.5	68.0	64.0	63.3	70.8	60.5	60.3	72.9
					B	78.0	71.9	70.2	72.5	76.4	59.5	60.4	82.6	66.5	59.2	68.0	58.7	66.3	65.0	59.9	59.5	59.3	59.4	55.6	65.7
			GC-MS/MS	Green tea	A	81.9	69.9	63.3	92.2	59.9	68.1	81.0	66.4	73.5	57.6	81.6	50.1	59.1	62.4	63.3	53.8	50.8	56.3	64.3	66.1
				Woolong tea	B	80.5	70.9	69.8	91.8	66.3	66.5	73.4	60.2	71.7	59.8	71.0	51.4	59.8	64.3	59.9	53.2	53.6	58.0	57.6	65.2
				tea	A*	105.7	88.1	77.7	86.4	74.8	74.8	67.6	125.4	79.8	62.8	86.5	63.9	78.3	66.6	63.6	53.5	53.6	64.9	43.4	75.6
					B	91.4	84.0	65.9	78.8	68.9	61.3	60.6	96.7	67.3	64.2	70.1	60.7	67.8	63.0	58.0	56.1	71.3	59.5	53.9	67.6
		Envi-Carb +PSA	GC-MS	Green tea	A	84.9	78.2	66.9	76.9	76.2	75.0	71.0	62.1	73.9	68.1	69.8	78.4	56.7	68.7	49.2	66.8	56.3	53.7	75.8	69.6
				Woolong tea	B	86.6	71.5	65.2	73.6	76.1	71.7	72.8	58.9	71.5	66.5	60.8	78.0	53.5	64.0	53.3	56.8	63.8	55.0	75.6	67.1
				tea	A	72.8	65.7	74.8	70.9	74.4	67.2	53.4	59.7	67.0	62.8	64.6	61.1	51.6	68.0	62.8	58.0	53.3	52.2	51.9	62.7
					B	67.3	58.3	74.6	63.0	66.5	64.6	63.9	52.7	53.7	51.8	54.3	60.1	50.7	64.6	54.7	53.2	51.1	51.6	50.4	58.3
			GC-MS/MS	Green tea	A	106.4	79.7	81.7	82.2	86.4	95.9	90.4	56.9	75.9	56.7	69.2	76.3	63.4	82.7	51.8	69.5	58.1	62.0	71.7	74.6
				Woolong tea	B	121.2	82.9	78.7	85.2	79.8	75.4	74.5	56.9	67.0	60.8	67.9	70.3	60.5	60.7	45.7	66.3	63.6	66.7	74.1	71.5
				tea	A	109.5	62.4	78.2	61.7	76.6	62.6	92.6	61.5	67.0	58.8	62.6	40.2	59.5	59.5	62.9	63.5	48.0	51.0	52.8	64.8
					B	86.2	59.3	64.3	56.9	66.8	60.9	67.4	51.5	53.7	54.3	60.9	38.2	55.0	55.7	47.7	56.1	56.3	45.9	52.8	57.4
107	isofenphos	Cleanert TPT	GC-MS	Green tea	A*	98.4	77.3	88.9	100.7	76.0	62.1	77.3	72.8	70.0	63.4	108.5	78.2	58.6	58.5	61.7	49.5	46.0	45.8	63.2	72.5
				Woolong tea	B	98.3	75.8	77.9	90.4	66.0	60.5	71.0	68.5	78.3	55.7	90.1	71.8	65.9	58.9	60.6	51.0	46.7	45.9	56.2	67.9
				tea	A	89.7	76.0	73.4	101.9	91.9	80.9	74.8	99.3	84.1	85.6	86.9	77.8	75.4	58.8	60.0	58.5	56.9	56.9	51.6	75.8
					B	85.9	79.7	70.0	98.1	87.3	73.7	65.8	93.9	67.4	69.9	68.0	63.8	70.1	57.6	54.8	55.5	48.6	57.0	50.2	69.3
			GC-MS/MS	Green tea	A	96.3	77.0	77.1	93.3	69.2	61.3	68.9	64.7	66.9	60.1	64.6	54.1	58.3	61.6	61.4	51.3	49.6	50.8	66.9	66.0
				Woolong tea	B	92.3	75.5	68.9	86.1	62.4	56.5	68.7	59.9	66.7	57.8	63.6	52.3	58.0	55.5	57.7	51.2	48.6	50.7	59.2	62.7
				tea	A	87.2	89.9	77.4	91.7	68.2	70.6	65.3	96.6	72.5	65.4	71.5	68.7	70.8	61.7	58.1	62.1	58.1	59.2	49.1	70.7
					B	83.9	83.6	69.3	84.8	67.4	62.5	55.6	88.8	60.3	60.4	57.5	59.5	58.6	60.5	51.2	59.6	51.9	57.4	49.2	64.3
		Envi-Carb +PSA	GC-MS	Green tea	A	98.0	107.4	87.5	81.9	91.2	99.3	70.2	69.5	90.2	107.4	85.0	100.0	72.6	70.8	50.4	54.5	60.7	50.3	57.7	79.2
				Woolong tea	B	100.7	93.6	83.0	70.9	86.0	86.5	69.6	63.1	93.1	101.1	72.7	83.8	64.8	68.8	43.9	47.9	55.0	55.3	58.0	73.6
				tea	A	98.4	72.2	87.1	78.9	98.9	73.7	71.7	65.3	67.4	70.1	69.1	67.3	65.2	58.4	64.4	51.9	48.8	54.9	51.6	69.1
					B	89.3	68.0	78.1	74.1	75.4	67.6	68.1	51.3	56.7	59.0	54.4	69.8	60.1	55.3	56.1	45.0	45.5	48.0	46.6	61.5
			GC-MS/MS	Green tea	A	99.1	92.0	76.7	73.6	80.2	85.1	72.0	64.1	66.5	65.6	67.5	59.6	67.5	70.4	54.9	65.0	62.3	59.2	61.4	70.7
				Woolong tea	B	93.1	83.3	74.7	68.6	75.9	75.3	67.8	60.5	63.6	66.5	59.6	57.0	58.5	64.6	49.2	56.4	56.7	56.8	64.6	65.9
				tea	A	92.0	69.1	78.3	68.3	79.5	65.5	65.2	57.2	67.4	62.4	59.9	57.8	53.9	54.1	63.2	62.1	48.2	53.6	49.5	63.5
					B	85.8	63.5	67.8	64.6	66.7	60.9	62.7	49.4	56.7	54.2	60.9	55.3	50.6	49.7	51.2	59.6	47.8	46.0	47.6	57.9
108	isopeopalin	Cleanert TPT	GC-MS	Green tea	A	104.5	75.4	76.5	97.5	64.6	56.0	69.3	65.9	63.5	61.1	93.6	66.9	66.6	60.6	67.3	54.7	49.7	52.2	71.0	69.3
				Woolong tea	B	102.5	74.6	66.8	85.3	59.1	54.2	67.9	63.3	73.4	52.5	78.6	61.5	61.8	58.8	62.9	54.8	50.9	51.6	62.5	65.4
				tea	A*	90.2	73.3	70.5	88.5	83.9	74.7	69.5	87.2	76.5	77.0	70.6	58.1	71.1	64.1	66.6	68.1	68.2	67.1	58.4	73.8
					B	86.0	73.5	66.4	84.5	77.7	67.1	52.5	68.6	59.3	64.5	57.3	55.9	64.1	62.9	59.4	63.5	57.3	65.8	57.0	66.4
			GC-MS/MS	Green tea	A	92.8	78.5	76.5	99.9	68.5	60.1	68.9	62.0	69.7	58.8	76.1	78.7	63.7	65.4	59.4	57.0	53.9	52.5	72.7	70.2
				Woolong tea	B	92.3	77.1	68.2	90.0	61.8	56.3	70.2	107.1	69.8	56.0	75.6	77.3	59.4	60.8	63.6	56.1	55.6	52.9	65.0	66.8
				tea	A*	90.4	78.1	73.7	94.1	69.1	76.4	68.6	64.1	75.8	71.7	79.5	71.2	70.5	64.3	66.1	68.4	66.0	65.3	54.4	74.2
					B*	83.5	73.7	67.1	89.4	69.8	66.7	51.0	60.6	62.5	65.3	61.4	57.5	58.9	62.3	59.1	62.9	56.3	62.0	52.1	66.2
		Envi-Carb +PSA	GC-MS	Green tea	A	95.7	94.1	76.2	71.5	80.6	87.7	85.2	63.5	64.7	86.1	73.9	82.8	68.0	68.8	51.1	64.3	62.0	62.4	64.5	73.9
				Woolong tea	B	91.1	85.8	73.6	62.3	76.5	77.1	73.9	55.1	61.2	81.3	68.4	71.2	63.9	63.2	48.5	54.6	56.2	55.5	62.9	67.7
				tea	A	85.3	60.6	76.3	70.0	85.4	69.6	60.7	56.9	66.0	68.0	65.4	66.2	64.4	64.1	69.6	61.1	53.7	60.0	55.4	66.9
					B	98.5	87.1	67.3	66.5	63.0	63.2	60.4	47.6	56.5	57.4	50.6	66.2	56.7	62.9	64.2	50.8	62.4	54.3	50.9	59.6
			GC-MS/MS	Green tea	A	100.4	82.3	76.1	65.2	80.3	97.3	79.3	64.1	59.1	72.4	75.6	67.0	76.3	71.4	57.9	70.1	49.3	63.7	50.9	73.3
				Woolong tea	B	96.7	67.7	73.8	72.3	75.7	79.2	59.4	60.6	57.0	74.1	66.5	65.2	67.3	66.1	51.9	58.3	62.4	62.8	66.8	67.9
				tea	A	84.0	62.5	64.4	66.7	68.8	62.7	64.5	47.2	56.5	56.1	59.4	61.5	56.9	54.3	56.8	62.9	51.8	49.6	48.2	59.7

(Continued)

Annex 1: Content change for 201 pesticides in incurred tea Youden pair samples under 8 conditions with two SPE cartridge cleanup in three months (Nov.9, 2009-Feb.7, 2010) (cont.)

No	Pesticides	SPE	Method	Sample	Youden pair	Nov.9, 2009 (n=5)	Nov.14, 2009 (n=3)	Nov.19, 2009 (n=3)	Nov.24, 2009 (n=3)	Nov.29, 2009 (n=3)	Dec.4, 2009 (n=3)	Dec.9, 2009 (n=3)	Dec.14, 2009 (n=3)	Dec.19, 2009 (n=3)	Dec.24, 2009 (n=3)	Dec.14, 2009 (n=3)	Jan.3, 2010 (n=3)	Jan.8, 2010 (n=3)	Jan.13, 2010 (n=3)	Jan.18, 2010 (n=3)	Jan.23, 2010 (n=3)	Jan.28, 2010 (n=3)	Feb.2, 2010 (n=3)	Feb.7, 2010 (n=3)	AVE, μg/kg.
109	methoprene	Cleanert TPT	GC-MS	Green tea	A	177.9	124.8	136.9	166.5	120.7	101.1	127.0	113.4	123.1	99.8	137.0	109.2	107.5	108.0	111.8	91.9	90.7	96.1	116.6	119.0
				Green tea	B	172.4	126.2	121.9	150.7	110.0	96.6	117.1	106.1	121.8	89.8	117.0	102.1	100.9	104.4	102.8	90.5	88.7	93.5	103.2	111.4
				Woolong tea	A*	163.3	131.7	135.1	152.4	357.1	125.4	106.9	140.6	120.3	119.7	114.5	125.8	114.3	106.3	103.9	109.9	100.2	104.6	92.2	132.8
				Woolong tea	B*	158.6	128.0	126.4	145.1	235.5	113.1	92.8	128.8	100.3	102.3	88.6	102.2	99.9	99.7	94.3	97.9	87.6	102.9	87.9	115.4
			GC-MS/MS	Green tea	A	173.8	131.2	138.0	160.1	113.8	106.9	125.7	114.7	125.3	110.4	112.6	130.5	103.8	107.1	114.1	86.7	87.0	84.9	114.4	117.9
				Green tea	B	163.4	128.4	119.1	147.8	105.2	100.8	132.4	108.6	116.2	94.9	112.1	125.2	100.9	97.6	103.1	80.8	88.6	88.6	102.3	111.4
				Woolong tea	A*	140.3	137.3	113.6	153.8	113.5	119.6	114.0	158.3	140.0	130.8	125.6	117.4	110.6	111.0	97.4	105.2	93.3	105.5	89.3	120.9
				Woolong tea	B*	136.2	122.3	123.1	141.7	116.2	106.3	99.2	150.6	117.7	118.1	103.4	97.6	95.6	103.7	88.5	98.1	84.7	107.2	87.0	110.4
		Envi-Carb+PSA	GC-MS	Green tea	A	188.3	166.2	133.8	128.8	135.1	146.0	187.0	111.9	110.9	138.9	112.2	135.5	112.9	120.5	105.9	128.7	103.2	107.3	105.5	130.4
				Green tea	B	175.3	139.1	131.1	120.2	132.8	125.8	139.0	106.9	105.8	131.5	105.4	115.2	105.6	109.2	97.6	103.8	94.2	94.5	104.5	117.8
				Woolong tea	A*	183.7	127.1	138.0	127.2	394.1	121.4	441.9	487.4	122.1	116.9	102.1	94.2	106.4	104.7	115.5	96.0	91.0	96.9	89.3	166.1
				Woolong tea	B*	162.2	115.9	128.3	123.4	185.7	107.0	191.0	196.4	106.9	95.4	88.3	89.3	98.4	99.2	96.6	85.2	82.6	84.3	83.0	116.8
			GC-MS/MS	Green tea	A*	181.3	177.5	131.9	210.3	136.2	119.2	126.4	122.9	119.3	127.7	114.4	179.4	114.8	116.7	104.2	119.1	95.6	109.8	110.2	132.5
				Green tea	B*	163.9	164.0	127.3	205.1	136.8	126.6	125.9	115.7	116.6	125.8	102.8	162.0	101.0	102.0	93.3	102.0	92.0	99.5	111.8	125.0
				Woolong tea	A	161.1	118.9	140.5	120.5	131.9	114.1	108.3	101.8	122.1	122.9	101.2	98.0	104.1	97.6	109.1	105.2	83.5	92.0	88.2	111.6
				Woolong tea	B	144.0	116.3	116.6	116.6	121.9	106.0	105.0	93.5	106.9	101.9	104.9	92.8	95.0	89.3	88.5	98.1	82.0	79.6	84.8	102.3
110	methoprotryne	Cleanert TPT	GC-MS	Green tea	A*	148.6	113.0	137.4	139.3	116.5	91.7	113.1	105.2	99.3	76.2	112.3	83.7	92.7	89.9	95.8	78.6	77.2	73.8	96.3	102.1
				Green tea	B	144.0	110.9	122.4	128.4	100.2	88.7	104.8	97.5	99.0	67.8	89.5	80.5	87.8	86.7	87.2	78.5	75.6	71.2	85.5	95.1
				Woolong tea	A	135.8	114.3	112.1	138.9	117.9	111.7	93.5	121.1	82.9	107.7	75.8	98.3	103.8	91.6	89.3	90.6	89.1	87.4	77.3	102.1
				Woolong tea	B	127.4	119.0	107.7	131.2	121.0	105.1	89.2	116.8	61.7	89.1	67.8	88.4	91.6	83.0	79.2	78.3	74.6	83.2	71.4	94.0
			GC-MS/MS	Green tea	A*	148.5	132.0	133.8	148.5	107.8	92.2	108.2	106.2	107.5	89.9	108.5	91.8	96.1	98.8	96.6	78.9	73.1	73.9	100.5	104.9
				Green tea	B	141.3	123.1	116.3	138.8	95.4	87.1	110.9	98.4	100.2	91.8	98.3	89.2	87.8	95.6	87.8	77.3	72.9	74.4	87.9	98.8
				Woolong tea	A*	118.9	120.6	116.5	138.7	103.9	108.1	94.7	135.1	112.3	99.5	108.0	102.3	98.0	95.6	87.6	94.7	88.0	85.2	80.5	104.8
				Woolong tea	B	118.0	114.6	109.7	125.6	109.7	98.5	87.3	130.4	94.7	91.7	92.2	92.2	84.7	89.2	74.7	81.2	77.6	80.8	72.7	96.1
		Envi-Carb+PSA	GC-MS	Green tea	A*	167.7	155.6	132.3	128.0	127.2	142.1	104.5	101.2	95.3	103.4	85.1	93.4	95.3	104.5	90.4	108.4	87.7	88.7	88.1	110.4
				Green tea	B*	161.7	132.6	129.6	113.8	126.3	119.7	115.8	102.7	86.4	96.9	85.3	79.4	88.2	91.7	85.3	78.2	78.9	75.2	88.1	101.9
				Woolong tea	A	154.9	111.7	127.1	117.5	134.0	103.2	103.0	80.5	99.3	92.7	79.1	75.2	88.9	89.9	95.6	81.9	78.9	84.6	74.3	98.5
				Woolong tea	B	138.1	104.7	127.4	111.8	109.8	93.3	88.9	73.9	86.0	76.1	63.5	76.3	80.9	83.5	76.6	71.9	68.9	72.4	69.1	88.1
			GC-MS/MS	Green tea	A*	149.8	145.8	122.1	118.7	119.8	133.3	107.4	102.4	103.2	106.7	108.1	119.6	113.6	98.5	89.2	102.0	87.6	89.9	88.2	110.8
				Green tea	B	145.1	151.6	118.0	109.2	119.8	113.8	104.5	93.0	93.7	102.7	95.7	105.9	99.0	81.4	81.1	86.8	77.5	81.6	88.0	102.5
				Woolong tea	A*	143.4	109.3	113.5	103.9	120.6	99.3	96.5	82.6	99.3	94.3	87.1	86.2	81.4	81.9	91.8	94.7	74.0	80.7	76.4	95.6
				Woolong tea	B	135.0	104.6	108.9	101.7	105.2	90.6	92.7	75.4	86.0	80.9	89.0	79.7	72.5	73.0	72.1	81.2	68.9	69.4	73.5	87.4
111	methoxychlor	Cleanert TPT	GC-MS	Green tea	A*	47.5	46.3	41.6	76.4	36.0	34.6	35.9	36.7	30.6	29.1	42.2	32.0	32.4	31.7	34.5	29.2	22.6	23.2	34.8	36.7
				Green tea	B*	45.7	43.5	35.8	72.2	34.4	31.2	36.0	35.8	39.3	28.6	34.5	29.3	31.0	31.9	37.5	30.7	24.9	27.2	31.4	35.8
				Woolong tea	A	39.4	50.7	38.2	56.1	43.3	33.1	35.2	46.9	32.0	32.5	39.2	33.1	39.3	30.9	32.1	34.3	38.2	29.2	28.1	37.5
				Woolong tea	B	36.8	43.5	38.6	48.2	42.1	29.5	32.7	48.9	28.8	32.0	33.7	26.7	36.6	33.6	30.3	34.1	30.7	26.3	28.1	34.7
			GC-MS/MS	Green tea	A*	47.4	43.5	31.2	51.6	34.5	31.6	39.9	35.5	35.4	29.8	44.3	47.0	40.7	30.2	31.9	29.4	25.9	30.2	37.7	36.7
				Green tea	B	47.1	43.0	33.9	47.9	35.5	30.5	39.7	32.4	35.9	28.2	38.4	43.9	35.5	30.3	31.9	31.0	24.9	30.2	31.9	35.3
				Woolong tea	A*	52.6	44.5	34.2	41.9	31.6	37.1	35.5	73.3	39.2	37.9	39.9	38.3	36.8	36.0	34.0	33.4	27.2	30.8	27.0	38.5
				Woolong tea	B*	45.3	42.4	34.3	37.4	31.7	29.5	34.6	70.9	33.2	34.4	31.9	36.5	33.9	33.5	28.9	31.9	28.1	28.7	28.2	35.6
		Envi-Carb+PSA	GC-MS	Green tea	A	36.4	40.6	40.7	39.0	38.0	45.8	36.4	33.0	39.1	35.3	36.4	41.1	30.6	30.3	20.2	33.1	38.9	28.3	40.2	36.0
				Green tea	B	43.1	47.8	39.3	32.5	37.7	46.6	38.6	24.5	36.5	34.7	30.4	40.1	30.3	32.3	20.4	26.7	34.0	28.3	41.1	35.0
				Woolong tea	A*	42.8	25.0	33.6	28.5	54.7	34.5	25.0	28.7	36.5	25.6	27.3	32.4	28.0	30.8	30.8	30.8	23.4	23.1	23.6	35.0
				Woolong tea	B	47.0	22.0	35.0	30.3	42.0	35.1	30.9	23.8	47.1	27.4	23.7	35.2	25.1	33.8	34.0	23.5	23.6	25.9	20.9	30.9
			GC-MS/MS	Green tea	A*	51.3	75.7	41.0	37.1	41.2	25.5	45.9	36.7	39.1	42.7	44.3	32.6	46.8	33.0	31.8	39.4	36.2	36.3	38.3	40.8
				Green tea	B	48.1	62.6	41.1	36.8	38.4	39.6	30.5	32.7	35.1	37.4	34.9	31.3	44.2	30.7	29.3	29.1	36.6	33.0	39.4	37.4
				Woolong tea	A	40.7	26.9	26.9	36.7	36.6	41.5	30.7	31.3	45.5	38.3	34.8	35.3	28.6	27.4	29.5	43.2	18.3	36.0	29.8	33.6
				Woolong tea	B	43.9	25.1	32.3	36.9	35.3	35.1	31.1	29.9	47.1	30.3	35.9	33.9	26.3	27.2	29.1	38.0	18.2	29.3	30.4	32.4

No.	Compound	Cartridge	Method	Tea	Rep																				
112	methyl-parathion	Cleanert TPT	GC-MS	Green tea	A	203.0	178.7	152.4	123.3	227.8	251.0	152.0	157.2	125.7	130.7	183.1	131.7	149.8	143.8	142.5	118.5	118.1	99.0	140.1	154.1
					B	215.9	170.9	136.5	143.4	214.5	233.0	150.1	144.1	175.7	120.0	151.8	135.7	154.7	141.4	165.6	123.8	129.2	97.6	130.1	154.4
				Woolong tea	A*	178.7	174.8	240.3	307.8	425.0	220.5	158.9	189.3	133.9	144.3	165.1	127.7	178.9	132.1	133.3	145.4	157.6	121.9	115.8	181.7
					B*	169.9	157.5	219.1	276.3	358.2	191.5	114.5	168.3	110.8	127.0	135.1	109.8	147.1	139.9	122.8	149.0	133.2	119.0	109.8	161.0
			GC-MS/MS	Green tea	A	215.3	164.4	128.9	214.9	141.3	134.0	154.3	155.2	159.6	118.7	187.4	186.7	121.9	125.5	129.7	127.5	104.8	123.3	151.8	149.8
					B	212.0	163.0	139.6	207.4	143.1	129.6	152.7	137.3	154.1	124.7	159.7	191.7	126.6	132.8	125.8	133.9	106.4	129.4	131.6	147.4
				Woolong tea	A	200.1	186.2	142.1	206.3	196.7	147.9	220.0	245.0	233.6	126.1	231.7	140.2	154.6	137.8	137.5	225.9	207.3	225.7	224.4	188.9
					B	230.2	183.4	132.2	209.1	211.9	124.5	216.6	245.4	232.3	133.8	223.3	121.1	132.4	131.2	125.7	230.1	211.7	228.1	212.1	186.1
		Envi-Carb +PSA	GC-MS	Green tea	A	181.6	158.8	151.9	252.1	241.7	333.2	136.3	142.0	195.0	184.9	159.2	189.5	122.2	131.6	89.7	155.0	153.2	117.1	151.8	170.9
					B	198.2	161.7	146.8	216.5	237.5	301.8	129.8	118.8	183.4	166.0	123.9	187.2	120.2	133.7	97.3	128.0	130.8	107.8	155.5	160.3
				Woolong tea	A	153.2	119.8	233.3	231.4	334.2	238.5	113.0	121.1	127.9	124.5	122.8	123.5	105.4	132.1	137.8	126.4	114.1	108.7	101.4	151.0
					B	155.4	110.0	196.4	216.5	251.1	219.1	120.1	100.7	108.7	98.2	97.8	124.5	102.3	139.8	144.1	104.5	109.8	109.3	95.5	137.0
			GC-MS/MS	Green tea	A	227.1	185.0	165.1	166.6	182.0	243.2	184.1	129.7	179.9	137.9	181.9	161.7	151.2	160.3	112.3	132.0	137.2	149.8	171.1	166.2
					B	269.9	196.3	162.3	163.7	166.8	168.9	113.2	128.9	161.9	148.4	149.3	148.0	129.9	124.9	106.0	119.5	134.7	148.1	166.7	153.0
				Woolong tea	A	237.7	128.4	162.5	162.6	222.8	137.0	219.3	216.2	127.9	118.3	217.9	77.6	223.0	121.4	127.1	139.0	98.4	114.1	222.7	161.8
					B	197.2	123.1	128.5	176.4	225.6	129.4	215.4	215.0	108.7	111.4	234.2	79.6	213.8	116.7	108.3	125.8	95.3	105.0	203.2	153.3
113	metolachlor	Cleanert TPT	GC-MS	Green tea	A*	37.3	37.6	37.5	40.5	31.2	31.0	36.9	33.8	35.4	30.8	32.3	30.3	31.8	29.7	33.5	29.4	25.8	28.7	35.4	33.1
					B	37.5	36.4	33.3	38.2	29.6	27.9	33.8	31.5	35.4	29.2	27.5	28.9	30.8	27.6	30.6	28.7	26.7	29.5	31.5	31.3
				Woolong tea	A	43.0	42.4	38.6	42.2	33.5	36.6	33.2	49.1	39.6	39.0	38.6	42.5	41.0	37.7	35.6	36.2	37.0	30.2	26.7	38.0
					B	42.1	36.9	32.3	40.0	34.0	33.3	27.5	41.6	35.0	34.5	31.2	35.1	33.7	33.3	29.8	32.4	29.7	33.1	26.3	33.8
			GC-MS/MS	Green tea	A	42.8	36.2	35.5	45.9	32.1	29.9	34.8	34.0	32.8	29.0	36.9	26.8	31.4	28.6	31.4	26.2	26.6	27.3	34.1	32.8
					B	41.9	35.4	34.5	42.5	30.8	28.3	34.5	30.8	37.8	28.6	31.4	26.3	29.1	28.1	30.8	26.3	26.5	27.4	32.3	31.4
				Woolong tea	A	43.8	41.8	38.1	43.0	33.3	36.3	34.5	50.1	32.9	32.4	38.3	34.0	35.1	34.9	27.1	33.4	33.4	32.2	28.9	36.4
					B	40.8	36.5	33.3	39.0	33.2	31.3	26.6	42.9	36.6	30.8	30.2	29.9	28.6	31.0	27.9	29.6	29.6	30.3	27.1	32.1
		Envi-Carb +PSA	GC-MS	Green tea	A	42.6	41.4	32.1	34.2	33.7	36.1	55.1	33.8	36.1	32.0	35.4	37.4	30.7	36.6	27.9	35.5	28.2	30.5	33.0	35.5
					B	40.5	35.6	32.6	32.7	33.6	32.4	37.3	31.3	33.6	32.7	32.5	34.6	31.9	33.9	28.6	30.3	29.3	30.5	32.9	33.1
				Woolong tea	A	42.8	32.5	35.3	34.9	37.7	35.1	30.6	31.4	27.2	37.0	35.6	33.6	33.1	34.9	36.8	32.5	28.2	28.7	26.2	33.8
					B	40.9	28.9	34.9	31.5	31.4	32.5	32.8	28.8	31.3	28.9	29.7	28.6	29.2	31.2	29.6	27.6	29.1	27.1	25.7	30.1
			GC-MS/MS	Green tea	A	52.2	39.9	35.9	34.4	38.8	36.6	36.0	31.0	29.8	30.8	35.7	47.8	35.7	33.9	30.0	34.9	25.3	30.7	31.3	35.6
					B	58.6	35.2	35.5	34.6	37.6	35.2	38.2	30.2	33.6	32.1	31.4	40.2	29.8	26.2	27.7	29.9	29.4	29.2	32.7	33.8
				Woolong tea	A	55.8	35.9	47.1	34.9	38.9	32.5	37.8	28.7	29.6	30.1	29.9	25.9	31.3	29.5	32.4	33.3	27.6	28.5	27.4	33.6
					B	40.8	32.3	32.5	31.3	33.1	30.1	31.6	25.3	27.2	25.8	30.9	24.6	27.7	26.3	26.0	29.6	24.2	24.3	26.4	29.0
114	nitrapyrin	Cleanert TPT	GC-MS	Green tea	A	115.1	114.6	106.2	139.2	106.7	102.2	133.6	117.7	119.7	99.6	101.9	92.7	100.0	103.3	108.6	89.4	85.1	96.4	105.9	107.6
					B	114.4	113.6	96.7	142.7	103.1	92.7	123.1	113.7	122.0	89.9	93.4	97.4	103.1	110.8	108.1	85.2	88.8	100.2	100.2	105.2
				Woolong tea	A	94.6	129.3	109.0	133.9	95.6	91.7	102.3	122.6	96.3	93.1	110.5	100.6	104.0	106.1	100.4	95.6	92.4	85.3	95.3	103.1
					B	88.6	107.4	102.4	110.3	96.6	83.1	91.8	113.9	86.5	85.7	84.1	82.5	94.4	97.4	86.9	95.3	82.9	73.4	76.3	91.6
			GC-MS/MS	Green tea	A	117.3	110.9	81.2	132.9	94.7	90.5	107.3	101.9	100.1	92.7	108.0	98.3	88.9	88.3	96.2	80.1	78.6	83.3	101.5	97.5
					B	123.2	109.7	97.2	133.3	97.2	86.7	97.8	87.2	98.6	86.2	97.9	91.5	91.6	93.8	93.8	87.3	75.9	95.7	84.5	96.3
				Woolong tea	A	120.0	139.4	91.5	146.1	89.3	93.0	89.7	136.6	108.7	82.1	99.0	99.2	102.4	98.9	88.6	88.4	94.4	99.5	64.9	101.7
					B	108.3	122.9	87.3	120.5	73.7	69.6	75.6	99.3	99.8	84.1	81.8	81.9	92.2	84.7	77.0	79.5	80.3	79.3	60.6	86.8
		Envi-Carb +PSA	GC-MS	Green tea	A	126.3	102.7	107.1	112.8	113.7	126.9	128.6	118.2	145.6	112.0	116.6	105.6	94.3	106.9	75.3	121.4	106.2	102.6	117.5	112.6
					B	133.8	107.5	108.6	95.6	117.6	120.3	106.9	101.1	125.4	110.4	86.4	103.6	87.5	101.4	81.0	103.2	103.0	100.6	116.8	105.8
				Woolong tea	A	102.4	67.9	93.4	90.9	108.9	100.5	100.4	85.2	76.1	79.1	78.7	90.5	68.9	104.9	89.2	85.2	97.7	70.6	66.3	79.0
					B	89.7	60.1	92.5	86.7	93.7	95.2	99.7	76.5	60.2	69.1	68.4	87.0	67.9	96.1	75.4	74.2	78.1	68.9	61.1	108.8
			GC-MS/MS	Green tea	A	130.9	103.9	106.3	115.8	142.4	114.8	118.8	101.3	117.6	100.1	119.0	124.5	99.0	92.1	83.1	86.5	100.6	93.1	117.2	106.2
					B	160.4	77.7	109.9	105.3	138.7	107.7	117.8	99.1	103.5	106.1	97.9	110.8	93.7	85.0	70.4	85.8	102.3	101.8	124.8	90.4
				Woolong tea	A	106.2	73.5	128.6	81.1	119.2	98.1	125.7	94.1	76.1	80.2	87.0	51.6	83.7	84.7	87.1	88.4	61.0	63.5	67.9	76.5
					B	64.2	64.2	83.1	74.7	97.3	94.3	79.4	74.5	60.2	74.3	86.0	50.0	78.4	84.2	79.3	79.5	62.2	58.8	66.4	76.5

(Continued)

Annex 1: Content change for 201 pesticides in incurred tea Youden pair samples under 8 conditions with two SPE cartridge cleanup in three months (Nov.9, 2009-Feb.7, 2010) (cont.)

No	Pesticides	SPE	Method	Sample	Youden pair	Nov.9 2009 (n=5)	Nov.14 2009 (n=3)	Nov.19 2009 (n=3)	Nov.24 2009 (n=3)	Nov.29 2009 (n=3)	Dec.4 2009 (n=3)	Dec.9 2009 (n=3)	Dec.14 2009 (n=3)	Dec.19 2009 (n=3)	Dec.24 2009 (n=3)	Dec.14 2009 (n=3)	Jan.3 2010 (n=3)	Jan.8 2010 (n=3)	Jan.13 2010 (n=3)	Jan.18 2010 (n=3)	Jan.23 2010 (n=3)	Jan.28 2010 (n=3)	Feb.2 2010 (n=3)	Feb.7 2010 (n=3)	AVE, µg/kg
115	oxyfluorfen	Cleanert TPT	GC-MS	Green tea	A*	201.4	182.2	170.2	361.2	149.5	121.1	142.2	144.4	120.9	114.4	205.1	143.2	153.3	125.4	150.8	120.9	99.5	96.3	157.8	155.8
					B*	215.9	176.8	142.0	301.5	129.9	115.4	148.6	143.9	172.5	103.3	172.5	130.6	134.5	132.1	156.3	127.0	111.4	99.8	146.4	150.6
				Woolong tea	A	177.7	151.4	133.8	296.9	264.3	152.2	155.1	229.5	147.4	159.3	145.6	126.6	167.2	120.2	140.4	163.5	169.4	132.6	117.0	165.8
					B	168.9	174.7	145.5	197.2	213.8	139.4	119.7	228.2	112.6	136.0	125.1	106.0	154.0	134.4	129.7	164.8	130.8	134.0	119.2	149.1
			GC-MS/MS	Green tea	A	230.9	171.0	151.6	215.0	150.1	109.9	135.5	142.9	159.7	120.1	238.6	99.0	142.7	135.0	129.7	127.8	107.1	103.6	114.1	149.7
					B	220.5	166.2	148.4	196.6	126.9	105.6	145.2	135.1	165.0	113.3	206.0	95.3	120.3	137.7	123.4	126.9	111.2	105.6	159.9	142.6
				Woolong tea	A*	254.1	191.5	132.0	205.3	155.5	157.5	141.8	519.0	151.6	142.9	172.4	141.3	154.0	133.7	144.7	161.5	159.4	131.7	121.1	177.4
					B*	200.8	189.6	142.2	224.2	159.1	142.3	141.7	455.7	115.9	146.7	131.8	116.6	136.2	126.5	127.4	133.1	127.4	119.1	136.7	160.7
		Envi-Carb +PSA	GC-MS	Green tea	A	156.8	189.2	177.8	163.4	163.5	224.2	138.1	105.5	173.0	182.5	159.3	183.0	143.8	126.6	78.0	126.2	172.4	123.4	140.5	156.0
					B	204.8	109.2	168.4	117.6	158.0	210.2	149.8	141.7	157.7	173.8	132.1	166.4	145.0	136.3	73.4	106.1	144.8	115.1	147.8	149.1
				Woolong tea	A*	174.3	112.5	162.2	146.5	229.0	154.5	96.6	124.8	130.3	132.8	114.7	128.5	127.8	120.2	145.1	134.0	113.7	121.1	111.2	135.6
					B	199.5	215.0	150.1	154.3	163.2	147.4	109.4	106.4	121.9	119.5	89.5	143.3	113.1	134.4	168.3	94.9	106.7	116.2	102.0	129.1
			GC-MS/MS	Green tea	A	206.7	261.8	165.9	169.3	167.0	279.4	178.8	129.7	162.6	166.6	188.1	145.0	166.2	144.9	104.4	131.9	117.7	142.2	162.7	165.5
					B	279.0	110.5	159.4	143.0	137.1	190.5	76.3	122.4	154.6	176.2	146.8	139.4	151.9	139.0	98.2	115.8	107.8	153.6	173.9	155.1
				Woolong tea	A	220.4	113.8	161.6	158.3	200.9	147.2	117.3	114.5	130.3	126.9	115.8	116.8	122.4	130.1	129.1	161.5	99.7	107.8	108.4	134.9
					B	196.8	113.8	139.0	158.1	144.6	133.0	118.9	85.6	121.9	126.0	108.0	116.8	103.1	114.2	124.7	133.1	99.7	104.8	103.9	123.5
116	pendimethalin	Cleanert TPT	GC-MS	Green tea	A*	199.3	151.1	129.0	285.9	121.8	107.7	131.3	124.0	158.9	87.3	192.7	108.4	115.9	112.7	126.1	108.5	70.8	78.5	120.2	135.9
					B*	201.9	149.8	129.3	183.0	320.6	134.2	145.3	197.8	135.4	136.9	164.4	124.9	146.4	104.5	134.3	110.1	80.9	86.8	102.3	133.3
				Woolong tea	A*	170.0	152.0	127.5	174.9	242.6	116.6	110.2	193.2	98.2	113.5	127.0	95.9	130.7	119.7	128.3	164.3	129.5	120.7	102.3	150.0
					B	161.9	152.1	134.5	208.1	132.5	110.9	127.6	128.1	126.3	104.3	102.4	125.4	119.2	121.5	113.3	169.5	107.7	120.0	104.2	134.4
			GC-MS/MS	Green tea	A	199.1	151.1	127.9	188.8	117.9	106.8	129.4	120.1	144.5	98.8	179.0	122.6	110.3	120.0	125.2	102.3	84.9	95.1	131.0	132.0
					B	195.6	187.7	134.6	181.0	130.9	135.2	133.5	310.1	140.3	125.3	159.6	124.8	144.3	125.4	114.9	107.3	93.4	99.9	121.8	127.9
				Woolong tea	A*	216.6	185.2	125.0	177.3	128.9	116.1	98.9	288.1	100.8	122.1	153.7	100.1	122.1	124.4	121.2	134.8	126.9	125.3	102.5	150.2
					B*	185.3	172.5	151.6	146.0	155.6	197.2	163.8	94.5	166.3	122.9	118.9	179.6	131.1	102.2	107.1	122.2	107.8	114.4	131.1	134.6
		Envi-Carb +PSA	GC-MS	Green tea	A	153.3	177.9	145.5	107.1	155.6	185.4	153.2	117.1	164.9	163.2	152.1	153.1	122.6	124.2	59.4	96.1	168.7	102.7	132.6	143.4
					B	170.7	116.9	145.9	132.5	210.1	135.9	97.6	94.5	164.9	163.2	119.4	114.6	122.6	124.5	54.2	80.7	140.7	107.7	133.3	133.9
				Woolong tea	A*	161.4	116.9	125.2	129.4	133.5	128.5	110.3	117.1	122.7	117.8	108.6	124.1	116.5	104.5	115.3	117.9	93.1	107.0	99.2	122.9
					B*	158.0	108.7	147.0	143.0	158.5	168.2	110.3	114.1	105.6	132.8	84.9	131.7	161.9	122.6	140.5	77.5	88.5	118.0	89.8	111.4
			GC-MS/MS	Green tea	A	198.0	187.9	147.0	143.0	144.9	165.9	62.1	107.3	136.3	139.9	159.2	131.6	145.8	116.7	103.1	119.1	124.3	124.8	131.0	143.4
					B	231.2	265.9	145.1	130.9	167.5	128.5	110.8	109.4	139.1	139.9	128.9	131.6	109.6	113.4	93.2	103.4	113.8	152.4	93.1	139.1
				Woolong tea	A	212.8	119.7	134.5	122.7	126.5	117.2	115.0	81.3	122.7	100.3	102.8	114.9	122.1	124.4	120.6	134.8	91.5	98.3	86.7	123.4
					B*	177.0	113.2	121.9	122.7	126.5	116.1	98.9	56.1	100.8	100.3	118.9	100.1	122.1	102.2	109.1	122.2	91.4	91.5	86.7	110.2
117	picoxystrobin	Cleanert TPT	GC-MS	Green tea	A	95.2	77.7	85.9	91.9	71.6	64.8	79.1	68.6	74.4	61.4	75.3	66.5	67.0	60.8	68.0	57.5	57.8	56.5	73.8	71.3
					B	91.5	74.5	76.5	84.1	65.3	60.9	72.5	63.5	69.7	59.1	66.0	62.3	63.0	62.5	61.3	57.9	56.5	54.5	65.0	66.7
				Woolong tea	A	86.7	81.5	78.0	97.4	69.3	76.4	70.3	83.3	75.1	70.8	75.5	73.6	75.8	63.2	66.6	68.7	67.0	66.8	59.1	73.9
					B	83.3	75.9	72.0	88.2	70.4	70.0	63.7	78.1	66.8	65.6	62.8	65.1	66.3	62.8	60.7	62.0	57.9	67.2	58.0	68.3
			GC-MS/MS	Green tea	A	97.0	82.3	84.4	96.3	70.7	64.1	74.9	70.7	71.7	61.9	75.2	46.9	70.0	71.6	68.8	54.6	55.2	56.0	73.5	70.8
					B	93.5	79.0	73.6	88.0	65.2	60.5	73.4	64.1	70.4	61.7	66.8	45.5	60.2	70.3	62.7	54.6	53.6	54.9	64.6	66.5
				Woolong tea	A	79.2	78.7	80.3	91.8	70.5	75.0	66.8	88.5	76.4	68.7	74.8	76.0	73.6	69.3	64.6	71.0	65.7	64.8	60.2	73.5
					B	79.7	76.1	72.9	84.1	71.2	67.6	59.8	82.9	68.0	64.7	62.2	68.0	58.6	64.4	57.0	64.4	59.2	65.1	61.1	67.7
		Envi-Carb +PSA	GC-MS	Green tea	A	107.0	92.9	82.0	77.9	82.1	84.8	60.7	69.5	69.7	70.3	67.6	76.0	69.2	76.3	64.4	78.4	65.8	67.2	66.6	75.2
					B	100.8	79.9	78.3	70.6	80.6	76.7	57.6	66.1	66.7	68.9	63.8	68.6	63.8	67.1	58.7	63.1	58.5	57.9	67.1	69.2
				Woolong tea	A	89.6	76.0	75.3	70.8	79.3	70.0	71.6	59.1	71.5	67.6	65.3	64.7	64.0	61.8	58.0	61.5	57.7	64.1	56.6	62.5
					B	81.3	71.9	75.0	67.9	67.5	64.6	65.7	55.2	60.2	57.2	60.0	60.7	59.8	61.9	61.5	55.4	53.9	55.5	54.4	75.3
			GC-MS/MS	Green tea	A	99.5	89.7	81.2	79.4	80.3	92.4	69.5	64.2	66.8	72.3	67.4	60.4	74.5	62.0	54.7	61.6	62.6	62.5	67.1	69.4
					B	94.4	73.6	78.8	73.4	78.1	75.8	69.5	60.1	71.5	64.7	63.2	64.7	60.0	61.4	66.4	71.0	55.2	68.9	71.4	67.8
				Woolong tea	A	92.4	73.6	81.1	68.7	80.5	69.5	68.0	60.1	71.5	64.7	63.2	64.7	61.4	61.4	66.4	61.6	52.8	59.9	57.6	69.4
					B	84.3	69.8	72.4	67.1	70.5	66.7	66.7	56.1	60.2	55.4	66.1	62.4	56.8	56.5	53.8	64.4	53.7	53.1	55.1	62.6

No.	Compound	Cartridge	Method	Tea																					
118	piperophos	Cleanert TPT	GC-MS	Green tea	A*	114.7	123.3	125.7	195.4	108.2	88.8	109.5	113.5	98.1	82.6	136.5	97.0	101.5	97.7	101.1	83.4	71.0	72.0	98.3	106.2
				Woolong tea	B*	118.2	117.2	112.3	183.5	91.3	87.7	106.7	109.2	118.6	71.2	108.6	97.6	96.9	98.0	98.0	84.9	76.2	78.1	91.0	102.4
				Green tea	A	127.7	126.2	108.4	143.6	143.8	106.8	108.5	154.0	107.7	104.8	98.3	93.9	115.2	92.0	92.4	98.8	104.1	94.7	86.0	110.9
				Woolong tea	B	121.9	114.5	112.2	137.9	136.8	101.0	102.2	141.5	92.9	88.4	79.0	86.9	102.9	94.3	86.2	88.7	82.2	90.7	79.9	102.1
			GC-MS/MS	Green tea	A*	143.6	125.1	107.3	149.7	104.2	86.9	112.0	106.8	103.9	89.0	131.7	123.6	88.8	95.1	96.0	85.8	75.4	76.6	118.8	106.3
				Woolong tea	B*	140.0	121.2	108.0	141.1	99.4	85.4	109.3	94.5	104.3	86.9	108.6	123.8	91.0	100.0	90.9	87.0	76.8	79.2	73.8	102.7
				Green tea	A	186.2	151.7	105.9	148.4	107.2	111.3	106.4	335.6	103.8	100.9	123.1	97.2	110.8	99.4	93.5	97.3	103.9	86.9	72.6	123.3
				Woolong tea	B	162.6	160.6	106.6	139.0	107.4	96.5	91.0	307.6	72.1	101.7	101.8	92.8	94.5	90.0	81.1	84.9	79.0	77.6	104.1	111.6
		Envi-Carb +PSA	GC-MS	Green tea	A	122.0	138.4	117.3	117.1	113.2	127.4	140.3	103.5	117.5	120.6	119.2	123.4	100.4	97.2	69.1	84.5	108.8	81.6	106.0	110.8
				Woolong tea	B	138.7	138.1	114.4	103.0	116.1	119.9	123.7	83.5	112.2	111.1	99.7	113.6	88.1	95.5	62.5	74.5	99.1	88.9	80.1	104.6
				Green tea	A	137.0	87.6	120.9	114.8	142.5	106.3	82.7	87.0	98.2	97.7	90.1	86.2	90.6	92.0	95.8	84.4	77.6	85.6	75.1	97.7
				Woolong tea	B	137.4	85.9	115.1	111.8	105.9	96.6	88.7	68.7	83.0	82.1	67.4	103.1	81.8	94.3	85.6	71.4	70.0	80.0	99.8	89.7
			GC-MS/MS	Green tea	A	158.3	124.2	117.0	112.7	119.9	142.5	121.5	92.4	111.7	92.5	129.9	114.9	97.0	105.2	83.3	93.5	77.0	90.0	107.7	109.6
				Woolong tea	B	205.2	139.3	114.1	112.3	113.2	115.9	95.3	91.6	98.0	99.8	105.7	100.6	84.4	81.8	82.4	84.6	80.8	89.9	80.5	105.4
				Green tea	A	160.8	97.6	120.0	105.8	137.4	101.8	115.9	87.5	98.2	92.9	93.7	72.6	90.2	82.3	93.1	97.3	77.0	79.7	68.2	99.2
				Woolong tea	B	134.6	97.4	101.2	98.9	107.8	95.1	93.7	66.2	83.0	78.7	101.3	72.2	82.6	75.5	79.9	84.9	73.3	76.1	67.1	87.9
119	pirimiphoe-ethyl	Cleanert TPT	GC-MS	Green tea	A	93.4	68.2	71.9	79.8	60.2	50.0	72.3	66.0	65.7	55.5	86.8	63.5	62.5	59.8	63.2	53.2	52.3	52.5	59.7	65.5
				Woolong tea	B	91.9	67.2	63.8	71.6	54.7	48.2	66.8	61.3	69.0	48.6	72.1	58.6	58.9	56.8	57.8	52.6	51.1	51.2	54.5	61.2
				Green tea	A	85.6	64.6	63.9	78.3	62.5	66.9	64.2	82.2	70.0	72.8	67.4	65.7	70.1	64.4	63.0	63.2	60.8	61.7	51.9	67.5
				Woolong tea	B	81.5	67.0	61.2	75.9	61.5	62.0	52.8	74.1	54.8	59.9	53.2	58.4	59.0	59.9	56.4	57.0	52.5	60.3	64.1	61.0
			GC-MS/MS	Green tea	A	84.6	70.3	73.7	86.7	63.2	56.7	65.0	61.5	65.8	56.9	63.0	42.9	60.7	57.3	61.6	48.5	46.1	49.9	55.7	62.0
				Woolong tea	B	83.5	69.5	64.1	79.8	57.6	54.1	65.0	57.8	62.9	52.0	58.4	40.7	55.5	53.7	57.7	47.7	46.2	50.9	48.7	58.6
				Green tea	A*	75.3	72.1	74.9	83.0	62.7	68.0	59.9	89.4	72.1	61.2	71.5	66.7	64.6	62.2	66.8	59.9	57.7	57.9	47.4	66.6
				Woolong tea	B	72.8	69.1	62.9	77.4	63.3	58.6	49.4	79.0	62.4	56.4	56.7	57.9	53.0	56.8	50.6	53.4	51.1	55.5	55.5	59.7
		Envi-Carb +PSA	GC-MS	Green tea	A	98.0	88.3	70.7	67.2	73.7	74.0	68.4	65.3	63.3	85.1	69.0	84.1	65.1	70.2	60.3	71.9	59.6	61.2	61.0	71.4
				Woolong tea	B	92.4	73.9	68.7	61.5	70.5	63.7	62.6	60.8	60.3	78.7	65.8	68.7	60.4	63.5	56.7	59.4	54.3	54.1	60.7	65.1
				Green tea	A	89.2	63.6	68.4	63.0	72.2	65.0	66.5	55.4	61.7	65.5	60.0	56.2	60.0	64.4	67.8	57.8	53.6	58.1	52.5	63.2
				Woolong tea	B	79.6	58.1	61.3	60.3	55.7	58.6	60.3	47.9	51.6	53.0	48.7	55.0	55.6	59.9	56.4	51.4	48.3	50.0	49.4	55.8
			GC-MS/MS	Green tea	A	94.9	80.8	71.6	66.2	73.9	97.5	68.4	60.8	58.6	61.9	64.2	74.3	73.6	65.2	59.4	67.8	55.6	60.8	60.3	69.2
				Woolong tea	B	90.6	66.5	69.0	62.4	71.0	69.5	65.6	59.1	56.5	64.1	60.2	71.1	65.3	54.8	54.6	55.8	52.4	56.5	57.8	63.3
				Green tea	A	80.9	68.6	76.1	65.8	75.3	63.5	64.0	53.6	61.7	58.4	59.5	58.3	56.5	53.2	63.6	59.9	46.7	52.4	49.4	61.4
				Woolong tea	B	72.5	62.5	63.0	60.3	63.3	58.4	58.5	47.3	51.6	50.2	60.7	54.0	51.9	49.1	51.6	53.4	45.3	46.5	50.1	55.3
120	pirimiphos-methyl	Cleanert TPT	GC-MS	Green tea	A	45.8	36.2	38.1	43.6	33.6	30.5	37.3	34.1	32.3	30.4	41.9	33.0	30.7	31.1	31.5	26.7	26.2	20.7	33.4	33.5
				Woolong tea	B	45.8	35.4	34.3	40.0	31.2	29.1	34.2	31.5	35.9	27.8	34.9	30.3	28.9	31.7	32.2	26.3	26.8	20.5	29.8	31.9
				Green tea	A	42.9	37.1	36.3	44.8	36.3	35.1	33.0	43.0	35.2	35.7	35.5	31.4	35.2	31.0	30.4	31.2	30.0	29.1	27.8	34.8
				Woolong tea	B	41.0	35.1	32.8	40.5	35.4	31.7	27.1	38.5	28.6	30.6	29.8	28.2	30.0	29.4	27.7	28.7	26.1	28.0	26.1	31.4
			GC-MS/MS	Green tea	A	44.5	35.8	36.9	47.3	33.2	31.0	36.2	33.8	33.9	29.2	43.3	40.9	29.2	29.6	31.8	24.7	24.2	26.7	32.9	34.0
				Woolong tea	B	43.2	34.7	34.0	43.6	31.6	28.9	34.9	31.2	33.7	27.3	34.3	39.3	29.6	30.6	29.0	24.8	23.8	31.3	28.4	32.1
				Green tea	A*	41.3	40.7	37.9	46.4	33.0	35.1	32.1	46.9	36.4	31.2	40.0	34.2	34.5	32.6	32.4	30.6	29.5	29.9	28.5	35.5
				Woolong tea	B	39.1	36.2	32.4	42.1	32.5	30.7	26.6	42.7	33.7	31.8	31.8	29.3	29.2	30.6	27.1	27.3	27.9	30.0	26.2	31.8
		Envi-Carb +PSA	GC-MS	Green tea	A	46.1	42.8	35.7	35.8	39.8	40.3	35.6	32.8	31.6	42.6	36.9	42.0	32.4	33.2	27.5	35.1	30.7	27.3	31.6	36.0
				Woolong tea	B	44.4	36.3	36.2	33.1	38.6	35.5	36.9	29.7	32.5	40.2	35.3	36.2	30.2	31.0	26.9	29.5	28.4	30.0	31.5	33.6
				Green tea	A	43.7	33.2	38.2	34.6	39.2	32.5	31.0	30.0	27.2	32.9	31.1	30.6	25.8	31.0	32.7	28.6	26.2	27.3	25.2	32.1
				Woolong tea	B	38.4	29.6	33.8	32.1	31.6	30.1	30.8	26.2	33.4	27.7	26.2	31.5	33.0	29.3	27.6	25.2	24.1	28.1	23.7	28.7
			GC-MS/MS	Green tea	A	48.2	40.9	36.6	34.4	40.0	45.2	37.6	31.7	28.6	32.4	39.0	42.7	29.8	34.7	30.9	33.9	30.7	29.8	31.7	36.1
				Woolong tea	B	50.2	34.1	36.2	33.7	38.1	35.4	33.4	30.2	32.5	32.5	33.7	38.1	31.4	28.1	28.6	30.1	27.8	27.8	32.0	33.1
				Green tea	A	49.6	34.4	44.4	33.6	39.2	32.0	34.4	28.6	27.2	29.5	29.5	28.1	28.0	27.3	31.2	30.6	23.8	27.6	26.3	32.3
				Woolong tea	B	39.9	31.6	33.2	31.1	33.5	29.2	31.3	25.3	27.2	27.8	29.8	25.6	25.8	25.2	25.8	27.3	25.4	24.0	25.1	28.8

(Continued)

Annex 1: Content change for 201 pesticides in incurred tea Youden pair samples under 8 conditions with two SPE cartridge cleanup in three months (Nov.9, 2009-Feb.7, 2010) (cont.)

No	Pesticides	SPE	Method	Sample	Youden pair	Nov.9, 2009 (n=5)	Nov.14, 2009 (n=3)	Nov.19, 2009 (n=3)	Nov.24, 2009 (n=3)	Nov.29, 2009 (n=3)	Dec.4, 2009 (n=3)	Dec.9, 2009 (n=3)	Dec.14, 2009 (n=3)	Dec.14, 2009 (n=3)	Dec.19, 2009 (n=3)	Dec.24, 2009 (n=3)	Jan.3, 2010 (n=3)	Jan.8, 2010 (n=3)	Jan.13, 2010 (n=3)	Jan.18, 2010 (n=3)	Jan.23, 2010 (n=3)	Jan.28, 2010 (n=3)	Feb.2, 2010 (n=3)	Feb.7, 2010 (n=3)	AVE, μg/kg.
121	profenofos	Cleaner TPT	GC-MS	Green tea	A*	266.5	322.2	214.5	427.2	175.9	209.1	207.6	209.3	233.1	177.3	154.1	175.0	173.4	193.5	178.0	137.0	124.9	131.1	164.0	203.9
					B	262.3	291.1	192.7	375.1	180.5	188.4	201.5	205.1	206.0	228.0	146.5	163.2	185.9	189.1	211.9	148.5	135.0	150.6	145.7	200.4
				Woolong tea	A	285.7	353.8	219.8	304.5	307.2	155.0	196.7	217.3	245.9	180.7	168.0	190.0	255.0	179.7	158.3	188.6	220.9	165.2	132.3	217.1
					B	268.7	287.7	211.6	245.6	253.0	136.2	147.5	221.8	189.6	182.0	167.2	137.6	203.8	195.1	151.2	191.4	173.2	146.1	128.0	191.4
			GC-MS/MS	Green tea	A	331.0	341.3	175.0	327.9	191.7	203.8	210.4	165.2	266.8	185.6	122.7	168.0	150.6	134.3	148.7	141.8	98.9	168.0	223.0	197.6
					B	268.2	290.1	169.3	343.2	243.0	199.5	181.2	182.9	254.2	188.9	157.1	155.2	165.0	170.2	148.1	160.4	137.9	150.8	153.1	195.7
				Woolong tea	A*	384.3	475.0	188.5	307.4	160.4	220.7	169.2	387.3	238.5	178.7	152.2	171.5	219.4	178.6	195.8	187.9	173.9	181.7	79.9	223.7
					B	339.1	401.7	184.6	249.6	145.2	155.9	169.9	238.5	176.4	144.1	181.7	167.5	137.4	169.6	151.8	172.5	126.2	148.2	142.8	192.0
		Envi-Carb +PSA	GC-MS	Green tea	A	308.7	245.8	213.7	254.8	235.9	244.8	233.4	162.2	237.2	238.3	196.9	227.4	129.0	157.8	127.9	143.3	215.9	155.6	225.3	214.0
					B	303.8	243.7	192.9	218.9	223.1	222.9	215.9	142.5	158.4	215.0	171.6	181.4	132.0	183.0	169.5	174.0	157.2	133.7	126.2	195.6
				Woolong tea	A	240.8	138.4	200.5	180.3	257.1	211.8	162.0	160.8	178.5	187.8	134.4	181.0	133.4	194.9	159.7	143.9	153.4	142.3	117.0	176.1
					B	229.9	123.5	200.7	178.7	204.5	215.0	173.8	140.5	146.4	122.9	111.0	109.3	128.8	112.0	110.4	107.7	85.8	106.8	130.3	162.9
			GC-MS/MS	Green tea	B*	329.0	546.5	233.7	276.1	291.6	297.3	291.6	143.4	270.4	323.5	212.6	197.9	181.3	234.0	128.5	181.0	284.3	285.4	284.6	262.8
					A	332.4	124.7	197.2	264.1	240.9	226.1	210.1	125.9	223.2	248.9	209.4	194.2	146.5	186.0	141.8	146.3	254.7	254.5	278.3	239.8
				Woolong tea	A*	287.3	114.0	172.5	157.2	271.5	182.4	165.5	155.5	135.3	122.9	155.9	141.5	149.5	168.1	142.5	172.5	111.2	151.3	116.4	175.8
					B*	324.3	225.8	233.4	676.1	225.4	223.0	262.4	135.3	164.5	108.6	111.0	122.6	128.8	112.0	143.1	172.5	106.8	130.3	130.3	159.4
122	profluralin	Cleaner TPT	GC-MS	Green tea	A*	316.0	232.5	218.5	310.5	646.4	282.6	140.7	180.4	145.8	145.0	136.9	136.0	150.2	123.4	144.2	167.7	126.7	126.4	107.1	199.3
					B*	263.9	221.7	195.0	276.6	504.2	234.7	110.4	162.2	124.0	113.4	121.2	107.8	134.1	130.6	122.4	167.7	107.7	98.1	106.8	174.0
				Woolong tea	A	246.2	153.5	146.4	198.8	131.5	131.2	133.3	124.0	163.2	122.9	108.3	138.3	116.7	114.4	119.3	98.6	89.2	98.1	135.5	132.4
					B	192.0	150.8	128.3	183.3	126.5	130.8	128.6	118.9	152.8	141.7	106.0	132.6	105.5	122.7	110.4	102.4	123.8	119.5	119.4	128.1
			GC-MS/MS	Green tea	A*	187.4	176.4	145.7	190.0	131.8	117.8	123.6	180.4	144.7	141.7	123.2	144.8	142.5	128.8	129.3	127.7	106.4	113.5	101.0	149.9
					B*	161.7	166.6	135.1	179.0	127.7	117.8	98.1	292.8	114.4	115.3	123.2	111.4	125.1	120.3	112.6	127.7	106.4	113.5	101.0	133.9
		Envi-Carb +PSA	GC-MS	Green tea	A	331.3	276.6	243.7	268.7	257.8	322.6	234.8	116.4	145.0	250.6	144.8	185.2	132.5	133.5	83.4	117.3	156.1	126.7	155.1	193.8
					B	295.9	260.9	232.4	233.9	245.9	298.8	135.7	113.8	120.2	233.4	130.8	159.8	127.6	131.4	80.0	99.2	127.9	120.2	148.8	173.6
				Woolong tea	A*	265.2	195.2	224.4	209.8	363.2	249.5	127.0	120.2	124.9	125.6	127.8	123.0	124.1	123.4	132.0	124.6	104.5	112.8	104.7	162.2
					B	225.9	169.7	191.8	194.5	255.6	226.2	117.2	120.0	103.5	105.7	107.8	122.1	107.4	130.6	143.6	84.5	100.1	103.8	96.1	141.4
			GC-MS/MS	Green tea	A*	185.2	169.1	146.6	147.8	164.9	151.8	190.5	110.5	154.7	143.3	127.6	127.9	165.6	153.0	105.2	130.6	134.2	122.5	169.5	147.5
					B*	184.2	179.4	143.5	141.9	147.2	158.0	49.3	106.3	129.4	131.1	128.5	124.5	159.9	108.5	99.4	106.5	124.1	130.6	165.3	132.5
				Woolong tea	A*	184.6	129.2	165.6	134.6	165.1	130.1	88.5	110.0	117.6	125.6	119.6	116.4	123.2	118.4	124.2	132.4	101.8	107.8	98.9	126.2
					B*	157.3	114.9	126.5	125.4	130.5	121.2	119.9	90.1	114.6	105.7	108.8	119.5	108.0	103.5	108.5	127.7	98.6	101.2	96.2	114.6
123	propachlor	Cleaner TPT	GC-MS	Green tea	A	114.5	111.1	120.6	113.6	103.6	97.8	118.3	112.8	95.8	96.1	94.3	86.7	89.6	99.6	104.9	86.3	70.8	72.7	96.3	99.2
					B	112.3	106.0	106.8	116.8	97.7	88.3	110.8	105.9	82.7	105.7	91.2	87.9	92.1	99.0	102.2	85.1	76.5	80.2	89.7	96.8
				Woolong tea	A	111.8	132.0	111.0	132.2	102.0	103.0	115.9	113.8	109.6	105.5	106.0	118.3	117.8	94.8	96.9	95.9	106.6	91.4	87.1	109.6
					B	111.3	103.8	112.9	117.3	106.1	97.7	96.4	113.8	92.6	101.8	92.9	100.5	103.4	91.2	88.8	89.2	92.3	89.7	75.3	98.8
			GC-MS/MS	Green tea	A	117.9	109.8	90.5	135.4	92.2	89.8	107.3	107.9	99.2	102.4	86.2	54.7	88.3	77.5	87.2	78.6	70.5	84.3	93.5	93.3
					B	120.1	106.2	105.4	129.8	95.3	85.0	102.7	94.5	86.0	98.5	91.3	53.4	90.2	84.7	82.1	80.1	71.2	87.6	84.2	92.0
				Woolong tea	B*	137.7	129.5	109.9	120.7	98.7	109.8	106.7	158.7	105.4	111.3	88.5	99.5	106.3	99.5	88.1	91.9	90.0	103.0	77.6	107.0
					A*	128.1	106.6	95.0	105.6	94.6	90.4	83.3	124.6	86.9	94.3	87.1	88.5	87.5	89.2	78.7	81.0	81.0	93.4	74.8	93.2
		Envi-Carb +PSA	GC-MS	Green tea	A	118.6	113.3	106.2	113.3	106.5	110.9	128.8	97.1	97.8	106.1	87.8	95.7	89.8	98.8	74.9	110.5	97.0	90.7	100.1	102.3
					B	119.2	101.5	111.0	106.8	112.7	105.8	107.6	77.4	87.2	100.6	89.1	98.8	85.7	94.1	76.0	89.4	87.5	86.5	98.5	96.6
				Woolong tea	A	101.0	109.3	115.9	110.3	109.3	107.0	82.4	87.7	93.5	92.6	98.1	96.7	89.9	96.2	99.7	87.6	88.4	82.3	69.7	95.6
					B	84.6	96.5	107.4	98.7	96.0	99.3	99.2	82.0	87.5	76.7	80.5	89.6	88.7	91.8	85.3	79.9	82.2	70.8	70.6	87.8
			GC-MS/MS	Green tea	B*	157.1	115.0	107.8	110.0	120.1	151.7	109.6	89.4	106.5	91.5	88.0	187.9	99.1	107.0	85.3	97.3	84.1	87.9	96.9	110.6
					A*	199.0	105.6	111.1	111.5	120.8	101.1	122.2	88.7	95.5	91.5	88.0	177.4	99.1	79.4	88.4	91.0	83.4	87.4	74.7	107.2
				Woolong tea	A	193.1	100.6	97.3	97.3	115.5	93.9	134.2	88.7	86.4	92.6	87.2	60.2	79.8	82.3	85.7	91.9	62.8	75.6	74.7	97.9
					B	117.8	91.2	93.3	84.9	96.4	87.0	92.5	76.2	89.4	76.7	75.2	56.4	79.8	79.4	70.5	81.0	66.9	65.1	72.6	81.7

No.	Pesticide	Cleanup	Detection	Tea																						
124	propiconazole	Cleanert TPT	GC-MS	Green tea	A	176.7	127.7	123.7	132.6	108.9	86.1	105.8	103.0	96.9	82.4	120.9	94.3	86.7	90.2	89.0	71.6	72.2	66.2	87.6	101.2
				Green tea	B	168.8	119.6	114.3	121.9	94.3	85.4	97.9	93.1	101.7	79.6	109.8	90.7	81.4	94.2	80.0	69.0	68.0	63.7	75.6	95.2
				Woolong tea	A	149.5	101.8	126.4	102.6	116.3	106.1	92.7	115.3	110.4	107.5	84.0	88.7	96.6	87.5	82.7	82.9	81.3	82.6	72.5	99.3
				Woolong tea	B	136.2	114.6	112.9	99.2	114.9	99.5	89.4	113.4	83.9	95.3	75.1	82.2	88.4	85.6	73.6	73.1	68.8	77.8	66.9	92.1
			GC-MS/MS	Green tea	A	142.7	127.6	136.8	130.1	96.0	92.5	102.4	94.6	99.9	82.8	99.1	72.2	91.1	89.8	87.8	72.3	70.2	70.1	93.4	97.4
				Green tea	B	138.3	119.7	128.9	113.1	82.8	82.5	101.9	87.5	95.1	86.1	107.8	67.5	81.7	91.0	81.0	82.7	65.7	69.2	80.9	92.8
				Woolong tea	A	115.1	120.4	133.0	113.8	93.0	103.2	90.3	120.4	95.3	92.7	101.0	95.3	93.5	87.6	80.0	84.3	77.2	81.5	73.4	97.4
				Woolong tea	B	114.1	112.5	118.7	102.6	91.5	91.3	84.5	114.6	83.2	87.4	84.9	87.1	81.1	78.7	68.8	76.1	68.7	74.6	65.1	88.7
		Envi-Carb +PSA	GC-MS	Green tea	A	174.8	156.4	113.2	129.4	117.0	125.9	95.5	91.0	105.3	125.4	94.2	107.5	83.1	104.2	89.0	104.9	82.2	85.2	81.1	108.7
				Green tea	B	170.6	133.2	102.4	118.3	115.1	104.7	104.3	89.3	94.3	107.5	93.1	89.0	77.6	89.8	83.2	84.8	70.4	71.2	80.9	98.9
				Woolong tea	A	162.3	100.1	96.1	111.2	122.0	89.2	102.3	86.9	94.5	103.9	73.2	72.7	83.3	89.1	89.1	75.6	73.6	76.4	66.8	92.6
				Woolong tea	A*	151.8	96.5	95.9	107.2	84.1	81.2	89.5	72.1	78.8	85.4	55.4	76.9	79.2	79.5	68.6	68.2	65.2	66.7	63.7	82.4
			GC-MS/MS	Green tea	B*	148.5	139.1	113.9	140.2	109.3	113.9	96.6	94.9	96.7	95.0	92.6	287.4	103.7	93.0	83.2	98.6	78.5	84.3	84.1	113.3
				Green tea	A	140.9	109.4	107.1	110.0	108.2	100.1	94.0	86.8	86.7	88.9	86.0	279.4	92.7	78.6	76.4	83.7	67.8	73.5	83.2	102.8
				Woolong tea	B	139.5	106.5	97.9	109.7	111.0	92.1	91.2	76.8	94.5	88.3	87.7	82.4	73.4	78.1	83.3	84.3	68.5	74.3	71.4	90.0
				Woolong tea	B	130.8	100.5	97.9	106.3	89.9	87.3	86.5	72.0	78.8	73.5	87.9	76.2	68.4	71.5	67.8	75.9	66.5	63.2	68.4	82.6
125	propyzamide	Cleanert TPT	GC-MS	Green tea	A	161.5	133.4	167.4	143.3	118.8	106.0	127.2	123.7	118.5	110.7	140.9	117.1	113.2	106.5	114.3	98.6	99.2	90.6	132.5	122.3
				Green tea	B	159.7	129.9	147.6	121.0	107.7	100.5	117.8	108.7	122.4	101.9	121.0	108.7	106.0	109.1	105.6	98.1	98.0	88.9	117.0	114.2
				Woolong tea	A	139.9	124.3	158.0	127.5	127.9	125.7	106.9	144.8	127.0	117.0	129.0	118.0	123.1	116.1	110.2	112.4	112.2	110.8	102.1	122.8
				Woolong tea	B	128.7	116.5	140.1	116.3	120.0	108.1	92.9	121.7	104.7	108.1	103.6	97.3	107.0	106.6	97.1	100.1	96.3	104.7	96.8	108.8
			GC-MS/MS	Green tea	A	160.8	134.3	156.5	140.7	118.9	109.3	122.6	120.3	113.6	109.6	121.4	111.4	113.9	101.7	116.3	97.0	95.0	96.2	127.9	119.3
				Green tea	B	154.5	130.6	142.6	122.3	108.8	101.0	122.2	106.3	111.7	107.0	115.2	112.5	103.5	101.7	104.6	95.8	93.2	108.9	112.2	112.3
				Woolong tea	A	123.9	133.5	169.3	134.1	114.1	122.2	109.3	151.1	136.0	112.3	123.8	127.2	122.7	125.8	111.7	109.6	106.9	109.3	99.7	123.4
				Woolong tea	B	121.1	121.7	147.3	116.2	111.9	102.5	93.4	133.4	115.6	105.7	100.6	103.1	108.7	109.8	95.0	99.7	95.8	100.3	94.6	109.3
		Envi-Carb +PSA	GC-MS	Green tea	A	169.6	152.2	134.9	133.7	137.1	151.7	106.7	115.6	125.1	133.9	124.2	139.6	116.0	124.5	103.9	131.4	112.8	112.7	117.5	128.6
				Green tea	B	159.8	135.2	124.5	130.6	131.9	135.7	87.6	109.2	121.0	129.4	114.1	125.5	105.7	111.2	95.5	105.8	101.0	104.1	122.9	118.5
				Woolong tea	A	163.7	114.5	109.5	117.5	146.0	111.5	116.4	102.4	110.1	101.9	105.5	113.4	103.5	116.1	121.0	104.5	86.4	97.2	95.6	112.5
				Woolong tea	B	134.6	97.8	101.6	110.1	116.5	103.1	109.1	88.5	79.5	89.6	93.0	102.3	94.4	106.6	101.3	89.3	81.9	84.7	82.2	98.2
			GC-MS/MS	Green tea	A	160.4	147.7	130.8	132.9	135.6	111.9	120.8	113.9	116.7	123.1	121.3	69.3	119.9	118.2	104.2	125.9	111.5	114.6	119.5	121.0
				Green tea	B	152.7	126.8	121.3	128.9	129.4	136.1	111.2	106.7	111.2	124.1	108.4	67.0	103.4	99.3	93.9	104.1	99.6	108.9	121.9	111.4
				Woolong tea	A	159.5	117.3	112.8	147.2	146.4	112.1	109.7	99.6	110.1	95.3	104.7	102.8	106.5	117.3	109.4	109.4	83.0	93.8	90.9	111.4
				Woolong tea	B	126.8	98.7	103.7	109.1	121.3	102.0	109.3	85.8	79.5	88.4	104.0	93.9	106.2	90.8	92.8	99.6	85.1	81.6	82.6	97.4
126	ronnel	Cleanert TPT	GC-MS	Green tea	A	238.7	211.9	306.0	216.3	199.9	193.8	226.0	222.1	193.5	194.8	220.0	192.5	183.2	180.3	187.4	162.9	150.5	144.8	185.7	200.9
				Green tea	B	240.7	211.9	285.8	197.4	190.6	179.2	206.9	203.1	217.5	181.0	190.1	185.2	181.2	174.3	184.8	159.6	149.6	149.5	168.9	192.5
				Woolong tea	A	250.9	250.7	265.8	236.7	229.6	206.9	207.5	268.1	205.9	212.4	230.6	205.3	226.2	201.7	187.5	192.3	193.8	185.9	175.3	217.5
				Woolong tea	B	232.5	204.8	228.9	197.7	209.2	177.9	157.5	215.6	179.2	171.6	186.7	174.7	174.2	181.1	162.9	170.0	160.8	172.4	153.9	184.8
			GC-MS/MS	Green tea	A	228.1	208.3	277.7	198.3	193.7	198.8	227.8	202.2	191.8	182.7	209.6	172.7	180.1	168.3	184.7	160.6	145.1	159.0	190.7	193.7
				Green tea	B	224.8	206.0	266.0	206.5	197.3	188.9	213.2	184.2	191.9	179.4	179.8	163.7	180.9	169.7	177.6	158.6	146.3	166.7	173.8	188.2
				Woolong tea	A*	302.1	266.3	265.7	230.7	203.9	217.6	203.1	331.5	239.1	181.6	228.0	210.2	212.0	227.3	192.3	190.6	184.6	205.0	174.0	224.5
				Woolong tea	A*	258.7	220.0	224.9	190.0	187.8	178.6	161.8	249.8	193.2	176.8	186.6	184.6	169.5	159.4	153.1	166.2	193.2	177.5	157.8	189.9
		Envi-Carb +PSA	GC-MS	Green tea	A*	248.2	202.9	217.4	206.9	231.2	229.4	114.3	249.4	224.0	217.3	213.9	217.3	179.8	200.9	153.1	194.1	193.2	168.9	202.5	203.1
				Green tea	B	240.9	202.9	206.4	206.6	228.4	213.9	120.4	160.3	213.6	207.4	190.2	231.5	169.8	186.6	162.3	167.7	176.4	165.5	199.3	192.1
				Woolong tea	A*	243.6	200.2	213.2	225.1	230.6	197.4	172.7	177.9	201.4	193.0	193.7	181.6	168.4	201.7	189.8	170.4	161.8	162.4	156.9	191.7
				Woolong tea	B	201.7	170.3	182.1	201.3	188.1	176.0	180.1	151.1	159.2	151.9	160.9	169.0	153.3	181.1	158.3	147.4	144.4	145.8	144.2	166.6
			GC-MS/MS	Green tea	A*	306.8	239.1	231.9	215.1	248.1	429.4	246.6	182.5	207.5	178.4	200.5	112.0	195.9	206.2	175.7	199.3	178.2	173.7	196.2	217.0
				Green tea	B	336.1	212.7	228.6	212.9	234.1	220.5	213.1	177.9	190.7	185.8	182.2	109.4	176.2	160.7	169.2	175.5	167.4	177.4	202.4	196.5
				Woolong tea	A	322.7	210.1	198.7	289.5	246.4	194.5	250.8	175.6	201.4	179.0	190.5	134.8	183.5	174.8	186.8	190.6	144.9	168.1	163.3	200.3
				Woolong tea	B	223.8	183.9	170.5	183.2	198.4	174.6	187.1	148.0	159.2	157.3	185.1	125.5	160.6	157.2	143.2	166.2	138.7	143.5	154.4	166.3

(Continued)

Annex 1: Content change for 201 pesticides in incurred tea Youden pair samples under 8 conditions with two SPE cartridge cleanup in three months (Nov.9, 2009-Feb.7, 2010) (cont.)

No	Pesticides	SPE	Method	Sample	Youden pair	Nov.9, 2009 (n=5)	Nov.14, 2009 (n=3)	Nov.19, 2009 (n=3)	Nov.24, 2009 (n=3)	Nov.29, 2009 (n=3)	Dec.4, 2009 (n=3)	Dec.9, 2009 (n=3)	Dec.14, 2009 (n=3)	Dec.19, 2009 (n=3)	Dec.24, 2009 (n=3)	Dec.14, 2009 (n=3)	Jan.3, 2010 (n=3)	Jan.8, 2010 (n=3)	Jan.13, 2010 (n=3)	Jan.18, 2010 (n=3)	Jan.23, 2010 (n=3)	Jan.28, 2010 (n=3)	Feb.2, 2010 (n=3)	Feb.7, 2010 (n=3)	AVE. µg/kg.
127	sulfotep	Cleanert TPT	GC-MS	Green tea	A	47.4	35.8	37.0	40.4	33.4	30.4	35.8	35.8	32.5	30.0	38.5	32.3	31.9	30.0	31.6	27.0	26.8	28.0	33.9	33.6
					B	47.7	35.2	34.5	36.9	30.9	28.5	32.9	32.7	36.5	29.7	32.3	30.1	30.5	28.6	30.0	26.1	26.0	28.1	29.3	31.9
				Woolong tea	A*	43.7	34.1	36.7	42.6	35.5	35.1	32.0	39.1	35.3	34.1	32.9	32.2	34.5	32.5	31.4	33.0	31.8	31.1	27.5	34.5
					B	41.1	32.6	32.8	38.4	34.0	31.2	26.9	34.8	30.5	30.0	28.2	29.4	29.0	30.8	27.8	29.8	26.3	29.7	25.2	31.0
			GC-MS/MS	Green tea	A	43.3	36.4	39.1	44.4	32.7	31.1	35.1	33.1	33.9	30.0	37.4	20.5	33.1	28.4	31.7	25.9	24.3	27.2	32.1	32.6
					B	42.2	35.9	34.7	41.6	30.2	28.9	34.7	30.9	33.1	28.7	31.7	19.6	29.6	27.3	29.0	25.8	23.8	27.8	27.8	30.7
				Woolong tea	A*	36.9	39.2	36.7	43.7	33.9	34.0	31.2	46.3	36.9	31.4	34.7	34.4	33.3	34.1	30.9	29.4	28.2	31.1	26.0	34.3
					B	36.5	34.9	32.1	40.1	32.7	29.6	25.4	40.9	31.8	29.5	28.8	30.0	29.0	29.9	26.6	26.2	26.7	29.6	24.1	30.8
		Envi-Carb +PSA	GC-MS	Green tea	A	49.5	44.9	36.5	36.1	39.9	43.8	48.3	33.6	33.9	39.4	35.1	39.2	33.3	35.2	30.3	34.9	32.1	30.6	32.5	37.3
					B	46.5	38.0	36.2	33.1	39.0	37.4	39.6	30.9	32.1	36.2	33.1	34.0	30.9	32.3	29.5	30.9	30.0	27.9	31.7	34.2
				Woolong tea	A	44.5	33.4	35.5	33.2	37.5	32.0	31.4	28.7	32.0	32.6	31.5	28.5	29.6	32.6	33.6	27.9	25.7	27.9	25.7	31.8
					B	36.3	30.2	31.7	30.6	31.4	28.3	29.6	26.8	27.1	27.5	26.5	26.7	27.3	30.9	27.5	24.3	22.6	24.6	24.2	28.1
			GC-MS/MS	Green tea	A	48.4	41.0	36.5	37.0	40.7	54.2	37.4	31.7	32.1	34.6	32.8	32.8	40.1	34.2	30.2	35.6	31.6	30.3	31.9	36.5
					B	48.0	33.6	36.3	34.6	39.1	36.8	33.1	30.7	29.7	33.9	30.6	32.2	36.8	30.7	27.6	30.9	28.6	30.1	31.8	33.4
				Woolong tea	A	46.0	33.5	44.5	35.9	37.3	31.7	31.5	27.7	32.0	29.3	30.1	26.9	29.3	28.3	31.7	29.4	24.5	26.3	25.6	31.7
					B	35.0	29.7	30.7	33.5	31.3	29.5	29.7	24.7	27.1	25.3	30.4	25.2	27.1	26.2	25.1	26.2	23.8	22.3	25.4	27.8
128	tebufenpyrad	Cleanert TPT	GC-MS	Green tea	A	75.7	62.5	66.5	72.8	57.5	50.9	60.6	55.9	60.4	48.3	55.0	50.8	51.3	50.8	53.1	44.2	43.6	43.9	55.6	55.7
					B	71.1	60.1	59.8	67.6	50.9	47.0	56.1	52.1	56.5	47.6	48.4	47.1	48.3	48.2	47.8	43.9	42.3	42.3	49.0	51.9
				Woolong tea	A	75.2	64.7	61.4	72.1	57.5	59.9	51.4	64.6	59.5	55.5	58.4	56.3	58.2	52.2	52.3	52.6	50.5	52.4	47.1	58.0
					B	69.9	59.2	56.1	64.1	58.5	55.2	49.8	62.0	51.6	50.6	48.4	49.6	52.7	48.9	46.0	45.6	42.5	49.0	42.6	52.7
			GC-MS/MS	Green tea	A*	78.7	64.0	71.4	75.8	56.8	51.5	56.9	54.1	57.2	47.8	56.1	39.3	52.4	51.5	55.8	41.5	42.5	44.6	58.5	55.6
					B	74.0	61.7	61.4	69.6	51.2	47.5	58.3	50.8	54.7	47.2	52.4	37.4	46.8	50.2	49.7	41.6	42.6	43.8	50.1	52.2
				Woolong tea	A*	67.2	64.0	63.0	74.2	58.6	59.5	52.1	74.0	62.0	54.8	60.2	57.4	56.1	55.5	50.8	50.4	50.5	54.2	46.0	58.4
					B	64.5	60.9	56.4	65.6	59.8	53.4	48.0	71.4	53.4	50.6	49.9	52.3	48.9	50.7	43.6	44.1	45.8	50.2	40.4	53.2
		Envi-Carb +PSA	GC-MS	Green tea	A	86.4	76.0	63.1	63.1	62.9	64.9	46.6	54.6	53.4	53.5	51.3	55.2	54.1	58.7	51.2	61.5	44.5	52.2	50.9	58.5
					B	81.9	65.8	61.6	55.7	62.7	58.5	36.5	50.3	54.2	51.5	50.1	52.0	50.2	51.7	47.7	50.6	50.5	43.6	50.3	53.4
				Woolong tea	A	77.9	58.5	59.2	58.6	64.4	55.3	57.0	45.8	49.1	53.6	50.8	49.5	50.5	52.2	56.2	47.9	47.1	50.6	44.7	54.5
					B	72.4	53.4	59.6	54.0	53.8	48.4	48.0	43.3	55.4	44.7	44.3	45.0	45.5	48.9	45.7	41.2	40.1	44.1	41.2	48.5
			GC-MS/MS	Green tea	A	79.4	47.1	63.2	58.9	6.5	69.4	58.9	53.7	47.2	55.5	56.0	37.8	61.9	56.3	49.3	59.0	50.1	53.3	51.6	56.8
					B	77.6	51.9	61.6	54.8	6.8	59.9	55.3	49.7	56.0	55.1	50.6	35.9	55.6	51.1	45.1	50.0	43.3	48.3	51.6	53.1
				Woolong tea	A	76.1	58.5	61.9	58.0	63.7	55.5	51.5	45.7	55.4	52.3	48.7	48.9	50.2	47.4	52.3	50.4	42.9	48.7	44.7	53.3
					B	70.3	54.3	58.2	54.0	57.1	49.5	47.9	41.7	47.2	45.2	48.1	43.4	44.8	40.7	42.0	44.1	40.8	42.5	42.4	48.1
129	terbutryn	Cleanert TPT	GC-MS	Green tea	A	102.0	79.5	82.8	89.2	68.3	58.4	72.4	68.8	64.3	57.7	78.2	60.7	58.7	67.8	65.3	50.1	64.4	21.0	62.3	66.9
					B*	99.1	77.6	72.3	80.0	61.7	56.7	66.7	63.0	66.2	52.7	65.8	56.0	54.9	61.1	60.5	50.3	55.1	20.3	55.4	61.9
				Woolong tea	A	93.7	75.6	73.9	88.3	71.4	72.0	64.2	79.9	70.4	70.0	67.2	66.7	67.4	59.5	61.1	61.8	58.4	56.7	51.1	68.9
					B	89.5	75.3	69.4	83.4	70.0	65.2	54.4	70.5	57.2	57.6	52.4	57.2	57.1	55.2	52.9	54.5	48.9	56.7	49.0	61.9
			GC-MS/MS	Green tea	A	99.9	81.7	84.5	95.6	66.8	61.5	66.9	65.4	65.9	60.1	58.6	49.8	59.8	56.4	61.0	48.9	45.1	45.9	60.5	65.5
					B	93.6	80.1	86.9	86.9	65.0	57.4	68.6	61.5	63.3	60.1	59.8	48.5	53.8	54.4	54.6	47.4	48.0	48.5	54.3	61.9
				Woolong tea	A	76.2	75.6	81.5	92.9	66.6	70.0	64.8	80.8	71.8	63.9	67.9	66.8	63.8	62.2	57.4	57.1	61.3	57.6	54.0	68.0
					B	77.2	71.9	71.4	83.7	66.1	60.5	52.5	73.8	60.5	59.4	56.6	55.2	53.7	57.5	50.2	50.7	52.7	55.1	54.0	61.1
		Envi-Carb +PSA	GC-MS	Green tea	A	110.2	98.3	78.5	77.6	80.3	83.9	77.9	64.0	65.3	75.7	63.9	76.3	59.8	66.2	56.2	67.9	58.0	59.3	56.5	72.4
					B	103.7	81.6	77.1	69.9	77.5	73.3	75.0	60.9	63.3	70.9	60.6	62.6	54.7	58.9	51.3	54.3	50.7	50.0	56.5	65.9
				Woolong tea	A	98.6	71.5	76.3	71.0	83.6	67.8	67.4	56.1	65.2	64.6	59.0	55.7	57.2	58.8	64.7	54.3	54.4	55.7	47.7	64.7
					B	89.9	66.0	70.1	66.6	65.2	61.1	61.9	50.1	54.1	52.8	48.7	53.4	52.9	54.1	53.6	49.7	48.0	46.2	44.8	57.3
			GC-MS/MS	Green tea	A	105.7	90.0	80.0	75.2	77.4	83.6	71.7	62.9	63.0	63.1	66.1	75.9	70.9	69.3	55.9	64.6	53.3	55.5	56.5	70.6
					B	98.9	71.5	77.1	69.9	76.4	66.7	66.7	59.7	60.1	64.5	58.6	73.6	58.3	53.5	48.7	52.8	47.3	51.6	58.6	64.3
				Woolong tea	A	92.1	73.4	80.6	70.3	78.2	67.5	63.2	54.1	65.2	61.6	59.9	61.3	58.3	56.4	62.5	57.1	46.4	53.0	47.2	63.6
					B	84.4	68.3	70.4	65.2	65.2	62.1	60.9	49.1	54.1	51.9	60.8	57.1	53.8	48.8	48.2	50.7	45.6	42.2	46.0	57.1

No.	Compound	Cartridge	Method	Tea																						
130	thiobencarb	Cleanert TPT	GC-MS	Green tea	A	91.3	73.6	81.3	100.0	72.4	64.9	78.6	74.0	76.1	64.2	74.0	66.6	67.5	64.5	68.9	58.9	58.6	61.1	74.2	72.2
					B	87.8	72.6	72.9	91.1	66.7	61.4	72.8	68.2	74.0	60.5	66.7	62.7	64.4	62.0	63.2	57.9	56.9	58.7	65.0	67.7
				Woolong tea	A	86.6	76.0	76.1	85.8	73.1	73.8	65.7	81.8	74.1	69.8	72.0	74.5	72.7	65.4	66.9	65.1	63.0	65.5	58.1	71.9
					B	82.5	70.7	68.2	80.9	71.5	67.4	56.6	73.3	64.7	63.7	59.7	63.7	62.9	61.5	59.5	60.9	55.9	64.6	55.4	65.4
			GC-MS/MS	Green tea	A	95.3	77.1	82.8	96.0	68.2	65.4	74.1	71.3	72.0	66.6	71.8	61.0	65.8	62.9	69.6	55.8	54.3	57.8	73.9	70.6
					B	91.7	74.6	69.9	89.4	64.3	61.2	74.5	66.8	70.8	61.6	68.6	58.1	61.7	59.6	64.6	55.1	53.9	59.5	66.8	67.0
				Woolong tea	A	73.9	77.1	75.7	90.6	68.4	70.8	65.1	85.4	76.4	69.7	71.6	73.4	68.7	72.2	63.7	67.1	61.2	65.4	55.1	71.1
					B	73.8	70.3	66.5	81.8	69.0	62.3	55.5	78.0	67.4	63.8	59.6	62.9	60.1	64.0	55.5	63.6	55.5	62.0	53.4	64.5
		Envi-Carb +PSA	GC-MS	Green tea	A	101.0	89.8	76.5	78.4	82.5	85.2	60.8	72.2	74.5	73.0	67.5	76.9	72.0	75.9	67.8	82.3	65.2	68.9	67.3	75.7
					B	93.6	74.7	78.1	72.5	80.9	76.2	69.0	69.1	69.8	71.2	64.6	70.2	68.1	68.6	63.0	68.0	59.4	59.7	66.6	70.7
				Woolong tea	A	93.5	72.0	73.8	69.5	80.0	70.6	69.5	58.7	65.9	66.1	63.2	60.3	62.1	65.4	69.1	60.1	56.4	59.4	53.0	66.8
					B	77.7	64.7	68.7	65.6	66.9	62.7	64.0	54.6	58.2	55.6	55.4	57.7	57.9	61.2	58.8	53.8	51.3	51.6	50.1	59.8
			GC-MS/MS	Green tea	A	98.3	86.3	77.1	73.9	81.0	87.1	78.3	71.1	69.0	73.7	72.1	68.0	82.8	75.8	65.2	78.3	65.7	70.8	67.1	75.9
					B	92.5	70.3	74.8	71.7	77.4	78.5	73.7	67.1	64.5	73.2	64.9	64.2	71.6	62.6	60.6	67.9	61.2	66.3	64.1	69.9
				Woolong tea	A	86.1	70.6	82.2	66.9	77.6	67.6	62.8	58.7	65.9	65.0	62.2	63.4	64.1	60.0	64.1	67.1	54.0	57.9	55.9	65.9
					B	76.2	63.3	66.4	62.9	65.1	62.8	62.8	52.4	58.2	56.5	63.2	59.2	58.9	55.9	53.1	63.6	52.5	51.9	54.5	60.0
131	tralkoxydim	Cleanert TPT	GC-MS	Green tea	A	406.7	308.4	327.0	438.6	282.7	252.6	285.5	272.3	289.1	223.0	264.5	249.6	247.4	237.0	260.7	218.3	206.1	216.1	247.9	275.4
					B	345.3	300.1	289.6	389.0	249.1	230.3	259.4	256.9	271.0	227.4	226.0	229.5	223.2	232.6	248.2	215.6	207.7	195.4	200.2	253.5
				Woolong tea	A	309.9	261.6	298.0	411.6	276.2	301.8	221.9	325.7	241.4	245.8	283.1	275.8	309.0	274.7	287.9	270.1	235.6	259.6	201.1	278.5
					B	302.3	239.7	286.9	367.1	279.4	250.8	221.2	305.4	209.9	244.3	251.7	228.3	258.0	271.6	258.2	268.8	206.6	215.4	187.0	256.5
			GC-MS/MS	Green tea	A*	361.5	316.6	333.5	384.4	279.0	262.6	277.8	273.8	264.9	244.9	259.0	333.5	258.2	251.3	254.0	201.8	198.4	187.6	265.0	274.1
					B*	341.1	300.3	299.1	348.2	242.5	232.2	269.1	259.9	252.3	239.0	233.1	325.3	225.1	238.0	233.6	190.8	193.3	188.2	243.2	255.5
				Woolong tea	A	352.4	296.0	317.8	460.8	246.5	270.2	220.4	490.7	282.6	222.8	287.6	275.6	254.4	290.5	255.5	238.7	204.9	225.0	212.5	284.5
					B	308.8	279.5	281.7	418.5	258.5	222.5	239.2	467.8	243.8	240.3	265.0	229.0	249.8	246.8	225.9	219.4	193.5	193.4	209.8	262.8
		Envi-Carb +PSA	GC-MS	Green tea	A	436.9	386.7	334.4	299.1	314.1	330.8	330.7	252.7	284.2	326.7	267.0	323.6	269.8	228.1	215.4	258.9	263.6	265.7	201.3	294.2
					B	378.4	348.1	308.1	223.5	314.9	316.0	296.3	232.5	276.3	310.8	183.6	308.7	243.6	249.2	193.3	242.8	239.9	239.2	220.9	269.8
				Woolong tea	A	520.0	269.9	237.7	229.1	360.8	244.4	302.4	206.6	198.2	189.0	220.6	253.2	242.5	274.7	326.7	233.9	201.2	194.1	176.9	256.9
					B	428.5	213.5	256.8	226.8	299.0	238.8	252.5	187.1	134.0	187.6	204.6	225.9	212.7	271.6	292.6	198.7	180.3	193.2	156.6	229.5
			GC-MS/MS	Green tea	A	492.3	373.7	317.9	261.6	290.4	230.3	245.9	246.9	306.5	377.7	293.4	279.5	290.2	200.9	231.8	230.3	220.1	270.6	216.8	283.0
					B	552.5	328.6	307.0	217.3	293.2	306.3	260.7	237.9	293.8	356.0	180.9	268.2	254.2	223.4	217.5	232.5	196.9	251.8	243.1	274.8
				Woolong tea	A	267.5	282.2	473.9	249.5	388.6	245.3	283.8	212.7	198.2	187.2	209.3	232.8	209.0	209.3	289.8	238.7	181.8	217.5	179.9	250.4
					B	139.9	229.3	260.8	237.5	324.1	244.1	244.9	180.9	134.0	186.7	207.6	214.5	181.3	188.9	235.9	219.4	171.1	194.3	157.1	208.0
132	trans-chlordane	Cleanert TPT	GC-MS	Green tea	A	44.4	35.8	37.0	45.4	32.3	30.2	35.3	32.6	33.3	29.5	33.3	30.2	30.9	28.6	31.7	26.8	26.2	27.9	34.4	32.9
					B	42.7	34.8	33.5	42.1	30.4	28.2	32.8	30.2	33.9	28.3	30.2	28.3	29.1	27.9	29.6	26.5	25.9	27.7	30.5	31.2
				Woolong tea	A	40.0	39.6	38.2	41.5	33.5	35.6	33.5	41.4	36.3	33.8	36.3	35.7	35.9	32.2	32.6	33.9	32.9	32.6	28.9	35.5
					B	39.0	35.3	33.1	38.3	32.7	32.0	26.8	37.4	32.1	31.1	29.5	30.6	30.6	30.5	29.2	30.6	28.8	32.5	27.5	32.0
			GC-MS/MS	Green tea	A	46.7	35.4	38.4	43.9	33.3	28.2	35.8	31.0	34.0	29.5	37.5	27.9	29.9	29.5	30.7	25.2	24.8	26.0	34.0	32.7
					B	44.4	36.2	33.6	40.4	30.8	26.6	32.8	28.9	33.0	26.7	31.3	26.7	29.6	27.4	28.3	25.0	23.9	27.7	30.1	30.7
				Woolong tea	A	40.8	40.2	39.3	42.3	33.4	34.9	33.9	41.6	37.2	33.2	40.2	37.8	35.6	33.7	30.0	31.0	32.1	35.6	30.7	35.7
					B	39.6	36.7	33.2	39.3	33.2	31.2	28.3	36.3	35.1	31.9	32.3	31.7	29.3	30.0	26.5	28.8	28.3	32.6	24.5	32.0
		Envi-Carb +PSA	GC-MS	Green tea	A	46.1	42.1	36.6	36.2	38.6	39.0	30.2	31.3	31.5	29.8	29.7	32.8	31.5	34.2	27.6	34.2	30.1	29.2	30.8	33.8
					B	44.7	35.9	35.9	33.8	37.1	36.1	37.4	29.6	30.0	30.0	28.2	31.3	29.2	30.8	25.5	28.2	27.5	27.7	31.1	32.1
				Woolong tea	A	42.0	35.7	35.7	33.3	38.6	32.9	33.1	29.8	35.3	32.3	31.7	31.2	30.8	32.2	35.1	30.4	28.7	30.0	27.4	33.0
					B	36.1	31.9	33.2	31.0	32.1	30.5	31.5	26.3	29.2	27.8	28.9	28.3	28.9	30.3	29.8	27.1	25.9	26.5	26.5	29.6
			GC-MS/MS	Green tea	A	47.8	43.6	38.3	38.3	36.6	40.6	36.5	31.8	30.6	33.0	34.1	49.5	32.7	35.7	29.6	36.0	29.0	29.1	30.4	36.0
					B	46.9	36.3	36.9	35.4	37.1	38.6	34.5	30.2	30.1	33.4	30.5	38.4	30.3	33.0	26.7	29.1	27.1	30.6	30.7	33.5
				Woolong tea	A	44.6	34.6	42.2	33.2	40.5	33.6	34.2	28.8	35.3	31.8	31.8	30.4	28.8	29.4	33.5	31.0	26.0	30.2	27.2	33.0
					B	38.7	30.5	34.5	29.6	34.9	31.7	32.4	26.5	29.2	27.4	31.1	27.9	27.4	26.4	26.1	28.8	25.9	24.8	27.2	29.5

(Continued)

Annex 1: Content change for 201 pesticides in incurred tea Youden pair samples under 8 conditions with two SPE cartridge cleanup in three months (Nov.9, 2009-Feb.7, 2010) (cont.)

No	Pesticides	SPE	Method	Sample	Youden pair	Nov.9, 2009 (n=5)	Nov.14, 2009 (n=3)	Nov.19, 2009 (n=3)	Nov.24, 2009 (n=3)	Nov.29, 2009 (n=3)	Dec.4, 2009 (n=3)	Dec.9, 2009 (n=3)	Dec.14, 2009 (n=3)	Dec.19, 2009 (n=3)	Dec.24, 2009 (n=3)	Dec.29, 2009 (n=3)	Jan.3, 2010 (n=3)	Jan.8, 2010 (n=3)	Jan.13, 2010 (n=3)	Jan.18, 2010 (n=3)	Jan.23, 2010 (n=3)	Jan.28, 2010 (n=3)	Feb.2, 2010 (n=3)	Feb.7, 2010 (n=3)	AVE, μg/kg
133	trans-diallate	Cleanert TPT	GC-MS	Green tea	A	87.3	66.6	68.6	79.8	55.7	60.4	71.6	66.3	69.0	63.7	84.0	77.0	77.2	58.2	80.1	72.6	76.7	33.3	88.7	70.9
					B*	83.5	66.3	62.5	74.6	52.2	56.4	66.5	60.4	68.0	59.3	77.4	71.9	74.2	51.2	66.6	72.6	74.6	33.6	88.7	66.3
				Woolong tea	A	78.2	70.6	68.7	78.9	63.8	68.1	61.7	78.4	73.2	83.5	86.4	88.4	85.8	73.1	72.8	85.2	72.7	86.5	74.9	76.4
					B	75.6	62.8	61.2	70.0	63.2	61.8	53.6	70.1	64.8	73.8	69.5	76.3	74.1	73.1	62.9	51.7	64.4	80.7	68.5	68.3
			GC-MS/MS	Green tea	A	85.7	70.6	75.4	89.4	63.2	61.6	68.9	61.7	63.0	57.1	62.9	50.3	65.0	55.0	59.5	50.4	48.7	51.6	64.2	63.8
					B	82.2	68.6	67.4	83.3	62.0	57.9	68.2	59.0	64.5	54.6	71.9	47.0	55.9	54.3	58.8	51.7	48.1	52.5	57.6	60.8
				Woolong tea	A	69.6	72.9	72.9	85.2	60.9	65.2	62.4	77.7	71.7	66.3	60.3	68.9	66.6	64.3	61.5	63.3	55.9	62.4	55.1	67.1
					B	69.4	64.2	64.0	74.1	59.3	56.3	51.8	73.1	65.8	60.5	60.3	60.2	56.4	60.7	51.0	58.4	52.5	59.7	48.9	60.3
		Envi-Carb +PSA	GC-MS	Green tea	A	98.3	82.3	68.6	67.8	78.2	76.6	68.6	65.0	66.1	82.8	82.4	91.7	81.9	80.6	78.2	93.4	80.7	83.8	86.2	79.6
					B	88.6	69.6	68.1	62.9	74.4	68.8	83.0	61.3	62.5	79.7	77.8	81.9	80.0	75.4	74.6	78.2	73.6	75.3	84.4	74.7
				Woolong tea	A	80.5	62.7	65.2	60.5	69.5	56.7	66.6	53.7	65.5	79.2	78.1	74.0	74.4	79.3	90.3	78.1	73.2	80.9	72.3	71.6
					B	66.8	55.4	60.1	56.1	57.9	51.9	58.2	52.3	55.5	66.6	69.1	69.0	68.0	70.1	75.8	64.7	64.9	69.3	67.1	63.1
			GC-MS/MS	Green tea	A	92.2	78.8	71.4	69.8	77.2	76.8	75.8	62.1	64.7	70.6	68.8	77.9	82.8	72.6	63.1	71.7	64.1	60.4	66.1	71.9
					B	85.4	66.0	69.6	66.4	74.6	70.2	57.7	59.2	59.2	68.7	60.6	77.5	75.1	65.5	57.4	60.3	56.6	58.4	66.1	66.0
				Woolong tea	A	81.1	63.9	79.2	61.9	71.6	61.9	58.9	126.9	65.5	62.1	58.1	56.1	58.6	56.2	63.2	66.0	51.6	53.8	51.3	65.7
					B*	67.1	57.2	61.0	57.9	61.1	57.5	56.5	113.4	55.5	53.2	59.0	53.5	52.7	51.1	50.8	60.2	49.5	46.3	50.3	58.6
134	trifloxystrobin	Cleanert TPT	GC-MS	Green tea	A*	224.3	198.3	158.9	398.0	140.9	139.6	150.3	133.0	127.2	109.6	214.4	138.0	130.2	110.8	124.3	117.5	92.5	119.5	133.1	155.8
					B*	226.1	186.1	142.3	360.8	137.8	131.5	136.6	125.4	160.3	104.1	199.0	120.4	121.3	115.3	131.0	117.5	91.1	120.6	118.4	149.8
				Woolong tea	A	184.4	175.3	142.9	205.2	199.5	144.8	125.5	167.8	142.4	138.5	148.6	140.6	153.3	132.7	137.1	152.4	129.5	132.8	110.0	150.7
					B	176.9	161.7	146.3	181.4	78.5	126.6	131.2	160.7	115.4	127.9	121.7	123.1	142.2	132.8	122.3	137.7	111.3	126.5	107.1	138.5
			GC-MS/MS	Green tea	A	223.6	192.8	159.2	228.7	41.5	147.0	154.0	119.1	111.2	112.2	194.6	154.7	148.2	118.3	126.8	112.5	87.6	105.9	128.5	145.6
					B	215.3	178.5	143.7	215.2	40.8	136.1	139.8	115.4	137.6	116.7	177.4	143.9	125.5	121.6	123.9	108.9	93.4	117.3	105.7	139.8
				Woolong tea	A*	210.9	228.8	151.8	211.6	138.5	141.3	121.0	211.6	142.6	134.2	159.5	143.8	147.6	141.6	131.3	138.5	136.8	120.8	98.3	153.2
					B	188.0	220.6	145.7	180.0	133.2	121.8	124.6	182.4	118.2	131.2	135.5	129.3	130.0	133.2	116.1	130.7	116.0	108.6	93.0	138.9
		Envi-Carb +PSA	GC-MS	Green tea	A	232.5	177.0	181.3	179.2	168.0	162.6	134.9	127.6	179.3	183.6	191.7	238.4	138.9	152.8	86.2	119.7	200.7	124.7	163.3	165.4
					B	235.3	197.9	153.0	155.9	157.1	168.6	174.5	118.9	164.6	152.6	138.0	188.5	134.1	138.7	90.2	92.1	172.1	151.0	166.9	155.3
				Woolong tea	A	211.6	118.3	145.1	132.2	195.0	130.4	151.3	128.9	144.4	127.8	125.9	125.4	126.8	132.7	129.7	124.6	110.0	110.3	108.5	135.7
					B	190.9	114.3	143.3	134.1	156.0	126.7	128.0	108.2	116.3	117.2	105.4	132.5	110.0	132.8	123.6	94.3	103.7	109.3	103.2	123.7
			GC-MS/MS	Green tea	A*	228.0	187.8	174.3	164.2	167.6	119.7	263.6	118.9	189.3	151.7	183.0	0.0	153.4	166.0	91.2	127.2	177.9	127.3	167.0	155.7
					B	205.1	217.0	152.8	161.1	149.2	169.1	93.3	103.7	164.3	142.9	143.4	0.0	170.0	119.7	100.1	94.4	158.0	156.5	183.0	141.2
				Woolong tea	A	186.7	126.6	133.4	132.9	185.9	124.9	122.6	119.3	144.4	122.8	124.9	119.5	112.1	113.0	122.2	138.5	95.6	108.9	94.5	127.8
					B	185.7	121.6	141.6	136.7	157.8	126.6	126.0	102.1	116.3	118.9	125.9	118.7	99.9	109.7	105.4	130.7	98.3	103.7	89.1	121.8
135	zoxamide	Cleanert TPT	GC-MS	Green tea	A*	93.5	84.1	95.7	126.3	81.4	70.7	90.0	82.6	82.3	77.6	78.9	68.7	71.5	77.9	75.7	65.3	73.0	60.8	80.4	80.9
					B*	93.5	81.4	88.3	123.6	73.4	65.6	84.8	73.9	78.3	73.4	70.1	66.5	68.4	76.5	66.9	64.1	68.8	57.4	73.1	76.2
				Woolong tea	A	91.2	82.7	86.0	90.6	76.7	80.4	68.9	88.2	81.6	75.5	80.6	79.8	79.4	73.4	70.1	70.1	66.7	69.8	66.7	77.8
					B	88.1	74.3	81.0	86.3	83.2	78.0	69.9	86.2	77.1	68.8	70.1	73.7	72.5	69.4	63.9	61.3	59.3	72.1	59.6	73.4
			GC-MS/MS	Green tea	A	94.0	83.5	92.4	98.8	81.5	73.9	83.6	84.1	80.1	76.4	80.0	84.3	73.6	75.8	73.2	64.8	64.8	63.9	74.4	79.1
					B	93.7	81.6	84.8	92.9	73.5	68.1	86.1	74.1	75.9	74.7	71.9	77.9	70.6	66.8	64.2	63.7	63.1	62.1	67.7	74.4
				Woolong tea	A	85.6	84.1	86.7	94.4	78.3	84.3	67.6	88.9	85.2	75.1	82.2	76.7	70.6	75.8	66.9	62.3	76.7	75.3	67.9	78.5
					B	85.3	76.3	77.8	86.0	85.2	79.1	68.8	85.4	81.9	71.9	71.2	76.1	66.1	70.6	59.7	54.2	67.5	76.3	58.4	73.6
		Envi-Carb +PSA	GC-MS	Green tea	A	109.4	91.4	90.8	97.6	84.0	87.3	59.4	78.2	83.0	71.5	70.8	79.7	77.8	78.6	62.4	74.6	74.9	76.3	59.3	79.3
					B	111.1	82.9	89.8	87.1	87.5	80.4	66.6	67.5	75.5	69.6	70.8	78.8	67.6	73.7	58.9	66.2	68.1	63.1	75.4	75.8
				Woolong tea	A	105.4	83.3	81.8	82.1	80.1	71.3	67.8	61.7	67.1	72.4	67.1	62.0	67.1	73.4	77.5	60.4	61.2	66.0	60.2	72.0
					B	92.6	79.9	96.0	78.9	71.9	62.5	66.9	62.1	64.0	56.0	64.9	56.0	63.8	69.4	60.5	58.3	57.8	60.0	62.9	67.6
			GC-MS/MS	Green tea	A	105.9	86.6	88.8	96.0	86.7	80.7	79.7	74.9	87.5	74.9	82.9	78.5	79.5	75.8	63.6	78.0	64.5	75.9	81.0	81.1
					B	99.9	80.8	88.9	87.0	87.8	82.1	68.1	69.0	76.2	73.2	75.4	77.4	69.7	71.2	61.0	69.5	61.2	69.5	83.5	76.4
				Woolong tea	A	98.9	80.7	82.9	79.5	82.1	70.5	75.1	60.2	67.1	71.2	63.6	59.0	61.5	56.2	70.9	62.3	56.6	62.6	62.7	69.7
					B	94.3	77.7	89.8	74.1	78.0	62.4	70.7	60.4	64.0	56.8	68.9	54.8	57.8	55.1	50.6	54.2	58.5	56.5	68.3	65.9

No.	Compound	Cartridge	Method	Tea	Rep																					
136	2,4'-DDE	Cleanert TPT	GC-MS	Green tea	A	45.1	38.9	41.1	40.3	36.0	36.3	33.6	36.9	35.8	36.0	33.6	36.3	35.2	36.1	33.0	34.9	28.1	29.1	28.7	28.7	34.9	35.5
					B	43.2	35.4	35.5	36.3	31.9	34.1	30.9	32.8	33.5	31.9	30.9	31.9	33.4	35.0	30.9	30.0	27.5	27.9	27.3	27.3	31.1	32.6
				Woolong tea	A	42.0	37.2	37.9	42.3	36.1	38.9	38.5	33.7	37.0	36.1	38.5	38.0	42.9	38.4	35.9	35.7	35.4	34.9	33.5	33.5	30.7	37.3
					B	39.2	30.9	30.9	36.7	33.8	33.0	33.1	26.8	32.1	33.8	33.1	32.3	34.7	30.8	31.4	30.5	30.4	28.9	31.8	31.8	27.1	31.9
			GC-MS/MS	Green tea	A	45.5	36.2	43.8	37.8	37.6	33.6	33.0	37.9	35.1	37.6	33.0	33.3	34.8	32.5	34.0	35.8	26.6	28.8	28.2	28.2	35.7	35.1
					B	43.3	36.2	38.8	37.8	33.7	33.3	31.3	33.9	32.1	33.7	31.3	33.3	30.9	30.9	31.8	32.7	24.6	28.1	28.1	28.1	30.0	32.7
				Woolong tea	A	39.9	35.0	37.7	43.4	33.4	39.0	40.6	37.4	37.0	33.4	40.6	39.0	42.2	38.2	33.9	37.2	32.8	36.6	35.3	35.3	29.4	37.3
					B	37.5	30.3	30.5	36.8	40.7	33.8	34.6	28.5	30.6	40.7	34.6	33.8	32.9	29.3	29.0	31.2	28.1	28.7	29.6	29.6	26.6	31.4
		Envi-Carb +PSA	GC-MS	Green tea	A	56.8	42.8	41.7	42.8	40.2	35.7	43.4	33.5	50.2	40.2	43.4	35.7	32.2	35.2	42.3	39.2	42.7	36.0	37.9	37.9	35.6	39.8
					B	57.0	34.2	41.4	40.1	38.8	34.2	42.8	37.3	46.3	38.8	42.8	34.2	33.3	33.1	36.1	35.2	36.0	32.3	31.3	31.3	35.4	37.6
				Woolong tea	A	45.4	44.8	41.8	38.4	30.9	37.5	31.2	36.7	34.9	30.9	31.2	37.5	34.1	35.1	39.1	40.5	33.5	30.0	39.6	39.6	30.7	37.2
					B	36.9	35.2	36.9	33.2	42.1	29.9	27.9	32.4	29.3	42.1	27.9	29.9	28.8	32.0	31.6	31.6	29.5	25.9	31.6	31.6	27.5	31.2
			GC-MS/MS	Green tea	A	47.7	39.8	42.0	42.1	40.7	39.1	43.1	35.6	41.6	40.7	43.1	39.1	33.8	36.3	41.7	38.3	44.6	37.3	37.4	37.4	37.9	39.5
					B	45.4	30.8	42.4	38.0	38.4	38.5	42.5	42.4	36.1	38.4	42.5	38.5	32.3	33.9	41.2	35.3	40.1	35.7	32.0	32.0	38.5	37.6
				Woolong tea	A	43.2	39.2	48.1	40.3	31.0	29.9	32.6	35.0	35.6	31.0	32.6	29.9	34.4	32.7	36.8	37.8	27.1	24.7	38.0	38.0	29.4	35.8
					B	35.7	34.1	36.9	34.6	26.3	27.4	27.9	27.9	29.8	23.0	27.9	27.4	29.4	29.3	31.5	27.4	23.8	24.0	30.3	30.3	26.6	30.2
137	ametryn	Cleanert TPT	GC-MS	Green tea	A*	143.3	107.3	144.1	126.8	102.1	121.9	83.4	62.0	69.1	85.0	81.0	66.9	88.2	84.6	93.4	96.0	77.6	74.8	70.3	88.9	87.0	92.7
					B*	141.4	108.7	130.9	115.2	94.4	128.2	73.6	85.0	81.3	81.0	84.6	99.6	81.0	91.1	96.4	88.0	77.0	74.8	68.9	86.9	76.4	90.4
				Woolong tea	A*	129.0	104.2	112.8	136.6	94.4	106.7	60.2	45.6	110.2	78.9	101.5	84.9	101.7	84.6	90.0	97.1	94.5	93.6	86.9	79.8	69.9	99.4
					B*	116.8	95.1	103.3	128.4	103.0	77.9	79.3	40.2	95.3	80.5	101.5	66.9	79.2	80.9	78.9	80.9	78.0	73.0	79.6	69.9	82.7	86.5
			GC-MS/MS	Green tea	A	146.2	112.8	134.8	123.1	100.9	93.3	98.8	104.4	107.9	86.4	80.5	99.6	86.9	86.4	93.8	74.0	79.6	74.0	70.3	82.7	100.4	94.1
					B	137.6	113.3	116.3	108.5	93.9	93.4	90.6	92.0	98.5	77.5	86.4	93.3	83.8	77.5	88.8	76.4	79.6	74.0	67.5	72.2	72.2	94.1
				Woolong tea	A	124.4	102.8	107.0	131.0	123.5	111.5	105.8	103.0	108.2	110.3	77.5	93.4	119.1	110.3	88.8	93.4	78.3	76.4	88.3	73.9	73.9	104.6
					B	114.8	90.9	89.4	109.5	115.4	94.3	89.4	78.2	86.9	85.8	110.3	117.0	119.1	85.8	94.1	93.4	89.0	74.2	89.8	65.2	65.2	87.4
		Envi-Carb +PSA	GC-MS	Green tea	A*	158.4	113.5	122.1	123.8	118.3	63.7	143.6	123.3	135.1	90.7	85.8	82.9	90.5	85.8	87.8	91.1	77.5	74.2	88.8	83.4	83.4	110.9
					B*	148.8	113.1	118.6	114.3	89.0	24.9	98.5	158.6	100.5	81.5	90.7	82.1	169.4	90.7	81.5	80.9	102.0	91.1	77.2	84.2	84.2	97.6
				Woolong tea	A*	223.1	199.1	125.8	113.4	119.7	113.9	66.3	43.6	137.9	105.5	96.6	80.6	115.0	96.6	103.3	79.8	89.4	80.9	102.4	76.7	76.7	111.0
					B*	182.1	104.7	110.4	101.5	115.8	88.0	49.4	81.2	123.3	86.5	81.3	121.0	90.7	81.3	103.3	79.8	72.2	66.6	79.9	67.0	67.0	91.2
			GC-MS/MS	Green tea	A	140.5	122.4	124.8	123.4	112.5	102.5	118.5	101.1	116.4	104.3	91.2	87.9	116.6	91.2	78.3	66.6	106.3	92.6	85.2	86.3	86.3	106.1
					B	129.6	98.9	117.2	112.3	89.7	96.8	109.9	97.4	104.6	85.6	84.4	89.2	109.6	84.4	87.0	80.4	87.1	80.7	78.6	86.9	86.9	97.0
				Woolong tea	A	129.4	125.2	143.4	121.2	112.5	113.9	89.3	100.9	110.4	102.2	95.1	97.0	74.8	95.1	100.0	68.3	80.7	60.3	96.3	73.9	73.9	101.9
					B	113.8	107.7	112.4	101.8	89.7	88.0	72.4	89.1	88.3	84.0	85.1	80.8	62.5	85.1	73.0	60.3	80.7	65.2	77.3	65.2	65.2	84.6
138	bifenthrin	Cleanert TPT	GC-MS	Green tea	A	44.5	43.1	35.6	36.7	35.6	34.5	35.7	37.3	37.7	30.9	34.0	33.5	35.7	34.0	34.7	26.9	28.0	26.9	28.3	33.0	33.0	34.9
					B	43.3	39.4	31.2	33.5	31.3	33.4	33.6	33.3	35.3	29.1	32.1	32.5	34.4	32.1	31.3	26.4	27.5	26.4	27.4	27.0	27.0	32.3
				Woolong tea	A	49.2	43.8	45.3	54.0	49.4	54.3	49.7	41.5	54.8	45.0	54.3	53.2	53.7	54.3	50.6	43.4	42.3	43.4	46.8	42.3	42.3	48.6
					B	51.7	42.7	44.4	48.6	51.6	50.6	48.4	39.0	50.8	47.0	50.6	45.4	49.0	53.5	45.0	40.9	40.5	43.4	48.3	41.8	41.8	46.7
			GC-MS/MS	Green tea	A	46.9	35.0	41.9	38.1	32.7	30.7	31.8	34.2	35.1	34.0	31.4	30.7	32.5	26.8	34.0	24.8	25.6	24.8	25.2	31.3	31.3	32.6
					B	44.1	34.7	36.9	33.7	29.9	30.8	29.8	30.8	31.7	32.9	29.5	29.8	29.6	26.0	32.9	25.1	24.7	25.1	24.9	25.1	25.1	30.4
				Woolong tea	A	49.6	40.1	43.5	55.2	45.4	42.4	45.3	42.4	54.7	43.3	50.3	54.5	51.2	58.6	43.3	42.9	40.2	42.9	46.5	43.9	43.9	47.8
					B	52.3	41.0	42.5	49.2	48.9	39.6	45.9	39.6	48.8	44.8	48.2	50.7	45.5	58.3	44.8	40.4	38.7	40.4	47.3	43.9	43.9	45.9
		Envi-Carb +PSA	GC-MS	Green tea	A*	51.2	42.2	37.2	38.5	39.9	54.5	40.2	48.0	39.7	36.5	31.6	35.7	36.3	32.8	36.5	35.1	37.7	35.1	34.4	29.8	29.8	36.6
					B*	47.5	35.5	38.2	35.2	52.2	50.7	37.6	43.4	35.1	32.8	29.7	32.7	35.9	30.6	32.8	31.2	30.6	31.2	30.8	29.8	29.8	34.0
				Woolong tea	A*	54.0	53.1	56.3	50.3	48.2	49.9	73.1	34.8	48.4	47.5	53.8	49.9	44.4	50.2	47.5	43.6	43.7	43.6	57.4	44.4	44.4	51.4
					B*	45.7	43.5	57.1	48.8	38.3	46.8	71.9	33.0	45.3	43.3	50.9	46.8	44.2	47.4	43.3	40.8	39.0	40.8	50.6	43.2	43.2	47.7
			GC-MS/MS	Green tea	A*	43.6	39.2	38.0	38.3	37.3	34.3	38.7	34.8	36.5	24.6	29.4	34.3	57.1	33.4	29.4	31.9	35.6	31.9	28.9	29.8	32.0	35.9
					B*	40.3	32.6	34.8	37.3	49.2	31.3	35.4	33.0	34.2	42.4	27.6	30.9	47.3	28.9	24.6	27.8	28.6	27.8	26.4	31.2	31.2	32.5
				Woolong tea	A*	48.7	48.1	61.5	50.9	49.2	49.9	70.3	43.8	47.4	42.4	44.4	50.9	33.2	43.9	42.4	35.9	37.8	35.9	54.0	43.9	43.9	47.7
					B*	42.3	42.6	56.5	48.4	45.3	46.8	67.9	41.7	42.7	40.3	45.6	46.0	33.5	44.6	40.3	36.5	33.5	36.5	48.9	43.9	44.7	44.7

(Continued)

Annex 1: Content change for 201 pesticides in incurred tea Youden pair samples under 8 conditions with two SPE cartridge cleanup in three months (Nov.9, 2009–Feb.7, 2010) (cont.)

No	Pesticides	SPE	Method	Sample	Youden pair	Nov.9, 2009 (n=5)	Nov.14, 2009 (n=3)	Nov.19, 2009 (n=3)	Nov.24, 2009 (n=3)	Nov.29, 2009 (n=3)	Dec.4, 2009 (n=3)	Dec.9, 2009 (n=3)	Dec.14, 2009 (n=3)	Dec.14, 2009 (n=3)	Dec.19, 2009 (n=3)	Dec.24, 2009 (n=3)	Dec.14, 2009 (n=3)	Jan.3, 2010 (n=3)	Jan.8, 2010 (n=3)	Jan.13, 2010 (n=3)	Jan.18, 2010 (n=3)	Jan.23, 2010 (n=3)	Jan.28, 2010 (n=3)	Feb.2, 2010 (n=3)	Feb.7, 2010 (n=3)	AVE µg/kg.
139	bitertanol	Cleanert TPT	GC-MS	Green tea	A	161.4	132.1	149.6	123.6	117.9	93.8	112.7	119.8	118.1	98.4	83.2	118.1	108.1	94.4	138.2	99.4	105.5	96.5	88.9	101.0	112.8
					B	159.8	129.9	144.6	116.8	130.3	95.2	107.2	114.8	101.3	100.8	79.5	101.3	97.4	93.0	136.9	96.1	99.7	91.1	86.5	88.2	107.3
				Woolong tea	A*	130.6	93.9	11.9	127.0	103.8	120.6	93.4	97.2	86.5	119.8	160.8	86.5	103.4	99.8	97.6	110.1	107.4	100.9	101.4	86.2	108.0
					B*	113.5	101.2	118.6	119.9	98.5	103.1	91.7	99.0	68.9	85.5	132.5	68.9	86.7	92.2	87.2	93.1	84.9	82.0	92.5	73.2	96.0
			GC-MS/MS	Green tea	A*	148.0	122.5	153.3	145.9	116.8	107.9	112.9	117.6	111.2	105.4	107.2	111.2	116.1	86.2	146.7	104.1	90.2	82.1	83.0	98.2	113.3
					B*	143.5	122.8	135.3	117.5	99.9	102.1	105.0	107.0	98.7	104.0	101.3	98.7	102.9	79.7	154.0	94.2	86.6	79.8	79.1	88.0	105.3
				Woolong tea	A	168.6	108.1	110.8	139.8	106.0	116.2	98.2	108.3	117.4	122.0	130.5	117.4	119.7	112.8	94.8	97.6	95.4	95.9	91.9	80.3	111.3
					B	136.2	96.2	97.6	121.8	103.1	96.2	92.0	103.9	87.8	94.0	114.5	87.8	96.2	101.6	83.4	84.5	78.6	76.7	81.6	71.8	95.7
		Envi-Carb+PSA	GC-MS	Green tea	A*	150.0	133.9	138.2	124.7	133.3	104.2	86.3	138.2	98.9	86.2	104.1	98.9	258.9	103.3	113.9	105.1	120.4	103.3	106.8	96.0	121.4
					B*	161.1	153.4	128.4	111.4	129.1	92.8	105.5	121.3	95.1	81.5	104.1	95.1	164.4	96.3	108.1	93.2	103.2	90.4	98.6	98.1	112.1
				Woolong tea	A*	149.9	124.1	122.7	119.9	120.3	108.1	116.3	75.0	115.9	109.5	137.1	115.9	85.5	101.0	98.6	113.0	92.7	85.4	111.6	87.6	109.2
					B	137.4	106.8	120.0	115.0	86.4	91.0	90.6	59.5	90.4	91.7	100.4	90.4	82.8	85.7	93.0	86.5	76.7	73.7	90.9	76.1	92.1
			GC-MS/MS	Green tea	A*	156.1	126.7	129.7	134.6	135.2	125.6	94.4	135.8	98.7	114.5	103.3	98.7	206.4	114.2	93.0	96.0	116.2	106.6	95.9	104.5	119.9
					B*	164.1	118.9	125.9	115.5	127.5	116.1	121.2	117.4	90.1	99.0	103.3	90.1	162.5	96.2	93.0	87.4	99.8	92.4	82.7	102.6	110.8
				Woolong tea	A	127.5	125.3	124.4	122.8	117.2	101.9	109.0	88.8	95.9	109.5	99.9	95.9	78.6	94.4	94.1	110.5	81.3	72.9	102.0	80.3	102.3
					B	123.2	109.8	123.3	110.4	93.0	85.2	91.0	73.1	89.8	91.7	106.7	89.8	68.2	103.6	82.9	76.5	65.3	65.7	83.8	71.0	88.9
140	boscalid	Cleanert TPT	GC-MS	Green tea	A	205.6	174.4	90.5	160.4	125.7	206.0	165.1	214.3	162.7	141.9	175.9	162.7	176.2	131.6	213.6	134.1	177.2	117.4	124.0	143.0	165.2
					B*	211.2	158.3	180.1	158.2	132.3	185.9	152.4	209.6	152.7	177.6	173.6	152.7	146.6	125.3	235.4	156.9	175.1	115.0	125.3	124.3	162.9
				Woolong tea	A	205.6	143.8	165.9	222.2	207.2	210.4	122.6	159.9	182.9	147.6	169.9	182.9	149.4	138.4	114.1	187.9	180.0	146.5	150.8	130.7	165.0
					B	175.5	135.1	169.5	170.6	177.5	158.6	138.2	155.5	133.1	127.9	180.4	133.1	140.0	145.7	118.2	154.4	156.3	121.7	120.3	120.8	146.0
			GC-MS/MS	Green tea	A*	240.9	235.0	184.7	178.0	128.1	194.2	159.7	192.0	166.0	132.1	192.0	166.0	212.1	108.6	200.3	147.0	153.8	122.3	120.2	147.3	169.2
					B*	241.2	213.4	179.7	171.0	135.8	176.6	159.9	180.7	156.9	175.4	194.3	156.9	160.2	104.9	240.2	140.6	150.0	128.8	123.7	132.6	166.4
				Woolong tea	A	338.8	152.6	147.2	204.9	169.6	193.8	125.8	157.5	168.4	151.7	177.7	168.4	141.7	145.4	116.4	171.1	160.6	135.6	163.9	153.4	167.2
					B	236.1	134.9	138.8	167.3	155.2	145.2	133.2	150.5	121.9	114.8	184.6	121.9	111.5	149.9	113.4	144.7	133.5	112.1	122.9	123.1	141.8
		Envi-Carb+PSA	GC-MS	Green tea	A*	195.3	199.3	182.0	188.1	171.8	166.6	171.6	171.6	187.8	178.6	180.7	187.8	442.8	126.5	151.3	153.8	204.3	190.4	129.6	154.4	186.7
					B*	217.5	263.5	160.7	150.8	162.0	185.5	141.4	149.1	127.6	163.4	155.6	127.6	332.3	127.9	148.6	143.4	156.9	158.6	156.0	163.8	171.8
				Woolong tea	A	216.8	113.8	171.4	146.3	257.2	223.1	118.4	154.7	181.9	149.7	139.2	181.9	160.5	152.3	112.1	170.3	142.0	117.3	139.3	152.1	158.9
					B	204.8	98.6	171.7	153.1	204.5	224.0	134.9	119.7	159.9	113.0	126.4	159.9	158.2	117.8	132.5	171.7	86.6	104.5	135.0	126.8	144.4
			GC-MS/MS	Green tea	A*	226.1	209.7	178.4	195.0	159.2	194.0	198.3	165.1	167.4	192.2	176.1	167.4	183.2	127.2	145.4	129.1	156.7	165.2	120.5	162.8	171.1
					B*	258.6	288.1	159.8	159.5	159.0	213.6	143.7	133.7	126.1	164.8	150.4	126.1	177.4	132.1	130.5	128.4	119.4	135.0	139.4	177.6	163.0
				Woolong tea	A	221.9	122.0	179.6	151.2	260.3	239.4	141.3	145.5	157.5	149.7	130.7	157.5	154.9	156.9	119.4	162.9	122.7	90.6	139.4	153.4	157.3
					B	199.8	108.4	170.1	146.0	200.1	232.2	121.3	116.5	149.1	113.0	119.0	149.1	156.6	143.8	125.0	138.2	80.5	84.6	120.4	123.1	139.4
141	butafenacil	Cleanert TPT	GC-MS	Green tea	A*	59.4	71.9	83.7	41.2	33.5	54.9	38.6	52.2	40.7	37.4	44.1	40.7	46.3	34.5	53.9	35.1	45.5	30.2	32.2	35.6	45.8
					B*	59.0	64.4	87.7	40.7	35.0	49.6	36.1	48.6	39.8	45.8	42.0	39.8	38.0	31.8	62.1	40.6	45.0	30.7	33.4	29.9	45.3
				Woolong tea	A	48.3	38.4	40.0	53.7	52.0	45.5	30.4	39.6	40.5	39.9	44.9	40.5	38.3	37.8	29.5	43.5	44.6	38.2	40.1	32.2	40.9
					B	42.0	35.3	42.2	41.8	43.0	34.2	33.8	38.5	30.4	32.3	44.1	30.4	30.3	35.6	30.3	36.5	38.0	31.9	34.4	28.8	36.0
			GC-MS/MS	Green tea	A	72.2	63.4	49.8	38.6	31.2	53.9	39.6	47.5	40.1	31.2	54.6	40.1	45.2	25.9	54.3	37.0	40.0	32.0	32.0	37.5	43.5
					B	68.5	60.0	44.8	42.6	35.0	48.9	37.2	45.9	39.4	45.0	53.7	39.4	35.2	26.3	70.9	36.8	37.8	32.1	32.9	33.8	43.5
				Woolong tea	A	62.9	40.2	38.5	51.1	43.7	46.5	29.5	39.4	44.4	37.7	42.4	44.4	36.6	38.1	32.6	41.2	39.3	35.8	41.8	41.8	41.2
					B	47.7	36.5	35.8	41.4	38.3	35.4	33.7	35.4	32.1	29.9	42.1	32.1	30.4	36.4	29.8	35.4	33.7	29.6	42.5	35.7	35.4
		Envi-Carb+PSA	GC-MS	Green tea	A*	43.1	55.5	50.9	50.9	40.0	28.9	63.4	39.2	46.3	50.4	50.8	46.3	253.8	29.7	41.0	35.8	44.0	53.5	29.0	37.6	54.9
					B*	50.8	80.0	40.7	42.4	36.1	34.6	40.1	38.5	31.1	44.8	42.0	31.1	165.8	30.6	38.6	36.0	32.7	45.2	41.6	40.1	48.0
				Woolong tea	A	52.2	28.9	41.3	35.6	63.1	60.1	31.0	37.7	46.7	41.4	34.7	46.7	37.7	38.4	26.1	38.3	36.4	26.1	35.7	38.0	39.9
					B	46.8	24.9	42.1	35.5	50.6	58.6	77.9	30.6	41.8	31.9	29.5	41.8	38.4	30.2	33.2	37.9	26.6	24.7	34.3	32.6	35.9
			GC-MS/MS	Green tea	A*	40.9	60.0	43.3	51.7	40.8	47.6	34.4	39.1	37.9	61.2	48.0	37.9	51.2	28.9	44.8	29.6	32.7	52.2	25.5	41.0	45.2
					B	40.7	87.1	39.0	44.6	38.4	54.1	36.2	31.1	30.2	49.6	39.0	30.2	44.1	35.2	30.9	35.3	22.3	40.4	36.5	44.6	40.9
				Woolong tea	A	47.8	31.9	39.7	37.6	58.3	65.7	34.4	36.0	40.1	41.4	31.9	40.1	38.2	37.1	28.3	38.9	30.7	21.0	33.9	41.8	38.8
					B	47.6	28.0	41.8	36.8	48.0	62.5	30.9	31.0	38.2	31.9	27.1	38.2	36.6	28.2	31.1	33.1	19.8	21.8	31.2	35.7	34.8

No.	Compound	Cartridge	Method	Tea	Rep																				
142	carbaryl	Cleanert TPT	GC-MS	Green tea	A*	111.6	106.9	95.8	92.7	90.5	105.3	115.3	120.6	103.6	104.2	110.0	105.7	125.9	115.2	130.5	145.7	126.5	80.8	124.8	120.0
				Woolong tea	B	101.8	85.5	92.9	84.7	83.1	100.3	124.6	109.3	100.1	87.0	99.6	91.7	116.5	92.5	113.6	152.5	117.7	66.7	108.1	106.9
			GC-MS/MS	Green tea	A	119.4	103.4	102.0	108.3	107.3	114.3	134.5	121.8	139.7	115.7	104.1	113.1	122.3	99.3	123.9	127.4	123.9	107.3	175.8	125.6
				Woolong tea	B	98.9	87.9	95.4	91.9	93.7	93.2	107.8	94.8	118.0	100.6	87.5	104.5	101.7	114.5	107.6	110.0	83.8	77.1	123.4	117.3
			GC-MS	Green tea	A	108.0	112.9	98.0	94.0	80.2	103.9	113.1	104.3	74.7	80.9	138.8	95.6	115.0	83.5	136.9	156.9	151.1	104.4	70.1	88.8
				Woolong tea	B	101.1	94.7	107.8	102.8	104.1	107.4	122.7	88.7	70.7	80.9	148.7	84.4	100.7	101.2	115.4	113.5	128.7	99.0	68.4	74.7
			GC-MS/MS	Green tea	A	114.7	102.9	96.8	92.7	88.4	106.9	132.1	129.5	136.6	121.5	98.8	120.6	116.4	89.2	107.6	110.3	77.8	69.7	140.0	103.5
				Woolong tea	B	94.9	97.4	114.8	113.9	129.6	88.0	101.2	90.8	111.5	102.9	88.7	100.7	94.1	95.7	89.4	139.8	128.9	123.6	109.2	103.7
		Envi-Carb+PSA	GC-MS	Green tea	A*	127.6	104.5	114.8	113.9	129.6	113.7	121.6	113.5	181.5	95.8	117.2	114.5	137.7	122.6	168.0	139.8	128.9	123.6	140.1	170.4
				Woolong tea	B*	121.1	111.4	110.0	100.7	114.4	96.0	106.6	103.3	156.4	96.5	114.7	109.1	131.6	122.6	170.8	135.1	118.1	117.2	126.3	160.9
			GC-MS/MS	Green tea	A	119.9	105.3	134.4	93.8	109.1	123.3	103.6	113.8	108.6	135.2	103.9	126.8	103.6	123.6	119.5	131.7	114.8	130.5	142.8	153.3
				Woolong tea	B	103.1	87.8	105.1	85.1	83.7	97.5	100.3	99.2	93.8	122.3	89.0	96.9	87.7	108.5	114.1	114.1	114.4	123.6	110.3	125.3
			GC-MS	Green tea	A	124.5	119.9	104.1	122.4	128.6	115.4	130.2	106.2	166.1	83.4	122.1	112.0	136.6	98.6	124.8	119.8	135.7	129.6	150.9	158.8
				Woolong tea	B	117.5	127.4	104.5	106.0	113.4	99.8	103.3	90.9	156.0	87.3	122.5	108.1	116.3	116.7	117.6	116.7	127.3	121.7	142.6	155.0
			GC-MS/MS	Green tea	A	111.1	102.9	109.0	76.4	93.2	117.6	96.2	125.8	102.6	115.9	85.4	126.8	101.3	111.6	114.6	128.0	122.8	131.6	120.0	129.3
				Woolong tea	B	97.9	95.2	93.7	73.5	69.3	91.7	93.3	127.6	90.9	111.9	72.9	96.9	78.8	99.1	105.5	110.4	106.7	122.3	100.9	119.7
143	chlorobenzilate	Cleanert TPT	GC-MS	Green tea	A*	36.2	34.1	26.2	27.0	38.8	34.0	40.5	31.7	37.7	36.6	33.6	31.7	45.0	36.4	40.1	30.4	39.3	38.0	36.8	49.2
				Woolong tea	B*	35.7	29.1	27.2	29.1	38.2	36.5	43.0	29.3	32.8	34.1	33.5	39.8	44.5	34.6	37.7	30.1	37.7	34.0	39.2	48.4
			GC-MS/MS	Green tea	A	38.6	31.1	34.3	37.1	40.9	40.8	29.5	35.5	36.7	35.3	49.2	37.9	38.1	35.7	44.2	40.0	47.2	39.3	35.7	44.1
				Woolong tea	B	34.2	30.4	31.5	30.0	36.1	34.3	30.0	32.5	32.4	27.6	45.0	30.5	38.1	30.3	36.6	35.1	39.6	37.8	32.5	39.9
			GC-MS	Green tea	A	37.8	33.9	28.0	27.3	32.2	35.4	36.3	28.1	47.0	37.1	39.0	31.2	42.1	37.4	40.7	31.6	46.3	42.5	46.2	56.1
				Woolong tea	B	35.9	30.0	28.8	28.9	30.7	32.9	38.5	26.7	38.0	34.6	38.7	38.4	40.0	34.0	37.3	31.5	38.3	39.4	42.7	53.4
			GC-MS/MS	Green tea	A	39.4	34.2	37.9	36.4	37.2	37.1	33.2	39.5	38.7	42.0	43.7	36.5	38.3	34.0	42.8	37.7	47.3	36.2	35.4	61.0
				Woolong tea	B	33.4	29.9	32.0	29.1	31.9	31.7	30.5	34.2	29.7	29.4	40.5	29.2	35.6	30.8	34.0	34.2	38.9	32.3	32.1	49.1
		Envi-Carb+PSA	GC-MS	Green tea	A*	44.9	32.5	23.5	41.3	53.6	55.1	29.8	29.2	104.3	38.3	37.5	30.2	37.4	68.0	44.1	41.0	41.8	41.4	50.6	52.5
				Woolong tea	B*	40.6	36.1	37.9	38.6	43.1	47.8	35.1	25.6	75.2	25.5	35.7	31.8	32.0	46.1	46.5	37.5	31.8	37.8	53.6	53.2
			GC-MS/MS	Green tea	A	38.4	33.2	36.1	28.7	35.3	40.9	33.9	35.1	32.7	42.9	37.4	38.7	36.5	33.2	51.4	51.4	37.3	43.5	35.3	46.5
				Woolong tea	B	33.5	28.6	32.8	25.6	24.3	40.6	32.4	29.0	34.5	36.3	33.4	30.1	27.8	29.9	46.9	36.9	35.6	40.4	29.8	41.9
			GC-MS	Green tea	A	34.2	34.4	25.5	40.8	37.9	34.1	35.8	32.2	30.3	38.9	37.5	54.9	43.4	44.2	38.0	35.7	37.2	41.2	71.0	102.2
				Woolong tea	B	41.5	39.5	32.3	35.2	30.8	30.8	35.4	27.6	28.3	26.6	34.7	33.0	35.2	35.0	36.3	38.0	35.7	37.2	96.6	114.4
			GC-MS/MS	Green tea	A	38.1	34.2	33.7	24.2	30.0	38.6	34.2	35.2	35.5	37.5	34.0	38.5	35.2	33.0	45.6	38.0	39.6	49.0	35.9	60.5
				Woolong tea	B	33.2	29.9	29.7	23.2	21.8	31.9	36.9	34.8	34.2	34.8	30.7	30.1	26.2	30.7	40.6	36.3	36.0	40.2	31.7	51.1
144	chlorthal-dimethyl	Cleanert TPT	GC-MS	Green tea	A*	107.2	94.2	80.0	81.5	92.2	85.0	101.1	101.9	103.5	115.0	99.4	102.4	102.1	104.0	107.6	95.6	116.5	105.6	123.5	128.9
				Woolong tea	B*	101.7	83.5	78.9	81.5	83.1	84.6	94.3	93.3	95.3	86.9	95.4	105.8	94.8	97.4	100.1	91.4	108.4	115.2	117.3	126.3
			GC-MS/MS	Green tea	A	111.4	89.0	96.8	104.2	89.0	98.3	100.8	89.0	117.3	115.0	107.3	109.0	112.6	96.8	109.2	113.4	125.2	110.3	107.9	191.5
				Woolong tea	B	94.2	78.9	88.8	85.0	80.1	99.1	85.5	85.5	94.2	91.4	93.5	92.0	98.8	72.6	90.2	93.7	105.5	91.3	90.0	174.6
			GC-MS	Green tea	A	102.7	92.1	83.1	78.8	84.6	98.8	93.1	101.2	126.6	103.3	98.4	96.1	101.2	107.3	109.3	99.3	116.8	118.1	103.6	132.3
				Woolong tea	B	95.9	79.2	79.8	79.7	80.1	90.8	85.5	93.3	108.0	93.9	94.1	99.5	93.3	96.8	98.9	95.0	106.1	109.3	99.9	125.8
			GC-MS/MS	Green tea	A	107.7	84.4	99.0	100.1	96.5	100.2	99.3	93.4	118.6	117.6	116.6	110.9	110.6	101.3	109.7	99.8	129.8	103.2	100.9	145.0
				Woolong tea	B	89.5	76.1	87.5	81.5	82.3	84.4	94.8	117.1	99.8	84.7	97.8	98.4	99.3	75.3	99.2	90.2	111.8	85.6	87.0	123.4
		Envi-Carb+PSA	GC-MS	Green tea	A*	108.0	93.6	92.7	103.4	113.2	104.9	108.1	101.3	101.7	94.8	96.2	100.7	109.6	102.6	116.4	119.0	102.3	109.2	112.7	135.2
				Woolong tea	B*	101.3	93.6	91.6	94.0	95.9	98.3	101.5	109.6	100.7	86.9	101.4	97.7	94.7	102.6	111.3	114.0	110.1	108.9	96.0	128.3
			GC-MS/MS	Green tea	A	104.5	85.0	109.9	86.0	97.3	113.2	107.7	101.4	100.7	115.0	101.4	112.3	94.7	102.6	113.3	113.4	105.2	121.1	106.3	125.4
				Woolong tea	B	89.6	75.5	90.4	73.3	77.8	92.5	89.3	91.8	83.9	115.0	82.1	87.0	83.3	91.4	94.9	90.4	96.7	105.6	94.8	101.5
			GC-MS	Green tea	A	108.3	99.4	94.7	103.0	111.1	108.2	99.1	99.6	87.7	94.7	98.2	100.4	122.1	106.5	116.4	116.9	114.4	115.2	118.7	152.3
				Woolong tea	B	101.5	99.8	90.3	92.0	94.3	91.4	89.6	99.1	79.9	87.5	97.3	95.0	113.3	104.9	108.3	113.4	105.2	111.9	106.3	148.6
			GC-MS/MS	Green tea	A	103.7	84.4	109.4	71.2	85.0	100.5	96.4	100.5	86.9	101.5	102.7	112.3	91.7	103.0	103.7	116.1	112.2	149.3	111.2	123.5
				Woolong tea	B	85.9	76.1	88.1	64.8	69.0	82.0	88.7	84.3	75.8	93.0	82.9	87.0	74.0	89.4	86.9	90.6	96.7	107.2	94.3	100.3

(Continued)

Annex 1: Content change for 201 pesticides in incurred tea Youden pair samples under 8 conditions with two SPE cartridge cleanup in three months (Nov.9, 2009-Feb.7, 2010) (cont.)

No	Pesticides	SPE	Method	Sample	Youden pair	Nov.9 2009 (n=5)	Nov.14 2009 (n=3)	Nov.19 2009 (n=3)	Nov.24 2009 (n=3)	Nov.29 2009 (n=3)	Dec.4 2009 (n=3)	Dec.9 2009 (n=3)	Dec.14 2009 (n=3)	Dec.19 2009 (n=3)	Dec.24 2009 (n=3)	Dec.29 2009 (n=3)	Jan.3 2010 (n=3)	Jan.8 2010 (n=3)	Jan.13 2010 (n=3)	Jan.18 2010 (n=3)	Jan.23 2010 (n=3)	Jan.28 2010 (n=3)	Feb.2 2010 (n=3)	Feb.7 2010 (n=3)	AVE µg/kg.
145	dibutylsuccinate	Cleanert TPT	GC-MS	Green tea	A	78.8	64.0	67.5	69.7	60.0	64.8	61.3	57.3	58.7	54.1	59.4	54.9	56.0	56.5	51.9	43.1	47.3	47.8	48.1	57.9
					B	73.2	59.1	52.4	64.4	53.8	58.9	57.9	52.6	56.6	50.1	52.9	51.0	51.8	53.2	47.1	41.6	39.6	46.5	42.0	52.9
				Woolong tea	A	75.9	56.6	55.4	65.5	52.6	54.9	45.4	54.3	51.3	50.0	48.3	52.1	45.8	41.1	42.8	41.0	39.8	40.1	38.1	50.1
					B	68.1	46.9	47.4	55.8	47.2	44.5	36.0	45.4	43.0	42.5	37.1	41.3	36.8	35.8	34.4	34.7	29.7	34.1	30.1	41.6
			GC-MS/MS	Green tea	A	85.2	62.3	76.3	67.7	57.9	63.2	64.6	56.9	56.6	55.0	56.3	66.0	52.0	59.3	53.7	45.2	42.9	44.0	48.9	58.6
					B	78.5	62.0	66.1	61.5	54.5	57.0	57.7	52.6	56.3	50.4	52.7	57.2	47.2	59.5	48.7	42.9	42.7	42.6	43.2	54.5
				Woolong tea	A	71.0	52.3	52.8	65.8	48.4	52.1	45.6	49.5	51.3	52.2	52.0	49.8	45.6	41.2	41.5	39.2	38.1	39.5	32.2	48.4
					B	64.7	44.7	44.9	55.2	44.1	41.5	35.9	42.5	42.1	44.8	37.7	38.4	38.8	34.9	33.9	33.3	31.3	33.3	27.9	40.5
		Envi-Carb +PSA	GC-MS	Green tea	A	99.8	73.1	65.0	68.3	74.4	76.9	60.2	72.5	58.2	60.2	54.2	89.9	55.3	59.7	54.4	61.0	49.1	52.4	53.0	65.9
					B	88.3	61.9	67.5	61.9	72.6	66.6	60.8	68.7	53.2	56.5	50.5	71.5	50.5	54.6	48.4	51.2	44.1	44.7	30.4	59.7
				Woolong tea	A	78.2	62.6	64.6	58.1	56.4	50.7	54.9	41.1	49.7	50.6	51.7	38.0	42.7	46.1	44.8	39.0	35.0	33.6	24.5	49.4
					B	58.8	51.6	56.3	50.9	44.2	42.0	43.3	34.3	39.5	39.2	41.9	33.2	35.7	37.2	34.1	31.2	27.7	32.2	27.9	39.9
			GC-MS/MS	Green tea	A	81.4	68.6	70.5	68.3	70.5	66.9	63.5	73.2	60.9	60.1	57.0	85.9	56.7	66.6	54.4	61.3	55.5	53.9	54.7	64.7
					B	73.7	54.6	66.9	61.6	70.6	61.9	62.5	65.9	54.4	57.2	51.0	76.7	48.9	55.2	48.6	52.8	49.5	48.4	51.0	58.5
				Woolong tea	A	75.8	59.6	63.0	58.6	55.6	53.4	48.0	41.2	49.7	46.3	43.2	32.9	41.8	42.4	45.2	34.2	29.8	40.3	32.2	48.1
					B	56.7	49.8	55.9	49.9	45.0	43.4	40.6	33.4	39.5	36.3	39.0	29.2	35.8	32.4	31.9	26.9	25.5	32.3	27.9	38.6
146	diethofencarb	Cleanert TPT	GC-MS	Green tea	A*	270.7	217.1	273.9	270.0	216.5	195.7	225.6	253.4	218.9	201.6	248.7	244.1	205.5	253.8	223.4	180.7	180.7	167.9	229.7	227.0
					B*	295.0	248.6	282.0	250.0	204.1	198.6	223.6	234.9	223.6	178.8	223.9	229.0	212.9	256.8	211.0	187.4	187.4	166.6	202.5	222.9
				Woolong tea	A	299.9	226.6	236.9	260.7	220.2	266.7	227.7	202.2	199.8	424.2	303.5	331.9	302.5	198.3	237.7	253.2	253.2	223.5	208.5	256.1
					B	267.2	201.2	221.7	230.9	196.7	218.1	176.7	202.2	210.1	272.7	165.1	215.2	214.1	194.9	208.9	189.8	189.8	219.7	187.6	210.6
			GC-MS/MS	Green tea	A	295.5	245.5	271.3	255.3	222.9	236.8	236.9	242.8	210.9	223.3	224.4	306.4	217.1	265.4	223.3	180.8	180.8	181.4	214.2	234.1
					B	291.7	234.5	247.6	237.9	202.4	218.8	214.6	219.9	219.1	211.7	199.1	254.9	196.7	277.2	200.5	187.6	187.6	181.2	192.2	219.7
				Woolong tea	A	346.6	210.2	222.8	284.5	222.5	245.8	225.3	241.7	250.2	285.3	268.2	261.9	250.3	217.3	225.3	232.3	232.3	230.7	193.4	244.2
					B	291.8	194.9	197.5	242.7	206.0	202.3	184.6	218.9	187.0	245.0	189.1	197.5	213.7	193.3	194.4	184.8	184.8	206.1	175.9	206.1
		Envi-Carb +PSA	GC-MS	Green tea	A*	294.1	267.1	261.2	248.4	247.6	298.3	254.5	260.9	193.8	218.5	239.5	472.0	235.8	222.4	272.9	289.1	255.1	175.9	213.5	259.0
					B*	341.2	236.5	253.7	213.7	244.5	280.7	223.8	226.9	200.9	215.7	192.7	341.4	193.5	223.4	246.5	243.4	220.9	215.2	227.5	241.2
				Woolong tea	A	268.4	203.6	245.0	233.7	262.6	263.2	216.0	214.4	246.7	314.5	330.2	284.7	276.0	290.2	235.7	225.6	198.7	288.9	191.6	250.0
					B	242.5	215.8	232.0	217.8	195.3	217.1	188.7	156.3	193.9	197.1	212.5	192.8	192.2	200.5	203.0	172.0	171.9	203.0	171.4	198.7
			GC-MS/MS	Green tea	A	468.1	332.7	260.0	261.3	264.0	276.9	225.0	269.1	223.3	213.2	218.8	210.4	216.3	239.9	228.6	253.7	235.7	193.0	219.5	253.1
					B	523.4	332.6	246.0	228.1	253.4	258.6	234.3	240.3	208.3	210.9	185.9	197.2	185.7	216.8	204.4	213.7	204.7	197.7	231.2	240.7
				Woolong tea	A	306.2	242.1	288.8	252.2	260.9	241.6	220.6	205.9	246.7	231.7	228.2	202.0	218.4	213.8	243.8	190.1	160.0	238.2	193.4	230.8
					B	269.1	214.1	243.5	226.4	203.0	198.6	196.8	153.6	193.9	182.0	206.0	182.7	195.9	192.9	185.7	151.5	151.8	193.6	175.9	195.6
147	diflufenican	Cleanert TPT	GC-MS	Green tea	A	46.7	45.1	51.8	39.9	33.8	38.1	36.1	40.7	28.5	35.3	36.1	54.0	34.0	35.6	44.2	46.8	40.1	31.7	36.6	37.7
					B	46.2	38.9	44.9	37.2	31.0	35.1	35.0	35.9	28.2	34.9	31.3	48.0	28.5	36.1	40.1	39.3	35.4	35.9	40.0	35.9
				Woolong tea	A	43.9	32.6	35.6	46.9	37.5	36.6	32.3	34.4	37.6	36.9	44.9	33.3	37.1	39.4	43.8	33.8	33.7	40.5	31.7	37.7
					B	39.1	29.4	34.1	39.1	36.3	33.3	30.7	27.7	30.7	30.5	35.0	33.0	31.7	33.7	36.1	26.6	28.2	36.6	30.5	33.5
			GC-MS/MS	Green tea	A	50.9	41.5	43.8	41.4	34.6	39.4	37.7	39.0	33.5	35.4	36.4	48.0	28.8	41.3	34.5	30.4	28.1	27.3	32.6	37.1
					B	48.6	39.4	40.1	38.3	31.7	37.7	35.3	37.0	35.3	35.6	33.8	39.5	27.5	39.1	31.6	29.2	29.2	27.1	29.5	35.0
				Woolong tea	A	59.4	35.5	35.7	45.1	36.6	35.4	33.9	37.5	39.3	45.0	40.1	41.4	39.5	33.8	35.4	36.5	34.3	35.3	30.9	38.7
					B	48.2	32.1	31.5	37.2	34.6	40.1	30.9	35.5	30.5	41.1	29.2	32.9	35.4	30.0	30.6	30.8	28.0	29.6	27.9	33.1
		Envi-Carb +PSA	GC-MS	Green tea	A	54.1	49.3	41.0	40.5	41.6	33.5	46.9	40.7	33.9	35.3	36.1	54.0	34.0	35.6	44.2	46.8	40.1	31.7	36.6	41.4
					B	57.5	47.4	39.2	33.7	40.4	48.8	42.8	35.9	36.4	34.9	31.3	48.0	28.5	36.1	40.1	39.3	35.4	35.9	40.0	39.2
				Woolong tea	A	45.0	36.2	43.5	39.4	43.8	49.6	35.0	34.4	36.9	36.9	44.9	33.3	37.1	39.4	43.8	33.8	33.7	40.5	31.7	38.5
					B	40.3	32.2	40.9	37.1	33.7	41.2	29.3	27.7	30.1	30.5	35.0	33.0	31.7	33.7	36.1	26.6	28.2	36.6	30.5	33.1
			GC-MS/MS	Green tea	A	78.4	56.4	41.9	43.0	40.4	35.9	40.6	41.9	37.0	34.7	37.6	32.0	34.9	29.9	35.1	39.5	36.2	30.6	33.8	40.4
					B	86.1	58.9	39.5	36.2	40.9	43.1	39.9	36.5	37.6	33.2	30.6	30.8	29.0	27.1	32.0	32.5	32.4	30.7	35.6	38.2
				Woolong tea	A	50.2	37.6	36.2	39.9	42.1	40.5	34.5	32.4	34.0	35.8	36.5	28.2	35.2	34.1	39.7	29.9	25.9	36.4	30.9	36.3
					B	46.6	33.8	40.0	35.8	33.1	33.2	29.6	26.2	30.7	30.7	32.8	26.0	31.6	29.8	30.2	22.9	23.2	31.7	27.9	31.4

No.	Compound	Cartridge	Method	Tea																						
148	dimepiperate	Cleanert TPT	GC-MS	Green tea	A	86.6	78.2	70.9	77.7	62.5	60.1	65.6	68.0	65.2	52.7	72.9	71.5	65.9	67.1	64.1	61.4	53.3	53.7	62.3	66.3
					B	88.8	77.6	66.2	70.6	56.5	57.6	63.2	62.6	64.9	48.6	65.1	67.2	60.9	69.7	59.9	61.6	56.2	51.5	54.4	63.3
				Woolong tea	A*	80.1	62.7	71.5	80.3	63.7	69.6	63.6	78.9	74.0	96.9	66.5	80.0	73.3	65.7	71.2	68.5	67.9	62.6	57.5	71.3
					B*	72.2	55.4	63.6	73.4	59.1	57.8	45.1	70.9	59.8	82.9	51.4	66.0	61.6	60.8	61.8	60.8	58.2	58.3	52.7	61.7
			GC-MS/MS	Green tea	A	89.8	67.7	76.4	74.6	62.5	66.1	69.7	64.5	61.7	63.7	63.5	90.1	57.1	69.5	62.2	53.8	51.5	51.3	59.3	66.1
					B	84.6	65.8	69.7	70.5	57.7	61.0	63.8	59.9	63.5	57.7	60.2	77.0	53.9	67.4	55.7	51.7	52.3	50.1	52.0	61.8
				Woolong tea	A*	99.9	64.9	65.3	80.7	61.7	68.4	65.0	72.7	75.1	77.3	77.4	75.0	77.8	61.9	64.5	62.3	65.7	61.9	46.0	69.7
					B*	82.3	56.1	55.8	67.7	57.5	55.5	48.7	60.8	60.6	68.4	55.7	58.9	61.9	54.8	54.9	53.1	53.0	55.3	43.5	58.1
		Envi-Carb +PSA	GC-MS	Green tea	A	91.7	95.6	73.3	68.3	77.0	77.8	66.7	79.1	58.6	62.6	62.8	108.0	61.8	54.8	68.9	78.6	65.8	63.8	60.7	73.3
					B	102.7	79.1	72.1	62.8	73.4	72.3	65.0	72.4	66.5	62.8	60.3	74.3	57.0	70.8	64.7	64.4	59.9	58.1	61.0	67.7
				Woolong tea	A*	80.8	69.5	83.0	71.7	71.5	68.9	63.3	66.6	74.1	62.8	79.0	61.5	67.9	67.7	76.6	64.5	58.5	68.1	54.9	69.9
					B*	69.5	61.8	67.4	63.4	54.6	57.9	54.5	54.2	56.6	77.5	65.8	60.8	60.6	70.9	62.9	54.4	50.0	58.0	49.3	59.2
			GC-MS/MS	Green tea	A	99.9	74.8	74.0	73.4	78.2	76.1	68.2	78.7	74.1	62.3	61.5	79.7	66.0	61.1	65.2	73.8	65.1	59.6	61.9	71.2
					B	101.0	65.3	72.2	66.6	76.0	69.9	74.2	73.4	56.6	62.2	57.3	75.7	57.1	68.9	58.2	61.6	58.0	55.9	63.2	67.0
				Woolong tea	A*	80.8	72.8	93.8	76.6	72.7	73.3	65.9	59.3	60.3	63.1	64.9	62.8	63.2	64.5	73.2	53.9	47.0	68.9	45.1	67.3
					B*	65.8	62.1	67.6	65.1	56.6	59.0	57.5	49.0	56.6	66.8	60.3	56.5	53.5	64.5	55.1	43.4	42.9	56.4	42.1	55.8
149	dimethametryn	Cleanert TPT	GC-MS	Green tea	A	50.5	38.7	18.5	40.4	33.1	31.1	33.4	32.5	26.6	27.2	34.2	32.7	30.4	28.8	31.0	25.6	25.0	22.6	28.1	31.1
					B	49.3	36.6	12.4	36.2	29.4	31.6	30.8	30.0	26.5	26.5	32.0	30.5	28.6	28.9	27.4	25.3	25.0	22.1	25.3	29.2
				Woolong tea	A*	44.0	33.5	36.3	39.2	32.6	35.6	32.2	34.1	36.2	50.6	26.7	32.6	33.7	31.1	30.4	30.8	31.4	26.4	26.2	34.0
					B*	40.0	30.6	32.6	35.8	31.4	31.0	26.5	29.8	26.8	40.0	21.0	26.6	27.7	26.9	26.3	25.2	24.8	22.6	23.4	29.1
			GC-MS/MS	Green tea	A	49.1	37.5	45.8	43.0	34.9	34.3	30.5	31.6	30.5	31.5	33.6	27.4	26.5	32.0	31.0	25.2	24.1	23.1	29.3	32.9
					B	45.6	37.4	39.5	35.5	31.3	31.7	30.9	29.3	30.1	31.5	29.9	27.2	26.1	32.5	28.0	24.3	24.2	23.1	24.8	30.7
				Woolong tea	A*	41.6	33.1	36.0	43.5	32.7	35.4	32.9	34.6	36.8	37.9	38.3	37.1	35.3	29.6	32.0	28.5	32.4	28.1	24.2	34.2
					B*	38.5	30.5	29.6	36.5	31.2	29.3	26.7	30.3	30.8	32.2	27.6	29.1	28.0	25.5	26.8	24.8	26.1	25.7	20.8	28.9
		Envi-Carb +PSA	GC-MS	Green tea	A	54.7	43.8	40.0	39.0	40.4	40.4	31.7	38.5	26.7	31.7	29.6	61.1	28.4	31.9	27.8	33.2	29.6	29.3	27.8	36.1
					B	53.3	37.3	38.5	34.9	39.1	34.0	32.5	36.1	27.3	32.1	28.0	39.9	25.8	28.9	26.6	26.7	25.7	25.1	27.6	32.6
				Woolong tea	A*	51.0	42.1	41.3	38.2	37.9	33.9	34.5	24.7	36.0	37.3	32.5	25.0	32.9	34.3	34.9	29.7	26.7	33.9	24.9	34.3
					B*	45.3	33.1	37.3	33.7	28.5	28.6	29.5	20.7	27.0	29.8	28.1	24.2	28.0	27.5	27.3	24.5	22.5	26.6	22.2	28.6
			GC-MS/MS	Green tea	A	46.9	43.4	41.7	40.1	39.6	40.5	34.4	40.7	33.9	30.2	28.6	37.3	29.4	35.1	28.7	33.9	31.8	29.1	27.4	35.4
					B	44.0	35.1	39.7	35.3	38.2	36.2	34.0	36.7	31.9	30.3	25.5	35.3	27.0	30.2	25.6	28.0	26.8	25.3	27.4	32.2
				Woolong tea	A*	43.4	38.5	47.2	40.2	37.4	34.0	33.1	28.7	36.0	32.0	31.0	30.8	29.7	28.5	33.7	25.7	22.0	29.5	24.2	32.9
					B*	38.4	34.2	39.0	34.9	30.3	27.9	28.8	24.1	27.0	26.0	29.3	26.5	30.8	24.8	24.0	21.1	19.9	25.2	20.8	28.1
150	dimethomorph	Cleanert TPT	GC-MS	Green tea	A	247.4	187.6	276.0	85.3	82.1	81.5	80.6	77.9	78.2	66.3	85.5	72.7	71.5	97.4	75.3	67.5	65.4	63.9	76.5	102.1
					B	320.4	239.6	353.4	78.9	71.7	74.7	76.6	69.9	71.4	57.7	69.4	70.5	70.9	103.6	64.7	67.3	61.0	58.0	66.8	107.7
				Woolong tea	A*	117.1	76.1	80.3	89.5	80.1	80.3	69.7	75.1	85.0	84.3	73.3	84.6	79.4	72.5	74.0	72.9	72.9	72.8	65.9	79.3
					B*	88.4	70.1	72.4	77.2	72.9	69.7	63.2	70.5	67.1	72.0	58.9	68.9	69.2	65.4	63.2	57.1	60.8	65.2	54.3	67.7
			GC-MS/MS	Green tea	A*	101.4	77.6	102.2	90.8	85.7	77.1	81.1	78.5	76.5	70.8	81.6	85.4	61.3	113.7	77.1	66.2	62.5	62.9	74.6	80.4
					B*	105.0	79.6	93.6	80.5	74.5	71.4	77.7	70.2	72.2	63.1	68.7	76.7	63.8	130.3	68.4	65.5	60.2	58.2	66.0	76.1
				Woolong tea	A	111.7	77.7	77.7	94.7	72.5	77.8	66.2	76.4	92.8	93.0	83.1	87.7	79.4	69.8	73.6	72.1	73.0	73.7	60.9	79.2
					B	91.8	69.1	91.2	83.1	71.7	67.9	61.1	71.6	71.1	63.1	62.2	68.7	73.5	60.9	64.0	58.9	59.7	65.9	55.5	68.5
		Envi-Carb +PSA	GC-MS	Green tea	A	117.2	106.8	92.1	87.3	85.8	39.7	57.3	91.4	70.5	73.9	65.4	124.4	78.1	84.3	74.4	88.1	80.7	75.6	72.4	82.4
					B*	129.2	100.3	103.4	80.0	86.9	29.7	78.4	82.8	63.8	70.8	64.5	95.2	72.7	71.6	69.3	74.5	67.0	61.8	71.7	77.6
				Woolong tea	A	95.5	62.9	82.7	85.6	83.5	68.9	87.5	62.0	76.1	82.4	81.2	65.5	70.0	71.4	85.9	72.0	64.8	85.5	60.7	76.0
					B	87.9	71.6	80.8	79.7	67.4	58.5	68.4	53.1	66.6	55.3	66.1	56.0	60.7	64.8	63.9	61.0	57.4	66.7	55.1	65.3
			GC-MS/MS	Green tea	A	89.9	79.1	88.1	95.7	81.4	88.1	57.2	100.8	84.0	72.9	67.9	69.2	78.3	87.9	73.9	87.5	82.0	73.0	78.3	80.8
					B	102.9	72.3	85.8	83.7	87.6	80.9	83.2	88.0	71.3	71.3	66.6	64.0	70.0	74.6	68.2	74.8	69.0	63.9	76.3	76.5
				Woolong tea	A	88.5	82.2	91.8	86.0	87.9	74.8	80.0	62.2	76.1	80.4	66.0	65.5	73.9	67.6	85.3	61.9	57.6	81.8	61.1	74.8
					B	77.2	73.0	85.3	77.2	68.5	60.6	67.8	54.9	66.6	54.7	64.1	54.1	68.7	61.8	56.9	52.3	53.8	64.9	53.7	64.1

(Continued)

Annex 1: Content change for 201 pesticides in incurred tea Youden pair samples under 8 conditions with two SPE cartridge cleanup in three months (Nov.9, 2009-Feb.7, 2010) (cont.)

No	Pesticides	SPE	Method	Sample	Youden pair	Nov.9, 2009 (n=5)	Nov.14, 2009 (n=3)	Nov.19, 2009 (n=3)	Nov.24, 2009 (n=3)	Nov.29, 2009 (n=3)	Dec.4, 2009 (n=3)	Dec.9, 2009 (n=3)	Dec.14, 2009 (n=3)	Dec.19, 2009 (n=3)	Dec.24, 2009 (n=3)	Dec.14, 2009 (n=3)	Jan.3, 2010 (n=3)	Jan.8, 2010 (n=3)	Jan.13, 2010 (n=3)	Jan.18, 2010 (n=3)	Jan.23, 2010 (n=3)	Jan.28, 2010 (n=3)	Feb.2, 2010 (n=3)	Feb.7, 2010 (n=3)	AVE μg/kg
151	dimethylphthalate	Cleanert TPT	GC-MS	Green tea	A*	179.4	53.4	135.1	148.3	136.9	145.2	142.5	134.0	139.0	135.7	143.1	140.8	134.6	160.1	135.3	116.3	113.1	116.4	132.0	133.7
				Woolong tea	B*	171.1	24.8	125.0	138.8	129.9	134.9	133.3	127.7	137.3	129.8	132.0	131.0	128.4	152.6	128.2	116.3	111.6	113.6	118.8	125.5
			GC-MS/MS	Green tea	A	175.3	139.0	144.1	169.0	139.9	145.6	118.8	137.1	140.3	136.7	143.4	149.2	134.4	127.8	129.2	126.3	128.0	130.6	114.8	138.4
				Woolong tea	B	153.8	113.3	120.0	139.1	123.0	116.6	97.8	116.1	116.0	118.3	110.8	121.1	111.2	111.1	106.2	106.9	107.4	115.5	96.2	115.8
			GC-MS	Green tea	A	186.6	143.3	167.9	149.8	133.9	150.8	146.4	132.9	131.0	137.7	138.1	164.6	126.4	163.6	138.2	117.5	110.4	115.3	130.2	141.3
				Woolong tea	B	175.5	139.5	153.8	140.8	126.8	135.6	132.4	121.5	133.6	133.2	129.7	142.7	116.7	172.4	129.3	115.3	112.2	112.9	116.6	133.7
			GC-MS/MS	Green tea	A	165.9	129.0	134.2	172.6	124.6	140.3	123.3	133.3	138.7	142.9	145.7	144.8	136.9	123.6	128.0	121.8	125.7	128.9	108.9	135.2
				Woolong tea	B	145.7	107.3	111.5	140.6	111.9	110.5	97.2	112.6	114.9	123.7	107.8	113.0	116.0	105.9	107.0	102.9	106.4	114.6	97.8	113.0
		Envi-Carb +PSA	GC-MS	Green tea	A*	215.8	161.3	156.5	154.3	163.9	169.5	133.3	169.5	141.9	144.8	138.2	148.6	135.1	150.1	139.7	155.3	136.7	146.1	132.3	152.3
				Woolong tea	B*	195.7	137.1	151.1	142.3	163.5	151.7	134.4	157.3	129.7	136.1	119.6	141.5	122.4	138.7	123.3	133.2	122.4	130.1	130.5	140.0
			GC-MS/MS	Green tea	A	180.9	147.6	161.5	145.2	156.6	138.2	144.4	123.1	143.6	133.2	150.4	116.7	125.5	134.9	144.0	126.1	115.8	147.3	109.3	139.2
				Woolong tea	B	132.5	120.8	141.6	125.6	122.7	116.4	119.5	103.2	112.4	106.9	128.4	100.4	107.9	113.3	114.6	102.5	97.4	119.4	96.0	114.8
			GC-MS	Green tea	A	174.9	150.8	157.7	161.2	154.8	152.8	137.9	173.1	146.5	139.4	139.4	143.4	134.1	163.0	137.3	158.9	136.5	138.6	142.4	149.6
				Woolong tea	B	166.8	122.9	151.2	146.2	158.2	144.6	135.8	154.2	131.9	133.7	120.9	136.3	123.3	134.5	123.0	135.8	124.5	128.9	136.5	137.3
			GC-MS/MS	Green tea	A	175.8	138.3	213.3	145.8	152.7	139.8	130.3	122.4	143.6	128.9	122.7	106.4	120.8	122.4	142.2	109.1	99.1	137.7	108.9	134.7
				Woolong tea	B	129.5	114.3	140.4	123.0	123.3	113.3	112.5	99.6	112.4	101.9	113.5	95.3	106.0	105.6	106.2	87.7	89.8	114.5	97.8	109.8
152	diniconazole	Cleanert TPT	GC-MS	Green tea	A*	165.9	135.8	169.2	128.8	105.5	78.9	102.6	105.2	89.9	83.7	91.3	89.4	97.0	106.5	106.2	81.7	74.0	71.9	84.3	103.2
				Woolong tea	B*	159.1	130.3	152.8	115.8	91.5	81.9	95.4	97.2	89.3	78.6	84.3	92.3	92.7	103.3	90.0	85.1	73.0	68.5	72.4	97.5
			GC-MS/MS	Green tea	A	146.3	91.6	119.5	127.9	91.0	109.8	94.2	97.4	122.4	173.0	88.7	87.0	104.1	99.7	102.8	97.0	96.9	79.9	83.5	105.9
				Woolong tea	B	129.7	95.5	119.6	119.9	88.3	94.8	85.7	97.7	81.5	134.8	81.0	70.3	89.3	91.2	82.8	76.9	73.6	74.7	74.0	92.7
			GC-MS	Green tea	A	147.1	121.8	141.4	113.6	106.4	106.1	104.5	103.6	95.6	98.4	102.1	96.9	82.1	113.4	93.7	77.4	74.1	70.3	86.4	101.8
				Woolong tea	B	139.3	118.8	122.8	110.8	90.9	97.0	96.8	94.0	92.9	96.9	91.4	87.6	77.1	118.3	84.3	75.1	74.7	69.1	75.0	95.4
			GC-MS/MS	Green tea	A	147.0	106.4	107.7	139.2	97.7	107.1	94.7	103.4	111.8	124.8	112.1	113.8	102.5	87.5	90.3	90.8	92.7	87.6	76.6	104.9
				Woolong tea	B	125.7	96.8	95.4	114.4	92.7	89.6	82.4	96.2	85.9	109.0	82.9	87.8	88.3	76.8	75.9	75.1	75.2	78.5	68.2	89.3
		Envi-Carb +PSA	GC-MS	Green tea	A*	169.2	139.7	125.7	122.0	124.4	122.3	99.9	121.1	72.8	94.5	89.7	209.1	93.4	102.7	96.4	104.4	93.1	85.8	87.1	113.3
				Woolong tea	B*	185.2	127.0	119.9	103.5	116.9	114.6	108.1	113.3	68.6	99.3	85.7	139.6	86.5	94.4	86.1	87.0	87.0	80.7	86.6	104.7
			GC-MS/MS	Green tea	A	154.1	103.0	121.4	117.2	114.5	101.8	110.1	71.1	108.3	123.6	112.6	91.4	98.1	105.8	109.0	98.0	84.7	94.9	70.2	104.7
				Woolong tea	B	137.5	93.7	114.1	112.5	87.9	88.6	100.2	58.9	86.3	97.7	86.7	83.6	83.6	78.3	83.4	78.3	70.5	78.1	68.5	88.8
			GC-MS	Green tea	A	156.9	131.3	122.0	126.2	121.5	132.3	100.2	117.1	102.6	97.1	117.0	105.2	95.7	92.5	88.1	103.6	97.4	84.7	81.6	108.3
				Woolong tea	B	160.8	114.5	116.4	108.0	115.5	120.2	105.7	105.6	94.2	93.6	105.6	98.4	83.3	79.3	80.9	103.6	85.4	76.8	91.4	99.8
			GC-MS/MS	Green tea	A	131.3	112.7	129.4	120.3	112.1	100.9	98.9	83.0	108.3	100.9	83.0	89.4	94.6	91.5	103.6	79.6	67.3	101.7	76.6	99.8
				Woolong tea	B	119.3	100.6	117.0	108.1	86.4	84.0	85.6	69.2	86.3	81.2	88.1	78.1	74.7	80.7	76.7	62.5	61.2	84.1	68.2	84.8
153	diphenamid	Cleanert TPT	GC-MS	Green tea	A*	49.5	38.0	42.0	40.6	24.2	30.4	32.0	27.4	33.5	33.5	38.3	36.5	37.9	42.1	38.9	32.0	32.4	31.3	36.4	35.6
				Woolong tea	B*	47.5	36.5	37.7	37.1	25.9	31.6	32.0	32.2	36.0	31.6	34.4	34.5	34.9	43.0	33.4	32.4	32.0	30.0	32.2	34.5
			GC-MS/MS	Green tea	A	49.4	39.1	38.8	47.1	41.1	40.1	33.7	34.3	40.6	41.8	38.7	45.4	39.2	38.6	37.9	38.3	38.7	36.0	33.4	39.6
				Woolong tea	B	44.4	33.4	31.4	40.5	35.0	35.5	27.9	32.8	33.3	36.2	29.1	35.3	32.6	33.8	31.5	32.0	32.5	34.2	30.0	33.8
			GC-MS	Green tea	A	47.3	36.9	45.0	41.4	35.2	36.6	36.2	35.1	33.1	34.4	35.5	35.5	30.1	33.8	35.9	29.8	28.0	27.9	32.4	35.5
				Woolong tea	B	45.5	36.8	38.7	36.3	32.4	33.4	32.6	31.1	32.3	33.5	31.1	32.5	30.9	42.4	32.6	28.5	28.5	27.0	28.7	33.4
			GC-MS/MS	Green tea	A	40.5	36.5	35.8	42.8	33.7	37.5	34.5	36.8	39.1	39.5	41.2	42.2	40.6	34.0	33.0	34.4	35.8	34.3	28.8	36.9
				Woolong tea	B	37.7	31.9	29.9	35.4	31.0	30.0	26.6	31.3	33.0	33.6	28.7	33.0	31.5	29.4	28.5	28.9	28.4	31.5	25.9	30.8
		Envi-Carb +PSA	GC-MS	Green tea	A*	54.0	44.3	43.1	41.8	43.3	51.8	24.4	45.1	18.5	34.6	33.6	48.6	36.9	41.2	35.7	41.1	37.9	38.1	68.2	39.4
				Woolong tea	B*	51.5	38.2	41.7	39.9	42.5	43.1	60.1	37.6	18.0	34.0	29.8	37.9	33.5	36.1	32.2	32.9	33.3	33.1	36.4	37.4
			GC-MS/MS	Green tea	A	47.2	39.9	46.3	40.0	42.0	41.0	23.6	39.3	37.5	39.7	41.6	33.9	38.4	39.6	40.1	36.3	32.7	42.4	32.2	38.6
				Woolong tea	B	40.5	36.8	41.7	36.0	33.7	34.3	37.5	29.0	29.7	31.2	36.1	29.4	34.1	33.3	32.1	30.2	28.6	34.6	31.8	33.6
			GC-MS	Green tea	A	45.9	39.4	41.8	42.2	40.6	40.4	32.5	42.6	35.0	32.3	32.9	41.5	33.6	42.6	34.1	39.8	35.5	33.5	29.4	37.9
				Woolong tea	B	42.1	32.8	40.1	38.4	39.8	36.6	35.0	38.9	33.5	32.4	29.4	36.9	30.6	34.1	30.3	32.5	31.0	34.2	33.4	33.8
			GC-MS/MS	Green tea	A	43.3	39.7	48.0	38.7	37.5	35.7	34.9	30.3	37.5	34.6	33.4	29.3	32.3	39.1	35.2	28.4	24.5	37.6	28.8	34.9
				Woolong tea	B	37.0	33.8	37.9	33.4	30.8	29.5	31.3	25.0	29.7	26.8	32.0	24.6	29.1	28.5	26.2	24.1	23.4	30.4	25.9	29.4

No.	Pesticide	Cartridge	Method	Tea	Rep																					
154	dipropetryn	Cleanert TPT	GC-MS	Green tea	A	53.3	41.8	48.0	41.9	33.8	31.3	35.8	34.8	32.6	29.1	34.7	34.1	33.1	29.9	31.3	25.2	26.7	23.3	29.4	34.2
					B	51.9	40.8	43.4	36.5	29.5	31.4	32.8	31.7	32.9	25.5	30.7	33.0	31.6	29.4	28.4	25.0	25.8	22.6	25.9	32.0
				Woolong tea	A	53.2	44.6	36.0	39.6	33.8	38.2	33.5	37.1	45.5	45.9	28.3	43.7	36.4	34.2	30.0	30.9	32.0	28.6	26.4	36.7
					B	47.7	29.0	30.6	36.7	33.2	31.3	28.0	32.2	37.8	34.5	23.3	37.2	29.0	30.8	25.5	25.5	25.4	27.2	23.6	31.0
			GC-MS/MS	Green tea	A	48.9	36.2	45.4	38.5	36.9	35.9	35.6	32.6	30.8	32.1	32.7	32.7	29.8	35.1	31.8	26.4	25.4	23.9	27.7	33.6
					B	45.7	36.5	38.4	36.7	32.9	32.8	31.7	30.3	30.3	30.2	28.6	31.0	27.8	36.5	29.3	25.8	26.1	23.9	24.7	31.5
				Woolong tea	A	40.2	33.3	37.0	45.6	32.6	35.2	35.3	36.0	36.5	39.1	35.6	40.0	38.5	38.7	31.3	32.5	31.9	27.5	25.7	35.4
					B	37.5	27.0	31.5	38.3	30.8	28.9	27.1	31.8	31.1	33.5	27.4	29.8	31.2	34.9	27.8	26.9	23.7	25.8	23.6	29.9
		Envi-Carb +PSA	GC-MS	Green tea	A*	51.2	41.0	37.2	37.1	40.3	36.7	34.6	40.1	29.8	32.3	27.8	64.0	31.7	34.9	27.4	34.1	30.4	30.3	27.9	36.2
					B	52.0	34.1	36.4	34.4	39.4	31.6	34.2	37.6	29.5	31.6	28.7	44.3	29.3	32.0	25.1	27.0	26.8	26.2	28.1	33.1
				Woolong tea	A*	44.7	63.0	46.5	35.0	35.7	35.2	39.1	27.0	37.4	43.4	35.0	33.0	33.3	37.3	35.1	31.7	27.2	36.2	22.2	36.7
					B*	41.7	71.5	44.1	31.1	30.6	30.3	33.2	23.3	29.8	39.8	29.7	27.3	27.3	30.7	26.9	25.2	22.5	26.9	22.5	31.8
			GC-MS/MS	Green tea	A*	46.7	40.8	42.9	39.7	41.8	42.1	35.2	40.7	34.7	30.1	29.8	56.1	30.5	34.6	28.9	35.3	32.5	28.8	30.5	36.9
					B*	43.4	33.0	40.1	36.2	39.6	38.1	34.8	36.6	32.8	30.5	27.1	66.5	27.7	28.8	27.8	29.3	27.9	25.0	30.8	34.5
				Woolong tea	A	43.2	59.2	44.9	39.4	33.7	36.5	32.3	28.7	37.4	28.3	27.0	30.3	31.0	36.7	35.0	37.3	24.1	32.7	25.7	34.9
					B*	37.6	53.2	37.4	29.3	28.7	29.0	28.2	23.9	29.8	20.9	24.5	25.5	23.8	25.3	19.9	29.6	22.1	27.5	23.6	28.4
155	ethalfluralin	Cleanert TPT	GC-MS	Green tea	A	190.4	136.7	142.0	152.5	103.0	138.7	129.5	136.9	110.5	117.6	132.9	126.1	120.5	132.8	129.4	123.4	104.5	95.5	120.2	128.6
					B	183.9	128.2	126.9	138.0	107.6	131.4	123.6	132.7	132.6	112.4	124.6	114.7	111.3	133.6	124.0	119.8	106.3	96.9	109.3	124.1
				Woolong tea	A	167.5	126.2	136.7	174.3	150.5	169.6	121.4	136.4	141.5	150.3	136.2	144.8	141.2	117.7	156.1	157.8	133.5	136.7	121.5	143.2
					B	151.7	111.9	120.3	144.6	126.1	117.2	92.6	121.0	110.3	128.7	101.9	110.9	119.0	112.9	127.7	149.9	111.0	122.7	111.1	120.6
			GC-MS/MS	Green tea	B	191.2	153.5	160.8	108.2	121.7	150.9	131.3	129.7	120.1	126.9	137.1	154.2	99.3	137.2	126.5	110.1	99.0	96.6	111.7	130.3
					B	187.0	146.6	144.2	138.6	117.7	132.3	121.4	123.8	141.7	121.5	127.9	127.7	91.7	150.3	112.7	107.7	103.0	96.3	102.1	126.0
				Woolong tea	A	173.8	126.9	137.0	181.4	133.4	140.8	119.1	138.1	143.2	172.4	165.8	153.9	147.8	120.3	150.3	133.4	132.7	132.3	117.0	143.1
					B	153.9	109.9	115.4	182.6	118.4	109.2	90.6	118.1	112.7	145.4	112.2	110.4	121.5	106.4	124.5	115.6	110.0	113.6	101.0	118.0
		Envi-Carb +PSA	GC-MS	Green tea	A*	194.2	158.8	145.2	148.9	153.5	172.8	165.6	133.0	132.5	130.1	120.9	464.1	117.8	129.3	141.5	145.7	146.2	137.5	133.3	161.6
					B*	174.4	158.5	137.9	131.6	142.5	127.5	122.8	129.0	122.4	121.7	106.3	339.2	111.5	129.4	130.2	120.9	125.4	131.0	129.5	141.7
				Woolong tea	A*	164.1	126.3	161.5	139.9	176.6	182.4	125.1	104.5	145.8	134.0	154.6	119.4	138.1	119.3	160.7	134.2	105.1	140.4	121.7	139.7
					B*	131.8	110.7	129.1	126.7	118.5	155.1	109.0	84.7	108.9	110.6	125.3	110.5	112.0	127.4	160.5	86.1	93.1	120.5	104.4	117.1
			GC-MS/MS	Green tea	A*	176.0	161.3	152.8	154.2	150.8	230.2	162.4	134.6	133.5	129.4	114.5	151.9	113.4	153.6	109.2	140.3	138.7	110.1	147.2	145.5
					B*	175.1	157.0	141.5	138.7	143.2	219.8	88.1	124.8	126.4	124.7	104.4	140.5	116.7	128.6	106.8	109.3	121.7	114.5	141.6	132.8
				Woolong tea	A	172.3	139.1	208.9	154.2	159.4	167.8	116.1	103.2	145.8	134.7	143.4	123.2	175.8	142.6	159.6	116.4	100.7	100.2	117.0	141.6
					B*	132.4	113.9	134.4	133.5	124.2	134.2	103.7	81.3	108.9	117.3	123.8	111.2	150.1	117.8	123.7	86.4	88.1	100.9	101.0	114.8
156	etofenprox	Cleanert TPT	GC-MS	Green tea	A*	56.9	45.7	72.6	40.7	45.4	42.1	46.2	39.0	40.6	37.3	39.6	40.3	40.1	43.1	42.5	36.1	37.3	40.1	40.5	43.5
					B*	55.4	43.7	73.1	40.7	40.0	41.0	43.5	39.8	41.8	39.4	36.4	40.0	40.3	43.3	40.2	35.1	37.7	41.1	38.8	42.7
				Woolong tea	A	45.1	35.7	36.9	43.0	36.8	38.1	30.1	35.4	37.6	35.4	37.4	40.7	35.3	36.5	36.6	35.6	33.3	28.5	29.2	36.2
					B	40.0	31.7	34.2	35.4	34.9	32.6	27.9	32.2	32.0	31.3	30.4	32.8	31.3	37.0	32.8	29.9	28.1	24.7	25.6	31.8
			GC-MS/MS	Green tea	A	56.6	42.3	53.7	43.5	44.3	42.8	44.6	38.4	35.5	38.5	38.7	35.4	36.6	48.5	40.7	33.3	34.4	36.7	40.9	41.3
					B	54.0	42.0	48.0	41.2	39.8	40.4	39.9	39.9	36.3	38.3	36.1	35.2	34.6	49.4	38.2	34.0	36.3	37.6	38.6	40.0
				Woolong tea	A	42.4	33.5	34.8	42.6	34.5	37.5	29.7	34.6	39.9	40.0	37.7	39.3	36.8	32.1	32.0	30.7	32.6	31.4	28.0	35.3
					B	38.2	30.9	30.6	35.8	33.6	31.3	27.3	32.0	32.8	35.2	29.3	31.6	36.8	28.4	27.5	25.8	26.3	27.7	24.8	30.6
		Envi-Carb +PSA	GC-MS	Green tea	A	50.8	51.1	45.7	47.6	47.2	35.3	37.8	50.9	43.1	41.2	39.3	36.6	39.2	40.9	41.8	45.7	40.1	46.6	42.8	43.4
					B	49.0	43.4	46.5	46.3	49.7	33.6	43.2	52.6	40.3	41.7	40.0	38.7	38.3	40.0	40.6	41.9	38.9	41.9	42.4	42.6
				Woolong tea	A	46.8	38.9	41.4	36.8	38.7	34.4	35.8	28.5	35.9	33.9	38.3	32.1	32.1	35.6	32.6	30.6	29.0	38.1	28.2	35.3
					B	39.9	33.0	39.0	31.5	31.5	29.5	27.7	25.5	29.0	27.1	31.9	28.7	28.1	31.7	27.1	26.0	23.8	31.7	24.9	30.2
			GC-MS/MS	Green tea	A	46.0	43.3	47.4	63.6	42.0	43.2	40.4	63.3	68.2	39.5	39.3	31.9	46.6	35.4	43.1	64.0	56.5	55.5	62.8	49.0
					B	46.4	36.8	51.8	61.9	47.1	42.6	42.2	66.5	62.8	40.9	38.3	30.2	34.4	41.6	46.0	56.3	57.4	54.6	63.4	48.5
				Woolong tea	A	48.8	51.2	42.7	58.3	53.6	49.3	55.4	28.9	35.9	33.8	31.2	27.2	31.8	33.4	36.3	43.7	25.5	37.1	38.8	40.2
					B*	46.0	49.4	40.6	56.3	54.3	53.8	62.0	24.8	29.0	27.6	28.6	23.6	30.4	27.7	26.4	50.0	21.6	35.1	39.2	38.2

(Continued)

Annex 1: Content change for 201 pesticides in incurred tea Youden pair samples under 8 conditions with two SPE cartridge cleanup in three months (Nov.9, 2009-Feb.7, 2010) (cont.)

No	Pesticides	SPE	Method	Sample	Youden pair	Nov.9 2009 (n=5)	Nov.14 2009 (n=3)	Nov.19 2009 (n=3)	Nov.24 2009 (n=3)	Nov.29 2009 (n=3)	Dec.4 2009 (n=3)	Dec.9 2009 (n=3)	Dec.14 2009 (n=3)	Dec.19 2009 (n=3)	Dec.24 2009 (n=3)	Dec.14 2009 (n=3)	Jan.3 2010 (n=3)	Jan.8 2010 (n=3)	Jan.13 2010 (n=3)	Jan.18 2010 (n=3)	Jan.23 2010 (n=3)	Jan.28 2010 (n=3)	Feb.2 2010 (n=3)	Feb.7 2010 (n=3)	AVE, µg/kg
157	etridiazol	Cleanert TPT	GC-MS	Green tea	A	112.0	91.3	94.3	96.8	79.8	103.4	92.3	85.4	80.4	88.5	93.4	83.6	85.4	96.8	73.7	63.9	55.6	64.3	82.0	85.4
					B	106.0	78.1	85.3	92.5	80.6	90.6	83.0	84.1	83.7	80.2	89.2	75.5	78.3	98.0	70.1	62.3	56.1	67.8	80.5	81.2
				Woolong tea	A	101.4	87.0	90.6	118.4	86.8	100.1	82.4	89.6	80.3	89.4	103.3	81.7	88.4	72.7	81.7	77.0	87.6	87.7	81.3	88.8
					B	88.5	71.1	80.5	90.8	73.5	78.4	66.1	76.8	63.3	80.0	73.9	60.6	74.2	64.9	66.2	74.2	71.9	69.4	69.3	73.3
			GC-MS/MS	Green tea	A	114.2	102.5	92.1	114.8	79.0	104.5	91.2	88.7	84.1	98.0	94.4	113.6	72.8	107.1	86.9	77.7	71.2	75.7	77.1	91.9
					B	115.9	89.0	98.2	91.6	79.0	95.7	81.9	83.4	95.9	84.6	85.2	95.6	73.4	118.6	80.5	77.1	77.7	79.1	69.4	88.0
				Woolong tea	A	121.4	83.6	83.8	127.6	97.2	96.0	86.0	93.2	86.2	94.9	106.6	88.6	87.2	78.4	94.8	83.5	84.1	95.0	75.3	92.8
					B	100.3	68.0	72.5	93.8	75.9	72.9	72.5	77.1	66.0	87.9	72.7	67.2	75.5	66.2	77.1	69.7	66.2	74.5	68.2	75.0
		Envi-Carb +PSA	GC-MS	Green tea	A*	128.8	105.9	98.4	104.9	108.1	144.0	101.8	97.6	97.5	93.9	106.3	97.4	80.5	63.5	72.5	69.7	70.4	99.0	84.7	96.0
					B*	128.6	103.1	97.2	89.7	105.1	190.9	89.3	86.1	85.5	91.0	81.5	108.8	69.3	66.8	64.2	63.7	62.3	96.5	89.9	93.1
				Woolong tea	A	117.3	76.8	96.9	88.6	102.7	112.3	84.2	81.9	82.0	82.1	97.4	86.8	85.2	82.4	86.1	88.2	66.5	88.6	77.0	88.6
					B	90.0	53.0	92.7	79.9	76.2	95.3	85.0	64.4	54.3	66.5	87.0	76.3	70.4	69.3	70.5	60.7	59.5	76.7	64.8	73.3
			GC-MS/MS	Green tea	A*	136.1	95.5	101.2	107.4	103.1	126.2	108.1	106.2	92.6	92.3	101.7	166.1	87.1	92.6	101.3	109.1	76.5	81.0	92.2	104.0
					B	183.4	86.2	97.4	91.3	98.9	126.0	80.3	102.2	87.9	91.3	78.6	160.0	76.9	89.3	84.5	96.8	65.4	81.3	93.5	98.5
				Woolong tea	A	154.4	70.7	165.9	90.4	135.9	111.6	66.9	77.6	82.0	87.5	81.2	69.1	76.6	78.6	97.3	72.8	56.2	92.6	75.3	91.7
					B	87.5	55.7	88.8	79.3	98.7	89.2	62.6	59.5	54.3	73.9	76.3	61.3	57.2	71.7	76.3	47.6	51.6	70.6	68.2	70.0
158	fenazaquin	Cleanert TPT	GC-MS	Green tea	A	46.3	44.6	31.9	37.6	35.9	32.5	33.5	32.5	30.7	23.3	31.8	28.9	30.3	30.2	30.7	25.9	30.1	30.0	29.7	32.4
					B	43.9	42.7	27.3	34.0	30.6	30.1	30.4	29.7	27.8	24.8	28.8	26.1	29.0	28.6	28.0	24.6	29.4	30.4	26.0	30.1
				Woolong tea	A	44.2	32.5	34.2	38.6	31.5	34.6	28.2	32.5	29.5	35.0	24.2	32.7	32.8	30.5	30.7	29.8	39.0	38.6	33.0	33.3
					B	39.3	28.4	30.2	32.8	29.6	29.4	26.4	29.8	22.5	29.8	19.2	27.4	27.5	26.6	24.5	26.3	30.7	33.8	28.9	28.6
			GC-MS/MS	Green tea	A	47.3	35.6	42.4	40.2	31.7	33.1	33.5	30.3	29.4	31.0	31.5	24.9	27.1	31.1	29.9	24.4	23.4	24.5	27.1	31.5
					B	44.1	35.3	35.7	33.2	28.3	29.9	30.1	27.8	29.0	29.8	28.6	24.3	25.5	33.3	27.0	23.4	23.6	22.4	23.8	29.3
				Woolong tea	A	39.9	33.1	32.4	40.3	30.8	33.9	29.2	30.4	33.1	36.5	35.4	35.5	35.6	28.2	28.8	28.0	28.6	27.8	28.6	32.4
					B	36.2	29.8	27.8	32.9	30.2	28.0	25.3	28.2	27.5	31.7	26.5	27.7	28.9	24.7	24.4	23.2	23.3	24.2	24.7	27.7
		Envi-Carb +PSA	GC-MS	Green tea	A	49.2	38.8	33.1	35.2	39.2	36.2	30.7	37.8	27.2	27.7	24.3	45.1	30.9	32.5	30.2	33.0	37.0	40.1	35.9	35.2
					B	45.9	33.6	35.4	32.1	37.4	29.5	31.1	35.8	25.7	27.6	25.0	28.9	28.1	29.1	27.8	27.8	33.6	33.6	35.7	31.8
				Woolong tea	A	47.6	39.8	35.6	35.6	34.3	32.6	34.1	25.0	33.1	31.3	34.8	21.5	31.0	32.7	31.3	34.3	34.8	43.0	34.4	34.2
					B	40.6	32.6	37.0	31.9	26.9	26.6	27.8	22.1	26.3	24.7	27.3	21.0	26.9	27.1	26.0	27.2	27.9	36.1	31.3	28.8
			GC-MS/MS	Green tea	A	44.2	37.3	35.9	38.8	37.7	36.6	32.7	38.0	32.0	30.1	28.6	38.1	32.7	38.5	28.1	33.4	30.5	28.8	29.5	34.3
					B	40.4	30.7	34.4	33.8	36.8	33.4	31.3	33.7	29.1	29.5	26.3	36.5	27.5	31.2	26.8	27.3	26.7	25.5	28.8	31.0
				Woolong tea	A	41.3	36.5	42.0	36.8	34.4	31.2	31.1	26.1	33.1	30.2	30.3	22.6	30.7	29.4	32.1	24.9	21.6	30.4	28.6	31.2
					B	36.5	32.1	36.7	32.3	27.7	25.6	25.8	21.7	26.3	24.7	27.3	19.9	32.6	25.0	23.0	19.9	18.9	27.5	25.7	26.8
159	fenchlorphos	Cleanert TPT	GC-MS	Green tea	A*	247.9	196.7	336.5	240.4	193.3	207.0	208.6	206.3	197.4	189.6	210.5	196.6	202.0	187.1	197.5	172.2	168.7	154.9	174.0	204.6
					B*	245.3	201.0	309.2	219.8	189.0	201.0	188.9	192.5	207.6	181.8	190.5	182.2	189.8	174.2	188.9	164.8	163.9	156.0	162.3	195.2
				Woolong tea	A	225.5	218.6	220.6	248.4	200.3	209.6	202.3	236.6	200.0	224.8	201.8	207.2	201.1	196.2	184.2	193.5	206.2	187.4	171.0	207.1
					B	211.8	188.0	185.1	216.2	185.6	177.9	150.6	199.5	173.6	196.0	168.8	176.1	164.0	177.3	160.6	172.3	173.6	170.2	151.8	178.9
			GC-MS/MS	Green tea	A	226.4	219.8	206.6	202.9	189.0	235.7	219.6	205.5	190.5	228.4	212.8	258.5	181.4	186.5	196.5	157.2	158.8	165.0	169.1	200.5
					B	229.7	190.9	2'0.4	213.7	198.1	213.6	191.8	192.0	199.9	223.1	180.9	224.2	170.9	189.2	181.4	146.2	163.5	167.1	158.8	191.9
				Woolong tea	A	303.5	220.8	200.7	254.8	211.6	211.1	197.3	215.8	208.8	200.6	229.6	215.6	208.0	197.9	190.5	187.0	204.2	197.2	169.9	211.8
					B	263.9	188.1	168.9	209.1	185.8	170.6	156.2	187.1	177.6	185.0	184.4	182.6	167.4	172.7	166.7	159.5	171.2	174.6	157.2	180.5
		Envi-Carb +PSA	GC-MS	Green tea	A*	234.2	240.9	221.8	234.6	237.8	235.8	214.4	232.2	198.7	205.5	184.4	352.4	198.2	209.9	183.1	199.7	203.6	184.2	190.0	219.0
					B*	214.3	208.1	217.2	225.6	228.1	239.5	190.7	206.3	185.0	205.4	171.8	299.9	182.5	196.3	183.6	173.8	185.9	185.8	188.7	204.6
				Woolong tea	A	240.9	237.8	254.4	224.7	221.2	212.4	183.6	179.9	215.7	192.8	208.9	169.7	190.8	198.2	195.7	184.8	158.5	207.1	162.2	201.0
					B	198.4	191.3	208.5	204.2	177.0	187.8	188.0	153.3	171.5	157.2	187.3	158.6	175.1	176.2	160.5	156.7	139.2	177.8	148.3	174.6
			GC-MS/MS	Green tea	A	278.0	236.8	232.7	240.2	233.4	220.0	229.4	229.1	209.3	207.9	184.7	128.0	187.2	223.0	185.7	205.1	197.0	174.1	203.2	210.8
					B	340.1	220.6	220.6	230.1	221.2	169.9	169.2	220.1	194.9	202.9	174.8	123.4	179.9	173.6	181.8	170.0	172.0	179.6	206.8	199.8
				Woolong tea	A	322.2	223.1	349.8	225.4	226.9	209.7	202.0	183.1	215.7	176.4	188.4	163.2	171.1	181.1	195.7	161.8	137.0	212.0	169.9	206.0
					B	211.4	192.2	218.4	193.7	187.2	180.9	181.9	151.2	171.5	150.2	173.9	145.2	146.7	168.8	150.5	130.9	127.0	172.6	157.2	169.0

No.	Compound	Cartridge	Method	Tea	Level																				
160	fenoxanil	Cleanert TPT	GC-MS	Green tea	A	91.8	84.3	87.4	84.6	82.6	72.3	70.7	76.3	72.5	59.5	75.3	71.6	69.8	76.5	72.3	64.0	62.5	58.9	71.1	73.9
				Green tea	B	90.8	82.3	81.8	77.3	71.8	67.2	69.3	67.8	69.3	54.9	64.9	69.6	69.7	74.3	65.6	63.2	60.6	55.7	60.7	69.3
				Woolong tea	A	95.8	81.1	79.0	81.8	74.5	83.2	70.4	80.0	92.7	90.3	70.3	89.1	77.4	72.0	74.5	75.8	75.3	62.1	64.9	78.4
				Woolong tea	B	86.1	67.0	66.4	75.4	68.9	70.6	63.7	72.3	78.5	74.6	57.9	74.3	63.3	62.5	65.0	60.6	60.7	64.9	57.5	67.9
			GC-MS/MS	Green tea	A	100.0	64.0	99.4	72.3	82.7	64.1	75.2	59.1	67.9	52.2	70.1	83.7	78.6	92.1	66.1	59.2	58.2	60.9	70.0	72.4
				Green tea	B	100.4	72.6	79.3	72.3	70.2	58.9	66.2	57.1	64.1	44.8	61.5	73.4	70.9	77.0	58.1	56.7	58.1	58.5	60.5	66.3
				Woolong tea	A	96.2	76.2	72.8	83.2	70.8	64.7	64.9	71.9	82.8	81.5	82.9	65.2	81.1	62.8	66.0	71.5	73.2	69.1	63.0	73.7
				Woolong tea	B	79.0	62.1	61.3	72.6	69.8	53.5	46.5	66.7	69.7	67.6	61.4	51.4	66.7	51.5	56.7	56.9	62.1	63.2	56.6	61.9
		Envi-Carb +PSA	GC-MS	Green tea	A	114.1	95.2	83.8	87.0	85.8	90.5	63.9	89.3	72.2	80.8	68.3	109.3	77.9	83.8	75.7	81.8	72.3	73.7	64.5	82.6
				Green tea	B	111.3	77.3	81.0	77.7	86.5	70.1	73.3	82.7	66.3	72.9	65.2	86.3	67.5	75.9	68.4	67.2	62.8	63.7	67.9	74.9
				Woolong tea	A	92.0	111.4	80.2	82.4	81.1	71.2	80.3	62.0	79.8	82.7	80.0	66.3	76.6	85.2	81.1	66.3	62.4	83.4	58.9	78.1
				Woolong tea	B	76.2	105.1	80.3	70.8	65.3	59.6	67.1	53.1	70.8	61.2	67.1	59.5	61.4	66.8	61.1	57.5	55.1	65.3	53.5	66.1
			GC-MS/MS	Green tea	A	105.5	72.6	103.4	75.9	97.9	102.0	66.4	89.0	64.4	89.0	56.3	99.1	73.2	76.9	69.8	86.4	70.9	71.2	74.5	80.3
				Green tea	B	97.5	60.0	99.3	65.9	85.8	80.1	84.5	75.3	67.2	70.5	50.4	87.1	59.4	74.7	64.9	72.2	63.4	64.7	73.6	73.6
				Woolong tea	A	87.0	125.9	87.9	97.7	67.8	77.6	80.7	56.0	79.8	71.6	63.0	59.0	72.1	72.6	84.0	61.4	55.4	75.3	63.0	76.5
				Woolong tea	B	81.5	111.0	79.1	72.6	61.5	60.0	69.4	47.4	70.8	88.1	60.0	49.9	64.3	60.2	53.4	54.8	52.4	63.7	56.6	64.4
161	fenpropidin	Cleanert TPT	GC-MS	Green tea	A	99.7	59.2	86.9	73.0	64.8	53.6	64.4	75.3	61.4	55.6	69.4	66.0	59.7	80.3	64.5	54.2	54.2	48.1	62.3	65.6
				Green tea	B	95.0	56.7	74.1	67.0	54.7	52.2	59.2	62.0	54.7	48.7	60.5	64.5	56.6	83.3	53.5	51.0	50.3	44.9	55.3	59.9
				Woolong tea	A*	88.9	73.1	102.3	116.0	95.7	101.8	90.1	100.9	172.2	209.8	209.9	130.8	126.9	124.0	126.4	123.7	123.6	120.9	118.9	124.0
				Woolong tea	B	80.3	58.0	65.9	75.7	57.0	62.2	53.3	57.1	73.5	70.4	51.0	57.1	57.3	56.0	53.9	54.0	51.3	60.3	51.0	60.3
			GC-MS/MS	Green tea	A	90.6	68.8	88.9	67.7	71.1	66.5	65.1	64.5	60.2	61.0	70.4	56.8	62.1	85.8	65.1	54.7	53.6	50.4	63.0	66.6
				Green tea	B*	81.8	69.4	80.0	67.2	57.9	59.8	58.9	57.8	56.1	54.7	60.8	56.9	56.9	52.3	55.0	52.3	51.2	47.3	56.3	61.8
				Woolong tea	A	80.2	75.3	100.7	126.2	103.3	118.8	109.0	129.9	156.5	156.3	149.3	167.2	181.8	148.8	147.1	148.0	155.1	155.1	134.5	133.8
				Woolong tea	B	78.4	58.5	66.2	83.9	62.0	62.4	53.3	66.6	70.4	70.9	52.7	71.2	67.9	59.3	55.2	53.6	52.3	60.7	46.7	62.8
		Envi-Carb +PSA	GC-MS	Green tea	A	106.6	69.6	74.0	71.9	65.5	68.6	47.4	79.0	73.0	57.3	63.3	83.7	57.2	69.8	57.5	70.2	62.0	60.2	57.3	68.1
				Green tea	B	101.5	64.2	68.9	62.3	68.3	51.7	29.8	76.4	70.5	54.6	47.4	62.6	50.3	56.8	51.8	54.6	52.5	50.2	57.3	59.6
				Woolong tea	A	98.8	92.8	103.4	103.5	110.6	89.8	113.1	91.7	0.0	188.0	221.9	238.1	120.2	130.3	140.9	119.6	110.5	150.5	108.7	122.8
				Woolong tea	B	74.9	65.9	65.6	62.2	49.0	51.3	54.4	54.5	0.0	52.4	61.2	56.2	56.4	53.6	54.8	49.0	47.9	56.9	45.2	53.2
			GC-MS/MS	Green tea	A	94.6	67.3	72.9	77.6	75.8	73.9	50.2	76.8	63.2	58.2	75.1	84.2	59.6	81.2	58.1	69.7	63.0	56.1	61.9	69.4
				Green tea	B	86.0	60.6	70.1	66.7	79.5	67.5	63.5	67.1	58.3	57.4	54.3	80.0	50.3	55.5	51.9	55.2	53.8	50.0	61.0	62.6
				Woolong tea	A*	82.3	99.3	138.9	118.5	117.6	111.0	117.1	109.8	128.9	139.9	124.4	94.3	153.1	147.8	178.9	133.9	128.3	178.9	134.5	128.3
				Woolong tea	B*	62.7	73.4	72.1	65.1	53.3	51.7	57.8	48.3	59.0	49.6	59.8	39.3	49.3	51.1	50.1	40.1	45.7	55.7	46.7	54.6
162	fenson	Cleanert TPT	GC-MS	Green tea	A	47.8	36.0	56.6	39.3	32.4	42.8	38.9	42.2	36.4	39.8	40.2	41.6	37.2	40.3	33.6	39.8	30.3	32.5	36.0	39.1
				Green tea	B	47.6	34.7	52.7	38.7	33.0	40.7	37.8	38.5	40.9	40.2	37.4	36.6	35.9	36.5	39.1	30.8	30.8	31.6	32.5	38.2
				Woolong tea	A	38.8	37.5	43.0	48.6	46.1	44.8	38.3	32.4	37.0	38.9	41.6	39.3	34.9	32.7	41.3	42.1	38.9	34.7	33.1	39.2
				Woolong tea	B	35.5	33.4	38.8	40.4	39.8	37.2	27.2	32.5	29.8	37.2	31.4	32.1	32.0	30.8	33.9	35.7	31.8	30.7	30.1	33.7
			GC-MS/MS	Green tea	A	54.9	44.5	42.1	40.9	32.9	43.9	39.4	40.3	33.7	39.7	39.7	56.8	29.6	38.3	36.8	34.7	28.6	30.3	39.0	39.2
				Green tea	B	52.4	40.7	40.2	40.3	33.0	38.8	36.5	38.3	40.5	40.0	37.6	44.0	28.9	41.8	35.9	33.5	29.8	29.8	34.4	37.7
				Woolong tea	A	64.5	36.7	36.6	51.7	38.1	43.5	36.7	40.5	36.2	42.3	43.0	41.6	38.2	34.1	38.9	38.7	36.2	40.9	32.9	40.6
				Woolong tea	B	49.4	31.9	31.7	42.8	33.3	32.6	30.0	34.9	28.4	41.1	30.5	31.2	32.7	29.9	31.8	33.8	29.3	34.3	29.3	43.8
		Envi-Carb +PSA	GC-MS	Green tea	A*	56.2	47.3	41.9	43.9	42.9	50.5	52.7	42.6	40.6	40.8	40.3	62.3	38.7	39.7	38.9	47.1	41.0	30.5	34.9	40.6
				Green tea	B*	53.2	47.4	39.4	38.5	40.6	44.6	36.7	39.7	38.2	39.2	32.4	58.2	35.6	39.3	39.9	39.5	36.9	34.9	36.5	40.1
				Woolong tea	A	48.1	43.4	47.2	37.5	46.8	55.1	34.8	40.2	40.1	36.1	45.1	38.9	39.3	35.0	39.6	34.9	29.9	36.5	32.7	34.3
				Woolong tea	B	40.3	30.9	42.3	36.1	37.2	51.0	29.6	33.8	31.2	32.6	38.5	34.8	33.3	34.1	35.4	24.5	25.5	37.5	27.9	42.3
			GC-MS/MS	Green tea	A	63.3	48.3	43.9	44.2	43.2	44.4	51.4	42.3	42.4	40.6	42.3	34.4	33.4	40.3	35.7	41.0	39.6	33.1	39.4	39.5
				Green tea	B	66.0	53.8	39.4	38.5	41.0	45.3	40.0	37.7	38.1	38.0	32.5	31.8	32.2	37.6	35.4	32.5	35.1	32.7	40.0	37.8
				Woolong tea	A	42.8	36.2	47.4	39.6	46.7	46.0	37.7	35.0	40.1	33.6	37.3	38.0	36.4	35.0	40.0	31.8	26.0	35.9	32.9	32.1
				Woolong tea	B	37.4	31.1	38.0	35.6	36.2	40.0	32.2	27.7	31.2	30.0	34.1	34.7	29.4	33.2	32.7	21.8	23.1	36.2	29.3	

(Continued)

Annex 1: Content change for 201 pesticides in incurred tea Youden pair samples under 8 conditions with two SPE cartridge cleanup in three months (Nov.9, 2009-Feb.7, 2010) (cont.)

No	Pesticides	SPE	Method	Sample	Youden pair	Nov.9, 2009 (n=5)	Nov.14, 2009 (n=3)	Nov.19, 2009 (n=3)	Nov.24, 2009 (n=3)	Nov.29, 2009 (n=3)	Dec.4, 2009 (n=3)	Dec.9, 2009 (n=3)	Dec.14, 2009 (n=3)	Dec.19, 2009 (n=3)	Dec.24, 2009 (n=3)	Jan.3, 2010 (n=3)	Jan.8, 2010 (n=3)	Jan.13, 2010 (n=3)	Jan.18, 2010 (n=3)	Jan.23, 2010 (n=3)	Jan.28, 2010 (n=3)	Feb.2, 2010 (n=3)	Feb.7, 2010 (n=3)	AVE, μg/kg.
163	flufenacet	Cleanert TPT	GC-MS	Green tea	A	327.7	305.2	339.7	326.1	247.6	289.5	253.5	311.8	214.7	287.7	273.8	270.3	272.8	274.0	216.9	215.6	184.6	219.4	268.6
				Green tea	B	325.4	284.7	311.0	285.8	264.8	278.1	233.8	302.1	226.1	293.5	229.7	240.3	260.4	272.3	227.7	213.7	192.8	218.2	259.4
				Woolong tea	A	289.6	349.9	330.8	379.0	316.4	314.4	276.2	282.5	233.4	245.2	337.1	287.6	263.9	306.0	298.9	307.1	276.8	223.1	297.5
				Woolong tea	B*	264.3	296.7	280.6	297.1	279.7	269.2	190.9	272.8	258.1	233.3	260.9	263.1	245.2	251.6	303.4	273.6	236.9	207.9	259.5
			GC-MS/MS	Green tea	A	332.6	402.7	247.2	316.6	252.5	323.3	295.9	286.9	246.3	349.0	292.8	188.0	210.0	265.1	198.2	195.6	213.0	236.6	271.1
				Green tea	B	359.6	312.6	258.2	295.3	280.5	291.3	237.9	273.5	275.5	374.0	249.3	191.7	256.7	241.6	182.8	209.7	221.0	211.3	262.2
				Woolong tea	A	388.1	385.7	297.4	370.6	351.7	299.6	297.8	341.0	273.5	242.9	302.4	312.7	277.5	273.9	231.5	237.1	267.3	304.2	305.7
				Woolong tea	B	372.6	296.1	223.7	257.3	280.5	226.2	261.6	284.5	201.5	227.1	249.4	227.3	241.4	234.4	196.4	224.5	215.7	289.4	316.4
		Envi-Carb+PSA	GC-MS	Green tea	A*	310.3	358.6	307.0	334.5	305.3	282.1	267.7	267.6	300.9	268.1	700.0	277.0	272.5	308.1	315.3	340.0	316.6	244.1	295.3
				Green tea	B*	346.7	369.9	257.3	309.1	284.9	352.7	257.2	234.5	267.3	254.2	611.0	233.8	275.7	282.5	232.6	187.0	269.6	228.7	297.6
				Woolong tea	A	289.0	241.5	339.6	290.9	364.6	379.5	272.2	292.0	317.9	207.2	290.0	296.5	287.7	312.6	300.9	161.8	218.2	206.8	255.4
				Woolong tea	B	387.4	326.3	327.9	289.8	283.1	341.5	258.6	228.7	214.6	143.2	277.5	275.3	271.0	269.2	259.7	346.7	146.8	221.1	295.7
			GC-MS/MS	Green tea	A	410.1	275.1	322.5	347.5	333.1	368.8	354.8	310.9	237.6	286.4	280.2	228.4	328.6	250.7	276.5	259.6	200.6	281.2	273.5
				Green tea	B	372.4	322.3	297.9	317.5	289.2	367.2	276.2	299.5	241.6	285.3	257.0	202.1	248.6	234.5	223.3	196.0	325.6	304.1	282.4
				Woolong tea	A	269.3	279.0	406.8	284.2	412.2	340.9	247.5	270.6	317.9	211.1	201.2	217.1	199.5	271.3	218.1	151.2	242.1	289.4	229.3
				Woolong tea	B	296.0	270.0	296.1	246.7	285.7	296.0	240.3	212.8	214.6	178.3	178.4	178.1	206.5	198.6	158.5	151.2	180.0	200.0	225.0
164	furalaxyl	Cleanert TPT	GC-MS	Green tea	A	89.1	125.1	95.5	77.8	69.3	67.6	68.5	67.3	66.6	60.1	65.8	62.1	74.5	67.3	56.6	55.6	53.2	62.7	71.4
				Green tea	B	88.1	114.3	91.7	70.6	62.8	63.8	63.8	59.9	64.1	55.5	62.8	59.0	76.0	58.8	56.9	54.9	50.9	55.2	66.9
				Woolong tea	A	89.6	72.0	73.7	83.2	69.5	71.8	66.2	74.2	74.7	81.3	80.8	72.8	69.0	66.6	69.5	71.3	63.6	58.9	72.5
				Woolong tea	B	79.2	62.1	62.9	71.4	63.2	60.9	51.9	63.2	61.5	68.0	63.4	57.3	60.1	56.3	58.0	56.4	60.4	52.9	61.2
			GC-MS/MS	Green tea	A	94.6	72.6	86.5	76.5	71.0	71.5	71.4	66.9	62.6	65.5	87.6	62.1	72.4	65.7	54.5	53.7	52.7	60.2	69.2
				Green tea	B	90.4	71.6	76.3	71.1	64.7	64.7	63.8	59.5	61.8	64.3	75.6	60.5	78.7	58.8	52.6	54.5	51.3	53.3	64.9
				Woolong tea	A	87.1	67.5	69.9	83.5	65.8	72.3	68.0	71.2	73.3	77.5	80.4	78.1	65.8	63.9	64.5	68.7	63.8	53.4	71.3
				Woolong tea	B	77.2	60.3	59.3	70.7	61.9	58.0	53.0	61.0	62.9	67.1	61.2	58.8	57.0	53.5	53.0	54.9	59.8	49.7	59.8
		Envi-Carb+PSA	GC-MS	Green tea	A	100.1	79.5	78.5	79.2	78.7	83.2	58.0	79.3	66.4	61.8	77.7	64.0	70.2	63.0	74.2	66.8	66.1	61.9	72.0
				Green tea	B	94.6	69.8	75.4	73.5	76.6	73.6	65.2	73.9	62.9	61.8	65.1	58.2	63.4	58.4	59.5	58.6	58.0	62.5	66.7
				Woolong tea	A	87.9	78.5	81.9	73.7	75.9	68.4	71.0	60.1	79.1	71.8	60.3	67.7	70.3	74.4	64.2	56.8	75.9	54.7	70.9
				Woolong tea	B	74.9	69.7	74.2	66.2	60.4	58.6	63.5	52.0	59.4	54.2	52.9	57.9	60.3	57.5	54.8	50.2	59.7	49.5	60.0
			GC-MS/MS	Green tea	A	92.9	77.5	81.0	80.6	79.2	78.5	60.3	81.7	68.7	61.7	64.0	62.1	79.7	62.2	74.0	63.9	62.1	66.1	71.6
				Green tea	B	90.2	65.4	77.0	73.0	77.7	82.0	81.7	75.0	65.1	62.2	59.5	55.8	66.3	57.3	59.6	56.1	55.4	65.3	66.2
				Woolong tea	A	87.9	78.5	94.0	77.8	73.5	74.7	67.1	59.9	79.1	68.8	59.3	63.9	61.9	72.2	54.3	48.3	73.4	53.4	69.1
				Woolong tea	B	74.4	68.4	75.1	68.3	59.9	59.5	69.2	50.2	59.4	52.1	51.0	58.8	55.7	51.4	45.8	45.2	58.0	49.7	58.2
165	heptachlor	Cleanert TPT	GC-MS	Green tea	A	120.3	98.9	90.2	105.3	82.8	95.1	93.0	93.3	89.0	86.8	91.2	86.9	91.2	99.7	101.2	92.7	82.6	82.0	90.0
				Green tea	B	110.7	88.9	76.1	100.1	79.9	88.9	86.0	90.3	97.2	81.7	83.9	82.0	83.9	85.3	76.7	72.7	85.4	83.2	85.2
				Woolong tea	A	104.3	97.3	107.3	115.7	87.9	99.2	93.1	111.5	96.6	99.3	102.5	94.3	99.2	96.4	96.6	97.1	98.0	78.7	97.8
				Woolong tea	B	95.7	81.0	84.9	98.8	79.7	82.0	69.0	94.2	81.3	88.3	78.6	78.7	80.3	80.8	88.8	79.5	82.9	69.8	92.9
			GC-MS/MS	Green tea	A	121.6	98.2	102.9	114.2	85.8	98.7	93.7	90.4	87.6	89.6	111.8	79.4	89.0	87.7	73.3	70.2	81.9	72.7	91.8
				Green tea	B	117.8	90.1	98.1	95.0	82.4	90.3	85.9	85.0	93.3	81.9	97.1	72.6	88.8	80.8	70.7	70.9	74.0	79.9	86.0
				Woolong tea	A	123.7	91.9	94.6	117.5	92.4	98.0	91.8	101.9	101.3	113.9	105.4	93.3	88.9	97.3	90.4	92.3	96.4	72.2	98.8
				Woolong tea	B	109.8	78.2	80.0	100.9	81.9	79.9	72.3	91.2	80.3	97.9	81.1	75.1	77.7	82.0	77.7	78.1	83.2	70.3	82.9
		Envi-Carb+PSA	GC-MS	Green tea	A	124.7	106.9	100.9	107.1	104.5	121.3	102.8	98.5	87.8	88.5	81.0	86.9	91.2	99.7	101.2	92.7	82.6	82.0	97.5
				Green tea	B	117.6	97.7	99.6	95.6	105.2	132.7	93.8	89.9	83.3	87.5	88.0	81.3	91.8	95.4	86.7	86.3	85.4	83.2	93.8
				Woolong tea	A	110.1	107.0	112.1	105.4	111.2	106.1	75.7	85.4	98.4	91.6	88.4	92.5	99.2	106.3	91.1	75.5	98.0	78.7	96.7
				Woolong tea	B	87.5	82.5	96.1	92.0	81.1	90.7	82.6	69.3	73.8	75.0	79.0	81.1	84.9	93.5	69.3	65.3	82.9	69.8	81.4
			GC-MS/MS	Green tea	A	138.0	108.8	101.2	105.5	107.3	104.3	94.3	102.7	76.5	89.8	104.1	84.8	92.2	91.8	102.5	84.0	77.3	85.4	96.8
				Green tea	B	173.8	101.1	97.9	93.4	111.4	104.1	66.6	99.6	66.6	89.8	103.1	78.8	78.1	85.7	89.7	74.1	77.0	87.2	92.8
				Woolong tea	A	156.2	101.9	167.5	105.0	111.4	98.5	86.4	83.9	98.4	95.3	73.5	82.4	87.7	104.9	78.3	65.2	106.0	77.4	98.4
				Woolong tea	B	102.9	82.9	97.7	90.3	84.0	82.4	77.0	66.3	73.8	79.5	64.5	64.4	78.1	81.3	60.8	59.5	80.4	70.3	77.9

No.	Compound	Cartridge	Method	Tea	Rep	1	2	3	4	5	6	7	8	9	10	11	12	13	14	15	16	17	18	19	20
166	iprobenfos	Cleanert TPT	GC-MS	Green tea	A	154.0	128.5	128.3	112.1	87.6	75.4	98.7	92.5	81.2	63.7	92.7	85.3	89.9	109.2	95.0	83.8	78.1	65.3	83.3	95.0
					B	146.0	130.2	116.4	106.8	74.2	77.0	93.3	85.3	86.8	58.9	84.4	83.5	82.7	120.0	92.8	81.7	82.6	63.2	74.9	91.6
				Woolong tea	A*	122.7	82.1	101.4	109.5	79.0	103.1	105.7	89.9	96.4	136.1	79.2	76.7	97.3	85.8	93.8	94.2	98.4	87.3	80.8	96.3
					B	115.1	79.2	93.8	105.3	74.6	87.0	71.2	93.0	70.3	100.4	61.3	64.5	79.0	77.0	77.9	82.0	76.0	80.8	72.1	81.8
			GC-MS/MS	Green tea	A	125.0	98.9	111.0	97.1	87.3	89.0	96.5	83.7	86.6	86.8	93.8	127.1	75.6	94.6	87.3	73.0	70.9	68.2	82.5	92.2
					B	123.3	95.7	101.1	123.5	78.3	83.0	87.3	99.2	87.9	82.6	82.7	107.6	69.3	97.7	77.9	70.0	72.5	66.4	73.9	86.2
				Woolong tea	A	143.6	92.0	98.4	106.9	88.8	96.0	97.2	87.2	99.6	116.4	105.4	109.5	103.2	87.8	91.3	88.1	91.6	90.0	75.7	99.9
					B	123.0	81.5	82.8	101.7	77.6	78.3	66.3	115.9	77.1	96.1	73.9	79.9	80.5	74.9	77.3	73.7	71.5	81.2	65.5	81.9
		Envi-Carb +PSA	GC-MS	Green tea	A	141.6	112.1	108.0	84.2	120.5	114.3	90.9	101.0	72.5	88.4	90.6	135.4	87.8	96.9	94.6	103.7	90.4	82.1	84.8	101.7
					B	139.5	121.3	109.7	101.8	110.6	95.2	89.8	66.4	67.8	88.1	82.1	103.1	80.3	90.9	89.7	86.0	83.4	80.3	83.5	94.0
				Woolong tea	A	124.9	93.9	112.3	95.1	100.5	106.1	81.2	53.4	99.4	111.0	98.9	71.8	93.4	98.5	106.4	89.5	80.1	106.2	78.0	95.8
					B	104.3	84.1	88.0	102.3	69.3	87.7	79.1	108.2	75.6	85.4	80.6	65.9	80.3	82.2	87.4	73.4	67.6	84.0	65.6	79.4
			GC-MS/MS	Green tea	A	152.3	113.7	104.6	91.8	114.8	116.9	91.1	97.8	87.8	85.8	81.9	104.1	85.9	107.3	84.0	100.8	86.7	79.1	83.7	99.5
					B	183.6	104.1	99.6	114.9	108.9	106.5	89.4	78.5	81.7	86.3	76.8	97.4	75.6	82.5	76.7	82.2	75.5	75.3	84.9	93.5
				Woolong tea	A	150.3	105.8	168.9	96.5	107.3	97.5	88.6	60.8	99.4	94.8	90.4	72.7	87.3	90.4	100.2	76.1	66.7	101.7	75.7	98.3
					B	104.2	87.8	98.7	92.0	77.4	75.8	78.8	76.6	75.6	75.4	81.5	62.8	72.1	74.8	73.8	61.4	60.0	76.3	65.5	76.8
167	isazofos	Cleanert TPT	GC-MS	Green tea	A	97.7	67.7	89.6	81.7	65.9	61.7	70.3	70.7	74.3	66.7	80.0	75.0	74.3	68.2	65.1	59.2	58.8	52.4	65.2	71.6
					B	97.1	67.0	84.7	84.6	61.5	61.6	64.2	87.5	76.6	62.8	73.7	74.3	71.6	62.6	60.9	57.3	58.1	51.1	59.0	68.2
				Woolong tea	A	80.4	63.5	74.7	78.6	66.0	74.7	67.4	79.0	82.4	96.2	78.9	75.9	81.5	69.2	55.0	72.3	71.0	65.5	60.7	74.1
					B	72.2	56.8	65.2	82.8	60.7	62.2	52.0	69.3	71.5	91.8	59.9	65.6	71.1	62.9	46.4	65.1	58.4	59.4	52.6	64.8
			GC-MS/MS	Green tea	A	86.3	72.2	80.5	71.6	64.7	72.3	71.6	62.2	64.8	66.7	69.8	61.3	64.7	68.7	67.6	56.4	54.9	54.0	61.5	67.9
					B	85.1	71.4	73.5	88.5	60.6	67.9	63.9	71.9	65.0	64.2	61.4	59.4	57.7	68.0	62.7	55.2	55.3	54.5	53.7	63.9
				Woolong tea	A	90.4	64.9	68.3	74.9	66.3	69.4	66.2	62.7	71.5	75.3	75.9	76.7	76.2	66.0	64.1	60.0	72.3	63.0	55.0	70.6
					B	79.5	55.6	58.7	86.9	59.3	54.3	49.6	71.7	59.2	65.0	54.9	60.4	61.2	55.3	55.9	54.4	57.4	54.9	50.8	59.2
		Envi-Carb +PSA	GC-MS	Green tea	A	96.2	75.6	77.8	78.5	77.6	76.4	71.5	64.5	77.2	78.1	76.5	95.4	72.6	79.0	74.2	76.5	70.6	71.0	63.4	77.3
					B	89.6	84.5	78.0	76.7	78.1	60.0	68.4	77.6	75.5	76.6	72.5	83.9	71.9	69.1	68.4	65.3	66.4	63.6	63.0	72.5
				Woolong tea	A	86.2	72.4	81.1	68.7	78.1	71.7	96.1	72.7	72.7	89.5	85.5	77.5	75.0	81.5	82.2	67.2	58.4	75.8	56.1	76.9
					B	69.6	58.9	70.7	77.4	58.0	61.0	61.0	81.3	57.3	76.5	81.0	77.5	67.9	69.4	64.5	56.8	50.3	59.7	48.5	64.7
			GC-MS/MS	Green tea	A	96.4	73.1	76.8	70.1	81.5	76.4	71.5	76.0	72.7	64.4	64.8	89.4	64.7	76.4	60.6	69.6	63.7	63.9	62.7	73.0
					B	105.9	65.5	71.9	79.7	80.7	70.6	63.7	61.0	66.7	62.7	57.8	77.3	61.7	59.4	58.5	58.6	59.6	64.0	61.9	68.0
				Woolong tea	A	91.7	75.4	108.7	68.2	74.3	72.3	65.2	48.1	72.7	63.1	64.1	52.8	68.9	62.6	65.0	52.2	49.2	70.6	55.0	68.7
					B	66.7	62.1	67.1	76.7	58.3	57.6	57.7	65.5	57.3	52.7	59.4	45.2	59.2	55.7	48.2	43.8	44.3	56.0	50.8	55.7
168	isoprothiolane	Cleanert TPT	GC-MS	Green tea	A	90.4	65.0	42.8	69.3	67.5	67.1	68.2	59.8	65.3	60.9	70.1	66.3	63.3	77.0	66.4	54.3	52.3	52.2	60.6	64.8
					B	88.7	61.8	33.5	84.6	61.3	62.4	64.2	73.4	63.5	58.5	63.3	62.6	59.1	74.0	59.0	54.2	51.8	50.0	52.6	60.5
				Woolong tea	A	80.9	71.6	73.4	73.5	69.6	67.9	62.6	64.5	74.3	80.4	66.4	77.8	81.1	53.0	66.7	66.0	64.7	60.9	54.5	70.0
					B	73.1	63.1	64.0	78.2	63.5	56.8	50.6	66.8	61.2	67.5	53.2	61.6	57.4	52.1	55.1	54.9	52.4	56.1	49.0	59.5
			GC-MS/MS	Green tea	A	95.9	73.0	89.6	73.5	68.6	72.6	70.9	61.0	62.6	65.3	68.1	51.9	62.9	72.7	64.9	52.8	50.8	52.0	59.3	67.3
					B	91.1	72.2	75.5	87.9	61.2	65.7	63.8	70.8	61.7	63.9	60.4	52.0	57.4	74.7	57.7	50.3	51.0	50.4	51.9	62.9
				Woolong tea	A	88.6	67.0	69.6	73.1	67.1	71.1	66.0	60.5	74.5	80.7	77.4	79.1	71.1	65.9	63.0	65.0	64.0	61.5	53.0	70.7
					B	79.4	60.3	59.1	76.8	62.7	58.6	52.1	72.8	62.0	64.4	56.5	59.8	58.0	56.8	54.7	54.7	52.3	55.3	47.9	59.6
		Envi-Carb +PSA	GC-MS	Green tea	A	105.2	79.7	79.4	70.4	78.9	86.0	58.6	68.7	70.9	64.5	60.0	96.3	65.5	74.2	63.7	71.6	54.7	63.3	60.0	73.6
					B	98.6	67.8	76.8	74.0	76.8	73.0	67.5	57.5	67.6	67.7	57.1	72.7	61.0	65.3	58.6	63.7	71.6	56.0	59.7	67.2
				Woolong tea	A	92.6	85.4	82.9	66.5	76.3	70.7	68.7	50.1	74.1	54.6	76.0	58.6	70.1	70.4	72.7	57.2	57.2	70.9	52.8	70.3
					B	77.7	67.6	74.3	78.6	73.1	58.9	69.9	81.4	58.0	62.3	63.4	54.0	60.1	62.6	57.2	61.4	57.4	56.9	47.2	59.1
			GC-MS/MS	Green tea	A	93.5	80.1	80.3	71.4	80.2	81.0	68.6	74.7	69.1	63.0	62.5	97.3	66.2	75.5	63.5	50.5	54.8	59.3	64.0	73.7
					B	86.8	67.7	75.8	77.5	76.8	73.4	68.6	59.2	64.8	65.6	56.8	84.7	59.3	61.5	58.6	73.1	46.4	54.9	62.9	67.2
				Woolong tea	A	83.1	77.0	91.3	67.7	75.3	69.7	67.3	49.2	74.1	53.5	64.7	62.5	62.5	63.0	69.5	58.7	64.3	68.6	53.0	67.7
					B	74.1	66.8	75.0	67.7	60.1	57.6	59.8	49.2	58.0	53.5	59.4	53.5	61.8	54.7	50.6	54.6	56.7	55.7	47.9	57.6

(Continued)

Annex 1: Content change for 201 pesticides in incurred tea Youden pair samples under 8 conditions with two SPE cartridge cleanup in three months (Nov.9, 2009-Feb.7, 2010) (cont.)

No	Pesticides	SPE	Method	Sample	Youden pair	Nov.9 2009 (n=5)	Nov.14 2009 (n=3)	Nov.19 2009 (n=3)	Nov.24 2009 (n=3)	Nov.29 2009 (n=3)	Dec.4 2009 (n=3)	Dec.9 2009 (n=3)	Dec.14 2009 (n=3)	Dec.19 2009 (n=3)	Dec.24 2009 (n=3)	Dec.14 2009 (n=3)	Jan.3 2010 (n=3)	Jan.8 2010 (n=3)	Jan.13 2010 (n=3)	Jan.18 2010 (n=3)	Jan.23 2010 (n=3)	Jan.28 2010 (n=3)	Feb.2 2010 (n=3)	Feb.7 2010 (n=3)	AVE. μg/kg.
169	kresoxim-methyl	Cleanert TPT	GC-MS	Green tea	A	90.3	73.1	108.4	37.6	48.5	32.3	41.8	47.8	39.0	33.1	35.8	34.3	33.4	37.1	30.6	32.5	29.0	33.4	35.4	44.9
					B	100.5	78.9	95.0	33.6	41.5	33.1	39.5	43.8	36.2	28.3	33.8	33.3	33.7	36.1	34.0	30.6	29.6	33.5	36.9	43.8
				Woolong tea	A	45.0	34.2	36.3	34.0	28.2	44.7	37.6	45.2	41.1	52.2	31.9	43.7	34.2	35.4	33.9	36.4	33.6	31.8	29.5	37.3
					B*	40.1	30.6	30.7	33.0	31.6	49.8	28.7	38.5	35.6	42.1	31.5	35.1	30.3	32.1	28.5	31.3	26.9	29.7	26.2	33.3
			GC-MS/MS	Green tea	A	60.1	48.3	40.5	43.6	41.5	38.4	42.6	41.2	34.3	34.6	37.0	34.8	25.7	38.1	33.6	30.1	28.2	30.7	31.4	37.6
					B	57.6	46.8	37.4	37.3	38.3	36.3	34.8	36.0	35.9	34.0	33.3	30.3	25.4	40.2	31.9	28.7	27.6	27.9	25.8	35.0
				Woolong tea	A	49.3	35.1	43.4	40.2	34.9	40.7	35.2	39.2	34.7	41.1	40.4	35.1	36.3	32.1	34.1	34.8	35.3	34.4	32.2	37.3
					B	41.7	31.7	28.9	34.7	37.1	33.6	30.7	37.6	31.5	36.6	32.0	28.3	29.9	29.0	30.4	29.6	31.1	29.8	28.3	32.2
		Envi-Carb +PSA	GC-MS	Green tea	A	65.8	58.6	39.4	40.9	40.0	39.8	28.4	55.4	36.1	29.6	23.2	12.0	30.0	36.0	30.9	39.0	42.6	26.3	37.8	37.5
					B	50.2	58.6	34.0	40.2	48.2	19.9	37.0	43.3	33.5	35.6	26.6	8.1	28.2	34.3	28.7	31.1	39.4	33.7	35.7	35.1
				Woolong tea	A	47.5	30.5	41.5	45.1	25.7	28.1	29.9	40.3	38.2	40.2	38.6	34.8	36.6	38.7	36.3	31.7	24.6	33.1	31.8	35.4
					B	41.0	25.6	37.9	29.2	24.8	24.9	29.8	31.9	30.1	34.3	29.1	32.5	31.3	35.5	32.3	23.7	21.7	29.7	25.9	30.1
			GC-MS/MS	Green tea	A*	43.2	54.9	39.4	47.5	37.6	34.8	27.3	50.7	51.1	32.7	31.5	40.9	28.3	34.3	26.8	32.7	19.4	28.3	48.7	37.4
					B*	36.9	64.4	34.0	56.0	42.9	29.8	29.0	40.8	42.3	36.9	28.5	38.6	30.8	31.8	28.5	26.2	18.8	31.4	47.7	36.6
				Woolong tea	A	47.2	33.3	45.2	41.7	28.1	19.9	62.1	37.3	38.2	30.9	27.9	29.5	31.2	31.0	32.2	28.4	22.5	32.4	32.2	34.2
					B	42.4	28.9	41.0	36.2	27.6	16.9	51.0	32.8	30.1	27.8	26.0	28.6	32.5	28.9	26.8	20.4	20.0	31.4	28.3	30.4
170	mefenacet	Cleanert TPT	GC-MS	Green tea	A*	111.5	61.7	120.1	114.3	95.6	119.5	104.4	126.7	101.8	113.0	121.7	123.5	81.1	138.9	111.5	122.9	135.6	143.2	18.6	109.0
					B*	126.1	59.5	122.7	118.1	93.1	114.3	107.1	118.8	114.2	115.4	110.2	111.2	83.6	151.8	111.3	134.2	119.1	120.5	15.0	108.2
				Woolong tea	A	94.2	102.2	136.6	170.1	116.2	128.3	99.6	120.8	103.3	136.6	121.6	116.8	106.9	89.6	119.6	121.4	121.1	102.0	110.8	117.1
					B*	87.2	102.2	129.3	138.7	108.8	115.2	88.5	114.8	95.1	129.7	93.1	84.6	105.8	86.6	99.4	109.2	94.2	85.9	88.1	103.0
			GC-MS/MS	Green tea	A	127.6	142.2	98.2	160.5	92.8	116.5	170.7	145.0	105.2	124.6	129.1	189.5	76.8	160.2	184.8	150.6	133.1	129.6	159.6	136.7
					B	139.5	130.1	119.9	140.7	96.2	111.2	143.9	132.7	119.7	131.7	122.4	140.2	76.7	156.9	171.1	151.8	138.8	138.2	147.9	132.1
				Woolong tea	A	302.7	106.9	104.5	146.0	114.0	128.1	103.9	127.6	107.8	139.2	123.5	113.2	132.8	94.6	114.8	108.1	108.8	129.2	111.0	127.2
					B	200.3	97.4	101.2	122.6	102.7	105.2	93.3	123.9	69.7	131.1	90.8	93.4	113.2	86.8	97.3	87.1	85.5	104.5	90.4	105.1
		Envi-Carb +PSA	GC-MS	Green tea	A*	104.7	145.3	121.6	111.7	120.5	150.6	148.1	135.1	161.2	103.4	124.5	142.1	84.8	118.7	144.7	215.1	114.4	135.8	0.0	125.4
					B*	126.0	171.5	119.5	94.3	114.4	196.3	161.2	117.6	125.6	116.3	93.2	148.9	78.3	135.7	152.6	168.7	110.8	122.8	0.0	123.4
				Woolong tea	A*	110.0	107.2	123.6	116.1	140.5	144.7	78.0	335.5	104.0	112.4	120.0	103.4	110.8	90.5	122.9	110.0	86.6	108.3	107.0	122.7
					B*	110.0	89.0	124.1	120.7	101.9	126.8	85.8	345.2	83.8	98.3	111.7	107.4	98.6	127.2	127.2	69.1	70.9	100.2	93.9	113.8
			GC-MS/MS	Green tea	A	147.9	131.5	124.9	141.7	103.9	117.3	152.5	126.1	112.2	109.5	122.0	73.6	99.0	98.3	129.5	166.1	147.4	147.6	162.6	128.3
					B	141.0	94.5	119.0	116.8	120.1	125.7	118.5	109.5	100.5	102.5	97.9	65.1	85.0	121.9	130.9	148.6	148.8	151.7	170.2	119.4
				Woolong tea	A	160.8	87.5	134.4	111.7	165.7	137.3	118.7	112.3	104.0	106.9	118.7	85.0	108.6	90.9	118.4	91.9	80.0	120.8	110.8	113.9
					B	135.9	85.3	122.0	104.2	119.4	121.0	96.9	85.2	83.8	88.8	104.7	80.0	86.0	89.3	100.7	63.0	68.1	85.6	89.8	95.2
171	mepronil	Cleanert TPT	GC-MS	Green tea	A	49.0	73.1	48.0	41.9	44.7	41.3	39.8	36.4	40.0	31.2	40.6	40.1	37.9	49.9	41.5	34.5	32.9	33.7	40.4	41.9
					B	50.2	63.0	45.8	39.2	38.8	38.3	37.3	32.3	35.8	31.0	35.8	36.7	36.7	48.7	41.7	33.6	31.8	32.1	34.9	39.1
				Woolong tea	A	44.0	40.3	42.4	42.0	40.0	40.4	33.0	39.8	37.4	37.3	35.6	45.4	39.5	39.1	37.6	38.5	39.3	36.6	34.6	39.1
					B	39.9	35.0	37.8	37.2	38.9	36.4	30.9	36.2	32.2	33.8	36.2	37.8	35.9	36.8	32.1	31.9	32.6	33.0	30.1	34.5
			GC-MS/MS	Green tea	A	52.6	40.7	50.5	41.4	39.1	40.1	40.6	37.8	38.4	35.9	37.8	50.0	31.1	44.7	37.7	33.9	30.6	32.4	37.8	39.6
					B	50.0	40.6	45.2	39.3	34.9	36.5	37.0	33.2	35.2	34.2	34.3	42.9	30.2	45.2	34.4	32.2	30.4	30.6	33.2	36.8
				Woolong tea	A	44.9	35.6	37.9	45.7	36.5	42.8	35.3	38.8	41.2	43.4	43.0	44.2	46.1	36.5	34.4	36.6	39.0	36.7	32.5	39.6
					B	42.5	32.7	33.1	40.4	37.4	36.5	32.0	37.1	34.1	39.3	32.4	36.0	37.6	32.1	31.3	30.0	31.7	32.4	31.6	34.8
		Envi-Carb +PSA	GC-MS	Green tea	A*	53.2	45.1	46.4	47.3	44.7	42.3	29.0	45.2	38.6	42.7	37.9	69.3	37.2	42.4	41.1	44.6	40.7	42.3	39.2	43.6
					B	51.3	55.8	45.4	42.8	47.8	35.6	31.1	40.3	32.9	43.4	34.1	52.0	32.8	41.7	34.1	37.1	38.7	36.3	39.6	40.7
				Woolong tea	A	53.6	57.7	48.0	41.2	45.0	38.2	42.4	29.3	40.5	42.3	41.0	31.3	37.5	39.4	42.5	37.6	34.3	44.6	32.6	41.0
					B	44.2	47.2	52.1	39.8	38.9	32.3	35.4	27.7	35.5	33.8	35.3	29.5	32.9	34.3	34.1	31.1	29.3	37.5	29.4	35.8
			GC-MS/MS	Green tea	A	45.4	46.0	45.0	36.7	41.8	30.5	35.2	48.1	43.2	38.7	37.1	33.2	44.5	38.0	37.6	44.6	38.8	38.4	40.8	40.2
					B	41.9	36.7	44.1	33.3	42.5	26.8	39.0	41.7	39.7	37.0	33.1	31.7	35.8	33.2	35.2	37.8	35.8	35.2	40.1	38.4
				Woolong tea	A	49.7	44.8	52.0	41.6	42.0	37.8	38.8	32.8	40.5	37.2	35.9	26.0	38.6	35.9	40.9	32.2	29.4	40.3	32.5	38.4
					B	42.2	41.1	46.5	36.9	36.7	31.1	32.5	29.7	35.5	30.2	34.5	23.8	37.1	31.8	29.6	26.6	26.8	34.3	30.3	33.5

No.	Compound	Cartridge	Method	Tea	A/B	1	2	3	4	5	6	7	8	9	10	11	12	13	14	15	16	17	18	19	20
172	metribuzin	Cleanert TPT	GC-MS	Green tea	A	147.3	122.0	116.5	87.1	53.8	59.9	61.6	68.6	49.6	45.2	53.3	49.9	52.3	61.6	49.5	50.8	43.3	39.9	44.7	66.1
					B	146.5	115.0	106.2	77.0	55.3	58.7	60.1	63.5	57.9	44.3	47.6	47.7	48.7	54.8	52.4	49.5	42.7	40.1	38.8	63.5
				Woolong tea	A	123.5	89.0	75.4	86.5	66.1	87.1	66.5	65.5	61.2	81.1	59.2	50.2	61.3	55.8	65.7	61.7	58.0	55.6	48.0	69.3
					B	110.3	80.5	66.6	71.6	55.4	66.0	51.0	57.6	43.2	63.1	43.1	39.0	47.8	47.9	50.1	49.1	42.9	45.4	41.4	56.4
			GC-MS/MS	Green tea	A	134.9	89.3	85.9	81.6	65.3	73.3	63.4	64.2	53.6	67.3	57.2	74.4	45.4	67.2	52.6	44.9	42.2	40.9	47.4	65.8
					B	129.1	83.8	80.8	83.1	67.1	64.5	55.5	57.7	61.1	68.8	48.6	74.0	41.9	73.6	50.3	42.2	42.5	41.0	41.4	62.7
				Woolong tea	A	151.3	97.6	80.8	99.8	76.4	82.1	50.3	72.9	66.7	71.4	80.7	52.5	70.5	61.4	59.9	56.0	54.1	54.7	50.0	75.0
					B	122.8	81.1	61.4	78.7	63.7	59.9	64.3	59.1	48.2	59.5	55.4	129.7	52.2	48.3	47.8	43.2	41.2	43.3	41.4	58.4
		Envi-Carb+PSA	GC-MS	Green tea	A*	146.9	91.1	81.5	75.3	71.5	75.8	74.8	67.4	49.9	55.4	48.1	92.1	53.6	56.2	51.2	61.8	56.3	39.5	46.2	70.1
					B	153.8	108.5	72.2	66.7	66.9	52.4	64.9	63.4	50.0	54.4	42.6	54.7	45.7	55.9	46.9	49.5	49.7	47.7	48.2	64.8
				Woolong tea	A	130.0	79.8	81.9	73.9	88.2	95.2	73.5	50.0	77.5	66.1	68.9	47.9	63.6	55.1	64.2	56.8	45.3	56.0	49.9	70.0
					B	118.8	68.1	70.1	69.8	63.7	80.7	61.4	37.9	55.1	50.0	54.2	106.2	49.1	50.4	51.7	38.0	37.7	43.1	41.0	57.3
			GC-MS/MS	Green tea	A*	138.8	116.0	91.5	86.2	76.9	80.4	75.6	66.7	69.4	62.0	48.3	98.8	47.8	70.4	44.8	54.3	57.2	40.8	52.9	72.9
					B	134.2	125.1	81.9	79.5	71.7	80.4	57.3	58.5	64.3	59.8	45.8	59.8	46.5	52.3	43.1	40.6	48.5	44.0	55.5	67.8
				Woolong tea	A	119.2	91.8	98.3	89.1	95.7	90.4	72.8	60.4	77.5	56.5	63.6	59.7	61.4	53.8	65.2	48.5	37.7	54.1	50.0	70.8
					B	108.9	76.4	82.7	74.7	67.8	72.9	60.8	43.9	55.1	44.2	57.7	47.3	47.3	46.8	44.6	33.7	29.6	41.5	41.4	57.1
173	molinate	Cleanert TPT	GC-MS	Green tea	A	38.7	26.1	32.9	34.1	32.1	32.8	33.4	27.7	31.8	29.8	32.3	30.1	33.6	31.5	33.3	27.0	27.5	28.7	30.5	31.4
					B*	37.6	24.4	29.6	31.8	29.4	30.5	31.2	25.1	30.5	28.3	29.6	28.1	30.6	30.1	28.3	26.3	32.5	27.5	27.7	29.2
				Woolong tea	A	37.8	33.4	26.3	38.9	30.9	33.0	28.2	32.0	33.2	34.2	36.3	37.5	35.1	34.1	34.5	34.6	27.8	33.2	28.7	33.4
					B	32.9	25.7	21.1	31.7	27.0	26.2	22.7	26.0	26.9	27.9	28.4	31.2	29.3	30.6	28.6	29.6	25.4	31.3	25.3	27.9
			GC-MS/MS	Green tea	A	42.9	31.6	39.5	35.3	31.3	34.2	34.0	29.6	29.9	31.7	31.9	26.4	31.6	35.2	31.5	26.3	25.7	24.9	29.4	31.7
					B	39.7	31.2	35.2	32.4	29.0	30.9	30.5	27.9	30.2	28.6	29.5	26.1	27.9	37.2	28.5	25.3	28.8	25.3	26.0	29.9
				Woolong tea	A	37.1	28.2	30.2	39.6	29.3	32.4	28.7	31.3	34.2	35.7	35.4	35.7	33.1	29.9	30.5	29.7	23.8	29.3	25.0	31.8
					B	33.6	23.9	25.2	32.4	26.5	25.5	22.2	27.1	27.5	30.0	25.4	26.8	28.2	25.9	25.0	24.7	32.2	26.1	22.6	26.4
		Envi-Carb+PSA	GC-MS	Green tea	A	48.6	38.7	35.6	35.4	38.3	41.9	31.0	34.7	31.4	31.4	29.3	36.7	30.7	33.5	30.6	35.8	28.9	34.8	31.5	34.9
					B	43.5	30.3	34.9	32.2	37.8	35.1	31.4	33.5	30.2	29.8	26.9	31.9	28.4	30.9	26.4	30.8	31.3	30.7	30.2	31.8
				Woolong tea	A	41.3	35.8	35.4	31.8	34.0	32.2	32.6	24.5	33.2	32.0	33.7	28.7	32.8	35.9	38.3	31.2	26.2	37.1	27.4	33.1
					B	31.3	27.2	32.0	27.5	25.1	26.0	25.4	19.9	25.8	25.0	28.5	25.5	28.7	30.2	29.4	26.1	31.5	30.6	25.3	27.1
			GC-MS/MS	Green tea	A	40.3	34.3	36.7	37.1	36.9	35.4	33.8	40.3	33.4	31.6	32.2	34.7	31.4	42.7	31.1	36.1	28.8	31.7	32.2	34.9
					B*	37.4	27.1	34.9	33.9	37.3	33.1	32.7	36.7	29.8	30.7	28.1	32.2	28.0	42.8	27.3	31.1	22.9	28.1	30.8	32.1
				Woolong tea	A	40.0	32.1	47.6	34.3	34.1	32.2	29.5	26.8	33.2	31.4	28.7	25.0	28.9	29.9	33.9	25.3	26.1	30.6	25.0	31.1
					B	29.1	26.0	31.9	28.8	27.5	25.3	25.7	21.7	25.8	24.7	26.5	22.5	26.2	25.0	24.2	19.9	20.0	25.6	22.6	25.2
174	napropamide	Cleanert TPT	GC-MS	Green tea	A	135.4	110.4	119.8	118.1	103.8	102.6	105.6	103.6	103.6	93.3	112.2	100.6	94.1	98.6	104.9	88.9	79.3	79.2	101.4	102.9
					B	133.0	107.8	106.8	107.0	91.1	97.0	97.3	92.7	98.0	87.4	98.3	97.4	82.8	95.4	90.8	87.1	76.2	76.3	85.5	95.1
				Woolong tea	A	133.9	111.9	114.1	125.3	108.2	110.5	96.9	111.8	114.2	121.9	107.7	125.4	110.7	105.8	99.0	103.7	107.9	110.4	90.5	111.0
					B	120.3	96.4	97.1	107.0	97.8	98.4	83.9	94.6	101.3	105.6	83.9	98.1	84.1	90.7	81.6	86.2	86.4	98.4	80.8	94.4
			GC-MS/MS	Green tea	A	144.0	112.1	137.8	113.9	107.9	110.1	107.8	100.5	96.8	97.9	103.6	120.8	103.6	115.0	102.0	83.1	77.7	79.4	93.5	104.7
					B	135.4	113.1	113.9	106.9	94.5	97.5	95.6	90.5	94.3	95.0	92.1	105.7	84.5	122.9	92.1	78.8	78.9	77.7	81.4	97.2
				Woolong tea	A	123.7	100.1	105.4	131.6	100.9	108.8	102.5	108.0	112.5	119.3	117.7	123.1	92.1	100.8	97.5	97.2	104.5	94.5	81.3	107.9
					B	113.1	91.5	88.7	110.5	93.4	87.1	79.4	92.5	95.4	101.5	85.4	92.7	121.2	87.4	83.9	81.7	83.2	87.4	75.4	90.7
		Envi-Carb+PSA	GC-MS	Green tea	A	162.1	121.1	121.7	118.0	122.6	139.4	91.8	126.1	98.3	97.7	91.8	158.8	93.0	105.5	84.5	110.9	99.9	102.4	99.1	113.0
					B	148.0	103.6	116.5	108.2	116.5	114.5	101.9	114.5	95.7	95.8	86.8	116.3	95.1	93.3	78.6	89.0	89.4	87.7	91.4	102.1
				Woolong tea	A	142.8	123.3	125.9	116.4	120.7	105.7	110.6	93.0	116.4	110.7	118.1	93.5	92.9	109.0	115.4	101.4	88.8	116.8	87.6	110.4
					B	119.6	103.5	115.3	101.8	91.0	91.3	93.8	78.3	88.6	88.6	101.3	85.7	100.8	90.7	89.8	85.0	73.8	89.0	74.8	112.0
			GC-MS/MS	Green tea	A	140.9	117.6	122.6	120.4	119.8	123.0	96.3	125.3	102.5	95.4	93.0	109.5	86.7	110.0	94.0	115.0	101.5	94.7	102.2	110.0
					B	129.6	97.2	115.7	107.9	115.9	110.3	102.3	112.9	97.0	94.6	83.5	105.1	105.7	93.3	85.6	91.9	89.1	85.5	98.8	100.4
				Woolong tea	A	128.7	116.8	141.1	119.7	112.9	107.2	103.2	89.8	116.4	103.7	98.4	100.9	91.1	104.5	110.2	86.2	72.8	108.1	81.3	105.1
					B	112.9	100.6	112.7	102.3	90.1	86.0	92.4	74.5	88.6	82.9	91.1	86.4	89.7	87.9	80.9	71.6	67.4	89.4	75.4	88.6

(Continued)

Annex 1: Content change for 201 pesticides in incurred tea Youden pair samples under 8 conditions with two SPE cartridge cleanup in three months (Nov.9, 2009-Feb.7, 2010) (cont.)

No	Pesticides	SPE	Method	Sample	Youden pair	Nov.9, 2009 (n=5)	Nov.14, 2009 (n=3)	Nov.19, 2009 (n=3)	Nov.24, 2009 (n=3)	Nov.29, 2009 (n=3)	Dec.4, 2009 (n=3)	Dec.9, 2009 (n=3)	Dec.14, 2009 (n=3)	Dec.19, 2009 (n=3)	Dec.24, 2009 (n=3)	Dec.29, 2009 (n=3)	Jan.3, 2010 (n=3)	Jan.8, 2010 (n=3)	Jan.13, 2010 (n=3)	Jan.18, 2010 (n=3)	Jan.23, 2010 (n=3)	Jan.28, 2010 (n=3)	Feb.2, 2010 (n=3)	Feb.7, 2010 (n=3)	AVE μg/kg
175	nuarimol	Cleanert TPT	GC-MS	Green tea	A	98.1	79.4	99.8	77.7	71.9	65.8	70.1	69.5	65.6	53.5	68.3	60.2	59.5	68.8	63.1	53.5	51.7	51.0	57.7	67.6
					B	95.2	74.7	91.1	70.0	65.6	64.5	63.3	66.7	61.0	49.4	62.4	59.5	57.0	58.7	58.6	52.5	49.4	48.1	48.4	63.0
				Woolong tea	A	98.5	80.6	84.4	83.0	70.5	75.9	62.6	66.7	79.8	83.4	59.5	67.1	64.6	64.7	65.8	62.7	64.1	61.9	53.7	71.0
					B	86.9	68.2	74.0	73.3	68.8	63.1	54.9	60.9	68.1	73.6	47.5	59.1	61.5	59.1	53.5	51.1	51.6	53.5	45.2	61.8
			GC-MS/MS	Green tea	A	100.9	82.5	99.4	88.5	71.2	76.8	70.5	67.1	63.7	68.4	66.8	71.8	54.8	77.7	62.6	51.2	49.3	50.0	57.7	70.0
					B	96.0	81.5	82.8	74.3	64.7	67.3	64.7	61.1	61.6	66.7	61.5	63.1	52.6	87.3	56.4	49.7	49.4	48.5	50.3	65.2
				Woolong tea	A	89.6	74.8	71.2	91.9	72.0	72.0	63.6	66.1	75.6	76.9	76.9	77.1	72.5	59.0	62.0	59.1	63.9	60.6	52.4	70.4
					B	78.9	66.4	62.0	75.4	69.3	60.2	54.4	61.5	63.6	68.6	57.2	60.2	59.9	51.9	51.3	47.4	52.1	54.1	47.3	60.1
		Envi-Carb +PSA	GC-MS	Green tea	A*	102.9	47.9	85.1	86.6	78.1	72.9	54.4	79.1	65.2	66.3	55.6	113.5	60.8	70.9	58.6	68.9	61.9	62.8	55.9	70.9
					B	96.8	66.3	74.7	78.8	81.7	65.6	66.5	72.6	53.4	66.2	54.6	75.3	57.8	60.5	53.8	56.7	53.4	53.4	56.3	65.5
				Woolong tea	A	100.9	76.2	75.1	75.7	69.3	69.8	79.1	54.0	72.8	68.5	65.5	49.7	61.2	64.9	64.6	59.5	52.9	67.9	51.4	67.3
					B	84.2	70.3	76.0	71.3	60.5	57.2	62.6	46.8	59.2	50.5	58.2	46.1	53.6	55.4	48.4	50.4	45.0	56.1	47.2	57.8
			GC-MS/MS	Green tea	A	92.8	79.9	84.2	87.2	80.3	80.7	59.6	81.9	71.4	67.3	61.0	67.5	64.1	89.6	57.4	68.8	65.8	58.8	60.4	72.6
					B	91.9	66.7	79.1	76.9	78.7	74.2	66.3	73.2	64.6	64.1	56.1	64.2	58.0	67.5	53.0	56.2	58.5	53.1	58.1	66.3
				Woolong tea	A	91.8	78.0	78.6	77.7	76.3	65.4	68.4	56.1	72.8	65.5	59.7	55.8	58.5	59.1	65.8	52.5	44.5	70.7	58.1	66.1
					B	82.5	67.8	70.5	70.5	63.3	56.4	58.8	48.8	59.2	51.5	57.1	47.2	65.5	54.3	46.6	42.9	42.0	57.4	52.4	57.8
176	permethrin	Cleanert TPT	GC-MS	Green tea	A	95.2	88.1	75.3	84.7	78.1	83.3	64.1	75.3	77.0	79.1	81.1	77.3	73.7	70.1	75.1	72.4	70.8	68.6	61.5	73.9
					B	89.7	97.0	73.4	71.6	72.6	70.8	59.7	71.4	63.8	70.7	63.8	64.4	66.9	63.9	63.6	60.9	58.3	59.6	53.1	69.5
				Woolong tea	A	78.3	72.6	90.1	81.1	66.7	74.0	70.9	67.2	60.7	68.5	68.8	69.9	54.8	75.6	63.1	51.8	53.3	52.6	60.8	75.0
					B	95.9	63.0	78.5	85.0	61.2	67.0	64.7	63.5	64.5	64.7	69.2	62.6	51.2	71.6	59.0	50.3	54.1	53.3	52.3	65.4
			GC-MS/MS	Green tea	A	90.0	81.4	69.7	69.5	70.2	74.4	62.1	66.8	80.6	80.2	77.0	74.3	74.4	61.8	67.9	62.8	66.6	64.8	57.1	68.5
					B	89.4	70.7	69.4	75.6	67.2	72.8	56.4	63.0	64.6	72.2	58.9	62.3	62.9	56.4	58.3	53.8	54.5	55.8	51.5	64.4
				Woolong tea	A	77.6	74.6	69.3	71.3	70.2	73.5	56.4	66.8	80.6	73.0	77.0	74.3	74.4	57.9	67.9	43.1	41.7	58.1	67.9	71.3
					B	95.4	66.5	71.3	78.9	53.8	60.5	56.4	63.0	64.6	71.7	69.9	62.3	62.9	56.4	58.3	53.8	54.5	55.8	67.9	61.6
		Envi-Carb +PSA	GC-MS	Green tea	A	99.4	72.3	81.9	73.0	78.5	83.4	77.3	81.7	80.6	73.0	79.6	68.4	71.9	76.7	68.1	84.6	76.7	74.0	67.9	80.1
					B	94.5	74.9	78.4	81.6	37.5	71.4	60.9	77.6	70.3	71.7	66.8	62.3	61.3	67.7	65.7	67.7	67.4	65.5	67.2	74.8
				Woolong tea	A	82.3	75.2	78.3	71.5	68.9	71.4	79.4	79.5	77.6	68.5	68.8	67.1	68.3	64.4	77.4	67.2	62.7	78.1	62.1	74.8
					B	86.6	68.4	76.3	77.3	77.4	77.8	69.5	70.1	68.8	63.1	66.5	63.6	61.6	63.6	62.2	60.3	51.3	67.4	54.5	63.8
			GC-MS/MS	Green tea	A	79.7	62.9	83.1	67.5	76.3	72.8	67.6	57.8	69.0	67.0	65.2	57.9	59.6	64.5	58.9	57.7	60.4	57.9	68.1	72.5
					B	82.9	72.6	78.9	76.7	77.4	73.5	67.6	49.9	56.1	55.5	59.1	51.7	59.4	64.2	74.0	55.2	49.9	57.9	67.2	66.7
				Woolong tea	A	78.0	63.0	71.3	43.8	53.8	60.5	56.0	32.1	34.9	55.5	59.1	51.7	59.4	57.9	53.9	43.1	41.7	58.1	62.1	67.4
					B	77.6	81.4	45.2	35.5	34.2	34.6	34.7	29.7	34.9	29.8	34.9	32.5	30.3	32.3	34.4	30.6	30.6	74.0	51.5	58.4
177	phenothrin	Cleanert TPT	GC-MS	Green tea	A	49.7	39.8	40.0	35.5	34.2	34.6	34.7	32.1	31.9	29.7	34.9	32.7	32.5	32.9	34.4	28.0	26.7	27.0	30.6	33.8
					B	44.8	33.6	36.8	32.5	30.3	32.0	32.1	29.7	31.0	29.8	30.5	32.5	30.3	32.3	30.3	27.2	24.7	26.0	27.1	31.2
				Woolong tea	A	40.8	35.1	35.9	43.8	34.5	35.2	27.9	33.3	36.0	36.5	33.1	35.9	33.0	29.4	30.7	30.4	31.4	29.9	27.3	33.7
					B	38.7	31.9	31.4	35.9	32.2	30.9	25.9	30.1	30.0	32.4	26.3	30.0	27.4	25.5	24.6	26.4	25.9	26.9	22.9	29.2
			GC-MS/MS	Green tea	A*	56.4	33.1	45.2	33.8	33.4	34.9	33.4	31.1	29.8	31.6	41.4	37.2	44.5	32.0	29.3	24.8	26.5	29.2	25.9	34.4
					B*	46.2	34.2	41.2	32.1	30.1	32.0	30.6	29.3	29.0	30.3	41.6	32.8	51.4	36.8	36.9	25.3	24.2	27.4	22.2	33.3
				Woolong tea	A	41.1	32.2	33.0	41.4	32.4	34.7	29.6	45.4	56.4	40.8	66.0	41.9	99.7	29.2	56.9	47.9	46.8	35.5	25.5	44.0
					B	40.0	27.3	28.8	35.4	31.8	29.2	26.6	47.6	51.3	50.6	58.3	38.4	70.8	25.0	45.7	45.4	57.6	38.7	22.7	40.6
		Envi-Carb +PSA	GC-MS	Green tea	A*	54.6	34.2	33.0	37.3	42.4	35.6	31.9	38.9	32.2	33.2	29.7	58.4	31.2	36.1	31.0	35.9	33.2	31.7	29.3	36.6
					B	51.5	32.4	36.5	33.0	38.6	31.6	31.8	36.5	30.1	32.1	29.0	50.3	28.5	31.2	28.6	28.6	29.6	27.9	29.9	33.6
				Woolong tea	A*	35.4	29.6	38.5	37.5	36.4	33.3	34.1	27.2	49.7	33.0	34.9	27.2	30.9	33.9	33.2	30.3	25.7	34.4	24.4	33.3
					B	31.7	30.9	36.7	34.4	29.7	27.9	27.0	23.6	45.3	27.5	29.8	24.0	26.7	28.9	26.3	24.4	20.8	29.6	24.1	28.9
			GC-MS/MS	Green tea	A*	47.5	40.6	38.7	37.6	55.3	40.5	36.2	38.2	33.1	32.4	28.4	34.3	30.9	38.6	31.2	38.9	31.7	30.6	30.2	36.6
					B	41.7	33.1	35.3	34.6	45.4	36.3	34.8	34.8	30.1	31.4	27.6	32.7	27.6	36.9	29.3	29.0	28.1	29.6	24.1	33.0
				Woolong tea	A	35.4	47.9	38.9	70.3	78.2	45.1	52.2	26.0	49.7	77.2	59.1	45.0	42.9	60.3	45.5	36.0	34.7	34.4	33.4	48.0
					B	32.8	50.0	36.3	62.3	56.1	45.2	41.7	21.7	45.3	70.5	78.0	48.1	39.5	53.5	59.2	54.8	38.2	51.1	56.5	49.5

#	Compound	Cartridge	Method	Tea																						
178	piperonylbutoxide	Cleanert TPT	GC-MS	Green tea	A	48.5	37.5	48.3	42.3	36.2	33.3	37.4	34.3	31.0	29.1	31.8	32.1	30.9	31.1	31.7	27.0	26.5	24.4	30.3	33.9
					B	45.8	38.1	43.2	38.5	31.6	32.8	34.6	32.2	32.9	27.8	30.9	29.8	28.7	30.2	28.3	26.7	26.1	23.1	26.1	32.0
				Woolong tea	A*	44.1	32.5	40.4	43.8	35.2	38.3	31.9	35.9	40.1	50.3	24.0	31.2	33.0	30.2	31.0	31.5	30.0	29.9	25.2	34.7
					B*	38.7	32.6	38.9	40.3	34.6	33.2	27.6	32.2	29.9	43.7	18.3	26.9	28.3	26.6	26.0	26.3	24.2	27.4	22.2	30.4
			GC-MS/MS	Green tea	A	47.8	35.1	43.4	43.7	33.2	34.9	34.9	32.9	31.3	31.3	32.7	29.5	30.4	31.7	31.7	26.6	25.8	25.6	29.7	33.3
					B	44.1	34.8	37.2	34.3	29.3	31.5	31.8	30.7	31.1	29.5	29.5	27.9	28.5	31.7	28.8	25.6	26.0	24.4	25.7	30.7
				Woolong tea	A	37.9	32.1	33.6	42.6	33.7	34.8	30.5	32.8	36.4	38.9	36.8	37.2	36.5	30.5	30.7	30.6	31.0	29.9	25.7	34.0
					B	61.1	29.2	29.6	35.7	30.8	28.8	25.5	30.1	30.2	33.8	27.3	29.1	30.1	26.7	26.0	25.8	25.3	26.8	25.7	29.0
		Envi-Carb +PSA	GC-MS	Green tea	A*	54.0	45.5	41.7	40.0	44.2	47.1	34.2	45.2	29.1	32.4	30.5	58.4	32.5	35.4	32.2	35.4	31.9	33.2	23.0	38.9
					B	45.4	41.0	40.5	35.3	43.4	36.3	35.2	40.1	26.3	32.0	26.6	34.9	29.2	31.1	28.9	28.5	28.3	28.1	29.4	34.2
				Woolong tea	A	40.8	36.3	42.4	39.8	39.6	35.8	36.7	26.9	33.9	37.1	35.9	24.0	34.0	34.0	35.6	29.8	27.6	33.8	25.7	34.3
					B	45.2	33.4	39.3	36.6	29.6	30.1	30.2	23.1	27.3	30.6	26.8	23.4	27.8	27.8	26.8	25.3	21.9	28.2	22.3	29.0
			GC-MS/MS	Green tea	A	41.3	38.7	38.7	38.2	39.6	41.2	35.3	40.5	33.6	31.8	30.2	36.4	32.3	32.3	31.8	36.1	32.4	31.3	32.6	35.9
					B	42.0	31.8	37.4	34.1	38.2	36.9	35.7	36.6	30.6	31.7	27.6	34.2	30.2	30.6	28.5	29.4	28.4	27.8	31.3	32.7
				Woolong tea	A	37.7	37.3	42.1	38.3	35.3	32.0	31.9	26.8	33.9	32.8	30.7	23.0	30.8	31.1	34.3	26.3	22.7	32.4	25.7	32.1
					B	72.1	32.7	37.1	33.4	27.9	26.4	27.2	22.4	27.3	26.7	27.9	20.1	26.2	26.2	25.0	21.5	20.1	27.3	23.0	27.5
179	pretilachlor	Cleanert TPT	GC-MS	Green tea	A	70.1	68.6	80.5	76.5	65.4	63.9	66.5	69.6	67.0	60.1	67.9	63.4	60.8	59.6	65.1	51.9	54.1	49.3	55.0	64.1
					B	70.1	66.1	74.2	65.6	61.2	62.3	60.7	63.4	68.2	57.5	59.8	57.1	57.4	54.5	58.1	48.7	54.0	48.9	52.1	60.0
				Woolong tea	A*	75.9	80.0	73.3	66.4	66.4	70.8	71.9	77.4	70.3	66.2	73.6	75.7	66.1	64.0	59.9	61.5	73.0	60.6	60.1	69.9
					B*	69.9	65.3	59.7	61.7	61.7	62.2	49.4	69.9	62.6	60.4	60.6	59.8	52.2	56.7	53.0	54.8	60.8	56.8	51.3	59.8
			GC-MS/MS	Green tea	A	75.6	67.2	63.2	63.2	63.2	71.6	71.2	65.2	59.5	68.8	63.8	92.4	51.0	60.2	62.4	50.1	49.2	52.0	55.3	66.0
					B	75.1	55.3	62.3	63.2	63.2	63.2	60.9	58.7	60.5	68.5	52.7	81.7	49.3	64.9	55.1	44.7	49.0	51.8	49.2	59.5
				Woolong tea	A	89.5	68.1	64.5	71.0	71.0	67.9	69.4	70.0	69.1	63.3	75.6	73.0	76.7	65.5	57.7	61.3	64.9	61.7	51.3	68.5
					B	81.3	59.6	51.5	64.7	63.2	53.7	52.5	59.1	55.8	57.1	57.5	61.4	54.5	55.6	51.1	50.1	52.4	56.2	49.0	57.2
		Envi-Carb +PSA	GC-MS	Green tea	A*	81.9	77.3	70.5	72.2	72.9	92.2	63.9	72.8	61.6	56.8	52.9	73.4	61.1	67.7	55.1	65.0	68.8	66.4	61.3	68.1
					B	72.6	62.2	70.4	70.7	71.2	99.2	61.7	63.2	57.5	58.9	50.2	67.5	56.2	60.7	56.5	54.3	64.4	64.6	59.5	64.3
				Woolong tea	A	76.8	88.1	79.9	76.9	76.9	72.3	57.2	63.1	72.3	66.3	65.7	62.7	61.8	67.0	61.4	60.1	53.1	69.9	49.8	67.2
					B	61.0	73.0	74.5	61.5	61.5	62.9	65.0	51.4	53.0	51.9	59.7	52.4	56.1	56.5	51.3	51.2	47.3	58.5	48.0	58.0
			GC-MS/MS	Green tea	A*	105.2	70.2	75.4	74.4	74.4	61.9	63.1	71.4	58.9	57.8	53.8	266.3	58.7	63.9	56.1	65.1	55.3	49.7	62.8	76.2
					B*	140.3	57.8	70.4	74.6	74.6	65.3	45.3	69.9	57.4	59.5	51.0	250.4	51.6	49.9	53.2	51.9	46.2	50.2	61.1	72.7
				Woolong tea	A	137.2	76.9	116.0	74.5	59.6	69.3	64.4	60.6	72.3	54.5	62.2	54.9	50.5	58.3	66.9	52.0	43.3	78.9	51.3	69.3
					B	86.0	66.6	72.7	60.3	57.1	57.1	58.7	49.2	53.0	44.1	55.7	45.5	49.1	53.1	49.0	41.7	40.1	60.2	49.0	55.3
180	prometon	Cleanert TPT	GC-MS	Green tea	A	160.3	124.4	148.1	135.0	110.9	100.9	119.5	108.1	108.0	94.8	127.4	113.5	108.6	110.4	117.4	93.7	95.6	87.6	107.9	114.3
					B	158.6	118.5	132.3	122.4	98.5	99.7	109.8	97.4	105.0	88.0	113.4	107.7	104.3	108.6	102.2	91.4	94.3	84.7	94.3	106.9
				Woolong tea	A*	150.9	114.5	130.1	135.3	106.5	119.7	112.6	125.8	125.8	157.0	92.6	128.0	119.8	112.3	115.0	113.1	118.3	105.8	95.1	119.9
					B*	139.4	105.6	116.2	121.4	99.0	99.6	91.2	103.5	92.9	127.0	71.5	96.8	94.1	96.3	93.4	91.4	97.0	98.6	87.0	101.2
			GC-MS/MS	Green tea	A	143.9	114.5	140.7	121.4	109.4	111.8	110.5	104.1	99.4	104.5	107.5	120.1	83.2	111.3	104.9	81.5	87.8	79.9	97.6	107.0
					B	136.1	116.2	119.5	110.4	96.7	101.8	97.7	95.6	97.0	101.5	95.1	109.4	77.8	115.6	96.0	83.4	83.5	78.1	86.6	99.9
				Woolong tea	A	126.6	101.1	108.3	136.1	100.7	109.3	104.8	110.8	112.7	120.1	122.9	123.9	130.3	100.6	100.9	85.1	99.6	96.9	85.9	110.4
					B	115.6	91.6	90.4	114.8	94.0	88.6	79.0	94.2	99.0	103.6	87.7	93.9	98.0	86.4	87.1	83.6	85.1	89.8	77.4	92.6
		Envi-Carb +PSA	GC-MS	Green tea	A*	181.0	144.1	133.0	137.0	138.1	138.1	107.3	131.6	97.9	104.8	99.8	271.9	101.3	117.4	104.7	113.3	124.6	109.6	102.6	129.1
					B	163.3	120.2	129.2	132.1	123.2	107.7	111.6	124.0	99.8	103.4	92.9	148.7	93.4	102.9	94.9	100.8	98.2	96.4	103.0	121.2
				Woolong tea	A	174.9	153.7	144.4	124.4	123.5	116.4	119.5	92.0	118.7	126.4	128.5	91.0	111.4	126.3	125.0	99.4	108.2	126.9	92.2	121.2
					B	133.9	123.7	130.0	90.0	104.7	99.3	102.1	77.0	89.3	99.3	101.2	89.4	97.8	102.5	99.1	82.8	88.5	99.0	82.5	115.4
			GC-MS/MS	Green tea	A	144.2	122.7	125.9	126.2	126.2	126.2	105.9	127.7	109.7	96.7	95.8	133.2	100.0	130.4	96.6	104.9	116.9	99.0	104.5	104.6
					B	132.2	99.9	118.9	122.2	112.9	114.8	108.5	114.3	103.0	95.4	86.4	121.1	89.3	104.1	87.0	91.8	93.4	89.1	102.9	108.6
				Woolong tea	A	130.3	118.1	154.5	117.1	124.5	114.4	103.2	90.3	118.7	104.1	102.2	107.4	98.9	107.8	113.5	77.3	87.7	108.4	85.9	104.6
					B	107.7	100.6	113.3	104.9	91.5	90.9	92.2	74.6	89.3	85.0	96.2	94.0	82.5	88.4	82.5	69.1	71.8	87.4	77.4	89.4

(Continued)

Annex 1: Content change for 201 pesticides in incurred tea Youden pair samples under 8 conditions with two SPE cartridge cleanup in three months (Nov.9, 2009-Feb.7, 2010) (cont.)

No	Pesticides	SPE	Method	Sample	Youden pair	Nov.9, 2009 (n=5)	Nov.14, 2009 (n=3)	Nov.19, 2009 (n=3)	Nov.24, 2009 (n=3)	Nov.29, 2009 (n=3)	Dec.4, 2009 (n=3)	Dec.9, 2009 (n=3)	Dec.14, 2009 (n=3)	Dec.19, 2009 (n=3)	Dec.24, 2009 (n=3)	Dec.29, 2009 (n=3)	Jan.3, 2010 (n=3)	Jan.8, 2010 (n=3)	Jan.13, 2010 (n=3)	Jan.18, 2010 (n=3)	Jan.23, 2010 (n=3)	Jan.28, 2010 (n=3)	Feb.2, 2010 (n=3)	Feb.7, 2010 (n=3)	AVE, μg/kg
181	pronamide	Cleanert TPT	GC-MS	Green tea	A	170.7	139.3	179.7	148.9	121.6	121.2	122.3	123.9	122.0	112.3	136.8	127.3	117.7	135.4	120.1	105.4	104.2	98.3	129.2	128.2
		Cleanert TPT	GC-MS	Green tea	B	167.6	136.7	144.6	145.9	111.6	117.3	114.6	109.5	123.7	107.4	118.3	119.9	110.8	135.5	110.3	104.9	101.4	95.2	114.0	120.5
		Cleanert TPT	GC-MS	Woolong tea	A	173.3	131.2	139.7	143.6	126.4	127.2	104.9	123.7	125.3	152.3	124.3	124.2	119.2	112.1	121.7	114.2	108.6	111.1	101.5	125.5
		Cleanert TPT	GC-MS	Woolong tea	B	164.4	120.1	118.0	123.0	123.4	106.1	90.0	108.6	103.9	141.9	100.9	99.7	104.9	102.0	107.0	100.4	92.3	95.6	95.7	110.4
		Cleanert TPT	GC-MS/MS	Green tea	A*	161.3	134.0	156.2	164.3	124.7	137.5	127.7	127.1	115.7	133.3	126.7	144.6	97.6	144.5	121.2	106.0	101.5	101.9	115.7	128.0
		Cleanert TPT	GC-MS/MS	Green tea	B	154.2	134.7	135.3	132.8	115.2	125.8	115.9	123.0	117.0	118.3	112.2	128.9	92.1	145.6	111.6	102.2	103.3	99.6	115.7	118.9
		Cleanert TPT	GC-MS/MS	Woolong tea	A	153.3	118.9	126.1	151.2	128.8	122.9	112.7	124.0	130.4	133.2	138.0	132.5	125.5	111.0	125.9	105.0	112.6	108.9	93.3	123.9
		Cleanert TPT	GC-MS/MS	Woolong tea	B	144.4	108.1	112.3	131.1	125.3	101.8	96.3	110.9	113.4	129.3	107.9	103.5	105.9	96.9	110.7	94.8	98.4	101.6	83.6	109.3
		Envi-Carb+PSA	GC-MS	Green tea	A*	195.3	115.9	141.1	151.3	143.1	147.4	111.1	143.1	121.5	127.0	116.7	192.5	116.1	127.8	114.7	133.4	124.2	121.7	113.7	134.6
		Envi-Carb+PSA	GC-MS	Green tea	B	187.0	122.5	135.0	136.8	146.9	113.0	116.3	135.2	115.0	128.5	109.4	139.6	105.9	115.6	106.3	109.2	110.3	111.7	119.8	124.4
		Envi-Carb+PSA	GC-MS	Woolong tea	A	193.2	144.6	129.9	122.7	154.1	121.1	117.8	95.6	122.6	116.1	131.2	120.6	109.5	123.4	131.6	115.0	86.9	115.1	98.1	123.6
		Envi-Carb+PSA	GC-MS	Woolong tea	B	149.2	112.5	128.1	115.3	123.9	111.4	110.7	84.4	91.0	102.9	119.2	110.4	100.5	113.5	114.4	99.5	83.2	102.3	86.9	108.2
		Envi-Carb+PSA	GC-MS/MS	Green tea	A*	182.2	146.5	138.9	146.3	153.1	154.4	122.8	153.7	131.2	121.1	114.9	37.1	121.5	140.0	119.8	140.7	123.9	116.8	125.5	131.1
		Envi-Carb+PSA	GC-MS/MS	Green tea	B*	173.9	125.5	133.9	131.7	146.1	142.4	124.0	140.7	124.6	120.6	105.0	35.5	110.0	112.3	110.1	113.0	109.1	110.1	124.3	120.7
		Envi-Carb+PSA	GC-MS/MS	Woolong tea	A	178.5	129.8	165.6	135.8	152.3	118.7	113.3	100.0	122.6	105.1	107.6	95.0	113.1	110.8	130.0	96.0	76.7	110.4	93.3	118.7
		Envi-Carb+PSA	GC-MS/MS	Woolong tea	B	143.6	108.6	128.9	124.0	128.9	106.6	109.6	85.5	91.0	105.3	105.3	86.4	125.7	102.1	101.4	82.1	75.8	98.9	83.5	104.4
182	propetamphos	Cleanert TPT	GC-MS	Green tea	A	61.6	45.4	44.8	41.9	25.3	32.7	36.4	38.7	28.9	29.6	34.8	37.7	33.9	39.8	33.7	33.0	28.1	27.1	31.4	36.0
		Cleanert TPT	GC-MS	Green tea	B*	58.6	43.6	39.1	40.0	25.7	34.2	33.9	33.1	36.2	26.1	35.2	36.1	30.8	42.9	35.6	32.5	27.7	27.5	27.8	35.1
		Cleanert TPT	GC-MS	Woolong tea	A*	48.2	31.5	37.7	44.1	36.4	41.4	34.4	45.7	55.7	72.7	35.7	23.3	36.2	32.1	37.4	37.1	35.5	33.2	29.5	39.4
		Cleanert TPT	GC-MS	Woolong tea	B	43.6	30.2	33.9	38.5	33.5	33.6	27.2	34.0	32.8	38.5	31.6	23.6	30.0	29.2	30.5	31.5	28.6	29.9	26.4	31.9
		Cleanert TPT	GC-MS/MS	Green tea	A	52.2	43.2	43.4	40.5	32.6	41.8	37.6	37.9	31.6	38.4	36.8	39.3	30.9	38.0	33.2	29.7	27.4	27.7	32.4	36.6
		Cleanert TPT	GC-MS/MS	Green tea	B	50.4	42.1	38.6	37.6	32.2	38.5	33.3	35.5	36.7	38.3	34.7	33.3	27.7	42.0	31.8	28.4	28.8	28.9	29.1	35.2
		Cleanert TPT	GC-MS/MS	Woolong tea	A	49.8	35.9	37.5	49.1	37.9	39.4	34.6	38.5	40.0	41.9	41.9	40.2	41.1	34.4	36.0	35.2	33.3	35.2	29.8	38.5
		Cleanert TPT	GC-MS/MS	Woolong tea	B	42.6	32.0	31.5	40.8	32.6	30.2	27.7	31.6	29.8	36.7	29.3	30.7	31.3	29.3	30.0	30.3	26.7	30.4	26.3	31.6
		Envi-Carb+PSA	GC-MS	Green tea	A*	54.5	48.5	40.3	40.0	39.9	32.8	46.1	33.5	32.1	38.5	40.6	79.6	33.2	34.4	36.4	39.0	37.5	27.4	33.8	40.4
		Envi-Carb+PSA	GC-MS	Green tea	B*	50.7	46.6	36.5	34.3	37.7	27.3	33.6	30.3	31.8	36.8	32.4	56.8	30.3	34.0	34.0	31.5	33.6	33.6	34.0	36.1
		Envi-Carb+PSA	GC-MS	Woolong tea	A	49.5	31.2	42.8	36.8	44.1	43.8	36.7	34.1	39.7	48.6	35.3	36.6	34.2	34.8	37.7	33.5	28.1	35.3	29.7	37.5
		Envi-Carb+PSA	GC-MS	Woolong tea	B	41.2	28.6	34.6	34.1	32.2	39.6	29.7	18.7	29.8	27.1	33.2	29.9	28.5	31.1	32.4	24.5	24.2	30.3	25.6	30.3
		Envi-Carb+PSA	GC-MS/MS	Green tea	A*	58.6	53.4	40.5	42.8	42.9	42.4	50.5	37.3	40.5	38.0	34.4	76.6	29.4	40.5	31.0	35.6	37.7	28.0	36.6	41.9
		Envi-Carb+PSA	GC-MS/MS	Green tea	B*	58.7	54.9	36.9	38.3	39.5	42.8	31.6	34.1	37.6	35.1	28.3	69.4	30.8	30.3	30.5	28.1	31.4	31.7	37.1	38.3
		Envi-Carb+PSA	GC-MS/MS	Woolong tea	A	48.8	37.4	52.8	41.4	44.9	44.4	34.9	32.6	39.7	33.5	35.6	27.0	36.1	33.4	35.1	29.8	23.6	33.4	29.9	36.5
		Envi-Carb+PSA	GC-MS/MS	Woolong tea	B	38.9	30.9	36.0	35.5	32.8	37.4	30.2	25.1	29.8	28.0	32.0	25.4	32.4	29.7	26.9	22.0	21.9	27.6	26.4	29.9
183	propoxur-1	Cleanert TPT	GC-MS	Green tea	A	341.9	245.9	257.0	304.3	267.7	248.2	217.8	273.1	283.1	274.8	251.0	301.8	247.4	262.7	241.7	240.2	244.8	262.7	226.0	248.8
		Cleanert TPT	GC-MS	Green tea	B	377.6	279.2	367.1	323.6	286.0	297.3	307.5	286.9	283.0	287.2	291.1	291.3	261.0	413.6	282.5	247.3	234.6	233.3	271.7	295.9
		Cleanert TPT	GC-MS	Woolong tea	A	358.1	288.5	325.0	300.3	264.0	270.2	279.2	249.6	278.3	288.1	264.2	273.1	258.2	437.0	256.7	239.7	232.9	222.9	238.4	280.2
		Cleanert TPT	GC-MS	Woolong tea	B	324.6	265.4	282.5	353.4	261.8	300.7	280.3	295.3	242.1	337.1	210.3	342.0	304.9	288.4	275.8	285.7	288.2	274.8	228.2	286.4
		Cleanert TPT	GC-MS/MS	Green tea	A*	432.7	313.0	332.7	334.6	342.2	376.0	244.7	360.3	294.3	292.5	272.7	268.5	286.6	323.1	288.3	333.0	285.0	295.5	271.0	313.1
		Cleanert TPT	GC-MS/MS	Green tea	B*	400.5	266.4	332.7	308.7	342.3	323.6	289.8	344.4	277.4	283.8	251.0	262.1	284.8	290.4	267.1	279.2	256.1	255.2	269.3	293.2
		Cleanert TPT	GC-MS/MS	Woolong tea	A	348.6	317.5	346.1	319.1	316.7	281.5	315.2	257.6	318.8	320.9	321.2	271.4	285.1	310.8	321.0	272.1	255.3	327.6	236.9	302.4
		Cleanert TPT	GC-MS/MS	Woolong tea	B	273.2	281.6	296.7	275.4	250.6	228.8	267.3	228.1	252.0	241.3	281.9	225.0	251.5	258.1	242.6	233.1	219.5	233.1	260.7	251.9
		Envi-Carb+PSA	GC-MS	Green tea	A	365.3	307.6	332.2	336.1	323.4	313.2	249.5	367.7	294.7	275.5	276.0	284.5	284.4	372.9	289.9	335.9	303.1	281.8	289.1	309.6
		Envi-Carb+PSA	GC-MS	Green tea	B	332.7	247.2	321.3	310.5	337.6	289.2	287.9	337.3	272.8	272.2	250.7	264.5	253.0	279.9	266.2	284.2	275.6	253.6	282.5	285.2
		Envi-Carb+PSA	GC-MS	Woolong tea	A	334.2	340.2	415.0	315.1	300.7	283.8	305.6	255.8	318.8	286.4	273.2	200.5	279.6	281.5	309.3	235.3	216.2	307.2	228.2	288.8
		Envi-Carb+PSA	GC-MS	Woolong tea	B	275.4	280.0	308.9	266.7	287.9	229.2	259.6	211.9	252.0	214.8	251.9	178.8	252.5	241.5	219.5	201.1	199.9	250.7	208.5	239.5

No.	Compound	Cartridge	Method	Tea	Sample																				
184	propoxur-2	Cleanert TPT	GC-MS	Green tea	A	429.3	410.3	385.8	319.3	224.0	386.6	269.2	399.8	344.5	297.1	394.3	383.2	366.2	522.4	457.4	401.2	352.2	404.1	453.9	379.0
				Green tea	B	415.3	375.6	365.0	277.4	278.5	351.1	248.9	386.7	325.0	310.4	391.9	390.4	380.4	484.2	396.9	418.5	348.9	374.1	417.9	365.1
				Woolong tea	A	271.0	345.0	312.0	341.0	487.0	531.2	352.7	424.2	234.9	255.2	332.3	269.1	470.2	419.1	426.3	412.2	440.9	424.4	227.3	367.2
				Woolong tea	B	223.0	295.1	279.4	262.3	548.7	531.2	374.6	395.1	224.1	245.1	253.0	196.8	460.1	437.2	432.1	418.0	446.8	459.1	211.2	352.3
			GC-MS/MS	Green tea	A	439.5	527.5	262.1	337.2	235.0	410.1	281.9	365.6	211.2	318.6	304.9	321.2	176.4	261.4	261.3	203.1	227.5	243.8	211.2	295.0
				Green tea	B	461.6	440.6	273.3	311.8	274.9	363.1	211.5	397.9	359.9	365.0	275.5	274.3	161.2	341.6	283.7	192.7	232.4	269.6	190.3	299.0
				Woolong tea	A	371.9	381.7	336.8	387.6	412.5	336.2	313.0	321.4	230.2	235.6	412.1	300.0	289.5	261.6	352.1	225.2	259.2	323.7	351.5	321.2
				Woolong tea	B*	349.8	314.8	264.6	248.7	276.9	240.3	279.1	281.5	216.5	231.1	298.7	237.0	217.9	235.4	294.8	182.4	225.2	245.8	314.0	260.8
		Envi-Carb +PSA	GC-MS	Green tea	A	293.9	513.1	299.2	298.0	320.9	257.3	366.6	271.6	375.8	325.7	490.2	307.1	346.5	363.6	466.2	474.0	397.8	382.5	278.0	359.4
				Green tea	B*	229.6	623.6	267.6	332.8	286.6	184.3	464.4	234.4	390.3	278.8	377.8	333.8	342.6	366.9	471.2	463.5	403.7	412.5	292.5	355.6
				Woolong tea	A*	357.2	372.6	372.0	320.7	394.9	435.7	304.3	360.3	325.6	232.1	301.2	348.0	300.3	427.5	317.7	320.6	184.8	327.7	265.5	329.9
				Woolong tea	B*	289.6	355.6	355.6	287.5	361.8	430.2	1024.1	264.8	226.1	163.5	274.2	341.8	238.9	438.1	296.0	219.5	180.8	271.1	234.5	281.5
			GC-MS/MS	Green tea	A*	578.0	698.9	331.7	367.5	325.0	339.6	1024.1	229.8	339.3	345.8	278.2	467.6	239.0	407.1	167.7	260.5	230.0	137.4	298.7	371.9
				Green tea	B*	574.9	1062.2	298.1	328.0	256.3	366.8	226.5	210.7	391.0	350.1	197.4	360.4	239.0	322.1	179.8	168.9	158.2	269.0	484.4	335.8
				Woolong tea	A*	421.4	245.7	436.3	313.2	445.5	406.7	244.2	279.1	325.6	215.0	253.5	141.2	200.3	186.1	263.1	249.5	152.6	262.1	351.5	283.8
				Woolong tea	B*	308.5	197.3	292.2	303.0	325.0	314.5	229.8	204.6	226.1	173.5	241.5	122.0	187.6	217.0	209.3	168.4	156.2	254.2	314.0	233.9
185	prothiophos	Cleanert TPT	GC-MS	Green tea	A	42.9	35.2	40.7	36.8	30.0	30.3	33.9	33.0	30.6	25.8	32.5	29.6	32.8	31.7	32.1	27.7	26.8	26.2	30.0	32.0
				Green tea	B	42.2	34.1	36.7	33.7	28.3	29.9	31.7	30.5	31.2	23.6	29.1	27.8	29.9	30.3	30.3	27.2	26.5	25.7	26.4	30.3
				Woolong tea	A*	38.4	32.1	35.8	39.2	31.6	35.1	31.0	34.6	34.3	41.9	28.2	35.2	34.2	32.4	33.2	33.0	33.6	31.2	28.3	33.9
				Woolong tea	B	35.0	28.7	31.1	34.5	29.2	29.3	23.6	30.9	26.9	34.1	23.1	28.3	27.6	28.5	28.1	28.2	27.4	28.6	25.5	28.9
			GC-MS/MS	Green tea	A	43.6	34.8	37.6	36.4	30.6	35.6	34.7	31.9	30.4	33.1	34.3	32.9	27.7	34.2	31.1	26.8	24.7	28.0	28.8	32.5
				Green tea	B	41.9	34.0	34.0	34.5	30.1	33.2	34.7	30.1	31.2	30.8	30.6	29.3	26.3	34.2	28.1	25.6	24.1	26.7	24.8	30.5
				Woolong tea	A*	47.1	34.2	33.3	41.9	32.0	34.2	32.3	32.0	35.7	38.5	38.6	38.5	34.3	33.4	32.1	30.3	33.5	32.8	24.8	35.0
				Woolong tea	B	40.7	28.8	28.0	35.2	29.7	28.1	25.1	30.7	29.9	32.7	28.9	30.1	27.4	27.8	26.5	26.9	26.6	30.2	28.4	29.3
		Envi-Carb +PSA	GC-MS	Green tea	A*	41.6	39.3	36.0	35.2	38.4	35.9	35.8	37.9	30.6	31.2	27.0	61.9	31.1	34.7	29.2	33.8	34.0	30.5	23.9	35.5
				Green tea	B	40.0	34.8	35.1	33.9	36.5	31.7	30.4	35.2	27.4	31.2	24.7	40.7	28.8	31.1	28.1	27.8	30.3	30.0	30.6	32.0
				Woolong tea	A*	41.8	38.4	37.7	36.2	36.3	35.2	33.2	27.3	35.7	35.0	36.3	26.8	33.0	34.4	35.7	31.5	27.6	34.8	30.3	33.9
				Woolong tea	B	35.4	31.2	33.4	32.0	28.2	29.7	29.2	22.7	27.5	28.4	29.3	25.1	28.7	29.2	28.7	25.8	23.4	29.2	27.4	28.5
			GC-MS/MS	Green tea	A*	44.0	40.0	37.3	37.5	39.8	36.9	36.9	37.3	35.7	32.0	30.9	41.4	31.0	32.2	30.7	33.7	30.5	27.1	34.0	35.3
				Green tea	B	45.4	34.3	34.9	35.1	37.9	36.3	29.7	34.4	31.7	31.2	27.6	36.9	29.8	26.4	29.5	26.3	25.9	26.9	33.3	32.3
				Woolong tea	A*	46.2	35.7	47.3	38.2	35.6	33.8	33.0	28.3	35.7	31.3	31.5	30.2	30.5	31.6	35.6	26.9	22.9	34.1	28.4	33.5
				Woolong tea	B	36.8	30.5	34.2	33.0	27.9	28.8	28.3	23.5	27.5	26.8	29.6	26.1	23.8	26.8	27.1	21.2	20.9	28.7	23.9	27.6
186	pyridaben	Cleanert TPT	GC-MS	Green tea	A*	76.1	65.0	52.1	53.0	27.7	45.6	36.2	47.0	34.9	36.9	39.3	41.2	35.0	45.8	37.2	45.4	25.0	28.0	32.8	42.3
				Green tea	B*	53.5	53.3	54.9	52.9	28.0	41.2	34.9	46.6	41.5	40.6	37.6	34.4	32.4	44.5	41.4	44.4	28.4	27.9	30.0	40.4
				Woolong tea	A	54.8	33.6	38.0	52.2	43.3	46.6	32.9	39.2	38.0	44.8	40.4	37.0	34.7	26.4	41.2	40.8	35.1	34.4	29.6	39.1
				Woolong tea	B	66.3	32.3	36.7	41.1	38.4	37.0	33.8	36.1	31.0	44.6	30.0	28.7	33.7	28.1	32.7	35.9	28.0	29.2	27.5	35.3
			GC-MS/MS	Green tea	A*	65.6	43.9	56.6	46.3	49.4	50.8	56.1	47.2	42.5	52.7	41.8	34.3	43.7	68.1	61.4	48.7	49.9	48.5	61.1	51.0
				Green tea	B	48.9	44.0	55.2	47.8	52.2	48.1	49.0	49.6	60.5	60.5	42.4	36.0	44.2	68.5	56.8	49.5	56.0	52.5	55.7	51.6
				Woolong tea	A*	55.2	43.1	57.4	47.7	51.8	68.4	45.3	53.3	63.8	54.9	50.7	57.0	60.5	48.6	57.0	51.8	55.0	55.5	46.1	53.5
				Woolong tea	B	96.7	42.9	49.3	57.2	57.2	65.7	51.2	50.1	57.0	54.1	46.4	60.4	43.5	50.0	62.0	51.3	54.3	56.1	53.8	53.7
		Envi-Carb +PSA	GC-MS	Green tea	A*	52.4	27.3	38.0	46.8	42.9	57.6	51.1	44.1	32.1	42.5	38.4	55.8	34.4	33.5	40.4	42.1	42.8	32.0	35.4	46.4
				Green tea	B*	45.0	47.7	32.2	47.1	38.4	43.1	37.4	39.3	26.6	36.7	28.4	102.6	32.0	40.4	37.5	39.3	39.1	37.4	35.4	40.7
				Woolong tea	A	44.4	22.6	42.7	35.0	55.3	52.9	33.5	52.5	38.3	33.9	42.8	87.2	36.9	33.5	43.9	34.2	26.1	34.0	37.5	38.3
				Woolong tea	B	52.5	27.4	41.9	36.7	40.0	49.1	29.0	33.4	29.3	32.6	36.1	33.9	29.0	34.8	45.1	21.4	21.9	32.6	34.2	33.9
			GC-MS/MS	Green tea	A*	75.3	44.6	40.2	44.3	39.4	45.7	38.3	40.9	36.8	34.4	31.9	34.9	32.7	29.8	30.4	35.3	32.7	24.9	27.4	38.4
				Green tea	B	43.9	47.6	37.7	39.4	40.6	48.0	31.9	35.5	32.5	33.8	28.7	51.7	27.2	32.6	29.5	28.4	25.0	27.2	34.3	37.4
				Woolong tea	A*	24.2	40.1	74.6	29.7	48.6	48.4	53.5	33.8	38.3	60.7	52.0	54.5	49.1	39.0	62.3	45.2	47.4	53.5	37.0	47.3
				Woolong tea	B	38.6	38.6	41.3	27.1	45.9	51.1	54.3	28.6	29.3	56.1	58.9	43.2	63.1	30.1	54.2	43.2	47.4	58.3	53.9	44.6

(Continued)

Annex 1: Content change for 201 pesticides in incurred tea Youden pair samples under 8 conditions with two SPE cartridge cleanup in three months (Nov.9, 2009-Feb.7, 2010) (cont.)

No	Pesticides	SPE	Method	Sample	Youden pair	Nov.9 2009 (n=5)	Nov.14 2009 (n=3)	Nov.19 2009 (n=3)	Nov.24 2009 (n=3)	Nov.29 2009 (n=3)	Dec.4 2009 (n=3)	Dec.9 2009 (n=3)	Dec.14 2009 (n=3)	Dec.19 2009 (n=3)	Dec.24 2009 (n=3)	Dec.14 2009 (n=3)	Jan.3 2010 (n=3)	Jan.8 2010 (n=3)	Jan.13 2010 (n=3)	Jan.18 2010 (n=3)	Jan.23 2010 (n=3)	Jan.28 2010 (n=3)	Feb.2 2010 (n=3)	Feb.7 2010 (n=3)	AVE, μg/kg
187	pyridaphenthion	Cleanert TPT	GC-MS	Green tea	A	40.8	48.2	39.3	53.1	32.8	33.4	34.4	39.7	31.5	31.4	40.3	37.5	33.5	41.9	34.9	31.8	28.5	31.2	30.7	36.6
					B	41.2	47.0	36.8	49.6	31.2	33.9	32.6	36.4	35.7	30.2	35.2	32.8	31.2	42.3	32.9	31.6	27.1	30.7	28.9	35.1
				Woolong tea	A*	36.8	35.2	49.9	42.2	36.9	38.1	34.3	38.1	36.3	51.8	34.4	31.3	32.8	30.6	34.1	34.0	42.0	30.2	35.9	37.1
					B	33.4	32.8	43.1	42.7	36.4	34.0	27.3	38.6	29.0	42.8	27.7	24.4	29.1	28.2	28.7	29.2	34.9	26.0	28.9	32.5
			GC-MS/MS	Green tea	A	45.4	43.4	47.0	34.5	31.6	38.1	40.0	41.7	33.7	41.8	38.1	68.5	23.7	39.7	31.7	29.1	25.8	26.2	33.9	37.6
					B	45.3	34.4	37.7	36.7	31.5	36.6	37.5	36.7	36.3	44.1	32.4	51.7	24.6	44.5	29.7	27.0	26.7	25.9	31.0	35.3
				Woolong tea	A	75.4	37.9	48.1	35.0	40.5	38.1	35.3	39.6	39.2	39.7	39.9	37.3	40.3	31.2	34.4	34.7	34.8	37.4	32.8	39.6
					B	57.9	33.8	39.7	32.0	36.9	31.4	30.7	36.7	25.7	36.9	31.6	30.5	35.5	28.1	29.5	27.0	26.7	31.2	27.2	33.0
		Envi-Carb +PSA	GC-MS	Green tea	A*	44.3	47.4	39.7	39.7	41.6	52.7	32.6	40.2	34.7	32.5	35.4	146.7	35.5	34.9	33.3	38.3	35.8	35.6	36.4	44.1
					B*	46.3	49.4	34.7	38.6	40.9	64.2	33.4	33.0	30.8	32.5	30.5	102.9	31.1	33.6	32.6	35.2	32.5	37.1	37.1	40.9
				Woolong tea	A	40.2	34.7	39.6	42.4	39.5	41.2	26.9	30.5	36.7	39.8	37.5	29.2	34.2	29.7	35.5	34.3	29.0	34.6	31.0	35.1
					B	39.1	31.0	37.8	41.4	32.4	36.2	29.3	23.3	28.3	29.9	32.1	30.3	28.8	29.6	30.2	26.1	24.9	33.4	29.2	31.2
			GC-MS/MS	Green tea	A	47.1	39.5	40.9	40.8	41.5	41.0	35.1	42.7	35.4	32.4	32.3	37.3	32.1	40.0	32.3	38.0	33.9	31.5	33.3	37.2
					B	43.2	32.5	37.1	38.7	40.4	37.4	35.4	39.0	33.3	32.5	29.6	35.5	30.1	33.4	29.8	31.3	30.7	28.8	32.2	34.3
				Woolong tea	A	59.3	34.4	38.7	53.3	49.1	41.9	33.4	33.2	36.7	32.8	35.3	25.5	29.8	31.3	38.2	27.9	22.6	38.7	27.3	36.6
					B	42.2	32.2	34.9	39.8	36.7	34.6	29.5	25.2	28.3	26.2	31.3	21.1	19.6	29.6	29.5	20.4	19.8	27.3	27.2	29.2
188	pyrimethanil	Cleanert TPT	GC-MS	Green tea	A*	48.1	37.3	40.9	44.2	34.0	33.3	34.5	34.1	33.7	30.0	36.1	32.8	34.7	34.0	33.4	28.5	28.1	27.4	31.7	34.6
					B	46.6	36.0	36.9	38.6	34.7	33.1	31.9	30.3	33.4	28.7	32.3	31.6	32.5	32.9	30.4	28.1	26.9	25.7	28.0	32.3
				Woolong tea	A	44.5	34.6	40.8	37.0	31.6	37.9	32.0	38.4	35.4	40.8	31.6	36.4	34.9	33.1	35.9	33.9	34.1	32.0	29.7	35.7
					B	40.0	30.2	31.4	31.4	34.5	31.6	25.1	31.8	29.4	34.6	25.9	28.0	29.1	29.8	29.7	28.4	28.0	29.6	26.3	30.3
			GC-MS/MS	Green tea	A	47.3	36.7	45.2	42.1	31.6	37.5	37.0	32.5	31.8	33.8	33.1	42.3	28.3	33.6	34.1	27.7	27.2	25.2	31.2	34.8
					B	44.6	37.4	38.2	37.1	32.4	34.1	33.1	30.6	32.1	31.9	29.6	37.8	25.7	33.9	31.3	26.4	26.9	24.7	27.2	32.3
				Woolong tea	A*	41.8	32.7	35.3	44.2	29.2	36.5	33.4	35.8	34.0	39.1	37.4	39.5	37.9	30.7	32.8	31.6	32.6	32.8	26.6	35.1
					B*	38.6	29.0	29.6	37.2	39.6	29.8	25.4	30.8	29.8	34.0	27.8	30.4	30.6	28.3	27.0	28.0	27.1	28.6	25.3	29.8
		Envi-Carb +PSA	GC-MS	Green tea	A*	57.2	42.5	41.4	41.4	39.6	47.1	32.7	41.6	31.9	33.6	31.8	67.3	34.7	37.6	33.9	38.5	33.8	34.5	31.5	39.5
					B	51.5	37.0	39.4	39.4	38.9	37.5	34.8	39.2	30.5	32.8	29.3	40.7	31.5	34.4	31.5	31.7	30.6	29.8	31.2	35.2
				Woolong tea	A	46.3	39.8	42.6	35.3	36.6	35.2	34.2	26.0	38.4	36.8	38.1	25.5	32.1	38.7	37.0	32.3	28.2	37.8	27.9	35.3
					B	38.1	33.2	35.3	33.0	23.0	29.5	29.6	22.7	30.0	29.5	31.7	24.3	29.1	30.8	28.4	25.6	24.2	30.1	24.7	29.4
			GC-MS/MS	Green tea	A	51.4	45.1	51.0	56.7	46.3	42.6	52.1	37.0	68.0	44.8	51.6	36.8	43.1	42.1	53.3	62.1	55.0	55.6	59.0	50.2
					B	49.1	33.3	54.1	58.7	51.1	44.1	43.9	31.8	63.7	47.9	50.9	34.9	47.7	44.6	52.6	55.6	57.2	57.0	61.1	49.4
				Woolong tea	A	42.2	39.0	50.4	39.8	33.6	36.3	33.7	29.7	38.4	33.9	33.6	29.6	33.2	34.7	36.2	28.0	26.0	34.3	26.6	35.0
					B	35.2	32.4	37.2	33.4	30.9	29.2	29.1	24.0	30.0	28.1	30.6	26.6	26.0	29.0	27.2	23.0	23.2	26.4	25.3	28.8
189	pyriproxyfen	Cleanert TPT	GC-MS	Green tea	A	45.6	40.4	42.3	48.2	38.5	35.0	37.0	36.8	34.9	32.0	35.8	33.7	34.2	33.8	34.3	29.3	29.9	29.2	32.5	36.0
					B	44.7	38.7	38.3	43.1	34.2	32.5	34.8	34.9	33.3	31.2	32.3	31.6	32.3	32.1	31.1	29.5	28.4	27.0	28.4	33.6
				Woolong tea	A	48.7	38.6	46.1	38.1	36.2	34.7	33.5	37.1	37.7	39.6	32.4	38.6	35.4	33.3	33.1	34.8	33.8	28.4	29.8	36.6
					B	42.9	34.6	39.2	34.1	34.1	33.6	29.6	33.8	31.1	34.8	26.3	31.7	34.8	29.8	26.6	28.6	27.6	29.8	25.2	31.7
			GC-MS/MS	Green tea	A*	53.2	48.6	43.5	43.5	42.0	39.9	31.0	41.1	32.2	33.4	30.6	69.9	34.1	38.3	26.6	40.6	35.9	36.1	61.2	50.3
					B*	48.5	37.9	41.4	41.4	41.0	31.9	32.4	37.9	29.6	32.9	29.5	55.8	32.9	34.4	33.1	32.9	27.2	53.2	55.5	50.1
				Woolong tea	A	50.3	39.6	41.5	43.7	33.3	34.4	36.7	28.3	36.5	35.6	39.3	30.6	33.8	37.1	37.6	32.1	30.9	33.2	28.4	36.0
					B	44.2	35.3	38.3	41.5	30.3	28.6	29.1	25.2	30.0	29.6	31.6	26.5	37.2	26.8	29.1	26.8	21.6	31.9	25.4	31.2
		Envi-Carb +PSA	GC-MS	Green tea	A																				40.0
					B																				35.9
				Woolong tea	A																				36.2
					B																				30.8
			GC-MS/MS	Green tea	A	53.5	45.4	56.7	51.9	47.8	45.4	36.7	59.2	67.6	45.6	49.1	35.3	44.6	42.0	52.8	61.9	55.3	56.0	58.9	50.8
					B	51.3	34.0	55.0	55.0	52.2	46.3	37.0	61.3	62.9	48.7	50.3	33.3	48.8	45.0	52.1	55.6	57.6	57.3	61.0	51.0
				Woolong tea	A	41.8	39.4	40.9	41.4	37.9	35.1	35.0	28.7	36.5	34.4	31.8	22.1	36.1	34.0	36.3	28.2	25.6	34.7	61.0	34.1
					B	39.6	34.8	35.3	39.8	31.5	28.7	28.8	25.3	30.0	27.9	29.2	19.5	33.6	28.8	26.3	22.5	22.3	29.6	28.4	29.4

No.	Compound	Cleanup	Method	Tea	Run																			
190	quinalphos	Cleanert TPT	GC-MS	Green tea	A	36.8	40.0	43.1	38.2	31.3	35.6	35.2	36.9	36.0	33.5	36.0	42.4	37.4	39.3	34.1	33.7	29.0	27.3	31.4	35.6
					B	37.4	39.5	39.6	35.6	30.5	34.4	34.4	34.4	38.3	31.7	33.5	40.4	37.2	39.1	34.9	31.5	29.0	27.3	28.9	34.6
				Woolong tea	A*	50.0	37.3	40.1	48.0	37.3	38.8	32.4	38.2	40.9	55.4	38.2	39.5	37.4	35.1	31.6	30.2	35.5	31.7	30.4	38.4
					B*	45.2	32.4	37.9	40.3	34.9	34.6	27.3	35.0	34.5	47.6	35.0	30.3	31.4	34.9	27.3	28.1	30.2	28.8	29.0	34.0
			GC-MS/MS	Green tea	A	44.2	37.7	35.9	44.3	32.2	39.1	36.4	36.0	32.2	36.1	36.0	41.6	32.0	33.0	33.5	28.6	25.8	29.1	30.9	35.1
				Woolong tea	B	43.6	33.9	34.3	37.1	32.3	34.4	33.1	30.8	35.0	36.1	30.8	35.7	30.8	33.7	34.4	26.3	26.3	28.0	27.6	33.1
				Green tea	A	57.8	34.8	28.9	45.1	34.6	38.4	34.9	40.3	38.0	33.5	40.6	39.2	40.3	33.4	32.9	33.5	34.7	33.6	30.7	37.5
					B	46.9	30.4	38.5	37.4	30.7	29.9	28.2	32.6	29.2	35.8	32.6	30.3	30.2	29.1	28.3	27.7	27.5	30.8	26.6	31.1
		Envi-Carb +PSA	GC-MS	Green tea	A*	40.3	47.2	37.2	38.8	38.8	44.9	40.9	32.9	31.9	35.8	32.9	82.4	34.3	36.5	35.9	33.9	37.9	32.2	34.1	39.8
					B*	41.4	41.3	42.2	35.1	36.5	50.2	32.5	30.6	31.0	37.5	30.6	47.7	30.1	35.1	31.9	34.0	35.2	32.8	33.3	36.2
				Woolong tea	A*	43.2	34.4	37.9	40.0	38.6	36.6	30.4	35.3	37.0	30.7	50.0	33.9	32.5	34.3	38.2	32.7	27.7	34.9	28.6	36.1
					B	36.9	35.2	37.9	36.5	29.3	32.2	29.3	34.4	28.6	33.9	35.3	32.6	28.7	30.1	30.8	27.0	24.4	29.4	26.6	30.9
			GC-MS/MS	Green tea	A	53.5	41.8	39.3	40.7	41.0	39.9	41.5	37.5	35.3	33.4	34.4	39.7	32.9	38.7	29.3	37.5	30.8	28.4	31.9	37.4
				Woolong tea	B	61.2	38.7	36.7	38.4	39.7	38.5	33.9	30.8	33.5	26.5	28.8	36.9	30.7	35.3	28.4	30.8	27.0	31.0	32.0	35.4
				Green tea	A	53.5	35.9	54.6	34.3	41.0	32.7	34.5	29.3	37.0	27.0	32.6	27.8	33.0	33.7	37.1	29.3	23.7	37.9	30.7	35.9
					B	38.9	30.9	36.5	37.0	31.3	34.8	30.6	23.6	28.6	25.5	29.2	25.9	27.5	28.4	27.4	23.6	27.7	23.6	26.6	29.3
191	quinoxyphen	Cleanert TPT	GC-MS	Green tea	A	45.5	34.9	38.1	34.6	30.6	30.6	31.6	34.5	32.7	34.9	28.4	32.8	28.4	31.6	33.4	29.6	27.4	24.0	28.8	33.1
					B	44.5	33.5	34.6	34.6	31.0	31.0	30.3	32.1	36.6	30.0	30.3	31.6	30.3	30.3	30.6	29.0	27.4	23.1	24.9	31.1
				Woolong tea	A*	40.0	31.9	40.7	36.9	37.3	31.6	29.2	34.0	30.2	34.2	32.8	38.6	28.7	32.8	32.5	30.6	28.7	28.2	29.1	33.6
					B*	36.0	27.5	35.0	31.0	34.9	31.9	29.0	26.2	31.4	27.6	23.8	31.6	23.8	29.0	25.8	25.2	23.1	25.5	25.7	28.7
			GC-MS/MS	Green tea	A	45.7	36.7	43.8	29.2	36.1	35.5	35.7	33.2	31.2	31.7	33.2	41.5	33.4	35.7	29.3	27.2	26.0	26.2	30.5	34.0
				Woolong tea	B	42.8	36.3	34.3	35.1	31.9	29.4	36.2	36.9	36.1	31.2	29.3	35.3	29.8	36.2	31.8	26.0	26.1	25.5	26.6	31.6
				Green tea	A	41.6	33.2	28.9	39.7	35.5	41.6	31.3	32.2	30.2	32.2	31.9	38.1	37.1	32.2	37.1	31.7	31.3	31.1	27.4	34.6
					B	38.5	30.0	35.4	35.4	29.4	36.6	26.4	32.6	29.8	26.3	27.5	29.5	21.3	28.1	26.2	26.5	26.4	27.6	24.4	29.7
		Envi-Carb +PSA	GC-MS	Green tea	A*	52.7	37.5	39.8	39.6	41.6	44.9	40.9	32.0	28.1	30.0	32.0	44.4	33.1	37.0	33.5	36.7	33.9	31.3	31.3	36.5
					B*	50.1	36.8	38.1	36.1	36.6	50.2	32.5	34.7	36.0	34.2	29.0	39.1	31.1	33.4	30.4	30.3	29.9	27.5	30.4	34.0
				Woolong tea	A*	47.8	46.9	39.9	36.9	38.6	36.6	36.0	36.0	28.2	27.6	36.0	31.3	31.6	36.5	35.7	31.5	28.8	37.1	31.1	35.5
					B	40.7	33.8	38.3	32.9	29.3	32.2	29.3	30.0	34.3	31.7	30.0	27.5	32.0	30.0	28.1	25.6	23.8	30.5	24.5	29.6
			GC-MS/MS	Green tea	A	45.2	39.4	39.3	39.9	41.0	39.9	41.5	30.7	31.7	30.7	30.7	33.6	32.7	38.7	32.1	37.3	33.4	30.7	33.1	35.8
				Woolong tea	B	42.3	32.4	37.8	35.6	39.7	38.5	33.9	28.0	36.0	28.0	28.0	32.5	34.7	35.3	29.0	30.3	29.8	34.2	32.4	32.8
				Green tea	A	40.5	37.7	42.7	38.8	37.7	32.7	34.5	31.4	31.4	31.4	31.4	30.6	32.7	32.3	34.7	27.3	23.7	27.5	27.4	33.4
					B	37.1	33.0	34.0	34.0	35.6	34.8	30.6	28.8	28.8	26.5	28.8	26.5	32.7	27.4	26.2	21.8	21.3	27.6	24.4	28.2
192	telodrin	Cleanert TPT	GC-MS	Green tea	A	171.2	152.0	150.5	155.9	153.5	129.2	131.1	122.0	131.6	134.6	131.6	131.5	119.8	121.1	123.2	108.8	101.4	100.4	115.8	129.8
					B	162.2	147.7	144.3	140.8	135.7	117.0	125.1	142.2	122.0	131.5	122.0	117.0	110.5	125.3	124.4	105.5	102.6	102.7	103.9	125.0
				Woolong tea	A*	164.3	135.9	163.4	138.9	146.8	126.3	151.0	135.3	153.9	134.5	153.9	140.9	144.4	128.8	138.4	130.8	133.7	126.5	104.5	138.5
					B*	149.6	115.4	139.4	120.2	115.3	97.0	122.8	117.7	114.9	122.6	114.9	114.3	111.4	120.1	114.1	115.6	109.8	118.0	95.1	117.9
			GC-MS/MS	Green tea	A	195.9	154.4	136.8	148.9	153.0	126.5	125.0	106.3	133.2	141.1	133.2	127.1	108.9	128.1	124.1	111.9	100.2	101.1	108.7	128.5
				Woolong tea	B	186.2	149.4	132.8	135.3	140.3	112.3	118.9	125.1	122.9	137.8	122.9	107.0	96.8	132.7	115.1	109.3	103.0	106.0	96.4	123.4
				Green tea	A	175.3	138.6	171.0	132.0	145.9	124.7	138.3	129.3	151.5	147.7	151.5	142.3	132.1	128.6	126.5	123.9	133.6	138.0	117.7	137.3
					B	149.3	114.9	138.3	109.2	110.4	99.3	114.6	107.8	108.9	131.8	108.9	111.1	104.7	111.2	109.5	105.5	110.4	120.8	101.5	113.6
		Envi-Carb +PSA	GC-MS	Green tea	A*	157.7	159.3	150.0	129.9	124.8	167.7	135.1	146.1	119.7	141.5	119.7	176.6	102.3	136.0	105.5	119.6	135.6	111.2	133.2	136.8
					B*	146.2	165.4	142.8	120.9	118.2	110.9	124.9	132.6	88.8	131.1	88.8	162.5	96.9	124.3	109.1	93.5	121.5	120.1	126.2	125.5
				Woolong tea	A*	164.3	146.9	142.9	160.8	160.2	132.3	123.3	147.9	153.7	131.1	153.7	130.3	114.6	119.9	140.4	127.0	133.7	139.1	117.2	139.0
					B	131.7	161.3	123.2	133.1	140.8	118.3	105.2	123.3	132.7	104.3	132.7	115.1	109.0	152.1	120.2	101.4	92.0	116.6	98.6	138.1
			GC-MS/MS	Green tea	A	159.8	158.7	149.2	137.5	140.4	131.6	131.6	108.6	113.5	138.8	113.5	146.0	116.6	113.7	104.8	125.0	139.4	116.6	136.6	124.6
				Woolong tea	B	155.5	137.2	137.6	128.7	144.2	118.7	118.7	132.3	99.3	126.2	99.3	132.3	131.3	123.5	104.1	91.5	124.0	116.5	130.7	135.7
				Green tea	A	169.6	113.6	141.3	196.2	148.2	121.2	121.2	147.9	136.5	116.6	136.5	123.5	97.4	113.2	142.8	112.1	89.5	129.6	117.7	135.7
					B	134.2	113.6	119.2	136.3	130.5	99.3	99.3	108.6	128.2	100.0	128.2	110.7	97.4	113.2	109.2	83.3	81.3	109.6	101.5	111.1

(Continued)

Annex 1: Content change for 201 pesticides in incurred tea Youden pair samples under 8 conditions with two SPE cartridge cleanup in three months (Nov.9, 2009-Feb.7, 2010) (cont.)

No	Pesticides	SPE	Method	Sample	Youden pair	Nov.9, 2009 (n=5)	Nov.14, 2009 (n=3)	Nov.19, 2009 (n=3)	Nov.24, 2009 (n=3)	Nov.29, 2009 (n=3)	Dec.4, 2009 (n=3)	Dec.9, 2009 (n=3)	Dec.14, 2009 (n=3)	Dec.19, 2009 (n=3)	Dec.24, 2009 (n=3)	Dec.29, 2009 (n=3)	Jan.3, 2010 (n=3)	Jan.8, 2010 (n=3)	Jan.13, 2010 (n=3)	Jan.18, 2010 (n=3)	Jan.23, 2010 (n=3)	Jan.28, 2010 (n=3)	Feb.2, 2010 (n=3)	Feb.7, 2010 (n=3)	AVE, μg/kg
193	tetrasul	Cleanert TPT	GC-MS	Green tea	A	41.0	32.6	39.1	36.4	35.0	34.0	33.5	30.5	31.9	31.0	33.5	32.3	34.2	31.0	33.4	33.2	32.7	32.3	37.3	33.8
					B	41.9	31.1	35.2	33.4	30.0	31.9	31.3	28.8	30.5	29.5	29.8	29.8	30.7	29.5	29.5	32.5	32.0	31.2	32.0	31.6
				Woolong tea	A	37.9	34.5	34.4	39.6	35.9	34.4	30.8	35.1	35.8	34.5	36.4	39.0	34.2	33.0	33.4	40.5	40.7	37.7	34.7	35.8
					B	33.4	28.9	28.9	33.6	31.3	29.5	25.6	30.1	29.8	30.2	28.4	31.1	28.4	29.2	32.2	34.7	33.8	35.3	31.8	30.9
			GC-MS/MS	Green tea	A	41.5	33.2	42.1	32.9	32.5	35.9	33.2	31.7	31.1	31.0	29.1	30.8	29.1	34.2	32.9	27.2	25.7	27.6	31.5	32.5
					B	39.6	33.8	36.0	34.3	30.2	32.4	30.0	29.6	30.9	28.6	27.4	27.6	27.4	32.0	30.0	25.5	27.2	26.9	27.6	30.5
				Woolong tea	A	39.2	33.0	33.5	39.6	32.3	34.7	30.6	33.2	36.1	36.4	32.1	37.4	32.1	31.3	32.7	31.4	32.7	31.4	26.5	33.7
					B	36.3	28.2	28.5	33.7	30.0	28.4	24.6	29.0	30.5	32.0	26.4	29.6	26.4	28.5	28.0	27.3	26.6	29.2	25.6	28.9
		Envi-Carb +PSA	GC-MS	Green tea	A	48.8	37.1	36.7	37.0	37.9	41.8	32.0	38.6	31.9	32.0	29.0	30.3	32.9	36.8	36.2	40.9	39.3	37.2	37.0	36.5
					B	45.0	31.3	36.7	34.0	36.9	35.9	32.3	37.0	30.1	31.8	27.8	29.1	30.9	31.8	32.2	31.5	35.1	33.3	36.5	33.6
				Woolong tea	A	45.5	46.8	38.4	36.4	36.0	32.5	33.6	28.1	35.3	33.3	36.2	30.3	32.2	36.4	43.6	38.5	34.3	42.2	33.8	36.5
					B	37.3	33.1	36.4	31.8	29.1	28.4	28.9	25.3	27.9	26.9	30.7	26.1	28.0	29.9	33.3	32.0	29.7	35.7	30.7	30.6
			GC-MS/MS	Green tea	A	40.9	38.6	38.1	37.3	39.9	38.6	34.0	39.7	32.6	31.5	30.5	39.2	35.2	34.4	33.2	38.6	32.8	33.1	33.0	35.9
					B	37.5	30.5	37.1	33.9	38.2	34.4	32.3	36.9	31.3	31.9	28.3	36.9	30.1	31.8	30.3	31.2	29.3	30.3	32.7	32.9
				Woolong tea	A	41.0	36.7	42.1	36.2	34.8	31.6	33.0	28.3	35.3	32.8	30.0	30.5	30.8	32.2	35.2	28.6	23.3	34.0	26.5	32.8
					B	36.0	31.6	35.0	31.8	27.9	26.4	28.9	23.4	27.9	26.8	28.6	26.4	27.9	27.5	25.9	23.3	21.5	28.4	25.6	27.9
194	thiazopyr	Cleanert TPT	GC-MS	Green tea	A	102.1	69.9	83.1	78.5	66.6	68.1	69.9	66.9	66.7	63.8	71.8	67.9	63.5	66.6	68.4	60.9	57.5	53.8	61.2	68.6
					B	89.9	69.1	74.2	71.2	61.6	63.9	63.9	60.5	65.6	60.8	62.7	63.5	60.2	66.6	60.9	56.1	54.3	52.4	54.6	63.8
				Woolong tea	A	85.8	75.8	74.5	84.1	71.1	74.4	68.3	73.9	76.5	77.0	77.1	81.3	73.5	69.5	70.9	70.3	69.9	67.1	59.6	73.7
					B	77.7	64.5	64.2	70.6	63.4	62.0	50.2	61.9	62.1	62.5	56.9	64.1	57.4	60.5	58.7	58.6	56.0	59.7	52.2	61.2
			GC-MS/MS	Green tea	A	105.1	72.7	88.4	51.8	73.1	67.2	68.7	68.9	66.7	70.2	62.8	83.2	64.5	78.9	68.0	54.8	51.4	54.1	64.5	69.2
					B	94.8	71.3	75.4	68.0	63.8	60.4	60.5	62.7	65.7	65.7	55.6	69.6	62.9	82.8	58.4	52.9	52.4	50.0	55.7	64.7
				Woolong tea	A	88.4	65.6	72.8	90.0	66.1	74.9	70.3	67.2	78.5	90.6	85.7	84.4	75.7	67.7	63.7	63.2	66.5	70.0	59.3	73.7
					B	81.5	60.5	59.3	76.7	60.5	58.6	52.6	57.0	63.9	76.7	62.4	64.2	55.6	55.4	54.3	56.0	56.6	59.2	53.4	61.3
		Envi-Carb+PSA	GC-MS	Green tea	A	100.5	89.0	75.0	76.5	80.9	73.5	68.0	79.2	68.0	77.0	56.4	76.4	63.7	69.1	66.4	74.2	67.0	64.8	61.3	72.4
					B	93.6	72.9	87.3	72.6	78.0	70.3	76.5	74.1	61.7	64.1	56.4	66.9	59.3	64.2	60.9	60.8	60.2	58.7	63.1	67.4
				Woolong tea	A	88.8	83.9	72.3	75.4	76.7	72.6	76.5	60.6	79.3	74.1	80.2	66.6	68.2	74.6	75.3	66.9	58.9	74.7	55.8	73.5
					B	72.3	68.4	76.3	68.6	59.2	60.3	64.3	50.1	58.8	57.2	66.1	55.3	59.3	61.8	60.2	54.9	50.1	60.1	50.1	60.5
			GC-MS/MS	Green tea	A	96.4	74.1	76.3	76.1	87.1	79.1	67.0	81.1	67.5	63.7	58.6	73.4	65.9	71.8	62.3	72.2	71.5	64.3	70.0	72.6
					B	92.9	61.1	71.3	73.4	84.0	72.4	67.4	77.0	63.8	63.1	55.1	67.2	61.1	64.9	59.2	57.0	61.8	58.2	64.3	67.1
				Woolong tea	A	83.9	79.8	99.3	80.2	72.0	71.1	68.8	60.2	79.3	71.3	72.2	69.3	65.7	69.6	69.8	54.6	46.9	66.3	59.3	70.5
					B	69.4	67.7	74.5	65.5	56.2	58.7	58.8	49.1	58.8	55.1	62.9	59.8	71.5	63.8	54.8	45.1	43.0	58.6	53.4	59.3
195	tolclofos-methyl	Cleanert TPT	GC-MS	Green tea	A	42.9	36.9	42.5	41.5	33.4	33.7	36.1	35.4	34.5	31.9	37.5	35.8	35.6	34.5	32.4	30.3	29.5	28.2	31.9	35.1
					B	42.3	35.9	38.7	38.1	31.3	32.8	33.1	32.5	34.9	30.3	33.8	33.8	34.8	33.2	33.6	29.2	28.9	27.7	29.0	33.3
				Woolong tea	A	39.8	35.3	37.8	40.9	34.2	36.5	35.3	41.1	35.2	41.8	36.0	38.0	40.2	38.2	33.6	35.3	37.3	33.8	31.4	36.9
					B	36.6	30.5	31.9	36.0	31.1	30.5	26.6	33.5	28.6	35.7	29.7	31.7	30.5	34.1	29.3	30.1	30.7	30.7	27.3	31.3
			GC-MS/MS	Green tea	A	39.1	36.2	36.8	37.4	33.8	38.7	38.0	36.3	33.3	36.9	37.3	38.3	31.5	35.2	34.5	28.6	27.2	29.2	30.6	34.7
					B	39.4	33.3	33.3	35.6	33.5	34.8	33.5	33.2	34.1	32.0	32.8	33.9	29.6	34.7	30.7	27.2	27.7	29.8	27.2	32.6
				Woolong tea	A	49.7	34.1	37.0	44.5	35.8	37.2	36.0	37.6	38.0	38.2	40.5	41.4	39.9	33.5	36.0	32.9	35.2	36.3	30.1	37.6
					B	42.1	29.9	29.7	36.3	32.3	29.2	27.1	31.8	32.8	34.5	31.2	31.6	29.3	28.5	30.7	28.7	29.8	31.9	27.3	31.3
		Envi-Carb+PSA	GC-MS	Green tea	A*	41.8	39.6	38.5	38.1	40.7	40.8	35.6	41.6	32.6	36.2	33.2	41.4	39.9	39.7	34.3	38.4	35.2	36.3	30.1	38.3
					B*	39.7	33.9	38.1	36.5	39.2	39.5	33.2	37.5	30.8	34.8	33.2	31.6	29.3	36.0	34.3	34.4	35.5	31.9	27.3	35.4
				Woolong tea	A	41.6	40.2	41.5	38.5	38.2	35.8	33.6	30.7	38.2	36.2	31.1	45.3	33.0	38.3	34.3	33.1	32.4	33.6	29.4	35.4
					B	33.6	33.0	35.5	34.2	29.5	30.5	32.0	25.6	29.9	34.8	31.1	30.9	34.6	33.4	27.7	28.0	25.3	31.6	26.3	30.6
			GC-MS/MS	Green tea	A	46.6	38.7	39.3	39.8	41.0	39.1	36.4	42.6	35.8	34.7	33.2	35.5	32.0	36.2	33.4	36.8	36.0	35.4	36.6	37.5
					B	55.6	32.0	38.1	37.6	40.0	36.9	31.1	39.9	34.0	33.4	30.8	32.1	33.1	29.4	31.3	31.6	30.9	32.2	37.0	35.1
				Woolong tea	A	54.6	36.7	59.9	38.7	40.1	35.3	35.1	32.5	38.2	32.1	31.8	30.3	32.5	32.7	37.1	28.7	25.1	37.5	30.1	36.3
					B	37.1	30.5	35.8	32.4	31.7	28.9	30.9	26.8	29.9	26.7	31.1	25.4	25.4	27.9	27.2	24.6	22.4	30.2	27.3	29.1

No.	Compound	Cartridge	Method	Tea	Rep	1	2	3	4	5	6	7	8	9	10	11	12	13	14	15	16	17	18	19	20
196	transfluthrin	Cleanert TPT	GC-MS	Green tea	A	49.3	36.1	39.5	41.4	31.7	34.6	34.5	33.1	33.7	35.4	31.4	33.2	33.3	34.4	36.0	28.3	32.6	26.8	32.2	34.6
				Green tea	B	46.4	34.6	37.5	38.1	29.8	32.2	32.5	30.6	33.5	31.9	29.7	31.1	31.1	38.3	33.8	27.5	27.2	25.6	27.9	32.6
				Woolong tea	A	45.1	34.9	36.0	41.1	34.7	36.9	32.3	38.7	39.7	34.3	36.6	40.5	35.8	33.5	34.7	34.7	33.6	32.5	28.9	36.0
				Woolong tea	B	42.1	29.9	31.5	35.7	31.4	30.7	25.5	31.4	34.4	27.2	31.8	31.5	29.5	30.0	28.5	28.6	27.5	29.5	26.1	30.7
			GC-MS/MS	Green tea	A	44.8	33.8	40.8	38.5	32.4	36.1	35.7	32.2	31.1	33.1	32.1	33.9	32.5	35.5	31.9	28.4	27.3	27.6	30.7	33.6
				Green tea	B	42.2	34.5	35.5	34.8	30.1	31.6	31.9	29.9	31.7	30.5	30.3	30.9	29.8	34.4	30.8	26.1	27.4	27.1	27.0	31.4
				Woolong tea	A	40.1	31.8	33.9	42.7	31.7	35.7	33.1	35.8	36.9	38.3	39.1	39.7	37.1	32.7	33.2	31.7	32.6	33.4	27.9	35.1
				Woolong tea	B	37.0	28.4	28.3	35.2	29.5	28.8	25.0	30.2	31.3	27.7	33.2	30.5	28.2	27.4	27.4	26.6	26.3	30.2	25.7	29.3
		Envi-Carb +PSA	GC-MS	Green tea	A	48.7	39.4	36.3	37.8	39.7	34.7	33.6	37.6	31.6	29.6	34.2	50.8	32.8	37.7	29.7	37.0	33.5	33.3	31.2	36.3
				Green tea	B	42.1	41.7	33.6	39.5	38.5	26.6	33.0	36.2	29.5	27.1	33.8	40.5	29.8	32.9	28.0	30.6	29.9	29.9	30.6	33.4
				Woolong tea	A	43.4	35.9	42.1	36.7	37.4	35.7	36.8	30.6	37.7	38.5	35.1	30.6	34.3	37.3	37.2	33.1	29.1	36.4	28.4	35.6
				Woolong tea	B	36.5	31.3	36.3	33.8	29.4	29.5	31.2	26.0	29.3	32.8	28.1	27.4	29.9	30.6	29.9	27.6	24.7	29.9	25.1	30.0
			GC-MS/MS	Green tea	A*	44.4	38.1	37.8	37.3	38.3	39.3	35.9	40.1	35.6	30.1	33.9	35.7	32.7	39.3	32.1	37.6	34.8	31.8	34.0	36.2
				Green tea	B*	39.5	31.9	36.3	34.5	37.3	36.7	33.2	36.9	32.4	28.0	33.0	33.2	30.2	30.5	30.3	31.7	31.2	29.4	32.9	33.1
				Woolong tea	A	41.4	36.9	50.0	38.6	36.3	35.2	33.2	29.4	37.7	32.6	32.9	24.9	33.7	32.6	36.7	28.4	24.6	34.3	27.9	34.1
				Woolong tea	B	33.6	31.1	35.7	33.2	29.0	28.4	28.9	24.2	29.3	29.9	27.0	22.3	29.1	28.0	27.2	22.7	22.5	29.2	25.7	28.3
197	triadimefon	Cleanert TPT	GC-MS	Green tea	A	94.0	74.6	85.3	76.8	61.3	57.0	69.8	72.2	68.9	74.3	58.2	75.5	66.8	75.2	66.3	60.5	62.7	51.3	60.5	69.0
				Green tea	B	94.5	70.3	75.8	69.1	58.8	57.1	65.6	65.3	66.2	69.7	54.4	72.1	61.0	77.5	66.0	60.0	63.0	50.2	54.4	65.8
				Woolong tea	A*	84.7	63.2	74.5	82.9	69.4	76.6	67.0	69.3	75.9	67.3	116.4	76.6	74.2	75.0	72.3	73.1	70.7	65.9	54.3	74.2
				Woolong tea	B*	76.9	57.6	64.9	72.4	64.3	63.3	53.9	60.1	61.4	77.4	89.4	61.0	66.9	73.7	60.9	64.2	63.5	61.5	48.7	65.4
			GC-MS/MS	Green tea	A	133.2	139.5	120.2	117.3	97.9	101.2	101.9	101.4	96.0	104.1	98.6	110.6	83.7	108.0	96.6	78.6	78.9	76.7	87.7	101.7
				Green tea	B	89.1	97.1	70.6	73.8	60.8	64.5	63.4	62.3	63.7	64.3	65.4	67.5	54.6	77.0	59.7	51.6	52.4	50.8	51.2	65.2
				Woolong tea	A*	89.0	71.0	71.8	90.0	65.3	73.4	69.0	71.6	76.0	83.1	80.8	81.4	73.4	69.6	69.0	59.9	60.9	60.0	50.2	71.9
				Woolong tea	B*	78.6	62.3	59.2	76.9	60.4	58.1	51.6	61.6	65.2	56.1	71.7	61.6	58.4	59.3	54.6	49.1	50.0	51.3	47.6	59.7
		Envi-Carb +PSA	GC-MS	Green tea	A	100.0	107.2	75.1	72.3	74.2	80.7	66.9	79.2	60.6	65.5	64.6	127.6	63.2	83.8	72.5	70.4	73.5	66.3	60.3	77.0
				Green tea	B	98.3	100.0	72.6	65.1	72.5	65.3	69.5	76.6	61.2	62.7	66.0	97.9	57.1	73.2	64.4	56.9	66.7	63.2	60.0	71.0
				Woolong tea	A*	92.2	76.6	83.5	75.0	77.5	71.8	72.0	59.4	75.1	86.2	84.1	77.6	78.7	74.6	84.0	70.4	66.1	70.1	53.8	75.2
				Woolong tea	B*	79.0	67.0	70.5	68.5	57.5	60.2	63.7	46.4	57.9	84.7	65.4	79.6	66.6	71.5	68.7	59.8	57.4	53.4	48.9	64.6
			GC-MS/MS	Green tea	A	149.3	120.9	121.5	121.5	126.7	127.7	98.7	119.3	101.8	96.2	98.9	107.2	97.5	116.6	91.7	112.0	93.7	85.8	100.9	109.9
				Green tea	B	141.9	102.6	117.3	109.2	119.7	115.1	106.4	113.6	98.6	85.1	99.5	99.5	87.6	104.5	84.6	88.6	78.6	84.6	103.6	102.1
				Woolong tea	A*	85.4	77.2	94.0	79.7	74.5	76.2	68.7	60.0	75.1	61.1	63.6	63.3	67.6	64.1	77.0	57.3	50.4	69.5	50.2	69.2
				Woolong tea	B*	75.9	66.8	76.3	70.6	60.9	61.6	60.9	49.6	57.9	60.2	51.5	55.0	53.7	54.3	55.7	46.6	46.3	57.6	47.6	58.4
198	triadimenol	Cleanert TPT	GC-MS	Green tea	A	162.2	115.1	144.5	121.2	99.5	78.1	103.1	103.7	98.1	89.7	67.3	94.4	98.0	125.1	100.0	86.6	84.1	83.7	96.5	102.7
				Green tea	B	155.6	109.0	130.5	110.4	91.2	82.6	97.4	94.0	102.2	83.3	61.5	95.5	93.0	131.6	90.7	85.6	81.3	78.3	83.9	97.8
				Woolong tea	A*	158.0	111.8	127.7	136.6	81.4	116.2	98.5	90.7	108.9	83.7	154.1	86.6	110.2	103.1	98.7	99.3	96.4	89.3	81.4	107.0
				Woolong tea	B*	156.9	118.6	131.5	138.4	85.5	92.9	82.3	91.3	86.2	71.0	121.4	70.1	91.7	90.4	84.9	76.0	75.5	82.3	71.0	95.7
			GC-MS/MS	Green tea	A	136.8	114.8	130.0	107.0	102.2	97.7	99.7	97.3	90.9	96.7	89.8	82.3	74.3	125.0	90.2	76.2	74.0	69.4	82.5	96.7
				Green tea	B	143.5	124.2	119.4	111.8	95.9	94.4	100.2	96.5	95.4	94.4	100.8	86.9	79.4	144.1	91.3	80.4	79.8	72.2	78.6	99.4
				Woolong tea	A*	140.0	106.1	118.7	137.6	102.5	109.7	100.0	106.6	114.4	119.1	118.4	118.1	116.8	95.4	96.8	94.3	100.9	94.5	79.9	108.4
				Woolong tea	B*	124.5	96.5	93.9	118.5	97.3	91.4	81.2	96.4	95.3	88.2	104.7	94.3	92.3	83.7	82.9	77.9	80.6	85.2	74.1	92.6
		Envi-Carb +PSA	GC-MS	Green tea	A	188.4	129.0	122.5	119.0	123.2	125.4	82.8	103.7	82.7	79.6	82.5	94.4	96.9	110.4	99.1	115.0	100.2	97.8	89.8	113.0
				Green tea	B	210.6	133.9	121.1	102.4	123.2	116.6	95.0	96.0	81.0	85.6	78.0	118.5	88.0	110.6	91.0	92.4	87.6	86.3	91.6	105.1
				Woolong tea	A*	151.0	113.9	144.4	128.1	102.8	108.0	112.4	76.7	111.3	117.1	103.2	97.2	97.4	104.4	112.6	90.8	84.1	116.9	84.3	108.2
				Woolong tea	B*	139.0	117.9	146.9	133.7	103.6	91.2	98.2	56.4	91.1	94.3	74.1	100.7	87.4	89.4	82.5	75.8	72.9	91.7	75.5	95.9
			GC-MS/MS	Green tea	A	100.1	84.9	84.5	85.6	85.4	85.7	63.4	85.0	71.6	63.2	64.6	129.6	68.1	63.1	62.2	61.8	67.1	60.1	64.3	77.1
				Green tea	B	99.7	73.7	80.7	74.5	81.8	77.9	76.9	75.9	66.5	56.5	63.2	117.4	58.3	53.7	56.9	75.8	58.4	53.6	64.7	71.2
				Woolong tea	A*	133.7	120.2	130.8	122.5	113.2	104.6	106.0	87.1	111.3	93.4	103.3	85.2	95.4	92.0	106.3	80.4	71.1	105.6	79.9	102.2
				Woolong tea	B*	120.7	104.8	120.3	111.6	90.3	85.2	93.0	75.1	78.9	89.4	78.9	74.7	73.7	82.6	76.4	68.0	66.2	74.1	74.1	87.5

(Continued)

Annex 1: Content change for 201 pesticides in incurred tea Youden pair samples under 8 conditions with two SPE cartridge cleanup in three months (Nov.9, 2009–Feb.7, 2010) (cont.)

No	Pesticides	SPE	Method	Sample	Youden pair	Nov.9, 2009 (n=5)	Nov.14, 2009 (n=3)	Nov.19, 2009 (n=3)	Nov.24, 2009 (n=3)	Nov.29, 2009 (n=3)	Dec.4, 2009 (n=3)	Dec.9, 2009 (n=3)	Dec.14, 2009 (n=3)	Dec.14, 2009 (n=3)	Dec.19, 2009 (n=3)	Dec.24, 2009 (n=3)	Dec.14, 2009 (n=3)	Jan.3, 2010 (n=3)	Jan.8, 2010 (n=3)	Jan.13, 2010 (n=3)	Jan.18, 2010 (n=3)	Jan.23, 2010 (n=3)	Jan.28, 2010 (n=3)	Feb.2, 2010 (n=3)	Feb.7, 2010 (n=3)	AVE, μg/kg.
199	triallate	Cleanert TPT	GC-MS	Green tea	A	86.7	69.6	76.4	67.4	57.4	67.1	64.5	62.5	64.6	59.8	57.2	62.8	61.6	58.8	60.4	59.4	53.4	53.2	52.5	58.2	62.7
					B	83.7	67.1	68.8	63.8	56.2	62.3	59.1	59.4	60.3	63.9	52.5	59.8	57.6	55.3	59.3	58.3	52.3	51.9	52.8	50.8	59.8
				Woolong tea	A	78.1	63.2	65.3	77.0	64.9	68.0	56.7	69.1	66.7	65.5	69.3	66.4	68.7	68.0	64.1	65.4	65.5	62.6	62.1	50.2	65.8
					B	71.0	53.6	55.7	65.0	57.8	54.9	45.4	57.8	52.6	53.8	61.6	54.2	53.1	54.2	57.9	54.7	55.8	53.6	55.9	44.0	55.7
			GC-MS/MS	Green tea	A*	99.7	78.6	95.8	71.7	63.4	78.4	73.6	68.2	69.5	63.3	69.0	67.9	100.3	56.8	67.3	67.7	57.5	56.1	57.6	60.8	71.3
					B	84.1	68.2	73.2	65.3	56.5	63.9	58.4	58.1	59.9	63.0	55.7	57.8	70.5	47.9	62.0	56.1	49.0	51.1	49.6	47.8	60.0
				Woolong tea	A*	75.9	61.1	62.3	81.0	60.5	65.6	57.3	64.8	72.7	63.9	58.6	64.4	70.6	66.0	58.7	61.1	60.4	59.1	62.3	53.1	64.5
					B*	68.6	52.8	52.3	66.8	54.8	51.7	45.8	54.3	53.1	55.1	61.2	53.1	55.3	54.3	52.0	50.2	51.5	50.5	54.5	46.7	54.3
		Envi-Carb +PSA	GC-MS	Green tea	A*	90.2	76.8	70.2	69.7	73.2	67.2	75.3	68.8	59.8	64.4	66.4	66.4	126.2	56.5	67.5	57.3	64.0	68.6	62.0	60.8	70.8
					B*	78.7	70.2	66.0	65.2	70.8	57.5	57.8	64.7	51.2	59.0	61.8	61.8	90.1	54.9	61.2	55.2	51.4	62.8	60.1	61.1	63.1
				Woolong tea	A*	80.9	66.4	74.2	66.3	72.0	69.2	64.6	53.9	71.5	67.9	62.4	66.7	56.1	64.0	66.7	70.2	60.0	52.4	64.0	52.7	65.0
					B*	62.8	53.8	62.9	58.8	55.5	59.8	54.0	45.9	61.1	51.7	52.5	56.4	50.8	54.2	58.9	57.2	48.2	43.9	56.4	45.8	54.4
			GC-MS/MS	Green tea	A	39.4	37.3	34.9	35.0	36.6	35.6	38.6	34.7	27.7	34.5	32.5	27.7	44.9	27.8	37.7	28.2	31.9	32.0	27.1	31.8	34.1
					B	36.0	33.4	33.1	32.1	35.0	34.9	26.2	31.2	24.7	31.0	30.6	27.4	39.7	28.1	29.6	27.4	24.7	28.0	27.4	30.1	30.7
				Woolong tea	A	76.9	65.6	92.3	70.5	70.9	68.7	60.8	54.0	62.8	67.9	57.5	59.6	60.4	60.8	61.1	66.2	50.7	44.0	59.6	53.1	63.4
					B	60.1	54.1	64.3	60.3	57.2	56.5	52.5	45.1	58.8	51.7	48.4	52.4	53.8	51.7	52.5	49.5	40.9	39.8	52.4	46.7	52.4
200	tribenuron-methyl	Cleanert TPT	GC-MS	Green tea	A	65.0	80.7	59.8	45.4	33.6	51.6	43.8	56.0	35.9	40.1	35.9	36.0	40.7	45.8	31.2	27.5	24.7	17.8	39.2	46.6	43.2
					B	62.7	77.3	54.8	44.3	32.4	48.3	44.5	49.5	33.6	49.5	33.6	34.2	33.5	43.1	34.9	27.9	24.9	18.7	40.3	40.3	41.8
				Woolong tea	A	49.5	30.1	47.9	65.6	55.2	74.1	37.0	42.3	38.7	35.5	51.2	38.7	39.6	42.3	27.8	27.9	38.6	37.8	36.1	39.1	43.7
					B	43.8	32.4	57.3	50.6	45.8	57.4	34.9	40.6	30.0	30.0	51.1	30.0	31.4	41.0	28.6	41.7	36.5	32.2	31.3	39.6	39.5
			GC-MS/MS	Green tea	A*	44.8	35.3	37.9	28.7	23.8	35.7	28.4	34.1	27.9	26.0	26.8	27.9	24.6	22.3	47.2	35.4	34.2	23.5	24.7	33.4	30.9
					B	65.5	54.7	52.5	41.8	35.0	50.3	41.6	48.3	38.3	45.4	38.3	38.3	34.6	32.6	91.3	39.8	52.1	36.0	35.1	42.3	46.1
				Woolong tea	A*	59.5	30.8	40.4	61.3	43.0	59.8	33.9	42.4	44.1	43.6	52.9	44.1	41.5	42.0	28.9	45.2	50.3	38.9	53.1	42.7	45.0
					B*	52.1	31.7	44.9	49.4	37.3	46.7	32.8	39.9	31.9	36.6	53.3	31.9	31.1	39.9	29.0	38.9	45.2	30.9	43.0	35.9	39.5
		Envi-Carb +PSA	GC-MS	Green tea	A*	70.2	69.9	52.0	55.8	47.4	46.6	125.4	54.1	29.4	43.4	54.8	29.4	17.0	52.1	17.2	24.2	23.9	22.6	48.3	45.4	47.4
					B*	58.1	87.6	43.4	45.9	45.4	36.6	61.7	49.1	18.1	40.4	45.7	18.1	12.9	49.9	16.5	19.5	19.6	21.1	47.1	47.0	40.3
				Woolong tea	A*	57.6	28.2	44.2	36.2	62.6	65.4	39.4	42.8	40.2	39.8	34.5	40.2	38.1	42.7	32.5	41.2	37.7	31.0	36.9	36.4	41.4
					B*	54.6	22.2	42.0	39.0	45.1	61.3	29.5	35.8	38.2	31.6	28.2	38.2	37.6	39.0	34.3	39.1	25.9	29.8	31.8	32.3	36.7
			GC-MS/MS	Green tea	A	76.7	132.0	91.1	116.7	77.6	112.6	143.2	109.1	63.0	108.4	94.4	63.0	0.0	78.1	107.1	59.1	98.3	114.0	84.5	106.9	93.3
					B	75.3	172.2	83.9	90.4	76.3	116.4	162.4	85.1	42.6	95.4	82.1	42.6	0.0	78.1	88.4	46.9	70.4	96.2	90.4	109.1	87.3
				Woolong tea	A	64.8	26.5	41.4	38.8	76.9	70.4	35.2	39.2	37.7	39.8	36.6	37.7	36.4	40.3	30.5	49.9	28.9	29.7	38.0	42.7	42.3
					B	60.5	23.1	41.5	40.2	55.1	68.0	29.5	30.8	36.8	31.6	28.6	36.8	34.4	37.9	31.8	42.7	20.5	29.8	32.7	35.9	37.4
201	vinclozolin	Cleanert TPT	GC-MS	Green tea	A	45.8	40.0	45.3	43.0	34.4	40.1	38.4	38.5	38.5	36.9	37.5	38.5	37.7	35.3	35.5	36.4	32.5	37.6	30.7	35.5	37.9
					B	45.6	39.1	41.1	39.9	32.9	37.5	36.2	36.0	41.7	39.2	38.3	41.7	34.8	33.4	35.9	33.2	31.8	34.4	31.1	30.8	36.1
				Woolong tea	A	44.9	37.0	39.1	46.9	40.6	40.1	35.4	42.0	31.6	38.7	39.8	31.6	41.4	38.5	38.2	42.4	38.0	38.6	36.4	32.5	39.5
					B	40.4	31.8	32.7	40.9	35.3	32.1	29.3	34.8	37.7	32.6	34.6	37.7	34.3	31.6	33.6	40.1	32.7	32.7	33.9	32.5	33.5
			GC-MS/MS	Green tea	A*	50.0	42.8	44.7	41.5	32.7	42.0	38.4	37.8	34.0	32.1	38.3	34.0	45.1	33.1	40.4	32.4	32.5	31.0	30.2	32.8	38.0
					B	47.8	42.9	33.9	39.0	32.1	37.8	34.2	34.5	43.4	35.4	37.6	43.4	37.5	31.6	41.0	38.0	31.0	31.7	30.8	28.9	35.9
				Woolong tea	A*	50.8	36.6	37.9	47.7	36.1	38.1	34.7	39.2	30.1	39.5	44.7	30.1	42.5	39.3	37.2	35.7	34.9	37.8	37.1	31.9	39.2
					B*	44.2	32.7	31.5	38.3	33.0	30.7	27.5	34.4	35.7	34.3	38.2	35.7	33.3	32.6	34.1	34.6	28.9	29.3	33.3	28.7	32.9
		Envi-Carb +PSA	GC-MS	Green tea	A*	52.3	46.7	45.3	44.0	42.8	44.7	44.6	44.0	30.7	40.9	39.6	30.7	52.4	33.6	42.7	30.4	39.8	40.7	37.9	35.8	41.9
					B*	45.5	41.9	41.3	41.7	40.1	37.6	38.8	40.8	42.4	38.0	37.1	42.4	45.9	33.6	36.2	33.3	32.5	36.4	36.1	35.7	37.9
				Woolong tea	A*	47.0	40.8	45.4	39.9	44.4	43.2	39.3	36.2	37.1	42.1	35.7	37.1	35.2	35.4	38.5	32.4	37.1	31.6	41.4	29.1	39.4
					B*	39.3	33.0	33.9	36.9	34.5	37.4	33.1	30.2	33.9	31.1	30.1	33.9	30.8	30.9	32.7	40.3	29.5	27.3	33.7	34.6	33.1
			GC-MS/MS	Green tea	A	43.6	44.8	43.3	41.6	45.4	43.6	46.1	41.4	29.8	43.0	38.3	29.8	0.0	34.7	41.2	33.6	36.7	38.0	31.5	35.9	37.5
					B	42.1	42.9	39.6	38.1	42.0	42.2	32.9	38.4	36.5	38.4	35.7	36.5	0.0	34.2	32.5	31.7	29.7	35.9	31.4	31.9	34.4
				Woolong tea	A	40.5	40.9	49.9	39.6	42.3	40.8	36.0	32.8	33.9	42.1	33.8	33.9	36.1	35.6	35.3	41.2	33.4	26.2	37.2	28.7	37.5
					B	35.4	34.4	38.4	34.8	34.8	34.7	30.7	27.3	33.9	31.1	28.9	33.9	32.6	27.3	30.2	30.1	25.8	25.0	32.2	28.7	31.4

Annex 2: RSD data for 201 pesticides in incurred tea Youden pair samples under 8 conditions with two SPE cartridge cleanup in three months (Nov.9, 2009-Feb.7, 2010)

No	Pesticides	SPE	Method	Sample	Youden pair	Nov.9, 2009 (n=5)	Nov.14, 2009 (n=3)	Nov.19, 2009 (n=3)	Nov.24, 2009 (n=3)	Nov.29, 2009 (n=3)	Dec.4, 2009 (n=3)	Dec.9, 2009 (n=3)	Dec.14, 2009 (n=3)	Dec.19, 2009 (n=3)	Dec.24, 2009 (n=3)	Dec.14, 2009 (n=3)	Jan.3, 2010 (n=3)	Jan.8, 2010 (n=3)	Jan.13, 2010 (n=3)	Jan.18, 2010 (n=3)	Jan.23, 2010 (n=3)	Jan.28, 2010 (n=3)	Feb.2, 2010 (n=3)	Feb.7, 2010 (n=3)
1	2,3,4,5-tetrachloroaniline	Cleanert TPT	GC-MS	Green tea	A	8.0	3.3	3.8	2.3	6.1	4.8	6.1	4.0	9.9	3.9	4.3	1.7	8.0	9.8	4.5	3.2	15.4	2.3	6.8
					B	10.0	8.7	7.3	9.2	3.5	11.1	1.4	4.1	5.8	12.3	0.5	8.2	3.4	9.9	5.2	0.8	9.2	4.7	7.6
				Woolong tea	A*	3.5	3.5	2.3	3.4	4.2	3.3	6.9	5.3	3.1	1.6	2.5	7.9	6.2	7.2	4.6	3.4	2.0	2.8	1.9
					B*	8.8	3.7	6.7	12.1	2.0	4.1	7.2	14.5	4.5	6.1	0.2	3.6	3.9	2.2	6.3	1.8	3.4	2.6	5.9
			GC-MS/MS	Green tea	A	5.9	2.1	7.7	1.8	5.8	5.5	12.2	8.7	4.2	7.6	13.8	3.5	1.7	6.7	8.0	13.8	8.5	17.3	13.2
					B	5.4	4.8	3.3	5.4	1.6	6.9	2.6	4.8	5.6	17.9	2.9	4.7	13.8	4.9	6.2	12.8	4.2	10.1	12.0
				Woolong tea	A*	2.9	6.4	6.3	4.0	5.2	2.4	9.9	2.3	11.9	10.1	9.9	2.5	8.0	14.1	33.7	10.5	13.7	0.4	12.3
					B*	7.8	12.8	17.5	2.3	5.4	8.3	5.3	16.6	8.6	6.0	7.8	11.8	5.9	6.7	15.9	5.6	19.9	9.8	18.5
		Envi-Carb+PSA	GC-MS	Green tea	A	8.1	14.1	19.6	7.4	6.0	3.1	9.8	1.7	5.3	5.3	3.5	8.4	5.6	1.4	10.3	5.5	6.7	7.5	5.6
					B	12.4	20.6	10.3	7.0	2.0	3.5	2.2	7.4	10.2	3.2	20.2	5.4	1.7	4.9	10.6	9.2	7.1	8.3	5.8
				Woolong tea	A*	0.9	2.0	1.2	4.8	10.4	2.0	6.6	3.2	4.2	4.0	7.2	11.8	15.7	7.0	7.5	2.1	5.8	4.3	5.0
					B*	4.6	4.0	12.7	6.5	6.5	5.6	2.1	6.3	11.7	5.9	8.1	2.4	7.2	8.5	7.0	5.3	25.6	2.0	4.7
			GC-MS/MS	Green tea	A	6.2	2.8	2.9	2.8	10.1	8.8	8.1	4.9	13.9	6.1	20.3	6.4	13.6	15.4	5.0	15.3	5.8	13.5	15.3
					B	7.6	8.3	5.7	4.5	2.6	1.1	1.4	4.9	10.6	6.4	25.4	1.8	9.1	3.5	3.7	6.8	16.8	16.3	16.0
				Woolong tea	A*	5.1	6.0	1.3	14.7	4.4	4.7	0.4	7.5	7.0	5.0	2.0	13.1	6.2	5.8	6.6	9.5	3.8	10.8	11.5
					B*	7.6	5.1	14.0	9.5	10.0	2.4	7.1	3.3	19.2	5.1	11.1	4.4	4.1	9.4	21.3	4.0	3.9	10.7	8.1
2	2,3,5,6-tetrachloroaniline	Cleanert TPT	GC-MS	Green tea	A	4.3	5.2	1.3	0.6	6.9	4.6	5.3	3.4	6.0	2.8	4.5	2.3	5.9	4.9	5.2	2.1	4.5	6.1	4.8
					B	6.6	8.8	3.8	8.3	2.7	9.5	3.0	1.8	3.1	8.5	1.9	6.8	2.7	4.5	3.7	1.5	11.9	2.2	5.1
				Woolong tea	A*	3.9	2.1	0.4	2.9	4.5	1.8	7.8	5.0	3.3	2.4	2.3	1.8	3.6	1.7	7.2	3.2	4.0	1.3	2.0
					B*	6.7	2.6	4.9	5.4	2.0	3.7	5.0	1.8	5.8	4.1	0.6	0.7	3.2	2.5	3.9	0.8	0.6	1.6	1.5
			GC-MS/MS	Green tea	A	3.4	6.2	2.6	2.0	6.3	6.2	6.8	4.0	5.6	4.5	4.9	1.6	4.2	7.3	1.4	0.7	4.5	3.8	2.9
					B	4.2	8.9	4.1	8.7	6.2	3.5	5.1	2.3	1.4	9.8	4.1	2.8	3.5	3.8	1.8	1.2	11.9	4.8	8.0
				Woolong tea	A*	4.1	1.7	2.6	5.7	4.1	2.5	8.4	1.6	1.1	2.0	1.7	7.9	0.3	0.9	3.8	0.7	6.6	2.5	6.4
					B*	4.9	2.2	5.8	2.6	1.6	3.4	3.9	2.8	6.5	2.0	2.8	4.6	2.6	2.9	5.2	3.1	2.4	1.6	2.9
		Envi-Carb+PSA	GC-MS	Green tea	A	5.1	8.1	2.0	3.1	1.6	3.2	11.6	5.4	2.7	1.6	7.4	3.7	4.4	3.0	5.0	5.1	11.8	8.7	5.4
					B	9.7	2.9	4.0	3.5	8.0	1.9	0.6	1.5	1.6	0.2	19.9	0.3	1.2	2.2	14.5	2.3	7.5	5.3	5.1
				Woolong tea	A*	3.7	3.2	2.5	3.9	8.1	2.9	2.3	5.8	10.7	2.7	4.2	3.1	2.8	7.1	2.7	2.7	5.2	4.1	4.1
					B*	4.9	5.0	8.5	6.3	4.3	1.1	1.7	0.3	6.2	3.1	6.7	4.2	5.0	7.8	8.4	8.5	12.4	4.6	4.3
			GC-MS/MS	Green tea	A	4.1	11.2	2.5	4.4	3.4	4.0	1.7	1.7	2.4	2.2	21.3	2.3	0.6	7.6	17.1	3.4	4.2	18.3	5.0
					B	1.9	5.3	4.6	6.4	5.7	2.4	3.0	3.3	2.8	1.2	2.6	7.5	0.6	3.5	7.9	2.3	7.4	2.0	3.8
				Woolong tea	A*	4.4	4.0	5.3	3.7	5.9	2.8	3.4	5.5	9.7	1.6	2.9	5.2	1.8	9.5	8.1	1.8	6.8	2.9	2.3
					B*	6.3	5.9	9.2	3.5	4.3	8.2	8.8	4.9	6.7	4.6	6.2	4.3	6.7	2.7	5.6	0.8	6.1	11.3	10.3
3	4,4-dibromobenzophenone	Cleanert TPT	GC-MS	Green tea	A	5.5	8.2	3.4	12.5	6.3	7.0	2.5	3.5	4.6	10.3	2.8	7.9	4.1	5.6	7.9	3.7	6.2	6.2	9.1
					B	12.3	10.6	4.8	0.8	7.2	6.1	9.6	10.6	4.4	5.4	3.0	7.3	8.6	6.7	7.6	8.6	5.8	0.4	11.0
				Woolong tea	A*	7.4	4.2	1.8	4.2	5.4	2.6	5.0	1.4	4.8	2.3	1.3	3.8	2.7	2.8	3.8	2.7	3.3	10.0	14.4
					B*	12.6	6.7	1.5	3.8	6.0	7.8	9.1	4.9	2.8	2.9	8.0	3.8	6.4	2.4	0.4	3.9	4.6	13.8	2.5
			GC-MS/MS	Green tea	A*	4.5	9.2	2.1	6.2	1.8	5.4	4.6	2.6	4.4	10.1	5.6	8.0	3.4	2.7	1.5	1.3	7.5	5.2	5.2
					B*	4.9	7.0	7.2	6.0	8.3	2.1	11.1	5.3	4.4	10.1	2.9	2.4	0.7	3.6	7.7	3.7	2.4	7.9	3.6
		Envi-Carb+PSA	GC-MS	Green tea	A	3.6	2.2	3.9	7.8	1.7	3.2	8.3	1.8	5.3	3.4	2.9	3.5	9.9	3.9	5.4	0.6	4.0	3.1	5.4
					B	5.0	4.4	2.7	5.7	8.7	15.9	8.4	1.6	1.4	3.3	9.5	8.2	3.0	4.5	7.7	2.2	8.3	5.2	5.2
				Woolong tea	A*	7.9	7.6	1.9	10.3	4.4	9.2	4.1	11.3	7.7	4.9	4.9	5.5	3.0	4.4	12.9	6.4	11.5	3.1	14.8
					B*	10.1	4.5	5.6	6.5	11.2	2.2	3.9	5.9	7.2	3.6	6.0	7.5	10.0	8.1	4.0	5.7	6.2	1.4	11.3
			GC-MS/MS	Green tea	A	7.0	1.9	6.5	7.2	10.9	4.4	2.5	9.0	3.2	2.2	10.2	2.5	5.5	7.1	16.8	6.6	4.9	10.0	3.1
					B	4.9	6.3	1.9	7.7	8.3	3.9	7.4	2.2	9.2	11.8	12.1	5.9	3.0	11.2	6.4	2.7	10.4	1.9	3.7
				Woolong tea	A*	10.3	8.5	4.7	9.0	4.3	1.4	3.0	7.0	10.1	1.7	16.2	6.5	10.1	7.3	14.1	1.9	5.4	8.2	3.3
					B	8.8	2.6	4.9	5.1	8.8	2.2	4.3	2.9	3.0	3.2	7.3	5.6	7.7	5.3	7.2	2.2	10.9	11.9	12.2
					B	8.0	2.5	11.4	1.7	8.7	2.9	3.6	9.7	10.7	2.5	2.5	3.3	4.8	5.9	13.1	2.8	4.0	8.9	5.8

(Continued)

Annex 2: RSD data for 201 pesticides in incurred tea Youden pair samples under 8 conditions with two SPE cartridge cleanup in three months (Nov.9, 2009-Feb.7, 2010) (cont.)

No	Pesticides	SPE	Method	Sample	Youden pair	Nov.9 2009 (n=5)	Nov.14 2009 (n=3)	Nov.19 2009 (n=3)	Nov.24 2009 (n=3)	Nov.29 2009 (n=3)	Dec.4 2009 (n=3)	Dec.9 2009 (n=3)	Dec.14 2009 (n=3)	Dec.19 2009 (n=3)	Dec.24 2009 (n=3)	Jan.3 2010 (n=3)	Jan.8 2010 (n=3)	Jan.13 2010 (n=3)	Jan.18 2010 (n=3)	Jan.23 2010 (n=3)	Jan.28 2010 (n=3)	Feb.2 2010 (n=3)	Feb.7 2010 (n=3)
4	4,4-dichlorobenzophenone	Cleanert TPT	GC-MS	Green tea	A	8.2	5.3	3.1	0.8	5.5	5.4	8.9	3.8	3.6	4.9	1.9	7.0	2.6	3.6	5.1	2.7	12.1	3.5
				Green tea	B	7.2	7.3	5.8	9.3	4.0	4.7	3.5	4.0	5.1	6.7	5.3	4.7	5.2	5.5	2.1	7.3	4.5	4.9
				Woolong tea	A	10.3	4.8	8.1	6.5	4.0	7.1	10.3	2.4	2.5	2.0	4.0	20.7	3.3	2.7	3.4	6.9	9.0	0.8
				Woolong tea	B	8.8	12.2	7.7	3.2	3.5	4.6	10.5	8.0	8.6	1.1	7.0	4.3	3.1	11.7	0.9	9.0	6.9	5.0
			GC-MS/MS	Green tea	A	7.2	4.9	1.1	2.4	5.0	4.7	8.3	4.6	5.8	3.9	2.6	6.3	3.0	3.0	0.8	3.0	11.2	4.6
				Green tea	B	5.2	7.9	6.1	7.9	4.8	4.3	4.7	3.8	2.4	9.7	6.1	1.5	0.1	4.0	2.9	9.0	8.4	4.4
				Woolong tea	A	9.8	4.2	8.1	8.8	4.2	6.5	10.1	5.6	7.9	4.5	2.2	19.6	3.9	1.5	4.2	5.1	6.7	2.6
				Woolong tea	B	5.9	9.8	10.1	4.4	4.9	3.7	8.9	6.0	9.1	5.1	6.2	6.1	3.7	14.0	1.7	6.9	7.2	6.6
		Envi-Carb+PSA	GC-MS	Green tea	A	4.2	4.6	0.9	10.5	7.9	7.4	17.0	4.2	3.1	4.4	9.0	3.6	9.0	5.7	1.9	13.4	2.2	6.7
				Green tea	B*	6.3	7.4	4.3	9.0	4.3	2.5	4.1	10.7	5.8	7.8	3.4	11.7	3.4	9.9	3.4	4.7	3.6	5.1
				Woolong tea	A*	10.9	11.2	5.9	4.5	7.3	10.2	11.8	3.1	15.2	6.6	2.6	3.3	4.7	2.0	5.2	2.7	4.6	7.9
				Woolong tea	B*	4.2	2.1	7.6	5.3	7.3	7.0	10.8	4.8	9.6	3.9	7.4	2.2	7.9	16.3	6.0	2.9	16.9	4.1
			GC-MS/MS	Green tea	A*	5.6	3.5	0.1	7.0	7.0	5.0	7.7	3.8	5.0	2.4	3.9	4.7	5.4	5.7	4.2	10.8	4.0	3.3
				Green tea	B*	8.2	9.5	4.7	8.4	6.9	2.8	2.2	8.0	3.3	9.9	10.1	8.7	5.7	12.7	4.1	4.9	11.8	3.0
				Woolong tea	A*	14.1	11.2	4.1	2.4	9.0	10.3	12.8	4.2	15.3	1.3	5.8	3.7	5.9	8.4	5.2	8.3	12.5	6.5
				Woolong tea	B*	5.6	0.7	8.1	3.5	7.1	7.3	13.2	3.7	4.9	5.8	6.7	10.0	8.4	13.8	4.5	4.2	11.4	5.1
5	acetochlor	Cleanert TPT	GC-MS	Green tea	A	5.7	6.0	3.9	3.9	7.6	5.0	6.0	4.9	8.8	3.1	0.8	7.9	12.3	11.1	3.5	18.1	8.5	5.3
				Green tea	B	7.2	9.6	6.3	9.7	4.9	3.3	2.6	2.3	1.5	4.0	7.9	1.0	9.9	15.7	4.8	23.7	2.4	6.2
				Woolong tea	A	5.8	4.5	6.2	3.7	3.6	3.5	7.0	1.2	4.5	4.1	5.1	1.3	4.3	7.8	2.4	3.6	2.0	1.6
				Woolong tea	B	6.5	6.1	0.4	2.5	3.7	3.4	6.4	5.5	3.0	2.6	2.4	1.2	4.5	6.0	4.7	2.5	1.3	4.5
			GC-MS/MS	Green tea	A	9.4	8.6	4.1	3.9	4.9	7.4	3.9	3.2	4.1	2.2	2.8	9.9	11.1	15.6	2.2	15.7	7.6	5.2
				Green tea	B	6.7	7.7	4.8	8.2	4.1	3.8	4.6	3.2	0.6	9.9	5.1	5.5	4.3	17.6	6.7	23.9	5.9	9.6
				Woolong tea	A	5.1	6.3	0.3	5.8	10.9	8.2	8.0	0.7	2.8	1.3	4.3	8.8	2.7	6.2	4.4	0.3	7.8	5.7
				Woolong tea	B	3.7	2.8	8.1	5.0	4.2	4.9	6.2	0.7	9.6	6.8	0.1	9.2	2.1	2.7	4.2	0.5	3.1	5.8
		Envi-Carb+PSA	GC-MS	Green tea	A	5.9	1.1	1.3	3.1	11.0	4.9	3.9	6.0	4.7	10.1	3.6	2.7	3.1	2.6	1.5	7.5	3.9	8.4
				Green tea	B*	8.1	5.8	3.4	6.4	3.7	5.1	2.2	1.2	14.8	1.6	4.8	3.7	4.5	10.5	4.1	4.7	8.1	13.3
				Woolong tea	A*	5.6	7.8	1.9	7.2	6.5	2.1	1.6	4.1	9.3	1.2	4.8	5.5	6.1	5.3	4.9	3.9	2.1	4.6
				Woolong tea	B*	8.6	5.5	11.1	5.7	9.7	3.0	4.5	8.7	64.0	2.8	3.0	2.7	8.4	10.2	4.9	4.7	1.4	3.0
			GC-MS/MS	Green tea	A*	4.3	5.4	0.2	3.8	6.8	3.8	8.5	1.6	4.8	2.5	6.1	9.7	7.3	5.3	7.8	7.9	4.3	2.5
				Green tea	B*	19.7	9.6	1.4	4.0	2.9	3.5	2.3	2.9	5.8	2.2	3.3	6.6	3.9	9.3	8.8	4.7	13.2	6.8
				Woolong tea	A*	3.7	5.1	0.3	3.2	7.1	7.1	3.8	8.6	9.2	1.8	6.4	7.4	2.3	10.2	4.9	6.1	3.6	5.8
				Woolong tea	B*	8.8	2.7	11.4	3.2	11.6	7.1	10.1	6.1	6.4	2.3	7.9	6.8	5.8	7.1	1.5	2.6	5.4	5.3
6	alachlor	Cleanert TPT	GC-MS	Green tea	A	5.9	4.4	3.7	6.8	6.4	4.5	5.5	7.5	6.6	10.2	2.4	3.1	2.0	3.5	1.0	6.8	8.8	4.7
				Green tea	B	7.7	12.0	8.6	6.0	5.4	2.3	2.2	0.3	5.0	8.4	9.6	1.8	3.5	5.2	4.2	4.1	6.5	13.6
				Woolong tea	A	4.1	1.1	4.0	5.9	2.3	2.1	12.8	3.9	9.2	6.3	4.8	3.0	3.0	8.6	1.1	5.1	2.0	2.0
				Woolong tea	B	6.7	8.2	11.6	2.5	5.7	4.5	5.0	2.1	3.0	4.1	0.9	2.0	1.9	7.7	2.2	0.9	2.0	4.4
			GC-MS/MS	Green tea	A	40.5	6.3	5.0	6.3	6.3	5.0	6.3	4.5	5.7	10.2	5.8	10.2	5.9	5.3	2.5	2.5	6.3	5.7
				Green tea	B	46.5	9.1	4.2	6.7	4.8	5.8	3.3	5.8	2.1	0.8	7.0	3.9	3.3	7.1	1.5	5.8	3.7	6.5
				Woolong tea	A*	4.6	3.6	1.9	6.2	7.7	3.8	11.9	3.0	8.7	9.4	4.5	10.6	2.7	7.6	3.5	2.4	7.7	8.0
				Woolong tea	B*	1.5	3.8	4.5	4.6	1.6	1.0	5.9	1.8	2.9	1.4	2.6	8.8	4.7	5.2	6.2	0.7	6.8	5.6
		Envi-Carb+PSA	GC-MS	Green tea	A	6.9	4.1	2.2	2.6	6.4	5.1	4.2	3.3	4.8	4.6	2.3	5.6	4.3	2.1	2.0	9.3	8.7	8.5
				Green tea	B	10.0	8.8	3.1	3.3	2.0	7.1	3.0	1.4	14.3	1.6	1.9	6.9	2.5	6.5	4.7	3.5	16.4	5.4
				Woolong tea	A*	9.8	4.3	5.4	4.0	2.4	2.5	1.8	7.8	8.1	10.6	4.3	2.2	4.2	4.2	3.2	3.2	4.4	3.7
				Woolong tea	B	5.3	2.3	10.4	4.6	10.4	3.0	1.4	1.4	0.4	4.0	2.3	4.6	8.7	11.5	1.3	7.6	1.9	4.3
			GC-MS/MS	Green tea	A*	8.1	6.3	1.1	2.6	6.1	5.0	6.8	3.1	3.7	5.3	23.8	9.0	9.8	4.3	5.8	7.1	11.7	3.4
				Green tea	B	29.6	10.6	1.5	7.1	3.5	3.2	3.8	3.8	4.4	1.0	12.6	2.7	6.3	12.3	6.2	7.7	12.8	4.4
				Woolong tea	A	3.8	2.1	2.9	3.7	7.6	3.1	2.7	6.8	12.5	1.9	8.4	4.9	5.8	8.3	2.2	9.9	8.7	5.3
				Woolong tea	B	7.2	2.8	10.5	3.7	7.8	5.5	2.9	4.2	3.7	3.0	3.8	4.7	10.5	12.2	2.2	4.5	4.1	7.2

No.	Compound	Cartridge	Method	Tea		1	2	3	4	5	6	7	8	9	10	11	12	13	14	15	16	17	18	19
7	atratone	Cleanert TPT	GC-MS	Green tea	A	5.1	5.1	2.5	1.9	9.1	4.8	8.1	5.4	4.2	4.7	7.5	5.4	9.0	2.7	2.9	5.6	4.9	6.4	4.7
					B	4.7	6.1	6.1	8.9	3.2	4.0	4.8	3.5	5.0	11.6	11.7	13.1	2.2	1.4	3.0	3.6	3.3	1.1	2.0
				Woolong tea	A	3.4	3.3	2.8	2.5	2.9	5.4	10.5	4.8	4.1	6.6	11.2	2.7	3.1	3.9	8.3	5.7	4.0	5.0	1.7
					B	7.2	2.8	3.3	3.4	2.7	2.8	5.4	3.4	7.6	5.6	10.4	1.1	3.0	5.7	6.4	0.3	1.9	1.4	2.2
			GC-MS/MS	Green tea	A*	5.0	5.4	2.1	2.8	9.1	7.7	7.1	6.4	4.7	0.6	5.1	5.2	8.2	4.6	2.3	2.1	7.7	6.6	4.8
					B*	4.7	7.1	8.8	7.0	7.3	15.0	4.6	5.5	3.8	10.7	6.8	6.8	3.1	4.7	7.0	6.2	6.7	6.9	5.0
				Woolong tea	A*	4.7	1.2	0.6	6.1	1.4	2.4	10.6	1.7	0.5	6.2	2.1	5.9	4.4	2.9	8.4	4.2	1.5	1.9	6.4
					B*	4.6	2.1	5.3	6.4	0.3	2.9	4.2	4.1	5.4	4.0	5.6	1.6	7.5	4.3	6.6	1.7	2.5	2.2	1.4
		Envi-Carb+PSA	GC-MS	Green tea	A	5.5	5.3	2.0	9.0	8.9	5.5	4.1	0.5	4.0	14.2	5.4	4.9	3.4	6.4	7.5	7.4	9.9	5.3	5.9
					B	4.1	8.2	4.7	5.8	3.3	5.1	4.5	5.6	7.4	17.6	29.1	7.9	5.1	6.5	14.7	2.7	2.9	12.6	7.7
				Woolong tea	A	4.2	4.0	2.9	2.4	2.5	2.4	3.8	0.5	30.4	3.2	8.3	8.4	2.4	5.4	5.0	3.2	6.1	3.7	5.7
					B	4.2	3.4	9.1	1.8	14.7	6.0	2.8	6.5	12.5	5.6	9.8	3.1	5.7	8.0	10.8	3.2	5.9	5.3	6.3
			GC-MS/MS	Green tea	A*	4.8	6.3	2.6	7.9	7.2	3.6	5.1	0.4	0.7	2.4	6.5	12.3	4.4	11.7	9.1	5.7	8.9	2.4	1.8
					B*	3.2	10.8	5.5	6.7	2.9	1.2	2.5	3.5	4.8	2.3	22.2	9.7	1.8	11.5	18.6	1.8	1.7	12.4	7.0
				Woolong tea	A*	3.7	4.7	4.5	4.9	5.1	2.3	5.8	3.7	4.9	1.6	6.6	9.7	3.9	6.6	10.7	3.3	5.9	5.4	5.5
					B*	5.6	3.9	9.8	3.8	9.6	6.3	3.2	5.9	12.1	4.8	1.2	6.8	8.5	8.9	11.3	0.8	4.7	6.3	6.7
8	benodanil	Cleanert TPT	GC-MS	Green tea	A	6.4	15.0	3.3	1.5	14.4	3.9	16.7	10.1	7.5	5.5	4.3	7.1	13.6	10.2	2.4	8.6	4.3	14.2	2.1
					B*	10.3	11.5	2.3	12.5	7.8	9.8	8.1	11.1	10.4	14.8	8.1	8.1	14.0	10.1	8.7	15.4	8.8	14.6	7.9
				Woolong tea	A*	40.5	10.2	7.9	30.1	19.6	32.7	49.0	16.8	8.5	7.2	9.5	39.0	10.6	5.2	6.8	9.1	8.3	51.2	13.6
					B*	52.1	5.2	18.9	17.5	4.3	12.7	21.5	45.8	5.5	28.2	3.7	5.6	9.3	29.0	20.4	9.6	13.9	37.2	29.4
			GC-MS/MS	Green tea	A	9.0	18.5	4.3	3.9	11.2	7.2	18.9	14.5	20.1	5.5	3.4	5.0	11.8	9.8	7.6	3.8	6.1	21.6	3.7
					B	14.7	6.2	3.6	13.1	5.0	7.2	17.0	24.2	2.5	16.4	7.7	16.0	16.5	7.5	10.9	12.5	12.8	11.7	15.3
				Woolong tea	A*	20.9	6.7	1.2	3.8	23.3	34.2	5.9	22.0	42.8	7.0	11.5	1.7	16.5	15.8	13.4	18.9	18.6	56.4	15.8
					B*	54.4	15.8	31.5	28.4	20.6	32.1	18.0	17.2	22.0	51.3	12.6	5.4	9.9	7.8	29.5	12.8	17.1	26.9	29.6
		Envi-Carb+PSA	GC-MS	Green tea	A*	16.7	16.0	24.3	11.4	8.0	24.1	7.1	9.4	9.3	9.4	2.7	30.4	6.7	3.2	5.7	27.0	16.9	10.1	12.2
					B*	18.1	3.0	11.2	17.7	15.8	14.8	14.4	20.0	13.7	8.0	31.4	16.7	44.5	15.8	21.9	8.7	7.2	24.3	17.9
				Woolong tea	A	39.7	36.3	6.5	39.6	19.6	13.5	76.0	8.1	22.2	21.6	8.2	33.5	24.3	8.9	7.8	27.0	3.9	7.0	14.3
					B	37.6	34.3	17.1	23.0	26.8	23.6	23.2	5.1	10.3	36.9	15.2	8.3	4.1	8.6	20.4	8.0	14.2	11.5	8.7
			GC-MS/MS	Green tea	A	17.2	16.6	20.5	12.8	7.1	22.5	44.6	25.4	13.9	10.5	11.8	4.7	4.6	5.1	4.6	6.3	13.4	1.3	10.4
					B	8.5	6.5	8.2	19.9	9.6	7.7	24.3	32.2	14.5	7.2	25.0	7.3	8.3	11.6	8.3	20.0	2.9	20.9	18.1
				Woolong tea	A*	32.0	47.9	2.9	45.1	20.7	18.6	4.7	44.0	11.5	1.8	18.3	41.1	14.3	2.0	14.3	49.0	16.8	37.0	29.3
					B*	34.1	11.5	23.1	30.7	40.3	26.0	39.9	9.4	35.8	59.0	7.3	5.5	31.1	10.2	21.6	10.6	21.7	67.4	6.5
9	benoxacor	Cleanert TPT	GC-MS	Green tea	A*	5.3	6.4	2.7	10.0	11.7	5.8	6.8	11.2	11.9	2.9	6.4	4.8	3.4	14.1	9.9	8.8	6.1	13.5	5.2
					B	8.6	11.6	7.9	13.4	7.2	2.3	5.6	14.3	7.7	9.7	4.5	9.3	13.6	3.2	12.3	8.6	6.5	11.3	7.4
				Woolong tea	A*	7.1	0.7	7.2	6.9	14.9	3.9	10.5	8.9	8.0	4.6	15.0	8.5	12.0	10.0	7.1	10.7	10.2	4.9	5.3
					B	11.1	2.6	5.6	2.5	10.8	14.8	5.9	13.5	9.4	3.2	5.8	5.6	8.5	3.9	9.8	7.6	3.7	1.4	14.4
			GC-MS/MS	Green tea	A*	8.6	9.5	5.6	10.1	14.2	3.6	13.5	12.0	3.1	1.1	6.1	8.6	17.6	4.0	5.2	5.2	9.2	15.0	8.1
					B	6.6	9.3	4.3	7.5	7.7	10.2	4.8	11.2	7.7	6.6	2.6	11.8	9.2	12.0	5.1	3.8	14.2	7.1	16.3
				Woolong tea	A	11.1	11.3	6.2	7.1	2.3	9.4	1.6	3.4	5.1	2.4	17.4	9.5	17.0	3.6	2.9	12.2	5.9	23.1	9.9
					B	25.1	6.1	6.6	2.9	3.5	11.9	2.9	2.5	14.3	4.5	11.3	6.8	7.4	3.7	11.2	4.9	1.7	13.2	1.6
		Envi-Carb+PSA	GC-MS	Green tea	A*	17.9	10.6	13.4	8.0	10.4	18.8	11.4	11.0	9.4	7.0	3.7	20.2	1.6	6.8	3.8	10.5	3.9	7.0	3.5
					B	18.4	8.5	4.0	11.9	11.8	12.3	10.5	9.4	7.8	6.4	30.2	12.2	40.5	11.0	11.1	4.5	8.1	22.4	9.9
				Woolong tea	A*	11.5	7.9	6.4	10.4	4.5	4.6	5.4	10.7	4.8	4.3	4.3	14.6	11.9	5.9	3.6	17.0	10.6	15.9	4.7
					B	19.3	13.5	4.9	3.0	15.7	8.4	9.5	10.1	15.4	16.0	15.4	6.8	11.1	8.5	12.4	13.3	16.7	22.7	6.8
			GC-MS/MS	Green tea	A	20.1	11.2	12.2	3.6	7.1	22.0	12.0	13.9	13.1	6.2	18.2	23.3	9.8	24.2	7.6	14.6	14.8	16.1	4.0
					B	38.1	15.5	2.1	6.2	11.8	13.4	7.1	13.8	11.0	5.6	16.0	11.9	33.5	9.2	18.7	3.6	11.7	21.1	9.6
				Woolong tea	A*	8.7	5.5	9.2	19.0	3.7	6.0	9.5	6.6	6.2	5.2	5.3	16.7	22.7	8.1	11.1	16.0	3.5	24.0	7.6
					B*	15.5	5.2	11.0	7.6	14.6	14.5	10.2	5.6	20.4	15.0	5.8	1.0	4.3	13.8	14.1	8.8	12.1	33.2	2.0

(Continued)

Annex 2: RSD data for 201 pesticides in incurred tea Youden pair samples under 8 conditions with two SPE cartridge cleanup in three months (Nov.9, 2009-Feb.7, 2010) (cont.)

No	Pesticides	SPE	Method	Sample	Youden pair	Nov.9, 2009 (n=5)	Nov.14, 2009 (n=3)	Nov.19, 2009 (n=3)	Nov.24, 2009 (n=3)	Nov.29, 2009 (n=3)	Dec.4, 2009 (n=3)	Dec.9, 2009 (n=3)	Dec.14, 2009 (n=3)	Dec.19, 2009 (n=3)	Dec.24, 2009 (n=3)	Dec.14, 2009 (n=3)	Jan.3, 2010 (n=3)	Jan.8, 2010 (n=3)	Jan.13, 2010 (n=3)	Jan.18, 2010 (n=3)	Jan.23, 2010 (n=3)	Jan.28, 2010 (n=3)	Feb.2, 2010 (n=3)	Feb.7, 2010 (n=3)
10	bromophos-ethyl	Cleanert TPT	GC-MS	Green tea	A*	4.3	5.5	4.0	3.5	7.8	3.7	6.2	4.4	5.2	4.1	8.0	5.6	6.2	2.6	1.7	2.3	4.0	5.4	4.0
					B	8.3	8.4	7.0	9.3	4.2	2.5	1.4	1.7	4.8	10.2	6.2	11.9	6.1	1.6	5.3	2.8	5.6	0.5	6.4
				Woolong tea	A	5.0	2.2	2.2	6.1	1.7	2.5	10.4	5.1	2.9	5.3	13.0	2.4	6.1	3.3	8.3	2.8	4.3	1.4	1.0
					B	7.8	0.8	2.9	4.4	2.9	3.3	3.8	4.0	6.2	2.8	10.8	1.4	3.2	3.6	3.2	0.5	1.5	1.2	2.7
			GC-MS/MS	Green tea	A*	4.6	7.7	5.3	3.3	7.6	5.6	8.1	3.4	6.1	3.2	3.8	4.3	4.6	2.2	1.9	1.6	0.3	6.6	3.8
					B	4.7	9.4	5.9	6.3	3.9	8.8	3.6	3.7	4.8	9.6	4.4	7.9	2.3	2.0	1.3	3.1	7.5	3.0	7.4
				Woolong tea	A	5.3	3.2	1.3	4.4	2.0	3.7	7.1	1.0	2.8	1.7	3.8	1.0	7.6	2.9	7.3	2.7	1.0	3.0	1.7
					B	1.4	2.7	4.4	5.8	3.1	2.8	3.6	3.8	6.3	2.1	7.1	2.8	3.5	0.5	5.3	1.5	2.5	2.1	3.4
		Envi-Carb+PSA	GC-MS	Green tea	A*	10.0	3.9	4.2	4.6	6.9	1.7	5.3	3.6	2.9	2.5	5.7	0.5	4.3	3.0	2.1	6.3	9.2	5.6	5.8
					B*	6.6	9.7	4.2	2.4	2.8	1.4	5.2	1.9	5.3	13.4	27.5	5.6	3.6	3.9	8.0	3.0	4.4	11.7	6.7
				Woolong tea	A	1.5	3.5	3.5	3.3	2.2	1.9	3.7	3.2	13.8	1.9	5.2	3.1	1.4	3.5	3.1	1.4	5.6	4.3	4.2
					B	7.3	3.7	8.9	3.7	11.0	3.9	1.8	6.6	10.6	3.8	9.4	0.8	4.4	8.4	9.8	2.6	7.8	1.3	4.5
			GC-MS/MS	Green tea	A	9.9	4.2	3.3	3.0	6.4	3.8	8.9	3.1	1.7	0.7	13.7	3.4	2.3	6.6	3.2	3.1	7.9	4.9	2.4
					B	10.9	11.7	2.1	1.5	3.3	2.4	5.1	0.7	4.4	4.2	16.0	3.2	5.9	4.2	13.5	2.7	3.5	12.9	7.1
				Woolong tea	A	4.2	4.0	3.9	5.8	6.1	2.7	4.5	5.9	3.8	1.9	5.6	1.9	1.0	4.7	6.7	0.5	5.2	4.2	7.7
					B	8.1	3.0	11.7	3.9	8.6	2.7	2.5	6.5	11.5	1.5	2.2	0.5	1.9	10.3	9.9	1.8	3.5	1.0	2.5
11	butralin	Cleanert TPT	GC-MS	Green tea	A*	4.6	5.2	4.8	3.9	15.2	5.0	13.2	13.2	6.8	10.4	10.4	12.7	12.8	9.1	9.9	7.8	5.1	11.9	3.7
					B	5.7	7.2	5.8	15.8	6.6	5.3	7.0	19.6	10.6	10.2	9.3	13.6	13.2	16.4	10.1	11.6	22.8	11.3	13.7
				Woolong tea	A*	5.6	2.7	5.4	4.8	25.0	11.4	7.6	7.5	6.9	13.0	11.8	10.9	5.6	9.8	3.5	19.3	10.2	1.2	2.7
					B	9.0	5.3	5.6	6.8	20.8	13.8	10.7	8.8	8.1	11.9	11.8	5.6	6.6	9.8	12.1	14.4	1.4	2.6	5.2
			GC-MS/MS	Green tea	A*	7.9	6.4	4.1	6.2	10.3	8.9	9.5	13.9	23.5	1.8	9.8	9.4	16.2	9.9	4.3	1.3	4.1	11.0	7.9
					B	6.1	10.2	6.5	8.5	7.6	19.2	10.0	14.4	6.5	10.0	7.1	17.0	10.4	7.8	3.7	6.8	16.3	5.5	15.0
				Woolong tea	A	10.0	7.7	1.9	6.3	10.3	7.5	15.4	3.9	11.2	6.9	16.7	10.9	3.9	4.0	6.5	4.9	4.0	9.5	4.5
					B	41.1	15.3	1.9	4.6	4.6	10.4	15.4	15.5	8.9	5.0	13.7	5.1	2.7	2.5	2.6	3.7	3.2	7.7	18.4
		Envi-Carb+PSA	GC-MS	Green tea	A*	21.0	3.4	20.2	12.9	8.7	32.3	14.9	6.8	5.2	6.9	10.5	24.5	4.5	12.1	12.6	12.9	6.5	20.1	8.5
					B	15.4	11.3	8.3	12.1	12.6	20.4	3.8	9.5	32.4	8.7	31.0	13.1	42.9	13.8	20.7	6.8	16.4	23.2	18.6
				Woolong tea	A*	8.4	5.9	3.7	7.0	26.4	5.2	9.0	10.3	14.2	8.8	3.3	13.2	21.7	20.4	8.0	48.1	19.9	11.4	1.5
					B*	9.5	7.0	10.7	5.2	27.5	18.5	25.2	15.6	12.2	6.2	15.7	4.0	13.9	30.9	26.6	11.9	30.5	18.1	3.9
			GC-MS/MS	Green tea	A	26.6	10.1	13.2	7.1	9.5	16.2	28.1	38.0	13.2	5.9	8.1	14.2	7.2	5.9	5.0	13.4	22.7	18.3	3.8
					B	27.9	15.6	4.2	7.8	12.7	12.0	22.7	38.0	22.4	2.5	18.5	14.8	32.2	6.1	6.4	8.5	11.1	23.8	11.1
				Woolong tea	A*	13.9	3.0	4.0	8.6	5.4	4.2	8.0	30.6	4.0	2.0	5.9	8.2	10.0	3.8	6.9	11.7	17.0	40.6	4.7
					B*	15.6	4.6	7.5	5.2	12.4	10.8	7.3	6.4	13.5	6.1	5.8	7.6	12.1	8.5	10.7	11.7	11.6	29.2	4.5
12	chlorfenapyr	Cleanert TPT	GC-MS	Green tea	A	6.1	6.2	1.1	4.2	6.8	5.0	7.8	4.2	5.0	4.0	4.3	3.5	6.8	2.4	1.3	1.1	3.0	7.1	4.0
					B	6.7	8.0	5.6	8.7	3.8	3.3	1.1	3.7	3.7	6.7	3.5	6.9	0.7	3.5	4.2	0.7	4.5	3.2	3.7
				Woolong tea	A	6.7	2.9	0.5	1.4	1.6	2.8	8.4	5.0	1.3	3.1	1.8	2.8	3.7	3.4	7.2	4.5	3.4	1.9	1.7
					B	7.3	3.3	4.3	8.6	2.0	5.7	5.9	5.1	6.3	5.4	0.2	1.5	1.8	4.2	2.0	0.8	3.4	0.6	3.3
			GC-MS/MS	Green tea	A	5.6	4.5	2.1	4.2	4.5	6.1	6.9	5.5	3.1	3.0	4.8	0.5	7.5	2.7	3.8	4.5	3.6	5.6	5.3
					B	5.7	7.9	6.9	5.2	4.2	2.7	2.5	6.9	4.0	9.9	4.0	6.7	4.2	1.9	1.1	2.4	5.8	2.7	2.4
				Woolong tea	A	5.2	2.1	3.5	6.0	5.5	3.8	6.7	1.5	2.2	4.3	3.0	3.8	10.5	2.3	9.3	5.4	4.8	3.6	1.3
					B	4.8	3.9	6.9	4.9	0.9	2.8	5.2	8.3	3.1	1.5	5.6	3.0	2.4	3.7	7.8	1.5	5.9	3.6	7.9
		Envi-Carb+PSA	GC-MS	Green tea	A	6.4	2.9	2.5	6.2	9.1	1.6	6.2	0.5	6.6	1.3	8.8	2.2	2.2	5.0	5.3	6.3	10.2	2.8	6.1
					B	3.7	9.0	1.8	4.9	3.9	1.9	4.9	5.3	4.6	4.7	19.4	2.2	1.5	6.6	10.6	1.6	4.6	9.1	4.6
				Woolong tea	A	4.6	1.8	3.6	1.9	6.8	1.1	4.1	1.1	2.2	0.7	3.7	1.7	3.0	4.1	3.6	1.5	3.4	4.4	4.7
					B	6.1	3.3	8.9	5.9	7.1	2.4	1.9	8.9	11.0	3.4	7.2	2.9	2.7	8.4	10.3	3.8	5.6	4.2	3.8
			GC-MS/MS	Green tea	A	5.8	3.4	1.7	4.5	7.4	1.6	6.0	2.5	3.0	1.8	13.3	2.7	3.1	3.6	7.0	7.0	4.8	7.3	2.3
					B	9.6	9.6	2.9	6.3	1.5	3.6	2.2	4.8	2.7	3.8	16.2	8.0	1.0	8.0	12.4	3.1	2.9	13.3	3.8
				Woolong tea	A	6.3	2.4	4.8	1.4	4.9	2.5	5.7	3.2	1.1	2.9	7.6	4.4	4.5	4.0	5.8	2.2	5.8	5.3	7.2
					B	7.5	5.1	10.3	4.7	6.4	2.5	3.3	9.3	10.2	3.8	3.5	3.8	7.7	11.1	9.9	0.6	7.8	5.6	3.1

No.	Compound	Cartridge	Method	Tea	±	1	2	3	4	5	6	7	8	9	10	11	12	13	14	15	16	17	18	19
13	clomazone	Cleanert TPT	GC-MS	Green tea	A	5.9	6.3	2.5	8.7	14.3	2.0	10.0	13.1	8.8	0.8	5.5	4.0	9.7	14.9	3.7	3.3	9.4	8.9	3.0
				Green tea	B	5.1	8.8	2.7	12.3	6.0	3.2	4.5	9.1	8.4	7.8	5.1	7.2	9.1	1.1	11.1	5.0	5.1	5.6	4.3
				Woolong tea	A*	4.4	1.5	1.7	9.0	2.4	1.9	16.2	6.5	7.5	3.9	13.8	1.5	3.0	0.8	6.6	4.4	3.6	1.1	1.8
				Woolong tea	B	7.4	2.9	3.6	7.8	3.9	7.1	5.3	8.9	9.4	7.0	3.7	3.0	1.0	5.1	8.8	6.6	1.0	1.5	6.6
			GC-MS/MS	Green tea	A*	5.2	6.8	2.7	6.9	14.3	4.0	11.0	13.5	11.1	5.3	4.7	3.9	6.5	6.8	3.5	4.4	1.2	9.0	3.3
				Green tea	B*	5.7	8.8	5.3	9.1	4.9	6.4	4.1	9.6	1.9	9.5	2.7	8.6	14.2	4.4	5.6	0.3	12.5	4.7	5.6
				Woolong tea	A*	8.3	3.1	2.4	4.9	9.1	2.7	4.4	6.3	10.0	1.5	6.1	1.1	3.7	3.4	8.3	0.7	2.8	7.0	6.7
				Woolong tea	B	5.1	2.0	7.4	6.2	2.2	7.1	6.6	1.7	9.9	3.1	5.7	5.4	5.7	2.8	8.5	5.7	3.7	5.4	2.0
		Envi-Carb+PSA	GC-MS	Green tea	A	17.3	0.1	11.9	5.5	7.4	19.1	12.2	13.8	11.6	3.7	1.9	16.1	2.6	13.8	9.7	11.0	11.5	12.1	6.1
				Green tea	B	17.2	9.0	5.3	9.8	8.3	5.2	11.9	12.2	12.6	12.6	6.4	14.8	20.9	6.2	4.7	14.5	3.1	22.0	11.6
				Woolong tea	A	6.4	2.2	3.3	7.5	3.9	0.9	4.2	12.9	7.6	1.6	3.7	6.2	7.9	1.1	6.7	5.3	7.3	10.4	3.2
				Woolong tea	B*	8.5	7.5	6.9	2.4	13.3	6.4	5.8	5.8	10.1	3.2	8.2	6.5	5.9	7.6	10.2	10.7	12.4	11.8	8.2
			GC-MS/MS	Green tea	A	19.5	6.9	8.9	3.5	4.9	18.0	11.7	15.9	10.0	3.7	2.3	12.2	9.8	2.1	4.1	19.1	5.8	9.7	1.1
				Green tea	B	25.7	13.3	3.9	6.9	5.7	3.4	12.7	14.7	10.4	9.8	8.6	10.5	12.5	13.0	7.6	5.6	5.2	25.9	2.6
				Woolong tea	A	8.0	1.6	4.1	7.0	4.0	2.0	3.8	9.1	2.3	0.8	3.8	11.1	4.8	5.2	9.6	2.5	9.7	10.0	4.7
				Woolong tea	B	10.1	4.1	8.8	1.8	10.6	6.7	4.4	2.5	14.0	3.9	4.8	7.4	5.8	7.7	14.0	2.0	8.2	16.7	6.5
14	cycloate	Cleanert TPT	GC-MS	Green tea	A	4.7	5.2	2.6	0.7	9.0	6.5	5.8	2.8	6.0	2.5	7.1	1.6	7.4	2.1	1.9	1.1	7.0	7.6	5.5
				Green tea	B	6.7	8.4	4.1	8.4	2.9	4.1	3.7	1.2	4.5	9.5	1.8	7.4	3.1	1.1	3.7	2.0	8.0	3.3	5.4
				Woolong tea	A*	7.3	3.4	3.9	0.9	2.4	3.2	8.8	2.4	4.9	5.1	11.8	5.9	4.5	2.1	3.6	0.9	2.9	1.6	1.3
				Woolong tea	B	9.6	1.1	3.5	7.1	4.1	4.5	2.1	3.7	9.4	3.8	9.9	1.5	1.8	5.6	1.9	1.1	1.6	1.6	5.0
			GC-MS/MS	Green tea	A*	3.5	5.6	2.4	1.3	7.9	6.2	6.5	4.0	5.6	1.9	6.4	2.6	4.9	6.4	2.6	1.5	3.0	7.3	3.5
				Green tea	B*	5.0	7.6	4.7	8.3	4.7	6.7	4.8	3.1	3.5	11.2	5.8	6.9	0.9	4.1	4.5	1.9	2.1	4.2	9.7
				Woolong tea	A*	7.6	4.6	3.7	3.1	2.2	2.6	10.1	3.3	7.9	3.3	5.2	3.8	5.5	2.2	4.9	3.4	2.4	1.6	1.6
				Woolong tea	B*	6.6	4.9	4.7	6.9	2.9	4.3	5.2	3.3	8.2	5.9	4.1	1.5	10.5	3.1	4.6	12.1	12.5	1.7	2.1
		Envi-Carb+PSA	GC-MS	Green tea	A	5.1	4.6	1.6	7.0	4.7	3.7	5.2	0.9	28.2	2.5	9.9	2.3	7.1	13.5	8.9	4.9	18.7	5.3	5.0
				Green tea	B	3.7	5.3	4.7	3.5	2.6	1.8	4.6	5.0	2.8	5.8	17.3	4.2	6.7	4.8	18.8	8.8	4.6	8.4	6.3
				Woolong tea	A	4.4	2.0	6.4	3.3	4.1	2.8	4.6	10.3	6.5	3.7	4.5	7.0	4.7	13.0	1.5	2.7	7.0	8.6	4.7
				Woolong tea	B	8.6	8.5	7.5	2.1	11.7	3.8	5.3	5.8	10.2	6.1	8.0	4.7	2.2	4.6	6.2	10.5	11.1	6.9	4.9
			GC-MS/MS	Green tea	A	3.9	6.1	1.4	6.6	4.6	0.3	4.8	0.3	6.0	2.2	10.5	13.4	3.5	6.4	6.2	2.3	4.4	6.1	2.4
				Green tea	B	2.4	8.1	5.9	5.5	3.1	2.4	2.7	2.4	2.4	2.3	22.2	14.0	2.0	10.3	18.3	7.6	9.6	11.7	2.1
				Woolong tea	A	5.4	2.9	6.3	4.7	4.5	1.1	2.8	12.3	2.2	4.7	6.0	8.0	1.2	3.8	6.0	0.4	6.2	19.6	1.6
				Woolong tea	B	8.6	4.7	8.2	0.5	11.2	5.2	5.1	4.6	11.5	6.3	1.3	7.2	4.6	2.9	11.5	4.8	3.9	19.3	4.9
15	cycluron	Cleanert TPT	GC-MS	Green tea	A	14.6	6.8	0.3	4.6	5.4	5.2	9.0	5.4	7.2	0.6	8.8	3.6	7.8	5.6	2.2	2.9	3.9	7.6	2.3
				Green tea	B	11.5	9.8	6.3	11.8	5.7	7.5	1.8	6.8	3.7	6.6	4.5	7.9	2.0	14.6	11.5	4.0	4.2	3.1	8.2
				Woolong tea	A*	3.9	2.0	1.9	2.7	1.4	1.0	8.1	5.0	2.2	2.9	2.9	3.7	4.1	2.8	8.3	0.7	1.3	3.1	1.0
				Woolong tea	B*	7.3	2.6	4.9	6.2	3.3	3.9	5.6	0.5	6.2	3.3	1.0	0.5	4.4	3.7	3.1	12.1	3.9	4.3	2.2
			GC-MS/MS	Green tea	A*	15.1	5.7	6.1	9.8	5.0	1.5	7.3	4.5	1.1	7.7	5.6	3.9	8.7	6.9	7.0	6.3	9.4	8.2	3.1
				Green tea	B*	7.2	12.5	5.6	14.5	6.0	17.6	7.5	8.6	12.1	13.3	12.6	4.9	2.5	20.4	6.1	2.9	3.4	10.9	6.8
				Woolong tea	A	3.6	2.5	1.8	5.7	1.7	1.8	9.6	1.1	3.4	4.2	1.5	4.9	3.4	3.4	8.9	0.2	2.6	2.5	1.5
				Woolong tea	B	4.8	2.3	5.3	5.6	1.3	2.9	4.2	2.1	4.8	3.8	1.9	1.9	5.7	3.8	6.3	4.6	9.2	0.5	2.3
		Envi-Carb+PSA	GC-MS	Green tea	A*	10.7	8.3	8.8	6.6	6.4	7.7	7.2	2.8	7.0	5.0	12.8	7.5	2.6	15.2	2.3	6.7	1.3	2.7	6.7
				Green tea	B*	12.8	8.1	1.6	4.7	6.7	8.5	7.4	9.4	3.4	7.4	26.2	2.4	7.7	3.5	16.7	2.9	11.6	8.0	3.2
				Woolong tea	A	4.7	3.9	2.7	4.2	4.6	1.1	0.9	0.6	13.0	1.1	3.1	2.4	3.9	6.9	3.5	4.2	4.2	4.8	5.9
				Woolong tea	B	5.4	3.9	6.5	3.7	10.4	3.7	2.5	5.7	9.6	4.3	4.7	2.4	1.7	4.1	7.0	5.9	2.1	7.6	3.0
			GC-MS/MS	Green tea	A	6.5	5.9	13.0	6.1	7.9	1.5	22.5	8.4	4.9	13.2	18.9	3.1	8.7	27.5	6.3	8.5	5.4	1.0	5.4
				Green tea	B	16.6	10.2	6.5	15.9	18.3	15.8	5.6	11.4	4.2	12.9	5.6	1.0	10.5	1.5	9.6	4.5	4.6	11.0	1.6
				Woolong tea	A	4.8	4.1	4.1	5.9	5.8	1.5	3.5	1.5	3.6	1.5	4.8	7.3	9.5	9.5	8.0			5.2	5.6
				Woolong tea	B	6.5	3.9	7.7	2.9	9.7	4.3	0.9	7.0	11.7	2.3	0.5	4.9	3.2	8.5	7.8			3.6	2.5

(Continued)

Annex 2: RSD data for 201 pesticides in incurred tea Youden pair samples under 8 conditions with two SPE cartridge cleanup in three months (Nov.9, 2009-Feb.7, 2010) (cont.)

No	Pesticides	SPE	Method	Sample	Youden pair	Nov.9, 2009 (n=5)	Nov.14, 2009 (n=3)	Nov.19, 2009 (n=3)	Nov.24, 2009 (n=3)	Nov.29, 2009 (n=3)	Dec.4, 2009 (n=3)	Dec.9, 2009 (n=3)	Dec.14, 2009 (n=3)	Dec.19, 2009 (n=3)	Dec.24, 2009 (n=3)	Dec.14, 2009 (n=3)	Dec.19, 2009 (n=3)	Jan.3, 2010 (n=3)	Jan.8, 2010 (n=3)	Jan.13, 2010 (n=3)	Jan.18, 2010 (n=3)	Jan.23, 2010 (n=3)	Jan.28, 2010 (n=3)	Feb.2, 2010 (n=3)	Feb.7, 2010 (n=3)
16	cyhalofop-butyl	Cleanert TPT	GC-MS	Green tea	A	5.2	8.1	2.3	11.5	13.6	3.0	7.0	11.5	8.0	3.6	9.9	6.6	9.9	7.8	7.5	7.5	0.6	5.3	6.9	3.8
				Woolong tea	B	5.0	7.0	5.6	10.7	4.1	3.2	4.6	7.6	4.9	11.3	9.1	7.0	10.6	10.6	4.0	7.5	1.7	14.6	7.5	1.4
			GC-MS/MS	Green tea	A*	7.1	4.7	3.0	3.5	1.4	2.9	9.9	5.6	5.5	5.5	16.9	10.7	10.7	9.0	2.0	0.6	7.5	9.0	13.0	1.6
				Woolong tea	B*	13.0	1.8	2.6	7.2	0.8	3.2	12.2	6.8	3.0	1.5	4.9	2.4	2.4	5.8	2.1	4.2	1.8	3.5	13.9	7.1
		Envi-Carb+PSA	GC-MS	Green tea	A*	6.1	6.6	0.3	5.9	12.1	3.5	10.6	13.4	10.1	4.3	8.2	6.4	6.4	7.1	8.8	1.8	4.0	4.6	6.0	2.4
				Woolong tea	B*	5.4	5.8	4.8	9.2	5.9	4.9	4.2	9.5	17.2	2.9	3.1	6.7	6.7	9.3	3.3	3.3	5.4	10.3	3.5	3.4
			GC-MS/MS	Green tea	A*	4.9	7.3	2.7	7.9	10.9	5.3	14.9	5.5	4.0	2.9	9.8	6.7	5.7	13.0	13.0	10.7	10.4	10.6	4.2	5.4
				Woolong tea	B	13.0	4.8	8.9	9.3	4.1	2.3	7.2	0.8	5.5	9.8	6.7	3.2	5.9	3.2	3.2	6.9	2.5	2.7	8.3	4.4
		Cleanert TPT	GC-MS	Green tea	A	14.7	5.3	14.8	5.5	8.9	17.6	2.8	5.0	10.2	0.8	7.8	8.9	8.9	12.0	3.0	7.1	4.4	8.2	8.2	5.2
				Woolong tea	B	20.2	7.1	5.8	7.3	6.4	3.7	10.4	4.5	9.5	4.1	15.0	12.3	12.3	8.5	7.2	12.8	8.7	5.7	14.8	6.6
			GC-MS/MS	Green tea	A*	13.3	5.8	4.3	11.6	6.8	4.4	7.5	5.1	11.7	11.6	9.7	9.2	12.3	7.2	11.2	10.1	6.4	2.7	8.2	6.0
				Woolong tea	B*	6.0	1.1	13.8	13.5	12.5	8.3	5.7	7.5	4.3	5.9	9.9	2.3	0.9	11.2	9.3	5.0	9.3	4.6	15.9	3.4
		Envi-Carb+PSA	GC-MS	Green tea	A*	16.3	8.9	11.9	6.8	5.9	16.6	3.5	8.8	43.0	23.5	0.5	0.9	0.9	12.0	5.0	5.0	12.4	9.6	7.2	5.2
				Woolong tea	B*	20.6	12.8	3.7	11.1	5.4	2.3	33.1	2.6	6.3	33.1	13.6	2.8	15.8	16.3	13.8	10.0	17.7	1.2	21.7	9.7
			GC-MS/MS	Green tea	A*	7.6	1.8	1.9	13.9	12.8	5.1	4.2	9.2	6.8	10.9	9.2	15.8	15.2	16.3	13.8	17.6	20.7	7.1	14.3	11.2
				Woolong tea	B*	7.2	2.5	21.4	12.2	6.0	0.9	7.6	4.6	8.9	7.1	9.5	8.2	12.3	15.5	9.7	11.5	5.2	10.5	7.3	3.9
17	cyprodinil	Cleanert TPT	GC-MS	Green tea	A	4.5	5.6	2.3	1.2	11.2	8.0	7.1	6.1	5.8	2.0	6.5	11.9	1.3	6.7	5.6	0.4	2.1	3.8	1.6	2.6
				Woolong tea	B	4.2	7.8	5.5	5.5	7.8	2.4	2.9	3.3	4.9	10.3	7.8	12.7	4.3	7.8	4.3	5.7	2.7	5.4	4.3	10.1
			GC-MS/MS	Green tea	A	5.6	2.5	1.5	5.3	1.8	3.0	6.2	17.5	5.7	6.1	18.5	1.3	1.8	5.1	3.4	2.6	2.0	3.3	2.2	0.9
				Woolong tea	B	8.7	1.7	3.6	2.9	0.8	0.3	15.4	4.6	9.3	3.9	18.0	1.8	1.3	7.8	5.9	1.5	2.0	10.6	7.8	6.3
		Envi-Carb+PSA	GC-MS	Green tea	A	6.0	7.1	1.7	3.3	9.4	4.6	7.2	5.9	6.1	2.5	6.8	3.7	3.7	6.4	5.9	2.6	4.0	6.6	6.4	8.5
				Woolong tea	B	4.0	7.3	5.8	6.4	9.0	8.5	5.1	4.6	5.5	9.4	6.2	10.1	10.1	2.2	2.6	9.5	2.4	1.0	4.2	5.6
			GC-MS/MS	Green tea	A	7.2	0.9	2.8	5.8	1.4	2.9	8.8	2.7	1.9	4.8	5.0	2.9	2.9	3.5	1.6	5.8	1.9	1.9	1.2	8.7
				Woolong tea	B	5.5	2.5	6.1	7.0	1.4	2.8	7.6	3.0	8.3	4.4	5.1	2.8	2.8	5.3	1.5	5.8	7.0	11.7	0.1	10.2
		Cleanert TPT	GC-MS	Green tea	A*	6.5	6.2	3.7	5.4	11.5	3.3	4.8	1.9	7.8	7.1	5.1	6.7	6.7	2.5	6.0	8.2	5.2	4.0	8.7	7.3
				Woolong tea	B*	3.3	9.0	4.7	9.5	1.5	3.5	3.9	5.8	6.2	10.5	34.6	6.3	6.3	5.2	5.2	17.0	4.2	1.7	14.7	7.1
			GC-MS/MS	Green tea	A	3.9	1.3	5.4	10.0	7.2	3.3	10.9	17.4	15.7	13.0	2.8	5.1	5.0	6.5	8.2	2.6	2.2	5.1	5.5	6.3
				Woolong tea	B	5.2	4.5	19.1	3.7	4.0	8.3	9.5	7.2	33.1	4.2	13.3	4.2	4.1	3.4	4.1	9.7	2.2	5.1	0.4	4.4
		Envi-Carb+PSA	GC-MS	Green tea	A	4.1	3.2	2.5	5.8	6.7	1.4	4.7	0.6	5.5	2.8	10.2	3.9	2.1	3.3	4.4	5.3	5.8	11.1	1.9	4.4
				Woolong tea	B	2.3	6.4	3.0	6.2	1.9	1.7	3.1	3.4	4.7	2.2	23.0	2.1	2.1	1.4	4.4	14.0	3.4	3.1	1.8	5.0
			GC-MS/MS	Green tea	A	6.3	2.7	3.7	7.0	5.8	0.9	5.4	2.2	5.8	3.4	5.2	8.1	8.1	8.2	2.7	9.7	9.5	3.4	7.1	5.0
				Woolong tea	B	5.7	2.1	10.4	3.3	5.2	5.6	5.2	6.4	12.3	1.9	2.3	5.6	8.1	4.9	7.6	9.1	13.8	9.5	7.0	3.5
18	dacthal	Cleanert TPT	GC-MS	Green tea	A	4.3	5.8	4.3	3.5	8.5	3.9	6.6	8.0	5.6	2.6	4.5	2.5	5.6	6.7	3.6	2.3	1.9	1.8	2.6	4.6
				Woolong tea	B	6.6	8.9	5.1	4.4	4.0	2.5	1.9	4.3	4.1	7.1	2.3	6.4	2.5	1.3	1.4	2.9	6.0	3.7	5.5	5.9
			GC-MS/MS	Green tea	A	5.0	2.7	1.3	2.7	2.7	2.2	12.7	6.5	3.4	3.2	3.7	1.5	0.9	3.5	2.5	6.8	2.4	4.3	0.4	1.0
				Woolong tea	B	6.9	2.3	4.6	2.2	2.4	2.4	3.2	5.5	6.1	3.8	1.9	0.9	0.9	2.0	3.1	4.5	0.9	1.0	1.9	2.8
		Envi-Carb+PSA	GC-MS	Green tea	A	3.9	5.7	3.6	6.2	6.2	3.3	7.0	6.5	5.6	2.7	5.1	3.2	3.2	6.4	4.4	0.5	1.6	3.8	7.1	3.9
				Woolong tea	B	6.0	7.9	7.4	6.1	4.1	5.0	2.7	4.9	3.6	9.2	4.3	8.9	8.9	3.0	2.1	2.0	3.4	10.0	2.8	7.4
			GC-MS/MS	Green tea	A	5.2	2.1	1.0	6.3	0.7	3.3	9.9	0.3	5.4	2.7	6.3	1.7	4.8	2.7	3.4	9.1	3.4	0.8	0.8	5.5
				Woolong tea	B	2.2	1.8	6.0	4.6	1.5	2.0	3.6	5.6	6.2	2.5	4.7	7.7	7.7	2.6	2.9	6.9	2.9	2.3	1.6	0.9
		Cleanert TPT	GC-MS	Green tea	A	8.9	3.2	2.4	4.8	7.5	3.9	8.1	3.4	3.7	1.7	8.1	3.2	1.8	4.3	5.4	5.4	12.5	10.4	10.4	4.8
				Woolong tea	B	6.1	7.7	3.1	3.5	1.2	1.6	4.1	1.2	5.5	4.8	19.9	2.8	2.8	1.3	1.2	15.7	1.2	4.8	12.3	6.4
			GC-MS/MS	Green tea	A	1.4	3.4	2.4	4.4	1.2	2.5	2.8	5.5	6.4	0.9	4.1	3.2	3.2	3.4	3.4	2.8	2.8	6.3	4.5	3.4
				Woolong tea	B	5.3	3.0	8.8	2.7	4.8	4.6	2.3	7.4	12.0	2.2	8.4	4.4	4.4	7.7	10.4	3.1	7.7	8.3	2.7	5.2
		Envi-Carb+PSA	GC-MS	Green tea	A*	8.4	4.3	2.4	5.3	8.2	3.5	4.3	3.3	2.9	0.6	12.0	3.2	1.5	2.1	3.5	6.8	2.1	8.0	7.0	1.6
				Woolong tea	B*	3.7	9.0	3.5	3.4	7.1	2.7	3.4	0.9	4.8	0.2	20.5	1.5	6.8	6.8	13.8	3.5	5.2	4.3	12.9	3.6
			GC-MS/MS	Green tea	A	4.7	2.5	3.4	6.5	2.4	2.5	4.7	1.3	2.7	0.5	3.7	4.4	4.4	4.1	9.6	1.7	5.2	5.4	5.1	6.1
				Woolong tea	B	4.5	2.1	10.2	3.1	7.5	5.6	2.4	6.6	11.7	2.7	3.7	3.7	9.2	12.0	9.2	12.0	1.5	5.0	0.4	3.5

#	Compound	Cartridge	Method	Tea	A/B	V1	V2	V3	V4	V5	V6	V7	V8	V9	V10	V11	V12	V13	V14	V15	V16	V17	V18	V19
19	DE-PCB101	Cleanert TPT	GC-MS	Green tea	A	5.2	5.1	3.5	1.5	7.3	4.8	6.1	3.7	5.7	3.0	8.0	3.2	9.8	2.0	4.0	4.1	7.1	7.1	3.8
					B	6.3	8.4	4.7	8.5	3.5	4.0	2.8	1.4	4.8	8.7	3.3	4.8	2.4	2.2	2.4	1.8	5.8	1.5	4.9
				Woolong tea	A	5.7	3.9	1.0	3.7	4.7	4.2	10.0	5.2	3.4	6.3	1.2	6.6	1.4	6.1	10.9	2.9	4.9	1.9	3.9
					B	7.3	3.4	4.8	4.4	3.0	2.6	4.9	1.7	5.1	6.0	2.2	1.7	2.8	6.2	4.8	0.5	1.4	0.8	3.7
			GC-MS/MS	Green tea	A	4.2	6.8	4.2	4.3	6.6	4.4	7.7	3.3	5.9	3.2	5.7	2.2	6.2	4.4	1.0	5.4	4.0	7.4	4.4
					B	5.5	9.3	1.8	9.5	2.7	2.6	4.8	3.4	3.1	10.3	3.0	7.1	2.5	0.4	5.5	5.6	8.0	5.9	7.8
				Woolong tea	A	6.0	3.9	6.3	5.6	4.8	3.2	9.9	1.0	1.9	2.7	2.8	3.9	4.7	2.4	7.7	4.1	3.2	5.2	4.2
					B	6.0	2.7	1.3	5.3	2.2	1.3	4.5	2.1	3.2	3.3	12.6	0.9	4.6	2.9	4.3	2.3	3.3	5.3	3.4
		Envi-Carb+PSA	GC-MS	Green tea	A	5.5	5.8	3.7	5.2	8.2	1.3	5.7	3.4	7.2	3.9	20.7	3.4	6.0	6.9	7.5	8.5	9.9	3.0	6.7
					B	3.9	8.2	1.7	4.7	1.8	1.6	3.8	5.3	3.3	4.8	5.2	0.7	3.0	4.1	13.1	5.0	5.6	7.6	4.4
				Woolong tea	A	1.8	3.5	9.9	4.3	5.1	1.8	2.6	0.9	6.4	3.9	9.2	3.3	2.0	4.2	1.4	2.9	5.2	3.1	6.2
					B	4.8	2.2	1.4	4.2	8.3	4.7	3.0	7.0	9.9	1.9	11.6	4.3	5.7	8.2	8.7	2.6	4.1	3.7	5.3
			GC-MS/MS	Green tea	A	4.7	5.4	4.3	5.7	6.9	0.9	6.2	1.2	4.9	0.9	22.2	5.5	4.0	11.6	4.3	7.5	10.8	2.3	2.7
					B	4.5	10.0	3.9	6.9	1.9	3.4	3.5	3.5	4.5	2.1	5.8	3.2	6.2	3.9	13.0	2.2	4.8	12.4	4.5
				Woolong tea	A	5.0	2.7	10.0	5.3	6.2	3.1	4.8	1.9	2.9	1.7	3.1	3.0	5.8	3.1	10.9	1.6	4.9	4.8	5.4
					B	4.4	2.7	2.6	1.6	5.9	6.8	2.3	6.7	11.6	3.2	5.4	4.9	0.9	6.1	9.5	3.6	4.3	0.3	4.5
20	DE-PCB118	Cleanert TPT	GC-MS	Green tea	A	5.2	5.1	5.2	0.8	7.7	4.5	9.7	3.9	4.4	3.8	3.4	2.5	9.6	3.9	3.4	2.3	6.8	9.5	4.7
					B	5.5	9.3	0.2	8.5	2.3	3.4	4.8	3.8	2.1	10.1	2.9	5.6	4.7	7.6	6.2	3.4	9.6	4.7	4.4
				Woolong tea	A	6.4	2.3	4.6	3.9	4.2	2.7	7.3	4.6	2.5	1.0	4.8	4.8	4.1	2.7	10.4	3.8	2.8	0.2	4.3
					B	7.9	3.2	3.3	1.7	1.4	2.3	3.6	1.0	4.6	4.2	6.0	1.3	3.1	0.9	1.9	3.3	4.4	1.7	2.4
			GC-MS/MS	Green tea	A	4.4	3.8	5.1	2.9	5.5	5.3	7.7	2.7	5.5	3.8	4.4	3.7	5.0	4.8	3.0	2.4	4.8	9.0	4.0
					B	5.1	9.3	2.9	6.5	4.6	3.9	3.7	2.1	1.6	9.6	1.9	7.2	2.2	2.3	5.3	3.8	6.2	3.0	6.1
				Woolong tea	A	6.2	2.1	5.2	5.5	3.2	2.6	7.8	0.5	1.3	2.5	3.4	2.0	0.7	1.6	6.3	3.3	2.3	5.5	3.1
					B	6.2	0.2	0.5	6.0	0.8	3.1	4.1	2.9	4.7	3.8	13.0	0.4	3.7	3.7	5.0	0.4	2.7	1.8	3.9
		Envi-Carb+PSA	GC-MS	Green tea	A	5.5	4.5	2.2	6.1	8.6	1.2	6.8	1.5	5.7	1.5	19.9	2.9	4.1	7.3	7.8	3.6	10.7	4.6	6.1
					B	4.1	8.9	2.6	5.2	2.1	4.0	4.2	7.0	2.4	4.0	4.8	0.7	3.6	4.2	12.6	2.5	6.8	8.6	6.9
				Woolong tea	A	4.3	2.7	10.0	4.3	5.2	2.0	2.1	1.6	4.4	3.1	10.6	4.0	2.1	5.4	3.5	3.0	6.5	1.3	9.0
					B	5.1	2.1	1.0	4.8	8.0	4.7	4.4	6.8	9.4	3.7	11.1	1.9	5.0	9.3	9.6	3.8	4.4	5.3	2.3
			GC-MS/MS	Green tea	A	4.5	4.6	4.6	6.6	7.5	1.0	5.5	1.8	5.7	1.1	21.5	5.4	2.2	6.4	6.8	8.7	6.1	3.3	3.8
					B	2.7	9.8	3.8	6.7	2.6	1.3	2.7	2.4	4.6	0.9	3.8	3.7	1.2	5.7	14.2	2.0	7.9	9.7	4.7
				Woolong tea	A	5.9	2.3	10.4	5.4	6.3	2.5	5.3	0.7	4.1	1.4	2.4	4.0	3.9	3.4	9.0	2.2	5.8	2.8	4.1
					B	6.0	1.3	2.0	2.9	8.1	5.8	3.9	5.8	10.1	2.6	5.4	3.2	3.1	7.2	11.5	1.8	3.8	2.7	1.9
21	DE-PCB138	Cleanert TPT	GC-MS	Green tea	A	5.5	4.8	5.2	0.8	7.5	5.2	6.4	4.3	5.4	2.3	3.5	1.9	7.3	4.2	2.2	1.9	3.6	6.6	3.8
					B	5.2	8.2	0.7	8.7	3.3	3.9	3.1	5.8	5.1	9.3	3.0	7.2	0.8	1.2	6.2	2.7	4.2	2.3	4.3
				Woolong tea	A	6.4	2.7	5.4	2.9	1.4	2.3	8.3	2.0	1.9	3.3	0.1	3.0	4.0	2.9	7.6	3.4	3.3	2.3	1.1
					B	8.1	3.1	1.7	4.3	1.0	3.4	4.2	5.5	5.3	4.1	7.6	1.8	2.6	3.7	3.1	0.7	1.6	1.3	2.1
			GC-MS/MS	Green tea	A	3.5	4.2	5.7	0.6	5.0	2.3	8.9	2.3	7.3	1.9	5.7	3.5	8.3	1.4	3.3	2.4	6.4	7.5	3.6
					B	6.6	6.0	1.3	8.5	7.3	1.0	3.6	4.3	8.2	9.4	2.6	7.5	1.4	2.8	2.8	3.8	7.5	7.0	9.6
				Woolong tea	A	8.2	2.2	5.2	3.3	5.1	4.4	8.2	1.8	3.0	4.4	8.0	1.0	5.9	3.9	8.6	2.4	3.2	2.8	0.5
					B	9.0	5.2	1.3	3.6	1.5	3.2	4.2	1.4	3.6	4.6	10.2	3.0	2.8	3.3	6.6	2.8	4.0	3.8	1.2
		Envi-Carb+PSA	GC-MS	Green tea	A	5.3	4.9	2.9	5.6	9.0	1.6	6.0	1.2	6.6	1.5	20.3	2.5	3.4	5.9	7.4	7.4	10.7	3.4	5.6
					B	3.5	8.3	2.4	5.5	1.9	1.8	3.5	1.1	3.6	4.7	4.7	0.9	1.1	4.9	12.7	1.9	5.8	8.4	6.6
				Woolong tea	A	4.8	2.6	9.8	2.7	6.6	1.8	3.0	5.8	5.3	2.2	7.3	2.6	1.9	4.9	2.5	1.9	4.3	4.6	3.8
					B	5.3	2.3	1.4	4.5	8.8	3.8	2.5	2.0	10.8	2.8	14.6	3.0	3.1	7.6	10.4	4.3	6.8	5.3	5.0
			GC-MS/MS	Green tea	A	5.0	4.5	4.8	5.3	6.1	2.8	7.9	7.5	6.1	1.9	19.4	3.9	3.8	13.9	8.4	6.1	6.9	7.7	1.7
					B	5.2	12.7	5.3	8.0	2.2	4.0	5.6	0.9	4.5	1.6	4.0	1.0	3.0	6.5	10.9	3.5	4.3	10.7	5.2
				Woolong tea	A	6.4	4.6	9.1	4.6	5.8	1.2	4.3	4.0	5.2	0.9	7.4	6.7	2.6	5.7	6.3	0.7	6.4	3.7	4.5
					B	5.5	2.6	2.1	2.1	4.8	5.3	4.2	5.0	10.4	2.2	3.8	6.6	6.7	10.3	7.5	0.9	1.8	4.8	2.3

(Continued)

Annex 2: RSD data for 201 pesticides in incurred tea Youden pair samples under 8 conditions with two SPE cartridge cleanup in three months (Nov.9, 2009-Feb.7, 2010) (cont.)

No	Pesticides	SPE	Method	Sample	Youden pair	Nov.9, 2009 (n=5)	Nov.14, 2009 (n=3)	Nov.19, 2009 (n=3)	Nov.24, 2009 (n=3)	Nov.29, 2009 (n=3)	Dec.4, 2009 (n=3)	Dec.9, 2009 (n=3)	Dec.14, 2009 (n=3)	Dec.19, 2009 (n=3)	Dec.24, 2009 (n=3)	Dec.29, 2009 (n=3)	Jan.3, 2010 (n=3)	Jan.8, 2010 (n=3)	Jan.13, 2010 (n=3)	Jan.18, 2010 (n=3)	Jan.23, 2010 (n=3)	Jan.28, 2010 (n=3)	Feb.2, 2010 (n=3)	Feb.7, 2010 (n=3)
22	DE-PCB180	Cleanert TPT	GC-MS	Green tea	A	5.4	4.7	1.1	9.5	6.5	4.0	6.3	4.6	5.8	1.5	5.3	2.4	5.4	2.8	2.7	2.6	4.4	7.6	2.8
					B	5.0	7.1	4.8	8.3	3.1	3.2	1.7	3.3	4.4	8.6	3.8	7.6	0.7	1.8	4.9	3.5	5.2	2.0	2.4
				Woolong tea	A	6.9	1.6	1.1	2.8	1.3	3.1	4.9	4.6	2.3	2.5	1.8	5.5	4.2	2.1	6.1	3.2	3.3	4.6	3.2
					B	8.7	2.4	5.0	4.7	0.8	3.4	3.5	0.5	6.0	4.8	0.4	2.8	1.4	3.1	1.9	1.1	1.8	1.8	2.5
			GC-MS/MS	Green tea	A	3.1	7.0	2.0	4.0	6.6	5.0	9.7	5.0	4.2	5.1	7.5	4.0	1.1	3.5	2.3	4.7	11.0	13.1	2.6
					B	4.6	9.3	4.9	8.5	0.7	4.9	2.5	1.8	4.5	11.0	5.5	8.7	4.5	2.4	6.3	7.3	5.5	7.6	6.5
				Woolong tea	A	6.0	6.8	4.3	3.1	3.4	3.8	4.8	4.1	6.6	4.5	1.8	6.9	4.6	1.7	10.1	1.8	9.6	8.2	5.8
					B	4.3	5.2	5.4	7.0	0.6	1.6	5.1	3.6	7.7	6.7	6.5	3.9	4.5	4.8	3.5	5.4	0.7	4.0	2.5
		Envi-Carb+PSA	GC-MS	Green tea	A	6.1	4.8	3.1	7.6	8.9	1.2	17.5	0.7	6.5	1.8	9.5	3.0	4.0	9.4	5.7	10.5	10.5	3.9	6.4
					B	3.5	7.4	4.6	6.3	1.7	1.9	2.4	3.9	3.1	4.2	19.9	2.0	1.1	9.0	13.0	1.7	5.6	9.2	5.4
				Woolong tea	A	5.2	2.6	4.6	1.8	5.8	1.4	2.5	3.1	4.3	3.3	5.6	2.7	1.7	5.8	2.5	1.9	4.8	5.3	4.3
					B	5.1	3.0	9.0	3.3	6.7	3.0	2.1	7.1	9.7	3.3	8.4	2.6	3.4	7.1	9.4	4.4	7.6	5.6	3.3
			GC-MS/MS	Green tea	A	5.3	3.2	1.9	6.2	9.4	2.5	8.2	2.3	5.7	13.8	13.8	3.9	0.5	1.3	7.9	9.2	2.3	5.5	4.1
					B	2.7	11.3	4.2	7.0	2.2	0.9	2.7	1.9	1.5	25.6	25.6	1.0	8.1	8.0	9.8	1.1	4.0	14.0	1.8
				Woolong tea	A	7.6	0.4	3.0	3.3	6.6	2.6	7.5	7.0	3.7	5.0	7.5	6.2	4.3	6.5	7.1	0.5	6.3	10.1	8.7
					B	5.8	2.5	11.0	3.9	10.9	3.6	4.4	4.1	9.3	2.4	1.4	0.3	7.1	9.6	8.6	2.8	0.5	1.4	4.8
23	DE-PCB28	Cleanert TPT	GC-MS	Green tea	A	4.7	4.6	2.3	2.2	12.1	4.5	4.9	3.1	5.7	2.1	5.7	1.4	6.2	2.6	2.8	2.1	2.6	7.2	4.4
					B	8.5	11.1	3.5	8.8	2.1	6.2	2.4	1.4	4.7	7.4	2.5	6.1	2.2	1.7	5.4	3.3	4.0	2.9	3.6
				Woolong tea	A	4.2	3.3	0.2	3.5	2.5	2.6	8.9	4.6	2.1	2.2	3.2	3.8	2.4	2.7	8.6	2.2	3.1	1.1	1.6
					B	6.5	3.3	5.1	3.7	1.4	2.8	3.8	4.3	5.9	4.7	0.6	1.3	2.3	2.8	4.1	1.0	1.9	2.6	0.9
			GC-MS/MS	Green tea	A	4.6	5.3	2.6	2.6	6.8	3.5	6.9	3.3	5.6	3.3	6.2	2.1	5.6	6.2	0.8	1.1	4.6	6.6	5.5
					B	5.5	9.1	5.7	7.4	5.5	3.1	4.1	2.0	3.2	10.0	5.8	7.2	1.5	2.7	2.7	1.8	2.3	6.6	7.0
				Woolong tea	A	6.3	3.9	0.9	5.4	3.0	2.4	7.7	1.5	1.7	3.5	2.0	4.5	4.4	1.9	8.4	2.8	9.9	2.9	3.9
					B	8.8	3.7	5.0	4.7	0.4	2.8	4.3	1.6	4.0	2.8	2.7	0.5	3.7	2.4	5.0	1.5	3.4	2.9	1.9
		Envi-Carb+PSA	GC-MS	Green tea	A	5.6	4.0	5.2	7.0	6.2	3.1	5.8	1.2	8.2	1.6	9.4	4.5	2.5	4.2	4.8	6.6	9.8	2.7	3.3
					B	3.8	9.0	2.9	3.4	1.6	1.3	8.4	4.7	3.3	3.7	17.2	2.5	1.4	4.0	12.6	3.2	6.4	7.8	6.4
				Woolong tea	A	3.9	3.1	2.6	4.0	6.5	1.9	2.8	0.7	3.1	0.8	4.5	2.7	1.6	3.6	2.5	2.7	6.4	5.1	4.4
					B	5.2	2.8	9.0	4.5	7.2	4.3	2.6	5.3	10.1	3.4	6.6	4.1	4.0	6.8	8.7	4.0	7.6	3.9	4.3
			GC-MS/MS	Green tea	A*	4.9	6.0	1.6	5.8	5.7	1.7	5.3	1.8	5.6	1.6	8.9	33.8	4.9	6.6	4.5	8.1	8.6	5.4	1.6
					B	3.4	10.6	5.1	4.8	2.9	2.0	1.8	2.5	1.9	1.2	21.8	21.7	1.1	6.6	13.4	1.7	2.2	11.2	2.1
				Woolong tea	A	4.1	3.3	4.3	5.9	5.7	2.8	4.1	1.6	2.8	0.7	4.4	6.9	3.3	3.0	7.4	3.2	5.0	4.6	5.3
					B	5.5	3.0	9.3	3.0	7.8	2.3	2.9	5.0	5.0	1.7	1.8	5.0	5.5	7.0	10.3	0.9	4.8	3.3	2.0
24	DE-PCB31	Cleanert TPT	GC-MS	Green tea	A	4.9	4.8	3.1	0.6	5.2	4.9	6.3	2.6	5.4	2.6	5.5	1.5	7.4	2.4	1.4	4.0	2.6	7.1	3.8
					B	8.1	8.8	3.0	8.6	7.2	6.3	3.0	2.3	5.8	8.4	1.9	6.7	3.0	3.0	4.9	1.5	5.1	3.1	4.4
				Woolong tea	A	4.2	3.2	0.4	3.6	2.5	2.9	9.3	3.6	2.3	2.5	3.1	5.1	3.8	2.9	6.8	4.3	2.8	2.3	1.5
					B	6.6	3.0	4.8	4.7	1.4	3.4	5.1	0.6	6.3	3.1	1.0	1.7	2.9	3.4	3.9	1.2	2.1	1.8	1.3
			GC-MS/MS	Green tea	A	4.6	5.3	2.6	2.6	6.8	3.5	6.9	2.2	5.6	3.3	6.2	2.1	5.6	6.2	0.8	1.1	4.6	6.6	5.5
					B	5.5	9.1	5.7	7.4	5.5	3.1	4.1	2.0	3.2	10.0	5.8	7.2	1.5	2.7	2.7	1.8	9.9	2.9	7.0
				Woolong tea	A	5.3	6.0	0.9	5.4	3.0	2.8	4.3	1.6	4.0	3.5	2.0	6.9	3.7	4.1	5.2	3.2	5.0	1.1	1.9
					B	8.8	1.1	5.0	4.7	0.4	2.5	3.0	1.7	7.7	2.8	2.7	5.0	3.3	3.0	10.3	0.9	4.8	4.1	3.8
		Envi-Carb+PSA	GC-MS	Green tea	A	5.5	4.1	3.5	6.9	6.6	1.2	20.8	5.5	3.3	3.8	8.9	3.8	4.2	1.6	5.2	6.6	10.3	7.5	5.0
					B	3.8	8.3	3.1	3.5	2.0	2.4	1.8	0.3	2.8	2.6	17.2	1.5	1.7	3.1	2.5	3.6	5.7	3.2	5.1
				Woolong tea	A	2.9	3.3	2.6	3.7	6.7	3.4	2.3	5.8	2.8	2.3	3.6	3.7	3.3	6.5	8.2	3.2	6.2	4.1	4.3
					B	5.0	2.8	4.8	4.5	7.5	1.7	5.3	1.8	5.6	1.6	8.1	4.4	4.9	6.6	4.5	8.1	8.6	5.4	1.6
			GC-MS/MS	Green tea	A	4.9	6.0	1.6	5.8	5.7	2.0	1.8	2.5	1.9	1.2	8.9	3.2	1.1	6.6	13.4	1.7	2.2	11.2	2.1
					B	3.4	10.6	5.1	4.8	2.9	2.0	4.1	1.6	2.8	0.7	21.8	5.1	1.1	6.6	7.4	3.2	5.0	4.6	5.3
				Woolong tea	A	3.7	3.3	4.3	5.9	5.7	2.3	2.9	5.0	10.7	1.7	4.4	6.9	3.3	3.0	10.3	3.2	5.0	4.6	5.3
					B	11.2	3.0	9.3	3.0	7.8	4.6	2.9	5.0	10.7	1.7	1.8	5.0	5.5	7.0	10.3	0.9	4.8	3.3	2.0

No.	Compound	Cartridge	Method	Tea	Rep																		
25	dichlorofop-methyl	Cleanert TPT	GC-MS	Green tea	A*	5.3	4.8	4.2	13.8	14.2	2.8	14.9	17.1	8.7	3.4	10.7	8.4	13.0	15.4	3.4	0.5	3.4	14.4	4.3
		Cleanert TPT	GC-MS	Woolong tea	B*	6.9	9.4	12.9	13.9	2.7	2.9	11.0	12.9	2.9	6.0	3.1	3.0	16.2	11.5	3.2	6.4	18.2	11.6	3.3
		Cleanert TPT	GC-MS	Green tea	A*	5.5	2.4	5.7	5.1	4.2	8.1	7.4	7.4	11.8	15.7	16.0	6.3	4.3	1.3	3.8	8.0	1.9	4.0	6.6
		Cleanert TPT	GC-MS	Woolong tea	B*	8.8	0.8	11.5	3.8	3.2	7.4	10.2	4.8	11.2	5.7	17.1	4.3	2.3	5.9	13.7	5.7	0.9	4.0	9.2
		Cleanert TPT	GC-MS/MS	Green tea	A	12.7	7.8	3.6	13.9	18.0	5.1	18.3	11.0	17.0	7.2	14.8	5.9	14.7	12.0	5.0	5.3	9.3	17.9	0.6
		Cleanert TPT	GC-MS/MS	Woolong tea	B	11.2	6.0	7.9	13.2	5.3	6.3	7.4	13.4	4.7	10.2	2.3	12.7	23.9	9.9	5.8	7.4	17.5	11.1	7.3
		Cleanert TPT	GC-MS/MS	Green tea	A	12.0	7.9	5.0	3.5	8.7	5.1	3.2	13.0	6.4	5.4	7.9	4.0	2.4	6.9	7.0	1.7	4.4	14.0	5.9
		Cleanert TPT	GC-MS/MS	Woolong tea	B	14.8	3.7	7.0	5.9	2.2	7.1	9.2	1.7	15.7	7.5	7.7	5.3	7.2	3.8	8.9	5.2	0.7	7.5	12.3
		Envi-Carb+PSA	GC-MS	Green tea	A*	28.7	15.8	26.5	6.0	5.9	32.6	18.4	21.0	17.0	12.0	12.8	33.5	2.3	21.8	3.5	10.6	7.0	14.4	12.7
		Envi-Carb+PSA	GC-MS	Woolong tea	B*	38.1	16.1	5.8	21.7	14.2	5.3	6.9	3.9	12.2	24.4	15.6	24.9	34.5	8.6	17.3	22.4	8.7	25.7	15.9
		Envi-Carb+PSA	GC-MS	Green tea	A*	7.6	5.6	3.3	7.5	8.2	3.8	12.2	15.7	6.2	1.2	3.2	5.6	9.8	2.2	1.2	15.5	8.0	10.0	7.1
		Envi-Carb+PSA	GC-MS	Woolong tea	B*	16.0	9.3	8.2	4.9	13.9	7.5	9.8	5.9	12.4	9.2	9.6	14.0	3.4	15.8	18.4	1.7	12.7	17.1	7.4
		Envi-Carb+PSA	GC-MS/MS	Green tea	A	33.1	19.1	21.6	5.2	5.7	34.7	19.9	29.1	16.3	10.5	7.1	2.4	15.5	19.6	13.0	34.7	7.3	5.2	5.8
		Envi-Carb+PSA	GC-MS/MS	Woolong tea	B	42.9	18.5	3.1	22.5	12.3	3.9	20.2	21.5	17.3	17.9	24.2	3.0	27.0	20.8	24.6	27.3	3.6	32.6	11.4
		Envi-Carb+PSA	GC-MS/MS	Green tea	A	8.7	7.1	7.0	10.6	5.5	2.8	5.9	20.1	3.5	3.7	2.6	14.9	7.6	4.2	6.4	18.1	9.8	28.1	9.7
		Envi-Carb+PSA	GC-MS/MS	Woolong tea	B	17.1	4.9	10.9	3.2	14.3	10.3	9.3	4.2	13.5	11.2	4.9	8.3	3.1	8.7	12.6	1.1	10.2	33.6	6.6
26	dimethenamid	Cleanert TPT	GC-MS	Green tea	A*	6.1	6.3	3.5	3.2	7.2	5.3	5.6	5.6	5.0	0.7	6.2	5.6	10.1	6.9	6.8	3.2	5.4	5.4	4.7
		Cleanert TPT	GC-MS	Woolong tea	B*	6.4	9.8	8.0	8.3	3.8	3.3	1.6	2.4	4.8	8.2	6.0	9.8	1.7	2.4	8.9	2.1	2.6	0.5	5.9
		Cleanert TPT	GC-MS	Green tea	A*	3.9	4.9	1.8	5.5	4.5	2.5	14.0	2.9	3.3	4.9	5.0	2.7	4.6	0.8	10.2	4.5	5.3	2.2	1.8
		Cleanert TPT	GC-MS	Woolong tea	B*	7.5	2.8	3.2	2.9	3.3	4.5	6.0	7.9	8.1	4.6	3.9	0.9	1.5	5.3	3.7	0.8	2.3	2.6	4.2
		Cleanert TPT	GC-MS/MS	Green tea	A	5.3	5.7	3.5	3.0	4.9	4.2	5.9	5.8	2.4	1.8	3.1	2.3	7.3	7.0	1.2	1.5	0.7	7.3	3.4
		Cleanert TPT	GC-MS/MS	Woolong tea	B	5.3	10.2	6.0	6.2	4.7	6.3	4.0	5.4	4.5	9.8	6.5	7.1	0.9	2.5	3.3	3.0	6.9	6.0	9.1
		Cleanert TPT	GC-MS/MS	Green tea	A	2.7	5.2	2.5	6.0	5.7	5.7	11.6	3.4	1.9	3.9	2.2	3.1	9.4	2.5	8.8	1.8	1.9	2.8	5.4
		Cleanert TPT	GC-MS/MS	Woolong tea	B	2.3	2.5	4.2	6.0	1.9	0.4	5.4	6.6	7.1	3.2	2.9	2.2	4.5	2.0	6.1	3.8	1.2	5.4	3.8
		Envi-Carb+PSA	GC-MS	Green tea	A*	7.3	1.2	0.1	4.5	6.8	4.4	4.0	1.9	1.2	1.2	8.4	0.1	3.8	6.3	5.3	9.2	7.6	8.1	6.0
		Envi-Carb+PSA	GC-MS	Woolong tea	B*	7.6	6.6	2.2	3.1	1.0	4.0	3.5	2.9	4.3	10.0	18.1	2.7	3.1	3.4	8.9	2.3	3.6	11.7	7.4
		Envi-Carb+PSA	GC-MS	Green tea	A*	2.8	4.5	1.9	4.9	3.5	2.3	2.5	2.8	12.8	1.1	5.1	2.6	3.9	5.9	8.6	4.0	4.7	3.7	4.3
		Envi-Carb+PSA	GC-MS	Woolong tea	B*	6.1	2.2	9.1	6.6	10.0	4.1	3.7	7.0	9.7	4.4	7.1	2.2	2.5	11.0	8.8	2.6	6.7	5.0	5.0
		Envi-Carb+PSA	GC-MS/MS	Green tea	A	8.1	7.3	1.6	4.6	7.1	3.0	3.6	1.8	0.7	1.9	15.3	3.0	8.3	6.8	3.5	2.5	7.5	5.6	2.7
		Envi-Carb+PSA	GC-MS/MS	Woolong tea	B	25.2	9.4	1.3	2.5	2.8	3.0	2.4	1.9	1.6	1.6	13.1	2.3	1.7	8.4	12.7	3.4	7.6	13.1	5.1
		Envi-Carb+PSA	GC-MS/MS	Green tea	A	4.7	4.4	4.1	7.4	7.0	2.3	6.6	1.9	3.5	1.6	5.1	8.7	4.0	4.5	11.4	5.9	4.3	4.6	7.0
		Envi-Carb+PSA	GC-MS/MS	Woolong tea	B	7.4	2.3	10.3	4.0	7.9	5.2	2.2	6.5	10.7	2.6	2.0	1.0	4.4	8.7	12.2	1.4	2.0	3.9	5.3
27	diofenolan-1	Cleanert TPT	GC-MS	Green tea	A*	4.7	5.6	0.8	0.9	5.9	7.0	8.1	3.9	5.7	2.1	5.9	4.0	8.1	4.3	3.5	2.2	2.9	8.3	4.5
		Cleanert TPT	GC-MS	Woolong tea	B*	7.6	6.9	4.5	8.8	3.8	4.0	3.4	1.2	3.9	7.8	4.1	8.6	1.6	2.9	9.0	1.6	4.1	2.1	4.8
		Cleanert TPT	GC-MS	Green tea	A*	4.6	2.2	1.0	2.2	2.8	2.9	6.6	6.2	1.2	4.4	6.9	5.8	4.5	2.8	8.5	4.6	1.3	6.5	2.2
		Cleanert TPT	GC-MS	Woolong tea	B*	7.9	1.4	2.9	4.2	3.0	6.9	4.5	2.8	3.3	4.0	5.6	2.0	4.0	4.2	1.1	0.3	2.5	1.0	1.0
		Cleanert TPT	GC-MS/MS	Green tea	A	4.7	5.9	2.2	3.0	5.7	5.2	7.5	5.0	5.9	2.4	6.1	3.2	4.1	4.6	5.2	0.5	3.1	4.5	4.4
		Cleanert TPT	GC-MS/MS	Woolong tea	B	3.8	7.5	6.0	7.7	4.0	5.3	3.2	3.6	4.0	8.2	5.8	8.3	0.8	3.7	7.8	2.9	7.7	4.4	6.6
		Cleanert TPT	GC-MS/MS	Green tea	A	6.6	2.5	3.3	4.3	2.9	1.9	5.7	2.7	0.6	3.1	1.6	1.6	2.3	2.8	6.7	3.7	1.9	3.4	3.7
		Cleanert TPT	GC-MS/MS	Woolong tea	B	3.8	1.5	2.0	5.7	1.2	1.2	4.2	3.5	4.7	4.1	3.5	1.4	7.4	3.4	5.5	1.2	0.6	2.9	2.7
		Envi-Carb+PSA	GC-MS	Green tea	A*	6.1	9.5	4.0	6.0	8.7	2.9	3.4	0.7	7.0	3.0	6.6	2.1	3.8	3.0	4.2	8.4	10.9	2.7	5.5
		Envi-Carb+PSA	GC-MS	Woolong tea	B*	2.7	9.1	2.4	6.6	2.0	5.7	1.8	8.6	4.3	5.6	26.9	3.4	1.8	5.3	14.2	3.1	3.8	8.2	6.3
		Envi-Carb+PSA	GC-MS	Green tea	A*	5.4	2.0	3.7	2.0	6.5	1.9	3.6	2.2	4.8	1.9	4.0	1.9	3.0	5.8	2.8	1.9	5.0	4.2	4.4
		Envi-Carb+PSA	GC-MS	Woolong tea	B*	4.4	2.0	9.0	4.2	9.9	3.7	3.7	6.6	11.8	3.3	6.4	0.6	1.7	9.4	10.5	5.6	5.3	5.2	5.2
		Envi-Carb+PSA	GC-MS/MS	Green tea	A	4.6	5.0	1.7	5.8	6.1	1.2	3.7	2.2	6.1	1.5	8.9	3.3	5.2	12.4	5.0	8.5	8.6	6.2	4.7
		Envi-Carb+PSA	GC-MS/MS	Woolong tea	B	1.5	11.3	3.9	5.4	1.7	2.1	2.4	4.0	3.3	2.3	6.2	4.0	3.9	7.8	12.3	3.0	3.8	11.6	2.9
		Envi-Carb+PSA	GC-MS/MS	Green tea	A	7.9	2.6	6.4	1.3	6.0	1.8	4.0	1.9	1.3	2.2	6.2	9.1	2.6	3.6	7.3	3.9	6.5	7.0	3.9
		Envi-Carb+PSA	GC-MS/MS	Woolong tea	B	6.5	2.5	9.3	3.0	8.2	4.0	3.2	6.4	11.7	3.2	1.3	4.3	6.6	6.9	10.8	3.3	2.9	1.5	4.8

Annex 2: RSD data for 201 pesticides in incurred tea Youden pair samples under 8 conditions with two SPE cartridge cleanup in three months (Nov.9, 2009-Feb.7, 2010) (cont.)

No	Pesticides	SPE	Method	Sample	Youden pair	Nov.9 2009 (n=5)	Nov.14 2009 (n=3)	Nov.19 2009 (n=3)	Nov.24 2009 (n=3)	Nov.29 2009 (n=3)	Dec.4 2009 (n=3)	Dec.9 2009 (n=3)	Dec.14 2009 (n=3)	Dec.19 2009 (n=3)	Dec.24 2009 (n=3)	Dec.29 2009 (n=3)	Jan.3 2010 (n=3)	Jan.8 2010 (n=3)	Jan.13 2010 (n=3)	Jan.18 2010 (n=3)	Jan.23 2010 (n=3)	Jan.28 2010 (n=3)	Feb.2 2010 (n=3)	Feb.7 2010 (n=3)
28	diofenolan-2	Cleanert TPT	GC-MS	Green tea	A	5.1	5.2	1.2	1.3	7.0	4.5	6.4	4.1	6.1	2.7	5.3	2.8	6.4	2.4	3.1	1.2	4.3	6.3	4.6
					B	5.9	7.2	4.6	8.0	4.1	3.7	2.1	1.9	3.5	7.5	3.7	7.1	1.6	1.6	5.0	1.8	4.6	2.6	4.1
				Woolong tea	A	6.5	1.8	1.2	1.9	2.7	1.5	7.2	6.4	1.7	3.3	5.9	4.7	4.4	2.5	6.1	3.1	2.3	3.3	1.8
					B	7.6	2.3	3.5	3.8	0.5	4.3	5.0	3.7	7.0	3.8	3.6	1.8	2.7	3.7	1.0	1.1	2.4	1.5	4.5
			GC-MS/MS	Green tea	A	9.5	7.3	1.6	2.2	6.8	5.7	7.8	3.6	4.9	3.4	5.2	3.7	6.7	5.1	1.7	3.8	4.5	8.6	7.2
					B	15.5	5.8	5.9	7.9	3.2	3.5	4.0	3.6	2.4	8.9	6.5	8.0	2.0	1.3	2.3	5.2	5.0	5.2	6.4
				Woolong tea	A	7.4	3.5	1.8	4.3	3.3	2.0	3.0	2.3	2.3	3.0	0.5	1.7	3.4	2.7	8.7	3.0	3.2	1.5	3.7
					B	4.6	2.0	4.6	5.7	1.1	1.2	4.1	4.4	3.9	6.4	4.3	1.2	7.4	1.8	5.7	4.9	1.0	3.7	2.7
		Envi-Carb+PSA	GC-MS	Green tea	A	5.6	4.6	3.0	6.5	8.9	1.2	6.2	1.2	7.5	4.1	6.2	3.4	1.7	5.4	5.6	8.0	9.6	2.8	5.3
					B*	3.5	8.1	3.2	4.3	1.9	2.5	2.7	5.5	3.7	4.0	21.4	3.2	2.0	4.8	12.4	3.5	4.4	8.3	6.0
				Woolong tea	A	5.9	2.2	4.4	1.7	4.3	4.0	5.4	2.2	2.4	2.2	4.2	1.4	2.3	6.2	0.8	2.2	5.2	4.4	3.7
					B	5.6	2.8	8.8	5.1	8.6	2.3	2.6	6.2	10.2	2.9	8.1	1.7	2.8	6.0	9.5	3.9	5.7	4.7	4.2
			GC-MS/MS	Green tea	A*	9.4	4.4	1.8	7.5	7.4	1.2	2.6	1.1	6.4	1.9	7.2	24.6	5.6	38.3	4.4	7.4	10.1	3.7	2.9
					B	9.3	12.1	4.0	4.8	4.6	1.6	1.1	4.1	2.6	3.6	22.4	12.9	2.0	18.6	15.8	1.2	5.3	13.4	4.6
				Woolong tea	A	7.1	2.6	6.4	1.4		2.5	5.3	1.9	3.3	0.7	4.3		1.6		7.3	3.9	5.0	7.9	4.9
					B	7.5	2.4	9.4	3.0		5.1	2.2	6.4	10.6	1.8	1.1		7.6		10.7	3.3	3.3	3.7	7.4
29	fenbuconazole	Cleanert TPT	GC-MS	Green tea	A	8.6	15.0	0.6	2.8	4.4	5.6	8.5	1.8	6.4	5.3	9.8	2.5	6.2	8.6	11.5	8.7	3.6	0.2	3.5
					B	4.2	12.7	5.2	6.5	5.0	3.4	1.9	9.2	2.7	11.3	5.3	13.2	6.7	6.8	11.2	2.9	11.0	5.9	7.0
				Woolong tea	A	4.7	10.0	6.2	4.6	2.0	2.5	7.4	3.0	2.3	5.3	8.4	5.0	5.7	3.5	6.2	8.1	6.1	4.4	5.0
					B*	11.0	1.3	2.3	8.4	4.1	4.6	7.5	2.0	9.7	5.7	7.5	1.2	2.5	1.8	3.0	1.7	3.3	2.2	1.3
			GC-MS/MS	Green tea	A	8.2	14.2	2.4	7.7	3.9	7.4	9.0	1.1	7.1	7.6	7.2	0.6	7.6	7.6	5.6	6.1	4.9	7.1	2.6
					B	4.0	12.8	5.4	5.7	3.7	9.8	2.5	10.3	0.2	21.0	7.1	8.3	1.0	8.3	13.4	5.0	12.6	8.6	6.6
				Woolong tea	A	6.9	3.0	2.4	4.6	1.7	4.9	9.5	3.8	2.5	6.4	5.5	0.9	3.3	3.7	7.5	5.7	5.2	9.7	2.8
					B	3.3	2.4	8.0	8.9	3.1	2.5	5.9	3.1	1.9	4.4	5.2	2.3	3.8	3.0	6.0	3.6	1.9	5.3	2.8
		Envi-Carb+PSA	GC-MS	Green tea	A*	13.0	29.9	4.8	7.7	4.6	6.4	11.5	1.7	9.5	6.2	3.8	2.7	9.1	8.3	8.0	8.9	12.7	5.2	2.5
					B*	8.6	4.7	4.2	9.7	5.5	4.0	14.9	6.9	5.7	10.5	30.2	5.1	6.2	6.0	15.6	6.1	1.9	6.9	1.6
				Woolong tea	A	7.2	4.1	9.2	3.3	4.3	9.6	6.7	6.2	16.9	19.3	6.2	10.3	4.3	14.8	9.5	6.5	1.8	4.6	5.8
					B	10.4	9.6	20.3	9.2	14.9	16.7	4.1	10.8	9.4	3.4	10.6	1.1	2.1	30.9	7.9	7.5	15.8	8.0	3.4
			GC-MS/MS	Green tea	A	13.5	29.9	4.0	7.9	2.9	7.1	54.1	1.5	7.4	6.5	10.2	1.7	6.1	7.9	4.2	9.9	8.9	5.6	2.9
					B	7.7	5.0	4.5	9.0	7.8	2.7	2.9	2.4	6.0	3.3	21.5	6.0	10.9	13.6	14.5	2.3	4.2	11.4	6.2
				Woolong tea	A	10.1	3.1	9.4	3.4	5.8	4.8	7.7	7.2	2.2	24.3	10.7	16.6	11.0	8.2	15.4	6.5	6.4	6.4	6.6
					B	8.8	3.1	21.7	8.1	17.2	13.3	3.2	6.5	6.4	3.2	10.0	10.3	4.8	30.8	7.9	5.7	18.2	7.9	1.7
30	fenpyroximate	Cleanert TPT	GC-MS	Green tea	A	8.1	5.3	3.7	11.4	6.3	10.1	19.5	8.7	6.5	6.7	13.9	5.9	14.0	12.1	8.3	12.4	0.5	20.8	8.8
					B	9.9	12.8	7.7	15.8	8.3	12.5	9.1	14.5	8.1	2.9	14.1	9.2	15.6	15.8	14.8	12.9	20.3	14.9	2.6
				Woolong tea	A	8.2	14.0	3.8	2.0	11.8	2.2	30.7	13.9	11.2	3.3	12.3	16.3	11.4	9.5	11.9	12.1	12.7	19.6	4.7
					B	7.8	10.2	7.2	13.3	10.3	14.9	7.2	1.0	5.2	5.7	13.0	10.6	5.4	3.9	9.3	9.7	1.1	9.0	10.1
			GC-MS/MS	Green tea	A	18.2	2.9	2.0	17.8	7.6	8.4	24.6	12.8	12.2	9.1	18.7	2.7	11.4	7.4	5.9	11.0	1.8	23.5	6.8
					B	13.6	16.3	14.2	15.5	9.4	5.9	12.0	14.9	4.5	6.3	12.1	4.4	13.3	14.3	6.8	4.5	18.6	13.7	14.2
				Woolong tea	A	8.0	20.1	11.7	7.6	21.4	3.4	20.2	11.7	4.6	6.0	14.1	9.1	9.8	14.0	11.7	11.6	10.0	26.4	2.9
					B	36.4	11.3	2.8	11.9	8.7	22.2	9.5	1.5	4.4	6.9	12.3	8.8	9.0	7.5	3.8	6.0	7.5	8.4	11.9
		Envi-Carb+PSA	GC-MS	Green tea	A	14.1	12.3	12.7	10.4	12.4	36.6	28.2	6.2	11.1	13.7	23.6	29.1	10.3	12.5	13.7	9.7	15.6	3.3	2.3
					B	26.4	12.8	5.1	14.0	20.2	14.2	3.3	12.5	7.3	7.7	23.9	13.9	32.7	13.6	14.1	8.2	7.1	2.3	3.4
				Woolong tea	A	11.8	8.9	10.6	19.0	17.8	4.6	12.2	12.2	12.0	2.6	14.0	12.8	23.1	9.0	0.8	26.6	20.3	13.5	15.2
					B	18.7	8.3	14.5	19.0	18.7	6.6	19.0	9.4	12.0	19.3	8.8	4.8	4.4	14.1	18.0	12.5	5.3	29.8	4.6
			GC-MS/MS	Green tea	A	16.3	6.2	12.5	8.6	13.5	44.2	18.1	18.5	6.5	8.2	17.0	6.8	8.6	35.4	14.6	24.4	17.2	3.4	2.6
					B	18.6	12.4	11.8	17.9	30.9	14.8	6.5	19.5	7.8	12.7	27.7	13.5	31.1	22.6	19.3	16.2	5.4	2.6	8.2
				Woolong tea	A	23.4	13.5	12.3	18.4	16.0	8.0	12.6	19.9	7.2	5.0	6.9	20.6	31.4	14.4	4.0	25.2	7.5	32.9	15.1
					B	16.5	9.4	11.9	16.0	13.7	9.2	14.3	12.5	8.1	14.0	10.0	9.8	13.0	20.6	11.4	10.0	5.8	42.5	5.3

No.	Compound	Cartridge	Method	Tea	Col.																		
31	fluotrimazole	Cleanert TPT	GC-MS	Green tea	A	5.5	5.9	0.7	3.5	5.7	3.1	8.0	7.8	5.3	9.6	14.9	6.3	0.9	2.7	5.2	5.2	6.5	3.8
					B	4.9	6.7	6.4	5.7	5.1	4.1	1.7	11.1	11.5	7.6	8.2	3.5	2.6	2.4	3.1	5.8	2.5	3.7
				Woolong tea	A	4.2	4.7	3.2	1.3	2.5	4.0	8.5	5.5	8.8	16.0	5.3	5.6	5.5	7.5	3.6	3.5	0.4	0.7
					B	8.1	1.1	4.2	9.6	3.6	3.7	5.0	8.6	13.4	7.2	2.9	1.9	8.4	2.1	1.2	2.0	0.6	1.6
			GC-MS/MS	Green tea	A*	4.8	5.2	2.1	3.8	12.4	12.4	8.3	7.8	3.4	4.2	4.0	8.0	1.4	5.4	6.4	9.1	7.0	6.0
					B	4.7	6.9	7.9	8.6	6.7	15.2	5.5	4.1	8.6	7.7	5.9	1.7	2.6	10.0	5.5	5.8	2.4	6.7
				Woolong	A*	8.5	1.8	2.4	3.9	0.8	4.1	7.7	5.7	5.4	3.4	2.5	4.4	4.3	2.2	5.6	0.8	1.0	4.2
				tea	B	4.0	5.3	2.7	9.9	7.8	6.5	4.3	7.5	1.8	3.2	1.8	13.4	2.8	4.3	5.6	5.4	3.1	6.3
		Envi-Carb+PSA	GC-MS	Green tea	A	6.2	5.0	4.1	7.4	5.9	4.1	6.5	4.4	1.5	1.2	4.0	3.3	5.5	9.0	7.0	5.4	3.1	4.7
					B	4.3	7.6	3.0	6.8	5.0	2.8	2.3	11.6	5.4	16.0	7.2	1.3	6.0	13.8	1.4	6.5	8.6	7.5
				Woolong	A	4.4	3.6	5.1	1.2	12.4	1.2	3.0	9.5	1.4	5.9	3.3	2.8	4.2	2.2	1.4	4.2	3.9	3.8
				tea	B*	5.4	4.9	5.3	7.0	13.7	2.4	2.7	9.5	10.9	7.7	3.7	2.2	10.6	8.9	4.8	6.6	3.3	4.5
32	fluoxypr-1-methylhep-tyl ester	Cleanert TPT	GC-MS	Green tea	A	3.9	4.8	2.8	6.6	6.1	3.6	3.7	5.1	3.8	8.7	10.6	4.6	6.7	4.4	6.1	11.2	5.2	3.5
					B	2.2	8.1	3.5	5.1	1.1	14.8	3.4	6.2	5.6	23.2	9.3	4.6	20.4	12.5	0.7	5.8	16.0	6.0
				Woolong	A	7.4	1.4	5.8	7.6	9.7	1.2	2.9	6.2	4.9	7.9	8.5	9.5	4.9	11.0	2.6	5.7	4.5	6.1
				tea	B*	6.2	2.8	8.9	2.4	11.5	3.9	4.3	8.1	4.1	3.6	3.3	7.3	6.9	10.8	5.0	7.3	1.4	3.6
			GC-MS/MS	Green tea	A*	7.0	7.7	0.9	6.4	10.4	0.8	7.3	9.2	2.3	4.6	4.8	5.0	4.0	2.1	5.5	6.4	25.5	3.2
					B	5.0	6.7	1.8	9.1	4.5	3.6	1.5	4.7	7.7	2.2	4.7	7.0	5.7	5.9	3.9	7.9	3.1	4.7
				Woolong	A*	6.7	4.1	7.0	7.1	1.1	3.4	11.7	4.4	3.7	9.4	5.1	5.4	0.9	3.8	2.5	2.6	1.3	2.3
				tea	B	9.1	4.8	5.9	7.1	2.3	0.8	3.5	7.2	3.2	3.7	1.0	0.6	4.6	5.8	0.4	2.1	1.3	6.3
		Envi-Carb+PSA	GC-MS	Green tea	A	3.2	6.9	6.4	9.0	17.1	6.7	11.9	11.6	2.7	7.7	5.0	8.1	2.8	8.5	4.3	2.6	14.2	7.9
					B	7.2	5.7	5.9	7.0	0.3	11.8	3.1	4.4	6.6	3.1	14.8	7.2	5.6	4.0	5.3	18.4	13.1	6.9
				Woolong	A*	9.2	0.6	3.2	0.6	4.8	3.7	9.8	9.7	4.5	2.8	6.7	6.2	4.8	6.7	9.4	10.3	2.2	5.7
				tea	B	2.5	11.0	9.1	6.8	7.6	7.1	9.3	6.4	2.4	10.0	2.0	9.1	7.9	0.3	1.5	10.7	1.6	13.1
			GC-MS	Green tea	A	17.7	6.4	11.4	4.6	7.6	12.7	7.2	7.9	2.3	2.6	11.9	2.7	8.8	3.4	5.1	10.7	10.0	4.4
					B	16.7	9.6	5.0	4.8	4.4	2.0	7.2	15.8	8.5	11.0	12.4	13.7	1.6	4.7	13.9	4.7	18.1	10.7
				Woolong	A	4.7	2.6	5.6	1.3	2.9	2.9	4.2	0.9	3.1	5.2	4.4	2.7	5.3	1.7	8.6	4.4	8.3	1.0
				tea	B	6.7	5.3	8.5	0.5	8.6	2.8	4.7	9.7	2.7	7.4	5.1	2.2	5.7	9.6	1.8	8.7	4.4	3.9
33	iprovali-carb-1	Cleanert TPT	GC-MS/MS	Green tea	A*	19.5	7.1	8.0	2.1	7.8	9.0	9.0	9.8	1.3	15.1	5.4	16.7	6.0	10.5	14.7	6.2	11.0	5.0
					B	19.9	10.0	1.1	6.8	6.4	5.2	15.8	14.4	6.3	11.5	7.6	15.1	17.0	15.2	9.4	11.7	23.1	33.5
				Woolong	A	10.4	1.3	11.2	3.4	5.3	5.3	10.0	1.4	8.2	4.6	4.6	5.2	9.3	19.5	9.0	11.7	5.0	16.1
				tea	B*	7.8	7.5	10.3	3.5	7.5	1.9	6.1	9.5	8.5	9.8	0.4	1.7	5.2	4.4	2.4	18.4	11.4	6.4
			GC-MS	Green tea	A*	6.1	3.8	1.3	2.6	14.8	4.9	11.0	5.7	1.5	5.4	5.8	9.9	4.4	2.1	8.4	6.1	3.8	4.7
					B*	6.1	9.2	5.0	9.1	6.7	7.3	3.4	4.9	6.6	6.1	8.7	2.5	2.8	6.0	6.7	3.8	4.0
				Woolong	A	4.7	4.3	7.9	2.6	6.6	4.3	3.3	7.2	8.7	10.5	2.3	3.9	5.9	5.6	2.1	0.9	2.6
					B*	9.4	1.6	6.8	27.4	11.0	16.6	12.2	6.4	3.0	7.7	5.2	4.0	4.1	0.7	4.5	2.8	8.1
		Envi-Carb+PSA	GC-MS/MS	Green tea	A*	4.6	7.2	0.7	1.8	6.3	8.5	8.8	8.2	2.4	7.9	9.0	9.1	5.4	1.3	4.6	3.0	7.2
					B	4.2	6.2	4.0	5.7	3.3	10.9	1.1	2.8	7.3	4.0	14.2	12.6	2.8	8.0	7.4	1.4	6.4
				Woolong	A	8.6	13.1	2.3	2.7	2.3	4.7	0.8	9.1	5.5	5.6	6.3	11.1	4.3	4.0	2.3	2.5	3.0
				tea	B	9.1	11.7	5.7	7.9	4.2	5.3	9.0	6.2	4.5	7.2	3.3	2.3	2.6	1.2	2.2	4.6	3.9
			GC-MS	Green tea	A*	12.1	4.1	5.9	8.0	8.7	4.1	21.6	4.4	3.1	1.9	5.0	17.9	6.5	34.4	9.7	30.8	7.7
					B*	7.2	2.0	9.6	6.5	10.3	6.8	9.7	11.4	22.3	29.5	6.1	1.9	34.4	35.8	2.7	27.0	14.3
				Woolong	A	4.2	3.9	6.4	3.3	3.7	4.6	33.1	6.6	5.0	4.1	1.4	5.9	2.9	3.0	6.9	9.2	1.6
					B	12.2	4.1	10.3	4.9	14.3	9.7	8.7	8.7	7.3	2.7	3.8	8.7	14.2	4.0	8.9	5.5	6.3
		Envi-Carb+PSA	GC-MS/MS	Green tea	A	17.0	6.6	4.3	6.8	6.2	2.5	12.2	11.1	2.9	8.0	4.5	8.1	4.2	12.8	8.5	13.1
					B	9.7	11.7	3.2	6.5	3.0	3.9	5.3	23.9	3.3	24.7	7.6	19.0	4.2	19.0	3.8	4.6	8.8
				Woolong	A*	6.9	1.7	4.9	1.9	5.5	1.7	4.1	4.2	3.4	4.8	9.2	1.9	3.1	1.9	5.1	6.6	4.5
				tea	B*	11.1	3.6	11.6	2.2	11.1	6.5	4.3	10.7	0.3	3.8	2.1	5.1	7.3	10.8	0.5	4.3	3.5

(Continued)

Annex 2: RSD data for 201 pesticides in incurred tea Youden pair samples under 8 conditions with two SPE cartridge cleanup in three months (Nov.9, 2009-Feb.7, 2010) (cont.)

No	Pesticides	SPE	Method	Sample	Youden pair	Nov.9 2009 (n=5)	Nov.14 2009 (n=3)	Nov.19 2009 (n=3)	Nov.24 2009 (n=3)	Nov.29 2009 (n=3)	Dec.4 2009 (n=3)	Dec.9 2009 (n=3)	Dec.14 2009 (n=3)	Dec.19 2009 (n=3)	Dec.24 2009 (n=3)	Dec.14 2009 (n=3)	Jan.3 2010 (n=3)	Jan.8 2010 (n=3)	Jan.13 2010 (n=3)	Jan.18 2010 (n=3)	Jan.23 2010 (n=3)	Jan.28 2010 (n=3)	Feb.2 2010 (n=3)	Feb.7 2010 (n=3)
34	iprovali-carb-2	Cleanert TPT	GC-MS	Green tea	A*	4.1	7.3	1.7	2.3	9.1	2.6	7.8	11.1	11.6	9.5	7.8	12.1	7.4	5.1	2.3	2.5	8.8	11.7	4.8
					B*	4.8	6.2	7.9	11.8	6.0	6.1	3.1	10.0	3.5	4.9	8.9	12.4	4.4	5.0	5.1	4.8	5.4	6.0	5.4
				Woolong tea	A	4.3	5.5	4.5	3.4	5.6	4.0	2.9	12.9	5.4	8.0	16.4	7.0	5.4	2.5	5.8	0.7	3.1	3.4	0.6
					B*	5.9	5.0	3.7	5.3	7.1	7.8	14.6	9.5	7.1	8.3	14.9	0.8	2.8	3.6	3.4	0.9	2.5	0.4	5.6
			GC-MS/MS	Green tea	A*	4.5	7.2	1.6	1.0	6.9	7.1	8.8	12.9	7.2	1.9	5.3	5.8	7.0	7.8	2.1	0.1	4.6	5.4	4.9
					B	4.8	6.9	5.3	5.7	3.4	10.0	0.8	8.0	2.5	8.2	5.7	11.1	3.3	4.2	8.1	3.8	9.5	0.1	3.8
				Woolong tea	A	8.6	13.1	2.3	2.7	0.8	5.3	1.3	5.1	7.8	4.9	5.6	4.1	12.3	2.0	5.8	2.0	0.4	4.6	1.3
					B*	9.1	11.7	5.7	7.9	3.1	4.8	9.1	9.1	8.5	3.7	9.0	1.8	8.2	1.3	2.2	2.0	10.5	0.3	4.8
		Envi-Carb+PSA	GC-MS	Green tea	A*	13.7	7.2	7.4	7.2	7.8	3.5	19.9	13.5	17.0	3.9	3.3	5.8	1.8	19.7	4.1	42.0	4.5	33.0	3.9
					B*	5.3	8.7	5.1	5.0	2.2	7.7	7.3	6.7	20.6	20.1	47.2	12.2	30.0	1.6	34.7	1.0	5.4	33.6	10.8
				Woolong tea	A	4.2	3.9	7.9	3.7	4.5	2.2	3.1	19.2	21.0	0.4	7.3	3.6	3.7	7.0	0.8	6.1	8.1	3.3	3.9
					B	4.1	5.5	11.1	0.1	15.8	5.6	6.9	10.7	12.0	5.1	9.4	7.3	5.3	3.8	9.9	2.5	7.5	0.7	4.0
			GC-MS/MS	Green tea	A*	17.2	7.0	5.7	6.6	6.0	5.0	14.3	8.2	12.9	2.9	7.7	5.0	6.6	6.6	3.6	16.4	3.5	12.6	10.9
					B*	10.8	10.2	2.6	5.4	2.1	4.5	6.1	9.2	10.4	4.3	29.3	2.3	21.2	5.6	18.9	1.2	5.7	17.5	9.9
				Woolong tea	A	6.9	1.7	4.9	1.9	5.5	3.8	3.3	18.4	3.2	4.6	5.1	6.9	3.6	3.1	5.0	5.4	3.5	10.2	3.5
					B*	11.1	3.6	11.6	2.2	11.1	6.0	5.3	5.6	11.1	2.8	2.4	3.3	8.3	7.3	8.0	1.3	5.7	7.0	3.8
35	isodrin	Cleanert TPT	GC-MS	Green tea	A*	7.1	9.1	5.1	2.1	9.0	4.7	8.6	3.2	12.2	5.8	2.7	5.3	7.6	4.1	4.7	7.4	6.5	12.2	5.8
					B	8.2	9.3	2.7	9.3	4.7	1.4	3.2	1.4	5.3	11.7	1.4	4.9	3.4	8.2	5.7	4.6	8.0	3.2	5.3
				Woolong tea	A*	5.0	2.6	1.6	3.8	2.1	1.2	9.9	3.7	6.7	9.0	8.8	10.7	2.8	5.7	8.1	11.8	1.3	9.3	2.6
					B	5.3	1.9	3.9	9.9	1.4	2.8	3.4	2.6	1.7	8.4	3.1	4.2	2.5	2.4	2.6	2.3	6.9	4.8	2.7
			GC-MS/MS	Green tea	A*	6.1	6.8	5.3	5.0	8.3	4.9	5.3	3.7	7.9	3.4	6.0	6.3	7.3	3.1	6.1	2.7	12.0	5.6	6.7
					B	5.4	7.8	5.9	8.7	7.6	2.6	5.1	5.2	4.4	10.8	7.4	7.4	2.7	6.6	3.4	3.7	6.5	5.2	10.5
				Woolong tea	A	2.6	3.4	1.1	5.2	2.7	3.5	8.5	1.1	2.7	2.0	3.3	0.1	10.0	2.3	8.6	19.4	2.0	6.3	6.2
					B	5.8	4.5	4.0	8.4	4.7	6.0	6.6	1.5	6.6	2.4	9.8	1.3	4.2	5.6	8.7	3.2	6.7	3.4	7.5
		Envi-Carb+PSA	GC-MS	Green tea	A*	4.7	8.0	1.5	7.6	7.5	2.8	12.0	6.8	10.0	9.0	7.2	3.5	6.3	3.6	9.0	6.1	5.0	4.2	8.7
					B	4.6	4.3	5.1	4.0	5.1	1.3	8.3	1.3	6.0	2.8	15.1	2.3	5.7	3.1	13.4	0.6	4.7	8.7	9.6
				Woolong tea	A	4.5	4.1	2.6	2.9	7.3	3.7	6.3	5.2	16.7	3.8	18.0	7.0	5.0	2.4	4.8	2.1	6.4	0.6	9.2
					B	9.2	4.6	8.9	2.3	9.6	4.8	1.2	3.7	10.6	7.9	8.8	9.1	2.3	8.0	9.9	6.4	8.3	14.2	2.5
			GC-MS/MS	Green tea	A*	5.9	5.4	2.6	5.2	4.5	2.6	7.7	1.9	7.3	2.6	16.3	4.8	2.2	2.3	9.5	11.2	4.6	5.7	5.7
					B*	8.1	7.9	4.4	5.3	5.0	7.2	3.5	5.7	3.3	2.3	21.9	3.6	4.2	4.9	13.9	2.9	6.6	13.7	2.5
				Woolong tea	A	4.6	4.4	5.8	19.7	0.2	4.3	6.8	1.5	4.4	1.4	4.1	8.1	11.6	7.4	11.8	9.4	7.2	13.2	9.4
					B	4.8	1.9	11.1	4.7	6.1	5.3	5.2	5.2	12.4	1.6	5.1	3.8	5.2	6.8	10.6	7.5	4.4	2.1	2.2
36	isoprocarb-1	Cleanert TPT	GC-MS	Green tea	A*	8.5	6.4	2.0	0.8	7.1	6.6	7.3	3.5	5.0	6.2	7.1	1.3	9.6	1.4	5.1	4.0	4.5	9.1	4.1
					B	9.3	10.6	4.0	8.8	1.7	4.9	4.8	7.9	3.0	3.0	5.2	6.8	5.8	5.4	4.0	2.8	4.1	4.7	5.4
				Woolong tea	A	3.6	3.1	2.7	4.0	3.8	2.8	11.9	7.3	2.4	3.4	0.5	3.5	3.2	2.7	8.0	3.5	0.3	10.6	1.4
					B	10.8	3.2	3.6	7.6	2.0	3.0	7.9	0.7	4.5	6.0	2.6	1.5	2.8	2.6	3.1	1.7	8.1	5.6	4.6
			GC-MS/MS	Green tea	A*	6.2	7.3	1.8	6.6	6.0	4.3	7.7	3.1	7.2	1.8	7.5	3.3	6.8	5.1	2.1	2.6	9.0	10.2	3.3
					B	4.6	9.4	6.5	8.0	4.8	4.0	5.1	6.1	2.3	8.9	7.5	8.2	1.6	4.1	4.3	3.2	2.9	4.8	7.3
				Woolong tea	A	6.3	1.9	4.5	2.6	2.5	2.1	9.0	0.6	5.5	3.8	1.6	3.9	10.4	1.7	9.1	1.7	1.3	3.8	2.6
					B	5.4	1.5	5.5	3.4	0.6	3.2	4.2	1.8	3.3	4.5	2.1	1.6	5.9	1.5	5.5	7.6	13.7	2.8	3.1
		Envi-Carb+PSA	GC-MS	Green tea	A*	4.1	6.1	1.6	7.2	5.9	4.8	15.2	5.3	8.6	5.4	10.0	9.5	4.5	4.0	6.7	4.3	6.2	2.7	5.6
					B	3.6	4.3	3.3	6.7	4.1	3.2	2.1	11.1	6.5	4.1	19.2	4.5	4.6	4.3	15.0	3.6	6.6	4.7	4.1
				Woolong tea	A	4.7	4.5	3.4	3.8	5.9	3.9	4.5	6.0	1.0	2.9	7.0	6.8	3.3	6.8	3.1	4.6	2.9	2.0	6.1
					B	7.3	1.7	10.7	5.6	9.3	5.6	4.5	5.9	8.0	7.8	7.3	6.0	2.4	10.6	9.4	12.5	11.9	9.7	3.4
			GC-MS/MS	Green tea	A*	4.0	7.2	0.8	6.6	5.8	6.2	10.7	2.2	6.7	4.1	9.4	17.1	5.7	9.5	5.6	0.9	4.0	3.8	2.7
					B*	4.9	7.3	5.1	8.6	5.1	4.8	1.3	7.5	5.6	2.8	23.1	9.6	2.7	9.3	19.1	3.1	3.8	10.8	1.4
				Woolong tea	A	4.6	2.0	2.6	7.3	5.2	0.6	2.8	1.1	2.9	0.7	4.9	7.4	6.2	1.9	5.9	2.6	4.6	3.8	6.4
					B	7.0	3.9	8.4	4.2	8.7	3.4	2.1	6.7	11.6	2.3	1.5	4.4	1.6	6.8	7.9	2.7	4.6	4.1	1.7

No.	Compound	Cartridge	Method	Tea	A/B																			
37	isoprocarb-2	Cleanert TPT	GC-MS	Green tea	A*	12.4	12.4	2.1	5.4	13.0	1.7	9.0	14.9	8.3	2.3	14.7	6.4	10.6	3.3	16.9	6.0	7.4	9.1	9.3
				Green tea	B*	15.5	9.3	7.7	7.6	6.1	3.4	5.9	7.3	7.9	11.3	3.6	8.3	15.3	12.0	4.3	2.9	8.4	7.2	11.0
				Woolong tea	A	13.5	5.3	4.3	5.2	3.6	1.8	8.6	4.2	12.1	8.6	12.0	4.4	3.3	6.3	9.0	4.8	6.5	23.2	9.9
				Woolong tea	B	12.3	2.4	11.7	7.8	3.4	14.3	4.7	14.1	15.7	8.7	1.5	1.3	3.8	5.2	11.3	6.8	2.4	8.9	14.8
			GC-MS/MS	Green tea	A	11.7	6.2	5.8	15.1	14.2	4.5	14.4	14.7	12.2	10.2	7.1	5.2	7.0	9.7	6.1	10.7	12.9	7.4	10.0
				Green tea	B	6.1	6.2	9.8	12.5	5.9	9.7	2.3	8.4	11.7	9.5	3.8	1.8	10.6	10.4	9.9	5.5	17.9	4.3	6.3
				Woolong tea	A	13.5	6.9	4.5	7.2	5.1	8.6	2.1	6.2	2.1	9.9	1.5	7.5	5.1	3.8	14.2	16.5	3.0	4.7	2.0
				Woolong tea	B	10.2	5.0	17.1	3.3	1.2	3.1	1.5	2.2	4.5	11.3	3.6	5.1	5.4	8.2	23.2	3.3	2.6	2.4	2.7
		Envi-Carb+PSA	GC-MS	Green tea	A*	19.2	28.5	9.7	19.5	9.8	6.6	34.4	20.0	35.2	12.4	2.7	35.0	6.1	18.8	11.6	24.8	5.8	7.6	1.7
				Green tea	B*	25.8	3.2	6.9	14.7	6.7	6.8	19.7	2.7	42.0	22.9	32.1	28.1	43.3	5.1	17.8	12.7	2.6	37.4	10.5
				Woolong tea	A	14.6	6.0	13.1	4.7	8.4	4.7	2.1	20.2	5.6	3.2	14.4	14.2	3.9	12.2	8.3	4.7	4.6	12.4	5.1
				Woolong tea	B	20.1	5.4	11.2	5.5	9.0	5.8	5.2	6.6	14.0	16.2	9.0	4.1	8.3	9.9	8.1	7.2	11.2	19.8	7.1
			GC-MS/MS	Green tea	A	11.9	6.0	13.2	4.4	3.9	21.7	14.9	16.3	24.1	11.9	11.1	0.8	3.2	8.1	21.6	10.3	22.2	7.3	21.2
				Green tea	B	24.9	12.2	9.7	14.5	9.4	9.2	15.7	34.6	34.8	7.2	7.8	3.3	1.0	7.0	13.7	2.0	10.4	7.1	26.6
				Woolong tea	A	8.4	5.2	17.2	8.0	7.2	5.3	7.1	18.6	4.3	9.8	5.6	8.7	8.3	3.9	14.3	15.7	5.4	3.0	10.1
				Woolong tea	B	16.0	6.7	15.3	8.2	10.8	4.2	5.4	6.6	27.0	11.9	2.6	4.4	3.0	10.1	14.0	5.3	2.5	4.7	3.0
38	lenacil	Cleanert TPT	GC-MS	Green tea	A*	5.8	6.9	2.4	17.4	18.4	4.4	18.2	23.6	24.1	6.1	5.5	10.0	12.4	14.8	8.2	14.2	12.9	30.8	2.1
				Green tea	B*	6.9	6.3	3.2	15.9	8.9	7.6	5.8	17.4	7.5	5.6	2.1	10.3	12.8	13.3	8.3	16.5	29.8	22.5	6.4
				Woolong tea	A	7.0	9.2	6.4	3.9	26.0	9.6	35.1	7.3	4.7	4.9	16.1	19.0	4.8	4.8	4.3	14.9	6.4	4.7	6.8
				Woolong tea	B	10.1	7.2	3.9	3.4	12.9	24.9	2.0	23.0	5.1	5.2	14.2	10.4	4.4	8.6	15.1	4.7	3.6	1.3	26.6
			GC-MS/MS	Green tea	A	11.2	7.8	0.4	7.3	13.8	21.3	21.1	35.7	22.6	4.1	9.2	10.5	18.0	8.9	4.3	23.6	12.9	29.0	14.2
				Green tea	B	10.4	5.3	3.3	12.0	6.2	8.5	7.6	21.1	6.0	8.9	2.5	17.0	20.9	4.4	7.2	4.3	6.5	16.2	20.9
				Woolong tea	A	18.5	32.5	4.2	2.9	25.8	13.0	35.8	17.1	36.2	4.4	28.2	4.7	7.6	2.1	7.8	15.0	0.2	24.3	7.1
				Woolong tea	B	36.6	29.8	14.4	3.4	8.9	31.7	0.1	17.1	24.9	5.0	23.2	7.6	7.0	4.2	13.7	9.6	4.1	24.0	25.4
		Envi-Carb+PSA	GC-MS	Green tea	A*	21.7	7.7	18.6	11.9	11.8	5.6	44.6	10.2	10.5	8.6	11.2	24.8	5.0	27.2	8.0	41.7	25.4	28.8	6.6
				Green tea	B*	17.8	7.1	5.8	2.1	11.5	3.6	12.6	40.5	25.6	6.5	42.6	19.5	50.5	10.4	41.4	3.8	5.7	34.8	18.4
				Woolong tea	A	8.7	3.8	5.5	7.4	18.4	11.3	11.9	41.1	6.5	4.4	5.9	10.8	15.3	3.7	3.0	29.2	10.9	10.6	6.6
				Woolong tea	B	10.2	12.0	7.7	7.8	15.2	22.7	23.3	12.9	12.5	1.5	5.3	8.2	4.2	12.9	18.6	2.6	20.2	17.2	7.5
			GC-MS/MS	Green tea	A	28.4	6.9	14.3	10.1	6.4	1.8	5.5	5.2	13.7	4.4	37.8	4.4	1.1	9.6	24.1	13.3	12.6	9.3	1.5
				Green tea	B	17.4	16.7	4.3	9.9	6.3	1.3	5.2	6.4	1.9	8.7	26.7	10.3	16.1	17.2	31.1	2.8	9.2	11.8	1.7
				Woolong tea	A	22.8	1.9	5.4	5.2	9.0	9.8	4.2	45.2	0.9	3.1	5.9	12.0	6.1	3.8	4.4	35.9	11.3	9.6	12.7
				Woolong tea	B	29.5	5.9	10.8	5.9	19.9	7.8	16.5	8.1	12.2	7.2	3.5	4.7	13.0	6.5	12.6	3.9	9.0	21.3	5.3
39	metalaxyl	Cleanert TPT	GC-MS	Green tea	A*	7.1	13.1	2.3	2.4	6.2	3.6	6.8	7.8	6.0	10.8	5.9	6.0	9.2	7.4	1.1	3.9	4.5	4.3	4.1
				Green tea	B*	4.2	9.4	4.3	7.5	6.0	7.9	2.7	11.0	6.2	10.6	6.5	13.5	9.5	11.0	6.5	3.8	5.2	1.0	5.1
				Woolong tea	A	3.8	3.9	7.6	6.3	1.8	4.4	10.3	4.2	3.1	3.1	10.0	3.6	3.7	3.2	8.4	3.3	4.7	2.6	4.2
				Woolong tea	B	7.8	4.0	2.7	8.3	2.2	6.7	2.4	6.4	9.3	9.3	7.5	1.1	2.1	5.6	3.1	0.3	0.7	1.7	3.2
			GC-MS/MS	Green tea	A	5.8	8.3	4.5	4.5	5.0	6.1	8.5	6.1	4.5	3.8	6.2	4.6	7.8	10.9	3.7	4.8	5.4	4.5	3.0
				Green tea	B	4.0	10.0	5.6	6.9	5.2	4.4	4.8	7.7	5.5	11.1	8.7	10.1	3.2	3.4	5.4	5.7	6.6	3.3	7.7
				Woolong tea	A	4.6	4.5	0.6	8.4	2.6	0.7	8.5	0.7	3.4	7.0	2.9	1.3	3.7	2.4	9.5	4.8	2.6	4.5	7.8
				Woolong tea	B	6.8	0.8	5.5	5.7	1.4	8.4	4.5	5.0	3.0	2.1	5.6	3.2	8.1	4.0	4.5	3.1	2.2	5.0	6.1
		Envi-Carb+PSA	GC-MS	Green tea	A*	9.0	9.4	8.1	4.9	7.4	3.7	27.3	3.9	4.6	5.6	12.1	2.2	6.8	5.5	4.8	11.0	9.7	6.2	7.1
				Green tea	B*	4.6	3.3	6.0	6.1	5.9	4.9	1.1	11.3	11.9	11.9	23.1	7.1	6.6	4.0	13.0	1.2	3.9	13.5	7.6
				Woolong tea	A	6.7	4.2	3.3	5.1	3.5	8.2	1.7	3.7	22.6	10.1	4.5	11.0	3.9	2.5	5.2	3.4	5.6	3.0	3.2
				Woolong tea	B	5.9	2.6	14.2	5.2	12.3	1.6	2.4	6.4	7.1	7.1	11.8	3.3	5.6	13.6	10.7	4.2	4.0	3.3	9.1
			GC-MS/MS	Green tea	A	7.4	11.6	3.2	5.9	7.2	2.9	20.9	2.9	3.2	1.6	12.7	0.8	6.1	2.3	6.9	8.4	6.9	4.0	5.8
				Green tea	B	3.7	5.2	3.4	4.0	2.8	5.0	3.5	1.2	5.5	2.6	20.7	4.0	1.2	13.4	13.4	1.9	1.6	12.9	5.4
				Woolong tea	A	3.7	3.7	3.3	5.9	5.1	8.3	6.5	3.8	4.6	12.4	7.1	11.8	7.0	4.1	13.0	6.2	9.6	5.1	3.3
				Woolong tea	B	6.5	3.2	16.3	4.3	10.7	8.3	3.4	5.7	10.5	3.1	0.8	4.5	3.5	16.5	10.7	3.2	7.2	0.2	7.0

(Continued)

Annex 2: RSD data for 201 pesticides in incurred tea Youden pair samples under 8 conditions with two SPE cartridge cleanup in three months (Nov.9, 2009-Feb.7, 2010) (cont.)

No	Pesticides	SPE	Method	Sample	Youden pair	Nov.9, 2009 (n=5)	Nov.14, 2009 (n=3)	Nov.19, 2009 (n=3)	Nov.24, 2009 (n=3)	Nov.29, 2009 (n=3)	Dec.4, 2009 (n=3)	Dec.9, 2009 (n=3)	Dec.14, 2009 (n=3)	Dec.19, 2009 (n=3)	Dec.24, 2009 (n=3)	Dec.29, 2009 (n=3)	Jan.3, 2010 (n=3)	Jan.8, 2010 (n=3)	Jan.13, 2010 (n=3)	Jan.18, 2010 (n=3)	Jan.23, 2010 (n=3)	Jan.28, 2010 (n=3)	Feb.2, 2010 (n=3)	Feb.7, 2010 (n=3)
40	metazachlor	Cleanert TPT	GC-MS	Green tea	A	7.3	7.8	6.1	3.3	6.4	4.8	6.4	8.3	5.8	2.6	9.5	6.0	8.0	3.1	8.5	6.1	5.8	10.0	1.5
					B*	8.2	10.4	9.6	5.6	6.0	7.1	1.6	6.2	3.2	10.7	7.1	7.5	3.2	7.9	9.0	1.7	10.3	8.0	5.3
				Woolong tea	A*	8.3	6.6	6.2	3.6	9.3	2.1	15.9	4.5	4.1	7.6	9.3	2.4	5.9	3.0	10.3	3.5	12.0	4.0	3.0
					B*	7.3	2.9	5.6	5.0	6.1	6.7	7.7	5.3	6.2	7.6	10.0	1.1	6.5	4.2	9.2	7.0	2.5	1.5	8.2
			GC-MS/MS	Green tea	A	7.2	7.9	4.4	5.9	4.2	1.9	2.9	10.1	3.6	3.8	4.6	4.0	10.2	7.1	3.5	2.5	6.2	3.9	6.4
					B	5.8	9.6	4.3	7.3	4.5	4.9	3.5	8.4	7.1	7.8	4.6	3.2	4.0	6.3	5.2	2.2	7.0	4.4	5.4
				Woolong tea	A	7.9	4.6	4.1	5.5	9.7	6.1	14.8	5.2	11.1	3.8	4.1	2.6	9.6	5.3	4.7	3.5	1.9	1.4	5.9
					B	6.8	6.8	8.2	5.0	3.1	1.5	8.2	6.5	8.6	5.6	4.5	6.4	6.4	4.6	7.3	3.5	3.5	2.3	2.8
		Envi-Carb+PSA	GC-MS	Green tea	A*	7.2	6.2	5.2	4.9	6.3	7.9	13.7	6.5	4.8	1.2	6.2	8.8	2.8	5.0	5.2	9.1	5.9	20.9	6.0
					B*	9.7	8.8	2.1	3.6	4.4	7.3	13.7	6.6	7.3	12.9	25.5	7.5	12.6	4.4	12.7	4.1	5.2	24.4	8.7
				Woolong tea	A*	7.2	7.4	4.6	7.6	1.7	5.4	3.2	4.5	7.6	6.4	9.5	5.1	4.6	5.4	5.2	5.4	6.5	4.7	5.5
					B*	8.1	7.8	12.8	8.8	8.3	4.6	4.8	13.2	10.3	8.1	8.8	5.9	6.5	7.7	16.1	2.8	11.9	5.7	1.6
			GC-MS/MS	Green tea	A	12.7	10.7	2.6	4.4	5.1	6.4	14.1	2.6	8.1	2.3	23.0	13.9	9.1	2.5	0.8	4.1	8.6	8.7	4.4
					B	41.6	11.5	5.6	2.5	5.7	6.2	4.9	5.2	5.7	6.5	12.6	9.0	11.2	13.5	6.6	6.0	0.8	13.0	10.2
				Woolong tea	A*	3.1	1.7	6.2	8.0	7.7	6.7	5.0	2.8	6.8	9.6	4.4	12.2	0.9	5.2	13.7	3.3	4.5	6.2	4.4
					B*	10.5	2.1	15.2	8.3	7.0	7.9	1.5	6.0	13.3	2.0	1.9	1.0	3.5	11.7	13.1	4.2	5.1	5.8	6.0
41	methabenz-thiazuron	Cleanert TPT	GC-MS	Green tea	A	5.9	7.0	3.0	2.6	3.5	6.4	7.9	5.8	4.6	1.6	1.2	5.6	7.9	4.1	1.5	3.2	7.1	5.5	4.5
					B	5.2	6.9	6.7	8.5	2.9	6.9	2.0	5.8	5.5	12.2	6.1	10.4	2.4	1.5	13.1	3.7	12.1	3.5	4.6
				Woolong tea	A*	10.2	2.2	3.1	2.8	4.9	2.2	8.0	7.4	2.0	7.7	10.3	3.2	5.6	3.9	4.5	2.5	0.8	3.1	1.2
					B*	11.3	1.7	4.0	3.9	5.6	4.0	2.2	3.3	6.0	2.3	6.3	3.2	4.3	5.2	3.8	1.8	3.5	4.7	5.4
			GC-MS/MS	Green tea	A	1.5	5.0	2.4	5.0	2.0	4.7	9.0	4.3	4.7	3.3	3.3	4.8	5.1	3.6	2.2	2.9	5.1	5.3	6.8
					B	5.2	3.4	6.3	5.8	4.7	1.4	5.9	5.4	2.7	5.3	6.6	9.1	3.3	4.5	9.8	4.6	8.8	15.5	6.4
				Woolong tea	A*	4.8	10.3	0.8	7.0	4.5	7.7	4.3	6.0	11.0	11.0	5.8	4.4	7.5	10.5	12.0	1.4	3.9	2.8	5.8
					B*	6.1	0.5	5.6	5.0	4.4	2.9	10.9	4.1	1.9	4.4	4.5	2.3	4.5	3.0	8.5	1.5	1.2	3.5	6.5
		Envi-Carb+PSA	GC-MS	Green tea	A	2.0	5.8	5.6	9.0	2.2	1.6	4.2	5.1	5.6	4.3	7.3	1.2	6.0	3.3	6.9	4.7	3.4	3.0	3.7
					B	7.1	4.1	8.3	6.8	9.5	6.8	11.1	2.1	4.6	8.3	10.0	8.0	2.4	8.1	8.3	8.1	13.8	1.2	6.6
				Woolong tea	A*	4.5	1.9	6.3	3.4	4.0	4.9	7.7	2.9	6.1	11.2	31.2	7.9	9.5	4.7	19.0	6.9	4.0	8.2	8.1
					B*	3.8	6.3	3.1	8.2	4.9	3.3	5.0	8.5	19.2	6.2	11.7	5.6	2.4	8.0	8.3	4.4	6.3	7.6	4.3
			GC-MS/MS	Green tea	A	5.1	3.2	11.1	8.0	13.5	10.5	4.9	7.0	8.8	1.8	8.4	1.1	4.4	9.2	6.7	5.0	5.1	4.8	6.2
					B	5.6	9.2	5.0	8.0	7.7	3.0	23.8	3.0	3.7	3.3	11.8	0.9	4.2	11.0	5.8	11.7	9.7	8.2	8.2
				Woolong tea	A*	3.3	7.4	4.1	8.4	2.9	1.5	4.4	1.5	6.4	2.2	27.5	1.8	7.7	11.5	17.2	2.5	5.4	16.5	7.5
					B*	5.2	2.9	3.5	3.3	1.1	1.4	5.9	5.4	2.7	3.3	6.6	4.4	3.2	4.5	9.8	1.0	8.8	15.5	5.8
42	mirex	Cleanert TPT	GC-MS	Green tea	A	5.1	2.9	4.8	7.0	15.0	4.9	12.8	10.1	6.8	2.7	2.9	6.8	10.5	8.1	1.1	5.1	9.6	14.9	3.7
					B	3.7	4.8	2.2	8.6	2.7	1.5	4.3	10.4	6.0	9.5	4.7	4.4	10.6	11.2	4.3	8.3	16.1	12.4	6.3
				Woolong tea	A*	6.3	7.5	2.1	2.2	9.2	4.0	4.1	7.6	5.2	4.1	7.0	11.6	2.9	1.3	7.5	8.6	5.9	4.1	2.8
					B*	4.9	2.4	6.2	3.7	6.7	9.3	6.9	12.3	7.1	5.9	3.4	2.3	2.2	7.3	10.4	4.2	2.4	1.9	9.8
			GC-MS/MS	Green tea	A	7.6	0.5	4.5	3.1	9.1	2.6	10.8	9.5	9.8	2.6	6.4	8.5	9.6	4.9	1.2	5.0	6.1	11.5	2.9
					B	5.7	5.1	3.5	8.5	2.1	1.4	2.9	5.9	5.0	10.4	2.1	9.9	11.9	3.2	1.8	6.5	10.2	3.8	7.8
				Woolong tea	A	5.9	7.6	3.5	3.7	1.1	6.6	3.6	2.9	12.9	3.1	9.6	1.1	3.4	3.4	12.1	5.4	2.3	8.5	6.9
					B	8.4	6.3	6.1	4.5	0.6	4.4	7.6	7.4	9.8	3.4	6.2	4.7	4.2	5.0	7.7	1.0	2.8	6.1	8.1
		Envi-Carb+PSA	GC-MS	Green tea	A*	12.8	2.1	9.6	6.3	8.0	15.4	18.3	8.2	11.4	4.6	11.8	17.4	4.2	14.3	4.7	16.4	16.1	11.8	9.4
					B	18.4	8.6	3.8	2.0	8.0	4.3	29.4	13.4	11.8	6.1	20.8	10.0	23.2	5.3	15.4	7.0	6.2	20.9	13.4
				Woolong tea	A	21.0	2.4	3.9	4.2	3.5	7.2	5.4	14.1	8.0	4.3	2.9	7.7	10.1	5.2	3.4	18.2	7.2	11.8	6.1
					B	3.9	5.6	7.0	2.3	2.9	10.3	14.9	6.2	14.8	5.0	11.9	5.8	5.4	16.4	16.8	5.8	14.2	16.9	8.8
			GC-MS/MS	Green tea	A	7.0	5.7	8.3	3.8	14.8	16.0	14.6	10.5	5.6	2.3	17.4	14.4	5.7	9.5	0.2	9.1	11.9	11.3	3.2
					B	19.9	13.4	2.6	4.4	9.9	2.8	11.8	11.4	8.3	5.6	16.6	10.1	29.2	10.2	9.1	9.9	6.9	19.7	8.2
				Woolong tea	A	27.6	2.8	6.7	9.7	7.5	2.2	3.8	15.2	6.0	2.5	4.8	10.6	9.7	4.1	9.1	10.6	8.0	16.0	5.6
					B	11.8	4.0	10.5	3.9	10.1	8.3	6.5	8.6	16.8	9.1	8.6	3.4	8.6	12.9	13.4	3.0	12.7	22.2	7.3

No.	Compound	Sorbent	Method	Tea	A/B	C1	C2	C3	C4	C5	C6	C7	C8	C9	C10	C11	C12	C13	C14	C15	C16	C17	C18	C19
43	monalide	Cleanert TPT	GC-MS	Green tea	A	1.8	6.8	3.3	6.1	12.8	1.5	5.5	4.9	10.8	3.9	7.8	10.8	4.4	10.3	5.7	2.8	7.6	10.6	5.9
				Woolong tea	B	6.8	8.2	3.2	7.5	8.0	4.7	5.4	5.7	7.2	8.3	5.5	0.7	11.5	14.3	5.5	5.4	12.6	3.4	8.4
					B	11.3	3.5	9.6	5.3	6.8	2.5	2.0	8.3	15.1	10.2	14.3	2.7	4.2	2.7	6.7	11.8	1.5	23.3	10.5
			GC-MS/MS	Green tea	A	11.9	5.8	18.7	15.7	3.5	6.4	7.8	0.4	10.5	10.3	3.8	6.9	8.0	5.9	16.3	5.3	9.4	11.5	18.0
				Woolong tea	B	5.6	5.7	1.4	4.8	7.0	2.2	7.2	4.2	5.3	3.4	6.5	4.9	8.8	5.1	1.9	5.8	4.4	12.8	5.7
					A	4.6	8.0	5.8	6.7	3.7	3.6	5.2	4.0	5.8	8.8	7.7	6.5	1.1	3.9	3.3	3.7	11.5	3.8	10.2
				Green tea	B	5.7	3.7	3.0	6.8	3.0	1.8	10.7	0.6	0.1	2.4	2.7	1.0	11.5	2.4	7.2	3.7	7.1	3.0	6.1
					A	6.0	2.1	5.0	5.4	0.8	3.2	5.2	1.7	4.7	1.2	3.7	1.6	4.1	1.5	6.0	0.8	1.7	1.7	3.2
		Envi-Carb+PSA	GC-MS	Woolong tea	A*	8.8	6.8	1.8	1.6	5.1	18.5	6.6	3.9	13.3	2.6	9.6	18.9	3.7	1.1	2.6	9.0	10.9	4.5	11.8
					B*	10.0	12.5	5.7	6.9	6.8	7.9	6.1	5.0	6.4	6.1	14.4	12.2	29.6	3.6	2.2	7.4	5.1	8.0	9.9
				Green tea	B	4.7	5.0	5.3	15.2	1.9	4.8	7.4	15.4	5.3	3.3	7.9	14.9	9.9	5.9	10.7	26.8	6.0	7.1	8.4
					A	10.8	14.9	24.4	10.8	8.9	4.5	7.2	6.7	18.3	6.2	1.5	5.2	6.0	4.2	13.9	6.4	8.9	6.8	8.2
			GC-MS/MS	Woolong tea	B	5.7	5.1	1.3	6.2	5.3	2.3	2.3	1.1	4.6	2.2	11.0	2.9	6.6	2.7	5.0	6.9	9.0	7.7	4.7
					A	4.0	8.8	3.7	4.6	2.8	3.1	2.4	5.8	4.3	2.8	23.1	1.2	3.3	4.8	12.2	2.8	2.9	13.8	8.0
				Green tea	B	3.6	4.9	5.2	6.1	7.8	3.8	5.0	3.5	5.7	3.6	6.5	11.5	6.1	4.0	10.3	3.9	6.5	3.0	4.2
					A	5.7	1.8	9.5	5.8	7.7	5.0	2.5	6.0	11.9	2.5	2.5	7.6	6.6	10.0	9.2	1.5	4.3	3.8	4.1
44	paraoxon-ethyl	Cleanert TPT	GC-MS	Woolong tea	B	12.4	2.4	3.1	4.3	7.2	3.7	5.1	4.8	2.8	2.6	0.3	1.4	2.1	7.8	2.8	2.1	10.2	11.8	4.6
					A*	7.8	0.7	8.9	1.0	4.5	2.9	2.3	3.4	3.1	6.2	7.0	3.5	5.4	10.6	0.3	5.7	3.2	5.1	13.5
				Green tea	B	4.9	7.6	4.1	8.9	3.6	9.3	4.1	6.7	9.0	9.2	4.9	8.3	6.0	4.9	7.4	8.9	5.3	7.2	5.8
					A	4.4	4.7	4.1	6.9	6.9	3.9	4.5	1.9	2.4	3.1	1.6	13.0	7.5	11.0	12.6	3.7	1.3	13.3	11.7
			GC-MS/MS	Woolong tea	B	8.1	11.6	9.0	20.9	10.4	11.3	4.9	11.8	16.1	15.7	9.9	28.5	2.6	26.1	9.1	3.7	23.1	19.6	
					A	10.1	9.9	6.9	10.7	9.7	9.6	6.9	14.1	17.2	16.6	3.0	21.3	2.6	16.4	5.2	7.2	24.4	27.4	
				Green tea	A*	2.3	13.1	9.0	12.4	6.2	7.0	3.7	6.8	6.8	2.2	4.2	1.9	16.6	3.6	2.4				
					B	16.4	15.7	11.5	3.2	5.5	9.4	10.0	3.6	3.6	9.4	19.7	5.6	8.5	3.3	15.5				4.5
		Envi-Carb+PSA	GC-MS	Woolong tea	A	14.1	5.0	3.3	8.9	3.9	2.0		7.7	7.4	4.4	5.5	9.4	9.6	4.8	2.1	6.9	6.9	9.4	6.8
					B	21.5	2.1	0.7	4.5	8.0	0.8	2.3	10.3	1.3	5.2	6.5	7.4	10.2	1.8	2.6	1.9	6.3	10.3	6.8
				Green tea	A	6.2	6.3	9.3	7.6	17.1	2.1	9.7	3.6	7.4	9.2	6.9	6.9	12.2	7.6	8.7	12.1	17.3	8.5	8.4
					B	8.8	12.8	2.5	13.2	1.6	1.0	38.3	8.1	4.2	3.2	5.7	5.7	11.7	2.9	16.9	3.0	18.5	4.4	3.7
			GC-MS/MS	Woolong tea	A*	25.8	21.4	19.4	5.9	10.3	9.5	11.1	17.9	13.6	7.4	17.8	4.3	30.1	19.4	21.4	10.3	20.5		
					B	29.2	20.8	7.5	9.0	38.2	20.6	3.6	23.9	16.9	8.3	24.6	3.2	26.6	13.8	9.7	14.7	20.4		
				Green tea	A*	15.4	7.8	7.5	18.4	28.7	9.8	4.2	8.9	3.1	21.9	12.6	11.1	25.9	4.7	14.0	32.6			
					B	20.8	9.5	15.7	15.2	15.8	14.9	4.2	12.4	15.1	14.2	12.0	11.4	20.9	17.2	8.8	18.2			
					A	3.5	4.3	2.6	2.5	9.6	9.0	4.4	2.4	7.1	5.4	6.6	2.2	5.1	3.9	2.8	4.1	3.1	10.5	4.8
					B	7.9	7.3	4.2	6.5	2.3	5.3	4.0	3.3	4.7	10.9	2.7	8.8	8.9	4.9	5.7	2.9	7.3	6.9	4.5
45	pebulate	Cleanert TPT	GC-MS		A	3.8	2.1	1.0	2.3	3.4	0.6	9.1	5.2	6.8	2.9	4.2	3.7	4.9	2.7	5.8	11.1	14.5	8.5	4.4
					B	8.8	4.4	5.3	8.7	3.8	5.3	6.0	0.3	5.7	4.2	2.7	1.1	4.6	2.1	3.6	11.9	9.3	13.3	3.5
			GC-MS/MS		A							5.5	2.1	6.0	5.4	8.0	4.1	5.3	5.7	5.9	6.6	2.0	2.8	4.0
					B							6.3	3.7	2.9	13.6	5.0	9.0	2.1	5.1	11.1	4.8	12.0	12.4	7.1
		Envi-Carb+PSA	GC-MS		A	5.6	4.7	2.1	5.9	6.3	3.6	11.4	0.6	5.0	6.4	4.3	7.5	16.4	2.3	10.8	13.8	19.7	12.6	8.1
					B	3.5	3.4	4.4	3.9	2.2	2.0	5.4	2.8	3.7	28.0	3.8	0.5	4.1	7.9	10.0	11.6	10.4	23.2	8.4
			GC-MS/MS		A*	4.4	3.5	3.2	5.3	4.2	1.5	8.7	4.1	2.6	2.5	8.5	3.7	4.6	4.9	5.6	10.8	8.2	15.4	14.4
					B*	7.8	5.2	5.0	5.0	9.8	3.5	3.0	1.7	9.7	3.1	23.7	2.9	12.8	13.4	12.1	10.8	8.6	8.7	2.5

(Continued)

Annex 2: RSD data for 201 pesticides in incurred tea Youden pair samples under 8 conditions with two SPE cartridge cleanup in three months (Nov.9, 2009-Feb.7, 2010) (cont.)

No	Pesticides	SPE	Method	Sample	Youden pair	Nov.9, 2009 (n=5)	Nov.14, 2009 (n=3)	Nov.19, 2009 (n=3)	Nov.24, 2009 (n=3)	Nov.29, 2009 (n=3)	Dec.4, 2009 (n=3)	Dec.9, 2009 (n=3)	Dec.14, 2009 (n=3)	Dec.19, 2009 (n=3)	Dec.24, 2009 (n=3)	Dec.14, 2009 (n=3)	Jan.3, 2010 (n=3)	Jan.8, 2010 (n=3)	Jan.13, 2010 (n=3)	Jan.18, 2010 (n=3)	Jan.23, 2010 (n=3)	Jan.28, 2010 (n=3)	Feb.2, 2010 (n=3)	Feb.7, 2010 (n=3)
46	pentachloro-aniline	Cleanert TPT	GC-MS	Green tea	A	4.0	5.1	2.4	1.5	8.6	3.8	6.7	3.6	4.3	2.5	2.1	3.3	4.6	4.8	1.0	0.5	6.0	6.9	5.1
					B	24.0	9.0	5.4	7.4	2.2	4.5	2.0	2.4	3.0	9.7	2.8	11.8	3.6	1.6	3.5	3.1	7.2	1.7	5.5
				Woolong tea	A	4.5	1.4	0.2	3.1	2.0	2.1	21.4	13.3	3.4	4.1	4.1	2.2	4.1	2.8	7.9	2.9	3.8	1.3	1.5
					B	6.9	2.5	5.2	3.5	2.3	3.7	9.5	5.2	5.4	5.1	3.1	0.4	2.8	2.8	4.1	0.5	1.0	1.4	1.6
			GC-MS/MS	Green tea	A	3.7	5.8	5.1	0.9	5.4	2.9	8.4	3.8	5.1	4.1	4.4	3.8	4.1	7.2	6.2	2.6	6.5	3.1	11.8
					B	4.8	9.1	3.6	8.2	4.3	1.6	4.3	4.0	7.0	10.6	5.4	13.0	5.1	4.1	3.8	1.9	5.8	5.4	3.7
				Woolong tea	A*	2.6	0.7	3.3	5.7	1.5	1.4	9.8	1.4	1.9	3.6	2.1	2.3	3.0	2.1	5.3	2.0	8.0	3.7	3.9
					B	4.3	1.8	4.9	4.2	1.1	2.2	4.6	4.8	4.6	1.8	3.4	3.0	0.9	3.8	6.9	3.1	4.6	5.9	6.1
		Envi-Carb+PSA	GC-MS	Green tea	A	7.9	1.1	3.5	5.3	9.5	3.9	6.6	0.7	4.7	1.1	13.1	3.1	4.6	4.8	6.3	4.2	13.5	10.0	5.2
					B	5.3	5.2	2.8	4.2	0.7	3.6	3.2	2.4	3.1	3.8	18.8	6.2	8.0	3.2	13.5	6.7	4.9	8.5	8.4
				Woolong tea	A	4.2	3.0	3.3	4.3	5.0	0.2	7.0	14.6	0.8	0.4	3.8	3.9	0.5	3.2	2.6	3.0	5.2	4.9	4.8
					B	5.4	3.9	7.6	4.3	8.8	4.0	11.6	13.7	9.8	7.0	8.3	4.3	2.3	6.9	7.9	3.7	3.1	3.1	3.5
			GC-MS/MS	Green tea	A	5.4	5.1	2.8	6.5	6.1	2.7	1.0	2.0	5.3	3.1	9.4	3.0	4.2	12.7	11.1	5.2	17.2	6.4	3.6
					B	5.6	7.3	4.3	6.2	0.7	1.2	1.7	3.7	4.4	2.7	22.0	0.5	9.3	2.7	10.4	11.3	5.0	9.4	10.1
				Woolong tea	A	4.1	6.9	6.9	4.9	5.4	0.3	3.7	3.7	1.5	2.1	3.6	6.6	4.5	5.2	8.1	4.6	8.9	4.6	7.5
					B	5.5	2.0	8.7	4.2	8.9	2.5	2.4	3.4	11.0	2.7	2.2	2.5	3.7	10.0	5.5	1.3	1.8	1.4	4.5
47	pentachloro-anisole	Cleanert TPT	GC-MS	Green tea	A	5.3	5.3	2.8	1.6	8.7	6.1	6.2	4.0	6.9	3.1	4.8	3.1	6.2	4.2	2.7	1.0	2.5	6.8	4.5
					B	6.5	8.5	3.7	8.6	2.4	5.3	3.5	0.8	5.4	9.8	2.1	8.1	2.5	3.4	4.8	2.0	3.4	1.4	5.7
				Woolong tea	A*	4.1	1.8	0.3	2.0	2.1	0.3	9.0	4.6	6.6	3.1	3.3	3.4	4.7	2.8	7.7	3.2	5.1	1.0	3.6
					B*	8.0	5.0	6.6	6.7	3.5	5.1	5.8	1.6	6.9	6.2	2.5	0.4	4.1	3.2	4.5	1.1	2.4	1.3	3.0
			GC-MS/MS	Green tea	A	4.1	7.3	3.4	3.0	10.1	7.8	6.4	2.4	4.2	3.3	5.6	0.6	5.3	9.0	1.3	2.0	5.5	7.4	5.2
					B	5.7	8.9	4.2	5.4	6.4	5.1	4.6	4.7	1.3	10.8	6.6	6.3	1.5	4.6	0.9	2.7	7.4	3.4	6.5
				Woolong tea	A	5.3	1.5	1.6	4.2	2.2	1.7	9.8	1.8	5.7	5.0	3.6	5.8	9.0	4.7	9.0	5.2	3.0	3.7	2.9
					B	6.1	6.8	5.5	5.2	2.2	5.3	4.9	0.2	6.0	3.7	6.0	1.0	2.1	1.9	9.6	2.6	2.6	2.9	2.8
		Envi-Carb+PSA	GC-MS	Green tea	A	6.7	5.1	1.3	8.1	6.8	2.5	7.8	1.4	9.9	1.2	9.2	3.4	3.6	3.0	5.9	8.0	11.3	4.6	4.9
					B	5.0	7.3	3.8	3.8	1.9	1.0	2.0	4.3	2.3	2.6	19.6	0.9	3.9	2.3	14.3	4.1	5.8	8.6	5.4
				Woolong tea	A	4.0	2.7	2.6	5.3	4.8	1.9	1.7	2.8	1.1	1.2	5.7	3.9	1.0	3.2	2.8	5.2	5.9	4.8	7.1
					B*	7.4	4.6	5.3	5.4	8.7	2.5	3.5	8.0	11.7	3.0	7.6	4.9	2.7	8.7	6.6	3.6	6.3	3.9	4.0
			GC-MS/MS	Green tea	A	6.3	5.5	1.9	6.1	6.1	4.9	5.4	1.9	4.9	0.7	9.1	3.7	2.1	11.7	8.1	8.0	10.3	6.2	4.8
					B	2.5	12.2	4.9	3.0	4.3	2.9	1.3	2.6	3.4	3.6	19.2	5.7	6.6	8.3	16.3	0.9	1.5	13.7	8.0
				Woolong tea	A	4.1	3.2	5.0	7.2	6.8	1.6	4.4	2.1	2.6	1.6	4.5	5.4	5.0	3.7	7.7	5.8	6.5	5.7	6.0
					B	6.6	3.6	6.2	4.2	7.8	3.1	3.3	5.5	12.8	2.0	5.1	2.5	2.3	11.8	8.2	2.7	5.9	2.2	2.3
48	pentachloro-benzene	Cleanert TPT	GC-MS	Green tea	A	3.9	3.6	2.4	1.7	8.5	7.8	4.1	1.3	5.9	4.2	7.0	1.6	5.5	1.8	2.0	2.7	3.4	7.0	3.4
					B	7.0	8.6	4.6	7.8	2.6	5.4	4.1	2.7	5.3	12.2	1.7	6.8	5.1	1.2	5.5	3.5	6.0	4.5	4.0
				Woolong tea	A	7.4	1.9	0.5	2.8	2.7	5.7	5.2	4.6	2.7	4.6	4.3	2.3	6.2	4.7	7.4	2.1	2.9	2.5	0.8
					B	9.3	2.3	4.3	2.0	1.1	3.0	7.4	4.2	7.6	2.2	2.3	0.7	1.6	5.1	3.8	0.7	1.8	1.0	3.3
			GC-MS/MS	Green tea	A	3.0	3.8	2.2	1.1	8.0	7.6	4.3	0.8	5.9	5.9	8.0	4.4	4.8	7.9	0.5	5.6	4.1	6.3	2.6
					B	7.5	7.9	4.7	7.3	6.7	5.5	5.8	2.2	5.4	13.4	3.9	7.4	4.8	7.2	4.8	4.0	6.7	6.3	6.5
				Woolong tea	A	5.9	0.6	2.6	5.2	2.0	3.0	8.9	2.3	2.8	4.4	5.4	2.2	9.8	3.0	9.4	2.1	0.2	1.7	2.8
					B	2.6	1.5	5.8	5.8	0.8	3.0	3.7	5.8	4.7	2.2	4.4	1.0	8.0	2.4	4.9	1.3	4.1	1.2	3.5
		Envi-Carb+PSA	GC-MS	Green tea	A	5.1	5.9	1.1	6.5	6.2	3.6	5.4	1.6	8.8	1.3	10.1	5.1	3.7	2.5	5.8	9.5	13.0	4.1	4.4
					B	3.0	4.5	4.4	4.3	2.0	7.9	5.3	4.5	2.1	2.1	22.0	1.6	8.5	3.0	23.3	4.7	7.2	7.9	4.6
				Woolong tea	A	3.4	1.8	2.7	2.9	4.6	1.5	3.5	6.2	1.9	2.4	4.7	2.7	2.4	2.6	2.7	1.7	6.1	5.3	3.5
					B	5.5	3.8	9.7	2.9	7.1	4.6	3.2	7.0	11.1	5.5	10.7	2.7	3.7	8.3	9.0	4.4	7.0	1.7	5.7
			GC-MS/MS	Green tea	A	5.2	6.7	1.5	5.1	3.8	2.6	5.2	1.4	9.2	0.7	10.5	5.0	4.7	4.2	7.0	11.2	11.8	6.4	1.0
					B	2.6	6.3	5.8	4.9	3.3	2.8	5.5	3.0	2.4	0.8	24.3	8.6	9.6	11.9	26.3	1.2	3.9	11.2	3.2
				Woolong tea	A	3.2	1.6	3.5	3.5	5.9	2.4	4.1	2.9	4.1	1.3	5.3	8.5	3.1	3.4	8.2	2.3	3.7	4.5	3.9
					B	6.7	2.7	10.6	3.0	8.1	6.0	3.3	6.8	12.3	3.2	1.7	5.0	1.9	9.0	10.0	1.6	6.3	5.9	5.9

#	Compound	Cartridge	Method	Tea	A/B																			
49	perthane	Cleanert TPT	GC-MS	Green tea	A	6.3	4.6	3.1	1.6	7.9	7.3	8.2	5.7	6.1	3.7	5.7	3.9	7.0	3.5	6.6	1.0	5.1	7.9	1.8
				Green tea	B	6.3	7.6	4.3	7.9	4.7	6.7	6.3	3.1	5.2	8.8	3.9	6.5	2.7	7.1	3.6	4.2	4.2	2.4	9.0
				Woolong tea	A	4.0	0.6	1.5	3.3	1.3	6.5	6.1	4.4	4.0	6.6	4.3	3.3	3.2	6.1	4.7	4.0	3.5	2.3	2.8
				Woolong tea	B	6.7	1.7	3.8	3.0	2.1	6.2	5.4	4.8	6.4	5.0	2.4	1.6	1.9	9.1	3.6	2.6	2.2	6.8	4.4
			GC-MS/MS	Green tea	A	3.8	4.6	2.6	2.3	7.4	4.3	8.0	6.4	4.8	2.3	5.7	2.0	9.3	3.4	1.2	2.0	4.1	4.6	4.6
				Green tea	B	3.7	7.9	5.6	7.0	5.0	3.1	3.3	5.7	4.7	9.7	5.6	8.7	3.2	1.4	2.5	5.0	8.9	3.2	5.7
				Woolong tea	A	4.9	3.8	1.3	3.5	5.5	2.9	9.8	5.0	1.1	2.3	5.0	5.3	3.6	4.4	9.3	5.5	7.6	9.1	9.3
				Woolong tea	B	3.2	8.1	7.6	8.5	1.8	3.7	2.4	1.2	7.8	3.6	2.8	3.2	4.4	3.1	7.1	1.4	6.8	2.7	1.2
		Envi-Carb+PSA	GC-MS	Green tea	A	13.7	4.2	4.4	4.7	8.5	2.6	6.9	4.4	5.7	0.8	7.9	2.1	1.6	3.0	3.1	5.5	8.8	5.0	7.1
				Green tea	B	6.1	9.5	3.0	3.4	0.5	4.7	7.2	6.7	4.1	2.7	19.7	2.7	4.0	3.5	10.6	7.9	4.3	9.2	8.2
				Woolong tea	A	1.5	4.8	2.9	3.2	2.3	0.8	2.2	1.6	2.7	1.0	6.8	6.3	0.5	2.9	1.0	4.1	8.1	6.7	5.7
				Woolong tea	B	4.9	5.3	10.1	5.7	7.4	4.0	2.0	5.0	12.6	2.7	9.0	5.8	3.6	9.6	9.9	2.8	8.4	5.6	7.0
			GC-MS/MS	Green tea	A	10.1	5.2	4.2	6.2	7.2	2.9	8.4	4.0	5.2	2.1	10.6	1.5	7.2	3.5	5.1	4.1	6.8	5.4	3.9
				Green tea	B	6.8	10.9	5.0	5.2	1.2	2.4	5.6	2.6	5.6	2.5	17.3	3.1	6.7	8.2	12.2	5.5	4.1	13.8	3.6
				Woolong tea	A	3.2	2.0	6.1	6.5	5.7	2.8	4.4	1.3	2.5	1.5	5.8	6.2	3.4	4.0	8.3	4.6	5.3	6.7	12.1
				Woolong tea	B	5.9	1.1	11.1	2.5	7.2	6.4	1.5	5.1	13.3	3.5	5.7	5.2	4.1	7.2	9.8	0.9	3.8	2.3	3.5
50	phenanthrene	Cleanert TPT	GC-MS	Green tea	A	3.9	5.0	2.6	1.3	7.8	5.1	5.1	2.4	6.1	3.4	4.4	1.1	6.4	2.7	2.6	2.3	3.8	6.7	4.8
				Green tea	B	6.4	8.5	3.8	7.7	4.3	6.1	2.3	2.5	3.6	9.1	3.4	6.7	2.3	2.9	3.9	1.3	4.4	1.5	5.7
				Woolong tea	A	4.0	2.3	0.4	2.9	2.3	1.3	8.3	4.1	2.0	3.3	2.4	2.7	3.2	4.1	7.3	2.5	3.6	0.9	0.3
				Woolong tea	B	6.2	2.3	5.0	4.2	1.9	3.7	4.3	3.0	6.6	3.3	0.4	0.7	3.3	3.5	3.4	1.1	0.7	1.4	1.7
			GC-MS/MS	Green tea	A	3.8	7.4	1.6	1.8	3.8	1.2	1.8	6.2	7.8	4.5	4.2	9.3	4.2	11.3	5.9	13.0	13.1	8.0	7.1
				Green tea	B	5.4	9.4	5.0	3.9	9.2	2.0	5.6	4.5	5.3	1.8	3.9	16.7	8.7	4.4	4.8	7.3	7.4	6.6	5.0
				Woolong tea	A	5.6	1.1	1.8	5.2	4.1	2.1	7.1	2.9	3.4	6.0	7.0	6.8	13.2	0.3	15.8	1.2	6.3	7.5	8.7
				Woolong tea	B	4.1	5.1	3.5	3.4	2.8	4.3	4.2	2.9	12.4	3.7	6.9	4.2	6.0	6.5	9.3	5.8	4.3	1.1	4.9
		Envi-Carb+PSA	GC-MS	Green tea	A*	6.3	4.8	1.7	8.8	6.5	2.8	4.8	1.0	9.2	2.7	7.0	3.0	3.0	1.8	4.0	6.2	14.3	3.6	4.3
				Green tea	B*	3.3	6.4	1.7	3.7	1.9	1.3	2.7	4.4	2.4	1.4	21.5	1.9	2.2	3.3	12.3	4.8	5.7	9.2	5.3
				Woolong tea	A	3.9	2.8	2.8	4.3	4.2	2.7	2.4	1.9	2.5	1.7	4.2	3.4	1.2	3.8	3.0	2.1	5.0	5.1	6.5
				Woolong tea	B	5.4	2.8	8.3	4.5	9.0	2.6	3.1	5.7	10.7	3.4	7.1	4.1	2.8	7.0	8.0	2.7	5.7	4.3	3.3
			GC-MS/MS	Green tea	A*	5.5	3.7	1.2	3.9	2.5	3.1	2.6	1.4	5.9	4.7	8.5	1.9	6.0	18.0	4.5	11.3	16.4	8.8	0.9
				Green tea	B*	3.6	7.9	5.6	2.1	2.8	4.3	1.4	3.4	0.9	3.3	24.0	2.7	2.6	10.2	13.7	5.9	2.1	14.7	1.4
				Woolong tea	A	6.2	4.9	5.8	3.3	6.3	2.1	4.8	4.3	6.9	5.0	11.4	9.9	10.2	7.0	7.7	4.3	9.4	15.6	9.4
				Woolong tea	B	6.1	4.8	8.6	5.0	10.4	3.4	3.3	7.0	10.5	0.3	3.2	16.5	8.0	11.1	12.0	7.0	5.7	3.5	4.7
51	pirimicarb	Cleanert TPT	GC-MS	Green tea	A	6.6	2.2	3.6	1.4	9.0	4.0	6.8	6.2	5.8	3.6	8.3	7.7	7.5	3.8	3.3	1.8	5.4	4.9	5.0
				Green tea	B	3.4	13.7	5.0	8.9	4.3	6.1	1.4	4.1	4.2	7.2	8.6	14.0	3.0	1.4	4.7	3.4	5.0	0.5	4.7
				Woolong tea	A	5.6	1.6	3.4	4.8	9.7	4.9	13.1	0.6	3.3	6.9	14.5	2.7	1.5	3.3	8.0	2.6	3.6	6.2	1.5
				Woolong tea	B	9.8	1.4	4.2	2.2	8.4	3.8	2.8	1.3	8.2	5.5	14.7	1.3	2.8	3.8	3.4	0.1	1.0	1.2	3.6
			GC-MS/MS	Green tea	A*	4.2	5.3	3.1	1.7	7.0	4.3	7.9	7.8	4.3	3.3	5.5	2.8	6.3	8.0	3.0	1.0	4.1	6.1	3.2
				Green tea	B*	4.2	8.6	7.3	7.1	6.2	10.3	3.6	8.3	3.2	9.9	6.3	6.3	3.2	0.7	5.1	1.7	8.9	3.3	5.6
				Woolong tea	A	5.7	0.6	0.3	6.5	0.8	4.1	9.5	1.1	2.8	5.1	5.1	9.4	6.1	1.8	9.9	1.2	4.1	1.2	8.5
				Woolong tea	B	1.2	3.7	4.2	5.5	1.3	3.8	3.5	6.5	5.1	2.6	7.2	2.1	7.4	2.9	6.0	1.0	2.1	1.2	4.2
		Envi-Carb+PSA	GC-MS	Green tea	A*	8.6	4.7	3.8	5.2	7.8	4.3	3.3	4.4	2.3	5.9	3.8	1.6	2.8	5.9	4.7	10.7	9.9	6.7	5.2
				Green tea	B*	4.4	8.0	5.7	3.2	2.3	9.9	5.2	2.8	5.9	18.2	29.4	4.5	8.6	2.8	15.1	2.1	4.4	12.0	8.3
				Woolong tea	A	6.3	6.6	3.6	3.7	1.9	2.1	26.6	8.4	33.0	2.2	9.2	7.4	1.8	3.4	3.9	2.9	4.7	5.5	3.1
				Woolong tea	B	3.7	4.2	13.0	2.8	14.9	5.5	4.0	8.7	11.0	3.4	8.1	4.9	4.5	8.1	11.0	3.4	7.5	3.7	2.9
			GC-MS/MS	Green tea	A	8.5	5.9	4.0	7.6	7.5	1.8	3.2	3.8	3.5	3.7	8.3	8.0	4.7	10.8	4.5	8.3	8.1	4.2	6.3
				Green tea	B	4.7	9.6	4.7	5.4	3.0	2.8	3.6	1.2	4.0	4.2	22.1	0.3	2.0	16.3	16.8	3.1	2.7	14.4	4.9
				Woolong tea	A	2.7	3.4	2.6	7.1	5.9	2.5	5.6	1.5	3.6	2.8	6.2	1.3	8.4	4.4	10.3	2.2	7.6	4.4	3.8
				Woolong tea	B	5.9	3.3	9.3	3.3	11.4	6.6	4.5	3.9	11.6	4.0	2.3	7.0	7.6	7.1	11.7	3.5	6.3	2.7	4.1

(Continued)

Annex 2: RSD data for 201 pesticides in incurred tea Youden pair samples under 8 conditions with two SPE cartridge cleanup in three months (Nov.9, 2009-Feb.7, 2010) (cont.)

No	Pesticides	SPE	Method	Sample	Youden pair	Nov.9, 2009 (n=5)	Nov.14, 2009 (n=3)	Nov.19, 2009 (n=3)	Nov.24, 2009 (n=3)	Nov.29, 2009 (n=3)	Dec.4, 2009 (n=3)	Dec.9, 2009 (n=3)	Dec.14, 2009 (n=3)	Dec.14, 2009 (n=3)	Dec.19, 2009 (n=3)	Dec.24, 2009 (n=3)	Dec.14, 2009 (n=3)	Jan.3, 2010 (n=3)	Jan.8, 2010 (n=3)	Jan.13, 2010 (n=3)	Jan.18, 2010 (n=3)	Jan.23, 2010 (n=3)	Jan.28, 2010 (n=3)	Feb.2, 2010 (n=3)	Feb.7, 2010 (n=3)
		Cleanert TPT	GC-MS	Green tea	A	5.7	5.8	2.3	1.1	6.9	4.5	6.1	4.6	5.2	4.6	3.9	5.1	11.4	9.0	5.2	2.1	1.7	4.2	7.0	4.4
				Woolong tea	B	5.9	6.6	5.5	7.4	3.7	3.9	3.7	5.9	3.0	5.9	6.0	3.7	7.1	2.1	8.6	3.4	1.6	4.2	2.4	4.9
			GC-MS/MS	Green tea	A	6.2	3.5	0.8	2.9	2.0	2.7	8.9	4.7	1.3	4.7	3.6	20.0	9.4	12.1	6.8	8.3	2.4	3.4	3.0	1.3
				Woolong tea	B*	8.3	3.8	5.4	3.6	1.9	4.7	5.2	0.8	6.7	0.8	2.9	12.1	12.1	4.7	4.0	2.0	0.8	2.2	0.6	2.6
		Envi-Carb+PSA	GC-MS	Green tea	A	4.4	7.2	1.5	2.4	5.5	4.7	7.8	6.4	5.7	6.4	2.0	5.1	1.8	6.6	0.9	1.7	1.7	7.4	5.4	4.6
				Woolong tea	B	4.2	7.1	5.7	7.6	3.6	3.0	4.2	4.8	3.4	4.8	7.4	3.8	7.0	1.4	3.9	3.4	1.7	7.1	3.4	6.9
			GC-MS/MS	Green tea	A	7.7	2.7	1.8	4.6	3.7	3.6	9.2	2.1	0.9	2.1	3.4	3.2	2.2	4.9	2.3	7.9	4.1	2.5	4.0	3.6
				Woolong tea	B	6.1	2.7	3.9	4.1	0.6	1.1	4.4	4.1	6.2	4.1	2.9	4.1	0.4	2.1	4.5	6.5	1.6	0.9	1.1	2.5
52	procymidone	Cleanert TPT	GC-MS	Green tea	A	5.3	4.2	1.0	5.9	8.5	1.5	6.3	0.1	5.8	0.1	3.0	7.9	11.7	5.7	2.3	14.5	7.6	10.5	3.9	5.0
				Woolong tea	B	4.4	7.0	2.8	4.1	1.4	1.3	3.8	6.7	3.3	3.8	2.3	19.4	5.0	9.0	9.2	8.6	1.8	4.6	8.8	5.5
			GC-MS/MS	Green tea	A*	4.1	3.8	3.0	3.8	6.3	2.7	2.6	0.8	2.8	2.6	0.4	9.1	2.1	4.5	26.3	4.8	2.7	6.5	3.8	4.8
				Woolong tea	B*	5.7	2.0	9.0	5.4	7.9	3.9	3.0	6.1	10.8	6.1	2.8	8.8	7.0	13.6	8.9	9.0	3.8	6.0	4.8	5.4
		Envi-Carb+PSA	GC-MS	Green tea	A	4.5	4.3	1.7	6.5	8.1	0.8	2.9	0.5	5.4	0.5	1.7	10.4	0.6	4.2	11.9	6.8	5.9	8.1	3.1	3.2
				Woolong tea	B	2.6	8.9	3.1	4.8	2.9	2.1	2.7	2.7	4.4	2.7	1.5	20.9	2.1	1.7	7.8	13.2	2.0	4.2	12.8	3.9
			GC-MS/MS	Green tea	A	5.5	1.6	3.5	4.6	7.1	2.1	5.7	2.5	4.3	2.5	1.3	3.8	6.4	3.9	3.5	10.5	2.8	2.9	4.2	5.8
				Woolong tea	B	5.9	0.8	9.4	2.6	7.9	5.6	3.1	5.9	4.3	5.9	3.7	2.2	5.3	1.5	8.7	11.1	1.3	3.1	4.0	3.8
53	prometrye	Cleanert TPT	GC-MS	Green tea	A	8.8	10.5	5.0	2.6	9.6	3.1	7.5	4.4	3.6	4.4	3.7	10.4	7.3	7.6	0.7	3.7	2.5	5.0	6.2	4.6
				Woolong tea	B	6.9	7.0	11.6	0.9	6.6	4.5	2.5	1.5	5.3	1.5	9.3	14.1	14.9	2.6	6.3	3.0	7.5	3.1	1.7	4.4
			GC-MS/MS	Green tea	A*	5.9	2.8	0.7	4.0	2.6	3.3	5.4	6.0	3.7	6.0	5.1	12.7	4.2	6.1	1.8	8.2	6.1	3.6	1.3	1.8
				Woolong tea	B*	9.3	0.9	1.1	2.2	4.8	3.3	7.0	6.5	7.7	6.5	4.7	13.7	1.5	2.4	3.3	4.8	1.4	1.8	0.4	3.4
		Envi-Carb+PSA	GC-MS	Green tea	A*	5.0	4.5	2.4	2.3	7.1	8.1	8.5	6.4	6.8	6.4	0.9	4.5	2.8	5.1	4.9	3.0	3.9	6.8	8.3	6.5
				Woolong tea	B*	4.6	7.0	7.7	6.7	5.6	12.8	3.9	4.6	6.5	4.6	13.3	4.6	4.0	1.5	4.6	5.6	5.1	6.3	3.6	7.9
			GC-MS/MS	Green tea	A	10.0	1.8	2.5	5.7	0.6	3.5	9.8	3.4	1.7	3.4	3.6	3.1	2.4	4.5	4.1	8.3	2.9	3.5	3.5	4.2
				Woolong tea	B	7.3	6.7	4.5	6.2	1.9	0.5	5.0	7.0	5.8	7.0	2.8	6.2	0.6	5.5	1.9	6.7	0.5	1.9	1.9	7.3
54	propham	Cleanert TPT	GC-MS	Green tea	A*	6.3	6.3	3.3	6.7	8.3	5.0	4.0	4.1	5.6	6.1	5.0	5.8	6.9	2.9	7.3	12.7	5.3	4.9	8.9	6.2
				Woolong tea	B*	7.3	9.5	6.5	6.8	4.4	1.3	3.5	3.7	6.1	32.0	18.7	30.5	2.9	1.5	7.3	4.6	5.2	5.6	2.1	7.5
			GC-MS/MS	Green tea	A	6.6	3.2	3.3	1.5	0.4	4.6	3.3	7.7	32.0	13.1	2.5	8.9	8.2	5.7	8.1	11.6	7.6	6.4	5.8	3.3
				Woolong tea	B	4.4	2.7	11.4	3.6	14.0	2.0	6.0	0.9	13.1	3.8	6.0	9.8	4.9	6.7	11.1	7.5	7.6	8.9	3.6	6.0
		Envi-Carb+PSA	GC-MS	Green tea	A*	4.5	4.3	2.1	5.4	8.0	6.4	3.1	1.3	3.8	4.0	2.9	11.3	3.9	1.8	8.3	7.5	2.9	8.9	3.6	6.1
				Woolong tea	B*	2.8	10.9	3.1	6.4	2.4	1.3	6.9	2.0	4.0	2.7	2.9	22.6	2.6	5.0	8.3	23.9	13.0	1.8	12.4	5.0
			GC-MS/MS	Green tea	A	4.8	2.5	3.0	4.0	8.5	5.6	5.1	2.3	2.7	3.0	0.7	4.8	8.5	8.0	1.5	11.7	11.7	4.2	1.9	3.9
				Woolong tea	B	5.7	4.0	11.2	2.3	8.4	5.6	5.8	6.2	12.3	3.6	3.0	0.9	3.6	8.0	9.9	10.3	6.9	8.3	3.8	5.9

54	propham	Cleanert TPT	GC-MS	Green tea	A	6.6	6.3	3.4	4.8	2.2	3.2	8.6	7.2	9.0	8.1	8.1	6.6	1.8	3.5	13.4	6.3	23.2	12.0	10.9	4.9
				Woolong tea	B	6.6	4.0	1.7	4.8	2.6	2.0	5.7	5.1	9.2	5.1	1.4	25.3	2.7	23.7	0.5	29.3	3.3	4.8	12.2	5.2
			GC-MS/MS	Green tea	A*	4.0	4.9	7.5	3.7	6.7	1.3	14.8	14.3	9.2	14.3	7.4	7.2	6.9	1.4	4.0	3.5	2.5	7.8	5.7	1.0
				Woolong tea	B*	5.2	6.2	3.7	6.2	4.0	5.9	3.5	9.2	9.2	9.2	6.0	5.0	7.9	2.8	6.6	8.4	14.8	3.8	2.7	14.8
		Envi-Carb+PSA	GC-MS	Green tea	A	9.4	10.3	4.0	5.4	1.7	10.0	3.5	4.8	5.5	4.8	3.4	6.7	0.5	3.5	4.8	4.3	14.8	7.3	11.1	7.2
				Woolong tea	B	6.1	2.8	4.4	4.3	5.3	1.5	2.2	10.0	2.9	10.0	4.2	28.6	2.3	14.2	9.3	24.1	2.3	1.7	14.5	7.2
			GC-MS/MS	Green tea	A	8.5	2.8	4.4	4.3	5.3	1.5	2.2	10.0	2.9	4.6	2.3	4.4	8.9	4.6	1.7	8.3	3.4	8.1	8.7	5.2
				Woolong tea	B	10.5	3.6	8.7	4.4	10.0	4.3	3.1	5.4	12.2	5.4	1.2	1.2	5.8	2.2	7.2	8.1	1.0	4.7	2.4	2.4

No.	Compound	Cartridge	Method	Tea																			
55	prosulfocarb	Cleanert TPT	GC-MS	Green tea	A	5.8	6.4	3.4	1.6	7.9	4.4	6.9	3.4	4.9	2.3	7.1	2.6	9.8	2.8	2.0	1.9	3.4	7.7	1.9
					B	5.8	5.5	5.3	9.2	2.3	3.0	2.8	2.1	5.6	9.1	6.2	11.0	2.4	2.8	4.2	3.3	1.9	5.0	2.1
				Woolcng tea	A	4.8	2.6	2.9	2.9	2.9	1.3	8.1	6.3	1.9	3.8	9.9	3.1	3.7	2.9	8.0	8.5	3.0	5.9	4.0
					B	7.5	4.0	6.2	5.3	3.3	2.5	1.8	2.8	7.9	2.2	8.1	1.6	1.1	1.7	2.7	0.7	1.5	1.3	1.7
			GC-MS/MS	Green tea	A	4.9	5.6	4.2	2.3	6.6	4.1	6.4	2.8	6.8	2.7	5.9	3.4	5.9	7.4	1.8	2.4	3.9	8.5	4.1
					B	5.3	8.0	6.0	7.0	5.4	5.9	3.6	3.8	2.3	11.6	5.9	9.7	1.9	0.5	5.1	4.0	7.7	3.6	9.1
				Woolcng tea	A	4.1	2.6	1.9	6.1	3.5	3.9	7.1	1.3	2.1	4.7	2.2	1.9	4.9	3.2	10.4	7.1	4.8	8.8	0.8
					B	4.2	2.0	4.0	6.8	0.6	2.3	3.0	3.7	4.9	4.4	4.6	1.0	5.3	3.5	3.5	2.6	3.7	3.4	4.0
		Envi-Carb+PSA	GC-MS	Green tea	A*	6.5	4.7	2.6	6.1	6.7	3.0	14.8	8.2	5.7	4.4	7.4	3.8	3.4	4.4	5.8	5.7	3.7	6.3	4.0
					B*	4.5	4.9	6.3	4.6	3.2	2.7	3.8	4.2	3.4	12.7	25.2	4.6	6.5	2.7	10.7	7.5	11.5	9.8	5.6
				Woolcng tea	A	5.0	7.6	4.9	3.7	2.5	2.0	3.2	0.2	22.7	1.9	6.3	2.9	2.1	3.4	3.5	2.8	4.8	3.6	9.5
					B	6.0	3.2	6.6	3.9	11.4	5.1	2.3	6.6	11.3	3.8	9.1	1.3	6.1	16.9	11.0	3.5	5.7	5.6	7.9
			GC-MS/MS	Green tea	A	4.6	4.6	2.8	6.3	5.9	0.7	4.4	2.0	3.2	2.4	8.5	3.9	4.6	8.9	1.3	8.0	6.3	4.1	2.6
					B	1.9	10.3	4.7	6.0	3.7	2.1	2.3	3.2	1.1	3.7	21.2	4.9	3.1	4.5	14.3	2.6	8.3	10.1	3.9
				Woolcng tea	A	4.6	4.5	5.2	4.7	6.8	2.0	5.8	3.1	2.9	1.2	4.1	9.6	4.7	1.9	9.1	3.7	6.9	2.9	3.7
					B	6.2	4.4	10.0	2.7	10.3	6.1	2.9	4.8	11.3	3.3	1.0	3.3	7.3	7.3	11.2	2.8	5.9	2.6	2.9
56	secbumeton	Cleanert TPT	GC-MS	Green tea	A	4.3	6.3	4.1	3.5	12.5	4.8	5.9	5.7	4.8	6.3	11.1	6.7	8.6	3.7	3.9	3.2	5.8	8.8	4.4
					B	4.3	5.1	4.8	8.8	6.9	4.9	3.2	4.4	5.8	8.1	14.2	11.9	3.3	3.1	7.7	0.6	7.0	1.7	3.8
				Woolcng tea	A	5.9	5.5	5.4	8.0	1.4	3.3	10.0	6.1	2.4	9.4	19.4	0.8	3.7	3.7	7.4	3.2	3.8	1.5	1.3
					B	10.9	2.3	3.8	1.8	3.9	2.5	4.0	3.0	8.7	10.1	10.8	1.7	4.6	3.0	4.2	0.1	2.0	1.3	2.8
			GC-MS/MS	Green tea	A*		5.4	0.8	1.8	9.8	9.7	7.5	5.8	2.6	2.7	6.4	4.0	8.2	8.3	1.3	3.4	8.7	8.7	7.9
					B*		9.1	6.1	7.2	6.6	13.8	3.2	7.0	5.3	9.7	6.8	7.0	4.1	1.5	3.8	5.6	8.7	4.0	4.7
				Woolcng tea	A	4.2	0.7	1.3	6.5	0.9	3.4	9.0	2.3	4.6	4.7	5.8	1.9	6.4	6.6	7.8	0.8	2.2	3.9	6.1
					B	2.2	3.4	2.9	7.2	4.5	2.4	3.2	5.2	4.9	3.0	7.0	1.9	11.9	2.2	6.1	2.8	1.7	4.0	3.7
		Envi-Carb+PSA	GC-MS	Green tea	A*	9.6	4.0	2.1	8.8	12.5	5.2	4.6	1.1	5.6	9.9	3.5	7.4	3.7	4.4	8.3	7.8	11.5	3.8	6.5
					B*	4.7	8.9	6.8	12.3	9.7	8.5	4.7	5.4	8.0	15.6	34.7	2.5	5.1	5.1	15.0	3.0	7.1	9.6	7.9
				Woolcng tea	A	4.0	6.2	7.7	5.3	3.0	3.1	1.9	5.8	26.0	2.5	18.4	8.3	3.2	4.0	2.9	3.2	7.0	2.9	2.8
					B	3.6	6.3	11.3	6.2	18.7	7.2	4.2	7.9	8.5	5.2	6.0	5.7	7.5	10.3	10.7	3.8	7.0	1.5	6.5
			GC-MS/MS	Green tea	A	4.4	3.8	1.2	8.1	5.9	4.6	4.5	0.4	3.6	1.6	6.9	3.0	3.1	4.4	12.0	9.5	7.8	4.6	4.1
					B	1.8	9.8	5.4	8.3	3.1	6.5	1.6	1.6	5.7	2.6	22.0	3.5	3.7	11.9	17.5	4.6	2.6	12.3	9.7
				Woolcng tea	A	4.4	4.4	3.4	4.5	7.4	2.0	5.7	3.3	2.8	3.1	6.2	12.3	3.1	2.9	13.5	2.7	6.9	5.7	5.5
					B	4.3	2.3	10.0	1.9	13.4	5.9	3.7	6.1	13.6	2.4	1.1	6.9	7.8	10.3	12.5	1.5	4.4	3.9	2.2
57	silafluofen	Cleanert TPT	GC-MS	Green tea	A	12.2	7.7	8.7	0.2	7.2	10.5	13.4	3.1	7.1	2.9	1.9	2.1	10.2	13.9	7.5	6.1	2.4	12.9	3.9
					B	8.4	2.1	1.7	5.8	4.5	8.0	3.4	1.7	10.3	8.9	9.3	3.1	9.8	9.3	3.4	6.6	13.8	3.7	4.3
				Woolcng tea	A	4.5	7.5	3.1	5.9	8.3	6.9	5.7	5.6	7.8	5.3	3.7		1.8	2.6	3.9	10.1	5.0	8.1	2.4
					B	7.3	4.4	2.1	11.4	12.4	11.2	4.3	8.5	2.7	11.9	0.1	1.6	2.3	5.6	4.4	2.4	7.3	2.6	2.4
			GC-MS/MS	Green tea	A	5.5	5.5	0.7	2.8	5.1	6.8	8.1	2.9	5.5	4.5	7.5	9.1	3.3	9.5	5.4	1.9	4.0	7.0	2.3
					B	3.2	6.0	5.1	8.6	4.9	0.6	1.7	1.7	2.9	7.5	7.1	3.2	3.0	11.4	5.6	3.9	7.7	7.0	6.1
				Woolcng tea	A	8.8	2.3	4.9	6.3	3.5	2.1	3.5	0.7	1.5	4.4	2.2	3.8	2.3	4.5	4.7	6.4	10.3	3.4	3.4
					B	6.9	0.8	4.2	5.2	1.3	2.8	4.2	3.8	2.8	4.3	6.0	1.0	3.8	2.6	4.3	7.4	4.0	3.5	9.4
		Envi-Carb+PSA	GC-MS	Green tea	A	5.9	21.4	6.8	12.8	11.4	3.9	0.4	5.1	8.7	7.0	11.2	5.5	2.3	4.5	6.8	6.2	13.2	1.6	2.2
					B	9.2	7.5	6.3	11.5	1.1	3.5	10.2	6.1	6.3	3.8	9.9	9.2	32.0	3.6	15.3	3.8	6.7	5.6	11.0
				Woolang tea	A*	6.0	3.4	11.2	10.1	6.0	1.5	3.5	7.7	1.1	10.4	5.7	8.7	38.4	2.9	4.6	2.3	16.7	11.5	2.4
					B	9.1	3.1	7.1	3.9	14.4	4.9	4.1	6.1	9.4	1.4	5.4	11.7	13.7	6.0	16.0	1.5	10.8	8.1	3.5
			GC-MS/MS	Green tea	A	5.0	3.4	1.5	7.8	7.3	2.7	6.4	1.7	8.1	2.7	5.4	5.7	14.6	2.0	9.1	11.3	10.8	8.1	5.1
					B	1.5	10.7	5.5	10.7	3.5	2.2	2.1	4.1	5.4	2.4	22.0	1.1	4.6	15.7	14.4	4.2	5.6	7.3	9.2
				Woolang tea	A	7.4	0.3	5.9	1.5	7.1	7.0	4.2	7.0	1.6	3.1	4.0	7.4	10.9	7.1	2.7	2.7	3.2	14.2	7.5
					B	7.3	2.0	8.6	4.4	9.6	4.5	4.4	7.0	12.1	1.5	1.0	4.4	11.0	2.5	8.0	2.1	1.6	5.0	1.6

(Continued)

Annex 2: RSD data for 201 pesticides in incurred tea Youden pair samples under 8 conditions with two SPE cartridge cleanup in three months (Nov.9, 2009-Feb.7, 2010) (cont.)

No	Pesticides	SPE	Method	Sample	Youden pair	Nov.9, 2009 (n=5)	Nov.14, 2009 (n=3)	Nov.19, 2009 (n=3)	Nov.24, 2009 (n=3)	Nov.29, 2009 (n=3)	Dec.4, 2009 (n=3)	Dec.9, 2009 (n=3)	Dec.14, 2009 (n=3)	Dec.19, 2009 (n=3)	Dec.24, 2009 (n=3)	Dec.14, 2009 (n=3)	Jan.3, 2010 (n=3)	Jan.8, 2010 (n=3)	Jan.13, 2010 (n=3)	Jan.18, 2010 (n=3)	Jan.23, 2010 (n=3)	Jan.28, 2010 (n=3)	Feb.2, 2010 (n=3)	Feb.7, 2010 (n=3)
58	tebupirimfos	Cleanert TPT	GC-MS	Green tea	A	3.7	4.6	4.0	3.0	11.9	3.7	5.2	0.6	4.8	4.7	9.3	4.7	7.0	0.7	1.9	2.2	3.5	7.8	4.4
					B	5.4	8.1	3.9	11.5	3.4	3.5	1.9	2.4	5.5	10.3	9.5	12.7	2.5	1.9	3.3	1.3	4.3	1.0	5.8
				Woolong tea	A	4.7	1.6	3.7	5.1	8.3	3.8	7.7	4.7	3.8	6.1	10.7	3.9	4.0	2.2	7.8	3.1	3.4	2.5	1.2
					B	7.3	2.9	4.1	5.3	7.1	5.6	2.3	0.9	9.1	3.5	7.1	1.6	2.7	1.3	2.3	0.8	1.6	2.2	3.2
			GC-MS/MS	Green tea	A*		4.1	5.7	4.0	9.2	6.2	7.5	4.9	6.4	2.7	6.1	4.5	8.6	4.1	1.7	8.7	1.9	3.8	2.0
					B*		8.6	7.3	8.5	5.0	11.4	3.5	5.9	8.9	13.2	6.0	10.0	1.7	2.7	6.0	5.7	9.3	5.3	9.5
				Woolong tea	A	4.1	1.2	3.2	6.8	3.0	4.0	10.1	3.5	2.6	3.2	5.1	5.5	1.7	3.4	7.6	3.2	3.8	13.7	0.7
					B	3.8	1.5	6.1	7.0	1.0	1.7	4.4	6.0	1.9	1.8	8.2	3.2	1.4	7.8	6.1	5.9	4.3	7.3	4.9
		Envi-Carb+PSA	GC-MS	Green tea	A	11.5	2.9	9.1	6.0	6.7	10.2	7.9	2.5	5.4	5.6	8.2	2.6	4.1	3.6	5.9	7.0	11.1	4.3	5.2
					B	8.3	7.1	6.1	1.0	4.9	10.6	25.0	6.0	2.9	20.2	8.9	5.0	1.8	4.7	12.9	3.3	5.8	9.6	5.6
				Woolong tea	A	7.1	4.0	3.2	1.8	13.0	1.4	5.9	1.0	30.4	1.8	26.7	4.3	1.6	4.8	2.4	2.9	5.1	3.4	5.3
					B	6.1	4.8	8.8	1.4	18.3	7.8	4.7	7.9	9.7	4.0	7.3	4.0	3.8	7.6	8.7	4.2	7.0	3.5	4.7
			GC-MS/MS	Green tea	A	6.8	5.9	1.8	7.7	5.4	1.7	2.2	1.2	2.1	4.7	7.8	2.1	5.7	11.5	5.3	9.1	13.5	9.6	1.2
					B	2.6	7.2	5.6	4.5	3.7	5.8	2.2	0.6	4.5	4.0	11.1	1.9	3.7	6.7	13.4	1.9	6.1	11.5	4.2
				Woolong tea	A	6.6	4.0	6.1	8.2	7.2	2.8	5.1	4.1	0.7	1.8	19.8	9.3	5.1	4.0	8.8	2.4	3.7	3.9	8.0
					B	6.2	4.7	8.0	4.4	11.0	5.0	3.4	6.0	10.3	1.9	4.7	6.9	3.1	8.7	8.0	2.8	7.9	9.4	4.8
59	tebutam	Cleanert TPT	GC-MS	Green tea	A	4.9	6.0	3.4	1.8	8.4	5.4	6.5	3.6	5.2	2.4	6.1	1.3	7.0	3.3	5.1	2.3	4.9	8.8	5.1
					B*	5.9	8.4	4.8	8.7	1.4	2.7	3.3	3.4	2.7	7.7	5.4	7.0	1.9	2.4	7.3	1.8	5.1	2.4	5.4
				Woolong tea	A	5.3	2.1	1.0	4.6	5.5	2.6	8.2	3.4	2.7	3.2	4.8	3.7	3.8	2.9	7.7	2.9	4.4	2.6	0.5
					B	5.0	3.2	5.3	9.3	2.8	4.3	5.6	0.4	5.5	2.3	3.0	1.6	2.8	3.5	3.2	0.3	0.3	2.2	2.5
			GC-MS/MS	Green tea	A	3.7	5.6	2.1	3.8	6.8	3.8	8.1	5.6	5.2	2.6	5.2	3.2	7.6	4.5	2.9	2.5	6.1	6.3	7.2
					B	4.2	8.3	6.5	7.4	6.9	5.0	4.0	4.4	3.3	8.1	7.8	8.8	2.3	3.3	3.9	6.0	5.6	4.6	7.1
				Woolong tea	A	3.5	0.8	2.1	7.0	0.9	2.6	9.8	5.0	3.6	5.1	2.8	2.7	3.4	3.7	8.2	2.1	5.6	1.7	5.5
					B	4.7	1.5	4.4	6.6	2.1	1.8	4.5	2.8	5.4	3.9	3.5	1.2	6.7	2.1	6.3	2.7	1.9	0.1	2.4
		Envi-Carb+PSA	GC-MS	Green tea	A	8.4	3.3	2.5	6.7	7.6	2.5	4.8	1.5	6.0	2.7	9.4	2.9	4.2	5.7	5.8	7.3	10.7	3.4	6.1
					B*	4.0	6.8	4.4	4.2	3.3	3.0	4.5	4.6	3.4	6.6	20.8	1.5	2.6	6.0	13.4	3.2	5.1	7.4	5.4
				Woolong tea	A	4.5	4.2	2.3	3.7	3.5	1.5	3.4	1.3	15.9	1.0	5.0	3.2	1.6	2.6	2.9	3.4	5.6	4.3	4.8
					B	4.9	2.6	8.4	4.0	10.1	4.1	3.2	5.5	9.3	3.8	8.9	1.7	3.4	7.2	9.4	3.3	6.1	4.9	6.4
			GC-MS/MS	Green tea	A	4.7	5.0	2.9	6.3	5.4	1.0	3.0	1.2	3.5	1.7	11.5	1.3	3.9	12.9	4.7	8.3	7.0	4.5	4.0
					B	2.8	7.7	4.4	6.4	2.2	1.2	1.9	1.6	0.9	3.2	21.9	0.4	1.9	6.9	17.1	1.2	3.4	12.5	3.5
				Woolong tea	A	5.1	3.4	4.0	5.8	7.2	1.1	4.0	2.0	1.8	0.4	5.7	8.8	6.1	7.0	8.5	2.4	3.9	4.6	7.3
					B	5.8	3.0	9.5	2.3	9.0	4.7	1.8	6.0	13.3	2.4	1.7	6.3	6.0	7.5	9.8	1.5	5.0	2.9	1.7
60	tebuthiuron	Cleanert TPT	GC-MS	Green tea	A*	7.9	7.6	2.8	9.2	2.0	2.7	8.0	3.4	4.7	5.8	9.0	3.7	8.4	3.9	3.1	6.6	6.1	6.7	3.9
					B*	8.9	8.7	4.5	6.8	3.1	4.9	4.3	9.4	6.5	13.4	9.8	7.9	4.4	3.2	8.9	3.8	8.2	1.5	2.6
				Woolong tea	A	7.4	4.4	5.4	5.5	8.7	6.9	9.7	11.7	0.7	2.9	4.4	4.3	3.3	7.0	7.7	5.1	3.7	5.0	1.7
					B	13.0	5.4	2.0	9.1	5.4	4.7	5.4	4.4	3.5	1.2	5.9	3.3	1.7	5.8	1.0	1.1	2.6	1.9	2.4
			GC-MS/MS	Green tea	A*	6.4	1.8	2.2	0.6	3.5	2.9	10.5	6.3	6.7	7.7	3.5	0.3	8.4	6.1	5.4	2.2	7.0	7.5	4.1
					B*	2.9	9.8	4.8	4.4	6.0	7.1	3.8	12.6	6.1	15.8	8.5	8.5	2.8	3.5	11.0	7.5	10.9	0.9	4.6
				Woolong tea	A	5.9	11.2	4.6	3.9	6.6	3.1	9.6	5.9	5.1	9.2	3.4	2.9	5.8	3.0	7.6	2.8	3.6	7.7	5.8
					B	4.9	3.2	6.5	2.6	2.8	1.1	5.3	4.7	3.9	2.9	6.3	5.1	6.2	3.2	4.6	1.9	1.9	2.5	2.0
		Envi-Carb+PSA	GC-MS	Green tea	A*	7.9	12.2	7.7	6.8	15.5	7.2	34.4	1.9	4.3	5.9	9.8	4.1	3.1	2.2	8.4	9.9	12.2	3.8	7.2
					B*	12.0	5.0	8.0	5.9	11.9	2.9	5.8	2.8	4.9	16.6	27.0	7.7	8.0	2.4	18.5	3.0	5.0	11.0	8.6
				Woolong tea	A	7.3	8.8	5.7	3.0	2.6	5.1	5.0	4.7	18.1	12.6	7.7	11.9	4.3	5.6	7.0	3.6	4.8	0.3	4.7
					B	4.7	4.9	2.8	9.9	19.2	9.8	2.7	8.5	17.9	6.7	9.2	1.3	5.8	22.1	11.6	4.6	6.5	8.0	6.4
			GC-MS/MS	Green tea	A*	8.7	16.3	14.8	7.7	8.7	5.5	36.0	2.4	1.7	3.6	12.7	9.3	2.7	10.0	6.0	9.9	9.8	9.0	5.1
					B*	4.3	6.7	8.4	8.7	3.4	6.9	3.3	1.4	6.1	3.3	24.4	13.6	3.0	13.5	19.0	0.6	2.4	14.3	7.9
				Woolong tea	A*	4.4	3.8	5.0	5.6	9.3	3.5	6.9	1.0	3.8	12.6	7.9	13.4	4.7	3.2	13.2	4.2	5.5	1.3	5.8
					B	5.9	3.5	4.7	5.0	12.7	10.1	2.6	7.4	9.8	4.4	2.8	7.3	4.3	20.4	11.5	3.2	6.0	4.9	5.1

No.	Pesticide	Cartridge	Method	Tea																					
61	tefluthrin	Cleanert TPT	GC-MS	Green tea	A	5.7	5.3	4.1	1.8	8.5	5.8	6.4	2.9	6.7	1.7	5.8	1.1	7.4	14.2	8.8	2.5	12.9	7.8	4.1
					B	6.9	7.8	5.8	8.6	3.3	2.2	1.5	0.6	4.9	11.0	4.5	7.8	1.1	1.5	10.3	2.1	9.1	1.7	4.8
				Woolong tea	A	4.0	3.3	2.3	4.9	3.9	3.0	6.9	3.3	2.8	2.5	4.1	3.9	3.8	3.5	5.9	3.7	4.8	2.8	4.4
					B	7.0	2.4	4.1	5.5	2.3	3.8	3.9	1.7	6.1	2.2	3.6	0.7	2.8	3.1	2.8	0.7	0.4	2.5	2.9
			GC-MS/MS	Green tea	A	4.4	6.6	3.2	2.9	7.2	3.8	8.6	4.8	6.8	3.7	5.7	3.1	7.4	6.4	5.1	5.2	3.4	11.0	6.8
					B	5.6	9.3	5.9	7.8	4.7	5.3	5.3	2.4	2.6	9.9	4.8	8.2	0.9	3.9	9.4	1.3	9.4	5.1	9.1
				Woolong tea	A	3.4	5.1	2.1	5.0	1.8	3.9	7.6	0.8	3.1	5.2	2.7	3.9	1.5	4.2	3.4	2.5	0.4	1.2	5.1
					B	5.1	1.9	3.8	5.4	1.9	2.3	3.0	4.6	5.0	3.1	3.2	1.5	4.3	1.7	5.4	1.9	2.3	1.1	4.9
		Envi-Carb+PSA	GC-MS	Green tea	A	6.8	4.5	2.0	5.7	7.6	2.1	5.0	2.0	7.0	1.8	10.4	6.6	5.1	5.5	13.1	7.3	10.9	3.9	5.2
					B	5.2	6.0	5.4	4.3	2.8	3.1	5.1	4.5	2.8	5.6	19.0	1.4	4.4	5.8	2.5	3.1	5.6	8.2	8.9
				Woolong tea	A	3.8	2.5	2.3	4.1	5.1	1.3	3.0	0.2	6.6	2.2	7.8	3.1	0.7	3.8	10.1	3.1	5.5	3.8	9.3
					B	5.1	2.4	8.4	3.6	10.3	4.8	3.8	7.5	9.9	2.7	7.0	3.4	5.2	7.2	6.2	3.8	7.2	7.7	10.3
			GC-MS/MS	Green tea	A	6.8	6.1	2.7	5.7	6.1	0.5	4.3	1.1	5.3	1.8	13.3	1.3	3.3	13.3	13.2	8.7	8.6	5.3	2.2
					B	3.1	8.8	4.4	5.3	3.1	1.7	3.0	1.9	1.8	1.8	20.4	2.8	1.8	10.0	8.7	2.2	5.6	12.5	4.3
				Woolong tea	A	4.5	5.0	3.6	4.9	6.6	1.7	6.0	2.9	1.6	1.3	5.1	8.5	8.0	4.7	8.8	2.4	5.5	6.4	5.4
					B	6.6	2.0	8.8	1.8	9.4	6.8	3.3	6.6	12.3	2.1	3.2	5.1	7.6	10.6	2.7	1.4	5.8	3.0	4.8
62	thenylchlor	Cleanert TPT	GC-MS	Green tea	A	9.9	6.9	2.2	8.2	6.5	5.9	9.9	8.3	9.3	2.2	5.1	7.0	8.3	9.6	10.2	4.9	4.4	13.4	4.7
					B	8.9	8.4	8.6	6.3	7.0	0.4	2.9	4.0	4.2	8.2	12.2	7.1	7.1	11.1	8.2	3.2	5.7	2.2	7.9
				Woolong tea	A	8.7	6.5	1.2	6.8	7.6	2.1	15.2	1.7	2.1	6.5	1.9	5.1	5.4	3.9	4.6	5.3	10.1	2.9	4.2
					B	8.1	6.2	3.8	1.1	4.6	6.4	8.5	9.8	4.5	4.4	3.4	0.7	7.9	5.6	3.6	2.2	4.8	1.2	5.8
			GC-MS/MS	Green tea	A	9.2	7.5	3.7	5.6	3.0	3.2	4.9	5.4	3.1	2.8	5.8	3.4	13.2	10.9	2.5	8.5	12.6	4.7	1.6
					B	6.3	7.6	2.7	5.1	3.5	3.5	1.1	7.6	4.4	9.7	5.6	3.9	2.9	6.8	6.5	2.4	5.0	8.7	8.5
				Woolong tea	A*	7.3	6.5	8.4	1.8	16.1	6.6	14.4	10.3	5.3	3.7	9.6	3.8	16.1	7.0	8.5	4.7	5.0	2.8	10.8
					B	6.2	4.7	8.8	7.0	3.9	4.4	9.6	5.3	9.7	8.4	1.7	6.1	4.4	4.2	6.0	2.0	3.0	3.6	10.9
		Envi-Carb+PSA	GC-MS	Green tea	A*	10.0	13.7	5.4	6.7	7.3	5.6	2.8	5.4	2.0	4.6	6.0	2.6	6.1	3.3	8.9	10.0	4.4	2.9	10.7
					B	11.1	11.4	3.9	2.0	10.2	3.7	4.7	4.6	5.9	10.2	17.4	3.5	6.1	1.1	6.6	5.2	2.5	18.8	5.8
				Woolong tea	A*	7.8	5.6	6.5	3.9	5.3	6.4	3.4	5.1	3.3	2.7	6.6	2.0	6.6	10.3	14.1	6.9	0.7	3.0	12.3
					B	12.9	5.1	12.5	7.0	2.8	0.5	3.2	9.2	14.5	10.6	9.1	4.8	3.9	10.1	1.3	2.4	8.1	6.0	18.0
			GC-MS/MS	Green tea	A*	15.3	10.2	5.7	4.3	6.8	9.4	15.5	2.3	5.0	4.7	25.7	4.1	16.0	1.9	9.4	8.7	15.3	0.8	2.1
					B*	48.4	14.5	6.4	2.3	7.3	7.4	3.5	5.8	8.3	9.4	6.1	7.9	15.8	17.6	11.7	11.9	2.0	10.0	9.6
				Woolong tea	A	7.5	1.8	6.6	6.1	5.4	6.2	6.3	4.5	8.5	9.1	5.8	13.6	9.2	6.7	8.3	3.0	6.2	5.7	3.2
					B	12.1	4.0	14.8	9.0	8.2	5.8	1.7	8.1	14.9	1.8	3.5	5.2	9.1	6.5	8.3	1.9	8.2	8.3	6.8
63	thionazin	Cleanert TPT	GC-MS	Green tea	A	5.5	5.1	0.7	3.3	12.0	5.1	6.2	3.5	5.3	1.2	7.2	4.7	5.3	3.1	3.4	4.4	7.1	5.7	4.2
					B	7.1	7.8	7.4	5.1	6.8	1.2	1.3	2.8	2.9	6.2	5.8	10.5	1.9	5.2	8.8	4.1	5.9	1.9	5.3
				Woolong tea	A*	5.2	1.6	2.8	8.3	4.1	2.3	8.1	2.8	3.8	4.0	5.1	2.8	6.5	3.1	4.0	1.7	5.4	1.1	2.5
					B*	8.6	4.8	8.9	9.1	2.6	2.2	4.4	6.6	8.3	2.3	5.3	1.3	4.5	3.2	3.1	0.9	3.4	0.7	3.5
			GC-MS/MS	Green tea	A	3.1	5.8	1.8	2.9	7.6	5.6	5.7	5.0	4.0	1.8	4.9	3.8	3.2	4.6	2.9	1.0	3.4	6.7	3.9
					B	4.8	8.6	5.5	6.0	5.2	13.1	4.1	5.2	2.0	7.5	5.5	6.4	4.6	1.9	5.8	2.5	8.5	2.1	7.5
				Woolong tea	A*	4.1	1.1	0.9	5.6	2.0	2.8	10.1	4.8	2.0	5.0	3.4	3.4	0.6	1.5	7.5	4.7	2.7	1.6	3.8
					B*	2.4	2.7	4.2	3.6	1.4	2.4	3.3	0.8	1.8	3.4	3.7	1.3	4.0	1.5	2.6	1.5	2.5	3.0	1.7
		Envi-Carb+PSA	GC-MS	Green tea	A*	7.6	5.8	3.5	6.2	3.5	2.9	9.8	2.2	4.4	3.9	4.6	1.9	1.5	2.7	17.6	9.7	10.9	3.0	1.7
					B*	7.5	4.9	6.3	4.9	6.7	4.2	15.1	0.8	5.7	15.6	23.9	1.7	2.7	0.7	4.4	2.0	5.4	6.2	4.1
				Woolong tea	A	8.4	4.1	3.9	7.1	5.0	5.8	2.2	1.7	5.8	2.8	7.3	5.0	5.8	8.0	5.5	2.5	3.4	8.5	4.8
					B	3.3	4.1	7.6	1.7	12.6	4.2	2.3	7.9	10.5	3.3	7.8	0.7	3.2	6.2	5.6	3.8	4.6	6.0	4.8
			GC-MS/MS	Green tea	A	6.6	5.4	2.9	5.9	4.9	0.7	1.1	2.0	4.8	1.1	10.6	7.4	3.8	5.6	16.8	7.8	9.9	0.9	1.5
					B	7.4	9.6	4.5	4.1	3.7	2.6	3.1	1.1	2.5	2.4	18.2	7.5	4.3	6.4	8.4	2.2	4.0	5.8	2.8
				Woolong tea	A	4.7	3.4	5.7	5.2	5.5	1.7	4.1	2.0	2.2	1.7	4.6	6.1	7.1	2.7	9.5	2.6	4.7	4.6	5.4
					B	6.6	3.2	8.2	3.6	8.7	4.6	1.9	5.1	11.5	0.9	0.7	3.7	2.2	6.8		1.1	4.4	2.3	5.2

(Continued)

Annex 2: RSD data for 201 pesticides in incurred tea Youden pair samples under 8 conditions with two SPE cartridge cleanup in three months (Nov.9, 2009–Feb.7 2010) (cont.)

No	Pesticides	SPE	Method	Sample	Youden pair	Nov.9 2009 (n=5)	Nov.14 2009 (n=3)	Nov.19 2009 (n=3)	Nov.24 2009 (n=3)	Nov.29 2009 (n=3)	Dec.4 2009 (n=3)	Dec.9 2009 (n=3)	Dec.14 2009 (n=3)	Dec.14 2009 (n=3)	Dec.19 2009 (n=3)	Dec.24 2009 (n=3)	Jan.3 2010 (n=3)	Jan.8 2010 (n=3)	Jan.13 2010 (n=3)	Jan.18 2010 (n=3)	Jan.23 2010 (n=3)	Jan.28 2010 (n=3)	Feb.2 2010 (n=3)	Feb.7 2010 (n=3)
64	trichloronat	Cleanert TPT	GC-MS	Green tea	A	3.8	4.8	3.6	1.3	9.9	3.5	6.0	6.3	5.6	4.2	7.2	5.9	6.9	4.9	2.5	1.4	2.4	4.8	4.1
				Green tea	B	5.7	8.0	5.7	10.1	4.8	2.8	1.0	3.2	4.4	3.9	6.7	10.8	1.3	1.6	3.6	2.3	4.8	0.5	5.5
				Woolong tea	A*	4.5	1.7	2.6	4.8	1.7	2.8	11.3	1.7	0.9	3.7	5.9	1.8	2.9	2.3	7.4	3.0	3.6	1.8	0.5
				Woolong tea	B	7.4	0.5	3.8	3.1	5.7	4.1	3.1	5.5	5.3	6.2	10.7	0.3	1.9	2.4	5.1	0.2	1.4	1.2	3.6
			GC-MS/MS	Green tea	A*	3.2	6.4	3.8	0.7	8.5	5.3	8.0	5.7	5.7	7.3	4.4	3.1	5.2	4.9	0.8	2.1	3.5	4.4	4.7
				Green tea	B*	6.4	8.7	5.5	6.3	4.5	7.4	2.2	3.5	3.5	3.0	4.1	8.1	4.2	4.0	3.1	3.7	7.4	1.6	7.3
				Woolong tea	A*	5.3	3.3	1.1	7.4	2.9	2.9	9.6	2.3	3.4	1.0	4.3	2.0	5.5	2.1	9.5	3.9	3.8	0.2	2.5
				Woolong tea	B	3.0	2.2	3.2	5.6	2.4	2.2	3.5	4.0	4.0	6.0	4.8	2.0	6.6	3.6	6.2	0.7	4.6	0.8	3.5
		Envi-Carb+PSA	GC-MS	Green tea	A*	11.7	3.9	6.6	4.4	6.7	4.2	7.0	6.3	3.2	3.4	4.1	4.2	1.2	3.8	1.5	5.8	9.0	7.5	5.2
				Green tea	B*	8.1	9.7	6.4	0.9	3.2	5.5	5.8	3.2	3.6	3.4	4.1	5.8	7.6	3.8	1.8	3.2	3.8	13.6	7.0
				Woolong tea	A	4.6	6.7	3.0	3.4	1.7	1.0	1.8	3.6	3.6	23.0	2.8	4.3	0.8	3.3	9.1	4.3	6.5	6.1	4.1
				Woolong tea	B	5.5	3.0	9.4	0.8	13.5	1.0	1.8	7.5	7.5	10.2	9.1	2.1	3.9	7.2	1.8	3.8	8.0	0.6	4.9
			GC-MS/MS	Green tea	A	12.3	5.6	3.6	4.7	6.3	4.5	7.2	5.6	5.6	4.6	9.3	0.0	4.5	2.6	4.0	4.3	5.7	8.3	5.1
				Green tea	B	9.6	10.8	3.5	3.5	4.3	2.3	6.5	4.1	4.1	6.1	13.5	0.0	6.1	9.0	13.4	4.7	6.4	14.4	2.3
				Woolong tea	A	3.2	3.3	4.0	6.5	4.3	1.3	4.9	4.8	4.8	2.3	3.7	5.2	3.5	3.2	8.9	2.4	4.9	5.3	4.2
				Woolong tea	B	6.9	2.7	8.9	1.7	9.9	6.9	2.5	6.2	6.2	11.0	2.8	4.9	2.1	10.5	10.7	1.7	6.1	3.7	3.5
65	trifluralin	Cleanert TPT	GC-MS	Green tea	A*	3.8	5.3	4.0	3.8	13.0	3.6	6.3	10.3	10.3	5.0	6.2	4.4	10.3	4.8	3.6	2.3	4.2	14.4	3.9
				Green tea	B*	5.3	8.3	4.5	4.7	3.9	3.8	2.2	6.2	6.2	5.8	5.8	9.3	8.1	6.9	2.6	11.9	13.7	24.0	7.3
				Woolong tea	A	3.8	1.6	3.3	4.7	12.6	5.6	8.6	4.2	4.2	4.8	14.2	5.1	4.5	2.9	1.0	10.4	6.9	2.7	0.8
				Woolong tea	B	7.7	1.8	3.8	5.3	8.0	6.7	4.2	5.7	5.7	6.8	5.9	3.3	3.6	4.2	11.5	5.8	0.7	2.1	4.3
			GC-MS/MS	Green tea	A*	2.8	7.8	4.0	4.8	16.4	10.0	8.4	8.7	8.7	9.4	4.3	4.9	7.8	6.1	1.1	4.8	0.9	6.7	3.6
				Green tea	B*	3.0	8.3	5.0	7.4	3.6	10.5	1.5	7.8	7.8	4.0	5.2	10.4	4.6	11.3	2.5	4.5	13.1	4.6	13.2
				Woolong tea	A*	2.5	4.4	1.1	7.3	8.7	4.1	8.7	0.2	0.2	6.3	7.2	4.7	5.7	4.2	3.3	3.9	5.1	7.4	6.3
				Woolong tea	B*	17.7	8.6	2.0	3.7	6.9	2.7	4.9	10.1	10.1	4.9	9.4	3.5	5.5	1.8	2.4	4.3	3.9	6.0	1.8
		Envi-Carb+PSA	GC-MS	Green tea	A*	12.4	3.2	9.1	5.6	7.0	15.6	6.4	12.8	12.8	8.1	6.1	6.3	1.7		2.9	13.9	16.8	8.2	7.6
				Green tea	B*	11.1	7.6	6.0	2.5	5.5	10.4	11.6	10.6	10.6	10.0	6.1	7.0	19.0	7.5	13.7	5.0	21.6	16.8	9.0
				Woolong tea	A	7.5	3.4	3.2	2.4	21.1	1.5	2.7	14.3	14.3	15.8	9.3	3.5	9.9	13.0	3.1	21.9	10.2	7.2	2.7
				Woolong tea	B	6.2	5.0	8.3	1.6	18.3	8.7	6.7	10.5	10.5	8.8	1.0	1.0	5.8	8.4	14.4	5.5	15.1	5.4	4.8
			GC-MS/MS	Green tea	A	11.6	2.9	6.6	6.4	9.8	6.0	9.1	15.4	15.4	11.3	2.1	12.9	3.5	4.4	7.3	5.8	7.2	8.1	5.0
				Green tea	B	13.5	9.9	4.6	6.6	7.2	5.0	17.2	11.8	11.8	10.8	3.2	4.0	8.4	16.1	7.5	5.8	7.1	16.9	8.0
				Woolong tea	A	7.7	2.5	6.1	3.9	5.6	1.3	4.7	9.6	9.6	0.6	2.6	7.4	5.2	0.7	10.2	5.9	13.8	29.0	10.6
				Woolong tea	B	6.0	2.7	9.3	5.3	12.5	8.8	0.8	8.1	8.1	13.9	3.0	1.7	7.1	7.6	1.8	4.2	11.4	10.4	7.6
66	2,4'-DDT	Cleanert TPT	GC-MS	Green tea	A*	2.5	3.9	5.3	11.0	13.4	1.4	9.0	8.5	8.5	9.0	3.2	4.5	8.1	8.3	1.8	2.3	6.0	12.5	4.3
				Green tea	B*	8.4	7.2	3.5	13.6	7.7	3.5	5.0	8.9	8.9	1.5	7.8	4.4	7.6	10.6	4.3	4.7	14.3	5.0	6.8
				Woolong tea	A	7.5	3.9	3.5	4.9	7.4	6.6	12.5	14.8	14.8	9.8	3.6	12.9	5.1	1.6	6.1	10.0	4.9	3.2	6.3
				Woolong tea	B	6.0	2.4	7.9	4.6	2.3	4.2	7.0	2.6	2.6	7.9	9.1	0.6	2.3	10.6	12.7	0.3	6.0	6.8	8.6
			GC-MS/MS	Green tea	A*	9.5	6.1	3.8	7.6	11.5	2.7	14.1	11.0	11.0	14.8	2.0	7.6	7.5	6.3	4.0	8.0	9.0	11.0	1.5
				Green tea	B*	8.2	8.2	2.9	12.1	7.5	2.3	3.7	11.2	11.2	3.9	7.7	5.5	4.8	4.6	4.6	9.8	12.5	2.0	6.4
				Woolong tea	A*	2.7	8.8	3.6	6.9	8.9	4.2	0.9	1.9	1.9	5.9	4.0	11.5	8.7	3.1	5.7	7.5	14.2	9.5	6.0
				Woolong tea	B*	4.5	9.2	7.2	10.8	4.2	1.4	12.8	11.9	11.9	17.2	11.0	3.9	7.0	13.8	13.8	1.7	5.9	6.3	8.4
		Envi-Carb+PSA	GC-MS	Green tea	A	18.7	7.8	13.1	5.0	9.5	20.8	3.4	13.7	13.7	5.5	19.2	18.8	0.4	7.5	1.4	6.9	12.1	12.7	8.8
				Green tea	B	19.1	6.9	4.8	8.7	12.3	7.7	14.4	12.6	12.6	13.7	15.1	14.3	28.3	4.0	10.3	6.3	4.4	18.0	14.0
				Woolong tea	A	15.9	10.1	7.5	20.4	5.5	15.5	11.1	17.5	17.5	10.2	7.1	17.1	19.6	3.5	4.1	23.7	5.9	21.0	8.6
				Woolong tea	B	15.7	13.3	6.1	4.3	18.5	17.3	17.2	8.1	8.1	18.5	18.6	5.0	6.3	10.6	19.1	6.0	23.3	39.3	4.9
			GC-MS/MS	Green tea	A	21.9	5.3	14.1	5.0	4.6	18.4	13.8	16.3	16.3	7.4	18.7	7.3	4.8	26.0	3.7	4.8	6.1	4.9	2.3
				Green tea	B	36.9	8.2	4.4	17.4	5.0	7.3	21.9	13.4	13.4	13.5	17.5	2.3	15.8	12.1	10.6	6.1	11.9	9.6	15.1
				Woolong tea	A	2.7	8.8	3.6	14.7	17.4	16.5	8.8	20.6	20.6	6.9	16.3	15.2	19.1	9.1	9.8	7.5	0.9	17.2	1.3
				Woolong tea	B	4.5	9.2	7.2	10.8	16.4	19.4	34.9	5.5	5.5	23.7	8.9	1.3	5.8	19.9	19.0	1.7	20.1	28.6	8.3

No.	Compound	Cartridge	Method	Tea																				
67	4,4'-DDE	Cleanert TPT	GC-MS	Green tea	A	6.2	3.9	2.9	3.3	7.3	4.7	6.8	3.2	3.5	2.8	6.0	2.0	6.4	2.7	2.6	2.1	3.6	7.5	4.0
					B	5.5	8.5	5.0	8.4	3.3	4.0	3.1	1.7	5.7	8.6	3.3	7.1	1.4	1.8	5.3	2.2	4.7	2.4	5.5
				Woolong tea	A	7.0	3.3	0.5	3.6	3.3	2.2	10.7	4.8	1.8	2.1	2.1	2.7	6.1	2.1	8.6	3.0	3.7	3.2	0.2
					B	4.8	4.2	3.3	3.6	0.7	2.8	4.7	1.2	5.3	3.9	2.4	2.7	2.3	3.2	4.3	1.2	1.6	2.4	1.2
			GC-MS/MS	Green tea	A	4.8	3.4	3.7	1.6	9.0	2.1	6.2	4.0	3.7	0.8	5.0	3.2	10.6	2.2	3.9	3.4	2.7	5.5	4.3
					B	7.6	8.9	7.7	8.6	7.0	5.5	8.7	3.5	6.1	9.2	6.3	7.9	1.0	4.4	2.0	6.4	4.9	3.6	5.6
				Woolong tea	A*	7.0	5.0	2.2	3.6	3.3	0.4	7.0	0.6	2.9	4.7	3.0	3.0	9.6	4.0	8.8	6.1	4.3	3.8	1.5
					B*	7.5	4.7	7.5	5.6	4.7	4.3	5.1	5.1	3.7	3.5	5.9	1.4	2.9	2.4	5.9	2.8	6.1	3.4	7.4
		Envi-Carb+PSA	GC-MS	Green tea	A	5.5	5.2	0.6	5.4	8.2	2.0	127.5	1.2	6.0	1.9	10.1	4.3	3.3	6.2	6.2	6.9	10.9	1.5	5.4
					B	3.1	7.1	3.9	4.8	2.4	0.7	27.0	5.6	3.9	1.4	21.4	1.7	1.3	3.6	11.7	1.8	5.3	6.3	5.7
				Woolong tea	A	6.5	4.1	1.8	2.6	7.1	0.6	4.3	6.8	1.4	3.4	4.7	3.5	3.6	3.5	2.5	1.0	5.6	4.0	4.8
					B	5.0	1.7	10.0	4.3	9.4	4.2	1.4	3.5	11.1	3.6	7.8	3.0	1.6	3.2	9.4	3.4	5.9	8.3	3.2
			GC-MS/MS	Green tea	A	6.7	8.1	3.2	5.9	8.7	1.4	3.1	4.1	6.6	1.0	13.2	5.3	3.4	13.0	6.2	9.3	10.7	2.2	9.3
					B	4.2	9.3	7.9	1.5	3.4	1.9	1.3	0.3	3.3	0.5	19.3	2.5	4.6	7.9	13.9	2.9	2.7	8.7	6.9
				Woolong tea	A	4.4	5.1	3.8	3.0	8.9	1.9	1.0	6.2	1.9	3.3	5.3	3.5	2.0	1.8	9.2	6.1	8.5	1.9	6.1
					B	6.6	1.5	9.3	4.3	5.7	6.5	5.6	8.9	9.8	3.3	0.2	4.1	5.2	9.0	9.9	2.8	3.8	1.7	3.0
68	benalaxyl	Cleanert TPT	GC-MS	Green tea	A	5.9	4.6	0.8	1.3	9.0	5.5	6.6	5.3	5.2	2.9	4.6	3.6	5.7	5.4	1.7	2.1	1.5	12.8	2.6
					B	5.9	9.8	6.7	8.8	2.6	3.7	2.3	6.5	3.8	5.3	4.1	6.8	7.6	1.6	5.5	9.8	10.2	15.0	4.9
				Woolong tea	A	10.4	1.2	1.0	2.9	1.9	4.1	11.2	5.1	7.2	4.5	7.1	1.7	4.1	2.7	0.8	2.4	7.5	3.6	6.3
					B	4.6	1.5	1.7	4.7	1.8	3.8	3.5	6.9	6.5	4.2	5.7	1.1	3.7	5.9	8.1	8.6	5.2	8.7	8.6
			GC-MS/MS	Green tea	A	4.2	7.8	3.2	4.0	14.1	3.8	9.0	5.4	8.2	3.0	7.9	4.9	1.4	3.9	4.0	8.9	9.5	6.1	3.9
					B	7.5	7.6	7.0	7.6	2.3	7.3	6.9	1.9	1.6	7.2	5.1	1.8	2.9	6.8	6.0	6.0	6.4	0.8	3.6
				Woolong tea	A	5.8	1.8	3.6	2.9	3.2	1.4	8.9	4.9	8.3	6.2	3.4	2.0	4.8	4.6	4.2	3.5	11.7	4.5	5.4
					B	2.2	3.2	3.9	5.9	3.9	1.7	0.8	7.3	2.1	5.2	6.2	3.5	6.8	6.8	6.1	3.7	2.0	7.4	8.6
		Envi-Carb+PSA	GC-MS	Green tea	A	17.9	5.7	3.4	5.9	8.2	3.5	5.4	4.5	8.6	2.1	5.8	6.2	4.6	7.0	6.1	6.4	5.5	22.9	5.7
					B	5.9	7.2	6.0	2.4	2.4	1.3	12.3	10.0	1.8	5.2	21.0	5.5	7.2	1.3	3.0	12.2	9.4	14.5	11.3
				Woolong tea	A	4.3	1.2	3.2	1.2	5.2	2.6	4.0	8.8	12.2	1.2	4.7	0.8	10.3	2.7	13.5	4.0	4.1	11.1	7.6
					B	4.8	3.3	9.0	3.3	11.7	6.2	8.4	5.3	5.3	2.4	8.4	2.1	0.6	4.3	7.3	4.8	11.9	5.4	5.8
			GC-MS/MS	Green tea	A	13.5	7.9	3.8	2.6	9.3	10.1	7.1	5.6	7.8	3.5	8.2	6.9	3.9	9.8	19.4	6.0	8.9	1.7	1.8
					B	16.1	8.8	3.9	6.3	2.4	2.7	18.5	7.1	0.8	5.3	11.7	11.6	6.3	13.2	8.0	10.3	7.7	20.4	16.0
				Woolong tea	A	5.2	3.3	4.9	1.1	5.1	3.9	1.2	6.9	11.7	2.6	9.8	4.2	3.5	4.8	10.6	3.4	11.0	3.1	3.5
					B	8.8	4.5	10.5	5.8	4.2	4.5	5.3	6.0	4.2	3.7	2.0	5.3	1.7	8.9	9.2	3.7	6.6	2.8	6.8
69	benzoylprop-ethyl	Cleanert TPT	GC-MS	Green tea	A	5.0	6.2	1.5	2.2	7.4	3.8	7.1	4.0	2.5	1.6	3.8	2.3	2.6	2.2	16.9	0.4	4.8	3.3	3.1
					B	4.9	5.5	3.5	9.7	3.2	2.5	0.3	4.6	2.0	3.3	3.2	6.1	5.6	5.9	1.7	3.7	5.8	1.5	4.0
				Woolong tea	A	6.9	1.5	1.3	1.3	2.5	2.9	7.0	2.4	8.1	0.4	3.5	2.7	3.3	1.4	4.8	3.1	2.9	4.3	0.4
					B	4.4	2.1	3.2	7.1	1.6	2.6	5.3	8.6	5.3	4.5	0.9	2.5	0.9	5.8	7.9	1.6	2.7	1.2	1.0
			GC-MS/MS	Green tea	A	4.7	4.8	1.5	1.6	9.0	3.7	9.1	6.4	2.3	1.8	6.5	4.4	3.8	8.7	1.8	2.8	3.5	2.0	4.1
					B	5.7	5.1	5.6	8.7	6.7	4.4	6.7	1.0	3.0	4.3	5.6	4.1	2.6	3.8	4.1	4.1	6.5	2.5	3.2
				Woolong tea	A	8.2	2.1	2.6	1.4	2.8	3.2	6.1	4.6	2.3	5.3	1.7	3.2	4.8	1.7	2.5	3.1	1.3	6.6	4.6
					B	3.0	3.4	4.7	6.9	5.6	3.8	1.9	3.3	7.2	4.4	6.9	2.7	3.7	9.0	4.7	2.5	1.7	1.6	1.5
		Envi-Carb+PSA	GC-MS	Green tea	A	8.9	3.5	5.6	7.0	8.0	4.3	5.4	1.8	5.9	1.8	6.3	2.7	1.6	2.7	4.4	5.4	10.3	4.2	5.3
					B	6.6	6.3	5.0	3.7	1.4	2.9	10.7	6.3	1.3	4.8	15.7	5.2	4.9	4.8	3.0	5.3	3.1	10.8	8.5
				Woolong tea	A	4.5	1.8	4.4	0.2	5.9	2.0	2.1	5.6	10.4	0.8	6.2	0.6	7.4	3.6	8.7	2.3	5.0	5.6	3.4
					B	5.4	3.6	8.9	4.3	10.9	4.6	4.1	3.9	8.6	1.1	8.7	2.9	5.4	8.2	2.0	3.5	5.5	1.0	4.1
			GC-MS/MS	Green tea	A	9.1	3.9	6.8	7.6	7.8	6.7	4.0	1.5	8.2	1.9	6.2	15.8	6.7	8.4	11.3	5.4	9.8	2.2	3.7
					B	9.9	6.7	6.4	4.5	2.0	1.9	12.1	6.2	1.0	3.3	14.4	5.2	1.0	4.1	5.9	4.8	7.6	12.9	9.0
				Woolong tea	A	6.4	2.6	5.0	1.6	4.6	1.5	0.0	6.2	9.9	2.2	8.4	3.1	6.5	5.6	11.4	3.1	6.7	4.4	2.7
					B	8.4	3.9	8.9	5.0	6.1	3.6	2.0	3.2	3.2	3.2	1.5	3.1	2.1	6.6	9.1	2.5	5.8	3.6	6.1

(Continued)

Annex 2: RSD data for 201 pesticides in incurred tea Youden pair samples under 8 conditions with two SPE cartridge cleanup in three months (Nov.9, 2009-Feb.7, 2010) (cont.)

No	Pesticides	SPE	Method	Sample	Youden pair	Nov.9, 2009 (n=5)	Nov.14, 2009 (n=3)	Nov.19, 2009 (n=3)	Nov.24, 2009 (n=3)	Nov.29, 2009 (n=3)	Dec.4, 2009 (n=3)	Dec.9, 2009 (n=3)	Dec.14, 2009 (n=3)	Dec.19, 2009 (n=3)	Dec.24, 2009 (n=3)	Dec.29, 2009 (n=3)	Jan.3, 2010 (n=3)	Jan.8, 2010 (n=3)	Jan.13, 2010 (n=3)	Jan.18, 2010 (n=3)	Jan.23, 2010 (n=3)	Jan.28, 2010 (n=3)	Feb.2, 2010 (n=3)	Feb.7, 2010 (n=3)
70	bromofos	Cleanert TPT	GC-MS	Green tea	A*	5.9	3.6	3.5	3.0	9.3	3.0	6.1	5.5	5.8	4.3	3.6	3.8	3.8	5.8	4.9	3.9	3.8	5.9	3.0
				Green tea	B*	8.7	9.6	4.4	5.5	5.7	4.3	1.4	0.7	2.5	6.0	2.2	4.2	3.8	4.7	9.3	2.4	6.6	6.8	9.5
				Woolong tea	A*	7.7	6.3	2.4	3.8	3.2	4.6	13.5	11.2	5.9	8.4	5.1	1.2	3.2	2.4	9.4	2.5	6.8	4.0	4.3
				Woolong tea	B	5.5	2.5	4.7	4.2	1.2	0.9	3.4	1.8	10.8	5.4	5.6	0.5	2.5	4.4	9.6	4.5	2.8	2.8	3.8
			GC-MS/MS	Green tea	A	5.1	4.7	9.9	4.2	11.5	9.4	12.1	6.2	6.2	6.9	5.1	7.9	5.5	6.7	5.2	6.5	17.7	9.4	9.0
				Green tea	B	8.4	13.1	7.6	8.5	1.2	8.0	9.6	3.3	3.2	6.6	5.8	1.5	2.4	1.8	4.7	9.0	5.3	6.0	9.6
				Woolong tea	A*	11.6	12.4	9.8	4.5	9.5	4.2	16.9	5.1	14.7	12.4	5.9	7.4	4.4	1.6	9.7	3.4	10.2	0.5	5.8
				Woolong tea	B	7.0	4.9	10.7	7.9	11.8	3.6	0.9	0.2	3.5	10.1	3.7	2.2	11.1	5.6	4.6	3.4	10.8	17.8	5.9
		Envi-Carb+PSA	GC-MS	Green tea	A	11.1	6.2	3.7	2.6	7.9	8.3	7.8	5.3	5.9	1.4	15.1	7.8	2.0	2.1	4.9	3.4	9.5	11.3	4.4
				Green tea	B	13.9	7.5	2.2	8.4	2.0	0.8	9.6	1.0	5.8	9.0	14.5	9.8	8.3	3.1	4.4	9.9	2.5	13.6	8.0
				Woolong tea	A	5.8	0.9	4.0	6.4	4.1	2.8	10.0	4.1	11.6	3.7	2.0	2.2	3.1	2.4	6.2	2.2	3.4	7.7	4.5
				Woolong tea	B	9.0	5.8	9.2	6.4	7.9	5.5	5.7	6.5	11.1	0.4	9.6	3.2	6.9	4.4	12.5	4.2	8.2	6.8	4.9
			GC-MS/MS	Green tea	A	16.0	0.6	3.4	7.9	2.1	17.8	6.6	7.3	9.1	4.4	10.1	0.5	7.3	9.0	4.8	5.0	3.3	9.1	12.6
				Green tea	B	32.6	13.1	2.3	13.1	5.2	5.9	13.2	6.5	6.2	15.7	9.8	7.3	11.0	7.8	4.6	7.7	11.9	19.0	12.2
				Woolong tea	A*	3.2	3.4	6.4	0.6	7.0	5.0	3.8	3.5	15.5	2.8	8.5	2.3	8.6	7.5	7.3	9.0	10.5	8.9	8.4
				Woolong tea	B*	6.1	5.8	11.9	11.9	14.1	4.5	8.9	5.1	11.2	4.0	5.3	3.2	9.5	16.2	16.2	3.4	3.6	6.4	3.7
71	bromopropylate	Cleanert TPT	GC-MS	Green tea	A*	4.6	5.3	1.3	5.7	11.9	1.7	13.2	5.1	1.9	2.6	6.9	7.2	8.3	16.2	11.3	3.4	8.6	7.7	3.7
				Green tea	B*	5.6	6.4	2.2	11.9	4.5	3.5	2.9	12.9	6.1	5.9	3.0	10.2	6.1	12.2	5.2	7.3	13.1	12.6	3.6
				Woolong tea	A*	7.3	5.7	4.0	1.8	9.1	4.1	3.5	12.0	7.9	7.6	14.3	10.8	3.3	9.5	4.2	8.4	4.0	2.8	4.0
				Woolong tea	B	4.1	3.1	3.2	4.0	5.4	8.6	15.2	19.1	12.7	4.5	11.6	2.3	0.9	0.2	6.2	6.1	5.5	4.5	12.5
			GC-MS/MS	Green tea	A	6.9	6.4	1.4	7.0	10.8	3.3	13.5	6.5	1.4	1.5	10.3	6.7	1.1	5.5	4.8	1.0	5.8	10.1	5.6
				Green tea	B	7.1	6.5	3.7	9.1	6.8	3.6	2.9	11.4	11.9	6.3	3.8	7.7	5.8	6.5	4.6	4.0	7.7	2.9	5.4
				Woolong tea	A*	8.9	10.9	2.2	1.6	7.4	5.0	3.0	9.8	9.7	6.9	9.8	6.2	2.1	0.4	3.1	5.5	4.5	7.6	3.0
				Woolong tea	B*	6.3	13.9	4.2	6.3	6.8	7.7	9.3	13.1	20.3	2.0	11.1	5.6	4.7	3.0	4.9	6.3	3.1	4.3	16.9
		Envi-Carb+PSA	GC-MS	Green tea	A	21.1	4.5	12.6	9.4	9.4	12.9	49.8	17.2	22.9	4.6	0.9	15.3	1.8	1.6	5.9	2.7	12.6	31.7	6.9
				Green tea	B	15.0	8.7	5.2	5.1	5.1	5.5	6.8	25.0	0.1	3.9	36.5	11.5	42.1	33.6	6.1	46.4	7.3	29.5	14.7
				Woolong tea	A	4.3	3.2	5.6	6.1	14.6	2.1	3.0	31.7	12.5	2.6	5.5	6.9	4.9	5.9	46.1	1.3	5.4	14.3	6.0
				Woolong tea	B	7.2	6.4	8.0	3.8	7.5	7.7	12.5	4.8	7.7	4.5	9.9	1.5	4.3	0.2	0.7	14.4	15.2	8.8	6.7
			GC-MS/MS	Green tea	A	24.5	7.5	11.5	7.0	3.5	10.9	27.5	16.4	19.4	1.7	1.5	2.4	8.7	5.5	12.3	2.5	7.5	14.1	12.2
				Green tea	B	22.7	10.3	7.3	6.5	3.7	5.0	21.8	16.7	2.4	4.6	31.0	1.8	30.1	3.6	4.6	18.7	8.9	17.6	17.3
				Woolong tea	A*	12.2	0.3	4.8	3.7	11.4	1.7	2.3	24.5	8.3	1.5	7.7	7.8	1.4	9.6	17.5	2.3	6.3	8.7	6.3
				Woolong tea	B*	14.5	5.6	7.0	1.0	9.5	4.7	11.0	5.3	2.2	3.8	3.3	2.2	2.4	4.3	8.5	6.3	9.5	12.9	8.2
72	buprofenzin	Cleanert TPT	GC-MS	Green tea	A*	6.1	5.3	4.8	3.6	9.5	6.6	6.4	4.8	6.6	11.5	7.1	7.6	3.5	6.8	3.4	4.3	3.8	23.8	4.6
				Green tea	B*	7.5	8.9	3.5	7.3	3.9	4.8	5.0	2.5	2.2	7.4	2.7	7.1	2.9	5.8	5.5	5.5	14.2	13.8	7.9
				Woolong tea	A*	4.6	6.4	3.6	2.8	3.4	0.5	11.9	4.2	6.6	2.0	2.6	12.8	5.4	10.3	13.9	13.8	5.8	5.2	1.3
				Woolong tea	B	13.4	0.5	3.2	3.1	3.1	2.4	4.8	4.5	1.3	1.3	3.0	12.9	3.6	5.2	2.7	2.4	2.0	3.7	5.2
			GC-MS/MS	Green tea	A	4.9	4.5	2.3	3.7	12.5	3.9	8.3	8.9	3.8	7.6	2.0	8.5	8.5	4.0	1.1	1.2	4.2	8.6	1.8
				Green tea	B	7.7	8.6	9.7	8.9	7.1	5.7	7.8	3.0	5.0	4.4	3.6	2.2	6.0	7.9	5.1	3.7	14.3	6.2	4.2
				Woolong tea	A	7.0	2.6	1.7	2.9	1.4	4.4	9.7	4.3	4.1	5.2	2.0	9.4	8.8	6.3	6.7	5.2	5.8	9.5	2.9
				Woolong tea	B	4.3	1.4	3.5	4.2	2.8	4.2	3.0	7.0	2.5	3.5	7.1	4.5	1.8	14.6	8.3	0.9	2.9	1.2	6.7
		Envi-Carb+PSA	GC-MS	Green tea	A	6.7	7.2	2.8	5.5	8.9	1.3	3.2	2.7	4.5	0.4	9.5	6.7	10.6	16.5	8.0	19.4	23.5	11.3	15.9
				Green tea	B	3.6	4.8	3.6	3.5	3.1	2.1	3.6	7.9	5.2	3.4	33.7	7.1	22.5	5.2	27.0	3.1	4.7	4.5	0.9
				Woolong tea	A	2.4	6.4	6.8	2.5	5.1	0.2	5.8	4.6	2.3	5.5	5.8	9.3	19.8	5.8	8.2	19.5	4.7	6.2	23.2
				Woolong tea	B	3.5	3.8	9.9	6.0	5.3	4.5	5.4	5.9	2.0	2.9	8.3	10.5	4.6	10.3	14.0	4.2	2.9	11.9	5.0
			GC-MS/MS	Green tea	A	5.2	6.5	2.6	5.9	11.6	4.5	3.9	1.8	9.3	0.7	8.6	0.8	11.0	6.6	9.6	11.0	18.6	1.5	6.2
				Green tea	B	5.1	8.6	5.7	1.3	9.2	2.7	0.8	4.9	2.5	2.5	18.2	1.7	9.4	8.1	21.1	2.9	8.1	1.1	9.3
				Woolong tea	A	2.5	3.9	3.2	4.7	8.3	1.6	1.8	4.5	9.7	6.0	5.9	1.6	8.8	12.6	13.0	5.2	7.3	12.2	7.6
				Woolong tea	B	5.6	4.8	8.7	3.7	6.4	9.7	1.9	5.0	10.3	2.8	2.8	5.3	7.2	12.6	10.7	0.9	3.2	14.7	4.9

(Continued)

No.	Pesticide	Cartridge	Method	Tea	Sample																			
73	butachlor	Cleanert TPT	GC-MS	Green tea	A	5.4	3.4	2.6	1.0	6.9	6.2	6.1	5.8	10.8	9.0	5.7	2.5	5.7	3.6	3.5	6.0	2.5	7.9	4.6
				Green tea	B	5.5	8.0	5.3	7.7	2.3	5.5	1.5	1.6	1.2	11.5	0.8	2.3	2.9	3.0	10.3	1.0	6.0	5.2	7.2
				Woolong tea	A	8.0	4.4	0.7	2.1	3.3	2.5	11.1	16.1	1.2	4.0	2.0	1.6	1.3	2.6	8.9	3.4	7.7	3.0	1.8
				Woolong tea	B	4.8	5.3	6.9	5.9	1.0	2.5	3.9	0.9	2.1	5.9	3.4	1.9	2.1	3.0	2.3	0.8	2.5	1.3	5.8
			GC-MS/MS	Green tea	A	5.5	2.4	3.6	1.2	11.1	2.3	0.4	4.6	5.2	4.1	10.0	6.7	6.6	3.9	3.7	3.4	4.0	10.1	1.8
				Green tea	B	7.1	9.4	5.6	8.2	6.4	6.5	8.5	1.9	7.3	6.2	2.5	5.3	1.0	6.0	4.1	0.5	1.8	5.8	7.0
				Woolong tea	A	7.9	10.0	3.3	5.5	3.6	4.7	8.9	1.5	2.6	1.9	2.3	1.0	7.3	2.9	8.0	5.1	8.6	4.0	4.4
				Woolong tea	B	2.5	1.6	4.6	5.3	3.9	4.1	4.8	1.7	2.2	7.0	9.1	3.8	2.6	2.2	9.4	6.0	3.8	3.2	7.6
		Envi-Carb+PSA	GC-MS	Green tea	A	7.7	2.9	3.0	5.7	6.6	3.5	4.4	7.0	3.4	2.2	14.6	2.5	2.4	3.4	2.1	9.0	11.0	12.1	9.5
				Green tea	B	11.6	9.2	5.1	4.6	1.5	2.7	2.9	4.5	2.7	5.7	22.9	4.1	4.4	5.1	10.1	4.7	4.3	12.0	10.3
				Woolong tea	A	4.8	1.5	3.7	3.0	4.6	3.8	4.2	16.8	8.5	4.8	0.5	2.8	2.9	3.3	5.1	4.2	3.8	7.8	2.3
				Woolong tea	B	12.6	1.9	5.6	5.3	10.2	5.4	2.5	7.0	9.3	3.5	8.2	2.5	7.0	4.1	10.7	2.5	9.9	1.8	10.5
			GC-MS/MS	Green tea	A	4.7	6.8	1.2	8.2	6.8	5.2	4.4	3.6	5.6	3.0	8.7	6.3	3.8	14.2	5.7	9.7	10.6	4.0	8.9
				Green tea	B	5.6	8.8	6.7	3.6	3.3	2.2	0.2	8.6	5.8	2.2	20.4	4.6	2.4	12.3	13.0	3.9	8.0	8.6	6.4
				Woolong tea	A	5.4	7.4	2.1	2.2	9.8	1.5	3.8	2.8	2.1	2.5	9.2	6.9	2.9	9.8	13.2	5.3	7.9	2.9	5.2
				Woolong tea	B	7.5	0.7	9.4	4.8	3.2	5.0	4.7	10.8	11.3	5.8	2.6	5.2	5.6	9.4	10.1	6.0	5.5	2.7	4.3
74	butylate	Cleanert TPT	GC-MS	Green tea	A	9.9	5.7	3.2	3.0	12.9	9.3	5.6	3.7	6.4	8.2	6.4	1.8	13.1	5.6	0.4	14.3	5.7	7.1	3.8
				Green tea	B	8.4	9.3	6.1	7.7	3.0	5.1	2.6	14.1	4.2	15.7	0.6	7.9	4.7	10.7	4.6	4.2	4.5	7.6	3.6
				Woolong tea	A	6.2	5.0	21.9	3.2	27.9	7.2	8.3	2.4	34.6	6.2	3.0	7.0	14.8	3.6	12.7	5.8	4.7	2.1	2.8
				Woolong tea	B	5.7	5.9	4.1	6.4	5.0	10.3	7.1	0.9	6.3	4.2	1.7	0.7	12.4	5.5	27.9	1.9	2.4	2.8	37.7
			GC-MS/MS	Green tea	A	2.6	2.3	2.9	2.7	11.0	7.2	6.8	3.3	4.5	6.7	8.5	3.3	5.4	4.7	2.2	5.7	3.0	4.8	4.0
				Green tea	B	9.2	7.7	5.3	6.9	4.9	9.6	5.7	3.2	8.6	14.2	3.2	5.9	7.9	2.5	2.2	4.3	5.9	5.1	4.6
				Woolong tea	A	7.0	1.5	1.0	1.7	4.8	7.6	11.1	0.7	8.0	5.2	3.1	5.8	14.0	3.4	8.5	4.1	6.1	2.2	5.6
				Woolong tea	B	4.3	4.8	6.9	6.7	6.4	0.9	3.6	6.2	1.4	5.4	6.7	3.0	1.9	6.2	9.0	2.7	4.1	1.7	6.7
		Envi-Carb+PSA	GC-MS	Green tea	A	20.2	48.1	31.8	14.2	15.9	5.2	4.3	2.9	7.6	3.5	46.8	19.8	18.7	128.0	28.6	16.3	20.5	7.2	4.4
				Green tea	B	17.9	49.0	55.0	37.6	7.7	20.2	5.8	3.0	1.7	12.6	30.6	6.5	15.9	28.2	33.9	7.4	19.8	7.8	9.8
				Woolong tea	A	33.6	46.1	10.4	6.2	17.1	41.4	1.0	6.9	1.8	1.9	6.9	4.2	6.9	3.6	4.5	12.5	18.0	20.5	10.9
				Woolong tea	B	16.7	16.4	7.9	4.4	13.2	5.9	1.1	46.2	11.6	0.6	9.7	2.8	22.2	5.5	5.7	5.0	7.3	10.1	4.9
			GC-MS/MS	Green tea	A	5.8	10.3	2.7	6.8	5.4	5.1	4.1	1.5	5.8	1.0	9.9	32.3	13.4	14.5	8.7	14.6	11.6	5.9	1.8
				Green tea	B	5.2	6.1	6.9	1.5	1.0	6.3	10.5	2.4	3.0	0.9	25.2	15.5	19.6	20.2	35.5	3.1	10.8	9.4	7.6
				Woolong tea	A	3.5	2.8	4.1	4.8	7.0	0.9	2.3	4.1	2.7	3.6	8.2	2.7	4.6	6.4	7.5	4.1	7.3	4.7	3.8
				Woolong tea	B	10.8	6.9	6.4	1.6	7.0	2.0	6.4	11.4	10.6	2.6	3.2	2.8	2.5	11.6	7.4	2.7	3.2	2.7	4.3
75	carbofenothion	Cleanert TPT	GC-MS	Green tea	A	6.1	4.7	2.5	0.7	5.7	3.4	8.0	6.0	3.1	3.3	9.6	6.3	9.0	2.7	2.6	1.8	3.9	14.5	5.7
				Green tea	B	7.2	7.4	2.4	10.1	7.0	1.0	1.7	2.1	2.7	3.3	8.1	10.3	1.8	0.5	3.5	4.1	5.3	7.7	5.1
				Woolong tea	A	7.7	4.1	5.6	1.9	5.7	3.1	6.0	13.2	3.9	5.0	12.5	0.7	5.1	3.8	5.4	3.9	7.0	2.0	2.6
				Woolong tea	B	3.9	0.7	1.2	3.9	7.8	4.6	5.2	0.7	6.9	10.2	15.9	0.3	3.1	6.8	1.1	6.1	6.1	0.5	17.8
			GC-MS/MS	Green tea	A	5.8	13.2	3.3	1.1	10.0	3.8	8.8	3.6	3.4	8.1	8.9	9.1	0.7	11.5	12.8	9.7	9.7	15.6	3.8
				Green tea	B	8.3	8.5	2.1	4.5	7.5	8.9	5.6	2.6	5.1	1.6	9.3	6.2	9.6	6.7	17.5	12.7	12.7	9.5	27.7
				Woolong tea	A	5.2	2.5	1.6	9.6	2.5	5.8	5.5	8.8	26.3	9.8	8.0	4.3	11.5	22.2	4.8	4.9	6.1	20.7	29.8
				Woolong tea	B	3.9	9.7	8.1	13.6	13.1	3.0	7.3	3.9	3.0	7.5	4.9	1.7	5.5	7.1	25.4	3.0	14.5	13.6	18.5
		Envi-Carb+PSA	GC-MS	Green tea	A	11.9	3.0	6.4	3.0	7.1	2.8	4.7	4.1	1.0	9.3	4.0	9.8	0.3	6.8	6.5	8.5	12.5	11.6	7.2
				Green tea	B	5.5	8.1	6.0	2.8	2.9	4.1	5.9	1.6	7.1	3.0	31.7	12.7	6.1	2.5	15.1	2.8	6.8	12.7	8.7
				Woolong tea	A	5.8	3.4	7.4	0.3	3.3	2.6	7.9	1.8	2.4	5.1	4.7	4.9	4.8	2.8	3.7	5.6	2.8	8.2	6.4
				Woolong tea	B	3.6	3.7	8.1	1.5	15.7	4.8	5.8	6.1	11.4	2.9	12.3	11.3	9.2	2.6	12.0	5.6	8.2	3.7	6.9
			GC-MS/MS	Green tea	A	7.7	7.5	6.3	4.4	5.7	4.3	9.5	4.0	7.6	1.0	8.5	9.8	15.6	12.1	30.4	15.0	14.2	9.8	24.5
				Green tea	B	13.7	11.2	3.4	4.2	7.9	3.3	7.8	5.9	6.7	4.8	19.3	8.3	12.1	13.9	19.5	9.1	19.3	15.1	15.3
				Woolong tea	A	11.2	4.1	5.5	4.7	4.6	4.9	2.1	5.6	5.9	3.0	5.5	6.3	10.8	4.9	22.8	4.9	16.5	8.4	23.4
				Woolong tea	B	7.6	6.5	11.9	2.8	8.2	9.2	10.0	0.8	12.3	3.3	12.2	6.4	9.9	10.7	28.7	3.0	17.8	17.2	23.8

Annex 2: RSD data for 201 pesticides in incurred tea Youden pair samples under 8 conditions with two SPE cartridge cleanup in three months (Nov.9, 2009-Feb.7, 2010) (cont.)

No	Pesticides	SPE	Method	Sample	Youden pair	Nov.9, 2009 (n=5)	Nov.14, 2009 (n=3)	Nov.19, 2009 (n=3)	Nov.24, 2009 (n=3)	Nov.29, 2009 (n=3)	Dec.4, 2009 (n=3)	Dec.9, 2009 (n=3)	Dec.14, 2009 (n=3)	Dec.19, 2009 (n=3)	Dec.14, 2009 (n=3)	Dec.24, 2009 (n=3)	Jan.3, 2010 (n=3)	Jan.8, 2010 (n=3)	Jan.13, 2010 (n=3)	Jan.18, 2010 (n=3)	Jan.23, 2010 (n=3)	Jan.28, 2010 (n=3)	Feb.2, 2010 (n=3)	Feb.7, 2010 (n=3)
76	chlorfenson	Cleanert TPT	GC-MS	Green tea	A*	8.9	5.9	1.7	15.8	12.9	2.4	15.8	12.2	12.4	2.3	9.6	5.0	11.3	15.6	6.5	11.4	8.1	6.1	5.2
					B*	8.5	8.5	7.3	21.2	7.3	5.7	7.1	11.0	2.1	5.4	2.9	4.5	14.1	9.8	6.3	7.7	16.1	7.9	5.9
				Woolong tea	A	8.0	4.5	5.6	4.9	12.8	2.6	5.4	13.9	11.3	3.0	13.1	10.9	5.5	1.5	4.5	8.2	5.9	3.9	4.7
					B	3.2	4.1	5.1	0.7	4.6	9.7	13.0	3.2	8.9	4.5	11.0	4.7	0.4	9.7	11.5	3.0	3.1	2.4	9.6
			GC-MS/MS	Green tea	A*	16.4	6.9	3.3	14.2	14.8	1.5	21.8	16.9	20.7	1.9	16.4	10.7	11.6	8.6	2.7	11.0	4.2	15.5	6.4
					B	11.3	7.6	4.6	17.2	5.4	5.5	4.8	12.2	4.3	9.0	2.7	4.8	15.6	6.5	7.2	5.9	17.9	8.4	6.8
				Woolong tea	A*	10.8	8.4	5.9	2.7	12.3	1.4	8.2	0.2	9.8	5.3	13.1	4.9	3.9	2.7	7.2	7.0	11.8	16.2	12.0
					B*	9.3	8.5	6.9	3.2	8.2	12.7	12.7	6.9	17.4	3.3	9.6	7.5	5.7	4.5	10.5	8.7	4.1	6.7	15.6
		Envi-Carb+PSA	GC-MS	Green tea	A*	27.9	8.8	20.9	5.3	10.0	32.9	8.9	22.7	12.1	5.6	3.8	30.8	4.4	24.4	11.5	13.3	13.7	15.7	8.8
					B	36.8	10.4	4.8	10.6	15.5	8.2	8.9	19.0	16.8	13.9	11.9	22.4	45.4	9.7	8.4	12.7	8.0	27.3	16.6
				Woolong tea	A	6.6	3.0	7.6	14.5	3.8	3.7	10.4	19.1	3.4	1.0	2.9	11.4	16.6	1.5	2.8	22.4	5.2	21.9	8.6
					B	13.9	10.5	7.1	10.0	16.1	12.9	16.0	9.9	14.7	8.5	10.4	9.2	6.6	9.7	15.8	4.0	12.6	27.8	9.3
			GC-MS/MS	Green tea	A*	29.7	4.4	20.6	6.2	8.4	24.8	22.4	28.3	8.7	5.1	4.5	3.2	4.6	16.0	7.1	12.0	14.0	15.3	8.9
					B	39.6	11.9	4.4	21.0	17.2	7.5	32.1	23.3	23.6	14.9	6.3	0.6	34.5	8.6	6.4	14.9	7.1	25.8	16.0
				Woolong tea	A	12.7	1.3	7.0	11.4	8.7	2.2	7.4	23.6	3.5	2.1	3.7	15.7	13.2	3.6	6.3	7.0	7.9	12.5	9.9
					B	13.8	9.8	7.5	4.0	14.1	10.5	21.4	7.5	12.7	9.3	7.1	6.1	2.6	11.7	16.7	8.7	10.4	24.1	9.5
77	chlorfenvin-phos	Cleanert TPT	GC-MS	Green tea	A*	5.9	4.8	1.7	11.4	12.5	4.0	8.0	7.5	5.3	3.9	6.5	3.9	7.1	6.1	12.5	3.7	6.5	6.5	8.6
					B*	9.0	9.6	3.9	8.5	5.5	4.8	2.0	3.1	1.2	2.4	7.0	6.0	10.2	8.6	12.8	6.7	9.8	8.2	6.7
				Woolong tea	A	7.7	1.1	1.8	3.6	11.0	2.7	21.9	18.3	8.1	8.4	13.2	9.0	3.6	6.2	10.3	1.3	12.7	3.4	4.7
					B	5.0	3.0	5.0	2.8	4.2	3.9	3.6	8.2	4.2	5.4	15.6	3.8	8.3	7.1	5.9	3.4	5.1	1.7	3.3
			GC-MS/MS	Green tea	A	6.2	6.8	4.7	5.1	9.7	5.1	7.2	15.8	14.8	1.7	10.2	8.0	11.3	6.6	4.6	6.4	6.2	9.7	1.6
					B	6.1	9.4	5.4	5.3	5.8	1.7	6.8	6.1	5.4	10.8	3.4	3.3	5.9	3.5	7.1	9.9	11.6	0.7	12.9
				Woolong tea	A*	6.8	1.8	5.3	4.0	8.7	3.1	15.1	1.4	10.6	9.1	8.7	6.2	6.8	5.7	8.7	7.5	14.2	5.7	6.1
					B*	2.6	7.3	4.5	8.0	5.6	0.6	6.7	11.8	14.4	7.6	4.3	5.8	12.3	0.7	6.1	5.8	5.4	9.1	11.7
		Envi-Carb+PSA	GC-MS	Green tea	A	14.9	4.8	11.4	7.8	10.7	12.2	2.1	13.8	9.8	4.3	14.3	14.4	4.4	13.2	7.4	5.0	6.0	16.5	6.5
					B	16.5	7.8	4.9	5.7	9.0	1.9	10.3	5.9	13.8	10.4	22.0	11.7	19.9	4.0	22.6	11.5	1.5	20.8	8.9
				Woolong tea	A	4.6	1.7	4.5	5.7	5.3	3.6	11.3	14.8	5.2	6.4	4.9	2.8	7.3	5.2	7.2	12.4	2.4	14.8	5.5
					B	9.9	9.2	7.3	4.6	14.0	10.9	7.4	7.4	13.5	0.1	11.5	3.4	7.2	3.1	17.7	5.0	16.6	6.7	6.7
			GC-MS/MS	Green tea	A*	18.5	6.8	12.9	0.8	3.7	28.7	9.7	12.8	8.4	3.6	9.9	2.8	4.4	11.9	7.2	5.1	11.9	7.8	12.1
					B	46.8	14.7	2.9	15.5	7.8	7.8	18.1	12.4	14.2	16.6	9.2	0.7	9.9	18.3	1.1	1.8	11.3	18.7	9.8
				Woolong tea	A	10.6	1.0	5.2	2.0	2.5	1.6	1.8	13.2	3.3	3.5	5.1	0.3	6.3	5.0	9.8	7.5	5.1	3.2	3.0
					B	9.6	5.6	7.6	6.2	14.1	7.0	15.5	2.2	12.7	0.7	4.1	2.3	2.1	14.7	17.4	5.8	9.8	7.3	6.2
78	chlormephos	Cleanert TPT	GC-MS	Green tea	A	5.2	6.4	5.8	2.9	8.6	8.1	4.4	4.0	4.2	6.2	6.1	0.8	4.9	1.0	3.2	3.2	6.2	4.5	3.9
					B	5.7	8.2	3.6	6.8	4.6	5.1	5.3	2.5	3.8	10.8	0.4	5.4	5.6	3.5	3.4	5.8	7.6	5.2	4.5
				Woolong tea	A	3.7	2.0	1.3	5.9	5.2	1.3	10.7	2.9	8.6	2.0	3.6	1.1	6.2	2.4	8.5	3.6	4.8	3.6	2.5
					B	7.7	1.3	7.8	5.7	4.2	4.8	5.4	0.8	6.0	6.1	1.3	2.7	4.3	4.3	6.4	3.4	4.8	2.0	4.1
			GC-MS/MS	Green tea	A	1.8	3.6	3.9	2.0	12.4	7.9	8.0	6.1	4.9	4.9	7.7	1.9	7.2	6.2	4.7	7.8	5.6	4.9	6.5
					B	8.8	6.6	5.6	6.9	4.5	6.8	6.7	3.6	4.8	12.1	4.6	3.7	7.3	5.0	2.4	1.0	5.6	3.1	5.3
				Woolong tea	A	5.9	0.7	1.5	3.6	2.5	0.7	7.7	2.4	5.8	4.3	2.2	2.8	10.8	0.4	6.4	5.0	5.6	2.5	4.1
					B	4.2	3.9	5.9	6.8	5.9	5.0	2.8	6.3	5.7	7.1	6.3	4.3	2.6	5.8	10.4	4.7	3.2	5.4	3.6
		Envi-Carb+PSA	GC-MS	Green tea	A	5.7	5.5	2.3	11.1	2.4	3.4	10.2	1.1	5.6	3.7	5.2	1.3	4.8	6.5	7.7	11.3	12.1	5.3	3.2
					B	3.4	5.1	5.1	3.7	3.3	2.1	6.1	2.2	1.8	1.7	26.6	2.1	18.1	0.6	32.3	1.4	6.1	7.6	5.9
				Woolong tea	A	6.2	8.3	4.5	7.3	4.0	1.8	1.5	8.1	1.3	1.3	5.3	3.8	1.7	2.6	3.4	5.4	5.0	5.9	6.1
					B	10.6	1.3	2.9	4.1	12.0	4.0	4.8	7.3	12.2	1.3	9.4	3.6	3.2	2.7	9.5	3.8	3.3	4.0	2.1
			GC-MS/MS	Green tea	A	7.8	7.5	1.4	6.7	5.2	3.1	4.7	0.7	5.0	2.3	8.8	5.8	10.6	12.8	6.5	12.5	12.9	5.6	2.1
					B	9.2	3.3	4.4	0.7	2.0	1.5	10.5	2.7	4.2	1.2	1.2	1.9	20.1	13.2	31.9	3.8	9.2	10.1	9.7
				Woolong tea	A	3.7	2.7	4.6	5.5	5.6	1.1	3.8	5.0	1.0	0.7	8.6	3.7	4.9	7.8	9.5	5.0	7.2	4.1	3.6
					B	9.1	7.4	6.8	0.3	7.5	3.6	6.7	9.4	9.5	1.1	2.7	4.7	1.8	10.7	11.8	4.7	7.1	2.4	3.5

79	chloroneb	Cleanert TPT	GC-MS	Green tea	A	6.3	4.7	3.3	2.2	8.3	3.3	4.9	5.5	4.2	1.4	5.7	0.7	5.3	4.3	3.5	4.5	1.5	5.0	4.2
					B	5.6	6.3	5.1	7.5	2.5	4.6	4.7	2.9	7.4	8.3	1.1	5.5	4.1	1.7	3.5	2.5	5.4	2.4	7.9
				Woolong tea	A	4.1	1.7	2.1	1.9	2.4	1.6	7.5	4.4	4.0	2.2	1.6	4.3	2.1	0.9	6.9	4.2	3.0	0.9	2.3
					B	3.5	1.1	3.8	5.2	1.7	3.4	2.2	0.5	6.0	6.2	2.8	1.2	2.4	3.0	4.8	1.1	3.5	1.6	8.6
			GC-MS/MS	Green tea	A	7.5	5.0	2.1	2.0	10.1	3.6	4.1	4.5	4.9	3.0	7.7	4.3	6.6	7.0	1.7	5.0	1.7	4.1	6.9
					B	8.0	6.6	3.9	8.6	5.1	8.1	6.9	3.0	7.1	7.5	4.1	5.0	3.3	3.3	2.9	4.9	9.2	4.8	7.5
				Woolong tea	A	4.1	2.1	1.9	2.8	2.0	1.1	8.7	2.1	1.2	5.0	4.1	2.6	7.9	3.1	6.9	5.7	6.2	5.1	6.0
					B	3.2	1.4	7.8	5.4	3.9	5.5	3.9	5.9	2.0	2.8	6.4	1.9	2.7	3.1	8.0	0.6	3.7	0.2	3.4
		Envi-Carb+PSA	GC-MS	Green tea	A	4.0	5.9	4.0	7.9	6.7	3.4	26.0	1.8	9.4	3.5	6.4	4.4	2.5	1.9	4.6	6.3	11.0	0.7	3.9
					B	5.1	2.0	5.4	3.9	3.6	1.7	5.0	5.2	3.1	2.5	5.9	2.6	4.4	1.3	15.5	3.3	6.9	5.0	5.3
				Woolong tea	A	8.5	5.7	9.2	3.5	1.0	1.1	2.6	7.4	2.0	15.7	18.7	3.7	1.8	2.9	1.4	1.6	6.2	3.5	4.7
					B	4.9	5.2	1.7	3.1	8.0	2.3	1.2	5.0	14.7	13.3	4.7	10.6	1.9	2.5	8.5	3.1	6.6	4.0	2.7
			GC-MS/MS	Green tea	A	5.0	8.2	6.6	5.9	5.7	3.9	1.9	3.0	6.8	1.2	5.1	9.0	18.3	7.0	7.4	8.8	8.1	4.2	3.1
					B	3.1	2.3	2.9	1.7	1.5	2.4	4.3	4.5	1.7	1.7	7.8	3.3	12.0	13.5	21.6	6.3	7.0	4.8	8.7
				Woolong tea	A	7.7	5.7	9.8	4.5	5.2	1.5	3.1	3.3	2.7	1.7	17.0	2.2	3.6	3.1	9.3	5.7	5.0	4.0	6.3
					B	4.1	10.0	5.9	2.6	5.3	4.1	4.3	7.5	8.3	2.9	6.3	4.2	2.3	7.5	8.9	0.6	5.8	5.0	3.2
80	chloropropylate	Cleanert TPT	GC-MS	Green tea	A	6.2	7.8	4.1	6.4	5.3	2.0	9.8	7.1	4.9	5.8	5.1	4.8	9.9	9.0	7.7	4.2	7.0	7.8	5.6
					B	4.4	1.3	3.0	13.0	10.1	4.8	2.8	9.7	3.3	13.2	4.5	13.2	6.5	4.5	4.8	6.5	7.7	3.0	5.7
				Woolong tea	A	4.3	2.6	3.1	3.2	0.7	3.3	0.2	11.9	3.9	4.3	6.5	3.1	3.0	2.4	7.3	2.5	3.2	2.2	1.7
					B	3.0	5.7	3.1	4.7	3.0	4.4	9.9	6.9	6.1	3.1	2.3	0.2	1.7	2.8	5.1	0.8	0.6	1.7	3.8
			GC-MS/MS	Green tea	A	6.8	8.6	4.8	2.3	9.5	2.8	6.7	11.1	6.0	2.4	6.1	4.1	7.2	4.3	0.8	3.5	3.8	7.2	6.0
					B	9.0	5.7	2.9	7.2	6.1	4.5	4.3	5.6	4.1	7.7	4.9	5.9	3.2	2.1	2.2	6.5	10.1	1.2	6.9
				Woolong tea	A*	2.8	7.1	4.3	0.7	2.6	4.1	3.9	0.2	5.8	5.2	6.1	3.4	5.6	2.4	5.4	3.8	6.0	4.9	10.5
					B*	13.8	5.7	6.8	7.1	5.3	5.4	5.0	8.6	6.6	3.8	8.9	1.9	2.1	5.9	7.1	2.1	2.9	5.0	5.9
		Envi-Carb+PSA	GC-MS	Green tea	A	13.8	5.7	6.8	6.8	5.4	8.7	2.8	12.3	12.4	2.7	7.1	1.3	27.7	16.7	8.8	37.8	8.4	33.6	5.9
					B	7.4	7.9	4.4	9.9	5.5	5.1	2.9	12.1	16.7	2.7	37.2	27.7	4.4	16.7	35.3	5.5	5.4	27.4	9.4
				Woolong tea	A	4.3	3.6	3.0	5.2	3.3	1.8	2.3	19.7	1.6	0.3	4.3	1.5	2.4	4.4	1.1	3.6	5.9	7.9	4.1
					B	5.9	4.4	9.3	1.2	10.8	6.4	3.7	7.4	10.4	1.5	7.2	4.6	2.8	2.4	10.1	3.1	7.4	1.6	4.8
			GC-MS/MS	Green tea	A*	12.9	7.0	4.0	4.9	6.9	2.5	15.3	6.8	3.0	1.4	6.5	17.1	19.4	2.8	4.7	17.6	9.1	6.6	10.1
					B	11.1	8.8	6.0	1.5	1.5	1.5	12.6	4.8	9.4	1.2	26.2	20.1	14.9	8.7	21.5	4.0	9.0	12.4	11.9
				Woolong tea	A	6.8	2.5	5.6	2.8	2.6	1.4	0.5	13.2	0.8	1.6	6.7	3.5	5.2	3.0	9.0	4.7	7.6	11.2	3.1
					B	10.1	3.8	8.8	1.2	11.2	5.5	7.4	6.7	10.0	2.7	3.1	3.0	8.9	4.0	15.0	2.1	4.5	19.7	8.6
81	chlorpropham	Cleanert TPT	GC-MS	Green tea	A	5.0	4.7	2.8	1.1	9.3	2.8	7.2	10.9	6.2	1.8	5.3	7.1	7.1	2.1	5.6	1.4	10.0	9.6	4.9
					B	6.2	9.4	4.2	8.8	3.9	2.4	1.7	4.0	2.9	5.9	4.0	1.3	4.9	7.9	7.9	3.8	5.0	2.1	5.9
				Woolong tea	A	4.6	2.5	2.4	2.8	3.9	5.2	0.3	16.7	3.9	6.1	13.2	5.7	5.3	6.3	9.8	2.4	2.4	2.9	2.2
					B	3.3	2.0	4.7	7.9	4.2	2.7	9.2	28.4	7.8	3.6	8.6	2.2	2.2	5.3	4.0	0.7	1.5	2.8	4.2
			GC-MS/MS	Green tea	A	4.0	5.5	2.0	3.2	9.6	4.4	10.3	12.8	7.9	1.4	6.6	6.3	8.3	4.3	2.6	5.2	6.5	4.9	3.7
					B	6.7	7.3	5.0	7.5	5.8	4.7	5.2	5.6	5.1	6.8	3.7	3.4	4.3	5.9	1.6	4.5	5.6	3.4	2.7
				Woolong tea	A*	5.4	2.0	1.4	3.5	1.6	2.6	4.4	0.7	5.4	3.0	8.9	3.9	5.9	3.4	8.2	8.5	6.1	2.9	2.6
					B*	2.5	7.5	5.1	6.3	4.5	6.3	4.2	10.6	7.2	3.7	7.3	4.0	3.4	3.9	6.7	2.9	1.3	2.1	7.3
		Envi-Carb+PSA	GC-MS	Green tea	A	10.8	5.7	4.9	5.7	7.9	4.3	11.5	9.1	11.3	3.2	5.5	2.5	6.1	4.0	5.8	27.6	11.8	27.0	5.4
					B	5.8	8.3	4.7	7.3	1.5	2.0	32.1	7.8	10.2	1.2	27.9	22.4	16.1	2.3	29.9	3.0	5.0	21.6	8.7
				Woolong tea	A	12.6	4.8	3.7	13.5	5.2	4.4	3.6	16.9	1.4	4.9	7.4	5.9	2.3	2.5	1.2	6.6	5.8	9.2	5.6
					B	26.6	3.2	9.7	6.1	11.9	8.0	6.2	5.7	9.7	2.9	14.4	14.4	2.8	5.9	13.4	3.3	9.8	0.5	11.3
			GC-MS/MS	Green tea	A*	18.6	3.2	7.0	5.5	1.4	6.2	31.7	9.8	4.2	1.4	4.9	9.0	10.6	2.8	3.9	17.8	8.6	10.4	13.0
					B	15.0	14.6	7.0	1.7	2.3	4.3	23.0	8.2	14.3	2.5	28.0	18.2	5.5	2.0	18.5	1.0	7.1	12.6	2.8
				Woolong tea	A*	6.3	4.9	4.0	4.3	2.3	0.7	2.7	10.0	1.2	2.7	6.9	0.4	3.4	4.7	9.5	8.5	6.0	4.0	7.0
					B*	10.5	3.3	10.2	2.3	7.4	5.3	10.5	6.7	9.5	3.3	3.6	1.9	9.5	2.9	11.0	2.9	6.8	4.8	

(Continued)

Annex 2: RSD data for 201 pesticides in incurred tea Youden pair samples under 8 conditions with two two SPE cartridge cleanup in three months (Nov.9, 2009-Feb.7, 2010) (cont.)

No	Pesticides	SPE	Method	Sample	Youden pair	Nov.9, 2009 (n=5)	Nov.14, 2009 (n=3)	Nov.19, 2009 (n=3)	Nov.24, 2009 (n=3)	Nov.29, 2009 (n=3)	Dec.4, 2009 (n=3)	Dec.9, 2009 (n=3)	Dec.14, 2009 (n=3)	Dec.19, 2009 (n=3)	Dec.24, 2009 (n=3)	Dec.14, 2009 (n=3)	Jan.3, 2010 (n=3)	Jan.8, 2010 (n=3)	Jan.13, 2010 (n=3)	Jan.18, 2010 (n=3)	Jan.23, 2010 (n=3)	Jan.28, 2010 (n=3)	Feb.2, 2010 (n=3)	Feb.7, 2010 (n=3)
82	chlorpyrifos (ethyl)	Cleanert TPT	GC-MS	Green tea	A*	6.5	4.0	4.4	3.0	10.7	2.9	6.5	6.8	5.1	4.8	5.8	2.1	5.6	3.8	4.1	1.0	3.3	4.3	3.4
					B	6.1	8.1	2.8	9.4	4.9	2.8	1.4	3.7	2.5	7.6	3.4	5.5	3.4	2.2	5.6	2.3	3.1	2.1	6.3
				Woolong tea	A	5.3	1.0	2.2	3.1	2.6	3.5	14.6	11.1	5.0	7.6	9.4	1.6	2.7	1.5	8.2	3.2	4.0	0.8	1.4
					B	3.5	3.3	3.7	4.3	3.7	3.5	4.3	3.0	6.3	3.6	6.5	1.6	2.3	3.5	6.4	0.6	1.3	1.0	2.4
			GC-MS/MS	Green tea	A	3.7	6.5	5.2	1.1	12.3	2.6	7.6	9.0	11.5	2.2	7.2	6.5	8.0	1.2	3.5	4.3	6.5	5.0	8.1
					B	7.0	9.9	5.4	7.7	7.3	4.7	6.8	3.1	6.7	9.2	2.2	2.0	3.7	3.5	0.9	3.8	4.8	4.1	5.1
				Woolong tea	A*	4.7	5.5	2.0	2.3	3.2	2.4	10.3	1.5	1.7	4.9	6.0	2.1	3.2	3.5	7.2	6.0	8.1	5.3	3.4
					B*	2.4	3.4	6.2	7.6	8.3	5.8	2.7	10.9	5.9	2.7	5.0	0.4	3.1	2.2	6.9	5.0	2.2	5.2	6.3
		Envi-Carb+PSA	GC-MS	Green tea	A	13.4	8.2	6.0	5.3	7.6	8.7	2.4	9.3	3.4	2.1	7.3	6.9	1.3	4.2	2.9	6.0	8.7	12.2	5.0
					B	12.0	7.3	4.4	4.6	4.9	4.6	4.6	7.6	9.6	4.2	16.0	7.2	12.2	2.7	7.0	7.3	3.3	13.7	8.2
				Woolong tea	A	4.8	3.6	3.7	3.6	2.9	1.0	5.7	6.0	0.3	0.5	4.0	2.0	1.8	2.4	2.8	5.1	4.6	7.1	4.2
					B	6.4	4.1	8.5	0.7	14.1	6.6	2.3	4.8	12.1	1.8	8.6	2.0	4.5	2.4	11.2	3.3	8.8	4.4	4.8
			GC-MS/MS	Green tea	A*	15.1	5.8	6.0	5.1	5.6	11.0	4.5	8.2	11.6	2.1	8.3	6.5	2.9	7.4	3.9	3.6	9.2	1.7	4.5
					B	26.7	10.2	3.0	8.6	1.9	5.3	18.7	4.1	8.4	7.8	10.3	8.1	7.9	5.6	9.9	4.0	5.1	17.0	12.2
				Woolong tea	A	5.5	3.5	5.6	3.8	2.1	1.2	0.5	7.2	1.3	3.8	9.2	1.6	6.0	8.5	9.7	6.0	5.0	9.1	4.4
					B	10.7	4.5	7.7	2.9	10.7	5.7	6.6	2.3	10.1	1.6	5.5	2.1	2.8	13.1	12.3	5.0	8.4	8.6	10.9
83	chlorthiophos	Cleanert TPT	GC-MS	Green tea	A	5.4	4.2	2.0	0.6	8.2	3.9	7.4	5.5	3.1	1.6	6.2	3.9	6.6	3.1	6.9	4.0	1.7	10.6	5.7
					B	5.5	7.8	3.7	7.7	4.0	2.8	1.6	2.0	2.3	2.7	5.6	7.0	6.6	6.5	6.2	13.5	10.0	3.4	6.7
				Woolong tea	A	7.3	1.3	1.5	2.8	0.8	2.4	8.7	10.7	3.6	5.1	7.5	2.1	10.3	6.0	2.6	3.1	10.6	2.9	5.9
					B	4.3	2.6	3.6	4.5	2.1	2.8	4.4	2.9	5.6	6.7	8.4	9.4	1.7	4.8	11.9	11.2	7.5	1.2	2.5
			GC-MS/MS	Green tea	A	5.1	7.5	3.7	4.1	8.2	4.8	10.4	6.5	5.3	6.4	2.7	0.4	2.8	5.0	4.3	0.9	5.6	4.4	1.9
					B	8.2	7.5	3.9	7.5	5.3	4.1	8.9	2.3	6.5	9.5	4.3	2.9	3.9	5.9	3.7	3.5	2.1	3.8	6.7
				Woolong tea	A*	5.7	0.8	1.8	1.4	3.7	1.6	11.1	2.5	1.1	3.8	3.0	4.2	3.7	4.6	8.6	6.7	8.0	0.2	11.2
					B*	3.4	0.4	6.7	5.2	4.0	3.2	4.5	6.2	6.1	3.8	4.4	2.7	5.5	8.5	5.2	2.4	5.3	5.3	5.6
		Envi-Carb+PSA	GC-MS	Green tea	A	9.6	2.7	4.2	4.8	9.0	2.9	17.1	4.0	3.7	3.0	8.7	0.7	4.9	2.3	8.9	2.4	7.6	8.6	8.7
					B*	6.8	7.9	4.3	3.0	1.5	0.9	11.0	0.4	3.2	4.5	24.1	4.0	10.9	6.1	17.5	11.8	9.5	10.2	10.4
				Woolong tea	A*	3.7	2.4	5.1	1.2	3.8	2.4	4.0	1.5	0.2	3.2	4.4	16.6	2.0	6.0	7.8	2.9	7.7	5.7	3.8
					B	5.2	3.1	8.9	11.4	11.4	4.8	2.8	6.2	11.6	3.2	8.2	3.1	12.4	3.5	17.9	6.6	6.8	2.4	4.9
			GC-MS/MS	Green tea	A	9.4	4.9	4.6	7.4	7.5	5.0	6.2	1.1	7.1	0.9	5.7	5.7	4.0	14.4	4.8	3.2	9.8	2.4	5.6
					B	15.5	9.8	3.6	7.7	2.9	3.0	11.1	1.7	8.2	4.5	16.5	3.9	3.1	3.3	11.4	3.8	9.6	11.7	12.9
				Woolong tea	A	7.4	0.5	4.1	3.4	5.2	0.4	1.0	5.3	2.3	1.6	7.7	3.5	1.3	2.9	10.7	6.7	4.5	3.5	2.7
					B	6.9	2.9	10.8	4.5	10.1	4.9	1.8	5.1	11.3	1.1	1.8	2.2	1.1	10.9	12.8	2.4	5.2	3.6	8.1
84	cis-chlordane	Cleanert TPT	GC-MS	Green tea	A	5.3	4.8	3.6	1.5	9.2	4.0	6.8	5.6	4.9	2.3	4.5	2.3	6.7	3.3	1.6	1.2	4.8	4.8	4.1
					B	5.7	8.3	4.4	8.7	3.4	4.0	2.6	2.9	5.4	8.1	2.4	6.2	2.2	1.2	3.7	3.6	4.8	0.7	5.7
				Woolong tea	A	5.3	3.6	0.7	4.2	2.3	3.7	8.0	6.3	3.4	3.1	4.2	1.0	3.5	3.0	8.6	2.4	3.4	2.0	1.4
					B	4.8	2.0	3.8	3.0	1.0	2.4	3.2	3.7	5.5	3.2	1.0	0.9	1.9	3.6	4.9	0.4	1.3	1.0	2.1
			GC-MS/MS	Green tea	A	4.9	5.3	4.9	2.0	11.7	2.9	12.1	8.6	5.3	5.2	11.1	5.9	7.6	1.5	3.5	5.6	6.2	5.2	3.3
					B	9.8	9.8	6.7	11.0	3.3	5.8	7.1	2.4	7.6	8.2	3.1	8.0	3.3	4.3	4.3	4.9	3.7	4.1	4.2
				Woolong tea	A	4.4	6.7	1.6	3.1	2.2	1.7	1.0	2.1	2.2	4.1	2.6	2.8	4.5	4.6	8.2	9.8	1.7	5.9	5.1
					B	6.5	0.3	3.1	7.4	3.6	5.6	1.4	0.5	2.7	4.4	12.6	2.3	3.3	2.0	8.2	5.5	6.0	10.3	6.8
		Envi-Carb+PSA	GC-MS	Green tea	A*	10.3	4.8	3.5	5.1	8.8	3.6	10.8	4.1	4.9	0.6	10.4	1.6	2.5	3.3	2.1	6.1	10.1	6.8	5.4
					B*	6.5	7.7	3.2	2.0	0.6	1.4	5.7	1.1	4.8	4.3	17.7	4.3	4.9	1.2	9.2	3.8	5.0	11.3	7.3
				Woolong tea	A	0.8	3.2	2.4	4.0	5.9	1.2	3.1	3.1	2.1	2.4	4.7	3.2	1.0	3.0	3.3	2.8	5.7	4.8	4.6
					B	5.5	3.1	9.7	2.8	8.9	5.6	3.0	7.3	12.1	3.2	8.4	4.0	5.1	3.6	11.0	2.7	7.4	2.1	5.4
			GC-MS/MS	Green tea	A*	12.0	4.9	4.1	6.3	11.0	4.4	6.0	3.0	7.1	4.0	9.9	24.7	9.4	6.8	9.4	4.9	10.2	4.2	8.1
					B	10.9	9.1	6.1	4.4	1.0	4.7	14.5	7.6	4.5	5.5	18.3	14.7	5.7	12.5	7.7	5.4	15.4	13.3	4.3
				Woolong tea	A	2.8	4.0	2.7	3.7	4.5	3.3	2.7	0.6	4.1	2.5	5.9	8.5	6.2	4.0	4.7	9.8	11.1	8.0	5.0
					B	9.0	4.5	7.8	5.0	10.5	5.5	4.5	9.0	7.7	2.5	4.1	8.4	6.6	15.4	16.9	5.5	3.9	3.0	9.3

85	cis-diallate	Cleanert TPT	GC-MS	Green tea	A	6.2	4.7	2.8	1.1	9.3	4.8	6.4	5.6	5.0	2.9	5.5	1.4	6.9	5.0	2.9	7.7	5.1	3.9
					B	6.4	8.9	4.6	8.6	3.3	2.4	2.1	2.9	4.1	9.5	3.1	6.8	0.5	5.3	9.5	8.2	0.6	5.5
				Woolong tea	A	3.2	2.1	1.0	3.7	2.9	2.0	7.1	4.5	3.7	3.1	3.8	0.5	3.9	2.3	3.1	3.0	0.8	2.1
					B	4.1	3.1	3.8	2.0	1.6	3.5	4.1	3.8	6.3	2.5	2.8	0.5	4.3	3.3	2.5	1.1	1.7	2.8
			GC-MS/MS	Green tea	A	3.7	7.8	3.2	9.6	9.9	4.2	8.1	6.8	5.5	0.7	7.5	3.5	5.2	6.8	0.7	3.7	3.0	4.7
					B	9.0	2.1	4.3	3.3	4.9	7.8	6.3	2.1	4.3	9.7	3.5	5.0	2.1	1.4	9.7	7.2	6.1	3.4
				Woolong tea	A	5.2	0.7	0.8	6.9	2.6	3.0	8.6	0.6	0.8	4.3	3.3	1.3	9.7	3.5	4.3	6.0	3.5	5.2
					B	2.5	3.3	5.5	5.3	3.3	3.8	1.2	5.9	3.2	4.3	4.1	2.7	1.7	5.1	4.3	4.0	0.7	4.2
		Envi-Carb+PSA	GC-MS	Green tea	A	8.8	3.3	3.5	0.8	6.2	4.2	10.3	4.3	6.0	0.7	7.5	2.1	3.1	1.9	0.7	9.7	4.8	4.9
					B	6.6	8.5	5.0	5.3	2.6	3.4	3.6	4.1	3.4	3.7	17.5	2.2	5.0	0.6	3.7	4.7	9.9	6.1
				Woolong tea	A	5.3	2.1	3.8	0.8	3.7	0.4	1.5	3.2	0.5	2.6	4.7	3.0	2.8	3.4	2.6	5.3	8.7	4.0
					B	5.8	3.0	7.9	3.2	11.6	4.7	0.7	6.8	10.7	1.7	7.6	1.8	2.5	3.3	1.7	5.7	4.1	4.5
			GC-MS/MS	Green tea	A	7.3	6.3	1.1	3.4	6.7	2.2	2.8	4.2	6.6	2.7	9.6	4.8	18.1	18.5	2.7	13.9	4.5	4.1
					B	6.0	6.1	5.4	5.9	1.6	2.3	7.3	0.6	2.6	1.5	15.8	0.8	6.6	11.2	1.5	10.2	10.9	5.9
				Woolong tea	A	4.0	2.8	4.2	3.5	5.0	1.8	1.0	2.6	1.6	2.0	7.2	3.6	2.8	4.6	2.0	6.2	4.5	4.2
					B	6.9	4.1	9.1	3.3	7.3	3.8	3.5	8.4	9.1	2.6	2.8	2.1	1.8	8.4	2.6	5.8	3.7	7.5
86	cyanofenphos	Cleanert TPT	GC-MS	Green tea	A*	5.2	6.1	2.2	1.4	13.6	5.1	7.4	11.8	7.9	3.1	6.0	5.0	7.8	8.8	3.1	6.0	10.4	4.2
					B*	6.2	6.8	4.3	12.2	5.0	3.4	1.4	4.7	1.7	4.2	4.0	4.1	12.4	4.9	4.2	11.6	2.0	5.6
				Woolong tea	A*	6.8	4.5	1.3	14.6	10.1	2.1	11.0	10.0	6.3	6.3	10.1	10.2	1.7	0.6	6.3	2.2	2.6	7.1
					B	3.2	1.8	4.0	2.0	6.1	3.2	5.1	2.2	3.6	1.9	8.7	1.6	1.7	2.0	1.9	3.0	2.9	3.9
			GC-MS/MS	Green tea	A	8.1	7.8	1.7	4.9	13.9	2.4	14.6	16.4	14.9	6.8	11.2	5.3	10.1	9.7	6.8	7.1	13.3	2.3
					B	7.9	7.8	7.9	10.1	6.7	3.1	5.3	10.9	3.4	5.9	2.3	2.4	14.4	2.7	5.9	11.5	3.1	3.5
				Woolong tea	A*	9.3	3.8	7.1	12.0	6.7	1.5	1.7	5.2	5.8	6.9	8.0	3.7	3.3	5.6	6.9	7.1	9.5	5.2
					B	6.3	5.7	6.3	2.4	2.8	3.2	7.4	10.4	11.5	6.7	5.8	5.5	4.4	3.9	6.7	1.6	4.6	9.9
		Envi-Carb+PSA	GC-MS	Green tea	A*	20.0	4.6	17.9	3.7	9.4	11.1	4.3	8.9	6.6	2.1	2.2	27.4	4.5	6.6	2.1	9.4	19.3	6.8
					B*	25.5	8.5	5.3	8.1	10.6	1.7	5.6	10.6	15.9	13.6	17.4	21.9	24.8	13.4	13.6	4.2	14.5	14.2
				Woolong tea	A	4.3	4.4	5.9	7.9	3.9	5.2	2.9	14.2	2.5	2.4	4.0	4.6	4.0	2.5	2.4	3.4	16.2	3.8
					B	10.0	10.3	7.5	3.3	15.7	2.8	6.3	12.5	15.1	2.1	16.8	3.9	3.4	1.5	2.1	8.0	14.4	7.3
			GC-MS/MS	Green tea	A	24.7	9.6	17.2	2.9	6.8	25.1	6.3	20.0	11.3	4.6	2.7	1.5	5.6	21.5	4.6	13.9	6.7	6.3
					B	36.6	8.7	3.0	23.9	10.1	3.6	27.4	18.6	17.6	17.1	11.1	2.4	18.3	13.3	17.1	1.7	18.8	13.2
				Woolong tea	A	10.8	5.0	6.3	6.4	1.3	3.8	4.2	18.7	3.4	3.2	12.0	9.9	9.8	4.6	3.2	7.9	12.1	5.7
					B	11.9	8.9	9.8	1.2	11.7	8.0	15.9	5.7	12.2	6.9	3.7	5.0	9.4	9.6	6.9	8.5	20.8	8.6
87	desmetryn	Cleanert TPT	GC-MS	Green tea	A	6.7	4.8	2.4	2.8	8.1	3.5	9.7	6.0	2.6	3.6	4.0	5.8	7.6	4.1	3.6	5.8	5.7	4.7
					B	4.3	8.1	7.6	8.3	3.9	3.4	2.8	2.2	2.6	8.3	11.4	10.4	0.9	3.2	8.3	8.1	1.0	4.7
				Woolong tea	A	6.6	2.4	3.2	3.4	2.2	3.5	5.2	7.6	5.6	13.0	9.1	2.5	2.6	4.5	13.0	3.3	3.7	1.6
					B	4.5	2.9	4.0	7.3	2.4	2.3	9.0	4.1	6.7	4.9	6.6	1.2	2.0	7.7	4.9	2.5	0.8	3.8
			GC-MS/MS	Green tea	A	4.5	4.3	3.0	2.2	9.4	6.8	6.4	8.1	7.2	1.9	7.8	7.7	7.4	3.9	1.9	5.3	3.3	4.9
					B	7.6	8.0	6.3	9.1	5.6	6.5	8.2	5.8	4.8	9.7	4.8	6.1	3.2	2.9	9.7	4.7	1.4	5.0
				Woolong tea	A	7.4	1.9	1.7	3.7	0.8	4.1	9.2	4.6	1.9	3.4	3.4	1.2	2.3	4.5	3.4	5.2	4.6	5.1
					B	4.3	1.4	4.9	7.1	4.3	1.6	1.9	6.5	5.4	2.7	6.1	0.5	3.4	7.6	2.7	4.0	3.7	5.7
		Envi-Carb+PSA	GC-MS	Green tea	A	5.7	5.7	3.5	8.5	9.3	2.7	13.5	1.6	3.6	4.2	8.5	2.4	3.4	6.5	4.2	10.6	3.3	5.4
					B	3.6	8.1	2.5	6.2	1.7	2.8	4.9	4.3	5.8	16.1	26.8	9.8	2.8	6.8	16.1	4.7	7.7	7.1
				Woolong tea	A	3.8	4.5	2.8	1.1	4.0	0.9	3.0	1.7	5.3	6.2	4.5	5.6	4.1	13.3	6.2	6.3	3.1	4.4
					B	4.1	3.2	10.0	2.6	15.0	5.6	2.1	7.1	11.5	5.7	9.5	5.0	4.6	3.5	5.7	5.9	6.2	5.2
			GC-MS/MS	Green tea	A	5.2	7.0	2.1	5.9	9.7	1.2	1.3	1.5	4.7	0.8	9.2	15.9	9.9	12.0	0.8	7.8	1.0	6.6
					B	5.8	8.4	7.3	2.5	0.4	3.5	4.8	4.6	5.5	1.4	16.0	3.6	5.3	10.4	1.4	3.5	10.6	13.4
				Woolong tea	A	4.1	5.6	1.9	3.0	5.4	1.6	2.1	2.8	1.0	2.2	7.1	5.1	4.3	16.2	2.2	5.0	4.6	3.2
					B	4.9	2.4	10.3	1.7	8.3	5.7	0.4	7.9	9.6	4.1	3.1	3.5	3.2	11.7	4.1	0.8	0.8	8.5

(Continued)

Annex 2: RSD data for 201 pesticides in incurred tea Youden pair samples under 8 conditions with two SPE cartridge cleanup in three months (Nov.9, 2009–Feb.7, 2010) (cont.)

No	Pesticides	SPE	Method	Sample	Youden pair	Nov.9, 2009 (n=5)	Nov.14, 2009 (n=3)	Nov.19, 2009 (n=3)	Nov.24, 2009 (n=3)	Nov.29, 2009 (n=3)	Dec.4, 2009 (n=3)	Dec.9, 2009 (n=3)	Dec.14, 2009 (n=3)	Dec.19, 2009 (n=3)	Dec.24, 2009 (n=3)	Dec.29, 2009 (n=3)	Jan.3, 2010 (n=3)	Jan.8, 2010 (n=3)	Jan.13, 2010 (n=3)	Jan.18, 2010 (n=3)	Jan.23, 2010 (n=3)	Jan.28, 2010 (n=3)	Feb.2, 2010 (n=3)	Feb.7, 2010 (n=3)
88	dichlobenil	Cleanert TPT	GC-MS	Green tea	A	4.7	5.2	3.2	2.4	7.9	8.6	4.4	3.0	2.0	6.4	6.1	2.4	4.1	1.7	1.3	1.9	4.6	12.1	3.8
					B	5.5	7.8	3.4	5.6	3.4	6.2	5.4	6.0	2.2	8.5	1.9	3.7	6.1	2.3	4.2	3.8	6.6	8.9	5.1
				Woolong tea	A	5.1	2.5	0.9	1.2	1.7	5.3	8.0	4.2	12.6	2.2	3.4	2.7	6.8	1.8	7.5	3.1	6.9	1.5	4.7
					B	8.6	6.5	5.6	5.5	2.5	5.6	6.6	9.3	5.3	5.6	4.2	5.1	2.5	4.0	9.6	5.8	2.3	1.2	5.1
			GC-MS/MS	Green tea	A	3.4	5.3	3.1	3.1	13.0	8.1	9.5	8.3	4.0	5.4	5.4	2.1	6.7	2.4	1.3	5.8	5.6	5.3	2.3
					B	9.0	5.8	3.4	6.5	8.2	7.1	5.7	5.4	3.8	3.2	3.2	6.7	12.5	3.3	2.2	1.7	6.9	3.7	0.7
				Woolong tea	A*	7.4	2.8	5.3	2.3	4.6	2.3	5.0	3.0	9.5	3.3	3.3	4.7	3.0	1.8	7.0	7.5	3.9	1.8	9.5
					B*	6.7	7.0	9.2	3.6	5.9	2.6	2.7	3.8	1.8	9.0	9.0	4.9	3.0	3.9	11.9	5.6	8.8	6.1	5.5
		Envi-Carb+PSA	GC-MS	Green tea	A	6.6	6.3	2.6	5.3	6.0	6.4	4.8	7.9	9.6	7.4	7.4	8.4	3.0	6.3	3.8	16.5	14.8	6.8	7.2
					B	3.8	5.6	2.2	4.4	4.1	2.3	3.9	9.8	2.7	31.3	31.3	2.1	4.2	5.5	35.4	5.6	5.0	3.9	6.3
				Woolong tea	A	5.1	7.6	4.8	3.6	5.0	1.6	3.1	10.2	1.0	5.2	5.2	3.4	2.9	1.8	2.8	7.9	4.6	6.7	4.9
					B	10.5	6.8	7.1	4.4	8.9	4.1	7.3	11.1	11.0	7.0	7.0	2.6	2.9	4.0	7.3	3.2	5.4	2.5	5.3
			GC-MS/MS	Green tea	A*	6.7	9.3	1.7	6.0	6.6	7.3	2.7	1.5	3.8	4.1	4.1	6.4	11.3	12.7	12.5	19.8	13.4	4.4	2.8
					B*	9.8	6.0	4.6	2.1	2.7	0.9	16.6	2.5	5.1	25.9	25.9	1.4	23.8	14.7	33.5	2.5	5.9	7.1	9.0
				Woolong tea	A	3.2	3.4	7.2	5.0	7.7	5.4	2.1	6.7	2.9	10.5	10.5	4.9	5.4	10.0	11.6	7.5	10.7	8.8	8.3
					B	10.5	8.2	7.9	3.4	8.4	5.3	11.8	9.3	4.6	4.0	4.0	3.3	1.7	9.1	11.1	5.6	9.1	8.0	9.6
89	dicloran	Cleanert TPT	GC-MS	Green tea	A	14.8	4.9	1.8	11.8	11.5	4.0	9.2	7.5	12.3	5.2	5.2	5.4	10.6	4.8	12.1	11.8	19.7	22.4	6.2
					B	12.0	12.6	8.1	17.5	11.0	6.1	5.4	12.4	2.6	3.9	3.9	6.3	9.4	6.6	13.8	5.8	8.2	12.6	10.8
				Woolong tea	B*	4.1	5.7	2.6	6.7	5.8	4.6	12.1	8.8	10.1	14.6	14.6	13.3	9.0	3.0	8.5	2.9	7.2	4.4	6.9
					B*	3.5	3.2	0.8	7.9	11.1	8.4	21.8	1.8	5.5	4.4	4.4	2.1	4.2	7.7	5.5	5.0	6.6	4.1	9.8
			GC-MS/MS	Green tea	A*	8.6	6.3	6.2	6.2	9.4	0.9	9.2	9.8	14.1	14.9	14.9	7.5	7.2	5.0	2.2	8.5	7.1	10.7	2.7
					A*	6.2	10.7	3.3	13.1	4.1	1.7	3.2	8.5	0.4	3.9	3.9	2.5	7.2	9.1	5.5	12.1	13.4	3.8	10.6
				Woolong tea	A*	8.2	8.8	3.0	1.9	7.5	6.3	2.7	14.5	11.0	13.0	13.0	8.7	7.4	2.4	9.4	4.0	8.8	4.9	7.5
					B*	3.8	2.8	1.6	4.5	12.0	4.2	4.2	7.6	14.4	14.2	14.2	6.9	10.8	7.4	8.9	3.8	3.8	7.7	5.9
		Envi-Carb+PSA	GC-MS	Green tea	A*	13.9	23.0	14.7	15.3	6.1	16.9	3.4	11.1	9.4	3.0	3.0	14.0	6.7	0.9	5.6	8.6	11.9	8.3	5.4
					B	8.3	8.5	3.6	10.7	10.9	4.2	3.0	18.4	8.9	29.3	29.3	16.0	19.4	9.7	18.2	4.7	18.8	22.2	10.7
				Woolong tea	A	6.2	3.2	2.5	4.8	2.1	2.4	11.5	0.6	3.7	3.8	3.8	2.8	14.8	11.4	4.4	14.0	9.2	18.2	6.5
					B	7.6	7.6	7.0	2.3	21.6	13.2	16.6	8.5	13.6	10.7	10.7	6.5	4.9	12.0	10.5	6.0	11.4	8.1	7.1
			GC-MS/MS	Green tea	A	10.3	11.9	14.2	5.1	0.7	11.1	26.9	12.8	9.8	4.5	4.5	4.9	11.3	5.0	4.6	11.4	10.9	8.1	7.7
					B	18.6	11.0	3.5	5.4	3.1	13.0	17.6	3.9	9.2	22.1	22.1	6.2	27.2	19.5	9.8	6.9	15.3	12.5	9.0
				Woolong tea	A	2.1	2.7	8.2	7.5	10.2	1.0	1.1	4.6	2.3	6.4	6.4	12.1	9.8	4.2	8.0	4.0	6.7	10.2	6.8
					B	2.6	9.0	7.7	7.4	13.1	12.1	19.1	3.0	12.8	7.7	7.7	5.4	2.2	9.9	9.4	3.8	7.6	5.2	8.6
90	dicofol	Cleanert TPT	GC-MS	Green tea	A	11.3	2.8	3.2	4.9	5.3	5.6	10.0	2.8	0.7	4.3	6.4	0.5	6.7	4.7	6.6	2.6	8.0	5.7	1.8
					B	9.5	7.8	5.4	9.0	3.2	4.7	4.3	1.7	5.3	5.8	4.8	9.5	4.4	9.9	6.6	4.1	7.2	9.0	5.6
				Woolong tea	A	9.9	5.0	8.1	7.5	3.2	6.5	9.0	6.1	2.6	2.8	2.8	2.6	3.3	2.9	2.7	1.5	7.8	8.1	1.5
					B	6.5	12.2	6.9	2.5	3.8	4.3	0.0	7.0	5.9	1.7	1.7	10.6	3.3	2.4	9.7	3.6	3.5	6.1	5.3
			GC-MS/MS	Green tea	A	8.2	5.0	2.3	3.8	8.5	4.5	1.5	3.2	2.5	4.0	4.0	3.6	7.5	2.2	4.3	5.3	6.0	2.8	3.6
					B	8.7	8.0	5.4	9.2	5.0	5.3	8.6	3.2	13.3	7.5	7.5	5.0	4.3	2.2	2.4	7.8	6.5	2.8	3.1
				Woolong tea	A	11.6	5.8	8.0	6.4	3.6	5.6	9.5	4.1	9.9	3.6	3.6	1.4	22.1	3.5	0.7	7.8	6.5	5.8	1.5
					B	4.6	8.8	9.0	3.5	1.6	4.2	8.8	3.1	1.6	10.5	10.5	6.1	5.2	10.5	14.9	1.8	10.8	6.0	6.0
		Envi-Carb+PSA	GC-MS	Green tea	A*	11.5	6.2	2.2	6.3	7.3	8.2	4.8	8.6	4.4	8.5	8.5	15.0	4.4	6.2	6.2	1.6	12.3	3.0	5.8
					B*	10.3	7.0	2.8	6.7	5.1	3.3	4.1	9.8	4.3	5.9	5.9	4.5	14.9	6.6	8.5	2.3	4.5	2.5	3.3
				Woolong tea	A	10.9	11.1	5.0	3.5	9.5	10.4	10.1	2.5	2.5	8.5	5.1	4.3	5.3	1.7	1.5	4.8	2.3	4.3	7.6
					B	4.2	2.7	7.5	4.9	8.2	7.0	11.7	6.9	13.3	6.1	6.1	3.7	4.2	2.9	14.2	5.7	2.6	16.7	2.9
			GC-MS/MS	Green tea	A	6.4	4.8	1.6	7.3	6.8	5.8	12.0	4.9	9.9	7.9	7.9	8.5	4.8	20.1	8.6	3.6	9.4	1.0	10.4
					B	9.0	7.2	6.0	2.0	5.5	2.2	7.1	10.1	1.6	14.0	14.0	4.2	7.8	16.5	9.9	4.8	7.1	9.0	6.8
				Woolong tea	A	11.3	9.8	5.3	3.8	9.1	10.7	10.2	2.2	3.9	3.7	3.7	1.9	8.3	7.5	8.0	7.8	3.9	3.8	9.5
					B	6.9	1.9	7.6	4.8	5.1	8.4	21.1	6.3	12.8	7.5	0.4	5.5	6.9	10.1	11.7	1.8	2.3	9.2	3.7

#	Pesticide	Cartridge	Method	Tea	Rep																			
91	dimethachlor	Cleanert TPT	GC-MS	Green tea	A	7.8	3.8	2.8	2.8	8.3	4.6	6.4	4.9	4.3	2.8	4.7	3.0	6.4	5.6	4.5	5.0	5.4	7.9	0.8
					B	7.4	10.0	8.6	7.0	3.6	5.4	5.2	3.8	2.7	3.4	4.2	3.7	4.9	4.0	10.4	2.0	7.3	6.4	7.9
				Woolong tea	A	5.6	5.6	1.4	4.2	4.5	3.4	13.5	9.8	4.2	7.0	3.4	2.1	4.6	6.1	10.9	1.6	7.0	1.5	2.9
					B	4.5	3.1	3.3	2.7	2.1	2.7	4.5	7.8	3.4	5.5	2.4	2.0	3.6	3.3	4.4	2.8	0.4	1.2	2.9
			GC-MS/MS	Green tea	A	4.1	6.7	6.2	6.2	6.6	2.0	2.8	6.1	6.4	2.0	11.1	5.2	9.4	2.4	3.6	5.5	2.9	3.3	2.3
					B	7.9	9.0	5.5	6.6	5.4	4.2	9.5	6.2	4.9	6.4	3.3	3.4	2.1	1.1	3.9	4.2	4.5	2.3	8.0
				Woolong tea	A	6.1	6.5	3.5	3.7	5.7	2.8	20.4	3.7	2.1	4.8	2.7	4.5	2.1	1.5	7.7	2.6	8.8	7.1	5.1
					B	6.7	1.1	5.5	7.4	2.6	4.2	5.9	8.2	9.1	9.3	2.4	3.4	9.8	6.9	4.3	3.2	3.3	1.9	4.4
		Envi-Carb+PSA	GC-MS	Green tea	A	7.3	5.2	3.2	7.7	8.6	2.7	4.9	3.4	3.2	0.5	14.0	3.2	3.4	6.5	1.8	9.9	9.9	11.9	5.5
					B	11.3	8.6	5.0	0.9	1.5	0.9	6.8	1.2	5.5	6.3	14.2	5.0	3.9	3.0	15.5	9.0	6.2	10.9	6.6
				Woolong tea	A	4.1	3.7	2.4	7.2	5.0	2.1	7.9	3.3	3.9	4.8	2.0	4.5	2.5	0.9	7.5	4.1	4.5	6.8	4.8
					B	7.3	3.7	8.0	7.7	10.3	8.2	4.5	5.2	11.2	2.3	9.1	4.8	6.9	3.4	13.7	4.4	8.2	2.0	4.2
			GC-MS/MS	Green tea	A*	9.5	5.4	3.1	3.4	3.6	9.9	11.8	2.3	7.8	2.5	12.5	0.5	2.3	6.6	4.6	1.7	10.9	3.6	4.6
					B	39.8	10.1	1.7	7.2	12.1	2.6	3.9	3.4	4.1	8.8	10.5	3.9	3.2	15.1	10.6	4.7	7.7	6.9	10.3
				Woolong tea	A	2.4	3.7	2.7	4.6	4.4	1.4	1.9	2.2	2.8	5.9	6.8	6.5	3.1	6.9	12.4	2.6	6.8	3.5	6.6
					B	7.3	3.8	7.9	8.7	7.7	5.0	2.3	6.0	9.5	3.3	1.8	3.2	2.0	12.2	10.7	3.2	7.1	1.8	6.4
92	dioxacarb	Cleanert TPT	GC-MS	Green tea	A*	10.6	12.4	3.8	11.6	4.5	5.6	6.9	4.0	4.2	8.1	7.2	3.2	11.0	14.5	6.5	10.1	3.7	2.5	3.9
					B*	7.9	11.2	6.4	13.4	2.9	9.3	3.0	10.5	4.9	15.7	7.8	6.0	2.1	6.8	15.7	6.4	10.0	4.8	6.3
				Woolong tea	A*	12.7	5.1	3.2	9.0	4.6	4.8	5.3	3.7	2.4	6.3	0.6	2.1	5.3	2.6	8.2	6.2	6.9	4.0	2.5
					B*	11.7	2.8	7.0	11.4	5.0	5.5	1.1	4.2	4.4	0.7	3.8	4.4	1.0	3.8	3.7	4.0	4.2	4.7	1.6
			GC-MS/MS	Green tea	A	8.4	18.7	3.9	10.8	7.2	7.2	3.4	4.7	5.2	9.7	7.0	6.3	9.9	22.7	7.8	12.6	0.8	1.6	5.2
					B	7.4	16.9	6.1	10.7	5.2	11.9	4.2	10.9	7.2	22.7	10.4	3.4	3.5	11.2	10.4	5.3	11.5	2.9	7.9
				Woolong tea	A*	4.7	6.1	1.9	9.2	7.7	5.9	7.1	2.2	4.6	6.6	3.3	2.8	10.1	2.5	4.5	8.0	11.0	3.7	3.6
					B*	4.9	6.3	12.6	12.3	6.0	5.6	2.2	14.5	6.9	2.4	10.8	2.6	3.0	1.5	3.8	11.3	3.0	1.1	14.1
		Envi-Carb+PSA	GC-MS	Green tea	A	11.2	29.2	21.1	4.8	6.6	6.0	15.3	6.4	5.0	4.6	10.5	2.9	3.1	5.6	14.4	8.3	10.9	6.2	7.7
					B	8.4	4.2	4.7	4.6	2.7	1.8	9.4	2.7	6.3	4.5	24.5	5.9	13.0	2.1	13.7	3.0	4.1	11.1	8.0
				Woolong tea	A	7.5	6.3	7.5	5.4	5.9	7.5	6.3	10.0	1.5	19.1	10.3	13.5	3.0	4.2	10.3	5.2	3.6	5.9	2.1
					B	7.6	5.2	16.8	7.8	17.7	17.0	1.4	8.3	3.5	3.5	9.4	12.1	8.6	4.3	6.5	10.2	16.9	4.6	7.1
			GC-MS/MS	Green tea	A	16.2	27.3	20.4	6.5	6.7	3.3	53.9	8.3	10.3	2.7	9.3	5.6	11.5	19.1	20.4	10.0	10.2	8.8	9.5
					B	7.3	6.4	4.7	0.5	2.5	2.3	19.4	4.0	10.3	3.4	23.0	16.7	19.9	12.9	22.3	3.9	4.2	16.0	11.8
				Woolong tea	A	6.4	4.8	6.9	2.7	5.1	8.5	7.7	13.4	1.9	28.1	12.1	11.9	5.8	9.6	8.1	8.0	5.2	3.5	7.1
					B	10.1	3.5	26.5	10.1	13.8	19.9	2.4	9.2	0.2	3.9	7.6	11.9	4.6	34.9	4.7	11.4	19.4	4.4	9.7
93	endrin	Cleanert TPT	GC-MS	Green tea	A	4.8	4.6	3.2	2.2	8.9	4.8	6.9	5.2	4.4	3.0	7.1	2.5	6.9	3.8	6.2	1.0	4.7	7.1	6.3
					B	5.9	8.0	5.0	8.9	3.9	3.7	3.3	2.5	4.5	8.3	3.5	7.2	1.1	1.1	5.0	3.0	5.3	1.9	5.7
				Woolong tea	A	4.6	2.4	2.1	4.2	2.3	3.6	8.0	9.1	2.6	5.5	3.8	1.7	3.5	4.9	8.4	5.0	3.6	6.7	0.5
					B	4.7	2.6	2.7	4.0	2.2	2.3	4.3	3.6	6.4	2.1	0.3	2.8	1.9	3.9	3.1	3.6	1.3	3.9	2.4
			GC-MS/MS	Green tea	A	5.3	4.5	3.1	2.8	10.5	7.2	6.6	7.0	5.6	4.0	4.2	2.0	3.1	3.8	1.5	3.0	4.5	3.5	3.0
					B	6.9	8.8	5.2	6.8	5.5	5.7	8.6	2.4	6.6	8.6	2.7	6.4	4.2	2.5	1.3	6.2	5.7	0.9	4.4
				Woolong tea	A*	7.3	3.7	1.3	3.1	2.9	5.6	10.4	3.5	3.7	4.3	5.8	3.8	3.8	3.8	7.8	4.4	5.0	3.9	4.5
					B*	4.3	2.3	4.1	5.1	5.0	3.1	3.7	6.3	3.3	2.8	4.8	2.5	5.6	0.4	4.3	1.9	2.2	1.3	4.1
		Envi-Carb+PSA	GC-MS	Green tea	A	8.3	4.8	2.1	5.6	9.2	1.6	13.9	1.6	4.0	1.9	8.1	2.2	2.4	2.7	7.3	10.3	8.8	7.0	5.4
					B	2.6	8.1	4.4	5.3	1.3	2.2	3.0	2.6	4.1	3.2	26.6	2.4	4.8	3.0	14.9	2.0	3.6	9.4	6.6
				Woolong tea	A	1.4	4.2	2.2	3.1	5.6	0.8	4.3	1.4	1.5	2.9	5.0	2.9	1.4	2.6	2.6	2.3	6.1	5.1	1.2
					B	4.8	3.0	9.0	2.3	11.1	6.0	3.0	7.0	11.8	4.8	9.0	1.4	4.7	3.9	10.3	3.5	7.1	3.8	2.6
			GC-MS/MS	Green tea	A*	8.5	6.4	5.0	5.0	8.0	2.2	2.5	3.0	5.2	1.5	12.7	2.2	7.0	14.1	8.9	7.8	10.7	3.8	4.1
					B*	3.7	7.5	6.4	2.6	1.5	1.8	1.5	4.8	2.7	1.2	21.3	1.0	5.0	6.7	14.4	3.5	7.7	9.7	6.6
				Woolong tea	A	3.1	5.0	4.1	5.5	4.9	0.8	1.7	1.6	3.4	1.8	5.8	3.0	3.1	7.0	9.9	4.4	7.1	4.0	3.9
					B	6.8	1.2	9.5	1.7	8.9	5.2	2.3	7.8	10.2	4.2	5.7	4.1	4.4	11.6	14.2	1.9	5.7	1.5	8.5

(Continued)

Annex 2: RSD data for 201 pesticides in incurred tea Youden pair samples under 8 conditions with two SPE cartridge cleanup in three months (Nov.9, 2009-Feb.7, 2010) (cont.)

No	Pesticides	SPE	Method	Sample	Youden pair	Nov.9 2009 (n=5)	Nov.14 2009 (n=3)	Nov.19 2009 (n=3)	Nov.24 2009 (n=3)	Nov.29 2009 (n=3)	Dec.4 2009 (n=3)	Dec.9 2009 (n=3)	Dec.14 2009 (n=3)	Dec.19 2009 (n=3)	Dec.24 2009 (n=3)	Dec.14 2009 (n=3)	Jan.3 2010 (n=3)	Jan.8 2010 (n=3)	Jan.13 2010 (n=3)	Jan.18 2010 (n=3)	Jan.23 2010 (n=3)	Jan.28 2010 (n=3)	Feb.2 2010 (n=3)	Feb.7 2010 (n=3)
94	epoxiconazole-2	Cleanert TPT	GC-MS	Green tea	A	5.1	8.2	0.9	1.6	6.1	3.4	7.7	4.9	4.5	5.3	9.2	7.3	7.1	3.3	2.0	1.1	4.5	3.3	3.1
					B	3.5	6.7	3.6	8.6	3.4	1.6	0.4	8.1	2.7	10.5	8.4	9.4	2.8	1.4	7.7	3.5	5.4	2.2	4.0
				Woolong tea	A	7.2	5.0	3.9	1.1	1.5	2.9	10.1	12.6	4.1	11.1	13.0	4.6	5.2	1.7	5.1	3.0	2.1	1.8	2.4
					B	5.1	2.9	4.2	7.7	2.0	3.3	5.1	3.7	7.8	6.3	12.9	3.5	3.1	4.5	3.7	2.4	3.6	1.9	1.2
			GC-MS/MS	Green tea	A	4.1	9.1	0.5	1.9	9.0	5.6	6.2	7.4	5.1	4.3	10.4	4.5	2.7	7.2	3.2	4.2	3.3	2.7	4.3
					B	6.1	6.7	5.3	7.5	4.7	3.9	5.4	8.5	2.6	9.3	4.7	2.6	2.9	2.3	5.9	5.9	4.3	1.8	3.4
				Woolong tea	A*	7.5	1.3	3.2	0.5	0.8	3.0	10.9	3.1	2.4	4.5	4.0	1.5	4.7	1.7	3.0	1.2	1.4	3.9	2.9
					B*	3.2	3.9	5.5	2.4	6.3	3.2	2.1	10.4	5.1	4.8	9.0	3.0	3.4	8.6	6.5	1.4	2.3	2.4	4.0
		Envi-Carb+PSA	GC-MS	Green tea	A	11.2	11.8	7.1	7.3	7.1	2.2	7.7	1.5	4.6	5.3	6.5	7.0	2.1	4.1	6.2	5.6	10.4	5.1	4.5
					B	4.3	5.6	3.9	6.2	1.4	2.6	2.9	1.2	8.0	2.7	28.9	2.9	6.7	4.4	7.4	3.8	2.4	9.7	7.7
				Woolong tea	A	5.5	2.3	6.4	2.2	4.1	3.7	1.5	10.4	5.2	14.6	7.7	6.5	1.6	1.3	3.5	2.5	2.4	5.1	3.7
					B	3.9	2.6	12.5	5.6	16.4	5.8	3.6	7.1	8.7	3.1	10.3	5.8	3.1	4.5	7.8	2.6	7.6	0.8	2.7
			GC-MS/MS	Green tea	A	9.9	12.1	4.8	7.4	6.3	4.4	31.5	3.2	8.1	1.2	5.4	3.2	8.3	7.1	11.0	5.7	10.5	3.4	2.5
					B	7.8	5.1	6.0	5.2	0.7	5.1	6.4	0.8	6.3	4.1	18.4	6.4	2.2	7.7	10.5	2.1	4.8	8.6	10.1
				Woolong tea	A	6.3	1.9	5.0	2.4	4.1	3.5	3.8	6.2	1.9	12.2	7.7	4.1	3.3	5.7	11.0	3.4	5.6	3.0	3.9
					B*	6.2	3.4	12.6	5.8	12.8	6.2	2.1	7.9	4.9	0.4	2.7	1.8	2.0	14.6	2.9	2.8	8.6	2.3	4.4
95	EPTC	Cleanert TPT	GC-MS	Green tea	A	3.8	3.3	3.3	2.6	9.7	11.4	5.3	5.5	6.7	9.7	10.0	7.5	7.0	5.1	4.4	1.4	6.6	4.4	4.1
					B*	7.0	9.4	4.0	6.4	2.8	6.2	5.3	3.1	5.9	13.9	1.7	4.7	4.0	4.8	8.4	3.9	3.1	3.8	4.4
				Woolong tea	A	5.3	9.4	1.1	1.9	8.1	2.1	9.0	3.6	14.2	4.1	3.8	4.6	7.2	3.1	8.0	4.3	7.6	0.7	4.3
					B	8.6	8.8	4.4	4.5	4.6	7.1	7.4	2.3	6.6	8.6	16.8	1.9	3.3	5.9	2.0	1.3	4.8	2.1	6.2
			GC-MS/MS	Green tea	A	3.2	2.8	4.1	6.3	9.4	11.3	7.3	2.4	7.5	8.6	8.2	1.3	4.8	4.7	2.6	2.5	4.5	4.4	5.3
					B	9.4	10.4	3.4	2.6	4.5	8.1	6.7	2.7	7.2	17.7	4.7	1.7	6.3	7.5	9.5	10.7	7.5	1.7	4.0
				Woolong tea	A	9.4	4.5	1.0	6.1	8.3	3.7	11.2	1.6	14.9	7.4	2.0	9.2	13.5	11.3	7.9	1.6	14.3	5.1	7.5
					B	6.7	6.2	10.7	2.6	4.8	8.7	3.9	2.6	4.6	4.4	7.8	5.6	1.5	8.0	5.6	3.4	2.1	8.7	6.4
		Envi-Carb+PSA	GC-MS	Green tea	A	6.4	8.2	5.1	6.1	5.9	4.1	3.9	3.0	7.5	8.0	7.8	1.2	3.7	12.4	37.9	2.8	11.3	5.9	9.4
					B	2.0	4.4	3.7	8.0	2.5	3.9	3.5	8.6	3.4	2.7	33.4	2.2	19.1	3.7	5.5	17.2	6.2	6.4	7.5
				Woolong tea	A	5.9	5.6	5.5	4.5	4.6	3.7	3.5	6.9	5.5	1.8	6.5	3.6	4.8	3.1	5.3	3.6	7.1	1.3	4.7
					B	13.9	8.3	7.4	6.3	12.5	4.5	10.8	10.5	13.4	2.2	9.3	2.6	2.0	5.9	13.0	9.0	7.6	8.9	6.1
			GC-MS/MS	Green tea	A	6.8	12.6	2.6	6.0	5.2	9.7	5.2	0.7	8.6	3.0	10.6	4.2	12.3	8.8	43.5	4.2	14.1	9.3	1.0
					B	6.2	9.2	6.8	4.8	1.3	3.5	13.2	3.9	4.5	1.0	28.4	2.4	25.1	21.0	9.7	13.7	8.5	6.5	8.9
				Woolong tea	A*	4.2	3.9	4.3	4.5	4.2	4.1	3.5	4.4	0.6	5.1	11.0	1.7	6.0	8.6	7.0	1.7	9.3	8.5	4.9
					B	13.1	9.0	12.0	1.1	6.7	4.5	11.5	14.0	12.0	2.9	4.0	1.7	2.2	14.7	1.9	3.4	4.9	3.6	7.6
96	ethiumesate	Cleanert TPT	GC-MS	Green tea	A	7.0	8.3	2.9	1.8	7.0	6.8	15.8	3.9	10.8	3.8	9.0	5.0	9.1	0.8	8.2	2.8	3.0	0.7	5.0
					B	5.3	7.7	2.5	9.9	9.7	7.6	5.1	9.4	8.6	4.9	2.4	10.7	1.8	3.2	7.1	2.7	3.5	2.4	6.4
				Woolong tea	A	6.1	2.7	3.3	4.1	13.9	4.4	29.0	24.9	4.9	9.1	5.7	2.6	5.6	3.5	4.7	5.0	6.3	2.1	1.0
					B	2.8	2.1	2.9	6.1	7.6	3.8	12.0	18.6	1.7	9.4	6.6	2.8	4.4	5.7	4.3	0.9	10.8	5.3	1.4
			GC-MS/MS	Green tea	A*	3.8	5.9	3.1	5.9	9.1	3.7	6.9	8.1	5.5	2.5	11.9	6.1	5.7	3.9	2.1	4.2	3.1	3.0	5.9
					B	8.1	7.5	6.3	0.8	5.2	4.4	6.7	6.9	5.8	5.9	3.6	3.6	1.3	5.4	7.4	2.4	4.6	4.8	4.2
				Woolong tea	A	6.9	1.8	1.8	8.8	2.6	4.9	17.1	1.7	0.8	4.5	3.2	2.5	8.2	4.0	7.9	3.4	4.0	4.1	4.3
					B	2.8	1.0	3.9	3.2	4.1	2.0	6.8	7.7	2.9	4.3	5.2	2.0	5.1	0.8	4.9	3.5	4.7	8.9	4.0
		Envi-Carb+PSA	GC-MS	Green tea	A	9.1	6.1	3.0	5.4	4.2	3.8	3.7	10.5	20.0	0.9	7.7	3.8	5.8	5.5	13.0	12.5	9.1	11.4	5.2
					B*	5.0	8.3	3.0	5.1	3.6	4.0	3.7	15.4	7.9	5.3	17.4	5.0	5.8	2.6	4.3	4.4	5.2	5.6	5.8
				Woolong tea	A	3.1	6.0	3.8	1.4	12.7	2.3	4.6	47.5	22.0	9.1	6.7	2.6	1.1	3.1	13.2	2.8	5.3	2.3	3.3
					B	4.9	5.2	7.7	4.1	11.2	14.8	18.4	14.2	24.7	9.4	5.2	0.4	1.8	3.2	4.0	3.3	3.8	1.4	7.2
			GC-MS/MS	Green tea	A	8.8	6.7	3.3	6.1	7.2	5.5	11.0	4.1	4.7	1.6	6.8	4.1	4.6	15.4	12.3	7.6	10.4	11.7	1.2
					B	9.1	6.4	6.3	5.9	0.9	1.3	8.8	1.0	6.2	2.4	18.1	5.5	3.6	7.7	11.8	2.2	6.1	6.0	9.5
				Woolong tea	A	2.2	4.6	4.0	0.4	7.2	2.8	8.9	2.3	0.5	6.9	5.1	4.8	2.1	6.8	9.2	3.4	7.3	2.2	2.9
					B	8.0	2.2	11.3	5.3	5.9	4.1	2.7	6.6	7.3	2.5	3.7	4.1	6.5	8.2		3.5	3.3		8.2

(Continued)

97	ethoprophos	Cleanert TPT	GC-MS	Green tea	A	8.0	4.8	2.9	1.2	9.7	4.1	6.8	6.3	5.3	2.7	6.2	1.6	6.0	3.1	4.8	1.8	3.3	4.7	4.1
					B	7.2	8.4	4.1	8.0	4.2	2.1	2.3	1.2	3.7	6.8	2.0	7.4	0.3	3.4	5.2	1.8	3.5	1.9	5.9
				Woolong tea	A	4.3	0.9	2.1	3.7	2.3	2.2	13.7	8.8	3.2	5.9	4.9	0.8	3.1	2.5	8.5	3.0	3.9	2.6	1.2
					B	4.2	3.6	4.3	5.9	3.2	3.6	4.2	2.7	5.6	2.8	3.1	1.6	2.0	3.3	4.5	0.3	0.9	2.6	1.7
			GC-MS/MS	Green tea	A*	3.8	5.8	3.2	2.4	10.3	4.5	6.1	6.7	5.4	1.6	6.1	2.6	6.0	3.4	1.9	3.6	3.5	2.1	1.9
					B*	7.4	8.6	5.0	7.0	5.3	4.8	6.2	3.3	5.0	9.0	4.2	4.0	1.7	1.3	1.0	3.0	6.1	0.1	4.8
				Woolong tea	A	4.3	2.2	1.2	3.9	2.0	3.2	11.5	3.5	2.2	3.3	3.1	1.3	9.0	0.8	8.9	5.9	4.8	2.9	4.2
					B	3.2	2.5	4.8	5.7	5.3	3.6	2.7	7.1	5.7	3.1	4.3	2.4	3.0	8.2	6.7	2.3	1.8	2.8	4.3
		Envi-Carb+PSA	GC-MS	Green tea	A	8.4	3.2	6.1	5.3	6.7	3.4	4.4	4.2	2.8	2.0	7.6	0.6	1.8	6.0	3.1	10.7	10.2	10.1	4.8
					B	8.1	8.2	8.0	0.6	2.4	3.3	6.4	1.8	4.4	1.4	20.6	1.7	10.1	0.4	14.2	2.4	4.9	11.8	6.2
				Woolong tea	A	7.7	3.7	3.4	3.0	3.7	1.0	2.9	4.5	1.2	1.8	4.1	1.8	0.8	2.9	3.0	2.9	4.4	6.3	4.4
					B	6.4	5.9	8.3	1.9	12.8	5.4	1.3	6.2	11.9	3.0	8.6	2.0	4.9	3.3	10.0	3.4	6.3	2.4	4.3
			GC-MS/MS	Green tea	A*	10.5	5.7	5.4	6.3	5.4	5.0	4.4	3.7	4.9	0.3	7.7	1.3	15.7	16.4	4.5	7.7	9.7	5.5	2.1
					B*	17.2	7.4	4.6	3.6	1.0	1.9	10.5	0.9	5.1	1.1	15.3	5.5	8.5	9.3	13.8	2.7	7.8	11.4	9.1
				Woolong tea	A	4.5	4.6	8.4	2.5	3.6	0.9	1.9	5.0	0.7	2.6	7.7	2.7	2.9	5.7	10.0	5.9	6.7	2.4	2.0
					B	6.6	3.8	2.5	1.9	7.8	4.9	6.0	8.2	9.4	3.1	3.2	2.3	3.0	8.7	10.4	2.3	6.0	1.7	7.4
98	etrimfos	Cleanert TPT	GC-MS	Green tea	A*	10.4	3.9	7.1	1.2	9.1	3.6	6.5	4.9	4.5	3.0	7.1	2.5	5.8	8.0	9.6	2.8	18.0	7.6	3.4
					B*	5.7	8.4	2.8	7.5	4.3	2.9	1.5	1.0	3.5	8.0	5.7	6.1	1.0	6.4	11.7	1.8	14.8	3.7	6.4
				Woolong tea	A	4.3	1.0	4.4	3.7	3.2	2.5	9.9	9.3	3.4	5.4	6.5	2.1	2.6	2.2	7.7	6.4	2.6	2.5	1.6
					B	4.2	2.4	3.6	5.2	4.6	2.3	5.0	2.4	6.3	5.1	4.0	1.8	2.4	5.8	5.3	0.9	4.3	0.7	2.2
			GC-MS/MS	Green tea	A*	4.5	5.6	2.9	2.2	9.6	6.1	6.9	5.9	5.8	1.5	6.4	6.6	6.2	5.7	2.9	6.7	2.8	4.6	2.6
					B*	9.0	7.8	9.1	8.3	5.0	5.6	6.6	2.2	5.1	8.7	3.7	3.4	4.4	1.4	5.2	6.5	5.9	1.0	6.6
				Woolong tea	A	5.1	3.4	3.5	0.8	1.9	3.5	10.9	4.9	0.5	1.9	3.0	2.2	1.2	3.4	8.7	7.2	4.3	7.1	6.2
					B	4.4	2.5	3.8	5.3	4.4	2.7	1.9	8.1	5.9	6.2	2.4	4.9	3.1	2.3	6.3	2.9	3.9	1.8	8.0
		Envi-Carb+PSA	GC-MS	Green tea	A	8.9	4.2	3.5	7.1	6.9	3.4	25.5	1.5	2.6	2.1	8.0	4.9	3.0	4.8	9.0	5.2	3.9	7.6	4.6
					B	6.4	7.6	3.8	1.5	1.9	0.9	4.5	2.9	4.6	6.6	18.9	1.7	3.7	2.0	9.5	4.8	1.5	8.9	5.9
				Woolong tea	A	6.0	4.5	2.9	3.6	5.6	4.3	2.8	2.2	2.1	1.3	4.8	1.5	0.9	2.2	2.2	6.2	4.2	4.7	4.7
					B	6.2	2.6	9.1	2.8	14.4	4.8	2.8	5.9	11.0	2.1	7.3	1.6	4.4	2.7	10.0	5.4	6.6	1.4	4.5
			GC-MS/MS	Green tea	A	9.5	5.6	2.5	3.9	4.9	4.8	2.7	2.7	6.0	2.0	9.8	3.2	11.0	15.3	5.7	5.5	8.8	4.1	3.4
					B	21.5	10.3	4.1	4.4	3.1	1.5	9.7	1.6	4.4	4.8	12.3	4.4	2.8	7.2	8.6	1.4	6.4	12.1	8.2
				Woolong tea	A*	4.1	3.2	3.8	3.5	2.2	1.3	2.4	0.8	2.0	2.7	8.4	1.6	5.8	6.5	6.3	7.2	7.6	4.3	1.3
					B*	6.7	3.7	8.3	1.8	8.1	5.6	5.4	6.6	10.2	1.7	2.4	3.8	1.6	11.6	10.1	2.9	7.5	4.6	6.3
99	fenamidone	Cleanert TPT	GC-MS	Green tea	A	5.8	6.9	1.0	3.3	5.0	4.8	6.1	6.0	4.4	3.6	7.3	3.8	6.1	2.0	3.3	1.1	4.3	11.7	3.3
					B	3.7	9.1	6.3	8.6	1.7	2.7	1.5	6.6	3.3	6.0	5.5	6.0	0.4	0.7	5.0	3.6	6.1	2.9	5.6
				Woolong tea	A	6.8	2.0	3.5	1.0	1.7	3.3	3.4	6.6	1.9	5.9	12.3	9.7	6.6	2.1	6.8	1.4	2.0	4.0	3.8
					B	4.8	1.7	5.9	7.4	4.5	4.6	6.9	11.1	6.2	4.2	8.8	3.5	1.8	3.8	3.3	1.8	1.2	2.2	1.1
			GC-MS/MS	Green tea	A*	5.2	8.3	1.1	1.2	7.6	4.4	7.7	2.3	6.1	3.3	7.7	0.9	4.4	8.6	5.8	3.9	2.4	5.4	2.0
					B*	5.3	5.8	3.7	8.6	5.1	4.6	6.2	7.1	4.9	6.1	7.4	5.4	4.3	15.0	1.9	4.8	8.3	0.5	2.9
				Woolong tea	A	7.4	1.5	4.3	0.8	3.3	3.6	7.1	3.7	3.5	4.4	4.7	3.5	9.9	1.7	4.4	2.5	2.1	3.0	1.1
					B	2.8	2.6	3.4	3.7	5.8	2.5	3.5	7.6	1.8	2.5	8.9	3.8	1.6	1.3	7.6	4.6	3.5	4.7	5.2
		Envi-Carb+PSA	GC-MS	Green tea	A	7.8	8.6	3.9	9.6	6.0	2.9	23.3	7.6	4.2	4.9	3.2	2.9	4.9	6.5	4.7	10.1	12.1	5.8	3.3
					B	3.9	4.5	3.5	14.0	4.2	2.3	6.9	0.3	2.3	2.8	29.4	1.2	2.3	4.7	11.9	2.1	4.0	7.8	6.8
				Woolong tea	A	6.6	2.6	4.4	1.6	4.0	2.5	3.4	5.5	5.9	5.9	7.4	7.4	1.7	2.1	2.8	2.6	2.1	3.6	3.9
					B	5.0	3.5	10.9	4.9	14.2	5.3	2.5	8.9	1.5	3.2	9.7	2.2	2.1	3.8	8.3	3.1	4.3	3.2	2.9
			GC-MS/MS	Green tea	A	8.5	9.8	3.5	8.4	7.1	1.3	18.6	7.4	9.6	3.4	5.2	5.1	8.5	3.7	7.1	9.3	9.3	3.7	2.8
					B	4.3	3.9	6.4	5.0	2.9	2.6	2.8	1.4	7.3	1.1	19.1	9.0	4.5	5.6	10.2	2.0	7.9	8.9	11.7
				Woolong tea	A	6.5	1.8	5.7	2.0	4.8	1.5	3.1	3.0	5.6	11.3	8.6	4.7	5.3	5.9	9.0	2.5	7.6	4.5	3.1
					B	6.1	3.1	11.6	5.1	9.6	4.8	0.6	8.5	7.0	2.6	0.8	3.7	8.6	12.5	10.0	4.7	6.7	3.8	6.2

Annex 2: RSD data for 201 pesticides in incurred tea Youden pair samples under 8 conditions with two SPE cartridge cleanup in three months (Nov.9, 2009-Feb.7, 2010) (cont.)

No	Pesticides	SPE	Method	Sample	Youden pair	Nov.9 2009 (n=5)	Nov.14 2009 (n=3)	Nov.19 2009 (n=3)	Nov.24 2009 (n=3)	Nov.29 2009 (n=3)	Dec.4 2009 (n=3)	Dec.9 2009 (n=3)	Dec.14 2009 (n=3)	Dec.19 2009 (n=3)	Dec.24 2009 (n=3)	Dec.29 2009 (n=3)	Jan.3 2010 (n=3)	Jan.8 2010 (n=3)	Jan.13 2010 (n=3)	Jan.18 2010 (n=3)	Jan.23 2010 (n=3)	Jan.28 2010 (n=3)	Feb.2 2010 (n=3)	Feb.7 2010 (n=3)
100	fenaromol	Cleanert TPT	GC-MS	Green tea	A	8.1	4.6	0.5	2.4	4.6	7.1	4.9	10.8	4.5	5.7	11.0	15.0	10.1	10.8	3.3	1.4	5.2	1.8	5.2
					B	4.5	10.4	5.3	11.0	5.7	2.5	1.9	2.3	3.6	10.0	7.4	10.1	3.1	1.3	2.2	1.7	10.9	6.8	7.7
				Woolong tea	A	7.3	4.0	2.0	1.5	1.4	3.9	10.8	8.9	7.5	9.0	13.0	2.1	14.3	2.7	8.0	5.3	2.6	6.4	2.6
					B	4.3	13.9	6.2	11.8	1.5	3.2	4.9	0.9	11.4	3.7	1.2	3.9	3.7	5.4	1.6	2.5	1.2	10.4	2.0
			GC-MS/MS	Green tea	A	5.7	6.4	0.5	1.9	9.5	6.5	6.4	5.3	7.1	3.0	6.9	3.7	1.6	6.4	2.0	6.2	2.8	7.5	2.7
					B	6.8	6.6	5.5	7.1	5.1	3.8	4.5	7.5	4.5	5.8	5.4	7.8	3.9	5.2	2.1	4.6	6.6	4.2	2.4
				Woolong tea	A*	7.2	0.9	3.9	0.9	1.4	3.3	3.6	4.1	5.6	5.2	7.1	4.1	8.5	2.0	5.5	2.7	5.1	4.5	0.8
					B*	4.4	2.8	2.9	7.6	6.5	4.7	2.1	9.5	6.1	4.0	5.0	1.7	1.2	5.2	7.2	4.0	4.5	5.3	3.5
		Envi-Carb+PSA	GC-MS	Green tea	A	12.9	12.7	4.7	7.5	2.4	5.8	4.6	7.9	23.8	5.2	2.6	11.3	4.9	8.5	14.1	10.2	7.5	10.7	9.1
					B	5.1	7.1	4.7	8.1	4.4	2.7	4.5	5.0	15.2	5.9	16.7	8.7	4.8	6.9	5.8	10.1	5.3	7.2	7.9
				Woolong tea	A	7.1	8.5	6.1	2.6	3.0	4.5	4.9	7.1	3.7	13.1	8.8	6.2	7.4	1.5	1.0	3.6	5.4	7.8	3.7
					B	4.1	3.5	7.9	12.0	11.5	5.5	4.9	10.0	6.8	6.6	8.6	6.2	6.4	5.0	7.5	6.3	2.1	1.7	7.2
			GC-MS/MS	Green tea	A	8.1	10.2	4.1	6.6	7.1	1.5	18.7	2.6	6.0	0.8	5.1	1.6	7.9	9.4	6.8	9.5	10.1	7.6	2.8
					B	6.5	5.4	7.2	6.9	0.4	3.8	3.3	0.1	6.7	1.6	21.7	1.6	3.3	8.4	18.6	2.5	8.1	10.3	10.8
				Woolong tea	A	6.3	2.0	6.1	0.9	6.1	1.3	3.9	7.2	4.6	9.3	8.9	6.4	5.0	7.5	9.3	2.7	6.5	3.4	3.2
					B	6.9	4.3	13.2	5.0	9.6	3.2	1.3	7.7	7.3	3.2	3.7	2.8	5.7	10.6	10.6	4.0	1.2	3.3	6.9
101	flamprop-isopropyl	Cleanert TPT	GC-MS	Green tea	A	6.6	6.7	1.8	2.1	7.9	5.3	7.9	6.4	4.0	3.0	4.5	4.6	7.3	5.6	4.6	1.7	3.3	6.9	4.3
					B	5.3	6.3	3.8	7.6	4.3	2.9	2.3	2.4	6.5	5.3	4.3	9.4	1.6	3.3	8.0	3.8	4.4	1.1	5.4
				Woolong tea	A	9.4	5.8	3.3	3.2	1.8	4.2	9.7	5.8	9.4	0.9	4.2	2.4	5.0	2.5	7.9	5.1	2.5	4.9	1.1
					B	6.7	2.6	4.6	7.5	3.1	3.4	5.6	5.6	0.8	6.9	1.4	5.2	2.6	4.4	1.4	1.7	3.4	0.9	2.3
			GC-MS/MS	Green tea	A	5.2	5.1	2.5	1.7	8.6	3.2	5.4	6.9	4.5	1.8	6.4	5.6	2.1	3.0	3.6	4.8	2.1	4.1	3.7
					B	6.6	6.4	6.1	8.3	6.8	4.6	7.7	4.3	6.6	6.7	4.2	3.8	2.6	6.2	1.3	4.2	3.3	1.4	3.3
				Woolong tea	A	7.3	1.9	3.6	1.8	2.8	2.7	8.2	2.7	2.4	4.1	1.8	1.9	5.8	1.4	7.9	4.9	2.9	4.7	4.0
					B	3.6	2.9	5.5	6.2	4.0	3.5	0.5	7.0	4.8	4.2	6.5	3.2	1.2	5.6	6.3	1.3	4.7	2.4	3.3
		Envi-Carb+PSA	GC-MS	Green tea	A	5.7	3.4	1.9	7.1	8.8	1.6	27.0	0.9	7.8	2.5	7.3	2.2	2.5	4.5	6.7	6.9	11.2	2.3	6.4
					B	2.9	6.8	2.5	3.9	3.5	1.3	8.7	4.3	3.1	4.4	22.9	3.2	1.2	5.0	10.4	1.2	3.9	6.9	6.7
				Woolong tea	A	4.4	1.4	2.8	3.7	6.2	1.2	4.1	1.1	1.5	2.8	8.1	6.8	1.2	2.5	3.9	4.5	5.3	5.2	4.0
					B	4.7	1.7	7.5	5.9	9.6	4.2	1.5	5.7	10.3	5.2	5.8	3.1	3.1	5.1	9.2	4.5	5.5	5.9	4.8
			GC-MS/MS	Green tea	A	6.4	5.2	2.6	6.4	10.4	3.1	4.5	7.1	5.9	0.1	9.6	1.5	10.9	15.2	6.7	5.8	9.1	3.2	4.1
					B	4.8	8.0	6.7	1.2	1.3	1.4	2.9	1.6	4.5	6.0	17.3	1.0	1.8	3.1	11.4	2.6	5.5	14.9	10.1
				Woolong tea	A	6.3	3.5	4.2	1.9	7.6	1.6	2.0	2.3	0.5	2.4	9.8	2.0	2.0	6.5	9.7	3.2	5.5	3.4	4.9
					B	6.6	3.2	10.4	4.0	8.6	5.1	0.7	7.1	9.7	3.9	3.0	2.6	3.4	8.6	10.2	2.2	3.7	3.5	7.9
102	flamprop-methyl	Cleanert TPT	GC-MS	Green tea	A*	7.2	4.9	3.2	2.7	12.2	4.1	11.3	2.2	7.7	3.7	7.5	7.1	6.6	5.0	1.6	2.4	5.4	3.2	5.7
					B*	4.1	6.6	5.9	10.9	6.5	8.1	2.9	10.3	4.1	8.9	0.5	4.2	4.6	3.0	4.3	5.2	4.1	0.7	5.2
				Woolong tea	A	6.1	3.2	0.6	4.1	6.1	4.0	12.6	2.0	6.5	5.1	7.6	10.1	2.7	2.1	8.6	2.1	3.1	2.7	3.0
					B	5.2	4.6	6.5	11.4	5.7	9.9	5.3	16.8	8.5	3.1	6.7	3.8	1.4	3.8	1.8	0.8	2.0	2.1	1.9
			GC-MS/MS	Green tea	A	4.7	5.7	2.7	2.9	10.0	2.7	7.3	9.9	5.1	2.8	7.0	4.3	5.3	2.9	4.6	4.9	3.9	3.5	3.7
					B	6.1	7.2	5.0	9.1	5.7	4.8	6.5	6.8	5.0	7.7	5.2	2.4	6.2	7.6	4.7	5.7	5.1	1.6	5.9
				Woolong tea	A	7.3	1.9	3.6	1.8	2.8	2.7	7.4	0.0	0.8	4.1	3.7	1.3	4.4	0.8	7.1	3.2	5.1	4.5	7.1
					B	3.6	2.9	5.5	6.2	4.0	3.5	2.8	5.9	3.1	5.9	5.2	1.8	1.8	6.7	7.2	2.0	0.4	1.7	7.9
		Envi-Carb+PSA	GC-MS	Green tea	A	11.7	4.7	8.0	7.4	7.8	1.0	17.7	11.7	12.2	1.2	3.9	4.5	2.8	3.4	2.2	3.9	9.1	6.9	5.9
					B	10.9	5.2	3.8	5.1	6.1	5.3	5.2	12.6	7.2	5.9	20.6	6.0	6.9	0.3	7.6	7.9	1.6	12.3	7.1
				Woolong tea	A	2.2	23.2	5.1	2.7	7.9	2.2	5.6	4.1	1.0	6.2	5.5	6.5	0.9	2.1	3.7	5.2	5.0	6.7	2.1
					B	5.2	6.6	11.7	5.3	13.3	7.8	1.2	6.4	10.7	4.1	17.7	3.5	2.6	3.8	9.9	1.9	5.4	1.4	6.5
			GC-MS/MS	Green tea	A	12.1	4.6	5.9	5.8	7.5	9.9	11.9	1.5	7.3	2.5	6.8	2.8	4.5	13.7	4.6	2.8	9.8	2.1	6.1
					B	13.6	6.9	4.9	6.0	3.7	0.6	15.4	5.5	9.7	0.7	11.7	2.3	2.1	4.2	8.4	4.4	2.4	9.7	10.1
				Woolong tea	A	6.3	3.5	4.2	1.9	7.6	1.6	2.0	2.4	0.5	2.4	9.8	2.0	2.0	6.5	12.1	4.9	3.4	3.4	3.1
					B	6.6	3.2	10.4	4.0	8.6	5.1	0.7	7.1	9.7	3.9	3.0	2.6	3.4	8.6	10.1	1.4	3.7	3.5	8.3

No.	Compound	Cartridge	Method	Tea	Level	1	2	3	4	5	6	7	8	9	10	11	12	13	14	15	16	17	18	19
103	fonofos	Cleanert TPT	GC-MS	Green tea	A	9.1	5.4	5.6	1.9	8.5	4.1	6.2	4.5	4.3	1.6	7.0	2.8	7.1	17.4	20.2	1.6	6.3	6.7	4.4
					B*	5.7	8.3	6.9	7.9	3.5	3.5	2.1	0.6	4.4	8.4	8.2	7.4	1.0	6.0	22.5	2.5	10.3	2.2	5.8
				Woolong tea	A	4.2	1.2	2.4	3.3	1.0	2.6	6.6	5.5	0.2	8.0	6.8	3.4	3.2	2.6	9.3	3.5	3.8	3.0	0.5
					B	4.2	4.8	4.1	1.4	2.5	3.1	4.4	3.0	5.2	3.9	4.6	0.3	3.0	3.2	3.7	1.4	2.2	2.2	1.0
			GC-MS/MS	Green tea	A	7.9	9.5	3.6	2.2	10.1	3.4	5.2	4.5	5.3	3.9	6.5	4.2	8.3	3.7	2.3	4.0	2.3	3.1	3.9
					B	2.8	3.4	5.1	9.1	5.2	6.0	6.2	3.0	5.4	1.3	4.9	3.1	2.6	3.9	1.0	0.4	5.1	2.0	4.5
				Woolong tea	A	2.9	0.5	1.9	3.0	1.9	3.4	2.4	3.1	5.4	3.7	3.4	1.4	5.4	1.1	7.6	4.6	3.7	3.6	5.3
					B	5.5	1.7	4.5	6.0	5.9	2.6	4.0	5.3	1.4	4.5	2.9	4.3	1.8	1.4	5.8	2.3	2.8	4.0	4.3
		Envi-Carb+PSA	GC-MS	Green tea	A	4.9	5.6	4.6	8.4	3.7	1.5	4.4	1.4	4.9	6.0	5.5	4.3	3.8	2.2	4.4	6.6	9.6	3.4	4.2
					B	6.4	4.4	1.1	5.2	2.6	5.7	2.8	2.2	2.9	5.6	18.9	4.4	4.8	2.0	11.8	2.5	6.4	7.9	5.3
				Woolong tea	A	5.0	3.0	3.4	2.2	3.4	0.8	1.9	1.3	2.3	2.9	5.2	1.8	1.0	2.6	0.9	2.3	5.8	4.5	4.3
					B	6.7	6.6	9.2	2.7	13.4	4.2	2.3	6.5	11.3	3.7	8.3	1.1	4.3	3.2	8.5	3.7	6.7	4.2	3.3
			GC-MS/MS	Green tea	A	4.5	6.6	2.3	5.8	6.0	0.2	5.4	2.7	5.1	3.7	7.7	7.4	15.6	14.5	5.9	8.5	9.9	2.7	2.1
					B	4.1	4.0	7.1	1.0	1.1	3.1	2.4	2.9	2.4	1.0	18.3	5.5	6.1	3.8	13.8	2.5	7.6	10.0	6.2
				Woolong tea	A	6.7	3.2	4.6	1.6	5.0	1.0	1.6	1.4	1.1	1.5	7.3	1.8	3.4	5.7	8.5	4.6	6.7	3.6	4.8
					B	3.4	5.0	9.8	1.9	7.6	4.3	1.6	7.4	8.2	3.6	0.7	2.1	1.9	8.8	10.8	2.3	5.0	2.3	4.6
104	hexachloro-benzene	Cleanert TPT	GC-MS	Green tea	A	6.2	5.0	3.1	0.7	14.7	5.2	6.9	9.1	5.3	3.6	4.1	4.0	7.3	5.2	2.4	3.1	2.6	6.9	4.4
					B	4.1	7.7	3.1	9.6	12.5	3.8	3.1	2.4	3.2	4.1	1.0	12.3	2.2	6.2	3.1	4.3	6.1	6.8	6.1
				Woolong tea	A	15.0	4.9	9.9	6.9	7.7	10.6	8.6	5.7	1.7	8.9	1.0	3.8	6.7	3.1	6.7	25.0	1.9	1.5	18.2
					B	4.1	6.4	14.5	4.4	7.8	14.1	5.8	0.4	18.9	8.6	2.8	3.8	9.4	1.5	3.8	6.2	15.2	1.9	7.9
			GC-MS/MS	Green tea	A	8.7	5.0	1.8	2.3	17.4	4.6	7.8	8.9	4.7	2.5	1.5	3.0	8.8	4.8	1.0	5.7	3.3	7.0	3.4
					B	5.1	7.6	3.6	8.1	13.0	5.5	5.6	5.9	5.7	4.2	6.3	5.5	4.6	4.5	0.2	3.5	8.4	9.6	3.3
				Woolong tea	A	15.5	4.2	9.6	6.4	6.2	10.7	7.2	0.1	3.8	7.9	4.3	9.2	5.7	2.5	6.9	23.6	2.4	6.6	19.5
					B	6.9	11.3	6.4	8.8	16.6	5.0	4.8	4.6	16.7	10.6	6.0	6.3	13.1	2.5	7.0	10.3	13.8	3.4	4.8
		Envi-Carb+PSA	GC-MS	Green tea	A	2.7	4.9	6.5	6.5	6.6	2.3	8.6	2.2	6.0	16.2	10.1	5.6	2.9	3.0	4.5	9.4	12.5	7.4	4.4
					B	5.0	4.0	5.8	5.6	3.0	2.2	8.6	2.4	2.4	0.5	21.5	1.3	6.4	0.6	14.1	4.8	6.6	10.9	7.0
				Woolong tea	A	6.2	4.0	4.3	3.0	4.1	2.2	0.7	3.0	1.2	1.4	4.3	1.6	1.8	3.1	1.8	4.3	6.1	6.0	4.6
					B	7.5	3.9	7.4	3.4	9.0	1.0	1.3	6.8	11.6	2.5	7.4	3.9	3.1	1.5	2.5	3.4	7.5	2.7	4.4
			GC-MS/MS	Green tea	A	4.6	5.6	4.1	7.3	6.2	4.9	2.7	1.9	6.3	3.6	9.7	3.0	16.2	20.4	5.8	10.6	14.2	7.1	3.2
					B	4.3	2.5	6.6	3.2	1.0	1.6	2.9	1.9	3.7	1.1	21.4	2.3	6.3	12.1	15.8	6.3	8.8	10.5	6.7
				Woolong tea	A	8.0	3.3	5.8	4.1	3.8	1.8	1.5	5.4	3.7	6.0	8.6	13.1	4.4	6.1	8.6	24.6	7.8	6.9	1.1
					B	11.0	1.5	8.2	1.6	2.5	1.8	0.6	5.5	0.5	7.7	11.2	5.0	2.8	10.3	11.8	7.0	7.3	3.8	6.7
105	hexazinone	Cleanert TPT	GC-MS	Green tea	A*	4.9	17.6	4.7	7.3	2.5	2.5	10.1	3.1	9.2	7.7	7.6	0.7	3.0	15.2	18.4	5.2	12.9	1.0	2.8
					B*	8.2	11.3	2.8	11.9	3.2	7.8	1.8	8.5	4.3	13.5	3.0	8.8	3.0	8.7	5.6	7.9	8.9	5.9	5.1
				Woolong tea	A	11.7	2.8	2.2	7.9	9.0	4.2	7.8	8.9	4.8	10.0	7.5	3.8	2.5	3.0	9.5	4.5	3.1	4.4	7.2
					B	10.2	19.7	9.6	14.2	4.7	5.3	8.1	3.0	6.4	3.0	7.5	5.4	4.0	4.6	9.9	13.3	1.0	2.7	8.5
			GC-MS/MS	Green tea	A	7.8	16.5	3.3	9.4	4.7	6.6	0.4	1.1	3.9	17.3	13.7	2.8	1.7	18.3	12.3	5.6	15.1	6.9	1.0
					B	9.3	5.8	5.0	13.3	5.0	10.8	5.2	8.6	6.3	3.1	6.8	3.0	6.0	20.3	7.8	10.8	7.9	7.2	5.1
				Woolong tea	A	9.9	3.6	6.3	8.1	6.8	4.4	7.8	3.8	1.0	5.8	4.5	3.7	3.6	1.0	7.7	7.3	3.8	4.4	4.9
					B	13.4	15.2	14.5	6.8	4.0	5.4	8.7	11.1	4.3	7.4	10.2	2.1	1.0	2.0	6.4	7.3	9.8	4.8	2.3
		Envi-Carb+PSA	GC-MS	Green tea	A	7.6	13.1	6.8	4.3	3.1	4.3	5.1	1.5	12.3	9.1	15.0	12.5	9.4	3.2	8.4	3.5	2.1	9.1	8.3
					B	9.0	4.0	5.9	4.3	5.2	3.9	6.5	5.0	3.1	19.4	9.8	7.7	7.2	3.4	10.9	3.7	1.8	3.4	9.5
				Woolong tea	A	8.2	2.4	11.8	8.4	7.0	6.9	6.2	7.9	4.2	4.0	9.8	7.7	4.7	2.7	6.5	11.2	20.6	2.3	3.5
					B	15.0	4.1	18.4	7.0	16.9	15.1	5.5	8.8	2.8	5.6	10.8	0.6	6.9	17.8	6.6	4.5	11.5	1.3	3.8
			GC-MS/MS	Green tea	A	8.5	26.0	11.7	7.5	2.5	3.4	54.2	3.9	12.6	10.4	11.4	0.8	12.3	16.5	6.7	5.3	5.1	7.8	5.6
					B	10.7	4.5	2.7	2.9	6.3	5.6	15.8	8.8	4.6	2.0	11.4	12.0	6.8	6.4	18.5	10.7	8.1	5.5	10.2
				Woolong tea	A	9.6	3.3	28.2	8.8	14.2	14.9	13.7	9.0	3.3	2.0	7.7	9.8	5.8	31.6	7.5	7.3	19.7	4.6	4.6

(Continued)

Annex 2: RSD data for 201 pesticides in incurred tea Youden pair samples under 8 conditions with two SPE cartridge cleanup in three months (Nov.9, 2009-Feb.7, 2010) (cont.)

No	Pesticides	SPE	Method	Sample	Youden pair	Nov.9, 2009 (n=5)	Nov.14, 2009 (n=3)	Nov.19, 2009 (n=3)	Nov.24, 2009 (n=3)	Nov.29, 2009 (n=3)	Dec.4, 2009 (n=3)	Dec.9, 2009 (n=3)	Dec.14, 2009 (n=3)	Dec.19, 2009 (n=3)	Dec.24, 2009 (n=3)	Dec.14, 2009 (n=3)	Jan.3, 2010 (n=3)	Jan.8, 2010 (n=3)	Jan.13, 2010 (n=3)	Jan.18, 2010 (n=3)	Jan.23, 2010 (n=3)	Jan.28, 2010 (n=3)	Feb.2, 2010 (n=3)	Feb.7, 2010 (n=3)
106	iodofenphos	Cleanert TPT	GC-MS	Green tea	A*	6.9	2.3	3.8	9.0	7.3	2.0	7.3	5.2	4.1	3.0	4.3	4.9	3.5	8.2	6.5	3.0	5.1	7.0	1.8
				Green tea	B*	9.2	11.5	4.8	6.3	5.0	6.0	3.3	0.3	1.8	5.3	4.1	6.8	6.8	6.5	11.3	4.4	8.6	7.4	12.4
				Woolong tea	A*	10.6	4.2	2.7	4.0	4.7	3.3	12.8	13.9	7.7	8.7	5.6	3.4	4.2	2.6	10.1	1.7	8.6	7.4	6.4
				Woolong tea	B	5.6	2.0	2.4	3.4	2.3	1.4	11.9	1.4	4.9	7.5	7.0	1.2	4.7	5.6	8.9	2.0	5.3	4.9	4.1
			GC-MS/MS	Green tea	A	6.3	9.5	10.4	5.9	8.3	2.5	2.6	10.4	10.9	6.1	5.6	6.5	8.9	6.3	11.1	6.0	12.9	6.3	4.2
				Green tea	B	4.6	10.5	6.8	6.6	7.8	8.7	12.1	5.0	7.7	12.1	15.1	1.8	7.1	4.2	3.9	7.7	9.9	3.4	10.5
				Woolong tea	A*	12.5	9.5	12.1	1.6	4.8	4.9	6.1	4.7	9.8	8.3	4.6	4.6	9.3	9.4	13.3	4.3	7.3	10.3	8.7
				Woolong tea	B	4.3	1.9	10.3	6.5	2.5	3.0	4.4	5.2	18.1	19.6	6.9	1.0	12.5	7.6	13.4	1.9	5.2	7.2	19.0
		Envi-Carb+PSA	GC-MS	Green tea	A	12.1	5.7	4.8	6.3	9.0	9.4	5.5	8.1	4.1	2.4	16.7	9.2	2.7	3.0	4.9	4.0	10.8	13.4	7.8
				Green tea	B	15.0	6.7	3.3	8.8	5.0	1.6	3.0	11.8	7.0	11.0	16.6	9.5	9.9	4.8	3.4	11.4	0.5	18.6	9.0
				Woolong tea	A	8.1	0.3	2.7	5.2	3.6	3.3	12.3	2.0	6.9	7.2	1.5	1.5	7.0	3.5	7.4	4.3	1.8	8.9	2.9
				Woolong tea	B	9.6	9.9	7.8	6.0	7.4	7.4	5.9	5.9	12.2	2.5	11.8	4.7	6.6	5.6	14.3	5.1	10.1	10.3	6.1
			GC-MS/MS	Green tea	A	21.0	5.1	10.3	3.3	8.1	19.8	2.8	11.0	17.1	8.7	13.8	5.3	6.5	18.4	11.2	3.0	13.6	14.3	16.5
				Green tea	B	42.0	10.6	2.3	16.6	10.4	8.3	16.6	4.2	7.4	18.4	8.7	2.6	2.8	10.9	1.5	15.2	10.2	21.4	12.1
				Woolong tea	A*	5.4	0.4	5.0	11.1	6.8	4.6	2.8	7.0	4.0	3.4	8.2	9.2	4.6	7.6	20.4	4.3	5.7	1.0	2.4
				Woolong tea	B	4.2	9.0	11.1	8.4	2.1	4.4	12.2	4.6	12.3	2.4	1.8	11.0	5.9	19.6	16.8	1.9	6.3	9.2	11.4
107	isofenphos	Cleanert TPT	GC-MS	Green tea	A*	7.7	4.0	4.1	1.5	7.2	5.0	12.5	7.7	3.8	6.6	11.0	7.2	5.6	8.4	4.6	1.8	5.2	3.4	4.4
				Green tea	B	5.4	8.5	4.8	8.4	5.3	2.8	7.2	3.5	5.7	5.5	9.2	13.8	1.2	3.5	3.0	5.8	5.4	0.5	5.3
				Woolong tea	A	5.7	4.3	5.3	3.3	1.5	3.5	7.2	13.2	4.7	12.8	13.1	3.3	4.0	2.1	3.0	10.3	3.0	2.9	0.9
				Woolong tea	B	3.2	7.2	8.0	5.2	3.7	3.1	4.5	10.6	7.5	9.7	15.4	0.4	1.3	3.8	9.4	1.5	4.8	0.7	2.7
			GC-MS/MS	Green tea	A	4.2	2.3	3.6	1.9	10.7	5.3	6.2	8.7	7.2	1.7	5.8	4.1	9.7	1.4	3.6	5.9	4.8	7.3	4.2
				Green tea	B	7.6	6.8	5.5	9.4	7.3	4.7	7.0	4.1	5.5	7.4	4.7	2.5	0.4	2.5	0.2	3.2	4.8	0.6	4.1
				Woolong tea	A	6.8	3.2	2.7	3.1	1.3	5.4	10.1	2.4	5.3	4.4	4.7	1.9	2.1	2.1	10.0	11.6	7.1	2.5	5.0
				Woolong tea	B	3.2	5.9	6.2	9.0	6.1	2.5	1.4	7.4	5.2	3.0	6.1	1.3	6.1	5.0	6.8	2.7	0.6	1.8	6.4
		Envi-Carb+PSA	GC-MS	Green tea	A	9.8	3.9	6.5	7.2	9.1	2.8	6.3	5.3	13.8	8.3	4.9	3.4	2.2	11.3	8.6	28.3	9.6	25.3	4.9
				Green tea	B	3.2	9.5	6.6	4.9	0.7	5.4	6.1	1.8	9.9	4.5	38.1	9.3	13.4	1.5	26.1	2.3	6.9	20.5	7.3
				Woolong tea	A	5.7	3.3	1.2	3.0	5.1	2.1	4.6	12.0	6.9	5.0	5.2	2.8	2.7	1.5	4.9	2.6	7.8	5.2	3.6
				Woolong tea	B	5.3	8.0	11.2	2.4	17.4	6.8	3.8	5.6	14.8	7.6	9.3	8.7	4.7	1.7	12.9	0.1	7.3	4.0	5.1
			GC-MS/MS	Green tea	A	10.7	6.8	3.6	5.5	8.9	1.6	8.7	4.6	2.6	2.1	10.2	2.6	5.7	3.0	8.4	12.1	9.2	4.3	12.6
				Green tea	B	4.4	8.2	6.6	2.7	1.6	5.2	6.3	0.8	7.3	0.5	21.5	0.5	8.4	3.6	16.3	1.4	5.7	12.1	3.8
				Woolong tea	A*	3.9	2.4	3.7	1.6	3.8	1.6	3.0	5.5	0.9	2.7	6.1	3.1	2.4	2.7	10.4	11.6	5.2	4.5	8.7
				Woolong tea	B	7.1	3.3	9.4	1.3	10.5	6.2	2.5	7.9	10.2	4.3	2.6	2.0	4.9	8.1	11.0	2.7	5.7	3.7	8.7
108	isopeopalin	Cleanert TPT	GC-MS	Green tea	A	4.4	4.7	4.1	0.9	10.4	4.4	7.2	6.7	4.4	7.5	11.4	7.8	6.7	7.8	4.0	0.9	4.2	8.0	5.4
				Green tea	B	4.8	8.9	5.6	10.6	4.3	3.2	1.7	4.4	4.8	9.0	9.7	10.3	1.2	5.4	2.2	4.0	7.0	1.6	4.0
				Woolong tea	A*	5.9	1.6	5.5	3.6	5.8	4.0	7.2	15.5	4.5	12.1	14.0	3.1	8.7	2.8	8.2	1.5	4.0	3.1	2.4
				Woolong tea	B	3.2	4.5	4.2	6.5	5.0	2.1	5.6	3.0	7.8	7.2	16.0	0.4	2.4	4.0	4.4	1.0	1.1	4.4	2.3
			GC-MS/MS	Green tea	A	2.7	4.6	2.0	3.1	9.7	6.2	5.6	8.9	6.1	1.8	10.2	4.2	12.1	4.8	3.6	3.4	7.0	4.7	4.4
				Green tea	B	8.0	10.7	5.5	10.6	7.5	2.5	4.8	4.9	6.2	11.4	5.0	8.9	4.8	6.1	1.6	6.7	5.1	2.2	5.9
				Woolong tea	A*	4.2	1.9	5.2	2.2	1.7	4.8	8.5	5.4	3.9	5.8	8.0	4.1	5.7	1.7	8.6	8.7	2.7	1.8	2.4
				Woolong tea	B*	2.8	2.1	3.1	7.6	7.5	3.3	4.0	9.9	5.2	3.1	13.2	2.3	3.3	8.2	3.0	2.5	2.6	0.6	3.5
		Envi-Carb+PSA	GC-MS	Green tea	A	10.9	7.8	5.9	5.8	9.9	8.8	8.1	6.9	3.5	5.6	8.2	2.0	0.9	6.1	7.1	9.3	11.0	8.8	6.9
				Green tea	B	2.9	7.8	6.2	2.5	2.7	1.3	7.9	3.9	6.5	5.5	30.6	9.6	6.6	1.1	13.3	0.7	6.0	10.9	7.8
				Woolong tea	A	5.5	5.9	3.8	1.1	3.9	8.0	4.8	7.4	4.6	6.0	4.9	5.9	2.7	2.8	0.7	9.3	6.9	6.5	5.8
				Woolong tea	B	4.2	3.4	9.3	5.5	17.7	1.1	4.8	7.4	12.9	5.7	14.5	6.6	8.2	4.0	12.2	3.8	10.0	3.3	5.4
			GC-MS/MS	Green tea	A	11.3	5.3	6.0	2.3	7.8	6.3	19.2	8.0	6.4	3.2	7.3	3.4	5.5	2.5	5.9	7.6	10.1	5.0	9.2
				Green tea	B	8.6	7.5	7.8	4.3	1.3	1.1	4.8	5.3	6.8	3.8	20.6	1.7	4.9	5.0	8.7	5.0	5.0	9.2	8.9
				Woolong tea	A	4.8	4.2	3.9	4.3	4.3	2.3	4.8	3.1	0.5	4.3	6.2	1.7	5.3	7.9	8.2	8.7	7.6	3.9	3.7
				Woolong tea	B	5.4	3.8	8.5	2.4	8.9	7.4	11.9	7.4	10.6	3.9	4.5	1.8	1.6	11.2	12.1	2.5	7.3	1.6	6.4

#	Analyte	Cartridge	Detection	Tea	Sample																			
109	methoprene	Cleanert TPT	GC-MS	Green tea	A	5.4	5.6	5.4	2.7	6.9	4.0	7.1	4.0	3.8	4.5	9.0	6.4	7.2	4.7	4.3	1.7	3.8	7.1	3.8
				Woolong tea	B	7.0	8.8	6.7	8.5	3.3	4.7	3.6	1.9	5.6	8.9	7.8	11.7	0.7	1.0	7.7	2.1	4.2	1.0	6.1
			GC-MS/MS	Green tea	A*	8.0	1.1	1.0	4.0	8.1	3.4	9.6	8.7	1.3	7.4	10.0	1.5	4.7	3.2	6.3	2.6	4.2	2.0	1.0
				Woolong tea	B*	2.5	3.2	4.1	3.9	4.6	2.8	6.0	6.8	5.4	5.4	9.9	3.0	3.9	2.1	3.8	1.2	1.8	1.6	3.7
		Envi-Carb+PSA	GC-MS	Green tea	A	5.1	5.5	4.9	0.8	11.2	6.2	3.5	4.6	7.1	2.8	5.6	5.1	10.2	6.0	5.6	3.7	0.2	5.6	4.5
				Woolong tea	B	9.6	9.3	6.8	10.5	5.9	6.0	9.6	2.2	5.8	8.5	5.3	7.9	1.5	5.8	6.0	4.3	5.9	1.9	5.3
			GC-MS/MS	Green tea	A*	5.5	0.6	4.9	4.6	2.4	5.3	11.4	6.8	3.2	5.0	6.1	1.2	5.9	2.8	9.4	4.9	4.8	4.6	4.4
				Woolong tea	B*	5.0	0.4	2.4	8.8	6.3	2.6	4.0	6.8	1.4	4.8	8.5	1.5	2.3	4.5	5.0	2.1	8.5	1.4	7.6
				Green tea	A*	8.2	6.6	2.1	4.0	8.4	2.3	7.6	1.6	5.0	3.8	8.8	6.6	2.8	5.3	6.2	8.4	11.2	3.3	5.7
				Woolong tea	B*	2.5	8.5	8.2	5.8	2.2	4.7	14.7	3.0	3.1	2.9	33.1	8.0	4.3	2.8	13.6	2.2	5.9	8.0	6.0
				Green tea	A	3.3	3.8	6.0	4.2	13.9	0.9	12.3	6.5	3.0	6.2	7.3	6.9	0.6	3.1	2.4	1.8	5.3	4.7	4.8
				Woolong tea	B	5.0	4.8	11.9	3.2	8.0	2.7	4.3	9.4	12.0	5.9	10.7	2.7	3.0	4.4	8.8	3.6	8.0	5.4	7.2
110	methoprotryne	Cleanert TPT	GC-MS	Green tea	A	6.7	37.7	3.6	9.9	7.6	4.0	5.2	3.3	2.4	2.1	10.2	13.0	4.8	13.5	7.7	9.0	10.7	2.0	8.1
				Woolong tea	B	4.7	9.2	6.3	8.0	5.7	5.3	4.0	2.0	7.6	1.4	20.6	9.0	4.3	4.3	14.8	1.6	11.0	10.8	8.3
			GC-MS/MS	Green tea	A*	4.5	5.5	3.4	6.1	8.8	1.6	6.1	2.5	1.1	5.3	6.1	2.2	4.7	6.7	9.1	4.9	6.7	3.6	4.6
				Woolong tea	B*	7.1	2.6	8.8	1.6	7.9	7.0	2.9	12.9	8.2	5.5	4.0	2.1	6.1	8.7	13.6	2.1	8.4	3.7	7.1
		Envi-Carb+PSA	GC-MS	Green tea	A	6.0	4.1	0.4	2.7	6.5	4.9	8.4	3.9	4.5	8.6	11.6	10.1	5.5	2.6	2.4	1.6	5.0	5.8	3.6
				Woolong tea	B	4.0	1.6	5.6	7.2	3.6	3.3	4.0	2.6	3.8	8.4	14.4	13.0	0.5	0.9	4.4	3.1	5.5	2.3	4.2
			GC-MS/MS	Green tea	A*	7.4	4.3	6.6	1.6	5.1	2.6	7.8	9.3	3.2	11.9	8.2	3.2	5.5	2.2	6.9	3.1	2.7	3.6	0.6
				Woolong tea	B*	4.3	2.3	11.3	8.3	1.6	2.8	5.3	3.0	3.8	7.9	11.4	0.4	2.0	4.5	1.8	1.7	2.5	1.5	1.9
				Green tea	A	5.0	5.2	2.0	1.9	9.5	5.7	4.3	7.2	4.3	2.0	8.2	5.5	5.0	5.8	3.3	3.2	2.9	4.9	5.0
				Woolong tea	B	6.6	4.7	4.4	8.7	6.1	4.3	7.7	4.0	3.9	9.0	4.8	3.6	3.1	6.6	3.7	4.9	6.8	0.8	2.1
				Green tea	A*	9.0	1.8	1.6	1.3	0.6	3.4	7.8	2.9	1.5	4.0	4.0	3.5	5.7	1.6	5.3	3.4	3.2	5.9	2.8
				Woolong tea	B*	4.2	3.2	3.8	6.5	3.8	2.5	2.6	4.5	4.4	3.8	5.3	1.5	1.1	1.6	5.5	1.8	2.4	3.2	3.1
				Green tea	A	2.9	1.4	2.6	6.2	8.6	7.7	9.3	2.9	6.9	8.8	5.6	12.2	2.6	5.0	5.9	6.4	10.5	2.3	5.4
				Woolong tea	B	4.5	7.8	3.5	7.1	7.0	3.0	3.9	5.7	5.8	8.7	36.2	16.9	1.0	5.9	10.0	2.1	4.3	7.7	7.6
				Green tea	A*	6.2	3.0	4.7	1.7	7.7	2.4	3.6	3.4	6.6	6.9	7.9	4.2	2.6	3.2	2.1	1.9	4.2	4.8	3.5
				Woolong tea	B*	3.3	2.9	10.8	4.4	15.7	4.0	1.6	6.8	10.0	4.0	8.5	9.9	1.8	12.9	8.4	3.3	4.3	4.3	3.6
				Green tea	A	5.7	6.3	2.5	7.0	9.3	2.9	2.0	1.3	7.3	2.6	7.0	9.9	9.7	5.5	7.9	6.6	11.3	1.0	3.2
				Woolong tea	B	5.4	8.6	6.1	0.7	2.7	3.5	2.1	4.6	5.1	2.5	16.5	6.6	5.9	6.0	13.9	1.0	6.6	6.2	9.9
				Green tea	A*	6.8	3.5	3.9	3.9	5.0	0.4	2.5	3.8	1.4	1.2	8.0	1.7	3.1	10.2	8.0	3.4	4.2	3.7	4.0
				Woolong tea	B*	7.2	5.3	9.8	0.8	10.2	4.3	2.4	7.2	9.1	4.2	1.3	1.4	3.2	8.4	10.8	1.8	4.9	1.3	6.0
111	methoxychlor	Cleanert TPT	GC-MS	Green tea	A*	4.8	4.6	2.9	14.6	12.8	1.0	9.8	13.6	9.0	2.9	6.2	4.7	9.7	2.1	2.1	2.2	6.3	13.2	6.2
				Woolong tea	B*	9.1	6.4	1.6	15.8	7.0	3.7	6.8	8.6	7.5	5.3	6.9	4.2	8.5	5.6	4.7	5.4	12.8	9.5	5.8
			GC-MS/MS	Green tea	A	8.1	6.5	6.7	5.6	6.7	8.4	3.3	20.4	11.2	6.5	11.6	15.0	4.2	6.9	6.9	9.7	7.6	2.1	4.5
				Woolong tea	B	5.3	2.5	3.5	1.5	4.2	4.5	10.6	0.9	9.0	6.4	13.2	2.1	4.1	10.6	10.3	0.8	10.3	8.8	10.0
		Envi-Carb+PSA	GC-MS	Green tea	A*	4.1	7.3	3.4	5.7	10.4	4.2	8.9	13.6	6.6	4.0	6.8	8.7	8.7	8.3	8.3	13.1	3.9	8.2	5.4
				Woolong tea	B*	6.4	6.0	2.2	11.8	6.6	5.9	4.3	5.3	3.3	8.2	4.8	9.1	9.1	7.7	7.7	13.0	7.3	5.8	1.0
			GC-MS	Green tea	A	11.4	3.7	6.5	5.0	10.7	7.5	7.1	5.8	4.8	2.7	4.2	7.7	4.7	10.3	9.7	8.1	5.8	7.7	3.1
				Woolong tea	B	12.7	7.4	8.1	8.9	6.5	3.5	3.3	11.4	4.5	7.1	7.8	5.3	3.0	9.7	9.7	2.4	13.6	6.6	1.3
				Green tea	A	21.5	1.4	16.4	9.4	10.9	22.0	5.1	17.6	4.8	6.6	16.5	24.4	32.0	3.1	11.2	8.6	4.5	23.3	15.0
				Woolong tea	B	18.3	8.9	5.3	7.3	13.1	9.5	7.4	12.9	9.2	7.8	20.0	16.8	18.5	11.2	3.2	6.2	2.5	24.2	6.9
				Green tea	A	17.8	9.4	6.1	16.7	9.9	14.9	12.5	27.1	9.4	9.0	5.5	17.0	6.4	16.6	3.2	24.3	22.0	33.2	5.8
				Woolong tea	B	16.7	13.0	3.3	5.4	21.2	18.9	16.7	7.7	20.9	15.9	18.8	1.7	14.6	10.6	4.0	6.7	14.0	11.5	6.0
				Green tea	A	15.2	3.8	16.1	6.6	11.7	8.8	12.0	11.9	7.4	3.0	7.5	4.0	23.4	23.4	4.0	7.7	14.0	13.2	10.2
				Woolong tea	B	11.0	8.7	4.0	19.6	3.0	2.0	25.8	7.1	13.3	9.5	15.2	1.0	9.9	1.0	4.6	16.4	7.5	2.1	7.2
				Green tea	A	8.4	10.2	6.8	9.0	8.2	15.8	6.2	9.3	2.2	6.6	6.7	1.6	8.8	1.6	5.9	13.0	13.3	2.1	7.2
				Woolong tea	B	9.3	12.4	7.4	9.0	9.1	15.2	25.8	9.4	2.2	12.4	3.0	5.6	6.3	11.1	13.4	7.5	26.3	4.4	8.0

(Continued)

Annex 2: RSD data for 201 pesticides in incurred tea Youden pair samples under 8 conditions with two SPE cartridge cleanup in three months (Nov.9, 2009-Feb.7, 2010) (cont.)

No	Pesticides	SPE	Method	Sample	Youden pair	Nov.9, 2009 (n=5)	Nov.14, 2009 (n=3)	Nov.19, 2009 (n=3)	Nov.24, 2009 (n=3)	Nov.29, 2009 (n=3)	Dec.4, 2009 (n=3)	Dec.9, 2009 (n=3)	Dec.14, 2009 (n=3)	Dec.19, 2009 (n=3)	Dec.24, 2009 (n=3)	Dec.14, 2009 (n=3)	Jan.3, 2010 (n=3)	Jan.8, 2010 (n=3)	Jan.13, 2010 (n=3)	Jan.18, 2010 (n=3)	Jan.23, 2010 (n=3)	Jan.28, 2010 (n=3)	Feb.2, 2010 (n=3)	Feb.7, 2010 (n=3)
112	methyl-parathion	Cleanert TPT	GC-MS	Green tea	A	11.8	5.1	1.7	4.1	10.8	2.7	5.4	4.3	5.8	6.8	6.5	2.5	4.7	14.7	9.9	0.8	12.9	17.3	1.2
					B	5.1	9.4	2.6	21.1	5.3	4.0	3.4	3.0	3.9	4.4	7.8	11.7	10.2	5.3	9.7	4.7	11.3	10.0	9.2
				Woolong tea	A*	6.0	2.6	2.6	6.5	8.7	2.9	14.7	18.4	9.0	14.4	9.0	14.5	6.5	6.9	9.9	3.3	9.9	1.6	6.0
					B*	5.1	4.2	5.9	6.8	4.8	1.9	4.6	1.9	6.0	6.6	9.8	3.4	10.7	2.1	11.7	3.2	8.1	4.0	4.8
			GC-MS/MS	Green tea	A	2.9	7.7	7.2	5.2	6.1	0.7	7.7	11.8	15.1	3.7	7.3	5.7	10.9	4.4	3.7	5.9	9.5	6.3	0.8
					B	6.2	10.1	5.3	6.8	4.4	1.1	5.9	8.8	1.7	9.9	2.9	0.6	2.4	5.7	7.1	8.8	11.8	3.8	6.9
				Woolong tea	A	9.2	2.0	4.0	3.7	7.2	6.4	5.0	0.9	1.3	8.7	1.8	7.4	2.7	2.8	8.2	2.1	8.4	8.6	5.3
					B	3.7	4.6	2.5	2.8	2.8	0.5	3.0	1.0	1.8	6.3	6.7	3.6	13.6	2.8	4.3	1.6	0.4	3.4	1.6
		Envi-Carb+PSA	GC-MS	Green tea	A	9.2	7.3	9.6	6.5	8.3	7.8	4.4	14.3	7.5	4.3	11.9	14.7	1.8	5.4	6.2	6.0	15.1	18.6	2.9
					B	6.9	3.8	5.6	2.1	5.4	5.6	3.2	4.2	1.8	5.8	24.7	13.5	23.9	7.6	12.8	13.5	2.6	20.8	10.9
				Woolong tea	A	10.9	4.7	0.4	5.1	1.1	0.9	10.1	7.2	1.8	11.1	1.9	0.9	15.8	7.1	3.7	15.4	3.6	11.7	4.6
					B	8.8	8.2	8.7	7.7	14.7	11.5	7.1	11.7	10.9	3.9	10.7	2.6	8.7	2.1	14.9	4.6	13.5	9.2	7.0
			GC-MS/MS	Green tea	A	15.8	7.7	11.2	4.0	5.1	20.2	11.8	12.2	9.1	0.7	6.2	11.4	3.9	5.1	6.4	4.3	12.4	6.5	4.6
					B	31.1	11.6	1.7	10.9	9.1	12.6	25.3	5.9	12.4	11.8	14.6	14.2	13.0	16.6	12.0	5.5	7.4	16.7	12.3
				Woolong tea	A	7.7	3.6	5.1	17.9	12.2	2.8	3.1	4.1	2.5	6.4	11.8	5.6	5.4	6.9	8.6	2.1	3.8	10.8	11.3
					B	8.3	7.4	5.0	15.5	6.3	8.3	2.5	10.0	11.4	1.2	3.3	2.6	6.6	11.8	12.5	5.1	7.0	4.1	3.7
113	metolachlor	Cleanert TPT	GC-MS	Green tea	A*	5.4	1.7	3.8	2.8	7.9	9.5	3.3	6.7	1.2	4.0	5.9	4.7	6.5	4.6	4.3	3.5	1.8	12.2	4.9
					B	6.2	9.6	5.1	6.9	3.5	2.9	7.3	2.0	2.7	6.2	6.0	8.2	1.5	2.6	2.3	1.6	4.7	3.2	5.6
				Woolong tea	A	7.8	3.5	2.5	6.4	3.8	3.2	9.9	12.4	3.5	11.7	7.1	1.9	1.1	2.7	6.6	7.1	2.0	4.7	10.6
					B	4.5	3.2	10.8	5.2	2.4	2.5	4.4	1.3	7.4	6.2	3.2	5.8	5.0	2.9	4.9	1.1	3.8	11.1	4.1
			GC-MS/MS	Green tea	A	5.9	6.1	4.0	3.5	7.7	2.9	3.8	6.3	5.9	1.7	3.5	1.1	8.2	3.1	1.2	6.5	3.8	5.9	2.9
					B	7.7	8.0	6.3	9.9	6.4	5.1	7.5	4.4	5.6	8.8	4.0	4.2	3.8	5.9	1.3	4.7	4.9	1.8	9.3
				Woolong tea	A	5.6	4.6	1.9	5.5	2.8	3.6	10.9	3.4	0.4	3.9	4.5	2.6	2.8	3.0	10.0	6.4	7.4	4.2	4.1
					B	4.7	0.8	5.3	7.1	4.1	2.0	3.7	9.2	6.3	5.8	4.0	5.4	6.4	4.2	7.1	2.4	3.5	1.4	3.3
		Envi-Carb+PSA	GC-MS	Green tea	A*	7.6	5.0	6.6	3.9	6.6	1.3	21.6	1.5	2.6	12.4	11.1	2.5	7.8	4.6	4.2	6.2	9.3	11.4	5.5
					B	4.1	7.1	2.0	0.4	0.9	1.2	9.7	3.2	3.3	4.9	18.4	5.1	5.8	2.3	9.7	4.5	10.0	9.9	6.2
				Woolong tea	A	2.5	22.7	6.4	4.2	4.0	1.3	8.4	1.9	1.7	1.1	3.4	2.6	1.3	3.8	4.6	4.2	4.6	2.1	7.0
					B	10.6	3.5	6.6	2.8	9.6	5.3	5.5	5.7	10.9	3.6	8.7	12.5	5.5	2.1	7.8	6.1	4.1	5.9	1.8
			GC-MS/MS	Green tea	A	8.3	5.7	1.6	7.1	5.5	4.8	3.3	2.4	4.5	1.6	11.1	9.7	0.9	15.6	6.9	4.5	6.3	1.6	2.9
					B	22.1	7.5	3.7	1.8	5.8	1.4	2.3	3.6	4.0	3.1	16.2	2.0	3.5	8.1	11.5	2.0	5.6	10.1	8.7
				Woolong tea	A	1.5	4.6	3.2	2.6	6.0	1.2	2.0	1.2	1.2	3.4	10.2	2.9	4.5	5.4	12.7	6.4	4.3	1.1	5.9
					B	7.0	1.8	9.7	3.9	7.7	6.0	0.6	8.3	9.8	4.8	1.3	5.0	3.5	11.1	13.4	2.4	5.0	3.5	8.3
114	nitrapyrin	Cleanert TPT	GC-MS	Green tea	A	2.9	3.6	2.1	10.6	11.4	7.9	7.0	2.9	8.6	6.6	1.6	3.8	8.5	7.6	8.6	7.1	14.1	10.7	1.7
					B	9.6	7.8	1.9	8.2	3.4	11.2	2.2	7.8	8.7	8.9	3.4	10.5	4.2	13.0	7.6	3.9	5.3	7.4	8.2
				Woolong tea	A	8.7	0.5	2.0	5.0	10.5	5.5	6.8	3.7	11.3	6.9	9.9	3.9	4.7	2.7	4.1	4.1	3.8	4.5	6.8
					B	11.4	9.1	1.0	5.4	4.4	3.2	5.6	3.9	8.1	4.6	6.8	6.4	3.6	9.2	19.9	2.5	11.4	4.5	17.6
			GC-MS/MS	Green tea	A	5.3	4.1	6.0	8.7	14.4	6.4	8.6	12.3	8.1	4.4	4.4	6.4	10.7	3.5	5.8	9.6	12.4	16.1	6.1
					B	8.7	8.0	5.5	11.0	9.9	4.9	8.9	7.0	11.7	4.8	4.8	1.6	4.8	5.7	5.7	7.7	5.5	5.0	5.2
				Woolong tea	A	11.6	4.0	4.8	9.3	5.9	8.6	1.2	3.7	5.8	9.3	9.3	5.3	4.1	3.6	7.3	3.1	2.7	7.5	8.0
					B	6.0	11.0	8.7	0.2	6.4	8.9	4.7	13.9	18.4	15.9	5.2	5.6	10.6	4.0	9.2	7.1	14.0	9.5	11.7
		Envi-Carb+PSA	GC-MS	Green tea	A	11.8	4.6	8.1	6.9	10.8	5.6	11.8	8.9	11.3	1.3	9.8	12.0	1.3	4.6	3.3	4.2	0.5	8.4	3.1
					B	11.7	5.7	3.6	12.4	5.5	4.2	9.4	5.1	3.6	5.6	18.7	9.7	13.3	4.6	12.1	5.8	2.8	13.6	4.7
				Woolong tea	A	13.1	2.4	7.5	9.8	4.2	4.6	11.5	6.9	6.6	0.6	5.0	8.0	9.3	3.0	6.0	8.9	9.6	16.5	10.7
					B	14.8	6.1	10.3	3.2	4.3	6.7	2.0	2.9	15.2	8.9	12.2	5.2	4.5	8.5	9.6	3.5	13.3	12.3	3.3
			GC-MS/MS	Green tea	A	13.6	10.6	7.3	5.7	6.1	15.9	8.8	7.5	12.8	2.5	5.4	11.6	6.3	15.4	6.8	17.3	13.3	14.8	16.1
					B	39.8	13.7	2.2	14.4	10.7	6.7	13.4	2.8	11.5	6.1	27.2	0.7	26.2	23.0	36.0	7.3	13.8	19.2	8.3
				Woolong tea	A	11.8	8.7	12.1	8.5	6.3	7.2	5.1	9.6	3.1	3.2	9.6	3.1	7.8	13.3	10.9	3.1	13.3	14.0	6.0
					B	16.1	10.2	4.8	10.9	13.0	11.1	16.7	2.7	17.5	8.6	8.1	4.9	10.5	15.9	11.4	7.1	4.9	14.0	14.5

No.	Compound	Cartridge	Method	Tea	Rep	1	2	3	4	5	6	7	8	9	10	11	12	13	14	15	16	17	18	19
115	oxyfluorfen	Cleanert TPT	GC-MS	Green tea	A*	9.8	5.7	2.0	8.9	13.4	2.5	7.6	12.3	3.8	7.7	12.9	8.8	12.8	11.8	2.9	4.4	9.7	12.3	7.0
				Green tea	B*	5.6	6.7	4.5	18.5	6.7	7.2	6.1	7.3	6.1	4.7	10.9	11.9	7.6	13.7	1.3	8.5	10.8	5.7	4.9
				Woolong tea	A	8.9	10.4	9.3	2.6	26.0	4.4	11.6	27.8	7.5	11.7	13.0	13.5	5.5	3.7	8.9	4.9	6.6	8.8	2.3
				Woolong tea	B	3.0	8.7	10.7	61.5	14.0	2.3	7.1	0.6	14.4	10.8	14.4	2.8	4.6	8.5	7.0	4.4	7.2	2.2	3.2
			GC-MS/MS	Green tea	A*	10.6	6.6	3.8	10.5	9.2	6.0	14.3	10.0	11.0	2.0	14.7	7.2	14.1	5.7	2.2	7.0	7.8	7.4	3.0
				Green tea	B*	7.7	10.7	7.7	9.3	5.4	5.4	1.2	12.6	5.3	11.7	6.4	14.3	0.6	3.8	7.6	10.9	13.4	2.1	7.1
				Woolong tea	A	5.8	5.7	7.2	3.7	4.1	3.2	1.8	14.9	7.0	8.3	16.0	12.0	2.7	4.8	7.8	4.0	6.4	8.5	7.9
				Woolong tea	B	8.1	2.5	3.0	11.9	14.3	3.9	10.7	14.2	5.1	5.2	19.6	5.1	4.2	7.3	1.2	4.1	10.7	6.2	8.4
		Envi-Carb+PSA	GC-MS	Green tea	A*	19.3	6.8	17.5	8.7	13.5	14.9	5.4	18.6	8.9	8.9	4.6	15.0	2.3	13.6	4.0	12.7	16.9	15.0	5.4
				Green tea	B*	11.3	10.0	8.0	2.9	8.9	12.8	2.4	11.7	8.1	6.9	41.0	8.3	29.8	6.4	17.2	6.6	1.1	24.2	14.4
				Woolong tea	A	9.1	7.8	9.0	9.2	14.9	0.4	3.2	7.7	16.2	7.4	4.9	3.4	11.5	3.7	6.9	25.7	7.6	13.3	2.8
				Woolong tea	B	6.9	6.7	13.3	3.6	28.3	17.4	11.5	10.5	4.4	7.1	15.6	6.6	10.8	8.5	14.8	6.2	14.8	8.2	5.6
			GC-MS/MS	Green tea	A*	17.2	9.8	10.8	6.6	8.4	10.4	13.0	18.8	18.9	8.8	9.3	2.5	12.6	6.1	5.0	12.3	7.1	6.1	11.4
				Green tea	B*	19.2	9.7	10.6	8.6	5.7	12.4	39.5	14.5	5.6	4.5	23.3	0.9	21.1	11.0	12.9	7.4	22.9	10.5	18.8
				Woolong tea	A	8.6	4.8	5.9	11.1	8.7	3.7	6.9	19.3	12.1	3.4	5.3	2.8	15.1	7.2	8.1	4.0	4.9	11.6	8.3
				Woolong tea	B	8.9	11.8	12.8	6.9	19.7	10.5	26.2	9.9	12.4	7.4	9.3	5.5	14.4	14.2	10.6	4.1	7.7	5.1	9.9
116	pendimeth-alin	Cleanert TPT	GC-MS	Green tea	A*	3.5	3.8	4.2	6.0	14.7	2.8	11.0	18.3	5.8	10.2	12.5	8.1	13.2	11.1	4.5	4.8	10.8	24.5	5.4
				Green tea	B*	5.9	9.5	5.3	17.5	5.7	5.1	4.2	14.4	7.5	8.7	8.8	13.5	5.4	12.0	2.1	6.8	17.6	16.7	5.2
				Woolong tea	A	5.7	3.4	5.8	4.9	30.1	3.6	0.3	21.8	5.8	14.3	18.5	10.7	4.6	3.1	6.8	11.6	4.8	6.0	0.5
				Woolong tea	B	2.5	6.5	6.5	5.1	13.5	9.5	11.6	4.1	17.4	10.9	17.2	2.0	3.4	6.4	7.6	6.7	3.4	2.4	2.2
			GC-MS/MS	Green tea	A*	5.4	3.8	2.8	4.6	12.0	6.2	10.8	17.5	2.7	10.9	10.6	6.1	13.1	6.7	4.9	7.1	7.5	5.3	7.2
				Green tea	B*	7.8	10.0	7.3	10.8	4.1	3.8	2.6	12.4	13.9	9.4	0.9	10.0	1.9	4.9	5.3	6.2	8.6	5.0	6.8
				Woolong tea	A	7.2	3.3	4.1	2.6	5.2	4.6	0.3	9.7	11.9	6.9	14.0	11.8	6.1	2.4	9.8	5.2	8.4	5.2	6.2
				Woolong tea	B	6.7	11.0	2.6	6.8	8.4	4.4	11.3	9.7	16.4	3.4	15.4	2.3	5.4	4.3	1.4	2.2	3.9	7.5	6.3
		Envi-Carb+PSA	GC-MS	Green tea	A*	17.1	6.2	13.1	8.3	11.2	19.2	24.8	27.8	19.9	8.9	5.7	14.5	0.7	28.2	4.1	23.4	15.9	22.1	4.8
				Green tea	B*	9.4	10.3	5.9	5.7	9.9	15.8	8.0	27.3	6.0	7.0	39.6	7.5	41.0	9.6	31.9	7.4	3.4	26.6	12.4
				Woolong tea	A	7.8	6.4	5.2	5.4	16.4	1.4	3.6	22.8	15.4	7.1	5.4	4.6	11.2	3.1	4.4	32.2	12.4	9.5	3.5
				Woolong tea	B	6.3	4.5	9.0	2.4	25.3	15.3	13.3	10.4	8.3	6.4	14.1	4.7	10.5	6.4	18.3	6.4	19.9	4.5	5.8
			GC-MS/MS	Green tea	A*	19.4	7.0	10.8	6.6	6.0	9.8	22.0	20.2	20.4	4.2	5.8	1.4	3.3	4.7	3.2	13.0	8.5	13.9	11.3
				Green tea	B*	23.3	13.7	7.9	4.8	1.0	10.5	36.9	20.6	3.0	2.3	23.2	4.9	14.8	6.1	14.1	4.2	13.6	23.5	18.0
				Woolong tea	A	8.4	4.2	5.8	4.7	2.6	2.7	5.6	22.0	13.4	7.9	5.7	6.2	8.8	4.8	5.9	5.2	6.8	10.7	6.2
				Woolong tea	B	12.0	5.2	8.1	1.0	13.9	10.5	30.0	6.2	4.1	4.9	6.0	3.1	8.7	11.4	9.9	2.2	10.4	9.6	9.6
117	picoxystrobin	Cleanert TPT	GC-MS	Green tea	A*	5.1	5.9	1.3	2.2	7.1	3.9	7.3	4.8	7.9	1.2	4.8	4.1	6.4	4.0	1.9	1.4	5.0	5.7	4.1
				Green tea	B*	4.7	7.3	6.4	7.3	3.2	3.1	3.0	9.2	2.5	2.7	5.7	8.5	1.7	3.9	4.3	2.5	5.4	1.0	4.8
				Woolong tea	A	6.5	1.4	0.7	2.8	1.2	2.0	8.0	8.4	6.1	2.8	4.0	2.3	3.9	4.6	7.9	3.3	2.9	3.4	0.4
				Woolong tea	B	3.5	1.7	2.2	6.0	0.8	2.8	4.1	2.9	5.4	2.7	5.7	1.1	1.9	5.3	2.1	1.2	2.2	0.7	2.3
			GC-MS/MS	Green tea	A*	4.4	6.7	2.9	1.7	10.0	4.2	6.0	6.9	4.4	3.4	1.9	5.5	8.3	2.4	3.8	3.9	5.6	3.0	2.8
				Green tea	B*	7.2	5.7	6.8	8.3	5.0	6.0	7.6	6.7	1.9	6.1	7.6	2.6	2.2	6.7	4.0	5.9	5.0	2.8	4.0
				Woolong tea	A	7.9	2.0	0.5	1.7	2.5	2.3	8.6	3.3	2.0	4.9	4.1	4.9	10.4	2.2	8.4	5.5	5.7	1.7	5.7
				Woolong tea	B	4.7	2.2	5.1	4.8	5.7	4.0	1.9	5.8	4.0	3.4	2.7	1.5	3.0	5.0	6.4	3.0	10.7	6.0	3.8
		Envi-Carb+PSA	GC-MS	Green tea	A*	6.4	5.1	2.2	5.9	8.6	1.7	8.5	1.0	3.2	3.2	5.8	2.6	1.7	3.9	5.3	6.1	4.6	3.1	6.1
				Green tea	B*	3.8	6.7	2.8	4.1	2.2	1.2	1.6	4.6	1.8	1.6	7.6	3.8	1.5	2.9	11.1	0.5	6.3	7.3	7.0
				Woolong tea	A	3.8	2.4	4.0	1.5	5.6	2.4	3.8	2.3	11.9	2.5	23.1	3.7	1.1	6.7	2.6	1.8	4.0	4.2	3.3
				Woolong tea	B	4.8	6.4	9.4	5.0	10.5	5.0	2.0	5.8	6.9	3.9	8.8	1.3	4.0	14.6	10.4	4.0	11.0	5.5	5.2
			GC-MS/MS	Green tea	A*	6.5	6.8	4.4	5.4	9.1	1.9	8.3	1.9	4.9	1.0	10.1	3.1	7.8	2.6	6.0	7.3	7.8	2.2	6.8
				Green tea	B*	6.1	5.5	6.5	1.2	3.9	0.8	3.8	4.5	0.6	1.4	10.5	1.6	5.0	6.5	10.8	1.7	5.8	7.6	5.8
				Woolong tea	A	5.9	2.4	4.0	2.1	7.3	0.5	2.5	3.7	9.4	5.1	17.5	2.0	2.4	10.9	7.7	5.5	5.4	2.2	6.5
				Woolong tea	B	7.2	2.3	9.5	4.4	7.7	6.1	1.6	6.5	9.4	3.2	2.2	1.8	2.9	10.9	9.0	3.0	5.4	2.0	9.0

(Continued)

Annex 2: RSD data for 201 pesticides in incurred tea Youden pair samples under 8 conditions with two SPE cartridge cleanup in three months (Nov.9, 2009–Feb.7, 2010) (cont.)

No	Pesticides	SPE	Method	Sample	Youden pair	Nov.9, 2009 (n=5)	Nov.14, 2009 (n=3)	Nov.19, 2009 (n=3)	Nov.24, 2009 (n=3)	Nov.29, 2009 (n=3)	Dec.4, 2009 (n=3)	Dec.9, 2009 (n=3)	Dec.14, 2009 (n=3)	Dec.19, 2009 (n=3)	Dec.24, 2009 (n=3)	Dec.29, 2009 (n=3)	Jan.3, 2010 (n=3)	Jan.8, 2010 (n=3)	Jan.13, 2010 (n=3)	Jan.18, 2010 (n=3)	Jan.23, 2010 (n=3)	Jan.28, 2010 (n=3)	Feb.2, 2010 (n=3)	Feb.7, 2010 (n=3)
118	piperophos	Cleanert TPT	GC-MS	Green tea	A*	5.4	4.4	1.3	5.1	10.6	1.8	8.5	11.0	10.6	4.8	3.7	9.3	9.6	7.1	3.2	1.7	7.2	3.8	2.7
				Woolong tea	B*	5.9	9.0	4.1	10.7	5.9	6.9	4.5	4.1	5.0	1.9	14.6	8.5	10.4	3.5	3.3	3.2	10.8	3.6	5.9
				Green tea	A	7.4	5.5	3.7	2.7	8.6	4.7	9.1	22.2	6.8	13.2	15.6	7.9	2.2	2.5	9.5	2.8	6.5	1.6	2.2
				Woolong tea	B	3.4	10.8	9.0	13.2	9.2	5.9	6.2	2.5	11.6	8.4	20.1	3.3	2.2	6.8	1.5	1.6	5.1	2.6	2.1
			GC-MS/MS	Green tea	A	3.7	7.7	1.3	7.0	9.9	4.4	10.7	12.1	10.5	2.8	7.3	6.4	6.9	10.2	2.3	4.4	2.8	10.0	3.6
				Woolong tea	B	6.8	6.7	4.9	7.7	5.7	3.4	4.9	9.4	6.2	6.2	8.8	6.6	2.2	7.8	6.4	7.6	4.9	3.3	6.4
				Green tea	A	7.2	6.1	6.6	1.9	6.9	5.9	6.3	7.0	7.2	5.0	8.2	5.6	2.3	2.9	9.6	2.1	14.6	10.1	5.0
				Woolong tea	B	2.3	7.0	3.9	5.0	5.1	1.3	8.2	6.3	13.6	6.2	6.3	2.2	7.6	1.6	4.9	3.3	6.7	4.3	17.5
		Envi-Carb+PSA	GC-MS	Green tea	A	14.0	2.6	10.3	8.5	12.2	8.1	1.4	9.5	1.3	4.3	6.4	8.6	2.2	14.2	3.2	22.4	4.0	28.2	4.9
				Woolong tea	B	10.5	6.7	5.6	7.4	2.3	7.5	4.9	8.0	11.3	2.4	33.7	13.3	18.5	1.4	21.9	1.2	3.2	22.9	10.0
				Green tea	A	5.5	4.4	6.2	2.3	10.8	1.6	3.4	21.9	4.2	1.0	6.6	9.4	1.2	2.5	1.5	6.3	8.6	6.6	3.0
				Woolong tea	B	8.1	10.2	9.2	1.1	24.6	7.5	7.5	2.0	14.5	6.0	14.4	2.9	8.3	6.8	11.6	4.4	3.6	4.7	4.6
			GC-MS/MS	Green tea	A	14.2	6.0	8.8	9.8	4.1	11.9	11.5	9.2	0.8	1.9	7.9	5.5	2.7	10.2	6.6	11.2	9.8	2.0	5.1
				Woolong tea	B	31.9	6.8	5.0	11.4	6.6	5.7	24.2	6.6	14.9	12.2	15.8	6.2	4.2	7.0	8.1	2.9	3.3	25.8	14.4
				Green tea	A	8.4	5.9	7.4	4.5	5.3	3.4	0.4	15.0	2.4	1.0	9.3	2.8	3.2	10.0	12.3	2.1	4.5	10.1	4.8
				Woolong tea	B	11.4	6.4	7.0	3.8	18.9	6.5	9.7	6.1	10.6	0.2	1.7	3.6	5.5	13.8	8.9	3.3	4.4	8.2	8.2
119	pirimiphoe-ethyl	Cleanert TPT	GC-MS	Green tea	A	7.6	4.5	4.1	2.5	8.3	4.6	7.2	4.0	4.3	5.8	10.3	6.7	6.2	2.4	2.5	1.5	4.4	6.8	4.0
				Woolong tea	B	4.9	8.2	4.7	7.9	4.3	2.7	3.2	1.9	5.1	8.3	10.1	12.4	0.8	0.8	3.8	2.7	5.3	1.6	6.0
				Green tea	A	6.0	1.9	4.7	3.6	1.6	1.9	8.5	9.6	2.9	12.3	12.8	3.5	3.6	2.7	8.9	3.2	3.5	2.7	0.8
				Woolong tea	B	3.8	4.7	4.5	5.8	4.0	5.6	4.6	3.5	7.8	8.2	11.8	0.7	2.4	2.7	2.9	0.6	1.2	1.2	2.3
			GC-MS/MS	Green tea	A	4.5	4.4	3.8	1.2	10.4	2.5	5.1	6.8	5.0	1.8	8.1	4.9	4.0	1.0	3.8	3.3	2.2	4.9	2.7
				Woolong tea	B	7.2	7.4	7.3	9.5	6.9	6.2	9.3	5.7	5.1	6.9	3.3	8.5	5.1	4.5	2.3	2.5	5.2	2.4	5.6
				Green tea	A	7.6	1.7	3.4	2.5	2.4	6.4	8.4	2.9	1.6	3.4	5.4	1.2	2.1	3.3	5.4	5.9	4.9	7.7	4.0
				Woolong tea	B	5.3	0.9	5.9	4.4	7.7	2.0	3.9	3.4	1.0	3.3	3.9	0.9	3.2	4.2	5.2	2.0	2.7	4.0	6.3
		Envi-Carb+PSA	GC-MS	Green tea	A	7.7	5.8	2.4	5.6	8.5	3.6	5.4	0.8	4.5	5.0	7.8	6.9	2.6	5.3	4.8	6.7	10.8	3.1	5.4
				Woolong tea	B	3.9	8.6	5.7	5.2	2.2	5.9	4.4	3.7	3.9	5.7	31.2	11.5	1.4	3.2	11.3	1.6	5.6	7.8	6.1
				Green tea	A	4.5	5.8	3.7	2.4	3.8	3.1	4.1	1.8	4.3	3.9	5.7	3.9	1.6	2.7	2.6	2.1	6.3	4.1	4.5
				Woolong tea	B	4.8	2.7	10.3	1.9	17.5	2.0	2.1	6.8	12.3	5.5	10.0	5.1	3.7	2.7	9.8	4.4	6.1	5.5	4.7
			GC-MS/MS	Green tea	A*	6.3	7.3	3.3	5.9	6.7	1.5	4.1	2.6	4.9	1.0	10.1	11.3	7.1	16.3	7.5	6.4	8.4	5.7	11.9
				Woolong tea	B	5.2	8.0	5.6	2.8	0.7	4.4	5.8	2.2	3.5	1.4	19.4	19.0	6.4	12.8	10.1	2.5	13.2	5.3	11.7
				Green tea	A	2.2	5.7	3.7	2.9	7.7	3.2	5.4	2.2	1.8	2.2	7.4	2.3	6.4	9.0	10.8	5.9	9.3	4.5	5.0
				Woolong tea	B	6.0	3.5	9.2	1.5	8.4	5.6	0.9	7.8	10.0	6.3	2.4	4.2	1.5	7.8	15.0	2.0	2.8	3.3	7.4
120	pirimiphos-methyl	Cleanert TPT	GC-MS	Green tea	A	7.8	4.4	3.6	2.2	8.3	3.7	6.2	4.5	5.4	3.8	7.1	3.9	6.3	7.7	2.0	1.3	9.7	6.5	3.1
				Woolong tea	B	5.2	8.8	5.4	8.0	3.8	2.8	2.2	1.7	3.9	6.7	6.8	9.1	1.9	7.0	7.2	1.7	4.0	2.8	6.3
				Green tea	A	4.7	2.3	2.2	3.6	1.5	3.1	9.1	11.3	7.0	9.9	7.8	3.8	2.0	2.4	8.3	3.2	4.1	0.3	1.6
				Woolong tea	B	3.9	3.4	3.0	10.4	1.9	1.8	4.0	3.5	6.9	6.4	11.4	1.4	1.7	3.4	4.4	0.2	1.2	2.1	2.2
			GC-MS/MS	Green tea	A	3.9	6.1	3.4	2.7	8.8	4.4	7.2	4.3	5.9	1.9	7.8	6.9	8.4	3.9	2.0	8.2	2.2	3.4	3.7
				Woolong tea	B	8.2	8.2	6.1	8.3	5.7	7.9	8.8	3.8	4.2	8.5	1.2	2.3	1.8	5.9	2.3	2.4	7.0	4.8	4.1
				Green tea	A	6.5	4.9	1.4	3.5	4.1	3.0	10.0	3.0	2.6	2.7	6.4	1.6	2.0	3.1	7.1	7.1	2.1	6.0	8.1
				Woolong tea	B	3.0	3.3	6.6	6.6	5.0	1.2	2.7	5.8	5.8	5.7	9.8	2.9	2.5	4.8	3.3	3.3	3.9	2.7	7.6
		Envi-Carb+PSA	GC-MS	Green tea	A	8.2	7.0	4.4	6.7	6.7	1.8	7.4	1.9	3.9	3.3	7.8	2.0	2.5	3.4	3.1	4.2	10.0	4.7	5.3
				Woolong tea	B	4.8	9.3	3.1	1.7	0.3	2.3	4.0	2.4	4.7	3.1	22.7	4.5	4.5	0.6	8.8	3.0	3.3	8.5	6.5
				Green tea	A	4.5	3.8	2.6	2.7	3.7	1.2	3.6	7.2	1.7	2.4	4.4	1.9	2.3	2.7	3.5	3.1	5.3	5.4	3.8
				Woolong tea	B	5.4	3.6	8.9	2.9	13.5	4.7	2.4	6.2	11.3	3.2	9.8	5.4	3.7	3.4	9.8	3.6	7.3	2.1	4.5
			GC-MS/MS	Green tea	A	9.5	4.3	2.4	6.3	4.8	5.0	4.6	1.5	7.3	2.8	9.8	12.5	6.5	14.8	6.4	5.3	8.9	2.6	4.3
				Woolong tea	B	16.9	9.3	3.8	1.6	2.9	2.4	7.0	1.9	5.8	4.6	19.1	8.2	5.9	4.2	9.4	5.1	7.2	8.9	6.6
				Green tea	A	2.4	3.6	2.8	3.9	3.7	1.3	2.5	4.7	1.2	2.2	5.6	2.7	5.2	4.4	7.9	7.1	9.6	4.8	5.6
				Woolong tea	B	6.6	4.7	9.1	3.2	6.9	5.6	1.6	5.6	9.6	2.8	2.9	3.6	3.8	10.8	10.7	3.3	5.3	3.0	7.1

(Continued)

No.	Compound	Cartridge	Method	Tea																					
121	profenofos	Cleanert TPT	GC-MS	Green tea	A*	10.5	3.3	1.3	14.4	15.4	1.6	8.1	12.2	11.2	1.2	7.1	11.1	5.4	17.1	11.4	3.2	14.6	13.6	2.1
					B	11.9	12.8	4.5	21.4	12.4	8.2	4.7	5.7	0.8	4.4	5.4	11.1	18.3	10.3	13.9	9.2	13.1	13.9	19.4
				Woolong tea	A	15.2	3.7	2.3	5.4	16.3	5.2	14.4	14.1	17.3	10.4	30.0	12.6	12.3	13.6	10.2	10.7	12.8	27.8	14.1
					B	13.6	5.0	1.4	3.0	3.8	6.2	9.0	6.4	9.8	13.1	30.2	2.1	13.2	6.2	13.3	9.2	13.8	10.8	16.7
			GC-MS/MS	Green tea	A*	15.0	7.5	12.8	4.6	6.7	11.6	30.3	16.9	19.5	8.8	20.0	4.6	18.8	17.1	6.1	19.2	5.4	22.5	14.8
					B	11.4	9.9	19.9	17.4	14.2	18.7	13.5	18.8	12.1	15.1	11.0	19.3	14.1	8.9	16.1	5.1	24.2	9.5	7.8
				Woolong tea	A	24.0	8.2	20.6	4.4	12.2	16.8	3.2	4.2	14.5	11.2	20.9	16.2	15.8	15.7	6.1	18.9	21.8	4.2	23.2
					B	9.9	13.7	16.2	12.6	2.1	11.9	3.7	0.2	28.9	20.9	8.1	10.1	10.6	4.7	11.2	10.5	8.2	23.4	42.0
		Envi-Carb+PSA	GC-MS	Green tea	A	16.5	18.6	11.3	13.4	8.2	21.3	10.3	19.6	5.2	7.6	9.1	35.8	4.2	12.0	8.5	7.3	14.3	20.1	1.6
					B	25.0	7.7	9.6	12.4	17.0	3.7	4.8	8.7	12.2	19.7	5.1	23.2	32.9	10.1	8.2	21.3	11.4	39.8	16.0
				Woolong tea	A	15.4	5.2	10.1	9.3	4.4	7.2	22.0	19.6	10.0	18.5	4.6	6.6	24.0	13.2	9.7	20.9	3.5	18.7	9.2
					B	25.4	9.6	10.7	2.3	9.3	16.8	10.1	10.5	17.6	15.2	21.9	10.0	8.4	8.0	15.7	6.0	24.8	18.6	13.7
			GC-MS/MS	Green tea	B*	30.0	6.7	25.8	17.3	11.4	32.0	23.9	26.2	24.7	2.2	9.2	4.6	4.9	37.9	6.4	6.3	21.0	2.8	42.7
					A	51.7	24.8	6.0	18.7	5.9	21.2	19.2	1.5	19.8	29.9	0.7	8.2	28.1	12.2	10.2	12.7	31.9	26.3	27.3
				Woolong tea	B*	12.7	15.5	12.8	5.3	33.5	8.1	10.7	39.2	11.3	3.4	4.2	18.3	15.0	7.2	5.5	18.9	16.5	16.4	11.5
					A	14.4	16.0	16.8	18.5	15.5	10.2	21.2	18.9	24.4	11.0	10.2	19.1	4.5	20.9	19.4	10.5	25.1	11.1	7.7
122	profluralin	Cleanert TPT	GC-MS	Green tea	A*	6.0	4.5	3.9	3.8	11.2	3.1	6.4	7.8	9.1	3.2	7.7	3.8	10.7	7.6	2.1	3.2	6.0	14.0	3.2
					B*	5.2	9.0	5.5	16.9	4.4	2.2	0.6	9.3	4.3	9.7	4.2	7.9	8.3	8.0	1.4	4.7	11.0	12.6	6.5
				Woolong tea	A*	4.5	1.1	3.4	3.6	15.9	3.2	9.9	9.3	9.1	9.7	11.5	7.1	5.0	2.7	4.2	11.2	1.7	3.7	1.4
					B	3.4	3.3	3.4	4.5	7.4	3.5	4.9	2.0	6.1	4.3	6.9	2.8	2.6	5.4	13.2	3.5	2.1	1.9	2.7
			GC-MS/MS	Green tea	A	4.3	5.0	3.7	4.7	13.3	6.3	12.2	10.7	16.1	0.5	8.9	5.2	8.8	2.5	3.0	5.9	3.6	9.5	4.2
					B	7.7	8.3	4.3	10.0	5.6	5.0	4.6	9.9	5.7	11.2	3.8	8.1	8.7	8.4	3.2	6.0	10.8	3.8	4.5
				Woolong tea	A*	6.7	0.9	1.6	0.7	8.2	5.3	3.8	4.5	9.5	5.8	10.1	5.1	6.5	3.9	9.8	2.9	6.6	1.2	6.1
					B*	8.1	8.7	2.9	6.4	6.9	4.3	7.8	10.0	7.2	1.9	11.1	5.2	5.2	3.4	3.2	4.6	3.7	3.2	9.6
		Envi-Carb+PSA	GC-MS	Green tea	A	12.0	6.0	8.6	5.8	9.6	9.3	12.6	20.3	6.6	3.5	6.6	10.1	0.4	9.9	4.2	10.0	15.2	9.6	7.9
					B	7.6	8.5	4.3	0.5	5.7	8.0	13.2	8.3	2.9	2.8	13.9	7.8	24.0	5.8	13.2	6.2	1.5	19.5	10.3
				Woolong tea	A	6.2	3.4	3.7	3.8	2.7	0.8	4.7	14.1	3.9	2.9	4.4	3.1	2.7	2.7	3.2	24.9	9.5	8.1	3.1
					B	15.9	3.7	8.3	1.0	17.5	8.5	11.0	10.6	11.6	3.6	9.9	3.1	5.4	5.4	14.8	6.0	16.7	8.3	5.4
			GC-MS/MS	Green tea	B*	23.0	3.9	8.5	3.4	8.1	17.1	8.7	17.0	8.6	2.5	6.0	2.3	16.6	16.6	5.3	2.1	9.6	5.1	5.7
					A	7.0	11.4	7.9	9.4	2.7	6.5	48.6	14.4	12.7	4.6	13.1	3.3	7.9	11.2	7.2	7.2	7.8	19.0	13.9
				Woolong tea	A*	10.5	3.8	3.8	5.8	3.9	1.3	2.4	17.5	0.7	3.9	5.3	7.1	11.0	6.6	5.6	2.9	11.2	8.8	4.0
					B	8.2	4.5	8.7	2.2	10.6	9.0	43.0	10.0	11.0	3.5	5.5	1.1	8.8	11.8	10.4	4.6	9.1	11.8	5.6
123	propachlor	Cleanert TPT	GC-MS	Green tea	A	7.8	4.6	0.7	5.2	8.3	4.7	7.1	7.9	6.3	5.4	4.6	3.9	3.1	1.4	8.4	4.1	5.2	8.5	4.7
					B	5.6	10.0	5.2	5.7	2.7	6.4	0.9	6.4	1.6	5.6	2.0	1.9	8.1	13.8	3.8	9.0	11.8	9.6	7.1
				Woolong tea	A*	4.4	7.9	1.7	4.6	10.3	6.3	11.6	6.4	5.5	3.4	2.2	3.8	6.9	6.0	9.8	4.2	4.2	2.4	7.9
					B	4.9	2.5	5.0	6.9	3.5	3.2	4.3	2.4	5.0	3.5	0.7	2.2	6.5	1.3	2.1	1.2	4.4	4.2	8.4
			GC-MS/MS	Green tea	A*	7.7	5.2	4.1	4.4	8.0	3.7	4.7	6.6	7.6	3.4	6.6	5.4	5.5	4.0	2.3	3.9	6.8	5.5	5.0
					B	7.0	9.2	4.1	6.7	8.1	4.9	7.7	4.4	4.1	8.4	4.1	1.3	7.8	5.1	2.8	5.2	6.9	2.4	6.8
				Woolong tea	A*	4.9	11.5	4.5	3.5	5.4	3.1	19.3	4.0	2.7	5.4	2.2	2.3	1.2	0.7	9.4	3.7	11.7	6.4	5.0
					B	6.5	0.8	5.9	5.5	3.9	0.6	5.5	4.3	9.1	8.7	1.6	3.1	2.2	2.3	8.3	0.7	5.8	6.7	9.0
		Envi-Carb+PSA	GC-MS	Green tea	B*	10.1	3.0	1.6	6.4	7.7	4.9	9.1	4.1	4.5	0.9	12.0	4.8	7.9	2.7	6.7	14.7	13.6	8.9	4.9
					A	6.6	6.7	5.6	4.5	0.7	1.4	7.0	3.4	5.4	5.6	14.9	9.2	1.2	4.3	20.2	5.9	4.4	14.9	5.2
				Woolong tea	B	7.7	2.9	3.4	7.0	5.8	1.3	5.5	3.1	4.6	2.7	4.0	2.2	8.2	4.9	7.5	4.2	4.1	6.6	3.6
					A	11.9	3.7	8.4	6.2	7.6	6.3	4.5	5.2	10.9	2.9	7.6	1.4	2.5	2.3	12.4	3.5	10.3	0.7	4.7
			GC-MS/MS	Green tea	B*	45.8	6.3	2.8	5.6	4.5	11.5	3.4	2.7	8.6	2.5	10.9	3.8	7.0	17.6	3.1	4.8	14.7	5.6	7.3
					A	2.3	11.2	2.3	7.9	11.7	2.3	4.6	3.7	7.0	9.4	10.2	1.5	5.3	16.0	12.5	4.0	6.9	10.2	7.4
				Woolong tea	A	6.4	5.1	3.1	5.5	5.6	1.4	0.9	4.8	4.4	5.0	7.6	2.4	3.5	6.3	12.9	3.7	5.0	3.6	8.9
					B		4.6	8.8	6.9	3.4	4.6	4.4	4.4	10.9	4.2	1.8	3.2	6.1	12.8	13.8	0.7	5.1	7.8	8.4

Annex 2: RSD data for 201 pesticides in incurred tea Youden pair samples under 8 conditions with two SPE cartridge cleanup in three months (Nov.9, 2009-Feb.7, 2010) (cont.)

No	Pesticides	SPE	Method	Sample	Youden pair	Nov.9, 2009 (n=5)	Nov.14, 2009 (n=3)	Nov.19, 2009 (n=3)	Nov.24, 2009 (n=3)	Nov.29, 2009 (n=3)	Dec.4, 2009 (n=3)	Dec.9, 2009 (n=3)	Dec.14, 2009 (n=3)	Dec.14, 2009 (n=3)	Dec.19, 2009 (n=3)	Dec.24, 2009 (n=3)	Dec.14, 2009 (n=3)	Jan.3, 2010 (n=3)	Jan.8, 2010 (n=3)	Jan.13, 2010 (n=3)	Jan.18, 2010 (n=3)	Jan.23, 2010 (n=3)	Jan.28, 2010 (n=3)	Feb.2, 2010 (n=3)	Feb.7, 2010 (n=3)
124	propiconazole	Cleanert TPT	GC-MS	Green tea	A	8.5	4.9	7.1	4.2	6.1	3.9	11.2	3.3	2.7	3.9	6.1	3.9	7.9	6.3	9.1	4.0	3.6	4.8	5.1	2.8
					B	3.7	2.9	8.5	2.7	6.3	3.8	4.1	2.8	5.1	5.1	10.6	7.4	10.7	0.7	4.5	5.8	4.6	5.5	1.0	4.4
				Woolong tea	A	6.0	5.7	2.2	2.0	5.6	2.5	9.5	8.9	8.1	10.3	9.3	10.3	5.5	4.8	3.6	6.3	3.8	2.8	2.2	1.0
					B	5.1	4.2	2.5	9.0	2.7	2.8	4.6	4.9	10.7	13.0	7.4	13.0	1.5	1.4	6.2	2.5	2.5	2.6	1.8	1.6
			GC-MS/MS	Green tea	A	4.9	7.7	1.3	3.0	5.7	11.3	6.0	7.0	4.7	6.3	3.6	6.3	2.1	3.6	2.4	4.6	4.7	2.0	1.8	5.2
					B	6.6	6.0	6.0	9.0	1.9	7.0	7.2	5.2	6.1	26.1	6.3	6.3	3.1	2.7	6.5	4.5	32.1	4.4	2.3	4.3
				Woolong tea	A	7.3	1.4	4.2	5.1	3.3	5.2	8.3	2.2	4.4	2.1	4.5	4.5	1.4	6.4	2.6	5.4	3.2	2.7	3.1	0.6
					B*	4.0	3.2	4.8	3.9	10.3	2.4	3.7	6.9	6.0	3.2	4.9	3.2	3.6	2.4	3.1	7.5	2.3	4.4	3.0	5.4
		Envi-Carb+PSA	GC-MS	Green tea	A	7.6	1.3	2.6	11.9	5.6	8.0	8.8	2.4	9.8	2.8	1.6	2.8	6.2	3.6	4.4	6.4	6.1	11.9	2.8	5.5
					B	4.3	8.2	8.4	8.0	5.6	2.3	3.4	13.0	2.7	32.1	6.8	32.1	8.9	4.0	5.0	9.8	2.6	4.1	7.7	6.2
				Woolong tea	A	6.1	2.3	8.2	4.0	3.2	2.6	3.1	3.2	7.9	7.4	6.4	7.4	5.2	9.0	2.5	6.3	2.7	2.5	6.2	2.9
					B	4.9	7.9	14.2	4.1	18.7	1.1	1.9	11.2	6.0	10.8	6.6	10.8	13.0	1.9	4.3	7.4	4.6	4.4	0.9	3.8
			GC-MS/MS	Green tea	A*	6.2	4.9	32.2	7.6	9.7	4.3	8.0	2.2	7.2	7.4	1.7	7.4	2.6	8.2	2.8	7.0	5.1	11.4	1.7	4.0
					B*	6.1	9.3	6.1	4.3	0.4	3.7	1.6	5.9	3.4	15.9	2.0	15.9	7.7	2.4	5.1	12.7	1.1	8.1	8.3	7.8
				Woolong tea	A	7.0	2.8	5.6	2.9	6.3	6.0	3.6	5.1	1.9	8.7	4.2	8.7	3.0	5.6	5.7	8.0	3.2	6.4	6.3	5.1
					B	7.6	0.8	9.0	3.9	12.4	3.0	3.2	10.3	8.3	1.4	2.3	1.4	2.9	6.1	8.2	12.3	2.3	3.9	3.1	4.8
125	propyzamide	Cleanert TPT	GC-MS	Green tea	A	2.5	8.1	10.2	8.4	3.6	2.1	3.3	5.2	6.4	4.4	6.4	4.4	8.0	2.3	1.7	6.4	3.3	7.0	1.2	0.6
					B	6.6	5.4	3.1	7.1	2.9	8.3	4.6	8.6	4.9	4.8	8.4	4.8	2.0	1.0	2.6	7.6	3.4	4.1	1.4	4.9
				Woolong tea	A	4.4	2.5	1.8	0.8	0.9	1.5	5.0	6.2	3.1	0.3	3.3	0.3	2.9	0.9	7.1	6.0	3.9	4.0	3.2	3.3
					B	4.7	6.7	2.7	2.0	10.0	4.1	7.0	7.0	5.0	6.0	1.9	6.0	5.4	8.3	3.0	1.7	5.6	3.5	3.1	3.3
			GC-MS/MS	Green tea	A	3.3	8.6	6.1	8.1	4.9	4.9	6.7	7.4	6.8	5.1	7.6	5.1	4.6	5.2	3.7	1.0	4.3	6.3	1.4	5.3
					B	7.0	5.5	1.4	9.4	4.9	7.9	5.3	3.4	2.3	3.1	3.2	3.1	3.0	2.9	1.7	6.9	1.1	1.2	4.3	6.7
				Woolong tea	A	4.8	2.4	5.0	1.0	1.4	2.0	4.7	8.4	4.0	3.4	5.4	3.4	3.3	2.0	7.9	10.2	4.6	7.3	5.2	6.4
					B	7.1	6.3	4.2	9.4	8.7	3.4	7.1	3.2	3.1	6.5	4.2	6.5	3.5	3.6	3.8	3.3	6.9	10.2	4.5	2.9
		Envi-Carb+PSA	GC-MS	Green tea	A	3.9	1.8	1.6	3.7	1.0	1.5	11.3	1.2	6.1	23.4	1.4	23.4	4.4	3.1	2.8	11.2	0.3	4.8	11.1	7.0
					B	13.8	3.5	1.5	11.9	5.0	5.0	4.1	7.0	6.3	3.7	2.7	3.7	3.9	1.7	2.6	4.2	2.2	2.4	6.6	7.4
				Woolong tea	A	7.0	5.4	9.2	4.5	12.4	5.9	5.8	5.6	12.6	10.6	4.8	10.6	1.5	4.7	7.0	11.1	2.3	8.8	8.8	3.0
					B	7.9	5.8	2.4	5.8	8.7	1.4	10.4	3.8	5.1	7.5	0.8	7.5	1.6	13.2	15.1	4.7	6.6	9.9	2.1	4.9
			GC-MS/MS	Green tea	A*	6.6	2.8	6.2	2.6	0.9	2.1	11.5	1.4	5.3	17.7	1.4	17.7	3.3	4.6	5.5	12.7	0.3	7.1	9.1	10.3
					B	16.1	3.1	1.3	8.6	4.0	2.8	3.3	4.3	5.6	8.4	1.7	8.4	3.4	0.9	6.4	12.1	1.2	5.4	4.4	8.0
				Woolong tea	A	9.2	2.4	12.2	1.5	8.7	5.9	5.4	5.6	11.9	1.1	11.2	1.1	2.4	2.9	11.7	11.7	4.6	7.3	12.3	7.0
					B	6.2	3.5	3.7	2.7	8.9	3.2	6.2	4.2	4.2	4.5	2.7	4.5	3.5	4.9	5.2	4.1	3.3	3.5	6.4	2.3
126	ronnel	Cleanert TPT	GC-MS	Green tea	A	6.2	3.5	3.7	2.7	8.9	3.2	6.2	4.2	2.7	4.5	2.7	4.5	3.5	4.9	5.2	4.1	3.3	3.5	6.4	8.1
					B	7.5	9.2	4.8	6.7	4.9	3.7	1.2	1.3	3.4	2.0	6.9	2.0	3.1	2.4	4.4	6.8	1.8	4.2	5.4	8.1
				Woolong tea	A	5.6	3.7	1.0	3.2	2.6	2.9	10.4	6.0	5.0	4.9	6.7	4.9	1.0	2.4	1.5	8.5	3.2	6.1	2.4	3.3
					B	4.5	2.7	3.4	3.9	1.1	1.9	3.3	0.9	5.1	4.4	4.9	4.4	0.8	2.1	3.3	6.9	0.8	2.3	1.7	2.8
			GC-MS/MS	Green tea	A	3.9	6.2	4.7	3.8	7.6	1.6	5.0	6.4	8.5	8.2	3.0	8.2	6.4	7.7	3.7	3.3	6.7	3.8	6.1	2.4
					B	7.4	9.1	4.6	6.2	5.3	5.7	8.5	2.4	5.4	3.0	10.0	3.0	1.6	1.1	1.2	3.5	7.1	6.6	3.5	8.1
				Woolong tea	A*	6.6	5.0	4.1	1.9	5.5	1.8	15.0	1.9	3.1	4.4	6.0	4.4	3.9	4.7	1.2	6.7	6.2	7.4	3.6	6.7
					B	4.8	1.8	5.0	6.0	2.7	2.9	2.3	7.6	9.8	0.8	9.9	0.8	2.1	5.2	4.1	8.2	1.4	3.5	2.0	7.3
		Envi-Carb+PSA	GC-MS	Green tea	A	10.4	6.1	3.7	7.5	7.3	5.5	5.0	5.1	2.2	11.6	0.3	11.6	5.4	2.4	1.5	4.7	2.7	10.0	9.4	4.5
					B	12.2	8.2	3.6	4.2	1.2	0.2	2.2	2.4	6.1	13.3	6.5	13.3	6.9	6.7	2.7	5.4	8.3	2.7	12.0	7.0
				Woolong tea	A	5.1	2.7	3.0	5.0	3.8	1.5	6.8	2.2	3.5	2.5	2.9	2.5	1.9	2.0	1.5	4.5	1.9	3.5	7.2	4.9
					B	7.2	4.2	8.7	4.6	9.2	4.8	4.2	5.6	11.0	7.8	1.1	7.8	3.4	5.9	3.3	11.5	3.8	7.6	3.6	4.1
			GC-MS/MS	Green tea	A*	12.8	5.1	5.9	3.1	4.6	11.9	6.0	2.6	9.3	9.1	4.2	9.1	1.9	2.9	7.4	0.9	3.2	10.0	0.8	1.7
					B	28.7	9.7	2.6	10.6	7.7	2.0	11.9	3.7	7.3	10.2	11.5	10.2	2.7	4.3	10.2	5.8	5.9	9.1	12.5	8.5
				Woolong tea	A	3.5	2.3	3.0	5.8	3.7	0.5	0.3	3.8	3.9	8.9	2.5	8.9	1.6	2.0	7.6	9.8	6.2	5.8	3.6	6.0
					B	5.7	4.9	9.2	5.7	5.7	4.3	5.5	3.7	7.9	6.5	1.0	6.5	4.4	6.0	11.1	10.2	1.4	5.1	4.3	6.8

127	sulfotep	Cleanert TPT	GC-MS	Green tea	A	10.8	5.1	2.9	1.6	8.1	2.9	6.7	5.1	4.5	1.8	3.8	2.1	5.0	0.9	1.2	1.7	5.3	9.8	5.4
					B	5.1	8.4	2.5	7.4	4.4	5.8	5.0	2.2	5.4	2.8	5.4	5.9	1.4	0.5	6.9	2.4	10.7	1.4	8.2
				Woolong	A	4.9	3.0	2.4	3.3	1.2	3.1	7.7	5.5	3.3	1.6	2.4	3.0	4.2	3.1	8.7	3.0	1.4	2.8	1.4
					B	4.0	2.8	3.0	6.1	1.0	3.1	3.7	2.8	6.7	1.8	5.2	1.0	2.6	2.4	6.0	1.3	0.4	3.1	1.7
				tea	A	4.1	5.6	2.3	2.1	12.9	5.1	4.1	6.3	4.5	3.7	6.7	2.3	4.8	6.1	3.8	3.3	3.0	5.7	2.6
					B*	7.5	7.5	5.3	7.5	7.0	6.6	6.0	3.4	5.4	9.6	3.5	4.3	4.6	2.2	1.4	4.9	6.2	2.0	3.8
			GC-MS/MS	Green tea	A*	6.1	8.1	2.6	3.1	1.9	4.0	8.0	2.7	1.0	5.8	3.3	1.5	3.1	2.3	1.4	3.4	2.5	4.2	4.1
					B	3.3	2.4	5.7	7.1	2.7	4.4	4.0	4.2	3.8	4.0	3.6	1.7	2.7	2.3	7.7	4.9	2.8	4.9	5.0
				Woolong	A	6.7	5.3	3.2	5.2	7.0	1.4	9.3	1.7	5.4	3.1	5.8	6.0	3.3	1.8	3.8	6.4	8.2	6.2	4.8
					B	6.2	7.5	4.6	1.7	1.6	1.8	9.9	2.8	2.4	3.3	22.7	7.7	3.3	1.3	9.8	1.9	9.9	8.7	6.1
				tea	A	7.8	1.4	4.0	2.3	6.3	2.3	2.4	1.7	3.6	1.6	7.4	3.0	1.6	2.3	2.8	3.8	7.4	4.8	4.2
					B	5.3	2.3	8.8	2.8	14.3	4.0	1.8	7.0	4.8	3.1	7.0	2.7	2.4	3.3	8.8	3.8	4.5	3.6	3.6
		Envi-Carb+PSA	GC-MS/MS	Green tea	A	7.5	5.7	0.7	5.1	4.9	2.2	3.2	3.4	6.3	1.4	8.9	3.0	11.1	17.7	6.4	7.9	9.3	7.1	3.3
					B	12.1	7.6	5.7	2.9	0.3	6.8	7.5	4.0	3.4	1.1	17.1	3.7	4.5	14.6	12.8	2.0	4.4	11.3	8.5
				Woolong	A	3.8	3.9	4.5	2.8	4.9	1.8	0.8	0.7	2.5	2.1	7.0	1.4	2.4	4.5	8.7	3.4	8.7	5.4	2.9
					B	7.9	1.7	9.7	1.3	10.1	4.4	2.6	8.2	6.9	2.9	4.2	2.9	3.2	10.9	10.0	4.9	5.9	1.5	5.4
				tea	A	5.1	5.2	0.9	1.4	6.8	4.4	7.7	4.0	3.8	1.5	5.7	2.8	5.6	2.9	3.4	1.0	3.7	5.3	3.1
					B	4.6	7.1	4.2	8.0	3.4	3.3	1.2	1.2	3.8	6.8	4.4	7.9	0.1	0.7	5.3	3.2	5.8	3.6	4.8
			GC-MS	Green tea	A	7.2	2.5	1.5	1.7	1.4	2.6	6.6	6.0	2.0	3.0	2.7	3.4	5.7	2.8	6.9	3.1	1.9	4.4	1.4
					B	4.1	1.6	2.2	6.9	1.3	3.7	5.6	2.3	6.6	3.0	3.2	1.5	1.7	4.1	1.0	2.7	2.4	1.7	1.3
				Woolong	A	5.3	4.9	1.4	0.9	9.5	3.7	7.1	6.4	4.9	1.5	9.4	1.2	4.5	3.6	4.2	4.7	3.0	4.5	3.7
					B	7.9	6.5	4.6	9.4	7.3	5.1	7.3	1.5	3.2	9.0	6.1	3.3	2.7	4.0	3.4	2.8	6.7	2.9	3.0
				tea	A*	8.2	3.8	2.7	2.0	2.8	4.6	7.0	2.7	2.1	3.9	2.2	3.6	4.4	1.6	6.5	3.1	2.6	7.5	6.0
					B	4.1	2.6	3.1	6.1	3.1	3.6	3.2	5.8	3.2	4.0	7.9	0.5	2.0	5.6	5.2	1.9	5.2	1.4	6.5
128	tebufenpyrad	Cleanert TPT	GC-MS	Green tea	A	5.6	2.6	2.9	7.9	8.6	2.0	46.9	1.7	8.3	3.5	7.2	2.6	2.6	6.3	6.0	6.8	10.8	2.3	5.4
					B	3.1	8.8	4.1	7.2	2.3	1.7	11.2	5.3	4.1	1.2	21.6	3.5	2.3	5.3	10.4	2.2	4.6	7.3	6.9
				Woolong	A	4.0	1.3	4.5	3.4	6.7	1.0	3.6	5.5	2.2	2.6	6.3	2.2	3.6	2.8	1.6	1.3	4.1	5.9	3.6
					B	3.4	3.0	8.9	6.0	10.6	3.4	1.8	6.7	10.9	2.8	7.5	1.1	1.2	4.1	7.9	3.3	5.0	5.0	2.8
				tea	A	5.3	10.4	3.3	8.9	7.8	2.2	2.8	2.5	9.3	1.4	4.8	2.9	14.9	8.4	6.6	9.5	11.2	3.5	3.1
					B	5.1	12.5	6.7	3.4	3.6	2.1	2.9	4.8	6.5	1.3	18.9	0.5	4.7	7.0	9.5	2.5	8.0	8.2	9.4
		Envi-Carb+PSA	GC-MS/MS	Green tea	A	7.9	2.9	5.0	3.0	8.8	0.8	5.5	5.6	1.9	0.9	10.8	0.5	3.7	5.8	6.0	3.1	4.7	3.1	4.2
					B	6.8	4.2	11.7	3.3	8.5	3.4	1.2	6.8	8.5	4.4	1.4	4.1	2.7	6.9	9.8	1.9	5.2	1.7	6.8
				Woolong	A	6.0	5.1	2.4	3.2	8.1	5.7	7.6	5.1	2.9	5.3	7.7	6.9	6.6	11.0	8.5	3.3	19.7	12.5	4.7
					B*	4.6	7.7	5.7	8.5	3.1	2.5	4.1	2.2	5.4	7.4	7.7	10.7	1.3	2.9	12.1	3.9	10.8	2.9	4.9
				tea	A	5.4	1.7	3.7	3.6	0.8	3.9	9.1	5.1	3.2	9.4	11.1	2.9	3.9	3.8	8.5	5.3	2.9	4.4	1.5
					B	3.2	3.6	2.6	5.9	2.9	2.8	4.9	6.5	7.3	4.7	9.5	0.3	1.8	5.7	3.6	2.4	2.3	0.3	2.2
			GC-MS	Green tea	A	4.7	4.5	4.6	0.5	8.4	6.4	6.7	2.7	7.9	1.9	5.3	5.7	1.3	5.7	2.7	5.1	5.0	2.2	5.8
					B	7.4	8.3	6.0	9.2	4.3	4.9	7.4	3.4	7.9	10.9	9.2	4.9	7.8	2.1	5.7	6.3	4.1	2.7	8.0
				Woolong	A	5.6	2.6	2.4	4.1	2.3	4.6	10.5	3.8	2.3	6.4	1.5	3.1	3.1	3.8	7.9	3.1	4.1	4.0	4.7
					B	3.6	0.5	4.9	7.2	4.9	1.2	3.8	4.0	5.6	1.2	3.7	1.9	0.5	3.6	6.4	6.3	1.7	5.7	6.5
				tea	A	6.1	6.5	2.7	6.9	9.7	3.1	5.1	0.5	3.8	4.9	7.4	7.1	2.3	5.3	7.3	1.3	10.4	8.1	3.9
					B	3.2	9.0	4.0	5.6	0.6	3.1	4.9	2.5	4.7	3.9	28.0	7.7	1.5	3.7	12.1	4.5	5.1	7.4	5.4
129	terbutryn	Cleanert TPT	GC-MS	Green tea	A	3.2	4.8	3.7	1.3	4.9	1.2	2.8	2.8	2.6	4.0	6.1	2.9	1.5	5.6	3.4	4.5	3.8	3.7	3.9
					B	4.3	3.0	9.9	2.7	15.7	5.3	2.3	8.5	11.5	4.7	9.5	3.4	2.9	2.3	14.4	3.6	3.5	6.5	5.4
				Woolong	A	5.9	6.7	3.3	7.4	7.5	1.5	4.4	1.2	5.1	2.6	9.0	16.7	9.9	16.3	10.5	8.1	10.1	1.3	1.7
					B	5.8	7.9	7.0	2.3	2.3	3.8	3.6	5.7	3.8	0.8	17.0	3.8	4.6	5.7	14.5	4.0	9.4	9.2	6.9
		Envi-Carb+PSA	GC-MS/MS	Woolong	A	3.5	3.1	3.9	2.6	7.6	1.3	1.6	3.7	0.6	1.9	5.9	6.1	5.8	8.3	6.0	6.3	7.5	3.2	7.9
				tea	B	6.4	2.2	11.0	4.7	6.0	4.5	1.7	7.1	10.0	4.3	3.7	1.7	0.4	11.6	14.1	3.1	4.3	1.0	4.1

(Continued)

Annex 2: RSD data for 201 pesticides in incurred tea Youden pair samples under 8 conditions with two SPE cartridge cleanup in three months (Nov.9, 2009–Feb.7, 2010) (cont.)

No	Pesticides	SPE	Method	Sample	Youden pair	Nov.9, 2009 (n=5)	Nov.14, 2009 (n=3)	Nov.19, 2009 (n=3)	Nov.24, 2009 (n=3)	Nov.29, 2009 (n=3)	Dec.4, 2009 (n=3)	Dec.9, 2009 (n=3)	Dec.14, 2009 (n=3)	Dec.19, 2009 (n=3)	Dec.24, 2009 (n=3)	Dec.29, 2009 (n=3)	Jan.3, 2010 (n=3)	Jan.8, 2010 (n=3)	Jan.13, 2010 (n=3)	Jan.18, 2010 (n=3)	Jan.23, 2010 (n=3)	Jan.28, 2010 (n=3)	Feb.2, 2010 (n=3)	Feb.7, 2010 (n=3)
130	thiobencarb	Cleanert TPT	GC-MS	Green tea	A	5.9	6.2	2.9	2.7	8.0	4.7	7.0	4.1	3.9	2.1	6.7	2.3	6.9	3.1	3.7	2.1	3.6	6.5	4.6
					B	5.7	9.2	5.7	9.2	3.5	3.2	3.0	1.7	4.5	7.9	3.6	7.0	1.9	0.8	3.4	1.7	5.1	0.4	6.6
				Woolong tea	A	5.2	2.2	1.5	3.6	2.0	3.3	7.3	3.3	2.6	4.8	6.2	3.3	3.9	3.0	8.7	5.2	3.3	2.8	0.4
					B	2.8	1.9	2.8	2.5	1.4	2.7	4.4	3.5	6.5	1.5	5.4	1.6	2.4	3.2	4.6	0.5	1.6	2.4	1.4
			GC-MS/MS	Green tea	A	5.5	5.3	3.9	1.6	8.8	2.0	5.3	3.8	5.0	2.4	7.8	5.5	6.2	2.2	4.3	3.6	4.5	5.8	2.8
					B	8.4	9.2	4.9	10.2	6.0	4.6	7.8	2.8	6.1	9.1	4.4	4.5	2.8	1.6	2.0	4.5	3.2	3.2	6.3
				Woolong tea	A	6.1	2.8	3.6	4.0	2.3	3.7	6.6	2.8	1.8	3.4	2.9	2.3	5.3	1.2	5.6	7.0	2.0	1.1	5.1
					B	4.3	2.4	3.9	6.7	3.7	2.2	3.8	2.5	4.3	3.8	4.2	0.7	3.2	5.5	5.0	1.7	2.5	2.4	6.3
		Envi-Carb+PSA	GC-MS	Green tea	A	5.9	10.8	4.0	6.2	7.7	3.2	29.6	3.4	5.2	3.7	7.0	3.8	2.7	4.3	5.2	5.9	11.1	2.2	5.1
					B	2.6	8.7	4.1	3.5	2.2	1.2	3.1	1.9	2.2	0.7	20.3	4.3	2.3	2.3	10.5	3.5	4.0	6.8	5.5
				Woolong tea	A	2.5	5.3	2.1	3.8	5.2	0.7	3.2	3.4	1.1	1.6	3.2	1.9	1.7	3.8	1.2	1.8	8.0	5.2	3.3
					B	4.0	3.4	9.3	3.9	10.7	4.7	2.7	6.8	10.7	3.3	6.6	2.1	3.4	3.2	9.7	3.5	5.4	7.5	4.5
			GC-MS/MS	Green tea	A	5.9	5.3	0.8	5.0	7.4	0.7	1.9	2.3	6.6	1.1	9.6	1.3	9.3	17.8	7.6	9.5	8.1	2.3	6.2
					B	4.3	7.7	7.3	0.6	1.1	2.4	3.8	5.7	2.7	1.6	18.7	4.7	5.3	3.5	13.3	3.7	6.3	7.3	8.0
				Woolong tea	A	2.1	5.8	3.8	3.2	5.7	1.5	2.4	1.8	1.9	1.8	7.3	3.5	3.3	7.5	9.8	7.0	5.0	2.7	3.9
					B	6.2	3.3	10.5	3.0	5.6	4.1	2.6	6.9	9.9	3.8	3.0	2.9	2.1	10.3	8.2	1.7	4.6	1.4	6.2
131	tralkoxydim	Cleanert TPT	GC-MS	Green tea	A	7.1	4.9	2.4	0.9	6.8	4.2	7.7	4.8	5.0	1.0	3.3	2.0	5.4	4.1	3.7	3.2	3.7	4.3	5.7
					B	5.2	5.9	3.5	10.3	3.4	3.8	1.9	2.7	2.8	8.3	5.5	5.8	1.7	2.0	3.1	9.6	8.5	7.3	4.1
				Woolong tea	A	16.8	9.5	3.3	11.3	4.6	12.6	3.7	14.3	7.9	3.7	2.6	1.3	6.8	1.3	1.1	3.8	1.6	12.6	3.8
					B	22.2	0.9	4.5	3.5	5.0	6.5	9.6	2.1	7.8	14.6	6.6	9.8	9.3	9.8	11.8	11.7	1.3	12.6	6.0
			GC-MS/MS	Green tea	A	3.6	5.2	1.3	2.2	8.2	4.3	8.0	3.4	3.4	0.8	6.0	5.5	5.6	9.7	6.8	5.6	4.4	4.6	2.3
					B	6.7	4.5	4.6	8.9	6.7	4.2	6.9	3.1	3.7	7.5	4.2	7.0	1.4	5.9	2.0	8.2	3.2	4.7	3.1
				Woolong tea	A*	4.7	7.1	3.7	13.0	3.9	13.7	3.1	6.2	8.3	5.2	4.1	5.3	8.7	3.8	1.4	7.0	7.4	9.7	3.7
					B*	12.7	6.6	10.5	8.1	7.9	7.5	10.4	7.9	4.0	5.9	4.0	4.6	0.6	14.7	14.0	3.7	16.2	6.6	3.3
		Envi-Carb+PSA	GC-MS	Green tea	A	11.2	3.4	4.6	10.3	9.7	8.0	13.1	15.1	3.0	5.0	4.3	4.4	5.2	23.1	8.3	22.9	5.2	14.7	6.6
					B	6.8	10.7	2.4	18.4	3.4	6.1	2.6	6.5	8.1	3.9	58.4	15.7	27.1	2.7	31.2	6.2	0.4	13.9	11.1
				Woolong tea	A	20.3	9.0	4.5	17.0	6.6	9.2	2.3	12.8	5.8	7.9	8.3	11.5	4.2	1.3	4.5	5.4	1.9	12.5	10.6
					B	11.9	10.6	4.6	2.1	12.8	10.2	11.9	3.3	15.3	8.4	11.1	1.7	5.4	9.8	14.0	4.5	13.8	17.6	3.6
			GC-MS/MS	Green tea	A	5.9	4.7	7.9	10.9	11.5	1.9	42.8	14.4	2.8	4.4	2.1	2.2	4.9	21.0	8.8	26.4	5.8	8.0	9.9
					B	21.5	11.4	4.0	13.0	3.1	1.8	8.9	6.2	10.1	2.7	62.8	1.7	12.7	19.6	16.2	6.6	3.7	11.1	13.2
				Woolong tea	A	65.7	11.6	1.4	9.8	9.2	10.9	6.9	8.8	11.1	15.0	18.3	7.1	4.2	7.1	12.8	7.0	9.6	4.9	6.4
					B	23.0	9.2	15.7	8.9	10.9	10.1	9.1	8.8	18.3	18.6	1.9	1.0	4.9	9.1	15.2	3.7	10.7	17.7	5.9
132	trans-chlor-dane	Cleanert TPT	GC-MS	Green tea	A	5.3	4.9	3.7	2.2	8.4	4.4	6.4	5.5	6.1	2.7	4.8	2.7	6.9	2.6	2.0	0.5	4.4	6.1	4.0
					B	4.9	8.7	4.8	8.9	3.3	3.6	2.5	2.0	6.9	9.2	2.3	6.7	0.5	3.1	1.8	2.4	9.0	4.1	4.5
				Woolong tea	A	4.8	4.8	3.5	3.5	2.0	3.7	8.2	6.8	2.4	3.7	4.7	1.6	3.1	3.9	7.6	2.0	4.9	1.3	0.9
					B	6.9	5.1	4.0	0.8	0.6	1.9	4.0	0.6	5.6	3.2	1.2	1.2	2.2	1.1	5.6	2.3	2.5	2.3	2.4
			GC-MS/MS	Green tea	A	7.1	8.7	3.6	0.6	12.7	10.7	6.2	4.0	5.5	3.4	4.9	4.7	7.0	3.9	3.8	7.0	5.0	3.2	9.1
					B	9.4	7.4	3.0	6.0	8.6	7.7	7.4	5.1	10.8	6.9	2.5	2.6	3.3	0.8	2.4	5.9	5.3	9.8	2.2
				Woolong tea	A	6.2	1.5	6.6	0.4	6.1	3.1	10.5	0.2	4.2	4.8	4.3	2.6	3.1	3.0	6.5	4.0	3.0	1.6	6.7
					B	8.0	4.8	2.5	5.3	5.5	6.7	1.3	3.0	7.8	2.7	6.1	5.1	1.6	4.5	9.2	10.7	7.0	1.3	13.8
		Envi-Carb+PSA	GC-MS	Green tea	A	4.0	7.6	3.7	4.0	8.7	1.7	52.5	1.5	4.1	1.2	12.6	0.5	2.9	5.2	6.1	1.3	4.1	9.5	5.3
					B	1.3	3.1	2.4	4.0	3.5	1.6	2.4	1.2	5.6	1.9	20.5	1.7	4.0	3.1	13.2	3.7	6.1	10.1	6.7
				Woolong tea	A	5.1	2.3	9.5	2.9	6.3	1.1	3.8	5.6	2.2	3.1	3.6	3.1	2.1	4.2	2.6	3.7	9.9	6.0	5.1
					B	8.0	7.7	3.8	7.6	9.8	5.6	2.1	8.8	12.3	3.4	8.9	3.4	4.7	3.9	9.6	4.2	9.9	4.0	4.9
			GC-MS/MS	Green tea	A	5.2	8.5	7.6	3.6	5.8	1.7	5.8	3.9	4.7	0.5	9.1	34.6	9.0	10.1	4.1	6.9	8.9	6.2	3.7
					B	3.9	1.9	3.6	3.6	2.4	2.2	3.0	1.7	3.6	3.8	22.1	20.6	3.4	3.4	17.3	3.1	6.0	11.9	11.5
				Woolong tea	A	3.9	1.9	1.9	4.2	6.0	1.9	1.6	2.0	2.4	2.1	3.2	0.8	2.5	6.3	5.9	5.9	5.5	1.2	0.5
					B	6.0	2.5	12.0	2.3	10.1	6.1	4.2	7.8	9.8	6.3	4.1	5.6	6.3	10.7	8.6	4.0	4.0	0.8	11.7

133	trans-diallate	Cleanert TPT	GC-MS	Green tea	A	5.4	4.9	2.4	2.0	8.4	3.3	4.6	4.5	5.1	5.9	4.6	2.3	8.3	4.4	8.3	3.4	6.3	8.4	4.4
					B*	5.8	8.7	4.5	8.3	2.6	3.4	0.9	1.9	6.1	9.6	3.1	6.6	1.3	3.8	10.4	3.5	11.7	0.8	6.2
				Woolong tea	A	4.4	3.0	1.0	2.7	2.4	5.0	8.8	6.2	5.8	4.1	3.1	3.8	6.1	8.1	7.9	6.1	3.9	1.3	3.4
					B	3.1	2.2	3.4	12.3	6.8	2.4	2.2	3.0	9.5	2.6	2.0	2.6	1.7	4.3	3.4	1.1	2.7	3.2	6.7
			GC-MS/MS	Green tea	A	3.1	3.8	3.8	2.3	12.1	4.1	6.9	5.9	5.6	2.2	7.7	4.1	4.8	5.8	4.2	3.7	1.1	2.7	3.2
					B	8.4	7.4	5.7	8.8	5.0	6.4	5.7	4.0	6.0	8.3	5.4	5.4	1.9	2.0	1.9	1.8	5.2	2.7	3.5
				Woolong tea	A	5.2	2.1	0.8	3.3	2.6	3.0	8.1	1.4	2.9	3.3	1.9	1.9	3.7	4.5	6.3	4.3	6.9	3.1	2.5
					B	2.5	0.7	5.5	6.9	4.5	3.8	1.7	6.6	1.1	4.3	4.1	1.4	0.8	5.7	8.3	5.1	3.4	0.9	5.4
		Envi-Carb+PSA	GC-MS	Green tea	A	7.2	3.9	2.0	6.7	7.6	3.0	18.0	1.6	4.5	2.6	7.6	2.1	1.2	4.1	2.8	5.2	10.0	6.3	5.1
					B	5.5	8.0	4.5	3.0	6.0	1.0	2.9	2.6	1.9	2.7	15.9	2.1	8.1	0.5	11.8	2.0	5.6	8.8	5.9
				Woolong tea	A	5.2	3.5	2.7	2.8	3.7	2.6	4.2	7.8	12.5	3.3	5.4	1.6	5.9	7.5	0.8	1.6	5.9	5.0	6.3
					B	5.8	2.7	7.9	3.4	7.3	3.6	3.6	9.0	3.8	1.2	7.5	4.6	2.8	2.3	8.4	1.9	6.6	3.4	2.4
			GC-MS/MS	Green tea	A*	9.1	7.6	4.1	5.2	5.6	5.8	3.8	4.3	6.7	1.0	6.2	7.4	11.6	14.9	10.6	8.3	10.2	5.8	2.4
					B*	10.4	7.3	6.7	4.4	2.3	2.9	12.7	0.6	4.8	0.7	14.7	7.1	6.6	15.9	14.4	4.5	7.3	8.4	8.8
				Woolong tea	A	4.0	2.8	4.2	3.3	5.0	1.8	1.0	2.6	2.6	1.2	7.2	3.6	2.8	4.7	8.3	5.7	3.0	3.3	4.2
					B	6.9	4.1	9.1	1.4	7.3	3.8	3.5	8.4	10.0	3.2	2.8	2.1	1.8	10.1	8.2	6.0	6.3	3.9	7.5
134	trifloxystrobin	Cleanert TPT	GC-MS	Green tea	A	9.7	6.0	2.7	13.1	13.0	2.3	12.7	14.2	12.0	6.3	8.5	8.7	9.8	11.5	3.3	5.6	6.6	14.7	4.8
					B	8.7	4.4	3.7	19.8	5.0	5.7	6.1	12.6	2.8	3.1	6.7	9.3	13.6	8.0	3.9	6.4	14.0	6.7	5.0
				Woolong tea	A*	7.7	7.0	2.4	1.5	11.9	3.0	1.5	9.8	9.2	8.6	15.2	6.5	3.7	0.9	5.7	5.5	0.7	1.6	2.4
					B	4.1	4.2	3.6	4.5	3.8	4.4	9.1	6.6	8.9	5.4	18.5	1.3	1.6	7.3	7.8	1.4	2.1	3.4	5.2
			GC-MS/MS	Green tea	A*	10.8	9.9	0.4	13.4	17.0	3.5	16.0	21.6	17.1	3.2	12.8	6.6	7.7	8.5	2.6	8.8	9.5	14.1	1.0
					B	8.7	4.8	7.2	11.4	4.3	3.8	4.9	13.0	3.7	5.8	1.4	7.4	19.0	2.8	6.4	5.5	12.8	10.8	7.3
				Woolong tea	A	11.6	3.7	5.2	0.6	8.2	1.9	1.2	3.0	7.4	4.4	5.4	7.3	1.6	1.3	5.3	12.5	3.4	6.4	5.9
					B	6.1	3.7	3.5	6.4	6.7	5.9	7.6	10.8	8.1	7.4	8.5	8.0	2.8	8.3	10.9	0.4	4.5	6.1	10.6
		Envi-Carb+PSA	GC-MS	Green tea	A*	29.4	9.7	21.3	8.0	8.0	27.3	15.0	21.9	7.8	6.5	6.6	27.7	5.8	19.1	14.2	9.0	12.2	19.1	6.3
					B*	39.1	6.5	6.8	8.8	14.2	6.9	5.3	19.5	22.1	14.9	8.5	21.5	30.9	6.6	18.1	21.3	2.2	26.1	14.8
				Woolong tea	A	5.3	7.3	4.7	6.7	4.3	4.0	9.6	22.1	2.0	1.5	4.4	5.3	6.3	0.9	2.8	16.8	6.2	11.1	1.7
					B	10.1	9.7	5.2	2.9	16.7	7.3	7.6	6.2	13.0	6.9	10.5	2.8	3.0	7.3	13.0	2.1	10.4	17.0	6.2
			GC-MS/MS	Green tea	A*	32.6	11.1	19.2	1.0	9.5	30.0	7.8	27.8	15.0	9.3	8.1	0.0	17.8	23.1	9.5	28.0	11.2	11.3	9.5
					B	44.8	13.6	5.7	25.4	14.9	6.7	26.9	20.3	23.5	19.1	23.9	0.0	10.1	9.7	14.3	23.3	5.3	23.8	10.2
				Woolong tea	A	10.5	4.3	6.9	7.7	5.0	3.4	7.4	19.0	2.8	3.5	7.3	10.6	4.5	4.7	4.9	12.5	8.3	13.0	3.4
					B	13.3	6.8	7.4	2.1	15.4	8.6	18.8	6.0	9.4	8.9	4.3	8.8	5.0	10.0	18.2	0.4	7.8	27.6	6.3
135	zoxamide	Cleanert TPT	GC-MS	Green tea	A	11.4	6.3	1.5	1.9	5.1	5.5	7.8	3.3	4.3	3.4	5.0	3.8	0.9	5.6	3.0	8.3	3.1	6.3	6.4
					B	5.9	10.3	5.5	6.7	1.1	4.8	1.2	6.5	2.5	5.1	5.7	6.5	9.2	7.2	10.0	1.8	11.7	5.5	3.8
				Woolong tea	A*	9.5	7.5	2.1	3.8	5.5	5.6	9.4	5.1	0.9	3.6	6.4	2.4	2.4	3.7	8.1	5.6	5.6	10.8	2.4
					B	10.2	1.8	4.4	7.5	0.4	13.5	0.6	2.3	4.4	5.0	7.8	3.6	7.5	3.3	2.2	2.9	1.4	2.0	3.3
			GC-MS/MS	Green tea	A	6.5	9.2	1.9	4.5	4.9	4.9	6.8	3.3	6.0	3.8	6.6	6.2	7.5	3.5	5.6	7.3	4.8	0.7	3.4
					B	8.6	7.8	4.7	5.9	4.9	7.4	15.5	6.6	4.2	7.9	5.1	3.0	2.4	3.7	2.4	3.2	4.2	4.9	3.8
				Woolong tea	A	10.0	8.0	5.1	3.4	5.7	3.9	3.0	1.0	0.7	2.9	1.5	3.3	10.2	3.8	5.7	5.3	0.2	14.7	3.8
					B	11.4	4.3	6.8	4.5	4.9	3.1	43.7	7.7	1.4	8.0	2.2	1.1	2.4	1.8	9.3	9.5	4.3	6.1	2.2
		Envi-Carb+PSA	GC-MS	Green tea	A	9.5	6.9	4.9	8.8	4.2	3.6	7.7	2.5	9.6	1.7	7.8	4.2	2.0	2.1	2.6	14.9	9.8	14.7	4.0
					B	5.9	6.5	4.0	1.4	6.9	3.9	5.3	6.0	6.3	5.8	16.6	3.9	1.7	2.2	10.5	8.0	2.1	14.7	6.0
				Woolong tea	A	5.8	2.9	6.6	8.3	8.1	3.8	6.0	5.5	0.5	8.6	7.2	6.6	10.7	3.7	3.9	6.4	1.5	4.7	8.2
					B	9.9	7.6	14.5	9.8	9.1	6.1	25.8	1.7	5.4	4.2	4.2	4.0	1.9	3.3	10.6	5.9	7.9	8.8	5.5
			GC-MS/MS	Green tea	A	15.4	5.6	5.3	5.6	2.1	4.0	13.1	4.2	8.8	2.1	7.5	10.7	10.4	0.7	4.6	6.2	10.4	1.9	4.5
					B	9.4	2.5	5.4	2.6	7.5	4.6	6.6	4.8	7.0	7.4	14.2	8.1	2.6	3.3	5.5	6.3	3.5	10.2	11.4
				Woolong tea	A	7.7	1.9	16.0	12.8	10.7	6.6	1.6	9.5	6.3	1.9	7.7	5.3	6.0	15.5	9.5	9.5	10.7	2.7	6.1

(Continued)

Annex 2: RSD data for 201 pesticides in incurred tea Youden pair samples under 8 conditions with two SPE cartridge cleanup in three months (Nov.9, 2009-Feb.7, 2010) (cont.)

No	Pesticides	SPE	Method	Sample	Youden pair	Nov.9 2009 (n=5)	Nov.14 2009 (n=3)	Nov.19 2009 (n=3)	Nov.24 2009 (n=3)	Nov.29 2009 (n=3)	Dec.4 2009 (n=3)	Dec.9 2009 (n=3)	Dec.14 2009 (n=3)	Dec.19 2009 (n=3)	Dec.24 2009 (n=3)	Jan.3 2010 (n=3)	Jan.8 2010 (n=3)	Jan.13 2010 (n=3)	Jan.18 2010 (n=3)	Jan.23 2010 (n=3)	Jan.28 2010 (n=3)	Feb.2 2010 (n=3)	Feb.7 2010 (n=3)
136	2,4'-DDE	Cleanert TPT	GC-MS	Green tea	A	6.3	7.3	3.3	8.3	5.5	5.8	11.9	2.8	5.2	3.5	1.4	10.0	4.8	4.1	3.7	2.9	10.4	5.5
					B	6.0	6.6	6.7	10.5	3.4	4.5	4.1	2.6	4.5	9.5	6.0	2.1	3.5	3.9	1.7	6.6	2.8	7.6
				Woolong tea	A	4.2	4.2	1.8	4.1	5.5	5.9	8.7	8.1	0.1	3.6	3.2	5.3	3.3	9.1	3.6	4.5	7.5	0.7
					B	4.8	3.6	3.7	5.7	1.1	4.4	6.0	2.6	3.3	2.0	3.1	4.6	3.5	2.0	1.0	1.1	4.5	1.1
			GC-MS/MS	Green tea	A	6.7	3.2	2.4	11.0	5.4	7.3	8.3	2.5	4.3	2.0	3.1	10.6	2.2	5.7	5.0	3.7	13.7	3.5
					B	6.6	8.9	9.9	9.3	3.7	6.6	5.5	5.2	4.7	7.8	11.6	3.9	1.4	7.4	4.6	8.9	6.8	6.3
				Woolong tea	A	5.5	4.0	3.5	9.6	4.9	6.4	7.7	5.7	1.8	2.9	3.0	9.7	8.6	12.9	7.9	8.4	3.1	1.1
					B	7.7	0.7	3.7	11.2	2.8	4.6	8.1	8.3	6.5	3.5	4.7	3.0	9.1	8.1	2.0	3.8	5.7	9.4
		Envi-Carb+PSA	GC-MS	Green tea	A	11.3	5.2	2.9	5.1	8.5	8.0	5.6	5.2	4.5	5.0	7.6	4.1	9.6	13.6	12.9	13.6	0.3	6.5
					B	16.9	7.3	4.6	10.3	4.2	4.4	2.1	13.8	6.7	3.7	5.1	1.5	2.6	16.0	9.4	6.3	4.4	5.4
				Woolong tea	A	2.0	4.0	2.2	3.0	10.7	1.9	4.5	3.5	1.9	4.6	4.2	8.2	5.1	2.1	7.4	5.5	3.2	7.6
					B	4.9	2.5	10.3	4.7	9.0	5.0	2.6	8.5	10.9	4.2	3.1	2.8	8.2	12.0	5.7	5.8	10.3	4.7
			GC-MS/MS	Green tea	A	7.1	0.7	1.5	9.3	3.1	8.8	1.9	4.4	4.8	4.6	2.1	6.6	10.9	12.3	15.7	10.4	2.7	7.1
					B	8.9	12.7	3.1	8.8	4.4	1.7	0.9	13.6	5.6	6.4	2.5	6.5	8.1	21.2	10.8	3.0	8.6	9.1
				Woolong tea	A	4.3	9.7	5.0	2.6	8.8	5.3	8.6	4.1	0.3	2.0	5.9	4.6	2.8	9.3	6.6	4.1	2.6	1.1
					B	5.8	4.8	14.4	8.9	4.2	5.1	6.6	5.9	7.5	4.2	4.9	8.9	6.6	11.3	5.8	7.5	4.5	9.4
137	ametryn	Cleanert TPT	GC-MS	Green tea	A*	1.8	4.2	2.5	2.4	23.3	4.2	33.8	7.0	14.3	7.1	6.8	7.3	4.7	1.2	2.4	5.6	5.9	3.7
					B*	6.5	7.7	5.6	7.8	21.9	17.3	10.7	14.8	5.3	10.1	14.0	2.4	6.3	4.1	3.5	4.7	0.6	4.4
				Woolong tea	A*	8.2	2.1	3.2	5.6	2.3	7.2	24.0	29.5	9.2	11.3	4.3	3.1	2.6	7.2	2.8	2.8	2.5	1.8
					B*	6.1	6.3	3.6	5.4	2.5	11.7	16.1	19.5	7.6	8.5	2.6	1.3	2.8	3.8	1.0	3.0	1.0	3.1
			GC-MS/MS	Green tea	A	4.3	3.2	2.0	3.6	10.4	5.2	7.9	5.4	4.2	2.9	1.6	9.7	11.0	1.1	2.2	8.2	6.3	4.7
					B	5.5	9.4	8.2	7.9	2.9	3.5	4.2	2.5	6.2	9.7	10.6	3.4	6.5	4.7	7.5	8.5	4.2	6.7
				Woolong tea	A	5.7	4.4	2.5	5.2	4.4	3.6	10.3	3.0	4.7	4.0	1.5	8.2	1.0	6.4	1.9	0.2	4.5	5.3
					B	5.1	4.1	4.3	7.4	2.2	0.8	7.1	7.4	7.6	3.2	3.8	8.9	1.9	6.2	3.5	3.9	3.7	4.0
		Envi-Carb+PSA	GC-MS	Green tea	A*	6.7	4.1	2.8	7.4	8.8	0.3	42.9	3.7	17.1	5.6	6.6	3.9	6.1	9.7	7.2	11.7	2.8	6.3
					B*	2.0	8.5	3.6	6.1	2.1	23.6	4.9	30.4	37.4	5.3	13.7	4.7	2.7	11.8	1.5	4.6	8.7	8.1
				Woolong tea	A*	8.8	2.7	2.1	0.5	5.1	2.7	9.9	22.8	20.8	3.4	10.5	1.3	2.5	3.2	3.4	5.7	4.4	3.2
					B	5.7	2.5	9.0	3.3	13.3	14.3	81.5	13.1	22.9	4.8	7.2	5.0	7.9	9.3	1.8	6.6	4.8	6.2
			GC-MS/MS	Green tea	A	7.5	3.6	1.3	7.6	7.8	1.8	4.6	2.0	5.5	0.6	0.8	1.9	3.8	7.7	6.1	8.4	2.8	6.2
					B	4.5	9.8	5.6	3.6	0.7	2.4	2.2	8.8	2.4	2.3	2.4	4.1	19.7	12.6	0.8	4.1	9.7	7.4
				Woolong tea	A	4.4	4.8	4.2	3.3	5.7	3.1	3.1	4.1	2.8	3.8	5.4	5.6	2.3	12.9	3.4	4.5	4.5	5.3
					B	5.4	3.6	10.2	5.1	8.7	5.1	4.2	8.2	7.3	4.5	2.6	3.7	8.2	13.6	2.7	2.8	5.0	4.0
138	bifenthrin	Cleanert TPT	GC-MS	Green tea	A	4.5	5.8	1.9	4.4	7.4	3.3	5.4	5.2	5.0	2.2	2.7	6.2	2.2	2.3	6.7	4.6	6.1	8.5
					B	5.1	7.6	4.4	8.2	5.3	6.1	3.6	0.6	4.5	7.4	7.5	1.1	1.4	6.7	1.6	5.7	2.0	4.1
				Woolong tea	A	7.0	1.4	3.6	6.7	3.3	6.1	6.2	8.4	8.6	3.3	2.4	10.6	2.9	4.8	3.2	1.9	3.0	1.3
					B	5.6	2.5	5.9	7.5	3.2	2.5	8.1	1.2	4.4	4.0	2.1	1.0	5.0	10.1	5.2	1.8	11.3	8.1
			GC-MS/MS	Green tea	A	5.2	4.1	2.1	3.1	9.8	4.1	7.1	3.7	3.8	1.5	1.7	11.0	4.5	4.4	1.7	7.5	4.9	1.7
					B	5.0	7.6	5.8	8.2	3.5	4.3	1.2	2.4	5.6	7.1	5.0	9.0	1.6	3.0	3.8	6.1	4.2	6.2
				Woolong tea	A	7.5	1.2	4.6	8.0	2.9	5.0	6.1	7.8	8.0	3.7	0.8	13.8	5.9	9.1	0.4	2.4	3.2	9.9
					B	5.8	4.3	5.7	7.9	3.3	3.9	9.9	6.8	11.0	3.5	2.9	5.2	3.3	9.9	7.7	1.3	8.8	4.2
		Envi-Carb+PSA	GC-MS	Green tea	A	6.8	4.7	3.8	3.9	8.9	2.5	4.1	1.5	7.4	3.1	1.2	1.0	3.4	2.6	5.3	10.0	4.5	5.5
					B	3.5	6.0	2.9	2.9	3.3	1.9	4.7	1.7	6.0	3.9	2.4	2.3	0.3	9.8	4.1	5.0	10.4	7.4
				Woolong tea	A*	4.5	3.9	3.8	5.2	13.4	1.7	2.5	8.1	2.7	1.0	5.6	1.8	8.4	2.8	6.3	6.3	8.5	10.7
					B*	7.7	8.8	11.2	5.9	9.0	3.2	7.5	5.2	12.0	5.1	2.6	4.2	8.9	6.2	5.9	1.6	11.7	4.2
			GC-MS/MS	Green tea	A*	5.0	9.4	4.9	1.0	0.7	1.0	3.9	4.5	3.6	2.2	33.5	3.6	15.3	8.6	1.9	2.6	4.1	5.3
					B*	4.8	4.2	11.7	5.3	11.5	3.0	4.6	6.3	1.8	2.5	19.9	8.5	11.9	4.4	8.6	8.0	5.3	7.9
				Woolong tea	A*	8.5	8.1	14.1	6.6	7.3	2.4	4.2	6.4	10.6	5.3	4.5	6.9	9.9	7.0	5.2	1.9	10.0	9.9
					B*	8.1	8.1	14.1	6.6	7.3	2.4	4.2	6.4	10.6	5.3	4.5	6.9	9.9	7.0	5.2	1.9	8.1	4.2

139	bitertanol	Cleanert TPT	GC-MS	Green tea	A	5.6	6.5	1.7	1.5	2.9	6.9	12.4	3.6	7.5	6.1	10.3	9.0	5.3	4.0	3.9	2.9	2.8	2.8	3.9
				Green tea	B	3.0	7.6	4.4	9.2	4.8	8.4	3.7	7.0	2.6	7.4	6.4	12.7	0.6	2.8	8.0	2.8	9.6	1.8	7.7
				Woolong tea	A*	5.7	2.2	8.5	2.7	10.1	4.4	0.4	14.2	5.5	11.8	12.7	10.8	6.3	3.0	5.7	3.3	2.4	3.7	8.7
				Woolong tea	B*	4.2	5.2	13.1	10.4	8.7	5.8	5.5	3.5	6.4	7.6	9.8	3.0	1.1	2.8	1.8	1.9	3.5	3.3	6.3
			GC-MS/MS	Green tea	A	5.3	6.9	0.9	12.3	9.7	1.1	9.5	5.3	4.5	3.8	6.3	0.9	11.1	16.0	1.8	1.7	6.3	3.4	3.1
				Green tea	B	4.5	10.6	6.4	7.5	4.6	3.3	1.2	8.8	1.9	11.4	6.4	5.7	4.6	3.0	7.0	3.8	5.5	5.7	6.8
				Woolong tea	A*	5.4	2.4	0.3	2.7	3.0	3.4	5.5	0.3	4.9	4.5	4.0	5.5	7.0	3.3	8.6	0.3	6.8	4.7	4.8
				Woolong tea	B*	4.6	3.8	3.7	7.3	5.5	3.5	4.6	9.9	10.8	2.7	6.0	3.0	7.1	4.1	5.1	3.6	1.7	6.0	1.5
		Envi-Carb+PSA	GC-MS	Green tea	A	7.2	13.0	6.4	10.0	6.9	1.8	30.7	1.6	3.5	6.7	1.7	11.4	3.8	7.1	11.7	14.1	7.4	15.8	3.8
				Green tea	B	6.5	15.1	0.7	5.9	1.9	7.1	1.0	6.7	7.5	3.5	39.6	10.4	9.7	4.7	18.1	3.1	1.9	9.0	17.0
				Woolong tea	A*	7.2	6.0	8.8	2.0	11.7	1.2	1.7	17.8	11.6	6.9	11.4	6.4	2.1	8.4	3.6	2.9	4.1	3.5	2.5
				Woolong tea	B*	4.6	3.3	15.2	3.6	21.7	8.5	1.7	6.5	5.9	1.6	10.6	2.7	0.8	14.2	6.1	1.8	8.8	5.0	1.3
			GC-MS/MS	Green tea	A	8.3	13.3	6.6	9.4	5.0	3.0	27.1	3.2	6.8	2.2	2.6	51.8	6.0	11.8	2.1	10.7	12.0	7.7	4.3
				Green tea	B	4.4	6.3	6.2	7.8	1.0	1.9	2.0	1.6	2.0	0.8	26.1	30.0	7.9	18.5	10.9	4.5	1.1	9.8	10.1
				Woolong tea	A*	4.7	3.5	3.7	1.1	8.3	1.2	2.1	9.6	5.4	10.8	7.0	4.2	22.9	8.7	5.7	3.4	4.1	6.6	4.2
				Woolong tea	B*	6.6	3.5	14.3	8.0	10.1	5.5	3.0	9.6	13.4	2.3	5.9	4.4	7.2	15.6	5.7	4.4	8.1	4.7	1.5
140	boscalid	Cleanert TPT	GC-MS	Green tea	A	8.6	11.2	1.6	17.4	14.7	2.1	11.1	8.5	5.6	4.1	11.4	8.7	12.6	12.5	5.1	11.7	7.2	23.4	4.2
				Green tea	B	7.7	5.0	5.6	13.1	7.4	3.1	4.6	6.5	7.5	10.0	4.3	12.6	12.5	12.7	12.6	9.7	13.6	10.8	5.2
				Woolong tea	A*	9.4	9.1	3.8	0.8	18.0	7.2	8.7	12.7	12.4	6.5	13.8	14.5	3.4	2.4	12.8	10.0	3.9	7.9	3.6
				Woolong tea	B*	5.8	2.9	5.1	2.5	8.5	14.4	1.1	3.8	16.1	3.5	12.9	8.2	1.5	11.7	1.6	0.3	3.9	6.2	12.0
			GC-MS/MS	Green tea	A	12.8	9.1	0.7	5.1	20.9	3.6	13.7	20.3	5.7	7.9	10.5	1.7	21.5	13.1	4.0	7.6	8.5	18.8	0.6
				Green tea	B	9.6	5.6	3.4	18.6	4.9	3.5	9.3	15.3	8.7	11.7	6.1	9.5	20.3	7.2	7.4	7.3	14.4	7.9	7.6
				Woolong tea	A*	12.1	3.7	0.5	3.5	15.0	7.8	15.9	16.4	22.6	6.0	10.9	18.2	4.8	5.1	5.4	5.8	5.8	19.0	12.0
				Woolong tea	B*	16.2	4.8	8.5	3.8	6.7	10.6	1.1	13.3	12.2	1.3	8.8	3.8	3.9	6.4	10.7	2.0	2.6	10.0	8.4
		Envi-Carb+PSA	GC-MS	Green tea	A	20.8	7.1	25.8	9.4	8.1	21.3	13.1	14.8	20.3	9.1	11.0	33.3	7.1	18.9	12.9	13.4	15.9	16.8	2.1
				Green tea	B	36.0	9.3	6.0	6.9	15.4	4.8	2.6	20.7	2.9	13.9	16.4	16.4	44.0	12.2	4.2	5.8	6.8	23.6	17.2
				Woolong tea	A*	12.0	9.6	7.2	16.6	9.1	9.8	10.3	32.8	15.0	12.7	6.3	15.7	13.4	4.2	1.4	33.6	7.2	23.6	13.2
				Woolong tea	B*	16.9	12.8	10.5	8.8	19.1	10.4	16.7	3.8	12.2	18.7	8.6	7.7	0.5	12.0	18.4	5.1	11.6	31.9	8.2
			GC-MS/MS	Green tea	A	22.2	7.7	21.9	7.8	9.1	34.4	31.2	28.0	22.3	2.1	8.1	5.7	4.5	7.9	9.6	7.9	17.0	15.0	3.9
				Green tea	B	37.9	12.4	7.5	13.4	18.3	9.7	14.8	28.7	3.5	13.3	17.5	6.6	40.9	20.7	3.0	18.1	6.0	27.1	16.3
				Woolong tea	A*	14.3	7.2	9.0	11.7	9.0	4.3	11.1	30.3	13.7	13.8	4.5	14.7	13.4	7.7	10.7	23.3	4.5	28.9	12.0
				Woolong tea	B*	17.8	11.1	12.9	2.1	17.2	14.4	15.3	4.3	12.9	17.5	10.6	9.5	4.9	8.9	14.6	1.1	14.6	41.5	8.4
141	butafenacil	Cleanert TPT	GC-MS	Green tea	A	10.2	9.6	2.9	21.1	17.1	4.7	16.4	14.1	6.5	7.2	12.0	7.0	12.1	14.9	3.6	8.8	6.5	25.8	1.0
				Green tea	B	9.0	7.7	8.5	14.3	9.9	10.2	11.2	12.9	11.9	13.3	2.2	12.5	18.9	11.8	12.8	8.3	19.6	13.4	9.0
				Woolong tea	A*	8.6	4.1	3.0	3.8	16.7	3.0	11.0	11.9	10.9	6.8	14.5	14.0	2.7	4.4	1.5	6.3	1.1	6.4	2.1
				Woolong tea	B*	4.9	4.5	4.1	3.6	10.2	13.1	13.6	6.2	17.8	2.5	13.7	10.5	3.2	10.8	9.9	1.9	2.6	4.1	11.0
			GC-MS/MS	Green tea	A	15.9	10.7	1.0	20.8	24.6	4.7	18.8	24.0	4.6	10.8	17.1	3.3	25.6	17.0	3.0	5.3	7.3	26.1	1.7
				Green tea	B	11.5	6.4	6.4	18.5	6.2	4.7	10.8	19.9	3.9	14.4	5.0	14.6	27.1	1.8	9.0	7.3	13.1	8.7	10.6
				Woolong tea	A*	12.9	2.8	0.6	4.1	13.6	4.9	8.5	18.3	20.7	5.8	7.0	12.6	7.9	4.6	6.3	6.7	4.5	14.5	10.9
				Woolong tea	B*	13.8	2.5	6.9	5.9	6.1	11.7	13.1	12.5	20.3	2.8	6.9	5.6	12.9	9.9	9.9	4.6	2.9	10.9	8.4
		Envi-Carb+PSA	GC-MS	Green tea	A	26.2	7.1	34.2	10.9	10.4	42.2	27.6	31.4	33.7	13.9	13.4	51.9	49.4	28.1	24.3	10.0	23.9	31.6	6.3
				Green tea	B	45.8	13.3	7.0	20.2	21.5	11.5	8.1	16.7	2.8	22.0	20.1	24.4	14.7	13.9	20.2	17.1	7.8	33.5	19.7
				Woolong tea	A*	9.7	10.2	3.9	15.3	7.8	10.8	16.2	34.0	12.1	21.4	7.4	17.3	0.4	5.0	4.8	20.3	9.8	15.4	9.9
				Woolong tea	B*	16.5	14.8	14.7	5.3	21.7	9.7	15.9	5.4	13.7	17.3	9.5	7.1	12.4	12.2	19.1	4.1	15.3	30.8	6.9
			GC-MS/MS	Green tea	A	31.0	3.3	26.7	3.6	10.3	45.5	19.3	39.0	29.7	1.3	17.5	21.8	37.7	30.7	19.5	27.3	20.4	25.2	7.7
				Green tea	B	58.1	13.2	5.0	28.1	23.5	9.4	17.5	35.9	1.2	22.5	27.9	15.3	5.3	19.2	32.0	33.4	10.1	33.7	22.3
				Woolong tea	A*	12.8	7.1	8.9	13.0	6.8	6.5	16.5	32.2	9.4	23.8	6.6	17.1	8.2	6.9	11.9	25.2	5.7	28.2	10.9
				Woolong tea	B*	21.4	10.9	16.8	1.6	16.3	13.5	15.9	7.0		16.1	9.7	15.2		18.3	18.0	2.1	11.1	42.5	8.4

(Continued)

Annex 2: RSD data for 201 pesticides in incurred tea Youden pair samples under 8 conditions with two SPE cartridge cleanup in three months (Nov.9, 2009–Feb.7, 2010) (cont.)

No	Pesticides	SPE	Method	Sample	Youden pair	Nov.9, 2009 (n=5)	Nov.14, 2009 (n=3)	Nov.19, 2009 (n=3)	Nov.24, 2009 (n=3)	Nov.29, 2009 (n=3)	Dec.4, 2009 (n=3)	Dec.9, 2009 (n=3)	Dec.14, 2009 (n=3)	Dec.14, 2009 (n=3)	Dec.19, 2009 (n=3)	Dec.24, 2009 (n=3)	Dec.14, 2009 (n=3)	Jan.3, 2010 (n=3)	Jan.8, 2010 (n=3)	Jan.13, 2010 (n=3)	Jan.18, 2010 (n=3)	Jan.23, 2010 (n=3)	Jan.28, 2010 (n=3)	Feb.2, 2010 (n=3)	Feb.7, 2010 (n=3)
142	carbaryl	Cleanert TPT	GC-MS	Green tea	A*	14.4	13.3	10.6	28.6	21.0	2.0	12.4	2.7	4.8	10.9	10.2	7.3	7.6	8.8	7.8	7.4	14.9	7.8	3.3	11.8
				Green tea	B	18.4	5.9	11.0	6.7	2.8	8.3	7.8	8.8	7.0	6.7	6.7	10.9	2.8	5.6	5.5	12.2	5.2	10.3	4.2	8.4
				Woolong tea	A	9.4	10.8	7.3	16.7	25.9	5.9	11.4	12.9	5.4	13.5	7.0	13.5	4.9	5.1	10.9	2.0	3.6	6.7	13.3	2.5
				Woolong tea	B	14.0	5.3	9.2	8.5	12.1	2.5	4.7	7.2	4.1	11.5	15.3	17.3	11.9	3.8	13.5	3.7	9.7	1.5	2.6	10.8
		Envi-Carb+PSA	GC-MS	Green tea	A	24.6	15.9	13.3	32.5	27.8	1.4	14.5	3.1	5.0	7.6	7.0	11.5	2.1	10.1	7.7	6.5	4.4	8.7	9.5	1.0
				Green tea	B	34.3	6.5	8.6	14.8	4.5	7.7	13.5	13.8	10.4	13.8	10.8	13.8	9.5	8.8	8.0	10.1	6.5	3.9	3.8	7.3
				Woolong tea	A	13.4	16.0	15.3	18.2	35.0	4.9	17.9	20.5	9.5	8.9	8.9	19.9	10.8	10.8	5.9	11.3	4.7	3.5	7.3	2.0
				Woolong tea	B	28.3	11.4	19.7	7.5	17.4	5.2	11.5	4.8	6.6	20.9	20.9	27.4	5.5	4.0	4.9	17.9	16.5	9.7	3.3	7.3
		Cleanert TPT	GC-MS/MS	Green tea	A	5.8	6.5	4.7	2.5	5.9	5.8	31.6	5.0	6.4	12.6	3.1	19.4	3.3	11.0	2.3	6.4	5.5	3.4	10.7	5.1
				Green tea	B*	13.4	3.5	0.9	3.0	2.8	3.1	6.4	7.6	12.6	29.0	5.0	29.0	8.2	3.8	3.9	18.9	7.0	4.8	18.0	8.9
				Woolong tea	A	5.7	3.7	4.6	10.0	4.1	6.0	7.2	14.3	2.7	1.4	9.6	1.4	5.5	5.5	15.4	6.6	1.4	9.3	5.9	2.3
				Woolong tea	B	6.2	6.2	15.3	4.1	12.9	4.4	5.5	5.6	4.8	5.4	6.6	5.4	5.6	1.8	9.9	9.8	15.9	4.8	5.1	10.2
		Envi-Carb+PSA	GC-MS/MS	Green tea	A	9.6	11.5	5.5	5.3	5.0	8.6	33.9	8.7	11.4	11.4	1.1	23.4	1.3	24.5	18.5	6.1	5.1	10.2	12.1	9.6
				Green tea	B	17.0	4.1	3.3	3.0	7.6	1.1	9.9	11.9	12.4	33.7	5.7	33.7	2.8	11.5	1.5	27.4	27.4	3.9	19.5	12.1
				Woolong tea	A	5.0	8.7	6.5	13.7	2.0	6.9	6.8	20.7	2.7	4.7	9.0	4.7	8.9	6.0	16.9	17.0	7.9	8.9	9.1	1.5
				Woolong tea	B	10.5	1.0	14.0	9.5	6.3	6.0	5.5	2.4	6.1	10.9	3.6	10.9	8.7	11.4	12.6	10.5	1.4	14.4	17.7	6.6
143	chlorobenzilate	Cleanert TPT	GC-MS	Green tea	A*	6.6	12.1	2.5	11.4	14.4	1.3	14.5	18.8	13.7	6.2	4.5	6.2	14.3	10.3	12.1	6.5	7.8	3.2	19.8	5.4
				Green tea	B	5.9	10.0	1.8	10.2	5.4	6.1	5.0	14.1	4.8	6.8	6.8	1.7	4.0	2.3	1.5	5.2	9.1	1.5	11.1	4.6
				Woolong tea	A	7.3	3.4	2.1	3.2	12.0	5.1	11.3	17.3	6.6	13.7	7.2	13.7	7.8	12.5	7.7	5.7	6.8	8.2	1.8	0.9
				Woolong tea	B	3.8	2.5	3.0	2.6	7.4	13.1	4.6	11.4	7.7	14.5	3.2	14.5	2.1	14.9	5.9	6.7	1.5	13.1	1.7	7.4
		Envi-Carb+PSA	GC-MS	Green tea	A	9.2	4.6	4.3	6.6	16.0	1.3	11.5	17.4	11.3	11.3	3.9	7.0	3.0	7.0	5.2	2.1	5.7	3.9	14.3	5.0
				Green tea	B	7.3	6.3	6.9	11.0	4.3	2.2	4.0	12.5	5.0	9.1	9.1	1.3	9.3	6.4	1.0	1.8	6.7	1.1	3.8	7.7
				Woolong tea	A	10.0	1.8	9.9	3.9	6.7	4.1	0.1	10.0	5.5	7.2	4.2	7.2	2.6	3.6	3.9	7.8	4.5	14.6	9.9	4.8
				Woolong tea	B	7.5	3.9	1.4	6.1	5.4	8.4	8.7	11.9	13.8	13.8	1.9	7.3	6.4	48.5	31.0	8.1	2.1	6.2	5.6	7.9
		Cleanert TPT	GC-MS/MS	Green tea	A	22.9	5.3	16.7	9.4	10.4	21.7	52.6	24.3	24.9	1.7	6.8	1.7	21.0	6.5	9.7	7.4	47.7	8.4	51.0	5.7
				Green tea	B*	24.9	10.7	5.2	12.7	9.4	3.9	11.4	29.1	30.2	41.8	8.1	41.8	1.7	3.8	1.2	46.7	9.0	14.2	32.3	12.6
				Woolong tea	A	4.5	3.2	4.2	7.4	8.9	1.9	5.0	28.2	2.1	5.3	2.1	5.3	7.4	4.6	11.8	1.1	20.0	9.0	13.6	5.6
				Woolong tea	B	8.2	7.5	7.1	5.5	17.2	9.8	13.0	6.0	12.7	6.6	6.6	11.2	1.9	38.1	7.9	15.1	2.5	1.5	14.8	6.8
		Envi-Carb+PSA	GC-MS/MS	Green tea	A	30.6	9.8	13.2	4.7	8.8	21.9	34.2	24.5	7.0	7.9	1.9	7.9	2.8	12.9	7.9	6.2	13.3	7.3	20.6	12.8
				Green tea	B	29.5	13.4	5.1	8.1	10.8	5.0	12.0	24.6	20.2	27.4	9.6	27.4	0.9	8.5	27.4	14.0	14.9	10.3	24.9	18.4
				Woolong tea	A	12.1	2.5	6.9	1.9	1.9	1.9	5.3	27.1	10.9	4.5	2.5	4.5	5.9	6.6	7.7	10.0	10.3	13.7	15.4	4.8
				Woolong tea	B	16.6	6.0	9.9	1.9	11.3	8.2	5.7	5.9	4.9	8.4	5.8	8.4	7.4	0.8	1.0	13.7	1.4	2.5	21.5	7.9
144	chlorthal-dimethyl	Cleanert TPT	GC-MS	Green tea	A*	4.2	6.0	3.9	1.0	8.4	3.7	6.3	7.7	4.4	4.4	2.9	4.4	2.5	2.7	1.9	2.5	2.2	2.6	5.4	3.8
				Green tea	B	5.7	6.1	4.7	7.9	4.1	3.9	1.2	2.6	3.2	7.0	7.0	2.5	6.4	2.6	2.7	2.6	4.0	3.4	0.3	5.9
				Woolong tea	A	4.9	2.5	1.8	3.8	1.6	2.9	10.0	5.8	5.6	4.4	2.1	4.4	2.1	9.4	3.6	7.3	2.8	0.7	2.2	0.5
				Woolong tea	B	4.1	2.4	4.5	4.2	2.5	3.5	3.8	4.1	4.5	4.6	4.6	0.7	1.0	1.1	3.6	4.6	0.5	4.1	1.3	2.5
		Envi-Carb+PSA	GC-MS	Green tea	A	4.6	4.2	2.7	2.9	9.6	2.7	6.3	5.7	6.5	8.2	2.1	8.2	0.2	4.8	2.6	3.8	1.1	6.0	4.3	3.6
				Green tea	B	5.5	6.9	6.7	7.5	4.7	2.9	3.1	4.4	3.5	9.8	9.8	2.0	7.3	4.1	0.7	4.3	4.0	5.8	3.1	8.1
				Woolong tea	A	4.7	4.0	1.1	4.3	5.4	3.1	10.2	0.1	5.7	5.7	4.6	2.1	3.0	2.1	2.8	7.1	2.7	1.2	3.3	6.5
				Woolong tea	B	4.5	4.9	4.9	5.1	3.0	3.3	3.1	7.0	6.2	3.9	2.0	3.9	4.1	7.5	4.8	7.8	2.7	9.9	0.9	3.5
		Cleanert TPT	GC-MS/MS	Green tea	A	8.5	6.8	4.6	4.9	7.9	2.1	8.1	3.1	2.3	8.9	2.0	8.9	2.8	0.4	1.8	7.2	13.0	5.2	11.9	5.1
				Green tea	B*	4.7	7.6	3.4	3.1	0.9	1.7	4.0	1.7	11.2	19.7	2.6	19.7	2.3	4.4	2.8	16.1	2.8	6.5	12.9	6.5
				Woolong tea	A	1.8	3.2	3.4	3.1	5.0	0.8	3.8	4.8	3.5	4.6	0.9	4.6	2.5	2.6	3.1	1.6	3.1	7.6	4.7	4.2
				Woolong tea	B	5.2	3.1	8.7	2.4	9.3	5.3	3.8	6.7	3.3	7.1	2.0	7.1	4.0	6.0	7.2	9.7	2.9	6.4	3.0	4.4
		Envi-Carb+PSA	GC-MS/MS	Green tea	A	9.4	3.7	3.7	5.9	7.0	2.4	4.1	3.8	2.4	4.6	1.8	4.6	3.1	1.6	2.2	6.1	7.8	2.5	5.4	3.9
				Green tea	B	5.3	8.6	5.0	0.8	1.2	1.8	2.8	3.0	8.9	21.1	1.0	21.1	3.8	7.4	12.2	11.1	1.0	5.4	10.6	6.6
				Woolong tea	A	3.2	4.7	6.7	5.0	4.1	1.6	2.2	5.7	4.7	5.2	1.8	5.2	3.2	3.1	3.1	8.9	2.6	3.4	3.4	6.5
				Woolong tea	B	7.0	3.1	11.2	3.8	7.3	5.7	4.2	7.5	3.1	6.2	3.5	6.2	4.6	9.4	9.4	10.0	1.7	5.4	3.1	3.5

145	dibutylsuccinate	Cleanert TPT	GC-MS	Green tea	A	3.3	5.6	3.1	1.5	8.6	3.1	5.6	1.9	6.1	3.1	7.2	2.2	6.8	1.5	1.1	0.5	1.1	3.0	9.1	4.0
					B	5.6	8.7	4.1	9.2	4.1	3.9	3.1	2.4	3.0	9.3	2.9	7.2	1.1	1.9	3.1	1.0	3.1	5.9	3.8	6.2
				Woolong tea	A	3.9	2.7	1.1	3.0	0.9	1.0	8.4	5.6	3.5	2.4	3.0	2.5	4.4	3.4	7.9	8.4	7.9	4.5	2.6	0.1
					B	4.0	5.0	5.9	6.9	3.5	4.6	5.0	2.2	4.8	2.4	4.3	1.4	3.3	3.0	2.5	1.2	2.5	4.9	2.9	1.9
			GC-MS/MS	Green tea	A	5.0	8.8	1.9	1.1	11.7	5.1	3.6	1.8	3.9	0.9	5.5	1.9	14.0	4.8	1.7	1.2	1.7	3.8	4.6	2.6
					B	4.7	1.3	2.4	8.0	3.3	3.8	3.6	3.5	4.1	9.0	3.3	9.4	17.7	3.3	4.2	10.0	4.2	4.6	5.0	6.6
				Woolong tea	A	4.8	3.9	3.9	4.2	4.8	0.9	9.0	4.6	6.1	3.6	2.0	3.7	8.1	1.7	8.5	2.6	8.5	1.7	0.9	5.3
					B	5.7	5.0	2.5	6.7	2.4	3.6	3.7	1.1	9.6	2.9	3.2	1.2	10.0	1.3	7.8	9.4	7.8	2.0	1.0	1.5
		Envi-Carb+PSA	GC-MS	Green tea	A	3.0	6.8	5.8	5.9	6.2	3.9	2.6	2.2	6.3	2.0	8.5	3.1	6.3	2.3	6.3	9.4	16.6	11.6	6.4	3.7
					B	4.0	3.4	3.0	4.6	5.0	3.3	5.6	3.7	3.8	1.2	19.9	5.5	9.3	1.4	16.6	3.4	16.6	5.8	6.4	3.8
				Woolong tea	A	5.9	3.3	7.4	1.9	4.5	3.9	2.6	8.3	1.5	4.5	5.9	0.2	2.5	2.7	1.6	4.9	1.6	6.2	2.9	3.6
			GC-MS/MS	Green tea	A	8.0	9.0	0.5	2.6	12.3	2.4	1.9	1.6	9.3	3.5	7.0	1.7	1.8	7.2	7.0	4.0	7.0	6.0	4.7	5.6
					B	2.9	7.5	6.8	8.1	3.1	0.9	4.3	1.6	6.5	2.2	8.1	6.8	1.7	12.4	3.4	8.8	3.4	9.7	6.3	4.1
				Woolong tea	A	3.7	4.3	4.4	2.2	2.3	2.7	2.9	5.1	1.4	0.8	22.0	3.8	14.1	13.8	18.3	4.0	18.3	3.5	10.9	4.8
					B	6.5	3.7	8.0	2.4	4.1	8.5	1.6	3.8	2.0	2.1	6.3	5.0	0.7	4.4	6.9	4.8	6.9	5.4	6.4	5.3
					B	4.9	5.6	2.2	5.5	7.6	4.6	2.6	8.1	7.7	2.1	3.5	2.4	8.1	8.1	6.7	1.7	6.7	5.5	2.6	1.5
146	diethofencarb	Cleanert TPT	GC-MS	Green tea	A	7.1	5.6	5.3	0.8	8.0	4.6	7.4	13.5	8.1	8.5	8.7	6.1	9.1	7.2	2.4	2.2	2.4	7.4	5.3	4.9
					B	6.4	10.0	3.7	8.0	6.5	5.6	2.0	8.0	3.3	5.8	4.7	14.2	4.9	2.3	4.1	4.5	4.1	5.8	2.8	11.0
				Woolong tea	A*	4.2	2.0	4.3	4.7	1.8	4.7	0.1	14.7	4.0	9.2	18.2	3.4	5.0	5.6	7.9	2.9	7.9	3.7	1.5	0.7
					B*	3.7	1.9	1.3	4.8	8.0	12.7	2.8	9.6	7.5	2.6	6.9	0.9	2.1	7.7	5.2	1.2	5.2	3.2	5.3	2.4
			GC-MS/MS	Green tea	A	3.7	4.2	7.0	4.8	9.3	2.4	7.3	8.5	4.6	1.2	5.8	1.5	9.9	4.2	0.1	1.5	0.1	7.5	5.2	4.2
					B	4.8	6.3	3.0	7.6	3.8	2.1	3.4	8.4	6.0	7.9	2.2	8.0	2.6	2.7	4.4	4.6	4.4	6.5	0.1	5.8
				Woolong tea	A	7.1	1.1	3.0	2.8	2.7	4.3	6.8	0.3	8.2	5.0	5.4	4.9	6.8	1.7	8.7	2.6	8.7	2.6	5.4	2.1
					B	3.6	3.1	3.5	8.5	4.0	3.9	3.7	9.3	8.5	0.6	9.3	1.5	1.3	2.7	5.2	3.3	5.2	2.1	4.6	5.1
					A*	10.9	5.3	10.7	7.7	10.0	3.3	28.8	13.1	14.1	4.7	2.8	4.9	3.0	13.7	13.9	34.8	13.9	14.2	32.9	3.2
		Envi-Carb+PSA	GC-MS	Green tea	B*	8.0	9.6	2.0	4.8	6.5	3.1	6.0	8.1	15.4	2.9	37.7	15.7	27.0	2.9	40.0	14.1	40.0	6.4	21.9	5.4
					A	3.1	3.7	7.9	4.9	8.0	2.4	2.2	16.6	6.3	3.7	6.5	5.1	3.5	4.9	5.0	4.5	5.0	7.7	5.2	2.6
				Woolong tea	B	4.4	8.1	8.2	2.0	14.8	7.7	4.3	7.6	7.1	3.1	8.5	0.4	5.0	4.2	9.8	3.3	9.8	5.9	2.9	4.3
			GC-MS/MS	Green tea	A	13.4	5.9	4.4	7.8	8.2	3.9	11.7	8.6	3.8	3.8	5.8	0.7	2.1	6.1	5.5	13.8	5.5	6.1	9.8	8.4
					B	5.9	9.7	4.5	2.6	2.0	2.9	4.5	3.6	10.1	2.7	30.3	1.8	19.0	12.9	18.2	2.7	18.2	4.2	14.8	11.7
				Woolong tea	B	5.5	5.1	5.2	2.9	3.4	0.7	3.3	12.2	2.2	3.2	5.7	6.4	2.4	1.9	12.0	3.7	12.0	8.1	5.9	2.1
					B	8.7	3.5	11.3	6.0	9.6	6.7	3.5	5.7	9.0	3.6	7.2	5.6	9.9	8.4	12.0	1.0	12.0	3.0	4.0	5.1
147	diflufenican	Cleanert TPT	GC-MS	Green tea	A	3.9	6.8	1.2	4.2	9.7	3.0	9.8	14.2	7.9	8.5	5.5	10.9	8.1	7.0	3.3	3.9	3.3	6.2	1.6	4.1
					B	3.8	6.9	3.1	7.7	3.2	3.9	2.1	8.6	3.9	6.2	1.8	9.9	6.4	4.8	3.2	1.7	3.2	11.3	9.7	9.7
				Woolong tea	A	6.3	4.0	9.4	1.6	5.3	4.3	5.7	14.6	4.4	4.0	9.6	5.5	6.8	2.3	6.8	3.9	6.8	2.8	3.1	11.5
					B	3.4	0.4	2.9	5.9	9.2	7.4	11.3	8.7	3.6	1.6	7.0	0.4	2.1	3.2	2.6	2.8	2.6	1.4	2.1	4.6
			GC-MS/MS	Green tea	A	6.3	3.5	0.5	2.7	11.9	4.7	10.4	10.4	7.2	0.8	6.5	1.9	10.0	1.8	2.6	2.3	2.6	8.8	2.4	3.7
					B	4.9	5.8	4.5	8.2	4.0	2.3	1.6	8.0	3.6	7.4	3.8	2.5	8.5	1.3	1.8	5.3	1.8	7.5	1.5	5.1
				Woolong tea	B	9.3	3.5	1.6	3.8	4.6	4.6	4.1	2.6	10.0	4.3	5.3	6.2	9.8	3.1	4.8	6.1	4.8	0.9	2.8	1.6
					B	5.6	4.5	2.2	6.9	3.3	6.8	5.1	12.4	10.1	1.7	6.5	0.7	5.1	0.7	5.9	1.6	5.9	4.5	5.0	5.5
		Envi-Carb+PSA	GC-MS	Green tea	A	13.8	9.9	8.4	7.5	8.9	9.8	35.8	10.5	14.5	2.2	2.9	12.8	2.1	14.1	9.0	31.1	9.0	10.8	25.9	9.7
					B	9.0	8.3	4.9	6.1	2.4	7.1	6.7	12.1	18.1	3.1	33.6	10.4	29.5	3.1	30.5	5.2	30.5	7.6	19.9	4.1
				Woolong tea	A	3.6	2.6	6.1	3.9	2.9	1.7	0.5	20.2	0.7	0.2	7.3	14.3	1.0	6.7	0.7	3.9	0.7	4.7	14.3	6.6
					B	5.2	4.2	7.9	1.2	11.5	6.3	8.2	7.2	9.5	2.5	14.5	2.6	2.2	4.2	9.2	3.6	9.2	7.6	7.9	3.6
			GC-MS/MS	Green tea	A	16.8	4.3	7.9	7.8	9.7	6.9	19.1	11.5	3.9	3.2	3.0	1.7	4.4	5.5	2.3	14.1	2.3	4.5	9.5	2.6
					B	9.4	9.6	5.8	1.8	4.1	1.8	5.4	8.1	10.1	3.2	29.3	0.8	17.8	6.3	15.4	2.6	15.4	2.7	16.7	13.1
				Woolong tea	A	6.2	2.5	5.7	2.8	2.9	4.3	1.0	16.9	0.5	2.3	8.0	2.6	7.4	4.3	4.7	4.4	4.7	2.1	11.2	1.6
					B	10.4	3.9	8.7	4.8	8.6	5.0	3.7	5.9	8.9	2.7	9.0	4.8	5.8	6.1	13.9	1.8	13.9	3.8	5.7	5.5

(Continued)

Annex 2: RSD data for 201 pesticides in incurred tea Youden pair samples under 8 conditions with two SPE cartridge cleanup in three months (Nov.9, 2009-Feb.7, 2010) (cont.)

No	Pesticides	SPE	Method	Sample	Youden pair	Nov.9, 2009 (n=5)	Nov.14, 2009 (n=3)	Nov.19, 2009 (n=3)	Nov.24, 2009 (n=3)	Nov.29, 2009 (n=3)	Dec.4, 2009 (n=3)	Dec.9, 2009 (n=3)	Dec.14, 2009 (n=3)	Dec.19, 2009 (n=3)	Dec.24, 2009 (n=3)	Dec.29, 2009 (n=3)	Jan.3, 2010 (n=3)	Jan.8, 2010 (n=3)	Jan.13, 2010 (n=3)	Jan.18, 2010 (n=3)	Jan.23, 2010 (n=3)	Jan.28, 2010 (n=3)	Feb.2, 2010 (n=3)	Feb.7, 2010 (n=3)
148	dimepiperate	Cleanert TPT	GC-MS	Green tea	A	5.5	6.2	6.5	2.7	9.4	4.1	7.2	6.9	6.8	4.7	7.5	5.3	10.0	6.1	1.4	2.4	6.9	6.2	5.3
				Woolong tea	B	7.9	5.4	7.8	8.1	5.3	2.2	1.9	2.6	2.5	10.2	5.0	13.9	1.5	2.2	3.5	1.4	7.0	2.3	4.9
				Green tea	A*	3.6	2.3	8.4	3.3	0.9	2.8	8.6	10.0	1.9	8.9	9.2	2.2	5.8	5.2	9.3	2.6	8.2	1.3	2.3
				Woolong tea	B*	4.5	0.4	7.1	5.2	2.8	4.2	8.3	5.6	3.3	7.3	9.8	2.1	3.1	3.9	1.9	1.8	2.6	2.0	0.2
			GC-MS/MS	Green tea	A	3.7	3.1	3.0	5.3	11.3	4.3	6.1	5.5	3.3	1.4	3.1	12.5	13.0	4.9	2.7	4.2	6.4	5.8	0.9
				Woolong tea	B	5.1	8.3	8.2	6.8	6.2	3.7	1.5	4.5	4.4	9.2	3.1	1.9	8.1	4.8	4.1	2.3	6.5	3.9	8.8
				Green tea	A*	4.2	1.9	2.5	5.5	3.2	4.8	11.4	4.2	2.4	4.2	3.8	1.9	8.4	1.5	7.1	2.3	6.2	5.9	3.9
				Woolong tea	B*	3.3	4.5	3.1	11.4	4.3	1.8	5.4	7.0	9.2	1.4	5.6	1.8	1.2	1.7	6.0	3.9	0.6	3.9	4.0
		Envi-Carb+PSA	GC-MS	Green tea	A	15.5	2.8	3.9	5.9	8.4	3.5	9.0	2.9	3.2	6.8	4.5	6.2	2.2	5.5	14.0	10.3	9.0	9.2	4.7
				Woolong tea	B	13.7	7.6	3.4	4.6	1.5	4.3	2.4	1.6	5.7	5.1	39.1	12.5	7.4	3.8	15.1	3.8	7.9	8.4	5.9
				Green tea	A*	6.4	5.0	2.5	0.6	6.1	0.9	4.1	5.0	3.7	4.7	7.6	4.0	2.8	5.7	3.2	5.9	6.1	6.0	4.7
				Woolong tea	B*	5.1	2.9	6.2	0.9	12.7	9.0	3.7	3.9	12.4	5.4	10.0	7.6	4.1	6.4	9.2	0.3	8.4	3.7	4.6
			GC-MS/MS	Green tea	A	8.6	5.6	2.0	7.9	7.0	1.7	6.3	2.4	3.5	1.6	10.7	2.7	3.4	3.2	4.4	5.2	6.2	6.8	3.5
				Woolong tea	B	3.0	7.3	4.5	2.2	2.4	1.5	2.2	3.9	3.5	0.9	24.7	4.5	5.8	6.3	12.7	8.8	3.3	10.1	7.6
				Green tea	A*	3.0	5.8	7.1	4.2	4.0	4.8	4.9	5.0	0.9	3.0	5.4	5.6	1.8	3.4	8.3	5.2	6.0	2.3	6.9
				Woolong tea	B*	4.6	3.3	10.5	7.6	8.9	6.2	5.2	8.1	8.0	2.9	7.1	4.8	2.5	8.8	9.5	1.9	6.6	5.0	1.4
149	dimethametryn	Cleanert TPT	GC-MS	Green tea	A	4.3	5.8	1.4	4.3	7.5	3.8	8.4	4.9	4.5	9.1	6.1	7.9	7.7	3.7	5.1	1.3	4.4	7.0	4.4
				Woolong tea	B	4.8	7.4	5.4	9.0	3.6	2.7	2.0	3.8	1.7	9.8	5.1	14.3	1.1	8.9	7.6	2.6	4.4	0.3	4.1
				Green tea	A*	6.2	2.8	1.6	3.1	2.2	5.8	8.2	8.3	9.4	9.6	8.8	5.6	5.2	4.3	6.7	5.0	3.4	4.8	2.1
				Woolong tea	B*	4.0	1.9	5.3	7.6	3.6	4.2	4.9	2.8	7.6	5.2	23.4	1.4	3.1	3.2	3.2	1.2	2.1	0.3	2.7
			GC-MS/MS	Green tea	A	3.9	4.0	2.3	7.9	11.0	4.2	6.7	5.9	2.9	1.7	8.2	2.0	5.5	1.8	5.6	3.4	7.0	8.9	1.9
				Woolong tea	B	6.5	10.1	8.4	8.1	4.5	3.7	3.5	4.0	6.3	8.5	5.2	13.4	10.1	5.2	2.6	1.1	7.0	5.1	5.4
				Green tea	A*	7.5	2.6	2.5	2.8	3.2	4.2	9.4	1.5	3.1	3.5	2.5	2.7	5.6	3.1	10.4	9.6	1.4	8.0	5.6
				Woolong tea	B*	4.4	4.8	7.2	7.1	1.3	3.5	4.8	10.0	8.8	3.1	6.3	5.2	8.2	2.6	3.4	9.4	2.8	6.1	6.3
		Envi-Carb+PSA	GC-MS	Green tea	A	4.3	2.4	1.1	6.2	8.7	2.1	5.4	1.0	2.2	8.7	4.4	10.0	1.0	5.0	11.1	7.3	12.2	2.5	6.2
				Woolong tea	B	3.0	9.2	4.1	6.5	2.8	8.2	3.4	5.4	4.9	3.0	38.2	14.7	3.1	2.3	13.0	1.6	4.7	8.5	7.8
				Green tea	A*	4.1	6.0	2.9	2.0	10.0	0.8	3.8	6.7	2.7	4.2	6.9	5.8	5.8	3.5	5.0	3.4	4.9	2.2	3.1
				Woolong tea	B*	5.1	4.0	10.3	1.5	16.3	4.3	2.9	6.9	10.7	5.0	10.2	5.3	4.5	8.2	11.1	4.9	6.3	5.0	4.0
			GC-MS/MS	Green tea	A	7.0	1.9	3.4	9.5	8.7	2.9	7.4	2.3	4.5	2.4	5.3	2.2	1.6	2.8	7.0	8.1	9.8	3.9	4.4
				Woolong tea	B	3.6	10.3	3.2	5.5	1.3	1.4	4.2	6.7	4.2	1.6	21.8	2.2	3.6	11.5	9.0	4.1	4.6	8.9	1.6
				Green tea	A*	6.6	5.4	5.8	0.8	7.5	6.3	2.3	3.4	0.8	3.3	5.0	3.0	15.6	5.3	12.2	1.8	5.5	5.3	5.6
				Woolong tea	B*	5.3	3.8	11.0	6.3	7.3	5.5	4.5	5.4	8.6	4.3	7.1	5.7	7.5	9.3	10.2	0.7	2.3	3.9	6.3
150	dimethomorph	Cleanert TPT	GC-MS	Green tea	A	48.9	10.4	2.8	5.2	1.5	4.0	5.0	1.1	9.3	6.2	6.3	4.6	1.4	6.0	6.6	6.3	11.2	3.2	7.0
				Woolong tea	B	18.8	26.8	9.3	3.3	6.9	4.3	3.9	22.2	6.9	4.8	11.3	8.8	10.3	2.3	8.4	8.4	1.6	6.6	11.4
				Green tea	A*	6.7	2.5	2.8	3.3	6.9	7.7	3.3	5.6	2.3	22.0	11.3	15.1	7.6	10.3	7.6	7.7	2.7	1.8	2.3
				Woolong tea	B*	33.1	8.1	17.1	8.7	18.5	12.5	51.4	10.8	5.6	4.0	12.4	2.4	8.9	28.0	8.4	6.3	19.5	8.2	3.6
			GC-MS/MS	Green tea	A	12.6	26.1	4.2	6.0	3.5	4.7	3.6	0.3	7.3	4.7	6.1	1.9	2.6	12.6	2.2	5.1	10.1	7.0	5.6
				Woolong tea	B	6.4	4.3	6.4	2.1	2.9	1.9	4.2	5.9	5.8	0.2	22.3	1.9	8.9	17.0	12.3	4.5	4.8	9.2	7.1
				Green tea	A*	5.8	10.4	2.7	6.0	5.9	5.7	5.0	5.9	10.0	4.8	9.4	12.2	12.2	12.1	14.7	4.6	4.8	7.7	7.0
				Woolong tea	B*	6.6	4.9	6.2	10.1	12.0	13.2	49.5	8.1	4.0	2.8	5.4	5.0	6.8	31.2	16.8	7.2	3.8	5.5	1.6
		Envi-Carb+PSA	GC-MS	Green tea	A	6.4	6.4	6.4	10.1	2.9	4.2	4.2	5.9	4.2	1.8	9.4	11.6	14.7	12.1	12.3	4.9	3.8	6.2	7.0
				Woolong tea	B	8.5	2.8	9.3	6.2	6.9	4.3	5.0	3.9	6.9	4.8	6.3	4.6	1.4	6.0	6.6	6.3	11.2	3.2	7.0
				Green tea	A*	13.0	9.3	2.5	3.3	6.9	7.7	3.3	5.6	2.3	22.0	11.3	15.1	7.6	10.3	7.6	7.7	2.7	1.8	2.3
				Woolong tea	B*	11.4	8.1	17.1	8.2	18.5	12.5	51.4	10.8	5.6	4.0	12.4	2.4	8.9	28.0	8.4	6.3	19.5	8.2	3.6
			GC-MS/MS	Green tea	A	12.0	26.1	4.2	8.7	3.5	4.7	3.6	0.3	7.3	4.7	6.1	1.9	2.6	12.6	2.2	5.1	10.1	7.0	5.6
				Woolong tea	B	5.8	4.3	6.4	6.0	2.9	1.9	4.2	5.9	5.8	0.2	22.3	1.9	8.9	17.0	12.3	4.5	4.8	9.2	7.1
				Green tea	A*	9.1	4.8	6.4	6.0	5.9	4.2	4.2	5.9	4.2	1.8	9.4	11.6	14.7	12.1	12.3	4.9	3.8	6.2	7.0
				Woolong tea	B*	7.4	4.8	19.8	10.9	12.0	13.2	4.0	8.1	2.8	2.8	5.4	6.8	5.0	31.2	16.8	7.2	16.8	5.5	1.6

No.	Compound	Cartridge	Method	Tea	Rep																		
151	dimethyl-phthalate	Cleanert TPT	GC-MS	Green tea	A*	3.3	31.5	1.6	1.1	8.9	4.2	5.3	4.6	5.4	3.4	4.5	1.8	6.1	0.5	0.5	1.5	3.5	4.4	3.3
				Green tea	B*	5.8	32.6	5.4	6.6	2.9	3.6	2.9	12.2	2.7	5.2	2.0	4.9	2.1	0.3	4.4	1.5	5.8	2.3	5.7
				Woolong tea	A	4.4	1.4	1.0	3.3	2.0	1.3	8.9	3.3	4.7	2.1	2.3	2.1	4.0	2.4	8.0	3.6	4.1	2.3	1.4
				Woolong tea	B	4.3	3.0	3.9	6.0	3.1	4.5	5.0	1.3	6.0	5.1	0.2	0.7	3.6	2.9	4.9	0.3	1.0	1.6	3.8
			GC-MS/MS	Green tea	A	3.9	5.1	2.2	1.6	11.5	5.2	5.2	5.4	3.6	1.8	5.3	1.7	8.9	4.4	3.1	2.2	5.9	3.5	2.0
				Green tea	B	5.4	8.0	4.9	7.0	4.1	3.2	3.8	6.3	4.2	7.1	3.0	9.4	8.0	2.2	4.1	2.7	4.9	5.4	7.0
				Woolong tea	A	5.1	2.3	1.7	4.9	5.0	0.8	9.5	0.9	5.8	3.3	1.3	3.3	6.4	0.7	9.0	5.0	0.9	0.5	4.9
				Woolong tea	B	4.7	4.0	4.1	6.9	2.5	3.9	4.4	4.1	10.0	3.9	2.3	1.5	13.6	1.6	9.7	2.8	1.7	0.2	2.6
		Envi-Carb+PSA	GC-MS	Green tea	A	5.8	4.9	3.1	3.9	5.4	1.7	8.2	4.3	6.1	3.7	5.8	1.9	2.3	3.5	3.3	7.4	11.4	3.9	3.9
				Green tea	B	2.7	5.5	4.5	4.5	1.4	2.5	6.2	1.9	5.1	1.9	19.6	1.6	8.5	1.8	18.3	2.7	4.6	8.6	5.0
				Woolong tea	A	3.9	3.4	1.3	1.6	5.1	1.2	1.7	5.0	2.5	2.2	4.0	3.9	0.9	2.7	4.6	2.9	4.6	4.8	3.9
				Woolong tea	B	6.3	3.0	7.1	3.9	10.4	4.2	0.9	6.6	10.0	2.3	5.7	4.9	1.3	7.9	7.9	2.3	4.1	2.4	2.8
			GC-MS/MS	Green tea	A	8.0	8.1	1.6	7.3	3.4	2.5	4.7	3.4	7.5	1.9	5.1	2.9	2.2	3.4	3.2	6.5	8.7	5.1	4.2
				Green tea	B	5.0	6.4	5.9	2.9	1.9	1.3	5.4	3.1	3.1	1.0	21.6	1.1	10.9	10.9	21.0	3.3	2.9	4.1	4.9
				Woolong tea	A	3.2	3.2	3.6	3.5	4.2	1.8	1.5	7.4	3.5	2.7	4.8	4.3	1.3	3.5	5.4	4.2	4.9	4.1	2.6
				Woolong tea	B	7.2	4.9	8.1	7.1	7.9	4.6	2.6	7.0	7.9	1.1	5.6	3.3	8.6	9.7	11.2	1.5	2.8	2.9	4.5
152	diniconazole	Cleanert TPT	GC-MS	Green tea	A*	5.3	4.7	0.3	0.5	5.0	6.1	8.2	6.9	6.4	7.8	12.8	10.5	8.7	5.6	4.4	1.4	4.6	6.1	5.9
				Green tea	B	5.5	7.8	3.3	7.8	8.4	5.7	1.3	6.5	2.0	5.2	7.8	8.9	4.7	5.7	6.8	7.4	9.3	3.4	4.3
				Woolong tea	A	7.4	6.7	8.2	9.1	1.7	5.2	2.6	14.8	3.5	9.2	11.3	11.7	3.3	7.5	7.8	1.2	2.9	4.7	2.3
				Woolong tea	B	4.5	9.2	10.9	10.6	4.0	1.8	6.3	3.0	12.7	7.8	9.6	2.5	2.8	7.9	3.6	4.3	5.3	5.6	1.3
			GC-MS/MS	Green tea	A	4.4	6.6	0.4	7.7	9.1	3.7	8.5	4.9	3.2	2.2	4.8	4.0	6.6	3.0	4.3	0.5	6.4	4.4	6.1
				Green tea	B	4.7	9.3	6.4	7.6	3.5	3.0	1.9	6.0	4.0	9.2	3.3	9.5	6.4	0.4	4.8	5.0	4.7	2.7	3.3
				Woolong tea	A	6.5	2.1	0.7	3.7	1.6	5.3	8.4	1.8	6.1	4.8	0.9	4.2	6.6	0.2	5.9	3.3	1.4	5.4	4.2
				Woolong tea	B	5.3	2.1	3.8	10.1	2.9	3.5	4.1	7.8	6.1	2.9	5.2	3.1	0.8	3.5	7.6	2.3	2.2	1.7	12.1
		Envi-Carb+PSA	GC-MS	Green tea	A	10.6	5.8	4.8	6.8	9.2	0.9	8.7	1.6	1.9	5.3	8.2	7.8	3.4	9.2	13.7	10.9	11.0	9.9	10.9
				Green tea	B	4.0	7.6	5.5	6.9	2.2	9.9	3.4	4.7	10.2	5.9	41.6	12.0	6.7	2.8	20.8	9.4	6.1	10.1	8.6
				Woolong tea	A	4.3	2.5	2.2	6.8	7.5	1.1	3.3	12.2	5.9	2.8	6.5	3.8	1.1	3.3	1.0	0.9	5.8	1.9	4.2
				Woolong tea	B	3.5	3.9	16.7	6.9	14.3	4.3	2.7	6.1	14.1	6.9	10.9	7.2	1.0	4.0	8.2	8.5	5.4	2.4	2.9
			GC-MS/MS	Green tea	A	7.8	4.6	6.2	8.2	7.3	1.8	7.9	3.5	5.0	2.2	3.7	7.0	4.7	4.7	6.1	6.3	10.2	5.2	7.1
				Green tea	B	4.4	8.3	4.8	5.2	3.1	1.8	2.6	4.4	5.3	2.2	23.5	2.9	6.0	7.8	14.8	1.5	2.8	11.4	3.3
				Woolong tea	A	3.9	3.6	5.5	3.3	3.9	1.7	2.5	5.1	0.5	4.3	7.7	3.0	4.3	2.9	12.1	3.6	5.6	5.3	4.2
				Woolong tea	B	5.4	3.7	13.1	4.3	12.8	5.3	2.3	7.6	8.8	2.2	6.7	4.8	5.1	9.3	9.8	1.6	1.7	4.2	4.0
153	diphenamid	Cleanert TPT	GC-MS	Green tea	A*	5.9	6.3	2.3	1.1	14.6	4.5	13.3	10.0	12.5	4.4	6.2	0.7	7.9	2.5	1.9	4.7	5.1	6.2	5.6
				Green tea	B*	3.5	5.9	7.7	8.9	11.1	9.9	5.9	3.6	6.3	6.4	6.1	12.1	3.8	4.9	5.4	3.6	5.3	1.3	1.1
				Woolong tea	A	4.2	1.8	1.6	7.9	2.3	11.2	44.5	10.9	3.8	7.3	7.3	0.3	4.7	1.8	8.9	3.8	4.2	3.3	2.0
				Woolong tea	B	4.5	3.1	6.4	13.1	3.7	5.1	6.8	4.8	5.5	1.6	5.8	1.5	3.7	2.7	3.6	1.0	5.4	0.8	2.9
			GC-MS/MS	Green tea	A	5.7	3.0	1.9	3.1	10.0	4.7	7.8	3.2	5.8	3.2	3.3	1.3	13.7	2.3	2.3	5.4	10.4	2.7	5.3
				Green tea	B	5.1	10.5	6.7	7.4	2.8	3.8	4.3	7.8	7.0	7.5	5.1	5.4	5.7	1.7	7.0	2.6	6.4	5.9	7.7
				Woolong tea	A	5.9	6.7	1.6	4.1	5.3	2.8	7.9	3.2	3.4	3.9	3.0	1.2	7.0	1.6	9.2	4.4	1.6	0.8	4.7
				Woolong tea	B	5.2	4.9	3.8	6.8	1.3	2.5	5.5	8.2	8.8	3.1	6.3	4.3	1.6	2.8	4.8	0.3	4.1	2.7	6.6
		Envi-Carb+PSA	GC-MS	Green tea	A*	6.8	14.1	2.2	3.4	8.5	6.1	14.9	6.9	22.0	3.5	6.6	1.6	1.4	5.1	11.4	7.4	11.2	3.1	7.7
				Green tea	B*	5.4	12.9	2.9	2.3	5.5	16.5	3.6	2.8	17.5	0.4	25.5	7.5	1.4	3.5	13.1	1.8	4.3	8.7	4.4
				Woolong tea	A	4.0	3.3	1.3	2.4	7.8	3.0	6.6	19.4	9.1	4.0	4.8	1.4	1.3	8.0	5.1	0.7	4.8	2.4	6.4
				Woolong tea	B	6.2	5.0	12.3	5.8	9.1	4.9	18.0	18.9	18.6	6.7	9.8	4.7	3.3	7.6	11.5	5.1	3.0	7.8	7.3
			GC-MS/MS	Green tea	A	5.6	5.9	1.1	6.9	8.5	1.9	13.6	1.3	1.9	2.8	5.4	10.0	3.7	2.1	8.1	6.6	6.4	1.6	5.9
				Green tea	B	4.6	5.3	4.6	4.2	1.8	1.1	3.1	6.1	3.9	2.9	20.7	9.7	2.0	12.6	12.9	1.5	2.7	11.5	7.7
				Woolong tea	A	4.5	4.5	4.0	1.3	7.2	0.8	3.3	0.7	4.1	6.6	7.0	9.8	5.4	1.6	10.9	5.4	7.6	4.0	4.7
				Woolong tea	B	5.9	1.9	11.6	8.7	7.7	6.1	2.9	6.4	7.9	4.4	8.1	3.8	11.7	8.3	10.3	4.7	0.6	5.3	4.7

(Continued)

Annex 2: RSD data for 201 pesticides in incurred tea Youden pair samples under 8 conditions with two SPE cartridge cleanup in three months (Nov.9, 2009-Feb.7, 2010) (*cont.*)

No	Pesticides	SPE	Method	Sample	Youden pair	Nov.9, 2009 (n=5)	Nov.14, 2009 (n=3)	Nov.19, 2009 (n=3)	Nov.24, 2009 (n=3)	Nov.29, 2009 (n=3)	Dec.4, 2009 (n=3)	Dec.9, 2009 (n=3)	Dec.14, 2009 (n=3)	Dec.19, 2009 (n=3)	Dec.24, 2009 (n=3)	Dec.14, 2009 (n=3)	Jan.3, 2010 (n=3)	Jan.8, 2010 (n=3)	Jan.13, 2010 (n=3)	Jan.18, 2010 (n=3)	Jan.23, 2010 (n=3)	Jan.28, 2010 (n=3)	Feb.2, 2010 (n=3)	Feb.7, 2010 (n=3)
154	dipropetryn	Cleanert TPT	GC-MS	Green tea	A	4.6	6.3	2.7	4.5	7.2	7.3	7.1	5.1	3.0	4.2	13.7	7.5	7.9	4.3	1.6	4.8	4.5	8.2	4.4
					B	4.8	9.8	6.6	9.0	4.3	3.9	2.4	2.1	3.0	11.1	6.9	11.9	2.0	6.1	7.4	3.2	5.2	0.3	4.6
				Woolong tea	A	5.6	4.9	4.2	3.4	2.9	4.3	9.9	7.6	4.9	19.0	3.3	6.6	3.9	3.0	14.7	5.4	4.0	4.8	1.8
					B	4.5	2.8	8.2	8.3	4.5	5.1	3.0	1.8	7.7	5.6	8.7	1.6	2.6	4.1	6.5	0.2	3.1	1.5	1.2
			GC-MS/MS	Green tea	A	4.4	6.8	1.1	5.3	9.4	3.2	6.9	4.6	3.1	2.5	5.4	2.9	11.4	2.9	5.1	2.6	5.1	3.3	5.0
					B	4.7	10.1	7.2	9.6	3.0	3.9	4.9	3.8	6.5	10.7	4.7	9.0	2.8	3.8	5.6	3.8	4.1	4.7	4.0
				Woolong tea	A	4.7	3.4	3.5	2.2	3.4	4.3	7.9	2.9	4.4	4.0	0.9	4.7	9.5	3.3	9.7	4.3	4.1	3.4	3.6
					B	6.4	5.5	6.1	7.1	3.5	3.8	6.0	5.6	4.8	2.8	11.6	4.1	2.1	5.1	7.8	4.8	5.7	4.5	6.4
		Envi-Carb+PSA	GC-MS	Green tea	A*	6.0	4.5	1.9	6.5	8.0	1.7	5.3	3.5	5.8	4.2	8.3	5.2	1.1	4.6	9.2	9.4	10.3	3.3	6.9
					B	5.9	7.8	2.9	3.7	4.1	8.1	1.9	4.0	3.7	5.4	28.7	12.2	1.7	2.3	11.0	6.3	5.3	8.5	7.2
				Woolong tea	A	9.0	6.0	4.6	5.2	10.9	2.9	3.7	4.2	2.2	7.2	7.1	21.0	20.0	4.2	9.4	2.9	4.5	19.0	21.4
					B*	12.7	3.6	8.4	11.8	11.1	1.7	2.0	7.4	4.5	17.3	7.8	2.9	7.4	7.5	7.7	8.3	5.3	20.2	8.1
			GC-MS/MS	Green tea	A*	7.1	3.7	3.0	6.4	7.5	2.4	6.6	4.6	4.1	2.2	10.7	14.8	3.7	4.0	9.0	7.0	10.7	3.1	5.4
					B*	3.6	10.8	5.1	5.5	1.8	1.9	3.1	8.8	4.6	2.0	21.4	16.3	5.0	11.9	14.8	0.9	1.7	5.7	7.0
				Woolong tea	A	3.9	2.4	10.0	10.5	3.1	1.0	3.8	0.5	0.7	0.3	9.2	9.5	11.0	12.7	13.0	3.5	7.6	6.2	3.6
					B*	5.4	1.8	13.8	17.6	7.4	4.9	4.9	7.4	8.3	16.8	4.3	5.0	5.8	16.3	15.4	2.8	4.9	2.6	6.4
155	ethalfluralin	Cleanert TPT	GC-MS	Green tea	A	3.4	5.4	3.8	5.3	13.6	2.6	8.4	14.2	15.3	2.5	6.0	6.4	11.1	8.6	4.3	4.1	6.0	10.7	1.9
					B	5.7	7.5	2.8	9.5	5.5	4.9	2.9	7.2	1.9	9.2	1.4	9.4	7.7	8.8	0.4	4.5	11.7	7.3	8.0
				Woolong tea	A	4.1	1.7	3.6	4.1	13.4	6.1	6.8	5.6	4.9	4.9	7.3	4.9	3.2	1.4	5.0	9.2	3.5	2.8	0.4
					B	2.8	3.6	3.2	4.0	9.0	3.7	4.8	4.4	6.9	1.0	5.9	4.5	2.3	5.4	10.4	1.1	2.4	1.8	3.6
			GC-MS/MS	Green tea	A	3.5	4.7	2.4	37.8	15.1	4.2	9.2	12.5	13.5	2.0	5.4	1.3	13.6	7.8	1.4	4.7	10.0	6.8	3.6
					B	5.5	7.8	5.0	10.0	2.4	2.1	3.0	11.1	6.9	11.0	3.0	4.7	3.7	4.9	2.6	4.5	11.4	0.4	10.0
				Woolong tea	A	7.0	1.0	4.9	3.3	8.3	3.3	6.8	4.5	9.7	7.6	7.6	5.0	9.0	5.0	9.2	3.0	3.5	6.3	1.4
					B	8.9	4.1	3.5	8.4	5.5	5.4	7.9	7.5	9.6	0.6	7.7	3.1	7.8	3.6	6.4	1.9	5.9	2.9	2.3
		Envi-Carb+PSA	GC-MS	Green tea	A*	14.6	6.4	12.1	5.9	8.3	23.4	7.8	10.6	5.1	0.9	4.9	14.6	2.1	9.3	5.7	10.7	13.0	11.7	7.4
					B*	13.0	9.7	4.0	3.2	8.5	32.4	11.8	14.8	8.1	5.0	15.0	8.1	21.4	4.2	16.4	4.3	6.9	11.9	8.3
				Woolong tea	A	6.9	4.5	4.6	6.2	5.1	4.1	3.1	19.0	1.4	4.6	3.8	3.6	7.7	2.0	1.4	20.1	8.9	7.0	0.7
					B*	6.8	4.3	7.7	3.4	19.3	10.5	7.0	8.1	11.2	2.1	9.4	0.9	6.3	30.4	13.4	5.2	12.6	9.0	5.5
			GC-MS/MS	Green tea	A*	13.3	7.5	7.7	6.1	5.2	16.8	10.5	15.4	11.0	2.5	0.9	6.2	2.4	6.3	5.8	2.4	12.9	7.6	3.2
					B*	16.6	11.3	5.2	5.3	5.5	15.3	14.7	13.5	10.0	4.7	14.5	4.0	17.8	13.5	9.8	10.1	6.5	19.6	6.5
				Woolong tea	A	7.4	2.9	5.0	6.0	2.6	5.3	1.3	20.7	2.2	1.6	7.6	7.2	4.3	2.1	7.7	5.2	8.8	20.1	1.4
					B*	10.7	3.4	9.3	3.1	10.9	10.7	5.6	5.6	9.1	2.4	9.6	5.0	9.0	8.7	9.2	1.2	9.0	31.7	2.3
156	etofenprox	Cleanert TPT	GC-MS	Green tea	A*	8.1	5.6	2.9	6.4	9.2	8.9	9.3	6.0	5.9	0.7	10.5	3.9	7.6	4.2	4.2	1.9	2.3	7.6	1.2
					B*	5.7	9.1	6.6	6.9	3.6	6.2	0.4	11.0	2.2	11.6	4.7	9.0	6.4	1.5	7.6	4.1	7.7	7.3	6.8
				Woolong tea	A	7.0	3.6	2.7	3.8	2.8	5.5	3.9	5.1	2.0	3.4	1.5	2.7	4.0	2.4	3.7	4.6	3.5	6.4	1.9
					B	2.6	2.5	3.4	11.1	0.9	5.9	3.8	6.5	6.6	3.8	3.4	2.1	4.7	6.1	13.5	3.6	0.3	2.9	7.1
			GC-MS/MS	Green tea	A	5.4	5.9	4.2	6.6	10.1	8.5	7.3	6.2	5.6	1.4	12.6	1.9	10.9	8.4	3.4	5.0	6.4	9.1	3.4
					B	4.1	8.2	7.8	5.7	5.4	7.1	1.5	8.5	2.8	11.4	6.0	10.0	1.2	6.5	5.6	1.4	6.4	5.4	7.3
				Woolong tea	A	6.2	3.4	1.0	2.5	3.7	4.4	5.4	1.4	0.8	3.8	2.6	4.9	7.4	2.5	4.5	4.3	3.7	6.1	6.8
					B	4.1	4.0	5.3	6.4	1.4	3.2	5.9	5.8	11.8	3.7	3.2	2.9	10.7	1.9	4.0	4.2	3.1	1.0	1.5
		Envi-Carb+PSA	GC-MS	Green tea	A	10.3	7.5	1.9	6.7	5.7	3.9	4.3	2.7	8.2	1.5	8.5	2.3	1.4	8.1	5.7	7.6	4.8	4.4	4.8
					B	10.6	11.1	3.1	2.8	4.6	7.8	3.3	6.3	9.3	2.9	21.8	7.4	2.9	8.5	14.7	3.3	8.4	4.8	11.5
				Woolong tea	A	5.1	5.4	6.1	6.1	5.6	5.2	3.0	4.6	0.6	4.4	4.6	1.9	6.7	5.1	18.3	7.1	5.9	3.0	9.9
					B	5.0	6.6	11.5	2.4	8.0	6.8	3.0	7.2	10.9	3.7	4.7	2.7	1.8	3.8	14.2	3.8	8.0	2.6	2.6
			GC-MS/MS	Green tea	A	8.7	7.3	4.6	7.9	7.0	3.5	4.0	7.9	11.2	3.0	7.8	1.6	2.8	26.1	29.8	11.7	12.9	7.2	6.0
					B	3.5	1.3	6.5	5.4	5.9	13.1	1.5	6.8	1.5	5.3	26.0	1.0	28.6	12.7	32.4	2.0	8.3	7.9	4.7
				Woolong tea	A	7.8	7.3	5.4	16.1	19.7	1.7	3.1	7.2	0.4	3.0	7.5	2.9	12.0	6.7	6.5	24.1	5.7	20.1	21.5
					B*	16.8	9.7	9.1	13.9	4.9	4.8	8.5	8.5	8.7	1.9	4.3	2.8	8.5	5.5	7.5	6.0	5.2	31.0	1.5

No.	Compound	Cartridge	Method	Tea	Rep																			
157	etridiazol	Cleanert TPT	GC-MS	Green tea	A	3.4	6.4	3.3	9.2	13.2	8.7	8.5	9.2	9.9	5.4	5.9	6.4	10.9	4.5	5.4	4.8	9.2	9.5	2.0
				Green tea	B	8.1	6.4	4.1	7.4	6.3	7.3	3.3	6.9	3.9	13.7	3.8	6.1	6.3	9.1	3.4	1.7	2.2	2.7	7.9
				Woolong tea	A	7.1	4.4	3.5	2.4	11.6	5.3	2.4	0.4	12.0	2.0	8.7	7.9	3.3	6.1	8.1	7.4	4.3	5.6	4.7
				Woolong tea	B	8.5	6.5	8.0	2.4	8.6	9.9	2.8	8.5	9.0	6.3	8.7	8.3	4.5	8.1	13.8	3.7	1.6	3.4	10.2
			GC-MS/MS	Green tea	A	6.3	2.8	3.7	13.4	15.7	5.8	12.5	7.4	11.9	8.6	5.1	10.8	10.1	5.7	0.7	8.9	12.1	9.0	5.7
				Green tea	B	7.8	6.5	2.6	9.3	4.0	4.0	2.6	9.1	7.4	12.0	3.5	10.3	16.5	3.6	4.1	4.5	10.4	6.1	7.1
				Woolong tea	A*	7.7	4.9	1.5	3.7	3.6	1.3	6.0	10.0	9.8	5.4	9.1	3.7	4.6	2.2	9.2	7.9	1.1	5.8	2.8
				Woolong tea	B*	13.1	5.4	9.7	0.8	2.6	10.9	9.8	10.6	16.7	4.1	9.3	12.5	11.2	6.6	13.2	3.4	1.5	4.2	5.5
		Envi-Carb+PSA	GC-MS	Green tea	A	11.3	3.7	9.4	7.2	5.0	9.5	5.0	3.0	13.4	1.6	7.8	9.8	6.4	10.1	1.5	12.9	10.3	15.4	2.9
				Green tea	B	14.2	6.3	3.6	9.6	7.2	8.3	4.5	5.1	4.1	4.1	26.1	6.9	28.5	1.6	34.8	3.8	9.1	8.2	5.4
				Woolong tea	A*	12.7	10.1	3.6	5.4	3.8	3.4	2.2	16.6	3.3	3.3	3.5	5.0	8.6	5.9	4.1	6.1	3.9	14.3	9.2
				Woolong tea	B	15.3	10.6	6.0	7.3	13.9	6.6	7.7	1.6	17.3	11.2	10.0	2.5	7.4	12.0	10.6	0.2	7.1	18.1	4.3
			GC-MS/MS	Green tea	A	9.9	6.4	5.1	5.3	7.7	9.6	13.7	4.1	15.6	3.8	5.9	10.8	2.4	7.3	1.8	20.5	15.3	3.4	3.4
				Green tea	B	21.5	4.0	5.2	9.2	5.1	9.5	10.9	19.8	7.5	1.2	25.3	8.9	37.6	13.9	39.9	1.8	5.2	13.8	12.0
				Woolong tea	A	13.1	8.7	8.8	12.9	2.3	6.5	0.1	6.1	3.6	4.6	1.7	9.3	33.7	6.4	11.0	13.5	10.9	16.2	2.8
				Woolong tea	B	16.5	9.9	6.7	6.1	14.9	10.0	12.8	4.2	14.7	11.7	9.6	5.5	15.7	9.7	13.1	7.3	8.6	25.7	5.5
158	fenazaquin	Cleanert TPT	GC-MS	Green tea	A	3.7	4.7	1.7	5.1	6.7	6.1	9.1	2.1	5.8	0.6	6.8	9.4	8.6	5.8	4.7	1.9	10.0	11.3	6.5
				Green tea	B	6.0	8.4	5.1	10.7	3.6	3.8	3.7	5.5	5.6	9.3	8.2	4.2	1.5	2.7	7.6	6.3	5.7	7.7	6.0
				Woolong tea	A	6.9	3.0	2.2	2.2	3.7	3.2	10.2	0.8	3.3	7.6	8.0	1.2	5.2	7.4	6.2	6.8	1.2	2.5	0.6
				Woolong tea	B	4.1	3.8	1.4	9.8	3.9	5.0	5.3	4.3	4.5	7.2	9.7	4.2	2.4	4.0	2.6	6.7	2.2	3.1	1.0
			GC-MS/MS	Green tea	A	7.6	6.8	0.9	7.5	9.9	5.5	7.0	2.6	4.8	2.0	3.8	14.3	5.2	6.3	3.9	1.5	5.4	11.9	3.4
				Green tea	B	5.3	10.7	4.3	8.9	3.7	2.7	2.6	1.6	6.2	10.8	4.4	3.9	9.0	6.0	3.6	4.6	4.2	5.3	6.0
				Woolong tea	A	6.2	3.4	0.8	1.4	3.7	3.4	9.0	7.3	2.7	3.8	2.3	3.3	5.4	1.8	5.9	5.5	2.9	4.9	13.0
				Woolong tea	B	6.0	4.4	3.7	8.0	0.4	4.0	4.0	3.9	8.1	2.9	3.6	9.8	6.4	4.1	5.2	2.9	2.1	1.8	3.4
		Envi-Carb+PSA	GC-MS	Green tea	A*	3.8	8.6	2.4	3.5	8.3	10.9	7.2	6.5	6.3	5.2	12.3	12.4	5.5	6.9	3.5	10.9	6.1	2.7	2.3
				Green tea	B	4.6	12.6	2.9	5.4	5.4	2.1	2.7	8.4	8.6	1.6	30.1	3.8	6.6	6.9	6.6	3.9	15.7	7.6	7.6
				Woolong tea	A	5.7	1.9	3.4	2.7	2.5	2.9	2.3	7.4	4.7	4.8	8.3	1.4	7.3	8.1	7.3	10.9	4.9	5.2	4.3
				Woolong tea	B	5.7	5.3	7.9	5.7	8.3	2.4	4.8	3.6	11.8	4.4	14.9	2.1	11.6	7.2	11.6	12.2	5.0	2.3	2.8
			GC-MS/MS	Green tea	A	6.5	1.0	3.1	13.2	8.9	2.7	3.2	7.8	6.3	0.2	3.3	5.4	3.7	12.1	3.7	4.7	9.4	4.5	5.6
				Green tea	B	4.6	5.7	4.5	4.9	0.4	1.4	1.2	7.8	3.3	3.4	17.3	1.4	8.2	16.0	11.4	3.3	3.6	8.5	7.4
				Woolong tea	A	6.5	2.6	4.8	0.5	5.4	4.6	2.3	6.5	0.8	4.5	8.7	1.6	2.4	6.0	8.0	3.2	5.2	12.2	6.2
				Woolong tea	B	5.8	4.4	10.9	6.9	10.6	4.7	3.0	7.3	8.9	1.1	7.2	3.2	23.2	7.5	10.3	2.1	3.2	2.8	3.4
159	fenchlorphos	Cleanert TPT	GC-MS	Green tea	A*	4.8	5.4	4.9	2.0	7.3	3.7	2.8	4.5	4.4	3.1	6.5	8.6	6.1	4.3	3.8	3.2	4.9	4.2	3.3
				Green tea	B	6.7	6.5	5.2	6.4	4.7	4.4	2.7	2.9	3.7	9.7	0.8	2.8	7.4	1.3	4.1	1.0	3.0	2.7	9.2
				Woolong tea	A	5.7	2.9	4.1	3.1	2.6	2.4	9.2	3.4	4.7	3.1	4.2	4.0	3.2	5.6	8.1	3.2	4.1	1.6	0.5
				Woolong tea	B	3.8	2.0	4.0	4.2	2.6	3.3	1.4	7.2	7.2	3.2	6.1	7.3	1.4	4.1	5.8	1.0	2.0	1.5	3.6
			GC-MS/MS	Green tea	A	5.1	3.4	3.1	12.0	10.3	1.0	4.9	3.1	5.7	8.4	7.9	8.5	1.3	5.3	3.8	1.9	6.3	3.3	3.8
				Green tea	B	5.6	7.6	3.5	8.4	4.8	4.4	4.4	4.2	10.7	12.1	2.1	3.5	10.7	5.1	4.7	3.0	5.2	2.4	5.7
				Woolong tea	A	4.0	3.8	2.3	4.9	6.7	1.7	10.8	7.0	6.4	2.9	2.6	2.2	8.5	2.2	10.3	1.4	5.7	1.5	6.8
				Woolong tea	B	6.6	5.4	5.3	6.7	1.7	2.4	3.8	1.3	9.5	5.9	5.8	3.5	4.4	2.2	9.8	4.8	1.0	1.8	2.6
		Envi-Carb+PSA	GC-MS	Green tea	A*	9.1	3.8	1.6	6.6	8.7	9.6	5.7	0.9	5.1	2.6	6.5	3.5	13.6	2.3	5.0	4.0	9.4	11.4	6.9
				Green tea	B*	13.2	9.5	1.4	0.5	1.1	0.2	3.9	3.0	6.3	6.7	17.6	2.9	4.5	2.2	4.5	3.4	4.4	9.1	5.3
				Woolong tea	A	4.9	3.3	4.7	2.6	5.3	0.3	2.8	4.2	4.0	0.7	8.2	5.1	5.0	3.4	3.1	2.4	3.8	6.5	2.5
				Woolong tea	B	5.6	3.1	7.2	3.6	9.5	3.7	1.8	2.9	10.2	3.6	8.1	0.8	3.1	8.3	5.1	2.6	3.0	5.4	4.5
			GC-MS/MS	Green tea	A	11.6	6.9	0.9	6.3	7.4	10.8	7.5	3.1	5.4	2.4	11.0	10.3	5.1	3.7	5.2	1.3	4.1	5.4	3.5
				Green tea	B	16.3	8.7	1.9	4.4	2.3	3.4	6.1	7.2	7.3	9.8	11.6	11.1	5.2	13.2	7.2	9.4	5.2	15.3	6.1
				Woolong tea	A	3.5	2.8	7.4	8.6	4.6	2.8	3.1	5.2	5.5	4.1	0.8	10.3	3.3	1.1	13.0	1.5	7.5	5.8	6.8
				Woolong tea	B	7.8	3.5	10.7	6.6	7.9	4.2	3.1	5.2	10.5	3.6	6.6	9.7	4.4	6.7	10.1	3.1	5.0	7.7	2.6

(Continued)

Annex 2: RSD data for 201 pesticides in incurred tea Youden pair samples under 8 conditions with two SPE cartridge cleanup in three months (Nov.9, 2009-Feb.7, 2010) (cont.)

No	Pesticides	SPE	Method	Sample	Youden pair	Nov.9, 2009 (n=5)	Nov.14, 2009 (n=3)	Nov.19, 2009 (n=3)	Nov.24, 2009 (n=3)	Nov.29, 2009 (n=3)	Dec.4, 2009 (n=3)	Dec.9, 2009 (n=3)	Dec.14, 2009 (n=3)	Dec.19, 2009 (n=3)	Dec.24, 2009 (n=3)	Dec.29, 2009 (n=3)	Jan.3, 2010 (n=3)	Jan.8, 2010 (n=3)	Jan.13, 2010 (n=3)	Jan.18, 2010 (n=3)	Jan.23, 2010 (n=3)	Jan.28, 2010 (n=3)	Feb.2, 2010 (n=3)	Feb.7, 2010 (n=3)
160	fenoxanil	Cleanert TPT	GC-MS	Green tea	A	10.5	6.1	1.4	3.7	3.2	4.9	7.4	4.5	4.8	4.4	7.9	0.6	9.8	7.9	4.6	4.0	5.1	6.3	3.9
					B	7.1	8.0	9.2	10.4	3.6	4.6	3.3	6.6	3.3	12.8	7.1	10.3	9.0	2.0	9.9	2.7	5.2	1.7	2.9
				Woolong tea	A	7.0	7.6	5.3	3.3	2.1	3.8	7.4	11.4	5.2	8.2	3.3	1.1	5.0	1.4	4.5	5.4	2.7	17.6	2.2
					B	16.7	4.3	4.7	4.3	2.1	5.8	3.6	8.7	1.9	5.2	0.1	4.6	0.8	8.8	0.4	2.3	3.1	11.2	0.8
			GC-MS/MS	Green tea	A	9.2	4.0	3.0	10.3	6.2	1.9	7.5	5.3	2.7	16.3	6.1	2.9	24.7	9.0	6.9	8.4	6.0	7.4	3.6
					B	12.1	16.0	9.4	9.0	7.1	7.4	3.2	15.0	3.6	7.0	6.4	9.3	5.5	11.5	9.5	0.8	7.0	6.0	4.3
				Woolong tea	A	7.5	12.2	4.7	2.7	6.8	6.5	7.9	7.3	4.1	6.1	3.3	3.4	6.1	3.2	9.0	7.6	4.0	10.0	7.3
					B	8.5	7.0	2.8	11.7	1.2	5.0	9.2	7.7	10.2	5.3	6.2	3.0	9.0	2.6	2.0	7.9	2.9	4.1	4.6
		Envi-Carb+PSA	GC-MS	Green tea	A	6.5	9.9	0.8	8.7	7.1	8.9	13.1	2.6	7.9	9.5	7.2	4.5	6.9	4.1	8.3	11.4	11.0	0.7	5.9
					B	5.2	4.5	1.9	5.8	9.0	2.5	1.2	2.5	5.8	10.3	28.1	14.2	11.6	4.1	15.7	2.1	2.7	9.8	8.4
				Woolong tea	A	12.5	6.2	7.0	4.5	7.0	2.8	3.2	3.3	2.7	4.3	2.6	4.4	5.9	4.7	4.3	4.8	5.4	2.3	7.0
					B	10.5	5.9	14.1	7.3	11.0	4.1	5.2	9.6	4.8	11.8	5.2	2.8	4.1	6.9	9.2	4.4	4.2	10.3	2.4
			GC-MS/MS	Green tea	A	7.5	17.2	18.5	12.1	16.8	5.8	12.5	8.4	2.6	4.1	5.4	22.9	3.7	5.5	4.7	11.6	5.6	3.1	5.4
					B	12.5	18.8	7.7	13.5	12.9	13.0	1.9	19.7	9.9	1.4	24.6	12.5	7.5	8.1	14.1	4.8	4.4	10.0	4.7
				Woolong tea	A	6.4	10.8	2.9	6.8	7.8	3.7	5.5	5.3	1.1	4.3	9.3	5.9	4.4	4.0	14.9	8.0	2.5	2.9	7.3
					B	5.9	4.9	14.7	10.8	7.2	3.4	10.4	7.9	7.6	15.9	4.0	1.1	3.8	12.1	11.9	6.0	5.4	6.9	4.7
161	fenpropidin	Cleanert TPT	GC-MS	Green tea	A	5.2	5.0	0.7	4.8	12.9	5.2	8.9	7.6	6.2	4.4	10.3	0.4	11.0	4.1	1.1	5.3	4.6	6.9	4.8
					B	6.4	9.5	5.6	9.0	4.0	1.4	5.2	13.8	12.4	14.3	9.3	11.7	0.3	4.0	9.8	5.4	8.1	1.4	4.1
				Woolong tea	A	6.0	2.1	6.4	4.8	2.0	5.8	16.2	12.8	5.9	6.6	23.8	8.3	9.4	4.8	11.9	7.0	6.2	13.0	3.8
					B	6.3	4.0	10.4	3.9	4.1	3.4	6.3	12.5	2.7	9.7	0.3	1.3	2.7	0.9	1.8	0.5	0.3	3.6	2.1
			GC-MS/MS	Green tea	A*	9.2	5.7	1.8	5.4	14.1	4.0	7.8	6.1	4.4	5.3	7.0	0.7	17.2	16.1	3.6	4.2	9.2	5.5	1.7
					B*	6.4	10.2	8.4	8.2	3.5	4.2	2.6	11.2	4.7	15.6	8.8	14.4	2.9	6.6	10.6	3.5	4.4	3.1	5.8
				Woolong tea	A	8.8	2.7	2.9	6.1	8.9	4.6	18.7	9.3	5.3	9.3	11.8	4.4	14.8	2.5	12.3	7.1	6.5	12.4	7.5
					B	13.3	4.6	6.8	10.3	3.4	4.4	6.0	4.7	6.7	3.1	4.4	3.4	8.7	1.7	4.2	1.8	1.4	2.5	3.7
		Envi-Carb+PSA	GC-MS	Green tea	A	11.3	22.8	8.4	9.9	11.3	5.3	40.0	9.7	25.9	7.3	7.6	6.3	2.1	9.5	8.1	8.5	12.0	3.2	7.6
					B	4.7	6.6	4.4	12.4	3.4	2.3	8.9	7.2	11.7	1.5	29.7	10.4	3.8	4.1	13.3	2.5	3.9	8.1	8.9
				Woolong tea	A	6.5	4.8	5.8	2.2	8.3	4.9	1.4	3.8		13.7	8.4	17.0	5.8	7.5	7.3	6.5	5.4	1.2	6.2
					B	7.5	1.4	16.9	6.1	15.2	8.5	2.9	11.7		10.0	4.1	2.4	1.3	10.4	7.5	4.1	3.2	8.1	3.7
			GC-MS/MS	Green tea	A		7.3	4.9	9.6	5.6	4.4	32.7	4.7	3.0	2.6	0.9	8.0	2.4	12.2	8.7	8.3	9.8	3.7	6.5
					B		6.3	5.0	5.2	4.4	1.5	2.3	3.2	3.2	3.2	3.9	16.3	7.8	20.7	15.7	4.3	2.0	9.6	8.9
				Woolong tea	A		3.9	4.4	6.0	8.2	0.7	7.2	8.1	2.7	0.3	1.7	13.5	5.1	10.6	15.8	10.5	4.0	7.2	7.5
					B		2.7	11.6	8.7	8.9	8.4	4.9	0.7	6.7	14.2	7.5	3.9	3.1	11.4	7.8	2.4	2.2	3.8	3.7
162	femson	Cleanert TPT	GC-MS	Green tea	A		5.6	2.4	7.2	8.9	2.1	9.2	10.9	10.7	4.2	2.6	8.0	10.2	11.7	10.4	7.2	8.3	11.8	3.1
					B		9.3	3.4	10.6	13.9	5.9	3.8	12.6	5.1	2.4	16.7	4.2	6.6	12.4	8.3	7.5	12.2	9.2	4.1
				Woolong tea	A		2.2	1.0	5.2	5.0	5.4	10.3	9.5	7.4	6.6	6.9	7.6	3.4	4.1	12.5	4.5	7.6	1.4	0.3
					B		5.0	4.3	6.0	4.8	13.1	5.8	14.0	6.7	3.6	3.6	5.2	4.0	6.7	8.5	3.1	2.2	2.8	12.3
			GC-MS/MS	Green tea	A		2.0	2.3	7.2	4.1	3.0	10.5	0.9	7.8	5.7	16.2	2.1	11.1	3.8	2.4	5.9	6.0	12.3	7.4
					B		11.0	4.6	4.6	15.0	3.2	7.0	13.8	5.0	2.4	4.6	7.6	12.6	5.8	1.6	5.9	13.7	6.0	8.2
				Woolong tea	A		3.4	1.4	5.2	3.9	2.9	1.5	12.9	14.5	8.0	13.1	5.3	5.5	5.1	7.1	3.8	5.4	9.4	1.3
					B		6.5	6.1	6.8	7.7	8.6	3.9	8.6	12.5	4.0	9.4	2.0	3.8	5.1	11.4	6.5	1.7	4.7	7.2
		Envi-Carb+PSA	GC-MS	Green tea	A*	18.2	6.9	13.2	4.8	4.9	15.2	27.6	17.0	7.7	1.9	19.3	16.1	3.3	13.2	8.9	16.9	14.6	22.9	5.0
					B*	21.1	12.5	5.7	6.3	7.2	1.2	3.1	21.9	12.1	3.9	6.9	10.6	21.8	7.8	15.7	7.3	6.1	18.4	15.3
				Woolong tea	A	3.2	5.2	4.4	6.4	0.4	8.1	11.4	5.5	8.9	9.7	8.0	13.6	9.5	1.0	5.2	10.4	10.7	16.7	4.7
					B	9.4	7.8	7.1	2.1	12.0	9.1	14.2	2.6	13.9	2.9		5.0	5.4	8.2	16.3	3.4	11.5	19.4	10.2
			GC-MS/MS	Green tea	A	24.5		11.8	7.8	7.7	19.4	17.9	19.6	10.4	3.5		1.2	6.0	10.2	11.0	4.4	11.9	5.2	4.5
					B	25.0		8.8	8.0	10.4	1.8	11.7	18.2	15.9	0.2		0.5	27.0	12.6	11.0	16.2	1.8	21.1	9.7
				Woolong tea	A	11.4		7.2	7.2	0.4	0.4	4.1	8.6	2.4	9.0		1.7	2.3	0.9	10.6	15.2	11.9	14.4	1.3
					B	13.6		7.2	1.9	13.6	9.2	7.4	5.6	11.4	6.0		8.6	5.6	9.3	14.0	3.2	6.4	17.0	7.2

No.	Compound	Cartridge	Method	Tea	Rep	1	2	3	4	5	6	7	8	9	10	11	12	13	14	15	16	17	18	19
163	flufenacet	Cleanert TPT	GC-MS	Green tea	A	7.1	7.0	5.5	5.7	10.8	8.7	8.0	9.8	10.8	2.6	14.1	7.7	13.1	9.7	15.1	4.1	11.6	8.8	3.6
		Cleanert TPT	GC-MS	Green tea	B	4.7	1.1	9.5	4.4	5.5	6.5	5.7	12.8	6.8	12.2	3.9	8.6	13.2	13.8	6.2	6.6	9.4	11.7	15.0
		Cleanert TPT	GC-MS	Woolong tea	A	11.0	4.4	9.1	1.4	7.1	1.0	14.4	0.4	11.9	1.1	14.8	5.4	7.1	15.6	4.8	3.6	11.8	13.7	3.2
		Cleanert TPT	GC-MS	Woolong tea	B*	3.0	0.9	5.4	3.8	5.1	10.7	10.1	12.6	7.3	6.4	9.5	11.6	8.7	12.5	5.3	6.5	2.4	7.1	14.4
		Cleanert TPT	GC-MS/MS	Green tea	A	8.0	7.5	3.0	1.0	16.8	4.0	8.7	11.6	13.0	14.5	4.8	5.1	12.3	10.1	0.4	5.3	11.6	8.1	18.3
		Cleanert TPT	GC-MS/MS	Green tea	B	9.6	11.4	7.8	10.5	4.7	5.4	5.4	13.2	19.2	13.5	4.0	13.3	25.7	7.7	8.8	9.6	5.3	7.2	3.0
		Cleanert TPT	GC-MS/MS	Woolong tea	A	3.5	7.2	10.0	3.3	7.6	2.5	10.2	22.2	14.3	7.7	4.6	7.3	4.2	3.5	7.7	6.4	7.0	6.3	13.6
		Cleanert TPT	GC-MS/MS	Woolong tea	A*	5.4	12.1	15.1	1.2	7.4	12.1	9.2	2.6	17.8	9.3	3.8	4.2	1.3	7.8	14.3	6.8	9.4	9.0	9.1
		Envi-Carb+PSA	GC-MS	Green tea	B*	11.9	8.0	3.9	6.3	12.3	13.9	9.0	13.6	13.3	8.2	3.3	5.7	6.7	4.4	9.3	6.5	12.2	32.2	9.6
		Envi-Carb+PSA	GC-MS	Green tea	A	21.4	10.7	3.2	7.1	9.1	7.0	5.3	14.3	9.0	8.5	16.3	10.7	13.7	2.0	12.3	12.4	6.0	13.9	6.9
		Envi-Carb+PSA	GC-MS	Woolong tea	B	10.6	6.6	9.4	6.1	5.3	3.8	5.1	14.1	8.4	6.3	12.9	3.7	9.9	3.8	12.4	5.7	15.8	14.2	18.8
		Envi-Carb+PSA	GC-MS	Woolong tea	A	15.8	12.5	6.4	2.9	11.4	6.2	7.8	5.9	19.3	13.2	11.8	7.9	11.5	10.2	13.6	6.9	17.4	17.4	6.9
		Envi-Carb+PSA	GC-MS/MS	Green tea	B	9.3	7.3	0.7	13.9	15.6	13.7	5.7	7.3	3.7	6.8	3.4	3.7	7.5	19.4	12.0	11.3	14.2	15.2	11.8
		Envi-Carb+PSA	GC-MS/MS	Green tea	A	23.4	2.9	3.5	10.3	8.7	10.5	12.4	9.2	13.8	21.8	5.5	3.7	10.7	15.3	8.0	7.0	21.2	24.0	19.8
		Envi-Carb+PSA	GC-MS/MS	Woolong tea	B	10.1	5.5	10.4	20.8	6.6	7.7	10.2	16.3	7.3	10.2	2.3	18.2	16.7	13.2	15.2	4.4	14.8	9.1	13.6
		Envi-Carb+PSA	GC-MS/MS	Woolong tea	A	18.7	8.8	18.5	7.4	8.5	11.4	10.4	7.8	17.1	4.1	12.4	24.4	13.2	3.1	13.9	5.4	43.0	15.1	9.1
164	furalaxyl	Cleanert TPT	GC-MS	Green tea	A	5.4	6.8	1.9	3.6	6.7	4.8	7.3	5.0	5.0	3.7	6.0	3.4	7.6	4.3	2.1	4.0	5.3	6.3	4.3
		Cleanert TPT	GC-MS	Green tea	B	4.5	8.7	5.4	8.1	3.0	3.5	3.4	8.2	4.6	8.5	4.7	9.1	1.7	3.9	4.8	3.3	5.6	1.2	4.8
		Cleanert TPT	GC-MS	Woolong tea	A	5.8	1.3	2.5	4.9	3.6	4.0	8.4	7.4	3.0	6.1	5.8	1.5	3.8	3.1	8.7	4.4	4.5	3.5	1.4
		Cleanert TPT	GC-MS	Woolong tea	B	6.2	2.1	2.6	4.9	2.0	3.9	5.0	3.7	5.4	1.3	4.5	1.9	2.5	2.2	1.0	0.3	1.4	0.8	2.1
		Cleanert TPT	GC-MS/MS	Green tea	A	5.9	4.7	1.9	5.5	8.7	4.8	7.0	4.8	5.5	4.5	5.1	2.3	9.4	4.8	1.4	3.7	7.8	6.2	2.4
		Cleanert TPT	GC-MS/MS	Green tea	B	4.4	10.3	7.8	7.1	4.3	3.8	4.4	9.1	6.7	11.4	3.5	12.2	7.7	2.0	7.5	5.0	3.3	4.5	6.4
		Cleanert TPT	GC-MS/MS	Woolong tea	A	6.1	2.4	1.7	4.5	5.7	3.2	10.4	1.2	2.7	4.8	0.5	4.3	6.4	1.5	7.2	4.6	2.9	5.2	5.9
		Cleanert TPT	GC-MS/MS	Woolong tea	B	7.3	4.8	6.0	7.8	1.6	1.8	5.4	7.5	6.3	3.0	6.6	3.8	9.9	2.0	6.5	2.1	1.9	3.1	6.4
		Envi-Carb+PSA	GC-MS	Green tea	A	6.3	7.8	2.1	4.5	7.9	1.7	19.3	0.7	8.1	3.0	10.2	2.6	2.5	6.4	6.7	7.0	11.1	2.9	6.5
		Envi-Carb+PSA	GC-MS	Green tea	B	3.4	5.2	3.0	5.0	2.5	1.1	3.8	4.5	5.4	1.6	24.9	12.2	0.4	2.9	12.8	1.7	3.8	8.8	7.8
		Envi-Carb+PSA	GC-MS	Woolong tea	A	2.7	3.5	3.7	2.0	6.3	3.1	4.7	2.5	1.9	6.6	6.7	6.3	0.9	4.7	5.4	4.2	4.4	1.9	3.4
		Envi-Carb+PSA	GC-MS	Woolong tea	B	5.1	3.4	12.3	6.8	11.1	7.7	1.9	5.8	10.2	5.1	9.6	3.1	5.5	12.7	10.0	4.2	3.5	5.7	5.5
		Envi-Carb+PSA	GC-MS/MS	Green tea	A	7.3	6.0	2.4	7.0	6.8	2.0	19.3	2.3	3.0	2.0	9.3	1.9	2.0	4.7	6.2	7.2	6.7	1.8	4.9
		Envi-Carb+PSA	GC-MS/MS	Green tea	B	4.1	5.1	4.2	3.9	1.6	1.3	2.7	6.3	4.8	1.2	21.4	2.6	3.1	27.2	13.0	1.1	1.1	9.1	6.3
		Envi-Carb+PSA	GC-MS/MS	Woolong tea	A	4.0	5.2	3.5	3.7	6.0	1.9	3.3	2.4	1.3	5.7	7.4	8.5	10.4	2.6	14.4	4.7	5.0	1.5	5.9
		Envi-Carb+PSA	GC-MS/MS	Woolong tea	B	5.8	3.7	13.5	8.9	8.4	8.1	2.5	8.1	8.8	3.7	6.4	4.0	6.4	13.1	10.5	4.3	3.3	5.8	6.4
165	heptachlor	Cleanert TPT	GC-MS	Green tea	A	4.2	7.1	3.0	1.8	9.4	5.2	6.1	7.6	6.1	4.2	4.5	3.5	6.7	4.2	5.6	0.4	6.8	6.7	3.2
		Cleanert TPT	GC-MS	Green tea	B	6.2	8.3	4.0	10.8	3.4	7.5	0.3	4.9	3.8	9.4	0.9	8.9	2.7	1.5	3.5	2.3	1.8	2.3	7.3
		Cleanert TPT	GC-MS	Woolong tea	A	3.5	2.3	2.8	3.9	5.0	3.2	5.9	5.3	4.6	2.8	6.6	3.9	1.5	5.7	8.1	4.3	5.0	0.9	1.0
		Cleanert TPT	GC-MS	Woolong tea	B	3.0	2.3	4.2	3.3	5.2	5.4	3.2	5.1	5.7	3.0	2.9	2.9	1.3	5.1	5.1	1.6	1.1	2.0	3.7
		Cleanert TPT	GC-MS/MS	Green tea	A	2.5	1.8	3.6	7.9	12.3	2.6	8.4	6.4	5.5	3.1	7.5	1.3	10.0	3.0	0.3	3.2	8.1	2.5	6.2
		Cleanert TPT	GC-MS/MS	Green tea	B	5.9	6.8	4.9	9.0	4.9	3.1	2.1	6.3	9.6	11.1	1.9	7.1	3.7	3.5	2.4	3.8	4.5	1.6	7.8
		Cleanert TPT	GC-MS/MS	Woolong tea	A	4.5	3.4	2.7	4.3	5.7	1.7	8.6	6.7	8.9	3.3	6.2	4.3	3.3	2.8	7.7	7.4	4.1	4.5	5.3
		Cleanert TPT	GC-MS/MS	Woolong tea	B	5.6	4.7	5.1	5.4	5.3	4.3	2.3	9.7	10.0	1.2	5.5	1.5	9.7	4.0	7.5	1.1	1.7	1.3	5.3
		Envi-Carb+PSA	GC-MS	Green tea	A	10.3	4.5	4.3	8.5	11.2	7.9	7.6	5.7	9.4	1.1	11.2	3.9	5.0	7.7	6.0	14.9	10.5	17.2	5.2
		Envi-Carb+PSA	GC-MS	Green tea	B	9.6	7.6	2.6	8.5	3.7	6.6	6.7	8.4	5.9	2.1	18.7	5.3	13.5	0.9	20.0	6.3	8.6	11.0	6.8
		Envi-Carb+PSA	GC-MS	Woolong tea	A	6.4	5.0	5.5	3.3	3.7	1.7	4.1	10.0	3.0	2.0	5.8	1.6	1.9	3.8	1.4	6.6	5.8	8.3	3.1
		Envi-Carb+PSA	GC-MS	Woolong tea	B	5.6	3.7	6.7	2.8	12.0	8.2	4.4	7.5	12.2	2.1	8.9	3.2	6.3	8.2	10.4	2.7	10.6	3.1	5.5
		Envi-Carb+PSA	GC-MS/MS	Green tea	A	10.9	6.4	4.0	6.1	7.0	7.4	8.8	4.8	3.4	3.7	9.9	5.2	3.7	2.1	4.5	10.7	3.2	6.0	8.4
		Envi-Carb+PSA	GC-MS/MS	Green tea	B	12.8	10.2	4.4	1.1	2.7	4.3	4.7	3.8	4.3	2.4	22.0	8.1	11.6	6.3	9.8	1.9	4.8	15.1	8.4
		Envi-Carb+PSA	GC-MS/MS	Woolong tea	A	6.5	4.9	6.9	6.3	4.1	2.7	2.0	12.1	3.6	2.8	4.5	—	15.7	2.2	7.7	3.1	10.1	3.3	5.3
		Envi-Carb+PSA	GC-MS/MS	Woolong tea	B	9.1	2.7	10.2	3.2	9.1	7.3	4.2	7.6	9.9	2.6	8.1	7.3	8.3	9.1	12.1	4.2	5.1	6.9	5.3

Annex 2: RSD data for 201 pesticides in incurred tea Youden pair samples under 8 conditions with two SPE cartridge cleanup in three months (Nov.9, 2009–Feb.7, 2010) (cont.)

No	Pesticides	SPE	Method	Sample	Youden pair	Nov.9, 2009 (n=5)	Nov.14, 2009 (n=3)	Nov.19, 2009 (n=3)	Nov.24, 2009 (n=3)	Nov.29, 2009 (n=3)	Dec.4, 2009 (n=3)	Dec.9, 2009 (n=3)	Dec.14, 2009 (n=3)	Dec.19, 2009 (n=3)	Dec.24, 2009 (n=3)	Dec.29, 2009 (n=3)	Jan.3, 2010 (n=3)	Jan.8, 2010 (n=3)	Jan.13, 2010 (n=3)	Jan.18, 2010 (n=3)	Jan.23, 2010 (n=3)	Jan.28, 2010 (n=3)	Feb.2, 2010 (n=3)	Feb.7, 2010 (n=3)
166	iprobenfos	Cleanert TPT	GC-MS	Green tea	A	14.0	4.8	3.7	5.6	13.5	5.7	6.2	7.0	3.6	7.4	10.6	5.9	9.2	10.8	6.7	1.9	8.3	7.6	4.0
				Woolong tea	B*	9.8	8.2	13.8	12.2	6.0	2.5	3.6	1.8	7.5	7.6	6.3	13.8	1.0	5.9	11.0	3.6	12.0	0.5	6.3
			GC-MS/MS	Green tea	A	5.9	3.1	10.7	5.9	2.3	4.8	1.9	11.9	3.0	13.2	8.4	14.8	3.0	5.7	9.5	5.3	5.2	5.0	1.2
				Woolong tea	B	4.8	7.1	11.1	10.3	8.0	5.2	4.8	5.2	10.0	11.2	9.5	6.4	1.1	3.3	2.2	1.4	2.1	2.4	1.0
		Envi-Carb+PSA	GC-MS	Green tea	A	3.6	3.3	3.3	2.8	10.9	5.9	7.6	5.0	3.7	3.5	6.2	1.3	9.3	8.2	3.5	0.6	6.3	5.5	4.3
				Woolong tea	B	5.2	8.2	7.7	7.8	4.6	3.7	4.8	6.0	6.7	11.0	2.4	14.6	0.7	0.4	3.7	3.6	4.9	2.4	7.5
			GC-MS/MS	Green tea	A	4.9	2.6	4.4	5.3	3.5	1.8	13.5	0.7	4.5	4.9	3.1	3.9	5.8	0.8	8.2	4.8	4.5	2.6	5.6
				Woolong tea	B	4.0	3.3	2.4	9.3	5.0	4.7	3.0	6.9	7.2	0.7	6.8	2.3	8.5	2.5	4.2	2.3	2.9	1.8	4.6
167	isazofos	Cleanert TPT	GC-MS	Green tea	A	10.6	3.0	3.2	10.6	6.8	4.3	5.1	8.6	9.0	10.4	1.3	8.2	3.3	5.4	13.2	15.1	10.9	13.1	6.2
				Woolong tea	B	4.6	29.7	7.7	3.1	1.7	14.5	3.0	8.2	12.5	5.3	39.0	6.5	10.7	3.3	19.4	6.6	8.5	10.6	7.2
			GC-MS/MS	Green tea	A	8.9	8.9	6.0	4.8	9.6	0.9	5.6	10.3	6.2	5.0	10.2	2.2	1.4	5.1	1.8	4.7	6.4	4.7	4.6
				Woolong tea	B*	5.7	4.6	13.4	0.9	24.1	7.3	2.8	2.9	14.9	4.6	15.2	4.1	7.1	7.8	10.7	2.3	5.9	5.4	5.1
		Envi-Carb+PSA	GC-MS	Green tea	A	10.7	5.8	2.1	7.2	7.8	2.9	2.8	3.0	3.7	3.0	10.5	1.5	3.0	1.7	6.3	8.5	5.0	6.9	4.9
				Woolong tea	B	3.4	8.8	5.8	3.8	3.9	3.9	3.1	1.8	4.5	1.0	23.5	1.8	6.4	15.8	13.9	3.7	4.9	10.4	9.5
			GC-MS/MS	Green tea	A	5.3	6.0	5.3	2.4	3.7	2.5	4.4	6.8	4.0	1.6	4.7	8.4	3.9	2.5	10.1	4.5	7.3	4.5	5.6
				Woolong tea	B	6.9	2.9	10.6	4.3	13.2	6.9	6.3	6.6	9.9	4.0	6.8	5.5	6.1	8.8	12.2	1.0	2.6	3.2	4.6
168	isoprothiolane	Cleanert TPT	GC-MS	Green tea	A*	4.6	8.5	4.7	7.6	2.5	3.0	2.2	4.8	1.7	7.4	5.1	4.3	8.6	14.0	2.8	3.4	4.5	5.6	3.6
				Woolong tea	B*	6.1	1.7	2.1	3.6	2.2	3.3	10.6	7.7	1.7	7.2	2.9	9.5	2.1	8.4	4.0	4.7	4.7	1.7	4.9
			GC-MS/MS	Green tea	A	4.5	2.2	2.5	5.3	2.6	3.1	4.5	5.6	6.0	1.5	4.5	1.4	4.9	8.3	7.7	2.7	3.2	3.1	0.4
				Woolong tea	B	4.2	4.8	2.6	3.7	9.8	3.5	7.4	6.2	2.8	2.4	6.1	0.2	8.8	7.3	1.7	2.7	2.3	5.8	2.1
		Envi-Carb+PSA	GC-MS	Green tea	A	4.3	8.9	7.0	8.6	3.6	3.0	3.0	4.2	5.6	9.3	4.0	3.5	7.7	2.9	5.3	3.8	7.0	4.9	5.1
				Woolong tea	B	6.5	2.8	0.9	3.0	4.4	4.1	8.5	0.8	5.2	5.1	0.9	2.9	6.6	1.1	7.5	2.8	4.9	3.8	5.6
			GC-MS/MS	Green tea	A	4.2	5.6	4.7	7.8	2.0	2.4	4.1	6.6	7.6	2.4	5.6	2.2	7.5	5.1	7.1	3.1	2.0	3.4	6.3
				Woolong tea	B	6.5	4.2	2.2	4.8	8.7	2.1	2.9	3.4	4.2	3.1	6.5	2.5	0.3	8.7	5.6	9.4	10.6	3.1	5.9

No.	Compound	Cartridge	Method	Tea	Rep																			
169	kresoxim-methyl	Cleanert TPT	GC-MS	Green tea	A	9.7	7.1	4.2	12.1	15.6	9.3	13.3	11.5	3.5	13.9	5.0	2.1	7.8	8.3	5.0	3.8	3.4	1.1	2.7
				Woolong tea	B	21.1	5.0	9.5	14.7	9.3	10.0	13.8	4.0	3.4	13.4	9.0	7.8	9.5	4.1	1.3	8.4	9.3	5.1	7.4
				Green tea	A	8.1	4.5	5.4	12.7	28.4	1.9	11.3	9.2	2.8	8.9	10.4	3.5	2.6	4.4	6.1	0.8	3.5	3.0	3.8
				Woolong tea	B*	4.6	0.8	7.0	23.5	7.0	13.4	8.9	1.4	8.2	0.3	0.2	3.7	3.0	4.7	5.3	4.5	2.8	5.1	5.2
			GC-MS/MS	Green tea	A	11.4	3.9	2.8	14.4	8.7	10.5	13.7	2.9	10.7	2.5	4.1	4.1	15.0	4.4	5.7	4.4	10.4	4.9	3.9
				Woolong tea	B	10.5	5.5	3.4	6.0	9.9	14.2	8.7	3.4	7.4	3.5	2.7	9.3	22.5	5.9	5.7	10.2	10.5	2.1	15.7
				Green tea	A	8.4	1.5	4.3	1.6	23.7	10.4	2.0	6.8	6.1	4.3	7.7	6.5	6.6	6.5	9.2	8.1	4.7	9.7	2.8
				Woolong tea	B	3.6	6.0	10.6	14.4	3.6	18.6	8.3	4.8	7.7	9.0	5.0	2.9	15.7	3.2	1.5	5.6	4.2	3.1	8.6
		Envi-Carb+PSA	GC-MS	Green tea	A	8.7	8.0	18.8	8.9	21.6	57.4	37.0	38.6	11.9	6.5	14.1	11.4	5.7	4.9	11.7	10.0	6.3	26.9	2.2
				Woolong tea	B	21.7	13.2	2.9	7.2	42.6	28.2	8.0	42.2	14.3	3.5	45.7	7.5	12.6	5.2	7.2	9.4	1.8	19.5	6.5
				Green tea	A	7.4	7.7	4.0	14.1	16.5	11.6	17.8	7.4	11.6	9.3	5.5	2.7	3.7	7.6	7.8	8.1	9.3	7.7	1.6
				Woolong tea	B	11.3	6.8	6.8	19.7	14.7	12.0	9.5	5.9	7.9	10.4	10.8	5.1	2.9	22.1	9.5	4.3	8.4	16.9	7.2
			GC-MS/MS	Green tea	A	31.0	16.7	13.4	6.6	8.5	15.0	44.3	7.1	16.5	7.9	7.9	1.2	14.4	4.2	8.5	18.0	15.5	1.8	12.0
				Woolong tea	B	43.4	14.8	4.1	15.4	27.0	7.1	7.1	4.2	8.5	3.1	14.4	0.9	21.8	1.9	8.4	20.9	13.9	17.9	5.4
				Green tea	A*	9.0	2.0	2.9	15.1	18.7	4.9	14.4	8.4	2.5	7.2	5.0	7.1	22.0	10.1	12.0	6.4	6.3	16.6	2.1
				Woolong tea	B*	16.4	7.0	10.1	16.2	10.0	6.6	29.4	9.2	9.9	9.4	7.2	11.8	5.4	12.4	7.7	2.8	6.7	20.2	8.6
170	mefenacet	Cleanert TPT	GC-MS	Green tea	A*	9.4	5.6	3.3	9.0	14.8	2.9	12.1	9.3	9.6	7.9	7.5	12.2	4.1	9.8	4.5	2.6	9.9	2.8	10.3
				Woolong tea	B*	7.2	7.4	3.1	8.8	7.0	8.9	5.8	5.4	7.5	7.9	4.7	16.3	14.8	7.1	9.1	1.0	1.5	2.8	8.8
				Green tea	A	8.0	8.0	10.7	3.9	14.7	6.6	5.0	1.6	11.0	3.6	13.5	7.9	2.6	3.3	3.1	2.3	4.2	4.3	8.4
				Woolong tea	B	3.4	2.4	10.8	3.3	8.2	11.5	11.0	16.0	9.8	8.7	14.8	10.4	4.0	5.0	3.5	2.1	5.8	3.8	10.9
			GC-MS/MS	Green tea	A	5.9	4.1	4.4	14.7	14.5	2.8	5.2	12.3	11.6	11.6	9.8	4.9	12.7	3.0	3.1	2.5	8.1	4.7	2.8
				Woolong tea	B	6.5	2.6	1.7	9.7	8.2	2.8	3.8	7.8	11.3	5.4	8.1	1.5	13.6	1.7	5.7	1.4	2.8	2.4	5.9
				Green tea	A	8.8	3.1	1.0	5.3	7.7	2.7	5.6	8.3	10.2	2.6	15.8	17.5	7.9	3.3	4.3	2.9	6.5	8.6	3.3
				Woolong tea	B	8.3	3.0	4.2	6.9	11.9	4.6	8.8	16.8	22.7	11.3	5.6	2.9	13.8	8.0	3.3	4.1	5.5	6.9	5.0
		Envi-Carb+PSA	GC-MS	Green tea	A*	18.1	3.5	13.8	6.8	12.3	16.6	8.5	6.8	20.9		7.0	2.0	9.0		19.4	14.4	14.7	4.9	
				Woolong tea	B	15.4	6.6	8.3	9.1	13.1	6.1	1.4	14.2	6.6	7.4	30.6	6.9	28.5	5.6	3.1	7.2	25.1	6.0	4.9
				Green tea	A*	8.8	5.7	7.7	11.3	13.2	4.1	3.7	11.8	3.0	3.9	10.4	10.1	5.2	5.0	11.5	14.9	3.0	15.1	11.5
				Woolong tea	B*	10.9	11.5	7.7	4.7	13.2	14.2	9.6	14.5	11.6	15.0	11.8	4.5	8.1	5.0	6.1	3.0	6.1	5.6	5.0
			GC-MS/MS	Green tea	A	8.4	3.2	8.8	13.6	7.5	10.7	11.9	18.7	7.2	5.8	3.0	13.4	14.5	8.8	9.9	9.1	6.6	5.6	5.7
				Woolong tea	B	16.6	4.6	6.0	4.1	8.8	7.2	6.8	22.5	20.3	10.2	33.5	12.7	16.0	5.2	6.3	4.8	7.4	8.9	2.9
				Green tea	A	9.4	4.0	12.6	5.3	7.9	4.4	4.7	22.3	3.7	12.5	6.0	18.7	5.7	7.4	13.5	10.5	7.5	34.8	5.2
				Woolong tea	B	7.4	9.9	10.3	3.0	12.4	11.8	6.2	6.1	11.8	7.5	7.0	15.4	6.8	8.4	4.9	4.9	19.0	20.1	4.0
171	mepronil	Cleanert TPT	GC-MS	Green tea	A	6.1	8.1	0.7	4.3	7.7	4.9	8.1	5.4	5.5	6.5	6.3	3.1	8.1	2.6	2.8	2.6	3.9	5.6	3.4
				Woolong tea	B	6.3	7.4	5.2	8.7	2.5	3.4	1.1	5.2	1.5	4.5	8.1	9.7	1.4	1.2	3.9	3.9	7.1	1.1	2.4
				Green tea	A	7.5	2.5	1.6	4.4	1.8	3.7	8.6	10.8	7.2	2.5	10.8	5.2	3.9	6.3	5.1	4.6	2.1	5.0	1.3
				Woolong tea	B	5.3	2.9	1.9	2.8	0.8	3.7	4.5	7.2	8.4	1.4	6.9	1.2	5.1	6.5	1.6	1.6	1.2	1.8	1.8
			GC-MS/MS	Green tea	A	6.5	6.7	3.9	3.9	6.9	3.8	7.3	8.4	8.1	8.1	3.8	3.2	13.2	3.5	2.0	5.3	5.3	5.8	1.8
				Woolong tea	B	5.2	11.5	5.6	5.6	3.7	2.9	1.7	8.1	12.9	3.6	8.3	9.4	8.5	6.5	5.5	3.1	3.6	4.4	6.0
				Green tea	A	6.1	2.1	3.8	4.0	7.0	3.8	10.7	7.3	3.4	1.0	2.7	4.6	6.5	2.1	5.5	3.9	2.2	6.2	12.0
				Woolong tea	B	6.4	3.7	3.3	7.1	3.6	4.1	4.1	1.5	8.6	16.9	4.7	3.9	4.7	4.1	4.6	1.9	1.4	1.2	2.9
		Envi-Carb+PSA	GC-MS	Green tea	A*	5.1	8.0	2.2	7.1	6.6	3.6	16.2	9.0	5.0	13.7	2.1	3.0	2.1	5.4	27.6	8.7	13.3	7.9	3.8
				Woolong tea	B	7.2	7.5	4.1	5.9	3.2	3.0	1.6	9.2	15.6	6.0	30.4	13.8	9.0	4.7	14.8	2.3	5.3	10.4	7.6
				Green tea	A	9.2	1.1	3.6	6.5	7.8	2.1	3.6	5.7	5.2	3.1	4.4	4.9	1.2	6.7	1.1	2.5	3.3	3.8	3.4
				Woolong tea	B	7.6	4.4	14.6	4.3	9.5	4.1	3.7	5.8	10.9	2.3	9.8	4.6	1.7	7.6	9.5	2.8	4.2	3.8	2.3
			GC-MS/MS	Green tea	A	6.9	3.2	3.5	7.9	6.1	16.2	19.8	4.3	9.7	1.5	6.3	3.3	10.1	6.7	4.0	7.1	9.3	6.4	3.5
				Woolong tea	B	6.0	8.4	4.1	2.2	4.9	27.3	1.5	5.6	9.0	4.3	22.7	2.7	4.5	12.1	12.7	3.1	10.1	10.5	7.9
				Green tea	A	5.8	1.8	3.4	2.6	6.9	1.7	8.2	4.6	4.6	4.3	7.4	4.9	15.2	4.0	8.8	3.2	4.8	9.4	6.3
				Woolong tea	B	9.0	4.4	9.0	11.0	8.8	5.1	7.5	5.6	3.8	3.4	3.8	3.3	3.4	11.2	9.5	3.2	4.6	3.8	2.9

(Continued)

Annex 2: RSD data for 201 pesticides in incurred tea Youden pair samples under 8 conditions with two SPE cartridge cleanup in three months (Nov.9, 2009-Feb.7, 2010) (cont.)

No	Pesticides	SPE	Method	Sample	Youden pair	Nov.9, 2009 (n=5)	Nov.14, 2009 (n=3)	Nov.19, 2009 (n=3)	Nov.24, 2009 (n=3)	Nov.29, 2009 (n=3)	Dec.4, 2009 (n=3)	Dec.9, 2009 (n=3)	Dec.14, 2009 (n=3)	Dec.19, 2009 (n=3)	Dec.24, 2009 (n=3)	Dec.29, 2009 (n=3)	Jan.3, 2010 (n=3)	Jan.8, 2010 (n=3)	Jan.13, 2010 (n=3)	Jan.18, 2010 (n=3)	Jan.23, 2010 (n=3)	Jan.28, 2010 (n=3)	Feb.2, 2010 (n=3)	Feb.7, 2010 (n=3)
172	metribuzin	Cleanert TPT	GC-MS	Green tea	A	3.7	5.7	5.8	2.4	6.2	0.7	9.4	9.2	11.0	3.9	2.9	2.7	6.7	6.1	3.4	2.8	7.0	8.5	4.0
					B	7.2	8.3	4.3	8.3	6.0	4.1	6.1	9.3	3.8	10.5	3.2	12.6	8.7	6.2	6.0	6.2	4.2	2.7	6.5
				Woolong tea	A	5.2	0.9	6.3	5.3	4.7	5.2	1.2	10.4	2.7	5.8	5.9	11.5	1.8	1.0	6.8	3.6	3.4	1.4	3.3
					B	5.8	2.7	1.3	4.3	5.1	6.0	12.0	4.9	5.0	3.2	3.5	3.8	1.0	15.4	7.8	0.8	1.6	6.9	7.7
			GC-MS/MS	Green tea	A	4.8	6.8	3.7	11.7	7.2	1.6	7.6	9.2	10.1	5.8	2.0	6.3	7.6	4.6	3.6	1.7	6.1	10.8	3.0
					B	5.0	4.8	6.0	6.9	3.6	3.5	7.6	11.3	9.7	10.9	4.2	11.6	11.0	3.6	4.8	5.7	5.6	2.2	6.1
				Woolong tea	A	7.0	6.4	5.2	3.2	4.8	5.3	3.3	3.0	11.3	3.3	3.1	1.4	10.9	5.0	8.7	2.0	3.5	2.9	3.0
					B	4.5	5.6	10.3	6.3	0.6	4.3	3.0	10.3	9.7	7.7	2.8	1.4	9.6	11.6	9.4	4.4	4.6	3.2	7.0
		Envi-Carb+PSA	GC-MS	Green tea	A*	15.8	6.8	15.2	10.2	8.3	19.0	17.0	8.2	14.3	4.4	12.0	20.9	6.4	5.4	7.8	10.8	10.5	21.6	7.3
					B	20.2	11.3	1.6	8.0	5.3	24.2	12.7	15.0	15.4	8.5	21.0	2.6	25.0	1.7	9.8	7.3	3.0	21.5	8.2
				Woolong tea	A	3.9	4.3	4.7	9.7	8.3	3.2	3.7	20.0	3.1	6.9	6.6	9.1	7.6	2.4	6.6	11.9	7.9	8.5	3.6
					B	4.2	2.5	13.7	2.6	17.8	5.9	6.5	7.2	10.7	6.2	12.7	4.8	3.9	4.5	13.4	3.3	6.5	13.5	5.8
			GC-MS/MS	Green tea	A*	18.9	1.6	11.0	3.3	7.0	20.2	16.6	14.9	8.5	1.3	10.6	2.8	5.6	5.6	6.7	7.5	7.7	6.5	7.2
					B	27.4	8.3	4.7	11.9	6.7	2.0	14.0	14.1	12.5	12.7	13.0	5.6	19.8	19.7	5.3	17.5	0.2	18.3	11.0
				Woolong tea	A	7.1	3.9	6.9	8.2	2.2	3.2	5.2	15.7	1.0	7.0	5.2	1.3	2.6	2.3	15.2	4.3	9.1	12.1	3.0
					B	13.5	4.8	15.0	5.2	12.2	8.1	4.5	6.8	8.3	3.8	6.9	7.4	6.9	12.7	13.0	2.8	4.1	18.3	7.0
173	molinate	Cleanert TPT	GC-MS	Green tea	A	6.1	9.6	1.6	0.8	8.8	6.3	6.2	4.2	7.0	4.0	9.4	12.0	5.0	3.7	3.6	4.8	13.3	5.6	4.7
					B*	6.4	9.1	4.0	8.2	2.9	4.2	3.0	1.8	3.3	10.4	4.0	1.3	4.2	3.8	4.3	1.8	6.4	1.9	2.1
				Woolong tea	A	4.3	0.7	2.3	2.7	3.8	1.8	10.0	4.0	5.1	3.4	3.5	10.2	3.4	4.7	9.2	3.1	2.2	1.9	0.3
					B	4.8	3.0	7.8	6.0	3.6	5.2	5.3	3.1	3.8	5.0	0.9	1.9	4.6	4.3	3.1	0.6	1.1	1.6	2.6
			GC-MS/MS	Green tea	A	3.6	3.1	2.7	1.3	12.0	6.4	4.7	2.4	2.9	2.2	5.3	1.2	7.9	4.3	1.2	1.7	6.2	5.6	1.7
					B	6.0	8.3	5.1	8.3	3.1	4.3	3.8	3.2	5.3	11.1	3.5	13.4	8.6	1.9	3.7	2.6	5.2	8.9	5.1
				Woolong tea	A	4.5	0.6	1.8	3.1	4.0	1.0	8.4	0.2	5.3	2.1	3.7	4.4	7.4	0.4	10.1	4.3	2.1	0.5	4.1
					B	5.1	3.7	5.1	6.1	2.5	2.5	4.7	4.3	11.0	3.8	2.7	1.7	8.5	0.1	9.0	2.4	1.5	2.9	1.9
		Envi-Carb+PSA	GC-MS	Green tea	A	5.1	1.0	2.1	5.0	5.1	0.6	1.4	1.5	6.3	2.5	9.9	4.6	1.5	3.0	3.9	8.6	13.4	3.3	3.2
					B	5.1	6.4	7.1	4.1	2.4	2.9	7.6	5.9	3.2	3.3	21.1	3.3	6.6	1.7	19.9	3.7	6.6	9.1	6.3
				Woolong tea	A	4.6	0.3	4.1	6.9	5.9	1.3	5.4	0.5	1.7	4.0	1.9	2.2	1.6	7.2	0.3	6.5	7.9	5.7	5.8
					B	5.8	4.0	9.1	5.7	13.0	5.3	2.6	6.9	11.8	2.9	7.7	3.9	4.2	5.9	4.7	2.6	2.8	4.2	3.3
			GC-MS/MS	Green tea	A	8.6	9.6	1.1	8.2	4.2	3.8	4.5	2.2	6.2	2.8	6.8	4.4	2.2	53.1	7.4	9.7	10.2	6.0	4.2
					B	3.7	6.0	6.7	2.8	1.6	1.0	4.6	4.5	1.0	1.1	21.7	2.0	9.6	9.6	21.5	2.5	3.8	10.4	4.3
				Woolong tea	A	2.9	4.5	4.0	4.2	4.7	2.0	1.1	5.2	1.6	2.5	6.2	4.8	3.5	4.5	8.8	5.6	4.7	4.7	4.1
					B	7.0	4.6	7.6	6.1	8.3	5.0	2.4	7.8	9.2	1.8	5.2	3.0	8.1	8.5	6.6	2.2	6.3	3.7	1.9
174	napropamide	Cleanert TPT	GC-MS	Green tea	A	4.6	4.7	1.7	2.6	8.6	4.0	8.3	6.7	4.9	4.7	6.1	2.9	14.8	3.2	2.9	3.1	1.4	7.7	4.4
					B	4.2	8.3	5.4	7.9	4.6	3.6	3.8	5.9	4.6	6.6	2.4	13.0	2.4	1.9	6.4	1.9	5.2	1.0	4.5
				Woolong tea	A	5.7	2.3	1.9	2.6	1.5	7.2	8.0	9.0	8.8	2.7	7.4	2.0	3.6	2.1	7.2	2.9	3.0	4.8	1.6
					B	3.7	1.3	2.2	1.5	1.1	1.3	3.5	2.4	6.4	0.9	8.2	1.6	4.2	3.6	5.0	0.8	1.1	2.4	5.1
			GC-MS/MS	Green tea	A	4.1	4.8	1.7	5.1	8.7	5.1	7.5	5.5	2.9	9.4	5.6	1.7	5.1	6.1	2.7	1.2	7.8	4.6	2.8
					B	5.3	1.7	6.8	7.9	4.4	4.1	4.4	6.9	6.6	4.8	4.5	8.8	6.3	2.0	4.3	2.3	5.0	4.3	7.3
				Woolong tea	A	5.6	11.6	1.2	3.1	5.1	4.4	9.5	2.1	2.9	1.3	1.3	3.0	7.9	1.9	9.8	4.2	0.5	3.7	4.3
					B	5.6	3.4	5.3	7.4	2.1	2.2	5.4	7.6	6.9	1.9	1.9	3.7	4.1	4.1	6.2	1.3	2.3	0.7	4.0
		Envi-Carb+PSA	GC-MS	Green tea	A	5.3	5.4	1.5	4.7	8.1	7.4	9.1	7.8	5.0	6.2	5.5	4.8	4.8	3.6	8.9	5.5	11.7	1.8	11.2
					B	3.5	4.6	2.4	4.6	0.8	3.2	4.1	4.5	6.1	1.0	8.5	3.2	3.2	2.6	9.5	2.5	9.6	7.5	12.4
				Woolong tea	A	4.9	8.5	2.0	1.6	7.0	2.3	4.7	1.4	2.2	0.8	27.2	20.6	1.8	4.8	2.3	4.8	5.2	2.1	4.5
					B	6.8	2.4	9.1	2.1	9.5	2.3	1.8	5.5	12.6	8.9	5.1	6.0	3.2	8.6	9.1	5.6	5.8	9.9	3.1
			GC-MS/MS	Green tea	A	6.6	4.6	1.7	7.7	7.5	7.4	8.1	1.9	3.1	3.3	10.8	6.0	3.0	9.1	8.2	6.3	9.2	1.5	6.1
					B	3.6	5.1	5.6	3.5	1.1	2.1	2.4	7.8	3.3	1.5	8.4	1.8	5.4	26.1	14.0	3.1	2.9	8.2	6.0
				Woolong tea	A	3.6	8.5	4.0	4.0	5.5	1.0	4.0	2.7	1.2	1.6	21.5	2.2	2.9	3.8	12.2	4.6	5.6	4.2	4.3
					B	5.8	3.3	11.6	6.7	7.8	5.7	3.6	7.7	10.1	5.4	5.4	6.6	5.6	8.1	11.6	2.0	4.0	4.6	4.0

No.	Compound	Cartridge	Method	Tea	Sample	1	2	3	4	5	6	7	8	9	10	11	12	13	14	15	16	17	18	19
175	nuarimol	Cleanert TPT	GC-MS	Green tea	A	7.2	4.7	3.5	3.4	7.3	12.9	6.5	5.8	0.6	3.5	6.1	4.9	4.8	12.5	3.5	2.9	3.9	4.5	0.8
					B	6.1	6.6	7.2	7.2	6.1	9.5	1.7	2.6	6.6	6.9	3.1	13.5	2.4	2.5	7.1	3.1	8.0	3.6	2.5
				Woolong tea	A	9.5	5.0	6.4	0.6	7.2	8.9	14.5	9.3	5.8	3.8	8.1	6.3	6.9	2.6	5.2	4.3	3.6	1.5	2.0
					B	7.1	3.1	5.2	7.2	3.0	2.8	6.6	4.4	8.1	6.3	6.2	0.8	0.9	3.1	9.5	1.1	1.5	3.0	0.8
			GC-MS/MS	Green tea	A*	5.3	8.0	0.6	7.8	9.2	5.2	8.1	4.2	3.2	2.6	5.4	3.8	7.0	9.7	3.0	3.0	6.8	4.3	0.6
					B	4.2	11.0	5.9	7.5	3.8	3.8	2.2	7.8	3.6	10.3	3.0	2.6	5.2	6.9	7.5	4.5	5.8	4.6	6.3
				Woolong tea	A	5.9	1.9	1.9	3.1	4.3	4.0	12.6	1.5	2.7	3.8	3.5	5.5	6.2	3.9	6.4	6.7	0.2	2.1	5.8
					B	6.3	4.4	5.2	9.6	1.3	1.8	4.6	8.2	8.6	2.1	2.7	3.3	5.4	5.7	7.0	1.4	4.7	4.0	2.5
		Envi-Carb+PSA	GC-MS	Green tea	A	7.0	12.0	2.1	6.4	3.2	2.7	25.2	1.2	19.0	7.8	4.3	6.9	2.1	6.9	5.1	5.1	11.3	2.9	4.3
					B	5.6	27.4	5.9	9.0	4.4	4.0	2.2	2.3	4.7	6.3	23.3	11.5	11.0	3.6	8.5	1.8	1.3	5.9	7.0
				Woolong tea	A	7.3	4.0	6.0	3.1	12.1	2.0	3.9	4.9	3.4	8.6	8.2	6.9	7.2	4.7	4.9	6.1	3.5	4.0	2.3
					B	7.5	5.8	11.0	7.8	9.3	8.3	0.2	10.2	6.9	1.8	8.7	4.9	2.8	20.3	9.8	3.6	7.2	2.0	2.6
			GC-MS/MS	Green tea	A*	7.0	8.6	5.4	7.1	6.4	4.0	26.6	2.3	7.2	1.4	5.1	1.8	5.5	9.7	7.4	3.4	9.0	2.2	3.7
					B*	6.6	5.9	4.0	5.0	1.0	1.8	1.4	4.9	3.5	5.8	18.8	5.4	3.3	19.2	16.5	1.9	2.5	8.3	4.8
				Woolong tea	A	7.1	2.0	5.0	0.9	5.2	5.6	2.2	5.5	1.2	13.1	8.5	4.5	20.3	5.2	10.6	4.5	5.9	5.4	5.8
					B	5.3	3.8	14.8	7.9	8.6	6.0	3.2	9.8	8.0	2.1	5.6	6.8	10.3	14.7	9.4	3.8	6.0	7.7	2.5
176	permethrin	Cleanert TPT	GC-MS	Green tea	A	3.9	6.1	2.0	1.4	7.1	3.3	7.2	5.1	6.1	3.8	5.3	2.6	4.6	3.8	3.5	1.6	1.1	2.7	3.0
					B	3.5	7.0	3.7	7.0	2.5	4.0	1.9	2.9	2.0	8.5	2.9	9.1	5.1	5.1	4.2	4.3	5.2	1.1	2.1
				Woolong tea	A	6.1	2.9	3.2	2.6	1.7	5.0	7.0	7.4	3.1	2.7	7.4	2.7	5.5	1.4	5.1	3.9	3.1	4.3	1.8
					B	3.2	2.0	2.5	6.3	1.4	5.6	4.3	2.1	7.8	3.8	3.5	4.6	1.4	5.0	1.2	2.6	2.0	2.3	2.2
			GC-MS/MS	Green tea	A*	4.8	3.3	0.6	3.4	10.7	2.9	7.7	3.9	5.4	2.0	6.0	2.6	3.8	1.9	0.1	1.8	6.9	3.5	0.2
					B*	3.5	8.1	4.5	8.5	4.3	5.2	0.6	3.4	6.1	8.8	4.9	3.5	6.4	3.0	5.4	5.5	5.4	8.5	6.0
				Woolong tea	A	6.5	3.0	1.7	1.5	3.3	3.5	8.1	2.7	1.5	7.6	2.6	5.9	8.1	3.0	6.1	4.3	4.5	4.0	3.3
					B	3.7	3.8	5.3	7.1	0.5	8.7	4.5	7.4	12.6	5.0	5.0	2.6	12.7	2.5	4.2	3.5	1.8	4.0	4.8
		Envi-Carb+PSA	GC-MS	Green tea	A	9.5	4.3	6.4	7.9	8.9	4.0	13.3	3.1	9.4	1.6	1.3	3.4	14.3	2.1	2.6	3.5	9.5	4.9	4.8
					B	16.5	11.4	4.1	10.3	2.7	2.5	3.9	2.2	8.2	3.8	16.2	5.2	5.6	1.6	8.3	4.0	3.5	10.7	9.0
				Woolong tea	A	5.3	1.4	5.5	11.1	4.0	3.5	1.1	10.9	2.6	3.9	10.6	2.3	2.3	6.2	2.4	3.1	5.0	6.3	3.1
					B	6.5	5.0	7.3	2.9	10.3	4.4	5.4	8.7	8.2	2.5	9.9	4.6	1.4	4.1	6.4	2.6	6.0	0.3	2.0
			GC-MS/MS	Green tea	A*	8.6	3.5	7.2	8.0	5.5	2.6	9.1	4.9	8.9	0.9	2.7	2.0	3.8	5.5	2.3	4.4	7.6	2.4	3.1
					B	9.3	8.5	5.3	5.8	3.3	5.1	4.1	1.5	6.2	3.0	16.9	1.5	4.6	14.1	2.6	2.4	7.6	9.5	4.6
				Woolong tea	A	6.2	4.3	6.6	1.9	3.9	3.7	1.0	6.4	1.0	3.8	7.8	2.4	16.9	5.5	7.8	0.8	5.0	9.4	3.3
					B	6.6	5.4	10.2	5.4	7.5	3.7	3.2	5.8	7.1	3.8	4.0	5.2	9.3	7.0	9.1	4.1	3.8	1.7	4.8
177	phenothrin	Cleanert TPT	GC-MS	Green tea	A	3.2	6.5	2.0	1.7	5.5	7.7	6.8	4.9	6.3	3.8	3.2	5.6	6.7	13.5	10.8	7.7	3.1	6.3	4.3
					B	3.7	6.9	2.8	9.5	2.7	3.7	1.3	1.0	2.1	10.2	4.8	5.1	6.1	8.0	8.6	6.6	8.1	2.2	5.4
				Woolong tea	A	5.4	5.2	1.1	3.8	1.5	2.3	8.6	9.7	1.2	7.4	6.0	1.2	5.6	1.0	6.3	0.0	4.7	6.3	1.2
					B	3.9	1.7	1.3	5.7	4.0	2.0	5.0	3.3	6.5	3.1	6.7	6.5	4.8	3.2	0.8	1.3	4.0	2.2	4.6
			GC-MS/MS	Green tea	A*	9.2	4.5	1.8	16.9	10.7	2.0	7.1	4.4	5.6	3.7	8.9	3.7	23.3	4.6	6.6	5.7	4.8	6.7	1.5
					B*	2.3	5.8	8.2	9.0	3.4	4.2	2.5	2.6	6.0	9.7	6.8	9.8	16.9	14.5	8.9	9.8	8.4	1.4	3.1
				Woolong tea	A	8.4	5.0	1.1	7.7	2.2	3.6	2.9	2.8	25.7	50.3	13.9	39.3	26.3	2.4	42.5	17.2	8.1	33.7	10.3
					B	4.4	6.1	5.0	3.7	2.1	4.9	6.4	14.3	27.3	20.6	24.4	30.9	42.0	6.0	40.3	28.3	26.2	14.7	1.1
		Envi-Carb+PSA	GC-MS	Green tea	A	6.7	21.3	8.4	4.4	5.5	1.6	5.3	3.9	8.1	2.8	7.7	2.3	4.0	5.6	6.3	6.8	11.3	4.0	6.3
					B	9.5	11.4	7.7	6.6	6.1	2.3	2.6	2.4	4.2	3.1	21.6	8.2	1.2	5.7	13.7	2.8	5.4	9.0	7.7
				Woolong tea	A	4.8	3.7	8.5	3.6	6.7	3.0	2.8	2.5	4.8	7.5	7.7	5.9	3.9	6.7	1.1	6.6	5.9	7.4	1.9
					B	6.5	3.1	8.1	8.7	11.2	5.0	3.4	8.3	10.5	3.4	8.5	3.7	2.8	6.4	8.4	4.1	4.1	0.4	1.1
			GC-MS/MS	Green tea	A*	7.1	1.9	11.7	5.7	23.5	3.3	6.7	3.1	6.0	1.6	8.3	1.4	11.8	4.1	10.3	18.9	6.6	5.9	3.9
					B	5.7	7.3	10.9	11.0	27.5	3.1	6.2	5.1	5.4	0.6	19.8	10.0	1.5	32.4	8.2	2.8	2.6	10.8	8.0
				Woolong tea	A	5.7	29.1	4.9	18.7	22.1	12.2	11.9	7.0	4.8	13.5	28.3	0.8	7.1	47.0	40.2	30.2	8.8	31.1	46.8
					B	6.0	28.0	12.2	19.3	27.4	14.3	19.4	9.5	17.0	37.4	18.2	16.8	35.0	11.3	42.9	14.6	25.2	25.7	44.9

(Continued)

Annex 2: RSD data for 201 pesticides in incurred tea Youden pair samples under 8 conditions with two SPE cartridge cleanup in three months (Nov.9, 2009-Feb.7, 2010) (cont.)

No	Pesticides	SPE	Method	Sample	Youden pair	Nov.9 2009 (n=5)	Nov.14 2009 (n=3)	Nov.19 2009 (n=3)	Nov.24 2009 (n=3)	Nov.29 2009 (n=3)	Dec.4 2009 (n=3)	Dec.9 2009 (n=3)	Dec.14 2009 (n=3)	Dec.19 2009 (n=3)	Dec.24 2009 (n=3)	Dec.14 2009 (n=3)	Jan.3 2010 (n=3)	Jan.8 2010 (n=3)	Jan.13 2010 (n=3)	Jan.18 2010 (n=3)	Jan.23 2010 (n=3)	Jan.28 2010 (n=3)	Feb.2 2010 (n=3)	Feb.7 2010 (n=3)
178	piperonylbutoxide	Cleanert TPT	GC-MS	Green tea	A	3.8	4.9	0.8	1.8	8.6	4.5	7.8	7.5	12.2	8.3	9.1	9.9	7.4	4.2	4.2	3.0	3.9	6.0	4.8
					B	4.8	10.5	5.6	8.8	3.5	4.0	2.4	3.1	1.7	7.5	2.7	15.1	3.6	1.0	6.0	3.8	4.8	2.0	3.7
				Woolong tea	A*	6.1	6.4	4.5	2.6	2.5	4.0	7.6	6.9	5.6	13.4	19.1	9.2	4.9	2.3	6.9	3.7	1.9	8.5	1.7
					B*	3.7	6.0	9.1	7.8	6.8	4.9	5.5	4.2	8.2	13.0	23.2	2.8	1.6	4.3	1.4	0.9	1.0	2.1	0.9
			GC-MS/MS	Green tea	A	5.3	3.3	1.0	11.7	8.9	4.7	8.3	4.0	2.9	0.9	5.7	10.5	10.3	5.5	5.8	0.6	5.7	5.2	2.6
					B	4.7	9.1	7.0	8.1	3.0	3.2	2.2	2.9	4.1	8.5	3.6	4.4	6.1	1.6	3.8	3.8	2.9	3.3	5.4
				Woolong tea	A*	6.9	3.3	0.8	2.5	2.9	3.5	9.0	2.6	2.6	5.0	1.1	2.5	6.4	1.7	7.5	5.5	2.1	5.0	4.9
					B*	5.1	4.0	3.1	7.6	3.4	4.3	4.9	7.6	7.3	1.4	5.9	12.1	3.5	3.5	5.0	1.6	1.3	1.4	3.2
		Envi-Carb+PSA	GC-MS	Green tea	A	9.3	6.2	2.8	7.6	9.4	2.6	3.4	1.6	11.9	2.6	4.9	10.8	2.7	1.0	6.9	7.7	13.2	3.2	6.9
					B	6.3	9.5	4.7	6.2	3.8	7.3	5.0	6.9	5.0	6.8	35.7	3.9	1.8	4.5	11.5	3.6	3.8	7.3	6.8
				Woolong tea	A*	4.8	4.6	6.5	1.9	8.0	0.8	3.3	6.6	3.6	2.7	10.2	7.0	4.2	6.4	1.0	4.5	4.5	4.7	4.7
					B*	3.9	0.9	10.6	1.5	18.8	4.7	1.7	8.2	12.5	2.9	16.7	2.3	2.4	4.6	10.5	3.6	5.4	5.4	3.4
			GC-MS/MS	Green tea	A	6.5	2.9	3.9	8.3	7.9	0.8	5.1	2.5	5.1	2.4	5.9	0.4	7.7	4.6	5.5	3.4	1.5	2.8	6.3
					B	3.2	9.4	4.3	4.8	1.7	0.6	2.1	6.3	3.7	0.6	21.0	3.4	3.4	2.3	12.0	4.0	4.5	9.0	6.3
				Woolong tea	A*	6.0	2.9	6.0	1.9	3.7	2.2	3.1	3.7	1.8	2.5	7.2	3.9	19.9	0.8	8.1	3.3	3.0	5.6	4.9
					B*	5.5	3.7	9.8	5.0	10.1	4.3	4.2	9.6	8.7	3.9	5.4	7.5	4.6	7.2	9.1	7.0	6.3	5.1	3.2
179	pretilachlor	Cleanert TPT	GC-MS	Green tea	A	6.8	5.7	3.8	4.6	5.6	6.0	4.6	11.1	4.5	3.4	1.4	10.9	11.0	5.0	8.5	5.0	9.0	5.6	6.4
					B	4.8	6.2	5.9	10.0	4.0	7.2	3.8	5.9	5.8	2.6	0.7	10.9	5.2	3.1	9.7	3.8	2.7	2.3	4.9
				Woolong tea	A*	6.4	5.2	2.3	2.4	2.7	2.6	13.9	11.9	2.7	4.5	3.0	8.6	6.3	6.1	9.4	2.8	4.1	7.2	7.8
					B*	4.8	3.1	4.9	1.8	3.0	4.4	2.8	3.7	5.7	10.5	8.2	3.5	6.6	7.0	2.3	6.5	2.5	13.1	12.6
			GC-MS/MS	Green tea	A	8.8	3.8	0.7	32.2	7.3	3.9	6.4	5.9	4.2	5.8	2.3	18.4	8.9	7.3	2.0	7.0	7.9	1.1	5.7
					B	4.6	10.1	4.5	7.4	2.9	5.9	5.6	4.0	16.4	4.4	1.5	5.2	14.6	2.3	6.9	3.0	1.9	3.1	4.2
				Woolong tea	A*	5.5	5.1	3.4	3.3	6.0	0.4	13.5	8.3	5.8	6.0	7.4	3.2	6.3	4.0	10.0	6.3	5.0	1.1	4.4
					B*	6.6	6.0	8.6	1.1	2.8	2.0	6.4	6.2	12.0	1.8	9.8	9.5	12.5	0.8	8.1	1.4	2.2	13.8	9.6
		Envi-Carb+PSA	GC-MS	Green tea	A	7.1	1.0	2.6	2.5	10.3	5.9	9.2	5.4	4.9	5.9	15.6	8.7	4.5	6.9	12.6	3.8	8.9	7.0	3.8
					B	11.1	8.9	2.4	5.2	4.9	4.1	3.6	1.7	8.0	3.4	8.3	4.9	1.0	2.8	8.0	5.0	2.7	5.6	6.4
				Woolong tea	A*	5.7	1.1	5.5	3.0	5.7	1.5	6.9	5.4	2.3	2.6	8.8	2.7	7.8	6.5	12.6	3.8	4.1	2.3	4.9
					B*	6.5	3.8	7.5	5.7	6.6	4.9	4.1	6.5	10.4	4.5	13.1	1.7	6.9	13.4	10.7	2.8	2.5	7.2	7.8
			GC-MS/MS	Green tea	A	11.7	4.6	4.7	4.7	7.2	7.7	9.8	1.2	6.4	10.5	10.9	1.3	6.3	19.2	4.7	6.5	9.7	13.1	12.6
					B	17.0	6.0	0.8	4.4	4.8	7.6	14.6	1.8	7.1	5.8	4.7	17.0	6.3	28.6	3.2	7.0	6.7	7.9	5.7
				Woolong tea	A*	3.7	1.5	6.4	10.7	6.9	3.5	3.1	8.0	4.2	2.9	11.4	10.8	2.0	0.4	14.5	3.0	5.6	1.9	4.2
					B*	8.7	3.2	14.7	9.6	4.1	4.0	3.0	7.2	8.3	2.9	10.1	4.9	15.3	10.4	11.8	0.1	5.6	6.9	4.4
180	prometon	Cleanert TPT	GC-MS	Green tea	A	5.2	5.0	3.5	4.0	8.4	4.1	8.3	5.0	4.6	4.6	10.1	4.9	7.3	3.9	1.9	3.0	4.8	3.3	4.2
					B	5.9	9.0	7.0	7.6	4.2	2.5	5.6	2.5	3.7	10.6	8.7	11.0	2.4	1.2	4.8	5.3	3.1	5.2	1.7
				Woolong tea	A*	6.8	1.0	1.8	4.0	2.1	3.9	21.7	6.3	2.4	8.9	9.3	4.0	3.4	5.5	8.7	1.5	6.4	1.5	0.8
					B*	3.6	2.3	3.5	6.8	4.0	4.3	5.3	4.3	7.0	7.5	10.4	3.1	2.0	1.4	4.5	0.8	9.2	4.0	2.8
			GC-MS/MS	Green tea	A	4.8	9.6	8.6	8.1	3.4	3.6	4.9	5.6	3.2	3.8	3.9	8.7	2.4	5.5	3.6	4.0	5.6	3.4	5.6
					B	5.0	3.0	3.1	3.6	3.4	1.8	10.4	1.0	2.0	2.0	0.4	1.9	8.7	1.4	6.1	2.8	2.6	3.4	5.9
				Woolong tea	A*	4.3	4.6	4.5	7.6	1.7	1.5	5.3	6.6	4.9	5.9	3.9	3.1	9.7	3.3	3.3	1.9	3.3	2.2	5.3
					B*	5.1	3.4	0.6	6.0	9.9	4.3	4.5	1.0	4.0	6.2	7.8	16.0	2.1	2.6	8.6	7.0	9.2	4.8	5.4
		Envi-Carb+PSA	GC-MS	Green tea	A	5.4	9.5	4.2	8.3	1.9	3.0	5.2	4.2	8.4	4.9	29.5	16.6	1.8	3.9	12.9	1.9	4.8	10.5	6.2
					B	3.9	6.5	4.0	4.1	6.6	6.8	4.1	3.3	5.2	4.9	4.3	5.0	2.2	9.0	3.0	3.0	4.4	2.1	4.4
				Woolong tea	A*	1.4	2.9	9.2	2.8	15.6	1.8	2.2	8.3	13.5	6.8	6.5	2.8	6.4	7.1	9.3	6.5	5.2	5.2	4.1
					B*	6.4	4.5	1.6	7.6	7.9	0.6	4.7	2.6	3.1	1.0	7.6	16.8	2.0	9.5	7.5	7.5	1.5	1.7	4.3
			GC-MS/MS	Green tea	A	7.3	10.2	4.9	3.4	1.9	3.5	3.3	6.9	4.5	2.4	22.3	16.4	4.9	30.2	14.6	2.5	11.0	2.5	6.7
					B	4.5	5.2	5.3	3.9	5.8	3.7	3.7	4.0	2.5	1.8	5.3	7.2	5.1	3.6	12.6	4.3	2.5	5.9	5.9
				Woolong tea	A*	2.6	5.2	5.3	3.9	5.8	3.5	3.7	4.0	2.5	1.8	5.3	7.2	5.1	3.6	12.6	4.3	2.3	11.0	5.9
					B*	5.7	3.9	10.8	5.1	8.9	6.4	3.9	8.0	10.0	4.1	7.0	4.6	5.2	10.1	11.4	2.1	3.5	3.5	5.3

No.	Compound	Cartridge	Method	Tea	Rep																			
181	pronamide	Cleanert TPT	GC-MS	Green tea	A	6.0	3.3	2.1	2.6	7.7	2.9	6.4	5.4	5.2	2.1	6.9	2.5	7.8	3.8	2.3	3.4	5.4	5.4	2.8
					B	4.2	8.5	9.4	11.9	3.7	1.3	3.1	5.4	3.1	6.9	4.1	10.0	2.1	4.0	5.5	3.0	4.4	1.3	5.4
				Woolong tea	A	11.1	5.5	2.9	5.0	3.1	7.6	4.2	8.9	4.7	7.8	1.3	3.7	0.6	3.4	6.7	1.1	0.8	6.0	2.4
					B	8.4	8.3	11.4	12.4	2.7	2.1	5.3	6.9	3.4	3.3	3.1	2.0	0.6	8.0	5.4	3.9	4.8	3.6	4.5
			GC-MS/MS	Green tea	A	4.7	4.7	2.0	9.9	9.9	3.7	6.7	5.3	4.1	1.7	4.6	3.1	13.4	3.6	1.7	2.1	6.6	4.4	3.0
					B	3.0	10.3	7.9	7.3	3.6	3.4	3.7	8.6	6.5	8.5	1.9	5.8	5.6	2.6	4.5	3.6	5.2	3.2	6.8
				Woolong tea	A	8.0	6.5	2.6	7.8	7.6	8.8	5.9	0.6	4.4	5.3	3.6	1.9	4.8	0.5	6.8	1.6	3.7	5.1	7.4
					B	5.2	3.5	4.5	2.4	0.8	0.4	1.6	9.2	6.5	6.8	4.9	3.2	6.4	7.5	10.6	5.8	5.3	5.6	4.8
		Envi-Carb+PSA	GC-MS	Green tea	A*	7.4	5.5	3.6	14.9	9.8	0.6	10.2	2.9	4.8	5.3	6.9	8.5	2.0	2.8	4.3	6.2	10.2	3.8	6.2
					B	4.1	10.8	4.2	14.4	6.9	8.8	5.2	2.6	5.9	1.5	24.0	10.6	3.3	2.7	11.7	1.8	4.2	9.4	10.1
				Woolong tea	A	15.7	11.1	7.5	11.4	5.6	4.1	4.6	9.1	2.6	3.4	4.5	3.5	1.4	4.7	2.6	4.7	4.3	9.3	7.7
					B	7.1	5.0	8.1	2.9	12.9	6.0	5.5	4.3	13.1	9.0	8.7	2.0	3.0	9.1	9.6	3.7	7.1	11.5	4.8
			GC-MS/MS	Green tea	A*	8.1	3.4	3.8	6.3	6.7	2.8	8.5	4.0	3.1	2.9	6.5	4.8	1.6	3.3	7.2	4.4	7.8	1.3	5.3
					B*	5.5	7.6	4.8	2.3	1.8	1.2	5.0	3.0	4.4	0.5	20.6	3.0	3.4	11.7	10.9	1.0	1.4	10.6	7.3
				Woolong tea	A	15.5	1.5	2.3	9.5	4.5	2.7	4.7	8.2	4.5	2.0	3.9	5.9	15.6	1.6	12.2	4.3	4.3	4.6	7.2
					B	8.6	3.3	14.6	2.7	8.5	6.2	4.2	5.7	12.4	12.6	4.9	4.4	6.1	10.4	12.1	1.0	4.7	13.5	4.8
182	propetam-phos	Cleanert TPT	GC-MS	Green tea	A	4.6	4.8	6.7	7.6	13.0	3.0	10.2	14.4	7.6	3.0	10.7	9.8	8.9	11.5	6.2	0.9	9.4	8.7	2.5
					B	7.1	7.8	4.6	14.5	7.3	10.2	4.5	6.9	2.6	1.1	8.8	10.3	8.6	6.4	6.9	3.3	11.2	4.0	5.9
				Woolong tea	A*	5.6	4.4	9.8	7.3	6.1	5.4	3.6	7.5	2.7	10.5	9.7	11.6	2.7	3.6	6.2	4.2	3.3	2.1	1.5
					B	4.9	6.3	6.8	6.8	7.4	6.1	6.5	7.9	5.3	4.8	2.3	11.6	2.6	6.5	5.4	3.8	1.1	0.9	4.0
			GC-MS/MS	Green tea	A*	5.0	5.0	4.3	5.3	15.6	1.3	9.8	11.4	6.7	2.7	6.0	3.8	10.4	11.9	4.6	0.4	8.8	9.2	0.9
					B*	5.1	7.9	5.6	9.3	4.0	1.6	4.1	9.3	6.6	8.7	1.0	2.4	10.6	2.1	4.0	2.8	10.0	2.8	9.3
				Woolong tea	A	5.0	0.2	3.1	2.2	4.5	3.7	5.0	6.8	11.7	3.5	7.0	4.9	4.8	3.0	11.5	3.2	3.4	5.9	2.4
					B	3.4	1.5	5.2	7.2	6.9	7.1	1.9	6.2	10.3	2.5	1.6	1.1	6.5	2.3	9.0	3.4	2.8	4.7	4.3
		Envi-Carb+PSA	GC-MS	Green tea	A*	18.0	10.6	13.7	6.3	9.9	17.0	26.5	12.1	12.5	10.5	8.2	17.8	5.1	14.5	7.8	4.1	11.3	25.0	5.9
					B*	20.0	28.2	1.4	3.0	11.0	15.1	9.8	14.0	8.7	2.9	25.3	3.5	22.2	5.5	18.4	18.5	4.3	22.8	8.6
				Woolong tea	A	9.4	5.7	4.1	5.7	7.3	1.8	4.3	22.2	12.6	3.1	8.6	4.1	5.1	2.5	1.8	4.8	7.3	8.2	1.8
					B*	12.8	2.4	11.3	3.1	21.4	6.9	5.7	6.2	12.4	2.0	8.2	3.5	3.5	7.7	11.0	10.8	9.7	9.9	5.4
			GC-MS/MS	Green tea	A*	21.6	7.9	9.9	4.0	7.5	20.3	14.2	16.1	4.8	1.6	3.4	33.5	5.4	11.1	6.7	2.2	9.3	9.9	6.5
					B*	25.5	10.9	3.2	12.0	8.9	3.6	11.0	12.6	11.2	10.4	11.4	39.7	18.6	14.4	5.1	7.0	6.9	17.0	10.9
				Woolong tea	A	7.6	2.0	6.0	7.4	3.1	4.5	4.4	13.3	2.8	3.8	6.8	7.1	4.7	3.8	10.3	17.9	7.5	9.5	2.5
					B	11.3	3.4	8.8	3.9	12.0	7.7	3.2	5.5	9.6	2.9	8.7	6.2	5.6	8.9	10.6	5.3	5.8	17.6	3.8
183	propoxur-1	Cleanert TPT	GC-MS	Green tea	A	6.2	7.7	2.2	6.7	5.7	4.5	8.2	3.6	6.5	6.8	5.3	0.8	7.6	4.2	2.4	2.0	2.6	8.6	4.4
					B	5.7	11.7	5.6	8.7	2.6	3.7	3.4	11.4	5.0	6.3	5.6	5.8	2.9	2.4	5.1	4.9	7.6	2.6	5.8
				Woolong tea	A	2.0	2.5	2.4	5.7	3.8	3.3	10.1	8.8	4.4	4.2	2.0	4.4	5.4	2.0	9.3	1.7	4.0	4.9	1.0
					B	7.1	3.3	2.9	6.6	1.9	5.0	8.3	3.2	5.0	4.2	0.7	3.2	3.9	2.1	1.5	4.4	1.2	2.6	1.3
			GC-MS/MS	Green tea	A	5.6	7.0	1.3	2.3	9.2	5.3	7.2	2.7	4.5	4.0	5.2	1.6	10.3	10.2	2.0	0.1	4.9	6.4	2.2
					B*	6.0	13.6	8.9	6.8	4.5	4.5	3.9	9.6	5.5	10.7	4.6	6.0	5.1	4.6	6.0	3.5	4.5	5.7	6.5
				Woolong tea	A	3.9	6.8	3.3	6.5	6.6	2.4	10.1	5.5	2.7	4.8	0.5	3.9	6.2	1.3	8.6	3.5	2.0	3.4	5.0
					B	6.7	5.5	3.0	8.7	1.1	3.1	7.4	8.4	15.4	2.3	7.8	2.6	11.8	1.5	5.8	3.7	1.8	0.9	4.0
		Envi-Carb+PSA	GC-MS	Green tea	A	4.8	10.7	0.6	6.4	7.0	5.7	23.8	3.0	5.0	4.9	7.9	5.8	2.9	3.3	5.6	7.7	12.9	3.1	4.1
					B	3.9	5.4	4.5	8.1	3.7	4.0	3.4	7.0	3.9	0.5	21.4	1.4	2.6	2.7	13.9	3.2	3.4	7.3	4.8
				Woolong tea	A	5.8	4.2	1.3	3.5	6.8	4.0	4.0	5.3	2.2	8.8	3.4	7.1	0.9	3.1	2.6	5.4	5.2	0.7	5.5
					B	9.1	3.0	12.1	5.8	11.4	8.1	0.8	9.4	7.7	5.7	5.5	5.7	1.0	17.9	8.1	3.2	3.7	6.6	4.1
			GC-MS/MS	Green tea	A	4.5	10.0	3.3	8.0	3.9	5.5	20.1	1.3	3.8	1.9	7.0	5.3	7.8	7.3	5.9	7.6	13.5	4.1	2.9
					B	3.0	5.4	6.3	5.6	1.0	1.9	1.1	7.8	2.7	1.7	24.6	11.1	3.6	16.6	17.0	4.6	2.9	7.9	4.4
				Woolong tea	A	6.0	6.0	3.3	1.7	6.4	1.2	4.3	2.8	2.1	8.8	6.5	12.4	2.9	6.8	11.5	6.4	4.2	1.7	5.0
					B	6.6	2.7	11.9	9.5	6.7	8.6	4.2	8.3	7.5	3.2	5.6	2.3	8.7	15.7	9.6	4.7	4.2	5.7	4.0

(Continued)

Annex 2: RSD data for 201 pesticides in incurred tea Youden pair samples under 8 conditions with two SPE cartridge cleanup in three months (Nov.9, 2009-Feb.7, 2010) (cont.)

No	Pesticides	SPE	Method	Sample	Youden pair	Nov.9, 2009 (n=5)	Nov.14, 2009 (n=3)	Nov.19, 2009 (n=3)	Nov.24, 2009 (n=3)	Nov.29, 2009 (n=3)	Dec.4, 2009 (n=3)	Dec.9, 2009 (n=3)	Dec.14, 2009 (n=3)	Dec.19, 2009 (n=3)	Dec.24, 2009 (n=3)	Dec.29, 2009 (n=3)	Jan.3, 2010 (n=3)	Jan.8, 2010 (n=3)	Jan.13, 2010 (n=3)	Jan.18, 2010 (n=3)	Jan.23, 2010 (n=3)	Jan.28, 2010 (n=3)	Feb.2, 2010 (n=3)	Feb.7, 2010 (n=3)
184	propoxur-2	Cleanert TPT	GC-MS	Green tea	A	15.5	4.8	2.7	6.9	14.7	7.7	19.1	4.3	6.8	11.6	7.5	4.3	8.8	5.1	13.5	6.8	3.5	5.3	5.5
				Green tea	B	14.3	12.7	3.1	11.6	10.8	9.6	13.2	8.1	5.0	12.9	4.1	13.1	1.1	7.4	6.8	3.8	9.8	10.2	5.8
				Woolong tea	A	16.4	1.1	8.3	1.0	4.7	3.5	20.9	13.0	14.9	6.3	19.1	14.2	9.3	3.4	9.9	6.2	5.3	3.7	5.6
				Woolong tea	B	14.7	2.1	11.6	6.3	4.9	3.7	12.0	0.9	12.8	7.4	1.2	17.1	2.5	1.6	1.1	1.5	2.0	5.4	11.2
			GC-MS/MS	Green tea	A*	6.8	9.7	9.8	4.3	15.9	8.1	22.4	37.9	31.8	5.4	4.8	8.4	14.9	15.2	0.8	8.5	14.4	28.9	7.3
				Green tea	B*	9.1	5.1	11.2	17.3	5.5	3.9	4.6	24.9	16.9	10.8	3.7	8.2	21.4	7.0	8.9	3.4	19.6	16.2	12.8
				Woolong tea	A*	8.4	17.7	12.4	6.1	5.6	6.9	4.6	39.3	4.5	8.0	15.4	6.3	14.2	8.8	8.7	11.3	8.4	18.2	10.0
				Woolong tea	B*	8.9	7.5	20.9	6.3	5.2	26.7	2.5	3.5	28.2	11.3	4.2	6.1	8.9	10.2	18.2	5.5	3.6	8.9	9.9
		Envi-Carb+PSA	GC-MS	Green tea	A	28.1	10.8	19.2	28.3	4.0	4.8	24.3	51.5	3.8	6.7	10.0	12.8	2.0	10.2	6.5	13.4	14.4	17.7	15.2
				Green tea	B	60.4	13.6	1.3	35.5	13.7	11.8	4.5	37.0	2.8	21.6	22.1	11.3	11.1	0.8	21.7	5.6	5.7	11.3	14.0
				Woolong tea	A	8.3	6.8	5.9	19.4	8.3	3.4	6.1	41.7	10.8	10.2	11.5	4.3	14.0	7.5	15.4	17.8	10.8	14.1	23.9
				Woolong tea	B	13.4	8.1	9.1	10.9	10.4	11.1	10.4	7.8	30.7	13.5	6.9	6.1	18.2	18.0	24.2	7.4	18.7	33.8	16.4
			GC-MS/MS	Green tea	A*	21.0	23.8	9.5	8.6	17.7	26.2	40.9	56.2	39.0	1.4	10.0	38.8	13.1	28.0	58.1	19.5	58.1	24.8	23.3
				Green tea	B*	43.9	10.0	11.6	33.0	5.3	12.9	40.1	60.9	50.9	26.9	25.3	21.7	43.6	40.3	35.8	43.2	28.0	53.6	37.4
				Woolong tea	A*	17.2	9.2	11.5	23.9	4.5	5.9	13.5	47.3	5.7	12.1	4.1	8.5	10.8	16.7	19.2	14.1	16.0	18.7	10.0
				Woolong tea	B*	30.5	5.0	18.2	4.3	16.7	11.0	11.6	8.3	18.2	3.4	8.1	21.4	18.6	18.5	15.2	11.9	4.5	48.3	9.9
185	prothiophos	Cleanert TPT	GC-MS	Green tea	A	4.2	5.5	4.5	0.9	8.2	3.7	6.0	8.3	4.9	6.0	7.6	11.2	7.1	5.3	1.0	1.5	3.5	5.1	2.6
				Green tea	B	5.7	7.7	3.4	7.6	5.0	2.1	1.8	3.9	3.1	10.9	2.5	13.5	0.8	1.1	5.5	1.9	3.0	1.9	5.8
				Woolong tea	A	6.0	2.2	6.4	3.6	0.6	4.0	8.1	8.3	2.1	9.4	6.9	6.3	1.7	3.9	9.6	3.0	3.7	2.2	1.1
				Woolong tea	B	3.7	1.7	5.5	6.7	4.0	3.6	4.1	2.1	7.1	4.6	12.6	1.9	2.3	3.7	4.9	1.0	0.9	0.8	3.0
			GC-MS/MS	Green tea	A*	4.2	2.5	5.2	4.1	9.7	2.4	9.9	5.2	4.7	2.7	5.1	1.4	7.3	6.6	8.8	2.8	7.0	4.2	5.0
				Green tea	B*	6.1	7.2	7.8	9.4	5.0	6.2	4.7	3.3	7.4	11.7	4.9	6.2	4.9	3.5	4.3	3.5	7.4	4.5	1.5
				Woolong tea	A*	6.8	1.9	2.7	4.5	4.2	2.6	11.3	2.4	5.9	3.6	1.8	2.7	8.2	5.2	2.2	9.8	3.7	2.2	5.3
				Woolong tea	B*	5.2	6.2	6.8	7.3	2.2	3.8	3.3	7.7	7.4	2.6	4.7	3.4	10.1	3.7	8.7	2.2	3.6	1.7	4.7
		Envi-Carb+PSA	GC-MS	Green tea	A	10.2	3.2	2.6	4.1	9.1	6.5	5.4	3.3	6.5	4.0	4.9	5.1	0.8	1.8	2.9	3.0	8.3	8.7	7.2
				Green tea	B	8.8	10.7	3.8	3.8	1.6	6.0	6.6	2.8	8.7	3.9	29.0	14.1	4.1	1.2	9.2	7.4	4.3	11.6	8.8
				Woolong tea	A	1.5	2.9	4.4	2.0	4.8	1.1	3.6	4.8	2.7	1.6	5.9	5.5	0.4	3.5	3.0	1.6	5.3	4.7	4.1
				Woolong tea	B	5.0	2.7	9.8	3.9	14.9	4.5	2.4	5.5	11.5	1.6	12.3	8.5	4.3	7.1	10.4	1.8	6.5	1.4	3.7
			GC-MS/MS	Green tea	A*	13.0	4.9	2.7	5.0	6.5	7.9	8.5	5.7	8.2	0.3	8.6	11.4	10.0	10.4	2.7	2.7	8.3	1.6	3.9
				Green tea	B*	11.5	11.8	3.5	7.0	5.8	5.5	7.6	1.1	7.4	4.6	12.9	18.2	3.7	18.7	8.9	7.8	5.8	9.5	9.6
				Woolong tea	A*	2.9	1.8	5.8	4.3	1.9	0.7	2.6	6.6	1.7	3.3	2.6	3.9	8.6	3.7	11.7	0.4	5.4	7.3	5.3
				Woolong tea	B*	8.8	6.0	11.3	2.7	8.6	5.0	4.5	7.8	10.0	3.5	7.9	9.1	4.7	8.8	7.8	4.0	3.8	2.6	4.7
186	pyridaben	Cleanert TPT	GC-MS	Green tea	A	18.4	11.1	37.8	10.4	13.2	3.1	12.1	13.2	14.6	2.0	11.2	7.4	8.1	6.5	2.0	13.9	5.5	21.7	3.3
				Green tea	B	12.3	10.9	23.6	9.9	10.2	8.3	3.3	12.6	1.8	10.7	3.8	14.9	11.7	13.4	9.0	14.0	13.6	13.1	5.7
				Woolong tea	A	34.5	7.6	7.5	6.2	16.5	7.3	10.1	16.4	9.3	8.8	13.2	11.9	5.3	2.4	3.2	9.3	5.3	3.0	3.3
				Woolong tea	B	22.2	4.7	5.7	3.4	9.6	12.1	7.1	2.3	9.2	0.5	15.1	7.1	2.4	7.4	6.7	1.6	5.4	6.1	8.2
			GC-MS/MS	Green tea	A*	7.8	11.2	3.2	3.6	13.8	9.1	5.9	7.6	3.3	6.6	1.9	9.9	1.6	3.5	3.6	1.1	9.2	10.2	2.2
				Green tea	B*	8.3	7.2	8.3	9.9	5.9	8.8	3.1	3.7	13.4	8.0	0.4	10.8	2.1	5.1	5.3	2.2	8.0	4.1	6.8
				Woolong tea	A*	10.3	1.2	4.6	4.6	8.6	15.9	19.8	0.3	20.4	7.2	4.4	5.1	8.4	4.7	7.6	0.6	6.6	2.6	19.5
				Woolong tea	B*	11.2	5.1	9.0	7.0	10.6	2.3	9.9	1.6	7.4	2.0	7.4	3.2	13.3	5.6	7.3	3.9	4.1	5.9	6.3
		Envi-Carb+PSA	GC-MS	Green tea	A	42.3	11.8	15.5	5.4	7.3	27.1	13.0	21.2	25.4	8.3	5.3	26.6	18.0	14.3	9.7	11.3	13.5	18.5	5.7
				Green tea	B	13.1	10.0	7.7	6.2	14.3	11.5	26.9	12.4	10.1	15.6	23.0	19.3	35.3	7.0	11.8	2.3	7.3	12.8	13.7
				Woolong tea	A	10.1	13.1	11.7	8.0	8.8	8.2	11.5	17.3	6.5	3.1	2.4	19.0	11.4	3.9	4.1	30.6	12.0	19.6	11.1
				Woolong tea	B	12.4	7.6	4.2	11.9	18.3	13.2	19.7	27.5	14.2	8.0	11.1	6.8	3.8	27.7	15.7	3.5	17.7	18.4	8.7
			GC-MS/MS	Green tea	A*	7.4	4.0	2.6	10.0	5.8	9.2	9.5	23.1	11.0	16.7	14.2	4.1	0.7	9.2	3.3	8.8	9.2	3.8	7.1
				Green tea	B*	13.1	5.3	11.0	0.8	2.6	4.1	6.5	26.5	1.4	11.9	18.3	3.0	8.9	14.1	11.1	3.9	8.3	7.5	3.6
				Woolong tea	A*	67.8	8.8	8.6	23.5	10.4	7.8	8.5	16.1	2.3	2.6	7.0	6.0	17.3	5.7	5.0	10.8	5.0	7.9	19.7
				Woolong tea	B*	24.1	6.9	10.8	16.6	8.5	3.4	4.4	6.3	12.9	6.5	6.1	7.6	9.5	7.7	4.3	5.9	6.3	7.0	6.3

No.	Compound	Cartridge	Method	Tea																				
187	pyridaphen-thion	Cleanert TPT	GC-MS	Green tea	A	6.8	8.2	0.9	4.6	6.3	3.7	5.3	11.1	8.7	5.8	9.2	12.1	8.1	8.0	4.7	2.5	14.5	14.9	4.6
					B	5.4	5.3	5.9	7.5	2.3	2.6	1.0	4.4	6.4	8.5	3.5	15.1	10.3	3.1	3.2	1.4	5.7	14.4	3.6
				Woolong tea	A*	9.1	5.6	11.8	5.3	7.0	4.1	0.7	14.9	11.5	10.4	16.6	13.6	3.1	7.7	2.4	10.0	4.5	—	8.4
					B	4.6	1.8	9.5	8.1	8.2	8.5	4.8	5.8	8.7	2.4	18.8	8.9	5.1	5.3	2.4	4.9	7.2	4.9	11.5
			GC-MS/MS	Green tea	A	6.5	5.2	1.7	12.4	12.2	1.5	9.1	11.4	10.4	11.8	9.9	4.2	13.7	10.8	3.9	1.6	6.9	7.6	9.9
					B	7.4	13.6	3.8	6.5	5.1	1.4	2.9	12.1	12.4	11.3	4.8	19.9	10.4	4.1	9.0	5.6	1.9	2.1	4.3
				Woolong tea	A*	7.4	4.1	7.5	2.2	3.0	4.2	6.0	7.1	8.2	3.1	7.2	14.4	10.7	4.4	5.0	3.2	0.3	3.6	1.1
					B*	4.4	3.7	7.5	7.3	8.8	3.3	5.5	16.5	11.7	2.7	4.7	2.9	13.4	4.3	7.0	5.5	6.6	5.8	3.2
		Envi-Carb+PSA	GC-MS	Green tea	A*	10.7	5.1	10.1	6.1	12.6	13.9	13.6	7.0	19.9	8.9	4.3	7.9	5.4	4.7	5.0	9.1	13.6	29.4	9.5
					B*	13.8	7.4	2.9	5.4	0.3	5.4	3.2	10.8	8.1	8.1	33.0	13.8	15.6	6.0	7.0	2.4	4.1	6.0	6.8
				Woolong tea	A	6.3	5.0	7.2	1.2	10.0	3.4	5.0	17.3	1.4	7.5	14.7	4.3	1.0	5.7	5.0	7.5	0.9	12.9	3.4
					B	7.8	4.9	10.1	4.8	10.8	9.2	2.8	0.9	11.3	11.9	12.9	14.3	8.8	3.5	7.0	4.9	6.1	8.3	7.4
			GC-MS/MS	Green tea	A	12.5	8.8	4.3	7.5	9.9	15.1	15.4	11.3	8.1	1.6	8.4	1.0	1.4	15.7	4.9	7.1	7.7	15.3	7.4
					B	15.6	8.8	3.9	9.3	3.5	8.8	7.7	13.6	17.1	12.9	20.1	10.3	9.3	24.1	12.2	6.2	5.6	18.0	17.4
				Woolong tea	A	4.9	2.0	9.2	5.1	2.3	3.0	2.8	20.0	1.9	12.4	7.2	18.6	25.4	3.1	7.6	4.4	10.7	16.8	1.1
					B	11.7	6.2	11.7	5.2	12.6	8.7	5.6	5.5	9.3	7.8	6.8	16.8	9.5	7.6	7.8	5.5	15.5	13.2	3.2
188	pyrimethanil	Cleanert TPT	GC-MS	Green tea	A	4.8	5.2	4.5	3.1	9.5	3.6	7.0	3.2	5.0	3.5	7.3	3.4	7.5	3.2	4.3	3.3	6.5	7.0	3.7
					B	5.1	8.8	5.6	8.7	3.4	2.8	3.0	3.0	3.2	9.4	3.0	10.3	1.8	4.7	7.5	2.0	7.2	1.2	4.7
				Woolong tea	A	3.5	1.3	3.5	3.9	2.5	3.6	8.2	6.4	1.4	6.9	13.5	3.7	2.8	5.1	2.4	4.7	3.5	3.8	0.9
					B	4.1	2.2	2.9	6.2	2.2	3.7	3.2	3.5	2.9	7.0	6.4	0.9	5.1	7.3	4.9	1.7	0.8	1.6	2.0
			GC-MS/MS	Green tea	A	5.8	3.9	3.8	2.8	11.7	4.9	6.5	3.6	5.2	0.4	3.0	10.4	7.3	3.6	2.4	1.9	3.4	2.7	8.3
					B	4.8	10.6	7.7	8.2	4.8	4.4	3.7	4.6	4.2	9.7	6.4	3.2	1.6	2.8	4.9	1.9	6.7	3.8	8.0
				Woolong tea	A	5.4	1.4	1.9	3.9	5.8	3.2	8.9	2.6	2.8	4.1	3.0	3.9	7.1	5.9	5.8	2.3	4.4	3.6	3.6
					B	5.5	1.8	2.4	6.0	2.3	2.1	4.6	8.6	7.3	1.6	2.4	3.9	10.9	4.1	11.5	1.9	11.7	3.8	5.5
		Envi-Carb+PSA	GC-MS	Green tea	A*	4.4	8.0	4.8	7.1	9.3	5.4	3.8	4.4	4.2	5.3	25.0	8.1	2.4	3.3	6.4	3.7	6.1	8.1	7.2
					B	2.9	5.0	3.1	1.4	3.0	3.8	3.9	4.4	2.2	6.4	6.4	11.2	4.0	2.4	3.7	4.3	9.1	3.5	3.8
				Woolong tea	A	4.1	3.0	10.3	3.5	5.5	4.7	2.0	6.7	7.5	7.4	2.5	1.3	2.9	3.9	6.9	6.9	4.6	5.8	5.5
					B	7.6	5.2	1.8	8.8	13.6	1.8	5.5	1.7	4.1	1.4	8.4	1.3	1.8	8.0	6.5	6.5	7.9	4.8	4.6
			GC-MS/MS	Green tea	A	4.1	7.2	4.5	3.3	6.5	0.9	2.7	5.9	1.1	0.2	21.4	13.4	4.1	3.5	7.1	2.5	7.1	6.2	6.8
					B	2.2	5.3	5.0	1.8	1.2	1.8	6.8	3.4	3.6	7.6	5.8	4.3	5.3	2.9	20.0	3.2	5.3	6.9	8.0
				Woolong tea	A	4.0	0.4	11.0	6.7	4.4	2.6	5.8	8.7	7.2	1.9	5.8	2.7	4.2	3.6	8.5	2.1	3.2	2.4	5.1
					B	5.0	5.1	0.5	1.2	8.7	6.3	9.9	3.4	4.3	1.9	8.6	1.8	10.9	3.9	14.5	3.2	2.1	7.8	3.8
189	pyriproxyfen	Cleanert TPT	GC-MS	Green tea	A	4.1	7.7	4.3	7.3	5.6	3.3	5.3	4.4	4.4	9.7	6.2	1.8	5.2	1.6	3.6	0.5	4.3	2.3	3.7
					B	7.3	4.5	2.0	4.2	2.4	2.9	5.1	1.1	1.3	5.8	5.5	10.0	2.6	4.9	3.9	2.2	7.8	5.5	0.3
				Woolong tea	A	1.9	2.4	1.9	8.4	3.0	4.0	0.0	5.9	6.4	3.9	3.5	4.7	4.9	0.3	5.1	4.7	1.7	4.8	1.5
					B	7.4	10.7	3.1	4.3	13.2	10.3	6.2	0.2	11.5	3.5	2.6	1.0	0.3	3.5	3.6	3.7	1.5	6.2	2.3
			GC-MS/MS	Green tea	A*	2.2	8.5	8.6	9.9	4.8	9.0	4.4	2.3	9.4	2.6	1.8	10.0	3.5	2.9	0.8	0.8	5.8	3.8	0.3
					B*	4.0	2.0	1.2	2.7	2.7	4.0	7.9	0.9	1.4	1.8	3.3	10.3	0.7	4.2	2.4	2.4	3.7	5.6	1.5
				Woolong tea	A	5.0	3.9	4.3	5.7	0.5	3.2	5.7	7.4	11.1	4.0	3.9	6.0	10.1	2.7	4.7	7.1	4.5	2.1	2.3
					B	5.7	5.0	4.3	7.9	8.2	5.6	3.9	1.5	6.8	5.2	5.2	2.7	16.5	4.4	3.7	7.1	8.3	3.9	6.8
		Envi-Carb+PSA	GC-MS	Green tea	A	3.4	6.4	4.6	6.8	2.3	4.7	2.1	3.5	5.1	3.9	5.9	9.9	1.5	4.5	0.8	0.7	1.7	2.1	5.6
					B	6.5	3.6	4.2	3.1	5.4	0.5	1.4	5.7	0.6	1.0	8.7	1.2	3.7	2.7	2.4	1.2	8.3	7.5	2.1
				Woolong tea	A	5.0	0.6	11.0	2.6	15.3	2.5	2.3	5.4	12.2	3.3	15.4	4.7	4.5	1.8	5.6	4.1	5.0	4.0	6.3
					B	7.6	4.7	1.4	10.2	6.6	5.6	4.6	5.7	10.4	13.6	18.9	2.6	2.7	7.8	11.3	8.9	5.4	5.1	2.1
			GC-MS/MS	Green tea	A	9.8	5.4	11.9	1.7	1.4	2.0	1.4	10.6	10.4	15.4	6.9	0.7	1.8	2.8	2.8	4.3	9.4	4.2	7.3
					B	8.0	2.9	5.7	0.9	5.1	2.5	2.0	6.6	1.3	2.5	6.9	2.9	7.8	10.6	10.6	8.9	8.6	7.5	4.0
				Woolong tea	A	6.6	3.9	10.0	6.2	7.2	3.7	2.7	9.2	6.9	2.7	3.9	2.9	3.7	6.3	13.8	4.3	4.8	14.2	5.6
					B	—	—	—	—	—	—	—	6.9	—	—	—	—	—	—	6.6	2.8	2.9	3.8	2.1

(Continued)

Annex 2: RSD data for 201 pesticides in incurred tea Youden pair samples under 8 conditions with two SPE cartridge cleanup in three months (Nov.9, 2009-Feb.7, 2010) (cont.)

No	Pesticides	SPE	Method	Sample	Youden pair	Nov.9, 2009 (n=5)	Nov.14, 2009 (n=3)	Nov.19, 2009 (n=3)	Nov.24, 2009 (n=3)	Nov.29, 2009 (n=3)	Dec.4, 2009 (n=3)	Dec.9, 2009 (n=3)	Dec.14, 2009 (n=3)	Dec.19, 2009 (n=3)	Dec.24, 2009 (n=3)	Dec.14, 2009 (n=3)	Jan.3, 2010 (n=3)	Jan.8, 2010 (n=3)	Jan.13, 2010 (n=3)	Jan.18, 2010 (n=3)	Jan.23, 2010 (n=3)	Jan.28, 2010 (n=3)	Feb.2, 2010 (n=3)	Feb.7, 2010 (n=3)
190	quinalphos	Cleanert TPT	GC-MS	Green tea	A	5.8	6.8	3.7	1.5	9.9	5.5	7.6	12.3	11.6	7.0	10.2	6.2	9.1	4.6	12.8	1.7	4.3	5.2	3.8
					B	5.4	5.4	5.0	9.0	4.9	3.9	2.2	3.7	8.3	11.5	3.8	13.3	2.8	3.1	4.3	5.3	2.8	1.3	5.0
				Woolong tea	A*	5.5	2.1	14.0	4.2	1.9	2.4	4.0	7.4	5.8	4.3	3.6	3.9	6.9	8.8	10.4	3.6	5.3	1.6	4.3
					B*	2.7	0.7	4.8	6.3	9.4	5.6	8.1	2.2	6.4	4.4	6.4	2.8	9.5	4.5	5.2	3.5	2.6	2.1	3.5
			GC-MS/MS	Green tea	A	4.7	5.1	1.4	9.7	12.7	2.5	8.1	8.7	5.3	4.3	8.1	2.1	12.5	2.9	7.5	4.6	6.0	6.1	9.2
					B	5.7	5.2	2.5	6.9	5.6	2.8	2.3	5.6	8.4	10.7	4.8	5.0	13.1	5.5	6.4	0.7	2.1	3.7	9.1
				Woolong tea	A	4.8	5.2	1.6	1.7	3.3	0.6	12.9	2.0	6.9	4.3	7.5	4.4	7.5	0.6	5.8	12.0	5.5	2.0	7.1
					B	2.6	3.7	6.2	10.5	5.5	1.5	3.6	6.0	14.0	1.7	6.6	2.0	3.9	2.8	6.8	1.0	3.0	3.9	6.6
		Envi-Carb+PSA	GC-MS	Green tea	A*	12.5	6.5	4.7	7.8	8.5	9.9	14.7	6.0	5.4	6.0	6.6	10.3	2.5	4.0	12.7	7.3	7.9	14.8	6.3
					B*	17.4	11.4	2.4	3.0	3.4	5.8	3.5	7.8	6.4	5.1	41.1	9.3	8.2	0.5	10.5	3.7	4.9	9.4	6.2
				Woolong tea	A*	2.2	3.7	2.2	2.2	5.9	4.1	5.1	10.1	4.4	1.9	10.8	5.2	3.2	7.4	7.9	0.8	6.7	7.9	11.0
					B	6.9	3.4	5.7	1.7	15.3	2.6	2.9	6.5	10.4	3.2	10.5	6.2	6.2	2.5	11.5	4.3	6.7	5.6	6.7
			GC-MS/MS	Green tea	A	15.7	5.2	1.7	5.5	6.4	10.7	6.1	8.4	2.9	0.1	7.4	17.9	4.4	5.3	3.8	3.1	6.4	12.6	10.0
					B	12.2	10.0	3.6	5.1	2.7	4.1	9.1	7.0	9.1	8.1	17.7	16.8	9.2	11.1	9.9	6.0	8.7	20.2	17.2
				Woolong tea	A	2.0	0.8	5.9	5.4	1.3	2.7	3.2	7.2	3.3	2.2	4.7	7.4	1.7	2.9	7.0	5.0	10.2	17.3	7.1
					B	9.3	4.1	11.0	3.8	9.0	5.4	3.3	6.6	9.9	4.4	8.0	9.9	8.9	6.0	13.2	1.3	7.3	9.5	6.6
191	quinoxyphen	Cleanert TPT	GC-MS	Green tea	A	5.6	6.0	0.8	3.5	9.0	4.4	9.8	4.1	5.2	5.2	5.5	2.2	6.2	4.9	3.2	1.7	3.6	6.4	1.8
					B	5.1	6.8	4.1	8.0	3.3	3.6	0.6	4.3	2.5	8.3	7.4	6.9	2.1	1.3	4.8	2.7	4.1	1.7	3.0
				Woolong tea	A	7.0	11.8	10.2	3.6	2.9	6.4	7.8	8.0	4.3	4.3	3.9	1.6	4.8	3.6	6.9	4.5	4.1	3.6	1.4
					B	5.1	9.3	10.7	5.5	2.3	3.7	11.3	3.1	6.2	0.6	1.8	0.6	3.4	4.5	3.7	3.7	1.6	2.1	1.4
			GC-MS/MS	Green tea	A	4.2	4.0	1.3	3.9	9.4	5.0	7.7	5.3	3.1	1.9	5.8	1.9	8.5	3.9	3.2	0.8	6.3	5.4	3.1
					B	7.4	10.0	6.1	8.2	4.3	3.0	1.7	3.1	5.6	10.4	1.2	5.0	5.9	0.6	3.4	3.5	4.0	5.3	5.1
				Woolong tea	A	7.0	2.6	0.6	3.4	4.3	4.5	7.1	1.6	2.9	3.9	1.2	3.4	6.1	4.0	7.9	2.6	1.5	3.8	4.5
					B	4.3	4.1	4.5	6.8	0.9	3.1	5.0	7.4	7.0	2.4	5.6	3.2	3.0	3.9	6.7	2.7	2.3	2.2	4.8
		Envi-Carb+PSA	GC-MS	Green tea	A	4.7	11.2	2.2	5.1	7.3	1.6	3.2	2.4	5.7	3.4	6.5	4.9	2.1	6.2	3.3	3.3	12.0	4.1	4.0
					B	4.3	14.3	1.3	7.7	1.6	2.8	2.4	2.4	5.5	10.9	24.1	5.0	2.3	5.1	1.2	1.2	5.5	6.6	9.4
				Woolong tea	A	7.8	2.4	4.2	1.0	6.4	3.9	3.1	4.2	2.9	0.7	4.5	5.1	3.2	6.8	1.2	1.5	6.1	3.4	4.2
					B	4.6	0.5	8.8	3.1	9.5	4.8	2.2	5.8	10.0	4.1	9.6	3.4	4.5	5.0	1.5	2.9	6.6	2.8	5.1
			GC-MS/MS	Green tea	A	6.2	2.2	3.2	7.7	7.5	1.4	2.3	2.3	5.3	1.8	6.8	1.6	5.4	8.1	2.9	6.0	10.1	2.3	3.3
					B	4.3	7.8	4.4	3.7	0.5	1.3	1.7	6.5	3.5	0.8	22.5	1.6	4.0	5.0	8.6	1.7	1.8	9.3	5.8
				Woolong tea	A	5.8	3.2	4.6	2.1	5.9	2.9	1.9	3.8	0.6	2.6	6.3	5.8	6.2	8.1	2.4	2.4	4.2	8.9	4.5
					B	5.8	3.4	10.3	5.5	7.6	4.9	3.5	7.6	8.8	4.1	7.4	4.2	4.2	7.8	9.7	2.9	1.1	5.0	4.8
192	telodrin	Cleanert TPT	GC-MS	Green tea	A	3.3	5.0	4.9	8.3	10.1	6.7	8.9	10.6	8.4	2.8	3.8	4.2	8.1	4.6	4.4	7.4	7.4	6.8	2.2
					B*	6.1	11.0	4.1	7.2	3.8	3.4	3.7	8.0	4.2	8.8	2.0	6.1	5.2	5.9	5.1	2.8	5.1	2.3	5.8
				Woolong tea	A	5.4	2.9	5.0	5.1	7.5	2.6	8.1	3.5	10.9	2.4	7.7	2.8	2.8	3.2	6.0	4.1	7.9	3.3	8.2
					B	3.8	6.8	6.2	1.8	3.0	5.3	5.2	5.4	7.8	5.8	2.7	4.0	3.1	10.1	8.5	1.1	2.8	1.2	4.5
			GC-MS/MS	Green tea	A	7.5	3.9	3.3	7.0	16.0	3.4	10.8	10.7	7.5	5.0	8.2	4.5	11.1	6.5	0.3	3.9	0.3	10.0	5.3
					B	5.9	7.3	4.6	10.5	4.0	4.9	4.2	8.0	5.7	10.7	2.4	7.2	11.2	1.4	3.9	5.8	7.5	3.8	7.6
				Woolong tea	A	8.1	3.0	1.5	7.2	6.1	3.7	8.1	9.1	7.9	5.1	2.0	2.6	2.5	2.8	3.1	7.3	9.8	4.9	1.9
					B*	8.3	6.7	8.6	8.4	7.0	4.6	3.6	6.1	8.3	5.4	6.9	3.1	13.9	7.7	7.7	5.5	4.0	4.2	3.8
		Envi-Carb+PSA	GC-MS	Green tea	A	20.3	9.7	12.8	6.0	7.6	16.7	9.7	10.6	9.0	5.0	3.9	13.4	4.5	7.6	15.3	6.1	5.7	12.5	10.0
					B*	23.4	11.7	0.4	12.2	6.1	4.1	12.0	14.4	14.4	11.7	10.0	16.4	19.7	2.7	12.9	15.4	10.4	18.9	9.2
				Woolong tea	A	5.2	3.2	4.4	6.3	2.9	2.7	6.1	11.8	2.8	1.6	4.4	4.8	4.2	8.9	3.1	11.9	4.0	7.8	4.4
					B	8.4	1.2	11.7	0.8	11.9	6.0	5.7	6.4	13.0	4.0	8.4	7.3	6.8	13.0	12.9	2.2	6.8	11.4	5.8
			GC-MS/MS	Green tea	A	27.3	11.5	10.6	12.8	7.6	21.5	11.6	11.8	6.9	2.6	5.8	6.3	4.2	15.7	8.3	10.5	9.5	8.8	7.3
					B	29.0	10.9	2.6	7.7	11.2	4.9	16.9	13.4	11.9	11.4	13.4	1.6	15.1	1.0	4.6	21.9	6.2	15.3	16.4
				Woolong tea	A	7.8	2.6	4.8	7.7	2.9	7.9	3.5	12.7	3.1	2.2	4.8	4.9	6.4	7.0	9.2	3.4	11.7	12.1	1.9
					B	11.1	5.4	10.7	4.7	11.9	9.3	1.7	5.0	10.8	5.0	8.1	7.8	6.2		13.5	1.1	7.1	15.9	3.8

#	Pesticide	Cartridge	Method	Tea		1	2	3	4	5	6	7	8	9	10	11	12	13	14	15	16	17	18	19	20	21
193	tetrasul	Cleanert TPT	GC-MS	Green tea	A	5.1	4.5	2.4	3.2	7.3	4.6	7.0	3.1	5.3	2.7	6.0	1.7	8.6	2.3		2.7		1.7	5.1	6.0	7.3
					B	16.6	7.9	3.5	9.0	3.5	5.5	1.7	3.2	3.3	8.9	4.4	6.4	2.6	1.5		4.0		3.1	5.5	1.4	5.4
				Woolong tea	A	6.5	1.5	1.3	2.4	0.6	1.5	6.6	7.7	2.6	2.6	3.6	2.4	5.1	2.9		8.9		3.3	2.3	6.6	1.3
					B	8.4	2.6	3.7	4.2	1.8	4.0	5.3	4.4	7.0	3.5	0.6	3.0	3.1	5.0		0.7		0.8	1.1	5.7	1.8
			GC-MS/MS	Green tea	A	5.2	4.1	2.3	12.8	10.0	2.4	7.9	2.9	3.8	4.5	6.9	3.1	9.6	3.8		5.9		3.3	4.3	7.1	0.6
					B	4.6	9.4	5.5	9.6	5.0	2.8	2.0	4.5	5.5	10.2	6.1	10.0	7.1	1.0		4.1		2.5	7.3	4.3	4.2
				Woolong tea	A	6.5	0.7	2.7	4.3	3.3	3.2	6.9	2.2	2.4	2.3	1.8	5.1	6.2	2.6		9.4		6.2	4.6	2.5	5.1
					B	5.2	5.9	5.1	8.4	0.8	0.9	4.0	9.5	8.4	1.2	4.3	1.3	7.9	4.4		5.7		4.4	0.9	3.3	3.8
		Envi-Carb+PSA	GC-MS	Green tea	A	5.4	2.3	0.2	6.1	8.3	2.2	6.3	1.1	6.8	2.8	9.0	1.5	2.2	6.5		8.1		10.0	10.3	5.7	5.5
					B	3.0	8.8	5.5	5.8	1.8	0.9	2.4	2.6	5.1	0.8	20.6	2.6	3.2	3.9		9.0		1.8	4.9	10.9	7.0
				Woolong tea	A	5.4	3.2	7.0	1.4	6.1	3.1	3.6	1.8	2.3	2.1	4.5	3.3	1.6	4.1		8.8		2.0	5.4	5.2	3.8
					B	6.0	4.3	9.8	4.2	9.2	4.2	1.3	7.4	11.0	4.5	7.3	3.6	3.4	7.1		8.9		3.0	5.7	4.2	4.0
			GC-MS/MS	Green tea	A	5.5	3.8	1.4	8.6	8.1	1.0	6.3	0.8	4.7	0.7	6.4	3.7	7.4	10.7		3.1		6.7	6.2	7.6	8.1
					B	3.7	7.4	5.2	2.8	1.7	2.3	1.2	5.4	1.2	2.9	23.9	5.5	0.9	12.5		12.2		3.6	0.5	12.6	8.8
				Woolong tea	A	6.0	4.2	5.5	3.4	3.7	2.5	2.7	0.4	0.9	2.2	4.0	6.8	2.1	4.3		10.2		2.5	3.8	5.0	5.1
					B	6.6	3.0	11.0	6.5	6.8	5.7	4.5	5.7	9.3	4.0	6.4	0.2	7.9	8.6		12.6		3.4	3.5	5.6	3.8
194	thiazopyr	Cleanert TPT	GC-MS	Green tea	A	6.5	5.7	3.4	3.5	9.1	2.7	7.4	6.3	5.6	4.6	5.6	2.4	7.0	3.8		1.8		1.8	4.3	6.3	4.3
					B	5.6	7.8	4.9	8.3	3.1	5.0	1.6	7.2	5.6	4.3	3.7	7.1	1.0	3.1		3.8		3.4	5.5	1.0	7.2
				Woolong tea	A	4.7	7.8	4.2	3.3	0.7	4.3	8.7	7.5	5.0	4.3	4.1	3.1	4.8	2.9		7.7		3.4	5.0	3.9	1.7
					B	3.6	3.2	3.5	3.8	3.8	4.7	3.9	2.8	5.3	2.3	1.6	1.2	3.3	4.7		2.9		1.8	1.3	4.5	2.1
			GC-MS/MS	Green tea	A	5.9	6.5	5.1	43.6	14.2	2.7	7.8	6.2	4.3	8.0	3.5	8.0	10.1	1.7		5.0		1.1	15.5	3.7	4.6
					B	7.4	7.9	6.7	8.5	7.9	4.9	5.0	7.3	5.4	6.8	2.8	16.5	4.2	4.2		6.8		4.4	3.2	4.7	9.2
				Woolong tea	A	6.9	5.5	4.6	8.3	8.4	3.9	7.4	2.6	8.0	4.9	3.4	8.8	10.8	3.3		9.9		2.2	5.1	0.3	6.3
					B	5.9	5.8	3.5	17.0	7.5	2.6	3.8	11.9	7.1	8.4	3.8	4.6	8.1	3.7		8.8		2.8	3.7	7.7	10.7
		Envi-Carb+PSA	GC-MS	Green tea	A	12.7	3.7	1.0	6.6	6.1	1.8	1.8	2.6	1.8	2.0	10.2	1.9	2.2	6.7		7.3		10.0	10.3	4.9	6.3
					B	9.0	6.8	6.6	3.6	1.6	1.6	4.2	3.8	5.8	1.7	22.7	5.1	1.4	1.6		14.0		2.5	4.4	10.1	9.9
				Woolong tea	A	4.0	3.9	3.3	1.3	5.5	0.6	3.2	1.2	2.5	0.6	5.7	4.6	3.5	2.8		3.7		3.0	6.3	2.9	4.1
					B	4.4	0.5	7.6	4.2	11.9	6.1	3.3	6.6	11.5	4.0	9.7	2.1	6.2	9.5		8.6		1.6	6.4	5.7	6.0
			GC-MS/MS	Green tea	A	7.1	5.6	4.9	5.6	7.7	2.9	3.9	1.2	5.4	4.9	10.3	1.6	1.6	4.3		5.2		11.5	15.0	7.2	4.1
					B	5.3	7.7	5.2	7.4	7.5	3.8	2.0	5.4	2.2	2.3	21.6	1.9	7.1	16.1		13.3		3.9	5.9	13.3	8.6
				Woolong tea	A	3.5	8.9	5.5	5.6	9.6	2.4	0.4	5.8	1.5	1.5	7.0	1.6	17.3	0.8		15.5		2.1	6.8	4.0	6.3
					B	6.9	1.2	12.3	8.4	10.6	8.7	6.2	8.1	9.1	6.0	9.0	9.7	13.7	7.7		14.2		5.0	3.3	8.0	10.7
195	tolclofos-methyl	Cleanert TPT	GC-MS	Green tea	A*	4.6	5.3	3.6	0.8	7.3	3.8	4.8	6.0	5.1	1.9	8.1	0.6	7.6	3.7		0.4		0.6	3.5	6.3	2.9
					B*	5.9	8.2	5.2	8.3	4.2	2.9	2.6	0.8	3.2	7.7	1.1	8.4	2.0	3.4		13.9		0.8	3.9	2.4	7.4
				Woolong tea	A	4.2	2.2	5.0	3.5	1.7	2.5	8.5	5.3	3.7	2.1	2.5	4.5	3.0	2.1		11.7		4.1	4.1	2.4	1.0
					B	4.1	2.2	3.9	2.9	4.6	3.8	3.4	2.9	7.4	1.8	2.0	1.6	6.7	3.8		3.0		1.0	1.0	2.4	2.0
			GC-MS/MS	Green tea	A	5.4	5.0	2.2	3.9	8.9	3.0	5.6	4.9	5.5	5.5	7.9	3.5	6.8	2.9		1.1		4.1	9.1	1.5	6.2
					B	6.7	4.9	4.9	8.7	5.6	2.5	4.3	5.6	9.3	5.5	0.6	3.8	2.9	4.4		5.2		4.1	0.8	2.0	6.3
				Woolong tea	A	3.8	2.1	2.7	3.9	5.6	3.0	9.9	2.2	1.8	2.9	3.8	4.2	2.2	2.1		9.3		4.9	3.9	4.3	10.8
					B	6.6	5.6	7.9	7.4	3.0	2.6	4.9	4.3	8.7	2.8	2.1	4.5	2.2	5.9		9.5		6.1	7.7	8.3	6.8
		Envi-Carb+PSA	GC-MS	Green tea	A	8.1	5.4	2.3	8.4	8.4	4.0	4.2	1.0	2.2	2.4	7.3	3.2	15.7	2.8		1.8		4.9	10.1	5.8	7.2
					B	6.6	9.3	3.2	3.3	1.0	1.1	3.2	3.0	4.5	2.5	20.3	6.4	3.1	2.8		10.8		0.9	4.5	7.7	7.7
				Woolong tea	A	4.0	4.4	3.8	1.6	5.7	0.9	2.7	1.3	2.5	1.0	6.8	4.4	3.9	8.4		3.9		2.5	4.6	5.0	5.0
					B	5.4	3.3	8.9	3.7	12.3	4.1	1.9	5.2	12.0	0.5	8.7	0.5	3.4	8.5		7.2		2.9	5.8	2.0	2.0
			GC-MS/MS	Green tea	A	9.9	4.8	4.1	6.3	8.1	6.6	4.9	2.1	4.4	0.7	8.2	2.4	5.0	3.8		8.7		6.7	5.3	5.0	7.3
					B	10.1	9.5	2.6	1.0	1.5	1.9	4.9	5.3	4.1	4.2	17.3	3.4	1.5	10.7		12.8		5.1	7.3	5.5	8.5
				Woolong tea	A	3.4	4.8	5.1	2.7	7.4	2.5	4.5	3.7	3.1	4.4	3.5	8.7	6.0	3.1		8.7		4.3	6.8	2.5	10.8
					B	9.9	4.1	10.6	4.3	4.8	4.2	3.7	5.3	8.1	2.9	4.0	5.6	10.5	10.0		13.0		3.9	5.0	2.2	6.8

(Continued)

Annex 2: RSD data for 201 pesticides in incurred tea Youden pair samples under 8 conditions with two SPE cartridge cleanup in three months (Nov.9, 2009-Feb.7, 2010) (cont.)

No	Pesticides	SPE	Method	Sample	Youden pair	Nov.9, 2009 (n=5)	Nov.14, 2009 (n=3)	Nov.19, 2009 (n=3)	Nov.24, 2009 (n=3)	Nov.29, 2009 (n=3)	Dec.4, 2009 (n=3)	Dec.9, 2009 (n=3)	Dec.14, 2009 (n=3)	Dec.19, 2009 (n=3)	Dec.24, 2009 (n=3)	Dec.29, 2009 (n=3)	Jan.3, 2010 (n=3)	Jan.8, 2010 (n=3)	Jan.13, 2010 (n=3)	Jan.18, 2010 (n=3)	Jan.23, 2010 (n=3)	Jan.28, 2010 (n=3)	Feb.2, 2010 (n=3)	Feb.7, 2010 (n=3)
196	transfluthrin	Cleanert TPT	GC-MS	Green tea	A	9.9	5.0	4.8	1.8	6.9	4.6	6.3	2.7	6.0	2.0	5.3	2.5	8.1	9.5	13.0	4.4	10.4	7.2	2.6
					B	7.1	8.2	8.2	9.6	2.4	0.9	0.5	2.7	4.5	9.3	4.0	8.0	1.4	8.3	13.1	2.9	2.7	2.1	6.3
				Woolong tea	A	4.9	0.3	9.3	3.0	2.1	4.4	8.6	6.0	1.6	4.6	4.8	2.0	4.0	1.3	9.2	3.1	3.2	4.1	1.5
					B	4.7	3.2	3.1	1.6	2.1	5.1	2.2	3.1	4.2	3.4	3.0	2.8	2.6	3.4	2.0	2.5	1.0	0.6	0.3
			GC-MS/MS	Green tea	A*	3.6	3.8	3.1	2.9	10.4	3.2	6.5	4.2	4.1	2.1	4.4	2.6	10.2	4.9	2.4	2.2	6.3	8.1	5.5
					B*	6.1	10.3	7.9	8.7	5.6	3.0	4.2	3.4	7.7	10.2	3.0	6.4	3.1	2.7	6.3	4.1	6.6	5.6	6.0
				Woolong tea	A	4.6	3.5	3.6	4.5	7.4	3.6	8.2	0.6	4.5	6.0	0.8	4.2	8.2	2.0	6.0	3.1	4.1	2.1	5.2
					B	4.7	2.8	4.1	12.1	0.9	3.3	4.6	6.1	8.4	1.9	3.3	2.7	10.5	2.6	6.9	3.5	2.8	1.1	3.0
		Envi-Carb+PSA	GC-MS	Green tea	A	8.3	5.5	5.1	9.3	6.7	7.3	5.6	2.7	6.1	5.0	8.5	6.9	4.1	1.8	4.6	3.1	9.0	2.9	4.5
					B	7.3	39.0	6.1	6.9	3.0	1.4	5.0	5.0	3.5	1.1	17.8	6.9	2.5	3.8	11.9	2.5	4.2	7.7	5.8
				Woolong tea	A	8.5	4.9	6.0	1.6	2.9	1.8	2.1	2.4	4.2	2.6	4.0	2.8	1.4	3.4	3.8	5.0	6.0	2.1	7.5
					B	5.3	2.0	9.0	3.4	11.9	5.0	1.6	8.6	11.2	5.7	8.6	1.8	3.8	7.6	11.0	1.9	3.5	2.8	3.2
			GC-MS/MS	Green tea	A*	9.1	5.2	2.6	8.3	6.8	1.5	5.9	3.6	3.7	1.0	9.3	1.1	1.2	13.9	1.8	4.8	7.1	2.8	5.6
					B*	4.8	9.0	5.2	1.7	2.2	2.3	3.3	6.1	4.0	0.3	19.6	1.6	4.3	9.4	11.0	4.0	1.7	10.1	5.2
				Woolong tea	A	2.2	4.7	5.4	3.0	4.0	3.7	3.1	4.5	2.8	1.9	5.9	4.1	3.6	2.2	10.7	5.5	6.4	4.5	5.2
					B	5.9	2.1	10.6	5.4	8.1	7.0	3.7	6.4	8.0	2.6	6.3	4.1	6.0	10.4	12.4	1.3	4.9	2.6	3.0
197	triadimefon	Cleanert TPT	GC-MS	Green tea	A	3.9	5.6	1.5	3.7	10.3	5.3	7.8	1.9	4.4	3.9	11.8	4.2	8.7	1.2	2.1	1.0	2.1	8.7	9.5
					B	7.4	6.3	4.4	8.0	3.3	5.9	1.6	9.0	3.8	6.1	8.8	5.9	0.2	3.6	7.7	6.9	4.3	9.5	11.1
				Woolong tea	A	5.3	2.4	2.5	6.0	4.2	2.8	6.4	10.0	8.2	13.8	2.5	6.8	6.8	2.1	4.1	4.2	5.2	4.6	0.9
					B	4.8	6.4	1.8	6.6	3.0	4.7	2.3	7.1	8.8	13.4	8.3	6.5	8.5	4.9	5.0	5.5	4.1	3.9	2.0
			GC-MS/MS	Green tea	A*	8.8	14.9	2.1	5.2	12.6	4.7	8.0	6.4	5.9	3.5	6.1	1.4	10.4	5.2	2.0	1.6	6.1	7.4	7.3
					B*	7.5	16.4	3.4	6.7	6.9	4.4	3.7	7.5	6.2	8.1	3.7	8.7	1.7	4.6	3.3	5.9	9.2	6.5	9.0
				Woolong tea	A	2.8	2.0	3.7	3.7	4.9	5.4	9.1	7.6	1.6	1.9	2.2	5.1	7.6	1.4	5.6	4.6	1.8	1.4	8.0
					B	6.1	4.7	3.3	6.1	2.9	2.4	4.7	7.2	13.3	2.6	6.0	4.2	7.7	4.7	8.5	4.5	2.9	5.1	6.8
		Envi-Carb+PSA	GC-MS	Green tea	A	9.0	6.1	3.6	5.8	7.3	1.6	8.4	3.6	3.9	10.6	6.0	0.9	2.1	3.4	5.9	6.1	7.4	14.6	6.2
					B	3.2	13.8	2.3	8.2	4.0	6.8	0.6	5.0	15.2	5.6	29.1	14.3	1.3	2.2	18.7	2.2	9.1	7.5	9.1
				Woolong tea	A	3.3	2.6	3.5	6.4	14.0	1.5	3.4	6.3	8.3	0.8	4.3	11.9	1.0	13.1	2.6	3.1	7.0	3.7	6.5
					B	5.4	4.3	7.1	4.5	12.8	7.3	2.0	7.5	11.3	7.8	9.9	2.7	4.3	7.4	11.1	2.0	6.3	10.1	5.9
			GC-MS/MS	Green tea	A*	6.6	1.7	3.9	9.4	7.1	3.2	6.8	2.5	2.8	0.3	3.2	12.0	3.0	10.2	6.1	9.5	12.2	3.8	6.4
					B*	4.4	6.2	3.8	3.5	3.5	1.3	2.0	7.3	1.7	4.8	19.3	13.9	3.7	10.0	11.9	3.5	3.2	9.8	8.5
				Woolong tea	A	1.2	4.4	3.8	2.2	1.8	4.0	4.8	6.9	0.9	2.6	7.5	6.4	5.6	7.0	11.6	3.2	12.5	3.7	8.0
					B	4.8	1.1	11.4	9.4	9.9	4.3	6.0	9.5	8.6	0.0	5.5	6.7	5.6	9.5	14.1	1.4	3.4	3.5	6.8
198	triadimenol	Cleanert TPT	GC-MS	Green tea	A	10.8	4.5	0.8	3.0	6.8	4.6	7.7	5.1	4.4	8.1	11.5	9.8	7.6	6.7	2.5	4.4	4.6	7.5	4.5
					B	7.0	9.3	7.4	6.7	5.5	4.1	3.1	6.2	3.1	11.6	7.3	10.0	0.6	4.7	5.4	3.7	6.4	4.1	3.9
				Woolong tea	A	5.8	4.3	2.4	3.2	4.2	6.7	11.6	12.0	0.3	13.8	14.1	10.2	4.8	1.2	7.1	4.9	4.3	7.8	0.8
					B	2.9	3.6	7.2	4.7	19.0	7.2	1.5	1.2	8.4	12.1	6.8	7.7	1.9	5.8	2.9	3.5	3.0	8.3	2.6
			GC-MS/MS	Green tea	A*	5.1	4.1	2.5	4.5	7.2	1.6	7.4	4.8	4.5	7.1	5.4	1.5	10.7	2.5	1.2	3.8	7.0	5.2	1.7
					B*	3.9	9.4	6.8	6.4	4.6	4.5	2.6	7.8	5.2	13.5	2.7	12.5	7.5	0.8	7.5	4.5	4.4	4.8	7.5
				Woolong tea	A	5.7	2.2	0.7	3.2	4.9	5.0	12.8	2.2	2.7	4.2	3.8	4.5	6.5	4.1	6.4	2.1	1.5	6.7	5.9
					B	6.9	3.8	5.7	8.7	3.5	3.1	4.7	8.8	7.1	3.6	5.6	2.9	7.9	5.7	6.5	1.6	2.3	0.2	3.4
		Envi-Carb+PSA	GC-MS	Green tea	A	9.5	12.8	5.5	8.0	6.8	2.6	19.7	1.7	3.2	3.6	6.2	10.3	1.6	5.7	8.9	11.0	11.6	5.7	5.3
					B	5.4	5.6	1.1	0.7	3.9	9.4	3.2	3.8	6.2	7.9	22.4	12.1	3.2	5.5	15.5	2.8	3.6	9.2	7.9
				Woolong tea	A	3.1	2.6	5.9	1.1	15.3	7.5	3.0	16.4	5.6	4.4	6.8	5.8	3.9	4.5	9.9	2.8	4.2	7.4	1.0
					B	6.4	2.9	7.7	8.6	14.6	8.1	8.4	19.5	17.4	3.3	20.7	18.6	3.3	15.1	6.3	8.7	6.9	10.2	7.1
			GC-MS/MS	Green tea	A*	8.3	9.2	4.2	4.7	6.7	1.9	18.9	2.9	4.1	2.8	4.6	15.7	4.0	12.0	6.2	6.8	10.7	3.3	3.1
					B*	4.3	6.0	5.1	1.0	4.4	1.0	1.3	5.5	4.8	1.4	22.5	7.9	5.7	20.6	13.0	2.7	0.4	10.1	6.7
				Woolong tea	A	5.8	2.5	4.6	1.5	3.6	1.5	3.3	3.5	2.4	7.5	6.3	6.8	8.7	3.9	12.5	3.3	4.6	4.7	5.9
					B	5.6	2.7	14.5	2.3	13.1	7.0	3.5	7.8	7.3	3.2	4.8	5.1	5.3	13.2	9.3	3.3	5.3	6.8	3.4

199	triallate	Cleanert TPT	GC-MS	Green tea	A	3.6	5.5	4.1	3.6	11.4	4.2	6.7	6.0	8.0	3.8	6.5	1.4	7.3	5.1	2.2	1.3	1.6	3.8	2.0
				Woolong tea	B	5.6	9.1	3.8	9.0	4.6	3.4	1.7	4.7	4.1	9.2	0.8	8.4	3.3	1.1	5.1	4.9	5.6	5.5	3.3
				Green tea	A	4.1	1.6	2.1	4.1	1.9	3.2	9.0	1.7	0.9	2.3	3.0	2.1	3.2	1.3	6.1	3.1	3.9	1.3	8.5
				Woolong tea	B*	3.5	2.8	4.2	4.3	3.6	4.4	3.5	4.8	8.7	4.0	2.4	3.5	5.3	3.9	6.9	1.7	1.9	1.8	5.4
			GC-MS/MS	Green tea	A*	5.0	7.1	0.7	10.9	14.0	3.6	7.9	6.5	6.3	3.1	5.5	2.8	9.6	1.3	3.1	1.8	7.0	7.8	0.9
				Woolong tea	B	3.9	9.7	6.0	9.7	3.4	4.9	2.0	5.8	6.6	7.9	2.6	17.1	2.6	1.8	1.2	3.9	8.7	4.2	8.3
				Green tea	A*	5.3	2.7	1.9	4.8	5.4	2.9	7.6	3.7	5.1	3.4	2.2	4.3	5.7	0.9	7.6	3.1	2.6	0.3	1.7
				Woolong tea	B*	3.2	4.6	4.0	7.4	3.2	5.1	2.7	3.7	7.4	2.6	1.4	1.0	8.0	1.6	9.4	5.9	1.7	1.2	5.9
		Envi-Carb+PSA	GC-MS	Green tea	A	14.6	5.7	8.3	3.5	7.0	12.5	8.2	7.4	5.3	0.8	5.1	10.6	6.4	4.2	9.5	3.4	7.3	9.6	5.0
				Woolong tea	B*	15.4	9.0	4.0	3.6	4.0	4.1	8.8	7.8	9.2	6.2	12.3	4.0	13.5	3.3	6.8	12.5	4.7	13.9	7.8
				Green tea	A	5.3	2.8	3.3	4.2	3.8	1.0	3.1	7.5	1.9	1.8	4.9	2.8	1.8	6.5	3.1	4.4	5.6	7.1	3.3
				Woolong tea	B	7.2	4.0	8.5	2.2	11.9	5.0	2.9	6.5	11.6	2.2	8.4	5.8	3.4	6.5	11.6	3.3	8.5	4.1	4.9
			GC-MS/MS	Green tea	A*	17.4	8.2	6.3	6.1	4.8	13.0	8.0	10.2	5.9	1.5	3.7	12.6	2.2	16.9	3.2	5.9	10.2	6.4	6.1
				Woolong tea	B*	18.4	10.4	5.8	8.8	6.7	3.9	9.1	5.5	6.5	6.5	12.5	12.1	9.3	20.0	6.3	11.1	6.3	14.7	4.9
				Green tea	A	5.3	3.7	6.0	6.8	3.2	2.1	2.0	8.0	2.6	1.3	3.7	5.2	2.5	3.0	8.2	2.7	4.8	6.0	1.7
				Woolong tea	B	8.7	4.4	10.1	5.0	8.2	6.0	2.8	5.9	9.2	1.7	7.6	5.0	6.4	7.0	9.9	1.1	6.9	11.1	5.9
200	tribenuron-methyl	Cleanert TPT	GC-MS	Green tea	A	8.0	17.6	2.8	12.5	12.8	1.8	21.3	11.4	8.9	7.8	11.3	4.8	10.6	8.7	4.8	6.9	5.2	12.8	7.2
				Woolong tea	B	5.6	13.7	1.2	14.8	10.9	10.0	8.0	13.5	6.3	17.2	2.4	8.7	8.5	3.6	13.4	9.9	11.7	6.4	4.6
				Green tea	A	11.6	19.2	7.8	10.0	16.7	8.4	10.3	12.1	1.3	4.0	16.0	13.2	5.5	5.9	3.9	5.2	2.5	5.5	5.7
				Woolong tea	B*	12.6	13.4	10.3	5.7	13.9	15.4	10.9	10.2	12.6	3.6	20.1	6.7	8.2	8.1	11.0	3.6	2.1	1.0	13.8
			GC-MS/MS	Green tea	A	3.5	4.5	3.9	2.4	16.1	5.3	19.3	13.9	8.3	6.6	13.4	3.3	19.6	13.7	2.7	7.3	4.3	12.7	2.9
				Woolong tea	B*	5.9	9.7	5.5	12.2	8.9	12.6	5.4	15.4	3.0	21.9	3.0	22.2	15.6	6.5	8.8	8.5	7.2	10.2	4.6
				Green tea	A	12.2	11.5	6.3	10.4	8.4	11.2	3.6	2.6	2.5	6.0	10.0	6.9	6.4	3.4	10.8	12.0	7.6	15.7	13.9
				Woolong tea	B*	17.3	7.1	10.4	11.0	9.4	16.5	13.1	17.8	16.5	3.6	14.4	4.4	9.0	2.6	8.6	7.9	7.7	4.0	4.4
		Envi-Carb+PSA	GC-MS	Green tea	A	20.4	7.9	35.0	13.3	12.1	14.1	67.3	20.2	14.9	10.9	5.6	23.4	3.4	4.2	7.2	5.6	10.7	8.2	6.0
				Woolong tea	B	19.2	13.4	14.6	14.8	24.3	13.4	16.3	24.0	18.5	9.3	53.1	14.1	22.3	4.1	26.9	4.2	5.4	13.5	9.4
				Green tea	A	8.1	21.2	10.1	19.5	12.6	7.1	7.2	15.1	1.7	16.6	5.2	10.9	8.6	7.3	7.4	11.8	6.7	8.6	8.0
				Woolong tea	B*	21.1	14.5	15.5	15.2	22.5	16.7	19.0	6.0	12.6	9.7	6.4	9.7	11.8	6.0	16.3	1.0	8.7	15.9	9.9
			GC-MS/MS	Green tea	A	21.2	25.9	29.9	14.4	5.0	13.1	27.4	15.5	8.3	10.1	3.0	0.0	2.6	17.0	4.9	4.7	8.7	15.1	6.1
				Woolong tea	B	20.6	5.5	10.6	12.5	20.3	14.5	20.3	25.6	3.0	4.6	58.1	0.0	41.4	28.0	38.5	8.7	2.7	22.6	16.9
				Green tea	A*	14.6	14.5	6.3	14.4	20.8	5.0	9.5	19.7	2.5	18.2	10.2	10.5	5.9	9.4	15.8	20.6	2.0	18.9	13.9
				Woolong tea	B*	19.1	17.4	21.4	4.6	23.7	21.7	19.5	4.7	16.5	5.9	6.2	5.1	5.6	14.8	14.1	5.8	6.8	20.1	4.4
201	vinclozolin	Cleanert TPT	GC-MS	Green tea	A	3.5	5.5	3.9	3.1	10.0	2.3	4.5	8.4	8.3	1.9	3.1	2.9	7.8	4.7	5.5	3.4	11.9	4.4	3.2
				Woolong tea	B	4.9	9.0	2.0	8.8	3.9	3.0	2.4	3.9	7.4	12.0	4.1	5.2	4.9	13.3	4.0	2.4	8.4	1.1	7.3
				Green tea	A	8.1	1.2	1.3	6.4	1.3	3.3	7.5	2.6	2.8	3.4	5.3	2.1	3.5	3.7	6.2	3.6	2.5	3.2	1.8
				Woolong tea	B	3.9	2.7	5.0	3.3	1.4	5.8	2.5	0.1	2.7	5.0	2.0	1.4	3.3	4.6	6.4	0.6	0.7	0.6	2.2
			GC-MS/MS	Green tea	A	4.8	5.7	2.3	5.4	15.6	2.6	7.5	9.9	6.8	3.1	6.3	5.1	10.4	6.4	0.2	4.8	8.4	6.9	4.7
				Woolong tea	B	5.0	12.3	3.0	8.7	4.3	3.6	4.5	3.4	3.8	7.0	0.6	8.1	4.1	5.5	1.8	4.8	6.2	2.4	10.9
				Green tea	A	8.1	2.0	2.2	6.2	5.2	4.2	8.5	5.5	6.9	2.6	2.0	2.7	6.6	1.6	3.2	5.4	1.2	5.9	2.1
				Woolong tea	B	6.3	5.6	6.3	6.9	4.5	6.4	3.6	8.2	9.3	4.6	3.4	1.4	7.4	4.6	6.1	6.9	2.1	8.1	1.6
		Envi-Carb+PSA	GC-MS	Green tea	A	15.1	3.2	8.5	7.5	7.5	11.5	5.7	8.9	3.2	1.4	3.4	10.8	6.7	2.5	10.2	2.2	8.6	6.4	4.6
				Woolong tea	B	12.7	7.4	6.1	4.4	3.2	2.1	8.0	7.0	9.6	7.2	9.9	8.2	13.8	1.8	4.6	13.1	3.8	10.8	7.1
				Green tea	A	2.7	1.7	1.9	8.3	2.9	1.6	4.7	11.2	2.0	2.3	5.2	4.6	4.3	2.7	2.2	2.3	5.6	7.4	4.2
				Woolong tea	B	6.3	3.7	7.8	1.4	9.8	3.4	2.0	7.4	11.7	3.1	8.3	4.2	4.4	6.0	11.2	1.6	5.9	3.0	4.6
			GC-MS/MS	Green tea	A	21.1	8.6	5.6	1.6	8.3	12.3	7.6	8.2	5.7	1.9	1.2	0.0	4.3	3.1	5.0	2.9	5.1	4.1	7.1
				Woolong tea	B	20.0	9.0	7.6	7.8	9.0	3.4	10.2	4.6	8.0	8.4	14.9	0.0	12.4	13.0	7.6	11.5	6.8	13.0	9.4
				Green tea	A	5.3	4.8	8.2	4.6	4.9	1.7	3.0	5.4	2.3	1.4	4.3	3.4	8.6	8.3	8.5	5.4	6.3	3.4	2.1
				Woolong tea	B	7.9	4.5	7.6	0.8	8.6	5.6	3.8	4.4	9.4	1.6	10.5	3.4	8.2	8.4	17.5	1.7	4.9	13.0	1.6

REFERENCES

[1] J.M. Telepchak, F.T. August, G. Chaney, Humana Press, Totowa, NJ, 2004.

[2] Hennion, M.C., 1999. J. Chromatogr. A 856, 3–54.

[3] Park, Y.S., Abd El-Aty, A.M., Choi, J.H., Cho, S.K., Shin, D.H., Shim, J.H., 2007. Biomed. Chromatogr. 21 (1), 29–39.

[4] Yahya, R.T., Mohammad, F.Z., Thaer, A.B., 2006. Biomed. Chromatogr. 558 (1–2), 2–68.

[5] Baugros, J.B., Cren-Olivé, C., Giroud, B., Gauvrit, J.Y., Lantéri, P., Grenier-Loustalot, M.F., 2009. J. Chromatogr. A 1216 (25), 4941–4949.

[6] Albero, B., Sánchez-Brunete, C., Tadeo, J.L., 2005. Talanta 66 (4), 917–924.

[7] Chen, S.B., Yu, X.J., He, X.Y., Xie, D.H., Fan, Y.M., Peng, J.F., 2009. Food Chemistry 113 (4), 1297–1300.

[8] Gervais, G., Brosillon, S., Laplanche, A., Helen, C., 2008. J. Chromatogr. A 1202 (2), 163–172.

[9] Hernández, F., Pozo, O.J., Sancho, J.V., Bijlsma, L., Barreda, M., Pitarch, E., 2006. J. Chromatogr. A 1109 (2), 242–252.

[10] Yagüe, C., Herrera, A., Ariño, A., Lázaro, R., Bayarri, S., Conchello, P., 2002. J. AOAC Int. 85 (5), 1181–1186.

[11] Kitagawa, Y., Okihashi, M., Takatori, S., Okamoto, Y., Fukui, N., Murata, H., Sumimoto, T., Obana, H., 2009. Shokuhin Eiseigaku Zasshi 50 (5), 198–207.

[12] Okihashi, M., Takatori, S., Kitagawa, Y., Tanaka, Y., 2007. J. AOAC Int. 90 (4), 1165–1179.

[13] Lou, Z.Y., Chen, Z.M., Luo, F.J., Tang, F.B., Liu, G.M., 2008. Chin. J. Chromatogr. 5, 568–576.

[14] Huang, Z., Zhang, Y., Wang, L., Ding, L., Wang, M., Yan, H., Li, Y., Zhu, S., 2009. J. Sep. Sci. 32 (9), 1294–1301.

[15] Fillion, J., Sauvé, F., Selwyn, J., 2000. J. AOAC Int. 83 (3), 698–713.

[16] Wong, J.W., Webster, M.G., Halverson, C.A., Hengel, M.J., Ngim, K.K., Ebeler, S.E., 2003. J. Agric. Food Chem. 51 (5), 1148–1161.

[17] Wong, J.W., Zhang, K., Tech, K., Hayward, D.G., Krynitsky, A.J., Cassias, I., Schenck, F.J., Banerjee, K., Dasgupta, S., Brown, D., 2010. J. Agric. Food Chem. 58 (10), 5884–5896.

[18] Schenck, F.J., Lehotay, S.J., Vega, V., 2002. J. High Resolut. Chromatogr. 25 (14), 883–890.

[19] Amvrazi, E.G., Albanis, T.A., 2006. J. Agric. Food Chem. 54 (26), 9642–9651.

[20] Standardization Administration of P.R. China, 2008. GB/T 19426-2006, Method for the Determination of 497 Pesticides and Related Chemicals Residues in Honey, Fruit Juice and Wine By Gas Chromatography–Mass Spectrometry Method. Standards Press, Beijing, China.

[21] Standardization Administration of P.R. China, 2008. GB/T 23216-2008, Determination of 503 Pesticides and Related Chemicals Residues in Mushrooms—GC-MS Method. Standards Press, Beijing, China.

[22] Standardization Administration of P.R. China, 2008. GB/T 19648-2006, Method for Determination of 500 Pesticides and Related Chemicals Residues in Fruits and Vegetables by Gas Chromatography-Mass Spectrometry Method. Standards Press, Beijing, China.

[23] Standardization Administration of P.R. China, 2008. GB/T 19649-2006, Method for Determination of 475 Pesticides and Related Chemicals Residues in Grains By Gas Chromatography-Mass Spectrometry Method. Standards Press, Beijing, China.

[24] Standardization Administration of P.R. China, 2008. GB/T 23210-2008, Determination of 511 Pesticides and Related Chemicals Residues in Milk and Milk Powder—GC-MS Method. Standards Press, Beijing, China.

[25] Pang, G.F., Fan, C.L., Zhang, F., Li, Y., Chang, Q.Y., Cao, Y.Z., Wang, Q.J., Hu, X.Y., Liang, P., 2010. J. AOAC Int. 94 (4), 1253–1296.

3 Study on the Influences of Tea Hydration for the Method Efficiency and Uncertainty Evaluation of the Determination of Pesticide Multiresidues in Tea Using Three Sample Preparation Methods/GC–MS/MS

Chapter 3.1

A Comparative Study on the Influences of Tea Hydration for the Method Efficiency of Pesticide Multiresidues Using Three Sample Preparation Methods/GC–MS/MS

3.1.1 INTRODUCTION

Tea is rich in physiologically active compounds, such as polyphenols, alkaloids, tea pigments, aromatic substances, amino acids, and vitamins, and it possesses such functions as reducing blood sugar, blood pressure, antithrombosis and antiatherosclerosis, bacteriostat, improving immunity, and is antitumor, in other words, a naturally healthy drink. More than 2 billion people from 150-plus countries and regions drink tea, making tea one of the three most popular drinks in the world. Tea is mostly planted in warm temperate zones and subtropical areas which are susceptible to threats from plant diseases and insect pests year-round. For the purpose of plant disease and insect pest prevention, chemical pesticides are widely used, leading to contamination with pesticide residues. To hold this in check, countries worldwide have prescribed pesticide maximum residue limits in tea (MRL). For instance, the EU has stipulated 453 varieties, Japan 268, and Germany 530. Hence, the acceleration of the development of techniques for pesticide multiresidues in tea from GC [1] to GC–MS [2], GC–MS/MS [3], LC–MS/MS [4], UPLC–MS/MS [5], and GC×GC-ToF [6], and so on.

Tea has a complex matrix with pigments, polyphenols, alkaloids, and some lipids that act as interferences for determination of pesticide residues in tea. The question of what pretreatment techniques should be adopted to sufficiently extract pesticide multiresidues in tea matrices, lowering interference from foreign matter to the greatest extent, is one of the most difficult projects in the tea residue analytical field.

For pesticide residues in tea, two modes of extraction are usually adopted, one of which is using common organic solvents or their composite mixing solvents, such as acetonitrile, hexane, cyclohexane, ethyl acetate, acetone, dichloromethane, or methanol. Alternatively, making an extraction in a mode such as homogeneous [7], oscillating [8], ultrasonic [4], vortex [9], accelerated solvent extraction [10], matrix solid-phase dispersion [11], head space solid-phase microextractions

[6], pressurized solvent extraction, or dispersive liquid–liquid microextraction [12]. Acetonitrile extraction, among those listed earlier, possesses a wide range of pesticide polarities and less interference from coextracting matters, making it the first choice of solvents for pesticide multiresidue extraction in years past.

The most widely used cleanup techniques for multiresidues in tea is solid phase extraction (SPE), which features lesser loss of pesticide residue and good cleanup results. The cleanup fillers mainly used include Carb-NH$_2$ [13], Carb-PSA, and Florisil [1]. Pang et al. [14,15] employed acetonitrile homogeneous extraction and SPE cleanup and established a simultaneous determination method for 653 pesticides in tea, with an appraisal study of different kinds of analytical conditions, among which the limit of detection (LOD) by GC–MS for 490 pesticides was 1.0–500 µg/kg, while the LOD by LC–MS/MS for 448 pesticides was 0.03–4820 µg/kg.

For GC–MS determination of 490 pesticides at the low fortification level of 0.01 –100 µg/kg, the average recoveries for 94% of pesticides fall within 60%–120%, with RSD for 77% pesticides below 20%; for LC–MS/MS determination of 448 pesticides, the average recoveries for 91% of pesticides fall within 60%–120%, with RSD for 76% pesticides below 20%. The new-type SPE cartridge Cleanert TPT, composed of three ingredients: graphitized carbon black (PestiCarb), polyamine silica, and amide polystyrene, is high in cleanup efficiency and is of good reproducibility and repeatability in analytical results. Other common cleanup techniques include liquid–liquid extraction [8], DSPE [3], GPC [10], DLLME [12]. Xchurek et al. [6] adopted head space solid phase microextraction in conjunction with the GC×GC-TOF approach for determination of 36 pesticide residues in tea. Ana Lozanoa et al. [3] adopted GC–MS/MS and LC–MS/MS for respective determination of 86 pesticide residues in green tea, black tea, dark tea, and jasmine flower tea, with a much lower limit of detection with LC–MS/MS than GC–MS/MS.

Another way of extraction is first hydrating samples before adopting organic solvents for extraction. For hydration of tea leaves, adding water quantities varies from 1 mL/g to 10 mL/g as different hydration methods are used. The varieties of pesticides determined by these analytical methods vary from more than a dozen to more than 100. For example, Hong-Ping Li et al. [8] used acetone to extract the hydrated tea leaves, employed 5% NaCl water solutions (1:1:5, v/v) liquid–liquid partitioning for cleanup, and made a GC determination of 84 pesticide residues, with recoveries 65%–120% and RSD 0.34%–16% at the fortification concentration of 0.02–3.0 mg/kg. Zhiqiang Huang et al. [2] used acetone, ethyl acetate, and hexane mixing solutions for extraction and GPC in conjunction with SPEC for cleanup, making a GC–MS determination of 102 pesticide residues in tea, with recoveries 59.7%–120.9% and RSD 3.0%–20.8% at the fortification concentrations of 0.01–2.5 µg/m. Ana Lozano et al. [3] used acetone oscillation for extraction and PSA and DSPE for cleanup after salting out, making a GC–MS/MS and LC–MS/MS determination of 86 pesticide residues in black tea, green tea, dark tea, and jasmine flower tea, with recoveries 70%–120% and RSD less than 20% at the fortification concentration of 10–100 µg/kg. LOD is 0.1–210 µg/kg.

For comparison of the superiority of these two modes of extraction, there are only two papers that have been found for appraisal study, one of which is Pang et al. [16], who use hydration (4 mL/g) homogeneous extraction and SPE cleanup for determination of 201 pesticide residues in tea. The comparison of the hydration method with the other two has found that both accuracy and precision with the hydration method is poor. Tomas Cajka et al. [17] proposed that tea samples be added with water and made to sit still for 30 min, which effectively helps improve the extraction efficiency of pesticide residues in tea.

For a further probe into the influences of hydration on the method efficacy of tea multiresidues, we chose M1 to be equivalent to Cajka's method [17] and used acetonitrile for extraction after tea hydration, then used part of the extraction solution and hexane liquid–liquid partitioning for cleanup. The extraction of M2 is identical to that of M1, but it takes all the extraction solution and cleans up with SPE, with cleanup procedures the same as those of M3. M3 is equivalent to Pang's method [14]: sample extraction through pure acetonitrile and SPE for cleanup. There only exist cleanup differences with M1 and M2, while extraction differences only exist with M2 and M3. Through recovery experiments at three fortification levels, aged sample determination experiments, and field incurred sample determination, aspects such as extraction efficiencies, cleanup results, and the applicability of the process to different methods have been studied and compared. These results reflect the concrete influences of hydration and nonhydration on the method efficiency.

Recoveries from 456 pesticides at three fortification levels and the log Kow function correlation diagram of them with pesticide polarities have both been proven. First, tea leaf hydration is indeed capable of increasing extraction efficiency of certain polarity pesticides—for instance, 24 pesticides of strong polarities, with recoveries of M2 higher than those of M3. Hydration also lowers the extraction efficiency of some nonpolar pesticides, however. Taking the 28 nonpolar pesticides for an example, recoveries with M3 are higher than M2. The appraisal of overall extraction efficiencies of 456 pesticides has found that the hydration method does more harm than good. In addition, interference from coextraction after tea hydration increases greatly, leading to a decrease of method efficiency. Fortification experiments of uniform limit 0.010 mg/kg confirm that for the percentage of pesticides that comply with EU SANCO/12495/2011 technical requirements of 70%–120%

and RSD <20%, only 5.0% are accounted for with M1 hydration method, but 50% with M3 nonhydration method. (M1 fails to meet the requirement.)

Although the influence is not obvious for determined values of high-concentration fortification, it has a relatively big influence on RSD of the method. For instance, with 456 pesticide aged samples, there are 158 pesticides with their added RSD <10% for M1, accounting for 35%; there are 381 for M3, making up 84%, and M3 is 2.4 times superior to M1. Additionally, tea leaf hydration greatly lowers the sensitivity of the method. In a study of 456 pesticide aged tea samples, the signal/noise ratio of each pesticide was statistically calculated and the adding of signal/noise ratios of 456 pesticides was averaged, with M1 being 940 and M3 6781, showing that hydration lowered the sensitivity 7 times. This is also the main cause of failure of detection at 0.010 mg/kg to meet EU SANCO/12495/2011 technical requirements.

The log Kow values of 329 pesticides have been found in this study of 456 pesticides, among which there are only 39 polarity pesticides and strong polarity pesticides with log Kow values <2.0, accounting for 12%. There are 290 other pesticides of medium polarity, weak polarity, and nonpolarity, accounting for 88%. Therefore, hydration methods only increased the extraction efficiency of certain polarity pesticides while losing extraction efficiency for the majority of pesticides of medium polarity and nonpolarity, which is not worthwhile.

3.1.2   EXPERIMENTAL METHOD

3.1.2.1   Three Different Sample Preparation Methods

Method 1(M1): hydration + oscillating extraction + hexane liquid/liquid partitioning cleanup for part of extraction solution. It is equivalent to Cajka's method [17]; method 2 (M2): hydration + oscillating extraction + overall extraction solution SPE cleanup, that is, the extraction is identical with M1, and cleanup is identical with M3; method 3 (M3): pure acetonitrile homogeneous extraction + overall extraction solution SPE cleanup. It is equivalent to Pang's method [14]. The differences in the three methods are demonstrated in Fig. 3.1.1.

Fig. 3.1.1 shows that M1 and M3 are two totally different methods, while M2's extraction is identical with M1 and its cleanup is identical with M3, so the test results of M2 can testify to the advantages and disadvantages of M1 and M3.

3.1.2.2   Aged Sample Preparation

Four hundred and fifty-six pesticide mixing solutions are uniformly sprayed onto blank oolong tea powders, which are stored in the dark and aged 30 days. Quarter sampling is adopted, and a parallel determination of 456 pesticide concentrations is made for the sample. When RSD <4% ($n = 10$) is determined for each pesticide, it is considered to be a uniformly prepared aged tea samples.

3.1.2.3   Incurred Sample Preparation

Market-available 18 pesticides were sprayed in accordance with spraying procedures onto tea trees grown in the field. A parallel spray was conducted on two experimental fields. The first picking began 24 h after spraying, with picking

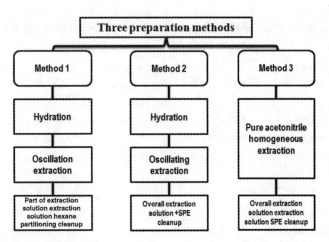

FIGURE 3.1.1   Sketch of the differences of M1, M2 and M3.

continuing once every day, lasting 1 month. On the 10th day after spraying the tea is picked to be an incurred sample and is prepared into uniform (RSD < 4%, $n = 10$) incurred samples.

3.1.3 EXPERIMENTAL RESULTS AND DISCUSSION

This experiment used three sample preparation methods, M1, M2, and M3. (1) For 456 pesticides in oolong tea, a fortification experiment was conducted at three levels of high, medium, and low residues, with recoveries and RSD data tabulated in Supplemental Tables 1–3; (2) for 456 pesticides, S/N of three fortification levels for oolong tea matrix standards were tested, with S/N raw data tabulated in Supplemental Tables 4–6; (3) for 456 pesticides in oolong tea aged samples, the concentration values and S/N were tested, with finding values, RSD ($n = 5$), and S/N data tabulated in Supplemental Table 7; (4) for 160 pesticides determined by Cajka's method [17], using two methods for sample preparation, recoveries determined at three fortification levels and RSD data were tabulated in Supplemental Table 8; (5) for 18 pesticide green tea incurred samples, three methods were used for sample preparation and their true content was tested and tabulated in Supplemental Table 9. Due to the huge quantity of data, supplementary materials have been provided for the reader to reference or trace the source when necessary. Supplemental tables are available in digital form on the *J. AOAC Int.* website, http://aoac.publisher.ingentaconnect.com/content/aoac/jaoac. The majority of data discussed in the next sections are derived from these supplemental tables.

3.1.3.1 Comparison of Accuracy and Precision for Fortification Recovery Experiments by the Three Methods

The three methods of M1, M2, and M3 are used to conduct a comparative experiment on the fortification recoveries at three levels (0.01, 0.1 and 1.0 mg/kg) for 456 pesticides determined in Pang's method [14], with the test raw data listed in Supplemental Tables 1–3. The three methods of M1, M2, and M3 are used to conduct a comparative experiment on fortification recoveries at three levels for 160 pesticides determined by Cajka's method [17], with data listed in Supplemental Table 8. Pesticide quantities and ratios with their fortified recoveries 70%–120% and RSD ≤ 20% in conformance with EU SANCO/12495/2011 technical requirements tabulated in Supplemental Tables 1–3 and Supplemental Table 8 are listed in Table 3.1.1.

Table 3.1.1 shows that in terms of EU SANCO/12495/2011 technical requirements with recoveries (REC) 70%–120% and RSD ≤20%, two items of REC and RSD at three fortification levels with M3 meet the technical standards of EU SANCO/12495/2011, with pesticide quantities determined in excess of those with M2 and M1.

In a practical experiment, one notable difference is also discovered. For M1 and M2, take 2 g oolong tea samples, with supernatant about 8–8.5 mL obtained through final centrifuging. For M1, take 1 mL acetonitrile supernatant (equivalent to 0.2 g tea) for cleanup; for M2, take overall acetonitrile supernatant (equivalent to 2 g tea) for cleanup. In fact, for 2 g samples, 8–8.5 mL supernatants are actually obtained, and based on this calculation, 1 mL acetonitrile is equivalent to 0.235–0.250 g tea; on the contrary, if 1 mL acetonitrile supernatant is equivalent to 0.2 g tea, 8–8.5 mL supernatant is equivalent to 1.6–1.7 g tea. Based on this calculation, due in part to extraction solutions adopted with M1, the analytical results from the pesticides are falsely higher than those of M2 at 17.5%–25.0%, with an average of 21.25%.

TABLE 3.1.1 Recoveries, RSD and Ratios (%) That Meet EU SANCO/12495/2011 Technical Requirements

| | 1.00 mg/kg | | 0.10 mg/kg | | 0.01 mg/kg | |
| | AVE REC | | AVE REC | | AVE REC | |
Fortified concentrations	70%–120%	RSD ≤ 20%	70%–120%	RSD ≤ 20%	70%–120%	RSD ≤ 20%
For 160 target pesticides						
M1	121(75.5)	159(99.4)	109(67.9)	144(89.9)	23(14.5)	24(15.1)
M2	134(84.3)	155(97.5)	99(62.3)	128(80.5)	18(11.3)	52(32.7)
M3	145(90.6)	160(100)	121(75.5)	156(97.5)	80(49.7)	112(69.8)
For 456 target pesticides						
M1	335(73.5)	387 (84.9)	313 (68.6)	287 (62.9)	121 (26.5)	156(34.2)
M2	353 (77.4)	313 (68.6)	358 (78.5)	345 (75.7)	350 (76.8)	287 (62.9)
M3	404 (88.6)	399 (87.5)	417(91.4)	413 (90.6)	325 (71.3)	303 (66.4)

TABLE 3.1.2 Total Average Recoveries for Pesticides at Three Fortification Levels by Three Methods

Fortification levels, mg/kg	M1[a]	M2	M3
1.0	61.8%	85.3%	88.0%
0.1	69.2%	84.9%	89.3%
0.01	52.7%	86.3%	83.4%
Average	61.2%	85.5%	86.8%

[a]Deducting falsely high number 20.0%.

The same thing also happened in comparison with M1 and M3, where the acetonitrile extraction solutions added with M3 are 30 mL and the supernatants obtained through centrifuging are 25 mL. Theoretically, when M1 and M3 are compared, the analytical results with M1 should be falsely higher by 20.0% than those with M3. See Table 3.1.2.

If, in deference to the deduction about the falsely high number 20.0%, it is deducted from the analytical results brought about by the procedure differences of M1, the average extraction efficiency ranking for the three methods is M3 > M2 > M1.

3.1.3.2   Correlation Comparison of the Three Methods' Extraction Efficiency with Pesticide log Kow Values

3.1.3.2.1   Correlation Comparison of 456 Pesticide Fortified Recoveries With Pesticide log Kow Values

For the purpose of evaluating the correlations of the recoveries and pesticide polarities of the three methods, log Kow values have been found for 329 pesticides out of 456 pesticides, with their range: −0.77 to +8.20. Take the log Kow values as horizontal ordinate and recoveries as vertical ordinate to draw the scatter diagram (eliminate recoveries being zero and the outliers exceeding 150%), and the correlation diagram of the recoveries of the three methods at three fortification levels and pesticide polarities is determined (see Fig. 3.1.2). To further prove that hydration influences the extraction efficiency

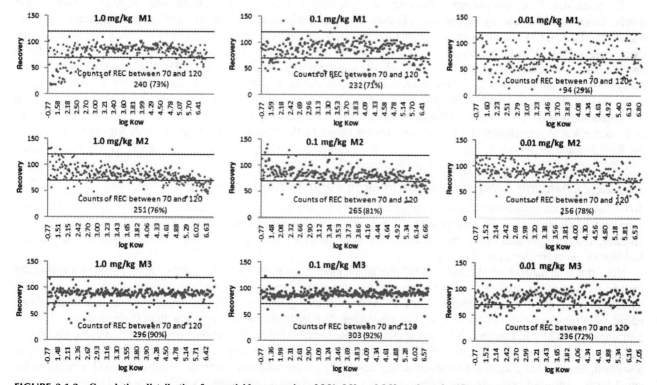

FIGURE 3.1.2   Correlation distribution for pesticide recoveries of M1, M2 and M3 at three fortification levels with log Kow in the range of −0.77–8.20.

TABLE 3.1.3 Distribution of Recoveries for 329 Pesticides by log Kow Sector

log Kow range	Pesticide polarities and number (Total:329)		1.0 mg/kg			0.1 mg/kg			0.01 mg/kg		
			M1	M2	M3	M1	M2	M3	M1	M2	M3
<1	Strong polarity	14	29.0	103.3	80.9	42.2	91.1	85.8	37.9	64.0	51.8
1–2	Polarity	25	38.4	101.3	82.0	65.3	87.2	82.7	75.0	87.3	71.1
2–3	Medium polarity	65	68.8	86.6	85.9	78.7	83.3	85.1	60.5	86.9	81.2
3–4		95	82.7	80.7	86.5	82.9	80.7	86.4	55.0	81.8	77.7
4–5		72	83.1	77.6	85.6	84.4	79.3	86.7	57.0	79.9	76.5
5–6	Weak polarity	27	73.5	71.7	84.4	74.9	72.0	81.1	47.3	64.1	70.1
>6	Nonpolar	31	70.2	62.7	87.2	59.4	63.5	88.8	48.8	61.2	64.8
AVE			63.7	83.4	84.6	69.7	79.6	85.2	54.5	75.0	70.5

of pesticides of different polarities, 329 pesticideswere divided into 7 sectors per log Kow values <1, 1–2, 2–3, 3–4, 4–5, 5–6, and >6 to statistically calculate the corresponding pesticide number, average recoveries, RSD, and mean S/N values for each sector (see Table 3.1.3). Recoveries, RSD, and S/N values can be traced to Supplemental Tables 1–6.

Fig. 3.1.2 shows that the number and ratios (%) of pesticides with recoveries 70%–120% at three fortification levels are 240 (73%), 232 (71%), and 94 (29%) for M1; 251 (76%), 265 (81%), and 256 (78%) for M2; and 296 (90%), 303 (92%), and 236 (72%) with M3. There are more varieties with M3 than M1 and M2. Moreover, the absolute majority of pesticide recovery values are relatively concentrated, which demonstrates that the M3 method has relatively wide application for a sufficient and balanced pesticide extraction of different polarities.

Table 3.1.3 shows: (1) there are 39 pesticides of relatively strong polarity in the range of log Kow <1 and 1–2, and at the three fortification levels. The recoveries for the overall ranking of the three methods is M2 > M3 > M1 because hydration has increased the extraction efficiency of these pesticides, enabling M2 > M3; the hexane liquid–liquid partitioning adopted for M1 has lowered the partitioning ratios of these polarity pesticides, causing recoveries to drop greatly; (2) there are 232 pesticides of medium polarity in the range of log Kow of 2–3, 3–4, and 4–5. At the high and medium fortification levels, the extraction efficiency for pesticides in these three ranges is balanced with M3, and their average recoveries are higher than M1 and M2, while with M2 pesticide recoveries within these three ranges drop regularly because with the drop of pesticide polarity hydration extraction efficiency obviously decreases. (3) There are 58 pesticides of weak polarity and nonpolarity in the range of log Kow 5–6 and >6, and recoveries for these pesticides with M1 and M2 are all obviously lower for than M3. This is because hydration has led to the marked drop of extraction efficiency in pesticides with weak polarity. The aforementioned analysis indicates that the hexane liquid–liquid partitioning adopted for cleanup with M1 sequentially and markedly lowers the pesticide extraction efficiency in the four sectors of 3–4, 2–3, 1–2, and <1.The hydration effect also gradually and sequentially lowers the extraction efficiency of nonpolar pesticides in the sectors of log Kow 5–6 and >6.

3.1.3.2.2 Changing Trend of Pesticide Polarities and Method Recoveries

To further prove the correlation of method recoveries and pesticide polarities, as per Table 3.1.3, choose 0.1 mg/kg fortified concentrated samples. Take the average value of recoveries for each sector obtained by the three methods as longitudinal ordinate, and the midvalues of the corresponding log Kow values as horizontal ordinate, to draw a histogram and draw a fortification recovery trend line (Fig. 3.1.3).

Next, the three methods of M1, M2, and M3 are independently analyzed. Fig. 3.1.3 shows that for the M1 trend line, the method recoveries increase—before decreasing with the pesticide polarities changing from strong to weak. This is because hydration lowers the extraction efficiency of pesticides with weak polarities, while the adoption of hexane liquid–liquid partitioning for cleanup causes the loss of polar pesticides, thus leading to an arc trend line high in the middle and low at both ends. For M2, hydration procedures are likewise adopted, increasing the extraction efficiency of pesticides with strong polarities and lowering the extraction efficiency of pesticides with weak polarities. Recoveries obviously take a high-to-low trend, demonstrating M2 has a relatively strong selectivity for pesticides of different polarities. For M3, deviations are small for recoveries of pesticides of different polarities without obvious fluctuations, which takes a balanced, stable, and similar-to-straight trend line.

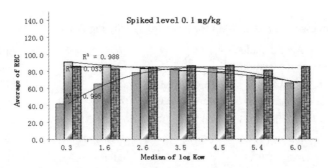

FIGURE 3.1.3   The recovery change diagram for 7 polarity sectors by the three methods.

We then make another comparative analysis of M1 and M2, M2 and M3, and M1 and M3, respectively. Likewise Fig. 3.1.3 shows that when comparing M1 and M2, the extraction efficiency of polarity pesticides (log Kow < 3.5) presents opposite trends, that is, M1 recoveries increase with the increase of log Kow and M2 recoveries decrease with the increase of log Kow. M1 recoveries begin at pesticides with log Kow > 3.5 and gradually increase until approximating those of M2. Owing to the identical extraction procedures of M1 and M2, such deviations are traced to M1's adoption of hexane liquid–liquid partitioning for cleanup, which lowers the recoveries of polarity pesticides. Average recoveries for 14 pesticides of strongest polarity were about one-half that of M2. M2's adoption of the SPE cleanup method, identical to M3, has shown balanced and good recoveries for pesticides of different polarities. Comparing M2 with M3, their trend lines cross at the point of log Kow = 2.6, that is, for 74 polarity pesticides (log Kow < 2.6), M2's recoveries were obviously higher than M3's. Beginning at log Kow > 2.6, M2's recoveries gradually decrease and are lower than M3's for 255 pesticides of medium and weak polarities. With similar cleanup procedures for M2 and M3, in comparison with M3's pure acetonitrile extraction, M2's hydration is capable of increasing the extraction efficiency of compounds of strong polarities while decreasing the extraction efficiency of compounds of weak polarities. Comparing M1 with M3, as well as M1's trend line overall under the M3 trend line, M1 has lower recoveries than M3 for the absolute majority of pesticides. Therefore, we conclude that hydration increases the extraction efficiency of polarity pesticides but, in the meantime, decreases the extraction efficiency of pesticides of weak polarity and nonpolarity—owing to the hexane partitioning design of the M1 procedures. Moreover, the adoption of hexane liquid–liquid partitioning for cleanup causes lower recovery of polarity pesticides. Therefore, M1's extraction and cleanup are self-contradictory, which results in an overall extraction efficiency lower than M3's.

The aforementioned analysis shows that M3's pure acetonitrile extraction achieves relatively good results for pesticides of different polarities, while M1's hydration increases the extraction efficiency of certain polarity water-soluble pesticides. However, the M1 adoption of hexane liquid–liquid partitioning causes certain loss of extracted polarity pesticides. Overall speaking, M3's extraction efficiency is superior to M1's.

3.1.3.2.3   Correlation Comparison of Determination Results of Pesticide Aged Samples With Pesticide log Kow Values

The correlation of extraction efficiency of three preparation methods and pesticide polarities regarding aged samples are shown in Table 3.1.4. Table 3.1.4 indicates that there are 23 out of 26 water-soluble pesticides with log Kow values less than 1.8, with the M2 determination concentration higher than M3, accounting for 92.3%; there are 28 fat-soluble pesticides with log Kow values higher than 6.0, each with the M3 determination concentration higher than M2.

A typical example for hydration decreasing the extraction efficiency of pesticides of nonpolarity is demonstrated in Table 3.1.5, which compares two kinds of specially composed pesticides DDT and long-lasting organic pollutants (PCB) fortification recoveries and aged samples determination results. Due to the very weak polarity of these two kinds of compounds, the content tested with M2 is far lower than with M3.

It has also been found in the experimental process that while carrying out hexane liquid–liquid partitioning with M1, acetonitrile extraction solution, hexane, and 5% NaCl water solutions (1:1:5,v/v) are statically layered after manual oscillation, and there is an emulsion layer between the upper layer of hexane phase and the lower layer of water phase, with their content varying as the determined lots and categories of tea vary.

After high-speed centrifuging at 10,000 rpm, this emulsified layer is usually compressed a black oily drop existing between the upper and lower phases. Research demonstrates that when the fat content in the samples exceeds 1% (tea contains 2% phospholipids, glycolipids, sulfatide, triglyceride, etc.), nonpolar pesticides will enter into the third phase, in addition to water and organic phases: an emulsified layer, leading to low recoveries, which is in conformity with the low recoveries

TABLE 3.1.4 Extraction Comparison of M2 and M3 for Pesticides of Strong Polarity and Nonpolarity in Aged Samples

No.	Log Kow < 1.8	Aged sample concentrations (µg/kg)		Log Kow	M3/ M2	Log Kow > 6.0	Aged sample concentration (µg/kg)		Log Kow	M3/ M2
		M2	M3				M2	M3		
1.	Sulfallate	153.3	186.3	−0.77	1.2	o,p′-DDE	263.9	329.0	6.47	1.2
2.	Dimethipin	290.2	116.1	−0.17	0.4	cis-Chlordane	186.3	390.4	6.10	2.1
3.	Dicrotophos	224.8	93.5	0.00	0.4	Aldrin	277.5	364.7	6.41	1.3
4.	Oxadixyl	344.2	160.7	0.73	0.5	Bioresmethrin	321.8	423.3	6.14	1.3
5.	2,6-Dichlorodenzamide	941.0	575.7	0.77	0.6	DE-PCB 118	226.5	378.4	6.57	1.7
6.	Phosphamidon-2	290.6	135.4	0.79	0.5	Bromophos-ethyl	286.8	341.7	6.15	1.2
7.	Phosphamidon-1	320.0	162.9	0.79	0.5	Flucythrinate	363.7	494.7	6.20	1.4
8.	Metamitron	225.7	218.5	0.98	1.0	Fenvalerate	483.0	1106.4	6.20	2.3
9.	Demeton-S-methyl	21.1	3.8	1.02	0.2	DE-PCB 101	252.3	355.5	6.16	1.4
10.	Mephosfolan	380.5	253.6	1.04	0.7	alpha-Cypermethrin	842.6	890.9	6.94	1.1
11.	Phthalimide	418.3	157.0	1.15	0.4	Flufenoxuron	242.1	432.8	6.16	1.8
12.	Desisopropyl-Atrazine	183.1	108.5	1.15	0.6	Heptachlor	477.4	735.1	6.66	1.5
13.	Paraoxon-methyl	2417.7	2842.6	1.33	1.2	Cypermethrin	520.1	1215.1	6.00	2.3
14.	Hexazinone	329.5	187.8	1.36	0.6	p,p′-DDD	387.8	683.9	6.02	1.8
15.	Formothion	1079.2	424.6	1.48	0.4	DE-PCB 153	198.5	357.0	6.80	1.8
16.	Atrazine-desethyl	212.4	147.9	1.51	0.7	Lambda-Cyhalothrin	441.3	691.9	7.00	1.6
17.	Propoxur-1	342.6	181.4	1.52	0.5	Bifenthrin	236.7	378.7	6.00	1.6
18.	Propoxur-2	440.4	280.6	1.52	0.6	Esfenvalerate	491.9	930.7	6.22	1.9
19.	Dimethyl Phthalate	387.9	250.1	1.56	0.6	DE-PCB 180	156.6	367.3	6.89	2.3
20.	Thionazin	453.3	306.9	1.58	0.7	o,p′-DDT	572.1	1578.9	6.53	2.8
21.	Metalaxyl	369.4	193.7	1.59	0.5	o,p′-DDD	182.3	389.3	6.42	2.1
22.	Pyroquilon	326.7	166.7	1.60	0.5	Chlorfluazuron	138.0	258.8	6.63	1.9
23.	Metribuzin	552.1	401.5	1.70	0.7	Octachlorostyrene	223.0	321.1	6.29	1.4
24.	Bendiocarb	357.7	280.4	1.70	0.8	Etofenprox	232.6	368.1	7.05	1.6
25.	Methabenzthiazuron	344.1	264.5	1.77	0.8	p,p′-DDE	133.4	346.9	6.96	2.6
26.	Tebuthiuron	376.7	195.7	1.79	0.5	p,p′-DDT	572.1	1578.9	6.36	2.8
27.						Permethrin	237.4	381.9	6.50	1.6
28.						Pyridaben	245.6	396.5	6.37	1.6
29.						Silafluofen	196.3	396.0	8.20	2.0
	M3/M2 < 1				24	M3/M2 > 1				29
	Accounting for				92.3%	Accounting for				100.0%

of these two kinds of specially composed DDT and PCB in the fortified recovery experiment. It illustrates that the water-soluble matrix impurities introduced by hydration interfere withe method efficiency.

3.1.3.3 General Analysis of Method Applicability

3.1.3.3.1 Comparison of the Sensitivity of the Three Methods

S/N test results for 456 pesticides in the tea matrices of three fortification levels by three methods are shown in Supplemental Tables 4–6. Average S/N values of these 456 pesticides are listed in Table 3.1.6.

TABLE 3.1.5 Hydration Decreases the Extraction Efficiency of DDT and PCB

Compounds		log Kow	Fortified recoveries/0.1 mg/kg				Aged sample determination values (µg/kg)			
			M1	M2	M3	M3/M1	M1	M2	M3	M3/M1
DDT	p,p′-DDD	6.02	76.6	61.5	87.4	1.1	552.0	387.8	683.9	1.2
	p,p′-DDT	6.36	70.0	51.7	78.5	1.1	1181.5	572.1	1578.9	1.3
	o,p′-DDD	6.42	73.0	50.0	87.0	1.2	351.5	182.3	389.3	1.1
	o,p′-DDE	6.47	69.8	64.5	86.8	1.2	320.5	263.9	329.0	1.0
	o,p′-DDT	6.53	71.3	56.3	88.7	1.2	1176.3	572.1	1578.9	1.3
	p,p′-DDE	6.96	68.3	45.6	87.6	1.3	278.7	133.4	346.9	1.2
PCB	DE-PCB 28	5.71	73.9	76.0	94.9	1.3	330.6	316.1	381.6	1.2
	DE-PCB 52	5.79	74.1	71.9	94.7	1.3	314.3	311.4	371.7	1.2
	DE-PCB 31	5.81	73.9	73.0	94.9	1.3	330.6	316.1	381.6	1.2
	DE-PCB 101	6.16	66.4	66.8	94.0	1.4	263.4	252.3	355.5	1.3
	DE-PCB 118	6.57	63.1	54.3	94.7	1.5	242.0	226.5	378.4	1.6
	DE-PCB 153	6.80	57.9	64.6	92.8	1.6	206.9	198.5	357.0	1.7
	DE-PCB180	6.89	53.0	46.3	92.7	1.7	172.1	156.6	367.3	2.1

Regarding aged samples, there are 379 pesticides that can be determined by all three sample preparation methods. S/N ratios in these 379 pesticides in tea matrices are shown in Supplemental Table 7. The average S/N for these 379 pesticides, taken with M1, M2, and M3, respectively, are sequentially 949, 6096, and 7027. The comprehensive analysis of the aforementioned results shows that S/N ratios obtained from the three sample preparation methods whether for 456 pesticides in the fortification samples or for 379 pesticides in the aged samples are S/N M3 > S/N M2 > S/N M1.

Because of transferring of only 1 mL of acetonitrile extraction solution for cleanup with M1, pesticides of low sensitivity cannot be detected or noises incurred by impurities in the matrices covered, causing decreasing method sensitivity. Figs. 3.1.4 and 3.1.5 show extraction ion chromatogram for GC–MS/MS determination of fluoroglycofen-ethyl and oxadixyl, respectively, with adoption of M1 and M3 for sample preparation. In comparison with M3, M1 is low in detection responses, has relatively large deviations for ion abundance ratio, poor peaks, and more interfering peaks close by.

In conclusion, from what is described earlier, despite transferring only 1 mL extraction solution for cleanup with M1 and taking the overall quantity (about 8 mL) for cleanup with M3, the concentrations of tea matrix that are finally injected into the instrument with M1 are only one-eighth that of M3, while the average S/N ratios with M1 in aged samples are only one-seventh of M3. This also demonstrates that hydration has increased the interfering matters of coextraction greatly, seriously affecting the method sensitivity.

3.1.3.3.2 Comparison of the Extraction Efficiencies of the Three Methods

Aged Tea Samples

The single-point quantification method is adopted for determination of the content of 456 pesticides in aged tea samples, with results tabulated in Supplemental Table 7. The number of pesticides detected with the three sample preparation methods respectively are 382 with M1, 404 with M2, and 410 with M3. The content ratio is calculated for the same pesticide

TABLE 3.1.6 Comparison of Average S/N Values From Three Pretreatment Methods and Three Fortification Concentrations

Fortification levels/mg/kg	M1	M2	M3
1.00	6234	11127	14224
0.10	432	1656	3048
0.01	73	240	229

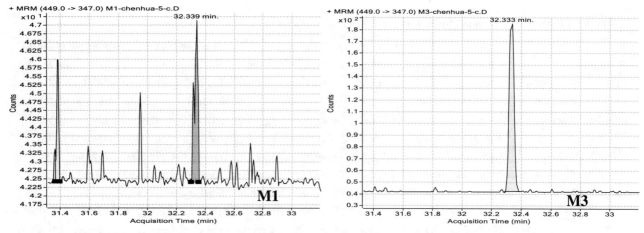

FIGURE 3.1.4 Fluoroglycofen-ethyl with respective M1 and M3 sample preparation and GC–MS/MS determination.

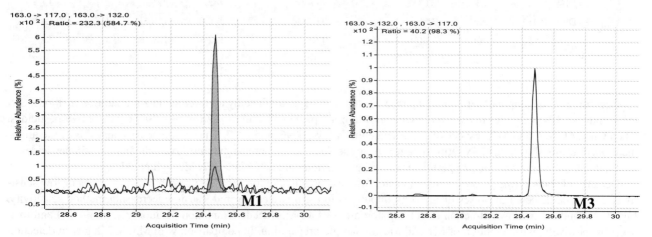

FIGURE 3.1.5 Oxadixyl with respective M1 and M3 sample preparation and GC–MS/MS determination.

with M3 and M1 to obtain the content ratio of M3/M1, and with M3 and M2 to obtain the content ratio of M3/M2, with the pesticide number statistically computed in the four ratio sectors listed in Table 3.1.7.

M3/M1 in Table 3.1.7 shows that there are 201 pesticides with content ratio 0.8–1.2, accounting for 53% in which pesticides can be detected overall by M1 and M3. There are 116 pesticides with ratios greater than 1.2, many more than the 65 pesticides with ratios less than 0.8. It is evident that the extraction efficiency for aged samples with M3 is higher than with M1. M3/M2 shows that there are 176 pesticides with a content ratio 0.8–1.2, accounting for 46% overall in which pesticides can be detected by M2 and M3, and there are 137 pesticides with ratios greater than 1.2, many more than the 87 pesticides with ratios less than 0.8. Thus, it reveals that the extraction efficiency for aged samples with M3 is higher than with M2.

In addition, compared with M3, there are 28 pesticides that fail detection with M1; compared with M2, there are 25 pesticides that fail detection with M1. To follow through on what was discussed earlier, nonhydration with M3 is superior

TABLE 3.1.7 Distribution of Pesticide Content Ratios Determined by the Three Sample Preparation Methods for Aged Samples

Ratiosector	≤0.6	0.6–0.8	0.8–1.2	≥1.2	The average ratio of overall pesticides that can be determined
M3/M1 pesticide number	13	52	201	116	1.17
M3/M2 pesticide number	36	51	176	137	1.09

to hydration with M1 and M2, no matter the extraction efficiency of pesticides or for the application ranges of pesticides. That is, M3 is high in extraction efficiency for aged samples and shows a larger range of pesticides that are applicable.

Incurred Tea Samples

The content of 18 pesticides for determination in incurred tea samples is tabulated in Supplemental Table 9, and the pesticide content ratios determined by the three sample preparation methods are respectively computed, with the pesticide numbers of M3/M1 and M3/M2 in the four ratio sectors listed, in Table 3.1.8.

TABLE 3.1.8 Distribution of Pesticide Content Ratios Determined by the Three Sample Preparation Methods for Incurred Tea Samples

Ratio range	≤0.6	0.6–0.8	0.8–1.2	≥1.2	Average ratios of 18 pesticides
M3/M1 pesticide number	0	4	10	4	1.01
M3/M2 pesticide number	0	5	7	6	1.09

Table 3.1.8 shows that M3/M1 has found 10 out of 18 pesticides with content ratios 0.8–1.2, 4 pesticides with ratios greater than 1.2 and less than 0.8, and no pesticides with ratios less than 0.6. The average ratio of the 18 pesticides is 1.01. Therefore, it can be said that for the majority of pesticides the extraction efficiencies of these two methods are identical, with certain differences among specific pesticides. M3/M2 has found that there are 7 out of 18 pesticides with content ratios 0.8–1.2, and there are 6 and 5 pesticides, respectively, with ratios greater than 1.2 and less than 0.8, and no pesticides with atios less than 0.6. The average ratio of the 18 pesticides is 1.09.

There exist certain differences when considering the extraction efficiencies of various pesticides with M3 and M2, but the overall extraction efficiencies are basically identical. Based on what was described earlier, the overall extraction efficiencies are approximately the same although there exist certain differences in extraction efficiencies for different pesticides with incurred samples when comparing the hydration of M1 and M2 with the nonhydration of M3. There are only 18 pesticides for the appraisal of incurred samples. Therefore, it is still beyond our capacity at present to judge which of these three methods is good or bad for the other hundreds of pesticides. This is an issue that needs further study.

Comparison of Fortified Recoveries for the Three Sample Preparation Methods Under "Uniform Limit Concentrations"

Fortified recovery experimental results from 456 pesticides by the three preparation methods are listed in Supplemental Tables 1–3; the fortified recovery experiment results from testing 160 pesticides by the two methods are tabulated in Supplemental Table 8. Also, pesticide ratios for 10 μg/kg fortification levels and at the same time conforming to recoveries 70%–120% and RSD < 20% are listed in Table 3.1.9.

In the fortification experiment of 456 pesticides, pesticides that conform to EU SANCO/12495/2011 technical requirements with M3 account for 50.2%, M2 45.0%, but withM1 only 5%.The method applicability is M3 > M2 > M1: in the fortification experiment of 160 pesticides, M3 is also superior to M1.

Comparison of RSD for Pesticides in Aged Tea Samples Determined by the Three Sample Preparation Methods

The analytical results from the aged oolong tea samples sprayed with 456 pesticides are shown in Supplemental Table 7, and the RSD data for the pesticides discovered are now divided into four sectors, with distribution shown in Table 3.1.10.

TABLE 3.1.9 Pesticide Ratios for 0.010 mg/kg "Uniform Limit Concentrations" and Conforming to Rec70%–120% and RSD < 20%

	M1		M2		M3	
Pesticide varieties	Pesticide number	Ratios/%	Pesticide number	Ratios/%	Pesticide number	Ratios/%
456	23	5.0	205	45.0	229	50.2
160	17	10.6			50	31.2

TABLE 3.1.10 Pesticide Number Discovered in Aged Samples and Distributions and Ratios (%) of 4 RSD Sectors

RSD%	M1	M2	M3
<10	158(41%)	112(28%)	381(93%)
10–15	136(36%)	209(52%)	14(3%)
15–20	58(15%)	56(14%)	7(2%)
>20	30(8%)	27(7%)	8(2%)
Total	382	404	410

From the aged samples sprayed with 456 pesticides, the pesticide number actually detected is, respectively, 382 with M1, 404 with M2, and 410 with M3. It can seen that M3 > M2 > M1 in terms of sample preparation efficacy. In terms of method ruggedness, there are 381 pesticides with RSD < 10% with M3, accounting for 93%, much more than the ratios with M1 and M2—41% and 28%, respectively. There are 30 and 27 pesticides with pesticide number RSD > 20%, respectively, for M1 and M2—much more than M3 with 8 pesticides. Therefore, when considering precision, M3 is better than M1 and M2 in applicability.

3.1.3.4 Comparison of the Cleanup Efficiency for the Three Methods

S/N results obtained from GC–MS/MS determination of 456 pesticide aged samples using the three sample preparation methods are shown in Supplemental Tables 4–6, with their added average S/N values for M1, M2, and M3, respectively, of 942, 5753, and 6781. It is evident that M3 method sensitivity is superior to M1. The color of interfering matters left on the SPE columns can testify this point as proof. See Fig. 3.1.6.

In addition, the oolong tea blank matrix extraction solutions prepared by the three methods are fully scanned for their total ion chromatogram (Scan-TIC) (Fig. 3.1.7), among which the contents of M1 (black), M2 (red), and M3(green) are all 0.2 g/mL tea.

Fig. 3.1.7 also shows that there are lower baselines and fewer interfering peaks in TIC of M3 than those in TIC of M1 and M2, which also proves that M3 acetonitrile extraction with SPE cleanup sample preparation techniques are superior to hexane liquid–liquid portioning technique with tea hydration extraction.

Based on the discussion of the analytical results from the three sample preparation methods, the comprehensive statistical analysis indexes are shown in Tables 3.1.11 and Table 3.1.12.

3.1.4 CONCLUSIONS

This experiment has adopted three sample preparation methods: M1, tea hydration + acetonitirle oscillating extraction + hexane liquid/liquid partitioning cleanup for part of extraction solution; M2, tea hydration + acetonitirle oscillating extraction + SPE cleanup; M3, pure acetonitrile homogeneous extraction + SPE cleanup—for a GC–MS/MS determination of 456 pesticide fortified samples, 456 pesticide aged samples, and 18 pesticide field-incurred samples. In this research of 456 pesticides, the log Kow values of 329 pesticides have been determined. Taking the pesticide recoveries as the

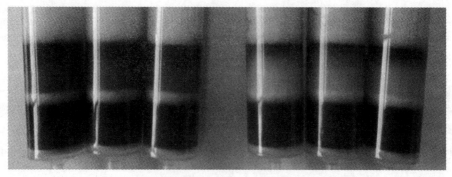

FIGURE 3.1.6   Sketch of Cleanert-TPT SPE cleanup effects with samples prepared by the three methods.

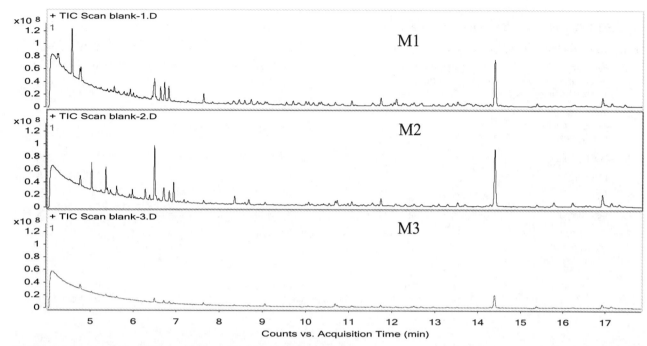

FIGURE 3.1.7   Scan-TIC of blank tea matrix prepared by the three methods.

TABLE 3.1.11 Comparison of the Analytical Results by GC–MS/MS in Four Kinds of Samples Prepared With M1, M2 and M3

Method	Ratio of REC 70%–120%			Ratio of RSD < 20%			Ratio of REC 70%–120% and RSD < 20%			S/N		
	1.0 mg/kg	0.1 mg/kg	0.01 mg/kg	1.0 mg/kg	0.1 mg/kg	0.01 mg/kg	1.0 mg/kg	0.1 mg/kg	0.01 mg/kg	1.0 mg/kg	0.1 mg/kg	0.01 mg/kg
(1) The fortified recoveries of 456 pesticides												
M1	73.5	68.6	26.5	84.9	62.9	34.2	63.8	45.8	5.0	6234	432	73
M2	77.4	78.5	76.8	68.6	75.7	62.9	53.8	59.2	45.0	11127	1656	240
M3	88.6	91.4	71.3	87.5	90.6	66.4	80.9	83.8	50.2	14224	3048	229
(2) The fortified recoveries of 159 pesticides												
M1	75.5	67.9	14.5	99.4	89.9	15.1	75.6	76.3	10.6	1234	145	
M3	90.6	75.5	49.7	100.0	97.5	69.8	90.6	80.0	33.1	36410	2459	

Method	No.s of peaking pesticides	S/N	RSD	the ratios of content		M3/M1	M3/M2	The ratios of content		M3/M1	M3/M2
	(3) The analytical results of 456 pesticides in aged samples (n = 5)							**(4) The analytical results of 456 pesticides in incurred samples (n = 4)**			
M1	382	942	12.2	≤0.8	Pesticide No.s	65	87	≤0.8	Pesticide No.s	6	6
M2	404	5753	12.8	0.8–1.2		201	176	0.8–1.2		8	8
M3	410	6781	4.1	≥1.2		116	137	≥1.2		4	4

TABLE 3.1.12 Technical Indexes for M1, M2 and M3

No.	Technical indexes	Sources	M1	M2	M3	Conclusions
1.	Percentage of pesticides with 70%–120% recoveries	0.01 mg/kg Fortified recovery experiment	26.54%	76.75%	71.27%	M2 > M3 > M1
2.	Percentage of pesticides with RSD < 20%		34.21%	62.94%	66.45%	M3 > M2 > M1
3.	LOD, µg/kg		3.2	1.9	1.5	M3 < M2 < M1
4.	LOQ, µg/kg		10.6	6.2	5.0	M3 < M2 < M1
5.	S/N	Aged sample testing	942	5753	6781	M3 > M2 > M1
6.	Peak height	Aged sample testing	7.6E+03	1.4E+05	1.2E+05	M2 > M3 > M1
7.	Correlation of recoveries VS pesticide polarities		log Kow↑, recovery: firstly ↑finally ↓	log Kow↑, recovery↓	log Kow↑, recovery stale and balanced	(1) Hydration and hexane LLE affect pesticide recoveries; (2) acetonitrile homogeneous and SPE have a sufficient and balanced extraction for pesticides of different polarities
8.	Calibration linear coefficient (R^2)	Incurred sample testing	0.9765	0.9995	0.9987	M2 > M3 > M1
9.	Extraction results		Low recoveries, and results tend to be falsely higher for aged sample and incurred sample determinations	Relatively high recoveries, and have good results in extracting water-soluble pesticides of polarities, with extraction matrices of high content of foreign materials	Recoveries fall within the scope of good practices, and have a stale and well-balanced extraction of pesticides	M3 > M2 > M1
10.	Cleanup results	0.01mg/kg Fortified recovery test	S/N = 73	S/N = 7240	S/N = 7229	M3 > M2 > M1
		Incurred sample testing				Regarding color of the pesticide extraction solution, (1) darkest with M2, M1 next to it, and relatively transparent and clear with M3; (2) with M1 and M2 the extraction solution will have flocculent precipitate after being stored away from exposure to light for a certain amount of time
		Oolong tea matrix Scan-TIC	High baselines	Relatively high baselines	Low baselines	
		Aged samples testing	$\delta_{RT} = 0.009$	$\delta_{RT} = 0.004$	$\delta_{RT} = 0.004$	M3 < M2 < M1

TABLE 3.1.12 Technical Indexes for M1, M2 and M3 (*cont.*)

No.	Technical indexes	Sources	M1	M2	M3	Conclusions
11.	Method fitness (number of peaking pestitcides)	Aged samples testing	382	404	410	M3 > M2 > M1
12.	Method selectivity/ special effect	Aged samples testing	Poor extraction efficiency for pesticides within the range of log Kow < 3.5 and log Kow > 5.4	Poor extraction efficiency for pesticides within the range of log Kow > 3.5	Relatively good extraction efficiency for pesticides within the range of log Kow (−0.77 to 8.20)	Have sufficient extraction for all pesticides with different polarities with M3

longitudinal axis and the log Kow as horizontal axis to establish a mathematical regression equation, these findings are obtained: the extraction efficiency of hydration method M1 presents obvious correlation with pesticide log Kow values and hydration increases extraction efficiencies of certain polarity pesticides, such as the 24 pesticides of strong polarity, but in addition, this has decreased the extraction efficiencies of certain pesticides of weak polarity, such as the 29 pesticides of nonpolarity. The M3 nonhydration method has almost nothing to do with the polarities of pesticides in extraction efficiencies, presenting a straight line nearly parallel to the log Kow horizontal axis, with its extraction efficiencies balanced and sufficient for pesticides of different polarities and with a wider range for application.

The statistics of the log Kow values of 329 pesticides also find that there are 14 pesticides with log Kow value <1.0, 25 with log Kow value 1.0–2.0, 65 with log Kow value 2.0–3.0, 95 with log Kow value 3.0–4.0, 72 with log Kow value 4.0–5.0, 27 with log Kow value 5.0–6.0, and 31 with log Kow value >6.0. If pesticides with log Kow values <1.0 are considered to be pesticides of strong polarity and those of 1.0–2.0 polarity pesticides, these two items total 39 pesticides, accounting for 12%. Other medium polarity, weak polarity, and nonpolarity pesticides total 290, accounting for 88%. The ratio of polarity pesticides is far smaller than that of nonpolarity pesticides, proving that hydration does more harm than good, which is the first point. Second, interfering matters from coextraction after tea hydration have increased, leading to an obvious drop in method efficiency. Only part of the extraction solution (1 mL) is transferred for cleanup with M1, which is equivalent to diluting the sample matrix 10 times—while the entire quantity is taken for cleanup with M3, and the target matter concentration is about 9 times larger than that with M1. S/N with M3 is 6781, 940 with M1, and the sensitivity of M3 is 7 times greater than that of M1.

The fortification experiment of 0.010 mg/kg (uniform limit) demonstrates that there are 23 pesticides with hydration method M1 that meet the EU SANCO/12495/2011 technical requirements of recoveries 70%–120% and conform to RSD ≤20%, accounting for 5.0%, while there are 229 pesticides with nonhydration M3, accounting for 50%, from which it is evident that M1 has failed to meet the good practice requirement of the EU. For aged samples with residual concentrations above 380 μg/kg, there are 158 pesticides that meet RSD ≤ 10% with M1, accounting for 35%, but 381 with M3, making up 84%, testifying that M3 is 2.4 times superior to M1.

In addition, the experiment also found that there are two technical defects with M1's design: on the one hand, tea hydration increases the extraction efficiencies of polarity pesticides, but on the other hand, hexane liquid–liquid partitioning is adopted for its cleanup, which increases the loss of polarity pesticides extracted in the hexane partitioning. Both M1 and M2 use hydration, but the cleanup adopted by M2 is identical to M3, and M2's recoveries for polarity pesticides are higher than those with M3, which is a good proof of M1's defect. Therefore, M1 equipped with hexane liquid–liquid partitioning is self-contradictory; only 1 mL extraction solution transferred with M1 is considered to be equivalent to 0.2 g sample, but the fact is that 10 mL acetonitile extraction solution is finally left to 8.0–8.5 mL, so 1 mL extraction solution taken is approximately equivalent to 2.35 g and computation results with M1 are about 20% falsely higher, which should be deducted from its results. For M1, the extraction solutions left after hydration extraction of different teas also change in volume. If taking part of it for cleanup, the variability is great, but if taking the overall, it is difficult for cleanup, so the variability with M1 is far greater than that with M3.

Owing to the reasons stated in the aforementioned three aspects, using tea hydration to increase extraction efficiencies of pesticides has more disadvantages than advantages for the determination of hundreds of multiclasses and multikinds of residue pesticides in tea, which proves it to be unworthy and unadvisable.

REFERENCES

[1] Oh, C.H., 2007. Purification method for multi-residual pesticides in green tea. Nat. Prod. Commun. 2 (10), 1025–1030.

[2] Huang, Z., Li, Y., Chen, B., 2007. Simultaneous determination of 102 pesticide residues in Chinese teas by gas chromatography–mass spectrometry. J. Chromatogr. B 853, 154–162.

[3] Lozano, A., Rajskia, Ł., Belmonte-Valles, N., Ucles, A., Ucles, S., Mezcua, M., Fernandez-Alba, A., 2012. Pesticide analysis in teas and chamomile by liquid chromatography and gas chromatography tandem mass spectrometry using a modified QuEChERS method: validation and pilot survey in real samples. J. Chromatogr. A 1268, 109–122.

[4] Huang, Z., Zhang, Y., Wang, L., 2009. Simultaneous determination of 103 pesticide residues in tea samples by LC–MS/MS. J. Sep. Sci. 32 (9), 1294–1301.

[5] Chen, G., Cao, P., Liu, R., 2011. A multi-residue method for fast determination of pesticides in tea by ultraperformance liquid chromatography–electrospray tandem mass spectrometry combined with modified QuEChERS sample preparation procedure. Food Chem. 125, 1406–1411.

[6] Schurek, J., Portol′e, T., Hajslova, J., 2008. Application of head-space solid-phase microextraction coupled to comprehensive two-dimensional gas chromatography–time-of-flight mass spectrometry for the determination of multiple pesticide residues in tea samples. Anal. Chim. Acta 611, 163–172.

[7] Wu, C.-C., Chu, C., Wang, Y.-S., 2009. Multiresidue method for high-performance liquid chromatography determination of carbamate pesticides residues in tea samples. J. Environ. Sci. Health B 44 (1), 58–68.

[8] Li, H.-P., Li, G.-C., Jen, J.-F., 2004. Fast multi-residue screening for 84 pesticides in tea by gas chromatography with dual-tower auto-sampler, dual-column and dual detectors. J. Chin. Chem. Soc. 51, 531–542.

[9] Bappaditya, K., Sudeb, M., Anjan, B., 2010. Validation and uncertainty analysis of a multiresidue method for 42 pesticides in made tea, tea infusion and spent leaves using ethyl acetate extraction and liquid chromatography-tandem mass spectrometry. J. Chromatogr. A 1217, 1926–1933.

[10] Hu, B., Song, W., Xie, L., 2008. Determination of 33 pesticides in tea by accelerated solvent extraction-gel permeation and solid-phase extraction purification-gas chromatography-mass spectrometry. Chin. J. Chromatogr. 1, 22–28.

[11] Hu, Y.-Y., Zheng, P., He, Y.-Z., 2005. Response surface optimization for determination of pesticide multiresidues by matrix solid-phase dispersion and gas chromatography. J. Chromatogr. A 1098, 188–193.

[12] Moinfara, S., Hosseinia, M.-R, 2009. Development of dispersive liquid–liquid microextraction method for the analysis of organophosphorus pesticides in tea. J. Hazard. Mater. 169, 907–911.

[13] Lou, Z., Chen, Z., Luo, F., 2008. Determination of 92 pesticide residues in tea by gas chromatography with solid-phase extraction. Chin. J. Chromatogr. 26 (5), 568–576.

[14] Pang, G.-F., Fan, C.-L., Zhang, F., 2011. High-throughput GC/MS and HPLC/MS/MS techniques for the multiclass, multiresidue determination of 653 pesticides and chemical pollutants in tea. J. AOAC Int. 94 (4), 1253–1263.

[15] Pang, G.-F., Fan, C.-L., Chang, Q.-Y., 2013. High-throughput analytical techniques for multiresidue, multiclass determination of 653 pesticides and chemical pollutants in tea—Part III: evaluation of the cleanup efficiency of an SPE cartridge newly developed for multiresidues in tea. J. AOAC Int. 96 (4), 887–896.

[16] Fan, C.-L., Chang, Q.-Y., Pang, G.-F., 2013. High-throughput analytical techniques for determination of residues of 653 multiclass pesticides and chemical pollutants in tea, Part II: comparative study of extraction efficiencies of three sample preparation techniques. J. AOAC Int. 96 (2), 432–440.

[17] Cajka, T., Sandy, C., Bachanova, V., 2012. Streamlining sample preparation and gas chromatography-tandem mass spectrometry analysis of multiple pesticide residues in tea. Anal. Chim. Acta 743, 51–60.

Chapter 3.2

Uncertainty Evaluation of the Determination of Multipesticide Residues in Tea by Gas Chromatography– Tandem Mass Spectrometry Coupled with Three Different Pretreatment Methods

3.2.1 INTRODUCTION

As one of the main beverages in China with its antioxidant properties, tea plays an important role in the exporting of traditional agricultural products. However, to ensure high-quality tea crop production, pesticides, such as carbamate and organophosphorus, are widely used in the period of cultivation or postharvest processing procedure to minimize the problems caused by weeds, disease, and pests. Many residual pesticides may be present in tea and thus cause potential health risks to tea consumers and impose great pressure on the environment. Some developed countries and international organizations have issued maximum residue limits (MRLs) for these pesticides in tea. Along with more strict demand on the content of pesticide residues, the export of tea will be greatly affected. Due to the high risk for consumer health and the guarantee of international trade of tea, rapid and cost-effective multiple pesticide residues analysis represents an important task for both the tea producers and regulatory agencies.

As one of the main parameters for the confirmation of analysis method [1], uncertainty is "a parameter associated with the measurement," or it has the ability of charactering the dispersion of the measurement [1]. Uncertainty is a token of the scope of the true value measured, and the real value of the measurement was decided by its magnitude. The smaller the uncertainty, the more precise the measurement result.

Analysis of Pesticide in Tea. http://dx.doi.org/10.1016/B978-0-12-812727-8.00009-2

The suggestion of quantitative uncertainty in a metrological calibration system was first proposed by Youden in 1962 [2]. In the "Guidance to Expression of Uncertainty in Measurement," which was published by ISO jointly with BIPM, IEC, IFCC, IUPAC, IUPAP, and OIML in 1993, the general principles were formally set up for the evaluation and expression of uncertainty in measurement fields [3]. In the second edition of the guide, it is emphasized that when the uncertainty evaluation is introduced into the quality control system of laboratory, it should be combined with existing quality assurance measures, for the quality assurance measures can provide a lot of information needed to evaluate the uncertainty of measurement, and meanwhile offer the examples of the source of uncertainty in chemical experiment [4]. As the application of measurement uncertainty in analytical chemistry is more and more wide [5–10], so the correct expression and evaluation of measurement uncertainty has gradually become the international demand.

At present, the two approaches, bottom-up approach [11] and top-down approach [12], have been widely used for the evaluation of measurement uncertainty, while the bottom-up approach was more common [13]. The first step of this approach is the assignment of the components of uncertainty or the establishment of the mathematical model of the measurement, then the identification of the belonging and the quantitative characterization of each component (both type A and type B evaluation method can be adopted), finally the combination of uncertainty. The advantage of this approach lies in its guarantee of analyzing the comprehensive understanding of the analysis method and the contribution of each uncertainty component to total uncertainty, so the critical control points can be mastered when the approach was applied in the determination of practical samples, thus the uncertainty can be decreased or controlled [9,14,15]. Until now, the bottom-up approach has been applied in the evaluation of measurement uncertainty of pesticides and contaminants in many works of literature [8,14,16,17], but the selection of uncertainty as evaluation criteria of determination method was seldom reported.

The research for the determination of 18 pesticide residues in tea by gas chromatography-tandem mass spectrometry (GC–MS/MS) coupled with different sample preparation techniques: (1) hydration expansion + oscillation extraction with acetonitrile + liquid-liquid extraction and cleanup with hexane; (2) hydration expansion + oscillation extraction with acetonitrile + solid-phase extraction and cleanup; (3) homogeneous extraction with acetonitrile + solid-phase extraction and cleanup, has been carried out by our team. According to Eurachem/Citac guide [4], the components of uncertainty arising from the process of determination of pesticide residue employing the aforementioned three sample preparation methods combined with GC–MS/MS by the "bottom-up" approach was discussed in detail on the basis of prior research. As the final model of the three methods consistent with each other, so the components of uncertainty of the three methods were the same and they were repeatability, weighing, preparation of standard solution, the concentration of internal standard solution, the volume of internal standard solution, the preparing process of quantitative using standard curve, the quantitative process using standard curve, the concentration of pesticides (this component is related to LOD), and recovery.

The main purpose of this chapter is to combine the relative standard uncertainty and calculate the contribution of each component to the combined relative standard uncertainty by the evaluation of the above nine uncertainty components, and to set uncertainty as the index for the comparison of the above three methods. Besides, one of the great contributions of this chapter is the introduction of the variation of the concentration of pesticides in solvents as a portion of uncertainty arising from the preparation process of quantitative using standard curve and it has not been reported in the existing literature. Meanwhile, the introduction of the concentration of pesticide residue in an actual sample as one of the uncertainty component was seldom reported.

3.2.2 REAGENTS AND MATERIALS

Solvents. Acetonitrile, toluene, hexane (HPLC grade), acetonitrile-toluene (3:1, v/v).

Anhydrous sodium sulfate. Analytically pure, the anhydrous sodium sulfate were baked for 4 h at 650°C in a muffle furnace.

Solid phase extraction column. Cleanert-TPT (12 mL, 2000 mg, Agela, China).

SPE cartridge. Suitable for 12 mL solid phase extraction column (57267), suitable for 6 mL solid phase extraction column (57020-U) (Sigma Aldrich Trading Co., Ltd.).

Heart-shaped bottle. 80 mL (Z680346-1EA, Sigma Aldrich Trading Co., Ltd.).

Liquid receiver. 30 mL (A82030, Agela, China).

Centrifuge tube. 80 mL.

Millipore filter (Nylon). 13 mm × 0.2 μm.

3.2.3 APPARATUS

GC–MS/MS system. Model 7890A gas chromatograph connected to a Model 7000A and equipped with electron impact (EI) ion source, a Model 7693 autosampler and Mass Hunter dataprocessing software (Agilent Technologies, Wilmington, DE). The column used was a DB-1701 capillary column (30 m × 0.25 mm × 0.25 μm, J&W Scientific, Folsom, CA, USA).

Homogenizer. Speed more than 13,500 r/min, T-25B (Janke & Kunkel, Staufen, Germany).

Rotary evaporator. Buchi EL131 (Flawil, Switzerland).

Centrifuge. Speed more than 4200 r/min, Z320 (B. HermLe AG, Gosheim, Germany).

Nitrogen evaporator. EVAP 112 (Organomation Associates, Inc. New Berlin, MA). Visiprep 5-port flask vacuum manifold: RS-SUPELCO 57101-U (Sigma Aldrich Trading Co., Ltd.).

3.2.4 EXPERIMENTAL METHOD

3.2.4.1 Method 1 (M1)

The tea samples were grinded to powder. Weigh 2 g of sample (accurate to 0.01 g) into a 50 mL centrifuge tube, add 10 mL of distilled water and shake for 30 s, then keep still for 30 min for the hydration expansion of the sample. Add 10 mL of acetonitrile, shake for 1 min. Add 1 g of NaCl and 4 g of MgSO$_4$, mix thoroughly, shake for 1 min, add 40 μL of internal standard (TPP), homogenize at 13,000 rpm for 5 min. Transfer 1 mL of the supernatants to a 15 mL pear-shaped flask, which equipped with 1 mL hexane and 5 mL 20% NaCl (w/w) water solution, shake for 1 min, homogenize at 10000 rpm for 5 min. Finally, the upper layer (hexane layer) was moved into the tiny sample bottle for GC–MS/MS analysis. Fig. 3.2.1 was the flow chart of the method for the determination of multipesticide residues in teas by GC–MS/MS coupled with M1 method.

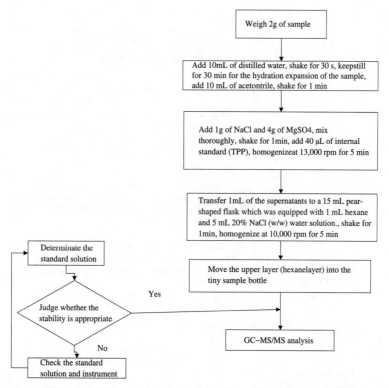

FIGURE 3.2.1 Flow chart of the method for the determination of multi pesticide residues in tea by GC–MS/MS coupled with M1 method.

3.2.4.2 Method 2 (M2)

The tea samples were grinded to powder. Weigh 2 g of sample (accurate to 0.01 g) into a 50 mL centrifuge tube, add 10 mL of distilled water and shake for 30 s, then keep still for 30 min for the hydration expansion of the sample. Add 10 mL of acetontrile, shake for 1 min. Add 1 g of NaCl and 4 g of MgSO$_4$, mix thoroughly, shake for 1 min, homogenize at 13,000 rpm for 5 min. The supernatants were flowed through 15 g of anhydrous sodium sulfate, eluted with 3 × 5 mL acetontrile, then the effluents were collected into pear-shaped flask and rotary evaporated (40°C water bath, 80 rpm/min) to about 2 mL. The concentrated solution was eluted with anhydrous sodium sulfate (ca 2 cm) TPT SPE cartridge, the flask was rinsed with 2 mL of acetonitrile-toluene (3:1, v/v) three times, and the washings were eluted with 25 mL of acetontrile-toluene (3:1, v/v). The eluted portion was concentrated to ca 0.5 mL by rotary evaporation at 40°C 120 rpmin. Add 20 µL of internal standard solution (heptachlor-epoxide), the residue was evaporated to dryness with nitrogen gas and it was dissolved in 1.5 mL of hexane and mixed thoroughly. Then pass the 0.20 µm filter membrane for GC–MS/MS determination. Fig. 3.2.2 was the flow chart of the method for the determination of multipesticide residues in teas by GC–MS/MS coupled with M2 method.

3.2.4.3 Method 3 (M3)

The tea samples were grinded to powder. Weigh 2 g of sample (accurate to 0.01 g) into an 80 mL centrifuge tube, add 15 mL of acetontrile, homogenize at 13,000 rpm for 1 min, then centrifuge for 5 min at 4200 rpm/min. Transfer the supernatants into a glass funnel, then the remaining sample was reextracted and the supernatants were combined and carefully evaporated to just 2 mL at 45°C with a vacuum evaporator in a water bath for the following cleanup procedure. Before the cleanup of the sample, anhydrous sodium sulfate (ca 2 cm) was placed on the top of the Cleanet-TPT cartridge, which connected a pear-shaped flask. The Cleanet-TPT was conditioned with 25 mL of acetontrile-toluene (3:1, v/v). When the conditioning solution reached the top of the sodium sulfate, the concentrated extract will be added to the Cleanet-TPT, the flask was rinsed with 2 mL of acetonitrile-toleune three times, and the washings were also applied to the Cleanet-TPT. The pesticides were eluted with 25 mL of acetontrile-toluene (3:1,v/v) and the eluted portion was concentrated to ca 0.5 mL by rotary evaporation at

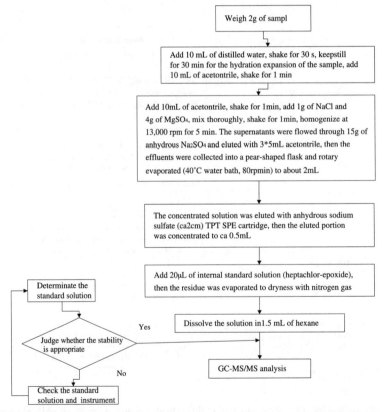

FIGURE 3.2.2 **Flow chart of the method for the determination of multi pesticide residues in teas by GC–MS/MS coupled with M2 method.**

45°C. Add 20 µL of internal standard solution (heptachlor-epoxide), then the residue was evaporated to dryness with nitrogen gas, and then it was dissolved in 1.5 mL of hexane and mixed thoroughly. Then pass the 0.20 µm filter membrane for GC–MS/MS determination. Fig. 3.2.3 was the flow chart of the method for the determination of multipesticide residues in teas by GC–MS/MS coupled with M3 method.

3.2.5 ESTIMATION OF UNCERTAINTY

This chapter presents a methodology for estimating the uncertainty associated with a multiresidue analytical method using gas chromatography–tandem mass spectrometry in tea through the bottom-up approach. The steps involved are as follows:

1. Specify the measured amount;
2. Identify the source of uncertainty;
3. Quantify uncertainty components;
4. Calculate combined uncertainty.

3.2.5.1 Identification of Uncertainty Sources

The critical step of "Bottom-up" approach is the identification of uncertainty sources while the evaluation of uncertainty was carried out on the basis of it.

3.2.5.1.1 Establishment M1 Model

The expression used for the quantification of pesticide residue (w) in tea on the basis of the experimental procedure provided by Section 3.2.4.1 is given by Eq. (3.2.1), where c is the concentration of pesticide residue in sample solution expressed in µg/mL, v_{hexane} is the volume of sample solution in tiny sample bottle (mL), m_{hexane} is the sample weight equivalent to the volume of sample solution in tiny sample bottle (g), R represents recovery rate.

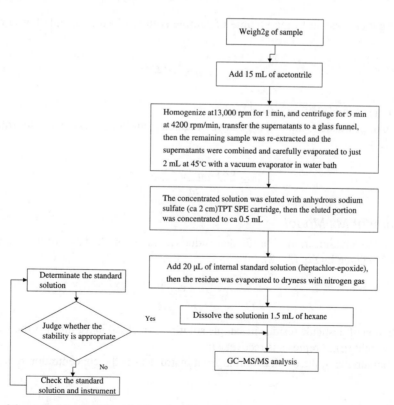

FIGURE 3.2.3 Flow chart of the method for the determination of multi pesticide residues in teas by GC–MS/MS coupled with M3 method.

$$w = \frac{c \times v_{hexane}}{m_{hexane}} \times \frac{1}{R} \qquad (3.2.1)$$

During the actual measurement, the result acquired from instrument is the concentration ratio of sample solution with internal standard, so

$$x = \frac{c}{c_{ISTD}} \qquad (3.2.2)$$

where x is the concentration ratio of sample solution with internal standard, c_{ISTD} is concentration of internal standard in tiny sample bottle (μg/mL). The following expression can be achieved from Eq. (3.2.2):

$$c = x \times c_{ISTD} \qquad (3.2.3)$$

So Eq. (3.2.1) can be changed to:

$$w = \frac{x \times c_{ISTD} \times v_{hexane}}{m_{hexane}} \times \frac{1}{R} \qquad (3.2.4)$$

According to the experimental procedure of M1, the concentration of internal standard in tiny sample bottle can be calculated using Eq. (3.2.5):

$$c_{ISTD} = \frac{c_{ISTD-add} \times v_{add}}{v_{ACN}} \times v_{supernatant} / v_{hexane} \qquad (3.2.5)$$

where $c_{ISTD\text{-}add}$ is the concentration of internal standard stock solution (μg/mL), v_{add} is the volume of internal standard stock solution spiked to extract solution (mL), v_{ACN} is the volume of solution after oscillation extraction with 10 mL acetonitrile (mL), $v_{supernatant}$ is the volume of supernatant transferred, v_{hexane} is the volume of sample solution in tiny sample bottle (mL).

The tea sample weight equivalent to the volume of sample solution in tiny sample bottle can be calculated using Eq. (3.2.6):

$$m_{hexane} = m \times \frac{v_{supernatant}}{v_{ACN}} \qquad (3.2.6)$$

where m is the weight of tea sample before treated (g).

So the concentration of pesticide residue in tea can be obtained from Eq. (3.2.7) by the substitution of the Eqs. (3.2.5) and (3.2.6) into Eq. (3.2.4).

$$w = \frac{x \times c_{ISTD-add} \times v_{add}}{m} \times \frac{1}{R} \qquad (3.2.7)$$

3.2.5.1.2 Establishment M2 Model

The expression used for the quantification of pesticide residue (w) in tea on the basis of the experimental procedure provided by Section 3.2.4.2 is given by Eq. (3.2.8).

$$w = \frac{c \times v}{m} \times \frac{1}{R} \qquad (3.2.8)$$

where c is the concentration of pesticide residue in sample solution expressed in μg/mL, v is the volume of sample solution (mL), m is the sample weight (g), R represents recovery rate.

During the actual measurement, the result acquired from instrument is the concentration ratio of sample solution with internal standard:

$$x = \frac{c}{c_{ISTD}} \qquad (3.2.2)$$

where x is the concentration ratio of sample solution with internal standard, c_{ISTD} is concentration of internal standard in tiny sample bottle (μg/mL). The following expression can be achieved from Eq. (3.2.2):

$$c = x \times c_{ISTD} \tag{3.2.3}$$

Meanwhile, Eq. (3.2.8) can be changed to:

$$w = \frac{x \times c_{ISTD} \times v}{m} \times \frac{1}{R} \tag{3.2.9}$$

According to the principle of mass conservation, the mass of internal standard in a certain volume of sample solution is equivalent to the mass added to sample solution:

$$c_{ISTD} \times v = c_{ISTD-add} \times v_{add} \tag{3.2.10}$$

where $c_{ISTD-add}$ is the concentration of internal standard stock solution (μg/mL), v_{add} is the volume of internal standard stock solution added to extract solution (mL).

So the concentration of pesticide residue in tea can be obtained from Eq. (3.2.7):

$$w = \frac{x \times c_{ISTD-add} \times v_{add}}{m} \times \frac{1}{R} \tag{3.2.7}$$

3.2.5.1.3 Establishment M3 Model

According to the experimental procedure provided by Section 3.2.4.3, the calculation principle of the concentration of pesticide residue in tea is the same with M2, so it can be obtained by Eq. (3.2.7), and it can be concluded that Eq. (3.2.7) can be adopted as model for determination of the concentration of pesticide residue in tea by GC–MS/MS using M1, M2, and M3 as the pretreatment method.

3.2.5.2 Calculation of Standard Uncertainty

It can be concluded from Eq. (3.2.7) that the main factors that contribute to the overall uncertainty of the concentration of pesticide residue in tea were: sample weight, concentration ratio of the sample and internal standard solution obtained from standard curve, concentration of internal standard stock solution, amount of internal standard stock solution added, and recovery. In addition, repeatability is of great importance. The uncertainty induced from the preparation and quantification of standard curve must be considered while the standard curve was selected for the calculation of the concentration ratio of the sample and internal standard solution. For convenient calculation, the estimation of uncertainty of pesticide residues in tea was performed in two aspects: type A uncertainty and type B uncertainty.

3.2.5.2.1 Type A Uncertainty

For obtaining accurate and reliable results, the determination was generally performed repeatedly and the results achieved was average value, therefore the uncertainty derived from the above process belongs to type A uncertainty and it can be calculated using Eq. (3.2.11).

$$u(A) = \frac{SD}{\sqrt{n}} \tag{3.2.11}$$

where n is determination times, SD is the standard deviation.

Relative standard uncertainty:

$$u_{rel}(A) = u(A)/\overline{w} \tag{3.2.12}$$

where $\overline{w}$ is the average value.

3.2.5.2.2   Type B Uncertainty

3.2.5.2.2.1   Uncertainty Arising From Sampling

The balance with 0.1 mg accuracy was selected for the weighing of tea sample. According to Eurachem/Citac guide [4], the uncertainty arising from the sampling is caused by: repeatability, readability (digital resolution), and the uncertainty derived from the calibration of balance. The uncertainty caused by the repeatability of the sampling has been included in type A uncertainty, so only the uncertainty caused by readability (digital resolution) should be considered.

The balance selected has a resolution of 0.1 mg, which is assumed to show a uniform distribution and is converted to standard uncertainty:

$$u_1(m) = {0.1}/{\sqrt{3}} = 0.058 \, \text{mg} \tag{3.2.13}$$

The linearity of balance demonstrates its ability in maintaining the linear relationship between load and display value. The balance verification certificate quotes ±0.2 mg for the linearity contribution, which is assumed to show a uniform distribution and is converted to standard uncertainty:

$$u_2(m) = 0.2 / \sqrt{3} = 0.12 \, \text{mg} \tag{3.2.14}$$

The standard uncertainty arising from sampling is:

$$
\begin{aligned}
u(m) &= \sqrt{2 \times \left(u_1(m)\right)^2 + 2 \times \left(u_2(m)\right)^2} \\
&= \sqrt{2 \times (0.12)^2 + 2 \times (0.058)^2} = 0.18 \, \text{mg}
\end{aligned} \tag{3.2.15}
$$

As the sample weight is 2 g, so the relative uncertainty arising from sampling is:

$$u_{rel,1} = u(m) / m_s = 0.18 \, \text{mg} / 2000 \, \text{mg} = 0.0091\% \tag{3.2.16}$$

3.2.5.2.2.2   Uncertainty Arising From Preparation of Standard Solution

The concentration of pesticide will be affected by the process of preparation of standard solution, so as the determination of x, which is the mathematical model established in this chapter, will be influenced, thus the uncertainty occurs. As many steps were involved in this part, so their contributions to uncertainty were regarded as a branch of total uncertainty. The uncertainty arising from preparation of standard solution was mainly due to two sides: uncertainty originated from standard substance, which is provided by production firm and uncertainty introduced by the preparation of standard solution.

1. Uncertainty originated from standard substance
   The purity and tolerance/uncertainty of the standard substance, which was provided by manufacturer, can be obtained from the certificate of standard substance. The uncertainty, originated from standard substance can be calculated using Eqs. (3.2.17) and (3.2.18).

   If the calculation is conducted by the adopting the purity of the standard substance, then:

$$u_{rel-std} = \frac{100 - P\%}{\sqrt{3}} \tag{3.2.17}$$

   If the calculation is conducted adopting the uncertainty of the standard substance, as the uncertainty contains the expanded uncertainty at $k = 2$ and 95% confidence interval, then:

$$u_{rel-std} = \frac{T\%}{2} \tag{3.2.18}$$

   The uncertainty originated from standard substance itself can be calculated using Eq. (3.2.18).

2. Uncertainty arising from preparation of standard solution

The process of preparation of standard solution: weigh a certain amount of pesticide standard substance to be tested (accurate to 0.01 g) into a 10 mL brown volumetric flask, dilute to volume with methanol and mix well, and then the stock solution was obtained. The preparation of intermediate solution was conducted by pipe a certain volume of stock solution into a 100 mL brown volumetric flask and dilute to volume with methanol. As these processes consisted of the weighing of standard substance, the fixing and dilution of standard solution, so these three aspects should be considered in the calculation of uncertainty arising from preparation of standard solution.

a. Uncertainty arising from the weighing of standard substance

The balance with 0.01 mg accuracy was selected for the weighing of standard substance. According to Eurachem/Citac guide [4], the uncertainty arising from the sampling is caused by: repeatability, readability (digital resolution), and the uncertainty derived from the calibration of balance. The calculation can be carried out with reference to Section 3.2.7.1.

The uncertainty caused by the repeatability of sampling of standard substance should not be included in type A uncertainty while the balance with 0.01 mg accuracy is adopted. The balance verification certificate quotes ±0.2 mg for the linearity contribution, which is assumed to show a uniform distribution:

$$u_{1(std|m)} = \frac{0.2}{\sqrt{3}} = 0.12\,\text{mg} \tag{3.2.19}$$

The balance selected has a resolution of 0.01 mg, which is assumed to show a uniform distribution and is converted to standard uncertainty:

$$u_{2(std-m)} = \frac{0.01}{\sqrt{3}} = 0.0058\,\text{mg} \tag{3.2.20}$$

The balance verification certificate quotes ±0.1 mg for the linearity contribution, which is assumed to show a uniform distribution and is converted to standard uncertainty:

$$u_{3(std|m)} = \frac{0.1}{\sqrt{3}} = 0.058\,\text{mg} \tag{3.2.21}$$

The uncertainty arising from the weighing of pesticide standard substance is:

$$\begin{aligned} u_{std-m} &= \sqrt{2 \times \left(u_{1(std-m)}\right)^2 + 2 \times \left(u_{2(std-m)}\right)^2 + 2 \times \left(u_{3(std-m)}\right)} \\ &= \sqrt{2 \times (0.12)^2 + 2 \times (0.0058)^2 + 2 \times (0.058)^2} = 0.18\,\text{mg} \end{aligned} \tag{3.2.22}$$

As the weight of pesticide standard substance is m_{std}, so the relative uncertainty arising from the weighing of pesticide standard substance:

$$u_{rel-std-m} = \frac{u_{std-m}}{m_{std}} \tag{3.2.23}$$

b. Uncertainty arising from the dilution of stock standard solution

The temperature labeled in the volumetric flask is 20°C and according to JJG196-2006 ≪Verification Regulation of Working Glass Containers≫, the volume is calibrated according to the specification of grade A within the range of ±0.02 mL for a 10 mL volumetric flask. As the confidence level and distributing information was not provided, so the standard uncertainty is obtained assuming a triangular distribution:

$$u_{f-10}(v) = \frac{0.02}{\sqrt{6}} 0.008\,\text{mL} \tag{3.2.24}$$

The uncertainty arising from the variation of the volume of liquid level to scale mark must be considered. The deviation of the operation of diluting to volume should not exceed one drop of solution; meanwhile, a little solution was usually left on the inner wall of volumetric flask. For the conservative estimate this uncertainty component, the deviation of the operation of diluting to volume was estimated to be two drops of solution, while one drop of solution was assumed to be 0.03 mL, so this uncertainty component is obtained assuming a uniform distribution:

$$u_{precision-10}(v) = \frac{0.06}{\sqrt{3}} = 0.035 \, \text{mL} \tag{3.2.25}$$

The coefficient of volume expansion of methanol is $1.10 \times 10^{-3}/°C$ and the temperature labeled in the volumetric flask is 20°C, so the temperature effect, which leads to an uncertainty in the determined volume due to the expansion of the liquid is:

$$\pm(10 \times 4 \times 1.10 \times 10^{-3}) = 0.044 \, \text{mL} \tag{3.2.26}$$

The standard uncertainty arising from the variation of the volume of volumetric flask is obtained assuming a uniform distribution:

$$u_{temp-10}(v) = \frac{0.044}{\sqrt{3}} = 0.025 \, \text{mL} \tag{3.2.27}$$

The uncertainty arising from the dilution of stock standard solution using 10 mL volumetric flask is:

$$u_{s-10}(v) = \sqrt{(u_{f-10}(v))^2 + (u_{precision-10}(v))^2 + (u_{temp-10}(v))^2} = 0.044 \, \text{mL} \tag{3.2.28}$$

The standard uncertainty arising from the dilution of stock standard solution using 10 mL volumetric flask is:

$$u_{rel-s-10}(v) = \frac{u_{s-10}(v)}{v_{s-10}} = 0.44\% \tag{3.2.29}$$

c. Uncertainty Arising From the Dilution of Standard Solution

As the concentration of pesticides differs from each other, so the volumes of stock standard solution fetched for the preparation of mixed standard solution varied greatly. In this experiment, the volume of solution fetched ranges from 113 to 9620 μL. Locomotive pipettes with various measurement ranges were applied for the dilution of standard solution, and the mixed solution achieved was diluted to volume into a brown volumetric flask.

The permitted precision and repeatability precision of the locomotive pipettes with various measurement ranges at room temperature 20°C can be obtained from JJG 646-2006 《Verification Regulation of Locomotive Pipette》, the uncertainty arising from locomotive pipettes was then calculated.

– Uncertainty arising from the volume of solution fetched

$SD_{allow}\%$ represents the permitted precision of the locomotive pipettes which is provided by JJG 646-2006 《Verification Regulation of Locomotive Pipette》, the results were obtained assuming a uniform distribution:

$$u_{rel-std-x} = \frac{SD_{allow}\%}{\sqrt{3}} \tag{3.2.30}$$

– Uncertainty arising from the variation of solution fetched

$SD_{repeat}\%$ represents the repeatability precision of the locomotive pipettes, which is provided by JJG 646-2006 《Verification Regulation of Locomotive Pipette》, the results were obtained assuming a uniform distribution:

$$u_{rel-std-x_{repeat}} = \frac{SD_{repeat}\%}{\sqrt{3}} \tag{3.2.31}$$

– Uncertainty arising from the temperature effect

The possible temperature variation is within the limits of 4°C (95% confidence level), the coefficient of volume expansion of methanol is $1.10 \times 10^{-3}/°C$, the volume variation due to the expansion of solution is:

$$\pm\left(v_{std}\times4\times1.10\times10^{-3}\right)\mu L \tag{3.2.32}$$

The standard uncertainty arising from the temperature effect is obtained assuming a uniform distribution:

$$u_{std-temp}=\frac{v_{std}\times4\times1.10\times10^{-3}}{\sqrt{3}}\mu L \tag{3.2.33}$$

The relative standard uncertainty is:

$$u_{rel-std-temp}=\frac{u_{std-temp}}{v_{std}} \tag{3.2.34}$$

The relative standard uncertainty arising from the fetching of stock solution during the dilution process is:

$$u_{rel-std-v}=\sqrt{\left(u_{rel-std-x}\right)^2+\left(u_{rel-std-x_{repeat}}\right)^2+\left(u_{rel-std-temp}\right)^2} \tag{3.2.35}$$

The temperature labeled in the volumetric flask is 20°C, according to JJG196-2006 ≪Verification Regulation of Working Glass Containers≫, the volume is calibrated according to the specification of grade A within the range of ±0.1 mL for a 100 mL volumetric flask. The uncertainty induced from 100 mL volumetric flask can be calculated with reference to:

$$u_{s-100}\left(v\right)=\sqrt{\left(u_{f-100}\left(v\right)\right)^2+\left(u_{precision-10}\left(v\right)\right)^2+\left(u_{temp-100}\left(v\right)\right)^2}=0.26\,\text{mL} \tag{3.2.36}$$

The standard uncertainty induced from 100 mL volumetric flask is:

$$u_{rel-s-100}\left(v\right)=\frac{u_{s-100}\left(v\right)}{v_{s-100}}=0.26\% \tag{3.2.37}$$

The relative standard uncertainty arising from the dilution of standard solution is:

$$u_{rel-std-dil}=\sqrt{\left(u_{rel-std-v}\right)^2+\left(u_{rel-s-100}\left(v\right)\right)^2} \tag{3.2.38}$$

The relative standard uncertainty arising from the preparation of standard solution becomes:

$$u_{rel-pre}=\sqrt{\left(u_{rel-std-m}\right)^2+\left(u_{rel-s-10}\left(v\right)\right)^2+\left(u_{rel-std-dil}\right)^2} \tag{3.2.39}$$

Above all, the relative standard uncertainty arising from the dilution of standard solution:

$$u_{rel,2}=\sqrt{\left(u_{rel-std}\right)^2+\left(u_{rel-pre}\right)^2} \tag{3.2.40}$$

3.2.5.2.2.3 Uncertainty Arising From the Concentration of Internal Standard Solution

The process of preparation of internal standard solution: weigh a certain amount of internal standard (TPP or heptachlor epoxide) into a 10 mL brown volumetric flask, dilute to volume with methanol and mix well, then the stock solution was obtained. The standard solution was acquired by the dilution of stock solution with methanol. The process of the evaluation of uncertainty arising from the concentration of internal standard solution is similar to the dilution of standard solution.

The calculation of the uncertainty arising from internal standard itself can be conducted with reference to "Uncertainty arising from preparation of standard solution" and the result can be called $u_{rel-ISTD}$. The calculation of the uncertainty arising from the preparation of internal standard can be carried out with reference to "Uncertainty arising from preparation of standard solution" and the result can be called $u_{rel-pre-ISTD}$. The uncertainty arising from the concentration of internal standard solution is then:

$$u_{rel,3} = \sqrt{\left(u_{rel-ISTD}\right)^2 + \left(u_{rel-pre-ISTD}\right)^2} \tag{3.2.41}$$

3.2.5.2.2.4 Uncertainty arising from the adding of internal standard solution

Take M1, for example, the calculation of the uncertainty arising from the fetching of 40 μL internal standard solution using locomotive pipette can be conducted with reference to "Uncertainty arising from preparation of standard solution." As the uncertainty arising from the variation of solution fetched is already included in type A uncertainty, so only the uncertainty arising from the volume of solution fetched and the uncertainty arising from the temperature effect should be considered.

a. Uncertainty arising from the volume of solution fetched

According to JJG 646-2006≪Verification Regulation of Locomotive Pipette≫, the permitted precision of the locomotive pipette when it was carried out for the fetching of 40 μL heptachlor epoxide solution at 20°C is 3.0%, so the relative standard uncertainty arising from the volume of solution fetched can be obtained assuming a uniform distribution:

$$u_{rel-ISTD-40} = \frac{3.0\%}{\sqrt{3}} = 0.017 \tag{3.2.42}$$

b. Uncertainty arising from the temperature effect

The possible temperature variation is within the limits of 4°C (95% confidence level), the coefficient of volume expansion of methanol is 1.10×10^{-3}/°C, the volume variation due to the expansion of solution is:

$$\pm \left(40 \times 4 \times 1.10 \times 10^{-3}\right) \mu L \tag{3.2.43}$$

The standard uncertainty arising from the temperature effect is obtained assuming a uniform distribution:

$$u_{ISTD-temp} = \frac{40 \times 4 \times 1.10 \times 10^{-3}}{\sqrt{3}} = 0.10 \, \mu L \tag{3.2.44}$$

So the relative standard uncertainty is:

$$u_{rel-ISTD-temp} = \frac{u_{ISTD-temp}(v)}{40} = 0.25\% \tag{3.2.45}$$

Then, the relative standard uncertainty arising from the adding of internal standard solution becomes:

$$u_{rel,4} = \sqrt{\left(u_{rel-ISTD-40}\right)^2 + \left(u_{rel-ISTD-temp}\right)^2} = 0.018 \tag{3.2.46}$$

3.2.5.2.2.5 Uncertainty arising from the prepare process of quantitative using standard curve

The quantitative method adopted in this experiment is five points matrix-matched standard calibration curve, the steps are: prepare 5 pieces of matrix blank sample, and the pretreatment was conducted using corresponding method, then 20, 50, 100, 200, and 500 μL mixed standard solution were spiked. The uncertainty will occur during the process of adding standard solution.

Meanwhile, the research result of our team demonstrated that the concentration of pesticide in mixed standard solutions will be influenced by external conditions, such as light, temperature, moisture, and volatilization of solvent, so as the quantitative of the concentration of pesticide will be exerted certain influence. In this paper, the uncertainty arising from the quantitative process using standard curve can be regarded as part of uncertainty induced from the quantitative process.

Uncertainty arising from the quantitative process using standard curve contains two parts: uncertainty arising from the volume of the mixed standard solution added and uncertainty arising from the variation of concentration of pesticide in solvent.

a. Uncertainty arising from the volume of the standard solution added

In this experiment, the volume of mixed standard solution added to matrix standard solution is 20, 50, 100, 200, and 500 μL, the uncertainty induced from this process can calculated with reference to "Uncertainty Arising From the Dilution of Standard Solution".

$$u_{rel-add-v} = \sqrt{\left(u_{rel-add-x}\right)^2 + \left(u_{rel-add-x_{repeat}}\right)^2 + \left(u_{rel-add-temp}\right)^2}$$

(3.2.47)

$$u_{rel-add-total} = \sqrt{\sum_{i=1}^{5} \left(u_{rel-add-v_i}\right)^2}$$

(3.2.48)

b. Uncertainty arising from the variation of concentration of pesticide in solvent

For the analysis of the variation of concentration of pesticide in solvent, the investigation of the stability of pesticide in solvent has been conducted in the previous research of our team. The processes were: by the premise of guarantee the stability of the instrument, the real response data of pesticide was acquired every 3 days for 3 months. This uncertainty can be calculated using Eq. (3.2.49):

$$u_{std-sol} = \frac{SD_{std-sol}}{\sqrt{n_{std-sol}}}$$

(3.2.49)

where $SD_{std-sol}$ is the standard deviation of the response data of pesticide during the 3 months, and $n_{std-sol}$ is the times of testing the response data during the 3 months.

Therefore, the relative standard uncertainty:

$$u_{rel-std-sol} = \frac{u_{std-sol}}{S}$$

(3.2.50)

where S is the average value of the response data of pesticides.

Then, the relative standard uncertainty arising from the prepare process of quantitative using standard curve can be calculated using the Eq. (3.2.51):

$$u_{rel,S} = \sqrt{\left(u_{rel-add-total}\right)^2 + \left(u_{rel-std-sol}\right)^2}$$

(3.2.51)

3.2.5.2.2.6 Uncertainty arising from the quantitative process using standard curve

According to Eurachem/Citac Guide [4], the uncertainty will occur when the fittings of standard curve were selected for the quantitative process. According to the related literature, the uncertainty arising from the quantitative process using standard calibration curve contributed to the total uncertainty, so it was regarded as an independent component in the evaluation of uncertainty.

The fitting of standard curve was conducted by assuming the relative concentration as horizontal ordinate and relative response as longitudinal ordinate, then the linear equation can be obtained, the mathematical model is:

$$A_j = x_i B_1 + B_0$$

(3.5.52)

where A_j is the relative response, x_i is the relative concentration, B_1 is the slope, B_0 is the intercept. According to Eurachem/Citac guide [4], the uncertainty arising from the fitting of standard curve can be calculated as follows:

$$u(x) = \frac{S}{B_1}\sqrt{\frac{1}{P} + \frac{1}{n} + \frac{\left(x-\bar{x}\right)^2}{S_{xx}}}$$

(3.2.53)

Where S is the residual standard deviation of standard curve, B_1 is the slope, P is repetition times, n is the determination times of standard curve, x is the relative concentration of pesticide, $\bar{X}$ is the average value of the relative concentration of standard curve.

$$S = \sqrt{\frac{\sum_{j=1}^{n}\left[A_j - \left(B_0 + B_1 x_j\right)\right]^2}{n-2}}$$

(3.2.54)

$$S_{xx} = \sum_{j=1}^{n}\left(x_j - \bar{x}\right)^2$$

(3.2.55)

Uncertainty arising from the quantitative process using standard curve is then:

$$u_{rel,6} = u_{rel}(x) = \frac{u(x)}{x}$$

(3.2.56)

3.2.5.2.2.7   Uncertainty arising from the concentration of pesticide

The blank value of the sample must be considered in the determination of pesticide residues in tea. The value of limits of detection (LOD) is estimated by the spiking of standard solution with different concentration levels into matrix blank sample, and it is defined as the corresponding concentration when the signal-to-noise ratio is three. According to the definition of LOD, the uncertainty of measurement should be 100% when the concentration of pesticide is near LOD, so as the result detected is unreliable. The uncertainty of measurement is relatively small when the concentration of pesticide is much higher than LOD. So the uncertainty arising from the concentration of pesticide can be calculated as:

$$u_{rel,7} = u_{LOD} = \frac{LOD}{c_{det}}$$

(3.2.57)

where LOD is the detection limit, c_{det} is the concentration of pesticide in sample solution.

3.2.5.2.2.8   Uncertainty arising from recovery

The extraction and cleanup of tea samples may cause the loss of pesticides, while the response of pesticides will be affected by the matrix, so as to exert influence on the recovery. The recovery has certain correction actions for the determination results when the results were at appropriate concentration levels. When the function of correction acts, any uncertainty related to recovery will contribute to total uncertainty. In this experiment, for the research of the fortified recoveries of M1, M2, and M3, weigh three samples, the pretreatment was conducted according to their respective flow chart, the evaluation of the uncertainty arising from the recovery was carried out according to JCGM 100:2008 [18].

Assuming the fortified recoveries were: R1%, R2%, R3%, R4%, R5%, and R6%, among these the maximum value was $R_{max}\%$, the minimum value was $R_{min}\%$, the average value was $\bar{R}\%$

$$b_+ = R_{max}\% - 100\%$$

(3.2.58)

$$b_- = 100\% - R_{min}\%$$

(3.2.59)

Then the standard uncertainty arising from the recovery is:

$$u(R) = \sqrt{\frac{\left(b_+ + b_-\right)^2}{12}}$$

(3.2.60)

The relative standard uncertainty arising from the recovery becomes:

$$u_{rel,8} = \frac{u(R)}{\bar{R}\%}$$

(3.2.61)

Above all, the type B relative standard uncertainty can be calculated using Eq. (3.2.62):

$$u_{rel}(B) = \sqrt{\left(u_{rel,1}\right)^2 + \left(u_{rel,2}\right)^2 + \left(u_{rel,3}\right)^2 + \left(u_{rel,4}\right)^2 + \left(u_{rel,5}\right)^2 + \left(u_{rel,6}\right)^2 + \left(u_{rel,7}\right)^2 + \left(u_{rel,8}\right)^2} \tag{3.2.62}$$

3.2.6 COMBINED UNCERTAINTY

The combined relative standard uncertainty is:

$$u_{rel} = \sqrt{\left(u_{rel}(A)\right)^2 + \left(u_{rel}(B)^2\right)} \tag{3.2.63}$$

So as the combined standard uncertainty:

$$u_{rel} \times w \tag{3.2.64}$$

3.2.7 EXPANDED UNCERTAINTY

While the coverage factor $k = 2$, the expanded uncertainty is:

$$U = u \times 2 \tag{3.2.65}$$

3.2.8 RESULT AND DISCUSSION

The uncertainty plays an important role in the expressing the determination results. Meanwhile, the importance of uncertainty lies in its capability of assessing the analysis process by the calculation of the contribution of each uncertainty component to the combined uncertainty. It can be concluded from the establishment of model M1, M2, and M3, and the process of evaluation of uncertainty that $u_{rel,1}$, $u_{rel,2}$, and $u_{rel,5}$ were the common components of the nine uncertainty components ($u_{rel,1}$, $u_{rel,2}$, $u_{rel,3}$, $u_{rel,4}$, $u_{rel,5}$, $u_{rel,6}$, $u_{rel,7}$, and $u_{rel,8}$), and these three components had the same numerical value in the three methods. Due to different methods, the numerical value of each uncertainty components varied with each other except $u_{rel,3}$ and $u_{rel,4}$ in M2 and M3.The combined uncertainty of each method is different from each other for the difference in uncertainty components and their contribution to total uncertainty. In this part, the uncertainty of M1, M2, and M3 was discussed to acquire the critical control point and then compared with each other.

3.2.8.1 Uncertainty of M1

Each of the relative standard uncertainty components and the combined relative standard uncertainty of M1 coupled with GC–MS/MS for the determination 18 pesticide residues are shown in Table 3.2.1. As $u_{rel,1}$, $u_{rel,3}$, and $u_{rel,4}$ were the components related to the dilution of standard solution, the concentration of interstandard solution and the volume of interstandard solution, respectively, so the three components of uncertainty of the three methods were the same while the other uncertainty components were different with kinds of pesticides.

For the assessment of the analysis procedure of M1, the research of the calculation the proportion of each uncertainty components to the algebraic sum of all uncertainty components and their respective contribution to the combined uncertainty was conducted. The distribution chart of the percentage of each uncertainty components of different kinds of pesticide was shown in Fig. 3.2.4 and it can be seen that the component of least contribution is $u_{rel,7}$ for it was related to LOD. As the concentration of pesticide detected in these experiments was much higher than LOD, so the percentage of the component related to LOD was relative tiny. The component of next least contribution is $u_{rel,1}$ for it was related to tea sample weight, as the balance with 0.1 mg accuracy was selected for the weighing of tea sample, so the percentage of this component was relative small.

The $u_{rel,2}$ was the uncertainty component related with the dilution of standard solution, thus it was concerned with the most steps. As the determination of x in the model was affected by the dilution of standard solution directly, so this uncertainty component played an important role during the experiment. As the $u_{rel,3}$ and $u_{rel,4}$ were the uncertainty components related with the concentration of inter standard solution and the volume of interstandard solution, respectively, so they had the same value for different pesticides. It can be seen from Fig. 3.2.4 that there was no significant difference among the percentage of the three uncertainty components, $u_{rel,2}$, $u_{rel,3}$, and $u_{rel,4}$, of the same pesticide for most of the pesticides.

TABLE 3.2.1 The Relative Standard Uncertainty Components and the Combined Relative Standard Uncertainty of M1 Coupled With GC–MS/MS for the Determination 18 Pesticide Residues

No.	Pesticide	$u(A)$	$u_{rel,1}$	$u_{rel,2}$	$u_{rel,3}$	$u_{rel,4}$	$u_{rel,5}$	$u_{rel,6}$	$u_{rel,7}$	$u_{rel,8}$	u_{rel}
1	Acetochlor	0.035	0.000091	0.020	0.020	0.018	0.082	0.62	0.0001892	0.047	0.632
2	Ametryn	0.018	0.000091	0.021	0.020	0.018	0.066	0.02	0.0000026	0.045	0.091
3	Bifenthrin	0.024	0.000091	0.019	0.020	0.018	0.091	0.18	0.0000035	0.039	0.210
4	Boscalid	0.011	0.000091	0.017	0.020	0.018	0.063	0.05	0.00000004	0.031	0.093
5	Butralin	0.019	0.000091	0.029	0.020	0.018	0.063	0.02	0.0000054	0.044	0.091
6	Cyprodinil	0.030	0.000091	0.020	0.020	0.018	0.068	0.02	0.0000001	0.061	0.105
7	Dimetho-morph	0.031	0.000091	0.022	0.020	0.018	0.070	0.02	0.0000149	0.134	0.160
8	Diniconazole	0.043	0.000091	0.017	0.020	0.018	0.075	0.34	0.0000042	0.043	0.350
9	Endosulfan	0.038	0.000091	0.043	0.020	0.018	0.074	0.03	0.0000013	0.053	0.115
10	Epoxicon-azole	0.021	0.000091	0.020	0.020	0.018	0.065	0.03	0.0000006	0.030	0.087
11	Metalaxyl	0.012	0.000091	0.018	0.020	0.018	0.074	0.07	0.0000036	0.030	0.109
12	Napropamide	0.024	0.000091	0.023	0.020	0.018	0.068	0.03	0.0000025	0.066	0.107
13	Pendimeth-alin	0.022	0.000091	0.018	0.020	0.018	0.068	0.02	0.0000007	0.007	0.081
14	Propicon-azole	0.018	0.000091	0.018	0.020	0.018	0.094	0.03	0.0000030	0.045	0.115
15	Pyridaben	0.026	0.000091	0.023	0.020	0.018	0.080	0.04	0.0000034	0.033	0.104
16	Pyrimethanil	0.016	0.000091	0.015	0.020	0.018	0.070	0.03	0.0000005	0.057	0.101
17	Triadimefon	0.018	0.000091	0.023	0.020	0.018	0.070	0.10	0.0000001	0.048	0.134
18	Trifloxystrobin	0.031	0.000091	0.020	0.020	0.018	0.070	0.11	0.0000094	0.054	0.149

In general, the percentage of the contribution of the above three uncertainty components to the combined uncertainty was less than 10% in addition to individual pesticides, this may be because the analysis procedure related to the three uncertainty components was mainly accomplished by means of balance, pipette, and volumetric flask, so the uncertainty induced by these three components was relative small as long as careful operation was conducted.

The $u(A)$ was the uncertainty component related with repeatability. As can be seen from Fig. 3.2.4, the contribution of this uncertainty component to the combined uncertainty was not high and it was slightly higher than the components of $u_{rel,2}$, $u_{rel,3}$, and $u_{rel,4}$.

It can be seen from Fig. 3.2.4 that the contribution of the three uncertainty components, $u_{rel,5}$, $u_{rel,6}$, and $u_{rel,8}$, to the combined uncertainty was relative high. The $u_{rel,5}$ was the uncertainty component related with the volume of mixed standard solution added and the variation of concentration of pesticide in solvent and according to the evaluation results, the variation of concentration of pesticide in solvent was the main factor of this component. The $u_{rel,5}$ made the greatest contribution to the combined uncertainty for all pesticides besides acetochlor, bifenthrin, diniconazole, and trifloxystrobin, this may be due to the uncertainty evaluation was carried out on the basis of the 32 repeated experimental results within 3 months for the relative conservative estimation the variation of concentration of pesticide in solvent.

The $u_{rel,6}$ and $u_{rel,8}$ were the uncertainty components related with quantitative process using standard curve and recovery. As can be seen from Fig. 3.2.4, the contribution of $u_{rel,6}$ to the combined uncertainty was inversely proportional to $u_{rel,8}$ for No.1 and No. 8 pesticide, and it demonstrated the importance of these two components in all the uncertainty components of M1. As the $u_{rel,6}$ will be affected by the concentration of pesticide residue in tea sample to some extent, so the relative standard uncertainty of this uncertainty component of some pesticides listed in Table 3.2.1 was probably related to $u_{rel,6}$. It can seen from Table 3.2.1 that the values of $u_{rel,8}$ of all pesticides were lower than 0.07 and were of uniform distribution except dimethomorph.

So it can be concluded that the $u_{rel,6}$ and $u_{rel,8}$ were the key components for M1 and attention must be paid to the operating steps concerned in the analysis process.

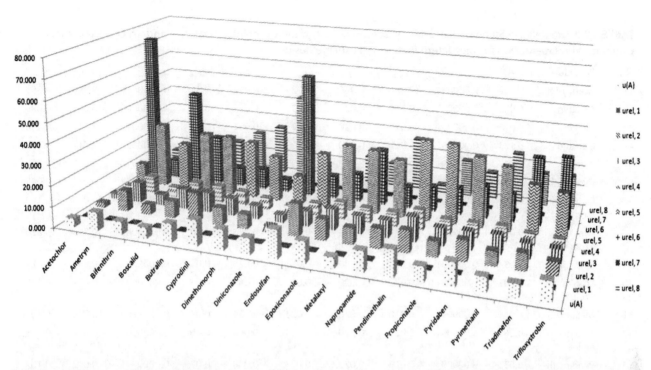

FIGURE 3.2.4 The percentage of each relative standard uncertainty components to the combined relative standard uncertainty of M1 coupled with GC–MS/MS for the determination 18 pesticide residue (%).

3.2.8.2 Uncertainty of M2

Each of the relative standard uncertainty components and the combined relative standard uncertainty of M2 coupled with GC–MS/MS for the determination 18 pesticide residues were shown in Table 3.2.2. As the model established in M2 was the same with the model established in M2, the evaluation results of the uncertainty components, $u_{rel,1}$, $u_{rel,3}$, $u_{rel,4}$, $u_{rel,7}$, and $u_{rel,2}$ were similarly the same with M1. It can be seen from Fig. 3.2.5 that the contribution of the above five uncertainty components to combined uncertainty in M2 was analogous to it in M1.

The percentage of the contribution of $u_{rel,6}$ to combined uncertainty was the highest among the remaining four uncertainty components. The values of $u_{rel,6}$ obtained from M2 ranged 3.09%–69.31%, among these the values ranged 20.04%–69.31% for pyridaben, napropamide, boscalid, pendimethalin, butralin, trifloxystrobin, diniconazole, acetochlor, and bifenthrin. The component of least contribution was $u_{rel,5}$ and the reason has been discussed in M1. The $u(A)$ was the uncertainty component related with repeatability and it was affected by the determination of parallel samples. The contribution of the $u(A)$ to the combined uncertainty, which was different from M1, was relatively high. As can be seen from Fig. 3.2.5 that the $u_{rel,8}$ contributed relative low to the total uncertainty.

So it can be concluded that the $u_{rel,6}$ and $u(A)$ were the key components for M2 and attention must be paid to the operating steps concerned in the analysis process.

3.2.8.3 Uncertainty of M3

Each of the relative standard uncertainty components and the combined relative standard uncertainty of M3 coupled with GC–MS/MS for the determination 18 pesticide residues were shown in Table 3.2.3, the evaluation results of the uncertainty components, $u_{rel,1}$, $u_{rel,3}$, $u_{rel,4}$, $u_{rel,7}$, and $u_{rel,2}$ were the same with M1 and M2, similarly, the contribution of them to combined uncertainty were similarly the same, so no more tautology here.

It can be seen from Fig. 3.2.6 that the contribution of the three uncertainty components, $u_{rel,5}$, $u_{rel,6}$, and $u_{rel,8}$, to the combined uncertainty was similar to that in M2. The $u_{rel,6}$ made the greatest contribution to the combined uncertainty, and the percentage of the $u_{rel,6}$ was relatively high for the values of them were 84.38%, 83.48%, 66.83%, 64.36%, and 52.61% fordiniconazole, acetochlor, trifloxystrobin, bifenthrin, and triadimefon, respectively. As can be seen from Table 3.2.3, the component of next least contribution is $u_{rel,i}$ and the values of $u_{rel,5}$ obtained from M3 ranged 0.063–0.094. It can be concluded by the comparison of Figs. 3.2.4–3.2.6 that the contribution of $u(A)$ to the combined uncertainty in M3 was higher

TABLE 3.2.2 The Relative Standard Uncertainty Components and the Combined Relative Standard Uncertainty of M2 Coupled With GC–MS/MS for the Determination 18 Pesticide Residues

No.	Pesticide	$u(A)$	$u_{rel,1}$	$u_{rel,2}$	$u_{rel,3}$	$u_{rel,4}$	$u_{rel,5}$	$u_{rel,6}$	$u_{rel,7}$	$u_{rel,8}$	u_{rel}
1	Acetochlor	0.012	0.000091	0.020	0.020	0.023	0.082	0.35	0.0000016	0.005	0.361
2	Ametryn	0.055	0.000091	0.021	0.020	0.023	0.066	0.02	0.0000026	0.013	0.096
3	Bifenthrin	0.018	0.000091	0.019	0.020	0.023	0.091	0.42	0.0000012	0.017	0.436
4	Boscalid	0.123	0.000091	0.017	0.020	0.023	0.063	0.09	0.00000001	0.017	0.172
5	Butralin	0.047	0.000091	0.029	0.020	0.023	0.063	0.09	0.0000014	0.028	0.131
6	Cyprodinil	0.025	0.000091	0.020	0.020	0.023	0.068	0.02	0.0000002	0.019	0.085
7	Dimetho-morph	0.114	0.000091	0.022	0.020	0.023	0.070	0.01	0.0000038	0.015	0.140
8	Dinicon-azole	0.051	0.000091	0.017	0.020	0.023	0.075	0.20	0.0000005	0.015	0.220
9	Endosulfan	0.034	0.000091	0.043	0.020	0.023	0.074	0.05	0.0000007	0.032	0.113
10	Epoxicon-azole	0.091	0.000091	0.020	0.020	0.023	0.065	0.03	0.0000003	0.014	0.121
11	Metalaxyl	0.055	0.000091	0.018	0.020	0.023	0.074	0.03	0.0000003	0.004	0.102
12	Naprop-amide	0.061	0.000091	0.023	0.020	0.023	0.068	0.06	0.0000005	0.012	0.115
13	Pendimeth-alin	0.031	0.000091	0.018	0.020	0.023	0.068	0.07	0.0000001	0.012	0.107
14	Propicon-azole	0.054	0.000091	0.018	0.020	0.023	0.094	0.01	0.0000012	0.002	0.114
15	Pyridaben	0.054	0.000091	0.023	0.020	0.023	0.080	0.05	0.0000010	0.019	0.119
16	Pyrimethanil	0.025	0.000091	0.015	0.020	0.023	0.070	0.01	0.0000002	0.015	0.083
17	Triadimefon	0.051	0.000091	0.023	0.020	0.023	0.070	0.02	0.0000001	0.009	0.098
18	Trifloxys-trobin	0.037	0.000091	0.020	0.020	0.023	0.070	0.18	0.0000006	0.017	0.203

than that in Fig. 3.2.4, but it was lower than that in Fig. 3.2.5, this may be due to the concentration of pesticide residues in tea samples or matrix effects. The values of $u_{rel,8}$ of all pesticides were lower than 9% except cyprodinil. So it can be concluded that the $u_{rel,6}$ and $u(A)$ were the key components for M3 and attention must be paid to the operating steps concerned in the analysis process.

3.2.8.4   Comparison of the Three Methods for Uncertainty Analysis

According to the analyses above, the five uncertainty components, $u_{rel,1}$, $u_{rel,3}$, $u_{rel,4}$, $u_{rel,7}$, and $u_{rel,2}$, contribute little to the total combined uncertainty, while other four uncertainty components, $u(A)$, $u_{rel,5}$, $u_{rel,6}$, and $u_{rel,8}$, contribute to the total combined uncertainty. Among the four components, $u(A)$ is related to the concentration of pesticide and repeatability, $u_{rel,5}$ is the component interrelated with the volume of standard solutions added and the variation of concentration of pesticides in the solvent, it is mainly affected by the variation of concentration of pesticide in solvent and it has the same value in different analytical methods. $u_{rel,6}$ is the uncertainty arising from the quantitative process using standard curve and it is induced by the deviation during the fitting standard curve. $u_{rel,8}$ is the component interrelated with the recovery and its values vary with method. So methods can be compared by the investigation of uncertainty component arising from the recovery.

As can be seen from Fig. 3.2.7, the values of $u_{rel,5}$ obtained from M1, ranged 0.030–0.134, were larger than M2 and M3 except pendimethalin, among them dimethomorph had the maximum value of $u_{rel,8}$. Meanwhile, M1 showed larger fluctuation than M2 (range 0.002–0.032) and M3 (range 0.01–0.023) in the values of $u_{rel,5}$ for different pesticides. M3 had the best stability among the three methods.

The comparison of the values of the combined relative standard uncertainty were shown in Fig. 3.2.8 and it can be seen from Fig. 3.2.8 that the values of them were fairly close to each other and of the same trend. It was obvious that the values of u_{rel} for acetochlor, bifenthrin, and diniconazole were higher than other pesticides in the three methods. It can be

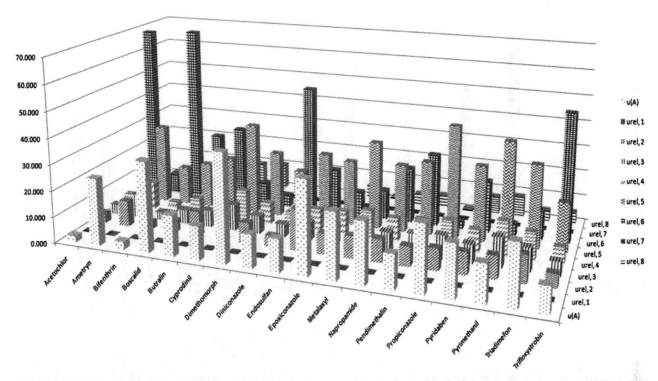

FIGURE 3.2.5 The percentage of each relative standard uncertainty components to the combined relative standard uncertainty of M2 coupled with GC–MS/MS for the determination 18 pesticide residue (%).

TABLE 3.2.3 The Relative Standard Uncertainty Components and the Combined Relative Standard Uncertainty of M3 Coupled With GC–MS/MS for the Determination 18 Pesticide Residues

No.	Pesticide	u(A)	$u_{rel,1}$	$u_{rel,2}$	$u_{rel,3}$	$u_{rel,4}$	$u_{rel,5}$	$u_{rel,6}$	$u_{rel,7}$	$u_{rel,8}$	u_{rel}
1	Acetochlor	0.011	0.000091	0.020	0.020	0.023	0.082	0.90	0.0000804	0.023	0.907
2	Ametryn	0.056	0.000091	0.021	0.020	0.023	0.066	0.09	0.0000020	0.020	0.131
3	Bifenthrin	0.033	0.000091	0.019	0.020	0.023	0.091	0.36	0.0000006	0.016	0.378
4	Boscalid	0.066	0.000091	0.017	0.020	0.023	0.063	0.02	0.00000003	0.020	0.101
5	Butralin	0.088	0.000091	0.029	0.020	0.023	0.063	0.08	0.0000020	0.014	0.143
6	Cyprodinil	0.025	0.000091	0.020	0.020	0.023	0.068	0.01	0.0000001	0.020	0.084
7	Dimetho-morph	0.101	0.000091	0.022	0.020	0.023	0.070	0.03	0.0000085	0.017	0.133
8	Dinicon-azole	0.007	0.000091	0.017	0.020	0.023	0.075	0.86	0.0000012	0.018	0.865
9	Endosulfan	0.046	0.000091	0.043	0.020	0.023	0.074	0.03	0.0000008	0.022	0.108
10	Epoxicon-azole	0.041	0.000091	0.020	0.020	0.023	0.065	0.04	0.0000010	0.014	0.095
11	Metalaxyl	0.025	0.000091	0.018	0.020	0.023	0.074	0.10	0.0000011	0.017	0.135
12	Naprop-amide	0.067	0.000091	0.023	0.020	0.023	0.068	0.16	0.0000010	0.016	0.190
13	Pendimeth-alin	0.078	0.000091	0.018	0.020	0.023	0.068	0.04	0.0000003	0.017	0.118
14	Propicon-azole	0.032	0.000091	0.018	0.020	0.023	0.094	0.02	0.0000008	0.015	0.108
15	Pyridaben	0.029	0.000091	0.023	0.020	0.023	0.080	0.01	0.0000066	0.017	0.096
16	Pyrimethanil	0.016	0.000091	0.015	0.020	0.023	0.070	0.04	0.0000005	0.014	0.088
17	Triadimefon	0.008	0.000091	0.023	0.020	0.023	0.070	0.18	0.0000001	0.019	0.200
18	Trifloxystrobin	0.035	0.000091	0.020	0.020	0.023	0.070	0.36	0.0000064	0.010	0.370

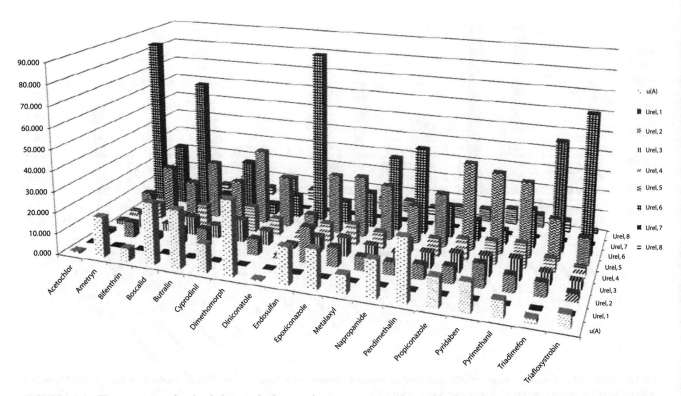

FIGURE 3.2.6 **The percentage of each relative standard uncertainty components to the combined relative standard uncertainty of M3 coupled with GC–MS/MS for the determination 18 pesticide residue (%).**

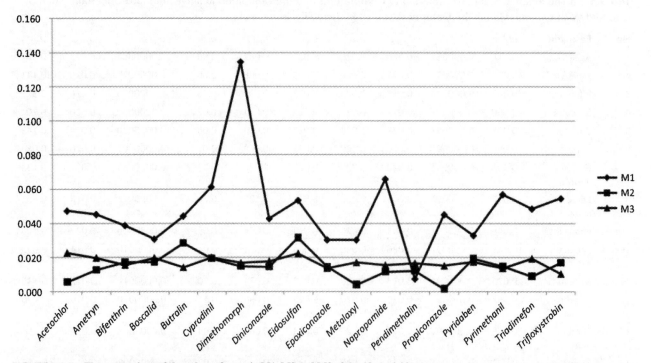

FIGURE 3.2.7 **The comparison of the values of $u_{rel,\ 8}$ in M1, M2 and M3 of the 18 pesticides.**

concluded from the front analysis that component of the largest contribution to the combined relative standard uncertainty was the uncertainty arising from the quantitative process using standard curve, and it was found that there is great difference between the theoretical response and actual response of the lowest point of the standard curve for the three pesticides in different methods by the analysis the initial data of the uncertainty arising from the quantitative process using standard curve, this may be due to etochlor, bifenthrin, and diniconazole were significantly affected by matrix effect while the

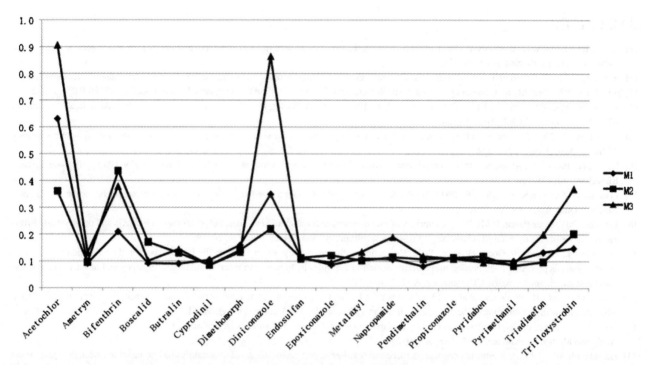

FIGURE 3.2.8   The comparison of the values of u_{rel} in M1, M2 and M3 of the 18 pesticides.

concentrations of them were relative low. So the concentration of the lowest point of the standard carve should be adjusted to decrease the uncertainty arising from the quantitative process using standard curve in further experiments.

So preliminary conclusion can be concluded on the basis of the earlier analysis: according to the uncertainty component of recovery of the three methods and the combined recovery, M3 was superior to M1 and M2.

3.2.9   CONCLUSION

As one of the main parameters for the confirmation of analysis method, uncertainty can be regarded as part of the results, but it was always ignored when it was used as criteria for the assessment of analysis method. In this paper, the research was carried out on the basis of the determination of 18 pesticide residues in tea by gas chromatography-tandem mass spectrometry (GC–MS/MS) coupled with different sample preparation technique, (1) hydration expansion + oscillation extraction with acetonitrile + liquid–liquid extraction and cleanup with hexane, (2) hydration expansion + oscillation extraction with acetonitrile + solid-phase extraction and cleanup, (3) homogeneous extraction with acetonitrile + solid-phase extraction and cleanup. The nine components of uncertainty arising from the determination of pesticide residue adopting the above three sample preparation methods combined with GC–MS/MS applying the "bottom-up" approach according to Eurachem/Citac guide [4] was discussed in detail and the nine components of uncertainty were repeatability, weighing, preparation of standard solution, the concentration of internal standard solution, the volume of internal standard solution, the preparation process of quantitative using standard curve, the quantitative process using standard curve, the concentration of pesticide, and recovery. By the comparison the contribution of each uncertainty component in different methods, it was found that the five uncertainty components, weighing, preparation of standard solution, the concentration of internal standard solution, the volume of internal standard solution, and the concentration of pesticide were secondary uncertainty components for their contributions to combined uncertainty was relative small, while the four uncertainty components, repeatability, the preparation process of quantitative using standard curve, the quantitative process using standard curve, the concentration of pesticide, and recovery were key uncertainty components for their contributions to combined uncertainty was relative great. By the quantitative analysis of the different uncertainty components of each method, it was found that the uncertainty component arising from recovery can be regarded as criteria for the appraising of the three methods. For the uncertainty arising from recovery, M2 and M3 were superior to M1, meanwhile relative to M2, M3 has the advantage of low variation and excellent data stabilization. The combined recovery of all methods was approaching except individual pesticides. It was notable that the conclusion of this chapter was from the angle of uncertainty and uncertainty was only one aspect of the numerous factors for the evaluation of methods. To obtain more accurate and reliable conclusion, other factors must be combined for the comprehensive and organic evaluation of the methods.

REFERENCES

[1] Rozet, E., Marini, R.D., Ziemons, E., Boulanger, B., Hubert, P., 2011. Advances in validation, risk and uncertainty assessment of bioanalytical methods. J. Pharm. Biomed. Anal. 55, 848.

[2] Youden, W., 1962. Uncertainties in calibration. In: Proceedings of 1962 International Conference on Precision Electromagnetic Measurements.

[3] Eurachem, 1995. Quantifying Uncertainty in Analytical Measurement. Laboratory of the Government Chemist, London, ISBN 0-948926-08-2.

[4] EURACHEM/CITAC Guide CG 4, 2000. In: Ellison, S.L.R., Rosslein, M., Williams, A. (Eds.), Quantifying Uncertainty in Analytical Measurement (QUAM). second ed. EURACHEM/CITAC.

[5] Armishaw, P., 2003. Estimating measurement uncertainty in an afternoon. A case study in the practical application of measurement uncertainty. Accredit. Qual. Assur. 8, 218–224.

[6] Ratola, N., Faria, J.L., Alves, A., 2004. Analysis and quantification of trans-resveratrol in wines from Alentejo region (Portugal). Food Technol. Biotechnol. 42, 125–130.

[7] Fisicaro, P., Amarouche, S., Lalere, B., 2008. Approaches to uncertainty evaluation based on proficiency testing schemes in chemical measurements. Accredit. Qual. Assur. 13, 361–366.

[8] Jiménez, O.P., Pérez Pastor, R.M., 2012. Estimation of measurement uncertainty of pesticides, polychlorinated biphenyls and polyaromatic hydrocarbons in sediments by using gas chromatography–mass spectrometry. Anal. Chim. Acta 724, 20–29.

[9] Quintela, M., Báguena, J., Gotor, G., et al., 2012. Estimation of the uncertainty associated with the results based on the validation of chromatographic analysis procedures: application to the determination of chlorides by high performance liquid chromatography and of fatty acids by high resolution gas chromatography. J. Chromatogr. A 1223, 107–117.

[10] Pelit, F.O., Ertas,, H., Seyrani, I., et al., 2013. Assessment of DFG-S19 method for the determination of common endocrine disruptor pesticides in wine samples with an estimation of the uncertainty of the analytical results. Food Chem. 138 (1), 54–61.

[11] ISO, 1993. Guide to the Expression of Uncertainty in Measurement. International Standards Organisation, Geneva.

[12] Analytical Methods Committee, 1995. Analyst 120, 2303.

[13] Hässelbarth, W., 1998. Measurement uncertainty procedures revisited: direct determination of uncertainty and bias handling. Accredit. Qual. Assur. 3, 418–422.

[14] Cuadros-RodríGuez, L., Hernández Torres, M.E., Almansa López, E., et al., 2002. Assessment of uncertainty in pesticide multiresidue analytical methods: main sources and estimation. Anal. Chim. Acta 454 (2), 297–314.

[15] Štěpán, R., Hajšlová, J., Kocourek, V., et al., 2004. Uncertainties of gas chromatographic measurement of troublesome pesticide residues in apples employing conventional and mass spectrometric detectors. Anal. Chim. Acta 520 (1/2), 245–255.

[16] Díaz, A., Vàzquez, L., Ventura, F., Galceran S M.T., 2004. Estimation of measurement uncertainty for the determination of nonylphenol in water using solid-phase extraction and solid-phase microextraction procedures. Anal. Chim. Acta 506, 71–80.

[17] Kanrar, B., Mandal, S., Bhattacharyya, A., 2010. Validation and uncertainty analysis of a multiresidue method for 42 pesticides in made tea, tea infusion and spent leaves using ethyl acetate extraction and liquid chromatography–tandem mass spectrometry. J. Chromatogr. A 1217 (12), 1926–1933.

[18] Evaluation of measurement data—Guide to the expression of uncertainty in measurement. JCGM 100, 2008.

FURTHER READING

[19] International Bureau of Weights and Measures, ISO, 1995. Guide to the Expression of Uncertainty in Measurement. International Organization for the Standardization, Geneva, pp. 2, 3.

4 Matrix Effect for Determination of Pesticide Residues in Tea

Chapter Outline

4.1 REVIEW

4.1.1 Current Situation of Matrix Effect

4.1.1.1 Matrix Effect Definition

Matrix usually refers to another component in a sample, excluding the target analyte, and matrix effect has to do with the influence of other components on the determination of analytes: that is, the quantitative and qualitative errors that arise from the other components. The definition for matrix effect is from the American Clinical and Laboratory Standards Institute: The effect of other components other than the analyte in the sample on the measured value of the analyte, namely, the interference of the matrix to the accuracy of the analytical method [1]. The definition of quality control procedures for pesticide residue analysis by the European Union is: matrix effect refers to the influence of one or several other components in a sample on the determination accuracy of the concentration or mass concentration of the analyte [2].

The influence of the matrix effect on the concentration and determination accuracy of the analyte is a very common consideration, such as that in gas chromatography (GC), liquid chromatography (LC), mass spectrometry (MS), tandem mass spectrometry (MS/MS), and inductively coupled plasma mass spectrometry (ICP-MS). In the analysis of pesticide residues, Gillespie et al. [3] established a method for the determination of organophosphorus pesticides in high fat samples in 1990, and found that the recoveries for vast majority of pesticides with P=O group exceeded 130% when using the solvent standard for quantitation, but the recoveries remained about 103% if the sample was corrected using the standard solution containing the sample matrix. In 1993, Erney et al. [4] first named this phenomenon as a matrix-induced response enhancement. In pesticide analysis with GC, most of the pesticides showed different degrees of matrix enhancement effects. However, with LC–MS, the common phenomenon is the matrix weakening effect; this is also one of the main disadvantages of LC–MS.

4.1.1.2 Classification of Matrix Effect

According to the different influences of matrix composition on the response of detection signals, matrix effects can be classified into matrix enhancement effects and matrix weakening effects. The matrix enhancement effect indicates that the presence of the sample matrix component reduces the chance that the active site of the chromatographic system interacts with the target molecule, which leads to the response of the target analyte higher than that of the pure solvent. The weakening effect refers to the phenomenon that the presence of the matrix component weakens the detection signal. Most of the pesticides in GC showed different degrees of matrix enhancement effects.

4.1.1.3 Production of the Matrix Effect and Its Influencing Factors

In the analysis of GC, it is generally believed that the mechanism of the matrix effect is that in the actual sample detection, the molecules of impurities from the sample matrix tend to block active sites (silanols, metal ions, and other active sites produced by thermally decomposed components of samples) in the GC inlet and column, thus reducing losses of susceptible analytes caused by adsorption or degradation on these active sites. Thus, the signal of the same content of analyte in the real sample is higher than that in the standard sample [5,6].

The factors that may affect the matrix effect mainly include:

1. *Type of extraction reagent and purification material*: The matrix effects will be different if there exists a difference among the types of the extraction reagent and the purification material [7–10]. Schenck et al. [9] found that the matrix effect was not improved when using GCB as an SPE column material for purification of fillers, but the matrix effect would be greatly reduced if a weak amino group, such as PSA or an NH_2 column, was used to connect with GCB for purification.

2. *Concentration of analytes*: The matrix effect has a close relationship with the concentration of the analyte. The matrix effect is obvious at low concentrations of target compounds. However, the matrix effect shows a weakening trend with the increase of concentration [11–13]. Liu Li et al. [13] found that in the determination of methamidophos, acephate and omethoate in cucumber, the matrix effects of three pesticides were decreased with the increase of pesticide concentration.

3. *The chemical structure and property of the analyte*: In general, organic ester, hydroxyl, imidazole, and amino compounds with thermal instability, polarity, and hydrogen bonding capacity are easy to produce matrix effects in GC analysis. In the analysis of modern pesticides, the pesticide with the following functional groups or characteristic structures: $(CHO)_3$—P,—O—CO—NH—,—OH,—N≡, R—NH—,—NH—CO—NH—, and so on, have the obvious matrix enhancement effect [5]. Liu Li et al. [13] found that, in the determination of organophosphorus pesticides in cucumbers, three kinds of pesticides, including methamidophos, acephate, and omethoate, were greatly affected by the matrix, in which the influence to acephate was the most significant, while the other nine pesticides had no obvious difference.

4. *Concentration, type, and character of matrix*: The concentration, type, and character of a sample matrix have a great influence on the matrix effect [12,14,15–21]. Guosheng et al. [18] found that the higher the water content, the weaker the matrix effect. The higher the carbohydrate, lipid, and protein content, the stronger the matrix effect. Pinheiro et al. [20] found that the matrix effect was relatively weak in some water samples with simple matrix composition, and the matrix effect was gradually increased with the complexity of the sample composition [21].

5. *Sampling technique*: Temperature program [22], direct sample introduction technique [23–25], and pulse splitless injection techniques [26] can reduce the matrix effect to some extent.

6. *Type of detector*: Different types of detectors also have an effect on the matrix effect [4,9,27]. Schenck et al. [9] found that the matrix effect of pesticides with gas chromatography-ion trap mass spectrometry was better than that of gas chromatography-flame photometric detector. Souverain et al. [27] investigated that the effects of two ionization modes of electrospray ionization (ESI) and atmospheric pressure chemical ionization (APCI) on the tolerance of matrix effect, and found that the ESI mode was more likely to be affected by the matrix effect than that of the APCI mode.

Moreover, the structure of inlet, the number and type of the activity point, the analysis condition (e.g., temperature, carrier gas flow rate, and pressure) [17,19], as well as the determination conditions of instruments (e.g., liner tube, column type, and pollution) also affect the matrix effect. Jufang et al. [28] studied systematically the effect of the matrix effect on pesticide residue determination in vegetables, and found that the adding of a certain amount of glass wool in the liner would effectively improve the matrix effect. After using the chromatogramic column for a long period of time, it is necessary to intercept a long column to improve the matrix effect. Also, the increasing of inlet temperature could effectively improve the matrix effect.

4.1.1.4 The Compensation Method of the Matrix Effect

In recent years, efforts to eliminate and compensate for the matrix effects of GC have been increasing. The main work includes:

1. *Improved sample preparation technique*: In the analysis process, the increase of purification steps could reduce the matrix effect. Although the improved sample preparation technique could improve the matrix effect, it cannot completely eliminate the matrix effect [9]. Furthermore, the increase of the preparation procedure may lead to the loss of some pesticides, which could result in the reduction of pesticide recovery.

2. *Optimization analysis condition*: Proper sampling strategy and correct operation and maintenance for the instrument system are beneficial to reduce the matrix effect. Cold column sampling, pulse nonshunt sampling, temperature programmed vaporization, and other injection methods could reduce the matrix effect. But they still cannot completely eliminate the matrix effect.

   Zhenlin et al. [29] used GC and GC–MS methods for the determination of acetochlor residue in 10 kinds of plant- and animal-derived foods. They found that the matrix effects have a great influence on acetochlor. The effect of the matrix effect could be greatly reduced if the sample is injected in a certain sequence, but this method cannot completely solve the matrix effect.

3. *Standard addition method*: The standard addition method is an effective method to compensate the matrix effect [30–32]. A known amount of analyte is added to a sample extract containing the target analyte, while the spiked concentrations of the target analyte are different. The chromatographic response values of samples spiked with standard and without standard were analyzed, and the standard curve was plotted. The concentration of analytes could be obtained according to the slope and intercept of the standard curve.

   Frenich et al. [30] established the method for the analysis of 12 pesticide residues in cucumbers and citrus by a single point standard addition method, and good results were achieved. The method was also successful in the analysis of a real sample. But the method is more complicated, which makes it not suitable for the determination of multiresidues and a large number of samples.

4. *Labeled internal standard method*: For the sensitive compounds, the addition of an isotope labeled internal standard or deuterated internal standard could effectively eliminate the matrix effect [33]. However, it is difficult to obtain the labeled internal standard (especially certified reference material, CRM). Moreover, each sensitive compound needs the corresponding internal standard, which increases the difficulty of the analysis process. At the same time, this method is also not suitable for the analysis of pesticide multiresidues in samples.

5. *Statistical correction factor*: Statistical correction factor is a method of correcting and evaluating the matrix effect by using mathematical statistics method. The application of this method requires that the sample analysis should be performed in the same system environment and the instrument operation conditions. When the operation conditions change, the system needs recalibration [34].

6. *Matrix-matched standardardization method*: At present, a lot of pesticide residue detection methods are based on the matrix-matched standard to calibrate [35–39], in order to make the standard sample in the same matrix environment as the sample. The method is usually used to prepare a pesticide-free blank matrix-matched standard solution, then add the standard solution. In this method, the sample is corrected by the matrix-matched standard, and the sample matrix in the standard compensated the response of the standard and sample solutions to the same extent, to improve the accuracy of the results.

   The standardization is usually adopted with the official laboratory in the United States for pesticide residue detection [40,41], and is also permitted by EU standards [42], but the Environmental Protection Agency (EPA) and the Food and Drug Administration (FDA) do not permit its use for food commodities [43]. It is not practical to apply this method in a routine analysis, especially for a large number of samples to be tested. The reasons are: (1) this procedure involves the preparation of calibration standards in blank extracts, but it is difficult to find pure blank samples for multiresidue analysis; (2) the matrix effect is commodity-dependent; it is too onerous for analysts to prepare matrix-matched standards for each commodity in a sequence with different types of samples; (3) along with the increasing amount of matrix material injected onto the column, GC maintenance will also have to increase.

7. *Method for analyte protectants (APs)*: In the analysis of GC, it is generally believed that the mechanism of the matrix effect is that in the actual sample detection, the molecules of impurities from the sample matrix tend to block active

sites (silanols, metal ions, and other active sites produced by thermally decomposed components of samples) in the GC inlet and column, thus reducing losses of susceptible analytes caused by adsorption or degradation on these active sites. Thus, the signal of the same content of analyte in the real sample is higher than that in the standard sample.

Based on this phenomenon, Erney et al. [44] believed that if the interaction between the active site and the analyte was blocked through the use of appropriate reagents, and the reagents in the sample before joining, to protect the object to be measured by the inlet near the active site of adsorption, the reagent is called analyte of protection agent. In 1993, Erney et al. [44] selected eight kinds of compounds with hydrogen bonding ability (including formic acid, formamide, glycerol, polyethylene glycol, *N,N,N',N'*-4[2-hydroxypropyl] ethylenediamine, hexamethylphosphoric triamide, dioctyl phosphate, 1,2,3-*tris*[2-cyanoethoxy] propane) as APs, and added it into the standard solution to simulate the presence of the matrix. But the experiment did not get the desired stability effect.

In 2003, Anastassiades et al. [43] introduced the idea of APs again. The author thought that the ideal protective agent should have good hydrogen bonding ability, good volatility, and convenient application. First of all, it was necessary to consider the fact that it could produce a strong matrix-induced enhancement effect, and Anastassiades considered that the compound with hydrogen bonding ability was the key factor for producing this phenomenon. Therefore, Anastassiades et al. selected 93 different compounds as APs to evaluate the effect of these compounds on the determination of 40 pesticides in a tomato matrix, in which the compounds with good hydrogen bonding ability were mainly studied, including sugar, sugar alcohols, sugar derivatives, and glycols. If these compounds were added before the sample, it was found that the compounds containing a lot of hydroxyl groups showed a good effect on the response of various pesticides in the GC analysis. The results showed that it was very important to keep the APs and pesticide in the same part of the chromatographic column when the temperature reached to make the analyte in the chromatographic column a small partition coefficient. The study also confirmed that 3-ethoxy-1,2-propanediol was a suitable AP for volatile pesticides, gulonic acid-(-lactone was suitable for a semivolatile pesticide, and sorbitol was suitable for a low volatile pesticide. In addition, the combination of the three AP mixtures could be realized to the quantitative analysis of pesticides with different properties.

On the basis of the previous study, in 2005, Mastovska et al. [45] evaluated various combinations of APs to find a combination that would effectively compensate for the matrix-induced response and "a mixture of ethylglycerol, L-gulonic acid-(-lactone, and sorbitol was found to be the most promising combination." APs also have the advantage of reducing the matrix-induced response diminishment effect, improving the robustness of the method and reducing system maintenance.

On the basis of the previous study, in 2005, Mastovska et al. [45] chose different types of pesticides, such as those easily affected by matrix effects of pesticides, not easily affected by matrix effects of pesticides, and pesticides with different polarity, size, and thermal stability, and chose eight kinds of APs with better effect, including 3-ethoxy-1,2-propanediol, 1,2,3,4-erythritol, 4,6-*O*-ethylidene-D-glucose, caffeine, L-gulonic acid-lactone, polyethylene glycol, triglyceride, D-sorbitol, to study the effects of eight AP combinations for quantitative analysis of fruits and vegetables. The results show that a certain concentration combination of gulonolatone, sobitol, and ethylglycerol could reduce pesticide loss and improve peak shape. At the same time, the quantitative analysis results with AP combinations were compared with the results from the matrix-matched standard solution. It was found that there was little difference between the two quantitative results, and the addition of APs made the quantitative operation more convenient. Therefore, the author believes that the use of APs could reduce the matrix-induced response diminishment effect, improve the robustness of the method, and reduce system maintenance.

The same year, Sánchez Brunete et al. [46] studied the matrix effects of 25 types of pesticides in soil, honey, and juice by GC–MS and evaluated four APs (2,3-butanediol, L-gulonic acid-(-lactone, corn oil, and olive oil) to counteract the matrix effects. The result indicated that L-gulonic acid-(-lactone was suitable for the compensation of matrix effect in soil and honey, and olive oil was suitable for the compensation of matrix effect in the juice samples.

González-Rodríguez et al. developed methods using APs to compensate for matrix-induced enhancement effect to detect 23 fungicides and insecticides residues in 75 green and leafy vegetables [47] and 11 new fungicides in grapes and wines [48], the results showed that the APs mixtures provide the best results in terms of effective compensation for matrix-induced enhancement effect.

In 2011, Wang et al. [49] established a method for simultaneously determining 195 pesticide residues in Chinese herbs by GC–MS using a APs combination of D-ribonic acid-c-lactone and sorbitol. The authors compared 7 kinds of APs, and then established a simple and rapid method for the determination of pesticide residues in Chinese herbs by the appropriate APs combination. Through the experiment, the author found that the pesticides of 2-toluene-4-phenoxy propionic acid, dichlofluanid, omethoate, fenthion sulfoxide, and fluoroglycofen have been greatly affected by the matrix. So the author concluded that the compounds with P=O, —O—CO—NH—, —OH, —N=, R—NH—, —NH—CO—NH— and other functional groups or characteristic structures have the obvious matrix enhancement effect.

Domestic researches on APs mainly include: in 2006, Huang Baoyong et al. [50] established a GC–MS method for determination of 52 kinds of pesticides in vegetables, and investigated the application of the matrix effects and APs (e.g., 3-ethoxy-1,2-propanediol, sorbitol, L-gulonic acid-(-lactone) in the method, then evaluated the effects of 3 kinds of AP combinations on the compensation of the matrix effect, and determined the optimum proportion of AP combinations. The same year, the author established a GC–MS method under a selected ion-monitoring mode for the determination of 45 kinds of pesticides in vegetables and fruits, and evaluated the effect of addition of APs (3-ethoxy-1,2-propanediol and D-sorbitol) on analytical results reliability and compensation matrix effects of pesticide residues in the process of analysis. The result showed that for most pesticides (especially methamidophos, acephate, omethoate, paraoxon), the addition of the above APs could significantly compensate the quantitative error resulting from the matrix enhancement effect and reducing the detection limit. The author also found that *o,p*′-DDT and *p,p*′-DDT displayed the matrix weakening effect, and the AP has no compensation effect on it.

4.1.2 Cluster Analysis and Its Application in Chemical Analysis

Chemometrics was founded in 1970s by a Swedish chemist Wold and an American chemist Kowalski [52]. Since 1980s, the chemometrics has been applied in the field of chemical industry as a good analytical tool. Chemometrics mainly uses statistics, mathematics, modern computer science, as well as other scientific theories and methods to optimize the chemical measurement process, and to maximize the extraction of useful chemical information according to the chemical measurement data. It is an emerging discipline based on the basis of interdiscipline, and is an important branch of chemistry. Chemometrics provide theory and methods for chemical measurements, and data analysis for various spectral and chemical data. It also provides new solutions and new ways of thinking about mechanism research and optimization of chemicals and the chemical processes.

According to the classification of the *Journal of the American Chemical* (Analytical Chemistry) biennial review [53–56], chemometrics can be divided into 11 research directions, including signal processing, optimization methods, factor analysis, multivariate calibration, chemical image analysis, chemical pattern recognition, parameter estimation, relationship of structure-active and structure-property of a compound, artificial intelligence, chemical database retrieval, and multivariate curve resolution. Since the 1980s, chemometrics have been widely applied in food analysis, by combining with atomic absorption spectrometry, atomic emission spectrometry, infrared spectrometry, chromatography, mass spectrometry [57].

Chemical pattern recognition methods include principal component analysis (PCA), cluster analysis (CA), and discriminant analysis (DA).

CA is an unsupervised pattern recognition method, which only depends on the properties of the object itself to distinguish the similarity between different objects [58]. According to the principle of "Like attracts like," it gathers the samples into different classes or clusters. First, in the classification process, the distances between different objects are calculated according to certain mathematical methods, and the distance from the nearest sample is clustered into a new class. Then, the differences between the new class and the rest are calculated, and two categories with smaller differences are combined. The calculation is repeated until all samples are aggregated into a class [59]. The purpose of CA is that after clustering, the samples with higher similarities are clustered into one class, and the samples with differences become different categories, to find the internal structure of the data set. CA is a basic method of data analysis, and it has been widely applied in the field of chemistry. As a branch of statistics, the statistical methods of CA are included in many statistical software, such as SPSS and SAS [60,61].

For different substances, the contents of trace elements, chromatographic and spectroscopic properties have some differences. Based on these differences, the CA method can establish a corresponding model to achieve the purpose of classification of different substances. For example, da Silva Torres et al. [62] applied hierarchical cluster analysis (HCA) to categorize foods according to their nutritive values. The result showed that French fries were the highest caloric preparation, with a considerable total fiber content. Milled white rice is rich in carbohydrates. Arugula offers the highest protein and total fiber content, whereas lettuce presented the smallest amount of these two nutrients. To better understand the occurrence of high F and As in the groundwater of the Datong basin, Li et al. [63] collected a total of 486 groundwater samples for the HCA of 18 hydrochemical parameters. Groundwater samples were divided into 36 and 19 groups for shallow and deep groundwater, respectively. The contents of F and As in the groundwater were determined, and the groundwater distribution was sketched. Jin et al. [64] clustered 8 h ozone levels in California area with the CA method, and the corresponding study provided a theoretical support for the establishment of a meteorologically representative pollution regimes model.

4.2 STUDY ON THE MATRIX EFFECTS OF DIFFERENT TEA VARIETIES FROM DIFFERENT PRODUCING AREAS

4.2.1 Introduction

Matrix effects are affected by the origin and matrix type. Different matrices have different matrix effects, and the same matrix sometimes also has different matrix effects. Kocourek et al. [65] evaluated the stability of pesticides in three kinds of plant extracts, for example, cabbage, citrus, and wheat, and they found that the stability of wheat extract was significantly higher than that of the other two plants. Kittlaus et al. [66] found that the matrix effects of green tea and oolong tea have great differences in LC–MS/MS analysis.

The common strategy for avoiding matrix effects is the use of matrix-matched standards, but this procedure involves the preparation of the corresponding matrix-matched standards for each matrix. However, the whole process of the operation is cumbersome, while reducing the sample processing speed. Therefore, in order to ensure the accuracy of the experiment and to simplify the experimental operation, different groupings have been proposed for a wide variety of matrices based on similarities of major constituents, such as water, fat/oil, sugar, and acid content [42,67]. A large number of possible representative matrices for each group also have been proposed [42,68,69]. Romero-González et al. [68] measured more than 90 pesticides in peaches, oranges, pineapples, apples, and the several fruit-mixed matrices. The authors evaluated the matrix effects of these fruits in the determination process, and found that many fruits matrices could be used as a representative matrix for verifying the whole experiment. The results obtained were within the range of the relevant standards. The EU is also listed as the representative matrix on behalf of similar substances in the relevant laws and regulations, such as apples and pears, using as the representative of some substances, and the representative of the high water content of fruit materials [42]. However, in the process of tea sample analysis, many scholars still use the single matrix-matched standard solution [35,70].

Hierarchical clustering analysis is one of the most popular multivariate exploratory methods used in analytical chemistry [71]. In HCA, samples are grouped on the basis of their similarities without taking into account information about their class membership. The tool allows the definition of the sample as a point in the variable space. Then the distances among the points are calculated by a certain calculation method, and the samples are classified according to the distance. First, the nearest sample is classified into a new class, and then the distance between the new class and the rest of the class is recalculated. Last, the samples are merged according to the distance, until all the samples are classified into one class. There are many different ways to calculate the distance, such as Euclidean distance, absolute distance, Chebyshev distance, Minkowski distance, variance weighted distance, and Mahalanobis distance. The similarity between the samples are represented by a two-dimensional tree diagram, in which the vertical line shows the distance between two objects, and the line connecting the two objects indicates the similarity between the two objects.

In this chapter, the matrix effects of 28 kinds of tea samples with different origins and varities were studied based on instruments of GC–MS and GC–MS/MS, and HCA was used to classify the matrix effects of tea samples detected by GC–MS and GC–MS/MS methods. It is hoped that on the basis of these results, representative tea matrixes can be selected to prepare the matrix-matched standards for the 28 tea matrixes, which can simplify the operation process.

4.2.2 Experimental Materials

4.2.2.1 Origin and Classification of 28 Teas

The origin and classification of the 28 teas are shown in Table 4.1.

4.2.2.2 Reagents and Materials

Acetone, acetonitrile, hexane, toluene, and methanol (LC grade), purchased from Fisher Scientific (Fair Lawn, NJ); Clean-ert TPT solid phase extraction column (2000 mg/12 mL) purchased from Tianjin agela company; pesticide standard, Purity, ≥95% (LGC Promochem, Wesel, Germany) are shown in Appendix Tables 4.1 and 4.2.

Preparation of pesticide standard solutions: Weigh 5–10 mg of individual pesticide standards (accurate to 0.1 mg) into a 10 mL volumetric flask. Dissolve and dilute to volume with toluene, toluene-acetone combination, methanol, acetonitrile, isooctane, depending on each individual compound solubility. Stock standard solutions should be stored in the dark below 4, and can be used for 1 month.

Depending on the properties and the retention times of compounds, the compounds are divided into three groups of A, B, and C. The mixed standard solution concentration is determined by its sensitivity on the instrument used for analysis. Depending on the group number, mixed standard solution concentration, and stock standard solution concentration, appropriate amounts of individual stock standard solution were pipetted into a 100 mL valumetric flask, diluted to volume with methanol. Mixed standerd solutions should be stored in the dark below 4, and can be used for 1 month.

TABLE 4.1 Variety, Species, Degree of Fermentation, and Origin of the 28 Teas Investigated in This Work

No.	Variety	Species of Tea	Degree of Fermentation of Tea	Origin
1	Huangshan Mao Feng	Green	Unfermented	Huizhou District, Anhui Province
2	Deng Village green tea	Green	Unfermented	Deng Cun Town, Yiling District, Hubei Province
3	Longjing	Green	Unfermented	Longjing Village, Zhejiang Province
4	Lushan Yunwu	Green	Unfermented	Han Pokou, Jiangxi Province
5	Taiping Houkui tea	Green	Unfermented	Huangshan District, Anhui Province
6	Enshi Yulu tea	Green	Unfermented	Bajiaodongzu Town, Hubei Province
7	Baisha green tea	Green	Unfermented	Baisha Li Autonomous County, Hainan Province
8	Guzhang Maojian tea	Green	Unfermented	Guzhang County, Hunan Province
9	Dongting Biluochun	Green	Unfermented	Dongshan Town, Jiangsu Province
10	Yu Hua tea	Green	Unfermented	Nanjing, Jiangsu Province
11	An Ji Bai tea	Green	Unfermented	Anji County, Zhejiang Province
12	Liuan Guapian green tea	Green	Unfermented	Yu'an District, Anhui Province
13	Mengshan tea	Green	Unfermented	Mengding Mountain, Sichuan Province
14	Laoshan green tea	Green	Unfermented	Wang Ge Zhuang, Shandong Province
15	Rizhao green tea	Green	Unfermented	Donggang District, Shandong Province
16	Junshan Yinzhen tea	Yellow	Partially fermented	Qingluo Island, Hunan Province
17	White Pony	White	Partially fermented	Zhenghe County, Fujian Province
18	Bai Hao Yin Zhen	White	Partially fermented	Tailao Mountain, Fujian Province
19	Anxi Tie Guan Yin	Oolong	Partially fermented	Western Anxi, Quanzhou, Fujian Anxi County
20	YongChun Fo Shou	Oolong	Partially fermented	Yongchun County, Fujian Province
21	Huang JinGui	Oolong	Partially fermented	Meizhuang Village, Fujian Province
22	Fenghuang Dancong	Oolong	Partially fermented	Fenghuang Town, Guangdong Province
23	Keemun black tea	Black	Completely fermented	Keemun County, Anhui Province
24	Dian black tea	Black	Completely fermented	Fengqing County, Yunnan Province
25	Min black tea	Black	Completely fermented	Gutian County, Fujian Province
26	Anhua dark tea	Dark	Post-fermented	Anhua County, Hunan Province
27	Liu Pao tea	Dark	Post-fermented	Liupao town, Guangxi Zhuang Autonomous Region
28	Pu'er tea	Pu'er	Post-fermented	Simao District, Yunnan Province

Working standard mixed solution in matrix: Working standard mixed solutions with different concentrations in a matrix are prepared with a blank extract to plot standard work curve. Working standard mixed solutions in the matrix must be prepared fresh.

Preparation of solvent standard solution: Place appropriate amounts of mixed standard solutions and 40 μL internal standard into a 1 mL valumetric flask. Under a stream of nitrogen after drying, dilute to volume with acetonitrile.

Internal standard solution: Weigh 3.5 mg heptachlor epoxide into a 100 mL volumetric flask. Dissolve and dilute to volume with toluene. Before each determination, place 40 μL internal standard solution into the test solution to correct the errors in the measurement of constant volume and instrument.

4.2.2.3 Apparatus

The main instruments used in the paper and their origins are shown in Table 4.2.

4.2.3 Preparation of a Test Sample

Tea samples extracted with the method of Pang et al. [35]: Weigh 5 g of the test sample (accurate to 0.01 g) into an 80 mL centrifuge tube, add 15 mL acetonitrile, and homogenize at 15,000 r/min for 1 min, then centrifuge at 4200 r/min for 5 min.

TABLE 4.2 Name and Place of the Instrument Using in the Experiment

Name	Place
Medical glass syringe(1 mL)	Shinva Ande Healthcare Apparatus Co., Ltd
Color single channel liquid gun(5–200 μL)	Eppendorf Ltd., German
Color single channel liquid gun(100–1000 μL)	Eppendorf Ltd., German
FJ200-S **homogenize**	IKA, German
Mettler PM6400 **Electronic Balance**(0.01 g)	Mettler, Switzerland
SC-60 vertical transparent door type cold box	Qingdao Aucma Co., Ltd
8893 type ultrasonic cleaner	Cole parmer, USA
VORTEX-2 **Vortex Mixer**	Scientific Industries, USA
SHB-B95 Type Vacuum Pump	Zhengzhou Greatwall Scientific Industrial and Trade Co., Ltd
Agilent6890/5973N GC–MSD with electron ionization (EI) source,and equipped with a Model 7683 autosampler	Agilent Technologies, USA
Agilent7890/7000 GC–MS/MSwith electron ionization (EI) source,and equipped with a Model CTC autosampler	Agilent Technologies, USA
Agilent1290/6460A LC–MS/MS	Agilent Technologies, USA
Milli-Q Ultrapure Water System	Millipore, USA
SA31 rotary evaporator	Büchi, Switzerland
Nitrogen evaporator	Organomation Associates, Jnc, USA

Other equipment: 10 mL pipette, 0.2 μm membrane, etc.

Pipette the acetonitrile layer of the extracts into a pear-shaped flask. The residue is once again extracted with 15 mL acetonitrile and the aforementioned procedure is repeated. The two portions collected are combined and concentrate the extract to ca 1 mL with a rotary evaporator at 40 for cleanup.

Add sodium sulfate into the Cleanert-TPT cartridge to ca 2 cm. Prewash the cartridge with 10 mL acetonitrile-toluent(3 + 1) before adding the sample. Fix the cartridge into a support to which a pear-shaped flask is connected. Once the solution gets to the top of sodium sulfate, pipette the eluate into the cartridge immediately. Rinse the pear-shaped flask with 3 × 2 mL acetonitrile-toluent (3 + 1) and decant it into the cartridge. Insert a 50 mL reservoir into the cartridges. Elute the pesticides with 25 mL acetonitrile-toluene (3 + 1). Collect the eluate into a pear-shaped flask. Evaporate the eluate to ca 0.5 mL using a rotary evaporator at 40. Add 40 μL internal standard solution and mix thoroughly. After drying under a stream of nitrogen, it is diluted to 1 mL with acetonitrile. Filter through a 0.2 μm membrane; the solution is ready for GC–MS determination.

Spiked experiment: Weigh a 5 g test sample, add mixed standard solution. After 30 min, the sample is processed as earlier described in this section.

4.2.4 Instrumental Analysis Condition

4.2.4.1 GC–MS and GC–MS/MS Instrumental Condition

1. GC condition.
   Column: DB-1701 capillary column (30 m × 0.25 mm × 0.25 μm) (J&W Scientific, Folsom, CA); Column temperature: 40°C hold 1 min, at 30°C/min to 130°C, at 5°C/min to 250°C, at 10°C/min to 300°C, hold 5 min; Carrier gas: helium, purity ≥99.999%, flow rate: 1.2 mL/min; injection port temperature: 290°C; injection volume: 1 μL; injection mode: splitless, purge on after 1.5 min.

2. MS/MS operating condition.
   Ion source: EI (electron impact ionization source); ion source temperature: 230; ionization voltage: 70 eV; GC–MS interface temperature: 280; scan mode: selected ion monitoring mode: each compound selects 1 quantifying ion and 2–3 qualifying ions. The retention times, quantifying ions, qualifying ions, and the abundance of ratios of quantifying ions and qualifying ions for each compound are listed in Appendix Tables 4.1 and 4.2.

TABLE 4.3 The Conditions of Mobile Phase and Gradient Elution

Time/min	Flow Rate/(μL/min)	Mobile Phase A(0.1% formic acid)/%	Mobile Phase B(acetonitrile)/%
0.00	400	99.0	1.0
3.00	400	70.0	30.0
6.00	400	60.0	40.0
9.00	400	60.0	40.0
15.00	400	40.0	60.0
19.00	400	1.0	99.0
23.00	400	1.0	99.0
23.01	400	99.0	1.0

4.2.4.2 LC–MS/MS

1. LC condition.
   Column: ZORBAX SB-C_{18} column, 3.5 μm, 100 mm × 2.1 mm (inside diameter) or equivalent. Mobile phase and gradient elution conditions are listed in Table 4.3. Column temperature: 40; injection volume: 10 μL.
2. MS/MS operating condition.
   Ion source: ESI; scan mode: positive ion scan; nebulizer gas: nitrogen; nebulizer gas pressure: 0.28 MPa; ion spray voltage: 4 000 V; dry gas temperature: 350; dry gas flow rate: 10 L/min; MRM transitions for precursor/product ion, quantifying for precursor/product ion, declustering potential, collision energy, and the voltage-source fragmentation (Appendix Table 4.3).

4.2.5 Qualitative Analysis of the Sample and Data Processing

4.2.5.1 Qualitative Analysis of the Sample

For the samples determined, if the retention times of peaks found in the sample solution chromatogram are the same as the peaks in the standard in blank matrix extract chromatogram, and the aboundance ratios of MRM transtions for precursor/product ion are within the expected limits, (relative abundance >50%, permission ±20% deviation; relative abundance >20% to 50%, permission ±25% deviation; relative abundance >10% to 20%, permission ±30% deviation; relative abundance ≤10%, permission ±50% deviation), the sample is confirmed to contain this pesticide compound.

4.2.5.2 Calculation of the Matrix Effect

The matrix effect of each pesticide was obtained by comparing the response of the pesticide in the standard solution of the tea matrix with that in the pure acetonitrile solvent. The average value of each sample was measured three times. The matrix effect was calculated according to the following formula: matrix effect (%) = (A2 − A1)/A1 × 100 (%). A1 was the peak response of the pesticide in acetonitrile solution; A2 was the peak response of the pesticide in the blank sample, and the spiked concentration was the MRL value of each pesticide.

4.2.5.3 Data Processing Method

In the experiment, the hierarchical clustering method is processed by SPSS 19 software, clustered with the Euclidean distance method, and classified with the in-group connection method.

4.2.6 Matrix Effect Evaluation of GC–MS

4.2.6.1 Matrix Effect Evaluation of Pesticides

Through the matrix effect of 186 pesticides in the 28 kinds of tea matrix, it can be seen that the matrix effects of the 186 pesticides in tea are all positive, that is to say, the matrix enhancement effect.

The matrix effect in 0–20% is a weak matrix enhancement effect (soft), 20%–50% is a medium matrix enhancement effect (medium), greater than 50% is a strong matrix enhancement effect (high) [73]. The 186 pesticides were classified according to the matrix effect, and the results are listed in Fig. 4.1.

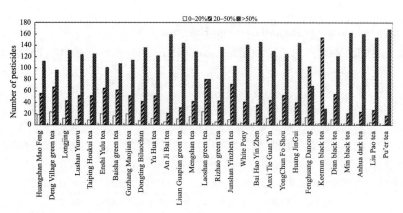

FIGURE 4.1   **Distribution of matrix effects on 186 pesticides in the 28 types of tea matrix.**

It can be obsreved from Fig. 4.1 that most of the pesticides demonstrated moderate and strong signal enhancement, but the matrix effect of different tea varieties had a great difference. For example, in Keemun black tea, only 29 kinds of pesticides (15.6% of the total) showed a strong enhancement matrix effect, but there were 69 (37.1%), 81 (43.5%), and 96 (51.6%) kinds of pesticides to strong enhancement matrix effects in Phoenix Dancong, Laoshan green tea, and den Village green tea, respectively. In the other 24 kinds of tea, the pesticides with strong matrix enhancement were greater than 100; Pu'er tea is even as high as 168 (accounting for 90.3% of the total). On the contrary, there were 154 pesticides in Keemun black tea and 103 pesticides in Phoenix Dancong to show moderate matrix enhancement effects.

The analysis shows that in 186 kinds of pesticides in tea, at least 162 pesticides (Laoshan green tea) show moderate and strong matrix enhancement effects, accounting for 87.1% of the total, and 185 kinds of pesticides in Pu'er tea showed moderate or strong matrix enhancement, accounting for 99.5% of the total. Therefore, it was concluded that the tea had strong matrix effects.

The current thinking suggests that the matrix enhancement effects of pesticides with the following functional groups or characteristic structure are obvious: P=O, —O—CO—NH—, —OH, —N=, R—NH—, —NH—CO—NH—, and so on. This kind of pesticide was usually polar, thermal unstable, and has good hydrogen bonding ability, such as methamidophos, acephate, omethoate, dicrotophos, malaoxon, carbaryl, chlorothalonil, captan, and deltamethrin. It was also found that the same phenomenon in the paper, the matrix effects of 2-phenylphenol, difenoconazole, flumioxazin, fluquinconazole, and bromopropylate were 126%, 257%, 258%, 156%, and 136%, respectively. Some literatures have also mentioned that organic chlorine pesticides have little effect on the matrix because they contain few functional groups acting with the active site [36]. But in this experiment, these organic chlorine pesticides, such as acifluorfen, dicloran, endosulfan sulfate, endrin, heptachlor, methoxychlor, oxyfluorfen, *p,p'*- DDT, quinoxyfen, quintozene and tetradifon, showed strong matrix enhancement effects, which indicated that the tea was a strong matrix effect material.

4.2.6.2   *Evaluation of the Matrix Effects of Different Origins and Varieties of Tea*

The same kind of pesticide showed different matrix effects in different tea matrices. For example, oxygen fenchlorphos in Deng village green tea and Guzhang Maojian tea showed weak matrix enhancement effect (the matrix effects of two teas are both 17%), but showed moderate matrix enhancement effect (the matrix effects are in the range of 23%–46%) in 16 kinds of teas, including Phenghuang Dancong, Laoshan green tea, Enshi Yulu tea, Huangshan Maofeng tea, Taiping Houkui tea, Yu Hua tea, Longjing tea, Dian black tea, Baisha green tea, Min black tea, Yongchun Fo Shou, Anhua dark tea, Dongting Biluochun, Rizhao green tea, Anxi Tie Guan Yin, Huang Jin Gui. Strong matrix enhancement effect (matrix effect in the range of 50%–111%), was shown in 10 kinds of teas, including Lushan Yunwu, Junshan Yizhen tea, Liu Pao tea, Bai Hao Yin Zhen, white peony, An Ji Bai tea, Mengshan tea, Keemun black tea, Pu'er tea, and Luan Guapian green tea.

Selection of 7 kinds of representative tea matrices based on 28 kinds of tea in the matrix: Mount Huangshan Maofeng tea, Junshan Yizhen tea, white peony, Anxi Tie Guan Yin, Keemun black tea, Anhua dark tea, and Pu'er tea, representing green tea, yellow tea, white tea, black tea, and oolong tea, black tea, Pu'er tea categories, respectively, and also randomly selected 8 kinds of pesticides (alpha-HCH, bromuconazole, dimethomorph, fenchlorphos-oxon, fenthion-sulfoxide, methoprene, *p,p'*-DDE, methyl parathion-methyl) to plot the distribution of matrix effects of 8 pesticides in the 7 representative matrices (Fig. 4.2). Fig. 4.2 shows that the different types of tea had different matrix effects, some of them even had big differences. The results are similar to those obtained from Kittlaus [66]. The matrix effects of teas from different origins were clustered with SPSS software, and the systematic differences of the matrix effects among different tea varieties were

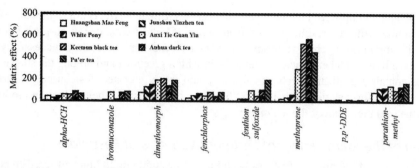

FIGURE 4.2   **Distribution of matrix effects of 8 pesticides in the 7 representative matrices.**

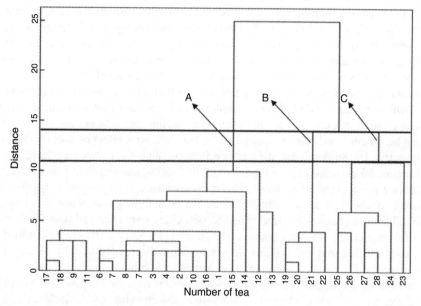

FIGURE 4.3   **Dendrogram of the cluster analysis of the 28 tea matrices.**

investigated. The result is shown with the dendrogram in Fig. 4.3, in which the number of the origin of the tea is in the same order as in Table 4.1.

Fig. 4.3 shows that the 28 tea matrices are separated into three clusters, A, B, and C, at linkage distances between 12 and 14. Group A has 18 samples, including Huangshan Mao Feng, Deng Village green tea, Longjing, Lushan Yunwu tea, Taiping Houkui tea, Enshi Yulu tea, Baisha green tea, Guzhang Maojian tea, Dongting Biluochun tea, Yu Hua tea, An Ji Bai tea, Liuan Guapian green tea, Mengshan tea, Laoshan green tea, Rizhao green tea, Junshan Yinzhen tea, White Peony, and Bai Hao Yin Zhen, which belong to green tea, yellow tea, and white tea kinds of tea varieties; category B consists 4 samples, including Anxi Tie Guan Yin, Yongchun Fo Shou tea, Huang Jin Gui and Fenghuang Dancong tea, belonging to oolong tea species; category C consists 6 samples, including Keemun black tea, Dian black tea, Min black tea, Anhua dark tea, Liu Pao tea, and Pu'er Tea, belonging to black tea and Pu'er tea species. At the same time, Fig. 4.2 shows that the same variety of tea samples from different origins do not have long distance, which also indicates that the differences between different origins are very small.

According to the different processing technology, the tea can be divided into six varieties: green tea, yellow tea, dark tea, white tea, oolong tea, black tea (including Pu'er tea) [74]. Green tea is unfermented tea (fermentation degree is zero), yellow tea is slightly fermented tea (fermentation degrees are 10–20 m), white tea is mild fermented tea (fermentation degrees are 20–30 m), oolong tea is partially fermented tea (fermentation degrees are 30–60 m), black tea is completely fermented tea (fermentation degrees are 80–90 m), dark tea is post-fermented tea (fermentation degree is 100 m).

Based on the clustering results of 28 kinds of tea, we can group the matrix effects of tea into three clusters according to the fermentation degree: the first one is unfermented tea and slightly fermented tea, including green tea, yellow tea, and white tea, and their matrix effects are in the range of 62%–99%; the second cluster is partially fermented teas, including oolong teas, their matrix effects are in the range of 109%–129%; the third one is black tea, dark tea, and Pu'er tea, belonging to the completely fermented and postfermented tea varieties, and their matrix effects are in the range of 146%–168%. The clustering results are similar with the result obtained from Kim [75] and Fernandez et al. [76]. Fernandez et al. made a separation for catechol compounds by chromatographic methods. At the same time, they successfully separated green tea, oolong tea, and instant tea based on chemometrics methods, in which green tea and oolong tea showed obvious difference.

Green tea samples contain significant levels of catechins, and these tea catechins can be converted into aflavins and thearubigins as a result of tea fermentation process [75,76]. Thus, catechins and their oxidation products were supposedly the substance that influenced the effect of the tea matrix.

4.2.6.3 Evaluation of the Matrix Effect of Different Varieties of Pesticides

Calculating the average value of the matrix effect for each kind of pesticide in 28 kinds of tea, 186 pesticides showed different matrix effects, in which the minimum is p,p'-DDE (11%) and the maximum is dirthofencarb (287%). Among 186 pesticides, except p,p'-DDE is weak matrix enhancement effect, and 37 pesticides, for example, aldrin, alpha-HCH, benalaxyl, benfluralin, ethyl-bromophos, bromuconazole, chlorbenside, chlorthal-dimethyl, diazinon, dieldrin, dioxathion, disulfoton sulfoxide, ethalfluralin, ethion, ethofumesate, fenchlorphos-oxon, fenpropimorph, fenpyroximate, fenthion, Lindane, lambda-cyhalothrin, permethrin-1, pirimicarb, procymidone, propoxur-1, prosulfocarb, tolclofos-methyl, terbuthylazine, tetraconazole, tetradifon, thiobencarb, *trans*-chlordane, triadimefon, triallate, and propyzamide showed moderate matrix enhancement effect; the other 148 pesticides showed strong matrix enhancement effect.

The matrix effect data of eight pesticides (e.g., chlorpyrifos-methyl, diphenylamine, fenpropimorph, fenthion sulfoxide, flurtamone, p,p'-DDE, parathion-methyl, and procymidone) in 7 representative tea matrices were plotted (Fig. 4.4). It can see from Fig. 4.4, that the matrix effect of fenthion sulfoxide in Huangshan Mao Feng is 0%, which indicates that the pesticide in Huangshan Mao Feng nearly has no matrix effect, but the matrix effect of flurtamone is as high as 394%. For Huangshan Mao Feng matrix, in 186 pesticides, the matrix effects of 18 pesticides are less than 20%, showed weak matrix effect; The matrix effects of 56 pesticides are in the range of 20%–50%, showed moderate matrix effect; and the matrix effects of 112 pesticides are greater than 50%, showed strong matrix effect. The result indicates that the tea varieties have a great effect on the matrix effects of pesticides in the same tea matrix, and the physical and chemical properties of the pesticide also have some influence on the matrix effect. The 186 pesticides were clustered with SPSS software, and the result is shown with the dendrogram in Fig. 4.5 (the number in the figure is the same as that in Appendix Table 4.1). The matrix effects of 186 pesticides were clustered with SPSS software, and the result is shown with the dendrogram in Fig. 4.3, in which the number of the pesticide is in the same order as in Appendix Table 4.1.

The dendrogram in Fig. 4.5 shows that when the distance is greater than 14 and less than 19, 29 pesticides, including azinphos-ethyl, bifenazate, bioresmethrin, bitertanol, chlorbufam, chlorfenvinphos, diethofencarb, difenoconazole, ethiofencarb, phenamiphos, fenamiphos-sulfone, fenvalerate-2,flucythrinate-1, flumioxazin, fluquinconazole, flurtamone, imazalil, iprovalicarb-2, iprovalicarb-1 metamitron, methoprene, mevinphos, oxadixyl, paclobutrazol, profenofos, pyraclofos,

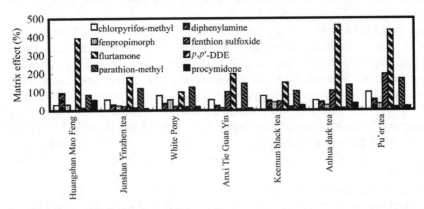

FIGURE 4.4 Distribution of matrix effects in the 7 representative matrices.

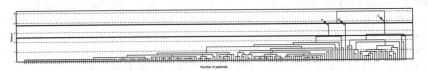

FIGURE 4.5 Dendrogram of the cluster analysis of the 186 pesticides.

pyrimidifen, spiromesifen, and triazophos, are together as category A, in which the average matrix effects of these pesticides are 156%–287%, belonging to the strong matrix-enhanced response of the pesticide. Malathion is as a separate category, and its average matrix effect is 158%. The other pesticides belong to category C, statistically, which also show that the remaining 156 pesticides (accounting for about 84.4% of the total) have similar matrix effects. In Section 4.2.6.1, we discussed that in the same tea matrix, physical and chemical properties of the pesticides have certain effects on their matrix effect. However, through the CA, we found that the differences of physical and chemical properties were not evident on the matrix effect. This phenomenon may be due to the complex matrix of tea and the strong matrix effects of the vast majority of pesticides, resulting in similar clustering results. Stahnke et al. [77] also found similar results in the evaluation of matrix effects in tea by using postcolumn injection.

4.2.6.4 Confirmation Experiments on Classification of Tea Varieties

According to the result from Section 4.2.6.2, teas can be divided into three categories according to their matrix effects, so we can choose any type of tea in each cluster to make a matrix-matched solution. Here, we chose the commercial green, white, oolong, and black teas to study the spiked recovery experiments of 186 pesticides at the MRL level The single-point matrix-matched standard was used for quantitative analysis, and at the same time, a single-point matrix-matched standard of white tea was used for quantitative analysis for each matrix.

The results show that the number of pesticides for which 70%–120% recoveries were obtained from the matrix-matched standard of the green, white, oolong, and black tea matrixes was 186. But the numbers of the green, white, oolong, and black tea with the matrix-matched standard of black tea were 185, 186, 104, and 64, respectively. This shows that the green tea has a similar recovery range when using the single-point matrix-matched standard of white tea and the matrix-matched standard for quantitative analysis. Relative standard deviation (RSD) values of spiked recoveries for the two quantitative methods are in the range of 0.1%–19.8%, which can be considered that two values are similar. The results show that white and green tea have the same matrix effects, and white tea can be used as the representative matrix for green and white tea to prepare the matrix-matched standard solution. The results also showed that the recoveries of oolong and black tea were high by using white tea as a representative matrix for quantitative; for example, the spiked recoveries of difenoconazole in oolong and black tea were 155.3% and 318.5%, respectively, which is due to the complex matrix of oolong and black tea, and the matrix enhancement effect of matrix composition on the pesticide is more obvious than that of white tea. Therefore, the recoveries of oolong and black tea were significantly high when using the single-point matrix-matched standard of white tea. The RSDs of the two recoveries with two matrix-matched standard methods were in the range of 0.1%–60.2% and 0.4%–74.5%, respectively, which indicates that the results of two methods differ greatly. Thus, the white tea cannot be used as the representative matrix for oolong and black tea. This observation agreed well with the clustering results of Section 4.2.6.2.

4.2.7 Evaluation of the Matrix Effect in Determination of GC–MS/MS

4.2.7.1 Evaluation of the Matrix Effect for Pesticide

The matrix effects of 205 pesticides in 28 teas matrices were evaluated by GC–MS/MS. The results showed that the matrix effects of 205 pesticides in tea were all positive, a matrix enhancement effect. The distribution of matrix effects on 205 pesticides in the 28 types of tea matrix are shown in Fig. 4.6.

Fig. 4.6 shows that, similar to GC–MS, the majority of pesticides in 28 tea matrices showed moderate and strong matrix enhancement effects in determination of GC–MS/MS. In the total pesticide with the medium and strong matrix enhancement effect, the number of pesticides in An Ji Bai tea matrix was the smallest, in which 167 pesticides showed moderate or strong matrix enhancement, accounting for 81.5% of the total; however, the numbers in Anhua dark tea and Keemun black tea are the largest, and 205 pesticides all showed moderate and strong matrix enhancement, accounting for 100%.

4.2.7.2 Evaluation of the Matrix Effect for Different Varieties of Tea

Similar to GC–MS, when using GC–MS/MS for determination, pesticides showed different matrix effects in different varieties of tea. For example, matrix effects on *p,p'*-DDE were between 2% and 55%. Weak matrix enhancement effects (with matrix effects of 2%–19%) were exhibited in 14 teas, including Luan Guapian green tea, Deng village green tea, Mengshan tea, Dongting Biluochun, Junshan Yinzhen tea, Rizhao green tea, Enshi Yulu tea, Min black tea, Bai Hao Yin Zhen, An Ji Bai tea, Lushan Yunwu tea, Yongchun Fo Shou, Laoshan green tea, and Huangshan Mao Feng tea. Medium matrix enhancement effects (22%–47%) were exhibited in 11 teas, including Liu Pao tea, Baisha green tea, Yu Hua tea, Pu'er tea, white Peony, Longjing tea, Keemun black tea, Guzhang Maojian tea, Anxi Tie Guan Yin tea, Taiping Houkui tea,

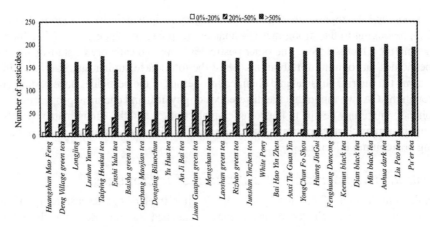

FIGURE 4.6 Distribution of matrix effects of 28 representative matrices.

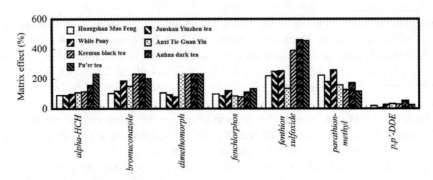

FIGURE 4.7 Distribution of matrix effects of 7 pesticides in the 7 representative matrices.

and Huang Jin Gui, The other 3 tea varieties (e.g., Fenghuang Dancong, Anhua dark tea, and Dian black tea) showed strong matrix enhancement and the values of matrix effect ranged from 52% to 55%. Phorate showed a strong matrix enhancement effect in all 28 tea matrices, and the values ranged between 52% and 208%, but the different tea matrix has big differences. Fig. 4.7 shows the distribution of matrix effects of 7 pesticides, for example, alpha-HCH, bromuconazole, dimethomorph, fenchlorphos, fenthion sulfoxide, parathion-methyl, and p,p'-DDE in the 7 representative matrices, including Huangshan Mao feng (green tea), Junshan Yinzhen tea (yellow tea), White Peony (white tea), Anxi Tie Guan Yin (oolong tea), Keemun black tea (black tea), Anhua dark tea (dark tea), and Pu'er tea (Pu'er tea). Fig. 4.7 shows that for the same kind of pesticide, the matrix effects of different tea matrices are different. The matrix effects of teas from different origins were clustered with SPSS software, and the result was shown with the dendrogram in Fig. 4.8, in which the number of the tea is in the same order as in Table 4.1.

Fig. 4.8 shows that the 28 tea matrices were separated into three clusters, A, B, and C, at linkage distances between 7 and 10. Group A has 6 sample, including Keemun black tea, Dian black tea, Min black tea, Anhua dark tea, Liu Pao tea, and Pu'er tea, belonging to completely fermented and postfermented tea species. Category B consisted of 18 samples: Huangshan Mao Feng, Deng Village green tea, Longjing, Lushan Yunwu tea, Taiping Houkui tea, Enshi Yulu tea, Baisha green tea, Guzhang Maojian tea, Dongting Biluochun tea, Yu Hua tea, An Ji Bai tea, Liuan Guapian green tea, Mengshan tea, Laoshan green tea, Rizhao green tea, Junshan Yinzhen tea, White Peony, and Bai Hao Yin Zhen, which belong to green tea, yellow tea, and white tea, 3 kinds of tea varieties, and also are unfermented and sightly fermented teas. Category C consisted of 4 samples, including Anxi Tie Guan Yin, Yongchun Fo Shou tea, Huang Jin Gui, and Fenghuang Dancong tea, belonging to oolong tea species, and are partially fermented teas. At the same time, Fig. 4.2 shows that the difference in the same samples obtained from different origins is very small. This observation agreed well with the results of the GC–MS.

Therefore, the clustering results obtained from GC–MS/MS were the same as those obtained from GC–MS: the first cluster is unfermented and sightly fermented teas, including green tea, yellow tea, and white tea, and their matrix effect values are in the range of 72%–120%. The second one is partially fermented teas, including oolong tea, and their values are 141%–153%. The third cluster is black tea and dark tea, which belongs to completely fermented teas, and their matrix effect values are 194%–208%.

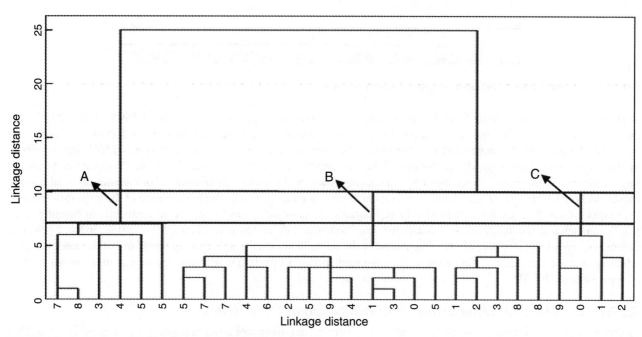

FIGURE 4.8   Dendrogram of the cluster analysis of the 28 tea matrices.

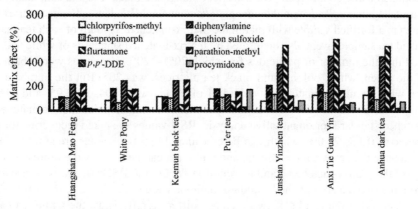

FIGURE 4.9   Distribution of matrix effects in the 7 representative matrices of different pesticides.

4.2.7.3   Evaluation of the Matrix Effect of Different Pesticides

To calculate the average value of the matrix effect for each kind of pesticides in 28 teas, it can be seen from the average values, that 205 pesticides showed the different average matrix effects, in which the minimum value is p,p'-DDE (with matrix effect value of 21%), the largest is fenamiphos (with matrix effect value of 305%). The average matrix effects of 205 pesticides showed all moderate and strong matrix enhancement effects.

In the Huangshan Mao Feng tea matrix, the values of the matrix effect for 205 pesticides ranged from 7% to 420%, in which the matrix effect of dichlobenil is only 7%, and showed a weak matrix enhancement effect. Therefore, the effect of enemy on pesticides can be ignored in the actual analysis, but the matrix effects of fenamiphos are as high as 420%. Therefore, it can be concluded that the different pesticides on the same tea had different matrix effects. In order to evaluate the effect of pesticide varieties on matrix effects, we chose the matrix effect values of 8 pesticides, for example, chlorpyrifos-methyl, diphenylamine, fenpropimorph, fenthion sulfoxide, flurtamone, parathion-methyl, p,p'-DDE, and procymidone in 7 representative tea matrices to draw the distribution of matrix effects (Fig. 4.9). Fig. 4.9 shows that the matrix effects were dependent on the pesticide varieties,

Fig. 4.10 shows the dendrogram of the CA for the matrix effects of 205 pesticides obtained with the SPSS software (the number of pesticide is in the same order as in Appendix Table 4.2). Fig. 4.10 shows that the 205 pesticides were separated into three clusters, A, B, and C, at linkage distances between 12 and 19. Cluster A was composed of 4 pesticides, including

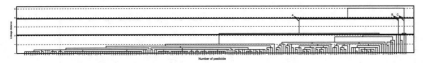

FIGURE 4.10   Dendrogram of the cluster analysis of the 205 pesticides.

flurtamone, monolinuron, dichlofluanid, and fenamiphos, and their average matrix effects were 259%, 270%, 275%, and 305%, respectively, which show strong matrix effects. At the same time, we found that the matrix effect of some pesticides near with the mentioned four pesticides, for example, the average matrix effect of metamitron was 256%, but it is far in distance from these four pesticides in the tree. This is mainly because the four pesticides in the 28 substrates showed a large difference in the matrix effect, together as a cluster. However, the average matrix effects of metamitron in 28 teas are closer. As Fig. 4.10 shows, cluster B only contained one pesticide (acetochlor), which is because of the large difference of the matrix effects of acetochlor in the different tea matrices, for example, the matrix effects of acetochlor in Keemun black tea, Dian black tea, Min black tea, Anhua dark tea, Liu Pao tea, and Pu'er Tea were 371%, 390%, 191%, 363%, 131%, and 165%, respectively. In addition, cluster C consisted of the other 200 pesticides (accounting for 97.6% of the total number), which means that the vast majority of pesticides have similar matrix effects. These results are similar to the previous clustering results of GC–MS.

4.2.7.4   Confirmation Experiments on Classification of Tea Varieties

According to the result from Section 4.2.7.2, teas can be divided into three categories according to their matrix effects. This result means that an analyst can choose any type of tea in each cluster to make matrix-matched solutions. The validation of the groupings was studied using commercial green, white, oolong, and black teas and estimated by means of recovery experiments. Each tea variety was spiked with the 186 pesticides at the MRL level. The recoveries were determined by comparing the peak area in a fortified sample with that in the matrix-matched standard of each tea matrix. In order to validate the grouping, spiked tea samples were also quantified by matrix-matched standards of white tea.

The results showed that the number of pesticides for which 70%–120% recoveries were obtained from the matrix-matched standard of the green, white, oolong, and black tea matrices, was 205. But the numbers of the green, white, oolong, and black tea with the matrix-matched standard of black tea were 204, 205, 95, and 116, respectively. This shows that the green tea has a similar recovery range when using the single-point matrix-matched standard of white tea and the matrix-matched standard for quantitative analysis. RSD values of spiked recoveries for the two quantitative methods are in the range of 0.1%–20.0%, which can be considered that two values are similar. The results show that the white and green tea have the same matrix effects, and white tea can be used as the representative matrix for green and white tea to prepare the matrix-matched standard solution. While the RSDs of the two recoveries for oolong and black tea obtained from the two matrix-matched standard methods were in the range of 0.1%–69.6% and 0.2%–67.4%, respectively, which indicates that the results of two methods differ greatly. Thus, the white tea cannot be used as the representative matrix for oolong and black tea. This observation agreed with the clustering results of Section 4.2.7.2. The results were also similar to the result of GC–MS, which proved the validity of the hierarchical clustering method for classification of tea.

4.2.8   Evaluation of the Matrix Effect in Determination of LC–MS/MS

Matrix effects of 110 pesticides in 28 teas matrices were evaluated by LC–MS/MS. As the results showed that the matrix effects of 110 pesticides in tea had positive or negative effect matrix, which means that some pesticides showed a matrix enhancement effect, and some pesticides showed a weakened effect matrix.

4.2.8.1   Evaluation of the Pesticide Matrix Effect

Strong suppression was exhibited by 180 data points, accounting for 6%.

Among 3080 data points of matrix effects obtained, 223 (in the range of 0% to 20%) accounted for 7% of the total and showed a matrix enhancement effect; the remaining 2857 data points accounted for 93% and presented signal suppression. In total, 1843 data points showed soft signal suppression, accounting for 60% of the total number; 834 data points presented medium signal suppression, accounting for 27%; and 180 exhibited strong suppression, accounting for 6%. The distribution of matrix effects on 110 pesticides in the 28 types of tea matrix are shown in Fig. 4.11. It can be observed that most of the pesticides demonstrated soft and medium signal suppression.

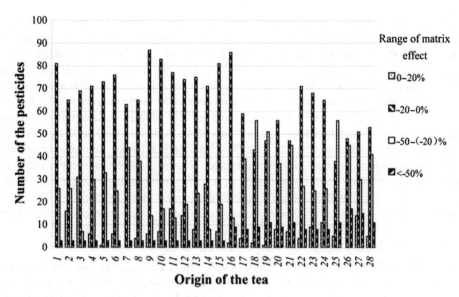

FIGURE 4.11   Distribution of matrix effects of 28 representative matrices.

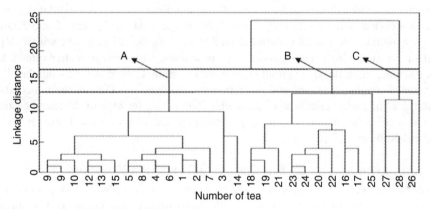

FIGURE 4.12   Dendrogram of the cluster analysis of the 28 tea matrices.

Dimethoate, resmethrin, and bioresmethrin exhibited strong signal suppression effects in all 28 types of tea, and the matrix effects ranged from −51% to −78%, −54% to −95%, and −55% to −92%, respectively. However, the other pesticides exhibited different matrix effects in different varieties of tea. For example, matrix effects on diphenylamine were between −24% and 19%. Matrix enhancement effects were exhibited in Liuan Guapian green tea, An Ji Bai tea, Guzhang Maojian tea, Rizhao green tea, Pu'er tea, Deng Village green tea, Lushan Yunwu, Longjing, Huang JinGui, Keemun black tea, Laoshan green tea, Yu Hua tea, and Liu Pao tea varieties, but the other 15 tea varieties showed matrix suppression effects. Napropamide exhibited signal suppression effects in all 28 tea matrices, and the values ranged between −2% and −77%. Rizhao green tea, Laoshan green tea, An Ji Bai tea, Yu Hua tea, Longjing green tea, Liuan Guapian green tea, Mengshan tea, Dongting Biluochun, Enshi Yulu tea, Guzhang Maojian tea, Taiping Houkui tea, Lushan Yunwu, and Deng Village green tea exhibited soft signal suppression. Baisha green tea and Huangshan Mao Feng showed medium signal suppression, and the other 13 varieties showed strong suppression.

4.2.8.2   Evaluation of the Matrix Effect of Different Tea Varieties From Different Producing Areas

Similar to GC–MS and GC–MS/MS determination, different varieties of teas showed different matrix effects. Using the same CA method of GC–MS, the matrix effects of 110 pesticides were classified, and the dendrogram is shown in Fig. 4.12 (the number of the tea is in the same order as in Table 4.1). Fig. 4.12 shows that the 28 tea matrices were separated into three clusters, A, B, and C, at linkage distances between 12.5 and 17.0: cluster A consisted of 15 samples, including Huangshan Mao Feng, Deng Village green tea, Longjing, Lushan Yunwu tea, Taiping Houkui tea, Enshi Yulu tea, Baisha green tea, Guzhang Maojian tea, Dongting Biluochun tea, Yu Hua tea, An Ji Bai tea, Liuan Guapian green tea, Mengshan tea, Laoshan green tea, Rizhao green tea, which belong to the green tea variety, and also are unfermented tea. Cluster B contained 10 samples, including Junshan Yinzhen tea, White Peony and Bai Hao Yin Zhen, Anxi Tie Guan Yin, Yongchun Fo Shou tea, Huang Jin Gui and Fenghuang Dancong tea, Keemun black tea, Dian black tea and Min black tea, belonging to slightly fermented, partially fermented and completely fermented teas, and this type of tea could be defined as the fermentation tea. Cluster C consisted of 3 samples, Anhua dark tea, Liu Pao tea, and Pu'er Tea, belonging to the postfermented tea. At the same time, as seen in Fig. 4.12, the differences between the same samples from different origins is very small. The results of CA based on matrix effects were different from those obtained from GC–MS and GC–MS/MS. Although the tea samples were divided into three categories, there were subtle differences in three kinds of tea. This may be caused by their own characteristics of instruments or the difference of pesticide types. LC–MS/MS used the electrospray ionization method for ionization of samples, and multiple reaction detection modes for the confirmation of samples. Therefore, compared to GC–MS and GC–MS/MS, its ionization mode was softer, and the application of multiple reaction detection modes attains high sensitivity. Furthermore, the pesticides determined with LC method was usually more difficult to vaporize, which may also be a cause of the difference between the two classifications. The related research will be carried out in the following experiments.

4.2.8.3 Evaluation of the Matrix Effect of Different Tea Varieties

Similar to GC–MS and GC–MS/MS determination, different varieties of pesticides showed different matrix effects. Using the same CA method of GC–MS, the matrix effects of 110 pesticides were classified, and the dendrogram is shown in Fig. 4.13 (the number of the tea is in the same order as in Appendix Table 4.3). Fig. 4.13 shows that the 110 pesticides were separated into two clusters, A and B, at linkage distances larger than 10. Group A was composed of 107 pesticides and group B consisted of three pesticides (bioresmethrin, resmethrin, and dimethoate). They exhibited a strong effect of signal suppression. Group A was divided into two clusters, C and D, at linkage distances between 5 and 9. Group C contained 95 pesticides whose average matrix effects of 28 varieties ranged from −1% to −27%, among which 78 pesticides exhibited soft effects of signal suppression. Group D consisted of 12 pesticides, with average matrix effects between −31% and −49%. Higher absolute values of pesticides in group D compared to group C were derived from the higher absolute values of matrix effects obtained from the fermented and postfermented tea matrices. Despite the remarkable physicochemical diversities of the pesticides, the matrix effects of 78 pesticides (accounting for 86% of the tested pesticides) were surprisingly similar. Thus, matrix effects should not be primarily considered analyte-dependent. These results verified the findings of Stahnke et al. [77] and Kittlaus et al. [66].

4.2.8.4 Confirmation Experiments on the Classification of Tea Varieties

According to the result from Section 4.2.8.2, teas can be divided into three categories according to their matrix effects. This result means that an analyst can choose any type of tea in each cluster to make matrix-matched solutions. The validation of the groupings was studied using commercial green, oolong, black, and dark teas and estimated by means of recovery experiments. Each tea variety was spiked with the 110 pesticides at the MRL level. The recoveries were determined by comparing the peak area in a fortified sample with that in the matrix-matched standard of each tea matrix. In order to validate the grouping, spiked tea samples were also quantified by matrix-matched standard of black tea.

The results showed that the numbers of pecitides for which 70%–120% recoveries were obtained from the matrix-matched standard of the green, oolong, black, and dark tea matrixes were 109, 109, 110, 110, respectively. But the numbers of the green, white, oolong, and black tea with the matrix-matched standard of black tea were 85, 109, 110, and 77, respectively. This shows that the oolong tea had a similar recovery range when using the single-point matrix-matched standard of black tea and the matrix-matched standard for quantitative analysis. RSD values of the two spiked recoveries were in the range of 0.5%–19.3%, which can be considered that two values are similar. The results showed that the black and oolong tea had the same matrix effects, and black tea could be used as the representative matrix for oolong tea to prepare the matrix-matched standard solution. While the RSDs of the two recoveries for green and dark tea obtained from the two matrix-matched standard methods were in the range of 0.2%–77.7% and 0.3%–60.0%, respectively, which indicated that

FIGURE 4.13 **Dendrogram of the cluster analysis of the 110 pesticides.**

the results of two methods differ greatly. Thus, the black tea could not be used as the representative matrix for green and dack Tea. This observation agreed well with the clustering results of Section 4.2.8.2. The results were also similar to the result of GC–MS, which proved the validity of the hierarchical clustering method for classification of tea.

4.3 COMPENSATION FOR THE MATRIX EFFECTS IN THE GAS CHROMATOGRAPHY–MASS SPECTROMETRY ANALYSIS OF 186 PESTICIDES IN TEA MATRICES USING ANALYTE PROTECTANTS

4.3.1 Introduction

It is generally believed that the formation mechanism of the GC matrix effect for the detection of actual sample is because the molecules of impurities from the sample matrix tend to block active sites (metal ions, silanols, and other active sites produced by thermally decomposed components of samples) in the GC inlet and column, thus reducing losses of susceptible analytes caused by adsorption or degradation on these active sites. Thus, the signal of the same content of analyte in the real sample is higher than that in the standard sample.

Based on this phenomenon, Erney et al. [44] thought that if the interaction between the active site in the GC inlet and column and the analyte could be blocked by adding an appropriate reagent in the sample before sampling, to reduce losses of susceptible analytes caused by adsorption or degradation on these active sites. Thus, the reagent is called analyte of protectant (APs). In 1993, Erney et al. [44] studied the compensation for matrix effects of APs, but did not get the desired results. In 2003, Anastassiades et al. [43] reintroduced the concept of the APs, and obtained good results. Since then, compensation of matrix effects using APs has been widely used.

4.3.2 Experiment and Material

4.3.2.1 Origin and Classification of Tea

Tea samples were obtained from Wumart supermarket, Chaoyang District, Beijing City.

4.3.2.2 Instruments, Reagents, Consumables, and Pesticide Standards

See Section 4.2.2.2. Full-scan total ion current chromatograms of 186 pesticides added APs of triglyceride and D(+)-gulonic acid-(-lactone (the concentration of APt was 2 mg/mL) are shown in Fig. 4.14. Among them, group A contains 69 pesticides, group B contains 60 pesticides, and group C contains 57 pesticides. The corresponding names of pesticides are shown in Appendix Table 4.1. Fig. 4.14 shows that each group of pesticide has been successfully separated with symmetry peak and better peak time.

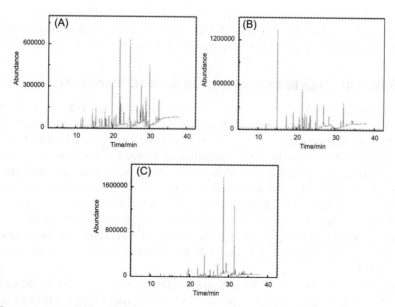

FIGURE 4.14   Total Ion Chromatography (TIC) of pesticides.

4.3.2.3 Analyte Protectant

Table 4.4 lists the 11 substances used as APs. The 11 substances purchased from Aldrich (Milwaukee, WI), Supelco (Bellefonte, PA, USA), Sigma (St. Louis, MO, USA), and Fluka (Buchs, Germany). Olive oil (10 mg/mL) was prepared with ethyl acetate. 3-ethoxy-1,2-propanediol (100 mg/mL), poly(ethylene glycol) (PEG) 300 (10 mg/mL), and 2,3-butanediol (10 mg/mL) in acetonitrile. D-sorbitol (10 mg/mL), 1,2,3,4-butanetetrol (8.5 mg/mL), triglycerol (10 mg/mL), and D-ribonic acid-γ-lactone (40 mg/mL) in acetonitrile-water (85:15, v/v). L-gulonic acid-γ-lactone (20 mg/mL), 4,6-O-ethylidene-α-D-glucopyranose(10 mg/mL) and 1-O-methyl-β-D-xylopyranoside (10 mg/mL) in acetonitril-water (80:20, v/v).

4.3.3 Experimental Method

4.3.3.1 The Preparation of Pesticide Standard Solution, the Preparation of Matrix Mixture Standard Working Solution, the Preparation of Solvent Standard Solution, the Preparation of Internal Standard Solution, Sample Pretreatment Method, the Spiked Experiment

See Section 4.2.2.2.

4.3.3.2 Application of Analyte Protectant

Application of the AP: the sample was dried in a stream of nitrogen, then a certain volume of AP was added, and using acetonitrile to volume 1 mL.

4.3.3.3 Analytical Conditions of Instruments

See Section 4.2.4.

4.3.3.4 Evaluation of the Effect of Analyte Protectants on the Long-Term Operation of the System

In order to evaluate the effect of APs combination on the GC–MS system, the 5 needle solvent standard was inserted into the 95 needle sample.

The samples were tested in the following sequence: (1) solvent standards; (2–16) black tea samples; (17–21) black tea matrix-matched standard solution; (22) solvent standards; (23–37) oolong tea samples; (38–42) oolong tea matrix-matched standard solution; (43) solvent standards; (44–58) green tea samples; (59–63) green tea matrix-matched standard solution; (64) solvent standards; (65–94) determination of commercially available tea samples; (95) solvent standards. After 95 times of determination, 5 times solvent standards were detected, and the stability of the whole system was evaluated by evaluating the peak area (area), peak height (height), peak height/peak area (H/A), and retention time (t_R) of the 5 solvent standards.

After the sample is determined, the effect of the solvent standard added AP on the stability of the instrument was evaluated with the same sample test sequence. The difference is that when the sample is determined, adding the APs of 2 mg/mL triglyceride and D-ribonic acid-γ-lactone combination in the solvent standards and samples are needed.

4.3.4 Effect of Different Analyte Protectants on Solvent Standards

In order to evaluate the influence of 11 kinds of APs with different concentrations (0.5, 1, 2, and 5 mg/mL, 3-ethyoxyl-1,2-propylene glycol also includes two concentrations of 10 and 20 mg/mL) on the chromatographic behaviors of 186 kinds of pesticides in acetonitrile, the pesticide responses of solvent standard adding AP and without adding AP were compared. The impact of the APs on pesticides was evaluated by the following formula: ratio(%) = (A2 – A1)/A1 × 100(%). Herein, A1 is the peak response of pesticides in acetonitrile solution, A2 is the peak response of pesticides in acetonitrile solution containing different concentrations of APs, and the spiked concentration of AP is the MRL of each pesticide.

Fig. 4.15 shows the impact of the 11 APs on the peak shape of paclobutrazol. The right side shows that the chromatograms from the high to the low order were the extraction ion chromatograms of paclobutrazol adding AP at concentrations of 5, 2, 1, 0.5 mg/mL and paclobutrazol solvent standard, respectlvely. AP 6 is extraction ion chromatograms of paclobutrazol adding AP at concentrations of 20, 10, 5, 2, 1, 0.5 mg/mL and paclobutrazol solvent standard, and the names of the AP are the same as that in Table 4.4.

The vast majority of 186 pesticides (87.1%) exhibited moderate or strong matrix effects in the matrix tea, with ME >20%. Therefore, in order to get a greater degree of compensation effect matrix, we figured out the number of pesticides showing

TABLE 4.4 Compounds Evaluated as Aps

No.	Compound Name	CAS Number	Molecular Structure
1	Olive oil	8001-25-0	
2	2,3-Butanediol	513-85-9	
3	1-O-Methyl-β-D-xylopyranoside	612-05-5	
4	D-Sorbitol	50-70-4	
5	L-Gulonic acid-γ-lactone	1128-23-0	
6	3-Ethoxy-1,2-propane-diol	1874-62-0	
7	D-Ribonic acid-γ-lactone	5336-08-3	

(Continued)

TABLE 4.4 Compounds Evaluated as Aps (*cont.*)

No.	Compound Name	CAS Number	Molecular Structure
8	1,2,3,4-Butanetetrol	149-32-6	
9	4,6-O-ethylidene-α-D-glucopyranose	13224-99-2	
10	Poly(ethylene glycol) (PEG) 300	25322-68-3	
11	Triglycerol	20411-31-8	

strong matrix enhancement (ME > 50%) in the matrix tea, and the results are listed in Table 4.5. Fig. 4.15 and Table 4.5, show that the matrix enhancement effect of No. 1 and No. 2 APs on 186 pesticides is not apparent. At four spiked concentrations, responses of pesticide added APs and without APs have no obvious difference (the ratios are around 1.0), while the peak response in chromatogram have also no obvious enhancement phenomenon. However, the other 9 kinds of APs had different degree of matrix enhancement effects to each pesticide, and the different concentrations of APs also showed different matrix enhancement effects for pesticide.

As Table 4.5 shows that, with the exception of AP1 and AP2, with the increase of APs concentration, the number of pesticides with ratios above 50% increases. However, APs were dissolved with some amount of water, which can harm the GC system. Furthermore, an ideal AP is a compound that gives the full response enhancement effect at a lower concentration. The reported concentration of APs was 1 mg/mL in most cases and sometimes 2 mg/mL. As Table 4.5 shows, No. 4, No. 5, No. 7, No. 10, and No. 11 APs give the largest number of ratios greater than 50% at concentrations of 0.5, 1, and 2 mg/mL. However, in the case of No. 10 APs, "the long-term injections of PEG solutions deteriorated column performance to such an extent that even cutting a large portion from the front part of the column did not cover it" [31]. Therefore, APs 4, 5, 7, and No. 11 were used in the next stage.

Fig. 4.16 shows the influence of APs (including No. 4, 5, 7, and 11) at concentrations of 1 and 2 mg/mL on chromatographic behaviors of the two isomers of phenothrin (0.05 mg/kg) as well as fenthion sulfoxide (0.1 mg/kg). Among them, Fig. 4.16A–B are extracted ion chromatograms of phenothrin and fenthion sulfoxide, respectively, with AP concentration of 1 mg/mL. C and D were extracted ion chromatograms of phenothrin and fenthion sulfoxide, respectively, with AP concentration of 2 mg/mL. It can be seen that, from the right side, the chromatograms from high to low are the extracted ion chromatograms of pesticide added APs 11, 7, 5, and 4, as well as pesticide solvent standard, respectively. As shown in Fig. 4.3, after adding APs, the peak responses of pesticides increased, and the peak shapes were also improved. At the same time, the four kinds of APs with different concentrations yielded different degrees of enhancement on the peaks and intensities of the pesticides.

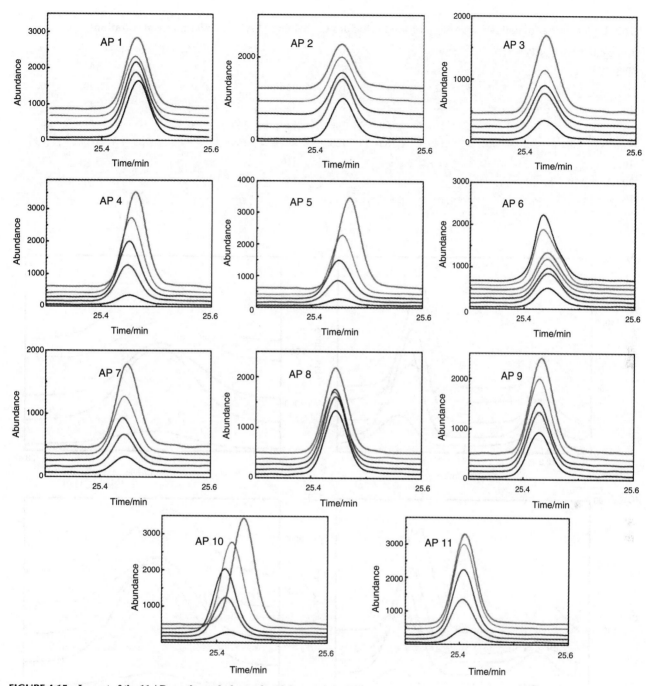

FIGURE 4.15 Impact of the 11 APs on the peak shape of paclobutrazol.

4.3.5   Effects of Different Analyte Protectant Combinations on Tea

4.3.5.1   Experimental Design of the Evaluation of Different Analyte Protectant Combinations

The selected AP combination was designed to be used in the tea matrix, and the method in the paper used the internal standardization for quantitation, therefore the evaluation of the AP combinations was based on the internal standard calibration. First, different AP combinations were selected, the peak area of the matrix-matched standard was assigned to 100%, and the solvent standard was quantified using Eq. (4.1):

$$\text{Normalized ratio (\%)} = (A_{std} / A_{stdis}) / (A_{mstd} / A_{mstdis}) \times 100 \tag{4.1}$$

TABLE 4.5 Number of Ratios Above 50% for 186 Pestticides Containing 11 APs at Different Concentrations

Concentration/ (mg/mL)	AP 1	2	3	4	5	6	7	8	9	10	11
0.5	3	4	62	113	109	33	75	36	32	120	100
1	4	4	86	150	153	43	131	44	47	147	154
2	10	4	116	163	168	62	149	48	97	157	164
5	14	10	149	165	169	74	166	79	133	160	173
10	–	–	–	–	–	116	–	–	–	–	–
20	–	–	–	–	–	145	–	–	–	–	–

The numbers of APs in the table are in the same order as in Table 4.4.

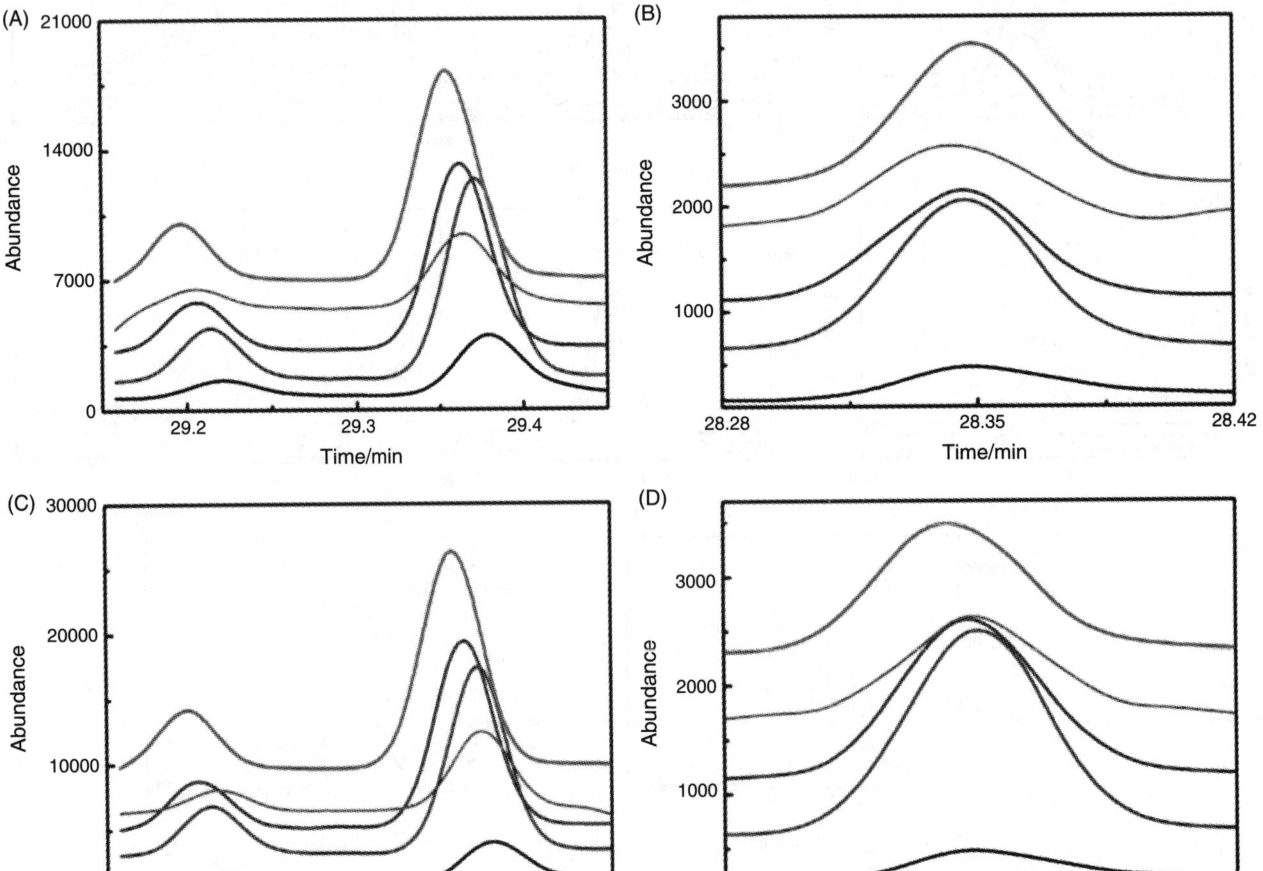

FIGURE 4.16 Effects of different APs on the peak shapes and intensities of phenothrin and fenthion sulfoxide.

where A_{std} is the peak area of the analytes from the standard solution with APs, A_{stdis} is the peak area of corresponding internal standard from the standard solution with APs, A_{mstd} is the peak area of analyte from the matrix-matched standard solution, A_{mstdis} is the peak area of the corresponding internal standard from matrix-matched standard solution.

On the basis of these experiments, the primary AP combinations were evaluated by means of recovery experiment. The selected AP combiantions were added to both solvent standards and the spiked samples extracts. The peak area of the solvent standard was assigned to 100% and the spiked sample was quantified using Eq. (4.2):

TABLE 4.6 Composition of AP Combinations and the Numbers of Pesticides With Normalized Ratio Above 70%

AP	Combination of APs[a]															
	1	2	3	4	5	6	7	8	9	10	11	12	13	14	15	16
4	–	–	1	1	–	–	1	1	–	–	2	2	–	–	–	1
5	2	–	1	–	1	–	2	–	–	2	2	–	2	–	1	1
7	–	2	–	1	–	1	–	2	2	–	–	2	–	2	1	1
11	–	–	–	–	1	1	–	–	1	1	–	–	2	2	1	–
Matrix	Numbers of pesticides with normalized ratio above 70%															
Green tea	124	121	131	117	123	126	174	143	135	130	151	149	172	169	127	145
Oolong tea	87	91	137	111	137	137	161	132	171	171	160	151	174	178	172	139
Black tea	107	88	102	114	106	120	126	124	130	136	136	117	162	169	137	127

aThe concentration is the actual concentration in the vial injected to the GC system. The unit is mg/mL.

$$\text{Recovery (\%)} = (A / A_{is}) / (A_{std} / A_{stdis}) \times 100 \tag{4.2}$$

where A is the peak area of the analyte from the spiked sample with APs, A_{is} is the peak area of the corresponding internal standard from the spiked sample with APs, and the terms A_{std} and A_{stdis} are the same as Eq. (4.1).

4.3.5.2   Evaluation of the Impact of Different Analyte Protectants on the Tea Matrice

In Section 4.2.6.2 we found that, according to the matrix effect, tea can be classified into black, green, and oolong tea: three types of teas. So, in this phase of the experiment, the commercially available black, green, and oolong teas were selected to carried out the experiment of APs.

No. 4 and No. 11 AP combinations showed better enhancement effect on some pesticides with small polarity [36], while No. 5 and No. 7 AP combinations had better protective effects on most of the pesticides. Therefore, the effect of the four kinds of AP combinations on the pesticides were evaluated according to the concentrations listed in Table 4.6.

According to Table 4.6, remove a certain volume of APs into the solvent standard containing 180 kinds of pesticides at MRLs and with an ISTD concentration of 280 g/kg. At the same time, black, green, and oolong tea matrix-matched standard solution were prepared and the normalized ratios were calculated with the matrix-matched standard quantitative standard.

An ideal AP is a compound that interacts with active sites in the GS system to reduce the interaction of the pesticide with those sites and increases the response of pesticides. Therefore, the response of the pesticides in the standard solutions with AP should be similar to that of the matrix-matched standards, and the normalized ratios should be near or more than 100%. We list the numbers of the pesticides whose normalized ratios are more than 70% in Table 4.6.

The combinations with the top nine amounts of pesticides with high-normalized ratios in black, green, and oolong tea in Table 4.6 are merged and 10 AP combinations composed of the 7–16 are selected. The 10 AP combinations are used to evaluate the matrix effect of the three kinds of tea. The 10 AP combinations listed in Table 4.6 are added to the spiked sample extracts, then quantified using the solvent standards with the same AP combinations. The recoveries are calculated according to the second formula. The average recoveries of each AP combination are listed in Table 4.7, and the number of pesticides whose recoveries ranged between 70% and 120% are also included.

When using the combinations 11 and 12, the average recoveries in three tea matrices were all more than 120%, even up to 149.7% in black tea. Therefore, neither the combination 11 nor 12 is suitable for the tea matrices. The average recoveries of 186 pesticides using the other 8 AP combinations exhibit little difference in green (84.8%–100.3%) and oolong tea (88.8%–116.8%), but prominent differences are observed in black tea (96.9%–161.4%). The phenomenon can be rationalized, because the black tea is completely fermented and has a more complex matrix, than green or oolong tea.

The average recoveries obtained by using the combination of 7 and 8, 9, and 10, 11, and 12, 13, and 14 were similar in the same tea matrix. This meant that the No. 4 (or No. 11) AP combined with No. 5 and No. 7 AP in the same proportion, caused little difference in the average recoveries, so APs of No. 5 and No. 7 showed similar compensations in the same tea matrix. Conversely, the average recoveries of combination 15 and 16 showed prominent differences in black tea matrix with combination 16 exhibiting higher recovery than those of combination 15. Therefore, No. 11 was more suitable for the black tea matrix than No. 4.

TABLE 4.7 Average Recoveries and the Number of Pesticides with Recoveries within the Range of 70%–120%

Matrix	AP combinations	7	8	9	10	11	12	13	14	15	16
Average recovery	Green tea	93.0	99.7	87.0	89.1	129.3	146.7	100.3	84.8	90.4	94.5
	Oolong tea	115.2	116.8	96.1	104.8	125.5	128.6	99.3	91.1	88.8	110.7
	Black tea	161.4	145.1	94.0	96.9	135.2	149.7	104.7	99.2	102.0	142.3
Number of pesticides with recoveries of 70%–120%	Green tea	156	142	142	158	89	75	153	180	146	139
	Oolong tea	115	113	133	145	92	90	172	180	151	128
	Black tea	46	65	159	152	70	57	153	184	155	67

The sequence of the AP combination is the same as in Table 4.6.

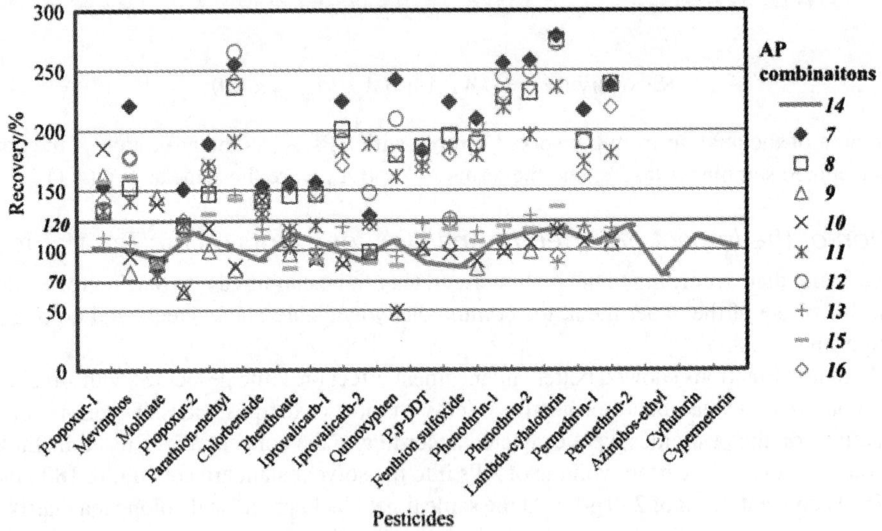

FIGURE 4.17 Recoveries of 20 pesticides obtained by 10 AP combinations in black tea matrix.

Twenty pesticides representing organophosphorus (azinphos-ethyl, fenthion sulfoxide, parathion-methyl, phenthoate, meviphos), organochlorine (chlorbenside, p,p'-DDT, quinoxyphen), carbamate (iprovalicarb-1, iprovalicarb-2, molinate, propoxur-1, propoxur-2), pyrethroid pesticides (cyfluthrin, cypermethrin, lambda-cyhalothrin, permethrin-1, permethrin-2, phenothrin-1, phenothrin-2) (which mostly had RTs over 25 min) were used to figure the recoveries obtained by using the 10 AP combinations in black tea, Fig. 4.17 schematically shows the recoveries of these pesticides. As seen from Fig. 4.17, the combination 14 gives the best results for the selected 20 pesticides.

4.3.6 Effectiveness Evaluation of Analyte Protectant Compensation Matrix Effect Method

By evaluating the sensitivity, accuracy, precision, the standard curve, and the linear range of the analytical method, it could be judged whether an analytical method meets the requirements of the analysis.

4.3.6.1 Linear Range of Standard Curve

Preparing the series of different concentration of various pesticides mixed standard solution containing APs (including 2 mg/mL triglycerol and D-(+)-ribonic acid-γ-lactone) in a sampling bottle were prepared and determined in the selected conditions. The standard curve of each pesticide was plotted with each peak area (y) of compounds versus mass concentration (x). The linear ranges, linear equations, and correlation coefficients of 186 pesticides were listed in Appendix Table 4.4. Appendix Table 4.4 shows that the calibration curves of the vast majority of pesticides were linear in the certain concentration ranges, and the correlation coefficients of 182 pesticides (accounting for 97.8% of the detected pesticides) were above 0.99.

4.3.6.2 Sensitivity of the Method

In the pesticide residue analysis method, the detection limit (LOD) is the most commonly used indicator to measure the sensitivity of pesticide residues. The detection limit refers to the minimum concentration or the minimum amount of the analytes detected from the sample with the specific analysis method in a given confidence level. Usually the signal value is 3 times the noise (S/N = 3) when the concentration of the analyte is the method of detection limit (LOD) and the signal value is 10 times the noise (S/N = 10) when the concentration of analyte is the method of limit quantitation (LOQ). The LOD and LOQ of the method are listed in Appendix Table 4.4. Appendix Table 4.4 shows that the LODs of 186 pesticides are in the range of 1.67–833.33 g/kg, and the LOQs are 5–2500.00 g/kg. The LOQs are lower than the maximum residue levels (MRLs) established by the European Union (EU) legislation or Japan, so this method could be applied in the detection of pesticide residues in tea.

4.3.6.3 Accuracy and Precision of the Method

The accuracy of the method is referred to the degree of coincidence between the measured value and the true value of the experiment, and the reliability of the method is often measured with the spiked recovery in the detection of pesticide residues. Here in, the recoveries experiment of 186 pesticides from three blank tea samples (e.g., black, green, and oolong teas) at spiked levels of 0.8MRL, MRL, and 2MRL are carried out. The spiked samples are left to stand for 30 min, so that the pesticide is fully absorbed by the sample. Then the spiked samples are extracted, purified, and determined with the selected method. Each level of the experiments repeats for 5 times, and the spiked recoveries are calculated.

The precision of the method refers to the degree of approximation of the measured values of a sample by a specific procedure, which reflects the random error of the method or the measurement system. Standard deviation, relative standard deviation, range, average deviation, and other parameters could be used to measure the precision, and the RSD is often used in pesticide residue analysis. The RSD is the ratio of the standard deviation to the arithmetic mean value of measurement results. The RSD can be used to measure the deviation degree of spiked recovery of the pesticide, so as to measure the repeatability of the test method of pesticide residue.

The experimental results show that the recoveries of 186 pesticide are in the range of 59.6%–127.0%, which indicate that the recoveries of 92%–100% of the tested pesticides could reach the ideal result (with recoveries of 70%–120%). The RSDs of 186 pesticides are all less than 20%. The method shows good accuracy.

4.3.6.4 Analysis of a Real Sample

The established method is used to analyze 28 tea samples for sale in the Beijing market, and the results are shown in Table 4.8. In 28 tea samples, 6 samples were not detected in any pesticide. The other 22 samples were detected in 14 pesticides, in which 8 pesticides, including fenobucarb, methabenzthiazuron, pirimiphos-methyl, propiconazol-1, tebufenpyrad, cyfluthrin, difenoconazole, and dimethomorph were all detected in 22 samples, but did not reach the method of LOQ. Among the remaining 6 pesticides, the highest detection rate of lambda-cyhalothrin was found, and there were 10 samples (36% of the total samples) to be detected, and their residual amounts were in the range of 16.0–170.0 µg/kg. Bifenthrin was detected in 8 tea samples, in which four of them did not reach the LOD, and the residual amounts in the rest of sample were 12.7–95.5 µg/kg. Chlorpyrifos was detected in three teas, one was detected only, and the residues amounts in the remaining two teas were between 25.6 and 53.1 g/kg. Diphenylamine was detected in eight teas, in which six of them did not reach the LOD, and the residue amounts of the remaining two teas were 8.5 and 36.2 µg/kg, respectively. Malathion was detected in two teas, one of which did not reach the LOD, the residue amount of another tea was 85.8 µg/kg. Triadimenol-2 was detected in two teas, and the residual amounts were 20.5 and 88.8 g/kg, respectively. It could be concluded that, although there were certain pesticide residues in tea, their amounts did not exceed the MRLs established by the EU and Japan.

4.3.6.5 Effect of Long-Running Time on Instruments

The effect of the APs on the stability of the instrument in the order of Section 4.3.3.4 was investigated by adding triglycerides and D-(+)-ribonic acid-γ-lactone as APs into the tea sample and solvent standard. The peaks of five needle-solvent standards (adding AP) in Section 4.3.3.4 were overlapped, and the results shown in Fig. 4.18 were obtained, in which A and B were the overlay of extracted ion chromatograms of phenyl phenol and fenvalerate in the absence of APs, and C and D were the overlay of extracted ion chromatograms of the two pesticides in the presence of APs. Fig. 4.18 shows that the increase of sample numbers, reproducibility of o-phenylphenol and flucythrinate (including isomers) solvent standard without APs was not an improvement. But adding APs, the RSDs for the peak area and the peak height of five needle-solvent standards were greatly reduced, and good reproducibility was obtained.

TABLE 4.8 Summary of the Analysis Results from 28 Batch of Teas

Concentration/μg/kg

Sample	Bifenthrin	Chlorpyrifos	Cyfluthrin	Difenoconazole	Dimethomorph	Diphenylamine	Fenobucarb	Lambda-Cyhalothrin	Malathion	Methabenzthiazuron	Pirimiphos-Methyl	Propiconazole-1	Tebufenpyrad	Triadimenol-2
1	Detected	–	–	–	–	8.5	–	–	–	–	–	–	Detected	–
2	Detected	–	–	–	–	Detected	–	–	–	Detected	–	Detected	–	–
3	Detected	Detected	–	–	–	Detected	Detected	16.0	–	–	–	–	–	–
4	–	–	–	–	–	Detected	–	–	–	–	–	–	–	–
5	12.7	–	–	–	–	–	–	–	–	–	–	–	–	–
6	–	–	–	–	–	Detected	–	–	–	–	–	–	–	–
7	–	–	–	–	–	–	–	–	Detected	–	–	–	–	–
8	–	–	–	–	–	Detected	–	–	–	–	–	–	–	–
9	–	–	–	–	–	–	–	60.0	–	–	–	–	–	–
10	–	–	–	–	–	–	–	–	–	–	–	–	–	–
11	–	–	–	–	–	–	–	–	–	–	–	–	–	–
12	–	–	–	–	–	Detected	–	–	–	–	–	–	–	–
13	–	–	–	–	–	–	–	–	–	–	–	–	–	20.5
14	–	–	–	–	–	–	–	–	–	–	–	–	–	88.8
15	–	25.6	Detected	–	–	–	–	120.0	–	–	–	–	–	–
16	27.8	–	–	–	–	–	–	–	–	–	–	–	–	–
17	–	–	–	–	–	–	–	–	–	–	–	–	–	–
18	95.5	–	–	–	–	–	–	89.0	–	–	–	–	–	–
19	–	–	–	–	–	–	–	76.0	–	–	–	–	–	–
20	–	–	–	–	–	–	–	57.0	–	–	–	–	–	–
21	–	–	–	–	–	36.2	–	59.0	–	–	–	–	–	–
22	–	–	–	–	–	–	–	–	–	–	–	–	–	–
23	–	53.1	–	–	Detected	–	–	18.0	–	–	–	–	–	–
24	–	–	–	–	–	–	–	–	–	–	–	–	–	–
25	Detected	–	–	–	–	–	–	–	–	–	–	–	–	–
26	–	–	–	–	–	–	–	37.0	–	–	–	–	–	–
27	–	–	–	–	–	–	–	–	–	–	–	–	–	–
28	32.0	–	–	Detected	–	–	–	170.0	85.8	–	Detected	–	–	–

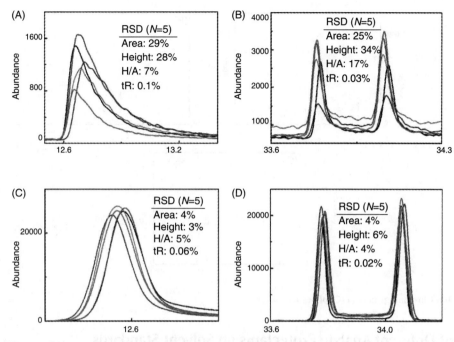

FIGURE 4.18   Overlay of extracted ion chromatograms of 2-phenylphenol and flucythrinate.

4.4   COMPENSATION FOR THE MATRIX EFFECTS IN THE GAS CHROMATOGRAPHY– MASS SPECTROMETRY ANALYSIS OF 205 PESTICIDES IN TEA MATRICES

4.4.1   Introduction

It is generally believed that the matrix effect of GC–MS is caused by the active sites in the GC system. At the same time, some scholars suggest that the different detector types also affect the matrix effect. Therefore, based on the previous report, according to the study of the matrix effects in the GC–MS/MS analysis of pesticides and the compensation for matrix effects using APs, the quantitative method for the determination of the compensation matrix effect of GC–MS/MS was established. The difference of the compensation effect between two kinds of different detectors (the quadrupole and the tandem quadrupole) were also investigated.

4.4.2   Experimental Materials

4.4.2.1   Origin and Classification of Tea

The sample is the same as that determined with GC–MS method.

4.4.2.2   Instruments, Reagents, Consumables, and Analyte Protectants of Pesticide Standards

See Section 4.2.2.2 and Section 4.3.2.3. Full-scan total ion current chromatograms of 205 pesticides added analyte protectant of triglycerides and D-(+)-RNA-γ-lactone (concentration of analyte protectant were 2 mg/mL) as shown in Fig. 4.19. Among them, group A contained 71 pesticides, group B contained 68 pesticides, and group C contained 66 pesticides. The corresponding names of pesticides are shown in Appendix Table 4.2. Fig. 4.19 shows that each group of pesticides is successfully separated with symmetry peak and reasonable peak time. The addition of APs did not adversely affect the whole chromatogram.

4.4.3   Experimental Method

See Section 4.3.3.

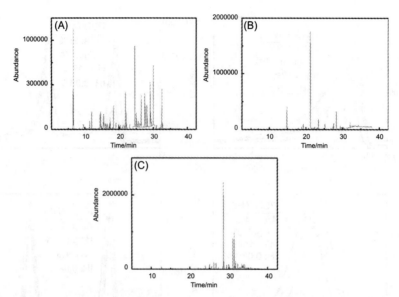

FIGURE 4.19 Total Ion Chromatography (TIC) of pesticides.

4.4.4 Effect of Different Analyte Protectants on Solvent Standards

According to the evaluation method of GC–MS, in order to study the influence of 11 APs in different spiked concentrations (0.5, 1, 2, and 5 mg/mL, 3-ethyoxyl-1,2-propylene glycol also includes two concentrations of 10 and 20 mg/mL) on the chromatographic behaviors of 186 pesticide in acetonitrile solvent, the pesticide responses of solvent standard adding AP and without adding AP are compared. Fig. 4.20 shows the impact of 11 APs on the peak shape of paclobutrazol. The right side shows that the chromatograms from the high to the low order were the extraction ion chromatograms of paclobutrazol adding AP at concentration of 5, 2, 1, 0.5 mg/mL, and paclobutrazol solvent standard, respectively. AP 6 is an extraction ion chromatogram of paclobutrazol adding AP at concentration of 20, 10, 5, 2, 1, 0.5 mg/mL, and paclobutrazol solvent standard. The names of the APs are the same as those in Table 4.4.

The vast majority of 205 pesticides (83.4%) exhibited moderate or strong matrix effects in the matrix tea, with ME > 20%. Referring to the selection criteria of GC–MS, we determined the number of pesticides showing strong matrix enhancement (ME > 50%) in the matrix tea, and the results are listed in Table 4.9. Table 4.9 and Fig. 4.20 show that, similar to the GC–MS method, the matrix enhancement effect of No. 1 and No. 2 APs on 205 pesticides is not apparent. At four spiked concentrations, responses of pesticide-added APs and APs without have no obvious difference (the difference between the two is about 20%). However, the other nine kinds of APs had different degree of matrix enhancement effects for most of 205 pesticides, and the number of pesticides with matrix enhancement effects increased with the increasing of APs concentrations.

Table 4.9 shows that the same as that of the GC–MS method, APs 4, 5, 7, 10, and 11 give the largest number of ratios greater than 50% at concentrations of 0.5, 1, and 2 mg/mL. Therefore, APs 4, 5, 7, and No. 11 were used in the next stage.

Fig. 4.21 shows the effects of APs (including No. 4, 5, 7, and 11) at concentrations of 1 and 2 mg/mL on the peak shapes and intensities of propyzamide and fenpropidin. Among them, Fig.4.21A–B are extracted ion chromatograms of propyzamide and fenpropidin, respectively, with AP concentration of 1 mg/mL. C and D are extracted ion chromatograms of propyzamide and fenpropidin, respectively, with an AP concentration of 2 mg/mL. The right side shows the chromatograms from high to low that are the extracted ion chromatograms of pesticide-added APs 11, 7, 5, and 4, as well as the pesticide solvent standard, respectively. Fig. 4.21 shows that after adding APs, the peak responses of pesticides are increased, and the peak shapes are also improved. At the same time, the four kinds of APs with different concentrations yielded different degrees of enhancement on the peaks and intensities of the pesticides, in which AP 11 is especially prominent.

4.4.5 Effects of Different Analyte Protectant Combinations on Tea

In Section 4.2.7.2, it was determined that according to the matrix effect, tea can be classified into black, green, and oolong tea: three types of teas. In this phase of the experiment, the commercially available black, green, and oolong teas have been selected to carry out the experiment of APs.

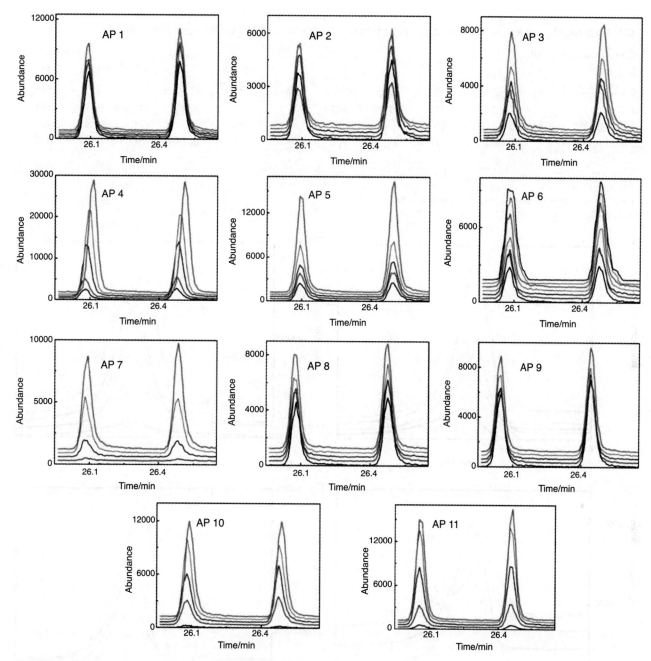

FIGURE 4.20   Impact of 11 APs on the peak shape of iprovalicarb.

The concentration and proportion of the APs as well as the calculation of the results are also referred to in Section 4.3.5.

Per Table 4.10, remove a certain volume of APs into the solvent standard containing 205 kinds of pesticides at MRLs and with an ISTD concentration of 280 g/kg. At the same time, black, green, and oolong tea matrix-matched standard solutions are prepared and the normalized ratios are calculated with the matrix-matched standard quantitative standard.

The obtained normalized ratios show big differences when using different concentrations of AP or an AP combination, for example, the normalized ratios are in the range of 28.2%–151.1% for o-phenylphenol and 0.4%–389.4% for tebuconazole. The results show that different APs and AP combinations had different degree of matrix enhancement effects for different pesticides. The normalized ratios of o-phenylphenol are in the range of 39.5%–151.1% in the green tea matrix, and are 48.2%–92.4% and 28.2%–122.0% in oolong tea and black tea matrices, respectively. Which indicate that the different of APs and AP combinations have different matrix enhancement effects for different teas.

TABLE 4.9 Number of Ratios Above 50% for 186 Pesticides Containing 11 APs at Different Concentrations

Concentration/ (mg/mL)	AP										
	1	**2**	**3**	**4**	**5**	**6**	**7**	**8**	**9**	**10**	**11**
0.5	4	25	38	47	97	19	113	28	57	145	129
1	38	51	41	103	119	36	140	35	63	157	161
2	57	6	98	162	135	56	154	66	113	158	165
5	68	38	149	118	180	82	157	86	106	164	171
10						106					
20						125					

The numbers of APs in the table are in the same order as in Table 4.4.

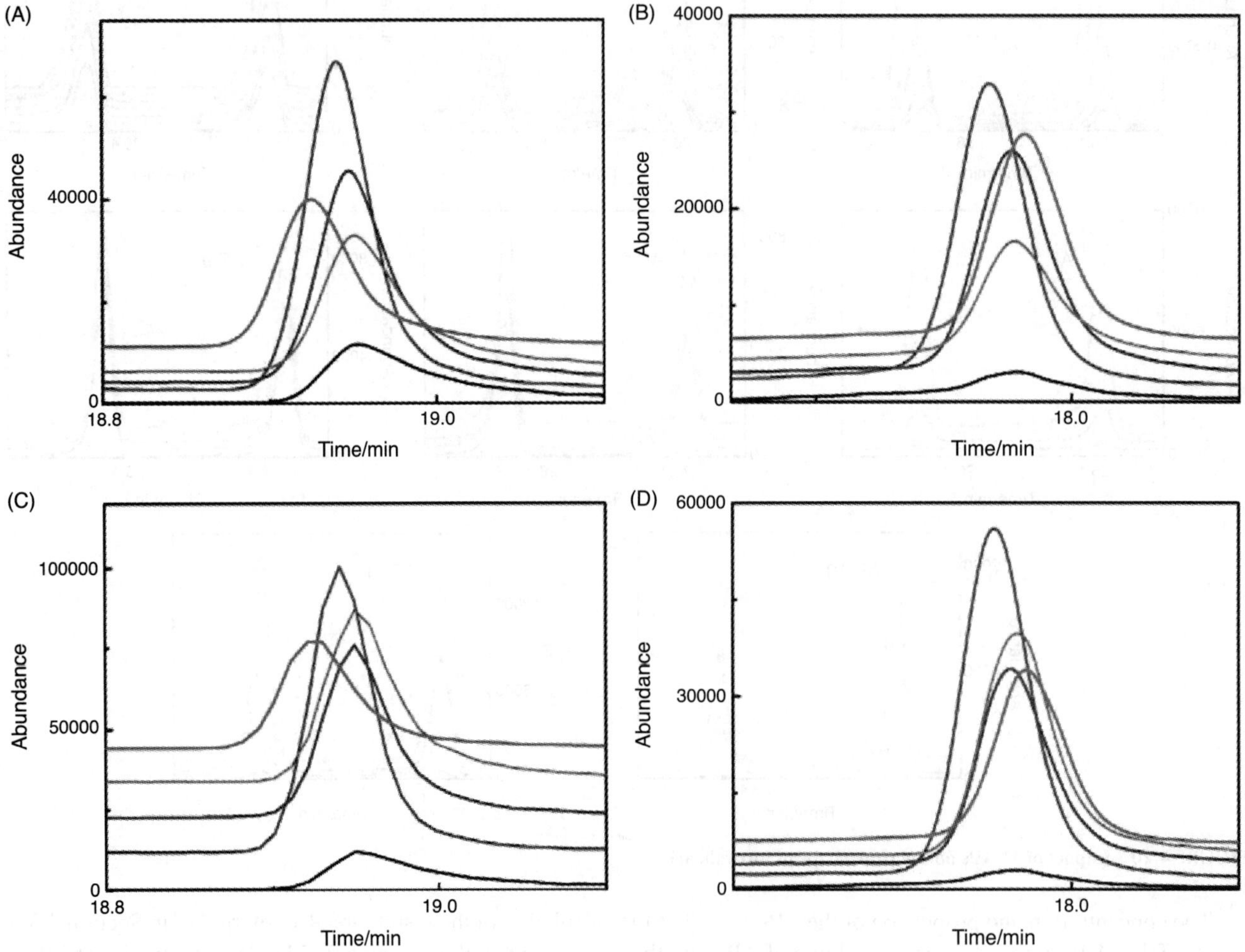

FIGURE 4.21   Effects of different APs on the peak shapes and intensities of propyzamide and fenpropidin.

An ideal AP is a compound that interacts with active sites to reduce the interaction of the pesticide with those sites, and to increase the response of pesticides. Therefore, the response of the pesticides in the standard solutions with AP should be similar to that of the matrix-matched standards, and the normalized ratios should be near or more than 100%. We list the numbers of the pesticides whose normalized ratios are more than 70% in Table 4.10.

The combinations with the top nine amounts of pesticides with high normalized ratios in black, green, and oolong tea (Table 4.10) are merged and 10 AP combinations composed of the 7–16 selected. The 10 AP combinations were used to evaluate the matrix effect of the three kinds of tea. The 10 AP combinations listed in Table 4.10 were added to the spiked sample extracts, then quantified using the solvent standards with the same AP combinations. The recoveries were calculated

TABLE 4.10 Composition of AP Combinations and the Numbers of Pesticides With Normalized Ratio Above 70%

APs	Combination of APs[a]															
	1	2	3	4	5	6	7	8	9	10	11	12	13	14	15	16
4	–	–	1	1	–	–	1	1	–	–	2	2	–	–	–	1
5	2	–	1	–	1	–	2	–	–	2	2	–	2	–	1	1
7	–	2	–	1	–	1	–	2	2	–	–	2	–	2	1	1
11	–	–	–	–	1	1	–	–	1	1	–	–	2	2	1	–
Matrix	Numbers of pesticides with normalized ratio above 70%															
Green tea	122	147	104	147	100	146	159	148	155	150	161	174	173	184	165	171
Oolong tea	90	108	129	140	120	145	186	178	184	162	170	171	189	192	181	171
Black tea	104	131	151	145	166	167	185	168	180	180	181	172	190	189	181	188

[a]The concentration is the actual concentration in the vial injected to the GC system. The unit is mg/mL.

TABLE 4.11 Average Recoveries and the Number of Pesticides With Recoveries Within the Range of 70%–120%

		AP Combinations									
		7	8	9	10	11	12	13	14	15	16
Average recovery	Green tea	113.24	101	105	113	108	107.8	103	89.4	101.1	106
	Oolong tea	103.64	102	91.9	100	96.9	87.45	86.05	91.2	102	96.2
	Black tea	106.39	105	93.4	107	93.5	98.75	94.14	89	98.65	104
Number of pesticides with re-coveries of 70%–120%	Green tea	109	124	149	128	145	143	159	188	131	132
	Oolong tea	119	132	134	135	167	166	175	191	147	122
	Black tea	135	133	161	157	148	160	159	187	154	140

The sequence of the AP combination is the same as in Table 4.10.

according to the second formula mentioned earlier. The average recoveries of each AP combination are listed in Table 4.11, and the number of pesticides whose recoveries ranged between 70% and 120% are also included.

When using the 10 AP combinations, the average recoveries in three tea matrices have no obvious difference, but the numbers of pesticides with recoveries of 70%–120% are the highest with combinations 14.

Twenty pesticides representing organophosphorus (phosalone, pyraclofos, pyrazophos), organochlorine (dicloran, dicofol, Tetradifon), pyrethroid pesticides (etofenprox, fenvalerate-1, fenvalerate-2, flucythrinate-1, flucythrinate-2, permethrin-1, permethrin-2), and organic nitrogen (acibenzolar, nicotinic acid, bromuconazole, diflufenican, flumioxazin, pyriproxyfen, thiobencarb) (which mostly had RTs over 30 min) are used to figure the recoveries obtained by using the 10 AP combinations in black tea. Fig. 4.22 schematically shows the recoveries of these pesticides. As seen in Fig. 4.22, the combination 14 gives ideal results for the selected 20 pesticides. Except for the recoveries of fenvalerate-2 and phoxim that were 126% and 64.3%, respectively, the recoveries of the other pesticides are all in the range of 70%–120%. However, the recoveries of the pesticides are higher when using the other combinations.

4.4.6 Effectiveness Evaluation of Analyte Protectant Compensation Matrix Effect Method

By evaluating the sensitivity, accuracy, precision, the standard curve, and the linear range of the analytical method, it could be judged whether an analytical method meets the requirements of the analysis.

The evaluation methods refer to Section 4.3.6.

4.4.6.1 Linear Range of Standard Curves

The linear ranges, linear equations, and correlation coefficients of 205 pesticides are listed in Appendix Table 4.5, which shows that the calibration curves of the vast majority of pesticides are linear in certain concentration ranges, and the correlation coefficients of 202 pesticides (accounting for 98.5% of the detected pesticides) are above 0.99.

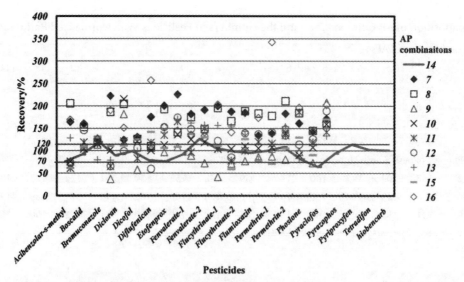

FIGURE 4.22 Recoveries of 20 pesticides obtained by 10 AP combinations in black tea matrix.

4.4.6.2 Sensitivity of the Method

The LOD and LOQ of the method are listed in Appendix Table 4.5, which shows that the LODs of 205 pesticides are in the range of 1.67–50.00 g/kg, and the LOQs are 5.00–150.00 g/kg. The LOQs are all lower than the MRLs established by the EU legislation or Japan, so this method ccould be applied in the detection of pesticide residues in tea.

4.4.6.3 Accuracy and Precision of the Method

Selecting three blank tea samples (e.g., black, green, and oolong teas) as matrices, the recoveries experiment of 205 pesticides at spiked levels of 0.8 MRL, MRL, and 2 MRL were carried out. The spiked samples are left to stand for 30 min, so that the pesticide is fully absorbed by the sample. Then the spiked samples are extracted, purified, and determined with the selected method. Each level of experiment repeats 5 times. The spiked recoveries and relative standard deviations are calculated.

The experimental results show that the recoveries of 194–202 types of pesticide (accounting for 95%–99%) could reach the ideal result (with recoveries of 70%–120%), and all RSDs are less than 20%. The method showed good accuracy.

4.4.6.4 Analysis of a Real Sample

The established method is used to analyze 28 tea samples for sale in the Beijing market, and the results are shown in Table 4.12. In 28 tea samples, 2 samples are not detected in any pesticide. The other 26 samples are detected in 21 pesticides, in which 5 pesticides, including methabenzthiazuron, pirimiphos-methyl, propiconazol-1, cyfluthrin, and difenoconazole are all detected in the 26 samples, but do not reach the method of LOQ. Among the remaining 16 pesticides, the highest detection rate of lambda-cyhalothrin is found, and there are 19 samples (68% of the total) to be detected, in which, only 16 samples reach the method of LOQ and their residual amounts are in the range of 12.2–151.7 μg/kg. Bifenthrin is detected in eight teas, in which three of them do not reach the LOD, and the residual amounts of the remaining five teas are 11.8–99.2 μg/kg. Chlorpyrifos is detected in 10 teas, in which six of them are detected only, and the residual amounts of the remaining four teas are between 10.0 g/kg and 53.3 g/kg. Dimethoate is detected in two teas, in which one does not reach the LOD, and the residue amount of another tea is 59.9 μg/kg (MRL was 50.0 μg/kg). Dimethomorph is detected in 1 tea, and its amount is 26.4 μg/kg. o-Tolidine is detected in five tea leaves, and their amounts are 9.1–59.6 μg/kg (MRL was 50.0 μg/kg). Fenobucarb is detected in four teas, in which two of them do not reach the LOQ, and the residual amounts of the other two teas are 51.0 and 56.9 μg/kg, respectively. Lenacil is detected in six teas, in which four of them do not reach the LOQ, and the residual amounts of the other two teas are 28.1 and 33.6 μg/kg, respectively. Malathion is detected in three teas, in which, two of them do not reach the LOD, and the residual amount of another tea is 134.6 μg/kg. Parathion-methyl is detected in three teas, in which one of them do not reach the LOD, and the residual amounts of the other two teas are 29.4 and 38.0 μg/kg, respectively. Procymidone is detected in one tea with an amount of 17.1 μg/kg. Propham is detected in four teas, in which two of them do not reach the LOD, and the residual amounts of the other two teas are 92.2 and 123.0 μg/kg (MRL was 50.0 μg/kg), respectively. Propisochlor is detected in four teas, in which one of them do not reach the LOD, and

TABLE 4.12 Summary of the Analysis Results from 28 Batch of Teas

Concentration/(μg/kg)

Sample	Bifenthrin	Chlorpyrifos	Cyfluthrin	Dichlofluanid	Difenoconazole	dimethoate	Dimethomorph	Diphenylamine	Fenobucarb	Lambda-Cyhalothrin	Lenacil	Malathion	Metha-benzthiazuron	Parathion-methyl	Pirimiphos-Methyl	Procymidone	Propham	Propiconazole-1	Propisochlor	Tebufenpyrad	Triadimenol-2
1	Detected	-	-	-	-	-	-	23.2	-	-	-	-	-	-	-	-	-	-	检出	4.2	检出
2	11.8	10.0	-	-	-	-	-	12.1	-	15.3	-	-	检出	-	-	-	-	检出	-	-	检出
3	Detected	12.8	-	-	-	-	-	-	-	37.2	-	-	-	-	-	-	检出	-	-	-	Detected
4	-	-	-	-	-	-	-	9.1	-	Detected	-	-	-	-	-	-	-	-	-	-	Detected
5	13.5	-	-	-	-	-	-	9.8	56.9	-	-	-	-	-	-	-	-	-	6.2	-	-
6	-	-	-	-	-	-	-	-	-	-	-	-	-	-	-	-	-	-	-	-	-
7	-	-	-	64.3	-	-	-	-	-	-	-	Detected	-	38.0	-	-	-	-	-	-	-
8	-	-	-	-	-	-	-	-	-	-	-	-	-	-	-	-	-	-	-	-	Detected
9	-	Detected	-	-	-	-	-	-	-	62.1	-	-	-	Detected	-	-	-	-	-	-	-
10	-	-	-	-	-	-	-	-	-	-	-	-	-	-	-	-	-	-	-	-	-
11	-	Detected	-	-	-	-	-	-	-	Detected	-	-	-	-	-	-	Detected	-	5.1	-	Detected
12	-	-	-	Detected	-	-	-	-	-	12.2	-	-	-	-	-	-	-	-	-	-	-
13	-	-	-	-	-	-	-	-	-	34.6	-	-	-	-	-	-	-	-	-	-	22.0
14	-	-	-	-	-	-	-	-	-	84.7	-	Detected	-	29.4	-	-	-	Detected	-	-	144.5
15	-	30.8	-	-	-	-	-	-	-	110.1	-	-	-	-	-	-	-	Detected	-	-	Detected
16	29.0	Detected	-	-	-	-	-	-	-	Detected	Detected	-	-	-	-	-	-	-	-	-	-
17	-	-	-	-	-	-	-	-	-	-	Detected	-	-	-	-	-	-	-	-	-	Detected
18	99.2	-	-	-	-	Detected	-	-	-	94.8	-	-	-	-	-	-	-	-	-	-	Detected
19	-	Detected	-	-	-	-	-	-	Detected	97.4	-	-	-	-	-	-	-	-	-	-	Detected
20	-	-	-	-	-	-	-	-	Detected	40.8	-	-	-	-	-	-	123.0	-	-	-	-
21	-	-	-	-	-	-	-	59.6	Detected	44.7	-	-	-	-	-	-	92.2	-	-	-	Detected
22	-	-	-	-	-	-	-	-	-	-	-	-	-	-	-	-	-	-	-	-	-
23	-	53.3	Detected	-	-	-	26.4	-	-	26.5	-	-	-	-	-	17.1	-	-	-	-	-
24	Detected	Detected	-	-	-	-	-	-	-	-	-	-	-	-	-	-	-	-	-	-	Detected
25	-	-	-	-	-	-	-	-	-	23.9	28.1	-	-	-	-	-	-	-	-	-	-
26	-	Detected	-	-	-	-	-	-	51.0	51.3	33.6	-	-	-	-	-	-	-	7.3	-	Detected
27	-	-	-	-	-	59.9	-	-	-	69.9	Detected	-	-	-	-	-	-	-	-	-	-
28	37.5	-	Detected	-	Detected	-	-	-	-	151.7	Detected	134.6	-	-	Detected	-	-	-	-	-	-

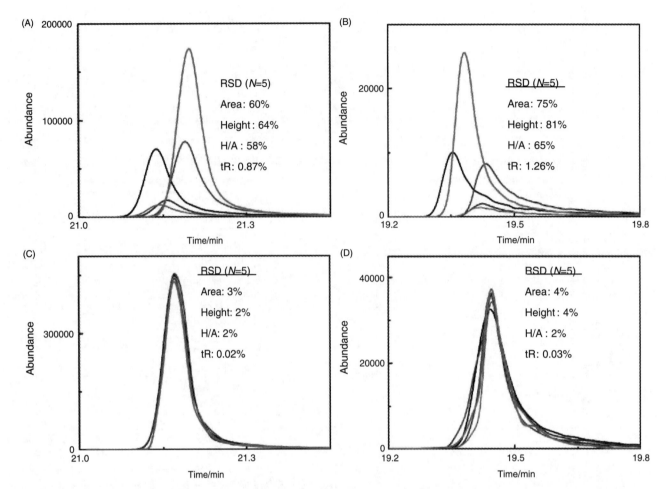

FIGURE 4.23 Overlay of extracted ion chromatograms of simeconzaole and fenpropidin obtained in the five standard solutions without (A,B) and with (C,D) the addition of APs(triglycerol and D-ribonic acid-(-lactone both at 2 mg/mL in the injected sample).

the residual amounts of the other three teas are 5.1, 6.2, and 7.3 µg/kg, respectively. Tebufenpyrad is detected in one tea with an amount of 4.2 µg/kg. Triadimenol-2 is detected in 13 teas, in which 11 of them do not reach the LOD, and the residual amounts of the other two teas are 22.0 and 144.5 µg/kg, respectively. The result indicates, although certain pesticide residues are detected in teas, compared with the MRL limit we find that only three samples exceed the limit: the residue of dimethoate in sample No. 27, and the residues of o-tolidine in sample No. 20 and propham in sample No. 21.

The sample analysis of GC–MS/MS is the same as what of the GC–MS, so we compared the results obtained from two kinds of instruments. It could be found that the two results are similar in the residue amount of pesticides, for example, the lambda-cyhalothrin amounts of No. 28 tea are 170 and 151.7 g/kg by GC–MS and GC–MS/MS, respectively. However, there are still some differences in the detection of the number of pesticides. The results indicated 14 kinds of pesticides detected by GC–MS and 21 kinds of pesticides detected by GC–MS/MS. For the same kind of pesticide, such as lambda-cyhalothrin, there is also some difference in the determination results of the two instruments; for example, lambda-cyhalothrin residue is not detected in No. 14 tea by GC–MS, but its residue amount is as high as 84.7 g/kg with GC–MS/MS. This kind of phenomenon is mainly due to the greater sensitivity of GC–MS/MS as compared to GC–MS.

4.4.6.5 Evaluation of the Long-Term Performance

An evaluation of long-term performance was performed using the description in Section 4.3.3.4. Fig. 4.23 demonstrates the long-term repeatability of the equipment by overlaying the five standard solutions. Fig. 4.5 shows that the RSDs of the peak area, height, H/A, and t_R of simeconzaole and fenpropidin are lower with APs in the injected samples than those without APs. The addition of APs minimized the gradual decrease in peak intensity with the long-term system operation and reduced the need for maintenance of the equipment.

REFERENCES

[1] NCCLS, Evaluation of matrix effects, Proposed Guideline, NCCLS document EPI4-P. NCCLS, Wayne, PA, United States, 1998.

[2] EU, Quality control procedures for pesticide residues analysis, Guidelines for Residues Monitoring in the European Union. EU Document No. SANCO/10476/2003.

[3] Gillespie, M., Walters, S., 1991. Rapid clean-up of fat extracts for organophosphorus pesticide residue determination using C18 solid-phase extraction cartridges. Anal. Chim. Acta 245, 259–265.

[4] Erney, D.R., Gillespie, A.M., Gilvydis, D.M., et al., 1993. Explanation of the matrix-induced chromatographic response enhancement of organophosphorus pesticides during open tubular column gas chromatography with splitless or hot on-column injection flame photometric detection. J. Chromatogr. A 638 (1), 57–63.

[5] Hajlova, J., Zrostlíková, J., 2003. Matrix effects in (ultra)trace analysis of pesticide residues in food biotic matrices. J. Chromatogr. A 1000 (1-2), 181–197.

[6] Arumugam, R., Ragupathi Raja Kannan, R., Jayalakshmi, J., et al., 2012. Determination of element contents in herbal drugs: chemometric approach. J. Food Chem. 135 (4), 2372–2377.

[7] Gao, H.J., Jiang, X., Wang, F., et al., 2004. Extraction solvent and solid-phase material in determination organochlorine pesticides in leaf vegetables. Environ. Chem. 23 (5), 587–590.

[8] Schenck, F.J., Lehotay, S.J., Vega, V., 2002. Comparision of solid-phase extraction sorbents for cleanup in pesticide residue analysis of fresh fruits and vegetables. J. Sep. Sci. 25 (14), 883–890.

[9] Schenck, F.J., Lehotay, S.J., 2000. Does further clean-up reduce the matrix enhancement effect in gas chromatographic analysis of pesticide residues in food. J. Chromatogr. A 868 (1), 51–61.

[10] Zhou, L., Wang, X.Q., Xu, H., et al., 2011. On the matrix effects in the LC–MS/MS determination of 16 pesticides in leaf vegetables. Phys. Test. Chem. Anal. 47 (12), 1398–1401.

[11] Hajslova, J., Holadov, K., Kocourek, V., et al., 1998. Matrix-induced effects: a critical point in the gas chromatographic analysis of pesticide residues. J. Chromatogr. A 800 (2), 283–295.

[12] Lehotay, S.J., Katerina, M., Yun S, 2005. Evaluation of two fast and easy methods for pesticide residue analysis in fatty food matrixes. J. AOAC Int. 88 (2), 615–629.

[13] Liu, L., Luo, J., 2009. Solutions to matrix-induced response enhancement in organic pesticide residue analysis by gas chromatography. Acta Agr. Jiangxi 21 (7), 146–148.

[14] Chai, L.K., Elie, F., 2013. A rapid multi-residue method for pesticide residues determination in white and black pepper (*Piper nigrum* L). Food Control 32 (1), 322–326.

[15] Lehotay, S.J., Maštovská, K., Lightfield A, 2005. Use of buffering and other means to improve results of problematic pesticides in a fast and easy method for residue analysis of fruits and vegetables. J. AOAC Int. 88 (2), 630–638.

[16] Lehotay S, 2003. Application of gas chromatography in food analysis. TrAC Trends Anal. Chem. 21 (9–10), 686–697.

[17] Molinar, G.P., Cavanna, S., Fornara, L., 1998. Determination of fenthion and its oxidative metabolites in olives and olive oil: errors caused by matrix effects food. Addit. Contam. 15 (6), 661–670.

[18] Yi, S.G., Hou, X., Han, M., et al., 2012. Study on matrix effects and matrix classification of pesticide residues in vegetable by gas chromatography-tandem mass spectrometry. Southwest Chin. J. Agric. Sci. 25 (2), 537–543.

[19] Godula, M., Hajslová, J., Alterova, K., 1999. Pulsed splitless injection and the extent of matrix effects in the analysis of pesticides. J. High Resolut. Chromatogr. 22, 395–402.

[20] Souza Pinheiro, de.A., Rocha, da.G.O., Andrade, de.J.B., 2011. A SDME/GC–MS methodology for determination of organophosphate and pyrethroid pesticides in water. Microchem. J. 99 (2), 303–308.

[21] Demeestere, K., Dewulf, J., Witte De, B., et al., 2007. Sample preparation for the analysis of volatile organic compounds in air water matrices. J. Chromatogr. A 1153 (1-2), 130–144.

[22] Zrostikova, J., Hajslova, J., Godula, M., et al., 2001. Performance of programmed temperature vaporizer pulsed splitless on-column injection techniques in analysis of pesticide residues in plant matrices. J. Chromatogr. A 937 (1–2), 73–86.

[23] Lehotay, S., 2000. J. Analysis of pesticide residues in mixed fruit and vegetable extracts by direct sample introduction/gas chromatography/tandem mass spectrometry. J. AOAC Int. 83 (3), 680 697.

[24] Jing, H.W., Amirav, A., 1997. Pesticide analysis with the pulsed-flame photometer detector and a direct sample introduction device. Anal. Chem. 69 (7), 1426–1435.

[25] Poole, C.F., 2007. Matrix-induced response enhancement in pesticide residue analysis by gas chromatography. J. Chromatogr. A 1158 (1–2), 241–250.

[26] Wylie, P.L., Uchiyama, K., 1996. Improved gas chromatographic analysis of organophosphorus pesticides with pulsed splitless injection. J. AOAC Int. 79 (2), 571–577.

[27] Souverain, S., Rudaz, S., Veuthey, J.L., 2004. Matrix effect in LC-ESI-MS LC-APCI-MS with off-line on-line extraction procedures. J. Chromatogr. A 1058 (1–2), 61–66.

[28] Ou, J.F., 2008. Matrix Effects in Determination of Multiresidues of Pesticides in Vegetables Using Gas Chromatography-Mass Spectrometry. Chinese Academy of Agricultural Sciences, Beijing, China, 6.

[29] Dong, Z.L., Yang, C.G., Xiao, S., et al., 2009. Determination of acetochlor residue in food by gas chromatography and gas chromatography-mass spectrometry. Chin. J. Anal. Chem. 37 (5), 698–702.

[30] Garrido Frenich, A., Martínez Vidal, J.L., Fernández Moreno, J.L., et al., 2009. Compensation for matrix effects in gas chromatography-tandem mass spectrometry using a single point standard addition. J. Chromatogr. A 1216 (23), 4798–4808.

[31] Sandra, P., Tienpont, B., Vercammen, J., et al., 2001. Stir bar sorptive extraction applied to the determination of dicarboximide fungicides in wine. J. Chromatogr. A 928 (1), 117–126.

[32] Sandra, P., Tienpont, B., David, F., 2003. Multi-residue screening of pesticides in vegetables fruits baby food by stir bar sorptive extraction-thermal desorption-capillary gas chromatography-mass spectrometry. J. Chromatogr. A 1000 (1–2), 299–309.

[33] Ueno, E., Oshim, H., Saito, I., et al., 2004. Multiresidue analysis of pesticides in vegetables and fruits by gas chromatography/mass spectrometry after gel permeation chromatography and graphitized carbon column cleanup. J. AOAC Int. 87 (13), 1003–1015.

[34] Sotiropoulou, S., Chaniotakis, N., 2003. A carbon nanotube array-based biosensor. Anal. Bioanal. Chem. 375 (1), 103–105.

[35] Pang, G.F., Fan, C.L., Zhang, F., et al., 2011. High-throughput GC/MS and HPLC/MS/MS techniques for the multiclass, multiresidue determination of 653 pesticides and chemical pollutants in tea. J. AOAC Int. 94 (4), 1253–1296.

[36] Pizzutti, I.R., Kok de, A., Dickow, C.C., et al., 2012. Multi-residue method for pesticides analysis in green coffee beans using gas chromatography–negative chemical ionization mass spectrometry in selective ion monitoring. J. Chromatogr. A 1251, 16–26.

[37] Yang, X., Zhang, H., Liu, Y., et al., 2011. Multiresidue method for determination of 88 pesticides in berry fruits using solid-phase extraction and gas chromatography–mass spectrometry: determination of 88 pesticides in berries using SPE and GC–MS. Food Chem. 127 (2), 855–865.

[38] Banerjee, K., Mujawar, S., Utture, S.C., et al., 2013. Optimization of gas chromatography–single quadrupole mass spectrometry conditions for multiresidue analysis of pesticides in grapes in compliance to EU-MRLs. Food Chem. 138 (1), 600–607.

[39] Wu, G., Bao, X.X., Zhao, S.H., et al., 2011. Analysis of multi-pesticide residues in the foods of animal origin by GC–MS coupled with accelerated solvent extraction and gel permeation chromatography cleanup. Food Chem. 126 (2), 646–654.

[40] Mercer G, 2005. Determination of 112 halogenated pesticides using gas chromatography/mass spectrometry with selected ion monitoring. J. AOAC Int. 88 (5), 1452–1462.

[41] Lehotay, S., 2002. Determination of pesticide residues in nonfatty fooda by percritical extraction and gas chromatography/mass spectrometry: collaborative study. J. AOAC Int. 85 (5), 1148–1166.

[42] EU, Method Validation And Quality Control Procedures For Pesticide Residues Analysis In Food And Feed. EU Document No. SANCO/12495/2011.

[43] Anastassiades, M., Mastovská, K., Lehotay, S.J., 2003. Evaluation of analyte protectants to improve gas chromatographic analysis of pesticides. J. Chromatogr. A 1015 (1-2), 163–184.

[44] Erney, D.R., Poole, C., 1993. A study of single compound additives to minimize the matrix induced chromatographic response enhancement observed in the gas chromatography of pesticide residues. J. High Resolut. Chromatogr. 16 (8), 501–503.

[45] Mastovská, K., Lehotay, S.J., Anastassiades, M., 2005. Combination of analyte protectants to overcome matrix effects in routine GC analysis of pesticide residues in food matrixes. Anal. Chem. 77 (24), 8129–8137.

[46] Sánchez Brunete, C., Albero, B., Martín, G., et al., 2005. Determination of pesticide residues by GC–MS using analyte protectants to counteract the matrix effect. Anal. Sci. 21 (11), 1291–1296.

[47] González-Rodríguez, R.M., Rial-Otero, R., Cancho-Grande, B., et al., 2008. Occurrence of fungicide and insecticide residues in trade samples of leafy vegetables. Food Chem. 107 (3), 1342–1347.

[48] González-Rodríguez, R.M., Cancho-Grande, B., Simal-Gándara, J., 2009. Multiresidue determination of 11 new fungicides in grapes wines by liquid-liquid extraction/clean-up programmable temperature vaporization injection with analyte protectants/gas chromatography/ion trap mass spectrometry. J. Chromatogr. A 1216 (32), 6033–6042.

[49] Wang, Y., Jin, H.Y., Ma, S.C., Lu, J., et al., 2011. Determination of 195 pesticide residues in chinese herbs by gas chromatography-mass spectrometry using analyte protectants. J. Chromatogr. A 1218 (2), 334–342.

[50] Huang, B.Y., Pan, C.P., WAN, G.Y.R., et al., 2006. Rapid determination of pesticide multiresidues in vegetable by gas chromatography-mass spectrometry and compensation for matrix effect with protectants. Chem. J. Chin. Univ. 27 (2), 227–232.

[51] Huang, B.Y., Pan, C.P., Zhang, W., et al., 2006. Rapid determination of 45 pesticide residues in fruits and vegetables by dispersive PSA cleanup and gas chromatography–mass spectrometry with correction of matrix effects. J. Instrum. Anal. 25 (3), 11–16.

[52] Mei, M.H., 2011. Application of Chemometrics on the Fingerprint Analysis of Several Traditional Chinese Medicine or Food. Nanchang University, Jiangxi, China, 6.

[53] Tran, C.D., Grishko, V.I., Oliveira, D., 2003. Determination of enantiomeric compositions of amino acids by near-infrared spectrometry through complexation with carbohydrate. Anal. Chem. 75 (23), 6455–6462.

[54] Lavine, B.K., 1998. Chemometrics. Anal. Chem. 70 (12), 209–228.

[55] Lavine, B.K., 2000. Chemometrics. Anal. Chem. 72 (12), 91–98.

[56] Lavine, B.K., Workman, J., 2002. Chemometrics. Anal. Chem. 74 (12), 2763–2770.

[57] Reid, L.M., O'Donnell, C.P., Downey, G., 2006. Recent technological advances for the determination of food authenticity. Trends Food Sci. Technol. 17 (7), 344–353.

[58] Liang, Y.Z., Yu, R.Q., 2003. Chemometrics. Higher Education Press, Beijing, China, pp. 67–69.

[59] Kaufman, L., Rousseeuw, P., 1990. Finding Groups in Data: An Introduction to Cluster Analysis. John Wiley & Sons, New York, NY, United States.

[60] Ng, R.T., Han, J, 1994. Efficient and Effective Clustering Method for Spatial Data Mining. Proceeding VLDB '94 Proceedings of the 20th International Conference on Very Large Data bases San Francisco, CA, United States. pp. 144–155.

[61] Pei, X.D., 1991. Multivariate Statistic Analysis and its Application. Beijing Agricultural University Press, Beijing, China, pp. 36–38.

[62] Silva Torres, da.E.A.F., Garbelotti, M.L., Neto, J.M.M., 2006. The Application of hierarchical clusters analysis to the study of the composition of foods. Food Chem. 99 (3), 622–629.

[63] Li, J.X., Wang, Y.X., Xie, X.J., et al., 2012. Hierarchical cluster analysis of arsenic and fluoride enrichments in groundwater from the datong basin, Northern China. J. Geochem. Explor. 118, 77–89.

[64] Jin, L., Harley, R.A., Brown, N.J., 2011. Ozone pollution regimes modeled for a summer season in California's San Joaquin Valley: a cluster analysis. Atmos. Environ. 45 (27), 4707–4718.

[65] Kocourek, V., Hajslova, J., Holadova, K., et al., 1998. Stability of pesticides in plant extracts used as calibrants in the gas chromatographic analysis of residues. J. Chromatogr. A 800 (2), 297–304.

[66] Kittlaus, S., Schimanke, J., Kempe, G., et al., 2012. Assessment of sample cleanup matrix effects in the pesticide residue analysis of foods using postcolumn infusion in liquid chromatography-tandem mass spectrometry. J. Chromatogr. A 1218 (46), 8399–8410.

[67] Codex Alimentarius Committee, Guidelines on Good Laboratory Practice in Residue Analysis, CAC/GL 40-1993, Rev.1, FAO, 2003.

[68] Romero-González, R., Frenich, A.G., Vidal, J.L.M., 2008. Multiresidue method for fast determination of pesticides in fruit juices by ultra performance liquid chromatography coupled to tandem mass spectrometry. Talanta 76 (1), 211–225.

[69] Martínez Vidal, J.L., Garrido Frenich, A., López López, T., et al., 2005. Selection of a representative matrix for calibration in multianalyte determination of pesticides in vegetables by liquid chromatography-electrospray tandem mass spectrometry. Chromatographia 61 (3–4), 127–131.

[70] Wu, F., Lu, W., Chen, J., et al., 2010. Single-walled carbon nanotubes coated fibers for solid-phase microextraction and gas chromatography–mass spectrometric determination of pesticides in tea samples. Talanta 82 (3), 1038–1043.

[71] Lima, D.C., Santos dos, A.M.P., Araujo, R.G.O., et al., 2010. Principal component analysis and hierarchical cluster analysis for homogeneity evaluation during the preparation of a wheat flour laboratory reference material for inorganic analysis. Microchem. J. 95 (2), 222–226.

[72] Matuszewski, B.K., Constanzer, M.L., Chavez-Eng, C.M., 2003. Strategies for the assessment of matrix effect in quantitative bioanalytical methods based on HPLC-MS/MS. Anal. Chem. 75 (13), 3019–3030.

[73] Kmellar, B., Fodor, P., Pareja, L., et al., 2008. Validation and uncertainty study of a comprehensive list of 160 pesticide residues in multi-class vegetables by liquid chromatography-tandem mass spectrometry. J. Chromatogr. A 1215 (1–2), 37–50.

[74] Peng, S.P., Gu, Z.L., 2005. The classification of commercial tea and comparison of content of tea polyphenols. Fujian Chaye 2, 32–33.

[75] Kim, Y., Goodner, K.L., Park, J.D., et al., 2011. Changes in antioxidant phytochemicals and volatile composition of *Camellia sinensis* by oxidation during tea fermentation. Food Chem. 129 (4), 1331–1342.

[76] Fernández, P.L., Martín, M.J., González, A.G., et al., 2000. HPLC determination of catechins and caffeine in tea: differentiation of green, black and instant teas. Analyst 125, 421–425.

[77] Stahnke, H., Reemtsma, T., Alder, L., 2009. Compensation of matrix effects by postcolumn infusion of a monitor substance in multiresidue analysis with LC–MS/MS. Anal. Chem. 81 (6), 2185–2192.

APPENDIX TABLE 4.1 Parameters of GC–MS for 186 Pesticides

Group	No.	Name	Rention time/min	MRL/ (mg/kg)[a]	Quantitative Ion	Qualitative Ion 1	Qualitative Ion 2	Qualitative Ion 3
ISTD		Heptachlor Epoxide	22.2	–	353(100)	355(79)	351(52)	
A	1	Acetochlor	19.7	0.01	146(100)	162(59)	223(59)	
A	2	Acibenzolar-S-Methyl	20.4	0.05	182(100)	135(64)	153(34)	
A	3	Aclonifen	27.1	0.05	264(100)	212(65)	194(57)	
A	4	Azinphos-Ethyl	32.1	0.05	160(100)	132(103)	77(51)	
A	5	Benalaxyl	27.6	0.10	148(100)	206(32)	325(8)	
A	6	Bifenazate	30.2	0.02	300(100)	258(99)	199(100)	
A	7	Bioresmethrin[b]	27.6	0.10	123(100)	171(54)	143(31)	
A	8	Bitertanol	32.4	0.10	170(100)	112(8)	141(6)	
A	9	Bromophos-Ethyl	23.1	0.10	359(100)	303(77)	357(74)	
A	10	Bromuconazole	30.8	0.05	173(100)	175(65)	214(15)	
A	11	Buprofezin	24.9	0.05	105(100)	172(54)	305(24)	
A	12	Butralin	22.1	0.02	266(100)	224(16)	295(9)	
A	13	Carbaryl	14.3	0.10	144(100)	115(100)	116(43)	
A	14	Carfentrazone-Ethyl	28.0	0.02	312(100)	330(52)	290(53)	
A	15	Chlorbufam	17.9	0.10	223(100)	153(53)	164(64)	
A	16	Chlorpyrifos	21.0	0.10	314(100)	258(57)	286(42)	
A	17	Clodinafop-Propargyl[b]	27.9	0.02	349(100)	238(96)	266(83)	
A	18	Clomazone	17.0	0.02	204(100)	138(4)	205(13)	
A	19	Cyproconazole	27.4	0.05	222(100)	224(35)	223(11)	
A	20	Cyprodinil	22.0	0.05	224(100)	225(62)	210(9)	
A	21	Diazinon	17.2	0.02	304(100)	179(192)	137(172)	
A	22	Dichlofluanid[b]	21.7	5(0.50)	224(100)	226(74)	167(120)	
A	23	Dieldrin	24.4	0.02	263(100)	277(82)	380(30)	345(35)
A	24	Dimethachlor	19.9	0.02	134(100)	197(47)	210(16)	
A	25	Dimethenamid	19.7	0.02	154(100)	230(43)	203(21)	
A	26	Diphenylamine	14.6	0.05	169(100)	168(58)	167(29)	
A	27	Disulfoton Sulfone	26.4	0.05	213(100)	229(4)	185(11)	
A	28	Ethofumesate	21.9	0.10	207(100)	161(54)	286(27)	
A	29	Etoxazole	29.5	0.05	300(100)	330(69)	359(65)	
A	30	Fenarimol	31.7	0.05	139(100)	219(70)	330(42)	
A	31	Fenazaquin	29.1	10(0.10)	145(100)	160(46)	117(10)	
A	32	Fenitrothion	21.7	0.50	277(100)	260(52)	247(60)	
A	33	Fenthion Sulfoxide	28.2	0.05	278(100)	279(290)	294(145)	
A	34	Flufenoxuron	18.9	15(0.15)	305(100)	126(67)	307(32)	
A	35	Flurtamone	32.4	0.05	333(100)	199(63)	247(25)	
A	36	Flusilazole	26.4	0.05	233(100)	206(33)	315(9)	
A	37	Flutolanil	26.3	0.05	173(100)	145(25)	323(14)	
A	38	Flutriafol	25.4	0.05	219(100)	164(96)	201(7)	
A	39	Iprovalicarb-1	25.8	0.10	119(100)	134(126)	158(62)	
A	40	Iprovalicarb-2	26.4	0.10	134(100)	119(75)	158(48)	
A	41	Methabenzthiazuron	16.4	0.05	164(100)	136(81)	108(27)	
A	42	Methidathion	24.5	0.10	145(100)	157(2)	302(4)	

(Continued)

APPENDIX TABLE 4.1 Parameters of GC–MS for 186 Pesticides (*cont.*)

Group	No.	Name	Rention time/min	MRL/ (mg/kg)[a]	Quantitative Ion	Qualitative Ion 1	Qualitative Ion 2	Qualitative Ion 3
A	43	Mevinphos	11.3	0.02	127(100)	192(39)	164(29)	
A	44	Molinate	12.1	0.10	126(100)	187(24)	158(2)	
A	45	Napropamide	24.9	0.05	271(100)	128(111)	171(34)	
A	46	Paclobutrazol	25.3	0.02	236(100)	238(37)	167(39)	
A	47	Phorate	15.5	0.10	260(100)	121(160)	231(56)	153(3)
A	48	Pirimicarb	19.0	0.05	166(100)	238(23)	138(8)	
A	49	Pirimiphos-Methyl	20.4	0.05	290(100)	276(86)	305(74)	
A	50	Procymidone	24.4	0.10	283(100)	285(70)	255(15)	
A	51	Propachlor	14.8	0.05	120(100)	176(45)	211(11)	
A	52	Propargite	28.1	5.00	135(100)	350(7)	173(16)	
A	53	Propham	11.4	0.10	179(100)	137(66)	120(51)	
A	54	Propiconazole-1	28.0	0.10	259(100)	173(97)	261(65)	
A	55	Propiconazole-2	28.2	0.10	259(100)	173(97)	261(65)	
A	56	Propisochlor	20.1	0.01	162(100)	223(200)	146(17)	
A	57	Propoxur-1	6.7	0.10	110(100)	152(16)	111(9)	
A	58	Propoxur-2	15.3	0.10	110(100)	152(19)	111(8)	
A	59	Pyriproxyfen	30.0	0.05	136(100)	226(8)	185(10)	
A	60	Tebufenpyrad	29.2	0.10	318(100)	333(78)	276(44)	
A	61	Terbuthylazine	18.2	0.05	214(100)	229(33)	173(35)	
A	62	Tolclofos-Methyl	19.8	0.10	265(100)	267(36)	250(10)	
A	63	Triadimenol-1	24.5	0.20	112(100)	168(81)	130(15)	
A	64	Triadimenol-2	24.9	0.20	112(100)	168(71)	130(10)	
A	65	Triallate	17.3	0.10	268(100)	270(73)	143(19)	
A	66	Triazophos	28.3	0.02	161(100)	172(47)	257(38)	
A	67	Trifloxystrobin	27.5	0.05	116(100)	131(40)	222(30)	
A	68	Trifluralin	15.4	0.10	306(100)	264(72)	335(7)	
A	69	Zoxamide	22.4	0.05	187(100)	242(68)	299(9)	
B	70	Alachlor	20.1	0.05	188(100)	237(35)	269(15)	
B	71	Boscalid	34.0	0.5(0.05)	342(100)	140(229)	112(71)	
B	72	Butylate	9.6	0.05	156(100)	146(115)	217(27)	
B	73	Diethofencarb	21.6	0.05	267(100)	225(98)	151(31)	
B	74	Dimethomorph	37.1	0.05	301(100)	387(32)	165(28)	
B	75	Diniconazole	27.2	0.05	268(100)	270(65)	232(13)	
B	76	Disulfoton	17.8	0.05	88(100)	274(15)	186(18)	
B	77	Disulfoton-Sulfoxide	8.6	0.05	212(100)	153(61)	184(20)	
B	78	Epoxiconazole	29.5	0.05	192(100)	183(24)	138(35)	
B	79	Ethiofencarb[b]	11.0	0.05	107(100)	168(34)	77(26)	
B	80	Ethion	26.7	3(0.30)	231(100)	384(13)	199(9)	
B	81	Ethoprophos	14.5	0.02	158(100)	200(40)	242(23)	168(15)
B	82	Etofenprox	32.9	0.01	163(100)	376(4)	183(6)	
B	83	Fenamiphos	25.3	0.05	303(100)	154(56)	288(31)	217(22)
B	84	Fenamiphos Sulfone	31.5	0.05	320(100)	292(57)	335(7)	
B	85	Fenamiphos Sulfoxide	31.1	0.05	304(100)	319(29)	196(22)	

(Continued)

APPENDIX TABLE 4.1 Parameters of GC–MS for 186 Pesticides (*cont.*)

Group	No.	Name	Rention time/min	MRL/ (mg/kg)[a]	Quantitative Ion	Qualitative Ion 1	Qualitative Ion 2	Qualitative Ion 3
B	86	Fenbuconazole	34.2	0.05	129(100)	198(51)	125(31)	
B	87	Fenobucarb[b]	14.8	0.50	121(100)	150(32)	107(8)	
B	88	Fenpropathrin	29.7	2(0.02)	265(100)	181(237)	349(25)	
B	89	Fenpropimorph	19.4	0.10	128(100)	303(5)	129(9)	
B	90	Fenpyroximate	17.3	0.10	213(100)	142(21)	198(9)	
B	91	Fenthion	21.5	0.05	278(100)	169(16)	153(9)	
B	92	Flurochloridone	24.5	0.10	311(100)	187(74)	313(66)	
B	93	Fuberidazole	22.2	0.05	184(100)	155(21)	129(12)	
B	94	Hexythiazox	26.6	0.05	227(100)	156(158)	184(93)	
B	95	Imazalil	25.5	0.10	215(100)	173(66)	296(5)	
B	96	Kresoxim-Methyl	25.1	0.10	116(100)	206(25)	131(66)	
B	97	Linuron	22.4	0.10	61(100)	248(30)	160(12)	
B	98	Malathion	21.6	0.50	173(100)	158(36)	143(15)	
B	99	Mepronil	28.1	0.10	119(100)	269(26)	120(9)	
B	100	Metalaxyl	20.7	0.10	206(100)	249(53)	234(38)	
B	101	Metamitron	28.7	0.10	202(100)	174(52)	186(12)	
B	102	Metazachlor	23.3	0.20	209(100)	133(120)	211(32)	
B	103	Methacrifos	12.0	0.10	125(100)	208(74)	240(44)	
B	104	Methoprene	21.4	0.05	73(100)	191(29)	153(29)	
B	105	Metribuzin	20.5	0.10	198(100)	199(21)	144(12)	
B	106	Oxyfluorfen	26.3	0.05	252(100)	361(35)	300(35)	
B	107	Phenothrin-1	29.2	0.05	123(100)	183(74)	350(6)	
B	108	Phenothrin-2	29.4	0.05	123(100)	183(74)	350(6)	
B	109	Phenthoate[b]	23.3	0.10	274(100)	246(24)	320(5)	
B	110	Phosalone	31.3	0.10	182(100)	367(30)	154(20)	
B	111	Picoxystrobin	24.7	0.10	335(100)	303(43)	367(9)	
B	112	Profenofos	24.7	0.10	339(100)	374(39)	297(37)	
B	113	Propyzamide	18.8	0.05	173(100)	255(23)	240(9)	
B	114	Prosulfocarb	19.6	0.05	251(100)	252(14)	162(10)	
B	115	Pyraclofos[b]	32.0	5(0.50)	360(100)	194(79)	362(38)	
B	116	Pyridaben	32.0	0.05	147(100)	117(11)	364(7)	
B	117	Pyrifenox-1[b]	22.4	5(0.50)	262(100)	294(18)	227(15)	
B	118	Pyrifenox-2[b]	23.4	5(0.50)	262(100)	294(18)	227(15)	
B	119	Pyrimethanil	17.4	0.10	198(100)	199(45)	200(5)	
B	120	Pyrimidifen[b]	33.6	5(0.50)	184(100)	186(32)	185(10)	
B	121	Quinalphos	23.1	0.10	146(100)	298(28)	157(66)	
B	122	Quinoxyphen	27.1	0.05	237(100)	272(37)	307(29)	
B	123	Tebuconazole	29.6	0.05	250(100)	163(55)	252(36)	
B	124	Tebuthiuron[b]	14.0	0.02	156(100)	171(30)	157(9)	
B	125	Terbufos	16.9	0.01	231(100)	153(25)	288(10)	186(13)
B	126	Tetraconazole	23.6	0.02	336(100)	338(33)	171(10)	
B	127	Thiobencarb	20.7	0.10	100(100)	257(25)	259(9)	
B	128	Triadimefon	22.3	0.20	208(100)	210(50)	181(74)	

(*Continued*)

APPENDIX TABLE 4.1 Parameters of GC–MS for 186 Pesticides (*cont.*)

Group	No.	Name	Rention time/min	MRL/ (mg/kg)ᵃ	Quantitative Ion	Qualitative Ion 1	Qualitative Ion 2	Qualitative Ion 3
B	129	Vinclozolin	20.4	0.10	285(100)	212(109)	198(96)	
C	130	2-Phenylphenol	12.6	0.10	170(100)	169(72)	141(31)	
C	131	Aldrin	19.6	0.02	263(100)	265(65)	293(40)	329(8)
C	132	alpha-HCH	16.1	0.02	219(100)	183(98)	221(47)	254(6)
C	133	Benfluralin	15.5	0.05	292(100)	264(20)	276(13)	
C	134	Bifenox	30.7	0.05	341(100)	189(30)	310(27)	
C	135	Bifenthrin	28.6	5(0.50)	181(100)	166(25)	165(23)	
C	136	Bromopropylate	29.4	0.05	341(100)	183(34)	339(49)	
C	137	Chlorbenside	23.0	0.10	268(100)	270(41)	143(11)	
C	138	Chlorfenapyr	27.2	50(0.05)	247(100)	328(54)	408(51)	
C	139	Chlorfenson	25.1	0.10	302(100)	175(282)	177(103)	
C	140	Chlorfenvinphos	23.3	0.05	323(100)	267(139)	269(92)	
C	141	Chlorobenzilate	26.1	0.10	251(100)	253(65)	152(5)	
C	142	Chlorpropham	15.6	0.10	213(100)	171(59)	153(24)	
C	143	Chlorpyrifos-Methyl	19.4	0.10	286(100)	288(70)	197(5)	
C	144	Chlorthal-Dimethyl	21.3	0.01	301(100)	332(27)	221(17)	
C	145	Cyfluthrin	33.0	0.10	206(100)	199(63)	226(72)	
C	146	Cypermethrin	33.2	0.50	181(100)	152(23)	180(16)	
C	147	Dichlobenil	9.8	0.05	171(100)	173(68)	136(15)	
C	148	Diclofop-Methyl	28.1	0.05	253(100)	281(50)	342(82)	
C	149	Dicloran	17.9	0.01	206(100)	176(128)	160(52)	
C	150	Difenoconazole	35.5	0.05	323(100)	325(66)	265(83)	
C	151	Diflufenican	28.6	0.05	266(100)	394(25)	267(14)	
C	152	Dioxathion	17.6	0.10	270(100)	197(43)	169(19)	
C	153	Endosulfan-Sulfate	29.1	30(0.03)	387(100)	272(165)	389(64)	
C	154	Endrin	25.1	0.01	263(100)	317(30)	345(26)	
C	155	Ethalfluralin	15.1	0.02	276(100)	316(81)	292(42)	
C	156	Fenamidone	30.4	0.05	268(100)	238(111)	206(32)	
C	157	Fenchlorphos	19.8	0.10	285(100)	287(69)	270(6)	
C	158	Fenvalerate-1	34.5	0.05	167(100)	225(53)	419(37)	181(41)
C	159	Fenvalerate-2	34.9	0.05	167(101)	225(54)	419(38)	181(42)
C	160	Flucythrinate-1	33.8	0.10	199(100)	157(90)	451(22)	
C	161	Flucythrinate-2	34.1	0.10	199(101)	157(91)	451(23)	
C	162	Fludioxonil	29.0	0.05	248(100)	127(24)	154(21)	
C	163	Flumioxazine	35.6	0.10	354(100)	287(24)	259(15)	
C	164	Fluquinconazole	32.7	0.05	340(100)	342(37)	341(20)	
C	165	gamma-HCH	17.7	0.05	183(100)	219(93)	254(13)	221(40)
C	166	Heptachlor	18.4	0.02	272(100)	237(40)	337(27)	
C	167	Hexaconazole	25.0	0.05	214(100)	231(62)	256(26)	
C	168	Lambda-Cyhalothrin	31.3	1.00	181(100)	197(100)	141(20)	
C	169	Methiocarb Sulfone	25.1	0.10	200(100)	185(40)	137(16)	
C	170	Methoxychlor	29.3	0.10	227(100)	228(16)	212(4)	
C	171	Oxadiazon	25.1	0.05	175(100)	258(62)	302(37)	

(Continued)

APPENDIX TABLE 4.1 Parameters of GC–MS for 186 Pesticides (*cont.*)

Group	No.	Name	Rention time/min	MRL/(mg/kg)[a]	Quantitative Ion	Qualitative Ion 1	Qualitative Ion 2	Qualitative Ion 3
C	172	Oxadixyl	29.1	0.02	163(100)	233(18)	278(11)	
C	173	*p,p'*-DDE	27.3	0.20	235(100)	237(65)	246(7)	165(34)
C	174	*p,p'*-DDT	24	0.20	318(100)	316(80)	246(139)	248(70)
C	175	Parathion-Methyl	20.8	0.05	263(100)	233(66)	246(8)	200(6)
C	176	Pendimethalin	22.6	0.10	252(100)	220(22)	162(12)	
C	177	Permethrin-1	31.4	0.10	183(100)	184(14)	255(1)	
C	178	Permethrin-2	31.7	0.10	183(100)	184(14)	255(1)	
C	179	Pyraflufen Ethyl	28.6	0.05	412(100)	349(41)	339(34)	
C	180	Pyrazophos	31.7	0.10	221(100)	232(35)	373(19)	
C	181	Quintozene	16.8	0.05	295(100)	237(159)	249(114)	
C	182	Spiromesifen	29.4	0.02	272(100)	254(27)	370(14)	
C	183	Tecnazene	13.6	0.10	261(100)	203(135)	215(113)	
C	184	Tetradifon	30.7	0.05	227(100)	356(70)	159(196)	
C	185	Tolyfluanid	23.5	0.10	238(100)	240(71)	137(210)	
C	186	Trans-Chlordane	23.3	0.02	373(100)	375(96)	377(51)	

[a]Because of the large maximum residue limits (MRLs) for these pesticides, the amount of pesticide needed would be too high, so we reduced the concentration of the sample. The numbers in the brackets are the actual concentrations injected to the system, and the numbers outside of the brackets are the MRLs set by Japan or the EU.
[b]The MRLs of the pesticides listed are set by Japan, and the others without the given symbol are set by the EU.

APPENDIX TABLE 4.2 Parameters of GC–MS/MS for 205 Pesticides

Group	No.	Name	Rention time/min	MRL/(mg/kg)[a]	MRM Quantitative Ion Pair and Qualitative Ion Pair	Collision energy/V
ISTD		Heptachlor Epoxide	22.1	–	353.0/263.0; 353.0/282.0	17;17
A	1	Acetochlor	19.8	0.01	146.0/131.0; 146.0/118.0	10;10
A	2	Acibenzolar-S-Methyl	20.5	0.05	182.0/153.0; 182.0/107.0	25;25
A	3	Azinphos-Ethyl	32.3	0.05	132.0/77.0;132.0/104.0	15;5
A	4	Benalaxyl	27.8	0.10	148.0/79.0;148.0/105.0	25; 15
A	5	Bioresmethrin[b]	27.5	0.10	171.0/143.0; 171.0/128.0	14; 14
A	6	Bromophos-Ethyl	23.1	0.10	359.0/303.0; 359.0/331.0	10; 10
A	7	Bromuconazole	31.0	0.05	173.0/109.0; 173.0/145.0	25; 15
A	8	Buprofezin	25.0	0.05	105.0/77.0; 172.0/116.0	18; 7
A	9	Butralin	22.2	0.02	266.0/190.0; 266.0/174.0	12; 12
A	10	Carbaryl	14.5	0.10	144.0/115.0; 144.0/116.0	20; 20
A	11	carboxin	26.6	0.05	235.0/143.0; 235.0/87.0	15; 15
A	12	Carfentrazone-Ethyl	28.3	0.02	330.0/310.0; 330.0/241.0	18; 18
A	13	Chlorpyrifos	21.0	0.10	314.0/286.0; 314.0/258.0	5; 5
A	14	Clodinafop-Propargyl[b]	28.1	0.02	349.0/266.0; 349.0/238.0	10; 15
A	15	Clomazone	17.1	0.02	204.0/107.0; 204.0/78.0	25; 25
A	16	Cyproconazole	27.8	0.05	222.0/125.0; 222.0/82.0	15; 10
A	17	Cyprodinil	22.0	0.05	224.0/208.0; 224.0/222.0	15; 15
A	18	Diazinon	17.2	0.02	304.0/179.0; 304.0/162.0	8; 8

(Continued)

APPENDIX TABLE 4.2 Parameters of GC–MS/MS for 205 Pesticides (*cont.*)

Group	No.	Name	Rention time/min	MRL/ (mg/kg)[a]	MRM Quantitative Ion Pair and Qualitative Ion Pair	Collision energy/V
A	19	Dichlofluanid[b]	22.0	5(0.50)	224.0/123.0; 224.0/77.0	10; 25
A	20	dicofol	21.4	20.00	250.0/139.0; 250.0/215.0	15; 10
A	21	Dieldrin	22.2	0.02	263.0/193.0,262.0/191.0	30; 30
A	22	Dimethachlor	20.1	0.02	197.0/148.0; 197.0/120.0	10; 15
A	23	Dimethenamid	19.9	0.02	230.0/154.0; 230.0/111.0	8; 25
A	24	Diphenylamine	14.6	0.05	169.0/168.0; 169.0/167.0	15; 25
A	25	Disulfoton Sulfone	27.0	0.05	213.0/153.0; 213.0/125.0	5; 10
A	26	Ethofumesate	22.3	0.10	207.0/161.0; 207.0/137.0	5; 15
A	27	Etoxazole	29.5	0.05	141.0/113.0,141.0/63.1	15; 25
A	28	Fenazaquin	29.1	10(0.10)	145.0/117.0; 145.0/91.0	10; 25
A	29	Fenitrothion	22.0	0.50	277.0/260.0; 277.0/109.0	5; 15
A	30	Fenthion Sulfoxide	28.3	0.05	278.9/108.9,278.9/168.9	20; 15
A	31	Flufenoxuron	19.0	15(0.15)	307.0/126.0; 307.0/98.0	25; 25
A	32	Flurtamone	32.8	0.05	333.0/120.0; 199.0/157.0	10; 15
A	33	Flusilazole	26.7	0.05	233.0/165.0; 233.0/152.0	15; 15
A	34	Flutolanil	26.8	0.05	173.0/145.0; 173.0/95.0	10; 25
A	35	Flutriafol	25.8	0.05	219.0/123.0; 219.0/95.0	15; 25
A	36	Hexaflumuron[b]	16.7	15.00	176.0/148.0; 176.0/121.0	15; 25
A	37	Iprovalicarb-1	26.7	0.10	134.0/93.0; 134.0/91.0	14; 14
A	38	Iprovalicarb-2	26.7	0.10	134.0/93.0; 134.0/91.0	14; 14
A	39	malaoxon	19.9	0.50	173.0/99.0; 173.0/127.0	10; 5
A	40	Methabenzthiazuron	16.6	0.05	164.0/136.0; 164.0/108.0	10; 25
A	41	Methidathion	24.8	0.10	145.0/85.0; 145.0/58.0	10; 25
A	42	Mevinphos	11.3	0.02	127.0/109.0,192.0/127.0	10; 10
A	43	Molinate	12.0	0.10	126.0/55.0; 126.0/83.0	10; 5
A	44	Monolinuron	18.5	0.10	126.0/99.0; 214.0/61.0	10; 10
A	45	Myclobutanil[b]	27.8	20.00	179.0/125.0; 179.0/90.0	15; 25
A	46	Napropamide	25.1	0.05	271.0/72.0; 271.0/128.0	10; 5
A	47	Paclobutrazol	25.7	0.02	236.0/125.0; 236.0/167.0	15; 10
A	48	Phorate	15.5	0.10	121.0/65.0,260.0/75.2	5; 15
A	49	Phosphamidon -1	19.7	0.02	264.0/127.0; 264.0/193.0	15; 5
A	50	Pirimicarb	19.1	0.05	238.0/166.0; 238.0/96.0	10; 10
A	51	Pirimiphos-Methyl	21.7	0.05	333.0/168.0; 333.0/180.0	25; 15
A	52	Procymidone	24.7	0.10	283.0/96.0; 283.0/255.0	10; 10
A	53	Propachlor	15.0	0.05	176.0/77.0; 176.0/120.0	25; 10
A	54	Propargite	28.1	5.00	135.1/107.1,135.1/77.1	15; 25
A	55	Propham	11.5	0.10	179.0/93.0; 179.0/137.0	15; 10
A	56	Propiconazole-1	28.4	0.10	259.0/69.0; 259.0/173.0	10; 15
A	57	Propiconazole-2	28.6	0.10	259.0/69.0; 259.0/173.0	10; 15
A	58	Propisochlor	20.1	0.01	222.6/132.0; 222.6/147.1	25; 10
A	59	Propoxur-1	6.5	0.10	110.0/63.0; 110.0/64.0	25; 15
A	60	Propoxur-2	15.5	0.10	110.0/63.0; 110.0/64.0	25; 15
A	61	Pyriproxyfen	30.0	0.05	136.0/78.0; 136.0/96.0	25; 15

(Continued)

APPENDIX TABLE 4.2 Parameters of GC–MS/MS for 205 Pesticides (*cont.*)

Group	No.	Name	Rention time/min	MRL/ (mg/kg)[a]	MRM Quantitative Ion Pair and Qualitative Ion Pair	Collision energy/V
A	62	Tebufenpyrad	29.2	0.10	333.0/171.0; 333.0/276.0	15; 5
A	63	Terbuthylazine	18.5	0.05	214.0/71.0; 214.0/132.0	15; 10
A	64	Tolclofos-Methyl	19.9	0.10	267.0/252.0; 267.0/93.0	15; 25
A	65	Triadimenol-1	24.8	0.20	168.0/70.0; 128.0/100.0	10; 15
A	66	Triadimenol-2	25.0	0.20	168.0/70.0; 128.0/100.0	10; 15
A	67	Triallate	17.2	0.10	270.0/186.0; 270.0/228.0	15; 10
A	68	Triazophos	28.4	0.02	161.0/134.0; 161.0/106.0	5; 10
A	69	Trifloxystrobin	27.7	0.05	222.0/162.0; 222.0/190.0	5; 10
A	70	Trifluralin	15.5	0.10	306.0/264.0; 306.0/206.0	12; 15
A	71	Zoxamide	22.2	0.05	242.0/214.0; 242.0/187.0	10; 15
B	72	Alachlor	20.3	0.05	237.0/160.0; 237.0/146.0	8; 20
B	73	Boscalid	34.3	0.5(0.05)	342.0/140.0; 342.0/112.0	18; 18
B	74	bupirimate	26.3	0.05	273.0/108.0; 273.0/193.0	15; 15
B	75	dichlorvos	8.0	0.02	109.0/79.0,185.0/93.0	5; 10
B	76	Diethofencarb	21.8	0.05	225.0/96.0; 225.0/168.0	25; 10
B	77	dimethoate	19.8	0.05	125.0/79.0; 143.0/111.0	8; 12
B	78	Dimethomorph	37.6	0.05	301.0/165.0; 301.0/139.0	100; 100
B	79	Diniconazole	27.5	0.05	268.0/232.0; 268.0/136.0	10; 25
B	80	Disulfoton	17.9	0.05	88.0/60.0; 88.0/59.0	10; 25
B	81	Disulfoton-Sulfoxide	8.4	0.05	212.0/97.0; 212.0/174.0	15; 25
B	82	Epoxiconazole	29.7	0.05	192.0/138.0; 192.0/111.0	10; 25
B	83	Ethiofencarb[b]	11.0	0.05	168.0/107.0; 168.0/77.0	20; 20
B	84	Ethion	27.0	3(0.30)	384.0/129.0; 384.0/203.0	25; 5
B	85	Ethoprophos	14.5	0.02	158.0/97.0; 158.0/114.0	12; 7
B	86	Etofenprox	32.9	0.01	163.0/107.0; 163.0/135.0	15; 10
B	87	Fenamiphos	25.6	0.05	303.0/195.0; 303.0/288.0	10; 10
B	88	Fenamiphos Sulfone	31.9	0.05	320.0/292.0; 320.0/79.0	10; 25
B	89	Fenbuconazole	34.5	0.05	198.0/129.0; 198.0/102.0	15; 25
B	90	Fenobucarb[b]	14.8	0.50	120.7/77.1; 120.7/103.1	20; 15
B	91	Fenpropidin	17.7	0.05	98.0/70.0; 98.0/69.0	10; 15
B	92	Fenpropimorph	19.1	0.10	128.0/70.0; 128.0/110.0	15; 15
B	93	Fenpyroximate	17.5	0.10	213.0/77.0; 213.0/212.0	25; 10
B	94	Fenthion	21.7	0.05	278.0/109.0; 278.0/169.0	15; 15
B	95	Flufenacet	23.1	0.05	211.0/96.0; 211.0/123.0	12; 12
B	96	Flurochloridone	25.0	0.10	311.0/174.0; 311.0/311.0/103.0	10; 10
B	97	Fuberidazole	22.6	0.05	184.0/156.0; 184.0/129.0	10; 15
B	98	Hexythiazox	26.9	0.05	226.6/148.8; 155.6/112.1	10; 15
B	99	Imazalil	25.7	0.10	215.0/173.0; 215.0/145.0	5; 25
B	100	Kresoxim-Methyl	25.3	0.10	131.0/89.0; 131.0/130.0	25; 10
B	101	Malathion	21.9	0.50	173.0/99.0; 173.0/127.0	10; 5
B	102	mecarbam	24.0	0.10	296.0/196.0; 296.0/168.0	10; 25
B	103	Mepronil	28.4	0.10	119.0/91.0; 119.0/65.0	10; 25
B	104	Metalaxyl	20.9	0.10	206.0/132.0; 206.0/105.0	15; 15

(Continued)

APPENDIX TABLE 4.2 Parameters of GC–MS/MS for 205 Pesticides (*cont.*)

Group	No.	Name	Rention time/min	MRL/ (mg/kg)[a]	MRM Quantitative Ion Pair and Qualitative Ion Pair	Collision energy/V
B	105	Metamitron	29.2	0.10	202.0/174.0; 202.0/104.0	5; 25
B	106	Metazachlor	23.6	0.20	209.0/132.0; 209.0/133.0	15; 5
B	107	Methacrifos	12.1	0.10	208.0/180.0; 208.0/110.0	5; 15
B	108	Metribuzin	20.8	0.10	198.0/82.0; 198.0/110.0	15; 15
B	109	Oxyfluorfen	26.6	0.05	361.0/317.0; 361.0/300.0	5; 10
B	110	pencycuron	14.3	0.05	209.0/180.0; 209.0/125.0	5; 25
B	111	Phenothrin-1	29.4	0.05	123.0/81.0; 123.0/79.0	10; 12
B	112	Phenothrin-2	29.6	0.05	123.0/81.0; 123.0/79.0	10; 12
B	113	Phenthoate[b]	23.3	0.10	274.0/121.0,274.0/125.0	10; 20
B	114	Phosalone	31.5	0.10	182.0/111.0; 182.0/75.0	15; 25
B	115	Picoxystrobin	24.9	0.10	335.0/173.0; 335.0/303.0	10; 10
B	116	prochloraz	33.3	0.10	180.0/138.0; 308.0/70.0	15; 15
B	117	Profenofos	24.8	0.10	374.0/339.0; 374.0/337.0	5; 10
B	118	propanil	23.3	0.10	163.0/90.0; 163.0/99.0	25; 25
B	119	Propyzamide	19.0	0.05	173.0/145.0; 173.0/109.0	15; 25
B	120	Prosulfocarb	19.6	0.05	251.0/128.0; 251.0/86.0	5; 10
B	121	Pyraclofos[b]	32.1	5(0.50)	360.0/97.0; 360.0/194.0	25; 15
B	122	Pyridaben	32.1	0.05	147.0/117.0; 147.0/132.0	25; 15
B	123	Pyrifenox-1[b]	23.5	5(0.50)	262.0/200.0; 262.0/192.0	12; 12
B	124	Pyrifenox-2[b]	24.5	5(0.50)	262.0/200.0; 262.0/192.0	12; 12
B	125	Pyrimethanil	17.4	0.10	200.0/199.0; 183.0/102.0	10; 30
B	126	Pyrimidifen[b]	33.5	5(0.50)	184.0/169.0; 184.0/141.0	15; 25
B	127	Quinalphos	23.3	0.10	157.0/102.0; 157.0/129.0	25; 15
B	128	Quinoxyphen	27.1	0.05	273.0/208.0; 273.0/182.0	25; 25
B	129	Resmethrin-1	27.5	0.20	171.0/143.0; 171.0/128.0	5; 10
B	130	Resmethrin-2	27.5	0.20	171.0/143.0; 171.0/128.0	5; 10
B	131	Simeconazole[b]	21.4	10.00	121.0/101.0; 121.0/75.0	10; 25
B	132	Spirodiclofen	32.4	0.05	312.0/259.0; 312.0/294.0	10; 5
B	133	Tebuconazole	29.9	0.05	250.0/125.0; 250.0/153.0	15; 10
B	134	Tebuthiuron[b]	14.2	0.02	156.0/74.0; 156.0/89.0	15; 10
B	135	Terbufos	17.0	0.01	231.0/129.0; 231.0/175.0	25; 15
B	136	Tetraconazole	23.7	0.02	336.0/204.0; 336.0/156.0	25; 25
B	137	Thiobencarb	20.7	0.10	257.0/100.0; 257.0/72.0	5; 25
B	138	Triadimefon	22.5	0.20	210.0/183.0; 210.0/129.0	5; 10
B	139	Vinclozolin	20.6	0.10	285.0/212.0; 285.0/178.0	10; 10
C	140	2-Phenylphenol	12.6	0.10	169.0/141.0; 141.0/115.0	15; 10
C	141	Aldrin	19.7	0.02	263.0/193.0,263.0/191.0	30; 30
C	142	alpha-HCH	16.2	0.02	219.0/183.0; 219.0/147.0	5; 15
C	143	Benfluralin	15.6	0.05	292.0/264.0; 292.0/160.0	10; 15
C	144	*beta*-HCH	21.6	0.02	219.0/183.0; 219.0/147.0	10; 20
C	145	Bifenthrin	28.6	5(0.50)	181.0/165.0; 181.0/166.0	15; 25
C	146	Bromopropylate	29.5	0.05	341.0/183.0; 341.0/185.0	15; 15
C	147	Chlorbenside	23.0	0.10	270.0/125.0; 270.0/127.0	10; 10

(Continued)

APPENDIX TABLE 4.2 Parameters of GC–MS/MS for 205 Pesticides (*cont.*)

Group	No.	Name	Rention time/min	MRL/ (mg/kg)[a]	MRM Quantitative Ion Pair and Qualitative Ion Pair	Collision energy/V
C	148	Chlorfenapyr	27.6	50(0.05)	408.0/59.0; 408.0/363.0	14; 14
C	149	Chlorfenson	25.4	0.10	302.0/111.0; 302.0/175.0	25; 10
C	150	Chlorfenvinphos	23.5	0.05	323.0/267.0; 323.0/159.0	15; 25
C	151	Chlorobenzilate	26.2	0.10	251.0/139.0; 251.0/111.0	15; 25
C	152	Chlorpropham	15.7	0.10	213.0/171.0; 213.0/127.0	5; 15
C	153	Chlorpyrifos-Methyl	19.5	0.10	286.0/93.0; 286.0/271.0	15; 15
C	154	Chlorthal-Dimethyl	21.4	0.01	301.0/223.0; 301.0/273.0	25; 15
C	155	Cyfluthrin	33.4	0.10	206.0/151.0; 206.0/177.0	25; 25
C	156	cyhalofop-butyl	31.5	0.05	357.0/256.0; 357.0/229.0	18; 18
C	157	Cypermethrin	36.1	0.50	181.0/152.0; 181.0/87.0	25; 40
C	158	Deltamethrin	35.8	5.00	181.0/152.0,253.0/172.0	25; 10
C	159	Dichlobenil	9.8	0.05	171.0/136.0; 171.0/100.0	15; 15
C	160	Diclofop-Methyl	28.2	0.05	342.0/255.0; 342.0/184.0	18; 18
C	161	Dicloran	18.2	0.01	206.0/176.0; 206.0/124.0	15; 25
C	162	Difenoconazole	35.8	0.05	323.0/265.0; 323.0/202.0	15; 25
C	163	Diflufenican	28.8	0.05	266.0/218.0; 266.0/246.0	25; 10
C	164	Dioxathion	17.8	0.10	270.0/197.0; 270.0/141.0	5; 15
C	165	Endosulfan -1	23.0	30.00	241.0/206.0; 241.0/170.0	25; 25
C	166	Endosulfan-Sulfate	29.4	30(0.03)	387.0/289.0; 387.0/253.0	5; 5
C	167	Endrin	25.1	0.01	263.0/191.0; 263.0/193.0	20; 12
C	168	Ethalfluralin	15.2	0.02	316.0/202.0; 316.0/279.0	25; 10
C	169	Fenamidone	30.7	0.05	268.0/180.0; 268.0/77.0	15; 25
C	170	Fenchlorphos	19.8	0.10	287.0/272.0; 287.0/242.0	10; 10
C	171	Fenvalerate-1	34.6	0.05	167.0/125.0; 225.0/119.0	10; 15
C	172	Fenvalerate-2	35.0	0.05	167.0/125.0; 225.0/119.0	10; 15
C	173	Flucythrinate-1	32.3	0.10	199.0/107.0; 199.0/157.0	25; 5
C	174	Flucythrinate-2	32.5	0.10	199.0/107.0; 199.0/157.0	25; 5
C	175	Fludioxonil	29.6	0.05	248.0/154.0; 248.0/182.0	15; 8
C	176	Flumioxazine	36.1	0.10	354.0/176.0; 354.0/326.0	15; 10
C	177	Fluquinconazole	32.8	0.05	340.0/298.0; 340.0/286.0	15; 25
C	178	gamma-HCH	17.8	0.05	219.0/183.0; 219.0/147.0	5; 15
C	179	halfenprox[b]	32.6	10.00	262.6/234.9,262.6/115.0	15; 15
C	180	Heptachlor	18.4	0.02	272.0/237.0; 272.0/235.0	10; 10
C	181	hexachlorobenzene	14.3	0.02	284.0/249.0; 284.0/214.0	18; 25
C	182	Hexaconazole	25.1	0.05	213.9/159.0,213.9/172.0	20; 20
C	183	Lambda-Cyhalothrin	31.6	1.00	197.0/141.0; 197.0/91.0	10; 25
C	184	Lenacil	31.5	0.10	153.0/110.0; 153.0/136.0	15; 15
C	185	Methoxychlor	29.4	0.10	227.0/169.0; 227.0/212.0	15; 15
C	186	Monocrotophos	18.7	0.10	127.0/109.0,127.0/95.0	5; 15
C	187	Nitrofen	26.4	0.02	283.0/162.0; 283.0/253.0	25; 10
C	188	o,p′-DDT	26.7	0.20	235.0/165.0; 235.0/199.0	25; 25
C	189	Oxadiazon	25.3	0.05	258.0/175.0; 258.0/112.0	10; 25
C	190	Oxadixyl	29.7	0.02	163.0/132.0; 163.0/117.0	10; 25

(Continued)

APPENDIX TABLE 4.2 Parameters of GC–MS/MS for 205 Pesticides (*cont.*)

Group	No.	Name	Rention time/min	MRL/ (mg/kg)[a]	MRM Quantitative Ion Pair and Qualitative Ion Pair	Collision energy/V
C	191	p,p'-DDE	23.9	0.20	318.0/248.0; 318.0/246.0	25; 25
C	192	p,p'-DDT	26.7	0.20	235.0/165.0; 235.0/199.0	25; 25
C	193	Parathion-Methyl	21.2	0.05	263.0/109.0; 263.0/246.0	12; 5
C	194	Pendimethalin	22.7	0.10	252.0/162.0; 252.0/161.0	10; 25
C	195	Permethrin-1	31.7	0.10	183.0/168.0; 183.0/153.0	15; 15
C	196	Permethrin-2	32.0	0.10	183.0/168.0; 183.0/153.0	15; 15
C	197	prothiophos[b]	24.1	5.00	309.0/239.0; 309.0/221.0	15; 25
C	198	Pyraclostrobin	31.9	0.05	132.0/77.0; 132.0/104.0	15; 10
C	199	Pyraflufen Ethyl	28.9	0.05	412.0/349.0; 412.0/307.0	15; 25
C	200	Pyrazophos	31.8	0.10	221.0/193.0; 221.0/149.0	10; 15
C	201	Quintozene	16.6	0.05	295.0/237.0; 295.0/265.0	15; 10
C	202	Spiromesifen	29.7	0.02	272.0/254.0; 272.0/209.0	25; 25
C	203	Tecnazene	13.5	0.10	203.0/83.0; 203.0/143.0	10; 10
C	204	Tetradifon	30.9	0.05	356.0/159.0; 356.0/229.0	10; 10
C	205	Trans-Chlordane	23.3	0.02	375.0/266.0; 375.0/303.0	15; 10

[a]*Because of the large maximum residue limits (MRLs) for these pesticides, the amount of pesticide needed would be too high, so we reduced the concentration of the sample. The numbers in the brackets are the actual concentrations injected to the system, and the numbers outside of the brackets are the MRLs set by Japan or the EU.*
[b]*The MRLs of the pesticides listed are set by Japan, and the others without the given symbol are set by the EU.*

APPENDIX TABLE 4.3 Parameters of LC–MS/MS for 110 Pesticides

Group	No.	Pesticide	RT/min	MRM Transitions/(m/z)	Fragmentor/v	Collison Energy/v	MRL/ (mg/kg)[a]
A	1	Acetochlor	12.66	270.2/224; 270.2/148.2	80	5,20	0.01
A	2	Acibenzolar-S-Methyl	9.21	211.1/91; 211.1/136.0	120	20,30	0.05
A	3	Azinphos-Ethyl	13.30	346.0/233; 346.0/261.1	120	10,5	0.05
A	4	Benalyxyl	14.15	326.2/148.1; 326.2/294	120	1,5	0.1
A	5	Bioresmethrin[b]	18.66	339.2/171.1; 339.2/143.1	100	15,25	0.1
A	6	Bitertanol	12.78	338.2/70; 338.2/269.2	60	5,1	0.1
A	7	Bromuconazole	10.47	376.0/159.0; 376/70	80	20,20	0.05
A	8	Butralin	17.93	296.1/240.1; 296.1/222.1	100	10,20	0.02
A	9	Carbaryl	6.31	202.1/145.1; 202.1/127.1	80	10,5	0.1
A	10	Carboxin	6.57	236.1/143.1; 236.1/87	120	15,20	0.05
A	11	Carfentrazone-Ethyl	14.32	412/346; 412/366	140	25; 15	0.02
A	12	Chlorpyrifos	17.57	350/198; 350.0/79.0	100	20,35	0.1
A	13	Clodinafop-Propargyl[b]	15.16	350.1/266.1; 350.1/238.1	120	15,20	0.02
A	14	Clomazone	8.04	240.1/125.0; 240.1/89.1	100	20,50	0.02
A	15	Cyproconazole	9.31	292.1/70.0; 292.1/125	120	15,15	0.05
A	16	Cyprodinil	8.18	226.0/93.0; 226.0/ 108	120	40,30	0.05
A	17	Diazinon	14.98	305.0/169.1; 305.0/153.2	160	20,20	0.02
A	18	Dimethachloro	7.75	256.1/224.2; 256.1/148.2	120	10,20	0.02
A	19	Dimethenamid	9.74	276.1/244.1; 276.1/168.1	120	10,15	0.02
A	20	Diphenylamin	12.39	170.2/93.1; 170.2/152	120	30,30	0.05

(*Continued*)

APPENDIX TABLE 4.3 Parameters of LC–MS/MS for 110 Pesticides (*cont.*)

Group	No.	Pesticide	RT/min	MRM Transitions/(m/z)	Fragmentor/v	Collison Energy/v	MRL/ (mg/kg)[a]
A	21	Disulfoton Sulfone	8.57	307.0/97.0; 307/125	100	30,10	0.05
A	22	Etoxazole	17.89	360.2/141.1; 360.2/304	100	30,15	0.05
A	23	Fenarimol	10.70	331.0/268.1; 331.0/81	120	25,30	0.05
A	24	Fenazaquin	18.08	307.2/57.1; 307.2/161.2	120	20,15	10(0.10)
A	25	Fenitrothion	12.88	278.1/125.0; 278.1/246	140	15,15	0.5
A	26	Fenthion Sulfoxide	6.08	295.1/109; 295.1/280.0	140	35,20	0.05
A	27	Flufenoxuron	17.59	489.0/158.1; 489.0/141.1	80	10,15	15(0.15)
A	28	Flurtamone	10.00	334.1/247.1; 334.1/303	120	30,20	0.05
A	29	Flusilazole	12.42	316.1/247.1; 316.1/165.1	120	15,20	0.05
A	30	Flutolanil	12.99	324.2/262.1; 324.2/282.1	120	20,10	0.05
A	31	Flutriafol	6.45	302.1/70; 302.1/123.0	120	15,20	0.05
A	32	Hexaflumuron[b]	16.13	461/141.1; 461.0/158.1	120	35,35	15
A	33	Iprovalicarb	10.64	321.1/119.0; 321.1/203.2	100	25,5	0.1
A	34	Lactofen	17.55	479.1/344.0; 479.1/223	120	15,35	0.02
A	35	Methabenzthazurop	6.02	222.2/165.1; 222.2/149.9	100	15,35	0.05
A	36	Methidathion	9.23	303.0/145.1; 303.0/85	80	5,10	0.1
A	37	Mevinphos	3.62	225.0/127.0; 225.0/193	80	15,1	0.02
A	38	Monolinuron	6.66	215.1/126.0; 215.1/148.1	100	15,10	0.1
A	39	Myclobutanil	10.68	289.1/125; 289.1/70.0	120	20,15	0.05
A	40	Napropamide	11.76	272.2/171.1; 272.2/129.2	120	15,15	0.05
A	41	Paclobutrazol	8.78	294.2/70.0; 294.2/125	100	15,25	0.02
A	42	Phorate	15.75	261.0/75.0; 261/199	80	10,5	0.1
A	43	Phosphamidon	4.75	300.1/174.1; 300.1/127.0	120	10,20	0.02
A	44	Pirimicarb	3.47	239.2/72.0; 239.2/182.2	120	20,15	0.05
A	45	Pirimiphos-Methyl	14.83	306.2/164; 306.2/108.1	120	20,30	0.05
A	46	Propachlor	7.48	212.1/170.1; 212.1/94.1	100	10,30	0.05
A	47	Propham	7.56	180.1/138.0; 180.1/120	80	5,15	0.1
A	48	Propiconazole	13.28	342.1/159.1; 342.1/69	120	20,20	0.1
A	49	Propisochlor	14.41	284/224; 284/212	80	5,15	0.01
A	50	Propoxur	5.75	210.1/111; 210.1/168.1	80	10,5	0.1
A	51	Pyriproxyfen	17.37	322.1/96.0; 322.1/227.1	120	15,10	0.05
A	52	Tebufenpyrad	16.68	334.3/147; 334.3/117.1	160	25,40	0.1
A	53	Terbuthylazine	8.93	230.1/174.1; 230.1/132	120	15,20	0.05
A	54	Tolclofos-Methyl	15.74	301.2/269; 301.2/125.2	120	15,20	0.1
A	55	Triadimenol	8.59	296.1/70.0; 296.1/99.1	80	10,10	0.2
A	56	Trifloxystrobin	16.69	409.3/186.1; 409.3/206.2	120	15,10	0.05
A	57	Zoxamide	15.02	336/187; 336/156	80	20; 40	0.05
B	58	Alachlor	13.81	270.2/238.2; 270.2/162.2	80	10,20	0.05
B	59	Azinphos-Methyl	9.52	318.1/125; 318.1/160	80	15,10	0.1
B	60	Bupirimate	9.11	317.2/166; 317.2/272	120	25,20	0.05
B	61	Chlorpyrifos-Methyl	15.03	322/290; 322/125	80	15; 15	0.1
B	62	Diethofencarb	9.67	268.1/226.2; 268.1/152.1	80	5,20	0.05
B	63	Dimethoate	3.85	230.0/199.0; 230/171	80	5,10	0.05

(Continued)

APPENDIX TABLE 4.3 Parameters of LC–MS/MS for 110 Pesticide (*cont.*)

Group	No.	Pesticide	RT/min	MRM Transitions/(*m/z*)	Fragmentor/v	Collison Energy/v	MRL/ (mg/kg)[a]
B	64	Dimethomorph	8.90	388.1/165.1; 388.1/301.1	120	25,20	0.05
B	65	Diniconazole	13.07	326.1/70.0; 326.1/159	120	25,30	0.05
B	66	Disulfoton-Sulfoxide	6.40	291.0/185; 291/157	80	10,20	0.05
B	67	Epoxiconazole	11.29	330.1/141.1; 330.1/121.1	120	20,20	0.05
B	68	Ethiofencarb[b]	6.65	227/107; 227/164	80	5,5	0.05
B	69	Ethoprophos	11.04	243.1/173; 243.1/215.0	120	10,10	0.02
B	70	Fenamiphos	10.64	304.0/216.9; 304.0/202	100	20,35	0.05
B	71	Fenamiphos Sulfone	5.67	336.1/188.2; 336.1/266.2	120	30,20	0.05
B	72	Fenamiphos Sulfoxide	4.66	320.1/171.1; 320.1/292.1	140	25,15	0.05
B	73	Fenbuconazole	12.49	337.1/70; 337.1/125.0	120	20,20	0.05
B	74	Fenobucarb[b]	8.94	208.2/95.0; 208.2/152.1	80	10,5	0.5
B	75	Fenpropimorph	7.35	304.0/147.2; 304.0/130	120	30,30	0.1
B	76	Fenpyroximate	17.95	422.2/366.2; 422.2/135	120	10,35	0.1
B	77	Fenthion	14.73	279.0/169.1; 279/247	120	15,10	0.05
B	78	Flufenacet	13.16	364.0/194.0; 364/152	80	5,10	0.05
B	79	Flurochloridone	13.04	312.2/292.2; 312.2/53.1	140	25,30	0.1
B	80	Hexythiazox	17.59	353.1/168.1; 353.1/228.1	120	20,10	0.05
B	81	Imazalil	5.44	297.0/159.0; 297/255	120	20,20	0.1
B	82	Kresoxim-Methyl	14.36	314.1/267; 314.1/206	80	5,5	0.1
B	83	Linuron	9.24	249.0/160.1; 249/182.1	100	15,15	0.1
B	84	Mecarbam	13.82	330/227; 330.0/199.0	80	5,10	0.1
B	85	Mepronil	12.31	270.2/119.1; 270.2/228.2	100	30,15	0.1
B	86	Metalaxyl	6.79	280.1/192.2; 280.1/220.2	120	15,20	0.1
B	87	Metamitron	3.51	203.1/175.1; 203.1/104	120	15,20	0.1
B	88	Metazachlor	7.59	278.1/134.1; 278.1/210.1	80	20,5	0.2
B	89	Metribuzin	5.36	215.1/187.2; 215.1/131.1	120	15,20	0.1
B	90	Pencycuron	15.80	329.2/125.0; 329.2/218.1	120	20,15	0.05
B	91	Phenthoate[b]	15.03	321.1/247; 321.1/163.1	80	5,10	0.1
B	92	Phosalone	16.06	368.1/182.0; 368.1/322	80	10,5	0.1
B	93	Picoxystrobin	14.76	368.1/145.0; 368.1/205.0	80	20,5	0.1
B	94	Prochloraz	10.67	376.1/308.0; 376.1/266	80	10,10	0.1
B	95	Profenefos	16.21	373.0/302.9; 373/345	120	15,10	0.1
B	96	Propanil	8.13	218.0/162.1; 218/127	120	15,20	0.1
B	97	Propyzamide	11.15	256/190; 256/173	80	10; 15	0.05
B	98	Prosulfocarb	16.57	252.1/91.0; 252.1/128.1	120	15,10	0.05
B	99	Pyraclofos[b]	14.76	361.1/257.0; 361.1/138	120	25,35	5(0.50)
B	100	Pyrifenox[b]	6.52	295.0/93.1; 295.0/163.0	120	15,15	5(0.50)
B	101	Pyrimethanil	5.95	200.2/107.0; 200.2/183.1	120	25,25	0.1
B	102	Quinaphos	14.08	299.1/147.1; 299.1/163.1	120	20,20	0.1
B	103	Quinoxyphen	16.44	308.0/197.0; 308.0/272.0	180	35,35	0.05
B	104	Resmethrin	18.66	339.2/171.1; 339.2/143.1	80	10,25	0.2
B	105	Simeconazole[b]	10.40	294.2/70.1; 294.2/135.1	120	15,15	10

(Continued)

APPENDIX TABLE 4.3 Parameters of LC–MS/MS for 110 Pesticides (*cont.*)

Group	No.	Pesticide	RT/min	MRM Transitions/(*m/z*)	Fragmentor/v	Collison Energy/v	MRL/(mg/kg)[a]
B	106	Tebuconazole	11.82	308.2/70.0; 308.2/125	100	25,25	0.05
B	107	Tebuthiuron[b]	4.61	229.2/172.2; 229.2/116	120	15,20	0.02
B	108	Tetraconazole	11.93	372.0/159.0; 372/70	120	35,35	0.02
B	109	Thiobencarb	15.27	258.1/125.0; 258.1/89	80	20,55	0.1
B	110	Triadimefon	11.30	294.2/69; 294.2/197.1	100	20,15	0.2

[a]*Because of the large maximum residue limits (MRLs) for these pesticides, the amount of pesticide needed would be too high, so we reduced the concentration of the sample. The numbers in the brackets are the actual concentrations injected to the system, and the numbers outside of the brackets are the MRLs set by Japan or the EU.*
[b]*The MRLs of the pesticides listed are set by Japan, and the others without the given symbol are set by the EU.*

APPENDIX TABLE 4.4 The Linear Range, Linear Equation, Correlation Coefficient and Limit Detect Limit of 186 Pesticides

Pesticides	Linear Equation	Correlation Coefficient/R^2	Linear Range	LOD/(μg/kg)	LOQ/(μg/kg)
Acetochlor	$y = 0.243477x - 0.00287968$	0.999607	1.00–100.00	1.67	5.00
Acibenzolar-S-Methyl	$y = 0.324629x - 0.0219021$	0.998973	5.00–500.00	8.33	25.00
Aclonifen	$y = 0.117638x - 0.022126$	0.997293	5.00–500.00	8.33	25.00
Azinphos-Ethyl	$y = 0.625343x - 0.0302604$	0.999322	5.00–500.00	8.33	25.00
Benalaxyl	$y = 3.12038x + 0.0595937$	0.999755	10.00–1000.00	3.33	10.00
Bifenazate	$y = 0.20663x + 0.00126451$	0.999631	2.00–200.00	3.33	10.00
Bioresmethrin[b]	$y = 1.85888x - 0.0362327$	0.999754	10.00–1000.00	6.67	20.00
Bitertanol	$y = 3.08548x - 0.089465$	0.999697	10.00–1000.00	16.67	50.00
Bromophos-Ethyl	$y = 0.62053x - 0.0325951$	0.999479	10.00–1000.00	3.33	10.00
Bromuconazole	$y = 0.210974x + 0.0141142$	0.999135	5.00–500.00	8.33	25.00
Buprofezin	$y = 0.60058x - 0.0058473$	0.999417	5.00–500.00	8.33	25.00
Butralin	$y = 0.106153x + 0.0021868$	0.995396	2.00–200.00	3.33	10.00
Carbaryl	$y = 1.36119x - 0.0608191$	0.999317	10.00–1000.00	16.67	50.00
Carfentrazone-Ethyl	$y = 0.089555x - 0.00125853$	0.999843	2.00–200.00	3.33	10.00
Chlorbufam	$y = 0.234991x + 0.00252651$	0.999508	10.00–1000.00	16.67	50.00
Chlorpyrifos	$y = 0.392483x - 0.0101337$	0.999788	10.00–1000.00	3.33	10.00
Clodinafop-Propargyl	$y = 0.0801005x - 0.00711062$	0.997871	2.00–200.00	3.33	10.00
Clomazone	$y = 0.357311x + 0.0068728$	0.999056	2.00–200.00	3.33	10.00
Cyproconazole	$y = 2.0524x + 0.0906933$	0.998767	5.00–500.00	8.33	25.00
Cyprodinil	$y = 1.74919x - 0.0481225$	0.999715	5.00–500.00	8.33	25.00
Diazinon	$y = 0.0843778x + 0.00353355$	0.99907	2.00–200.00	3.33	10.00
Dichlofluanid	$y = 0.411668x - 0.0113484$	0.998947	50.00–5000.00	16.67	50.00
Dieldrin	$y = 0.0370636x + 0.00314647$	0.999713	2.00–200.00	3.33	10.00
Dimethachlor	$y = 0.412155x - 0.00072803$	0.999948	2.00–200.00	3.33	10.00
Dimethenamid	$y = 0.39018x + 0.00974809$	0.999793	2.00–200.00	3.33	10.00
Diphenylamine	$y = 1.63976x - 0.0297536$	0.999489	5.00–500.00	8.33	25.00
DisulfotonSulfone	$y = 0.293521x - 0.0131231$	0.999311	5.00–500.00	8.33	25.00
Ethofumesate	$y = 1.38756x + 0.0278572$	0.999865	10.00–1000.00	6.67	20.00

(*Continued*)

APPENDIX TABLE 4.4 The Linear Range, Linear Equation, Correlation Coefficient and Limit Detect Limit of 186 Pesticides (*cont.*)

Pesticides	Linear Equation	Correlation Coefficient/R^2	Linear Range	LOD/(μg/kg)	LOQ/(μg/kg)
Etoxazole	$y = 0.597587x-0.0136917$	0.999652	5.00–500.00	8.33	25.00
Fenarimol	$y = 0.302443x-0.000720652$	0.999554	5.00–500.00	8.33	25.00
Fenazaquin	$y = 3.16394x-0.110145$	0.999562	10.00–1000.00	6.67	20.00
Fenitrothion	$y = 1.62443x-0.289081$	0.995463	50.00–5000.00	16.67	50.00
FenthionSulfoxide	$y = 0.0351038x + 0.00856193$	0.998535	5.00–500.00	8.33	25.00
Flufenoxuron	$y = 0.178618x-0.0022442$	0.998787	15.00–600.00	10.00	30.00
Flurtamone	$y = 0.602918x-0.0161275$	0.999696	5.00–500.00	8.33	25.00
Flusilazole	$y = 1.33607x-0.0252591$	0.999659	10.00–500.00	8.33	25.00
Flutolanil	$y = 2.21346x + 0.069371$	0.999256	5.00–500.00	8.33	25.00
Flutriafol	$y = 0.41959x-0.00896769$	0.999527	5.00–500.00	8.33	25.00
Iprovalicarb-1	$y = 1.30184x + 6.81777$	0.962502	10.00–1000.00	16.67	50.00
Iprovalicarb-2	$y = 2.83603x-0.0259138$	0.999608	10.00–1000.00	16.67	50.00
Methabenzthiazuron	$y = 3.62275x-0.0743011$	0.999768	5.00–500.00	8.33	25.00
Methidathion	$y = 3.21933x-0.23402$	0.999231	10.00–1000.00	16.67	50.00
Mevinphos	$y = 0.0822241x-0.00673095$	0.996171	4.00–200.00	3.33	10.00
Molinate	$y = 1.29782x-0.14882$	0.994074	10.00–1000.00	3.33	10.00
Napropamide	$y = 0.258938x-0.0138842$	0.998647	5.00–500.00	8.33	25.00
Paclobutrazol	$y = 0.275553x-0.00367183$	0.999424	2.00–200.00	3.33	10.00
Phorate	$y = 0.337764x-0.0244917$	0.998833	10.00–1000.00	6.67	20.00
Pirimicarb	$y = 1.56531x-0.00578965$	0.9999	5.00–500.00	8.33	25.00
Pirimiphos-Methyl	$y = 0.559172x-0.0136974$	0.999883	5.00–500.00	8.33	25.00
Procymidone	$y = 0.794793x + 9.45643e-005$	0.99988	10.00–1000.00	6.67	20.00
Propachlor	$y = 0.583892x + 0.0115542$	0.999614	5.00–500.00	8.33	25.00
Propargite	$y = 1.01671x + 0.0954889$	0.997256	50.00–5000.00	16.67	50.00
Propham	$y = 0.500908x-0.0410306$	0.997354	10.00–1000.00	6.67	20.00
Propiconazole-1	$y = 1.73438x-0.029936$	0.999166	10.00–1000.00	6.67	20.00
Propiconazole-2	$y = 1.75397x + 0.0126585$	0.999829	10.00–1000.00	6.67	20.00
Propisochlor	$y = 0.395281x + 0.0126905$	0.999815	1.00–100.00	1.67	5.00
Propoxur-1	$y = 1.91429x + 0.173937$	0.997108	10.00–1000.00	6.67	20.00
Propoxur-2	$y = 1.1872x-0.0572777$	0.999503	10.00–1000.00	6.67	20.00
Pyriproxyfen	$y = 6.15713x-0.198117$	0.999611	5.00–500.00	8.33	25.00
Tebufenpyrad	$y = 0.902878x-0.0247762$	0.999555	10.00–1000.00	6.67	20.00
Terbuthylazine	$y = 2.87534x + 0.0689208$	0.999538	5.00–500.00	8.33	25.00
Tolclofos-Methyl	$y = 2.44057x + 0.00337552$	0.999873	10.00–1000.00	6.67	20.00
Triadimenol-1	$y = 11.3162x-0.473727$	0.999615	20.00–2000.00	10.00	30.00
Triadimenol-2	$y = 2.38613x-0.0828164$	0.998992	20.00–2000.00	6.67	20.00
Triallate	$y = 0.612077x + 0.00266439$	0.999727	10.00–1000.00	6.67	20.00
Triazophos	$y = 0.0959229x + 0.00518835$	0.998676	2.00–200.00	3.33	10.00
Trifloxystrobin	$y = 0.556335x + 0.215369$	0.998179	10.00–500.00	8.33	25.00
Trifluralin	$y = 0.627033x-0.0809194$	0.996451	10.00–1000.00	16.67	50.00
Zoxamide	$y = 0.272353x-0.0410538$	0.993323	5.00–500.00	8.33	25.00

(Continued)

APPENDIX TABLE 4.4 The Linear Range, Linear Equation, Correlation Coefficient and Limit Detect Limit of 186 Pesticides (cont.)

Pesticides	Linear Equation	Correlation Coefficient/R^2	Linear Range	LOD/($\mu g/kg$)	LOQ/($\mu g/kg$)
Alachlor	$y = 0.325195x - 0.00417005$	0.999916	5.00–500.00	8.33	25.00
Boscalid	$y = 0.166137x + 0.00414857$	0.999592	5.00–500.00	8.33	25.00
Butylate	$y = 0.0411048x + 0.00352921$	0.797508	5.00–500.00	8.33	25.00
Diethofencarb	$y = 0.32232x - 0.00577103$	0.999903	5.00–500.00	8.33	25.00
Dimethomorph	$y = 0.178338x + 0.00145976$	0.99954	5.00–500.00	8.33	25.00
Diniconazole	$y = 0.715077x + 0.0110414$	0.999571	5.00–500.00	8.33	25.00
Disulfoton	$y = 0.773548x - 0.00815407$	0.999862	5.00–500.00	8.33	25.00
Disulfoton-Sulfoxide	$y = 0.134691x + 0.00169302$	0.999942	5.00–500.00	8.33	25.00
Epoxiconazole	$y = 0.323794x + 0.0038501$	0.999626	5.00–500.00	8.33	25.00
Ethiofencarb[b]	$y = 3.41569x - 0.104838$	0.999695	5.00–500.00	8.33	25.00
Ethion	$y = 4.40242x - 0.301633$	0.999611	30.00–3000.00	10.00	30.00
Ethoprophos	$y = 0.144821x + 0.0029424$	0.999356	4.00–200.00	3.33	10.00
Etofenprox	$y = 0.27019x + 0.00808777$	0.999314	1.00–100.00	1.67	5.00
Fenamiphos	$y = 1.43666x - 0.162416$	0.997248	5.00–500.00	8.33	25.00
FenamiphosSulfone	$y = 0.195033x + 0.000699509$	0.999884	5.00–500.00	8.33	25.00
FenamiphosSulfoxide	$y = 0.0454506x - 0.00241671$	0.999714	5.00–500.00	8.33	25.00
Fenbuconazole	$y = 0.616689x + 0.00604147$	0.999726	5.00–500.00	8.33	25.00
Fenobucarb	$y = 13.158x + 0.146441$	0.999747	50.00–5000.00	16.67	50.00
Fenpropathrin	$y = 0.0500269x + 0.00300387$	0.997761	2.00–200.00	3.33	10.00
Fenpropimorph	$y = 6.48781x - 0.232584$	0.999821	10.00–1000.00	6.67	20.00
Fenpyroximate	$y = 0.302328x + 0.0069848$	0.999624	10.00–1000.00	6.67	20.00
Fenthion	$y = 1.61502x - 0.0391857$	0.999913	5.00–500.00	8.33	25.00
Flurochloridone	$y = 0.388428x - 0.015557$	0.999799	10.00–1000.00	6.67	20.00
Fuberidazole	$y = 0.693947x - 0.0796524$	0.998684	5.00–500.00	8.33	25.00
Hexythiazox	$y = 0.0896625x + 0.00124884$	0.999943	5.00–500.00	8.33	25.00
Imazalil	$y = 0.409616x - 0.0451656$	0.997103	10.00–1000.00	16.67	50.00
Kresoxim-Methyl	$y = 4.35062x - 0.159947$	0.999798	10.00–1000.00	6.67	20.00
Linuron	$y = 0.0182895x - 0.00590316$	0.999084	10.00–1000.00	6.67	20.00
Malathion	$y = 0.1446x - 0.00607535$	0.999918	50.00–5000.00	16.67	50.00
Mepronil	$y = 4.4617x - 0.0738565$	0.999892	10.00–1000.00	16.67	50.00
Metalaxyl	$y = 0.7791x - 0.0120014$	0.999741	10.00–1000.00	16.67	50.00
Metamitron	$y = 0.171121x - 0.00853191$	0.999237	10.00–1000.00	16.67	50.00
Metazachlor	$y = 1.14791x - 0.0350553$	0.999829	20.00–2000.00	10.00	30.00
Methacrifos	$y = 2.17719x - 0.107395$	0.992539	10.00–1000.00	16.67	50.00
Methoprene	$y = -0.0303012x + 0.766626$	0.942978	5.00–500.00	8.33	25.00
Metribuzin	$y = 4.02025x - 0.273466$	0.999467	10.00–1000.00	10.00	30.00
Oxyfluorfen	$y = 0.162725x - 0.0206276$	0.99639	5.00–500.00	8.33	25.00
Phenothrin-1	$y = 2.85686x - 0.111794$	0.999724	5.00–500.00	8.33	25.00
Phenothrin-2	$y = 0.687938x + 0.0102534$	0.999862	5.00–500.00	8.33	25.00
Phenthoate	$y = 0.486922x - 0.0358188$	0.999529	10.00–1000.00	6.67	20.00
Phosalone	$y = 0.38093x - 0.00514638$	0.999836	10.00–1000.00	6.67	20.00

(Continued)

APPENDIX TABLE 4.4 The Linear Range, Linear Equation, Correlation Coefficient and Limit Detect Limit of 186 Pesticides (*cont.*)

Pesticides	Linear Equation	Correlation Coefficient/R^2	Linear Range	LOD/(μg/kg)	LOQ/(μg/kg)
Picoxystrobin	$y = 0.669272x - 0.0325167$	0.99955	10.00–1000.00	6.67	20.00
Profenofos	$y = 0.118584x - 0.00275472$	0.999402	10.00–1000.00	16.67	50.00
Propyzamide	$y = 0.851874x - 0.0232845$	0.999837	5.00–500.00	8.33	25.00
Prosulfocarb	$y = 0.324623x - 0.00485576$	0.999922	5.00–500.00	8.33	25.00
Pyraclofos	$y = 0.69966x - 0.0262114$	0.999839	50.00–5000.00	33.33	100.00
Pyridaben	$y = 1.40949x - 0.0348727$	0.999683	5.00–500.00	8.33	25.00
Pyrifenox-1	$y = 2.4516x - 0.272185$	0.999001	50.00–5000.00	33.33	100.00
Pyrifenox-2	$y = 1.08383x - 0.126907$	0.998265	50.00–5000.00	33.33	100.00
Pyrimethanil	$y = 18.0053x + 0.350283$	0.999788	10.00–1000.00	6.67	20.00
Pyrimidifen	$y = 0.100443x + 0.00254231$	0.999833	50.00–5000.00	16.67	50.00
Quinalphos	$y = 1.21369x - 0.10384$	0.999139	10.00–1000.00	6.67	20.00
Quinoxyphen	$y = 0.770074x - 0.0257063$	0.999784	5.00–500.00	8.33	25.00
Tebuconazole	$y = 0.365077x - 0.00492377$	0.99971	5.00–500.00	8.33	25.00
Tebuthiuron	$y = 0.255677x + 0.000818772$	0.998873	2.00–200.00	3.33	10.00
Terbufos	$y = 0.10157x + 0.000184582$	0.999382	1.00–100.00	1.67	5.00
Tetraconazole	$y = 0.352888x - 0.0114929$	0.999816	2.00–200.00	3.33	10.00
Thiobencarb	$y = 3.04487x - 0.0879727$	0.999902	10.00–1000.00	6.67	20.00
Triadimefon	$y = 1.61124x - 0.0620809$	0.999831	20.00–2000.00	6.67	20.00
Vinclozolin	$y = 0.358327x - 0.00774347$	0.999903	10.00–1000.00	6.67	20.00
2-Phenylphenol	$y = 2.25229x - 0.0745689$	0.999754	10.00–1000.00	16.67	50.00
Aldrin	$y = 0.0807499x + 0.00425193$	0.998824	2.00–200.00	3.33	10.00
alpha-HCH	$y = 0.114717x - 0.00152848$	0.999936	2.00–200.00	3.33	10.00
Benfluralin	$y = 2.24657x - 0.283652$	0.995125	5.00–500.00	8.33	25.00
Bifenox	$y = 0.0834065x - 0.0107562$	0.998343	5.00–500.00	8.33	25.00
Bifenthrin	$y = 33.2025x - 0.517676$	0.999605	10.00–1000.00	3.33	10.00
Bromopropylate	$y = 1.93158x - 0.111489$	0.999278	5.00–500.00	8.33	25.00
Chlorbenside	$y = 0.478x - 0.0427448$	0.998525	10.00–1000.00	6.67	20.00
Chlorfenapyr	$y = 0.0717662x + 0.00659795$	0.999742	5.00–500.00	8.33	25.00
Chlorfenson	$y = 2.02872x - 0.0353055$	0.99988	10.00–1000.00	6.67	20.00
Chlorfenvinphos	$y = 0.135335x - 0.0093328$	0.999368	5.00–500.00	8.33	25.00
Chlorobenzilate	$y = 2.39154x - 0.0281453$	0.999892	10.00–1000.00	6.67	20.00
Chlorpropham	$y = 0.506693x + 0.00311291$	0.999894	10.00–1000.00	6.67	20.00
Chlorpyrifos-Methyl	$y = 0.918049x - 0.0226158$	0.999537	10.00–1000.00	3.33	10.00
Chlorthal-Dimethyl	$y = 0.184471x + 0.00344368$	0.99993	1.00–100.00	1.67	5.00
Cyfluthrin	$y = 0.115184x + 0.00196516$	0.999951	10.00–1000.00	16.67	50.00
Cypermethrin	$y = 0.878216x - 0.0528531$	0.999366	50.00–5000.00	16.67	50.00
Dichlobenil	$y = 0.747991x - 0.0391099$	0.994266	10.00–500.00	8.33	25.00
Diclofop-Methyl	$y = 0.408981x + 0.017467$	0.999624	5.00–500.00	8.33	25.00
Dicloran	$y = 0.0245974x - 0.00220262$	0.999342	2.00–100.00	1.67	5.00
Difenoconazole	$y = 0.301656x - 0.0189551$	0.999689	10.00–500.00	8.33	25.00
Diflufenican	$y = 1.99433x - 0.0715995$	0.999849	5.00–500.00	8.33	25.00

(Continued)

APPENDIX TABLE 4.4 The Linear Range, Linear Equation, Correlation Coefficient and Limit Detect Limit of 186 Pesticides (*cont.*)

Pesticides	Linear Equation	Correlation Coefficient/R^2	Linear Range	LOD/(μg/kg)	LOQ/(μg/kg)
Dioxathion	$y = 1.36144x - 0.0146534$	0.999919	10.00–1000.00	16.67	50.00
Endosulfan-Sulfate	$y = 0.0323688x + 0.00282669$	0.998396	6.00–300.00	5.00	15.00
Endrin	$y = 0.0106593x - 0.00103649$	0.996651	2.00–100.00	1.67	5.00
Ethalfluralin	$y = 0.113033x - 0.00346621$	0.997577	4.00–200.00	3.33	10.00
Fenamidone	$y = 0.499974x - 0.0237113$	0.999277	5.00–500.00	8.33	25.00
Fenchlorphos	$y = 1.30156x - 0.0262318$	0.999814	10.00–1000.00	16.67	50.00
Fenvalerate-1	$y = 0.145068x - 0.00404652$	0.999503	10.00–500.00	8.33	25.00
Fenvalerate-2	$y = 0.0695794x + 0.00504717$	0.999525	10.00–500.00	8.33	25.00
Flucythrinate-1	$y = 1.99581x - 0.127283$	0.998804	10.00–1000.00	6.67	20.00
Flucythrinate-2	$y = 2.18207x - 0.16117$	0.998335	10.00–1000.00	6.67	20.00
Fludioxonil	$y = 0.687749x - 0.0166624$	0.999813	5.00–500.00	8.33	25.00
Flumioxazine	$y = 0.293395x - 0.0266918$	0.998507	10.00–1000.00	6.67	20.00
Fluquinconazole	$y = 0.755807x - 0.024921$	0.999503	5.00–500.00	8.33	25.00
gamma-HCH	$y = 0.17934x - 0.0112942$	0.999176	5.00–500.00	8.33	25.00
Heptachlor	$y = 0.0143465x - 0.0005402$	0.999895	4.00–200.00	3.33	10.00
Hexaconazole	$y = 0.40651x - 0.0186033$	0.999635	5.00–500.00	8.33	25.00
Lambda-Cyhalothrin	$y = 14.8432x - 1.17241$	0.999462	10.00–1000.00	3.33	10.00
MethiocarbSulfone	$y = 0.194317x - 0.0121428$	0.999624	10.00–1000.00	6.67	20.00
Methoxychlor	$y = 0.309124x - 0.0537547$	0.998061	20.00–400.00	6.67	20.00
Oxadiazon	$y = 7.35157x - 0.100172$	0.999721	5.00–500.00	8.33	25.00
Oxadixyl	$y = 0.127866x - 0.0152541$	0.999994	2.00–200.00	3.33	10.00
p,p'-DDE	$y = 2.79471x - 0.0703144$	0.999857	20.00–2000.00	6.67	20.00
p,p'-DDT	$y = 0.24091x - 0.039343$	0.978365	20.00–2000.00	6.67	20.00
Parathion-Methyl	$y = 0.0746467x - 0.0117323$	0.994519	10.00–500.00	8.33	25.00
Pendimethalin	$y = 0.563779x - 0.0463309$	0.995097	10.00–400.00	16.67	50.00
Permethrin-1	$y = 3.60799x - 0.0827044$	0.999946	10.00–1000.00	6.67	20.00
Permethrin-2	$y = 7.27294x - 0.202223$	0.999899	10.00–1000.00	6.67	20.00
PyraflufenEthyl	$y = 0.388462x - 0.0127018$	0.999618	5.00–500.00	8.33	25.00
Pyrazophos	$y = 0.961908x - 0.0371656$	0.999511	10.00–1000.00	6.67	20.00
Quintozene	$y = 0.0742249x - 0.00512027$	0.999021	5.00–500.00	8.33	25.00
Spiromesifen	$y = 0.494727x - 0.00669344$	0.999683	2.00–200.00	3.33	10.00
Tecnazene	$y = 0.240024x - 0.0189064$	0.999252	10.00–1000.00	6.67	20.00
Tetradifon	$y = 0.211233x - 0.00021588$	0.999969	10.00–500.00	8.33	25.00
Tolyfluanid	$y = 0.0767107x - 0.00209674$	0.998294	10.00–400.00	6.67	20.00
Trans-Chlordane	$y = 0.16351x - 0.00437265$	0.999587	2.00–200.00	3.33	10.00

APPENDIX TABLE 4.5 The Linear Range, Linear Equation, Correlation Coefficient and Limit Detect Limit of 205 Pesticides

Pesticides	Linear Equation	Correlation Coefficient/R^2	Linear Range	LOD/(μg/kg)	LOQ/(μg/kg)
Acetochlor	$y = 0.142706x - 0.980656$	0.9952	1.00–100.00	1.67	5.00
Acibenzolar-S-Methyl	$y = 0.005553x - 0.023614$	0.9973	5.00–500.00	8.33	25.00
Azinphos-Ethyl	$y = 0.013708x - 0.034916$	0.9967	5.00–500.00	8.33	25.00
Benalaxyl	$y = 0.113096x + 0.045981$	0.9993	10.00–1000.00	3.33	10.00
Bioresmethrin	$y = 0.144904x - 0.179271$	0.9979	10.00–1000.00	6.67	20.00
Bromophos-Ethyl	$y = 0.033699x - 0.215104$	0.9955	10.00–1000.00	3.33	10.00
Bromuconazole	$y = 0.015596x - 0.013204$	0.9982	5.00–500.00	8.33	25.00
Buprofezin	$y = 0.030766x + 1.296196$	0.9976	5.00–500.00	8.33	25.00
Butralin	$y = 0.002096x - 0.013148$	0.9960	2.00–200.00	3.33	10.00
Cabaryl	$y = 0.080358x - 0.375158$	0.9976	10.00–1000.00	16.67	50.00
Carboxin	$y = 0.033345x + 0.062243$	0.9977	5.00–500.00	8.33	25.00
Carfentrazone-Ethyl	$y = 0.002098x + 0.002627$	0.9979	2.00–200.00	3.33	10.00
Chlorpyrifos(-Ethyl)	$y = 0.015922x - 0.085601$	0.9970	10.00–1000.00	3.33	10.00
Clodinafop-Propargyl	$y = 0.002092x - 9.943911E-004$	0.9964	2.00–200.00	3.33	10.00
Clomazone	$y = 0.024922x - 0.011653$	0.9999	2.00–200.00	3.33	10.00
Cyproconazole	$y = 0.061153x - 0.009838$	0.9991	5.00–500.00	8.33	25.00
Cyprodinil	$y = 0.029465x - 0.140893$	0.9982	5.00–500.00	8.33	25.00
Diazinon	$y = 0.009485x - 0.015004$	0.9999	2.00–200.00	3.33	10.00
Dichloflunid	$y = 0.144573x - 0.630657$	0.9999	50.00–5000.00	16.67	50.00
Dicofol	$y = 0.015856x - 0.121882$	0.9989	20.00–2000.00	6.67	20.00
Dieldrin	$y = 0.006088x - 0.045820$	0.9960	2.00–200.00	3.33	10.00
Dimethachlor	$y = 0.020140x - 0.074584$	0.9977	2.00–200.00	3.33	10.00
Dimethenamid	$y = 0.030310x - 0.120272$	0.9990	2.00–200.00	3.33	10.00
Diphenylamine	$y = 0.254851x - 1.949361$	0.9989	5.00–500.00	8.33	25.00
Disulfotonsulfone	$y = 0.011846x + 0.006298$	0.9965	5.00–500.00	8.33	25.00
Ethofumesate	$y = 0.069057x - 0.408414$	0.9981	10.00–1000.00	6.67	20.00
Etoxazole	$y = 0.077423x - 0.013124$	0.9990	5.00–500.00	8.33	25.00
Fenazaquin	$y = 0.264733x - 0.626998$	0.9978	10.00–1000.00	6.67	20.00
Fenitrothion	$y = 0.084314x - 0.862127$	0.9948	50.00–200.00	16.67	50.00
Fenthionsulfoxide	$y = 0.002112x + 0.004117$	0.9946	5.00–500.00	8.33	25.00
Flufenoxuron	$y = 0.010152x - 0.055663$	0.9952	15.00–1500.00	10.00	30.00
Flurtamone	$y = 0.004167x - 0.008647$	0.9972	5.00–500.00	8.33	25.00
Flusilazole	$y = 0.016251x - 8.578358E-004$	0.9976	5.00–500.00	8.33	25.00
Flutolanil	$y = 0.130491x - 0.025559$	0.9982	5.00–500.00	8.33	25.00
Flutriafol	$y = 0.039921x - 0.017981$	0.9973	5.00–500.00	8.33	25.00
Hexaflumuron	$y = 0.030949x - 0.026410$	0.9989	15.00–1500.00	20.00	60.00
Iprovalicarb-1	$y = 0.017570x + 0.015372$	0.9954	10.00–1000.00	16.67	50.00
Iprovalicarb-2	$y = 0.020805x - 0.033902$	0.9980	10.00–1000.00	16.67	50.00
Malaoxon	$y = 0.006443x - 0.041558$	0.9970	50.00–5000.00	16.67	50.00
Methabenzthiazuron	$y = 0.062198x - 0.308813$	0.9994	5.00–500.00	8.33	25.00
Methidathion	$y = 0.293079x - 0.937489$	0.9967	10.00–1000.00	16.67	50.00

(Continued)

APPENDIX TABLE 4.5 The Linear Range, Linear Equation, Correlation Coefficient and Limit Detect Limit of 205 Pesticides (*cont.*)

Pesticides	Linear Equation	Correlation Coefficient/R^2	Linear Range	LOD/(µg/kg)	LOQ/(µg/kg)
Mevinphos	$y = 0.016624x - 0.042462$	0.9991	2.00–200.00	3.33	10.00
Molinate	$y = 0.241112x - 0.642192$	0.9985	10.00–1000.00	3.33	10.00
Monolinuron	$y = 0.010035x - 0.119980$	0.9979	10.00–1000.00	16.67	50.00
Myclobutanil	$y = 0.290141x - 0.180668$	0.9992	2.00–200.00	3.33	10.00
Napropamide	$y = 0.015771x - 0.013469$	0.9957	5.00–500.00	8.33	25.00
Paclobutrazol	$y = 0.014151x - 0.015263$	0.9967	2.00–200.00	3.33	10.00
Phorate	$y = 0.126607x - 0.753766$	0.9987	10.00–1000.00	3.33	10.00
Phosphamidon-1	$y = 6.547714E{-}004x - 0.002473$	0.9989	2.00–200.00	3.33	10.00
Pirimicarb	$y = 0.056032x - 0.287951$	0.9974	5.00–500.00	8.33	25.00
Pirimiphos-Methyl	$y = 0.011892x - 0.057497$	0.9986	5.00–500.00	8.33	25.00
Procymidone	$y = 0.075991x - 0.061303$	0.9979	10.00–1000.00	3.33	10.00
Propachlor	$y = 0.018811x - 0.075960$	0.9988	5.00–500.00	8.33	25.00
Propargite	$y = 0.153049x + 1.005685$	0.9985	50.00–5000.00	16.67	50.00
Propham	$y = 0.081086x - 0.572363$	0.9972	10.00–1000.00	3.33	10.00
Propiconazole-1	$y = 0.060186x - 0.026298$	0.9982	10.00–1000.00	6.67	20.00
Propiconazole-2	$y = 0.062495x + 0.012193$	0.9985	10.00–1000.00	6.67	20.00
Propisochlor	$y = 0.002221x - 0.008737$	0.9972	1.00–100.00	1.67	5.00
Propoxur-1	$y = 0.352927x - 0.709904$	0.9997	10.00–1000.00	6.67	20.00
Propoxur-2	$y = 0.353409x - 0.663997$	0.9999	10.00–1000.00	6.67	20.00
Pyriproxyfen	$y = 0.172555x - 0.450338$	0.9971	5.00–500.00	8.33	25.00
Tebufenpyrad	$y = 0.033960x - 0.074862$	0.9966	10.00–1000.00	3.33	10.00
Terbuthylazine	$y = 0.024726x - 0.092429$	0.9986	5.00–500.00	8.33	25.00
Tolclofos-Methyl	$y = 0.031930x - 0.132382$	0.9987	10.00–1000.00	3.33	10.00
Triadimenol-1	$y = 0.277734x - 1.294600$	0.9957	20.00–2000.00	10.00	30.00
Triadimenol-2	$y = 0.048408x - 0.068162$	0.9959	20.00–2000.00	6.67	20.00
Triallate	$y = 0.035705x - 0.068750$	0.9999	10.00–1000.00	6.67	20.00
Triazophos	$y = 0.004901x - 0.005784$	0.9952	2.00–200.00	3.33	10.00
Trifloxystrobin	$y = 0.006590x + 0.009191$	0.9981	5.00–500.00	8.33	25.00
Trifluralin	$y = 0.043890x - 0.362895$	0.9957	10.00–1000.00	16.67	50.00
Zoxamide	$y = 0.008310x - 0.059562$	0.9966	5.00–500.00	8.33	25.00
Alachlor	$y = 0.009967x - 0.041637$	0.9983	5.00–500.00	8.33	25.00
Boscalid	$y = 0.012445x - 0.031024$	0.9976	5.00–500.00	8.33	25.00
Bupirimate	$y = 0.011383x - 0.014900$	0.9982	5.00–500.00	8.33	25.00
Dichlorvos	$y = 0.014408x + 0.007446$	0.9981	2.00–200.00	3.33	10.00
Diethofencarb	$y = 0.011198x - 0.073085$	0.9986	5.00–500.00	8.33	25.00
Dimethoate	$y = 0.062664x - 0.234168$	0.9973	5.00–500.00	8.33	25.00
Dimethomorph	$y = 0.007485x + 0.005841$	0.9986	5.00–500.00	8.33	25.00
Diniconazole	$y = 0.029015x + 0.044529$	0.9994	5.00–500.00	8.33	25.00
Disulfoton	$y = 0.075509x - 0.287507$	0.9983	5.00–500.00	8.33	25.00
Disulfoton-Sulfoxide	$y = 0.006706x + 0.001842$	0.9985	5.00–500.00	8.33	25.00
Epoxiconazole	$y = 0.036567x - 0.101698$	0.9968	5.00–500.00	8.33	25.00

(*Continued*)

APPENDIX TABLE 4.5 The Linear Range, Linear Equation, Correlation Coefficient and Limit Detect Limit of 205 Pesticides (*cont.*)

Pesticides	Linear Equation	Correlation Coefficient/R^2	Linear Range	LOD/(μg/kg)	LOQ/(μg/kg)
Ethiofencarb	$y = 0.012761x + 0.056111$	0.9974	5.00–500.00	8.33	25.00
Ethion	$y = 0.002048x - 0.010825$	0.9979	30.00–3000.00	10.00	30.00
Ethoprophos	$y = 0.013843x - 0.056499$	0.9988	2.00–200.00	3.33	10.00
Etofenprox	$y = 0.014853x + 0.013508$	0.9981	1.00–100.00	1.67	5.00
Fenamiphos	$y = 0.003961x - 0.017009$	0.9989	5.00–200.00	8.33	25.00
Fenamiphossulfone	$y = 0.006689x - 0.007691$	0.9979	5.00–500.00	8.33	25.00
Fenbuconazole	$y = 0.040939x + 0.004061$	0.9992	5.00–500.00	8.33	25.00
Fenobucarb	$y = 2.105211x - 13.125309$	0.9975	50.00–5000.00	16.67	50.00
Fenpropidin	$y = 0.040933x - 0.153077$	0.9987	5.00–500.00	8.33	25.00
Fenpropimorph	$y = 0.184706x - 1.629563$	0.9953	10.00–1000.00	6.67	20.00
Fenpyroximate	$y = 0.016876x + 0.065896$	0.9975	10.00–1000.00	3.33	10.00
Fenthion	$y = 0.041806x - 0.064446$	0.9972	5.00–500.00	8.33	25.00
Flufenacet	$y = 0.003233x - 0.021228$	0.9954	5.00–500.00	8.33	25.00
Flurochloridone	$y = 0.016972x - 0.084360$	0.9973	10.00–1000.00	6.67	20.00
Fuberidazole	$y = 0.013893x - 0.069432$	0.9960	5.00–500.00	8.33	25.00
Hexythiazox	$y = 0.004284x - 0.045807$	0.9968	5.00–500.00	8.33	25.00
Imazalil	$y = 0.024081x - 0.046623$	0.9954	10.00–1000.00	16.67	50.00
Kresoxim-Methyl	$y = 0.094331x - 0.551772$	0.9974	10.00–1000.00	6.67	20.00
Malathion	$y = 0.009620x - 0.050400$	0.9987	50.00–5000.00	16.67	50.00
Mecarbam	$y = 8.858172E-004x - 0.009214$	0.9955	10.00–1000.00	16.67	50.00
Mepronil	$y = 0.632518x - 0.060150$	0.9982	10.00–1000.00	16.67	50.00
Metalaxyl	$y = 0.019446x - 0.110057$	0.9957	10.00–1000.00	16.67	50.00
Metamitron	$y = 0.011042x - 0.070024$	0.9991	10.00–1000.00	16.67	50.00
Metazachlor	$y = 0.097175x - 0.587904$	0.9971	20.00–2000.00	10.00	30.00
Methacrifos	$y = 0.079232x - 0.431486$	0.9995	10.00–1000.00	16.67	50.00
Metribuzin	$y = 0.059336x - 0.355733$	0.9974	10.00–1000.00	10.00	30.00
Oxyfluorfen	$y = 0.001146x - 0.009650$	0.9951	5.00–500.00	8.33	25.00
Pencycuron	$y = 0.008273x + 0.024527$	0.9981	5.00–500.00	8.33	25.00
Phenothrin-1	$y = 0.018199x - 0.009249$	0.9975	5.00–500.00	8.33	25.00
Phenothrin-2	$y = 0.072008x - 0.222139$	0.9971	5.00–500.00	8.33	25.00
Phenthoate	$y = 0.023926x - 0.202987$	0.9950	10.00–1000.00	6.67	20.00
Phoslone	$y = 0.021818x - 0.057863$	0.9972	10.00–1000.00	6.67	20.00
Picoxystrobin	$y = 0.017749x - 0.086893$	0.9956	10.00–1000.00	6.67	20.00
Prochloraz	$y = 0.007637x - 0.028578$	0.9953	10.00–1000.00	16.67	50.00
Profenofos	$y = 0.001360x - 0.010675$	0.9924	10.00–1000.00	16.67	50.00
Propanil	$y = 0.019676x - 0.052814$	0.9994	10.00–1000.00	16.67	50.00
Propyzamide	$y = 0.069118x - 0.267981$	0.9981	5.00–500.00	8.33	25.00
Prosulfocarb	$y = 0.021141x - 0.111633$	0.9993	5.00–500.00	8.33	25.00
Pyraclofos	$y = 0.010989x - 0.050904$	0.9963	50.00–200.00	33.33	100.00
Pyridaben	$y = 0.077044x - 0.209123$	0.9974	5.00–500.00	8.33	25.00
Pyrifenox-1	$y = 0.020751x - 0.135164$	0.9994	50.00–5000.00	33.33	100.00

(Continued)

APPENDIX TABLE 4.5 The Linear Range, Linear Equation, Correlation Coefficient and Limit Detect Limit of 205 Pesticides (*cont.*)

Pesticides	Linear Equation	Correlation Coefficient/R^2	Linear Range	LOD/(μg/kg)	LOQ/(μg/kg)
Pyrifenox-2	$y = 0.073897x - 0.728561$	0.9952	50.00–5000.00	33.33	100.00
Pyrimethanil	$y = 0.042818x - 0.206738$	0.9962	10.00–1000.00	3.33	10.00
Pyrimidifen	$y = 0.004769x + 0.084232$	0.9999	50.00–5000.00	16.67	50.00
Quinalphos	$y = 0.031584x - 0.325226$	0.9951	10.00–1000.00	3.33	10.00
Quinoxyphen	$y = 0.039100x + 0.015162$	0.9992	5.00–500.00	8.33	25.00
Resmethrin-1	$y = 0.054600x + 0.101945$	0.9991	20.00–2000.00	16.67	50.00
Resmethrin-2	$y = 0.292679x - 2.853469$	0.9348	20.00–2000.00	16.67	50.00
Simeconzaole	$y = 2.475381x - 21.127155$	0.9964	10.00–1000.00	6.67	20.00
Spirodiclofen	$y = 2.112822E{-}007x - 3.274923E{-}004$	0.9005	5.00–500.00	8.33	25.00
Tebuconazole	$y = 0.014157x - 0.022761$	0.9978	5.00–500.00	8.33	25.00
Tebuthiuron	$y = 0.006566x - 9.945701E{-}004$	0.9983	2.00–200.00	3.33	10.00
Terbufos	$y = 0.007188x - 0.006215$	0.9996	1.00–100.00	1.67	5.00
Tetraconazole	$y = 0.002320x - 0.013782$	0.9981	2.00–200.00	3.33	10.00
Thiobencarb	$y = 0.042532x - 0.466272$	0.9967	10.00–1000.00	6.67	20.00
Triadimefon	$y = 0.028330x - 0.202684$	0.9973	20.00–2000.00	6.67	20.00
Vinclozolin	$y = 0.017751x - 0.104369$	0.9960	10.00–1000.00	3.33	10.00
2-Phenylphenol	$y = 0.205359x - 1.353870$	0.9960	10.00–1000.00	16.67	50.00
Aldrin	$y = 0.004259x - 0.016833$	0.9996	2.00–200.00	3.33	10.00
Alpha-HCH	$y = 0.022230x - 0.136286$	0.9970	2.00–200.00	3.33	10.00
Benfluralin	$y = 0.029966x - 0.180607$	0.9965	5.00–500.00	8.33	25.00
Beta-HCH	$y = 0.071476x - 0.403593$	0.9985	2.00–200.00	3.33	10.00
Bifenthrin	$y = 2.325491x - 7.685865$	0.9995	10.00–1000.00	3.33	10.00
Bromopropylate	$y = 0.163829x - 0.546602$	0.9999	5.00–500.00	8.33	25.00
Chlorbenside	$y = 0.031042x - 0.229748$	0.9995	10.00–1000.00	6.67	20.00
Chlorfenapyr	$y = 0.001497x - 0.002615$	0.9996	5.00–500.00	8.33	25.00
Chlorfenson	$y = 0.097104x - 0.176074$	0.9994	10.00–1000.00	6.67	20.00
Chlorfenvinphos	$y = 0.017029x - 0.084111$	0.9991	5.00–500.00	8.33	25.00
Chlorobenzilate	$y = 0.382145x - 0.945344$	0.9999	10.00–1000.00	6.67	20.00
Chlorpropham	$y = 0.032680x + 0.466607$	0.9849	10.00–1000.00	6.67	20.00
Chlorpyrifos-Methyl	$y = 0.025284x - 0.146925$	0.9994	10.00–1000.00	3.33	10.00
Chlorthal-Dimethyl	$y = 0.007410x - 0.011822$	0.9999	1.00–100.00	1.67	5.00
Cyfluthrin-1	$y = 0.013095x - 0.055969$	0.9993	10.00–400.00	16.67	50.00
Cyhalofop-Butyl	$y = 0.028059x - 0.158475$	0.9997	5.00–500.00	8.33	25.00
Cypermethrin-1	$y = 0.444036x - 3.422555$	0.9961	50.00–5000.00	16.67	50.00
Deltamethrin	$y = 4.300280x - 29.818791$	0.9995	50.00–5000.00	50.00	150.00
Dichlobenil	$y = 0.085922x - 0.827815$	0.9952	5.00–500.00	8.33	25.00
Diclofop-Methyl	$y = 0.016542x - 0.041846$	0.9996	5.00–500.00	8.33	25.00
Dicloran	$y = 0.005614x - 0.036360$	0.9991	1.00–100.00	1.67	5.00
Difenconazole	$y = 0.190568x - 1.560202$	0.9983	5.00–500.00	8.33	25.00
Diflufenican	$y = 0.050664x - 0.186496$	0.9999	5.00–500.00	8.33	25.00
Dioxathion	$y = 0.017331x - 0.053212$	0.9997	10.00–1000.00	16.67	50.00

(*Continued*)

APPENDIX TABLE 4.5 The Linear Range, Linear Equation, Correlation Coefficient and Limit Detect Limit of 205 Pesticides (*cont.*)

Pesticides	Linear Equation	Correlation Coefficient/R^2	Linear Range	LOD/(µg/kg)	LOQ/(µg/kg)
Endosulfan-1	$y = 0.001827x - 0.008728$	0.9966	30.00–3000.00	20.00	60.00
Endosulfan-Sulfate	$y = 7.976857E - 004x - 0.002184$	0.9994	3.00–300.00	5.00	15.00
Endrin	$y = 0.002762x - 0.012733$	0.9992	1.00–100.00	1.67	5.00
Ethalfluralin	$y = 9.445113E - 004x - 0.003996$	0.9972	2.00–200.00	3.33	10.00
Fenamidone	$y = 0.022195x - 0.118509$	0.9998	5.00–500.00	8.33	25.00
Fenchlorphos	$y = 0.027369x - 0.139707$	0.9996	10.00–1000.00	16.67	50.00
Fenvalerate-1	$y = 0.115549x - 0.858440$	0.9993	5.00–500.00	8.33	25.00
Fenvalerate-2	$y = 0.063404x - 0.258348$	0.9999	5.00–500.00	8.33	25.00
Flucythrinate-1	$y = 0.226453x - 1.561030$	0.9997	10.00–1000.00	6.67	20.00
Flucythrinate-2	$y = 0.206765x - 1.442237$	0.9995	10.00–1000.00	6.67	20.00
Fludioxonil	$y = 0.058432x - 0.338051$	0.9998	5.00–500.00	8.33	25.00
Flumioxazin	$y = 0.012761x - 0.112480$	0.9987	10.00–1000.00	6.67	20.00
Fluquinconazole	$y = 0.029346x - 0.175305$	0.9995	5.00–500.00	8.33	25.00
Gamma-HCH	$y = 0.071417x - 0.409223$	0.9981	5.00–500.00	8.33	25.00
Halfenprox	$y = 0.003206x - 0.007399$	0.9993	10.00–1000.00	16.67	50.00
Heptachlor	$y = 0.010566x - 0.057577$	0.9978	2.00–200.00	3.33	10.00
Hexachlorobenzene	$y = 0.014900x - 0.095200$	0.9955	2.00–200.00	3.33	10.00
Hexaconazole	$y = 0.009699x - 0.051299$	0.9993	5.00–500.00	8.33	25.00
Lamda-Cyhalothrin	$y = 0.941595x - 6.533547$	0.9995	10.00–1000.00	3.33	10.00
Lenacil	$y = 0.267829x - 1.471553$	0.9998	10.00–1000.00	3.33	10.00
Methoxychlor	$y = 0.095325x - 0.126221$	0.9996	10.00–1000.00	3.33	10.00
Monocrotophos	$y = 0.004182x - 0.033939$	0.9998	10.00–1000.00	16.67	50.00
Nitrofen	$y = 0.004335x - 0.037662$	0.9987	2.00–200.00	3.33	10.00
o,p-DDT	$y = 0.393376x - 0.376215$	0.9992	20.00–2000.00	6.67	20.00
Oxadiazone	$y = 0.030410x - 0.065796$	0.9998	5.00–500.00	8.33	25.00
Oxadixyl	$y = 0.024441x - 0.102268$	0.9997	2.00–200.00	3.33	10.00
p,p-DDE	$y = 0.147765x - 0.490977$	0.9999	20.00–2000.00	6.67	20.00
p,p-DDT	$y = 0.136960x - 1.664583$	0.9940	20.00–2000.00	6.67	20.00
Parathion-Methyl	$y = 0.007894x - 0.055873$	0.9993	5.00–500.00	8.33	25.00
Pendimethalin	$y = 0.023651x - 0.191210$	0.9955	10.00–1000.00	16.67	50.00
Permethrin-1	$y = 0.072413x - 0.324682$	0.9998	10.00–1000.00	6.67	20.00
Permethrin-2	$y = 0.155898x - 0.940434$	0.9997	10.00–1000.00	6.67	20.00
Prothiophos	$y = 0.028066x - 0.096588$	0.9998	5.00–500.00	3.33	10.00
Pyraclostrobin	$y = 0.002671x + 0.005887$	0.9964	5.00–500.00	8.33	25.00
Pyraflufenethyl	$y = 0.014354x - 0.087499$	0.9987	5.00–500.00	8.33	25.00
Pyrazophos	$y = 0.127383x - 0.874606$	0.9996	10.00–1000.00	6.67	20.00
Quintozene	$y = 0.008832x - 0.051713$	0.9990	5.00–500.00	8.33	25.00
Spiromesifen	$y = 0.007735x - 0.025419$	0.9999	2.00–200.00	3.33	10.00
Tecnazene	$y = 0.006040x - 0.040031$	0.9967	10.00–1000.00	6.67	20.00
Tetradifon	$y = 0.015762x - 0.061954$	0.9995	5.00–500.00	8.33	25.00
Trans-Chlordane	$y = 0.003451x - 0.012882$	0.9999	2.00–200.00	3.33	10.00

5 The Evaluation of the Ruggedness of the Method, Error Analysis, and the Key Control Points of the Method

5.1 INTRODUCTION

Food is of paramount necessity to people, and food safety is an eternal theme in human life. Food safety standards or regulations can be traced to the first English food law in 1202 called "the Assize of Bread"[1] and the first pure food and drug act [2] proclaimed in the United States in 1906. In development over more than 100 years, food safety has caused increasing concern all over the globe. In 2000, FAO/WHO proposed WHO global strategy for food safety, appealing for the implementation of a food safety administrative strategy for the entire process from "farm to table" [3]. In 2001, the 53rd World Health Assembly passed the Food Safety Resolution, rendering food safety the first priority in public health [4]. From that point on, countries worldwide have adopted a series of important strategic measures for food safety. In particular, an important study topic concerning food safety is the pesticide residue limit, which has been one of the focuses for food safety. After the ban of the use of dichlorodiphenyltrichloroethane (DDT) [5] by the United States in the 1970s, the United States and European Union released a pesticide residue standard for fruits. Up till the present, the European Union has published 145,000 pesticide residue limit standards [6], the Codex Alimentarius Commission 3338 standards [7], the United States more than 10,000 standards [8], Japan more than 50,000 standards [9], and China 478 standards [10]. Pesticide residue standards have already become an important technical index of food safety as well as the threshold of access for international trade. In terms of tea, there are at present 17 countries with more than 800 pesticide residue limit standards, among which Germany has 81 standards [11], Japan 269 [12], and the EU 438 [13]. Due to increasing limits on pesticide residues in food safety determinations by countries across the globe and generally stricter tendencies, a single analytical technique for determining pesticide residue has been rendered incapable of meeting present detection requirements.

For that reason, the development and application of simultaneous determination techniques for pesticide multiresidues is presently the focus of the study. The study work of pesticide multiresidue analytical techniques by Lehotay et al. [14], Wong et al. [15], Kitagawa et al. [16], Okihashi et al. [17], Huang et al. [18], Fillion et al. [19], Walorczyk [20], Nguyen

Analysis of Pesticide in Tea. http://dx.doi.org/10.1016/B978-0-12-812727-8.00011-0

et al. [21], and Przybylski and Segard [22]. In addition, Koesukwiwat et al. [23] adopted the QuEChERS approach, PSA, C18, and graphite carbon filler cleanup, making a LP-GC/TOFMS determination of 150 pesticide residues in tomatoes, strawberries, potatoes, oranges, and lettuces under three spiked levels with recoveries falling within 70%–120%, and more than 126 pesticides having RSD < 20%. Schenck et al. [24] used the improved QuEChERS method, PSA, and graphite carbon for cleanup, making a determination of 102 pesticide residues in grapes, oranges, spinach, and tomatoes, with recoveries 63%–125% and the recoveries for the majority of pesticides >80%. Nguyen et al. [25] employed QuEChERS approach, making a GC–MS determination of 107 pesticide residues in cabbages and turnips with recoveries falling 80%–115% and RSD < 15%. Saito et al. [26] adopted PSA and graphite carbon for cleanup, making GC–MS detection of 114 pesticide residues in tomatoes, spinach, Japanese pears, grapes, and brown rice under the spiked level of 0.02–0.4 μg/g, with recoveries for 108 pesticides >60%. Hirahara et al. [27] used ion-trap GC–MS/MS to determine 199 pesticide residues in potatoes, spinach, cabbages, apples, oranges, soybeans, and brown rice, with recoveries for 194 pesticides falling 50%–150%. Pizzutti et al. [28] omitted the cleanup process after sample extraction, making a LC–MS/MS determination of 169 pesticide residues in soybeans, with recoveries for majority pesticides of 60%–120% and RSD ≤ 20%. Nguyen et al. [29] used matrix solid phase dispersion, making a GC–MS determination of 234 pesticides in South Korean herbal medicines, with recoveries falling 62%–119% and RSD < 21%. Hu et al. [30] adopted matrix solid phase dispersion for extraction and cleanup, making a GC–MS determination of 106 pesticide residues in apple juices under the spiked level of 0.01–0.2 mg/kg, with recoveries ranging 70%–110%. Chu et al. [31] employed matrix solid phase dispersion method, making a GC–MS determination of 266 pesticide residues in vegetable juices. Among these, the recoveries for 258 pesticides fell 70.8%–116.8%, with RSD < 24%. Nguyen et al. [32] described a method for measuring 118 pesticides in vegetable juice using a matrix solid-phase dispersion preparation followed by GC–MS and LC–MS/MS analysis, under the spiked level of 10–120 μg/kg, with recoveries 77%–114%. Matsumoto et al. [33] used a minicolumn for cleanup, making a GC/MS and LC/MS/MS determination of 235 pesticide residues in eight foods, such as sweet potato jam, tomato jam, and pea jam, under the spiked levels of 0.05 or 0.10 μg/g (method GC) and 0.025 or 0.05 μg/g (method LC), with recoveries for 214 pesticides ranging 50%–100% and RSD < 20%.

In the past 10 years, the author's team has also systematically studied high throughput analytical techniques for multigrouped residues from more than 1000 pesticides and environmental pollutants and has established a series of China Official Methods for simultaneous determinations of 400–500 pesticide residues, which are applied to honey, fruit and vegetable juices, fruit wines (orange juices, pear juices, apple juices, grape juices, cabbage juices, carrot juices, dry wines, half-dry wines, half-sweet wines, and sweet wines) [34]; fruits and vegetables (apples, oranges, cabbages, celeries, tangerines, grapes, and tomatoes) [35]; grains and cereals (barley, wheat, oats, rice, and corn) [36]; animal muscles (pork, beef, mutton, rabbit meat, and chicken meat) [37]; teas (green tea, black tea, Pu"(tm)er tea, and oolong tea) [38]; edible fungi (nameko, needle mushrooms, Jew's ear, and champignon) [39]; aquatic products (puffer fish, eels, and prawns) [40]; Chinese medicinal herbs (mulberry twigs, honeysuckles, Barbary Wolfberry fruit, and lotus leaves) [41]; and milk and milk powders [42].

For the purpose of having counterparts in residual analysis share these multigrouped pesticide residual analytical techniques, 653 pesticides were selected from tea in preparation for the AOAC intercollaborative study. In keeping with this objective, this chapter deals especially with the study of ruggedness of method. Two teas (green and oolong teas), two cartridges (Cleanert TPT and ENVI-Carb + PSA), and two kinds of instruments (GC–MS and GC–MS/MS) were chosen, along with eight different experimental schemes designed to run a 3-month circulative test. There was one test every 5 days, 3 parallel samples at each time (5 parallel samples for the first time), and 19 circulative determinations conducted over 3 months, from which a total of 61,101 test data were obtained (2 teas × 2 cartridges × 2 kinds of instruments × 2 Youden pair samples, once every 5 days, 3 parallel samples each time, and 19 determinations conducted over 3 months × 201 pesticides). The classification of these data and multiclasses analysis find that the ruggedness of the method is satisfactory.

At the same time, concrete analysis was made to error types from different analytical stages and their causes, while chromatographic resolutions were carried out to the seven types of errors, one by one, especially from the chromatographic peaks. Elaborated discussions were conducted on key control points in different stages of the method.

5.2 EXPERIMENT

5.2.1 Reagents and Materials

Solvents: acetonitrile, toluene, hexane, acetone (LC grade), purchased from Dikma Co. (Beijing, China). Anhydrous sodium sulfate: analytically pure. Baked at 650 C for 4 h and stored in a desiccator. Pesticides standard and internal standard:

purity, $\geq 95\%$ (LGC Promochem, Wesel, Germany). Stock standard solutions: weigh 5–10 mg of individual pesticide standards (accurate to 0.1 mg) into a 10 mL volumetric flask. Dissolve and dilute to volume with toluene, toluene-acetone combination, cyclohexane, depending on each individual compound solubility. Stock standard solutions should be stored in the dark below 4 C. Mixed standard solution: depending on properties and retention time of each pesticide, all 201 pesticides for GC/MS and GC/MS/MS analysis are divided into three groups, A–C. The concentration of mixed standard solutions was dependent on the sensitivity of each compound for the instrument used for analysis. Mixed standard solutions should be stored in the dark below 4 C. SPE: Cleanert-TPT (Agela, Tianjin, China); ENVI-Carb (Supelco, American); and PSA (Varian, USA).

5.2.2 Apparatus

GC/MS system: Model 6890N gas chromatograph connected to a Model 5973N MSD and equipped with a Model 7683 autosampler (Agilent Technologies, Wilmington, DE, USA). The column used was a DB-1701 capillary column (30 m $\times$ 0.25 mm $\times$ 0.25 μm, J&W Scientific, Folsom, CA, USA). GC/MS/MS system: Model 7890 gas chromatograph connected to a Model 7000A and equipped with a Model 7693 autosampler (Agilent Technologies, Wilmington, DE, USA). The column used was a DB-1701 capillary column (30 m $\times$ 0.25 mm $\times$ 0.25 μm, J&W Scientific, Folsom, CA, USA). Visiprep 5-port flask vacuum manifold: RS-SUPELCO 57101-U (Sigma Aldrich Trading Co., Ltd, Shanghai, China). Homogenizer: T-25B (Janke & Kunkel, Staufen, Germany). Rotary evaporator: Buchi EL131 (Flawil, Switzerland). Centrifuge: Z 320 (B. HermLe AG, Gosheim, Germany). Nitrogen evaporator: EVAP 112 (Organomation Associates, Inc. New Berlin, MA, USA).

5.2.3 Preparing Pesticides-Aged Tea Samples and the Evaluation of Uniformity for Preparation

(1) Take green tea and oolong tea that are found to be free from the target pesticide after testing and pass them respectively through 10 mesh and 16 mesh sieves after initial blending via blender. (2) Take 10–16 mesh 500 g oolong tea and green tea each, and spread them uniformly over the bottom of a stainless steel vessel with 40 cm diameter and wait for spraying. (3) Accurately transfer a certain amount of pesticide mixed standard solutions into the full-glass sprayer and spray the tea leaves. Spray while stirring the tea leaves with a glass rod so as to make a uniform spray. (4) After completion of the spraying, continue to stir the tea leaves for half an hour to wait for the volatile solvents in the tea leaves, after which, place the sprayed tea leaves into a 4 L brown bottle, and avoid exposure to light for storage at room temperature. (5) Spread the aged tea samples onto the bottom of a flat-bottomed vessel, draw a cross, and weigh a total of five portions of aged tea samples located at the symmetrical four points of the cross and the central part. Submit them for GC/MS and GC/MS/MS determination, and calculate the average value of the pesticide content of the aged samples and RSD. When RSD < 4% for GC–MS and GC–MS/MS, and RSD < 5% for LC–MS/MS, the conclusion can be drawn that tea samples have been sprayed and mixed homogeneously, and the average value of the content obtained in the five instances serves as the fixed value concentration of the pesticides in the aged samples.

5.2.4 Experimental Method

Weigh 5 g of sample (accurate to 0.01 g) into an 80 mL centrifuge tube. Add 15 mL of acetontrile, homogenize at 15,000 r/min for 1 min, and centrifuge for 5 min at 4200 r/min. Transfer the supernatants to a pear-shaped flask. The sample is then reextracted in the same way and the supernatants combined. Carefully evaporate just to ca 1 mL at 45 C with a vacuum evaporator in a water bath for the following cleanup procedure.

Before the sample application, anhydrous sodium sulfate (ca 2 cm high) was placed on the top of the SPE cartridge, which connected a pear-shaped flask. The cartridge was conditioned with 10 mL of acetontrile-toluene (3:1). When the conditioning solution reached the top of the sodium sulfate, the concentrated extract was added to the cartridge, the flask was rinsed with 3 mL of acetonitrile-toleune twice, and the washings were also applied to the cartridge. A 50 mL of reservoir was attached to the cartridge and the pesticides were eluted with 25 mL of acetontrile-toluene (3:1). The eluted portion was concentrated to ca 0.5 mL by rotary evaporation at 40 C. For determination of pesticides by GC/MS (GC–MS/MS), add 5 mL of hexane for solvent exchange on a rotary evaporator in a water bath of 40 C and repeat it twice and make up the final solution volume to ca 1 mL. Add 40 μL of internal standard solution and mix thoroughly for the determination of GC–MS and GC–MS/MS.

5.3 EVALUATION OF THE RUGGEDNESS OF THE METHOD

To investigate the ruggedness of the method, two teas, two Youden pair samples, two cartridges, and two instruments are used to design eight testing procedures (Fig. 5.1). Determination is made once every 5 days, with 3 parallel samples each time (the first time with 5 parallel samples) over a successive period of 3 months and 19 circulative determinations of 944 samples. A total of 189,744 original data have been obtained, and the average content and RSD of each pesticide for every sample has been calculated three times, with the results tabulated in Appendix Tables 5.1 and 5.2. The Youden pair ratios of each average content from 19 circulative determinations of 944 samples appear in Appendix Table 5.3.

Based on the eight analytical procedures, both single-class analysis and multiclass analysis can be undertaken for the evaluation of the ruggedness of the method at different levels and in different angles.

5.3.1 Reproducibility Comparison of the First Determination RSD of Parallel Samples from Pesticide-Aged Youden Pair Samples with 18 Circulative Determination Values

After preparing the 201 pesticide aged samples, take five each from the aged Youden pair samples A and B, make a determination of the 201 pesticides content, and calculate the average content and RSD values. At that point, make 18 circulative determinations over 3 months with three parallel samples each time, calculating the average content and RSD, with a total of 57,888 RSD data obtained (201 pesticides × 18 circulative determinations × 2 teas × 2 SPE cartridges × 2 instruments × 2 concentrations) (Appendix Table 5.13). To conduct the single-class analysis of the results from the eight analytical procedures, classify the RSD (Appendix Table 5.13) into four levels and tabulate them (Table 5.1) for single-class analysis. The percentage of individual pesticides making up the total number and falling within the optimum range of RSD ≤ 15% (Table 5.1) is expressed in the bar chart shown in Fig. 5.2.

Table 5.1 and Fig. 5.2 show that the percentage of the number of each pesticide RSD of the fixed values falling within the optimum range (RSD ≤ 15%) from the other analytical procedures except green tea + ENVI-Carb + PSA cartridges differs <1.5% with the percentage of 18 circulative determinations. In addition, the percentage of pesticides falling within the optimum range is >93.1%, demonstrating good reproducibility and ruggedness of these analytical procedures.

Comparing the two cartridges applied, the percentage of detected values falling within the optimum range (RSD ≤ 15%) from TPT cartridge is higher than that from ENVI-Carb + PSA cartridges by 7.7%, on the average. For multiclasses analysis, the distributions of RSD results from fixed values and 18 circulative determinations of parallel samples are listed in Table 5.2, in accordance with different teas, different cartridges, and different instruments. The percentage of individual pesticides falling within the optimum range of RSD ≤ 15% over the total number of pesticides (Table 5.5) is expressed in the bar chart in Fig. 5.3. Table 5.2 and Fig. 5.3 show that the percentage of fixed values of pesticide numbers from two teas, two cartridges, and two instruments, with RSD falling within the optimum range (RSD ≤ 15%), has a difference not exceeding 6% from the percentage of circulative determinations, demonstrating an excellent reproducibility.

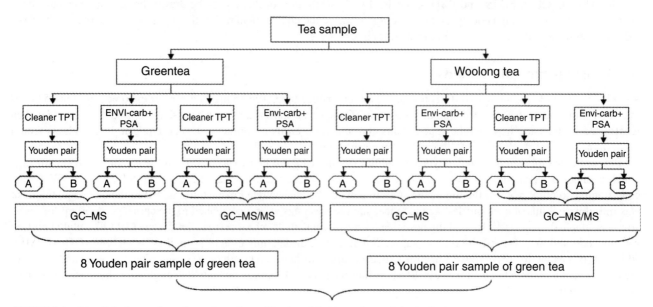

FIGURE 5.1   8 Analytical procedures for each tea for verification of the Ruggedness of the method.

TABLE 5.1 RSD Data Distribution Range (*n* = 3) of Green Tea and Oolong Tea "Parallel Samples"

RSD (%)	Cleanert TPT GC–MS		Cleanert TPT GC–MSMS		ENVI-Carb + PSA GC–MS		ENVI-Carb + PSA GC–MSMS	
	First fixed value	18 Determination values	First fixed value	18 Determination values	First fixed value	18 Determination values	First fixed value	18 Determination values
Green tea								
≤10	368 (91.5)	6348 (87.7)	370 (92)	6374 (88.1)	256 (63.7)	5544 (76.6)	242 (60.2)	5392 (74.5)
10.1–15	25 (6.2)	801 (11.1)	21 (5.2)	638 (8.8)	77 (19.2)	908 (12.5)	48 (11.9)	1020 (14.1)
≤15	393 (97.7)	7149 (98.8)	391 (97.2)	7012 (96.7)	333 (82.9)	6452 (89.1)	290 (72.1)	6412 (88.6)
15.1–20	6 (1.5)	63 (0.9)	5 (1.2)	147 (2)	35 (8.7)	308 (4.3)	43 (10.7)	396 (5.5)
>20	3 (0.7)	24 (0.3)	4 (1)	65 (0.9)	34 (8.5)	470 (6.5)	67 (16.7)	404 (5.6)
No data	0 (0)	0 (0)	2 (0.5)	12 (0.2)	0 (0)	6 (0.1)	2 (0.5)	24 (0.3)
Total	402 (100)	7236 (100)	402 (100)	7236 (100)	402 (100)	7236 (100)	402 (100)	7236 (100)
Oolong tea								
≤10	362 (90)	6537 (90.3)	355 (88.3)	6576 (90.9)	340 (84.6)	5848 (80.8)	319 (79.4)	5947 (82.2)
10.1–15	29 (7.2)	545 (7.5)	29 (7.2)	441 (6.1)	40 (10)	956 (13.2)	52 (12.9)	874 (12.1)
≤15	391 (97.2)	7082 (97.8)	384 (95.5)	7017 (97.0)	380 (94.6)	6804 (94.0)	371 (93.1)	6821 (94.3)
15.1–20	5 (1.2)	93 (1.3)	6 (1.5)	116 (1.6)	14 (3.5)	283 (3.9)	17 (4.2)	233 (3.2)
>20	6 (1.5)	59 (0.8)	8 (2)	81 (1.1)	8 (2)	147 (2)	12 (3)	160 (2.2)
No data	0 (0)	2 (0)	4 (1)	22 (0.3)	0 (0)	2 (0)	2 (0.5)	22 (0.3)
Total	402 (100)	7236 (100)	402 (100)	7236 (100)	402 (100)	7236 (100)	402 (100)	7236 (100)

Data in parentheses are the percentage.

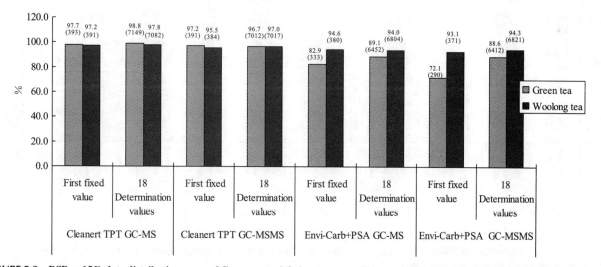

FIGURE 5.2 **RSD ≤ 15% data distribution range of Green tea and Oolong tea "parallel samples" (*n* = 3).** Data in brackets are the numbers of the first determination and consequent 18 determinations.

5.3.2 Evaluation of the Ruggedness of the Method Based on the RSD of the Youden Pair Ratios from the 3-Month Circulative Experiment

In the 3-month circulative experiment, calculate the Youden pair ratios from each determination before working out the RSD values over the ratios from 19 determinations (Appendix Table 5.14). To conduct the single-class analysis of the results from the eight analytical procedures, classify the RSD (Appendix Table 5.14) into four levels and tabulate them

TABLE 5.2 Statistical Analysis of RSD Data Per Class for Green and Oolong Tea Parallel Samples

	Two cartridges				Two tea samples				Two instruments			
	Cleanert TPT		ENVI-Carb + PSA		Green tea		Oolong tea		GC–MS		GC–MS/MS	
RSD (%)	First fixed value	18 Determination values	First fixed value	First fixed value	18 Determination values	First fixed value	First fixed value	18 Determination values	First fixed value	First fixed value	18 Determination values	First fixed value
≤10	1455 (90.5)	25,835 (89.3)	1157 (72)	22,731 (78.5)	1236 (76.9)	23,658 (81.7)	1376 (85.6)	24,908 (86.1)	1326 (82.5)	24,277 (83.9)	1286 (80)	24,289 (83.9)
10.1–15	104 (6.5)	2425 (8.4)	217 (13.5)	3758 (13)	171 (10.6)	3367 (11.6)	150 (9.3)	2816 (9.7)	171 (10.6)	3210 (11.1)	150 (9.3)	2973 (10.3)
≤15	1559 (97.0)	28,260 (97.7)	1374 (85.5)	26,489 (91.5)	1407 (87.5)	27,025 (93.3)	1526 (94.9)	27,724 (95.8)	1497 (93.1)	27,487 (95)	1436 (89.3)	27,262 (94.2)
15.1–20	22 (1.4)	419 (1.4)	109 (6.8)	1220 (4.2)	89 (5.5)	914 (3.2)	42 (2.6)	725 (2.5)	60 (3.7)	747 (2.6)	71 (4.4)	892 (3.1)
>20	21 (1.3)	229 (0.8)	121 (7.5)	1181 (4.1)	108 (6.7)	963 (3.3)	34 (2.1)	447 (1.5)	51 (3.2)	700 (2.4)	91 (5.7)	710 (2.5)
No data	6 (0.4)	36 (0.1)	4 (0.2)	54 (0.2)	4 (0.2)	42 (0.1)	6 (0.4)	48 (0.2)	0 (0)	10 (0)	10 (0.6)	80 (0.3)
Total	1608 (100)	28,944 (100)	1608 (100)	28,944 (100)	1608 (100)	28,944 (100)	1608 (100)	28,944 (100)	1608 (100)	28,944 (100)	1608 (100)	28,944 (100)

Data in parentheses are the percentage.

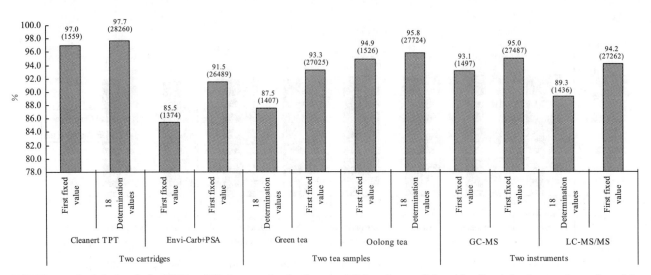

FIGURE 5.3 **Statistical analysis of RSD ≤ 15% data per class for Green and Oolong tea parallel samples.** Data in brackets are the numbers of the first determination and consequent 18 determinations.

(Table 5.3) for single-class analysis. The percentage of individual pesticides falling with the optimum range of RSD ≤ 15% (Table 5.6) making up the total number is expressed in the bar chart in Fig. 5.4.

Table 5.3 and Fig. 5.4 show that the average percentage of RSD value ≤15% of green tea Youden pair samples is 92.7%, and, moreover, no matter whether it is two SPE cartridges or two instruments this percentage is >85%. For oolong tea samples the percentage of RSD value ≤15% of Youden pair sample ratios is 93.9%, and no matter whether it is two SPE cartridges or two instruments this percentage is >87%. Such data indicate that Youden pair sample ratios from 19 circulative determinations under different conditions are constant and RSD ≤ 15% for the majority of pesticides.

TABLE 5.3 RSD Distribution Range of Green and Oolong Tea Youden Pair Sample Ratios (19 Determinations)

RSD (%)	Green tea				Oolong tea			
	Cleanert TPT GC–MS	Cleanert TPT GC–MS/MS	ENVI-Carb + PSA GC–MS	ENVI-Carb + PSA GC–MS/MS	Cleanert TPT GC–MS	Cleanert TPT GC–MS/MS	ENVI-Carb + PSA GC–MS	ENVI-Carb + PSA GC–MS/MS
≤5	39.3 (79)	17.4 (35)	10 (20)	3.5 (7)	15.4 (31)	9 (18)	11.4 (23)	0.5 (1)
5–10	53.2 (107)	65.7 (132)	58.7 (118)	59.7 (120)	71.6 (144)	76.6 (154)	54.7 (110)	75.1 (151)
10–15	7.5 (15)	12.4 (25)	21.4 (43)	21.9 (44)	10 (20)	10.9 (22)	20.9 (42)	19.4 (39)
≤15	100 (201)	95.5 (192)	90.1 (181)	85.1 (171)	97 (195)	96.5 (194)	87.0 (175)	95.0 (191)
15–20	0 (0)	4 (8)	7 (14)	8.5 (17)	2 (4)	1.5 (3)	10 (20)	4 (8)
>20	0 (0)	0.5 (1)	3 (6)	6.5 (13)	1 (2)	2 (4)	3 (6)	1 (2)
Total	100 (201)	100 (201)	100 (201)	100 (201)	100 (201)	100 (201)	100 (201)	100 (201)

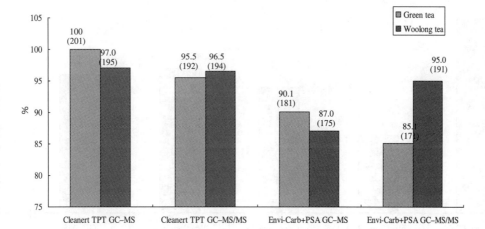

FIGURE 5.4   Bar chart of RSD distribution range of Green and Oolong tea Youden pair sample ratios (19 determinations).

See Table 5.4 and Fig. 5.5 for the results for multiclass analysis from two teas, two cartridges, and two instruments. In terms of multiclass analysis, Table 5.4 and Fig. 5.5 shows that the percentage of RSD ≤ 15% for two teas and two analytical methods runs neck-to-neck with one another; in terms of two cartridges, the TPT cartridge is higher than the cartridge in tandem by 8%.

Comparing the two different SPE cleaning methods, the percentage of RSD falling within the optimum range for oolong tea differs only by 1.5% (96.5% for TPT versus 95% for the ENVI-Carb + PSA), except for oolong tea with GC–MS/MS determination. For other samples using TPT cleaning, the percentages of pesticides with RSD ≤ 15% are 10% higher than those using ENVI-Carb + PSA cleaning. The results demonstrate that TPT cleaning is more stable and robust than ENVI-Carb + PSA.

In summary, the thinking is totally different although both Evaluation I and Evaluation II calculate RSD. The former works out RSD by determining the content of target pesticides of three parallel samples each time during a circulative experiment, with a total of 61,104 data obtained, while the latter computes RSD by Youden pair ratios from 19 circulative determinations, with a total of 1608 data obtained""yet the conclusions from these two approaches are absolutely identical. Therefore, it can be said that the earlier described Evaluations I and II prove the ruggedness of the method.

5.4   ERROR ANALYSIS AND KEY CONTROL POINTS

5.4.1   Error Analysis of Sample Pretreatment and Key Control Points

This chapter further used a recovery test to investigate the influences of rotary evaporation temperatures on recoveries, with test results tabulated in Table 5.5. The statistical analysis in Table 5.5 found that the average recovery under

TABLE 5.4 RSD Statistical Analysis of Youden Pair Ratio Per Class

RSD %	Two cartridges		Two tea samples		Two instruments	
	Cleanert TPT	ENVI-Carb + PSA	Green tea	Oolong tea	GC–MS	GC–MS/MS
≤5	20.3 (163)	6.3 (51)	17.5 (141)	9.1 (73)	19 (153)	7.6 (61)
5–10	66.8 (537)	62.1 (499)	59.3 (477)	69.5 (559)	59.6 (479)	69.3 (557)
10–15	10.2 (82)	20.9 (168)	15.8 (127)	15.3 (123)	14.9 (120)	16.2 (130)
≤15	97.3 (782)	89.3 (718)	92.6 (745)	93.9 (755)	93.5 (752)	93.1 (748)
15–20	1.9 (15)	7.3 (59)	4.9 (39)	4.4 (35)	4.7 (38)	4.5 (36)
>20	0.9 (7)	3.4 (27)	2.5 (20)	1.7 (14)	1.7 (14)	2.5 (20)
Total	100 (804)	100 (804)	100 (804)	100 (804)	100 (804)	100 (804)

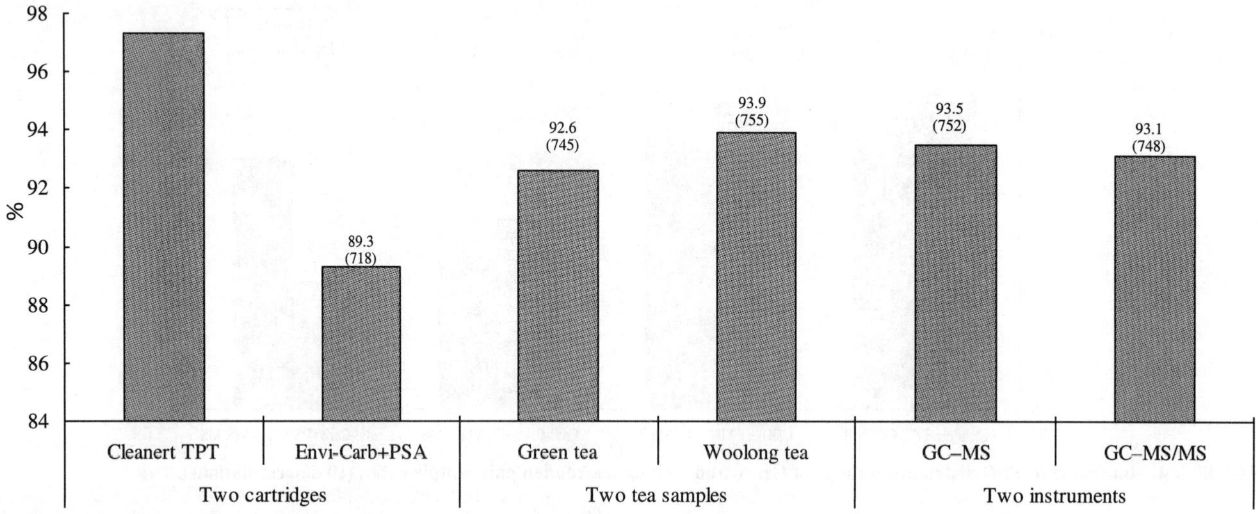

FIGURE 5.5   RSD statistical analysis of Green tea and Oolong tea Youden pair ratios.

just dryness conditions is 103.6%, with recovery <80% for 6.2% pesticides; the continuation of evaporation for 2 min after dryness found that average recovery of pesticides under this condition is 96.5%, with recovery <80% for 14.5%. It further proved that during the process of sample pretreatment, evaporation temperature is one of key control points in influencing recoveries.

The aforementioned experiments found that in the process of sample preparation two key control points must be grasped to ensure accurate quantification results. First, the evaporation temperature should not exceed 45 C when using rotary evaporators for concentration, with the final dryness to 0.3–0.5 mL being suitable. Excessively high temperatures will cause part of the drugs to dissolve, and evaporation to dryness will also lead to pesticide loss. Second, solvent exchange is made twice and is carried out once again if the liquid surface still has a separating layer after the two times exchange. Precautions are taken in the exchange with hexane that the revolutions of the rotary evaporator are reduced properly, if evaporation is going very rapidly, so as to prevent pesticide loss from evaporation to dryness.

5.4.2   The Effects of Equipment Status on Test Results (Error Analysis) and Critical Control Point

5.4.2.1   Effect of Equipment Status on Instrumental Reproducibility, Sensitivity and Peak Shape

Evaluations were made to peak asymmetric factors, instrumental reproducibility, and instrumental sensitivity on two instruments of GC–MS and GC–MS/MS in our laboratory, with the results tabulated in Table 5.6 and Table 5.7.

TABLE 5.5 The Influence of Dryness of Rotary Evaporation on the Recoveries of 227 Pesticides

No.	Pesticides	Just dryness conditions	After evaporation to dryness 2 min	No.	Pesticides	Just dryness conditions	After evaporation to dryness 2 min
1	Isoprocarb-1	114.4	108.2	9	2,3,5,6-tetrachloroani...	103.0	94.8
2	Pebulate	101.6	79.0	10	Pentachloroanisole	106.3	99.8
3	Pentachlorobenzene	101.8	83.1	11	Trifluralin	93.3	79.4
4	Propham	109.6	99.0	12	Tebutam	106.3	95.7
5	Cycloate	103.5	93.7	13	Methabenzthiazuron	117.2	109.8
6	Isoprocarb-2	86.6	91.3	14	Simeton	111.3	103.6
7	Thionazin	93.6	86.5	15	Atratone	116.9	104.0
8	Tebuthiuron	115.9	109.7	16	Phenanthrene	104.8	100.1
17	Clomazone	85.5	88.1	86	Ethoprophos	108.2	102.4
18	Tefluthrin	94.3	98.9	87	cis-Diallate	115.1	95.7
19	Fenpyroximate	120.4	130.1	88	Propachlor	90.9	81.3
20	Tebupirimfos	96.5	84.4	89	trans-Diallate	100.6	92.0
21	Cycluron	116.8	123.6	90	Sulfotep	105.3	97.7
22	DE-PCB28	104.0	100.3	91	Chlorpropham	106.9	96.4
23	DE-PCB31	104.0	99.8	92	Quintozen	54.7	46.5
24	Secbumeton	114.5	103.8	93	Fonofos	106.5	100.0
25	Pyroquilon	105.4	99.8	94	Profluralin	95.8	84.3
26	2,3,4,5-Tetrachloroani...	115.8	114.6	95	Propazine	98.8	89.4
27	Pentachloroaniline	95.1	93.1	96	Etrimfos	96.4	84.4
28	Pirimicarb	106.4	99.3	97	Dicloran	89.1	103.5
29	Benoxacor	87.1	71.4	98	Propyzamide	114.5	104.9
30	Prosulfocarb	112.3	97.5	99	Chlorprifos-methyl	89.3	79.8
31	Acetochlor	103.4	97.0	100	Desmetryn	107.0	100.1
32	Dimethenamid	105.2	98.4	101	Ronnel	96.8	91.7
33	DE-PCB52	106.1	101.2	102	Dimethachlor	100.1	92.1
34	Alachlor	104.3	95.4	103	Pirimiphos-methyl	110.4	102.1
35	Prometrye	103.3	95.1	104	Mefenoxam	112.3	102.9
36	Monalide	96.6	80.5	105	Endosulfan-1	57.9	52.7
37	Beta-HCH	N.D.	N.D.	106	Thiobencarb	110.5	103.2
38	Metalaxyl	110.0	102.7	107	Terbutryn	111.7	104.4
39	Isodrin	114.4	101.8	108	Chlorpyifos(ethyl)	102.3	93.8
40	Trichloronat	106.6	100.1	109	Methyl-parathion	88.8	74.5
41	Dacthal	107.0	101.6	110	Dicofol	103.9	102.8
42	4,4-dichlorobenzophenone	101.7	101.8	111	Metolachlor	110.4	99.5
43	Anthraquinone	139.3	128.1	112	Pirimiphoe-ethyl	110.4	103.4
44	Paraoxon-ethyl	101.2	97.7	113	Methoprene	108.8	104.6
45	Cyprodinil	112.5	108.1	114	Bromofos	95.3	90.5

(Continued)

TABLE 5.5 The Influence of Dryness of Rotary Evaporation on the Recoveries of 227 Pesticides (*cont.*)

No.	Pesticides	Just dryness conditions	After evaporation to dryness 2 min	No.	Pesticides	Just dryness conditions	After evaporation to dryness 2 min
46	Butralin	83.7	71.3	115	Ethfumesate	97.4	138.6
47	De-pcb101	108.3	102.6	116	Zoxamide	106.5	107.0
48	Bromophos-ethyl	108.9	104.9	117	Isopeopalin	110.9	98.9
49	Metazachlor	97.4	90.6	118	Pendimethalin	100.5	84.9
50	*trans*-Nonachlor	78.4	69.2	119	Isofenphos	110.0	100.9
51	DEF	102.9	90.5	120	*trans*-Chlordane	107.7	100.8
52	Procymidone	107.9	101.2	121	Chlorfenvinphos	88.8	74.4
53	Perthane	115.1	104.5	122	*cis*-Chlordane	107.9	100.1
54	DE-PCB118	108.4	107.4	123	4,4″(tm)-DDE	112.1	107.4
55	4,4-Dibromobenzo-phenone	127.1	125.3	124	Butachlor	114.9	102.3
56	Iprovalicarb-1	132.0	119.9	125	Chlorfurenol	68.4	54.5
57	DE-PCB153	108.9	103.8	126	Iodofenphos	89.8	82.9
58	4,4-DDD	63.3	58.3	127	Profenofos	46.9	38.2
59	Pyraflufenethyl	65.9	62.4	128	Picoxystrobin	110.3	103.4
60	Tetradifon	106.1	105.3	129	Buprofenzin	112.9	110.2
61	Iprovalicarb-2	109.2	100.7	130	Endrin	111.7	104.0
62	Diofenolan-1	109.1	101.0	131	Chlorfenson	88.4	76.7
63	Diofenolan-2	109.2	104.5	132	2,4″(tm)-DDT	95.4	80.4
64	DE-PCB138	108.3	103.0	133	Methoprotryne	110.5	110.3
65	Chlorfenapyr	106.6	103.5	134	Chloropropylate	105.8	96.1
66	Dichlorofop-methyl	75.9	82.2	135	Flamprop-methyl	124.4	106.9
67	Fluotrimazole	114.0	109.0	136	Oxyfluorfen	100.3	81.9
68	Fluroxypr-1-methyl-hept...	98.5	93.4	137	Chlorthiophos	105.4	100.5
69	Mirex	95.7	84.2	138	Flamprop-isopro-pyl	109.7	101.2
70	Benodanil	112.4	105.6	139	Carbofenothion	113.4	105.8
71	Thenylchlor	94.4	88.8	140	Trifloxystrobin	94.1	85.2
72	DE-PCB180	107.5	104.8	141	Benalaxyl	106.7	100.5
73	Lenacil	75.7	64.3	142	Propiconazole	104.2	101.1
74	Cyhalofop-butyl	93.8	89.7	143	Cyanofenphos	103.2	82.7
75	Silafluofen	132.0	118.8	144	Tebufenpyrad	107.9	101.4
76	Fenbuconazole	109.5	110.4	145	Methoxychlor	91.9	75.5
77	EPTC	85.4	57.4	146	Bromopropylate	90.0	86.8
78	Butylate	88.4	68.8	147	Benzoylprop-ethyl	112.9	111.6
79	Dichlobenil	95.4	78.5	148	Epoxiconazole-2	105.3	104.5
80	Chlormephos	99.2	73.2	149	Piperophos	102.7	93.0
81	Nitrapyrin	91.7	59.6	150	Hexazinone	113.9	110.6
82	Dioxacarb	123.8	113.7	151	Fenamidone	105.4	108.9
83	Chloroneb	112.2	93.9	152	Fenaromol	111.6	104.1

TABLE 5.5 The Influence of Dryness of Rotary Evaporation on the Recoveries of 227 Pesticides (*cont.*)

No.	Pesticides	Just dryness conditions	After evaporation to dryness 2 min	No.	Pesticides	Just dryness conditions	After evaporation to dryness 2 min
84	Tecnazene	56.1	36.8	153	Tralkoxydim	103.5	104.8
85	Hexachlorobenzene	117.9	106.6	154	Fluridone	179.8	181.8
155	Propoxur-1	108.4	105.3	192	Fenson	96.1	97.0
156	Tribenuron-methyl	70.7	69.7	193	Dimethametryn	128.5	119.5
157	Etridiazol	78.4	53.9	194	Flufenacet	73.6	63.4
158	Dimethylphthalate	86.0	73.0	195	Quinalphos	98.4	92.3
159	Molinate	101.0	89.8	196	Diphenamid	111.6	105.1
160	Dibutylsuccinate	102.4	91.6	197	Fenoxanil	126.7	126.2
161	Carbaryl	101.8	115.2	198	Furalaxyl	111.2	103.3
162	2,4-D	91.6	88.2	199	Prothiophos	109.7	101.5
163	Ethalfluralin	88.6	75.2	200	Triadimenol	114.1	111.0
164	Propoxur-2	106.4	105.8	201	Pretilachlor	108.0	99.5
165	Prometon	114.3	108.6	202	Napropamide	117.5	112.8
166	Diazinon	106.8	99.0	203	Kresoxim-methyl	214.7	248.1
167	Triallate	97.2	92.7	204	Tetrasul	109.9	102.5
168	Pyrimethanil	108.2	102.0	205	Isoprothiolane	109.0	103.6
169	Fenpropidin	114.3	107.1	206	Chlorobenzilate	99.8	86.0
170	Propetamphos	122.4	97.5	207	Aclonifen	85.0	53.4
171	Heptachlor	99.7	87.0	208	Metoconazole	108.8	109.5
172	Iprobenfos	113.5	103.1	209	Diniconazole	113.4	106.2
173	Isazofos	102.3	91.7	210	Quinoxyphen	107.5	105.3
174	Pronamide	110.5	103.3	211	Piperonylbutoxide	108.5	101.7
175	Transfluthrin	104.2	101.5	212	Mepronil	103.3	109.5
176	Dinitramine	95.9	77.0	213	Bifenthrin	108.6	105.0
177	Fenpropimorph	108.5	97.7	214	Diflufenican	104.4	98.0
178	Tolclofos-methyl	102.4	97.9	215	Fenazaquin	111.7	109.5
179	Fenchlorphos	96.4	91.1	216	Nuarimol	106.0	104.0
180	Ametryn	101.8	96.7	217	Phenothrin	112.2	101.5
181	Telodrin	93.0	87.6	218	Pyriproxyfen	97.3	93.0
182	Vinclozolin	96.4	91.1	219	Pyridaphenthion	87.5	81.0
183	Metribuzin	96.4	91.2	220	Mefenacet	86.5	72.6
184	Dipropetryn	111.9	104.0	221	Permethrin	105.0	104.5
185	Chlorthal-dimethyl	105.8	100.6	222	Pyridaben	92.0	75.0
186	Fenitrothion	117.4	94.1	223	Bitertanol	110.3	100.0
187	Diethofencarb	108.8	101.0	224	Etofenprox	102.9	103.0
188	Thiazopyr	109.8	104.4	225	Butafenacil	75.3	68.6
189	Triadimefon	104.3	101.3	226	Boscalid	81.0	91.4
190	Dimepiperate	107.0	103.9	227	Dimethomorph	104.8	108.2
191	2,4″(tm)-DDE	119.1	116.4				

TABLE 5.6 Evaluation Results of GC–MS Using Pesticide Quality Control Standard

		Chromatographic peak asymmetry factor (calculate internal standard) 1.00 (±0.10)	Instrument repeatability	Instrument sensitivity (internal standard signal to noise ratio) >1000
[0,1-2]Evaluation and technical target		Retention time RSD% (n = 3) <0.1	Peak area RSD% (n = 3) <5	
The results before maintain	1.11 (unacceptable)	0.01 (acceptable)	5.5 (unacceptable)	386 (unacceptable)
The results after maintain	0.94 (acceptable)	0.003 (acceptable)	0.8 (acceptable)	1824.3 (acceptable)

TABLE 5.7 Evaluation Results of GC–MS/MS Using Pesticide Quality Control Standard

		Chromatographic peak asymmetry factor (calculate internal standard) 1.00 (±0.10)	Instrument repeatability	Instrument sensitivity (internal standard signal to noise ratio) >2000
Evaluation and technical target		Retention time RSD% (n = 3) <0.1	Peak area RSD% (n = 3) <5	
The results before maintain	1.29 (unacceptable)	0.054 (acceptable)	17.8 (unacceptable)	36.7 (unacceptable)
The results after maintain	0.91 (acceptable)	0.008 (acceptable)	1.4 (acceptable)	2904.6 (acceptable)

The evaluation is capable of discovering problems existing with the instrument, which can be restored into a comparatively ideal state through related maintenance, with all indexes satisfying the requirements for collaborative tests. See Table 5.8 for technical indexes and maintenance measures for problems discovered.

As the saying goes, you can"(tm)t make bricks without straw. Therefore, the instruments must be checked for their technical indexes and conditions to see if they meet the technical conditions required. This is the first key control point before starting determinations.

5.4.2.2 Effect of Equipment Status on Test Results of Pesticide Aged Tea Sample

Statistical analysis is conducted twice on the test data from 227 pesticide aged tea samples, before and after cleaning the ion source, with results tabulated in Table 5.9. The statistical analysis of data in Table 5.9 finds that pesticides with RSD < 5% for parallel determination only account for 16.3% of the total number of pesticides prior to the cleaning of the ion source, but 74% after the cleaning of the ion source. This illustrates that the parallelism of the analytical results from after cleaning the ion source is evidently better than those from before cleaning the ion source (Fig. 5.6). Therefore, the evaluation of the state of the instruments and instrumental maintenance is regarded as one of key control points in the process of determination.

5.4.3 Chromatographic Resolution (Error Analysis) and Key Control Points

In the process of studying the ruggedness of the method, the statistical analysis of the integration results from GC–MS and GC–MS/MS determination for 271 pesticides in green tea and oolong tea found that there were 12.5%–28.4% automatic integration errors for GC–MS, and 4.0%–9.7% for GC–MS/MS, which needed to be corrected manually. The frequency of the integration errors was closely related to the extent of contamination of the instrument. A comparative study conducted on the integration errors before and after cleaning the ion source (including replacing the inlet lining tube), shows the statistical results tabulated in Table 5.10 for GC–MS and Table 5.11 for GC–MS/MS.

TABLE 5.8 Evaluation Technical Indexes and Maintenance Measures of GC–MS and GC–MS/MS

Evaluation	Chromatographic peak asymmetry factor (calculate internal standard)	Instrument repeatability Retention time RSD% ($n = 3$)	Instrument sensitivity (internal standard signal to noise ratio) Peak area RSD% ($n = 3$)
GC–MS technical target	1.00 (±0.10)	<0.1	<5	>1000
GC–MS/MS technical target	1.00 (±0.10)	<0.1	<5	>2000
Maintain measure for unacceptable results	1. Cut off a section of the contaminated capillary column 2. Replace the inlet liner 3. Replace capillary column	Adjust the flow (flow should not be <0.8 mL/min)	1. System leak check 2. Replacement Inlet Septa 3. Replace the inlet liner	1. Replace the inlet liner 2. Cut off a section of the contaminated capillary column 3. Cleaning the ion source

TABLE 5.9 Comparison of the Two Test Results From Before and After Cleaning the Ion Source for 227 Pesticide Aged Tea Samples ($n=3$)

No.	Pesticides	RSD (%) Before cleaning	After cleaning	No.	Pesticides	RSD (%) Before cleaning	After cleaning
1	Isoprocarb-1	11.0	0.2	5	Cycloate	6.7	2.6
2	Pebulate	9.7	6.1	6	Isoprocarb-2	11.2	4.9
3	Pentachlorobenzene	9.0	4.6	7	Thionazin	7.3	0.6
4	Propham	6.3	5.1	8	Tebuthiuron	2.4	0.6
9	2,3,5,6-Tetrachloroani…	10.6	7.8	77	EPTC	1.8	9.0
10	Pentachloroanisole	9.3	7.0	78	Butylate	10.1	8.6
11	Trifluralin	9.7	1.4	79	Dichlobenil	4.6	8.1
12	Tebutam	10.7	2.5	80	Chlormephos	4.5	6.8
13	Methabenzthiazuron	9.2	0.3	81	Nitrapyrin	3.3	9.1
14	Simeton	9.9	1.7	82	Dioxacarb	5.7	6.1
15	Atratone	4.8	2.1	83	Chloroneb	11.1	2.4
16	Phenanthrene	14.6	1.9	84	Tecnazene	2.1	9.4
17	Clomazone	12.2	4.0	85	Hexachlorobenzene	8.2	19.8
18	Tefluthrin	11.5	1.2	86	Ethoprophos	5.8	0.6
19	Fenpyroximate	10.6	5.4	87	cis-Diallate	4.0	3.4
20	Tebupirimfos	6.9	1.9	88	Propachlor	5.9	6.6
21	Cycluron	11.9	5.2	89	trans-Diallate	6.8	4.2
22	DE-PCB28	6.0	2.0	90	Sulfotep	4.8	4.6
23	DE-PCB31	9.6	1.1	91	Chlorpropham	6.1	2.5
24	Secbumeton	3.0	2.9	92	Quintozen	7.5	7.0
25	Pyroquilon	1.7	2.3	93	Fonofos	7.0	2.7

(Continued)

TABLE 5.9 Comparison of the Two Test Results From Before and After Cleaning the Ion Source for 227 Pesticide Aged Tea Samples (*n*=3) (*cont.*)

| No. | Pesticides | RSD (%) | | No. | Pesticides | RSD (%) | |
		Before cleaning	After cleaning			Before cleaning	After cleaning
26	2,3,4,5-Tetra-chloroani...	5.5	1.0	94	Profluralin	4.7	2.5
27	Pentachloro-aniline	7.4	6.2	95	Propazine	6.5	1.7
28	Pirimicarb	18.0	1.5	96	Etrimfos	6.1	4.0
29	Benoxacor	7.4	5.8	97	Dicloran	3.1	2.2
30	Prosulfocarb	8.6	0.6	98	Propyzamide	5.9	0.9
31	Acetochlor	12.2	1.7	99	Chlorprifos-methyl	5.0	8.0
32	Dimethenamid	9.3	0.0	100	Desmetryn	2.2	2.3
33	DE-PCB52	10.5	2.2	101	Ronnel	6.6	7.2
34	Alachlor	9.5	1.0	102	Dimethachlor	14.0	6.0
35	Prometrye	3.7	4.6	103	Pirimiphos-methyl	9.5	3.6
36	Monalide	0.4	0.8	104	Mefenoxam	4.9	1.4
37	Beta-HCH	6.9	6.9	105	Endosulfan-1	1.7	7.3
38	Metalaxyl	7.6	1.3	106	Thiobencarb	9.0	0.9
39	Isodrin	10.1	3.4	107	Terbutryn	5.5	1.8
40	Trichloronat	0.8	3.5	108	Chlorpyifos (ethyl)	4.1	4.1
41	Dacthal	8.5	1.9	109	Methyl-para-thion	10.9	8.5
42	4,4-Dichloro-benzophenone	8.5	6.1	110	Dicofol	6.9	5.3
43	Anthraquinone	9.5	5.3	111	Metolachlor	7.7	3.1
44	Paraoxon-ethyl	0.4	4.3	112	Pirimiphoe-ethyl	7.5	2.3
45	Cyprodinil	14.4	5.1	113	Methoprene	4.2	1.4
46	Butralin	11.5	2.7	114	Bromofos	5.8	9.1
47	De-pcb101	11.5	2.1	115	Ethfumesate	8.8	1.0
48	Bromophos-ethyl	15.5	0.4	116	Zoxamide	1.2	6.8
49	Metazachlor	5.0	0.2	117	Isopeopalin	4.9	1.5
50	*trans*-Non-achlor	14.8	9.0	118	Pendimethalin	4.6	2.7
51	DEF	12.0	2.7	119	Isofenphos	8.0	1.8
52	Procymidone	9.0	1.1	120	*trans*-Chlor-dane	10.3	2.9
53	Perthane	18.7	7.8	121	Chlorfenvin-phos	10.8	8.4
54	DE-PCB118	8.9	1.9	122	*cis*-Chlordane	6.2	3.0
55	4,4-Dibromo-benzophenone	9.3	6.1	123	4,4″(tm)-DDE	9.1	1.6
56	Iprovalicarb-1	12.6	0.3	124	Butachlor	7.4	7.4
57	DE-PCB153	17.5	1.5	125	Chlorfurenol	3.2	9.0

TABLE 5.9 Comparison of the Two Test Results From Before and After Cleaning the Ion Source for 227 Pesticide Aged Tea Samples (*n*=3) (*cont.*)

| No. | Pesticides | RSD (%) | | No. | Pesticides | RSD (%) | |
		Before cleaning	After cleaning			Before cleaning	After cleaning
58	4,4-ddd	14.4	8.4	126	Iodofenphos	13.7	9.6
59	Pyraflufenethyl	15.2	8.4	127	Profenofos	3.6	14.9
60	Tetradifon	1.9	8.4	128	Picoxystrobin	7.1	0.9
61	Iprovalicarb-2	12.7	1.7	129	Buprofenzin	5.8	2.1
62	Diofenolan-1	15.9	0.4	130	Endrin	2.3	2.9
63	Diofenolan-2	16.1	1.0	131	Chlorfenson	1.7	7.7
64	DE-PCB138	22.9	0.8	132	2,4″(tm)-DDT	7.8	2.9
65	Chlorfenapyr	13.1	0.3	133	Methoprotryne	10.1	0.3
66	Dichlorofop-methyl	7.7	0.9	134	Chloropropylate	13.9	1.0
67	Fluotrimazole	6.0	0.5	135	Flamprop-methyl	10.9	2.3
68	Fluroxypr-1-methylhept...	10.2	0.3	136	Oxyfluorfen	8.9	0.4
69	Mirex	11.1	4.3	137	Chlorthiophos	1.5	2.4
70	Benodanil	3.7	1.2	138	Flamprop-isopropyl	2.4	0.7
71	Thenylchlor	19.5	4.8	139	Carbofenothion	10.9	1.5
72	DE-PCB180	12.6	0.4	140	Trifloxystrobin	6.4	3.7
73	Lenacil	14.5	3.8	141	Benalaxyl	12.2	2.2
74	Cyhalofop-butyl	18.0	0.1	142	Propiconazole	12.7	0.6
75	Silafluofen	22.4	3.6	143	Cyanofenphos	16.4	4.8
76	Fenbuconazole	13.9	3.0	144	Tebufenpyrad	10.3	0.8
145	Methoxychlor	2.1	4.4	187	Diethofencarb	19.0	4.5
146	Bromopropylate	8.9	2.5	188	Thiazopyr	27.8	2.3
147	Benzoylprop-ethyl	10.2	0.5	189	Triadimefon	6.0	5.7
148	Epoxiconazole-2	9.8	1.0	190	Dimepiperate	14.8	2.2
149	Piperophos	10.1	3.0	191	2,4″(tm)-dde	21.4	0.5
150	Hexazinone	14.0	7.2	192	Fenson	5.5	3.8
151	Fenamidone	11.4	1.0	193	Dimethametryn	18.5	0.5
152	Fenaromol	13.5	1.4	194	Flufenacet	5.2	2.5
153	Tralkoxydim	14.2	0.2	195	Quinalphos	18.9	0.7
154	Fluridone	0.3	19.0	196	Diphenamid	21.6	1.2
155	Propoxur-1	15.6	1.7	197	Fenoxanil	31.4	6.9
156	Tribenuron-methyl	16.4	0.2	198	Furalaxyl	23.4	1.3
157	Etridiazol	0.9	9.1	199	Prothiophos	12.9	1.7
158	Dimethyl-phthalate	11.5	4.1	200	Triadimenol	21.1	1.3

(Continued)

TABLE 5.9 Comparison of the Two Test Results From Before and After Cleaning the Ion Source for 227 Pesticide Aged Tea Samples (*n*=3) (*cont.*)

No.	Pesticides	RSD (%) Before cleaning	After cleaning	No.	Pesticides	RSD (%) Before cleaning	After cleaning
159	Molinate	18.9	7.8	201	Pretilachlor	14.1	0.8
160	Dibutylsuccinate	17.6	3.5	202	Napropamide	22.8	1.0
161	Carbaryl	9.3	7.9	203	Kresoxim-methyl	21.9	7.8
162	2,4-D	9.6	7.1	204	Tetrasul	14.8	4.4
163	Ethalfluralin	12.2	3.2	205	Isoprothiolane	23.4	0.4
164	Propoxur-2	26.6	1.6	206	Chlorobenzilate	25.8	4.1
165	Prometon	9.8	1.7	207	Aclonifen	1.4	7.4
166	Diazinon	22.2	2.0	208	Metoconazole	14.9	8.7
167	Triallate	15.4	3.8	209	Diniconazole	16.7	0.0
168	Pyrimethanil	13.9	1.7	210	Quinoxyphen	21.3	3.0
169	Fenpropidin	12.5	1.3	211	Piperonylbutoxide	23.6	6.7
170	Propetamphos	12.4	4.1	212	Mepronil	19.2	4.0
171	Heptachlor	17.6	1.6	213	Bifenthrin	21.1	5.2
172	Iprobenfos	19.2	0.8	214	Diflufenican	18.3	2.4
173	Isazofos	16.0	2.5	215	Fenazaquin	18.3	1.0
174	Pronamide	21.7	0.5	216	Nuarimol	14.6	0.3
175	Transfluthrin	16.2	0.6	217	Phenothrin	14.0	5.0
176	Dinitramine	10.2	1.6	218	Pyriproxyfen	21.2	3.8
177	Fenpropimorph	14.6	1.9	219	Pyridaphenthion	16.1	1.8
178	Tolclofos-methyl	14.1	3.1	220	Mefenacet	9.1	5.3
179	Fenchlorphos	17.5	3.6	221	Permethrin	13.5	0.2
180	Ametryn	16.2	0.6	222	Pyridaben	17.1	7.1
181	Telodrin	29.1	6.5	223	Bitertanol	12.4	2.0
182	Vinclozolin	17.6	1.7	224	Etofenprox	11.0	2.5
183	Metribuzin	6.9	0.8	225	Butafenacil	5.9	4.9
184	Dipropetryn	21.2	2.0	226	Boscalid	13.4	3.9
185	Chlorthal-dimethyl	15.1	2.7	227	Dimethomorph	14.6	4.8
186	Fenitrothion	11.9	2.8				

Table 5.10 finds that there are 146 pesticides with mistaken peaks or without integration owing to deviation of retention time, accounting for 34.7%; there are 121 pesticides with mistaken peaks or with integration, accounting for 28.7%; and there are 124 pesticides that need to be manually corrected for their integration due to the drifting of baselines and/or unreasonable choice of integration lines. The sum of the three makes up 92.8% of the total number of manual integrations. Obviously, deviation of retention time, drifting of baselines, and dropping of sensitivity of the instrument constitute the major causes of automatic integration errors from the instrument.

The comparison of before and after cleaning the ion source finds that there are 114 pesticide peaks with mistaken peaks or without integration due to low sensitivity of instrument and weak response before cleaning the ion source, while there

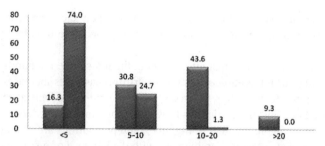

FIGURE 5.6   RSD distribution for GC–MS analytical results from before and after cleaning the ion source.

TABLE 5.10 Comparison of Manual Integration of Pesticide Peaks in Tea Samples Using GC–MS Determination

| | | Oolong tea sample I | | Oolong tea sample II | | Green tea sample I | | Green tea sample II | | | Percentage (manual integration, %) |
| | | Ion source | | Ion source | | Ion source | | Ion source | | | |
The reasons of manual integration		Before cleaning	After cleaning	Before cleaning	After cleaning	Before cleaning	After cleaning	Before cleaning	After cleaning	Total	
1	Retention time migrate, no integration	6	6	5	5	1	4	2	2	31	7.4
2	Retention time migrate, error integration	17	21	3	9	12	14	22	17	115	27.3
	Subtotal	23	27	8	14	13	18	24	19	146	34.7
3	Low response, no integration	19	2	57	2	17	1	11	0	109	25.9
4	Low response, error integration	3	1	0	0	7	1	0	0	12	2.8
	Subtotal	22	3	57	2	24	2	11	0	121	28.7
5	Baseline drift	18	20	7	5	19	23	12	20	124	29.4
6	Isomers, error integration	1	1	2	2	2	1	0	1	10	2.4
7	Qualitative and quantitative ion interfere by matrix, error integration	1	5	3	11	0	0	0	0	20	4.8
	Total	65	56	77	34	58	44	47	40	421	100
	Percentage (271 pesticides, %)	24.0	20.7	28.4	12.5	21.4	16.2	17.3	14.8		

are seven peaks after cleaning the ion source. Thus, it can be seen that timely cleaning of the ion source and changing the inlet lining tube increases the instrument sensitivity and contributes to accurate qualitative and quantitative analysis of pesticide peaks.

The comparison also finds, at the same time, that after cleaning the ion source the number of retention time deviations increases from 68 to 78; the number of integration errors due to drifting baselines increases from 56 to 68, and qualification and quantification of ions are interfered with by matrixes, while the number of mistaken peaks increases from 4 to 16. Laboratory personnel are required to adjust the carrier gas flow of the instrument to reduce deviation of retention time after cleaning the ion source (including changing the inlet lining tube), and precautions are also taken because baselines will be elevated and ion interference in the matrixes will also increase with the increase of instrumental sensitivity.

Tables 5.10 and 5.11 show that the precision rate of automatic integration for GC–MS/MS is higher than that for GC–MS, with 93.6% and 80.6%, respectively, on the average. When the instrument is in a sound state (after cleaning the ion source), the precision rate of automatic integration for GC–MS and GC–MS/MS can reach above 95% and 84%, respectively. To ensure that the analytical results are correct, automatic integration results in the experimental process are checked one by one. Twenty-six examples are given for concrete error analysis and chromatographic resolution based on seven manual integrations:

1. Failure of integration due to low instrumental sensitivity. Instrumental sensitivity increases after cleaning the ion source, and automatic integration starts (Figs. 5.7–5.10).
2. Integration errors caused by mistaken peaks from low instrumental sensitivity (instrumental automatic integration gives priority to the relatively high responsive peaks in time windows and identifies them as the target peaks, which, however, are not the target pesticides after comparing them with the abundance ratio of ions) (Figs. 5.11–5.14).
3. Failure of integration due to retention time deviation: the retention time of pesticide peaks deviates from the time window, failing to be identified. After adjusting the carrier gas flow, the instrument starts automatic integration (Figs. 5.15–5.18).
4. Mistaken peaks caused by retention time deviations: retention time deviates and target pesticides fail to integrate, leading to mistaken peaks. The retention time is restored and the instrument correctly identifies pesticide peaks after cleaning the ion source and adjusting the carrier gas flow rate (Figs. 5.19–5.22).
5. Isomer peaks are easy to misidentify because their retention times are similar. The correct retention time and the identical monitoring ions should be listed and double checked through their retention time and peaking sequences (Figs. 5.23 and 5.24).
6. Baseline drifting: it will cause certain integration lines of chromatic peaks to be wrongly chosen, leading to erroneous integration. Incorrect choice of automatic integration lines due to drifting of baselines causes the automatic integration to give an erroneous result (Figs. 5.25–5.28).
7. Interfering ions existing in the matrixes, if any, will affect the automatic integration results. During automatic integration interfering ion peaks are mistaken for target pesticide peaks, which need to be corrected manually (Figs. 5.29–5.32).

TABLE 5.11 Comparison of Manual Integration of Pesticide Peaks in Tea Samples Using GC–MS/MS Determination

| The reasons of manual integration | | Oolong tea sample | | Green tea sample | | Total | Percentage (manual integration, %) |
| | | Ion source | | Ion source | | | |
		Before cleaning	After cleaning	Before cleaning	After cleaning		
1	Low response, error integration	13	2	10	3	28	48.3
2	Isomers, error integration	5	5	5	5	20	34.5
3	Qualitative and quantitative ion interfere by matrix, error integration	4	2	2	2	10	17.2
Total		22	9	17	10	58	100
Percentage (271 pesticides, %)		9.7	4.0	7.5	4.4		

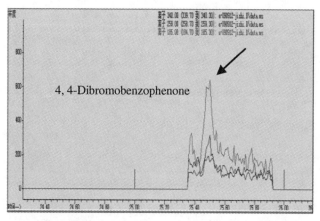

FIGURE 5.7   Before cleaning ion source, no integration.

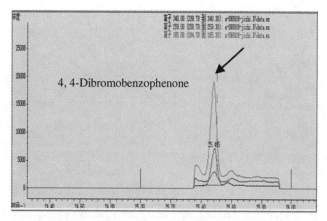

FIGURE 5.8   After cleaning ion source, automatic integration.

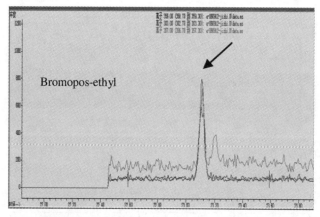

FIGURE 5.9   Before cleaning ion source, no integration.

The error analysis and chromatographic resolution of the aforementioned 26 cases show that the following three key control points are kept in mind to ensure qualitative and quantitative results free from errors: first, the instrument is always be maintained in a sound working state by carrying out maintenance work, such as timely cleaning of ion sources, changing the lining tube or chromatic column; second, the retention time of internal standards are checked for each determination to ensure that each pesticide peak falls within the integration window; third, the retention time and ion abundance ratio of each

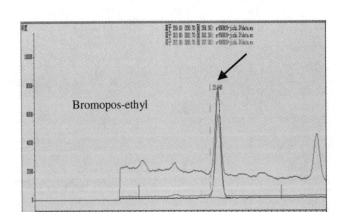

FIGURE 5.10　After cleaning ion source, automatic integration.

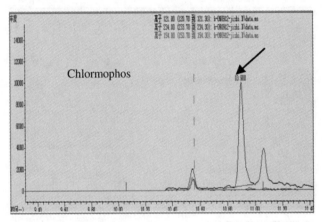

FIGURE 5.11　Before cleaning ion source, error integration.

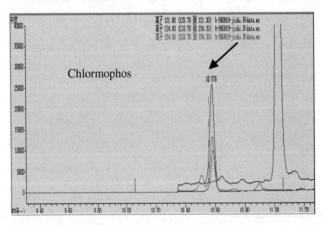

FIGURE 5.12　After cleaning ion source, correct integration.

pesticide peak are checked according to the qualitative and quantitative requirements of the method; fourth, the integration line of each pesticide peak are checked to see if it is correctly chosen. For pesticides with integration line problems, a manual integration mode is adopted from peak valley to peak valley. Only when the three key control points are followed and executed can accurate qualitative and quantitative analysis be achieved.

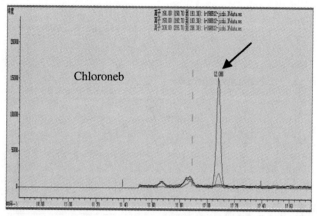

FIGURE 5.13 Before cleaning ion source, error integration.

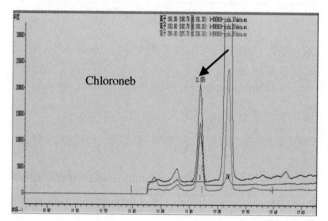

FIGURE 5.14 After cleaning ion source, correct integration.

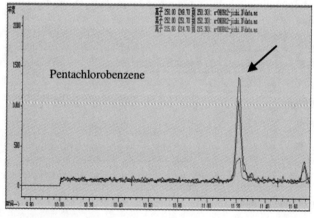

FIGURE 5.15 Retention time migrate, no integration.

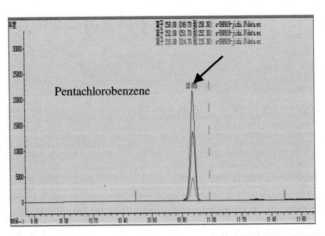

FIGURE 5.16 After adjust flow, correct integration.

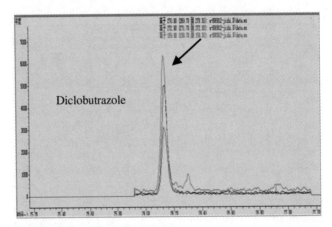

FIGURE 5.17 Retention time migrate, no integration.

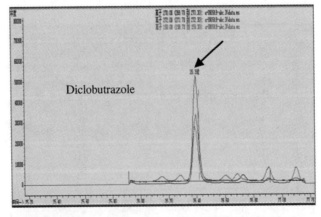

FIGURE 5.18 After adjust flow, correct integration.

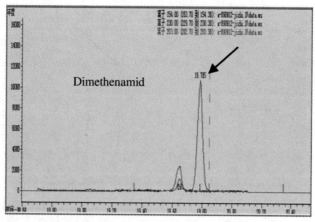

FIGURE 5.19   Retention time migrate, error integration.

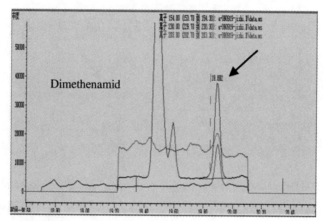

FIGURE 5.20   After adjust flow, correct integration.

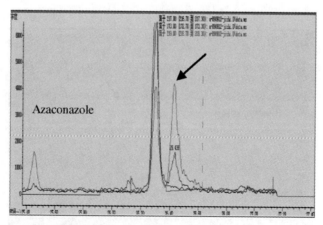

FIGURE 5.21   Retention time migrate, error integration.

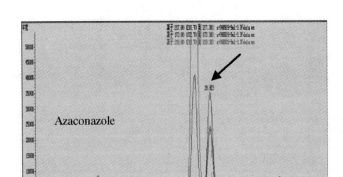

FIGURE 5.22 After adjust flow, correct integration.

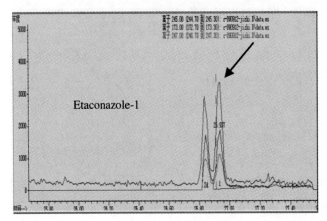

FIGURE 5.23 Automatic integration error.

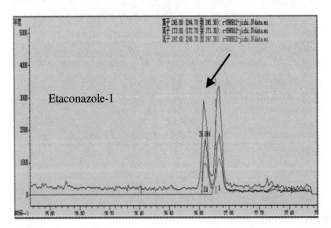

FIGURE 5.24 Manual integration correct.

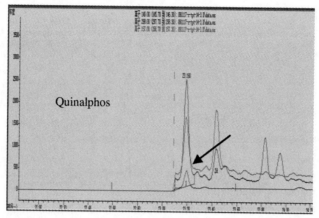

FIGURE 5.25   **Automatic integration error.**

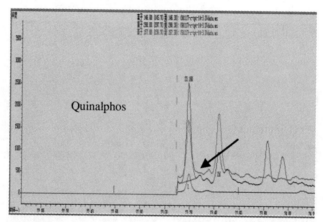

FIGURE 5.26   **Manual integration correct.**

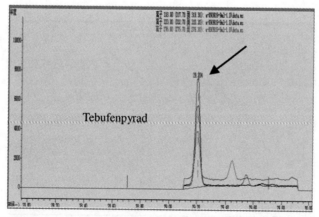

FIGURE 5.27   **Automatic integration error.**

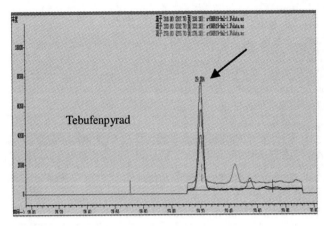

FIGURE 5.28   **Manual integration correct.**

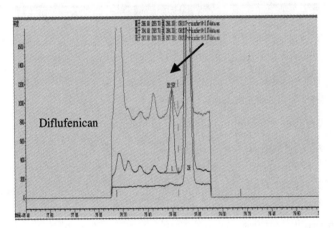

FIGURE 5.29   **Automatic integration error.**

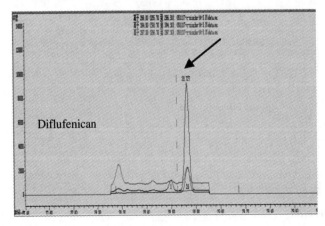

FIGURE 5.30   **Manual integration correct.**

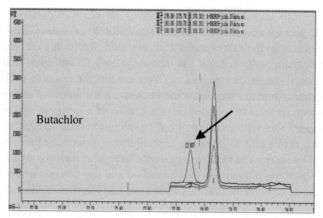

FIGURE 5.31 Automatic integration error.

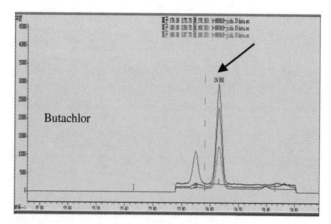

FIGURE 5.32 Manual integration correct.

5.5 CONCLUSIONS

Based on the systematic evaluations of the ruggedness of the method, the causes of the errors were analyzed and the corresponding key control points were established, which provided a technical warrantee for the method efficacy. The study of ruggedness also discovers that certain errors exist with specific data, 26 examples are given for concrete chromatographic resolution based on the seven manual integrations and conducted chromatographic resolution; it has been found that there are some deviations for certain samples, for which mass spectra have to be checked. The causes of the errors have been analyzed and traced, and key control points have been formulated accordingly. Errors mainly consist of the following three aspects: (1) errors from the pretreatment of samples, which chiefly occur during the stage of rotary evaporation. Tests find that the recoveries of certain pesticides tend to drop with the increase of rotary evaporation temperature and speeding of the rotary evaporation, which agree with the errors revealed during tests of practical aged samples; (2) errors from the instrumental conditions, which chiefly influences the parallelism of the test results. Tests find that pesticides with RSD < 5% for parallel determination only account for 16.3% of the total number of pesticides prior to the cleaning of the ion source, but 74% after the cleaning of the ion source; and (3) errors from chromatogram resolutions. The automatic integration results from GC–MS and GC–MS/MS have been compared to find that there are 12.5%–28.4% automatic integration errors for GC–MS, and 4.0%–9.7% for GC–MS/MS, which need to be corrected manually.

The corresponding key control points have been formulated against the errors of the aforementioned three aspects: (1) concentration temperature in the first place shall not exceed 40 C when using rotary evaporation as over high temperature will cause part of the drugs to dissolve and evaporation to dryness will also lead to pesticide loss. Plus, solvent exchange is made twice and is carried out once again if the liquid surface still has a separating layer after the two times exchange. Precautions are taken to the exchange with hexane that the revolutions of the rotary evaporator are reduced properly if evaporation is going very rapidly so as to prevent pesticide loss from evaporation to dryness. (2) Before the samples are submitted for analysis, the instrument is checked for their three states and conditions, including chromatographic asymmetric factors, instrumental reproducibility, and sensitivity to see if it has met the technical conditions required. (3) For each determination, the retention time of the internal standards are checked to ensure that the peaks of each target pesticide are all within the integration windows; the retention time of every pesticide peak and ion abundance ratio is verified strictly according to the qualitative and quantitative requirements of the method; the choice of every pesticide peak integration line is verified to see if it is correct. Regarding pesticides with integration line problems, manual integration of peak valley to peak valley are uniformly adopted. In addition, errors tend to get worse if problems exist with the performance of the instrument. At the same time, the technical skills of operators and lack of experience in residue analysis will also lead to errors of analytical results. The systematic evaluations of the ruggedness of the method and the analysis of the causes of errors have resulted in the formulations of corresponding key control points, which have laid the foundation for the AOAC collaborative study of multiresidues in tea.

Appendix Table 5.1 The 201 Pesticides Average Content for 19 Circulative Determinations on 3 Months Under 16 Factors (Two Cartridges, Two Instruments, Two Tea Samples, and Two Youden Pair Samples)

Average content (μg/kg)

No.	Pesticides	SPE	Method	Sample	Youden pair	Nov 9, 2009 (n=5)	Nov 14, 2009 (n=3)	Nov 19, 2009 (n=3)	Nov 24, 2009 (n=3)	Nov 29, 2009 (n=3)	Dec 4, 2009 (n=3)	Dec 9, 2009 (n=3)	Dec 14, 2009 (n=3)	Dec 14, 2009 (n=3)	Dec 19, 2009 (n=3)	Dec 24, 2009 (n=3)	Dec 14, 2009 (n=3)	Jan 3, 2010 (n=3)	Jan 8, 2010 (n=3)	Jan 13, 2010 (n=3)	Jan 18, 2010 (n=3)	Jan 23, 2010 (n=3)	Jan 28, 2010 (n=3)	Feb 2, 2010 (n=3)	Feb 7, 2010 (n=3)	Ave (μg/kg)
1	2,3,4,5-Tetrachloroaniline	Cleanert TPT	GC–MS	Green tea	A	10.0	5.7	4.6	3.4	5.1	4.2	4.7	4.5	4.8	5.1	4.3	4.8	4.3	3.6	4.0	3.2	2.7	4.1	3.4	4.2	4.5
					B	7.1	5.4	4.8	3.2	4.7	4.2	4.2	3.9	4.3	4.4	4.2	3.9	4.0	3.5	4.1	3.2	2.7	3.6	3.3	3.8	4.1
				Oolong tea	A	7.3	6.6	5.7	6.1	5.8	4.6	4.8	4.4	5.2	5.4	5.0	4.4	5.2	5.1	4.8	6.5	4.9	3.9	5.0	4.7	5.3
					B	6.9	5.4	4.4	4.8	5.1	3.7	3.7	4.0	4.1	4.5	4.2	4.0	4.4	4.0	3.9	5.3	3.8	3.1	4.4	3.9	4.4
			GC–MS/MS	Green tea	A	6.6	5.5	6.5	6.2	3.9	5.6	4.6	4.1	3.5	4.5	3.6	4.1	4.0	3.7	3.9	3.5	2.5	3.4	2.9	4.0	4.3
					B	6.2	5.2	5.6	5.6	4.1	5.2	4.1	3.8	3.4	4.3	3.9	3.8	4.1	3.4	4.2	3.5	2.6	3.3	2.7	3.6	4.1
				Oolong tea	A	6.8	6.7	5.9	5.5	6.2	4.5	4.9	5.5	5.3	5.4	5.2	5.5	5.3	4.6	4.9	4.3	4.7	3.2	3.4	3.8	5.1
					B	6.3	5.8	4.6	4.8	4.8	3.9	3.5	4.2	3.7	4.3	3.9	4.2	4.1	3.5	3.7	3.2	3.4	2.5	2.8	3.6	4.0
		ENVI-Carb + PSA	GC–MS	Green tea	A	5.6	7.7	7.5	6.1	5.9	5.6	4.3	4.8	4.6	4.7	4.8	4.8	4.7	3.9	4.2	3.8	4.7	4.3	4.3	4.1	5.0
					B	5.4	6.9	6.0	5.5	5.5	4.9	4.4	4.8	4.3	4.7	5.0	4.8	4.6	3.6	4.2	3.2	4.1	3.9	3.9	4.1	4.7
				Oolong tea	A	9.0	6.9	6.4	4.5	7.0	5.5	6.2	4.8	4.7	5.1	5.7	4.0	4.4	3.1	3.9	4.3	3.4	3.9	3.9	4.1	5.1
					B	7.3	5.9	5.5	3.9	5.6	4.3	5.2	4.0	4.2	3.8	4.4	4.2	3.6	3.3	3.1	3.0	2.8	2.9	2.9	3.7	4.2
			GC–MS/MS	Green tea	A	5.9	6.7	5.2	6.0	5.8	5.0	4.8	4.5	4.8	4.6	4.8	4.5	4.8	3.4	4.0	3.4	4.3	3.2	3.3	2.9	4.9
					B	5.8	5.4	5.0	4.9	5.6	4.9	4.4	4.1	3.6	4.5	4.8	4.1	3.6	3.9	4.1	2.6	3.7	3.2	3.0	3.9	4.6
				Oolong tea	A	8.5	8.1	7.4	5.0	6.2	5.6	5.4	5.0	4.8	5.1	5.0	3.8	4.3	3.9	3.9	4.2	3.7	3.1	3.8	3.8	5.1
					B	6.8	6.6	5.4	5.0	5.1	4.3	4.7	3.8	4.4	3.8	3.9	3.1	4.3	3.1	3.2	2.9	2.5	2.6	3.7	3.7	4.1
2	2,3,5,6-Tetrachloroaniline	Cleanert TPT	GC–MS	Green tea	A	44.3	36.2	38.9	39.8	33.2	37.3	34.2	34.1	35.0	35.8	34.4	34.1	34.7	35.3	36.3	33.8	23.7	31.5	29.4	38.2	35.0
					B	41.3	34.8	34.8	37.3	29.7	35.4	31.3	31.2	31.5	35.7	32.8	31.2	32.4	33.4	33.5	30.3	23.2	30.5	27.5	33.4	32.6
				Oolong tea	A	39.8	40.9	35.0	37.8	37.6	31.6	31.8	38.8	38.7	39.5	38.5	38.8	39.2	38.5	38.2	35.8	36.5	32.8	34.9	37.1	37.0
					B	37.0	34.4	28.9	32.0	34.6	26.4	26.6	32.8	30.7	33.4	33.0	32.8	33.0	31.3	33.8	29.6	31.5	28.1	31.3	30.9	31.5
			GC–MS/MS	Green tea	A	43.3	38.7	41.9	42.2	30.0	34.9	35.4	31.4	35.8	36.8	32.6	31.4	35.3	34.9	31.9	35.2	28.1	31.1	27.8	39.6	35.1
					B	42.7	36.4	36.7	37.2	28.3	33.9	32.4	29.8	32.5	36.3	31.9	29.8	33.4	33.0	31.9	33.3	27.2	30.4	27.4	35.5	33.2
				Oolong tea	A	37.2	40.9	35.3	36.2	36.4	33.0	32.7	40.4	37.6	38.3	37.6	40.4	37.7	37.3	39.0	33.0	35.4	32.4	32.2	32.9	36.3
					B	35.2	33.7	29.7	30.6	33.5	28.0	26.7	35.0	31.7	31.2	32.8	35.0	30.9	32.6	35.8	27.2	30.7	29.1	29.6	27.7	31.1
		ENVI-Carb + PSA	GC–MS	Green tea	A	41.1	41.0	37.6	41.0	42.4	39.7	35.5	37.2	35.4	37.8	34.6	37.2	38.6	37.8	42.7	37.0	40.1	36.7	36.7	35.0	38.3
					B	38.2	34.8	37.8	37.0	40.9	36.4	37.1	36.1	32.4	34.4	33.8	36.1	37.0	34.1	37.9	33.4	34.1	34.7	31.3	32.7	35.5
				Oolong tea	A	46.0	39.0	37.3	35.5	40.8	36.4	41.4	32.0	35.3	41.5	37.3	32.0	31.6	32.5	35.0	36.1	29.9	29.6	31.9	31.3	35.8
					B	35.7	33.1	33.5	31.4	33.3	30.8	35.2	28.7	30.4	32.9	30.1	28.7	27.5	28.5	29.4	28.8	24.8	25.2	26.4	28.0	30.2
			GC–MS/MS	Green tea	A	33.5	42.0	38.9	40.5	42.5	41.1	36.2	37.5	32.5	36.7	39.7	37.5	116.2	36.7	43.5	28.4	39.1	41.0	33.7	35.0	42.5
					B	48.2	41.2	46.3	35.7	42.0	38.0	33.9	34.1	30.8	34.5	37.6	34.1	112.4	35.8	38.8	33.4	33.7	35.4	29.7	39.3	39.3
				Oolong tea	A	38.2	35.1	33.6	35.6	40.5	36.6	37.7	34.8	31.5	31.5	36.8	34.8	30.7	31.5	33.3	35.7	27.5	25.6	33.0	26.5	35.7
					B	42.7	34.8	31.1	30.7	33.5	30.7	33.9	29.7	28.8	29.8	28.8	30.0	27.4	28.8	29.8	26.8	22.9	23.2	26.7	25.6	29.9
3	4,4-Dibromobenzophenone	Cleanert TPT	GC–MS	Green tea	A	47.2	44.7	46.6	44.7	40.7	44.7	51.7	49.0	52.7	66.3	39.3	49.0	59.8	56.3	50.9	56.8	81.9	51.5	48.1	43.8	51.4
					B	45.6	40.7	44.2	38.4	36.8	42.6	46.8	45.8	47.2	52.7	37.2	45.8	51.4	52.0	49.7	50.4	79.6	48.5	43.9	38.5	46.9
				Oolong tea	A	43.6	41.3	48.3	49.1	55.2	39.7	64.6	67.7	44.7	43.1	44.7	67.7	40.8	51.6	48.3	56.7	59.0	61.6	37.2	40.4	50.5
					B	39.4	37.1	42.0	42.0	47.6	35.0	59.3	64.0	39.5	38.1	39.5	64.0	35.7	43.2	41.2	45.0	50.7	51.0	36.4	35.3	43.8
			GC–MS/MS	Green tea	A	52.6	41.0	49.4	44.1	38.6	47.6	42.9	42.1	52.1	51.6	36.0	42.1	45.4	60.2	40.5	42.6	36.8	39.3	33.8	35.3	47.5
					B	55.3	40.9	43.7	44.1	34.3	42.5	39.2	41.0	47.0	51.6	33.4	41.0	45.1	53.3	40.2	40.4	36.1	39.1	33.0	79.3	43.9
				Oolong tea	A	64.2	67.2	43.6	46.5	41.6	40.0	50.7	63.5	41.5	57.6	41.5	63.5	44.2	51.9	51.9	54.7	56.3	47.0	47.4	48.1	52.2
					B	63.1	58.5	37.1	39.3	41.5	37.0	42.4	60.7	36.0	51.1	36.0	60.7	37.2	42.8	44.4	43.9	49.0	38.2	47.1	42.6	45.6
		ENVI-Carb + PSA	GC–MS	Green tea	A	55.3	46.8	53.6	53.4	49.1	51.1	62.4	57.7	48.2	67.3	47.9	57.7	58.9	51.9	51.3	63.4	57.2	92.0	38.2	51.0	56.9
					B	54.2	43.5	54.5	49.5	48.1	45.2	58.4	59.0	52.5	59.9	55.0	59.0	51.4	50.3	44.9	60.2	45.5	73.7	52.0	50.4	53.1
				Oolong tea	A	56.0	44.3	42.9	46.0	58.1	43.5	52.4	93.6	42.0	42.5	45.0	93.6	44.2	58.3	65.6	44.5	38.6	49.6	36.8	41.4	49.8
					B	47.9	38.9	42.5	40.4	47.5	35.4	40.6	76.9	36.6	34.7	36.6	76.9	36.9	50.6	46.5	35.4	33.0	37.8	31.3	38.7	41.3
			GC–MS/MS	Green tea	A	63.2	67.4	47.8	52.9	47.8	50.2	57.2	52.0	45.9	59.3	45.9	52.0	37.4	49.8	55.4	53.5	56.1	50.3	53.5	58.7	53.0
					B	66.6	47.2	47.1	46.7	45.3	43.7	72.8	48.8	44.0	51.7	44.0	48.8	39.0	44.6	45.5	49.2	46.2	47.0	42.4	55.9	48.7
				Oolong tea	A	76.6	43.4	40.4	37.4	44.0	40.2	42.8	47.5	43.5	42.5	43.5	47.5	38.5	38.4	36.6	40.0	33.1	66.0	95.8	55.2	47.5
					B	74.3	38.3	36.8	34.0	39.2	33.8	39.7	42.5	34.7	34.7	33.1	32.1	32.1	33.4	31.3	30.9	27.0	54.7	64.5	50.2	40.3

(Continued)

Appendix Table 5.1 The 201 Pesticides Average Content for 19 Circulative Determinations on 3 Months Under 16 Factors (Two Cartridges, Two Instruments, Two Tea Samples, and Two Youden Pair Samples) (cont.)

No.	Pesticides	SPE	Method	Sample	Youden pair	Nov 9, 2009 (n=5)	Nov 14, 2009 (n=3)	Nov 19, 2009 (n=3)	Nov 24, 2009 (n=3)	Nov 29, 2009 (n=3)	Dec 4, 2009 (n=3)	Dec 9, 2009 (n=3)	Dec 14, 2009 (n=3)	Dec 19, 2009 (n=3)	Dec 24, 2009 (n=3)	Dec 14, 2009 (n=3)	Jan 3, 2010 (n=3)	Jan 8, 2010 (n=3)	Jan 13, 2010 (n=3)	Jan 18, 2010 (n=3)	Jan 23, 2010 (n=3)	Jan 28, 2010 (n=3)	Feb 2, 2010 (n=3)	Feb 7, 2010 (n=3)	Ave (µg/kg)
4	4,4-Dichlorobenzophenone	Cleanert TPT	GC-MS	Green tea	A	116.8	96.7	114.6	107.2	90.7	103.1	104.8	104.2	122.8	90.9	105.5	108.9	105.5	102.2	105.2	87.6	102.4	96.7	103.7	103.7
					B	111.0	94.1	109.2	101.6	81.0	93.5	95.6	94.9	112.5	87.4	92.3	103.1	98.6	96.7	89.9	85.0	96.0	87.4	90.0	95.8
				Oolong tea	A	59.0	69.0	51.0	52.9	60.4	53.7	52.4	67.1	67.2	53.7	65.8	53.7	49.5	64.0	72.6	69.3	71.3	60.4	57.7	60.6
					B	57.7	62.0	49.3	51.5	64.5	48.4	52.4	60.9	64.4	56.3	62.1	57.7	49.3	63.2	67.0	62.4	61.6	63.6	57.4	58.5
			GC-MS/MS	Green tea	A	128.3	100.5	119.5	112.4	88.2	101.1	97.4	92.2	115.7	87.5	107.2	110.4	106.5	95.8	96.7	79.8	89.2	75.6	156.8	103.2
					B	125.2	97.5	98.8	102.8	79.8	93.1	89.4	87.6	104.3	83.5	94.5	104.7	96.5	91.8	89.0	79.7	88.0	74.4	127.7	95.2
				Oolong tea	A	57.5	74.0	52.4	53.8	54.9	54.3	51.6	66.7	74.6	54.2	66.5	53.0	47.2	65.5	65.5	67.4	61.1	60.7	54.3	59.7
					B	59.7	67.2	51.3	52.5	61.0	49.6	51.6	60.6	69.5	57.3	59.0	56.9	51.1	64.4	62.6	63.4	57.1	65.4	54.4	58.7
		ENVI-Carb + PSA	GC-MS	Green tea	A	112.7	112.0	115.0	121.7	111.9	109.2	126.1	116.0	135.3	96.2	102.9	116.9	107.0	130.6	123.1	143.0	138.5	133.5	102.4	118.6
					B	108.3	98.9	116.4	111.7	111.0	96.4	126.5	115.7	120.4	96.5	106.1	113.0	103.7	113.6	115.0	114.4	120.7	103.8	100.2	110.1
				Oolong tea	A	70.2	48.5	55.2	60.1	67.5	55.6	62.0	73.2	61.7	58.5	58.2	55.7	53.8	70.1	69.4	49.5	53.5	55.8	67.6	60.3
					B	56.1	54.0	59.7	53.4	51.7	54.4	54.3	73.9	59.8	55.6	52.8	53.3	51.3	70.3	59.7	46.1	49.2	51.9	72.9	56.9
			GC-MS/MS	Green tea	A	123.4	136.2	112.3	120.4	113.8	115.1	112.3	110.5	124.6	105.3	97.5	37.0	91.2	118.5	112.9	125.7	119.0	112.2	110.1	110.5
					B	114.7	104.2	110.1	107.5	111.8	100.0	112.7	102.6	109.1	101.3	102.7	34.4	92.0	105.1	105.7	104.1	106.2	91.7	110.1	101.4
				Oolong tea	A	82.8	47.4	66.8	57.4	59.5	54.8	58.3	62.4	61.7	58.8	59.8	51.5	49.9	51.1	64.6	46.2	60.3	90.4	73.8	60.9
					B	66.8	54.1	60.0	50.9	48.1	52.7	55.8	61.1	59.8	54.6	53.5	53.0	47.1	56.4	53.4	43.4	58.2	75.5	79.6	57.1
5	Acetochlor	Cleanert TPT	GC-MS	Green tea	A	92.2	76.6	83.9	94.6	58.5	72.3	63.9	73.2	85.4	56.4	74.8	61.7	62.4	89.2	82.3	60.1	83.6	52.1	59.2	72.8
					B	88.9	73.2	71.0	90.3	56.7	67.1	57.2	66.8	73.7	54.1	65.3	58.9	61.2	104.2	74.2	58.5	73.3	51.2	53.5	68.4
				Oolong tea	A	86.6	92.7	73.1	70.7	63.0	57.9	67.1	73.1	78.9	72.6	77.2	76.4	72.7	75.7	65.6	70.1	70.3	56.8	65.1	71.9
					B	83.0	72.7	58.9	61.3	57.8	49.4	50.8	60.8	66.6	62.4	56.3	62.9	57.8	63.4	52.9	59.8	60.1	52.7	57.9	60.4
			GC-MS/MS	Green tea	A	87.0	71.8	65.6	110.4	57.2	78.2	68.3	68.6	70.0	55.5	68.2	67.5	64.4	64.8	67.6	56.4	71.7	56.7	69.6	69.4
					B	87.4	68.7	63.6	72.6	57.0	74.0	59.2	64.5	68.9	59.5	62.7	62.9	63.7	65.1	66.6	55.1	64.4	54.4	65.9	65.1
				Oolong tea	A	82.0	95.5	72.1	74.4	65.1	64.2	72.7	96.0	80.3	75.3	90.7	73.8	67.1	73.7	66.5	64.7	64.3	62.4	68.9	74.2
					B	75.8	76.0	59.2	62.4	61.1	54.0	67.6	77.7	70.3	64.6	68.8	62.2	61.8	64.7	55.7	62.3	57.5	56.5	63.6	64.3
		ENVI-Carb + PSA	GC-MS	Green tea	A	69.3	96.5	73.8	70.1	83.7	78.9	80.2	70.1	75.8	97.8	75.9	68.9	63.0	83.4	69.0	79.7	85.2	68.0	61.0	75.9
					B	69.0	78.9	75.5	69.3	82.9	71.0	71.1	61.3	71.2	88.2	67.6	69.5	57.2	71.7	73.0	69.2	81.5	61.6	61.1	71.1
				Oolong tea	A	86.2	83.0	78.1	75.4	63.0	67.7	68.7	76.2	79.3	71.8	71.6	59.5	61.8	60.4	71.5	50.0	52.2	54.8	56.2	69.3
					B	77.4	70.7	64.6	62.3	69.4	58.4	68.8	70.4	60.8	59.0	57.1	52.6	55.7	53.0	53.7	44.4	46.2	46.5	52.1	59.1
			GC-MS/MS	Green tea	A	76.4	85.6	69.9	73.6	77.9	77.6	72.0	66.8	100.3	65.7	68.2	85.9	53.8	72.7	71.0	69.9	73.6	61.0	60.3	72.8
					B	82.0	70.2	70.4	69.6	78.5	69.3	62.1	63.1	70.7	68.4	67.6	79.8	60.5	62.3	68.0	58.3	69.3	55.6	62.8	67.8
				Oolong tea	A	80.9	80.0	112.9	68.0	81.9	71.1	76.0	65.0	79.3	67.7	65.4	53.7	57.5	57.4	65.1	42.1	48.0	67.1	56.2	68.2
					B	55.1	70.3	70.2	63.4	66.3	60.0	62.2	57.0	60.8	54.8	59.7	49.1	54.9	51.7	49.3	38.7	45.7	56.8	58.7	57.1
6	Alachlor	Cleanert TPT	GC-MS	Green tea	A	127.3	132.9	120.0	98.4	120.5	100.6	98.3	100.7	110.9	92.2	109.7	83.6	96.4	90.0	95.3	97.5	83.0	74.8	94.5	100.6
					B	125.6	108.6	102.4	95.4	115.0	94.5	87.7	90.1	104.0	85.7	98.0	81.0	90.3	85.5	87.8	91.5	82.1	68.6	82.2	93.4
				Oolong tea	A	113.7	132.0	109.8	110.9	133.1	85.0	103.2	121.2	115.8	116.0	114.3	117.7	110.8	104.4	92.6	102.9	95.1	87.5	98.7	106.9
					B	107.7	105.2	86.0	93.6	92.0	73.2	74.1	94.0	105.2	94.2	87.9	95.5	88.3	95.7	77.1	90.5	76.5	87.5	86.2	89.6
			GC-MS/MS	Green tea	A	193.5	103.3	94.8	149.5	83.7	123.2	97.7	88.2	105.3	83.0	102.3	87.3	96.6	86.3	91.4	83.1	79.7	77.6	109.4	101.8
					B	154.2	101.4	99.4	97.7	81.5	115.0	84.6	83.2	104.0	87.1	93.1	81.3	92.6	87.7	88.2	79.4	73.7	74.7	95.1	93.3
				Oolong tea	A	131.9	139.3	100.5	106.0	80.3	93.5	106.3	130.7	108.6	107.2	128.5	105.7	101.1	112.6	96.4	107.1	91.3	88.4	88.7	107.2
					B	115.3	113.0	85.8	87.8	86.3	77.4	78.1	108.0	91.3	94.3	91.3	89.3	86.8	98.4	79.7	87.7	68.2	88.8	81.4	90.2
		ENVI-Carb + PSA	GC-MS	Green tea	A	88.2	132.9	108.1	98.4	120.5	118.1	124.2	106.7	97.0	164.2	100.1	93.3	95.3	112.5	88.0	98.8	117.8	73.5	86.9	106.6
					B	97.2	108.6	108.9	95.4	115.0	106.7	114.7	80.6	88.6	146.8	100.4	95.9	87.8	99.1	89.0	87.2	106.6	76.9	85.4	99.5
				Oolong tea	A	118.0	114.0	112.1	110.9	133.1	99.5	97.6	117.9	118.5	105.2	100.7	85.2	85.4	95.8	101.4	79.3	75.7	80.9	83.5	100.8
					B	107.3	99.2	94.2	92.6	100.1	86.8	103.4	92.1	92.7	83.6	82.0	76.0	76.7	82.3	82.2	67.7	66.1	67.8	76.8	85.8
			GC-MS/MS	Green tea	A	119.9	120.8	98.8	108.0	107.8	115.8	104.3	91.4	78.4	96.5	97.0	168.9	77.8	113.8	96.4	96.5	95.7	67.8	89.3	103.2
					B	151.3	104.6	100.3	103.7	111.4	104.8	91.2	87.5	91.8	99.2	91.9	131.9	83.2	93.7	93.5	79.4	87.7	83.3	95.7	99.0
				Oolong tea	A	108.6	122.5	171.5	99.4	113.0	96.7	115.8	96.4	118.5	99.8	99.1	75.4	90.3	88.3	95.2	78.2	70.8	94.5	78.3	100.6
					B	67.8	103.6	97.4	89.5	91.9	83.9	100.8	80.5	92.7	82.1	84.4	70.2	76.7	78.9	73.5	63.6	68.0	74.0	75.4	81.8

7	Atratone	Cleanert TPT	GC–MS	Green tea	A	60.2	42.8	43.0	43.8	33.5	33.7	32.8	33.7	40.7	41.0	48.0	31.7	35.8	31.9	30.6	22.9	29.6	25.3	36.0	36.7
					B	54.2	41.2	36.8	39.5	29.7	32.9	30.6	32.9	38.7	35.8	38.9	28.8	32.5	30.4	27.1	23.3	28.8	25.2	31.0	33.5
				Oolong tea	A	47.7	42.2	37.0	38.0	35.6	31.0	32.9	38.4	40.9	38.1	45.9	38.8	37.2	34.1	35.5	34.7	33.0	32.7	30.5	37.1
					B	45.0	37.0	32.1	32.4	33.0	26.6	26.1	31.6	30.5	31.8	32.7	30.9	29.2	31.0	28.9	29.4	26.6	30.2	27.6	31.2
			GC–MS/MS	Green tea	A	56.8	48.3	53.9	52.9	35.2	89.6	40.1	35.8	41.2	38.1	46.8	41.0	41.6	36.9	41.4	31.8	33.3	28.7	45.6	44.2
					B	55.1	46.4	45.9	47.2	32.6	72.9	36.9	33.4	40.7	40.3	40.7	37.4	38.8	38.4	38.7	32.0	33.1	29.4	39.4	41.0
				Oolong tea	A	48.2	50.2	45.5	41.7	39.6	39.3	40.6	49.8	43.2	47.3	51.8	45.8	42.2	45.4	40.7	40.4	38.9	39.9	36.1	43.5
					B	45.8	43.5	39.2	37.3	37.1	33.8	31.1	41.5	36.3	38.9	38.0	38.3	36.4	38.9	31.9	35.0	31.5	38.4	30.9	37.0
		ENVI-Carb + PSA	GC–MS	Green tea	A	48.2	41.9	44.8	41.9	49.3	52.6	34.0	35.2	33.8	85.0	36.9	37.1	32.1	37.4	32.8	36.1	32.8	32.2	30.5	40.8
					B	41.6	39.7	43.0	37.3	45.6	44.3	34.5	33.4	33.0	59.8	35.0	33.9	28.6	32.5	29.0	28.5	28.9	28.1	31.0	36.2
				Oolong tea	A	54.2	42.6	41.1	40.4	65.0	36.6	40.0	39.4	55.1	43.1	48.8	29.2	31.6	37.2	35.6	28.2	28.9	31.4	29.5	39.9
					B	44.8	36.5	35.5	34.9	43.8	30.8	35.4	32.1	42.7	32.5	35.0	27.6	27.9	30.6	28.2	22.2	24.4	25.4	25.3	32.4
			GC–MS/MS	Green tea	A	47.5	55.9	48.7	52.4	53.0	57.5	41.9	41.4	40.9	45.7	36.5	46.5	37.1	48.6	39.8	44.9	44.6	35.9	40.6	45.2
					B	44.4	44.2	46.7	45.8	51.9	49.8	42.0	37.5	41.2	45.3	33.2	41.8	37.8	37.8	35.9	36.3	40.0	31.2	40.9	41.2
				Oolong tea	A	60.3	50.2	57.2	43.4	47.4	43.9	46.4	34.4	55.1	46.3	41.8	38.6	36.4	38.9	43.1	34.6	28.1	38.9	33.6	43.4
					B	50.1	43.0	42.0	38.6	37.1	37.5	42.7	34.4	42.7	36.3	36.0	34.2	32.5	33.0	31.4	28.1	26.7	29.7	29.9	36.1
8	Benodanil	Cleanert TPT	GC–MS	Green tea	A	140.2	149.0	122.7	118.2	84.0	111.2	103.0	88.9	209.4	96.8	200.8	209.4	190.2	101.7	133.0	337.2	103.4	121.8	119.0	140.8
					B	154.1	140.0	107.5	102.4	83.8	105.2	85.4	85.4	169.0	94.0	173.9	169.0	166.4	122.2	128.5	322.6	109.7	98.8	82.9	131.2
				Oolong tea	A	238.9	120.6	113.2	226.8	352.0	252.0	80.8	80.8	80.8	133.4	399.7	287.2	97.3	91.5	120.4	120.4	155.8	77.9	77.9	155.1
					B	162.6	120.9	127.2	157.9	263.1	142.7	93.6	80.8	114.9	111.5	287.2	201.2	130.9	89.8	103.4	103.1	131.5	45.5	78.9	128.5
			GC–MS/MS	Green tea	A	161.7	158.5	133.1	126.1	109.6	290.4	98.2	93.2	135.7	113.0	201.2	93.3	221.4	111.4	111.8	125.3	92.3	80.8	354.2	146.6
					B	173.2	148.6	129.9	113.3	99.6	225.3	93.2	91.4	155.7	100.3	168.8	86.3	181.6	115.1	101.5	124.0	88.7	76.7	224.0	130.4
				Oolong tea	A	336.1	154.9	131.0	241.1	415.8	181.1	109.0	76.2	155.1	68.8	313.4	81.3	88.3	65.3	128.1	130.3	106.9	118.6	72.4	162.3
					B	217.6	159.8	119.6	178.1	368.5	109.0	137.7	184.8	118.4	113.4	273.6	59.9	112.8	70.9	100.3	118.2	92.9	72.3	70.3	138.2
		ENVI-Carb + PSA	GC–MS	Green tea	A	113.9	119.5	157.8	84.8	155.8	138.2	155.0	116.3	211.1	155.6	161.2	160.8	118.5	92.7	127.2	109.7	300.4	111.8	120.8	140.9
					B	146.7	150.3	155.2	73.5	141.8	158.9	103.4	103.4	128.4	154.2	122.9	139.6	114.8	96.1	103.9	88.3	233.1	111.8	124.9	131.7
				Oolong tea	A	114.6	141.4	120.2	70.1	210.3	76.4	61.2	388.2	56.1	56.1	81.9	67.3	166.2	161.1	116.0	93.6	91.2	99.4	99.2	128.9
					B	124.4	104.2	131.2	87.3	144.0	88.1	139.1	193.6	136.3	119.1	153.9	80.2	117.3	167.6	101.9	61.4	102.1	82.1	65.0	103.6
			GC–MS/MS	Green tea	A	125.9	127.4	132.2	101.1	141.7	146.1	140.1	127.1	127.1	123.8	153.9	119.1	110.1	125.5	108.4	105.7	102.1	97.0	133.7	124.0
					B	134.7	174.3	137.6	86.0	141.7	155.0	112.3	183.2	127.1	112.3	112.6	123.7	118.7	125.5	118.7	105.7	90.5	116.6	123.7	121.9
				Oolong tea	A	103.8	158.2	147.0	61.8	123.9	70.1	124.2	80.2	67.3	63.7	71.6	84.7	102.9	65.0	97.0	82.3	131.1	279.9	203.0	111.5
					B	91.2	119.1	137.9	74.7	86.6	80.1	79.7	83.4	72.7	72.6	73.2	85.3	84.7	80.7	97.4	46.4	115.7	211.7	116.7	91.1
9	Benoxacor	Cleanert TPT	GC–MS	Green tea	A	112.4	104.0	90.2	114.6	65.2	73.0	73.1	88.9	91.4	77.0	94.8	87.4	82.1	91.4	93.3	90.0	80.5	60.9	51.1	84.9
					B	110.4	96.2	77.3	113.9	66.6	66.9	72.6	85.4	83.6	65.7	79.1	80.1	77.8	101.6	94.1	87.6	83.1	53.1	50.2	82.0
				Oolong tea	A	87.4	106.3	88.4	99.5	71.0	68.2	95.8	84.7	87.9	83.3	121.9	76.8	88.7	77.1	62.1	69.9	75.0	59.4	71.0	82.9
					B	80.0	89.0	76.2	76.7	64.9	56.2	78.9	56.0	70.5	78.8	84.7	60.9	77.8	75.1	56.3	66.8	54.6	70.8	70.8	70.6
			GC–MS/MS	Green tea	A	109.8	85.6	55.2	98.2	61.7	104.4	65.7	65.7	78.9	58.0	108.9	61.3	94.8	61.0	76.9	86.3	49.2	51.9	102.3	77.9
					B	103.5	82.1	65.1	70.8	62.0	105.2	58.2	58.2	89.9	63.1	103.2	54.3	87.9	67.5	74.3	89.3	51.5	51.5	88.1	75.4
				Oolong tea	A	107.1	94.4	69.3	90.3	80.7	71.2	95.2	93.7	71.6	67.3	127.8	72.6	73.1	78.5	84.0	89.6	77.2	71.1	71.9	83.5
					B	92.7	86.1	63.6	71.3	71.5	53.5	83.1	72.0	64.0	68.6	91.2	53.4	72.5	70.5	63.6	76.1	60.8	58.5	64.5	70.4
		ENVI-Carb + PSA	GC–MS	Green tea	A	84.5	111.0	93.0	60.1	98.2	109.5	82.3	90.3	95.9	77.0	94.8	104.8	88.2	66.4	72.4	76.2	112.4	82.7	74.0	88.1
					B	92.4	100.5	97.2	53.7	89.0	110.8	74.3	83.9	92.5	70.5	79.1	100.9	84.9	70.8	75.9	66.8	99.8	78.8	75.8	84.1
				Oolong tea	A	69.9	79.8	90.8	85.5	111.1	88.1	95.8	72.4	71.1	73.4	76.1	84.8	92.4	78.1	79.6	66.0	59.1	59.7	66.2	80.6
					B	73.7	71.0	77.4	80.9	78.4	78.5	78.6	73.8	67.7	53.1	64.0	82.3	78.9	82.9	57.0	50.9	57.0	59.6	60.2	71.9
			GC–MS/MS	Green tea	A	78.5	75.7	77.1	65.9	80.4	100.1	70.7	62.7	57.8	52.4	89.1	88.8	69.3	69.8	74.7	54.8	69.9	56.0	70.1	72.4
					B	97.7	91.1	79.3	60.9	77.9	96.4	69.0	62.8	65.0	62.8	68.4	75.5	75.2	68.8	68.1	55.7	62.8	59.2	78.9	72.4
				Oolong tea	A	80.3	80.0	107.2	57.3	74.3	67.3	82.9	80.6	73.4	63.2	66.5	50.4	66.2	59.4	59.6	56.9	88.5	111.1	98.0	74.9
					B	57.7	68.1	68.1	57.3	57.4	66.1	66.8	70.0	65.7	53.1	58.8	53.7	54.9	60.2	55.2	40.5	87.4	100.5	84.0	64.5

(Continued)

Appendix Table 5.1 The 201 Pesticides Average Content for 19 Circulative Determinations on 3 Months Under 16 Factors (Two Cartridges, Two Instruments, Two Tea Samples, and Two Youden Pair Samples) (cont.)

Average content (μg/kg)

No.	Pesticides	SPE	Method	Sample	Youden pair	Nov 9, 2009 (n=5)	Nov 14, 2009 (n=3)	Nov 19, 2009 (n=3)	Nov 24, 2009 (n=3)	Nov 29, 2009 (n=3)	Dec 4, 2009 (n=3)	Dec 9, 2009 (n=3)	Dec 14, 2009 (n=3)	Dec 19, 2009 (n=3)	Dec 24, 2009 (n=3)	Jan 3, 2010 (n=3)	Jan 8, 2010 (n=3)	Jan 13, 2010 (n=3)	Jan 18, 2010 (n=3)	Jan 23, 2010 (n=3)	Jan 28, 2010 (n=3)	Feb 2, 2010 (n=3)	Feb 7, 2010 (n=3)	Ave (μg/kg) (n=3)
10	Bromophos-ethyl	Cleaner TPT	GC-MS	Green tea	A	43.9	35.0	34.2	39.4	28.8	30.8	30.4	29.7	37.6	31.5	36.6	32.7	31.3	29.3	25.7	27.5	25.7	31.7	33.3
					B	42.3	33.9	30.5	35.5	27.0	29.5	27.8	27.7	36.0	28.9	33.3	30.5	29.9	26.8	25.3	26.6	24.5	28.0	30.9
				Oolong tea	A	38.0	37.8	32.9	34.9	33.0	29.0	31.7	37.1	39.1	37.0	37.0	34.9	34.5	32.1	33.5	31.2	30.3	30.2	34.7
					B	35.8	32.9	28.1	29.4	30.2	24.5	24.1	32.2	31.6	32.1	30.0	28.3	30.5	26.7	28.6	25.7	27.5	27.1	29.4
			GC-MS/MS	Green tea	A	40.6	34.5	33.4	36.1	28.9	49.8	29.6	28.3	33.7	31.1	29.0	34.9	31.7	30.1	26.1	25.8	24.4	37.5	32.7
					B	40.8	33.2	32.3	32.0	27.2	42.7	26.9	27.5	33.5	29.6	27.4	32.4	32.1	28.9	25.5	24.4	24.2	31.9	30.8
				Oolong tea	A	41.5	41.1	35.9	36.2	32.2	29.6	31.0	39.5	36.7	33.8	36.6	36.1	36.7	32.8	34.1	28.5	29.2	29.6	34.8
					B	35.8	34.8	29.7	31.8	29.3	25.5	24.2	34.1	30.5	30.2	29.5	28.7	31.8	26.8	29.3	23.1	27.3	26.3	29.4
		ENVI-Carb + PSA	GC-MS	Green tea	A	34.8	41.6	38.8	33.0	42.7	41.3	35.5	33.8	34.9	90.7	38.8	30.9	35.8	32.1	34.1	35.9	29.5	29.4	38.5
					B	34.2	35.4	37.6	30.6	39.5	37.0	31.8	30.3	32.7	75.8	35.4	28.4	33.2	31.4	28.4	32.0	27.2	29.1	35.0
				Oolong tea	A	42.3	38.2	35.8	34.3	47.3	32.6	36.3	37.7	40.1	36.2	30.2	30.2	33.8	34.7	26.7	26.2	27.5	27.3	34.5
					B	35.4	32.5	35.1	30.1	34.7	28.5	31.4	29.8	31.4	29.4	27.4	26.6	28.4	27.4	22.6	22.7	23.4	24.9	28.9
			GC-MS/MS	Green tea	A	37.8	40.4	35.2	33.9	38.7	40.7	35.0	33.5	32.8	33.8	37.6	31.9	37.8	33.1	32.9	33.3	27.6	31.6	34.8
					B	38.0	34.4	34.3	32.4	37.8	36.3	31.3	31.4	33.2	33.6	35.9	30.6	32.3	30.5	27.9	29.4	25.8	31.2	32.5
				Oolong tea	A	39.4	40.5	45.5	33.1	34.5	32.8	36.5	32.4	40.1	33.8	29.3	29.2	29.8	31.8	25.3	25.9	32.5	29.3	33.3
					B	30.4	34.9	31.6	29.3	27.7	28.4	32.7	26.6	31.4	27.6	26.2	25.2	26.4	24.6	21.0	23.8	25.7	27.5	27.9
11	Butralin	Cleaner TPT	GC-MS	Green tea	A	219.9	154.2	139.7	145.7	92.4	125.2	134.9	115.1	193.5	109.3	165.0	197.3	140.4	181.5	389.7	112.7	151.8	107.2	163.5
					B	203.3	149.3	118.5	127.4	95.9	121.9	126.2	123.6	183.5	98.2	148.3	170.0	132.0	184.0	364.3	122.8	123.8	97.4	152.9
				Oolong tea	A	164.3	132.6	140.1	170.5	235.8	145.5	199.9	187.2	161.0	138.3	135.8	151.4	108.2	158.0	171.2	228.8	132.1	123.2	161.9
					B	153.0	123.7	133.7	145.3	152.1	104.6	137.1	169.2	122.7	120.9	102.9	138.7	133.8	119.4	157.0	177.1	120.3	123.6	134.9
			GC-MS/MS	Green tea	A	227.0	144.3	121.7	187.7	99.4	582.2	108.0	114.8	148.8	100.3	107.1	148.3	125.6	129.7	186.9	89.7	71.7	283.5	167.4
					B	206.2	140.4	122.1	131.6	91.1	455.8	104.7	108.8	192.0	101.1	100.9	127.3	131.7	127.3	187.8	92.4	72.4	223.7	153.0
				Oolong tea	A	186.1	226.5	128.4	132.6	116.4	154.5	176.6	212.5	172.8	130.9	114.6	123.1	138.8	136.0	136.3	134.2	138.9	109.9	157.5
					B	177.8	201.7	117.2	122.9	124.5	117.3	190.9	185.2	140.1	112.2	93.8	118.6	125.1	111.4	123.6	101.4	113.2	111.9	133.2
		ENVI-Carb + PSA	GC-MS	Green tea	A	137.2	162.1	192.2	85.3	189.5	222.2	244.9	180.6	146.2	193.3	206.3	158.7	88.0	130.1	139.1	292.6	149.7	116.3	171.3
					B	149.0	166.8	176.3	65.5	159.3	223.5	162.2	146.7	156.8	156.8	191.1	147.8	102.0	124.4	117.3	218.9	122.1	115.9	151.4
				Oolong tea	A	148.8	130.4	136.3	131.7	487.8	169.4	103.5	310.0	155.8	133.9	145.8	196.7	313.3	80.1	117.6	114.0	114.6	114.3	172.1
					B	137.9	113.5	108.3	120.5	210.0	153.6	100.7	192.4	125.8	119.2	141.0	149.2	282.4	109.7	60.6	112.0	95.7	100.8	134.4
			GC-MS/MS	Green tea	A	162.4	135.3	134.7	122.9	144.3	176.1	199.6	145.2	110.5	217.7	191.5	111.0	135.0	124.5	107.2	112.4	76.3	131.6	139.8
					B	184.0	216.0	135.7	105.2	138.1	163.0	69.8	126.0	119.1	168.2	163.3	128.6	113.3	119.0	92.6	101.8	92.6	149.0	131.7
				Oolong tea	A	130.4	142.5	167.5	118.3	135.1	130.3	138.9	184.1	155.8	125.4	101.0	116.1	107.3	106.8	105.3	256.4	393.4	172.7	153.2
					B	101.8	121.0	111.6	113.1	135.1	117.6	123.4	115.6	125.8	105.6	101.6	96.7	91.3	92.8	74.7	219.7	294.5	148.7	123.3
12	Chlorfenapyr	Cleaner TPT	GC-MS	Green tea	A	352.3	304.4	328.8	363.7	267.4	297.4	278.9	276.4	297.0	284.1	302.4	296.6	286.4	268.4	207.2	256.5	225.4	295.0	287.5
					B	354.4	289.2	298.7	344.9	241.2	269.1	252.6	250.6	291.5	279.7	281.3	256.7	274.2	241.1	201.1	248.6	220.6	264.9	269.8
				Oolong tea	A	316.5	377.1	314.0	328.9	297.0	253.7	285.1	336.4	328.5	338.2	339.5	357.2	327.5	297.0	315.4	297.1	282.8	293.9	316.2
					B	303.7	318.6	270.8	276.2	294.5	236.3	238.9	301.0	280.9	298.9	290.7	280.7	291.4	254.6	269.1	248.1	270.6	262.6	276.8
			GC-MS/MS	Green tea	A	361.1	306.2	325.2	361.5	269.6	295.1	278.9	257.1	307.8	279.3	257.4	297.1	295.3	278.4	227.4	231.6	229.7	301.8	287.8
					B	358.5	297.5	301.0	303.6	244.4	276.5	253.4	239.5	300.7	284.7	249.4	281.4	302.8	252.8	214.2	231.7	219.6	275.8	271.8
				Oolong tea	A	377.4	386.8	313.4	315.4	306.9	258.3	276.1	345.4	330.4	300.3	337.9	318.5	345.2	309.1	307.5	265.1	268.6	267.6	315.6
					B	324.3	333.1	283.5	268.4	298.3	230.0	236.5	309.7	282.5	281.1	293.9	263.0	299.7	263.5	260.9	219.7	260.2	239.9	275.6
		ENVI-Carb + PSA	GC-MS	Green tea	A	298.4	371.0	312.3	332.9	310.8	321.4	310.6	295.5	303.3	275.6	308.0	276.0	333.3	274.9	274.5	311.8	295.0	272.6	304.7
					B	308.4	307.9	313.7	310.8	327.9	297.5	300.0	260.6	278.3	260.6	284.6	250.3	289.6	257.9	264.0	274.7	236.1	272.5	283.6
				Oolong tea	A	394.8	348.7	323.5	329.1	350.6	297.6	336.6	294.8	357.1	317.1	272.7	274.9	292.8	327.1	250.0	248.7	262.2	260.1	307.1
					B	347.2	308.8	308.7	289.5	297.4	262.8	296.7	260.1	289.0	262.3	240.0	253.1	246.0	258.0	219.4	216.3	223.3	244.5	267.6
			GC-MS/MS	Green tea	A	352.4	349.6	310.9	339.8	319.2	347.0	292.5	284.0	282.7	290.9	265.6	293.3	325.0	288.1	302.8	304.0	258.5	278.6	302.7
					B	374.6	289.0	306.9	305.4	325.6	311.3	297.4	269.2	273.3	294.3	251.5	264.6	276.5	266.7	249.6	264.9	231.4	261.2	282.4
				Oolong tea	A	360.2	350.1	424.5	312.6	338.7	293.6	333.0	294.7	357.1	319.9	256.1	254.4	277.2	300.8	236.3	225.8	306.6	257.8	304.0
					B	283.1	312.3	327.3	286.3	287.5	264.0	298.3	257.7	289.0	264.4	231.3	237.4	244.1	230.5	201.2	216.0	243.5	244.4	262.1

No.	Compound	Cleanup	Instrument	Tea	A/B	1	2	3	4	5	6	7	8	9	10	11	12	13	14	15	16	17	18	19	20
13	Clomazone	Cleanert TPT	GC-MS	Green tea	A	56.1	39.9	34.1	37.5	30.2	36.5	31.5	30.8	34.4	32.1	48.6	39.6	37.0	37.6	35.2	47.0	32.5	27.7	37.1	36.4
					B	52.3	37.7	31.9	33.9	30.6	34.0	28.7	29.8	38.6	31.9	44.3	35.0	33.2	41.0	38.4	45.3	29.6	26.9	35.5	31.0
				Oolong tea	A	42.0	41.1	36.8	40.9	41.9	31.6	35.2	39.6	38.4	33.7	51.7	35.3	36.4	33.6	34.2	35.8	36.0	33.1	37.3	30.8
					B	38.8	35.4	32.1	32.0	34.6	25.4	27.8	33.9	32.1	30.0	38.8	29.3	30.7	30.7	27.9	31.0	30.4	29.6	31.5	28.3
			GC-MS/MS	Green tea	A	55.5	40.5	37.2	39.1	29.4	66.0	31.4	30.6	33.6	34.1	46.3	35.5	35.4	30.7	33.4	41.6	29.0	26.2	37.8	42.7
					B	52.6	36.6	33.9	34.6	30.8	59.3	28.4	29.8	38.1	35.0	44.2	32.1	33.2	34.6	34.9	40.2	28.8	27.5	36.3	35.1
				Oolong tea	A	43.8	45.7	37.3	38.3	36.8	34.3	34.2	42.0	42.9	36.5	52.2	37.6	34.3	38.5	33.9	35.1	33.1	33.5	37.9	29.9
					B	38.3	38.5	31.2	31.6	32.5	26.7	29.2	35.7	34.2	32.6	38.4	29.6	29.4	35.0	28.4	32.3	29.2	29.4	32.0	26.0
		ENVI-Carb + PSA	GC-MS	Green tea	A	44.7	46.5	47.7	29.6	43.3	41.0	43.3	39.0	38.0	54.1	55.0	51.0	36.6	39.2	41.2	37.6	41.8	29.3	41.6	31.9
					B	37.8	45.5	42.9	26.3	39.7	41.4	31.3	37.3	35.7	43.5	45.1	45.2	33.7	36.8	41.0	29.9	34.9	31.5	37.5	32.2
				Oolong tea	A	43.9	41.9	41.3	34.3	51.5	35.8	35.8	41.7	40.0	35.9	33.1	33.1	33.3	41.9	30.3	27.2	27.8	28.1	36.2	30.2
					B	35.0	35.2	35.7	31.5	37.1	31.5	33.1	33.4	31.3	30.3	28.0	30.6	27.7	37.9	27.0	20.6	24.8	24.7	30.6	25.9
			GC-MS/MS	Green tea	A	42.4	50.6	43.2	35.1	42.4	41.0	43.7	36.8	43.5	38.3	52.8	32.2	28.4	40.6	41.5	34.7	35.4	27.6	39.3	37.1
					B	37.2	49.9	39.9	31.0	39.2	41.3	28.0	33.1	40.3	35.3	43.3	32.0	34.5	32.7	40.9	29.1	32.1	28.3	36.0	35.6
				Oolong tea	A	42.9	43.1	48.5	34.5	37.6	36.1	35.4	36.9	40.0	37.6	35.0	29.5	33.2	30.0	31.9	28.7	29.1	33.2	35.5	32.1
					B	35.0	35.6	34.4	31.6	30.2	31.5	33.0	30.2	31.3	31.6	30.8	28.3	26.9	27.3	25.6	21.8	26.0	27.9	29.9	29.1
14	Cycloate	Cleanert TPT	GC-MS	Green tea	A	41.6	31.6	33.0	34.0	27.8	30.7	29.8	29.8	30.8	29.4	31.8	27.9	31.0	29.0	27.4	19.2	25.2	23.0	29.7	31.9
					B	38.5	30.4	30.2	32.4	25.4	28.1	27.2	27.1	30.6	27.4	28.9	26.3	29.1	28.5	24.6	18.9	24.1	22.4	27.8	27.4
				Oolong tea	A	90.7	80.0	71.6	90.3	102.2	73.6	58.3	70.1	80.4	66.5	138.9	64.9	69.9	68.7	78.3	74.3	68.1	64.0	77.3	57.5
					B	84.3	73.9	64.7	63.4	78.2	53.0	58.0	67.6	68.0	65.7	101.9	59.1	64.5	68.4	68.4	64.7	59.7	59.0	67.2	54.6
			GC-MS/MS	Green tea	A	39.2	31.8	35.1	30.0	26.0	36.2	30.3	26.1	31.1	26.9	31.9	29.7	28.4	25.0	28.0	23.2	25.6	22.4	29.3	29.3
					B	38.3	30.7	31.3	32.3	24.6	32.8	27.6	25.4	30.9	25.8	29.0	28.1	27.2	26.1	26.6	23.0	24.6	22.2	28.0	26.1
				Oolong tea	A	81.9	89.6	83.7	76.0	79.9	63.2	60.0	75.2	80.0	68.5	103.6	68.3	64.4	72.7	67.9	68.8	57.1	57.2	72.0	50.3
					B	78.0	82.7	75.4	64.2	75.6	56.0	59.5	70.9	67.9	65.1	82.6	60.3	64.9	70.1	59.1	60.8	50.5	51.0	65.4	47.3
		ENVI-Carb + PSA	GC-MS	Green tea	A	36.7	36.1	33.6	34.1	37.7	35.5	31.2	31.5	39.0	41.0	30.3	35.1	30.7	36.6	32.4	35.3	30.0	31.4	34.0	28.5
					B	32.7	30.6	33.3	30.9	36.9	32.2	30.8	30.9	28.5	37.2	27.8	32.9	26.8	32.9	28.0	28.1	29.7	26.9	30.8	28.3
				Oolong tea	A	98.4	86.1	76.8	69.3	110.2	67.7	81.7	97.0	89.5	69.3	66.8	60.9	78.5	112.9	62.7	55.6	53.3	54.0	76.3	58.1
					B	87.3	77.7	76.2	67.2	86.1	61.3	64.5	79.4	71.4	60.9	58.3	57.9	68.1	103.9	57.4	44.2	48.6	53.0	67.2	53.5
			GC-MS/MS	Green tea	A	33.2	36.5	33.3	35.0	37.3	35.7	31.9	30.8	30.3	33.6	27.3	52.4	27.4	36.0	31.2	32.4	31.7	27.5	33.3	28.6
					B	31.0	29.3	33.2	31.0	37.0	32.1	30.5	28.6	29.3	32.6	25.8	45.3	28.0	34.0	28.4	27.7	29.8	23.7	30.8	27.6
				Oolong tea	A	106.0	90.3	86.9	69.5	72.4	67.5	71.4	67.2	89.5	72.4	67.8	57.3	64.6	56.2	59.1	49.0	73.8	107.3	73.6	69.7
					B	96.6	82.4	80.5	66.4	63.2	61.3	64.7	58.9	71.4	62.2	59.9	56.0	52.7	53.3	50.4	38.6	66.9	93.6	65.4	64.4
15	Cycluron	Cleanert TPT	GC-MS	Green tea	A	140.3	113.4	128.5	150.2	121.8	116.4	110.4	107.6	129.9	114.5	136.8	111.2	129.4	122.6	108.6	80.2	96.3	88.8	116.6	108.5
					B	143.7	113.2	122.2	144.1	111.8	112.5	99.9	96.2	112.6	101.4	123.8	112.2	126.8	117.0	96.6	73.0	87.9	84.2	109.3	96.8
				Oolong tea	A	36.3	35.7	29.5	26.2	24.1	21.3	27.3	33.3	33.7	32.5	36.3	32.5	31.9	31.1	29.8	28.4	26.1	28.7	30.2	28.6
					B	33.8	29.4	24.7	22.2	21.9	18.7	22.3	27.8	27.8	27.6	27.7	27.0	26.1	27.1	24.3	24.6	22.3	25.4	25.5	24.7
			GC-MS/MS	Green tea	A	113.6	95.7	110.3	98.9	88.4	199.2	92.1	86.9	100.0	83.5	80.1	91.1	94.8	66.8	84.5	64.9	78.2	88.8	96.3	111.1
					B	122.2	93.2	111.3	109.7	83.7	167.1	79.8	77.4	92.5	98.2	65.8	87.6	91.6	75.8	77.3	60.7	71.6	85.8	92.1	99.5
				Oolong tea	A	33.0	34.9	29.2	32.3	29.7	28.4	28.4	33.1	34.3	33.8	35.5	31.0	29.0	32.2	28.5	27.7	24.8	27.6	30.5	26.5
					B	30.9	29.0	25.1	27.0	27.0	23.3	22.4	28.4	28.9	28.5	27.0	26.1	27.4	27.4	23.2	24.5	21.5	24.9	26.0	22.0
		ENVI-Carb + PSA	GC-MS	Green tea	A	113.0	132.7	123.7	133.9	144.0	122.7	115.7	104.2	119.9	187.5	176.4	116.3	109.1	120.1	110.5	116.3	105.2	100.7	123.9	102.0
					B	107.9	106.1	125.6	117.8	147.1	109.9	110.6	96.8	111.3	164.3	109.5	113.1	93.1	116.7	101.1	97.9	94.5	89.2	111.2	100.3
				Oolong tea	A	41.7	33.8	31.9	32.3	37.7	31.0	34.9	27.0	32.9	31.4	30.3	26.9	26.9	28.9	31.4	25.2	25.5	26.0	30.6	26.3
					B	32.1	28.2	28.1	28.1	28.6	25.8	28.8	23.7	26.7	24.0	25.0	23.5	23.0	23.7	24.2	20.4	20.1	21.9	25.2	23.2
			GC-MS/MS	Green tea	A	102.0	123.7	97.6	131.6	98.4	108.6	83.5	84.2	93.9	89.2	82.0	102.9	73.1	126.6	94.2	101.6	99.9	92.9	99.3	96.0
					B	89.7	87.1	92.9	115.4	101.4	95.1	83.5	81.1	102.0	96.0	80.8	102.1	76.0	86.3	85.1	92.7	85.9	76.7	91.1	101.1
				Oolong tea	A	42.3	33.5	43.3	30.6	34.6	30.2	32.1	28.9	32.9	31.6	29.5	24.5	26.2	27.6	29.6	22.8	19.4	23.6	29.8	22.7
					B	31.3	28.3	28.5	26.4	27.6	25.6	28.3	24.2	26.7	24.5	25.4	22.4	25.2	23.6	21.8	18.6	17.3	19.1	24.5	20.8

(Continued)

Appendix Table 5.1 The 201 Pesticides Average Content for 19 Circulative Determinations on 3 Months Under 16 Factors (Two Cartridges, Two Instruments, Two Tea Samples, and Two Youden Pair Samples) (cont.)

No.	Pesticides	SPE	Method	Sample	Youden pair	Nov 9, 2009 (n=5)	Nov 14, 2009 (n=3)	Nov 19, 2009 (n=3)	Nov 24, 2009 (n=3)	Nov 29, 2009 (n=3)	Dec 4, 2009 (n=3)	Dec 9, 2009 (n=3)	Dec 14, 2009 (n=3)	Dec 19, 2009 (n=3)	Dec 24, 2009 (n=3)	Jan 3, 2010 (n=3)	Jan 8, 2010 (n=3)	Jan 13, 2010 (n=3)	Jan 18, 2010 (n=3)	Jan 23, 2010 (n=3)	Jan 28, 2010 (n=3)	Feb 2, 2010 (n=3)	Feb 7, 2010 (n=3)	Ave (µg/kg)
16	Cyhalofop-butyl	Cleanert TPT	GC-MS	Green tea	A	114.2	84.0	76.6	73.0	64.7	76.0	65.5	63.4	91.1	77.9	96.2	91.0	76.9	82.3	84.3	64.9	70.1	79.9	80.2
					B	106.4	80.1	69.1	69.5	63.1	69.0	59.3	60.8	98.4	74.3	94.4	78.7	84.5	82.6	78.2	62.4	59.5	71.0	76.4
				Oolong tea	A	119.4	161.8	120.7	125.5	118.5	101.3	102.9	118.8	105.7	276.1	323.3	254.0	120.5	125.2	120.2	102.4	120.9	112.9	159.4
					B	131.9	131.1	99.8	133.0	127.3	109.2	82.6	106.2	108.3	115.5	166.2	115.2	130.3	108.7	96.6	83.6	138.4	108.6	116.9
			GC-MS/MS	Green tea	A	121.7	86.8	87.3	57.1	68.6	179.5	65.2	58.1	41.4	69.9	75.7	78.2	69.0	69.6	73.9	56.4	50.5	107.5	78.5
					B	117.2	81.8	74.0	73.1	65.0	148.2	58.3	57.0	49.6	144.1	66.0	71.7	80.0	69.7	72.3	54.6	51.7	86.4	78.5
				Oolong tea	A	113.1	142.9	106.4	101.4	92.0	92.1	98.2	117.7	110.8	123.9	112.1	106.6	123.2	106.0	112.6	80.7	84.1	94.5	108.3
					B	102.8	114.5	84.8	82.5	86.5	75.4	75.0	93.1	91.1	100.2	98.4	88.5	94.5	85.2	88.8	70.0	83.1	78.1	89.4
		ENVI-Carb + PSA	GC-MS	Green tea	A	94.2	91.7	99.4	67.3	86.8	79.2	84.3	76.0	99.8	88.3	109.4	97.0	117.5	105.0	89.6	97.8	62.8	65.4	90.2
					B	79.7	100.8	89.1	59.0	83.3	81.1	70.8	73.6	92.7	83.8	104.5	96.2	97.9	86.7	68.2	66.4	66.4	69.9	83.6
				Oolong tea	A	173.0	143.6	132.2	137.8	173.1	111.6	132.3	109.7	120.2	317.6	242.5	236.2	124.3	141.4	96.3	93.5	110.3	93.8	158.5
					B	151.5	129.8	127.3	122.4	134.8	97.7	120.5	99.0	94.6	110.7	103.2	110.2	111.3	111.3	82.8	85.2	90.4	87.9	110.7
			GC-MS/MS	Green tea	A	80.5	107.7	88.6	76.0	87.6	85.5	87.8	71.6	148.9	40.9	212.9	71.2	99.7	79.4	75.2	66.4	63.2	81.2	90.8
					B	69.5	105.9	81.0	64.0	84.1	83.6	20.9	65.8	124.3	36.8	199.8	82.7	76.5	79.1	58.4	61.4	60.0	78.8	80.1
				Oolong tea	A	151.4	157.0	147.3	114.4	126.2	93.3	119.6	91.3	120.2	113.6	86.6	84.3	93.8	114.3	73.5	66.3	95.4	75.0	105.6
					B	113.4	131.9	108.7	93.5	113.1	79.2	106.4	79.3	94.6	77.0	66.5	80.9	84.7	67.1	73.1	63.8	69.3	73.0	87.2
17	Cyprodinil	Cleanert TPT	GC-MS	Green tea	A	48.8	37.5	40.5	39.3	30.5	31.3	30.1	31.3	35.2	30.7	30.5	35.4	29.8	29.8	23.7	27.4	24.6	32.8	33.3
					B	45.1	35.9	34.1	34.6	26.5	30.5	27.3	29.6	35.6	27.4	30.4	32.1	28.5	26.8	23.3	26.6	22.7	28.1	30.7
				Oolong tea	A	41.1	41.8	34.8	37.5	33.2	30.3	31.8	40.1	39.4	40.7	42.2	39.3	37.5	39.6	36.0	33.2	34.3	30.6	36.9
					B	39.1	37.7	29.5	31.5	30.8	24.5	25.2	37.5	33.0	33.0	34.1	33.5	35.5	32.1	31.6	26.2	31.1	29.3	31.8
			GC-MS/MS	Green tea	A	46.2	36.9	40.7	40.2	27.6	44.7	32.1	28.3	34.9	31.6	32.5	30.9	31.7	29.9	24.7	26.6	23.0	34.3	33.2
					B	44.3	35.5	34.7	35.6	25.8	38.4	29.3	27.2	33.3	30.2	31.0	29.0	31.7	28.4	24.6	25.8	22.9	28.9	30.9
				Oolong tea	A	37.6	39.8	34.3	35.3	33.8	30.6	29.5	39.9	39.4	39.1	36.9	31.9	37.4	32.6	32.7	30.0	31.0	28.6	34.8
					B	36.6	34.5	29.7	30.6	31.5	26.3	25.9	31.1	33.2	32.0	31.3	28.0	31.8	27.6	28.1	25.3	28.7	26.4	29.9
		ENVI-Carb + PSA	GC-MS	Green tea	A	40.3	41.2	40.7	41.8	43.2	41.9	34.6	34.1	35.1	71.9	35.1	33.1	35.2	29.8	37.1	30.5	31.1	27.9	37.7
					B	39.2	36.6	38.9	36.1	39.7	36.8	33.8	32.6	33.0	56.7	32.2	29.9	30.6	26.5	29.4	27.0	26.8	29.6	33.9
				Oolong tea	A	48.8	47.3	36.5	33.1	42.6	41.2	36.6	40.8	35.3	41.0	31.4	35.8	37.8	37.4	30.9	31.2	27.7	30.1	37.3
					B	41.7	40.1	35.7	30.7	34.8	34.3	31.2	28.7	28.7	30.5	29.8	31.2	31.7	30.7	24.7	26.1	21.1	27.8	31.2
			GC-MS/MS	Green tea	A	37.8	42.8	38.0	42.6	41.3	43.3	33.6	33.2	33.9	35.2	34.4	24.5	34.2	31.2	33.8	34.8	29.1	31.3	34.9
					B	35.8	35.2	36.8	37.4	39.9	37.3	32.7	30.8	32.9	34.7	32.1	26.7	31.6	27.8	28.3	30.6	24.8	31.2	32.3
				Oolong tea	A	53.7	41.8	46.3	36.1	39.9	34.4	37.2	33.6	35.3	36.9	28.2	29.8	30.8	33.5	23.9	23.7	30.1	25.9	34.5
					B	44.1	36.4	34.8	30.3	32.1	29.2	33.9	29.2	28.7	28.7	25.8	27.6	27.3	25.6	20.6	21.4	23.9	24.9	29.2
18	Dacthal	Cleanert TPT	GC-MS	Green tea	A	126.3	102.1	104.6	113.6	86.3	100.6	90.3	88.0	103.3	94.9	101.0	96.3	94.5	95.8	104.4	82.1	76.4	98.0	97.9
					B	123.1	98.9	94.4	104.3	81.5	92.7	83.4	83.4	104.0	92.1	93.8	90.4	91.2	90.6	100.1	81.6	73.8	86.7	92.5
				Oolong tea	A	107.7	119.9	101.7	106.8	99.9	87.8	105.8	116.2	109.8	105.9	110.9	106.3	102.2	96.3	101.9	107.7	93.4	90.9	105.0
					B	101.2	99.6	84.2	90.5	89.6	73.3	77.3	101.3	92.6	91.5	88.6	86.0	91.1	78.9	86.7	87.1	85.1	81.5	88.4
			GC-MS/MS	Green tea	A	131.5	105.6	103.1	113.4	86.4	96.4	93.5	84.9	106.5	90.0	88.2	98.9	95.7	90.7	80.0	77.7	73.3	110.3	96.6
					B	130.9	122.2	96.9	99.4	82.3	89.7	86.4	81.8	106.2	87.8	85.2	91.0	95.9	86.7	79.8	73.9	72.1	91.7	91.5
				Oolong tea	A	132.5	122.8	100.1	107.4	103.1	91.5	103.0	119.5	113.0	107.6	109.5	107.8	108.0	97.7	98.9	88.7	89.7	87.2	106.0
					B	113.6	104.0	84.2	88.7	90.7	76.8	76.2	102.9	90.9	91.5	87.3	85.6	94.7	79.1	86.6	72.7	82.8	76.8	88.2
		ENVI-Carb + PSA	GC-MS	Green tea	A	101.7	119.0	119.0	102.5	115.6	111.4	104.9	100.5	96.9	82.6	107.9	95.0	102.0	99.0	99.3	109.6	81.3	87.8	101.2
					B	100.2	99.8	106.3	94.5	110.9	103.4	96.0	89.2	92.8	84.5	102.5	87.5	96.8	91.7	85.3	97.7	80.5	87.4	94.6
				Oolong tea	A	127.4	113.5	107.9	101.5	116.1	100.6	109.0	108.5	110.0	104.8	92.1	94.2	97.7	100.9	81.6	84.0	84.5	85.6	100.9
					B	103.6	96.2	94.3	90.5	90.9	86.5	94.2	90.6	87.0	84.7	78.0	80.9	82.3	81.1	66.5	71.5	70.0	76.1	84.6
			GC-MS/MS	Green tea	A	112.4	113.1	107.2	107.5	115.9	115.9	102.4	98.5	97.4	100.3	34.2	91.3	108.1	93.9	100.8	97.8	84.8	92.2	98.5
					B	106.3	99.5	106.6	98.2	113.5	106.3	103.0	91.8	98.5	98.6	31.8	85.5	97.4	87.0	87.3	91.2	76.8	93.5	92.6
				Oolong tea	A	120.1	118.4	129.8	101.7	110.9	102.0	107.6	101.2	110.0	103.5	87.6	94.1	85.8	100.6	80.0	80.8	89.8	87.8	99.9
					B	101.8	100.6	96.4	88.3	89.0	86.4	95.2	83.3	87.0	82.2	76.6	78.4	74.7	77.0	64.5	65.8	70.8	80.5	83.4

No.	Compound	Cleanup	Instrument	Tea	Rep	1	2	3	4	5	6	7	8	9	10	11	12	13	14	15	16	17	18	19	20
19	DE-PCB101	Cleanert TPT	GC-MS	Green tea	A	42.0	31.7	34.0	35.9	28.1	32.4	29.8	28.2	30.8	30.3	31.5	31.5	32.0	30.2	29.3	20.4	27.0	24.9	32.6	30.7
					B	39.3	30.7	30.6	33.5	26.2	29.6	27.3	26.2	31.3	28.6	27.7	29.7	30.4	28.5	25.7	20.6	26.2	24.2	28.5	28.7
				Oolong tea	A	35.0	40.0	32.3	34.2	31.4	28.9	29.1	35.6	35.8	36.4	37.3	35.8	36.1	35.0	31.4	33.4	30.2	31.3	31.7	33.7
					B	33.1	33.1	26.3	29.0	28.9	24.7	23.8	30.1	30.7	31.0	29.4	29.7	27.8	31.2	26.4	28.8	25.6	28.6	27.4	28.7
			GC-MS/MS	Green tea	A	40.5	33.2	36.2	31.1	26.6	28.9	30.2	27.4	31.9	29.4	32.4	30.8	30.0	30.4	28.8	24.4	26.4	24.0	31.2	30.2
					B	39.8	32.3	31.2	33.2	25.4	27.3	27.3	25.7	30.5	27.7	28.9	28.7	28.5	29.1	27.5	23.5	25.4	23.9	28.2	28.6
				Oolong tea	A	34.5	38.5	32.9	34.4	31.8	28.4	30.1	35.1	37.4	34.3	37.2	37.4	34.0	36.5	29.7	31.3	27.6	30.3	30.5	33.3
					B	33.0	31.5	27.5	29.6	29.3	24.1	23.9	30.0	31.0	29.8	27.4	29.6	27.2	32.0	25.3	27.4	24.5	28.0	26.9	28.3
		ENVI-Carb + PSA	GC-MS	Green tea	A	37.1	37.2	33.9	35.1	36.9	35.1	31.4	32.0	32.0	25.3	29.2	29.2	31.1	35.8	32.0	35.9	31.1	32.1	29.3	33.0
					B	33.3	31.3	33.5	32.4	35.9	33.2	30.7	30.5	30.3	26.9	27.3	27.3	29.3	31.6	29.5	29.8	28.2	27.7	28.8	30.7
				Oolong tea	A	44.3	38.4	35.1	34.3	39.1	32.5	37.2	28.0	30.3	35.5	32.8	28.2	28.8	32.9	33.7	27.3	25.2	28.2	26.4	33.0
					B	36.0	32.1	31.3	29.6	31.4	27.9	32.2	25.1	27.7	28.7	27.4	27.2	25.5	26.1	25.8	23.6	21.6	23.0	24.1	27.6
			GC-MS/MS	Green tea	A	33.6	37.5	34.3	34.0	39.2	35.9	31.9	31.1	31.6	32.8	28.2	33.1	33.1	35.3	32.2	33.2	31.8	28.5	30.0	32.9
					B	30.2	30.4	33.2	31.3	38.3	32.7	29.6	29.1	30.7	32.3	27.2	30.8	30.3	31.0	28.5	27.9	30.4	25.0	30.0	30.3
				Oolong tea	A	49.0	38.2	42.2	33.6	37.2	32.4	35.0	30.9	37.7	34.1	31.7	29.3	28.9	30.3	32.5	24.5	21.2	27.7	23.7	32.6
					B	41.2	32.6	31.9	29.1	29.9	27.4	32.0	26.5	28.4	26.8	27.9	25.1	25.1	26.5	24.5	20.4	19.4	22.0	22.6	27.3
20	DE-PCB118	Cleanert TPT	GC-MS	Green tea	A	42.9	32.7	35.0	36.4	29.4	33.2	30.4	28.4	31.9	31.8	31.8	32.2	32.5	28.6	28.6	21.6	28.3	26.5	34.4	31.4
					B	40.3	31.4	31.3	34.2	26.6	30.1	27.3	26.6	33.2	30.5	28.5	30.4	29.6	28.4	27.5	21.4	26.7	25.3	29.8	29.4
				Oolong tea	A	35.5	40.6	32.9	34.7	30.5	30.4	30.8	35.1	37.2	35.8	38.5	37.4	33.7	35.3	31.1	34.4	31.0	31.7	31.8	34.3
					B	33.8	33.9	27.4	29.6	30.5	25.4	25.2	29.9	31.9	32.1	30.8	30.8	26.7	31.4	27.1	29.7	26.0	29.7	28.5	29.5
			GC-MS/MS	Green tea	A	40.6	35.0	37.3	32.3	27.8	31.0	30.2	27.2	33.1	30.2	33.0	31.1	32.3	34.1	29.8	24.3	27.2	23.9	34.9	31.3
					B	39.4	33.5	32.0	32.7	26.1	28.7	27.8	25.7	31.8	28.0	29.8	28.8	30.5	33.2	28.8	23.7	26.8	23.8	30.2	29.5
				Oolong tea	A	36.7	39.3	33.0	35.6	33.1	29.0	30.2	35.2	37.1	35.3	39.1	36.9	35.1	35.7	31.6	32.5	27.9	29.3	29.9	33.8
					B	34.9	33.6	28.5	31.1	31.1	24.8	25.1	31.5	30.5	30.0	29.4	29.9	27.8	31.9	26.6	28.9	23.3	27.9	26.5	29.1
		ENVI-Carb + PSA	GC-MS	Green tea	A	37.9	37.0	34.3	36.3	37.2	36.2	31.6	32.4	32.9	26.4	29.9	29.1	32.4	35.7	32.7	36.6	32.6	32.7	30.0	33.6
					B	34.7	31.8	34.0	33.4	36.1	33.8	31.1	31.1	31.5	27.6	29.1	28.0	31.1	33.4	29.7	31.5	29.5	27.8	28.8	31.4
				Oolong tea	A	46.9	38.9	33.8	35.8	38.2	33.1	37.6	27.0	36.7	35.4	33.0	28.9	27.0	33.1	34.6	28.2	26.6	27.8	27.7	33.2
					B	38.8	33.2	31.5	31.1	30.9	28.7	32.5	24.3	29.1	27.6	27.7	25.8	24.3	27.2	26.6	23.5	22.8	23.7	24.4	28.1
			GC-MS/MS	Green tea	A	34.2	38.7	35.0	35.3	38.8	37.1	32.7	32.3	33.7	33.7	28.2	31.7	31.7	38.4	32.5	35.0	32.8	30.0	31.3	33.9
					B	31.2	32.0	35.1	32.2	37.6	33.5	30.4	30.1	31.3	33.3	27.3	28.3	30.7	34.1	30.4	28.4	29.4	25.8	30.7	31.1
				Oolong tea	A	48.9	38.5	40.5	34.1	37.4	33.4	35.8	30.9	36.7	34.8	31.2	28.8	28.8	30.0	33.3	23.7	23.2	28.9	26.4	33.0
					B	42.8	33.4	33.1	30.0	31.6	28.2	32.6	27.0	29.1	27.6	26.8	24.7	24.7	25.9	25.5	19.9	20.8	23.2	24.7	26.4
21	DE-PCB138	Cleanert TPT	GC-MS	Green tea	A	42.1	32.3	34.9	35.4	28.6	32.5	29.6	27.8	30.8	30.0	31.1	30.8	30.8	30.0	28.5	20.4	26.9	24.8	32.2	30.5
					B	40.0	31.0	31.4	33.0	25.8	29.3	27.0	26.1	31.0	28.5	27.6	28.9	28.9	28.3	25.9	20.2	26.5	23.6	28.4	28.5
				Oolong tea	A	35.6	40.1	32.5	34.1	31.7	28.7	29.0	35.2	35.7	34.4	36.8	34.1	34.1	33.8	31.4	33.0	30.2	30.7	29.6	33.2
					B	33.9	33.3	27.2	29.0	30.2	24.8	25.0	30.3	30.4	29.9	29.1	27.9	27.9	30.1	26.3	28.0	24.8	28.5	26.7	28.7
			GC-MS/MS	Green tea	A	39.8	34.5	38.3	32.5	26.4	29.7	29.4	27.2	32.7	29.8	32.2	31.0	31.0	33.3	28.5	23.4	25.8	24.3	33.3	30.3
					B	38.2	33.4	32.6	34.9	24.7	28.3	27.2	26.1	31.7	27.5	26.1	28.2	28.2	34.4	27.2	23.2	26.0	23.6	28.6	29.0
				Oolong tea	A	37.3	37.0	33.0	29.0	32.7	29.9	29.1	34.4	37.6	32.5	34.4	31.9	31.9	35.2	30.7	30.7	28.4	29.1	29.1	33.0
					B	35.5	32.2	27.9	35.7	32.7	25.7	24.8	31.2	37.6	28.6	35.2	26.9	26.9	29.7	26.6	26.8	24.2	28.2	28.2	28.5
		ENVI-Carb + PSA	GC-MS	Green tea	A	37.6	37.1	33.8	35.7	36.0	35.1	31.4	31.0	31.5	24.5	29.9	28.5	30.0	34.9	31.1	34.9	31.1	31.0	28.5	32.5
					B	34.6	32.1	33.4	32.4	35.3	32.1	30.6	30.2	29.9	25.6	27.4	27.4	27.1	31.1	28.3	28.4	27.8	26.1	28.2	30.1
				Oolong tea	A	47.0	38.0	34.4	35.1	35.9	32.4	36.7	26.6	35.5	33.8	31.5	29.2	29.2	31.0	34.3	26.5	26.3	28.2	28.0	32.6
					B	39.5	33.0	32.6	30.7	28.7	28.0	31.9	25.5	28.2	27.6	27.2	25.5	25.5	25.8	26.8	22.2	22.3	23.4	25.3	27.8
			GC-MS/MS	Green tea	A	32.0	36.9	35.8	34.8	37.7	34.4	32.8	32.1	32.3	31.4	27.2	29.0	29.0	35.2	29.8	33.8	32.0	29.5	28.8	32.7
					B	30.1	30.4	35.1	31.9	36.8	32.8	30.1	29.7	30.1	31.1	26.4	26.4	26.4	31.1	28.6	27.3	29.0	25.2	28.4	30.2
				Oolong tea	A	48.9	37.7	39.3	31.9	35.7	32.0	35.5	30.5	35.5	34.9	29.5	26.5	26.5	29.2	32.2	24.9	21.6	28.0	26.3	32.0
					B	42.8	34.0	33.8	28.5	29.9	27.7	31.1	26.8	28.2	27.4	26.4	24.0	24.0	26.1	24.0	20.9	20.4	21.9	24.2	27.5

(Continued)

Appendix Table 5.1 The 201 Pesticides Average Content for 19 Circulative Determinations on 3 Months Under 16 Factors (Two Cartridges, Two Instruments, Two Tea Samples, and Two Youden Pair Samples) (cont.)

No.	Pesticides	SPE	Method	Sample	Youden pair	Nov 9, 2009 (n=5)	Nov 14, 2009 (n=3)	Nov 19, 2009 (n=3)	Nov 24, 2009 (n=3)	Nov 29, 2009 (n=3)	Dec 4, 2009 (n=3)	Dec 9, 2009 (n=3)	Dec 14, 2009 (n=3)	Dec 19, 2009 (n=3)	Dec 24, 2009 (n=3)	Jan 3, 2010 (n=3)	Jan 8, 2010 (n=3)	Jan 13, 2010 (n=3)	Jan 18, 2010 (n=3)	Jan 23, 2010 (n=3)	Jan 28, 2010 (n=3)	Feb 2, 2010 (n=3)	Feb 7, 2010 (n=3)	Ave (µg/kg) (n=3)
22	DE-PCB180	Cleanert TPT	GC-MS	Green tea	A	40.0	32.3	34.1	35.2	27.5	29.9	28.6	26.4	30.4	28.1	31.2	30.8	28.2	27.9	21.7	25.6	24.1	31.1	29.7
					B	39.3	31.0	30.6	31.2	24.7	27.4	26.5	25.3	30.3	27.4	28.6	28.4	27.0	25.3	20.9	24.9	22.4	27.9	27.7
				Oolong tea	A	35.3	37.7	32.0	34.0	31.5	28.4	27.9	33.6	35.0	33.3	33.3	32.9	32.4	31.1	32.7	31.0	29.2	28.4	32.5
					B	33.5	32.3	27.5	29.0	30.2	24.8	24.8	30.4	29.9	29.4	28.7	27.5	28.8	26.1	27.6	25.0	26.9	25.6	28.3
			GC-MS/MS	Green tea	A	40.6	33.9	36.3	35.7	28.4	29.0	27.5	25.9	31.3	26.7	28.6	29.1	30.8	27.2	22.1	25.1	23.0	31.9	29.7
					B	40.1	32.5	31.8	31.7	25.1	26.2	25.4	25.3	30.1	26.4	28.5	27.3	30.6	25.2	21.5	24.3	22.4	27.1	27.9
				Oolong tea	A	36.2	35.5	31.2	35.7	31.8	27.2	28.6	34.1	36.6	32.4	34.0	30.7	33.7	28.7	29.6	27.0	28.5	27.6	31.8
					B	34.6	32.4	26.4	29.7	30.0	23.9	25.0	32.8	30.1	28.4	29.2	26.9	29.1	24.6	25.3	22.7	26.2	24.9	27.9
		ENVI-Carb + PSA	GC-MS	Green tea	A	36.7	36.6	33.8	34.1	35.3	34.3	37.7	30.2	30.5	23.7	32.3	28.4	32.2	30.3	30.0	31.5	29.6	27.4	31.7
					B	34.9	32.6	33.8	30.5	34.6	31.1	29.5	29.7	28.9	24.9	30.5	26.5	30.9	27.3	23.9	27.6	25.1	27.6	29.3
				Oolong tea	A	46.8	35.7	33.5	33.8	37.4	31.3	35.6	30.7	34.8	32.9	28.9	29.7	30.5	33.7	26.4	26.3	28.3	25.4	32.3
					B	40.0	31.8	32.4	30.2	31.2	27.2	30.6	27.3	28.2	26.9	24.6	25.6	25.2	26.2	21.6	22.0	23.7	23.0	27.6
			GC-MS/MS	Green tea	A	33.7	37.4	34.6	34.0	37.1	35.4	32.2	30.5	31.4	31.1	36.1	28.8	38.5	30.6	33.2	31.7	27.1	28.2	32.6
					B	30.4	31.1	34.1	31.8	36.6	31.6	30.3	27.7	29.7	30.0	34.4	26.8	32.7	28.2	27.2	28.1	22.9	28.8	30.0
				Oolong tea	A	48.4	37.2	37.9	31.7	33.9	30.0	31.7	30.7	34.8	33.6	26.4	28.3	28.0	32.9	23.6	21.2	27.4	24.7	31.2
					B	43.8	33.3	33.7	28.1	28.4	25.8	29.5	28.2	28.2	26.0	24.1	23.2	24.4	23.3	19.5	19.3	21.3	22.8	26.8
23	DE-PCB28	Cleanert TPT	GC-MS	Green tea	A	41.9	33.9	38.6	41.4	32.6	37.2	32.6	30.9	33.8	33.8	33.6	33.2	33.3	30.8	22.7	30.7	27.1	35.5	33.5
					B	40.0	33.4	32.8	38.4	26.8	33.3	29.5	29.0	35.0	32.2	31.1	31.3	32.1	28.4	22.2	28.9	26.0	30.9	31.1
				Oolong tea	A	37.0	40.3	33.2	35.3	34.4	29.9	30.9	37.2	38.5	37.3	36.6	37.0	36.1	33.1	34.5	30.6	33.0	33.3	35.2
					B	34.8	33.6	27.5	30.6	31.5	25.3	25.4	31.6	32.8	31.8	30.8	30.1	32.3	27.8	30.0	25.8	29.9	29.3	30.2
			GC-MS/MS	Green tea	A	42.1	34.8	39.3	39.6	28.8	32.8	32.7	29.6	34.7	31.5	33.5	32.7	30.4	32.1	25.8	29.2	26.7	35.7	33.0
					B	40.8	33.6	34.1	35.1	27.6	30.9	30.1	28.6	34.4	30.4	31.5	31.1	30.1	30.6	25.3	28.6	26.4	30.8	31.2
				Oolong tea	A	34.0	38.0	33.9	35.7	34.0	31.3	31.8	36.9	37.1	36.2	35.8	33.7	37.6	32.2	33.3	28.6	31.6	31.4	34.3
					B	30.6	33.6	28.5	30.5	31.1	26.6	25.5	32.4	31.7	31.7	30.2	31.3	32.5	26.7	29.5	28.7	31.6	27.0	29.6
		ENVI-Carb + PSA	GC-MS	Green tea	A	39.7	37.5	34.7	37.4	39.9	36.7	32.9	34.8	35.6	31.5	36.3	34.1	38.9	35.0	38.4	34.4	35.0	31.7	35.7
					B	34.8	32.1	35.3	34.4	38.7	33.5	33.4	33.4	32.6	31.4	35.1	31.5	35.2	32.3	32.6	31.2	29.5	31.1	33.1
				Oolong tea	A	45.6	38.4	36.1	38.6	36.1	34.9	39.2	28.9	38.7	35.5	31.1	30.5	32.9	36.3	29.2	28.8	30.8	29.3	34.4
					B	36.0	32.3	32.2	34.2	29.8	29.3	33.2	27.4	31.4	29.1	31.1	26.8	27.9	26.3	24.3	24.8	25.7	26.3	29.2
			GC-MS/MS	Green tea	A	34.5	38.9	36.9	37.8	41.1	38.4	35.2	34.6	33.9	37.8	57.0	29.1	39.0	34.8	37.2	36.5	32.6	32.7	36.8
					B	31.5	31.9	36.1	34.1	39.8	34.4	33.0	32.4	32.9	36.3	44.8	29.2	35.5	32.4	31.6	33.4	28.1	32.1	33.7
				Oolong tea	A	59.6	39.2	43.1	34.4	38.3	34.2	36.7	32.2	38.7	35.7	28.6	29.7	31.1	34.3	26.8	23.1	29.8	27.1	34.5
					B	50.1	33.3	32.0	30.2	31.6	29.0	32.6	28.2	31.4	28.7	26.3	26.3	27.1	25.8	22.6	21.3	24.1	25.3	29.2
24	DE-PCB31	Cleanert TPT	GC-MS	Green tea	A	43.2	33.8	38.3	40.5	30.9	35.3	32.7	31.0	33.9	32.4	34.2	33.7	33.0	31.0	22.3	30.4	27.7	35.4	33.4
					B	41.1	32.7	32.9	38.2	27.9	31.7	29.8	28.8	34.9	31.0	32.1	32.2	32.1	28.6	21.8	28.5	25.9	30.9	31.1
				Oolong tea	A	36.3	39.2	33.3	35.5	34.1	30.1	30.4	36.7	37.9	36.9	37.8	36.1	36.0	33.8	34.2	31.2	32.9	32.6	35.0
					B	34.2	32.8	27.5	30.6	31.3	25.3	25.1	31.1	32.8	32.0	31.7	29.2	31.9	28.1	29.7	26.4	29.7	29.0	30.0
			GC-MS/MS	Green tea	A	42.1	34.8	39.3	39.6	28.8	32.8	32.7	29.6	34.7	31.5	33.5	32.7	30.4	32.1	25.8	29.2	26.7	35.7	33.0
					B	40.8	33.6	34.1	35.1	27.6	30.9	30.1	28.6	34.4	30.4	31.5	31.1	30.1	30.6	25.3	28.6	26.4	30.8	31.2
				Oolong tea	A	35.3	38.0	33.9	35.7	34.0	31.3	31.8	36.9	37.1	36.2	35.8	33.7	37.6	32.2	33.3	28.6	31.6	31.4	34.4
					B	32.8	33.6	28.5	30.5	31.1	26.6	25.5	32.4	31.7	31.7	30.2	31.3	32.5	26.7	29.5	25.6	28.9	27.0	29.8
		ENVI-Carb + PSA	GC-MS	Green tea	A	39.4	37.6	35.1	37.3	39.9	37.3	28.7	35.0	35.4	32.4	37.6	33.4	39.3	35.1	38.8	33.7	34.8	31.6	35.5
					B	34.5	31.8	35.5	34.4	38.9	34.2	27.9	33.7	32.4	32.8	35.5	31.2	35.7	32.6	33.0	30.8	29.0	31.0	32.9
				Oolong tea	A	44.2	38.6	36.0	38.5	35.2	34.4	38.5	29.3	38.7	35.5	31.5	30.2	32.9	35.9	28.9	28.9	30.8	29.5	34.2
					B	35.1	32.3	32.1	34.2	29.1	29.2	33.3	27.7	31.4	28.7	26.7	26.3	27.7	28.3	24.3	24.3	25.0	26.6	29.0
			GC-MS/MS	Green tea	A	34.5	38.9	36.9	37.8	41.1	38.4	34.6	34.6	33.9	37.8	36.0	29.1	39.0	34.8	37.2	36.5	32.6	32.7	35.7
					B	31.5	31.9	36.1	34.1	39.8	34.4	33.0	32.4	32.9	36.3	34.7	29.2	35.5	32.4	31.6	33.4	28.1	32.1	33.1
				Oolong tea	A	37.0	39.2	42.8	30.2	31.6	34.2	36.7	32.2	38.7	35.7	28.6	29.7	31.1	34.3	26.8	23.1	29.8	27.1	33.3
					B	31.3	33.3	32.0	30.2	31.6	29.0	32.6	28.2	31.4	28.7	26.3	26.3	27.1	25.8	22.6	21.3	24.1	25.3	28.2

No.	Pesticide	Cleanup	Method	Tea	Rep	1	2	3	4	5	6	7	8	9	10	11	12	13	14	15	16	17	18	19	20
25	Dichlorofop-methyl	Cleanert TPT	GC-MS	Green tea	A	37.7	32.8	29.0	24.9	68.5	33.5	32.6	35.3	48.6	52.0	30.8	30.9	28.8	27.4	42.4	30.9	30.5	30.1	44.3	62.8
					B	35.9	28.2	27.4	25.0	63.5	36.4	32.0	31.9	41.0	48.5	29.2	38.9	26.9	25.4	36.8	34.4	28.0	28.7	41.4	58.7
				Oolong tea	A	38.8	26.1	32.5	36.5	36.8	33.6	32.3	34.5	28.3	82.4	31.5	46.0	36.5	30.1	31.6	52.3	41.2	39.0	46.2	39.2
					B	31.5	22.4	25.1	26.9	27.8	24.5	27.8	29.9	23.7	53.7	33.0	36.2	31.9	32.3	22.9	41.5	31.0	34.3	38.9	35.6
			GC-MS/MS	Green tea	A	39.2	51.2	22.9	24.5	47.3	32.3	33.0	38.5	32.0	51.3	36.0	28.8	28.6	28.1	78.6	34.0	25.8	36.0	44.9	71.9
					B	38.2	39.1	23.3	25.7	46.9	34.9	38.5	34.1	27.6	50.2	38.8	35.8	26.9	24.7	68.6	35.3	32.3	32.5	40.5	69.4
				Oolong tea	A	38.5	26.0	30.4	31.3	36.8	35.1	37.3	32.4	33.0	62.0	32.3	45.1	41.6	31.0	32.6	44.1	44.0	42.3	48.5	45.3
					B	31.4	20.8	22.4	23.4	27.3	26.9	30.6	27.2	25.3	42.6	30.7	37.7	34.9	30.5	25.2	37.2	33.0	36.9	43.0	40.6
		ENVI-Carb + PSA	GC-MS	Green tea	A	42.4	32.0	27.0	47.1	33.1	39.8	40.1	38.5	55.5	60.6	46.8	45.4	42.2	39.9	38.7	40.9	25.5	56.9	46.4	48.8
					B	38.1	33.1	30.0	33.4	25.2	38.4	34.7	35.0	47.6	48.8	41.4	42.1	42.2	28.3	42.9	37.6	21.9	42.4	59.2	39.4
				Oolong tea	A	38.4	29.6	25.4	27.3	28.4	28.5	52.1	34.0	36.1	34.6	32.6	43.6	46.5	41.4	34.4	59.7	36.5	41.6	51.6	46.4
					B	31.1	23.1	20.9	21.7	17.6	23.8	44.8	24.2	30.5	27.4	27.4	30.1	37.7	30.2	31.8	44.8	33.0	39.2	42.4	40.9
			GC-MS/MS	Green tea	A	42.1	44.9	26.5	27.1	34.9	40.4	42.1	34.8	37.1	65.5	36.3	55.1	42.3	54.3	38.2	42.4	28.4	45.1	60.4	44.8
					B	37.0	41.9	30.1	25.1	24.5	44.3	27.3	43.9	34.4	53.2	32.1	44.9	35.6	20.1	42.0	38.7	25.5	38.2	66.9	34.6
				Oolong tea	A	36.3	35.1	51.8	36.6	25.0	28.7	26.6	29.6	31.9	36.3	37.3	43.6	34.0	34.1	32.2	35.8	33.0	41.4	52.1	44.3
					B	30.5	27.4	39.6	29.7	15.6	21.8	24.2	22.1	28.1	29.4	32.7	30.1	27.9	30.5	29.2	28.7	31.0	41.9	44.0	44.6
26	Dimethenamid	Cleanert TPT	GC-MS	Green tea	A	34.5	30.9	24.6	30.0	28.9	33.5	35.5	32.5	31.2	37.7	29.6	37.2	34.2	32.4	34.6	30.2	48.7	41.5	37.8	43.8
					B	32.4	27.6	24.1	29.6	28.4	31.1	36.0	31.3	29.2	33.3	28.6	35.3	30.3	29.4	32.6	28.1	45.7	35.4	35.9	42.7
				Oolong tea	A	36.0	33.2	31.5	31.9	34.3	31.9	35.1	35.9	38.6	38.7	38.5	37.9	40.8	34.4	27.6	31.5	37.8	35.7	46.6	41.4
					B	30.0	29.2	29.3	25.8	29.9	28.4	31.8	27.7	30.4	28.2	31.6	31.1	32.4	25.1	23.5	29.2	31.9	29.0	37.4	38.8
			GC-MS/MS	Green tea	A	33.0	36.8	25.9	26.0	27.0	30.8	28.2	31.7	30.2	34.6	28.8	35.5	30.1	31.7	41.8	27.9	52.3	32.4	35.7	40.4
					B	30.9	33.0	25.6	24.8	26.2	28.9	29.6	31.3	27.8	31.8	29.8	35.0	27.7	28.3	38.2	27.0	34.4	33.3	34.8	40.2
				Oolong tea	A	36.5	33.8	31.0	31.7	35.1	32.1	38.2	34.9	37.0	43.7	36.4	36.7	44.2	36.6	30.3	31.8	36.1	34.6	45.1	43.5
					B	30.1	28.0	27.2	23.7	29.6	25.8	32.8	30.8	30.6	32.7	31.1	29.9	35.9	26.5	25.5	29.6	30.1	29.3	35.7	37.7
		ENVI-Carb + PSA	GC-MS	Green tea	A	36.0	30.4	27.0	39.6	34.5	31.0	36.5	32.4	35.3	31.9	53.3	32.2	33.1	37.6	40.2	40.8	35.0	36.0	46.2	31.5
					B	33.6	30.3	26.7	35.8	29.8	30.8	33.6	29.4	33.2	31.5	49.8	30.9	26.1	35.5	35.9	39.0	33.7	37.3	36.6	33.4
				Oolong tea	A	34.7	27.6	28.6	27.6	25.3	33.5	32.6	30.4	31.2	35.0	37.3	41.7	39.6	33.3	34.1	43.2	38.8	37.6	39.5	41.6
					B	29.1	25.0	23.6	23.9	21.8	26.8	27.7	25.9	27.1	28.8	28.7	32.2	31.9	34.6	29.2	32.6	31.8	32.3	33.6	35.0
			GC-MS/MS	Green tea	A	33.7	30.0	28.1	33.1	33.0	32.7	35.6	26.3	36.9	32.4	33.0	28.1	31.2	34.0	40.2	37.6	36.0	34.6	39.0	38.5
					B	32.6	31.6	25.5	29.7	27.9	30.7	30.3	27.5	35.2	29.4	34.1	32.6	29.7	31.4	35.6	37.4	34.2	34.6	34.7	47.7
				Oolong tea	A	34.1	28.3	31.4	23.4	25.4	32.6	29.1	31.2	26.0	30.9	34.6	41.7	32.9	38.1	33.4	38.7	33.1	34.6	41.1	37.9
					B	27.6	27.2	25.5	21.9	21.4	24.7	25.5	27.2	23.6	27.2	27.6	32.2	27.0	33.7	28.8	31.8	29.1	57.7	35.1	23.6
27	Diofeno-lan-1	Cleanert TPT	GC-MS	Green tea	A	68.7	72.3	56.8	59.1	47.6	61.6	67.3	70.9	73.7	78.3	62.8	71.6	62.8	67.8	71.3	64.0	79.5	71.8	72.6	93.7
					B	64.3	63.2	52.9	57.4	46.7	57.8	65.4	66.9	67.2	70.6	60.5	70.0	58.4	62.2	65.2	55.9	74.0	65.0	73.1	88.5
				Oolong tea	A	70.9	62.9	68.2	64.8	69.7	71.6	72.9	71.7	70.1	82.9	69.8	75.8	73.7	61.9	60.1	72.8	73.0	68.5	75.3	81.7
					B	61.4	56.6	62.1	52.6	58.9	59.6	63.0	61.5	60.7	63.7	60.6	60.8	65.7	53.6	53.6	67.8	63.2	59.6	65.1	76.5
			GC-MS/MS	Green tea	A	71.2	82.1	53.2	57.0	53.0	65.7	77.0	70.4	66.3	75.9	65.5	72.5	65.7	66.4	87.9	62.1	87.2	84.5	75.1	93.5
					B	66.2	82.1	56.9	56.9	52.5	63.3	76.7	64.0	63.0	66.4	62.7	68.8	55.6	60.8	78.7	58.0	72.9	72.7	72.5	88.7
				Oolong tea	A	71.1	81.4	63.0	58.4	68.8	68.0	78.0	72.9	78.1	83.2	76.2	78.8	65.7	62.7	61.8	69.9	75.4	70.1	78.7	80.9
					B	62.1	72.3	56.5	49.1	59.0	67.1	64.7	64.2	69.9	63.0	64.7	72.7	57.4	54.4	54.9	67.1	65.7	61.8	70.8	76.7
		ENVI-Carb + PSA	GC-MS	Green tea	A	75.8	82.4	69.2	74.8	78.7	81.7	94.2	72.7	75.3	72.7	71.9	81.7	71.9	65.2	84.4	81.7	78.4	82.4	81.8	81.6
					B	69.6	63.4	58.1	66.0	64.7	79.8	90.1	73.9	72.7	66.4	69.5	79.8	69.5	65.1	73.9	79.8	71.1	76.9	71.0	76.9
				Oolong tea	A	71.9	60.7	60.0	58.1	55.1	94.4	72.8	67.6	78.8	80.2	63.7	72.8	63.7	79.6	67.6	94.4	74.6	72.0	79.9	93.6
					B	61.0	53.7	50.6	48.4	45.9	73.6	59.4	57.2	72.7	66.5	59.0	73.6	59.0	66.2	57.2	73.6	68.5	70.0	69.6	82.3
			GC-MS/MS	Green tea	A	74.2	66.0	62.5	72.0	74.9	82.5	76.6	72.9	73.2	72.9	73.2	82.5	73.2	70.6	84.6	82.5	82.8	78.4	84.8	79.1
					B	68.6	67.1	54.6	64.9	63.3	80.9	72.5	74.7	67.5	62.2	67.5	80.9	67.5	67.1	74.7	80.9	72.9	76.0	71.2	73.9
				Oolong tea	A	70.1	59.1	78.6	78.6	49.8	76.7	58.2	64.9	65.1	72.5	65.1	59.2	73.2	74.9	67.2	76.7	72.1	86.4	79.9	103.5
					B	60.1	54.0	62.0	62.0	42.2	64.7	57.4	57.4	57.4	58.1	57.4	58.2	65.1	66.4	57.4	64.7	64.0	72.1	72.1	88.1

(Continued)

Appendix Table 5.1 The 201 Pesticides Average Content for 19 Circulative Determinations on 3 Months Under 16 Factors (Two Cartridges, Two Instruments, Two Tea Samples, and Two Youden Pair Samples) (cont.)

No.	Pesticides	SPE	Method	Sample	Youden pair	Average content (µg/kg) Nov 9, 2009 (n=5)	Nov 14, 2009 (n=3)	Nov 19, 2009 (n=3)	Nov 24, 2009 (n=3)	Nov 29, 2009 (n=3)	Dec 4, 2009 (n=3)	Dec 9, 2009 (n=3)	Dec 14, 2009 (n=3)	Dec 19, 2009 (n=3)	Dec 24, 2009 (n=3)	Dec 14, 2009 (n=3)	Jan 3, 2010 (n=3)	Jan 8, 2010 (n=3)	Jan 13, 2010 (n=3)	Jan 18, 2010 (n=3)	Jan 23, 2010 (n=3)	Jan 28, 2010 (n=3)	Feb 2, 2010 (n=3)	Feb 7, 2010 (n=3)	Ave (µg/kg)
28	Diofeno-lan-2	Cleanert TPT	GC-MS	Green tea	A	93.4	73.3	77.6	79.9	65.9	74.0	68.5	64.5	73.2	65.7	77.6	77.1	72.5	69.0	64.2	48.6	61.2	58.4	74.5	70.5
					B	87.7	71.2	70.2	74.9	59.1	66.2	62.5	60.2	70.6	63.9	68.5	70.7	67.4	65.8	58.5	47.5	59.3	54.4	64.6	65.4
				Oolong tea	A	81.7	85.4	70.8	75.6	76.1	62.4	62.7	76.2	76.2	75.4	86.2	74.8	73.8	73.3	74.0	71.9	66.3	68.0	64.9	73.5
					B	77.4	72.6	61.1	64.7	71.9	54.7	56.5	66.5	63.7	65.1	65.4	63.2	62.4	64.6	61.4	59.8	54.2	61.7	56.4	63.3
			GC-MS/MS	Green tea	A	95.4	78.2	88.2	90.4	66.2	90.1	68.9	54.3	75.7	67.8	79.2	70.2	71.6	78.6	67.3	55.4	58.9	56.9	88.8	73.8
					B	95.0	77.9	79.6	80.0	62.8	84.3	66.1	55.5	76.2	66.8	72.6	68.4	69.1	82.0	67.1	60.2	60.3	58.0	78.1	71.6
				Oolong tea	A	75.9	79.0	70.3	75.4	71.7	61.8	60.1	79.1	77.5	76.8	86.2	79.1	65.1	77.3	69.3	73.8	58.8	62.3	61.5	71.6
					B	72.9	69.1	62.9	65.7	68.0	54.8	54.3	69.8	62.9	64.8	64.9	66.6	63.6	68.5	59.1	60.9	50.7	54.5	53.1	62.5
		ENVI-Carb + PSA	GC-MS	Green tea	A	82.6	82.7	79.1	82.1	82.4	78.7	69.9	73.9	75.0	102.6	65.2	76.8	71.3	80.8	73.4	79.4	77.0	70.1	65.1	77.3
					B	77.8	73.1	77.2	73.5	80.9	72.3	70.2	70.0	68.6	98.5	62.1	71.5	64.0	71.7	65.7	65.3	66.8	59.2	64.5	71.2
				Oolong tea	A	99.9	82.0	73.6	76.3	91.0	69.8	80.8	65.1	79.6	73.1	70.2	63.8	66.3	82.6	71.9	57.7	59.8	61.3	61.8	73.0
					B	87.6	71.7	72.0	67.5	73.0	60.1	66.2	61.1	66.1	59.5	58.7	54.6	57.2	65.9	56.2	48.0	49.7	51.8	55.1	62.2
			GC-MS/MS	Green tea	A	77.4	89.2	80.2	83.5	82.4	86.6	73.8	77.0	76.8	78.5	63.4	52.2	63.0	134.2	69.3	74.7	73.1	67.5	67.9	77.4
					B	67.9	73.7	77.6	74.0	80.2	75.2	68.4	70.7	69.7	77.4	59.7	59.3	60.9	104.9	63.6	62.9	64.9	57.2	68.0	70.3
				Oolong tea	A	102.2	80.0	88.2	72.1	0.0	66.7	74.9	65.1	79.6	74.0	67.7	0.0	57.6	0.0	70.5	49.8	62.4	83.9	62.2	60.9
					B	85.1	72.2	72.1	64.0	0.0	57.6	66.8	57.4	66.1	60.0	58.5	0.0	52.1	60.0	54.2	42.2	55.5	64.5	58.1	51.9
29	Fenbucon-azole	Cleanert TPT	GC-MS	Green tea	A	101.4	81.5	89.3	80.9	73.6	69.5	64.8	67.5	79.0	62.3	104.0	74.1	80.2	60.0	61.6	46.8	61.9	58.6	73.5	73.2
					B	93.5	79.2	81.5	77.8	62.2	67.6	62.0	60.1	75.0	58.6	78.9	69.3	77.9	63.9	51.7	45.4	56.3	51.8	67.2	67.4
				Oolong tea	A	95.2	77.6	74.3	76.5	74.3	72.2	58.7	75.8	78.6	68.1	92.6	77.7	73.7	69.2	68.4	67.8	61.1	62.7	60.0	72.9
					B	84.8	78.1	70.5	74.7	75.5	63.3	58.6	74.1	61.6	60.6	69.9	67.1	61.8	61.4	56.5	51.7	50.6	54.8	47.7	64.4
			GC-MS/MS	Green tea	A	89.3	77.5	96.8	116.8	74.6	97.4	68.3	64.4	78.4	63.3	81.3	69.6	73.6	63.2	63.6	51.3	60.4	55.9	79.8	75.0
					B	91.9	77.3	86.1	81.8	61.4	84.2	66.0	58.9	71.3	57.4	68.1	66.3	73.5	80.1	55.6	51.8	54.2	52.2	69.8	68.8
				Oolong tea	A	89.6	85.8	79.7	79.0	66.7	66.6	62.3	76.1	72.6	77.8	82.6	80.1	63.5	76.9	68.4	66.9	55.3	58.5	57.6	71.9
					B	78.0	76.3	70.1	72.5	69.1	59.2	56.9	72.3	59.7	64.1	63.7	68.8	65.4	65.6	56.6	53.3	46.4	52.4	45.8	63.0
		ENVI-Carb + PSA	GC-MS	Green tea	A	81.3	76.1	87.9	39.4	92.3	88.5	56.6	77.4	82.7	138.9	76.6	88.8	72.4	76.0	70.8	87.4	76.0	68.3	65.0	81.2
					B	83.1	75.9	84.0	78.5	95.2	81.3	56.7	72.4	80.8	109.0	71.5	75.6	58.8	70.1	65.6	67.5	64.8	57.2	64.4	73.8
				Oolong tea	A	111.6	94.7	76.6	31.2	136.8	68.8	87.7	82.7	93.3	80.0	80.3	46.7	61.7	64.1	87.5	55.7	52.8	65.2	58.4	78.2
					B	96.1	82.8	84.6	30.3	96.4	57.3	69.4	67.5	79.6	50.2	66.7	41.5	55.2	54.3	59.3	47.5	44.8	50.2	53.4	65.1
			GC-MS/MS	Green tea	A	73.5	80.1	85.9	94.2	87.5	88.7	54.8	75.7	79.9	85.5	61.6	63.7	67.4	78.0	68.2	74.8	73.1	63.0	65.4	74.8
					B	76.1	73.4	84.3	79.8	91.7	81.4	76.5	67.5	70.2	77.4	61.2	60.1	68.4	63.9	63.8	64.4	63.1	51.8	64.3	70.5
				Oolong tea	A	117.7	83.8	75.7	76.8	81.3	72.0	73.4	67.9	93.3	75.4	60.7	52.4	58.4	54.7	70.1	49.3	48.9	69.3	59.0	70.5
					B	105.2	75.2	75.2	71.5	72.5	60.4	65.0	60.4	79.6	47.5	55.1	45.8	44.6	51.0	46.8	42.1	44.6	51.6	54.0	60.4
30	Fenpyroxi-mate	Cleanert TPT	GC-MS	Green tea	A	331.5	236.3	395.6	350.3	310.2	285.2	302.1	279.7	271.4	269.1	205.0	162.5	219.3	286.1	212.1	81.4	247.1	200.4	282.2	259.3
					B	282.6	240.5	350.3	346.7	232.4	254.1	271.3	234.0	235.0	219.5	162.8	171.6	218.2	268.7	160.9	81.3	214.9	200.1	259.3	231.8
				Oolong tea	A	336.5	392.7	234.7	235.3	234.6	239.9	174.4	285.0	241.8	359.5	176.2	375.4	282.7	344.2	234.9	230.8	164.5	277.3	298.1	269.4
					B	337.8	306.4	171.8	217.8	260.3	218.9	179.1	226.0	214.8	263.5	173.3	350.5	209.6	244.7	214.7	193.0	146.3	268.0	231.1	233.8
			GC-MS/MS	Green tea	A	256.0	233.3	472.2	466.2	229.0	148.0	316.2	240.6	321.1	218.0	223.6	279.5	193.0	244.0	244.8	114.8	239.1	267.6	189.4	257.7
					B	250.5	244.0	370.3	233.1	173.6	142.5	299.9	218.6	224.5	209.5	174.2	267.8	201.3	210.7	196.0	107.3	213.0	244.5	186.0	226.6
				Oolong tea	A	231.7	298.0	219.3	230.1	211.0	213.1	182.0	278.6	260.6	333.1	241.7	320.4	258.9	304.1	222.9	214.9	190.0	214.7	265.0	247.0
					B	226.0	225.5	202.2	230.1	237.0	204.0	260.5	231.3	219.8	257.0	215.2	303.6	220.0	253.2	194.2	180.0	164.7	217.9	193.7	223.0
		ENVI-Carb + PSA	GC-MS	Green tea	A	287.9	293.1	275.8	499.7	301.0	308.7	217.3	288.4	296.1	317.9	158.4	220.2	237.4	325.6	245.5	324.6	195.3	301.6	236.5	280.6
					B	304.4	222.3	290.6	432.6	323.5	213.9	268.8	242.5	254.8	299.5	197.1	229.0	216.8	271.9	227.5	278.9	186.4	220.6	231.5	258.6
				Oolong tea	A	572.6	319.9	259.5	389.0	213.4	279.9	397.7	160.2	250.6	320.5	258.5	214.2	181.1	185.2	389.2	221.7	197.7	277.1	226.7	279.7
					B	440.3	273.0	255.6	304.1	200.6	201.1	314.7	186.9	238.8	216.3	209.8	150.7	180.7	142.3	221.5	258.4	159.8	216.3	231.1	231.3
			GC-MS/MS	Green tea	A	285.6	307.4	295.8	415.6	323.0	328.4	205.3	253.4	275.0	272.7	147.5	339.7	197.9	249.6	249.1	306.2	311.6	276.5	219.2	264.5
					B	321.6	200.6	326.8	362.3	363.4	204.0	330.9	240.2	241.3	278.4	184.8	290.6	193.3	287.0	232.5	283.9	274.5	196.6	212.3	264.5
				Oolong tea	A	535.6		329.3	340.9	378.1	259.9	368.3	215.9	250.6	288.2	234.4	218.6	223.3	304.4	334.7	208.9	113.5	148.4	129.2	268.3
					B	389.8	211.2	229.3	275.9	350.6	187.4	307.8	227.5	238.8	191.6	204.5	172.4	218.9	217.5	207.9	219.4	105.6	113.3	146.6	221.9

31	Fluotrimazole	Cleanert TPT	GC-MS	Green tea	A	50.2	38.5	38.5	35.2	32.2	29.3	30.6	28.9	30.1	24.9	34.4	27.2	28.5	29.8	28.2	20.0	25.9	23.2	31.2	30.9
					B	45.4	36.3	34.8	31.9	27.8	26.6	27.6	26.4	30.7	24.1	31.7	24.4	25.7	28.5	25.3	19.7	24.8	22.0	26.8	28.4
				Oolong tea	A	43.5	36.2	32.1	34.8	33.0	30.0	29.1	32.8	36.9	31.1	33.5	34.4	34.1	32.3	30.7	31.3	28.5	29.9	28.7	32.8
					B	40.5	34.2	30.1	31.0	32.1	26.1	26.3	29.6	29.4	29.1	28.2	29.5	28.3	28.2	25.9	26.1	23.2	26.8	25.0	28.9
			GC-MS/MS	Green tea	A	45.8	39.0	46.0	21.2	28.9	60.5	30.4	26.2	34.0	31.3	36.2	29.8	27.1	29.8	28.6	22.5	25.8	23.0	32.1	32.5
					B	44.6	37.5	37.1	34.3	24.4	43.3	27.9	25.4	32.0	29.5	30.8	30.3	25.7	33.5	26.8	22.9	25.2	23.0	27.1	30.6
				Oolong tea	A	37.1	37.8	37.4	31.9	26.5	28.5	30.0	36.3	30.5	33.6	39.4	35.3	30.5	33.9	28.4	24.0	24.9	27.2	25.1	31.7
					B	35.7	35.0	34.2	30.0	29.0	25.2	26.3	32.6	27.3	29.4	30.4	28.8	29.2	29.2	24.9	24.0	21.4	25.9	21.7	28.5
		ENVI-Carb + PSA	GC-MS	Green tea	A	44.1	40.8	40.4	39.7	39.3	36.6	30.8	31.8	35.0	40.6	25.8	31.6	30.0	35.9	30.3	34.0	30.6	29.0	27.4	34.4
					B	40.1	38.5	36.3	34.9	40.3	32.0	31.3	30.1	35.0	36.1	27.9	28.6	26.7	31.2	27.7	27.5	26.6	24.4	27.6	31.7
				Oolong tea	A	50.4	43.7	35.5	36.4	50.8	33.1	39.1	33.4	53.7	39.4	35.0	26.9	29.1	28.3	33.4	25.9	25.6	27.4	26.5	35.5
					B	43.2	38.2	33.8	32.9	39.5	28.3	31.7	28.1	41.2	28.9	25.7	24.6	24.8	24.3	24.9	21.8	21.3	22.6	23.9	29.5
			GC-MS/MS	Green tea	A	36.7	43.4	39.1	40.5	38.3	48.5	30.9	31.2	33.4	37.7	27.0	44.1	24.8	35.9	30.4	32.8	29.6	26.9	29.0	34.8
					B	34.7	35.9	37.3	35.5	37.2	36.0	29.5	28.2	31.2	35.9	25.7	39.2	27.4	26.8	26.1	25.8	26.4	21.8	29.0	31.0
				Oolong tea	A	55.3	41.7	44.0	31.1	30.6	32.5	34.0	29.3	53.7	35.2	31.8	26.1	28.5	28.3	30.8	23.9	21.2	25.6	24.3	33.0
					B	46.5	37.7	36.3	29.7	25.7	27.3	30.8	25.9	41.2	27.0	27.8	23.2	24.6	25.4	24.0	19.8	18.6	20.9	22.3	28.1
32	Fluoxypr-1-methyl-heptyl ester	Cleanert TPT	GC-MS	Green tea	A	54.1	41.5	38.6	36.3	32.8	37.3	31.9	31.5	35.3	34.0	44.9	43.1	36.9	32.6	34.3	43.7	30.0	33.0	37.0	37.3
					B	49.5	39.2	34.4	33.8	31.6	34.7	29.1	30.4	40.2	34.2	40.6	39.5	33.9	32.2	34.2	42.3	29.1	27.2	31.7	35.2
				Oolong tea	A	42.9	44.0	37.9	41.5	41.8	31.0	30.3	37.2	41.1	37.8	56.4	36.5	38.8	35.3	38.0	37.8	36.6	34.6	31.5	38.5
					B	40.3	38.1	35.9	33.3	38.2	26.7	31.3	35.7	34.2	34.5	42.1	31.0	34.5	32.9	30.5	31.5	29.7	31.0	28.0	33.7
			GC-MS/MS	Green tea	A	52.3	41.5	45.9	36.7	31.7	50.0	32.6	31.9	32.5	31.9	44.2	32.6	40.1	37.9	33.4	38.7	28.8	25.7	52.6	37.9
					B	52.6	37.3	41.9	36.8	30.1	47.1	32.0	32.1	36.1	31.8	40.7	29.6	37.6	36.4	33.3	38.0	25.7	25.5	40.6	36.1
				Oolong tea	A	38.9	46.5	41.4	43.7	38.1	32.6	34.8	39.3	41.1	36.1	52.2	34.8	32.0	41.5	39.8	30.2	33.4	30.6	25.5	37.5
					B	36.7	42.6	38.0	34.9	34.2	28.0	30.2	33.0	36.2	33.0	37.4	31.4	30.2	30.2	35.3	26.3	25.1	29.1	20.2	32.2
		ENVI-Carb + PSA	GC-MS	Green tea	A	42.8	46.2	47.9	35.7	42.6	40.5	40.0	39.5	41.2	40.7	47.2	48.3	35.9	35.0	40.2	36.8	46.4	30.8	32.5	40.8
					B	39.3	47.4	43.9	30.5	41.7	40.5	29.5	39.0	37.7	37.5	40.7	44.7	32.8	35.0	40.7	29.8	38.5	30.5	33.3	37.5
				Oolong tea	A	47.9	41.6	39.5	37.0	47.2	33.9	42.2	44.9	45.4	36.3	35.6	33.4	35.4	43.7	35.8	30.7	31.1	30.3	32.7	38.1
					B	43.9	36.9	38.5	34.3	38.5	29.6	32.9	38.8	36.5	30.4	30.4	29.6	29.3	37.7	29.4	23.6	26.5	27.0	28.9	32.8
			GC-MS/MS	Green tea	A	43.0	65.3	43.6	36.5	36.6	41.1	44.6	39.7	49.8	35.5	39.8	44.3	31.5	37.8	41.1	35.4	31.1	26.8	28.6	39.6
					B	39.8	62.0	40.9	31.0	37.7	37.6	31.3	34.2	43.4	34.1	32.1	40.4	33.8	28.3	38.1	25.8	31.5	28.8	26.7	35.7
				Oolong tea	A	50.4	45.9	46.1	35.7	43.5	36.1	35.2	36.3	45.4	37.9	32.8	30.7	32.7	30.5	29.1	27.9	36.7	44.8	40.0	37.8
					B	44.6	41.2	41.6	34.6	34.6	32.4	30.8	30.5	36.5	33.8	29.0	28.8	24.9	27.8	26.1	20.3	31.5	37.9	32.0	32.6
33	Iprovalicarb-1	Cleanert TPT	GC-MS	Green tea	A	199.0	173.2	180.4	182.7	152.7	140.3	165.4	177.7	159.5	135.3	215.3	173.2	163.4	151.2	166.0	330.9	123.4	114.5	150.7	171.3
					B	187.8	164.1	175.3	166.9	139.3	139.6	173.6	181.5	178.9	133.0	190.7	154.3	152.7	150.1	165.8	321.6	127.3	111.4	138.9	165.9
				Oolong tea	A	193.5	137.8	188.0	185.5	172.3	169.8	167.3	168.6	168.1	150.9	154.1	178.0	170.7	156.5	180.2	181.5	191.8	152.2	134.9	168.5
					B	203.6	145.7	184.3	170.0	162.8	161.5	165.8	177.1	127.7	136.3	144.0	161.0	156.4	149.5	154.6	153.5	160.9	136.6	132.0	157.0
			GC-MS/MS	Green tea	A	211.1	173.7	174.8	177.8	144.9	284.1	137.2	117.8	142.9	145.8	230.3	128.8	161.2	191.3	145.0	144.1	122.0	103.5	213.2	165.8
					B	206.2	164.7	167.1	157.5	127.1	241.3	128.9	124.2	160.4	145.3	224.3	125.0	149.2	206.6	147.5	144.7	116.7	102.1	181.9	159.0
				Oolong tea	A	235.4	161.6	158.8	167.0	147.1	154.1	226.3	194.3	169.9	159.6	199.6	154.2	123.1	166.4	152.8	158.2	147.9	144.8	116.4	165.1
					B	201.3	165.4	143.9	146.6	142.2	134.6	214.4	204.2	157.5	140.7	148.1	130.2	154.8	148.7	127.0	133.0	119.4	127.3	120.5	150.5
		ENVI-Carb + PSA	GC-MS	Green tea	A	156.1	231.8	201.4	179.9	191.1	203.8	159.8	185.4	211.3	256.3	161.0	175.5	150.0	138.7	165.8	134.2	251.6	101.8	138.8	178.6
					B	163.2	191.5	209.6	151.5	194.3	180.4	136.3	155.1	187.8	255.7	135.4	164.5	132.9	145.6	141.5	120.2	207.5	128.7	151.9	166.0
				Oolong tea	A	218.4	148.7	162.0	156.3	257.1	170.4	151.6	244.3	212.8	148.3	145.6	133.1	157.3	214.5	180.9	131.8	161.4	138.1	141.7	172.3
					B	200.6	146.1	164.7	163.4	190.7	158.1	168.9	188.3	176.2	123.3	127.8	119.7	141.7	173.1	144.5	106.0	135.5	126.0	129.3	151.8
			GC-MS/MS	Green tea	A	174.9	184.6	169.5	173.0	177.6	206.2	185.6	165.4	204.7	170.6	170.8	140.4	131.0	141.2	143.6	145.6	146.7	114.7	138.4	162.3
					B	194.7	208.5	166.0	147.8	177.1	181.9	170.7	144.5	196.8	164.1	137.7	129.2	127.8	138.6	132.5	132.4	125.0	114.4	153.1	154.9
				Oolong tea	A	161.4	174.1	177.8	148.1	162.4	146.0	156.8	152.2	212.8	150.6	143.9	115.2	130.9	130.2	142.3	119.3	145.2	194.1	164.8	154.1
					B	143.5	160.9	161.6	139.0	132.6	126.8	141.0	119.4	176.2	123.7	129.7	128.9	101.5	119.6	115.4	94.5	129.5	159.5	145.5	134.1

(Continued)

Appendix Table 5.1 The 201 Pesticides Average Content for 19 Circulative Determinations on 3 Months Under 16 Factors (Two Cartridges, Two Instruments, Two Tea Samples, and Two Youden Pair Samples) (cont.)

No.	Pesticides	Method	SPE	Sample	Youden pair	Nov 9, 2009 (n=5)	Nov 9, 2009 (n=3)	Nov 14, 2009 (n=3)	Nov 19, 2009 (n=3)	Nov 24, 2009 (n=3)	Nov 29, 2009 (n=3)	Dec 4, 2009 (n=3)	Dec 9, 2009 (n=3)	Dec 14, 2009 (n=3)	Dec 19, 2009 (n=3)	Dec 24, 2009 (n=3)	Jan 3, 2010 (n=3)	Jan 8, 2010 (n=3)	Jan 13, 2010 (n=3)	Jan 18, 2010 (n=3)	Jan 23, 2010 (n=3)	Jan 28, 2010 (n=3)	Feb 2, 2010 (n=3)	Feb 7, 2010 (n=3)	Ave (µg/kg)
34	Iprovalicarb-2	GC-MS	Cleanert TPT	Green tea	A	216.2	178.6	163.0	185.3	135.9	133.4	137.5	143.9	197.4	130.5	284.2	185.4	169.8	145.7	166.8	338.1	129.6	116.1	147.2	173.9
				Green tea	B	208.8	179.1	141.9	161.9	121.8	133.3	131.9	136.7	170.6	117.6	243.2	161.3	155.1	150.0	163.7	325.8	130.5	107.6	133.0	161.8
				Oolong tea	A	195.2	156.0	147.0	167.2	159.5	133.4	221.4	203.4	168.8	144.6	192.0	157.3	149.7	142.7	151.4	158.0	188.6	140.8	136.5	164.4
				Oolong tea	B	192.5	151.5	142.7	155.9	148.8	119.9	211.1	206.7	126.5	131.5	141.6	129.7	149.3	139.1	127.4	133.0	148.7	128.8	126.7	148.0
		GC-MS/MS		Green tea	A	212.3	177.4	182.0	181.1	147.9	318.3	141.6	133.9	153.1	140.7	202.3	140.8	163.8	159.5	148.6	145.3	126.2	105.5	234.6	169.2
				Green tea	B	207.1	165.4	174.8	159.6	130.6	267.2	132.1	130.2	157.4	144.9	182.9	129.7	150.8	174.4	143.0	143.2	118.9	104.7	200.5	158.8
				Oolong tea	A	235.4	161.6	158.8	167.0	146.3	148.2	240.9	198.6	191.0	161.1	202.3	158.3	148.6	163.8	158.5	162.0	156.0	146.9	126.5	170.1
				Oolong tea	B	201.3	165.4	143.9	146.6	144.6	129.2	223.6	206.3	146.8	144.1	149.7	133.1	152.9	153.3	132.6	138.2	127.9	128.6	125.6	152.3
		GC-MS	ENVI-Carb + PSA	Green tea	A	158.2	216.8	197.9	154.0	201.4	203.2	176.8	180.3	142.7	395.9	158.3	171.6	157.3	140.9	169.9	128.2	256.4	95.1	135.4	181.1
				Green tea	B	177.5	192.9	188.8	132.7	190.5	182.3	161.4	151.1	136.5	278.8	143.7	147.7	132.3	148.3	143.9	120.1	214.8	122.0	139.8	163.4
				Oolong tea	A	200.6	158.0	158.3	153.0	300.8	149.1	160.6	301.9	212.0	158.9	181.0	128.9	156.5	176.6	156.2	122.1	137.6	129.7	135.5	172.5
				Oolong tea	B	184.6	146.4	148.2	146.2	197.7	132.3	140.2	214.6	177.8	161.2	135.6	130.8	132.6	149.4	130.9	97.4	119.0	112.9	121.7	144.7
		GC-MS/MS		Green tea	A	174.4	201.4	171.1	177.3	177.9	212.1	158.7	170.0	150.5	155.9	166.5	149.4	124.1	145.5	153.1	144.3	151.3	119.1	141.8	160.5
				Green tea	B	195.2	210.1	167.9	150.7	176.1	188.9	150.5	149.0	156.9	154.7	132.7	142.6	126.6	134.1	140.5	130.8	131.9	119.8	153.6	153.4
				Oolong tea	A	161.4	174.1	177.8	148.1	162.4	148.0	157.9	152.2	212.0	141.6	141.6	123.0	114.4	130.2	140.3	119.9	137.8	199.6	168.3	153.9
				Oolong tea	B	143.5	160.9	161.6	139.0	132.6	128.7	140.3	119.4	177.8	127.7	127.5	117.6	109.3	119.6	128.8	95.3	123.8	161.0	150.2	135.0
35	Isodrin	GC-MS	Cleanert TPT	Green tea	A	46.0	35.0	33.7	38.5	32.4	34.0	32.1	31.3	33.4	27.9	30.8	36.1	33.2	32.1	32.6	23.6	29.6	27.9	34.4	32.9
				Green tea	B	43.9	34.2	31.2	35.1	29.8	32.4	28.9	28.7	32.4	27.5	29.0	34.5	31.4	30.7	30.6	23.0	29.8	24.2	30.0	30.9
				Oolong tea	A	40.7	41.4	34.0	37.8	34.3	30.3	34.8	40.7	37.6	39.5	36.8	39.0	36.5	37.1	33.9	35.6	32.9	32.2	32.2	36.2
				Oolong tea	B	39.6	35.1	28.0	30.2	31.0	25.5	26.9	32.8	35.4	33.6	31.6	32.3	29.5	31.8	28.1	32.8	27.7	27.9	27.9	30.9
		GC-MS/MS		Green tea	A	46.1	34.3	36.9	40.3	27.9	30.5	31.4	28.0	33.1	29.3	34.9	33.0	33.6	29.8	30.7	24.7	24.4	23.9	32.5	32.0
				Green tea	B	43.7	32.4	33.4	35.4	27.4	28.8	29.1	28.0	31.9	26.8	30.3	30.4	30.2	28.7	27.8	24.1	31.7	23.4	27.6	29.7
				Oolong tea	A	38.0	40.5	34.0	36.2	32.8	31.5	30.7	36.5	39.4	36.9	42.5	39.1	33.8	36.4	32.4	34.3	28.7	33.6	31.5	35.4
				Oolong tea	B	36.1	33.2	29.0	30.0	30.2	26.1	23.7	32.0	32.9	31.2	31.4	30.5	28.6	31.7	24.9	34.7	30.5	30.0	26.6	30.1
		GC-MS	ENVI-Carb + PSA	Green tea	A	44.6	40.7	33.7	35.8	40.9	37.4	30.5	29.1	37.7	22.1	32.8	30.0	32.0	37.1	33.8	35.8	30.3	28.7	25.4	33.6
				Green tea	B	37.3	34.6	33.7	34.5	37.5	35.5	26.2	27.6	34.5	25.5	29.2	29.4	31.4	33.8	31.3	30.2	28.1	26.4	27.3	31.4
				Oolong tea	A	46.5	39.1	36.8	36.7	38.4	35.1	38.2	31.6	37.6	34.6	27.3	35.1	31.4	31.5	39.4	27.9	24.9	28.6	27.1	34.3
				Oolong tea	B	43.7	32.7	33.7	30.8	29.8	30.7	35.4	27.7	30.6	28.5	29.3	30.6	28.5	26.7	31.5	24.1	31.9	25.1	23.7	29.9
		GC-MS/MS		Green tea	A	37.4	38.2	35.0	36.5	39.3	36.5	31.9	31.3	32.7	34.6	29.1	57.6	28.4	39.5	33.7	34.8	29.7	29.5	29.5	35.1
				Green tea	B	36.6	31.2	35.3	35.1	39.5	33.2	31.3	30.2	31.7	34.2	27.8	56.3	28.1	35.4	30.6	30.0	23.5	26.8	27.4	33.2
				Oolong tea	A	46.2	42.9	48.0	35.2	37.9	33.8	36.3	32.9	37.6	38.7	33.3	30.1	32.7	30.7	37.4	27.6	20.5	28.0	26.7	34.7
				Oolong tea	B	37.0	35.9	34.0	34.2	31.4	29.2	32.9	28.4	30.6	29.4	29.1	27.0	27.8	27.2	27.0	23.9		22.5	25.7	29.1
36	Isoprocarb-1	GC-MS	Cleanert TPT	Green tea	A	80.2	70.1	82.0	82.9	65.8	74.0	70.3	69.9	73.8	65.9	65.6	69.9	75.3	66.3	62.4	42.2	61.3	57.8	78.0	69.1
				Green tea	B	78.4	70.6	74.4	79.5	58.8	65.8	64.9	60.4	72.4	59.5	56.4	64.6	70.7	63.1	56.9	41.3	59.0	55.4	67.3	64.2
				Oolong tea	A	72.3	86.0	72.2	76.1	68.9	63.8	63.1	77.0	74.4	78.2	75.9	81.8	77.1	77.1	76.3	75.7	68.0	69.2	73.0	74.0
				Oolong tea	B	69.9	70.0	59.2	66.3	64.1	54.5	51.8	63.4	63.3	63.1	59.2	68.1	62.2	66.9	62.5	64.2	56.1	64.0	59.9	62.6
		GC-MS/MS		Green tea	A	87.7	71.5	92.1	84.8	63.9	72.3	69.7	61.1	73.6	63.3	73.8	71.8	66.2	53.5	66.4	50.2	64.4	55.5	75.3	69.3
				Green tea	B	85.7	70.8	79.1	79.3	58.8	67.1	64.3	57.1	70.6	62.0	66.2	68.1	65.4	59.5	59.5	49.2	61.7	53.9	53.9	65.3
				Oolong tea	A	91.1	101.6	90.9	90.2	109.1	116.1	98.3	108.0	131.1	115.9	120.7	125.5	131.2	136.8	138.2	140.4	98.1	103.1	111.7	113.6
				Oolong tea	B	101.7	97.8	92.7	90.6	119.5	113.1	100.0	107.5	131.8	113.7	121.1	124.3	140.1	142.3	135.6	140.8	103.1	111.5	112.2	115.8
		GC-MS	ENVI-Carb + PSA	Green tea	A	80.5	77.5	72.9	87.4	81.0	77.3	63.7	71.3	72.9	63.6	65.1	76.7	66.9	80.1	71.2	82.9	71.2	73.7	65.0	73.7
				Green tea	B	74.0	67.7	73.9	79.0	80.7	71.9	70.3	71.9	66.6	65.6	61.4	73.5	61.7	71.1	65.5	68.2	63.9	60.6	64.8	69.1
				Oolong tea	A	100.9	82.6	73.8	77.3	81.2	74.4	84.4	60.0	130.7	76.2	67.9	64.6	66.0	67.5	73.3	60.6	60.8	66.2	62.8	75.3
				Oolong tea	B	77.6	67.3	64.8	67.4	63.4	60.6	69.1	54.7	128.7	57.0	58.1	53.1	56.3	54.6	57.3	50.8	50.6	51.4	55.2	63.0
		GC-MS/MS		Green tea	A	75.5	83.7	77.6	87.2	87.4	80.4	68.4	68.1	72.6	88.1	61.7	96.6	66.5	87.4	71.5	78.8	76.2	67.3	67.8	77.0
				Green tea	B	70.4	66.8	76.5	76.4	86.8	75.8	73.4	63.9	68.5	82.7	59.9	108.9	66.3	73.9	64.5	66.3	70.1	56.2	66.3	72.3
				Oolong tea	A	132.4	94.9	126.7	92.8	102.2	103.2	109.8	110.1	130.7	127.6	116.3	100.2	108.9	110.5	143.7	112.8	74.2	93.7	100.0	110.0
				Oolong tea	B	112.9	95.9	121.2	96.0	98.5	106.4	114.5	114.0	128.7	118.9	119.1	109.8	125.7	111.7	127.7	111.1	79.6	91.9	108.0	110.1

37	Isoprocarb-2	Cleanert TPT	GC–MS	Green tea	A	69.4	105.0	67.7	66.7	69.5	79.4	69.1	60.0	64.3	72.8	99.0	68.1	52.1	63.7	78.3	172.4	67.2	48.4	63.4	75.6
					B	86.6	88.3	66.0	65.5	70.4	72.3	63.5	57.8	75.1	85.3	104.7	68.8	49.8	65.6	72.4	174.8	64.8	48.0	62.0	75.9
				Oolong tea	A	104.0	98.6	80.0	87.5	90.3	71.0	84.2	93.3	94.9	79.2	122.4	67.2	77.5	77.5	59.4	69.1	55.1	67.8	55.8	80.8
					B	88.0	83.8	63.8	67.2	83.6	62.9	77.1	82.5	83.5	77.4	103.0	58.1	62.7	70.7	49.3	58.4	54.3	57.4	56.5	70.5
			GC–MS/MS	Green tea	A	118.0	102.1	66.2	70.9	58.2	96.4	69.2	67.8	66.9	59.9	110.7	61.6	62.6	48.2	68.5	125.8	48.6	68.1	75.8	76.1
					B	110.7	87.4	62.8	66.2	67.8	97.0	59.1	63.5	79.1	77.2	100.8	59.7	60.1	65.0	83.2	115.8	47.3	69.5	70.7	75.9
				Oolong tea	A	81.7	94.9	74.0	85.5	98.3	59.3	83.3	106.7	123.8	64.6	123.8	67.3	116.6	76.0	65.6	74.6	103.2	106.5	116.6	90.7
					B	73.0	78.6	68.2	64.0	114.5	49.3	74.6	118.1	128.5	73.2	117.4	57.1	138.1	75.9	53.1	72.2	112.1	118.7	122.8	90.0
		ENVI-Carb + PSA	GC–MS	Green tea	A	93.0	142.6	101.5	64.1	86.3	83.4	87.2	95.3	91.0	82.5	132.8	133.8	75.8	65.3	92.0	67.9	66.3	57.1	65.2	88.6
					B	89.7	122.1	101.7	57.2	90.9	81.3	58.1	87.0	91.6	70.1	103.2	125.2	69.3	67.5	78.7	60.7	66.3	77.5	64.0	82.2
				Oolong tea	A	74.5	83.4	95.3	83.2	108.9	78.0	75.9	75.5	91.8	73.1	80.4	79.8	62.7	74.2	79.6	65.4	48.6	52.1	58.7	75.8
					B	64.4	73.0	88.6	71.1	91.9	72.3	78.6	63.3	62.7	65.3	72.1	76.8	66.0	76.9	64.2	54.6	48.0	53.1	58.5	68.5
			GC–MS/MS	Green tea	A	80.2	126.9	106.1	76.9	94.7	109.0	92.1	104.5	74.6	74.2	130.3	82.3	112.1	125.0	112.5	142.8	80.5	120.7	87.5	101.8
					B	88.2	134.1	109.0	74.8	103.3	95.6	57.1	98.8	91.9	79.6	106.6	80.7	126.0	120.2	115.7	130.9	76.4	115.9	110.1	100.8
				Oolong tea	A	92.8	86.9	134.0	86.1	88.4	79.3	81.7	73.5	91.8	78.4	107.3	95.3	96.2	94.4	52.4	56.7	78.3	96.0	100.8	87.9
					B	66.1	68.9	98.0	85.5	85.1	74.0	71.6	63.0	62.7	75.5	115.7	106.7	113.3	101.0	46.3	40.7	87.8	94.5	116.3	82.8
38	Lenacil	Cleanert TPT	GC–MS	Green tea	A	262.9	235.7	189.6	179.0	158.3	197.6	151.2	190.6	170.6	169.8	359.1	368.8	223.1	176.2	237.5	798.1	132.4	139.2	191.6	238.5
					B	258.0	215.7	172.8	166.4	159.5	183.0	150.5	186.3	295.8	174.9	337.1	310.7	194.5	187.3	270.7	795.3	152.7	134.7	157.3	237.0
				Oolong tea	A	230.3	177.1	228.4	261.0	266.3	198.8	456.6	246.1	207.6	142.8	376.4	162.1	187.1	141.0	201.3	194.8	349.5	174.4	122.5	227.6
					B	210.9	178.3	259.5	226.2	199.6	152.3	450.6	263.8	172.1	144.2	263.9	135.0	207.1	163.8	154.9	175.1	279.2	150.0	142.1	206.8
			GC–MS/MS	Green tea	A	327.1	233.6	206.4	257.2	185.0	208.6	148.0	163.4	185.9	200.8	340.9	195.0	263.3	208.7	188.2	279.2	95.9	66.7	466.3	225.5
					B	326.9	216.4	193.0	186.8	185.4	200.3	148.6	175.4	262.5	209.9	321.0	168.9	227.8	224.5	217.2	387.2	123.1	84.7	405.5	224.5
				Oolong tea	A	502.0	183.0	226.4	239.0	246.0	249.9	371.5	273.7	310.4	175.9	376.6	167.0	166.4	153.2	180.1	309.6	142.8	129.1	198.7	242.2
					B	354.0	269.6	211.7	191.3	203.1	195.0	351.1	267.5	193.1	169.5	239.4	139.0	210.1	161.7	155.1	223.4	176.5	127.5	286.7	217.1
		ENVI-Carb + PSA	GC–MS	Green tea	A	207.5	235.4	262.7	151.9	222.7	217.5	289.2	210.7	202.9	328.8	305.2	286.0	225.2	147.7	241.1	166.2	427.5	118.4	144.0	231.1
					B	223.6	270.1	250.0	117.0	202.7	238.8	220.3	212.1	209.4	291.0	195.8	249.6	189.3	166.4	184.5	134.3	323.0	155.3	152.4	209.8
				Oolong tea	A	228.5	205.0	215.6	173.5	428.5	187.8	180.7	684.3	245.5	178.2	181.3	182.0	206.8	375.6	154.1	159.4	207.8	144.5	187.0	238.2
					B	219.7	176.7	209.1	177.5	276.1	175.6	143.1	381.4	191.3	149.6	156.1	184.4	158.9	355.6	160.7	102.6	178.7	134.6	159.1	194.2
			GC–MS/MS	Green tea	A	229.3	273.9	229.8	185.1	224.1	228.6	235.9	213.0	292.6	222.7	145.4	181.5	206.8	222.4	358.1	296.4	332.0	270.9	363.5	248.0
					B	236.7	383.7	217.8	147.7	214.6	241.2	196.3	219.1	281.0	188.2	171.6	174.6	186.1	168.5	384.2	258.4	335.7	247.4	367.1	243.2
				Oolong tea	A	159.0	217.1	178.3	169.5	187.4	186.2	163.3	242.3	245.5	186.3	180.3	175.5	183.4	141.6	141.9	132.8	198.4	219.3	267.9	188.2
					B	175.6	193.6	197.8	164.1	137.6	169.0	154.3	164.3	191.3	162.9	160.4	196.2	128.7	137.2	138.6	82.6	209.0	207.2	277.3	170.9
39	Metalaxyl	Cleanert TPT	GC–MS	Green tea	A	135.5	108.2	108.8	115.6	89.5	91.3	90.8	97.2	103.0	84.6	122.5	86.0	116.5	108.3	104.5	86.4	96.5	88.3	116.7	102.6
					B	123.9	103.8	98.9	105.8	80.2	81.2	85.1	80.0	103.0	70.7	103.5	82.5	104.9	107.0	93.9	85.3	93.3	85.9	101.6	94.2
				Oolong tea	A	123.8	116.4	102.2	112.9	99.7	91.5	92.0	118.0	109.3	100.3	118.5	129.6	119.1	109.1	114.4	121.9	117.6	109.9	119.1	111.9
					B	118.1	103.7	92.6	97.2	83.5	78.8	71.3	96.9	79.7	77.3	88.9	107.3	97.7	99.5	94.1	101.4	94.0	103.0	100.3	94.0
			GC–MS/MS	Green tea	A	139.2	110.6	119.4	154.4	88.4	121.0	93.0	82.3	97.9	87.6	111.0	95.9	93.0	80.1	91.9	73.1	81.1	70.9	107.4	99.9
					B	131.5	105.2	104.8	106.3	81.3	109.7	84.1	76.1	96.6	85.7	96.1	90.3	88.7	86.3	84.5	71.3	77.2	72.6	94.6	91.7
				Oolong tea	A	117.1	128.2	102.1	105.5	98.8	94.4	99.7	120.9	113.2	114.5	127.0	114.4	99.5	113.7	90.7	95.7	89.5	85.5	93.5	105.5
					B	107.5	108.0	86.5	91.6	90.5	77.7	74.7	98.3	96.1	93.9	92.9	86.7	85.8	96.9	73.5	80.6	71.3	81.5	81.1	88.2
		ENVI-Carb + PSA	GC–MS	Green tea	A	111.4	121.8	123.3	110.7	121.2	116.1	89.7	91.2	91.6	175.7	102.3	128.3	107.9	122.0	106.9	116.0	116.5	98.9	100.8	113.3
					B	110.2	108.2	116.0	99.0	117.3	98.3	92.3	88.9	94.2	145.5	97.7	106.3	97.7	109.8	96.9	93.8	102.0	91.7	101.4	103.5
				Oolong tea	A	136.8	119.0	109.6	103.7	138.9	103.7	115.3	108.7	125.3	113.7	101.8	84.9	103.8	106.8	125.2	97.0	93.2	105.0	94.1	109.8
					B	121.0	97.6	96.5	94.0	101.3	87.9	99.4	88.7	98.1	79.5	81.9	77.0	88.2	91.5	97.6	80.9	82.7	83.3	86.3	91.2
			GC–MS/MS	Green tea	A	112.4	121.0	111.9	121.3	120.2	123.7	86.9	97.2	95.7	102.5	89.3	110.2	78.4	103.5	95.1	106.1	95.7	88.5	88.2	102.5
					B	107.4	103.5	109.4	107.0	115.5	110.9	93.9	90.7	96.7	100.4	83.2	86.1	78.2	87.4	84.5	86.7	87.8	77.3	91.7	95.5
				Oolong tea	A	142.6	120.1	145.6	103.8	116.8	103.3	107.8	96.7	125.3	109.2	97.8	74.9	91.1	86.6	102.7	76.2	66.0	87.6	82.1	102.5
					B	110.7	102.9	103.0	93.5	94.5	86.0	100.4	80.5	98.1	78.4	84.3		81.0	79.6	73.8	63.6	60.2	65.2	77.4	84.6

(Continued)

Appendix Table 5.1 The 201 Pesticides Average Content for 19 Circulative Determinations on 3 Months Under 16 Factors (Two Cartridges, Two Instruments, Two Tea Samples, and Two Youden Pair Samples) (cont.)

No.	Pesticides	SPE	Method	Sample	Youden pair	Nov 9, 2009 (n=5)	Nov 14, 2009 (n=3)	Nov 19, 2009 (n=3)	Nov 24, 2009 (n=3)	Nov 29, 2009 (n=3)	Dec 4, 2009 (n=3)	Dec 9, 2009 (n=3)	Dec 14, 2009 (n=3)	Dec 19, 2009 (n=3)	Dec 24, 2009 (n=3)	Dec 14, 2009 (n=3)	Jan 3, 2010 (n=3)	Jan 8, 2010 (n=3)	Jan 13, 2010 (n=3)	Jan 18, 2010 (n=3)	Jan 23, 2010 (n=3)	Jan 28, 2010 (n=3)	Feb 2, 2010 (n=3)	Feb 7, 2010 (n=3)	Ave (µg/kg)
40	Metazachlor	Cleanert TPT	GC-MS	Green tea	A	120.9	120.5	126.0	177.0	99.0	117.9	122.2	117.9	132.8	89.8	129.8	94.6	95.4	101.9	105.4	108.4	85.8	76.3	85.5	110.9
					B	125.8	109.9	115.4	171.6	96.4	106.6	108.5	107.1	127.2	86.4	114.2	85.5	94.3	100.8	92.7	107.2	89.3	78.0	81.2	105.2
				Oolong tea	A	109.1	160.2	117.1	116.0	87.8	88.0	162.3	210.9	116.8	130.7	131.7	110.5	102.0	98.2	85.6	95.6	103.1	83.2	102.4	116.4
					B	106.1	117.6	99.5	103.6	86.3	77.1	123.6	179.4	96.6	109.7	99.5	87.9	84.9	87.6	77.3	87.1	84.8	82.6	89.6	99.0
			GC-MS/MS	Green tea	A	127.0	111.8	76.4	163.4	90.7	120.8	97.9	98.1	116.5	82.4	110.8	83.4	104.1	77.0	97.2	92.1	76.6	78.2	132.3	101.9
					B	128.0	106.1	97.0	102.5	89.0	117.4	85.0	86.7	114.8	95.6	100.9	77.3	105.2	91.3	88.9	90.8	69.2	76.4	121.9	97.0
				Oolong tea	A	154.4	159.2	98.9	113.9	105.8	102.2	137.7	143.3	118.3	112.8	152.7	116.2	115.8	120.3	103.2	121.8	106.1	81.9	110.0	119.7
					B	131.8	128.7	84.9	89.5	100.1	80.6	101.4	119.2	97.5	100.1	115.5	96.3	94.4	102.2	87.6	97.7	74.8	75.2	96.1	98.6
		ENVI-Carb + PSA	GC-MS	Green tea	A	79.7	165.7	103.1	98.4	121.7	125.4	95.0	120.2	139.5	198.9	112.7	93.3	93.4	93.2	89.0	92.4	153.0	91.0	93.2	113.6
					B	100.2	125.0	109.0	98.1	117.7	123.3	106.9	91.8	132.6	190.9	113.7	92.8	87.1	92.1	91.0	81.6	133.5	93.5	93.5	109.2
				Oolong tea	A	113.8	123.9	116.6	132.8	144.9	108.3	95.9	230.1	105.1	114.3	106.5	81.7	90.3	84.5	102.4	80.4	74.7	78.4	83.1	108.8
					B	104.8	106.2	104.5	108.0	113.4	98.0	117.5	169.8	82.6	85.2	106.5	73.1	79.9	83.5	88.4	66.6	68.0	73.1	82.3	94.3
			GC-MS/MS	Green tea	A	117.3	123.7	95.3	106.8	108.6	123.8	97.9	89.9	66.6	77.9	108.7	107.7	84.1	122.8	102.4	85.7	101.9	75.6	80.9	98.7
					B	168.3	125.5	97.6	106.0	112.0	113.7	89.7	89.7	93.0	88.4	93.2	93.9	92.6	91.6	93.7	94.6	84.6	75.2	95.7	99.2
				Oolong tea	A	106.2	130.4	168.0	103.1	121.1	96.3	120.2	113.2	105.1	100.1	93.2	93.7	89.7	79.9	95.5	75.2	80.7	126.9	102.2	104.3
					B	67.8	113.6	105.5	92.2	98.3	87.0	102.5	88.4	82.6	80.4	86.5	71.0	77.4	82.4	73.6	63.1	80.9	107.8	96.5	87.2
41	Methabenz-thiazuron	Cleanert TPT	GC-MS	Green tea	A	478.9	358.1	379.5	390.2	308.8	354.6	322.8	321.4	354.8	274.1	398.0	306.7	365.1	314.5	311.8	299.0	284.5	279.6	373.9	340.9
					B	438.6	347.2	337.8	349.7	280.5	324.0	293.0	284.1	336.2	258.4	345.4	280.6	327.5	298.8	264.7	290.9	275.8	259.1	322.0	311.3
				Oolong tea	A	457.0	421.3	345.1	379.4	374.9	329.0	333.6	379.4	366.2	366.2	382.0	368.2	357.4	345.7	368.4	361.5	324.2	323.2	303.4	362.4
					B	417.3	370.4	296.0	315.9	317.2	277.4	263.0	320.4	275.8	313.3	269.4	304.5	281.1	307.7	288.8	296.1	261.8	305.5	281.0	303.3
			GC-MS/MS	Green tea	A	417.4	365.7	428.9	492.3	301.4	529.5	325.0	263.1	358.5	307.0	363.1	342.9	361.3	275.6	312.7	269.0	281.2	239.7	550.5	358.2
					B	415.7	351.1	365.3	362.6	280.6	453.7	289.0	263.1	346.4	307.4	315.9	319.9	325.2	298.6	288.5	262.9	276.8	247.7	434.3	326.6
				Oolong tea	A	379.3	433.3	377.1	366.5	328.0	315.1	344.5	392.0	369.3	382.4	427.9	391.8	328.5	376.0	324.1	354.8	303.8	310.9	278.5	357.0
					B	369.2	377.7	318.9	310.7	309.3	275.6	266.9	345.6	300.0	316.3	303.8	303.8	291.6	316.8	262.2	304.2	253.4	290.3	264.0	304.3
		ENVI-Carb + PSA	GC-MS	Green tea	A	441.6	405.1	431.5	396.6	451.0	436.2	309.9	359.3	339.5	621.5	292.1	332.6	342.3	383.6	354.9	365.7	430.0	333.8	299.1	385.6
					B	410.0	370.0	429.0	347.4	440.3	367.5	354.4	345.7	329.1	524.4	300.8	317.5	301.0	349.8	305.2	300.0	358.7	281.2	304.3	354.5
				Oolong tea	A	528.9	445.1	380.9	369.0	547.9	371.9	416.3	450.6	458.8	372.2	363.1	296.5	329.6	397.3	349.3	294.4	281.6	308.5	292.0	381.8
					B	458.4	377.8	341.1	330.6	390.3	300.7	336.9	354.6	351.1	283.7	298.6	262.8	276.0	321.8	284.1	258.1	239.9	261.2	261.3	315.2
			GC-MS/MS	Green tea	A	369.2	421.1	384.1	411.2	412.7	424.3	300.9	358.9	328.7	389.3	297.8	411.7	324.9	385.7	338.4	357.5	333.0	294.9	340.8	362.4
					B	350.8	369.0	383.5	351.8	393.1	375.4	339.2	327.7	333.8	372.2	279.3	379.9	301.6	330.3	294.3	295.4	315.6	270.1	353.0	337.5
				Oolong tea	A	508.5	413.2	448.8	350.0	384.0	372.5	359.0	346.5	458.8	373.4	335.4	298.2	325.3	298.9	328.0	269.5	351.5	496.0	353.2	372.1
					B	431.3	363.1	346.8	316.2	315.4	303.4	331.0	277.9	351.1	285.6	295.1	272.8	285.6	266.7	256.1	217.6	309.7	354.8	307.5	309.9
42	Mirex	Cleanert TPT	GC-MS	Green tea	A	41.9	31.3	30.2	32.6	23.3	32.1	25.1	24.8	27.8	26.0	32.4	33.1	27.8	28.1	30.7	42.6	20.3	20.2	24.8	29.2
					B	42.0	30.6	26.6	30.4	22.4	27.6	22.4	24.3	28.8	25.6	30.5	29.2	25.9	27.7	30.6	39.2	22.0	19.9	22.6	27.8
				Oolong tea	A	35.8	38.0	32.4	36.3	34.9	27.2	38.1	40.3	34.7	30.9	42.0	29.7	30.9	25.1	27.5	31.1	36.0	26.9	26.5	32.9
					B	33.1	31.2	28.4	29.4	29.8	22.3	30.8	32.6	29.1	28.3	31.8	23.9	26.5	25.8	23.3	28.3	28.8	25.4	24.7	28.1
			GC-MS/MS	Green tea	A	46.5	33.7	26.1	37.0	24.6	32.4	24.1	25.3	30.2	23.8	35.4	25.6	27.6	27.7	25.8	28.3	20.9	18.4	32.3	28.7
					B	44.2	31.8	25.7	28.0	24.3	30.1	22.0	24.5	34.0	23.4	32.7	22.9	25.8	28.1	25.3	26.9	20.9	18.5	29.1	27.3
				Oolong tea	A	42.5	42.7	28.8	35.6	34.2	32.2	36.5	42.3	40.6	29.3	42.9	28.8	34.0	31.4	28.8	29.9	30.6	27.8	25.3	33.9
					B	38.5	36.0	25.3	28.3	31.3	24.4	30.5	35.8	31.2	27.4	31.2	24.0	26.7	26.5	24.3	26.8	23.8	24.3	23.4	28.4
		ENVI-Carb + PSA	GC-MS	Green tea	A	30.2	36.6	32.0	23.6	34.3	32.5	24.9	28.5	24.9	20.3	34.6	35.8	27.8	24.9	26.0	23.0	42.4	19.1	26.0	28.8
					B	32.2	34.0	31.2	21.3	31.9	32.8	21.9	22.4	25.4	21.0	28.0	32.4	26.0	26.1	25.6	19.9	35.3	21.6	26.4	27.1
				Oolong tea	A	39.3	33.9	32.8	30.3	40.9	30.9	27.6	40.7	28.0	29.0	28.4	28.4	29.5	33.0	26.2	24.3	23.7	23.3	24.3	30.3
					B	34.2	29.3	29.5	27.9	30.4	28.7	28.9	31.1	21.0	26.3	25.8	26.4	25.2	30.6	25.2	17.5	22.4	21.4	22.2	26.5
			GC-MS/MS	Green tea	A	35.9	30.6	30.4	26.3	33.5	30.7	33.3	25.5	22.7	23.2	34.5	45.1	28.4	35.2	26.1	24.7	25.7	21.4	19.8	29.3
					B	37.7	35.0	30.4	24.3	32.7	31.0	21.1	24.0	24.8	24.2	29.8	38.5	29.2	28.1	25.2	20.7	23.3	19.4	27.3	27.7
				Oolong tea	A	35.1	35.8	40.1	29.2	33.1	28.9	32.1	32.5	28.0	28.9	27.6	25.1	26.2	25.6	26.0	22.9	22.3	30.1	26.1	29.2
					B	28.4	30.8	30.8	26.7	25.9	26.7	28.4	25.9	21.0	26.5	25.7	24.4	21.0	23.9	22.4	17.7	21.4	26.3	23.9	25.1

No.	Compound	Cleanup	Instrument	Tea	Rep	1	2	3	4	5	6	7	8	9	10	11	12	13	14	15	16	17	18	19	20
43	Monalide	Cleanert TPT	GC-MS	Green tea	A	85.3	89.1	67.8	70.0	84.7	79.9	81.2	76.6	96.7	94.3	94.1	86.5	90.4	79.0	103.2	77.4	75.0	85.3	97.9	90.9
					B	83.8	78.4	65.9	73.7	90.1	81.8	80.5	74.6	89.6	93.1	111.7	89.5	87.5	69.1	92.7	84.1	75.3	79.1	85.6	90.5
				Oolong tea	A	93.5	78.1	82.6	93.3	90.7	82.5	87.1	97.2	84.4	127.6	81.5	105.1	101.8	69.3	86.3	98.5	97.3	89.4	118.6	99.4
					B	86.4	71.9	78.0	89.6	91.8	75.2	85.9	93.8	78.0	110.8	91.9	97.2	91.5	73.3	69.2	93.0	72.6	76.0	103.1	98.4
			GC-MS/MS	Green tea	A	76.3	78.3	65.3	66.2	61.4	73.1	67.2	71.1	78.3	81.4	69.2	76.5	68.8	72.4	72.4	66.2	84.2	88.3	82.6	94.0
					B	70.6	68.4	59.9	64.2	60.4	69.6	69.6	69.8	72.4	72.8	68.9	74.9	63.5	65.0	67.3	62.1	81.7	75.8	79.2	80.4
				Oolong tea	A	78.7	69.0	68.8	72.1	77.0	70.9	84.9	76.9	82.2	93.2	85.0	87.5	87.2	70.1	68.9	76.5	69.1	75.6	86.8	75.0
					B	67.6	60.8	66.8	62.0	66.6	59.2	73.8	67.7	69.8	71.8	72.3	72.0	74.3	56.2	59.3	69.3	82.9	64.9	74.4	83.3
		ENVI-Carb + PSA	GC-MS	Green tea	A	89.0	73.7	86.1	82.2	96.1	93.8	95.5	79.4	117.3	100.2	71.9	107.2	104.1	70.5	82.4	105.7	78.1	63.2	96.4	80.9
					B	82.6	74.4	83.3	76.1	82.6	86.5	99.7	84.6	98.3	78.5	71.4	78.9	90.8	57.7	97.9	106.9	86.4	62.8	80.6	109.6
				Oolong tea	A	92.0	84.7	70.4	64.3	168.0	80.8	87.0	82.1	88.1	80.2	94.9	85.1	85.0	88.7	91.8	90.2	81.0	97.0	113.1	90.6
					B	85.3	82.7	63.0	63.6	121.2	75.8	95.7	72.3	88.6	100.9	91.0	69.7	81.9	80.8	87.3	84.7	90.3	100.6	88.8	80.4
			GC-MS/MS	Green tea	A	79.4	74.7	77.1	78.3	84.5	75.6	81.7	63.4	64.7	68.0	80.0	80.9	73.9	75.7	90.8	90.2	79.3	87.6	91.4	73.6
					B	73.6	74.0	68.4	71.7	70.0	70.5	75.2	64.3	60.3	64.2	79.0	77.5	69.5	73.0	82.0	87.8	81.0	84.8	74.2	105.5
				Oolong tea	A	77.0	63.5	66.5	52.0	62.9	75.0	68.4	70.8	66.6	73.0	81.7	85.1	75.1	86.2	73.5	90.5	70.8	98.4	87.5	88.2
					B	65.7	63.6	55.2	48.5	51.8	58.4	60.2	60.0	60.4	64.0	66.0	69.7	64.6	78.0	64.5	74.1		75.6	75.0	
44	Paraoxon-ethyl	Cleanert TPT	GC-MS	Green tea	A	194.6	201.9	226.4	197.4	192.7	206.5	189.2	198.4	211.8	179.8	154.9	212.5	202.7	152.7	205.4	198.2	184.1	179.8	210.2	193.7
					B	193.3	216.7	212.8	202.2	197.4	206.0	190.9	206.6	210.1	182.6	164.2	201.2	194.8	129.7	197.4	192.8	184.8	186.0	205.9	190.4
				Oolong tea	A	198.0	204.7	154.2	188.2	203.6	180.5	189.7	189.2	247.8	186.3	207.9	206.2	199.3	194.3	208.8	175.0	215.0	223.3	201.0	186.5
					B	196.5	183.0	172.8	186.6	190.0	186.5	211.5	185.5	207.6	196.2	201.0	211.4	207.3	217.8	214.3	169.3	216.8	202.2	185.4	189.1
			GC-MS/MS	Green tea	A	137.4	0.0	65.2	57.8	143.8	101.5	122.0	216.1	92.1	200.0	110.5	164.5	105.6	136.4	321.2	133.8	170.0	85.2	178.8	206.8
					B	135.1	0.0	63.5	46.2	143.7	106.9	158.0	165.1	77.8	196.3	156.4	185.9	104.2	123.2	274.5	142.1	131.7	115.2	168.5	207.1
				Oolong tea	A	114.9	0.0	0.0	0.0	0.0	149.5	131.6	121.9	116.9	259.2	115.1	0.0	174.7	176.3	97.6	118.3	118.6	178.1	242.1	183.6
					B	98.3	0.0	0.0	0.0	0.0	111.1	115.6	122.4	89.9	183.9	123.0	0.0	143.3	135.2	98.8	102.8	96.9	158.0	228.3	158.9
		ENVI-Carb + PSA	GC-MS	Green tea	A	190.9	202.4	203.0	190.0	213.1	205.5	207.8	228.0	186.9	197.1	153.1	211.7	219.3	0.0	195.6	179.2	191.6	184.3	188.6	269.7
					B	189.6	209.2	208.2	191.8	197.9	205.2	202.6	220.0	188.9	201.7	162.0	212.7	205.2	0.0	196.7	208.5	197.3	182.1	204.7	207.6
				Oolong tea	A	200.5	216.1	221.5	225.5	203.9	206.1	172.9	196.2	178.6	190.9	209.6	196.6	171.8	211.6	191.3	190.3	218.4	196.7	204.2	207.7
					B	195.2	226.0	217.0	216.1	184.8	181.9	180.6	187.4	176.4	193.7	191.9	137.8	179.7	204.6	187.3	212.5	199.7	210.4	215.9	204.7
			GC-MS/MS	Green tea	A	142.3			145.9	85.2	169.5	131.4	114.6	146.4	189.6	115.2	116.2	157.2	98.3	217.4	100.3	115.4	148.0	215.3	153.9
					B	133.5			100.8	106.0	145.6	113.1	112.7	140.3	138.5	124.7	122.6	126.2	89.6	178.6	117.5	112.8	159.3	219.5	162.5
				Oolong tea	A	105.6				90.1	89.8	66.6	108.2	86.9	114.2	114.2	137.8	116.9	161.5	141.9	76.0	90.0	195.7	174.6	184.0
					B	87.1				60.8	86.1	90.4	98.2	78.8	104.7	97.3	84.6	78.5	133.3	123.0	69.1	110.3	123.8	145.7	117.6
45	Pebulate	Cleanert TPT	GC-MS	Green tea	A	81.2	82.1	85.7	98.7	99.2	98.6	66.7	67.6	77.0	86.7	98.8	87.1	89.6	87.3	98.5	107.8	99.1	94.3	104.1	112.9
					B	76.4	85.1	83.1	91.3	77.3	84.4	60.8	58.9	70.7	76.0	93.7	79.5	86.9	83.5	92.6	108.1	89.8	95.7	86.3	103.0
				Oolong tea	A	88.7	83.0	82.0	78.0	72.2	98.5	92.6	79.4	72.6	81.1	86.2	97.2	71.1	94.3	89.0	102.0	88.7	89.6	90.5	107.1
					B	74.8	67.3	70.7	64.5	67.5	72.5	77.1	56.9	59.6	67.3	64.9	75.2	59.0	76.6	72.4	77.4	76.9	82.4	74.6	96.0
			GC-MS/MS	Green tea	A											0.0									
					B																				
				Oolong tea	A																				
					B																				
		ENVI-Carb + PSA	GC-MS	Green tea	A	91.1																			103.0
					B	84.1																			95.2
				Oolong tea	A	87.9																			121.4
					B	71.1																			88.6
			GC-MS/MS	Green tea	A	0.0																			
					B																				
				Oolong tea	A																				
					B																				

(Continued)

Appendix Table 5.1 The 201 Pesticides Average Content for 19 Circulative Determinations on 3 Months Under 16 Factors (Two Cartridges, Two Instruments, Two Tea Samples, and Two Youden Pair Samples) (cont.)

No.	Pesticides	SPE	Method	Sample	Youden pair	Nov 9, 2009 (n=5)	Nov 14, 2009 (n=3)	Nov 19, 2009 (n=3)	Nov 24, 2009 (n=3)	Nov 29, 2009 (n=3)	Dec 4, 2009 (n=3)	Dec 9, 2009 (n=3)	Dec 14, 2009 (n=3)	Dec 19, 2009 (n=3)	Dec 24, 2009 (n=3)	Jan 3, 2010 (n=3)	Jan 8, 2010 (n=3)	Jan 13, 2010 (n=3)	Jan 18, 2010 (n=3)	Jan 23, 2010 (n=3)	Jan 28, 2010 (n=3)	Feb 2, 2010 (n=3)	Feb 7, 2010 (n=3)	Ave (µg/kg)
46	Pentachloro-aniline	Cleanert TPT	GC-MS	Green tea	A	97.8	36.3	41.3	38.9	30.3	37.2	29.3	32.7	37.2	33.6	35.5	34.7	32.3	30.1	24.7	29.0	27.8	35.5	36.7
					B	86.3	35.2	35.3	36.8	27.0	34.2	27.1	30.0	34.8	32.1	30.5	31.1	30.8	28.9	22.8	27.7	25.8	32.2	33.7
				Oolong tea	A	35.2	36.8	30.9	33.2	32.4	31.2	32.8	45.0	36.5	39.5	33.7	33.5	33.4	31.8	31.8	28.8	31.1	29.5	34.0
					B	32.6	30.3	25.7	28.1	29.3	26.7	29.8	39.9	32.0	34.0	28.3	27.4	29.7	26.2	27.5	24.8	27.5	26.1	29.2
			GC-MS/MS	Green tea	A	42.0	35.7	39.8	39.8	29.3	35.2	31.0	30.3	33.8	31.1	33.9	33.7	29.6	30.9	26.5	29.3	24.5	38.5	33.1
					B	41.2	35.4	35.3	35.1	28.7	32.0	28.5	28.5	34.3	31.3	32.6	31.6	30.9	29.1	25.3	28.3	24.4	32.5	31.3
				Oolong tea	A	36.2	34.6	31.4	32.1	31.8	29.2	28.2	32.7	35.2	32.8	34.7	31.9	32.7	31.8	29.8	27.3	28.4	28.5	31.9
					B	33.1	29.6	27.2	27.1	30.3	24.0	23.1	30.3	30.0	28.9	28.3	27.1	28.9	25.8	27.1	24.8	25.1	24.6	27.5
		ENVI-Carb + PSA	GC-MS	Green tea	A	39.7	38.8	36.5	39.2	42.0	38.8	33.7	35.2	32.1	35.0	38.3	37.0	39.2	32.9	38.4	36.7	33.4	32.5	36.5
					B	35.9	33.5	36.6	35.2	39.7	35.9	32.5	31.7	29.9	34.0	36.1	32.5	36.4	27.8	33.1	34.2	30.0	31.9	33.7
				Oolong tea	A	42.3	34.6	33.2	33.9	35.0	35.6	46.2	35.3	36.9	40.0	27.8	29.5	30.8	32.3	26.4	25.9	27.7	24.2	33.4
					B	32.0	29.3	29.6	29.8	27.9	30.7	45.6	39.9	29.7	31.3	24.2	25.6	26.0	25.6	21.8	21.8	23.2	24.2	28.8
			GC-MS/MS	Green tea	A	37.3	41.3	36.1	38.9	42.3	41.0	34.1	35.9	34.7	36.6	33.5	32.1	39.6	35.8	38.0	39.0	31.7	37.1	36.6
					B	34.6	34.5	35.7	34.8	39.2	38.1	33.0	33.0	32.5	35.6	32.0	29.5	36.9	30.2	32.6	35.8	27.4	33.0	33.6
				Oolong tea	A	43.1	35.4	43.7	31.7	35.0	32.3	33.7	29.2	36.9	33.4	28.5	28.5	28.7	30.9	23.9	21.4	27.7	26.3	31.6
					B	32.4	30.2	29.0	26.5	28.6	26.9	30.1	25.5	29.7	26.5	25.2	25.2	25.5	22.9	19.7	18.8	22.0	23.3	26.1
47	Pentachloro-anisole	Cleanert TPT	GC-MS	Green tea	A	43.9	32.0	33.3	34.9	28.6	34.1	33.5	33.3	34.0	30.8	31.1	32.0	31.0	29.0	22.0	27.6	24.7	32.9	31.6
					B	41.0	30.8	30.6	33.2	26.5	30.7	30.7	30.8	35.1	29.3	29.4	30.5	30.8	26.7	21.5	26.8	23.8	28.9	29.8
				Oolong tea	A	33.9	33.0	28.0	32.2	30.9	27.8	29.6	36.3	33.0	36.2	32.8	32.5	31.1	30.7	30.6	27.5	28.7	27.9	31.3
					B	30.9	26.9	23.5	26.3	27.6	22.7	24.6	30.3	24.7	27.7	27.5	26.4	27.4	24.8	26.5	23.5	25.4	23.9	26.3
			GC-MS/MS	Green tea	A	37.8	32.4	32.7	26.0	27.5	33.1	30.5	27.9	32.4	30.1	31.0	32.6	28.3	29.9	23.4	27.2	23.9	34.2	30.2
					B	37.6	31.0	29.9	31.9	26.2	30.2	28.1	26.3	31.8	27.8	28.9	30.3	26.7	28.0	22.9	26.5	23.4	29.6	28.8
				Oolong tea	A	33.8	32.6	27.4	30.6	31.7	29.4	27.3	33.9	33.9	31.4	31.3	31.8	31.0	28.8	29.5	25.4	29.0	28.0	30.7
					B	31.9	27.0	23.8	25.6	28.1	24.0	22.3	29.6	28.3	28.1	25.1	25.6	26.4	24.0	25.8	22.1	25.4	23.9	26.0
		ENVI-Carb + PSA	GC-MS	Green tea	A	31.5	36.3	33.3	34.2	37.2	35.7	36.4	35.7	38.4	32.9	30.0	32.0	35.2	32.1	34.5	32.5	31.3	29.0	33.6
					B	35.5	30.4	33.4	31.6	36.7	33.3	37.2	36.3	35.4	30.6	28.4	28.9	32.6	29.1	29.3	29.7	27.1	28.4	31.8
				Oolong tea	A	40.6	30.3	29.5	30.7	32.6	31.1	36.1	29.4	32.3	30.1	26.4	28.8	29.0	30.5	25.1	24.7	25.4	25.3	29.9
					B	29.3	25.6	27.9	27.0	25.7	25.9	30.7	25.7	25.6	24.1	22.5	24.7	24.0	23.3	19.7	20.4	20.7	22.5	24.8
			GC-MS/MS	Green tea	A	33.5	36.5	34.2	34.1	38.8	34.8	32.0	32.2	31.1	35.0	38.5	33.2	39.4	31.3	32.1	32.1	29.7	30.5	33.6
					B	31.7	30.0	33.6	30.6	38.7	32.4	30.9	29.4	30.3	33.2	34.8	29.9	33.7	29.2	27.4	29.9	25.9	30.3	31.0
				Oolong tea	A	46.6	31.0	40.4	29.0	33.3	30.0	31.5	27.4	32.3	31.3	26.5	26.5	26.5	29.1	23.2	20.3	25.4	24.3	29.6
					B	33.8	26.1	27.2	25.5	26.9	24.9	28.1	23.7	25.6	23.6	22.5	24.7	22.1	20.5	18.0	18.3	20.3	22.2	24.2
48	Pentachloro-benzene	Cleanert TPT	GC-MS	Green tea	A	37.5	29.8	32.4	31.9	27.3	30.9	30.7	30.6	33.5	29.5	29.6	32.9	31.8	29.4	21.4	26.9	23.5	31.7	30.1
					B	34.8	28.5	29.7	30.6	24.8	27.8	28.1	28.3	32.3	26.8	27.7	31.1	29.5	26.1	21.5	26.2	23.2	27.6	28.0
				Oolong tea	A	37.4	41.8	35.5	37.4	36.4	30.8	35.8	38.6	39.0	36.8	37.5	36.2	35.6	34.3	35.4	34.7	32.2	30.9	36.2
					B	35.7	35.2	30.3	31.7	34.4	26.4	26.8	34.1	31.9	32.6	30.7	30.1	32.0	28.9	30.0	28.1	30.3	28.7	31.1
			GC-MS/MS	Green tea	A	36.4	29.8	34.2	36.7	25.5	30.8	30.3	28.5	33.1	28.6	30.1	30.4	25.4	28.9	23.8	25.9	23.0	31.9	29.7
					B	34.8	28.4	28.6	30.6	24.4	28.0	27.8	27.5	31.6	25.7	28.5	28.5	24.7	27.4	23.5	25.4	23.4	28.8	27.7
				Oolong tea	A	44.6	41.3	33.0	36.7	35.3	31.6	34.1	40.6	40.1	36.5	36.0	33.1	40.2	34.0	33.5	31.5	32.4	30.3	36.3
					B	39.0	36.6	29.5	31.5	33.3	27.1	27.2	36.4	32.9	32.3	30.4	31.2	35.0	27.8	29.5	26.5	30.2	27.0	31.4
		ENVI-Carb + PSA	GC-MS	Green tea	A	36.1	34.7	31.7	33.2	36.6	33.4	34.9	34.4	34.5	26.6	34.6	33.1	37.0	33.7	35.0	33.4	32.5	29.1	33.5
					B	32.4	28.7	32.0	30.2	36.5	32.3	33.5	34.1	31.7	26.3	32.4	28.2	34.3	28.2	29.6	30.7	27.2	28.2	30.8
				Oolong tea	A	46.4	41.0	37.1	36.6	39.9	35.6	39.2	33.2	39.0	36.8	32.5	32.0	35.8	34.8	28.7	28.6	28.7	30.1	35.3
					B	40.3	34.9	35.0	32.6	31.8	30.9	32.2	29.1	31.5	30.1	28.6	27.8	29.1	28.0	23.8	24.5	24.2	27.2	30.1
			GC-MS/MS	Green tea	A	32.8	36.0	32.4	33.5	37.8	34.7	33.4	32.4	32.0	32.9	38.0	33.8	36.3	32.4	32.8	34.8	28.7	30.1	33.4
					B	30.9	28.3	32.8	29.9	37.8	32.3	28.8	30.1	29.7	31.1	38.5	29.6	34.5	27.4	28.3	32.3	24.9	29.0	30.7
				Oolong tea	A	43.2	40.4	41.8	34.9	38.6	34.4	36.9	34.4	39.0	35.7	28.3	28.9	30.7	33.2	27.2	24.9	31.3	28.4	34.0
					B	38.7	36.1	35.1	31.4	31.5	29.6	33.1	29.0	31.5	26.8	26.8	24.2	27.6	26.2	22.1	23.3	25.9	26.3	29.3

49	Perthane	Cleanert TPT	GC-MS	Green tea	A	53.1	35.4	35.9	37.3	30.7	35.8	27.5	30.1	33.4	32.3	40.3	36.7	36.4	32.1	32.4	26.8	28.5	26.9	34.6	34.0
				Green tea	B	45.3	33.8	32.6	34.6	28.4	32.7	26.2	28.5	35.1	30.7	36.1	33.4	32.4	32.3	29.2	26.1	28.1	25.5	31.1	31.7
				Oolong tea	A	191.8	43.1	33.9	37.3	39.1	34.7	32.9	38.2	38.0	37.3	38.6	35.8	37.7	35.1	34.9	36.6	33.6	33.5	31.8	44.4
				Oolong tea	B	180.0	37.1	29.1	32.8	33.0	26.5	26.2	31.5	33.4	32.1	29.7	31.6	31.3	33.1	28.7	32.1	28.2	28.7	29.5	38.7
			GC-MS/MS	Green tea	A	48.2	36.9	49.6	43.3	28.6	38.4	31.3	27.6	33.0	31.0	40.4	31.6	32.0	34.9	31.8	27.5	27.1	24.6	37.0	34.5
				Green tea	B	46.3	35.6	45.0	34.9	26.7	34.8	28.4	26.6	33.6	29.2	35.7	29.8	30.0	35.0	30.3	27.0	26.8	24.4	31.9	32.2
				Oolong tea	A	36.9	39.0	37.9	37.7	39.8	31.9	32.1	37.9	36.4	36.4	44.9	35.5	35.9	34.9	31.6	33.1	28.4	30.0	33.1	35.5
				Oolong tea	B	35.3	33.9	31.1	30.4	35.3	26.1	25.8	32.0	30.3	31.6	34.0	29.3	30.6	30.9	25.8	28.1	26.5	27.8	30.7	30.3
		ENVI-Carb + PSA	GC-MS	Green tea	A	40.7	42.5	38.7	36.8	39.7	39.6	35.2	35.3	35.7	41.8	35.9	38.3	33.1	39.1	33.8	38.0	34.8	31.6	29.1	36.8
				Green tea	B	37.6	36.9	37.6	32.9	38.0	37.1	34.4	30.8	33.2	41.0	32.9	35.0	29.8	33.9	31.4	30.1	30.9	28.4	29.1	33.7
				Oolong tea	A	47.6	38.6	38.3	39.1	39.7	36.4	39.5	35.9	40.6	37.2	34.6	31.5	33.2	34.9	35.0	28.6	28.4	31.4	29.1	35.8
				Oolong tea	B	37.9	34.6	32.5	35.2	30.5	31.2	34.4	30.0	32.8	31.2	30.8	27.5	28.9	29.9	29.0	23.1	24.7	25.2	26.2	30.3
			GC-MS/MS	Green tea	A	40.3	44.0	37.9	39.0	41.0	39.9	36.6	34.4	34.3	35.8	33.9	34.7	26.4	39.5	34.1	34.6	33.3	28.8	32.3	35.8
				Green tea	B	37.6	38.0	36.8	34.0	39.4	35.6	34.2	31.3	33.7	35.7	30.4	32.5	27.8	34.3	31.7	28.7	29.2	25.6	32.2	33.1
				Oolong tea	A	48.4	38.9	47.7	34.5	37.5	36.5	37.4	33.6	40.6	37.3	33.7	29.6	31.4	31.5	34.8	27.6	27.5	33.3	26.1	35.2
				Oolong tea	B	38.9	35.0	33.2	30.4	31.4	31.6	34.5	29.6	32.8	30.2	30.5	27.2	27.3	27.6	26.9	22.4	24.4	26.0	23.7	29.7
50	Phenan-threne	Cleanert TPT	GC-MS	Green tea	A	43.2	36.0	38.9	42.8	34.1	39.3	34.9	32.7	35.8	34.4	35.5	34.4	37.1	35.5	33.0	24.6	32.3	28.7	38.5	35.3
				Green tea	B	41.0	34.7	34.7	40.8	30.8	35.1	32.0	30.1	36.4	32.8	31.9	32.0	35.1	33.6	30.0	24.2	30.8	26.9	33.7	33.0
				Oolong tea	A	39.1	42.2	34.5	37.0	38.3	31.9	31.6	39.1	39.0	38.4	40.3	39.3	38.4	38.1	36.4	37.1	32.1	35.1	34.0	36.9
				Oolong tea	B	36.7	34.9	28.5	34.2	34.2	27.0	26.7	32.6	33.4	33.1	31.9	33.3	31.8	33.9	29.7	32.1	27.9	31.8	30.6	31.7
			GC-MS/MS	Green tea	A	43.2	38.6	37.2	46.3	30.0	36.4	36.2	31.2	37.5	33.6	38.5	37.4	36.9	29.1	35.8	26.7	32.3	28.9	40.0	35.6
				Green tea	B	43.0	37.3	33.9	37.6	28.9	34.2	34.6	30.9	36.8	33.1	36.6	35.0	34.6	31.3	33.7	27.4	32.8	27.6	34.7	34.0
				Oolong tea	A	37.0	43.0	31.1	38.0	36.0	33.0	34.1	39.9	42.9	39.0	42.7	38.9	36.4	41.2	28.0	32.7	32.5	37.9	30.4	36.9
				Oolong tea	B	35.5	35.8	28.7	33.6	35.0	28.8	28.1	34.2	35.2	35.0	33.5	33.0	31.0	37.6	37.6	30.1	30.1	35.1	27.2	32.4
		ENVI-Carb + PSA	GC-MS	Green tea	A	41.2	39.5	37.4	40.3	42.2	40.2	35.5	38.1	38.4	34.3	35.6	38.2	38.4	43.6	37.4	41.3	36.4	35.4	34.3	38.3
				Green tea	B	36.8	34.1	37.6	37.0	41.1	37.4	35.5	35.6	35.1	34.2	33.1	36.6	34.7	39.2	34.2	35.3	33.1	30.6	33.4	35.5
				Oolong tea	A	45.4	39.4	37.3	39.3	41.0	36.3	40.0	31.8	42.0	37.4	34.2	33.1	32.6	36.9	36.3	30.4	30.2	31.9	31.9	36.2
				Oolong tea	B	35.6	32.9	33.3	34.7	42.7	30.9	33.9	29.2	35.3	30.5	29.9	28.7	28.9	30.5	29.5	25.5	25.3	26.6	28.4	30.7
			GC-MS/MS	Green tea	A	36.6	41.9	39.8	44.3	41.8	43.9	37.9	38.1	37.4	42.4	34.8	38.1	29.6	45.3	39.1	38.7	38.0	37.4	37.6	39.1
				Green tea	B	34.8	32.9	39.4	40.5	40.1	39.6	36.9	33.7	35.6	41.4	30.7	36.0	31.8	38.9	36.6	33.6	35.3	31.2	36.7	36.2
				Oolong tea	A	46.7	39.5	48.7	35.3	34.6	37.1	38.7	34.3	42.0	40.3	36.3	27.5	30.8	32.5	34.4	27.3	24.7	34.3	27.4	35.7
				Oolong tea	B	37.4	34.9	35.7	32.2	34.6	30.7	34.9	28.8	35.3	32.4	32.2	27.0	30.1	29.9	25.2	24.2	22.8	26.5	25.4	30.5
51	Pirimicarb	Cleanert TPT	GC-MS	Green tea	A	103.8	74.4	83.5	73.3	57.7	63.7	59.8	56.8	67.6	52.7	91.5	52.3	66.1	62.6	61.7	54.8	54.6	47.5	65.4	65.8
				Green tea	B	95.9	74.2	65.8	66.0	53.6	60.6	58.5	53.8	69.3	48.6	76.9	49.3	60.2	62.2	57.2	53.7	52.3	46.1	57.5	61.1
				Oolong tea	A	99.1	75.4	67.3	73.7	64.6	55.1	63.6	69.0	72.7	64.3	79.2	69.8	69.2	65.6	63.4	65.2	63.2	60.3	56.8	68.3
				Oolong tea	B	91.5	67.4	58.9	64.3	55.6	47.0	51.0	61.0	55.0	54.6	53.6	55.4	55.5	57.9	51.5	54.9	50.3	56.5	51.6	57.6
			GC-MS/MS	Green tea	A	93.8	72.7	86.2	86.2	53.2	110.7	60.3	51.6	64.9	57.0	72.5	62.4	58.9	54.2	61.5	51.7	53.4	46.3	72.7	67.0
				Green tea	B	90.4	70.4	73.1	70.3	49.8	94.9	55.2	50.1	65.4	59.4	64.7	58.2	55.0	57.2	59.2	51.1	52.4	45.4	61.7	62.3
				Oolong tea	A	82.7	79.7	68.0	68.9	62.6	59.9	63.2	75.1	68.8	70.4	84.3	70.8	58.6	70.4	61.9	61.8	56.4	57.3	53.4	67.1
				Oolong tea	B	74.6	67.9	57.9	59.0	56.1	50.4	47.0	64.4	55.9	58.8	59.2	55.7	54.2	61.6	61.4	53.8	44.9	54.2	46.9	56.5
		ENVI-Carb + PSA	GC-MS	Green tea	A	83.8	79.0	77.8	70.7	85.7	79.5	58.8	65.4	66.7	164.9	65.0	54.7	62.9	67.2	64.9	67.0	67.8	57.2	55.3	73.4
				Green tea	B	75.9	66.8	81.2	61.9	76.9	69.7	62.0	61.5	62.8	119.5	58.6	49.8	57.5	62.4	58.6	54.5	59.6	52.1	56.1	65.6
				Oolong tea	A	93.4	83.7	77.4	73.3	110.6	68.1	72.9	74.1	83.6	69.7	76.5	52.5	60.8	65.0	63.2	52.1	53.5	56.3	55.3	70.6
				Oolong tea	B	79.4	66.2	60.9	66.7	72.0	57.4	55.3	57.5	63.7	53.0	56.5	51.7	51.8	53.5	50.9	42.2	44.5	44.6	48.4	56.6
			GC-MS/MS	Green tea	A	77.3	83.4	74.8	76.9	77.3	81.4	64.0	65.0	64.4	72.2	57.7	69.8	48.5	75.5	62.2	64.8	63.2	54.2	59.5	68.0
				Green tea	B	71.6	71.4	72.6	67.3	75.2	71.9	65.8	59.3	65.0	69.8	53.2	66.9	51.8	59.9	58.1	53.4	56.9	49.4	59.8	63.1
				Oolong tea	A	89.4	79.7	86.7	65.1	70.3	65.8	68.3	62.0	83.6	68.7	65.3	57.2	58.5	57.5	62.5	49.4	47.2	60.9	52.2	65.8
				Oolong tea	B	75.5	66.4	63.4	58.5	56.0	56.0	61.9	51.0	63.7	53.5	54.9	52.2	50.5	49.8	46.9	39.6	42.0	47.1	46.3	54.5

(Continued)

Appendix Table 5.1 The 201 Pesticides Average Content for 19 Circulative Determinations on 3 Months Under 16 Factors (Two Cartridges, Two Instruments, Two Tea Samples, and Two Youden Pair Samples) (cont.)

No.	Pesticides	SPE	Method	Sample	Youden pair	Nov 9, 2009 (n=5)	Nov 14, 2009 (n=3)	Nov 19, 2009 (n=3)	Nov 24, 2009 (n=3)	Nov 29, 2009 (n=3)	Dec 4, 2009 (n=3)	Dec 9, 2009 (n=3)	Dec 14, 2009 (n=3)	Dec 19, 2009 (n=3)	Dec 24, 2009 (n=3)	Dec 14, 2009 (n=3)	Jan 3, 2010 (n=3)	Jan 8, 2010 (n=3)	Jan 13, 2010 (n=3)	Jan 18, 2010 (n=3)	Jan 23, 2010 (n=3)	Jan 28, 2010 (n=3)	Feb 2, 2010 (n=3)	Feb 7, 2010 (n=3)	Ave (μg/kg)
52	Procymidone	Cleanert TPT	GC-MS	Green tea	A	47.9	36.8	40.4	41.7	33.5	37.6	34.7	33.8	36.5	35.6	31.0	38.6	39.5	42.3	32.4	24.7	32.2	29.3	38.8	36.2
					B	45.0	35.5	36.2	39.0	30.6	33.8	31.6	30.2	35.6	34.7	27.6	36.1	42.9	43.5	29.2	24.1	31.1	27.9	33.8	34.1
				Oolong tea	A	39.5	47.1	36.4	38.8	36.7	33.5	33.7	39.9	40.5	40.4	40.1	47.3	46.1	48.3	36.5	38.3	35.3	35.3	35.5	39.4
					B	37.2	39.3	30.6	33.5	34.3	28.4	27.9	33.7	34.2	35.1	36.0	39.0	37.6	48.1	31.2	32.8	29.9	33.2	32.1	34.4
			GC-MS/MS	Green tea	A	46.9	39.2	42.8	44.1	31.9	34.6	34.7	30.2	37.1	34.0	39.1	35.7	35.1	32.6	33.5	27.6	30.5	28.3	38.0	35.6
					B	45.2	37.1	36.3	38.3	29.8	32.3	31.7	28.7	36.3	31.7	34.8	34.7	33.0	34.0	32.5	27.4	29.9	28.1	32.5	33.4
				Oolong tea	A	38.8	42.2	36.3	40.5	36.7	32.7	33.9	42.4	42.3	39.8	44.7	41.1	34.7	41.1	34.9	36.5	32.3	34.0	32.7	37.8
					B	36.8	36.5	31.2	34.9	34.8	28.6	28.1	35.9	36.0	34.2	33.3	34.3	32.5	36.0	29.1	31.7	27.5	32.9	29.4	32.8
		ENVI-Carb + PSA	GC-MS	Green tea	A	41.8	43.0	40.7	41.7	40.9	41.3	28.9	36.4	37.8	32.8	33.7	40.0	37.6	47.8	41.1	41.4	36.3	37.3	34.0	38.7
					B	38.6	36.7	39.2	37.9	40.5	37.8	29.1	35.0	36.7	33.3	31.7	35.6	39.5	41.2	33.9	34.1	32.2	31.8	34.1	34.1
				Oolong tea	A	50.9	43.6	39.3	39.3	41.8	37.4	43.0	32.0	41.8	39.3	34.9	31.2	40.4	46.2	38.9	30.6	30.4	33.3	32.5	38.3
					B	41.9	36.6	36.7	34.7	34.1	32.3	37.6	29.6	33.8	31.4	31.3	28.2	33.9	42.0	27.8	26.3	26.4	27.4	29.5	32.7
			GC-MS/MS	Green tea	A	39.0	43.8	40.3	41.9	43.5	42.7	35.3	36.7	36.6	38.6	32.2	43.0	30.3	46.2	36.0	39.6	38.9	33.4	36.4	38.6
					B	36.6	36.3	39.0	37.8	42.1	39.0	34.8	33.8	35.5	37.9	30.8	40.2	30.5	39.1	33.7	32.9	33.9	31.7	36.8	35.8
				Oolong tea	A	53.1	43.1	49.2	37.7	42.4	37.1	39.7	36.4	41.8	39.6	35.3	31.8	32.3	33.1	36.3	29.2	24.2	31.7	30.7	37.1
					B	43.7	37.0	38.0	33.4	34.8	31.5	36.6	31.6	33.8	31.5	31.7	28.7	26.9	33.1	28.0	24.5	23.0	25.6	27.4	31.4
53	Prometrye	Cleanert TPT	GC-MS	Green tea	A	53.6	41.1	45.4	39.8	30.4	32.7	31.7	31.2	35.2	27.8	47.9	26.5	33.7	29.8	27.2	20.5	26.7	22.6	31.7	33.6
					B	43.6	37.6	37.4	36.2	34.5	33.0	30.3	28.0	34.1	26.0	35.2	27.8	34.7	28.0	25.0	24.5	26.2	30.7	27.6	30.6
				Oolong tea	A	41.6	43.6	35.6	37.4	31.5	43.0	24.5	37.7	37.8	32.8	25.4	26.0	27.6	35.5	31.6	21.6	23.0	23.1	27.6	34.7
					B	40.2	39.0	30.6	33.5	27.8	37.6	31.1	31.1	27.8	28.3	28.3	24.8	26.4	30.7	26.9	20.5	22.9	22.6	29.3	29.4
			GC-MS/MS	Green tea	A	46.3	38.6	43.1	38.7	26.0	54.2	28.5	27.4	33.9	29.8	35.3	35.7	30.2	28.0	31.4	22.6	27.6	28.4	26.3	33.5
					B	45.1	37.5	36.6	35.7	31.3	45.1	33.1	26.3	33.1	29.7	31.3	33.5	28.2	29.4	28.8	30.7	23.0	23.0	32.5	31.3
				Oolong tea	A	42.7	46.7	35.4	35.4	30.0	30.6	26.4	38.4	35.8	37.0	42.3	28.5	31.4	36.6	30.4	25.1	27.5	22.9	28.0	34.9
					B	42.9	41.5	31.4	31.1	44.8	26.1	33.6	33.0	30.4	32.0	29.9	25.5	28.9	31.6	25.4	24.6	30.5	29.8	28.0	30.3
		ENVI-Carb + PSA	GC-MS	Green tea	A	46.1	44.0	44.6	40.2	41.5	46.1	32.5	34.3	33.0	86.7	31.2	40.0	30.2	36.5	31.5	27.2	26.9	29.4	24.8	38.9
					B	39.0	38.5	41.7	35.5	54.7	38.9	39.6	32.5	30.7	61.9	29.5	37.5	28.1	32.3	26.7	34.9	25.7	30.2	28.1	34.0
				Oolong tea	A	52.6	45.3	39.3	38.8	36.7	34.9	33.8	33.9	43.2	36.9	38.5	42.1	26.4	32.4	35.7	27.8	22.0	26.1	28.4	36.4
					B	45.8	38.0	34.1	33.8	41.2	29.9	33.4	26.0	33.2	28.4	28.5	34.7	23.9	25.7	26.3	28.1	20.4	28.4	27.7	29.7
			GC-MS/MS	Green tea	A	38.4	46.1	40.2	42.6	39.3	45.0	31.5	33.0	32.3	35.4	28.3	34.8	24.9	35.1	31.4	23.8	33.4	22.8	24.4	35.2
					B	35.9	36.9	39.0	37.2	38.8	39.1	36.6	30.6	31.5	34.9	27.0	40.6	30.5	29.7	27.1	34.9	30.4	29.1	29.8	32.3
				Oolong tea	A	53.1	42.2	47.9	34.8	31.5	35.6	33.7	32.6	43.2	36.8	34.3	32.7	26.8	33.1	33.0	28.6	22.0	25.5	30.6	34.5
					B	43.0	36.5	35.4	32.0	33.8	29.8	37.2	27.7	33.2	28.8	28.4	45.3	25.2	25.2	25.0	23.1	20.4	28.2	25.4	28.9
54	Propham	Cleanert TPT	GC-MS	Green tea	A	44.3	41.2	40.7	34.7	34.4	38.6	37.2	37.3	40.3	38.5	44.0	40.0	43.1	38.1	40.0	58.6	34.0	30.9	40.4	39.8
					B	42.4	39.3	37.0	32.0	32.5	35.6	33.7	35.2	41.3	40.3	40.3	37.5	40.9	37.9	38.4	54.9	33.5	29.9	36.5	37.7
				Oolong tea	A	44.2	41.5	37.7	42.3	40.4	33.1	45.0	46.4	42.2	39.5	42.8	42.1	42.8	40.8	42.9	42.2	45.9	40.7	42.6	42.3
					B	40.8	35.5	33.3	35.9	35.9	28.2	40.5	43.4	36.2	34.9	36.4	42.8	37.5	38.6	36.1	37.9	39.8	38.0	35.2	36.9
			GC-MS/MS	Green tea	A	47.4	40.7	44.6	44.7	33.6	53.8	36.0	32.9	39.6	35.1	38.3	34.8	40.5	31.1	37.3	34.3	32.6	28.7	54.9	39.4
					B	47.2	38.4	41.9	39.9	31.7	48.3	33.5	32.1	40.4	34.8	40.1	40.6	38.3	33.0	36.3	33.8	32.2	29.0	46.7	37.5
				Oolong tea	A	47.6	44.5	39.1	39.5	38.0	36.3	47.5	43.7	42.9	42.3	52.7	32.7	38.1	42.8	38.7	40.9	35.8	37.1	33.7	41.1
					B	41.7	38.7	32.9	33.5	33.9	30.6	38.6	40.3	34.9	35.7	37.6	45.3	36.6	37.8	31.3	36.5	32.2	32.7	31.1	35.2
		ENVI-Carb + PSA	GC-MS	Green tea	A	38.1	44.6	47.0	41.5	45.3	43.8	41.4	46.0	41.0	55.0	44.4	42.8	41.4	36.8	43.2	38.4	52.0	32.6	35.2	42.8
					B	37.3	39.3	45.6	35.7	44.1	38.0	40.4	41.5	39.8	49.6	38.0	34.7	36.4	37.7	37.3	34.9	45.9	32.4	37.0	39.7
				Oolong tea	A	46.3	39.8	41.4	38.6	51.8	41.3	43.8	51.6	48.7	40.8	34.0	35.6	37.9	43.1	38.9	34.3	51.9	55.1	45.0	43.4
					B	37.1	33.8	36.6	34.7	40.4	33.4	35.9	41.1	39.8	33.2	34.6	33.4	33.6	37.5	33.1	28.5	50.9	55.1	40.6	37.4
			GC-MS/MS	Green tea	A	40.4	46.5	42.5	44.7	44.0	44.1	39.1	40.9	36.8	46.5	36.1	34.8	37.5	46.2	40.5	38.3	39.8	29.0	38.5	40.8
					B	41.9	41.2	41.7	38.0	43.7	40.5	41.8	37.2	37.8	43.6	32.4	40.6	46.1	42.7	36.2	34.2	37.0	31.5	38.6	38.4
				Oolong tea	A	42.6	42.0	50.9	36.5	41.9	38.9	38.9	38.2	48.7	40.0	36.2	31.7	42.7	33.7	36.7	30.4	35.0	46.2	38.6	39.0
					B	33.7	35.5	35.9	32.7	33.8	32.5	35.3	31.2	39.8	31.9	31.5	30.2	34.6	30.5	28.4	24.5	31.3	36.2	37.1	32.6

No.	Pesticide	Cleanup	Instrument	Tea	A/B																				
55	Prosulfocarb	Cleanert TPT	GC–MS	Green tea	A	46.8	35.2	39.3	36.9	29.9	33.4	32.0	31.0	34.4	28.5	39.0	29.7	35.2	33.0	30.2	20.4	28.8	25.2	33.5	32.8
					B	42.1	34.4	32.9	34.9	27.3	31.5	29.2	28.2	33.9	26.1	33.7	27.4	31.3	29.7	26.5	20.3	27.7	24.1	29.6	30.0
				Oolong tea	A	39.7	39.0	32.4	35.6	31.9	29.5	28.2	35.5	36.4	35.7	35.9	37.0	37.1	35.6	33.3	33.1	30.7	32.9	30.8	34.2
					B	37.3	33.5	27.9	30.7	29.3	25.0	24.1	29.5	29.0	29.2	25.4	29.6	27.8	30.7	26.7	28.7	24.7	28.4	27.4	28.7
			GC–MS/MS	Green tea	A	42.8	34.4	36.4	47.3	27.1	42.8	31.6	27.1	33.1	28.4	35.4	31.7	31.7	31.0	31.0	25.2	27.7	25.5	24.4	32.9
					B	41.4	33.0	30.2	35.1	25.4	37.2	28.7	26.4	32.7	27.5	32.0	29.9	29.9	29.9	29.2	24.0	26.0	24.8	29.3	30.1
				Oolong tea	A	36.1	37.7	33.8	34.2	31.3	28.5	30.0	36.9	34.2	36.7	38.7	36.1	29.9	36.6	31.6	30.5	28.5	28.2	27.9	33.0
					B	33.5	31.6	29.1	29.5	28.9	24.4	23.9	31.7	29.0	30.5	29.1	30.1	29.0	32.5	25.6	27.9	24.3	25.7	25.0	28.5
		ENVI-Carb + PSA	GC–MS	Green tea	A	40.2	38.7	37.5	37.4	41.7	40.1	37.5	34.0	33.7	69.0	30.6	34.4	33.3	37.9	32.4	34.8	32.0	33.0	28.6	37.2
					B	35.7	33.5	37.4	34.0	39.3	35.6	32.8	31.2	30.8	58.3	29.1	31.5	30.8	33.6	30.6	30.1	29.1	27.5	29.1	33.7
				Oolong tea	A	46.2	39.5	35.9	35.5	44.0	33.6	38.3	32.0	39.7	36.5	36.2	29.1	30.5	34.9	34.3	27.7	28.4	29.1	26.9	34.6
					B	36.1	32.4	30.6	31.1	32.3	28.7	32.4	25.9	31.0	27.9	27.4	25.1	25.7	26.3	26.3	22.5	22.9	23.8	24.5	28.1
			GC–MS/MS	Green tea	A	35.5	39.9	37.1	39.6	40.4	40.1	34.3	33.4	29.0	36.7	29.0	33.4	27.1	34.3	31.8	34.5	34.2	30.4	31.0	34.5
					B	32.9	32.4	36.4	34.5	39.6	34.8	34.1	30.7	28.2	36.4	33.3	33.2	27.6	30.9	29.1	29.8	30.3	25.5	30.0	32.0
				Oolong tea	A	45.7	38.5	47.9	33.3	35.5	34.2	35.8	30.9	33.3	34.9	28.2	27.0	29.0	30.5	33.0	25.3	22.8	27.2	25.4	33.2
					B	35.9	32.6	32.7	29.5	28.7	28.5	31.8	26.3	28.3	27.2	28.3	24.7	26.0	25.9	24.3	21.0	21.2	21.3	23.9	27.4
56	Secbumeton	Cleanert TPT	GC–MS	Green tea	A	62.1	43.1	43.8	40.9	30.7	27.8	32.6	31.0	35.7	26.5	52.0	26.1	36.8	35.0	33.4	23.1	30.0	24.9	36.2	35.3
					B	54.9	41.0	36.1	37.5	24.9	27.9	29.8	27.9	35.8	23.6	40.6	25.2	33.2	33.2	30.4	23.1	28.7	24.2	31.7	32.1
				Oolong tea	A	50.7	37.5	34.7	35.3	31.5	29.5	31.4	33.4	38.3	34.4	43.7	34.9	34.3	34.9	35.0	35.8	33.7	34.5	32.3	35.6
					B	46.4	35.9	31.2	33.0	29.3	25.5	24.9	28.4	28.8	29.1	29.0	28.0	27.5	31.4	29.5	30.3	26.8	32.6	29.3	30.4
			GC–MS/MS	Green tea	A	0.0	0.0	0.0	0.0	0.0	0.0	0.0	0.0	0.0	0.0	0.0	0.0	0.0	0.0	0.0	0.0	0.0	0.0	0.0	0.0
					B	45.5	39.1	42.4	37.1	24.5	62.4	30.0	26.7	33.4	29.7	33.3	32.9	32.0	30.9	30.2	27.0	27.8	23.5	32.0	33.7
				Oolong tea	A	39.1	40.9	38.5	34.7	28.4	31.4	32.9	38.0	33.0	39.2	42.7	35.4	31.9	37.0	32.2	33.7	30.1	31.8	29.1	34.8
					B	38.0	35.7	34.0	32.5	28.8	27.1	26.0	33.6	28.8	32.4	32.0	29.3	31.0	32.2	27.0	28.2	25.6	30.1	27.0	30.5
		ENVI-Carb + PSA	GC–MS	Green tea	A	45.2	46.4	48.7	39.4	44.3	52.2	33.6	34.5	32.9	71.3	29.8	33.1	31.0	37.5	33.0	37.3	32.7	31.5	29.7	39.2
					B	41.5	40.1	45.0	32.4	38.6	42.2	33.5	32.6	30.8	48.4	28.0	27.9	32.0	32.3	28.8	29.3	29.1	27.2	30.2	34.0
				Oolong tea	A	55.5	41.4	40.3	36.9	70.2	35.3	39.0	32.1	49.5	39.4	45.4	25.1	28.5	33.6	35.2	28.1	29.0	31.5	31.1	38.4
					B	46.2	35.4	33.8	35.9	42.9	30.1	34.1	26.7	37.7	29.0	34.5	25.7	27.9	26.3	28.5	23.0	24.6	25.9	27.3	31.3
			GC–MS/MS	Green tea	A	39.4	45.6	40.9	43.2	41.1	48.5	34.7	35.1	34.4	37.1	29.3	30.8	24.3	37.4	32.1	35.9	32.9	29.8	31.7	36.0
					B	37.4	36.8	39.2	36.7	39.5	39.2	35.0	31.8	35.0	36.2	27.3	29.2	24.8	31.3	27.9	29.2	30.1	25.7	31.7	32.8
				Oolong tea	A	53.1	40.0	46.1	33.1	35.0	36.0	37.8	31.8	49.5	37.4	35.4	29.5	31.2	32.9	35.9	27.0	23.6	31.3	27.9	35.5
					B	44.1	35.5	35.2	30.6	29.3	30.3	34.7	27.1	37.7	29.8	30.2	27.0	27.3	28.1	26.6	22.5	21.6	24.2	26.6	29.9
57	Silafluofen	Cleanert TPT	GC–MS	Green tea	A	67.6	57.7	35.7	41.5	32.3	43.5	34.6	27.2	28.0	33.9	51.6	44.2	46.6	31.4	34.9	23.1	27.6	25.1	33.5	37.8
					B	50.4	62.7	31.0	40.6	26.2	37.4	29.2	24.9	31.9	32.3	44.6	38.9	38.8	29.1	33.8	21.1	28.8	24.5	31.6	34.6
				Oolong tea	A	43.6	33.1	35.0	30.7	37.3	31.9	24.7	34.1	39.6	35.1	42.4	35.1	36.0	32.3	34.4	31.2	30.9	31.4	31.5	32.4
					B	42.6	28.7	33.2	23.8	34.5	29.1	23.8	37.0	33.0	31.8	33.9	0.0	36.5	31.3	27.9	27.4	24.6	29.6	26.6	29.2
			GC–MS/MS	Green tea	A	46.1	36.5	41.9	28.8	29.5	32.1	32.8	27.9	34.9	29.6	35.8	32.7	31.4	35.7	31.4	24.0	27.0	23.6	35.2	32.5
					B	44.3	35.6	34.5	35.6	26.6	29.0	29.8	28.5	34.7	29.0	31.4	29.5	29.7	36.5	29.8	24.1	25.2	24.5	30.4	31.0
				Oolong tea	A	38.5	39.7	30.5	34.9	38.7	30.5	38.2	31.6	31.8	35.7	32.8	34.4	28.3	38.3	30.2	28.0	29.9	26.5	26.8	33.5
					B	36.7	34.4	30.4	31.9	34.1	29.8	31.6	26.3	27.6	30.8	29.6	30.2	25.4	33.7	28.2	28.0	23.7	26.5	23.2	29.8
		ENVI-Carb + PSA	GC–MS	Green tea	A	49.5	64.9	37.7	40.6	38.7	32.1	34.0	29.8	41.8	29.6	30.1	35.8	36.0	34.4	48.8	34.7	28.2	31.3	30.3	37.4
					B	43.3	40.7	37.9	37.7	33.1	29.0	31.6	31.2	30.1	29.0	21.3	35.2	35.2	30.5	38.1	27.4	28.7	25.6	31.4	34.1
				Oolong tea	A	75.2	34.7	32.8	32.8	43.9	33.9	38.2	33.8	32.8	35.7	32.8	30.0	51.4	41.2	36.3	27.5	30.1	25.6	31.4	37.2
					B	47.3	40.7	34.4	30.5	33.1	27.6	32.0	27.6	29.6	27.4	28.5	27.4	37.1	32.6	28.2	22.2	26.5	22.8	30.3	30.4
			GC–MS/MS	Green tea	A	35.9	40.9	38.2	40.6	39.0	40.1	34.0	34.7	28.5	34.7	30.1	34.5	28.2	38.7	32.0	34.9	34.1	25.6	31.2	35.1
					B	34.7	34.8	37.7	35.1	39.2	36.2	37.5	31.6	34.7	33.8	28.3	33.0	29.9	34.8	29.7	29.1	31.1	25.5	31.3	32.6
				Oolong tea	A	54.6	38.4	43.2	36.6	39.8	32.2	37.5	31.8	36.6	36.6	32.9	28.0	27.4	34.0	34.0	25.0	24.2	29.3	27.6	33.9
					B	44.8	35.0	35.1	33.2	33.0	27.1	31.2	28.6	28.0	28.0	27.1	25.8	23.6	25.7	25.1	20.8	20.4	24.5	25.7	28.7

(Continued)

Appendix Table 5.1 The 201 Pesticides Average Content for 19 Circulative Determinations on 3 Months Under 16 Factors (Two Cartridges, Two Instruments, Two Tea Samples, and Two Youden Pair Samples) (cont.)

No.	Pesticides	SPE	Method	Sample	Youden pair	Nov 9, 2009 (n=5)	Nov 14, 2009 (n=3)	Nov 19, 2009 (n=3)	Nov 24, 2009 (n=3)	Nov 29, 2009 (n=3)	Dec 4, 2009 (n=3)	Dec 9, 2009 (n=3)	Dec 14, 2009 (n=3)	Dec 19, 2009 (n=3)	Dec 24, 2009 (n=3)	Dec 14, 2009 (n=3)	Jan 3, 2010 (n=3)	Jan 8, 2010 (n=3)	Jan 13, 2010 (n=3)	Jan 18, 2010 (n=3)	Jan 23, 2010 (n=3)	Jan 28, 2010 (n=3)	Feb 2, 2010 (n=3)	Feb 7, 2010 (n=3)	Ave (µg/kg)
58	Tebupirimfos	Cleanert TPT	GC-MS	Green tea	A	275.4	203.1	192.7	191.6	126.3	125.9	119.3	54.3	69.1	53.0	79.4	48.3	63.3	57.1	54.4	39.2	51.4	44.5	60.6	100.5
					B	248.2	194.6	168.0	169.9	121.6	122.2	110.2	50.8	65.1	48.3	66.3	46.1	58.1	54.1	49.0	38.4	49.2	42.9	53.0	92.4
				Oolong tea	A	237.9	193.0	168.2	182.6	132.3	103.6	111.7	67.5	69.9	62.1	76.1	66.1	62.8	63.5	59.2	60.3	53.9	58.5	54.7	99.1
					B	225.1	182.1	155.2	162.2	108.4	83.1	86.7	56.2	53.4	53.1	54.4	54.5	50.4	56.0	48.9	51.4	45.3	52.4	49.0	85.7
			GC-MS/MS	Green tea	A	0.0	0.0	0.0	0.0	0.0	0.0	0.0	0.0	0.0	0.0	0.0	0.0	0.0	0.0	0.0	0.0	0.0	0.0	0.0	0.0
					B	81.3	61.4	58.2	67.3	47.3	107.2	51.8	48.7	61.0	50.6	58.8	52.4	56.1	55.0	49.4	44.5	50.1	46.5	53.7	58.0
				Oolong tea	A	70.2	69.1	65.7	61.9	56.2	54.3	57.8	70.4	61.4	61.3	74.2	70.5	53.1	68.3	57.8	56.9	48.4	54.3	51.5	61.2
					B	67.9	58.7	55.8	55.5	54.7	46.1	45.6	61.0	52.9	51.7	55.9	54.9	48.5	62.4	46.9	50.0	43.2	50.4	44.9	53.0
		ENVI-Carb + PSA	GC-MS	Green tea	A	184.3	195.8	204.2	154.2	229.5	187.2	56.3	61.2	62.4	157.1	58.3	61.2	58.2	75.4	59.6	65.8	59.4	58.1	52.7	107.4
					B	165.4	190.1	192.5	136.7	208.1	169.8	72.9	58.1	57.8	113.3	56.9	55.5	52.4	66.7	55.3	54.6	54.5	50.5	51.6	98.0
				Oolong tea	A	245.4	202.6	207.4	177.5	240.2	137.3	69.0	65.7	82.9	67.3	76.1	49.1	54.8	58.8	64.3	49.9	49.8	51.8	52.3	105.4
					B	206.3	173.3	175.0	164.3	134.2	122.5	55.8	53.1	67.2	51.7	55.1	45.9	47.7	48.6	50.2	41.5	41.6	42.8	45.3	85.4
			GC-MS/MS	Green tea	A	69.0	77.2	69.8	70.6	80.4	81.7	61.4	60.8	66.5	68.6	53.7	73.8	49.0	65.4	61.9	60.8	64.3	59.3	56.3	65.3
					B	65.1	61.0	68.7	64.8	78.8	70.5	58.3	55.7	55.8	64.6	53.8	67.8	49.2	58.9	57.9	52.5	60.3	49.7	55.9	60.5
				Oolong tea	A	89.4	74.7	88.2	61.6	63.5	64.7	65.9	54.2	82.9	67.9	60.6	50.0	55.1	55.4	62.4	47.5	41.3	55.0	51.6	62.7
					B	66.4	62.6	56.6	54.1	52.9	53.6	58.4	48.4	67.2	54.8	51.3	47.3	49.8	48.0	46.7	40.2	37.0	43.0	43.6	51.7
59	Tebutam	Cleanert TPT	GC-MS	Green tea	A	90.7	68.5	70.9	73.8	58.1	66.0	60.5	62.5	65.5	59.8	69.6	57.9	65.8	64.4	60.9	41.7	57.7	48.9	66.5	63.7
					B	84.6	66.3	64.1	68.7	53.4	61.2	55.8	55.4	64.8	56.7	60.1	54.8	61.4	59.8	56.4	40.9	55.2	47.5	58.5	59.2
				Oolong tea	A	81.7	77.9	66.1	73.3	64.9	57.5	59.9	72.6	72.3	70.2	73.8	72.9	69.0	69.5	64.2	65.8	58.8	62.3	60.8	68.1
					B	76.9	65.4	54.3	60.7	59.0	47.5	47.7	59.1	59.3	59.5	57.3	59.3	54.5	60.6	52.7	56.6	49.0	56.3	53.5	57.3
			GC-MS/MS	Green tea	A	87.3	69.1	81.9	77.6	55.2	72.8	62.2	54.4	68.2	57.5	69.7	64.8	60.9	52.7	60.6	50.3	55.9	48.3	65.2	63.9
					B	83.3	66.9	69.5	69.4	52.4	66.4	56.6	52.3	66.1	56.5	62.3	61.0	58.2	53.8	57.2	50.1	53.9	47.7	56.7	60.0
				Oolong tea	A	69.5	79.5	64.3	69.1	63.2	61.6	60.5	73.2	71.5	74.0	78.9	71.9	63.8	72.5	60.7	61.3	57.9	63.1	58.1	67.1
					B	65.5	64.6	54.3	58.9	54.3	50.3	47.3	61.5	61.1	60.8	57.6	57.7	58.6	61.2	50.4	54.2	49.9	56.6	49.8	56.7
		ENVI-Carb + PSA	GC-MS	Green tea	A	82.1	77.9	73.9	73.2	57.2	77.7	63.7	64.4	66.3	93.6	61.1	68.2	63.4	74.6	64.8	72.1	61.9	63.5	58.7	70.6
					B	70.7	67.0	72.6	66.2	76.3	70.6	64.7	61.6	61.4	84.3	58.0	63.3	57.2	65.7	59.3	59.3	57.3	54.1	57.4	64.6
				Oolong tea	A	90.1	78.1	71.8	70.3	80.8	68.1	73.3	58.9	76.5	69.0	65.7	58.2	59.1	62.8	66.8	54.2	53.5	57.1	54.7	66.8
					B	68.8	63.3	61.9	61.1	60.1	57.1	63.2	51.3	60.7	54.6	54.7	50.3	51.5	52.5	52.4	45.4	44.9	46.3	48.9	55.2
			GC-MS/MS	Green tea	A	71.8	80.1	71.7	77.2	80.0	76.5	65.2	63.2	63.1	71.6	58.0	65.7	58.2	77.9	64.2	70.0	67.2	60.7	61.5	68.6
					B	67.3	65.2	70.7	68.5	77.8	67.3	63.4	57.8	62.4	68.1	54.6	60.5	58.3	66.1	60.6	59.1	62.2	52.2	61.4	63.3
				Oolong tea	A	94.1	77.3	96.1	66.7	73.1	66.5	69.7	61.2	76.5	70.9	67.0	55.4	60.3	60.3	64.2	50.8	41.4	52.4	48.9	65.9
					B	70.4	65.5	61.6	58.4	58.8	62.6	62.6	51.0	60.7	55.5	55.7	50.3	53.0	51.8	48.5	41.7	37.8	42.4	46.3	54.1
60	Tebuthiuron	Cleanert TPT	GC-MS	Green tea	A	214.5	142.8	168.7	158.0	128.1	113.2	129.0	129.8	152.6	109.8	123.9	152.4	109.8	122.5	124.9	89.5	117.3	106.1	149.6	133.1
					B	196.4	133.0	139.9	126.2	114.7	114.0	119.1	112.2	130.9	98.5	126.4	138.3	105.6	114.7	107.0	88.5	111.1	101.5	132.3	122.0
				Oolong tea	A	191.9	153.1	129.0	154.4	123.2	131.0	127.5	155.3	140.5	153.6	187.3	146.1	144.9	114.7	141.1	141.7	131.9	129.8	124.1	143.2
					B	175.4	142.7	110.7	131.1	115.6	108.1	102.7	131.0	117.7	132.5	143.8	122.3	116.9	117.5	115.6	114.7	106.0	123.2	114.6	123.3
			GC-MS/MS	Green tea	A	180.5	154.1	203.3	158.7	112.8	272.9	128.1	110.1	135.3	108.4	144.3	131.5	140.8	104.7	127.6	99.2	116.1	99.1	162.8	141.6
					B	175.1	148.2	165.8	151.0	104.9	224.4	119.7	101.3	129.8	106.0	126.5	127.1	131.7	124.1	112.3	95.8	112.1	98.7	135.7	131.1
				Oolong tea	A	143.5	179.2	158.8	139.7	115.3	122.6	129.3	153.1	126.3	168.8	157.6	149.4	133.7	155.0	133.5	135.7	125.8	119.9	121.0	140.4
					B	137.2	155.5	133.4	125.9	106.2	106.9	99.1	131.5	107.2	136.4	118.0	116.7	113.1	129.4	108.8	114.6	103.0	114.0	104.4	119.0
		ENVI-Carb + PSA	GC-MS	Green tea	A	176.9	158.9	166.1	160.2	173.4	168.5	111.0	138.5	126.5	276.0	122.1	128.6	121.7	130.4	137.8	153.1	142.1	128.6	118.4	149.9
					B	169.5	143.7	161.0	140.3	159.1	140.1	137.4	134.1	127.1	200.4	117.5	115.0	115.0	119.5	117.2	119.0	122.1	110.8	121.4	135.3
				Oolong tea	A	215.3	174.6	149.9	149.9	258.1	136.7	161.5	147.0	235.3	153.7	154.3	108.5	125.8	157.7	158.2	115.0	110.6	126.2	117.8	155.6
					B	179.2	146.2	136.1	130.1	177.6	117.4	136.6	119.2	185.1	105.3	128.4	97.8	103.8	133.4	124.0	98.1	97.6	96.7	107.5	127.4
			GC-MS/MS	Green tea	A	146.0	171.2	155.5	171.8	164.7	181.9	110.1	140.1	133.5	164.4	117.6	147.2	122.3	155.7	131.5	145.2	139.5	121.3	128.0	144.6
					B	142.8	143.0	154.4	147.5	157.8	145.9	134.7	125.1	133.3	157.0	113.4	138.6	116.8	127.2	115.6	115.3	124.0	106.2	132.0	133.2
				Oolong tea	A	211.0	178.7	178.3	135.4	143.7	144.6	142.4	130.3	235.3	149.4	134.7	120.8	121.6	117.2	138.2	109.0	104.8	146.3	124.8	145.6
					B	175.7	152.7	137.2	126.2	121.0	123.1	134.2	107.8	185.1	105.4	116.9	106.0	111.6	107.4	99.7	89.4	97.0	108.3	111.5	121.9

No.	Compound	Cleanup	Instrument	Tea	A/B	1	2	3	4	5	6	7	8	9	10	11	12	13	14	15	16	17	18	19	20
61	Tefluthrin	Cleanert TPT	GC-MS	Green tea	A	43.7	32.3	33.1	34.3	26.8	29.8	28.9	28.5	29.9	26.8	31.2	28.7	30.3	34.9	27.8	19.7	29.4	22.8	30.9	30.0
					B	39.5	31.2	29.5	31.9	24.7	27.8	26.8	26.5	30.8	25.4	27.9	27.4	28.5	36.8	26.2	19.8	25.9	22.0	26.8	28.2
				Oolong tea	A	37.3	37.3	32.1	33.6	31.0	26.4	28.3	36.6	34.4	34.1	35.2	36.1	32.8	30.9	30.8	30.0	28.3	29.6	28.3	32.3
					B	35.5	31.6	27.0	29.6	28.7	22.5	23.5	30.2	29.4	28.9	27.6	30.8	27.1	27.3	25.9	26.3	23.7	27.4	25.2	27.8
			GC-MS/MS	Green tea	A	41.0	32.4	33.0	41.5	25.8	31.6	29.2	25.1	31.5	26.2	32.4	29.2	28.0	27.6	27.7	22.8	24.2	22.2	29.0	29.5
					B	39.4	31.2	27.9	32.0	24.1	29.0	26.7	24.7	30.5	24.3	28.8	27.5	25.9	26.7	25.9	22.3	23.3	22.2	25.5	27.3
				Oolong tea	A	34.7	35.6	31.2	33.7	30.1	28.3	28.5	34.9	35.1	34.9	35.9	34.2	29.6	33.6	28.1	27.8	26.4	28.2	27.2	31.5
					B	32.8	30.3	26.6	29.3	27.7	23.8	22.1	30.4	29.4	28.9	26.5	27.6	25.0	28.8	23.5	23.8	22.3	26.0	22.9	26.7
		ENVI-Carb + PSA	GC-MS	Green tea	A	36.9	36.3	34.4	33.1	36.9	35.2	31.1	30.4	31.6	39.3	29.7	33.3	29.9	33.1	30.5	33.3	28.4	28.7	27.1	26.7
					B	32.6	31.4	33.8	30.4	35.8	31.9	29.7	30.5	29.4	37.1	28.1	30.3	27.3	30.4	28.1	26.7	26.2	25.7	26.2	30.1
				Oolong tea	A	45.0	36.1	34.7	34.1	35.7	33.1	35.5	28.7	37.4	33.0	33.6	29.3	28.6	30.6	32.1	24.1	25.7	27.0	26.4	32.1
					B	34.7	30.8	30.4	30.2	27.6	27.9	30.9	25.8	28.9	26.6	29.2	25.5	25.3	25.1	25.7	20.2	21.8	23.0	22.4	26.9
			GC-MS/MS	Green tea	A	34.2	37.5	33.5	33.9	37.2	35.3	30.9	30.3	30.4	34.2	27.0	36.3	25.1	32.9	29.3	31.9	31.0	26.0	27.8	31.8
					B	30.9	30.5	32.6	30.8	36.3	31.8	28.5	27.9	29.1	32.8	26.0	33.8	24.6	29.9	27.1	26.0	28.9	22.7	28.7	29.4
				Oolong tea	A	43.9	36.2	46.1	32.4	35.4	32.8	32.7	28.9	37.4	33.1	30.6	26.4	28.2	29.1	30.1	23.7	20.3	25.3	24.2	31.4
					B	33.1	30.1	30.1	27.9	28.5	27.0	29.3	24.5	28.9	25.7	25.9	24.4	24.4	24.5	22.9	19.4	18.5	21.0	21.7	25.7
62	Thenylchlor	Cleanert TPT	GC-MS	Green tea	A	73.6	90.2	85.2	125.8	65.2	72.5	70.6	62.5	81.2	60.4	70.4	71.9	62.9	80.7	67.8	78.5	58.7	66.0	63.9	74.1
					B	79.6	79.5	73.2	126.5	62.4	66.5	62.8	58.5	72.3	62.3	67.1	69.3	61.3	82.9	72.5	78.2	61.2	59.0	62.3	71.4
				Oolong tea	A	73.5	115.9	78.4	70.3	59.2	54.3	71.9	80.2	77.5	85.8	89.4	79.3	70.3	70.6	58.3	69.5	70.5	57.1	69.5	73.8
					B	75.8	85.0	61.9	64.2	62.4	49.4	52.3	72.3	66.5	72.2	66.5	66.1	59.3	66.0	52.3	60.2	55.6	55.6	65.0	63.8
			GC-MS/MS	Green tea	A	81.3	77.9	44.4	126.5	61.5	79.6	67.0	68.2	76.8	52.2	75.0	49.8	69.9	57.0	62.0	61.0	51.8	57.7	85.8	68.7
					B	84.9	73.5	69.9	70.8	61.9	81.1	57.3	60.7	73.1	63.9	69.9	46.4	72.7	69.2	59.6	61.4	45.3	53.1	77.9	65.9
				Oolong tea	A	106.8	99.7	67.0	77.4	71.2	68.0	87.7	100.4	81.6	72.0	101.1	73.9	76.0	80.7	75.6	82.2	63.1	66.8	80.1	80.6
					B	90.3	79.6	59.5	59.1	70.4	56.5	64.4	85.1	67.8	67.1	79.4	64.6	62.7	69.4	63.9	64.2	44.2	56.5	69.2	67.0
		ENVI-Carb + PSA	GC-MS	Green tea	A	54.9	130.8	65.3	65.7	78.3	73.2	81.7	72.2	66.1	83.6	69.7	67.3	65.2	71.6	59.9	66.2	113.0	58.0	67.7	74.2
					B	69.8	95.8	73.6	68.8	78.5	71.0	73.6	46.0	61.0	87.0	69.4	68.3	64.2	66.3	60.8	59.4	100.4	56.3	69.3	70.5
				Oolong tea	A	79.2	76.0	72.6	85.2	82.7	69.8	54.2	111.4	64.0	73.6	67.9	67.1	62.1	63.0	67.4	60.8	53.9	56.1	62.0	69.9
					B	76.7	71.0	72.3	68.0	74.4	72.9	72.0	83.5	52.8	56.5	58.5	54.6	57.4	62.1	54.2	53.2	46.3	50.7	60.2	62.5
			GC-MS/MS	Green tea	A	77.0	91.5	61.1	71.1	69.4	79.7	74.4	60.3	44.4	47.8	55.5	76.0	60.4	94.7	73.9	63.6	68.4	51.0	55.2	68.2
					B	121.9	88.8	65.0	72.7	75.9	76.1	61.8	59.4	63.3	56.5	70.1	69.7	71.3	67.4	70.8	57.3	58.4	48.3	64.3	69.4
				Oolong tea	A	76.3	89.1	114.4	69.9	76.9	62.1	82.0	72.7	64.0	64.0	60.6	46.8	62.5	54.4	64.9	47.4	57.4	77.2	68.5	69.0
					B	45.3	81.6	74.3	65.5	65.3	57.9	68.2	56.6	52.8	53.3	58.4	42.5	53.3	55.3	49.5	40.4	55.3	68.0	64.3	58.3
63	Thionazin	Cleanert TPT	GC-MS	Green tea	A	49.6	36.3	37.7	37.4	31.5	30.6	31.8	31.3	34.8	27.1	34.6	27.7	34.2	29.2	29.9	24.4	27.6	26.0	31.7	32.3
					B	46.0	35.3	35.1	35.0	27.4	28.9	29.9	28.2	33.6	26.4	30.4	25.8	32.3	29.2	26.1	23.0	26.2	24.7	28.3	30.1
				Oolong tea	A	42.5	42.0	39.9	42.3	30.7	32.3	33.6	38.0	35.4	33.2	41.3	37.9	36.6	36.1	35.8	38.1	34.3	35.0	35.2	36.8
					B	38.0	33.8	31.9	34.4	32.8	27.7	27.8	34.1	28.5	29.6	31.5	32.1	30.1	34.0	30.6	33.8	29.3	33.2	33.5	31.9
			GC-MS/MS	Green tea	A	42.0	35.2	38.6	34.8	27.7	62.3	31.8	28.7	33.9	27.1	34.8	31.2	31.7	25.7	30.2	25.8	27.2	23.3	35.0	33.0
					B	42.1	33.9	34.4	34.8	25.9	49.6	29.2	27.0	34.1	27.8	31.9	29.4	30.8	28.9	28.9	25.0	25.9	23.4	30.8	31.2
				Oolong tea	A	40.7	40.6	33.4	34.1	30.8	30.5	31.7	31.5	36.4	37.0	40.9	33.8	32.2	33.9	31.2	31.5	28.4	29.0	29.3	33.8
					B	35.4	33.0	28.2	29.1	28.1	25.3	24.7	31.8	30.5	30.7	30.4	28.2	30.3	30.1	25.4	26.9	23.8	25.9	24.5	28.5
		ENVI-Carb + PSA	GC-MS	Green tea	A	36.1	42.4	39.5	36.8	45.6	38.3	42.0	33.8	30.8	66.7	32.8	32.7	34.2	34.7	32.7	32.8	33.8	29.1	26.8	36.9
					B	34.7	35.3	39.2	33.6	41.9	34.8	36.6	31.0	29.0	49.9	30.1	31.5	30.8	32.2	31.0	28.7	30.0	25.8	27.0	33.3
				Oolong tea	A	48.1	40.2	42.0	42.9	48.7	36.2	39.0	37.1	40.2	36.1	36.4	27.5	30.5	30.9	37.9	31.0	29.0	31.8	32.0	36.7
					B	37.8	32.4	33.7	34.3	34.3	31.3	33.4	31.9	32.1	29.0	30.2	25.7	27.1	27.1	30.3	26.5	25.1	27.8	29.7	30.5
			GC-MS/MS	Green tea	A	37.0	39.4	36.6	36.8	39.2	39.0	33.8	33.6	32.0	36.1	32.8	41.8	30.2	37.5	33.9	32.7	33.2	28.1	30.3	34.9
					B	37.4	33.5	35.9	33.2	39.0	34.7	34.0	31.2	31.8	34.4	30.0	37.3	30.4	34.7	31.4	28.6	30.3	25.5	30.2	32.8
				Oolong tea	A	40.3	38.4	47.5	32.2	36.3	33.3	35.8	32.1	40.2	34.0	31.0	27.0	27.5	28.7	31.2	24.6	24.3	28.7	27.0	32.6
					B	29.9	32.6	31.2	28.8	28.7	28.3	31.1	26.5	32.1	27.0	27.2	24.6	27.1	25.0	23.5	20.4	21.3	23.0	23.8	27.0

(Continued)

Appendix Table 5.1 The 201 Pesticides Average Content for 19 Circulative Determinations on 3 Months Under 16 Factors (Two Cartridges, Two Instruments, Two Tea Samples, and Two Youden Pair Samples) (cont.)

No.	Pesticides	SPE	Method	Sample	Youden pair	Nov 9, 2009 (n=5)	Nov 14, 2009 (n=3)	Nov 19, 2009 (n=3)	Nov 24, 2009 (n=3)	Nov 29, 2009 (n=3)	Dec 4, 2009 (n=3)	Dec 9, 2009 (n=3)	Dec 14, 2009 (n=3)	Dec 19, 2009 (n=3)	Dec 24, 2009 (n=3)	Dec 14, 2009 (n=3)	Jan 3, 2010 (n=3)	Jan 8, 2010 (n=3)	Jan 13, 2010 (n=3)	Jan 18, 2010 (n=3)	Jan 23, 2010 (n=3)	Jan 28, 2010 (n=3)	Feb 2, 2010 (n=3)	Feb 7, 2010 (n=3)	Ave (µg/kg)
64	Trichloronat	Cleanert TPT	GC-MS	Green tea	A	53.3	36.4	35.0	36.3	28.4	30.3	30.1	29.9	36.2	29.5	48.3	32.0	33.6	31.2	29.9	27.4	27.3	25.1	32.0	33.3
					B	49.9	35.9	31.3	32.2	26.7	29.4	27.9	27.9	36.2	27.3	42.0	29.3	31.1	30.2	28.2	26.8	26.4	24.2	28.1	31.1
				Oolong tea	A	42.6	36.6	35.3	37.0	36.4	31.7	31.2	36.6	38.4	33.0	43.7	35.4	34.3	33.9	31.4	32.8	31.1	31.0	28.8	34.8
					B	40.2	32.5	30.0	32.1	31.0	25.6	23.6	31.1	30.4	29.3	31.8	28.9	27.9	30.3	26.0	28.1	25.8	28.0	26.1	29.4
			GC-MS/MS	Green tea	A	43.6	35.0	37.9	25.8	28.9	51.9	29.7	27.7	32.3	29.9	36.8	30.6	32.8	31.0	30.4	27.0	26.6	23.0	34.3	32.4
					B	43.7	33.5	34.7	33.1	27.0	46.1	27.2	26.5	33.7	28.6	34.3	29.0	30.2	30.5	29.2	27.0	25.4	23.3	28.9	31.1
				Oolong tea	A	40.1	38.9	36.1	35.9	31.0	29.2	32.2	37.0	35.1	37.0	41.7	36.0	33.7	35.9	31.5	33.2	28.5	30.3	28.8	34.1
					B	35.6	32.9	30.4	31.0	28.9	25.1	25.1	32.7	29.4	29.5	30.7	28.7	28.1	31.4	25.8	27.7	24.4	26.8	25.6	28.9
		ENVI-Carb + PSA	GC-MS	Green tea	A	38.1	41.0	43.1	32.3	44.1	41.0	35.5	34.2	35.1	73.2	39.0	37.6	31.6	32.2	34.2	33.6	34.0	29.1	28.8	37.8
					B	35.6	37.5	40.6	28.7	40.5	37.1	30.3	31.8	32.3	55.8	35.6	33.9	29.3	32.5	33.0	27.6	30.4	27.4	28.7	34.1
				Oolong tea	A	46.5	39.7	38.7	34.6	57.1	34.1	36.6	36.9	43.3	36.3	39.3	27.1	30.2	33.8	33.0	26.8	26.8	27.1	28.0	35.6
					B	37.4	33.1	30.9	29.9	37.2	28.8	32.0	29.2	33.2	29.1	29.7	25.8	26.0	29.1	26.9	21.8	22.8	23.2	24.8	29.0
			GC-MS/MS	Green tea	A	37.8	40.9	37.7	32.7	39.6	40.6	35.5	33.4	35.0	34.6	36.1	0.0	29.1	37.7	32.9	32.2	32.0	28.2	31.2	33.0
					B	34.9	36.5	35.9	30.1	37.7	37.3	28.3	31.1	33.3	33.8	31.6	0.0	28.9	34.2	31.4	27.2	28.8	26.5	31.5	30.5
				Oolong tea	A	40.2	39.8	43.7	33.5	35.2	33.2	34.6	31.1	43.3	34.2	31.6	29.4	29.4	29.6	32.0	26.0	23.4	29.5	26.5	33.0
					B	32.8	33.8	30.8	29.6	28.5	27.9	31.2	26.3	33.2	27.7	27.3	26.5	25.8	26.7	24.5	20.7	21.6	24.9	24.7	27.6
65	Trifluralin	Cleanert TPT	GC-MS	Green tea	A	107.2	74.0	70.2	71.2	53.2	64.3	65.2	59.0	81.9	59.9	80.2	66.9	102.8	65.8	74.1	107.2	55.2	84.6	65.7	74.1
					B	96.2	70.0	61.5	63.3	52.4	61.0	60.5	56.8	76.5	56.7	70.1	60.6	86.9	64.0	68.4	95.6	55.0	66.9	57.7	67.4
				Oolong tea	A	84.0	74.2	67.4	78.8	103.8	65.8	75.9	82.3	80.5	71.3	111.1	71.0	77.4	68.8	84.3	88.5	103.5	75.1	63.4	80.4
					B	76.9	65.4	58.6	65.6	76.7	48.4	55.2	66.8	62.1	61.3	77.4	56.6	63.0	65.6	61.9	76.6	81.4	67.3	58.4	65.5
			GC-MS/MS	Green tea	A	96.5	75.7	87.9	54.1	52.3	188.7	61.4	57.8	71.9	56.2	84.9	63.8	79.1	61.2	68.8	66.0	54.4	45.4	119.0	76.1
					B	92.1	71.9	74.0	70.4	50.0	147.7	56.8	57.1	79.2	58.4	79.7	59.5	75.2	65.5	66.1	66.8	55.5	45.5	91.1	71.7
				Oolong tea	A	73.1	108.1	78.0	71.4	63.9	73.1	79.8	83.2	77.7	86.6	102.4	80.2	75.1	85.1	72.2	77.6	73.0	72.3	59.3	78.5
					B	75.1	87.0	67.2	64.8	59.9	56.9	57.3	72.2	63.8	71.6	72.6	61.1	66.3	74.8	61.5	61.5	55.2	65.3	55.2	66.0
		ENVI-Carb + PSA	GC-MS	Green tea	A	77.3	76.3	82.8	59.0	87.2	87.8	83.8	64.3	68.1	93.1	80.6	81.8	70.5	0.0	75.9	62.3	208.5	52.9	62.8	77.6
					B	68.6	73.4	77.4	51.9	79.3	82.6	63.2	61.3	64.2	76.9	71.6	73.2	64.8	0.0	72.9	53.3	127.6	51.7	60.5	67.1
				Oolong tea	A	86.9	77.4	74.6	70.0	70.0	76.2	68.5	183.2	106.8	73.4	75.7	61.1	90.1	117.7	61.7	61.2	64.5	61.7	64.1	87.4
					B	69.1	62.7	60.6	61.7	61.7	65.7	62.5	118.6	82.2	58.2	58.6	56.4	72.6	102.3	60.3	39.3	56.5	52.4	54.7	67.9
			GC-MS/MS	Green tea	A	74.1	75.2	73.7	70.5	76.9	83.8	83.5	71.3	72.0	72.7	89.1	71.4	69.2	84.9	70.9	71.1	70.3	55.6	85.0	74.8
					B	71.5	76.7	71.6	61.7	72.7	78.6	46.9	63.1	63.9	69.8	78.3	62.7	69.5	69.8	67.5	58.5	66.6	48.9	82.5	67.4
				Oolong tea	A	85.6	83.7	96.3	63.6	69.8	70.9	74.6	73.4	106.8	74.3	73.9	64.9	70.3	67.8	71.1	57.9	98.5	126.1	82.2	79.6
					B	65.8	69.2	61.2	58.6	53.9	61.3	65.9	56.1	82.2	59.8	60.1	61.4	59.5	57.2	56.7	42.1	79.8	92.7	70.5	63.9
66	2,4'-DDT	Cleanert TPT	GC-MS	Green tea	A	89.8	77.8	73.5	128.4	62.8	64.2	61.7	64.4	57.0	56.2	74.8	59.3	59.2	56.3	64.5	56.6	41.4	44.0	64.6	66.1
					B	84.3	73.8	64.0	119.8	63.6	57.4	62.1	63.2	69.2	53.9	61.6	54.9	55.8	56.8	69.1	55.5	45.4	51.5	56.5	64.1
				Oolong tea	A	77.7	92.9	73.9	107.8	75.7	55.6	70.4	98.9	61.9	64.7	75.0	69.2	76.7	65.4	62.5	68.2	71.5	58.5	54.1	73.2
					B	73.2	78.1	71.3	91.5	72.1	54.9	62.0	92.1	57.1	63.9	64.7	54.3	68.8	68.9	58.2	68.4	59.5	54.2	54.0	66.7
			GC-MS/MS	Green tea	A	101.4	75.7	61.2	101.7	60.2	59.5	65.9	61.2	57.1	56.5	80.9	45.9	61.7	59.7	58.0	53.4	42.9	48.8	64.1	64.0
					B	93.3	76.0	61.8	97.7	62.3	56.1	60.0	56.5	62.9	52.5	73.3	45.4	54.5	58.9	56.3	55.1	44.9	52.9	57.0	62.0
				Oolong tea	A												0.0	0.0	0.0	0.0	0.0	0.0	0.0	0.0	0.0
					B												0.0	0.0	0.0	0.0	0.0	0.0	0.0	0.0	0.0
		ENVI-Carb + PSA	GC-MS	Green tea	A	72.2	70.0	71.7	69.7	72.7	79.9	58.8	56.3	69.5	59.3	62.7	73.5	58.1	62.9	36.6	58.0	69.3	53.2	69.5	64.4
					B	74.8	77.3	70.3	62.8	69.5	81.9	54.6	50.4	69.1	60.6	55.8	71.0	58.3	62.0	37.0	49.1	62.3	55.6	72.0	62.9
				Oolong tea	A	84.0	54.9	66.9	54.8	101.0	66.8	54.4	58.2	49.9	48.4	54.4	64.9	53.7	65.4	62.6	60.1	44.7	44.0	47.4	59.8
					B	82.2	47.7	67.2	55.5	77.2	68.6	64.5	49.6	34.6	53.3	49.7	68.4	48.1	68.2	66.8	46.3	46.9	49.0	42.9	57.2
			GC-MS/MS	Green tea	A	104.6	94.0	72.7	67.8	74.7	77.0	80.8	54.1	67.3	52.1	78.1	72.8	65.6	59.6	44.3	57.2	62.0	57.7	71.4	69.1
					B	123.6	78.4	71.9	69.1	67.8	76.6	66.3	52.3	63.7	60.6	67.0	66.3	57.1	54.5	42.8	48.6	58.3	60.8	74.7	66.3
				Oolong tea	A	109.4	55.8	54.2	54.2	111.0	57.5	92.7	59.2	49.9	45.7	52.8	44.8	52.9	55.5	61.3	65.3	34.9	45.8	45.6	61.5
					B	91.8	50.0	62.9	51.3	84.0	62.3	67.5	47.3	34.6	55.7	53.1	50.5	45.2	55.8	58.4	62.8	37.4	45.5	41.5	55.7

No.	Compound	Cleanup	Method	Tea	A/B	1	2	3	4	5	6	7	8	9	10	11	12	13	14	15	16	17	18	19	20
67	4,4'-DDE	Cleanert TPT	GC–MS	Green tea	A	33.6	35.8	27.3	28.3	27.6	33.0	30.9	32.1	31.1	32.7	31.0	35.5	33.4	37.0	30.8	33.8	42.9	37.8	35.1	42.6
					B	31.7	31.6	26.5	28.1	27.4	30.1	30.0	30.4	29.2	29.4	29.6	34.3	31.4	33.9	28.9	31.3	40.2	34.2	34.4	40.5
				Oolong tea	A	37.0	31.5	34.5	33.7	34.5	34.6	35.2	38.2	38.3	38.1	36.4	37.8	41.7	34.6	37.7	34.4	44.0	39.2	38.7	40.8
					B	34.3	30.5	34.7	30.0	31.9	31.4	33.7	33.9	33.8	32.1	33.4	34.4	38.7	31.7	34.9	35.3	40.7	35.0	34.9	40.1
			GC–MS/MS	Green tea	A	33.8	35.0	28.3	25.7	26.1	33.4	34.7	30.8	33.9	33.5	31.2	33.8	33.0	35.7	30.2	32.4	45.7	39.7	35.7	44.1
					B	31.9	31.9	28.3	25.8	25.6	30.1	31.5	30.4	32.1	30.0	28.9	32.8	30.6	34.9	28.7	30.3	41.9	34.8	35.0	42.2
				Oolong tea	A	37.2	31.9	34.0	30.8	37.0	33.5	35.6	38.0	38.3	37.3	36.3	39.4	43.7	35.4	38.0	35.5	45.2	40.2	38.3	38.8
					B	34.4	28.6	33.9	30.5	33.5	30.2	35.1	32.8	33.9	32.4	32.6	35.7	40.7	30.7	34.2	37.0	40.9	36.4	35.3	39.2
		ENVI-Carb + PSA	GC–MS	Green tea	A	39.2	31.7	32.7	30.9	38.8	32.4	36.3	33.3	33.3	30.9	31.4	32.3	33.0	103.3	39.2	38.9	37.1	37.0	43.0	49.9
					B	34.6	31.5	28.1	28.1	32.2	30.2	32.4	31.1	31.8	30.0	31.9	31.0	31.8	60.4	35.7	38.2	34.6	36.5	36.0	46.5
				Oolong tea	A	35.0	29.8	32.7	30.6	31.9	38.1	35.2	32.9	32.9	33.3	34.0	36.6	30.1	35.3	35.3	39.9	36.1	37.6	37.2	46.0
					B	32.0	29.1	29.0	28.3	29.1	32.1	33.5	31.0	30.7	30.8	29.6	31.1	29.0	33.4	32.9	33.9	33.4	37.1	35.0	39.7
			GC–MS/MS	Green tea	A	35.7	33.7	34.0	29.5	36.1	31.7	31.5	33.7	37.2	33.7	33.1	33.5	33.5	36.5	39.7	37.2	36.5	37.0	43.0	47.1
					B	33.3	33.2	31.0	28.8	31.1	28.7	30.0	30.0	34.9	32.3	33.1	31.6	32.4	35.5	35.2	36.5	34.6	36.3	33.6	43.7
				Oolong tea	A	34.3	30.9	30.0	28.7	37.0	35.9	30.3	29.8	31.9	33.9	33.1	36.6	29.5	35.2	34.5	38.9	34.7	41.7	37.2	42.7
					B	31.6	31.1	26.1	28.7	33.5	28.2	29.2	28.8	29.5	35.1	28.9	31.1	26.7	34.0	32.2	33.7	33.2	36.5	34.9	38.4
68	Benalaxyl	Cleanert TPT	GC–MS	Green tea	A	35.8	37.0	32.7	27.6	28.7	31.0	28.6	29.8	32.0	38.1	29.4	35.6	34.2	39.5	33.4	36.9	55.4	42.3	39.6	49.0
					B	33.7	32.9	32.3	28.5	29.8	28.5	28.0	27.8	29.7	33.7	28.9	35.7	31.6	35.1	30.9	34.0	51.3	39.3	37.3	44.3
				Oolong tea	A	36.7	28.9	33.4	32.7	35.8	35.1	33.7	33.9	33.0	36.8	34.3	36.3	40.7	34.0	38.3	38.5	45.7	41.1	42.9	43.0
					B	34.5	26.6	34.0	29.3	31.6	31.2	30.3	29.5	28.1	30.0	32.0	31.7	38.7	30.0	35.4	38.3	42.0	37.0	39.4	41.3
			GC–MS/MS	Green tea	A	32.7	35.8	26.8	24.1	26.3	31.4	32.4	34.6	33.4	35.8	31.2	33.2	32.2	35.4	32.0	33.9	46.0	41.9	39.8	48.7
					B	36.3	30.8	27.2	24.6	25.5	29.1	30.0	28.6	32.6	33.9	31.0	33.7	29.8	35.0	30.0	32.7	43.2	36.5	39.5	46.9
				Oolong tea	A	32.8	28.1	31.6	31.9	36.4	30.6	35.3	35.2	34.0	37.0	33.6	35.6	41.3	32.2	36.4	33.7	46.9	39.4	43.9	45.9
					B	37.3	27.4	30.3	27.9	32.6	27.3	30.5	30.8	30.7	31.1	31.9	30.8	37.9	29.3	31.8	33.9	41.4	35.0	39.7	43.2
		ENVI-Carb + PSA	GC–MS	Green tea	A	34.2	33.8	29.3	32.9	35.9	25.3	32.7	31.2	38.0	34.7	33.3	36.1	33.1	34.3	42.0	41.1	42.5	42.5	49.3	59.9
					B	34.1	35.2	29.0	31.3	27.4	23.0	29.8	28.1	34.7	30.0	32.7	34.2	29.7	35.0	39.7	40.3	37.5	41.2	43.2	48.4
				Oolong tea	A	31.5	29.3	28.7	28.9	29.1	37.4	31.3	28.5	31.0	32.2	33.0	33.1	30.1	35.4	36.0	44.1	35.7	39.2	37.9	46.8
					B	39.0	29.4	28.5	26.1	27.5	37.0	30.3	26.1	28.0	29.2	28.3	28.5	26.5	33.5	33.3	36.5	34.2	39.2	34.6	42.0
			GC–MS/MS	Green tea	A	35.5	36.4	29.6	32.8	35.0	26.7	36.7	37.9	31.7	28.5	33.0	34.0	30.9	41.8	30.3	38.9	59.5	65.0	45.6	56.4
					B	32.5	36.2	29.1	27.8	28.1	24.7	29.5	32.6	31.9	30.6	33.3	32.8	28.9	30.1	37.5	37.4	51.0	59.0	42.4	51.0
				Oolong tea	A	32.2	26.0	28.3	25.2	36.4	31.9	29.0	25.2	27.3	30.4	31.4	33.1	28.3	31.8	33.7	39.7	33.1	38.3	36.0	46.9
					B	29.8	26.2	25.0	26.0	32.8	25.9	26.0	24.1	26.8	30.5	27.7	28.5	25.6	32.0	31.4	34.9	31.3	35.5	33.5	43.3
69	Benzoyl-prop-ethyl	Cleanert TPT	GC–MS	Green tea	A	109.4	108.7	87.5	83.4	86.0	96.5	93.7	100.0	103.0	114.4	94.0	112.0	107.9	118.6	101.3	111.7	165.7	129.0	123.1	142.4
					B	102.9	95.8	85.6	81.1	85.3	90.7	92.4	96.2	95.7	102.3	95.5	110.2	100.1	109.0	92.6	101.8	151.1	115.9	117.1	136.0
				Oolong tea	A	111.0	89.8	101.8	97.3	100.4	99.6	107.8	106.9	110.6	116.9	101.9	114.2	118.6	106.5	112.5	112.9	136.8	117.0	125.1	131.9
					B	103.2	84.2	96.8	85.2	89.5	91.4	96.0	100.8	99.1	98.1	100.2	103.0	117.1	104.2	104.5	116.0	124.7	109.0	116.4	124.4
			GC–MS/MS	Green tea	A	109.2	115.7	85.3	79.8	85.2	100.0	95.3	104.9	110.6	116.7	94.9	109.8	106.4	111.7	104.0	106.5	145.8	128.8	124.2	148.2
					B	103.3	102.6	86.4	79.7	83.9	93.7	95.9	94.3	105.0	106.9	93.7	107.6	98.7	111.2	94.3	99.6	137.4	113.5	116.9	142.4
				Oolong tea	A	109.8	85.0	97.6	95.5	104.9	98.2	97.8	107.5	108.7	114.3	103.8	116.9	124.2	99.2	114.5	106.6	139.1	120.5	125.5	126.8
					B	101.8	80.4	93.5	86.4	91.3	86.6	94.0	96.1	97.7	97.0	98.5	105.4	119.0	96.6	104.5	108.4	125.7	110.7	120.2	122.7
		ENVI-Carb + PSA	GC–MS	Green tea	A	114.2	102.1	100.1	105.3	111.8	87.1	111.3	104.1	113.8	104.6	103.3	110.3	101.9	108.1	124.8	121.4	126.0	125.2	147.1	161.4
					B	106.0	103.2	89.4	92.4	90.7	83.7	96.4	96.6	105.6	97.4	98.8	101.0	96.1	102.7	118.7	121.3	112.4	120.6	133.4	153.2
				Oolong tea	A	102.9	86.5	94.3	87.8	91.9	100.8	105.0	92.6	95.4	97.4	101.4	102.4	90.6	107.1	104.3	125.4	111.9	111.6	105.0	143.8
					B	94.4	81.5	84.9	79.6	78.6	84.9	99.9	82.3	89.9	89.8	87.8	90.8	87.0	96.4	96.6	107.2	109.2	115.4	99.8	131.7
			GC–MS/MS	Green tea	A	113.6	109.4	97.5	99.3	110.1	90.2	112.6	111.9	122.4	108.1	108.0	114.6	103.2	121.5	116.1	123.0	108.3	111.5	139.0	152.4
					B	105.0	110.7	93.2	88.6	90.4	83.5	97.3	103.2	108.9	98.2	103.1	102.0	94.9	94.8	117.3	117.7	107.0	114.3	123.1	147.2
				Oolong tea	A	100.6	85.2	90.7	83.0	104.9	98.8	89.2	85.1	91.2	97.1	97.5	102.4	87.5	99.8	100.8	121.9	106.7	119.9	109.3	140.6
					B	92.8	82.5	80.2	80.6	91.3	80.0	82.2	77.9	84.9	98.2	84.8	90.8	83.0	96.7	91.9	108.5	101.8	112.8	104.3	130.9

(Continued)

Appendix Table 5.1 The 201 Pesticides Average Content for 19 Circulative Determinations on 3 Months Under 16 Factors (Two Cartridges, Two Instruments, Two Tea Samples, and Two Youden Pair Samples) (cont.)

No.	Pesticides	SPE	Method	Sample	Youden pair	Nov 9, 2009 (n=5)	Nov 14, 2009 (n=3)	Nov 19, 2009 (n=3)	Nov 24, 2009 (n=3)	Nov 29, 2009 (n=3)	Dec 4, 2009 (n=3)	Dec 9, 2009 (n=3)	Dec 14, 2009 (n=3)	Dec 19, 2009 (n=3)	Dec 24, 2009 (n=3)	Dec 14, 2009 (n=3)	Jan 3, 2010 (n=3)	Jan 8, 2010 (n=3)	Jan 13, 2010 (n=3)	Jan 18, 2010 (n=3)	Jan 23, 2010 (n=3)	Jan 28, 2010 (n=3)	Feb 2, 2010 (n=3)	Feb 7, 2010 (n=3)	Ave (µg/kg)
70	Bromofos	Cleanert TPT	GC-MS	Green tea	A	80.2	76.3	75.0	114.3	66.7	65.9	78.0	71.3	64.8	64.4	70.4	61.4	61.2	63.3	64.5	54.9	50.6	53.2	62.1	68.3
					B	79.9	73.1	70.4	108.5	65.8	61.0	71.5	66.9	74.8	59.6	60.7	60.7	62.1	60.2	66.0	54.7	51.7	56.5	57.3	66.4
				Oolong tea	A	79.6	90.4	82.0	88.5	73.6	65.7	68.4	88.0	64.2	66.5	72.2	66.2	76.0	66.6	60.3	62.8	66.0	59.8	58.8	71.3
					B	78.1	72.3	70.6	78.4	70.8	60.0	53.8	76.3	59.6	57.0	62.4	59.5	60.4	63.2	54.9	58.6	56.9	58.0	54.0	63.4
			GC-MS/MS	Green tea	A	85.6	70.9	63.9	92.1	70.3	67.4	77.7	71.0	66.2	59.7	79.7	53.7	55.9	60.6	63.7	60.3	48.0	53.9	65.3	66.6
					B	85.9	70.7	67.6	88.0	69.0	64.5	67.6	66.9	70.6	57.9	61.3	45.4	63.5	53.0	61.0	62.3	52.7	55.5	60.2	64.4
				Oolong tea	A	94.2	92.9	79.3	82.2	68.6	70.0	64.5	122.6	80.8	62.3	67.9	64.6	79.5	71.7	62.0	68.2	52.7	62.5	52.7	74.3
					B	86.2	81.9	69.4	68.4	66.8	59.7	51.5	92.3	59.8	68.4	60.2	57.8	63.6	64.4	58.0	57.9	68.4	59.0	53.7	64.7
		ENVI-Carb + PSA	GC-MS	Green tea	A	88.9	79.2	73.5	76.7	76.8	74.5	88.0	63.3	71.6	68.4	67.8	76.9	58.7	67.9	50.7	65.9	67.3	54.8	71.2	70.6
					B	85.8	69.8	72.4	73.1	78.1	70.6	82.7	52.9	69.3	66.1	60.1	77.0	55.9	63.7	54.5	56.9	61.9	55.7	71.4	67.3
				Oolong tea	A	77.4	67.5	74.8	69.9	72.8	64.1	53.7	58.3	64.9	61.6	63.2	57.3	54.6	66.6	62.1	56.2	52.5	52.3	51.6	62.2
					B	67.8	60.0	72.0	61.7	64.4	61.2	62.5	51.4	55.3	51.1	54.6	56.6	52.7	63.2	53.6	51.0	49.7	50.1	49.8	57.3
			GC-MS/MS	Green tea	A	110.2	82.9	80.3	87.9	80.1	105.2	89.3	63.4	77.4	55.1	65.6	65.3	68.0	74.7	52.8	74.2	59.0	61.8	75.6	75.2
					B	122.3	75.7	87.8	97.7	78.2	75.0	69.3	63.3	68.9	60.8	62.4	61.7	61.2	56.7	50.0	62.6	55.8	69.1	69.8	71.0
				Oolong tea	A	110.9	66.8	89.3	63.9	83.2	64.2	79.9	54.8	64.9	56.1	62.0	38.0	57.9	67.4	63.7	68.2	51.3	52.8	63.1	66.2
					B	81.4	60.3	65.4	56.6	67.7	60.2	60.8	49.9	55.3	51.4	63.1	38.5	51.6	62.9	52.1	57.9	51.9	45.4	55.9	57.3
71	Bromopropylate	Cleanert TPT	GC-MS	Green tea	A	93.2	87.0	82.8	146.0	73.0	67.1	74.7	66.6	66.6	59.5	89.6	68.3	66.6	58.6	65.9	62.1	42.7	62.7	77.2	74.2
					B	92.2	83.1	74.5	132.1	69.4	63.1	73.0	66.0	81.3	58.3	82.1	61.8	62.9	59.9	69.0	62.3	45.3	66.3	69.4	72.2
				Oolong tea	A	87.4	86.2	73.9	100.7	94.0	72.1	72.4	96.3	74.8	72.2	76.2	70.1	78.1	61.8	69.9	79.1	69.1	70.6	55.4	76.9
					B	83.4	86.7	76.5	91.5	90.5	67.8	74.6	104.2	63.8	69.3	61.4	60.2	77.3	66.5	63.5	72.9	61.3	69.4	52.9	73.4
			GC-MS/MS	Green tea	A	100.2	87.5	81.9	103.3	71.5	68.6	75.0	70.2	64.1	61.9	100.0	52.2	75.3	63.1	68.0	59.3	49.4	54.7	75.5	72.7
					B	96.8	83.6	75.8	98.6	66.7	64.1	74.6	64.9	73.5	61.3	81.4	46.0	61.6	64.3	64.3	59.5	51.2	56.8	66.6	69.0
				Oolong tea	A	119.2	110.5	77.1	99.0	68.6	72.9	68.3	157.0	74.7	67.3	80.0	72.8	79.3	68.4	68.3	74.4	68.0	61.2	49.6	80.9
					B	103.9	116.7	73.9	90.4	69.6	65.6	69.8	155.2	60.0	69.0	64.8	66.0	73.8	66.9	60.6	69.1	62.2	54.7	50.0	75.9
		ENVI-Carb + PSA	GC-MS	Green tea	A	88.4	94.6	84.1	82.4	82.3	87.9	53.9	68.3	89.8	79.5	81.4	88.0	72.2	57.2	41.1	47.5	91.2	53.3	72.8	72.8
					B	99.8	94.1	80.3	63.0	79.9	85.7	73.9	56.2	95.1	75.8	60.0	79.1	61.5	66.9	32.9	42.4	81.8	75.4	78.6	68.5
				Oolong tea	A	90.9	62.5	73.9	66.7	95.5	68.2	68.0	68.8	70.6	67.4	64.4	66.7	64.8	61.8	68.8	64.1	59.9	58.5	60.8	68.5
					B	89.9	61.4	75.0	68.7	75.3	65.7	63.7	54.4	60.5	62.3	56.6	70.4	57.3	66.5	67.4	49.4	54.7	60.0	59.7	64.2
			GC-MS/MS	Green tea	A	102.6	104.3	78.0	75.8	82.0	74.6	88.0	66.8	81.9	73.6	86.2	70.3	84.4	70.3	50.9	62.8	84.0	67.4	71.5	77.6
					B	116.1	123.2	76.5	68.1	77.4	81.2	60.9	60.8	77.0	73.3	63.7	69.0	73.8	64.5	45.7	53.5	74.9	73.4	80.0	74.4
				Oolong tea	A	109.5	67.1	71.8	70.7	89.2	68.2	68.7	61.5	70.6	64.6	63.7	63.5	57.2	60.3	64.4	74.4	54.9	56.9	52.4	67.9
					B	109.3	65.1	73.0	69.2	74.9	64.5	64.9	52.2	60.5	60.6	63.7	66.2	51.7	56.2	57.8	69.1	53.8	54.1	50.9	64.1
72	Buprofenzin	Cleanert TPT	GC-MS	Green tea	A	94.2	75.5	80.0	93.4	70.2	63.7	77.1	69.0	66.2	60.3	73.2	56.6	45.8	66.0	66.7	52.2	61.9	70.0	68.8	69.0
					B	87.7	71.3	70.5	85.2	63.5	60.8	72.3	63.9	65.9	53.8	64.3	50.2	45.5	62.3	58.8	50.6	62.1	68.0	61.5	64.1
				Oolong tea	A	90.4	86.6	83.8	92.8	84.5	86.7	70.8	89.1	83.0	78.4	78.9	85.0	81.7	78.0	66.7	57.7	68.9	64.5	63.6	78.5
					B	92.2	77.8	78.1	88.4	82.5	83.0	63.0	79.9	74.1	74.4	72.0	90.3	65.9	67.1	60.7	51.1	61.6	68.6	61.5	73.3
			GC-MS/MS	Green tea	A	94.3	76.7	78.5	94.7	67.8	63.0	70.4	69.1	68.3	66.0	63.0	54.0	70.2	71.9	69.6	53.1	59.5	58.5	69.6	69.4
					B	89.6	73.8	69.6	85.2	64.2	58.7	70.7	62.7	70.4	65.1	68.1	49.2	66.6	74.3	60.4	51.5	55.6	59.0	61.6	66.1
				Oolong tea	A	83.4	83.2	85.6	95.0	70.5	79.2	69.8	85.1	75.4	76.3	76.1	78.2	73.6	76.4	67.6	67.7	69.0	67.5	69.1	76.2
					B	84.1	76.2	73.5	86.6	70.7	70.1	57.3	82.9	65.7	62.4	65.9	71.7	63.8	70.8	64.9	62.3	65.4	72.0	63.3	70.0
		ENVI-Carb + PSA	GC-MS	Green tea	A	106.1	91.8	73.5	73.3	79.4	90.4	77.5	65.2	57.0	63.5	57.9	66.8	54.5	76.5	74.4	82.8	53.9	62.7	65.2	72.2
					B	96.3	75.8	71.1	67.6	78.5	81.5	79.3	61.1	56.9	61.5	55.4	56.7	41.2	65.6	70.9	73.5	57.1	49.1	63.2	66.4
				Oolong tea	A	99.4	83.3	86.1	81.2	85.9	81.4	74.2	63.6	70.9	74.2	70.7	75.9	65.8	75.0	82.8	56.7	65.1	62.0	60.4	74.4
					B	84.2	77.7	87.4	73.3	79.7	73.9	71.7	58.0	61.6	65.3	65.5	64.7	75.7	68.9	56.4	56.9	62.3	56.9	75.4	69.8
			GC-MS/MS	Green tea	A	100.1	86.6	74.1	72.2	78.9	70.2	74.2	65.6	66.2	68.2	63.0	55.9	72.8	62.6	71.7	78.4	51.3	66.9	62.9	70.6
					B	91.3	70.0	58.7	55.5	81.8	76.2	71.8	63.0	62.5	70.4	57.6	53.9	59.6	53.6	62.9	71.7	51.9	58.7	66.0	65.1
				Oolong tea	A	97.7	73.8	92.0	70.9	80.7	76.7	71.4	61.6	70.9	70.9	63.4	58.7	53.6	63.1	69.7	67.7	56.7	72.3	61.0	70.1
					B	84.6	69.3	74.9	64.7	72.2	67.7	68.6	55.5	61.6	59.0	64.1	53.8	53.7	57.5	56.2	62.3	61.5	61.7	65.2	63.9

#	Compound	Cleanup	Instrument	Tea	Rep	1	2	3	4	5	6	7	8	9	10	11	12	13	14	15	16	17	18	19	20
73	Butachlor	Cleanert TPT	GC-MS	Green tea	A	76.5	73.0	73.5	83.7	64.6	62.2	72.7	64.4	64.9	56.8	63.3	54.1	53.8	58.8	63.3	54.4	48.7	48.8	63.3	63.2
					B	74.5	71.3	65.9	80.5	59.8	56.5	69.0	59.3	66.9	53.9	53.0	52.8	53.4	55.3	59.3	52.9	50.8	52.0	57.0	60.2
				Oolong tea	A	76.7	80.6	75.0	83.0	64.8	71.5	68.1	81.2	62.7	68.5	66.6	65.1	69.0	61.8	57.2	62.3	64.4	52.8	56.0	67.8
					B	74.9	66.6	64.5	75.8	66.7	68.0	57.7	82.2	57.7	56.0	59.2	57.6	55.4	56.8	52.8	56.4	52.2	54.0	51.6	61.4
			GC-MS/MS	Green tea	A	88.4	71.7	77.4	92.4	68.2	60.6	67.4	65.1	69.0	61.3	66.3	66.0	65.2	67.4	68.3	54.8	52.3	57.8	67.3	67.7
					B	84.6	72.3	68.0	84.2	63.5	58.2	69.0	61.3	68.0	59.7	63.6	62.8	59.3	62.0	62.0	55.5	50.4	60.1	62.8	64.6
				Oolong tea	A	77.4	72.9	82.7	94.2	67.8	78.9	69.0	83.4	78.8	71.6	74.2	77.3	73.5	70.8	68.9	69.7	68.5	69.4	54.8	73.9
					B	77.2	65.1	72.0	85.2	67.1	70.2	58.3	75.9	73.6	64.9	62.7	68.9	66.1	67.2	58.5	65.9	66.2	63.4	58.6	67.7
		ENVI-Carb + PSA	GC-MS	Green tea	A	82.8	79.2	71.4	74.7	71.5	75.5	67.6	61.5	59.7	57.5	57.7	63.7	54.4	63.0	48.4	61.9	60.3	54.8	65.2	64.8
					B	77.5	70.1	72.8	72.3	70.7	70.7	62.7	52.7	58.5	56.1	56.0	62.7	53.1	55.7	49.4	53.8	55.0	52.3	63.0	61.3
				Oolong tea	A	96.4	69.4	71.5	70.5	72.4	68.6	57.3	51.2	75.1	58.2	60.4	54.9	51.0	59.1	64.5	55.6	51.3	56.8	49.5	62.8
					B	68.1	63.4	71.1	64.4	63.0	63.5	65.1	44.2	62.6	49.2	53.6	53.7	48.5	56.3	53.8	50.1	49.4	51.1	45.6	56.7
			GC-MS/MS	Green tea	A	93.2	82.4	68.8	66.5	77.1	50.1	72.6	67.7	67.1	66.8	68.3	62.9	75.7	64.3	63.5	73.5	58.6	65.8	67.0	69.1
					B	86.5	66.4	68.1	66.3	73.8	70.9	71.4	62.5	65.4	67.9	62.1	59.0	64.1	61.7	57.8	63.9	57.5	62.3	62.6	65.8
				Oolong tea	A	90.6	72.4	81.9	71.4	77.9	71.7	69.5	60.7	75.1	66.1	68.4	63.7	56.1	65.6	73.8	69.8	55.7	64.5	58.1	69.1
					B	81.3	69.1	71.8	63.7	68.0	66.9	66.6	57.7	62.6	58.9	72.3	59.9	55.1	59.4	57.4	65.8	56.7	56.9	60.3	63.7
74	Butylate	Cleanert TPT	GC-MS	Green tea	A	114.4	92.7	99.3	108.6	97.7	79.5	78.6	58.1	91.9	46.1	36.3	37.8	32.2	28.9	25.9	24.7	24.0	60.9	71.0	63.6
					B	112.5	88.7	92.0	109.0	85.1	73.0	73.7	58.2	88.9	42.6	34.5	35.3	29.7	32.1	24.5	24.2	24.7	71.5	64.8	60.8
				Oolong tea	A	110.4	86.6	88.0	107.3	73.7	81.6	77.3	83.8	93.3	68.3	73.1	75.9	74.3	67.3	80.6	64.1	67.0	64.9	79.3	79.8
					B	105.5	76.7	98.7	92.3	85.1	74.1	68.1	79.3	87.7	64.8	58.9	64.1	69.4	63.7	60.2	63.9	60.3	58.8	57.9	73.1
			GC-MS/MS	Green tea	A	106.3	80.2	88.8	98.0	78.5	72.4	83.8	73.5	79.2	74.2	78.3	53.2	79.6	75.5	74.3	57.9	59.7	57.5	77.8	76.3
					B	98.1	78.4	80.0	95.0	73.6	68.1	82.5	68.2	78.8	65.9	73.2	50.9	70.4	74.6	68.3	60.6	60.4	61.8	70.8	72.6
				Oolong tea	A	88.9	79.7	84.0	102.1	70.8	77.9	73.2	82.9	87.0	72.7	77.2	73.7	76.1	75.8	68.9	67.9	66.5	71.1	58.0	76.6
					B	84.3	65.5	72.1	81.5	63.1	61.8	59.9	72.5	72.0	62.8	61.6	59.3	63.7	64.4	56.5	60.0	57.1	61.7	48.3	64.6
		ENVI-Carb + PSA	GC-MS	Green tea	A	396.7	154.4	92.6	70.7	86.8	121.2	105.3	64.5	46.0	52.4	263.2	36.8	28.3	75.4	27.5	30.2	30.4	83.1	68.0	96.5
					B	218.4	139.1	118.3	93.5	107.7	131.7	101.0	62.2	43.4	92.7	30.9	30.6	23.1	24.8	18.8	25.7	29.3	75.7	66.7	75.5
				Oolong tea	A	80.8	94.1	86.4	78.1	94.6	85.0	79.6	85.5	74.1	76.7	66.7	64.1	85.3	67.3	77.3	61.1	55.7	63.8	58.9	75.5
					B	53.8	103.8	105.7	77.3	83.7	79.1	69.2	59.8	58.2	61.5	66.4	62.8	57.3	63.7	65.4	49.2	57.4	70.2	57.8	68.5
			GC-MS/MS	Green tea	A	109.5	93.8	66.4	66.1	97.4	71.7	87.7	82.0	74.0	85.8	83.9	154.8	96.8	80.5	68.1	79.2	72.0	75.5	79.8	85.5
					B	105.9	71.5	72.4	65.7	98.2	86.4	78.7	77.4	71.0	83.8	69.6	128.5	79.4	74.6	55.9	69.8	66.6	71.0	78.8	79.3
				Oolong tea	A	109.5	69.0	103.5	69.2	80.0	77.7	66.4	61.3	74.1	74.1	68.1	65.3	79.4	65.7	74.9	69.8	55.4	59.2	58.0	71.8
					B	79.3	58.9	70.5	62.8	63.9	67.1	61.8	50.7	58.2	57.1	65.8	58.5	63.7	55.4	54.2	67.9	49.0	49.0	48.3	59.6
75	Carbofeno-thion	Cleanert TPT	GC-MS	Green tea	A	78.2	74.8	71.5	96.7	64.3	56.7	69.8	67.5	67.3	50.5	68.8	55.3	53.6	48.8	50.2	41.2	35.6	39.4	45.4	59.8
					B	80.8	72.3	66.1	88.2	58.4	53.0	66.6	63.9	73.5	46.6	59.2	54.4	51.7	48.7	51.6	41.0	38.0	35.7	40.2	59.8
				Oolong tea	A	80.9	75.7	68.8	86.3	79.5	64.0	69.7	94.9	60.4	63.6	63.8	58.6	61.0	52.5	50.2	47.0	47.8	41.4	37.8	57.4
					B	77.1	75.3	65.1	80.4	75.4	60.9	57.9	88.6	49.3	51.9	48.9	47.8	50.7	47.5	50.2	40.3	38.4	39.7	31.4	63.4
			GC-MS/MS	Green tea	A	91.5	76.3	67.6	89.3	62.5	55.2	60.5	65.9	58.2	54.4	63.7	64.6	47.0	51.9	45.7	37.7	36.3	39.1	42.4	56.4
					B	89.2	75.6	68.0	81.3	64.5	51.5	60.8	56.6	58.9	51.4	56.3	60.9	44.8	46.8	40.7	33.6	37.3	34.1	41.9	58.4
				Oolong tea	A	115.5	88.3	70.2	87.0	64.9	72.3	59.2	146.0	63.9	58.6	64.4	53.9	64.6	48.8	50.9	57.4	55.2	50.0	32.1	55.5
					B	99.5	83.7	66.6	79.7	64.5	62.7	49.1	112.9	44.2	50.9	51.0	49.6	54.4	46.4	43.5	36.6	48.5	38.6	47.7	68.6
		ENVI-Carb + PSA	GC-MS	Green tea	A	82.6	85.9	75.6	71.0	75.4	78.1	73.8	63.9	64.3	67.2	64.4	61.9	54.4	54.3	41.5	49.9	44.8	38.6	41.8	59.8
					B	87.2	76.3	72.2	62.9	72.9	71.0	68.2	53.3	62.2	66.7	53.5	56.2	51.7	50.8	47.2	41.2	40.7	37.7	45.3	62.6
				Oolong tea	A	76.9	62.0	72.6	65.0	75.3	63.7	55.8	57.0	60.8	55.7	52.6	51.5	48.5	50.1	40.5	41.0	37.7	38.3	35.2	58.5
					B	77.0	62.8	71.4	61.6	61.6	58.0	55.3	47.4	60.8	47.9	42.3	51.0	44.7	46.9	34.1	35.2	33.4	38.1	35.2	55.1
			GC-MS/MS	Green tea	A	117.8	84.0	113.7	118.1	70.0	59.6	71.2	53.5	56.7	60.7	63.7	66.4	53.4	55.5	28.8	48.8	46.0	33.0	31.0	66.3
					B	135.9	79.0	141.0	135.7	71.9	63.3	79.7	53.3	53.0	60.4	63.6	61.7	46.9	49.2	38.6	45.6	35.5	46.5	40.2	67.8
				Oolong tea	A	92.5	62.9	62.9	59.2	81.8	62.0	66.2	51.9	60.8	52.8	53.6	39.2	42.6	48.2	57.4	35.5	30.8	46.7	47.0	53.9
					B	85.3	62.4	60.6	55.4	63.1	57.7	53.4	42.5	49.3	46.2	49.3	34.9	41.1	44.7	36.6	36.4	27.7	31.7	34.3	47.6

(Continued)

Appendix Table 5.1 The 201 Pesticides Average Content for 19 Circulative Determinations on 3 Months Under 16 Factors (Two Cartridges, Two Instruments, Two Tea Samples, and Two Youden Pair Samples) (cont.)

No.	Pesticides	SPE	Method	Sample	Youden pair	Nov 9, 2009 (n=5)	Nov 14, 2009 (n=3)	Nov 19, 2009 (n=3)	Nov 24, 2009 (n=3)	Nov 29, 2009 (n=3)	Dec 4, 2009 (n=3)	Dec 9, 2009 (n=3)	Dec 14, 2009 (n=3)	Dec 14, 2009 (n=3)	Dec 19, 2009 (n=3)	Dec 24, 2009 (n=3)	Jan 3, 2010 (n=3)	Jan 8, 2010 (n=3)	Jan 13, 2010 (n=3)	Jan 18, 2010 (n=3)	Jan 23, 2010 (n=3)	Jan 28, 2010 (n=3)	Feb 2, 2010 (n=3)	Feb 7, 2010 (n=3)	Ave (µg/kg)
76	Chlorfenson	Cleanert TPT	GC-MS	Green tea	A	94.0	100.5	73.8	244.3	76.0	79.1	76.7	69.7	73.0	97.5	61.5	74.6	72.5	64.1	68.1	69.0	45.8	59.5	74.0	82.8
					B	97.1	91.0	70.1	215.5	78.1	73.6	75.4	66.2	89.1	94.0	64.1	66.2	68.7	63.0	82.1	68.0	47.5	62.3	68.9	81.1
				Oolong tea	A	88.6	104.1	76.5	112.8	109.4	71.3	74.9	92.1	72.0	91.0	68.3	74.3	86.4	66.2	76.8	83.5	71.4	69.4	49.3	80.9
					B	85.9	99.9	77.6	92.9	95.0	61.0	70.1	91.4	63.8	72.7	70.0	61.1	81.8	71.1	67.7	79.5	62.5	67.2	51.0	74.8
			GC-MS/MS	Green tea	A	105.8	97.3	71.5	124.0	74.5	80.8	84.4	70.3	65.6	120.8	65.2	146.3	71.5	74.4	67.0	64.0	48.6	62.7	78.9	82.8
					B	105.2	94.6	73.6	124.5	81.8	75.0	75.0	64.0	84.2	107.7	68.0	131.8	67.7	78.5	68.0	64.3	50.0	68.1	68.4	81.6
				Oolong tea	A	144.2	129.4	80.0	110.1	65.9	69.0	70.3	153.1	73.5	90.2	68.6	75.4	83.5	71.8	74.7	87.0	70.3	61.2	50.7	85.7
					B	109.1	127.3	75.0	93.9	60.5	56.4	68.9	121.7	58.3	69.2	75.1	64.4	79.3	71.3	64.0	77.8	65.1	53.3	54.0	76.0
		ENVI-Carb + PSA	GC-MS	Green tea	A	92.8	87.6	90.1	92.8	87.7	91.0	79.5	70.1	87.4	95.1	83.2	109.3	78.1	68.9	41.0	57.7	100.3	63.6	96.9	82.8
					B	98.4	95.7	79.2	77.4	82.1	100.1	77.5	61.0	89.4	64.4	76.6	99.0	74.5	73.1	40.6	48.4	89.9	73.5	98.8	78.9
				Oolong tea	A	85.0	57.0	72.6	57.8	104.7	72.0	67.9	69.1	76.7	68.3	65.8	70.9	66.1	66.2	67.1	69.8	57.3	55.0	53.4	68.6
					B	82.7	53.6	70.8	62.2	82.3	74.0	63.4	59.4	62.3	61.3	65.0	77.0	57.4	71.1	73.1	49.7	56.2	56.9	48.6	64.6
			GC-MS/MS	Green tea	A	115.6	96.0	90.8	87.9	86.1	69.4	124.0	66.8	101.6	104.6	77.6	68.9	73.0	79.6	53.5	64.0	96.2	76.8	94.0	85.6
					B	121.7	118.8	75.1	76.6	75.6	96.5	68.2	62.5	94.5	72.3	74.1	66.3	75.2	63.3	52.7	51.5	88.7	87.7	101.6	80.1
				Oolong tea	A	98.4	62.1	71.5	62.7	108.7	66.2	76.6	68.9	76.7	67.3	65.1	60.2	68.9	60.3	64.6	87.0	55.3	63.9	52.1	70.3
					B	95.6	60.3	68.2	63.9	83.3	68.0	69.7	56.5	62.3	66.4	68.0	66.5	57.1	61.2	60.9	77.8	56.2	57.4	47.3	65.6
77	Chlorfenvinphos	Cleanert TPT	GC-MS	Green tea	A	126.4	121.6	114.8	203.8	98.9	92.2	106.5	102.5	82.5	110.6	87.3	87.2	84.5	102.4	78.4	83.5	70.6	71.8	92.8	102.0
					B	120.2	116.4	103.1	198.5	92.7	85.2	103.1	97.5	109.7	91.7	79.3	86.7	87.1	96.8	107.1	101.2	77.6	79.6	78.1	99.7
				Oolong tea	A	115.8	141.0	114.5	139.2	125.2	92.4	108.3	141.8	95.4	98.4	100.6	100.0	120.2	94.6	89.5	95.4	109.5	90.0	89.7	108.8
					B	115.0	118.6	105.8	127.3	120.2	88.0	86.6	127.3	87.7	83.0	83.8	75.6	92.8	93.9	87.4	78.4	90.7	88.1	81.9	97.3
			GC-MS/MS	Green tea	A	131.2	113.3	93.3	145.2	93.3	87.4	106.7	99.5	99.1	116.7	80.2	92.4	86.1	78.8	91.3	85.3	69.8	75.7	100.8	96.8
					B	131.3	114.9	103.1	142.0	96.0	85.0	101.8	87.9	98.1	96.5	85.2	92.9	83.8	83.0	85.4	101.1	74.1	78.5	88.5	95.4
				Oolong tea	A	164.6	162.5	108.6	136.5	100.7	108.8	106.1	224.1	113.1	121.1	89.4	96.7	116.9	99.6	92.1	90.2	119.5	116.1	67.2	118.1
					B	149.8	159.1	99.1	121.7	99.1	93.5	83.1	168.5	86.3	94.6	89.8	85.5	94.5	89.8	81.5	104.0	92.3	97.0	73.3	102.6
		ENVI-Carb + PSA	GC-MS	Green tea	A	119.3	115.2	108.3	116.0	103.3	115.8	105.4	88.4	109.5	108.4	99.0	109.3	80.5	88.5	62.5	84.0	110.8	67.3	103.2	100.8
					B	121.1	121.8	103.3	99.3	103.8	108.9	104.0	69.0	107.8	82.0	95.8	109.9	71.5	87.2	65.3	86.0	103.5	67.3	111.4	95.8
				Oolong tea	A	117.6	87.9	114.0	109.6	120.6	106.0	66.2	86.9	99.2	91.1	94.1	85.5	77.6	92.3	95.4	75.5	75.9	74.7	74.7	92.4
					B	112.8	80.5	108.5	98.6	98.2	100.2	92.8	71.0	78.5	73.8	73.7	90.2	73.2	94.6	91.7	94.7	73.0	75.2	72.2	86.0
			GC-MS/MS	Green tea	A	148.5	126.4	96.7	97.3	121.8	239.0	133.1	79.4	115.9	113.2	80.6	117.7	86.4	101.1	70.0	82.5	93.3	93.8	103.3	111.2
					B	196.3	136.1	118.2	119.2	114.5	107.3	89.0	78.6	108.2	98.1	84.8	115.5	80.4	80.5	74.4	101.1	91.3	96.0	109.2	104.2
				Oolong tea	A	209.8	95.7	141.2	101.6	132.3	92.3	133.4	92.6	99.2	88.3	83.5	58.0	84.2	86.7	85.5	90.2	66.2	77.7	82.0	100.6
					B	149.5	92.1	102.8	90.6	105.3	89.1	95.2	72.9	78.5	90.3	76.1	56.4	73.7	84.3	72.5	90.2	68.1	70.7	70.9	85.8
78	Chlormephos	Cleanert TPT	GC-MS	Green tea	A	79.8	64.6	72.9	76.6	66.4	59.9	70.9	63.2	68.3	62.6	60.0	58.4	64.8	64.3	67.2	51.9	49.3	51.2	64.9	64.1
					B	75.9	62.9	64.2	74.4	62.0	55.5	67.4	58.2	67.1	57.6	54.5	55.4	62.7	59.8	61.8	53.0	49.6	52.0	58.2	60.6
				Oolong tea	A	80.3	69.2	70.7	83.3	63.0	65.3	63.5	74.5	70.7	66.7	64.6	67.7	69.4	60.0	63.1	60.0	58.3	60.8	52.8	66.5
					B	75.4	61.4	62.0	70.5	59.2	56.1	52.8	71.1	59.9	51.8	55.4	56.6	60.0	56.4	55.3	55.5	49.5	54.1	46.3	58.4
			GC-MS/MS	Green tea	A	80.4	66.2	70.1	79.5	66.8	59.5	68.5	62.2	64.7	70.0	61.6	72.2	60.6	65.0	61.1	47.2	48.2	52.3	66.8	64.4
					B	75.7	64.9	64.0	76.3	62.0	55.2	68.2	56.2	63.4	60.4	54.6	69.0	59.3	61.9	57.1	49.5	49.2	55.7	60.4	61.2
				Oolong tea	A	78.3	69.0	68.9	81.7	60.1	64.0	61.7	81.1	72.6	66.0	61.8	62.3	63.9	59.0	57.0	58.7	54.9	59.2	48.4	64.7
					B	72.6	58.7	59.9	68.0	55.1	52.4	50.3	69.6	59.8	52.7	54.0	50.3	55.2	53.4	46.3	51.7	51.7	54.0	40.9	55.6
		ENVI-Carb + PSA	GC-MS	Green tea	A	92.4	79.8	70.2	73.8	74.3	75.3	74.1	69.5	64.7	64.1	68.1	66.0	69.2	69.8	55.2	68.1	61.9	62.6	64.4	69.7
					B	89.0	65.4	71.9	68.1	74.1	70.7	70.3	64.2	61.0	54.0	67.0	62.1	59.4	68.0	46.9	60.5	58.4	56.7	64.5	64.8
				Oolong tea	A	79.7	61.5	62.2	59.6	68.9	65.6	60.2	57.0	61.3	56.8	62.7	54.9	57.6	58.8	64.2	54.3	50.0	53.3	46.7	59.7
					B	64.0	51.2	59.9	56.2	55.4	57.4	56.3	50.5	50.1	51.9	50.6	51.0	51.6	54.8	52.2	44.3	43.5	46.5	44.7	52.2
			GC-MS/MS	Green tea	A	86.5	76.6	60.0	60.9	77.1	62.5	70.4	65.5	64.1	65.9	73.1	86.8	70.8	70.9	56.8	64.2	63.3	63.9	67.4	68.8
					B	91.1	60.6	48.7	45.1	76.0	70.6	67.6	61.6	60.4	54.1	71.1	82.7	61.1	57.6	45.4	57.8	58.6	61.3	66.9	63.1
				Oolong tea	A	91.3	57.7	85.7	57.0	69.7	61.9	59.6	54.6	61.3	55.7	59.2	53.9	56.5	53.7	58.5	58.7	47.5	49.7	49.5	60.1
					B	65.2	50.8	57.0	52.4	56.0	56.6	53.9	47.3	50.1	56.9	47.6	49.6	50.3	48.2	45.8	51.7	43.0	42.3	47.5	51.2

79	Chloroneb	Cleanert TPT	GC–MS	Green tea	A	48.0	38.6	41.8	45.4	38.9	34.7	41.7	37.8	41.8	34.8	34.3	34.3	36.7	35.8	35.8	32.7	31.9	33.7	40.2	37.8
					B	45.4	37.2	38.7	43.8	35.4	32.3	38.9	34.2	39.7	32.7	32.7	32.5	34.4	34.3	33.3	32.4	31.1	32.4	35.4	35.6
				Oolong tea	A	42.3	38.9	40.6	44.3	36.7	37.2	34.6	42.5	43.0	40.4	43.5	45.2	43.6	40.3	38.9	38.9	35.4	38.0	32.7	39.8
					B	40.3	34.1	36.4	39.4	34.6	32.2	30.8	38.6	39.0	39.4	35.9	39.2	38.5	38.1	35.5	36.1	33.8	36.7	32.4	36.4
			GC–MS/MS	Green tea	A	45.8	39.6	41.9	48.0	36.4	33.9	40.5	37.5	38.1	35.7	37.0	45.5	34.9	34.3	35.5	31.6	29.6	31.4	39.7	37.7
					B	43.3	37.6	37.7	45.1	34.8	32.2	39.6	34.5	37.8	33.1	35.5	42.9	35.3	32.7	32.5	31.5	29.2	31.7	35.7	35.9
				Oolong tea	A	37.2	38.9	41.2	46.4	34.9	36.2	34.0	39.7	39.0	36.1	39.6	37.4	37.4	33.9	34.8	35.2	34.0	35.4	30.2	36.9
					B	37.4	33.0	35.1	40.6	33.1	31.1	29.4	36.0	35.4	33.1	32.2	33.3	32.1	33.1	30.6	32.8	33.8	32.3	26.9	33.2
		ENVI-Carb + PSA	GC–MS	Green tea	A	52.0	47.3	41.4	41.2	43.1	43.3	40.6	40.0	38.2	35.1	36.6	37.9	40.2	41.4	35.9	41.9	35.9	38.0	38.0	40.4
					B	47.2	38.7	41.1	37.4	43.0	39.8	40.2	37.9	34.5	34.8	33.5	36.3	36.3	38.1	32.1	36.4	33.2	32.8	37.4	37.4
				Oolong tea	A	42.7	34.6	33.2	33.9	40.4	34.1	35.8	31.3	34.9	37.6	38.3	37.9	36.3	40.3	41.0	35.2	33.7	35.7	32.1	36.3
					B	33.5	31.4	32.9	32.3	33.8	31.3	32.5	27.7	30.0	32.3	34.6	34.6	34.7	39.2	34.6	31.6	30.6	31.7	31.2	32.7
			GC–MS/MS	Green tea	A	47.3	44.1	43.8	43.1	42.8	28.0	40.1	38.7	37.4	40.9	39.6	39.7	44.1	38.1	36.1	40.3	35.2	38.0	39.5	39.8
					B	45.6	34.7	39.4	37.4	41.9	39.3	36.2	36.3	34.4	38.9	35.2	35.8	39.2	35.7	31.2	35.6	33.6	34.7	39.6	37.1
				Oolong tea	A	43.9	35.6	45.4	33.0	39.2	34.9	32.9	30.8	34.9	34.9	32.1	32.0	33.1	31.4	36.1	35.2	30.0	30.4	29.7	34.5
					B	34.7	31.4	33.7	31.6	34.5	32.4	32.5	28.3	30.0	29.8	31.3	29.7	30.3	29.8	29.5	32.8	28.5	26.5	30.2	30.9
80	Chloropropylate	Cleanert TPT	GC–MS	Green tea	A	229.5	42.6	35.7	73.1	23.3	35.3	35.7	48.0	32.2	29.7	43.0	35.2	32.1	31.5	33.9	31.5	25.9	26.1	38.7	46.5
					B	224.4	37.4	39.4	59.6	25.7	34.7	34.5	31.3	37.3	30.4	40.7	33.7	31.0	33.2	35.7	31.8	27.4	26.8	34.3	44.7
				Oolong tea	A	85.3	80.8	75.9	96.3	82.5	72.8	66.8	96.5	76.9	75.2	73.9	75.4	74.3	67.3	73.4	79.3	72.1	73.3	61.5	76.8
					B	83.4	82.2	75.5	87.2	81.2	67.5	70.0	92.5	66.7	68.1	62.8	62.7	70.0	68.4	67.2	73.0	62.1	71.1	61.0	72.3
			GC–MS/MS	Green tea	A	50.2	41.3	41.1	49.5	35.0	32.2	36.2	33.9	32.8	31.1	39.9	35.6	34.0	34.1	34.2	28.6	25.3	28.6	38.2	35.9
					B	48.4	39.9	37.0	46.3	32.7	30.6	36.1	32.3	35.3	30.2	37.4	32.4	31.0	33.0	32.0	28.4	26.0	29.8	33.8	34.3
				Oolong tea	A	113.0	105.7	86.7	104.5	76.4	78.2	75.8	144.8	84.3	77.7	89.1	81.6	81.6	80.3	76.4	81.8	75.0	75.3	60.4	86.8
					B	99.6	104.3	76.9	93.8	73.5	69.4	64.0	129.7	71.4	73.0	71.6	70.1	75.1	71.7	65.9	76.4	69.4	69.8	63.5	78.4
		ENVI-Carb + PSA	GC–MS	Green tea	A	47.4	45.6	41.9	42.1	37.4	42.5	36.7	35.2	43.2	39.1	40.1	41.2	36.5	29.8	24.5	29.4	42.0	31.5	36.3	38.0
					B	49.8	41.5	40.8	33.5	37.0	40.8	36.4	31.2	43.7	38.1	32.4	36.0	29.2	34.5	20.8	26.0	37.4	39.7	38.1	36.2
				Oolong tea	A	130.0	69.1	68.8	41.9	88.9	71.3	69.8	65.5	76.5	65.2	68.3	68.8	63.4	66.6	73.6	66.4	61.0	64.1	58.9	70.4
					B	104.9	63.9	75.6	52.2	72.0	68.8	67.5	58.2	67.0	58.8	57.6	75.3	57.5	69.8	67.4	55.8	57.4	57.8	56.9	65.5
			GC–MS/MS	Green tea	A	51.0	37.2	38.5	37.6	40.6	33.4	67.7	33.4	37.4	36.5	39.4	27.8	39.1	33.6	28.7	33.1	37.6	31.8	35.4	37.9
					B	53.3	50.8	36.2	33.5	38.6	40.1	36.8	31.1	36.7	37.5	30.7	25.8	32.5	29.7	25.7	28.3	35.0	33.3	38.7	35.5
				Oolong tea	A	98.7	74.9	88.2	73.2	92.9	77.1	72.4	68.7	76.5	71.9	72.4	68.4	73.6	69.1	79.3	81.8	63.6	71.3	64.2	75.7
					B	89.1	69.0	73.1	70.1	77.7	71.8	72.1	59.8	67.0	63.0	73.3	67.2	67.6	66.1	66.2	76.4	64.0	62.1	61.1	69.3
81	Chlor-propham	Cleanert TPT	GC–MS	Green tea	A	92.2	85.5	86.7	126.8	79.7	72.1	85.2	82.4	82.1	71.8	87.0	75.3	77.2	72.8	78.4	69.6	67.7	40.6	85.1	79.9
					B	94.1	84.0	80.1	114.1	74.5	68.5	81.3	76.0	86.0	70.0	88.8	71.1	74.3	85.3	77.6	69.4	66.0	39.3	77.7	77.3
				Oolong tea	A	44.4	43.7	41.4	51.6	45.4	40.2	38.3	45.5	41.4	39.7	42.9	42.4	43.2	37.5	39.9	41.0	37.2	38.4	31.0	41.3
					B	43.1	41.4	38.6	46.9	42.4	36.6	33.6	43.1	36.5	36.5	35.2	35.5	38.5	36.9	35.1	37.9	32.9	38.4	37.9	37.9
			GC–MS/MS	Green tea	A	93.1	86.8	84.0	105.7	78.0	72.0	82.7	81.0	77.0	71.4	105.6	89.5	74.4	69.0	75.5	67.3	60.6	70.3	92.3	80.8
					B	94.8	85.4	79.6	100.7	73.5	68.1	82.0	74.4	80.5	71.3	82.5	83.9	71.7	68.2	72.8	66.8	60.7	71.3	80.5	77.3
				Oolong tea	A	52.3	48.6	38.8	48.5	35.3	36.8	34.9	64.7	37.6	36.1	40.4	37.2	39.0	33.7	33.3	36.1	33.8	32.0	26.3	39.2
					B	47.2	49.3	35.7	44.4	35.5	33.0	32.0	61.9	30.8	34.2	32.4	32.2	35.0	33.2	29.9	34.0	30.2	31.0	26.0	36.2
		ENVI-Carb + PSA	GC–MS	Green tea	A	99.5	101.0	89.2	88.0	89.7	93.1	92.0	78.3	87.9	83.8	81.9	86.9	82.7	73.0	61.4	71.2	86.1	70.4	77.0	83.9
					B	102.6	89.9	87.1	75.0	88.3	87.6	103.6	70.4	89.6	81.7	69.2	83.5	72.3	78.1	54.5	65.4	79.0	78.1	80.9	80.9
				Oolong tea	A	43.1	36.9	40.6	36.5	45.9	38.4	37.2	35.6	35.9	37.0	36.1	35.4	35.4	37.5	39.3	35.1	32.8	34.7	31.7	37.1
					B	39.5	32.9	37.1	35.5	37.3	35.4	34.7	30.1	30.5	31.9	32.7	34.9	32.2	36.9	35.1	30.6	30.7	31.4	30.1	33.7
			GC–MS/MS	Green tea	A	106.3	100.7	99.9	95.7	91.7	83.4	84.0	77.9	84.6	85.7	90.5	71.1	94.0	77.3	69.7	79.5	82.8	79.5	80.3	86.0
					B	114.2	95.4	109.5	98.0	87.8	89.9	89.4	73.3	80.8	86.5	71.3	68.1	80.9	72.9	60.9	73.6	76.3	82.6	86.3	84.1
				Oolong tea	A	54.3	34.5	37.8	34.7	92.2	34.5	33.7	30.6	35.9	32.6	32.4	32.3	29.1	30.9	68.0	36.1	27.1	56.2	27.5	40.0
					B	51.0	33.2	35.3	33.8	70.9	32.6	33.4	25.9	30.5	29.3	33.2	32.2	26.0	28.4	57.2	33.3	27.2	49.6	27.0	36.3

(Continued)

Appendix Table 5.1 The 201 Pesticides Average Content for 19 Circulative Determinations on 3 Months Under 16 Factors (Two Cartridges, Two Instruments, Two Tea Samples, and Two Youden Pair Samples) (cont.)

No.	Pesticides	SPE	Method	Sample	Youden pair	Nov 9, 2009 (n = 5)	Nov 14, 2009 (n = 3)	Nov 19, 2009 (n = 3)	Nov 24, 2009 (n = 3)	Nov 29, 2009 (n = 3)	Dec 4, 2009 (n = 3)	Dec 9, 2009 (n = 3)	Dec 14, 2009 (n = 3)	Dec 19, 2009 (n = 3)	Dec 24, 2009 (n = 3)	Jan 3, 2010 (n = 3)	Jan 8, 2010 (n = 3)	Jan 13, 2010 (n = 3)	Jan 18, 2010 (n = 3)	Jan 23, 2010 (n = 3)	Jan 28, 2010 (n = 3)	Feb 2, 2010 (n = 3)	Feb 7, 2010 (n = 3)	Ave (μg/kg)
82	Chlorpyifos (ethyl)	Cleanert TPT	GC-MS	Green tea	A	43.8	37.0	36.4	54.7	32.5	30.0	35.9	34.7	30.1	28.0	30.6	31.3	29.3	31.8	27.5	25.0	23.6	32.5	33.4
					B	44.4	36.1	33.3	49.9	30.7	28.6	33.8	32.7	35.8	25.9	28.8	30.1	29.5	31.0	27.2	24.7	24.1	29.2	32.1
				Oolong tea	A	41.5	37.9	35.3	43.8	36.9	33.4	32.7	44.1	32.7	33.3	33.1	36.0	31.3	31.1	32.1	30.8	30.5	27.4	34.6
					B	40.2	36.1	32.7	40.2	34.7	30.2	25.9	38.1	27.7	29.0	29.0	30.4	30.4	28.2	29.7	26.7	29.6	26.1	31.2
			GC-MS/MS	Green tea	A	42.9	36.1	34.1	45.8	32.5	30.6	35.0	33.5	33.2	29.2	32.3	29.7	29.4	31.8	26.0	23.4	26.8	32.0	32.8
					B	42.3	35.9	32.7	43.7	32.3	28.6	33.1	30.2	35.2	28.0	30.6	29.7	28.7	29.7	26.2	23.9	26.8	30.2	31.7
				Oolong tea	A	46.2	40.5	35.9	44.6	32.9	33.7	32.7	55.1	37.1	31.1	34.0	36.0	32.1	30.8	33.8	31.2	30.9	25.0	35.8
					B	42.4	36.5	32.1	40.5	31.1	29.3	26.5	45.8	29.4	30.0	29.5	30.6	30.9	27.9	30.6	26.6	29.8	24.3	31.7
		ENVI-Carb + PSA	GC-MS	Green tea	A	43.3	41.1	35.8	36.2	38.9	38.8	38.2	32.2	36.9	38.1	41.4	31.5	33.0	24.0	30.6	34.3	28.5	33.5	35.4
					B	42.5	36.9	35.0	32.2	37.4	36.3	35.9	28.0	35.6	35.1	37.0	29.6	31.8	24.2	26.2	31.4	29.2	33.4	33.1
				Oolong tea	A	42.1	30.7	35.5	33.0	39.6	31.6	30.4	28.0	31.9	30.8	28.9	28.9	31.1	31.8	28.3	25.9	27.0	25.6	31.1
					B	37.4	27.8	32.2	30.9	31.3	29.6	30.9	24.0	26.3	26.3	28.2	26.7	30.0	28.6	24.5	24.2	24.9	24.2	28.1
			GC-MS/MS	Green tea	A	47.1	39.1	34.4	34.2	41.1	59.1	40.9	29.8	34.9	31.6	35.4	34.0	37.1	26.5	31.5	31.4	30.4	34.3	36.2
					B	51.4	38.4	33.9	34.1	37.9	36.9	30.0	28.4	31.9	31.4	33.2	31.4	30.2	26.8	27.3	30.5	30.4	34.8	33.2
				Oolong tea	A	51.5	31.9	43.8	32.3	39.3	30.9	34.6	28.4	31.9	29.0	34.7	31.6	28.3	30.0	33.8	25.5	26.6	25.4	32.1
					B	40.3	29.7	31.1	30.6	31.6	29.2	31.1	24.7	26.3	26.1	24.0	28.5	26.6	25.3	29.8	24.9	24.0	24.7	28.4
83	Chlorthiophos	Cleanert TPT	GC-MS	Green tea	A	126.8	110.8	107.4	138.7	93.4	80.5	98.0	90.3	85.7	69.9	70.4	80.9	79.4	83.9	69.6	70.0	58.2	67.0	87.7
					B	126.7	105.1	98.5	128.3	85.4	76.4	91.1	84.7	91.8	65.7	67.9	85.6	81.4	87.7	72.4	72.4	59.1	60.0	84.9
				Oolong tea	A	124.4	112.4	105.2	120.4	100.5	94.5	90.4	114.2	86.8	83.7	94.2	101.5	86.3	83.6	80.5	77.6	62.9	59.0	92.6
					B	119.7	105.2	97.2	110.3	99.1	86.9	75.3	101.6	73.3	71.2	78.8	84.1	85.9	72.8	67.7	72.3	60.4	52.6	83.2
			GC-MS/MS	Green tea	A	127.2	108.8	107.0	131.8	93.1	82.0	96.2	86.0	83.5	69.7	48.4	74.8	73.0	69.2	56.4	52.1	56.3	68.8	82.8
					B	122.6	109.0	99.8	120.3	85.8	77.7	95.1	79.5	83.1	70.6	47.6	66.6	70.1	65.9	55.6	54.1	57.2	61.4	79.1
				Oolong tea	A	140.6	121.5	108.0	122.0	91.6	94.5	86.5	141.2	93.6	80.2	81.7	88.1	81.8	69.7	67.9	69.7	63.8	55.6	92.1
					B	128.0	109.2	95.9	112.2	91.8	84.1	75.3	124.9	77.3	77.2	72.0	69.6	68.5	57.6	60.6	57.0	58.4	47.4	81.1
		ENVI-Carb + PSA	GC-MS	Green tea	A	131.5	127.7	108.9	102.2	105.4	113.7	121.6	84.7	82.3	82.3	90.5	83.9	91.4	74.7	90.3	77.2	57.9	69.5	92.8
					B	136.4	112.9	106.4	93.8	104.2	95.4	152.8	76.7	81.4	81.4	82.4	86.7	82.0	75.1	74.9	75.6	56.3	68.5	90.2
				Oolong tea	A	129.9	98.5	107.5	95.9	108.8	87.2	82.8	73.9	81.4	78.9	78.3	84.4	86.8	102.8	67.9	51.0	59.2	50.6	84.2
					B	121.2	92.7	105.9	89.2	89.3	80.5	79.4	66.2	69.9	65.9	71.7	74.2	85.9	85.5	66.5	46.4	51.3	46.8	76.3
			GC-MS/MS	Green tea	A	153.3	122.8	117.9	117.2	111.1	168.7	104.2	80.5	86.1	77.8	108.1	77.9	80.4	63.0	73.9	61.2	63.1	67.1	95.7
					B	162.4	106.6	108.4	108.4	103.4	97.8	95.3	77.3	79.8	81.3	101.8	70.0	67.0	60.4	61.8	58.1	60.9	69.2	86.6
				Oolong tea	A	149.9	96.7	117.4	97.7	111.7	84.0	96.5	72.2	81.4	78.0	64.4	68.5	64.0	71.2	67.9	50.7	56.3	51.2	81.8
					B	124.7	93.3	96.5	90.5	94.7	79.3	82.9	61.4	69.9	67.8	60.4	62.4	56.4	54.7	60.6	50.1	48.4	50.3	72.5
84	cis-Chlordane	Cleanert TPT	GC-MS	Green tea	A	89.1	73.6	88.8	83.1	68.5	60.7	52.2	73.0	65.3	65.5	63.1	63.1	64.3	53.0	61.6	56.0	59.3	52.8	67.4
					B	86.2	71.2	68.3	95.1	63.2	58.4	66.7	62.2	65.8	61.8	58.0	59.8	56.4	63.7	54.8	51.5	52.2	61.1	64.0
				Oolong tea	A	80.9	81.0	77.0	87.2	68.8	72.0	67.7	82.3	73.0	67.4	72.6	73.4	66.5	59.6	68.4	64.6	65.7	58.0	72.0
					B	78.7	72.2	68.0	78.8	66.3	64.1	55.6	74.4	65.1	62.2	62.1	62.3	62.9	58.7	62.1	56.2	65.6	56.3	64.9
			GC-MS/MS	Green tea	A	89.6	71.3	75.3	88.4	65.1	63.6	68.5	64.4	63.5	62.1	61.6	66.9	60.2	65.4	50.9	50.7	53.7	65.7	66.3
					B	87.0	72.8	72.8	85.0	67.1	61.1	68.3	61.9	65.2	59.0	54.2	62.0	55.6	63.3	52.5	49.0	58.2	56.6	63.9
				Oolong tea	A	80.1	73.7	81.3	91.0	71.1	69.8	63.4	82.4	78.5	70.8	73.3	76.4	70.6	61.5	67.3	60.2	63.0	56.3	71.9
					B	75.6	66.5	68.8	83.1	68.5	60.7	52.2	73.0	65.3	65.5	63.1	59.6	64.3	53.0	61.6	56.0	59.3	52.8	63.6
		ENVI-Carb + PSA	GC-MS	Green tea	A	93.9	85.8	74.3	74.8	78.4	78.5	117.2	63.4	65.8	61.8	70.5	64.3	70.2	55.0	68.9	64.8	60.9	66.1	72.5
					B	89.7	74.0	72.2	69.7	76.0	73.8	122.9	60.4	63.6	61.8	67.0	59.8	63.8	52.7	56.3	58.8	57.7	66.5	68.7
				Oolong tea	A	84.3	70.1	73.5	67.5	77.8	66.4	68.4	61.3	63.6	65.5	63.3	60.6	66.5	70.2	60.7	56.5	60.5	55.1	66.5
					B	73.2	62.7	69.4	62.6	65.1	61.7	65.2	54.5	68.8	56.6	58.9	60.1	62.9	60.4	53.0	52.3	53.0	60.0	60.0
			GC-MS/MS	Green tea	A	94.9	88.2	69.1	69.4	77.3	79.3	74.1	64.0	63.5	65.2	96.8	66.7	65.4	56.6	69.0	61.2	66.3	66.3	71.7
					B	90.9	75.8	61.3	58.3	71.8	74.6	57.6	62.3	68.8	64.5	84.3	59.6	56.7	53.0	56.0	57.0	61.4	70.3	65.2
				Oolong tea	A	91.3	68.3	87.6	68.3	80.9	63.3	66.5	62.8	61.1	63.4	61.1	55.4	59.3	61.4	67.3	52.8	61.9	57.2	66.3
					B	75.9	60.8	67.7	64.3	67.7	61.3	63.4	51.8	57.9	55.6	57.0	50.4	52.3	48.7	61.6	53.4	51.4	53.5	58.9

85	cis-Diallate	Cleanert TPT	GC–MS	Green tea	A	85.7	67.9	72.7	88.2	64.9	61.2	71.6	63.9	68.2	58.0	67.3	59.6	62.0	59.1	62.4	51.4	56.1	34.0	66.3	64.2
					B	82.5	66.5	67.1	82.4	61.5	57.8	65.7	60.9	68.7	55.4	61.3	55.6	58.9	55.1	57.4	51.4	53.7	33.5	58.9	60.7
				Oolong tea	A	77.2	71.8	71.1	81.0	65.7	67.1	61.0	75.5	69.7	63.9	70.1	67.9	67.5	62.4	63.5	63.4	56.5	63.1	52.4	66.9
					B	74.7	64.2	63.9	74.3	62.9	59.8	51.0	67.4	60.8	58.4	56.4	58.5	58.2	60.8	55.7	57.0	51.5	59.7	50.0	60.3
			GC–MS/MS	Green tea	A	86.2	71.0	76.1	90.8	63.6	62.2	71.2	64.4	67.8	60.6	68.7	48.5	65.8	59.8	65.1	52.6	50.0	56.0	69.4	65.8
					B	84.2	69.2	68.3	85.6	60.5	58.6	70.0	60.7	67.7	56.9	63.4	47.9	58.6	56.8	60.9	51.2	49.4	57.6	61.5	62.6
				Oolong tea	A	69.6	72.8	72.9	85.2	60.3	65.3	59.5	75.9	69.4	62.3	68.1	67.0	64.1	64.7	60.0	62.0	55.3	59.3	51.9	65.6
					B	69.4	64.2	64.0	74.1	58.7	56.3	50.4	67.8	64.5	58.4	57.8	57.3	55.3	58.1	52.7	55.9	51.8	57.1	47.4	59.0
		ENVI-Carb + PSA	GC–MS	Green tea	A	100.0	82.6	71.7	71.4	77.6	74.6	72.4	63.5	64.7	66.6	65.1	73.1	62.9	68.8	58.0	67.1	63.9	62.7	65.4	70.1
					B	88.4	71.2	70.2	66.3	76.0	69.6	73.6	61.5	60.5	63.3	59.4	66.7	58.5	61.4	54.5	56.7	58.8	58.0	64.4	65.2
				Oolong tea	A	78.3	63.6	67.2	64.2	73.0	63.1	67.8	54.1	62.9	61.5	60.2	57.0	58.6	62.4	62.8	53.6	51.6	57.4	51.3	61.6
					B	68.7	56.3	62.2	59.5	59.6	56.6	59.7	49.6	53.3	52.6	53.9	53.3	54.1	60.4	53.6	46.8	48.3	50.7	47.4	55.1
			GC–MS/MS	Green tea	A	94.9	84.7	72.3	71.0	78.5	76.7	73.8	62.1	65.2	66.8	70.2	65.9	87.7	71.4	58.4	67.3	63.1	63.9	69.1	71.7
					B	89.3	68.7	69.7	66.9	76.8	74.6	63.8	59.1	61.2	65.9	63.0	62.3	74.9	59.8	54.0	57.7	56.2	60.9	68.0	65.9
				Oolong tea	A	81.1	63.9	79.2	61.9	71.6	61.9	58.9	53.2	62.9	59.2	58.1	56.1	58.6	54.8	61.3	62.0	50.3	54.2	51.3	61.1
					B	67.1	57.2	61.0	57.9	61.1	57.6	56.5	47.6	53.3	51.2	59.0	53.5	52.7	51.3	50.7	55.9	47.8	47.6	50.3	54.7
86	Cyanofen-phos	Cleanert TPT	GC–MS	Green tea	A	48.9	47.8	41.7	102.7	38.5	39.2	44.9	39.4	41.7	32.8	46.9	34.5	37.8	32.8	34.0	31.8	24.7	29.6	36.7	41.4
					B	50.6	45.0	38.4	88.3	38.5	37.4	43.2	38.1	44.0	32.1	42.3	30.8	34.9	33.1	35.8	32.5	26.1	32.3	32.7	39.8
				Oolong tea	A	46.8	46.4	41.9	53.7	49.8	45.3	40.0	56.0	40.8	43.6	44.7	38.6	43.7	36.6	35.8	44.9	35.1	33.1	29.8	42.5
					B	44.9	45.5	42.5	47.4	48.2	43.3	38.9	54.1	38.4	40.3	39.5	34.1	39.5	38.5	34.4	40.8	30.1	32.7	30.7	40.2
			GC–MS/MS	Green tea	A	51.2	46.5	38.1	57.3	35.4	39.3	41.3	35.4	31.7	30.0	48.2	40.3	38.9	31.8	34.5	31.3	25.0	30.1	35.6	38.0
					B	50.3	44.6	37.3	56.3	37.4	36.4	37.2	33.2	38.1	32.6	45.3	37.1	32.3	32.8	33.8	31.9	26.0	31.6	31.7	37.2
				Oolong tea	A	61.1	53.9	39.9	51.7	35.7	38.0	34.0	64.0	39.2	32.2	44.0	36.4	41.0	36.9	36.0	38.6	35.1	32.9	28.2	41.0
					B	51.8	51.7	37.7	43.5	34.7	31.9	34.9	52.6	32.6	35.1	36.5	33.9	35.6	34.1	31.8	35.2	31.1	29.7	27.7	37.0
		ENVI-Carb + PSA	GC–MS	Green tea	A	51.1	49.3	45.8	45.7	44.1	44.0	119.6	41.0	52.9	44.7	71.3	56.4	35.8	49.4	23.5	57.8	47.2	49.1	45.6	51.3
					B	53.7	51.7	42.3	39.6	42.4	43.9	106.1	38.2	49.0	41.8	48.2	49.3	33.9	46.1	25.4	47.4	41.4	51.7	47.6	47.4
				Oolong tea	A	48.4	31.1	40.6	35.8	50.7	43.2	40.8	40.4	36.1	41.0	31.3	33.8	34.3	36.7	35.2	31.0	28.0	31.6	29.5	36.8
					B	46.1	30.1	42.3	37.2	42.2	43.0	37.9	35.1	29.5	36.5	25.5	36.0	29.9	38.5	34.7	24.7	25.8	30.9	28.5	34.4
			GC–MS/MS	Green tea	A	56.7	51.2	52.0	51.2	42.8	30.3	59.5	32.9	47.9	37.0	47.2	37.6	39.2	42.7	27.5	34.8	43.4	35.8	45.9	42.9
					B	56.4	42.9	50.3	52.3	38.3	43.1	28.3	30.5	42.4	35.9	37.8	35.8	38.8	32.4	27.8	26.5	41.4	41.1	47.9	39.5
				Oolong tea	A	50.0	31.8	36.1	33.0	47.5	32.8	38.8	32.6	36.1	31.1	33.1	27.3	29.5	30.6	33.9	38.6	25.4	29.3	27.0	33.9
					B	48.8	31.1	36.9	33.0	40.7	33.4	33.6	28.9	29.5	30.6	33.3	28.3	24.5	30.5	29.2	35.2	26.2	26.9	24.4	31.8
87	Desmetryn	Cleanert TPT	GC–MS	Green tea	A	53.5	41.3	42.7	46.1	34.9	30.1	36.9	35.0	32.6	30.5	43.7	31.8	30.5	28.0	29.5	25.7	26.4	16.7	31.9	34.1
					B	51.5	40.2	37.3	41.3	31.4	29.0	33.9	30.8	33.9	28.2	36.5	29.8	28.6	26.9	27.9	26.0	26.1	16.6	28.3	31.8
				Oolong tea	A	49.6	38.7	38.3	45.6	38.4	37.2	33.1	41.3	36.9	36.7	35.0	33.5	36.9	32.6	33.2	30.0	29.1	28.6	26.7	35.9
					B	46.5	37.9	35.6	42.8	37.0	33.4	30.1	36.7	30.4	32.3	32.5	28.7	31.1	29.5	28.3	25.5	24.1	28.0	25.8	32.2
			GC–MS/MS	Green tea	A	48.5	40.8	42.4	47.9	34.2	31.7	33.9	33.1	32.0	29.1	30.8	21.7	29.9	26.1	30.6	24.7	23.9	20.0	30.4	32.3
					B	46.4	39.2	37.1	43.8	31.4	29.3	34.1	30.4	31.7	29.0	35.8	21.0	27.3	28.2	29.3	24.5	24.3	21.3	27.3	30.9
				Oolong tea	A	38.5	40.3	39.8	46.4	35.7	35.8	32.0	40.8	35.4	31.3	28.0	34.1	33.4	32.1	27.6	31.1	28.4	28.1	24.5	34.3
					B	39.6	37.1	35.1	41.6	35.1	31.2	26.2	36.4	31.3	28.9	35.6	29.0	26.7	28.3	24.5	27.2	23.9	27.7	22.7	30.6
		ENVI-Carb + PSA	GC–MS	Green tea	A	57.7	49.8	39.8	39.4	40.8	44.3	37.3	33.0	34.1	42.3	31.2	41.0	31.0	34.0	28.2	39.5	28.2	28.1	28.0	37.5
					B	52.5	42.0	39.3	35.6	39.3	34.7	37.2	31.1	32.9	43.3	28.3	34.6	28.7	30.9	25.7	29.4	25.4	24.4	28.5	34.2
				Oolong tea	A	50.2	36.7	39.1	35.3	43.1	30.7	33.9	28.2	33.4	33.1	24.4	38.6	29.4	31.3	32.9	29.4	25.6	27.7	24.2	32.8
					B	45.9	33.7	35.9	33.1	32.9	43.3	31.4	25.1	27.2	27.7	33.2	29.6	27.8	27.8	26.5	24.1	22.8	22.7	22.4	29.0
			GC–MS/MS	Green tea	A	51.4	46.9	35.5	34.1	39.4	38.8	34.7	32.8	31.0	33.3	29.5	29.6	38.0	29.1	27.5	32.3	27.5	28.4	28.4	35.3
					B	47.9	37.6	32.3	29.7	37.6	33.8	34.0	30.4	30.8	33.1	29.3	26.7	31.1	24.6	24.6	27.2	25.0	26.3	28.9	32.0
				Oolong tea	A	47.3	36.8	41.4	34.9	39.5	30.5	30.8	27.6	33.4	31.3	29.4		29.1	27.8	30.3	31.0	23.8	26.3	24.4	32.0
					B	42.6	33.9	35.0	33.4	32.6		31.2	24.6	27.2	26.1		26.7	27.0	24.9	24.2	27.2	23.0	21.4	23.9	28.7

(Continued)

Appendix Table 5.1 The 201 Pesticides Average Content for 19 Circulative Determinations on 3 Months Under 16 Factors (Two Cartridges, Two Instruments, Two Tea Samples, and Two Youden Pair Samples) (cont.)

No.	Pesticides	SPE	Method	Sample	Youden pair	Nov 9, 2009 (n=5)	Nov 14, 2009 (n=3)	Nov 19, 2009 (n=3)	Nov 24, 2009 (n=3)	Nov 29, 2009 (n=3)	Dec 4, 2009 (n=3)	Dec 9, 2009 (n=3)	Dec 14, 2009 (n=3)	Dec 19, 2009 (n=3)	Dec 24, 2009 (n=3)	Dec 14, 2009 (n=3)	Jan 3, 2010 (n=3)	Jan 8, 2010 (n=3)	Jan 13, 2010 (n=3)	Jan 18, 2010 (n=3)	Jan 23, 2010 (n=3)	Jan 28, 2010 (n=3)	Feb 2, 2010 (n=3)	Feb 7, 2010 (n=3)	Ave (µg/kg)
88	Dichlobenil	Cleanert TPT	GC–MS	Green tea	A	8.5	6.6	7.3	7.2	7.1	6.2	7.6	6.5	7.1	6.3	6.7	7.0	6.4	5.5	6.0	5.0	5.2	5.9	6.8	6.6
					B	7.8	6.4	6.7	6.9	6.5	5.7	7.0	5.8	6.6	5.7	6.3	6.6	6.2	5.2	6.2	5.0	5.2	5.8	6.1	6.2
				Oolong tea	A	8.3	6.7	7.2	8.3	5.9	6.9	6.3	6.3	7.0	6.0	7.0	6.9	7.0	6.4	7.0	5.8	5.7	6.2	5.3	6.6
					B	7.6	5.9	6.2	6.9	5.5	6.0	5.3	6.1	5.9	5.4	5.9	6.0	6.2	5.7	5.4	5.3	5.1	5.6	4.7	5.8
			GC–MS/MS	Green tea	A	8.7	7.0	7.3	8.1	6.6	6.4	7.1	6.1	6.9	6.2	6.1	6.2	6.9	6.8	6.5	5.1	4.7	5.6	7.2	6.6
					B	7.9	6.5	6.5	7.8	6.3	6.0	6.7	5.9	7.0	5.9	6.1	5.7	6.2	6.3	6.1	5.3	4.9	6.0	6.4	6.3
				Oolong tea	A	7.8	7.1	7.2	8.4	6.1	6.6	6.0	7.9	7.3	6.0	5.9	6.0	6.8	6.0	5.9	5.9	5.8	5.4	4.6	6.5
					B	7.3	6.2	6.2	6.7	5.4	5.3	5.4	6.9	6.3	5.5	7.9	5.2	6.1	5.5	4.9	5.3	5.4	5.1	4.6	5.7
		ENVI-Carb + PSA	GC–MS	Green tea	A	9.6	8.3	7.4	7.9	8.0	7.6	7.5	7.5	7.0	6.9	6.6	7.0	6.4	5.7	5.4	6.5	5.8	7.1	7.4	7.2
					B	9.3	6.5	7.7	7.3	8.2	7.2	7.0	7.2	6.5	6.7	5.5	6.8	5.3	6.4	4.5	5.9	5.6	6.3	7.1	6.6
				Oolong tea	A	8.5	5.8	6.2	5.9	6.7	6.6	6.7	4.6	6.6	6.2	5.8	5.7	5.7	5.7	6.8	5.5	5.2	5.6	4.8	6.1
					B	6.7	5.1	6.3	5.5	5.6	5.7	5.7	4.7	5.4	5.1	5.5	5.1	5.1	6.4	5.1	4.8	4.6	4.7	4.6	5.3
			GC–MS/MS	Green tea	A	8.5	7.8	6.1	6.0	7.9	5.5	7.0	7.1	6.9	8.0	5.8	15.4	8.0	5.7	5.5	6.5	6.4	6.6	7.9	7.4
					B	8.5	5.9	6.8	5.8	8.0	7.4	6.0	6.8	6.4	7.7	7.5	15.4	6.9	6.8	4.7	5.7	5.9	6.1	7.9	7.1
				Oolong tea	A	8.3	5.6	7.9	5.6	7.2	6.5	5.7	6.0	6.6	6.1	5.8	5.3	5.6	6.3	5.9	5.9	5.4	5.4	4.8	6.0
					B	6.5	4.8	6.1	5.5	5.9	6.0	5.6	5.0	5.4	5.0	5.6	5.1	5.2	4.7	4.7	5.3	4.3	4.5	4.5	5.2
89	Dicloran	Cleanert TPT	GC–MS	Green tea	A	117.2	107.9	95.0	165.5	104.9	110.5	90.5	109.3	138.1	76.7	118.6	86.4	85.7	69.3	79.9	69.6	74.2	25.3	79.9	95.0
					B	144.6	104.6	84.4	145.0	97.5	101.4	82.7	95.7	128.6	72.5	104.1	85.7	81.4	74.4	72.2	65.3	67.0	24.2	79.5	90.0
				Oolong tea	A	92.0	104.1	102.4	130.0	137.9	85.8	104.6	114.4	97.8	115.3	82.5	78.7	97.9	71.4	89.7	83.5	107.5	64.9	64.5	96.0
					B	89.6	100.6	93.3	113.1	117.3	83.8	73.0	79.7	75.4	87.5	75.8	69.6	90.2	70.3	75.5	82.1	108.9	65.6	59.1	84.8
			GC–MS/MS	Green tea	A	103.1	87.0	59.3	111.2	74.9	64.5	77.7	78.1	74.8	57.0	117.7	68.5	67.4	59.5	66.3	63.9	51.8	62.9	88.1	75.5
					B	102.2	87.4	73.8	108.7	75.8	65.0	75.4	69.9	76.5	61.8	92.0	69.0	63.7	69.3	67.5	69.1	57.5	65.7	73.5	74.9
				Oolong tea	A	120.7	96.1	72.0	106.6	69.5	74.1	69.0	147.7	84.2	63.5	91.0	71.6	88.4	71.6	80.0	71.6	77.9	71.3	60.4	83.5
					B	99.7	83.2	70.1	101.5	70.0	61.3	54.7	107.1	64.5	69.5	66.6	60.2	80.0	68.9	67.0	71.3	67.7	59.1	60.1	72.8
		ENVI-Carb + PSA	GC–MS	Green tea	A	127.0	113.5	107.5	133.7	100.8	109.3	75.0	85.3	155.2	106.2	115.1	118.6	66.4	91.8	55.4	76.5	88.3	58.6	78.0	98.0
					B	126.3	111.8	107.4	104.8	100.6	98.9	77.6	70.5	130.2	101.0	82.7	101.7	66.1	92.4	59.0	70.1	85.3	63.7	78.3	91.0
				Oolong tea	A	82.0	67.1	104.7	88.2	135.6	99.3	74.3	110.9	65.7	91.7	84.0	90.6	62.9	74.5	69.1	79.2	78.8	65.4	57.9	83.3
					B	85.1	62.8	102.1	90.9	106.0	92.3	70.6	86.6	59.2	69.4	65.2	87.7	60.5	73.7	76.2	71.0	79.7	64.7	55.0	76.8
			GC–MS/MS	Green tea	A	111.7	87.8	109.1	115.7	84.9	96.3	85.6	68.1	88.5	73.2	90.7	64.4	81.1	66.3	55.8	63.8	58.0	74.8	80.7	81.9
					B	130.1	88.9	128.9	120.1	81.1	93.5	74.6	68.5	80.9	82.7	72.8	62.8	77.6	64.4	54.4	62.2	63.8	70.1	83.5	82.1
				Oolong tea	A	97.7	56.6	82.6	68.5	103.0	88.9	85.0	66.9	65.7	60.2	61.3	38.8	67.1	60.5	61.5	71.6	52.0	65.4	59.0	68.0
					B	76.9	55.7	62.5	67.3	80.6	68.2	63.3	55.2	59.2	58.1	60.0	43.7	57.7	59.7	58.9	71.3	55.1	57.5	55.2	61.4
90	Dicofol	Cleanert TPT	GC–MS	Green tea	A	104.5	145.4	168.7	209.6	149.0	134.5	170.1	160.5	166.9	129.8	135.2	128.4	132.8	137.5	154.9	123.0	132.3	114.3	155.0	144.9
					B	113.3	144.1	151.1	185.5	132.8	124.2	155.0	146.4	147.6	125.2	120.5	121.5	126.2	127.4	134.2	121.3	125.8	110.1	130.4	133.8
				Oolong tea	A	107.4	101.6	105.3	135.0	107.9	114.6	136.2	134.7	112.2	106.6	97.4	117.3	138.7	117.1	122.7	120.8	123.8	107.1	123.8	117.4
					B	106.5	91.9	103.8	132.6	117.0	103.5	135.2	120.1	104.6	115.1	93.3	112.8	127.5	106.1	108.1	103.5	103.1	105.1	113.4	110.7
			GC–MS/MS	Green tea	A	182.8	150.6	172.9	185.9	143.1	128.6	152.1	151.3	152.8	130.7	155.2	189.9	132.6	127.6	140.9	116.2	121.9	114.5	171.0	148.5
					B	174.4	145.8	146.5	169.4	129.0	117.2	155.8	139.3	141.6	123.3	131.5	182.3	129.4	115.9	125.0	113.3	116.3	113.2	148.8	137.8
				Oolong tea	A	104.9	104.9	107.1	142.9	96.9	111.7	136.4	161.9	115.0	111.4	99.4	108.5	123.2	107.0	105.3	110.5	114.5	97.8	107.0	114.2
					B	103.0	95.2	102.1	136.1	107.6	98.8	134.6	141.4	107.9	118.8	94.5	113.9	125.9	109.7	105.3	105.4	107.1	104.4	108.7	111.6
		ENVI-Carb + PSA	GC–MS	Green tea	A	126.1	189.5	157.1	168.7	160.4	165.5	69.2	160.4	163.5	135.8	124.6	134.0	149.5	172.1	161.0	189.4	139.4	155.3	146.2	150.9
					B	136.4	150.3	155.3	161.3	162.0	143.5	71.2	159.4	144.1	135.5	130.7	131.3	147.4	144.4	152.6	152.7	125.0	125.7	145.2	140.7
				Oolong tea	A	125.9	104.7	122.0	118.5	144.1	117.1	122.7	89.8	112.5	102.9	105.6	122.4	112.9	117.1	137.0	118.2	139.9	114.2	125.7	118.6
					B	100.8	116.3	134.0	105.4	113.8	113.9	109.6	91.4	108.5	95.3	91.8	113.7	103.2	106.1	111.9	105.1	122.2	95.3	121.8	108.4
			GC–MS/MS	Green tea	A	204.2	199.2	296.3	294.6	151.0	145.8	160.4	151.6	155.3	147.5	136.9	75.1	156.0	153.2	139.5	170.3	137.2	154.7	152.2	167.4
					B	191.3	141.1	231.7	223.2	153.8	147.6	148.0	141.3	137.1	142.2	140.2	70.6	143.9	123.4	130.5	141.3	123.4	135.5	147.3	148.2
				Oolong tea	A	128.2	103.8	143.5	119.3	142.4	114.4	106.3	91.4	112.5	96.3	100.4	106.6	107.0	91.8	120.8	110.5	119.2	99.6	115.2	112.1
					B	109.7	118.0	132.0	106.4	114.4	113.5	109.4	91.1	108.5	90.8	99.4	107.9	102.3	99.9	101.3	105.4	122.4	91.2	129.5	108.1

No.	Compound	Cleanup	Instrument	Tea	A/B	1	2	3	4	5	6	7	8	9	10	11	12	13	14	15	16	17	18	19	20
91	Dimethachlor	Cleanert TPT	GC–MS	Green tea	A	117.0	111.5	120.3	119.8	103.2	97.2	108.0	113.3	93.7	93.8	105.7	91.9	91.6	95.8	102.8	85.7	78.0	69.7	99.2	99.9
					B	116.4	107.1	106.3	119.1	96.2	89.7	105.0	100.8	106.0	88.6	89.2	90.8	95.6	90.4	102.2	84.1	81.2	75.7	90.7	96.6
				Oolong tea	A	119.8	130.9	120.3	128.5	103.0	105.6	109.5	141.0	102.3	106.9	106.7	109.1	112.1	96.4	89.9	95.2	104.8	92.7	94.8	108.9
					B	117.3	105.7	102.9	119.5	104.0	98.4	81.7	117.4	93.8	89.5	91.8	94.6	90.3	92.7	84.0	88.5	84.9	92.0	87.1	96.6
			GC–MS/MS	Green tea	A	121.5	109.6	99.6	139.5	96.8	91.6	113.1	107.4	108.6	88.5	105.8	69.5	94.1	81.9	98.3	81.3	79.4	84.0	101.5	98.5
					B	122.4	107.7	106.7	129.8	96.3	88.2	106.5	93.5	101.3	90.5	96.6	68.4	90.7	86.0	90.4	81.4	78.5	86.9	89.8	95.4
				Oolong tea	A	143.8	133.0	112.7	129.4	104.1	111.9	111.2	178.0	113.9	96.4	111.7	103.3	111.7	100.4	92.3	99.4	108.5	108.8	83.3	113.4
					B	132.8	114.1	96.8	116.8	102.4	94.4	83.3	140.4	96.2	94.4	91.0	92.5	88.7	93.0	82.8	86.8	88.4	97.9	80.1	98.6
		ENVI-Carb + PSA	GC–MS	Green tea	A	116.2	112.3	100.4	112.3	111.4	118.5	122.6	94.2	107.8	93.0	97.6	108.9	86.5	105.1	71.4	104.6	93.1	81.6	100.6	102.0
					B	119.0	99.1	105.2	108.7	114.5	108.8	102.3	72.2	105.2	94.4	91.3	107.1	83.9	91.9	77.1	92.6	89.6	83.1	100.8	97.2
				Oolong tea	A	111.1	105.5	114.4	112.6	110.1	105.4	73.1	87.0	99.4	98.3	94.2	92.7	85.4	94.0	90.2	82.5	79.2	81.7	76.9	94.4
					B	96.9	93.6	108.7	99.1	94.0	96.4	94.8	78.3	79.6	77.2	82.6	84.4	80.4	90.0	79.4	75.7	74.0	73.6	74.1	85.9
			GC–MS/MS	Green tea	A	161.8	116.7	58.2	57.9	122.7	126.5	110.5	90.6	96.2	82.6	109.7	103.5	107.2	101.8	87.9	101.5	89.3	88.2	96.0	100.5
					B	205.5	109.2	79.3	80.5	119.1	108.5	124.7	91.0	92.7	89.9	97.1	97.9	90.7	78.3	85.1	91.0	84.1	89.7	100.1	100.7
				Oolong tea	A	203.7	105.5	165.0	102.5	122.9	99.3	137.6	91.2	99.4	89.4	90.7	68.5	90.7	83.9	92.8	99.4	70.7	81.5	84.7	104.2
					B	127.5	96.2	100.8	90.5	102.4	91.2	98.1	75.8	79.6	77.9	95.2	63.4	85.5	81.6	73.9	86.8	74.2	70.2	83.3	87.0
92	Dioxacarb	Cleanert TPT	GC–MS	Green tea	A	342.8	273.1	343.4	447.7	294.4	248.9	318.9	277.6	303.0	274.6	303.8	248.9	244.5	229.6	271.9	224.0	262.7	245.8	298.9	287.1
					B	354.5	277.8	310.2	411.2	271.2	235.7	316.3	245.8	288.1	232.2	252.8	246.1	256.3	262.6	223.2	234.7	247.7	234.6	262.6	271.8
				Oolong tea	A	359.0	308.9	300.2	351.1	364.9	302.6	250.3	476.6	292.6	281.7	296.9	302.3	290.3	258.2	258.2	267.3	249.9	242.9	213.0	298.3
					B	364.3	296.4	290.0	312.9	342.9	288.6	222.9	520.2	265.2	250.6	235.7	253.5	253.3	241.7	238.2	230.5	221.0	256.6	201.7	278.3
			GC–MS/MS	Green tea	A	317.3	284.2	336.4	340.0	278.6	245.6	283.3	265.8	266.2	274.2	299.8	336.4	232.5	262.2	250.9	193.5	240.3	242.7	289.5	276.0
					B	340.8	290.4	303.5	325.9	257.7	232.7	311.4	235.6	259.7	230.7	254.2	328.0	225.5	303.6	194.8	204.2	226.9	249.6	260.5	265.0
				Oolong tea	A	360.9	365.4	306.2	371.5	296.8	288.1	259.6	705.7	279.6	272.8	292.0	284.5	277.2	264.2	226.4	259.2	239.0	219.5	171.4	302.1
					B	350.5	375.0	274.6	336.3	291.6	268.3	221.2	656.2	236.7	246.7	224.5	244.6	244.6	242.0	205.7	233.7	215.3	216.2	171.8	276.6
		ENVI-Carb + PSA	GC–MS	Green tea	A	414.9	327.5	299.2	341.5	310.4	303.6	379.1	276.3	303.5	273.4	264.6	304.3	280.7	274.2	231.7	283.4	274.0	240.0	248.8	296.4
					B	414.5	302.2	326.3	292.8	326.9	309.6	379.4	263.6	299.9	273.0	237.2	286.5	245.9	266.0	210.0	226.5	238.3	237.4	267.1	284.4
				Oolong tea	A	346.0	298.6	270.6	270.7	293.7	267.9	272.7	249.8	258.4	280.2	250.8	251.6	231.6	246.6	262.4	218.0	194.8	248.4	210.6	259.1
					B	322.5	267.2	287.5	285.7	271.1	255.2	250.8	244.1	241.5	195.9	257.7	215.9	226.9	238.3	205.4	206.0	202.1	186.7	211.0	239.5
			GC–MS/MS	Green tea	A	367.9	323.4	244.8	277.8	305.2	209.4	216.0	273.1	279.0	288.3	269.7	287.6	305.7	259.9	208.7	247.5	221.8	262.1	272.1	269.5
					B	382.9	303.8	224.6	206.7	316.0	309.9	238.0	254.8	279.7	286.8	226.7	271.1	259.8	192.3	179.8	209.3	198.9	265.5	273.3	257.9
				Oolong tea	A	360.1	306.4	270.6	268.5	281.7	258.1	219.6	236.1	258.4	263.4	238.9	223.1	243.8	215.0	248.1	259.1	184.1	245.8	214.6	252.4
					B	359.4	281.2	296.5	271.6	261.3	244.3	242.5	207.1	241.5	178.7	273.0	203.8	230.1	226.1	196.3	233.7	202.3	185.7	219.0	239.7
93	Endrin	Cleanert TPT	GC–MS	Green tea	A	534.0	426.7	441.1	512.2	388.5	348.6	413.8	385.8	393.3	347.6	389.8	359.8	369.1	338.3	380.8	314.0	293.6	324.9	402.5	387.6
					B	514.8	413.4	395.5	471.7	359.6	330.1	387.9	360.7	397.6	327.0	350.4	339.7	348.4	326.6	355.0	312.3	297.9	314.2	351.7	366.0
				Oolong tea	A	482.9	451.5	436.8	503.6	416.7	423.9	393.6	494.6	418.3	387.2	423.6	410.8	418.7	375.0	375.8	387.9	371.1	373.4	328.2	414.4
					B	469.5	416.5	392.2	463.2	411.4	381.1	316.9	452.5	364.6	347.9	344.8	343.4	376.0	361.1	343.8	358.3	317.7	382.8	315.5	376.8
			GC–MS/MS	Green tea	A	532.7	437.1	445.4	527.0	372.6	352.2	392.1	379.4	377.2	346.4	426.2	232.9	390.0	394.3	368.6	297.6	298.3	310.9	400.8	383.2
					B	512.5	431.5	398.9	489.2	349.5	328.0	395.7	354.3	378.0	331.5	373.2	217.6	335.7	376.2	341.1	291.7	291.9	315.6	362.4	361.8
				Oolong tea	A	493.6	461.8	438.7	512.1	387.5	417.0	387.7	499.9	439.7	396.9	429.8	427.6	431.8	398.4	359.5	379.0	364.8	358.9	326.9	416.4
					B	473.8	420.7	393.5	472.6	383.8	360.0	315.9	440.6	380.2	361.8	351.8	357.1	357.1	372.1	316.1	351.3	318.8	355.2	310.9	373.9
		ENVI-Carb + PSA	GC–MS	Green tea	A	536.2	508.5	432.0	422.5	449.9	470.4	985.1	372.9	363.0	370.5	352.7	394.3	370.5	405.3	331.5	401.3	343.7	345.8	359.1	432.4
					B	545.0	429.2	426.3	388.7	440.0	424.8	1065.0	348.1	357.1	366.6	330.1	367.9	346.1	359.7	303.0	332.0	319.3	321.6	361.4	412.2
				Oolong tea	A	500.9	407.9	435.2	396.1	452.5	394.4	378.5	341.9	389.8	373.5	366.6	356.4	354.9	375.0	386.7	349.1	311.3	336.7	309.8	379.9
					B	445.1	373.5	402.9	372.1	369.4	361.6	363.5	302.5	332.6	309.8	325.6	335.4	332.7	354.1	331.8	306.1	287.1	309.3	292.1	342.5
			GC–MS/MS	Green tea	A	578.7	492.4	444.0	440.9	460.3	428.8	405.6	365.6	362.7	371.9	387.8	346.5	396.6	378.1	339.3	412.3	317.5	359.2	354.4	402.2
					B	576.9	420.6	420.6	412.2	440.1	422.9	412.4	351.2	346.6	384.2	350.2	329.4	374.0	328.7	303.7	355.7	308.4	343.5	354.2	380.8
				Oolong tea	A	521.6	407.0	471.3	392.9	456.1	392.8	414.2	336.1	389.8	367.4	366.8	354.2	343.6	344.3	373.4	379.0	307.7	333.2	319.9	382.7
					B	456.5	370.1	389.1	359.9	383.8	361.9	373.8	297.1	332.6	316.0	366.0	338.5	308.9	313.4	306.6	351.3	302.7	289.0	315.4	343.8

(Continued)

Appendix Table 5.1 The 201 Pesticides Average Content for 19 Circulative Determinations on 3 Months Under 16 Factors (Two Cartridges, Two Instruments, Two Tea Samples, and Two Youden Pair Samples) (cont.)

No.	Pesticides	SPE	Method	Sample	Youden pair	Nov 9, 2009 (n=5)	Nov 14, 2009 (n=3)	Nov 19, 2009 (n=3)	Nov 24, 2009 (n=3)	Nov 29, 2009 (n=3)	Dec 4, 2009 (n=3)	Dec 9, 2009 (n=3)	Dec 14, 2009 (n=3)	Dec 14, 2009 (n=3)	Dec 19, 2009 (n=3)	Dec 24, 2009 (n=3)	Jan 3, 2010 (n=3)	Jan 8, 2010 (n=3)	Jan 13, 2010 (n=3)	Jan 18, 2010 (n=3)	Jan 23, 2010 (n=3)	Jan 28, 2010 (n=3)	Feb 2, 2010 (n=3)	Feb 7, 2010 (n=3)	Ave (μg/kg)
94	Epoxiconazole-2	Cleanert TPT	GC-MS	Green tea	A	401.2	336.4	344.3	393.9	293.9	238.6	301.5	276.7	337.7	271.5	210.8	240.6	246.8	227.3	251.9	210.9	201.7	204.5	255.4	276.1
					B	390.3	323.9	308.1	360.1	256.9	226.3	283.5	251.8	282.6	278.5	185.5	219.2	235.2	235.4	233.0	210.0	197.3	196.1	224.0	257.8
				Oolong tea	A	373.4	296.8	276.7	352.0	302.3	289.5	252.1	327.7	230.2	269.7	259.1	264.3	278.7	241.3	239.2	250.3	230.7	232.5	200.7	272.0
					B	346.2	312.8	277.3	331.7	303.4	268.5	237.0	330.0	193.8	213.4	222.1	232.0	252.7	227.8	212.7	216.5	194.4	215.1	181.3	251.0
			GC-MS/MS	Green tea	A	397.5	341.0	345.2	374.2	288.5	247.9	287.9	276.8	327.6	275.2	236.8	268.2	262.3	225.8	250.3	208.8	196.2	199.1	262.7	277.5
					B	386.2	327.4	310.5	356.0	257.0	231.1	290.7	250.8	285.6	270.8	238.9	255.1	230.0	248.3	228.6	206.4	192.9	199.0	232.0	263.0
				Oolong tea	A	381.0	338.0	305.9	357.4	274.6	289.9	249.8	390.4	291.1	286.6	252.7	265.7	272.3	237.2	231.7	242.0	229.7	225.6	200.9	280.1
					B	343.5	325.7	280.4	331.9	284.2	262.8	232.3	275.7	237.9	243.9	239.2	239.9	235.4	232.8	200.8	212.5	200.5	206.3	184.1	256.8
		ENVI-Carb + PSA	GC-MS	Green tea	A	392.9	377.2	336.3	331.7	312.5	330.8	295.7	267.1	249.7	263.7	304.3	294.9	259.5	269.2	226.3	277.5	244.5	231.7	240.9	289.8
					B	423.2	360.4	316.9	292.4	316.7	298.0	290.2	244.5	235.4	239.3	269.1	251.7	240.4	241.5	213.3	231.4	214.4	204.9	243.4	269.8
				Oolong tea	A	393.3	268.2	279.3	275.6	329.5	254.7	269.3	220.8	214.9	258.8	248.2	202.5	229.9	241.3	251.6	210.3	194.6	217.6	188.8	250.0
					B	385.0	251.7	287.9	275.7	270.9	232.1	231.8	189.6	176.2	223.8	188.4	198.5	207.3	228.6	205.3	182.3	180.2	191.2	180.5	225.6
			GC-MS/MS	Green tea	A	400.2	352.4	342.4	343.8	308.9	260.7	255.2	269.4	281.3	280.7	271.5	316.2	283.7	265.7	223.6	270.2	232.1	236.1	250.2	286.5
					B	416.1	319.7	344.5	325.4	307.7	310.2	269.3	248.8	284.8	250.5	262.9	297.1	250.1	224.5	205.8	226.3	202.8	219.8	253.0	272.6
				Oolong tea	A	366.3	283.2	276.6	272.6	311.5	250.7	263.9	219.4	229.6	258.8	245.5	222.4	194.4	202.2	234.6	212.0	181.8	215.0	191.3	245.4
					B	359.9	269.7	288.0	267.7	273.1	233.2	239.6	199.8	230.9	223.8	196.9	206.6	183.0	196.1	189.8	212.3	181.6	184.4	187.6	227.6
95	EPTC	Cleanert TPT	GC-MS	Green tea	A	101.6	76.6	86.5	123.0	86.9	75.8	87.5	81.9	64.8	89.2	75.7	61.3	60.7	60.3	59.1	47.9	50.2	55.2	75.7	74.7
					B	92.3	76.7	80.9	124.9	79.6	70.9	85.5	72.6	62.1	87.0	68.2	57.7	58.4	59.9	52.7	50.4	51.8	58.2	69.5	71.5
				Oolong tea	A	102.3	83.2	87.3	104.7	75.3	85.9	80.3	92.6	80.8	95.5	78.6	83.9	82.6	71.1	75.1	73.4	70.0	76.9	52.8	81.7
					B	89.2	70.6	78.3	85.3	66.9	70.4	70.1	81.0	63.2	77.0	70.4	65.2	70.0	62.6	61.2	63.1	59.3	66.3	44.2	69.2
			GC-MS/MS	Green tea	A	103.4	80.5	85.9	97.2	79.3	74.3	84.1	74.2	67.0	78.1	78.2	51.9	89.2	73.3	75.7	60.2	60.7	56.5	81.4	76.4
					B	94.1	79.5	79.9	97.2	74.7	67.6	81.7	67.7	72.8	79.6	68.3	51.3	78.4	74.0	71.1	60.2	61.0	64.9	72.0	73.5
				Oolong tea	A	95.5	78.0	81.3	100.1	69.2	77.8	74.8	84.8	74.1	89.0	71.4	71.6	74.1	78.7	67.4	70.4	63.2	73.8	58.4	76.8
					B	86.9	63.3	68.8	75.4	58.6	60.0	61.3	70.8	62.1	68.9	61.5	59.4	65.2	62.7	54.1	59.3	58.2	62.4	46.7	63.4
		ENVI-Carb + PSA	GC-MS	Green tea	A	106.9	100.7	89.7	92.4	101.8	96.9	105.0	91.2	68.0	76.2	90.4	91.1	56.5	53.0	51.6	70.6	60.0	77.9	81.3	82.2
					B	104.8	78.7	96.3	84.2	107.0	93.6	105.8	84.2	56.0	74.6	84.0	85.7	46.8	52.3	40.9	59.5	56.1	73.1	75.0	76.8
				Oolong tea	A	102.6	64.5	78.8	74.3	90.4	87.3	79.5	73.2	71.9	77.3	84.7	71.3	69.0	71.1	71.9	65.1	50.7	60.0	48.0	73.2
					B	78.2	53.9	81.5	67.2	70.2	74.9	68.4	60.8	63.5	60.0	61.7	60.5	57.7	62.6	54.4	47.2	42.5	54.6	44.0	61.2
			GC-MS/MS	Green tea	A	104.0	91.7	59.2	61.1	97.9	67.2	89.8	84.7	88.9	76.0	90.9	101.8	98.1	85.0	64.6	81.1	73.5	76.5	81.6	82.8
					B	102.6	69.0	59.6	54.3	102.8	89.3	80.4	80.8	68.9	72.4	91.2	96.9	81.5	69.7	51.6	69.6	69.2	76.2	82.0	77.3
				Oolong tea	A	114.9	63.8	103.7	63.7	82.0	78.9	66.4	67.3	66.4	77.3	75.4	62.7	67.0	68.7	74.9	70.4	56.6	58.5	55.5	72.3
					B	79.6	54.5	70.1	59.7	63.1	67.8	61.3	51.8	62.9	59.2	56.5	54.9	57.3	57.6	53.5	59.3	49.8	47.7	51.7	58.8
96	Ethumesate	Cleanert TPT	GC-MS	Green tea	A	99.5	82.3	86.2	102.6	63.1	86.2	79.4	94.6	72.5	87.3	78.5	68.6	73.2	66.6	68.9	61.2	59.5	56.9	76.6	77.2
					B	96.0	78.0	77.8	93.1	63.1	86.6	71.4	101.0	66.7	78.9	70.2	65.1	78.3	69.6	66.1	60.1	58.8	55.8	68.2	74.1
				Oolong tea	A	93.2	86.6	86.9	86.2	59.3	77.5	113.0	102.8	96.0	90.3	85.1	78.4	78.1	74.8	67.3	73.3	103.7	69.7	61.2	83.5
					B	88.8	76.9	75.6	80.1	60.6	70.3	113.4	97.6	74.7	78.1	72.4	71.5	68.5	69.5	63.2	68.6	110.7	69.1	60.2	77.9
			GC-MS/MS	Green tea	A	94.6	80.9	81.4	100.4	71.8	67.2	74.6	70.8	74.3	74.2	75.8	71.7	69.5	63.5	71.2	57.7	56.1	60.0	75.0	72.4
					B	91.8	78.8	73.5	92.8	67.8	62.1	73.3	65.1	71.1	69.6	68.4	68.2	62.5	61.1	65.9	57.1	56.6	60.8	67.7	69.0
				Oolong tea	A	90.3	85.4	81.0	91.3	71.4	74.8	65.3	76.5	79.7	79.2	72.6	76.8	71.1	71.0	65.9	69.1	62.7	65.9	57.3	73.6
					B	85.6	78.4	72.0	82.4	69.0	64.5	62.7	67.3	64.9	66.3	62.1	66.9	61.6	65.7	57.8	62.8	61.5	63.8	56.6	67.1
		ENVI-Carb + PSA	GC-MS	Green tea	A	102.1	93.6	83.9	84.6	86.2	88.4	61.3	100.5	70.9	85.4	86.9	75.5	68.8	79.5	63.9	74.2	73.1	68.4	67.5	80.5
					B	97.9	82.7	82.1	77.5	86.8	86.4	60.8	106.7	64.6	78.1	81.3	69.3	65.1	72.5	59.3	62.5	67.0	64.2	67.7	75.7
				Oolong tea	A	97.7	76.2	77.4	69.9	65.9	89.2	68.0	87.5	75.6	79.5	80.2	65.6	70.8	69.9	76.6	64.6	59.2	67.9	57.8	73.0
					B	85.4	69.5	75.6	63.3	61.5	77.8	92.6	34.0	69.8	61.2	59.6	60.3	65.3	66.9	64.3	56.9	56.4	59.5	57.4	65.0
			GC-MS/MS	Green tea	A	103.9	85.7	80.8	77.4	81.0	63.8	71.4	70.3	77.9	78.8	85.3	66.5	79.3	76.8	65.2	76.0	69.8	60.8	75.0	75.6
					B	99.4	75.6	79.1	72.8	77.8	76.5	74.2	66.6	65.3	75.3	83.7	64.3	72.8	65.1	58.9	65.4	63.4	67.0	78.9	71.6
				Oolong tea	A	93.5	73.6	82.6	67.2	81.5	68.2	70.5	61.4	62.8	72.1	71.5	65.3	68.9	58.9	68.3	69.1	52.8	60.3	55.9	68.6
					B	85.0	67.4	70.1	63.9	71.0	64.4	59.2	56.8	66.1	59.8	57.3	60.5	60.7	55.8	55.8	62.8	53.2	53.1	55.4	62.5

97	Ethoprophos	Cleanert TPT	GC-MS	Green tea	A	120.5	106.6	108.1	137.0	97.8	90.0	104.9	96.4	97.0	84.4	106.5	87.9	86.4	84.6	90.3	78.0	73.0	63.4	97.4	95.3
					B	125.4	104.7	99.3	127.5	90.8	85.2	99.3	89.1	102.8	79.9	94.9	83.0	82.2	87.5	85.7	76.6	71.7	65.3	88.2	91.5
				Oolong tea	A	121.8	112.4	106.3	125.2	104.3	98.9	93.5	115.6	99.6	96.9	99.7	98.0	94.0	85.9	88.2	87.5	83.6	88.9	80.9	99.0
					B	117.4	100.9	95.1	112.6	98.2	87.5	73.6	103.6	84.9	83.9	80.1	80.5	79.3	81.8	77.9	79.0	71.2	86.0	74.1	87.8
			GC-MS/MS	Green tea	A	123.4	107.4	105.3	130.8	95.1	88.1	102.6	97.3	95.6	83.2	105.4	74.2	91.8	79.2	89.8	75.2	73.1	80.3	96.9	94.5
					B	121.2	105.8	100.4	123.1	91.0	83.3	101.0	90.0	96.7	84.2	93.7	68.5	84.1	81.2	84.8	75.7	72.6	82.0	87.9	90.9
				Oolong tea	A	128.3	122.9	108.1	127.1	93.9	98.7	92.8	148.6	103.6	92.0	101.8	95.9	99.6	98.1	87.9	89.5	85.8	90.4	75.0	102.1
					B	118.9	108.3	95.0	114.1	89.8	84.0	75.9	129.9	86.4	84.5	82.6	81.0	85.6	85.5	76.2	82.7	76.5	83.3	70.0	90.0
		ENVI-Carb + PSA	GC-MS	Green tea	A	143.7	126.2	106.7	105.9	111.9	115.2	126.0	96.1	97.5	100.2	100.1	106.3	90.0	92.7	73.3	90.2	87.4	83.8	91.9	102.4
					B	136.1	113.4	107.8	94.9	112.7	105.9	111.5	86.0	94.7	95.0	89.0	97.8	83.1	88.1	70.5	78.0	81.5	82.2	91.5	95.8
				Oolong tea	A	119.0	94.5	102.9	97.4	110.5	94.7	85.9	79.8	90.7	89.7	85.9	79.9	80.4	85.9	87.7	79.5	73.3	81.6	73.7	89.1
					B	102.2	83.8	91.8	88.1	87.4	85.2	83.8	67.9	76.0	75.0	73.6	76.2	73.9	81.0	75.7	69.6	64.4	71.9	67.6	78.7
			GC-MS/MS	Green tea	A	143.3	122.0	107.5	105.4	114.5	118.3	108.5	91.8	94.7	97.8	100.7	111.4	111.5	100.0	84.2	96.0	90.6	88.5	94.9	104.3
					B	155.0	109.7	106.3	99.8	110.8	107.7	97.7	87.4	89.4	98.1	88.9	105.6	98.9	79.9	76.2	84.9	84.7	88.5	96.9	98.2
				Oolong tea	A	148.7	97.4	135.7	95.2	112.4	92.6	95.9	79.6	90.7	86.0	86.4	76.7	87.6	81.1	90.7	89.5	71.3	77.6	74.1	93.1
					B	109.7	86.4	90.9	87.2	91.3	85.1	84.9	68.1	76.0	73.5	86.0	72.4	78.1	73.4	72.6	82.7	69.9	67.2	69.9	80.3
98	Etrimfos	Cleanert TPT	GC-MS	Green tea	A	45.6	36.3	38.1	43.7	33.5	30.1	37.4	36.0	30.8	30.1	40.3	31.1	29.8	31.5	32.7	27.2	31.8	8.4	32.1	33.0
					B	46.5	35.3	34.4	40.4	31.1	28.8	35.1	33.0	36.1	28.0	33.6	29.5	28.8	32.9	32.7	26.6	27.8	8.4	28.9	31.5
				Oolong tea	A	43.0	36.3	36.4	42.6	35.6	33.8	32.7	42.1	33.5	34.6	34.5	31.1	34.9	31.7	30.6	30.4	30.6	30.2	27.3	34.3
					B	41.3	33.9	33.0	39.5	34.8	30.9	26.4	38.9	28.0	29.6	27.9	28.5	29.1	29.3	27.4	28.9	26.5	29.5	25.5	31.0
			GC-MS/MS	Green tea	A	42.1	36.4	35.9	46.0	33.1	31.5	35.4	33.4	33.8	29.5	35.2	25.2	30.3	28.8	31.7	26.3	25.6	26.1	33.9	32.6
					B	41.3	35.4	33.8	43.2	31.0	30.1	34.7	31.3	34.1	28.6	31.0	25.1	28.8	29.3	30.0	26.3	24.8	26.7	29.0	31.3
				Oolong tea	A	42.7	40.6	38.2	44.6	32.2	33.8	31.4	52.4	36.0	30.2	34.6	33.7	32.6	32.5	28.7	30.6	30.0	29.7	26.8	34.8
					B	39.9	36.1	32.9	40.7	31.6	29.3	25.4	45.0	31.2	29.9	28.6	30.1	27.8	29.6	24.8	28.0	27.4	28.7	24.0	31.1
		ENVI-Carb + PSA	GC-MS	Green tea	A	44.9	43.4	36.9	36.9	38.9	39.9	46.2	33.0	34.6	40.5	34.8	41.9	32.1	36.1	28.0	32.6	31.6	29.2	31.9	36.5
					B	42.9	37.3	36.3	33.9	38.4	35.9	48.9	28.6	32.3	37.4	31.3	36.5	29.9	34.0	27.2	28.3	28.9	27.6	31.3	34.1
				Oolong tea	A	43.5	33.8	36.3	34.4	36.5	31.8	28.8	27.5	31.0	31.2	31.0	27.9	28.0	31.1	31.3	27.2	26.3	27.1	25.1	31.0
					B	36.9	30.1	32.5	31.3	29.7	29.2	29.6	24.3	26.6	26.0	26.3	27.5	26.3	29.3	26.8	25.2	23.9	24.4	23.6	27.9
			GC-MS/MS	Green tea	A	48.7	40.3	36.6	34.9	39.1	58.8	37.5	31.0	32.5	32.4	36.4	36.0	41.1	36.0	28.9	33.9	28.8	31.0	31.6	36.6
					B	53.0	35.2	35.6	34.3	37.8	37.0	34.3	29.7	30.3	33.2	32.2	34.3	36.1	27.8	27.3	30.2	27.2	30.2	32.8	33.6
				Oolong tea	A	50.6	33.6	44.7	34.1	37.9	30.9	34.1	26.8	31.0	28.9	28.4	24.5	30.4	27.5	29.7	30.6	23.6	26.4	25.1	31.5
					B	37.3	30.2	30.1	31.2	31.8	28.7	30.6	24.1	26.6	25.4	29.6	23.7	27.4	25.2	24.5	28.0	23.6	23.1	24.4	27.6
99	Fenamidone	Cleanert TPT	GC-MS	Green tea	A	47.4	42.9	46.5	54.1	40.7	34.5	42.0	38.7	39.5	31.3	43.0	34.3	34.8	33.1	36.0	30.2	29.4	31.7	38.2	38.3
					B	46.4	42.7	41.3	50.7	37.1	34.1	40.0	35.2	39.3	30.5	36.9	32.5	33.3	33.6	32.6	30.3	28.8	27.9	33.9	36.2
				Oolong tea	A	46.1	39.7	40.2	47.4	41.2	40.2	32.7	43.0	38.6	37.3	36.1	36.7	38.4	34.6	34.8	35.1	33.1	34.6	30.4	37.9
					B	42.8	40.2	40.2	45.5	42.9	38.0	33.6	44.4	33.5	33.2	29.9	33.5	36.5	33.8	31.4	30.7	29.7	33.1	29.3	35.9
			GC-MS/MS	Green tea	A	49.0	44.0	46.8	50.5	39.5	34.9	39.6	38.6	40.3	34.2	41.0	34.2	40.2	36.0	37.4	28.8	28.4	29.1	38.6	38.5
					B	47.2	41.8	41.8	47.7	35.7	32.3	40.5	35.6	39.4	34.4	36.6	33.2	34.1	35.4	33.7	28.9	28.0	28.0	32.9	36.2
				Oolong tea	A	47.4	41.9	40.6	49.7	37.8	39.7	34.1	51.1	40.5	34.1	38.6	38.8	38.2	36.9	33.9	33.5	32.1	33.6	28.1	38.4
					B	44.3	40.7	39.0	46.6	40.7	36.4	34.7	51.9	34.9	34.2	33.0	36.8	33.6	34.8	30.9	29.5	32.9	32.9	26.5	36.4
		ENVI-Carb + PSA	GC-MS	Green tea	A	54.6	49.9	43.8	44.3	41.6	46.5	39.2	37.2	38.2	39.8	37.0	39.2	36.8	39.0	32.9	39.8	34.8	34.4	34.3	40.2
					B	55.4	45.4	42.7	37.9	42.1	42.7	40.9	34.6	34.5	37.3	34.8	36.0	34.2	35.8	30.8	33.6	30.8	30.2	34.3	37.6
				Oolong tea	A	49.9	36.4	37.5	36.3	43.7	36.2	37.1	30.4	35.0	35.4	31.9	31.1	32.6	34.6	37.9	31.0	29.5	33.1	29.9	35.2
					B	47.1	34.9	39.4	37.4	36.5	32.7	31.6	28.3	32.1	28.7	28.0	30.8	30.1	33.9	30.8	27.1	27.3	29.3	28.6	32.3
			GC-MS/MS	Green tea	A	50.4	48.4	43.3	41.7	41.3	40.6	36.0	37.7	37.4	38.3	38.7	38.7	45.5	35.9	32.5	39.3	33.8	35.7	35.5	39.5
					B	51.9	43.0	42.4	38.3	41.5	41.9	39.4	34.8	34.1	37.8	35.7	35.6	40.2	34.3	30.1	34.5	30.3	31.7	35.7	37.5
				Oolong tea	A	48.3	37.1	37.1	36.3	41.9	34.9	34.8	29.6	35.0	34.8	31.1	31.0	28.5	29.1	35.0	33.5	27.9	33.5	28.7	34.1
					B	47.5	36.6	39.6	36.3	37.0	32.5	33.3	27.7	32.1	28.8	32.6	28.6	27.4	28.2	27.1	29.5	27.4	30.2	28.4	32.1

(Continued)

Appendix Table 5.1 The 201 Pesticides Average Content for 19 Circulative Determinations on 3 Months Under 16 Factors (Two Cartridges, Two Instruments, Two Tea Samples, and Two Youden Pair Samples) (cont.)

Average content (μg/kg)

No.	Pesticides	SPE	Method	Sample	Youden pair	Nov 9, 2009 (n=5)	Nov 14, 2009 (n=3)	Nov 19, 2009 (n=3)	Nov 24, 2009 (n=3)	Nov 29, 2009 (n=3)	Dec 4, 2009 (n=3)	Dec 9, 2009 (n=3)	Dec 14, 2009 (n=3)	Dec 19, 2009 (n=3)	Dec 24, 2009 (n=3)	Dec 14, 2009 (n=3)	Jan 3, 2010 (n=3)	Jan 8, 2010 (n=3)	Jan 13, 2010 (n=3)	Jan 18, 2010 (n=3)	Jan 23, 2010 (n=3)	Jan 28, 2010 (n=3)	Feb 2, 2010 (n=3)	Feb 7, 2010 (n=3)	Ave (μg/kg)
100	Fenaromol	Cleanert TPT	GC-MS	Green tea	A	101.3	83.9	89.3	112.8	79.0	69.9	76.2	77.1	74.6	62.6	70.7	71.7	66.4	67.1	71.6	58.0	56.4	54.2	70.2	74.4
					B	96.4	81.1	82.8	105.8	70.8	66.7	69.7	67.0	70.2	58.2	62.0	57.3	65.1	74.5	67.8	61.7	56.9	53.8	61.6	70.0
				Oolong tea	A	96.1	77.9	78.4	94.4	77.5	78.3	63.4	84.4	71.3	67.4	68.6	70.1	75.6	66.8	66.9	65.1	61.1	71.2	61.6	73.5
					B	91.9	69.6	74.6	84.6	78.3	72.7	59.5	85.1	58.6	61.7	59.5	67.2	72.5	64.2	59.6	57.5	54.0	70.3	58.2	68.4
			GC-MS/MS	Green tea	A	97.0	87.9	93.0	94.9	77.4	62.0	73.9	70.8	69.3	63.5	79.4	70.9	65.8	61.4	66.2	53.5	52.1	49.8	70.5	71.5
					B	95.0	85.6	79.2	89.6	69.4	59.6	77.2	66.7	67.3	63.6	69.4	66.9	61.0	64.8	58.0	54.4	50.0	50.2	61.6	67.9
				Oolong tea	A	93.7	83.5	79.2	94.9	70.4	73.4	63.5	94.9	75.3	68.7	74.8	68.3	71.1	64.6	59.5	61.3	58.6	56.6	50.3	71.7
					B	88.0	82.0	73.4	87.9	73.3	67.3	63.7	101.8	63.7	65.5	61.9	60.6	64.0	61.0	51.6	54.5	51.7	50.0	46.8	66.8
		ENVI-Carb + PSA	GC-MS	Green tea	A	97.8	93.1	86.0	93.1	87.4	80.6	77.6	72.0	62.5	69.2	64.8	79.8	65.9	81.6	64.0	80.5	64.8	59.7	65.3	76.1
					B	101.0	87.2	86.5	76.8	92.8	71.6	71.9	65.3	58.2	65.8	77.1	65.6	68.9	65.4	60.5	64.2	59.2	55.0	70.3	71.8
				Oolong tea	A	105.8	72.2	67.2	71.4	75.7	70.5	68.2	62.1	67.4	63.9	59.2	51.2	64.5	67.4	67.2	56.0	59.3	69.3	56.6	67.1
					B	95.1	66.7	69.8	67.0	67.0	62.7	61.3	61.1	59.2	48.8	50.7	51.8	62.5	63.5	57.1	49.0	53.2	63.0	54.1	61.2
			GC-MS/MS	Green tea	A	97.8	96.6	84.5	83.2	78.0	75.7	65.6	70.1	73.4	73.9	70.7	69.7	71.2	68.6	58.6	69.0	59.6	61.9	61.1	73.1
					B	102.4	84.3	81.5	74.4	78.5	76.0	66.2	64.0	64.4	69.6	62.3	67.3	66.6	58.6	51.2	58.0	55.2	57.1	60.8	68.3
				Oolong tea	A	95.5	73.6	73.5	69.5	80.7	64.8	63.2	56.7	67.4	64.7	58.6	57.0	55.0	54.1	63.5	61.3	48.8	57.8	52.1	64.1
					B	92.1	70.9	74.9	69.8	69.3	61.8	60.8	52.2	59.2	52.6	58.9	54.6	47.2	50.1	48.8	54.5	48.5	52.0	49.2	59.3
101	Flamprop-isopropyl	Cleanert TPT	GC-MS	Green tea	A	45.8	38.7	40.7	45.6	35.3	32.4	38.2	35.4	37.4	30.5	36.1	32.4	32.5	32.0	34.6	28.9	28.2	29.2	36.3	35.3
					B	43.8	36.9	36.8	42.6	32.8	30.2	35.4	32.0	35.4	29.6	31.6	30.0	30.9	31.5	32.2	28.7	28.2	27.8	31.9	33.1
				Oolong tea	A	44.8	41.1	38.0	43.7	36.8	37.9	34.0	41.9	40.1	36.0	40.0	38.8	37.3	34.1	33.0	34.7	33.8	33.2	30.3	37.3
					B	42.9	35.5	35.2	40.4	36.8	35.8	31.1	39.9	34.1	33.3	30.9	32.1	32.6	31.9	30.1	31.4	28.4	32.7	28.5	33.9
			GC-MS/MS	Green tea	A	49.4	40.9	41.9	47.3	34.6	33.0	36.3	34.2	36.0	31.0	37.7	32.4	33.7	36.1	32.7	27.2	27.1	27.0	35.8	35.5
					B	46.8	39.3	36.8	43.5	32.2	30.8	36.3	31.3	34.4	30.8	36.6	31.4	29.6	33.4	30.1	27.4	27.2	27.9	32.2	33.6
				Oolong tea	A	39.4	40.8	39.6	45.7	34.9	37.0	33.8	42.3	39.2	34.8	37.5	36.7	35.3	34.1	31.6	34.4	32.7	33.6	29.4	36.5
					B	39.5	38.5	36.0	41.7	35.2	33.2	30.4	40.0	33.9	32.4	31.1	32.3	30.7	32.3	28.1	31.6	29.6	32.5	28.1	33.5
		ENVI-Carb + PSA	GC-MS	Green tea	A	52.3	45.6	40.0	39.6	39.1	40.8	47.5	34.1	32.9	32.6	32.2	36.2	33.4	37.3	32.1	39.6	31.0	29.2	32.5	37.5
					B	49.9	39.7	38.7	36.2	38.9	37.0	35.6	31.9	31.7	32.9	32.2	33.2	30.8	32.7	29.2	31.5	27.7	28.4	32.7	34.2
				Oolong tea	A	46.3	35.9	38.4	35.9	40.2	35.2	35.7	29.6	34.6	35.9	31.0	34.3	31.7	33.8	36.7	31.7	28.9	32.4	28.3	34.7
					B	41.7	33.6	38.1	34.1	34.0	31.7	32.9	28.1	30.3	28.8	30.2	30.7	29.3	31.9	30.2	27.9	26.4	27.3	26.4	31.2
			GC-MS/MS	Green tea	A	50.4	45.1	39.5	38.7	40.5	38.7	36.4	32.9	33.8	36.8	33.4	35.5	37.1	36.3	30.5	37.4	30.2	34.2	32.9	36.9
					B	47.4	37.3	38.4	36.3	39.1	37.3	34.9	30.9	32.4	36.1	31.2	34.2	33.0	28.2	27.7	30.3	27.4	34.5	33.6	34.2
				Oolong tea	A	45.9	36.3	38.9	34.7	39.8	33.9	34.2	29.5	34.6	33.0	31.6	31.5	30.4	30.4	34.7	34.7	26.4	30.2	29.3	33.7
					B	43.0	34.3	36.6	33.2	35.0	31.3	33.1	27.2	30.3	28.6	32.0	29.1	27.6	27.3	28.3	31.5	26.1	26.4	28.6	33.5
102	Flamprop-methyl	Cleanert TPT	GC-MS	Green tea	A	51.2	42.6	47.4	77.6	39.1	40.8	41.6	39.9	39.5	31.7	40.9	35.8	34.0	31.6	33.1	29.5	28.0	28.9	38.1	39.3
					B	49.8	42.4	44.2	71.5	38.9	37.0	42.7	35.3	39.5	35.8	39.7	35.9	32.5	36.7	32.2	29.4	27.4	28.5	33.8	38.2
				Oolong tea	A	44.2	37.8	43.5	45.8	44.7	42.5	35.7	42.1	37.5	33.8	33.8	34.0	39.7	34.6	35.0	35.9	33.9	35.0	30.8	38.1
					B	41.2	38.8	43.1	46.7	44.7	40.6	31.2	34.3	32.1	32.3	30.4	28.4	34.5	34.6	32.0	31.8	29.3	35.0	29.1	35.3
			GC-MS/MS	Green tea	A	49.7	42.6	41.7	49.3	45.8	40.6	36.8	34.4	34.3	31.9	36.2	37.8	34.2	33.9	32.6	27.9	25.9	28.4	37.3	36.0
					B	47.6	40.8	37.1	45.8	35.0	34.5	36.0	31.5	34.1	31.9	32.8	36.0	30.3	34.2	30.8	27.9	25.5	30.1	32.2	34.2
				Oolong tea	A	39.4	40.8	39.6	45.7	34.9	37.0	33.1	44.4	39.7	34.9	38.6	36.2	37.9	33.2	32.3	34.7	31.3	32.9	28.4	36.6
					B	39.5	38.5	36.0	41.7	35.2	33.2	30.0	40.7	34.7	33.3	32.0	31.4	32.2	31.8	28.0	31.5	28.6	32.9	27.0	33.6
		ENVI-Carb + PSA	GC-MS	Green tea	A	52.0	57.1	45.2	46.3	42.8	43.0	34.2	38.9	45.4	36.1	40.3	41.8	33.9	39.1	28.5	38.4	37.9	32.8	34.5	40.4
					B	49.2	48.6	49.3	40.9	45.7	42.0	34.2	36.1	40.1	34.8	33.1	38.1	31.4	35.0	27.3	30.1	32.7	31.7	36.0	37.7
				Oolong tea	A	47.2	37.9	42.2	38.8	44.1	41.6	26.8	29.3	34.6	31.8	35.3	34.9	33.3	36.7	37.6	32.9	29.0	32.5	29.7	35.6
					B	42.6	29.2	44.2	37.4	36.8	38.8	35.8	28.5	30.3	29.1	32.0	39.6	30.6	34.6	32.6	28.0	27.7	28.7	27.9	33.4
			GC-MS/MS	Green tea	A	52.0	44.7	44.2	40.6	41.5	36.7	40.9	33.5	36.9	34.9	34.9	32.1	37.2	34.6	34.6	36.1	34.2	33.0	37.5	37.5
					B	48.0	40.9	38.6	38.5	39.3	38.5	30.0	31.3	34.0	34.5	32.2	30.7	34.2	30.3	30.3	28.7	30.9	30.7	37.4	34.5
				Oolong tea	A	45.9	36.3	38.9	34.7	39.8	33.9	34.2	29.5	34.6	33.0	31.6	31.5	30.4	30.4	34.4	34.4	26.4	30.2	27.6	34.5
					B	43.0	34.3	36.6	33.2	35.0	31.3	33.1	27.2	30.3	28.6	32.0	29.1	27.6	27.3	26.9	31.7	26.1	26.4	28.2	30.9

No.	Compound	Cleanup	Detection	Tea	Rep	1	2	3	4	5	6	7	8	9	10	11	12	13	14	15	16	17	18	19	20
103	Fonofos	Cleanert TPT	GC–MS	Green tea	A	49.1	36.4	39.0	41.9	34.3	30.7	37.6	36.2	33.8	30.1	38.5	31.4	32.8	41.7	36.1	27.8	28.1	40.3	34.7	35.8
				Green tea	B	48.7	36.0	35.4	38.3	31.6	29.1	35.3	33.3	36.1	27.5	34.1	29.3	30.9	47.1	36.5	27.5	27.9	39.1	30.7	34.5
				Oolong tea	A	42.5	35.6	36.3	42.1	36.2	36.0	31.7	40.2	35.8	34.3	34.0	33.0	35.2	31.8	31.7	31.4	29.6	31.5	27.4	34.5
				Oolong tea	B	40.9	34.0	33.2	40.2	34.4	32.3	26.9	36.6	29.7	29.8	28.0	29.2	30.7	30.0	28.1	29.0	26.3	30.3	25.9	31.3
			GC–MS/MS	Green tea	A	44.8	37.7	39.5	46.2	34.4	31.5	36.0	34.2	35.6	31.6	36.9	36.8	31.1	29.9	32.1	26.4	26.0	28.4	33.9	34.4
				Green tea	B	43.4	36.3	35.4	42.9	31.6	29.4	35.9	31.8	34.6	29.1	32.8	35.1	30.0	29.6	29.7	26.3	25.4	28.2	30.3	32.5
				Oolong tea	A	39.3	38.2	38.3	45.5	33.0	35.1	31.4	41.8	37.2	33.0	35.2	34.1	33.1	33.1	31.1	30.4	29.4	30.5	26.5	34.5
				Oolong tea	B	38.2	33.9	33.5	41.2	32.8	30.9	26.2	37.9	32.8	29.9	29.2	29.5	28.7	30.7	27.0	28.5	27.3	29.2	24.6	31.2
		ENVI-Carb + PSA	GC–MS	Green tea	A	48.5	47.1	39.6	36.4	42.2	41.3	35.4	34.2	34.8	40.7	35.3	39.9	34.3	35.4	31.1	36.5	31.3	31.8	31.5	37.2
				Green tea	B	46.3	41.1	38.1	34.7	39.6	36.5	33.2	32.6	32.3	37.4	33.1	35.2	31.9	32.7	29.3	31.0	27.9	28.0	31.3	34.3
				Oolong tea	A	43.6	33.5	36.3	33.8	40.8	33.1	33.2	28.4	32.6	31.8	30.1	28.8	30.3	31.8	33.0	28.5	26.8	29.0	25.8	32.2
				Oolong tea	B	36.9	29.9	32.6	31.7	32.0	29.9	30.0	25.1	28.1	26.6	25.8	27.4	27.7	30.0	27.9	25.2	24.3	25.3	24.3	28.5
			GC–MS/MS	Green tea	A	48.6	42.4	37.6	36.3	40.3	47.0	36.9	33.5	33.9	36.4	34.7	35.2	38.4	36.2	31.7	35.1	30.2	32.8	33.1	36.8
				Green tea	B	47.1	35.4	37.5	34.7	38.9	37.7	35.3	31.6	31.2	35.6	31.3	34.8	34.4	29.4	28.6	30.9	28.3	30.0	32.8	34.0
				Oolong tea	A	42.3	34.1	42.3	33.4	38.5	32.2	32.3	27.4	32.6	30.5	31.2	29.2	31.0	25.0	32.0	30.4	25.0	26.9	25.5	31.8
				Oolong tea	B	35.4	30.5	32.1	31.4	31.5	29.7	30.8	24.9	28.1	26.5	32.6	27.5	28.2	26.1	25.7	28.5	24.8	23.5	25.2	28.5
104	Hexachloro-benzene	Cleanert TPT	GC–MS	Green tea	A	36.8	31.7	32.9	40.8	28.9	27.6	33.6	28.9	31.6	25.8	29.7	26.9	28.6	26.1	27.9	24.5	22.0	18.5	31.8	29.2
				Green tea	B	36.0	31.1	30.5	38.7	27.2	25.8	31.6	28.3	32.9	25.1	27.1	25.2	27.0	24.7	26.6	23.7	23.0	20.8	28.4	28.1
				Oolong tea	A	30.2	30.6	29.6	37.5	26.2	28.4	24.2	34.7	26.6	28.8	31.7	30.0	26.1	26.1	28.0	26.2	22.8	27.7	22.4	28.3
				Oolong tea	B	32.6	27.3	27.2	32.7	23.8	25.5	20.2	32.0	25.9	24.7	25.5	26.3	23.5	26.2	24.2	22.3	22.0	26.5	20.2	25.7
			GC–MS/MS	Green tea	A	35.6	31.6	32.6	39.3	28.3	28.3	31.9	28.1	29.9	27.0	29.5	32.1	28.1	26.2	29.0	23.5	21.5	18.7	32.6	29.1
				Green tea	B	35.0	31.3	31.0	36.5	27.5	26.4	31.5	28.0	30.8	25.4	28.4	28.7	26.5	24.5	27.5	22.1	22.4	21.6	28.6	28.1
				Oolong tea	A	31.7	31.5	31.9	38.1	24.8	28.0	24.2	38.2	27.9	26.8	31.8	29.7	25.8	25.9	26.9	24.5	21.9	27.0	22.1	28.4
				Oolong tea	B	32.7	27.9	28.0	32.2	22.6	24.6	19.8	33.9	27.0	24.9	25.2	26.6	23.1	26.4	23.6	20.8	22.0	26.3	19.5	25.6
		ENVI-Carb + PSA	GC–MS	Green tea	A	43.1	38.3	33.5	32.9	35.4	36.2	40.7	31.4	30.2	29.7	29.3	31.7	31.8	31.7	26.6	32.1	29.3	30.0	29.5	32.8
				Green tea	B	41.6	33.2	33.6	30.2	35.6	33.7	36.8	30.3	28.3	29.3	27.1	29.9	29.3	30.5	24.5	27.8	27.6	27.6	29.1	30.8
				Oolong tea	A	37.0	27.8	29.7	29.3	32.8	29.0	30.6	25.9	29.7	28.4	28.6	26.7	27.2	26.1	29.2	26.9	24.4	27.4	23.7	28.4
				Oolong tea	B	29.7	25.6	28.0	27.8	26.9	26.6	27.6	23.1	25.0	24.7	26.1	25.1	24.9	26.2	26.4	22.4	22.0	24.1	22.3	25.5
			GC–MS/MS	Green tea	A	41.8	38.0	33.8	33.3	36.8	34.4	34.1	30.8	29.8	30.9	30.3	32.1	34.3	33.1	27.7	32.2	29.1	32.0	29.8	32.9
				Green tea	B	40.0	32.2	33.8	30.2	35.5	33.1	34.7	28.7	28.0	30.5	27.9	31.1	30.8	29.6	24.5	28.4	25.7	30.1	29.6	30.8
				Oolong tea	A	36.2	28.1	36.5	29.1	31.9	29.5	28.9	25.2	29.7	28.5	28.2	26.4	28.4	25.2	29.0	24.8	22.9	26.1	23.2	28.3
				Oolong tea	B	29.9	25.4	28.2	27.2	26.5	27.4	26.6	22.3	25.0	24.5	27.8	25.3	25.8	24.2	24.6	21.9	22.4	22.4	22.3	25.2
105	Hexazinone	Cleanert TPT	GC–MS	Green tea	A	116.7	112.0	134.7	188.2	121.3	99.6	126.0	108.9	118.0	108.7	117.2	98.1	101.0	97.5	108.6	95.3	115.9	104.6	120.8	115.4
				Green tea	B	118.0	114.8	127.4	172.8	106.0	91.6	123.8	94.7	105.8	87.6	93.5	93.7	104.5	104.2	89.9	94.2	100.7	95.7	100.6	106.3
				Oolong tea	A	136.9	115.3	109.3	133.5	133.0	123.7	91.2	128.5	110.9	107.6	108.6	112.4	109.1	107.0	91.7	99.9	93.4	91.4	75.4	109.4
				Oolong tea	B	129.9	116.5	103.8	123.6	130.3	113.7	90.4	126.4	94.9	92.4	90.2	105.3	99.0	96.4	87.0	84.7	82.9	91.6	74.6	101.8
			GC–MS/MS	Green tea	A	126.9	114.7	136.4	132.5	113.2	106.1	117.0	111.8	112.3	109.1	115.2	110.0	109.7	92.7	107.3	86.4	109.0	99.1	117.9	112.0
				Green tea	B	130.9	113.7	124.9	129.8	104.8	97.3	125.8	97.6	102.4	98.6	93.2	113.3	103.7	114.1	89.1	89.2	98.7	93.1	102.0	106.4
				Oolong tea	A	137.6	120.7	121.0	136.2	116.0	118.5	92.2	193.1	111.8	102.0	111.2	112.8	108.7	102.2	88.5	92.8	92.7	89.1	87.2	112.3
				Oolong tea	B	127.1	115.5	105.2	119.2	117.0	107.7	88.8	175.6	100.1	94.8	94.8	106.6	91.4	89.8	78.3	79.9	84.0	88.2	73.9	102.0
		ENVI-Carb + PSA	GC–MS	Green tea	A	155.5	132.0	123.9	142.7	122.0	120.8	116.0	110.1	111.5	113.2	106.9	120.0	105.5	119.2	105.5	125.6	105.0	108.7	102.7	118.2
				Green tea	B	149.7	116.6	128.3	125.1	130.2	119.6	102.1	104.1	98.1	107.1	109.4	107.3	99.4	104.5	99.2	102.9	91.5	91.4	103.9	110.0
				Oolong tea	A	149.5	109.9	102.0	107.9	108.4	98.1	99.6	85.3	99.4	102.8	91.5	84.1	91.4	103.9	102.1	82.1	78.9	96.1	88.4	99.0
				Oolong tea	B	140.2	100.7	118.3	108.1	101.0	89.7	87.0	81.1	90.2	71.4	88.6	76.7	81.7	93.0	75.2	78.6	75.1	78.0	90.9	90.8
			GC–MS/MS	Green tea	A	150.5	123.1	121.0	134.1	122.4	75.1	97.8	107.0	125.2	101.1	107.1	104.7	120.4	126.9	104.3	128.9	100.2	108.0	111.6	114.2
				Green tea	B	151.5	104.6	121.3	127.0	127.7	118.4	88.3	100.9	104.6	98.3	109.3	99.4	117.4	88.4	100.8	107.8	87.3	95.2	110.3	108.3
				Oolong tea	A	134.2	114.9	96.2	101.7	107.3	95.5	84.5	87.4	99.4	98.7	90.4	82.8	81.2	84.6	95.5	93.0	73.8	94.2	82.8	94.6
				Oolong tea	B	133.8	105.9	115.3	102.6	105.0	90.1	93.2	85.7	90.2	69.9	98.6	73.0	73.5	84.9	68.5	80.0	75.6	72.8	86.6	89.7

Appendix Table 5.1 The 201 Pesticides Average Content for 19 Circulative Determinations on 3 Months Under 16 Factors (Two Cartridges, Two Instruments, Two Tea Samples, and Two Youden Pair Samples) (cont.)

No.	Pesticides	SPE	Method	Sample	Youden pair	Nov 9, 2009 (n=5)	Nov 14, 2009 (n=3)	Nov 19, 2009 (n=3)	Nov 24, 2009 (n=3)	Nov 29, 2009 (n=3)	Dec 4, 2009 (n=3)	Dec 9, 2009 (n=3)	Dec 14, 2009 (n=3)	Dec 19, 2009 (n=3)	Dec 24, 2009 (n=3)	Jan 3, 2010 (n=3)	Jan 8, 2010 (n=3)	Jan 13, 2010 (n=3)	Jan 18, 2010 (n=3)	Jan 23, 2010 (n=3)	Jan 28, 2010 (n=3)	Feb 2, 2010 (n=3)	Feb 7, 2010 (n=3)	Ave (µg/kg)
106	Iodofenphos	Cleanert TPT	GC-MS	Green tea	A	75.7	73.8	72.9	126.2	65.5	67.7	78.8	73.1	61.9	64.4	63.0	60.6	65.9	66.9	55.1	50.9	49.2	62.7	68.7
					B	77.2	72.3	68.6	121.3	63.6	62.7	74.3	69.6	74.8	60.0	63.1	62.9	62.6	69.1	55.6	53.6	53.8	56.5	67.4
				Oolong tea	A	79.7	89.5	82.3	82.8	79.1	64.1	65.6	93.3	67.9	68.5	69.2	80.5	68.0	64.0	63.3	70.8	60.5	60.3	72.9
					B	78.0	71.9	70.2	72.5	76.4	59.5	60.4	82.6	66.5	59.2	58.7	66.3	65.0	59.9	59.5	59.3	59.4	55.6	65.7
			GC-MS/MS	Green tea	A	81.9	69.9	63.3	92.2	59.9	68.1	81.0	66.4	73.5	57.6	50.1	59.1	62.4	63.3	53.8	50.8	56.3	64.3	66.1
					B	80.5	70.9	69.8	91.8	66.3	66.5	73.4	60.2	71.7	59.8	51.4	59.8	64.3	59.9	53.2	53.6	58.0	57.6	65.2
				Oolong tea	A	105.7	88.1	77.7	86.4	66.3	74.8	67.6	125.4	79.8	62.8	63.9	78.3	66.6	63.6	63.5	71.3	64.9	43.4	75.6
					B	91.4	84.0	65.9	78.8	68.9	61.3	60.6	96.7	67.3	64.2	60.7	67.8	63.0	58.0	56.1	56.3	59.5	53.9	67.6
		ENVI-Carb + PSA	GC-MS	Green tea	A	84.9	78.2	66.9	76.9	76.2	75.0	71.0	62.1	73.9	68.1	78.4	56.7	68.7	49.2	66.8	70.0	53.7	75.8	69.6
					B	86.6	71.5	65.2	73.6	76.1	71.7	72.8	58.9	71.5	66.5	78.0	53.5	64.0	53.3	56.8	63.8	55.0	75.6	67.1
				Oolong tea	A	72.8	65.7	74.8	70.9	74.4	67.2	53.4	59.7	67.0	62.8	61.1	51.6	68.0	62.8	58.0	53.3	52.2	51.9	62.7
					B	67.3	58.3	74.6	63.0	66.5	64.6	63.9	52.7	53.7	51.8	60.1	50.7	64.6	54.7	53.2	51.1	51.6	50.4	58.3
			GC-MS/MS	Green tea	A	106.4	79.7	81.7	82.2	86.4	95.9	90.4	56.9	75.9	56.7	76.3	63.4	82.7	51.8	69.5	58.1	62.0	71.7	74.6
					B	121.2	82.9	78.7	85.2	79.8	75.4	74.5	56.9	67.0	60.8	70.3	60.5	60.7	45.7	66.3	63.6	66.7	74.1	71.5
				Oolong tea	A	109.5	62.4	78.2	61.7	76.6	62.6	92.6	61.5	67.0	58.8	40.2	59.5	59.5	62.9	63.5	48.0	51.0	52.8	64.8
					B	86.2	59.3	64.3	56.9	66.8	60.9	67.4	51.5	53.7	54.3	38.2	55.0	55.7	47.7	56.1	56.3	45.9	52.8	57.4
107	Isofenphos	Cleanert TPT	GC-MS	Green tea	A	98.4	77.3	88.9	100.7	76.0	62.1	77.3	72.8	70.0	63.4	78.2	68.6	68.5	61.7	49.5	46.0	45.8	63.2	72.5
					B	98.3	75.8	77.9	90.4	66.0	60.5	71.0	68.5	78.3	55.7	71.8	65.9	58.9	60.6	51.0	46.7	45.9	56.2	67.9
				Oolong tea	A	89.7	76.0	73.4	101.9	91.9	80.9	74.8	99.3	84.1	85.6	77.8	75.4	58.8	60.0	58.5	56.9	56.9	51.6	75.8
					B	85.9	79.7	70.0	98.1	87.3	73.7	65.8	93.9	67.4	69.9	63.8	70.1	57.6	54.8	55.5	48.6	57.0	50.2	69.3
			GC-MS/MS	Green tea	A	96.3	77.0	77.1	93.3	69.2	61.3	68.9	64.7	66.9	60.1	54.1	58.3	61.6	61.4	51.3	49.6	50.8	66.9	66.0
					B	92.3	75.5	68.9	86.1	62.4	56.5	68.7	59.9	66.7	57.8	52.3	58.0	55.5	57.7	51.2	48.6	50.7	59.2	62.7
				Oolong tea	A	87.2	89.9	77.4	91.7	68.2	70.6	65.3	96.6	72.5	65.4	68.7	70.8	61.7	58.1	62.1	58.1	59.2	49.1	70.7
					B	83.9	83.6	69.3	84.8	67.4	62.5	55.6	88.8	60.3	60.4	59.5	58.6	60.5	51.2	59.6	51.9	57.4	49.2	64.3
		ENVI-Carb + PSA	GC-MS	Green tea	A	98.0	107.4	87.5	81.9	91.2	99.3	70.2	69.5	90.2	107.4	100.0	72.6	70.8	50.4	54.5	60.7	50.3	57.7	79.2
					B	100.7	93.6	83.0	70.9	86.0	86.5	69.6	63.1	93.1	101.1	83.8	64.8	68.8	43.9	47.9	55.0	55.3	58.0	73.6
				Oolong tea	A	98.4	72.2	87.1	78.9	98.9	73.7	71.7	65.3	67.4	70.1	67.3	65.2	58.4	64.4	51.9	48.8	54.9	49.0	69.1
					B	89.3	68.0	78.1	74.1	75.4	67.6	68.1	51.3	56.7	59.0	69.8	60.1	55.3	56.1	45.0	45.5	48.0	46.6	61.5
			GC-MS/MS	Green tea	A	99.1	92.0	76.7	73.6	80.2	85.1	72.0	64.1	66.5	65.6	59.6	67.5	70.4	54.9	65.0	62.3	59.2	61.4	70.7
					B	93.1	83.3	74.7	68.6	75.9	75.3	67.8	60.5	63.6	66.5	57.0	58.5	64.6	49.2	56.4	56.7	56.8	64.6	65.9
				Oolong tea	A	92.0	69.1	78.3	68.3	79.5	65.5	65.2	57.2	67.4	62.4	57.8	53.9	54.1	63.2	62.1	48.2	53.6	49.5	63.5
					B	85.8	63.5	67.8	64.6	66.7	60.9	62.7	49.4	56.7	54.2	55.3	50.6	49.7	51.2	59.6	47.8	46.0	47.6	57.9
108	Isopeopalin	Cleanert TPT	GC-MS	Green tea	A	104.5	75.4	76.5	97.5	64.6	56.0	69.3	65.9	63.5	61.1	66.9	66.6	60.6	67.3	54.7	49.7	52.2	71.0	69.3
					B	102.5	74.6	66.8	85.3	59.1	54.2	67.9	63.3	73.4	52.5	61.5	61.8	58.8	62.9	54.8	50.9	51.6	62.5	65.4
				Oolong tea	A	90.2	73.3	70.5	88.5	83.9	74.7	69.5	95.3	76.5	77.0	68.1	64.1	64.1	66.6	68.1	68.2	67.1	58.4	73.8
					B	86.0	73.5	66.4	84.5	77.7	67.1	52.5	87.2	59.3	64.5	55.9	56.7	62.9	59.4	63.5	57.3	65.8	57.0	66.4
			GC-MS/MS	Green tea	A	92.8	78.5	76.5	99.9	68.5	60.1	68.9	68.6	69.7	58.8	78.7	76.3	65.4	71.3	57.0	53.9	52.5	72.7	70.2
					B	92.3	77.1	68.2	90.0	61.8	56.3	70.2	62.0	69.8	56.0	77.3	67.3	60.8	63.6	56.1	55.6	52.9	65.0	66.8
				Oolong tea	A	90.4	78.1	73.7	94.1	69.1	76.4	68.6	107.1	75.8	71.7	71.2	59.4	64.3	66.1	68.4	66.0	65.3	54.4	74.2
					B	83.5	73.7	67.1	89.4	69.8	66.7	51.0	97.0	62.5	65.3	57.5	70.5	62.3	59.1	62.9	56.3	62.0	52.1	66.2
		ENVI-Carb + PSA	GC-MS	Green tea	A	95.7	94.1	76.2	71.5	80.6	87.7	85.2	63.5	64.7	86.1	82.8	68.0	68.8	51.1	64.3	62.0	62.4	64.5	73.9
					B	95.5	85.8	73.6	62.3	76.5	77.1	73.9	55.1	61.2	81.3	71.2	63.9	63.2	48.5	54.6	56.2	55.5	62.9	67.7
				Oolong tea	A	91.1	66.7	76.3	70.0	85.4	69.6	60.7	56.9	66.0	68.0	66.2	64.4	64.1	69.6	61.1	53.7	60.0	55.4	66.9
					B	85.3	60.6	67.3	66.5	63.0	63.2	60.4	47.6	56.5	57.4	64.2	56.7	62.9	64.2	50.8	49.3	54.3	50.9	59.6
			GC-MS/MS	Green tea	A	88.5	87.1	76.1	68.9	80.3	97.3	79.3	64.1	59.1	72.4	67.0	76.3	71.4	57.9	70.1	62.4	63.7	66.0	73.3
					B	100.4	82.3	73.8	65.2	75.7	79.2	59.4	60.6	57.0	74.1	65.2	67.3	66.1	51.9	58.3	58.3	62.8	66.8	67.9
				Oolong tea	A	96.7	67.7	84.3	72.3	84.4	69.9	60.3	55.2	66.0	65.0	64.2	65.7	62.1	68.3	68.4	52.5	58.0	50.9	67.0
					B	84.0	62.5	64.4	66.7	68.8	62.7	64.5	47.2	56.5	56.1	61.5	56.9	54.3	56.8	62.9	51.8	49.6	48.2	59.7

109	Methoprene	Cleanert TPT	GC–MS	Green tea	A	177.9	124.8	136.9	166.5	120.7	101.1	127.0	113.4	123.1	99.8	137.0	109.2	107.5	108.0	111.8	91.9	90.7	96.1	116.6	119.0
					B	172.4	126.2	121.9	150.7	110.0	96.6	117.1	106.1	121.8	89.8	117.0	102.1	100.9	104.4	102.8	90.5	88.7	93.5	103.2	111.4
				Oolong tea	A	163.3	131.7	135.1	152.4	357.1	125.4	106.9	140.6	120.3	119.7	114.5	125.8	114.3	106.3	103.9	109.9	100.2	104.6	92.2	132.8
					B	158.6	128.0	126.4	145.1	235.5	113.1	92.8	128.8	100.3	102.3	88.6	102.2	99.9	99.7	94.3	97.9	87.6	102.9	87.9	115.4
			GC–MS/MS	Green tea	A	173.8	131.2	138.0	160.1	113.8	106.9	125.7	114.7	125.3	110.4	112.6	130.5	103.8	107.1	114.1	86.7	87.0	84.9	114.4	117.9
					B	163.4	128.4	119.1	147.8	105.2	100.8	132.4	108.6	116.2	94.9	112.1	125.2	100.9	97.6	103.1	80.8	88.6	88.6	102.3	111.4
				Oolong tea	A	140.3	137.3	133.6	153.8	113.5	119.6	114.0	158.3	140.0	130.8	125.6	117.4	110.6	111.0	97.4	105.2	93.3	105.5	89.3	120.9
					B	136.2	122.3	123.1	141.7	116.2	106.3	99.2	150.6	117.7	118.1	103.4	97.6	95.6	103.7	88.5	98.1	84.7	107.2	87.0	110.4
		ENVI-Carb + PSA	GC–MS	Green tea	A	188.3	166.2	133.8	128.8	135.1	146.0	187.0	111.9	110.9	138.9	112.2	135.5	112.9	120.5	105.9	128.7	103.2	107.3	105.5	130.4
					B	175.3	139.1	131.1	120.2	132.8	125.8	139.0	106.9	105.8	131.5	105.4	115.2	105.6	109.2	97.6	103.8	94.2	94.5	104.5	117.8
				Oolong tea	A	183.7	127.1	138.0	127.2	394.1	121.4	441.9	487.4	122.1	116.9	102.1	94.2	106.4	104.7	115.5	96.0	91.0	96.9	89.3	166.1
					B	162.2	115.9	128.3	123.4	185.7	107.0	191.0	196.4	106.9	95.4	88.3	89.3	98.4	99.2	96.6	85.2	82.6	84.3	83.0	116.8
			GC–MS/MS	Green tea	A	181.3	177.5	131.9	210.3	136.2	119.2	126.4	122.9	119.3	127.7	114.4	179.4	114.8	116.7	104.2	119.1	95.6	109.8	110.2	132.5
					B	163.9	164.0	127.3	205.1	136.8	126.6	125.9	115.7	116.6	125.8	102.8	162.0	101.0	102.0	93.3	102.0	92.0	99.5	111.8	125.0
				Oolong tea	A	161.1	118.9	140.5	120.5	131.9	114.1	108.3	101.8	122.1	122.9	101.2	98.0	104.1	97.6	109.1	105.2	83.5	92.0	88.2	111.6
					B	144.0	116.3	116.0	116.6	121.9	106.0	105.0	93.5	106.9	101.9	104.9	92.8	95.0	89.3	88.5	98.1	82.0	79.6	84.8	102.3
110	Methoprotryne	Cleanert TPT	GC–MS	Green tea	A	148.0	113.0	137.4	139.3	116.5	91.7	113.1	105.2	99.3	76.2	112.3	83.7	92.7	89.3	95.8	98.1	77.2	73.8	96.3	102.1
					B	144.0	110.9	122.4	128.4	100.2	88.7	104.8	97.5	99.0	67.8	89.5	80.5	87.8	86.7	87.2	78.6	75.6	71.2	85.5	95.1
				Oolong tea	A	135.8	114.3	112.1	138.9	117.9	111.7	93.5	121.1	82.9	107.7	75.8	98.3	103.8	91.6	89.3	90.6	89.1	87.4	77.3	102.1
					B	127.4	119.0	107.7	131.2	121.0	105.1	89.2	116.8	61.7	89.1	67.8	88.4	91.6	83.0	79.2	78.3	74.6	83.2	71.4	94.0
			GC–MS/MS	Green tea	A	148.5	132.0	133.8	148.5	107.8	92.2	108.2	106.2	107.5	89.9	108.5	91.8	96.1	98.8	96.6	78.9	73.1	73.9	100.5	104.9
					B	141.3	123.1	118.3	138.8	95.4	87.1	110.9	98.4	100.2	91.8	98.3	89.2	87.8	95.6	87.8	77.3	72.9	74.4	87.9	98.8
				Oolong tea	A	118.9	120.6	118.5	138.7	103.9	108.1	94.7	135.1	112.3	99.5	108.0	102.3	98.0	95.6	87.6	94.7	88.0	85.2	80.5	104.8
					B	118.0	114.6	109.7	125.6	109.7	98.5	87.3	130.4	94.7	91.7	92.2	92.2	84.7	89.2	74.7	81.2	77.6	80.8	72.7	96.1
		ENVI-Carb + PSA	GC–MS	Green tea	A	167.7	155.6	132.3	128.0	127.2	142.1	104.5	101.2	95.3	103.4	85.1	93.4	95.3	89.2	90.4	108.4	87.7	88.7	87.0	110.4
					B	161.7	132.6	129.6	113.8	126.3	119.7	115.8	92.7	86.4	96.9	85.3	79.4	88.2	91.7	85.3	89.2	78.2	75.2	88.1	101.9
				Oolong tea	A	154.9	111.7	127.1	117.5	134.0	103.2	103.0	80.5	99.3	92.7	79.1	75.2	88.9	89.9	95.6	81.9	78.9	84.6	74.3	98.5
					B	138.1	104.7	127.4	111.8	109.8	93.3	88.9	73.9	86.0	76.1	63.5	76.3	80.9	83.5	76.6	71.9	68.9	72.4	69.1	88.1
			GC–MS/MS	Green tea	A	149.8	145.8	122.1	118.7	119.8	133.3	107.4	102.4	103.2	106.7	108.1	119.6	113.6	98.5	89.2	102.0	87.6	89.9	88.2	110.8
					B	145.1	151.6	118.0	109.2	119.8	113.8	104.5	93.0	93.7	102.7	95.7	105.9	99.0	81.4	81.1	86.8	77.5	81.6	88.0	102.5
				Oolong tea	A	143.4	109.3	113.5	103.9	120.6	99.3	96.5	82.6	99.3	94.3	87.1	86.2	81.4	81.9	91.8	94.7	74.0	80.7	76.4	95.6
					B	135.0	104.6	108.9	101.7	105.2	90.6	92.7	75.4	86.0	80.9	89.0	79.7	72.5	73.0	72.1	81.2	68.9	69.4	73.5	87.4
111	Methoxychlor	Cleanert TPT	GC–MS	Green tea	A	47.5	46.3	41.6	76.4	36.0	34.6	35.9	36.7	30.6	29.1	42.2	32.0	32.4	31.7	34.5	29.2	22.6	23.2	34.8	36.7
					B	45.7	43.5	35.8	72.2	34.4	31.2	36.0	35.8	39.3	28.6	34.5	29.3	31.0	31.9	37.5	30.7	24.9	27.2	31.4	35.8
				Oolong tea	A	39.4	50.7	38.2	56.1	43.3	33.1	35.2	46.9	32.0	32.5	39.2	33.1	39.3	30.9	32.1	34.3	38.2	29.2	28.1	37.5
					B	36.8	43.5	38.6	48.2	42.1	29.5	32.7	48.9	28.8	32.0	33.7	26.7	36.6	33.6	30.3	34.1	30.7	26.3	26.5	34.7
			GC–MS/MS	Green tea	A	47.4	43.5	31.2	51.6	34.5	31.6	39.9	35.5	35.4	29.8	44.3	47.0	40.7	30.2	31.9	29.4	25.9	30.2	37.7	36.7
					B	47.1	43.0	33.9	47.9	35.5	30.5	39.7	32.4	35.9	28.2	38.4	43.9	35.5	30.3	30.5	31.0	24.9	30.2	31.9	35.3
				Oolong tea	A	52.6	44.5	34.2	41.9	31.6	37.1	35.5	73.3	39.2	37.9	39.9	38.3	36.8	36.0	34.0	33.4	27.2	30.8	27.0	38.5
					B	45.3	42.4	34.3	37.4	31.7	29.5	34.6	70.9	33.2	34.4	31.9	36.5	33.9	33.5	28.9	31.9	28.1	28.7	28.2	35.6
		ENVI-Carb + PSA	GC–MS	Green tea	A	36.4	40.6	40.7	39.0	38.0	45.8	36.4	33.0	39.1	35.3	36.4	41.1	30.6	30.3	20.2	33.1	38.9	28.3	40.2	36.0
					B	43.1	47.8	39.3	32.5	37.7	46.6	38.6	24.5	36.5	34.7	30.4	40.1	30.3	32.3	20.4	26.7	34.0	28.3	41.1	35.0
				Oolong tea	A	42.8	22.0	33.6	28.5	54.7	34.5	25.0	28.7	45.5	25.6	27.3	32.4	28.0	30.9	30.8	30.8	23.4	23.1	23.6	31.3
					B	47.0	35.0	35.0	30.3	42.0	35.1	30.9	23.8	47.1	27.4	23.7	35.2	25.1	33.8	34.0	23.5	23.6	25.9	20.9	30.9
			GC–MS/MS	Green tea	A	51.3	75.7	41.0	37.1	41.2	25.5	45.9	36.7	39.1	42.7	44.3	32.6	44.2	33.0	31.8	36.2	39.4	36.3	20.9	30.9
					B	48.1	62.6	41.1	36.8	38.4	39.6	30.5	32.7	35.1	37.4	34.9	31.3	28.6	33.0	29.3	29.1	36.6	33.0	39.4	37.4
				Oolong tea	A	40.7	26.9	26.9	36.7	36.6	41.5	30.7	31.3	45.5	38.3	34.8	35.3	28.6	27.4	29.5	43.2	18.3	36.0	29.8	33.6
					B	43.9	25.1	32.3	36.9	35.3	35.1	31.1	29.9	47.1	30.3	35.9	33.9	26.3	27.2	29.1	38.0	18.2	29.3	30.4	32.4

(Continued)

Appendix Table 5.1 The 201 Pesticides Average Content for 19 Circulative Determinations on 3 Months Under 16 Factors (Two Cartridges, Two Instruments, Two Tea Samples, and Two Youden Pair Samples) (cont.)

Average content (μg/kg)

No.	Pesticides	Method	SPE	Sample	Youden pair	Nov 9, 2009 (n=5)	Nov 14, 2009 (n=3)	Nov 19, 2009 (n=3)	Nov 24, 2009 (n=3)	Nov 29, 2009 (n=3)	Dec 4, 2009 (n=3)	Dec 9, 2009 (n=3)	Dec 14, 2009 (n=3)	Dec 19, 2009 (n=3)	Dec 24, 2009 (n=3)	Dec 14, 2009 (n=3)	Jan 3, 2010 (n=3)	Jan 8, 2010 (n=3)	Jan 13, 2010 (n=3)	Jan 18, 2010 (n=3)	Jan 23, 2010 (n=3)	Jan 28, 2010 (n=3)	Feb 2, 2010 (n=3)	Feb 7, 2010 (n=3)	Ave (μg/kg)
112	Methyl-parathion	GC-MS	Cleanert TPT	Green tea	A	203.0	178.7	152.4	123.3	227.8	251.0	152.0	157.2	125.7	130.7	183.1	131.7	149.8	143.8	142.5	118.5	118.1	99.0	140.1	154.1
					B	215.9	170.9	136.5	143.4	214.5	233.0	150.1	144.1	175.7	120.0	151.8	135.7	154.7	141.4	165.6	123.8	129.2	97.6	130.1	154.4
				Oolong tea	A	178.7	174.8	240.3	307.8	425.0	220.5	158.9	189.3	133.9	144.3	165.1	127.7	178.9	132.1	133.3	145.4	157.6	121.9	115.8	181.7
					B	169.9	157.5	219.1	276.3	358.2	191.5	114.5	168.3	110.8	127.0	135.1	109.8	147.1	139.9	122.8	149.0	133.2	119.0	109.8	161.0
		GC-MS/MS		Green tea	A	215.3	164.4	128.9	214.9	141.3	134.0	154.3	155.2	159.6	118.7	187.4	186.7	121.9	125.5	129.7	127.5	104.8	123.3	151.8	149.8
					B	212.0	163.0	139.6	207.4	143.1	129.6	152.7	137.3	154.1	124.7	159.7	191.7	126.6	132.8	125.8	133.9	106.4	129.4	131.6	147.4
				Oolong tea	A	200.1	186.2	142.1	206.3	211.9	147.9	220.0	245.4	233.6	126.1	231.7	140.2	154.6	137.8	137.5	225.9	207.3	225.7	224.4	188.9
					B	230.2	183.4	132.2	209.1	241.7	124.5	216.6	142.0	232.3	133.8	223.3	121.1	132.4	131.2	125.7	230.1	211.7	228.1	212.1	186.1
		GC-MS	ENVI-Carb + PSA	Green tea	A	181.6	158.8	151.9	252.1	237.5	333.2	136.3	118.8	195.0	184.9	159.2	189.5	122.2	131.6	89.7	155.0	153.2	117.1	151.8	170.9
					B	198.2	161.7	146.8	216.5	334.2	301.8	129.8	121.1	183.4	166.0	123.9	187.2	120.2	133.7	97.3	128.0	130.8	107.8	155.5	160.3
				Oolong tea	A	153.2	119.8	233.3	231.4	251.1	238.5	113.0	121.1	127.9	124.5	122.8	123.5	105.4	132.1	137.8	104.5	114.1	108.7	101.4	151.0
					B	155.4	110.0	196.4	216.5	216.6	219.1	120.1	100.7	108.7	98.2	97.8	124.5	102.3	139.8	144.1	132.0	137.2	110.6	95.5	137.0
		GC-MS/MS		Green tea	A	227.1	185.0	165.1	166.6	182.0	243.2	184.1	129.7	179.9	137.9	181.9	161.7	151.2	160.3	112.3	119.5	134.7	149.8	171.1	166.2
					B	269.9	196.3	162.3	163.7	166.8	168.9	113.2	148.9	161.9	148.4	149.3	148.0	129.9	124.9	106.0	106.0	106.0	148.1	166.7	153.0
				Oolong tea	A	237.7	128.4	162.5	162.6	222.8	137.0	219.3	216.2	127.9	118.3	217.9	77.6	223.0	121.4	127.1	139.0	98.4	114.1	222.7	161.8
					B	197.2	123.1	128.5	176.4	225.6	129.4	215.4	215.0	108.7	111.4	234.2	79.6	213.8	116.7	108.3	125.8	95.3	105.0	203.2	153.3
113	Metolachlor	GC-MS	Cleanert TPT	Green tea	A	37.3	37.6	37.5	40.5	31.2	31.0	36.9	33.8	35.4	30.8	32.3	30.3	31.8	29.7	33.5	29.4	25.8	28.7	35.4	33.1
					B	37.5	36.4	33.3	38.2	29.6	27.9	33.8	31.5	35.4	29.2	27.5	28.9	30.8	27.6	30.6	28.7	26.7	29.5	31.5	31.3
				Oolong tea	A	43.0	42.4	38.6	42.2	33.5	36.6	33.2	49.1	39.6	39.0	38.6	42.5	41.0	37.7	35.6	36.2	37.0	30.2	26.7	38.0
					B	42.1	36.9	32.3	40.0	34.0	33.3	27.5	41.6	35.0	34.5	31.2	35.1	33.7	33.3	29.8	32.4	29.7	33.1	26.3	33.8
		GC-MS/MS		Green tea	A	42.8	36.2	35.5	45.9	32.1	29.9	34.8	34.0	34.0	29.0	36.9	26.8	31.4	28.6	31.4	26.2	26.6	27.3	34.1	32.8
					B	41.9	35.4	34.5	42.5	30.8	28.3	34.5	30.8	32.8	28.6	31.4	26.3	29.1	28.1	29.1	26.3	26.5	27.4	32.3	31.4
				Oolong tea	A	43.8	41.8	38.1	43.0	33.3	36.3	34.5	50.1	37.8	32.4	30.8	34.0	35.1	34.9	30.8	33.3	33.4	32.2	28.9	36.4
					B	40.8	36.5	33.3	39.0	33.2	31.3	26.6	42.9	32.9	30.8	30.2	29.9	30.8	31.0	27.1	29.6	28.2	30.3	27.1	32.1
		GC-MS	ENVI-Carb + PSA	Green tea	A	42.6	41.4	32.1	34.2	33.7	36.1	55.1	33.8	36.6	32.7	35.4	37.4	30.7	36.6	27.9	35.5	29.3	30.5	33.0	35.5
					B	40.5	35.6	32.6	32.7	33.6	32.4	37.3	31.3	36.1	37.0	32.5	34.6	31.9	33.9	28.6	30.3	28.2	30.5	32.9	33.1
				Oolong tea	A	42.8	32.5	35.3	34.9	37.7	35.1	30.6	31.4	33.6	28.9	35.6	33.6	33.1	34.9	36.8	32.5	29.1	28.7	26.2	33.8
					B	40.9	28.9	34.9	31.5	31.4	32.5	32.8	28.8	27.2	30.8	29.7	28.6	29.2	31.2	29.6	27.6	25.3	27.1	25.7	30.1
		GC-MS/MS		Green tea	A	52.2	39.9	35.9	34.4	38.8	36.6	36.0	31.0	31.3	32.1	35.7	47.8	35.7	33.9	30.0	34.9	29.4	30.7	31.3	35.6
					B	58.6	35.2	35.5	34.6	37.6	35.2	38.2	30.2	29.8	30.1	31.4	40.2	29.8	26.2	27.7	29.9	27.6	29.2	32.7	33.8
				Oolong tea	A	55.8	35.9	47.1	34.9	38.9	32.5	37.8	28.7	33.6	29.9	29.9	25.9	31.3	29.5	32.4	33.3	24.2	28.5	27.4	33.6
					B	40.8	32.3	32.5	31.3	33.1	30.1	31.6	25.3	27.2	25.8	30.9	24.6	27.7	26.3	26.0	29.6	26.0	24.3	26.4	29.0
114	Nitrapyrin	GC-MS	Cleanert TPT	Green tea	A	115.1	114.6	106.2	139.2	106.7	102.2	133.6	117.7	119.7	99.6	101.9	92.7	100.0	110.3	108.6	89.4	85.1	96.4	105.9	107.6
					B	114.4	113.6	96.7	142.7	103.1	92.7	123.1	113.7	122.0	89.9	93.4	89.9	103.1	110.8	108.1	85.2	88.8	100.2	100.2	105.2
				Oolong tea	A	94.6	129.3	109.0	133.9	95.6	91.7	102.3	122.6	96.3	93.1	110.5	100.6	104.0	106.1	100.4	95.6	92.4	85.3	95.3	103.1
					B	88.6	107.4	102.4	110.3	96.6	83.1	91.8	113.9	86.5	85.7	84.1	82.5	94.4	97.4	86.9	95.3	82.9	73.4	76.3	91.6
		GC-MS/MS		Green tea	A	117.3	110.9	81.2	132.9	94.7	90.5	107.3	101.9	100.1	92.7	108.0	98.3	88.9	88.3	96.2	80.1	78.6	83.3	101.5	97.5
					B	123.2	109.7	97.2	133.3	97.2	86.7	97.8	87.2	98.6	86.2	97.9	91.5	91.6	93.8	93.8	87.3	75.9	95.7	84.5	96.3
				Oolong tea	A	120.0	139.4	91.5	146.1	89.3	93.0	89.7	136.6	108.7	82.1	99.0	99.2	102.4	98.9	88.6	88.4	94.4	99.5	64.9	101.7
					B	108.3	122.9	87.3	120.5	73.7	69.6	75.6	99.3	89.8	84.1	81.8	81.9	92.2	84.7	77.0	79.5	80.3	79.3	60.6	86.8
		GC-MS	ENVI-Carb + PSA	Green tea	A	126.3	102.7	107.1	112.8	113.7	126.9	128.6	118.2	145.6	112.0	116.6	105.6	94.3	106.9	75.3	121.4	106.2	102.6	117.5	112.6
					B	133.8	107.5	108.6	95.6	117.6	120.3	106.9	101.1	125.4	110.4	106.4	103.6	87.5	101.4	81.0	103.2	103.0	100.6	116.8	105.8
				Oolong tea	A	102.4	67.9	93.4	90.9	108.9	100.5	100.4	85.2	76.1	79.1	78.7	90.5	68.9	104.9	89.2	85.2	79.7	70.6	66.3	86.3
					B	89.7	60.1	92.5	86.7	93.7	95.2	99.7	76.5	60.2	69.1	68.4	87.0	67.9	96.1	75.4	74.2	78.1	68.9	61.1	79.0
		GC-MS/MS		Green tea	A	130.9	103.9	106.3	115.8	142.4	114.8	118.8	101.3	117.6	100.1	119.0	124.5	99.0	92.1	83.1	86.5	100.6	93.1	117.2	108.8
					B	160.4	97.7	109.9	105.3	138.7	107.7	117.8	99.1	103.5	106.1	97.9	110.8	93.7	85.0	70.4	85.8	102.3	101.8	124.8	106.2
				Oolong tea	A	166.1	73.5	128.6	81.1	119.2	98.1	125.7	94.1	76.1	80.2	87.0	51.6	83.7	84.7	87.1	88.4	61.0	63.5	67.9	90.4
					B	106.2	64.2	83.1	74.7	97.3	94.3	79.4	74.5	60.2	74.3	86.0	50.0	78.4	84.2	79.3	79.5	62.2	58.8	66.4	76.5

115	Oxyfluorfen	Cleanert TPT	GC-MS	Green tea	A	201.4	182.2	361.2	170.2	149.5	121.1	142.2	144.4	120.9	114.4	205.5	143.2	153.3	125.4	150.8	120.7	99.5	96.3	157.8	155.8
					B	215.9	176.8	301.5	142.0	129.9	115.4	148.6	143.9	172.5	103.3	172.5	130.6	134.5	132.1	156.3	127.0	111.4	99.8	146.4	150.6
				Oolong tea	A	177.7	151.4	296.9	133.8	264.3	152.2	155.1	229.5	147.4	159.3	145.6	126.6	167.2	120.2	140.4	163.5	169.4	132.6	117.0	165.8
					B	168.9	174.7	197.2	145.5	213.8	139.4	119.7	228.2	112.6	136.0	125.1	106.0	154.0	134.4	128.6	164.8	130.8	134.0	119.2	149.1
			GC-MS/MS	Green tea	A	230.9	171.0	215.0	151.6	150.1	109.9	135.5	142.9	159.7	120.1	238.6	99.0	142.7	135.0	129.7	127.8	107.1	103.6	174.1	149.7
					B	220.5	166.2	196.6	148.4	126.9	105.6	145.2	135.1	165.0	113.3	206.0	95.3	120.3	137.7	123.4	126.9	111.2	105.6	159.9	142.6
				Oolong tea	A	254.1	191.5	205.3	132.0	155.5	157.5	141.8	519.0	151.6	142.9	172.4	141.3	154.0	133.7	144.7	161.5	159.4	131.7	121.1	177.4
					B	200.8	189.6	224.2	142.2	159.1	142.3	121.7	455.7	115.9	146.7	131.8	116.6	136.2	126.5	127.4	133.1	127.4	119.1	136.7	160.7
		ENVI-Carb + PSA	GC-MS	Green tea	A	156.8	189.2	163.4	177.8	163.5	224.2	138.1	141.7	173.0	182.5	159.3	183.0	143.8	126.6	78.0	126.2	172.4	123.4	140.5	156.0
					B	204.8	220.6	117.6	168.4	158.0	210.2	149.8	105.5	157.7	173.8	132.1	166.4	145.0	136.3	73.4	106.1	144.8	115.1	147.8	149.1
				Oolong tea	A	174.3	109.2	146.5	161.2	229.0	154.5	96.6	124.8	130.3	132.8	114.7	128.5	127.8	120.2	145.1	134.0	113.7	121.1	111.2	135.6
					B	199.5	112.5	154.3	150.1	163.2	147.4	109.4	106.4	121.9	119.5	89.5	143.3	113.1	134.4	168.3	94.9	106.7	116.2	102.0	129.1
			GC-MS/MS	Green tea	A	206.7	215.0	169.3	165.9	167.0	279.4	178.8	129.7	162.6	166.6	188.1	145.0	166.2	144.9	104.4	131.9	117.7	142.2	162.7	165.5
					B	279.0	261.8	143.0	159.4	157.1	190.5	76.3	122.4	154.6	176.2	146.8	139.4	151.9	139.0	98.2	115.8	107.8	153.6	173.9	155.1
				Oolong tea	A	220.4	110.5	158.3	161.6	200.9	147.2	117.3	114.5	130.3	126.9	115.8	104.3	122.4	130.1	129.1	161.5	96.1	107.8	108.4	134.9
					B	196.8	113.8	158.1	139.0	144.6	133.0	118.9	85.6	121.9	126.0	108.0	116.8	103.1	114.2	124.7	133.1	99.7	104.8	103.9	123.5
116	Pendimethalin	Cleanert TPT	GC-MS	Green tea	A	199.3	152.4	337.6	148.4	128.5	112.5	125.6	117.6	94.4	102.2	192.7	108.0	128.0	108.8	126.1	108.5	70.8	78.5	131.9	133.3
					B	201.9	151.1	285.9	129.0	121.8	107.7	131.3	124.0	158.9	87.3	164.4	108.4	115.9	112.7	134.3	110.1	80.9	86.8	120.2	150.0
				Oolong tea	A	170.0	149.8	183.0	129.3	320.6	134.2	145.3	197.8	135.4	136.9	127.0	124.9	146.4	104.5	128.3	164.3	129.5	120.7	102.3	134.4
					B	161.9	152.0	174.9	127.5	242.6	116.6	110.2	193.2	98.2	113.5	102.4	95.9	130.7	119.7	113.3	169.5	107.7	120.0	104.2	132.0
			GC-MS/MS	Green tea	A	199.1	152.1	208.1	134.5	132.5	110.9	127.6	128.1	126.3	104.3	179.0	125.4	119.2	121.5	125.2	102.3	84.9	95.1	131.0	127.9
					B	195.6	151.1	188.8	127.9	117.9	106.8	129.4	120.1	144.5	98.8	159.6	122.6	110.3	120.0	114.9	107.3	93.4	99.9	121.8	150.2
				Oolong tea	A	216.6	187.7	181.0	134.6	130.9	135.2	133.5	310.1	140.3	125.3	153.7	124.8	144.3	125.4	121.2	134.8	126.9	125.3	102.5	134.6
					B	185.3	185.2	177.3	125.0	128.9	116.1	98.9	288.1	100.8	122.1	118.9	100.1	122.1	124.4	107.1	122.2	107.8	114.4	113.1	143.4
		ENVI-Carb + PSA	GC-MS	Green tea	A	153.3	172.5	146.0	151.6	155.6	197.2	163.8	121.3	166.3	172.9	152.1	179.6	131.1	102.2	59.4	96.1	168.7	102.7	132.6	133.9
					B	170.7	177.9	107.1	146.5	210.1	185.4	153.2	94.5	164.9	163.2	119.4	153.1	122.6	124.2	54.2	80.7	140.7	107.7	133.3	122.9
				Oolong tea	A	161.4	116.9	132.5	145.9	133.5	135.9	97.6	117.1	122.7	117.8	108.6	114.6	116.5	104.5	115.3	117.9	93.1	107.0	99.2	111.4
					B	158.0	108.7	129.4	126.2	158.5	128.5	103.3	86.2	105.6	102.3	84.9	124.1	103.0	119.7	140.5	77.5	88.5	99.1	89.8	143.4
			GC-MS/MS	Green tea	A	198.0	187.9	143.0	147.0	144.9	168.2	168.2	114.1	136.3	132.8	159.2	131.7	161.9	122.6	103.1	119.1	124.3	118.0	131.0	139.1
					B	231.2	265.5	130.9	145.1	167.5	165.9	62.1	107.3	139.1	139.9	128.9	131.6	145.8	116.7	93.2	103.4	113.8	124.8	152.4	123.4
				Oolong tea	A	212.8	119.7	130.2	163.7	126.5	128.5	110.8	109.4	122.7	108.1	106.8	102.6	109.6	113.4	120.6	134.8	91.5	98.3	93.1	110.2
					B	177.0	113.2	122.7	121.9	115.0	117.2	115.0	81.3	105.6	100.3	102.8	114.9	97.3	96.6	109.1	122.2	91.4	91.5	86.7	110.2
117	Picoxystrobin	Cleanert TPT	GC-MS	Green tea	A	95.2	77.7	91.9	85.9	71.6	64.8	79.1	68.6	74.4	61.4	75.3	66.5	67.0	60.8	68.0	57.5	57.8	56.5	73.8	71.3
					B	91.5	74.5	84.1	76.5	65.3	60.9	72.5	63.5	69.7	59.1	66.0	62.3	63.0	62.5	61.3	57.9	56.5	54.5	65.0	66.7
				Oolong tea	A	86.7	81.5	97.4	78.0	69.3	76.4	70.3	83.3	75.1	70.8	75.5	73.6	75.8	63.2	66.6	68.7	67.0	66.8	59.1	73.9
					B	83.3	75.9	88.2	72.0	70.4	70.0	63.7	78.1	66.8	65.6	62.8	65.1	66.3	62.8	60.7	62.0	57.9	67.2	58.0	68.3
			GC-MS/MS	Green tea	A	97.0	82.3	96.3	84.4	70.7	64.1	74.9	70.7	71.7	61.9	75.2	46.9	70.0	71.6	68.8	54.6	55.2	56.0	70.8	70.8
					B	93.5	79.0	88.0	73.6	65.2	60.5	73.4	64.1	70.4	61.7	66.8	45.5	60.2	70.3	62.7	54.6	53.6	54.9	73.5	66.5
				Oolong tea	A	79.2	78.7	91.8	80.3	70.5	75.0	66.8	88.5	76.4	68.7	74.8	76.0	73.6	69.3	64.6	71.0	65.7	64.8	60.2	73.5
					B	79.7	76.1	84.1	72.9	71.2	67.6	59.8	82.9	68.0	64.7	62.2	68.0	58.6	64.4	57.0	64.4	59.2	65.1	61.1	67.7
		ENVI-Carb + PSA	GC-MS	Green tea	A	107.0	92.9	77.9	82.0	82.1	84.8	60.7	69.5	69.7	70.3	67.6	76.0	69.2	76.3	64.4	78.4	65.8	67.2	66.6	75.2
					B	100.8	79.9	70.6	78.3	80.6	76.7	57.6	66.1	66.7	68.9	63.8	68.6	63.8	67.1	58.7	63.1	58.5	57.9	67.1	69.2
				Oolong tea	A	89.6	76.0	70.8	75.3	79.3	70.0	71.6	59.1	71.5	67.6	65.3	64.7	64.0	61.8	71.9	61.5	57.7	64.1	56.6	68.3
					B	81.3	71.9	67.9	75.0	67.5	64.6	65.7	55.2	60.2	57.2	60.0	60.7	59.8	61.9	60.0	55.4	53.9	55.5	54.4	62.5
			GC-MS/MS	Green tea	A	99.5	89.7	79.4	81.2	80.3	92.4	70.7	68.6	70.7	71.9	72.4	61.9	81.7	73.6	61.5	74.5	62.6	68.9	69.3	75.3
					B	94.4	75.6	73.4	78.8	78.1	75.8	69.5	64.2	66.8	72.3	67.4	60.4	74.5	62.0	54.7	61.6	55.2	62.5	71.4	69.4
				Oolong tea	A	92.4	73.6	68.7	81.1	80.5	70.3	68.0	60.1	71.5	64.7	63.2	64.7	60.0	61.4	66.4	71.0	52.8	59.9	57.6	67.8
					B	84.3	69.8	67.1	72.4	70.5	64.9	66.7	56.1	60.2	55.4	66.1	62.4	56.8	56.5	53.8	64.4	53.7	53.1	55.1	62.6

(Continued)

Appendix Table 5.1 The 201 Pesticides Average Content for 19 Circulative Determinations on 3 Months Under 16 Factors (Two Cartridges, Two Instruments, Two Tea Samples, and Two Youden Pair Samples) (cont.)

No.	Pesticides	SPE	Method	Sample	Youden pair	Nov 9, 2009 (n = 5)	Nov 14, 2009 (n = 3)	Nov 19, 2009 (n = 3)	Nov 24, 2009 (n = 3)	Nov 29, 2009 (n = 3)	Dec 4, 2009 (n = 3)	Dec 9, 2009 (n = 3)	Dec 14, 2009 (n = 3)	Dec 19, 2009 (n = 3)	Dec 24, 2009 (n = 3)	Jan 3, 2010 (n = 3)	Jan 8, 2010 (n = 3)	Jan 13, 2010 (n = 3)	Jan 18, 2010 (n = 3)	Jan 23, 2010 (n = 3)	Jan 28, 2010 (n = 3)	Feb 2, 2010 (n = 3)	Feb 7, 2010 (n = 3)	Ave (µg/kg) (n = 3)
118	Piperophos	Cleanert TPT	GC-MS	Green tea	A	114.7	123.3	125.7	195.4	108.2	88.8	109.5	113.5	98.1	82.6	97.0	101.5	97.7	101.1	83.4	71.0	72.0	98.3	106.2
					B	118.2	117.2	112.3	183.5	91.3	87.7	106.7	109.2	118.6	71.2	97.6	96.9	98.0	98.0	84.9	76.2	78.1	91.0	102.4
				Oolong tea	A	127.7	126.2	108.4	143.6	143.8	106.8	108.5	154.0	107.7	104.8	93.9	115.2	92.0	92.4	98.8	104.1	94.7	86.0	110.9
					B	121.9	114.5	112.2	137.9	136.8	101.0	102.2	141.5	92.9	88.4	86.9	102.9	94.3	86.2	88.7	82.2	90.7	79.9	102.1
			GC-MS/MS	Green tea	A	143.6	125.1	107.3	149.7	104.2	86.9	112.0	106.8	103.9	89.0	123.6	88.8	95.1	96.0	85.8	75.4	76.6	118.8	106.3
					B	140.0	121.2	108.0	141.1	99.4	85.4	106.9	94.5	104.3	86.9	123.8	91.0	100.0	90.9	87.0	76.8	79.2	104.6	102.7
				Oolong tea	A	186.2	151.7	105.9	148.4	107.2	111.3	106.4	335.6	103.8	100.9	97.2	110.8	99.4	93.5	97.3	103.9	86.9	73.8	123.3
					B	162.6	160.6	106.6	139.0	107.4	96.5	91.0	307.6	72.1	101.7	92.8	94.5	90.0	81.1	84.9	79.0	77.6	72.6	111.6
		ENVI-Carb + PSA	GC-MS	Green tea	A	122.0	138.4	117.3	117.1	113.2	127.4	140.3	103.5	117.5	120.6	123.4	100.4	97.2	69.1	84.5	108.8	81.6	104.1	110.8
					B	138.7	138.1	114.4	103.0	116.1	119.9	123.7	83.5	112.2	111.1	113.6	88.1	95.5	62.5	74.5	99.1	88.9	106.0	104.6
				Oolong tea	A	137.0	87.6	120.9	114.8	142.5	106.3	82.7	87.0	98.2	97.7	86.2	90.6	92.0	95.8	84.4	77.6	85.6	80.1	97.7
					B	137.4	85.9	115.1	111.8	105.9	96.6	88.7	68.7	83.0	82.1	103.1	81.8	94.3	85.6	71.4	70.0	80.0	75.1	89.7
			GC-MS/MS	Green tea	A	158.3	124.2	117.0	112.7	119.9	142.5	121.5	92.4	111.7	92.5	114.9	97.0	105.2	83.3	93.5	77.0	90.0	99.8	109.6
					B	205.2	139.3	114.1	112.3	113.2	115.9	95.3	91.6	98.0	99.8	100.6	84.4	81.8	82.4	84.6	80.8	89.9	107.7	105.4
				Oolong tea	A	160.8	97.6	120.0	105.8	137.4	101.8	115.9	87.5	98.2	92.9	72.6	90.2	82.3	93.1	97.3	77.0	79.7	80.5	99.2
					B	134.6	97.4	101.2	98.9	107.8	95.1	93.7	66.2	83.0	78.7	72.2	82.6	75.5	79.9	84.9	73.3	76.1	68.2	87.9
119	Pirimiphoe-ethyl	Cleanert TPT	GC-MS	Green tea	A	93.4	68.2	71.9	79.8	60.2	50.0	72.3	66.0	65.7	55.5	63.5	62.5	59.8	63.2	53.2	52.3	52.5	67.1	65.5
					B	91.9	67.2	63.8	71.6	54.7	48.2	66.8	61.3	69.0	48.6	58.6	58.9	56.8	57.8	52.6	51.1	51.2	59.7	61.2
				Oolong tea	A	85.6	64.6	63.9	78.3	62.5	66.9	64.2	82.2	70.0	72.8	65.7	70.1	64.4	63.0	63.2	60.8	61.7	54.5	67.5
					B	81.5	67.0	61.2	75.9	61.5	62.0	52.8	74.1	54.8	59.9	58.4	59.0	59.9	56.4	57.0	52.5	60.3	51.9	61.0
			GC-MS/MS	Green tea	A	84.6	70.3	73.7	86.7	63.2	56.7	65.0	61.5	65.8	56.9	42.9	60.7	57.3	61.6	48.5	46.1	49.9	64.1	62.0
					B	83.5	69.5	64.1	79.8	57.6	54.1	65.0	57.8	62.9	52.0	40.7	55.5	53.7	57.7	47.7	46.2	50.9	55.7	58.6
				Oolong tea	A	75.3	72.1	74.9	83.0	62.7	68.0	59.9	89.4	72.1	61.2	66.7	64.6	62.2	56.8	59.9	57.7	57.9	48.7	66.6
					B	72.8	69.1	62.9	77.4	63.3	58.6	49.4	79.0	62.4	56.4	57.9	53.0	56.8	50.6	53.4	51.1	55.5	47.4	59.7
		ENVI-Carb + PSA	GC-MS	Green tea	A	98.0	88.3	70.7	67.2	73.7	74.0	68.4	65.3	63.3	85.1	84.1	65.1	70.2	60.3	71.9	59.6	61.2	61.0	71.4
					B	92.4	73.9	68.7	61.5	70.5	63.7	62.6	60.8	60.3	78.7	68.7	60.4	63.5	56.7	59.4	54.3	54.1	60.7	65.1
				Oolong tea	A	89.2	63.6	68.4	63.0	72.2	65.0	66.5	55.4	61.7	65.5	56.2	60.0	64.4	67.8	57.8	53.6	52.5	52.5	63.2
					B	79.6	58.1	61.3	60.3	55.7	58.6	60.3	47.9	51.6	53.0	55.0	55.6	59.9	56.4	51.4	48.3	58.1	49.4	55.8
			GC-MS/MS	Green tea	A	94.9	80.8	71.6	66.2	73.9	97.5	68.4	60.8	58.6	61.9	74.3	73.6	65.2	59.4	67.8	55.6	60.8	60.3	69.2
					B	90.6	66.5	69.0	62.4	71.0	69.5	65.6	59.1	56.5	64.1	71.1	65.3	54.8	54.6	55.8	52.4	56.5	57.8	63.3
				Oolong tea	A	80.9	68.6	76.1	65.8	75.3	63.5	64.0	53.6	61.7	58.4	58.3	56.5	53.2	63.6	59.9	46.7	52.4	49.4	61.4
					B	72.5	62.5	63.0	60.3	63.3	58.4	58.5	47.3	51.6	50.2	54.0	51.9	49.1	51.6	53.4	45.3	46.5	50.1	55.3
120	Pirimiphos-methyl	Cleanert TPT	GC-MS	Green tea	A	45.8	36.2	38.1	43.6	31.2	30.5	37.3	34.1	32.3	30.4	33.0	30.7	31.1	31.5	26.7	26.2	20.7	33.4	33.5
					B	45.8	35.4	34.3	40.0	31.2	29.1	34.2	31.5	35.9	27.8	30.3	28.9	31.7	32.2	26.3	26.8	20.5	29.8	31.9
				Oolong tea	A	42.9	37.1	36.3	44.8	36.3	35.1	33.0	43.0	35.2	35.7	31.4	35.2	31.0	30.4	31.2	30.0	29.8	27.8	34.8
					B	41.0	35.1	32.8	40.5	35.4	31.7	27.1	38.5	28.6	30.6	28.2	30.0	29.4	27.7	28.7	26.1	29.1	26.1	31.4
			GC-MS/MS	Green tea	A	44.5	35.8	36.9	47.3	33.2	31.0	36.2	33.8	33.9	29.2	40.9	29.2	29.6	31.8	24.7	24.2	28.0	32.9	34.0
					B	43.2	34.7	34.0	43.6	31.6	28.9	34.9	31.2	33.7	27.3	39.3	29.6	30.6	29.0	24.8	23.8	26.7	28.4	32.1
				Oolong tea	A	41.3	40.7	37.9	46.4	33.0	35.1	32.1	46.9	36.4	31.2	34.2	34.5	32.6	32.4	30.6	29.5	31.3	28.5	35.5
					B	39.1	36.2	32.4	42.1	32.5	30.7	26.6	42.7	31.1	31.8	29.3	29.2	30.6	27.1	27.3	27.9	29.9	26.2	31.8
		ENVI-Carb + PSA	GC-MS	Green tea	A	46.1	42.8	35.7	35.8	39.8	40.3	35.6	32.8	33.7	42.6	42.0	32.4	33.2	27.5	35.1	30.7	30.0	31.6	36.0
					B	44.4	36.3	36.2	33.1	38.6	35.5	36.9	29.7	31.6	40.2	36.2	30.2	31.0	26.9	29.5	28.4	27.3	31.5	33.6
				Oolong tea	A	43.7	33.2	38.2	34.6	39.2	32.5	31.0	30.0	32.5	32.9	30.6	28.2	31.0	32.7	28.6	26.2	28.1	25.2	32.1
					B	38.4	29.6	33.8	32.1	31.6	30.1	30.8	26.2	27.2	27.7	31.5	25.8	29.3	27.6	25.2	24.1	24.9	23.7	28.7
			GC-MS/MS	Green tea	A	48.2	40.9	36.6	34.4	40.0	45.2	37.6	31.7	33.4	32.4	42.7	33.0	34.7	30.9	33.9	30.7	29.8	31.7	36.1
					B	50.2	34.1	36.2	33.7	38.1	35.4	33.4	30.2	28.6	32.5	38.1	29.8	28.1	28.6	30.1	27.8	27.8	32.0	33.1
				Oolong tea	A	49.6	34.4	44.4	33.6	39.2	32.0	34.4	28.6	32.5	29.5	28.1	31.4	27.3	31.2	30.6	23.8	27.6	26.3	32.3
					B	39.9	31.6	33.2	31.1	33.5	29.2	31.3	25.3	27.2	27.8	25.6	28.0	25.2	25.8	27.3	25.4	24.0	25.1	28.8

121	Profenofos	Cleanert TPT	GC–MS	Green tea	A	266.5	322.2	214.5	427.2	175.9	209.1	207.6	209.3	177.3	154.1	233.1	175.0	173.4	193.5	178.0	137.0	124.9	131.1	164.0	203.9
					B	262.3	291.1	192.7	375.1	180.5	188.4	201.5	205.1	228.0	146.5	206.0	163.2	185.9	189.1	211.9	148.5	135.0	150.6	145.7	200.4
				Oolong tea	A	285.7	353.8	199.8	304.5	307.2	155.0	196.7	217.3	180.7	168.0	245.9	190.0	255.0	179.7	158.3	188.6	220.9	165.2	132.3	217.1
					B	268.7	287.7	211.6	245.6	253.0	136.2	147.5	221.8	182.0	167.2	189.6	137.6	203.8	195.1	151.2	191.4	173.2	146.1	128.0	191.4
			GC–MS/MS	Green tea	A	331.0	341.3	175.0	327.9	191.7	203.8	210.4	165.2	185.6	127.7	266.8	168.0	150.6	134.3	148.7	141.8	98.9	168.0	223.0	197.6
					B	268.2	290.1	169.3	343.2	243.0	199.5	181.2	182.9	188.9	157.1	254.2	155.2	165.0	170.2	148.1	160.4	137.9	150.8	153.1	195.7
				Oolong tea	A	384.3	475.0	188.5	307.4	160.4	220.7	169.2	387.3	178.7	152.2	238.5	171.5	219.4	178.6	195.8	187.9	173.9	181.7	79.9	223.7
					B	339.1	401.7	184.6	249.6	145.2	155.9	169.9	238.5	144.1	181.7	176.4	167.5	183.3	169.6	151.8	172.5	126.2	148.2	142.8	192.0
		ENVI-Carb + PSA	GC–MS	Green tea	A	308.7	245.8	213.7	254.8	235.9	244.8	233.4	162.2	238.3	237.3	237.2	240.8	137.4	160.0	117.0	181.5	259.0	146.7	212.4	214.0
					B	303.8	243.7	192.9	218.9	223.1	222.9	215.9	142.5	215.0	196.9	158.4	227.4	129.0	157.8	127.9	143.3	215.9	155.6	225.3	195.6
				Oolong tea	A	240.8	138.4	200.5	180.3	257.1	211.8	162.0	160.8	187.8	171.6	178.5	181.4	132.0	183.0	169.5	174.0	157.2	133.7	126.2	176.1
					B	229.9	123.5	200.7	178.7	204.5	215.0	173.8	140.5	122.9	134.4	146.4	181.0	133.4	194.9	159.7	143.9	153.4	142.3	117.0	162.9
			GC–MS/MS	Green tea	A	329.0	546.5	233.7	276.1	291.6	297.3	291.6	143.4	323.5	209.4	270.4	197.9	181.3	234.0	128.5	181.0	284.3	285.4	284.6	262.8
					B	341.0	667.5	197.2	264.1	240.9	226.1	210.1	125.9	248.9	223.2	223.2	194.2	146.5	186.0	141.8	146.3	254.7	254.5	278.3	239.8
				Oolong tea	A	332.4	124.7	188.9	153.6	339.8	148.1	222.1	146.6	187.8	163.3	177.4	75.5	149.5	168.1	142.5	187.9	111.2	163.6	156.8	175.8
					B	287.3	114.0	171.7	157.2	271.5	182.4	165.5	125.5	122.9	155.9	179.9	82.1	141.5	166.4	143.1	172.5	120.6	151.3	116.4	159.4
122	Profluralin	Cleanert TPT	GC–MS	Green tea	A	324.3	225.8	233.4	676.1	225.4	223.0	262.4	135.3	108.6	111.0	164.5	122.6	128.8	112.0	127.3	107.7	85.8	106.8	130.3	190.1
					B	316.0	222.3	207.9	573.8	216.7	210.0	247.6	123.4	146.1	104.2	148.7	109.3	118.6	111.2	125.4	107.1	86.0	103.4	115.3	178.6
				Oolong tea	A	263.9	232.5	218.5	310.5	646.4	282.6	140.7	180.4	145.0	136.9	145.8	136.0	150.2	123.4	144.2	167.7	126.7	128.7	107.1	199.3
					B	246.2	221.7	199.0	276.6	504.2	234.7	110.4	162.2	113.4	121.2	113.4	107.8	134.1	130.6	122.4	167.7	107.7	126.4	106.8	174.0
			GC–MS/MS	Green tea	A	192.0	153.5	146.4	198.8	131.5	131.2	133.3	124.0	122.9	108.3	163.2	138.3	116.7	114.4	119.3	98.6	89.2	98.1	135.5	132.4
					B	187.4	150.8	128.3	183.3	126.5	120.4	128.6	118.9	141.7	106.0	152.8	132.6	105.5	122.7	110.4	102.4	93.2	102.6	119.4	128.1
				Oolong tea	A	161.7	176.4	149.7	190.0	131.8	138.8	123.6	339.2	138.8	133.0	146.7	144.8	142.5	128.8	129.3	132.4	123.8	119.5	97.7	149.9
					B	156.1	166.6	135.1	179.0	127.7	117.8	98.1	292.8	115.3	132.2	114.4	111.4	125.1	120.3	112.6	127.7	106.4	113.5	101.0	133.9
		ENVI-Carb + PSA	GC–MS	Green tea	A	331.3	276.6	243.7	268.7	257.8	322.6	234.8	116.4	250.6	144.8	145.0	185.2	132.5	133.5	83.4	117.3	156.1	126.7	155.1	193.8
					B	295.9	260.9	232.4	233.9	245.9	298.8	135.7	113.8	233.4	130.8	123.0	159.8	127.6	131.4	80.0	99.2	127.9	120.2	148.8	173.6
				Oolong tea	A	265.2	195.2	224.4	209.8	363.2	249.5	127.0	120.2	125.6	127.4	124.9	123.0	124.1	123.4	132.0	124.6	104.5	112.8	104.7	162.2
					B	225.9	169.7	191.8	194.5	255.6	226.2	117.2	99.9	105.7	107.8	103.5	122.1	107.4	130.6	143.6	84.5	100.1	103.8	96.1	141.4
			GC–MS/MS	Green tea	A	185.2	169.1	148.6	147.8	164.9	151.8	190.5	110.5	143.3	127.6	154.7	127.9	165.6	153.0	105.2	130.6	134.2	122.5	169.5	147.5
					B	184.2	179.4	143.5	141.9	147.2	158.0	49.3	106.3	131.1	128.5	129.4	124.5	159.9	108.5	99.4	106.5	124.1	130.6	165.3	132.5
				Oolong tea	A	184.6	129.2	169.6	134.6	165.1	130.1	88.5	110.0	125.6	119.6	117.6	116.4	123.2	118.4	124.2	132.4	101.8	107.8	98.9	126.2
					B	157.3	114.9	126.5	125.4	130.5	121.2	119.9	90.1	105.7	108.8	114.6	119.5	108.0	103.5	108.5	127.7	98.6	101.2	96.2	114.6
123	Propachlor	Cleanert TPT	GC–MS	Green tea	A	114.5	111.1	120.6	113.6	103.6	97.8	112.8	96.1	96.1	94.3	95.8	86.7	89.6	99.6	104.9	86.3	72.7	89.2	96.3	99.2
					B	112.3	106.0	108.2	116.8	97.7	88.3	110.8	105.7	105.5	91.2	82.7	87.9	92.1	99.0	102.2	85.1	76.5	80.2	89.7	96.8
				Oolong tea	A	111.8	132.0	130.4	132.2	106.1	103.0	103.0	113.8	105.5	106.0	109.6	118.3	117.8	94.8	96.9	95.9	106.6	90.1	87.1	109.6
					B	111.3	103.8	112.9	117.3	106.1	97.7	96.4	113.8	101.8	92.9	92.6	100.5	103.4	91.2	100.5	89.2	92.3	89.7	75.3	98.8
			GC–MS/MS	Green tea	A	117.9	109.8	90.6	135.4	92.2	89.8	107.3	107.9	102.4	86.2	99.2	54.7	88.3	77.5	87.2	78.6	70.5	84.3	93.5	93.3
					B	120.1	106.2	105.4	129.8	95.3	85.0	102.7	94.5	98.5	91.3	86.0	53.4	90.2	84.7	82.1	80.1	71.2	87.6	84.2	92.0
				Oolong tea	A	137.7	129.5	109.9	120.7	98.7	109.8	106.7	158.7	111.3	88.5	105.4	99.5	106.3	89.5	88.1	91.9	90.0	103.0	77.6	107.0
					B	128.1	106.6	95.0	105.6	94.6	90.4	83.3	124.6	94.3	87.1	86.9	88.5	87.5	89.2	78.7	81.0	81.0	93.4	74.8	93.2
		ENVI-Carb + PSA	GC–MS	Green tea	A	118.6	113.3	106.2	113.3	106.5	110.9	128.8	97.1	106.1	87.8	97.8	95.7	89.8	98.8	74.9	110.5	97.0	90.7	100.1	102.3
					B	119.2	111.0	111.0	106.8	112.7	105.8	107.6	77.4	100.6	89.1	87.2	98.8	85.7	94.1	76.0	89.4	87.5	86.5	86.5	96.6
				Oolong tea	A	101.0	109.3	115.9	110.3	109.3	107.0	82.4	87.7	92.6	98.1	93.5	96.7	89.9	96.2	97.7	87.6	88.4	82.3	69.7	95.6
					B	84.6	96.5	107.4	98.7	96.0	99.3	99.2	82.0	76.7	80.5	87.5	89.6	88.7	91.8	85.3	79.9	82.2	70.8	70.6	87.8
			GC–MS/MS	Green tea	A	157.1	115.0	107.8	110.0	120.1	151.7	109.6	89.4	96.6	80.0	106.5	187.9	107.5	107.0	93.2	97.3	84.1	87.9	93.0	110.6
					B	199.0	105.6	110.1	111.5	120.8	101.1	122.2	89.8	91.5	88.0	95.5	177.4	99.1	79.4	88.4	91.0	83.4	87.4	96.9	107.2
				Oolong tea	A	193.1	100.6	151.8	97.3	115.5	93.9	134.2	88.7	92.6	85.8	86.4	60.2	87.2	82.3	85.7	91.9	62.8	75.6	74.7	97.9
					B	117.8	91.2	93.8	84.9	96.4	87.0	92.5	76.2	76.7	75.2	89.4	56.4	79.8	79.4	70.5	81.0	66.9	65.1	72.6	81.7

(Continued)

Appendix Table 5.1 The 201 Pesticides Average Content for 19 Circulative Determinations on 3 Months Under 16 Factors (Two Cartridges, Two Instruments, Two Tea Samples, and Two Youden Pair Samples) (cont.)

No.	Pesticides	SPE	Method	Sample	Youden pair	Nov 9, 2009 (n=5)	Nov 14, 2009 (n=3)	Nov 19, 2009 (n=3)	Nov 24, 2009 (n=3)	Nov 29, 2009 (n=3)	Dec 4, 2009 (n=3)	Dec 9, 2009 (n=3)	Dec 14, 2009 (n=3)	Dec 19, 2009 (n=3)	Dec 24, 2009 (n=3)	Jan 3, 2010 (n=3)	Jan 8, 2010 (n=3)	Jan 13, 2010 (n=3)	Jan 18, 2010 (n=3)	Jan 23, 2010 (n=3)	Jan 28, 2010 (n=3)	Feb 2, 2010 (n=3)	Feb 7, 2010 (n=3)	Ave (μg/kg)
124	Propiconazole	Cleanert TPT	GC-MS	Green tea	A	176.7	127.7	132.6	123.7	108.9	86.1	105.8	103.0	96.9	82.4	94.3	86.7	90.2	89.0	71.6	72.2	66.2	87.6	101.2
					B	168.8	119.6	121.9	114.3	94.3	85.4	97.9	93.1	101.7	79.6	90.7	81.4	94.2	80.0	69.0	68.0	63.7	75.6	95.2
				Oolong tea	A	149.5	101.8	102.6	126.4	116.3	106.1	92.7	115.3	110.4	107.5	88.7	96.6	87.5	82.7	82.9	81.3	82.6	72.5	99.3
					B	136.2	114.6	99.2	112.9	114.9	99.5	89.4	113.4	83.9	95.3	82.2	88.4	85.6	73.6	73.1	88.8	77.8	66.9	92.1
			GC-MS/MS	Green tea	A	142.7	127.6	130.1	136.8	96.0	92.5	102.4	94.6	99.9	82.8	72.2	91.1	89.8	87.8	82.3	70.2	70.1	93.4	97.4
					B	138.3	119.7	113.1	128.9	82.8	82.5	101.9	87.5	95.1	86.1	67.5	81.7	91.0	81.0	82.7	65.7	69.2	80.9	92.8
				Oolong tea	A	115.1	120.4	113.8	133.0	93.0	103.2	90.3	120.4	95.3	92.7	95.3	93.5	87.6	80.0	84.3	77.2	81.5	73.4	97.4
					B	114.1	112.5	102.6	118.7	91.5	91.3	84.5	114.6	83.2	87.4	87.1	81.1	78.7	68.8	76.1	68.7	74.6	65.1	88.7
		ENVI-Carb + PSA	GC-MS	Green tea	A	174.8	156.4	129.4	113.2	117.0	125.9	95.5	91.0	105.3	125.4	107.5	83.1	104.2	89.0	104.9	82.2	85.2	81.1	108.7
					B	170.6	133.2	118.3	102.4	115.1	104.7	104.3	89.3	94.3	107.5	89.0	77.6	89.8	83.2	84.8	70.4	71.2	80.9	98.9
				Oolong tea	A	162.3	100.1	111.2	96.1	122.0	89.2	102.3	86.9	94.5	103.9	72.7	83.3	81.1	89.1	75.6	73.6	76.4	66.8	92.6
					B	151.8	96.5	107.2	95.9	84.1	81.2	89.5	72.1	78.8	85.4	76.9	79.2	79.5	68.6	68.2	65.2	66.7	63.7	92.4
			GC-MS/MS	Green tea	A	148.5	139.1	140.2	113.9	109.3	113.9	96.6	94.9	96.7	95.0	287.4	103.7	93.0	83.2	98.6	78.5	84.3	84.1	113.3
					B	140.9	109.4	110.0	107.1	108.2	100.1	94.0	86.8	86.7	88.9	279.4	92.7	78.6	76.4	83.7	67.8	73.5	83.2	102.8
				Oolong tea	A	139.5	106.5	109.7	97.9	111.0	92.1	91.2	76.8	94.5	88.3	82.4	73.4	78.1	83.3	84.3	68.5	74.3	71.4	90.0
					B	130.8	100.5	106.3	97.9	89.9	87.3	86.5	72.0	78.8	73.5	76.2	68.4	71.5	67.8	75.9	66.5	63.2	68.4	82.6
125	Propyzamide	Cleanert TPT	GC-MS	Green tea	A	161.5	133.4	143.3	167.4	118.8	106.0	127.2	123.7	118.5	110.7	117.1	113.2	106.5	114.3	98.6	99.2	90.6	132.5	122.3
					B	159.7	129.9	121.0	147.6	107.7	100.5	117.8	108.7	122.4	101.9	108.7	106.0	109.1	105.6	98.1	98.0	88.9	117.0	114.2
				Oolong tea	A	139.9	124.3	127.5	158.0	127.9	125.7	106.9	144.8	127.0	117.0	118.0	123.1	116.1	110.2	112.4	112.2	110.8	102.1	122.8
					B	128.7	116.5	116.3	140.1	120.0	108.1	92.9	121.7	104.7	108.1	97.3	107.0	106.6	97.1	100.1	96.3	104.7	96.8	108.8
			GC-MS/MS	Green tea	A	160.8	134.3	140.7	156.5	118.9	109.3	122.6	120.3	113.6	109.6	111.4	113.9	101.7	116.3	97.0	95.0	96.2	127.9	119.3
					B	154.5	130.6	122.3	142.6	108.8	101.0	122.2	106.3	111.7	107.0	102.5	103.5	101.7	104.6	95.8	93.2	98.6	122.2	112.3
				Oolong tea	A	123.9	135.5	134.1	169.3	114.1	122.2	109.3	151.3	136.0	112.3	127.2	122.7	125.8	111.7	99.7	106.9	109.3	99.7	123.4
					B	121.1	121.7	116.2	147.3	111.9	102.5	93.4	133.4	115.6	105.7	103.1	108.7	109.8	95.0	99.7	95.8	100.3	94.6	109.3
		ENVI-Carb + PSA	GC-MS	Green tea	A	169.6	152.2	133.7	134.9	137.1	151.7	106.7	115.6	125.1	133.9	139.6	116.0	124.5	103.9	131.4	112.8	112.7	117.5	128.6
					B	159.8	135.2	130.6	124.5	131.9	135.7	87.6	109.2	121.0	129.4	125.5	105.7	111.2	95.5	105.8	101.0	104.1	122.9	118.5
				Oolong tea	A	163.7	114.5	117.5	109.5	146.0	111.5	116.4	102.4	110.1	101.9	113.4	103.5	116.1	121.0	104.5	86.4	97.2	95.6	112.5
					B	134.6	97.8	110.1	101.6	116.5	103.1	109.1	88.5	79.5	89.6	102.3	94.4	106.6	101.3	89.3	81.9	84.7	82.2	98.2
			GC-MS/MS	Green tea	A	160.4	147.7	132.9	130.8	135.6	111.9	120.8	113.9	116.7	123.1	69.3	119.9	118.2	104.2	125.9	111.5	114.6	119.5	121.0
					B	152.7	126.8	128.9	121.3	129.4	136.1	111.2	106.7	111.2	124.1	67.0	103.4	99.3	93.9	104.1	99.6	108.9	121.9	113.4
				Oolong tea	A	159.5	117.3	147.2	112.8	146.4	112.1	109.7	99.6	110.1	95.3	102.8	106.5	97.7	117.3	109.4	83.0	93.8	90.9	111.4
					B	126.8	98.7	109.1	103.7	121.3	102.0	109.3	85.8	79.5	88.4	93.9	96.2	90.8	92.8	99.6	85.1	81.6	82.6	97.4
126	Ronnel	Cleanert TPT	GC-MS	Green tea	A	238.7	219.4	216.3	306.0	199.9	193.8	226.0	222.1	193.5	194.8	192.5	183.2	180.3	187.4	162.9	150.5	144.8	185.7	200.9
					B	240.7	211.9	197.4	285.8	190.6	179.2	206.9	203.1	217.5	181.0	185.2	181.2	174.3	184.8	159.3	149.6	149.5	168.9	192.5
				Oolong tea	A	250.9	250.7	236.7	265.8	229.6	206.9	207.5	268.1	205.9	212.4	205.3	226.2	201.7	187.5	192.3	193.8	185.9	175.3	217.5
					B	232.5	204.8	197.7	228.9	209.2	177.9	157.5	215.6	179.2	171.6	174.7	174.2	181.1	162.9	170.0	160.8	172.4	153.9	184.8
			GC-MS/MS	Green tea	A	228.1	208.3	198.3	277.7	193.7	198.8	227.8	202.2	191.8	182.7	172.7	180.1	168.3	184.7	160.6	145.1	159.0	190.7	193.7
					B	224.8	206.0	206.5	266.0	197.3	188.9	213.2	184.2	191.9	179.4	163.7	180.9	169.7	177.6	158.6	146.3	166.7	173.8	188.2
				Oolong tea	A	302.1	266.3	230.7	265.7	203.9	217.6	203.1	331.5	239.1	181.6	210.2	212.0	227.3	192.3	190.6	184.6	205.0	174.0	224.5
					B	258.7	220.0	190.0	224.9	187.8	178.6	161.8	249.8	193.2	176.8	184.6	169.5	192.5	159.4	166.2	153.1	177.5	157.8	188.9
		ENVI-Carb + PSA	GC-MS	Green tea	A	248.2	231.3	206.9	217.4	231.2	229.4	114.3	190.1	224.0	217.3	242.0	179.8	200.9	153.1	194.1	193.2	168.9	202.5	203.1
					B	240.9	202.9	206.6	226.4	228.4	213.9	120.4	160.3	213.6	207.4	231.5	169.8	186.6	162.3	167.7	176.4	165.5	199.3	192.1
				Oolong tea	A	243.6	200.2	225.1	213.2	230.6	197.4	172.7	151.1	213.6	193.0	169.0	153.3	201.7	189.8	194.1	161.8	162.4	156.9	191.7
					B	201.7	170.3	201.3	182.1	188.1	176.0	180.1	182.5	213.6	185.8	112.0	195.9	181.6	153.1	167.7	144.4	145.8	144.2	166.6
			GC-MS/MS	Green tea	A	306.8	239.1	215.1	231.9	248.1	429.4	246.6	177.9	201.4	178.4	109.4	176.2	201.7	162.3	199.3	178.2	173.7	196.2	217.0
					B	336.1	212.7	212.9	228.6	234.1	220.5	213.1	175.6	207.5	185.8	134.8	183.5	160.7	169.2	175.5	167.4	177.4	202.4	196.5
				Oolong tea	A	322.7	210.1	289.5	198.7	246.4	194.5	250.8	190.5	190.7	179.0	125.5	183.5	174.8	186.8	190.6	144.9	168.1	163.3	200.3
					B	223.8	183.9	183.2	170.5	198.4	174.6	187.1	185.1	201.4	157.3	125.5	160.6	157.2	143.2	166.2	138.7	143.5	154.4	166.3

127	Sulfotep	Cleanert TPT	GC–MS	Green tea	A	47.4	35.8	37.0	40.4	33.4	30.4	35.8	35.8	32.5	30.0	38.5	32.3	31.9	30.0	30.0	31.6	27.0	26.8	28.0	33.9	33.6
					B	47.7	35.2	34.5	36.9	30.9	28.5	32.9	32.7	36.5	29.7	32.3	30.1	30.5	28.6	29.5	30.0	26.1	26.0	28.1	29.3	31.9
				Oolong tea	A	43.7	34.1	36.7	42.6	36.5	35.1	32.0	39.1	35.3	34.1	32.9	32.2	34.5	32.5	31.4	31.4	33.0	31.8	31.1	27.5	34.5
					B	41.1	32.6	32.8	38.4	34.0	31.2	26.9	34.8	30.5	30.0	28.2	29.4	29.0	30.8	27.8	27.8	29.8	26.3	29.7	25.2	31.0
			GC–MS/MS	Green tea	A	43.3	36.4	39.1	44.4	32.7	31.1	35.1	33.1	33.9	30.0	37.4	20.5	33.1	28.4	31.7	31.7	25.9	24.3	27.2	32.1	32.6
					B	42.2	35.9	34.7	41.6	30.2	28.9	34.7	30.9	33.1	28.7	31.7	19.6	29.6	27.3	29.0	29.0	25.8	23.8	27.8	27.8	30.7
				Oolong tea	A	36.9	39.2	36.7	43.7	33.9	34.0	31.2	46.3	36.9	31.4	34.7	34.4	33.3	34.1	30.9	30.9	29.4	28.2	31.1	26.0	34.3
					B	36.5	34.9	32.1	40.1	32.7	29.6	25.4	40.9	31.8	29.5	28.8	30.0	29.0	29.9	26.6	26.6	26.2	26.7	29.6	24.1	30.8
		ENVI-Carb + PSA	GC–MS	Green tea	A	49.5	44.9	36.5	36.1	39.9	43.8	48.3	33.6	33.9	39.4	35.1	39.2	33.3	35.2	30.3	30.3	34.9	32.1	30.6	32.5	37.3
					B	46.5	38.0	36.2	33.1	39.0	37.4	39.6	30.9	32.1	36.2	33.1	34.0	30.9	32.3	29.5	29.5	30.9	30.0	27.9	31.7	34.2
				Oolong tea	A	44.5	33.4	35.5	33.2	37.5	32.0	31.4	28.7	32.0	32.6	31.5	28.5	29.6	32.6	33.6	33.6	27.9	25.7	27.9	25.7	31.8
					B	36.3	30.2	31.7	30.6	31.4	28.3	29.6	26.8	27.1	27.5	26.5	26.7	27.3	30.9	27.5	27.5	24.3	22.6	24.6	24.2	28.1
			GC–MS/MS	Green tea	A	48.4	41.0	36.5	37.0	40.7	54.2	37.4	31.7	32.1	34.6	32.8	32.8	40.1	34.2	30.2	30.2	35.6	31.6	30.3	31.9	36.5
					B	48.0	33.6	36.3	34.6	39.1	36.8	33.1	30.7	29.7	33.9	30.6	32.2	36.8	30.7	27.6	27.6	30.9	30.9	30.1	31.8	33.4
				Oolong tea	A	46.0	33.5	44.5	33.9	37.3	31.7	31.5	27.7	32.0	29.3	30.1	26.9	29.3	28.3	31.7	31.7	29.4	24.5	26.3	25.6	31.7
					B	35.0	29.7	30.7	33.5	31.3	29.5	29.7	24.7	27.1	25.3	30.4	25.2	27.1	26.2	25.1	25.1	26.2	23.8	22.3	25.4	27.8
128	Tebufen-pyrad	Cleanert TPT	GC–MS	Green tea	A	75.7	62.5	66.5	72.8	57.5	50.9	60.6	55.9	60.4	48.3	55.0	50.8	51.3	50.8	53.1	53.1	44.2	43.6	43.9	55.6	55.7
					B	71.1	60.1	59.8	67.6	50.9	47.0	56.1	52.1	56.5	47.6	48.4	47.1	48.3	48.2	47.8	47.8	43.9	42.3	42.3	49.0	51.9
				Oolong tea	A	75.2	64.7	61.4	72.1	57.5	59.9	51.4	64.6	59.5	55.5	58.4	56.3	58.2	52.2	52.3	52.3	52.6	50.5	52.4	47.1	58.0
					B	69.9	59.2	56.1	64.1	58.5	55.2	49.8	62.0	51.6	50.6	48.4	49.6	52.7	48.9	46.0	46.0	45.6	42.5	49.0	42.6	52.7
			GC–MS/MS	Green tea	A	78.7	64.0	71.4	75.8	56.8	51.5	56.9	54.1	57.2	47.8	56.1	39.3	52.4	51.5	55.8	55.8	41.5	42.5	44.6	58.5	55.6
					B	74.0	61.7	61.4	69.6	51.2	47.5	58.3	50.8	54.7	47.2	52.4	37.4	46.8	50.2	49.7	49.7	41.6	42.6	43.8	50.1	52.2
				Oolong tea	A	67.2	64.0	63.0	74.2	58.6	59.5	52.1	74.0	62.0	54.8	60.2	57.4	56.1	55.5	50.8	50.8	50.4	50.5	54.2	46.0	58.4
					B	64.5	60.9	56.4	65.6	59.8	53.4	48.0	71.4	53.4	50.6	49.9	52.3	48.9	50.7	43.6	43.6	44.1	45.8	50.2	40.4	53.2
		ENVI-Carb + PSA	GC–MS	Green tea	A	86.4	76.0	63.1	63.1	62.9	64.9	46.6	54.6	54.2	53.5	51.3	55.2	54.1	58.7	51.2	51.2	61.5	50.5	52.2	50.9	58.5
					B	81.9	65.8	61.6	55.7	62.7	58.5	36.5	50.3	49.1	51.5	50.1	52.0	50.2	51.7	47.7	47.7	50.6	44.5	43.6	50.3	53.4
				Oolong tea	A	77.9	58.5	59.2	58.6	64.4	55.3	57.0	45.8	55.4	53.6	50.8	49.5	50.5	52.2	56.2	56.2	47.9	47.1	50.6	44.7	54.5
					B	72.4	53.4	59.6	54.0	53.8	48.4	48.0	43.3	47.2	44.7	44.3	45.0	45.5	48.9	45.7	45.7	41.2	40.1	44.1	41.2	48.5
			GC–MS/MS	Green tea	A	79.4	47.1	63.2	58.9	61.5	69.4	58.9	53.7	56.0	55.5	56.0	37.8	61.9	56.3	49.3	49.3	59.0	50.1	53.3	51.6	56.8
					B	77.6	51.9	61.6	54.8	61.8	59.9	55.3	49.7	50.1	55.1	50.6	35.9	55.6	51.1	45.1	45.1	50.0	43.3	48.3	51.6	53.1
				Oolong tea	A	76.1	58.5	61.9	58.0	63.7	55.5	51.5	45.7	55.4	52.3	48.7	48.9	50.2	47.4	52.3	52.3	50.4	42.9	48.7	44.7	53.3
					B	70.3	54.3	58.2	54.0	57.1	49.5	47.9	41.7	47.2	45.2	48.1	43.4	44.8	40.7	42.0	42.0	44.1	40.8	42.5	42.4	48.1
129	Terbutryn	Cleanert TPT	GC–MS	Green tea	A	102.0	79.5	82.8	89.2	68.3	58.4	72.4	68.8	64.3	57.7	78.2	60.7	58.7	67.8	65.3	65.3	50.1	64.4	21.0	62.3	56.9
					B	99.1	77.6	72.9	80.0	61.7	56.7	66.7	63.0	66.2	52.7	65.8	56.0	54.9	61.1	60.5	60.5	50.3	55.1	20.3	51.1	61.9
				Oolong tea	A	93.7	75.6	73.9	88.3	70.0	72.0	64.2	79.9	70.4	70.0	67.2	66.7	67.4	59.5	61.1	61.1	61.8	58.4	56.7	69.0	68.9
					B	89.5	75.3	69.4	83.4	66.8	65.2	54.4	70.5	57.2	57.6	52.4	57.2	57.1	55.2	52.9	52.9	54.5	48.9	56.7	61.9	61.9
			GC–MS/MS	Green tea	A	99.9	81.7	84.6	95.6	63.0	61.5	66.9	65.4	65.9	60.1	68.6	49.8	59.8	56.4	61.0	61.0	48.9	45.1	45.9	60.5	65.5
					B	93.6	80.1	72.5	86.9	66.6	57.4	68.6	61.5	63.3	60.1	59.8	48.5	53.8	54.4	54.6	54.6	47.4	48.0	48.5	54.3	61.9
				Oolong tea	A	76.2	75.6	81.5	92.9	66.1	70.0	64.8	80.8	71.8	63.9	67.9	66.8	63.8	62.2	57.4	57.4	57.1	61.3	57.6	54.0	68.0
					B	77.2	71.9	71.4	83.7	80.3	60.5	52.5	73.8	60.5	59.4	56.6	55.2	53.7	57.5	50.2	50.2	50.7	52.7	55.1	52.5	61.1
		ENVI-Carb + PSA	GC–MS	Green tea	A	110.2	98.3	78.5	77.6	77.5	83.9	77.9	64.0	65.3	75.7	63.9	76.3	59.8	66.2	56.2	56.2	67.9	58.0	59.3	56.5	72.4
					B	103.7	81.6	77.1	69.9	75.0	73.3	75.0	60.9	63.3	70.9	60.6	62.6	54.7	58.9	51.3	51.3	54.3	50.7	50.0	56.5	65.9
				Oolong tea	A	98.6	71.5	76.3	71.0	83.6	67.8	67.4	50.1	65.2	64.6	59.0	55.7	57.2	58.8	64.7	64.7	54.3	54.4	55.7	47.7	64.7
					B	89.9	66.0	70.1	66.6	77.5	61.1	61.9	50.1	54.1	52.8	48.7	53.4	52.9	54.1	53.6	53.6	49.7	48.0	46.2	44.8	57.3
			GC–MS/MS	Green tea	A	105.7	90.0	80.0	75.2	77.4	83.6	71.7	62.9	63.0	63.1	66.1	75.9	70.9	69.3	55.9	55.9	64.6	53.3	55.5	56.5	70.6
					B	98.9	71.5	77.1	69.9	76.4	73.8	66.7	59.7	60.1	64.5	58.6	73.6	58.3	53.5	48.7	48.7	52.8	47.3	51.6	58.6	64.3
				Oolong tea	A	92.1	73.4	80.6	70.3	78.2	67.5	63.2	54.1	65.2	61.6	59.9	61.3	58.3	56.4	62.5	62.5	57.1	46.4	53.0	47.2	63.6
					B	84.4	68.3	70.4	65.2	65.2	62.1	60.9	49.1	54.1	51.9	60.8	57.1	53.8	48.8	48.2	48.2	50.7	45.6	42.2	46.0	57.1

(Continued)

Appendix Table 5.1 The 201 Pesticides Average Content for 19 Circulative Determinations on 3 Months Under 16 Factors (Two Cartridges, Two Instruments, Two Tea Samples, and Two Youden Pair Samples) (cont.)

No.	Pesticides	SPE	Method	Sample	Youden pair	Nov 9, 2009 (n=5)	Nov 14, 2009 (n=3)	Nov 19, 2009 (n=3)	Nov 24, 2009 (n=3)	Nov 29, 2009 (n=3)	Dec 4, 2009 (n=3)	Dec 9, 2009 (n=3)	Dec 14, 2009 (n=3)	Dec 19, 2009 (n=3)	Dec 24, 2009 (n=3)	Jan 3, 2010 (n=3)	Jan 8, 2010 (n=3)	Jan 13, 2010 (n=3)	Jan 18, 2010 (n=3)	Jan 23, 2010 (n=3)	Jan 28, 2010 (n=3)	Feb 2, 2010 (n=3)	Feb 7, 2010 (n=3)	Ave (µg/kg)
130	Thiobencarb	Cleanert TPT	GC-MS	Green tea	A	91.3	73.6	81.1	100.0	72.4	64.9	78.6	74.0	76.1	64.2	66.6	67.5	64.5	68.9	58.9	58.6	61.1	74.2	72.2
					B	87.8	72.6	72.9	91.1	66.7	61.4	72.8	68.2	74.0	60.5	62.7	64.4	62.0	63.2	57.9	56.9	58.7	65.0	67.7
				Oolong tea	A	86.6	76.0	76.1	85.8	73.1	73.8	65.7	81.8	74.1	69.8	74.5	72.7	65.4	66.9	65.1	63.0	65.5	58.1	71.9
					B	82.5	70.7	68.2	80.9	71.5	67.4	56.6	73.3	64.7	63.7	63.7	62.9	61.5	59.7	60.9	55.9	64.6	55.4	65.4
			GC-MS/MS	Green tea	A	95.3	77.1	82.8	96.0	68.2	65.4	74.1	71.3	72.0	66.6	61.0	65.8	62.9	69.6	55.8	54.3	57.8	73.9	70.6
					B	91.7	74.6	69.9	89.4	64.3	61.2	74.5	66.8	70.8	61.6	58.1	61.7	59.6	64.6	55.1	53.9	59.5	66.8	67.0
				Oolong tea	A	73.9	77.1	75.7	90.6	68.4	70.8	65.1	85.4	76.4	69.7	73.4	68.7	72.2	63.7	67.1	61.2	65.4	55.1	71.1
					B	73.8	70.3	66.5	81.8	69.0	62.3	55.5	78.0	67.4	63.8	62.9	60.1	64.0	55.5	63.6	55.5	62.0	53.4	64.5
		ENVI-Carb + PSA	GC-MS	Green tea	A	101.0	89.8	76.5	78.4	82.5	85.2	60.8	72.2	74.5	73.0	76.9	72.0	75.9	67.8	82.3	65.2	68.9	67.3	75.7
					B	93.6	74.7	78.1	72.5	80.9	76.2	69.0	69.1	69.8	71.2	70.2	68.1	68.6	63.0	68.0	59.4	59.7	66.6	70.7
				Oolong tea	A	93.5	72.0	73.8	69.5	80.0	70.6	69.5	58.7	65.9	66.1	60.3	62.1	65.4	69.1	60.1	56.4	59.4	53.0	66.8
					B	77.7	64.7	68.7	65.6	66.9	62.7	64.0	54.6	58.2	55.6	57.7	57.9	61.2	58.8	53.8	51.3	51.6	50.1	59.8
			GC-MS/MS	Green tea	A	98.3	86.3	77.1	73.9	81.0	87.1	78.3	71.1	69.0	73.7	68.0	82.8	75.8	65.2	78.3	65.7	70.8	67.1	75.9
					B	92.5	70.3	74.8	71.7	77.4	78.5	73.7	67.1	64.5	73.2	64.2	71.6	62.6	60.6	67.9	61.2	66.3	64.1	69.9
				Oolong tea	A	86.1	70.6	82.2	66.9	77.6	67.6	62.8	58.7	65.9	65.0	63.4	64.1	60.0	64.1	67.1	54.0	57.9	55.9	65.0
					B	76.2	63.3	64.4	62.9	65.1	62.8	62.8	52.4	58.2	56.5	59.2	58.9	55.9	53.1	63.6	52.5	51.9	54.5	60.0
131	Tralkoxydim	Cleanert TPT	GC-MS	Green tea	A	406.2	308.4	327.0	438.6	282.7	252.6	285.5	272.3	289.1	264.5	249.6	247.4	237.0	260.7	218.3	206.1	216.1	247.9	275.4
					B	345.3	300.1	289.6	389.0	249.1	230.3	259.4	256.9	271.0	226.0	229.5	223.2	232.6	248.2	215.6	207.7	195.4	220.2	253.5
				Oolong tea	A	309.9	261.6	298.0	411.6	276.2	301.8	221.9	325.7	241.4	245.8	275.8	309.0	274.7	287.9	270.1	235.6	259.6	201.1	278.5
					B	302.3	239.7	286.9	367.1	279.4	250.8	241.2	305.4	209.9	249.6	228.3	286.6	271.6	258.2	236.8	206.6	215.4	187.0	256.5
			GC-MS/MS	Green tea	A	361.5	316.6	333.5	384.4	279.0	262.6	277.8	273.8	264.9	244.3	333.5	258.0	251.3	254.0	201.8	198.4	187.6	265.0	274.1
					B	341.1	300.3	299.1	348.2	242.5	232.2	269.1	259.9	252.3	239.0	325.3	225.1	238.0	233.6	190.8	193.3	188.2	243.2	255.5
				Oolong tea	A	352.4	296.0	317.8	460.8	246.5	270.2	220.4	490.7	282.6	222.8	275.6	254.4	290.5	255.5	238.7	204.9	225.0	212.5	284.5
					B	308.8	279.5	281.7	418.5	258.5	222.5	239.2	467.8	243.8	240.3	229.0	249.8	246.8	255.9	219.4	193.5	193.4	209.8	262.8
		ENVI-Carb + PSA	GC-MS	Green tea	A	436.9	386.7	334.4	299.1	314.1	330.8	330.7	252.7	284.2	267.0	323.6	269.8	228.1	215.4	258.9	263.6	265.7	201.3	294.2
					B	378.4	348.1	308.1	223.5	314.9	316.0	296.3	232.5	276.3	310.8	308.7	243.6	249.2	193.3	242.8	239.9	239.2	220.9	269.8
				Oolong tea	A	520.0	269.9	237.7	229.1	360.8	244.4	302.4	206.6	198.2	189.0	253.2	242.5	274.7	326.7	233.9	201.2	194.1	176.9	256.9
					B	428.5	213.5	256.8	226.8	299.0	238.8	252.5	187.1	134.0	187.6	225.9	212.7	271.6	292.6	198.7	180.3	193.2	156.6	229.5
			GC-MS/MS	Green tea	A	492.3	373.7	317.9	261.6	290.4	230.3	245.9	246.9	306.5	377.7	279.5	290.2	200.9	231.8	230.3	220.1	270.6	216.8	283.0
					B	552.5	328.6	307.0	217.3	293.2	306.3	260.7	237.9	293.8	356.0	268.2	254.2	223.4	217.5	232.5	196.9	251.8	243.1	274.8
				Oolong tea	A	267.5	282.2	473.9	249.5	388.6	245.3	283.8	212.7	198.2	187.2	232.8	209.0	209.3	289.8	238.7	181.8	217.5	179.9	250.4
					B	139.9	229.3	260.8	237.5	324.1	244.1	244.9	180.9	134.0	186.7	214.5	181.3	188.9	235.9	219.4	171.1	194.3	157.1	208.0
132	trans-Chlordane	Cleanert TPT	GC-MS	Green tea	A	44.4	35.8	37.0	45.4	32.3	30.2	35.3	32.6	33.3	29.5	30.2	30.9	28.6	31.7	26.8	26.2	27.9	34.4	32.9
					B	42.7	34.8	33.5	42.1	30.4	28.2	32.8	30.2	33.9	28.3	28.3	29.1	27.9	29.6	26.5	25.9	27.7	30.5	31.2
				Oolong tea	A	40.0	39.6	38.2	41.5	33.5	35.6	33.5	41.4	36.3	33.8	35.7	35.9	32.2	32.6	33.9	32.9	32.6	28.9	35.5
					B	39.0	35.3	33.1	38.3	32.7	32.0	26.8	37.4	32.1	31.1	30.6	30.6	30.5	29.2	30.6	28.8	32.5	27.5	32.0
			GC-MS/MS	Green tea	A	46.7	35.4	38.4	43.9	33.3	28.2	35.8	31.0	34.0	29.5	27.9	29.9	29.5	30.7	25.2	24.8	26.0	34.0	32.7
					B	44.4	36.2	33.6	40.4	30.8	26.6	32.8	28.9	33.0	26.7	26.7	29.6	27.4	28.3	25.0	23.9	27.7	30.1	30.7
				Oolong tea	A	40.8	40.2	39.3	42.3	33.4	34.9	33.9	41.6	37.2	33.2	37.8	35.6	33.7	30.0	31.0	32.1	35.6	26.0	35.7
					B	39.6	36.7	33.2	39.3	33.2	31.2	28.3	36.3	35.1	31.9	31.7	29.3	30.0	26.5	28.8	28.3	32.6	24.5	32.0
		ENVI-Carb + PSA	GC-MS	Green tea	A	46.1	42.1	36.6	36.2	38.6	39.0	30.2	31.3	31.5	29.8	32.8	31.5	34.2	27.6	34.2	30.1	29.2	30.8	33.8
					B	44.7	35.9	35.9	33.8	37.1	36.1	37.4	29.6	30.0	30.0	31.3	29.2	30.8	25.5	28.2	27.5	27.7	31.1	32.1
				Oolong tea	A	42.0	35.7	35.7	33.3	38.6	32.9	33.1	29.8	35.3	32.3	31.2	30.8	32.2	35.1	30.4	28.7	30.0	27.4	33.0
					B	36.1	31.9	33.2	31.0	32.1	30.5	31.5	26.3	29.2	27.8	28.3	28.9	30.3	29.8	27.1	25.9	26.5	25.6	29.6
			GC-MS/MS	Green tea	A	47.8	43.6	38.3	38.3	36.6	40.6	36.5	31.8	30.6	33.0	49.5	32.7	35.7	29.8	36.0	29.0	29.1	30.4	36.0
					B	46.9	36.3	36.9	35.4	37.1	38.6	34.5	30.2	30.1	33.4	30.4	30.3	33.0	26.7	29.1	27.1	30.6	30.7	33.5
				Oolong tea	A	44.6	34.6	42.2	33.2	40.5	33.6	34.2	28.8	35.3	31.8	30.4	28.8	29.4	33.5	31.0	26.0	30.2	27.2	33.0
					B	38.7	30.5	34.5	29.6	34.9	31.7	32.4	26.5	29.2	27.4	27.9	27.4	26.4	26.1	28.8	25.9	24.8	26.5	29.5

No.	Compound	Cleanup	Method	Matrix		C1	C2	C3	C4	C5	C6	C7	C8	C9	C10	C11	C12	C13	C14	C15	C16	C17	C18	C19	C20
133	trans-Diallate	Cleanert TPT	GC-MS	Green tea	A	87.3	66.6	68.6	79.8	65.7	60.4	71.6	66.3	69.0	63.7	84.0	77.0	77.2	58.2	80.1	72.6	33.3	76.7	88.7	70.9
				Green tea	B	83.5	66.3	62.5	74.6	62.2	56.4	66.5	60.4	68.0	59.3	77.4	71.9	74.2	51.2	66.6	72.6	33.6	74.6	78.7	66.3
				Oolong tea	A	78.2	70.6	68.7	78.9	63.8	68.1	61.7	78.4	73.2	83.5	86.4	88.4	85.8	73.1	72.8	85.2	86.5	72.7	74.9	76.4
				Oolong tea	B	75.6	62.8	61.2	70.0	63.2	61.8	53.6	70.1	64.8	73.8	69.5	76.3	74.1	73.1	62.9	71.5	80.7	64.4	68.5	68.3
			GC-MS/MS	Green tea	A	85.7	70.6	75.4	89.4	63.2	61.6	68.9	61.7	63.0	57.1	69.5	50.3	65.0	55.0	59.5	51.7	51.6	48.7	64.2	63.8
				Green tea	B	82.2	68.6	67.4	83.3	62.0	57.9	68.2	59.0	64.5	54.6	62.9	47.0	55.9	54.3	58.8	50.4	52.5	48.1	57.6	60.8
				Oolong tea	A	69.6	72.9	72.9	85.2	60.9	65.2	62.4	77.7	71.7	66.3	71.9	68.9	66.6	64.3	61.5	63.3	62.4	55.9	55.1	67.1
				Oolong tea	B	69.4	64.2	64.0	74.1	59.3	56.3	51.8	73.1	65.8	60.5	60.3	60.2	56.4	60.7	51.0	58.4	59.7	52.5	48.9	60.3
		ENVI-Carb + PSA	GC-MS	Green tea	A	98.3	82.3	68.6	67.8	78.2	76.6	68.6	65.0	66.1	82.8	82.4	91.7	81.9	80.6	78.2	93.4	83.8	80.7	86.2	79.6
				Green tea	B	88.6	69.6	68.1	62.9	74.4	68.8	83.0	61.3	62.5	79.7	77.8	81.9	80.0	75.4	74.6	78.2	75.3	73.6	84.4	74.7
				Oolong tea	A	80.5	62.7	65.2	60.5	69.5	56.7	66.6	53.7	65.5	79.2	78.1	74.0	74.4	79.3	90.3	78.1	80.9	73.2	72.3	71.6
				Oolong tea	B	66.8	55.4	60.1	56.1	57.9	51.9	58.2	52.3	55.5	66.6	69.1	69.0	68.0	70.1	75.8	64.7	69.3	64.9	67.1	63.1
			GC-MS/MS	Green tea	A	92.2	78.8	71.4	69.8	77.2	76.8	75.8	62.1	64.7	70.6	68.8	77.9	82.8	72.6	63.1	71.7	60.4	64.1	66.1	71.9
				Green tea	B	85.4	66.0	69.6	66.4	74.6	70.2	57.7	59.2	59.2	68.7	60.6	77.5	75.1	65.5	57.4	60.3	58.4	56.6	66.1	66.0
				Oolong tea	A	81.1	63.9	79.2	61.9	71.6	61.9	58.9	126.9	65.5	62.1	58.1	56.1	58.6	56.2	63.2	66.0	53.8	51.6	51.3	65.7
				Oolong tea	B	67.1	57.2	61.0	57.9	61.1	57.5	56.5	113.4	55.5	53.2	59.0	53.5	52.7	51.1	50.8	60.2	46.3	49.5	50.3	58.6
134	Trifloxy-strobin	Cleanert TPT	GC-MS	Green tea	A	224.3	198.3	158.9	398.0	140.9	139.6	150.3	133.0	127.2	109.6	214.4	138.0	130.2	110.8	124.3	117.5	119.5	92.5	133.1	155.8
				Green tea	B	226.1	186.1	142.3	360.8	137.8	131.5	136.6	125.4	160.3	104.1	199.0	120.4	121.3	115.3	131.0	117.5	120.6	91.1	118.4	149.8
				Oolong tea	A	184.4	175.3	142.9	205.2	199.5	144.8	125.5	167.8	142.4	138.5	148.6	140.6	153.3	132.7	137.1	152.4	132.8	129.5	110.0	150.7
				Oolong tea	B	176.9	161.7	146.3	181.4	178.5	126.6	131.2	160.7	115.4	127.9	121.7	123.1	142.2	132.8	122.3	137.7	126.5	111.3	107.1	138.5
			GC-MS/MS	Green tea	A	223.6	192.8	159.2	228.7	141.5	147.0	154.0	119.1	111.2	112.2	194.6	154.7	148.2	118.3	126.8	112.5	105.9	87.6	128.5	145.6
				Green tea	B	215.3	178.5	143.7	215.2	140.8	136.1	139.8	115.4	137.6	116.7	177.4	143.9	125.5	121.6	123.9	108.9	117.3	93.4	105.7	139.8
				Oolong tea	A	210.9	228.8	151.8	211.6	138.5	141.3	121.0	211.6	142.6	134.2	159.5	143.8	147.6	141.6	131.3	138.5	120.8	136.8	98.3	153.2
				Oolong tea	B	188.0	220.6	145.7	180.0	133.2	121.8	124.6	182.4	118.2	131.2	135.5	129.3	130.0	133.2	116.1	130.7	108.6	116.0	93.0	138.9
		ENVI-Carb + PSA	GC-MS	Green tea	A	232.5	177.0	181.3	179.2	168.0	162.6	134.9	127.6	179.3	183.6	191.7	238.4	138.9	132.8	86.2	119.7	124.7	200.7	163.3	165.4
				Green tea	B	235.3	197.9	153.0	155.9	157.1	168.6	174.5	118.9	144.6	152.6	152.6	188.5	134.1	138.7	90.2	92.1	151.0	172.1	166.9	135.3
				Oolong tea	A	211.6	118.3	145.1	132.2	195.0	130.4	151.3	128.9	144.4	127.8	125.9	125.4	126.8	132.7	129.7	124.6	110.3	110.0	108.5	135.7
				Oolong tea	B	190.9	114.3	143.3	134.1	156.0	126.7	128.0	108.2	116.3	117.2	105.4	132.5	110.0	132.8	123.6	94.3	109.3	103.7	103.2	123.7
			GC-MS/MS	Green tea	A	228.0	187.8	174.3	164.2	167.6	119.7	263.6	118.9	189.3	151.7	183.0	0.0	153.4	166.0	91.2	127.2	127.3	177.9	167.0	155.7
				Green tea	B	205.1	217.0	152.8	161.1	149.2	169.1	93.3	103.7	164.3	142.9	143.4	0.0	170.0	119.7	100.1	94.4	156.5	158.0	183.0	141.2
				Oolong tea	A	186.7	126.6	133.4	132.9	185.9	124.9	122.6	119.3	144.4	122.8	124.9	119.5	112.1	113.0	122.2	138.5	108.9	95.6	94.5	127.8
				Oolong tea	B	185.7	121.6	141.6	136.7	157.8	126.6	126.0	102.1	116.3	118.9	125.9	118.7	99.9	109.7	105.4	130.7	103.7	98.3	89.1	121.8
135	Zoxamide	Cleanert TPT	GC-MS	Green tea	A	93.5	84.1	95.7	126.3	81.4	70.7	90.0	82.6	82.3	77.6	78.9	68.7	71.5	77.9	75.7	65.3	60.8	73.0	80.4	80.9
				Green tea	B	93.5	81.4	88.3	123.6	73.4	65.6	84.8	73.9	78.3	73.4	70.1	66.5	68.4	76.5	66.9	64.1	57.4	68.8	73.1	76.2
				Oolong tea	A	91.2	82.7	86.0	90.6	76.7	80.4	88.9	88.2	81.6	75.5	80.6	79.8	79.4	73.4	70.1	70.1	69.8	66.7	66.7	77.8
				Oolong tea	B	88.1	74.3	81.0	86.3	83.2	78.0	69.9	86.2	77.1	68.8	70.1	73.7	72.5	69.4	63.9	61.3	72.1	59.3	59.6	73.4
			GC-MS/MS	Green tea	A	94.0	83.5	92.4	98.8	81.5	73.9	83.6	84.1	80.1	76.4	80.0	84.3	73.6	75.8	73.2	64.8	63.9	64.8	74.4	79.1
				Green tea	B	93.7	81.6	84.8	92.9	73.5	68.1	86.1	74.1	75.9	74.7	71.9	77.9	70.6	66.8	64.2	63.7	62.1	63.1	67.7	74.4
				Oolong tea	A	85.6	84.1	86.7	94.4	78.3	84.3	67.6	88.9	85.2	75.1	82.2	76.7	77.5	75.8	66.9	62.3	75.3	76.7	67.9	78.5
				Oolong tea	B	85.3	76.3	77.8	86.0	85.2	79.1	68.8	85.4	81.9	71.9	71.2	76.1	66.1	70.6	59.7	54.2	76.3	67.5	58.4	73.6
		ENVI-Carb + PSA	GC-MS	Green tea	A	109.4	91.4	90.8	97.6	84.0	87.3	59.4	78.2	83.0	71.5	70.8	79.7	77.8	78.6	62.4	74.6	63.1	74.9	72.6	79.3
				Green tea	B	111.1	82.9	89.8	87.1	87.5	80.4	66.6	67.5	75.5	69.6	70.8	78.8	67.6	73.7	58.9	66.2	63.0	68.1	75.4	75.8
				Oolong tea	A	105.4	83.3	81.8	82.1	80.1	71.3	67.8	61.7	72.4	72.4	67.1	62.0	67.1	73.4	57.5	60.4	66.0	61.2	60.2	72.0
				Oolong tea	B	92.6	79.9	96.0	78.9	71.9	62.5	66.9	62.1	64.0	56.0	64.9	56.0	63.8	69.4	60.5	58.3	60.0	57.8	62.9	67.6
			GC-MS/MS	Green tea	A	105.9	86.6	88.8	96.0	86.7	80.7	79.7	74.9	87.5	74.9	82.9	78.5	79.5	75.8	63.6	78.0	75.9	64.5	81.0	76.4
				Green tea	B	99.9	80.8	88.9	87.0	87.8	82.1	68.1	69.0	76.2	73.2	75.4	77.4	69.7	71.2	61.0	69.5	69.5	63.1	83.5	69.7
				Oolong tea	A	98.9	80.7	82.9	79.5	82.1	70.5	75.1	60.2	67.1	71.2	63.6	59.0	61.5	56.2	70.9	62.3	62.6	56.6	62.7	65.9
				Oolong tea	B	94.3	77.7	89.8	74.1	78.0	62.4	70.7	60.4	64.0	56.8	68.9	54.8	57.8	55.1	50.6	54.2	56.5	58.5	68.3	65.9

(Continued)

Appendix Table 5.1 The 201 Pesticides Average Content for 19 Circulative Determinations on 3 Months Under 16 Factors (Two Cartridges, Two Instruments, Two Tea Samples, and Two Youden Pair Samples) (cont.)

No.	Pesticides	SPE	Method	Sample	Youden pair	Nov 9, 2009 (n=5)	Nov 14, 2009 (n=3)	Nov 19, 2009 (n=3)	Nov 24, 2009 (n=3)	Nov 29, 2009 (n=3)	Dec 4, 2009 (n=3)	Dec 9, 2009 (n=3)	Dec 14, 2009 (n=3)	Dec 19, 2009 (n=3)	Dec 24, 2009 (n=3)	Dec 14, 2009 (n=3)	Jan 3, 2010 (n=3)	Jan 8, 2010 (n=3)	Jan 13, 2010 (n=3)	Jan 18, 2010 (n=3)	Jan 23, 2010 (n=3)	Jan 28, 2010 (n=3)	Feb 2, 2010 (n=3)	Feb 7, 2010 (n=3)	Ave (μg/kg)
136	2,4-dde	Cleanert TPT	GC-MS	Green tea	A	45.1	38.9	41.1	40.3	36.0	35.8	36.9	33.6	36.3	33.6	36.5	35.2	36.1	33.0	34.9	28.1	29.1	28.7	34.9	35.5
					B	43.2	35.4	35.5	36.3	31.9	33.5	32.8	30.9	34.1	31.2	31.9	33.4	35.0	30.9	30.0	27.5	27.9	27.3	31.1	32.6
				Oolong tea	A	42.0	37.2	37.9	42.3	36.1	37.0	33.7	38.5	38.9	38.0	40.0	42.9	38.4	35.9	35.7	35.4	34.9	33.5	30.7	37.3
					B	39.2	30.9	30.9	36.7	33.8	32.1	26.8	33.1	33.0	32.3	31.7	34.7	30.8	31.4	30.5	30.4	28.9	31.8	27.1	31.9
			GC-MS/MS	Green tea	A	45.5	36.2	43.8	37.8	37.6	35.1	37.9	33.0	33.6	33.3	37.2	34.8	32.5	34.0	35.8	26.6	28.8	28.2	35.7	35.1
					B	43.3	36.2	38.8	37.8	33.7	32.1	33.9	31.3	33.3	29.3	32.5	33.2	30.9	31.8	32.7	24.6	28.1	28.1	30.0	32.7
				Oolong tea	A	39.9	35.0	37.7	43.4	33.8	37.0	37.4	40.6	39.0	39.8	39.6	42.2	38.2	33.9	37.2	32.8	36.6	35.3	29.4	37.3
					B	37.5	30.3	30.5	36.8	33.4	30.6	28.5	34.6	33.8	34.1	30.5	32.9	29.2	29.0	31.2	28.1	28.7	29.6	26.6	31.4
		ENVI-Carb + PSA	GC-MS	Green tea	A	56.8	34.2	41.7	42.8	40.7	50.2	33.5	43.4	35.7	43.4	31.6	32.2	35.2	42.3	39.2	42.7	36.0	37.9	35.6	39.8
					B	57.0	44.8	41.4	40.1	40.2	46.3	37.3	42.8	34.2	37.0	31.4	33.3	33.1	36.1	35.2	36.0	32.3	31.3	35.4	37.6
				Oolong tea	A	45.4	35.2	41.8	38.4	38.8	34.9	36.7	31.2	37.5	35.6	39.3	34.1	35.1	39.1	40.5	33.5	30.0	39.6	30.7	37.2
					B	36.9	39.8	36.9	33.2	30.9	29.3	32.4	27.9	29.9	28.7	33.9	28.8	32.0	31.6	31.6	29.5	25.9	31.6	27.5	31.2
			GC-MS/MS	Green tea	A	47.7	30.8	42.0	42.1	42.1	41.6	35.6	43.1	39.1	34.1	35.8	33.8	36.3	41.7	38.3	44.6	37.3	37.4	37.9	39.5
					B	45.4	39.2	42.4	38.0	40.7	36.1	42.4	42.5	38.5	35.0	33.1	32.3	33.9	41.2	35.3	40.1	35.7	32.0	38.5	37.6
				Oolong tea	A	43.2	34.1	48.1	40.3	38.4	35.6	35.0	32.6	37.5	35.0	34.8	34.4	32.7	36.8	37.8	27.1	24.7	29.4	29.4	35.8
					B	35.7	36.9	36.9	34.6	31.0	29.8	31.2	27.9	29.9	27.4	32.9	29.4	29.3	31.5	27.4	23.8	24.0	30.3	26.6	30.2
137	Ametryn	Cleanert TPT	GC-MS	Green tea	A	143.3	107.3	144.1	126.8	26.3	69.1	62.0	83.4	121.9	84.8	106.2	94.9	91.1	93.4	96.0	77.6	76.3	70.3	87.0	92.7
					B	141.4	108.7	130.9	115.2	23.0	81.3	85.0	73.6	128.2	81.0	94.8	88.2	84.6	96.4	88.0	77.0	74.8	68.9	76.4	90.4
				Oolong tea	A	129.0	104.2	112.8	136.6	102.1	110.2	45.6	60.2	106.7	151.6	84.9	101.7	101.5	90.0	97.1	94.5	93.6	86.9	79.8	99.4
					B	116.8	95.1	103.3	128.4	94.4	95.3	40.2	79.3	77.9	125.1	66.9	79.2	80.5	78.9	80.9	78.0	73.0	79.6	69.9	86.5
			GC-MS/MS	Green tea	A	146.2	112.8	134.8	123.1	103.0	107.9	104.4	98.8	93.3	101.4	99.6	86.9	86.4	107.6	93.8	79.6	74.0	70.3	82.7	100.4
					B	137.6	113.3	116.3	108.5	93.1	98.5	92.0	90.6	93.4	98.5	87.9	83.8	77.5	113.9	88.8	78.3	76.4	67.5	72.2	94.1
				Oolong tea	A	124.4	102.8	107.0	131.0	100.9	108.2	103.0	105.8	111.5	116.3	117.0	119.1	110.3	92.2	94.1	89.0	93.4	88.3	73.9	104.6
					B	114.8	90.9	89.4	109.5	93.9	86.9	78.2	89.4	94.3	102.4	82.9	90.5	85.8	77.2	78.1	77.5	74.2	79.7	65.2	87.4
		ENVI-Carb + PSA	GC-MS	Green tea	A	158.4	133.5	122.1	123.8	123.5	135.1	123.3	143.6	63.7	90.9	82.1	169.4	90.7	94.5	87.8	102.0	91.1	88.8	83.4	110.9
					B	148.8	113.1	118.6	114.3	115.4	100.5	158.6	98.5	24.9	90.9	80.6	115.0	81.5	88.1	81.5	82.2	80.9	77.2	84.2	97.6
				Oolong tea	A	223.1	199.1	125.8	113.4	118.3	137.9	43.6	66.3	113.9	102.4	121.0	90.7	96.6	105.5	103.3	89.4	79.8	102.4	76.7	111.0
					B	182.1	109.7	110.4	101.5	89.0	123.3	81.2	49.4	88.0	87.9	96.3	87.3	81.3	86.5	78.3	72.2	66.6	79.9	67.0	91.2
			GC-MS/MS	Green tea	A	140.5	122.4	124.8	123.4	119.7	116.4	101.1	118.5	102.5	87.6	89.6	116.6	91.2	104.3	87.0	106.3	92.6	85.2	86.3	106.1
					B	129.6	98.9	117.2	112.3	115.8	104.6	97.4	109.9	96.8	89.2	81.0	109.6	84.4	85.6	87.0	87.1	80.4	78.6	86.9	97.0
				Oolong tea	A	129.4	125.2	143.4	121.2	112.5	110.4	100.9	89.3	113.9	97.0	102.0	74.8	95.1	102.2	100.0	80.7	68.3	96.3	73.9	101.9
					B	113.8	107.7	112.4	101.8	89.7	88.3	89.1	72.4	88.0	80.8	93.5	62.5	85.1	84.0	73.0	62.3	60.3	77.3	65.2	84.6
138	Bifenthrin	Cleanert TPT	GC-MS	Green tea	A	44.5	43.1	35.6	36.7	35.6	37.7	37.3	35.7	34.5	33.5	36.6	35.7	34.0	30.9	34.7	28.0	26.9	28.3	33.0	34.9
					B	43.3	39.4	31.2	33.5	31.3	35.3	33.3	33.6	33.4	32.5	32.6	33.4	32.1	29.1	31.3	27.5	26.4	27.4	27.0	32.3
				Oolong tea	A	49.2	43.8	45.3	54.0	49.4	54.8	41.5	49.7	54.3	49.2	53.2	53.7	54.3	45.0	50.6	42.3	43.4	46.8	42.3	48.6
					B	51.7	42.7	44.4	48.6	51.6	50.8	39.0	48.4	50.6	46.8	45.4	49.0	53.5	47.0	45.5	40.5	40.9	48.3	41.8	46.7
			GC-MS/MS	Green tea	A	46.9	35.0	41.9	38.1	32.7	35.1	34.2	31.8	30.7	31.4	32.2	32.5	26.8	34.0	31.6	25.6	24.8	25.2	28.7	32.6
					B	44.1	34.7	36.9	33.7	29.9	31.7	30.8	29.8	30.8	29.5	29.2	29.6	26.0	32.9	28.7	24.7	25.1	24.9	25.1	30.4
				Oolong tea	A	49.6	40.1	43.5	55.2	45.4	54.7	42.4	45.3	54.5	50.3	52.4	51.2	58.6	43.3	48.4	40.2	42.9	46.5	43.9	47.8
					B	52.3	41.0	42.5	49.2	48.9	48.8	39.6	45.9	50.7	48.2	43.4	45.5	58.3	44.8	43.3	38.7	40.4	47.3	43.9	45.9
		ENVI-Carb + PSA	GC-MS	Green tea	A	51.2	42.2	37.2	38.5	39.2	39.7	31.8	40.2	35.7	34.4	31.6	36.3	32.8	36.5	32.0	37.7	35.1	34.4	29.8	36.6
					B	47.5	35.5	38.2	35.2	39.9	35.1	29.5	37.6	32.7	33.3	29.7	35.9	30.6	32.8	30.6	30.6	31.2	30.8	29.8	34.0
				Oolong tea	A	54.0	53.1	56.3	50.3	52.2	48.4	48.0	73.1	49.9	54.0	53.8	44.4	50.2	47.5	51.5	43.7	43.6	57.4	44.4	51.4
					B	45.7	43.5	57.1	48.8	48.2	45.3	43.4	71.9	46.8	48.3	50.9	44.2	47.4	43.3	47.3	39.0	40.8	50.6	43.2	47.7
			GC-MS/MS	Green tea	A	43.6	39.2	38.0	38.3	38.3	34.2	34.8	38.7	34.3	30.9	29.4	57.1	33.4	29.4	30.7	35.6	31.9	28.9	32.0	35.9
					B	40.3	32.6	37.4	34.8	37.3	34.2	33.0	35.4	31.3	30.5	27.6	47.3	28.9	24.6	28.5	28.6	27.8	26.4	31.2	32.5
				Oolong tea	A	48.7	48.1	61.5	50.9	49.2	47.4	43.8	70.3	49.9	50.9	44.4	33.2	43.9	42.4	49.2	37.8	35.9	54.0	43.9	47.7
					B	42.3	42.6	56.5	48.4	45.3	42.7	41.7	67.9	46.8	46.0	45.6	33.4	44.6	40.3	42.5	33.5	36.5	48.9	43.9	44.7

No.	Compound	Cleanup	Detection	Tea	Rep	1	2	3	4	5	6	7	8	9	10	11	12	13	14	15	16	17	18	19	20
139	Bitertanol	Cleanert TPT	GC-MS	Green tea	A	161.4	132.1	149.6	123.6	117.9	93.8	112.7	119.8	98.4	83.2	118.1	108.1	94.4	138.2	99.4	105.5	96.5	88.9	101.0	112.8
					B	159.8	129.9	144.6	116.8	100.3	95.2	107.2	114.8	100.8	79.5	101.3	97.4	93.0	136.9	96.1	99.7	91.1	86.5	88.2	107.3
				Oolong tea	A	130.6	93.9	111.9	127.0	103.8	120.6	93.4	97.2	119.8	160.8	86.5	103.4	99.8	97.6	110.1	107.4	100.9	101.4	86.2	108.0
					B	113.5	101.2	118.6	119.9	98.5	103.1	91.7	99.0	85.5	132.5	68.9	86.7	92.2	87.2	93.1	84.9	82.0	92.5	73.2	96.0
			GC-MS/MS	Green tea	A	148.0	122.5	150.9	145.9	116.8	107.9	112.9	117.6	105.4	107.2	111.2	116.1	86.2	146.7	104.1	90.2	82.1	83.0	98.2	113.3
					B	143.5	122.8	135.3	117.5	99.9	102.1	105.0	107.0	104.0	101.3	98.7	102.9	79.7	154.0	94.2	86.6	79.8	79.1	88.0	105.3
				Oolong tea	A	168.6	108.1	110.8	139.8	106.0	116.2	98.2	108.3	122.0	130.5	117.4	119.7	112.8	94.8	97.6	95.4	95.9	91.9	80.3	111.3
					B	136.2	96.2	97.6	121.8	103.1	96.2	92.0	103.9	94.0	114.5	87.8	96.2	101.6	83.4	84.5	78.6	76.7	81.6	71.8	95.7
		ENVI-Carb + PSA	GC-MS	Green tea	A	150.0	133.9	138.2	124.7	133.3	104.2	86.3	138.2	86.2	104.1	98.9	258.9	103.3	113.9	105.1	120.4	103.3	106.8	96.0	121.4
					B	161.1	153.4	128.4	111.4	129.1	92.8	105.5	121.3	81.5	98.0	95.1	164.4	96.3	108.1	93.2	103.2	90.4	98.6	98.1	112.1
				Oolong tea	A	149.9	124.1	122.7	119.9	120.3	108.1	116.3	75.0	109.5	137.1	115.9	85.5	101.0	98.6	113.0	92.7	85.4	111.6	87.6	109.2
					B	137.4	106.8	120.0	115.0	86.4	91.0	90.6	59.5	91.7	100.4	90.4	82.8	85.7	88.5	86.5	76.7	73.7	90.9	76.1	92.1
			GC-MS/MS	Green tea	A	156.1	126.7	129.7	134.6	125.2	125.6	94.4	135.8	114.5	103.3	98.7	206.4	114.2	93.0	96.0	116.2	106.6	95.9	104.5	119.9
					B	164.1	118.9	125.9	115.5	127.5	116.1	121.2	117.4	99.0	99.9	90.1	162.5	96.2	86.3	87.4	99.8	92.4	82.7	102.6	110.8
				Oolong tea	A	127.5	125.3	124.4	122.8	117.2	101.9	109.0	88.8	109.5	106.7	95.9	78.6	94.4	94.1	87.4	81.3	72.9	102.0	80.3	102.3
					B	123.2	109.8	123.3	110.4	93.0	85.2	91.0	73.1	91.7	80.9	89.8	68.2	103.6	82.9	76.5	65.3	65.7	83.8	71.0	88.9
140	Boscalid	Cleanert TPT	GC-MS	Green tea	A	205.6	174.4	190.5	160.4	125.7	206.0	165.1	214.3	141.9	175.9	162.7	176.2	131.6	213.6	134.1	177.2	117.4	124.0	143.0	165.2
					B	211.2	158.3	180.1	158.2	132.3	185.9	152.4	209.6	177.6	173.6	152.7	146.6	125.3	235.4	156.9	175.1	115.0	125.3	124.3	162.9
				Oolong tea	A	205.6	143.8	165.9	222.2	207.2	210.4	122.6	159.9	147.6	169.9	182.9	149.4	138.4	114.1	187.9	180.0	146.5	150.8	130.7	165.0
					B	175.5	135.1	169.5	170.6	177.5	158.6	138.2	155.5	127.9	180.4	133.1	114.0	145.7	118.2	154.4	156.3	121.7	120.3	120.8	146.0
			GC-MS/MS	Green tea	A	240.9	235.0	184.7	178.0	128.1	194.2	159.7	192.0	132.1	192.0	166.0	212.1	108.6	200.3	147.0	153.8	122.3	120.2	147.3	169.2
					B	241.2	213.4	179.7	171.0	135.8	176.6	155.9	180.7	175.4	194.3	156.9	160.2	104.9	240.2	140.6	150.0	128.8	123.7	132.6	166.4
				Oolong tea	A	338.8	152.6	147.2	204.9	169.6	193.8	125.8	157.5	151.7	177.7	168.4	141.7	145.4	116.4	171.1	160.6	135.6	163.9	153.4	167.2
					B	236.1	134.9	138.8	167.3	155.2	145.2	133.2	150.5	114.8	184.6	121.9	111.5	149.9	113.4	147.7	133.5	112.1	122.9	123.1	141.8
		ENVI-Carb + PSA	GC-MS	Green tea	A	195.3	199.3	182.0	188.1	171.8	166.6	171.6	171.6	178.6	180.7	187.8	332.3	126.5	151.3	153.8	204.3	190.4	129.6	154.4	186.7
					B	217.5	263.5	160.7	150.8	162.0	185.5	141.4	149.1	163.4	155.6	127.6	160.5	127.9	148.6	143.4	156.9	158.6	156.0	163.8	171.8
				Oolong tea	A	216.8	113.8	171.4	146.3	257.2	223.1	118.4	154.7	149.7	139.2	181.9	181.9	152.3	112.1	170.3	142.0	117.3	139.3	152.1	158.9
					B	204.8	98.6	171.7	153.1	204.5	224.0	134.9	119.7	113.0	126.4	159.9	159.9	117.8	132.5	171.7	86.6	104.5	135.0	126.8	144.4
			GC-MS/MS	Green tea	A	226.1	209.7	178.4	195.0	159.2	194.0	198.3	165.1	192.2	176.1	167.4	183.2	127.2	145.4	129.1	156.7	165.2	120.5	162.8	171.1
					B	258.6	288.1	159.8	159.5	159.0	213.6	143.7	133.7	164.8	150.4	126.1	177.4	132.1	130.5	128.4	119.4	135.0	139.4	177.6	163.0
				Oolong tea	A	221.9	122.0	179.6	151.2	260.3	239.4	141.3	145.5	149.7	130.7	157.5	154.9	156.9	119.4	162.9	122.7	90.6	129.4	153.4	157.3
					B	199.8	108.4	170.1	146.0	200.1	232.2	121.3	116.5	113.0	119.0	149.1	156.6	143.8	125.0	138.2	80.5	84.6	120.4	123.1	139.4
141	Butafenacil	Cleanert TPT	GC-MS	Green tea	A	59.4	71.9	83.7	41.2	33.5	54.9	38.6	52.2	37.4	44.1	40.7	46.3	34.5	53.9	35.1	45.5	30.2	32.2	35.6	45.8
					B	59.0	64.4	87.7	40.7	35.0	49.6	36.1	48.6	45.8	42.0	39.8	38.0	31.8	62.1	40.6	45.0	30.7	33.4	29.9	45.3
				Oolong tea	A	48.3	38.4	40.0	53.7	52.0	45.5	30.4	39.9	39.9	44.9	40.5	30.3	37.8	29.5	43.5	44.6	38.2	40.1	32.2	40.9
					B	42.0	35.3	42.2	41.8	43.0	34.2	33.8	38.5	32.3	44.1	30.4	45.2	35.6	30.3	36.5	38.0	31.9	34.4	28.8	36.0
			GC-MS/MS	Green tea	A	72.2	63.4	49.8	38.6	31.2	53.9	39.6	47.5	31.2	54.6	40.1	35.2	25.9	54.3	37.0	40.0	32.0	32.0	37.5	43.5
					B	68.5	60.0	44.8	42.6	35.0	48.9	37.2	45.9	45.0	53.7	39.4	36.6	26.3	70.9	36.8	37.8	32.1	32.9	33.8	43.5
				Oolong tea	A	62.9	40.2	38.5	51.1	43.7	46.5	29.5	39.4	37.7	42.4	44.4	30.4	38.1	32.6	41.2	39.3	35.8	41.8	41.8	41.2
					B	47.7	36.5	35.8	41.4	38.3	35.4	33.7	35.4	29.9	42.1	32.1	32.1	36.4	29.8	35.4	33.7	29.6	32.5	35.7	35.4
		ENVI-Carb + PSA	GC-MS	Green tea	A	43.1	55.5	50.9	50.9	40.0	28.9	63.4	39.2	50.4	50.8	46.3	253.8	29.7	41.0	35.8	44.0	53.5	29.0	37.6	54.9
					B	50.8	80.0	40.7	42.4	36.1	34.6	40.1	38.5	44.8	34.7	31.1	165.8	30.6	38.6	36.0	32.7	45.2	53.5	40.1	48.0
				Oolong tea	A	52.2	28.9	41.3	35.6	63.1	60.1	40.1	37.7	41.4	29.5	46.7	31.1	38.4	26.1	38.3	36.4	26.1	45.2	38.0	39.9
					B	46.8	24.9	42.1	35.5	50.6	58.6	31.0	30.6	31.9	48.0	41.8	46.7	30.2	33.2	37.9	26.6	24.7	24.7	32.6	35.9
			GC-MS/MS	Green tea	A	40.9	60.0	48.3	51.7	40.8	47.6	77.9	39.1	61.2	39.0	37.9	38.4	28.9	44.8	29.6	32.7	52.2	52.2	41.0	45.2
					B	40.7	87.1	39.0	44.6	38.4	54.1	34.4	31.1	49.6	31.9	30.2	51.2	35.2	30.9	35.3	22.3	40.4	25.5	44.6	40.9
				Oolong tea	A	47.8	31.9	39.7	37.6	58.3	65.7	36.2	36.0	41.4	27.1	40.1	44.1	37.1	28.3	38.9	30.7	21.0	36.5	41.8	38.8
					B	47.6	28.0	41.8	36.8	48.0	62.5	30.9	31.0	31.9	27.1	38.2	38.2	28.2	31.1	33.1	19.8	21.8	31.2	35.7	34.8

(Continued)

Appendix Table 5.1 The 201 Pesticides Average Content for 19 Circulative Determinations on 3 Months Under 16 Factors (Two Cartridges, Two Instruments, Two Tea Samples, and Two Youden Pair Samples) (cont.)

No.	Pesticides	SPE	Method	Sample	Youden pair	Nov 9, 2009 (n=5)	Nov 14, 2009 (n=3)	Nov 19, 2009 (n=3)	Nov 24, 2009 (n=3)	Nov 29, 2009 (n=3)	Dec 4, 2009 (n=3)	Dec 9, 2009 (n=3)	Dec 14, 2009 (n=3)	Dec 19, 2009 (n=3)	Dec 14, 2009 (n=3)	Dec 24, 2009 (n=3)	Jan 3, 2010 (n=3)	Jan 8, 2010 (n=3)	Jan 13, 2010 (n=3)	Jan 18, 2010 (n=3)	Jan 23, 2010 (n=3)	Jan 28, 2010 (n=3)	Feb 2, 2010 (n=3)	Feb 7, 2010 (n=3)	Ave (μg/kg)
142	Carbaryl	Cleanert TPT	GC-MS	Green tea	A	120.0	124.8	80.8	126.5	145.7	130.5	115.2	125.9	105.7	104.2	110.0	103.6	120.6	115.3	105.3	90.5	92.7	95.8	106.9	111.6
					B	106.9	108.1	66.7	117.7	152.5	113.6	92.5	116.5	91.7	87.0	99.6	100.1	109.3	124.6	100.3	83.1	84.7	92.9	85.5	101.8
				Oolong tea	A	125.6	175.8	107.3	123.9	127.4	123.9	99.3	122.3	113.1	115.7	104.1	139.7	121.8	134.5	114.3	107.3	108.3	102.0	103.4	119.4
					B	117.3	123.4	77.1	83.8	110.0	107.6	92.0	101.7	104.5	100.0	87.5	118.0	94.8	107.8	93.2	85.7	91.9	95.4	87.9	98.9
			GC-MS/MS	Green tea	A	88.8	70.1	104.4	163.6	131.4	136.9	114.5	115.0	95.6	100.6	138.8	74.7	104.3	113.1	103.9	93.7	92.1	98.0	112.9	108.0
					B	74.7	68.4	99.0	151.1	156.9	115.4	83.5	100.7	84.4	80.9	148.7	70.7	88.7	122.7	107.4	80.2	94.0	98.0	94.7	101.1
				Oolong tea	A	103.5	140.0	104.8	128.7	113.5	107.6	101.2	116.4	120.6	121.5	98.8	136.6	129.5	132.1	106.9	104.1	102.8	107.8	102.9	114.7
					B	103.7	109.2	69.7	77.8	110.3	89.4	89.2	94.1	100.7	102.9	88.7	111.5	90.8	101.2	88.0	88.4	92.7	96.8	97.4	94.9
		ENVI-Carb + PSA	GC-MS	Green tea	A	170.4	140.1	123.6	128.9	139.8	168.0	95.7	137.7	114.5	95.8	117.2	181.5	113.5	121.6	113.7	129.6	113.9	114.8	104.5	127.6
					B	160.9	126.3	117.2	118.1	135.1	170.8	122.6	131.6	109.1	96.5	114.7	156.4	97.7	110.6	96.0	114.4	100.7	100.0	111.4	121.1
				Oolong tea	A	153.3	142.8	130.5	114.8	131.7	119.5	123.6	103.6	126.8	135.2	135.2	108.6	111.8	103.6	103.3	109.1	93.8	134.4	105.3	119.9
					B	125.3	110.3	123.6	114.4	114.1	114.1	108.5	87.7	96.9	123.3	89.0	93.8	99.2	100.3	97.5	83.7	85.1	105.1	87.8	103.1
			GC-MS/MS	Green tea	A	158.8	150.9	129.6	135.7	119.8	124.8	98.6	136.6	112.0	83.4	122.1	166.1	106.2	130.2	115.4	128.6	122.4	104.1	119.9	124.5
					B	155.0	142.6	121.7	127.3	116.7	117.6	116.7	116.3	108.1	87.3	122.5	156.0	90.9	103.3	99.8	113.4	106.0	104.5	127.4	117.5
				Oolong tea	A	129.3	120.0	131.6	122.8	128.0	114.6	111.6	101.3	126.8	115.9	85.4	102.6	125.8	96.2	117.6	93.2	76.4	109.0	102.9	111.1
					B	119.7	100.9	122.3	106.7	110.4	105.5	99.1	78.8	96.9	111.9	72.9	90.9	127.6	93.3	91.7	69.3	73.5	93.7	95.2	97.9
143	Chlorobenzilate	Cleanert TPT	GC-MS	Green tea	A	49.2	36.8	38.0	39.3	30.4	40.1	36.4	45.0	31.7	36.6	33.6	37.7	31.7	40.5	34.0	38.8	27.0	26.2	34.1	36.2
					B	48.4	39.2	34.0	37.7	30.1	37.7	34.6	44.5	39.8	34.1	33.5	32.8	29.3	43.0	36.5	38.2	29.1	27.2	29.1	35.7
				Oolong tea	A	44.1	35.7	39.3	47.2	40.0	44.2	35.7	38.1	37.9	35.3	49.2	36.7	35.5	29.5	40.8	40.9	37.1	34.3	31.1	38.6
					B	39.9	32.5	37.8	39.6	35.1	36.6	30.3	38.1	30.5	27.6	45.0	32.4	32.5	30.0	34.3	36.1	30.0	31.5	30.4	34.2
			GC-MS/MS	Green tea	A	56.1	46.2	42.5	46.3	31.6	40.7	37.4	42.1	31.2	37.1	39.0	47.0	28.1	36.3	35.4	32.2	27.3	28.0	33.9	37.8
					B	53.4	42.7	39.4	38.3	31.5	37.3	34.0	40.0	38.4	34.6	38.7	38.0	26.7	38.5	32.9	30.7	28.9	28.8	30.0	35.9
				Oolong tea	A	61.0	35.4	36.2	47.3	37.7	42.8	34.0	38.3	36.5	42.0	43.7	38.7	39.5	33.2	37.1	37.2	36.4	37.9	34.2	39.4
					B	49.1	32.1	32.3	38.9	34.2	34.0	30.8	35.6	29.2	29.4	40.5	29.7	34.2	30.5	31.7	31.9	29.1	32.0	29.9	33.4
		ENVI-Carb + PSA	GC-MS	Green tea	A	52.5	50.6	41.4	41.8	41.0	44.1	68.0	37.4	30.2	38.3	37.5	104.3	29.2	29.8	55.1	53.6	41.3	23.5	32.5	44.9
					B	53.2	53.6	37.8	31.8	37.5	46.5	46.1	32.0	31.8	25.5	35.7	75.2	25.6	35.1	47.8	43.1	38.6	37.9	36.1	40.6
				Oolong tea	A	46.5	35.3	43.5	37.3	51.4	51.4	33.2	36.5	38.7	42.9	37.4	32.7	35.1	33.9	40.9	35.3	28.7	36.1	33.2	38.4
					B	41.9	29.8	40.4	35.6	36.9	46.9	29.9	27.8	30.1	36.3	33.4	34.5	29.0	32.4	40.6	24.3	25.6	32.8	28.6	33.5
			GC-MS/MS	Green tea	A	102.2	71.0	41.2	42.9	40.5	43.4	54.9	37.5	38.5	38.9	37.5	30.3	32.2	35.8	34.1	37.9	40.8	32.8	34.4	43.1
					B	114.4	96.6	37.2	35.7	38.0	44.2	33.0	32.5	36.6	26.6	34.7	28.3	27.6	35.4	30.8	30.8	35.2	32.3	39.5	41.5
				Oolong tea	A	60.5	35.9	49.0	39.6	47.4	45.6	35.0	35.2	38.7	37.5	34.0	35.5	35.2	34.2	38.6	30.0	24.2	33.7	34.2	38.1
					B	51.1	31.7	40.2	36.0	36.3	40.6	30.7	26.2	30.1	34.8	30.7	34.2	34.8	36.9	31.9	21.8	23.2	29.7	29.9	33.2
144	Chlorthal-dimethyl	Cleanert TPT	GC-MS	Green tea	A	128.9	123.5	200.2	116.5	95.6	107.6	104.0	102.1	102.4	106.7	99.4	103.5	98.8	101.9	97.4	92.2	81.1	80.0	94.2	107.2
					B	126.3	117.3	190.2	108.4	91.4	100.1	97.4	94.8	105.8	95.9	95.4	95.3	92.2	94.3	93.3	90.1	81.5	78.9	83.5	101.7
				Oolong tea	A	191.5	107.9	110.3	125.2	103.7	109.2	98.2	112.6	109.0	119.3	107.3	117.3	107.2	100.8	104.0	103.8	104.2	96.8	89.0	111.4
					B	174.6	90.0	91.3	105.5	93.7	90.2	72.6	98.8	92.0	89.8	93.5	94.2	86.6	90.0	86.1	88.5	85.0	88.8	78.9	94.2
			GC-MS/MS	Green tea	A	132.3	103.6	118.1	116.8	99.3	109.3	107.3	101.2	96.1	103.3	98.4	126.6	93.1	107.6	98.8	84.6	78.8	83.1	92.1	102.7
					B	125.8	99.9	109.3	106.1	95.0	98.9	96.8	93.3	99.5	93.9	94.1	108.0	85.5	106.7	90.8	80.1	79.7	79.8	79.2	95.9
				Oolong tea	A	145.0	100.0	103.2	129.8	99.8	109.7	101.3	110.6	110.9	117.6	116.6	118.6	102.9	99.3	100.2	96.5	100.1	99.0	84.4	107.7
					B	123.4	87.0	85.6	107.8	90.2	87.5	75.3	93.4	90.6	84.7	99.8	92.3	85.4	86.3	84.4	82.3	81.5	87.5	76.1	89.5
		ENVI-Carb + PSA	GC-MS	Green tea	A	135.2	112.7	109.2	111.8	119.0	127.4	111.2	117.1	101.3	94.8	97.8	101.7	97.2	108.1	104.9	113.2	103.4	92.7	93.6	108.0
					B	128.3	96.0	108.9	102.3	114.0	116.4	102.6	109.6	97.7	86.9	96.2	100.7	90.4	101.5	98.3	95.9	94.0	91.6	93.4	101.3
				Oolong tea	A	125.4	73.6	121.1	110.1	117.3	111.3	101.9	94.7	112.3	115.0	101.4	98.8	103.2	107.7	113.2	97.3	86.0	109.9	85.0	104.5
					B	101.5	94.8	105.6	96.7	90.4	94.9	91.4	83.3	87.0	99.6	82.1	83.9	89.3	91.8	92.5	77.8	73.3	90.4	75.5	89.6
			GC-MS/MS	Green tea	A	152.3	118.7	115.2	114.4	116.9	116.4	106.5	122.1	100.4	94.7	98.2	87.7	97.5	108.2	99.7	111.1	103.0	94.7	99.4	108.3
					B	148.6	106.3	111.9	105.2	113.4	108.3	104.9	113.3	95.0	87.5	97.3	79.9	89.6	99.1	91.4	94.3	92.0	90.3	99.8	101.5
				Oolong tea	A	123.5	111.2	149.3	112.2	116.1	103.7	103.0	91.7	112.3	101.5	102.7	86.9	96.4	100.5	109.5	85.0	71.2	109.4	84.4	103.7
					B	100.3	94.3	107.2	96.7	90.6	86.9	89.4	74.0	87.0	93.0	82.9	75.8	84.3	88.7	82.0	69.0	64.8	88.1	76.1	85.9

Average content (μg/kg)

#	Compound	Instrument	Cleanup	Sample	Rep	1	2	3	4	5	6	7	8	9	10	11	12	13	14	15	16	17	18	19	20
145	Dibutylsuccinate	GC–MS	Cleanert TPT	Green tea	A	78.8	64.0	67.5	69.7	60.0	64.8	61.3	57.3	58.7	54.1	59.4	54.9	56.0	56.5	51.9	43.1	47.3	47.8	48.1	57.9
					B	73.2	59.1	52.4	64.4	53.8	58.9	57.9	52.6	56.6	50.1	52.9	51.0	51.8	53.2	47.1	41.6	39.6	46.5	42.0	52.9
				Oolong tea	A	75.9	56.6	55.4	65.5	52.6	54.9	45.4	54.3	51.3	50.0	48.3	52.1	45.8	41.1	42.8	41.0	39.8	40.1	38.1	50.1
					B	68.1	46.9	47.4	55.8	47.2	44.5	36.0	45.4	43.0	42.5	37.1	41.3	36.8	35.8	34.4	34.7	29.7	34.1	30.1	41.6
		GC–MS/MS		Green tea	A	85.2	62.3	76.3	67.7	57.9	63.2	64.6	56.9	56.0	55.0	56.3	66.0	52.0	59.3	53.7	45.2	42.9	44.0	48.9	58.6
					B	78.5	62.0	68.1	61.5	54.5	57.0	57.7	52.6	56.3	50.4	52.7	57.2	47.2	59.5	48.7	42.9	42.7	42.6	43.2	54.5
				Oolong tea	A	71.0	52.3	52.8	65.8	48.4	52.1	45.6	49.5	51.3	52.2	52.0	49.8	45.6	41.2	41.5	39.2	38.1	39.5	32.2	48.4
					B	64.7	44.7	44.9	55.2	44.1	41.5	35.9	42.5	42.1	44.8	37.7	38.4	38.8	34.9	33.9	33.3	31.3	33.3	27.9	40.5
		GC–MS	ENVI-Carb + PSA	Green tea	A	99.8	73.1	69.0	68.3	74.4	76.9	60.2	72.5	58.2	60.2	54.2	89.9	55.3	59.7	54.4	61.0	49.1	61.0	55.0	65.9
					B	88.3	61.9	67.5	61.9	72.6	66.6	60.8	68.7	53.2	56.5	50.5	71.5	50.5	54.6	48.4	51.2	44.1	52.4	53.0	59.7
				Oolong tea	A	78.2	62.6	64.6	58.1	56.4	50.7	54.9	41.1	49.7	50.6	51.7	38.0	42.7	46.1	44.8	39.0	35.0	44.7	30.4	49.4
					B	58.8	51.6	56.3	50.9	44.2	42.0	43.3	34.3	39.5	39.2	41.9	33.2	35.7	37.2	34.1	31.2	27.7	33.6	24.5	39.9
		GC–MS/MS		Green tea	A	81.4	68.6	70.5	68.3	70.5	66.9	63.5	73.2	60.9	60.1	57.0	85.9	56.7	66.6	54.4	61.3	55.5	53.9	54.7	64.7
					B	73.7	54.6	66.9	61.6	70.6	61.9	62.5	65.9	54.4	57.2	51.0	76.7	48.9	55.2	48.6	52.8	49.5	48.4	51.0	58.5
				Oolong tea	A	75.8	59.6	83.0	58.6	55.6	53.4	48.0	41.2	49.7	46.3	43.2	32.9	41.8	42.4	45.2	34.2	29.8	40.3	32.2	48.1
					B	56.7	49.8	55.9	49.9	45.0	43.4	40.6	33.4	39.5	36.3	39.0	29.2	35.8	35.2	31.9	26.9	25.5	32.3	27.9	38.6
146	Diethofencarb	GC–MS	Cleanert TPT	Green tea	A	270.7	217.1	273.9	270.0	216.5	195.7	225.6	253.4	218.9	201.6	248.7	244.1	205.5	253.8	223.4	216.3	180.7	167.9	229.7	227.0
					B	295.0	248.6	282.0	250.0	204.1	198.6	215.4	234.9	223.6	178.8	223.9	229.0	212.9	256.8	211.0	213.4	187.4	166.6	202.5	222.9
				Oolong tea	A	299.9	226.6	236.9	260.7	220.2	266.7	222.7	202.2	199.8	424.2	303.5	331.9	302.5	198.3	237.7	247.3	253.2	223.5	208.5	256.1
					B	267.2	201.2	221.7	230.9	196.7	218.1	176.7	202.2	210.1	272.7	165.1	215.2	214.1	194.9	208.9	209.3	189.8	219.7	187.6	210.6
		GC–MS/MS		Green tea	A	295.5	245.5	271.3	255.3	222.9	236.8	236.9	242.8	210.9	223.3	224.4	306.4	217.1	265.4	223.3	194.4	180.8	181.4	214.2	234.1
					B	291.7	234.5	247.6	237.9	202.4	218.8	214.6	199.9	219.1	211.7	199.1	254.9	196.7	277.2	200.5	187.0	187.6	181.2	192.2	219.7
				Oolong tea	A	346.6	210.2	222.8	284.5	222.5	245.8	225.3	241.7	250.2	285.3	268.2	261.9	250.3	217.3	225.5	187.0	232.3	230.7	193.4	244.2
					B	291.8	194.9	197.5	242.7	206.0	202.3	184.6	218.9	187.0	245.0	189.1	197.5	213.7	193.3	194.4	190.1	184.8	206.1	175.9	206.1
		GC–MS	ENVI-Carb + PSA	Green tea	A	294.1	267.1	261.2	248.4	247.6	298.3	254.5	260.9	193.8	218.5	239.5	472.0	235.8	222.4	272.9	289.1	255.1	175.9	213.5	259.0
					B	341.2	236.5	253.7	213.7	244.5	280.7	263.8	226.9	200.9	215.7	192.7	341.4	193.5	223.4	246.5	243.4	220.9	215.2	227.5	241.2
				Oolong tea	A	268.4	203.6	245.0	233.7	262.6	263.2	216.0	214.4	246.7	314.5	330.2	284.7	276.0	290.2	235.7	225.6	198.7	248.9	191.6	250.0
					B	242.5	215.8	232.0	217.8	195.3	217.1	188.7	156.3	193.9	197.1	212.5	192.8	192.2	200.5	203.0	172.0	171.9	203.5	171.4	198.7
		GC–MS/MS		Green tea	A	468.1	332.7	260.0	261.3	264.0	276.9	225.0	269.1	223.3	213.2	218.8	210.4	216.3	239.9	228.6	253.7	235.7	193.0	219.5	253.1
					B	523.4	332.6	246.0	228.1	253.4	258.6	234.3	240.3	208.3	210.9	185.9	197.2	185.7	216.8	204.4	213.7	204.7	197.7	231.2	240.7
				Oolong tea	A	306.2	242.1	288.8	252.2	260.9	241.6	220.6	205.9	246.7	231.7	228.2	202.0	218.4	213.8	243.8	190.1	160.0	238.2	193.4	230.8
					B	269.1	214.1	243.5	226.4	203.0	198.6	196.8	153.6	193.0	182.0	206.0	182.7	195.9	192.9	185.7	151.5	151.8	193.6	175.9	195.6
147	Diflufenican	GC–MS	Cleanert TPT	Green tea	A	46.7	45.1	51.8	39.9	33.8	38.1	36.1	41.2	33.9	35.5	37.5	38.9	34.4	41.9	35.8	35.3	26.9	27.6	35.7	37.7
					B	46.2	38.9	44.9	37.2	31.0	35.1	35.0	39.5	36.4	33.9	34.3	34.3	33.1	42.3	34.8	35.6	29.1	28.7	31.1	35.9
				Oolong tea	A	43.9	32.6	35.6	46.9	37.5	36.6	32.3	37.9	36.9	43.6	41.9	37.6	38.4	34.3	38.6	38.4	33.7	35.3	34.6	37.7
					B	39.1	29.4	34.1	39.1	36.3	33.3	30.7	37.5	30.1	40.1	30.0	30.6	35.3	32.6	33.0	33.0	28.3	32.2	32.2	33.5
		GC–MS/MS		Green tea	A	50.9	41.5	43.8	41.4	34.6	39.4	37.7	39.0	33.5	35.4	36.4	48.0	28.8	41.3	34.5	30.4	28.1	27.3	32.6	37.1
					B	48.6	39.4	40.1	38.3	31.7	35.4	35.3	37.0	36.2	35.6	33.8	39.5	27.5	39.1	31.6	29.2	29.2	27.1	29.5	35.0
				Oolong tea	A	59.4	35.5	35.7	45.1	36.6	40.1	33.9	37.5	39.3	45.0	40.1	41.4	39.5	33.8	35.4	36.5	34.3	35.3	30.9	38.7
					B	48.2	32.1	31.5	37.2	34.6	33.5	30.9	35.5	30.5	41.1	29.2	32.9	35.4	30.0	30.6	30.8	28.0	29.6	27.9	33.1
		GC–MS	ENVI-Carb + PSA	Green tea	A	54.1	49.3	41.0	40.5	41.6	48.8	46.9	40.7	28.5	35.3	36.1	54.0	35.4	35.6	46.8	30.8	40.1	31.7	36.6	33.1
					B	57.5	47.4	39.2	33.7	40.4	49.6	42.8	35.9	28.2	34.9	31.3	48.0	34.0	36.1	39.3	46.8	35.4	35.9	41.4	39.2
				Oolong tea	A	45.0	36.2	43.5	39.4	43.8	41.2	35.0	34.4	37.6	36.9	44.9	33.3	28.5	39.4	33.8	39.3	33.7	40.5	39.2	38.5
					B	40.3	32.2	40.9	37.1	33.7	35.9	29.3	27.7	30.7	30.5	35.0	33.0	37.1	33.7	26.6	33.8	28.2	36.6	31.7	33.1
		GC–MS/MS		Green tea	A	78.4	56.4	41.9	43.0	40.4	43.1	40.6	41.9	37.0	34.7	37.6	32.0	34.9	29.9	35.1	39.5	36.2	36.6	33.8	40.4
					B	86.1	58.9	39.5	36.2	40.9	40.5	39.9	36.5	34.0	33.2	30.6	30.8	29.0	27.1	32.0	32.5	32.4	30.6	35.6	38.2
				Oolong tea	A	50.2	37.6	43.5	39.9	42.1	39.7	34.5	32.4	37.6	35.8	36.5	28.2	35.2	34.1	39.7	29.9	25.9	36.4	30.9	36.3
					B	46.6	33.8	40.0	35.8	33.1	33.2	29.6	26.2	30.7	30.7	32.8	26.0	31.6	29.8	30.2	22.9	23.2	31.7	27.9	31.4

(Continued)

Appendix Table 5.1 The 201 Pesticides Average Content for 19 Circulative Determinations on 3 Months Under 16 Factors (Two Cartridges, Two Instruments, Two Tea Samples, and Two Youden Pair Samples) (cont.)

No	Pesticides	SPE	Method	Sample	Youden pair	Nov 9, 2009 (n=5)	Nov 14, 2009 (n=3)	Nov 19, 2009 (n=3)	Nov 24, 2009 (n=3)	Nov 29, 2009 (n=3)	Dec 4, 2009 (n=3)	Dec 9, 2009 (n=3)	Dec 14, 2009 (n=3)	Dec 19, 2009 (n=3)	Dec 24, 2009 (n=3)	Dec 14, 2009 (n=3)	Jan 3, 2010 (n=3)	Jan 8, 2010 (n=3)	Jan 13, 2010 (n=3)	Jan 18, 2010 (n=3)	Jan 23, 2010 (n=3)	Jan 28, 2010 (n=3)	Feb 2, 2010 (n=3)	Feb 7, 2010 (n=3)	Ave (µg/kg)
148	Dimepiperate	Cleanert TPT	GC-MS	Green tea	A	86.6	78.2	70.9	77.7	62.5	60.1	65.6	68.0	65.2	52.7	72.9	71.5	65.9	67.1	64.1	61.4	53.3	53.7	62.3	66.3
					B	88.8	77.6	66.2	70.6	56.5	57.6	63.2	62.6	64.9	48.6	65.1	67.2	60.9	69.7	59.9	61.6	56.2	51.5	54.4	63.3
				Oolong tea	A	80.1	62.7	71.5	80.3	63.7	69.6	63.6	78.9	74.0	96.9	66.5	80.0	73.3	65.7	71.2	68.5	67.9	62.6	57.5	71.3
					B	72.2	55.4	63.6	73.4	59.1	57.8	45.1	70.9	59.8	82.9	51.4	66.0	61.6	60.8	61.8	60.8	58.2	58.3	52.7	61.7
			GC-MS/MS	Green tea	A	89.8	67.7	76.4	74.6	62.5	66.1	69.7	64.5	61.7	63.7	63.5	90.1	57.1	69.5	62.2	53.8	51.5	51.3	59.3	66.1
					B	84.6	65.8	69.7	70.5	57.7	61.0	63.8	59.9	63.5	57.7	60.2	77.0	53.9	67.4	55.7	51.7	52.3	50.1	52.0	61.8
				Oolong tea	A	99.9	64.9	65.3	80.7	61.7	68.4	65.0	72.7	75.1	77.3	77.4	75.0	77.8	61.9	64.5	62.3	65.7	61.9	46.0	69.7
					B	82.3	56.1	55.8	67.7	57.5	55.5	48.7	60.8	60.6	68.4	55.7	58.9	61.9	54.8	54.9	53.1	53.0	55.3	43.5	58.1
		ENVI-Carb + PSA	GC-MS	Green tea	A	91.7	95.6	73.3	68.3	77.0	77.8	66.7	79.1	58.6	62.6	62.8	74.3	61.8	70.8	68.9	78.6	65.8	63.8	60.7	73.3
					B	102.7	79.1	72.1	62.8	73.4	72.3	65.0	72.4	56.5	62.8	60.3	108.0	57.0	67.7	64.7	64.4	59.9	58.1	61.0	67.7
				Oolong tea	A	80.8	69.5	83.0	71.7	71.5	68.9	63.3	66.6	74.1	77.5	79.0	61.5	67.9	70.9	76.6	64.5	58.5	68.1	54.9	69.9
					B	69.5	61.8	67.4	63.4	54.6	57.9	54.5	54.2	56.6	62.3	65.8	60.8	60.6	61.1	62.9	54.4	50.0	58.0	49.3	59.2
			GC-MS/MS	Green tea	A	99.9	74.8	74.0	73.4	78.2	76.1	68.2	78.7	64.8	62.2	61.5	79.7	66.0	68.9	65.2	73.8	65.1	59.6	61.9	71.2
					B	101.0	65.3	72.2	66.6	76.0	69.9	74.2	73.4	60.3	63.1	57.3	75.7	57.1	64.5	58.2	61.6	58.0	55.9	63.2	67.0
				Oolong tea	A	80.8	72.8	93.8	76.6	72.7	73.3	65.9	59.3	74.1	66.8	64.9	62.8	63.2	64.5	73.2	53.9	47.0	68.9	45.1	67.3
					B	65.8	62.1	67.6	65.1	56.6	59.0	57.5	49.0	56.6	55.1	60.3	56.5	53.5	56.3	55.1	43.4	42.9	56.4	42.1	55.8
149	Dimetham-etryn	Cleanert TPT	GC-MS	Green tea	A	50.5	38.7	18.5	40.4	33.1	31.1	33.4	32.5	26.6	27.2	34.2	32.7	30.4	28.8	31.0	25.6	25.0	22.6	28.1	31.1
					B	49.3	36.6	12.4	36.2	29.4	31.6	30.8	30.0	26.5	26.5	32.0	30.5	28.6	28.9	27.4	25.3	25.0	22.1	25.3	29.2
				Oolong tea	A	44.0	33.5	36.3	39.2	32.6	35.6	32.2	34.1	36.2	50.6	26.7	32.6	33.7	31.1	30.4	30.8	31.4	28.4	26.2	34.0
					B	40.0	30.6	32.6	35.8	31.4	31.0	26.5	29.8	26.8	40.0	21.0	26.6	27.7	26.9	26.3	25.2	24.8	26.4	23.4	29.1
			GC-MS/MS	Green tea	A	49.1	37.5	45.8	43.0	34.9	34.3	34.5	31.6	30.5	31.5	33.6	27.4	26.5	32.0	31.0	25.2	24.1	22.6	29.3	32.9
					B	45.6	37.4	39.5	35.5	31.3	31.7	30.9	29.3	30.1	31.5	29.9	27.2	26.1	32.5	28.0	24.3	24.2	23.1	24.8	30.7
				Oolong tea	A	41.6	33.1	36.0	43.5	32.7	35.4	32.9	34.6	36.8	37.9	38.3	37.1	35.3	29.6	32.0	24.8	32.4	28.1	24.2	34.2
					B	38.5	30.5	29.6	36.5	31.2	29.3	26.7	30.3	30.8	32.2	27.6	29.1	28.0	25.5	26.8	24.8	26.1	25.7	20.8	28.9
		ENVI-Carb + PSA	GC-MS	Green tea	A	54.7	43.8	40.0	39.0	40.4	40.4	31.7	38.5	26.7	31.7	29.6	61.1	28.4	31.9	27.8	33.2	29.6	29.3	27.8	36.1
					B	53.3	37.3	38.5	34.9	39.1	34.0	32.5	36.1	27.3	32.1	28.0	39.9	25.8	28.9	26.6	26.7	25.7	25.1	27.6	32.6
				Oolong tea	A	51.0	42.1	41.3	38.2	37.9	33.9	34.5	24.7	36.0	37.3	32.5	25.0	32.9	34.3	34.9	29.7	26.7	33.9	24.9	34.3
					B	45.3	33.1	37.3	33.7	28.5	28.6	29.5	20.7	27.0	29.8	28.1	24.2	28.0	27.5	27.3	24.5	22.5	26.6	22.2	28.6
			GC-MS/MS	Green tea	A	46.9	43.4	41.7	40.1	39.6	40.5	34.4	40.7	33.9	30.2	28.6	37.3	29.4	35.1	28.7	33.9	31.8	29.1	27.4	35.4
					B	44.0	35.1	39.7	35.3	38.2	36.2	34.0	36.7	31.9	30.3	25.5	35.3	27.0	30.2	25.6	28.0	26.8	25.3	27.4	32.2
				Oolong tea	A	43.4	38.5	47.2	40.2	37.4	34.0	33.1	28.7	36.0	32.0	31.0	30.8	29.7	28.5	33.7	25.7	22.0	29.5	24.2	32.9
					B	38.4	34.2	39.0	34.9	30.3	27.9	28.8	24.1	27.0	26.0	29.3	26.5	30.8	24.8	24.0	21.1	19.9	25.2	20.8	28.1

150	Dimetho-morph	Cleanert TPT	GC-MS	Green tea	A	247.4	187.6	276.7	85.3	82.1	81.5	80.6	77.9	78.2	66.3	85.5	72.7	71.5	97.4	75.3	67.5	65.4	63.9	76.5	102.1
					B	320.4	239.6	353.4	78.9	71.7	74.7	76.6	69.9	71.4	57.7	69.4	70.5	70.9	103.6	64.7	67.3	61.0	58.0	66.8	107.7
				Oolong tea	A	117.1	76.1	80.3	89.5	80.1	80.3	69.7	75.1	85.0	84.3	73.3	84.6	79.4	72.5	74.0	72.9	72.9	72.8	65.9	79.3
					B	88.4	70.1	72.4	77.2	72.9	69.7	63.2	70.5	67.1	72.0	58.9	68.9	69.2	65.4	63.2	57.1	60.8	65.2	54.3	67.7
			GC-MS/MS	Green tea	A	101.4	77.6	102.2	90.8	85.7	77.1	81.1	78.5	76.5	70.8	81.6	85.4	61.3	113.7	77.1	66.2	62.5	62.9	74.6	80.4
					B	105.0	79.6	93.6	80.5	74.5	71.4	77.7	70.2	72.2	63.1	68.7	76.7	63.8	130.3	68.4	65.5	60.2	58.2	66.0	76.1
				Oolong tea	A	111.7	69.1	77.7	94.7	72.5	77.8	66.2	76.4	92.8	93.0	83.1	87.7	79.4	69.8	73.6	72.1	73.0	73.7	60.9	79.2
					B	91.8	64.2	69.1	83.1	71.7	67.9	61.1	71.6	71.1	80.2	62.2	68.7	73.5	60.9	64.0	58.9	59.7	65.9	55.5	68.5
		ENVI-Carb + PSA	GC-MS	Green tea	A	117.2	106.8	92.1	87.3	85.8	39.7	57.3	91.4	70.5	73.9	65.4	124.4	78.1	84.3	74.4	88.1	80.7	75.6	72.4	82.4
					B	129.2	100.3	103.4	80.0	86.9	29.7	78.4	82.8	63.8	70.8	64.5	95.2	72.7	71.6	69.3	74.5	67.0	61.8	71.7	77.6
				Oolong tea	A	95.5	62.9	82.7	85.6	83.5	68.9	87.5	62.0	76.1	82.4	81.2	65.5	70.0	71.4	85.9	72.0	64.8	85.5	60.7	76.0
					B	87.9	71.6	80.8	79.7	67.4	58.5	68.4	53.1	66.6	55.3	66.1	56.0	60.7	64.8	63.9	61.0	57.4	66.7	55.1	65.3
			GC-MS/MS	Green tea	A	89.9	79.1	88.1	95.7	81.4	88.1	57.2	100.8	84.0	72.9	67.9	69.2	78.3	87.9	73.9	87.5	82.0	73.0	78.3	80.8
					B	102.9	72.3	85.8	83.7	87.6	80.9	83.2	88.0	71.3	71.3	66.6	64.0	70.0	74.6	68.2	74.8	69.0	63.9	76.3	76.5
				Oolong tea	A	88.5	82.2	91.8	86.0	77.9	74.8	80.0	62.2	76.1	80.4	66.0	65.5	73.9	77.6	85.3	61.9	57.6	81.8	61.1	74.8
					B	77.2	73.0	85.3	77.2	68.5	60.6	67.8	54.9	66.6	54.7	64.1	54.1	68.7	61.8	56.9	52.3	53.8	64.9	55.7	64.1
151	Dimethyl-phthalate	Cleanert TPT	GC-MS	Green tea	A	179.4	53.4	135.1	148.3	136.9	145.2	142.5	134.0	139.0	135.7	143.1	140.8	134.6	160.1	135.3	116.3	113.1	116.4	132.0	133.7
					B	171.1	24.8	125.0	138.8	129.9	134.9	133.3	127.7	137.3	129.8	132.0	131.0	128.4	152.6	128.2	116.3	111.6	113.6	118.8	125.5
				Oolong tea	A	175.3	139.0	144.1	169.0	139.9	145.6	118.8	137.1	140.3	136.7	143.4	149.2	134.4	127.8	129.2	126.3	128.0	130.6	114.8	138.4
					B	153.8	113.3	120.0	139.1	123.0	116.6	97.8	116.1	116.0	118.3	110.8	121.1	111.2	111.1	106.2	106.9	107.4	115.5	96.2	115.8
			GC-MS/MS	Green tea	A	186.6	143.3	167.9	149.8	133.9	150.8	146.4	132.9	131.0	137.7	138.1	164.6	126.4	163.6	138.2	117.5	110.4	115.3	130.2	141.3
					B	175.5	139.5	153.8	140.8	126.8	135.6	132.4	121.5	133.6	133.2	129.7	142.7	116.7	172.4	129.3	115.3	112.2	112.9	116.6	133.7
				Oolong tea	A	165.9	129.0	134.2	172.6	124.6	140.3	123.3	133.3	138.7	142.9	145.7	144.8	136.9	123.6	128.0	121.8	125.7	128.9	108.9	135.2
					B	145.7	107.3	111.5	140.6	111.9	110.5	97.2	112.6	114.9	123.7	107.8	113.0	116.0	105.9	107.0	102.9	106.4	114.6	97.8	113.0
		ENVI-Carb + PSA	GC-MS	Green tea	A	215.8	161.3	156.5	154.3	163.9	169.5	133.3	169.5	141.9	144.8	138.2	148.6	135.1	150.1	139.7	155.3	136.7	146.1	132.3	152.3
					B	195.7	137.1	151.1	142.3	163.5	151.7	134.4	157.3	129.7	136.1	119.6	141.5	122.4	138.7	123.3	133.2	122.4	130.1	130.5	140.0
				Oolong tea	A	180.9	147.6	161.5	145.2	156.6	138.2	144.4	123.1	143.6	133.2	150.4	116.7	125.5	134.9	144.0	126.1	115.8	147.3	109.3	139.2
					B	132.5	120.8	141.6	125.6	122.7	116.4	119.5	103.2	112.4	106.9	128.4	100.4	107.9	113.3	114.6	102.5	97.4	119.4	96.0	114.8
			GC-MS/MS	Green tea	A	174.9	150.8	157.7	161.2	154.8	152.8	137.9	173.1	146.5	139.4	139.4	143.4	134.1	163.0	137.3	158.9	136.5	138.6	142.4	149.6
					B	166.8	122.9	151.2	146.2	158.2	144.6	135.8	154.2	131.9	133.7	120.9	136.3	123.3	134.5	123.0	135.8	124.5	128.9	136.5	137.3
				Oolong tea	A	175.8	138.3	213.3	145.8	152.7	139.8	130.3	122.4	143.6	128.9	122.7	106.4	120.8	122.4	142.2	109.1	99.1	137.7	108.9	134.7
					B	129.5	114.3	140.4	123.0	123.3	113.3	112.5	99.6	112.4	101.9	113.5	95.3	106.0	105.6	106.2	87.7	89.8	114.5	97.8	109.8
152	Dinico-azole	Cleanert TPT	GC-MS	Green tea	A	165.9	135.8	169.2	128.8	105.5	78.9	102.6	105.2	89.9	83.7	91.3	89.4	97.0	106.5	98.3	81.7	74.0	71.9	84.3	103.2
					B	159.1	130.3	152.8	115.8	91.5	81.9	95.4	97.2	89.3	78.6	84.3	92.3	92.7	103.3	90.0	85.1	73.0	68.5	72.4	97.5
				Oolong tea	A	146.3	91.6	119.5	127.9	91.0	109.8	94.2	97.4	122.4	173.0	88.7	87.0	104.1	99.7	102.8	97.0	96.9	79.9	83.5	105.9
					B	129.7	95.5	119.6	119.9	88.3	94.8	85.7	97.7	81.5	134.8	81.0	70.3	89.3	91.2	82.8	76.9	73.6	74.7	74.0	92.7
			GC-MS/MS	Green tea	A	147.1	121.8	141.4	113.6	106.4	106.1	104.5	103.6	95.6	98.4	102.1	96.9	82.1	113.4	93.7	77.4	74.1	70.3	86.4	101.8
					B	139.3	118.8	122.8	110.8	90.9	97.0	96.8	94.0	92.9	96.9	91.4	87.6	77.1	118.3	84.3	75.1	74.7	69.1	75.0	95.4
				Oolong tea	A	147.0	106.4	107.7	139.2	97.7	107.1	94.7	103.4	111.8	124.8	112.1	113.8	102.5	87.5	90.3	90.8	92.7	87.6	76.6	104.9
					B	125.7	96.8	95.4	114.4	92.7	89.6	82.4	96.2	85.9	109.0	82.9	87.8	88.3	76.8	75.9	75.1	75.2	78.5	68.2	89.3
		ENVI-Carb + PSA	GC-MS	Green tea	A	169.2	139.7	125.7	122.0	124.4	122.3	99.9	121.1	72.8	94.5	89.7	209.1	93.4	102.7	96.4	104.4	93.1	85.8	87.1	113.3
					B	185.2	127.0	119.9	103.5	116.9	114.6	108.1	113.3	68.6	99.3	85.7	139.6	86.5	94.4	86.1	87.0	87.0	80.7	86.6	104.7
				Oolong tea	A	154.1	103.0	121.4	117.2	114.5	101.8	110.1	71.1	108.3	123.6	112.6	91.4	98.1	105.8	109.0	98.0	84.7	94.9	70.2	104.7
					B	137.5	93.7	114.1	112.5	87.9	88.6	91.5	58.9	86.3	97.7	86.7	83.6	95.7	86.6	83.4	78.3	70.5	78.1	68.5	88.8
			GC-MS/MS	Green tea	A	156.9	131.3	122.0	126.2	121.5	132.3	100.2	117.1	102.6	97.1	91.6	105.2	93.4	92.5	88.1	103.6	97.4	84.7	91.6	108.3
					B	160.8	114.5	116.4	108.0	115.5	120.2	105.7	105.6	94.2	93.6	82.6	98.4	83.3	79.3	80.9	83.5	85.4	76.8	91.4	99.8
				Oolong tea	A	131.3	112.7	129.4	120.3	112.1	100.9	98.9	83.0	108.3	100.9	94.2	89.4	94.6	91.5	103.6	79.6	67.3	101.7	76.6	99.8
					B	119.3	100.6	117.0	108.1	86.4	84.0	85.6	69.2	86.3	81.2	88.1	78.1	74.7	80.7	76.7	62.5	61.2	84.1	68.2	84.8

(Continued)

Appendix Table 5.1 The 201 Pesticides Average Content for 19 Circulative Determinations on 3 Months Under 16 Factors (Two Cartridges, Two Instruments, Two Tea Samples, and Two Youden Pair Samples) (cont.)

No.	Pesticides	SPE	Method	Sample	Youden pair	Nov 9, 2009 (n=5)	Nov 14, 2009 (n=3)	Nov 19, 2009 (n=3)	Nov 24, 2009 (n=3)	Nov 29, 2009 (n=3)	Dec 4, 2009 (n=3)	Dec 9, 2009 (n=3)	Dec 14, 2009 (n=3)	Dec 19, 2009 (n=3)	Dec 24, 2009 (n=3)	Dec 14, 2009 (n=3)	Jan 3, 2010 (n=3)	Jan 8, 2010 (n=3)	Jan 13, 2010 (n=3)	Jan 18, 2010 (n=3)	Jan 23, 2010 (n=3)	Jan 28, 2010 (n=3)	Feb 2, 2010 (n=3)	Feb 7, 2010 (n=3)	Ave (µg/kg)
153	Diphenamid	Cleanert TPT	GC-MS	Green tea	A	49.5	38.0	42.0	40.6	24.2	30.4	32.0	27.4	33.5	33.5	38.3	36.5	37.9	42.1	38.9	32.0	32.4	31.3	36.4	35.6
					B	47.5	36.5	37.7	37.1	26.9	31.6	32.0	32.2	36.0	31.6	34.4	34.5	34.9	43.0	33.4	32.4	32.0	30.0	32.2	34.5
				Oolong tea	A	49.4	39.1	38.8	47.1	41.1	40.1	33.7	34.3	40.6	41.8	38.7	45.4	39.2	38.6	37.9	38.3	38.7	36.0	33.4	39.6
					B	44.4	33.4	31.4	40.5	36.0	35.5	27.9	32.8	33.3	36.2	29.1	35.3	32.6	33.8	31.5	32.0	32.5	34.2	30.0	33.8
			GC-MS/MS	Green tea	A	47.3	36.9	45.0	41.4	35.2	36.6	36.2	35.1	33.1	34.4	35.5	35.5	30.1	39.1	35.9	29.8	28.0	27.9	32.4	35.5
					B	45.5	36.8	38.7	36.3	32.4	33.4	32.6	31.1	32.3	33.5	31.1	32.5	30.9	42.4	32.6	28.5	28.5	27.0	28.7	33.4
				Oolong tea	A	40.5	36.5	35.8	42.8	33.7	37.5	34.5	36.8	39.1	39.5	41.2	42.2	40.6	34.0	33.0	34.4	35.8	34.3	28.8	36.9
					B	37.7	31.9	29.9	35.4	31.0	30.0	26.6	31.3	33.0	33.6	28.7	33.0	31.5	29.4	28.5	28.9	28.4	31.5	25.9	30.8
		ENVI-Carb + PSA	GC-MS	Green tea	A	54.0	44.3	43.1	41.8	43.3	51.8	24.4	45.1	18.5	34.6	33.6	48.6	36.9	41.2	35.7	41.1	37.9	38.1	35.6	39.4
					B	51.5	38.2	41.7	39.9	42.5	43.1	60.1	37.6	18.0	34.0	29.8	37.9	33.5	36.1	32.2	32.9	33.3	33.1	35.5	37.4
				Oolong tea	A	47.2	39.9	46.3	40.0	42.0	41.0	23.6	39.3	37.5	39.7	41.6	33.9	38.4	39.6	40.1	36.3	32.7	42.4	31.8	38.6
					B	40.5	36.8	41.7	36.0	33.7	34.3	37.5	29.0	29.7	31.2	36.1	29.4	34.1	33.3	32.1	30.2	28.6	34.6	29.4	33.6
			GC-MS/MS	Green tea	A	45.9	39.4	41.8	42.2	40.6	40.4	32.5	42.6	35.0	32.3	32.9	41.5	33.6	42.6	34.1	39.8	35.5	33.5	33.4	37.9
					B	42.1	32.8	40.1	38.4	39.8	36.6	35.0	38.9	33.5	32.4	29.4	36.9	30.6	39.1	30.3	32.5	31.0	29.5	33.9	34.9
				Oolong tea	A	43.3	39.7	48.0	38.7	37.5	35.7	34.9	30.3	37.5	34.6	33.4	29.3	32.3	33.7	35.2	28.4	24.5	37.6	28.8	34.9
					B	37.0	33.8	37.9	33.4	30.8	29.5	31.3	25.0	29.7	26.8	32.0	24.6	29.1	28.5	26.2	24.1	23.4	30.4	25.9	29.4
154	Dipropetryn	Cleanert TPT	GC-MS	Green tea	A	53.3	41.8	48.0	41.9	33.8	31.3	35.8	34.8	32.6	29.1	34.7	34.1	33.1	29.9	31.3	25.2	26.7	23.3	29.4	34.2
					B	51.9	40.8	43.4	36.5	29.5	31.4	32.8	31.7	32.9	25.5	30.7	33.0	31.6	29.4	28.4	25.0	25.8	22.6	25.9	32.0
				Oolong tea	A	53.2	44.6	36.0	39.6	33.8	38.2	33.5	37.1	45.5	45.9	28.3	43.7	36.4	34.2	30.0	30.9	32.0	22.6	26.4	36.7
					B	47.7	29.0	30.6	36.7	33.2	31.3	28.0	32.2	37.8	34.5	23.3	37.2	29.0	30.8	25.5	25.5	25.4	28.6	23.6	31.0
			GC-MS/MS	Green tea	A	48.9	36.2	45.4	38.5	36.9	35.9	35.6	32.6	30.8	32.1	32.7	32.7	29.8	35.1	31.8	26.4	25.4	23.9	27.7	33.6
					B	45.7	36.5	38.4	36.7	32.9	32.8	31.7	30.3	30.3	30.2	28.6	31.0	27.8	36.5	29.3	25.8	26.1	23.9	24.7	31.5
				Oolong tea	A	40.2	33.3	37.0	45.6	32.6	35.2	35.3	36.0	36.5	39.1	35.6	40.0	38.5	38.7	31.3	32.5	31.9	27.5	25.7	35.4
					B	37.5	27.0	31.5	38.3	30.8	28.9	27.1	31.8	31.1	33.5	27.4	29.8	31.2	34.9	27.8	26.9	23.7	25.8	23.6	29.9
		ENVI-Carb + PSA	GC-MS	Green tea	A	51.2	41.0	37.2	37.1	40.3	36.7	34.6	40.1	29.8	32.3	27.8	64.0	31.7	34.9	27.4	34.1	30.4	30.3	27.9	36.2
					B	52.0	34.1	36.4	34.4	39.4	31.6	34.2	37.6	29.5	31.6	28.7	44.3	29.3	32.0	25.1	27.0	26.8	26.2	28.1	33.1
				Oolong tea	A	44.7	63.0	46.5	35.0	35.7	35.2	39.1	27.0	37.4	43.4	35.0	33.0	33.3	37.3	35.1	31.7	27.2	36.2	22.2	36.7
					B	41.7	71.5	44.1	31.1	30.6	30.3	33.2	23.3	29.8	29.8	29.7	27.3	27.3	30.7	26.9	25.2	22.5	26.9	22.5	31.8
			GC-MS/MS	Green tea	A	46.7	40.8	42.9	39.7	41.8	42.1	35.2	40.7	34.7	30.1	29.8	56.1	30.5	34.6	28.9	35.3	32.5	28.8	30.5	36.9
					B	43.4	33.0	40.1	36.2	39.6	38.1	34.8	36.6	32.8	30.5	27.7	66.5	27.7	28.8	27.8	29.3	27.9	25.0	30.8	34.5
				Oolong tea	A	43.2	59.2	44.9	39.4	33.7	36.5	32.3	28.7	37.4	28.3	31.0	30.3	31.0	36.7	35.0	37.3	24.1	32.7	25.7	34.9
					B	37.6	53.2	37.4	29.3	28.7	29.0	28.2	23.9	29.8	20.9	23.8	25.5	23.8	25.3	19.9	29.6	22.1	27.5	23.6	28.4
155	Ethalfluralin	Cleanert TPT	GC-MS	Green tea	A	190.4	136.9	142.0	152.5	103.0	138.7	129.5	136.9	110.5	117.6	132.9	126.1	120.5	132.8	129.4	123.4	104.5	95.5	120.2	128.6
					B	183.9	128.2	126.9	138.0	107.6	131.4	123.6	132.7	132.6	112.4	124.6	114.7	111.3	133.6	124.0	119.8	106.3	96.9	109.3	124.1
				Oolong tea	A	167.5	126.2	136.7	174.3	150.5	169.6	121.4	136.4	141.5	150.3	136.2	144.8	141.2	117.7	156.1	157.8	133.5	136.7	121.5	143.2
					B	151.7	111.9	120.3	144.6	126.1	117.2	92.6	121.0	110.3	128.7	101.9	110.9	119.0	112.9	127.7	149.9	111.0	122.7	111.1	120.6
			GC-MS/MS	Green tea	A	191.2	153.5	160.8	138.2	121.0	150.9	131.3	129.7	120.1	126.9	137.1	154.2	99.3	137.2	126.5	110.1	99.0	96.6	121.7	130.3
					B	187.0	146.6	144.2	138.6	117.7	123.3	121.4	123.8	141.7	121.5	127.9	127.7	91.7	150.3	112.7	107.7	103.0	96.3	102.1	126.0
				Oolong tea	A	173.8	126.9	137.0	181.4	133.4	140.8	119.1	138.1	143.2	172.4	165.8	153.9	147.8	120.3	150.3	133.4	132.7	132.3	117.0	143.1
					B	153.9	109.9	115.4	152.6	118.4	109.2	90.6	118.1	112.7	145.4	112.2	110.4	112.5	106.4	124.5	115.6	110.0	113.6	101.0	118.0
		ENVI-Carb + PSA	GC-MS	Green tea	A	194.2	158.8	145.2	148.9	153.5	172.8	165.6	133.0	132.5	130.1	120.9	464.1	117.8	129.3	141.5	145.7	146.2	137.5	133.3	161.6
					B	174.4	158.5	137.9	131.6	142.5	127.5	122.8	129.0	122.4	121.7	106.3	339.2	111.5	129.4	130.2	120.9	125.4	131.0	129.5	141.7
				Oolong tea	A	164.1	126.3	161.5	139.9	176.6	182.4	125.1	104.5	145.8	134.0	154.6	119.4	138.1	119.3	160.7	134.2	105.1	140.4	121.7	139.7
					B	131.8	110.7	129.1	126.7	118.5	155.1	109.0	84.7	108.9	110.6	125.3	110.5	112.0	127.4	160.5	86.1	93.1	120.5	104.4	117.1
			GC-MS/MS	Green tea	A	176.0	161.3	152.8	154.2	150.8	230.2	162.4	134.6	133.5	129.4	114.5	151.9	113.4	153.6	109.2	140.3	138.7	110.1	147.2	145.5
					B	175.1	157.0	141.5	138.7	143.2	219.8	88.1	124.8	126.4	124.7	104.4	140.5	116.7	128.6	106.8	109.3	121.7	114.5	141.6	132.8
				Oolong tea	A	172.3	139.1	208.9	154.2	159.4	167.8	116.1	103.2	145.8	134.7	143.4	123.2	175.8	142.6	159.6	116.4	100.7	110.2	117.0	141.6
					B	132.4	113.9	134.4	133.5	124.2	134.2	103.7	81.3	108.9	111.3	123.8	111.2	150.1	117.8	123.7	86.4	88.1	100.9	101.0	114.8

156	Etofenprox	Cleanert TPT	GC–MS	Green tea	A	56.9	45.7	72.6	40.7	45.4	42.1	39.0	40.6	37.3	39.6	40.3	40.1	43.1	42.5	36.1	37.3	40.1	40.5	43.5
					B	55.4	43.7	73.1	40.7	40.0	41.0	39.8	41.8	39.4	36.4	40.0	40.3	43.3	40.2	35.1	37.7	41.1	38.8	42.7
				Oolong tea	A	45.1	35.7	36.9	43.0	36.8	38.1	35.4	37.6	35.4	37.4	40.7	35.3	36.5	36.6	35.6	33.3	28.5	29.2	36.2
					B	40.0	31.7	34.2	35.4	34.9	32.6	32.2	32.0	31.3	30.4	32.8	31.3	37.0	32.8	29.9	28.1	24.7	25.6	31.8
			GC–MS/MS	Green tea	A	56.6	42.3	53.7	43.5	44.3	42.8	38.4	35.5	38.5	38.7	35.4	36.6	48.5	40.7	33.3	34.4	36.7	40.9	41.3
					B	54.0	42.0	48.0	41.2	39.8	40.4	39.9	36.3	38.3	36.1	35.2	34.6	49.4	38.2	34.0	36.3	37.6	38.6	40.0
				Oolong tea	A	42.4	33.5	34.8	42.6	34.5	37.5	34.6	39.9	40.0	37.7	39.3	36.8	32.1	32.0	30.7	32.6	31.4	28.0	35.3
					B	38.2	30.9	30.6	35.8	33.6	31.3	32.0	32.8	35.2	29.3	31.6	32.1	28.4	27.5	25.8	26.3	27.7	24.8	30.6
		ENVI-Carb + PSA	GC–MS	Green tea	A	50.8	51.1	45.7	47.6	47.2	37.8	50.9	43.1	41.2	39.3	36.6	39.2	40.9	41.8	45.7	40.1	46.6	42.8	43.4
					B	49.0	43.4	46.5	46.3	49.7	43.2	52.6	40.3	41.7	40.0	38.7	38.3	40.0	40.6	41.9	38.9	41.9	42.4	42.6
				Oolong tea	A	46.8	38.9	41.4	39.3	38.7	35.8	28.5	35.9	33.9	38.3	32.1	32.1	35.6	32.6	30.6	29.0	38.1	28.2	35.3
					B	39.9	33.0	39.0	36.8	31.5	27.7	25.5	29.0	27.1	31.9	28.7	28.1	31.7	27.1	26.0	23.8	31.7	24.9	30.2
			GC–MS/MS	Green tea	A	46.0	43.3	47.4	63.6	42.0	43.2	63.3	68.2	39.5	39.3	31.9	46.6	35.4	43.1	64.0	56.5	55.5	62.8	49.0
					B	46.4	36.8	51.8	61.9	47.1	42.6	66.5	62.8	40.9	38.3	30.2	34.4	41.6	46.0	56.3	57.4	54.6	63.4	48.5
				Oolong tea	A	48.8	51.2	42.7	58.3	53.6	49.3	28.9	35.9	33.8	31.2	27.2	31.8	33.4	36.3	43.7	25.5	37.1	38.8	40.2
					B	46.0	49.4	40.6	56.3	54.3	53.8	24.8	29.0	27.6	28.6	23.6	30.4	27.7	26.4	50.0	21.6	35.1	39.2	38.2
157	Etridiazol	Cleanert TPT	GC–MS	Green tea	A	112.0	91.3	94.3	96.8	79.8	103.4	85.4	80.4	88.5	93.4	83.6	85.4	96.8	73.7	63.9	55.6	64.3	82.0	85.4
					B	106.0	78.1	85.3	92.5	80.6	90.6	84.1	83.7	80.2	89.2	75.5	78.3	98.0	70.1	62.3	56.1	67.8	80.5	81.2
				Oolong tea	A	101.4	87.0	90.6	118.4	86.8	100.1	89.6	80.3	89.4	103.3	81.7	88.4	72.7	81.7	77.0	87.6	87.7	81.3	88.8
					B	88.5	71.1	80.5	90.8	73.5	78.4	76.8	63.3	80.0	73.9	60.6	74.2	64.9	66.2	74.2	71.9	69.4	69.3	73.3
			GC–MS/MS	Green tea	A	114.2	102.5	92.1	114.8	79.0	104.5	88.7	84.1	98.0	94.4	113.6	72.8	107.1	86.9	77.7	71.2	75.7	77.1	91.9
					B	115.9	89.0	98.2	91.6	79.0	95.7	83.4	95.9	84.6	85.2	95.6	73.4	118.6	80.5	77.1	77.7	79.1	69.4	88.0
				Oolong tea	A	121.4	83.6	83.8	127.6	97.2	96.0	93.2	86.2	94.9	106.6	88.6	87.2	78.4	94.8	83.5	84.1	95.0	75.3	92.8
					B	100.3	68.0	72.5	93.8	75.9	72.9	77.1	66.0	87.9	72.7	67.2	75.5	66.2	77.1	69.7	66.2	74.5	68.2	75.0
		ENVI-Carb + PSA	GC–MS	Green tea	A	128.8	105.9	98.4	104.9	108.1	144.0	97.6	97.5	93.9	106.3	97.4	80.5	63.5	72.5	69.7	70.4	99.0	84.7	96.0
					B	128.6	103.1	97.2	89.7	105.1	190.9	86.1	85.5	91.0	81.5	69.3	69.3	66.8	64.2	69.7	62.3	96.5	89.9	93.1
				Oolong tea	A	117.3	76.8	96.9	88.6	102.7	112.3	81.9	82.0	82.1	97.4	85.2	85.2	82.4	86.1	63.7	66.5	88.6	77.0	88.6
					B	90.0	53.0	92.7	79.9	76.2	95.3	64.4	54.3	66.5	87.0	76.3	70.4	69.3	70.5	88.2	59.5	76.7	64.8	73.3
			GC–MS/MS	Green tea	A	136.1	95.5	101.2	107.4	103.1	126.2	106.2	92.6	92.3	101.7	166.1	87.1	92.6	101.3	60.7	76.5	81.0	92.2	104.0
					B	183.4	86.2	97.4	91.3	98.9	126.0	102.2	87.9	91.3	78.6	160.0	76.9	89.3	84.5	109.1	65.4	81.3	93.5	98.5
				Oolong tea	A	154.4	70.7	165.9	90.4	135.9	111.6	77.6	82.0	87.5	81.2	69.1	76.6	78.6	97.3	96.8	56.2	92.6	75.3	91.7
					B	87.5	55.7	88.8	79.3	98.7	89.2	59.5	54.3	73.9	76.3	61.3	57.2	71.7	76.3	72.8	51.6	70.6	68.2	70.0
158	Fenazaquin	Cleanert TPT	GC–MS	Green tea	A	46.3	44.6	31.9	37.6	35.9	32.5	32.5	30.7	23.3	31.8	28.9	30.3	30.2	30.7	25.9	30.1	30.0	29.7	32.4
					B	43.9	42.7	27.3	34.0	30.6	30.1	29.7	27.8	24.8	28.8	26.1	29.0	28.6	28.0	24.6	29.4	30.4	26.0	30.1
				Oolong tea	A	44.2	32.5	34.2	38.6	31.5	34.6	32.5	29.5	35.0	24.2	32.7	32.8	30.5	30.7	29.8	39.0	38.6	33.0	33.3
					B	39.3	28.4	30.2	32.8	29.6	29.4	29.8	22.5	29.8	19.2	27.4	27.5	26.6	24.5	26.3	30.7	38.6	28.9	28.6
			GC–MS/MS	Green tea	A	47.3	35.6	42.4	40.2	31.7	33.5	30.3	29.4	31.0	31.5	24.9	27.1	31.1	29.9	24.4	23.4	24.5	27.1	31.5
					B	44.1	35.3	36.7	33.2	28.3	30.1	27.8	29.0	29.8	28.6	24.3	25.5	33.3	27.0	23.4	23.6	22.4	23.8	29.3
				Oolong tea	A	39.9	33.1	32.4	40.3	30.8	29.2	30.4	33.1	36.5	35.4	35.5	35.6	28.2	28.8	28.0	28.6	27.8	28.6	32.4
					B	36.2	29.8	27.8	32.9	30.2	25.3	28.2	27.5	31.7	26.5	27.7	28.9	24.7	24.4	23.2	23.3	24.2	24.7	27.7
		ENVI-Carb + PSA	GC–MS	Green tea	A	49.2	38.8	38.1	35.2	39.2	36.2	37.8	27.2	27.7	24.3	45.1	30.9	32.5	30.2	33.0	37.0	40.1	35.9	35.2
					B	45.9	33.6	35.4	32.1	37.4	29.5	35.8	25.7	27.6	25.0	28.9	28.1	29.1	27.8	27.8	33.6	33.6	35.7	31.8
				Oolong tea	A	47.6	39.4	39.8	35.6	34.3	32.6	25.0	33.1	31.3	34.8	21.5	31.0	32.7	34.3	34.3	34.8	43.0	35.9	34.2
					B	40.6	32.6	37.0	31.9	26.9	26.6	22.1	26.3	24.7	27.3	21.0	26.9	27.2	27.9	27.2	27.9	36.1	34.4	28.8
			GC–MS/MS	Green tea	A	44.2	37.3	35.9	38.8	37.7	36.6	38.0	32.0	30.1	28.6	38.1	32.7	38.5	28.1	33.4	30.5	28.8	31.3	34.3
					B	40.4	30.7	34.4	33.8	36.8	33.4	33.7	29.1	29.5	26.3	36.5	27.5	31.2	26.0	27.3	26.7	25.5	29.5	31.0
				Oolong tea	A	41.3	36.5	42.0	36.8	34.4	31.2	26.1	33.1	30.2	30.3	22.6	30.7	29.4	32.1	24.9	21.6	30.4	28.8	31.2
					B	36.5	32.1	36.7	32.3	27.7	25.8	21.7	26.3	24.7	27.3	19.9	32.6	25.0	23.0	19.9	18.9	27.5	25.7	26.8

(Continued)

Appendix Table 5.1 The 201 Pesticides Average Content for 19 Circulative Determinations on 3 Months Under 16 Factors (Two Cartridges, Two Instruments, Two Tea Samples, and Two Youden Pair Samples) (cont.)

No.	Pesticides	SPE	Method	Sample	Youden pair	Nov 9, 2009 (n=5)	Nov 14, 2009 (n=3)	Nov 19, 2009 (n=3)	Nov 24, 2009 (n=3)	Nov 29, 2009 (n=3)	Dec 4, 2009 (n=3)	Dec 9, 2009 (n=3)	Dec 14, 2009 (n=3)	Dec 19, 2009 (n=3)	Dec 24, 2009 (n=3)	Dec 29, 2009 (n=3)	Jan 3, 2010 (n=3)	Jan 8, 2010 (n=3)	Jan 13, 2010 (n=3)	Jan 18, 2010 (n=3)	Jan 23, 2010 (n=3)	Jan 28, 2010 (n=3)	Feb 2, 2010 (n=3)	Feb 7, 2010 (n=3)	Ave (μg/kg)
159	Fenchlorphos	Cleanert TPT	GC-MS	Green tea	A	247.9	196.7	336.5	240.4	193.3	207.0	208.6	206.3	197.4	189.6	210.5	196.6	202.0	187.1	197.5	172.2	168.7	154.9	174.0	204.6
					B	245.3	201.0	309.2	219.8	189.0	201.0	188.9	192.5	207.6	181.8	190.5	182.2	189.8	174.2	188.9	164.8	163.9	156.0	162.3	195.2
				Oolong tea	A	225.5	218.6	220.6	248.4	200.3	209.6	202.3	236.6	200.0	224.8	201.8	207.2	201.1	196.2	184.2	193.5	206.2	187.4	171.0	207.1
					B	211.8	188.0	185.1	216.2	185.6	177.9	150.6	199.5	173.6	196.0	168.8	176.1	164.0	177.3	160.6	172.3	173.6	170.2	151.8	178.9
			GC-MS/MS	Green tea	A	226.4	219.8	206.6	202.9	189.0	235.7	219.6	205.5	190.5	228.4	212.8	258.5	181.4	186.5	196.5	157.2	158.8	165.0	169.1	200.5
					B	229.7	190.9	210.4	213.7	198.1	213.6	191.8	192.0	199.9	223.1	180.9	224.2	170.9	189.2	181.4	146.2	163.5	167.1	158.8	191.9
				Oolong tea	A	303.5	220.8	200.7	254.8	211.6	211.1	197.3	215.8	208.8	200.6	229.6	215.6	208.0	197.9	190.5	187.0	204.2	197.2	169.9	211.8
					B	263.9	188.1	168.9	209.1	185.8	170.6	156.2	187.1	177.6	185.0	184.4	182.6	167.4	172.7	166.7	159.5	171.2	174.6	157.2	180.5
		ENVI-Carb + PSA	GC-MS	Green tea	A	234.2	240.9	221.8	234.6	237.8	235.8	214.4	232.2	198.7	205.5	184.4	352.4	198.2	209.9	183.1	199.7	203.6	184.2	190.0	219.0
					B	214.3	208.1	217.2	225.6	228.1	239.5	190.7	206.3	185.0	205.4	171.8	299.9	182.5	196.3	183.6	173.8	185.8	188.7	162.2	204.6
				Oolong tea	A	240.9	237.8	234.4	224.7	221.2	212.4	183.6	179.9	215.7	205.4	187.3	169.7	190.8	198.2	195.7	184.8	158.5	207.1	148.3	201.0
					B	198.4	191.3	208.5	204.2	177.0	177.8	188.0	153.3	171.5	192.8	153.2	158.6	175.1	176.2	160.5	156.7	139.2	177.8	148.3	174.6
			GC-MS/MS	Green tea	A	278.0	236.8	232.7	240.2	233.4	220.0	229.4	229.1	209.3	207.9	184.7	128.0	187.2	223.0	185.7	205.1	197.0	174.1	203.2	210.8
					B	340.1	205.1	220.6	228.8	230.1	221.2	169.9	220.1	194.9	202.9	174.8	123.4	179.9	173.6	181.8	170.0	172.0	179.6	206.8	199.8
				Oolong tea	A	322.2	223.1	349.8	225.4	226.9	209.7	202.0	183.1	215.7	176.4	188.4	163.2	171.1	181.1	195.7	161.8	137.0	212.0	169.9	206.0
					B	211.4	192.2	218.4	193.7	187.2	180.9	181.9	151.2	171.5	150.2	173.9	145.2	146.7	168.8	150.5	130.9	127.0	172.6	157.2	169.0
160	Fenoxanil	Cleanert TPT	GC-MS	Green tea	A	91.8	84.3	87.4	84.6	82.6	72.3	70.7	76.3	72.5	59.5	75.3	71.6	69.8	76.5	72.3	64.0	62.5	58.9	71.1	73.9
					B	90.8	82.3	81.8	77.3	71.8	67.2	69.3	67.8	69.3	54.9	64.9	69.6	69.7	74.3	65.6	63.2	60.6	55.7	60.7	69.3
				Oolong tea	A	95.8	81.1	79.0	81.8	74.5	83.2	70.4	80.0	92.7	90.3	70.3	89.1	77.4	72.0	74.5	75.8	75.3	62.1	64.9	78.4
					B	86.1	67.0	66.4	75.4	68.9	70.6	63.7	72.3	78.5	74.6	57.9	74.3	63.3	62.5	65.0	60.6	60.7	64.9	57.5	67.9
			GC-MS/MS	Green tea	A	100.0	64.0	99.4	72.3	82.7	64.1	75.2	59.1	67.9	52.2	70.1	83.7	78.6	92.1	66.1	59.2	58.2	60.9	70.0	72.4
					B	100.4	72.6	79.3	72.3	70.2	58.9	66.2	57.1	64.1	44.8	61.5	73.4	70.9	77.0	66.0	56.7	58.1	58.5	60.5	66.3
				Oolong tea	A	96.2	76.2	72.8	83.2	70.8	64.7	64.9	71.9	82.8	81.5	82.9	65.2	81.1	62.8	56.7	71.5	73.2	69.1	63.0	73.7
					B	79.0	62.1	61.3	72.6	69.8	53.5	46.5	66.7	69.7	67.6	61.4	51.4	66.7	51.5	75.7	56.9	62.1	63.2	56.6	61.9
		ENVI-Carb + PSA	GC-MS	Green tea	A	114.1	95.2	83.8	87.0	85.8	90.5	63.9	89.3	72.2	80.8	68.3	86.3	67.5	83.8	68.4	67.2	62.8	73.7	67.9	82.6
					B	111.3	77.3	81.0	77.7	86.5	71.1	73.3	82.7	66.3	72.9	65.2	66.3	76.6	75.9	81.1	66.3	62.4	83.4	58.9	74.9
				Oolong tea	A	92.0	111.4	80.2	82.4	81.1	71.2	80.3	62.0	79.8	82.7	80.0	59.5	61.4	85.2	61.1	57.5	55.1	63.7	58.9	78.1
					B	76.2	105.1	80.3	70.8	65.3	59.6	67.1	53.1	70.8	61.2	67.1	99.1	73.2	66.8	69.8	86.4	70.9	65.3	53.5	66.1
			GC-MS/MS	Green tea	A	105.5	72.6	103.4	75.9	97.9	102.0	66.4	89.0	64.4	70.5	56.3	87.1	59.4	76.9	64.9	72.2	63.4	71.2	74.5	80.3
					B	97.5	60.0	99.3	65.9	85.8	80.1	84.5	75.3	67.2	71.6	50.4	59.0	72.1	74.7	84.0	61.4	55.4	64.7	73.6	73.6
				Oolong tea	A	87.0	125.9	87.9	97.7	67.8	77.6	80.7	56.0	79.8	88.1	63.0	49.9	64.3	72.6	53.4	54.8	52.4	75.3	63.0	76.5
					B	81.5	111.0	79.1	72.6	61.5	60.0	69.4	47.4	70.8	55.6	60.0	66.0	59.7	60.2	64.5	54.2	54.2	63.7	56.6	64.4
161	Fenpropidin	Cleanert TPT	GC-MS	Green tea	A	99.7	59.2	86.9	73.0	64.8	53.6	64.4	75.3	54.7	48.7	60.5	66.0	59.7	80.3	64.5	51.0	50.3	44.9	55.3	65.6
					B	95.0	56.7	74.1	67.0	54.7	52.2	59.2	62.0	54.7	42.0	60.5	64.5	56.6	83.3	53.5	54.0	62.0	60.7	57.3	65.9
				Oolong tea	A	88.9	73.1	102.3	116.0	95.7	101.8	90.1	100.9	172.2	209.8	209.9	130.8	126.9	124.0	126.4	123.7	123.6	120.9	118.9	124.0
					B	80.3	58.0	65.9	75.7	57.0	62.2	53.3	57.1	73.5	70.4	51.0	57.1	57.3	56.0	53.9	54.0	51.3	51.3	51.0	60.3
			GC-MS/MS	Green tea	A	90.6	68.8	88.9	67.7	71.1	66.5	65.1	64.5	60.2	61.0	70.4	56.8	62.1	85.8	65.1	54.7	53.6	50.4	63.0	66.6
					B	81.8	69.4	80.0	67.2	57.9	59.8	58.9	57.8	56.1	54.7	60.8	56.9	57.5	93.4	55.0	52.3	51.2	47.3	56.3	61.8
				Oolong tea	A	80.2	75.3	100.7	126.2	103.3	118.8	109.0	129.9	156.5	156.3	149.3	167.2	181.8	148.8	147.1	148.0	155.1	155.1	134.5	133.8
					B	78.4	58.5	66.2	83.9	62.0	62.4	53.3	66.6	70.4	70.9	52.7	71.2	67.9	59.3	55.2	53.6	52.3	60.7	46.7	62.8
		ENVI-Carb + PSA	GC-MS	Green tea	A	106.6	69.6	74.0	71.9	65.5	68.6	47.4	79.0	73.0	57.3	63.3	83.7	57.2	69.8	57.5	70.2	62.0	60.2	57.3	68.1
					B	101.5	64.2	68.9	62.3	68.3	51.7	29.8	76.4	70.5	54.6	47.4	62.6	50.3	56.8	51.8	54.6	52.5	50.2	57.3	59.6
				Oolong tea	A	98.8	92.8	103.4	103.5	110.6	89.8	113.1	91.7	0.0	188.0	221.9	238.1	120.2	130.3	140.9	119.6	110.5	150.5	108.7	122.8
					B	74.9	65.9	65.6	62.2	49.0	51.3	54.4	54.5	0.0	52.4	61.2	56.2	56.4	53.6	54.8	49.0	47.9	56.9	45.2	53.2
			GC-MS/MS	Green tea	A	94.6	67.3	72.9	77.6	75.8	73.9	50.2	76.8	63.2	58.2	75.1	84.2	59.6	81.2	58.1	69.7	63.0	56.1	61.9	69.4
					B	86.0	60.6	70.1	66.7	79.5	67.5	63.5	67.1	58.3	57.4	54.3	80.0	50.3	55.5	51.9	55.2	53.8	50.0	61.0	62.6
				Oolong tea	A	82.3	99.3	138.9	118.5	117.6	111.0	117.1	109.8	128.9	139.9	124.4	94.3	153.1	147.8	178.9	133.9	128.3	178.9	134.5	128.3
					B	62.7	73.4	72.1	65.1	53.3	51.7	57.8	48.3	59.0	49.6	59.8	39.3	49.3	51.1	50.1	46.4	45.7	55.7	46.7	54.6

	Cleanup	Detection	Tea	A/B	1	2	3	4	5	6	7	8	9	10	11	12	13	14	15	16	17	18	19	20
162 Fenson	Cleanert TPT	GC-MS	Green tea	A	47.8	36.0	56.6	39.3	32.4	42.8	38.9	42.2	36.4	39.8	40.2	41.6	37.2	40.3	33.6	40.1	30.3	32.5	36.0	39.1
			Green tea	B	47.6	34.7	52.7	38.7	33.0	40.7	37.8	38.5	40.9	40.2	37.4	36.6	35.9	36.5	39.1	39.8	30.8	31.6	32.5	38.2
			Oolong tea	A	38.8	37.5	43.0	48.6	46.1	44.8	38.3	32.4	37.0	38.9	41.6	39.3	34.9	32.7	41.3	42.1	38.9	34.7	33.1	39.2
			Oolong tea	B	35.5	33.4	38.8	40.4	39.8	37.2	27.2	32.5	29.8	37.2	31.4	32.1	32.0	30.8	33.9	35.7	31.8	30.7	30.1	33.7
		GC-MS/MS	Green tea	A	54.9	44.5	42.1	40.9	32.9	43.9	39.4	40.3	33.7	39.7	39.7	56.8	29.6	38.3	36.8	34.7	28.6	29.8	39.0	39.2
			Green tea	B	52.4	40.7	40.2	40.3	33.0	38.8	36.5	38.3	40.5	40.0	37.6	44.0	28.9	41.8	35.9	33.5	29.8	30.3	34.4	37.7
			Oolong tea	A	64.5	36.7	36.6	51.7	38.1	43.5	36.7	40.5	36.2	42.3	43.0	41.6	38.2	34.1	38.9	38.7	36.2	40.9	32.9	40.6
			Oolong tea	B	49.4	31.9	31.7	42.8	33.3	32.6	30.0	34.9	28.4	41.1	30.5	31.2	32.7	29.9	31.8	33.8	29.3	34.3	29.3	33.6
	ENVI-Carb + PSA	GC-MS	Green tea	A	56.2	47.3	41.9	43.9	42.9	50.5	52.7	42.6	40.6	40.8	40.3	62.3	38.7	39.7	38.9	47.1	41.0	30.5	34.9	43.8
			Green tea	B	53.2	47.4	39.4	38.5	40.6	44.6	36.7	39.7	38.2	39.2	32.4	58.2	35.6	39.3	39.9	39.5	36.9	36.5	36.5	40.6
			Oolong tea	A	48.1	43.4	47.2	37.5	46.8	55.1	34.8	40.2	40.1	36.1	45.1	38.9	39.3	35.0	39.6	34.9	29.9	37.5	32.7	40.1
			Oolong tea	B	40.3	30.9	42.3	36.1	37.2	51.0	29.6	33.8	31.2	32.6	38.5	34.8	33.3	34.1	35.4	24.5	25.5	33.1	27.9	34.3
		GC-MS/MS	Green tea	A	63.3	48.3	43.9	44.2	43.2	44.4	51.4	42.3	42.4	40.6	42.3	34.4	33.4	40.3	35.7	41.0	39.6	32.7	39.4	42.3
			Green tea	B	66.0	53.8	39.4	38.5	41.0	45.3	40.0	37.7	38.1	38.0	32.5	31.8	32.2	37.6	35.4	32.5	35.1	35.9	40.0	39.5
			Oolong tea	A	42.8	36.2	47.4	39.6	46.7	46.0	37.7	35.0	40.1	33.6	37.3	38.0	36.4	35.0	40.0	31.8	26.0	36.2	32.9	37.8
			Oolong tea	B	37.4	31.1	38.0	35.6	36.2	40.0	32.2	27.7	31.2	30.0	34.1	34.7	29.4	33.2	32.7	21.8	23.1	31.9	29.3	32.1
163 Flufenacet	Cleanert TPT	GC-MS	Green tea	A	327.7	305.2	339.7	326.1	247.6	289.5	253.5	311.8	214.7	287.7	272.0	273.8	270.3	272.8	274.0	216.9	215.6	184.6	219.4	268.6
			Green tea	B	325.4	284.7	311.0	285.8	264.8	278.1	233.8	302.1	226.1	293.5	267.7	229.7	240.3	260.4	272.3	227.7	213.7	192.8	192.8	259.4
			Oolong tea	A	289.6	349.9	330.8	379.0	316.4	314.4	276.2	282.5	233.4	245.2	334.5	337.1	287.6	263.9	306.0	298.9	307.1	276.8	223.1	297.5
			Oolong tea	B	264.3	296.7	281.6	297.1	279.7	269.2	190.9	272.8	258.1	233.3	243.3	260.9	263.1	245.2	251.6	303.4	273.6	236.9	207.9	259.5
		GC-MS/MS	Green tea	A	332.6	402.7	241.2	316.6	252.5	323.3	295.9	286.9	246.3	349.0	303.7	292.8	188.0	210.0	265.1	198.2	195.6	213.0	236.6	271.1
			Green tea	B	359.6	312.6	258.2	295.3	280.5	291.3	237.9	273.5	275.5	374.0	260.3	249.3	191.7	256.7	241.6	182.8	209.7	221.0	211.3	262.2
			Oolong tea	A	388.1	385.7	291.4	370.6	351.7	299.6	297.8	341.0	273.5	242.9	358.9	302.4	312.7	277.5	273.9	231.5	237.1	267.3	304.2	305.7
			Oolong tea	B	372.6	296.1	223.7	257.3	280.5	226.2	261.6	284.5	201.5	227.1	267.7	249.4	227.3	241.4	234.4	196.4	224.5	215.7	289.4	251.4
	ENVI-Carb + PSA	GC-MS	Green tea	A	310.3	358.6	307.0	334.5	305.3	282.1	267.7	267.6	300.9	268.1	280.4	700.0	277.0	272.5	308.1	315.3	340.0	272.8	242.7	316.4
			Green tea	B	310.1	369.9	257.3	309.1	284.9	352.7	257.2	234.5	267.3	254.2	220.1	611.0	233.8	275.7	282.5	232.6	297.3	316.6	244.1	295.3
			Oolong tea	A	346.7	355.3	339.6	290.9	364.6	379.5	272.2	292.0	317.9	207.2	315.6	290.0	296.5	287.7	312.6	300.9	187.0	269.6	228.7	297.6
			Oolong tea	B	289.0	241.5	321.9	289.8	283.1	341.5	258.6	228.7	237.6	143.2	301.4	277.5	275.3	271.0	269.2	259.7	161.8	218.2	206.8	255.4
		GC-MS/MS	Green tea	A	387.4	326.3	322.5	347.5	333.1	368.8	354.8	310.9	241.6	286.4	264.1	280.2	228.4	328.6	250.7	276.5	346.7	146.8	221.1	295.7
			Green tea	B	410.1	275.1	297.9	317.5	289.2	367.2	276.2	299.5	317.9	285.3	229.9	257.0	202.1	248.6	234.5	223.3	259.6	200.6	281.2	273.5
			Oolong tea	A	372.4	322.3	406.8	284.2	412.2	340.9	247.5	270.6	214.6	211.1	245.8	201.2	217.1	199.5	271.3	218.1	196.0	325.6	304.1	282.4
			Oolong tea	B	269.3	279.0	296.1	246.7	285.7	296.0	240.3	212.8	214.6	178.3	235.7	178.4	178.1	206.5	198.6	158.5	151.2	242.1	289.4	229.3
164 Furalaxyl	Cleanert TPT	GC-MS	Green tea	A	89.1	125.1	95.5	77.8	69.3	67.6	68.5	67.3	66.6	60.1	71.1	65.8	62.1	74.5	67.3	56.6	55.6	53.2	62.7	71.4
			Green tea	B	88.1	114.3	91.7	70.6	62.8	63.8	63.8	59.9	64.1	55.5	61.7	62.8	59.0	76.0	58.8	56.9	54.9	50.9	55.2	66.9
			Oolong tea	A	89.6	72.0	73.7	83.2	69.5	71.8	66.2	74.2	74.7	81.3	69.5	80.8	72.8	69.0	66.6	69.5	71.3	63.6	58.9	72.5
			Oolong tea	B	79.2	62.1	62.9	71.4	63.2	60.9	51.9	63.2	61.5	68.0	53.7	63.4	57.3	60.1	56.3	58.0	56.4	60.4	52.9	61.2
		GC-MS/MS	Green tea	A	94.6	72.6	86.5	76.5	71.0	71.5	71.4	66.9	62.6	65.5	67.5	87.6	62.1	72.4	65.7	54.5	53.7	52.7	60.2	69.2
			Green tea	B	90.4	71.6	76.3	71.1	64.7	64.7	63.8	59.5	61.8	64.3	59.6	75.6	60.5	78.7	58.8	52.6	54.5	51.3	53.3	71.3
			Oolong tea	A	87.1	67.5	69.9	83.5	65.8	72.3	68.0	71.2	73.3	77.5	79.5	80.4	78.1	65.8	63.9	64.5	68.7	63.8	53.4	59.8
			Oolong tea	B	77.2	60.3	59.3	70.7	61.9	58.0	53.0	61.0	62.9	67.1	56.4	61.2	58.8	57.0	53.5	53.0	54.9	59.8	49.7	72.0
	ENVI-Carb + PSA	GC-MS	Green tea	A	100.1	79.5	78.5	79.2	78.7	83.2	58.0	79.3	66.4	61.8	59.3	77.7	64.0	70.2	63.0	74.2	66.8	66.1	61.9	66.7
			Green tea	B	94.6	69.8	75.4	73.5	76.6	73.6	65.2	73.9	62.9	61.8	55.6	65.1	58.2	63.4	58.4	59.5	58.6	58.0	62.5	70.9
			Oolong tea	A	87.9	78.5	81.9	73.7	75.9	68.4	71.0	60.1	79.1	71.8	74.2	60.3	67.7	70.3	74.4	64.2	56.8	75.9	54.7	60.0
			Oolong tea	B	74.9	69.7	74.2	66.2	60.4	58.6	63.5	52.0	59.4	54.2	64.2	52.9	57.9	60.3	57.5	54.8	50.2	59.7	49.5	71.6
		GC-MS/MS	Green tea	A	92.9	77.5	81.0	80.6	79.2	78.5	60.3	81.7	68.7	61.7	61.7	64.0	64.2	79.7	62.2	74.0	63.9	62.1	66.1	66.2
			Green tea	B	90.2	65.4	77.0	73.0	77.7	72.1	67.1	75.0	65.1	62.2	55.8	59.5	57.1	66.3	57.3	59.6	56.1	55.4	65.3	69.1
			Oolong tea	A	87.9	78.5	94.0	77.8	73.5	74.7	69.2	59.9	79.1	68.8	63.3	59.3	63.9	61.9	72.2	54.3	48.3	73.4	53.4	58.2
			Oolong tea	B	74.4	68.4	75.1	68.3	59.9	59.5	62.1	50.2	59.4	52.1	60.1	51.0	58.8	55.7	51.4	45.8	45.2	58.0	49.7	58.2

(Continued)

Appendix Table 5.1 The 201 Pesticides Average Content for 19 Circulative Determinations on 3 Months Under 16 Factors (Two Cartridges, Two Instruments, Two Tea Samples, and Two Youden Pair Samples) (cont.)

No.	Pesticides	SPE	Method	Sample	Youden pair	Nov 9, 2009 (n=5)	Nov 14, 2009 (n=3)	Nov 19, 2009 (n=3)	Nov 24, 2009 (n=3)	Nov 29, 2009 (n=3)	Dec 4, 2009 (n=3)	Dec 9, 2009 (n=3)	Dec 14, 2009 (n=3)	Dec 19, 2009 (n=3)	Dec 24, 2009 (n=3)	Dec 29, 2009 (n=3)	Jan 3, 2010 (n=3)	Jan 8, 2010 (n=3)	Jan 13, 2010 (n=3)	Jan 18, 2010 (n=3)	Jan 23, 2010 (n=3)	Jan 28, 2010 (n=3)	Feb 2, 2010 (n=3)	Feb 7, 2010 (n=3)	Ave (μg/kg)
165	Heptachlor	Cleanert TPT	GC-MS	Green tea	A	120.3	98.9	90.2	105.3	82.8	95.1	93.0	93.3	89.0	86.8	95.7	91.2	87.7	92.5	91.1	80.3	70.7	65.1	81.8	90.0
					B	110.7	88.9	76.1	100.1	79.9	88.9	86.0	90.3	97.2	81.7	87.2	83.9	82.0	88.5	85.3	76.7	72.7	66.4	76.2	85.2
				Oolong tea	A	104.3	97.3	101.3	115.7	87.9	99.2	93.1	111.5	96.6	99.3	105.0	102.5	94.3	87.5	96.4	96.6	97.1	91.2	81.4	97.8
					B	95.7	81.0	84.9	98.8	79.7	82.0	69.0	94.2	81.3	88.3	79.1	78.6	78.7	80.3	80.8	88.8	79.5	81.9	72.7	82.9
			GC-MS/MS	Green tea	A	121.6	98.2	102.9	114.2	85.8	98.7	93.7	90.4	87.6	89.6	98.4	111.8	79.4	89.0	87.7	73.3	70.2	71.8	79.9	91.8
					B	117.8	90.1	98.1	95.0	82.4	90.3	85.9	85.0	93.3	81.9	87.7	97.1	72.6	88.8	80.8	70.7	70.9	74.0	72.2	86.0
				Oolong tea	A	123.7	91.9	94.6	117.5	92.4	98.0	91.8	101.9	101.3	113.9	108.2	105.4	93.3	88.9	97.3	90.4	92.3	96.4	77.4	98.8
					B	109.8	78.2	80.0	100.9	81.9	79.9	72.3	91.2	80.3	97.9	77.2	81.1	75.1	77.7	82.0	77.7	78.1	77.4	70.3	82.9
		ENVI-Carb + PSA	GC-MS	Green tea	A	124.7	106.9	100.9	107.1	104.5	121.3	102.8	102.7	87.8	88.5	86.9	81.0	86.9	91.2	99.7	101.2	92.7	82.6	82.0	97.5
					B	117.6	97.7	99.6	95.6	105.2	132.7	93.8	99.6	83.3	87.5	81.3	88.0	81.3	91.8	95.4	86.7	86.3	85.4	83.2	93.8
				Oolong tea	A	110.1	107.0	112.1	105.4	111.2	106.1	75.7	85.4	98.4	91.6	104.1	88.4	92.5	99.2	106.3	91.1	75.5	98.0	78.7	96.7
					B	87.5	82.5	96.1	92.0	81.1	90.7	82.6	69.3	73.8	75.0	90.8	79.0	81.1	84.9	93.5	69.3	65.3	82.9	69.8	81.4
			GC-MS/MS	Green tea	A	138.0	108.8	101.2	105.5	107.3	104.3	94.3	102.7	76.5	89.8	89.5	104.1	84.8	92.2	91.8	102.5	84.0	77.3	85.4	96.8
					B	173.8	101.1	97.9	93.4	105.0	104.1	66.6	99.6	76.6	89.8	80.8	103.1	78.8	78.1	85.7	89.7	74.1	77.0	87.2	92.8
				Oolong tea	A	156.2	101.9	167.5	105.0	111.4	98.5	86.4	83.9	98.4	95.3	90.6	73.5	82.4	87.7	104.9	78.3	65.2	106.0	77.4	98.4
					B	102.9	82.9	97.7	90.3	84.0	82.4	77.0	66.3	73.8	79.5	83.3	64.5	64.4	78.1	81.3	60.8	59.5	80.4	70.3	77.9
166	Iprobenfos	Cleanert TPT	GC-MS	Green tea	A	154.0	128.5	128.3	112.1	87.6	75.4	98.7	81.2	86.8	63.7	81.2	85.3	89.9	109.2	95.0	83.8	70.9	65.3	83.3	95.0
					B	146.0	130.2	116.4	106.8	74.2	77.0	93.3	86.8	58.9	58.9	86.6	83.5	82.7	120.0	92.8	81.7	82.6	63.2	74.9	91.6
				Oolong tea	A	122.7	82.1	101.4	109.5	79.0	103.1	105.7	99.7	96.4	136.1	105.7	76.7	97.3	85.8	93.8	94.2	98.4	87.3	80.8	96.3
					B	115.1	79.2	93.8	102.8	74.6	87.0	71.2	70.3	70.3	100.4	70.3	64.5	79.0	77.0	77.9	82.0	76.0	80.8	72.1	81.8
			GC-MS/MS	Green tea	A	125.0	98.9	111.0	105.3	87.3	89.0	96.5	93.0	87.3	86.8	87.9	127.1	75.6	94.6	87.3	73.0	72.5	68.2	82.5	92.2
					B	123.3	95.7	101.1	97.1	78.3	83.0	87.3	83.7	86.8	82.6	87.9	107.6	69.3	97.7	77.9	70.0	91.6	66.4	73.9	86.2
				Oolong tea	A	143.6	92.0	98.4	123.5	88.8	96.0	97.2	99.2	99.2	116.4	99.6	109.5	103.2	87.8	91.3	88.1	71.5	90.0	75.7	99.9
					B	123.0	81.5	82.8	106.9	77.6	78.3	66.3	87.2	87.2	96.1	77.1	79.9	80.5	74.9	77.3	73.7	83.4	81.2	65.5	91.9
		ENVI-Carb + PSA	GC-MS	Green tea	A	141.6	112.1	108.0	101.7	120.5	114.3	90.9	86.6	87.9	88.4	99.6	135.4	87.8	96.9	94.6	103.7	83.5	81.2	84.8	101.7
					B	139.5	121.3	109.7	101.8	110.6	95.2	89.8	87.9	84.4	82.6	75.6	103.1	80.3	90.9	89.7	86.0	80.1	84.0	83.5	94.0
				Oolong tea	A	124.9	93.9	112.3	101.8	100.5	106.1	81.2	96.4	79.2	111.0	64.8	71.8	93.4	98.5	106.4	89.5	67.6	79.1	78.0	95.8
					B	104.3	84.1	88.0	95.1	69.3	74.7	67.4	70.3	61.3	85.4	65.0	65.9	71.6	82.2	87.4	73.4	86.7	75.3	65.6	79.4
			GC-MS/MS	Green tea	A	152.3	113.7	104.6	102.3	114.8	116.9	91.1	93.8	82.7	85.8	71.5	104.1	81.5	82.5	84.0	100.8	75.5	75.3	83.7	99.5
					B	183.6	104.1	99.6	91.8	108.9	106.5	89.4	82.7	84.4	94.8	59.2	97.4	71.1	90.4	76.7	82.2	66.7	101.7	84.9	93.5
				Oolong tea	A	150.3	105.8	168.9	114.9	107.3	97.5	88.6	105.4	73.9	75.4	72.7	72.7	64.7	74.8	100.2	76.1	60.0	76.3	75.7	98.3
					B	104.2	87.8	98.7	96.5	77.4	75.8	78.8	73.9	57.3	52.7	57.3	62.8	57.7	55.3	73.8	61.4	55.3	54.9	65.5	76.8
167	Isazofos	Cleanert TPT	GC-MS	Green tea	A	97.7	67.7	89.6	92.0	65.9	61.7	70.3	60.4	66.7	62.7	64.5	75.0	72.6	79.0	74.2	76.5	70.6	71.0	63.4	71.6
					B	97.1	67.0	84.7	81.7	61.5	61.6	64.2	66.7	62.7	57.8	59.0	74.3	71.9	69.1	68.4	65.3	66.4	63.6	63.0	68.2
				Oolong tea	A	80.4	63.5	74.7	84.6	66.0	74.7	67.4	67.4	80.7	64.1	63.7	75.9	75.0	81.5	82.2	67.2	58.4	75.8	56.1	64.8
					B	72.2	56.8	65.2	82.8	60.7	72.3	52.0	52.0	57.3	59.4	48.1	65.6	67.9	69.4	64.5	56.8	50.3	59.7	48.5	64.8
			GC-MS/MS	Green tea	A	86.3	72.2	80.5	71.6	64.7	62.2	71.6	71.6	62.7	66.7	62.7	65.6	74.3	59.4	60.6	69.6	63.7	63.9	62.7	67.9
					B	85.1	71.4	73.5	88.5	60.6	67.9	63.9	63.9	62.7	62.7	57.8	61.3	71.6	62.6	62.7	58.6	59.6	64.0	61.9	63.9
				Oolong tea	A	90.4	64.9	68.3	74.9	66.3	69.4	66.2	66.2	72.7	72.7	64.1	59.4	74.3	66.0	65.0	60.0	72.3	70.6	55.0	70.6
					B	79.5	55.6	58.7	86.9	59.3	54.3	49.6	49.6	57.3	57.3	48.1	76.7	64.5	55.7	48.2	54.4	57.4	56.0	50.8	59.2
		ENVI-Carb + PSA	GC-MS	Green tea	A	96.2	75.6	77.8	78.5	77.6	76.4	71.5	71.5	65.0	65.0	59.2	60.4	77.6	79.0	74.2	76.5	70.6	54.9	70.6	77.3
					B	89.6	84.5	78.0	76.7	78.1	60.0	68.4	68.4	66.7	62.2	57.8	95.4	72.7	69.1	68.4	65.3	66.4	71.0	65.5	72.5
				Oolong tea	A	86.2	72.4	81.1	68.7	78.1	71.7	96.1	96.1	72.7	75.9	73.3	83.9	81.3	81.5	82.2	67.2	58.4	63.6	65.2	76.9
					B	69.6	58.9	70.7	77.4	58.0	61.0	61.0	61.0	57.3	65.0	57.3	77.5	76.0	69.4	64.5	56.8	50.3	75.8	59.0	64.7
			GC-MS/MS	Green tea	A	96.4	73.1	76.8	70.1	81.5	76.4	63.7	63.7	66.7	66.0	52.8	89.4	68.9	59.4	60.6	69.6	63.7	59.7	60.7	73.0
					B	105.9	65.5	71.9	79.7	80.7	70.6	65.2	65.2	62.7	62.7	48.2	77.3	59.2	62.6	62.7	58.6	59.6	63.9	58.4	68.0
				Oolong tea	A	91.7	75.4	108.7	68.2	74.3	72.3	57.7	57.7	57.3	63.1	45.2	52.8	52.6	62.6	65.0	52.2	49.2	70.6	55.0	68.7
					B	66.7	62.1	67.1	58.3	58.3	57.6	57.7	57.3	57.3	52.7	44.3	45.2	48.2	55.7	48.2	43.8	44.3	56.0	53.7	55.7

168	Isoprothiolane	Cleanert TPT	GC-MS	Green tea	A	90.4	65.0	42.8	76.7	67.5	67.1	68.2	65.5	65.3	60.9	70.1	66.3	63.3	77.0	66.4	54.3	52.3	52.2	60.6	64.8
					B	88.7	61.8	33.5	69.3	61.3	62.4	64.2	59.8	63.5	58.5	63.3	62.6	59.1	74.0	59.0	54.2	51.8	50.0	52.6	60.5
				Oolong tea	A	80.9	71.6	73.4	84.6	69.6	67.9	62.6	73.4	74.3	80.4	66.4	77.8	81.1	53.0	66.7	66.0	64.7	60.9	54.5	70.0
					B	73.1	63.1	64.0	73.5	63.5	56.8	50.6	64.5	61.2	67.5	53.2	61.6	57.4	52.1	55.1	54.9	52.4	56.1	49.0	59.5
			GC-MS/MS	Green tea	A	95.9	73.0	89.6	78.2	68.6	72.6	70.9	66.8	62.6	63.9	68.1	51.9	62.9	72.7	64.9	52.8	50.8	52.0	59.3	67.3
					B	91.1	72.2	75.5	73.5	61.2	65.7	63.8	61.0	61.7	65.3	60.4	52.0	57.4	74.7	57.7	50.3	51.0	50.4	51.9	62.9
				Oolong tea	A	88.6	67.0	69.6	87.9	67.1	71.1	66.0	70.8	74.5	80.7	77.4	79.1	71.1	65.9	63.0	65.0	64.0	61.5	53.0	70.7
					B	79.4	60.3	59.1	73.1	62.7	58.6	52.1	60.5	62.0	68.8	56.5	59.8	58.0	56.8	54.2	54.7	52.3	55.3	47.9	59.6
		ENVI-Carb + PSA	GC-MS	Green tea	A	105.2	79.7	79.4	76.8	78.9	86.0	65.7	72.8	70.9	64.4	60.0	96.3	65.5	74.2	63.7	71.6	64.7	63.3	60.0	73.6
					B	98.6	67.8	76.1	70.4	76.3	73.0	68.7	68.7	67.6	64.5	57.1	72.7	61.0	65.3	58.6	57.2	57.4	56.0	59.7	67.2
				Oolong tea	A	92.6	85.4	82.9	74.0	73.1	70.7	69.9	57.5	74.1	67.7	76.0	58.6	70.1	70.4	72.7	61.4	54.8	70.9	52.8	70.3
					B	77.7	67.6	74.3	66.5	57.3	58.9	59.6	50.1	58.0	54.6	63.4	54.0	60.1	62.6	57.2	50.5	46.4	56.9	47.2	59.1
			GC-MS/MS	Green tea	A	93.5	80.1	80.3	78.6	80.2	81.0	68.6	71.4	69.1	62.3	62.5	97.3	66.2	75.5	63.5	73.1	64.3	59.3	64.0	73.7
					B	86.8	67.7	75.8	71.4	76.8	73.4	68.6	74.7	64.8	63.0	56.8	84.7	59.3	58.6	58.6	54.6	56.7	54.9	62.9	67.2
				Oolong tea	A	83.1	77.0	91.3	77.5	75.3	69.7	67.3	59.2	74.1	65.6	64.7	62.5	62.5	63.0	69.5	54.6	47.4	68.6	53.0	67.7
					B	74.1	66.8	75.0	67.7	60.1	57.6	59.8	49.2	58.0	53.5	59.4	53.5	61.8	54.7	50.6	44.9	44.2	55.7	47.9	57.6
169	Kresoximmethyl	Cleanert TPT	GC-MS	Green tea	A	90.3	73.1	108.4	37.6	48.5	32.3	41.8	47.8	39.0	33.1	35.8	34.3	33.4	37.1	30.6	32.5	29.0	33.4	35.4	44.9
					B	100.5	78.9	95.0	33.6	41.5	33.1	39.5	43.8	36.2	28.3	33.8	33.3	33.7	36.1	34.0	30.6	29.6	33.5	36.9	43.8
				Oolong tea	A	45.0	34.2	36.3	34.0	28.2	44.7	37.6	45.2	41.1	52.2	31.9	43.7	34.2	35.4	33.9	36.4	33.6	31.8	29.5	37.3
					B	40.1	30.6	30.7	33.0	31.6	49.8	28.7	38.5	35.6	42.1	31.5	35.1	30.3	32.1	28.5	31.3	26.9	29.7	26.2	33.3
			GC-MS/MS	Green tea	A	60.1	48.3	40.5	43.6	41.5	38.4	42.6	41.2	34.3	34.6	37.0	34.8	25.7	38.1	33.6	30.1	28.2	30.7	31.4	37.6
					B	57.6	46.8	37.4	37.3	38.3	36.3	34.8	36.0	35.9	34.0	33.3	30.3	25.4	40.2	31.9	28.7	27.6	27.9	25.8	35.0
				Oolong tea	A	49.3	35.1	43.4	40.2	34.9	40.7	35.2	39.2	34.7	41.1	40.4	35.1	36.3	32.1	34.1	34.8	35.3	34.4	32.2	37.3
					B	41.7	31.7	28.9	34.7	37.1	33.6	30.7	37.6	31.5	36.6	32.0	28.3	36.3	29.0	30.4	29.6	31.1	29.8	32.2	32.2
		ENVI-Carb + PSA	GC-MS	Green tea	A	65.8	58.6	39.4	40.9	40.0	39.8	28.4	55.4	36.1	36.6	23.2	12.0	29.9	36.0	30.9	39.0	42.6	26.3	37.8	37.5
					B	50.2	58.6	34.0	40.2	48.2	19.9	37.0	43.3	33.5	29.6	26.6	8.1	28.2	34.3	28.7	31.1	39.4	33.7	35.7	35.1
				Oolong tea	A	47.5	30.5	41.5	45.1	25.7	28.1	29.9	40.3	38.2	35.6	38.6	34.8	36.6	38.7	36.3	31.7	24.6	33.1	31.8	35.4
					B	41.0	25.6	37.9	29.2	24.8	24.9	29.8	31.9	30.1	40.2	29.1	32.5	31.3	35.5	32.3	23.7	21.7	29.7	25.9	30.1
			GC-MS/MS	Green tea	A	43.2	54.9	39.4	47.5	37.6	34.8	27.3	50.7	51.1	34.3	31.5	40.9	28.3	34.3	26.8	32.7	19.4	28.3	48.7	37.4
					B	36.9	64.4	34.0	56.0	42.9	29.8	29.0	40.8	42.3	32.7	28.5	38.6	30.8	31.8	28.5	26.2	18.8	31.4	47.7	36.6
				Oolong tea	A	47.2	33.3	43.2	41.7	28.1	19.9	62.1	37.3	38.1	36.9	27.9	29.5	31.2	31.0	32.2	28.4	22.5	32.4	32.2	34.2
					B	42.4	28.9	41.0	36.2	27.6	16.9	51.0	32.8	30.1	27.8	26.0	28.6	32.5	28.9	26.8	20.4	20.0	31.4	28.3	30.4
170	Mefenacet	Cleanert TPT	GC-MS	Green tea	A	111.5	61.7	120.1	114.1	95.6	119.5	110.4	126.7	101.8	113.0	121.7	123.5	81.1	138.9	111.5	122.9	135.6	143.2	18.6	109.0
					B	126.1	59.5	122.7	118.1	93.1	114.3	107.1	118.8	114.2	125.4	110.2	111.2	83.6	151.8	111.3	134.2	119.1	120.5	15.0	108.2
				Oolong tea	A	94.2	110.0	136.6	170.1	116.2	128.3	99.6	120.8	103.3	136.6	121.6	116.8	106.9	89.6	119.6	121.4	121.1	122.0	110.8	117.1
					B	87.2	102.2	129.3	138.7	108.8	115.2	88.5	114.8	95.1	129.7	93.1	84.6	105.8	86.6	99.4	109.2	94.2	85.9	88.1	103.0
			GC-MS/MS	Green tea	A	127.6	142.2	98.2	160.5	92.8	116.5	170.7	145.0	105.2	124.6	129.1	189.5	76.8	160.2	184.8	150.6	133.1	129.6	159.6	136.7
					B	139.5	130.1	119.9	140.7	96.2	111.2	143.9	132.7	119.7	131.7	122.4	140.2	76.7	156.9	171.1	151.8	138.8	138.2	147.9	132.1
				Oolong tea	A	302.0	106.9	104.5	146.0	114.0	128.1	103.9	127.6	107.8	139.2	123.5	113.2	132.8	94.6	114.8	108.1	108.8	129.2	111.0	127.2
					B	200.3	97.4	101.2	122.6	102.7	105.2	93.3	123.9	99.7	131.1	90.8	93.4	113.2	86.8	97.3	87.1	85.5	104.5	90.4	105.1
		ENVI-Carb + PSA	GC-MS	Green tea	A	104.7	145.3	121.6	111.7	120.5	150.6	148.1	135.1	161.2	103.4	124.5	142.1	84.8	118.7	144.7	215.1	114.4	135.8	0.0	125.4
					B	126.0	171.5	119.5	94.3	114.4	196.3	161.2	117.6	125.6	106.3	93.2	148.9	78.3	135.7	152.6	168.7	110.8	122.8	0.0	123.4
				Oolong tea	A	110.0	107.2	123.6	116.1	140.5	144.7	78.0	335.5	104.0	112.4	120.0	103.4	110.8	90.5	122.9	110.0	86.6	108.3	107.0	122.7
					B	110.0	89.0	124.1	120.7	101.9	126.8	85.8	345.2	83.8	98.3	111.7	107.4	98.6	98.3	127.2	69.1	70.9	100.2	93.9	113.8
			GC-MS/MS	Green tea	A	147.9	131.5	124.9	141.7	103.9	117.3	152.5	126.1	112.2	109.5	122.0	73.6	99.0	121.9	129.5	166.1	147.4	147.6	162.6	128.3
					B	141.0	94.5	119.0	116.8	120.1	125.7	118.5	109.5	100.5	102.5	97.9	65.1	85.0	122.1	130.9	148.6	148.8	151.7	170.2	119.4
				Oolong tea	A	160.8	87.5	134.4	111.7	165.7	137.3	118.7	112.3	104.0	106.9	118.7	85.0	108.6	90.9	118.4	91.9	80.0	120.8	110.8	113.9
					B	135.9	85.3	122.0	104.2	119.4	121.0	96.9	85.2	83.8	88.8	104.7	80.0	86.0	89.3	100.7	63.0	68.1	85.6	89.8	95.2

(Continued)

Appendix Table 5.1 The 201 Pesticides Average Content for 19 Circulative Determinations on 3 Months Under 16 Factors (Two Cartridges, Two Instruments, Two Tea Samples, and Two Youden Pair Samples) (cont.)

No.	Pesticides	SPE	Method	Sample	Youden pair	Nov 9, 2009 (n=5)	Nov 14, 2009 (n=3)	Nov 19, 2009 (n=3)	Nov 24, 2009 (n=3)	Nov 29, 2009 (n=3)	Dec 4, 2009 (n=3)	Dec 9, 2009 (n=3)	Dec 14, 2009 (n=3)	Dec 19, 2009 (n=3)	Dec 24, 2009 (n=3)	Jan 3, 2010 (n=3)	Jan 8, 2010 (n=3)	Jan 13, 2010 (n=3)	Jan 18, 2010 (n=3)	Jan 23, 2010 (n=3)	Jan 28, 2010 (n=3)	Feb 2, 2010 (n=3)	Feb 7, 2010 (n=3)	Ave (μg/kg)
171	Mepronil	Cleanert TPT	GC-MS	Green tea	A	49.0	73.1	48.0	41.9	44.7	41.3	39.8	36.4	40.0	31.2	40.1	37.9	49.9	41.5	34.5	32.9	33.7	40.4	41.9
					B	50.2	63.0	45.8	39.2	38.8	38.3	37.3	32.3	35.8	31.0	36.7	36.7	48.7	41.7	33.6	31.8	32.1	34.9	39.1
				Oolong tea	A	44.0	40.3	42.4	42.0	40.0	40.4	33.0	39.8	37.4	37.3	45.4	39.5	39.1	37.6	38.5	39.3	36.6	34.6	39.1
					B	39.9	35.0	37.8	37.2	38.9	36.4	30.9	36.2	32.2	33.8	37.8	35.9	36.8	32.1	31.9	32.6	33.0	30.1	34.5
			GC-MS/MS	Green tea	A	52.6	40.7	50.5	41.4	39.1	40.1	40.6	37.8	38.4	35.9	50.0	31.1	44.7	37.7	33.9	30.6	32.4	37.8	39.6
					B	50.0	40.6	45.2	39.3	34.9	36.5	37.0	33.2	35.2	34.3	42.9	30.2	45.2	34.4	32.2	30.4	30.6	33.2	36.8
				Oolong tea	A	44.9	35.6	37.9	45.7	36.5	42.8	35.3	38.8	41.2	43.4	44.2	46.1	36.5	36.3	36.6	39.0	36.7	32.5	39.6
					B	42.5	32.7	33.1	40.4	37.4	36.5	32.0	37.1	34.1	39.3	36.0	37.6	32.1	31.3	30.0	31.7	32.4	31.6	34.8
		ENVI-Carb + PSA	GC-MS	Green tea	A	53.2	45.1	46.4	47.3	44.7	42.3	29.0	45.2	38.6	42.7	69.3	37.6	42.4	41.1	44.6	40.7	42.3	39.2	43.6
					B	51.3	55.8	45.4	42.8	47.8	35.6	31.1	40.3	32.9	43.4	52.0	37.2	41.7	34.1	37.1	38.7	36.3	39.6	40.7
				Oolong tea	A	53.6	57.7	48.0	41.2	45.0	38.2	42.4	29.3	40.5	42.3	31.3	32.8	39.4	42.5	37.6	34.3	44.6	32.6	41.0
					B	44.2	47.2	52.1	39.8	38.9	32.3	35.4	27.7	35.5	33.8	29.5	32.9	34.3	34.1	31.1	29.3	37.5	29.4	35.8
			GC-MS/MS	Green tea	A	45.4	46.0	45.0	36.7	41.8	30.5	35.2	48.1	43.2	38.7	33.2	44.5	38.0	37.6	44.6	38.8	38.4	40.8	40.2
					B	41.9	36.7	44.1	33.3	42.5	26.8	39.0	41.7	39.7	37.0	31.7	35.8	33.2	35.2	37.8	35.8	35.2	40.1	36.9
				Oolong tea	A	49.7	44.8	52.0	41.6	42.0	37.8	38.8	32.8	40.5	37.2	26.0	38.6	35.9	40.9	32.2	29.4	40.3	32.5	38.4
					B	42.2	41.1	46.5	36.9	36.7	31.1	32.5	29.7	35.5	30.2	23.8	37.1	31.8	29.6	26.6	26.8	34.3	30.3	33.5
172	Metribuzin	Cleanert TPT	GC-MS	Green tea	A	147.3	122.0	116.5	87.1	53.8	59.9	61.6	68.6	49.6	45.2	49.9	52.3	61.6	49.5	50.8	43.3	39.9	44.7	66.1
					B	146.5	115.0	106.2	77.0	55.3	58.7	60.1	63.5	57.9	44.3	47.7	48.7	54.8	52.4	49.5	42.7	40.1	38.8	63.5
				Oolong tea	A	123.5	89.0	75.4	86.5	66.1	87.1	66.5	65.5	61.2	81.1	50.2	61.3	55.8	65.7	61.7	58.0	55.6	48.0	69.3
					B	110.3	80.5	66.6	71.6	55.4	66.0	51.0	57.6	43.2	63.1	39.0	47.8	47.9	50.1	49.1	42.9	45.4	41.4	56.4
			GC-MS/MS	Green tea	A	134.9	89.3	85.9	81.6	65.3	73.3	63.4	64.2	53.6	57.2	74.4	45.4	67.2	52.6	44.9	42.2	40.9	47.4	65.8
					B	129.1	83.8	80.8	83.1	67.1	64.5	55.5	57.7	61.1	67.3	58.3	41.9	73.6	52.3	42.2	42.5	41.0	41.4	62.7
				Oolong tea	A	151.3	97.6	80.8	99.8	76.4	82.1	64.3	72.9	66.7	68.8	74.0	70.5	61.4	59.9	56.0	54.1	54.7	50.0	75.0
					B	122.8	81.1	61.4	78.7	63.7	59.9	50.3	59.1	48.2	71.4	52.5	52.2	48.3	47.8	43.2	41.2	43.3	41.4	58.4
		ENVI-Carb + PSA	GC-MS	Green tea	A	146.9	91.1	81.5	75.3	71.5	75.8	74.8	67.4	49.9	59.5	129.7	53.6	56.2	51.2	61.8	56.3	39.5	46.2	70.1
					B	153.8	108.5	72.2	66.7	66.9	52.4	64.9	63.4	50.0	54.4	92.1	45.7	55.9	46.9	49.5	49.7	47.7	48.2	64.8
				Oolong tea	A	130.0	79.8	81.9	73.9	88.2	95.2	73.5	50.0	77.5	66.1	54.7	63.6	55.1	64.2	56.8	45.3	56.0	49.9	70.0
					B	118.8	68.1	70.1	69.8	63.7	80.7	61.4	37.9	55.1	50.0	47.9	49.1	50.4	51.7	38.0	37.7	43.1	41.0	57.3
			GC-MS/MS	Green tea	A	138.8	116.0	91.5	86.2	76.9	80.4	75.6	66.7	69.4	62.0	106.2	47.8	70.4	44.8	54.3	57.2	40.8	52.9	72.9
					B	134.2	125.1	81.9	79.5	71.7	80.4	57.3	58.5	64.3	59.8	98.8	46.5	52.3	43.1	40.6	48.5	44.0	55.5	67.8
				Oolong tea	A	119.2	91.8	98.3	89.1	95.7	90.4	72.8	60.4	77.5	56.5	59.8	61.4	53.8	65.2	48.5	37.7	54.1	50.0	70.8
					B	108.9	76.4	82.7	74.7	67.8	72.9	60.8	43.9	55.1	44.2	49.7	47.3	46.8	44.6	33.7	34.6	41.5	41.4	57.1
173	Molinate	Cleanert TPT	GC-MS	Green tea	A	38.7	26.1	32.9	34.1	32.1	32.8	33.4	27.7	31.8	29.8	30.1	33.6	31.5	33.3	27.0	29.6	28.7	30.5	31.4
					B	37.6	24.4	29.6	31.8	29.4	30.5	31.2	25.1	30.5	28.3	28.1	30.6	30.1	28.3	26.3	27.5	27.5	27.7	29.2
				Oolong tea	A	37.8	33.4	26.3	38.9	30.9	33.0	28.2	32.0	33.2	34.2	37.5	35.1	34.1	34.5	34.6	32.5	33.2	28.7	33.4
					B	32.9	25.7	21.1	31.7	27.0	26.2	22.7	26.0	26.9	27.9	31.2	29.3	30.6	28.6	29.6	27.8	31.3	25.3	27.9
			GC-MS/MS	Green tea	A	42.9	31.6	39.5	35.3	31.3	34.2	34.0	29.6	29.9	31.7	26.4	31.6	35.2	31.5	26.3	25.4	24.9	29.4	31.7
					B	39.7	31.2	35.2	32.4	29.0	30.9	30.5	27.9	30.2	28.6	26.1	27.9	37.2	28.5	25.3	25.7	25.3	26.0	29.9
				Oolong tea	A	37.1	28.2	30.2	39.6	29.3	32.4	28.7	31.3	34.2	35.7	35.7	33.1	29.9	30.5	29.7	28.8	29.3	25.0	31.8
					B	33.6	23.9	25.2	32.4	26.5	25.5	22.2	27.1	27.5	30.0	26.8	28.2	25.9	25.0	24.7	23.8	26.1	22.6	26.4
		ENVI-Carb + PSA	GC-MS	Green tea	A	48.6	38.7	35.6	35.4	38.3	41.9	31.0	34.7	31.4	31.4	36.7	30.7	33.5	30.6	35.8	32.2	34.8	31.5	34.9
					B	43.5	30.3	34.9	32.2	37.8	35.1	31.4	33.5	33.2	29.3	31.9	30.7	30.9	26.4	30.8	28.9	30.7	30.2	31.8
				Oolong tea	A	41.3	35.8	35.4	31.8	34.0	32.2	32.6	24.5	33.2	32.0	28.7	28.4	35.9	38.3	31.2	31.3	37.1	27.4	33.1
					B	31.3	27.2	32.0	27.5	25.1	26.0	25.4	19.9	25.8	25.0	25.5	28.7	30.2	29.4	26.1	26.2	30.6	25.3	27.1
			GC-MS/MS	Green tea	A	40.3	34.3	36.7	37.1	36.9	35.4	33.8	40.3	33.4	31.6	34.7	31.4	42.7	31.1	36.1	31.5	31.7	32.2	34.9
					B	37.4	27.1	34.9	33.9	37.3	33.1	32.7	36.7	29.8	30.7	32.2	28.0	42.8	27.3	31.1	28.8	28.1	30.8	32.1
				Oolong tea	A	40.0	32.1	47.6	34.3	34.1	32.2	29.5	26.8	33.2	31.4	25.0	28.9	29.9	33.9	25.3	22.9	30.6	25.0	31.1
					B	29.1	26.0	31.9	28.8	27.5	25.3	25.7	21.7	25.8	24.7	22.5	26.2	25.0	24.2	19.9	20.0	25.6	22.6	25.2

No.	Compound	Cleanup	Method	Tea	Rep	1	2	3	4	5	6	7	8	9	10	11	12	13	14	15	16	17	18	19	20	21
174	Napropamide	Cleanert TPT	GC–MS	Green tea	A	135.4	110.4	119.8	118.1	103.8	102.6	105.6	103.6	103.6	103.6	93.3	112.2	100.6	94.1	98.6	104.9	88.9	79.3	79.2	101.4	102.9
					B	133.0	107.8	106.8	107.0	91.1	97.0	97.3	92.7	98.0	92.7	87.4	98.3	97.4	82.8	95.4	90.8	87.1	76.2	76.3	85.5	95.1
				Oolong tea	A	133.9	111.9	114.1	125.3	108.2	110.5	96.9	111.8	114.2	114.2	121.9	107.7	125.4	110.7	105.8	99.0	103.7	107.9	110.4	90.5	111.0
					B	120.3	96.4	97.1	107.0	97.8	98.4	83.9	94.6	101.3	101.3	105.6	83.9	98.1	84.1	90.7	81.6	86.2	86.4	98.4	80.8	94.4
			GC–MS/MS	Green tea	A	144.0	112.1	137.8	113.9	107.9	110.1	107.8	100.5	96.8	96.8	97.9	103.6	120.8	84.5	115.0	102.0	83.1	77.7	79.4	93.5	104.7
					B	135.4	113.1	113.9	106.9	106.9	97.5	95.6	90.5	94.3	94.3	95.0	92.1	105.7	80.2	122.9	92.1	78.8	78.9	77.7	81.4	97.2
				Oolong tea	A	123.7	100.1	105.4	131.6	100.9	108.8	102.5	108.0	112.5	112.5	119.3	117.7	123.1	121.2	100.8	97.5	97.2	104.5	94.5	81.3	107.9
					B	113.1	91.5	88.7	110.5	93.4	87.1	79.4	92.5	95.4	95.4	101.5	85.4	92.7	93.0	87.4	83.9	81.7	83.2	87.4	75.4	90.7
		ENVI-Carb + PSA	GC–MS	Green tea	A	162.1	121.1	121.7	118.0	122.6	139.4	91.8	126.1	98.3	98.3	97.7	91.8	158.8	95.1	105.5	84.5	110.9	99.9	102.4	99.1	113.0
					B	148.0	103.6	116.5	108.2	116.5	114.5	101.9	114.5	95.7	95.7	95.8	86.8	116.3	92.9	93.3	78.6	89.0	89.4	87.7	91.4	102.1
				Oolong tea	A	142.8	123.3	125.9	116.4	120.7	105.7	110.6	93.0	116.4	116.4	110.7	118.1	93.5	100.8	109.0	115.4	101.4	88.8	116.8	87.6	110.4
					B	119.6	103.5	115.3	101.8	91.0	91.3	93.8	78.3	88.6	88.6	88.6	101.3	85.7	86.7	90.7	89.8	85.0	73.8	89.0	74.8	92.0
			GC–MS/MS	Green tea	A	140.9	117.6	122.6	120.4	119.8	123.0	96.3	125.3	102.5	102.5	95.4	93.0	109.5	105.7	110.0	94.0	115.0	101.5	94.7	102.2	110.0
					B	129.6	97.2	115.7	107.9	115.9	110.3	102.3	112.9	97.0	97.0	94.6	83.5	105.1	91.1	93.3	85.6	91.9	89.1	85.5	98.8	100.4
				Oolong tea	A	128.7	116.8	141.1	119.7	112.9	107.2	103.2	98.8	116.4	116.4	103.7	98.4	100.9	94.5	104.5	110.2	86.2	72.8	108.1	81.3	105.1
					B	112.9	100.6	112.7	102.3	90.1	86.0	92.4	74.5	88.6	88.6	82.9	91.1	86.4	89.7	87.9	80.9	71.6	67.4	89.4	75.4	88.6
175	Nuarimol	Cleanert TPT	GC–MS	Green tea	A	98.1	79.4	99.8	77.7	71.9	65.8	70.1	69.5	65.6	65.6	53.5	68.3	60.2	59.5	68.8	63.1	53.5	51.7	51.0	57.7	67.6
					B	95.2	74.7	91.1	70.0	65.6	64.5	63.3	66.7	61.0	61.0	49.4	62.4	59.5	57.0	58.7	58.6	52.5	49.4	48.1	48.4	63.0
				Oolong tea	A	98.5	80.6	84.4	83.0	70.5	75.9	62.6	66.7	79.8	79.8	83.4	59.5	67.1	64.6	64.7	65.8	62.7	64.1	61.9	53.7	71.0
					B	86.9	68.2	74.0	73.3	68.8	63.1	54.9	60.9	68.1	68.1	73.6	47.5	59.1	61.5	59.1	53.5	51.1	51.6	53.5	45.2	61.8
			GC–MS/MS	Green tea	A	100.9	82.5	99.4	88.5	71.2	76.8	70.5	67.1	63.7	63.7	68.4	66.8	71.8	54.8	77.7	62.6	51.2	49.3	50.0	57.7	70.0
					B	96.0	81.5	82.8	74.3	64.7	67.3	64.7	61.1	61.6	61.6	66.7	61.5	63.1	52.6	87.3	56.4	49.7	49.4	48.5	50.3	65.2
				Oolong tea	A	89.6	74.8	71.2	91.9	72.0	72.0	63.6	66.1	75.6	75.6	76.9	76.9	77.1	72.5	59.0	62.0	59.1	63.9	60.6	52.4	70.4
					B	78.9	66.4	62.0	75.4	69.3	60.2	54.4	61.5	63.6	63.6	68.6	57.2	60.2	59.9	51.9	51.3	47.4	52.1	54.1	47.3	60.1
		ENVI-Carb + PSA	GC–MS	Green tea	A	102.9	47.9	85.1	86.6	78.1	72.9	54.4	79.1	65.2	65.2	66.3	55.6	113.5	57.8	70.9	58.6	56.7	61.9	62.8	55.9	70.9
					B	96.8	66.3	74.7	78.8	81.7	65.6	66.5	72.6	53.4	53.4	66.2	54.6	75.3	57.8	60.5	53.8	53.8	53.4	53.4	56.3	65.5
				Oolong tea	A	100.9	76.2	75.1	75.7	69.3	69.8	79.1	54.0	72.8	72.8	68.5	65.5	49.7	61.2	64.9	64.6	59.5	52.9	67.9	51.4	67.3
					B	84.2	70.3	76.0	71.3	60.5	57.2	62.6	46.8	59.2	59.2	50.5	58.2	46.1	53.6	55.4	48.4	50.4	45.0	56.1	47.2	57.8
			GC–MS/MS	Green tea	A	92.8	79.9	84.2	87.2	80.3	80.7	59.6	81.9	71.4	71.4	67.3	61.0	67.5	64.1	89.6	57.4	68.8	65.8	58.8	60.4	72.6
					B	91.9	66.7	79.1	76.9	78.7	74.2	66.3	73.2	64.6	64.6	64.1	56.1	64.2	58.0	67.5	53.0	56.2	58.5	53.1	58.1	66.3
				Oolong tea	A	91.8	78.0	84.6	77.7	76.3	65.4	68.4	56.1	72.8	72.8	65.5	59.7	55.8	58.5	59.1	65.8	52.5	44.5	70.7	52.4	66.1
					B	82.5	67.8	78.7	70.5	63.3	56.4	58.8	48.8	59.2	59.2	51.5	57.1	47.2	65.5	54.3	46.6	42.9	42.0	57.4	47.3	57.8
176	Permethrin	Cleanert TPT	GC–MS	Green tea	A	95.2	78.7	92.5	78.6	70.9	79.2	76.9	74.2	76.1	76.1	70.3	75.8	73.4	70.9	70.9	71.8	63.1	59.9	58.3	67.9	73.9
					B	90.7	74.3	81.9	71.5	64.8	72.3	70.4	69.3	77.0	77.0	69.7	68.0	73.2	67.1	68.6	66.4	62.2	57.1	56.5	59.0	69.5
				Oolong tea	A	89.3	69.8	73.4	84.7	78.1	83.3	64.1	75.3	77.1	77.1	79.1	81.1	77.3	73.7	70.1	75.1	72.4	70.8	68.6	61.5	75.0
					B	78.3	62.3	67.3	71.6	72.6	70.8	59.7	71.4	63.8	63.8	68.5	63.8	64.4	66.9	63.9	63.6	60.9	58.3	59.6	53.1	65.4
			GC–MS/MS	Green tea	A	95.9	74.9	90.1	81.1	66.7	74.0	70.9	67.2	60.7	60.7	64.7	68.8	69.9	54.8	75.6	63.1	51.8	53.3	52.6	60.8	68.5
					B	90.0	75.2	78.5	70.5	61.2	67.0	64.7	63.5	64.5	64.5	80.2	69.2	62.6	51.2	71.6	59.0	50.3	54.1	53.3	52.3	64.4
				Oolong tea	A	89.4	68.4	69.7	85.0	70.2	74.4	62.1	66.8	80.6	80.6	72.2	77.0	74.3	74.4	61.8	67.9	62.8	66.6	64.8	57.1	71.3
					B	77.6	62.9	61.4	69.5	67.2	61.6	56.4	63.0	64.6	64.6	73.0	58.9	62.3	62.9	56.4	58.3	53.8	54.5	55.8	51.5	61.6
		ENVI-Carb + PSA	GC–MS	Green tea	A	95.4	88.1	79.3	75.6	80.6	71.0	81.4	81.7	80.6	80.6	71.7	69.9	120.1	77.3	75.8	68.1	84.6	76.7	74.0	67.9	80.1
					B	99.4	97.0	75.3	71.3	87.5	73.4	76.0	77.6	70.3	70.3	69.9	79.6	100.2	66.6	76.7	65.7	67.7	62.7	65.5	67.2	74.8
				Oolong tea	A	94.5	72.6	81.9	78.9	68.9	83.4	77.3	62.3	69.0	69.0	59.7	66.8	68.4	71.9	64.4	77.4	67.2	51.3	78.1	62.1	74.8
					B	82.3	63.0	78.4	73.0	77.4	71.4	60.9	55.3	56.1	56.1	66.1	61.1	62.3	61.3	63.6	62.2	53.3	69.2	62.2	54.5	63.8
			GC–MS/MS	Green tea	A	86.6	81.4	78.3	81.6	76.3	77.8	79.4	79.5	77.6	77.6	63.1	56.5	67.1	68.3	64.5	60.3	71.1	60.4	57.9	69.5	72.5
					B	79.7	70.7	76.3	71.5	77.4	72.8	69.5	70.1	68.8	68.8	67.0	65.2	63.6	61.6	64.2	58.9	57.7	49.9	66.6	68.1	66.7
				Oolong tea	A	82.9	74.6	83.1	77.3	63.8	73.5	67.6	57.8	69.0	69.0	55.5	59.1	57.9	59.6	57.9	74.0	55.2	41.7	58.1	57.1	67.4
					B	78.0	66.5	78.9	67.5	63.8	60.5	56.0	49.9	56.1	56.1	55.5	59.1	51.7	59.4	57.9	53.9	43.1	41.7	58.1	51.5	58.4

(Continued)

Appendix Table 5.1 The 201 Pesticides Average Content for 19 Circulative Determinations on 3 Months Under 16 Factors (Two Cartridges, Two Instruments, Two Tea Samples, and Two Youden Pair Samples) (cont.)

Average content (µg/kg)

No.	Pesticides	SPE	Method	Sample	Youden pair	Nov 9, 2009 (n=5)	Nov 14, 2009 (n=3)	Nov 19, 2009 (n=3)	Nov 24, 2009 (n=3)	Nov 29, 2009 (n=3)	Dec 4, 2009 (n=3)	Dec 9, 2009 (n=3)	Dec 14, 2009 (n=3)	Dec 19, 2009 (n=3)	Dec 24, 2009 (n=3)	Dec 14, 2009 (n=3)	Jan 3, 2010 (n=3)	Jan 8, 2010 (n=3)	Jan 13, 2010 (n=3)	Jan 18, 2010 (n=3)	Jan 23, 2010 (n=3)	Jan 28, 2010 (n=3)	Feb 2, 2010 (n=3)	Feb 7, 2010 (n=3)	Ave (µg/kg)
177	Phenothrin	Cleanert TPT	GC-MS	Green tea	A	49.7	39.8	40.0	35.5	34.2	34.6	34.7	32.1	31.9	29.7	34.9	32.7	32.5	32.9	34.4	28.0	26.7	27.0	30.6	33.8
					B	44.8	33.6	36.8	32.5	30.3	32.0	32.1	29.7	31.0	29.8	30.5	32.5	30.3	32.3	30.3	27.2	24.7	26.0	27.1	31.2
				Oolong tea	A	40.8	35.1	35.9	43.8	34.5	35.2	27.9	33.3	36.0	36.5	33.1	35.9	33.0	29.4	30.7	30.4	31.4	29.9	27.3	33.7
					B	38.7	31.9	31.4	35.9	32.2	30.9	25.9	30.1	30.0	32.4	26.3	30.0	27.4	25.5	24.6	26.4	25.9	26.9	22.9	29.2
			GC-MS/MS	Green tea	A	56.4	33.1	45.2	33.8	33.4	34.9	33.4	31.1	29.8	31.6	41.4	37.2	44.5	32.0	29.3	24.8	26.5	29.2	25.9	34.4
					B	46.2	34.2	41.2	32.1	30.1	32.0	30.6	29.3	29.0	30.3	41.6	32.8	51.4	36.8	36.9	25.3	24.2	27.4	22.2	33.3
				Oolong tea	A	41.1	32.2	33.0	41.4	32.4	34.7	29.6	45.4	56.4	40.8	66.0	41.9	99.7	29.2	56.9	47.9	46.8	35.5	25.5	44.0
					B	40.0	27.3	28.8	35.4	31.8	29.2	26.6	47.6	51.3	50.6	58.3	38.4	70.8	25.0	45.7	45.4	57.6	38.7	22.7	40.6
		ENVI-Carb + PSA	GC-MS	Green tea	A	54.6	34.2	38.0	37.3	42.4	35.6	31.9	38.9	32.2	33.2	29.7	58.4	31.2	36.1	31.0	35.9	33.2	31.7	29.3	36.6
					B	51.5	32.4	36.5	33.0	38.6	31.6	31.8	36.5	30.1	32.1	29.0	50.3	28.5	31.2	28.6	28.6	29.6	27.9	29.9	33.6
				Oolong tea	A	35.4	29.6	38.5	37.5	36.4	33.3	34.1	27.2	49.7	33.0	34.9	27.2	30.9	33.9	33.2	30.3	25.7	34.4	26.9	33.3
					B	31.7	30.9	36.7	34.4	29.7	27.9	27.0	23.6	45.3	27.5	29.8	24.0	26.7	28.9	26.3	24.4	20.8	29.6	24.1	28.9
			GC-MS/MS	Green tea	A	47.5	40.6	38.7	37.6	55.3	40.5	36.2	38.2	33.1	32.4	28.4	34.3	30.9	38.6	31.2	38.9	31.7	30.6	30.2	36.6
					B	41.7	33.1	35.3	34.6	45.4	36.3	34.8	34.8	30.1	31.4	27.6	32.7	27.6	36.9	29.3	29.0	28.1	28.1	29.4	33.0
				Oolong tea	A	35.4	47.9	38.9	70.3	78.2	45.1	52.2	26.0	49.7	77.2	59.1	45.0	42.9	60.3	45.5	36.0	34.7	34.3	33.4	48.0
					B	32.8	50.0	36.3	62.3	56.1	45.2	41.7	21.7	45.3	70.5	78.0	48.1	39.5	53.5	59.2	54.8	38.2	51.1	56.5	49.5
178	Piperonylbu-toxide	Cleanert TPT	GC-MS	Green tea	A	48.5	38.1	43.2	42.3	36.2	33.3	37.4	34.3	32.9	27.8	30.9	29.8	28.7	30.2	28.3	26.7	26.1	23.1	26.1	33.9
					B	45.8	32.5	40.4	38.5	35.2	38.3	31.9	35.9	40.1	50.3	24.0	31.2	33.0	30.2	31.0	31.5	30.0	29.9	25.2	32.0
				Oolong tea	A	44.1	32.6	38.9	43.8	34.6	33.2	27.6	32.2	29.9	43.7	18.3	26.9	28.3	26.6	26.0	26.3	24.2	27.4	22.2	34.7
					B	38.7	35.1	43.4	40.3	33.2	34.9	34.9	32.9	31.3	31.3	32.7	29.5	30.4	31.7	31.7	26.6	25.8	25.6	29.7	30.4
			GC-MS/MS	Green tea	A	47.8	34.8	37.2	38.2	29.3	33.2	35.3	30.7	31.3	31.3	32.7	27.9	30.4	30.6	31.0	25.6	26.0	24.4	25.7	33.3
					B	44.1	32.1	33.6	34.1	29.6	30.1	31.9	30.7	31.1	31.3	29.5	29.5	28.3	31.7	26.0	26.3	25.8	27.4	22.2	30.7
				Oolong tea	A	41.9	29.2	29.6	43.7	34.6	34.9	34.9	32.9	31.3	31.3	30.4	30.4	28.5	30.6	25.0	25.6	24.2	25.6	29.7	34.0
					B	37.9	32.7	41.7	40.3	33.2	30.1	30.5	30.7	30.2	30.7	36.5	36.5	30.2	31.1	34.3	26.3	21.5	29.9	25.7	29.0
		ENVI-Carb + PSA	GC-MS	Green tea	A	61.1	45.5	40.5	40.0	44.2	47.1	34.2	45.2	29.1	32.4	30.5	58.4	32.5	35.4	32.2	35.4	31.9	33.2	29.8	38.9
					B	54.0	41.0	42.4	35.3	43.4	36.3	35.2	40.1	26.3	32.0	26.6	34.9	29.2	31.1	28.9	28.5	28.3	28.1	29.4	34.2
				Oolong tea	A	45.4	36.3	39.3	39.8	39.6	35.8	36.7	26.9	33.9	37.1	35.9	24.0	31.1	34.0	35.6	29.8	27.6	33.8	25.7	34.3
					B	40.8	33.4	36.6	36.6	29.6	30.1	30.2	23.1	27.3	30.6	26.8	23.4	27.1	27.8	26.8	25.3	21.9	28.2	22.3	29.0
			GC-MS/MS	Green tea	A	45.2	38.7	38.2	38.2	39.6	41.2	35.3	40.5	33.6	31.8	30.2	36.4	36.5	32.3	31.8	36.1	32.4	31.3	32.6	35.9
					B	41.3	31.8	37.4	34.1	38.2	36.9	35.7	36.6	30.6	31.7	30.7	34.2	30.2	30.6	28.5	29.4	28.4	27.8	31.3	32.7
				Oolong tea	A	42.0	37.3	42.1	38.3	35.3	32.0	31.9	26.8	33.9	32.8	30.7	23.0	30.8	31.1	34.3	21.5	22.7	32.4	25.7	32.1
					B	37.7	32.7	37.1	33.4	27.9	26.4	27.2	22.4	27.3	26.7	27.9	20.1	32.6	26.2	25.0	21.5	20.1	27.3	23.0	27.5
179	Pretilachlor	Cleanert TPT	GC-MS	Green tea	A	72.1	68.6	80.5	76.5	65.4	63.9	66.5	69.6	67.0	60.1	67.9	63.4	60.8	59.6	65.1	51.9	54.1	49.3	55.0	64.1
					B	70.1	66.1	74.2	65.6	61.2	62.3	60.7	63.4	68.2	57.5	59.8	57.1	57.4	54.5	58.1	48.7	54.0	48.9	52.1	60.0
				Oolong tea	A	75.9	80.0	73.3	81.8	66.4	70.8	71.9	77.4	70.3	66.2	73.6	75.7	66.1	64.0	59.9	61.5	73.0	60.6	60.1	69.9
					B	69.9	65.3	59.7	69.1	61.7	62.2	49.4	69.9	62.6	60.4	60.6	59.8	52.2	56.7	53.0	54.8	60.8	56.8	51.3	59.8
			GC-MS/MS	Green tea	A	75.6	67.2	63.2	112.9	63.2	71.6	71.2	65.2	59.5	68.8	63.8	92.4	51.0	60.2	62.4	50.1	49.2	52.0	55.3	66.0
					B	75.1	55.3	62.3	65.4	63.2	63.2	60.9	58.7	60.5	68.5	52.7	81.7	49.3	64.9	55.1	44.7	49.0	51.8	49.2	59.5
				Oolong tea	A	89.5	68.1	64.5	81.0	71.0	67.9	69.4	70.0	69.1	63.3	75.6	73.0	76.7	65.5	57.7	61.3	64.9	61.7	51.3	68.5
					B	81.3	59.6	51.5	64.7	63.2	53.7	52.5	59.1	55.8	57.1	57.5	61.4	54.5	55.6	51.1	50.1	52.4	56.2	49.0	57.2
		ENVI-Carb + PSA	GC-MS	Green tea	A	81.9	77.3	70.5	72.2	72.9	92.2	63.9	72.8	61.6	56.8	52.9	73.4	61.1	67.7	55.1	65.0	68.8	66.4	61.3	68.1
					B	72.6	62.2	70.4	70.7	71.2	99.2	61.7	63.2	57.5	58.9	50.2	67.5	56.2	60.7	56.5	54.3	64.4	64.6	59.5	64.3
				Oolong tea	A	76.8	88.1	79.9	73.1	76.9	72.3	57.2	63.1	72.3	66.3	65.7	62.7	61.8	67.0	61.4	60.1	53.1	69.9	59.5	67.2
					B	61.0	73.0	74.5	66.0	61.5	62.9	65.0	51.4	53.0	51.9	59.7	52.4	61.8	56.5	51.3	51.2	47.3	58.5	48.0	58.0
			GC-MS/MS	Green tea	A	105.2	70.2	75.4	78.6	74.4	61.9	61.1	71.4	58.9	57.8	53.8	266.3	58.7	63.9	56.1	65.1	55.3	49.7	62.8	76.2
					B	140.3	57.8	70.4	76.1	74.6	65.3	45.3	69.9	57.4	59.5	51.0	250.4	51.6	49.9	53.2	51.9	46.2	50.2	61.1	72.7
				Oolong tea	A	137.2	76.9	116.0	72.5	74.5	69.3	64.4	60.6	72.3	54.5	62.2	54.9	50.5	58.3	66.9	52.0	43.3	78.9	51.3	69.3
					B	86.0	66.6	72.7	60.3	59.6	57.1	58.7	49.2	53.0	44.1	55.7	45.5	49.1	53.1	49.0	41.7	40.1	60.2	49.0	55.3

No.	Compound	Cleanup	Detection	Matrix	A/B	1	2	3	4	5	6	7	8	9	10	11	12	13	14	15	16	17	18	19	20
180	Prometon	Cleanert TPT	GC-MS	Green tea	A	160.3	124.4	148.1	135.0	110.9	100.9	119.5	108.1	108.0	94.8	127.4	113.5	108.6	110.4	117.4	95.6	93.7	87.6	107.9	114.3
					B	158.6	118.5	132.3	122.4	98.5	99.7	109.8	97.4	105.0	88.0	113.4	107.7	104.3	108.6	102.2	94.3	91.4	84.7	94.3	106.9
				Oolong tea	A	150.9	114.5	130.1	135.3	106.5	119.7	112.6	125.8	125.8	157.0	92.6	128.0	119.8	112.3	115.0	118.3	113.1	105.8	95.1	119.9
					B	139.4	105.6	116.2	121.4	99.0	99.6	91.2	103.5	92.9	127.0	71.5	96.8	94.1	96.3	93.4	97.0	91.4	98.6	87.0	101.2
			GC-MS/MS	Green tea	A	143.9	114.5	140.7	121.4	109.4	111.8	110.5	104.1	99.4	104.5	107.5	120.1	83.2	111.3	104.9	87.8	81.5	79.9	97.6	107.0
					B	136.1	116.2	119.5	110.4	96.7	101.8	97.7	95.6	97.0	101.5	95.1	109.4	77.8	115.6	96.0	83.5	83.4	78.1	86.6	99.9
				Oolong tea	A	126.6	101.1	108.3	136.1	100.7	109.3	104.8	110.8	112.7	120.1	122.9	123.9	130.3	100.6	100.9	99.6	106.6	96.9	85.9	110.4
					B	115.6	91.6	90.4	114.8	94.0	88.6	79.0	94.2	99.0	103.6	87.7	93.9	98.0	86.4	87.1	83.6	85.1	89.8	77.4	92.6
		ENVI-Carb + PSA	GC-MS	Green tea	A	181.0	144.1	133.0	132.5	137.0	138.1	107.3	131.6	97.9	104.8	99.8	271.9	101.3	117.4	104.7	124.6	113.3	109.6	102.6	129.1
					B	163.3	120.2	129.2	123.2	132.1	107.7	111.6	124.0	99.8	103.4	92.9	148.7	93.4	102.9	94.9	98.2	100.8	96.4	103.0	112.9
				Oolong tea	A	174.9	153.7	144.4	123.5	124.4	116.4	119.5	92.0	118.7	126.4	128.5	91.0	114.4	126.3	125.0	108.2	99.4	126.9	92.2	121.2
					B	133.9	123.7	130.0	104.7	90.0	99.3	102.1	77.0	89.3	99.3	101.2	89.4	97.8	102.5	99.1	88.5	82.8	99.0	82.5	99.6
			GC-MS/MS	Green tea	A	144.2	122.7	125.9	126.1	126.2	126.2	105.9	127.7	109.7	96.7	95.8	133.2	100.0	130.4	96.6	116.9	104.9	99.0	104.5	115.4
					B	132.2	99.9	118.9	112.9	122.2	114.8	108.5	114.3	103.0	95.4	86.4	121.1	89.3	104.1	87.0	93.4	91.8	89.1	102.9	104.6
				Oolong tea	A	130.3	118.1	154.5	124.5	117.1	114.4	103.2	90.3	118.7	104.1	102.2	107.4	98.9	107.8	113.5	87.7	77.3	108.4	85.9	108.6
					B	107.7	100.6	113.3	104.9	91.5	90.9	92.2	74.6	89.3	85.0	96.2	94.0	82.5	88.4	82.5	71.8	69.1	87.4	77.4	104.4
181	Pronamide	Cleanert TPT	GC-MS	Green tea	A	170.7	139.3	179.7	148.9	121.6	121.2	122.3	123.9	122.0	112.3	136.8	127.3	117.7	135.4	120.1	105.4	104.2	98.3	129.2	89.4
					B	167.6	136.7	144.6	145.9	116.6	117.3	114.6	109.5	123.7	107.4	118.3	119.9	110.8	135.5	110.3	104.9	101.4	95.2	114.0	128.2
				Oolong tea	A	173.3	131.2	139.7	143.6	126.4	127.2	104.9	123.7	125.3	152.3	124.3	124.2	119.2	112.1	121.7	114.2	108.6	111.1	101.5	120.5
					B	164.4	120.1	118.0	123.0	123.4	106.1	90.0	108.6	103.9	141.9	100.9	99.7	104.9	99.7	107.0	104.4	92.3	95.6	95.7	125.5
			GC-MS/MS	Green tea	A	161.3	134.0	156.2	164.3	124.7	137.5	127.7	127.1	115.7	123.3	126.7	144.6	97.6	145.6	121.2	106.0	101.5	101.9	115.7	110.4
					B	154.2	134.7	135.3	132.8	115.2	125.8	115.9	113.3	117.0	118.3	112.2	128.9	92.1	145.6	111.6	102.2	103.3	99.6	101.7	128.0
				Oolong tea	A	153.3	118.9	126.1	151.2	128.8	122.9	112.7	124.0	130.4	133.2	138.0	132.5	125.5	111.0	125.9	105.0	112.6	112.6	93.3	118.9
					B	144.4	108.1	112.3	131.3	125.3	101.8	96.3	110.9	113.4	129.3	107.9	103.5	105.9	96.9	110.7	94.8	98.4	101.6	83.6	123.9
		ENVI-Carb + PSA	GC-MS	Green tea	A	195.3	115.9	141.1	151.3	143.1	147.4	111.1	143.1	121.5	127.0	116.7	192.5	116.1	127.8	114.7	133.4	124.2	121.7	113.7	109.3
					B	187.0	122.5	135.0	136.8	146.9	113.0	116.3	135.2	115.0	128.5	109.4	139.6	105.9	115.6	106.3	109.2	110.3	111.7	119.8	134.6
				Oolong tea	A	193.2	144.6	129.9	122.7	154.1	121.1	117.8	95.6	122.6	116.1	131.2	120.6	109.5	123.4	131.6	115.0	86.9	115.1	98.1	124.4
					B	149.2	112.5	138.9	115.3	123.9	111.4	110.7	84.4	91.0	102.9	119.2	110.4	100.5	113.5	111.4	99.5	83.2	102.3	86.9	123.6
			GC-MS/MS	Green tea	A	182.2	146.5	146.3	146.3	153.1	154.4	122.8	153.7	131.2	121.1	114.9	37.1	121.5	140.0	119.8	140.7	123.9	116.8	125.5	108.2
					B	173.9	125.5	133.9	131.7	146.1	142.4	124.0	140.7	124.4	120.6	105.0	35.5	110.0	112.3	110.1	113.0	109.1	110.1	124.3	131.1
				Oolong tea	A	178.5	129.8	165.6	135.8	152.3	118.7	113.3	100.0	122.6	105.1	107.6	95.0	113.1	110.8	130.0	96.0	76.7	110.4	93.3	120.7
					B	143.6	108.6	128.9	124.0	128.9	106.6	109.6	85.5	91.0	95.2	105.3	86.4	125.7	102.1	101.4	82.1	75.8	98.9	83.5	118.7
182	Propetamphos	Cleanert TPT	GC-MS	Green tea	A	61.6	45.4	44.8	41.9	25.3	32.7	36.4	38.7	28.9	29.6	34.8	37.7	33.9	39.8	33.7	33.0	28.1	27.1	31.4	36.0
					B	58.6	43.6	39.1	40.0	25.7	34.2	33.9	33.1	36.2	26.1	35.2	36.1	30.8	42.9	35.6	32.5	27.7	27.5	27.8	35.1
				Oolong tea	A	48.2	31.5	37.7	44.1	36.4	41.4	34.4	45.7	55.7	72.7	35.7	23.3	36.2	32.1	37.4	37.1	35.5	33.2	29.5	39.4
					B	43.6	30.2	33.9	38.5	33.5	33.6	27.2	34.0	32.8	38.5	31.6	23.6	30.0	29.2	30.5	31.5	28.6	29.9	26.4	31.9
			GC-MS/MS	Green tea	A	52.2	43.2	43.4	40.5	32.6	41.8	37.6	37.9	31.6	38.4	36.8	39.3	30.9	38.0	33.2	29.7	27.4	27.7	32.4	36.6
					B	50.4	42.1	38.6	37.6	32.2	38.5	33.3	35.5	36.7	38.3	34.7	33.3	27.7	42.0	31.8	28.4	28.8	28.9	29.1	35.2
				Oolong tea	A	49.8	35.9	37.5	49.1	37.9	39.4	34.6	38.5	40.0	41.9	41.9	40.2	41.1	34.4	36.0	35.2	33.3	35.2	29.8	38.5
					B	42.6	32.0	31.5	40.8	32.6	30.2	27.7	31.6	29.8	36.7	29.3	30.7	31.3	29.3	30.0	30.3	26.7	30.4	26.3	31.6
		ENVI-Carb + PSA	GC-MS	Green tea	A	54.5	48.5	40.3	40.0	39.9	32.8	46.1	33.5	32.1	38.5	40.6	79.6	33.2	34.4	36.4	39.0	37.5	27.4	33.8	40.4
					B	50.7	46.6	36.5	34.3	37.7	27.3	33.6	30.3	31.8	36.8	32.4	56.8	30.3	34.0	34.0	31.5	33.6	28.9	34.0	36.1
				Oolong tea	A	49.5	31.2	42.8	36.8	44.1	43.8	36.7	34.1	39.7	48.6	35.3	36.6	34.2	34.8	37.7	33.5	28.1	35.3	29.7	37.5
					B	41.2	28.6	34.6	34.1	32.2	39.6	29.7	18.7	29.8	27.1	33.2	29.9	28.5	31.1	32.4	24.5	24.2	30.3	25.6	30.3
			GC-MS/MS	Green tea	A	58.6	53.4	40.5	42.8	42.9	42.4	50.5	37.3	40.5	38.0	34.4	76.6	29.4	40.5	31.0	35.6	37.7	28.0	36.6	41.9
					B	58.7	54.9	36.9	38.3	39.5	42.8	31.6	34.1	37.6	35.1	28.3	69.4	30.8	30.3	30.5	28.1	31.4	31.7	37.1	38.3
				Oolong tea	A	48.8	37.4	52.8	41.4	44.9	44.4	34.9	32.6	39.7	33.5	35.6	27.0	36.1	33.4	35.1	29.8	23.6	33.4	29.9	36.5
					B	38.9	30.9	36.0	35.5	32.8	37.4	30.2	25.1	29.8	28.0	32.0	25.4	32.4	29.7	26.9	22.0	21.9	27.6	26.4	29.9

(Continued)

Appendix Table 5.1 The 201 Pesticides Average Content for 19 Circulative Determinations on 3 Months Under 16 Factors (Two Cartridges, Two Instruments, Two Tea Samples, and Two Youden Pair Samples) (cont.)

Average content (µg/kg)

No.	Pesticides	SPE	Method	Sample	Youden pair	Nov 9, 2009 (n=5)	Nov 14, 2009 (n=3)	Nov 19, 2009 (n=3)	Nov 24, 2009 (n=3)	Nov 29, 2009 (n=3)	Dec 4, 2009 (n=3)	Dec 9, 2009 (n=3)	Dec 14, 2009 (n=3)	Dec 19, 2009 (n=3)	Dec 24, 2009 (n=3)	Dec 29, 2009 (n=3)	Jan 3, 2010 (n=3)	Jan 8, 2010 (n=3)	Jan 13, 2010 (n=3)	Jan 18, 2010 (n=3)	Jan 23, 2010 (n=3)	Jan 28, 2010 (n=3)	Feb 2, 2010 (n=3)	Feb 7, 2010 (n=3)	Ave (µg/kg)
183	Propoxur-1	Cleanert TPT	GC-MS	Green tea	A	336.4	294.8	342.3	325.9	310.1	297.1	300.0	279.3	310.9	271.5	304.9	287.1	272.3	340.6	282.8	245.1	242.4	250.0	286.4	293.7
				Green tea	B	338.4	285.3	301.6	310.2	274.6	271.9	279.1	243.6	286.8	250.5	267.6	276.7	266.7	335.4	250.7	240.7	235.0	233.9	251.2	273.7
				Oolong tea	A	372.6	298.2	316.4	354.1	298.1	295.1	277.3	333.7	286.6	328.4	333.6	372.2	310.6	303.8	290.1	293.9	293.8	281.3	260.7	313.3
				Oolong tea	B	341.9	245.9	257.0	304.3	267.7	248.2	217.8	273.1	283.1	274.8	251.0	301.8	247.4	262.7	241.7	240.2	244.8	262.7	226.0	262.7
			GC-MS/MS	Green tea	A	377.6	279.2	367.1	323.6	286.0	297.3	307.5	286.9	283.0	287.2	291.1	291.3	261.0	413.6	282.5	247.3	234.6	233.3	271.7	295.9
				Green tea	B	358.1	288.5	325.0	300.3	264.0	270.2	279.2	249.6	278.3	288.1	264.2	273.1	258.2	437.0	256.7	239.7	232.9	222.9	238.4	280.2
				Oolong tea	A	324.6	265.4	282.5	353.4	261.8	300.7	280.3	295.3	242.1	337.1	210.3	342.0	304.9	288.4	275.8	285.7	288.2	274.8	228.2	286.4
				Oolong tea	B	298.6	232.0	233.8	301.2	246.1	241.6	208.0	250.2	269.8	282.9	237.8	264.6	254.7	242.5	230.0	237.5	234.9	253.1	208.5	248.8
		ENVI-Carb + PSA	GC-MS	Green tea	A	432.7	313.0	334.9	334.6	342.2	376.0	244.7	360.3	294.3	292.5	272.7	268.5	286.6	323.1	288.3	333.0	285.0	295.5	270.0	313.1
				Green tea	B	400.5	266.4	332.7	308.7	347.3	323.6	289.8	344.4	277.4	283.8	251.0	262.1	264.8	290.4	267.1	279.2	256.1	255.2	269.3	293.2
				Oolong tea	A	348.6	317.5	348.1	319.1	316.7	281.5	315.2	257.6	318.8	320.9	321.2	271.4	285.1	310.8	321.0	272.1	255.3	327.6	236.9	302.4
				Oolong tea	B	273.2	281.6	298.7	275.4	250.6	228.8	267.3	228.1	252.0	241.3	281.9	225.0	251.5	258.1	242.6	233.1	219.5	264.0	213.4	251.9
			GC-MS/MS	Green tea	A	365.3	307.6	332.2	336.1	323.4	313.2	249.5	367.7	294.7	275.5	276.0	284.5	284.4	372.9	289.9	335.9	303.1	281.8	289.1	309.6
				Green tea	B	332.7	247.2	321.3	310.5	337.6	289.2	287.9	337.3	272.8	272.2	250.7	264.5	253.0	279.9	266.2	284.2	275.6	253.6	282.5	285.2
				Oolong tea	A	334.2	340.2	415.0	315.1	300.7	283.8	305.6	255.8	318.8	286.4	273.2	200.5	279.6	261.5	309.3	235.3	216.2	307.2	228.2	288.8
				Oolong tea	B	275.4	280.0	308.9	266.7	247.9	229.2	259.6	211.9	252.0	214.8	251.9	178.8	252.5	241.5	219.5	201.1	199.9	250.7	208.5	239.5
184	Propoxur-2	Cleanert TPT	GC-MS	Green tea	A	429.3	410.3	385.8	319.3	224.0	386.6	269.2	399.8	344.5	297.1	399.3	383.2	366.2	522.4	457.4	401.2	352.2	404.1	453.9	379.0
				Green tea	B	415.3	375.6	365.0	277.4	278.5	351.1	248.9	386.7	325.0	310.4	391.9	390.4	380.4	484.2	396.9	418.5	348.9	374.1	417.9	365.1
				Oolong tea	A	271.0	345.0	312.0	341.0	487.0	531.1	352.7	424.2	234.9	255.2	332.3	269.1	470.2	419.1	426.3	412.2	440.9	424.4	227.3	367.2
				Oolong tea	B	223.0	295.1	279.4	262.3	548.7	531.2	374.6	395.1	224.1	245.1	253.0	196.8	460.1	437.2	432.1	418.0	446.8	243.8	217.2	352.3
			GC-MS/MS	Green tea	A	439.5	527.5	262.1	337.2	235.0	410.1	281.9	365.6	211.2	318.6	304.9	321.2	176.4	261.4	283.7	203.1	227.5	243.8	190.3	295.0
				Green tea	B	461.6	440.6	273.3	311.8	274.9	363.1	211.5	397.9	359.9	365.0	275.5	274.3	161.2	341.6	352.1	192.7	259.2	269.6	351.5	299.0
				Oolong tea	A	371.9	381.7	336.8	387.6	412.5	336.2	313.0	321.4	230.2	235.6	412.1	300.0	289.5	321.6	294.8	225.2	259.2	323.7	351.5	321.2
				Oolong tea	B	349.8	314.8	264.6	248.7	276.9	240.3	279.1	281.5	216.5	231.1	298.7	237.0	217.9	235.4	294.8	182.4	225.2	245.8	314.0	260.8
		ENVI-Carb + PSA	GC-MS	Green tea	A	293.9	513.1	299.2	298.0	320.9	257.3	1024.1	271.6	375.8	325.7	490.2	307.1	346.5	363.6	466.2	474.0	397.8	382.5	278.0	359.4
				Green tea	B	574.9	623.6	267.6	332.8	286.6	184.3	226.5	234.4	390.3	278.8	477.8	333.8	342.6	366.9	471.2	463.5	403.7	412.5	292.5	355.6
				Oolong tea	A	421.4	245.7	436.3	313.2	445.5	406.7	244.2	279.1	325.6	215.0	253.5	141.2	200.3	427.5	317.7	320.6	184.8	327.7	265.5	329.9
				Oolong tea	B	308.5	197.3	292.2	303.0	325.0	314.5	229.8	204.6	226.1	173.5	241.5	122.0	187.6	438.1	296.0	219.5	180.8	271.1	234.5	281.5
			GC-MS/MS	Green tea	A	578.0	698.9	331.7	367.5	325.0	339.6	313.0	229.8	339.3	345.8	278.2	467.6	239.0	407.1	167.7	260.5	230.0	137.4	298.7	371.9
				Green tea	B	574.9	1062.2	298.1	328.0	256.3	366.8	226.5	210.7	391.0	350.1	197.4	360.4	231.1	322.1	179.8	168.9	158.2	269.0	428.4	335.8
				Oolong tea	A	421.4	245.7	436.3	313.2	445.5	406.7	244.2	279.1	325.6	215.0	253.5	141.2	200.3	186.1	263.1	249.5	152.6	262.1	351.5	283.8
				Oolong tea	B	308.5	197.3	292.2	303.0	325.0	314.5	229.8	204.6	226.1	173.5	241.5	122.0	187.6	217.0	209.3	168.4	156.2	254.2	314.0	233.9
185	Prothiophos	Cleanert TPT	GC-MS	Green tea	A	42.9	35.2	40.7	36.8	30.0	30.3	33.9	33.0	30.6	25.8	32.5	29.6	32.8	31.7	32.1	27.7	26.8	26.2	30.0	32.0
				Green tea	B	42.2	34.1	36.7	33.7	28.3	29.9	31.7	30.5	31.2	23.6	29.1	27.8	29.9	30.3	30.3	27.2	26.5	25.7	26.4	30.3
				Oolong tea	A	38.4	32.1	35.8	39.2	31.6	35.1	31.0	34.6	34.3	41.9	28.2	35.2	34.2	32.4	33.2	33.0	33.6	31.2	28.3	33.9
				Oolong tea	B	35.0	28.7	31.1	34.5	29.2	29.3	23.6	30.9	26.9	34.1	23.1	28.3	27.6	28.5	28.1	28.2	27.4	28.6	25.5	28.9
			GC-MS/MS	Green tea	A	43.6	34.8	37.6	36.4	30.6	35.6	34.7	31.9	30.4	34.1	34.3	32.9	27.7	34.2	31.1	26.8	24.7	28.0	26.7	32.5
				Green tea	B	41.9	34.0	34.0	34.5	30.1	33.2	30.2	30.1	31.2	30.8	33.1	29.3	26.3	34.2	28.1	25.6	24.1	26.7	24.8	30.5
				Oolong tea	A	47.1	34.2	33.3	41.9	32.0	34.2	32.3	34.4	35.7	38.5	34.4	38.5	34.3	33.4	32.1	30.3	33.5	32.8	28.4	35.0
				Oolong tea	B	40.7	28.8	28.0	35.2	29.7	28.1	25.1	30.7	29.9	32.7	30.6	30.1	27.4	27.8	26.5	26.9	26.6	30.2	23.9	29.3
		ENVI-Carb + PSA	GC-MS	Green tea	A	41.6	39.3	36.0	35.2	38.4	35.9	35.8	37.9	30.6	31.2	27.0	61.9	31.1	34.7	29.2	33.8	34.0	30.2	30.6	35.5
				Green tea	B	40.0	34.8	35.1	33.9	36.5	31.7	30.4	35.2	27.4	31.2	24.7	40.7	28.8	31.1	28.1	27.8	30.3	30.0	30.3	32.0
				Oolong tea	A	41.8	38.4	37.7	36.2	36.3	35.2	33.2	35.2	35.7	35.0	36.3	26.8	33.0	34.4	35.7	31.5	27.6	34.8	27.4	33.9
				Oolong tea	B	35.4	31.2	33.4	32.0	28.2	29.7	29.2	22.7	27.5	28.4	29.3	25.1	28.7	29.2	28.7	25.8	23.4	29.2	24.1	28.5
			GC-MS/MS	Green tea	A	44.0	40.0	37.3	37.5	39.8	38.3	36.9	37.3	35.7	32.0	30.9	41.4	31.0	32.2	30.7	33.7	30.5	27.1	34.0	35.3
				Green tea	B	45.4	34.3	34.9	35.1	37.9	36.3	29.7	34.4	31.7	31.2	29.8	36.9	29.8	26.4	29.5	26.3	25.9	26.9	33.3	32.3
				Oolong tea	A	46.2	35.7	47.3	38.2	35.6	33.8	33.0	28.3	35.7	31.3	28.3	30.2	30.5	31.6	35.6	26.9	22.9	34.1	28.4	33.5
				Oolong tea	B	36.8	30.5	34.2	33.0	27.9	28.8	28.3	23.5	27.5	26.8	23.5	26.1	23.8	26.8	27.1	21.2	20.9	28.7	23.9	27.6

186	Pyridaben	Cleanert TPT	GC–MS	Green tea	A	76.1	65.0	52.1	53.0	27.7	45.6	36.2	47.0	34.9	36.9	39.3	41.2	35.0	45.8	37.2	25.0	28.0	32.8	42.3
					B	53.5	53.3	54.9	52.9	28.0	41.2	34.9	46.6	41.5	40.6	37.6	34.4	32.4	44.5	41.4	28.4	27.9	30.0	40.4
				Oolong tea	A	54.8	33.6	38.0	52.2	43.3	46.6	32.9	39.2	38.0	44.8	40.4	37.0	34.7	26.4	41.2	35.1	34.4	29.6	39.1
					B	66.3	32.3	36.7	41.1	38.4	37.0	33.8	36.1	31.0	44.6	30.0	28.7	33.7	28.1	32.7	28.0	29.2	27.5	35.3
			GC–MS/MS	Green tea	A	66.9	43.9	56.6	46.3	49.4	50.8	56.1	47.2	42.5	52.7	41.8	34.3	43.7	68.1	61.4	49.9	48.5	61.1	51.0
					B	65.6	44.0	55.2	47.8	52.2	48.1	49.0	49.6	46.6	60.5	42.4	36.0	44.2	68.5	56.8	56.0	52.5	55.7	51.6
				Oolong tea	A	48.9	43.1	47.7	57.4	51.8	68.4	45.3	53.3	63.8	54.9	50.7	57.0	60.5	48.6	57.0	55.0	55.5	46.1	53.5
					B	55.2	42.9	47.3	49.3	57.2	65.7	51.2	50.1	57.0	54.1	46.4	55.8	60.4	50.0	62.0	54.3	56.1	53.8	53.7
		ENVI-Carb + PSA	GC–MS	Green tea	A	96.7	27.3	38.0	46.8	42.9	57.6	51.1	44.1	32.1	42.5	38.4	102.6	34.4	33.5	40.4	42.8	32.0	35.4	46.4
					B	52.4	47.7	32.2	47.1	38.4	43.1	37.4	39.3	26.6	36.7	28.4	87.2	32.0	34.8	37.5	39.1	37.4	37.5	40.7
				Oolong tea	A	45.0	22.6	42.7	35.0	55.3	52.9	33.5	52.5	38.3	33.9	42.8	33.9	36.9	29.8	43.9	26.1	34.0	34.2	38.3
					B	44.4	27.4	41.9	36.7	40.0	49.1	29.0	33.4	29.3	32.6	36.1	34.9	29.0	32.6	45.1	21.9	32.6	27.4	33.9
			GC–MS/MS	Green tea	A	52.5	44.6	40.2	44.3	39.4	45.7	38.3	40.9	36.8	34.4	31.9	51.7	32.7	39.0	30.4	32.7	24.9	34.3	38.4
					B	75.3	47.6	37.7	39.4	40.6	48.0	31.9	35.5	32.5	33.8	28.7	54.5	27.2	30.1	29.5	25.0	27.2	37.0	37.4
				Oolong tea	A	43.9	40.1	74.6	29.7	48.6	48.4	53.5	33.8	38.3	60.7	52.0	40.5	49.1	31.5	62.3	47.4	53.5	46.2	47.3
					B	24.2	38.6	41.3	27.1	45.9	51.1	54.3	28.6	29.3	56.1	58.9	43.2	63.1	29.2	54.2	47.4	58.3	53.9	44.6
187	Pyridaphen-thion	Cleanert TPT	GC–MS	Green tea	A	40.8	48.2	53.1	39.3	32.8	33.4	34.4	39.7	31.5	31.4	40.3	37.5	33.5	41.9	34.9	28.5	30.7	30.7	36.6
					B	41.2	47.0	49.6	36.8	31.2	33.9	32.6	36.4	35.7	30.2	35.2	32.8	31.2	42.3	32.9	27.1	30.2	28.9	35.1
				Oolong tea	A	36.8	35.2	42.2	49.9	36.9	38.1	34.3	38.1	36.3	51.8	34.4	31.3	32.8	30.6	34.1	42.0	26.0	35.9	37.1
					B	33.4	32.8	42.7	43.1	36.4	34.0	27.3	38.6	29.0	42.8	27.7	24.4	29.1	28.2	28.7	34.9	26.2	28.9	32.5
			GC–MS/MS	Green tea	A	45.4	43.4	34.5	47.0	31.6	38.1	40.0	41.7	33.7	41.8	38.1	68.5	23.7	39.7	31.7	25.8	25.9	33.9	37.6
					B	45.3	34.4	36.7	37.7	31.5	36.6	37.5	36.7	36.3	44.1	32.4	51.7	24.6	44.5	29.7	26.7	37.4	31.0	35.3
				Oolong tea	A	75.4	37.9	35.0	48.1	40.5	38.1	35.3	39.6	39.2	39.7	39.9	37.3	40.3	31.2	34.4	34.8	31.2	32.8	39.6
					B	57.9	33.8	32.0	39.7	36.9	31.4	30.7	36.7	25.7	36.9	31.6	30.5	34.3	28.1	29.5	26.7	35.6	27.2	33.0
		ENVI-Carb + PSA	GC–MS	Green tea	A	44.3	47.4	39.7	34.7	41.6	52.7	32.6	40.2	34.7	32.6	35.4	146.7	35.5	34.9	33.3	35.8	37.1	36.4	44.1
					B	46.3	49.4	38.6	39.6	40.9	64.2	33.4	33.0	30.8	32.5	30.5	102.9	31.1	33.6	32.6	32.5	34.6	37.1	40.9
				Oolong tea	A	40.2	34.7	42.4	37.8	39.5	41.2	26.9	30.5	36.7	39.8	37.5	29.2	34.2	29.7	35.5	29.0	33.4	31.0	35.1
					B	39.1	31.0	41.4	40.9	32.4	36.2	29.3	23.3	28.3	29.9	32.1	30.3	28.8	29.6	30.2	24.9	31.5	29.2	31.2
			GC–MS/MS	Green tea	A	47.1	39.5	40.8	40.9	41.5	41.0	35.1	42.7	35.4	32.4	32.3	37.3	32.1	40.0	32.3	33.9	28.8	33.3	37.2
					B	43.2	32.5	38.7	37.1	40.4	37.4	35.4	39.0	33.3	32.5	29.6	35.5	30.1	33.4	29.8	30.7	38.2	32.2	34.3
				Oolong tea	A	59.3	34.4	53.3	38.7	49.1	41.9	33.4	33.2	36.7	32.8	35.3	25.5	29.8	31.3	38.2	22.6	38.7	32.8	36.6
					B	42.2	32.2	39.8	34.9	36.7	34.6	29.5	25.2	28.3	26.2	31.3	21.1	19.6	29.6	29.5	19.8	27.3	27.2	29.2
188	Pyrimethanil	Cleanert TPT	GC–MS	Green tea	A	48.1	37.3	44.2	40.9	34.0	33.3	34.5	34.1	33.7	30.0	36.1	32.8	34.7	34.0	33.4	28.1	27.4	31.7	34.6
					B	46.6	36.0	38.6	36.9	30.4	33.1	31.9	30.3	33.4	28.7	32.3	31.6	32.5	32.9	30.4	28.1	25.7	28.0	32.3
				Oolong tea	A	44.5	34.6	37.0	40.8	34.7	37.9	32.0	38.4	35.4	40.8	31.6	36.4	34.9	33.1	35.9	34.1	32.0	29.7	35.7
					B	40.0	30.2	31.4	36.0	31.6	31.6	25.1	31.8	29.4	34.6	25.9	28.0	29.1	29.8	29.7	28.0	29.6	26.3	30.3
			GC–MS/MS	Green tea	A	47.3	36.7	45.2	41.2	34.5	37.5	37.0	32.5	31.8	33.8	33.1	42.3	28.3	33.6	34.1	27.2	25.2	31.2	34.8
					B	44.6	37.4	38.2	37.1	31.6	34.1	33.1	30.6	32.1	31.9	29.6	37.8	25.7	33.9	31.3	26.9	24.7	27.2	32.3
				Oolong tea	A	41.8	32.7	35.3	44.2	32.4	36.5	33.4	35.8	34.0	39.1	37.4	39.5	37.9	30.7	32.8	32.6	32.8	26.6	35.1
					B	38.6	29.0	29.6	37.2	29.2	29.8	25.4	30.8	29.8	34.0	27.8	30.4	30.6	28.3	27.0	27.1	28.6	25.3	29.8
		ENVI-Carb + PSA	GC–MS	Green tea	A	57.2	42.5	41.4	40.1	39.6	47.1	32.7	41.6	31.9	33.6	31.8	67.3	34.7	37.6	33.9	33.8	34.5	31.5	39.5
					B	51.5	37.0	39.4	37.0	38.9	37.5	34.8	39.2	30.5	32.8	29.3	40.7	31.5	34.4	31.5	30.6	29.8	31.2	35.2
				Oolong tea	A	46.3	39.8	42.6	37.4	36.6	35.2	34.2	26.0	38.4	36.8	38.1	25.5	32.1	38.7	37.0	28.2	37.8	27.9	35.3
					B	38.1	33.2	35.3	33.0	28.0	29.5	29.6	22.7	30.0	29.5	31.7	24.3	29.1	30.8	28.4	24.2	30.1	24.7	29.4
			GC–MS/MS	Green tea	A	51.4	45.1	51.0	56.7	46.3	42.6	52.1	37.0	68.0	44.8	51.6	36.8	43.1	42.1	53.3	55.0	55.6	59.0	50.2
					B	49.1	33.3	54.1	58.7	51.1	44.1	43.9	31.8	63.7	47.9	50.9	34.9	47.7	44.6	52.6	57.2	57.0	61.1	49.4
				Oolong tea	A	42.2	39.0	50.4	39.8	38.6	36.3	33.7	29.7	38.4	33.9	33.6	29.6	33.2	34.7	36.2	26.0	34.3	26.6	35.0
					B	35.2	32.4	37.2	33.4	30.9	29.2	29.1	24.0	30.0	28.1	30.6	26.6	26.0	29.0	27.2	23.2	26.4	25.3	28.8

(Continued)

Appendix Table 5.1 The 201 Pesticides Average Content for 19 Circulative Determinations on 3 Months Under 16 Factors (Two Cartridges, Two Instruments, Two Tea Samples, and Two Youden Pair Samples) (cont.)

No.	Pesticides	SPE	Method	Sample	Youden pair	Nov 9, 2009 (n=5)	Nov 14, 2009 (n=3)	Nov 19, 2009 (n=3)	Nov 24, 2009 (n=3)	Nov 29, 2009 (n=3)	Dec 4, 2009 (n=3)	Dec 9, 2009 (n=3)	Dec 14, 2009 (n=3)	Dec 19, 2009 (n=3)	Dec 24, 2009 (n=3)	Dec 29, 2009 (n=3)	Jan 3, 2010 (n=3)	Jan 8, 2010 (n=3)	Jan 13, 2010 (n=3)	Jan 18, 2010 (n=3)	Jan 23, 2010 (n=3)	Jan 28, 2010 (n=3)	Feb 2, 2010 (n=3)	Feb 7, 2010 (n=3)	Ave (µg/kg)
											Average content (µg/kg)													
189	Pyriproxyfen	Cleanert TPT	GC–MS	Green tea	A	45.6	40.4	48.2	42.3	38.5	35.0	37.0	36.8	34.9	32.0	35.8	33.7	34.2	33.8	34.3	29.3	29.9	29.2	32.5	36.0
					B	44.7	38.7	43.1	38.3	34.2	32.5	34.8	34.9	33.3	31.2	32.3	31.6	32.3	32.1	31.1	29.5	28.4	27.0	28.4	33.6
				Oolong tea	A	48.7	38.6	38.1	46.1	36.2	34.7	33.5	37.1	37.7	39.6	32.4	38.6	35.4	33.3	33.1	34.8	33.8	33.5	29.8	36.6
					B	42.9	34.6	34.1	39.2	34.1	33.6	29.6	33.8	31.1	34.8	26.3	31.7	31.0	29.8	26.6	28.6	27.6	28.6	25.2	31.7
			GC–MS/MS	Green tea	A	68.9	44.0	58.2	46.6	49.5	49.6	57.5	34.4	43.7	52.4	41.3	50.7	41.7	69.6	61.9	49.0	27.2	48.7	61.2	50.3
					B	67.4	45.4	55.8	48.4	51.6	47.2	51.1	32.4	47.5	59.5	42.6	47.5	43.5	69.9	56.8	50.2	27.3	53.2	55.5	50.1
				Oolong tea	A	42.0	34.1	35.2	43.4	35.5	37.5	32.9	35.2	39.7	40.4	38.5	39.2	37.2	32.6	33.2	33.0	33.4	33.2	28.4	36.0
					B	38.0	30.9	30.8	36.2	34.2	31.6	29.6	32.9	32.3	35.7	29.4	32.2	32.9	28.8	28.5	27.7	26.9	28.6	25.4	31.2
		ENVI-Carb + PSA	GC–MS	Green tea	A	53.2	48.6	43.5	43.0	42.0	39.9	31.0	41.1	32.2	33.4	30.6	69.9	34.1	38.3	34.1	40.6	35.9	36.1	32.2	40.0
					B	48.5	37.9	41.4	37.1	41.0	31.9	32.4	37.9	29.6	32.9	29.5	55.8	32.9	34.4	33.1	32.9	30.9	30.7	31.3	35.9
				Oolong tea	A	50.3	39.6	43.7	41.1	38.3	34.4	36.7	28.3	36.5	35.6	39.3	30.6	33.8	37.1	37.6	32.1	26.8	36.8	29.5	36.2
					B	44.2	35.3	41.5	38.3	30.3	28.6	29.1	25.2	30.0	29.6	31.6	26.5	29.6	30.5	29.1	32.1	21.6	31.9	25.7	30.8
			GC–MS/MS	Green tea	A	53.5	45.4	51.9	56.7	47.8	45.4	37.0	59.2	67.6	45.6	49.1	35.3	44.6	42.0	52.8	61.9	55.3	56.0	58.9	50.8
					B	51.3	34.0	55.0	58.6	52.2	46.3	37.0	61.3	62.9	48.7	50.3	33.3	48.8	45.0	52.1	55.6	57.6	57.3	61.0	51.0
				Oolong tea	A	41.8	39.4	41.4	40.9	37.9	35.1	35.0	28.7	36.5	34.4	31.8	22.1	36.1	34.0	36.3	28.2	25.6	34.7	28.4	34.1
					B	39.6	34.8	39.8	35.3	31.5	28.7	28.8	25.3	30.0	27.9	29.2	19.5	33.6	28.8	26.3	22.5	22.3	29.6	25.4	29.4
190	Quinalphos	Cleanert TPT	GC–MS	Green tea	A	36.8	40.0	43.1	38.2	31.3	35.6	35.2	36.9	36.0	33.5	36.0	42.4	37.4	39.3	34.1	33.7	29.0	27.3	31.4	35.6
					B	37.4	39.5	39.6	35.6	30.5	34.4	34.4	34.4	38.3	31.7	33.5	40.4	37.2	39.1	34.9	31.5	29.0	27.3	28.9	34.6
				Oolong tea	A	50.0	37.3	40.1	48.0	37.3	38.8	32.4	39.3	40.9	55.4	38.2	39.5	37.4	35.1	31.6	30.2	35.5	31.7	30.4	38.4
					B	45.2	32.4	37.9	40.3	34.9	34.6	27.3	36.7	34.5	47.6	35.0	30.3	31.4	34.9	27.3	28.1	30.2	28.8	29.0	34.0
			GC–MS/MS	Green tea	A	44.2	37.7	35.9	44.3	32.2	39.1	41.5	36.4	32.2	36.1	36.0	41.6	32.0	33.0	33.5	28.6	30.8	29.1	30.9	35.1
					B	43.6	33.9	35.7	37.1	32.3	38.0	33.9	33.1	34.9	36.1	30.8	35.7	30.8	33.7	30.0	26.3	26.3	28.0	27.6	33.1
				Oolong tea	A	57.8	34.8	34.3	45.1	34.6	38.4	34.5	38.6	38.0	37.9	40.6	39.2	40.3	33.4	32.9	33.5	34.7	33.6	30.7	37.5
					B	46.9	30.4	28.9	37.4	30.7	29.9	28.2	33.0	29.2	33.5	32.6	30.3	30.2	29.1	28.3	27.7	27.5	30.8	26.6	31.1
		ENVI-Carb + PSA	GC–MS	Green tea	A	40.3	47.2	38.5	38.8	38.8	44.9	40.9	39.5	31.9	35.8	32.9	82.4	34.3	36.5	35.9	33.9	37.9	32.2	34.1	39.8
					B	41.4	41.3	37.2	35.1	36.5	50.2	32.5	35.7	31.0	36.5	30.6	47.7	30.1	35.1	31.9	34.0	35.2	32.8	33.3	36.2
				Oolong tea	A	43.2	34.4	42.2	40.0	38.6	36.6	30.4	32.3	37.0	37.5	50.0	33.9	32.5	34.3	38.2	32.7	27.7	34.9	28.6	36.1
					B	36.9	35.2	37.9	36.5	29.3	32.2	29.3	25.5	28.6	30.7	35.3	32.6	28.7	30.1	30.8	27.0	24.4	29.4	26.6	30.9
			GC–MS/MS	Green tea	A	53.5	41.8	39.3	40.7	41.0	39.9	41.5	39.9	35.3	33.9	34.4	39.7	32.9	38.7	29.3	37.5	30.8	28.4	31.9	37.4
					B	61.2	38.7	36.7	38.4	39.7	38.5	33.9	36.7	33.5	33.4	28.8	36.9	30.7	35.3	28.4	30.8	27.0	31.0	32.0	35.4
				Oolong tea	A	53.5	35.9	34.3	38.1	41.0	38.0	34.5	31.9	37.0	31.3	32.6	27.8	33.0	33.7	37.1	29.3	23.7	37.9	30.7	35.9
					B	38.9	30.9	29.7	34.3	31.3	32.7	30.6	24.9	28.6	26.5	29.2	25.9	27.5	28.4	27.4	23.6	22.8	29.6	26.6	29.3
191	Quinoxy-phen	Cleanert TPT	GC–MS	Green tea	A	45.5	34.9	43.1	37.0	33.7	34.8	33.8	32.6	33.6	27.0	34.5	32.8	30.3	31.6	33.4	29.6	27.7	24.0	28.8	33.1
					B	44.5	33.5	38.1	34.6	30.6	32.0	31.0	30.8	32.7	25.5	32.1	31.6	28.7	30.3	30.6	29.0	27.4	23.1	24.9	31.1
				Oolong tea	A	40.0	31.9	40.7	36.9	31.0	35.7	28.5	36.3	36.6	34.9	34.0	38.6	31.9	32.8	32.5	30.6	28.7	28.2	29.1	33.6
					B	36.0	27.5	35.0	31.0	29.2	31.6	23.0	31.7	30.2	31.3	26.2	31.6	26.4	29.0	25.8	25.2	23.1	28.2	25.7	28.7
			GC–MS/MS	Green tea	A	45.7	36.7	43.8	39.7	33.4	35.7	34.9	32.7	31.4	32.7	33.2	41.5	27.4	35.7	32.1	27.2	26.0	25.5	30.5	34.0
					B	42.8	36.3	38.2	35.1	30.6	31.9	31.9	30.3	31.2	31.9	29.3	35.3	26.3	36.2	29.3	26.0	26.1	26.2	26.6	31.6
				Oolong tea	A	41.6	33.2	34.3	42.6	33.1	35.5	32.2	34.7	36.1	37.5	36.9	38.1	36.7	32.2	31.8	31.7	31.3	31.1	27.4	34.6
					B	38.5	30.0	29.7	35.4	31.5	29.4	27.2	31.0	30.2	33.0	27.5	29.5	31.3	28.1	27.1	26.5	26.4	27.6	24.4	29.7
		ENVI-Carb + PSA	GC–MS	Green tea	A	52.7	37.5	39.8	39.6	38.6	41.6	32.8	39.4	29.8	32.0	34.5	44.4	33.1	37.0	33.5	36.7	33.9	31.3	30.4	36.5
					B	50.1	36.8	38.3	36.1	38.3	36.6	33.1	37.0	28.1	30.0	32.0	39.1	31.6	33.4	30.4	30.3	29.9	27.5	31.1	34.0
				Oolong tea	A	47.8	46.9	39.9	36.9	39.0	34.0	34.0	29.8	36.0	34.2	36.0	31.3	32.0	36.5	35.7	31.5	28.8	37.1	27.7	35.5
					B	40.7	33.8	38.3	32.9	30.3	28.9	28.1	25.7	28.2	27.6	30.0	27.5	27.6	30.0	28.1	25.6	23.8	30.5	24.5	29.6
			GC–MS/MS	Green tea	A	45.2	39.4	39.4	39.9	38.5	39.9	34.0	40.3	34.3	31.7	30.7	33.6	34.0	32.2	32.1	37.3	33.4	30.7	33.1	35.8
					B	42.3	32.4	37.8	35.6	37.7	36.0	34.7	36.7	31.7	31.2	28.0	32.5	29.5	28.9	29.0	30.3	29.8	27.4	32.4	32.8
				Oolong tea	A	40.5	37.7	42.7	38.8	36.2	34.0	33.5	28.8	36.0	32.2	31.4	30.6	32.7	32.3	34.7	27.3	23.7	34.2	27.4	33.4
					B	37.1	33.0	37.4	34.0	29.5	27.8	28.6	24.3	28.2	26.3	28.8	26.5	25.7	27.4	26.2	21.8	21.3	28.2	24.4	28.2

No.	Compound	Cleanup	Instrument	Tea		1	2	3	4	5	6	7	8	9	10	11	12	13	14	15	16	17	18	19	20
192	Telodrin	Cleanert TPT	GC–MS	Green tea	A	171.2	152.0	155.9	150.5	113.4	153.0	129.2	131.1	122.0	134.6	131.6	131.5	119.8	121.1	123.2	108.8	101.4	100.4	115.8	129.8
					B	162.2	147.7	140.8	144.3	115.0	135.7	117.0	125.1	142.2	131.5	122.0	117.0	110.5	125.3	124.4	105.5	102.6	102.7	103.9	125.0
				Oolong tea	A	164.3	135.9	138.9	163.4	143.0	146.8	126.3	151.0	135.3	134.5	153.9	140.9	134.4	128.8	138.4	130.8	133.7	126.5	104.5	138.5
					B	149.6	115.4	120.2	139.4	126.8	115.3	97.0	122.8	117.7	122.6	114.9	114.3	111.4	120.1	114.1	115.6	109.8	118.0	95.1	117.9
			GC–MS/MS	Green tea	A	195.9	154.4	148.9	136.8	111.0	153.0	126.5	125.0	106.3	141.1	133.2	127.1	108.9	128.1	124.1	111.9	100.2	101.1	108.7	128.5
					B	186.2	149.4	135.3	132.8	117.6	140.3	112.3	118.9	125.1	137.8	122.9	107.0	96.8	132.7	115.1	109.3	103.0	106.0	96.4	123.4
				Oolong tea	A	175.3	138.6	132.0	171.0	111.9	145.9	124.7	138.3	129.3	147.7	151.5	142.3	132.1	128.6	126.5	123.9	133.6	138.0	117.7	137.3
					B	149.3	114.9	109.2	138.3	99.5	110.4	99.3	114.6	107.8	131.8	108.9	111.1	104.7	111.2	109.5	105.5	110.4	120.8	101.5	113.6
		ENVI-Carb + PSA	GC–MS	Green tea	A	157.7	159.3	129.9	150.0	147.8	124.8	167.7	135.1	146.1	141.5	119.7	176.6	102.3	136.0	104.1	119.6	135.6	111.2	133.2	136.8
					B	146.2	165.4	120.9	142.8	137.6	118.2	110.9	124.9	132.6	131.1	98.8	162.5	96.9	124.3	109.1	93.5	121.5	120.1	126.2	125.5
				Oolong tea	A	164.3	146.9	160.8	142.9	159.0	160.2	132.3	123.3	147.9	122.1	153.7	130.3	134.6	133.5	140.4	127.0	105.6	139.1	117.2	139.0
					B	131.7	115.5	133.1	123.2	122.2	140.8	118.3	105.2	108.6	104.3	132.7	115.1	111.2	119.9	120.2	101.4	92.0	116.6	98.6	116.3
			GC–MS/MS	Green tea	A	159.8	161.3	137.5	149.2	147.9	140.4	180.6	131.6	145.6	138.8	113.5	146.0	109.0	152.1	104.8	125.0	139.4	104.1	136.6	138.1
					B	155.5	158.7	128.7	137.6	137.5	144.2	98.9	118.7	132.3	126.2	99.3	132.3	116.6	113.7	104.1	91.5	124.0	116.5	130.7	124.6
				Oolong tea	A	169.6	137.2	136.2	141.3	164.1	148.2	130.0	121.2	147.9	116.6	136.5	123.5	131.3	123.5	142.8	112.1	89.5	129.6	117.7	135.7
					B	134.2	113.6	136.3	119.2	127.4	130.5	107.5	99.3	108.6	100.0	128.2	110.7	97.4	113.2	109.2	83.3	81.1	109.6	101.5	111.1
193	Tetrasul	Cleanert TPT	GC–MS	Green tea	A	41.0	32.6	39.1	36.4	33.0	34.0	33.5	30.5	31.9	31.0	33.5	32.3	34.2	31.0	33.4	33.2	32.7	32.3	37.3	33.8
					B	41.9	31.1	35.2	33.4	30.0	31.9	31.3	28.8	30.5	29.5	29.8	29.8	30.7	29.5	29.5	32.5	32.0	31.2	32.0	31.6
				Oolong tea	A	37.9	34.5	34.4	39.6	33.9	34.4	30.8	35.1	35.8	34.5	36.4	39.0	34.2	33.0	33.4	40.5	40.7	37.7	34.7	35.8
					B	33.4	28.9	28.9	33.6	31.3	29.5	25.6	30.1	29.8	30.2	28.4	31.1	28.4	29.2	32.2	34.7	33.8	35.3	31.8	30.9
			GC–MS/MS	Green tea	A	41.5	33.2	42.1	32.9	32.5	35.9	33.2	31.7	31.1	31.0	32.4	30.8	29.1	34.2	32.9	27.2	25.7	27.6	31.5	32.5
					B	39.6	33.8	36.0	34.3	30.2	32.4	30.0	29.6	30.9	28.6	29.5	27.6	27.4	32.0	30.0	25.5	27.2	26.9	27.6	30.5
				Oolong tea	A	39.2	33.0	33.5	39.6	32.3	34.7	30.6	33.2	36.1	36.4	36.6	37.4	32.1	31.3	32.7	31.4	32.7	31.4	26.5	33.7
					B	36.3	28.2	28.5	33.7	30.0	28.4	24.6	29.0	30.5	32.0	27.5	29.6	26.4	28.5	28.0	27.3	26.6	29.2	25.6	28.9
		ENVI-Carb + PSA	GC–MS	Green tea	A	48.8	37.1	36.7	37.0	37.9	41.8	32.0	38.6	31.9	32.0	29.0	30.3	32.9	36.8	36.2	40.9	39.3	37.2	37.0	36.5
					B	45.0	31.3	36.7	34.0	36.9	35.9	32.3	37.0	30.1	31.8	27.8	29.1	30.9	31.8	32.2	31.5	35.1	33.3	36.5	33.6
				Oolong tea	A	45.5	46.8	38.4	36.4	36.0	32.5	33.6	28.1	35.3	33.3	36.2	30.3	32.2	36.4	43.6	38.5	34.3	42.2	33.8	36.5
					B	37.3	33.1	36.4	31.8	29.1	28.4	28.9	25.3	27.9	26.9	30.7	26.1	28.0	29.9	33.3	32.0	29.7	35.7	30.7	30.6
			GC–MS/MS	Green tea	A	40.9	38.6	38.1	37.3	39.9	38.6	34.0	39.7	32.6	31.5	30.5	39.2	35.2	34.4	33.2	38.6	32.8	33.1	33.0	35.9
					B	37.5	30.5	37.1	33.9	38.2	34.4	32.3	36.9	31.3	31.9	28.3	36.9	30.1	31.8	30.3	31.2	29.3	30.3	32.7	32.9
				Oolong tea	A	41.0	36.7	42.1	36.2	34.8	31.6	33.0	28.3	35.3	32.8	30.0	30.5	30.8	32.2	35.2	28.6	23.3	34.0	26.5	32.8
					B	36.0	31.6	35.0	31.8	27.9	26.4	28.9	23.4	27.9	26.8	28.6	26.4	27.9	27.5	25.9	23.3	21.5	28.4	25.6	27.9
194	Thiazopyr	Cleanert TPT	GC–MS	Green tea	A	102.1	69.9	83.1	78.5	66.6	68.1	69.9	66.9	66.7	63.8	71.8	67.9	63.5	67.8	68.4	57.5	55.6	53.8	61.2	68.6
					B	89.9	69.1	74.2	71.2	61.6	63.9	63.9	60.5	65.6	60.8	62.7	63.5	60.2	66.6	60.9	56.1	54.3	52.4	54.6	63.8
				Oolong tea	A	85.8	75.8	74.5	84.1	71.1	74.4	68.3	73.9	76.5	77.0	77.1	81.3	73.5	69.5	70.9	70.3	69.9	67.1	59.6	73.7
					B	77.7	64.5	64.2	70.6	63.4	62.0	50.2	61.9	62.1	62.5	56.9	64.1	57.4	60.5	58.7	58.6	56.0	59.7	52.2	61.2
			GC–MS/MS	Green tea	A	105.1	72.7	88.4	51.8	73.1	67.2	68.7	68.9	66.7	70.2	62.8	83.2	64.5	78.9	68.0	54.8	59.7	54.1	64.5	69.2
					B	94.8	71.3	75.4	68.0	63.8	60.4	60.5	62.7	65.7	65.7	55.6	69.6	62.9	82.8	58.4	52.9	51.4	50.0	55.7	64.7
				Oolong tea	A	88.4	65.6	72.8	90.0	66.1	74.9	70.3	67.2	78.5	90.6	85.7	84.4	75.7	67.7	63.7	63.2	52.4	70.0	59.3	73.7
					B	81.5	60.5	59.8	76.7	60.5	58.6	52.6	57.0	63.9	76.7	62.4	64.2	55.6	55.4	54.3	56.0	66.5	59.2	53.4	61.3
		ENVI-Carb + PSA	GC–MS	Green tea	A	100.5	89.0	76.3	76.5	80.9	73.5	66.5	79.2	64.3	65.2	60.5	76.4	63.7	69.1	66.4	74.2	56.6	64.8	61.3	72.4
					B	93.6	72.9	75.0	72.6	78.0	70.3	68.0	74.1	61.7	64.1	56.4	66.9	59.3	64.2	60.9	60.8	67.0	58.7	63.1	67.4
				Oolong tea	A	88.8	83.9	87.8	75.4	76.7	72.6	76.5	60.6	79.3	74.1	80.2	66.6	68.2	74.6	75.3	66.9	60.2	74.7	55.8	73.5
					B	72.3	68.4	72.8	68.6	59.2	60.3	64.3	50.1	58.8	57.2	58.1	55.3	59.3	61.8	60.2	54.9	58.9	60.1	50.1	60.5
			GC–MS/MS	Green tea	A	96.4	74.1	76.8	76.1	87.1	79.1	67.0	81.1	67.5	63.7	58.6	73.4	65.9	71.8	62.3	72.2	54.1	64.3	70.0	72.6
					B	92.9	61.1	71.8	73.4	84.0	72.4	67.4	77.0	63.8	63.1	55.1	67.2	61.1	64.9	59.2	57.0	71.5	58.2	64.3	67.1
				Oolong tea	A	83.9	79.8	99.3	80.2	72.0	71.1	68.8	60.2	79.3	71.3	72.2	69.3	65.7	69.6	69.8	54.6	61.8	66.3	59.3	70.5
					B	69.4	67.7	74.5	65.5	56.2	58.7	58.8	49.1	58.8	55.1	62.9	59.8	71.5	63.8	54.8	45.1	46.9	58.6	53.4	59.3

(Continued)

Appendix Table 5.1 The 201 Pesticides Average Content for 19 Circulative Determinations on 3 Months Under 16 Factors (Two Cartridges, Two Instruments, Two Tea Samples, and Two Youden Pair Samples) (cont.)

No.	Pesticides	SPE	Method	Sample	Youden pair	Nov 9, 2009 (n = 5)	Nov 14, 2009 (n = 3)	Nov 19, 2009 (n = 3)	Nov 24, 2009 (n = 3)	Nov 29, 2009 (n = 3)	Dec 4, 2009 (n = 3)	Dec 9, 2009 (n = 3)	Dec 14, 2009 (n = 3)	Dec 19, 2009 (n = 3)	Dec 24, 2009 (n = 3)	Jan 3, 2010 (n = 3)	Jan 8, 2010 (n = 3)	Jan 13, 2010 (n = 3)	Jan 18, 2010 (n = 3)	Jan 23, 2010 (n = 3)	Jan 28, 2010 (n = 3)	Feb 2, 2010 (n = 3)	Feb 7, 2010 (n = 3)	Ave (µg/kg)
195	Tolclofos-methyl	Cleanert TPT	GC-MS	Green tea	A	42.9	36.9	42.6	41.5	33.4	33.7	36.1	35.4	34.5	31.9	35.8	35.6	34.5	34.7	30.3	29.5	28.2	31.9	35.1
					B	42.3	35.9	38.7	38.1	31.3	32.8	33.1	32.5	34.9	30.3	33.8	34.8	33.2	32.4	29.2	28.9	27.7	29.0	33.3
				Oolong tea	A	39.8	35.3	37.8	40.9	34.2	36.5	35.3	41.1	35.2	41.8	38.0	40.2	38.2	33.6	35.3	37.3	33.8	31.4	36.9
					B	36.6	30.5	31.9	36.0	31.1	30.5	26.6	33.5	28.6	35.7	31.7	30.5	34.1	29.3	30.1	30.7	30.7	27.3	31.3
			GC-MS/MS	Green tea	A	39.1	36.2	36.8	37.4	33.8	38.7	38.0	36.3	33.3	36.9	38.3	31.5	35.2	34.5	28.6	27.2	29.2	30.6	34.7
					B	39.4	33.3	36.2	35.6	33.5	34.8	33.5	33.2	34.1	32.0	33.9	29.6	34.7	30.7	27.2	27.7	29.8	27.2	32.6
				Oolong tea	A	49.7	34.1	37.0	44.5	35.8	37.2	36.0	37.6	38.0	38.2	41.4	39.9	33.5	36.0	32.9	35.2	36.3	30.1	37.6
					B	42.1	29.9	29.7	36.3	32.3	29.2	27.1	31.8	32.8	34.5	31.6	29.3	28.5	30.7	28.7	29.8	31.9	27.3	31.3
		ENVI-Carb + PSA	GC-MS	Green tea	A	41.8	39.6	38.5	38.1	40.7	40.8	35.6	41.6	32.6	36.2	59.5	34.3	39.7	34.3	38.4	35.5	33.6	33.3	38.3
					B	39.7	33.9	38.1	36.5	39.2	39.5	33.2	37.5	30.8	34.8	45.3	33.0	36.0	33.4	34.4	32.4	31.6	32.9	35.4
				Oolong tea	A	41.6	40.2	41.5	38.5	38.2	35.8	33.6	30.7	38.2	38.3	30.9	36.3	38.3	34.3	33.1	29.8	38.4	29.4	36.1
					B	33.6	33.0	35.5	34.2	29.5	30.5	32.0	25.6	29.9	30.9	28.5	32.0	33.4	27.7	28.0	25.3	31.6	26.3	30.6
			GC-MS/MS	Green tea	A	46.6	38.7	39.8	39.8	41.0	39.1	36.4	42.6	35.8	34.7	35.5	34.6	36.2	33.4	36.8	36.0	35.4	36.6	37.5
					B	55.6	32.0	38.1	37.6	40.0	36.9	31.1	39.9	34.0	33.4	32.1	33.1	29.4	31.3	31.6	30.9	32.2	37.0	35.1
				Oolong tea	A	54.6	36.7	59.9	38.7	40.1	35.3	35.1	32.5	38.2	32.1	30.3	32.5	32.7	37.1	28.7	25.1	37.5	30.1	36.3
					B	37.1	30.5	35.8	32.4	31.7	28.9	30.9	26.8	29.9	26.7	25.4	25.4	27.9	27.2	24.6	22.4	30.2	27.3	29.1
196	Transfluthrin	Cleanert TPT	GC-MS	Green tea	A	49.3	36.1	39.5	41.4	31.7	34.6	34.5	33.1	33.7	31.4	33.2	33.3	34.4	36.0	28.3	32.6	26.8	32.2	34.6
					B	46.4	34.6	37.5	38.1	29.8	32.2	32.5	30.6	33.5	29.7	31.1	31.1	38.3	33.8	27.5	27.2	25.6	27.9	32.6
				Oolong tea	A	45.1	34.9	36.0	41.1	34.7	36.9	32.3	38.7	39.7	36.6	40.5	35.8	33.5	34.7	34.7	33.6	32.5	28.9	36.0
					B	42.1	29.9	31.5	35.7	31.4	30.7	25.5	31.4	34.4	31.8	31.5	29.5	30.0	28.5	28.6	27.5	29.5	26.1	30.7
			GC-MS/MS	Green tea	A	44.8	33.8	40.8	38.5	32.4	36.1	35.7	32.2	31.1	32.1	33.9	32.5	35.5	31.9	26.1	27.3	27.6	30.7	33.6
					B	42.2	34.5	35.5	34.8	30.1	31.6	31.9	29.9	31.7	30.3	30.9	29.8	34.4	30.8	30.7	27.4	27.1	27.0	31.4
				Oolong tea	A	40.1	31.8	33.9	42.7	31.7	35.7	33.1	35.8	36.9	39.1	39.7	37.1	32.7	33.2	31.7	32.6	33.4	27.9	35.1
					B	37.0	28.4	28.3	35.2	29.5	28.8	25.0	30.2	31.3	33.2	30.5	28.2	27.4	27.4	26.6	26.3	30.2	25.7	29.3
		ENVI-Carb + PSA	GC-MS	Green tea	A	48.7	39.4	36.3	37.8	39.7	34.7	33.6	37.6	31.6	34.2	50.8	32.8	37.7	36.0	37.0	33.5	33.3	31.2	36.3
					B	42.1	41.7	33.6	39.5	38.5	26.6	33.0	36.2	29.5	33.8	40.5	29.8	32.9	29.7	30.6	29.9	33.3	30.6	33.4
				Oolong tea	A	43.4	35.9	42.1	36.7	37.4	35.7	36.8	30.6	37.7	35.1	30.6	34.3	37.3	37.2	33.1	29.1	36.4	28.4	35.6
					B	36.5	31.3	36.3	33.8	29.4	29.5	31.2	26.0	29.3	28.1	27.4	29.9	30.6	29.9	27.6	24.7	29.9	25.1	30.0
			GC-MS/MS	Green tea	A	44.4	38.1	37.8	37.3	38.3	39.3	35.9	40.1	35.6	33.9	35.7	32.7	39.3	32.1	37.6	34.8	31.8	34.0	36.2
					B	39.5	31.9	36.3	34.5	37.3	36.7	33.2	36.9	32.4	33.0	33.2	30.2	30.5	30.3	31.7	31.2	29.4	32.9	33.1
				Oolong tea	A	41.4	36.9	50.0	38.6	36.3	35.2	33.2	29.4	37.7	32.6	24.9	33.7	32.6	36.7	28.4	24.6	34.3	27.9	34.1
					B	33.6	31.1	35.7	33.2	29.0	28.4	28.9	24.2	29.3	27.0	22.3	29.1	28.0	27.2	22.7	22.5	29.2	25.7	28.3
197	Triadimefon	Cleanert TPT	GC-MS	Green tea	A	94.0	74.6	85.3	76.8	61.3	57.0	69.8	72.2	68.9	58.2	75.5	66.8	75.2	66.3	60.5	62.7	51.3	60.5	69.0
					B	94.5	70.3	75.8	69.1	58.8	57.1	65.6	65.3	66.2	54.4	72.1	61.0	77.5	66.0	60.0	63.0	50.2	54.4	65.8
				Oolong tea	A	84.7	63.2	74.5	82.9	69.4	76.6	67.0	69.3	75.9	116.4	76.6	74.2	75.0	72.3	73.1	70.7	65.9	54.3	74.2
					B	76.9	57.6	64.9	72.4	64.3	63.3	53.9	60.1	61.4	89.4	61.0	66.9	73.7	60.9	64.2	63.5	61.5	48.7	65.4
			GC-MS/MS	Green tea	A	133.2	139.5	120.2	117.3	97.9	101.2	101.9	101.4	96.0	98.6	110.6	83.7	108.0	96.6	78.6	78.9	76.7	87.7	101.7
					B	89.1	97.1	70.6	73.8	60.8	64.5	63.4	62.3	63.7	65.4	67.5	54.6	77.0	59.7	51.6	52.4	50.8	51.2	65.2
				Oolong tea	A	89.0	71.0	71.8	90.0	65.3	73.4	69.0	71.6	76.0	80.8	81.4	73.4	69.6	69.0	59.9	60.9	60.0	50.2	71.9
					B	78.6	62.3	59.2	76.9	60.4	58.1	51.6	61.6	65.2	71.7	61.6	58.4	59.3	54.6	49.1	50.0	51.3	47.6	59.7
		ENVI-Carb + PSA	GC-MS	Green tea	A	100.0	107.2	75.1	72.3	74.2	80.7	66.9	79.2	60.6	64.6	127.6	63.2	83.8	72.5	70.4	73.5	66.3	60.3	77.0
					B	98.3	100.0	72.6	65.1	72.5	65.3	69.5	76.6	61.2	66.0	97.9	57.1	73.2	64.4	56.9	66.7	63.2	60.0	71.0
				Oolong tea	A	92.2	76.6	83.5	75.0	77.5	71.8	72.0	59.4	75.1	84.1	77.6	86.2	74.6	84.0	70.4	66.1	70.1	53.8	75.2
					B	79.0	67.0	70.5	68.5	57.5	60.2	63.7	46.4	57.9	65.4	77.6	66.6	71.5	68.7	59.8	57.4	53.4	48.9	64.6
			GC-MS/MS	Green tea	A	149.3	120.9	121.5	121.5	126.7	127.7	98.7	119.3	101.8	98.9	107.2	97.5	116.6	91.7	112.0	93.7	85.8	100.9	109.9
					B	141.9	102.6	117.3	109.2	119.7	115.1	106.2	113.6	98.6	99.5	99.5	87.6	104.5	84.6	88.6	78.6	84.6	103.6	102.1
				Oolong tea	A	85.4	77.2	94.0	79.7	74.5	76.2	68.7	60.0	75.1	63.6	63.3	67.6	64.1	77.0	57.3	50.4	69.5	50.2	69.2
					B	75.9	66.8	76.3	70.6	60.9	61.6	60.9	49.6	57.9	51.5	55.0	53.7	54.3	55.7	46.6	46.3	57.6	47.6	58.4

198	Triadimenol	Cleaner TPT	GC-MS	Green tea	A	162.2	115.1	144.5	121.2	99.5	78.1	103.1	103.7	98.1	67.3	89.7	94.4	98.0	125.1	100.0	86.6	84.1	83.7	96.5	102.7
					B	155.6	109.0	130.5	110.4	91.2	82.6	97.4	94.0	102.2	61.5	83.3	95.5	93.0	131.6	90.7	85.6	81.3	78.3	83.9	97.8
				Oolong tea	A	158.0	111.8	127.7	136.6	81.4	116.2	98.5	90.7	108.9	154.1	83.7	86.6	110.2	103.1	98.7	99.3	96.4	89.3	81.4	107.0
					B	156.9	118.6	131.5	138.4	85.5	92.9	82.3	91.3	86.2	121.4	71.0	70.1	91.7	90.5	84.9	76.0	75.5	82.3	71.0	95.7
			GC-MS/MS	Green tea	A	136.8	114.8	130.0	107.0	102.2	97.7	99.7	97.3	90.9	89.8	96.7	82.3	74.3	125.0	90.2	76.2	74.0	69.4	82.5	96.7
					B	143.5	124.2	119.4	111.8	95.9	94.4	100.2	96.5	95.4	100.8	94.4	86.9	79.4	144.1	91.3	80.4	79.8	72.2	78.6	99.4
				Oolong tea	A	140.0	106.1	108.7	137.6	102.5	109.7	100.0	106.6	114.4	118.4	119.1	118.1	116.8	95.4	96.8	94.3	100.9	94.5	79.9	108.4
					B	124.5	96.5	93.9	118.5	97.3	91.4	81.2	96.4	95.3	104.7	88.2	94.3	92.3	83.7	82.9	77.9	80.6	85.2	74.1	92.6
		ENVI-Carb + PSA	GC-MS	Green tea	A	188.4	129.0	122.5	119.0	123.2	125.4	82.8	106.0	82.7	82.5	79.6	197.1	96.9	110.4	99.1	115.0	100.2	97.8	89.8	113.0
					B	210.6	133.9	121.1	102.4	123.2	116.6	95.0	96.0	81.0	78.0	85.6	118.5	88.0	98.6	91.0	92.4	87.6	86.3	91.6	105.1
				Oolong tea	A	151.0	113.9	144.4	128.1	102.8	108.0	112.4	76.7	111.3	103.2	117.1	97.2	97.5	104.4	112.6	90.8	84.1	116.9	84.3	108.2
					B	139.0	117.9	146.9	133.7	103.6	91.2	98.2	56.4	91.1	74.1	94.3	100.7	87.4	89.4	82.5	76.4	72.9	91.7	75.5	95.9
			GC-MS/MS	Green tea	A	100.1	84.9	84.5	85.6	85.4	85.7	63.4	85.0	71.6	64.6	63.2	129.6	68.1	63.1	62.2	75.8	67.1	60.1	64.3	77.1
					B	99.7	73.7	80.7	74.5	81.8	77.9	76.9	75.9	66.5	63.2	56.5	117.4	58.3	53.7	56.9	61.8	58.4	53.6	64.7	71.2
				Oolong tea	A	133.7	120.2	130.8	122.5	113.2	104.6	106.0	87.1	111.3	103.3	93.4	85.2	95.4	92.0	106.3	80.4	71.1	105.6	79.9	102.2
					B	120.7	104.8	120.3	111.6	90.3	85.2	93.0	75.1	91.1	78.9	89.4	74.7	73.7	82.6	76.4	68.0	66.2	86.5	74.1	87.5
199	Triallate	Cleaner TPT	GC-MS	Green tea	A	86.7	69.6	76.4	67.4	57.4	67.1	64.5	62.5	59.8	57.2	64.6	61.6	58.8	60.4	59.4	53.4	53.2	52.5	58.2	62.7
					B	83.7	67.1	68.8	63.8	56.2	62.3	59.1	59.4	63.9	52.5	60.3	57.6	55.3	59.3	58.3	52.3	51.9	52.8	50.8	59.8
				Oolong tea	A	78.1	63.2	65.3	77.0	64.9	68.0	56.7	59.1	65.5	69.3	66.7	68.7	68.0	64.1	65.4	65.5	62.6	62.1	50.2	65.8
					B	71.0	53.6	55.7	65.0	57.8	54.9	45.4	57.8	53.8	61.6	52.6	53.1	54.2	57.9	54.7	55.8	53.6	55.9	44.0	55.7
			GC-MS/MS	Green tea	A	99.7	78.6	95.8	71.7	63.4	78.4	73.6	68.2	63.3	69.0	69.5	100.3	56.8	67.3	67.7	57.5	56.1	57.6	60.8	71.3
					B	84.1	68.2	73.2	65.3	56.5	63.9	58.4	58.1	63.0	55.7	59.9	70.5	47.9	62.0	56.1	49.0	51.1	49.6	47.8	60.0
				Oolong tea	A	75.9	61.1	62.3	81.0	60.5	65.6	57.3	64.8	63.9	68.6	72.7	70.6	66.0	58.7	61.1	60.4	59.1	62.3	53.1	64.5
					B	68.6	52.8	52.3	66.8	54.8	51.7	45.8	54.3	55.1	61.2	53.1	55.3	54.3	52.0	50.2	51.5	50.5	54.5	46.7	54.3
		ENVI-Carb + PSA	GC-MS	Green tea	A	90.2	76.8	70.2	69.7	73.2	67.2	75.3	68.8	64.4	66.4	59.8	126.2	56.5	67.5	57.3	64.0	68.6	62.0	60.8	70.8
					B	78.7	70.2	66.0	65.2	70.8	57.5	57.8	64.7	59.0	61.8	51.2	90.1	54.9	61.2	55.2	51.4	62.8	60.1	61.1	63.1
				Oolong tea	A	80.9	66.4	74.2	66.3	72.0	69.2	64.6	53.9	67.9	62.4	71.5	56.1	64.0	66.7	70.2	60.0	52.4	64.0	52.7	65.0
					B	62.8	53.8	62.9	58.8	55.5	59.8	54.0	45.9	51.7	52.5	61.1	50.8	54.2	58.9	57.2	48.2	43.9	56.4	45.8	54.4
			GC-MS/MS	Green tea	A	39.4	37.3	34.9	35.0	36.6	35.6	38.6	34.7	34.5	32.5	27.7	44.9	27.8	37.7	28.2	31.9	32.0	27.1	31.8	34.1
					B	36.0	33.4	33.1	32.1	35.0	34.9	26.2	31.2	31.0	30.6	24.7	39.7	28.1	29.6	27.4	24.7	28.0	27.4	30.1	30.7
				Oolong tea	A	76.9	65.6	92.3	70.5	70.9	68.7	60.8	54.0	67.9	57.5	62.8	60.4	60.8	61.1	66.2	50.7	44.0	59.6	53.1	63.4
					B	60.1	54.1	64.3	60.3	57.2	56.5	52.5	45.1	51.7	48.4	58.8	53.8	51.7	52.5	49.5	40.9	39.8	52.4	46.7	52.4
200	Tribenuron-methyl	Cleaner TPT	GC-MS	Green tea	A	65.0	80.7	59.8	45.4	33.6	51.6	43.8	56.0	40.1	35.9	36.0	40.7	45.8	31.2	27.5	24.7	17.8	39.2	46.6	43.2
					B	62.7	77.3	54.8	44.3	32.4	48.3	44.5	49.5	49.5	33.6	34.2	33.5	43.1	34.9	27.9	24.9	18.7	40.3	40.3	41.8
				Oolong tea	A	49.5	30.1	47.9	65.6	55.2	74.1	37.0	42.3	35.5	51.2	38.7	39.6	42.3	27.8	41.7	38.6	37.8	36.1	39.1	43.7
					B	43.8	32.4	57.3	50.6	45.8	57.4	34.9	40.6	30.0	51.1	30.0	31.4	41.0	28.6	35.4	36.5	32.2	31.3	39.6	39.5
			GC-MS/MS	Green tea	A	44.8	35.3	37.9	28.7	23.8	35.7	28.4	34.1	26.0	26.8	27.9	24.6	22.3	47.2	28.3	34.2	23.5	24.7	33.4	30.9
					B	65.5	54.7	52.5	41.8	35.0	50.3	41.6	48.3	45.4	38.3	38.3	34.6	32.6	91.3	39.8	52.1	36.0	35.1	42.3	46.1
				Oolong tea	A	59.5	30.8	40.4	61.3	43.0	59.8	33.9	42.4	43.6	52.9	44.1	41.5	42.0	28.9	45.2	50.3	38.9	53.1	42.7	45.0
					B	52.1	31.7	44.9	49.4	37.3	46.7	32.8	39.9	36.6	53.3	31.9	31.1	39.9	29.0	38.9	45.2	30.9	43.0	35.9	39.5
		ENVI-Carb + PSA	GC-MS	Green tea	A	70.2	69.9	52.0	55.8	47.4	46.6	47.4	54.1	43.4	54.8	29.4	17.0	52.1	17.2	24.2	23.9	22.6	48.3	45.4	47.4
					B	58.1	87.6	43.4	45.9	45.4	36.6	61.7	49.1	40.4	45.7	18.1	12.9	49.9	16.5	19.5	19.6	21.1	47.1	47.0	40.3
				Oolong tea	A	57.6	28.2	44.2	36.2	62.6	65.4	39.4	42.8	39.8	34.5	40.2	38.1	42.7	32.5	41.2	37.7	31.0	36.9	36.4	41.4
					B	54.6	22.2	42.0	39.0	45.1	61.3	29.5	35.8	31.6	28.2	38.2	37.6	39.0	34.3	39.1	25.9	29.8	31.8	32.3	36.7
			GC-MS/MS	Green tea	A	76.7	132.0	91.1	116.7	77.6	112.6	143.2	109.1	108.4	94.4	63.0	0.0	78.1	107.1	59.1	98.3	114.0	84.5	106.9	93.3
					B	75.3	172.2	80.9	90.4	76.3	116.4	162.4	85.1	95.4	82.1	42.6	0.0	78.1	88.4	46.9	70.4	96.2	90.4	109.1	87.3
				Oolong tea	A	64.8	26.5	41.4	38.8	76.9	70.4	35.2	39.2	39.8	36.6	37.7	36.4	40.3	30.5	49.9	28.9	29.7	38.0	42.7	42.3
					B	60.5	23.1	41.5	40.2	55.1	68.0	29.5	30.8	31.6	28.6	36.8	34.4	37.9	31.8	42.7	20.5	29.8	32.7	35.9	37.4

(Continued)

Appendix Table 5.1 The 201 Pesticides Average Content for 19 Circulative Determinations on 3 Months Under 16 Factors (Two Cartridges, Two Instruments, Two Tea Samples, and Two Youden Pair Samples) (cont.)

No.	Pesticides	SPE	Method	Sample	Youden pair	Average content (µg/kg)																			Ave (µg/kg)
						Nov 9, 2009 (n=5)	Nov 14, 2009 (n=3)	Nov 19, 2009 (n=3)	Nov 24, 2009 (n=3)	Nov 29, 2009 (n=3)	Dec 4, 2009 (n=3)	Dec 9, 2009 (n=3)	Dec 14, 2009 (n=3)	Dec 19, 2009 (n=3)	Dec 24, 2009 (n=3)	Dec 29, 2009 (n=3)	Jan 3, 2010 (n=3)	Jan 8, 2010 (n=3)	Jan 13, 2010 (n=3)	Jan 18, 2010 (n=3)	Jan 23, 2010 (n=3)	Jan 28, 2010 (n=3)	Feb 2, 2010 (n=3)	Feb 7, 2010 (n=3)	
201	Vinclozolin	Cleanert TPT	GC–MS	Green tea	A	45.8	40.0	45.3	43.0	34.4	40.1	38.4	38.5	36.9	37.5	38.5	37.7	35.3	35.5	36.4	32.5	37.6	30.7	35.5	37.9
					B	45.6	39.1	41.1	39.9	32.9	37.5	36.2	36.0	39.2	38.3	35.8	34.8	33.4	35.9	33.2	31.8	34.4	31.1	30.8	36.1
				Oolong tea	A	44.9	37.0	39.1	46.9	40.6	40.1	35.4	42.0	38.7	39.8	41.7	41.4	38.5	38.2	40.1	38.0	38.6	36.4	32.5	39.5
					B	40.4	31.8	32.7	40.9	35.3	32.1	29.3	34.8	32.6	34.6	31.6	34.3	31.6	33.6	32.4	32.7	32.7	33.9	29.0	33.5
			GC–MS/MS	Green tea	A	50.0	42.8	44.7	41.5	32.7	42.0	38.4	37.8	32.1	38.3	37.7	45.1	33.1	40.4	38.0	32.5	31.0	30.2	32.8	38.0
					B	47.8	42.9	38.9	39.0	32.1	37.8	34.2	34.5	35.4	37.6	34.0	37.5	31.6	41.0	35.7	31.0	31.7	30.8	28.9	35.9
				Oolong tea	A	50.8	36.6	37.9	47.7	36.1	38.1	34.7	39.2	39.5	44.7	43.4	42.5	39.3	37.2	34.6	34.9	37.8	37.1	31.9	39.2
					B	44.2	32.7	31.5	38.3	33.0	30.7	27.5	34.4	34.3	38.2	30.1	33.3	32.6	34.1	30.4	28.9	29.3	33.3	28.7	32.9
		ENVI-Carb + PSA	GC–MS	Green tea	A	52.3	46.7	45.3	44.0	42.8	44.7	44.6	44.0	40.9	39.6	35.7	52.4	33.6	42.7	33.3	39.8	40.7	37.9	35.8	41.9
					B	45.5	41.9	41.3	41.7	40.1	37.6	38.8	40.8	38.0	37.1	30.7	45.9	31.6	36.2	32.4	32.5	36.4	36.1	35.7	37.9
				Oolong tea	A	47.0	40.8	45.4	39.9	44.4	43.2	39.3	36.2	42.1	35.7	42.4	35.2	35.4	38.5	40.3	37.1	31.6	41.4	33.4	39.4
					B	39.3	33.0	38.9	36.9	34.5	37.4	33.1	30.2	31.1	30.1	37.1	30.8	30.9	32.7	33.6	29.5	27.3	33.7	29.1	33.1
			GC–MS/MS	Green tea	A	43.6	44.8	43.3	41.6	45.4	43.6	46.1	41.4	43.0	38.3	33.9	0.0	34.7	41.2	31.7	36.7	38.0	31.5	34.6	37.5
					B	42.1	42.9	39.6	38.1	42.0	42.2	32.9	38.4	38.4	35.7	29.8	0.0	34.2	32.5	32.6	29.7	35.9	31.4	35.9	34.4
				Oolong tea	A	40.5	40.9	49.9	39.6	42.3	40.8	36.0	32.8	42.1	33.8	36.5	36.1	35.6	35.3	41.2	33.4	26.2	37.2	31.9	37.5
					B	35.4	34.4	38.4	34.8	34.8	34.7	30.7	27.3	31.1	28.9	33.9	32.6	27.3	30.2	30.1	25.8	25.0	32.2	28.7	31.4

Appendix Table 5.2 RSD of 201 Pesticides in Youden Pair Aged Tea Sample for Parallel Determinations (Nov 9, 2009–Feb 7, 2010)

No.	Pesticides	SPE	Method	Sample	Youden pair	RSD%																		
						Nov 9, 2009 (n = 5)	Nov 14, 2009 (n = 3)	Nov 19, 2009 (n = 3)	Nov 24, 2009 (n = 3)	Nov 29, 2009 (n = 3)	Dec 4, 2009 (n = 3)	Dec 9, 2009 (n = 3)	Dec 14, 2009 (n = 3)	Dec 19, 2009 (n = 3)	Dec 24, 2009 (n = 3)	Dec 29, 2009 (n = 3)	Jan 3, 2010 (n = 3)	Jan 8, 2010 (n = 3)	Jan 13, 2010 (n = 3)	Jan 18, 2010 (n = 3)	Jan 23, 2010 (n = 3)	Jan 28, 2010 (n = 3)	Feb 2, 2010 (n = 3)	Feb 7, 2010 (n = 3)
1	2,3,4,5-Tetrachloroaniline	TPT	GC-MS	Green tea	A	8.0	3.3	3.8	2.3	6.1	4.8	6.1	4.0	9.9	3.9	4.3	1.7	8.0	9.8	4.5	3.2	15.4	2.3	6.8
					B	10.0	8.7	7.3	9.2	3.5	11.1	1.4	4.1	5.8	12.3	0.5	8.2	3.4	9.9	5.2	0.8	9.2	4.7	7.6
				Oolong tea	A	3.5	3.5	2.3	3.4	4.2	3.3	6.9	5.3	3.1	1.6	2.5	7.9	6.2	7.2	4.6	3.4	2.0	2.8	1.9
					B	8.8	3.7	6.7	12.1	2.0	4.1	7.2	14.5	4.5	6.1	0.2	3.6	3.9	2.2	6.3	1.8	3.4	2.6	5.9
			GC-MS/MS	Green tea	A	5.9	2.1	7.7	1.8	5.8	5.5	12.2	8.7	4.2	7.6	13.8	3.5	1.7	6.7	8.0	13.8	8.5	17.3	13.2
					B	5.4	4.8	3.3	5.4	1.6	6.9	2.6	4.8	5.6	17.9	2.9	4.7	13.8	4.9	6.2	12.8	4.2	10.1	12.0
				Oolong tea	A	2.9	6.4	6.3	4.0	5.2	2.4	9.9	2.3	11.9	10.1	9.9	2.5	8.0	14.1	33.7	10.5	8.2	0.4	12.3
					B	7.8	12.8	17.5	2.3	5.4	8.3	5.3	16.6	8.6	6.0	7.8	11.8	5.9	6.7	15.9	5.6	13.7	9.8	18.5
		ENVI-Carb+PSA	GC-MS	Green tea	A	8.1	14.1	19.6	7.4	6.0	3.1	9.8	1.7	5.3	5.3	3.5	8.4	5.6	1.4	10.3	5.5	19.9	7.5	5.6
					B	12.4	20.6	10.3	7.0	2.0	3.5	2.2	7.4	10.2	3.2	20.2	5.4	1.7	4.9	10.6	9.2	6.7	8.3	5.8
				Oolong tea	A	0.9	2.0	1.2	4.8	10.4	2.0	6.6	3.2	4.2	4.0	7.2	11.8	15.7	7.0	7.5	2.1	7.1	4.3	5.0
					B	4.6	4.0	12.7	6.5	6.5	5.6	2.1	6.3	11.7	5.9	8.1	2.4	7.2	8.5	7.0	5.3	5.8	2.0	4.7
			GC-MS/MS	Green tea	A	6.2	2.8	2.9	2.8	10.1	8.8	8.1	4.9	13.9	6.1	20.3	6.4	13.6	15.4	5.0	15.3	25.6	13.5	15.3
					B	7.6	8.3	5.7	4.5	2.6	1.1	1.4	4.9	10.6	6.4	25.4	1.8	9.1	3.5	3.7	6.8	5.8	16.3	16.0
				Oolong tea	A	5.1	6.0	1.3	14.7	4.4	4.7	0.4	7.5	7.0	5.0	2.0	13.1	6.2	5.8	6.6	9.5	16.8	10.8	11.5
					B	7.6	5.1	14.0	9.5	10.0	2.4	7.1	3.3	19.2	5.1	11.1	4.4	4.1	9.4	21.3	4.0	3.8	10.7	8.1
2	2,3,5,6-Tetrachloroaniline	TPT	GC-MS	Green tea	A	4.5	5.2	1.3	0.6	6.9	4.6	5.3	3.4	6.0	2.8	4.5	2.3	5.9	4.9	5.2	2.1	3.9	6.1	4.8
					B	6.6	8.8	3.8	8.3	2.7	9.5	3.0	1.8	3.1	8.5	1.9	6.8	2.7	4.5	3.7	1.5	4.5	2.2	5.1
				Oolong tea	A	3.9	2.1	0.4	2.9	4.5	1.8	7.8	5.0	3.3	2.4	2.3	1.8	3.6	1.7	7.2	3.2	4.0	1.3	2.0
					B	6.7	2.6	4.9	5.4	2.0	3.7	5.0	1.8	5.8	4.1	0.6	0.7	3.2	2.5	3.9	0.8	0.6	1.6	1.5
			GC-MS/MS	Green tea	A	3.4	6.2	2.6	2.0	6.3	6.2	6.8	4.0	5.6	4.5	4.9	1.6	4.2	7.3	1.4	0.7	4.5	3.8	2.9
					B	4.2	8.9	4.1	8.7	6.2	3.5	5.1	2.3	1.4	9.8	4.1	2.8	3.5	3.8	1.8	1.2	11.9	4.8	8.0
				Oolong tea	A	4.1	1.7	2.6	5.7	4.1	2.5	8.4	1.6	1.1	2.0	1.7	7.9	0.3	0.9	3.8	0.7	6.6	2.5	6.4
					B	4.9	2.2	5.8	2.6	1.6	3.4	3.9	2.8	6.5	2.0	2.8	4.6	2.6	2.9	5.2	3.1	2.4	1.6	2.9
		ENVI-Carb+PSA	GC-MS	Green tea	A	5.1	5.6	2.0	5.2	3.7	3.2	5.9	1.4	7.0	1.6	7.4	3.7	2.9	4.2	5.0	6.6	11.8	2.9	9.2
					B	9.7	8.1	4.0	3.1	1.6	2.4	11.6	5.4	2.7	3.2	19.9	0.3	4.4	3.0	14.5	5.1	7.5	8.7	5.4
				Oolong tea	A	3.7	2.9	2.5	3.5	8.0	1.9	0.6	1.5	1.6	0.2	4.2	3.1	1.2	2.2	2.7	2.3	5.2	5.3	5.1
					B	4.9	3.2	8.5	3.9	8.1	2.9	2.3	5.8	10.7	2.7	6.7	4.2	2.8	7.1	8.4	2.7	6.7	4.1	4.1
			GC-MS/MS	Green tea	A	4.1	5.0	2.5	6.3	4.3	1.1	1.7	0.3	6.2	3.1	4.6	7.0	5.0	7.8	5.9	8.5	12.4	4.6	4.3
					B	1.5	11.2	4.6	4.4	3.4	4.0	1.7	3.3	2.4	2.2	21.3	2.3	0.6	7.6	17.1	3.4	4.2	18.3	5.0
				Oolong tea	A	4.4	5.3	5.3	6.4	5.7	2.4	3.0	1.7	2.8	1.2	2.6	7.5	0.6	3.5	7.9	2.3	7.4	2.0	3.8
					B	6.3	4.0	9.2	3.7	5.9	2.8	3.4	5.5	9.7	1.6	1.0	5.2	1.8	9.5	8.1	1.8	6.8	2.3	2.3

(Continued)

Appendix Table 5.2 RSD of 201 Pesticides in Youden Pair Aged Tea Sample for Parallel Determinations (Nov 9, 2009–Feb 7, 2010) (cont.)

RSD%

No.	Pesticides	SPE	Method	Sample	Youden pair	Nov 9, 2009 (n = 5)	Nov 14, 2009 (n = 3)	Nov 19, 2009 (n = 3)	Nov 24, 2009 (n = 3)	Nov 29, 2009 (n = 3)	Dec 4, 2009 (n = 3)	Dec 9, 2009 (n = 3)	Dec 14, 2009 (n = 3)	Dec 19, 2009 (n = 3)	Dec 24, 2009 (n = 3)	Dec 14, 2009 (n = 3)	Jan 3, 2010 (n = 3)	Jan 8, 2010 (n = 3)	Jan 13, 2010 (n = 3)	Jan 18, 2010 (n = 3)	Jan 23, 2010 (n = 3)	Jan 28, 2010 (n = 3)	Feb 2, 2010 (n = 3)	Feb 7, 2010 (n = 3)
3	4,4-Dibromobenzophenone	TPT	GC-MS	Green tea	A	5.5	5.9	3.4	3.5	4.3	8.2	8.8	4.9	6.7	4.6	6.2	4.3	6.7	2.7	7.9	0.8	6.1	11.3	10.3
					B	12.3	8.2	4.8	12.5	6.3	7.0	2.5	3.5	4.6	10.3	2.8	7.9	4.6	5.6	7.6	3.7	6.2	6.2	9.1
				Oolong tea	A	7.4	10.6	1.8	0.8	7.2	6.1	9.6	10.6	4.4	5.4	3.0	7.3	4.4	6.7	3.8	8.6	5.8	0.4	11.0
					B	12.6	4.2	1.5	4.2	5.4	2.6	5.0	1.4	4.8	2.3	1.3	3.8	4.8	2.8	0.4	2.7	3.3	10.0	14.4
			GC-MS/MS	Green tea	A	4.5	6.7	2.1	3.8	6.0	7.8	9.1	4.9	2.8	2.9	8.0	2.8	2.8	2.4	1.5	3.9	4.6	13.8	2.5
					B	4.9	9.2	7.2	6.2	1.8	5.4	4.6	2.6	4.4	10.1	5.6	8.0	4.4	2.7	4.7	1.3	7.5	5.2	5.2
				Oolong tea	A	6.4	7.0	3.9	6.0	8.3	2.1	11.1	5.3	5.3	3.4	2.9	2.4	5.3	3.6	7.7	3.7	2.4	7.9	3.6
					B	3.6	2.2	2.7	7.8	1.7	3.2	8.3	1.8	1.4	3.3	9.5	3.5	1.4	3.9	5.4	0.6	4.0	3.1	5.4
		ENVI-Carb +PSA	GC-MS	Green tea	A	5.0	4.4	1.9	5.7	8.7	15.9	8.4	1.6	7.7	4.9	4.9	8.2	7.7	4.5	7.7	2.2	8.3	5.2	5.2
					B	7.9	7.6	5.6	10.3	4.4	9.2	4.1	11.3	7.2	3.6	18.6	5.5	7.2	4.4	12.9	6.4	11.5	3.1	14.8
				Oolong tea	A	10.1	4.5	4.3	6.5	11.2	2.2	3.9	5.9	3.2	2.2	6.0	7.5	3.2	8.1	4.0	5.7	6.2	1.4	11.3
					B	7.0	1.9	6.5	7.2	10.9	4.4	2.5	9.0	9.2	11.8	10.2	2.5	9.2	7.1	16.8	6.6	4.9	10.0	3.1
			GC-MS/MS	Green tea	A	4.9	6.3	1.9	7.7	8.3	3.9	7.4	2.2	10.1	1.7	12.1	5.9	10.1	11.2	6.4	2.7	10.4	1.9	3.7
					B	10.3	8.5	4.7	9.0	4.3	1.4	3.0	7.0	2.9	3.2	16.2	6.5	2.9	7.3	14.1	1.9	5.4	8.2	3.3
				Oolong tea	A	8.8	2.6	4.9	5.1	8.8	2.2	4.3	9.5	3.0	0.2	7.3	5.6	3.0	5.3	7.2	2.2	10.9	11.9	12.2
					B	8.0	2.5	11.4	1.7	8.7	2.9	3.6	9.7	10.7	2.5	2.5	3.3	10.7	5.9	13.1	2.8	4.0	8.9	5.8
4	4,4-Dichlorobenzophenone	TPT	GC-MS	Green tea	A	8.2	5.3	3.1	0.8	5.5	5.4	8.9	3.8	3.6	4.9	5.2	1.9	3.6	7.0	3.6	5.1	2.7	12.1	3.5
					B	7.2	7.3	5.8	9.3	4.0	4.7	3.5	4.0	5.1	6.7	3.8	5.3	5.1	4.7	5.5	2.1	7.3	4.5	4.9
				Oolong tea	A	10.3	4.8	8.1	6.5	4.0	7.1	10.3	2.4	2.5	2.0	1.2	4.0	2.5	20.7	2.7	3.4	6.9	9.0	0.8
					B	8.8	12.2	7.7	3.2	3.5	4.6	10.5	8.0	8.6	1.1	0.6	7.0	8.6	4.3	11.7	0.9	9.0	6.9	5.0
			GC-MS/MS	Green tea	A	7.2	4.9	1.1	2.4	5.0	4.7	8.3	4.6	5.8	3.9	7.3	2.6	5.8	6.3	3.0	0.8	3.0	11.2	4.6
					B	5.2	7.9	6.1	7.9	4.8	4.3	4.7	3.8	2.4	9.7	7.8	6.1	2.4	1.5	4.0	2.9	9.0	8.4	4.4
				Oolong tea	A	9.8	4.2	8.1	8.8	4.2	6.5	10.1	5.6	2.2	7.5	2.4	2.2	2.2	19.6	1.5	4.2	5.1	6.7	2.6
					B	5.9	5.9	10.1	4.4	4.9	3.7	8.9	6.0	7.9	4.5	8.8	6.2	7.9	6.1	14.0	1.7	6.9	7.2	6.6
		ENVI-Carb +PSA	GC-MS	Green tea	A	4.2	4.6	0.9	10.5	7.9	7.4	17.0	4.2	9.1	5.1	9.9	9.0	9.1	3.6	5.7	1.9	13.4	2.2	6.7
					B	6.3	7.4	4.3	9.0	4.3	2.5	4.1	10.7	3.1	4.4	15.5	3.4	3.1	11.7	9.9	3.4	4.7	3.6	5.1
				Oolong- tea	A	10.9	11.2	5.9	4.5	7.3	10.2	11.8	3.1	5.8	7.8	6.5	2.6	5.8	3.3	2.0	5.2	2.7	4.6	7.9
					B	4.2	2.1	7.6	5.3	7.3	7.0	10.8	4.8	15.2	6.6	5.9	7.4	15.2	2.2	16.3	6.0	2.9	16.9	4.1
			GC-MS/MS	Green tea	A	5.6	3.5	0.1	7.0	7.0	5.0	7.7	3.8	9.6	3.9	10.8	3.9	9.6	4.7	5.7	4.2	10.8	4.0	3.3
					B	8.2	9.5	4.7	8.4	6.9	2.8	2.2	8.0	5.0	2.4	19.0	10.1	5.0	8.7	12.7	4.1	4.9	11.8	3.0
				Oolong tea	A	14.1	11.2	4.1	2.4	9.0	10.3	12.8	4.2	3.3	8.2	7.3	5.8	3.3	3.7	8.4	5.2	8.3	12.5	6.5
					B	5.6	0.7	8.1	3.5	7.1	7.3	13.2	3.7	15.3	5.8	0.6	6.7	15.3	5.9	13.8	4.5	4.2	11.4	5.1
5	Acetochlor	TPT	GC-MS	Green tea	A	5.7	6.0	3.9	3.9	7.6	5.0	6.0	4.9	4.9	3.1	3.1	0.8	4.9	9.0	11.1	6.1	18.1	8.5	5.3
					B	7.2	9.6	6.3	9.7	4.9	3.3	2.6	2.3	8.8	4.0	5.8	7.9	8.8	3.4	15.7	3.5	23.7	2.4	6.2
				Oolong tea	A	5.8	4.5	6.2	3.7	3.6	0.8	7.0	1.2	1.5	4.1	5.3	5.1	1.5	4.7	7.8	4.8	3.6	2.0	1.6
					B	6.5	6.1	0.4	2.5	3.7	3.5	6.4	5.5	4.5	2.6	5.5	2.4	4.5	1.2	6.0	2.4	2.5	1.3	4.5
			GC-MS/MS	Green tea	A	9.4	8.6	4.1	3.9	4.9	3.4	3.9	3.2	3.0	2.2	5.0	2.8	3.0	9.9	15.6	4.7	15.7	7.6	5.2
					B	6.7	7.7	4.8	8.2	4.1	7.4	4.6	3.2	4.1	9.9	5.2	5.1	4.1	5.5	17.6	2.2	23.9	5.9	9.6
				Oolong tea	A	5.1	6.3	0.3	5.8	10.9	8.2	8.0	0.7	0.6	1.3	4.5	4.3	0.6	8.8	6.2	6.7	0.3	7.8	5.7
					B	3.7	2.8	8.1	5.0	4.2	3.8	6.2	0.7	2.8	7.4	2.6	0.1	2.8	9.2	2.7	4.4	0.5	3.1	5.8
		ENVI-Carb +PSA	GC-MS	Green tea	A	5.9	1.1	1.3	11.0	4.2	4.9	3.9	6.0	9.6	6.8	6.8	3.6	9.6	2.7	2.6	4.2	7.5	3.9	8.4
					B	8.1	5.8	3.4	3.7	4.9	4.9	2.2	1.2	4.7	10.1	16.2	4.8	4.7	3.7	10.5	1.5	4.7	8.1	13.3
				Oolong tea	A	5.6	7.8	1.9	6.5	3.6	5.1	1.6	4.1	14.8	1.6	3.3	4.8	14.8	5.5	5.3	4.1	3.9	2.1	4.6
					B	8.6	5.5	11.1	9.7	3.7	2.1	4.5	8.7	9.3	1.2	6.0	3.0	9.3	2.7	10.2	4.9	4.7	1.4	3.0
			GC-MS/MS	Green tea	A	4.3	5.4	0.2	5.7	4.9	3.0	8.5	1.6	64.0	2.8	14.0	6.1	64.0	8.4	5.3	4.9	15.7	7.6	2.5
					B	19.7	9.6	1.4	3.8	4.1	3.8	2.3	2.9	4.8	2.5	6.1	3.3	4.8	7.3	9.3	4.9	23.9	4.3	6.8
				Oolong tea	A	3.7	5.1	0.3	4.0	7.1	3.5	3.8	8.6	5.8	2.2	3.2	6.4	5.8	3.9	10.2	7.8	4.7	13.2	5.8
					B	8.8	2.7	11.4	3.2	11.6	7.1	10.1	6.1	9.2	1.8	3.1	7.9	9.2	2.3	7.1	8.8	6.1	3.6	5.3

6	Alachlor	TPT	GC-MS	Green tea	A	5.9	4.4	3.7	3.2	6.4	4.5	5.5	7.5	6.4	2.3	5.7	2.4	3.1	2.0	3.5	1.5	6.8	8.8	4.7
					B	7.7	12.0	8.6	6.8	5.4	2.3	2.2	0.3	6.6	10.2	10.5	9.6	1.8	3.5	5.2	1.0	4.1	6.5	13.6
				Oolong tea	A	4.1	1.1	4.0	6.0	2.3	2.1	12.8	3.9	5.0	8.4	6.3	4.8	3.0	3.0	8.6	4.2	5.1	2.0	2.0
					B	6.7	8.2	11.6	5.9	5.7	4.5	5.0	8.6	9.2	4.1	10.2	0.9	2.0	1.9	7.7	1.1	0.9	2.0	4.4
			GC-MS/MS	Green tea	A	40.5	6.3	5.0	2.5	6.3	5.0	6.3	7.2	3.0	0.8	2.1	5.8	10.2	5.9	5.3	2.2	2.5	6.3	5.7
					B	46.5	9.1	4.2	6.3	4.8	5.8	3.3	3.5	5.7	9.4	3.2	7.0	3.9	3.3	7.1	2.5	5.8	3.7	6.5
				Oolong tea	A	4.5	3.6	1.9	6.7	7.7	3.8	11.9	9.4	2.1	1.4	4.1	4.5	10.6	2.7	7.6	1.5	2.4	7.7	8.0
					B	1.5	3.8	4.5	6.2	1.6	1.0	5.9	10.3	8.7	4.6	2.1	2.6	8.8	4.7	5.2	3.5	0.7	6.8	5.6
		ENVI-Carb+PSA	GC-MS	Green tea	A	6.3	4.1	2.2	4.6	6.4	5.1	4.2	2.2	2.9	1.6	5.8	2.3	5.6	4.3	2.1	6.2	9.3	8.7	8.5
					B	10.0	8.8	3.1	2.6	2.0	7.1	3.0	4.1	4.8	10.6	14.2	11.9	6.9	2.5	6.5	2.0	3.5	16.4	5.4
				Oolong tea	A	9.8	4.3	5.4	3.3	2.4	2.5	1.8	3.3	14.3	4.0	6.9	4.3	2.2	4.2	4.2	4.7	3.2	4.4	3.7
					B	5.3	2.3	10.4	4.0	10.4	3.0	1.4	7.8	8.1	5.3	6.7	2.3	4.6	8.7	11.5	3.2	7.6	1.9	4.3
			GC-MS/MS	Green tea	A	8.1	6.3	1.1	4.6	6.1	5.0	6.8	1.4	0.4	1.0	16.9	23.8	9.0	9.8	4.3	1.3	7.1	11.7	3.4
					B	29.6	10.6	1.5	2.6	3.5	3.2	3.8	3.1	3.7	1.9	11.2	12.6	2.7	6.3	12.3	5.8	7.7	12.8	4.4
				Oolong tea	A	3.8	2.1	2.9	7.1	7.6	3.1	2.7	3.8	4.4	3.0	1.9	8.4	4.9	5.8	8.3	6.2	9.9	8.7	5.3
					B	7.2	2.8	10.5	3.7	7.8	5.5	2.9	6.8	12.5	2.8	4.2	3.8	4.7	10.5	12.2	2.2	4.5	4.1	7.2
7	Atratone	TPT	GC-MS	Green tea	A	5.1	5.1	2.5	1.9	9.1	4.8	8.1	5.4	4.2	4.7	7.5	5.4	9.0	2.7	2.9	5.6	4.9	6.4	4.7
					B	4.7	6.1	6.1	8.9	3.2	4.0	4.8	3.5	5.0	11.6	11.7	13.1	2.2	1.4	3.0	3.6	3.3	1.1	2.0
				Oolong tea	A	3.4	3.3	2.8	2.5	2.9	5.4	10.5	4.8	4.1	6.6	11.2	2.7	3.1	3.9	8.3	5.7	4.0	5.0	1.7
					B	7.2	2.8	3.3	3.4	2.7	2.8	5.4	3.4	7.6	5.6	10.4	1.1	3.0	5.7	6.4	0.3	1.9	1.4	2.2
			GC-MS/MS	Green tea	A	5.0	5.4	2.1	2.8	9.1	7.7	7.1	6.4	4.7	0.6	5.1	5.2	8.2	4.6	2.3	2.1	7.7	6.6	4.8
					B	4.7	7.1	8.8	7.0	7.3	15.0	4.6	5.5	3.8	10.7	6.8	6.8	3.1	4.7	7.0	6.2	6.7	6.9	5.0
				Oolong tea	A	4.7	1.2	0.6	6.1	1.4	2.4	10.6	1.7	0.5	6.2	2.1	5.9	4.4	2.9	8.4	4.2	1.5	1.9	6.4
					B	4.6	2.1	5.3	6.4	0.3	2.9	4.2	4.1	5.4	4.0	5.6	1.6	7.5	4.3	6.6	1.7	2.5	2.2	1.4
		ENVI-Carb+PSA	GC-MS	Green tea	A	5.5	5.3	2.0	9.0	8.9	5.5	4.1	0.5	4.0	14.2	5.4	4.9	3.4	6.4	7.5	7.4	9.9	5.3	5.9
					B	4.1	8.2	4.7	5.8	3.3	5.1	4.5	5.6	7.4	17.6	29.1	7.9	5.1	6.5	14.7	2.7	2.9	12.6	7.7
				Oolong tea	A	4.2	4.0	2.9	2.4	2.5	2.4	3.8	0.5	30.4	3.2	8.3	8.4	2.4	5.4	5.0	3.2	6.1	3.7	5.7
					B	4.8	3.4	9.1	1.8	14.7	6.0	2.8	6.5	12.5	5.6	9.8	3.1	5.7	8.0	10.8	3.2	5.9	5.3	6.3
			GC-MS/MS	Green tea	A	4.8	6.3	2.6	7.9	7.2	3.6	5.1	0.4	0.7	2.4	6.5	12.3	4.4	11.7	9.1	5.7	8.9	2.4	1.8
					B	5.2	10.8	5.5	6.7	2.9	1.2	2.5	3.5	4.8	2.3	22.2	9.7	1.8	11.5	18.6	1.8	1.7	12.4	7.0
				Oolong tea	A	3.7	4.7	4.5	4.9	5.1	2.3	5.8	3.7	4.9	1.6	6.6	9.7	3.9	6.6	10.7	3.3	5.9	5.4	5.5
					B	5.6	3.9	9.8	3.8	9.6	6.3	3.2	5.9	12.1	4.8	1.2	6.8	8.5	8.9	11.3	0.8	4.7	6.3	6.7
8	Benodanil	TPT	GC-MS	Green tea	A	6.4	15.0	3.3	1.5	14.4	3.9	16.7	10.1	7.5	5.5	4.3	7.1	13.6	10.2	2.4	8.6	4.3	14.2	2.1
					B	-0.3	11.5	2.3	12.5	7.8	9.8	8.1	11.1	10.4	14.8	8.1	8.1	14.0	10.1	8.7	15.4	8.8	14.6	7.9
				Oolong tea	A	40.5	10.2	7.9	30.1	19.6	32.7	49.0	16.8	8.5	7.2	9.5	39.0	10.6	5.2	6.8	9.1	8.3	51.2	13.6
					B	52.1	5.2	18.9	17.5	4.3	12.7	21.5	45.8	5.5	28.2	3.7	5.6	10.6	29.0	20.4	9.6	13.9	37.2	29.4
			GC-MS/MS	Green tea	A	9.0	18.5	4.3	3.9	11.2	7.2	18.9	14.5	20.1	5.5	3.4	5.0	9.3	9.8	7.6	3.8	6.1	21.6	3.7
					B	14.7	6.2	3.6	13.1	5.0	7.2	17.0	24.2	2.5	16.4	7.7	16.0	11.8	7.5	10.9	12.5	12.8	11.7	15.3
				Oolong tea	A	20.9	16.7	1.2	33.8	23.3	34.2	5.9	22.0	42.8	7.0	11.5	1.7	16.5	15.8	13.4	18.9	18.6	56.4	15.8
					B	54.4	15.8	31.5	28.4	20.6	12.1	18.0	17.2	22.0	51.3	12.6	5.4	16.5	7.8	29.5	12.8	17.1	26.9	29.6
		ENVI-Carb+PSA	GC-MS	Green tea	A	16.7	16.0	24.3	11.4	8.0	24.1	7.1	9.4	9.3	9.4	2.7	30.4	9.9	15.8	5.7	27.0	16.9	10.1	12.2
					B	18.1	3.0	11.2	17.7	15.8	14.8	14.4	20.0	13.7	8.0	31.4	16.7	6.7	8.9	21.9	8.7	7.2	24.3	17.9
				Oolong tea	A	39.7	36.3	6.5	39.6	19.6	13.5	76.0	8.1	22.2	21.6	8.2	33.5	44.5	8.6	7.8	27.0	3.9	7.0	14.3
					B	37.6	34.3	17.1	23.0	26.8	23.6	23.2	5.1	10.3	36.9	15.2	8.3	24.3	5.1	20.4	8.0	14.2	11.5	8.7
			GC-MS/MS	Green tea	A	17.2	16.6	20.5	12.8	7.1	22.5	44.6	25.4	13.9	10.5	11.8	4.7	4.6	11.6	4.6	6.3	13.4	1.3	10.4
					B	8.5	6.5	8.2	19.9	9.6	7.7	24.3	32.2	14.5	7.2	25.0	7.3	57.0	2.0	8.3	20.0	2.9	20.9	18.1
				Oolong tea	A	32.0	47.9	2.9	45.1	20.7	18.6	4.7	44.0	11.5	1.8	18.3	41.1	31.1	10.2	14.3	49.0	16.8	37.0	29.3
					B	34.1	11.5	23.1	30.7	40.3	26.0	39.9	9.4	35.8	59.0	7.3	5.5	3.4	14.1	21.6	10.6	21.7	67.4	6.5

(Continued)

Appendix Table 5.2 RSD of 201 Pesticides in Youden Pair Aged Tea Sample for Parallel Determinations (Nov 9, 2009–Feb 7, 2010) (cont.)

No.	Pesticides	SPE	Method	Sample	Youden pair	RSD% Nov 9 2009 (n=5)	Nov 14 2009 (n=3)	Nov 19 2009 (n=3)	Nov 24 2009 (n=3)	Nov 29 2009 (n=3)	Dec 4 2009 (n=3)	Dec 9 2009 (n=3)	Dec 14 2009 (n=3)	Dec 19 2009 (n=3)	Dec 24 2009 (n=3)	Dec 29 2009 (n=3)	Jan 3 2010 (n=3)	Jan 8 2010 (n=3)	Jan 13 2010 (n=3)	Jan 18 2010 (n=3)	Jan 23 2010 (n=3)	Jan 28 2010 (n=3)	Feb 2 2010 (n=3)	Feb 7 2010 (n=3)
9	Benoxacor	TPT	GC-MS	Green tea	A	5.3	6.4	2.7	10.0	11.7	5.8	6.8	11.2	11.9	2.9	6.4	4.8	13.6	3.2	9.9	8.8	6.1	13.5	5.2
					B	8.6	11.6	7.9	13.4	7.2	2.3	5.6	14.3	7.7	9.7	4.5	9.3	12.0	10.0	12.3	8.6	6.5	11.3	7.4
				Oolong tea	A	7.1	0.7	7.2	6.9	14.9	3.9	10.5	8.9	8.0	4.6	15.0	8.5	4.2	3.9	7.1	10.7	10.2	4.9	5.3
					B	11.1	2.6	5.6	2.5	10.8	14.8	5.9	13.5	9.4	3.2	5.8	5.6	8.5	4.0	9.8	7.6	3.7	1.4	14.4
			GC-MS/MS	Green tea	A	8.6	9.5	5.6	10.1	14.2	3.6	13.5	12.0	3.1	1.1	6.1	8.6	17.6	12.0	5.2	5.2	9.2	15.0	8.1
					B	6.6	9.3	4.3	7.5	7.7	10.2	4.8	11.2	7.7	6.6	2.6	11.8	9.2	3.6	5.1	3.8	14.2	7.1	16.3
				Oolong tea	A	11.1	11.3	6.2	7.1	2.3	9.4	1.6	3.4	5.1	2.4	17.4	9.5	17.0	3.7	2.9	12.2	5.9	23.1	9.9
					B	25.1	6.1	6.6	2.9	3.5	11.9	2.9	2.5	14.3	4.5	11.3	6.8	7.4	6.8	11.2	4.9	1.7	13.2	1.6
		ENVI-Carb+PSA	GC-MS	Green tea	A	17.9	10.6	13.4	8.0	10.4	18.8	11.4	11.0	9.4	7.0	3.7	20.2	1.6	11.0	3.8	10.5	3.9	7.0	3.5
					B	18.4	8.5	4.0	11.9	11.8	12.3	10.5	9.4	7.8	6.4	30.2	12.2	40.5	5.9	11.1	4.5	8.1	22.4	9.9
				Oolong tea	A	11.5	7.9	6.4	10.4	4.5	4.6	5.4	10.7	4.8	4.3	15.4	14.6	11.9	8.5	3.6	17.0	10.6	15.9	4.7
					B	19.3	13.5	4.9	3.0	15.7	8.4	9.5	10.1	15.4	16.0	15.4	6.8	11.1	24.2	12.4	13.3	16.7	22.7	6.8
			GC-MS/MS	Green tea	A	20.1	11.2	12.2	3.6	7.1	22.0	12.0	13.9	13.1	6.2	18.2	23.3	9.8	9.2	7.6	14.6	14.8	16.1	4.0
					B	38.1	15.5	2.1	6.2	11.8	13.4	7.1	13.8	11.0	5.6	16.0	11.9	33.5	8.1	18.7	3.6	11.7	21.1	9.6
				Oolong tea	A	8.7	5.5	9.2	19.0	3.7	6.0	9.5	6.6	6.2	5.2	5.3	16.7	22.7	5.1	11.1	16.0	3.5	24.0	7.6
					B	15.5	5.2	11.0	7.6	14.6	14.5	10.2	5.6	20.4	15.0	5.8	1.0	4.3	13.8	14.1	8.8	12.1	33.2	2.0
10	Bromophos-ethyl	TPT	GC-MS	Green tea	A	4.3	5.5	4.0	3.5	7.8	3.7	6.2	4.4	5.2	4.1	8.0	5.6	6.2	2.6	1.7	2.3	4.0	5.4	4.0
					B	8.3	8.4	7.0	9.3	4.2	2.5	1.4	1.7	4.8	10.2	6.2	11.9	1.3	1.6	5.3	2.8	5.6	0.5	6.4
				Oolong tea	A	5.0	2.2	2.2	6.1	1.7	2.5	10.4	5.1	2.9	5.3	13.0	2.4	6.1	3.3	8.3	2.8	4.3	1.4	1.0
					B	7.8	0.8	2.9	4.4	2.9	3.3	3.8	4.0	6.2	2.8	10.8	1.4	3.2	3.6	3.2	0.5	1.5	1.2	2.7
			GC-MS/MS	Green tea	A	4.6	7.7	5.3	3.3	7.6	5.6	8.1	3.4	6.1	3.2	3.8	4.3	4.6	2.2	1.9	1.6	0.3	6.6	3.8
					B	4.7	9.4	6.9	6.3	3.9	8.8	3.6	3.7	4.8	9.6	4.4	7.9	2.3	2.0	1.3	3.1	7.5	3.0	7.4
				Oolong tea	A	5.3	3.2	1.3	4.4	2.0	3.7	7.1	1.0	2.8	1.7	3.8	1.0	7.6	2.9	7.3	2.7	1.0	3.0	1.7
					B	1.4	2.7	4.4	5.8	3.1	2.8	3.6	3.8	6.3	2.1	7.1	2.8	3.5	0.5	5.3	1.5	2.5	2.1	3.4
		ENVI-Carb+PSA	GC-MS	Green tea	A	10.0	3.9	4.2	4.6	6.9	1.7	5.3	3.6	2.9	2.5	5.7	0.5	4.3	3.0	2.1	6.3	9.2	5.6	5.8
					B	6.6	9.7	4.2	2.4	2.8	1.4	5.2	1.9	5.3	13.4	27.5	5.6	3.6	3.9	8.0	3.0	4.4	11.7	6.7
				Oolong tea	A	1.5	3.5	3.5	3.3	2.2	1.9	3.7	3.2	13.8	1.9	5.2	3.1	1.4	3.5	3.1	1.4	5.6	4.3	4.2
					B	7.3	3.7	8.9	3.7	11.0	3.9	1.8	6.6	10.6	3.8	9.4	0.8	4.4	8.4	9.8	2.6	7.8	1.3	4.5
			GC-MS/MS	Green tea	A	9.9	4.2	3.3	3.0	6.4	3.8	8.9	3.1	1.7	0.7	13.7	3.4	2.3	6.6	3.2	3.1	7.9	4.9	2.4
					B	10.9	11.7	2.1	1.5	3.3	2.4	5.1	0.7	4.4	4.2	16.0	3.2	5.9	4.2	13.5	2.7	3.5	12.9	7.1
				Oolong tea	A	4.2	4.0	3.9	5.8	6.1	2.7	4.5	5.9	3.8	1.9	5.6	1.9	1.0	4.7	6.7	1.5	5.2	4.2	7.7
					B	8.1	3.0	11.7	3.9	8.6	5.0	2.5	6.5	11.5	1.5	2.2	0.5	1.9	10.3	9.9	1.8	3.5	1.0	2.5
11	Butralin	TPT	GC-MS	Green tea	A	4.6	5.2	4.8	9.6	15.2	5.3	13.2	13.2	6.8	10.4	10.4	12.7	12.8	9.1	10.1	7.8	5.1	11.9	3.7
					B	5.7	7.2	5.8	15.8	6.6	11.4	7.0	19.6	10.6	10.2	9.3	13.6	13.2	16.4	9.5	11.6	22.8	11.3	13.7
				Oolong tea	A	5.6	2.7	6.4	4.8	25.0	3.7	7.6	7.5	6.9	13.0	11.8	10.9	5.6	9.8	3.5	19.3	10.2	1.2	2.7
					B	9.0	5.3	5.6	6.8	20.8	13.8	10.7	8.8	8.1	11.9	11.8	5.6	6.6	3.2	12.1	14.4	1.4	2.6	5.2
			GC-MS/MS	Green tea	A	7.9	6.4	4.1	6.2	10.3	8.9	9.5	13.9	23.5	1.8	9.8	9.4	16.2	9.9	4.3	1.3	4.1	11.0	7.9
					B	6.1	10.2	6.5	8.5	7.6	19.2	1.1	14.4	6.5	10.0	7.1	17.0	10.4	7.8	3.7	6.8	16.3	5.5	15.0
				Oolong tea	A	10.0	7.7	1.9	6.3	10.3	7.5	10.0	3.9	11.2	6.9	16.7	10.9	3.9	4.0	6.5	4.9	4.0	9.5	4.5
					B	41.1	15.3	1.9	4.6	4.6	10.4	15.4	15.5	8.9	5.0	13.7	5.1	2.7	4.0	2.6	3.7	3.2	7.7	18.4
		ENVI-Carb+PSA	GC-MS	Green tea	A	21.0	3.4	20.2	12.9	8.7	32.3	14.9	6.8	5.2	6.9	10.5	24.5	4.5	12.1	12.6	12.9	6.5	20.1	8.5
					B	15.4	11.3	8.3	12.1	12.6	20.4	3.8	9.5	32.4	8.7	31.0	13.1	42.9	13.8	20.7	6.8	16.4	23.2	18.6
				Oolong tea	A	8.4	5.9	3.7	7.0	26.4	5.2	9.0	10.3	14.2	8.8	3.3	13.2	21.7	20.4	8.0	48.1	19.9	11.4	1.5
					B	9.5	7.0	10.7	5.2	27.5	18.5	25.2	15.6	12.2	6.2	15.7	4.0	13.9	30.9	26.6	11.9	30.5	18.1	3.9
			GC-MS/MS	Green tea	A	26.6	10.1	13.2	7.1	9.5	16.2	28.1	38.0	13.2	5.9	8.1	14.2	7.2	5.9	5.0	13.4	22.7	18.3	3.8
					B	27.9	15.6	4.2	7.8	12.7	12.0	22.7	38.0	22.4	2.5	18.5	14.8	32.2	6.1	6.4	8.5	11.1	23.8	11.1
				Oolong tea	A	13.9	3.0	4.0	8.6	5.4	4.2	8.0	30.6	4.0	2.0	5.9	8.2	10.0	3.8	6.9	11.7	17.0	40.6	4.7
					B	15.6	4.6	7.5	5.2	12.4	10.8	7.3	6.4	13.5	6.1	5.8	7.6	12.1	8.5	10.7	0.5	11.6	29.2	4.5

12	Chlorfena-pyr	TPT	GC-MS	Green tea	A	6.1	6.2	1.1	4.2	6.8	5.0	7.8	4.2	5.0	4.0	4.3	3.5	6.8	2.4	1.3	1.1	3.0	7.1	4.0
				Green tea	B	6.7	8.0	5.6	8.7	3.8	3.3	1.1	3.7	3.3	6.7	3.5	6.9	0.7	3.5	4.2	0.7	4.5	3.2	3.7
				Oolong tea	A	6.7	2.9	0.5	1.4	1.6	2.8	8.4	5.0	2.8	3.1	1.8	2.8	3.7	3.4	7.2	4.5	3.4	1.9	1.7
				Oolong tea	B	7.3	3.3	4.3	8.6	2.0	5.7	5.9	5.1	5.7	5.4	0.2	1.5	1.8	4.2	2.0	0.8	3.4	0.6	3.3
			GC-MS/MS	Green tea	A	5.6	4.5	2.1	4.2	4.5	6.1	6.9	5.5	9.1	3.0	4.8	6.7	7.5	2.7	3.8	4.5	3.6	5.6	5.3
				Green tea	B	5.7	7.9	6.9	5.2	4.2	2.7	2.5	6.9	3.8	9.9	4.0	3.8	4.2	1.9	1.1	2.4	5.8	2.7	2.4
				Oolong tea	A	5.2	2.1	3.5	6.0	5.5	3.8	6.7	1.5	2.8	4.3	3.0	3.0	10.5	2.3	9.3	5.4	4.8	3.6	1.3
				Oolong tea	B	4.8	3.9	6.9	4.9	0.9	2.8	5.2	8.3	1.6	1.5	5.6	2.2	2.4	3.7	7.8	1.5	5.9	7.9	7.9
		ENVI-Carb+PSA	GC-MS	Green tea	A	6.4	2.9	2.5	6.2	9.1	1.6	6.2	0.5	6.6	1.3	8.8	2.2	2.2	5.0	5.3	6.3	10.2	2.8	6.1
				Green tea	B	3.7	9.0	1.8	4.9	3.9	1.9	4.9	5.3	4.6	4.7	19.4	1.7	1.5	6.6	10.6	1.6	4.6	9.1	4.6
				Oolong tea	A	4.6	1.8	3.6	1.9	6.8	2.7	4.1	1.1	2.2	0.7	3.7	2.9	3.0	4.1	3.6	1.5	3.4	4.4	4.7
				Oolong tea	B	6.1	3.3	8.9	5.9	7.1	2.4	1.9	8.9	11.0	3.4	7.2	2.7	2.7	8.4	10.3	3.8	5.6	4.2	3.8
			GC-MS/MS	Green tea	A	5.8	3.4	1.7	4.5	7.4	1.6	6.0	2.5	3.0	1.8	13.3	0.3	3.1	3.6	7.0	7.0	4.8	7.3	2.3
				Green tea	B	9.6	9.6	2.9	6.3	1.5	3.6	2.2	4.8	2.7	3.8	16.2	4.4	1.0	8.0	12.4	3.1	2.9	13.3	3.8
				Oolong tea	A	6.3	2.4	4.8	1.4	4.9	2.5	5.7	3.2	5.8	2.9	7.6	3.8	4.5	11.1	5.8	2.2	5.8	5.3	7.2
				Oolong tea	B	7.5	5.1	10.3	4.7	6.4	2.5	3.3	9.3	8.8	3.8	3.5	4.0	7.7	14.9	9.9	0.6	7.8	5.6	3.1
13	Clomazone	TPT	GC-MS	Green tea	A	5.9	6.3	2.5	8.7	14.3	2.0	10.0	13.1	8.8	0.8	5.5	7.2	9.7	1.1	3.7	3.3	9.4	8.9	3.0
				Green tea	B	5.1	8.8	2.7	12.3	6.0	3.2	4.5	9.1	8.4	7.8	5.1	1.5	9.1	0.8	11.1	5.0	5.1	5.6	4.3
				Oolong tea	A	4.4	1.5	1.7	9.0	2.4	1.9	16.2	6.5	7.5	3.9	13.8	3.0	3.0	5.1	6.6	4.4	3.6	1.1	1.8
				Oolong tea	B	7.4	2.9	3.6	7.8	3.9	7.1	5.3	8.9	9.4	7.0	3.7	3.9	1.0	6.8	8.8	2.1	1.0	1.5	6.6
			GC-MS/MS	Green tea	A	5.2	6.8	2.7	6.9	14.3	4.0	11.0	13.5	11.1	5.3	4.7	8.6	6.5	4.4	3.5	6.6	1.2	9.0	3.3
				Green tea	B	5.7	8.8	5.3	9.1	4.9	6.4	4.1	9.6	10.0	9.5	2.7	1.1	14.2	3.4	5.6	4.4	12.5	4.7	5.6
				Oolong tea	A	8.5	3.1	2.4	4.9	9.1	2.7	4.4	6.3	9.9	1.5	6.1	5.4	3.7	2.8	8.3	0.3	2.8	7.0	6.7
				Oolong tea	B	5.1	2.0	7.4	6.2	2.2	7.1	6.6	1.7	9.9	3.1	5.7	16.1	5.7	13.8	8.5	0.7	3.7	5.4	2.0
		ENVI-Carb+PSA	GC-MS	Green tea	A	17.3	0.1	11.9	5.5	7.4	19.1	12.2	13.8	11.6	3.7	1.9	14.8	2.6	6.2	9.7	5.7	11.5	12.1	6.1
				Green tea	B	17.2	9.0	5.3	9.8	8.3	5.2	11.9	12.2	12.6	12.6	6.4	6.2	20.9	1.1	4.7	11.0	3.1	22.0	11.6
				Oolong tea	A	6.4	2.2	3.3	7.5	3.9	0.9	4.2	12.9	7.6	1.6	3.7	6.5	7.9	7.6	6.7	14.5	7.3	10.4	3.2
				Oolong tea	B	8.5	7.5	6.9	2.4	13.3	6.4	5.8	5.8	10.1	3.2	8.2	12.2	5.9	2.1	10.2	5.3	12.4	11.8	8.2
			GC-MS/MS	Green tea	A	19.5	6.9	8.9	3.5	4.9	18.0	11.7	15.9	10.0	3.7	2.3	10.5	9.8	13.0	4.1	10.7	5.8	9.7	1.1
				Green tea	B	25.7	13.3	3.9	6.9	5.7	3.4	12.7	14.7	10.4	9.8	8.6	11.1	12.5	5.2	7.6	19.1	5.2	25.9	2.6
				Oolong tea	A	8.0	1.6	4.1	7.0	4.0	2.0	3.8	9.1	2.3	0.8	3.8	7.4	4.8	7.7	9.6	5.6	9.7	10.0	4.7
				Oolong tea	B	10.1	4.1	8.8	1.8	10.6	6.7	4.4	2.5	14.0	3.9	4.8	1.6	5.8	2.1	14.0	2.5	8.2	16.7	4.9
14	Cycloate	TPT	GC-MS	Green tea	A	4.7	5.2	2.6	0.7	9.0	6.5	5.8	2.8	6.0	2.5	7.1	7.4	7.4	1.1	1.9	2.0	7.0	7.6	6.5
				Green tea	B	6.7	8.4	4.1	8.4	2.9	4.1	3.7	1.2	4.5	9.5	1.8	5.9	3.1	2.1	3.7	1.1	8.0	3.3	5.5
				Oolong tea	A	7.3	3.4	3.9	0.9	2.4	3.2	8.8	2.4	4.9	5.1	11.8	1.5	4.5	5.6	3.6	2.0	2.9	1.6	5.4
				Oolong tea	B	9.6	1.1	3.5	7.1	4.1	4.5	2.1	3.7	9.4	3.8	9.9	2.6	1.8	6.4	1.9	0.9	1.6	1.6	1.3
			GC-MS/MS	Green tea	A	3.5	5.6	2.4	1.3	7.9	6.2	6.5	4.0	5.6	1.9	6.4	6.9	4.9	4.1	2.6	1.1	3.0	7.3	5.0
				Green tea	B	5.0	7.6	4.7	8.3	4.7	6.7	4.8	3.1	1.9	11.2	5.8	3.8	0.9	2.2	4.5	1.5	6.8	4.2	3.5
				Oolong tea	A	7.5	4.6	3.7	3.1	2.2	2.6	2.7	3.3	11.2	3.3	5.2	1.5	5.5	3.1	4.9	1.9	2.1	1.6	9.7
				Oolong tea	B	6.5	4.9	4.7	6.9	2.9	4.3	5.2	3.3	3.3	5.9	4.1	2.3	10.5	13.5	4.6	3.4	2.4	1.7	1.6
		ENVI-Carb+PSA	GC-MS	Green tea	B	5.1	4.6	1.6	7.0	4.7	3.7	5.2	0.9	2.5	2.5	9.9	4.2	7.1	4.8	8.9	12.1	12.5	5.3	2.1
				Green tea	B	3.7	5.3	4.7	3.5	2.6	1.8	4.6	5.0	5.8	5.8	17.3	7.0	6.7	13.0	18.8	4.9	18.7	8.4	5.0
				Oolong tea	A	4.4	2.0	6.4	3.3	4.1	2.8	4.6	10.3	5.8	3.7	4.5	4.7	4.7	4.6	1.5	8.8	4.6	8.6	6.3
				Oolong tea	B	8.5	8.5	7.5	2.1	11.7	3.8	5.3	5.8	9.4	6.1	8.0	13.4	2.2	6.4	6.2	2.7	7.0	6.9	4.7
			GC-MS/MS	Green tea	A	3.3	6.1	1.4	6.6	4.6	0.3	4.8	0.3	3.8	2.2	10.5	14.0	3.5	10.3	6.2	10.5	11.1	6.1	4.9
				Green tea	B	2.4	8.1	5.9	5.5	3.1	2.4	2.7	2.4	1.9	2.3	22.2	8.0	2.0	3.8	18.3	2.3	4.4	11.7	2.4
				Oolong tea	A	5.4	2.9	6.3	4.7	4.5	1.1	2.8	12.3	2.2	4.7	6.0	7.2	1.2	2.9	6.0	7.6	9.6	19.6	2.1
				Oolong tea	B	8.6	4.7	8.2	0.5	11.2	5.2	5.1	4.6	11.5	6.3	1.3	7.2	4.6	2.9	11.5	0.4	6.2	19.3	1.6

(Continued)

Appendix Table 5.2 RSD of 201 Pesticides in Youden Pair Aged Tea Sample for Parallel Determinations (Nov 9, 2009–Feb 7, 2010) (cont.)

No.	Pesticides	SPE	Method	Sample	Youden pair	Nov 9, 2009 (n=5)	Nov 14, 2009 (n=3)	Nov 19, 2009 (n=3)	Nov 24, 2009 (n=3)	Nov 29, 2009 (n=3)	Dec 4, 2009 (n=3)	Dec 9, 2009 (n=3)	Dec 14, 2009 (n=3)	Dec 19, 2009 (n=3)	Dec 24, 2009 (n=3)	Dec 14, 2009 (n=3)	Jan 3, 2010 (n=3)	Jan 8, 2010 (n=3)	Jan 13, 2010 (n=3)	Jan 18, 2010 (n=3)	Jan 23, 2010 (n=3)	Jan 28, 2010 (n=3)	Feb 2, 2010 (n=3)	Feb 7, 2010 (n=3)
15	Cycluron	TPT	GC-MS	Green tea	A	14.6	6.8	0.3	4.6	5.4	5.2	9.0	5.4	7.2	0.6	8.8	3.6	7.8	5.6	2.2	4.8	3.9	7.6	2.3
					B	11.5	9.8	6.3	11.8	5.7	7.5	1.8	6.8	3.7	6.6	4.5	7.9	2.0	14.6	11.5	2.9	6.9	3.1	8.2
				Oolong tea	A	3.9	2.0	1.9	2.7	1.4	1.0	8.1	5.0	2.2	2.9	2.9	3.7	4.1	2.8	8.3	4.0	4.2	3.1	1.0
					B	7.3	2.6	4.9	6.2	3.3	3.9	5.6	0.5	6.2	3.3	1.0	0.5	4.4	3.7	3.1	0.7	1.3	4.3	2.2
			GC-MS/MS	Green tea	A	15.1	5.7	6.1	9.8	5.0	1.5	7.3	4.5	1.1	7.7	5.6	3.9	8.7	6.9	7.0	12.1	3.9	8.2	3.1
					B	7.2	12.5	5.6	14.5	6.0	17.6	7.5	8.6	12.1	13.3	12.6	4.9	2.5	20.4	6.1	6.3	9.4	10.9	6.8
				Oolong tea	A	3.6	2.5	1.8	5.7	1.7	1.8	9.6	1.1	3.4	4.2	1.5	4.9	3.4	3.4	8.9	2.9	3.4	2.5	1.5
					B	4.8	2.3	5.3	5.6	1.3	2.9	4.2	2.1	4.8	3.8	1.9	1.9	5.7	3.8	6.3	0.2	2.6	0.5	2.3
		ENVI-Carb+PSA	GC-MS	Green tea	A	10.7	8.3	8.8	6.6	6.4	7.7	7.2	2.8	7.0	5.0	12.8	7.5	2.6	15.2	2.3	4.6	9.2	2.7	6.7
					B	12.8	8.1	1.6	4.7	6.7	8.5	7.4	9.4	3.4	7.4	26.2	2.4	7.7	2.0	16.7	6.7	1.3	8.0	3.2
				Oolong tea	A	4.7	3.9	2.7	4.2	4.6	1.1	7.4	0.6	13.0	1.1	3.1	2.4	3.9	3.5	3.5	2.9	3.6	4.8	5.9
					B	5.4	3.9	6.5	3.7	10.4	3.7	0.9	5.7	9.6	4.3	4.7	2.4	1.7	6.9	7.0	4.2	11.6	7.6	3.0
			GC-MS/MS	Green tea	A	6.5	5.9	13.0	6.1	7.9	1.5	2.5	8.4	4.9	13.2	18.9	3.1	8.7	4.1	6.3	5.9	4.2	1.0	5.4
					B	16.6	10.2	6.5	15.9	18.3	15.8	22.5	11.4	4.2	12.9	5.6	1.0	10.5	27.5	9.6	8.5	2.1	11.0	1.6
				Oolong tea	A	4.8	4.1	4.1	5.9	5.8	1.5	5.6	1.5	3.6	1.5	4.8	7.3	9.5	1.5	8.0	4.5	5.4	5.2	5.6
					B	6.5	3.9	7.7	2.9	9.7	4.3	0.9	7.0	11.7	2.3	0.5	4.9	3.2	8.5	7.8	2.6	4.6	3.6	2.5
16	Cyhalofop-butyl	TPT	GC-MS	Green tea	A	5.2	8.1	2.3	11.5	13.6	3.0	7.0	11.5	8.0	3.6	9.9	6.6	9.9	7.8	7.5	0.6	5.3	6.9	3.8
					B	5.0	7.0	5.6	10.7	4.1	3.2	4.6	7.6	4.9	11.3	9.1	7.0	10.6	4.0	7.5	1.7	14.6	7.5	1.4
				Oolong tea	A	7.1	4.7	3.0	3.5	1.4	2.9	9.9	5.6	5.5	5.5	16.9	10.7	9.0	2.0	0.6	7.5	9.0	13.0	1.6
					B	13.0	1.8	2.6	7.2	0.8	3.2	12.2	6.8	3.0	1.5	4.9	2.4	5.8	2.1	4.2	1.8	3.5	13.9	7.1
			GC-MS/MS	Green tea	A	6.1	6.6	0.3	5.9	12.1	3.5	10.6	13.4	10.1	4.3	8.2	6.4	7.1	8.8	1.8	4.0	4.6	6.0	2.4
					B	5.4	5.8	4.8	9.2	5.9	4.9	4.2	9.5	17.2	2.9	3.1	6.7	9.3	3.3	3.3	5.4	10.3	3.5	3.4
				Oolong tea	A	4.9	7.3	2.7	7.9	10.9	5.3	14.9	5.5	4.0	2.9	9.8	6.7	5.7	13.0	10.7	10.4	10.6	4.2	5.4
					B	13.0	4.8	8.9	9.3	4.1	2.3	7.2	0.8	5.5	9.8	6.7	3.2	5.9	3.2	6.9	2.5	2.7	8.3	4.4
		ENVI-Carb+PSA	GC-MS	Green tea	A	14.7	5.3	14.8	5.5	8.9	17.6	2.8	5.0	10.2	0.8	7.8	8.9	2.0	3.0	7.1	4.4	8.2	8.3	5.2
					B	20.2	7.1	5.8	7.3	6.4	3.7	10.4	4.5	9.5	4.1	15.0	12.3	12.3	8.5	12.8	8.7	5.7	14.8	6.6
				Oolong tea	A	13.3	5.8	4.3	11.6	6.8	4.4	7.5	5.1	11.7	11.6	9.7	12.3	9.2	7.2	10.1	6.4	2.7	8.2	6.0
					B	6.0	1.1	13.8	13.5	12.5	8.3	5.7	7.5	4.3	5.9	9.9	0.9	2.3	11.2	9.3	5.0	4.6	15.9	3.4
			GC-MS/MS	Green tea	A	16.3	8.9	11.9	6.8	5.9	16.6	3.5	8.8	43.0	23.5	0.5	0.9	7.9	12.0	5.0	12.4	9.6	7.2	5.2
					B	20.6	12.8	3.7	11.1	5.4	2.3	33.1	2.6	6.3	33.1	13.6	2.8	15.2	16.3	10.0	17.7	1.2	21.7	9.7
				Oolong tea	A	7.6	1.8	1.9	13.9	12.8	5.1	4.2	9.2	6.8	10.9	9.2	15.8	12.3	13.8	17.6	20.7	7.1	14.3	11.2
					B	7.2	2.5	21.4	12.2	6.0	0.9	7.6	4.6	8.9	7.1	9.5	8.2	10.7	15.5	9.7	11.5	5.2	10.5	3.9
17	Cyprodinil	TPT	GC-MS	Green tea	A	4.5	5.6	2.3	1.2	11.2	8.0	7.1	6.1	5.8	2.0	6.5	11.9	7.3	6.7	5.6	2.1	3.8	7.3	2.6
					B	4.2	7.8	5.5	5.5	7.8	2.4	2.9	3.3	4.9	10.3	7.8	12.7	0.5	1.3	0.4	2.7	5.4	1.6	10.1
				Oolong tea	A	5.6	2.5	1.5	5.3	1.8	3.0	6.2	17.5	5.7	6.1	18.5	1.3	7.8	4.3	5.7	6.0	3.4	4.3	2.1
					B	8.7	1.7	3.6	2.9	0.8	0.3	15.4	4.6	9.3	3.9	18.0	1.8	5.1	3.4	2.6	1.7	3.3	2.2	0.9
			GC-MS/MS	Green tea	A	6.0	7.1	1.7	3.3	9.4	4.6	7.2	5.9	6.1	2.5	6.8	3.7	6.4	5.9	1.5	2.0	10.6	7.8	6.3
					B	4.0	7.3	5.8	6.4	9.0	8.5	5.1	4.6	5.5	9.4	6.2	10.1	2.2	2.6	2.6	4.0	6.6	6.4	8.5
				Oolong tea	A	7.2	0.9	2.8	5.8	1.4	2.7	8.8	2.7	1.9	4.8	5.0	2.9	3.5	1.6	9.5	2.4	1.0	4.2	5.6
					B	5.5	2.5	6.1	7.0	1.4	2.8	7.6	3.0	8.3	4.4	5.0	2.8	5.3	1.5	5.8	1.9	1.9	1.2	8.7
		ENVI-Carb+PSA	GC-MS	Green tea	A	6.5	6.2	3.7	5.4	11.5	3.3	4.8	1.9	7.8	7.1	5.1	6.7	2.5	6.0	8.2	7.0	11.7	0.1	10.2
					B	3.3	9.0	4.7	9.5	1.5	3.5	3.9	5.8	6.2	10.5	34.6	6.3	6.2	5.2	17.0	5.2	4.0	8.7	7.3
				Oolong tea	A	3.9	1.3	5.4	10.0	7.2	3.3	10.9	17.4	15.7	13.0	2.8	5.1	5.0	6.5	2.6	4.2	1.7	14.7	7.1
					B	5.2	4.5	19.1	3.7	4.0	8.3	9.5	7.2	33.1	4.2	13.3	4.2	3.4	8.2	9.7	2.2	5.1	17.6	6.3
			GC-MS/MS	Green tea	A	4.1	3.2	2.5	5.8	6.7	1.4	4.7	0.6	5.5	2.8	10.2	3.9	3.3	4.1	5.3	5.8	11.1	2.6	4.4
					B	2.3	6.4	3.0	6.2	1.9	1.7	3.1	3.4	4.7	2.2	23.0	2.1	1.4	4.4	14.0	3.4	3.1	11.5	5.0
				Oolong tea	A	6.3	2.7	3.7	15.3	5.8	0.9	5.4	2.2	5.8	1.9	5.2	8.1	8.2	2.7	9.7	9.5	3.4	7.0	5.0
					B	5.7	2.1	10.4	3.3	5.2	5.6	5.2	6.4	12.3	2.3	2.3	5.6	4.9	7.6	9.1	13.8	1.8	2.6	3.5

No.	Compound	Method	Detection	Tea	A/B	1	2	3	4	5	6	7	8	9	10	11	12	13	14	15	16	17	18	19
18	Dacthal	TPT	GC-MS	Green tea	A	4.3	5.8	4.3	2.6	8.5	3.9	6.6	8.0	5.6	2.6	4.5	2.5	6.7	3.6	2.3	1.9	3.7	5.5	4.6
					B	6.6	8.9	5.1	8.2	4.0	2.5	1.9	4.3	4.1	7.1	2.3	6.4	1.3	1.4	2.9	6.0	5.6	0.4	5.9
				Oolong tea	A	5.0	2.7	1.3	3.9	2.7	2.2	12.7	6.5	3.4	3.2	3.7	1.5	3.5	2.5	6.8	2.4	4.3	1.9	1.0
					B	6.9	2.3	4.6	2.2	2.4	2.4	3.2	5.5	6.1	3.8	1.9	0.9	2.0	3.1	4.5	0.9	1.0	1.8	2.8
			GC-MS/MS	Green tea	A	3.9	5.7	3.6	6.1	6.2	3.3	7.0	6.5	5.6	2.7	5.1	3.2	6.4	4.4	0.5	1.6	3.8	7.1	3.9
					B	6.0	7.9	7.4	6.3	4.1	5.0	2.7	4.9	3.6	9.2	4.3	8.9	3.0	2.1	2.0	3.4	10.0	2.8	7.4
				Oolong tea	A	5.2	2.1	1.0	4.6	0.7	3.3	9.9	0.3	5.4	2.7	6.3	2.1	4.8	2.7	9.1	3.4	0.8	0.8	5.5
					B	2.2	1.8	6.0	4.8	1.5	2.0	3.6	5.6	6.2	2.5	4.7	1.7	7.7	2.6	6.9	2.9	2.3	1.6	0.9
		ENVI-Carb +PSA	GC-MS	Green tea	A	8.9	3.2	2.4	3.5	7.5	3.9	8.1	3.4	3.7	1.7	8.1	1.8	2.1	4.3	5.4	12.5	10.4	10.4	4.8
					B	6.1	7.7	3.1	4.4	1.2	1.6	4.1	1.2	5.5	4.8	19.9	2.8	6.9	1.3	15.7	1.2	4.8	12.3	6.4
				Oolong tea	A	1.4	3.4	2.4	2.7	4.8	2.5	2.8	5.5	6.4	0.9	4.1	3.2	1.1	3.4	2.8	2.8	6.3	4.5	3.4
					B	5.3	3.0	8.8	5.3	8.2	4.6	2.3	7.4	12.0	2.2	8.4	4.4	4.2	7.7	10.4	3.1	8.3	2.7	5.2
			GC-MS/MS	Green tea	A	8.4	4.3	2.4	3.4	7.1	3.5	4.3	3.3	2.9	0.6	12.0	3.2	2.4	2.1	3.5	6.8	8.0	7.0	1.6
					B	3.7	9.0	3.5	6.5	2.4	2.7	3.4	0.9	4.8	0.2	20.5	1.5	9.3	6.8	13.8	5.2	4.3	12.9	3.6
				Oolong tea	A	4.7	2.5	3.4	3.1	5.8	2.5	4.7	1.3	2.7	0.5	3.7	4.4	2.2	4.1	9.6	1.7	5.4	5.1	6.1
					B	4.5	2.1	10.2	1.5	7.5	5.6	2.4	6.6	11.7	2.7	3.7	3.7	2.5	9.2	12.0	1.5	5.0	0.4	3.5
19	DE-PCB101	TPT	GC-MS	Green tea	A	5.2	5.1	3.5	8.5	7.3	4.8	6.1	3.7	5.7	3.0	8.0	3.2	9.8	2.0	4.0	4.1	7.1	7.1	3.8
					B	6.3	8.4	4.7	3.7	3.5	4.0	2.8	1.4	4.8	8.7	3.3	4.8	2.4	2.2	2.4	1.8	5.8	1.5	4.9
				Oolong tea	A	5.7	3.9	1.0	4.4	4.7	4.2	10.0	5.2	3.4	6.3	1.2	6.6	1.4	6.1	10.9	2.9	4.9	1.9	3.9
					B	7.3	3.4	4.8	4.3	3.0	2.6	4.9	1.7	5.1	6.0	2.2	1.7	2.8	6.2	4.8	0.5	1.4	0.8	3.7
			GC-MS/MS	Green tea	A	4.2	6.8	4.2	9.5	6.6	4.4	7.7	3.3	5.9	3.2	5.7	2.2	6.2	4.4	1.0	5.4	4.0	7.4	4.4
					B	5.5	9.3	4.2	5.6	2.7	2.6	4.8	3.4	3.1	10.3	5.7	7.1	2.5	0.4	5.5	5.6	8.0	5.9	7.8
				Oolong tea	A	6.0	3.9	1.8	5.3	4.8	3.2	9.9	1.0	1.9	2.7	3.0	3.9	4.7	2.4	7.7	4.1	3.2	5.2	4.2
					B	6.0	2.7	6.3	5.2	2.2	1.3	4.5	2.1	3.2	3.3	2.8	0.9	4.6	2.9	4.3	2.3	3.3	5.3	3.4
		ENVI-Carb +PSA	GC-MS	Green tea	A	5.5	5.8	1.3	4.7	8.2	1.3	5.7	3.4	7.2	3.9	12.6	3.4	6.0	6.9	7.5	8.5	9.9	3.0	6.7
					B	3.9	8.2	3.7	4.3	1.8	1.6	3.8	5.3	3.3	4.8	20.7	0.7	3.0	4.1	13.1	5.0	5.6	7.6	4.4
				Oolong tea	A	1.8	3.5	1.7	4.2	5.1	1.8	2.6	0.9	6.4	3.9	5.2	3.3	2.0	4.2	1.4	2.9	5.2	3.1	6.2
					B	4.8	2.2	9.9	5.7	8.3	4.7	3.0	7.0	9.9	3.9	9.2	4.3	5.7	8.2	8.7	2.6	4.1	3.7	5.3
			GC-MS/MS	Green tea	A	4.7	5.4	1.4	6.9	6.9	0.9	6.2	1.2	4.9	4.8	11.6	5.5	4.0	11.6	4.3	7.5	10.8	2.3	2.7
					B	4.5	10.0	4.3	5.3	1.9	3.4	3.5	3.5	4.5	3.9	22.2	3.2	6.2	3.9	13.0	2.2	4.8	12.4	4.5
				Oolong tea	A	5.0	2.7	3.9	1.6	6.2	3.1	4.8	1.9	2.9	1.7	5.8	3.0	5.8	3.1	10.9	1.6	4.9	4.8	5.4
					B	4.4	2.7	10.0	0.8	5.9	6.8	2.3	6.7	11.6	3.2	3.1	4.9	0.9	6.1	9.5	3.6	4.3	0.3	4.5
20	DE-PCB118	TPT	GC-MS	Green tea	A	5.2	5.1	2.6	8.5	7.7	4.5	9.7	3.9	4.4	3.8	5.4	2.5	9.6	3.9	3.4	2.3	6.8	9.5	4.7
					B	5.5	9.3	5.2	3.9	2.3	3.4	4.8	3.8	2.1	10.1	3.4	5.6	4.7	7.6	6.2	3.4	9.6	4.7	4.4
				Oolong tea	A	6.4	2.3	0.2	1.7	4.2	2.7	7.3	4.6	2.5	1.0	2.9	4.8	4.1	2.7	10.4	3.8	2.8	0.2	4.3
					B	7.9	3.2	4.6	2.9	1.4	2.3	3.6	1.0	4.6	4.2	4.8	1.3	3.1	0.9	1.9	3.3	4.4	1.7	2.4
			GC-MS/MS	Green tea	A	4.4	3.8	3.3	6.5	5.5	5.3	7.7	2.7	5.5	3.8	6.0	3.7	5.0	4.8	3.0	2.4	4.8	9.0	4.0
					B	5.1	9.3	5.1	5.5	4.6	3.9	3.7	2.1	1.6	9.6	4.4	7.2	2.2	2.3	5.3	3.8	6.2	3.0	6.1
				Oolong tea	A	6.2	2.1	2.9	6.0	3.2	2.6	7.8	0.5	1.3	2.5	1.9	2.0	0.7	1.6	6.3	3.3	2.3	5.5	3.1
					B	6.2	0.2	5.2	6.1	0.8	3.1	4.1	2.9	4.7	3.8	3.4	0.4	3.7	3.7	5.0	0.4	2.7	1.8	3.9
		ENVI-Carb +PSA	GC-MS	Green tea	A	5.5	4.5	0.5	5.2	8.6	4.0	6.8	1.5	5.7	1.5	13.0	2.9	4.1	7.3	7.8	3.6	10.7	4.6	6.1
					B	4.1	8.9	2.2	4.3	2.1	2.0	4.2	7.0	2.4	4.0	19.9	0.7	3.6	4.2	12.6	2.5	6.8	8.6	6.9
				Oolong tea	A	4.3	2.7	2.6	4.8	5.2	4.7	2.1	1.6	4.4	3.1	4.8	4.0	2.1	5.4	3.5	3.0	6.5	1.3	9.0
					B	5.1	2.1	10.0	6.6	8.0	1.0	4.4	6.8	9.4	3.7	10.6	1.9	5.0	9.3	9.6	3.8	4.4	5.3	2.3
			GC-MS/MS	Green tea	A	4.5	4.6	1.0	6.7	7.5	1.3	5.5	1.8	5.7	1.1	11.1	5.4	2.2	6.4	6.8	8.7	6.1	3.3	3.8
					B	2.7	9.8	4.6	5.4	2.6	4.0	2.7	2.4	4.6	0.9	21.5	3.7	1.2	5.7	14.2	2.0	7.9	9.7	4.7
				Oolong tea	A	5.9	2.3	3.8	2.9	6.3	2.0	5.3	0.7	4.1	1.4	3.8	4.0	3.9	3.4	9.0	2.2	5.8	2.8	4.1
					B	6.0	1.3	10.4		8.1	5.8	3.9	5.8	10.1	2.6	2.4	3.2	3.1	7.2	11.5	1.8	3.8	2.7	1.9

(Continued)

Appendix Table 5.2 RSD of 201 Pesticides in Youden Pair Aged Tea Sample for Parallel Determinations (Nov 9, 2009–Feb 7, 2010) (cont.)

No.	Pesticides	SPE	Method	Sample	Youden pair	Nov 9, 2009 (n=5)	Nov 14, 2009 (n=3)	Nov 19, 2009 (n=3)	Nov 24, 2009 (n=3)	Nov 29, 2009 (n=3)	Dec 4, 2009 (n=3)	Dec 9, 2009 (n=3)	Dec 14, 2009 (n=3)	Dec 19, 2009 (n=3)	Dec 24, 2009 (n=3)	Dec 14, 2009 (n=3)	Jan 3, 2010 (n=3)	Jan 8, 2010 (n=3)	Jan 13, 2010 (n=3)	Jan 18, 2010 (n=3)	Jan 23, 2010 (n=3)	Jan 28, 2010 (n=3)	Feb 2, 2010 (n=3)	Feb 7, 2010 (n=3)
21	DE-PCB138	TPT	GC-MS	Green tea	A	5.5	4.8	2.0	0.8	7.5	5.2	6.4	4.3	5.4	2.3	5.4	1.9	7.3	4.2	2.2	1.9	3.6	6.6	3.8
					B	5.2	8.2	5.2	8.7	3.3	3.9	3.1	2.0	5.1	9.3	3.5	7.2	0.8	1.2	6.2	2.7	4.2	2.3	4.3
				Oolong tea	A	6.4	2.7	0.7	2.9	1.4	2.3	8.3	5.5	1.9	3.3	3.0	3.0	4.0	2.9	7.6	3.4	3.3	2.3	1.1
					B	8.1	3.1	5.4	4.3	1.0	3.4	4.2	2.3	5.3	4.1	0.1	1.8	2.6	3.7	3.1	0.7	1.6	1.3	2.1
			GC-MS/MS	Green tea	A	3.5	4.2	1.7	0.6	5.0	2.3	8.9	4.3	7.3	1.9	7.6	3.5	8.3	1.4	3.3	2.4	6.4	7.5	3.6
					B	6.6	6.0	5.7	8.5	7.3	1.0	3.6	1.8	8.2	9.4	5.7	7.5	1.4	2.8	2.8	3.8	7.5	7.0	9.6
				Oolong tea	A	8.2	2.2	1.3	3.3	5.1	4.4	8.2	1.4	3.0	2.6	2.6	1.0	5.9	3.9	8.6	2.4	3.2	2.8	0.5
					B	9.0	5.2	5.2	3.6	1.5	3.2	4.2	1.2	3.6	4.6	8.0	3.0	2.8	3.3	6.6	2.8	4.0	3.8	1.2
		ENVI-Carb+PSA	GC-MS	Green tea	A	5.3	4.9	1.3	5.6	9.0	1.6	6.0	1.1	6.6	1.5	10.2	2.5	3.4	5.9	7.4	7.4	10.7	3.4	5.6
					B	3.5	8.3	2.9	5.5	1.9	1.8	3.5	5.8	3.6	4.7	20.3	0.9	1.1	4.9	12.7	1.9	5.8	8.4	6.6
				Oolong tea	A	4.8	2.6	2.4	2.7	6.6	1.8	3.0	2.0	5.3	2.2	4.7	2.6	1.9	4.9	2.5	1.9	4.3	4.6	3.8
					B	5.3	2.3	9.8	4.5	8.8	3.8	2.5	7.5	10.8	2.8	7.3	3.0	3.1	7.6	10.4	4.3	6.8	5.3	5.0
			GC-MS/MS	Green tea	A	5.0	4.5	1.4	5.3	6.1	2.8	7.9	0.9	6.1	1.9	14.6	3.9	3.8	13.9	8.4	6.1	6.9	7.7	1.7
					B	5.2	12.7	4.8	8.0	2.2	4.0	5.6	4.0	4.5	1.6	19.4	1.0	3.0	6.5	10.9	3.5	4.3	10.7	5.2
				Oolong tea	A	6.4	4.6	5.3	4.6	5.8	1.2	4.3	5.0	5.2	0.9	7.4	6.7	2.6	5.7	6.3	0.7	6.4	3.7	4.5
					B	5.5	2.6	9.1	2.1	4.8	5.3	4.2	4.7	10.4	2.2	3.8	6.6	6.7	10.3	7.5	0.9	1.8	4.8	2.3
22	DE-PCB180	TPT	GC-MS	Green tea	A	5.4	4.7	1.1	9.5	6.5	4.0	6.3	4.6	5.8	1.5	5.3	2.4	5.4	2.8	2.7	2.6	4.4	7.6	2.8
					B	5.0	7.1	4.8	8.3	3.1	3.2	1.7	3.3	4.4	8.6	3.8	7.6	0.7	1.8	4.9	3.5	5.2	2.0	2.4
				Oolong tea	A	6.9	1.6	1.1	2.8	1.3	3.1	4.9	4.6	2.3	2.5	1.8	5.5	4.2	2.1	6.1	3.2	3.3	4.6	3.2
					B	8.7	2.4	5.0	4.7	0.8	3.4	3.5	0.5	6.0	4.8	0.4	2.8	1.4	3.1	1.9	1.1	1.8	1.8	2.5
			GC-MS/MS	Green tea	A	3.1	7.0	2.0	4.0	6.6	5.0	9.7	5.0	4.2	5.1	7.5	4.0	1.1	3.5	2.3	4.7	11.0	13.1	2.6
					B	4.6	9.3	4.9	8.5	0.7	4.9	2.5	1.8	4.5	11.0	5.5	8.7	4.5	2.4	6.3	7.3	5.5	7.6	6.5
				Oolong tea	A	6.0	6.8	4.3	3.1	3.4	3.8	4.8	4.1	6.6	4.5	1.8	6.9	4.6	1.7	10.1	1.8	9.6	8.2	5.8
					B	4.3	5.2	5.4	7.0	0.6	1.6	5.1	3.6	7.7	6.7	6.5	3.9	4.5	4.8	3.5	5.4	0.7	4.0	2.5
		ENVI-Carb+PSA	GC-MS	Green tea	A	6.1	4.8	3.1	7.6	8.9	1.2	17.5	0.7	6.5	1.8	9.5	3.0	4.0	9.4	5.7	10.5	10.5	3.9	6.4
					B	3.5	7.4	4.6	6.3	1.7	1.9	2.4	3.9	3.1	4.2	19.9	2.0	1.1	9.0	13.0	1.7	5.6	9.2	5.4
				Oolong tea	A	5.2	2.6	4.6	1.8	5.8	1.4	2.5	3.1	4.3	3.3	5.6	2.7	1.7	5.8	2.5	1.9	4.8	5.3	4.3
					B	5.1	3.0	9.0	3.3	6.7	3.0	2.1	7.1	5.9	4.7	8.4	2.6	3.4	7.1	9.4	4.4	7.6	5.6	3.3
			GC-MS/MS	Green tea	A	5.3	3.2	1.9	6.2	9.4	2.5	8.2	2.3	5.7	3.3	13.8	3.9	0.5	6.2	7.9	9.2	2.3	5.5	4.1
					B	2.7	11.3	4.2	7.0	2.2	0.9	2.7	1.9	1.5	1.4	25.6	1.0	8.1	2.7	9.8	1.1	4.0	14.0	1.8
				Oolong tea	A	7.6	0.4	3.0	3.3	6.6	2.6	7.5	7.0	3.7	5.0	7.5	6.2	4.3	1.9	7.1	0.5	6.3	10.1	8.7
					B	5.8	2.5	11.0	3.9	10.9	3.6	4.4	4.1	9.3	2.4	1.4	0.3	7.1	9.6	8.6	2.8	0.5	1.4	4.8
23	DE-PCB28	TPT	GC-MS	Green tea	A	4.7	4.6	2.3	2.2	12.1	4.5	4.9	3.1	5.7	2.1	5.7	1.4	6.2	2.6	2.8	2.1	2.6	7.2	4.4
					B	8.5	11.1	3.5	8.8	2.1	6.2	2.4	1.4	4.7	7.4	2.5	6.1	2.2	1.7	5.4	3.3	4.0	2.9	3.6
				Oolong tea	A	4.2	3.3	0.2	3.5	2.5	2.6	8.9	4.6	2.1	2.2	3.2	3.8	2.4	2.7	8.6	2.2	3.1	1.1	1.6
					B	6.5	3.3	5.1	3.7	1.4	2.8	3.8	4.3	5.9	4.7	0.6	1.3	2.3	2.8	4.1	1.0	1.9	2.6	0.9
			GC-MS/MS	Green tea	A	4.6	5.3	2.6	2.6	6.8	3.5	6.9	2.2	5.6	3.3	6.2	2.1	5.6	6.2	0.8	1.1	4.6	6.6	5.5
					B	5.5	9.1	5.7	7.4	5.5	3.1	4.1	2.0	3.2	10.0	5.8	7.2	1.5	2.7	2.7	1.8	9.9	2.9	7.0
				Oolong tea	A	6.3	3.9	0.9	5.4	3.0	2.4	7.7	1.5	1.7	3.5	2.0	4.5	4.4	1.9	8.4	2.8	3.4	2.9	3.9
					B	8.8	3.7	5.0	4.7	0.4	2.8	4.3	1.6	4.0	2.8	2.7	0.5	3.7	2.4	5.0	1.5	0.8	1.1	1.9
		ENVI-Carb+PSA	GC-MS	Green tea	A	5.6	4.0	5.2	7.0	6.2	3.1	5.8	1.2	8.2	1.6	9.4	4.5	2.5	4.2	4.8	6.6	9.8	1.1	3.3
					B	3.8	9.0	2.9	3.4	1.6	1.3	8.4	4.7	3.3	3.7	17.2	2.5	1.4	4.0	12.6	3.2	6.4	7.8	6.4
				Oolong tea	A	3.9	3.1	2.6	4.0	6.5	1.9	2.8	0.7	3.1	0.8	4.5	2.7	1.6	3.6	2.5	2.7	6.4	5.1	4.4
					B	5.2	2.8	9.0	4.5	7.2	4.3	2.6	5.3	10.1	3.4	6.6	4.1	4.0	6.8	8.7	4.0	7.6	3.9	4.3
			GC-MS/MS	Green tea	A	4.9	6.0	1.6	5.8	5.7	1.7	5.3	1.8	5.6	1.6	8.9	33.8	4.9	6.6	4.5	8.1	8.6	5.4	1.6
					B	3.4	10.6	5.1	4.8	2.9	2.0	1.8	2.5	1.9	1.2	21.8	21.7	1.1	6.6	13.4	2.2	2.2	11.2	2.1
				Oolong tea	A	4.1	3.3	4.3	5.9	5.7	2.3	4.1	1.6	2.8	0.7	4.4	6.9	3.3	3.0	7.4	5.0	5.0	4.6	5.3
					B	5.5	3.0	9.3	3.0	7.8	4.6	2.9	5.0	10.7	1.7	1.8	5.0	5.5	7.0	10.3	0.9	4.8	3.3	2.0

No.	Compound	Cleanup	Instrument	Tea	Group	1	2	3	4	5	6	7	8	9	10	11	12	13	14	15	16	17	18	19
24	DE-PCB31	TPT	GC–MS	Green tea	A	4.9	4.8	3.1	0.6	5.2	4.9	6.3	2.6	5.4	2.6	5.5	1.5	7.4	2.4	1.4	4.0	2.6	7.1	3.8
					B	8.1	8.8	3.0	8.6	7.2	6.3	3.0	2.3	5.8	8.4	1.9	6.7	3.0	3.0	4.9	1.5	5.1	3.1	4.4
				Oolong tea	A	4.2	3.2	0.4	3.6	2.5	2.9	9.3	3.6	2.3	2.5	3.1	5.1	3.8	2.9	6.8	4.3	2.8	2.3	1.5
					B	6.6	3.0	4.8	4.7	1.4	3.4	5.1	0.6	6.3	3.1	1.0	1.7	2.9	3.4	3.9	1.2	2.1	1.8	1.3
			GC–MS/MS	Green tea	A	4.6	5.3	2.6	2.6	6.8	3.5	6.9	2.2	5.6	3.3	6.2	2.1	5.6	6.2	0.8	1.8	4.6	6.6	5.5
					B	5.5	9.1	5.7	7.4	5.5	3.1	4.1	2.0	3.2	10.0	5.8	7.2	1.5	2.7	2.7	2.8	9.9	2.9	7.0
				Oolong tea	A	5.3	6.0	0.9	5.4	3.0	2.4	7.7	1.5	1.7	3.5	2.0	4.5	4.4	1.9	8.4	1.5	3.4	2.9	3.9
					B	8.8	1.1	5.0	4.7	0.4	2.8	4.3	1.6	4.0	2.8	2.7	0.5	3.7	2.4	5.0	6.6	0.8	1.1	1.9
		ENVI-Carb+PSA	GC–MS	Green tea	A	5.5	4.1	3.5	6.9	6.6	2.5	3.0	1.7	7.7	2.8	8.9	3.8	4.2	4.1	5.2	3.6	10.3	4.1	3.8
					B	3.8	8.3	3.1	3.5	2.0	1.2	20.8	5.5	3.3	3.8	17.2	1.5	1.7	1.6	11.9	2.4	5.7	7.5	5.0
				Oolong tea	A	2.9	3.3	2.6	3.7	6.7	2.4	1.8	0.3	2.8	2.6	3.6	3.7	1.0	3.1	2.5	3.2	5.1	3.2	5.1
					B	5.0	2.8	9.4	4.5	7.5	3.4	2.3	5.8	11.7	2.3	8.1	4.4	3.3	6.5	8.2	8.1	6.2	4.1	4.3
			GC–MS/MS	Green tea	A	4.9	6.0	1.6	5.8	5.7	1.7	5.3	1.8	5.6	1.6	8.9	3.2	4.9	6.6	4.5	1.7	8.6	5.4	1.6
					B	3.4	10.6	5.1	4.8	2.9	2.0	1.8	2.5	1.9	1.2	21.8	5.1	1.1	6.6	13.4	3.2	2.2	11.2	2.1
				Oolong tea	A	3.7	3.3	4.3	5.9	5.7	2.3	4.1	1.6	2.8	0.7	4.4	6.9	3.3	3.0	7.4	0.9	5.0	4.6	5.3
					B	11.2	3.0	9.3	3.0	7.8	4.6	2.9	5.0	10.7	1.7	1.8	5.0	5.5	7.0	10.3	0.5	4.8	3.3	2.0
25	Dichloro-fop-methyl	TPT	GC–MS	Green tea	A	5.3	4.8	4.2	13.8	14.2	2.8	14.9	17.1	8.7	3.4	10.7	8.4	13.0	15.4	3.4	6.4	3.4	14.4	4.3
					B	6.9	9.4	12.9	13.9	2.7	2.9	11.0	12.9	2.9	6.0	3.1	3.0	16.2	11.5	3.2	8.0	18.2	11.6	3.3
				Oolong tea	A	5.5	2.4	5.7	5.1	4.2	8.1	7.4	7.4	11.8	15.7	16.0	6.3	4.3	1.3	3.8	5.7	1.9	4.0	6.6
					B	8.8	0.8	11.5	3.8	3.2	7.4	10.2	4.8	11.2	5.7	17.1	4.3	2.3	5.9	13.7	5.3	0.9	4.0	9.2
			GC–MS/MS	Green tea	A	12.7	7.8	3.6	13.9	18.0	5.1	18.3	11.0	17.0	7.2	14.8	5.9	14.7	12.0	5.0	7.4	9.3	17.9	0.6
					B	11.2	6.0	7.9	13.2	5.3	6.3	7.4	13.4	4.7	10.2	2.3	12.7	23.9	9.9	5.8	1.7	17.5	11.1	7.3
				Oolong tea	A	12.0	7.9	5.0	3.5	8.7	5.1	3.2	13.0	6.4	5.4	7.9	4.0	2.4	6.9	7.0	5.2	4.4	14.0	5.9
					B	14.3	3.7	7.0	5.9	2.2	7.1	9.2	1.7	15.7	7.5	7.7	5.3	7.2	3.8	8.9	10.6	0.7	7.5	12.3
		ENVI-Carb+PSA	GC–MS	Green tea	A	28.7	15.8	26.5	6.0	5.9	32.6	18.4	21.0	17.0	12.0	12.8	33.5	2.3	21.8	3.5	22.4	7.0	14.4	12.7
					B	38.1	16.1	5.8	21.7	14.2	5.3	6.9	3.9	12.2	24.4	15.6	24.9	34.5	8.6	17.3	15.5	8.7	25.7	15.9
				Oolong tea	A	7.6	5.6	3.3	7.5	8.2	3.8	12.2	15.7	6.2	1.2	3.2	5.6	9.8	2.2	1.2	1.7	8.0	10.0	7.1
					B	16.3	9.3	8.2	4.9	13.9	7.5	9.8	5.9	12.4	9.2	9.6	14.0	3.4	15.8	18.4	34.7	12.7	17.1	7.4
			GC–MS/MS	Green tea	A	33.1	19.1	21.6	5.2	5.7	34.7	19.9	29.1	16.3	10.5	24.2	2.4	15.5	19.6	13.0	27.3	7.3	5.2	5.8
					B	42.9	18.5	3.1	22.5	12.3	3.9	20.2	21.5	17.3	17.9	2.6	3.0	27.0	20.8	24.6	18.1	3.6	32.6	11.4
				Oolong tea	A	8.7	7.1	7.0	10.6	5.5	2.8	5.9	20.1	3.5	3.7	4.9	14.9	7.6	4.2	6.4	1.1	10.2	28.1	9.7
					B	17.1	4.9	10.9	3.2	14.3	10.3	9.3	4.2	13.5	11.2	6.2	8.3	3.1	8.7	12.6	3.2	5.4	33.6	6.6
26	Dimethe-namid	TPT	GC–MS	Green tea	A	6.1	6.3	3.5	3.2	7.2	5.3	5.6	5.6	5.0	0.7	6.0	5.6	10.1	6.9	6.8	2.1	2.6	5.4	4.7
					B	6.4	9.8	8.0	8.3	3.8	3.3	1.6	2.4	4.8	8.2	6.0	9.8	1.7	2.4	8.9	4.5	5.3	0.5	5.9
				Oolong tea	A	3.9	4.9	1.8	5.5	4.5	2.5	14.0	2.9	3.3	4.9	5.0	2.7	4.6	0.8	10.2	0.8	2.3	2.2	1.8
					B	7.5	2.8	3.2	2.9	3.3	4.5	6.0	7.9	8.1	4.6	3.9	0.9	1.5	5.3	3.7	1.5	0.7	2.6	4.2
			GC–MS/MS	Green tea	A	5.3	5.7	3.5	3.0	4.9	4.2	5.9	5.8	2.4	1.8	3.1	2.3	7.3	7.0	1.2	3.0	6.9	7.3	3.4
					B	5.3	10.2	6.0	6.2	4.7	6.3	4.0	5.4	4.5	9.8	6.5	7.1	0.9	2.5	3.3	1.8	1.9	6.0	9.1
				Oolong tea	A	2.7	5.2	2.5	6.0	5.7	5.7	11.6	3.4	1.9	3.9	2.2	3.1	9.4	2.5	8.8	3.8	1.2	2.8	5.4
					B	2.3	2.5	4.2	6.0	1.9	0.4	5.4	6.6	7.1	3.2	2.9	2.2	4.5	2.0	6.1	9.2	7.6	5.4	3.8
		ENVI-Carb+PSA	GC–MS	Green tea	A	7.3	1.2	0.1	4.5	6.8	4.4	4.0	1.9	1.2	1.2	8.4	0.1	3.8	6.3	5.3	2.3	3.6	8.1	6.0
					B	7.6	6.6	2.2	3.1	1.0	4.0	3.5	2.9	4.3	10.0	18.1	2.7	3.1	3.4	8.9	4.0	4.7	11.7	7.4
				Oolong tea	A	2.8	4.5	1.9	4.9	3.5	2.3	2.5	2.8	12.8	1.1	5.1	2.6	3.9	5.9	8.6	2.6	6.7	3.7	4.3
					B	6.1	2.2	9.1	6.6	10.0	4.1	3.7	7.0	9.7	4.4	7.1	2.2	2.5	11.0	8.8	2.5	7.5	5.0	5.0
			GC–MS/MS	Green tea	A	8.1	7.3	1.6	4.6	7.1	3.0	3.6	1.8	0.7	1.9	15.3	3.0	8.3	6.8	3.5	3.4	7.6	5.6	2.7
					B	25.2	9.4	1.3	2.5	2.8	3.0	2.4	1.9	1.6	1.6	13.1	2.3	1.7	8.4	12.7	5.9	4.3	13.1	5.1
				Oolong tea	A	4.7	4.4	4.1	7.4	7.0	2.3	6.6	2.0	3.5	1.6	5.1	8.7	4.0	4.5	11.4	1.4	2.0	4.6	7.0
					B	7.4	2.3	10.3	4.0	7.9	5.2	2.2	6.5	10.7	2.6	2.0	1.0	4.4	8.7	12.2			3.9	5.3

(Continued)

Appendix Table 5.2 RSD of 201 Pesticides in Youden Pair Aged Tea Sample for Parallel Determinations (Nov 9, 2009–Feb 7, 2010) (cont.)

No.	Pesticides	SPE	Method	Sample	Youden pair	RSD%																		
						Nov 9, 2009 (n = 5)	Nov 14, 2009 (n = 3)	Nov 19, 2009 (n = 3)	Nov 24, 2009 (n = 3)	Nov 29, 2009 (n = 3)	Dec 4, 2009 (n = 3)	Dec 9, 2009 (n = 3)	Dec 14, 2009 (n = 3)	Dec 19, 2009 (n = 3)	Dec 24, 2009 (n = 3)	Dec 14, 2009 (n = 3)	Jan 3, 2010 (n = 3)	Jan 8, 2010 (n = 3)	Jan 13, 2010 (n = 3)	Jan 18, 2010 (n = 3)	Jan 23, 2010 (n = 3)	Jan 28, 2010 (n = 3)	Feb 2, 2010 (n = 3)	Feb 7, 2010 (n = 3)
27	Diofeno-lan-1	TPT	GC–MS	Green tea	A	4.7	5.6	0.8	0.9	5.9	7.0	8.1	3.9	5.7	2.1	5.9	4.0	8.1	4.3	3.5	2.2	2.9	8.3	4.5
					B	7.6	6.9	4.5	8.8	3.8	4.0	3.4	1.2	3.9	7.8	4.1	8.6	1.6	2.9	9.0	1.6	4.1	2.1	4.8
				Oolong tea	A	4.6	2.2	1.0	2.2	2.8	2.9	6.6	6.2	1.2	4.4	6.9	5.8	4.5	2.8	8.5	4.6	1.3	6.5	2.2
					B	7.9	1.4	2.9	4.2	3.0	6.9	4.5	2.8	3.3	4.0	5.6	2.0	4.0	4.2	1.1	0.3	2.5	1.0	1.0
			GC–MS/MS	Green tea	A	4.7	5.9	2.2	3.0	5.7	5.2	7.5	5.0	5.9	2.4	6.1	3.2	4.1	4.6	5.2	0.5	3.1	4.5	4.4
					B	3.8	7.5	6.0	7.7	4.0	5.3	3.2	3.6	4.0	8.2	5.8	8.3	0.8	3.7	7.8	2.9	7.7	4.4	6.6
				Oolong tea	A	6.6	2.5	3.3	4.3	2.9	1.9	5.7	2.7	0.6	3.1	1.6	1.6	2.3	2.8	6.7	3.7	1.9	3.4	3.7
					B	3.8	1.5	2.0	5.7	1.2	1.2	4.2	3.5	4.7	4.1	3.5	1.4	7.4	3.4	5.5	1.2	0.6	2.9	2.7
		ENVI-Carb+PSA	GC–MS	Green tea	A	6.1	9.5	4.0	6.0	8.7	2.9	3.4	0.7	7.0	3.0	6.6	2.1	3.8	3.0	4.2	8.4	10.9	2.7	5.5
					B	2.7	9.1	2.4	6.6	2.0	5.7	1.8	8.6	4.3	5.6	26.9	3.4	1.8	5.3	14.2	3.1	3.8	8.2	6.3
				Oolong tea	A	5.4	2.0	3.7	2.0	6.5	1.9	3.6	2.2	4.8	1.9	4.0	1.9	3.0	5.8	2.8	1.9	5.0	4.2	4.4
					B	4.4	2.0	9.0	4.2	9.9	3.7	3.7	6.6	11.8	3.3	6.4	0.6	1.7	9.4	10.5	5.6	5.3	5.2	5.2
			GC–MS/MS	Green tea	A	4.6	5.0	1.7	5.8	6.1	1.2	3.7	2.2	6.1	1.5	8.9	3.3	5.2	12.4	5.0	8.5	8.6	6.2	4.7
					B	1.5	11.3	3.9	5.4	1.7	2.1	2.4	4.0	3.3	2.3	20.8	4.0	3.9	7.8	12.3	3.0	3.8	11.6	2.9
				Oolong tea	A	7.9	2.6	6.4	1.3	6.0	1.3	4.0	1.9	1.3	2.2	6.2	9.1	2.6	3.6	7.3	3.9	6.5	7.0	3.9
					B	6.5	2.5	9.3	3.0	8.2	4.3	3.2	6.4	11.7	3.2	1.3	4.3	6.6	6.9	10.8	3.3	2.9	1.5	4.8
28	Diofeno-lan-2	TPT	GC–MS	Green tea	A	5.1	5.2	1.2	1.3	7.0	4.5	6.4	4.1	6.1	2.7	5.3	2.8	6.4	2.4	3.1	1.2	4.3	6.3	4.6
					B	5.9	7.2	4.6	8.0	4.1	3.7	2.1	1.9	3.5	7.5	3.7	7.1	1.6	1.6	5.0	1.8	4.6	2.6	4.1
				Oolong tea	A	6.5	1.8	1.2	1.9	2.7	1.5	7.2	6.4	1.7	3.3	5.9	4.7	4.4	2.5	6.1	3.1	2.3	3.3	1.8
					B	7.6	2.3	3.5	3.8	0.5	4.3	5.0	3.7	7.0	3.8	5.2	1.8	2.7	3.7	1.0	1.1	2.4	1.5	4.5
			GC–MS/MS	Green tea	A	9.5	7.3	1.6	2.2	6.8	5.7	7.8	3.6	4.9	3.4	6.5	3.7	6.7	5.1	1.7	3.8	4.5	8.6	7.2
					B	15.5	5.8	5.9	7.9	3.2	3.5	4.0	4.0	2.4	8.9	6.5	8.0	2.0	1.3	2.3	5.2	5.0	5.2	6.4
				Oolong tea	A	7.4	3.5	1.8	4.3	3.3	2.0	3.0	2.3	2.3	3.0	0.5	1.7	3.4	2.7	8.7	3.0	3.2	1.5	3.7
					B	4.6	2.0	4.6	5.7	1.1	1.2	4.1	4.4	3.9	6.4	4.3	1.2	7.4	1.8	5.7	4.9	1.0	3.7	2.7
		ENVI-Carb+PSA	GC–MS	Green tea	A	5.6	4.6	3.0	6.5	8.9	1.2	6.2	1.2	7.5	4.1	6.2	3.4	1.7	5.4	5.6	8.0	9.6	2.8	5.3
					B	3.5	8.1	3.2	4.3	1.9	2.5	2.7	5.5	3.7	4.0	21.4	3.2	2.0	4.8	12.4	3.5	4.4	8.3	6.0
				Oolong tea	A	5.9	2.2	4.4	1.7	4.3	4.0	5.4	2.2	2.4	2.2	4.2	1.4	2.3	6.2	0.8	2.2	5.2	4.4	3.7
					B	5.6	2.8	8.8	5.1	8.6	2.3	2.6	6.2	10.2	2.9	8.1	1.7	2.8	6.0	9.5	3.9	5.7	4.7	4.2
			GC–MS/MS	Green tea	A	9.4	4.4	1.8	7.5	7.4	1.2	2.6	1.1	6.4	1.9	7.2	24.6	5.6	38.3	4.4	7.4	10.1	3.7	2.9
					B	9.3	12.1	4.0	4.8	4.6	1.5	1.1	4.1	2.6	3.6	22.4	12.9	2.0	18.6	15.8	1.2	5.3	13.4	4.6
				Oolong tea	A	7.1	2.6	6.4	1.4	#DIV/0!	2.5	5.3	1.9	3.3	0.7	4.3		1.6		7.3	3.9	5.0	7.9	4.9
					B	7.5	2.4	9.4	3.0	#DIV/0!	5.1	2.2	6.4	10.6	1.8	1.1		7.6		10.7	3.3	3.3	3.7	7.4

29	Fenbuconazole	TPT	GC-MS	Green tea	A	8.6	15.0	0.6	2.8	4.4	5.6	8.5	1.8	6.4	5.3	9.8	2.5	6.2	8.6	11.5	8.7	3.6	0.2	3.5
					B	4.2	12.7	5.2	6.5	5.0	3.4	1.9	9.2	2.7	11.3	5.3	13.2	6.7	6.8	11.2	2.9	11.0	5.9	7.0
				Oolong tea	A	4.7	10.0	6.2	4.6	2.0	2.5	7.4	3.0	2.3	5.3	8.4	5.0	5.7	3.5	6.2	8.1	6.1	4.4	5.0
					B	11.0	1.3	2.3	8.4	4.1	4.6	7.5	2.0	9.7	5.7	7.5	1.2	2.5	1.8	3.0	1.7	3.3	2.2	1.3
			GC-MS/MS	Green tea	A	8.2	14.2	2.4	7.7	3.9	7.4	9.0	1.1	7.1	7.6	7.2	0.6	7.6	7.6	5.6	6.1	4.9	7.1	2.6
					B	4.0	12.8	5.4	5.7	3.7	9.8	2.5	10.3	0.2	21.0	7.1	8.3	1.0	8.3	13.4	5.0	12.6	8.6	6.6
				Oolong tea	A	6.9	3.0	2.4	4.6	1.7	4.9	9.5	3.8	2.5	6.4	5.5	0.9	3.3	3.7	7.5	5.7	5.2	9.7	2.2
					B	3.3	2.4	8.0	8.9	3.1	2.5	5.9	3.1	1.9	4.4	5.2	2.3	3.8	3.0	6.0	3.6	1.9	5.3	2.8
		ENVI-Carb+PSA	GC-MS	Green tea	A	13.3	29.9	4.8	7.7	4.6	6.4	11.5	1.7	9.5	6.2	3.8	2.7	9.1	8.3	8.0	8.9	12.7	5.2	2.5
					B	8.6	4.7	4.2	9.7	5.5	4.0	14.9	6.9	5.7	10.5	30.2	5.1	6.2	6.0	15.6	6.1	1.9	6.9	1.6
				Oolong tea	A	7.2	4.1	9.2	3.3	4.3	9.6	6.7	6.2	16.9	19.3	6.2	10.3	4.3	14.8	9.5	6.5	1.8	4.6	5.8
					B	10.4	9.6	20.3	9.2	14.9	16.7	4.1	10.8	9.4	3.4	10.6	1.1	2.1	30.9	7.9	7.5	15.8	8.0	3.4
			GC-MS/MS	Green tea	A	13.5	29.9	4.0	7.9	2.9	7.1	54.1	1.5	7.4	6.5	10.2	1.7	6.1	7.9	4.2	9.9	8.9	5.6	2.9
					B	7.7	5.0	4.5	9.0	7.8	2.7	2.9	2.4	6.0	3.3	21.5	6.0	10.9	13.6	14.5	2.3	4.2	11.4	6.2
				Oolong tea	A	10.1	3.1	9.4	3.4	5.8	4.8	7.7	7.2	2.2	24.3	10.7	16.6	11.0	8.2	15.4	6.5	6.4	6.4	6.6
					B	8.8	3.1	21.7	8.1	17.2	13.3	3.2	6.5	6.4	3.2	10.0	10.3	4.8	30.8	7.9	5.7	18.2	7.9	1.7
30	Fenpyroximate	TPT	GC-MS	Green tea	A	8.1	5.3	3.7	11.4	6.3	10.1	19.5	8.7	6.5	6.7	13.9	5.9	14.0	12.1	8.3	12.4	0.5	20.8	8.8
					B	9.9	9.9	7.7	15.8	8.3	12.5	9.1	14.5	8.1	2.9	14.1	9.2	15.6	15.8	14.8	12.9	20.3	14.9	2.6
				Oolong tea	A	8.2	14.0	3.8	2.0	11.8	2.2	30.7	13.9	11.2	3.3	12.3	16.3	11.4	9.5	11.9	12.1	12.7	19.6	4.7
					B	7.8	10.2	7.2	13.3	10.3	14.9	7.2	1.0	5.2	5.7	13.0	10.6	5.4	3.9	9.3	9.7	1.1	10.1	10.1
			GC-MS/MS	Green tea	A	18.2	2.9	2.0	17.8	7.6	8.4	24.6	12.8	12.2	9.1	18.7	2.7	11.4	7.4	5.9	11.0	1.8	23.5	6.8
					B	13.6	16.3	14.2	15.5	9.4	5.9	12.0	14.9	4.5	6.3	12.1	4.4	13.3	14.3	6.8	4.5	18.6	13.7	14.2
				Oolong tea	A	8.0	20.1	11.7	7.6	21.4	3.4	20.2	11.7	4.6	6.0	14.1	9.1	9.8	14.0	11.7	11.6	10.0	26.4	2.9
					B	36.4	11.3	2.8	11.9	8.7	22.2	9.5	1.5	4.4	6.9	12.3	8.8	9.0	7.5	3.8	6.0	7.5	8.4	11.9
		ENVI-Carb+PSA	GC-MS	Green tea	A	14.1	12.3	12.7	10.4	12.4	14.2	28.2	6.2	11.1	13.7	23.6	29.1	10.3	12.5	13.7	9.7	15.6	3.3	2.3
					B	26.4	12.8	5.1	14.0	20.2	36.6	3.3	12.5	7.3	7.7	23.9	13.9	32.7	13.6	14.1	8.2	7.1	2.3	3.4
				Oolong tea	A	11.8	8.9	10.6	21.4	17.8	14.2	12.2	12.2	12.0	2.6	14.0	12.8	23.1	9.0	0.8	26.6	2.4	13.5	15.2
					B	18.7	8.3	14.5	19.0	18.7	4.6	19.0	9.4	12.0	19.3	8.8	4.8	4.4	14.1	18.0	12.5	5.3	29.8	4.6
			GC-MS/MS	Green tea	A	16.3	6.2	12.5	8.6	13.5	44.2	18.1	18.5	6.6	8.2	17.0	6.8	8.6	35.4	14.6	24.4	17.2	3.4	2.6
					B	18.6	12.4	11.8	17.9	30.9	14.8	6.5	19.5	7.8	12.7	27.7	13.5	31.1	22.6	19.3	16.2	5.4	2.6	8.2
				Oolong tea	A	23.4	13.5	12.3	18.4	16.0	8.0	12.6	19.9	7.2	5.0	6.9	20.6	31.4	14.4	4.0	25.2	7.5	32.9	15.1
					B	16.5	9.4	11.9	16.0	13.7	9.2	14.3	12.5	8.1	14.0	10.0	9.8	13.0	20.6	11.4	10.0	5.8	42.5	5.3
31	Fluotrimazole	TPT	GC-MS	Green tea	A	5.5	5.9	0.7	3.5	5.7	3.1	8.0	3.9	7.8	5.3	9.6	14.9	6.3	0.9	2.7	5.2	5.2	6.5	3.8
					B	4.9	6.7	6.4	5.7	5.1	4.1	1.7	3.8	7.8	11.5	7.6	8.2	3.5	2.6	2.4	3.1	5.8	2.5	3.7
				Oolong tea	A	4.2	4.7	3.2	1.3	2.5	4.0	8.5	8.2	5.5	8.8	16.0	5.3	5.6	5.5	7.5	3.6	3.5	0.4	0.7
					B	8.1	1.1	4.2	9.6	3.6	3.7	5.0	3.8	8.6	13.4	7.2	2.9	1.9	8.4	2.1	1.2	2.0	0.6	1.6
			GC-MS/MS	Green tea	A	4.8	5.2	2.1	3.8	12.4	12.4	8.3	4.9	7.8	3.4	4.2	4.0	8.0	1.4	5.4	6.4	9.1	7.0	6.0
					B	4.7	6.9	7.9	8.6	6.7	15.2	5.5	4.0	4.1	8.6	7.7	5.9	1.7	2.6	10.0	5.5	5.8	2.4	6.7
				Oolong tea	A	8.5	1.8	2.4	3.9	0.8	4.1	7.7	4.2	5.7	5.4	3.4	2.5	4.4	4.3	2.2	5.6	0.8	1.0	4.2
					B	4.0	5.3	2.7	9.9	7.8	6.5	4.3	6.6	7.5	1.8	3.2	1.8	13.4	2.8	4.3	5.6	5.4	3.1	6.3
		ENVI-Carb+PSA	GC-MS	Green tea	A	6.2	5.0	4.1	7.4	5.9	4.1	6.5	1.4	4.4	1.5	1.2	4.0	3.3	5.5	9.0	7.0	10.5	2.5	4.7
					B	4.3	7.6	3.0	6.8	4.1	2.8	2.3	6.1	11.6	5.4	16.0	7.2	1.3	6.0	13.8	1.4	6.5	8.6	7.5
				Oolong tea	A	4.4	3.6	5.1	1.2	2.8	1.2	3.0	5.1	9.5	1.4	5.9	3.3	2.8	4.2	2.2	1.4	4.2	3.9	3.8
					B	5.1	4.9	5.3	7.0	2.4	2.4	2.7	6.8	9.5	10.9	7.7	3.7	2.2	10.6	8.9	4.8	6.6	3.3	4.5
			GC-MS/MS	Green tea	A	3.9	4.8	2.8	6.6	3.6	3.6	3.7	3.2	5.1	3.8	8.7	10.6	4.6	6.7	4.4	6.1	11.2	5.2	3.5
					B	2.2	8.1	3.5	5.1	6.1	14.8	3.4	4.4	6.2	5.6	23.2	9.3	4.6	20.4	12.5	0.7	5.8	16.0	6.0
				Oolong tea	A	7.4	1.4	5.8	7.6	9.7	1.2	2.9	4.2	6.2	4.9	7.9	8.5	9.5	4.9	11.0	2.6	5.7	4.5	6.1
					B	6.2	2.8	8.9	2.4	11.5	3.9	4.3	4.1	8.1	4.1	3.6	3.3	7.3	6.9	10.8	5.0	7.3	1.4	3.6

(Continued)

Appendix Table 5.2 RSD of 201 Pesticides in Youden Pair Aged Tea Sample for Parallel Determinations (Nov 9, 2009–Feb 7, 2010) (cont.)

RSD%

No.	Pesticides	SPE	Method	Sample	Youden pair	Nov 9, 2009 (n=5)	Nov 14, 2009 (n=3)	Nov 19, 2009 (n=3)	Nov 24, 2009 (n=3)	Nov 29, 2009 (n=3)	Dec 4, 2009 (n=3)	Dec 9, 2009 (n=3)	Dec 14, 2009 (n=3)	Dec 19, 2009 (n=3)	Dec 24, 2009 (n=3)	Dec 29, 2009 (n=3)	Jan 3, 2010 (n=3)	Jan 8, 2010 (n=3)	Jan 13, 2010 (n=3)	Jan 18, 2010 (n=3)	Jan 23, 2010 (n=3)	Jan 28, 2010 (n=3)	Feb 2, 2010 (n=3)	Feb 7, 2010 (n=3)
32	Fluoxypr-1-methylheptyl ester	TPT	GC-MS	Green tea	A	7.0	7.7	0.9	6.4	10.4	0.8	7.3	9.8	9.2	2.3	4.6	4.8	5.0	4.0	2.1	5.5	6.4	25.5	3.2
		TPT	GC-MS	Green tea	B	5.0	6.7	1.8	9.1	4.5	3.6	1.5	6.5	4.7	7.7	2.2	4.7	7.0	5.7	5.9	3.9	7.9	3.1	4.7
		TPT	GC-MS	Oolong tea	A	6.7	4.1	7.0	1.2	1.1	3.4	11.7	5.5	4.4	3.7	9.4	5.1	5.4	0.9	3.8	0.9	4.0	2.6	2.3
		TPT	GC-MS	Oolong tea	B	9.1	4.8	5.9	7.1	2.3	0.8	3.5	0.3	7.2	3.2	3.7	1.0	0.6	4.6	5.8	0.4	2.1	1.3	6.3
		TPT	GC-MS/MS	Green tea	A	3.2	6.9	6.4	9.0	17.1	6.7	11.9	11.0	11.6	2.7	7.7	5.0	8.1	2.8	8.5	8.9	4.3	14.2	7.9
		TPT	GC-MS/MS	Green tea	B	7.2	5.7	5.9	7.0	0.3	11.8	3.1	5.5	4.4	6.6	3.1	14.8	7.2	5.6	4.0	5.3	2.6	13.1	6.9
		TPT	GC-MS/MS	Oolong tea	A	9.2	0.6	3.2	0.6	4.8	3.7	9.8	6.2	9.7	4.5	2.8	6.7	6.2	4.8	6.7	9.4	18.4	2.2	5.7
		TPT	GC-MS/MS	Oolong tea	B	2.5	11.0	9.1	6.8	7.6	7.1	9.3	0.1	6.4	2.4	10.0	2.0	9.1	7.9	0.3	1.5	10.3	1.6	13.1
		ENVI-Carb+PSA	GC-MS	Green tea	A	17.7	6.4	11.4	4.6	7.6	12.7	7.2	11.3	7.9	2.3	2.6	11.9	2.7	8.8	3.4	5.1	10.7	10.0	4.4
		ENVI-Carb+PSA	GC-MS	Green tea	B	16.7	9.6	5.0	4.8	4.4	2.0	7.2	5.7	15.8	8.5	11.0	12.4	13.7	1.6	4.7	13.9	4.7	18.1	10.7
		ENVI-Carb+PSA	GC-MS	Oolong tea	A	4.7	2.6	5.6	1.3	2.9	2.9	4.2	15.0	0.9	3.1	5.2	4.4	2.7	5.3	1.7	8.6	4.4	8.3	1.0
		ENVI-Carb+PSA	GC-MS	Oolong tea	B	6.7	5.3	8.5	0.5	8.6	2.8	4.7	7.5	9.7	2.7	7.4	5.1	2.2	5.7	9.6	1.8	8.7	4.4	3.9
		ENVI-Carb+PSA	GC-MS/MS	Green tea	A	19.5	7.1	8.0	2.1	7.8	9.0	9.0	11.0	9.8	1.3	15.1	5.4	16.7	6.0	10.5	14.7	1.7	11.0	5.0
		ENVI-Carb+PSA	GC-MS/MS	Green tea	B	19.9	10.0	1.1	6.8	6.4	5.2	15.8	12.6	14.4	6.3	11.5	7.6	15.1	17.0	15.2	9.4	6.2	23.1	33.5
		ENVI-Carb+PSA	GC-MS/MS	Oolong tea	A	10.4	1.3	11.2	3.4	5.3	5.3	10.0	8.2	1.4	8.2	4.6	4.6	5.2	9.3	19.5	9.0	11.7	5.0	16.1
		ENVI-Carb+PSA	GC-MS/MS	Oolong tea	B	7.8	7.5	10.3	3.5	7.5	1.9	6.1	8.3	9.5	8.5	9.8	0.4	1.7	5.2	4.4	2.4	18.4	11.4	6.4
33	Iprovalicarb-1	TPT	GC-MS	Green tea	A	6.1	3.8	1.3	2.6	14.8	4.9	11.0	11.3	5.7	1.5	5.4	5.8	9.9	4.4	2.1	1.3	8.4	6.1	4.7
		TPT	GC-MS	Green tea	B	6.1	9.2	5.0	9.1	6.7	7.3	3.4	1.2	4.9	6.6	6.1	8.7	2.5	2.8	6.0	6.5	6.7	3.8	4.0
		TPT	GC-MS	Oolong tea	A	4.7	4.3	7.9	2.6	6.6	4.3	3.3	10.6	7.2	8.7	10.5	2.3	3.9	5.9	5.6	0.4	2.1	0.9	2.6
		TPT	GC-MS	Oolong tea	B	9.4	1.6	6.8	27.4	11.0	16.6	12.2	9.8	6.4	3.0	7.7	5.2	4.0	4.1	0.7	1.7	4.5	2.8	8.1
		TPT	GC-MS/MS	Green tea	A	4.6	7.2	0.7	1.8	6.3	8.5	8.8	8.5	8.2	2.4	7.9	9.0	9.1	12.9	1.3	1.9	4.6	3.0	7.2
		TPT	GC-MS/MS	Green tea	B	4.2	6.2	4.0	5.7	3.3	10.9	1.1	12.5	2.8	7.3	4.0	14.2	3.4	5.4	8.0	5.2	7.4	1.4	6.4
		TPT	GC-MS/MS	Oolong tea	A	8.6	13.1	2.3	2.7	2.3	4.7	0.8	1.7	9.1	5.5	5.6	6.3	12.6	2.8	4.3	3.0	2.3	2.5	3.0
		TPT	GC-MS/MS	Oolong tea	B	9.1	11.7	5.7	7.9	4.2	5.3	9.0	5.1	6.2	4.5	7.2	3.3	11.1	2.6	4.0	1.2	2.2	4.6	3.9
		ENVI-Carb+PSA	GC-MS	Green tea	A	12.1	4.1	5.9	8.0	8.7	4.1	21.6	14.4	4.4	3.1	1.9	5.0	2.3	17.9	6.5	35.8	9.7	30.8	7.7
		ENVI-Carb+PSA	GC-MS	Green tea	B	7.2	2.0	9.6	6.5	10.3	6.3	9.7	3.4	11.4	22.3	29.5	6.1	29.6	1.9	34.4	2.5	2.7	27.0	14.3
		ENVI-Carb+PSA	GC-MS	Oolong tea	A	4.2	3.9	6.4	3.3	3.7	4.6	33.1	17.6	6.6	5.0	4.1	1.4	5.4	5.9	2.9	3.0	6.9	9.2	1.6
		ENVI-Carb+PSA	GC-MS	Oolong tea	B	12.2	4.1	10.3	4.9	14.3	9.7	8.7	5.7	8.7	7.3	2.7	3.8	4.2	8.7	14.2	4.0	8.9	5.5	6.3
		ENVI-Carb+PSA	GC-MS/MS	Green tea	A	17.0	6.6	4.3	6.8	6.2	2.5	12.2	8.3	11.1	2.9	8.0	4.5	7.8	8.1	4.2	12.8	8.5	14.7	13.1
		ENVI-Carb+PSA	GC-MS/MS	Green tea	B	9.7	11.7	3.2	6.5	3.0	3.9	5.3	8.9	23.9	3.3	24.7	7.6	19.0	7.5	19.0	3.8	4.6	17.3	8.8
		ENVI-Carb+PSA	GC-MS/MS	Oolong tea	A	6.9	1.7	4.9	1.9	5.5	1.7	4.1	18.4	4.2	3.4	4.8	9.2	1.9	3.1	6.3	5.1	6.6	11.5	4.5
		ENVI-Carb+PSA	GC-MS/MS	Oolong tea	B	11.1	3.6	11.6	2.2	11.1	6.5	4.3	5.6	10.7	0.3	3.8	2.1	5.1	7.3	10.8	0.5	4.3	6.7	3.5
34	Iprovalicarb-2	TPT	GC-MS	Green tea	A	4.1	7.3	1.7	2.3	9.1	2.6	7.8	13.5	11.6	3.9	7.8	12.1	7.4	5.1	2.3	2.5	8.8	11.7	4.8
		TPT	GC-MS	Green tea	B	4.8	6.2	7.9	11.8	6.0	6.1	3.1	6.7	3.5	9.5	8.9	12.4	4.4	5.0	5.1	4.8	5.4	6.0	5.4
		TPT	GC-MS	Oolong tea	A	4.3	5.5	4.5	3.4	5.6	4.0	2.9	19.2	5.4	4.9	16.4	7.0	5.4	2.5	5.8	0.7	3.1	3.4	0.6
		TPT	GC-MS	Oolong tea	B	5.9	5.0	3.7	5.3	7.1	7.8	14.6	10.7	7.1	8.0	14.9	0.8	2.8	3.6	3.4	0.9	2.5	0.4	5.6
		TPT	GC-MS/MS	Green tea	A	4.5	7.2	1.6	1.0	6.9	7.1	8.8	12.9	7.2	8.3	5.3	5.8	7.0	7.8	2.1	0.1	4.6	5.4	4.9
		TPT	GC-MS/MS	Green tea	B	4.8	6.9	5.3	5.7	3.4	10.0	0.8	9.5	2.5	1.9	5.7	11.1	3.3	4.2	8.1	3.8	9.5	0.1	3.8
		TPT	GC-MS/MS	Oolong tea	A	8.6	13.1	2.3	2.7	0.8	5.3	1.3	8.0	7.8	8.2	5.6	4.1	12.3	2.0	5.8	2.0	0.4	4.6	1.3
		TPT	GC-MS/MS	Oolong tea	B	9.1	11.7	5.7	7.9	3.1	4.8	9.1	5.1	8.5	4.9	9.0	1.8	8.2	1.3	2.2	2.0	6.7	0.3	4.8
		ENVI-Carb+PSA	GC-MS	Green tea	A	13.7	7.2	7.4	7.2	7.8	3.5	19.9	9.1	17.0	3.7	3.3	5.8	1.8	19.7	4.1	42.0	10.5	33.0	3.9
		ENVI-Carb+PSA	GC-MS	Green tea	B	5.3	8.7	5.1	5.0	2.2	7.7	7.3	13.5	20.6	20.1	47.2	12.2	30.0	1.6	34.7	1.0	4.5	33.6	10.8
		ENVI-Carb+PSA	GC-MS	Oolong tea	A	4.2	3.9	7.9	3.7	4.5	2.2	3.1	6.7	21.0	0.4	7.3	3.6	3.7	7.0	0.8	6.1	5.4	3.3	3.9
		ENVI-Carb+PSA	GC-MS	Oolong tea	B	4.1	5.5	11.1	0.1	15.8	5.6	6.9	19.2	12.0	5.1	9.4	7.3	5.3	3.8	9.9	2.5	8.1	0.7	4.0
		ENVI-Carb+PSA	GC-MS/MS	Green tea	A	17.2	7.0	5.7	6.6	6.0	5.0	14.3	8.2	12.9	2.9	7.7	5.0	6.6	6.6	3.6	16.4	7.5	12.6	10.9
		ENVI-Carb+PSA	GC-MS/MS	Green tea	B	10.8	10.2	2.6	5.4	2.1	4.5	6.1	9.2	10.4	4.3	29.3	2.3	21.2	5.6	18.9	1.2	3.5	17.5	9.9
		ENVI-Carb+PSA	GC-MS/MS	Oolong tea	A	6.9	1.7	4.9	1.9	5.5	3.8	3.3	18.4	3.2	4.6	5.1	6.9	3.6	3.1	5.0	5.4	5.7	10.2	3.5
		ENVI-Carb+PSA	GC-MS/MS	Oolong tea	B	11.1	3.6	11.6	2.2	11.1	6.0	5.3	5.6	11.1	2.8	2.4	3.3	8.3	7.3	8.0	1.3	4.3	7.0	3.8

35	Isodrin	TPT	GC-MS	Green tea	A	7.1	9.1	5.1	2.1	9.0	4.7	8.6	3.2	12.2	5.8	2.7	5.3	7.6	4.1	4.7	7.4	3.6	12.2	5.8
					B	8.2	9.3	2.7	9.3	4.7	1.4	3.2	1.4	5.3	11.7	1.4	4.9	3.4	8.2	5.7	4.6	6.5	3.2	5.3
				Oolong tea	A	5.0	2.6	1.6	3.8	2.1	1.2	9.9	3.7	6.7	9.0	8.8	10.7	2.8	5.7	8.1	11.8	8.0	9.3	2.6
					B	5.3	1.9	3.9	9.9	1.4	2.8	3.4	2.6	1.7	8.4	3.1	4.2	2.5	2.4	2.6	2.3	1.3	4.8	2.7
			GC-MS/MS	Green tea	A	6.1	6.8	5.3	5.0	8.3	4.9	5.3	3.7	7.9	3.4	6.0	6.3	7.3	3.1	6.1	2.7	6.9	5.6	6.7
					B	5.4	7.8	5.9	8.7	7.6	2.6	5.1	5.2	4.4	10.8	7.4	7.4	2.7	6.6	3.4	3.7	12.0	5.2	10.5
				Oolong tea	A	2.6	3.4	1.1	5.2	2.7	3.5	8.5	1.1	2.7	2.0	3.3	0.1	10.0	2.3	8.6	19.4	6.5	6.3	6.2
					B	5.8	4.5	4.0	8.4	4.7	6.0	6.6	1.5	6.6	2.4	9.8	1.3	4.2	5.6	8.7	3.2	2.0	3.4	7.5
		ENVI-Carb+PSA	GC-MS	Green tea	A	4.7	8.0	1.5	7.6	7.5	2.8	12.0	6.8	10.0	9.0	7.2	3.5	6.3	3.6	9.0	6.1	6.7	4.2	8.7
					B	4.6	4.3	5.1	4.0	5.1	1.1	8.3	1.3	6.0	2.8	15.1	2.3	5.7	3.1	13.4	0.6	5.0	8.7	9.6
				Oolong tea	A	4.5	4.1	2.6	2.9	7.3	3.7	6.3	5.2	16.7	3.8	18.0	7.0	5.0	2.4	4.8	2.1	4.7	0.6	9.2
					B	9.2	4.6	8.9	2.3	9.6	4.8	1.2	3.7	10.6	7.9	8.8	9.1	2.3	8.0	9.9	6.4	6.4	14.2	2.5
			GC-MS/MS	Green tea	A	5.9	5.4	2.6	5.2	4.5	2.6	7.7	1.9	7.3	2.6	16.3	4.8	2.2	2.3	9.5	11.2	8.3	5.7	5.7
					B	8.1	7.9	4.4	5.3	5.0	7.2	3.5	5.7	3.3	2.3	21.9	3.6	4.2	4.9	13.9	2.9	4.6	13.7	2.5
				Oolong tea	A	4.6	4.4	5.8	19.7	0.2	4.3	6.8	1.5	4.4	1.4	4.1	8.1	11.6	7.4	11.8	9.4	6.6	13.2	9.4
					B	4.8	1.9	11.1	4.7	6.1	5.3	5.2	5.2	12.4	1.6	5.1	3.8	5.2	6.8	10.6	7.5	7.2	2.1	2.2
36	Isoprocarb-1	TPT	GC-MS	Green tea	A	8.5	6.4	2.0	0.8	7.1	6.6	7.3	3.5	5.0	6.2	7.1	1.3	9.6	1.4	5.1	4.0	4.4	9.1	4.1
					B	9.3	10.6	4.0	8.8	1.7	4.9	4.8	7.9	3.0	3.0	5.2	6.8	5.8	5.4	4.0	2.8	4.5	4.7	5.4
				Oolong tea	A	3.6	3.1	2.7	4.0	3.8	2.8	11.9	7.3	2.4	3.4	0.5	3.5	3.2	2.7	8.0	3.5	4.1	10.6	1.4
					B	10.8	3.2	3.6	7.6	2.0	3.0	7.9	0.7	4.5	6.0	2.6	1.5	2.8	2.6	3.1	1.7	0.3	5.6	4.6
			GC-MS/MS	Green tea	A	6.2	7.3	1.8	6.6	6.0	4.3	7.7	3.1	7.2	1.8	7.5	3.3	6.8	5.1	2.1	3.1	8.1	10.2	3.3
					B	4.6	9.4	6.5	8.0	4.8	4.0	5.1	6.1	2.3	8.9	1.6	4.1	1.6	4.1	4.3	2.6	9.0	4.8	7.3
				Oolong tea	A	6.3	1.9	4.5	2.6	2.5	2.1	9.0	0.6	5.5	3.8	7.5	3.9	10.4	1.7	9.1	3.2	2.9	3.8	2.6
					B	5.4	1.5	5.5	3.4	0.6	3.2	15.2	1.8	3.3	4.5	2.1	1.6	5.9	1.5	5.5	1.7	1.3	2.8	3.1
		ENVI-Carb+PSA	GC-MS	Green tea	A	4.1	6.1	1.6	7.2	5.9	4.8	2.1	5.3	8.6	5.4	10.0	9.5	4.5	4.0	6.7	7.6	13.7	2.7	5.6
					B	3.6	4.3	3.3	6.7	4.1	3.2	4.5	11.1	6.5	4.1	19.2	4.5	4.6	4.3	15.0	4.3	6.2	4.7	4.1
				Oolong tea	A	4.7	4.5	3.4	3.8	5.9	3.9	4.5	6.0	1.0	2.9	7.0	6.8	3.3	2.5	3.1	3.6	6.6	2.0	6.1
					B	7.3	1.7	10.7	5.6	9.3	5.6	10.7	5.9	8.0	7.8	7.3	6.0	2.4	10.6	9.4	4.6	2.9	9.7	3.4
			GC-MS/MS	Green tea	A	4.0	7.2	0.8	6.6	5.8	6.2	1.3	2.2	6.7	4.1	9.4	17.1	5.7	9.5	5.6	12.5	11.9	3.8	2.7
					B	4.9	7.3	5.1	8.6	5.1	4.8	2.8	7.5	5.6	2.8	23.1	9.6	2.7	9.3	19.1	0.9	4.0	10.8	1.4
				Oolong tea	A	4.6	2.0	2.6	7.3	5.2	0.6	2.1	1.1	2.9	0.7	4.9	7.4	6.2	1.9	5.9	3.1	3.8	3.8	6.4
					B	7.0	3.9	8.4	4.2	8.7	3.4	9.0	6.7	11.6	2.3	1.5	4.4	1.6	6.8	7.9	2.7	4.6	4.1	1.7
37	Isoprocarb-2	TPT	GC-MS	Green tea	A	12.4	12.4	2.1	5.4	13.0	1.7	9.0	14.9	8.3	2.3	14.7	6.4	10.6	3.3	16.9	6.0	7.4	9.1	9.3
					B	15.5	9.3	7.7	7.6	6.1	3.4	5.9	7.3	7.9	11.3	3.6	8.3	15.3	12.0	4.3	2.9	8.4	7.2	11.0
				Oolong tea	A	13.5	5.3	4.3	5.2	3.6	1.8	8.6	4.2	12.1	8.6	12.0	4.4	3.3	6.3	9.0	4.8	6.5	23.2	9.9
					B	12.3	2.4	11.7	7.8	3.4	14.3	4.7	14.1	15.7	8.7	1.5	1.3	3.8	5.2	11.3	6.8	2.4	8.9	14.8
			GC-MS/MS	Green tea	A	11.7	6.1	5.8	15.1	14.2	4.5	14.4	14.7	12.2	10.2	7.1	5.2	7.0	9.7	6.1	10.7	12.9	7.4	10.0
					B	6.1	6.2	9.8	12.5	5.9	9.7	2.3	8.4	11.7	9.5	3.8	1.8	10.6	10.4	9.9	5.5	17.9	4.3	6.3
				Oolong tea	A	13.5	6.9	4.5	7.2	5.1	8.6	2.1	6.2	2.1	9.9	1.5	7.5	5.1	3.8	14.2	16.5	3.0	4.7	2.0
					B	10.2	5.0	17.1	3.3	1.2	3.1	1.5	2.2	4.5	11.3	3.6	5.1	5.4	8.2	23.2	3.3	2.6	2.4	2.7
		ENVI-Carb+PSA	GC-MS	Green tea	A	19.2	28.5	9.7	19.5	9.8	6.6	34.4	20.0	35.2	12.4	2.7	35.0	6.1	18.8	11.6	24.8	5.8	7.6	1.7
					B	25.3	3.2	6.9	14.7	6.7	6.8	19.7	2.7	42.0	22.9	32.1	28.1	43.3	5.1	17.8	12.7	2.6	37.4	10.5
				Oolong tea	A	14.5	6.0	13.1	4.7	8.4	4.7	2.1	20.2	5.6	3.2	14.4	14.2	3.9	12.2	8.3	4.7	4.6	12.4	5.1
					B	20.1	5.4	11.2	5.5	9.0	5.8	5.2	6.6	14.0	16.2	9.0	4.1	8.3	9.9	8.1	7.2	11.2	19.8	7.1
			GC-MS/MS	Green tea	A	11.9	6.0	13.2	4.4	3.9	21.7	14.9	16.3	24.1	11.9	11.1	0.8	3.2	8.1	21.6	10.3	22.2	7.3	21.2
					B	24.9	12.2	9.7	14.5	9.4	9.2	15.7	34.6	34.8	7.2	7.8	3.3	1.0	7.0	13.7	2.0	10.4	7.1	26.6
				Oolong tea	A	8.4	5.2	17.2	8.0	7.2	5.3	7.1	18.6	4.3	9.8	5.6	8.7	8.3	3.9	14.3	15.7	5.4	3.0	10.1
					B	16.0	6.7	15.3	8.2	10.8	4.2	5.4	6.6	27.0	11.9	2.6	4.4	3.0	10.1	14.0	5.3	2.5	4.7	3.0

(Continued)

Appendix Table 5.2 RSD of 201 Pesticides in Youden Pair Aged Tea Sample for Parallel Determinations (Nov 9, 2009–Feb 7, 2010) (cont.)

No.	Pesticides	SPE	Method	Sample	Youden pair	RSD% Nov 9, 2009 (n=5)	Nov 14, 2009 (n=3)	Nov 19, 2009 (n=3)	Nov 24, 2009 (n=3)	Nov 29, 2009 (n=3)	Dec 4, 2009 (n=3)	Dec 9, 2009 (n=3)	Dec 14, 2009 (n=3)	Dec 19, 2009 (n=3)	Dec 24, 2009 (n=3)	Dec 14, 2009 (n=3)	Jan 3, 2010 (n=3)	Jan 8, 2010 (n=3)	Jan 13, 2010 (n=3)	Jan 18, 2010 (n=3)	Jan 23, 2010 (n=3)	Jan 28, 2010 (n=3)	Feb 2, 2010 (n=3)	Feb 7, 2010 (n=3)
38	Lenacil	TPT	GC-MS	Green tea	A	5.8	6.9	2.4	17.4	18.4	4.4	18.2	23.6	24.1	6.1	5.5	10.0	12.4	14.8	8.2	14.2	12.9	30.8	2.1
				Green tea	B	6.9	6.3	3.2	15.9	8.9	7.6	5.8	17.4	7.5	5.6	2.1	10.3	12.8	13.3	8.3	16.5	29.8	22.5	6.4
				Oolong tea	A	7.0	9.2	6.4	3.9	26.0	9.6	35.1	7.3	4.7	4.9	16.1	19.0	4.8	4.8	4.3	14.9	6.4	4.7	6.8
				Oolong tea	B	10.1	7.2	3.9	3.4	12.9	24.9	2.0	23.0	5.1	5.2	14.2	10.4	4.4	8.6	15.1	4.7	3.6	1.3	26.6
			GC-MS/MS	Green tea	A	11.2	7.8	0.4	7.3	13.8	8.5	21.1	35.7	22.6	4.1	14.2	10.5	18.0	8.9	4.3	23.6	12.9	29.0	14.2
				Green tea	B	10.4	5.3	3.3	12.0	25.8	13.0	7.6	21.1	6.0	8.9	2.5	17.0	20.9	4.4	7.2	4.3	6.5	16.2	20.9
				Oolong tea	A	18.5	32.5	4.2	2.9	8.9	29.1	35.8	17.1	36.2	4.4	28.2	4.7	7.6	2.1	7.8	15.0	0.2	24.3	7.1
				Oolong tea	B	36.6	29.8	14.4	3.4	8.9	31.7	0.1	17.1	24.9	5.0	23.2	7.6	7.0	4.2	13.7	9.6	4.1	24.0	25.4
		ENVI-Carb+PSA	GC-MS	Green tea	A	21.7	7.7	18.6	11.9	11.8	5.6	44.6	10.2	10.5	8.6	11.2	24.8	5.0	27.2	8.0	41.7	25.4	28.8	6.6
				Green tea	B	17.8	7.1	5.8	2.1	11.5	3.6	12.6	40.5	25.6	6.5	42.6	19.5	50.5	10.4	41.4	3.8	5.7	34.8	18.4
				Oolong tea	A	8.7	3.8	5.5	7.4	18.4	11.3	11.9	41.1	6.5	4.4	5.9	10.8	15.3	3.7	3.0	29.2	10.9	10.6	6.6
				Oolong tea	B	10.2	12.0	7.7	7.8	15.2	22.7	23.3	12.9	12.5	1.5	5.3	8.2	4.2	12.9	18.6	2.6	20.2	17.2	7.5
			GC-MS/MS	Green tea	A	28.4	6.9	14.3	10.1	6.4	1.8	5.5	5.2	13.7	4.4	37.8	4.4	1.1	9.6	24.1	13.3	12.6	9.3	1.5
				Green tea	B	17.4	16.7	4.3	9.9	6.3	1.3	5.2	6.4	1.9	8.7	26.7	10.3	16.1	17.2	31.1	2.8	9.2	11.8	1.7
				Oolong tea	A	22.8	1.9	5.4	5.2	9.0	9.8	4.2	8.1	0.9	3.1	5.9	12.0	6.1	3.8	4.4	35.9	11.3	9.6	12.7
				Oolong tea	B	29.5	5.9	10.8	5.9	19.9	7.8	16.5	8.1	12.2	7.2	3.5	4.7	13.0	6.5	12.6	3.9	9.0	21.3	5.3
39	Metalaxyl	TPT	GC-MS	Green tea	A	7.1	13.1	2.3	2.4	6.2	7.8	6.8	8.1	6.0	10.8	5.9	6.0	9.2	7.4	1.1	3.9	4.5	4.3	4.1
				Green tea	B	4.2	9.4	4.3	7.5	6.0	3.6	2.7	11.0	6.2	10.6	6.5	13.5	9.5	11.0	6.5	3.8	5.2	1.0	5.1
				Oolong tea	A	3.8	3.9	7.6	6.3	1.8	7.9	10.3	4.2	3.1	9.6	10.0	3.6	3.7	3.2	8.4	3.3	4.7	2.6	4.2
				Oolong tea	B	7.8	4.0	2.7	8.3	2.2	4.4	2.4	6.4	9.3	10.8	7.5	1.1	2.1	5.6	3.1	0.3	0.7	1.7	3.2
			GC-MS/MS	Green tea	A	5.8	8.3	2.7	4.5	5.0	6.7	8.5	6.1	4.5	3.8	6.2	4.6	7.8	10.9	3.7	4.8	5.4	4.5	3.0
				Green tea	B	4.0	10.0	5.6	6.9	5.2	6.1	4.8	7.7	5.5	11.1	8.7	10.1	3.2	3.4	5.4	5.7	6.6	3.3	7.7
				Oolong tea	A	4.6	4.5	0.6	8.4	2.6	4.4	8.5	0.7	3.4	7.0	2.9	1.3	3.7	2.4	9.5	4.8	2.6	4.5	7.8
				Oolong tea	B	6.8	0.8	5.5	5.7	1.4	0.7	4.5	5.0	3.0	2.1	5.6	3.2	8.1	4.0	4.5	3.1	2.2	5.0	6.1
		ENVI-Carb+PSA	GC-MS	Green tea	A	9.0	9.4	8.1	4.9	7.4	8.4	4.5	3.9	4.6	5.6	12.1	2.2	6.8	5.5	4.8	11.0	9.7	6.2	7.1
				Green tea	B	4.6	3.3	6.0	6.1	5.9	3.7	1.1	11.3	11.9	11.9	23.1	7.1	6.6	4.0	13.0	1.2	3.9	13.5	7.6
				Oolong tea	A	6.7	4.2	3.3	5.1	3.5	4.9	1.7	3.7	22.6	10.1	4.5	11.0	3.9	2.5	5.2	3.4	5.6	3.0	3.2
				Oolong tea	B	5.9	2.6	14.2	5.2	12.3	8.2	2.4	6.4	7.1	7.1	11.8	3.3	5.6	13.6	10.7	4.2	4.0	3.3	9.1
			GC-MS/MS	Green tea	A	7.4	11.6	3.2	5.9	7.2	1.6	20.9	2.9	3.2	1.6	12.7	0.8	6.1	2.3	6.9	8.4	6.9	4.0	5.8
				Green tea	B	3.7	5.2	3.4	4.0	2.8	2.9	3.5	1.2	5.5	2.6	20.7	4.0	1.2	13.4	13.4	1.9	1.6	12.9	5.4
				Oolong tea	A	3.7	3.7	3.3	5.9	5.1	5.0	6.5	3.8	4.6	12.4	7.1	11.8	7.0	4.1	13.0	6.2	9.6	5.1	3.3
				Oolong tea	B	6.5	3.2	16.3	4.3	10.7	8.3	3.4	5.7	10.5	3.1	0.8	4.5	3.5	16.5	10.7	3.2	7.2	0.2	7.0

No.	Compound	Cleanup	Method	Tea	Type	1	2	3	4	5	6	7	8	9	10	11	12	13	14	15	16	17	18	19
40	Metazachlor	TPT	GC-MS	Green tea	A	7.3	7.8	6.1	3.3	6.4	4.8	6.4	8.3	5.8	2.6	9.5	6.0	8.0	3.1	8.5	6.1	5.8	10.0	1.5
					B	8.2	10.4	9.6	5.6	6.0	7.1	1.6	6.2	3.2	10.7	7.1	7.5	3.2	7.9	9.0	1.7	10.3	8.0	5.3
				Oolong tea	A	8.3	6.6	6.2	3.6	9.3	2.1	15.9	4.5	4.1	7.6	9.3	2.4	5.9	3.0	10.3	3.5	12.0	4.0	3.0
					B	7.3	2.9	5.6	5.0	6.1	6.7	7.7	5.3	6.2	7.6	10.0	1.1	6.5	4.2	9.2	7.0	2.5	1.5	8.2
			GC-MS/MS	Green tea	A	7.2	7.9	4.4	5.9	4.2	1.9	2.9	10.1	3.6	3.8	4.6	4.0	10.2	7.1	3.5	2.5	6.2	3.9	6.4
					B	5.8	9.6	4.3	7.3	4.5	4.9	3.5	8.4	7.1	7.8	4.6	3.2	4.0	6.3	5.2	2.2	7.0	4.4	5.4
				Oolong tea	A	7.9	4.6	4.1	5.5	9.7	6.1	14.8	5.2	11.1	3.8	4.1	2.6	9.6	5.3	4.7	3.5	1.9	1.4	5.9
					B	6.8	6.8	8.2	5.0	3.1	1.5	8.2	6.5	8.6	5.6	4.5	6.4	6.4	4.6	7.3	3.5	3.5	2.3	2.8
		ENVI-Carb+PSA	GC-MS	Green tea	A	7.2	6.2	5.2	4.9	6.3	7.9	13.7	6.5	4.8	1.2	6.2	8.8	2.8	5.0	5.2	9.1	5.9	20.9	6.0
					B	9.7	8.8	2.1	3.6	4.4	7.3	13.7	6.6	7.3	12.9	25.5	7.5	12.6	4.4	12.7	4.1	5.2	24.4	8.7
				Oolong tea	A	7.2	7.4	4.6	7.6	1.7	5.4	3.2	4.5	7.6	6.4	9.5	5.1	4.6	5.4	5.2	5.4	6.5	4.7	5.5
					B	8.1	7.8	12.8	8.8	8.3	4.6	4.8	13.2	10.3	8.1	8.8	5.9	6.5	7.7	16.1	2.8	11.9	5.7	1.6
			GC-MS/MS	Green tea	A	12.7	10.7	2.6	4.4	5.1	6.4	14.1	2.6	8.1	2.3	23.0	13.9	9.1	2.5	0.8	4.1	8.6	8.7	4.4
					B	41.6	11.5	5.6	2.5	5.7	6.2	4.9	5.2	5.7	6.5	12.6	9.0	11.2	13.5	6.6	6.0	0.8	13.0	10.2
				Oolong tea	A	3.1	1.7	6.2	8.0	7.7	6.7	5.0	2.8	6.8	9.6	4.4	12.2	0.9	5.2	13.7	3.3	4.5	6.2	4.4
					B	10.5	2.1	15.2	8.3	7.0	7.9	1.5	6.0	13.3	2.0	1.9	1.0	3.5	11.7	13.1	4.2	5.1	5.8	6.0
41	Methabenzthiazuron	TPT	GC-MS	Green tea	A	5.9	7.0	3.0	2.6	3.5	6.4	7.9	5.8	4.6	1.6	1.2	3.5	7.9	4.1	1.5	3.2	7.1	5.5	4.5
					B	5.2	6.9	6.7	8.5	2.9	6.9	2.0	5.8	5.5	12.2	6.1	7.9	2.4	1.5	13.1	3.7	12.1	3.5	4.6
				Oolong tea	A	10.2	2.2	3.1	2.8	4.9	2.2	8.0	7.4	2.0	7.7	10.3	2.4	5.6	3.9	4.5	2.5	0.8	3.1	1.2
					B	11.3	1.7	4.0	3.9	5.6	4.0	2.2	3.3	6.0	2.3	6.3	5.6	4.3	5.2	3.8	1.8	3.5	4.7	5.4
			GC-MS/MS	Green tea	A	1.5	5.0	2.4	3.6	2.0	4.7	9.0	4.3	4.7	3.3	3.3	4.3	5.1	3.6	2.2	2.9	5.1	5.3	6.8
					B	4.8	10.3	6.3	5.0	4.5	11.4	3.4	8.6	5.3	11.0	5.8	5.1	3.3	5.3	5.1	2.5	9.6	2.0	6.4
				Oolong tea	A	6.1	0.5	0.8	7.0	4.4	2.9	10.9	4.1	1.9	4.4	4.5	1.0	4.5	3.0	8.5	1.5	1.2	3.5	6.5
					B	2.0	5.8	5.6	5.0	2.2	1.6	4.2	5.1	5.6	4.3	7.3	4.4	6.0	3.3	6.9	4.7	3.4	3.0	3.7
		ENVI-Carb+PSA	GC-MS	Green tea	A	7.1	4.1	5.6	9.0	9.5	6.8	11.1	2.1	4.6	8.3	10.0	6.8	2.4	8.1	8.3	8.1	13.8	1.2	6.6
					B	4.5	1.9	8.3	6.8	4.0	4.9	7.7	2.9	6.1	11.2	31.2	4.4	9.5	4.7	19.0	6.9	4.0	8.2	8.1
				Oolong tea	A	3.8	6.3	6.3	3.4	4.9	3.3	5.0	8.5	19.2	6.2	11.7	11.6	2.4	8.0	8.3	4.4	6.3	7.6	4.3
					B	5.1	3.2	11.1	8.2	13.5	10.5	4.9	7.0	8.8	1.8	8.4	2.3	4.4	9.2	6.7	5.0	5.1	4.8	6.2
			GC-MS/MS	Green tea	A	5.6	9.2	5.0	8.0	7.7	3.0	23.8	3.0	3.7	3.3	11.8	8.5	4.2	11.0	5.8	11.7	9.7	8.2	8.2
					B	3.3	7.4	4.1	8.4	2.9	1.5	4.4	1.5	6.4	2.2	27.5	9.9	7.7	11.5	17.2	2.5	5.4	16.5	7.5
				Oolong tea	A	5.2	3.4	3.8	5.8	4.7	1.4	5.9	5.4	2.7	5.3	6.6	1.1	3.2	4.5	9.8	4.6	8.8	15.5	7.0
					B	5.1	2.9	11.3	3.3	10.0	7.7	4.3	6.0	11.0	3.3	1.1	4.7	7.5	10.5	12.0	1.4	3.9	2.8	5.8
42	Mirex	TPT	GC-MS	Green tea	A	3.7	4.8	4.8	7.0	15.0	4.9	12.8	10.1	6.8	2.7	2.9	17.4	10.5	8.1	1.1	5.1	9.6	14.9	3.7
					B	6.3	7.5	2.2	8.6	2.7	1.5	4.3	10.4	6.0	9.5	4.7	10.0	10.6	11.2	4.3	8.3	16.1	12.4	6.3
				Oolong tea	A	4.9	2.4	2.1	2.2	9.2	4.0	4.1	7.6	5.2	4.1	7.0	7.7	2.9	1.3	7.5	8.6	5.9	4.1	2.8
					B	7.6	0.5	6.2	3.7	6.7	9.3	6.9	12.3	7.1	5.9	3.4	5.8	2.2	7.3	10.4	4.2	2.4	1.9	9.8
			GC-MS/MS	Green tea	A	5.7	5.1	4.5	3.1	9.1	2.6	10.8	9.5	9.8	2.6	6.4	14.4	9.6	4.9	1.2	5.0	6.1	11.5	2.9
					B	5.9	7.6	3.5	8.5	2.1	1.4	2.9	5.9	5.0	10.4	2.1	10.1	10.8	3.2	1.8	6.5	10.2	3.8	7.8
				Oolong tea	A	8.4	6.3	3.5	3.7	1.1	6.6	3.6	2.9	12.9	3.1	9.6	10.6	11.9	3.4	12.1	5.4	2.3	8.5	6.9
					B	12.3	4.8	6.1	4.5	0.6	4.4	7.6	7.4	9.8	3.4	6.2	3.4	3.4	5.0	7.7	1.0	2.8	6.1	8.1
		ENVI-Carb+PSA	GC-MS	Green tea	A	18.4	2.1	9.6	6.3	8.0	15.4	18.3	13.4	11.4	4.6	11.8	4.2	4.2	14.3	4.7	16.4	9.6	11.8	9.4
					B	21.0	8.6	3.8	2.0	3.5	4.3	29.4	14.1	11.8	6.1	20.8	6.1	23.2	5.3	15.4	7.0	6.2	20.9	13.4
				Oolong tea	A	3.9	2.4	3.9	4.2	2.9	7.2	5.4	6.2	8.0	4.3	2.9	4.3	10.1	5.2	3.4	18.2	7.2	11.8	6.1
					B	7.0	5.6	7.0	2.3	14.8	10.3	14.9	10.5	14.8	5.0	11.9	5.0	5.4	16.4	16.8	5.8	14.2	16.9	8.8
			GC-MS/MS	Green tea	A	19.3	5.7	8.3	3.8	9.9	16.0	14.6	11.4	5.6	2.3	17.4	2.3	5.7	9.5	0.2	9.1	11.9	11.3	3.2
					B	27.5	13.4	2.6	4.4	7.5	2.8	11.8	15.2	8.3	5.6	16.6	5.6	29.2	10.2	9.1	9.9	6.9	19.7	5.6
				Oolong tea	A	6.2	2.8	6.7	9.7	7.5	2.2	3.8	3.4	6.0	2.5	4.8	2.5	9.7	4.1	9.1	10.6	8.0	16.0	7.3
					B	11.3	4.0	10.5	3.9	10.1	8.3	6.5	8.6	16.8	9.1	8.6	9.1	8.6	12.9	13.4	3.0	12.7	22.2	

(Continued)

Appendix Table 5.2 RSD of 201 Pesticides in Youden Pair Aged Tea Sample for Parallel Determinations (Nov 9, 2009–Feb 7, 2010) (cont.)

RSD%

No.	Pesticides	SPE	Method	Sample	Youden pair	Nov 9, 2009 (n=5)	Nov 14, 2009 (n=3)	Nov 19, 2009 (n=3)	Nov 24, 2009 (n=3)	Nov 29, 2009 (n=3)	Dec 4, 2009 (n=3)	Dec 9, 2009 (n=3)	Dec 14, 2009 (n=3)	Dec 19, 2009 (n=3)	Dec 24, 2009 (n=3)	Dec 29, 2009 (n=3)	Jan 3, 2010 (n=3)	Jan 8, 2010 (n=3)	Jan 13, 2010 (n=3)	Jan 18, 2010 (n=3)	Jan 23, 2010 (n=3)	Jan 28, 2010 (n=3)	Feb 2, 2010 (n=3)	Feb 7, 2010 (n=3)
43	Monalide	TPT	GC-MS	Green tea	A	1.8	6.8	3.3	6.1	12.8	1.5	5.5	4.9	10.8	3.9	7.8	10.8	4.4	10.3	5.7	2.8	7.6	10.6	5.9
					B	6.8	8.2	3.2	7.5	8.0	4.7	5.4	5.7	7.2	8.3	5.5	0.7	11.5	14.3	5.5	5.4	12.6	3.4	8.4
				Oolong tea	A	11.3	3.5	9.6	5.3	6.8	2.5	2.0	8.3	15.1	10.2	14.3	2.7	4.2	2.7	6.7	11.8	1.5	23.3	10.5
					B	11.9	5.8	18.7	15.7	3.5	6.4	7.8	0.4	10.5	10.3	3.8	6.9	8.0	5.9	16.3	5.3	9.4	11.5	18.0
			GC-MS/MS	Green tea	A	5.6	5.7	1.4	4.8	7.0	2.2	7.2	4.2	5.3	3.4	6.5	4.9	8.8	5.1	1.9	0.5	4.4	12.8	5.7
					B	4.6	8.0	5.8	6.7	3.7	3.6	5.2	4.0	5.8	8.8	7.7	6.5	1.1	3.9	3.3	5.8	11.5	3.8	10.2
				Oolong tea	A	5.7	3.7	3.0	6.8	3.0	1.8	10.7	0.6	0.1	2.4	2.7	1.0	11.5	2.4	7.2	3.7	7.1	3.0	6.1
					B	6.0	2.1	5.0	5.4	0.8	3.2	5.2	1.7	4.7	1.2	3.7	1.6	4.1	1.5	6.0	0.8	1.7	1.7	3.2
		ENVI-Carb+PSA	GC-MS	Green tea	A	8.8	6.8	1.8	1.6	5.1	18.5	6.6	3.9	13.3	2.6	9.6	18.9	3.7	1.1	2.6	9.0	10.9	4.5	11.8
					B	10.0	12.5	5.7	6.9	6.8	7.9	6.1	5.0	6.4	6.1	14.4	12.2	29.6	3.6	2.2	7.4	5.1	8.0	9.9
				Oolong tea	A	4.7	5.0	5.3	15.2	1.9	4.8	7.4	15.4	6.4	3.3	7.9	14.9	9.9	5.9	10.7	26.8	6.0	7.1	8.4
					B	10.8	14.9	24.4	10.8	8.9	4.5	7.2	6.7	5.3	6.2	1.5	5.2	6.0	4.2	13.9	6.4	8.9	6.8	8.2
			GC-MS/MS	Green tea	A	5.7	5.1	1.3	6.2	5.3	2.3	2.3	1.1	18.3	2.2	11.0	2.9	6.6	2.7	5.0	6.9	9.0	7.7	4.7
					B	4.0	8.8	3.7	4.6	2.8	3.1	2.4	5.8	4.6	2.8	23.1	1.2	3.3	4.8	12.2	2.8	2.9	13.8	8.0
				Oolong tea	A	3.6	4.9	5.2	6.1	7.8	3.8	5.0	3.5	4.3	3.6	6.5	11.5	6.1	4.0	10.3	3.9	6.5	3.0	4.2
					B	5.7	1.8	9.5	5.8	7.7	5.0	2.5	6.0	5.7	2.5	2.5	7.6	6.6	10.0	9.2	1.5	4.3	3.8	4.1
44	Paraoxon-ethyl	TPT	GC-MS	Green tea	A	12.4	2.4	3.1	4.3	7.2	3.7	5.1	4.8	2.8	2.6	0.3	1.4	2.1	7.8	2.8	2.1	10.2	11.8	4.6
					B	7.8	0.7	8.9	1.0	4.5	2.9	2.3	3.4	3.1	6.2	7.0	3.5	5.4	10.6	0.3	5.7	3.2	5.1	13.5
				Oolong tea	A	4.9	7.6	4.1	8.9	3.6	9.3	4.1	6.7	9.0	9.2	4.9	8.3	6.0	4.9	7.4	8.9	5.3	7.2	5.8
					B	4.4	4.7	4.1	6.9	6.9	3.9	4.5	1.9	2.4	3.1	1.6	13.0	7.5	11.0	12.6	3.7	1.3	13.3	11.7
			GC-MS/MS	Green tea	A	8.1	11.6	9.0	20.9	10.4	11.3	4.9	11.8	16.1	15.7	9.9	28.5	2.6	26.1	9.1	3.7	23.1	19.6	
					B	10.1	9.9	6.9	10.7	9.7	9.6	6.9	14.1	17.2	16.6	3.0	21.3	2.6	16.4	5.2	7.2	24.4	27.4	
				Oolong tea	A	2.3	13.1	9.0	12.4	6.2	7.0	3.7	6.8		2.2	4.2	1.9	16.6	3.6	2.4				#DIV/0!
					B	16.4	15.7	11.5	3.2	5.5	9.4	10.0	3.6		9.4	19.7	5.6	8.5	3.3	15.5				#DIV/0!
		ENVI-Carb+PSA	GC-MS	Green tea	A	14.1	5.0	3.3	8.9	3.9	2.0		7.7	3.3	4.4	5.5	9.4	9.6	4.8	2.1	6.9	6.9	9.4	4.5
					B	21.5	2.1	0.7	4.5	8.0	0.8		10.3	1.3	5.2	6.5	7.4	10.2	1.8	2.6	1.9	6.3	10.3	6.8
				Oolong tea	A	6.2	6.3	9.3	7.6	17.1	2.1	2.3	3.6	7.4	9.2	3.4	6.9	12.2	7.6	8.7	12.1	17.3	8.5	8.4
					B	8.8	12.8	2.5	13.2	1.6	1.0	9.7	8.1	4.2	3.2	7.3	5.7	11.7	2.9	16.9	3.0	18.5	4.4	3.7
			GC-MS/MS	Green tea	A	25.8	21.4	19.4	5.9	10.3	9.5	38.3	17.9	13.6	7.4	17.8	4.3	30.1	19.4	21.4	10.3	20.5	12.4	4.0
					B	29.2	20.8	7.5	9.0	38.2	20.6	11.1	23.9	16.9	8.3	24.6	3.2	26.6	13.8	9.7	14.7	20.4	12.6	
				Oolong tea	A	15.4	7.8	7.5	18.4	28.7	9.8	3.6	8.9	3.1	21.9	12.6	11.1	25.9	4.7	14.0	32.6	#DIV/0!	#DIV/0!	#DIV/0!
					B	20.8	9.5	15.7	15.2	15.8	14.9	4.2	12.4	15.1	14.2	12.0	11.4	20.9	17.2	8.8	18.2	#DIV/0!	#DIV/0!	#DIV/0!
45	Pebulate	TPT	GC-MS	Green tea	A	3.5	4.3	2.6	2.5	9.6	9.0	4.4	2.4	7.1	5.4	6.6	2.2	5.1	3.9	2.8	4.1	3.1	10.5	4.8
					B	7.9	7.3	4.2	6.5	2.3	5.3	4.0	3.3	4.7	10.9	2.7	8.8	8.9	4.9	5.7	2.9	7.3	6.9	4.5
				Oolong tea	A	3.8	2.1	1.0	2.3	3.4	0.6	9.1	5.2	6.8	2.9	4.2	3.7	4.9	2.7	5.8	11.1	14.5	8.5	4.4
					B	8.8	4.4	5.3	8.7	3.8	5.3	6.0	0.3	5.7	4.2	2.7	1.1	4.6	2.1	3.6	11.9	9.3	13.3	3.5
			GC-MS/MS	Green tea	A							5.5	2.1	6.0	5.4	8.0	4.1	5.3	5.7	5.9	6.6	2.0	2.8	4.0
					B							6.3	3.7	2.9	13.6	5.0	9.0	2.1	5.1	11.1	4.8	12.0	12.4	7.1
				Oolong tea	A							11.4	0.6	5.0	6.4	4.3	7.5	21.7	2.3	10.8	13.8	19.7	12.6	8.1
					B							5.4	2.8	3.7	28.0	3.8	0.5	16.4	7.2	10.0	11.6	10.4	23.2	8.4
		ENVI-Carb+PSA	GC-MS	Green tea	A	5.6	4.7	2.1	5.9	6.3	3.6	3.6	0.4	8.8	2.5	8.5	3.7	4.1	19.9	3.4	19.2	17.0	15.4	14.4
					B	3.5	3.4	4.4	3.9	2.2	2.0	8.7	4.1	2.6	3.1	23.7	2.2	8.2	4.9	31.2	16.9	8.2	8.7	2.5
				Oolong tea	A	4.4	3.5	3.2	5.3	4.2	1.5	3.0	1.7	9.7	0.2	4.5	2.9	4.6	13.4	5.6	10.8	6.2	11.2	9.1
					B	7.8	5.2	5.3	5.0	9.8	3.5	1.0	7.4	10.0	2.0	8.9	4.3	12.8	16.5	12.1	7.6	8.6	12.3	19.0
			GC-MS/MS	Green tea	A							6.6	0.4	8.5	9.8	9.8	0.7	2.4	43.5	2.7	21.7	22.0	10.2	6.2
					B							7.7	1.3	3.0	24.9	9.8	1.7	24.5	20.6	30.1	31.6	4.6	14.0	0.9
				Oolong tea	A							1.1	0.9	1.9	2.0	6.5	5.8	41.0	18.8	18.4	23.6	11.0	3.7	12.8
					B							3.4	8.0	14.3	1.7	1.7	4.6	5.1	30.0	17.3	10.3	10.8	1.4	0.8

46	Pentachloroaniline	TPT	GC–MS	Green tea	A	4.0	5.1	2.4	1.5	8.6	3.8	6.7	3.6	4.3	2.5	2.1	3.3	4.6	4.8	1.0	0.5	6.0	6.9	5.1
					B	24.0	9.0	5.4	7.4	2.2	4.5	2.0	2.4	3.0	9.7	2.8	11.8	3.6	1.6	3.5	3.1	7.2	1.7	5.5
				Oolong tea	A	4.5	1.4	0.2	3.1	2.0	2.1	21.4	13.3	3.4	4.1	4.1	2.2	4.1	2.8	7.9	2.9	3.8	1.3	1.5
					B	6.9	2.5	5.2	3.5	2.3	3.7	9.5	5.2	5.4	5.1	3.1	0.4	2.8	2.8	4.1	0.5	1.0	1.4	1.6
			GC–MS/MS	Green tea	A	3.7	5.8	5.1	0.9	5.4	2.9	8.4	3.8	5.1	4.1	4.4	3.8	4.1	7.2	6.2	2.6	6.5	3.1	11.8
					B	4.8	9.1	3.6	8.2	4.3	1.6	4.3	4.0	7.0	10.6	5.4	13.0	5.1	4.1	3.8	1.9	5.8	5.4	3.7
				Oolong tea	A	2.6	0.7	3.3	5.7	1.5	1.4	9.8	1.4	1.9	3.6	2.1	2.3	3.0	2.1	5.3	2.0	8.0	3.7	3.9
					B	4.3	1.8	4.9	4.2	1.1	2.2	4.6	4.8	4.6	1.8	3.4	3.0	0.9	3.8	6.9	3.1	4.6	5.9	6.1
		ENVI-Carb+PSA	GC–MS	Green tea	A	7.9	1.1	3.5	5.3	9.5	3.9	6.6	0.7	4.7	1.1	13.1	3.1	4.6	4.8	6.3	4.2	13.5	10.0	5.2
					B	5.3	5.2	2.8	4.2	0.7	3.6	3.2	2.4	3.1	3.8	18.8	6.2	8.0	3.2	13.5	6.7	4.9	8.5	8.4
				Oolong tea	A	4.2	3.0	3.3	4.3	5.0	0.2	7.0	14.6	0.8	0.4	3.8	3.9	0.5	3.2	2.6	3.0	5.2	4.9	4.8
					B	5.4	3.9	7.6	4.2	8.8	4.0	11.6	13.7	9.8	7.0	8.3	4.3	2.3	6.9	7.9	3.7	6.7	3.1	3.5
			GC–MS/MS	Green tea	A	5.4	5.1	2.8	6.5	6.1	2.7	1.0	2.0	5.3	3.1	9.4	3.0	4.2	12.7	11.1	5.2	17.2	6.4	3.6
					B	5.6	7.3	4.3	6.2	0.7	1.2	1.7	3.7	4.4	2.7	22.0	0.5	9.3	2.7	10.4	11.3	5.0	9.4	10.1
				Oolong tea	A	4.1	6.9	6.9	4.9	5.4	0.3	3.7	3.7	1.5	2.1	3.6	6.6	4.5	5.2	8.1	4.6	8.9	4.6	7.5
					B	5.5	2.0	8.7	4.2	8.9	2.5	2.4	3.4	11.0	2.7	2.2	2.5	3.7	10.0	5.5	1.3	1.8	1.4	4.5
47	Pentachloroanisole	TPT	GC–MS	Green tea	A	5.3	5.3	2.8	1.6	8.7	6.1	6.2	4.0	6.9	3.1	4.8	3.1	6.2	4.2	2.7	1.0	2.5	6.8	4.5
					B	6.5	8.5	3.7	8.6	2.4	5.3	3.5	0.8	5.4	9.8	2.1	8.1	2.5	3.4	4.8	2.0	3.4	1.4	5.7
				Oolong tea	A	4.1	1.8	0.3	2.0	2.1	0.3	9.0	4.6	6.6	3.1	3.3	3.4	4.7	2.8	7.7	3.2	5.1	1.0	3.6
					B	8.0	5.0	6.6	6.7	3.5	5.1	5.8	1.6	6.9	6.2	2.5	0.4	4.1	3.2	4.5	1.1	2.4	1.3	3.0
			GC–MS/MS	Green tea	A	4.1	7.3	3.4	3.0	10.1	7.8	6.4	2.4	4.2	3.3	5.6	0.6	5.3	9.0	1.3	2.0	5.5	7.4	5.2
					B	5.7	8.9	4.2	5.4	6.4	5.1	4.6	4.7	1.3	10.8	6.6	6.3	1.5	4.6	0.9	2.7	7.4	3.4	6.5
				Oolong tea	A	5.3	1.5	1.6	4.2	2.2	1.7	9.8	1.8	5.7	5.0	3.6	5.8	9.0	4.7	9.0	5.2	3.0	3.7	2.9
					B	6.1	6.8	5.5	5.2	2.2	5.3	4.9	0.2	6.0	3.7	3.6	1.0	2.1	1.9	9.6	2.6	2.6	2.9	2.8
		ENVI-Carb+PSA	GC–MS	Green tea	A	6.7	5.1	1.3	8.1	6.8	2.5	7.8	1.4	9.9	1.2	9.2	3.4	3.6	3.0	9.6	8.0	11.3	4.6	4.9
					B	5.0	7.3	3.8	3.8	1.9	1.0	2.0	4.3	2.3	2.6	19.6	0.9	3.9	2.3	14.3	4.1	5.8	8.6	5.4
				Oolong tea	A	4.0	2.7	2.6	5.3	4.8	1.9	1.7	2.8	1.1	1.2	5.7	3.9	1.0	3.2	2.8	5.2	5.9	4.8	7.1
					B	7.4	4.6	5.3	5.4	8.7	2.5	3.5	8.0	11.7	3.0	7.6	4.9	2.7	8.7	6.6	3.6	6.3	3.9	4.0
			GC–MS/MS	Green tea	A	6.3	5.5	1.9	6.1	6.1	4.9	5.4	1.9	4.9	0.7	9.1	3.7	2.1	11.7	8.1	8.0	10.3	6.2	4.8
					B	2.5	12.2	4.9	3.0	4.3	2.9	1.3	2.6	3.4	3.6	19.2	5.7	6.6	8.3	16.3	0.9	1.5	13.7	8.0
				Oolong tea	A	4.1	3.2	5.0	7.2	6.8	1.6	4.4	2.1	2.6	1.6	4.5	5.4	5.0	3.7	7.7	5.8	6.5	5.7	6.0
					B	6.6	3.6	6.2	4.2	7.8	3.1	3.3	5.5	12.8	2.0	5.1	2.5	2.3	11.8	8.2	2.7	5.9	2.2	2.3
48	Pentachlorobenzene	TPT	GC–MS	Green tea	A	3.9	3.6	2.4	1.7	8.5	7.8	4.1	1.3	5.9	4.2	7.0	1.6	5.5	1.8	2.0	2.7	3.4	7.0	3.4
					B	7.0	8.6	4.6	7.8	2.6	5.4	4.1	2.7	5.3	12.2	1.7	6.8	5.1	1.2	5.5	3.5	6.0	4.5	4.0
				Oolong tea	A	7.4	1.9	0.5	2.8	2.7	5.7	5.2	4.6	2.7	4.6	4.3	2.3	6.2	4.7	7.4	2.1	2.9	2.5	0.8
					B	9.3	2.3	4.3	2.0	1.1	3.0	7.4	4.2	7.6	2.2	2.3	0.7	1.6	5.1	3.8	0.7	1.8	1.0	3.3
		ENVI-Carb+PSA	GC–MS	Green tea	A	3.0	3.8	2.2	1.1	8.0	7.6	4.3	0.8	5.9	5.9	8.0	4.4	4.8	7.9	0.5	5.6	4.1	6.3	2.6
					B	7.5	7.9	4.7	7.3	6.7	5.5	5.8	2.2	5.4	13.4	3.9	7.4	4.8	7.2	4.8	4.0	6.7	6.3	6.5
				Oolong tea	A	5.9	0.6	2.6	5.2	2.0	3.0	8.9	2.3	2.8	4.4	5.4	2.2	9.8	3.0	9.4	2.1	0.2	1.7	2.8
					B	2.6	1.5	5.8	5.8	0.8	3.0	3.7	5.8	4.7	2.2	4.4	1.0	8.0	2.4	4.9	1.3	4.1	1.2	3.5
			GC–MS/MS	Green tea	A	5.1	5.9	1.1	6.5	6.2	3.6	5.4	1.6	8.8	1.3	10.1	5.1	3.7	2.5	5.8	9.5	13.0	4.1	4.4
					B	3.0	4.5	4.4	4.3	2.0	7.9	5.3	4.5	2.1	2.1	22.0	1.6	8.5	3.0	23.3	4.7	7.2	7.9	4.6
				Oolong tea	A	3.4	1.8	2.7	2.9	4.6	1.5	3.5	6.2	1.9	2.4	4.7	2.7	2.4	2.6	2.7	1.7	6.1	5.3	3.5
					B	5.5	3.8	9.7	2.9	7.1	4.6	3.2	7.0	11.1	5.5	10.7	2.7	3.7	8.3	9.0	4.4	7.0	1.7	5.7
			GC–MS/MS	Green tea	A	5.2	6.7	1.5	5.1	3.8	2.6	5.2	1.4	9.2	0.7	10.5	5.0	4.7	4.2	7.0	11.2	11.8	6.4	1.0
					B	2.6	6.3	5.8	4.9	3.3	2.8	5.5	3.0	2.4	0.8	24.3	8.6	9.6	11.9	26.3	1.2	3.9	11.2	3.2
				Oolong tea	A	3.2	1.6	3.5	3.5	5.9	2.4	4.1	2.9	4.1	1.3	5.3	8.5	3.1	3.4	1.2	2.3	3.7	4.5	3.9
					B	6.7	2.7	10.6	3.0	8.1	6.0	3.3	6.8	12.3	3.2	1.7	5.0	1.9	9.0	10.0	1.6	6.3	1.0	5.9

(Continued)

Appendix Table 5.2 RSD of 201 Pesticides in Youden Pair Aged Tea Sample for Parallel Determinations (Nov 9, 2009–Feb 7, 2010) (cont.)

No.	Pesticides	SPE	Method	Sample	Youden pair	RSD% Nov 9, 2009 (n=5)	Nov 14, 2009 (n=3)	Nov 19, 2009 (n=3)	Nov 24, 2009 (n=3)	Nov 29, 2009 (n=3)	Dec 4, 2009 (n=3)	Dec 9, 2009 (n=3)	Dec 14, 2009 (n=3)	Dec 19, 2009 (n=3)	Dec 24, 2009 (n=3)	Dec 14, 2009 (n=3)	Jan 3, 2010 (n=3)	Jan 8, 2010 (n=3)	Jan 13, 2010 (n=3)	Jan 18, 2010 (n=3)	Jan 23, 2010 (n=3)	Jan 28, 2010 (n=3)	Feb 2, 2010 (n=3)	Feb 7, 2010 (n=3)
49	Perthane	TPT	GC-MS	Green tea	A	6.3	4.6	3.1	1.6	7.9	7.3	8.2	5.7	6.1	3.7	5.7	3.9	7.0	3.5	6.6	1.0	5.1	7.9	1.8
					B	6.3	7.6	4.3	7.9	4.7	6.7	6.3	3.1	5.2	8.8	3.9	6.5	2.7	7.1	3.6	4.2	4.2	2.4	9.0
				Oolong tea	A	4.0	0.6	1.5	3.3	1.3	6.5	6.1	4.4	4.0	6.6	4.3	3.3	3.2	6.1	4.7	4.0	3.5	2.3	2.8
					B	6.7	1.7	3.8	3.0	2.1	6.2	5.4	4.8	6.4	5.0	2.4	1.6	1.9	9.1	3.6	2.6	2.2	6.8	4.4
			GC-MS/MS	Green tea	A	3.8	4.6	2.6	2.3	7.4	4.3	8.0	6.4	4.8	2.3	5.7	2.0	9.3	3.4	1.2	2.0	4.1	4.6	4.6
					B	3.7	7.9	5.6	7.0	5.0	3.1	3.3	5.7	4.7	9.7	5.6	8.7	3.2	1.4	2.5	5.0	8.9	3.2	5.7
				Oolong tea	A	4.9	3.8	1.3	3.5	5.5	2.9	9.8	5.0	1.1	2.3	5.0	5.3	3.6	4.4	9.3	5.5	7.6	9.1	9.3
					B	3.2	8.1	7.6	8.5	1.8	3.7	2.4	1.2	7.8	3.6	2.8	3.2	4.4	3.1	7.1	1.4	6.8	2.7	1.2
		ENVI-Carb+PSA	GC-MS	Green tea	A	13.7	4.2	4.4	4.7	8.5	2.6	6.9	4.4	5.7	0.8	7.9	2.1	1.6	3.0	3.1	5.5	8.8	5.0	7.1
					B	6.1	9.5	3.0	3.4	0.5	4.7	7.2	6.7	4.1	2.7	19.7	2.7	4.0	3.5	10.6	7.9	4.3	9.2	8.2
				Oolong tea	A	1.5	4.8	2.9	3.2	2.3	0.8	2.2	1.6	2.7	1.0	6.8	6.3	0.5	2.9	1.0	4.1	8.1	6.7	5.7
					B	4.9	5.3	10.1	5.7	7.4	4.0	2.0	5.0	12.6	2.7	5.0	5.8	3.6	9.6	9.9	2.8	8.4	5.6	7.0
			GC-MS/MS	Green tea	A	10.1	5.2	4.2	6.2	7.2	2.9	8.4	4.0	5.2	2.1	10.6	3.1	7.2	3.5	5.1	4.1	6.8	5.4	3.9
					B	6.8	10.9	5.0	5.2	1.2	2.4	5.6	2.6	5.6	2.5	17.3	6.2	6.7	8.2	12.2	5.5	4.1	13.8	3.6
				Oolong tea	A	3.2	2.0	6.1	6.5	5.7	2.8	4.4	1.3	2.5	1.5	5.8	5.2	3.4	4.0	8.3	4.6	5.3	6.7	12.1
					B	5.9	1.1	11.1	2.5	7.2	6.4	1.5	5.1	13.3	3.5	5.7	1.1	4.1	7.2	9.8	0.9	3.8	2.3	3.5
50	Phenan-threne	TPT	GC-MS	Green tea	A	3.9	5.0	2.6	1.3	7.8	5.1	5.1	2.4	6.1	3.4	4.4	6.7	6.4	2.9	2.6	2.3	4.4	1.5	4.8
					B	6.4	8.5	3.8	7.7	4.3	6.1	2.3	2.5	3.6	9.1	3.4	2.7	2.3	4.1	3.9	1.3	3.6	0.9	5.7
			GC-MS/MS	Green tea	A	4.0	2.3	0.4	2.9	2.3	1.3	8.3	4.1	2.0	3.3	2.4	0.7	3.2	3.5	7.3	2.5	3.6	1.4	0.3
					B	6.2	2.3	5.0	4.2	1.9	3.7	4.3	3.0	6.6	3.5	0.4	9.3	3.3	11.3	3.4	1.1	0.7	0.9	1.7
				Oolong tea	A	3.8	7.4	1.6	1.8	3.8	1.2	1.8	6.2	7.8	4.5	4.2	16.7	4.2	4.4	5.9	13.0	13.1	14.7	7.1
					B	5.4	9.4	5.0	3.9	9.2	2.0	5.6	4.5	5.3	11.2	3.9	6.8	8.7	4.4	4.8	7.3	7.4	15.6	5.0
				Oolong tea	A	5.6	1.1	1.8	5.2	4.1	2.1	7.1	2.9	3.4	6.0	7.0	6.8	13.2	0.3	15.8	1.2	6.3	7.5	8.7
					B	4.1	5.1	3.5	3.4	2.8	4.3	4.2	2.9	12.4	3.7	6.9	4.2	6.0	6.5	9.3	5.8	4.3	1.1	4.9
		ENVI-Carb+PSA	GC-MS	Green tea	A	6.3	4.8	1.7	8.8	6.5	2.8	4.8	1.0	9.2	2.7	7.0	3.0	2.2	1.8	4.0	6.2	14.3	3.6	4.3
					B	3.3	6.4	1.7	3.7	1.9	1.3	2.7	4.4	2.4	1.4	21.5	1.9	2.2	3.3	12.3	4.8	5.7	9.2	5.3
				Oolong tea	A	3.9	2.8	2.8	4.3	4.2	2.7	2.4	1.9	2.5	1.7	4.2	3.4	1.2	2.8	3.0	2.1	5.0	5.1	6.5
					B	5.4	2.8	8.3	4.5	9.0	2.6	3.1	5.7	10.7	3.4	7.1	4.1	2.8	7.0	8.0	2.7	5.7	4.3	3.3
			GC-MS/MS	Green tea	A	5.5	3.7	1.2	3.9	2.5	3.1	2.6	1.4	5.9	4.7	8.5	1.9	6.0	18.0	4.5	11.3	16.4	8.8	0.9
					B	3.6	7.9	5.6	2.1	2.8	4.3	1.4	3.4	0.9	3.3	24.0	2.7	2.6	10.2	13.7	5.9	2.1	14.7	1.4
				Oolong tea	A	6.2	4.9	5.8	3.3	6.3	2.1	4.8	4.3	6.9	5.0	11.4	9.9	10.2	7.0	7.7	4.3	9.4	15.6	9.4
					B	6.1	4.8	8.6	5.0	10.4	3.4	3.3	7.0	10.5	0.3	3.2	16.5	8.0	11.1	12.0	7.0	5.7	3.5	4.7
51	Pirimicarb	TPT	GC-MS	Green tea	A	6.6	2.2	3.6	1.4	9.0	4.0	6.8	6.2	5.8	3.6	8.3	7.7	7.5	3.8	3.3	1.8	5.4	4.9	5.0
					B	3.4	13.7	5.0	8.9	4.3	6.1	1.4	4.1	4.2	7.2	8.6	14.0	3.0	1.4	4.7	3.4	5.0	0.5	4.7
				Oolong tea	A	5.6	1.6	3.4	4.8	9.7	4.9	13.1	0.6	3.3	6.9	14.5	2.7	1.5	3.3	8.0	2.6	3.6	6.2	1.5
					B	9.8	1.4	4.2	2.2	8.4	3.8	2.8	1.3	8.2	5.5	14.7	1.3	2.8	3.8	3.4	0.1	1.0	1.2	3.6
			GC-MS/MS	Green tea	A	4.2	5.3	3.1	1.7	7.0	4.3	7.9	7.8	4.3	3.3	5.5	5.2	6.3	8.0	3.0	1.0	4.1	6.1	3.2
					B	4.2	8.6	7.3	7.1	6.2	10.3	3.6	8.3	3.2	9.9	6.3	9.4	3.2	0.7	5.1	1.7	8.9	3.3	5.6
				Oolong tea	A	5.7	3.7	0.3	6.5	0.8	4.1	9.5	1.1	2.8	5.1	5.1	2.1	6.1	1.8	9.9	1.2	4.1	1.2	8.5
					B	1.2	4.7	4.2	5.5	1.3	3.8	3.5	6.5	5.1	2.6	7.2	1.6	7.4	2.9	6.0	1.0	2.1	1.2	4.2
		ENVI-Carb+PSA	GC-MS	Green tea	A	8.6	4.7	3.8	5.2	7.8	4.3	3.3	4.4	2.3	5.9	3.8	4.5	2.8	5.9	4.7	10.7	9.9	6.7	5.2
					B	4.4	8.0	5.7	3.2	2.3	9.9	5.2	2.8	5.9	18.2	29.4	7.4	8.6	2.8	15.1	2.1	4.4	12.0	8.3
				Oolong tea	A	6.3	6.6	3.6	3.7	1.9	2.1	26.6	8.4	33.0	2.2	9.2	4.9	1.8	3.4	3.9	2.9	4.7	5.5	3.1
					B	3.7	4.2	13.0	2.8	14.9	5.5	4.0	8.7	11.0	3.4	8.1	8.0	4.5	8.1	11.0	0.1	7.5	3.7	2.9
			GC-MS/MS	Green tea	A	8.5	5.9	4.0	7.6	7.5	1.8	3.2	3.8	3.5	3.7	8.3	0.3	4.7	10.8	4.5	8.3	8.1	4.2	6.3
					B	4.7	9.6	4.7	5.4	3.0	2.8	3.6	1.2	4.0	4.2	22.1	1.3	2.0	16.3	16.8	3.1	2.7	14.4	4.9
				Oolong tea	A	2.7	3.4	2.6	7.1	5.9	2.5	3.6	1.5	3.6	2.8	6.2	10.0	8.4	4.4	10.3	2.2	7.6	4.4	3.8
					B	5.9	3.3	9.3	3.3	11.4	6.6	4.5	3.9	11.6	4.0	2.3	7.0	7.6	7.1	11.7	3.5	6.3	2.7	4.1

52	Procymidone	TPT	GC–MS	Green tea	A	5.7	5.8	2.3	1.1	6.9	4.5	6.1	4.6	5.2	3.9	5.1	11.4	9.0	5.2	2.1	1.7	4.2	7.0	4.4
					B	5.9	6.6	5.5	7.4	3.7	3.9	3.7	5.9	3.0	6.0	3.7	7.1	2.1	8.6	3.4	1.6	4.2	2.4	4.9
				Oolong tea	A	6.2	3.5	0.8	2.9	2.0	2.7	8.9	4.7	1.3	3.6	20.0	9.4	12.1	6.8	8.3	2.4	3.4	3.0	1.3
					B	8.3	3.8	5.4	3.6	1.9	4.7	5.2	0.8	6.7	2.9	12.1	12.1	4.7	4.0	2.0	0.8	2.2	0.6	2.6
			GC–MS/MS	Green tea	A	4.4	7.2	1.5	2.4	5.5	3.0	7.8	6.4	5.7	2.0	5.1	7.0	6.6	0.9	1.7	1.7	7.4	5.4	4.6
					B	4.2	7.1	5.7	7.6	3.6	3.6	4.2	4.8	3.4	7.4	3.8	2.2	1.4	1.6	3.4	0.2	7.1	3.4	6.9
				Oolong tea	A	7.7	2.7	1.8	4.6	3.7	1.1	9.2	2.1	0.9	3.4	3.2	0.4	4.9	2.3	7.9	4.1	2.5	4.0	3.6
					B	6.1	2.7	3.9	4.1	0.6	1.5	4.4	4.1	6.2	2.9	4.1	11.7	2.1	4.5	6.5	1.6	0.9	1.1	2.5
		ENVI-Carb +PSA	GC–MS	Green tea	A	5.3	4.2	1.0	5.9	8.5	1.3	6.3	0.1	5.8	3.0	7.9	5.0	5.7	2.3	14.5	7.6	10.5	3.9	5.0
					B	4.4	7.0	2.8	4.1	1.4	2.7	3.8	6.7	3.3	2.3	19.4	2.1	9.0	9.2	8.6	1.8	4.6	8.8	5.5
				Oolong tea	A	4.1	3.8	3.0	3.8	6.3	3.9	2.6	0.8	2.8	0.4	9.1	7.0	4.5	26.3	4.8	2.7	6.5	3.8	4.8
					B	5.7	2.0	9.0	5.4	7.9	5.4	3.0	6.1	10.8	2.8	8.8	0.6	13.6	8.9	9.0	3.8	6.0	4.8	5.4
			GC–MS/MS	Green tea	A	4.5	4.3	1.7	6.5	8.1	0.8	2.9	0.5	5.4	1.7	10.4	7.0	4.2	11.9	6.8	5.9	8.1	3.1	3.2
					B	2.6	8.9	3.1	4.8	2.9	2.1	2.7	2.7	4.4	1.5	20.9	0.6	1.7	7.8	13.2	2.0	4.2	12.8	3.9
				Oolong tea	A	5.5	1.6	3.5	4.6	7.1	2.1	5.7	2.5	4.3	1.3	3.8	2.1	3.9	3.5	10.5	2.8	2.9	4.2	5.8
					B	5.9	0.8	9.4	2.6	7.9	5.6	3.1	5.9	12.6	3.7	2.2	6.4	1.5	8.7	11.1	1.3	3.1	4.0	3.8
53	Prometrye	TPT	GC–MS	Green tea	A	8.8	10.5	5.0	0.9	9.6	3.1	7.5	4.4	3.6	5.4	10.4	5.3	7.6	0.7	3.7	2.5	5.0	6.2	4.6
					B	6.9	7.0	11.6	9.2	6.6	4.5	2.5	1.5	5.3	9.3	14.1	7.3	2.6	6.3	3.0	7.5	3.1	1.7	4.4
				Oolong tea	A	5.9	2.8	0.7	4.0	2.6	3.3	5.4	6.0	3.7	5.1	12.7	14.9	6.1	1.8	8.2	6.1	3.6	1.3	1.8
					B	9.3	0.9	1.1	2.2	4.8	3.3	7.0	6.5	7.7	4.7	13.7	4.2	2.4	3.3	4.8	1.4	1.8	0.4	3.4
			GC–MS/MS	Green tea	A	5.0	4.5	2.4	2.3	7.1	8.1	8.5	6.4	6.8	0.9	4.5	1.5	5.1	4.9	3.0	3.9	6.8	8.3	6.5
					B	4.6	7.0	7.7	6.7	5.6	12.8	3.9	4.6	6.5	13.3	4.6	2.8	1.5	4.6	5.6	5.1	6.3	3.6	7.9
				Oolong tea	A	10.0	1.8	2.5	5.7	0.6	3.5	9.8	3.4	1.7	3.6	3.1	4.0	4.5	4.1	8.3	2.9	3.5	3.5	4.2
					B	7.3	6.7	4.5	6.2	1.9	0.5	5.0	7.0	5.8	2.8	6.2	2.4	5.5	1.9	6.7	0.5	1.9	1.9	7.3
		ENVI-Carb +PSA	GC–MS	Green tea	A	6.3	6.3	3.3	6.7	8.3	2.5	6.5	1.9	5.6	5.0	5.8	0.6	2.4	7.8	6.5	5.8	9.9	2.9	6.2
					B	7.3	9.5	6.5	6.8	4.4	5.0	4.0	4.1	6.1	18.7	30.5	0.9	2.9	7.3	12.7	5.3	4.9	8.9	7.5
				Oolong tea	A	6.6	3.2	3.3	1.5	0.4	1.3	3.5	3.7	32.0	2.5	8.9	6.9	1.5	2.7	4.6	5.2	5.6	2.1	3.3
					B	4.4	2.7	11.4	3.6	14.0	4.6	3.3	7.7	13.1	6.0	9.8	8.2	5.7	8.1	11.6	7.6	6.4	5.8	6.0
			GC–MS/MS	Green tea	A	4.5	4.3	2.1	5.4	8.0	2.0	6.0	0.9	3.8	2.9	11.3	4.9	6.7	11.1	7.5	7.5	8.9	3.6	6.1
					B	2.8	10.9	3.1	6.4	2.4	6.4	3.1	1.3	4.0	2.9	22.6	3.9	1.8	8.3	23.9	2.9	1.8	12.4	5.0
				Oolong tea	A	4.8	2.5	3.0	4.0	8.5	1.3	6.9	2.0	2.7	0.7	4.8	2.6	5.0	1.5	11.7	13.0	4.2	1.9	3.9
					B	5.7	4.0	11.2	2.3	8.4	5.6	5.1	6.2	12.3	3.0	0.9	8.5	8.0	9.9	10.3	11.7	8.3	3.8	5.9
54	Propham	TPT	GC–MS	Green tea	A	3.7	6.0	2.2	1.2	5.3	4.6	5.8	10.5	9.0	1.7	5.2	3.6	7.8	3.9	1.7	6.9	4.3	5.7	5.5
					B	4.6	9.2	3.3	7.1	3.2	2.0	4.0	5.2	4.5	7.4	3.7	2.0	1.1	1.0	2.9	9.1	7.6	2.2	6.7
				Oolong tea	A	3.6	2.3	2.0	4.3	1.2	1.3	6.1	6.8	3.0	6.0	13.3	6.8	4.4	2.1	6.3	2.8	6.7	4.1	12.0
					B	7.1	2.9	5.0	6.5	3.4	5.9	4.5	6.5	6.8	3.0	7.6	3.0	3.0	2.6	4.4	0.9	10.3	4.5	30.4
			GC–MS/MS	Green tea	A	1.5	6.6	1.2	1.0	7.9	3.8	5.6	8.3	5.8	2.3	5.3	1.6	7.0	6.3	2.0	3.9	2.8	6.2	5.2
					B	3.3	7.9	5.5	7.3	5.6	5.0	2.1	6.7	2.7	7.6	5.8	3.7	0.8	4.7	4.8	4.4	9.0	2.7	5.9
				Oolong tea	A	6.6	4.1	0.6	4.9	3.1	1.7	7.2	1.5	7.8	4.1	8.9	9.2	13.2	0.5	7.1	4.6	2.5	2.0	3.5
					B	5.5	4.3	3.0	4.5	0.5	6.1	5.3	7.3	6.8	3.1	6.8	4.8	5.0	0.7	5.7	1.4	2.8	2.5	5.2
		ENVI-Carb +PSA	GC–MS	Green tea	A	6.7	4.9	3.4	5.1	5.5	3.4	18.1	7.2	9.2	3.1	25.3	0.6	3.5	13.4	6.3	23.2	12.0	10.9	4.9
					B	6.6	6.3	4.9	4.8	2.2	3.2	8.6	5.1	8.1	8.1	7.2	1.8	23.7	0.5	29.3	3.4	4.8	12.2	5.2
				Oolong tea	A	4.0	4.0	1.7	3.7	2.6	2.0	5.7	14.3	10.5	1.4	5.0	2.7	1.4	4.0	3.5	3.3	7.8	5.7	1.0
					B	5.2	4.9	7.5	3.8	6.7	10.0	3.5	9.2	9.2	1.6	6.7	6.9	2.8	6.6	8.4	2.5	3.8	2.7	14.8
			GC–MS/MS	Green tea	A	9.4	6.2	3.7	6.2	4.0	2.5	14.8	4.8	9.2	3.4	28.6	7.9	3.5	4.8	4.3	14.8	7.3	11.1	7.2
					B	6.1	2.8	4.0	5.4	1.7	1.8	2.5	5.4	5.5	4.2	4.4	0.5	14.2	9.3	24.1	2.3	1.7	14.5	7.2
				Oolong tea	A	8.5	10.3	4.4	4.3	5.3	1.5	2.2	10.0	2.9	2.3	8.9	2.3	4.6	1.7	8.3	3.4	8.1	8.7	5.2
					B	10.5	3.6	8.7	4.4	10.0	4.3	3.1	5.4	12.2	1.2	1.2	5.8	2.2	7.2	8.1	1.0	4.7	2.4	2.4

(Continued)

Appendix Table 5.2 RSD of 201 Pesticides in Youden Pair Aged Tea Sample for Parallel Determinations (Nov 9, 2009–Feb 7, 2010) (cont.)

No.	Pesticides	SPE	Method	Sample	Youden pair	RSD% Nov 9, 2009 (n = 5)	Nov 14, 2009 (n = 3)	Nov 19, 2009 (n = 3)	Nov 24, 2009 (n = 3)	Nov 29, 2009 (n = 3)	Dec 4, 2009 (n = 3)	Dec 9, 2009 (n = 3)	Dec 14, 2009 (n = 3)	Dec 19, 2009 (n = 3)	Dec 24, 2009 (n = 3)	Dec 14, 2009 (n = 3)	Jan 3, 2010 (n = 3)	Jan 8, 2010 (n = 3)	Jan 13, 2010 (n = 3)	Jan 18, 2010 (n = 3)	Jan 23, 2010 (n = 3)	Jan 28, 2010 (n = 3)	Feb 2, 2010 (n = 3)	Feb 7, 2010 (n = 3)
55	Prosulfo-carb	TPT	GC-MS	Green tea	A	5.8	6.4	3.4	1.6	7.9	4.4	6.9	3.4	4.9	2.3	7.1	2.6	9.8	2.8	2.0	1.9	3.4	7.7	1.9
					B	5.8	5.5	5.3	9.2	2.3	3.0	2.8	2.1	5.6	9.1	6.2	11.0	2.4	2.8	4.2	3.3	1.9	5.0	2.1
				Oolong tea	A	4.8	2.6	2.9	2.9	2.9	1.3	8.1	6.3	1.9	3.8	9.9	3.1	3.7	2.9	8.0	8.5	3.0	5.9	4.0
					B	7.5	4.0	6.2	5.3	3.3	2.5	1.8	2.8	7.9	2.2	8.1	1.6	1.1	1.7	2.7	0.7	1.5	1.3	1.7
			GC-MS/MS	Green tea	A	4.9	5.6	4.2	2.3	6.6	4.1	6.4	2.8	6.8	2.7	5.9	3.4	5.9	7.4	1.8	2.4	3.9	8.5	4.1
					B	5.3	8.0	6.0	7.0	5.4	5.9	3.6	3.8	2.3	11.6	5.9	9.7	1.9	0.5	5.1	4.0	7.7	3.6	9.1
				Oolong tea	A	4.1	2.6	1.9	6.1	3.5	3.9	7.1	1.3	2.1	4.7	4.6	1.9	4.9	3.2	10.4	7.1	4.8	8.8	0.8
					B	4.2	2.0	4.0	6.8	0.6	2.3	3.0	3.7	4.9	4.4	1.0	1.0	5.3	3.5	3.5	2.6	3.7	3.4	4.0
		ENVI-Carb +PSA	GC-MS	Green tea	A	6.5	4.7	2.6	6.1	6.7	3.0	14.8	8.2	5.7	4.4	7.4	3.8	3.4	4.4	5.8	5.7	11.5	6.3	5.6
					B	4.5	4.9	6.3	4.6	3.2	2.7	3.8	4.2	3.4	12.7	25.2	4.6	6.5	2.7	10.7	7.5	4.8	9.8	9.5
				Oolong tea	A	5.0	7.6	4.9	3.7	2.5	2.0	3.2	0.2	22.7	1.9	6.3	2.9	2.1	3.4	3.5	2.8	5.7	3.6	7.9
					B	6.0	3.2	6.6	3.9	11.4	5.1	2.3	6.6	11.3	3.8	9.1	1.3	6.1	16.9	11.0	3.5	6.3	5.6	2.6
			GC-MS/MS	Green tea	A	4.6	4.6	2.8	6.3	5.9	0.7	4.4	2.0	3.2	2.4	8.5	3.9	4.6	8.9	1.3	2.6	3.3	4.1	2.7
					B	1.9	10.3	4.7	6.0	3.7	2.1	2.3	3.2	1.1	3.7	21.2	4.9	3.1	4.5	14.3	3.7	10.1	10.1	3.9
				Oolong tea	A	4.6	4.5	5.2	4.7	6.8	2.0	5.8	3.1	2.9	1.2	4.1	9.6	4.7	1.9	9.1	2.8	6.9	2.9	3.7
					B	6.2	4.4	10.0	2.7	10.3	6.1	2.9	4.8	11.3	3.3	1.0	3.3	7.3	7.3	11.2	2.8	5.9	2.6	2.9
56	Secbume-ton	TPT	GC-MS	Green tea	A	4.3	6.3	4.1	3.5	12.5	4.8	5.9	5.7	4.8	6.3	11.1	6.7	8.6	3.7	3.9	3.2	5.8	8.8	4.4
					B	4.3	5.1	4.8	8.8	6.9	4.9	3.2	4.4	5.8	8.1	14.2	11.9	3.3	3.1	7.7	0.6	7.0	1.7	3.8
				Oolong tea	A	5.9	5.5	3.8	8.0	1.4	3.3	10.0	6.1	2.4	9.4	19.4	0.8	3.7	3.7	7.4	3.2	3.8	1.5	1.3
					B	10.9	2.3	3.8	6.1	3.9	2.5	4.0	3.0	8.7	10.1	10.8	1.7	4.6	3.0	4.2	0.1	2.0	1.3	2.8
			GC-MS/MS	Green tea	A		5.4	0.8	1.8	9.8	9.7	7.5	5.8	2.6	2.7	6.4	4.0	8.2	8.3	1.3	3.4	7.2	8.7	7.9
					B		9.1	6.1	7.2	6.6	13.8	3.2	7.0	5.3	9.7	6.8	7.0	4.1	1.5	3.8	5.6	8.7	4.0	4.7
				Oolong tea	A	4.2	0.7	1.3	6.5	0.9	3.4	9.0	2.3	4.6	4.7	5.8	1.9	6.4	6.6	7.8	0.8	2.2	3.9	6.1
					B	2.2	3.4	2.9	7.2	4.5	2.4	3.2	5.2	4.9	3.0	7.0	1.9	11.9	2.2	6.1	2.8	1.7	4.0	3.7
		ENVI-Carb +PSA	GC-MS	Green tea	A	9.6	4.0	2.1	8.8	12.5	5.2	4.6	1.1	5.6	9.9	3.5	7.4	3.7	4.4	8.3	7.8	11.5	3.8	6.5
					B	4.7	8.9	6.8	12.3	9.7	8.5	4.7	5.4	8.0	15.6	34.7	2.5	5.1	5.1	15.0	3.0	7.1	9.6	7.9
				Oolong tea	A	4.0	6.2	7.7	5.3	3.0	3.1	1.9	5.8	26.0	2.5	18.4	8.3	3.2	4.0	2.9	3.2	7.0	2.9	2.8
					B	3.6	6.3	11.3	6.2	18.7	7.2	4.2	7.9	8.5	5.2	6.0	5.7	7.5	10.3	10.7	3.8	7.0	1.5	6.5
			GC-MS/MS	Green tea	A	4.4	3.8	1.2	8.1	5.9	4.6	4.5	0.4	3.6	1.6	6.9	3.0	3.1	4.4	12.0	9.5	7.8	4.6	4.1
					B	1.8	9.8	5.4	8.3	3.1	6.5	1.6	1.6	5.7	2.6	22.0	3.5	5.5	11.9	17.5	4.6	2.6	12.3	9.7
				Oolong tea	A	4.4	4.4	3.4	4.5	7.4	2.0	5.7	3.3	2.8	3.1	6.2	12.3	3.1	2.9	13.5	2.7	6.9	5.7	5.5
					B	4.3	2.3	10.0	1.9	13.4	5.9	3.7	6.1	13.6	2.4	1.1	6.9	7.8	10.3	12.5	1.5	4.4	12.9	2.2
57	Silafluofen	TPT	GC-MS	Green tea	A	12.2	7.7	8.7	0.2	7.2	10.5	13.4	3.1	7.1	2.9	1.9	2.1	10.2	13.9	7.5	6.1	2.4	3.9	3.9
					B	8.4	2.1	1.7	5.8	4.5	8.0	3.4	1.7	10.3	8.9	9.3	3.1	9.8	9.3	3.4	6.6	13.8	3.7	4.3
				Oolong tea	A	4.5	7.5	3.1	5.9	8.3	6.9	5.7	5.6	7.8	5.3	3.7		1.8	2.6	3.9	10.1	5.0	8.1	2.4
					B	7.3	4.4	2.1	11.4	12.4	11.2	4.3	8.5	2.7	11.9	0.1		2.3	5.6	4.4	2.4	7.3	2.6	2.4
			GC-MS/MS	Green tea	A	5.5	5.5	0.7	2.8	5.1	6.8	8.1	2.9	5.5	4.5	7.5	1.6	3.3	9.5	5.4	1.9	4.0	7.0	2.3
					B	3.2	6.0	5.1	8.6	4.9	0.6	1.7	1.7	2.9	7.5	7.1	9.1	3.0	11.4	5.6	3.9	7.7	7.0	6.1
				Oolong tea	A	8.8	2.3	4.9	6.3	3.5	2.1	3.5	0.7	1.5	4.4	2.2	3.2	3.8	4.5	4.7	6.4	10.3	3.4	3.4
					B	6.9	0.8	4.2	5.2	1.3	2.8	4.2	3.8	2.8	4.3	6.0	1.0	2.3	2.6	4.3	7.4	4.0	3.5	9.4
		ENVI-Carb +PSA	GC-MS	Green tea	A	5.9	21.4	6.8	12.8	11.4	3.9	0.4	5.1	8.7	7.0	11.2	5.5	32.0	4.5	6.8	6.2	8.0	1.6	2.2
					B	9.2	7.5	6.3	11.5	1.1	3.9	10.2	6.1	6.3	3.8	9.9	9.2	38.4	3.6	15.3	3.8	13.2	5.6	11.0
				Oolong tea	A	6.0	3.4	11.2	10.1	6.0	3.5	3.5	7.7	1.1	10.4	5.7	8.7	13.7	2.9	4.6	2.3	6.7	11.5	2.4
					B	9.1	3.1	7.1	3.9	14.4	1.5	4.1	6.1	9.4	1.4	5.4	11.7	14.6	6.0	16.0	1.5	16.7	8.1	3.5
			GC-MS/MS	Green tea	A	5.0	3.4	1.5	7.8	7.3	4.9	6.4	1.7	8.1	2.7	6.1	5.7	4.6	2.0	9.1	11.3	10.8	7.3	5.1
					B	1.5	10.7	5.5	10.7	3.5	2.7	2.1	4.1	5.4	2.4	22.0	1.1	10.9	15.7	14.4	4.2	5.6	14.2	9.2
				Oolong tea	A	7.4	0.3	5.9	1.5	7.1	2.2	4.2	7.0	1.6	3.1	4.0	7.4	7.5	7.1	7.4	2.7	3.2	5.7	7.5
					B	7.3	2.0	8.6	4.4	9.6	4.5	4.4	7.0	12.1	1.5	1.0	4.4	11.0	2.5	8.0	2.1	1.6	5.0	1.6

58	Tebupirimfos	TPT	GC–MS	Green tea	A	3.7	4.6	4.0	3.0	11.9	3.7	5.2	0.6	4.8	4.7	9.3	4.7	7.0	0.7	1.9	2.2	3.5	7.8	4.4
					B	5.4	8.1	3.9	11.5	3.4	3.5	1.9	2.4	5.5	10.3	9.5	12.7	2.5	1.9	3.3	1.3	4.3	1.0	5.8
				Oolong tea	A	4.7	1.6	3.7	5.1	8.3	3.8	7.7	4.7	3.8	6.1	10.7	3.9	4.0	2.2	7.8	3.1	3.4	2.5	1.2
					B	7.3	2.9	4.1	5.3	7.1	5.6	2.3	0.9	9.1	3.5	7.1	1.6	2.7	1.3	2.3	0.8	1.6	2.2	3.2
			GC–MS/MS	Green tea	A		4.1	5.7	4.0	9.2	6.2	7.5	4.9	6.4	2.7	6.1	4.5	8.6	4.1	1.7	8.7	1.9	3.8	2.0
					B	4.1	8.6	7.3	8.5	5.0	11.4	3.5	5.9	8.9	13.2	6.0	10.0	1.7	2.7	6.0	5.7	9.3	5.3	9.5
				Oolong tea	A	3.3	1.2	3.2	6.8	3.0	4.0	10.1	3.5	2.6	3.2	5.1	5.5	1.7	3.4	7.6	3.2	3.8	13.7	0.7
					B	11.5	1.5	6.1	7.0	1.0	1.7	4.4	6.0	1.9	1.8	8.2	3.2	1.4	3.4	6.1	5.9	4.3	7.3	4.9
		ENVI-Carb +PSA	GC–MS	Green tea	A	8.3	2.9	9.1	6.0	6.7	10.2	7.9	2.5	5.4	5.6	8.9	2.6	4.1	3.6	5.9	7.0	11.1	4.3	5.2
					B	7.1	7.1	6.1	1.0	4.9	10.6	25.0	6.0	2.9	20.2	26.7	5.0	1.8	4.7	12.9	3.3	5.8	9.6	5.6
				Oolong tea	A	6.1	4.0	3.2	1.8	13.0	1.4	5.9	1.0	30.4	1.8	7.3	4.3	1.6	4.8	2.4	2.9	5.1	3.4	5.3
					B	6.8	4.8	8.8	1.4	18.3	7.8	4.7	7.9	9.7	4.0	7.8	4.0	3.8	7.6	8.7	4.2	7.0	3.5	4.7
			GC–MS/MS	Green tea	A	2.6	5.9	1.8	7.7	5.4	1.7	2.2	1.2	2.1	4.7	11.1	2.1	5.7	11.5	5.3	9.1	13.5	9.6	1.2
					B	6.6	7.2	5.6	4.5	3.7	5.8	2.2	0.6	4.5	4.0	19.8	1.9	3.7	6.7	13.4	1.9	6.1	11.5	4.2
				Oolong tea	A	6.6	4.0	6.1	8.2	7.2	2.8	5.1	4.1	0.7	1.8	4.7	9.3	5.1	4.0	8.8	2.4	3.7	3.9	8.0
					B	6.2	4.7	8.0	4.4	11.0	5.0	3.4	6.0	10.3	1.9	4.5	6.9	3.1	8.7	8.0	2.8	7.9	9.4	4.8
59	Tebutam	TPT	GC–MS	Green tea	A	4.9	6.0	3.4	1.8	8.4	5.4	6.5	3.6	5.2	2.4	6.1	1.3	7.0	3.3	5.1	2.3	4.9	2.4	5.1
					B	5.9	8.4	4.8	8.7	1.4	2.7	3.3	3.4	2.7	7.7	5.4	7.0	1.9	2.4	7.3	1.8	5.1	2.6	5.4
				Oolong tea	A	5.3	2.1	1.0	4.6	5.5	2.6	8.2	3.4	2.7	3.2	4.8	3.7	3.8	2.9	7.7	2.9	4.4	2.2	0.5
					B	5.0	3.2	5.3	9.3	2.8	4.3	5.6	0.4	5.5	2.3	3.0	1.6	2.8	3.5	3.2	0.3	0.3	2.6	2.5
			GC–MS/MS	Green tea	A	3.7	5.6	2.1	3.8	6.8	3.8	8.1	5.6	5.2	2.6	5.2	3.2	7.6	4.5	2.9	2.5	6.1	6.3	7.2
					B	4.2	8.3	6.5	7.4	6.9	5.0	4.0	4.4	3.3	8.1	7.8	6.9	2.3	3.3	3.9	6.0	6.9	4.6	7.1
				Oolong tea	A	3.5	0.8	2.1	7.0	0.9	2.6	9.8	5.0	3.6	5.1	2.8	2.7	3.4	3.7	8.2	2.1	5.6	1.7	5.5
					B	4.7	1.5	4.4	6.6	2.1	1.8	4.5	2.8	5.4	3.9	3.5	1.2	6.7	2.1	6.3	2.7	1.9	0.1	2.4
		ENVI-Carb +PSA	GC–MS	Green tea	A	8.4	3.3	2.5	6.7	7.6	2.5	4.8	1.5	6.0	2.7	9.4	2.9	4.2	5.7	5.8	7.3	10.7	3.4	6.1
					B	4.0	6.8	4.4	4.2	3.3	3.0	4.5	4.6	3.4	6.6	20.8	1.5	2.6	6.0	13.4	3.2	5.1	7.4	5.4
				Oolong tea	A	4.5	4.2	2.3	3.7	3.5	1.5	3.4	1.3	15.9	1.0	5.0	3.2	1.6	2.6	2.9	3.4	5.6	4.3	4.8
					B	4.9	2.6	8.4	4.0	10.1	4.1	3.2	5.5	9.3	3.8	8.9	1.7	3.4	7.2	9.4	3.3	6.1	4.9	6.4
			GC–MS/MS	Green tea	A	4.7	5.0	2.9	6.3	5.4	1.0	3.0	1.2	3.5	1.7	11.5	1.3	3.9	12.9	4.7	8.3	7.0	4.5	4.0
					B	2.8	7.7	4.4	6.4	2.2	1.2	1.9	1.6	0.9	3.2	21.9	0.4	1.9	6.9	17.1	1.2	3.4	12.5	3.5
				Oolong tea	A	5.1	3.4	4.0	5.8	7.2	1.1	4.0	2.0	1.8	0.4	5.7	8.8	6.1	7.0	8.5	2.4	3.9	4.6	7.3
					B	5.8	3.0	9.5	2.3	9.0	4.7	1.8	6.0	13.3	2.4	1.7	6.3	6.0	7.5	9.8	1.5	5.0	2.9	1.7
60	Tebuthi-uron	TPT	GC–MS	Green tea	A	7.9	7.6	2.8	9.2	2.0	2.7	8.0	3.4	4.7	5.8	9.0	3.7	8.4	3.9	3.1	6.6	6.1	2.9	3.9
					B	8.9	8.7	4.5	6.8	3.1	4.9	4.3	9.4	6.5	4.4	9.8	7.9	4.4	3.2	8.9	3.8	8.2	6.7	2.6
				Oolong tea	A	7.2	4.4	5.4	5.5	8.7	6.9	9.7	11.7	0.7	3.3	27.0	4.3	3.3	7.0	7.7	5.1	3.7	1.5	1.7
					B	13.0	1.8	2.0	9.1	5.4	4.7	5.4	4.4	3.5	1.2	7.7	3.3	1.7	5.8	1.0	1.1	2.6	5.0	2.4
			GC–MS/MS	Green tea	A	6.4	9.8	2.2	0.6	3.5	2.9	10.5	6.3	6.7	7.7	9.2	0.3	8.4	6.1	5.4	2.2	7.0	1.9	4.1
					B	2.5	11.2	4.8	4.4	6.0	7.1	3.8	12.6	6.1	15.8	3.5	8.5	2.8	3.5	11.0	7.5	10.9	0.9	4.6
				Oolong tea	A	5.5	3.2	4.6	3.9	6.6	3.1	9.6	5.9	5.1	9.2	3.4	2.9	5.8	3.0	7.6	2.8	3.6	7.7	5.8
					B	4.5	2.3	6.5	2.6	2.8	1.1	5.3	4.7	3.9	2.9	6.3	5.1	6.2	3.2	4.6	1.9	1.9	2.5	2.0
		ENVI-Carb +PSA	GC–MS	Green tea	A	7.5	12.2	7.7	6.8	15.5	7.2	34.4	1.9	4.3	5.9	9.8	4.1	3.1	2.2	8.4	9.9	12.2	3.8	7.2
					B	12.0	5.0	8.0	5.9	11.9	2.9	5.8	2.8	6.9	16.6	27.0	7.7	8.0	2.4	18.5	3.0	5.0	11.0	8.6
				Oolong tea	A	7.3	8.8	5.7	3.0	2.6	5.1	5.0	4.7	18.1	12.6	7.7	11.9	4.3	5.6	7.0	3.6	4.8	0.3	4.7
					B	4.7	4.9	14.8	9.9	19.2	9.8	2.7	8.5	17.9	6.7	9.2	1.3	3.3	22.1	1.0	4.6	6.5	8.0	6.4
			GC–MS/MS	Green tea	A	8.7	16.3	8.4	7.7	8.7	5.5	36.0	2.4	1.7	3.6	12.7	9.3	5.8	10.0	11.6	9.9	9.8	9.0	5.1
					B	4.3	6.7	5.0	8.7	3.4	6.9	3.3	1.4	6.1	3.3	24.4	13.6	2.7	13.5	6.0	0.6	2.4	14.3	7.9
				Oolong tea	A	4.4	3.8	4.7	5.6	9.3	3.5	6.9	1.0	3.8	12.6	7.9	13.4	3.0	3.2	19.0	4.2	5.5	1.3	5.8
					B	5.9	3.5	15.0	5.0	12.7	10.1	2.6	7.4	9.8	4.4	2.8	7.3	4.3	20.4	11.5	3.2	6.0		5.1

(Continued)

Appendix Table 5.2 RSD of 201 Pesticides in Youden Pair Aged Tea Sample for Parallel Determinations (Nov 9, 2009–Feb 7, 2010) (cont.)

RSD%

No.	Pesticides	SPE	Method	Sample	Youden pair	Nov 9, 2009 (n=5)	Nov 14, 2009 (n=3)	Nov 19, 2009 (n=3)	Nov 24, 2009 (n=3)	Nov 29, 2009 (n=3)	Dec 4, 2009 (n=3)	Dec 9, 2009 (n=3)	Dec 14, 2009 (n=3)	Dec 19, 2009 (n=3)	Dec 24, 2009 (n=3)	Dec 14, 2009 (n=3)	Jan 3, 2010 (n=3)	Jan 8, 2010 (n=3)	Jan 13, 2010 (n=3)	Jan 18, 2010 (n=3)	Jan 23, 2010 (n=3)	Jan 28, 2010 (n=3)	Feb 2, 2010 (n=3)	Feb 7, 2010 (n=3)
61	Tefluthrin	TPT	GC-MS	Green tea	A	5.7	5.3	4.1	1.8	8.5	5.8	6.4	2.9	6.7	1.7	5.8	1.1	7.4	14.2	8.8	2.5	12.9	7.8	4.1
					B	6.9	7.8	5.8	8.6	3.3	2.2	1.5	0.6	4.9	11.0	4.5	7.8	1.1	1.5	10.3	2.1	9.1	1.7	4.8
				Oolong tea	A	4.0	3.3	2.3	4.9	3.9	3.0	6.9	3.3	2.8	2.5	4.1	3.9	3.8	3.5	5.9	3.7	4.8	2.8	4.4
					B	7.0	2.4	4.1	5.5	2.3	3.8	3.9	1.7	6.1	2.2	3.6	0.7	2.8	3.1	2.8	0.7	0.4	2.5	2.9
			GC-MS/MS	Green tea	A	4.4	6.6	3.2	2.9	7.2	3.8	8.6	4.8	6.8	3.7	5.7	3.1	7.4	6.4	2.4	5.2	3.4	11.0	6.8
					B	5.6	9.3	5.9	7.8	4.7	5.3	5.3	2.4	2.6	9.9	4.8	3.1	0.9	3.9	5.1	1.3	9.4	5.1	9.1
				Oolong tea	A	3.4	5.1	2.1	5.0	1.8	3.9	7.6	0.8	3.1	5.2	2.7	3.9	1.5	4.2	9.4	2.5	0.4	1.1	5.1
					B	5.1	1.9	3.8	5.4	1.9	2.3	3.0	4.6	5.0	3.1	3.2	1.5	4.3	1.7	3.4	1.9	2.3	3.9	4.9
		ENVI-Carb+PSA	GC-MS	Green tea	A	6.8	4.5	2.0	5.7	7.6	2.1	5.0	2.0	7.0	1.8	5.5	6.6	5.1	5.5	5.4	7.3	10.9	8.2	5.2
					B	5.2	6.0	5.4	4.3	2.8	3.1	5.1	4.5	6.6	5.6	2.8	1.4	4.4	5.8	13.1	3.1	5.6	3.8	8.9
				Oolong tea	A	3.8	2.5	2.3	4.1	5.1	1.3	3.0	0.2	9.9	2.2	6.6	3.1	0.7	3.8	2.5	3.1	5.5	7.7	9.3
					B	5.1	2.4	8.4	3.6	10.3	4.8	3.8	7.5	5.3	2.7	7.0	3.4	5.2	7.2	10.1	3.8	7.2	5.3	10.3
			GC-MS/MS	Green tea	A	6.8	6.1	2.7	5.7	6.1	0.5	4.3	1.1	1.8	1.8	13.3	1.3	3.3	13.3	6.2	8.7	8.6	12.5	2.2
					B	3.1	8.8	4.4	5.3	3.1	1.7	3.0	1.9	1.6	1.8	20.4	2.8	1.8	10.0	13.2	2.2	5.6	6.4	4.3
				Oolong tea	A	4.5	5.0	3.6	4.9	6.6	1.7	6.0	2.9	1.6	1.3	5.1	8.5	8.0	4.7	8.7	2.4	5.5	13.4	5.4
					B	6.6	2.0	8.8	1.8	9.4	6.8	3.3	6.6	12.3	2.1	3.2	5.1	7.6	10.6	8.8	1.4	5.8	2.2	4.8
62	Thenyl-chlor	TPT	GC-MS	Green tea	A	9.9	6.9	2.2	8.2	6.5	5.9	2.9	8.3	9.3	2.2	13.3	7.1	7.1	11.1	10.2	3.2	5.7	2.2	7.9
					B	8.9	8.4	8.6	6.3	7.0	0.4	15.2	4.0	4.2	8.2	1.9	5.1	5.4	3.9	8.2	5.3	10.1	2.9	4.2
				Oolong tea	A	8.7	6.5	1.2	6.8	7.6	2.1	8.5	1.7	4.5	4.4	3.4	0.7	7.9	5.6	4.6	2.2	4.8	1.2	5.8
					B	8.1	6.2	3.8	1.1	4.6	6.4	4.9	9.8	3.1	2.8	5.8	3.4	13.2	10.9	3.6	8.5	12.6	4.7	1.6
			GC-MS/MS	Green tea	A	9.2	7.5	3.7	5.6	3.0	3.2	1.1	5.4	4.4	2.8	5.6	3.9	2.9	6.8	2.5	2.4	5.0	8.7	8.5
					B	6.3	7.6	2.7	5.1	3.5	3.5	14.4	5.4	5.3	9.7	5.6	3.8	16.1	7.0	6.5	4.7	5.0	2.8	10.8
				Oolong tea	A	7.3	6.5	8.4	1.8	1.8	6.6	9.6	4.6	9.6	9.7	5.6	3.9	6.8	7.0	6.5	4.7	5.0	3.6	10.9
					B	6.2	4.7	8.8	7.0	16.1	4.4	1.1	5.1	9.7	3.7	9.6	3.8	16.1	4.2	8.5	2.0	5.0	3.6	10.7
		ENVI-Carb+PSA	GC-MS	Green tea	A	10.0	13.7	5.4	6.7	3.9	5.6	2.8	5.4	2.0	4.6	6.0	2.6	6.1	3.3	6.0	10.0	4.4	18.8	5.8
					B	11.1	11.4	3.9	2.0	7.3	3.7	4.7	4.6	5.9	10.2	17.4	3.5	6.1	1.1	8.9	5.2	2.5	3.0	12.3
				Oolong tea	A	7.8	5.6	6.5	3.9	10.2	6.4	3.4	5.1	3.3	2.7	6.6	2.0	6.6	10.3	6.6	6.9	0.7	6.0	18.0
					B	12.9	5.1	12.5	7.0	5.3	0.5	3.2	9.2	14.5	10.6	9.1	4.8	3.9	10.1	14.1	2.4	8.1	0.8	2.1
			GC-MS/MS	Green tea	A	15.3	10.2	5.7	4.3	2.8	9.4	15.5	2.3	5.0	4.7	25.7	4.1	16.0	1.9	1.3	8.7	15.3	10.0	9.6
					B	48.4	14.5	6.4	2.3	6.8	7.4	3.5	5.8	8.3	9.4	6.1	7.9	15.8	17.6	9.4	11.9	2.0	5.7	3.2
				Oolong tea	A	7.5	1.8	6.6	6.1	7.3	6.2	6.3	4.5	8.5	9.1	5.8	13.6	9.2	6.7	11.7	3.0	6.2	8.3	6.8
					B	12.1	4.0	14.8	9.0	5.4	5.8	1.7	8.1	14.9	9.1	3.5	5.2	9.1	6.5	8.3	1.9	8.2	5.7	4.2
63	Thionazin	TPT	GC-MS	Green tea	A	5.5	5.1	0.7	3.3	8.2	5.1	6.2	3.5	5.3	3.9	7.2	1.9	2.7	3.1	3.4	9.7	5.4	6.2	6.8
					B	7.1	7.8	7.4	5.1	12.0	1.2	1.3	2.8	2.9	15.6	5.8	1.7	5.8	5.2	8.8	2.0	5.9	8.5	4.2
				Oolong tea	A	5.2	1.6	2.8	8.3	6.8	2.3	8.1	6.6	3.8	2.8	5.1	5.0	3.2	3.1	4.4	2.5	3.4	6.0	3.5
					B	8.6	4.8	8.9	9.1	4.1	2.2	8.1	5.0	8.3	2.3	5.3	3.8	4.9	4.6	3.1	1.0	4.6	6.7	3.9
			GC-MS/MS	Green tea	A	3.1	5.8	1.8	2.9	2.6	5.6	4.4	5.2	4.0	1.8	4.9	3.8	6.1	1.9	2.9	2.5	8.5	2.1	7.5
					B	4.8	8.6	5.5	6.0	7.6	13.1	4.1	4.8	1.8	7.5	5.5	6.4	0.6	1.5	5.8	4.7	2.7	1.6	3.8
				Oolong tea	A	4.1	1.1	0.9	3.6	5.2	2.8	10.1	0.8	1.8	5.0	3.4	3.4	4.0	1.5	5.8	1.5	2.7	3.0	1.7
					B	2.4	2.7	4.2	3.6	2.0	2.4	3.3	2.4	4.4	3.4	3.7	1.3	1.5	1.5	7.5	1.5	2.5	3.0	1.7
		ENVI-Carb+PSA	GC-MS	Green tea	A	7.6	5.8	3.5	6.2	1.4	2.9	9.8	2.2	5.7	3.9	4.6	1.9	2.7	2.7	2.6	9.7	5.4	6.2	4.1
					B	7.5	4.9	6.3	4.9	3.5	4.2	15.1	0.8	5.8	15.6	23.9	1.7	5.8	0.7	17.6	2.0	5.4	8.5	4.8
				Oolong tea	A	8.4	4.1	3.9	7.1	6.7	5.8	2.2	1.7	20.2	2.8	7.3	5.0	3.2	8.0	4.4	2.5	3.4	6.0	4.8
					B	3.3	4.1	7.6	1.7	5.0	4.2	2.3	7.9	10.5	3.3	7.8	0.7	3.8	6.2	5.5	3.8	4.6	0.9	1.5
			GC-MS/MS	Green tea	A	6.6	5.4	2.9	5.9	12.6	1.1	1.1	2.0	4.8	1.1	10.6	7.4	1.8	5.6	5.6	7.8	9.9	5.8	2.8
					B	7.4	9.6	4.5	4.1	4.9	3.1	3.1	1.1	2.5	2.4	18.2	7.5	6.4	16.8	6.4	2.2	4.0	12.6	2.8
				Oolong tea	A	4.7	3.4	5.7	5.2	5.5	4.1	4.1	2.0	2.2	1.7	4.6	6.1	4.3	4.6	8.4	2.6	4.7	4.6	5.4
					B	6.6	3.2	8.2	3.6	8.7	4.6	1.9	5.1	11.5	0.9	0.7	3.7	2.2	6.8	9.5	1.1	4.4	2.3	5.2

No.	Compound	Cleanup	Instrument	Tea	Rep	1	2	3	4	5	6	7	8	9	10	11	12	13	14	15	16	17	18	19
64	Trichloronat	TPT	GC–MS	Green tea	A	3.8	4.8	3.6	1.3	9.9	3.5	6.0	6.3	5.6	4.2	7.2	5.9	6.9	4.9	2.5	1.4	2.4	4.8	4.1
					B	5.7	8.0	5.7	10.1	4.8	2.8	1.0	3.2	3.9	10.0	6.7	10.8	1.3	1.6	3.6	2.3	4.8	0.5	5.5
				Oolong tea	A	4.5	1.7	2.6	4.8	1.7	2.8	11.3	1.7	3.7	5.9	10.7	1.8	2.9	2.3	7.4	3.0	3.6	1.8	0.5
					B	7.4	0.5	3.8	3.1	5.7	4.1	3.1	5.5	7.4	3.7	11.5	0.3	1.9	2.4	5.1	0.2	1.4	1.2	3.6
			GC–MS/MS	Green tea	A	3.2	6.4	3.8	0.7	8.5	5.3	8.0	5.7	7.3	1.7	4.4	3.1	5.2	4.9	0.8	2.1	3.5	4.4	4.7
					B	6.4	8.7	5.5	6.3	4.5	7.4	2.2	3.5	3.0	9.7	4.1	8.1	4.2	4.0	3.1	3.7	7.4	1.6	7.3
				Oolong tea	A	5.3	3.3	1.1	7.4	2.9	2.9	9.6	2.3	1.0	3.4	4.3	2.0	5.5	2.1	9.5	3.9	3.8	0.2	2.5
					B	3.0	2.2	3.2	5.6	2.4	2.2	3.5	4.0	6.0	2.7	4.8	2.0	6.6	3.6	6.2	0.7	4.6	0.8	3.5
		ENVI-Carb +PSA	GC–MS	Green tea	A	11.7	3.9	6.6	4.4	6.7	4.2	7.0	6.3	3.4	2.3	4.1	4.2	1.2	9.7	1.5	5.8	9.0	7.5	5.2
					B	8.1	9.7	6.4	0.9	3.2	5.5	5.8	3.2	8.0	14.3	23.1	5.8	7.6	3.8	7.4	5.6	3.8	13.6	7.0
				Oolong tea	A	4.6	6.7	3.6	3.4	1.7	1.7	1.7	3.6	23.0	1.4	7.3	4.3	0.8	3.3	1.8	3.2	6.5	6.1	4.1
					B	5.5	3.0	9.4	0.8	13.5	1.0	1.8	7.5	10.2	2.8	9.1	2.1	3.9	7.2	9.1	3.8	8.0	0.6	4.9
			GC–MS/MS	Green tea	A	12.5	5.6	3.6	4.7	6.3	4.5	7.2	5.6	4.6	1.6	9.3	0.0	4.5	2.6	4.0	4.3	5.7	8.3	5.1
					B	9.6	10.8	3.5	3.5	6.3	2.3	6.5	4.1	6.1	2.6	13.5	0.0	6.1	9.0	13.4	4.7	6.4	14.4	2.3
				Oolong tea	A	3.2	3.3	4.0	6.5	4.3	1.3	4.9	4.8	2.3	1.9	3.7	5.2	3.5	3.2	8.9	2.4	4.9	5.3	4.2
					B	6.9	2.7	8.9	1.7	4.3	6.9	2.5	6.2	11.0	1.6	2.8	4.9	2.1	10.5	10.7	1.7	6.1	3.7	3.5
65	Trifluralin	TPT	GC–MS	Green tea	A	3.8	5.3	4.0	3.8	13.0	3.6	6.3	10.3	5.0	2.7	6.2	4.4	10.3	4.8	3.6	2.3	4.2	14.4	3.9
					B	5.3	8.3	4.5	11.4	3.9	3.8	2.2	6.2	5.8	9.1	5.8	9.3	8.1	6.9	2.6	11.9	13.7	24.0	7.3
				Oolong tea	A	3.8	1.6	3.3	4.7	12.6	5.6	8.6	4.2	4.8	5.0	14.2	5.1	4.5	2.9	1.0	10.4	6.9	2.7	0.8
					B	7.7	1.8	3.8	5.3	8.0	6.7	4.2	5.7	6.8	1.4	5.9	3.3	3.6	4.2	11.5	5.8	0.7	2.1	4.3
			GC–MS/MS	Green tea	A	2.8	7.8	4.0	4.8	16.4	10.0	8.4	8.7	9.4	3.2	4.3	4.9	7.8	4.2	1.1	4.8	0.9	6.7	3.6
					B	3.0	8.3	5.0	7.4	3.6	10.5	1.5	7.8	4.0	12.5	5.2	10.4	4.6	6.1	2.5	4.5	13.1	4.6	13.2
				Oolong tea	A	2.5	4.4	1.1	7.3	8.7	4.1	8.7	0.2	6.3	5.4	7.2	4.7	5.7	4.2	3.3	3.9	5.1	7.4	6.3
					B	17.2	8.6	2.0	3.7	6.9	2.7	4.9	10.1	4.9	4.7	9.4	3.5	5.5	1.8	2.4	4.3	3.9	6.0	1.8
		ENVI-Carb +PSA	GC–MS	Green tea	A	12.4	3.2	9.1	5.6	7.0	15.6	6.4	12.8	8.1	1.6	6.1	6.3	1.7		2.9	13.9	16.8	8.2	7.6
					B	11.?	7.6	6.0	2.5	5.5	10.4	11.6	10.6	10.0	9.7	17.3	7.0	19.0		13.7	5.0	21.6	16.8	9.0
				Oolong tea	A	7.5	3.4	3.2	2.4	21.1	1.5	2.7	14.3	15.8	2.6	6.1	3.5	9.9	7.5	3.1	21.9	10.2	7.2	2.7
					B	6.2	5.0	8.3	1.6	18.3	8.7	6.7	10.5	8.8	2.1	9.3	1.0	5.8	13.0	14.4	5.5	15.1	5.4	4.8
			GC–MS/MS	Green tea	A	11.6	2.9	6.6	6.4	9.8	6.0	9.1	15.4	11.3	2.1	7.9	12.9	3.5	8.4	7.3	5.8	7.2	8.1	5.0
					B	13.5	9.9	4.6	6.6	7.2	5.0	17.2	11.8	10.8	3.2	15.0	4.0	8.4	4.4	16.1	5.8	7.1	16.9	8.0
				Oolong tea	A	7.7	2.5	6.1	3.9	5.6	1.3	4.7	9.6	0.6	2.6	3.5	7.4	5.2	0.7	7.5	5.9	13.8	29.0	10.6
					B	6.0	2.7	9.3	5.3	12.5	8.8	0.8	8.1	13.9	3.0	4.0	1.7	7.1	7.6	10.2	4.2	11.4	10.4	7.6
66	2,4-DDT	TPT	GC–MS	Green tea	A	2.5	3.9	5.3	11.0	13.4	8.8	9.0	8.5	9.0	3.2	4.9	4.5	8.1	8.3	1.8	4.7	6.0	12.5	4.3
					B	8.4	7.2	3.5	13.6	7.7	1.4	5.0	8.9	1.5	7.8	1.5	4.4	7.6	10.6	4.3	2.3	14.3	5.0	6.8
				Oolong tea	A	7.5	3.9	3.5	4.9	7.4	6.6	12.5	14.8	9.8	3.6	6.2	12.9	5.1	1.6	6.1	4.7	4.9	5.0	6.3
					B	6.0	2.4	7.9	4.6	2.3	4.2	7.0	2.6	7.9	9.1	5.4	0.6	2.3	10.6	12.7	10.0	6.0	3.2	8.6
			GC–MS/MS	Green tea	A	9.5	6.1	3.8	7.6	11.5	2.7	14.1	11.0	14.8	2.0	8.7	7.6	7.5	6.3	4.0	0.3	6.0	11.0	6.3
					B	8.2	8.2	2.9	12.1	7.5	2.3	3.7	11.2	3.9	7.7	6.2	5.5	4.8	4.6	4.6	8.0	9.0	2.0	8.4
				Oolong tea	A			6.3	6.9	8.9	4.2	0.9	1.9	5.9	4.0	13.0	11.5	8.7	3.1	5.7	9.8	12.5	9.5	8.4
					B	18.7	7.8	11.1	2.8	4.2	1.4	12.8	11.9	17.2	11.0	5.7	3.9	7.0	13.8	13.8	7.5	14.2	6.3	8.8
		ENVI-Carb +PSA	GC–MS	Green tea	A	19.1	6.9	13.1	5.0	9.5	20.8	3.4	13.7	5.5	5.2	19.2	18.8	0.4	7.5	1.4	1.7	5.9	12.7	14.0
					B	15.9	10.1	4.8	8.7	12.3	7.7	14.4	12.6	11.1	6.5	15.1	14.3	28.3	4.0	10.3	6.9	12.1	18.0	8.6
				Oolong tea	A	15.7	13.3	7.5	20.4	5.5	15.5	11.1	17.5	10.2	8.3	7.1	17.1	19.6	3.5	4.1	6.3	4.4	21.0	14.0
					B	21.9	5.3	6.1	4.3	18.5	17.3	17.2	8.1	18.5	14.9	18.6	5.0	6.3	10.6	19.1	23.7	5.9	21.0	8.6
			GC–MS/MS	Green tea	A	36.9	8.2	14.1	5.0	4.6	18.4	13.8	16.3	7.4	2.9	18.7	7.3	4.8	26.0	3.7	6.0	23.3	39.3	4.9
					B	2.7	8.8	4.4	17.4	6.5	7.3	21.9	13.4	13.5	7.4	17.5	2.3	15.8	12.1	10.6	4.8	6.1	4.9	2.3
				Oolong tea	A	2.7	8.8	3.6	14.7	6.3	16.5	8.8	20.6	6.9	11.4	16.3	15.2	19.1	9.1	9.8	7.5	11.9	9.6	15.1
					B	4.5	9.2	7.2	10.8	16.4	19.4	34.9	5.5	23.7	25.4	8.9	1.3	5.8	19.9	19.0	1.7	20.1	28.6	8.3

(Continued)

Appendix Table 5.2 RSD of 201 Pesticides in Youden Pair Aged Tea Sample for Parallel Determinations (Nov 9, 2009–Feb 7, 2010) (cont.)

No.	Pesticides	SPE	Method	Sample	Youden pair	RSD%																		
						Nov 9, 2009 (n=5)	Nov 14, 2009 (n=3)	Nov 19, 2009 (n=3)	Nov 24, 2009 (n=3)	Nov 29, 2009 (n=3)	Dec 4, 2009 (n=3)	Dec 9, 2009 (n=3)	Dec 14, 2009 (n=3)	Dec 19, 2009 (n=3)	Dec 24, 2009 (n=3)	Dec 14, 2009 (n=3)	Jan 3, 2010 (n=3)	Jan 8, 2010 (n=3)	Jan 13, 2010 (n=3)	Jan 18, 2010 (n=3)	Jan 23, 2010 (n=3)	Jan 28, 2010 (n=3)	Feb 2, 2010 (n=3)	Feb 7, 2010 (n=3)
67	4,4-DDE	TPT	GC–MS	Green tea	A	6.2	3.9	2.9	3.3	7.3	4.7	6.8	3.2	3.5	2.8	6.0	2.0	6.4	2.7	2.6	2.1	3.6	7.5	4.0
					B	5.5	8.5	5.0	8.4	3.3	4.0	3.1	1.7	5.7	8.6	3.3	7.1	1.4	1.8	5.3	2.2	4.7	2.4	5.5
				Oolong tea	A	7.0	3.3	0.5	3.6	3.3	2.2	10.7	4.8	1.8	2.1	2.1	2.7	6.1	2.1	8.6	3.0	3.7	3.2	0.2
					B	4.8	4.2	3.3	3.6	0.7	2.8	4.7	1.2	5.3	3.9	2.4	2.7	2.3	3.2	4.3	1.2	1.6	2.4	1.2
			GC–MS/MS	Green tea	A	4.8	3.4	3.7	1.6	9.0	2.1	6.2	4.0	3.7	0.8	5.0	3.2	10.6	2.2	3.9	3.4	2.7	5.5	4.3
					B	7.6	8.9	7.7	8.6	7.0	5.5	8.7	3.5	6.1	9.2	6.3	7.9	1.0	4.4	2.0	6.4	4.9	3.6	5.6
				Oolong tea	A	7.0	5.0	2.2	3.6	3.3	0.4	7.0	0.6	2.9	4.7	3.0	3.0	9.6	4.0	8.8	6.1	4.3	3.8	1.5
					B	7.5	4.7	7.5	5.6	4.7	4.3	5.1	5.1	3.7	3.5	5.9	1.4	2.9	2.4	5.9	2.8	6.1	3.4	7.4
		ENVI-Carb+PSA	GC–MS	Green tea	A	5.5	5.2	0.6	5.4	8.2	2.0	127.5	1.2	6.0	1.9	10.1	4.3	3.3	6.2	6.2	6.9	10.9	1.5	5.4
					B	3.1	7.1	3.9	4.8	2.4	0.7	27.0	5.6	3.9	1.4	21.4	1.7	1.3	3.6	11.7	1.8	5.3	6.3	5.7
				Oolong tea	A	6.5	4.1	1.8	2.6	7.1	0.6	4.3	1.8	1.4	3.4	4.7	3.5	1.6	3.5	2.5	1.0	5.6	4.0	4.8
					B	5.0	1.7	10.0	4.3	9.4	4.2	1.4	6.8	11.1	3.6	7.8	3.0	3.4	3.2	9.4	3.4	5.9	8.3	3.2
			GC–MS/MS	Green tea	A	6.7	8.1	3.2	5.9	8.7	1.4	3.1	3.5	6.6	1.0	13.2	5.3	4.6	13.0	6.2	9.3	10.7	2.2	9.3
					B	4.2	9.3	7.9	1.5	3.4	1.9	1.3	4.1	3.3	0.5	19.3	2.5	2.0	7.9	13.9	2.9	2.7	8.7	6.9
				Oolong tea	A	4.4	5.1	3.8	3.0	8.9	1.9	1.0	0.3	1.9	3.3	5.3	3.5	5.2	1.8	9.2	6.1	8.5	8.7	6.1
					B	6.6	1.5	9.3	4.3	5.7	6.5	5.6	6.2	9.8	3.3	0.2	4.1	5.7	9.0	9.9	2.8	3.8	1.7	3.0
68	Benalaxyl	TPT	GC–MS	Green tea	A	5.9	4.6	0.8	1.3	9.0	5.5	6.6	8.9	5.2	2.9	4.6	3.6	7.6	5.4	1.7	2.1	1.5	12.8	2.6
					B	5.9	9.8	6.7	8.8	2.6	3.7	2.3	5.3	3.8	5.3	4.1	6.8	4.1	1.6	5.5	9.8	10.2	15.0	4.9
				Oolong tea	A	10.4	1.2	1.0	2.9	1.9	4.1	11.2	6.5	7.2	4.5	7.1	1.7	3.7	2.7	0.8	2.4	7.5	3.6	6.3
					B	4.6	1.5	1.7	4.7	1.8	3.8	3.5	5.1	6.5	4.2	5.7	1.1	1.4	5.9	8.1	8.6	5.2	8.7	8.6
			GC–MS/MS	Green tea	A	4.2	7.8	3.2	4.0	14.1	3.8	9.0	6.9	6.5	3.0	7.9	4.9	2.9	3.9	4.0	8.9	9.5	6.1	3.9
					B	7.5	7.6	7.0	7.6	2.3	7.3	6.9	5.4	8.2	7.2	5.1	1.8	4.8	6.8	4.2	6.0	6.4	0.8	3.6
				Oolong tea	A	5.8	1.8	3.6	2.9	3.2	1.4	8.9	1.9	1.6	6.2	3.4	2.0	7.2	4.6	6.1	3.5	11.7	4.5	5.4
					B	2.2	3.2	3.9	5.9	3.9	1.7	0.8	4.9	8.3	5.2	6.2	3.5	2.8	6.8	6.1	3.7	2.0	7.4	8.6
		ENVI-Carb+PSA	GC–MS	Green tea	A	17.9	5.7	3.4	5.9	8.2	3.5	5.4	7.3	2.1	2.1	5.8	6.2	1.9	7.0	3.0	6.4	5.5	22.9	5.7
					B	5.9	7.2	6.0	2.4	2.4	1.3	12.3	4.5	8.6	5.2	21.0	5.5	10.3	1.3	13.5	12.2	9.4	14.5	11.3
				Oolong tea	A	4.3	1.2	3.2	1.2	5.2	2.6	4.0	10.0	1.8	1.2	4.7	0.8	0.6	2.7	7.3	4.0	4.1	11.1	7.6
					B	4.8	3.3	9.0	3.3	11.7	6.2	8.4	8.8	12.2	2.4	8.4	2.1	3.9	4.3	19.4	4.8	11.9	5.4	5.8
			GC–MS/MS	Green tea	A	13.5	7.9	3.8	2.6	9.3	10.1	7.1	5.3	5.3	3.5	8.2	6.9	6.3	9.8	8.0	6.0	8.9	1.7	1.8
					B	16.1	8.8	3.9	6.3	2.4	2.7	18.5	5.6	7.8	5.3	11.7	11.6	3.5	13.2	10.6	10.3	7.7	20.4	16.0
				Oolong tea	A	5.2	3.3	4.9	1.1	5.1	3.9	1.2	7.1	0.8	2.6	9.8	4.2	1.7	4.8	9.2	3.4	11.0	3.1	3.5
					B	8.8	4.5	10.5	5.8	4.2	4.5	5.3	6.9	11.7	3.7	2.0	5.3	2.6	8.9	16.9	3.7	6.6	2.8	6.8
69	Benzoyl-prop-ethyl	TPT	GC–MS	Green tea	A	5.0	6.2	1.5	2.2	7.4	3.8	7.1	6.0	4.2	1.6	3.8	2.3	5.6	2.2	1.7	0.4	4.8	3.3	3.1
					B	4.9	5.5	3.5	9.7	3.2	2.5	0.3	4.0	2.5	3.3	3.2	6.1	3.3	5.9	4.8	3.7	5.8	1.5	4.0
				Oolong tea	A	6.9	1.5	1.3	1.3	2.5	2.9	7.0	4.6	2.0	0.4	3.5	2.7	0.9	1.4	7.9	3.1	2.9	4.3	0.4
					B	4.4	2.1	3.2	7.1	1.6	2.6	5.3	2.4	8.1	4.5	0.9	2.5	3.8	5.8	1.8	1.6	2.7	1.2	1.0
			GC–MS/MS	Green tea	A	4.7	4.8	1.5	1.6	9.0	3.7	9.1	8.6	5.3	1.8	6.5	4.4	2.6	8.7	4.1	2.8	3.5	2.0	4.1
					B	5.7	5.1	5.6	8.7	6.7	4.4	6.7	6.4	2.3	4.3	5.6	4.1	4.8	3.8	2.5	4.1	6.5	2.5	3.2
				Oolong tea	A	8.2	2.1	2.6	1.4	2.8	3.2	6.1	1.0	3.0	5.3	1.7	3.2	3.7	1.7	4.7	3.1	1.3	6.6	4.6
					B	3.0	3.4	4.7	6.9	5.6	3.8	1.9	4.6	2.3	4.4	6.9	2.7	1.6	9.0	4.4	2.5	1.7	1.6	1.5
		ENVI-Carb+PSA	GC–MS	Green tea	A	8.9	3.5	5.6	7.0	8.0	4.3	5.4	3.3	7.2	1.8	6.3	2.7	3.1	2.7	3.0	5.4	10.3	4.2	5.3
					B	6.6	6.3	5.0	3.7	1.4	2.9	10.7	1.8	5.9	4.8	15.7	5.2	4.9	4.8	8.7	5.3	3.1	10.8	8.5
				Oolong tea	A	4.5	1.8	4.4	0.2	5.9	2.0	2.1	6.3	1.3	0.8	6.2	0.6	0.8	3.6	2.0	2.3	5.0	5.6	3.4
					B	5.4	3.6	8.9	4.3	10.9	4.6	4.1	5.6	10.4	1.1	8.7	2.9	5.4	8.2	11.3	3.5	5.5	1.0	4.1
			GC–MS/MS	Green tea	A	9.1	3.9	6.8	7.6	7.8	6.7	4.0	3.9	8.6	1.9	6.2	15.8	6.7	8.4	5.9	5.4	9.8	2.2	3.7
					B	9.9	6.7	6.4	4.5	2.0	1.9	12.1	1.5	8.2	3.3	14.4	5.2	1.0	4.1	11.4	4.8	7.6	12.9	9.0
				Oolong tea	A	6.4	2.6	5.0	1.6	4.6	1.5	0.0	6.2	1.0	2.2	8.4	3.1	6.5	5.6	9.1	3.1	6.7	4.4	2.7
					B	8.4	3.9	8.9	5.0	6.1	3.6	2.0	6.2	9.9	3.2	1.5	3.1	2.1	6.6	11.1	2.5	5.8	3.6	6.1

70	Bromofos	TPT	GC-MS	Green tea	A	5.9	3.6	3.5	3.0	9.3	3.0	6.1	5.5	5.8	4.3	3.6	3.8	3.8	5.8	4.9	3.9	3.8	5.9	3.0
					B	8.7	9.6	4.4	5.5	5.7	4.3	1.4	0.7	2.5	6.0	2.2	4.2	3.8	4.7	9.3	2.4	6.6	6.8	9.5
				Oolong tea	A	7.7	6.3	2.4	3.8	3.2	4.6	13.5	11.2	5.9	8.4	5.1	1.2	3.2	2.4	9.4	2.5	6.8	4.0	4.3
					B	5.5	2.5	4.7	4.2	1.2	0.9	3.4	1.8	5.4	5.4	5.6	0.5	2.5	4.4	9.6	0.4	2.8	2.8	3.8
			GC-MS/MS	Green tea	A	5.1	4.7	9.9	8.5	11.5	9.4	12.1	6.2	10.8	6.9	5.1	7.9	5.5	6.7	5.2	4.5	17.7	9.4	9.0
					B	8.4	13.1	7.6	4.5	1.2	8.0	9.6	3.3	6.2	6.6	5.8	1.5	2.4	1.8	4.7	6.5	5.3	6.0	9.6
				Oolong tea	A	11.5	12.4	9.8	7.9	9.5	4.2	16.9	5.1	3.2	12.4	5.9	7.4	4.4	5.6	9.7	9.0	10.8	0.5	5.8
					B	7.0	4.9	10.7	2.6	11.8	3.6	0.9	0.2	14.7	10.1	3.7	2.2	11.1	2.1	4.6	3.4	10.8	17.8	5.9
		ENVI-Carb+PSA	GC-MS	Green tea	A	11.1	6.2	3.7	8.4	7.9	8.3	7.8	5.3	3.5	1.4	15.1	7.8	2.0	3.1	4.9	3.4	9.5	11.3	4.4
					B	13.9	7.5	2.2	6.4	2.0	0.8	9.6	1.0	5.9	9.0	14.5	9.8	8.3	2.4	4.4	9.9	2.5	13.6	8.0
				Oolong tea	A	5.8	0.9	4.0	6.4	4.1	2.8	10.0	4.1	5.8	3.7	2.0	2.2	3.1	4.4	6.2	2.2	3.4	7.7	4.5
					B	9.0	5.8	9.2	7.9	7.9	5.5	5.7	6.5	11.6	0.4	9.6	3.2	6.9	9.0	12.5	4.2	8.2	6.8	4.9
			GC-MS/MS	Green tea	A	16.0	0.6	3.4	13.1	2.1	17.8	6.6	7.3	11.1	4.4	10.1	0.5	7.3	7.8	4.1	5.0	3.3	9.1	12.6
					B	32.6	13.1	2.3	0.6	5.2	5.9	13.2	6.5	9.1	15.7	9.8	0.5	11.0	7.5	6.3	7.7	11.9	19.0	12.2
				Oolong tea	A	3.2	3.4	6.4	11.9	7.0	5.0	3.8	3.5	6.2	2.8	8.5	2.3	8.6	16.2	16.2	9.0	10.5	8.9	8.4
					B	6.1	5.8	11.9	5.7	14.1	4.5	8.9	5.1	15.5	4.0	5.3	3.2	9.5	12.2	11.3	3.4	3.6	6.4	3.7
71	Bromopro-pylate	TPT	GC-MS	Green tea	A	4.6	5.3	1.3	11.9	11.9	1.7	13.2	12.9	11.2	2.6	6.9	7.2	8.3	9.5	5.2	7.3	8.6	7.7	3.7
					B	5.6	6.4	2.2	4.5	4.5	3.5	2.9	12.0	1.9	5.9	3.0	10.2	6.1	0.2	4.2	8.4	13.1	12.6	3.6
				Oolong tea	A	7.3	5.7	4.0	1.8	9.1	4.1	3.5	19.1	6.1	7.6	14.3	10.8	3.3	5.5	6.2	6.1	4.0	2.8	4.0
					B	4.1	3.1	3.2	4.0	5.4	8.6	15.2	6.5	7.9	4.5	11.6	2.3	0.9	6.5	4.8	1.0	5.5	4.5	12.5
			GC-MS/MS	Green tea	A	6.9	6.4	1.4	7.0	10.8	3.3	13.5	11.4	12.7	1.5	10.3	6.7	1.1	0.4	4.6	4.0	5.8	10.1	5.6
					B	7.1	10.9	3.7	9.1	6.8	3.6	2.9	9.8	1.4	6.3	3.8	7.7	5.8	3.0	3.1	5.5	7.7	2.9	5.4
				Oolong tea	A	8.9	13.9	2.2	1.6	7.4	5.0	3.0	1.9	11.9	6.9	9.8	6.2	2.1	1.6	4.9	6.3	4.5	7.6	3.0
					B	6.3	4.5	4.2	6.3	6.8	7.7	9.3	13.1	9.7	2.0	11.1	5.6	4.7	33.6	5.9	2.7	3.1	4.3	16.9
		ENVI-Carb+PSA	GC-MS	Green tea	A	21.1	8.7	12.6	9.4	9.4	12.9	49.8	17.2	20.3	4.6	0.9	15.3	1.8	5.9	6.1	46.4	12.6	31.7	6.9
					B	15.0	3.2	5.2	5.1	4.9	5.5	6.8	25.0	22.9	3.9	36.5	11.5	42.1	0.2	46.1	1.3	7.3	29.5	14.7
				Oolong tea	A	4.3	6.4	5.6	6.1	5.1	2.1	3.0	31.7	0.1	2.6	5.5	6.9	4.9	5.5	0.7	14.4	5.4	14.3	6.0
					B	7.2	7.5	8.0	3.8	14.6	7.7	12.5	4.8	12.5	4.5	9.9	1.5	4.3	3.6	12.3	2.5	15.2	8.8	6.7
			GC-MS/MS	Green tea	A	24.5	10.3	11.5	7.0	7.5	10.9	27.5	16.4	7.7	1.7	1.5	2.4	8.7	9.6	4.6	18.7	7.5	14.1	12.2
					B	22.7	0.3	7.3	6.5	3.5	5.0	21.8	16.7	19.4	4.6	31.0	1.8	30.1	4.3	17.5	2.3	8.9	17.6	17.3
				Oolong tea	A	12.2	5.6	4.8	3.7	3.7	1.7	2.3	24.5	2.4	1.5	7.7	7.8	1.4	8.2	8.5	6.3	6.3	8.7	6.3
					B	14.5	5.3	7.0	1.0	11.4	4.7	11.0	5.3	8.3	3.8	3.3	2.2	2.4	3.4	13.6	2.7	9.5	12.9	8.2
72	Buprofen-zin	TPT	GC-MS	Green tea	A	6.1	8.9	4.8	3.6	9.5	6.6	6.4	4.8	2.2	5.1	7.1	7.6	3.5	6.8	3.4	4.3	3.8	23.8	4.6
					B	7.5	6.4	3.5	7.3	3.9	4.8	5.0	2.5	6.6	11.5	2.6	7.1	2.9	6.1	5.5	5.5	14.2	13.8	7.9
				Oolong tea	A	4.6	0.5	3.6	2.8	3.4	0.5	11.9	4.2	1.3	7.4	3.0	12.8	5.4	5.1	13.9	13.8	5.8	5.2	1.3
					B	13.4	4.5	3.2	3.1	3.1	2.4	4.8	4.5	3.8	2.0	2.0	12.9	3.6	1.2	2.7	2.4	2.0	8.6	5.2
			GC-MS/MS	Green tea	A	4.5	8.6	2.3	3.7	12.5	3.9	8.3	8.9	5.0	1.3	3.6	8.5	8.5	4.0	1.1	1.2	4.2	6.2	1.8
					B	7.7	2.6	9.7	8.9	7.1	5.7	7.8	3.0	4.1	7.6	2.0	2.2	6.0	7.9	5.1	3.7	14.3	9.5	4.2
				Oolong tea	A	7.0	1.4	1.7	2.9	1.4	4.4	9.7	4.3	2.5	4.4	7.1	9.4	8.8	6.3	6.7	5.2	5.8	1.2	2.9
					B	4.3	7.2	3.5	4.2	2.8	4.2	3.0	7.0	4.5	5.2	9.5	4.5	1.8	14.6	8.3	0.9	2.9	11.3	6.7
		ENVI-Carb+PSA	GC-MS	Green tea	A	6.7	4.8	2.8	5.5	8.9	1.3	3.2	2.7	5.2	3.5	33.7	6.7	10.6	16.5	8.0	19.4	23.5	4.5	15.9
					B	3.6	6.4	3.6	3.5	3.1	2.1	3.6	7.9	2.3	0.4	5.8	7.1	22.5	5.2	27.0	3.1	4.7	6.2	0.9
				Oolong tea	A	2.4	3.8	6.8	2.5	5.1	0.2	5.8	4.6	2.0	3.4	8.3	9.3	19.8	5.8	8.2	19.5	4.7	11.9	23.2
					B	3.5	6.5	9.9	6.0	5.3	4.5	5.4	5.9	9.3	5.5	8.6	10.5	4.6	10.3	14.0	4.2	2.9	1.5	5.0
			GC-MS/MS	Green tea	A	5.2	8.6	2.6	5.9	11.6	4.5	3.9	1.8	2.5	2.9	18.2	0.8	11.0	6.6	9.6	11.0	18.6	1.1	6.2
					B	5.1	3.9	5.7	1.3	9.2	2.7	0.8	4.9	9.7	0.7	5.9	1.7	9.4	8.1	21.1	2.9	8.1	12.2	9.3
				Oolong tea	A	2.5	4.8	3.2	4.7	8.3	1.6	1.8	4.5	3.3	2.5	2.8	1.6	8.8	12.6	13.0	5.2	7.3	14.7	7.6
					B	5.6		8.7	3.7	6.4	9.7	1.9	5.0	10.3	6.0		5.3	7.2		10.7	0.9	3.2		4.9

(Continued)

Appendix Table 5.2 RSD of 201 Pesticides in Youden Pair Aged Tea Sample for Parallel Determinations (Nov 9, 2009–Feb 7, 2010) (cont.)

No.	Pesticides	SPE	Method	Sample	Youden pair	RSD% Nov 9, 2009 (n=5)	Nov 14, 2009 (n=3)	Nov 19, 2009 (n=3)	Nov 24, 2009 (n=3)	Nov 29, 2009 (n=3)	Dec 4, 2009 (n=3)	Dec 9, 2009 (n=3)	Dec 14, 2009 (n=3)	Dec 19, 2009 (n=3)	Dec 24, 2009 (n=3)	Dec 14, 2009 (n=3)	Jan 3, 2010 (n=3)	Jan 8, 2010 (n=3)	Jan 13, 2010 (n=3)	Jan 18, 2010 (n=3)	Jan 23, 2010 (n=3)	Jan 28, 2010 (n=3)	Feb 2, 2010 (n=3)	Feb 7, 2010 (n=3)
73	Butachlor	TPT	GC-MS	Green tea	A	5.4	3.4	2.6	1.0	6.9	6.2	6.1	5.8	10.8	9.0	5.7	2.5	5.7	3.6	3.5	6.0	2.5	7.9	4.6
					B	5.5	8.0	5.3	7.7	2.3	5.5	1.5	1.6	1.2	11.5	0.8	2.3	2.9	3.0	10.3	1.0	6.0	5.2	7.2
				Oolong tea	A	8.0	4.4	0.7	2.1	3.3	2.5	11.1	16.1	1.2	4.0	2.0	1.6	1.3	2.6	8.9	3.4	7.7	3.0	1.8
					B	4.8	5.3	6.9	5.9	1.0	2.5	3.9	0.9	2.1	5.9	3.4	1.9	2.1	3.0	2.3	0.8	2.5	1.3	5.8
			GC-MS/MS	Green tea	A	5.5	2.4	3.6	1.2	11.1	2.3	0.4	4.6	5.2	4.1	10.0	6.7	6.6	3.9	3.7	3.4	4.0	10.1	1.8
					B	7.1	9.4	5.6	8.2	6.4	6.5	8.5	1.9	7.3	6.2	2.5	5.3	1.0	6.0	4.1	0.5	1.8	5.8	7.0
				Oolong tea	A	7.9	10.0	3.3	5.5	3.6	4.7	8.9	1.5	2.6	1.9	2.3	1.0	7.3	2.9	8.0	5.1	8.6	4.0	4.4
					B	2.5	1.6	4.6	5.3	3.9	4.1	4.8	1.7	2.2	7.0	9.1	3.8	2.6	2.2	9.4	6.0	3.8	3.2	7.6
		ENVI-Carb+PSA	GC-MS	Green tea	A	7.7	2.9	3.0	5.7	6.6	3.5	4.4	7.0	3.4	2.2	14.6	2.5	2.4	3.4	2.1	9.0	11.0	12.1	9.5
					B	11.6	9.2	5.1	4.6	1.5	2.7	2.9	4.5	2.7	5.7	22.9	4.1	4.4	5.1	10.1	4.7	4.3	12.0	10.3
				Oolong tea	A	4.8	1.5	3.7	3.0	4.6	3.8	4.2	16.8	8.5	4.8	0.5	2.8	2.9	3.3	5.1	4.2	3.8	7.8	2.3
					B	12.6	1.9	5.6	5.3	10.2	5.4	2.5	7.0	9.3	3.5	8.2	2.5	7.0	4.1	10.7	2.5	9.9	1.8	10.5
			GC-MS/MS	Green tea	A	4.7	6.8	1.2	8.2	6.8	5.2	4.4	3.6	5.6	3.0	8.7	6.3	3.8	14.2	5.7	9.7	10.6	4.0	8.9
					B	5.6	8.8	6.7	3.6	3.3	2.2	0.2	8.6	5.8	2.2	20.4	4.6	2.4	12.3	13.0	3.9	8.0	8.6	6.4
				Oolong tea	A	5.4	7.4	2.1	2.2	9.8	1.5	3.8	2.8	2.1	2.5	9.2	6.9	2.9	9.8	13.2	6.0	7.9	2.9	5.2
					B	7.5	0.7	9.4	4.8	3.2	5.0	4.7	10.8	11.3	5.8	2.6	5.2	5.6	9.4	10.1	14.3	5.7	7.1	4.3
74	Butylate	TPT	GC-MS	Green tea	A	9.9	5.7	3.2	3.0	12.9	9.3	5.6	3.7	6.4	8.2	0.6	1.8	13.1	5.6	0.4	4.2	4.5	7.6	3.6
					B	8.4	9.3	6.1	7.7	3.0	5.1	2.6	14.1	4.2	15.7	3.0	7.9	4.7	10.7	4.6	5.8	4.7	2.1	2.8
				Oolong tea	A	6.2	5.0	21.9	3.2	27.9	10.3	8.3	2.4	34.6	6.2	1.7	7.0	14.8	3.6	12.7	1.9	2.4	2.8	37.7
					B	5.7	5.9	4.1	6.4	5.0	7.2	7.1	0.9	6.3	4.2	8.5	0.7	12.4	5.5	27.9	5.7	3.0	4.8	4.0
			GC-MS/MS	Green tea	A	2.6	2.3	2.9	2.7	11.0	9.6	6.8	3.3	4.5	6.7	3.2	3.3	5.4	4.7	2.2	4.3	5.9	5.1	4.6
					B	9.2	7.7	5.3	6.9	4.9	7.6	5.7	3.2	8.6	14.2	3.1	5.9	7.9	2.5	2.2	4.1	6.1	2.2	5.6
				Oolong tea	A	7.0	1.5	1.0	1.7	4.8	0.9	11.1	0.7	8.0	5.2	6.7	5.8	14.0	3.4	8.5	2.7	4.1	1.7	6.7
					B	4.3	4.8	6.9	6.7	6.4	5.2	3.6	6.2	1.4	5.4	46.8	3.0	1.9	6.2	9.0	16.3	20.5	7.2	4.4
		ENVI-Carb+PSA	GC-MS	Green tea	A	20.2	48.1	31.8	14.2	15.9	20.2	4.3	2.9	7.6	3.5	30.6	19.8	18.7	128.0	28.6	7.4	19.8	7.8	9.8
					B	17.9	49.0	55.0	37.6	7.7	41.4	5.8	3.0	1.7	12.6	6.9	6.5	15.9	28.2	33.9	12.5	18.0	20.5	10.9
				Oolong tea	A	33.6	46.1	10.4	6.2	17.1	5.9	1.0	6.9	1.8	1.9	9.7	4.2	6.9	3.6	4.5	5.0	18.0	10.1	4.9
					B	16.7	16.4	7.9	4.4	13.2	5.1	1.1	46.2	11.6	0.6	9.9	2.8	22.2	5.5	5.7	14.6	7.3	5.9	1.8
			GC-MS/MS	Green tea	A	5.8	10.3	2.7	6.8	5.4	6.3	4.1	1.5	5.8	1.0	25.2	32.3	13.4	14.5	8.7	3.1	11.6	9.4	7.6
					B	5.2	6.1	6.9	1.5	1.0	0.9	10.5	2.4	3.0	0.9	8.2	15.5	19.6	20.2	35.5	4.1	10.8	4.7	3.8
				Oolong tea	A	3.5	2.8	4.1	4.8	7.0	2.0	2.3	4.1	2.7	3.6	3.2	2.7	4.6	6.4	7.5	2.7	7.3	2.7	4.3
					B	10.8	6.9	6.4	1.6	7.0	3.4	6.4	11.4	10.6	2.6	9.6	2.8	2.5	11.6	7.4	1.8	3.2	14.5	5.7
75	Carbofeno-thion	TPT	GC-MS	Green tea	A	6.1	4.7	2.5	0.7	5.7	1.0	8.0	6.0	3.1	3.3	8.1	6.3	9.0	2.7	2.6	4.1	5.3	7.7	5.1
					B	7.2	7.4	2.4	10.1	7.0	3.1	1.7	2.1	2.7	5.0	12.5	10.3	1.8	0.5	3.5	3.9	7.0	2.0	2.6
				Oolong tea	A	7.7	4.1	5.6	1.9	5.7	2.2	6.0	13.2	3.9	10.2	15.9	0.7	5.1	3.8	5.4	6.1	3.6	0.5	2.6
					B	3.9	0.7	1.2	3.9	7.8	4.6	5.2	0.7	6.9	8.1	8.9	0.3	3.1	6.8	1.1	9.7	21.3	15.6	17.8
			GC-MS/MS	Green tea	A	5.8	3.3	3.3	1.1	10.0	3.8	8.8	3.6	3.4	1.6	9.3	9.1	0.7	11.5	12.8	12.7	2.6	9.5	3.8
					B	8.3	8.5	2.1	4.5	7.5	8.9	5.6	2.6	5.1	9.8	8.0	6.2	9.6	6.7	17.5	4.9	6.1	20.7	27.7
				Oolong tea	A	5.2	2.5	1.6	9.6	2.5	5.8	5.5	8.8	26.3	7.5	7.5	4.3	11.5	22.2	4.8	3.0	14.5	13.6	29.8
					B	3.9	9.7	8.1	13.6	13.1	3.0	7.3	3.9	3.0	9.3	4.9	1.7	5.5	7.1	25.4	8.5	12.5	11.6	18.5
		ENVI-Carb+PSA	GC-MS	Green tea	A	11.9	3.0	6.4	3.0	7.1	2.8	4.7	4.1	1.0	3.0	4.0	9.8	0.3	6.8	6.5	2.8	6.8	12.7	7.2
					B	5.5	8.1	6.0	2.8	2.9	4.1	5.9	1.6	7.1	8.0	31.7	12.7	6.1	2.5	15.1	5.6	2.8	8.2	8.7
				Oolong tea	A	5.8	3.4	7.4	0.3	3.3	2.6	7.9	1.8	2.4	5.1	4.7	4.9	4.8	2.8	3.7	5.6	8.2	3.7	6.4
					B	3.6	3.7	8.1	1.5	15.7	4.8	5.8	6.1	11.4	2.9	12.3	11.3	9.2	2.6	12.0	15.0	14.2	3.7	6.9
			GC-MS/MS	Green tea	A	7.7	7.5	6.3	4.4	5.7	4.3	9.5	4.0	7.6	1.0	8.5	9.8	15.6	12.1	30.4	9.1	2.6	15.1	24.5
					B	13.7	11.2	3.4	4.2	7.9	3.3	7.8	5.9	6.7	19.3	19.3	8.3	12.1	13.9	19.5	4.9	6.1	8.4	15.3
				Oolong tea	A	11.2	4.1	5.5	4.7	4.6	4.9	2.1	5.6	5.9	5.5	5.5	6.3	10.8	4.9	22.8	3.0	19.3	17.2	23.4
					B	7.6	6.5	11.9	2.8	8.2	9.2	10.0	0.8	12.3	3.3	12.2	6.4	9.9	10.7	28.7	3.0	16.5	17.2	23.8

76	Chlorfenson	TPT	GC–MS	Green tea	A	8.9	5.9	1.7	15.8	12.9	2.4	15.8	12.2	12.4	2.3	9.6	5.0	11.3	15.6	6.5	11.4	8.1	6.1	5.2
					B	8.5	8.5	7.3	21.2	7.3	5.7	7.1	11.0	2.1	5.4	2.9	4.5	14.1	9.8	6.3	7.7	16.1	7.9	5.9
				Oolong tea	A	8.0	4.5	5.6	4.9	12.8	2.6	5.4	13.9	11.3	3.0	13.1	10.9	5.5	1.5	4.5	8.2	5.9	3.9	4.7
					B	3.2	4.1	5.1	0.7	4.6	9.7	13.0	3.2	8.9	4.5	11.0	4.7	0.4	9.7	11.5	3.0	3.1	2.4	9.6
			GC–MS/MS	Green tea	A	16.4	6.9	3.3	14.2	14.8	1.5	21.8	16.9	20.7	1.9	16.4	10.7	11.6	8.6	2.7	11.0	4.2	15.5	6.4
					B	11.3	7.6	4.6	17.2	5.4	5.5	4.8	12.2	4.3	9.0	2.7	4.8	15.6	6.5	7.2	5.9	17.9	8.4	6.8
				Oolong tea	A	10.8	8.4	5.9	2.7	12.3	1.4	8.2	0.2	9.8	5.3	13.1	4.9	3.9	2.7	7.2	7.0	11.8	16.2	12.0
					B	9.3	8.5	6.9	3.2	8.2	12.7	12.7	6.9	17.4	3.3	9.6	7.5	5.7	4.5	10.5	8.7	4.1	6.7	15.6
		ENVI-Carb +PSA	GC–MS	Green tea	A	27.9	8.8	20.9	5.3	10.0	32.9	8.9	22.7	12.1	5.6	3.8	30.8	4.4	24.4	11.5	13.3	13.7	15.7	8.8
					B	36.8	10.4	4.8	10.6	15.5	8.2	8.9	19.0	16.8	13.9	11.9	22.4	45.4	9.7	8.4	12.7	8.0	27.3	16.6
				Oolong tea	A	6.6	3.0	7.6	14.5	3.8	3.7	10.4	19.1	3.4	1.0	2.9	11.4	16.6	1.5	2.8	22.4	5.2	21.9	8.6
					B	13.9	10.5	7.1	10.0	16.1	12.9	16.0	9.9	14.7	8.5	10.4	9.2	6.6	9.7	15.8	4.0	12.6	27.8	9.3
			GC–MS/MS	Green tea	A	29.7	4.4	20.6	6.2	8.4	24.8	22.4	28.3	8.7	5.1	4.5	3.2	4.6	16.0	7.1	12.0	14.0	15.3	8.9
					B	39.6	11.9	4.4	21.0	17.2	7.5	32.1	23.3	23.6	14.9	6.3	0.6	34.5	8.6	6.4	14.9	7.1	25.8	16.0
				Oolong tea	A	12.7	1.3	7.0	11.4	8.7	2.2	7.4	23.6	3.5	2.1	3.7	15.7	13.2	3.6	6.3	7.0	7.9	12.5	9.9
					B	13.8	9.8	7.5	4.0	14.1	10.5	21.4	7.5	12.7	9.3	7.1	6.1	2.6	11.7	16.7	8.7	10.4	24.1	9.5
77	ChlorfEN-Vtnphos	TPT	GC–MS	Green tea	A	5.9	4.8	1.7	11.4	12.5	4.0	8.0	8.0	5.3	3.9	6.5	3.9	7.1	6.1	12.5	3.7	6.5	6.5	8.6
					B	9.0	9.6	3.9	8.5	5.5	4.8	2.0	3.1	1.2	2.4	7.0	6.0	10.2	8.6	12.8	6.7	9.8	8.2	6.7
				Oolong tea	A	7.7	1.1	1.8	3.6	11.0	2.7	21.9	18.3	8.1	8.4	13.2	9.0	3.6	6.2	10.3	1.3	12.7	3.4	4.7
					B	5.0	3.0	5.0	2.8	4.2	3.9	3.6	8.2	4.2	5.4	15.6	3.8	8.3	7.1	5.9	3.4	5.1	1.7	3.3
			GC–MS/MS	Green tea	A	6.2	6.8	4.7	5.1	9.7	5.1	7.2	15.8	14.8	1.7	10.2	8.0	11.3	6.6	4.6	6.4	6.2	9.7	1.6
					B	6.1	9.4	5.4	5.3	5.8	1.7	6.8	6.1	5.4	10.8	3.4	3.3	5.9	3.5	7.1	9.9	11.6	0.7	12.9
				Oolong tea	A	6.8	1.8	5.3	4.0	8.7	3.1	15.1	1.4	10.6	9.1	8.7	6.2	6.8	5.7	8.7	7.5	14.2	5.7	6.1
					B	2.6	7.3	4.5	8.0	5.6	0.6	6.7	11.8	14.4	7.6	4.3	5.8	12.3	0.7	6.1	5.8	5.4	9.1	11.7
		ENVI-Carb +PSA	GC–MS	Green tea	A	14.9	4.8	11.4	7.8	10.7	12.2	2.1	13.8	9.8	4.3	14.3	14.4	4.4	13.2	7.4	5.0	6.0	16.5	6.5
					B	16.5	7.8	4.9	5.7	9.0	1.9	10.3	5.9	13.8	10.4	22.0	11.7	19.9	4.0	22.6	11.5	1.5	20.8	8.9
				Oolong tea	A	4.6	1.7	4.5	5.7	5.3	3.6	11.3	14.8	5.2	6.4	4.9	2.8	7.3	5.2	7.2	12.4	2.4	14.8	5.5
					B	9.9	9.2	7.3	4.6	14.0	10.9	7.4	7.4	13.5	0.1	11.5	3.4	7.2	3.1	17.7	5.0	16.6	6.7	6.7
			GC–MS/MS	Green tea	A	18.5	6.8	12.9	0.8	3.7	28.7	9.7	12.8	8.4	3.6	9.9	2.8	4.4	11.9	7.2	5.1	11.9	7.8	12.1
					B	46.8	14.7	2.9	15.5	7.8	7.8	18.1	12.4	14.2	16.6	9.2	0.7	9.9	18.3	1.1	1.8	11.3	18.7	9.8
				Oolong tea	A	10.6	1.0	5.2	2.0	2.5	1.6	1.8	13.2	3.3	3.5	5.1	0.3	6.3	5.0	9.8	7.5	5.1	3.2	3.0
					B	9.6	5.6	7.6	6.2	14.1	7.0	15.5	2.2	12.7	0.7	4.1	2.3	2.1	14.7	17.4	5.8	9.8	7.3	6.2
78	Chlorme-phos	TPT	GC–MS	Green tea	A	5.2	6.4	5.8	2.9	8.6	8.1	4.4	4.0	4.2	6.2	6.1	0.8	4.9	1.0	3.2	3.2	6.2	4.5	3.9
					B	5.7	8.2	3.6	6.8	4.6	5.1	5.3	2.5	3.8	10.8	0.4	5.4	5.6	3.5	3.4	5.8	7.6	5.2	4.5
				Oolong tea	A	3.7	2.0	1.3	5.9	5.2	1.3	10.7	2.9	8.6	2.0	3.6	1.1	6.2	2.4	8.5	3.6	4.8	3.6	2.5
					B	7.7	1.3	7.8	5.7	4.2	4.8	5.4	0.8	6.0	6.1	1.3	2.7	4.3	4.3	6.4	3.4	5.9	2.0	4.1
			GC–MS/MS	Green tea	A	1.8	3.6	3.9	2.0	12.4	7.9	8.0	6.1	4.9	4.9	7.7	1.9	7.2	6.2	4.7	7.8	5.6	4.9	6.5
					B	8.8	6.6	5.6	6.9	4.5	6.8	6.7	3.6	4.8	12.1	4.6	3.7	7.3	5.0	2.4	1.0	7.8	3.1	5.3
				Oolong tea	A	5.9	0.7	1.5	3.6	2.5	0.7	7.7	2.4	5.8	4.3	2.2	2.8	10.8	0.4	6.4	5.0	5.6	2.5	4.1
					B	4.2	3.9	5.9	6.8	5.9	5.0	2.8	6.3	5.7	7.1	6.3	4.3	2.6	5.8	10.4	4.7	3.2	5.4	3.6
		ENVI-Carb +PSA	GC–MS	Green tea	A	5.7	5.5	2.3	11.1	2.4	3.4	10.2	1.1	5.2	3.7	5.2	1.3	4.8	6.5	7.7	11.3	12.1	5.3	3.2
					B	3.4	5.1	5.1	3.7	3.3	2.1	6.1	2.2	1.8	1.7	26.6	2.1	18.1	0.6	32.3	1.4	6.1	7.6	5.9
				Oolong tea	A	6.2	8.3	4.5	7.3	4.0	1.8	1.5	8.1	1.3	1.3	5.3	3.8	1.7	2.6	3.4	5.4	5.0	5.9	6.1
					B	10.6	1.3	2.9	4.1	12.0	4.0	4.8	7.3	12.2	1.3	9.4	3.6	3.2	2.7	9.5	3.8	3.3	4.0	2.1
			GC–MS/MS	Green tea	A	7.8	7.5	1.4	6.7	5.2	3.1	4.7	0.7	5.0	2.3	8.8	5.8	10.6	12.8	6.5	12.5	12.9	5.6	2.1
					B	9.2	3.3	4.4	0.7	2.0	1.5	10.5	2.7	4.2	1.2	26.0	1.9	20.1	13.2	31.9	3.8	9.2	10.1	9.7
				Oolong tea	A	3.7	2.7	4.6	5.5	5.6	1.1	3.8	5.0	1.0	0.7	8.6	3.7	4.9	7.8	9.5	5.0	7.2	4.1	3.6
					B	9.1	7.4	6.8	0.3	7.5	3.6	6.7	9.4	9.5	1.1	2.7	4.7	1.8	10.7	11.8	4.7	7.1	2.4	3.5

(Continued)

Appendix Table 5.2 RSD of 201 Pesticides in Youden Pair Aged Tea Sample for Parallel Determinations (Nov 9, 2009–Feb 7, 2010) (cont.)

No.	Pesticides	SPE	Method	Sample	Youden pair	Nov 9, 2009 (n=5)	Nov 14, 2009 (n=3)	Nov 19, 2009 (n=3)	Nov 24, 2009 (n=3)	Nov 29, 2009 (n=3)	Dec 4, 2009 (n=3)	Dec 9, 2009 (n=3)	Dec 14, 2009 (n=3)	Dec 19, 2009 (n=3)	Dec 24, 2009 (n=3)	Dec 29, 2009 (n=3)	Jan 3, 2010 (n=3)	Jan 8, 2010 (n=3)	Jan 13, 2010 (n=3)	Jan 18, 2010 (n=3)	Jan 23, 2010 (n=3)	Jan 28, 2010 (n=3)	Feb 2, 2010 (n=3)	Feb 7, 2010 (n=3)
79	Chloroneb	TPT	GC-MS	Green tea	A	6.3	4.7	3.3	2.2	8.3	3.3	4.9	5.5	4.2	1.4	5.7	0.7	5.3	4.3	3.5	4.5	1.5	5.0	4.2
					B	5.6	6.3	5.1	7.5	2.5	4.5	4.7	2.9	7.4	8.3	1.1	5.5	4.1	1.7	3.5	2.5	5.4	2.4	7.9
				Oolong tea	A	4.1	1.7	2.1	1.9	2.4	1.5	7.5	4.4	4.0	2.2	1.6	4.3	2.1	0.9	6.9	4.2	3.0	0.9	2.3
					B	4.1	1.1	3.8	5.2	1.7	3.4	2.2	0.5	6.0	6.2	2.8	1.2	2.4	3.0	4.8	1.1	3.5	1.6	8.6
			GC-MS/MS	Green tea	A	3.5	5.0	2.1	2.0	10.1	3.5	4.1	4.5	4.9	3.0	7.7	4.3	6.6	7.0	1.7	5.0	1.7	4.1	6.9
					B	7.5	6.6	3.9	8.6	5.1	8.1	6.9	3.0	7.1	7.5	4.1	5.0	3.3	3.3	2.9	4.9	9.2	4.8	7.5
				Oolong tea	A	8.0	2.1	1.9	2.8	2.0	1.1	8.7	2.1	1.2	5.0	4.1	2.6	7.9	8.3	6.9	5.7	6.2	5.1	6.0
					B	4.1	1.4	7.8	5.4	3.9	5.5	3.9	5.9	2.0	2.8	6.4	1.9	2.7	1.9	4.6	0.6	3.7	0.2	3.4
		ENVI-Carb+PSA	GC-MS	Green tea	A	3.2	5.9	2.0	7.9	6.7	3.4	26.0	1.8	9.4	3.5	5.9	4.4	2.5	1.3	15.5	6.3	11.0	5.0	3.9
					B	4.0	7.9	4.0	3.9	3.6	1.7	5.0	5.2	3.1	2.5	18.7	2.6	4.4	2.9	1.4	3.3	6.9	3.5	5.3
				Oolong tea	A	5.1	2.0	5.4	3.5	1.0	1.1	2.6	7.4	2.0	15.7	4.7	3.7	1.8	2.5	8.5	1.6	6.2	4.0	4.7
					B	8.5	5.7	9.2	3.1	8.0	2.3	1.2	5.0	14.7	13.3	5.1	10.6	1.9	7.0	7.4	3.1	6.6	4.2	2.7
			GC-MS/MS	Green tea	A	4.9	5.2	1.7	5.9	5.7	3.9	1.9	3.0	6.8	1.2	7.8	9.0	18.3	13.5	21.6	8.8	8.1	4.8	3.1
					B	5.0	8.2	6.6	1.7	1.5	2.4	4.3	4.5	1.7	1.7	17.0	3.3	12.0	3.1	9.3	6.3	7.0	4.0	8.7
				Oolong tea	A	3.1	2.3	2.9	4.5	5.2	1.3	3.1	3.3	2.7	1.7	6.3	2.2	3.6	7.5	8.9	5.7	5.0	5.8	6.3
					B	7.7	5.7	9.8	2.6	5.3	4.1	4.3	7.5	8.3	2.9	3.0	4.2	2.3	9.0	7.7	0.6	5.8	7.0	3.2
80	Chloropropylate	TPT	GC-MS	Green tea	A	4.1	10.0	5.9	6.4	5.3	2.0	9.8	7.1	4.9	5.8	4.5	4.8	9.9	4.5	4.8	4.2	7.0	3.0	5.6
					B	6.2	7.8	4.1	13.0	10.1	4.8	2.8	9.7	3.3	13.2	6.5	13.2	6.5	2.4	7.3	6.5	7.7	7.8	5.7
				Oolong tea	A	4.4	1.3	3.0	3.2	0.7	3.3	0.2	11.9	3.9	4.3	3.3	3.1	3.0	2.8	5.1	2.5	3.2	2.2	1.7
					B	3.0	2.6	3.1	4.7	3.0	4.4	9.9	6.9	6.1	3.1	6.1	0.2	1.7	4.3	0.8	0.8	0.6	1.7	3.8
			GC-MS/MS	Green tea	A	6.8	5.7	3.1	2.3	9.5	2.8	6.7	11.1	6.0	2.4	4.9	4.1	7.2	2.1	2.2	3.5	3.8	7.2	6.0
					B	9.0	8.6	4.8	7.2	6.1	4.5	4.3	5.6	4.1	7.7	6.1	5.9	3.2	2.4	5.4	6.5	10.1	1.2	6.9
				Oolong tea	A	2.8	5.7	2.9	0.7	2.6	4.1	3.9	0.2	5.8	5.2	4.9	5.9	5.6	5.9	7.1	3.8	6.0	4.9	4.9
					B	13.8	7.1	4.3	7.1	5.3	5.4	5.0	8.6	6.6	3.8	6.1	3.4	2.1	16.7	8.8	2.1	2.9	5.0	10.5
		ENVI-Carb+PSA	GC-MS	Green tea	A	7.4	5.7	6.8	6.8	5.4	8.7	2.8	12.3	12.4	2.7	7.1	6.0	1.3	4.4	35.3	37.8	8.4	33.6	5.9
					B	4.3	7.9	4.4	9.9	5.5	5.1	2.9	12.1	16.7	2.7	37.2	6.1	27.7	2.4	1.1	5.5	5.4	27.4	9.4
				Oolong tea	A	5.9	3.6	3.0	5.2	3.3	1.8	2.3	19.7	1.6	0.3	4.3	3.3	1.5	2.8	10.1	3.6	5.9	7.9	4.1
					B	12.9	4.4	9.3	1.2	10.8	6.4	3.7	7.4	10.4	1.5	7.2	3.7	4.6	19.4	4.7	3.1	7.4	1.6	4.8
			GC-MS/MS	Green tea	A	11.1	7.0	4.0	4.9	6.9	2.5	15.3	6.8	3.0	1.4	6.5	5.3	17.1	14.9	21.5	17.6	9.1	6.6	10.1
					B	6.8	8.8	6.0	1.5	1.5	1.5	12.6	4.8	9.4	1.2	26.2	8.7	20.1	5.2	9.0	4.0	9.0	12.4	11.9
				Oolong tea	A	10.1	2.5	5.6	2.8	2.6	1.4	0.5	13.2	0.8	1.6	6.7	3.0	3.5	8.9	15.0	4.7	7.6	11.2	3.1
					B	5.0	3.8	8.8	1.2	11.2	5.5	7.4	6.7	10.0	2.7	3.1	4.0	3.0	7.1	5.6	2.1	4.5	19.7	8.6
81	Chlorpropham	TPT	GC-MS	Green tea	A	6.2	4.7	2.8	1.1	9.3	2.8	7.2	10.9	6.2	1.8	5.3	2.1	7.1	4.9	7.9	1.4	10.0	9.6	4.9
					B	4.6	9.4	4.2	8.8	3.9	2.4	1.7	4.0	2.9	5.9	4.0	7.9	1.3	5.3	9.8	3.8	5.0	2.1	5.9
				Oolong tea	A	3.3	2.5	2.4	2.8	3.9	5.2	0.3	16.7	3.9	6.1	13.2	6.3	5.7	8.3	4.0	2.4	2.4	2.9	2.2
					B	4.0	2.0	4.7	7.9	4.2	2.7	9.2	28.4	7.8	3.6	8.6	5.3	2.2	4.3	2.6	0.7	1.5	2.8	4.2
			GC-MS/MS	Green tea	A	6.7	5.5	2.0	3.2	9.6	4.4	10.3	12.8	7.9	1.4	6.6	3.1	6.3	5.9	1.6	5.2	6.5	4.9	3.7
					B	5.4	7.3	5.0	7.5	5.8	4.7	5.2	5.6	5.1	6.8	3.7	5.4	3.4	3.4	8.2	4.5	5.6	3.4	2.7
				Oolong tea	A	2.5	7.5	1.4	3.5	1.6	2.6	4.4	0.7	5.4	3.0	8.9	2.8	3.9	6.1	6.7	8.5	6.1	2.9	2.6
					B	10.8	5.7	4.9	5.7	7.9	6.3	4.2	10.6	7.2	3.7	7.3	2.7	4.0	16.1	5.8	2.9	1.3	2.1	7.3
		ENVI-Carb+PSA	GC-MS	Green tea	A	5.8	8.3	4.7	7.3	1.5	2.0	11.5	9.1	11.3	3.2	5.5	2.9	2.5	2.3	29.9	27.6	11.8	27.0	5.4
					B	12.6	4.8	3.7	13.5	5.2	4.4	32.1	7.8	10.2	1.2	27.9	3.9	22.4	2.5	1.2	3.0	5.0	21.6	8.7
				Oolong tea	A	26.6	3.2	9.7	6.1	11.9	8.0	3.6	16.9	1.4	4.9	7.4	3.7	2.7	9.5	13.4	6.6	5.8	9.2	2.2
					B	18.6	3.2	7.0	5.5	9.6	6.2	6.2	5.7	9.7	2.9	14.4	2.8	5.9	10.6	3.9	3.3	9.8	0.5	5.6
			GC-MS/MS	Green tea	A	15.0	14.6	7.0	1.7	1.4	4.3	31.7	9.8	4.2	1.4	4.9	2.0	9.0	5.5	18.5	17.8	8.6	10.4	11.3
					B	6.3	4.9	4.0	4.3	2.3	0.7	23.0	8.2	14.3	2.5	28.0	4.7	18.2	3.4	9.5	1.0	7.1	12.6	13.0
				Oolong tea	A	10.5	3.3	10.2	2.3	7.4	5.3	2.7	10.0	1.2	2.7	6.9	3.9	0.4	9.5	11.0	8.5	6.0	4.0	2.8
					B							10.5	6.7	9.5	3.3	3.6	3.6	1.9			2.9	6.8	4.8	7.0

No.	Compound	Extraction	Instrument	Matrix		1	2	3	4	5	6	7	8	9	10	11	12	13	14	15	16	17
82	Chlorpyrifos (ethyl)	TPT	GC-MS	Green tea	A	6.5	4.0	4.4	3.0	10.7	2.9	6.5	6.8	5.1	4.8	5.8	3.8	4.1	1.0	3.3	4.3	3.4
					B	6.1	8.1	2.8	9.4	4.9	2.8	1.4	3.7	2.5	7.6	3.4	2.2	5.6	2.3	3.1	2.1	6.3
				Oolong tea	A	5.3	1.0	2.2	3.1	2.6	3.5	14.6	11.1	5.0	7.6	9.4	1.5	8.2	3.2	4.0	0.8	1.4
					B	3.5	3.3	3.7	4.3	3.7	3.5	4.3	3.0	11.5	3.6	6.5	3.5	6.4	0.6	1.3	1.0	2.4
			GC-MS/MS	Green tea	A	3.7	6.5	5.2	1.1	12.3	2.6	7.6	9.0	6.7	2.2	7.2	1.2	3.5	4.3	6.5	5.0	8.1
					B	7.0	9.9	5.4	7.7	7.3	4.7	6.8	3.1	1.7	9.2	2.2	3.5	0.9	3.8	4.8	4.1	5.1
				Oolong tea	A	4.7	5.5	2.0	2.3	3.2	2.4	10.3	1.5	5.9	4.9	6.0	3.5	7.2	6.0	8.1	5.3	3.4
					B	2.2	3.4	6.2	7.6	8.3	5.8	2.7	10.9	3.4	2.7	5.0	2.2	6.9	5.0	2.2	5.2	6.3
		ENVI-Carb+PSA	GC-MS	Green tea	A	13.4	8.2	6.0	5.3	7.6	8.7	2.4	9.3	9.6	2.1	7.3	4.2	2.9	6.0	8.7	12.2	5.0
					B	12.0	7.3	4.4	4.6	4.9	4.6	4.6	7.6	0.3	4.2	16.0	2.7	7.0	7.3	3.3	13.7	8.2
				Oolong tea	A	4.8	3.6	3.7	3.6	2.9	1.0	5.7	6.0	12.1	0.5	4.0	2.2	2.8	5.1	4.6	7.1	4.2
					B	6.-	4.1	8.5	0.7	14.1	6.6	2.3	4.8	11.6	1.8	8.6	2.4	11.2	3.3	8.8	4.4	4.8
			GC-MS/MS	Green tea	A	15.1	5.8	6.0	5.1	5.6	11.0	4.5	8.2	11.6	2.1	8.3	7.4	3.9	3.6	9.2	1.7	4.5
					B	26.7	10.2	3.0	8.6	1.9	5.3	18.7	4.1	8.4	7.8	10.3	5.6	9.9	4.0	5.1	17.0	12.2
				Oolong tea	A	5.5	3.5	5.6	3.8	2.1	1.2	0.5	7.2	1.3	3.8	9.2	8.5	9.7	6.0	5.0	9.1	4.4
					B	10.7	4.5	7.7	2.9	10.7	5.7	6.6	2.3	10.1	1.6	5.5	13.1	12.3	5.0	8.4	8.6	10.9
83	Chlorthiophos	TPT	GC-MS	Green tea	A	5.4	4.2	2.0	0.6	8.2	3.9	7.4	5.5	3.1	2.7	6.2	3.1	6.9	4.0	1.7	10.6	5.7
					B	5.5	7.8	3.7	7.7	4.0	2.8	1.6	2.0	2.3	5.1	5.6	6.5	6.2	13.5	10.0	3.4	6.7
				Oolong tea	A	7.3	1.3	1.5	2.8	0.8	2.4	8.7	10.7	3.6	6.7	7.5	6.0	2.6	3.1	10.6	2.9	5.9
					B	4.3	2.6	3.6	4.5	2.1	2.8	4.4	2.9	5.6	4.3	8.4	4.8	11.9	11.2	7.5	1.2	2.5
			GC-MS/MS	Green tea	A	5.1	7.5	3.7	4.1	8.2	4.8	10.4	6.5	5.3	6.4	2.7	5.0	4.3	0.9	5.6	4.4	1.9
					B	8.2	7.5	3.9	7.5	5.3	4.1	8.9	2.3	6.5	9.5	4.3	5.9	3.7	3.5	2.1	3.8	6.7
				Oolong tea	A	5.7	0.8	1.8	1.4	3.7	1.6	11.1	2.5	6.1	3.8	3.0	4.6	8.6	6.7	8.0	0.2	11.2
					B	3.4	0.4	6.7	5.2	4.0	3.2	4.5	6.2	3.7	3.8	4.4	8.5	5.2	2.4	5.3	5.3	5.6
		ENVI-Carb+PSA	GC-MS	Green tea	A	9.5	2.7	4.2	4.8	9.0	2.9	17.1	4.0	3.2	3.0	8.7	2.3	8.9	2.4	7.6	8.6	8.7
					B	6.3	7.9	4.3	3.0	1.5	0.9	11.0	0.4	0.2	4.5	24.1	6.1	17.5	11.8	9.5	10.2	10.4
				Oolong tea	A	3.7	2.4	5.1	1.2	3.8	2.4	4.0	1.5	11.6	4.0	4.4	6.0	7.8	2.9	7.7	5.7	3.8
					B	5.2	3.1	8.9	3.0	11.4	4.8	2.8	6.2	7.1	3.2	8.2	3.5	17.9	6.6	6.8	2.4	4.9
			GC-MS/MS	Green tea	A	9.4	4.9	4.6	7.4	7.5	5.0	6.2	1.1	8.2	0.9	5.7	14.4	4.8	3.2	9.8	2.4	5.6
					B	15.5	9.4	3.6	7.7	2.9	3.0	11.1	1.7	2.3	4.5	16.5	3.3	11.4	3.8	9.6	11.7	12.9
				Oolong tea	A	7.4	0.5	4.1	3.4	5.2	0.4	1.0	5.3	11.3	1.6	7.7	2.9	10.7	6.7	4.5	3.5	2.7
					B	6.9	2.9	10.8	4.5	10.1	4.9	1.8	5.1	4.9	1.1	1.8	10.9	12.8	2.4	5.2	3.6	8.1
84	Cis-Chlordane	TPT	GC-MS	Green tea	A	5.3	4.8	3.6	1.5	9.2	4.0	6.8	5.6	5.4	2.3	4.5	3.3	1.6	1.2	4.8	4.8	4.1
					B	5.7	8.3	4.4	8.7	3.4	4.0	2.6	2.9	3.4	6.2	2.4	1.2	3.7	3.6	4.8	0.7	5.7
				Oolong tea	A	5.3	3.6	0.7	4.2	2.3	3.7	8.0	6.3	5.5	1.0	4.2	3.0	8.6	2.4	3.4	2.0	1.4
					B	4.8	2.0	3.8	3.0	1.0	2.4	3.2	3.7	3.4	0.9	1.0	3.6	4.9	0.4	1.3	1.0	2.1
			GC-MS/MS	Green tea	A	4.9	5.3	4.9	2.0	11.7	2.9	12.1	8.6	5.5	5.9	11.1	1.5	3.5	5.6	6.2	5.2	3.3
					B	9.8	9.8	6.7	11.0	3.3	5.8	7.1	2.4	5.3	8.0	3.1	4.3	4.3	4.9	3.7	4.1	4.2
				Oolong tea	A	4.4	6.7	1.6	3.1	2.2	1.7	10.0	2.1	7.6	2.8	2.6	4.6	8.2	9.8	1.7	5.9	5.1
					B	6.5	0.3	3.1	7.4	3.6	5.6	1.4	0.5	2.2	2.3	12.6	2.0	8.2	5.5	6.0	10.3	6.8
		ENVI-Carb+PSA	GC-MS	Green tea	A	13.3	4.8	3.5	5.1	8.8	3.6	10.8	4.1	4.9	1.6	10.4	3.3	2.1	6.1	10.1	6.8	5.4
					B	6.5	7.7	3.2	2.0	0.6	1.4	5.7	1.1	4.8	4.3	17.7	1.2	9.2	3.8	5.0	11.3	7.3
				Oolong tea	A	0.8	3.2	2.4	4.0	5.9	1.2	3.1	3.1	2.1	3.2	4.7	3.0	3.3	2.8	5.7	4.8	4.6
					B	5.5	3.1	9.7	2.8	8.9	5.6	3.0	7.3	12.1	4.0	8.4	3.6	11.0	2.7	7.4	2.1	5.4
			GC-MS/MS	Green tea	A	12.0	4.9	4.1	6.3	11.0	4.4	6.0	3.0	7.1	3.2	9.9	6.8	9.4	4.9	10.2	4.2	8.1
					B	13.9	9.1	6.1	4.4	1.0	4.7	14.5	7.6	4.5	4.0	18.3	12.5	7.7	5.4	15.4	13.3	4.3
				Oolong tea	A	2.8	4.0	2.7	3.7	4.5	3.3	2.7	0.6	4.1	5.5	5.9	4.0	4.7	9.8	11.1	8.0	5.0
					B	9.0	4.5	7.8	5.0	10.5	5.5	4.5	9.0	7.7	2.5	4.1	15.4	16.9	5.5	3.9	3.0	9.3

(Continued)

Appendix Table 5.2 RSD of 201 Pesticides in Youden Pair Aged Tea Sample for Parallel Determinations (Nov 9, 2009–Feb 7, 2010) (cont.)

No.	Pesticides	SPE	Method	Sample	Youden pair	Nov 9, 2009 (n=5)	Nov 14, 2009 (n=3)	Nov 19, 2009 (n=3)	Nov 24, 2009 (n=3)	Nov 29, 2009 (n=3)	Dec 4, 2009 (n=3)	Dec 9, 2009 (n=3)	Dec 14, 2009 (n=3)	Dec 19, 2009 (n=3)	Dec 24, 2009 (n=3)	Dec 14, 2009 (n=3)	Jan 3, 2010 (n=3)	Jan 8, 2010 (n=3)	Jan 13, 2010 (n=3)	Jan 18, 2010 (n=3)	Jan 23, 2010 (n=3)	Jan 28, 2010 (n=3)	Feb 2, 2010 (n=3)	Feb 7, 2010 (n=3)
85	Cis-Dial-late	TPT	GC-MS	Green tea	A	6.2	4.7	2.8	1.1	9.3	4.8	6.4	5.6	5.0	2.9	5.5	1.4	6.9	5.0	2.9	2.0	7.7	5.1	3.9
					B	6.4	8.9	4.6	8.6	3.3	2.4	2.1	2.9	4.1	9.5	3.1	6.8	0.5	5.3	4.2	0.9	8.2	0.6	5.5
				Oolong tea	A	3.2	2.1	1.0	3.7	2.9	2.0	7.1	4.5	3.7	3.1	3.8	0.5	3.9	2.3	7.9	4.0	3.0	0.8	2.1
					B	4.1	3.1	3.8	3.2	1.6	3.5	4.1	3.8	6.3	2.5	2.8	0.5	4.3	3.3	4.8	0.4	1.1	1.7	2.8
			GC-MS/MS	Green tea	A	3.7	3.1	3.2	2.0	9.9	4.2	8.1	6.8	5.5	0.7	7.5	3.5	5.2	6.8	1.5	5.1	3.7	3.0	4.7
					B	9.0	7.8	4.3	9.6	4.9	7.8	6.3	2.1	4.3	9.7	3.5	5.0	2.1	1.4	2.8	4.9	7.2	6.1	3.4
				Oolong tea	A	5.2	2.1	0.8	3.3	2.6	3.0	8.6	0.6	0.8	4.3	3.3	1.3	9.7	3.5	9.8	6.9	6.0	3.5	5.2
					B	2.5	0.7	5.5	6.9	3.3	3.8	1.2	5.9	3.2	4.3	4.1	2.7	1.7	5.1	6.2	1.8	4.0	0.7	4.2
		ENVI-Carb+PSA	GC-MS	Green tea	A	8.8	3.3	3.5	5.3	6.2	4.2	10.3	4.3	6.0	0.7	7.5	2.1	3.1	1.9	1.1	5.1	9.7	4.8	4.9
					B	6.6	8.5	5.0	0.8	2.6	3.4	3.6	4.1	3.4	3.7	17.5	0.6	5.0	0.6	12.0	3.1	4.7	9.9	6.1
				Oolong tea	A	5.3	2.1	3.8	3.2	3.7	0.4	1.5	3.2	0.5	2.6	4.7	3.0	2.8	3.4	1.4	3.6	5.3	8.7	4.0
					B	5.8	3.0	7.9	3.4	11.6	4.7	0.7	6.8	10.7	1.7	7.6	1.8	2.5	3.3	8.9	3.5	5.7	4.1	4.5
			GC-MS/MS	Green tea	A	7.3	6.3	1.1	5.9	6.7	2.2	2.8	4.2	6.6	2.7	9.6	4.8	18.1	18.5	3.2	6.2	13.9	4.5	4.1
					B	6.0	6.1	5.4	3.5	1.6	2.3	7.3	0.6	2.6	1.5	15.8	0.8	6.6	11.2	13.1	3.6	10.2	10.9	5.9
				Oolong tea	A	4.0	2.8	4.2	3.3	5.0	1.8	1.0	2.6	1.6	2.0	7.2	3.6	2.8	4.6	9.9	6.9	6.2	4.5	4.2
					B	6.9	4.1	9.1	1.4	7.3	3.8	3.5	8.4	9.1	2.6	2.8	2.1	1.8	8.4	11.0	1.8	5.8	3.7	7.5
86	Cyanofen-phos	TPT	GC-MS	Green tea	A	5.2	6.1	2.2	12.2	13.6	5.1	7.4	11.8	7.9	3.1	6.0	5.0	7.8	8.8	3.7	3.4	6.0	10.4	4.2
					B	6.2	6.8	4.3	14.6	5.0	3.4	1.4	4.7	1.7	4.2	4.0	4.1	12.4	4.9	4.6	13.0	11.6	2.0	5.6
				Oolong tea	A	6.8	4.5	1.3	2.0	10.1	2.1	11.0	10.0	6.3	6.3	10.1	10.2	1.7	0.6	11.6	14.1	2.2	2.6	7.1
					B	3.2	1.8	4.0	4.9	6.1	3.2	5.1	2.2	3.6	1.9	8.7	1.6	1.7	2.0	5.2	9.6	3.0	2.9	3.9
			GC-MS/MS	Green tea	A	8.1	7.8	1.7	1.7	13.9	2.4	14.6	16.4	14.9	6.8	11.2	5.3	10.1	9.7	2.0	5.1	7.1	13.3	2.3
					B	7.9	3.8	7.9	10.1	6.7	3.1	5.3	10.9	3.4	5.9	2.3	2.4	14.4	2.7	3.4	5.5	11.5	3.1	3.5
				Oolong tea	A	9.3	5.7	7.1	12.0	6.7	1.5	1.7	5.2	5.8	6.9	8.0	3.7	3.3	5.6	6.4	1.6	7.1	9.5	5.2
					B	6.3	4.6	6.3	2.4	2.8	3.2	7.4	10.4	11.5	6.7	5.8	5.5	4.4	3.9	8.7	18.3	1.6	4.6	9.9
		ENVI-Carb+PSA	GC-MS	Green tea	A	20.0	8.5	17.9	3.7	9.4	11.1	4.3	8.9	6.6	2.1	2.2	27.4	4.5	6.6	12.7	11.8	9.4	19.3	6.8
					B	25.5	4.4	5.3	8.1	10.6	1.7	5.6	10.6	15.9	13.6	17.4	21.9	24.8	13.4	17.9	14.3	4.2	14.5	14.2
				Oolong tea	A	4.3	2.9	5.9	10.0	3.9	5.2	2.9	14.2	2.5	2.4	4.0	4.6	4.0	0.6	6.3	0.6	3.4	16.2	3.8
					B	10.0	4.3	7.5	7.9	15.7	2.8	6.3	12.5	15.1	2.1	16.8	3.9	3.4	1.5	13.6	13.7	8.0	14.4	7.3
			GC-MS/MS	Green tea	A	24.7	9.6	17.2	3.3	6.8	25.1	6.3	20.0	11.3	4.6	2.7	1.5	5.6	21.5	7.0	19.6	13.9	6.7	6.3
					B	36.6	8.7	3.0	2.9	10.1	3.6	27.4	18.6	17.6	17.1	11.1	2.4	18.3	13.3	11.3	5.5	1.7	18.8	13.2
				Oolong tea	A	10.8	5.0	6.3	23.9	1.3	3.8	4.2	18.7	3.4	3.2	12.0	9.9	9.8	4.6	10.5	1.5	7.9	12.1	5.7
					B	11.9	8.9	9.8	6.4	11.7	8.0	15.9	5.7	12.2	6.9	3.7	5.0	9.4	9.6	15.4	1.4	8.5	20.8	8.6
87	Desmetryn	TPT	GC-MS	Green tea	A	6.7	4.8	2.4	2.8	8.1	3.5	9.7	6.0	2.6	3.6	4.0	5.8	7.6	4.1	1.1	4.0	5.8	5.7	4.7
					B	4.3	8.1	7.6	8.3	3.9	3.4	2.8	2.2	5.6	8.3	11.4	10.4	0.9	3.2	4.5	0.5	8.1	1.0	4.7
				Oolong tea	A	6.6	2.4	3.2	3.4	2.2	3.5	5.2	7.6	6.7	13.0	9.1	2.5	2.6	4.3	7.7	1.2	3.3	3.7	1.6
					B	4.5	2.9	4.0	7.3	2.4	2.3	9.0	4.1	7.2	4.9	6.6	1.2	2.0	3.0	3.9	5.1	2.5	0.8	3.8
			GC-MS/MS	Green tea	A	4.5	4.3	3.0	2.2	9.4	6.8	6.4	8.1	4.1	1.9	7.8	7.7	7.4	5.5	2.9	4.0	5.3	3.3	4.9
					B	7.6	8.0	6.3	9.1	5.6	6.5	8.2	5.8	5.8	9.7	4.8	6.1	3.2	3.1	4.5	7.1	4.7	1.4	5.0
				Oolong tea	A	7.4	1.9	1.7	3.7	0.8	4.1	9.2	4.6	1.9	3.4	3.4	1.2	2.3	3.2	7.6	3.9	5.2	4.6	5.1
					B	4.3	1.4	4.9	7.1	4.3	1.6	1.9	6.5	5.4	2.7	6.1	0.5	3.4	3.7	6.5	6.3	4.0	3.7	5.7
		ENVI-Carb+PSA	GC-MS	Green tea	A	5.7	5.7	3.5	8.5	9.3	2.7	13.5	1.6	3.6	4.2	8.5	2.4	2.5	6.5	6.8	8.1	10.6	3.3	5.4
					B	3.6	8.1	2.5	6.2	1.7	2.8	4.9	4.3	5.8	16.1	26.8	9.8	2.8	2.5	13.3	2.8	4.7	7.7	7.1
				Oolong tea	A	3.8	4.5	2.8	1.1	4.0	0.9	3.0	1.7	5.3	6.2	4.5	5.6	4.1	4.9	3.5	4.2	6.3	3.1	4.4
					B	4.1	3.2	10.0	2.6	15.0	5.5	2.1	7.1	11.5	5.7	9.5	5.0	4.6	3.7	12.0	9.8	5.9	6.2	5.2
			GC-MS/MS	Green tea	A	5.2	7.0	2.1	5.9	9.7	1.2	1.3	1.5	4.7	0.8	9.2	15.9	9.9	16.2	10.4	1.3	7.8	1.0	6.6
					B	5.8	8.4	7.3	2.5	0.4	3.5	4.8	4.6	5.5	1.4	16.0	3.6	5.3	10.0	16.2	7.1	3.5	10.6	13.4
				Oolong tea	A	4.1	5.6	1.9	3.0	5.4	1.6	2.1	2.8	1.0	2.2	7.1	5.1	4.3	4.1	12.2	5.0	4.0	4.6	3.2
					B	4.9	2.4	10.3	1.7	8.3	5.7	0.4	7.9	9.6	4.1	3.1	3.5	3.2	9.9	11.7	3.3	0.8	0.8	8.5

No.	Compound	Sorbent	Instrument	Matrix	Rep	1	2	3	4	5	6	7	8	9	10	11	12	13	14	15	16	17	18	19
88	Dichlobenil	TPT	GC-MS	Green tea	A	4.7	5.2	3.2	2.4	7.9	8.6	4.4	3.0	2.0	6.4	6.1	2.4	4.1	1.7	1.3	1.9	4.6	12.1	3.8
					B	5.5	7.8	3.4	5.6	3.4	6.2	5.4	6.0	2.2	8.5	1.9	3.7	6.1	2.3	4.2	3.8	6.6	8.9	5.1
				Oolong tea	A	5.1	2.5	0.9	1.2	1.7	5.3	8.0	4.2	12.6	2.2	3.4	2.7	6.8	1.8	7.5	3.1	6.9	1.5	4.7
					B	8.6	6.5	5.6	5.5	2.5	5.6	6.6	9.3	5.3	5.6	4.2	2.7	2.5	4.0	9.6	1.4	2.3	1.2	5.1
			GC-MS/MS	Green tea	A	3.4	5.3	3.1	3.1	13.0	8.1	9.5	8.3	4.0	7.2	5.4	5.1	2.1	2.4	1.3	5.8	5.6	5.3	2.3
					B	9.0	5.8	3.4	6.5	8.2	7.1	5.7	5.4	3.8	13.0	3.2	4.1	6.7	3.3	2.2	1.7	6.9	3.7	0.7
				Oolong tea	A	7.4	2.8	5.3	2.3	4.6	2.3	5.0	3.0	9.5	3.3	3.3	4.7	12.5	1.8	7.0	7.5	3.9	1.8	9.5
					B	6.7	7.0	9.2	3.6	5.9	2.6	2.7	3.8	1.8	3.3	9.0	4.9	3.0	3.9	11.9	5.6	8.8	6.1	5.5
		ENVI-Carb +PSA	GC-MS	Green tea	A	6.5	6.3	2.6	5.3	6.0	6.4	4.8	7.9	9.6	6.0	7.4	8.4	4.2	6.3	3.8	16.5	14.8	6.8	7.2
					B	3.8	5.6	2.2	4.4	4.1	2.3	3.9	9.8	2.7	3.5	31.3	2.1	10.6	5.5	35.4	5.6	5.0	3.9	6.3
				Oolong tea	A	5.1	7.6	4.8	3.6	5.0	1.6	3.1	10.2	1.0	4.7	5.2	3.4	1.5	1.8	2.8	7.9	4.6	6.7	4.9
					B	10.5	6.8	7.1	4.4	8.9	4.1	7.3	11.1	13.0	2.9	7.0	2.6	2.9	4.0	7.3	3.2	5.4	2.5	5.3
			GC-MS/MS	Green tea	A	6.7	9.3	1.7	6.0	6.6	7.3	2.7	1.5	3.8	2.9	4.1	6.4	11.3	12.7	12.5	19.8	13.4	4.4	2.8
					B	9.8	6.0	4.6	2.1	2.7	0.9	16.6	2.5	5.1	1.2	25.9	1.4	23.8	14.7	33.5	2.5	5.9	7.1	9.0
				Oolong tea	A	3.2	3.4	7.2	5.0	7.7	5.4	2.1	6.7	2.9	4.6	10.5	4.9	5.4	10.0	11.6	7.5	10.7	8.8	8.3
					B	10.5	8.2	7.9	3.4	8.4	5.3	11.8	9.3	7.9	1.3	4.0	3.3	1.7	9.1	11.1	5.6	9.1	8.0	9.6
89	Dicloran	TPT	GC-MS	Green tea	A	14.8	4.9	1.8	11.8	11.5	4.0	9.2	7.5	12.3	4.0	5.2	5.4	10.6	4.8	12.1	11.8	19.7	22.4	6.2
					B	12.0	12.6	8.1	17.5	11.0	6.1	5.4	12.4	2.6	4.3	3.9	6.3	9.4	6.6	13.8	5.8	8.2	12.6	10.8
				Oolong tea	A	4.1	5.7	2.6	6.7	5.8	4.6	12.1	8.8	10.1	8.4	14.6	13.3	9.0	3.0	8.5	2.9	7.2	4.4	6.9
					B	3.5	3.2	0.8	7.9	11.1	8.4	21.8	1.8	5.5	1.2	4.4	2.1	4.2	7.7	5.5	5.0	6.6	4.1	9.8
			GC-MS/MS	Green tea	A	8.6	6.3	6.2	6.2	9.4	0.9	9.2	9.8	14.1	3.6	14.9	7.5	7.2	5.0	2.2	8.5	7.1	10.7	2.7
					B	6.2	10.7	3.3	13.1	4.1	1.7	3.2	8.5	0.4	12.6	3.9	2.5	7.4	9.1	5.5	12.1	13.4	3.8	10.6
				Oolong tea	A	8.2	8.8	3.0	1.9	7.5	6.3	2.7	14.5	11.0	8.2	13.0	8.7	6.6	2.4	9.4	4.0	8.8	4.9	7.5
					B	3.8	2.8	1.6	4.5	12.0	4.2	4.2	7.6	14.4	1.2	14.2	6.9	10.8	7.4	8.9	3.8	3.8	7.7	5.9
		ENVI-Carb +PSA	GC-MS	Green tea	A	13.9	23.0	14.7	15.3	6.1	16.9	3.4	11.1	9.4	2.6	3.0	14.0	6.7	0.9	5.6	8.6	11.9	8.3	5.4
					B	8.3	8.5	3.6	10.7	10.9	4.2	3.0	18.4	8.9	4.6	29.3	16.0	19.4	9.7	18.2	4.7	18.8	22.2	10.7
				Oolong tea	A	6.2	3.2	2.5	4.8	2.1	2.4	11.5	0.6	3.7	3.5	3.8	2.8	14.8	11.4	4.4	14.0	9.2	18.2	6.5
					B	7.6	7.6	7.0	2.3	21.6	13.2	16.6	8.5	13.6	3.5	10.7	6.5	4.9	12.0	10.5	6.0	11.4	8.1	7.1
			GC-MS/MS	Green tea	A	10.3	11.9	14.2	5.1	0.7	11.1	26.9	12.8	9.8	7.7	4.5	4.9	11.3	5.0	4.6	11.4	10.9	3.4	7.7
					B	18.6	11.0	3.5	5.4	3.1	13.0	17.6	3.9	9.2	7.1	22.1	6.2	27.2	19.5	9.8	6.9	15.3	12.5	9.0
				Oolong tea	A	2.1	2.7	8.2	7.5	10.2	1.0	1.1	4.6	2.3	2.2	6.4	12.1	9.8	4.2	8.0	4.0	6.7	10.2	6.8
					B	2.6	9.0	7.7	7.4	13.1	12.1	19.1	3.0	12.8	4.5	7.7	5.4	2.2	9.9	9.4	3.8	7.6	5.2	8.6
90	Dicofol	TPT	GC-MS	Green tea	A	11.3	2.8	3.2	4.9	5.3	5.6	10.0	2.8	0.7	4.3	6.4	4.3	7.3	1.6	3.8	1.4	1.5	10.2	1.8
					B	9.5	7.8	5.4	9.0	3.2	4.7	4.3	1.7	5.3	5.8	4.8	5.8	4.4	4.7	6.6	2.6	8.0	5.7	5.6
				Oolong tea	A	9.9	5.0	8.1	7.5	3.2	6.5	8.4	6.1	2.6	6.1	2.8	6.1	20.4	1.7	2.7	4.1	7.2	9.0	1.5
					B	6.5	12.2	6.9	2.5	3.8	4.3	9.0	7.0	7.6	2.8	1.7	2.8	3.3	2.9	9.7	1.5	7.8	8.1	5.3
			GC-MS/MS	Green tea	A	6.2	5.0	2.3	3.8	8.5	4.5	1.5	3.2	3.6	1.3	4.0	1.3	7.5	2.4	4.3	3.6	3.5	6.1	3.6
					B	8.7	8.0	5.4	9.2	5.0	5.3	8.6	3.2	6.6	6.9	7.5	6.9	4.3	2.2	2.4	5.3	6.0	2.8	3.1
				Oolong tea	A	11.6	5.8	8.0	6.4	3.6	5.6	9.5	4.1	0.9	9.1	3.6	9.1	22.1	3.5	0.7	7.8	6.5	5.8	1.5
					B	4.6	8.8	9.0	3.5	1.6	4.2	8.8	3.1	5.9	2.0	10.5	2.0	5.2	0.6	14.9	1.8	10.8	6.0	6.0
		ENVI-Carb +PSA	GC-MS	Green tea	A	1.5	6.2	2.2	6.3	7.3	8.2	4.8	8.6	11.7	8.5	8.5	15.0	4.4	10.5	6.2	1.6	12.3	3.0	5.8
					B	0.3	7.0	2.8	6.7	5.1	3.3	4.1	9.8	4.3	5.9	15.3	4.5	14.9	6.6	8.5	2.3	4.5	2.5	3.3
				Oolong tea	A	0.9	11.1	5.0	3.5	9.5	10.4	10.1	2.5	2.5	8.5	5.1	4.3	5.3	1.7	1.5	4.8	2.3	4.3	7.6
					B	4.2	2.7	7.5	4.9	8.2	7.0	11.7	6.9	13.3	6.1	6.1	3.7	4.2	2.9	14.2	5.7	2.6	16.7	2.9
			GC-MS/MS	Green tea	A	6.4	4.8	1.6	7.3	6.8	5.8	12.0	4.9	9.9	2.8	7.9	8.5	4.8	20.1	8.6	3.6	9.4	1.0	10.4
					B	9.0	7.2	6.0	2.0	5.5	2.2	7.1	10.1	1.6	1.3	14.0	4.2	7.8	16.5	9.9	4.8	7.1	9.0	6.8
				Oolong tea	A	11.3	9.8	5.3	3.8	9.1	10.7	10.2	2.2	3.9	10.0	3.7	1.9	8.3	7.5	8.0	7.8	3.9	3.8	9.5
					B	6.9	1.9	7.6	4.8	5.1	8.4	21.1	6.3	12.8	7.5	0.4	5.5	6.9	10.1	11.7	1.8	2.3	9.2	3.7

(Continued)

Appendix Table 5.2 RSD of 201 Pesticides in Youden Pair Aged Tea Sample for Parallel Determinations (Nov 9, 2009–Feb 7, 2010) (cont.)

No.	Pesticides	SPE	Method	Sample	Youden pair	Nov 9, 2009 (n = 5)	Nov 14, 2009 (n = 3)	Nov 19, 2009 (n = 3)	Nov 24, 2009 (n = 3)	Nov 29, 2009 (n = 3)	Dec 4, 2009 (n = 3)	Dec 9, 2009 (n = 3)	Dec 14, 2009 (n = 3)	Dec 19, 2009 (n = 3)	Dec 24, 2009 (n = 3)	Dec 14, 2009 (n = 3)	Jan 3, 2010 (n = 3)	Jan 8, 2010 (n = 3)	Jan 13, 2010 (n = 3)	Jan 18, 2010 (n = 3)	Jan 23, 2010 (n = 3)	Jan 28, 2010 (n = 3)	Feb 2, 2010 (n = 3)	Feb 7, 2010 (n = 3)
91	Dimetha-chlor	TPT	GC-MS	Green tea	A	7.8	3.8	2.8	2.6	8.3	4.6	6.4	4.9	4.3	2.8	4.7	3.0	6.4	5.6	4.5	5.0	5.4	7.9	0.8
					B	7.4	10.0	8.6	7.0	3.6	5.4	5.2	3.8	2.7	3.4	4.2	3.7	4.9	4.0	10.4	2.0	7.3	6.4	7.9
				Oolong tea	A	5.6	5.6	1.4	4.2	4.5	3.4	13.5	9.8	4.2	7.0	3.4	2.1	4.6	6.1	10.9	1.6	7.0	1.5	2.9
					B	4.5	3.1	3.3	2.7	2.1	2.7	4.5	7.8	3.4	5.5	2.4	2.0	3.6	3.3	4.4	2.8	0.4	1.2	2.9
			GC-MS/MS	Green tea	A	4.1	6.7	6.2	6.2	6.6	2.0	2.8	6.1	6.4	2.0	11.1	5.2	9.4	2.4	3.6	5.5	2.9	3.3	2.3
					B	7.9	9.0	5.5	6.6	5.4	4.2	9.5	6.2	4.9	6.4	3.3	3.4	2.1	1.1	3.9	4.2	4.5	2.3	8.0
				Oolong tea	A	6.1	6.5	3.5	3.7	5.7	2.8	20.4	3.7	2.1	4.8	2.7	4.5	2.1	1.5	7.7	2.6	8.8	7.1	5.1
					B	6.7	1.1	5.5	7.4	2.6	4.2	5.9	8.2	9.1	9.3	2.4	3.4	9.8	6.9	4.3	3.2	3.3	1.9	4.4
		ENVI-Carb+PSA	GC-MS	Green tea	A	7.3	5.2	3.2	7.7	8.6	2.7	4.9	3.4	3.2	0.5	14.0	3.2	3.4	6.5	1.8	9.9	9.9	11.9	5.5
					B	11.3	8.6	5.0	0.9	1.5	0.9	6.8	1.2	5.5	6.3	14.2	5.0	3.9	3.0	15.5	9.0	6.2	10.9	6.6
				Oolong tea	A	4.1	3.7	2.4	7.2	5.0	2.1	7.9	3.3	3.9	4.8	2.0	4.5	2.5	0.9	7.5	4.1	4.5	6.8	4.8
					B	7.3	3.7	8.0	7.7	10.3	8.2	4.5	5.2	11.2	2.3	9.1	4.8	6.9	3.4	13.7	4.4	8.2	2.0	4.2
			GC-MS/MS	Green tea	A	9.5	5.4	3.1	3.4	3.6	9.9	11.8	2.3	7.8	2.5	12.5	0.5	2.3	6.6	4.6	1.7	10.9	3.6	4.6
					B	39.8	10.1	1.7	7.2	12.1	2.6	3.9	3.4	4.1	8.8	10.5	3.9	3.2	15.1	10.6	4.7	7.7	6.9	10.3
				Oolong tea	A	2.4	3.7	2.7	4.6	4.4	1.4	1.9	2.2	2.8	5.9	6.8	6.5	3.1	6.9	12.4	2.6	6.8	3.5	6.6
					B	7.3	3.8	7.9	8.7	7.7	5.0	2.3	6.0	9.5	3.3	1.8	3.2	2.0	12.2	10.7	3.2	7.1	1.8	6.4
92	Dioxacarb	TPT	GC-MS	Green tea	A	10.6	12.4	3.8	11.6	4.5	5.6	6.9	4.0	4.2	8.1	7.2	3.2	11.0	14.5	6.5	10.1	3.7	2.5	3.9
					B	7.9	11.2	6.4	13.4	2.9	9.3	3.0	10.5	4.9	15.7	7.8	6.0	2.1	6.8	15.7	6.4	10.0	4.8	6.3
				Oolong tea	A	12.7	5.1	3.2	9.0	4.6	4.8	5.3	3.7	2.4	6.3	0.6	2.1	5.3	2.6	8.2	6.2	6.9	4.0	2.5
					B	11.7	2.8	7.0	11.4	5.0	5.5	1.1	4.2	4.4	0.7	3.8	4.4	1.0	3.8	3.7	4.0	4.2	4.7	1.6
			GC-MS/MS	Green tea	A	8.4	18.7	3.9	10.8	7.2	7.2	3.4	4.7	5.2	9.7	7.0	6.3	9.9	22.7	7.8	12.6	0.8	1.6	5.2
					B	7.4	16.9	6.1	10.7	5.2	11.9	4.2	10.9	7.2	22.7	10.4	3.4	3.5	11.2	10.4	5.3	11.5	2.9	7.9
				Oolong tea	A	4.7	6.1	1.9	9.2	7.7	5.9	7.1	2.2	4.6	6.6	3.3	2.8	10.1	2.5	7.8	8.0	11.0	3.7	3.6
					B	4.9	6.3	12.6	12.3	6.0	5.6	2.2	14.5	6.9	2.4	10.8	2.6	3.0	1.5	4.5	11.3	3.0	1.1	14.1
		ENVI-Carb+PSA	GC-MS	Green tea	A	11.2	29.2	21.1	4.8	6.6	6.0	15.3	6.4	5.0	4.6	10.5	2.9	3.1	5.6	3.8	8.3	10.9	6.2	7.7
					B	8.4	4.2	4.7	4.6	2.7	1.8	9.4	2.7	6.3	4.5	24.5	5.9	13.0	2.1	14.4	3.0	4.1	11.1	8.0
				Oolong tea	A	7.5	6.3	7.5	5.4	5.9	7.5	6.3	10.0	1.5	19.1	10.3	13.5	3.0	4.2	13.7	5.2	3.6	5.9	2.1
					B	7.6	5.2	16.8	7.8	17.7	17.0	1.4	8.3	3.5	3.5	9.4	12.1	8.6	4.3	10.3	10.2	16.9	4.6	7.1
			GC-MS/MS	Green tea	A	16.2	27.3	20.4	6.5	6.7	3.3	53.9	8.3	3.2	2.7	9.3	5.6	11.5	19.1	6.5	10.0	10.2	8.8	9.5
					B	7.3	6.4	4.7	0.5	2.5	2.3	19.4	4.0	10.3	3.4	23.0	2.5	19.9	12.9	20.4	3.9	4.2	16.0	11.8
				Oolong tea	A	6.4	4.8	6.9	2.7	5.1	8.5	7.7	13.4	1.9	28.1	12.1	16.7	5.8	9.6	22.3	8.0	5.2	7.1	7.1
					B	10.1	3.5	26.5	10.1	13.8	19.9	2.4	9.2	0.2	3.9	7.6	11.9	4.6	34.9	8.1	11.4	19.4	4.4	9.7
93	Endrin	TPT	GC-MS	Green tea	A	4.8	4.6	3.2	2.2	8.9	4.8	6.9	5.2	4.4	3.0	7.1	2.5	6.9	3.8	6.2	1.0	4.7	7.1	6.3
					B	5.9	8.0	5.0	8.9	3.9	3.7	3.3	2.5	4.5	8.3	3.5	7.2	1.1	1.1	5.0	3.0	5.3	1.9	5.7
				Oolong tea	A	4.6	2.4	2.1	4.2	2.3	3.6	8.0	9.1	2.6	5.5	3.8	1.7	3.5	4.9	8.4	5.0	3.6	6.7	0.5
					B	4.7	2.6	2.7	4.0	2.2	2.3	4.3	3.6	6.4	2.1	0.3	2.8	1.9	3.9	3.1	3.6	1.3	3.9	2.4
			GC-MS/MS	Green tea	A	5.3	4.5	3.1	2.8	10.5	7.2	6.6	7.0	5.6	4.0	4.2	2.0	3.1	3.8	1.5	3.0	4.5	3.5	3.0
					B	6.9	8.8	5.2	6.8	5.5	5.7	8.6	2.4	6.6	8.6	2.7	6.4	4.2	2.5	1.3	6.2	5.7	0.9	4.4
				Oolong tea	A	7.3	3.7	1.3	3.1	2.9	5.6	10.4	3.5	3.7	4.3	5.8	3.4	3.8	3.8	7.8	4.4	5.0	3.9	4.5
					B	4.3	2.3	4.1	5.1	5.0	3.1	3.7	6.3	3.3	2.8	4.8	2.5	5.6	0.4	4.3	1.9	2.2	1.3	4.1
		ENVI-Carb+PSA	GC-MS	Green tea	A	8.3	4.8	2.1	5.6	9.2	1.6	13.9	1.6	4.0	1.9	8.1	2.2	2.4	2.7	7.3	10.3	8.8	7.0	5.4
					B	2.6	8.1	4.4	5.3	1.3	2.2	3.0	2.6	4.1	3.2	26.6	2.4	4.8	3.0	14.9	2.0	3.6	9.4	6.6
				Oolong tea	A	1.4	4.2	2.2	3.1	5.6	0.8	4.3	1.4	1.5	2.9	5.0	2.9	1.4	2.6	2.6	2.3	6.1	5.1	1.2
					B	4.8	3.0	9.0	2.3	11.1	6.0	3.0	7.0	7.0	4.8	9.0	1.4	4.7	3.9	10.3	3.5	7.1	3.8	2.6
			GC-MS/MS	Green tea	A	8.5	6.4	5.0	5.0	8.0	2.2	2.5	3.0	5.2	1.5	12.7	2.2	7.0	14.1	8.9	7.8	10.7	3.8	4.1
					B	3.7	7.5	6.4	2.6	1.5	1.8	1.5	4.8	2.7	1.2	21.3	1.0	5.0	6.7	14.4	3.5	7.7	9.7	6.6
				Oolong tea	A	3.1	5.0	4.1	5.5	4.9	0.8	1.7	1.6	3.4	1.8	5.8	3.0	3.1	7.0	9.9	4.4	7.1	4.0	3.9
					B	6.8	1.2	9.5	1.7	8.9	5.2	2.3	7.8	10.2	4.2	5.7	4.1	4.4	11.6	14.2	1.9	5.7	1.5	8.5

94	Epoxiconazole-2	TPT	GC–MS	Green tea	A	5.1	8.2	0.9	1.6	6.1	3.4	7.7	4.9	4.5	5.3	9.2	7.3	7.1	3.3	2.0	1.1	4.5	3.3	3.1
					B	3.5	6.7	3.6	8.6	3.4	1.6	0.4	8.1	2.7	10.5	8.4	9.4	2.8	1.4	7.7	3.5	5.4	2.2	4.0
				Oolong tea	A	7.2	5.0	3.9	1.1	1.5	2.9	10.1	12.6	4.1	11.1	13.0	4.6	5.2	1.7	5.1	3.0	2.1	1.8	2.4
					B	5.1	2.9	4.2	7.7	2.0	3.3	5.1	3.7	7.8	6.3	12.9	3.5	3.1	4.5	3.7	2.4	3.6	1.9	1.2
			GC–MS/MS	Green tea	A	4.1	9.1	0.5	1.9	9.0	5.6	6.2	7.4	5.1	4.3	10.4	3.2	2.7	7.2	3.2	4.2	3.3	2.7	4.3
					B	6.1	6.7	5.3	7.5	4.7	3.9	5.4	8.5	2.6	9.3	4.7	4.5	2.9	2.3	5.9	5.9	4.3	1.8	3.4
				Oolong tea	A	7.5	1.3	3.2	0.5	0.8	3.0	10.9	3.1	2.4	4.5	4.0	2.6	4.7	1.7	3.0	1.2	1.4	3.9	2.9
					B	3.2	3.9	5.5	2.4	6.3	3.2	2.1	10.4	5.1	4.8	9.0	1.5	3.4	8.6	6.5	1.4	2.3	2.4	4.0
		ENVI-Carb +PSA	GC–MS	Green tea	A	11.2	11.8	7.1	7.3	7.1	2.2	7.7	1.5	4.6	5.3	6.5	3.0	2.1	4.1	6.2	5.6	10.4	5.1	4.5
					B	4.3	5.6	3.9	6.2	1.4	2.6	2.9	1.2	8.0	2.7	28.9	7.0	6.7	4.4	7.4	3.8	2.4	9.7	7.7
				Oolong tea	A	5.5	2.3	6.4	2.2	4.1	3.7	1.5	10.4	5.2	14.6	7.7	2.9	1.6	1.3	3.5	2.5	2.4	5.1	3.7
					B	3.9	2.6	12.5	5.6	16.4	5.8	3.6	7.1	8.7	3.1	10.3	6.5	3.1	4.5	7.8	2.6	7.6	0.8	2.7
			GC–MS/MS	Green tea	A	9.9	12.1	4.8	7.4	6.3	4.4	31.5	3.2	8.1	1.2	5.4	5.8	8.3	7.1	6.0	5.7	10.5	3.4	2.5
					B	7.8	5.1	6.0	5.2	0.7	5.1	6.4	0.8	6.3	4.1	18.4	3.2	2.2	7.7	11.0	2.1	4.8	8.6	10.1
				Oolong tea	A	6.3	1.9	5.0	2.4	4.1	3.5	3.8	6.2	1.9	12.2	7.7	6.4	3.3	5.7	10.5	1.2	5.6	3.0	3.9
					B	6.2	3.4	12.6	5.8	12.8	6.2	2.1	7.9	4.9	0.4	2.7	4.1	2.0	14.6	11.0	1.4	8.6	2.3	4.4
95	EPTC	TPT	GC–MS	Green tea	A	3.8	3.3	3.3	2.6	9.7	11.4	5.3	5.5	6.7	9.7	10.0	1.8	7.0	5.1	2.9	3.9	6.6	4.4	4.1
					B	7.0	9.4	4.0	6.4	2.8	6.2	5.3	3.1	5.9	13.9	1.7	7.5	4.0	4.8	4.4	4.3	3.1	3.8	4.4
				Oolong tea	A	5.3	9.4	1.1	1.9	8.1	2.1	9.0	3.6	14.2	4.1	3.8	4.7	7.2	3.1	8.4	1.3	7.6	0.7	4.3
					B	8.6	8.8	4.4	4.5	4.6	7.1	7.4	2.3	6.6	8.6	16.8	4.6	3.3	5.9	8.0	2.5	4.8	2.1	6.2
			GC–MS/MS	Green tea	A	3.2	2.8	4.1	4.6	9.4	11.3	7.3	2.4	7.5	8.6	8.2	1.9	4.8	4.7	2.0	10.7	4.5	4.4	5.3
					B	9.4	10.4	3.4	6.3	4.5	8.1	6.7	2.7	7.2	17.7	4.7	1.3	6.3	7.5	2.6	1.6	7.5	4.4	4.0
				Oolong tea	A	9.4	4.5	1.0	2.6	8.3	3.7	11.2	1.6	14.9	7.4	2.0	9.2	13.5	11.3	9.5	3.4	14.3	1.7	7.5
					B	6.7	6.2	10.7	6.1	4.8	8.7	3.9	2.6	4.6	4.4	7.8	5.6	1.5	8.0	7.9	2.8	2.1	5.1	6.4
		ENVI-Carb +PSA	GC–MS	Green tea	A	6.4	8.2	5.1	4.6	5.9	4.1	3.9	3.0	7.5	8.0	7.4	1.2	3.7	12.4	5.6	17.2	11.3	8.7	9.4
					B	2.0	4.4	3.7	8.0	2.5	3.9	3.5	8.6	3.4	2.7	33.4	2.2	19.1	3.7	37.9	3.6	6.2	5.9	7.5
				Oolong tea	A	5.9	5.6	5.5	4.5	4.6	3.7	3.5	6.9	5.5	1.8	6.5	3.6	4.8	3.1	5.5	9.0	7.1	6.4	4.7
					B	13.9	8.3	7.4	6.3	12.5	4.5	10.8	10.5	13.4	2.2	9.3	2.6	2.0	5.9	5.3	4.2	7.6	1.3	6.1
			GC–MS/MS	Green tea	A	6.8	12.6	2.6	6.0	5.2	9.7	5.2	0.7	8.6	3.0	10.6	4.2	12.3	8.8	13.0	13.7	14.1	8.9	1.0
					B	6.2	9.2	6.8	4.8	1.3	3.5	13.2	3.9	4.5	1.0	28.4	2.4	25.1	21.0	43.5	1.7	8.5	9.3	8.9
				Oolong tea	A	4.2	3.9	4.3	4.5	4.2	4.1	3.5	4.4	0.6	5.1	11.0	1.7	6.0	8.6	9.7	3.4	9.3	6.5	4.9
					B	13.1	9.0	12.0	1.1	6.7	4.5	11.5	14.0	12.0	2.9	4.0	1.7	2.2	14.7	7.0	2.8	4.9	8.5	7.6
96	Ethfumesate	TPT	GC–MS	Green tea	A	7.0	8.3	2.9	1.8	7.0	6.8	15.8	3.9	10.8	3.8	9.0	5.0	9.1	0.8	1.9	4.0	3.0	3.6	5.0
					B	5.3	7.7	2.5	9.9	9.7	7.6	5.1	9.4	8.6	4.9	2.4	10.7	1.8	3.2	8.2	2.7	3.5	0.7	6.4
				Oolong tea	A	6.1	2.7	3.3	6.1	13.9	4.4	29.0	24.9	4.9	5.2	5.7	2.6	5.6	3.5	7.1	5.0	6.3	2.4	1.0
					B	2.8	2.1	2.9	7.8	7.6	3.8	12.0	18.6	1.7	7.8	6.6	2.8	4.4	5.7	4.7	0.9	10.8	2.1	1.4
			GC–MS/MS	Green tea	A	3.8	5.9	3.1	0.8	9.1	3.7	6.9	8.1	5.5	2.5	11.9	6.1	5.7	3.9	4.3	4.2	3.1	5.3	5.9
					B	8.1	7.5	6.3	8.8	5.2	4.4	6.7	6.9	5.8	5.9	3.6	3.6	1.3	5.4	2.1	2.4	4.6	3.0	4.2
				Oolong tea	A	6.9	1.8	1.8	3.2	2.6	4.9	17.1	1.7	0.8	4.5	3.2	2.5	8.2	4.0	7.4	3.4	4.0	4.8	4.3
					B	2.8	1.0	3.9	5.4	4.1	2.0	6.8	7.7	2.9	4.3	5.2	2.0	5.1	0.8	7.9	3.5	4.7	4.1	4.0
		ENVI-Carb +PSA	GC–MS	Green tea	A	9.1	6.1	3.0	5.1	4.2	3.8	3.7	10.5	20.0	0.9	7.7	3.8	5.8	5.5	4.9	12.5	9.1	8.9	5.2
					B	5.0	8.3	3.0	1.4	3.6	4.0	3.7	15.4	7.9	5.3	17.4	5.0	5.8	2.6	13.0	4.4	5.2	11.4	5.8
				Oolong tea	A	3.1	6.0	3.8	4.1	12.7	2.3	4.6	47.5	22.0	9.1	6.7	2.6	1.1	3.1	4.3	2.8	5.3	5.6	3.3
					B	4.9	5.2	7.7	6.1	11.2	14.8	18.4	14.2	24.7	9.4	5.2	0.4	1.8	3.2	13.2	3.3	3.8	2.3	7.2
			GC–MS/MS	Green tea	A	8.8	6.7	3.3	5.9	7.2	5.5	11.0	4.1	4.7	1.6	6.8	4.1	4.6	15.4	4.0	7.6	10.4	1.4	1.2
					B	9.1	6.4	6.3	0.4	0.9	1.3	8.8	1.0	6.2	2.4	18.1	5.5	3.6	7.7	12.3	2.2	6.1	11.7	9.5
				Oolong tea	A	2.2	4.6	4.0	4.0	7.2	2.8	8.9	2.3	0.5	6.9	5.1	4.8	2.1	6.8	11.8	3.4	7.3	6.0	2.9
					B	8.0	2.2	11.3	5.3	5.9	4.1	2.7	6.6	7.3	2.5	3.7	4.1	6.5	8.2	9.2	3.5	3.3	2.2	8.2

(Continued)

Appendix Table 5.2 RSD of 201 Pesticides in Youden Pair Aged Tea Sample for Parallel Determinations (Nov 9, 2009–Feb 7, 2010) (cont.)

No.	Pesticides	SPE	Method	Sample	Youden pair	Nov 9, 2009 (n=5)	Nov 14, 2009 (n=3)	Nov 19, 2009 (n=3)	Nov 24, 2009 (n=3)	Nov 29, 2009 (n=3)	Dec 4, 2009 (n=3)	Dec 9, 2009 (n=3)	Dec 14, 2009 (n=3)	Dec 19, 2009 (n=3)	Dec 24, 2009 (n=3)	Jan 3, 2010 (n=3)	Jan 8, 2010 (n=3)	Jan 13, 2010 (n=3)	Jan 18, 2010 (n=3)	Jan 23, 2010 (n=3)	Jan 28, 2010 (n=3)	Feb 2, 2010 (n=3)	Feb 7, 2010 (n=3)
97	Ethoprophos	TPT	GC-MS	Green tea	A	8.0	4.8	2.9	1.2	9.7	4.1	6.8	6.3	5.3	2.7	1.6	6.0	3.1	4.8	1.8	3.3	4.7	4.1
					B	7.2	8.4	4.1	8.0	4.2	2.1	2.3	1.2	3.7	6.8	7.4	0.3	3.4	5.2	1.8	3.5	1.9	5.9
				Oolong tea	A	4.3	0.9	2.1	3.7	2.3	2.2	13.7	8.8	3.2	5.9	0.8	3.1	2.5	8.5	3.0	3.9	2.6	1.2
					B	4.2	3.6	4.3	5.9	3.2	3.6	4.2	2.7	5.6	2.8	1.6	2.0	3.3	4.5	0.3	0.9	2.6	1.7
			GC-MS/MS	Green tea	A	3.8	5.8	3.2	2.4	10.3	4.5	6.1	6.7	5.4	1.6	2.6	6.0	3.4	1.9	3.6	3.5	2.1	1.9
					B	7.4	8.6	5.0	7.0	5.3	4.8	6.2	3.3	5.0	9.0	4.0	1.7	1.3	1.0	3.0	6.1	0.1	4.8
				Oolong tea	A	4.3	2.2	1.2	3.9	2.0	3.2	11.5	3.5	2.2	3.3	1.3	9.0	0.8	8.9	5.9	4.8	2.9	4.2
					B	3.2	2.5	4.8	5.7	5.3	3.6	2.7	7.1	5.7	3.1	2.4	3.0	8.2	6.7	2.3	1.8	2.8	4.3
		ENVI-Carb+PSA	GC-MS	Green tea	A	8.4	3.2	6.1	5.3	6.7	3.4	4.4	4.2	2.8	2.0	0.6	1.8	6.0	3.1	10.7	10.2	10.1	4.8
					B	8.1	8.2	8.0	0.6	2.4	3.3	6.4	1.8	4.4	1.4	1.7	10.1	0.4	14.2	2.4	4.9	11.8	6.2
				Oolong tea	A	7.7	3.7	3.4	3.0	3.7	1.0	2.9	4.5	1.2	1.8	1.8	0.8	2.9	3.0	2.9	4.4	6.3	4.4
					B	6.4	5.9	8.3	1.9	12.8	5.4	1.3	6.2	3.0	3.0	2.0	4.9	3.3	10.0	3.4	6.3	2.4	4.3
			GC-MS/MS	Green tea	A	10.5	5.7	3.9	6.3	5.4	5.0	4.4	3.7	4.9	0.3	1.3	15.7	16.4	4.5	7.7	9.7	5.5	2.1
					B	17.2	7.4	5.4	3.6	1.0	1.9	10.5	0.9	5.1	1.1	5.5	8.5	9.3	13.8	2.7	7.8	11.4	9.1
				Oolong tea	A	4.5	4.6	4.6	2.5	3.6	0.9	1.9	5.0	0.7	2.6	2.7	2.9	5.7	10.0	5.9	6.7	2.4	2.0
					B	6.6	3.8	8.4	1.9	7.8	4.9	6.0	8.2	9.4	3.1	2.3	3.0	8.7	10.4	2.3	6.0	1.7	7.4
98	Etrimfos	TPT	GC-MS	Green tea	A	10.4	3.9	2.5	1.2	9.1	3.6	6.5	4.9	4.5	3.0	2.5	5.8	8.0	9.6	2.8	18.0	7.6	3.4
					B	5.7	8.4	7.1	7.5	4.3	2.9	1.5	1.0	3.5	8.0	6.1	1.0	6.4	11.7	1.8	14.8	3.7	6.4
				Oolong tea	A	4.3	1.0	2.8	3.7	3.2	2.5	9.9	9.3	3.4	5.4	2.1	2.6	2.2	7.7	6.4	2.6	2.5	1.6
					B	4.2	2.4	4.4	5.2	4.6	2.3	5.0	2.4	6.3	5.1	1.8	2.4	5.8	5.3	0.9	4.3	0.7	2.2
			GC-MS/MS	Green tea	A	4.5	5.6	3.6	2.2	9.6	6.1	6.9	5.9	5.8	1.5	6.6	6.2	5.7	2.9	6.7	2.8	4.6	2.6
					B	9.0	7.8	5.7	8.3	5.0	5.6	6.6	2.2	5.1	8.7	3.4	4.4	1.4	5.2	6.5	5.9	1.0	6.6
				Oolong tea	A	5.1	3.4	2.2	0.8	1.9	3.5	10.9	4.9	0.5	1.9	2.2	1.2	3.4	8.7	7.2	4.3	7.1	6.2
					B	4.4	2.5	5.1	5.3	4.4	3.5	1.9	8.1	5.9	6.2	4.9	3.1	2.3	6.3	2.9	3.9	1.8	8.0
		ENVI-Carb+PSA	GC-MS	Green tea	A	8.9	4.2	3.5	7.1	6.9	2.7	25.5	1.5	2.6	2.1	1.7	3.0	4.8	9.0	5.2	9.5	7.6	4.6
					B	6.4	7.6	3.8	1.5	1.9	3.4	4.5	2.9	4.6	6.6	1.5	3.7	2.0	9.5	4.8	1.5	8.9	5.9
				Oolong tea	A	6.0	4.5	2.9	3.6	5.6	0.9	2.8	2.2	2.1	1.3	1.6	0.9	2.2	2.2	6.2	4.2	4.7	4.7
					B	6.2	2.6	9.1	2.8	14.4	4.3	2.8	5.9	11.0	2.1	1.6	4.4	2.7	10.0	5.4	6.6	1.4	4.5
			GC-MS/MS	Green tea	A	9.5	5.6	2.5	3.9	4.9	4.8	2.7	2.7	6.0	2.0	3.2	11.0	15.3	5.7	5.5	8.8	4.1	3.4
					B	21.5	10.3	4.1	4.4	3.1	1.5	9.7	1.6	4.4	4.8	4.4	2.8	7.2	8.6	1.4	6.4	12.1	8.2
				Oolong tea	A	4.1	3.2	3.8	3.5	2.2	1.3	2.4	0.8	2.0	2.7	1.6	5.8	6.5	6.3	7.2	7.6	4.3	1.3
					B	6.7	3.7	8.3	1.8	8.1	5.5	5.4	6.6	10.2	1.7	3.8	1.6	11.6	10.1	2.9	7.5	4.6	6.3
99	Fenamidone	TPT	GC-MS	Green tea	A	5.8	6.9	1.0	3.3	5.0	4.8	6.1	6.0	4.4	3.6	6.0	6.1	2.0	3.3	1.1	4.3	11.7	3.3
					B	3.7	9.1	6.3	8.6	1.7	2.7	1.5	6.6	3.3	6.0	9.7	0.4	0.7	5.0	3.6	6.1	2.9	5.6
				Oolong tea	A	6.8	2.0	3.5	1.0	1.7	3.3	3.4	11.1	1.9	5.9	3.5	6.6	2.1	6.8	1.4	2.0	4.0	3.8
					B	4.8	1.7	5.9	7.4	4.5	4.6	6.9	2.3	6.2	4.2	0.9	1.8	3.8	3.3	1.8	1.2	2.2	1.1
			GC-MS/MS	Green tea	A	5.2	8.3	1.1	1.2	7.6	4.4	7.7	7.1	6.1	3.3	5.4	4.4	8.6	5.8	3.9	2.4	5.4	2.0
					B	5.3	5.8	3.7	8.6	5.1	4.6	6.2	7.6	4.9	6.1	3.5	4.3	15.0	1.9	4.8	8.3	0.5	2.9
				Oolong tea	A	7.4	1.5	4.3	0.8	3.3	3.6	7.1	3.7	3.5	4.4	3.8	9.9	1.7	4.4	2.5	2.1	3.0	1.1
					B	2.8	2.6	3.4	3.7	5.8	2.5	3.5	7.6	1.8	4.7	2.9	1.6	1.3	7.6	4.6	3.5	4.7	5.2
		ENVI-Carb+PSA	GC-MS	Green tea	A	7.8	8.6	3.9	9.6	6.0	2.9	23.3	0.3	4.2	2.5	1.2	4.9	6.5	4.7	10.1	12.1	5.8	3.3
					B	3.9	4.5	3.5	14.0	4.2	2.3	6.9	5.5	5.9	4.9	7.4	2.3	4.7	11.9	2.1	4.0	7.8	6.8
				Oolong tea	A	6.6	2.6	4.4	1.6	4.0	2.5	3.4	8.9	1.5	2.8	2.2	1.7	2.1	2.8	2.6	2.1	3.6	3.9
					B	5.0	3.5	10.9	4.9	14.2	5.3	2.5	7.4	9.6	5.9	5.1	2.1	3.8	8.3	3.1	4.3	3.2	2.9
			GC-MS/MS	Green tea	A	8.5	9.8	3.5	8.4	7.1	1.3	18.6	1.4	7.3	3.2	9.0	8.5	3.7	7.1	9.3	9.3	3.7	2.8
					B	4.3	3.9	6.4	5.0	2.9	2.6	2.8	3.0	5.6	3.4	4.7	4.5	5.6	10.2	2.0	7.9	8.9	11.7
				Oolong tea	A	6.5	1.8	5.7	2.0	4.8	1.5	3.1	7.0	2.3	1.1	6.0	5.3	5.9	9.0	2.5	7.6	4.5	3.1
					B	6.1	3.1	11.6	5.1	9.6	4.8	0.6	8.5	7.0	2.6	3.7	8.6	12.5	10.0	4.7	6.7	3.8	6.2

100	Fenaromol	TPT	GC-MS	Green tea	A	8.1	4.6	0.5	2.4	4.6	7.1	4.9	10.8	4.5	5.7	11.0	15.0	10.1	10.8	3.3	1.4	5.2	1.8	5.2
					B	4.5	10.4	5.3	11.0	5.7	2.5	1.9	2.3	3.6	10.0	7.4	10.1	3.1	1.3	2.2	1.7	10.9	6.8	7.7
				Oolong tea	A	7.3	4.0	2.0	1.5	1.4	3.9	10.8	8.9	7.5	9.0	13.0	2.1	14.3	2.7	8.0	5.3	2.6	6.4	2.6
					B	4.3	13.9	6.2	11.8	1.5	3.2	4.9	0.9	11.4	3.7	1.2	3.9	3.7	5.4	1.6	2.5	1.2	10.4	2.0
			GC-MS/MS	Green tea	A	5.7	6.4	0.5	1.9	9.5	6.5	6.4	5.3	7.1	3.0	6.9	3.7	1.6	6.4	2.0	6.2	2.8	7.5	2.7
					B	6.23	6.6	5.5	7.1	5.1	3.8	4.5	7.5	4.5	5.8	5.4	7.8	3.9	5.2	2.1	4.6	6.6	4.2	2.4
				Oolong tea	A	7.2	0.9	3.9	0.9	1.4	3.3	3.6	4.1	5.6	5.2	7.1	4.1	8.5	2.0	5.5	2.7	5.1	4.5	0.8
					B	4.4	2.8	2.9	7.6	6.5	4.7	2.1	9.5	6.1	4.0	5.0	1.7	1.2	5.2	7.2	4.0	4.5	5.3	3.5
		ENVI-Carb+PSA	GC-MS	Green tea	A	12.9	12.7	4.7	7.5	2.4	5.8	4.6	7.9	23.8	5.2	2.6	11.3	4.9	8.5	14.1	10.2	7.5	10.7	9.1
					B	5.1	7.1	4.7	8.1	4.4	2.7	4.5	5.0	15.2	5.9	16.7	8.7	4.8	6.9	5.8	10.1	5.3	7.2	7.9
				Oolong tea	A	7.1	8.5	6.1	2.6	3.0	4.5	4.9	7.1	3.7	13.1	8.8	1.9	7.4	1.5	1.0	3.6	5.4	7.8	3.7
					B	4.1	3.5	7.9	12.0	11.5	5.5	4.9	10.0	6.8	6.6	8.6	6.2	6.4	5.0	7.5	6.3	2.1	1.7	7.2
			GC-MS/MS	Green tea	A	8.1	10.2	4.1	6.6	7.1	1.5	18.7	2.6	6.0	0.8	5.1	1.6	7.9	9.4	6.8	9.5	10.1	7.6	2.8
					B	6.5	5.4	7.2	6.9	0.4	3.8	3.3	0.1	6.7	1.6	21.7	1.6	3.3	8.4	18.6	2.5	8.1	10.3	10.8
				Oolong tea	A	6.3	2.0	6.1	0.9	6.1	1.3	3.9	7.2	4.6	9.3	8.9	6.4	5.0	7.5	9.3	2.7	6.5	3.4	3.2
					B	6.9	4.3	13.2	5.0	9.6	3.2	1.3	7.7	7.3	3.2	3.7	2.8	5.7	10.6	10.6	4.0	1.2	3.3	6.9
101	Flamprop-isopropyl	TPT	GC-MS	Green tea	A	6.6	6.7	1.8	2.1	7.9	5.3	7.9	6.4	4.0	3.0	4.5	4.6	7.3	5.6	4.6	1.7	3.3	6.9	4.3
					B	5.3	6.3	3.8	7.6	4.3	2.9	2.3	2.4	6.5	5.3	4.3	2.8	1.6	3.3	8.0	3.8	4.4	1.1	5.4
				Oolong tea	A	9.4	5.8	3.3	3.2	1.8	4.2	9.7	5.8	9.4	0.9	4.2	2.4	5.0	2.5	7.9	5.1	2.5	4.9	1.1
					B	6.7	2.6	4.6	7.5	3.1	3.4	5.6	5.6	0.8	6.9	1.4	5.2	2.6	4.4	1.4	1.7	3.4	0.9	2.3
			GC-MS/MS	Green tea	A	5.2	5.1	2.5	1.7	8.6	3.2	5.4	6.9	4.5	1.8	6.4	5.6	2.1	3.0	3.6	4.8	2.1	4.1	3.7
					B	6.6	6.4	6.1	8.3	6.8	4.6	7.7	4.3	6.6	6.7	4.2	3.8	2.6	6.2	1.3	4.2	3.3	1.4	3.3
				Oolong tea	A	7.3	1.9	3.6	1.8	2.8	2.7	8.2	2.7	2.4	4.1	1.8	1.9	5.8	1.4	7.9	4.9	2.9	4.7	4.0
					B	3.6	2.9	5.5	6.2	4.0	3.5	0.5	7.0	4.8	4.2	6.5	3.2	1.2	5.6	6.3	1.3	4.7	2.4	3.3
		ENVI-Carb+PSA	GC-MS	Green tea	A	5.7	3.4	1.9	7.1	8.8	1.6	27.0	0.9	7.8	2.5	7.3	2.2	2.5	4.5	6.7	6.9	11.2	2.3	6.4
					B	2.9	6.8	2.5	3.9	3.5	1.3	8.7	4.3	3.1	4.4	22.9	3.2	1.2	5.0	10.4	1.2	3.9	6.9	6.7
				Oolong tea	A	4.4	1.4	2.8	3.7	6.2	1.2	4.1	1.1	1.5	2.8	8.1	6.8	3.1	2.5	3.9	4.5	5.3	5.2	4.0
					B	4.7	1.7	7.5	5.9	9.6	4.2	1.5	5.7	10.3	5.2	5.8	3.1	5.1	5.1	9.2	4.5	5.5	5.9	4.8
			GC-MS/MS	Green tea	A	6.4	5.2	2.6	6.4	10.4	3.1	4.5	7.1	5.9	0.1	9.6	1.5	10.9	15.2	6.7	5.8	9.1	3.2	4.1
					B	4.8	8.0	6.7	1.2	1.3	1.4	2.9	1.6	4.5	6.0	17.3	1.0	1.8	3.1	11.4	2.6	5.5	14.9	10.1
				Oolong tea	A	6.3	3.5	4.2	1.9	7.6	1.6	2.0	2.3	0.5	2.4	9.8	2.0	2.0	6.5	9.7	3.2	5.5	3.4	4.9
					B	6.6	3.2	10.4	4.0	8.6	5.1	0.7	7.1	9.7	3.9	3.0	2.6	3.4	8.6	10.2	2.2	3.7	3.5	7.9
102	Flamprop-methyl	TPT	GC-MS	Green tea	A	7.2	4.9	3.2	2.7	12.2	4.1	11.3	2.2	7.7	3.7	7.5	7.1	6.6	5.0	1.6	2.4	5.4	3.2	5.7
					B	4.1	6.6	5.9	10.9	6.5	8.1	2.9	10.3	4.1	8.9	0.5	4.2	4.6	3.0	4.3	5.2	4.1	0.7	5.2
				Oolong tea	A	6.1	3.2	0.6	4.1	6.1	4.0	12.6	2.0	6.5	5.1	7.6	10.1	2.7	2.1	8.6	5.2	3.1	2.7	3.0
					B	5.2	4.6	6.5	11.4	5.7	9.9	5.3	16.8	8.5	3.1	6.7	3.8	1.4	3.8	1.8	0.8	2.0	2.1	1.9
			GC-MS/MS	Green tea	A	6.1	5.7	2.7	2.9	10.0	2.7	7.3	9.9	5.1	2.8	7.0	4.3	5.3	2.9	4.6	4.9	3.9	3.5	3.7
					B	6.1	7.2	5.0	9.1	5.7	4.8	6.5	6.8	5.0	7.7	5.2	2.4	6.2	7.6	4.7	5.7	5.1	1.6	5.9
				Oolong tea	A	4.7	4.7	3.6	1.8	2.8	2.7	7.4	0.0	0.8	7.7	3.7	4.4	6.2	0.8	7.1	5.7	5.1	1.6	7.1
					B	6.1	7.2	5.0	9.1	4.0	3.5	2.8	5.9	3.1	4.1	5.2	1.8	4.4	6.7	7.2	3.2	5.1	4.5	7.9
		ENVI-Carb+PSA	GC-MS	Green tea	A	11.7	4.7	8.0	7.4	7.8	1.0	17.7	11.7	12.2	1.2	20.6	4.5	2.8	3.4	2.2	3.9	9.1	6.9	5.9
					B	10.9	5.2	3.8	5.1	6.1	5.3	5.2	12.6	7.2	5.9	5.5	6.0	6.9	0.3	7.6	7.9	1.6	12.3	7.1
				Oolong tea	A	2.2	23.2	5.1	2.7	7.9	2.2	5.6	4.1	1.0	6.2	17.7	3.5	0.9	2.1	3.7	5.2	5.0	6.7	2.1
					B	5.2	6.6	11.7	5.3	13.3	7.8	1.2	6.4	10.7	4.1	6.8	2.8	2.6	3.8	9.9	5.4	5.4	1.4	6.5
			GC-MS/MS	Green tea	A	-2.1	4.6	5.9	5.8	7.5	9.9	11.9	1.5	7.3	2.5	11.7	2.3	4.5	13.7	4.6	2.8	9.8	2.1	6.1
					B	-3.6	6.9	4.9	6.0	3.7	0.6	15.4	5.5	9.7	0.7	9.8	2.0	2.1	4.2	8.4	4.4	2.4	9.7	10.1
				Oolong tea	A	6.3	3.5	4.2	1.9	7.6	1.6	2.0	2.4	0.5	2.4	9.8	2.0	2.0	6.5	12.1	4.9	5.5	3.4	3.1
					B	6.6	3.2	10.4	4.0	8.6	5.1	0.7	7.1	9.7	3.9	3.0	2.6	3.4	8.6	10.1	1.4	3.7	3.5	8.3

(Continued)

Appendix Table 5.2 RSD of 201 Pesticides in Youden Pair Aged Tea Sample for Parallel Determinations (Nov 9, 2009–Feb 7, 2010) (cont.)

RSD%

No.	Pesticides	SPE	Method	Sample	Youden pair	Nov 9, 2009 (n=5)	Nov 14, 2009 (n=3)	Nov 19, 2009 (n=3)	Nov 24, 2009 (n=3)	Nov 29, 2009 (n=3)	Dec 4, 2009 (n=3)	Dec 9, 2009 (n=3)	Dec 14, 2009 (n=3)	Dec 19, 2009 (n=3)	Dec 14, 2009 (n=3)	Dec 24, 2009 (n=3)	Jan 3, 2010 (n=3)	Jan 8, 2010 (n=3)	Jan 13, 2010 (n=3)	Jan 18, 2010 (n=3)	Jan 23, 2010 (n=3)	Jan 28, 2010 (n=3)	Feb 2, 2010 (n=3)	Feb 7, 2010 (n=3)
103	Fonofos	TPT	GC-MS	Green tea	A	9.1	5.4	5.6	1.9	8.5	4.1	6.2	4.5	4.3	1.6	7.0	2.8	7.1	17.4	20.2	1.6	6.3	6.7	4.4
				Green tea	B	5.7	8.3	6.9	7.9	3.5	3.5	2.1	0.6	4.4	8.4	8.2	7.4	1.0	6.0	22.5	2.5	10.3	2.2	5.8
				Oolong tea	A	4.2	1.2	2.4	3.3	1.0	2.6	6.6	5.5	0.2	8.0	6.8	3.4	3.2	2.6	9.3	3.5	3.8	3.0	0.5
				Oolong tea	B	4.0	2.3	4.1	1.4	2.5	3.1	4.4	3.0	5.2	3.9	4.6	0.3	3.0	3.2	3.7	1.4	2.2	2.2	1.0
			GC-MS/MS	Green tea	A	4.2	4.8	3.6	2.2	10.1	3.4	5.2	4.5	5.3	1.3	6.5	3.9	8.3	3.7	2.3	4.0	2.3	3.1	3.9
				Green tea	B	7.9	9.5	5.1	9.1	5.2	6.0	6.2	3.0	5.4	6.3	4.9	4.2	2.6	3.9	1.0	0.4	5.1	2.0	4.5
				Oolong tea	A	2.8	3.4	1.9	3.0	1.9	3.4	7.2	3.1	1.3	3.7	3.4	3.1	5.4	1.1	7.6	4.6	3.7	3.6	5.3
				Oolong tea	B	2.9	0.5	4.5	6.0	5.9	2.6	2.4	5.3	3.9	4.5	2.9	2.4	1.8	1.4	5.8	2.3	2.8	4.0	4.3
		ENVI-Carb+PSA	GC-MS	Green tea	A	5.5	1.7	4.6	8.4	3.7	1.5	4.0	1.4	4.9	6.0	5.5	4.3	3.8	2.2	4.4	6.6	9.6	3.4	4.2
				Green tea	B	4.9	5.6	1.1	5.2	2.6	5.7	4.4	2.2	2.9	5.6	18.9	4.4	4.8	2.0	11.8	2.5	6.4	7.9	5.3
				Oolong tea	A	6.4	4.4	3.4	2.2	3.4	0.8	2.8	1.3	2.3	2.9	5.2	1.8	1.0	2.6	0.9	2.3	5.8	4.5	4.3
				Oolong tea	B	5.0	3.0	9.2	2.7	13.4	4.2	1.9	6.5	11.3	3.7	8.3	1.1	4.3	3.2	8.5	3.7	6.7	4.2	3.3
			GC-MS/MS	Green tea	A	6.7	6.6	2.3	5.8	6.0	0.2	2.3	2.7	5.1	1.0	7.7	7.4	15.6	14.5	5.9	8.5	9.9	2.7	2.1
				Green tea	B	4.5	6.6	7.1	1.0	1.1	3.1	5.4	2.9	2.4	1.5	18.3	5.5	6.1	3.8	13.8	2.5	7.6	10.0	6.2
				Oolong tea	A	4.1	4.0	4.6	1.6	5.0	1.0	2.4	1.4	1.1	1.6	7.3	1.8	3.4	5.7	8.5	4.6	6.7	3.6	4.8
				Oolong tea	B	6.7	3.2	9.8	1.9	7.6	4.3	1.6	7.4	8.2	3.6	0.7	2.1	1.9	8.8	10.8	3.1	5.0	2.3	4.6
104	Hexachlorobenzene	TPT	GC-MS	Green tea	A	3.4	5.0	3.1	0.7	14.7	5.2	6.9	9.1	5.3	4.1	4.1	4.0	7.3	5.2	2.4	4.3	2.6	6.9	4.4
				Green tea	B	6.2	7.7	3.8	9.6	12.5	3.8	3.1	2.4	3.2	8.9	1.0	12.3	2.2	6.2	3.1	6.1	6.1	6.8	6.1
				Oolong tea	A	4.1	4.9	9.9	6.9	7.7	10.6	8.6	5.7	1.7	8.6	2.8	3.8	6.7	3.1	6.7	1.9	1.9	1.5	18.2
				Oolong tea	B	15.0	6.4	14.5	4.4	7.8	14.1	5.8	0.4	18.9	2.5	1.5	3.0	9.4	1.5	3.8	15.2	15.2	1.9	7.9
			GC-MS/MS	Green tea	A	4.1	5.0	1.8	2.3	17.4	4.6	7.8	8.9	4.7	4.2	6.3	5.5	8.8	4.8	1.0	3.3	3.3	7.0	3.4
				Green tea	B	8.7	7.6	3.6	8.1	13.0	5.5	5.6	5.9	5.7	7.9	4.3	9.2	4.6	4.5	0.2	3.5	8.4	9.6	3.3
				Oolong tea	A	5.1	4.2	9.6	6.9	6.2	13.7	7.2	0.1	3.8	10.6	6.0	6.3	5.7	2.5	6.9	23.6	2.4	6.6	19.5
				Oolong tea	B	15.5	11.3	15.8	6.4	8.8	15.6	5.0	4.6	16.7	16.2	4.4	5.6	13.1	9.5	7.0	10.3	13.8	3.4	4.8
		ENVI-Carb+PSA	GC-MS	Green tea	A	6.9	4.9	2.4	6.5	6.6	2.3	4.8	2.2	6.0	0.5	10.1	1.3	2.9	3.0	4.5	9.4	12.5	7.4	4.4
				Green tea	B	2.7	4.0	5.8	5.6	3.0	2.2	8.6	2.4	2.4	1.4	21.5	1.6	6.4	0.6	14.1	4.8	6.6	10.9	7.0
				Oolong tea	A	5.0	4.0	4.3	3.0	4.1	1.0	0.7	3.0	1.2	2.5	4.3	3.9	1.8	3.1	1.8	4.3	6.1	6.0	4.6
				Oolong tea	B	6.2	3.9	7.4	3.4	9.0	4.0	1.3	6.8	11.6	3.6	7.4	3.0	3.1	1.5	2.5	3.4	7.5	2.7	4.4
			GC-MS/MS	Green tea	A	7.5	5.6	4.1	7.3	6.2	4.9	2.7	1.9	6.3	1.1	9.7	2.3	16.2	20.4	5.8	10.6	14.2	7.1	3.2
				Green tea	B	4.6	2.5	6.6	3.2	1.0	1.6	2.9	1.9	3.7	1.4	21.4	13.1	6.3	12.1	15.8	6.3	8.8	10.5	6.7
				Oolong tea	A	4.3	3.3	5.8	4.1	3.8	1.8	1.5	5.4	0.5	0.5	8.6	5.0	4.4	6.1	8.6	24.6	7.8	6.9	1.1
				Oolong tea	B	8.0	1.5	8.2	1.6	5.3	2.5	0.6	5.5	9.2	6.0	4.8	2.0	2.8	10.3	11.8	7.0	7.3	3.8	6.7
105	Hexazinone	TPT	GC-MS	Green tea	A	11.0	17.6	4.7	7.3	2.5	9.8	10.1	3.1	4.3	7.7	7.6	0.7	8.5	15.2	10.0	6.4	8.7	2.5	2.8
				Green tea	B	4.9	11.3	2.8	11.9	3.2	7.8	1.8	8.5	5.2	13.5	11.2	8.8	3.0	8.7	18.4	5.2	12.9	1.0	5.1
				Oolong tea	A	8.2	2.8	2.2	7.9	9.0	4.2	7.8	8.9	4.8	10.0	3.0	6.8	3.9	2.3	5.6	7.9	8.9	5.9	7.2
				Oolong tea	B	11.7	2.2	9.6	14.2	4.7	5.3	8.1	3.0	6.4	3.0	7.5	3.8	2.5	4.6	9.5	4.5	3.1	4.4	8.5
			GC-MS/MS	Green tea	A	10.2	19.7	3.3	9.4	4.7	6.6	0.4	1.1	3.9	10.0	13.7	5.4	4.0	18.3	9.9	13.3	9.8	2.7	1.0
				Green tea	B	7.8	16.5	5.0	13.3	5.0	10.8	5.2	8.6	6.3	17.3	6.8	2.8	1.7	20.3	12.3	5.6	2.1	6.9	5.1
				Oolong tea	A	9.3	5.8	6.3	8.1	6.8	4.4	7.8	3.8	1.0	3.1	4.5	3.0	6.0	1.0	7.8	10.8	7.9	7.2	4.9
				Oolong tea	B	9.9	3.6	15.2	14.5	4.0	5.4	8.7	11.1	4.3	5.8	10.2	3.7	3.6	2.0	7.7	7.3	3.8	4.4	2.3
		ENVI-Carb+PSA	GC-MS	Green tea	A	13.4	27.0	13.1	6.8	3.1	4.3	5.1	1.5	12.3	7.4	15.0	2.1	1.0	6.8	6.4	3.5	9.8	4.8	8.3
				Green tea	B	7.6	4.3	4.0	5.9	5.2	3.9	6.5	5.0	3.1	9.1	9.8	7.1	9.4	3.2	8.4	3.5	2.1	9.1	9.5
				Oolong tea	A	9.0	2.4	11.8	4.3	7.0	15.1	6.2	7.9	4.2	19.4	10.8	12.5	7.2	3.4	10.9	3.7	1.8	3.4	3.5
				Oolong tea	B	8.2	4.1	18.4	8.4	16.9	3.4	5.5	8.8	2.8	4.0	11.4	7.7	4.7	2.7	6.5	11.2	20.6	2.3	3.8
			GC-MS/MS	Green tea	A	15.0	26.0	11.7	11.7	2.5	5.6	15.8	3.9	12.6	5.6	12.4	0.6	6.9	17.8	6.5	4.5	11.5	1.3	5.6
				Green tea	B	8.5	4.5	2.7	7.5	6.3	6.3	4.4	8.8	4.6	10.4	7.7	0.8	12.3	16.5	6.7	5.3	5.1	7.8	10.2
				Oolong tea	A	10.7	0.5	8.3	2.9	6.9	5.6	4.4	7.2	3.1	26.7		12.0	6.8	6.4	18.5	10.7	8.1	5.5	5.5
				Oolong tea	B	9.6	3.3	28.2	8.8	14.2	14.9	13.7	9.0	3.3	2.0		9.8	5.8	31.6	7.5	7.3	19.7	4.6	4.6

106	Iodofenphos	TPT	GC–MS	Green tea	A	6.9	2.3	3.8	9.0	7.3	2.0	7.3	5.2	4.1	3.0	4.3	4.9	3.5	8.2	6.5	3.0	5.1	7.0	1.8
					B	9.2	11.5	4.8	6.3	5.0	6.0	3.3	0.3	1.8	5.3	4.1	6.8	6.8	6.5	11.3	4.4	8.6	7.4	12.4
				Oolong tea	A	10.6	4.2	2.7	4.0	4.7	3.3	12.8	13.9	7.7	8.7	5.6	3.4	4.2	2.6	10.1	1.7	8.6	7.4	6.4
					B	5.6	2.0	2.4	3.4	2.3	1.4	11.9	1.4	4.9	7.5	7.0	1.2	4.7	5.6	8.9	2.0	5.3	4.9	4.1
			GC–MS/MS	Green tea	A	6.3	9.5	10.4	5.9	8.3	2.5	2.6	10.4	10.9	6.1	5.6	6.5	8.9	6.3	11.1	6.0	12.9	6.3	4.2
					B	4.6	10.5	6.8	6.6	7.8	8.7	12.1	5.0	7.7	12.1	15.1	1.8	7.1	4.2	3.9	7.7	9.9	3.4	10.5
				Oolong tea	A	12.5	9.5	12.1	1.6	4.8	4.9	6.1	4.7	9.8	8.3	4.6	4.6	9.3	9.4	13.3	4.3	7.3	10.3	8.7
					B	4.3	1.9	10.3	6.5	2.5	3.0	4.4	5.2	18.1	19.6	6.9	1.0	12.5	7.6	13.4	1.9	5.2	7.2	19.0
		ENVI-Carb+PSA	GC–MS	Green tea	A	12.1	5.7	4.8	6.3	9.0	9.4	5.5	8.1	4.1	2.4	16.7	9.2	2.7	3.0	4.9	4.0	10.8	13.4	7.8
					B	15.0	6.7	3.3	8.8	5.0	1.6	3.0	11.8	7.0	11.0	16.6	9.5	9.9	4.8	3.4	11.4	0.5	18.6	9.0
				Oolong tea	A	8.1	0.3	2.7	5.2	3.6	3.3	12.3	2.0	6.9	7.2	1.5	1.5	7.0	3.5	7.4	4.3	1.8	8.9	2.9
					B	9.6	9.9	7.8	6.0	7.4	7.4	5.9	5.9	12.2	2.5	11.8	4.7	6.6	5.6	14.3	5.1	10.1	10.3	6.1
			GC–MS/MS	Green tea	A	21.0	5.1	10.3	3.3	8.1	19.8	2.8	11.0	17.1	8.7	13.8	5.3	6.5	18.4	11.2	3.0	13.6	14.3	16.5
					B	42.0	10.6	2.3	16.6	10.4	8.3	16.6	4.2	7.4	18.4	8.7	2.6	2.8	10.9	1.5	15.2	10.2	21.4	12.1
				Oolong tea	A	5.4	0.4	5.0	11.1	6.8	4.6	2.8	7.0	4.0	3.4	8.2	9.2	4.6	7.6	20.4	4.3	5.7	1.0	2.4
					B	4.2	9.0	11.1	8.4	2.1	4.4	12.2	4.6	12.3	2.4	1.8	11.0	5.9	19.6	16.8	1.9	6.3	9.2	11.4
107	Isofenphos	TPT	GC–MS	Green tea	A	7.7	4.0	4.1	1.5	7.2	5.0	12.5	7.7	3.8	6.6	11.0	7.2	5.6	8.4	4.6	1.8	5.2	3.4	4.4
					B	5.4	8.5	4.8	8.4	5.3	2.8	7.2	3.5	5.7	5.5	9.2	13.8	1.2	3.5	3.0	5.8	5.4	0.5	5.3
				Oolong tea	A	5.7	4.3	5.3	3.3	1.5	3.5	7.2	13.2	4.7	12.8	13.1	3.3	4.0	2.1	9.4	10.3	3.0	2.9	0.9
					B	3.2	7.2	8.0	5.2	3.7	3.1	4.5	10.6	7.5	9.7	15.4	0.4	1.3	3.8	3.9	1.5	1.0	0.7	2.7
			GC–MS/MS	Green tea	A	4.2	2.3	3.6	1.9	10.7	5.3	6.2	8.7	7.2	1.7	5.8	4.1	9.7	1.4	3.6	5.9	4.8	7.3	4.2
					B	7.6	6.8	5.5	9.4	7.3	4.7	7.0	4.1	5.5	7.4	4.7	2.5	0.4	2.5	0.2	3.2	4.8	0.6	4.1
				Oolong tea	A	6.8	3.2	2.7	3.1	1.3	5.4	10.1	2.4	5.3	4.4	4.7	1.9	2.1	2.1	10.0	11.6	7.1	2.5	5.0
					B	3.2	5.9	6.2	9.0	6.1	2.5	1.4	7.4	5.2	3.0	6.1	1.3	6.1	5.0	6.8	2.7	0.6	1.8	6.4
		ENVI-Carb+PSA	GC–MS	Green tea	A	9.8	3.9	6.5	7.2	9.1	2.8	6.3	5.3	13.8	8.3	4.9	3.4	2.2	11.3	8.6	28.3	9.6	25.3	4.9
					B	3.2	9.5	6.6	4.9	0.7	5.4	6.1	1.8	9.9	4.5	38.1	9.3	13.4	1.5	26.1	2.3	6.9	20.5	7.3
				Oolong tea	A	5.7	3.3	1.2	3.0	5.1	2.1	4.6	12.0	6.9	5.0	5.2	2.8	2.7	1.5	4.9	2.6	7.8	5.2	3.6
					B	5.3	8.0	11.2	2.4	17.4	6.8	3.8	5.6	14.8	7.6	9.3	8.7	4.7	1.7	12.9	0.1	7.3	4.0	5.1
			GC–MS/MS	Green tea	A	10.7	6.8	3.6	5.5	8.9	1.6	8.7	4.6	2.6	2.1	10.2	2.6	5.7	3.0	8.4	12.1	9.2	4.3	2.7
					B	4.4	8.2	6.6	2.7	1.6	5.2	6.3	0.8	7.3	0.5	21.5	0.5	8.4	3.6	16.3	1.4	5.7	12.1	12.6
				Oolong tea	A	3.9	2.4	3.7	1.6	3.8	1.6	3.0	5.5	0.9	2.7	6.1	3.1	2.4	2.7	10.4	11.6	5.2	4.5	3.8
					B	7.1	3.3	9.4	1.3	10.5	6.2	2.5	7.9	10.2	4.3	2.6	2.0	4.9	8.1	11.0	2.7	5.7	3.7	8.7
108	Isopeopalin	TPT	GC–MS	Green tea	A	4.4	4.7	4.1	0.9	10.4	4.4	7.2	6.7	4.4	7.5	11.4	7.8	6.7	7.8	4.0	0.9	4.2	8.0	5.4
					B	4.6	8.9	5.6	10.6	4.3	3.2	1.7	4.4	4.8	9.0	9.7	10.3	1.2	5.4	2.2	4.0	7.0	1.6	4.0
				Oolong tea	A	5.5	1.6	5.5	3.6	5.8	4.0	7.2	15.5	4.5	12.1	14.0	3.1	8.7	2.8	8.2	1.5	4.0	4.4	2.4
					B	3.2	4.5	4.2	6.5	5.0	2.1	5.6	3.0	7.8	7.2	16.0	0.4	2.4	4.0	4.4	1.0	1.1	4.7	2.3
			GC–MS/MS	Green tea	A	2.7	4.6	2.0	3.1	9.7	6.2	5.6	8.9	6.1	1.8	10.2	4.2	12.1	4.8	4.4	3.4	7.0	4.7	4.4
					B	8.0	10.7	5.5	10.6	7.5	2.5	4.8	4.9	6.2	11.4	5.0	8.9	4.8	6.1	3.6	6.7	5.1	2.2	5.9
				Oolong tea	A	4.2	1.9	5.2	2.2	1.7	4.8	8.5	5.4	3.9	5.8	8.0	4.1	5.7	1.7	1.6	8.7	2.7	1.8	2.4
					B	2.8	2.1	3.1	7.6	7.5	3.3	4.0	9.9	5.2	3.1	13.2	2.3	3.3	8.2	8.6	2.5	2.6	0.6	3.5
		ENVI-Carb+PSA	GC–MS	Green tea	A	10.9	5.2	5.9	5.8	9.9	4.0	8.1	6.9	3.5	5.6	8.2	2.3	0.9	6.6	7.1	9.3	11.0	8.8	6.9
					B	2.9	7.8	6.2	2.7	9.7	8.8	7.9	4.9	6.5	5.5	30.6	9.6	6.6	1.1	13.3	0.7	0.7	10.9	7.8
				Oolong tea	A	5.5	5.9	3.8	3.9	2.7	1.3	4.8	3.9	4.6	6.0	4.9	5.9	8.2	2.8	0.7	9.3	6.0	6.5	5.8
					B	4.2	3.4	9.3	1.1	17.7	8.0	4.8	7.4	12.9	5.7	14.5	6.6	5.5	4.0	12.2	3.8	6.9	3.3	5.4
			GC–MS/MS	Green tea	A	11.3	5.3	6.0	5.5	7.8	1.1	4.8	8.0	6.4	3.2	7.3	3.4	4.9	2.5	5.9	7.6	10.0	5.0	9.2
					B	8.6	7.5	7.8	2.3	1.3	6.3	19.2	5.3	6.8	3.8	20.6	1.7	5.3	5.0	8.7	5.0	10.1	9.2	8.9
				Oolong tea	A	4.8	4.2	3.9	4.3	4.3	2.3	4.8	3.1	0.5	4.3	6.2	1.7	5.3	7.9	8.2	8.7	5.0	6.5	3.7
					B	5.4	3.8	8.5	2.4	8.9	7.4	11.9	7.4	10.6	3.9	4.5	1.8	1.6	11.2	12.1	2.5	7.3	1.6	6.4

(Continued)

Appendix Table 5.2 RSD of 201 Pesticides in Youden Pair Aged Tea Sample for Parallel Determinations (Nov 9, 2009–Feb 7, 2010) (cont.)

No.	Pesticides	SPE	Method	Sample	Youden pair	Nov 9, 2009 (n=5)	Nov 14, 2009 (n=3)	Nov 19, 2009 (n=3)	Nov 24, 2009 (n=3)	Nov 29, 2009 (n=3)	Dec 4, 2009 (n=3)	Dec 9, 2009 (n=3)	Dec 14, 2009 (n=3)	Dec 19, 2009 (n=3)	Dec 24, 2009 (n=3)	Dec 14, 2009 (n=3)	Jan 3, 2010 (n=3)	Jan 8, 2010 (n=3)	Jan 13, 2010 (n=3)	Jan 18, 2010 (n=3)	Jan 23, 2010 (n=3)	Jan 28, 2010 (n=3)	Feb 2, 2010 (n=3)	Feb 7, 2010 (n=3)
109	Metho-prene	TPT	GC-MS	Green tea	A	5.4	5.6	5.4	2.7	6.9	4.0	7.1	4.0	3.8	4.5	9.0	6.4	7.2	4.7	4.3	1.7	3.8	7.1	3.8
					B	7.0	8.8	6.7	8.5	3.3	4.7	3.6	1.9	5.6	8.9	7.8	11.7	0.7	1.0	7.7	2.1	4.2	1.0	6.1
				Oolong tea	A	8.0	1.1	1.0	4.0	8.1	3.4	9.6	8.7	1.3	7.4	10.0	1.5	4.7	3.2	6.3	2.6	4.2	2.0	1.0
					B	2.5	3.2	4.1	3.9	4.6	2.8	6.0	6.8	5.4	5.4	9.9	3.0	3.9	2.1	3.8	1.2	1.8	1.6	3.7
			GC-MS/MS	Green tea	A	5.1	5.5	4.9	0.8	11.2	6.2	3.5	4.6	7.1	2.8	5.6	5.1	10.2	6.0	5.6	3.7	0.2	5.6	4.5
					B	9.6	9.3	6.8	10.5	5.9	6.0	9.6	2.2	5.8	8.5	5.3	7.9	1.5	5.8	6.0	4.3	5.9	1.9	5.3
				Oolong tea	A	5.5	0.6	4.9	4.6	2.4	5.3	11.4	5.3	3.2	5.0	6.1	1.2	5.9	2.8	9.4	4.9	4.8	4.6	4.4
					B	5.0	0.4	2.4	8.8	6.3	2.6	4.0	6.8	1.4	4.8	8.5	1.5	2.3	4.5	5.0	2.1	8.5	1.4	7.6
		ENVI-Carb+PSA	GC-MS	Green tea	A	8.2	6.6	2.1	4.0	8.4	2.3	7.6	1.6	5.0	3.8	8.8	6.6	2.8	5.3	6.2	8.4	11.2	3.3	5.7
					B	2.5	8.5	8.2	5.8	2.2	4.7	14.7	3.0	3.1	2.9	33.1	8.0	4.3	2.8	13.6	2.2	5.9	8.0	6.0
				Oolong tea	A	3.3	3.8	6.0	4.2	13.9	0.9	12.3	6.5	3.0	6.2	7.3	6.9	0.6	3.1	2.4	1.8	5.3	4.7	4.8
					B	5.0	4.8	11.9	3.2	8.0	2.7	4.3	9.4	12.0	5.9	10.7	2.7	3.0	4.4	8.8	3.6	8.0	5.4	7.2
			GC-MS/MS	Green tea	A	6.7	37.7	3.6	9.9	7.6	4.0	5.2	3.3	2.4	2.1	10.2	13.0	4.8	13.5	7.7	9.0	10.7	2.0	8.1
					B	4.7	9.2	6.3	8.0	5.7	5.3	4.0	2.0	7.6	1.4	20.6	9.0	4.3	4.3	14.8	1.6	11.0	10.8	8.3
				Oolong tea	A	4.5	5.5	3.4	6.1	8.8	1.6	6.1	2.5	1.1	5.3	6.1	2.2	4.7	6.7	9.1	4.9	6.7	3.6	4.6
					B	7.1	2.6	8.8	1.6	7.9	7.0	2.9	12.9	8.2	5.5	4.0	2.1	6.1	8.7	13.6	2.1	8.4	3.7	7.1
110	Methopro-tryne	TPT	GC-MS	Green tea	A	6.0	4.1	0.4	2.7	6.5	4.9	8.4	3.9	4.5	8.6	11.6	10.1	5.5	2.6	2.4	1.6	5.0	5.8	3.6
					B	4.0	1.6	5.6	7.2	3.6	3.3	4.0	2.6	3.8	8.4	14.4	13.0	0.5	0.9	4.4	3.1	5.5	2.3	4.2
				Oolong tea	A	7.4	4.3	6.6	1.6	5.1	2.6	7.8	9.3	3.2	11.9	8.2	3.2	5.5	2.2	6.9	3.1	2.7	3.6	0.6
					B	4.3	2.3	11.3	8.3	1.6	2.8	5.3	3.0	3.8	7.9	11.4	0.4	2.0	4.5	1.8	1.7	2.5	1.5	1.9
			GC-MS/MS	Green tea	A	5.0	5.2	2.0	1.9	9.5	5.7	4.3	7.2	4.3	2.0	8.2	5.5	5.0	5.8	3.3	3.2	2.9	4.9	5.0
					B	6.6	4.7	4.4	8.7	6.1	4.3	7.7	4.0	3.9	9.0	4.8	3.6	3.1	6.6	3.7	4.9	6.8	0.8	2.1
				Oolong tea	A	9.0	1.8	1.6	1.3	0.6	3.4	7.8	2.9	1.5	4.0	4.0	3.5	5.7	1.6	5.3	3.4	3.2	5.9	2.8
					B	4.2	3.2	3.8	6.5	3.8	2.5	2.6	4.5	4.4	3.8	5.3	1.5	1.1	1.6	5.5	1.8	2.4	3.2	3.1
		ENVI-Carb+PSA	GC-MS	Green tea	A	2.9	1.4	2.6	6.2	8.6	7.7	9.3	2.9	6.9	8.8	5.6	12.2	2.6	5.0	5.9	6.4	10.5	2.3	5.4
					B	4.5	7.8	3.5	7.1	7.0	3.0	3.9	5.7	5.8	8.7	36.2	16.9	1.0	5.9	10.0	2.1	4.3	7.7	7.6
				Oolong tea	A	6.2	3.0	4.7	1.7	7.7	2.4	3.6	3.4	6.6	6.9	7.9	4.2	2.6	3.2	2.1	1.9	4.2	4.8	3.5
					B	3.3	2.9	10.8	4.4	15.7	4.0	1.6	6.8	10.0	4.0	8.5	9.9	1.8	2.7	8.4	3.3	4.3	4.3	3.6
			GC-MS/MS	Green tea	A	5.7	6.3	2.5	7.0	9.3	2.9	2.0	1.3	7.3	2.6	7.0	9.9	9.7	12.9	7.9	6.6	11.3	1.0	3.2
					B	5.4	8.6	6.1	0.7	2.7	3.5	2.1	4.6	5.1	2.5	16.5	9.9	5.9	5.5	13.9	1.0	6.6	6.2	9.9
				Oolong tea	A	6.8	3.5	3.9	3.9	5.0	0.4	2.5	3.8	1.4	1.2	8.0	6.6	3.1	6.0	8.0	3.4	4.2	3.7	4.0
					B	7.2	5.3	9.8	0.8	10.2	4.3	2.4	7.2	9.1	4.2	1.3	1.7	3.2	10.2	10.8	1.8	4.9	1.3	6.0
111	Methoxy-chlor	TPT	GC-MS	Green tea	A	4.8	4.6	2.9	14.6	12.8	1.0	9.8	1.0	9.0	2.9	6.2	4.7	9.7	8.4	2.1	2.2	6.3	13.2	6.2
					B	9.1	6.4	1.6	15.8	7.0	3.7	6.8	8.6	7.5	5.3	6.9	4.2	8.5	9.5	5.6	5.4	12.8	9.5	5.8
				Oolong tea	A	8.1	6.5	6.7	5.6	10.9	8.4	3.3	20.4	11.2	6.5	11.6	15.0	4.2	4.1	6.9	9.7	7.6	2.1	4.5
					B	5.3	2.5	3.5	1.5	4.2	4.5	10.6	0.9	9.0	6.4	13.2	2.1	4.1	10.6	10.3	0.8	10.3	8.8	10.0
			GC-MS/MS	Green tea	A	4.1	7.3	3.4	5.7	10.4	4.2	8.9	13.6	6.6	4.0	6.8	6.6	8.7	10.6	8.3	13.1	3.9	8.2	5.4
					B	6.4	6.0	2.2	11.8	6.6	5.9	4.3	5.3	3.3	8.2	4.8	1.9	9.1	5.7	7.7	13.0	7.3	5.8	1.0
				Oolong tea	A	11.4	3.7	6.5	5.0	10.7	7.5	7.1	5.8	4.8	2.7	4.2	7.7	4.7	10.3	11.7	8.1	5.8	7.7	3.1
					B	12.7	7.4	8.1	8.9	6.5	3.5	3.3	11.4	4.5	7.1	7.8	5.3	1.4	14.9	9.7	2.4	13.6	6.6	1.3
		ENVI-Carb+PSA	GC-MS	Green tea	A	21.5	1.4	16.4	9.4	10.9	22.0	5.1	17.6	4.8	6.6	16.5	24.4	3.0	8.5	3.1	8.6	11.4	19.8	7.1
					B	18.3	8.9	5.3	7.3	13.1	9.5	7.4	12.9	9.2	7.8	20.0	16.8	32.0	3.9	11.2	6.2	4.5	23.3	15.0
				Oolong tea	A	17.8	9.4	6.1	16.7	9.9	14.9	12.5	27.1	9.4	9.0	5.5	17.0	18.5	4.8	3.2	24.3	2.5	24.2	6.9
					B	16.7	13.0	3.3	5.4	21.2	18.9	16.7	7.7	20.9	15.9	18.8	1.7	6.4	10.6	16.6	6.7	22.0	33.2	5.8
			GC-MS/MS	Green tea	A	15.2	3.8	16.1	6.6	11.7	8.8	12.0	11.9	7.4	3.0	7.5	4.0	14.6	23.4	4.0	7.7	14.0	11.5	6.0
					B	11.0	8.7	4.0	19.6	3.0	2.0	25.8	7.1	13.3	9.5	15.2	1.0	2.4	9.9	4.6	16.4	7.5	13.2	10.2
				Oolong tea	A	8.4	10.2	6.8	9.0	8.2	15.8	6.2	9.3	2.2	6.6	6.7	1.6	11.1	8.8	5.9	13.0	13.3	2.1	7.2
					B	9.3	12.4	7.4	9.0	9.1	15.2	25.8	9.4	2.2	12.4	3.0	5.6	6.3	20.7	13.4	7.5	26.3	4.4	8.0

#	Compound	Method	Instrument	Tea	Rep																			
112	Methyl-parathion	TPT	GC-MS	Green tea	A	11.8	5.1	1.7	4.1	10.8	2.7	5.4	4.3	5.8	6.8	6.5	2.5	4.7	14.7	9.9	0.8	12.9	17.3	1.2
					B	5.1	9.4	2.6	21.1	5.3	4.0	3.4	3.0	3.9	4.4	7.8	11.7	10.2	5.3	9.7	4.7	11.3	10.0	9.2
				Oolong tea	A	6.0	2.6	2.6	6.5	8.7	2.9	14.7	18.4	9.0	14.4	9.0	14.5	6.5	6.9	9.9	3.3	9.9	1.6	6.0
					B	5.1	4.2	5.9	6.8	4.8	1.9	4.6	1.9	6.0	6.6	9.8	3.4	10.7	2.1	11.7	3.2	8.1	4.0	4.8
			GC-MS/MS	Green tea	A	2.9	7.7	7.2	5.2	6.1	0.7	7.7	11.8	15.1	3.7	7.3	5.7	10.9	4.4	3.7	5.9	9.5	6.3	0.8
					B	6.2	10.1	5.3	6.8	4.4	1.1	5.9	8.8	1.7	9.9	2.9	0.6	2.4	5.7	7.1	8.8	11.8	3.8	6.9
				Oolong tea	A	9.2	2.0	4.0	3.7	7.2	6.4	5.0	0.9	1.3	8.7	1.8	7.4	2.7	2.8	8.2	2.1	8.4	8.6	5.3
					B	3.7	4.6	2.5	2.8	2.8	0.5	3.0	1.0	1.8	6.3	6.7	3.6	13.6	2.8	4.3	1.6	0.4	3.4	1.6
		ENVI-Carb+PSA	GC-MS	Green tea	A	9.2	7.3	9.6	6.5	8.3	7.8	4.4	14.3	7.5	4.3	11.9	14.7	1.8	5.4	6.2	6.0	15.1	18.6	2.9
					B	6.9	3.8	5.6	2.1	5.4	5.6	3.2	4.2	11.8	5.8	24.7	13.5	23.9	7.6	12.8	13.5	2.6	20.8	10.9
				Oolong tea	A	10.9	4.7	0.4	5.1	1.1	0.9	10.1	7.2	1.8	11.1	1.9	0.9	15.8	7.1	3.7	15.4	3.6	11.7	4.6
					B	8.8	8.2	8.7	4.2	14.7	11.5	7.1	11.7	10.9	3.9	10.7	2.6	8.7	2.1	14.9	4.6	13.5	9.2	7.0
			GC-MS/MS	Green tea	A	15.8	7.7	11.2	4.0	5.1	20.2	11.8	12.2	9.1	0.7	6.2	11.4	3.9	5.1	6.4	4.3	12.4	6.5	4.6
					B	31.1	11.6	1.7	10.9	9.1	12.6	25.3	5.9	12.4	11.8	14.6	14.2	13.0	16.6	12.0	5.5	7.4	16.7	12.3
				Oolong tea	A	7.7	3.6	5.1	17.9	12.2	2.8	3.1	4.1	2.5	6.4	11.8	5.6	5.4	6.9	8.6	2.1	3.8	10.8	11.3
					B	8.3	7.4	5.0	15.5	6.3	8.3	2.5	10.0	11.4	1.2	3.3	2.6	6.6	11.8	12.5	5.1	7.0	4.1	3.7
113	Metola-chlor	TPT	GC-MS	Green tea	A	5.1	1.7	3.8	2.8	7.9	9.5	3.3	6.7	1.2	4.0	5.9	4.7	6.5	4.6	4.3	3.5	1.8	12.2	4.9
					B	6.2	9.6	5.1	6.9	3.5	2.9	7.3	2.0	2.7	6.2	6.0	8.2	1.5	2.6	2.3	1.6	8.2	3.2	5.6
				Oolong tea	A	7.3	3.5	2.5	6.4	3.8	3.2	9.9	12.4	3.5	11.7	7.1	3.9	1.1	2.7	6.6	7.1	4.7	4.7	10.6
					B	4.5	3.2	10.8	5.2	2.4	2.5	4.4	1.3	7.4	6.2	3.2	1.9	5.0	2.9	4.9	1.1	2.0	11.1	4.1
			GC-MS/MS	Green tea	A	5.9	6.1	4.0	3.5	7.7	2.9	3.8	6.3	5.9	1.7	3.5	5.8	8.2	3.1	1.2	6.5	3.8	5.9	2.9
					B	7.7	8.0	6.3	9.9	6.4	5.1	7.5	4.4	5.6	8.8	4.0	1.1	3.8	5.9	1.3	4.7	4.9	1.8	9.3
				Oolong tea	A	5.6	4.6	1.9	5.5	2.8	3.6	10.9	3.4	0.4	3.9	4.5	4.2	2.8	3.0	10.0	6.4	7.4	4.2	4.1
					B	4.7	0.8	5.3	7.1	4.1	2.0	3.7	9.2	6.3	5.8	4.0	2.6	6.4	4.2	7.1	2.4	3.5	4.2	3.3
		ENVI-Carb+PSA	GC-MS	Green tea	A	7.6	5.0	6.6	3.9	6.6	1.3	21.6	1.5	2.6	12.4	11.1	5.4	7.8	4.6	4.2	6.2	9.3	11.4	5.5
					B	4.1	7.1	2.0	0.4	0.9	1.2	9.7	3.2	3.3	4.9	18.4	2.5	5.8	2.3	9.7	4.5	10.0	9.9	6.2
				Oolong tea	A	2.5	22.7	6.4	4.2	4.0	1.3	8.4	1.9	1.7	1.1	3.4	5.1	1.3	3.8	4.6	4.2	4.6	2.1	7.0
					B	10.6	3.5	6.6	2.8	9.6	5.3	5.5	5.7	10.9	3.6	8.7	2.6	5.5	2.1	7.8	6.1	4.1	5.9	1.8
			GC-MS/MS	Green tea	A	8.3	5.7	1.6	7.1	5.5	4.8	3.3	2.4	4.5	1.6	11.1	12.5	0.9	15.6	6.9	4.5	10.5	1.6	2.9
					B	22.1	7.5	3.7	1.8	5.8	1.4	2.3	3.6	4.0	3.1	16.2	9.7	3.5	8.1	11.5	2.0	6.3	10.1	8.7
				Oolong tea	A	1.5	4.6	3.2	2.6	6.0	1.2	2.0	1.2	1.2	3.4	10.2	2.0	4.5	5.4	12.7	6.4	5.6	1.1	5.9
					B	7.0	1.8	9.7	3.9	7.7	6.0	0.6	8.3	9.8	4.8	1.3	2.9	3.5	11.1	13.4	2.4	4.3	3.5	8.3
114	Nitrapyrin	TPT	GC-MS	Green tea	A	2.9	3.6	2.1	10.6	2.8	7.9	7.0	2.9	8.6	6.6	1.6	5.0	8.5	7.6	8.6	7.1	5.0	1.6	1.7
					B	9.6	7.8	1.9	8.2	3.4	11.2	2.2	7.8	8.7	8.9	3.4	3.8	4.2	13.0	7.6	3.9	14.1	10.7	8.2
				Oolong tea	A	11.4	0.5	2.0	5.0	10.5	5.5	6.8	3.7	7.1	6.9	9.9	3.9	4.7	2.7	4.1	2.5	5.3	7.4	6.8
					B	5.3	9.1	1.0	5.4	4.4	3.2	5.6	3.9	8.1	4.6	6.8	6.4	3.6	3.5	19.9	9.6	3.8	4.5	17.6
			GC-MS/MS	Green tea	A	8.7	4.1	6.0	8.7	14.4	6.4	8.6	12.3	10.5	0.9	4.4	1.6	10.7	5.7	5.8	7.7	11.4	16.1	6.1
					B	11.6	8.0	5.5	11.0	9.9	4.9	8.9	7.0	11.7	12.8	4.8	5.3	4.8	3.6	5.7	3.1	12.4	5.0	5.2
				Oolong tea	A	6.0	4.0	4.8	9.3	5.9	8.6	1.2	3.7	5.8	7.9	9.3	5.6	4.1	4.0	7.3	7.1	5.5	7.5	8.0
					B	11.8	11.0	8.7	0.2	6.4	8.9	4.7	13.9	18.4	15.9	5.2	10.6	13.3	4.6	9.2	4.2	2.7	9.5	11.7
		ENVI-Carb+PSA	GC-MS	Green tea	A	11.7	4.6	8.1	6.9	10.8	5.6	11.8	8.9	11.3	1.3	9.8	12.0	1.3	4.6	3.3	5.8	14.0	8.4	3.1
					B	13.1	5.7	3.6	12.4	5.5	4.2	9.4	5.1	3.6	5.6	18.7	9.7	13.3	4.6	12.1	8.9	0.5	13.6	4.7
				Oolong tea	A	14.8	2.4	7.5	9.8	4.2	4.6	11.5	6.9	6.6	0.6	5.0	8.0	9.3	3.0	6.0	3.5	2.8	16.5	10.7
					B	13.6	6.1	10.3	3.2	4.3	6.7	2.0	2.9	15.2	8.9	12.2	5.2	4.5	8.5	9.6	3.5	9.6	12.3	3.3
			GC-MS/MS	Green tea	A	39.8	10.6	7.3	5.7	6.1	15.9	8.8	7.5	12.8	2.5	5.4	11.6	6.3	15.4	6.8	17.3	13.3	10.2	16.1
					B	11.8	13.7	2.2	14.4	10.7	6.7	13.4	2.8	11.5	6.1	27.2	0.7	26.2	23.0	36.0	7.3	13.3	14.8	8.3
				Oolong tea	A	11.8	8.7	12.1	8.5	6.3	7.2	5.1	9.6	3.1	3.2	9.6	9.0	7.8	13.3	10.9	3.1	13.8	19.2	6.0
					B	16.1	10.2	4.8	10.9	13.0	11.1	16.7	2.7	17.5	8.6	8.1	4.9	10.5	15.9	11.4	7.1	4.9	14.0	14.5

(Continued)

Appendix Table 5.2 RSD of 201 Pesticides in Youden Pair Aged Tea Sample for Parallel Determinations (Nov 9, 2009–Feb 7, 2010) (cont.)

No.	Pesticides	SPE	Method	Sample	Youden pair	Nov 9, 2009 (n=5)	Nov 14, 2009 (n=3)	Nov 19, 2009 (n=3)	Nov 24, 2009 (n=3)	Nov 29, 2009 (n=3)	Dec 4, 2009 (n=3)	Dec 9, 2009 (n=3)	Dec 14, 2009 (n=3)	Dec 19, 2009 (n=3)	Dec 24, 2009 (n=3)	Dec 14, 2009 (n=3)	Jan 3, 2010 (n=3)	Jan 8, 2010 (n=3)	Jan 13, 2010 (n=3)	Jan 18, 2010 (n=3)	Jan 23, 2010 (n=3)	Jan 28, 2010 (n=3)	Feb 2, 2010 (n=3)	Feb 7, 2010 (n=3)
115	Oxyfluorfen	TPT	GC-MS	Green tea	A	9.8	5.7	2.0	8.9	13.4	2.5	7.6	12.3	3.8	7.7	12.9	8.8	12.8	11.8	2.9	4.4	9.7	12.3	7.0
					B	5.6	6.7	4.5	18.5	6.7	7.2	6.1	7.3	6.1	4.7	10.9	11.9	7.6	13.7	1.3	8.5	10.8	5.7	4.9
				Oolong tea	A	8.9	10.4	9.3	2.6	26.0	4.4	11.6	27.8	7.5	11.7	13.0	13.5	5.5	3.7	8.9	4.9	6.6	8.8	2.3
					B	3.0	8.7	10.7	61.5	14.0	2.3	7.1	0.6	8.0	10.8	14.4	2.8	4.6	8.5	7.0	4.4	7.2	2.2	3.2
			GC-MS/MS	Green tea	A	10.6	6.6	3.8	10.5	9.2	6.0	14.3	10.0	14.4	2.0	14.7	7.2	14.1	5.7	2.2	7.0	7.8	7.4	3.0
					B	7.7	10.7	7.7	9.3	5.4	5.4	1.2	12.6	11.0	11.7	6.4	14.3	0.6	3.8	7.6	10.9	13.4	2.1	7.1
				Oolong tea	A	5.8	5.7	7.2	3.7	4.1	3.2	1.8	14.9	5.3	8.3	16.0	12.0	2.7	4.8	7.8	4.0	6.4	8.5	7.9
					B	8.1	2.5	3.0	11.9	14.3	3.9	10.7	14.2	7.0	5.2	19.6	5.1	4.2	7.3	1.2	4.1	10.7	6.2	8.4
		ENVI-Carb+PSA	GC-MS	Green tea	A	19.3	6.8	17.5	8.7	13.5	14.9	5.4	18.6	5.1	8.9	4.6	15.0	2.3	13.6	4.0	12.7	16.9	15.0	5.4
					B	11.3	10.0	8.0	2.9	8.9	12.8	2.4	11.7	8.9	6.9	41.0	8.3	29.8	6.4	17.2	6.6	1.1	24.2	14.4
				Oolong tea	A	9.1	7.8	9.0	9.2	14.9	0.4	3.2	7.7	8.1	7.4	4.9	3.4	11.5	3.7	6.9	25.7	7.6	13.3	2.8
					B	6.9	6.7	13.3	3.6	28.3	17.4	11.5	10.5	16.2	7.1	15.6	6.6	10.8	8.5	14.8	6.2	14.8	8.2	5.6
			GC-MS/MS	Green tea	A	17.2	9.8	10.8	6.6	8.4	10.4	13.0	18.8	4.4	8.8	9.3	2.5	12.6	6.1	5.0	12.3	7.1	6.1	11.4
					B	19.2	9.7	10.6	8.6	5.7	12.4	39.5	14.5	18.9	4.5	23.3	0.9	21.1	11.0	12.9	7.4	22.9	10.5	18.8
				Oolong tea	A	8.6	4.8	5.9	11.1	8.7	3.7	6.9	19.3	5.6	3.4	5.3	2.8	15.1	7.2	8.1	4.0	4.9	11.6	8.3
					B	8.9	11.8	12.8	6.9	19.7	10.5	26.2	9.9	12.1	7.4	9.3	5.5	14.4	14.2	10.6	4.1	7.7	5.1	9.9
116	Pendimethalin	TPT	GC-MS	Green tea	A	3.5	3.8	4.2	6.0	14.7	2.8	11.0	18.3	12.4	10.2	12.5	8.1	13.2	11.1	4.5	4.8	10.8	24.5	5.4
					B	5.9	9.5	5.3	17.5	5.7	5.1	4.2	14.4	5.8	8.7	8.8	13.5	5.4	12.0	2.1	6.8	17.6	16.7	5.2
				Oolong tea	A	5.7	3.4	5.8	4.9	30.1	3.6	0.3	21.8	7.5	14.3	18.5	10.7	4.6	3.1	6.8	11.6	4.8	6.0	0.5
					B	2.5	6.5	6.5	5.1	13.5	9.5	11.6	4.1	5.8	10.9	17.2	2.0	3.4	6.4	7.6	6.7	3.4	2.4	2.2
			GC-MS/MS	Green tea	A	5.4	3.8	2.8	4.6	12.0	6.2	10.8	17.5	17.4	0.6	10.6	6.1	13.1	6.7	4.9	7.1	7.5	5.3	7.2
					B	7.8	10.0	7.3	10.8	4.1	3.8	2.6	12.4	2.7	9.4	0.9	10.0	1.9	4.9	5.3	6.2	8.6	5.0	6.8
				Oolong tea	A	7.2	3.3	4.1	2.6	5.2	4.6	0.3	9.7	13.9	6.9	14.0	11.8	6.1	2.4	9.8	5.2	8.4	5.2	6.2
					B	6.7	11.0	2.6	6.8	8.4	4.4	11.3	9.7	11.9	3.4	15.4	2.3	5.4	4.3	1.4	2.2	3.9	7.5	6.3
		ENVI-Carb+PSA	GC-MS	Green tea	A	17.1	6.2	13.1	8.3	11.2	19.2	24.8	27.8	16.4	8.9	5.7	14.5	0.7	28.2	4.1	23.4	15.9	22.1	4.8
					B	9.4	10.3	5.9	5.7	9.9	15.8	8.0	27.3	19.9	7.0	39.6	7.5	41.0	9.6	31.9	7.4	3.4	26.6	12.4
				Oolong tea	A	7.8	6.4	5.2	5.4	16.4	1.4	3.6	22.8	6.0	7.1	5.4	4.6	11.2	3.1	4.4	32.2	12.4	9.5	3.5
					B	6.3	4.5	9.0	2.4	25.3	15.3	13.3	10.4	15.4	6.4	14.1	4.7	10.5	6.4	18.3	6.4	19.9	4.5	5.8
			GC-MS/MS	Green tea	A	19.4	7.0	10.8	6.6	6.0	9.8	22.0	20.2	8.3	4.2	23.2	1.4	3.3	4.7	3.2	13.0	8.5	13.9	11.3
					B	23.3	13.7	7.9	4.8	1.0	10.5	36.9	20.6	20.4	2.3	5.7	4.9	14.8	6.1	14.1	4.2	13.6	23.5	18.0
				Oolong tea	A	8.4	4.2	5.8	4.7	2.6	2.7	5.6	22.0	3.0	7.9	6.0	6.2	8.8	4.8	5.9	5.2	6.8	10.7	6.2
					B	12.0	5.2	8.1	1.0	13.9	10.5	30.0	6.2	13.4	4.9	6.0	3.1	8.7	11.4	9.9	2.2	10.4	9.6	9.6
117	Picoxystrobin	TPT	GC-MS	Green tea	A	5.1	5.1	2.2	5.9	7.1	3.9	7.3	4.8	4.1	1.2	4.8	4.1	6.4	4.0	1.9	1.4	5.0	5.7	4.1
					B	4.7	6.7	2.8	4.1	3.2	3.1	3.0	9.2	7.9	2.7	5.7	8.5	1.7	3.9	4.3	2.5	5.4	1.0	4.8
				Oolong tea	A	6.5	1.4	4.0	1.5	5.6	2.0	8.0	8.4	2.5	2.8	4.0	2.3	3.9	4.6	7.9	3.3	2.9	3.4	0.4
					B	3.5	1.7	0.7	2.8	1.2	2.8	4.1	2.9	6.1	2.7	1.9	1.1	1.9	5.3	2.1	1.2	2.2	0.7	2.3
			GC-MS/MS	Green tea	A	4.4	5.7	2.9	6.0	10.0	4.2	6.0	6.9	5.4	3.4	7.6	5.5	8.3	2.4	3.8	3.9	5.6	3.0	4.0
					B	7.2	5.7	6.8	1.7	5.0	6.0	7.6	6.7	4.9	6.1	4.1	2.6	2.2	6.7	4.0	5.9	5.0	2.8	4.0
				Oolong tea	A	7.9	2.0	0.5	1.7	2.5	2.3	8.6	3.3	1.9	4.9	2.7	4.9	10.4	2.2	8.4	5.5	5.7	1.7	5.7
					B	4.7	2.2	5.1	4.8	5.7	4.0	1.9	5.8	2.0	3.4	5.8	1.5	3.0	1.1	6.4	3.0	2.2	6.0	3.8
		ENVI-Carb+PSA	GC-MS	Green tea	A	6.4	5.1	2.2	5.9	8.6	1.7	8.5	1.0	4.0	3.2	7.6	2.6	1.7	5.0	5.3	6.1	10.7	3.1	6.1
					B	3.8	6.7	2.8	4.1	2.2	1.2	1.6	4.6	4.0	1.6	23.1	3.8	1.5	3.9	11.1	0.5	4.6	7.3	7.0
				Oolong tea	A	3.8	2.4	4.0	1.5	5.6	2.4	3.8	2.3	1.8	2.5	8.8	3.7	1.1	2.9	2.6	1.8	6.3	4.2	3.3
					B	4.8	6.4	9.4	5.0	10.5	5.0	2.0	5.8	11.9	3.9	10.1	1.3	4.0	6.7	10.4	4.0	4.0	5.5	5.2
			GC-MS/MS	Green tea	A	6.5	6.8	4.4	5.4	9.1	1.9	8.3	1.9	6.9	1.0	14.6	3.1	7.8	14.6	6.0	7.3	11.0	2.2	6.8
					B	6.1	5.5	6.5	1.2	3.9	0.8	3.8	4.5	4.9	1.4	17.5	1.6	5.0	2.6	10.8	1.7	7.8	7.6	5.8
				Oolong tea	A	5.9	2.4	4.0	2.1	7.3	0.5	2.5	3.7	0.6	5.1	8.6	2.0	2.4	2.0	7.7	5.5	5.8	2.2	6.5
					B	7.2	2.3	9.5	4.4	7.7	6.1	1.6	6.5	9.4	3.2	2.2	1.8	2.9	10.9	9.0	3.0	5.4	2.0	9.0

(Continued)

No.	Compound	Cleanup	Method	Tea	A/B	1	2	3	4	5	6	7	8	9	10	11	12	13	14	15	16	17	18	19
118	Piperophos	TPT	GC-MS	Green tea	A	5.4	4.4	1.3	5.1	10.6	1.8	8.5	11.0	10.6	4.8	3.7	9.3	9.6	7.1	3.2	1.7	7.2	3.8	2.7
					B	5.9	9.0	4.1	10.7	5.9	6.9	4.5	4.1	5.0	1.9	14.6	8.5	10.4	3.5	3.3	3.2	10.8	3.6	5.9
				Oolong tea	A	7.4	5.5	3.7	2.7	8.6	4.7	9.1	22.2	6.8	13.2	15.6	7.9	2.2	2.5	9.5	2.8	6.5	1.6	2.2
					B	3.4	10.8	9.0	13.2	9.2	5.9	6.2	2.5	11.6	8.4	20.1	3.3	2.2	6.8	1.5	1.6	5.1	2.6	2.1
			GC-MS/MS	Green tea	A	3.7	7.7	1.3	7.0	9.9	4.4	10.7	12.1	10.5	2.8	7.3	6.4	6.9	10.2	2.3	4.4	2.8	10.0	3.6
					B	6.8	6.7	4.9	7.7	5.7	3.4	4.9	9.4	6.2	6.2	8.8	6.6	2.2	7.8	6.4	7.6	4.9	3.3	6.4
				Oolong tea	A	7.2	6.1	6.6	1.9	6.9	5.9	6.3	7.0	7.2	5.0	8.2	5.6	2.3	2.9	9.6	2.1	14.6	10.1	5.0
					B	2.3	7.0	3.9	5.0	5.1	1.3	8.2	6.3	13.6	6.2	6.3	2.2	7.6	1.6	4.9	3.3	6.7	4.3	17.5
		ENVI-Carb+PSA	GC-MS	Green tea	A	14.0	2.6	10.3	8.5	12.2	8.1	1.4	9.5	1.3	4.3	6.4	8.6	2.2	14.2	3.2	22.4	8.2	28.2	4.9
					B	10.5	6.7	5.6	7.4	2.3	7.5	4.9	8.0	11.3	2.4	33.7	13.3	18.5	1.4	21.9	1.2	4.0	22.9	10.0
				Oolong tea	A	5.5	4.4	6.2	2.3	10.8	1.6	3.4	21.9	4.2	1.0	6.6	9.4	1.2	2.5	1.5	6.3	3.2	6.6	3.0
					B	8.1	10.2	9.2	1.1	24.6	7.5	7.5	2.0	14.5	6.0	14.4	2.9	8.3	6.8	11.6	4.4	8.6	4.7	4.6
			GC-MS/MS	Green tea	A	14.2	6.0	8.8	9.8	4.1	11.9	11.5	9.2	0.8	12.2	7.9	5.5	2.7	10.2	6.6	11.2	3.6	2.0	5.1
					B	31.9	6.8	5.0	11.4	6.6	5.7	24.2	6.6	14.9	1.0	15.8	6.2	4.2	7.0	8.1	2.9	9.8	25.8	14.4
				Oolong tea	A	8.4	5.9	7.4	4.5	5.3	3.4	0.4	15.0	2.4	0.2	9.3	2.8	3.2	10.0	12.3	2.1	3.3	10.1	4.8
					B	11.4	6.4	7.0	3.8	18.9	6.5	9.7	6.1	10.6	5.8	1.7	3.6	5.5	13.8	8.9	3.3	4.5	8.2	8.2
119	Pirimi-phoe-ethyl	TPT	GC-MS	Green tea	A	7.6	4.5	4.1	2.5	8.3	4.6	7.2	4.0	4.3	8.3	10.3	6.7	6.2	2.4	2.5	1.5	4.4	6.8	4.0
					B	4.9	8.2	4.7	7.9	4.3	2.7	3.2	1.9	5.1	12.3	10.1	12.4	0.8	0.8	3.8	2.7	5.3	1.6	6.0
				Oolong tea	A	6.0	1.9	4.7	3.6	1.6	1.9	8.5	9.6	2.9	8.2	12.8	3.5	3.6	2.7	8.9	3.2	3.5	2.7	0.8
					B	3.8	4.7	4.5	5.8	4.0	5.6	4.6	3.5	7.8	1.8	11.8	0.7	2.4	2.7	2.9	0.6	1.2	1.2	2.3
			GC-MS/MS	Green tea	A	4.5	4.4	3.8	1.2	10.4	2.5	5.1	6.8	5.0	6.9	8.1	4.9	4.0	1.0	3.8	3.3	2.2	4.9	2.7
					B	7.2	7.4	7.3	9.5	6.9	6.2	9.3	5.7	5.1	3.4	3.3	8.5	5.1	4.5	2.3	2.5	5.2	2.4	5.6
				Oolong tea	A	7.6	1.7	3.4	2.5	2.4	6.4	8.4	2.9	1.6	3.3	5.4	1.2	2.1	3.3	5.4	5.9	4.9	7.7	4.0
					B	5.3	0.9	5.9	4.4	7.7	2.0	3.9	3.4	1.0	5.0	3.9	0.9	3.2	4.2	5.2	2.0	2.7	4.0	6.3
		ENVI-Carb+PSA	GC-MS	Green tea	A	7.7	5.8	2.4	5.6	8.5	3.6	5.4	0.8	4.5	5.7	7.8	6.9	2.6	5.3	4.8	6.7	10.8	3.1	5.4
					B	3.9	8.6	5.7	5.2	2.2	5.9	4.4	3.7	3.9	3.9	31.2	11.5	1.4	3.2	11.3	1.6	5.6	7.8	6.1
				Oolong tea	A	4.5	5.8	3.7	2.4	3.8	3.1	4.1	1.8	4.3	5.5	5.7	3.9	1.6	2.7	2.6	2.1	6.3	4.1	4.5
					B	4.8	2.7	10.3	1.9	17.5	2.0	2.1	6.8	12.3	1.0	10.0	5.1	3.7	2.7	9.8	4.4	6.1	5.5	4.7
			GC-MS/MS	Green tea	A	6.3	7.3	3.3	5.9	6.7	4.4	4.1	2.6	4.9	1.4	10.1	11.3	7.1	16.3	7.5	6.4	8.4	5.7	11.9
					B	5.2	8.0	5.6	2.8	0.7	3.2	5.8	4.4	3.5	2.2	19.4	19.0	6.4	12.8	10.1	2.5	13.2	5.3	11.7
				Oolong tea	A	2.2	5.7	3.7	2.9	7.7	5.6	5.4	2.2	1.8	6.3	7.4	2.3	6.4	9.0	10.8	5.9	9.3	4.5	5.0
					B	6.0	3.5	9.2	1.5	8.4	3.7	0.9	7.8	10.0	6.4	2.4	4.2	1.5	7.8	15.0	2.0	2.8	3.3	7.4
120	Pirimi-phos-methyl	TPT	GC-MS	Green tea	A	7.8	4.4	3.6	2.2	8.3	2.8	6.2	4.5	5.4	3.8	7.1	3.9	6.3	7.7	2.0	1.3	4.0	6.5	3.1
					B	5.2	8.8	3.1	8.0	3.8	3.7	2.2	1.7	3.9	6.7	6.8	9.1	1.9	7.0	7.2	1.7	4.1	2.8	6.3
				Oolong tea	A	4.7	2.3	2.6	3.6	1.5	2.8	9.1	11.3	7.0	9.9	7.8	3.8	2.0	2.4	8.3	3.2	3.2	0.3	1.6
					B	3.9	3.4	3.0	10.4	1.9	3.1	4.0	3.5	6.9	6.4	11.4	1.4	1.7	3.4	4.4	0.2	4.4	2.1	2.2
			GC-MS/MS	Green tea	A	3.9	6.1	3.4	2.7	8.8	1.8	7.2	4.3	5.9	1.9	7.8	6.9	8.4	3.9	2.0	8.2	2.2	3.4	3.7
					B	8.2	8.2	6.1	8.3	5.7	4.4	8.8	3.8	4.2	8.5	1.2	2.3	1.8	5.9	2.3	2.4	7.0	4.8	4.1
				Oolong tea	A	6.5	4.9	1.4	3.5	4.1	7.9	10.0	3.0	2.6	2.7	6.4	1.6	2.0	3.1	1.6	7.1	2.1	6.0	8.1
					B	3.0	3.3	6.6	6.6	5.0	3.0	2.7	5.8	5.8	5.7	9.8	2.9	2.5	4.8	2.9	3.3	3.9	2.7	7.6
		ENVI-Carb+PSA	GC-MS	Green tea	A	8.2	7.0	4.4	6.7	6.7	1.2	7.4	3.9	3.9	3.3	7.8	2.0	2.5	3.4	3.1	4.2	3.3	4.7	5.3
					B	4.8	9.3	3.1	1.7	0.3	1.8	4.0	4.7	4.7	3.1	22.7	4.5	4.5	0.6	8.8	3.0	5.3	8.5	6.5
				Oolong tea	A	4.5	3.8	2.6	2.7	3.7	2.3	3.6	1.7	1.7	2.4	4.4	1.9	2.3	2.7	3.5	3.1	7.3	5.4	3.8
					B	5.4	3.6	8.9	2.9	13.5	1.2	2.4	7.2	11.3	3.2	9.8	5.4	3.7	3.4	9.8	3.6	8.9	2.1	4.5
			GC-MS/MS	Green tea	A	9.5	4.3	2.4	6.3	4.8	4.7	4.6	6.2	7.3	2.8	9.8	12.5	6.5	14.8	6.4	5.3	7.2	2.6	4.3
					B	16.5	9.3	3.8	1.6	2.9	5.0	7.0	1.5	5.8	4.6	19.1	8.2	5.9	4.2	9.4	5.1	9.6	8.9	6.6
				Oolong tea	A	2.4	3.6	2.8	3.9	3.7	2.4	2.5	1.9	1.2	2.2	5.6	2.7	5.2	4.4	7.9	7.1	5.3	4.8	5.6
					B	6.6	4.7	9.1	3.2	6.9	1.3	1.6	4.7	9.6	2.8	2.9	3.6	3.8	10.8	10.7	3.3	5.3	3.0	7.1

Appendix Table 5.2 RSD of 201 Pesticides in Youden Pair Aged Tea Sample for Parallel Determinations (Nov 9, 2009–Feb 7, 2010) (cont.)

No.	Pesticides	SPE	Method	Sample	Youden pair	RSD% Nov 9, 2009 (n=5)	Nov 14, 2009 (n=3)	Nov 19, 2009 (n=3)	Nov 24, 2009 (n=3)	Nov 29, 2009 (n=3)	Dec 4, 2009 (n=3)	Dec 9, 2009 (n=3)	Dec 14, 2009 (n=3)	Dec 19, 2009 (n=3)	Dec 24, 2009 (n=3)	Jan 3, 2010 (n=3)	Jan 8, 2010 (n=3)	Jan 13, 2010 (n=3)	Jan 18, 2010 (n=3)	Jan 23, 2010 (n=3)	Jan 28, 2010 (n=3)	Feb 2, 2010 (n=3)	Feb 7, 2010 (n=3)
121	Profenofos	TPT	GC-MS	Green tea	A	10.5	3.3	1.3	14.4	15.4	1.6	8.1	12.2	11.2	1.2	11.1	5.4	17.1	11.4	3.2	14.6	13.6	2.1
					B	11.9	12.8	4.5	21.4	12.4	8.2	4.7	5.7	0.8	4.4	11.1	18.3	10.3	13.9	9.2	13.1	13.9	19.4
				Oolong tea	A	15.2	3.7	2.3	5.4	16.3	5.2	14.4	14.1	17.3	10.4	12.6	12.3	13.6	10.2	10.7	12.8	27.8	14.1
					B	13.6	5.0	1.4	3.0	3.8	6.2	9.0	6.4	9.8	13.1	2.1	13.2	6.2	13.3	9.2	13.8	10.8	16.7
			GC-MS/MS	Green tea	A	15.0	7.5	12.8	4.6	6.7	11.6	30.3	16.9	19.5	8.8	4.6	18.8	17.1	6.1	19.2	5.4	22.5	14.8
					B	11.4	9.9	19.9	17.4	14.2	18.7	13.5	18.8	12.1	15.1	19.3	14.1	8.9	16.1	5.1	24.2	9.5	7.8
				Oolong tea	A	24.0	8.2	20.6	4.4	12.2	16.8	3.2	4.2	14.5	11.2	16.2	15.8	15.7	6.1	18.9	21.8	4.2	23.2
					B	9.9	13.7	16.2	12.6	2.1	11.9	3.7	0.2	28.9	20.9	10.1	10.6	4.7	11.2	10.5	8.2	23.4	42.0
		ENVI-Carb+PSA	GC-MS	Green tea	A	16.5	18.6	11.3	13.4	8.2	21.3	10.3	19.6	5.2	7.6	35.8	9.1	12.0	8.5	7.3	14.3	20.1	1.6
					B	25.0	7.7	9.6	12.4	17.0	3.7	4.8	8.7	12.2	19.7	23.2	5.1	10.1	8.2	21.3	11.4	39.8	16.0
				Oolong tea	A	15.4	5.2	10.1	9.3	4.4	7.2	22.0	19.6	10.0	18.5	6.6	4.6	13.2	9.7	20.9	3.5	18.7	9.2
					B	25.4	9.6	10.7	2.3	9.3	16.8	10.1	10.5	17.6	15.2	10.0	21.9	8.0	15.7	6.0	24.8	18.6	13.7
			GC-MS/MS	Green tea	A	30.0	6.7	25.8	17.3	11.4	32.0	23.9	26.2	24.7	2.2	4.6	9.2	37.9	6.4	6.3	21.0	2.8	42.7
					B	51.7	24.8	6.0	18.7	5.9	21.2	19.2	1.5	19.8	29.9	8.2	0.7	12.2	10.2	12.7	31.9	26.3	27.3
				Oolong tea	A	12.7	15.5	12.8	5.3	33.5	8.1	10.7	39.2	11.3	3.4	18.3	4.2	7.2	5.5	18.9	16.5	16.4	11.5
					B	14.4	16.0	16.8	18.5	15.5	10.2	21.2	18.9	24.4	11.0	19.1	11.0	20.9	19.4	10.5	25.1	11.1	7.7
122	Profluralin	TPT	GC-MS	Green tea	A	6.0	4.5	3.9	3.8	11.2	3.1	6.4	7.8	9.1	3.2	3.8	10.7	7.6	2.1	3.2	6.0	14.0	3.2
					B	5.2	9.0	5.5	16.9	4.4	2.2	0.6	9.3	4.3	9.7	7.9	8.3	8.0	1.4	4.7	11.0	12.6	6.5
				Oolong tea	A	4.5	1.1	3.4	3.6	15.9	3.2	9.9	9.3	9.1	9.7	7.1	5.0	2.7	4.2	11.2	1.7	3.7	1.4
					B	3.4	3.3	3.4	4.5	7.4	3.5	4.9	2.0	6.1	4.3	2.8	2.6	5.4	13.2	3.5	2.1	1.9	2.7
			GC-MS/MS	Green tea	A	4.3	5.0	3.7	4.7	13.3	6.3	12.2	10.7	16.1	0.5	5.2	8.8	2.5	3.0	5.9	3.6	9.5	4.2
					B	7.7	8.3	4.3	10.0	5.6	5.0	4.6	9.9	5.7	11.2	8.1	8.7	8.4	3.2	6.0	10.8	3.8	4.5
				Oolong tea	A	6.7	0.9	1.6	0.7	8.2	5.3	3.8	4.5	9.5	5.8	5.1	6.5	3.9	9.8	2.9	6.6	1.2	6.1
					B	8.1	8.7	2.9	6.4	6.9	4.3	7.8	10.0	7.2	1.9	5.2	5.2	3.4	3.2	4.6	3.7	3.2	9.6
		ENVI-Carb+PSA	GC-MS	Green tea	A	12.0	6.0	8.6	5.8	9.6	9.3	12.6	20.3	6.6	3.5	10.1	0.4	9.9	4.2	10.0	15.2	9.6	7.9
					B	7.6	8.5	4.3	0.5	5.7	8.0	13.2	8.3	2.9	2.8	7.8	13.9	5.8	13.2	6.2	1.5	19.5	10.3
				Oolong tea	A	7.3	3.4	3.7	3.8	2.7	0.8	4.7	14.1	3.9	2.9	3.1	4.4	2.7	3.2	24.9	9.5	8.1	3.1
					B	6.2	3.7	8.3	1.0	17.5	8.5	11.0	10.6	11.6	3.6	3.1	9.9	5.4	14.8	6.0	16.7	8.3	5.4
			GC-MS/MS	Green tea	A	15.9	3.9	8.5	3.4	8.1	17.1	8.7	17.0	8.6	2.5	2.3	6.6	16.6	5.3	2.1	9.6	5.1	5.7
					B	23.0	11.4	7.9	9.4	2.7	6.5	48.6	14.4	12.7	4.6	3.3	7.9	11.2	7.2	7.2	7.8	19.0	13.9
				Oolong tea	A	7.0	3.8	3.8	5.8	3.9	1.3	2.4	17.5	0.7	3.9	7.1	8.8	6.6	5.6	2.9	11.2	8.8	4.0
					B	10.5	4.5	8.7	2.2	10.6	9.0	43.0	10.0	11.0	3.5	1.1	3.1	11.8	10.4	4.6	9.1	11.8	5.6
123	Propachlor	TPT	GC-MS	Green tea	A	8.2	4.6	0.7	5.2	8.3	4.7	7.1	7.9	6.3	5.4	3.9	8.1	1.4	8.4	4.1	5.2	8.5	4.7
					B	7.8	10.0	5.2	5.7	2.7	6.4	0.9	6.4	1.6	5.6	1.9	6.9	13.8	3.8	9.0	11.8	9.6	7.1
				Oolong tea	A	5.6	7.9	1.7	4.6	10.3	6.3	11.6	6.4	5.5	3.4	3.8	6.5	6.0	9.8	4.2	4.2	2.4	7.9
					B	4.4	2.5	5.0	6.9	3.5	3.2	4.3	2.4	5.0	3.5	2.2	5.5	1.3	2.1	1.2	4.4	4.2	8.4
			GC-MS/MS	Green tea	A	4.9	5.2	4.1	4.4	8.0	3.7	4.7	6.6	7.6	3.4	5.4	7.8	4.0	2.3	3.9	6.8	5.5	5.0
					B	7.7	9.2	4.1	6.7	8.1	4.9	7.7	4.4	4.1	8.4	1.3	1.2	5.1	2.8	5.2	6.9	2.4	6.8
				Oolong tea	A	7.0	11.5	4.5	3.5	5.4	3.1	19.3	4.0	2.7	5.4	2.3	2.2	0.7	9.4	3.7	11.7	6.4	5.0
					B	4.9	0.8	5.9	5.5	3.9	0.6	5.5	4.3	9.1	8.7	3.1	1.6	2.3	8.3	0.7	5.8	6.7	9.0
		ENVI-Carb+PSA	GC-MS	Green tea	A	6.5	3.0	1.6	6.4	7.7	4.9	9.1	4.1	4.5	0.9	4.8	1.2	2.7	6.7	14.7	13.6	8.9	4.9
					B	10.1	6.7	5.6	4.5	0.7	1.4	7.0	3.4	5.4	5.6	9.2	8.2	4.3	20.2	5.9	4.4	14.9	5.2
				Oolong tea	A	6.6	2.9	3.4	7.0	5.8	1.3	5.5	3.1	4.6	2.7	2.2	2.5	4.9	7.5	4.2	4.1	6.6	3.6
					B	7.7	3.7	8.4	6.2	7.6	6.3	4.5	5.2	10.9	2.9	1.4	7.0	2.3	12.4	3.5	10.3	0.7	4.7
			GC-MS/MS	Green tea	A	11.9	6.3	2.8	5.6	4.5	11.5	3.4	2.7	8.6	2.5	3.8	4.0	17.6	3.1	4.8	14.7	5.6	7.3
					B	45.8	11.2	2.3	7.9	11.7	2.3	4.6	3.7	7.0	9.4	1.5	5.3	16.0	12.5	4.0	6.9	10.2	7.4
				Oolong tea	A	2.3	5.1	3.1	5.5	5.6	1.4	0.9	4.8	4.4	5.0	2.4	7.6	6.3	12.9	3.7	5.0	3.6	8.9
					B	6.4	4.6	8.8	6.9	3.4	4.6	4.4	4.4	10.9	4.2	3.2	6.1	12.8	13.8	0.7	5.1	7.8	8.4

124	Propicon-azole	TPT	GC-MS	Green tea	A	8.5	4.9	7.1	4.2	6.1	3.9	11.2	3.3	2.7	6.1	3.9	7.9	6.3	9.1	4.0	3.6	4.8	5.1	2.8
					B	3.7	2.9	8.5	2.7	6.3	3.8	4.1	2.8	5.1	10.6	7.4	10.7	0.7	4.5	5.8	4.6	5.5	1.0	4.4
				Oolong tea	A	6.0	5.7	2.2	2.0	5.6	2.5	9.5	8.9	8.1	9.3	10.3	5.5	4.8	3.6	6.3	3.8	2.8	2.2	1.0
					B	5.1	4.2	2.5	9.0	2.7	2.8	4.6	4.9	10.7	7.4	13.0	1.5	1.4	6.2	2.5	2.5	2.6	1.8	1.6
			GC-MS/MS	Green tea	A	4.9	7.7	1.3	3.0	5.7	11.3	6.0	7.0	4.7	3.6	6.3	2.1	3.6	2.4	4.6	4.7	2.0	1.8	5.2
					B	6.6	6.0	6.0	9.0	1.9	7.0	7.2	5.2	6.1	6.3	26.1	3.1	2.7	6.5	4.5	32.1	4.4	2.3	4.3
				Oolong tea	A	7.3	1.4	4.2	5.1	3.3	5.2	8.3	2.2	4.4	4.5	2.1	1.4	6.4	2.6	5.4	3.2	2.7	3.1	0.6
					B	4.0	3.2	4.8	3.9	10.3	2.4	3.7	6.9	6.0	4.9	3.2	3.6	2.4	3.1	7.5	2.3	4.4	3.0	5.4
		ENVI-Carb+PSA	GC-MS	Green tea	A	7.6	1.3	2.6	11.9	5.6	8.0	8.8	2.4	9.8	1.6	2.8	6.2	3.6	4.4	6.4	6.1	11.9	2.8	5.5
					B	4.3	8.2	8.4	8.0	5.6	2.3	3.4	13.0	2.7	6.8	32.1	8.9	4.0	5.0	9.8	2.6	4.1	7.7	6.2
				Oolong tea	A	6.1	2.3	8.2	4.0	3.2	2.6	3.1	3.2	7.9	6.4	7.4	5.2	9.0	2.5	6.3	2.7	2.5	6.2	2.9
					B	4.9	7.9	14.2	4.1	18.7	1.1	1.9	11.2	6.0	6.6	10.8	13.0	1.9	4.3	7.4	4.6	4.4	0.9	3.8
			GC-MS/MS	Green tea	A	6.2	4.9	32.2	7.6	9.7	4.3	8.0	2.2	7.2	1.7	7.4	2.6	8.2	2.8	7.0	5.1	11.4	1.7	4.0
					B	6.1	9.3	6.1	4.3	0.4	3.7	1.6	5.9	3.4	2.0	15.9	7.7	2.4	5.1	12.7	1.1	8.1	8.3	7.8
				Oolong tea	A	7.0	2.8	5.6	2.9	6.3	6.0	3.6	5.1	1.9	4.2	8.7	3.0	5.6	5.7	8.0	3.2	6.4	6.3	5.1
					B	7.6	0.8	9.0	3.9	12.4	3.0	3.2	10.3	8.3	2.3	1.4	2.9	6.1	8.2	12.3	2.3	3.9	3.1	4.8
125	Propyza-mide	TPT	GC-MS	Green tea	A	6.9	5.9	1.8	2.8	7.4	4.3	6.9	5.5	4.9	3.2	5.7	3.3	7.5	4.1	2.6	2.7	4.3	4.8	4.2
					B	2.5	8.1	10.2	8.4	3.6	2.1	3.3	5.2	5.0	6.4	4.4	8.0	2.3	1.7	6.4	3.3	7.0	1.2	5.5
				Oolong tea	A	6.6	5.4	3.1	7.1	2.9	8.3	4.6	8.6	4.9	8.4	4.8	2.9	1.0	2.6	7.6	3.4	4.1	1.4	0.6
					B	4.4	2.5	1.8	0.8	0.9	1.5	5.0	6.2	3.1	3.3	0.3	2.9	0.9	7.1	6.0	3.9	4.0	3.2	4.9
			GC-MS/MS	Green tea	A	4.7	6.7	2.7	2.0	10.0	4.1	7.0	7.0	5.0	1.9	6.0	5.4	8.3	3.0	1.7	5.6	3.5	3.1	3.3
					B	3.3	8.6	6.1	8.1	4.9	4.9	6.7	7.4	6.8	7.6	5.1	4.6	5.2	3.7	1.0	4.3	6.3	1.4	5.3
				Oolong tea	A	7.0	5.5	1.4	9.4	4.9	7.9	5.3	3.4	2.3	3.2	3.1	3.0	2.9	1.7	6.9	1.1	1.2	4.3	6.7
					B	4.6	2.4	5.0	1.0	1.4	2.0	4.7	8.4	4.0	5.4	3.4	3.3	2.0	7.9	10.2	4.6	7.3	5.2	6.4
		ENVI-Carb+PSA	GC-MS	Green tea	A	7.1	6.3	4.2	9.4	8.7	3.4	7.1	3.2	3.1	4.2	6.5	0.5	3.6	3.8	3.3	6.9	10.2	4.5	2.9
					B	3.6	6.3	1.6	3.7	1.0	1.5	11.3	1.2	6.1	1.4	23.4	4.4	3.1	2.8	11.2	0.3	4.8	11.1	7.0
				Oolong tea	A	13.8	1.8	1.5	11.9	5.0	5.0	4.1	7.0	6.3	2.7	3.9	3.9	1.7	2.6	4.2	2.2	2.4	6.6	7.4
					B	7.0	3.5	9.2	4.5	12.4	5.9	5.8	5.6	12.6	4.8	10.6	1.5	4.7	7.0	11.1	2.3	8.8	8.8	3.0
			GC-MS/MS	Green tea	A	7.9	5.4	2.4	5.8	8.7	1.4	10.4	3.8	5.1	0.8	7.5	1.6	13.2	15.1	4.7	6.6	9.9	2.1	4.9
					B	6.6	5.8	6.2	2.6	0.9	2.1	11.5	1.4	5.3	1.4	17.7	3.3	4.6	5.5	12.7	0.3	7.1	9.1	10.3
				Oolong tea	A	16.1	2.8	1.3	8.6	4.0	2.8	3.3	4.3	5.6	1.7	8.4	3.4	0.9	6.4	12.1	1.2	5.4	4.4	8.0
					B	9.2	3.1	12.2	1.5	8.7	5.9	5.4	5.6	11.9	11.2	1.1	2.4	2.9	11.7	11.7	4.6	7.3	12.3	7.0
126	Ronnel	TPT	GC-MS	Green tea	A	6.2	3.5	3.7	2.7	8.9	3.2	6.2	4.2	4.2	2.7	4.5	3.5	4.9	5.2	4.1	3.3	3.5	6.4	2.3
					B	7.5	9.2	4.8	6.7	4.9	3.7	1.2	1.3	3.4	6.9	2.0	3.1	2.4	4.4	6.8	1.8	4.2	5.4	8.1
				Oolong tea	A	5.6	3.7	1.0	3.2	2.6	2.9	10.4	6.0	5.0	6.7	4.9	1.0	2.4	1.5	8.5	3.2	6.1	2.4	3.3
					B	4.5	2.7	3.4	3.9	1.1	1.9	3.3	0.9	5.1	4.9	4.4	0.8	2.1	3.3	6.9	0.8	2.3	1.7	2.8
			GC-MS/MS	Green tea	A	3.9	6.2	4.7	3.8	7.6	1.6	5.0	6.4	8.5	3.0	8.2	6.4	7.7	3.7	3.3	6.7	3.8	6.1	2.4
					B	7.4	9.1	4.6	6.2	5.3	5.7	8.5	2.4	5.4	10.0	3.0	1.6	1.1	1.2	3.5	7.1	6.6	3.5	8.1
				Oolong tea	A	6.5	5.0	4.1	1.9	5.5	1.8	15.0	1.9	3.1	6.0	4.4	3.9	4.7	1.2	6.7	6.2	7.4	3.6	6.7
					B	4.3	1.8	5.0	6.0	2.7	2.9	2.3	7.6	9.8	9.9	0.8	2.1	5.2	4.1	8.2	1.4	3.5	2.0	7.3
		ENVI-Carb+PSA	GC-MS	Green tea	A	10.4	6.1	3.7	7.5	7.3	5.5	5.0	5.1	2.2	0.3	11.6	5.4	2.4	1.5	4.7	2.7	10.0	9.4	4.5
					B	12.2	8.2	3.6	4.2	1.2	0.2	2.2	2.4	6.1	6.5	13.3	6.9	6.7	2.7	5.4	8.3	2.7	12.0	7.0
				Oolong tea	A	5.1	2.7	3.0	5.0	3.8	1.5	6.8	2.2	3.5	2.9	2.5	1.9	2.0	1.5	4.5	1.9	3.5	7.2	4.9
					B	12.8	4.2	8.7	4.6	9.2	4.8	4.2	5.6	11.0	1.1	7.8	3.4	5.9	3.3	11.5	3.8	7.6	3.6	4.1
			GC-MS/MS	Green tea	A	28.7	9.7	2.6	10.6	7.7	11.9	11.9	3.7	9.3	11.5	10.2	2.7	4.3	10.2	7.4	5.9	9.1	12.5	8.5
					B	3.5	2.3	3.0	5.8	3.7	0.5	0.3	3.8	3.9	2.5	8.9	1.6	2.0	7.6	8.9	6.2	5.8	3.6	6.0
				Oolong tea	A	5.7	4.9	9.2	5.7	5.7	4.3	5.5	3.7	7.9	1.0	6.5	4.4	6.0	11.1	10.2	1.4	5.1	4.3	6.8

(Continued)

Appendix Table 5.2 RSD of 201 Pesticides in Youden Pair Aged Tea Sample for Parallel Determinations (Nov 9, 2009–Feb 7, 2010) (cont.)

RSD%

No.	Pesticides	SPE	Method	Sample	Youden pair	Nov 9, 2009 (n = 5)	Nov 14, 2009 (n = 3)	Nov 19, 2009 (n = 3)	Nov 24, 2009 (n = 3)	Nov 29, 2009 (n = 3)	Dec 4, 2009 (n = 3)	Dec 9, 2009 (n = 3)	Dec 14, 2009 (n = 3)	Dec 19, 2009 (n = 3)	Dec 24, 2009 (n = 3)	Dec 29, 2009 (n = 3)	Jan 3, 2010 (n = 3)	Jan 8, 2010 (n = 3)	Jan 13, 2010 (n = 3)	Jan 18, 2010 (n = 3)	Jan 23, 2010 (n = 3)	Jan 28, 2010 (n = 3)	Feb 2, 2010 (n = 3)	Feb 7, 2010 (n = 3)
127	Sulfotep	TPT	GC-MS	Green tea	A	10.8	5.1	2.9	1.6	8.1	2.9	6.7	5.1	4.5	1.8	3.8	2.1	5.0	0.9	1.2	1.7	5.3	9.8	5.4
					B	5.1	8.4	2.5	7.4	4.4	5.8	5.0	2.2	5.4	2.8	5.4	5.9	1.4	0.5	6.9	2.4	10.7	1.4	8.2
				Oolong tea	A	4.9	3.0	2.4	3.3	1.2	3.1	7.7	5.5	3.3	1.6	2.4	3.0	4.2	3.1	8.7	3.0	1.4	2.8	1.4
					B	4.0	2.8	3.0	6.1	1.0	3.1	3.7	2.8	6.7	1.8	5.2	1.0	2.6	2.4	6.0	1.3	0.4	3.1	1.7
			GC-MS/MS	Green tea	A	4.1	5.6	2.3	2.1	12.9	5.1	4.1	6.3	4.5	3.7	6.7	2.3	4.8	6.1	3.8	3.3	3.0	5.7	2.6
					B	7.5	7.5	5.3	7.5	7.0	6.6	6.0	3.4	5.4	9.6	3.5	4.3	4.6	2.2	1.4	4.9	6.2	2.0	3.8
				Oolong tea	A	6.1	8.1	2.6	3.1	1.9	4.0	8.0	2.7	1.0	5.8	3.3	1.5	3.1	2.3	8.4	3.4	2.5	4.2	4.1
					B	3.3	2.4	5.7	7.1	2.7	4.4	4.0	4.2	3.8	4.0	3.6	1.7	2.7	7.6	7.7	4.9	2.8	4.9	5.0
		ENVI-Carb+PSA	GC-MS	Green tea	A	6.7	5.3	3.2	5.2	7.0	1.4	9.3	1.7	5.4	3.1	5.8	6.0	3.3	1.8	3.8	6.4	8.2	6.2	4.8
					B	6.2	7.5	4.6	1.7	1.6	1.8	9.9	2.8	2.4	3.3	22.7	7.7	3.3	1.3	9.8	1.9	9.9	8.7	6.1
				Oolong tea	A	7.8	1.4	4.0	2.3	6.3	2.3	2.4	1.7	3.6	1.6	7.4	3.0	1.6	2.3	2.8	3.8	7.4	4.8	4.2
					B	5.3	2.3	8.8	2.8	14.3	4.0	1.8	7.0	4.8	3.1	7.0	2.7	2.4	3.3	8.8	3.8	4.5	3.6	3.6
			GC-MS/MS	Green tea	A	7.5	5.7	0.7	5.1	4.9	2.2	3.2	3.4	6.3	1.4	8.9	3.0	11.1	17.7	6.4	7.9	9.3	7.1	3.3
					B	12.1	7.6	5.7	2.9	0.3	6.8	7.5	4.0	3.4	1.1	17.1	3.7	4.5	14.6	12.8	2.0	4.4	11.3	8.5
				Oolong tea	A	3.8	3.9	4.5	2.8	4.9	1.8	0.8	0.7	2.5	2.1	7.0	1.4	2.4	4.5	8.7	3.4	8.7	5.4	2.9
					B	7.9	1.7	9.7	1.3	10.1	4.4	2.6	8.2	6.9	2.9	4.2	2.9	3.2	10.9	10.0	4.9	5.9	1.5	5.4
128	Tebufen-pyrad	TPT	GC-MS	Green tea	A	5.1	5.2	0.9	1.4	6.8	4.4	7.7	4.0	3.8	1.5	5.7	2.8	5.6	2.9	3.4	1.0	3.7	5.3	3.1
					B	4.6	7.1	4.2	8.0	3.4	3.3	1.2	1.2	3.8	6.8	4.4	7.9	0.1	0.7	5.3	3.2	5.8	3.6	4.8
				Oolong tea	A	7.2	2.5	1.5	1.7	1.4	2.6	6.6	6.0	2.0	3.0	2.7	3.4	5.7	2.8	6.9	3.1	1.9	4.4	1.4
					B	4.1	1.6	2.2	6.9	1.3	3.7	5.6	2.3	6.6	1.5	3.2	1.5	1.7	4.1	1.0	2.7	2.4	1.7	1.3
			GC-MS/MS	Green tea	A	5.3	4.9	1.4	0.9	9.5	3.7	7.1	6.4	4.9	1.5	9.4	1.2	4.5	3.6	4.2	4.7	3.0	4.5	3.7
					B	7.9	6.5	4.6	9.4	7.3	5.1	7.3	1.5	3.2	9.0	6.1	3.3	2.7	4.0	3.4	2.8	6.7	2.9	3.0
				Oolong tea	A	8.2	3.8	2.7	2.0	2.8	4.6	7.0	2.7	2.1	3.9	2.2	3.6	4.4	1.6	6.5	3.1	2.6	7.5	6.0
					B	4.1	2.6	3.1	6.1	3.1	3.6	3.2	5.8	3.2	4.0	7.9	0.5	2.0	5.6	5.2	1.9	5.2	1.4	6.5
		ENVI-Carb+PSA	GC-MS	Green tea	A	5.6	2.6	2.9	7.9	8.6	2.0	46.9	1.7	8.3	3.5	7.2	2.6	2.6	6.3	6.0	6.8	10.8	2.3	5.4
					B	3.1	8.8	4.1	7.2	2.3	1.7	11.2	5.3	4.1	1.2	21.6	3.5	2.3	5.3	10.4	2.2	4.6	7.3	6.9
				Oolong tea	A	4.0	1.3	4.5	3.4	6.7	1.0	3.6	5.5	0.1	2.6	6.3	2.2	3.6	2.8	1.6	1.3	4.1	5.9	3.6
					B	3.4	3.0	8.9	6.0	10.6	3.4	1.8	6.7	10.9	2.8	7.5	1.1	1.2	4.1	7.9	3.3	5.0	5.0	2.8
			GC-MS/MS	Green tea	A	5.3	10.4	3.3	8.9	7.8	2.2	2.8	2.5	9.3	1.4	4.8	2.9	14.9	8.4	6.6	9.5	11.2	3.5	3.1
					B	5.1	12.5	6.7	3.4	3.6	2.1	2.9	4.8	6.5	1.3	18.9	0.5	4.7	7.0	9.5	2.5	8.0	8.2	9.4
				Oolong tea	A	7.9	2.9	5.0	3.0	8.8	0.8	5.5	5.6	1.9	0.9	10.8	0.5	3.7	5.8	6.0	3.1	4.7	3.1	4.2
					B	6.8	4.2	11.7	3.3	8.5	3.4	1.2	6.8	8.5	4.4	1.4	4.1	2.7	6.9	9.8	1.9	5.2	1.7	6.8

129	Terbutryn	TPT	GC-MS	Green tea	A	6.0	5.1	2.4	3.2	8.1	5.7	7.6	5.1	2.9	5.3	7.7	6.9	6.6	11.0	8.5	3.3	19.7	12.5	4.7
					B	4.6	7.7	5.7	8.5	3.1	2.5	4.1	2.2	5.4	7.4	7.7	10.7	1.3	2.9	12.1	3.9	10.8	2.9	4.9
				Oolong tea	A	5.4	1.7	3.7	3.6	0.8	3.9	9.1	5.1	3.2	9.4	11.1	2.9	3.9	3.8	8.5	5.3	2.9	4.4	1.5
					B	3.2	4.5	2.6	5.9	2.9	2.8	4.9	5.1	7.3	4.7	9.5	0.3	1.8	5.7	3.6	2.4	2.3	0.3	2.2
			GC-MS/MS	Green tea	A	4.7	8.3	4.6	0.5	8.4	6.4	6.7	6.5	5.7	1.9	5.3	4.9	9.3	1.3	2.7	5.1	3.6	2.2	6.5
					B	7.4	2.6	6.0	9.2	4.3	4.9	7.4	2.7	7.9	10.9	9.2	5.7	4.3	2.1	5.7	4.1	5.0	2.7	7.5
				Oolong tea	A	5.6	0.5	2.4	4.1	2.3	4.6	10.5	3.4	2.3	6.4	1.5	3.1	7.8	3.8	7.9	6.3	4.1	4.0	5.8
					B	3.6	6.5	4.9	7.2	4.9	1.2	3.8	4.0	5.6	1.2	3.7	1.9	0.5	3.6	6.4	3.1	1.7	5.7	8.0
		ENVI-Carb+PSA	GC-MS	Green tea	A	6.1	9.0	2.7	6.9	9.7	3.1	5.1	0.5	3.8	4.9	7.4	7.1	2.3	5.3	7.3	6.3	10.4	8.1	4.7
					B	3.2	4.8	4.0	5.6	0.6	3.1	4.9	2.5	4.7	3.9	28.0	7.7	1.5	3.7	12.1	1.3	5.1	7.4	6.5
				Oolong tea	A	3.2	3.0	3.7	1.3	4.9	1.2	2.8	2.8	2.6	4.0	6.1	2.9	1.5	5.6	3.4	4.5	3.8	3.7	3.9
					B	4.3	6.7	9.9	2.7	15.7	5.3	2.3	8.5	11.5	4.7	9.5	3.4	2.9	2.3	14.4	3.6	3.5	6.5	5.4
			GC-MS/MS	Green tea	A	5.9	7.9	3.3	7.4	7.5	1.5	4.4	1.2	5.1	2.6	9.0	16.7	9.9	16.3	10.5	8.1	10.1	1.3	1.7
					B	5.8	3.1	7.0	2.3	2.3	3.8	3.6	5.7	3.8	0.8	17.0	3.8	4.6	5.7	14.5	4.0	9.4	9.2	6.9
				Oolong tea	A	3.5	3.1	3.9	2.6	7.6	1.3	1.6	3.7	0.6	1.9	5.9	6.1	5.8	8.3	6.0	6.3	7.5	3.2	7.9
					B	6.4	2.2	11.0	4.7	6.0	4.5	1.7	7.1	10.0	4.3	3.7	1.7	0.4	11.6	14.1	3.1	4.3	1.0	4.1
130	Thiobencarb	TPT	GC-MS	Green tea	A	5.9	6.2	2.9	2.7	8.0	4.7	7.0	4.1	3.9	2.1	6.7	2.3	6.9	3.1	3.7	2.1	3.6	1.0	4.6
					B	5.7	9.2	5.7	9.2	3.5	3.2	3.0	1.7	4.5	7.9	3.6	7.0	1.9	0.8	3.4	1.7	5.1	6.5	6.6
				Oolong tea	A	5.2	2.2	1.5	3.6	2.0	3.3	7.3	3.3	2.6	4.8	6.2	3.3	3.9	3.0	8.7	5.2	3.3	0.4	0.4
					B	2.8	1.9	2.8	2.5	1.4	2.7	4.4	3.5	6.5	1.5	5.4	1.6	2.4	3.2	4.6	0.5	1.6	2.8	1.4
			GC-MS/MS	Green tea	A	5.5	5.3	3.9	1.6	8.8	2.0	5.3	3.8	5.0	2.4	7.8	5.5	6.2	2.2	4.3	3.6	4.5	2.4	2.8
					B	8.4	9.2	4.9	10.2	6.0	4.6	7.8	2.8	6.1	9.1	4.4	4.5	2.8	1.6	2.0	4.5	3.2	5.8	6.3
				Oolong tea	A	6.1	2.8	3.6	4.0	2.3	3.7	6.6	2.8	1.8	3.4	2.9	2.3	5.3	1.2	5.6	7.0	2.0	3.2	5.1
					B	4.3	2.4	3.9	6.7	3.7	2.2	3.8	2.5	4.3	3.8	4.2	0.7	3.2	5.5	5.0	1.7	2.5	1.1	6.3
		ENVI-Carb+PSA	GC-MS	Green tea	A	5.9	10.8	4.0	6.2	7.7	3.2	29.6	1.3	5.2	3.7	7.0	3.8	3.2	4.3	5.2	5.9	11.1	2.2	5.1
					B	2.6	8.7	4.1	3.5	2.2	1.2	3.1	3.4	2.2	0.7	20.3	4.3	2.3	2.3	10.5	3.5	4.0	6.8	5.5
				Oolong tea	A	2.5	5.3	2.1	3.8	5.2	0.7	3.2	1.9	1.1	1.6	3.2	1.9	1.7	3.8	1.2	1.8	8.0	5.2	3.3
					B	4.0	3.4	9.3	3.9	10.7	4.7	2.7	6.8	10.7	3.3	6.6	2.1	3.4	3.2	9.7	3.5	5.4	7.5	4.5
			GC-MS/MS	Green tea	A	5.9	5.3	0.8	5.0	7.4	0.7	1.9	2.3	6.6	1.1	9.6	1.3	9.3	3.2	7.6	9.5	8.1	2.3	6.2
					B	4.3	7.7	7.3	0.6	1.1	2.4	3.8	5.7	2.7	1.6	18.7	4.7	5.3	17.8	13.3	3.7	6.3	7.3	8.0
				Oolong tea	A	2.1	5.8	3.8	3.2	5.7	1.5	2.4	1.8	1.9	1.8	7.3	3.5	3.3	3.5	9.8	7.0	5.0	2.7	3.9
					B	6.2	3.3	10.5	3.0	5.6	4.1	2.6	6.9	9.9	3.8	3.0	2.9	4.4	7.5	8.2	1.7	4.6	1.4	6.2
131	Tralk-oxydim	TPT	GC-MS	Green tea	A	7.1	4.9	2.4	0.9	6.8	4.2	7.7	4.8	5.0	1.0	3.3	2.0	5.4	4.1	3.7	3.2	3.7	4.3	4.1
					B	5.2	5.9	3.5	10.3	3.4	3.8	1.9	14.3	2.8	8.3	5.5	5.8	1.7	2.0	1.1	9.6	8.5	7.3	3.8
				Oolong tea	A	16.3	9.5	3.3	11.3	4.6	12.6	3.7	2.1	7.9	3.7	2.6	5.9	6.8	1.3	11.8	3.8	1.6	12.6	6.0
					B	22.2	0.9	4.5	3.5	5.0	6.5	9.6	3.4	7.8	14.6	6.6	5.5	9.3	9.8	6.8	11.7	1.3	12.6	2.3
			GC-MS/MS	Green tea	A	3.6	5.2	1.3	2.2	8.2	4.3	8.0	3.1	3.4	0.8	6.0	7.0	5.6	9.7	3.8	5.6	4.4	4.6	3.1
					B	6.7	4.5	4.6	8.9	6.7	4.2	6.9	3.4	3.7	7.5	4.2	5.3	1.4	5.9	1.4	8.2	3.2	4.7	3.7
				Oolong tea	A	4.7	7.1	3.7	13.0	3.9	13.7	3.1	6.2	8.3	5.2	4.1	4.6	8.7	3.8	14.0	7.0	7.4	9.7	3.3
					B	12.7	6.6	10.5	8.1	7.9	7.5	10.4	7.9	4.0	5.9	4.0	4.4	0.6	14.7	8.3	3.7	16.2	6.6	6.6
		ENVI-Carb+PSA	GC-MS	Green tea	A	11.2	3.4	4.6	10.3	9.7	8.0	13.1	15.1	3.0	5.0	4.3	15.7	5.2	23.1	31.2	22.9	5.2	14.7	11.1
					B	6.8	10.7	2.4	18.4	3.4	6.1	2.6	6.5	8.1	0.8	58.4	4.5	27.1	2.7	4.5	6.2	0.4	13.9	10.6
				Oolong tea	A	20.3	9.0	4.5	17.0	6.6	9.2	2.3	12.8	5.8	7.9	8.3	11.5	4.2	1.3	14.0	5.4	1.9	12.5	3.6
					B	11.3	10.6	4.6	2.1	12.8	10.2	11.9	3.3	15.3	8.4	11.1	1.7	5.4	9.8	8.8	4.5	13.8	17.6	9.9
			GC-MS/MS	Green tea	A	5.9	4.7	7.9	10.9	11.5	1.9	42.8	14.4	2.8	4.4	2.1	2.2	4.9	21.0	16.2	26.4	5.8	8.0	13.2
					B	21.5	11.4	4.0	13.0	3.1	1.8	8.9	6.2	10.1	2.7	62.8	1.7	12.7	19.6	12.8	6.6	3.7	11.1	6.4
				Oolong tea	A	65.7	11.6	1.4	9.8	9.2	10.9	6.9	8.8	11.1	15.0	18.3	7.1	4.2	7.1	15.2	7.0	9.6	4.9	5.9
					B	23.0	9.2	15.7	8.9	10.9	10.1	9.1	8.8	18.3	18.6	1.9	1.0	4.9	9.1	15.2	3.7	10.7	17.7	5.9

(Continued)

Appendix Table 5.2 RSD of 201 Pesticides in Youden Pair Aged Tea Sample for Parallel Determinations (Nov 9, 2009–Feb 7, 2010) (cont.)

No.	Pesticides	SPE	Method	Sample	Youden pair	RSD% Nov 9, 2009 (n=5)	Nov 14, 2009 (n=3)	Nov 19, 2009 (n=3)	Nov 24, 2009 (n=3)	Nov 29, 2009 (n=3)	Dec 4, 2009 (n=3)	Dec 9, 2009 (n=3)	Dec 14, 2009 (n=3)	Dec 14, 2009 (n=3)	Dec 19, 2009 (n=3)	Dec 24, 2009 (n=3)	Dec 14, 2009 (n=3)	Jan 3, 2010 (n=3)	Jan 8, 2010 (n=3)	Jan 13, 2010 (n=3)	Jan 18, 2010 (n=3)	Jan 23, 2010 (n=3)	Jan 28, 2010 (n=3)	Feb 2, 2010 (n=3)	Feb 7, 2010 (n=3)
132	Trans-chlordane	TPT	GC-MS	Green tea	A	5.3	4.9	3.7	2.2	8.4	4.4	6.4	5.5	6.1	4.8	2.7	6.9	2.7	6.9	2.6	2.0	0.5	4.4	6.1	4.0
					B	4.9	8.7	4.8	8.9	3.3	3.6	2.5	2.0	6.9	2.3	9.2	0.5	6.7	0.5	3.1	1.8	2.4	9.0	4.1	4.5
				Oolong tea	A	5.1	4.8	3.5	3.5	2.0	3.7	8.2	6.8	2.4	4.7	3.7	3.1	1.6	3.1	3.1	7.6	2.4	4.9	1.3	0.9
					B	4.8	2.8	4.0	0.8	0.6	1.9	4.0	0.6	5.6	1.2	3.2	3.9	1.2	2.2	3.9	5.6	2.0	2.0	2.3	2.4
			GC-MS/MS	Green tea	A	6.9	5.1	3.6	0.6	12.7	10.7	6.2	4.0	5.5	4.9	3.4	4.7	2.2	7.0	3.9	3.8	11.1	2.5	2.3	9.1
					B	7.1	8.7	9.0	6.0	8.6	7.7	7.4	5.1	10.8	5.5	6.9	2.6	7.0	3.3	0.8	2.4	7.0	5.0	9.8	2.2
				Oolong tea	A	9.4	7.4	3.0	0.4	6.1	3.1	10.5	0.2	4.2	3.0	4.8	4.3	3.1	3.1	4.5	6.5	5.9	5.3	1.6	6.7
					B	6.2	1.5	6.6	5.3	5.5	6.7	1.3	3.0	7.8	7.8	2.7	6.1	1.6	1.6	5.2	9.2	4.0	3.0	1.3	13.8
		ENVI-Carb+PSA	GC-MS	Green tea	A	8.0	4.8	2.5	5.3	8.7	1.7	52.5	1.5	4.1	4.1	1.2	5.1	0.5	2.9	5.2	6.1	10.7	7.0	9.5	5.3
					B	4.0	7.6	3.7	4.0	3.5	1.6	2.4	1.2	5.6	5.6	1.9	0.5	1.7	4.0	3.1	13.2	1.3	4.1	10.1	6.7
				Oolong tea	A	1.3	3.1	2.4	4.0	6.3	1.1	3.8	5.6	2.2	2.2	3.1	1.2	3.1	2.1	4.2	2.6	3.7	6.1	6.0	5.1
					B	5.1	2.3	9.5	2.9	9.8	5.6	2.1	8.8	12.3	12.3	3.4	5.6	3.2	4.7	3.9	9.6	4.2	9.9	4.0	4.9
			GC-MS/MS	Green tea	A	8.0	7.7	3.8	7.6	5.8	1.7	5.8	3.9	4.7	4.7	0.5	3.8	34.6	9.0	10.1	4.1	6.9	8.9	6.2	3.7
					B	5.2	8.5	7.6	3.6	2.4	2.2	3.0	1.7	3.6	3.6	3.8	2.1	20.6	3.4	3.4	17.3	3.1	6.0	11.9	11.5
				Oolong tea	A	3.9	1.9	2.1	4.2	6.0	1.9	1.6	2.0	2.4	2.4	2.1	3.2	0.8	2.5	6.3	6.9	5.9	5.5	1.2	0.5
					B	6.0	2.5	12.0	2.3	10.1	6.1	4.2	7.8	9.8	9.8	6.3	4.1	5.6	4.0	10.7	8.6	4.0	4.0	0.8	11.7
133	Trans-diallate	TPT	GC-MS	Green tea	A	5.4	4.9	2.4	2.0	8.4	3.3	4.6	4.5	5.1	4.6	5.9	2.3	2.3	8.3	4.4	8.3	3.4	6.3	8.4	4.4
					B	5.8	8.7	4.5	8.3	2.6	3.4	0.9	1.9	6.1	3.1	9.6	1.9	6.6	1.3	3.8	10.4	3.5	11.7	0.8	6.2
				Oolong tea	A	4.4	3.0	1.0	2.7	2.4	5.0	8.8	6.2	5.8	3.1	4.1	6.2	3.8	6.1	8.1	7.9	6.1	3.9	1.3	3.4
					B	3.1	2.2	3.4	12.3	6.8	2.4	2.2	3.0	9.5	2.0	2.6	3.0	2.6	1.7	4.3	3.4	1.1	2.7	3.2	6.7
			GC-MS/MS	Green tea	A	3.1	3.8	3.8	2.3	12.1	4.1	6.9	5.9	5.6	7.7	2.2	5.9	4.1	4.8	5.8	4.2	3.7	1.1	2.7	3.2
					B	8.4	7.4	5.7	8.8	5.0	6.4	5.7	4.0	6.0	5.4	8.3	4.0	5.4	1.9	2.0	1.9	1.8	5.2	2.7	3.5
				Oolong tea	A	5.2	2.1	0.8	3.3	2.6	3.0	8.1	1.4	2.9	1.9	3.3	1.4	1.9	3.7	4.5	6.3	4.3	6.9	2.7	2.5
					B	2.5	0.7	5.5	6.9	4.5	3.8	1.7	6.6	1.1	4.1	4.3	6.6	1.4	0.8	5.7	8.3	5.1	3.4	0.9	5.4
		ENVI-Carb+PSA	GC-MS	Green tea	A	7.2	3.9	2.0	6.7	7.6	3.0	18.0	1.6	4.5	7.6	2.6	3.0	2.3	1.2	4.1	2.8	5.2	10.0	6.3	5.1
					B	5.5	8.0	4.5	3.0	6.0	1.0	2.9	2.6	1.9	2.6	2.7	1.6	2.1	8.1	0.5	11.8	2.0	5.6	8.8	5.9
				Oolong tea	A	5.2	3.5	2.7	2.8	3.7	2.6	4.2	7.8	12.5	7.8	3.3	7.8	1.6	5.9	7.5	0.8	1.6	5.9	5.0	6.3
					B	5.8	2.7	7.9	3.4	7.3	3.6	3.6	9.0	3.8	9.0	1.2	4.6	4.6	2.8	2.3	8.4	1.9	6.6	3.4	2.4
			GC-MS/MS	Green tea	A	9.1	7.6	4.1	5.2	5.6	5.8	3.8	4.3	6.7	4.3	1.0	6.2	7.4	11.6	14.9	10.6	8.3	10.2	5.8	2.4
					B	10.4	7.3	6.7	4.4	2.3	2.9	12.7	0.6	4.8	0.6	0.7	14.7	7.1	6.6	15.9	14.4	4.5	7.3	8.4	8.8
				Oolong tea	A	4.0	2.8	4.2	3.3	5.0	1.8	1.0	2.6	2.6	2.6	1.2	7.2	3.6	2.8	4.7	8.3	5.7	3.0	3.3	4.2
					B	6.9	4.1	9.1	1.4	7.3	3.8	3.5	8.4	10.0	8.4	3.2	2.8	2.1	1.8	10.1	8.2	6.0	6.3	3.9	7.5
134	Trifloxystrobin	TPT	GC-MS	Green tea	A	9.7	6.0	2.7	13.1	13.0	2.3	12.7	14.2	12.0	14.2	6.3	8.5	8.7	9.8	11.5	3.3	5.2	6.6	14.7	4.8
					B	8.7	4.4	3.7	19.8	5.0	5.7	6.1	12.6	2.8	12.6	3.1	6.7	9.3	13.6	8.0	3.9	6.4	14.0	6.7	5.0
				Oolong tea	A	7.7	7.0	2.4	1.5	11.9	3.0	1.5	9.8	9.2	9.8	8.6	15.2	6.5	3.7	0.9	5.7	5.5	0.7	1.6	2.4
					B	4.1	4.2	3.6	4.5	3.8	4.4	9.1	6.6	8.9	6.6	5.4	18.5	1.3	1.6	7.3	7.8	1.4	2.1	3.4	5.2
			GC-MS/MS	Green tea	A	10.8	9.9	0.4	13.4	17.0	3.5	16.0	21.6	17.1	21.6	3.2	12.8	6.6	7.7	8.5	2.6	8.8	9.5	14.1	1.0
					B	8.7	4.8	7.2	11.4	4.3	3.8	4.9	13.0	3.7	13.0	5.8	1.4	7.4	19.0	2.8	6.4	5.5	12.8	10.8	7.3
				Oolong tea	A	11.6	3.7	5.2	0.6	8.2	1.9	1.2	3.0	7.4	3.0	4.4	5.4	7.3	1.6	1.3	5.3	12.5	3.4	6.4	5.9
					B	6.1	3.7	3.5	6.4	6.7	5.9	7.6	10.8	8.1	10.8	7.4	8.5	8.0	2.8	8.3	10.9	0.4	4.5	6.1	10.6
		ENVI-Carb+PSA	GC-MS	Green tea	A	29.4	9.7	21.3	8.8	8.0	27.3	15.0	21.9	7.8	21.9	6.5	6.6	27.7	5.8	19.1	14.2	9.0	12.2	19.1	6.3
					B	39.1	6.5	6.8	14.5	14.2	6.9	5.3	19.5	22.1	19.5	14.9	8.5	21.5	30.9	6.6	18.1	21.3	2.2	26.1	14.8
				Oolong tea	A	5.3	7.3	4.7	6.7	4.3	4.0	9.6	22.1	2.0	22.1	1.5	4.4	5.3	6.3	0.9	2.8	16.8	6.2	11.1	1.7
					B	10.1	9.7	5.2	2.9	16.7	7.3	7.6	6.2	13.0	6.2	6.9	10.5	2.8	1.6	7.3	13.0	2.1	10.4	17.0	6.2
			GC-MS/MS	Green tea	A	32.6	11.1	19.2	1.0	9.5	30.0	7.8	27.8	15.0	27.8	9.3	8.1	0.0	0.7	23.1	9.5	28.0	11.2	11.3	9.5
					B	44.8	13.6	5.7	25.4	14.9	6.7	26.9	20.3	23.5	20.3	19.1	23.9	0.0	17.8	9.7	14.3	23.3	5.3	23.8	10.2
				Oolong tea	A	10.5	4.3	6.9	7.7	5.0	3.4	7.4	19.0	2.8	19.0	3.5	7.3	10.6	10.1	4.7	4.9	12.5	8.3	13.0	3.4
					B	13.3	6.8	7.4	2.1	15.4	8.6	18.8	6.0	9.4	6.0	8.9	4.3	8.8	4.5	10.0	18.2	0.4	7.8	27.6	6.3

135	Zoxamide	TPT	GC-MS	Green tea	A	11.4	6.3	1.5	1.9	5.1	5.5	7.8	3.3	4.3	3.4	5.0	3.8	5.0	5.6	3.0	8.3	3.1	6.3	6.4
					B	5.9	10.3	5.5	6.7	1.1	4.8	1.2	6.5	2.5	5.1	5.7	6.5	0.9	7.2	10.0	1.8	11.7	5.5	3.8
				Oolong tea	A	5.9	7.5	2.1	3.8	5.5	5.6	9.4	5.1	0.9	3.6	6.4	2.4	9.2	3.7	8.1	5.6	5.6	10.8	2.4
					B	9.5	1.8	4.4	7.5	0.4	13.5	6.5	2.3	4.4	5.0	7.8	3.6	2.4	3.3	2.2	2.9	1.4	2.0	3.3
			GC-MS/MS	Green tea	A	10.2	9.2	1.9	4.5	4.9	4.9	0.6	3.3	6.0	3.8	6.6	6.2	7.5	3.5	5.6	7.3	4.8	0.7	3.4
					B	6.5	7.8	4.7	5.9	4.9	7.4	6.8	6.6	4.2	7.9	5.1	3.0	0.7	3.7	2.4	3.2	4.2	4.9	3.8
				Oolong tea	A	8.6	8.0	5.1	3.4	5.7	3.9	15.5	1.0	0.7	2.9	1.5	3.3	10.2	3.8	5.7	5.3	0.2	14.7	3.8
					B	10.0	4.3	6.8	4.5	4.9	3.1	3.0	7.7	1.4	8.0	2.2	1.1	2.4	1.8	9.3	9.5	4.3	6.1	2.2
		ENVI-Carb +PSA	GC-MS	Green tea	A	11.4	6.9	4.9	8.8	4.2	3.6	43.7	1.7	9.6	1.7	7.8	4.2	2.0	2.1	2.6	14.9	9.8	14.7	4.0
					B	9.5	6.5	4.0	1.4	6.9	3.9	7.7	2.5	6.3	5.8	16.6	3.9	1.7	2.2	10.5	8.0	2.1	14.7	6.0
				Oolong tea	A	5.9	2.9	6.6	8.3	8.1	3.8	5.3	6.0	0.5	8.6	7.2	6.6	10.7	3.7	3.9	6.4	1.5	4.7	8.2
					B	5.8	3.0	14.5	9.8	9.1	6.1	6.0	5.5	5.4	4.2	4.2	4.0	1.9	3.3	10.6	5.9	7.9	8.8	5.5
			GC-MS/MS	Green tea	A	9.9	7.6	5.3	5.6	2.1	4.0	25.8	1.7	8.8	2.1	7.5	10.7	10.4	0.7	4.6	6.2	10.4	1.9	4.5
					B	15.4	5.6	5.4	2.6	7.5	4.6	13.1	4.2	7.0	7.4	14.2	8.1	2.6	3.3	5.5	6.3	3.5	10.2	11.4
				Oolong tea	A	9.4	2.5	6.9	4.1	7.6	2.9	6.6	4.8	1.8	13.1	8.9	4.2	7.9	3.6	7.9	5.3	7.1	2.0	7.2
					B	7.7	1.9	16.0	12.8	10.7	6.6	1.6	9.5	6.3	1.9	7.7	5.3	5.2	15.5	6.0	9.5	10.7	2.7	6.1
136	2,4'-DDE	TPT	GC-MS	Green tea	A	6.3	7.3	3.3	8.3	5.5	5.8	11.9	2.8	5.2	3.5	7.1	1.4	10.0	4.8	4.1	3.7	2.9	10.4	5.5
					B	6.0	6.6	6.7	10.5	3.4	4.5	4.1	2.6	4.5	9.5	6.0	6.0	2.1	3.5	3.9	1.7	6.6	2.8	7.6
				Oolong tea	A	4.2	4.2	1.8	4.1	5.5	5.9	8.7	8.1	0.1	3.6	1.4	3.2	5.3	3.3	9.1	3.6	4.5	7.5	0.7
					B	4.8	3.6	3.7	5.7	1.1	4.4	6.0	2.6	3.3	2.0	2.9	3.1	4.6	3.5	2.0	1.0	1.1	4.5	1.1
			GC-MS/MS	Green tea	A	6.7	3.2	2.4	11.0	5.4	7.3	8.3	2.5	4.3	2.0	7.6	1.0	10.6	2.2	5.7	5.0	3.7	13.7	3.5
					B	6.6	8.9	9.9	9.3	3.7	6.6	5.5	5.2	4.7	7.8	7.0	11.6	3.9	1.4	7.4	4.6	8.9	6.8	6.3
				Oolong tea	A	5.5	4.0	3.5	9.6	4.9	6.4	7.7	5.7	1.8	2.9	4.4	3.0	9.7	8.6	12.9	7.9	8.4	3.1	1.1
					B	7.7	0.7	3.7	11.2	2.8	4.6	8.1	8.3	6.5	3.5	8.3	4.7	3.0	9.1	8.1	2.0	3.8	5.7	9.4
		ENVI-Carb +PSA	GC-MS	Green tea	A	11.3	5.2	2.9	5.1	8.5	8.0	5.6	5.2	4.5	5.0	13.1	7.6	4.1	9.6	13.6	12.9	13.6	0.3	6.5
					B	16.9	7.3	4.6	10.3	4.2	4.4	2.1	13.8	6.7	3.7	24.4	5.1	1.5	2.6	16.0	9.4	6.3	4.4	5.4
				Oolong tea	A	2.0	4.0	2.2	3.0	10.7	1.9	4.5	3.5	1.9	4.6	3.9	4.2	8.2	5.1	2.1	7.4	5.5	3.2	7.6
					B	4.9	2.5	10.3	4.7	9.0	5.0	2.6	8.5	10.9	4.2	7.6	3.1	2.8	8.2	12.0	5.7	5.8	10.3	4.7
			GC-MS/MS	Green tea	A	7.1	0.7	1.5	9.3	3.1	8.8	1.9	4.4	4.8	4.6	11.8	2.1	6.6	10.9	12.3	15.7	10.4	2.7	7.1
					B	8.9	12.7	3.1	8.8	4.4	1.7	0.9	13.6	5.6	4.6	29.3	2.5	6.5	8.1	21.2	10.8	3.0	8.6	9.1
				Oolong tea	A	4.3	9.7	5.0	2.6	8.8	5.3	8.6	4.1	0.3	2.0	6.1	5.9	4.6	2.8	9.3	6.6	4.1	2.6	1.1
					B	5.8	4.8	14.4	8.9	4.2	5.1	6.6	5.9	7.5	4.2	5.0	4.9	8.9	6.6	11.3	5.8	7.5	4.5	9.4
137	Ametryn	TPT	GC-MS	Green tea	A	1.8	4.2	2.5	2.4	23.3	4.2	33.8	7.0	14.3	7.1	9.9	6.8	7.3	4.7	1.2	2.4	5.6	5.9	3.7
					B	6.5	7.7	5.6	7.8	21.9	17.3	10.7	14.8	5.3	10.1	3.7	14.0	2.4	6.3	4.1	3.5	4.7	0.6	4.4
				Oolong tea	A	8.2	2.1	3.2	5.6	2.3	7.2	24.0	29.5	9.2	11.3	7.5	4.3	3.1	2.6	7.2	2.8	2.8	2.5	1.8
					B	6.1	6.3	3.6	5.4	2.5	11.7	16.1	19.5	7.6	8.5	15.7	2.6	1.3	2.8	3.8	1.0	3.0	1.0	3.1
			GC-MS/MS	Green tea	A	4.3	3.2	2.0	3.6	10.4	5.2	7.9	5.4	4.2	2.9	5.5	1.6	9.7	11.0	1.1	2.2	8.2	6.3	4.7
					B	5.5	9.4	8.2	7.9	2.9	3.5	4.2	2.5	6.2	9.7	5.0	10.6	3.4	6.5	4.7	7.5	8.5	4.2	6.7
				Oolong tea	A	5.7	4.4	2.5	5.2	4.4	3.6	10.3	3.0	4.7	4.0	3.8	1.5	8.2	1.0	6.4	1.9	0.2	4.5	5.3
					B	5.1	4.1	4.3	7.4	2.2	2.2	7.1	7.4	7.6	3.2	9.7	3.8	8.9	1.9	6.2	3.5	3.9	3.7	4.0
		ENVI-Carb +PSA	GC-MS	Green tea	A	6.7	8.5	2.8	7.4	8.8	0.8	42.9	3.7	3.7	5.6	6.6	6.6	3.9	6.1	9.7	7.2	11.7	2.8	6.3
					B	2.0	2.7	3.6	6.1	2.1	0.3	4.9	30.4	17.1	5.3	28.7	13.7	4.7	2.7	11.8	1.5	4.6	8.7	8.1
				Oolong tea	A	8.8	2.7	2.1	0.5	5.1	23.6	9.9	22.8	37.4	3.4	6.0	10.5	1.3	2.5	3.2	3.4	5.7	4.4	3.2
					B	5.7	2.5	9.0	3.3	13.3	2.7	81.5	13.1	20.8	4.8	7.8	7.2	5.0	7.9	9.3	1.8	6.6	4.8	6.2
			GC-MS/MS	Green tea	A	7.5	3.6	1.3	7.6	7.8	14.3	4.6	2.0	22.9	0.6	5.3	0.8	1.9	3.8	7.7	6.1	8.4	2.8	6.2
					B	4.5	9.8	5.6	3.6	0.7	1.8	2.2	8.8	2.4	2.3	22.6	2.4	4.1	19.7	12.6	0.8	4.1	9.7	7.4
				Oolong tea	A	4.4	4.8	4.2	3.3	5.7	2.4	3.1	4.1	2.8	3.8	7.7	5.4	5.6	2.3	12.9	3.4	4.5	4.5	5.3
					B	5.4	3.6	10.2	5.1	8.7	3.1	4.2	8.2	7.3	4.5	6.8	2.6	3.7	8.2	13.6	2.7	2.8	5.0	4.0

(Continued)

Appendix Table 5.2 RSD of 201 Pesticides in Youden Pair Aged Tea Sample for Parallel Determinations (Nov 9, 2009–Feb 7, 2010) (cont.)

No.	Pesticides	SPE	Method	Sample	Youden pair	Nov 9, 2009 (n=5)	Nov 14, 2009 (n=3)	Nov 19, 2009 (n=3)	Nov 24, 2009 (n=3)	Nov 29, 2009 (n=3)	Dec 4, 2009 (n=3)	Dec 9, 2009 (n=3)	Dec 14, 2009 (n=3)	Dec 19, 2009 (n=3)	Dec 24, 2009 (n=3)	Dec 29, 2009 (n=3)	Jan 3, 2010 (n=3)	Jan 8, 2010 (n=3)	Jan 13, 2010 (n=3)	Jan 18, 2010 (n=3)	Jan 23, 2010 (n=3)	Jan 28, 2010 (n=3)	Feb 2, 2010 (n=3)	Feb 7, 2010 (n=3)
												RSD%												
138	Bifenthrin	TPT	GC-MS	Green tea	A	4.5	5.8	1.9	4.4	7.4	3.3	5.4	5.2	5.0	2.2	6.3	2.7	6.2	2.2	2.3	6.7	4.6	6.1	8.5
					B	5.1	7.6	4.4	8.2	5.3	6.1	3.6	0.6	4.5	7.4	3.9	7.5	1.1	1.4	4.8	1.6	5.7	2.0	4.1
				Oolong tea	A	7.0	1.4	3.6	6.7	3.3	6.1	6.2	8.4	8.6	3.3	4.7	2.4	10.6	2.9	10.1	3.2	1.9	3.0	1.3
					B	5.6	2.5	5.9	7.5	3.2	2.5	8.1	1.2	4.4	4.0	6.3	2.1	1.0	5.0	4.4	5.2	1.8	11.3	8.1
			GC-MS/MS	Green tea	A	5.2	4.1	2.1	3.1	9.8	4.1	7.1	3.7	3.8	1.5	6.1	1.7	11.0	4.5	5.3	1.7	7.5	4.9	1.7
					B	5.0	7.6	5.8	8.2	3.5	4.3	1.2	2.4	5.6	7.1	3.1	5.0	9.0	1.6	3.0	3.8	6.1	4.2	6.2
				Oolong tea	A	7.5	1.2	4.6	8.0	2.9	5.0	6.1	7.8	8.0	3.7	1.6	0.8	13.8	5.9	9.1	0.4	2.4	3.2	9.9
					B	5.8	4.3	5.7	7.9	3.3	3.9	9.9	6.8	11.0	3.5	5.6	2.9	5.2	3.3	9.9	7.7	1.3	8.8	4.2
		ENVI-Carb +PSA	GC-MS	Green tea	A	6.8	4.7	3.8	3.9	8.9	2.5	5.8	1.5	7.4	3.1	7.9	1.2	1.0	3.4	2.6	5.3	10.0	4.5	5.5
					B	3.5	6.0	3.8	2.9	3.3	1.9	4.1	1.7	6.0	3.9	17.1	2.4	2.3	0.3	9.8	4.1	5.0	10.4	7.4
				Oolong tea	A	4.5	3.9	11.2	5.2	13.4	1.7	4.7	8.1	2.7	1.0	6.5	5.6	1.8	8.4	2.8	6.3	6.3	8.5	10.7
					B	7.7	8.8	13.9	5.9	9.0	3.2	2.5	5.2	12.0	5.1	6.4	2.6	4.2	8.9	6.2	5.9	1.6	11.7	4.2
			GC-MS/MS	Green tea	A	8.5	3.1	3.1	8.4	7.7	2.8	7.5	2.2	7.3	1.1	7.9	33.5	5.8	15.3	3.5	4.2	7.9	4.1	5.3
					B	5.0	9.4	4.9	1.0	0.7	1.0	0.7	4.5	0.7	2.2	19.1	19.9	3.6	11.9	8.6	1.9	2.6	10.0	7.9
				Oolong tea	A	4.8	4.2	11.7	5.3	11.5	3.0	3.9	6.3	3.6	2.5	6.9	4.9	8.5	7.1	4.4	8.6	8.0	8.1	9.9
					B	8.5	8.1	14.1	6.6	7.3	2.4	4.6	6.4	4.6	5.3	2.9	4.5	6.9	9.9	7.0	5.2	1.9	7.2	4.2
139	Bitertanol	TPT	GC-MS	Green tea	A	5.6	6.5	1.7	1.5	2.9	6.9	4.2	3.6	7.5	6.1	10.3	9.0	5.3	4.0	3.9	2.9	2.8	2.8	3.9
					B	3.0	7.6	4.4	9.2	4.8	8.4	12.4	7.0	2.6	7.4	6.4	12.7	0.6	2.8	8.0	2.8	9.6	1.8	7.7
				Oolong tea	A	5.7	2.2	8.5	2.7	10.1	4.4	3.7	14.2	5.5	11.8	12.7	10.8	6.3	3.0	5.7	3.3	2.4	3.7	8.7
					B	4.2	5.2	13.1	10.4	8.7	5.8	0.4	3.5	6.4	7.6	9.8	3.0	1.1	2.8	1.8	1.9	3.5	3.3	6.3
			GC-MS/MS	Green tea	A	5.3	6.9	0.9	12.3	9.7	1.1	5.5	5.3	4.5	3.8	6.3	0.9	11.1	16.0	1.8	1.7	6.3	3.4	3.1
					B	4.5	10.6	6.4	7.5	4.6	3.3	9.5	8.8	1.9	11.4	6.4	5.7	4.6	3.0	7.0	3.8	5.5	5.7	6.8
				Oolong tea	A	5.4	2.4	0.3	2.7	3.0	3.4	1.2	0.3	4.9	4.5	4.0	5.5	7.0	3.3	8.6	0.3	6.8	4.7	4.8
					B	4.6	3.8	3.7	7.3	5.5	3.5	5.5	9.9	10.8	2.7	10.8	3.0	7.1	4.1	5.1	3.6	1.7	6.0	1.5
		ENVI-Carb +PSA	GC-MS	Green tea	A	7.2	13.0	6.4	10.0	6.9	1.8	4.6	1.6	3.5	6.7	1.7	11.4	3.8	7.1	11.7	14.1	7.4	15.8	3.8
					B	6.5	15.1	0.7	5.9	1.9	7.1	30.7	6.7	7.5	3.5	39.6	10.4	9.7	4.7	18.1	3.1	1.9	9.0	17.0
				Oolong tea	A	7.2	6.0	8.8	2.0	11.7	1.2	1.0	17.8	4.7	6.9	11.4	6.4	2.1	8.4	3.6	2.9	4.1	3.5	2.5
					B	4.6	3.3	15.2	3.6	21.7	8.5	1.7	6.5	11.6	1.6	10.6	2.7	0.8	14.2	6.1	1.8	8.8	5.0	1.3
			GC-MS/MS	Green tea	A	8.3	13.3	6.6	9.4	5.0	3.0	1.7	3.2	5.9	2.2	2.6	51.8	6.0	11.8	2.1	10.7	12.0	7.7	4.3
					B	4.4	6.3	6.2	7.8	1.0	1.9	27.1	1.6	6.8	0.8	26.1	30.0	7.9	18.5	10.9	4.5	1.1	9.8	10.1
				Oolong tea	A	4.7	3.5	3.7	3.8	8.3	1.2	2.0	9.6	2.0	10.8	7.0	4.2	22.9	8.7	5.7	3.4	4.1	6.6	4.2
					B	6.6	11.2	14.3	8.0	10.1	5.5	2.1	9.6	5.4	2.3	5.9	4.4	7.2	15.6	5.7	4.4	8.1	4.7	1.5
140	Boscalid	TPT	GC-MS	Green tea	A	8.6	7.1	1.6	17.4	14.7	2.1	3.0	8.5	13.4	4.1	11.4	8.7	12.6	12.5	5.1	11.7	7.2	23.4	4.2
					B	7.7	9.3	5.6	13.1	7.4	3.1	11.1	6.5	5.6	10.0	4.3	12.6	12.5	12.7	12.6	9.7	13.6	10.8	5.2
				Oolong tea	A	9.4	9.6	3.8	0.8	18.0	7.2	4.6	12.7	7.5	6.5	13.8	14.5	3.4	2.4	1.6	10.0	3.9	7.9	3.6
					B	5.8	12.8	5.1	2.5	8.5	14.4	8.7	3.8	12.4	3.5	12.9	8.2	1.5	11.7	12.8	0.3	3.9	6.2	12.0
			GC-MS/MS	Green tea	A	12.8	9.1	0.7	5.1	20.9	3.6	7.6	20.3	16.1	7.9	10.5	1.7	21.5	13.1	4.0	7.6	8.5	18.8	0.6
					B	9.6	5.6	3.4	18.6	4.9	3.5	13.7	15.3	5.7	11.7	6.1	9.5	20.3	7.2	7.4	7.3	14.4	7.9	7.6
				Oolong tea	A	12.1	3.7	0.5	3.5	15.0	7.8	9.3	16.4	8.7	6.0	10.9	18.2	4.8	5.1	5.4	5.8	5.8	19.0	12.0
					B	16.2	4.8	8.5	3.8	6.7	10.6	15.9	13.3	22.6	1.3	8.8	3.8	3.9	6.4	10.7	2.0	2.6	10.0	8.4
		ENVI-Carb +PSA	GC-MS	Green tea	A	20.8	7.1	25.8	9.4	8.1	21.3	1.1	14.8	12.2	9.1	11.0	33.3	7.1	18.9	12.9	13.4	15.9	16.8	2.1
					B	36.0	9.3	6.0	6.9	15.4	4.8	13.1	20.7	20.3	13.9	16.4	16.4	44.0	12.2	4.2	5.8	6.8	23.6	17.2
				Oolong tea	A	12.0	9.6	7.2	16.6	9.1	9.8	2.6	32.8	2.9	12.7	6.3	15.7	13.4	4.2	1.4	33.6	7.2	23.6	13.2
					B	16.9	12.8	10.5	8.8	19.1	10.4	10.3	3.8	15.0	18.7	8.6	7.7	0.5	12.0	18.4	5.1	11.6	31.9	8.2
			GC-MS/MS	Green tea	A	22.2	7.7	21.9	7.8	9.1	34.4	16.7	28.0	12.2	2.1	8.1	5.7	4.5	7.9	9.6	7.9	17.0	15.0	3.9
					B	37.9	12.4	7.5	13.4	18.3	9.7	31.2	28.7	22.3	13.3	17.5	6.6	40.9	20.7	3.0	18.1	6.0	27.1	16.3
				Oolong tea	A	14.3	7.2	9.0	11.7	9.0	4.3	14.8	30.3	3.5	13.8	4.5	14.7	13.4	7.7	10.7	23.3	4.5	28.9	12.0
					B	17.8	11.1	12.9	2.1	17.2	14.4	15.3	4.3	13.7	17.5	10.6	9.5	4.9	8.9	14.6	1.1	14.6	41.5	8.4

141	Butafenacil	TPT	GC–MS	Green tea	A	10.2	9.6	2.9	21.1	17.1	4.7	16.4	14.1	12.9	12.0	7.2	12.1	14.9	3.6	8.8	6.5	25.8	1.0
					B	9.0	7.7	8.5	14.3	9.9	10.2	11.2	12.9	6.5	2.2	13.3	18.9	11.8	12.8	8.3	19.6	13.4	9.0
				Oolong tea	A	8.6	4.1	3.0	3.8	16.7	3.0	11.0	11.9	11.9	14.5	6.8	2.7	4.4	1.5	6.3	1.1	6.4	2.1
					B	4.9	4.5	4.1	3.6	10.2	13.1	13.6	6.2	10.9	13.7	2.5	3.2	10.8	9.9	1.9	2.6	4.1	11.0
			GC–MS/MS	Green tea	A	15.6	10.7	1.0	20.8	24.6	4.7	18.8	24.0	17.8	17.1	10.8	25.6	17.0	3.0	5.3	7.3	26.1	1.7
					B	11.5	6.4	6.4	18.5	6.2	4.7	10.8	19.9	4.6	5.0	14.4	27.1	1.8	9.0	7.3	13.1	8.7	10.6
				Oolong tea	A	12.9	2.8	0.6	4.1	13.6	4.9	8.5	18.3	3.9	7.0	5.8	7.9	4.6	6.3	6.7	4.5	14.5	10.9
					B	13.8	2.5	6.9	5.9	6.1	11.7	13.1	12.5	20.7	6.9	2.8	3.9	5.5	9.9	4.6	2.9	10.9	8.4
		ENVI-Carb+PSA	GC–MS	Green tea	A	13.6	7.1	34.2	10.9	10.4	42.2	27.6	31.4	20.3	13.4	13.9	12.9	28.1	24.3	10.0	23.9	31.6	6.3
					B	2.5	13.3	7.0	20.2	21.5	11.5	8.1	16.7	33.7	20.1	22.0	49.4	13.9	20.2	17.1	7.8	33.5	19.7
				Oolong tea	A	26.2	10.2	3.9	15.3	7.8	10.8	16.2	34.0	2.8	7.4	21.4	14.7	5.0	4.8	20.3	9.8	15.4	9.9
					B	45.8	14.8	14.7	5.3	21.7	9.7	15.9	5.4	12.1	9.5	17.3	0.4	12.2	19.1	4.1	15.3	30.8	6.9
			GC–MS/MS	Green tea	A	9.7	3.3	26.7	3.6	10.3	45.5	19.3	39.0	13.7	17.5	1.3	12.4	30.7	19.5	27.3	20.4	25.2	7.7
					B	16.5	13.2	5.0	28.1	23.5	9.4	17.5	35.9	29.7	27.9	22.5	37.7	19.2	32.0	33.4	10.1	33.7	22.3
				Oolong tea	A	31.0	7.1	8.9	13.0	6.8	6.5	16.5	32.2	1.2	6.6	23.8	5.3	6.9	11.9	25.2	5.7	28.2	10.9
					B	58.7	10.9	16.8	1.6	16.3	13.5	15.9	7.0	9.4	9.7	16.1	8.2	18.3	18.0	2.1	11.1	42.5	8.4
142	Carbaryl	TPT	GC–MS	Green tea	A	12.8	13.3	10.6	28.6	21.0	2.0	12.4	2.7	4.8	7.3	10.2	8.8	7.8	7.4	14.9	7.8	3.3	11.8
					B	21.4	5.9	11.0	6.7	2.8	8.3	7.8	8.8	7.0	10.9	6.7	5.6	5.5	12.2	5.2	10.3	4.2	8.4
				Oolong tea	A	14.4	10.8	7.3	16.7	25.9	5.9	11.4	12.9	5.4	13.5	7.0	5.1	10.9	2.0	5.6	5.0	13.3	2.5
					B	18.4	5.3	9.2	8.5	12.1	2.5	4.7	7.2	4.1	17.3	15.3	3.8	8.4	3.7	3.6	6.7	2.6	10.8
			GC–MS/MS	Green tea	A	9.4	15.9	13.3	32.5	27.8	1.4	14.5	3.1	5.0	11.5	7.0	10.1	13.5	6.5	9.7	1.5	9.5	1.0
					B	14.0	6.5	8.6	14.8	4.5	7.7	13.5	13.8	10.4	13.8	10.8	3.8	7.7	10.1	4.4	8.7	3.8	7.3
				Oolong tea	A	24.6	16.0	15.3	18.2	35.0	4.9	17.9	20.5	9.5	19.9	8.9	8.8	8.0	11.3	6.5	3.9	7.3	2.0
					B	34.3	11.4	19.7	7.5	17.4	5.2	11.5	4.8	6.6	27.4	20.9	10.8	5.9	17.9	4.7	3.5	3.3	7.3
		ENVI-Carb+PSA	GC–MS	Green tea	A	13.4	6.5	4.7	2.5	5.9	5.8	31.6	5.0	6.4	19.4	3.1	4.0	4.9	6.4	16.5	9.7	10.7	5.1
					B	28.3	3.5	0.9	3.0	2.8	3.1	6.4	7.6	12.6	29.0	5.0	11.0	2.3	18.9	5.5	3.4	18.0	8.9
				Oolong tea	A	5.8	3.7	4.6	10.0	4.1	6.0	7.2	14.3	2.7	1.4	9.6	3.8	3.9	6.6	7.0	5.4	5.9	2.3
					B	13.4	6.2	15.3	4.1	12.9	4.4	5.5	5.6	4.8	5.4	6.6	5.5	15.4	9.8	1.4	4.8	5.1	10.2
			GC–MS/MS	Green tea	A	5.7	11.5	5.5	5.3	5.0	8.6	33.9	8.7	11.4	23.4	1.1	1.8	9.9	6.1	15.9	9.3	12.1	9.6
					B	6.2	4.1	3.3	3.0	7.6	1.1	9.9	11.9	12.4	33.7	5.7	24.5	18.5	27.4	5.1	4.8	19.5	12.1
				Oolong tea	A	9.6	8.7	6.5	13.7	2.0	6.9	6.8	20.7	2.7	4.7	9.0	11.5	1.5	17.0	7.9	10.2	9.1	1.5
					B	17.0	1.0	14.0	9.5	6.3	6.0	5.5	2.4	6.1	10.9	3.6	6.0	16.9	10.5	1.4	3.9	17.7	6.6
143	Chloroben-zilate	TPT	GC–MS	Green tea	A	5.0	12.1	2.5	11.4	14.4	1.3	14.5	18.8	13.7	6.2	4.5	11.4	12.6	6.5	7.8	8.9	19.8	5.4
					B	10.5	10.0	1.8	10.2	5.4	6.1	5.0	14.1	4.8	1.7	6.8	10.3	12.1	5.2	9.1	14.4	11.1	4.6
				Oolong tea	A	6.6	3.4	3.0	3.2	12.0	5.1	11.3	17.3	6.6	13.7	7.2	2.3	1.5	5.7	6.8	3.2	1.8	0.9
					B	5.9	2.5	1.3	2.6	7.4	13.1	4.6	11.4	7.7	14.5	3.2	2.2	7.7	6.7	1.5	1.5	1.7	7.4
			GC–MS/MS	Green tea	A	7.3	4.6	4.3	6.6	16.0	1.3	11.5	17.4	11.3	7.0	3.9	12.5	5.9	2.1	5.7	8.2	14.3	5.0
					B	3.8	6.3	1.4	11.0	4.3	2.2	4.0	12.5	5.0	1.3	9.1	14.9	5.2	1.8	6.7	13.1	3.8	7.7
				Oolong tea	A	9.2	1.8	5.7	3.9	6.7	4.1	0.1	10.0	5.5	7.2	4.2	7.0	1.0	7.8	4.5	3.9	9.9	4.8
					B	7.3	3.9	16.7	6.1	5.4	8.4	8.7	11.9	13.8	7.3	1.9	6.4	3.9	8.1	2.1	1.1	5.6	7.9
		ENVI-Carb+PSA	GC–MS	Green tea	A	10.0	5.3	5.2	9.4	10.4	21.7	52.6	24.3	24.9	1.7	6.8	3.6	31.0	7.4	47.7	14.6	51.0	5.7
					B	7.5	10.7	4.2	12.7	9.4	3.9	11.4	29.1	30.2	41.8	8.1	48.5	9.7	46.7	9.0	6.2	32.3	12.6
				Oolong tea	A	22.9	3.2	7.1	7.4	8.9	1.9	5.0	28.2	2.1	5.3	2.1	6.5	1.2	1.1	20.0	8.4	13.6	5.6
					B	24.9	7.5	13.2	5.5	17.2	9.8	13.0	6.0	12.7	11.2	6.6	3.8	11.8	15.1	2.5	14.2	14.8	6.8
			GC–MS/MS	Green tea	A	4.5	9.8	5.1	4.7	8.8	21.9	34.2	24.5	7.0	7.9	1.9	4.6	7.9	6.2	13.3	9.0	20.6	12.8
					B	8.2	13.4	6.9	8.1	10.8	5.0	12.0	24.6	20.2	27.4	2.8	38.1	14.0	14.9	14.9	1.5	24.9	18.4
				Oolong tea	A	30.5	2.5	7.1	7.1	1.9	1.9	5.3	27.1	2.0	4.5	2.7	12.9	10.0	10.0	10.3	7.3	15.4	4.8
					B	29.3	6.0	9.9	1.9	11.3	8.2	5.7	5.9	10.9	8.4	5.8	8.5	7.7	13.7	1.4	7.7	21.5	7.9

(Continued)

Appendix Table 5.2 RSD of 201 Pesticides in Youden Pair Aged Tea Sample for Parallel Determinations (Nov 9, 2009–Feb 7, 2010) (cont.)

No.	Pesticides	SPE	Method	Sample	Youden pair	RSD% Nov 9, 2009 (n = 5)	Nov 14, 2009 (n = 3)	Nov 19, 2009 (n = 3)	Nov 24, 2009 (n = 3)	Nov 29, 2009 (n = 3)	Dec 4, 2009 (n = 3)	Dec 9, 2009 (n = 3)	Dec 14, 2009 (n = 3)	Dec 19, 2009 (n = 3)	Dec 24, 2009 (n = 3)	Dec 14, 2009 (n = 3)	Jan 3, 2010 (n = 3)	Jan 8, 2010 (n = 3)	Jan 13, 2010 (n = 3)	Jan 18, 2010 (n = 3)	Jan 23, 2010 (n = 3)	Jan 28, 2010 (n = 3)	Feb 2, 2010 (n = 3)	Feb 7, 2010 (n = 3)
144	Chlorthal-dimethyl	TPT	GC-MS	Green tea	A	4.2	6.0	3.9	1.0	8.4	3.7	6.3	7.7	4.9	2.9	4.4	2.5	6.6	1.1	2.5	2.2	4.8	5.4	3.8
					B	5.7	6.1	4.7	7.9	4.1	3.9	1.2	2.6	4.4	7.0	2.5	6.4	0.8	1.9	2.6	4.0	5.6	0.3	5.9
				Oolong tea	A	4.9	2.5	1.8	3.8	1.6	2.9	10.0	5.8	3.2	2.1	4.4	2.1	2.7	2.7	7.3	2.8	3.4	2.2	0.5
					B	4.1	2.4	4.5	4.2	2.5	3.5	3.8	4.1	5.6	4.6	0.7	1.0	2.6	3.6	4.6	0.5	0.7	1.3	2.5
			GC-MS/MS	Green tea	A	4.6	4.2	2.7	2.9	9.6	2.7	6.3	5.7	4.5	2.1	8.2	0.2	9.4	3.6	3.8	1.1	4.1	4.3	3.6
					B	5.5	6.9	6.7	7.5	4.7	2.9	3.1	4.4	6.5	9.8	2.0	7.3	1.1	2.6	4.3	4.0	6.0	3.1	8.1
				Oolong tea	A	4.7	4.0	1.1	4.3	5.4	3.1	10.2	0.1	3.5	4.6	2.1	3.0	4.8	0.7	7.1	2.7	5.8	3.3	6.5
					B	4.5	4.9	4.9	5.1	3.0	3.3	3.1	7.0	7.8	2.0	3.9	1.0	4.1	2.8	7.8	2.7	1.2	0.9	3.5
		ENVI-Carb+PSA	GC-MS	Green tea	A	8.5	6.8	4.6	4.9	7.9	2.2	8.1	3.1	5.7	2.0	8.9	2.8	7.5	4.8	7.2	13.0	9.9	11.9	5.1
					B	4.7	7.6	3.4	3.1	0.9	0.8	4.0	1.7	6.2	2.6	19.7	2.3	0.4	1.8	16.1	2.8	5.2	12.9	6.5
				Oolong tea	A	1.8	3.2	3.4	3.1	5.0	1.7	3.8	4.8	2.3	0.9	4.6	2.5	4.4	2.8	1.6	3.1	7.6	4.7	4.2
					B	5.2	3.1	8.7	2.4	9.3	5.3	3.8	6.7	11.2	2.0	7.1	4.0	2.6	7.2	9.7	2.9	6.4	3.0	4.4
			GC-MS/MS	Green tea	A	9.4	3.7	3.7	5.9	7.0	2.4	4.1	3.8	3.3	1.8	4.6	3.1	2.2	2.2	6.1	7.8	6.4	5.4	3.9
					B	5.3	8.6	5.0	0.8	1.2	1.8	2.8	3.0	3.3	1.0	21.1	3.8	6.0	12.2	11.1	1.0	2.5	10.6	6.6
				Oolong tea	A	3.2	4.7	6.7	5.0	4.1	1.6	2.2	5.7	2.4	1.8	5.2	3.2	1.6	3.1	8.9	2.6	3.4	3.4	6.5
					B	7.0	3.1	11.2	3.8	7.3	5.7	4.2	7.5	8.9	3.5	6.2	4.6	7.4	9.4	10.0	1.7	3.0	3.1	3.5
145	Dibutyl-succinate	TPT	GC-MS	Green tea	A	3.3	5.6	3.1	1.5	8.6	3.1	5.6	1.9	6.1	3.1	7.2	2.2	1.1	1.5	3.1	0.5	3.0	9.1	6.2
					B	5.6	8.7	4.1	9.2	4.1	3.9	3.1	2.4	3.0	9.3	2.9	7.2	1.9	1.9	3.1	8.4	5.9	3.8	0.1
				Oolong tea	A	3.9	2.7	1.1	3.0	0.9	1.0	8.4	5.6	3.0	2.4	3.0	2.5	4.4	3.4	7.9	1.2	4.5	2.6	1.9
					B	4.2	5.0	5.9	6.9	3.5	4.6	5.0	2.2	4.8	2.4	4.3	1.4	3.3	3.0	2.5	2.1	4.9	2.9	2.6
			GC-MS/MS	Green tea	A	4.0	4.0	1.9	1.1	11.7	5.1	4.5	1.8	3.9	0.9	5.5	1.9	14.0	4.8	1.7	1.7	3.8	4.6	6.6
					B	5.0	8.8	5.1	8.0	3.3	3.8	3.6	3.5	4.1	9.0	3.3	9.4	17.7	3.3	4.2	1.7	4.6	5.0	5.3
				Oolong tea	A	4.7	1.3	2.4	1.1	3.3	0.9	3.6	0.9	6.1	3.6	2.0	3.7	8.1	1.7	8.5	10.0	1.7	0.9	3.7
					B	4.8	3.9	3.9	8.0	4.8	3.6	9.0	4.6	9.6	2.9	3.2	1.2	10.0	1.3	7.8	2.6	2.0	1.0	1.5
		ENVI-Carb+PSA	GC-MS	Green tea	A	5.7	5.0	2.5	5.9	6.2	1.1	5.6	1.1	6.3	2.0	8.5	3.1	9.3	2.3	6.3	9.4	11.6	4.6	3.7
					B	3.0	6.8	5.8	4.6	5.0	3.9	5.6	2.2	3.8	1.2	19.9	5.5	2.5	1.4	16.6	3.4	5.8	6.4	3.8
				Oolong tea	A	4.0	3.4	3.0	1.9	4.5	3.3	2.6	3.7	1.5	4.5	5.9	0.2	1.8	2.7	7.0	4.9	6.2	2.9	3.6
					B	5.9	3.3	7.4	2.6	12.3	3.9	1.9	8.3	9.3	3.5	7.0	1.7	1.7	12.4	3.4	8.8	6.0	4.7	4.1
			GC-MS/MS	Green tea	A	8.0	9.0	0.5	8.1	3.1	2.4	4.3	1.6	6.5	2.2	8.1	6.8	14.1	13.8	18.3	4.0	3.5	10.9	4.8
					B	2.9	7.5	6.8	2.2	2.3	0.9	2.9	5.1	1.4	0.8	22.0	3.8	0.7	4.4	6.9	4.8	5.4	6.4	5.3
				Oolong tea	A	3.7	4.3	4.4	2.4	4.1	2.7	1.6	3.8	2.0	2.1	6.3	5.0	8.1	8.1	6.7	1.7	5.5	2.6	1.5
					B	6.5	3.7	8.0	5.5	7.6	8.5	2.6	8.1	7.7	2.1	3.5	2.4	9.1	7.2	7.8	2.2	7.4	5.3	4.9
146	Diethofen-carb	TPT	GC-MS	Green tea	A	4.9	5.6	2.2	0.8	8.0	4.6	7.4	13.5	8.1	8.5	8.7	6.1	9.1	7.2	2.4	4.5	7.4	5.3	11.0
					B	7.1	10.0	5.3	8.0	9.3	5.6	2.0	8.0	3.3	5.8	4.7	14.2	4.9	2.3	4.1	2.9	5.8	2.8	0.7
				Oolong tea	A	6.4	9.6	2.0	4.8	6.5	3.1	3.4	8.1	15.4	2.9	2.9	15.7	2.7	2.7	7.9	1.2	6.5	1.5	5.4
					B	4.2	3.7	7.9	4.9	8.0	2.4	6.8	0.3	6.3	3.7	5.4	5.1	5.0	4.2	5.2	1.5	2.6	5.3	2.6
			GC-MS/MS	Green tea	A	3.7	3.1	3.5	8.5	4.0	3.9	3.7	9.3	8.5	0.6	9.3	1.5	2.1	2.7	0.1	3.3	2.1	5.2	5.1
					B	4.8	5.3	10.7	7.7	10.0	3.3	28.8	13.1	14.1	4.7	2.8	4.9	3.0	13.7	13.9	34.8	14.2	32.9	3.2
				Oolong tea	A	8.0	9.6	2.0	4.8	6.5	3.1	6.0	8.1	15.4	2.9	37.7	15.7	27.0	2.9	40.0	14.1	6.4	21.9	5.4
					B	3.1	3.7	7.9	4.9	3.8	2.4	2.2	16.6	6.3	3.7	6.5	5.1	3.5	4.9	5.0	4.5	7.7	5.2	2.6
		ENVI-Carb+PSA	GC-MS	Green tea	A	10.9	5.3	8.2	2.8	2.7	4.3	7.3	8.1	7.1	3.1	8.5	0.4	5.0	4.2	13.9	2.7	6.1	5.2	4.3
					B	8.0	9.6	4.4	7.8	2.0	3.9	3.4	8.6	3.8	3.8	5.8	0.7	2.1	6.1	40.0	4.6	4.2	14.8	8.4
				Oolong tea	A	3.1	9.7	7.0	4.5	6.5	2.1	6.8	3.6	10.1	3.2	30.3	8.0	19.0	12.9	5.0	3.3	8.1	5.9	11.7
					B	4.4	5.1	8.2	2.6	14.8	4.3	11.7	7.6	2.2	3.2	5.7	6.4	2.4	1.9	9.8	2.6	6.1	2.9	2.1
			GC-MS/MS	Green tea	A	13.4	5.9	4.5	7.8	8.2	3.9	4.5	3.6	3.8	2.7	30.3	0.7	19.0	12.9	18.2	2.7	4.2	14.8	11.7
					B	5.9	9.7	4.5	2.6	2.0	2.9	3.3	12.2	2.2	3.2	5.7	6.4	2.4	1.9	12.0	3.7	8.1	5.9	2.1
				Oolong tea	A	5.5	5.1	5.2	2.9	3.4	0.7	3.3	12.2	2.2	3.2	5.7	5.6	2.4	8.4	12.0	1.0	3.0	5.9	5.1
					B	8.7	3.5	11.3	6.0	9.6	6.7	3.5	5.7	9.0	3.6	7.2	5.6	9.9	8.4	12.0	1.0	3.0	4.0	5.1

147	Diflufeni-can	TPT	GC–MS	Green tea	A	3.9	6.8	1.2	4.2	9.7	3.0	9.8	14.2	7.9	8.5	5.5	10.9	8.1	7.0	3.3	3.9	6.2	1.6	4.1
					B	3.8	6.9	3.1	7.7	3.2	3.9	2.1	8.6	3.9	6.2	1.8	9.9	6.4	4.8	3.2	1.7	11.3	9.7	1.7
				Oolong tea	A	6.3	4.0	9.4	1.6	5.3	4.3	5.7	14.6	4.4	4.0	9.6	5.5	6.8	2.3	6.8	3.9	2.8	3.1	11.5
					B	3.4	0.4	2.9	5.9	9.2	7.4	11.3	8.7	3.6	1.6	7.0	0.4	2.1	3.2	2.6	2.8	1.4	2.1	4.6
			GC–MS/MS	Green tea	A	6.3	3.5	0.5	2.7	11.9	4.7	10.4	10.4	7.2	0.8	6.5	1.9	10.0	1.8	2.6	2.3	8.8	2.4	3.7
					B	4.9	5.8	4.5	8.2	4.0	2.3	1.6	8.0	3.6	7.4	3.8	2.5	8.5	1.3	1.8	5.3	7.5	1.5	5.1
				Oolong tea	A	9.3	3.5	1.6	3.8	4.6	4.6	4.1	2.6	10.0	4.3	5.3	6.2	9.8	3.1	4.8	6.1	0.9	2.8	1.6
					B	5.6	4.5	2.2	6.9	3.3	6.8	5.1	12.4	10.1	1.7	6.5	0.7	5.1	0.7	5.9	1.6	4.5	5.0	5.5
		ENVI-Carb +PSA	GC–MS	Green tea	A	13.8	9.9	8.4	7.5	8.9	9.8	35.8	10.5	14.5	2.2	2.9	12.8	2.1	14.1	9.0	31.1	10.8	25.9	9.7
					B	9.0	8.3	4.9	6.1	2.4	7.1	6.7	12.1	18.1	3.1	33.6	10.4	29.5	3.1	30.5	5.2	7.6	19.9	4.1
				Oolong tea	A	3.6	2.6	6.1	3.9	2.9	1.7	0.5	20.2	0.7	0.2	7.3	14.3	1.0	6.7	0.7	4.5	4.7	14.3	6.6
					B	5.2	4.2	7.9	1.2	11.5	6.3	8.2	7.2	9.5	2.5	14.5	2.6	2.2	4.2	9.2	3.6	7.6	7.9	3.6
			GC–MS/MS	Green tea	A	16.8	4.3	7.9	7.8	9.7	6.9	19.1	11.5	3.9	3.2	3.0	1.7	4.4	5.5	2.3	14.1	4.5	9.5	2.6
					B	9.4	9.6	5.8	1.8	4.1	1.8	5.4	8.1	10.1	3.2	29.3	0.8	17.8	6.3	15.4	2.6	2.7	16.7	13.1
				Oolong tea	A	6.2	2.5	5.7	2.8	2.9	4.3	1.0	16.9	0.5	2.3	8.0	2.6	7.4	4.3	4.7	4.4	2.1	11.2	1.6
					B	10.4	3.9	8.7	4.8	8.6	5.0	3.7	5.9	8.9	2.7	9.0	4.8	5.8	6.1	13.9	1.8	3.8	5.7	5.5
148	Dimepi-perate	TPT	GC–MS	Green tea	A	5.5	6.2	6.5	2.7	9.4	4.1	7.2	6.9	6.8	4.7	9.0	5.3	5.8	6.1	1.4	2.4	6.9	6.2	5.3
					B	7.9	5.4	7.8	8.1	5.3	2.2	1.9	2.6	2.5	10.2	7.5	13.9	10.0	6.1	3.5	1.4	7.0	2.3	4.9
				Oolong tea	A	3.6	2.3	8.4	3.3	0.9	2.8	8.6	10.0	1.9	8.9	5.0	2.2	1.5	2.2	9.3	2.6	8.2	1.3	2.3
					B	4.5	0.4	7.1	5.2	2.8	4.2	8.3	5.6	3.3	7.3	9.2	2.1	5.8	5.2	1.9	1.8	2.6	2.0	0.2
			GC–MS/MS	Green tea	A	3.7	3.1	3.0	5.3	11.3	4.3	6.1	5.5	4.4	1.4	9.8	2.2	3.1	3.9	2.7	1.3	6.4	5.8	0.9
					B	5.1	8.3	8.2	6.8	6.2	3.7	1.5	4.5	4.7	9.2	5.8	12.5	13.0	4.9	4.1	4.2	6.5	3.9	8.8
				Oolong tea	A	4.2	1.9	2.5	5.5	3.2	4.8	11.4	4.2	2.4	4.2	3.1	1.9	8.1	4.8	7.1	2.3	6.2	5.9	3.9
					B	3.3	4.5	3.1	11.4	4.3	1.8	5.4	7.0	9.2	1.4	3.8	1.8	8.4	1.5	6.0	3.9	0.6	3.9	4.0
		ENVI-Carb +PSA	GC–MS	Green tea	A	15.5	2.8	3.9	5.9	8.4	3.5	9.0	2.9	3.2	6.8	4.5	6.2	1.2	1.7	14.0	3.8	9.0	9.2	4.7
					B	13.7	7.6	3.4	4.6	1.5	4.3	2.4	1.6	5.7	5.1	39.1	12.5	2.2	5.5	15.1	10.3	7.9	8.4	5.9
				Oolong tea	A	6.4	5.0	2.5	0.6	6.1	0.9	4.1	5.0	3.7	4.7	7.6	4.0	7.4	3.8	3.2	3.8	6.1	6.0	4.7
					B	5.1	2.9	6.2	0.9	12.7	9.0	3.7	3.9	12.4	5.4	10.0	7.6	2.8	5.7	9.2	5.9	8.4	3.7	4.6
			GC–MS/MS	Green tea	A	8.6	5.6	2.0	7.9	7.0	1.7	6.3	2.4	3.5	1.6	10.7	2.7	4.1	6.4	4.4	0.3	6.2	6.8	3.5
					B	3.0	7.3	4.5	2.2	2.4	1.5	2.2	3.9	3.5	0.9	24.7	4.5	3.4	3.2	12.7	8.8	3.3	10.1	7.6
				Oolong tea	A	3.0	5.8	7.1	4.2	4.0	4.8	4.9	5.0	0.9	3.0	5.4	5.6	5.8	6.3	8.3	5.2	6.0	2.3	6.9
					B	4.6	3.3	10.5	7.6	8.9	6.2	5.2	8.1	8.0	2.9	7.1	4.8	1.8	3.4	9.5	1.9	6.6	5.0	1.4
149	Dimeth-ametryn	TPT	GC–MS	Green tea	A	4.3	5.8	1.4	4.3	7.5	3.8	8.4	4.9	4.5	9.1	6.1	7.9	2.5	8.8	5.1	1.3	4.4	5.0	4.4
					B	4.8	7.4	5.4	9.0	3.6	2.7	2.0	3.8	1.7	9.8	5.1	14.3	7.7	3.7	7.6	2.6	4.4	7.0	4.1
				Oolong tea	A	6.2	2.8	1.6	3.1	2.2	5.8	8.2	8.3	9.4	9.6	8.8	5.6	1.1	8.9	6.7	5.0	3.4	0.3	2.1
					B	4.0	1.9	5.3	7.6	3.6	4.2	4.9	2.8	7.6	5.2	23.4	1.4	5.2	4.3	3.2	1.2	2.1	4.8	2.7
			GC–MS/MS	Green tea	A	4.0	4.0	2.3	7.9	11.0	4.2	6.7	5.9	2.9	1.7	8.2	2.0	3.1	3.2	5.6	3.4	7.0	0.3	2.7
					B	3.9	10.1	8.4	8.1	4.5	3.7	3.5	4.0	6.3	8.5	5.2	13.4	5.5	1.8	2.6	1.1	7.0	8.9	1.9
				Oolong tea	A	6.5	2.6	2.5	2.8	3.2	4.2	9.4	1.5	3.1	3.5	2.5	2.7	10.1	5.2	10.4	9.6	1.4	5.1	5.4
					B	7.5	4.8	7.2	7.1	1.3	3.5	4.8	10.0	8.8	3.1	6.3	5.2	5.6	3.1	3.4	9.4	2.8	8.0	5.6
		ENVI-Carb +PSA	GC–MS	Green tea	A	4.4	9.2	1.1	6.2	8.7	2.1	5.4	5.4	3.0	8.7	4.4	10.0	8.2	2.6	11.1	9.4	2.8	6.1	6.3
					B	3.0	6.0	4.1	6.5	2.8	8.2	3.4	6.7	2.2	3.0	38.2	14.7	1.0	5.0	13.0	7.3	12.2	2.5	6.2
				Oolong tea	A	4.1	4.0	2.9	2.0	10.0	0.8	3.8	6.9	4.9	4.2	6.9	5.8	3.1	2.3	5.0	1.6	4.7	8.5	7.8
					B	5.1	1.9	10.3	1.5	16.3	4.3	2.9	2.3	2.7	5.0	10.2	5.3	4.5	3.5	11.1	3.4	4.9	2.2	3.1
			GC–MS/MS	Green tea	A	7.0	10.3	3.4	9.5	8.7	2.9	7.4	6.7	10.7	2.4	5.3	2.2	1.6	8.2	7.0	4.9	9.8	5.0	4.0
					B	3.6	5.4	3.2	5.5	1.3	1.4	4.2	3.4	4.5	1.6	21.8	2.2	3.6	2.8	9.0	8.1	4.6	3.9	4.4
				Oolong tea	A	6.6	10.3	5.8	0.8	7.5	6.3	0.8	6.7	0.8	3.3	5.0	3.0	15.6	11.5	12.2	4.1	5.5	8.9	1.6
					B	5.3	3.8	11.0	6.3	7.3	5.5	8.6	5.4	8.6	4.3	7.1	5.7	7.5	5.3	10.2	0.7	2.3	5.3	5.6
																								6.3

(Continued)

Appendix Table 5.2 RSD of 201 Pesticides in Youden Pair Aged Tea Sample for Parallel Determinations (Nov 9, 2009–Feb 7, 2010) (cont.)

No.	Pesticides	SPE	Method	Sample	Youden pair	Nov 9, 2009 (n=5)	Nov 14, 2009 (n=3)	Nov 19, 2009 (n=3)	Nov 24, 2009 (n=3)	Nov 29, 2009 (n=3)	Dec 4, 2009 (n=3)	Dec 9, 2009 (n=3)	Dec 14, 2009 (n=3)	Dec 19, 2009 (n=3)	Dec 24, 2009 (n=3)	Dec 14, 2009 (n=3)	Jan 3, 2010 (n=3)	Jan 8, 2010 (n=3)	Jan 13, 2010 (n=3)	Jan 18, 2010 (n=3)	Jan 23, 2010 (n=3)	Jan 28, 2010 (n=3)	Feb 2, 2010 (n=3)	Feb 7, 2010 (n=3)
150	Dimetho-morph	TPT	GC-MS	Green tea	A	48.9	5.4	1.5	4.6	3.0	2.7	9.1	2.4	6.7	7.1	5.0	1.0	10.2	11.9	5.4	6.9	2.9	4.5	4.2
					B	18.8	7.0	4.6	7.9	3.8	6.1	1.7	11.3	1.2	20.0	1.9	9.2	1.9	4.0	9.8	1.9	12.0	4.5	4.6
				Oolong tea	A	6.7	1.1	5.2	4.3	2.3	3.2	8.5	8.4	2.2	3.6	5.8	5.7	6.3	2.9	7.9	5.3	4.2	6.7	5.4
					B	33.1	3.0	2.6	9.5	5.6	4.8	6.3	0.1	7.6	3.9	8.8	2.9	2.4	1.6	3.4	1.7	2.0	2.4	2.6
			GC-MS/MS	Green tea	A	12.6	11.0	2.0	9.3	6.5	7.3	10.0	1.2	4.6	6.1	7.0	1.9	10.4	15.6	12.5	4.5	7.2	4.4	2.3
					B	6.4	15.2	6.8	7.5	4.0	4.4	1.5	12.2	4.5	16.9	6.6	4.2	5.4	4.5	0.6	2.5	6.6	5.9	4.9
				Oolong tea	A	5.8	0.9	2.7	4.9	4.9	3.5	9.1	4.4	10.0	5.5	3.8	8.4	7.4	3.5	8.4	4.6	1.5	7.7	7.1
					B	6.6	4.9	6.2	10.1	4.8	5.7	5.0	6.7	9.3	1.7	5.5	5.1	6.1	2.3	3.4	1.6	4.4	3.5	1.5
		ENVI-Carb+PSA	GC-MS	Green tea	A	6.4	10.4	6.4	5.2	1.5	4.0	49.5	1.1	6.9	6.2	6.3	4.6	1.4	6.0	6.6	6.3	11.2	3.2	7.0
					B	8.5	26.8	2.8	6.2	6.9	4.3	5.0	3.9	2.3	4.8	22.2	8.8	4.4	2.3	8.4	5.7	1.6	6.6	11.4
				Oolong tea	A	13.0	2.5	9.3	3.3	6.9	7.7	3.3	5.6	5.6	22.0	11.3	15.1	2.1	10.3	7.6	7.0	2.7	1.8	2.3
					B	11.4	8.1	17.1	8.2	18.5	12.5	0.7	10.8	7.3	4.0	12.4	2.4	1.6	28.0	8.4	7.7	19.5	8.2	3.6
			GC-MS/MS	Green tea	A	12.0	26.1	4.2	8.7	3.5	4.7	51.4	0.3	5.8	4.7	6.1	1.9	8.9	12.6	2.2	6.3	10.1	7.0	5.6
					B	5.8	4.3	6.4	6.0	2.9	0.9	3.6	5.9	1.8	0.2	22.3	1.9	2.6	17.0	12.3	5.1	4.8	9.2	7.1
				Oolong tea	A	9.1	4.8	6.4	2.1	5.9	1.9	4.2	5.9	2.9	22.6	9.4	11.6	12.2	12.1	14.7	4.9	3.8	6.2	7.0
					B	7.4	4.8	19.8	10.9	12.0	13.2	4.0	8.1	5.4	2.8	5.4	6.8	5.0	31.2	8.5	7.2	16.8	5.5	1.6
151	Dimethyl-phthalate	TPT	GC-MS	Green tea	A	3.3	31.5	1.6	1.1	8.9	4.2	5.3	4.6	2.7	3.4	4.5	1.8	6.1	0.5	0.5	1.5	3.5	4.4	3.3
					B	5.8	32.6	5.4	6.6	2.9	3.6	2.9	12.2	4.7	5.2	2.0	4.9	2.1	0.3	4.4	1.5	5.8	2.3	5.7
				Oolong tea	A	4.4	1.4	1.0	3.3	2.0	1.3	8.9	3.3	6.0	2.1	2.3	2.1	4.0	2.4	8.0	3.6	4.1	2.3	1.4
					B	4.3	3.0	3.9	6.0	3.1	4.5	5.0	1.3	3.6	5.1	0.2	0.7	3.6	2.9	4.9	0.3	1.0	1.6	3.8
			GC-MS/MS	Green tea	A	3.9	5.1	2.2	1.6	11.5	5.2	5.2	5.4	4.2	1.8	5.3	1.7	8.9	4.4	3.1	2.2	5.9	3.5	2.0
					B	5.4	8.0	4.9	7.0	4.1	3.2	3.8	6.3	5.8	7.1	3.0	9.4	8.0	2.2	4.1	2.7	4.9	5.4	7.0
				Oolong tea	A	5.1	2.3	1.7	4.9	5.0	0.8	9.5	0.9	10.0	3.3	1.3	3.3	6.4	0.7	9.0	5.0	0.9	0.5	4.9
					B	4.7	4.0	4.1	6.9	2.5	3.9	4.4	4.1	6.1	3.9	4.1	1.5	13.6	1.6	9.7	2.8	1.7	0.2	2.6
		ENVI-Carb+PSA	GC-MS	Green tea	A	5.8	4.9	3.1	3.9	5.4	1.7	8.2	4.3	5.1	3.7	5.8	1.9	2.3	2.3	3.3	7.4	11.4	3.9	3.9
					B	2.7	5.5	4.5	4.5	1.4	2.5	6.2	1.9	2.5	1.9	19.6	1.6	8.5	1.5	18.3	2.7	4.6	8.6	5.0
				Oolong tea	A	3.9	3.4	1.3	1.6	5.1	1.2	1.7	5.0	10.0	2.2	4.0	3.9	0.9	1.8	4.6	2.9	4.6	4.8	3.9
					B	6.3	3.0	7.1	3.9	10.4	4.2	0.9	6.6	7.5	2.3	5.7	4.9	1.3	2.7	7.9	2.3	4.1	2.4	2.8
			GC-MS/MS	Green tea	A	8.0	8.1	1.6	7.3	3.4	2.5	4.7	3.4	3.1	1.9	5.1	2.9	2.2	3.5	3.2	6.5	8.7	5.1	2.8
					B	5.0	6.4	5.9	2.9	1.9	1.3	5.4	3.1	3.5	1.0	21.6	1.1	10.9	1.8	21.0	3.3	2.9	9.4	4.2
				Oolong tea	A	3.2	3.2	3.6	3.5	4.2	1.8	1.5	7.4	7.9	2.7	4.8	4.3	1.3	9.7	5.4	4.2	4.9	4.1	4.9
					B	7.2	4.9	8.1	7.1	7.9	4.6	2.6	7.0	6.4	1.1	5.6	3.3	8.6	3.5	11.2	1.5	2.8	2.9	2.6
152	Dinicon-azole	TPT	GC-MS	Green tea	A	5.3	4.7	0.3	0.5	5.0	6.1	8.2	6.9	2.0	7.8	12.8	7.8	8.7	5.6	4.4	1.4	4.6	6.1	3.9
					B	5.5	7.8	3.3	7.8	8.4	5.7	1.3	6.5	3.5	5.2	7.8	8.9	4.7	5.7	6.8	7.4	9.3	3.4	5.0
				Oolong tea	A	7.4	6.7	8.2	9.1	1.7	1.8	2.6	14.8	12.7	9.2	11.3	11.7	3.3	7.5	7.8	1.2	2.9	4.7	4.3
					B	4.5	9.2	10.9	10.6	4.0	3.7	6.3	3.0	3.2	7.8	9.6	2.5	2.8	7.9	3.6	4.3	5.3	5.6	2.8
			GC-MS/MS	Green tea	A	4.4	6.6	0.4	7.7	9.1	3.0	8.5	4.9	4.0	2.2	4.8	4.0	6.6	3.0	4.3	0.5	6.4	4.4	2.8
					B	4.7	9.3	6.4	7.6	3.5	5.3	1.9	6.0	6.1	9.2	3.3	9.5	6.4	0.4	4.8	5.0	4.7	2.7	6.1
				Oolong tea	A	6.5	2.1	0.7	3.7	1.6	3.5	8.4	1.8	6.1	4.8	0.9	4.2	6.6	0.2	5.9	3.3	1.4	5.4	3.3
					B	5.3	2.1	3.8	10.1	2.9	0.9	4.1	7.8	1.9	2.9	5.2	3.1	0.8	3.5	7.6	2.3	2.2	1.7	4.2
		ENVI-Carb+PSA	GC-MS	Green tea	A	10.6	5.8	4.8	6.8	9.2	9.9	8.7	1.6	10.2	5.3	8.2	7.8	3.4	9.2	13.7	10.9	11.0	9.9	12.1
					B	4.0	7.6	5.5	6.9	2.2	1.1	3.4	4.7	5.9	5.9	41.6	12.0	6.7	2.8	20.8	9.4	6.1	10.1	10.9
				Oolong tea	A	4.3	2.5	2.2	6.8	7.5	4.3	3.3	12.2	14.1	2.8	6.5	3.8	1.1	3.3	1.0	0.9	5.8	1.9	8.6
					B	3.5	3.9	16.7	6.9	14.3	1.8	2.7	6.1	5.0	6.9	10.9	7.2	1.0	4.0	8.2	8.5	5.4	2.4	4.2
			GC-MS/MS	Green tea	A	7.8	4.6	6.2	8.2	7.3	1.8	7.9	3.5	5.3	3.7	3.7	7.0	4.7	4.7	6.1	6.3	10.2	5.2	2.9
					B	4.4	8.3	4.8	5.2	3.1	1.7	2.6	4.4	0.5	2.2	23.5	2.9	6.0	7.8	14.8	1.5	1.4	11.4	7.1
				Oolong tea	A	3.9	3.6	5.5	3.3	3.9	5.3	2.5	5.1	5.3	4.3	7.7	3.0	4.3	2.9	12.1	3.6	2.2	5.3	3.3
					B	5.4	3.7	13.1	4.3	12.8	5.3	2.3	7.6	8.8	2.2	6.7	4.8	5.1	9.3	9.8	1.6	1.7	4.2	4.2

#	Compound	Sorbent	Method	Tea																				
153	Diphenamid	TPT	GC–MS	Green tea	A	5.9	6.3	2.3	1.1	14.6	4.5	13.3	10.0	12.5	4.4	6.2	0.7	7.9	2.5	1.9	4.7	5.1	6.2	4.0
					B	3.5	5.9	7.7	8.9	11.1	9.9	5.9	3.6	6.3	6.4	6.1	12.1	3.8	4.9	5.4	3.6	5.3	1.3	5.6
				Oolong tea	A	4.2	1.8	1.6	7.9	2.3	11.2	44.5	10.9	3.8	7.3	7.3	0.3	4.7	1.8	8.9	3.8	4.2	3.3	1.1
					B	4.5	3.1	6.4	13.1	3.7	5.1	6.8	4.8	5.5	1.6	5.8	1.5	3.7	2.7	3.6	1.0	5.4	0.8	2.0
			GC–MS/MS	Green tea	A	5.7	3.0	1.9	3.1	10.0	4.7	7.8	3.2	5.8	3.2	3.3	1.3	13.7	2.3	2.3	5.4	10.4	2.7	2.9
					B	5.1	10.5	6.7	7.4	2.8	3.8	4.3	7.8	7.0	7.5	5.1	5.4	5.7	1.7	7.0	2.6	6.4	5.9	5.3
				Oolong tea	A	5.9	6.7	1.6	4.1	5.3	2.8	7.9	3.2	3.4	3.9	3.0	1.2	7.0	1.6	9.2	4.4	1.6	0.8	7.7
					B	5.2	4.9	3.8	6.8	1.3	2.5	5.5	8.2	8.8	3.1	6.3	4.3	1.6	2.8	4.8	0.3	4.1	2.7	4.7
		ENVI-Carb +PSA	GC–MS	Green tea	A	6.8	14.1	2.2	3.4	8.5	6.1	14.9	6.9	22.0	3.5	6.6	1.4	1.4	5.1	11.4	7.4	11.2	3.1	6.6
					B	5.4	12.9	2.9	2.3	5.5	16.5	3.6	2.8	17.5	0.4	25.5	7.5	1.4	3.5	13.1	1.8	4.3	8.7	7.7
				Oolong tea	A	4.0	3.3	1.3	2.4	7.8	3.0	6.6	19.4	9.1	4.0	4.8	1.4	1.3	8.0	5.1	0.7	4.8	2.4	4.4
					B	6.2	5.0	12.3	5.8	9.1	4.9	18.0	18.9	18.6	6.7	9.8	4.7	3.3	7.6	11.5	5.1	3.0	7.8	6.4
			GC–MS/MS	Green tea	A	5.6	5.9	1.1	6.9	8.5	1.9	13.6	1.3	1.9	2.8	5.4	10.0	3.7	2.1	8.1	6.6	6.4	1.6	7.3
					B	4.6	5.3	4.6	4.2	1.8	1.1	3.1	0.7	3.9	2.9	20.7	9.7	2.0	12.6	12.9	1.5	2.7	11.5	5.9
				Oolong tea	A	4.5	4.5	4.0	1.3	7.2	0.8	3.3	6.1	4.1	6.6	7.0	9.8	5.4	1.6	10.9	5.4	7.6	4.0	7.7
					B	5.9	1.9	11.6	8.7	7.7	6.1	2.9	6.4	7.9	4.4	8.1	3.8	11.7	8.3	10.3	4.7	0.6	5.3	4.7
154	Dipropetryn	TPT	GC–MS	Green tea	A	4.6	6.3	2.7	4.5	7.2	7.3	7.1	5.1	3.0	4.2	13.7	7.5	7.9	4.3	1.6	4.8	4.5	8.2	4.4
					B	4.8	9.8	6.6	9.0	4.3	3.9	2.4	2.1	3.0	11.1	6.9	11.9	2.0	6.1	7.4	3.2	5.2	0.3	4.6
				Oolong tea	A	5.6	4.9	4.2	3.4	2.9	4.3	9.9	7.6	4.9	19.0	3.3	6.6	3.9	3.0	14.7	5.4	4.0	4.8	1.8
					B	4.5	2.8	8.2	8.3	4.5	5.1	3.0	1.8	7.7	5.6	8.7	1.6	2.6	4.1	6.5	0.2	3.1	1.5	1.2
			GC–MS/MS	Green tea	A	4.4	6.8	1.1	5.3	9.4	3.2	6.9	4.6	3.1	2.5	5.4	2.9	11.4	2.9	5.1	2.6	5.1	3.3	5.0
					B	4.7	10.1	7.2	9.6	3.0	3.9	4.9	3.8	6.5	10.7	4.7	9.0	2.8	3.8	5.6	3.8	4.1	4.7	4.0
				Oolong tea	A	4.7	3.4	3.5	2.2	3.4	4.3	7.9	2.9	4.4	4.0	0.9	4.7	9.5	3.3	9.7	4.3	4.1	3.4	3.6
					B	6.4	5.5	6.1	7.1	3.5	3.8	6.0	5.6	4.8	2.8	11.6	4.1	2.1	5.1	7.8	4.8	5.7	4.5	6.4
		ENVI-Carb +PSA	GC–MS	Green tea	A	6.0	4.5	1.9	6.5	8.0	1.7	5.3	3.5	5.8	4.2	8.3	5.2	1.1	4.6	9.2	9.4	10.3	3.3	6.9
					B	5.9	7.8	2.9	3.7	4.1	8.1	1.9	4.0	3.7	5.4	28.7	12.2	1.1	2.3	11.0	6.3	5.3	8.5	7.2
				Oolong tea	A	9.0	6.0	4.6	5.2	10.9	2.9	3.7	4.2	2.2	7.2	7.1	21.0	20.0	4.2	9.4	2.9	4.5	19.0	21.4
					B	12.7	3.6	8.4	11.8	11.1	1.7	2.0	7.4	4.5	17.3	7.8	2.9	7.4	7.5	7.7	8.3	5.3	20.2	8.1
			GC–MS/MS	Green tea	A	7.1	3.7	3.0	6.4	7.5	2.4	6.6	4.6	4.1	2.2	10.7	14.8	3.7	4.0	9.0	7.0	10.7	3.1	5.4
					B	3.6	10.8	5.1	5.5	1.8	1.9	3.1	8.8	4.6	2.0	21.4	16.3	5.0	11.9	14.8	0.9	1.7	5.7	7.0
				Oolong tea	A	3.9	2.4	10.0	10.5	3.1	1.0	3.8	0.5	0.7	0.3	9.2	9.5	11.0	12.7	13.0	3.5	7.6	6.2	3.6
					B	5.4	1.8	13.8	17.6	7.4	4.9	4.9	7.4	8.3	16.8	4.3	5.0	5.8	16.3	15.4	2.8	4.9	2.6	6.4
155	Ethalfluralin	TPT	GC–MS	Green tea	A	3.4	5.4	3.8	5.3	13.6	2.6	8.4	14.2	15.3	2.5	6.0	6.4	11.1	8.6	4.3	4.1	6.0	2.6	1.9
					B	5.7	7.5	2.8	9.5	5.5	4.9	2.9	7.2	1.9	9.2	1.4	9.4	7.7	8.8	0.4	4.5	11.7	10.7	8.0
				Oolong tea	A	4.1	1.7	3.6	4.1	13.4	6.1	6.8	5.6	4.9	4.9	7.3	4.9	3.2	1.4	5.0	9.2	3.5	7.3	0.4
					B	2.8	3.6	3.2	4.0	9.0	3.7	4.8	4.4	6.9	1.0	5.9	4.5	13.6	5.4	10.4	1.1	2.4	2.8	3.6
			GC–MS/MS	Green tea	A	3.5	4.7	2.4	37.8	15.1	4.2	9.2	12.5	13.5	2.0	5.4	1.3	3.7	7.8	1.4	4.7	10.0	1.8	3.6
					B	5.5	7.8	5.0	10.0	2.4	2.1	3.0	11.1	6.9	11.0	3.0	4.7	9.0	4.9	2.6	4.5	11.4	6.8	10.0
				Oolong tea	A	7.0	1.0	4.9	3.3	8.3	3.3	6.8	4.5	9.7	7.6	7.6	5.0	7.8	5.0	9.2	3.0	3.5	0.4	1.4
					B	8.9	4.1	3.5	8.4	5.5	5.4	7.9	7.5	9.6	0.6	7.7	3.1	2.1	3.6	6.4	1.9	5.9	6.3	2.3
		ENVI-Carb +PSA	GC–MS	Green tea	A	14.6	6.4	12.1	5.9	8.3	23.4	7.8	10.6	5.1	0.9	4.9	14.6	2.1	9.3	5.7	1.9	13.0	11.7	7.4
					B	13.0	9.7	4.0	3.2	8.5	32.4	11.8	14.8	8.1	5.0	15.0	8.1	21.4	4.2	16.4	10.7	6.9	11.9	8.3
				Oolong tea	A	6.9	4.5	4.6	6.2	5.1	4.1	3.1	19.0	1.4	4.6	3.8	3.6	7.7	2.0	1.4	4.3	8.9	7.0	0.7
					B	6.8	4.3	7.7	3.4	19.3	10.5	7.0	8.1	11.2	2.1	9.4	0.9	6.3	30.4	13.4	20.1	12.6	9.0	5.5
			GC–MS/MS	Green tea	A	13.3	7.5	7.7	6.1	5.2	16.8	10.5	15.4	11.0	2.5	0.9	6.2	2.4	6.3	5.8	5.2	12.9	7.6	3.2
					B	16.6	11.3	5.2	5.3	5.5	15.3	14.7	13.5	10.0	4.7	14.5	4.0	17.8	13.5	9.8	2.4	6.5	19.6	6.5
				Oolong tea	A	7.4	2.9	5.0	6.0	2.6	5.3	1.3	20.7	2.2	1.6	7.6	7.2	4.3	2.1	7.7	10.1	8.8	20.1	1.4
					B	10.7	3.4	9.3	3.1	10.9	10.7	5.6	5.6	9.1	2.4	9.6	5.0	9.0	8.7	9.2	1.2	9.0	31.7	2.3

(Continued)

Appendix Table 5.2 RSD of 201 Pesticides in Youden Pair Aged Tea Sample for Parallel Determinations (Nov 9, 2009–Feb 7, 2010) (cont.)

No.	Pesticides	SPE	Method	Sample	Youden pair	Nov 9, 2009 (n=5)	Nov 14, 2009 (n=3)	Nov 19, 2009 (n=3)	Nov 24, 2009 (n=3)	Nov 29, 2009 (n=3)	Dec 4, 2009 (n=3)	Dec 9, 2009 (n=3)	Dec 14, 2009 (n=3)	Dec 19, 2009 (n=3)	Dec 24, 2009 (n=3)	Dec 29, 2009 (n=3)	Jan 3, 2010 (n=3)	Jan 8, 2010 (n=3)	Jan 13, 2010 (n=3)	Jan 18, 2010 (n=3)	Jan 23, 2010 (n=3)	Jan 28, 2010 (n=3)	Feb 2, 2010 (n=3)	Feb 7, 2010 (n=3)
						RSD%																		
156	Etofenprox	TPT	GC–MS	Green tea	A	8.1	5.6	2.9	6.4	9.2	8.9	9.3	6.0	5.9	0.7	10.5	3.9	7.6	4.2	4.2	1.9	2.3	7.6	1.2
				Green tea	B	5.7	9.1	6.6	6.9	3.6	6.2	0.4	11.0	2.2	11.6	4.7	9.0	6.4	1.5	7.6	4.1	7.7	7.3	6.8
				Oolong tea	A	7.0	3.6	2.7	3.8	2.8	5.5	3.9	5.1	2.0	3.4	1.5	2.7	4.0	2.4	3.7	4.6	3.5	6.4	1.9
				Oolong tea	B	2.6	2.5	3.4	11.1	0.9	5.9	3.8	6.5	6.6	3.8	3.4	2.1	4.7	6.1	13.5	3.6	0.3	2.9	7.1
			GC–MS/MS	Green tea	A	5.4	5.9	4.2	6.6	10.1	8.5	7.3	6.2	5.6	1.4	12.6	1.9	10.9	8.4	3.4	5.0	6.4	9.1	3.4
				Green tea	B	4.1	8.2	7.8	5.7	5.4	7.1	1.5	8.5	2.8	11.4	6.0	10.0	1.2	6.5	5.6	1.4	6.4	5.4	7.3
				Oolong tea	A	6.2	3.4	1.0	2.5	3.7	4.4	5.4	1.4	0.8	3.8	2.6	4.9	7.4	2.5	4.5	4.3	3.7	6.1	6.8
				Oolong tea	B	4.1	4.0	5.3	6.4	1.4	3.2	5.9	5.8	11.8	3.7	3.2	2.9	10.7	1.9	4.0	4.2	3.1	1.0	1.5
		ENVI-Carb +PSA	GC–MS	Green tea	A	10.3	7.5	1.9	6.7	5.7	3.9	4.3	2.7	8.2	1.5	8.5	2.3	1.4	8.1	5.7	7.6	4.8	4.4	4.8
				Green tea	B	10.6	11.1	3.1	2.8	4.6	7.8	3.3	6.3	9.3	2.9	21.8	7.4	2.9	8.5	14.7	3.3	8.4	4.8	11.5
				Oolong tea	A	5.1	5.4	6.1	6.1	5.6	5.2	3.0	4.6	0.6	4.4	4.6	1.9	6.7	5.1	18.3	7.1	5.9	3.0	9.9
				Oolong tea	B	5.0	6.6	11.5	2.4	8.0	6.8	3.0	7.2	10.9	3.7	4.7	2.7	1.8	3.8	14.2	3.8	8.0	2.6	2.6
			GC–MS/MS	Green tea	A	8.7	7.3	4.6	7.9	7.0	3.5	4.0	7.9	11.2	3.0	7.8	1.6	2.8	26.1	29.8	11.7	12.9	7.2	6.0
				Green tea	B	3.5	1.3	6.5	5.4	5.9	13.1	1.5	6.8	1.5	5.3	26.0	1.0	28.6	12.7	32.4	2.0	8.3	7.9	4.7
				Oolong tea	A	7.8	7.3	5.4	16.1	19.7	1.7	3.1	7.2	0.4	3.0	7.5	2.9	12.0	5.5	6.5	24.1	5.7	20.1	21.5
				Oolong tea	B	16.8	9.7	9.1	13.9	4.9	4.8	8.5	8.5	8.7	1.9	4.3	2.8	8.5	4.5	7.5	6.0	5.2	31.0	1.5
157	Etridiazol	TPT	GC–MS	Green tea	A	3.4	6.4	3.3	9.2	13.2	8.7	8.5	9.2	9.9	13.7	5.9	6.4	10.9	9.1	5.4	4.8	9.2	9.5	2.0
				Green tea	B	8.1	6.4	4.1	7.4	6.3	7.3	3.3	6.9	3.9	2.0	3.8	6.1	6.3	6.1	3.4	1.7	2.2	2.7	7.9
				Oolong tea	A	7.1	4.4	3.5	2.4	11.6	5.3	2.4	0.4	12.0	6.3	8.7	7.9	3.3	8.1	8.1	7.4	4.3	5.6	4.7
				Oolong tea	B	8.5	6.5	8.0	2.4	8.6	9.9	2.8	8.5	9.0	8.6	8.7	8.3	4.5	5.7	13.8	3.7	1.6	3.4	10.2
			GC–MS/MS	Green tea	A	6.3	2.8	3.7	13.4	15.7	5.8	12.5	7.4	7.4	12.0	5.1	3.0	10.1	5.7	0.7	8.9	12.1	9.0	5.7
				Green tea	B	7.8	6.5	2.6	9.3	4.0	4.0	2.6	9.1	11.9	5.4	3.5	10.8	16.5	3.6	4.1	4.5	10.4	6.1	7.1
				Oolong tea	A	7.7	4.9	1.5	3.7	3.6	1.3	6.0	10.0	9.8	4.1	9.1	10.3	4.6	2.2	13.2	7.9	1.1	5.8	2.8
				Oolong tea	B	13.1	5.4	9.7	0.8	2.6	10.9	9.8	10.6	16.7	1.6	9.3	3.7	11.2	6.6	2.2	3.4	1.5	4.2	5.5
		ENVI-Carb +PSA	GC–MS	Green tea	A	11.3	3.7	9.4	7.2	5.0	9.5	5.0	3.0	13.4	4.1	7.8	12.5	6.4	10.1	1.5	12.9	10.3	15.4	2.9
				Green tea	B	14.2	3.7	3.6	9.6	7.2	8.3	4.5	5.1	4.1	1.6	26.1	9.8	28.5	1.6	34.8	3.8	9.1	8.2	5.4
				Oolong tea	A	12.7	10.1	3.6	5.4	3.8	3.4	2.2	16.6	3.3	4.1	3.5	6.9	8.6	5.9	4.1	6.1	3.9	14.3	9.2
				Oolong tea	B	15.3	10.6	6.0	7.3	13.9	6.6	7.7	1.6	17.3	3.3	10.0	5.0	7.4	12.0	10.6	0.2	7.1	18.1	4.3
			GC–MS/MS	Green tea	A	9.9	6.4	5.1	5.3	7.7	9.6	13.7	4.1	15.6	3.8	5.9	2.5	2.4	7.3	1.8	20.5	15.3	3.4	6.4
				Green tea	B	21.5	4.0	5.2	9.2	5.1	9.5	10.9	6.9	7.5	1.2	25.3	10.8	37.6	13.9	39.9	1.8	5.2	13.8	12.0
				Oolong tea	A	13.1	8.7	8.8	12.9	2.3	6.5	0.1	19.8	3.6	4.6	1.7	8.9	33.7	6.4	11.0	13.5	10.9	16.2	2.8
				Oolong tea	B	16.5	9.9	6.7	6.1	14.9	10.0	12.8	6.1	14.7	11.7	9.6	9.3	15.7	9.7	13.1	7.3	8.6	25.7	5.5
158	Fenaza-quin	TPT	GC–MS	Green tea	A	3.7	4.7	1.7	5.1	6.7	6.1	9.1	4.2	5.8	0.6	6.8	5.5	8.6	5.8	4.7	1.9	10.0	11.3	6.5
				Green tea	B	6.0	8.4	5.1	10.7	3.6	3.8	3.7	2.1	5.6	9.3	8.2	9.4	1.5	2.7	7.6	6.3	5.7	7.7	6.0
				Oolong tea	A	6.9	3.0	2.2	2.2	3.7	3.2	10.2	5.5	3.3	7.6	8.0	4.2	5.2	7.4	6.2	6.8	1.2	2.5	0.6
				Oolong tea	B	4.1	3.8	1.4	9.8	3.9	5.0	5.3	0.8	4.5	7.2	9.7	1.2	2.4	4.0	2.6	6.7	2.2	3.1	1.0
			GC–MS/MS	Green tea	A	7.6	6.8	0.9	7.5	9.9	5.5	7.0	4.3	4.8	2.0	3.8	4.2	5.2	6.3	3.9	1.5	5.4	11.9	3.4
				Green tea	B	5.3	10.7	4.3	8.9	3.7	2.7	2.6	2.6	6.2	10.8	4.4	14.3	9.0	6.0	3.6	4.6	4.2	5.3	6.0
				Oolong tea	A	6.2	3.4	0.8	1.4	3.7	3.4	9.0	1.6	2.7	3.8	2.3	3.9	5.4	1.8	5.9	5.5	2.9	4.9	13.0
				Oolong tea	B	6.0	4.4	3.7	8.0	0.4	4.0	4.0	7.3	8.1	2.9	3.6	3.3	6.4	4.1	5.2	2.9	2.1	1.8	3.4
		ENVI-Carb +PSA	GC–MS	Green tea	A	3.8	8.6	2.4	3.5	8.3	10.9	7.2	3.9	6.3	5.2	12.3	9.8	5.5	6.9	3.5	10.9	6.1	2.7	2.3
				Green tea	B	4.6	12.6	2.9	5.4	5.4	2.1	2.7	6.5	8.6	1.6	30.1	12.4	3.4	6.9	6.6	3.9	15.7	7.6	7.6
				Oolong tea	A	5.7	1.9	3.4	2.7	2.5	2.9	2.3	8.4	4.7	4.8	8.3	3.8	4.8	8.1	7.3	10.9	4.9	5.2	4.3
				Oolong tea	B	5.7	5.3	7.9	5.7	8.3	2.4	4.8	7.4	11.8	4.4	14.9	1.4	2.6	7.2	11.6	12.2	5.0	2.3	2.8
			GC–MS/MS	Green tea	A	6.5	1.0	3.1	13.2	8.9	2.7	3.2	3.6	6.3	0.2	3.3	2.1	8.2	12.1	3.7	4.7	9.4	4.5	5.6
				Green tea	B	4.6	5.7	4.5	4.9	0.4	1.4	1.2	7.8	3.3	3.4	17.3	5.4	2.4	16.0	11.4	3.3	3.6	8.5	7.4
				Oolong tea	A	6.5	2.6	4.8	0.5	5.4	4.6	2.3	7.8	0.8	4.5	8.7	1.4	23.2	6.0	8.0	3.2	5.2	12.2	6.2
				Oolong tea	B	5.8	4.4	10.9	6.9	10.6	4.7	3.0	6.5	8.9	1.1	7.2	1.6	6.1	7.5	10.3	2.1	3.4	2.8	3.4

159	Fenchlorphos	TPT	GC-MS	Green tea	A	4.3	5.4	4.9	2.0	7.3	3.7	2.8	7.3	4.4	3.1	6.5	3.2	7.4	4.3	3.8	3.2	4.9	4.2	3.3
					B	6.7	6.5	5.2	6.4	4.7	4.4	2.7	4.5	3.7	9.7	0.8	8.6	3.2	1.3	4.1	1.0	3.0	2.7	9.2
				Oolong tea	A	5.7	2.9	4.1	3.1	2.6	2.4	9.2	2.9	4.7	3.1	4.2	2.8	1.4	5.6	8.1	3.2	4.1	1.6	0.5
					B	3.8	2.0	4.0	4.2	2.6	3.3	1.4	3.4	7.2	3.2	6.1	4.0	1.3	4.1	5.8	1.0	2.0	1.5	3.6
			GC–MS/MS	Green tea	A	5.1	3.4	3.1	12.0	10.3	1.0	4.9	7.2	5.7	8.4	7.9	7.3	10.7	5.3	3.8	1.9	6.3	3.3	3.8
					B	5.6	7.6	3.5	8.4	4.8	4.4	4.4	3.1	10.7	12.1	2.1	8.5	8.5	5.1	4.7	3.0	5.2	2.4	5.7
				Oolong tea	A	4.0	3.8	2.3	4.9	6.7	1.7	10.8	4.2	6.4	2.9	2.6	3.5	4.4	2.2	10.3	1.4	5.7	1.5	6.8
					B	6.6	5.4	5.3	6.7	1.7	2.4	3.8	7.0	9.5	5.9	5.8	2.2	13.6	2.2	9.8	4.8	1.0	1.8	2.6
		ENVI-Carb+PSA	GC-MS	Green tea	A	9.1	3.8	1.6	6.6	8.7	9.6	5.7	1.3	5.1	2.6	6.5	2.2	4.5	2.3	5.0	4.0	9.4	11.4	6.9
					B	13.2	9.5	1.4	0.5	1.1	0.2	3.9	0.9	6.3	6.7	17.6	3.5	5.0	2.2	5.8	3.4	4.4	9.1	5.3
				Oolong tea	A	4.9	3.3	4.7	2.6	5.3	0.3	2.8	3.0	4.0	0.7	8.2	3.5	3.1	3.4	4.5	2.4	3.8	6.5	2.5
					B	5.6	3.1	7.2	3.6	9.5	3.7	1.8	4.2	10.2	3.6	8.1	2.9	5.1	8.3	10.2	2.6	6.4	5.4	4.5
			GC–MS/MS	Green tea	A	11.6	6.9	0.9	6.3	7.4	10.8	7.5	2.9	5.4	2.4	11.0	5.1	5.2	3.7	5.5	1.3	5.2	4.0	3.5
					B	16.3	8.7	1.9	4.4	2.3	3.4	6.1	3.1	7.3	9.8	11.6	0.8	3.3	13.2	7.2	9.4	3.4	15.3	6.1
				Oolong tea	A	3.5	2.8	7.4	8.6	4.6	2.8	3.1	7.2	5.5	4.1	0.8	10.3	4.4	1.1	13.0	1.5	7.5	5.8	6.8
					B	7.8	3.5	10.7	6.6	7.9	4.2	3.1	5.2	10.5	3.6	6.6	11.1	9.7	6.7	10.1	3.1	5.0	7.7	2.6
160	Fenoxanil	TPT	GC-MS	Green tea	A	10.5	6.1	1.4	3.7	3.2	4.9	7.4	4.5	4.8	4.4	7.9	0.6	9.8	7.9	4.6	4.0	5.1	6.3	3.9
					B	7.0	8.0	9.2	10.4	3.6	4.6	3.3	6.6	3.3	12.8	7.1	10.3	9.0	2.0	9.9	2.7	5.2	1.7	2.9
				Oolong tea	A	7.0	7.6	5.3	3.3	2.1	3.8	7.4	11.4	5.2	8.2	3.3	1.1	5.0	1.4	4.5	5.4	2.7	17.6	2.2
					B	16.7	4.3	4.7	4.3	2.1	5.8	3.6	8.7	1.9	5.2	0.1	4.6	0.8	8.8	0.4	2.3	3.1	11.2	0.8
			GC–MS/MS	Green tea	A	9.2	4.0	3.0	10.3	6.2	1.9	7.5	5.3	2.7	16.3	6.1	2.9	24.7	9.0	6.9	8.4	6.0	7.4	3.6
					B	12.1	16.0	9.4	9.0	7.1	7.4	3.2	15.0	3.6	7.0	6.4	9.3	5.5	11.5	9.5	0.8	7.0	6.0	4.3
				Oolong tea	A	7.5	12.2	4.7	2.7	6.8	6.5	7.9	7.3	4.1	6.1	3.3	3.4	6.1	3.2	9.0	7.6	4.0	10.0	7.3
					B	8.5	7.0	2.8	11.7	1.2	5.0	9.2	7.7	10.2	5.3	6.2	3.0	9.0	2.6	2.0	7.9	2.9	4.1	4.6
		ENVI-Carb+PSA	GC-MS	Green tea	A	6.5	9.9	0.8	8.7	7.1	8.9	13.1	2.6	7.9	9.5	7.2	4.5	6.9	4.1	8.3	11.4	11.0	0.7	5.9
					B	5.2	4.5	1.9	5.8	9.0	2.5	1.2	2.5	5.8	10.3	28.1	14.2	11.6	4.1	15.7	2.1	2.7	9.8	8.4
				Oolong tea	A	12.5	6.2	7.0	4.5	7.0	2.8	3.2	3.3	2.7	4.3	2.6	4.4	5.9	4.7	4.3	4.8	5.4	2.3	7.0
					B	10.5	5.9	14.1	7.3	11.0	4.1	5.2	9.6	4.8	11.8	5.2	2.8	4.1	6.9	9.2	4.4	4.2	10.3	2.4
			GC–MS/MS	Green tea	A	7.5	17.2	18.5	12.1	16.8	5.8	12.5	8.4	2.6	4.1	5.4	22.9	3.7	5.5	4.7	11.6	5.6	3.1	5.4
					B	12.5	18.8	7.7	13.5	12.9	13.0	1.9	19.7	9.9	1.4	24.6	12.5	7.5	8.1	14.1	4.8	4.4	10.0	4.7
				Oolong tea	A	6.4	10.8	2.9	6.8	7.8	3.7	5.5	5.3	1.1	4.3	9.3	5.9	4.4	4.0	14.9	8.0	2.5	2.9	7.3
					B	5.9	4.9	14.7	10.8	7.2	3.4	10.4	7.9	7.6	15.9	4.0	1.1	3.8	12.1	11.9	6.0	5.4	6.9	4.7
161	Fenpropidin	TPT	GC-MS	Green tea	A	5.2	5.0	0.7	4.8	12.9	5.2	8.9	7.6	6.2	4.4	10.3	0.4	11.0	4.1	1.1	5.3	4.6	6.9	4.8
					B	6.4	9.5	5.6	9.0	4.0	1.4	5.2	13.8	12.4	14.3	9.3	11.7	0.3	4.0	9.8	5.4	8.1	1.4	4.1
				Oolong tea	A	6.0	2.1	6.4	4.8	2.0	5.8	16.2	12.8	5.9	6.6	23.8	8.3	9.4	4.8	11.9	7.0	6.2	13.0	3.8
					B	6.5	4.0	10.4	3.9	4.1	3.4	6.3	12.5	2.7	9.7	0.3	1.3	2.7	0.9	1.8	0.5	0.3	3.6	2.1
			GC–MS/MS	Green tea	A	9.2	5.7	1.8	5.4	14.1	4.0	7.8	6.1	4.4	5.3	7.0	0.7	17.2	16.1	3.6	4.2	9.2	5.5	1.7
					B	6.4	10.2	8.4	8.2	3.5	4.2	2.6	11.2	4.7	15.6	8.8	14.4	2.9	6.6	10.6	3.5	4.4	3.1	5.8
				Oolong tea	A	8.8	2.7	2.9	6.1	8.9	4.6	18.7	9.3	5.3	9.3	11.8	4.4	14.8	2.5	12.3	7.1	6.5	12.4	7.5
					B	13.3	4.6	6.8	10.3	3.4	4.4	6.0	4.7	6.7	3.1	4.4	3.4	8.7	1.7	4.2	1.8	1.4	2.5	3.7
		ENVI-Carb+PSA	GC-MS	Green tea	A	11.3	22.8	8.4	9.9	11.3	5.3	40.0	9.7	25.9	7.3	7.6	6.3	2.1	9.5	8.1	8.5	12.0	3.2	7.6
					B	4.7	6.6	4.4	12.4	3.4	2.3	8.9	7.2	11.7	1.5	29.7	10.4	3.8	4.1	13.3	2.5	3.9	8.1	8.9
				Oolong tea	A	6.5	4.8	5.8	2.2	8.3	4.9	1.4	3.8	#DIV/0!	13.7	8.4	17.0	5.8	7.5	7.3	6.5	5.4	1.2	6.2
					B	7.5	1.4	16.9	6.1	15.2	8.5	2.9	11.7	#DIV/0!	10.0	4.1	2.4	1.3	10.4	7.5	4.1	3.2	8.1	3.7
			GC–MS/MS	Green tea	A	12.1	27.7	4.9	9.6	5.6	4.4	32.7	4.7	3.0	2.6	7.7	8.0	2.4	12.2	8.7	8.3	9.8	3.7	6.5
					B	6.5	4.6	5.0	5.2	4.4	1.5	2.3	8.1	3.2	0.3	24.4	16.3	7.8	20.7	15.7	4.3	2.0	9.6	8.9
				Oolong tea	A	5.8	6.0	4.4	6.0	8.2	0.7	7.2	0.7	2.7	14.2	10.5	13.5	5.1	10.6	15.8	10.5	4.0	7.2	7.5
					B	7.1	1.1	11.6	8.7	8.9	8.4	4.9	10.9	6.7	4.2	3.8	3.9	3.1	11.4	7.8	2.4	2.2	3.8	3.7

(Continued)

Appendix Table 5.2 RSD of 201 Pesticides in Youden Pair Aged Tea Sample for Parallel Determinations (Nov 9, 2009–Feb 7, 2010) (cont.)

No.	Pesticides	SPE	Method	Sample	Youden pair	RSD% Nov 9, 2009 (n=5)	Nov 14, 2009 (n=3)	Nov 19, 2009 (n=3)	Nov 24, 2009 (n=3)	Nov 29, 2009 (n=3)	Dec 4, 2009 (n=3)	Dec 9, 2009 (n=3)	Dec 14, 2009 (n=3)	Dec 19, 2009 (n=3)	Dec 24, 2009 (n=3)	Dec 14, 2009 (n=3)	Jan 3, 2010 (n=3)	Jan 8, 2010 (n=3)	Jan 13, 2010 (n=3)	Jan 18, 2010 (n=3)	Jan 23, 2010 (n=3)	Jan 28, 2010 (n=3)	Feb 2, 2010 (n=3)	Feb 7, 2010 (n=3)
162	Fenson	TPT	GC-MS	Green tea	A	5.1	7.3	2.4	13.3	13.9	2.1	9.2	12.6	10.7	2.4	3.8	8.0	10.2	11.7	10.4	7.2	8.3	11.8	3.1
					B	5.8	6.3	3.4	10.9	5.0	5.9	3.8	9.5	5.1	6.6	0.9	4.2	6.6	12.4	8.3	7.5	12.2	9.2	4.1
				Oolong tea	A	9.2	3.9	1.0	7.7	4.8	5.4	10.3	14.0	7.4	3.6	3.9	7.6	3.4	4.1	12.5	4.5	7.6	1.4	0.3
					B	2.4	2.7	4.3	4.6	4.1	13.1	5.8	0.9	6.7	5.7	1.7	5.2	4.0	6.7	8.5	3.1	2.2	2.8	12.3
			GC-MS/MS	Green tea	A	8.9	5.6	2.3	7.2	15.0	3.0	10.5	13.8	7.8	2.4	7.5	2.1	11.1	3.8	2.4	5.9	6.0	12.3	7.4
					B	7.2	9.3	4.6	10.6	3.9	3.2	7.0	12.9	5.0	8.0	2.6	7.6	12.6	5.8	1.6	5.9	13.7	6.0	8.2
				Oolong tea	A	9.2	2.2	1.4	5.2	7.7	2.9	1.5	8.6	14.5	4.0	16.7	5.3	5.5	5.1	7.1	3.8	5.4	9.4	1.3
					B	6.4	5.0	6.1	6.8	4.9	8.6	3.9	5.6	12.5	1.9	6.9	2.0	3.8	5.1	11.4	6.5	1.7	4.7	7.2
		ENVI-Carb+PSA	GC-MS	Green tea	A	18.2	2.0	13.2	4.8	8.2	15.2	27.6	17.0	7.7	3.9	3.6	16.1	3.3	13.2	8.9	16.9	14.6	22.9	5.0
					B	21.1	11.0	5.7	6.3	7.2	1.2	3.1	21.9	12.1	9.7	16.2	10.6	21.8	7.8	15.7	7.3	6.1	18.4	15.3
				Oolong tea	A	3.2	3.4	4.4	6.4	0.4	8.1	11.4	5.5	8.9	2.9	4.6	13.6	9.5	1.0	5.2	10.4	10.7	16.7	4.7
					B	9.4	6.5	7.1	2.1	12.0	9.1	14.2	2.6	13.9	3.5	13.1	5.0	5.4	8.2	16.3	3.4	11.5	19.4	10.2
			GC-MS/MS	Green tea	A	24.5	6.9	11.8	7.8	7.7	19.4	17.9	19.6	10.4	0.2	9.4	1.2	6.0	10.2	11.0	4.4	11.9	5.2	4.5
					B	25.0	12.5	4.0	8.0	10.4	7.5	11.7	18.2	15.9	9.0	19.3	0.5	27.0	12.6	11.0	16.2	1.8	21.1	9.7
				Oolong tea	A	11.4	5.2	8.8	7.2	0.4	1.8	4.1	18.5	2.4	3.0	6.9	1.7	2.3	0.9	10.6	15.2	11.9	14.4	1.3
					B	13.6	7.8	7.2	1.9	13.6	9.2	7.4	6.9	11.4	6.0	8.0	8.6	5.6	9.3	14.0	3.2	6.4	17.0	7.2
163	Flufenacet	TPT	GC-MS	Green tea	A	7.1	7.0	5.5	5.7	10.8	8.7	8.0	9.8	10.8	2.6	14.1	7.7	13.1	9.7	15.1	4.1	11.6	8.8	3.6
					B	4.7	1.1	9.5	4.4	5.5	6.5	5.7	12.8	6.8	12.2	3.9	8.6	13.2	13.8	6.2	6.6	9.4	11.7	15.0
				Oolong tea	A	11.0	4.4	9.1	1.4	7.1	1.0	14.4	0.4	11.9	1.1	14.8	5.4	7.1	15.6	4.8	3.6	11.8	13.7	3.2
					B	3.0	0.9	5.4	3.8	5.1	10.7	10.1	12.6	7.3	6.4	9.5	11.6	8.7	12.5	5.3	6.5	2.4	7.1	14.4
			GC-MS/MS	Green tea	A	8.0	7.5	3.0	1.0	16.8	4.0	8.7	11.6	13.0	14.5	4.8	5.1	12.3	10.1	0.4	5.3	11.6	8.1	18.3
					B	9.6	11.4	7.8	10.5	4.7	5.4	5.4	13.2	19.2	13.5	4.0	13.3	25.7	7.7	8.8	9.6	5.3	7.2	3.0
				Oolong tea	A	3.5	7.2	10.0	3.3	7.6	2.5	10.2	22.2	14.3	7.7	4.6	7.3	4.2	3.5	7.7	6.4	7.0	6.3	13.6
					B	5.4	12.1	15.1	1.2	7.4	12.1	9.2	2.6	17.8	9.3	3.8	4.2	1.3	7.8	14.3	6.8	9.4	9.0	9.1
		ENVI-Carb+PSA	GC-MS	Green tea	A	11.9	8.0	3.9	6.3	12.3	13.9	9.0	13.6	13.3	8.2	3.3	5.7	6.7	4.4	9.3	6.5	12.2	32.2	9.6
					B	21.4	10.7	3.2	7.1	9.1	7.0	5.3	14.3	9.0	8.5	16.3	10.7	13.7	2.0	12.3	12.4	6.0	13.9	6.9
				Oolong tea	A	10.6	6.6	9.4	6.1	5.3	3.8	5.1	14.1	8.4	6.3	12.9	3.7	9.9	3.8	12.4	5.7	15.8	14.2	18.8
					B	15.8	12.5	6.4	2.9	11.4	6.2	7.8	5.9	19.3	13.2	11.8	7.9	11.5	10.2	13.6	6.9	17.4	17.4	6.9
			GC-MS/MS	Green tea	A	9.3	7.3	0.7	13.9	15.6	13.7	5.7	7.3	3.7	6.8	3.4	3.7	7.5	19.4	12.0	11.3	14.2	15.2	11.8
					B	23.4	2.9	3.5	10.3	8.7	10.5	12.4	9.2	13.8	21.8	5.5	3.7	10.7	15.3	8.0	7.0	21.2	24.0	19.8
				Oolong tea	A	10.1	5.5	10.4	20.8	6.6	7.7	10.2	16.3	7.3	10.2	2.3	18.2	16.7	13.2	15.2	4.4	14.8	9.1	13.6
					B	18.7	8.8	18.5	7.4	8.5	11.4	10.4	7.8	17.1	4.1	12.4	24.4	13.2	3.1	13.9	5.4	43.0	15.1	9.1
164	Furalaxyl	TPT	GC-MS	Green tea	A	5.4	6.8	1.9	3.6	6.7	4.8	7.3	5.0	5.0	3.7	6.0	3.4	7.6	4.3	2.1	4.0	5.3	6.3	4.3
					B	4.5	8.7	5.4	8.1	3.0	3.5	3.4	8.2	4.6	8.5	4.7	9.1	1.7	3.9	4.8	3.3	5.6	1.2	4.8
				Oolong tea	A	5.8	1.3	2.5	4.9	3.6	4.0	8.4	7.4	3.0	6.1	5.8	1.5	3.8	3.1	8.7	4.4	4.5	3.5	1.4
					B	6.2	2.1	2.6	4.9	2.0	3.9	5.0	3.7	5.4	1.3	4.5	1.9	2.5	2.2	1.0	0.3	1.4	0.8	2.1
			GC-MS/MS	Green tea	A	5.9	4.7	1.9	5.5	8.7	4.8	7.0	4.8	1.9	4.5	5.1	2.3	9.4	4.8	1.4	3.7	7.8	6.2	2.4
					B	4.4	10.3	7.8	7.1	4.3	3.8	4.4	9.1	6.7	11.4	3.5	12.2	7.7	2.0	7.5	5.0	3.3	4.5	6.4
				Oolong tea	A	6.1	2.4	1.7	4.5	5.7	3.2	10.4	1.2	2.7	4.8	0.5	4.3	6.4	1.5	7.2	4.6	2.9	5.2	5.9
					B	7.3	4.8	6.0	7.8	1.6	1.8	5.4	7.5	6.3	3.0	6.6	3.8	9.9	2.0	6.5	2.1	1.9	3.1	6.4
		ENVI-Carb+PSA	GC-MS	Green tea	A	6.3	7.8	2.1	4.5	7.9	1.7	19.3	0.7	8.1	3.0	10.2	2.6	2.5	6.4	6.7	7.0	11.1	2.9	6.5
					B	3.4	5.2	3.0	5.0	2.5	1.1	3.8	4.5	5.4	1.6	24.9	12.2	0.4	2.9	12.8	1.7	3.8	8.8	7.8
				Oolong tea	A	2.7	3.5	3.7	2.0	6.3	3.1	4.7	2.5	1.9	6.6	6.7	6.3	0.9	4.7	5.4	4.2	4.4	1.9	3.4
					B	5.1	3.4	12.3	6.8	11.1	7.7	1.9	5.8	10.2	5.1	9.6	3.1	5.5	12.7	10.0	4.2	3.5	5.7	5.5
			GC-MS/MS	Green tea	A	7.3	6.0	2.4	7.0	6.8	2.0	19.3	2.3	3.0	2.0	9.3	1.9	2.0	4.7	6.2	7.2	6.7	1.8	4.9
					B	4.1	5.1	4.2	3.9	1.6	1.3	2.7	6.3	4.8	1.2	21.4	2.6	3.1	27.2	13.0	1.1	1.1	9.1	6.3
				Oolong tea	A	4.0	5.2	3.5	3.7	6.0	1.9	3.3	2.4	1.3	5.7	7.4	8.5	10.4	2.6	14.4	4.7	5.0	1.5	5.9
					B	5.8	3.7	13.5	8.9	8.4	8.1	2.5	8.1	8.8	3.7	6.4	4.0	6.4	13.1	10.5	4.3	3.3	5.8	6.4

165	Heptachlor	TPT	GC-MS	Green tea	A	4.2	7.1	3.0	1.8	9.4	5.2	6.1	7.6	6.1	4.2	4.5	3.5	6.7	4.2	5.6	0.4	6.8	6.7	3.2
					B	6.2	8.3	4.0	10.8	3.4	7.5	0.3	4.9	3.8	9.4	0.9	8.9	2.7	1.5	3.5	2.3	1.8	2.3	7.3
				Oolong tea	A	3.5	2.3	2.8	3.9	5.0	3.2	5.9	5.3	4.6	2.8	6.6	3.9	1.5	5.7	8.1	4.3	5.0	0.9	1.0
					B	3.0	2.3	4.2	3.3	5.2	5.4	3.2	5.1	5.7	3.0	2.9	2.9	1.3	5.1	5.1	1.6	1.1	2.0	3.7
			GC-MS/MS	Green tea	A	2.5	1.8	3.6	7.9	12.3	2.6	8.4	6.4	5.5	3.1	7.5	1.3	10.0	3.0	0.3	3.2	8.1	2.5	6.2
					B	5.9	6.8	4.9	9.0	4.9	3.1	2.1	6.3	9.6	11.1	1.9	7.1	3.7	3.5	2.4	3.8	4.5	1.6	7.8
				Oolong tea	A	4.5	3.4	2.7	4.3	5.7	1.7	8.6	6.7	8.9	3.3	6.2	4.3	3.3	2.8	7.7	7.4	4.1	4.5	5.3
					B	5.6	4.7	5.1	5.4	5.3	4.3	2.3	9.7	10.0	1.2	5.5	1.5	9.7	4.0	7.5	1.1	1.7	1.3	5.3
		ENVI-Carb +PSA	GC-MS	Green tea	A	10.3	4.5	4.3	8.5	11.2	7.9	7.6	5.7	9.4	1.1	11.2	7.1	5.0	7.7	6.0	14.9	10.5	17.2	5.2
					B	9.6	7.6	2.6	8.5	3.7	6.6	6.7	8.4	5.9	2.1	18.7	3.9	13.5	0.9	20.0	6.3	8.6	11.0	6.8
				Oolong tea	A	6.4	5.0	5.5	3.3	3.7	1.7	4.1	10.0	3.0	2.0	5.8	5.3	1.9	3.8	1.4	6.6	5.8	8.3	3.1
					B	5.6	3.7	6.7	2.8	12.0	8.2	4.4	7.5	12.2	2.1	8.9	1.6	6.3	8.2	10.4	2.7	10.6	3.1	5.5
			GC-MS/MS	Green tea	A	10.9	6.4	4.0	6.1	7.0	7.4	8.8	4.8	3.4	3.7	9.9	3.2	3.7	2.1	4.5	10.7	3.2	6.0	8.4
					B	12.8	10.2	4.4	1.1	2.7	4.3	4.7	3.8	4.3	2.4	22.0	5.2	11.6	6.3	9.8	1.9	4.8	15.1	8.4
				Oolong tea	A	6.5	4.9	6.9	6.3	4.1	2.7	2.0	12.1	3.6	2.8	4.5	8.1	15.7	2.2	7.7	3.1	10.1	3.3	5.3
					B	9.1	2.7	10.2	3.2	9.1	7.3	4.2	7.6	9.9	2.6	8.1	7.3	8.3	9.1	12.1	4.2	5.1	6.9	5.3
166	Iprobenfos	TPT	GC-MS	Green tea	A	14.0	4.8	3.7	5.6	13.5	5.7	6.2	7.0	3.6	7.4	10.6	5.9	9.2	10.8	6.7	1.9	8.3	7.6	4.0
					B	9.8	8.2	13.8	12.2	6.0	2.5	3.6	1.8	7.5	7.6	6.3	13.8	1.0	5.9	11.0	3.6	12.0	0.5	6.3
				Oolong tea	A	5.9	3.1	10.7	5.9	2.3	4.8	1.9	11.9	3.0	13.2	8.4	14.8	3.0	5.7	9.5	5.3	5.2	5.0	1.2
					B	4.8	7.1	11.1	10.3	8.0	5.2	4.8	5.2	10.0	11.2	9.5	6.4	1.1	3.3	2.2	1.4	2.1	2.4	1.0
			GC-MS/MS	Green tea	A	3.6	3.3	3.3	2.8	10.9	5.9	7.6	5.0	3.7	3.5	6.2	1.3	9.3	8.2	3.5	0.6	6.3	5.5	4.3
					B	5.2	8.2	7.7	7.8	4.6	1.8	4.8	6.0	6.7	11.0	2.4	14.6	0.7	0.4	3.7	3.6	4.9	2.4	7.5
				Oolong tea	A	4.9	2.6	4.4	5.3	3.5	3.7	13.5	0.7	4.5	4.9	3.1	3.9	5.8	0.8	8.2	4.8	4.5	2.6	5.6
					B	4.0	3.3	2.4	9.3	5.0	4.7	3.0	6.9	7.2	0.7	6.8	2.3	8.5	2.5	4.2	2.3	2.9	1.8	4.6
		ENVI-Carb +PSA	GC-MS	Green tea	A	10.6	3.0	3.2	10.6	6.8	4.3	5.1	8.6	9.0	10.4	1.3	8.2	3.3	5.4	13.2	15.1	10.9	13.1	6.2
					B	4.6	29.7	7.7	3.1	1.7	14.5	3.0	8.2	12.5	5.3	39.0	6.5	10.7	3.3	19.4	6.6	8.5	10.6	7.2
				Oolong tea	A	8.9	8.9	6.0	4.8	9.6	0.9	5.6	10.3	6.2	5.0	10.2	2.2	1.4	5.1	1.8	4.7	6.4	4.7	4.6
					B	5.7	4.6	13.4	0.9	24.1	7.3	2.8	2.9	14.9	4.6	15.2	4.1	7.1	7.8	10.7	2.3	5.9	5.4	5.1
			GC-MS/MS	Green tea	A	10.7	5.8	2.1	7.2	7.8	2.9	2.8	3.0	3.7	3.0	10.5	1.5	3.0	1.7	6.3	8.5	5.0	6.9	4.9
					B	3.4	8.8	5.8	3.8	3.9	3.9	3.1	1.8	4.5	1.0	23.5	1.8	6.4	15.8	13.9	3.7	4.9	10.4	9.5
				Oolong tea	A	5.3	6.0	5.3	2.4	3.7	2.5	4.4	6.8	4.0	1.6	4.7	8.4	3.9	2.5	10.1	4.5	7.3	4.5	5.6
					B	6.9	2.9	10.6	4.3	13.2	6.9	6.3	6.6	9.9	4.0	6.8	5.5	6.1	8.8	12.2	1.0	2.6	3.2	4.6
167	Isazofos	TPT	GC-MS	Green tea	A	5.8	6.5	3.8	8.1	9.6	4.2	4.3	5.0	4.8	2.8	6.8	2.0	9.2	13.0	3.8	3.8	3.6	8.8	4.2
					B	6.5	5.9	8.8	14.1	5.5	2.3	1.1	2.7	2.7	8.6	6.0	6.2	3.0	3.3	2.4	2.0	10.9	2.3	6.0
				Oolong tea	A	5.8	2.5	6.2	6.5	1.5	4.4	9.3	6.0	3.1	9.0	2.3	6.3	1.7	10.5	11.2	9.4	3.6	4.7	1.4
					B	3.1	4.3	6.6	8.2	4.2	4.6	3.1	2.9	5.7	7.1	5.4	1.8	2.9	2.2	6.2	2.3	0.6	5.2	2.2
			GC-MS/MS	Green tea	A	6.5	2.2	3.6	4.0	11.7	1.4	7.3	5.1	2.4	3.1	10.4	1.0	4.7	3.7	0.9	5.6	9.0	6.3	5.1
					B	6.9	8.1	5.7	8.3	4.1	5.5	3.2	5.3	3.9	8.5	6.4	17.8	3.5	3.6	5.6	6.4	7.1	7.3	5.8
				Oolong tea	A	5.3	1.5	1.6	3.7	4.8	6.1	6.5	0.5	5.2	3.2	4.4	1.7	3.4	2.5	10.5	8.4	5.8	3.2	7.4
					B	3.9	1.7	4.4	6.7	4.4	3.0	2.9	5.7	5.2	1.9	4.2	1.8	10.9	5.2	4.4	3.8	2.0	3.9	6.5
		ENVI-Carb +PSA	GC-MS	Green tea	A	8.8	2.7	4.7	5.4	11.8	6.8	2.8	1.9	5.6	4.2	5.9	5.5	4.4	5.1	1.0	7.7	12.8	6.4	5.6
					B	7.8	9.8	6.9	8.2	2.4	6.3	5.2	4.0	4.3	3.3	25.0	9.7	1.3	2.2	8.0	2.8	6.6	7.4	6.0
				Oolong tea	A	6.7	6.5	4.9	0.8	3.8	2.1	3.1	4.8	2.2	3.5	8.9	2.6	1.0	4.9	1.6	10.3	7.2	3.5	5.5
					B	4.6	2.4	11.1	1.6	16.5	3.7	1.2	8.4	8.0	3.0	9.1	1.6	4.1	7.5	10.6	2.2	3.3	8.5	3.8
			GC-MS/MS	Green tea	A	10.5	7.1	1.2	6.1	6.1	4.9	2.0	2.0	6.4	0.8	6.3	16.5	2.7	2.9	7.4	4.1	5.6	7.5	4.1
					B	5.4	7.4	5.8	0.7	5.3	2.0	4.4	4.0	3.4	3.1	18.8	7.1	2.9	9.0	16.3	5.6	4.1	13.9	11.9
				Oolong tea	A	0.8	4.3	5.6	2.5	1.1	1.3	1.9	6.0	3.0	4.1	6.7	4.7	4.7	4.7	10.7	2.4	10.8	5.6	7.4
					B	7.3	5.6	11.1	5.5	9.0	3.6	0.4	9.3	8.0	2.5	7.1	5.4	10.7	9.8	11.6	3.2	3.5	4.2	6.5

(Continued)

Appendix Table 5.2 RSD of 201 Pesticides in Youden Pair Aged Tea Sample for Parallel Determinations (Nov 9, 2009–Feb 7, 2010) (cont.)

No.	Pesticides	SPE	Method	Sample	Youden pair	Nov 9, 2009 (n=5)	Nov 14, 2009 (n=3)	Nov 19, 2009 (n=3)	Nov 24, 2009 (n=3)	Nov 29, 2009 (n=3)	Dec 4, 2009 (n=3)	Dec 9, 2009 (n=3)	Dec 14, 2009 (n=3)	Dec 19, 2009 (n=3)	Dec 24, 2009 (n=3)	Dec 29, 2009 (n=3)	Jan 3, 2010 (n=3)	Jan 8, 2010 (n=3)	Jan 13, 2010 (n=3)	Jan 18, 2010 (n=3)	Jan 23, 2010 (n=3)	Jan 28, 2010 (n=3)	Feb 2, 2010 (n=3)	Feb 7, 2010 (n=3)
168	Isoprothiolane	TPT	GC-MS	Green tea	A	4.4	4.8	3.1	2.7	8.8	2.2	9.0	6.0	6.3	2.5	5.1	4.3	8.6	14.0	2.8	3.4	4.5	5.6	3.6
					B	4.6	8.5	4.7	7.6	2.5	3.0	2.2	4.8	1.7	7.4	2.9	9.5	2.1	8.4	4.0	4.7	4.7	1.7	4.9
				Oolong tea	A	6.1	1.7	2.1	3.6	2.2	3.3	10.6	7.7	2.6	7.2	4.5	1.4	4.9	8.3	7.7	2.7	3.2	3.1	0.4
					B	4.5	2.2	2.5	5.3	2.6	3.1	4.5	5.6	6.0	1.5	4.6	0.2	8.8	7.3	1.7	2.7	2.3	0.7	2.1
			GC-MS/MS	Green tea	A	4.2	4.8	2.6	3.7	9.8	3.5	7.4	6.2	2.8	2.4	6.1	3.5	7.7	2.9	1.9	0.4	7.5	5.8	1.0
					B	4.3	8.9	7.0	8.6	3.6	3.0	3.0	4.2	5.6	9.3	4.0	21.1	5.1	2.9	5.3	3.8	7.0	4.9	5.1
				Oolong tea	A	6.5	2.8	0.9	3.0	4.4	4.1	8.5	0.8	5.2	5.1	0.9	2.9	6.6	1.1	7.5	2.8	4.9	3.8	5.6
					B	4.2	5.6	4.7	7.8	2.0	2.4	4.1	6.6	7.6	2.4	5.6	2.2	7.5	5.1	7.1	3.1	2.0	3.4	6.3
		ENVI-Carb+PSA	GC-MS	Green tea	A	6.5	4.2	2.2	4.8	8.7	2.1	2.9	3.4	4.2	3.1	6.5	2.5	0.3	8.7	5.6	9.4	10.6	3.1	5.9
					B	3.9	0.8	3.6	4.6	2.4	1.9	4.4	0.9	6.5	2.7	24.1	13.7	4.7	2.0	11.8	1.2	4.3	9.0	7.4
				Oolong tea	A	3.6	3.4	4.5	1.4	4.8	1.1	3.8	1.8	3.5	1.6	5.6	3.3	4.2	2.8	2.5	2.9	5.6	4.1	3.5
					B	4.2	1.9	8.2	3.8	10.5	6.7	2.1	6.9	9.5	5.8	8.7	2.6	6.2	4.2	10.1	3.3	5.2	4.3	4.8
			GC-MS/MS	Green tea	A	6.6	2.2	4.5	7.2	7.1	0.8	1.9	3.5	4.3	3.4	7.8	19.0	3.7	9.9	6.8	7.1	9.0	2.4	4.5
					B	4.8	7.0	5.3	3.8	1.3	1.4	2.6	5.6	4.7	2.1	21.7	16.0	3.4	28.7	10.9	2.1	3.7	11.0	8.1
				Oolong tea	A	3.9	3.5	5.4	3.1	5.4	3.2	3.8	3.0	2.3	1.4	6.1	5.0	11.4	1.9	14.2	3.0	6.7	3.8	5.6
					B	5.4	1.7	10.9	5.5	7.9	5.4	3.0	6.6	8.7	3.6	6.2	4.3	10.4	8.2	10.3	3.9	1.2	3.7	6.3
169	Kresoxim-methyl	TPT	GC-MS	Green tea	A	9.7	7.1	4.2	12.1	15.6	9.3	13.3	11.5	3.5	13.9	5.0	2.1	7.8	8.3	5.0	3.8	3.4	1.1	2.7
					B	21.1	5.0	9.5	14.7	9.3	10.0	13.8	4.0	3.4	13.4	9.0	7.8	9.5	4.1	1.3	8.4	9.3	5.1	7.4
				Oolong tea	A	8.1	4.5	5.4	12.7	28.4	1.9	11.3	9.2	2.8	8.9	10.4	3.5	2.6	4.4	6.1	0.8	3.5	3.0	3.8
					B	4.6	0.8	7.0	23.5	7.0	13.4	8.9	1.4	8.2	0.3	0.2	3.7	3.0	4.7	5.3	4.5	2.8	5.1	5.2
			GC-MS/MS	Green tea	A	11.4	3.9	2.8	14.4	8.7	10.5	13.7	2.9	10.7	2.5	4.1	4.1	15.0	4.4	5.3	4.4	10.4	4.9	3.9
					B	10.5	5.5	3.4	6.0	9.9	14.2	8.7	3.4	7.4	8.8	2.7	9.3	22.5	5.9	5.7	10.2	10.5	2.1	15.7
				Oolong tea	A	8.4	1.5	4.3	1.6	23.7	10.4	2.0	6.8	6.1	3.5	5.0	6.5	6.6	2.2	9.2	8.1	4.7	9.7	2.8
					B	3.6	6.0	10.6	14.4	3.6	10.4	8.3	4.8	7.7	4.3	7.7	6.5	15.7	6.5	9.4	5.6	4.2	3.1	8.6
		ENVI-Carb+PSA	GC-MS	Green tea	A	8.7	8.0	18.8	8.9	21.6	18.6	37.0	38.6	11.9	9.0	14.1	11.4	5.7	3.2	1.5	10.0	6.3	26.9	2.2
					B	21.7	13.2	2.9	7.2	42.6	57.4	8.0	42.2	14.3	6.5	45.7	7.5	12.6	4.9	11.7	9.4	1.8	19.5	6.5
				Oolong tea	A	7.4	7.7	4.0	14.1	16.5	28.2	17.8	7.4	11.6	3.5	5.5	2.7	3.7	5.2	7.2	8.1	9.3	7.7	1.6
					B	11.3	6.8	6.8	19.7	14.7	11.6	9.5	5.9	7.9	9.3	10.8	5.1	2.9	7.6	7.8	4.3	8.4	16.9	7.2
			GC-MS/MS	Green tea	A	31.0	16.7	13.4	6.6	8.5	15.0	44.3	7.1	16.5	10.4	7.9	1.2	14.4	22.1	9.5	18.0	15.5	1.8	12.0
					B	43.4	14.8	4.1	15.4	27.0	7.1	7.1	4.2	8.5	7.9	14.4	0.9	21.8	4.2	8.5	20.9	13.9	17.9	5.4
				Oolong tea	A	9.0	2.0	2.9	15.1	18.7	4.9	14.4	9.2	2.5	3.1	5.0	7.1	22.0	1.9	8.4	6.4	6.3	16.6	2.1
					B	16.4	7.0	10.1	16.2	10.0	6.6	29.4	16.8	22.7	7.2	7.2	11.8	5.4	10.1	12.0	2.8	6.7	20.2	8.6
170	Mefenacet	TPT	GC-MS	Green tea	A	9.4	5.6	3.3	9.0	14.8	2.9	12.1	9.3	9.6	9.4	4.7	12.2	4.1	12.4	7.7	2.6	9.9	2.8	10.3
					B	7.2	7.4	3.1	8.8	7.0	8.9	5.8	5.4	7.5	7.9	4.7	16.3	14.8	9.8	4.5	1.0	1.5	4.3	8.8
				Oolong tea	A	8.0	8.0	10.7	3.9	14.7	6.6	5.0	1.6	11.0	3.6	13.5	7.9	2.6	7.1	9.1	2.3	4.2	3.8	8.4
					B	3.4	2.4	10.8	3.3	8.2	11.5	11.0	16.0	9.8	8.7	14.8	10.4	4.0	3.3	3.1	2.1	5.8	4.7	10.9
			GC-MS/MS	Green tea	A	5.9	4.1	4.4	14.7	14.5	2.8	5.2	12.3	11.6	8.7	9.8	4.9	12.7	5.0	3.5	2.5	8.1	8.8	2.8
					B	6.5	2.6	1.7	9.7	8.2	2.8	3.8	7.8	11.3	11.6	8.1	1.5	13.6	3.0	3.1	1.4	2.8	2.4	5.9
				Oolong tea	A	8.8	3.1	1.0	5.3	7.7	2.7	5.6	8.3	10.2	5.4	15.8	17.5	7.9	1.7	5.7	2.9	6.5	8.6	3.3
					B	8.3	3.0	4.2	6.9	11.9	4.6	8.8	16.8	22.7	2.6	5.6	2.9	13.8	3.0	4.3	4.1	5.5	6.9	5.0
		ENVI-Carb+PSA	GC-MS	Green tea	A	18.1	3.5	13.8	6.8	12.3	15.6	8.5	6.8	20.9	11.3	30.6	2.0	9.0	8.0	3.3	14.4	14.7	4.9	10.3
					B	15.4	6.6	8.3	9.1	13.1	6.1	1.4	14.2	6.6	7.4	7.4	6.9	28.5	5.6	19.4	7.2	25.1	6.0	8.8
				Oolong tea	A	8.8	5.7	7.7	11.3	13.2	4.1	3.7	11.8	3.0	3.9	10.4	10.1	5.2	5.0	3.1	14.9	3.0	15.1	11.5
					B	10.9	11.5	7.7	4.7	17.4	14.2	9.6	14.5	11.6	15.0	11.8	4.5	8.1	8.8	11.5	3.0	6.1	5.6	5.0
			GC-MS/MS	Green tea	A	8.4	3.2	8.8	13.6	7.5	10.7	11.9	18.7	7.2	5.8	3.0	13.4	14.5	4.9	6.1	9.1	6.6	5.1	11.5
					B	16.6	4.6	6.0	4.1	8.8	7.2	6.8	22.5	20.3	10.2	33.5	12.7	16.0	18.3	9.9	4.8	7.4	8.9	5.0
				Oolong tea	A	9.4	4.0	12.6	5.3	7.9	4.4	4.7	22.3	3.7	7.2	6.0	18.7	5.7	5.2	6.3	10.5	7.5	34.8	2.9
					B	7.4	9.9	10.3	3.0	12.4	11.8	6.2	6.1	11.8	12.5	7.0	15.4	6.8	7.4	13.5	4.9	19.0	20.1	5.2

171	Mepronil	TPT	GC–MS	Green tea	A	6.1	8.1	0.7	4.3	7.7	4.9	8.1	5.4	5.5	7.5	6.3	3.1	8.1	8.4	2.8	2.6	3.9	5.6	4.0
					B	6.3	7.4	5.2	8.7	2.5	3.4	1.1	5.2	1.5	6.5	8.1	9.7	1.4	2.6	6.5	3.9	7.1	1.1	3.4
				Oolong tea	A	7.5	2.5	1.6	4.4	1.8	3.4	8.6	10.8	7.2	4.5	10.8	5.2	3.9	1.2	3.6	4.6	2.1	5.0	2.4
					B	5.3	2.9	1.9	2.8	0.8	3.7	4.5	7.2	8.4	2.5	6.9	1.2	5.1	6.3	1.6	1.6	1.2	1.8	1.3
			GC–MS/MS	Green tea	A	6.5	6.7	0.5	3.9	6.9	3.8	7.3	4.5	8.1	1.4	3.8	3.2	13.2	6.5	1.5	2.0	5.3	5.8	1.8
					B	5.2	11.5	6.9	5.6	3.7	2.9	1.7	7.3	12.9	8.1	8.3	9.4	8.5	3.5	5.5	3.1	3.6	4.4	6.0
				Oolong tea	A	6.1	2.1	0.5	3.8	7.0	3.8	10.7	1.5	3.4	3.6	2.7	4.6	6.5	2.1	5.5	3.9	2.2	6.2	12.0
					B	6.4	3.7	3.3	4.0	3.6	4.1	4.1	9.0	3.4	1.0	4.7	3.9	4.7	4.1	4.6	1.9	1.4	1.2	2.9
		ENVI-Carb+PSA	GC–MS	Green tea	A	5.1	8.0	2.2	7.1	6.6	3.6	16.2	7.0	5.0	16.9	2.1	3.0	2.1	5.4	27.6	8.7	1.4	7.9	3.8
					B	7.2	7.5	4.1	5.9	3.2	3.0	1.6	9.2	15.6	13.7	30.4	13.8	9.0	4.7	14.8	2.3	13.3	10.4	7.6
				Oolong tea	A	9.2	1.1	3.6	6.5	7.8	2.1	3.6	5.7	5.2	6.0	4.4	4.9	1.2	7.6	1.1	2.5	5.3	3.8	3.4
					B	7.6	4.4	14.6	4.3	9.5	4.1	3.7	5.8	10.9	3.1	9.8	4.6	1.7	6.7	9.5	2.8	4.2	3.8	2.3
			GC–MS/MS	Green tea	A	6.9	3.2	3.5	7.9	6.1	16.2	19.8	4.3	9.7	2.3	6.3	3.3	10.1	12.1	4.0	7.1	9.3	6.4	3.5
					B	6.0	8.4	4.1	2.2	4.9	27.3	1.5	5.6	9.0	1.5	22.7	2.7	4.5	4.0	12.7	3.1	10.1	10.5	7.9
				Oolong tea	A	5.8	1.8	3.4	2.6	6.9	1.7	8.2	4.6	4.6	4.3	7.4	4.9	15.2	6.0	8.8	3.2	4.8	9.4	6.3
					B	9.0	4.4	9.0	11.0	8.8	5.1	7.5	5.6	3.8	3.4	3.8	3.3	3.4	11.2	9.5	3.2	4.6	3.8	2.9
172	Metribuzin	TPT	GC–MS	Green tea	A	3.7	5.7	5.8	2.4	6.2	4.1	6.1	9.3	3.8	10.5	3.2	12.6	8.7	6.1	3.4	2.8	7.0	8.5	4.0
					B	7.2	8.3	4.3	8.3	6.0	5.2	1.2	10.4	2.7	5.8	5.9	11.5	1.8	6.2	6.0	6.2	4.2	2.7	6.5
				Oolong tea	A	5.2	0.9	6.3	5.3	4.7	6.0	12.0	4.9	5.0	3.2	3.5	3.8	1.0	1.0	6.8	3.6	3.4	1.4	3.3
					B	5.8	2.7	1.3	4.3	5.1	1.6	7.6	9.2	10.1	5.8	2.0	6.3	7.6	6.0	7.8	0.8	1.6	6.9	7.7
			GC–MS/MS	Green tea	A	4.8	6.8	3.7	11.7	7.2	3.5	7.6	11.3	9.7	10.9	4.2	11.6	11.0	15.4	3.6	1.7	6.1	10.8	3.0
					B	5.0	4.8	6.0	6.9	3.6	5.3	3.3	3.0	11.3	3.3	3.1	1.4	10.9	4.6	4.8	5.7	5.6	2.2	6.1
				Oolong tea	A	7.0	6.4	5.2	3.2	4.8	4.3	3.0	10.3	9.7	7.7	2.8	1.4	9.6	3.6	8.7	2.0	3.5	2.9	3.0
					B	4.5	5.6	10.3	6.3	0.6	19.0	17.0	8.2	14.3	4.4	12.0	20.9	6.4	5.0	9.4	4.4	4.6	3.2	7.0
		ENVI-Carb+PSA	GC–MS	Green tea	A	15.8	6.8	15.2	10.2	8.3	24.2	12.7	15.0	15.4	8.5	21.0	2.6	25.0	11.6	7.8	10.8	10.5	21.6	7.3
					B	20.2	11.3	1.6	8.0	5.3	3.2	3.7	20.0	3.1	6.9	6.6	9.1	7.6	5.4	9.8	7.3	3.0	21.5	8.2
				Oolong tea	A	3.9	4.3	4.7	9.7	8.3	5.9	6.5	7.2	10.7	6.2	12.7	4.8	3.9	1.7	6.6	11.9	7.9	8.5	3.6
					B	4.2	2.5	13.7	2.6	17.8	20.2	16.6	14.9	8.5	1.3	10.6	2.8	5.6	2.4	13.4	3.3	6.5	13.5	5.8
			GC–MS/MS	Green tea	A	18.9	1.6	11.0	3.3	7.0	2.0	14.0	14.1	12.5	12.7	13.0	1.3	19.8	4.5	6.7	7.5	7.7	6.5	7.2
					B	27.4	8.3	4.7	11.9	6.7	3.2	5.2	15.7	1.0	7.0	5.2	7.4	2.6	19.7	5.3	17.5	0.2	18.3	11.0
				Oolong tea	A	7.1	3.9	6.9	8.2	2.2	8.1	4.5	6.8	8.3	3.8	6.9	12.0	6.9	2.3	15.2	4.3	9.1	12.1	3.0
					B	13.5	4.8	15.0	5.2	12.2	6.3	6.2	4.2	7.0	4.0	9.4	1.3	5.0	12.7	13.0	2.8	4.1	18.3	7.0
173	Molinate	TPT	GC–MS	Green tea	A	6.1	9.6	1.6	0.8	8.8	6.3	6.2	4.2	7.0	4.0	9.4	10.2	5.0	3.7	3.6	4.8	13.3	5.6	4.7
					B	6.4	9.1	4.0	8.2	2.9	4.2	3.0	1.8	3.3	10.4	4.0	1.9	4.2	3.8	4.3	1.8	6.4	1.9	2.1
				Oolong tea	A	4.3	0.7	2.3	2.7	3.8	1.8	10.0	4.0	3.8	5.0	3.5	1.2	3.4	4.7	9.2	3.1	2.2	1.9	0.3
					B	4.8	3.0	7.8	6.0	3.6	5.2	5.3	3.1	2.9	2.2	5.3	13.4	4.6	4.3	3.1	0.6	1.1	1.6	2.6
			GC–MS/MS	Green tea	A	3.6	3.1	2.7	1.3	12.0	6.4	4.7	2.4	5.3	2.8	5.3	4.4	7.9	4.3	3.7	1.7	6.2	5.6	1.7
					B	6.0	8.3	5.1	8.3	3.1	4.3	3.8	3.2	5.3	11.1	3.5	1.7	8.6	1.9	10.1	2.6	5.2	8.9	5.1
				Oolong tea	A	4.8	0.6	1.8	3.1	4.0	1.0	8.4	0.2	5.3	2.1	3.7	4.6	7.4	0.4	9.0	4.3	2.1	0.5	4.1
					B	5.1	3.7	5.1	6.1	2.5	2.5	4.7	4.3	11.0	3.8	2.7	3.3	8.5	0.1	3.9	2.4	1.5	2.9	1.9
		ENVI-Carb+PSA	GC–MS	Green tea	A	5.1	1.0	2.1	5.0	0.6	0.6	1.4	1.5	6.3	2.5	9.9	2.2	1.5	3.0	19.9	8.6	13.4	3.3	3.2
					B	4.6	6.4	7.1	4.1	2.9	2.9	7.6	5.9	3.2	3.3	3.3	3.9	6.6	1.7	0.3	3.7	6.6	9.1	6.3
				Oolong tea	A	5.8	0.3	0.3	6.9	1.3	1.3	5.4	0.5	1.7	4.0	21.1	4.4	1.6	7.2	4.7	6.5	7.9	5.7	5.8
					B	8.6	4.0	4.0	5.7	5.9	5.3	2.6	6.9	11.8	2.9	1.9	2.0	4.2	5.9	7.4	2.6	2.8	4.2	3.3
			GC–MS/MS	Green tea	A	3.7	9.6	9.6	8.2	13.0	3.8	4.5	2.2	6.2	2.8	7.7	4.8	2.2	53.1	21.5	9.7	10.2	6.0	4.2
					B	2.9	6.0	1.1	2.8	4.2	1.0	4.6	4.5	1.0	1.1	6.8	3.0	9.6	9.6	0.3	2.5	3.8	10.4	4.3
				Oolong tea	A	7.0	4.5	6.7	4.2	4.7	2.0	1.1	5.2	1.6	1.0	21.7	4.8	3.5	4.5	8.8	5.6	4.7	4.7	4.1
					B		4.6	4.0	6.1	8.3	5.0	2.4	7.8	9.2	2.5	5.2	3.0	8.1	8.5	6.6	2.2	6.3	3.7	1.9

(Continued)

Appendix Table 5.2 RSD of 201 Pesticides in Youden Pair Aged Tea Sample for Parallel Determinations (Nov 9, 2009–Feb 7, 2010) (cont.)

No.	Pesticides	SPE	Method	Sample	Youden pair	RSD% Nov 9, 2009	Nov 14, 2009	Nov 19, 2009	Nov 24, 2009	Nov 29, 2009	Dec 4, 2009	Dec 9, 2009	Dec 14, 2009	Dec 19, 2009	Dec 24, 2009	Dec 29, 2009	Jan 3, 2010	Jan 8, 2010	Jan 13, 2010	Jan 18, 2010	Jan 23, 2010	Jan 28, 2010	Feb 2, 2010	Feb 7, 2010
						(n = 5)	(n = 3)	(n = 3)	(n = 3)	(n = 3)	(n = 3)	(n = 3)	(n = 3)	(n = 3)	(n = 3)	(n = 3)	(n = 3)	(n = 3)	(n = 3)	(n = 3)	(n = 3)	(n = 3)	(n = 3)	(n = 3)
174	Napropamide	TPT	GC-MS	Green tea	A	4.6	4.7	1.7	2.6	8.6	4.0	8.3	6.7	4.9	1.8	6.1	2.9	14.8	3.2	2.9	3.1	1.4	7.7	4.4
					B	4.2	8.3	5.4	7.9	4.6	3.6	3.8	5.9	4.6	4.7	2.4	13.0	2.4	1.9	6.4	1.9	5.2	1.0	4.5
				Oolong tea	A	5.7	2.3	1.9	2.6	1.5	7.2	8.0	9.0	8.8	6.6	7.4	2.0	3.6	2.1	7.2	2.9	3.0	4.8	1.6
					B	3.7	1.3	2.2	0.6	1.1	1.3	3.5	2.4	6.4	2.7	8.2	1.6	4.2	3.6	5.0	0.8	1.1	2.4	5.1
			GC-MS/MS	Green tea	A	4.1	4.8	1.7	5.1	8.7	5.1	7.5	5.5	2.9	0.9	5.6	1.7	8.4	6.1	2.7	1.2	7.8	4.6	2.8
					B	5.3	11.6	6.8	7.9	4.4	4.1	4.4	6.9	6.6	9.4	4.5	8.8	5.1	2.0	4.3	2.3	5.0	4.3	7.3
				Oolong tea	A	5.6	3.4	1.2	3.1	5.1	4.4	9.5	2.1	2.9	4.8	1.3	3.0	7.9	1.9	9.8	4.2	0.5	3.7	4.3
					B	5.6	5.4	5.3	7.4	2.1	2.2	5.4	7.6	6.9	1.9	5.5	3.7	3.2	4.1	6.2	1.3	2.3	0.7	4.0
		ENVI-Carb+PSA	GC-MS	Green tea	A	5.3	4.6	1.5	4.7	8.1	7.4	9.1	7.8	5.0	6.2	27.2	4.8	1.8	5.1	8.9	5.5	11.7	1.8	11.2
					B	3.5	8.5	2.4	4.6	0.8	3.2	4.1	4.5	6.1	1.0	5.1	20.6	3.2	3.6	9.5	2.5	9.6	7.5	12.4
				Oolong tea	A	4.9	2.4	2.0	1.6	7.0	2.3	4.7	1.4	2.2	0.8	10.8	6.0	3.0	2.6	2.3	4.8	5.2	2.1	4.5
					B	6.8	4.6	9.1	2.1	9.5	7.4	1.8	5.5	12.6	8.9	8.4	6.0	5.4	8.6	9.1	5.6	5.8	9.9	3.1
			GC-MS/MS	Green tea	A	6.6	5.1	1.7	7.7	7.5	2.1	8.1	1.9	3.1	3.3	21.5	1.8	2.9	9.1	8.2	6.3	9.2	1.5	6.1
					B	3.6	8.5	5.6	3.5	1.1	1.0	2.4	7.8	3.3	1.5	5.4	2.2	5.6	26.1	14.0	3.1	2.9	8.2	6.0
				Oolong tea	A	3.6	4.2	4.0	4.0	5.5	2.3	4.0	2.2	1.2	1.6	5.9	6.6	12.5	3.8	12.2	4.6	5.6	4.2	4.3
					B	5.8	3.3	11.6	6.7	7.8	5.7	3.6	7.7	10.1	5.4	6.1	3.7	4.8	8.1	11.6	2.0	4.0	4.6	4.0
175	Nuarimol	TPT	GC-MS	Green tea	A	7.2	4.7	3.5	3.4	7.3	12.9	6.5	5.8	0.6	3.5	3.1	4.9	2.4	12.5	3.5	2.9	3.9	4.5	2.5
					B	6.1	6.6	7.2	7.2	6.1	9.5	1.7	2.6	6.6	6.9	8.1	13.5	6.9	2.5	7.1	3.1	8.0	3.6	2.0
				Oolong tea	A	9.5	5.0	6.4	0.6	7.2	8.9	14.5	9.3	5.8	3.8	6.2	6.3	0.9	2.6	5.2	4.3	3.6	1.5	0.8
					B	7.1	3.1	5.2	7.2	3.0	2.8	6.6	4.4	8.1	6.3	5.4	0.8	7.0	3.1	9.5	1.1	1.5	3.0	0.6
			GC-MS/MS	Green tea	A	5.3	8.0	0.6	7.8	9.2	5.2	8.1	4.2	3.2	2.6	3.0	3.8	5.2	9.7	3.0	3.0	6.8	4.3	6.3
					B	4.2	11.0	5.9	7.5	3.8	3.8	2.2	7.8	3.6	10.3	3.5	2.6	6.2	6.9	7.5	4.5	5.8	4.6	5.8
				Oolong tea	A	5.9	1.9	1.9	3.1	4.3	4.0	12.6	1.5	2.7	3.8	2.7	5.5	5.4	3.9	6.4	6.7	0.2	2.1	2.5
					B	6.3	4.4	5.2	9.6	1.3	1.8	4.6	8.2	8.6	2.1	4.3	3.3	5.4	5.7	7.0	1.4	4.7	4.0	4.3
		ENVI-Carb+PSA	GC-MS	Green tea	A	7.0	12.0	2.1	6.4	3.2	2.7	25.2	1.2	19.0	7.8	23.3	6.9	11.0	6.9	5.1	5.1	11.3	2.9	7.0
					B	5.6	27.4	5.9	9.0	4.4	4.0	2.2	2.3	4.7	6.3	8.2	11.5	7.2	3.6	8.5	1.8	1.3	5.9	2.3
				Oolong tea	A	7.3	4.0	6.0	3.1	12.1	2.0	3.9	4.9	3.4	8.6	8.7	6.9	2.8	4.7	4.9	6.1	3.5	4.0	2.6
					B	7.5	5.8	5.4	7.8	9.3	8.3	0.2	10.2	6.9	1.8	5.1	4.9	5.5	20.3	9.8	3.6	7.2	2.0	3.7
			GC-MS/MS	Green tea	A	7.0	8.6	4.0	7.1	6.4	4.0	26.6	2.3	7.2	1.4	18.8	1.8	3.3	9.7	7.4	3.4	9.0	2.2	4.8
					B	6.6	5.9	5.0	5.0	1.0	1.4	1.4	4.9	3.5	5.8	8.5	5.4	20.3	19.2	16.5	1.9	2.5	8.3	5.8
				Oolong tea	A	7.1	2.0	1.7	0.9	5.2	1.8	2.2	5.5	1.2	13.1	5.6	4.5	10.3	5.2	10.6	4.5	5.9	5.4	2.5
					B	5.3	3.8	5.3	7.9	8.6	5.6	3.2	9.8	8.0	2.1	5.3	6.8	14.3	14.7	9.4	3.8	6.0	7.7	4.3
176	Permethrin	TPT	GC-MS	Green tea	A	3.9	6.1	2.0	1.4	7.1	6.0	7.2	5.1	6.1	3.8	2.9	2.6	5.1	3.8	3.5	1.6	1.1	2.7	7.0
					B	3.5	7.0	3.7	7.0	2.5	3.3	1.9	2.9	2.0	8.5	7.4	9.1	5.5	5.1	4.2	4.3	5.2	1.1	2.3
				Oolong tea	A	6.1	2.9	3.2	2.6	1.7	4.0	7.0	7.4	3.1	2.7	7.4	2.7	1.4	1.4	5.1	3.9	3.1	4.3	2.6
					B	3.2	2.0	2.5	6.3	1.4	5.0	4.3	2.1	7.8	3.8	2.1	4.6	12.3	5.0	1.2	2.6	2.0	2.3	3.7
			GC-MS/MS	Green tea	A	4.8	3.3	0.6	3.4	10.7	5.6	7.7	3.9	5.4	2.0	6.0	2.6	6.4	1.9	0.1	1.8	6.9	3.5	4.8
					B	3.5	8.1	4.5	8.5	4.3	2.9	0.6	3.4	6.1	8.8	4.9	3.5	8.1	3.0	5.4	5.5	5.4	8.5	5.8
				Oolong tea	A	6.5	3.0	1.7	1.5	3.3	5.2	8.1	2.7	1.5	7.6	2.6	5.9	12.7	3.0	6.1	4.3	4.5	9.4	2.5
					B	3.7	3.8	5.3	7.1	0.5	3.5	4.5	7.4	12.6	5.0	5.0	2.6	14.3	2.5	4.2	3.5	1.8	4.0	3.0
		ENVI-Carb+PSA	GC-MS	Green tea	A	9.5	4.3	6.4	7.9	8.9	8.7	13.3	3.1	9.4	1.6	1.3	3.4	5.6	2.1	2.6	3.5	9.5	4.9	4.8
					B	16.5	11.4	4.1	10.3	2.7	4.0	3.9	2.2	8.2	3.8	16.2	5.2	2.3	1.6	8.3	3.5	3.5	10.7	9.0
				Oolong tea	A	5.3	1.4	5.5	11.1	4.0	2.5	1.1	10.9	2.6	3.9	10.6	2.3	1.4	6.2	2.4	3.1	5.0	6.3	3.1
					B	6.5	5.0	7.3	2.9	10.3	3.5	5.4	8.7	8.2	2.5	9.9	4.6	5.0	4.1	6.4	2.6	6.0	0.3	2.0
			GC-MS/MS	Green tea	A	8.6	3.5	7.2	8.0	5.5	4.4	9.1	4.9	8.9	0.9	16.9	2.0	3.8	5.5	2.3	4.4	7.6	2.4	3.1
					B	9.3	8.5	5.3	5.8	3.3	2.6	4.1	1.5	6.2	3.0	7.8	1.5	4.6	14.1	2.6	2.4	7.6	9.5	4.6
				Oolong tea	A	6.2	4.3	6.6	1.9	3.9	5.1	1.0	6.4	1.0	3.8	7.8	2.4	16.9	5.5	7.8	0.8	5.0	9.4	3.3
					B	6.6	5.4	10.2	5.4	7.5	3.7	3.2	5.8	7.1	1.1	4.0	5.2	9.3	7.0	9.1	4.1	3.8	1.7	4.8

No.	Compound	Cleanup	Detection	Tea	A/B	1	2	3	4	5	6	7	8	9	10	11	12	13	14	15	16	17	18	19
177	Phenothrin	TPT	GC-MS	Green tea	A	3.2	6.5	2.0	1.7	5.5	7.7	6.8	4.9	6.3	3.8	3.2	5.6	6.7	13.5	10.8	7.7	3.1	6.3	4.3
					B	3.7	6.9	2.8	9.5	2.7	3.7	1.3	1.0	2.1	10.2	4.8	5.1	6.1	8.0	8.6	6.6	8.1	2.2	5.4
				Oolong tea	A	5.4	5.2	1.1	3.8	1.5	2.3	8.6	9.7	1.2	7.4	6.0	4.9	5.6	1.0	6.3	0.0	4.7	6.7	1.2
					B	3.9	1.7	1.3	16.9	4.0	2.0	5.0	3.3	6.5	3.1	6.7	1.8	4.8	3.2	0.8	1.3	4.0	1.4	4.6
			GC-MS/MS	Green tea	A	9.2	4.5	1.8	9.0	10.7	2.0	7.1	4.4	5.6	3.7	8.9	0.6	23.3	4.6	6.6	5.7	4.8	5.2	1.5
					B	2.3	5.8	8.2	7.7	3.4	4.2	2.5	2.6	6.0	9.7	6.8	4.6	16.9	14.5	8.9	9.8	8.4	3.7	3.1
				Oolong tea	A	8.4	5.0	1.1	3.7	2.2	3.6	2.9	2.8	25.7	50.3	13.9	39.3	26.3	2.4	42.5	17.2	8.1	33.7	10.3
					B	4.4	6.1	5.0	4.4	2.1	4.9	6.4	14.3	27.3	20.6	24.4	30.9	42.0	6.0	40.3	28.3	26.2	14.7	1.1
		ENVI-Carb +PSA	GC-MS	Green tea	A	6.7	21.3	8.4	6.6	5.5	1.6	5.3	3.9	8.1	2.8	7.7	2.3	4.0	5.6	6.3	6.8	11.3	4.0	6.3
					B	9.5	11.4	7.7	3.6	6.1	2.3	2.6	2.4	4.2	3.1	21.6	8.2	1.2	5.7	13.7	2.8	5.4	9.0	7.7
				Oolong tea	A	4.8	3.7	8.5	8.7	6.7	3.0	2.8	2.5	4.8	7.5	7.7	5.9	3.9	6.7	1.1	6.6	5.9	7.4	1.9
					B	6.5	3.1	8.1	5.7	11.2	5.0	3.4	8.3	10.5	3.4	8.5	3.7	2.8	6.4	8.4	4.1	4.1	0.4	1.1
			GC-MS/MS	Green tea	A	7.1	1.9	11.7	11.0	23.5	3.3	6.7	3.1	6.0	1.6	8.3	1.4	11.8	4.1	10.3	18.9	6.6	5.9	3.9
					B	5.7	7.3	10.9	4.8	27.5	3.1	6.2	5.1	5.4	0.6	19.8	10.0	1.5	32.4	8.2	2.8	2.6	10.8	8.0
				Oolong tea	A	5.7	29.1	4.9	18.7	22.1	12.2	11.9	7.0	4.8	13.5	28.3	0.8	7.1	47.0	40.2	30.2	8.8	31.1	46.8
					B	6.0	28.0	12.2	19.3	27.4	14.3	19.4	9.5	17.0	37.4	18.2	16.8	35.0	11.3	42.9	14.6	25.2	25.7	44.9
178	Piperonyl-butoxide	TPT	GC-MS	Green tea	A	3.8	4.9	0.8	1.8	8.6	4.5	7.8	7.5	12.2	8.3	9.1	9.9	7.4	4.2	4.2	3.0	3.9	6.0	4.8
					B	4.8	10.5	5.6	8.8	3.5	4.0	2.4	3.1	1.7	7.5	2.7	15.1	3.6	1.0	6.0	3.8	4.8	2.0	3.7
				Oolong tea	A	6.1	6.4	4.5	2.6	2.5	4.0	7.6	6.9	5.6	13.4	19.1	9.2	4.9	2.3	6.9	3.7	1.9	8.5	1.7
					B	3.7	6.0	9.1	7.8	6.8	4.9	5.5	4.2	8.2	13.0	23.2	2.8	1.6	4.3	1.4	0.9	1.0	2.1	0.9
			GC-MS/MS	Green tea	A	5.3	3.3	1.0	11.7	8.9	4.7	8.3	4.0	2.9	0.9	5.7	0.4	10.3	5.5	5.8	0.6	5.7	5.2	2.6
					B	4.7	9.1	7.0	8.1	3.0	3.2	2.2	2.9	4.1	8.5	3.6	10.5	6.1	1.6	3.8	3.8	2.9	3.3	5.4
				Oolong tea	A	6.9	3.3	0.8	2.5	2.9	3.5	9.0	2.6	2.6	5.0	1.1	4.4	6.4	1.7	7.5	5.5	2.1	5.0	4.9
					B	5.2	4.0	3.1	7.6	3.4	4.3	4.9	7.6	7.3	1.4	5.9	2.5	3.5	3.5	5.0	1.6	1.3	1.4	3.2
		ENVI-Carb +PSA	GC-MS	Green tea	A	9.3	6.2	2.8	7.6	9.4	2.6	3.4	1.6	11.9	2.6	4.9	12.1	2.7	2.7	6.9	7.7	13.2	3.2	6.9
					B	6.3	9.5	4.7	6.2	3.8	7.3	5.0	6.9	5.0	6.8	35.7	10.8	1.8	1.0	11.5	3.6	3.8	7.3	6.8
				Oolong tea	A	4.8	4.6	6.5	1.9	8.0	0.8	3.3	6.6	3.6	2.7	10.2	3.9	4.2	4.5	1.0	4.5	4.5	4.7	4.7
					B	3.9	0.9	10.6	1.5	18.8	4.7	1.7	8.2	12.5	2.9	16.7	7.0	2.4	6.4	10.5	3.6	5.4	5.4	3.4
			GC-MS/MS	Green tea	A	6.5	2.9	3.9	8.3	7.9	0.8	5.1	2.5	5.1	2.4	5.9	2.3	7.7	4.6	5.5	7.5	8.6	2.8	6.3
					B	3.2	9.4	4.3	4.8	1.7	0.6	2.1	6.3	3.7	0.6	21.0	0.4	3.4	2.3	12.0	3.4	1.5	9.0	6.3
				Oolong tea	A	6.3	2.9	6.0	1.9	3.7	2.2	3.1	3.7	1.8	2.5	7.2	3.4	19.9	5.5	8.6	4.0	4.5	5.6	4.9
					B	5.5	3.7	9.8	5.0	10.1	4.3	4.2	9.6	8.7	3.9	5.4	3.9	4.6	7.2	9.1	3.3	3.0	5.1	3.2
179	Pretila-chlor	TPT	GC-MS	Green tea	A	6.3	5.7	3.8	4.6	5.6	6.0	4.6	11.1	4.5	0.7	7.0	7.5	11.0	5.0	8.5	7.0	6.3	12.2	5.4
					B	4.5	6.2	5.9	10.0	4.0	7.2	3.8	5.9	5.8	8.1	1.4	10.9	5.2	3.1	9.7	1.3	9.0	9.6	9.1
				Oolong tea	A	6.4	5.2	2.3	2.4	2.7	2.6	13.9	11.9	2.7	5.4	0.7	8.6	6.3	6.1	9.4	4.9	7.9	2.0	1.9
					B	4.8	3.1	4.9	1.8	3.0	4.4	2.8	3.7	5.7	6.4	8.1	3.5	6.6	7.0	2.3	2.8	5.6	1.5	5.6
			GC-MS/MS	Green tea	A	8.8	3.8	0.7	32.2	7.3	3.9	6.4	5.9	4.2	8.3	2.3	6.0	8.9	7.3	2.0	3.4	7.9	2.5	4.9
					B	4.6	10.1	4.5	7.4	2.9	5.9	5.6	4.0	16.4	11.5	2.3	18.4	14.6	2.3	6.9	6.1	5.0	1.1	7.1
				Oolong tea	A	5.5	5.1	3.4	3.3	6.0	0.4	13.5	8.3	5.8	4.4	1.5	5.2	6.3	0.8	10.0	6.3	5.0	3.1	5.7
					B	6.6	6.0	8.6	1.1	2.8	2.0	6.4	6.2	12.0	6.0	7.4	3.2	12.5	4.0	8.1	1.4	2.2	1.1	4.2
		ENVI-Carb +PSA	GC-MS	Green tea	A	7.1	1.0	2.6	2.5	10.3	5.9	9.2	5.4	4.9	1.8	9.8	9.5	4.5	6.9	4.0	7.1	12.2	13.8	9.6
					B	11.1	8.9	2.4	5.2	4.9	4.1	3.6	1.7	8.0	5.9	15.6	8.7	4.5	2.8	8.0	3.8	8.9	7.0	3.8
				Oolong tea	A	5.7	1.1	5.5	3.0	5.7	1.5	6.9	5.4	2.3	3.4	8.3	4.9	1.0	6.5	12.6	5.0	2.7	5.6	6.4
					B	6.5	3.8	7.5	5.7	6.6	4.9	4.1	6.5	10.4	2.6	8.8	2.7	7.8	13.4	10.7	3.8	4.1	2.3	4.9
			GC-MS/MS	Green tea	A	1.7	4.6	4.7	4.7	7.2	4.9	9.8	1.2	6.4	4.5	13.1	1.7	6.9	19.2	4.7	2.8	2.5	7.2	7.8
					B	17.0	6.0	0.8	4.4	4.8	7.7	14.6	1.8	7.1	10.5	10.9	1.3	6.3	28.6	3.2	6.5	9.7	13.1	12.6
				Oolong tea	A	3.7	1.5	6.4	10.7	6.9	3.5	3.1	8.0	4.2	5.8	4.7	17.0	2.0	0.4	14.5	7.0	6.7	7.9	5.7
					B	8.7	3.2	14.7	9.6	4.1	4.0	3.0	7.2	8.3	2.9	11.4	10.8	15.3	10.4	11.8	3.0	5.6	1.9	4.2

(Continued)

Appendix Table 5.2 RSD of 201 Pesticides in Youden Pair Aged Tea Sample for Parallel Determinations (Nov 9, 2009–Feb 7, 2010) (cont.)

RSD%

No.	Pesticides	SPE	Method	Sample	Youden pair	Nov 9, 2009 (n=5)	Nov 14, 2009 (n=3)	Nov 19, 2009 (n=3)	Nov 24, 2009 (n=3)	Nov 29, 2009 (n=3)	Dec 4, 2009 (n=3)	Dec 9, 2009 (n=3)	Dec 14, 2009 (n=3)	Dec 19, 2009 (n=3)	Dec 24, 2009 (n=3)	Dec 14, 2009 (n=3)	Jan 3, 2010 (n=3)	Jan 8, 2010 (n=3)	Jan 13, 2010 (n=3)	Jan 18, 2010 (n=3)	Jan 23, 2010 (n=3)	Jan 28, 2010 (n=3)	Feb 2, 2010 (n=3)	Feb 7, 2010 (n=3)
180	Prometon	TPT	GC-MS	Green tea	A	5.2	5.0	3.5	4.0	8.4	4.1	8.3	5.0	4.6	4.6	10.1	4.9	7.3	3.9	1.9	0.1	5.6	6.9	4.4
					B	5.9	9.0	7.0	7.6	4.2	2.5	5.6	2.5	3.7	10.6	8.7	11.0	2.4	1.2	4.8	3.0	3.1	3.3	4.2
				Oolong tea	A	6.8	1.0	1.8	4.0	2.1	3.9	21.7	6.3	2.4	8.9	9.3	4.0	3.4	5.5	8.7	5.3	6.4	5.2	1.7
					B	3.6	2.3	3.5	6.8	4.0	4.3	5.3	4.3	7.0	7.5	10.4	3.1	2.0	1.4	4.5	1.5	1.2	1.5	0.8
			GC-MS/MS	Green tea	A	4.8	3.2	2.6	2.7	11.3	7.2	7.7	5.6	3.2	1.3	6.2	2.9	13.7	3.6	0.5	0.8	9.2	4.0	2.8
					B	5.0	9.6	8.6	8.1	3.4	4.0	4.9	4.3	5.9	9.6	3.9	8.7	2.4	5.5	3.6	4.0	5.6	3.4	5.6
				Oolong tea	A	4.3	3.0	3.1	3.6	3.4	3.6	10.4	1.4	2.0	3.8	0.4	1.9	8.7	1.4	8.5	2.8	2.6	3.4	5.9
					B	5.1	4.6	4.5	7.6	1.7	1.8	5.3	6.6	4.9	2.0	3.9	3.1	9.7	3.3	6.1	1.9	3.3	2.2	5.3
		ENVI-Carb+PSA	GC-MS	Green tea	A	5.4	3.4	0.6	6.0	9.9	4.3	4.5	1.0	8.4	5.9	7.8	4.0	2.1	2.6	8.6	7.0	9.2	4.8	5.4
					B	3.9	9.5	4.2	8.3	1.9	3.0	5.2	4.2	5.2	6.2	29.5	16.0	1.8	3.9	12.9	1.9	4.8	10.5	6.2
				Oolong tea	A	1.4	6.5	4.0	4.1	6.6	6.8	4.1	3.3	13.5	4.9	4.3	16.6	2.2	9.0	3.0	2.1	4.4	2.1	4.4
					B	6.4	2.9	9.2	2.8	15.6	1.8	2.2	8.3	3.1	6.8	6.5	5.0	6.4	7.1	9.3	3.0	5.7	5.2	4.1
			GC-MS/MS	Green tea	A	7.3	4.5	1.6	7.6	7.9	0.6	4.7	2.6	4.5	1.0	7.6	2.8	2.0	9.5	7.5	6.5	7.5	1.7	4.3
					B	4.5	10.2	4.9	3.4	1.9	3.5	3.3	6.9	2.5	2.4	22.3	16.8	4.9	30.2	14.6	2.5	2.3	11.0	6.7
				Oolong tea	A	2.6	5.2	5.3	3.9	5.8	6.4	3.7	4.0	10.0	1.8	5.3	16.4	5.1	3.6	12.6	4.3	4.9	2.5	5.9
					B	5.7	3.9	10.8	5.1	8.9	2.9	3.9	8.0	4.1	4.1	7.0	7.2	5.2	10.1	11.4	2.1	3.6	3.5	5.3
181	Pronamide	TPT	GC-MS	Green tea	A	6.0	3.3	2.1	2.6	7.7	2.9	6.4	5.4	5.2	2.1	4.1	4.6	7.8	3.8	2.3	3.4	5.4	5.4	2.8
					B	4.2	8.5	9.4	11.9	3.7	1.3	3.1	5.4	3.1	6.9	1.3	2.5	2.1	4.0	5.5	3.0	4.4	1.3	5.4
				Oolong tea	A	11.1	5.5	2.9	5.0	3.1	7.5	4.2	8.9	4.7	7.8	3.1	10.0	0.6	3.4	6.7	1.1	0.8	6.0	2.4
					B	8.4	8.3	11.4	12.4	2.7	2.1	5.3	6.9	3.4	3.3	4.6	3.7	0.6	8.0	5.4	3.9	4.8	3.6	4.5
			GC-MS/MS	Green tea	A	4.7	4.7	2.0	9.9	9.9	3.7	6.7	5.3	4.1	1.7	1.9	2.0	13.4	3.6	1.7	2.1	6.6	4.4	3.0
					B	3.0	10.3	7.9	7.3	3.6	3.4	3.7	8.6	6.5	8.5	3.6	3.1	5.6	2.6	4.5	3.6	5.2	3.2	6.8
				Oolong tea	A	8.0	6.5	2.6	7.8	7.6	8.8	5.9	0.6	4.4	5.3	4.9	5.8	4.8	0.5	6.8	1.6	3.7	5.1	7.4
					B	5.2	3.5	4.5	2.4	0.8	0.4	1.6	9.2	6.5	6.8	6.9	1.9	6.4	7.5	10.6	5.8	5.3	5.6	4.8
		ENVI-Carb+PSA	GC-MS	Green tea	A	7.4	5.5	3.6	14.9	9.8	0.6	10.2	2.9	4.8	5.3	24.0	3.2	2.0	2.8	4.3	6.2	10.2	3.8	6.2
					B	4.1	10.8	4.2	14.4	6.9	8.8	5.2	2.6	5.9	1.5	4.5	8.5	3.3	2.7	11.7	1.8	4.2	9.4	10.1
				Oolong tea	A	15.7	11.1	7.5	11.4	5.6	4.1	4.6	9.1	2.6	3.4	8.7	10.6	1.4	4.7	2.6	4.7	4.3	9.3	7.7
					B	7.1	5.0	8.1	2.9	12.9	6.0	5.5	4.3	5.3	9.0	6.5	3.5	3.0	9.1	9.6	3.7	7.1	11.5	4.8
			GC-MS/MS	Green tea	A	8.1	3.4	3.8	6.3	6.7	2.8	8.5	4.0	6.7	2.9	20.6	4.7	1.6	3.3	7.2	4.4	7.8	1.3	5.3
					B	5.5	7.6	4.8	2.3	1.8	1.2	5.0	3.0	6.6	0.5	3.9	2.0	3.4	11.7	10.9	1.0	1.4	10.6	7.3
				Oolong tea	A	15.5	1.5	2.3	9.5	4.5	2.7	4.7	8.2	11.7	2.0	4.9	3.0	15.6	1.6	12.2	1.0	4.3	4.6	7.2
					B	8.6	3.3	14.6	2.7	8.5	6.2	4.2	5.7	10.3	12.6	1.6	1.6	6.1	10.4	12.1	0.9	4.7	13.5	4.8
182	Propetamphos	TPT	GC-MS	Green tea	A	4.6	4.8	6.7	7.6	13.0	3.0	10.2	14.4	12.5	3.0	10.7	9.8	8.9	11.5	6.2	3.3	9.4	8.7	2.5
					B	7.1	7.8	4.6	14.5	7.3	10.2	4.5	6.9	8.7	1.1	8.8	10.3	8.6	6.4	6.9	4.2	11.2	4.0	5.9
				Oolong tea	A	5.6	4.4	9.8	7.3	6.1	5.4	3.6	7.5	12.6	10.5	9.7	11.6	2.7	3.6	6.2	3.8	3.3	2.1	1.5
					B	4.9	6.3	6.8	6.8	7.4	6.1	6.5	7.9	12.4	4.8	2.3	3.8	2.3	6.5	5.4	0.4	1.1	0.9	4.0
			GC-MS/MS	Green tea	A	5.0	5.0	4.3	5.3	15.6	1.3	9.8	11.4	6.7	2.7	6.0	2.4	10.4	11.9	4.6	2.8	8.8	9.2	0.9
					B	5.1	7.9	5.6	9.3	4.0	1.6	4.1	9.3	4.8	8.7	1.0	4.9	10.6	2.1	4.0	3.2	10.0	2.8	9.3
				Oolong tea	A	5.0	0.2	3.1	2.2	4.5	3.7	5.0	6.8	11.2	3.5	7.0	1.1	4.8	3.0	11.5	3.4	3.4	5.9	2.4
					B	3.4	1.5	5.2	7.2	6.9	7.1	1.9	6.2	2.8	2.5	1.6	17.8	6.5	2.3	9.0	4.1	2.8	4.7	4.3
		ENVI-Carb+PSA	GC-MS	Green tea	A	18.0	10.6	13.7	6.3	9.9	17.0	26.5	12.1	9.8	10.5	8.2	3.5	5.1	14.5	7.8	18.5	11.3	25.0	5.9
					B	20.0	28.2	1.4	3.0	11.0	15.1	9.8	14.0	8.7	2.9	25.3	4.1	22.2	5.5	18.4	4.8	4.3	22.8	8.6
				Oolong tea	A	9.4	5.7	4.1	5.7	7.3	1.8	4.3	22.2	4.3	3.1	8.6	3.5	5.1	2.5	1.8	10.8	7.3	8.2	1.8
					B	12.8	2.4	11.3	3.1	21.4	6.9	5.7	6.2	5.7	2.0	8.2	33.5	5.4	7.7	11.0	2.2	9.7	9.9	5.4
			GC-MS/MS	Green tea	A	21.6	7.9	9.9	4.0	7.5	20.3	14.2	16.1	14.2	1.6	3.4	39.7	18.6	11.1	6.7	7.0	9.3	2.8	6.5
					B	25.5	10.9	3.2	12.0	8.9	3.6	11.0	12.6	11.0	10.4	11.4	7.1	4.7	14.4	5.1	17.9	6.9	17.0	10.9
				Oolong tea	A	7.6	2.0	6.0	7.4	3.1	4.5	4.4	13.3	4.4	3.8	6.8	3.8	5.6	3.8	10.3	5.3	7.5	9.5	2.5
					B	11.3	3.4	8.8	3.9	12.0	7.7	3.2	5.5	9.6	2.9	8.7	6.2	5.6	8.9	10.6	2.0	5.8	17.6	3.8

183	Propoxur-1	TPT	GC–MS	Green tea	A	6.2	7.7	2.2	6.7	5.7	4.5	8.2	3.6	6.5	6.8	5.3	0.8	7.6	4.2	2.4	4.9	2.6	8.6	4.4
					B	5.7	11.7	5.6	8.7	2.6	3.7	3.4	11.4	5.0	6.3	5.6	5.8	2.9	2.4	5.1	1.7	7.6	2.6	5.8
				Oolong tea	A	2.0	2.5	2.4	5.7	3.8	3.3	10.1	8.8	4.4	4.2	2.0	4.4	5.4	2.0	9.3	4.4	4.0	4.9	1.0
					B	7.1	3.3	2.9	6.6	1.9	5.0	8.3	3.2	5.0	4.2	0.7	3.2	3.9	2.1	1.5	0.1	1.2	2.6	1.3
			GC–MS/MS	Green tea	A	5.6	7.0	1.3	2.3	9.2	5.3	7.2	2.7	4.5	4.0	5.2	1.6	10.3	10.2	2.0	3.5	4.9	6.4	2.2
					B	6.0	13.6	8.9	6.8	4.5	4.5	3.9	9.6	5.5	10.7	4.6	6.0	5.1	4.6	6.0	3.5	4.5	5.7	6.5
				Oolong tea	A	3.9	6.8	3.3	6.5	6.6	2.4	10.1	5.5	2.7	4.8	0.5	3.9	6.2	1.3	8.6	3.7	2.0	3.4	5.0
					B	6.7	5.5	3.0	8.7	1.1	3.1	7.4	8.4	15.4	2.3	7.8	2.6	11.8	1.5	5.8	2.1	1.8	0.9	4.0
		ENVI-Carb+PSA	GC–MS	Green tea	A	6.7	10.7	0.6	6.4	7.0	5.7	23.8	3.0	5.0	4.9	7.9	5.8	2.9	3.3	5.6	7.7	12.9	3.1	4.1
					B	4.8	5.4	4.5	8.1	3.7	4.0	3.4	7.0	3.9	0.5	21.4	1.4	2.6	2.7	13.9	3.2	3.4	7.3	4.8
				Oolong tea	A	3.9	4.2	1.3	3.5	6.8	4.0	4.0	5.3	2.2	8.8	3.4	7.1	0.9	3.1	2.6	5.4	5.2	0.7	5.5
					B	5.8	3.0	12.1	5.8	11.4	8.1	0.8	9.4	7.7	5.7	5.5	5.7	1.0	17.9	8.1	3.2	3.7	6.6	4.1
			GC–MS/MS	Green tea	A	9.1	10.0	3.3	8.0	3.9	5.5	20.1	1.3	3.8	1.9	7.0	5.3	7.8	7.3	5.9	7.6	13.5	4.1	2.9
					B	4.5	5.4	6.3	5.6	1.0	1.9	1.1	7.8	2.7	1.7	24.6	11.1	3.6	16.6	17.0	4.6	2.9	7.9	4.4
				Oolong tea	A	3.0	6.0	3.3	1.7	6.4	1.2	4.3	2.8	2.1	8.8	6.5	12.4	2.9	6.8	11.5	6.4	4.2	1.7	5.0
					B	6.6	2.7	11.9	9.5	6.7	8.6	4.2	8.3	7.5	3.2	5.6	2.3	8.7	15.7	9.6	4.7	4.2	5.7	4.0
184	Propoxur-2	TPT	GC–MS	Green tea	A	15.5	4.8	2.7	6.9	14.7	7.7	19.1	4.3	6.8	11.6	7.5	4.3	8.8	5.1	13.5	6.8	3.5	5.3	5.5
					B	14.3	12.7	3.1	11.6	10.8	9.6	13.2	8.1	5.0	12.9	4.1	13.1	1.1	7.4	6.8	3.8	9.8	10.2	5.8
				Oolong tea	A	16.4	1.1	8.3	1.0	4.7	3.5	20.9	13.0	14.9	6.3	19.1	14.2	9.3	3.4	9.9	6.2	5.3	3.7	5.6
					B	14.7	2.1	11.6	6.3	4.9	3.7	12.0	0.9	12.8	7.4	1.2	17.1	2.5	1.6	1.1	1.5	2.0	5.4	11.2
			GC–MS/MS	Green tea	A	6.8	9.7	9.8	4.3	15.9	8.1	22.4	37.9	31.8	5.4	4.8	8.4	14.9	15.2	0.8	8.5	14.4	28.9	7.3
					B	9.1	5.1	11.2	17.3	5.5	3.9	4.6	24.9	16.9	10.8	3.7	8.2	21.4	7.0	8.9	3.4	19.6	16.2	12.8
				Oolong tea	A	8.4	17.7	12.4	6.1	5.6	6.9	4.6	39.3	4.5	8.0	15.4	6.3	14.2	8.8	8.7	11.3	8.4	18.2	10.0
					B	8.9	7.5	20.9	6.3	5.2	26.7	2.5	3.5	28.2	11.3	4.2	6.1	8.9	10.2	18.2	5.5	3.6	8.9	9.9
		ENVI-Carb+PSA	GC–MS	Green tea	A	28.1	10.8	19.2	28.3	4.0	4.8	24.3	51.5	3.8	6.7	10.0	12.8	2.0	10.2	6.5	13.4	14.4	17.7	15.2
					B	60.4	13.6	1.3	35.5	13.7	11.8	4.5	37.0	2.8	21.6	22.1	11.3	11.1	0.8	21.7	5.6	5.7	11.3	14.0
				Oolong tea	A	8.3	6.8	5.9	19.4	8.3	3.4	6.1	41.7	10.8	10.2	11.5	4.3	14.0	7.5	15.4	17.8	10.8	14.1	23.9
					B	13.4	8.1	9.1	10.9	10.4	11.1	10.4	7.8	30.7	13.5	6.9	6.1	18.2	18.0	24.2	7.4	18.7	33.8	16.4
			GC–MS/MS	Green tea	A	21.0	23.8	9.5	8.6	17.7	26.2	40.9	56.2	39.0	1.4	10.0	38.8	13.1	40.3	28.0	19.5	58.1	24.8	23.3
					B	43.9	10.0	11.6	33.0	5.3	12.9	40.1	60.9	50.9	26.9	25.3	21.7	43.6	16.7	35.8	43.2	28.0	53.6	37.4
				Oolong tea	A	17.2	9.2	11.5	23.9	4.5	5.9	13.5	47.3	5.7	12.1	4.1	8.5	10.8	18.5	19.2	14.1	16.0	18.7	10.0
					B	30.5	5.0	18.2	4.3	16.7	11.0	11.6	8.3	18.2	3.4	8.1	21.4	18.6	12.6	15.2	11.9	4.5	48.3	9.9
185	Prothio-phos	TPT	GC–MS	Green tea	A	4.2	5.5	4.5	0.9	8.2	3.7	6.0	8.3	4.9	6.0	7.6	11.2	7.1	5.3	1.0	1.5	3.5	5.1	2.6
					B	5.7	7.7	3.4	7.6	5.0	2.1	1.8	3.9	3.1	10.9	2.5	13.5	0.8	1.1	5.5	1.9	3.0	1.9	5.8
				Oolong tea	A	6.0	2.2	6.4	3.6	0.6	4.0	8.1	8.3	2.1	9.4	6.9	6.3	1.7	3.9	9.6	3.0	3.7	2.2	1.1
					B	3.7	1.7	5.5	6.7	4.0	3.6	4.1	2.1	7.1	4.6	12.6	1.9	2.3	3.7	4.9	1.0	0.9	0.8	3.0
			GC–MS/MS	Green tea	A	4.2	2.5	5.2	4.1	9.7	2.4	9.9	5.2	4.7	2.7	5.1	1.4	7.3	6.6	8.8	2.8	7.0	4.2	5.0
					B	6.1	7.2	7.8	9.4	5.0	6.2	4.7	3.3	7.4	11.7	4.9	6.2	4.9	3.5	4.3	3.5	7.4	4.5	1.5
				Oolong tea	A	6.8	1.9	2.7	4.5	4.2	2.6	11.3	2.4	5.9	3.6	1.8	2.7	8.2	5.2	2.2	9.8	3.7	2.2	5.3
					B	5.2	6.2	6.8	7.3	2.2	3.8	3.3	7.7	7.4	2.6	4.7	3.4	10.1	3.7	8.7	2.2	3.6	1.7	4.7
		ENVI-Carb+PSA	GC–MS	Green tea	A	10.2	3.2	2.6	4.1	9.1	6.5	5.4	3.3	6.5	4.0	4.9	5.1	0.8	1.8	2.9	3.0	8.3	8.7	7.2
					B	8.8	10.7	3.8	3.8	1.6	6.0	6.6	2.8	8.7	3.9	29.0	14.1	4.1	1.2	9.2	7.4	4.3	11.6	8.8
				Oolong tea	A	1.5	2.9	4.4	2.0	4.8	1.1	3.6	4.8	2.7	1.6	5.9	5.5	0.4	3.5	3.0	1.6	5.3	4.7	4.1
					B	5.0	2.7	9.8	3.9	14.9	4.5	2.4	5.5	11.5	1.6	12.3	8.5	4.3	7.1	10.4	1.8	6.5	1.4	3.7
			GC–MS/MS	Green tea	A	13.0	4.9	2.7	5.0	6.5	7.9	8.5	1.1	8.2	0.3	8.6	11.4	10.0	10.4	2.7	2.7	8.3	1.6	3.9
					B	11.5	11.8	3.5	7.0	5.8	5.5	7.6	6.6	7.4	4.6	12.9	18.2	3.7	18.7	8.9	7.8	5.8	9.5	9.6
				Oolong tea	A	2.9	1.8	5.8	4.3	1.9	0.7	2.6	7.8	1.7	3.3	2.6	3.9	8.6	3.7	11.7	0.4	5.4	7.3	5.3
					B	8.8	6.0	11.3	2.7	8.6	5.0	4.5	7.8	10.0	3.5	7.9	9.1	4.7	8.8	7.8	4.0	3.8	2.6	4.7

(Continued)

Appendix Table 5.2 RSD of 201 Pesticides in Youden Pair Aged Tea Sample for Parallel Determinations (Nov 9, 2009–Feb 7, 2010) (cont.)

No.	Pesticides	SPE	Method	Sample	Youden pair	Nov 9, 2009 (n = 5)	Nov 14, 2009 (n = 3)	Nov 19, 2009 (n = 3)	Nov 24, 2009 (n = 3)	Nov 29, 2009 (n = 3)	Dec 4, 2009 (n = 3)	Dec 9, 2009 (n = 3)	Dec 14, 2009 (n = 3)	Dec 19, 2009 (n = 3)	Dec 24, 2009 (n = 3)	Dec 29, 2009 (n = 3)	Jan 3, 2010 (n = 3)	Jan 8, 2010 (n = 3)	Jan 13, 2010 (n = 3)	Jan 18, 2010 (n = 3)	Jan 23, 2010 (n = 3)	Jan 28, 2010 (n = 3)	Feb 2, 2010 (n = 3)	Feb 7, 2010 (n = 3)
186	Pyridaben	TPT	GC-MS	Green tea	A	18.4	11.1	37.8	10.4	13.2	3.1	12.1	13.2	14.6	2.0	11.2	7.4	8.1	6.5	2.0	13.9	5.5	21.7	3.3
					B	12.3	10.9	23.6	9.9	10.2	8.3	3.3	12.6	1.8	10.7	3.8	14.9	11.7	13.4	9.0	14.0	13.6	13.1	5.7
				Oolong tea	A	34.5	7.6	7.5	6.2	16.5	7.3	10.1	16.4	9.3	8.8	13.2	11.9	5.3	2.4	3.2	9.3	5.3	3.0	3.3
					B	22.2	4.7	5.7	3.4	9.6	12.1	7.1	2.3	9.2	0.5	15.1	7.1	2.4	7.4	6.7	1.6	5.4	6.1	8.2
			GC-MS/MS	Green tea	A	7.8	11.2	3.2	3.6	13.8	9.1	5.9	7.6	3.3	6.6	1.9	9.9	1.6	3.5	3.6	1.1	9.2	10.2	2.2
					B	8.3	7.2	8.3	9.9	5.9	8.8	3.1	3.7	13.4	8.0	0.4	10.8	2.1	5.1	5.3	2.2	8.0	4.1	6.8
				Oolong tea	A	10.3	1.2	4.6	4.6	8.6	15.9	19.8	0.3	6.3	7.2	4.4	5.1	8.4	4.7	7.6	0.6	6.6	2.6	19.5
					B	11.2	5.1	9.0	17.7	10.6	2.3	9.9	1.6	20.4	2.0	7.4	3.2	13.3	5.6	7.3	3.9	4.1	5.9	6.3
		ENVI-Carb+PSA	GC-MS	Green tea	A	42.3	11.8	15.5	5.4	7.3	27.1	13.0	21.2	25.4	8.3	5.3	26.6	18.0	14.3	9.7	11.3	13.5	18.5	5.7
					B	13.1	10.0	7.7	6.2	14.3	11.5	26.9	12.4	10.1	15.6	23.0	19.3	35.3	7.0	11.8	2.3	7.3	12.8	13.7
				Oolong tea	A	10.1	13.1	11.7	8.0	8.8	8.2	11.5	17.3	6.5	3.1	2.4	19.0	11.4	3.9	4.1	30.6	12.0	19.6	11.1
					B	12.4	7.6	4.2	11.9	18.3	13.2	19.7	27.5	14.2	8.0	11.1	6.8	3.8	27.7	15.7	3.5	17.7	18.4	8.7
			GC-MS/MS	Green tea	A	7.4	4.0	2.6	10.0	5.8	9.2	9.5	23.1	11.0	16.7	14.2	4.1	0.7	9.2	3.3	8.8	9.2	3.8	7.1
					B	13.1	5.3	11.0	0.8	2.6	4.1	6.5	26.5	1.4	11.9	18.3	3.0	8.9	14.1	11.1	3.9	8.3	7.5	3.6
				Oolong tea	A	67.8	8.8	8.6	23.5	10.4	7.8	8.5	16.1	2.3	2.6	7.0	6.0	17.3	5.7	5.0	10.8	5.0	7.9	19.7
					B	24.1	6.9	10.8	16.6	8.5	3.4	4.4	6.3	12.9	6.5	6.1	7.6	9.5	7.7	4.3	5.9	1.7	7.0	6.3
187	Pyrida-phenthion	TPT	GC-MS	Green tea	A	6.8	8.2	0.9	4.6	6.3	3.7	5.3	11.1	8.7	5.8	9.2	12.2	8.1	8.0	4.7	2.5	14.5	14.9	4.6
					B	5.4	5.3	5.9	7.5	2.3	2.6	1.0	4.4	6.4	8.5	3.5	15.1	10.3	3.1	3.2	1.4	5.7	14.4	3.6
				Oolong tea	A	9.1	5.6	11.8	5.3	7.0	4.1	0.7	14.9	11.5	10.4	16.6	13.6	3.1	7.7	7.2	5.1	10.0	4.5	8.4
					B	4.6	1.8	9.5	8.1	8.2	8.5	4.8	5.8	8.7	2.4	18.8	8.9	5.1	10.8	2.4	4.9	7.2	4.9	11.5
			GC-MS/MS	Green tea	A	7.4	5.2	1.7	12.4	12.2	1.5	9.1	11.4	10.4	11.8	9.9	4.2	13.7	10.8	3.9	1.6	6.9	7.6	9.9
					B	6.5	13.6	3.8	6.5	5.1	1.4	2.9	12.1	12.4	11.3	4.8	19.9	10.4	4.1	9.0	5.6	1.9	2.1	4.3
				Oolong tea	A	7.4	4.1	3.2	2.2	3.0	4.2	6.0	7.1	8.2	3.1	7.2	14.4	10.7	4.4	5.0	3.2	0.3	3.6	1.1
					B	4.4	3.7	7.5	7.3	8.8	3.3	5.5	16.5	11.7	2.7	4.7	2.9	13.4	4.3	7.0	5.5	6.6	5.8	3.2
		ENVI-Carb+PSA	GC-MS	Green tea	A	10.7	5.1	10.1	6.1	12.6	13.9	13.6	7.0	19.9	8.9	4.3	7.9	5.4	4.7	5.0	9.1	13.6	29.4	9.5
					B	13.8	7.4	2.9	5.4	0.3	5.4	3.2	10.8	8.1	8.1	33.0	13.8	15.6	6.0	7.0	2.4	4.1	6.0	6.8
				Oolong tea	A	6.3	5.0	7.2	1.2	10.0	3.4	5.0	17.3	1.4	7.5	14.7	4.3	1.0	5.7	5.5	7.5	0.9	12.9	3.4
					B	7.8	4.9	10.1	4.8	10.8	9.2	2.8	0.9	11.3	11.9	12.9	14.3	8.8	3.5	12.2	4.9	6.1	8.3	7.4
			GC-MS/MS	Green tea	A	12.5	8.8	4.3	7.5	9.9	15.1	15.4	11.3	8.1	1.6	8.4	1.0	1.4	15.7	4.9	7.1	7.7	15.3	7.4
					B	15.6	8.8	3.9	9.3	3.5	8.8	7.7	13.6	17.1	12.9	20.1	10.3	9.3	24.1	7.8	6.2	5.6	18.0	17.4
				Oolong tea	A	4.9	2.0	9.2	5.1	2.3	3.0	2.8	20.0	1.9	12.4	7.2	18.6	25.4	3.1	7.6	4.4	10.7	16.8	1.1
					B	11.7	6.2	11.7	5.2	12.6	8.7	5.6	5.5	9.3	7.8	6.8	16.8	9.5	7.6	11.0	5.5	15.5	13.2	3.2
188	Pyrimeth-anil	TPT	GC-MS	Green tea	A	4.8	5.2	4.5	3.1	9.5	3.6	7.0	3.2	5.0	3.5	7.3	3.4	7.5	3.2	1.1	3.3	6.5	7.0	3.7
					B	5.1	8.8	5.6	8.7	3.4	2.8	3.0	3.0	3.2	9.4	3.0	10.3	1.8	4.7	4.3	2.0	7.2	1.2	4.7
				Oolong tea	A	5.1	1.3	3.5	3.9	2.5	3.7	8.2	6.4	1.4	6.9	6.0	3.7	2.8	4.4	7.5	4.7	3.5	3.8	0.9
					B	3.5	2.2	2.9	6.2	2.2	3.6	3.2	3.5	2.9	7.0	13.5	3.0	5.1	3.3	2.4	1.7	0.8	1.6	2.0
			GC-MS/MS	Green tea	A	4.1	3.9	3.8	2.8	11.7	3.7	6.5	3.6	5.2	0.4	6.4	3.6	7.3	3.6	4.9	1.1	3.4	10.8	1.5
					B	5.8	10.6	7.7	8.2	4.8	4.9	3.7	4.6	4.2	9.7	3.0	10.4	1.6	2.8	2.4	1.9	6.7	2.7	8.3
				Oolong tea	A	4.8	1.4	1.9	3.9	5.8	4.4	8.9	2.6	2.8	4.1	2.4	3.2	7.1	5.9	5.8	7.5	4.4	3.8	8.0
					B	5.4	1.8	2.9	6.0	2.3	3.2	4.6	8.6	7.3	1.6	4.0	3.9	10.9	4.1	11.5	2.3	1.9	3.6	5.1
		ENVI-Carb+PSA	GC-MS	Green tea	A	5.5	3.5	2.4	5.9	9.3	2.1	6.7	0.9	4.9	3.8	5.3	7.8	1.6	3.3	5.9	6.4	11.7	3.8	5.5
					B	4.4	8.0	4.8	7.1	3.0	5.4	3.8	4.4	4.2	1.6	1.6	8.1	2.4	2.4	12.2	3.7	6.1	8.1	7.2
				Oolong tea	A	2.9	5.0	3.1	1.4	5.5	3.8	3.9	4.4	2.2	6.4	6.4	11.2	4.0	3.9	2.8	4.3	9.1	3.5	3.8
					B	4.1	3.0	10.3	3.5	13.6	4.7	2.0	6.7	7.5	7.4	2.5	1.3	2.9	8.0	7.8	6.9	4.6	5.8	5.5
			GC-MS/MS	Green tea	A	7.6	5.2	1.8	8.8	6.5	1.8	5.5	1.7	4.1	1.4	8.4	1.3	1.8	9.7	7.1	6.5	7.9	4.8	4.6
					B	4.1	7.2	4.5	3.3	1.2	0.9	2.7	5.9	1.1	0.2	21.4	13.4	4.1	20.0	14.1	2.5	7.1	6.2	6.8
				Oolong tea	A	2.2	5.3	5.0	1.8	4.4	1.8	6.8	3.4	3.6	7.6	5.8	4.3	5.3	2.6	8.5	3.2	5.3	6.9	8.0
					B	4.0	0.4	11.0	6.7	8.7	2.6	5.8	8.7	7.2	1.9	7.7	2.7	4.2	10.9	14.5	2.1	2.1	2.4	5.1

No.	Compound	Pretreatment	Instrument	Tea	Rep	C1	C2	C3	C4	C5	C6	C7	C8	C9	C10	C11	C12	C13	C14	C15	C16	C17	C18	C19
189	Pyriproxy-fen	TPT	GC-MS	Green tea	A	5.0	5.1	0.5	1.2	5.6	6.3	9.9	3.4	4.3	1.9	8.6	1.8	5.2	1.6	3.6	0.5	4.3	7.8	3.8
					B	4.1	7.7	4.3	7.3	2.4	3.3	5.3	1.1	4.4	9.7	6.2	10.0	2.6	1.8	3.9	2.2	7.8	2.3	3.7
				Oolong tea	A	7.3	4.5	2.0	4.2	3.0	2.9	5.1	5.9	1.3	5.8	5.5	4.7	4.9	1.9	6.3	4.7	1.7	5.5	0.3
					B	1.9	2.4	1.9	8.4	2.8	4.0	0.0	0.2	6.4	3.9	3.5	1.0	0.3	4.3	5.1	3.7	1.5	1.8	1.5
			GC-MS/MS	Green tea	A	7.4	10.7	3.1	4.3	13.2	10.3	6.2	3.7	3.5	5.8	2.6	10.0	3.5	2.9	3.6	0.8	5.8	10.4	2.3
					B	3.6	8.5	8.6	9.9	4.8	9.0	4.4	2.3	11.5	9.4	1.8	10.3	0.7	5.9	5.5	2.4	3.7	3.8	6.8
				Oolong tea	A	7.3	2.0	1.2	2.7	2.7	4.0	7.9	0.9	1.4	3.5	3.3	6.0	10.1	3.2	5.5	4.7	4.5	5.6	5.6
					B	5.1	3.9	4.3	5.7	0.5	3.2	5.7	7.4	11.1	4.0	3.9	2.3	16.5	2.7	5.3	2.2	1.4	0.9	2.1
		ENVI-Carb+PSA	GC-MS	Green tea	A	5.7	5.0	4.3	7.9	8.2	5.6	3.9	1.5	6.8	5.6	5.2	2.7	1.5	4.4	7.1	7.1	8.3	2.1	3.9
					B	3.4	6.4	4.6	6.8	2.3	4.7	2.1	3.5	5.1	3.9	22.0	9.9	3.7	5.6	10.7	0.7	5.0	7.5	6.3
				Oolong tea	A	6.5	3.6	4.2	3.1	5.4	0.5	1.4	5.7	0.6	1.0	5.9	1.2	4.5	6.6	1.7	1.2	4.3	4.0	2.1
					B	5.0	0.6	11.0	2.6	15.3	2.5	2.3	5.4	12.2	3.3	8.7	4.7	2.7	3.9	7.0	4.1	5.4	5.1	1.5
			GC-MS/MS	Green tea	A	7.6	4.7	1.4	10.2	6.6	6.4	5.6	4.6	10.4	13.6	15.4	2.6	1.8	10.6	2.8	8.9	9.4	4.2	7.3
					B	9.8	5.4	11.9	1.7	1.4	2.0	1.4	10.6	1.3	8.3	18.9	0.7	7.8	13.8	11.3	4.3	8.6	7.5	4.0
				Oolong tea	A	8.0	2.9	5.7	0.9	5.1	2.5	2.7	6.6	0.6	2.5	6.9	2.9	15.7	6.6	7.2	2.8	4.8	14.2	5.6
					B	6.6	3.9	10.0	6.2	7.2	3.7	3.2	9.2	6.9	1.9	3.9	2.9	3.7	6.3	7.4	2.3	2.9	3.8	2.1
190	Quinal-phos	TPT	GC-MS	Green tea	A	5.8	6.8	3.7	1.5	9.9	5.5	7.6	12.3	11.6	7.0	10.2	6.2	9.1	4.6	12.8	1.7	4.3	5.2	3.8
					B	5.4	5.4	5.0	9.0	4.9	3.9	2.2	3.7	8.3	11.5	3.8	13.3	2.8	3.1	4.3	5.3	2.8	1.3	5.0
				Oolong tea	A	5.5	2.1	14.0	4.2	1.9	2.4	4.0	7.4	5.8	4.3	3.6	3.9	6.9	8.8	10.4	3.6	5.3	1.6	4.3
					B	2.7	0.7	4.8	6.3	9.4	5.6	8.1	2.2	6.4	4.4	6.4	2.8	9.5	4.5	5.2	3.5	2.6	2.1	3.5
			GC-MS/MS	Green tea	A	4.7	5.1	1.4	9.7	12.7	2.5	8.1	8.7	5.3	4.3	8.1	2.1	12.5	2.9	7.5	4.6	6.0	6.1	9.2
					B	5.7	5.2	2.5	6.9	5.6	2.8	2.3	5.6	8.4	10.7	4.8	5.0	13.1	5.5	6.4	0.7	2.1	3.7	9.1
				Oolong tea	A	4.8	3.7	1.6	1.7	3.3	0.6	12.9	2.0	6.9	4.3	7.5	4.4	7.5	0.6	5.8	12.0	5.5	2.0	7.1
					B	2.6	6.5	6.2	10.5	5.5	1.5	3.6	6.0	14.0	1.7	6.6	2.0	3.9	2.8	6.8	1.0	3.0	3.9	6.6
		ENVI-Carb+PSA	GC-MS	Green tea	A	12.5	11.4	4.7	7.8	8.5	9.9	14.7	6.0	5.4	6.0	6.6	10.3	2.5	4.0	12.7	7.3	7.9	14.8	6.3
					B	17.4	3.7	2.4	3.0	3.4	5.8	3.5	7.8	6.4	5.1	41.1	9.3	8.2	0.5	10.5	3.7	4.9	9.4	6.2
				Oolong tea	A	2.2	3.4	2.2	2.2	5.9	4.1	5.1	10.1	4.4	1.9	10.8	5.2	3.2	7.4	7.9	0.8	6.7	7.9	11.0
					B	6.9	5.2	5.7	1.7	15.3	2.6	2.9	6.5	10.4	3.2	10.5	6.2	6.2	2.5	11.5	4.3	6.7	5.6	6.7
			GC-MS/MS	Green tea	A	15.7	10.0	1.7	5.5	6.4	10.7	6.1	8.4	2.9	0.1	7.4	17.9	4.4	5.3	3.8	3.1	6.4	12.6	10.0
					B	12.2	0.8	3.6	5.1	2.7	4.1	9.1	7.0	9.1	8.1	17.7	16.8	9.2	11.1	9.9	6.0	8.7	20.2	17.2
				Oolong tea	A	2.0	4.1	5.9	5.4	1.3	2.7	3.2	7.2	3.3	2.2	4.7	7.4	1.7	2.9	7.0	5.0	10.2	17.3	7.1
					B	9.3	6.0	11.0	3.8	9.0	5.4	3.3	6.6	9.9	4.4	8.0	9.9	8.9	6.0	13.2	1.3	7.3	9.5	6.6
191	Quinoxy-phen	TPT	GC-MS	Green tea	A	5.6	6.8	0.8	3.5	3.3	4.4	9.8	4.1	5.2	5.2	5.5	2.2	6.2	4.9	3.2	1.7	3.6	6.4	1.8
					B	5.1	11.8	4.1	8.0	2.9	3.6	0.6	4.3	2.5	8.3	7.4	6.9	2.1	1.3	4.8	2.7	4.1	1.7	3.0
				Oolong tea	A	7.0	9.3	10.2	3.6	2.3	6.4	7.8	8.0	4.3	4.3	3.9	1.6	4.8	3.6	6.9	4.5	4.1	3.6	1.4
					B	5.1	4.0	10.7	5.5	9.4	3.7	11.3	3.1	6.2	0.6	1.8	0.6	3.4	4.5	3.7	3.7	1.6	2.1	1.4
			GC-MS/MS	Green tea	A	4.2	10.0	1.3	3.9	4.3	5.0	7.7	5.3	3.1	1.9	5.8	1.9	8.5	3.9	3.2	0.8	6.3	5.4	3.1
					B	7.4	2.6	6.1	8.2	4.3	3.0	1.7	3.1	5.6	10.4	1.2	5.0	5.9	0.6	3.4	3.5	4.0	5.3	5.1
				Oolong tea	A	7.0	4.1	0.6	3.4	0.9	4.5	7.1	1.6	2.9	3.9	1.2	3.4	6.1	0.3	7.9	2.6	1.5	3.8	4.5
					B	4.3	11.2	4.5	6.8	7.3	3.1	5.0	7.4	7.0	2.4	5.6	3.2	3.0	4.0	6.7	2.7	2.3	2.2	4.8
		ENVI-Carb+PSA	GC-MS	Green tea	A	4.7	14.3	2.2	5.1	1.6	1.6	3.2	2.4	5.7	3.4	6.5	4.9	2.1	3.9	3.3	3.3	12.0	4.1	4.0
					B	4.3	2.4	1.3	7.7	6.4	2.8	2.4	2.4	5.5	10.9	24.1	5.0	2.3	6.2	8.8	1.2	5.5	6.6	9.4
				Oolong tea	A	7.8	0.5	4.2	1.0	9.5	3.9	3.1	4.2	1.4	0.7	4.5	5.1	3.2	5.1	3.2	1.5	6.1	3.4	4.2
					B	4.6	2.2	8.8	3.1	7.5	4.8	2.2	5.8	10.0	4.1	9.6	3.4	4.5	6.8	8.6	2.9	6.6	2.8	5.1
			GC-MS/MS	Green tea	A	6.2	7.8	3.2	7.7	0.5	1.4	2.3	2.3	5.3	1.8	6.8	1.6	5.4	5.0	4.4	6.0	10.1	2.3	3.3
					B	4.3	3.2	4.4	3.7	5.9	1.3	1.7	6.5	3.5	0.8	22.5	1.6	4.0	8.1	10.1	1.7	1.8	9.3	5.8
				Oolong tea	A	5.8	3.4	4.6	2.1	7.6	2.9	1.9	3.8	0.6	2.6	6.3	5.8	6.2	2.4	8.5	2.4	4.2	8.9	4.5
					B	5.8	10.3	1.1	5.5	9.7	4.9	3.5	7.6	8.8	4.1	7.4	4.2	4.2	7.8	—	2.9	—	5.0	4.8

(Continued)

Appendix Table 5.2 RSD of 201 Pesticides in Youden Pair Aged Tea Sample for Parallel Determinations (Nov 9, 2009–Feb 7, 2010) (cont.)

No.	Pesticides	SPE	Method	Sample	Youden pair	RSD%																	
						Nov 9, 2009 (n=5)	Nov 14, 2009 (n=3)	Nov 19, 2009 (n=3)	Nov 24, 2009 (n=3)	Nov 29, 2009 (n=3)	Dec 4, 2009 (n=3)	Dec 9, 2009 (n=3)	Dec 14, 2009 (n=3)	Dec 19, 2009 (n=3)	Dec 24, 2009 (n=3)	Jan 3, 2010 (n=3)	Jan 8, 2010 (n=3)	Jan 13, 2010 (n=3)	Jan 18, 2010 (n=3)	Jan 23, 2010 (n=3)	Jan 28, 2010 (n=3)	Feb 2, 2010 (n=3)	Feb 7, 2010 (n=3)
192	Telodrin	TPT	GC-MS	Green tea	A	3.3	5.0	4.9	8.3	10.1	6.7	8.9	10.6	8.4	2.8	4.2	8.1	3.6	4.4	7.4	5.1	6.8	2.2
					B	6.1	11.0	4.1	7.2	3.8	3.4	3.7	8.0	4.2	8.8	6.1	5.2	4.6	5.1	2.8	7.9	2.3	5.8
				Oolong tea	A	5.4	2.9	5.0	5.1	7.5	2.6	8.1	3.5	10.9	2.4	2.8	2.8	5.9	6.0	4.1	2.8	3.3	8.2
					B	3.8	6.8	6.2	1.8	3.0	5.3	5.2	5.4	7.8	5.8	4.0	3.1	3.2	8.5	1.1	0.3	1.2	4.5
			GC-MS/MS	Green tea	A	7.5	3.9	3.3	7.0	16.0	3.4	10.8	10.7	7.5	5.0	4.5	11.1	10.1	3.9	3.9	7.5	10.0	5.3
					B	5.9	7.3	4.6	10.5	4.0	4.9	4.2	8.0	5.7	10.7	7.2	11.2	6.5	5.8	5.8	9.8	3.8	7.6
				Oolong tea	A	8.1	3.0	1.5	7.2	6.1	3.7	8.1	9.1	7.9	5.1	2.6	2.5	1.4	3.1	7.3	4.0	4.9	1.9
					B	8.3	6.7	8.6	8.4	7.0	4.6	3.6	6.1	8.3	5.4	3.1	13.9	2.8	7.7	5.5	5.7	4.2	3.8
		ENVI-Carb+PSA	GC-MS	Green tea	A	20.3	9.7	12.8	6.0	7.6	16.7	9.7	10.6	9.0	5.0	13.4	4.5	7.7	15.3	6.1	10.4	12.5	10.0
					B	23.4	11.7	0.4	12.2	6.1	4.1	12.0	14.4	14.4	11.7	16.4	19.7	7.6	12.9	15.4	4.0	18.9	9.2
				Oolong tea	A	5.2	3.2	4.4	6.3	2.9	2.7	6.1	11.8	2.8	1.6	4.8	4.2	2.7	3.1	11.9	6.8	7.8	4.4
					B	8.4	1.2	11.7	0.8	11.9	6.0	5.7	6.4	13.0	4.0	7.3	6.8	8.9	12.9	2.2	9.5	11.4	5.8
			GC-MS/MS	Green tea	A	27.3	11.5	10.6	4.5	7.6	21.5	11.6	11.8	6.9	2.6	6.3	4.2	13.0	8.3	10.5	6.2	8.8	7.3
					B	29.0	10.9	2.6	12.8	11.2	4.9	16.9	13.4	11.9	11.4	1.6	15.1	15.7	4.6	21.9	11.7	15.3	16.4
				Oolong tea	A	7.8	2.6	4.8	7.7	2.9	7.9	3.5	12.7	3.1	2.2	4.9	6.4	1.0	9.2	3.4	7.1	12.1	1.9
					B	11.1	5.4	10.7	4.7	11.9	9.3	1.7	5.0	10.8	5.0	7.8	6.2	7.0	13.5	1.1	8.1	15.9	3.8
193	Tetrasul	TPT	GC-MS	Green tea	A	5.1	4.5	2.4	3.2	7.3	4.6	7.0	3.1	5.3	2.7	1.7	8.6	2.3	2.7	1.7	5.1	6.0	7.3
					B	16.6	7.9	3.5	9.0	3.5	5.5	1.7	3.2	3.3	8.9	6.4	2.6	1.5	4.0	3.1	5.5	1.4	5.4
				Oolong tea	A	6.5	1.5	1.3	2.4	0.6	1.5	6.6	7.7	2.6	2.6	2.4	5.1	2.9	8.9	3.3	2.3	6.6	1.3
					B	8.4	2.6	3.7	4.2	1.8	4.0	5.3	4.4	7.0	3.5	3.0	3.1	5.0	0.7	0.8	1.1	5.7	1.8
			GC-MS/MS	Green tea	A	5.2	4.1	2.3	12.8	10.0	2.4	7.9	2.9	3.8	4.5	3.1	9.6	3.8	5.9	3.3	4.3	7.1	0.6
					B	4.6	9.4	5.5	9.6	5.0	2.3	2.0	4.5	5.5	10.2	10.0	7.1	1.0	4.1	2.5	7.3	4.3	4.2
				Oolong tea	A	6.5	0.7	2.7	4.3	3.3	3.2	6.9	2.2	2.4	2.3	5.1	6.2	2.6	9.4	6.2	4.6	2.5	5.1
					B	5.2	5.9	5.1	8.4	0.8	0.9	4.0	9.5	8.4	1.2	1.3	7.9	4.4	5.7	4.4	0.9	3.3	3.8
		ENVI-Carb+PSA	GC-MS	Green tea	A	5.4	2.3	0.2	6.1	8.3	2.2	6.3	1.1	6.8	2.8	1.5	2.2	6.5	8.1	10.0	10.3	5.7	5.5
					B	3.0	8.8	5.5	5.8	1.8	0.9	2.4	2.6	5.1	0.8	2.6	3.2	3.9	9.0	1.8	4.9	10.9	7.0
				Oolong tea	A	5.4	3.2	7.0	1.4	6.1	3.1	3.6	1.8	2.3	2.1	3.3	1.6	4.1	8.8	2.0	5.4	5.2	3.8
					B	6.0	4.3	9.8	4.2	9.2	4.2	1.3	7.4	11.0	4.5	3.6	3.4	7.1	8.9	3.0	5.7	4.2	4.0
			GC-MS/MS	Green tea	A	5.5	3.8	1.4	8.6	8.1	1.0	6.3	0.8	4.7	0.7	3.7	7.4	10.7	3.1	6.7	6.2	4.7	8.1
					B	3.7	7.4	5.2	2.8	1.7	2.3	1.2	5.4	1.2	2.9	5.5	0.9	12.5	12.2	3.6	0.5	7.6	8.8
				Oolong tea	A	6.0	4.2	5.5	3.4	3.7	2.5	2.7	0.4	0.9	2.2	6.8	2.1	4.3	10.2	2.5	3.8	12.6	5.1
					B	6.6	3.0	11.0	6.5	6.8	5.7	4.5	5.7	9.3	4.0	0.2	7.9	8.6	12.6	3.4	3.5	5.0	3.8
194	Thiazopyr	TPT	GC-MS	Green tea	A	6.5	5.7	3.4	3.5	9.1	2.7	7.4	6.3	5.6	4.6	2.4	7.0	3.8	1.8	1.8	4.3	6.3	4.3
					B	5.6	7.8	4.9	8.3	3.1	5.0	1.6	7.2	5.6	4.3	7.1	1.0	3.1	3.8	3.4	5.5	1.0	7.2
				Oolong tea	A	4.7	7.8	4.2	3.3	0.7	4.3	8.7	7.5	5.0	4.3	3.1	4.8	2.9	7.7	3.4	5.0	3.9	1.7
					B	3.6	3.2	3.5	3.8	3.8	4.7	3.9	2.8	5.3	2.3	1.2	3.3	4.7	2.9	1.8	1.3	4.5	2.1
			GC-MS/MS	Green tea	A	5.9	6.5	5.1	43.6	14.2	2.7	7.8	6.2	4.3	6.8	8.0	10.1	1.7	5.0	1.1	15.5	3.7	4.6
					B	7.4	7.9	6.7	8.5	7.9	4.9	5.0	7.3	5.4	4.9	16.5	4.2	4.2	6.8	4.4	3.2	4.7	9.2
				Oolong tea	A	6.9	5.5	4.6	8.3	8.4	3.9	7.4	2.6	8.0	8.4	8.8	10.8	3.7	9.9	2.2	5.1	0.3	6.3
					B	5.9	5.8	3.5	17.0	7.5	2.6	3.8	11.9	7.1	2.0	4.6	8.1	8.6	8.8	2.5	3.7	7.7	10.7
		ENVI-Carb+PSA	GC-MS	Green tea	A	12.7	3.7	1.0	3.6	6.1	1.8	1.8	2.6	1.8	1.7	1.9	2.2	6.7	7.3	10.0	10.3	4.9	6.3
					B	9.0	6.8	6.6	3.6	1.6	1.6	4.2	3.8	5.8	0.6	5.1	1.4	1.6	14.0	2.5	4.4	10.1	9.9
				Oolong tea	A	4.0	3.9	3.3	1.3	5.5	0.6	3.2	1.2	2.5	4.0	4.6	3.5	2.8	3.7	3.0	6.3	2.9	4.1
					B	4.4	0.5	7.6	4.2	11.9	6.1	3.3	6.6	11.5	4.9	2.1	6.2	9.5	8.6	1.6	6.4	5.7	6.0
			GC-MS/MS	Green tea	A	7.1	5.6	4.9	5.6	7.7	2.9	3.9	1.2	5.4	2.3	1.6	1.6	4.3	5.2	11.5	15.0	7.2	4.1
					B	5.3	7.7	5.2	7.4	7.5	3.8	2.0	5.4	2.2	1.5	1.9	7.1	16.1	13.3	3.9	5.9	13.3	8.6
				Oolong tea	A	3.5	8.9	5.5	5.6	9.6	2.4	0.4	5.8	1.5	7.0	1.6	17.3	0.8	15.5	2.1	6.8	4.0	6.3
					B	6.9	1.2	12.3	8.4	10.6	8.7	6.2	8.1	9.1	6.0	9.7	13.7	7.7	14.2	5.0	3.3	8.0	10.7

195	Tolclofos-methyl	TPT	GC–MS	Green tea	A	4.6	5.3	3.6	0.8	7.3	3.8	4.8	6.0	5.1	1.9	8.1	0.6	7.6	3.7	0.4	3.0	3.5	6.3	2.9
					B	5.9	8.2	5.2	8.3	4.2	2.9	2.6	0.8	3.2	7.7	1.1	8.4	2.0	3.4	13.9	0.8	3.9	2.4	7.4
				Oolong tea	A	4.2	2.2	5.0	3.5	1.7	2.5	8.5	5.3	3.7	2.1	2.5	4.5	3.0	2.1	11.7	4.1	4.1	2.4	1.0
					B	4.1	5.0	3.9	2.9	4.6	3.8	3.4	2.9	7.4	1.8	2.0	1.6	6.7	3.8	3.0	1.0	1.0	1.5	2.0
			GC–MS/MS	Green tea	A	5.4	4.9	2.2	3.9	8.9	3.0	5.6	4.9	5.5	5.5	7.9	3.5	6.8	2.9	1.1	4.1	9.1	2.0	6.2
					B	6.7	2.1	4.9	8.7	5.6	2.5	4.3	5.6	9.3	5.5	0.6	3.8	2.9	4.4	5.2	4.1	0.8	4.3	6.3
				Oolong tea	A	3.8	5.6	2.7	3.9	5.6	3.0	9.9	2.2	1.8	2.9	3.8	4.2	2.2	2.1	9.3	4.9	3.9	8.3	10.8
					B	6.6	5.4	7.9	7.4	3.0	2.6	4.9	4.3	8.7	2.8	2.1	4.5	15.7	5.9	9.5	6.1	7.7	5.8	6.8
		ENVI-Carb +PSA	GC–MS	Green tea	A	8.1	9.3	2.3	8.4	8.4	4.0	4.2	1.0	2.2	2.4	7.3	3.2	3.1	2.6	1.8	4.9	10.1	7.2	5.6
					B	6.6	4.4	3.2	3.3	1.0	1.1	3.2	3.0	4.5	2.5	20.3	6.4	3.9	2.6	10.8	0.9	4.5	7.7	5.5
				Oolong tea	A	4.0	3.3	3.8	1.6	5.7	0.9	2.7	1.3	2.5	1.0	6.8	4.4	3.4	8.4	3.9	2.5	4.6	5.0	3.6
					B	5.4	4.8	8.9	3.7	12.3	4.1	1.9	5.2	12.0	0.5	8.7	0.5	5.0	8.5	7.2	2.9	5.8	2.0	3.9
			GC–MS/MS	Green tea	A	9.9	9.5	4.1	6.3	8.1	6.6	4.9	2.1	4.4	0.7	8.2	2.4	1.5	3.8	8.7	6.7	5.3	5.0	7.3
					B	10.1	4.8	2.6	1.0	1.5	1.9	4.9	5.3	4.1	4.2	17.3	3.4	6.0	10.7	12.8	5.1	7.3	5.5	8.5
				Oolong tea	A	3.4	4.1	5.1	2.7	7.4	2.5	4.5	3.7	3.1	4.4	3.5	8.7	10.5	3.1	8.7	4.3	6.8	2.5	10.8
					B	9.9	5.0	10.6	4.3	4.8	4.2	3.7	5.3	8.1	2.9	4.0	5.6	6.2	10.0	13.0	3.9	5.0	2.2	6.8
196	Transfluthrin	TPT	GC–MS	Green tea	A	9.9	8.2	4.8	1.8	6.9	4.6	6.3	2.7	6.0	2.0	5.3	2.5	8.1	9.5	13.0	4.4	10.4	7.2	2.6
					B	7.1	0.3	8.2	9.6	2.4	0.9	0.5	2.7	6.0	9.3	4.0	8.0	1.4	8.3	13.1	2.9	2.7	2.1	6.3
				Oolong tea	A	4.9	3.2	9.3	3.0	2.1	4.4	8.6	6.0	1.6	4.6	4.8	2.0	4.0	1.3	9.2	3.1	3.2	4.1	1.5
					B	4.7	3.8	3.1	1.6	2.1	5.1	2.2	3.1	4.2	3.4	3.0	2.8	2.6	3.4	2.0	2.5	1.0	0.6	0.3
			GC–MS/MS	Green tea	A	3.6	10.3	3.1	2.9	10.4	3.2	6.5	4.2	4.1	2.1	4.4	2.6	10.2	4.9	2.4	2.2	6.3	8.1	5.5
					B	6.1	3.5	7.9	8.7	5.6	3.0	4.2	3.4	7.7	10.2	3.0	6.4	3.1	2.7	6.3	4.1	6.6	5.6	6.0
				Oolong tea	A	4.6	2.8	3.6	4.5	7.4	3.6	8.2	0.6	4.5	6.0	0.8	4.2	8.2	2.0	6.0	3.1	4.1	2.1	5.2
					B	4.7	5.5	4.1	12.1	0.9	3.3	4.6	6.1	8.4	1.9	3.3	2.7	10.5	2.6	6.9	3.5	2.8	1.1	3.0
		ENVI-Carb +PSA	GC–MS	Green tea	A	6.3	6.1	5.1	9.3	6.7	7.3	5.6	2.7	6.1	5.0	8.5	3.8	4.1	1.8	4.6	3.1	9.0	2.9	4.5
					B	7.3	39.0	6.1	6.9	3.0	1.4	5.0	5.0	3.5	1.1	17.8	6.9	2.5	3.8	11.9	2.5	4.2	7.7	5.8
				Oolong tea	A	6.5	4.9	6.0	1.6	2.9	1.8	2.1	2.4	4.2	2.6	4.0	2.8	1.4	3.4	3.8	5.0	6.0	2.1	7.5
					B	5.3	2.0	9.0	3.4	11.9	5.0	1.6	8.6	11.2	5.7	8.6	1.8	3.8	7.6	11.0	1.9	3.5	2.8	3.2
			GC–MS/MS	Green tea	A	9.1	5.2	2.6	8.3	6.8	1.5	5.9	3.6	3.7	1.0	9.3	1.1	1.2	13.9	1.8	4.8	7.1	2.8	5.6
					B	6.8	9.0	5.2	1.7	2.2	2.3	3.3	6.1	4.0	0.3	19.6	1.6	4.3	9.4	11.0	4.0	1.7	10.1	5.2
				Oolong tea	A	2.2	4.7	5.4	3.0	4.0	3.7	3.1	4.5	2.8	1.9	5.9	4.1	3.6	2.2	10.7	5.5	6.4	4.5	5.2
					B	5.9	2.1	10.6	5.4	8.1	7.0	3.7	6.4	8.0	2.6	6.3	4.1	6.0	10.4	12.4	1.3	4.9	2.6	3.0
197	Triadimefon	TPT	GC–MS	Green tea	A	5.9	5.6	1.5	3.7	10.3	5.3	7.8	1.9	4.4	3.9	11.8	4.2	8.7	1.2	2.1	1.0	2.1	8.7	9.5
					B	7.4	6.3	4.4	8.0	3.3	5.9	1.6	9.0	3.8	6.1	8.8	5.9	0.2	3.6	7.7	6.9	4.3	9.5	11.1
				Oolong tea	A	5.3	2.4	2.5	6.0	4.2	2.8	6.4	10.0	8.2	13.8	2.5	6.8	6.8	2.1	4.1	4.2	5.2	4.6	0.9
					B	4.8	6.4	1.8	6.6	3.0	4.7	2.3	7.1	8.8	13.4	8.3	6.5	8.5	4.9	5.0	5.5	4.1	3.9	2.0
			GC–MS/MS	Green tea	A	8.8	14.9	2.1	5.2	12.6	4.7	8.0	6.4	5.9	3.5	6.1	1.4	10.4	5.2	2.0	1.6	6.1	7.4	7.3
					B	7.5	16.4	3.4	6.7	6.9	4.4	3.7	7.5	6.2	8.1	3.7	8.7	1.7	4.5	3.3	5.9	9.2	6.5	9.0
				Oolong tea	A	2.8	2.0	3.7	3.7	4.9	5.4	9.1	7.6	1.6	1.9	2.2	5.1	7.6	1.4	5.6	4.6	1.8	1.4	8.0
					B	6.1	4.7	3.3	6.1	2.9	2.4	4.7	7.2	13.3	2.6	6.0	4.2	7.7	4.7	8.5	4.5	2.9	5.1	6.8
		ENVI-Carb +PSA	GC–MS	Green tea	A	9.0	6.1	3.6	5.8	7.3	1.6	8.4	3.6	3.9	10.6	6.0	0.9	2.1	3.4	5.9	6.1	7.4	14.6	6.2
					B	3.2	13.8	2.3	8.2	4.0	6.8	0.6	5.0	15.2	5.6	29.1	14.3	1.3	2.2	18.7	2.2	9.1	7.5	9.1
				Oolong tea	A	3.3	2.6	3.5	6.4	14.0	1.5	3.4	6.3	8.3	0.8	4.3	11.9	1.0	15.1	2.6	3.1	7.0	3.7	6.5
					B	5.4	4.3	7.1	4.5	12.8	7.3	2.0	7.5	11.3	7.8	9.9	2.7	4.3	7.4	11.1	2.0	6.3	10.1	5.9
			GC–MS/MS	Green tea	A	6.6	1.7	3.9	9.4	7.1	3.2	6.8	2.5	2.8	0.3	3.2	12.0	3.0	10.2	6.1	9.5	12.2	3.8	6.4
					B	4.4	6.2	3.8	3.5	3.5	1.3	2.0	7.3	1.7	4.8	19.3	13.9	3.7	10.0	11.9	3.5	3.2	9.8	8.5
				Oolong tea	A	1.2	4.4	3.8	2.2	1.8	4.0	4.8	6.9	0.9	2.6	7.5	6.4	5.6	7.0	11.6	3.2	12.5	3.7	8.0
					B	4.8	1.1	11.4	9.4	9.9	4.3	6.0	9.5	8.6	0.0	5.5	6.7	5.6	9.5	14.1	1.4	3.4	3.5	6.8

(Continued)

Appendix Table 5.2 RSD of 201 Pesticides in Youden Pair Aged Tea Sample for Parallel Determinations (Nov 9, 2009–Feb 7, 2010) (cont.)

RSD%

No.	Pesticides	SPE	Method	Sample	Youden pair	Nov 9, 2009 (n=5)	Nov 14, 2009 (n=3)	Nov 19, 2009 (n=3)	Nov 24, 2009 (n=3)	Nov 29, 2009 (n=3)	Dec 4, 2009 (n=3)	Dec 9, 2009 (n=3)	Dec 14, 2009 (n=3)	Dec 19, 2009 (n=3)	Dec 24, 2009 (n=3)	Dec 14, 2009 (n=3)	Jan 3, 2010 (n=3)	Jan 8, 2010 (n=3)	Jan 13, 2010 (n=3)	Jan 18, 2010 (n=3)	Jan 23, 2010 (n=3)	Jan 28, 2010 (n=3)	Feb 2, 2010 (n=3)	Feb 7, 2010 (n=3)
198	Triadimenol	TPT	GC-MS	Green tea	A	10.8	4.5	0.8	3.0	6.8	4.6	7.7	5.1	4.4	8.1	11.5	9.8	7.6	6.7	2.5	4.4	4.6	7.5	4.5
					B	7.0	9.3	7.4	6.7	5.5	4.1	3.1	6.2	3.1	11.6	7.3	10.0	0.6	4.7	5.4	3.7	6.4	4.1	3.9
				Oolong tea	A	5.8	4.3	2.4	3.2	4.2	6.7	11.6	12.0	0.3	13.8	14.1	10.2	4.8	1.2	7.1	4.9	4.3	7.8	0.8
					B	2.9	3.6	7.2	4.7	19.0	7.2	1.5	1.2	8.4	12.1	6.8	7.7	1.9	5.8	2.9	3.5	3.0	8.3	2.6
			GC-MS/MS	Green tea	A	5.1	4.1	2.5	4.5	7.2	1.6	7.4	4.8	4.5	7.1	5.4	1.5	10.7	10.1	1.2	3.8	7.0	5.2	1.7
					B	3.9	9.4	6.8	6.4	4.6	4.5	2.6	7.8	5.2	13.5	2.7	12.5	7.5	2.5	7.5	4.5	4.4	4.8	7.5
				Oolong tea	A	5.7	2.2	0.7	3.2	4.9	5.0	12.8	2.2	2.7	4.2	3.8	4.5	6.5	0.8	6.4	2.1	1.5	6.7	5.9
					B	6.9	3.8	5.7	8.7	3.5	3.1	4.7	8.8	7.1	3.6	5.6	2.9	7.9	4.1	6.5	1.6	2.3	0.2	3.4
		ENVI-Carb+PSA	GC-MS	Green tea	A	9.5	12.8	5.5	7.5	6.8	2.6	19.7	1.7	3.2	7.9	6.2	10.3	1.6	5.7	8.9	11.0	11.6	5.7	5.3
					B	5.4	5.6	1.1	8.0	3.9	9.4	3.2	3.8	6.2	4.4	22.4	12.1	3.2	5.5	15.5	2.8	3.6	9.2	7.9
				Oolong tea	A	3.1	2.6	5.9	0.7	15.3	7.5	3.0	16.4	5.6	3.3	6.8	5.8	3.9	4.5	9.9	2.8	4.2	7.4	1.0
					B	6.4	2.9	7.7	1.1	14.6	8.1	8.4	19.5	17.4	6.1	20.7	18.6	3.3	15.1	6.3	8.7	6.9	10.2	7.1
			GC-MS/MS	Green tea	A	8.3	9.2	4.2	8.6	6.7	1.9	18.9	2.9	4.1	2.8	4.6	15.7	4.0	12.0	6.2	6.8	10.7	3.3	3.1
					B	4.3	6.0	5.1	4.7	4.4	1.0	1.3	5.5	4.8	1.4	22.5	7.9	5.7	20.6	13.0	2.7	0.4	10.1	6.7
				Oolong tea	A	5.8	2.5	4.6	2.3	3.6	1.5	3.3	3.5	2.4	7.5	6.3	6.8	8.7	3.9	12.5	3.3	4.6	4.7	5.9
					B	5.6	2.7	14.5	7.7	13.1	7.0	3.5	7.8	7.3	3.2	4.8	5.1	5.3	13.2	9.3	3.3	5.3	6.8	3.4
199	Triallate	TPT	GC-MS	Green tea	A	3.6	5.5	4.1	3.6	11.4	4.2	6.7	6.0	8.0	3.8	6.5	1.4	7.3	5.1	2.2	1.3	1.6	3.8	2.0
					B	5.6	9.1	3.8	9.0	4.6	3.4	1.7	4.7	4.1	9.2	0.8	8.4	3.3	1.1	5.1	4.9	5.6	5.5	3.3
				Oolong tea	A	4.1	1.6	2.1	4.1	1.9	3.2	9.0	1.7	0.9	2.3	3.0	2.1	3.2	1.3	6.1	3.1	3.9	1.3	8.5
					B	3.5	2.8	4.2	4.3	3.6	4.4	3.5	4.8	8.7	4.0	2.4	3.5	5.3	3.9	6.9	1.7	1.9	1.8	5.4
			GC-MS/MS	Green tea	A	5.0	7.1	0.7	10.9	14.0	3.6	7.9	6.5	6.3	3.1	5.5	2.8	9.6	1.3	3.1	1.8	7.0	7.8	0.9
					B	3.9	9.7	6.0	9.7	3.4	4.9	2.0	5.8	6.6	7.9	2.6	17.1	2.6	1.8	1.2	3.9	8.7	4.2	8.3
				Oolong tea	A	5.3	2.7	1.9	4.8	5.4	2.9	7.6	3.7	5.1	3.4	2.2	4.3	5.7	0.9	7.6	3.1	2.6	0.3	1.7
					B	3.2	4.6	4.0	7.4	3.2	5.1	2.7	3.7	7.4	2.6	1.4	1.0	8.0	1.6	9.4	5.9	1.7	1.2	5.9
		ENVI-Carb+PSA	GC-MS	Green tea	A	14.6	5.7	8.3	3.5	7.0	2.5	8.2	7.4	5.3	0.8	5.1	10.6	6.4	4.2	9.5	3.4	7.3	9.6	5.0
					B	15.4	9.0	4.0	3.6	4.0	4.1	8.8	7.8	9.2	6.2	12.3	4.0	13.5	3.3	6.8	12.5	4.7	13.9	7.8
				Oolong tea	A	5.3	2.8	3.3	4.2	3.8	1.0	3.1	7.5	1.9	1.8	4.9	2.8	1.8	6.5	3.1	4.4	5.6	7.1	3.3
					B	7.2	4.0	8.5	2.2	11.9	5.0	2.9	6.5	11.6	2.2	8.4	5.8	3.4	6.5	11.6	3.3	8.5	4.1	4.9
			GC-MS/MS	Green tea	A	17.4	8.2	6.3	6.1	4.8	13.0	8.0	10.2	5.9	1.5	3.7	12.6	2.2	16.9	3.2	5.9	10.2	6.4	6.1
					B	18.4	10.4	5.8	8.8	6.7	3.9	9.1	5.5	6.5	6.5	12.5	12.1	9.3	20.0	6.3	11.1	6.3	14.7	4.9
				Oolong tea	A	5.3	3.7	6.0	6.8	3.2	2.1	2.0	8.0	2.6	1.3	3.7	5.2	2.5	3.0	8.2	2.7	4.8	6.0	1.7
					B	8.7	4.4	10.1	5.0	8.2	6.0	2.8	5.9	9.2	1.7	7.6	5.0	6.4	7.0	9.9	1.1	6.9	11.1	5.9

No.	Compound	Sorbent	Method	Tea																				
200	Tribenuron-methyl	TPT	GC–MS	Green tea	A	8.0	17.6	2.8	12.5	12.8	1.8	21.3	11.4	8.9	7.8	11.3	4.8	10.6	8.7	4.8	6.9	5.2	12.8	7.2
					B	5.6	13.7	1.2	14.8	10.9	10.0	8.0	13.5	6.3	17.2	2.4	8.7	8.5	3.6	13.4	9.9	11.7	6.4	4.6
				Oolong tea	A	11.6	19.2	7.8	10.0	16.7	8.4	10.3	12.1	1.3	4.0	16.0	13.2	5.5	5.9	3.9	5.2	2.5	5.5	5.7
					B	12.6	13.4	10.3	5.7	13.9	15.4	10.9	10.2	12.6	3.6	20.1	6.7	8.2	8.1	11.0	3.6	2.1	1.0	13.8
		GC–MS/MS		Green tea	A	3.5	4.5	3.9	2.4	16.1	5.3	19.3	13.9	8.3	6.6	13.4	3.3	19.6	13.7	2.7	7.3	4.3	12.7	2.9
					B	5.9	9.7	5.5	12.2	8.9	12.6	5.4	15.4	3.0	21.9	3.0	22.2	15.6	6.5	8.8	8.5	7.2	10.2	4.6
				Oolong tea	A	12.2	11.5	6.3	10.4	8.4	11.2	3.6	2.6	2.5	6.0	10.0	6.9	6.4	3.4	10.8	12.0	7.6	15.7	13.9
					B	17.3	7.1	10.4	11.0	9.4	16.5	13.1	17.8	16.5	3.6	14.4	4.4	9.0	2.6	8.6	7.9	7.7	4.0	4.4
		ENVI-Carb+PSA	GC–MS	Green tea	A	20.4	7.9	35.0	13.3	12.1	14.1	67.3	20.2	14.9	10.9	5.6	23.4	3.4	4.2	7.2	5.6	10.7	8.2	6.0
					B	19.2	13.4	14.6	14.8	24.3	13.4	16.3	24.0	18.5	9.3	53.1	14.1	22.3	4.1	26.9	4.2	5.4	13.5	9.4
				Oolong tea	A	8.1	21.2	10.1	19.5	12.6	7.1	7.2	15.1	1.7	16.6	5.2	10.9	8.6	7.3	7.4	11.8	6.7	8.6	8.0
					B	21.1	14.5	15.5	15.2	22.5	16.7	19.0	6.0	10.5	9.7	6.4	9.7	11.8	6.0	16.3	1.0	8.7	15.9	9.9
		GC–MS/MS		Green tea	A	21.2	25.9	29.9	14.4	5.0	13.1	27.4	15.5	18.2	10.1	3.0	0.0	2.6	17.0	4.9	4.7	8.7	15.1	6.1
					B	20.6	5.5	10.6	12.5	20.3	14.5	20.3	25.6	12.5	4.6	58.1	0.0	41.4	28.0	38.5	8.7	2.7	22.6	16.9
				Oolong tea	A	14.5	14.5	6.3	14.4	20.8	5.0	9.5	19.7	3.4	18.2	10.2	10.5	5.9	9.4	15.8	20.6	2.0	18.9	13.9
					B	19.1	17.4	21.4	4.6	23.7	21.7	19.5	4.7	8.3	5.9	6.2	5.1	5.6	14.8	14.1	5.8	6.8	20.1	4.4
201	Vinclozolin	TPT	GC–MS	Green tea	A	3.5	5.5	3.9	3.1	10.0	2.3	4.5	8.4	7.4	1.9	3.1	2.9	7.8	4.7	5.5	3.4	11.9	4.4	3.2
					B	4.9	9.0	2.0	8.8	3.9	3.0	2.4	3.9	2.8	12.0	4.1	5.2	4.9	13.3	4.0	2.4	8.4	1.1	7.3
				Oolong tea	A	8.1	1.2	1.3	6.4	1.3	3.3	7.5	2.6	2.7	3.4	5.3	2.1	3.5	3.7	6.2	3.6	2.5	3.2	1.8
					B	3.9	2.7	5.0	3.3	1.4	5.8	2.5	0.1	6.8	5.0	2.0	1.4	3.3	4.6	6.4	0.6	0.7	0.6	2.2
		GC–MS/MS		Green tea	A	4.8	5.7	2.3	5.4	15.6	2.6	7.5	9.9	3.8	3.1	6.3	5.1	10.4	6.4	0.2	4.8	8.4	6.9	4.7
					B	5.0	12.3	3.0	8.7	4.3	3.6	4.5	3.4	6.9	7.0	0.6	8.1	4.1	5.5	1.8	4.8	6.2	2.4	10.9
				Oolong tea	A	8.1	2.0	2.2	6.2	5.2	4.2	8.5	5.5	5.7	2.6	2.0	2.7	6.6	1.6	3.2	5.4	1.2	5.9	2.1
					B	6.3	5.6	6.3	6.9	4.5	6.4	3.6	8.2	9.3	4.6	3.4	1.4	7.4	4.6	6.1	6.9	2.1	8.1	1.6
		ENVI-Carb+PSA	GC–MS	Green tea	A	15.1	3.2	8.5	7.5	7.5	11.5	5.7	8.9	3.2	1.4	3.4	10.8	6.7	2.5	10.2	2.2	8.6	6.4	4.6
					B	12.7	7.4	6.1	4.4	3.2	2.1	8.0	7.0	9.6	7.2	9.9	8.2	13.8	1.8	4.6	13.1	3.8	10.8	7.1
				Oolong tea	A	2.7	1.7	1.9	8.3	2.9	1.6	4.7	11.2	2.0	2.3	5.2	4.6	4.3	2.7	2.2	2.3	5.6	7.4	4.2
					B	6.3	3.7	7.8	1.4	9.8	3.4	2.0	7.4	11.7	3.1	8.3	4.2	4.4	6.0	11.2	1.6	5.9	3.0	4.6
		GC–MS/MS		Green tea	A	21.1	8.6	5.6	1.6	8.3	12.3	7.6	8.2	5.7	1.9	1.2	0.0	12.4	3.1	5.0	2.9	5.1	4.1	7.1
					B	20.0	9.0	7.6	7.8	9.0	3.4	10.2	4.6	8.0	8.4	14.9	0.0	12.4	13.0	7.6	11.5	6.8	13.0	9.4
				Oolong tea	A	5.3	4.8	8.2	4.6	4.9	1.7	3.0	5.4	2.3	1.4	4.3	3.4	8.6	8.3	8.5	5.4	6.3	3.4	2.1
					B	7.5	4.5	7.6	0.8	8.6	5.6	3.8	4.4	9.4	1.6	10.5	3.4	8.2	8.4	17.5	1.7	4.9	13.0	1.6

Appendix Table 5.3 Youden Pair Ratios of Each Average Content From 201 Pesticides and RSD for 19 Circulative Determinations

No.	Pesticides	SPE	Sample	Method	Nov 9, 2009 (n = 5)	Nov 14, 2009 (n = 3)	Nov 19, 2009 (n = 3)	Nov 24, 2009 (n = 3)	Nov 29, 2009 (n = 3)	Dec 4, 2009 (n = 3)	Dec 9, 2009 (n = 3)	Dec 14, 2009 (n = 3)	Dec 19, 2009 (n = 3)	Dec 24, 2009 (n = 3)	Dec 14, 2009 (n = 3)	Jan 3, 2010 (n = 3)	Jan 8, 2010 (n = 3)	Jan 13, 2010 (n = 3)	Jan 18, 2010 (n = 3)	Jan 23, 2010 (n = 3)	Jan 28, 2010 (n = 3)	Feb 2, 2010 (n = 3)	Feb 7, 2010 (n = 3)	RSD%
					Youden pair ratio																			
1	2,3,4,5-Tetra-chloroaniline	TPT	Green tea	GC-MS	1.05	1.22	1.29	1.27	1.14	1.24	1.30	1.10	1.20	1.17	1.27	1.19	1.28	1.2	1.23	1.30	1.24	1.14	1.21	5.7
				GC-MS/MS	1.09	1.14	1.29	1.14	1.29	1.15	1.38	1.30	1.25	1.33	1.41	1.30	1.29	1.32	1.32	1.41	1.31	1.24	1.05	8.1
			Oolong tea	GC-MS	1.24	1.17	1.16	1.17	1.25	1.26	1.20	1.20	1.32	1.29	1.11	1.20	0.96	1.26	1.41	1.23	1.21	1.37	1.11	8.2
				GC-MS/MS	1.24	1.24	1.11	1.00	1.22	1.30	1.17	1.32	1.33	1.28	1.09	1.37	1.24	1.20	1.43	1.50	1.22	1.50	1.03	11.2
		ENVI-Carb + PSA	Green tea	GC-MS	1.41	1.05	0.97	1.07	1.08	1.00	1.14	1.17	1.15	1.03	1.11	1.08	1.02	0.97	0.99	1.00	1.14	1.03	1.10	9.4
				GC-MS/MS	1.07	1.06	1.16	1.11	0.94	1.08	1.11	1.08	1.06	0.91	1.04	0.97	1.10	0.92	1.01	0.97	1.03	1.07	1.11	6.8
			Oolong tea	GC-MS	1.05	1.11	1.24	1.13	1.07	1.14	0.97	1.01	1.00	0.97	1.06	1.03	1.07	1.00	1.20	1.16	1.01	1.11	1.02	7.3
				GC-MS/MS	1.02	1.24	1.04	1.21	1.02	1.03	1.08	1.09	1.04	0.99	0.97	1.06	1.03	0.97	1.29	1.15	1.03	1.09	0.74	11.2
2	2,3,5,6-Tetra-chloroaniline	TPT	Green tea	GC-MS	1.07	1.04	1.12	1.07	1.12	1.05	1.09	1.09	1.00	1.05	1.11	1.07	1.06	1.08	1.12	1.02	1.03	1.07	1.14	3.4
				GC-MS/MS	1.01	1.06	1.14	1.14	1.06	1.03	1.09	1.05	1.01	1.02	1.10	1.06	1.06	1.00	1.05	1.03	1.02	1.01	1.12	4.1
			Oolong tea	GC-MS	1.08	1.19	1.21	1.18	1.09	1.19	1.19	1.19	1.18	1.17	1.26	1.19	1.23	1.1	1.21	1.16	1.17	1.12	1.20	3.9
				GC-MS/MS	1.06	1.21	1.19	1.18	1.09	1.18	1.23	1.15	1.23	1.15	1.30	1.22	1.15	1.09	1.21	1.15	1.11	1.09	1.19	5.3
		ENVI-Carb + PSA	Green tea	GC-MS	1.08	1.18	0.99	1.1	1.04	1.09	0.96	1.03	1.10	1.02	1.09	1.04	1.11	1.13	1.11	1.17	1.06	1.17	1.07	5.5
				GC-MS/MS	1.07	1.27	1.01	1.13	1.01	1.08	1.07	1.10	1.06	1.06	1.05	1.03	1.03	1.12	1.15	1.16	1.16	1.14	1.00	6.2
			Oolong tea	GC-MS	1.29	1.18	1.11	1.13	1.22	1.18	1.18	1.12	1.24	1.24	1.16	1.15	1.14	1.19	1.25	1.21	1.17	1.21	1.12	4.2
				GC-MS/MS	1.26	1.17	1.09	1.14	1.21	1.19	1.11	1.17	1.26	1.28	1.15	1.12	1.09	1.12	1.33	1.20	1.10	1.24	1.03	6.5
3	4,4-Dibromo-benzophe-none	TPT	Green tea	GC-MS	1.03	1.10	1.05	1.16	1.11	1.05	1.10	1.07	1.26	1.06	1.12	1.16	1.08	1.03	1.13	1.03	1.06	1.10	1.14	5.2
				GC-MS/MS	0.95	1.00	1.13	1.00	1.12	1.12	1.09	1.03	1.10	1.08	1.14	1.01	1.13	1.01	1.05	1.02	1.01	1.02	1.33	7.9
			Oolong tea	GC-MS	1.11	1.11	1.15	1.17	1.16	1.13	1.09	1.06	1.13	1.13	1.35	1.14	1.19	1.2	1.26	1.16	1.21	1.02	1.14	6.2
				GC-MS/MS	1.02	1.15	1.18	1.18	1.00	1.08	1.20	1.05	1.13	1.15	1.36	1.19	1.20	1.17	1.25	1.15	1.23	1.01	1.13	7.9
		ENVI-Carb + PSA	Green tea	GC-MS	1.02	1.08	0.98	1.08	1.02	1.13	1.07	0.98	1.12	0.87	0.92	1.15	1.03	1.14	1.05	1.26	1.25	1.2	1.01	9.7
				GC-MS/MS	1.12	1.43	1.01	1.13	1.05	1.15	0.79	1.07	1.15	1.04	0.94	0.96	1.12	1.22	1.09	1.21	1.07	1.26	1.05	12.3
			Oolong tea	GC-MS	1.17	1.14	1.01	1.14	1.22	1.23	1.29	1.22	1.20	1.23	1.26	1.20	1.15	1.41	1.26	1.17	1.31	1.18	1.07	7.3
				GC-MS/MS	1.03	1.13	1.06	1.10	1.12	1.19	1.08	1.12	1.22	1.31	1.16	1.20	1.15	1.17	1.29	1.23	1.21	1.49	1.10	8.9
4	4,4-Dichlo-robenzophe-none	TPT	Green tea	GC-MS	1.05	1.03	1.05	1.06	1.12	1.10	1.10	1.10	1.09	1.04	1.14	1.06	1.07	1.06	1.17	1.03	1.07	1.11	1.15	3.8
				GC-MS/MS	1.02	1.03	1.21	1.09	1.11	1.09	1.09	1.05	1.11	1.05	1.13	1.05	1.10	1.04	1.09	1.00	1.01	1.02	1.23	5.7
			Oolong tea	GC-MS	1.02	1.11	1.03	1.03	0.94	1.11	1.00	1.10	1.04	0.95	1.06	0.93	1.00	1.0	1.08	1.11	1.16	0.95	1.01	6.4
				GC-MS/MS	0.96	1.10	1.02	1.02	0.90	1.09	1.00	1.10	1.07	0.95	1.13	0.93	0.92	1.02	1.05	1.06	1.07	0.93	1.00	6.8
		ENVI-Carb + PSA	Green tea	GC-MS	1.04	1.13	0.99	1.09	1.01	1.13	1.00	1.00	1.12	1.00	0.97	1.03	1.03	1.15	1.07	1.25	1.15	1.29	1.02	8.3
				GC-MS/MS	1.08	1.31	1.02	1.12	1.02	1.15	1.00	1.08	1.14	1.04	0.95	1.08	0.99	1.13	1.07	1.21	1.12	1.22	1.02	8.2
			Oolong tea	GC-MS	1.25	0.90	0.92	1.13	1.30	1.02	1.14	0.99	1.02	1.05	1.10	1.05	1.05	1.00	1.16	1.07	1.09	1.08	0.93	9.7
				GC-MS/MS	1.24	0.88	0.90	1.13	1.24	1.04	1.05	1.02	1.03	1.08	1.12	0.97	1.06	0.91	1.21	1.07	1.04	1.20	0.93	10.6

5	Acetochlor	TPT	Green tea	GC-MS	1.04	1.05	1.18	1.05	1.03	1.08	1.12	1.10	1.16	1.04	1.14	1.05	1.02	0.86	1.11	1.03	1.14	1.02	1.11	6.8
				GC-MS/MS	1.00	1.04	1.03	1.52	1.00	1.06	1.15	1.06	1.02	0.93	1.09	1.07	1.01	1.00	1.01	1.02	1.11	1.04	1.06	11.3
			Oolong tea	GC-MS	1.04	1.27	1.24	1.15	1.09	1.17	1.32	1.20	1.19	1.16	1.37	1.21	1.26	1.2	1.24	1.17	1.17	1.08	1.13	6.8
				GC-MS/MS	1.08	1.26	1.22	1.19	1.06	1.19	1.08	1.24	1.14	1.17	1.32	1.19	1.08	1.14	1.20	1.04	1.12	1.10	1.08	6.5
		ENVI-Carb + PSA	Green tea	GC-MS	1.01	1.22	0.98	1.01	1.01	1.11	1.13	1.14	1.06	1.11	1.01	0.99	1.10	1.16	0.94	1.15	1.05	1.10	1.00	7.1
				GC-MS/MS	0.93	1.22	0.99	1.06	0.99	1.12	1.16	1.06	1.42	0.96	1.01	1.08	0.89	1.17	1.04	1.20	1.06	1.10	0.96	11.5
			Oolong tea	GC-MS	1.11	1.17	1.21	1.21	1.32	1.16	1.00	1.08	1.52	1.22	1.25	1.13	1.11	1.14	1.33	1.13	1.13	1.18	1.08	9.8
				GC-MS/MS	1.47	1.14	1.10	1.07	1.23	1.18	1.22	1.14	1.30	1.23	1.10	1.09	1.05	1.11	1.32	1.09	1.05	1.18	0.96	10.2
6	Alachlor	TPT	Green tea	GC-MS	1.01	1.04	1.17	1.08	1.06	1.06	1.12	1.12	1.07	1.07	1.12	1.03	1.07	1.05	1.09	1.06	1.01	1.09	1.15	4.0
				GC-MS/MS	1.25	1.02	0.95	1.53	1.02	1.07	1.15	1.06	1.01	0.95	1.10	1.07	1.04	0.98	1.04	1.05	1.08	1.04	1.15	12.0
			Oolong tea	GC-MS	1.06	1.25	1.28	1.21	1.10	1.16	1.39	1.29	1.10	1.23	1.30	1.23	1.26	1.1	1.20	1.14	1.24	1.07	1.14	7.6
				GC-MS/MS	1.14	1.23	1.17	1.21	1.08	1.21	1.36	1.21	1.19	1.14	1.34	1.18	1.16	1.14	1.21	1.22	1.34	1.00	1.09	7.6
		ENVI-Carb + PSA	Green tea	GC-MS	0.91	1.22	0.99	1.03	1.05	1.11	1.08	1.32	1.09	1.12	1.00	0.97	1.09	1.14	0.99	1.13	1.11	0.96	1.02	9.2
				GC-MS/MS	0.79	1.16	0.98	1.04	0.97	1.11	1.14	1.04	0.85	0.97	1.05	1.28	0.94	1.22	1.03	1.22	1.09	1.06	0.93	12.0
			Oolong tea	GC-MS	1.10	1.15	1.19	1.20	1.33	1.15	0.94	1.28	1.51	1.26	1.23	1.12	1.11	1.16	1.23	1.17	1.15	1.19	1.09	9.6
				GC-MS/MS	1.60	1.18	1.10	1.11	1.23	1.15	1.15	1.20	1.28	1.22	1.17	1.07	1.18	1.12	1.30	1.23	1.04	1.28	1.04	10.5
7	Atratone	TPT	Green tea	GC-MS	1.11	1.04	1.17	1.11	1.13	1.02	1.07	1.11	1.05	1.15	1.23	1.10	1.10	1.05	1.13	0.98	1.03	1.01	1.16	5.8
				GC-MS/MS	1.03	1.04	1.18	1.12	1.08	1.23	1.09	1.07	1.01	0.95	1.15	1.09	1.07	0.96	1.07	0.99	1.00	0.98	1.16	7.2
			Oolong tea	GC-MS	1.06	1.14	1.15	1.17	1.08	1.17	1.26	1.22	1.34	1.20	1.40	1.25	1.27	1.1	1.23	1.18	1.24	1.08	1.11	7.6
				GC-MS/MS	1.05	1.15	1.16	1.12	1.07	1.16	1.31	1.20	1.19	1.22	1.36	1.20	1.16	1.17	1.28	1.15	1.23	1.04	1.17	6.9
		ENVI-Carb + PSA	Green tea	GC-MS	1.16	1.06	1.04	1.12	1.08	1.19	0.98	1.05	1.02	1.42	1.05	1.09	1.12	1.15	1.13	1.27	1.14	1.15	0.98	9.2
				GC-MS/MS	1.07	1.26	1.04	1.14	1.02	1.16	1.00	1.10	0.99	1.01	1.10	1.11	0.98	1.28	1.11	1.24	1.11	1.15	0.99	8.4
			Oolong tea	GC-MS	1.21	1.17	1.16	1.16	1.48	1.19	1.13	1.23	1.98	1.33	1.39	1.06	1.13	1.22	1.26	1.27	1.18	1.24	1.16	15.9
				GC-MS/MS	1.20	1.17	1.13	1.12	1.28	1.17	1.09	1.17	1.29	1.27	1.16	1.13	1.12	1.18	1.37	1.23	1.05	1.31	1.12	7.1
8	Benodanil	TPT	Green tea	GC-MS	0.91	1.06	1.14	1.15	1.00	1.06	0.94	1.04	1.05	1.03	1.15	1.24	1.14	0.83	1.04	1.05	0.94	1.23	1.43	12.7
				GC-MS/MS	0.93	1.07	1.03	1.11	1.10	1.29	1.00	1.07	0.80	1.13	1.19	1.08	1.22	0.97	1.10	1.01	1.04	1.05	1.58	14.5
			Oolong tea	GC-MS	1.47	1.00	0.89	1.44	1.34	1.77	0.78	1.00	1.38	0.66	1.39	1.44	0.74	1.0	1.23	1.17	1.18	1.71	0.99	26.5
				GC-MS/MS	1.54	0.97	1.10	1.35	1.13	1.66	1.47	0.99	1.31	0.61	1.15	1.36	0.72	0.92	1.28	1.10	1.15	1.64	1.03	24.2
		ENVI-Carb + PSA	Green tea	GC-MS	0.78	0.79	1.02	1.15	1.10	0.87	0.89	1.13	1.64	1.01	1.31	1.15	1.03	0.96	1.22	1.24	1.29	0.85	0.97	19.9
				GC-MS/MS	0.93	0.73	0.96	1.18	1.02	0.94	0.76	1.25	1.07	0.96	1.37	1.07	0.93	1.28	1.12	1.29	1.13	0.83	1.08	17.0
			Oolong tea	GC-MS	0.92	1.36	0.92	0.80	1.45	0.87	1.43	2.00	2.04	0.70	0.98	1.10	1.42	0.96	1.14	1.52	1.12	1.21	1.53	30.3
				GC-MS/MS	1.14	1.33	0.94	0.83	1.43	0.88	0.96	1.56	2.65	0.88	0.98	0.99	1.47	0.81	1.22	1.77	1.13	1.32	1.74	35.7

(Continued)

Appendix Table 5.3 Youden Pair Ratios of Each Average Content From 201 Pesticides and RSD for 19 Circulative Determinations (cont.)

No.	Pesticides	SPE	Sample	Method	Youden pair ratio Nov 9, 2009 (n = 5)	Nov 14, 2009 (n = 3)	Nov 19, 2009 (n = 3)	Nov 24, 2009 (n = 3)	Nov 29, 2009 (n = 3)	Dec 4, 2009 (n = 3)	Dec 9, 2009 (n = 3)	Dec 14, 2009 (n = 3)	Dec 19, 2009 (n = 3)	Dec 24, 2009 (n = 3)	Dec 29, 2009 (n = 3)	Jan 3, 2010 (n = 3)	Jan 8, 2010 (n = 3)	Jan 13, 2010 (n = 3)	Jan 18, 2010 (n = 3)	Jan 23, 2010 (n = 3)	Jan 28, 2010 (n = 3)	Feb 2, 2010 (n = 3)	Feb 7, 2010 (n = 3)	RSD%
9	Benoxacor	TPT	Green tea	GC–MS	1.02	1.08	1.17	1.01	0.98	1.09	1.11	1.01	1.09	0.97	1.04	1.09	1.06	0.90	0.99	1.03	0.97	1.15	1.02	6.5
				GC–MS/MS	1.06	1.04	0.85	1.39	0.99	0.99	1.13	1.07	0.88	0.92	1.06	1.13	1.08	0.90	1.04	0.97	0.95	1.00	1.16	11.8
			Oolong tea	GC–MS	1.09	1.19	1.16	1.30	1.09	1.21	1.51	1.21	1.25	1.06	1.44	1.26	1.14	1.0	1.10	1.05	1.11	1.09	1.00	11.5
				GC–MS/MS	1.15	1.10	1.09	1.27	1.13	1.33	1.30	1.15	1.12	0.98	1.40	1.36	1.01	1.11	1.32	1.18	1.27	1.22	1.12	10.0
		ENVI-Carb + PSA	Green tea	GC–MS	0.91	1.10	0.96	1.12	1.10	0.99	1.08	1.11	1.09	1.04	1.20	1.04	1.04	0.94	0.95	1.14	1.13	1.05	0.98	7.5
				GC–MS/MS	0.80	0.83	0.97	1.08	1.03	1.04	1.18	1.02	0.83	0.89	1.30	1.18	0.92	1.01	1.10	0.98	1.11	0.95	0.89	13.1
			Oolong tea	GC–MS	0.95	1.12	1.17	1.06	1.42	1.12	0.92	1.39	1.58	1.05	1.19	1.03	1.17	0.94	1.02	1.30	1.04	1.00	1.10	15.5
				GC–MS/MS	1.39	1.18	1.13	1.00	1.29	1.02	1.15	1.24	1.38	0.96	1.13	0.94	1.21	0.99	1.08	1.40	1.01	1.11	1.17	12.8
10	Bromophos-ethyl	TPT	Green tea	GC–MS	1.04	1.03	1.12	1.11	1.07	1.04	1.09	1.07	1.05	1.09	1.15	1.10	1.07	1.05	1.09	1.01	1.03	1.05	1.13	3.6
				GC–MS/MS	0.99	1.04	1.03	1.13	1.07	1.17	1.10	1.03	1.01	1.05	1.08	1.06	1.08	0.99	1.04	1.02	1.06	1.01	1.18	5.0
			Oolong tea	GC–MS	1.06	1.15	1.17	1.19	1.09	1.18	1.31	1.15	1.24	1.15	1.38	1.23	1.23	1.1	1.20	1.17	1.21	1.10	1.12	6.5
				GC–MS/MS	1.16	1.18	1.21	1.14	1.10	1.16	1.28	1.16	1.21	1.12	1.32	1.24	1.26	1.15	1.23	1.16	1.24	1.07	1.12	5.5
		ENVI-Carb + PSA	Green tea	GC–MS	1.02	1.18	1.03	1.08	1.08	1.12	1.12	1.12	1.07	1.20	1.07	1.09	1.09	1.08	1.02	1.20	1.12	1.08	1.01	5.0
				GC–MS/MS	0.99	1.18	1.03	1.05	1.03	1.12	1.12	1.07	0.99	1.01	1.09	1.05	1.04	1.17	1.09	1.18	1.13	1.07	1.01	5.7
			Oolong tea	GC–MS	1.20	1.17	1.15	1.14	1.36	1.15	1.16	1.26	1.46	1.23	1.25	1.10	1.13	1.19	1.26	1.19	1.16	1.17	1.10	7.4
				GC–MS/MS	1.30	1.16	1.11	1.13	1.25	1.16	1.12	1.22	1.28	1.22	1.07	1.12	1.16	1.13	1.29	1.21	1.09	1.26	1.07	6.5
11	Butralin	TPT	Green tea	GC–MS	1.08	1.03	1.18	1.14	0.96	1.03	1.07	0.93	1.05	1.11	1.08	1.11	1.16	1.06	0.99	1.07	0.92	1.23	1.10	7.6
				GC–MS/MS	1.10	1.03	1.00	1.43	1.09	1.28	1.03	1.06	0.78	0.99	1.07	1.06	1.17	0.95	1.02	1.00	0.97	0.99	1.27	13.2
			Oolong tea	GC–MS	1.07	1.07	1.05	1.17	1.55	1.39	1.46	1.11	1.31	1.14	1.50	1.32	1.09	0.8	1.32	1.09	1.29	1.10	1.00	16.0
				GC–MS/MS	1.05	1.12	1.10	1.08	1.10	1.32	1.40	1.15	1.23	1.17	1.50	1.22	1.04	1.11	1.22	1.10	1.32	1.23	0.98	11.2
		ENVI-Carb + PSA	Green tea	GC–MS	0.92	0.97	1.09	1.30	1.19	0.99	1.51	1.23	0.93	1.23	1.27	1.08	1.07	0.86	1.05	1.19	1.34	1.23	1.00	14.8
				GC–MS/MS	0.88	0.63	0.99	1.17	1.04	1.08	2.86	1.15	0.93	1.00	1.29	1.17	0.86	1.19	1.05	1.16	1.10	0.82	0.88	40.3
			Oolong tea	GC–MS	1.08	1.15	1.26	1.09	2.32	1.10	1.03	1.61	1.55	1.12	1.37	1.03	1.32	1.11	0.73	1.94	1.02	1.20	1.13	28.8
				GC–MS/MS	1.28	1.18	1.17	1.05	1.39	1.11	1.13	1.59	1.24	1.17	1.19	0.99	1.20	1.18	1.15	1.41	1.25	1.34	1.16	11.3
12	Chlorfenapyr	TPT	Green tea	GC–MS	0.99	1.05	1.10	1.05	1.10	1.11	1.10	1.10	1.02	1.02	1.16	1.07	1.04	1.04	1.11	1.03	1.03	1.02	1.11	4.1
				GC–MS/MS	1.01	1.03	1.08	1.19	1.10	1.07	1.10	1.07	1.02	0.98	1.11	1.03	1.06	0.97	1.10	1.06	1.00	1.05	1.09	5.0
			Oolong tea	GC–MS	1.04	1.18	1.16	1.19	1.01	1.07	1.19	1.12	1.17	1.13	1.27	1.17	1.18	1.1	1.17	1.17	1.20	1.05	1.12	5.6
				GC–MS/MS	1.16	1.16	1.11	1.18	1.03	1.12	1.17	1.12	1.17	1.07	1.27	1.15	1.21	1.15	1.17	1.18	1.21	1.03	1.12	5.2
		ENVI-Carb + PSA	Green tea	GC–MS	0.97	1.20	1.00	1.07	1.01	1.08	1.04	1.13	1.09	0.97	1.06	1.06	1.10	1.15	1.07	1.18	1.13	1.14	1.00	6.5
				GC–MS/MS	0.94	1.21	1.01	1.11	0.98	1.11	0.98	1.06	1.03	0.99	1.06	1.06	1.11	1.18	1.08	1.21	1.15	1.12	1.07	7.2
			Oolong tea	GC–MS	1.14	1.13	1.05	1.14	1.18	1.13	1.13	1.13	1.21	1.21	1.14	1.14	1.09	1.19	1.27	1.14	1.15	1.17	1.06	4.5
				GC–MS/MS	1.27	1.12	1.02	1.09	1.18	1.11	1.12	1.14	1.24	1.21	1.06	1.11	1.07	1.14	1.30	1.17	1.05	1.26	1.05	7.2

13	Clomazone	TPT	Green tea	GC-MS	1.07	1.06	1.07	1.11	0.99	1.07	1.10	1.03	0.89	1.01	1.10	1.13	1.11	0.92	0.92	1.04	1.10	1.10	1.03	1.17	7.3
				GC-MS/MS	1.05	1.11	1.10	1.13	0.95	1.11	1.11	1.02	0.88	0.97	1.05	1.11	1.07	0.89	0.96	1.04	1.01	1.01	0.95	1.22	8.5
			Oolong tea	GC-MS	1.08	1.16	1.14	1.28	1.21	1.24	1.27	1.17	1.19	1.12	1.33	1.21	1.19	1.1	1.23	1.16	1.18	1.18	1.12	1.09	5.8
				GC-MS/MS	1.14	1.19	1.20	1.21	1.13	1.29	1.17	1.18	1.25	1.12	1.36	1.27	1.17	1.10	1.19	1.08	1.13	1.13	1.14	1.15	5.8
		ENVI-Carb + PSA	Green tea	GC-MS	1.18	1.02	1.11	1.12	1.09	0.99	1.38	1.05	1.06	1.24	1.22	1.13	1.09	1.07	1.01	1.26	1.20	1.20	0.93	0.99	10.1
				GC-MS/MS	1.14	1.01	1.08	1.13	1.08	0.99	1.56	1.11	1.08	1.08	1.22	1.01	0.82	1.24	1.02	1.19	1.10	1.10	0.98	1.04	13.3
			Oolong tea	GC-MS	1.25	1.19	1.16	1.09	1.39	1.14	1.08	1.25	1.68	1.18	1.18	1.08	1.20	1.11	1.12	1.32	1.12	1.12	1.14	1.17	11.8
				GC-MS/MS	1.23	1.21	1.15	1.09	1.25	1.15	1.07	1.23	1.28	1.19	1.14	1.04	1.23	1.10	1.25	1.32	1.12	1.12	1.19	1.10	6.4
14	Cycloate	TPT	Green tea	GC-MS	1.08	1.04	1.09	1.05	1.09	1.09	1.09	1.10	1.00	1.07	1.10	1.06	1.07	1.02	1.11	1.01	1.05	1.05	1.03	1.16	3.7
				GC-MS/MS	1.02	1.04	1.12	0.93	1.05	1.10	1.10	1.03	1.01	1.04	1.10	1.05	1.04	0.96	1.05	1.01	1.04	1.04	1.01	1.12	4.9
			Oolong tea	GC-MS	1.12	1.18	1.01	1.10	1.02	1.10	1.01	1.02	1.37	1.10	1.09	1.07	1.14	1.11	1.16	1.26	1.01	1.01	1.17	1.01	8.4
				GC-MS/MS	1.07	1.24	1.00	1.13	1.01	1.11	1.05	1.08	1.03	1.03	1.06	1.16	0.98	1.06	1.10	1.17	1.06	1.06	1.16	1.04	6.2
		ENVI-Carb + PSA	Green tea	GC-MS	1.07	1.21	1.19	1.18	1.10	1.14	1.22	1.20	1.21	1.18	1.31	1.21	1.22	1.2	1.23	1.15	1.17	1.17	1.13	1.16	4.4
				GC-MS/MS	1.07	1.21	1.16	1.20	1.10	1.22	1.26	1.17	1.19	1.19	1.32	1.19	1.06	1.17	1.23	1.13	1.15	1.15	1.11	1.20	5.4
			Oolong tea	GC-MS	1.30	1.20	1.14	1.15	1.32	1.20	1.21	1.14	1.52	1.31	1.21	1.14	1.17	1.22	1.30	1.24	1.27	1.27	1.19	1.13	7.6
				GC-MS/MS	1.35	1.19	1.13	1.16	1.25	1.18	1.14	1.20	1.23	1.29	1.16	1.20	1.04	1.17	1.36	1.23	1.12	1.12	1.23	1.09	7.2
15	Cycluron	TPT	Green tea	GC-MS	0.98	1.00	1.05	1.04	1.09	1.03	1.10	1.12	1.15	1.13	1.10	0.99	1.02	1.05	1.12	1.10	1.10	1.10	1.05	1.12	4.8
				GC-MS/MS	0.93	1.03	0.99	0.90	1.06	1.19	1.15	1.12	1.08	0.85	1.22	1.04	1.03	0.88	1.09	1.07	1.09	1.09	1.03	1.12	9.6
			Oolong tea	GC-MS	1.05	1.25	0.98	1.14	0.98	1.12	1.05	1.08	1.08	1.14	1.61	1.03	1.17	1.03	1.09	1.19	1.11	1.11	1.13	1.02	12.4
				GC-MS/MS	1.14	1.42	1.05	1.14	0.97	1.14	1.00	1.04	0.92	0.93	1.01	1.06	0.96	1.47	1.11	1.10	1.16	1.16	1.21	0.95	13.7
		ENVI-Carb + PSA	Green tea	GC-MS	0.91	1.23	1.21	1.12	1.11	1.10	1.24	1.12	0.98	2.39	3.12	1.95	2.21	0.9	1.15	1.24	1.23	1.23	0.87	1.04	43.6
				GC-MS/MS	1.10	1.25	1.26	1.23	1.06	1.22	1.31	1.26	1.22	1.24	1.30	1.14	1.20	1.30	1.24	1.27	1.15	1.15	1.01	1.21	6.8
			Oolong tea	GC-MS	1.14	1.11	1.04	1.13	1.28	1.14	1.10	1.11	0.77	2.87	2.43	2.35	2.14	1.12	1.27	1.16	1.10	1.10	1.22	1.07	41.6
				GC-MS/MS	1.34	1.19	1.01	1.22	1.12	1.18	1.12	1.15	1.27	1.48	1.04	1.30	1.04	1.11	1.70	1.01	1.04	1.04	1.38	1.03	15.2
16	Cyhalofop-butyl	TPT	Green tea	GC-MS	1.07	1.05	1.11	1.05	1.03	1.10	1.11	1.04	0.93	1.05	1.02	1.02	1.16	0.91	1.00	1.08	1.04	1.04	1.18	1.13	6.5
				GC-MS/MS	1.04	1.06	1.18	0.78	1.06	1.21	1.12	1.02	0.83	0.48	1.07	1.15	1.09	0.86	1.00	1.02	1.03	1.03	0.98	1.24	17.3
			Oolong tea	GC-MS	1.18	0.91	1.12	1.14	1.04	0.98	1.19	1.03	1.08	1.05	1.06	1.05	1.01	1.20	1.21	1.31	1.11	1.11	0.94	0.94	9.9
				GC-MS/MS	1.16	1.02	1.09	1.19	1.04	1.02	4.20	1.09	1.20	1.11	1.14	1.07	0.86	1.30	1.00	1.29	1.08	1.08	1.05	1.03	57.1
		ENVI-Carb + PSA	Green tea	GC-MS	1.08	1.08	1.11	1.20	1.10	1.17	1.00	1.04	1.18	1.01	1.36	1.10	1.08	1.0	1.14	1.15	1.14	1.14	1.08	1.05	7.6
				GC-MS/MS	1.05	1.08	1.11	1.18	1.06	1.13	1.01	1.06	1.18	1.05	1.25	1.13	0.99	1.04	1.15	1.13	1.13	1.13	1.12	1.06	6.0
			Oolong tea	GC-MS	1.13	1.11	1.01	1.03	1.28	1.10	1.27	1.22	1.31	1.14	1.15	1.05	1.15	1.09	1.09	1.26	1.10	1.10	1.02	1.09	8.0
				GC-MS/MS	1.10	1.10	0.99	1.05	1.15	1.10	1.10	1.14	1.25	1.16	1.13	1.02	1.22	1.05	1.17	1.27	1.10	1.10	1.15	1.08	6.6

(Continued)

Appendix Table 5.3 Youden Pair Ratios of Each Average Content From 201 Pesticides and RSD for 19 Circulative Determinations (cont.)

| No. | Pesticides | SPE | Sample | Method | Youden pair ratio | | | | | | | | | | | | | | | | | | | RSD% |
					Nov 9, 2009 (n=5)	Nov 14, 2009 (n=3)	Nov 19, 2009 (n=3)	Nov 24, 2009 (n=3)	Nov 29, 2009 (n=3)	Dec 4, 2009 (n=3)	Dec 9, 2009 (n=3)	Dec 14, 2009 (n=3)	Dec 19, 2009 (n=3)	Dec 24, 2009 (n=3)	Dec 14, 2009 (n=3)	Jan 3, 2010 (n=3)	Jan 8, 2010 (n=3)	Jan 13, 2010 (n=3)	Jan 18, 2010 (n=3)	Jan 23, 2010 (n=3)	Jan 28, 2010 (n=3)	Feb 2, 2010 (n=3)	Feb 7, 2010 (n=3)	
17	Cyprodinil	TPT	Green tea	GC–MS	1.08	1.05	1.19	1.14	1.15	1.03	1.10	1.06	0.99	1.12	1.12	1.00	1.10	1.05	1.12	1.02	1.03	1.09	1.17	5.3
				GC–MS/MS	1.04	1.04	1.17	1.13	1.07	1.16	1.09	1.04	1.05	1.05	1.11	1.05	1.07	1.00	1.05	1.00	1.03	1.01	1.19	5.3
			Oolong tea	GC–MS	1.05	1.11	1.18	1.19	1.08	1.24	1.27	1.07	1.19	1.23	1.28	1.24	1.17	1.1	1.23	1.14	1.27	1.10	1.04	7.0
				GC–MS/MS	1.03	1.15	1.15	1.15	1.07	1.17	1.14	1.28	1.18	1.22	1.35	1.18	1.14	1.18	1.18	1.16	1.18	1.08	1.08	6.2
		ENVI-Carb + PSA	Green tea	GC–MS	1.03	1.13	1.05	1.16	1.09	1.14	1.02	1.05	1.06	1.27	1.11	1.09	1.11	1.15	1.13	1.26	1.13	1.16	0.94	7.0
				GC–MS/MS	1.06	1.22	1.03	1.14	1.04	1.16	1.03	1.08	1.03	1.02	1.05	1.07	0.92	1.08	1.12	1.19	1.14	1.17	1.00	7.0
			Oolong tea	GC–MS	1.17	1.18	1.02	1.08	1.22	1.20	1.17	1.42	2.14	1.34	1.29	1.05	1.15	1.19	1.22	1.25	1.20	1.31	1.08	19.2
				GC–MS/MS	1.22	1.15	1.09	1.19	1.24	1.18	1.10	1.15	1.23	1.29	1.14	1.09	1.08	1.13	1.31	1.16	1.11	1.26	1.04	6.5
18	Dacthal	TPT	Green tea	GC–MS	1.03	1.03	1.11	1.09	1.06	1.09	1.08	1.05	0.99	1.03	1.11	1.08	1.06	1.04	1.06	1.04	1.01	1.04	1.13	3.4
				GC–MS/MS	1.00	1.03	1.06	1.14	1.05	1.07	1.08	1.04	1.00	1.02	1.10	1.04	1.09	1.00	1.05	1.00	1.05	1.02	1.20	4.9
			Oolong tea	GC–MS	1.06	1.20	1.21	1.18	1.12	1.20	1.37	1.15	1.19	1.16	1.32	1.25	1.24	1.1	1.22	1.18	1.24	1.10	1.11	6.3
				GC–MS/MS	1.17	1.18	1.19	1.21	1.14	1.19	1.35	1.16	1.24	1.18	1.38	1.25	1.26	1.14	1.24	1.14	1.22	1.08	1.13	6.1
		ENVI-Carb + PSA	Green tea	GC–MS	1.01	1.19	1.01	1.08	1.04	1.08	1.09	1.13	1.04	0.98	1.09	1.05	1.09	1.05	1.08	1.16	1.12	1.01	1.00	5.2
				GC–MS/MS	1.06	1.14	1.01	1.10	1.02	1.09	0.99	1.07	0.99	1.02	1.11	1.07	1.07	1.11	1.08	1.15	1.07	1.10	0.99	4.7
			Oolong tea	GC–MS	1.23	1.18	1.14	1.12	1.28	1.16	1.16	1.20	1.24	1.24	1.17	1.18	1.16	1.19	1.24	1.23	1.18	1.21	1.12	3.6
				GC–MS/MS	1.18	1.18	1.14	1.15	1.25	1.18	1.13	1.21	1.26	1.26	1.12	1.14	1.20	1.15	1.31	1.24	1.08	1.27	1.09	5.4
19	DE-PCB 101	TPT	Green tea	GC–MS	1.06	1.21	1.23	1.18	1.09	1.17	1.22	1.19	1.17	1.17	1.27	1.20	1.30	1.1	1.19	1.16	1.18	1.09	1.16	5.0
				GC–MS/MS	1.04	1.22	1.19	1.16	1.08	1.18	1.26	1.17	1.21	1.15	1.36	1.26	1.25	1.14	1.17	1.14	1.13	1.08	1.13	6.3
			Oolong tea	GC–MS	1.23	1.20	1.12	1.16	1.25	1.16	1.16	1.12	1.23	1.24	1.19	1.17	1.13	1.26	1.31	1.15	1.17	1.22	1.10	4.7
				GC–MS/MS	1.19	1.17	1.11	1.16	1.24	1.18	1.10	1.17	1.33	1.27	1.13	1.17	1.15	1.14	1.33	1.20	1.09	1.26	1.05	6.5
		ENVI-Carb + PSA	Green tea	GC–MS	1.05	1.20	1.20	1.17	1.10	1.20	1.22	1.17	1.17	1.11	1.25	1.21	1.26	1.1	1.15	1.16	1.19	1.07	1.12	4.9
				GC–MS/MS	1.05	1.17	1.16	1.15	1.07	1.17	1.21	1.12	1.22	1.18	1.33	1.24	1.26	1.12	1.19	1.13	1.20	1.05	1.13	6.1
			Oolong tea	GC–MS	1.21	1.17	1.07	1.15	1.23	1.16	1.16	1.11	1.20	1.28	1.19	1.17	1.12	1.22	1.30	1.20	1.17	1.17	1.14	4.7
				GC–MS/MS	1.14	1.16	1.07	1.20	1.18	1.18	1.10	1.14	1.26	1.26	1.16	1.18	1.17	1.16	1.30	1.19	1.11	1.25	1.07	5.4
20	DE-PCB 138	TPT	Greentea	GC–MS	1.05	1.20	1.20	1.17	1.05	1.16	1.16	1.16	1.17	1.15	1.27	1.20	1.22	1.1	1.19	1.18	1.22	1.08	1.11	5.0
				GC–MS/MS	1.05	1.15	1.18	1.21	1.08	1.16	1.18	1.10	1.19	1.13	1.32	1.22	1.19	1.18	1.16	1.14	1.17	1.03	1.16	5.6
			Oolong tea	GC–MS	1.19	1.15	1.06	1.14	1.25	1.16	1.15	1.05	1.21	1.23	1.16	1.16	1.15	1.20	1.28	1.19	1.18	1.20	1.11	5.0
				GC–MS/MS	1.14	1.11	1.02	1.12	1.19	1.15	1.10	1.14	1.26	1.27	1.12	1.17	1.10	1.12	1.34	1.20	1.06	1.28	1.09	7.2
		ENVI-Carb + PSA	Green tea	GC–MS	1.05	1.17	1.16	1.17	1.04	1.15	1.13	1.10	1.17	1.13	1.26	1.16	1.20	1.1	1.19	1.18	1.24	1.09	1.14	5.0
				GC–MS/MS	1.05	1.09	1.18	1.20	1.06	1.14	1.15	1.04	1.22	1.14	1.28	1.17	1.14	1.16	1.17	1.17	1.19	1.09	1.07	5.4
			Oolong tea	GC–MS	1.17	1.12	1.03	1.12	1.20	1.15	1.17	1.13	1.18	1.22	1.17	1.17	1.16	1.21	1.29	1.22	1.20	1.19	1.10	4.6
				GC–MS/MS	1.11	1.12	1.02	1.13	1.19	1.16	1.08	1.09	1.24	1.29	1.16	1.09	1.22	1.15	1.41	1.21	1.10	1.28	1.09	8.1

21	DE-PCB 28	TPT	Green tea	GC-MS	1.06	1.20	1.21	1.16	1.09	1.18	1.22	1.18	1.17	1.17	1.26	1.19	1.23	1.1	1.19	1.15	1.19	1.10	1.13	4.2
				GC-MS/MS	1.11	1.20	1.19	1.17	1.09	1.18	1.25	1.14	1.17	1.14	1.30	1.19	1.08	1.15	1.21	1.13	1.12	1.09	1.16	4.8
			Oolong tea	GC-MS	1.27	1.19	1.12	1.13	1.21	1.19	1.18	1.06	1.24	1.16	1.18	1.16	1.14	1.18	1.27	1.20	1.16	1.20	1.11	4.5
				GC-MS/MS	1.19	1.18	1.13	1.14	1.21	1.18	1.13	1.14	1.23	1.09	1.12	1.09	1.13	1.15	1.33	1.19	1.08	1.24	1.07	5.5
		ENVI-Carb + PSA	Green tea	GC-MS	1.06	1.20	1.21	1.16	1.09	1.19	1.21	1.18	1.16	1.15	1.25	1.19	1.24	1.1	1.20	1.15	1.18	1.11	1.12	4.2
				GC-MS/MS	1.08	1.13	1.19	1.17	1.09	1.18	1.25	1.14	1.17	1.14	1.30	1.19	1.08	1.15	1.21	1.13	1.12	1.09	1.16	5.0
			Oolong tea	GC-MS	1.26	1.19	1.12	1.13	1.21	1.18	1.16	1.06	1.21	1.18	1.16	1.15	1.15	1.19	1.27	1.19	1.19	1.23	1.11	4.5
				GC-MS/MS	1.18	1.18	1.13	1.14	1.21	1.18	1.13	1.14	1.23	1.09	1.12	1.13	1.13	1.15	1.33	1.19	1.08	1.24	1.07	5.5
22	DE-PCB101	TPT	Green tea	GC-MS	1.07	1.03	1.11	1.07	1.08	1.10	1.09	1.08	0.98	1.06	1.14	1.06	1.06	1.06	1.14	0.99	1.03	1.03	1.14	4.3
				GC-MS/MS	1.02	1.03	1.16	0.93	1.05	1.06	1.10	1.07	1.04	1.07	1.12	1.05	1.05	1.04	1.05	1.04	1.00	1.00	1.11	4.6
			Oolong tea	GC-MS	1.11	1.19	1.01	1.09	1.03	1.06	1.02	1.05	0.94	1.06	1.07	1.05	1.06	1.13	1.09	1.20	1.16	1.16	1.02	6.0
				GC-MS/MS	1.11	1.23	1.03	1.09	1.03	1.10	1.08	1.07	1.01	1.03	1.04	1.07	1.13	1.14	1.13	1.19	1.14	1.14	1.00	5.7
		ENVI-Carb + PSA	Green tea	GC-MS	1.07	1.04	1.12	1.06	1.10	1.11	1.11	1.07	1.04	1.06	1.12	1.06	1.10	1.01	1.04	1.01	1.06	1.05	1.15	4.3
				GC-MS/MS	1.03	1.04	1.17	0.99	1.07	1.08	1.09	1.06	1.08	1.08	1.11	1.08	1.06	1.03	1.04	1.03	1.01	1.00	1.16	4.4
			Oolong tea	GC-MS	1.09	1.16	1.01	1.09	1.03	1.07	1.02	1.04	0.96	1.05	1.03	1.05	1.07	1.13	1.10	1.16	1.11	1.18	1.04	5.4
				GC-MS/MS	1.10	1.21	1.00	1.10	1.03	1.11	1.08	1.07	1.01	1.08	1.03	1.10	1.13	1.14	1.07	1.23	1.12	1.16	1.02	5.7
23	DE-PCB138	TPT	Green tea	GC-MS	1.05	1.04	1.11	1.07	1.11	1.11	1.09	1.07	0.99	1.06	1.13	1.08	1.10	1.06	1.10	1.01	1.05	1.05	1.14	3.7
				GC-MS/MS	1.04	1.03	1.18	0.82	1.07	1.05	1.08	1.04	1.03	1.08	1.13	1.03	0.97	1.03	1.05	1.01	1.00	1.03	1.16	7.3
			Oolong tea	GC-MS	1.09	1.15	1.01	1.10	1.02	1.09	1.03	1.03	0.96	1.05	1.04	1.06	1.12	1.07	1.10	1.23	1.19	1.12	1.01	6.2
				GC-MS/MS	1.07	1.21	1.02	1.09	1.02	1.05	1.11	1.08	1.01	1.10	1.03	1.05	1.13	1.13	1.04	1.24	1.17	1.11	1.01	6.0
		ENVI-Carb + PSA	Green tea	GC-MS	1.02	1.04	1.11	1.13	1.12	1.09	1.08	1.05	1.02	1.06	1.12	1.09	1.04	1.04	1.10	1.04	1.03	1.08	1.11	3.7
				GC-MS/MS	1.01	1.03	1.14	1.12	1.13	1.11	1.08	1.02	1.01	1.05	1.14	1.00	1.00	1.08	1.08	1.03	1.03	1.03	1.18	5.1
			Oolong tea	GC-MS	1.05	1.12	1.00	1.12	1.02	1.10	1.28	1.02	0.95	1.06	1.06	1.05	1.04	1.07	1.11	1.26	1.14	1.18	0.99	7.8
				GC-MS/MS	1.11	1.20	1.01	1.07	1.01	1.12	1.06	1.10	1.03	1.06	1.02	1.06	1.18	1.09	1.09	1.22	1.13	1.19	0.98	6.3
24	DE-PCB28	TPT	Green tea	GC-MS	1.05	1.02	1.18	1.08	1.22	1.12	1.11	1.07	0.97	1.08	1.11	1.08	1.04	1.04	1.09	1.02	1.06	1.04	1.15	5.5
				GC-MS/MS	1.03	1.04	1.15	1.13	1.05	1.06	1.09	1.04	1.01	1.06	1.10	1.05	1.01	1.01	1.05	1.02	1.02	1.01	1.16	4.4
			Oolong tea	GC-MS	1.14	1.17	0.98	1.09	1.03	1.10	0.99	1.04	1.09	1.00	1.08	1.04	1.11	1.09	1.09	1.18	1.10	1.19	1.02	5.7
				GC-MS/MS	1.10	1.22	1.02	1.11	1.03	1.12	1.07	1.07	1.03	1.04	1.03	1.27	1.10	1.01	1.08	1.18	1.16	1.16	1.02	6.7
		ENVI-Carb + PSA	Green tea	GC-MS	1.05	1.03	1.16	1.06	1.11	1.09	1.10	1.04	0.97	1.07	1.12	1.07	1.03	1.05	1.07	1.02	1.07	1.07	1.14	4.3
				GC-MS/MS	1.03	1.04	1.15	1.13	1.05	1.12	1.09	1.07	1.01	1.06	1.10	1.06	1.01	1.02	1.02	1.02	1.01	1.01	1.16	4.4
			Oolong tea	GC-MS	1.14	1.18	0.99	1.08	1.03	1.09	1.03	1.04	1.09	1.06	1.05	1.06	1.10	1.10	1.08	1.18	1.10	1.20	1.02	5.7
				GC-MS/MS	1.10	1.22	1.02	1.11	1.03	1.12	1.07	1.07	1.03	1.04	1.03	1.04	1.10	1.10	1.08	1.18	1.09	1.16	1.02	5.5

(Continued)

Appendix Table 5.3 Youden Pair Ratios of Each Average Content From 201 Pesticides and RSD for 19 Circulative Determinations (cont.)

No	Pesticides	SPE	Sample	Method	Youden pair ratio																			RSD%
					Nov 9, 2009 (n = 5)	Nov 14, 2009 (n = 3)	Nov 19, 2009 (n = 3)	Nov 24, 2009 (n = 3)	Nov 29, 2009 (n = 3)	Dec 4, 2009 (n = 3)	Dec 9, 2009 (n = 3)	Dec 14, 2009 (n = 3)	Dec 19, 2009 (n = 3)	Dec 24, 2009 (n = 3)	Dec 29, 2009 (n = 3)	Jan 3, 2010 (n = 3)	Jan 8, 2010 (n = 3)	Jan 13, 2010 (n = 3)	Jan 18, 2010 (n = 3)	Jan 23, 2010 (n = 3)	Jan 28, 2010 (n = 3)	Feb 2, 2010 (n = 3)	Feb 7, 2010 (n = 3)	
25	Dichlorofop-methyl	TPT	Green tea	GC-MS	1.07	1.07	1.05	1.09	0.90	1.15	1.08	1.07	0.80	1.06	1.07	1.19	1.11	1.02	0.92	1.08	1.00	1.06	1.16	9.0
				GC-MS/MS	1.04	1.11	1.11	0.80	0.96	1.15	1.14	1.06	0.80	0.93	1.02	1.16	1.13	0.86	0.92	1.01	0.95	0.98	1.31	12.9
			Oolong tea	GC-MS	1.10	1.19	1.14	1.33	1.26	1.38	0.93	1.14	1.27	0.95	1.53	1.20	1.15	1.2	1.37	1.32	1.36	1.29	1.17	12.0
				GC-MS/MS	1.12	1.13	1.15	1.33	1.19	1.29	1.02	1.19	1.20	1.05	1.45	1.31	1.19	1.22	1.31	1.35	1.34	1.36	1.25	9.2
		ENVI-Carb + PSA	Green tea	GC-MS	1.24	0.78	1.34	1.17	1.09	0.90	1.41	1.00	1.08	1.13	1.24	1.17	1.10	1.16	1.03	1.31	1.41	0.90	0.97	15.5
				GC-MS/MS	1.30	0.90	1.18	1.11	1.10	0.91	2.70	1.19	1.23	1.13	1.23	1.08	0.79	1.54	0.91	1.43	1.08	0.88	1.07	34.2
			Oolong tea	GC-MS	1.13	1.22	1.06	1.11	1.33	1.08	1.37	1.23	1.43	1.19	1.26	1.18	1.40	1.16	1.20	1.62	1.25	1.21	1.29	10.9
				GC-MS/MS	0.99	1.18	0.99	1.06	1.25	1.10	1.12	1.22	1.45	1.14	1.24	1.13	1.34	1.10	1.32	1.60	1.23	1.31	1.28	12.5
26	Dimethena-mid	TPT	Green tea	GC-MS	1.03	1.05	1.17	1.06	1.07	1.06	1.10	1.13	1.05	1.04	1.13	1.07	1.04	0.98	1.08	1.02	1.01	1.02	1.12	4.5
				GC-MS/MS	1.00	1.03	0.97	1.52	1.03	1.09	1.12	1.09	1.01	0.97	1.09	1.08	1.01	0.95	1.07	1.03	1.05	1.01	1.12	11.3
			Oolong tea	GC-MS	1.07	1.25	1.23	1.18	1.08	1.17	1.37	1.26	1.22	1.22	1.37	1.27	1.30	1.1	1.12	1.15	1.24	1.07	1.14	7.7
				GC-MS/MS	1.15	1.26	1.18	1.20	1.07	1.19	1.38	1.23	1.23	1.17	1.34	1.21	1.13	1.16	1.24	1.19	1.34	1.14	1.21	6.3
		ENVI-Carb + PSA	Green tea	GC-MS	0.94	1.26	0.97	1.04	1.05	1.12	1.06	1.27	1.04	1.07	1.01	1.07	1.10	1.09	1.00	1.16	1.11	1.01	1.01	8.0
				GC-MS/MS	0.81	1.12	1.00	1.05	1.01	1.13	1.08	1.05	0.86	0.97	1.10	1.05	0.96	1.17	1.07	1.18	1.11	1.10	0.95	9.6
			Oolong tea	GC-MS	1.19	1.18	1.16	1.22	1.33	1.17	0.96	1.24	1.53	1.30	1.21	1.15	1.17	1.18	1.25	1.16	1.15	1.21	1.11	9.0
				GC-MS/MS	1.60	1.17	1.12	1.14	1.22	1.16	1.13	1.22	1.29	1.25	1.14	1.10	1.15	1.14	1.32	1.19	1.07	1.23	1.04	10.3
27	Diofenolan-1	TPT	Green tea	GC-MS	1.06	0.99	1.11	1.07	1.14	1.09	1.09	1.08	1.02	1.04	1.11	1.10	1.06	1.03	1.06	1.02	1.03	1.07	1.14	3.9
				GC-MS/MS	1.05	1.04	1.16	1.20	1.07	1.12	1.09	1.02	1.05	1.05	1.14	1.05	1.10	1.00	1.04	1.01	1.00	1.00	1.17	5.7
			Oolong tea	GC-MS	1.07	1.16	1.15	1.15	1.07	1.12	1.12	1.12	1.25	1.15	1.30	1.15	1.17	1.2	1.20	1.18	1.23	1.10	1.11	5.1
				GC-MS/MS	1.04	1.11	1.13	1.15	1.04	1.13	1.15	1.12	1.23	1.18	1.32	1.22	1.01	1.15	1.22	1.17	1.19	1.12	1.16	6.4
		ENVI-Carb + PSA	Green tea	GC-MS	1.06	1.15	1.07	1.10	1.02	1.14	1.00	1.03	1.10	1.04	1.00	1.09	1.10	1.12	1.12	1.22	1.13	1.19	1.01	5.7
				GC-MS/MS	1.07	1.19	1.03	1.14	1.02	1.13	1.05	1.08	1.07	1.02	1.01	1.04	1.04	1.17	1.09	1.18	1.11	1.14	0.98	5.7
			Oolong tea	GC-MS	1.14	1.15	1.03	1.09	1.28	1.18	1.20	1.08	1.31	1.23	1.24	1.15	1.15	1.30	1.30	1.20	1.20	1.19	1.13	6.6
				GC-MS/MS	1.17	1.11	1.02	1.13	1.19	1.17	1.13	1.14	1.21	1.25	1.15	1.09	1.09	1.15	1.31	1.18	1.14	1.27	1.09	5.9
28	Diofenolan-2	TPT	Green tea	GC-MS	1.06	1.03	1.11	1.07	1.11	1.12	1.10	1.07	1.04	1.03	1.13	1.09	1.08	1.05	1.10	1.02	1.03	1.07	1.15	3.5
				GC-MS/MS	1.00	1.05	1.16	1.18	1.10	1.12	1.09	1.03	1.04	1.06	1.14	1.08	1.09	1.00	1.05	0.97	1.02	1.03	1.19	5.9
			Oolong tea	GC-MS	1.06	1.18	1.16	1.17	1.06	1.14	1.11	1.15	1.20	1.16	1.32	1.18	1.18	1.1	1.21	1.20	1.22	1.10	1.15	5.1
				GC-MS/MS	1.04	1.14	1.12	1.15	1.05	1.13	1.11	1.13	1.23	1.18	1.33	1.19	1.02	1.13	1.17	1.21	1.16	1.14	1.16	6.1
		ENVI-Carb + PSA	Green tea	GC-MS	1.06	1.13	1.03	1.12	1.02	1.09	1.00	1.06	1.09	1.04	1.05	1.07	1.11	1.13	1.12	1.22	1.15	1.18	1.01	5.5
				GC-MS/MS	1.14	1.21	1.03	1.13	1.03	1.15	1.08	1.09	1.10	1.01	1.06	0.88	1.03	1.28	1.09	1.19	1.13	1.18	1.00	8.2
			Oolong tea	GC-MS	1.14	1.14	1.02	1.13	1.25	1.16	1.22	1.07	1.23	1.23	1.20	1.17	1.16	1.25	1.28	1.20	1.20	1.18	1.12	5.5
				GC-MS/MS	1.20	1.11	1.02	1.13		1.16	1.12	1.13	1.20	1.23	1.16		1.11		1.30	1.18	1.12	1.30	1.07	6.6

No.	Compound	Sorbent	Tea	Method	V1	V2	V3	V4	V5	V6	V7	V8	V9	V10	V11	V12	V13	V14	V15	V16	V17	V18	V19	V20
29	Fenbuconazole	TPT	Green tea	GC-MS	1.08	1.03	1.10	1.04	1.18	1.03	1.04	1.12	1.05	1.06	1.32	1.07	1.03	0.94	1.19	1.03	1.10	1.13	1.09	7.5
				GC-MS/MS	0.97	1.00	1.12	1.43	1.21	1.16	1.04	1.09	1.10	1.10	1.19	1.05	1.00	0.79	1.14	0.99	1.12	1.07	1.14	11.6
			Oolong tea	GC-MS	1.12	0.99	1.05	1.02	0.98	1.14	1.00	1.02	1.28	1.12	1.32	1.16	1.19	1.1	1.21	1.31	1.21	1.14	1.26	9.4
				GC-MS/MS	1.15	1.12	1.14	1.09	0.97	1.13	1.10	1.05	1.22	1.21	1.30	1.16	0.97	1.17	1.21	1.26	1.19	1.12	1.26	7.8
		ENVI-Carb + PSA	Green tea	GC-MS	0.98	1.00	1.05	1.14	0.97	1.09	1.00	1.07	1.17	1.27	1.07	1.04	1.23	1.08	1.08	1.30	1.17	1.19	1.01	8.9
				GC-MS/MS	0.97	1.09	1.02	1.18	0.95	1.09	0.72	1.12	1.14	1.10	1.01	1.06	0.99	1.22	1.07	1.16	1.16	1.22	1.02	11.0
			Oolong tea	GC-MS	1.16	1.14	0.90	1.01	1.42	1.20	1.26	1.23	1.26	1.59	1.20	1.12	1.12	1.18	1.48	1.17	1.18	1.30	1.09	13.0
				GC-MS/MS	1.12	1.11	0.90	1.07	1.12	1.19	1.13	1.12	1.17	1.59	1.10	1.15	1.31	1.07	1.50	1.17	1.10	1.34	1.09	13.5
30	Fenpyroximate	TPT	Green tea	GC-MS	1.17	0.98	1.13	1.01	1.33	1.12	1.11	1.20	1.15	1.23	1.26	0.95	1.00	1.06	1.32	1.00	1.15	1.00	1.09	10.2
				GC-MS/MS	1.02	0.96	1.28	1.26	1.32	1.04	1.05	1.10	1.43	1.04	1.28	1.04	0.96	1.16	1.25	1.07	1.12	1.09	1.02	11.8
			Oolong tea	GC-MS	1.00	1.28	1.37	1.08	0.90	1.10	0.97	1.26	1.13	1.36	1.02	1.07	1.35	1.3	1.09	1.20	1.12	1.03	1.34	12.6
				GC-MS/MS	1.03	1.32	1.08	1.01	0.89	1.04	0.70	1.20	1.19	1.30	1.12	1.06	1.18	1.20	1.15	1.19	1.15	0.99	1.37	14.0
		ENVI-Carb + PSA	Green tea	GC-MS	0.95	1.32	0.95	1.16	0.93	1.44	0.81	1.19	1.16	1.06	0.80	0.96	1.10	1.20	1.08	1.16	1.05	1.37	1.02	15.9
				GC-MS/MS	0.89	1.53	0.91	1.15	0.89	1.61	0.62	1.05	1.14	0.98	0.80	1.17	1.02	0.87	1.07	1.08	1.14	1.41	1.03	22.7
			Oolong tea	GC-MS	1.30	1.17	1.02	1.28	1.06	1.39	1.26	0.86	1.22	1.48	1.23	1.42	1.00	1.30	1.76	0.86	1.24	1.28	1.02	18.1
				GC-MS/MS	1.37	1.02	1.05	1.24	1.08	1.39	1.20	0.95	1.05	1.50	1.15	1.27	1.02	1.40	1.61	0.95	1.08	1.31	0.88	17.2
31	Fluotrimazole	TPT	Green tea	GC-MS	1.11	1.06	1.10	1.10	1.16	1.10	1.11	1.09	0.98	1.04	1.08	1.11	1.11	1.05	1.11	1.02	1.04	1.06	1.16	4.3
				GC-MS/MS	1.03	1.04	1.24	0.62	1.18	1.40	1.09	1.03	1.06	1.06	1.17	0.98	1.05	0.89	1.07	0.98	1.02	1.00	1.18	14.7
			Oolong tea	GC-MS	1.07	1.06	1.07	1.12	1.03	1.15	1.10	1.11	1.25	1.07	1.19	1.17	1.21	1.1	1.19	1.20	1.23	1.11	1.15	5.6
				GC-MS/MS	1.04	1.08	1.09	1.06	0.92	1.13	1.14	1.11	1.12	1.14	1.30	1.23	1.04	1.13	1.14	1.17	1.17	1.05	1.15	7.1
		ENVI-Carb + PSA	Green tea	GC-MS	1.10	1.06	1.11	1.14	0.98	1.15	0.98	1.06	1.00	1.13	0.92	1.10	1.12	1.15	1.10	1.24	1.15	1.19	0.99	7.5
				GC-MS/MS	1.06	1.21	1.05	1.14	1.03	1.35	1.05	1.11	1.07	1.05	1.05	1.13	0.90	1.34	1.16	1.28	1.12	1.23	1.00	10.4
			Oolong tea	GC-MS	1.17	1.15	1.05	1.11	1.28	1.17	1.23	1.19	1.38	1.37	1.36	1.09	1.17	1.16	1.34	1.19	1.20	1.21	1.11	8.1
				GC-MS/MS	1.19	1.11	1.02	1.05	1.19	1.19	1.10	1.13	1.30	1.30	1.14	1.12	1.16	1.11	1.28	1.21	1.14	1.22	1.09	6.8
32	Fluoxypr-1-methylheptyl ester	TPT	Green tea	GC-MS	1.09	1.06	1.12	1.07	1.04	1.08	1.10	1.04	0.88	0.99	0.99	1.09	1.09	1.01	1.00	1.03	1.03	1.21	1.17	6.7
				GC-MS/MS	1.07	1.15	1.06	1.24	1.09	1.16	0.97	1.04	1.20	1.10	1.10	1.18	1.12	1.1	1.24	1.20	1.23	1.12	1.12	7.7
			Oolong tea	GC-MS	1.06	1.09	1.09	1.25	1.11	1.16	1.15	1.19	1.14	1.09	1.40	1.11	1.06	1.37	1.13	1.15	1.33	1.05	1.26	9.1
				GC-MS/MS	1.09	0.97	1.09	1.17	1.02	1.00	1.36	1.01	1.09	1.09	1.16	1.08	1.09	1.13	0.99	1.23	1.20	1.01	0.98	9.1
		ENVI-Carb + PSA	Green tea	GC-MS	1.09	1.13	1.03	1.08	1.22	1.16	1.28	1.16	1.21	1.19	1.17	1.13	1.21	1.16	1.22	1.31	1.17	1.12	1.13	5.8
				GC-MS/MS	1.13	1.11	0.98	1.03	1.26	1.00	1.14	1.19	1.24	1.12	1.13	1.06	1.31	1.10	1.12	1.38	1.17	1.18	1.25	8.3
			Oolong tea	GC-MS	0.99	1.11	1.10	1.00	1.05	1.06	1.02	0.99	0.90	1.00	1.09	1.10	1.07	1.04	1.00	1.02	1.12	1.01	1.29	7.6
				GC-MS/MS	1.08	1.05	1.07	1.18	0.97	1.09	1.43	1.16	1.15	1.04	1.24	1.10	0.93	1.33	1.08	1.37	0.99	0.93	1.07	12.5

(Continued)

Appendix Table 5.3 Youden Pair Ratios of Each Average Content From 201 Pesticides and RSD for 19 Circulative Determinations (cont.)

No.	Pesticides	SPE	Sample	Method	Nov 9, 2009 (n=5)	Nov 14, 2009 (n=3)	Nov 19, 2009 (n=3)	Nov 24, 2009 (n=3)	Nov 29, 2009 (n=3)	Dec 4, 2009 (n=3)	Dec 9, 2009 (n=3)	Dec 14, 2009 (n=3)	Dec 19, 2009 (n=3)	Dec 24, 2009 (n=3)	Dec 14, 2009 (n=3)	Jan 3, 2010 (n=3)	Jan 8, 2010 (n=3)	Jan 13, 2010 (n=3)	Jan 18, 2010 (n=3)	Jan 23, 2010 (n=3)	Jan 28, 2010 (n=3)	Feb 2, 2010 (n=3)	Feb 7, 2010 (n=3)	RSD%
33	Iprovali-carb-1	TPT	Green tea	GC-MS	1.06	1.06	1.03	1.09	1.10	1.00	0.95	0.98	0.89	1.02	1.13	1.12	1.07	1.01	1.00	1.03	0.97	1.03	1.08	5.9
				GC-MS/MS	1.02	1.05	1.05	1.13	1.14	1.18	1.06	0.95	0.89	1.00	1.03	1.03	1.08	0.93	0.98	1.00	1.05	1.01	1.17	7.5
			Oolong tea	GC-MS	0.95	0.95	1.02	1.09	1.06	1.05	1.01	0.95	1.32	1.11	1.07	1.11	1.09	1.0	1.17	1.18	1.19	1.11	1.02	8.6
				GC-MS/MS	1.17	0.98	1.10	1.14	1.03	1.14	1.06	0.95	1.08	1.13	1.35	1.18	0.80	1.12	1.20	1.19	1.24	1.14	0.97	11.1
		ENVI-Carb + PSA	Green tea	GC-MS	0.96	1.21	0.96	1.19	0.98	1.13	1.17	1.20	1.13	1.00	1.19	1.07	1.13	0.95	1.17	1.12	1.21	0.79	0.91	11.3
				GC-MS/MS	0.90	0.89	1.02	1.17	1.00	1.13	1.09	1.14	1.04	1.04	1.24	1.09	1.03	1.02	1.08	1.10	1.17	1.00	0.90	9.1
			Oolong tea	GC-MS	1.09	1.02	0.98	0.96	1.35	1.08	0.90	1.30	1.32	1.20	1.14	1.11	1.11	1.24	1.25	1.24	1.19	1.10	1.10	11.0
				GC-MS/MS	1.13	1.08	0.98	1.07	1.22	1.15	1.11	1.27	1.21	1.22	1.11	0.89	1.29	1.09	1.23	1.26	1.12	1.22	1.13	8.9
34	Iprovali-carb-2	TPT	Green tea	GC-MS	1.04	1.00	1.15	1.14	1.12	1.00	1.04	1.05	1.16	1.11	1.17	1.15	1.09	0.97	1.02	1.04	0.99	1.08	1.11	5.9
				GC-MS/MS	1.03	1.07	1.04	1.13	1.13	1.19	1.07	1.03	0.97	0.97	1.11	1.09	1.09	0.91	1.04	1.01	1.06	1.01	1.17	6.6
			Oolong tea	GC-MS	1.01	1.03	1.03	1.07	1.07	1.11	1.05	0.98	1.33	1.10	1.36	1.21	1.07	1.0	1.19	1.19	1.27	1.09	1.08	9.7
				GC-MS/MS	1.17	0.98	1.10	1.14	1.01	1.15	1.08	0.96	1.30	1.12	1.35	1.19	0.97	1.07	1.20	1.17	1.22	1.14	1.01	9.6
		ENVI-Carb + PSA	Green tea	GC-MS	0.89	1.12	1.05	1.15	1.06	1.11	1.10	1.19	1.05	1.42	1.10	1.16	1.19	0.95	1.18	1.07	1.19	0.78	0.97	12.5
				GC-MS/MS	0.89	0.96	1.02	1.18	1.01	1.12	1.05	1.14	0.96	1.03	1.25	1.05	0.98	1.09	1.09	1.10	1.15	0.99	0.92	8.8
			Oolong tea	GC-MS	1.09	1.08	1.07	1.05	1.52	1.13	1.15	1.41	1.60	1.21	1.33	0.99	1.18	1.18	1.19	1.25	1.16	1.15	1.11	13.3
				GC-MS/MS	1.13	1.08	0.98	1.07	1.22	1.15	1.13	1.27	1.19	1.21	1.11	1.05	1.05	1.09	1.09	1.26	1.11	1.24	1.12	7.1
35	Isodrin	TPT	Green tea	GC-MS	1.05	1.02	1.08	1.10	1.09	1.05	1.11	1.09	1.03	1.02	1.06	1.05	1.06	1.05	1.06	1.03	0.99	1.15	1.15	4.0
				GC-MS/MS	1.05	1.06	1.10	1.14	1.02	1.06	1.08	1.00	1.04	1.09	1.15	1.09	1.11	1.04	1.11	1.02	1.14	1.02	1.17	4.6
			Oolong tea	GC-MS	1.03	1.18	1.22	1.25	1.11	1.19	1.30	1.24	1.06	1.17	1.17	1.21	1.24	1.2	1.20	1.09	1.19	1.15	1.15	5.7
				GC-MS/MS	1.05	1.22	1.17	1.20	1.09	1.21	1.29	1.14	1.20	1.18	1.35	1.28	1.18	1.15	1.30	0.99	1.10	1.12	1.18	7.6
		ENVI-Carb + PSA	Green tea	GC-MS	1.20	1.18	1.00	1.04	1.09	1.06	1.16	1.05	1.09	0.87	1.12	1.02	1.02	1.10	1.08	1.18	1.01	1.09	0.93	7.9
				GC-MS/MS	1.02	1.22	0.99	1.04	1.00	1.10	1.02	1.04	1.03	1.01	1.05	1.02	1.01	1.12	1.10	1.16	1.07	1.10	1.03	5.7
			Oolong tea	GC-MS	1.07	1.20	1.09	1.19	1.29	1.14	1.08	1.14	1.13	1.21	0.93	1.15	1.10	1.18	1.25	1.16	1.13	1.14	1.15	6.6
				GC-MS/MS	1.25	1.20	1.13	1.03	1.21	1.16	1.10	1.16	1.23	1.32	1.14	1.12	1.18	1.13	1.38	1.15	1.15	1.24	1.04	7.3
36	Isoprocarb-1	TPT	Green tea	GC-MS	1.02	0.99	1.10	1.04	1.12	1.13	1.08	1.16	1.02	1.11	1.16	1.08	1.06	1.05	1.10	1.02	1.04	1.04	1.16	4.8
				GC-MS/MS	1.02	1.01	1.16	1.07	1.09	1.08	1.08	1.07	1.04	1.02	1.12	1.05	1.01	0.95	1.12	1.02	1.04	1.03	1.14	4.9
			Oolong tea	GC-MS	1.03	1.23	1.22	1.15	1.07	1.17	1.22	1.21	1.18	1.24	1.28	1.20	1.24	1.2	1.22	1.18	1.21	1.08	1.22	5.4
				GC-MS/MS	0.90	1.04	0.98	1.00	0.91	1.03	0.98	1.00	0.99	1.02	1.08	1.01	0.94	0.96	1.02	0.93	0.95	0.92	1.00	4.9
		ENVI-Carb + PSA	Green tea	GC-MS	1.09	1.15	0.99	1.11	1.00	1.08	0.91	0.99	1.09	0.97	1.06	1.04	1.08	1.13	1.09	1.21	1.11	1.22	1.00	7.5
				GC-MS/MS	1.07	1.25	1.01	1.14	1.01	1.06	0.93	1.07	1.06	1.07	1.03	0.89	1.00	1.18	1.11	1.19	1.09	1.20	1.02	8.6
			Oolong tea	GC-MS	1.30	1.23	1.14	1.15	1.28	1.23	1.22	1.10	1.21	1.34	1.17	1.22	1.17	1.24	1.28	1.19	1.20	1.29	1.14	5.2
				GC-MS/MS	1.17	0.99	0.89	0.97	1.04	0.97	0.96	0.97	1.02	1.07	0.98	0.91	0.87	0.99	1.13	1.02	0.93	1.02	0.93	7.7

(Continued)

37	Isoprocarb-2	TPT	Green tea	GC–MS	0.80	1.19	1.03	1.02	0.99	1.10	1.09	1.04	0.86	0.85	0.95	0.99	1.05	0.97	1.08	0.99	1.04	1.01	1.02	9.2
				GC–MS/MS	.07	1.17	1.05	1.07	0.86	0.99	1.17	1.07	0.85	0.78	1.10	1.03	1.04	0.74	0.82	1.09	1.03	0.98	1.07	12.8
			Oolong tea	GC–MS	.18	1.18	1.25	1.30	1.08	1.13	1.09	1.13	1.14	1.02	1.19	1.16	1.24	1.1	1.20	1.18	1.01	1.18	0.99	7.2
				GC–MS/MS	.12	1.21	1.08	1.33	0.86	1.20	1.12	0.90	0.96	0.88	1.05	1.18	0.84	1.00	1.23	1.03	0.92	0.90	0.95	13.9
		ENVI-Carb + PSA	Greentea	GC–MS	1.04	1.17	1.00	1.12	0.95	1.03	1.50	1.10	0.99	1.18	1.29	1.07	1.09	0.97	1.17	1.12	1.00	0.74	1.02	14.2
				GC–MS/MS	0.91	0.99	1.02	1.08	0.96	1.19	1.69	1.11	0.85	0.98	1.28	1.07	0.93	1.09	1.02	1.14	1.10	1.09	0.83	17.5
			Oolong tea	GC–MS	1.16	1.14	1.08	1.17	1.19	1.08	0.97	1.19	1.47	1.12	1.12	1.04	0.95	0.97	1.24	1.20	1.01	0.98	1.00	11.3
				GC–MS/MS	1.40	1.26	0.97	1.01	1.04	1.07	1.14	1.17	1.46	1.04	0.93	0.89	0.85	0.93	1.13	1.40	0.89	1.02	0.87	17.5
38	Lenacil	TPT	Green tea	GC–MS	1.02	1.09	1.10	1.08	0.99	1.08	1.00	1.02	0.58	0.97	1.07	1.19	1.15	0.94	0.88	1.00	0.87	1.03	1.22	13.8
				GC–MS/MS	1.00	1.08	1.07	1.38	1.00	1.04	1.00	0.93	0.71	0.96	1.06	1.15	1.16	0.93	0.87	0.89	0.78	0.79	1.15	15.9
			Oolong tea	GC–MS	1.09	0.99	0.88	1.15	1.33	1.31	1.01	0.93	1.21	0.99	1.43	1.20	0.90	0.9	1.30	1.11	1.25	1.16	0.86	15.9
				GC–MS/MS	1.42	0.68	1.07	1.25	1.21	1.28	1.06	1.02	1.61	1.04	1.57	1.20	0.79	0.95	1.16	1.39	0.81	1.01	0.69	24.1
		ENVI-Carb + PSA	Green tea	GC–MS	0.93	0.87	1.05	1.30	1.10	0.91	1.31	0.99	0.97	1.13	1.56	1.15	1.19	0.89	1.31	1.24	1.32	0.76	0.94	18.5
				GC–MS/MS	0.97	0.71	1.06	1.25	1.04	0.95	1.20	0.97	1.04	1.18	0.85	1.04	1.11	1.32	0.93	1.15	0.99	1.10	0.99	13.7
			Oolong tea	GC–MS	1.04	1.16	1.03	0.98	1.55	1.07	1.26	1.79	1.36	1.19	1.16	0.99	1.30	1.06	0.96	1.55	1.16	1.07	1.18	18.7
				GC–MS/MS	0.91	1.12	1.00	1.03	1.36	1.10	1.06	1.47	1.28	1.14	1.12	0.89	1.42	1.03	1.02	1.61	0.95	1.06	0.97	17.8
39	Metalaxyl	TPT	Green tea	GC–MS	1.09	1.04	1.10	1.09	1.12	1.12	1.07	1.21	1.00	1.20	1.18	1.04	1.11	1.01	1.11	1.01	1.03	1.03	1.15	5.9
				GC–MS/MS	1.06	1.05	1.14	1.45	1.09	1.10	1.11	1.08	1.01	1.02	1.16	1.06	1.05	0.93	1.09	1.03	1.05	0.98	1.14	9.7
			Oolong tea	GC–MS	1.05	1.12	1.10	1.16	1.19	1.16	1.29	1.22	1.37	1.30	1.33	1.21	1.22	1.1	1.22	1.20	1.25	1.07	1.19	7.3
				GC–MS/MS	1.09	1.19	1.18	1.15	1.09	1.21	1.33	1.23	1.18	1.22	1.37	1.32	1.16	1.17	1.23	1.19	1.25	1.05	1.15	6.8
		ENVI-Carb + PSA	Green tea	GC–MS	1.01	1.13	1.06	1.12	1.03	1.18	0.97	1.03	0.97	1.21	1.05	1.21	1.10	1.11	1.10	1.24	1.14	1.08	0.99	7.4
				GC–MS/MS	1.05	1.17	1.02	1.13	1.04	1.12	0.93	1.07	0.99	1.02	1.07	1.09	1.00	1.18	1.13	1.22	1.09	1.14	0.96	7.3
			Oolong tea	GC–MS	1.13	1.22	1.14	1.10	1.37	1.18	1.16	1.23	1.62	1.43	1.24	1.10	1.18	1.17	1.28	1.20	1.13	1.26	1.09	10.8
				GC–MS/MS	1.29	1.17	1.10	1.11	1.24	1.20	1.07	1.20	1.28	1.39	1.16	1.15	1.12	1.09	1.39	1.20	1.10	1.34	1.06	8.8
40	Metazachlor	TPT	Green tea	GC–MS	0.96	1.10	1.09	1.03	1.03	1.11	1.13	1.10	1.04	1.04	1.14	1.11	1.01	1.01	1.14	1.01	0.96	0.98	1.05	5.4
				GC–MS/MS	0.99	1.05	0.79	1.59	1.02	1.03	1.15	1.13	1.02	0.86	1.10	1.08	0.99	0.84	1.09	1.01	1.11	1.02	1.09	15.5
			Oolong tea	GC–MS	1.03	1.36	1.18	1.12	1.02	1.14	1.31	1.18	1.21	1.19	1.32	1.26	1.20	1.1	1.11	1.10	1.22	1.01	1.14	8.5
				GC–MS/MS	1.17	1.24	1.16	1.27	1.06	1.27	1.36	1.20	1.21	1.13	1.32	1.21	1.23	1.18	1.18	1.25	1.42	1.09	1.14	7.3
		ENVI-Carb + PSA	Green tea	GC–MS	0.80	1.33	0.95	1.00	1.03	1.02	0.89	1.31	1.05	1.04	0.99	1.01	1.07	1.01	0.98	1.13	1.15	0.97	1.00	12.1
				GC–MS/MS	0.70	0.99	0.98	1.01	0.97	1.09	1.09	1.00	0.72	0.88	1.17	1.15	0.91	1.34	1.09	1.07	1.20	0.98	0.85	15.8
			Oolong tea	GC–MS	1.09	1.17	1.12	1.23	1.28	1.11	0.82	1.36	1.42	1.34	1.23	1.12	1.13	1.01	1.16	1.21	1.10	1.07	1.01	12.1
				GC–MS/MS	1.57	1.15	1.02	1.12	1.23	1.11	1.17	1.28	1.27	1.25	1.08	1.04	1.16	0.97	1.30	1.19	1.00	1.18	1.06	11.9

Appendix Table 5.3 Youden Pair Ratios of Each Average Content From 201 Pesticides and RSD for 19 Circulative Determinations (cont.)

No.	Pesticides	SPE	Sample	Method	Youden pair ratio Nov 9, 2009 (n = 5)	Nov 14, 2009 (n = 3)	Nov 19, 2009 (n = 3)	Nov 24, 2009 (n = 3)	Nov 29, 2009 (n = 3)	Dec 4, 2009 (n = 3)	Dec 9, 2009 (n = 3)	Dec 14, 2009 (n = 3)	Dec 19, 2009 (n = 3)	Dec 24, 2009 (n = 3)	Dec 14, 2009 (n = 3)	Jan 3, 2010 (n = 3)	Jan 8, 2010 (n = 3)	Jan 13, 2010 (n = 3)	Jan 18, 2010 (n = 3)	Jan 23, 2010 (n = 3)	Jan 28, 2010 (n = 3)	Feb 2, 2010 (n = 3)	Feb 7, 2010 (n = 3)	RSD%
41	Methabenz-thiazuron	TPT	Green tea	GC–MS	1.09	1.03	1.12	1.12	1.10	1.09	1.10	1.13	1.06	1.06	1.15	1.09	1.12	1.05	1.18	1.03	1.03	1.08	1.16	4.0
				GC–MS/MS	1.00	1.04	1.17	1.36	1.07	1.17	1.12	1.08	1.03	1.00	1.15	1.07	1.11	0.92	1.08	1.02	1.02	0.97	1.27	9.6
			Oolong tea	GC–MS	1.10	1.14	1.17	1.20	1.18	1.19	1.27	1.18	1.33	1.17	1.42	1.21	1.27	1.1	1.28	1.22	1.24	1.06	1.08	7.3
				GC–MS/MS	1.03	1.15	1.18	1.18	1.06	1.14	1.29	1.13	1.23	1.21	1.41	1.29	1.13	1.19	1.24	1.17	1.20	1.07	1.06	7.9
		ENVI-Carb + PSA	Green tea	GC–MS	1.08	1.09	1.01	1.14	1.02	1.19	0.87	1.04	1.03	1.19	0.97	1.05	1.14	1.10	1.16	1.22	1.20	1.19	0.98	8.6
				GC–MS/MS	1.05	1.14	1.00	1.17	1.05	1.13	0.89	1.11	0.98	1.05	1.07	1.08	1.08	1.17	1.15	1.21	1.06	1.09	0.97	7.4
			Oolong tea	GC–MS	1.15	1.18	1.12	1.12	1.40	1.24	1.24	1.27	1.75	1.31	1.22	1.13	1.19	1.23	1.23	1.14	1.17	1.18	1.12	11.9
				GC–MS/MS	1.18	1.14	1.10	1.11	1.22	1.23	1.08	1.25	1.31	1.31	1.14	1.09	1.14	1.12	1.28	1.24	1.13	1.40	1.15	7.4
42	Mirex	TPT	`Green tea	GC–MS	1.00	1.02	1.14	1.07	1.02	1.16	1.12	1.02	0.96	1.02	1.06	1.13	1.07	1.02	1.00	1.09	0.92	1.02	1.10	6.0
				GC–MS/MS	1.05	1.06	1.01	1.32	1.01	1.08	1.09	1.03	0.89	1.01	1.08	1.12	1.07	0.99	1.02	1.05	1.00	0.99	1.11	7.9
			Oolong tea	GC–MS	1.08	1.22	1.14	1.23	1.17	1.22	1.24	1.23	1.19	1.09	1.32	1.24	1.17	1.0	1.18	1.10	1.25	1.06	1.08	7.3
				GC–MS/MS	1.10	1.19	1.14	1.26	1.09	1.32	1.20	1.18	1.30	1.07	1.37	1.20	1.27	1.18	1.18	1.11	1.29	1.14	1.08	7.3
		ENVI-Carb + PSA	Green tea	GC–MS	0.94	1.08	1.03	1.11	1.08	0.99	1.14	1.27	0.98	0.97	1.23	1.11	1.07	0.96	1.02	1.16	1.20	0.89	0.99	9.9
				GC–MS/MS	0.95	0.88	1.00	1.08	1.03	0.99	1.58	1.06	0.91	0.96	1.16	1.17	0.97	1.25	1.04	1.19	1.10	1.02	0.93	15.0
			Oolong tea	GC–MS	1.15	1.16	1.11	1.08	1.35	1.08	0.96	1.31	1.33	1.10	1.11	1.08	1.17	1.08	1.04	1.39	1.06	1.09	1.10	10.1
				GC–MS/MS	1.24	1.16	1.05	1.09	1.28	1.08	1.13	1.25	1.33	1.09	1.07	1.03	1.25	1.07	1.16	1.30	1.04	1.14	1.10	8.2
43	Monalide	TPT	Green tea	GC–MS	1.00	1.14	1.08	1.00	0.92	1.11	1.14	1.03	0.97	0.84	1.01	1.08	1.03	1.01	0.98	0.94	0.95	1.03	1.14	7.9
				GC–MS/MS	1.05	1.04	1.16	1.34	1.07	1.38	1.11	1.08	1.02	1.00	1.12	1.08	1.02	0.97	1.05	1.02	1.03	1.09	1.14	7.5
			Oolong tea	GC–MS	1.06	1.15	1.18	1.34	1.06	1.25	0.95	1.11	1.08	0.89	1.15	1.08	1.04	1.0	1.10	0.99	1.04	1.06	1.09	9.4
				GC–MS/MS	1.07	1.17	1.17	1.18	1.10	1.16	1.25	1.17	1.22	1.17	1.30	1.18	1.14	1.15	1.20	1.16	1.16	1.03	1.14	5.0
		ENVI-Carb + PSA	Green tea	GC–MS	1.03	1.20	1.01	1.06	0.99	0.34	1.22	1.15	1.36	1.01	1.28	1.19	0.94	0.96	1.08	1.16	1.08	1.03	0.99	11.9
				GC–MS/MS	1.09	1.23	1.03	1.14	1.03	1.11	1.04	1.06	1.04	1.01	1.06	1.07	0.99	1.09	1.07	1.21	1.09	1.13	1.01	5.9
			Oolong tea	GC–MS	1.21	1.27	0.96	1.07	1.06	1.05	1.10	1.04	1.31	1.04	0.79	0.99	1.14	0.91	1.07	1.39	1.01	1.12	1.02	12.9
				GC–MS/MS	1.20	1.17	1.09	1.14	1.22	1.14	1.11	1.16	1.22	1.24	1.14	1.10	1.18	1.14	1.28	1.21	1.07	1.21	1.00	5.8
44	Paraoxon-ethyl	TPT	Green tea	GC–MS	1.02	1.02	0.97	1.00	1.03	1.04	1.18	1.04	1.06	0.94	0.98	1.01	0.96	0.99	1.00	0.98	0.98	1.06	0.93	5.4
				GC–MS/MS	1.00	1.06	0.74	1.29	0.94	1.17	1.11	1.01	0.88	0.71	1.02	1.18	1.31	0.77	0.95	1.00	1.25	1.03		17.4
			Oolong tea	GC–MS	0.99	1.08	1.10	0.99	1.03	0.97	0.89	0.96	0.98	1.03	0.95	1.19	1.02	0.9	0.97	1.07	1.01	0.89	1.12	8.0
				GC–MS/MS	1.16	1.06	1.13	1.22	1.15	0.99	1.30	1.22		0.94	1.41	1.30	1.00	1.14	1.35					12.1
		ENVI-Carb + PSA	Green tea	GC–MS	1.30	0.92	1.01	0.97	0.86	0.99		1.07	1.00	0.94	0.98	0.99	1.04	1.03	1.00	1.08	0.99	0.97	0.97	8.8
				GC–MS/MS	0.95	0.98	0.93	1.02	0.85	1.22	1.10	1.25	0.95	0.92	1.37	1.04	1.02	1.16	1.16	0.80	1.45			16.6
			Oolong tea	GC–MS	1.01	0.95	0.94	1.09	0.90	1.02	1.03	0.96	0.95	1.09	0.99	1.01	1.05	0.96	1.13	1.10	1.04	1.02	0.96	6.5
				GC–MS/MS	1.56	1.20	1.01	0.82	1.10	1.15	1.21	1.49	1.43	1.17	1.09	1.10	1.10	0.74	1.04	1.48				19.7

No.	Compound	Cleanup	Matrix	Method																				RSD
45	Pebulate	TPT	Green tea	GC-MS	1.10	1.04	1.10	1.03	1.10	1.10	1.07	1.09	0.99	1.12	1.11	1.06	1.04	1.02	1.14	0.98	1.01	1.01	1.08	4.5
				GC-MS/MS							1.08	1.06	1.02	1.07	1.10	1.06	1.05	0.98	1.17	0.99	1.04	0.92	1.10	6.0
			Oolong tea	GC-MS	1.12	1.25	1.20	1.23	1.13	1.23	1.29	1.22	1.25	1.21	1.35	1.24	1.26	1.0	1.21	1.14	1.02	1.11	1.16	6.9
				GC-MS/MS							1.32	1.22	1.25	1.39	1.40	1.26	1.11	0.96	1.10	1.07	0.91	1.07	1.24	13.1
		ENVI-Carb + PSA	Green tea	GC-MS	1.08	1.21	0.99	1.10	1.00	1.06	1.05	1.03	1.10	1.05	1.14	1.09	1.15	1.10	1.17	1.28	1.08	1.03	0.96	7.2
				GC-MS/MS							1.06	1.08	1.05		1.12	1.08	1.14	1.10	1.13	1.31	1.02	1.16	1.01	7.3
			Oolong tea	GC-MS	1.37	1.21	1.09	1.15	1.32	1.23	1.23	1.21	1.45	1.33	1.20	1.22	1.39	1.20	1.36	1.07	1.21	1.16	1.23	8.2
				GC-MS/MS							1.12	1.29	1.29	1.32	1.18	1.14	0.82	1.30	1.65	0.88	1.20	1.26	1.15	17.1
46	Penrthane	TPT	Green tea	GC-MS	1.07	1.16	1.17	1.14	1.19	1.31	1.26	1.21	1.14	1.16	1.30	1.13	1.20	1.1	1.22	1.14	1.19	1.16	1.08	5.8
				GC-MS/MS	1.05	1.15	1.22	1.24	1.13	1.22	1.25	1.18	1.20	1.15	1.32	1.21	1.17	1.13	1.23	1.18	1.07	1.08	1.08	6.1
			Oolong tea	GC-MS	1.25	1.12	1.18	1.11	1.30	1.17	1.15	1.20	1.29	1.19	1.12	1.15	1.15	1.17	1.21	1.24	1.15	1.24	1.11	5.0
				GC-MS/MS	1.24	1.11	1.16	1.14	1.19	1.15	1.08	1.14	1.24	1.24	1.10	1.09	1.15	1.14	1.30	1.23	1.13	1.28	1.10	5.7
		ENVI-Carb + PSA	Green tea	GC-MS	1.13	1.03	1.17	1.06	1.12	1.09	1.08	1.13	1.07	1.05	1.07	1.16	1.12	1.05	1.04	1.08	1.05	1.16	1.10	3.7
				GC-MS/MS	1.02	1.01	1.13	1.13	1.02	1.10	1.09	1.08	0.99	0.99	1.09	1.04	1.07	0.96	1.06	1.05	1.04	1.13	1.19	5.4
			Oolong tea	GC-MS	1.11	1.16	1.00	1.12	1.06	1.08	1.04	1.11	1.07	1.03	1.05	1.06	1.14	1.08	1.18	1.16	1.07	1.11	1.02	4.7
				GC-MS/MS	1.08	1.20	1.01	1.12	1.08	1.08	1.03	1.09	1.07	1.03	1.03	1.05	1.09	1.07	1.19	1.17	1.09	1.16	1.13	5.0
47	Pentachloro-aniline	TPT	Green tea	GC-MS	1.08	1.22	1.20	1.18	1.11	1.17	1.10	1.13	1.14	1.16	1.34	1.19	1.22	1.1	1.21	1.15	1.16	1.13	1.13	5.0
				GC-MS/MS	1.09	1.17	1.15	1.18	1.05	1.22	1.22	1.08	1.17	1.13	1.30	1.23	1.18	1.13	1.23	1.10	1.10	1.13	1.16	5.4
			Oolong tea	GC-MS	1.32	1.18	1.12	1.14	1.26	1.16	1.01	0.88	1.25	1.28	1.20	1.15	1.15	1.18	1.26	1.21	1.19	1.19	1.12	8.4
				GC-MS/MS	1.33	1.18	1.14	1.19	1.22	1.20	1.12	1.14	1.24	1.26	1.12	1.13	1.13	1.13	1.35	1.21	1.14	1.26	1.13	6.0
		ENVI-Carb + PSA	Green tea	GC-MS	1.07	1.04	1.09	1.05	1.08	1.11	1.09	1.08	0.97	1.05	1.10	1.06	1.05	1.00	1.09	1.02	1.03	1.04	1.14	3.8
				GC-MS/MS	1.01	1.05	1.11	0.81	1.05	1.09	1.09	1.06	1.02	1.08	1.12	1.07	1.08	1.06	1.07	1.02	1.02	1.02	1.16	6.6
			Oolong tea	GC-MS	0.89	1.19	1.00	1.08	1.01	1.07	0.98	0.98	1.09	1.02	1.08	1.06	1.11	1.08	1.10	1.18	1.09	1.16	1.02	7.0
				GC-MS/MS	1.06	1.22	1.02	1.12	1.00	1.07	1.04	1.09	1.03	1.05	1.07	1.11	1.11	1.17	1.07	1.17	1.07	1.15	1.01	5.5
48	Pentachloro-anisole	TPT	Green tea	GC-MS	1.09	1.23	1.19	1.22	1.12	1.23	1.21	1.20	1.26	1.19	1.25	1.19	1.23	1.1	1.24	1.15	1.17	1.17	1.17	3.8
				GC-MS/MS	1.06	1.21	1.15	1.19	1.13	1.22	1.23	1.15	1.20	1.18	1.30	1.25	1.24	1.17	1.20	1.14	1.15	1.14	1.17	4.5
			Oolong tea	GC-MS	1.38	1.18	1.06	1.14	1.27	1.20	1.17	1.14	1.29	1.30	1.13	1.17	1.17	1.21	1.31	1.27	1.21	1.23	1.12	6.6
				GC-MS/MS	1.38	1.19	1.08	1.14	1.24	1.21	1.12	1.16	1.26	1.33	1.09	1.18	1.07	1.20	1.41	1.29	1.11	1.25	1.10	8.4
		ENVI-Carb + PSA	Green tea	GC-MS	1.08	1.05	1.09	1.04	1.10	1.11	1.09	1.08	1.04	1.10	1.10	1.07	1.05	1.08	1.13	0.99	1.03	1.01	1.15	3.7
				GC-MS/MS	1.05	1.05	1.15	1.20	1.04	1.10	1.09	1.04	1.05	1.11	1.09	1.05	1.07	1.03	1.05	1.01	1.02	0.98	1.11	4.8
			Oolong tea	GC-MS	1.11	1.21	0.99	1.10	1.00	1.03	1.04	1.01	1.09	1.01	1.14	1.07	1.17	1.08	1.19	1.18	1.09	1.19	1.12	6.6
				GC-MS/MS	1.06	1.27	0.99	1.12	1.00	1.08	1.16	1.08	1.08	1.06	1.10	0.99	1.14	1.05	1.18	1.16	1.08	1.15	1.04	6.6

(Continued)

Appendix Table 5.3 Youden Pair Ratios of Each Average Content From 201 Pesticides and RSD for 19 Circulative Determinations (cont.)

No.	Pesticides	SPE	Sample	Method	Youden pair ratio																			RSD%
					Nov 9, 2009 (n=5)	Nov 14, 2009 (n=3)	Nov 19, 2009 (n=3)	Nov 24, 2009 (n=3)	Nov 29, 2009 (n=3)	Dec 4, 2009 (n=3)	Dec 9, 2009 (n=3)	Dec 14, 2009 (n=3)	Dec 14, 2009 (n=3)	Dec 19, 2009 (n=3)	Dec 24, 2009 (n=3)	Jan 3, 2010 (n=3)	Jan 8, 2010 (n=3)	Jan 13, 2010 (n=3)	Jan 18, 2010 (n=3)	Jan 23, 2010 (n=3)	Jan 28, 2010 (n=3)	Feb 2, 2010 (n=3)	Feb 7, 2010 (n=3)	
49	Pentachlorobenzene	TPT	Green tea	GC-MS	1.05	1.19	1.17	1.18	1.06	1.17	1.34	1.13	1.22	1.13	1.13	1.22	1.20	1.1	1.19	1.18	1.24	1.06	1.07	6.8
				GC-MS/MS	1.14	1.13	1.12	1.16	1.06	1.17	1.26	1.11	1.22	1.13	1.13	1.18	1.06	1.15	1.22	1.13	1.19	1.07	1.12	6.4
			Oolong tea	GC-MS	1.15	1.17	1.06	1.12	1.25	1.15	1.22	1.14	1.26	1.22	1.22	1.14	1.15	1.23	1.24	1.21	1.17	1.19	1.11	4.6
				GC-MS/MS	1.12	1.12	1.07	1.11	1.22	1.16	1.11	1.19	1.24	1.20	1.20	1.05	1.20	1.11	1.27	1.23	1.07	1.21	1.08	5.7
		ENVI-Carb + PSA	Green tea	GC-MS	1.17	1.05	1.10	1.08	1.08	1.09	1.05	1.06	0.95	1.05	1.05	1.10	1.12	0.99	1.11	1.03	1.01	1.06	1.11	4.8
				GC-MS/MS	1.04	1.04	1.11	1.24	1.07	1.10	1.10	1.04	0.98	1.06	1.06	1.06	1.07	1.00	1.05	1.02	1.01	1.00	1.08	5.9
			Oolong tea	GC-MS	1.08	1.15	1.03	1.12	1.04	1.07	1.02	1.14	1.07	1.02	1.02	1.09	1.11	1.15	1.08	1.26	1.12	1.11	1.00	5.6
				GC-MS/MS	1.07	1.16	1.03	1.15	1.04	1.12	1.07	1.10	1.02	1.00	1.00	1.07	0.95	1.15	1.08	1.21	1.14	1.13	1.00	6.0
50	Phenanthrene	TPT	Green tea	GC-MS	1.05	1.04	1.12	1.05	1.11	1.12	1.09	1.08	1.05	0.99	1.11	1.07	1.06	1.06	1.10	1.02	1.05	1.07	1.14	3.7
				GC-MS/MS	1.00	1.03	1.14	1.23	1.04	1.06	1.05	1.01	1.02	1.02	1.05	1.07	1.07	0.93	1.02	0.97	0.98	1.05	1.15	6.5
			Oolong tea	GC-MS	1.07	1.21	1.21	1.16	1.12	1.18	1.19	1.20	1.16	1.17	1.26	1.18	1.21	1.1	1.23	1.16	1.15	1.10	1.11	4.2
				GC-MS/MS	1.04	1.20	1.08	1.13	1.03	1.14	1.21	1.17	1.11	1.22	1.28	1.18	1.17	1.10	1.21	1.09	1.08	1.08	1.12	5.9
		ENVI-Carb + PSA	Green tea	GC-MS	1.12	1.16	1.00	1.09	1.03	1.08	1.00	1.07	1.00	1.09	1.07	1.04	1.11	1.11	1.09	1.17	1.10	1.16	1.02	4.9
				GC-MS/MS	1.05	1.27	1.01	1.10	1.02	1.11	1.03	1.13	1.02	1.05	1.13	1.06	0.93	1.16	1.07	1.15	1.08	1.20	1.02	7.3
			Oolong tea	GC-MS	1.27	1.20	1.12	1.13	1.25	1.18	1.18	1.09	1.23	1.20	1.14	1.15	1.13	1.21	1.23	1.19	1.19	1.20	1.12	4.1
				GC-MS/MS	1.25	1.13	1.09	1.10	1.16	1.21	1.11	1.19	1.24	1.19	1.13	1.02	1.02	1.09	1.37	1.13	1.09	1.29	1.08	7.9
51	Pirimicarb	TPT	Green tea	GC-MS	1.08	1.00	1.27	1.11	1.08	1.05	1.02	1.05	0.98	1.09	1.13	1.06	1.10	1.01	1.08	1.02	1.04	1.03	1.14	6.5
				GC-MS/MS	1.04	1.03	1.17	1.23	1.07	1.17	1.09	1.03	0.99	1.12	1.16	1.07	1.07	0.95	1.04	1.01	1.02	1.02	1.18	7.2
			Oolong tea	GC-MS	1.08	1.12	1.14	1.15	1.16	1.17	1.25	1.13	1.32	1.18	1.26	1.26	1.25	1.1	1.23	1.19	1.26	1.07	1.10	8.1
				GC-MS/MS	1.11	1.17	1.18	1.17	1.12	1.19	1.35	1.17	1.23	1.20	1.08	1.27	1.08	1.14	1.20	1.15	1.26	1.06	1.14	7.5
		ENVI-Carb + PSA	Green tea	GC-MS	1.10	1.18	1.12	1.07	1.11	1.11	1.06	1.07	1.06	1.38	1.11	1.10	1.09	1.08	1.11	1.23	1.14	1.10	0.99	8.7
				GC-MS/MS	1.08	1.17	1.18	1.15	1.03	1.13	0.97	1.10	0.99	1.03	1.04	1.04	0.94	1.26	1.07	1.21	1.11	1.10	0.99	7.6
			Oolong tea	GC-MS	1.06	1.03	1.04	1.16	1.54	1.18	1.32	1.29	1.99	1.31	1.02	1.02	1.17	1.22	1.24	1.23	1.20	1.26	1.14	16.0
				GC-MS/MS	1.04	1.06	1.03	1.11	1.26	1.18	1.10	1.21	1.31	1.28	1.16	1.10	1.16	1.16	1.33	1.25	1.12	1.29	1.13	6.2
52	Procymidone	TPT	Green tea	GC-MS	1.06	1.03	1.12	1.07	1.10	1.11	1.10	1.12	1.13	1.02	1.03	1.07	0.92	0.97	1.11	1.02	1.04	1.05	1.15	5.4
				GC-MS/MS	1.04	1.06	1.18	1.15	1.07	1.07	1.10	1.05	1.07	1.02	1.07	1.03	1.06	0.96	1.03	1.01	1.02	1.01	1.17	5.5
			Oolong-tea	GC-MS	1.06	1.20	1.19	1.16	1.07	1.18	1.21	1.19	1.19	1.19	1.15	1.21	1.22	1.0	1.17	1.17	1.18	1.07	1.10	5.3
				GC-MS/MS	1.05	1.16	1.16	1.16	1.05	1.14	1.21	1.18	1.18	1.18	1.16	1.20	1.07	1.14	1.20	1.15	1.18	1.03	1.12	6.0
		ENVI-Carb + PSA	Green tea	GC-MS	1.08	1.17	1.04	1.10	1.01	1.09	0.99	1.04	1.03	1.03	0.99	1.12	0.95	1.16	1.21	1.22	1.13	1.17	1.00	7.4
				GC-MS/MS	1.06	1.21	1.03	1.11	1.03	1.10	1.02	1.08	1.03	1.03	1.02	1.07	0.99	1.18	1.07	1.20	1.15	1.14	0.99	6.3
			Oolong-tea	GC-MS	1.21	1.19	1.07	1.13	1.23	1.16	1.14	1.08	1.24	1.24	1.25	1.11	1.19	1.10	1.40	1.17	1.15	1.21	1.10	6.6
				GC-MS/MS	1.21	1.16	1.07	1.13	1.22	1.18	1.09	1.15	1.24	1.24	1.26	1.11	1.20	1.15	1.30	1.19	1.05	1.24	1.12	5.8

53	Prometrye	TPT	Green tea	GC-MS	1.23	1.10	1.21	1.10	1.09	0.99	1.08	1.11	1.03	1.07	1.21	1.00	1.06	1.06	1.09	1.05	1.04	1.02	1.15	6.3
				GC-MS/MS	1.03	1.03	1.19	1.08	1.07	1.20	1.09	1.04	1.03	1.00	1.13	1.09	1.07	0.95	1.09	1.02	1.01	1.00	1.16	6.3
			Oolong tea	GC-MS	1.04	1.12	1.16	1.12	1.09	1.20	1.24	1.21	1.36	1.16	1.39	1.24	1.26	1.2	1.17	1.19	1.25	1.08	1.12	7.6
				GC-MS/MS	0.99	1.12	1.13	1.14	1.04	1.17	1.26	1.16	1.18	1.16	1.41	1.20	1.08	1.16	1.20	1.18	1.28	1.01	1.13	8.3
		ENVI-Carb + PSA	Green tea	GC-MS	1.18	1.14	1.07	1.13	1.08	1.19	1.04	1.06	1.07	1.40	1.06	1.15	1.09	1.13	1.18	1.26	1.12	1.16	0.99	8.0
				GC-MS/MS	1.07	1.25	1.03	1.15	1.05	1.15	1.06	1.08	1.03	1.01	1.05	1.07	0.96	1.18	1.16	1.22	1.10	1.14	0.97	7.3
			Oolong tea	GC-MS	1.15	1.19	1.15	1.15	1.49	1.17	1.17	1.30	1.93	1.30	1.35	1.05	1.12	1.26	1.36	1.18	1.17	1.25	1.14	15.4
				GC-MS/MS	1.23	1.16	1.10	1.09	1.23	1.19	1.09	1.18	1.30	1.28	1.21	1.12	1.14	1.18	1.32	1.12	1.08	1.28	1.07	6.8
54	Propham	TPT	Green tea	GC-MS	1.04	1.05	1.10	1.09	1.06	1.09	1.10	1.06	0.98	1.03	1.09	1.07	1.05	1.01	1.04	1.07	1.02	1.03	1.11	3.3
				GC-MS/MS	1.00	1.06	1.09	1.12	1.06	1.11	1.07	1.03	0.98	1.01	1.08	1.08	1.06	0.94	1.03	1.01	1.01	0.99	1.17	5.3
			Oolong tea	GC-MS	1.08	1.17	1.13	1.18	1.13	1.17	1.11	1.07	1.17	1.13	1.36	1.19	1.14	1.1	1.19	1.12	1.15	1.07	1.21	5.9
				GC-MS/MS	1.14	1.15	1.19	1.18	1.12	1.19	1.23	1.08	1.23	1.18	1.40	1.24	1.04	1.13	1.24	1.12	1.11	1.13	1.08	6.8
		ENVI-Carb + PSA	Green tea	GC-MS	1.02	1.14	1.03	1.16	1.03	1.15	1.03	1.11	1.03	1.11	1.17	1.06	1.14	0.98	1.16	1.10	1.13	1.00	0.95	6.4
				GC-MS/MS	0.96	1.13	1.02	1.17	1.01	1.09	0.93	1.10	0.97	1.07	1.19	1.07	1.04	1.08	1.12	1.12	1.08	1.09	1.00	6.6
			Oolong tea	GC-MS	1.25	1.18	1.13	1.11	1.28	1.24	1.22	1.26	1.36	1.23	1.16	1.10	1.13	1.15	1.17	1.20	1.02	1.00	1.11	7.6
				GC-MS/MS	1.26	1.18	1.12	1.12	1.24	1.20	1.10	1.22	1.22	1.25	1.15	1.05	1.10	1.10	1.29	1.24	1.12	1.28	1.10	6.3
55	Prosulfocarb	TPT	Green tea	GC-MS	1.11	1.02	1.19	1.06	1.09	1.06	1.10	1.10	1.02	1.09	1.16	1.08	1.13	1.11	1.14	1.00	1.04	1.05	1.14	4.6
				GC-MS/MS	1.03	1.04	1.17	1.35	1.07	1.15	1.10	1.03	1.01	1.03	1.11	1.06	1.06	1.04	1.06	1.05	1.07	1.03	1.17	7.3
			Oolong tea	GC-MS	1.06	1.17	1.16	1.16	1.09	1.18	1.17	1.20	1.26	1.23	1.41	1.25	1.33	1.2	1.25	1.15	1.25	1.16	1.12	6.9
				GC-MS/MS	1.08	1.19	1.16	1.16	1.08	1.17	1.26	1.16	1.18	1.21	1.33	1.20	1.03	1.12	1.24	1.09	1.17	1.10	1.12	6.1
		ENVI-Carb + PSA	Green tea	GC-MS	1.12	1.16	1.00	1.10	1.06	1.13	1.14	1.09	1.09	1.18	1.05	1.09	1.08	1.13	1.06	1.16	1.10	1.20	0.99	5.1
				GC-MS/MS	1.08	1.23	1.02	1.15	1.02	1.15	1.01	1.09	1.02	1.01	1.03	1.01	0.98	1.11	1.09	1.16	1.13	1.19	1.03	6.8
			Oolong tea	GC-MS	1.28	1.22	1.17	1.14	1.36	1.17	1.18	1.23	1.60	1.31	1.32	1.16	1.19	1.33	1.30	1.23	1.24	1.22	1.10	8.8
				GC-MS/MS	1.28	1.18	1.15	1.13	1.24	1.20	1.13	1.18	1.28	1.28	1.18	1.09	1.11	1.17	1.36	1.20	1.08	1.28	1.06	6.8
56	Secbumeton	TPT	Green tea	GC-MS	1.13	1.05	1.21	1.09	1.24	1.00	1.10	1.11	1.00	1.12	1.28	1.03	1.11	1.05	1.10	1.00	1.05	1.03	1.14	7.3
				GC-MS/MS		1.04	1.22	1.16	1.14	1.42	1.08	1.04	1.05	1.00	1.15	1.04	1.04	0.95	1.08	0.99	1.02	1.00	1.15	10.0
			Oolong tea	GC-MS	1.09	1.05	1.11	1.07	1.08	1.16	1.26	1.18	1.33	1.18	1.51	1.24	1.25	1.1	1.19	1.18	1.26	1.06	1.10	9.6
				GC-MS/MS	1.03	1.15	1.13	1.07	0.99	1.16	1.27	1.13	1.15	1.21	1.34	1.21	1.03	1.15	1.19	1.20	1.18	1.06	1.08	7.6
		ENVI-Carb + PSA	Green tea	GC-MS	1.09	1.16	1.08	1.22	1.15	1.24	1.00	1.06	1.07	1.47	1.06	1.19	1.12	1.16	1.15	1.28	1.12	1.16	0.98	9.6
				GC-MS/MS	1.05	1.24	1.04	1.18	1.04	1.24	0.99	1.10	0.98	1.03	1.08	1.06	0.98	1.19	1.15	1.23	1.09	1.16	1.00	8.1
			Oolong tea	GC-MS	1.20	1.17	1.19	1.03	1.64	1.17	1.14	1.20	1.70	1.36	1.32	0.98	1.14	1.28	1.23	1.22	1.18	1.22	1.14	14.1
				GC-MS/MS	1.20	1.13	1.09	1.08	1.20	1.19	1.09	1.17	1.31	1.26	1.17	1.09	1.14	1.17	1.35	1.20	1.09	1.29	1.05	7.2

(Continued)

Appendix Table 5.3 Youden Pair Ratios of Each Average Content From 201 Pesticides and RSD for 19 Circulative Determinations (cont.)

No.	Pesticides	SPE	Sample	Method	Nov 9, 2009 (n=5)	Nov 14, 2009 (n=3)	Nov 19, 2009 (n=3)	Nov 24, 2009 (n=3)	Nov 29, 2009 (n=3)	Dec 4, 2009 (n=3)	Dec 9, 2009 (n=3)	Dec 14, 2009 (n=3)	Dec 19, 2009 (n=3)	Dec 24, 2009 (n=3)	Dec 29, 2009 (n=3)	Jan 3, 2010 (n=3)	Jan 8, 2010 (n=3)	Jan 13, 2010 (n=3)	Jan 18, 2010 (n=3)	Jan 23, 2010 (n=3)	Jan 28, 2010 (n=3)	Feb 2, 2010 (n=3)	Feb 7, 2010 (n=3)	RSD%
57	Silafluofen	TPT	Green tea	GC-MS	1.34	0.89	1.15	1.02	1.23	1.16	1.18	1.09	0.88	1.05	1.16	1.14	1.20	1.08	1.03	1.09	0.96	1.02	1.06	10.6
				GC-MS/MS	1.04	1.03	1.19	0.81	1.11	1.11	1.10	0.98	1.01	1.02	1.14	1.11	1.06	0.98	1.05	1.00	1.07	0.98	1.16	8.1
			Oolong tea	GC-MS	1.02	1.15	1.05	1.29	1.08	1.09	1.04	0.92	1.20	1.10	1.25		0.99	1.0	1.23	1.14	1.25	1.06	1.19	9.2
				GC-MS/MS	1.05	1.15	1.09	1.16	1.02	1.12	1.05	1.10	1.21	1.11	1.24	1.14	0.94	1.14	1.19	1.15	1.21	1.14	1.16	6.6
		ENVI-Carb + PSA	Green tea	GC-MS	1.14	0.96	0.99	1.10	0.98	1.06	1.01	1.09	1.10	1.07	1.41	1.02	1.32	1.13	1.28	1.26	1.09	1.22	0.97	11.6
				GC-MS/MS	1.03	1.17	1.01	1.15	1.00	1.11	1.07	1.10	1.10	1.02	1.01	1.05	0.94	1.11	1.08	1.20	1.10	1.19	1.00	6.6
			Oolong tea	GC-MS	1.59	1.17	0.95	1.02	1.33	1.23	1.19	1.22	1.30	1.30	1.11	1.09	1.38	1.26	1.29	1.24	1.18	1.12	1.09	11.7
				GC-MS/MS	1.22	1.10	1.01	1.10	1.21	1.19	1.14	1.11	1.23	1.31	1.22	1.09	1.16	1.15	1.36	1.20	1.19	1.20	1.07	7.0
58	Tebupirimfos	TPT	Green tea	GC-MS	1.11	1.04	1.15	1.13	1.04	1.03	1.08	1.07	1.06	1.10	1.20	1.05	1.09	1.06	1.11	1.02	1.05	1.04	1.11	4.4
				GC-MS/MS	1.06	1.06	1.17	1.00	1.06	1.20	1.12	1.05	1.00	1.05	1.11	1.06	1.03	1.03	1.11	1.02	1.01	1.00	1.15	5.9
			Oolong tea	GC-MS	1.06	1.06	1.08	1.13	1.22	1.25	1.29	1.20	1.31	1.17	1.40	1.21	1.25	1.1	1.21	1.17	1.19	1.12	1.02	7.5
				GC-MS/MS	1.06	1.18	1.18	1.12	1.03	1.18	1.27	1.15	1.16	1.19	1.33	1.29	1.10	1.10	1.23	1.14	1.12	1.08	1.01	6.9
		ENVI-Carb + PSA	Green tea	GC-MS	1.11	1.03	1.06	1.13	1.10	1.10	0.77	1.05	1.08	1.39	1.02	1.10	1.11	1.13	1.08	1.21	1.09	1.15	1.15	10.3
				GC-MS/MS	1.06	1.26	1.02	1.09	1.02	1.16	1.05	1.09	1.01	1.06	1.00	1.09	0.99	1.11	1.07	1.16	1.07	1.19	1.18	6.7
			Oolong tea	GC-MS	1.19	1.17	1.19	1.08	1.79	1.12	1.24	1.24	1.92	1.30	1.38	1.07	1.15	1.21	1.28	1.20	1.20	1.21	1.15	17.5
				GC-MS/MS	1.35	1.19	1.16	1.14	1.20	1.21	1.13	1.12	1.23	1.24	1.18	1.06	1.11	1.15	1.34	1.18	1.12	1.28	1.18	6.3
59	Tebutam	TPT	Green tea	GC-MS	1.07	1.03	1.11	1.07	1.09	1.08	1.08	1.13	1.01	1.06	1.16	1.06	1.07	1.08	1.08	1.02	1.05	1.03	1.14	3.6
				GC-MS/MS	1.05	1.03	1.15	1.12	1.05	1.10	1.10	1.04	1.03	1.02	1.12	1.06	1.05	0.98	1.06	1.00	1.04	1.01	1.15	4.6
			Oolong tea	GC-MS	1.06	1.19	1.22	1.21	1.10	1.21	1.26	1.23	1.22	1.18	1.29	1.23	1.27	1.1	1.22	1.16	1.20	1.11	1.14	5.0
				GC-MS/MS	1.06	1.23	1.18	1.17	1.10	1.22	1.28	1.19	1.17	1.22	1.37	1.25	1.09	1.18	1.20	1.13	1.16	1.12	1.17	6.0
		ENVI-Carb + PSA	Green tea	GC-MS	1.16	1.16	1.02	1.11	1.04	1.10	0.98	1.05	1.08	1.11	1.05	1.08	1.11	1.14	1.09	1.22	1.08	1.17	1.02	5.4
				GC-MS/MS	1.07	1.23	1.01	1.13	1.03	1.14	1.03	1.09	1.01	1.05	1.06	1.09	1.00	1.18	1.06	1.18	1.08	1.16	1.00	6.3
			Oolong tea	GC-MS	1.31	1.23	1.16	1.15	1.35	1.19	1.16	1.15	1.53	1.26	1.20	1.16	1.15	1.20	1.27	1.19	1.19	1.24	1.12	7.9
				GC-MS/MS	1.34	1.18	1.17	1.14	1.24	1.19	1.11	1.20	1.26	1.28	1.20	1.10	1.14	1.16	1.32	1.22	1.09	1.24	1.06	6.4
60	Tebuthiuron	TPT	Green tea	GC-MS	1.09	1.07	1.21	1.25	1.12	0.99	1.08	1.16	1.17	1.12	1.30	1.02	1.10	0.90	1.17	1.01	1.06	1.05	1.13	7.9
				GC-MS/MS	1.03	1.04	1.25	1.05	1.07	1.22	1.07	1.09	1.04	1.02	1.34	1.03	1.07	0.84	1.14	1.04	1.04	1.00	1.20	8.3
			Oolong tea	GC-MS	1.09	1.07	1.17	1.16	1.07	1.21	1.24	1.18	1.19	1.16	1.30	1.19	1.24	1.0	1.22	1.24	1.24	1.05	1.08	7.2
				GC-MS/MS	1.05	1.15	1.19	1.11	1.09	1.15	1.30	1.16	1.18	1.24	1.34	1.28	1.18	1.20	1.23	1.18	1.22	1.05	1.16	6.6
		ENVI-Carb + PSA	Green tea	GC-MS	1.04	1.11	1.03	1.14	1.09	1.20	0.81	1.03	1.00	1.38	1.04	1.12	1.15	1.09	1.18	1.29	1.16	1.16	0.98	11.0
				GC-MS/MS	1.02	1.20	1.01	1.16	1.04	1.25	0.82	1.12	1.00	1.05	1.04	1.06	1.05	1.22	1.14	1.26	1.12	1.14	0.97	10.0
			Oolong tea	GC-MS	1.20	1.19	1.10	1.15	1.45	1.16	1.18	1.23	1.90	1.46	1.20	1.11	1.21	1.18	1.28	1.17	1.13	1.30	1.10	15.0
				GC-MS/MS	1.20	1.17	1.08	1.07	1.19	1.37	1.06	1.21	1.27	1.42	1.15	1.14	1.09	1.09	1.39	1.22	1.08	1.35	1.12	9.0

#	Compound	Cartridge	Tea	Method																				
61	Tefluthrin	TPT	Green tea	GC-MS	1.11	1.04	1.12	1.07	1.09	1.07	1.08	1.07	0.97	1.06	1.12	1.05	1.06	0.95	1.06	0.99	1.13	1.03	1.15	5.0
				GC-MS/MS	1.04	1.04	1.17	1.29	1.07	1.09	1.09	1.01	1.03	1.08	1.12	1.06	1.08	1.03	1.07	1.02	1.04	1.00	1.14	6.3
			Oolong tea	GC-MS	1.05	1.18	1.19	1.14	1.08	1.17	1.21	1.21	1.17	1.18	1.28	1.17	1.21	1.1	1.19	1.14	1.19	1.08	1.12	4.6
				GC-MS/MS	1.06	1.17	1.18	1.15	1.08	1.19	1.29	1.15	1.19	1.21	1.36	1.24	1.18	1.17	1.20	1.17	1.18	1.08	1.19	5.8
		ENVI-Carb + PSA	Green tea	GC-MS	1.13	1.16	1.02	1.09	1.03	1.10	1.05	1.02	1.07	1.06	1.05	1.10	1.09	1.09	1.09	1.25	1.08	1.12	1.03	5.0
				GC-MS/MS	1.11	1.23	1.03	1.10	1.03	1.11	1.08	1.08	1.04	1.04	1.04	1.07	1.02	1.10	1.08	1.23	1.07	1.14	0.97	6.0
			Oolong tea	GC-MS	1.29	1.17	1.14	1.13	1.29	1.18	1.15	1.11	1.33	1.24	1.15	1.15	1.13	1.22	1.25	1.19	1.18	1.17	1.18	5.2
				GC-MS/MS	1.33	1.20	1.15	1.16	1.24	1.22	1.12	1.18	1.30	1.29	1.18	1.08	1.15	1.19	1.31	1.22	1.10	1.21	1.12	6.0
62	Thenylchlor	TPT	Green tea	GC-MS	0.92	1.13	1.16	0.99	1.04	1.09	1.12	1.07	1.12	0.97	1.05	1.04	1.03	0.97	0.94	1.00	0.96	1.12	1.03	6.9
				GC-MS/MS	0.96	1.06	0.69	1.79	0.99	0.98	1.17	1.12	1.05	0.82	1.07	1.07	0.96	0.82	1.04	0.99	1.14	1.09	1.10	20.6
			Oolong tea	GC-MS	0.97	1.36	1.27	1.09	0.95	1.10	1.38	1.11	1.17	1.19	1.34	1.20	1.19	1.1	1.12	1.16	1.20	1.03	1.07	10.5
				GC-MS/MS	1.18	1.25	1.13	1.31	1.01	1.20	1.36	1.18	1.20	1.07	1.27	1.14	1.21	1.16	1.18	1.28	1.43	1.18	1.16	8.0
		ENVI-Carb + PSA	Green tea	GC-MS	0.79	1.37	0.89	0.95	1.00	1.03	1.11	1.57	1.08	0.96	1.01	0.99	1.02	1.08	0.99	1.11	1.13	1.03	0.98	16.0
				GC-MS/MS	0.63	1.03	0.94	0.98	0.91	1.05	1.20	1.02	0.70	0.85	1.08	1.09	0.85	1.40	1.04	1.11	1.17	1.05	0.86	17.9
			Oolong tea	GC-MS	1.03	1.07	1.00	1.25	1.11	1.11	0.75	1.33	1.25	1.30	1.16	1.23	1.08	1.01	1.24	1.14	1.16	1.11	1.03	11.9
				GC-MS/MS	1.68	1.09	0.92	1.07	1.18	1.07	1.20	1.29	1.21	1.20	1.04	1.10	1.17	0.98	1.31	1.17	1.04	1.14	1.06	14.0
63	Thionazin	TPT	Green tea	GC-MS	1.08	1.03	1.07	1.07	1.15	1.06	1.06	1.11	1.03	1.03	1.14	1.07	1.06	1.00	1.15	1.06	1.06	1.05	1.12	3.9
				GC-MS/MS	1.00	1.04	1.08	1.00	1.07	1.26	1.09	1.06	0.99	0.97	1.09	1.06	1.03	0.92	1.05	1.03	1.05	0.99	1.14	6.8
			Oolong tea	GC-MS	1.04	1.20	1.01	1.09	1.09	1.10	1.15	1.09	1.06	1.34	1.09	1.04	1.11	1.08	1.05	1.15	1.13	1.13	0.99	6.9
				GC-MS/MS	0.99	1.18	1.02	1.11	1.01	1.12	1.00	1.08	1.01	1.05	1.10	1.12	0.99	1.08	1.08	1.15	1.10	1.10	1.00	5.4
		ENVI-Carb + PSA	Green tea	GC-MS	1.12	1.24	1.25	1.23	0.94	1.17	1.21	1.11	1.24	1.12	1.31	1.18	1.22	1.1	1.17	1.13	1.17	1.05	1.05	7.7
				GC-MS/MS	1.15	1.23	1.18	1.17	1.10	1.21	1.28	1.17	1.19	1.21	1.35	1.20	1.06	1.13	1.23	1.17	1.19	1.12	1.20	5.4
			Oolong tea	GC-MS	1.27	1.24	1.25	1.25	1.42	1.16	1.17	1.16	1.63	1.24	1.20	1.07	1.13	1.14	1.25	1.25	1.16	1.14	1.08	10.5
				GC-MS/MS	1.35	1.18	1.13	1.12	1.26	1.18	1.15	1.21	1.25	1.26	1.14	1.10	1.02	1.15	1.33	1.21	1.14	1.25	1.13	6.9
64	Trichloronat	TPT	Green tea	GC-MS	1.07	1.01	1.12	1.13	1.07	1.03	1.08	1.07	1.00	1.08	1.15	1.09	1.08	1.03	1.06	1.02	1.04	1.04	1.14	4.0
				GC-MS/MS	1.00	1.05	1.09	0.78	1.07	1.13	1.09	1.04	0.96	1.05	1.07	1.06	1.09	1.02	1.04	1.00	1.05	0.99	1.19	7.9
			Oolong tea	GC-MS	1.06	1.13	1.18	1.15	1.15	1.24	1.32	1.18	1.26	1.13	1.38	1.22	1.23	1.1	1.21	1.17	1.20	1.11	1.10	6.6
				GC-MS/MS	1.13	1.18	1.19	1.16	1.16	1.17	1.29	1.13	1.20	1.13	1.36	1.25	1.20	1.14	1.22	1.20	1.17	1.13	1.13	5.6
		ENVI-Carb + PSA	Green tea	GC-MS	1.07	1.09	1.06	1.13	1.09	1.10	1.17	1.08	1.09	1.31	1.10	1.11	1.08	0.99	1.03	1.21	1.12	1.06	1.01	6.6
				GC-MS/MS	1.08	1.12	1.05	1.09	1.05	1.09	1.25	1.07	1.05	1.02	1.14		1.01	1.10	1.05	1.18	1.11	1.07	0.99	5.8
			Oolong tea	GC-MS	1.25	1.20	1.25	1.16	1.53	1.19	1.14	1.26	1.68	1.25	1.32	1.05	1.16	1.16	1.23	1.23	1.17	1.17	1.13	11.7
				GC-MS/MS	1.23	1.18	1.16	1.13	1.23	1.19	1.11	1.18	1.30	1.23	1.16	1.11	1.14	1.11	1.30	1.25	1.08	1.19	1.07	5.8

(Continued)

Appendix Table 5.3 Youden Pair Ratios of Each Average Content From 201 Pesticides and RSD for 19 Circulative Determinations (cont.)

No.	Pesticides	SPE	Sample	Method	Youden pair ratio																			RSD%
					Nov 9, 2009 (n=5)	Nov 14, 2009 (n=3)	Nov 19, 2009 (n=3)	Nov 24, 2009 (n=3)	Nov 29, 2009 (n=3)	Dec 4, 2009 (n=3)	Dec 9, 2009 (n=3)	Dec 14, 2009 (n=3)	Dec 19, 2009 (n=3)	Dec 24, 2009 (n=3)	Dec 29, 2009 (n=3)	Jan 3, 2010 (n=3)	Jan 8, 2010 (n=3)	Jan 13, 2010 (n=3)	Jan 18, 2010 (n=3)	Jan 23, 2010 (n=3)	Jan 28, 2010 (n=3)	Feb 2, 2010 (n=3)	Feb 7, 2010 (n=3)	
65	Trifluralin	TPT	Green tea	GC-MS	1.11	1.06	1.14	1.13	1.02	1.05	1.08	1.04	1.14	1.07	1.06	1.10	1.18	1.03	1.08	1.12	1.00	1.26	1.14	5.8
				GC-MS/MS	1.05	1.05	1.13	0.77	1.05	1.28	1.08	1.01	1.07	1.04	0.96	1.07	1.05	0.93	1.04	0.99	0.98	1.00	1.31	11.5
			Oolong tea	GC-MS	1.09	1.13	1.15	1.20	1.35	1.36	1.38	1.23	1.30	1.16	1.43	1.25	1.23	1.0	1.36	1.16	1.27	1.12	1.09	9.2
				GC-MS/MS	0.97	1.24	1.16	1.10	1.07	1.29	1.39	1.15	1.22	1.21	1.41	1.31	1.13	1.14	1.17	1.18	1.32	1.11	1.07	9.6
		ENVI-Carb + PSA	Green tea	GC-MS	1.13	1.04	1.07	1.14	1.10	1.06	1.33	1.05	1.06	1.21	1.12	1.12	1.09	1.15	1.04	1.17	1.63	1.02	1.04	12.8
				GC-MS/MS	1.04	0.98	1.03	1.14	1.06	1.07	1.78	1.13	1.13	1.04	1.14	1.14	1.00	1.19	1.05	1.22	1.06	1.14	1.03	15.3
			Oolong tea	GC-MS	1.26	1.23	1.23	1.13	1.95	1.16	1.10	1.55	1.65	1.26	1.29	1.08	1.24	1.15	1.02	1.56	1.14	1.18	1.17	18.1
				GC-MS/MS	1.30	1.21	1.21	1.08	1.29	1.16	1.13	1.31	1.30	1.24	1.23	1.06	1.18	1.19	1.25	1.37	1.23	1.36	1.17	7.0
66	2,4'-DDT	TPT	Green tea	GC-MS	1.07	1.05	1.15	1.07	0.99	1.12	0.99	1.02	0.82	1.04	1.21	1.08	1.06	0.9	0.93	1.02	0.91	0.85	1.14	9.7
				GC-MS/MS	1.09	1.00	0.99	1.04	0.97	1.06	1.07	1.08	0.91	1.08	1.10	1.01	1.13	1.01	1.03	0.97	0.95	0.92	1.12	6.4
			Oolong tea	GC-MS	1.06	1.19	1.04	1.18	1.05	1.20	1.14	1.07	1.08	1.01	1.16	1.27	1.11	0.9	1.07	1.00	1.20	1.08	1.00	7.7
				GC-MS/MS		1.00		1.10	1.01	1.28	1.12	1.31	1.27	0.88	1.20	1.18	1.07	1.10	1.14	1.04	1.17	1.15	0.92	10.7
		ENVI-Carb + PSA	Green tea	GC-MS	0.97	0.91	1.02	1.11	1.05	0.98	1.08	1.12	1.01	0.98	1.12	1.04	1.00	1.01	0.99	1.18	1.11	0.96	0.96	7.0
				GC-MS/MS	0.85	1.20	1.01	0.98	1.10	1.01	1.22	1.03	1.06	0.86	1.17	1.10	1.15	1.09	1.03	1.18	1.06	0.95	0.96	10.1
			Oolong tea	GC-MS	1.02	1.15	1.00	0.99	1.31	0.97	0.84	1.17	1.59	0.91	1.09	0.95	1.11	0.96	0.94	1.30	0.95	0.90	1.10	17.0
				GC-MS/MS	1.19	1.12	0.99	1.06	1.32	0.92	1.37	1.25	1.44	0.82	0.99	0.89	1.17	0.99	1.05	1.04	0.93	1.00	1.10	15.4
67	4,4'-DDE	TPT	Green tea	GC-MS	1.05	1.02	1.11	1.07	1.08	1.07	1.09	1.06	1.03	1.05	1.11	1.07	1.06	1.03	1.10	1.01	1.01	1.03	1.13	3.4
				GC-MS/MS	1.04	1.02	1.14	1.09	1.07	1.05	0.98	1.08	1.03	1.08	1.12	1.05	1.01	1.10	1.11	1.02	1.00	1.00	1.10	4.3
			Oolong tea	GC-MS	1.02	1.11	1.12	1.08	0.97	1.08	1.09	1.08	1.10	1.09	1.19	1.13	1.13	1.0	1.10	1.08	1.12	0.99	1.03	4.8
				GC-MS/MS	0.99	1.08	1.10	1.10	0.96	1.11	1.16	1.08	1.10	1.11	1.15	1.13	1.16	1.01	1.11	1.10	1.01	1.00	1.11	5.4
		ENVI-Carb + PSA	Green tea	GC-MS	1.07	1.19	1.01	1.07	1.02	1.10	1.71	1.04	1.04	0.99	1.03	1.05	1.07	1.12	1.08	1.20	1.10	1.17	1.01	14.2
				GC-MS/MS	1.08	1.28	1.02	1.05	1.02	1.13	1.03	1.03	1.06	1.00	1.04	1.07	1.13	1.05	1.11	1.16	1.03	1.10	1.01	6.2
			Oolong tea	GC-MS	1.16	1.06	1.01	1.08	1.18	1.07	1.06	1.04	1.15	1.15	1.08	1.07	1.06	1.05	1.18	1.09	1.08	1.13	1.03	4.7
				GC-MS/MS	1.11	1.07	1.03	1.05	1.15	1.07	1.04	1.11	1.17	1.15	0.97	1.08	1.03	1.04	1.27	1.10	1.00	1.15	0.99	6.9
68	Benalaxyl	TPT	Green tea	GC-MS	1.11	1.06	1.08	1.08	1.09	1.08	1.13	1.08	1.00	1.02	1.13	1.08	1.07	1.02	1.09	1.09	0.97	1.01	1.12	4.8
				GC-MS/MS	1.04	1.01	1.15	1.06	1.04	1.07	0.98	1.08	0.99	1.01	1.06	1.02	1.21	1.08	1.08	1.08	0.98	0.98	1.16	6.1
			Oolong tea	GC-MS	1.04	1.09	1.11	1.09	1.00	1.08	1.13	1.05	1.14	1.07	1.23	1.17	1.15	1.1	1.12	1.13	1.12	0.98	1.08	5.2
				GC-MS/MS	1.06	1.10	1.13	1.13	1.00	1.14	1.10	1.09	1.16	1.05	1.19	1.11	1.14	1.16	1.12	1.12	1.14	1.05	1.03	4.5
		ENVI-Carb + PSA	Green tea	GC-MS	1.24	1.14	1.03	1.13	1.02	1.06	0.98	1.11	1.05	1.02	1.16	1.10	1.11	1.10	1.10	1.31	1.05	1.01	0.96	7.9
				GC-MS/MS	1.10	1.08	1.01	1.07	1.04	0.81	1.39	1.07	1.04	0.99	1.26	0.99	1.16	1.24	1.08	1.24	1.18	1.02	1.01	11.8
			Oolong tea	GC-MS	1.12	1.10	1.00	1.04	1.21	1.08	1.06	1.14	1.19	1.17	1.10	1.11	1.09	1.03	1.01	1.06	1.11	1.01	1.00	5.8
				GC-MS/MS	1.08	1.07	1.00	1.06	1.14	1.07	0.99	1.11	1.16	1.13	1.00	1.02	1.05	1.11	1.23	1.11	0.97	1.13	0.99	6.5

No.	Compound	Cleanup	Tea	Instrument																				
69	Benzoylprop-ethyl	TPT	Green tea	GC-MS	1.05	1.05	1.11	1.10	1.10	1.09	1.09	1.08	1.02	0.98	1.12	1.08	1.04	1.01	1.06	1.01	1.03	1.02	1.13	4.0
				GC-MS/MS	1.04	1.05	1.13	1.06	1.07	1.10	0.99	1.08	1.02	1.01	1.09	1.05	1.11	1.02	1.07	0.99	1.00	0.99	1.13	4.5
			Oolong tea	GC-MS	1.06	1.07	1.07	1.10	0.97	1.08	1.02	1.01	1.11	1.02	1.19	1.12	1.06	1.12	1.09	1.1	1.14	1.05	1.07	4.8
				GC-MS/MS	1.03	1.04	1.09	1.11	0.98	1.10	1.03	1.04	1.11	1.05	1.18	1.11	1.12	1.15	1.13	1.04	1.11	1.04	1.06	4.5
		ENVI-Carb + PSA	Green tea	GC-MS	1.05	1.10	1.04	1.12	1.00	1.05	1.05	1.06	1.09	1.05	1.07	1.08	1.08	1.23	1.04	1.15	1.14	1.12	0.99	5.3
				GC-MS/MS	1.03	1.13	1.04	1.08	1.04	0.99	1.28	1.09	1.12	1.05	1.10	1.12	1.08	1.22	1.08	1.16	1.12	1.05	0.99	6.6
			Oolong tea	GC-MS	1.09	1.05	0.97	1.03	1.17	1.08	1.11	1.04	1.16	1.15	1.08	1.06	1.13	1.17	1.19	1.05	1.10	1.11	1.06	5.2
				GC-MS/MS	1.07	1.05	0.99	1.05	1.12	1.10	1.03	1.05	1.13	1.15	0.99	1.07	1.09	1.15	1.23	1.09	1.03	1.13	1.03	5.6
70	Bromofos	TPT	Green tea	GC-MS	1.00	1.04	1.06	1.05	1.01	1.08	1.09	1.07	0.87	1.08	1.16	1.01	0.98	1.00	0.98	1.05	0.98	0.94	1.08	6.3
				GC-MS/MS	1.00	1.00	0.95	1.05	1.02	1.04	1.07	1.06	0.94	1.03	1.30	1.18	0.88	0.97	1.04	1.14	0.91	0.97	1.08	9.6
			Oolong tea	GC-MS	1.02	1.25	1.16	1.13	1.04	1.09	1.27	1.15	1.08	1.17	1.16	1.11	1.26	1.07	1.10	1.1	1.16	1.03	1.09	6.7
				GC-MS/MS	1.09	1.13	1.14	1.20	1.03	1.17	1.25	1.33	1.35	0.94	1.13	1.12	1.25	1.18	1.07	1.11	1.21	1.06	0.98	9.4
		ENVI-Carb + PSA	Green tea	GC-MS	1.04	1.13	1.02	1.05	0.98	1.05	1.06	1.20	1.03	1.04	1.13	1.00	1.05	1.16	0.93	1.07	1.09	0.98	1.00	6.3
				GC-MS/MS	0.90	1.10	1.05	1.03	1.03	1.40	1.29	1.00	1.12	0.91	1.05	1.06	1.11	1.18	1.06	1.32	1.06	0.89	1.08	12.4
			Oolong tea	GC-MS	1.14	1.12	1.04	1.13	1.13	1.05	0.86	1.14	1.23	1.21	1.16	1.01	1.04	1.10	1.16	1.05	1.06	1.04	1.04	7.7
				GC-MS/MS	1.36	1.11	1.00	1.13	1.23	1.07	1.31	1.10	1.17	1.09	0.98	0.99	1.12	1.18	1.22	1.07	0.99	1.16	1.13	9.3
71	Bromopro-pylate	TPT	Green tea	GC-MS	1.01	1.05	1.11	1.10	1.05	1.06	1.02	1.01	0.82	1.02	1.09	1.10	1.06	1.00	0.96	0.98	0.94	0.95	1.11	7.3
				GC-MS/MS	1.03	1.05	1.08	1.05	1.07	1.07	1.00	1.08	0.87	1.01	1.23	1.14	1.22	1.00	1.06	0.98	0.97	0.96	1.13	8.3
			Oolong tea	GC-MS	1.05	0.99	0.97	1.10	1.04	1.06	0.97	0.92	1.17	1.04	1.24	1.16	1.01	1.08	1.10	0.9	1.13	1.02	1.05	7.9
				GC-MS/MS	1.15	0.95	1.04	1.10	0.99	1.11	0.98	1.01	1.25	0.98	1.23	1.10	1.08	1.08	1.13	1.02	1.09	1.12	0.99	7.8
		ENVI-Carb + PSA	Green tea	GC-MS	0.88	1.00	1.05	1.31	1.03	1.03	0.73	1.22	0.94	1.05	1.36	1.11	1.18	1.12	1.25	0.86	1.11	0.71	0.93	17.0
				GC-MS/MS	0.88	0.85	1.04	1.13	1.06	0.92	1.44	1.10	1.06	1.00	1.35	1.02	1.14	1.17	1.11	1.09	1.12	0.92	0.89	14.2
			Oolong tea	GC-MS	1.01	1.02	0.99	0.97	1.27	1.04	1.07	1.26	1.16	1.08	1.14	0.95	1.13	1.30	1.02	0.93	1.10	0.98	1.02	10.3
				GC-MS/MS	1.00	1.03	0.98	1.02	1.19	1.06	1.06	1.18	1.17	1.07	1.00	0.96	1.11	1.08	1.11	1.07	1.02	1.05	1.03	6.1
72	Buprofenzin	TPT	Green tea	GC-MS	1.08	1.06	1.13	1.10	1.11	1.05	1.07	1.08	1.00	1.12	1.14	1.13	1.01	1.03	1.14	1.06	1.00	1.03	1.12	4.4
				GC-MS/MS	1.05	1.04	1.13	1.11	1.06	1.07	0.97	1.10	0.97	1.01	0.93	1.10	1.05	1.03	1.15	0.97	1.07	0.99	1.13	6.1
			Oolong tea	GC-MS	0.98	1.11	1.07	1.05	1.02	1.04	1.12	1.12	1.12	1.05	1.10	0.94	1.24	1.13	1.10	1.2	1.12	0.94	1.03	6.9
				GC-MS/MS	0.99	1.09	1.16	1.10	1.00	1.13	1.22	1.03	1.15	1.22	1.16	1.09	1.15	1.09	1.04	1.08	1.06	0.94	1.09	6.9
		ENVI-Carb + PSA	Green tea	GC-MS	1.10	1.07	1.03	1.08	1.08	1.11	0.98	1.07	1.00	1.03	1.05	1.18	1.32	1.17	1.05	1.13	0.94	1.28	1.03	9.2
				GC-MS/MS	1.16	1.07	1.02	1.05	1.12	0.92	1.03	1.04	1.06	0.97	1.09	1.04	1.22	1.17	1.14	1.09	0.99	1.14	0.95	8.3
			Oolong tea	GC-MS	1.18	1.07	0.98	1.11	1.08	1.10	1.10	1.10	1.15	1.14	1.08	1.17	0.87	1.09	1.47	0.85	1.04	1.09	0.80	13.3
				GC-MS/MS	1.16	1.07	1.06	1.10	1.12	1.13	1.04	1.11	1.15	1.20	0.99	1.09	1.00	1.10	1.24	1.09	0.92	1.17	0.94	7.7

(Continued)

Appendix Table 5.3 Youden Pair Ratios of Each Average Content From 201 Pesticides and RSD for 19 Circulative Determinations (cont.)

No.	Pesticides	SPE	Sample	Method	Nov 9, 2009 (n=5)	Nov 14, 2009 (n=3)	Nov 19, 2009 (n=3)	Nov 24, 2009 (n=3)	Nov 29, 2009 (n=3)	Dec 4, 2009 (n=3)	Dec 9, 2009 (n=3)	Dec 14, 2009 (n=3)	Dec 19, 2009 (n=3)	Dec 24, 2009 (n=3)	Dec 14, 2009 (n=3)	Jan 3, 2010 (n=3)	Jan 8, 2010 (n=3)	Jan 13, 2010 (n=3)	Jan 18, 2010 (n=3)	Jan 23, 2010 (n=3)	Jan 28, 2010 (n=3)	Feb 2, 2010 (n=3)	Feb 7, 2010 (n=3)	RSD%
73	Butachlor	TPT	Green tea	GC-MS	1.03	1.02	1.12	1.04	1.08	1.10	1.05	1.09	0.97	1.05	1.19	1.03	1.01	1.06	1.07	1.03	0.96	0.94	1.11	5.7
				GC-MS/MS	1.04	0.99	1.14	1.10	1.07	1.04	0.95	1.06	1.01	1.03	1.04	1.05	1.10	1.09	1.10	0.99	1.04	0.96	1.07	4.7
			Oolong tea	GC-MS	1.02	1.21	1.16	1.10	0.97	1.05	1.18	0.99	1.09	1.22	1.12	1.13	1.25	1.1	1.08	1.11	1.23	0.98	1.09	7.7
				GC-MS/MS	1.00	1.12	1.15	1.11	1.01	1.12	1.18	1.10	1.07	1.10	1.18	1.12	1.11	1.05	1.18	1.06	1.03	1.09	0.94	6.0
		ENVI-Carb + PSA	Green tea	GC-MS	1.07	1.13	0.98	1.03	1.01	1.07	1.08	1.17	1.02	1.03	1.03	1.02	1.03	1.13	0.98	1.15	1.09	1.05	1.03	5.2
				GC-MS/MS	1.08	1.24	1.03	1.02	1.05	0.71	1.02	1.08	1.03	0.98	1.10	1.06	1.18	1.04	1.10	1.15	1.02	1.06	1.07	10.0
			Oolong tea	GC-MS	1.42	1.09	1.01	1.09	1.15	1.08	0.88	1.16	1.18	1.18	1.13	1.02	1.05	1.05	1.20	1.11	1.04	1.11	1.09	9.5
				GC-MS/MS	1.11	1.05	1.02	1.12	1.15	1.07	1.04	1.05	1.20	1.12	0.95	1.06	1.02	1.10	1.29	1.06	0.98	1.13	0.96	7.6
74	Butylate	TPT	Green tea	GC-MS	1.12	1.05	1.08	1.00	1.15	1.09	1.07	1.00	1.03	1.08	1.05	1.07	1.08	0.90	1.06	1.02	0.97	0.85	1.10	7.0
				GC-MS/MS	1.08	1.02	1.11	1.03	1.07	1.06	0.99	1.08	1.00	1.12	1.07	1.05	1.13	1.01	1.09	0.96	0.99	0.93	1.10	5.4
			Oolong tea	GC-MS	1.05	1.13	0.89	1.16	0.87	1.10	1.13	1.06	1.06	1.05	1.24	1.18	1.18	1.1	1.34	1.00	1.11	1.11	1.37	11.3
				GC-MS/MS	1.06	1.22	1.17	1.25	1.12	1.26	1.22	1.14	1.21	1.16	1.25	1.24	1.19	1.18	1.22	1.13	1.16	1.15	1.20	4.5
		ENVI-Carb + PSA	Green tea	GC-MS	1.82	1.11	0.78	0.76	0.81	0.92	1.04	1.04	1.06	0.57	8.52	1.20	1.22	3.04	1.46	1.17	1.04	1.10	1.02	113.1
				GC-MS/MS	1.03	1.31	0.96	1.05	0.99	0.83	1.11	1.06	1.04	1.00	1.20	1.20	1.22	1.08	1.22	1.14	1.08	1.06	1.01	10.4
			Oolong tea	GC-MS	1.50	0.91	0.82	1.01	1.13	1.07	1.15	1.43	1.18	1.25	1.00	1.02	1.49	1.06	1.18	1.24	0.97	0.91	1.02	17.2
				GC-MS/MS	1.38	1.17	1.06	1.10	1.25	1.16	1.07	1.21	1.27	1.30	1.04	1.12	1.15	1.19	1.38	1.13	1.12	1.21	1.10	8.5
75	Carbofeno-thion	TPT	Green tea	GC-MS	0.97	1.03	1.08	1.10	1.10	1.07	1.05	1.06	0.92	1.08	1.16	1.02	1.04	1.00	0.97	1.00	0.94	1.10	1.13	6.3
				GC-MS/MS	1.03	1.01	0.99	1.10	0.97	1.07	0.99	1.17	0.99	1.06	1.13	1.06	1.05	1.11	1.12	1.12	0.97	1.15	1.01	6.0
			Oolong tea	GC-MS	1.05	1.01	1.06	1.07	1.05	1.05	1.20	1.07	1.23	1.22	1.30	1.23	1.20	1.1	1.13	1.17	1.25	1.04	1.20	7.8
				GC-MS/MS	1.16	1.06	1.05	1.09	1.01	1.15	1.21	1.29	1.45	1.15	1.26	1.09	1.17	1.05	1.06	1.57	1.14	1.29	0.67	16.0
		ENVI-Carb + PSA	Green tea	GC-MS	0.95	1.13	1.05	1.13	1.03	1.10	1.08	1.20	1.03	1.01	1.20	1.10	1.05	1.07	1.05	1.21	1.10	0.98	0.92	7.5
				GC-MS/MS	0.87	1.06	0.96	1.04	0.97	0.94	0.89	1.00	1.07	1.01	1.19	1.08	1.14	1.13	1.18	1.07	1.30	1.00	0.85	11.1
			Oolong tea	GC-MS	1.00	0.99	1.02	1.06	1.22	1.10	1.01	1.20	1.16	1.16	1.25	1.01	1.08	1.07	1.17	1.16	1.13	1.16	1.14	7.2
				GC-MS/MS	1.08	1.01	0.96	1.07	1.30	1.07	1.24	1.22	1.23	1.14	0.95	1.12	1.04	1.08	1.06	1.57	1.11	1.03	1.30	13.0
76	Chlorfenson	TPT	Green tea	GC-MS	0.97	1.10	1.05	1.13	0.97	1.08	1.02	1.05	0.82	0.96	1.04	1.13	1.06	1.02	0.83	1.02	0.96	0.96	1.07	8.5
				GC-MS/MS	1.01	1.03	0.97	1.00	0.91	1.08	1.09	1.10	0.78	0.96	1.12	1.11	1.06	0.95	0.98	1.00	0.97	0.92	1.15	8.9
			Oolong tea	GC-MS	1.03	1.04	0.99	1.21	1.15	1.17	1.07	1.01	1.13	0.98	1.25	1.22	1.06	0.9	1.13	1.05	1.14	1.03	0.97	8.6
				GC-MS/MS	1.32	1.02	1.07	1.17	1.09	1.22	1.02	1.26	1.26	0.91	1.30	1.17	1.05	1.01	1.17	1.12	1.08	1.15	0.94	10.5
		ENVI-Carb + PSA	Green tea	GC-MS	0.94	0.92	1.14	1.20	1.07	0.91	1.03	1.15	0.98	1.09	1.48	1.10	1.05	0.94	1.01	1.19	1.12	0.87	0.98	13.3
				GC-MS/MS	0.95	0.81	1.10	1.05	1.14	0.72	1.82	1.07	1.08	1.05	1.45	1.04	0.97	1.26	1.02	1.24	1.08	0.88	0.93	22.2
			Oolong tea	GC-MS	1.03	1.06	1.03	0.93	1.27	0.97	1.07	1.16	1.22	1.01	1.11	0.92	1.15	0.93	0.92	1.41	1.02	0.97	1.10	12.3
				GC-MS/MS	1.03	1.03	1.02	0.98	1.30	0.97	1.10	1.22	1.23	0.96	1.01	0.91	1.21	0.99	1.06	1.12	0.98	1.11	1.10	10.0

77	Chlorfenvin-phos	TPT	Green tea	GC-MS	1.05	1.04	1.11	1.03	1.07	1.08	1.03	1.05	0.75	1.10	1.21	1.00	0.97	1.06	0.96	0.94	0.91	0.90	1.19	10.2
				GC-MS/MS	0.99	0.95	1.05	1.17	1.00	1.06	1.01	1.28	1.02	1.03	1.32	0.99	1.12	1.01	0.96	1.24	1.07	0.94	0.93	10.9
			Oolong tea	GC-MS	1.00	0.99	0.90	1.02	0.97	1.03	1.01	1.13	1.01	0.94	1.21	0.99	1.03	0.95	1.07	0.92	0.94	0.97	1.14	7.8
				GC-MS/MS	1.01	1.19	1.08	1.09	1.04	1.05	1.25	1.11	1.09	1.20	1.19	1.32	1.30	1.0	1.02	1.06	1.21	1.02	1.10	8.8
		ENVI-Carb + PSA	Green tea	GC-MS	1.10	1.02	1.10	1.12	1.02	1.16	1.28	1.33	1.31	1.00	1.28	1.13	1.24	1.11	1.13	1.12	1.29	1.20	0.92	10.1
				GC-MS/MS	0.76	0.93	1.03	1.02	1.06	2.23	1.50	1.01	1.07	0.95	1.15	1.02	1.07	1.26	0.94	1.15	1.02	0.98	0.95	27.9
			Oolong tea	GC-MS	1.04	1.09	1.05	1.11	1.23	1.06	0.71	1.22	1.24	1.28	1.23	0.95	1.06	0.98	1.04	1.14	1.04	0.99	1.03	12.3
				GC-MS/MS	1.40	1.04	0.98	1.12	1.26	1.04	1.40	1.27	1.26	1.10	0.98	1.03	1.14	1.03	1.18	1.12	0.97	1.10	1.16	11.7
78	Chlormephos	TPT	Green tea	GC-MS	1.05	1.03	1.14	1.03	1.07	1.08	1.05	1.09	1.02	1.10	1.09	1.05	1.03	1.07	1.09	0.98	0.99	0.98	1.12	4.1
				GC-MS/MS	1.06	1.02	1.10	1.04	1.08	1.08	0.97	1.11	1.02	1.13	1.16	1.05	1.02	1.05	1.07	0.95	0.98	0.94	1.11	5.7
			Oolong tea	GC-MS	1.06	1.13	1.14	1.18	1.06	1.16	1.20	1.05	1.18	1.16	1.29	1.20	1.16	1.1	1.14	1.08	1.18	1.12	1.14	5.2
				GC-MS/MS	1.08	1.18	1.15	1.20	1.09	1.22	1.23	1.17	1.21	1.14	1.25	1.24	1.16	1.10	1.23	1.14	1.06	1.10	1.18	5.0
		ENVI-Carb + PSA	Green tea	GC-MS	1.04	1.22	0.98	1.08	1.00	1.07	1.05	1.08	1.06	1.02	1.19	1.06	1.17	1.03	1.18	1.13	1.06	1.10	1.00	6.3
				GC-MS/MS	0.95	1.26	0.98	1.08	1.01	0.89	1.04	1.06	1.06	1.03	1.22	1.05	1.16	1.23	1.25	1.11	1.08	1.04	1.01	9.6
			Oolong tea	GC-MS	1.25	1.20	1.04	1.06	1.24	1.14	1.07	1.13	1.24	1.24	1.10	1.08	1.12	1.07	1.23	1.22	1.15	1.15	1.04	6.6
				GC-MS/MS	1.40	1.14	1.07	1.09	1.24	1.09	1.10	1.15	1.22	1.24	0.98	1.09	1.12	1.11	1.28	1.14	1.10	1.17	1.04	8.4
79	Chloroneb	TPT	Green tea	GC-MS	1.06	1.04	1.08	1.04	1.10	1.08	1.07	1.11	1.05	1.07	1.05	1.06	1.07	1.04	1.08	1.01	1.02	1.04	1.13	2.8
				GC-MS/MS	1.06	1.05	1.11	1.06	1.05	1.05	0.99	1.09	1.01	1.08	1.04	1.06	0.99	1.05	1.09	1.00	1.01	0.99	1.11	3.8
			Oolong tea	GC-MS	1.05	1.14	1.12	1.12	1.06	1.16	1.12	1.10	1.10	1.03	1.21	1.15	1.13	1.1	1.10	1.08	1.05	1.04	1.01	4.8
				GC-MS/MS	1.00	1.18	1.17	1.14	1.04	1.16	1.16	1.10	1.10	1.09	1.23	1.12	1.17	1.02	1.14	1.07	1.00	1.10	1.12	5.7
		ENVI-Carb + PSA	Green tea	GC-MS	1.10	1.22	1.01	1.10	1.00	1.09	1.01	1.05	1.11	1.01	1.09	1.04	1.11	1.09	1.12	1.15	1.08	1.16	1.02	5.5
				GC-MS/MS	1.04	1.27	1.02	1.05	1.02	0.71	1.11	1.06	1.09	1.05	1.12	1.11	1.12	1.07	1.16	1.13	1.05	1.10	1.00	9.9
			Oolong tea	GC-MS	1.28	1.10	1.01	1.05	1.20	1.09	1.10	1.13	1.22	1.16	1.11	1.09	1.05	1.03	1.18	1.11	1.10	1.13	1.03	6.3
				GC-MS/MS	1.26	1.13	1.07	1.05	1.13	1.08	1.01	1.09	1.16	1.17	1.03	1.08	1.09	1.05	1.22	1.07	1.05	1.15	0.98	6.5
80	Chloropro-pylate	TPT	Green tea	GC-MS	1.02	1.14	0.91	1.23	0.91	1.02	1.04	1.53	0.86	0.98	1.06	1.05	1.03	0.95	0.95	0.99	0.95	0.97	1.13	14.3
				GC-MS/MS	0.95	1.10	1.03	1.26	1.01	1.04	1.01	1.13	0.99	1.03	1.24	1.15	1.25	0.86	1.18	1.13	1.12	0.79	0.95	11.9
			Oolong tea	GC-MS	0.98	1.02	1.06	1.05	1.06	1.06	0.99	1.09	0.96	1.00	1.28	1.07	1.04	1.01	1.04	1.01	1.00	0.99	1.15	6.9
				GC-MS/MS	1.02	0.98	1.01	1.10	1.02	1.08	0.95	1.04	1.15	1.10	1.18	1.20	1.06	1.0	1.09	1.09	1.16	1.03	1.01	6.7
		ENVI-Carb + PSA	Green tea	GC-MS	1.13	1.01	1.13	1.11	1.04	1.13	1.18	1.12	1.18	1.06	1.24	1.16	1.09	1.12	1.16	1.07	1.08	1.08	0.95	6.0
				GC-MS/MS	0.93	1.06	1.02	1.09	1.04	0.93	0.94	1.06	1.05	0.99	1.27	1.04	1.16	1.06	1.14	1.08	1.08	0.96	0.93	8.5
			Oolong tea	GC-MS	1.24	1.08	0.91	0.80	1.23	1.04	1.03	1.13	1.23	1.11	1.19	0.91	1.10	0.95	1.09	1.19	1.06	1.11	1.04	11.1
				GC-MS/MS	1.11	1.09	1.09	1.05	1.20	1.07	1.00	1.15	1.14	1.14	0.99	1.02	1.09	1.04	1.20	1.07	0.99	1.15	1.05	5.9

(Continued)

Appendix Table 5.3 Youden Pair Ratios of Each Average Content From 201 Pesticides and RSD for 19 Circulative Determinations (cont.)

No.	Pesticides	SPE	Sample	Method	Nov 9, 2009 (n=5)	Nov 14, 2009 (n=3)	Nov 19, 2009 (n=3)	Nov 24, 2009 (n=3)	Nov 29, 2009 (n=3)	Dec 4, 2009 (n=3)	Dec 9, 2009 (n=3)	Dec 14, 2009 (n=3)	Dec 19, 2009 (n=3)	Dec 24, 2009 (n=3)	Dec 14, 2009 (n=3)	Jan 3, 2010 (n=3)	Jan 8, 2010 (n=3)	Jan 13, 2010 (n=3)	Jan 18, 2010 (n=3)	Jan 23, 2010 (n=3)	Jan 28, 2010 (n=3)	Feb 2, 2010 (n=3)	Feb 7, 2010 (n=3)	RSD%
81	Chlorpropham	TPT	Green tea	GC-MS	0.98	1.02	1.08	1.11	1.07	1.05	1.05	1.08	0.95	1.03	1.10	1.06	1.04	0.85	1.01	1.00	1.03	1.03	1.09	5.8
				GC-MS/MS	0.97	1.12	1.02	1.17	1.02	1.06	0.89	1.11	0.98	1.03	1.18	1.04	1.14	0.94	1.13	1.09	1.09	0.90	0.95	8.6
			Oolong tea	GC-MS	1.04	1.03	1.11	1.07	1.07	1.05	0.99	1.05	0.93	1.03	1.07	1.10	1.10	1.03	1.07	1.00	0.97	0.96	1.13	5.1
				GC-MS/MS	1.03	1.06	1.07	1.10	1.07	1.10	1.14	1.06	1.14	1.09	1.22	1.20	1.12	1.0	1.13	1.08	1.13	1.00	0.99	5.6
		ENVI-Carb + PSA	Green tea	GC-MS	1.11	0.99	1.08	1.09	0.99	1.12	1.09	1.05	1.22	1.06	1.25	1.16	1.12	1.01	1.11	1.06	1.12	1.03	1.01	6.5
				GC-MS/MS	0.96	0.73	1.03	1.09	1.05	0.83	1.84	1.07	1.02	0.97	1.28	1.08	1.20	1.13	1.12	1.17	1.07	0.96	0.91	20.7
			Oolong tea	GC-MS	1.09	1.12	1.10	1.03	1.23	1.08	1.07	1.18	1.15	1.16	1.11	1.02	1.10	1.02	1.12	1.14	1.07	1.10	1.05	5.1
				GC-MS/MS	1.06	1.04	1.01	1.03	1.30	1.06	1.01	1.18	1.18	1.11	0.98	1.00	1.12	1.09	1.19	1.09	1.13	1.13	1.02	7.7
82	Chlorpyrifos(ethyl)	TPT	Green tea	GC-MS	0.99	1.02	1.09	1.10	1.06	1.05	1.06	1.06	0.84	1.08	1.06	1.06	1.04	0.99	1.03	1.01	1.01	0.98	1.11	6.3
				GC-MS/MS	1.01	1.01	1.04	1.05	1.01	1.07	1.01	1.11	0.94	1.04	1.11	1.06	1.00	1.03	1.07	0.99	0.98	1.00	1.06	4.4
			Oolong tea	GC-MS	1.03	1.05	1.08	1.09	1.06	1.11	1.26	1.16	1.18	1.15	1.22	1.14	1.18	1.0	1.10	1.08	1.15	1.03	1.05	6.0
				GC-MS/MS	1.09	1.11	1.12	1.10	1.06	1.15	1.23	1.20	1.26	1.04	1.20	1.15	1.18	1.04	1.10	1.13	1.17	1.04	1.03	6.2
		ENVI-Carb + PSA	Green tea	GC-MS	1.02	1.11	1.02	1.12	1.04	1.07	1.07	1.15	1.04	1.08	1.16	1.12	1.06	1.04	0.99	1.17	1.09	0.98	1.00	5.4
				GC-MS/MS	0.92	1.02	1.06	1.05	1.08	1.60	1.36	1.05	1.10	1.01	1.14	1.07	1.08	1.23	0.99	1.15	1.03	1.00	0.98	14.2
			Oolong tea	GC-MS	1.13	1.10	1.10	1.07	1.26	1.07	0.98	1.17	1.22	1.17	1.17	1.02	1.08	1.04	1.11	1.15	1.07	1.08	1.06	6.3
				GC-MS/MS	1.28	1.07	1.10	1.06	1.24	1.06	1.11	1.15	1.21	1.11	0.99	1.03	1.11	1.06	1.18	1.13	1.02	1.11	1.03	7.0
83	Chlorthiophos	TPT	Green tea	GC-MS	1.00	1.05	1.09	1.08	1.09	1.05	1.08	1.07	0.93	1.06	1.16	1.04	0.94	0.98	0.96	0.96	0.97	0.98	1.12	6.4
				GC-MS/MS	1.04	1.00	1.07	1.10	1.08	1.06	0.98	1.08	1.01	0.99	1.11	1.02	1.12	1.04	1.05	1.02	0.96	0.98	1.12	4.9
			Oolong tea	GC-MS	1.04	1.07	1.08	1.09	1.01	1.09	1.20	1.12	1.19	1.18	1.21	1.19	1.21	1.0	1.15	1.19	1.07	1.04	1.12	6.2
				GC-MS/MS	1.10	1.11	1.13	1.09	1.00	1.12	1.15	1.13	1.21	1.04	1.26	1.13	1.27	1.19	1.21	1.12	1.22	1.09	1.17	6.1
		ENVI-Carb + PSA	Green tea	GC-MS	0.96	1.13	1.02	1.09	1.01	1.09	0.80	1.11	1.03	1.01	1.07	1.10	0.97	1.11	0.99	1.21	1.02	1.03	1.01	8.2
				GC-MS/MS	0.94	1.15	1.05	1.05	1.07	1.73	1.09	1.04	1.08	0.96	1.12	1.06	1.11	1.20	1.04	1.20	1.05	1.04	0.97	15.0
			Oolong tea	GC-MS	1.07	1.06	1.02	1.07	1.22	1.08	1.04	1.12	1.19	1.20	1.19	1.09	1.14	1.01	1.20	1.02	1.10	1.15	1.08	6.1
				GC-MS/MS	1.20	1.04	1.01	1.08	1.18	1.06	1.16	1.18	1.16	1.15	1.04	1.07	1.10	1.13	1.30	1.12	1.01	1.16	1.02	7.0
84	cis-Chlordane	TPT	Green tea	GC-MS	1.03	1.03	1.11	1.06	1.05	1.07	1.08	1.06	0.97	1.04	1.10	1.07	1.06	1.03	1.07	1.00	1.00	1.01	1.13	3.8
				GC-MS/MS	1.03	0.98	1.12	1.04	0.97	1.04	0.98	1.04	0.97	1.05	1.08	1.14	1.08	1.08	1.03	0.97	1.04	0.92	1.16	6.1
			Oolong tea	GC-MS	1.03	1.12	1.13	1.11	1.04	1.12	1.22	1.11	1.12	1.08	1.23	1.17	1.18	1.1	1.12	1.10	1.15	1.00	1.03	5.7
				GC-MS/MS	1.06	1.11	1.18	1.10	1.04	1.15	1.21	1.13	1.20	1.08	1.29	1.16	1.28	1.10	1.16	1.09	1.07	1.06	1.07	6.4
		ENVI-Carb + PSA	Green tea	GC-MS	1.05	1.16	1.03	1.07	1.03	1.06	0.95	1.05	1.03	1.00	1.08	1.05	1.08	1.10	1.04	1.22	1.10	1.06	0.99	5.6
				GC-MS/MS	1.04	1.16	1.02	1.08	1.08	1.06	1.28	1.03	1.06	1.01	1.12	1.15	1.02	1.15	1.07	1.23	1.07	1.08	1.07	6.9
			Oolong tea	GC-MS	1.15	1.12	1.06	1.08	1.20	1.08	1.05	1.12	1.22	1.16	1.09	1.08	1.06	1.06	1.16	1.15	1.08	1.14	1.05	4.7
				GC-MS/MS	1.20	1.12	1.08	1.06	1.19	1.03	1.05	1.21	1.19	1.14	0.96	1.07	1.10	1.13	1.26	1.09	0.99	1.20	1.07	7.3

85	cis-Diallate	TPT	Green tea	GC-MS	1.04	1.02	1.08	1.07	1.06	1.06	1.09	1.05	0.99	1.05	1.10	1.07	1.05	1.07	1.09	1.00	1.05	1.02	1.13	3.2
				GC-MS/MS	1.02	1.03	1.12	1.06	1.05	1.06	0.99	1.06	1.00	1.06	1.08	1.01	1.12	1.05	1.07	1.03	1.01	0.97	1.13	4.1
			Oolong tea	GC-MS	1.03	1.12	1.11	1.09	1.04	1.12	1.20	1.12	1.15	1.09	1.24	1.16	1.16	1.0	1.14	1.11	1.10	1.06	1.05	5.1
				GC-MS/MS	1.00	1.13	1.14	1.15	1.03	1.16	1.18	1.12	1.08	1.07	1.18	1.17	1.16	1.11	1.14	1.11	1.07	1.04	1.09	4.7
		ENVI-Carb + PSA	Green tea	GC-MS	1.13	1.16	1.02	1.08	1.02	1.07	0.98	1.03	1.07	1.05	1.10	1.10	1.07	1.12	1.06	1.18	1.09	1.08	1.02	4.7
				GC-MS/MS	1.06	1.23	1.02	1.04	1.02	1.03	1.16	1.05	1.06	1.01	1.11	1.06	1.17	1.19	1.08	1.17	1.12	1.05	1.02	6.2
			Oolong tea	GC-MS	1.14	1.13	1.08	1.08	1.22	1.12	1.13	1.09	1.19	1.17	1.12	1.07	1.08	1.03	1.17	1.15	1.07	1.13	1.08	4.3
				GC-MS/MS	1.21	1.12	1.07	1.07	1.17	1.08	1.04	1.12	1.18	1.16	0.98	1.05	1.11	1.07	1.21	1.11	1.05	1.14	1.02	5.7
86	Cyanofen-phos	TPT	Green tea	GC-MS	0.96	1.06	1.09	1.16	1.00	1.05	1.04	1.03	0.95	1.02	1.11	1.12	1.08	0.99	0.95	0.98	0.95	0.92	1.12	6.9
				GC-MS/MS	1.02	1.04	1.02	1.02	0.95	1.08	1.07	1.07	0.83	0.92	1.07	1.09	1.20	0.97	1.02	0.98	0.96	0.95	1.12	8.1
			Oolong tea	GC-MS	1.04	1.02	0.99	1.13	1.03	1.04	1.03	1.04	1.06	1.08	1.13	1.13	1.11	0.9	1.04	1.10	1.16	1.01	0.97	5.6
				GC-MS/MS	1.18	1.04	1.06	1.19	1.03	1.19	0.97	1.22	1.20	0.92	1.21	1.07	1.15	1.08	1.13	1.10	1.13	1.11	1.02	7.7
		ENVI-Carb + PSA	Green tea	GC-MS	0.95	0.95	1.08	1.15	1.04	1.00	1.13	1.07	1.08	1.07	1.48	1.15	1.06	1.07	0.92	1.22	1.14	0.95	0.96	11.8
				GC-MS/MS	1.01	1.19	1.09	1.03	1.12	0.70	2.11	1.08	1.13	1.03	1.25	1.05	1.01	1.32	0.99	1.31	1.05	0.87	0.96	24.9
			Oolong tea	GC-MS	1.05	1.03	0.96	0.96	1.20	1.01	1.08	1.15	1.22	1.13	1.23	0.94	1.15	0.95	1.01	1.26	1.08	1.02	1.04	9.4
				GC-MS/MS	1.02	1.02	0.95	1.00	1.17	0.98	1.15	1.13	1.22	1.01	0.99	0.96	1.20	1.00	1.16	1.10	0.97	1.09	1.11	8.2
87	Desmetryn	TPT	Green tea	GC-MS	1.04	1.03	1.14	1.11	1.11	1.04	1.09	1.13	0.96	1.08	1.20	1.07	1.07	1.04	1.06	0.99	1.01	1.01	1.13	5.6
				GC-MS/MS	1.05	1.04	1.14	1.09	1.09	1.08	0.97	1.09	1.01	1.00	1.06	1.03	1.10	0.93	1.05	1.01	0.98	0.94	1.11	5.8
			Oolong tea	GC-MS	1.07	1.02	1.08	1.06	1.04	1.11	1.10	1.13	1.21	1.13	1.27	1.16	1.19	1.1	1.17	1.18	1.21	1.02	1.04	6.4
				GC-MS/MS	0.97	1.09	1.13	1.12	1.05	1.15	1.22	1.12	1.13	1.08	1.28	1.17	1.25	1.13	1.13	1.14	1.19	1.02	1.08	6.9
		ENVI-Carb + PSA	Green tea	GC-MS	1.10	1.18	1.01	1.11	1.04	1.15	1.00	1.06	1.04	0.98	1.14	1.18	1.08	1.10	1.10	1.34	1.11	1.15	0.98	7.9
				GC-MS/MS	1.07	1.25	1.03	1.08	1.05	1.12	1.02	1.08	1.01	1.00	1.13	1.16	1.22	1.12	1.13	1.19	1.10	1.08	0.98	6.9
			Oolong tea	GC-MS	1.09	1.09	1.09	1.06	1.31	1.13	1.08	1.13	1.22	1.19	1.16	1.03	1.05	1.13	1.24	1.14	1.12	1.22	1.08	6.4
				GC-MS/MS	1.11	1.08	1.06	1.04	1.21	1.11	0.99	1.12	1.23	1.20	1.00	1.11	1.08	1.11	1.25	1.14	1.03	1.23	1.02	7.3
88	Dichlobenil	TPT	Green tea	GC-MS	1.09	1.03	1.09	1.03	1.09	1.08	1.08	1.11	1.08	1.10	1.07	1.06	1.03	1.05	1.10	1.00	1.00	1.02	1.11	3.4
				GC-MS/MS	1.10	1.08	1.12	1.04	1.04	1.07	0.99	1.05	0.99	1.06	0.95	1.09	1.12	1.08	1.05	0.95	0.95	0.93	1.13	6.1
			Oolong tea	GC-MS	1.09	1.14	1.16	1.20	1.07	1.15	1.18	1.03	1.17	1.11	1.19	1.15	1.14	1.1	1.18	1.09	1.12	1.11	1.11	3.9
				GC-MS/MS	1.07	1.16	1.15	1.26	1.12	1.24	1.13	1.14	1.17	1.09	1.21	1.15	1.13	1.09	1.20	1.13	1.07	1.07	0.98	5.7
		ENVI-Carb + PSA	Green tea	GC-MS	1.03	1.28	0.96	1.08	0.97	1.06	1.06	1.04	1.08	1.03	1.20	1.04	1.21	1.05	1.20	1.10	1.04	1.13	1.04	7.7
				GC-MS/MS	1.00	1.32	0.96	1.10	1.00	0.74	1.15	1.05	1.09	1.04	1.30	1.09	1.16	1.09	1.16	1.15	1.09	1.08	1.00	11.7
			Oolong tea	GC-MS	1.26	1.14	0.98	1.07	1.20	1.15	1.18	0.99	1.23	1.21	1.06	1.11	1.11	1.13	1.34	1.14	1.13	1.19	1.05	7.8
				GC-MS/MS	1.27	1.17	1.02	1.01	1.20	1.08	1.02	1.22	1.22	1.24	1.01	1.05	1.08	1.13	1.27	1.13	1.08	1.21	1.08	8.0

(Continued)

Appendix Table 5.3 Youden Pair Ratios of Each Average Content From 201 Pesticides and RSD for 19 Circulative Determinations (cont.)

No.	Pesticides	SPE	Sample	Method	Nov 9, 2009 (n=5)	Nov 14, 2009 (n=3)	Nov 19, 2009 (n=3)	Nov 24, 2009 (n=3)	Nov 29, 2009 (n=3)	Dec 4, 2009 (n=3)	Dec 9, 2009 (n=3)	Dec 14, 2009 (n=3)	Dec 19, 2009 (n=3)	Dec 24, 2009 (n=3)	Dec 14, 2009 (n=3)	Jan 3, 2010 (n=3)	Jan 8, 2010 (n=3)	Jan 13, 2010 (n=3)	Jan 18, 2010 (n=3)	Jan 23, 2010 (n=3)	Jan 28, 2010 (n=3)	Feb 2, 2010 (n=3)	Feb 7, 2010 (n=3)	RSD%
89	Dicloran	TPT	Green tea	GC-MS	0.81	1.03	1.13	1.14	1.08	1.09	1.09	1.14	1.07	1.06	1.14	1.01	1.05	0.93	1.11	1.07	1.11	1.05	1.00	7.6
				GC-MS/MS	1.01	1.00	0.80	1.02	0.99	0.99	1.02	1.12	0.98	0.92	1.28	0.99	1.06	0.86	0.98	0.93	0.90	0.96	1.20	11.0
			Oolong tea	GC-MS	1.03	1.03	1.10	1.15	1.18	1.02	1.43	1.44	1.30	1.32	1.09	1.13	1.08	1.0	1.19	1.02	0.99	0.99	1.09	12.4
				GC-MS/MS	1.21	1.16	1.03	1.05	0.99	1.21	1.26	1.38	1.31	0.91	1.37	1.19	1.11	1.04	1.19	1.01	1.15	1.21	1.01	11.5
		ENVI-Carb + PSA	Green tea	GC-MS	1.01	1.02	1.00	1.28	1.00	1.10	0.97	1.21	1.19	1.05	1.39	1.17	1.00	0.99	0.94	1.09	1.03	0.92	1.00	11.5
				GC-MS/MS	0.86	0.99	0.98	1.12	1.05	1.03	1.15	0.99	1.09	0.89	1.25	1.02	1.06	1.03	1.03	1.02	0.91	1.07	0.97	8.8
			Oolong tea	GC-MS	0.96	1.07	1.03	0.97	1.28	1.08	1.05	1.28	1.20	1.32	1.29	1.03	1.04	1.01	0.91	1.12	0.99	1.01	1.05	11.4
				GC-MS/MS	1.27	1.02	1.04	1.02	1.28	1.01	1.34	1.21	1.11	1.04	1.02	0.89	1.16	1.01	1.04	1.01	0.94	1.14	1.07	11.0
90	Dicofol	TPT	Green tea	GC-MS	0.92	1.01	1.12	1.13	0.92	1.08	1.10	1.10	1.13	1.04	1.12	1.06	1.05	1.08	1.15	1.01	1.05	1.04	1.19	5.7
				GC-MS/MS	1.05	1.03	1.18	1.10	0.90	1.10	0.96	1.09	1.08	1.06	1.18	1.04	1.02	1.10	1.13	1.02	1.05	1.01	1.15	5.4
			Oolong tea	GC-MS	1.01	1.11	1.01	1.02	0.92	1.11	1.01	1.12	1.07	0.93	1.04	1.04	1.09	1.1	1.14	1.17	1.20	1.02	1.09	6.9
				GC-MS/MS	1.02	1.10	1.05	1.05	0.90	1.13	1.01	1.14	1.07	0.94	1.05	0.95	0.98	0.98	1.04	1.05	1.07	0.94	0.98	6.5
		ENVI-Carb + PSA	Green tea	GC-MS	0.92	1.26	1.01	1.05	0.99	1.15	0.97	1.01	1.13	1.00	0.95	1.02	1.01	1.19	1.06	1.24	1.12	1.23	1.01	9.7
				GC-MS/MS	1.07	1.38	1.03	1.05	0.98	0.99	1.08	1.07	1.13	1.04	0.98	1.06	1.08	1.24	1.07	1.21	1.11	1.14	1.03	9.0
			Oolong tea	GC-MS	1.25	0.90	0.91	1.12	1.27	1.03	1.12	0.98	1.01	1.08	1.15	1.08	1.09	1.10	1.22	1.12	1.15	1.20	1.03	9.4
				GC-MS/MS	1.17	0.88	0.93	1.12	1.25	1.01	0.97	1.00	1.04	1.06	1.01	0.99	1.05	0.92	1.19	1.05	0.97	1.09	0.89	9.7
91	Dimethachlor	TPT	Green tea	GC-MS	1.01	1.04	1.13	1.01	1.07	1.08	1.03	1.12	0.88	1.06	1.18	1.01	0.96	1.06	1.01	1.02	0.96	0.92	1.09	7.2
				GC-MS/MS	0.99	1.02	0.93	1.07	1.01	1.04	1.01	1.15	1.07	0.98	1.10	1.02	1.04	0.95	1.09	1.00	1.01	0.97	1.13	5.7
			Oolong tea	GC-MS	1.02	1.24	1.17	1.08	0.99	1.07	1.34	1.20	1.09	1.19	1.16	1.15	1.24	1.0	1.07	1.08	1.23	1.01	1.09	8.4
				GC-MS/MS	1.08	1.17	1.16	1.11	1.02	1.19	1.33	1.27	1.18	1.02	1.23	1.12	1.26	1.08	1.12	1.14	1.23	1.11	1.04	7.6
		ENVI-Carb + PSA	Green tea	GC-MS	0.98	1.13	0.95	1.03	0.97	1.09	1.20	1.30	1.02	0.99	1.07	1.02	1.03	1.14	0.93	1.13	1.04	0.98	1.00	9.0
				GC-MS/MS	0.79	1.07	0.99	0.97	1.03	1.17	0.89	1.00	1.04	0.92	1.13	1.06	1.18	1.30	1.03	1.11	1.06	0.98	0.96	11.1
			Oolong tea	GC-MS	1.15	1.13	1.05	1.14	1.17	1.09	0.77	1.11	1.22	1.27	1.14	1.10	1.06	1.04	1.14	1.09	1.07	1.11	1.04	9.0
				GC-MS/MS	1.60	1.10	1.03	1.13	1.20	1.09	1.40	1.20	1.25	1.15	0.95	1.08	1.06	1.03	1.26	1.14	0.95	1.16	1.02	13.6
92	Dioxacarb	TPT	Green tea	GC-MS	0.97	0.98	1.11	1.09	1.09	1.06	1.01	1.13	1.05	1.18	1.20	1.01	0.95	0.87	1.22	0.95	1.06	1.05	1.14	8.7
				GC-MS/MS	1.00	1.08	0.92	1.17	0.95	0.98	1.00	1.05	1.01	1.00	1.12	1.06	1.14	1.03	1.10	1.25	1.15	1.01	0.93	8.3
			Oolong tea	GC-MS	0.93	0.98	1.11	1.04	1.08	1.06	0.94	1.13	1.02	1.19	1.18	1.03	1.03	0.86	1.29	0.95	1.06	0.97	1.11	9.8
				GC-MS/MS	0.99	1.04	1.04	1.12	1.06	1.05	1.12	0.92	1.10	1.12	1.26	1.19	1.14	1.1	1.08	1.16	1.13	0.95	1.06	7.6
		ENVI-Carb + PSA	Green tea	GC-MS	1.03	0.97	1.12	1.10	1.02	1.07	1.17	1.08	1.18	1.11	1.30	1.16	1.13	1.09	1.10	1.11	1.11	1.02	1.00	6.9
				GC-MS/MS	0.96	1.06	0.89	1.10	0.97	0.68	0.91	1.07	1.00	1.01	1.19	1.06	1.18	1.35	1.16	1.18	1.11	0.99	0.93	14.1
			Oolong tea	GC-MS	1.07	1.12	0.94	0.95	1.08	1.05	1.09	1.11	1.05	1.43	0.97	1.17	1.02	1.03	1.28	1.06	0.96	1.33	1.00	12.0
				GC-MS/MS	1.00	1.09	0.91	0.99	1.08	1.06	0.91	1.14	1.07	1.47	0.88	1.09	1.06	0.95	1.26	1.11	0.91	1.32	0.98	14.3

93	Endrin	TPT	Green tea	GC-MS	1.04	1.03	1.12	1.09	1.08	1.06	1.07	1.07	0.99	1.06	1.11	1.06	1.06	1.04	1.07	1.01	0.99	1.03	1.14	3.9
				GC-MS/MS	1.04	1.01	1.12	1.08	1.07	1.07	0.97	1.07	1.00	1.04	1.14	1.07	1.16	1.05	1.08	1.02	1.02	0.99	1.11	4.9
			Oolong tea	GC-MS	1.03	1.08	1.11	1.09	1.01	1.11	1.24	1.09	1.15	1.11	1.23	1.20	1.11	1.0	1.09	1.08	1.17	0.98	1.04	6.4
				GC-MS/MS	1.34	1.10	1.11	1.08	1.01	1.16	1.23	1.13	1.16	1.10	1.22	1.16	1.21	1.07	1.14	1.08	1.14	1.01	1.05	5.9
		ENVI-Carb + PSA	Green tea	GC-MS	0.38	1.18	1.01	1.09	1.02	1.11	0.92	1.07	1.02	1.01	1.07	1.07	1.07	1.13	1.09	1.21	1.08	1.08	0.99	6.4
				GC-MS/MS	1.30	1.17	1.01	1.02	1.05	1.01	0.98	1.04	1.05	0.97	1.11	1.05	1.06	1.15	1.12	1.16	1.03	1.05	1.00	5.7
			Oolong tea	GC-MS	1.13	1.09	1.08	1.06	1.23	1.09	1.04	1.13	1.20	1.21	1.13	1.06	1.07	1.06	1.17	1.14	1.08	1.09	1.06	5.0
				GC-MS/MS	1.14	1.10	1.06	1.09	1.19	1.09	1.11	1.13	1.17	1.16	1.00	1.05	1.11	1.10	1.22	1.08	1.02	1.15	1.01	5.5
94	Epoxicon-azole-2	TPT	Green tea	GC-MS	1.03	1.04	1.12	1.09	1.14	1.05	1.06	1.10	0.97	1.14	1.19	1.10	1.05	0.97	1.08	1.00	1.02	1.04	1.14	5.6
				GC-MS/MS	1.03	1.04	1.11	1.05	1.12	1.07	0.99	1.10	1.02	0.99	1.15	1.05	1.14	0.91	1.10	1.01	1.02	1.00	1.13	6.0
			Oolong tea	GC-MS	1.08	0.95	1.00	1.06	1.00	1.08	1.06	0.99	1.26	1.17	1.19	1.14	1.10	1.1	1.12	1.16	1.19	1.08	1.11	7.2
				GC-MS/MS	1.11	1.04	1.09	1.05	0.97	1.10	1.08	1.04	1.18	1.06	1.22	1.11	1.16	1.02	1.15	1.14	1.15	1.09	1.09	5.6
		ENVI-Carb + PSA	Green tea	GC-MS	0.93	1.05	1.06	1.13	0.99	1.11	1.02	1.09	1.10	1.13	1.06	1.17	1.08	1.11	1.06	1.20	1.14	1.13	0.99	6.3
				GC-MS/MS	0.96	1.10	1.04	1.10	1.00	0.84	0.95	1.10	1.12	1.03	1.13	1.06	1.13	1.18	1.09	1.19	1.14	1.07	0.99	8.3
			Oolong tea	GC-MS	1.02	1.07	0.97	1.00	1.22	1.10	1.16	0.99	1.17	1.32	1.22	1.02	1.11	1.23	1.06	1.15	1.08	1.14	1.05	8.1
				GC-MS/MS	1.02	1.05	0.94	1.02	1.14	1.08	1.10	1.04	1.16	1.25	0.99	1.08	1.06	1.24	1.03	1.14	1.00	1.17	1.02	7.6
95	EPTC	TPT	Green tea	GC-MS	1.10	1.00	1.07	0.99	1.09	1.07	1.02	1.13	1.03	1.11	1.04	1.06	1.04	1.01	1.12	0.95	0.97	0.95	1.09	5.4
				GC-MS/MS	1.10	1.01	1.08	1.00	1.06	1.10	0.99	1.10	0.98	1.15	0.92	1.01	1.14	0.99	1.07	1.00	0.99	0.87	1.13	7.2
			Oolong tea	GC-MS	1.15	1.18	1.11	1.23	1.13	1.22	1.15	1.16	1.24	1.12	1.28	1.06	1.18	1.1	1.23	1.16	1.18	1.16	1.19	4.4
				GC-MS/MS	1.10	1.23	1.18	1.33	1.18	1.30	1.22	1.10	1.29	1.16	1.27	1.05	1.14	1.26	1.25	1.19	1.09	1.18	1.25	5.4
		ENVI-Carb + PSA	Green tea	GC-MS	1.02	1.28	0.93	1.10	0.95	1.03	0.99	1.08	1.02	1.08	1.21	1.06	1.21	1.01	1.26	1.19	1.07	1.06	1.08	9.2
				GC-MS/MS	1.01	1.33	0.94	1.07	0.95	0.75	1.12	1.05	1.05	1.00	1.29	1.05	1.20	1.22	1.25	1.17	1.06	1.00	0.99	12.9
			Oolong tea	GC-MS	1.31	1.20	0.97	1.11	1.29	1.17	1.16	1.20	1.26	1.37	1.13	1.18	1.20	1.14	1.32	1.38	1.19	1.10	1.09	8.8
				GC-MS/MS	1.44	1.17	1.03	1.07	1.30	1.16	1.08	1.30	1.29	1.36	1.06	1.14	1.17	1.19	1.40	1.19	1.14	1.23	1.07	10.1
96	Ethiumesate	TPT	Green tea	GC-MS	1.04	1.06	1.11	1.10	1.00	1.00	1.11	0.94	1.06	1.02	1.09	1.05	1.07	0.96	1.04	1.02	1.01	1.02	1.12	5.0
				GC-MS/MS	1.04	1.13	1.02	1.09	0.99	1.02	1.01	0.94	1.25	0.98	1.10	1.09	1.06	1.10	1.08	1.19	1.09	1.07	1.00	6.9
			Oolong tea	GC-MS	1.03	1.03	1.11	1.08	1.06	1.08	0.99	1.09	0.99	1.01	1.05	1.05	1.11	1.04	1.08	1.01	0.99	0.99	1.11	4.1
				GC-MS/MS	1.05	1.13	1.15	1.08	0.98	1.10	1.00	1.05	1.08	1.14	1.29	1.10	1.14	1.1	1.07	1.07	0.94	1.01	1.02	7.1
		ENVI-Carb + PSA	Green tea	GC-MS	1.05	1.09	1.13	1.11	1.03	1.16	1.04	1.14	1.09	1.07	1.23	1.15	1.15	1.08	1.14	1.10	1.02	1.03	1.01	5.2
				GC-MS/MS	1.05	1.13	1.02	1.06	1.04	0.83	1.05	1.06	1.04	0.98	1.19	1.03	1.09	1.18	1.11	1.16	1.10	1.06	0.95	7.9
			Oolong tea	GC-MS	1.14	1.10	1.02	1.10	1.07	1.15	0.73	2.57	1.66	1.28	1.08	1.09	1.08	1.04	1.19	1.14	1.05	1.14	1.01	31.4
				GC-MS/MS	1.10	1.09	1.07	1.05	1.15	1.06	1.05	1.08	1.21	1.18	0.95	1.08	1.14	1.05	1.22	1.10	0.99	1.14	1.01	6.5

(Continued)

Appendix Table 5.3 Youden Pair Ratios of Each Average Content From 201 Pesticides and RSD for 19 Circulative Determinations (cont.)

No.	Pesticides	SPE	Sample	Method	Nov 9, 2009 (n = 5)	Nov 14, 2009 (n = 3)	Nov 19, 2009 (n = 3)	Nov 24, 2009 (n = 3)	Nov 29, 2009 (n = 3)	Dec 4, 2009 (n = 3)	Dec 9, 2009 (n = 3)	Dec 14, 2009 (n = 3)	Dec 19, 2009 (n = 3)	Dec 24, 2009 (n = 3)	Dec 14, 2009 (n = 3)	Jan 3, 2010 (n = 3)	Jan 8, 2010 (n = 3)	Jan 13, 2010 (n = 3)	Jan 18, 2010 (n = 3)	Jan 23, 2010 (n = 3)	Jan 28, 2010 (n = 3)	Feb 2, 2010 (n = 3)	Feb 7, 2010 (n = 3)	RSD%
												Youden pair ratio												
97	Ethoprophos	TPT	Green tea	GC–MS	0.96	1.02	1.09	1.07	1.08	1.06	1.06	1.08	0.94	1.06	1.12	1.06	1.05	0.97	1.05	1.02	1.02	0.97	1.10	4.9
				GC–MS/MS	1.02	1.02	1.05	1.06	1.05	1.06	0.99	1.08	0.99	0.99	1.12	1.08	1.09	0.97	1.06	0.99	1.01	0.98	1.10	4.5
			Oolong tea	GC–MS	1.04	1.11	1.12	1.11	1.06	1.13	1.27	1.12	1.17	1.15	1.25	1.22	1.19	1.1	1.13	1.11	1.17	1.03	1.09	5.9
				GC–MS/MS	1.08	1.14	1.14	1.11	1.04	1.17	1.22	1.14	1.20	1.09	1.23	1.18	1.16	1.15	1.15	1.08	1.12	1.08	1.07	4.6
		ENVI-Carb + PSA	Green tea	GC–MS	1.06	1.11	0.99	1.12	0.99	1.09	1.13	1.12	1.03	1.05	1.12	1.09	1.08	1.05	1.04	1.16	1.07	1.02	1.00	4.6
				GC–MS/MS	0.92	1.11	1.01	1.06	1.03	1.10	1.11	1.05	1.06	1.00	1.13	1.05	1.13	1.25	1.11	1.13	1.07	1.00	0.98	6.8
			Oolong tea	GC–MS	1.16	1.13	1.12	1.11	1.26	1.11	1.03	1.18	1.20	1.20	1.17	1.05	1.09	1.06	1.16	1.14	1.14	1.14	1.09	5.1
				GC–MS/MS	1.36	1.13	1.10	1.09	1.23	1.09	1.13	1.17	1.19	1.17	1.00	1.06	1.12	1.11	1.25	1.08	1.02	1.15	1.06	7.4
98	Etrimfos	TPT	Green tea	GC–MS	0.98	1.03	1.11	1.08	1.08	1.04	1.07	1.09	0.85	1.07	1.20	1.05	1.04	0.96	1.00	1.02	1.14	1.00	1.11	7.1
				GC–MS/MS	1.02	1.03	1.06	1.06	1.07	1.05	0.99	1.07	0.99	1.03	1.14	1.01	1.05	0.98	1.05	1.00	1.03	0.98	1.17	4.8
			Oolong tea	GC–MS	1.04	1.07	1.10	1.08	1.02	1.09	1.24	1.08	1.20	1.17	1.24	1.09	1.20	1.1	1.12	1.05	1.16	1.02	1.07	6.2
				GC–MS/MS	1.07	1.12	1.16	1.10	1.02	1.15	1.24	1.16	1.15	1.01	1.21	1.12	1.17	1.10	1.16	1.09	1.10	1.03	1.12	5.4
		ENVI-Carb + PSA	Green tea	GC–MS	1.05	1.16	1.02	1.09	1.01	1.11	0.94	1.15	1.07	1.08	1.11	1.15	1.07	1.06	1.03	1.15	1.09	1.06	1.02	5.3
				GC–MS/MS	0.92	1.14	1.03	1.02	1.03	1.59	1.09	1.04	1.07	0.97	1.13	1.05	1.14	1.29	1.06	1.12	1.06	1.03	0.97	13.3
			Oolong tea	GC–MS	1.18	1.12	1.12	1.10	1.23	1.09	0.98	1.13	1.22	1.20	1.18	1.01	1.07	1.06	1.17	1.08	1.10	1.11	1.06	6.1
				GC–MS/MS	1.36	1.11	1.09	1.09	1.19	1.07	1.12	1.11	1.17	1.14	0.96	1.03	1.11	1.09	1.21	1.09	1.00	1.14	1.03	7.7
99	Fenamidone	TPT	Green tea	GC–MS	1.02	1.01	1.13	1.07	1.10	1.01	1.05	1.10	1.01	1.02	1.17	1.05	1.04	0.98	1.10	1.00	1.02	1.14	1.03	5.1
				GC–MS/MS	1.04	1.05	1.12	1.06	1.11	1.08	0.98	1.08	1.02	0.99	1.12	1.03	1.18	1.02	1.11	1.00	1.05	1.04	1.17	5.4
			Oolong tea	GC–MS	1.08	0.99	1.00	1.04	0.96	1.06	0.97	0.97	1.15	1.12	1.21	1.10	1.05	1.0	1.11	1.14	1.12	1.05	1.03	6.5
				GC–MS/MS	1.07	1.03	1.04	1.07	0.93	1.09	0.98	0.98	1.16	1.00	1.17	1.05	1.14	1.06	1.10	1.14	1.09	1.02	1.06	6.0
		ENVI-Carb + PSA	Green tea	GC–MS	0.98	1.10	1.03	1.17	0.99	1.09	0.96	1.07	1.11	1.07	1.06	1.09	1.07	1.09	1.07	1.18	1.13	1.14	1.00	5.7
				GC–MS/MS	0.97	1.13	1.02	1.09	0.99	0.97	0.91	1.08	1.10	1.01	1.08	1.09	1.13	1.04	1.08	1.14	1.12	1.13	0.99	6.2
			Oolong tea	GC–MS	1.06	1.04	0.95	0.97	1.20	1.11	1.17	1.07	1.08	1.23	1.14	1.01	1.08	1.02	1.23	1.15	1.08	1.13	1.05	7.3
				GC–MS/MS	1.02	1.01	0.92	1.00	1.13	1.07	1.05	1.07	1.09	1.21	0.95	1.08	1.04	1.03	1.29	1.14	1.02	1.11	1.01	8.1
100	Fenarimol	TPT	Green tea	GC–MS	1.02	1.03	1.18	1.06	1.12	1.04	0.96	1.06	1.03	1.00	1.14	1.06	1.08	0.95	1.14	0.98	1.04	0.99	1.14	6.2
				GC–MS/MS	1.05	1.12	1.05	1.12	0.99	1.03	1.07	0.99	1.22	1.09	1.15	1.04	1.04	1.0	1.12	1.13	1.13	1.01	1.06	5.4
			Oolong tea	GC–MS	1.06	1.02	1.08	1.08	0.96	1.09	1.00	0.93	1.18	1.05	1.21	1.13	1.11	1.06	1.15	1.13	1.13	1.13	1.07	6.6
				GC–MS/MS	0.96	1.15	1.04	1.12	0.99	1.00	0.99	1.10	1.14	1.06	1.13	1.04	1.07	1.17	1.15	1.19	1.08	1.08	1.01	6.4
		ENVI-Carb + PSA	Green tea	GC–MS	1.11	1.08	0.96	1.07	1.13	1.12	1.11	1.02	1.14	1.31	1.17	0.99	1.03	1.06	1.18	1.14	1.11	1.10	1.05	7.1
				GC–MS/MS	1.04	1.04	0.95	1.00	1.17	1.05	1.04	1.09	1.14	1.23	1.00	1.04	1.16	1.08	1.30	1.13	1.01	1.11	1.06	8.1
			Oolong tea	GC–MS	1.05	1.03	1.08	1.07	1.12	1.05	1.09	1.15	1.06	1.08	1.14	1.25	1.02	0.90	1.06	0.94	0.99	1.01	1.14	7.4
				GC–MS/MS	0.97	1.07	0.99	1.21	0.94	1.13	1.08	1.10	1.07	1.05	0.84	1.22	0.96	1.25	1.06	1.25	1.10	1.09	0.93	10.6

101	Flamprop-isopropyl	TPT	Green tea	GC-MS	1.05	1.05	1.11	1.07	1.08	1.07	1.08	1.10	1.06	1.03	1.14	1.08	1.05	1.02	1.07	1.01	1.00	1.05	1.14	3.6
				GC-MS/MS	1.06	1.04	1.14	1.09	1.07	1.07	0.98	1.09	1.04	1.01	1.03	1.03	1.14	1.08	1.09	0.99	0.99	0.97	1.11	4.9
			Oolong tea	GC-MS	1.04	1.16	1.08	1.08	1.00	1.06	1.09	1.05	1.18	1.08	1.30	1.21	1.14	1.1	1.10	1.19	1.19	1.01	1.06	6.6
				GC-MS/MS	1.00	1.06	1.10	1.09	0.99	1.11	1.11	1.06	1.15	1.08	1.21	1.14	1.15	1.06	1.13	1.11	1.11	1.03	1.05	5.0
		ENVI-Carb + PSA	Green tea	GC-MS	1.05	1.15	1.03	1.10	1.00	1.10	1.34	1.07	1.04	0.99	1.04	1.09	1.09	1.14	1.10	1.26	1.12	1.15	0.99	7.9
				GC-MS/MS	1.06	1.21	1.03	1.07	1.04	1.04	1.04	1.06	1.05	1.02	1.07	1.04	1.13	1.29	1.10	1.23	1.10	0.99	0.98	7.6
			Oolong tea	GC-MS	1.11	1.07	1.01	1.05	1.18	1.11	1.08	1.06	1.19	1.24	1.11	1.12	1.08	1.06	1.22	1.14	1.10	1.19	1.07	5.7
				GC-MS/MS	1.07	1.06	0.99	1.04	1.14	1.09	1.03	1.08	1.14	1.16	0.99	1.08	1.10	1.11	1.23	1.10	1.01	1.15	1.02	5.7
102	Flamprop-methyl	TPT	Green tea	GC-MS	1.03	1.01	1.07	1.09	1.09	1.01	0.98	1.13	1.00	0.89	1.03	1.00	1.04	0.97	1.03	1.01	1.02	1.01	1.13	5.6
				GC-MS/MS	1.04	1.04	1.12	1.08	1.04	1.09	1.00	1.09	1.00	1.00	1.10	1.05	1.12	0.99	1.06	1.00	1.02	0.94	1.16	5.3
			Oolong tea	GC-MS	1.07	0.97	1.01	0.98	1.14	1.05	1.14	1.23	1.17	1.05	1.17	1.19	1.15	1.1	1.09	1.13	1.16	0.98	1.06	7.2
				GC-MS/MS	1.00	1.06	1.10	1.09	1.10	1.11	1.10	1.09	1.14	1.05	1.21	1.15	1.18	1.04	1.15	1.10	1.09	1.00	1.05	5.6
		ENVI-Carb + PSA	Green tea	GC-MS	1.06	1.17	0.92	1.13	1.00	1.02	1.00	1.08	1.13	1.04	1.22	1.10	1.08	1.12	1.05	1.28	1.16	1.04	0.96	8.7
				GC-MS/MS	1.08	1.09	1.06	1.05	1.36	0.95	1.36	1.07	1.08	1.01	1.13	1.05	1.09	1.14	1.06	1.26	1.11	1.07	0.98	8.4
			Oolong tea	GC-MS	1.11	1.30	0.95	1.04	0.75	1.07	0.75	1.03	1.13	1.09	1.10	0.88	1.09	1.06	1.15	1.17	1.05	1.13	1.07	11.1
				GC-MS/MS	1.07	1.06	0.99	1.04	1.03	1.09	1.03	1.08	1.14	1.16	0.99	1.08	1.10	1.11	1.26	1.09	1.01	1.15	0.98	6.3
103	Fonofos	TPT	Green tea	GC-MS	1.01	1.01	1.10	1.09	1.06	1.05	1.06	1.09	0.94	1.09	1.13	1.07	1.06	0.89	0.99	1.01	1.01	1.03	1.13	6.1
				GC-MS/MS	1.05	1.15	1.04	1.05	1.07	1.13	1.07	1.05	1.08	1.09	1.06	1.14	1.07	1.08	1.06	1.18	1.12	1.13	1.01	4.0
			Oolong tea	GC-MS	1.03	1.04	1.12	1.08	1.09	1.07	1.09	1.07	1.03	1.08	1.12	1.05	1.04	1.01	1.08	1.00	1.03	1.01	1.12	4.0
				GC-MS/MS	1.04	1.05	1.09	1.05	1.05	1.11	1.05	1.10	1.21	1.15	1.22	1.13	1.15	1.1	1.13	1.09	1.13	1.04	1.06	5.1
		ENVI-Carb + PSA	Green tea	GC-MS	1.03	1.13	1.14	1.10	1.01	1.14	1.01	1.10	1.14	1.10	1.21	1.15	1.15	1.08	1.15	1.07	1.08	1.03	1.13	4.8
				GC-MS/MS	1.03	1.20	1.00	1.05	1.04	1.25	1.04	1.06	1.09	1.02	1.11	1.01	1.12	1.23	1.11	1.13	1.07	1.13	1.01	6.7
			Oolong tea	GC-MS	1.18	1.12	1.11	1.07	1.27	1.11	1.27	1.13	1.19	1.19	1.17	1.05	1.09	1.06	1.18	1.13	1.11	1.15	1.14	5.1
				GC-MS/MS	1.20	1.12	1.10	1.06	1.22	1.25	1.22	1.10	1.16	1.15	0.96	1.06	1.10	1.24	1.24	1.07	1.01	1.14	1.01	6.7
104	Hexachlorobenzene	TPT	Green tea	GC-MS	1.02	1.01	1.05	1.08	0.99	1.07	0.99	1.00	0.97	1.06	1.04	1.12	1.06	1.07	1.05	1.06	0.96	0.87	1.14	5.9
				GC-MS/MS	0.93	1.12	1.09	1.15	1.20	1.11	1.20	1.08	1.03	1.16	1.24	1.14	1.11	1.0	1.15	1.17	1.04	1.05	1.11	6.7
			Oolong tea	GC-MS	0.97	1.13	1.14	1.18	1.18	1.14	1.22	1.13	1.04	1.08	1.26	1.12	1.12	0.98	1.14	1.18	1.00	1.03	1.14	7.1
				GC-MS/MS	1.04	1.18	1.00	1.10	1.21	1.04	1.21	1.07	1.06	1.01	1.09	1.03	1.11	1.12	1.13	1.13	1.13	1.07	1.01	5.1
		ENVI-Carb + PSA	Green tea	GC-MS	1.24	1.08	1.06	1.05	1.22	1.09	1.11	1.12	1.17	1.15	1.10	1.06	1.10	1.00	1.10	1.20	1.11	1.14	1.14	5.5
				GC-MS/MS	1.21	1.11	1.07	1.07	1.21	1.08	1.09	1.13	1.19	1.16	1.01	1.05	1.10	1.04	1.18	1.13	1.02	1.16	1.01	5.7
			Oolong tea	GC-MS	1.02	1.02	1.08	1.06	1.06	1.07	1.06	1.02	0.96	1.03	1.10	1.07	1.06	1.06	1.05	1.04	0.96	0.89	1.12	5.2
				GC-MS/MS	1.04	1.15	1.00	1.09	1.11	1.07	1.11	1.04	1.07	1.02	1.08	1.06	1.09	1.04	1.08	1.15	1.06	1.09	1.01	4.2

(Continued)

Appendix Table 5.3 Youden Pair Ratios of Each Average Content From 201 Pesticides and RSD for 19 Circulative Determinations (cont.)

No.	Pesticides	SPE	Sample	Method	Nov 9, 2009 (n = 5)	Nov 14, 2009 (n = 3)	Nov 19, 2009 (n = 3)	Nov 24, 2009 (n = 3)	Nov 29, 2009 (n = 3)	Dec 4, 2009 (n = 3)	Dec 9, 2009 (n = 3)	Dec 14, 2009 (n = 3)	Dec 14, 2009 (n = 3)	Dec 19, 2009 (n = 3)	Dec 24, 2009 (n = 3)	Dec 14, 2009 (n = 3)	Jan 3, 2010 (n = 3)	Jan 8, 2010 (n = 3)	Jan 13, 2010 (n = 3)	Jan 18, 2010 (n = 3)	Jan 23, 2010 (n = 3)	Jan 28, 2010 (n = 3)	Feb 2, 2010 (n = 3)	Feb 7, 2010 (n = 3)	RSD%
													Youden pair ratio												
105	Hexazinone	TPT	Green tea	GC-MS	0.99	0.98	1.06	1.09	1.14	1.09	1.02	1.15	1.12	1.15	1.25	1.24	1.05	0.97	0.94	1.21	1.01	1.15	1.09	1.20	8.7
				GC-MS/MS	0.97	1.01	1.09	1.02	1.08	1.09	0.94	1.15	1.10	1.15	1.24	1.11	0.97	1.06	0.81	1.20	0.97	1.11	1.06	1.16	9.4
			Oolong tea	GC-MS	1.05	0.99	1.05	1.08	1.02	1.09	1.01	1.02	1.17	1.02	1.20	1.16	1.07	1.10	1.1	1.05	1.18	1.13	1.00	1.01	6.1
				GC-MS/MS	1.08	1.04	1.15	1.14	0.99	1.10	1.04	1.10	1.12	1.10	1.17	1.08	1.06	1.19	1.14	1.13	1.16	1.10	1.01	1.18	5.2
		ENVI-Carb + PSA	Green tea	GC-MS	1.04	1.13	0.97	1.14	0.94	1.01	1.14	1.06	1.14	1.06	0.98	1.06	1.12	1.06	1.14	1.06	1.22	1.15	1.19	0.99	7.4
				GC-MS/MS	0.99	1.18	1.00	1.06	0.96	0.63	1.11	1.06	1.20	1.06	0.98	1.03	1.05	1.03	1.43	1.03	1.20	1.15	1.13	1.01	14.3
			Oolong tea	GC-MS	1.07	1.09	0.86	1.00	1.07	1.09	1.14	1.05	1.07	1.05	1.03	1.44	1.10	1.12	1.12	1.36	1.04	1.05	1.23	0.97	11.7
				GC-MS/MS	1.00	1.09	0.83	0.99	1.02	1.06	0.91	1.02	1.10	1.02	0.92	1.41	1.13	1.10	1.00	1.39	1.16	0.98	1.29	0.96	14.5
106	Iodofenphos	TPT	Green tea	GC-MS	0.98	1.02	1.06	1.04	1.03	1.08	1.06	1.05	0.83	1.05	1.19	1.07	1.00	0.96	1.05	0.97	0.99	0.95	0.92	1.11	7.8
				GC-MS/MS	1.02	0.99	0.91	1.00	0.90	1.02	1.03	1.10	1.03	1.10	1.15	0.96	0.97	0.99	0.97	1.06	1.01	0.95	0.97	1.12	6.4
			Oolong tea	GC-MS	1.02	1.24	1.17	1.14	1.04	1.08	1.08	1.13	1.02	1.13	1.12	1.16	1.18	1.21	1.0	1.07	1.06	1.19	1.02	1.08	6.3
				GC-MS/MS	1.16	1.05	1.18	1.10	0.96	1.22	1.11	1.30	1.19	1.30	1.23	0.98	1.05	1.15	1.06	1.10	1.13	1.27	1.09	0.81	10.5
		ENVI-Carb + PSA	Green tea	GC-MS	0.98	1.09	1.03	1.04	1.00	1.05	0.98	1.05	1.03	1.05	1.15	1.02	1.01	1.06	1.07	0.92	1.18	1.10	0.98	1.00	5.9
				GC-MS/MS	0.88	0.96	1.04	0.96	1.08	1.27	1.21	1.00	1.13	1.00	1.02	0.93	1.09	1.05	1.36	1.13	1.05	0.91	0.93	0.97	12.1
			Oolong tea	GC-MS	1.08	1.13	1.00	1.13	1.12	1.04	0.84	1.13	1.25	1.13	1.19	1.21	1.02	1.02	1.05	1.15	1.09	1.04	1.01	1.03	8.7
				GC-MS/MS	1.27	1.05	0.96	1.08	1.15	1.03	1.37	1.20	1.25	1.20	1.03	1.08	1.05	1.08	1.07	1.32	1.13	0.85	1.11	1.00	11.5
107	Isofenphos	TPT	Green tea	GC-MS	1.00	1.02	1.14	1.11	1.15	1.03	1.09	1.06	0.89	1.06	1.20	1.14	1.09	1.04	1.16	1.02	0.97	0.99	1.00	1.12	7.5
				GC-MS/MS	1.04	1.02	1.12	1.08	1.11	1.09	0.98	1.08	1.00	1.08	1.02	1.04	1.04	1.01	1.11	1.06	1.00	1.02	1.00	1.13	4.4
			Oolong tea	GC-MS	1.04	0.95	1.05	1.04	1.05	1.10	1.14	1.06	1.25	1.06	1.28	1.22	1.22	1.08	1.0	1.10	1.06	1.17	1.00	1.03	8.3
				GC-MS/MS	1.04	1.08	1.12	1.08	1.01	1.13	1.17	1.09	1.20	1.09	1.24	1.08	1.16	1.21	1.02	1.14	1.04	1.12	1.03	1.00	6.5
		ENVI-Carb + PSA	Green tea	GC-MS	0.97	1.15	1.05	1.16	1.06	1.15	1.01	1.10	0.97	1.10	1.17	1.06	1.19	1.12	1.03	1.15	1.14	1.10	0.91	1.00	7.4
				GC-MS/MS	1.06	1.11	1.03	1.07	1.06	1.13	1.06	1.06	1.05	1.06	1.13	0.99	1.05	1.15	1.09	1.12	1.15	1.10	1.04	0.95	5.0
			Oolong tea	GC-MS	1.10	1.06	1.11	1.06	1.31	1.09	1.05	1.27	1.14	1.27	1.27	1.19	1.14	1.08	1.06	1.15	1.15	1.07	1.14	1.05	7.8
				GC-MS/MS	1.07	1.09	1.08	1.06	1.19	1.08	1.04	1.16	1.19	1.16	0.98	1.15	1.19	1.06	1.09	1.23	1.04	1.01	1.17	1.04	6.3
108	Isopropalin	TPT	Green tea	GC-MS	1.02	1.01	1.14	1.14	1.09	1.03	1.02	1.04	0.87	1.04	1.19	1.16	1.09	1.08	1.03	1.07	1.00	0.98	1.01	1.14	7.3
				GC-MS/MS	1.00	1.10	1.04	1.15	1.05	1.14	1.15	1.15	1.06	1.15	1.08	1.06	1.16	1.06	1.09	1.05	1.18	1.10	1.12	1.03	4.8
			Oolong tea	GC-MS	1.01	1.02	1.12	1.11	1.11	1.07	0.98	1.11	1.00	1.11	1.01	1.05	1.02	1.07	1.07	1.12	1.02	0.97	0.99	1.12	5.1
				GC-MS/MS	1.05	1.00	1.06	1.05	1.08	1.11	1.32	1.09	1.29	1.09	1.23	1.19	1.22	1.11	1.0	1.12	1.07	1.19	1.02	1.02	8.6
		ENVI-Carb + PSA	Green tea	GC-MS	1.08	1.06	1.10	1.05	0.99	1.14	1.35	1.10	1.21	1.10	1.29	1.10	1.24	1.20	1.03	1.12	1.09	1.17	1.05	1.05	8.3
				GC-MS/MS	0.98	1.06	1.03	1.06	1.06	1.23	1.33	1.06	1.04	1.06	1.14	0.98	1.03	1.13	1.08	1.12	1.20	1.07	1.01	0.99	8.5
			Oolong tea	GC-MS	1.07	1.10	1.13	1.05	1.35	1.10	1.01	1.20	1.19	1.20	1.29	1.19	1.03	1.13	1.02	1.08	1.20	1.09	1.11	1.09	8.0
				GC-MS/MS	1.15	1.08	1.14	1.08	1.23	1.11	0.94	1.17	1.17	1.17	1.02	1.16	1.04	1.15	1.14	1.20	1.09	1.01	1.17	1.06	6.6

109	Methoprene	TPT	Green tea	GC–MS	1.03	0.99	1.12	1.11	1.10	1.05	1.08	1.07	1.01	1.11	1.17	1.07	1.07	1.03	1.09	1.02	1.02	1.03	1.13	4.5
				GC–MS/MS	1.06	1.02	1.16	1.08	1.08	1.06	0.93	1.06	1.08	1.16	1.00	1.04	1.03	1.10	1.11	1.07	0.98	0.96	1.12	5.8
			Oolong tea	GC–MS	1.03	1.03	1.07	1.05	1.52	1.11	1.15	1.09	1.20	1.17	1.29	1.23	1.14	1.1	1.10	1.12	1.14	1.02	1.05	10.4
				GC–MS/MS	1.03	1.12	1.09	1.09	0.98	1.12	1.15	1.05	1.19	1.11	1.21	1.20	1.16	1.07	1.10	1.07	1.10	0.98	1.03	6.1
		ENVI-Carb + PSA	Green tea	GC–MS	1.07	1.20	1.02	1.07	1.02	1.16	1.34	1.05	1.05	1.06	1.06	1.18	1.07	1.10	1.08	1.24	1.10	1.14	1.01	7.7
				GC–MS/MS	1.11	1.08	1.04	1.03	1.00	0.94	1.00	1.06	1.02	1.01	1.11	1.11	1.14	1.14	1.12	1.17	1.04	1.10	0.99	5.8
			Oolong tea	GC–MS	1.13	1.10	1.08	1.03	2.12	1.13	2.31	2.48	1.16	1.23	1.16	1.06	1.08	1.06	1.20	1.13	1.10	1.15	1.08	34.7
				GC–MS/MS	1.12	1.02	1.08	1.03	1.08	1.08	1.03	1.09	1.14	1.21	0.96	1.06	1.10	1.09	1.23	1.07	1.02	1.16	1.04	6.1
110	Methoprotryne	TPT	Green tea	GC–MS	1.03	1.02	1.12	1.09	1.16	1.03	1.08	1.08	1.00	1.12	1.25	1.04	1.06	1.04	1.10	1.00	1.02	1.04	1.13	5.9
				GC–MS/MS	1.05	1.07	1.13	1.07	1.13	1.06	0.96	1.08	1.07	0.98	1.10	1.03	1.09	1.03	1.10	1.02	1.00	0.99	1.14	4.9
			Oolong tea	GC–MS	1.07	0.96	1.04	1.06	0.97	1.06	1.05	1.04	1.35	1.21	1.12	1.11	1.13	1.1	1.13	1.16	1.20	1.05	1.08	8.0
				GC–MS/MS	1.01	1.05	1.08	1.10	0.95	1.10	1.09	1.04	1.19	1.08	1.17	1.11	1.16	1.07	1.17	1.17	1.13	1.06	1.11	5.6
		ENVI-Carb + PSA	Green tea	GC–MS	1.04	1.17	1.02	1.12	1.01	1.19	0.90	1.09	1.10	1.07	1.00	1.18	1.08	1.14	1.06	1.21	1.12	1.18	0.99	7.6
				GC–MS/MS	1.03	0.96	1.03	1.09	1.00	1.17	1.03	1.10	1.10	1.04	1.13	1.13	1.15	1.21	1.10	1.18	1.13	1.10	1.00	6.2
			Oolong tea	GC–MS	1.12	1.07	1.00	1.05	1.22	1.11	1.16	1.09	1.14	1.22	1.25	0.99	1.10	1.08	1.25	1.14	1.15	1.17	1.08	6.7
				GC–MS/MS	1.06	1.04	0.98	1.02	1.15	1.10	1.04	1.09	1.15	1.17	0.98	1.08	1.12	1.12	1.27	1.17	1.07	1.16	1.04	6.7
111	Methoxychlor	TPT	Green tea	GC–MS	1.04	1.06	1.16	1.06	1.05	1.11	1.00	1.02	0.78	1.02	1.22	1.09	1.05	0.99	0.92	0.95	0.91	0.85	1.11	10.4
				GC–MS/MS	1.01	1.01	0.92	1.08	0.97	1.03	0.99	1.09	0.99	1.05	1.15	1.07	1.15	1.00	1.05	0.95	1.04	1.00	1.18	6.7
			Oolong tea	GC–MS	1.07	1.17	0.99	1.16	1.03	1.12	1.08	0.96	1.11	1.02	1.16	1.24	1.07	0.9	1.06	1.01	1.24	1.11	1.06	8.1
				GC–MS/MS	1.16	1.05	1.00	1.12	1.00	1.26	1.02	1.03	1.18	1.10	1.25	1.05	1.08	1.07	1.18	1.05	0.97	1.07	0.95	8.1
		ENVI-Carb + PSA	Green tea	GC–MS	0.85	0.85	1.04	1.20	1.04	0.98	0.94	1.35	1.07	1.02	1.20	1.02	1.01	0.94	0.99	1.24	1.15	1.00	0.98	12.5
				GC–MS/MS	1.07	1.21	1.00	1.01	1.00	0.64	1.50	1.12	1.11	1.14	1.27	1.04	1.06	1.07	1.08	1.25	1.08	1.10	0.97	14.9
			Oolong tea	GC–MS	0.91	1.13	0.96	0.94	1.30	0.98	0.81	1.21	1.56	0.94	1.15	0.92	1.11	0.91	0.91	1.31	0.99	0.89	1.13	17.8
				GC–MS/MS	0.93	1.07	0.90	1.00	1.04	1.18	0.98	1.05	0.97	1.26	0.97	1.04	1.09	1.01	1.01	1.14	1.01	1.23	0.98	9.4
112	Methyl-parathion	TPT	Green tea	GC–MS	0.94	1.05	1.12	0.86	1.06	1.08	1.01	1.09	0.72	1.09	1.21	0.97	0.97	1.02	0.86	0.96	0.91	1.01	1.08	11.2
				GC–MS/MS	1.02	1.01	0.92	1.04	0.99	1.03	0.99	1.13	1.04	0.95	1.17	0.97	0.96	0.94	1.03	0.95	0.98	0.95	1.15	7.0
			Oolong tea	GC–MS	1.05	1.11	1.10	1.11	1.19	1.15	1.39	1.12	1.21	1.14	1.22	1.16	1.22	0.9	1.09	0.98	1.18	1.02	1.08	8.9
				GC–MS/MS	0.87	1.02	1.07	0.99	0.93	1.19	1.02	1.00	1.01	0.94	1.04	1.16	1.17	1.05	1.09	0.98	0.98	0.99	1.06	8.0
		ENVI-Carb + PSA	Green tea	GC–MS	0.92	0.98	1.03	1.16	1.02	1.10	1.05	1.20	1.06	1.11	1.28	1.01	1.02	0.98	0.92	1.21	1.17	1.09	1.05	9.5
				GC–MS/MS	0.84	0.94	1.02	1.02	1.09	1.44	1.63	1.01	1.11	0.93	1.22	1.09	1.16	1.28	1.06	1.10	1.02	1.01	1.06	16.6
			Oolong tea	GC–MS	0.99	1.09	1.19	1.07	1.33	1.09	0.94	1.20	1.22	1.27	1.26	0.99	1.03	0.95	0.96	1.21	1.04	0.99	0.98	11.1
				GC–MS/MS	1.21	1.04	1.05	0.92	0.99	1.06	1.02	1.01	1.18	1.06	0.93	0.98	1.04	1.04	1.17	1.11	1.03	1.09	1.10	7.3

(Continued)

Appendix Table 5.3 Youden Pair Ratios of Each Average Content From 201 Pesticides and RSD for 19 Circulative Determinations (cont.)

No.	Pesticides	SPE	Sample	Method	Youden pair ratio																			RSD%
					Nov 9, 2009 (n=5)	Nov 14, 2009 (n=3)	Nov 19, 2009 (n=3)	Nov 24, 2009 (n=3)	Nov 29, 2009 (n=3)	Dec 4, 2009 (n=3)	Dec 9, 2009 (n=3)	Dec 14, 2009 (n=3)	Dec 19, 2009 (n=3)	Dec 24, 2009 (n=3)	Dec 14, 2009 (n=3)	Jan 3, 2010 (n=3)	Jan 8, 2010 (n=3)	Jan 13, 2010 (n=3)	Jan 18, 2010 (n=3)	Jan 23, 2010 (n=3)	Jan 28, 2010 (n=3)	Feb 2, 2010 (n=3)	Feb 7, 2010 (n=3)	
113	Metolachlor	TPT	Green tea	GC-MS	1.00	1.03	1.13	1.06	1.05	1.11	1.09	1.07	1.00	1.06	1.17	1.05	1.03	1.08	1.09	1.02	0.97	0.97	1.13	5.2
				GC-MS/MS	1.02	1.02	1.03	1.08	1.04	1.06	0.98	1.10	1.04	1.02	1.18	1.02	1.08	1.02	1.08	0.99	1.00	1.00	1.05	4.4
			Oolong tea	GC-MS	1.02	1.15	1.19	1.06	0.98	1.10	1.21	1.18	1.13	1.13	1.24	1.21	1.22	1.1	1.20	1.12	1.25	0.91	1.01	8.3
				GC-MS/MS	1.07	1.14	1.14	1.10	1.00	1.16	1.30	1.17	1.15	1.05	1.27	1.14	1.23	1.13	1.14	1.12	1.18	1.06	1.07	6.4
		ENVI-Carb + PSA	Green tea	GC-MS	1.05	1.16	0.98	1.05	1.00	1.11	1.48	1.08	1.01	0.98	1.09	1.08	0.96	1.08	0.98	1.17	1.04	1.00	1.00	10.9
				GC-MS/MS	0.89	1.13	1.01	1.00	1.03	1.04	0.94	1.03	1.05	0.96	1.14	1.19	1.20	1.29	1.08	1.17	1.06	1.05	0.96	9.6
			Oolong tea	GC-MS	1.05	1.13	1.01	1.11	1.20	1.08	0.93	1.09	1.24	1.28	1.20	1.17	1.13	1.12	1.24	1.18	1.15	1.06	1.02	7.9
				GC-MS/MS	1.37	1.11	1.06	1.12	1.17	1.08	1.20	1.13	1.23	1.17	0.97	1.05	1.13	1.12	1.25	1.12	0.99	1.17	1.04	8.3
114	Nitrapyrin	TPT	Green tea	GC-MS	1.01	1.01	1.10	0.98	1.04	1.10	1.09	1.03	0.98	1.11	1.09	0.95	0.97	1.00	1.00	1.05	0.96	0.96	1.06	5.2
				GC-MS/MS	0.95	1.01	0.84	1.00	0.97	1.04	1.03	1.17	1.01	1.07	1.10	1.07	0.97	0.94	1.03	0.92	1.03	0.87	1.20	9.0
			Oolong tea	GC-MS	1.07	1.20	1.06	1.21	0.99	1.10	1.11	1.08	1.11	1.09	1.31	1.22	1.10	1.1	1.16	1.00	1.11	1.16	1.25	7.3
				GC-MS/MS	1.11	1.13	1.05	1.21	1.21	1.34	1.19	1.38	1.21	0.98	1.21	1.21	1.11	1.17	1.15	1.11	1.18	1.25	1.07	8.1
		ENVI-Carb + PSA	Green tea	GC-MS	0.94	0.96	0.99	1.18	0.97	1.06	1.20	1.17	1.16	1.01	1.35	1.02	1.08	1.05	0.93	1.18	1.03	1.02	1.01	10.3
				GC-MS/MS	0.82	1.06	0.97	1.10	1.03	1.07	1.01	1.02	1.14	0.94	1.22	1.12	1.06	1.08	1.18	1.01	0.98	0.92	0.94	9.4
			Oolong tea	GC-MS	1.14	1.13	1.01	1.05	1.16	1.06	1.01	1.11	1.37	1.14	1.15	1.04	1.01	1.09	1.18	1.15	1.02	1.03	1.09	8.0
				GC-MS/MS	1.56	1.14	0.99	1.09	1.22	1.04	1.58	1.26	1.26	1.08	1.01	1.03	1.07	1.01	1.10	1.11	0.98	1.08	1.02	15.3
115	Oxyfluorfen	TPT	Green tea	GC-MS	0.93	1.03	1.20	1.20	1.15	1.05	0.96	1.00	0.70	1.11	1.19	1.10	1.14	0.95	0.96	0.95	0.89	0.97	1.08	12.2
				GC-MS/MS	0.77	0.86	1.06	1.39	1.03	1.07	0.92	1.34	1.10	1.05	1.21	1.10	0.99	0.93	1.06	1.19	1.19	1.07	0.95	14.4
			Oolong tea	GC-MS	1.05	1.03	1.02	1.09	1.18	1.04	0.96	1.06	0.97	1.06	1.16	1.04	1.19	0.98	1.05	1.01	0.96	0.98	1.09	6.7
				GC-MS/MS	1.05	0.87	0.92	1.51	1.24	1.09	1.30	1.01	1.31	1.17	1.16	1.19	1.09	0.9	1.09	0.99	1.29	0.99	0.98	14.9
		ENVI-Carb + PSA	Green tea	GC-MS	1.26	1.01	0.93	0.92	0.98	1.11	1.17	1.14	1.31	0.97	1.31	1.21	1.13	1.06	1.14	1.21	1.25	1.11	0.89	12.0
				GC-MS/MS	0.74	0.82	1.04	1.18	1.06	1.47	2.34	1.06	1.05	0.95	1.28	1.04	1.09	1.04	1.06	1.14	1.09	0.93	0.94	29.8
			Oolong tea	GC-MS	0.87	0.97	1.07	0.95	1.40	1.05	0.88	1.17	1.11	1.11	1.28	0.90	1.13	0.89	0.86	1.41	1.06	1.04	1.09	15.5
				GC-MS/MS	1.12	0.97	1.04	1.00	1.39	1.11	0.99	1.34	1.07	1.01	1.07	0.89	1.19	1.14	1.03	1.21	0.96	1.03	1.04	11.6
116	Pendimethalin	TPT	Green tea	GC-MS	0.99	1.01	1.15	1.18	1.05	1.04	0.96	0.95	0.59	1.17	1.17	1.09	1.10	0.96	0.94	0.99	0.88	0.90	1.10	13.7
				GC-MS/MS	1.02	1.01	1.05	1.10	1.12	1.04	0.98	1.07	0.87	1.06	1.12	1.02	1.08	1.01	1.09	0.95	0.91	0.95	1.08	6.8
			Oolong tea	GC-MS	1.05	0.99	1.01	1.05	1.32	1.15	1.32	1.02	1.38	1.21	1.24	1.30	1.12	0.9	1.13	0.97	1.20	1.01	0.98	12.9
				GC-MS/MS	1.17	1.01	1.08	1.02	1.02	1.16	1.35	1.08	1.39	1.03	1.29	1.25	1.18	1.01	1.13	1.10	1.18	1.10	0.91	11.2
		ENVI-Carb + PSA	Green tea	GC-MS	0.90	0.97	1.03	1.36	1.07	1.06	1.07	1.28	1.01	1.06	1.27	1.17	1.07	0.82	1.10	1.19	1.20	0.95	1.00	12.6
				GC-MS/MS	0.86	0.71	1.01	1.09	1.09	1.01	2.71	1.06	0.98	0.95	1.23	1.00	1.11	1.05	1.11	1.15	1.09	0.95	0.86	36.7
			Oolong tea	GC-MS	1.02	1.08	1.16	1.02	1.57	1.06	0.88	1.36	1.18	1.15	1.28	0.92	1.13	0.87	0.82	1.52	1.05	1.08	1.11	18.0
				GC-MS/MS	1.20	1.06	1.12	1.06	1.32	1.10	0.96	1.34	1.16	1.08	1.04	0.89	1.13	1.17	1.10	1.10	1.00	1.07	1.07	9.8

(Continued)

No.	Compound	Cleanup	Tea	Detection	1	2	3	4	5	6	7	8	9	10	11	12	13	14	15	16	17	18	19	RSD
117	Picoxystrobin	TPT	Green tea	GC-MS	1.04	1.04	1.12	1.09	1.10	1.06	1.09	1.08	1.07	1.04	1.14	1.07	1.06	0.97	1.11	0.99	1.02	1.04	1.14	4.2
				GC-MS/MS	1.04	1.04	1.15	1.09	1.09	1.06	0.99	1.10	1.02	1.00	1.12	1.03	1.16	1.02	1.10	1.00	1.03	1.02	1.14	5.1
			Oolong tea	GC-MS	1.04	1.07	1.08	1.10	0.98	1.09	1.10	1.07	1.12	1.08	1.20	1.13	1.14	1.0	1.10	1.11	1.16	0.99	1.02	5.3
				GC-MS/MS	0.99	1.04	1.10	1.09	0.99	1.11	1.12	1.07	1.12	1.06	1.20	1.12	1.26	1.08	1.13	1.10	1.11	1.00	0.99	6.5
		ENVI-Carb + PSA	Green tea	GC-MS	1.06	1.16	1.05	1.10	1.02	1.11	1.05	1.05	1.04	1.02	1.06	1.11	1.08	1.14	1.10	1.24	1.12	1.16	0.99	5.5
				GC-MS/MS	1.05	1.19	1.03	1.08	1.03	1.22	1.02	1.07	1.06	1.00	1.07	1.02	1.10	1.19	1.12	1.21	1.14	1.10	0.97	6.7
			Oolong tea	GC-MS	1.10	1.06	1.00	1.04	1.18	1.08	1.09	1.07	1.17	1.18	1.09	1.07	1.07	1.00	1.20	1.11	1.07	1.15	1.04	5.4
				GC-MS/MS	1.10	1.06	1.02	1.02	1.14	1.08	1.02	1.07	1.19	1.17	0.96	1.04	1.06	1.09	1.23	1.10	0.98	1.13	1.05	6.5
118	Piperophos	TPT	Green tea	GC-MS	0.97	1.05	1.12	1.06	1.18	1.01	1.03	1.04	0.83	1.16	1.26	0.99	1.05	1.00	1.03	0.98	0.93	0.92	1.08	9.5
				GC-MS/MS	1.03	1.03	0.99	1.06	1.05	1.02	1.01	1.13	1.00	1.02	1.21	1.00	0.98	0.95	1.06	0.99	0.98	0.97	1.13	6.4
			Oolong tea	GC-MS	1.05	1.10	0.97	1.04	1.05	1.06	1.06	1.09	1.16	1.19	1.24	1.08	1.12	1.0	1.07	1.11	1.27	1.04	1.08	7.1
				GC-MS/MS	1.15	0.94	0.99	1.07	1.00	1.15	1.17	1.09	1.44	0.99	1.21	1.05	1.17	1.10	1.15	1.15	1.31	1.12	1.02	10.6
		ENVI-Carb + PSA	Green tea	GC-MS	0.88	1.00	1.03	1.14	0.97	1.06	1.13	1.24	1.05	1.09	1.20	1.09	1.14	1.02	1.11	1.13	1.10	0.92	0.98	8.6
				GC-MS/MS	0.77	0.89	1.03	1.00	1.06	1.23	1.28	1.01	1.14	0.93	1.23	1.14	1.15	1.29	1.01	1.10	0.95	1.00	0.93	13.1
			Oolong tea	GC-MS	1.00	1.02	1.05	1.03	1.35	1.10	0.93	1.27	1.14	1.19	1.34	0.84	1.11	0.98	1.12	1.18	1.11	1.07	1.07	11.8
				GC-MS/MS	1.19	1.00	0.99	1.07	1.27	1.07	1.24	1.32	1.18	1.18	0.93	1.01	1.09	1.09	1.16	1.15	1.05	1.05	1.18	9.3
119	Pirimiphoe-ethyl	TPT	Green tea	GC-MS	1.02	1.01	1.13	1.11	1.10	1.04	1.08	1.08	0.95	1.14	1.20	1.08	1.06	1.05	1.09	1.01	1.02	1.03	1.12	5.5
				GC-MS/MS	1.06	1.20	1.03	1.09	1.05	1.16	1.09	1.07	1.05	1.08	1.05	1.22	1.08	1.11	1.06	1.21	1.10	1.13	1.01	5.6
			Oolong tea	GC-MS	1.01	1.01	1.15	1.09	1.10	1.05	0.97	1.06	1.05	1.10	1.08	1.06	1.09	1.07	1.07	1.02	1.00	0.98	1.15	4.8
				GC-MS/MS	1.05	0.97	1.04	1.03	1.02	1.08	1.22	1.11	1.28	1.22	1.26	1.13	1.19	1.1	1.12	1.11	1.16	1.02	1.05	7.9
		ENVI-Carb + PSA	Green tea	GC-MS	1.04	1.04	1.19	1.07	0.99	1.16	1.21	1.13	1.16	1.08	1.26	1.15	1.22	1.09	1.12	1.12	1.13	1.04	1.03	6.5
				GC-MS/MS	1.05	1.21	1.04	1.06	1.04	1.40	1.04	1.03	1.04	0.96	1.07	1.04	1.13	1.19	1.09	1.22	1.06	1.08	1.04	9.1
			Oolong tea	GC-MS	1.12	1.10	1.11	1.05	1.29	1.11	1.10	1.16	1.22	1.24	1.23	1.02	1.08	1.08	1.20	1.12	1.11	1.16	1.06	6.4
				GC-MS/MS	1.12	1.10	1.08	1.09	1.19	1.09	1.09	1.13	1.20	1.16	0.98	1.08	1.09	1.08	1.23	1.12	1.03	1.13	0.99	5.8
120	Pirimiphos-methyl	TPT	Green tea	GC-MS	1.00	1.02	1.11	1.09	1.08	1.05	1.09	1.08	0.90	1.09	1.20	1.09	1.06	0.98	0.98	1.02	1.01	1.01	1.12	6.5
				GC-MS/MS	1.04	1.18	0.99	1.08	1.03	1.13	0.96	1.10	1.07	1.06	1.05	1.16	1.07	1.07	1.02	1.19	1.10	1.10	1.00	5.7
			Oolong tea	GC-MS	1.03	1.03	1.09	1.09	1.05	1.07	1.00	1.08	1.01	1.07	1.26	1.04	0.99	0.97	1.10	0.99	1.05	1.05	1.05	6.4
				GC-MS/MS	1.05	1.06	1.11	1.11	1.03	1.11	1.22	1.12	1.23	1.17	1.19	1.11	1.17	1.1	1.10	1.09	1.02	1.02	1.06	5.5
		ENVI-Carb + PSA	Green tea	GC-MS	1.06	1.12	1.17	1.10	1.01	1.15	1.20	1.10	1.17	0.98	1.26	1.17	1.18	1.07	1.20	1.12	1.06	1.05	1.09	6.4
				GC-MS/MS	0.96	1.20	1.01	1.02	1.05	1.28	1.12	1.05	1.17	1.00	1.16	1.12	1.11	1.23	1.08	1.13	1.11	1.07	0.99	7.8
			Oolong tea	GC-MS	1.14	1.12	1.13	1.08	1.24	1.08	1.01	1.15	1.20	1.19	1.19	0.97	1.09	1.06	1.19	1.13	1.09	1.13	1.06	6.1
				GC-MS/MS	1.24	1.09	1.08	1.08	1.17	1.09	1.10	1.13	1.19	1.06	0.99	1.10	1.12	1.08	1.21	1.12	0.94	1.15	1.05	6.6

Appendix Table 5.3 Youden Pair Ratios of Each Average Content From 201 Pesticides and RSD for 19 Circulative Determinations (cont.)

No.	Pesticides	SPE	Sample	Method	Nov 9, 2009 (n = 5)	Nov 14, 2009 (n = 3)	Nov 19, 2009 (n = 3)	Nov 24, 2009 (n = 3)	Nov 29, 2009 (n = 3)	Dec 4, 2009 (n = 3)	Dec 9, 2009 (n = 3)	Dec 14, 2009 (n = 3)	Dec 19, 2009 (n = 3)	Dec 24, 2009 (n = 3)	Dec 14, 2009 (n = 3)	Jan 3, 2010 (n = 3)	Jan 8, 2010 (n = 3)	Jan 13, 2010 (n = 3)	Jan 18, 2010 (n = 3)	Jan 23, 2010 (n = 3)	Jan 28, 2010 (n = 3)	Feb 2, 2010 (n = 3)	Feb 7, 2010 (n = 3)	RSD%
121	Profenofos	TPT	Green tea	GC-MS	1.02	1.11	1.11	1.14	0.97	1.11	1.03	1.02	0.78	1.05	1.13	1.07	0.93	1.02	0.84	0.92	0.92	0.87	1.13	10.6
				GC-MS/MS	1.23	1.18	1.03	0.96	0.79	1.02	1.06	0.90	0.98	0.78	1.05	1.08	0.91	0.79	1.00	0.88	0.72	1.11	1.46	17.9
			Oolong tea	GC-MS	1.06	1.23	1.04	1.24	1.21	1.14	1.33	0.98	0.99	1.00	1.30	1.38	1.25	0.9	1.05	0.99	1.28	1.13	1.03	12.2
				GC-MS/MS	1.13	1.18	1.02	1.23	1.11	1.42	1.00	1.62	1.24	0.84	1.35	1.02	1.20	1.05	1.29	1.09	1.38	1.23	0.56	19.9
		ENVI-Carb + PSA	Green tea	GC-MS	1.02	1.01	1.11	1.16	1.06	1.10	1.08	1.14	1.11	1.21	1.50	1.06	1.06	1.01	0.92	1.27	1.20	0.94	0.94	12.2
				GC-MS/MS	0.96	0.82	1.19	1.05	1.21	1.31	1.39	1.14	1.30	1.02	1.21	1.02	1.24	1.26	0.91	1.24	1.12	1.12	1.02	13.2
			Oolong tea	GC-MS	1.05	1.12	1.00	1.01	1.26	0.99	0.93	1.14	1.42	1.28	1.22	1.00	0.99	0.94	1.06	1.21	1.02	0.94	1.08	12.3
				GC-MS/MS	1.16	1.09	0.95	0.98	1.25	0.81	1.34	1.17	1.53	1.05	0.99	0.92	1.06	1.01	1.00	1.09	0.92	1.08	1.35	16.1
122	Profluralin	TPT	Green tea	GC-MS	1.03	1.02	1.12	1.18	1.04	1.06	1.06	1.10	0.74	1.06	1.11	1.12	1.09	1.01	1.01	1.01	1.00	1.03	1.13	8.5
				GC-MS/MS	1.02	1.02	1.14	1.08	1.04	1.09	1.02	1.04	0.87	1.02	1.07	1.04	1.11	0.93	1.08	0.96	0.96	0.96	1.14	6.9
			Oolong tea	GC-MS	1.07	1.05	1.10	1.12	1.28	1.20	1.28	1.11	1.28	1.13	1.29	1.26	1.12	0.9	1.18	1.00	1.18	1.02	1.00	9.5
				GC-MS/MS	1.04	1.06	1.11	1.06	1.03	1.18	1.26	1.16	1.20	1.08	1.28	1.30	1.14	1.07	1.15	1.04	1.16	1.05	0.97	8.2
		ENVI-Carb + PSA	Green tea	GC-MS	1.12	1.06	1.05	1.15	1.05	1.08	1.73	1.02	1.07	1.11	1.18	1.16	1.04	1.02	1.04	1.18	1.22	1.05	1.04	14.1
				GC-MS/MS	1.01	0.94	1.04	1.04	1.12	0.96	3.87	1.04	1.09	0.99	1.20	1.03	1.04	1.41	1.06	1.23	1.08	0.94	1.03	53.6
			Oolong tea	GC-MS	1.17	1.15	1.17	1.08	1.42	1.10	1.08	1.20	1.26	1.18	1.21	1.01	1.16	0.95	0.92	1.47	1.04	1.09	1.09	12.1
				GC-MS/MS	1.17	1.12	1.14	1.07	1.27	1.07	0.74	1.22	1.19	1.10	1.03	0.97	1.14	1.14	1.15	1.04	1.03	1.07	1.03	10.4
123	Propachlor	TPT	Green tea	GC-MS	1.02	1.05	1.11	0.97	1.06	1.11	1.07	1.07	0.91	1.03	1.16	0.99	0.97	1.01	1.03	1.01	0.92	0.91	1.07	6.7
				GC-MS/MS	0.98	1.03	0.86	1.04	0.97	1.06	1.00	1.14	1.04	0.94	1.15	1.02	0.98	0.92	1.06	0.98	0.99	0.96	1.11	7.3
			Oolong tea	GC-MS	1.00	1.27	1.16	1.13	0.96	1.05	1.20	1.10	1.04	1.14	1.18	1.18	1.14	1.0	1.09	1.08	1.16	1.02	1.16	7.0
				GC-MS/MS	1.07	1.22	1.16	1.14	1.04	1.21	1.28	1.27	1.18	1.02	1.21	1.12	1.21	1.12	1.12	1.14	1.11	1.10	1.04	6.6
		ENVI-Carb + PSA	Green tea	GC-MS	0.99	1.12	0.96	1.06	0.94	1.05	1.20	1.25	1.05	0.99	1.12	0.97	1.05	1.05	0.98	1.24	1.11	1.05	1.02	8.5
				GC-MS/MS	0.79	1.09	0.98	0.99	0.99	1.50	0.90	1.00	1.06	0.91	1.12	1.06	1.09	1.35	1.05	1.07	1.01	1.01	0.96	14.9
			Oolong tea	GC-MS	1.19	1.13	1.08	1.12	1.14	1.08	0.83	1.07	1.17	1.22	1.07	1.08	1.01	1.05	1.15	1.10	1.08	1.16	0.99	7.9
				GC-MS/MS	1.64	1.10	0.99	1.15	1.20	1.08	1.45	1.17	1.21	1.14	0.97	1.07	1.09	1.04	1.22	1.14	0.94	1.16	1.03	14.4
124	Propicon-azole	TPT	Green tea	GC-MS	1.05	1.07	1.09	1.08	1.15	1.01	1.08	1.11	0.95	1.04	1.10	1.04	1.06	0.96	1.11	1.04	1.06	1.04	1.16	5.1
				GC-MS/MS	1.03	1.07	1.15	1.06	1.16	1.12	0.99	1.08	1.05	0.96	0.92	1.07	1.11	0.99	1.08	0.87	1.07	1.01	1.15	7.4
			Oolong tea	GC-MS	1.10	0.89	1.03	1.12	1.01	1.07	1.04	1.02	1.31	1.13	1.12	1.08	1.09	1.0	1.12	1.13	1.18	1.06	1.08	7.8
				GC-MS/MS	1.01	1.07	1.11	1.12	1.02	1.13	1.07	1.05	1.15	1.06	1.19	1.09	1.15	1.11	1.16	1.11	1.12	1.09	1.13	4.3
		ENVI-Carb + PSA	Green tea	GC-MS	1.02	1.17	1.09	1.11	1.02	1.20	0.92	1.02	1.12	1.17	1.01	1.21	1.07	1.16	1.07	1.24	1.17	1.20	1.00	8.0
				GC-MS/MS	1.05	1.27	1.27	1.06	1.01	1.14	1.03	1.09	1.12	1.07	1.08	1.03	1.12	1.18	1.09	1.18	1.16	1.15	1.01	7.0
			Oolong tea	GC-MS	1.07	1.04	1.04	1.00	1.45	1.10	1.14	1.21	1.21	1.22	1.32	0.95	1.05	1.02	1.30	1.11	1.13	1.15	1.05	11.1
				GC-MS/MS	1.07	1.06	0.97	1.00	1.24	1.05	1.05	1.07	1.20	1.20	1.00	1.08	1.07	1.09	1.23	1.11	1.03	1.17	1.04	7.3

125	Propyzamide	TPT	Green tea	GC-MS	1.01	1.03	1.18	1.13	1.10	1.05	1.08	1.14	0.97	1.09	1.16	1.08	1.07	0.98	1.08	1.01	1.01	1.02	1.13	5.9
				GC-MS/MS	1.04	1.03	1.15	1.10	1.09	1.08	0.98	1.13	1.02	1.02	1.05	1.09	1.10	1.00	1.11	1.01	1.02	0.98	1.14	5.1
			Oolong tea	GC-MS	1.09	1.07	1.10	1.13	1.07	1.16	1.15	1.19	1.21	1.08	1.24	1.21	1.15	1.1	1.13	1.12	1.17	1.06	1.05	5.1
				GC-MS/MS	1.02	1.11	1.15	1.15	1.02	1.19	1.17	1.13	1.18	1.06	1.23	1.23	1.13	1.15	1.17	1.10	1.12	1.09	1.05	5.5
		ENVI-Carb + PSA	Green tea	GC-MS	1.06	1.13	1.02	1.08	1.04	1.12	1.22	1.06	1.03	1.03	1.09	1.11	1.10	1.12	1.09	1.24	1.12	1.08	0.96	6.0
				GC-MS/MS	1.05	1.17	1.03	1.08	1.05	0.82	1.09	1.07	1.05	0.99	1.12	1.04	1.16	1.19	1.11	1.21	1.12	1.05	0.98	8.2
			Oolong tea	GC-MS	1.22	1.17	1.07	1.08	1.25	1.08	1.07	1.16	1.49	1.14	1.13	1.11	1.10	1.09	1.19	1.17	1.06	1.15	1.16	8.5
				GC-MS/MS	1.26	1.19	1.07	1.09	1.21	1.10	1.00	1.16	1.39	1.08	1.01	1.10	1.11	1.08	1.26	1.10	0.98	1.15	1.10	8.9
126	Ronnel	TPT	Green tea	GC-MS	0.99	1.04	1.10	1.07	1.05	1.08	1.09	1.09	0.89	1.08	1.16	1.04	1.01	1.03	1.01	1.02	1.01	0.97	1.10	5.7
				GC-MS/MS	1.01	1.01	0.96	1.04	0.98	1.05	1.03	1.10	1.00	1.02	1.17	1.06	1.00	0.99	1.04	1.01	0.99	0.95	1.10	5.0
			Oolong tea	GC-MS	1.08	1.22	1.20	1.16	1.10	1.16	1.32	1.24	1.15	1.24	1.23	1.18	1.30	1.1	1.15	1.13	1.21	1.08	1.14	5.8
				GC-MS/MS	1.17	1.21	1.21	1.18	1.09	1.22	1.26	1.33	1.24	1.03	1.22	1.14	1.25	1.18	1.21	1.15	1.21	1.16	1.10	5.7
		ENVI-Carb + PSA	Green tea	GC-MS	1.03	1.14	1.00	1.05	1.01	1.07	0.95	1.19	1.05	1.05	1.12	1.05	1.06	1.08	0.94	1.16	1.10	1.02	1.02	6.1
				GC-MS/MS	0.91	1.12	1.01	1.01	1.06	1.95	1.16	1.03	1.09	0.96	1.10	1.02	1.11	1.28	1.04	1.14	1.07	0.98	0.97	19.9
			Oolong tea	GC-MS	1.21	1.18	1.12	1.17	1.23	1.12	0.96	1.18	1.26	1.27	1.20	1.07	1.10	1.11	1.20	1.16	1.12	1.11	1.09	6.4
				GC-MS/MS	1.44	1.14	1.10	1.17	1.24	1.11	1.34	1.19	1.27	1.14	1.03	1.07	1.14	1.11	1.30	1.15	1.04	1.17	1.06	9.2
127	Sulfotep	TPT	Green tea	GC-MS	0.99	1.02	1.07	1.09	1.08	1.07	1.09	1.09	0.89	1.01	1.19	1.07	1.05	1.05	1.05	1.03	1.03	1.00	1.15	6.1
				GC-MS/MS	1.03	1.01	1.13	1.07	1.08	1.08	0.98	1.07	1.02	1.05	1.18	1.04	1.12	1.04	1.09	1.01	1.02	0.98	1.15	5.2
			Oolong tea	GC-MS	1.06	1.05	1.12	1.11	1.07	1.13	1.19	1.12	1.15	1.14	1.17	1.10	1.19	1.1	1.13	1.11	1.21	1.05	1.09	4.4
				GC-MS/MS	1.01	1.12	1.14	1.09	1.04	1.15	1.23	1.13	1.16	1.07	1.20	1.15	1.15	1.14	1.16	1.12	1.05	1.05	1.08	5.1
		ENVI-Carb + PSA	Green tea	GC-MS	1.06	1.18	1.01	1.09	1.02	1.17	1.22	1.09	1.06	1.09	1.06	1.15	1.08	1.09	1.03	1.13	1.07	1.10	1.03	5.3
				GC-MS/MS	1.01	1.22	1.01	1.07	1.04	1.47	1.13	1.03	1.08	1.02	1.07	1.02	1.09	1.11	1.09	1.15	1.10	1.01	1.00	10.0
			Oolong tea	GC-MS	1.23	1.11	1.12	1.09	1.19	1.13	1.06	1.07	1.28	1.19	1.19	1.07	1.09	1.06	1.22	1.15	1.13	1.14	1.06	5.8
				GC-MS/MS	1.31	1.13	1.10	1.07	1.19	1.08	1.06	1.12	1.18	1.16	0.99	1.07	1.08	1.08	1.26	1.12	1.03	1.18	1.01	7.4
128	Tebufenpyrad	TPT	Green tea	GC-MS	1.06	1.04	1.11	1.08	1.13	1.08	1.08	1.07	1.07	1.01	1.14	1.08	1.06	1.05	1.11	1.01	1.03	1.04	1.13	3.6
				GC-MS/MS	1.06	1.04	1.16	1.09	1.11	1.08	0.97	1.06	1.04	1.01	1.07	1.05	1.12	1.03	1.12	1.00	1.00	1.02	1.17	5.2
			Oolong tea	GC-MS	1.08	1.09	1.09	1.12	0.98	1.08	1.03	1.04	1.15	1.10	1.02	1.13	1.10	1.1	1.14	1.15	1.19	1.07	1.11	4.9
				GC-MS/MS	1.04	1.05	1.12	1.13	0.98	1.11	1.08	1.04	1.16	1.08	1.11	1.10	1.15	1.09	1.17	1.14	1.10	1.08	1.14	4.9
		ENVI-Carb + PSA	Green tea	GC-MS	1.06	1.15	1.02	1.13	1.20	1.11	1.28	1.06	1.10	1.20	1.02	1.06	1.11	1.13	1.07	1.22	1.14	1.20	1.01	6.7
				GC-MS/MS	1.02	0.91	1.03	1.07	1.12	1.16	1.07	1.10	1.12	1.16	1.11	1.05	1.12	1.10	1.09	1.18	1.16	1.10	1.14	6.3
			Oolong tea	GC-MS	1.08	1.10	0.99	1.09	1.20	1.14	1.19	1.06	1.17	1.20	1.15	1.10	1.11	1.07	1.23	1.16	1.17	1.15	1.09	5.3
				GC-MS/MS	1.08	1.08	0.98	1.07	1.12	1.12	1.08	1.10	1.17	1.16	1.01	1.13	1.12	1.16	1.25	1.14	1.05	1.15	1.05	5.5

(Continued)

Appendix Table 5.3 Youden Pair Ratios of Each Average Content From 201 Pesticides and RSD for 19 Circulative Determinations (cont.)

No.	Pesticides	SPE	Sample	Method	Nov 9, 2009 (n=5)	Nov 14, 2009 (n=3)	Nov 19, 2009 (n=3)	Nov 24, 2009 (n=3)	Nov 29, 2009 (n=3)	Dec 4, 2009 (n=3)	Dec 9, 2009 (n=3)	Dec 14, 2009 (n=3)	Dec 19, 2009 (n=3)	Dec 24, 2009 (n=3)	Dec 14, 2009 (n=3)	Jan 3, 2010 (n=3)	Jan 8, 2010 (n=3)	Jan 13, 2010 (n=3)	Jan 18, 2010 (n=3)	Jan 23, 2010 (n=3)	Jan 28, 2010 (n=3)	Feb 2, 2010 (n=3)	Feb 7, 2010 (n=3)	RSD%
129	Terbutryn	TPT	Green tea	GC-MS	1.03	1.02	1.14	1.11	1.11	1.03	1.08	1.09	0.97	1.09	1.19	1.08	1.07	1.11	1.08	1.00	1.17	1.04	1.12	5.2
				GC-MS/MS	1.07	1.02	1.17	1.10	1.06	1.07	0.96	1.06	1.04	1.00	1.15	1.03	1.11	1.04	1.12	1.03	0.94	0.95	1.11	6.1
			Oolong tea	GC-MS	1.05	1.00	1.06	1.06	1.02	1.10	1.18	1.13	1.23	1.21	1.28	1.17	1.18	1.1	1.16	1.13	1.19	1.00	1.04	7.3
				GC-MS/MS	0.99	1.05	1.14	1.11	1.01	1.16	1.23	1.09	1.19	1.07	1.20	1.21	1.19	1.08	1.14	1.13	1.16	1.05	1.03	6.5
		ENVI-Carb + PSA	Green tea	GC-MS	1.06	1.20	1.02	1.11	1.04	1.14	1.04	1.05	1.03	1.07	1.05	1.22	1.09	1.13	1.10	1.25	1.14	1.18	1.00	6.6
				GC-MS/MS	1.07	1.26	1.04	1.08	1.01	1.13	1.07	1.05	1.05	0.98	1.13	1.03	1.22	1.30	1.15	1.22	1.13	1.07	0.96	8.4
			Oolong tea	GC-MS	1.10	1.08	1.09	1.07	1.28	1.11	1.09	1.12	1.23	1.22	1.21	1.04	1.08	1.09	1.21	1.09	1.13	1.21	1.07	6.2
				GC-MS/MS	1.09	1.07	1.05	1.08	1.20	1.09	1.04	1.10	1.21	1.19	0.99	1.07	1.08	1.15	1.30	1.13	1.02	1.25	1.03	7.5
130	Thiobencarb	TPT	Green tea	GC-MS	1.04	1.01	1.12	1.10	1.09	1.06	1.08	1.09	1.03	1.06	1.11	1.06	1.05	1.04	1.09	1.02	1.03	1.04	1.14	3.4
				GC-MS/MS	1.04	1.03	1.19	1.07	1.06	1.07	0.97	1.07	1.02	1.08	1.05	1.05	1.07	1.06	1.08	1.01	1.01	0.97	1.11	4.6
			Oolong tea	GC-MS	1.05	1.08	1.12	1.06	1.02	1.09	1.16	1.11	1.14	1.10	1.21	1.17	1.15	1.1	1.12	1.07	1.13	1.01	1.05	4.8
				GC-MS/MS	1.00	1.10	1.14	1.11	0.99	1.14	1.17	1.10	1.13	1.09	1.20	1.17	1.14	1.13	1.15	1.05	1.10	1.06	1.03	5.2
		ENVI-Carb + PSA	Green tea	GC-MS	1.08	1.20	0.98	1.08	1.02	1.12	0.88	1.04	1.07	1.03	1.05	1.09	1.06	1.11	1.08	1.21	1.10	1.15	1.01	7.1
				GC-MS/MS	1.06	1.23	1.03	1.03	1.05	1.11	1.06	1.06	1.07	1.01	1.11	1.06	1.16	1.21	1.08	1.15	1.07	1.07	1.05	5.5
			Oolong tea	GC-MS	1.20	1.11	1.07	1.06	1.19	1.13	1.09	1.08	1.17	1.19	1.14	1.05	1.07	1.07	1.17	1.12	1.10	1.15	1.06	4.6
				GC-MS/MS	1.13	1.12	1.09	1.06	1.19	1.08	1.00	1.12	1.13	1.15	0.98	1.07	1.09	1.07	1.21	1.05	1.03	1.11	1.03	5.5
131	Tralkoxydim	TPT	Green tea	GC-MS	1.18	1.03	1.13	1.13	1.14	1.10	1.10	1.06	1.07	0.98	1.17	1.09	1.11	1.02	1.05	1.01	0.99	1.11	1.13	5.3
				GC-MS/MS	1.06	1.05	1.12	1.10	1.15	1.13	1.01	1.05	1.05	1.02	1.11	1.03	1.15	1.06	1.09	1.06	1.03	1.00	1.09	4.3
			Oolong tea	GC-MS	1.03	1.09	1.04	1.12	0.99	1.20	0.92	1.07	1.15	0.98	1.12	1.21	1.08	1.0	1.12	1.14	1.14	1.21	1.08	7.4
				GC-MS/MS	1.14	1.06	1.13	1.10	0.95	1.21	0.92	1.05	1.16	0.93	1.09	1.20	1.02	1.18	1.13	1.09	1.06	1.16	1.01	8.1
		ENVI-Carb + PSA	Green tea	GC-MS	1.15	1.11	1.09	1.34	1.00	1.05	1.12	1.09	1.03	1.05	1.45	1.05	1.11	0.92	1.11	1.07	1.10	1.11	0.91	11.4
				GC-MS/MS	0.89	1.14	1.04	1.20	0.99	0.75	0.94	1.04	1.04	1.06	1.62	1.04	1.14	0.90	1.07	0.99	1.12	1.07	0.89	16.7
			Oolong tea	GC-MS	1.21	1.26	1.07	1.01	1.21	1.02	1.20	1.10	1.58	1.01	1.08	1.12	1.14	1.01	1.12	1.18	1.12	1.00	1.13	12.5
				GC-MS/MS	1.91	1.23	1.09	1.05	1.20	1.01	1.16	1.18	1.48	1.00	1.01	1.09	1.15	1.11	1.23	1.09	1.06	1.12	1.14	18.4
132	trans-Chlor-dane	TPT	Green tea	GC-MS	1.05	0.98	1.14	1.09	1.08	1.06	1.04	1.07	1.03	1.10	1.20	1.05	1.01	1.07	1.09	1.01	1.04	0.94	1.13	5.6
				GC-MS/MS	1.03	1.12	1.16	1.08	1.02	1.11	1.25	1.11	1.13	1.09	1.23	1.16	1.17	1.1	1.12	1.11	1.14	1.00	1.05	5.9
			Oolong tea	GC-MS	1.03	1.10	1.18	1.08	1.01	1.12	1.19	1.15	1.06	1.04	1.24	1.19	1.21	1.12	1.13	1.08	1.13	1.09	1.06	5.9
				GC-MS/MS	1.02	1.20	1.04	1.08	0.99	1.05	1.06	1.05	1.02	0.99	1.12	1.29	1.08	1.08	1.11	1.24	1.07	0.95	0.99	8.2
		ENVI-Carb + PSA	Green tea	GC-MS	1.17	1.12	1.08	1.07	1.20	1.08	1.05	1.13	1.22	1.16	1.10	1.10	1.07	1.06	1.18	1.12	1.11	1.13	1.03	4.6
				GC-MS/MS	1.15	1.13	1.06	1.12	1.16	1.06	1.06	1.09	1.21	1.16	1.02	1.09	1.05	1.11	1.28	1.08	1.00	1.22	1.03	6.6
			Oolong tea	GC-MS	1.04	1.03	1.11	1.08	1.06	1.07	1.08	1.08	0.98	1.04	1.10	1.07	1.06	1.02	1.07	1.01	1.01	1.01	1.13	3.6
				GC-MS/MS	1.03	1.17	1.02	1.07	1.04	1.08	0.81	1.06	1.05	0.99	1.05	1.05	1.08	1.11	1.08	1.21	1.09	1.05	0.99	7.6

133	trans-Diallate	TPT	Green tea	GC–MS	1.05	1.00	1.10	1.07	1.06	1.07	1.08	1.10	1.01	1.07	1.09	1.07	1.04	1.14	1.20	1.00	1.03	0.99	1.13	4.9
				GC–MS/MS	1.04	1.03	1.12	1.07	1.02	1.06	0.99	1.04	0.98	1.05	1.10	1.07	1.16	1.01	1.01	1.03	1.01	0.98	1.12	4.8
			Oolong tea	GC–MS	1.04	1.12	1.12	1.13	1.01	1.10	1.15	1.12	1.13	1.13	1.24	1.16	1.16	1.0	1.16	1.19	1.13	1.07	1.09	5.3
				GC–MS/MS	1.00	1.14	1.14	1.15	1.03	1.16	1.21	1.06	1.09	1.10	1.19	1.14	1.18	1.06	1.21	1.08	1.06	1.05	1.13	5.5
		ENVI-Carb + PSA	Green tea	GC–MS	1.11	1.18	1.01	1.08	1.05	1.11	0.83	1.06	1.06	1.04	1.06	1.12	1.02	1.07	1.05	1.20	1.10	1.11	1.02	7.2
				GC–MS/MS	1.08	1.19	1.03	1.05	1.03	1.09	1.31	1.05	1.09	1.03	1.14	1.01	1.10	1.11	1.10	1.19	1.13	1.03	1.00	7.0
			Oolong tea	GC–MS	1.21	1.13	1.08	1.08	1.20	1.09	1.14	1.03	1.09	1.19	1.13	1.07	1.09	1.13	1.19	1.21	1.13	1.17	1.08	4.7
				GC–MS/MS	1.21	1.12	1.07	1.07	1.17	1.08	1.04	1.12	1.18	1.17	0.98	1.05	1.11	1.10	1.24	1.10	1.04	1.16	1.02	6.2
134	Trifloxys-trobin	TPT	Green tea	GC–MS	0.99	1.07	1.12	1.10	1.02	1.06	1.10	1.06	0.79	1.05	1.08	1.15	1.07	0.96	0.95	1.00	1.02	0.99	1.12	7.8
				GC–MS/MS	1.04	1.08	1.11	1.06	1.00	1.08	1.07	1.03	0.81	0.96	1.10	1.07	1.18	0.97	1.02	1.03	0.94	0.90	1.21	9.1
			Oolong tea	GC–MS	1.04	1.08	0.98	1.13	1.12	1.14	0.96	1.04	1.23	1.08	1.22	1.14	1.08	1.0	1.12	1.11	1.16	1.05	1.03	6.9
				GC–MS/MS	1.12	1.04	1.04	1.18	1.04	1.16	0.97	1.16	1.21	1.02	1.18	1.11	1.14	1.06	1.13	1.06	1.18	1.11	1.06	6.0
		ENVI-Carb + PSA	Green tea	GC–MS	0.99	0.89	1.19	1.15	1.07	0.96	0.77	1.07	1.09	1.20	1.39	1.26	1.04	1.10	0.96	1.30	1.17	0.83	0.98	14.9
				GC–MS/MS	1.11	0.87	1.14	1.02	1.12	0.71	2.82	1.15	1.15	1.06	1.28		0.90	1.39	0.91	1.35	1.13	0.81	0.91	39.3
			Oolong tea	GC–MS	1.11	1.04	1.01	0.99	1.25	1.03	1.18	1.19	1.24	1.09	1.19	0.95	1.15	1.00	1.05	1.32	1.06	1.01	1.05	9.5
				GC–MS/MS	1.01	1.04	0.94	0.97	1.18	0.99	0.97	1.17	1.24	1.03	0.99	1.01	1.12	1.03	1.16	1.06	0.97	1.05	1.06	8.0
135	Zoxamide	TPT	Green tea	GC–MS	1.00	1.03	1.08	1.02	1.11	1.08	1.06	1.12	1.05	1.06	1.13	1.03	1.05	1.02	1.13	1.02	1.06	1.06	1.10	3.7
				GC–MS/MS	1.00	1.02	1.09	1.06	1.11	1.08	0.97	1.14	1.06	1.02	1.11	1.08	1.04	1.13	1.14	1.02	1.03	1.03	1.10	4.7
			Oolong tea	GC–MS	1.03	1.11	1.06	1.05	0.92	1.03	0.99	1.02	1.06	1.10	1.15	1.08	1.10	1.1	1.10	1.14	1.12	0.97	1.12	5.7
				GC–MS/MS	1.00	1.10	1.11	1.10	0.92	1.07	0.98	1.04	1.04	1.04	1.15	1.01	1.17	1.07	1.12	1.15	1.14	0.99	1.16	6.6
		ENVI-Carb + PSA	Green tea	GC–MS	0.98	1.10	1.01	1.12	0.96	1.09	0.89	1.16	1.10	1.03	1.00	1.01	1.15	1.07	1.06	1.13	1.10	1.00	0.96	7.0
				GC–MS/MS	1.06	1.07	1.00	1.10	0.99	0.98	1.17	1.08	1.15	1.02	1.10	1.01	1.14	1.06	1.04	1.12	1.05	1.09	0.97	5.5
			Oolong tea	GC–MS	1.14	1.04	0.85	1.04	1.11	1.14	1.01	0.99	1.08	1.29	1.03	1.11	1.05	1.06	1.28	1.04	1.06	1.10	0.96	9.4
				GC–MS/MS	1.05	1.04	0.88	1.07	1.05	1.13	1.06	1.00	1.05	1.25	0.92	1.08	1.06	1.02	1.40	1.15	0.97	1.11	0.92	11.2
136	2,4′-DDE	TPT	Green tea	GC–MS	1.04	1.08	1.12	1.11	1.13	1.07	1.13	1.09	1.07	1.07	1.14	1.05	1.03	1.07	1.16	1.02	1.04	1.05	1.12	3.8
				GC–MS/MS	1.05	1.00	1.13	1.00	1.11	1.10	1.12	1.05	1.01	1.14	1.15	1.05	1.05	1.07	1.10	1.08	1.03	1.00	1.19	5.1
			Oolong tea	GC–MS	1.23	1.20	1.22	1.15	1.07	1.15	1.26	1.16	1.18	1.18	1.26	1.24	1.25	1.1	1.17	1.17	1.21	1.05	1.13	4.9
				GC–MS/MS	1.06	1.16	1.24	1.18	1.01	1.21	1.31	1.17	1.15	1.17	1.30	1.28	1.31	1.17	1.19	1.17	1.27	1.19	1.10	6.8
		ENVI-Carb + PSA	Green tea	GC–MS	1.00	1.25	1.01	1.07	1.01	1.13	0.90	1.01	1.04	0.98	1.01	0.97	1.07	1.17	1.11	1.19	1.12	1.21	1.01	8.8
				GC–MS/MS	1.05	1.29	0.99	1.11	1.03	1.15	0.84	1.01	1.02	0.97	1.08	1.05	1.07	1.01	1.09	1.11	1.05	1.17	0.99	8.7
			Oolong tea	GC–MS	1.23	1.18	1.13	1.16	1.26	1.19	1.13	1.12	1.28	1.24	1.16	1.18	1.10	1.24	1.28	1.14	1.16	1.25	1.12	4.9
				GC–MS/MS	1.21	1.15	1.08	1.16	1.24	1.20	1.12	1.17	1.26	1.28	1.06	1.17	1.12	1.17	1.38	1.14	1.03	1.26	1.10	7.2

(Continued)

Appendix Table 5.3 Youden Pair Ratios of Each Average Content From 201 Pesticides and RSD for 19 Circulative Determinations (cont.)

No.	Pesticides	SPE	Sample	Method	Nov 9, 2009 (n = 5)	Nov 14, 2009 (n = 3)	Nov 19, 2009 (n = 3)	Nov 24, 2009 (n = 3)	Nov 29, 2009 (n = 3)	Dec 4, 2009 (n = 3)	Dec 9, 2009 (n = 3)	Dec 14, 2009 (n = 3)	Dec 19, 2009 (n = 3)	Dec 24, 2009 (n = 3)	Dec 14, 2009 (n = 3)	Jan 3, 2010 (n = 3)	Jan 8, 2010 (n = 3)	Jan 13, 2010 (n = 3)	Jan 18, 2010 (n = 3)	Jan 23, 2010 (n = 3)	Jan 28, 2010 (n = 3)	Feb 2, 2010 (n = 3)	Feb 7, 2010 (n = 3)	RSD%
137	Ametryn	TPT	Green tea	GC-MS	1.01	1.02	1.12	1.10	1.15	0.85	0.73	1.13	0.95	1.05	1.12	1.08	1.08	0.97	1.09	1.01	1.02	1.02	1.14	10.2
				GC-MS/MS	1.06	1.00	1.16	1.13	1.11	1.10	1.14	1.09	1.00	1.03	1.13	1.04	1.11	0.94	1.06	1.02	0.97	1.04	1.15	6.0
			Oolong tea	GC-MS	1.22	1.10	1.09	1.06	1.08	1.16	1.13	0.76	1.37	1.21	1.27	1.28	1.26	1.1	1.20	1.21	1.28	1.09	1.14	11.1
				GC-MS/MS	1.08	1.13	1.20	1.20	1.08	1.24	1.32	1.18	1.18	1.14	1.41	1.32	1.29	1.19	1.20	1.15	1.26	1.11	1.13	7.3
		ENVI-Carb + PSA	Green tea	GC-MS	1.06	1.18	1.03	1.08	1.07	1.28	0.78	1.46	2.56	1.00	1.02	1.47	1.11	1.07	1.08	1.24	1.13	1.15	0.99	30.7
				GC-MS/MS	1.08	1.24	1.06	1.10	1.03	1.11	1.04	1.08	1.06	0.98	1.11	1.06	1.08	1.22	1.13	1.22	1.15	1.08	0.99	6.4
			Oolong tea	GC-MS	1.22	1.10	1.14	1.12	1.33	1.12	0.54	1.34	1.27	1.16	1.26	1.04	1.19	1.22	1.32	1.24	1.20	1.28	1.15	14.9
				GC-MS/MS	1.14	1.16	1.12	1.19	1.25	1.25	1.13	1.23	1.29	1.20	1.09	1.20	1.12	1.22	1.37	1.30	1.13	1.25	1.13	6.2
138	Bifenthrin	TPT	Green tea	GC-MS	1.03	1.08	1.13	1.10	1.14	1.07	1.12	1.06	1.03	1.03	1.12	1.07	1.06	1.06	1.11	1.02	1.02	1.03	1.22	4.8
				GC-MS/MS	1.06	1.01	1.13	1.13	1.10	1.11	1.11	1.07	1.00	1.07	1.10	1.10	1.03	1.03	1.10	1.04	0.99	1.01	1.15	4.6
			Oolong tea	GC-MS	1.18	1.03	1.02	1.11	0.96	1.08	1.06	1.03	1.07	1.05	1.17	1.10	1.02	1.0	1.11	1.04	1.06	0.97	1.01	5.9
				GC-MS/MS	0.95	0.98	1.02	1.12	0.93	1.12	1.07	0.99	1.07	1.04	1.21	1.13	1.00	0.97	1.12	1.04	1.06	0.98	1.00	7.0
		ENVI-Carb + PSA	Green tea	GC-MS	1.08	1.19	0.97	1.09	0.98	1.08	1.08	1.07	1.09	1.03	1.06	1.01	1.07	1.11	1.04	1.23	1.12	1.12	1.00	6.0
				GC-MS/MS	1.08	1.20	1.02	1.10	1.03	1.07	1.06	1.09	1.10	1.01	1.07	1.21	1.15	1.20	1.08	1.24	1.15	1.10	1.03	6.3
			Oolong tea	GC-MS	1.18	1.11	0.99	1.03	1.08	1.07	1.10	1.02	1.08	1.12	1.06	1.01	1.06	1.10	1.09	1.12	1.07	1.13	1.03	4.5
				GC-MS/MS	1.15	1.13	0.94	1.05	1.09	1.11	1.05	1.04	1.07	1.11	0.97	0.99	0.98	1.05	1.16	1.13	0.98	1.10	1.00	6.1
139	Bitertanol	TPT	Green tea	GC-MS	1.01	1.03	1.00	1.06	1.18	0.99	1.05	1.04	0.98	1.05	1.16	1.11	1.02	1.01	1.03	1.06	1.06	1.03	1.15	5.5
				GC-MS/MS	1.03	1.00	1.11	1.24	1.17	1.06	1.08	1.10	1.01	1.06	1.13	1.13	1.08	0.95	1.10	1.04	1.03	1.05	1.12	6.1
			Oolong tea	GC-MS	1.09	0.93	0.94	1.05	1.05	1.17	1.02	0.98	1.40	1.21	1.26	1.19	1.08	1.1	1.18	1.26	1.23	1.10	1.18	10.7
				GC-MS/MS	1.24	1.12	1.14	1.15	1.03	1.21	1.07	1.04	1.30	1.14	1.34	1.24	1.11	1.14	1.15	1.21	1.25	1.13	1.12	7.2
		ENVI-Carb + PSA	Green tea	GC-MS	0.93	0.87	1.08	1.12	1.03	1.22	0.82	1.14	1.06	1.06	1.04	1.58	1.07	1.05	1.13	1.17	1.14	1.08	0.98	14.3
				GC-MS/MS	0.95	1.07	1.03	1.16	0.98	1.08	0.78	1.16	1.16	1.03	1.09	1.27	1.19	1.08	1.10	1.16	1.15	1.16	1.02	10.0
			Oolong tea	GC-MS	1.09	1.01	1.02	1.04	1.39	1.19	1.28	1.26	1.13	1.36	1.28	1.03	1.18	1.11	1.31	1.21	1.16	1.23	1.15	9.7
				GC-MS/MS	1.03	1.14	0.98	1.11	1.26	1.20	1.20	1.22	1.19	1.32	1.07	1.15	0.91	1.14	1.44	1.25	1.11	1.22	1.13	10.3
140	Boscalid	TPT	Green tea	GC-MS	0.97	1.14	1.09	1.01	0.95	1.11	1.08	1.02	0.80	1.01	1.07	1.20	1.05	0.91	0.85	1.01	1.02	0.99	1.15	9.8
				GC-MS/MS	1.00	1.10	1.03	1.04	0.94	1.10	1.02	1.06	0.75	0.99	1.06	1.32	1.03	0.83	1.05	1.03	0.95	0.97	1.11	11.3
			Oolong tea	GC-MS	1.06	1.06	0.98	1.30	1.17	1.33	0.89	1.03	1.15	0.94	1.37	1.31	0.95	1.0	1.22	1.15	1.20	1.25	1.08	13.1
				GC-MS/MS	1.43	1.13	1.06	1.22	1.09	1.33	0.94	1.05	1.32	0.96	1.38	1.27	0.97	1.03	1.18	1.20	1.21	1.33	1.25	12.7
		ENVI-Carb + PSA	Green tea	GC-MS	0.90	0.76	1.13	1.25	1.06	0.97	1.21	1.15	1.09	1.16	1.47	1.33	0.99	1.02	1.07	1.30	1.20	0.83	0.94	16.3
				GC-MS/MS	0.87	0.73	1.12	1.22	1.00	0.91	1.38	1.24	1.17	1.17	1.33	1.03	0.96	1.11	1.01	1.31	1.22	0.86	0.92	16.6
			Oolong tea	GC-MS	1.06	1.09	1.00	0.96	1.26	1.00	0.88	1.29	1.42	1.10	1.14	1.01	1.29	0.85	0.99	1.64	1.12	1.03	1.20	17.3
				GC-MS/MS	1.11	1.13	0.95	1.04	1.30	1.03	1.16	1.25	1.32	1.10	1.06	0.99	1.09	0.96	1.18	1.52	1.07	1.07	1.25	12.6

141	Butafenacil	TPT	Green tea	GC-MS	1.01	1.08	1.06	0.96	1.11	1.07	1.07	0.82	1.05	1.02	1.22	1.09	0.87	0.86	1.01	0.98	0.96	1.19	10.1
				GC-MS/MS	1.05	1.06	1.11	0.89	1.10	1.06	1.04	0.69	1.02	1.02	1.28	0.98	0.77	1.00	1.06	0.99	0.97	1.11	12.9
			Oolong tea	GC-MS	1.12	1.09	0.95	1.21	1.33	0.90	1.03	1.23	1.02	1.33	1.26	1.06	1.0	1.19	1.17	1.20	1.17	1.12	11.2
				GC-MS/MS	1.32	1.10	1.07	1.14	1.31	0.87	1.11	1.26	1.01	1.38	1.20	1.05	1.09	1.16	1.17	1.21	1.28	1.17	10.5
		ENVI-Carb + PSA	Green tea	GC-MS	0.85	0.69	1.25	1.11	0.99	1.58	1.02	1.13	1.21	1.49	1.53	0.97	1.06	0.99	1.35	1.18	0.70	0.94	22.5
				GC-MS/MS	1.01	0.69	1.24	1.06	0.88	2.27	1.26	1.23	1.23	1.25	1.16	0.82	1.45	0.84	1.47	1.29	0.70	0.92	30.8
			Oolong tea	GC-MS	1.12	1.07	0.98	1.25	1.03	1.29	1.23	1.31	1.17	1.12	0.98	1.27	0.79	1.01	1.37	1.06	1.04	1.17	13.0
				GC-MS/MS	1.00	1.14	0.95	1.21	1.05	1.17	1.16	1.30	1.18	1.05	1.04	1.32	0.91	1.18	1.55	0.96	1.09	1.17	13.5
142	Carbaryl	TPT	Green tea	GC-MS	1.12	1.04	1.10	0.96	1.15	1.25	1.08	1.15	1.10	1.20	1.03	1.10	0.92	1.05	1.09	1.09	1.03	1.25	7.7
				GC-MS/MS	1.19	1.03	1.05	0.84	1.19	1.37	1.14	1.13	0.93	1.24	1.06	1.18	0.92	0.97	1.17	0.98	1.00	1.19	12.0
			Oolong tea	GC-MS	1.22	1.42	1.39	1.16	1.15	1.08	1.20	1.08	1.19	1.16	1.18	1.28	1.2	1.23	1.25	1.18	1.07	1.18	9.1
				GC-MS/MS	1.00	1.28	1.50	1.03	1.20	1.13	1.24	1.20	1.11	1.18	1.23	1.43	1.31	1.22	1.18	1.11	1.11	1.06	13.4
		ENVI-Carb + PSA	Green tea	GC-MS	1.06	1.11	1.05	1.04	1.04	0.78	1.05	1.37	1.02	0.99	1.16	1.16	1.10	1.18	1.13	1.13	1.04	0.94	8.7
				GC-MS/MS	1.02	1.06	1.07	1.03	1.06	0.84	1.17	1.31	1.00	0.95	1.06	1.17	1.26	1.16	1.13	1.16	1.00	0.94	9.2
			Oolong tea	GC-MS	1.22	1.06	1.06	1.15	1.05	1.14	1.18	1.37	1.17	1.11	1.16	1.15	1.03	1.26	1.30	1.10	1.28	1.20	8.5
				GC-MS/MS	1.08	1.19	1.00	1.16	1.09	1.13	1.29	1.31	1.17	1.04	1.13	0.99	1.03	1.28	1.35	1.04	1.16	1.08	9.3
143	Chlorobenzilate	TPT	Green tea	GC-MS	1.02	1.03	1.12	1.01	1.07	1.05	1.01	0.80	1.00	1.07	1.15	1.08	0.94	0.93	1.01	0.93	0.96	1.17	8.5
				GC-MS/MS	1.05	1.08	1.08	1.00	1.09	1.10	1.05	0.81	1.01	1.07	1.23	1.05	0.94	1.08	1.05	0.94	0.97	1.13	9.0
			Oolong tea	GC-MS	1.11	1.10	1.04	1.14	1.21	1.18	1.00	1.24	1.09	1.28	1.13	1.09	1.0	1.19	1.13	1.24	1.09	1.02	7.4
				GC-MS/MS	1.24	1.11	1.12	1.10	1.26	1.10	1.08	1.25	1.08	1.43	1.31	1.15	1.09	1.17	1.16	1.25	1.19	1.14	7.8
		ENVI-Carb + PSA	Green tea	GC-MS	0.99	0.95	1.09	1.09	0.98	1.47	1.17	0.95	1.05	1.50	1.39	1.14	0.85	1.15	1.24	1.07	0.62	0.90	20.0
				GC-MS/MS	0.89	0.74	1.11	1.07	0.98	1.66	1.15	1.05	1.08	1.46	1.07	1.17	1.01	1.11	1.23	1.16	0.79	0.87	19.7
			Oolong tea	GC-MS	1.11	1.13	1.07	1.39	1.09	1.11	1.31	1.27	1.12	1.18	0.95	1.21	1.05	1.01	1.45	1.12	1.10	1.16	11.1
				GC-MS/MS	1.18	1.13	1.03	1.31	1.12	1.14	1.34	1.28	1.11	1.08	1.04	1.01	0.93	1.21	1.38	1.05	1.13	1.14	10.4
144	Chlorthal-dimethyl	TPT	Green tea	GC-MS	1.02	1.05	1.11	1.05	1.07	1.07	1.07	0.97	1.04	1.11	1.08	1.07	1.08	1.04	1.02	1.00	1.01	1.13	3.9
				GC-MS/MS	1.05	1.04	1.08	1.05	1.10	1.11	1.08	0.97	1.05	1.10	1.17	1.09	1.01	1.09	1.06	0.99	1.04	1.16	4.9
			Oolong tea	GC-MS	1.24	1.20	1.21	1.11	1.21	1.35	1.14	1.19	1.15	1.33	1.24	1.24	1.1	1.21	1.17	1.23	1.09	1.13	5.7
				GC-MS/MS	1.17	1.16	1.21	1.11	1.25	1.34	1.18	1.22	1.17	1.39	1.28	1.21	1.15	1.19	1.17	1.23	1.13	1.16	6.1
		ENVI-Carb + PSA	Green tea	GC-MS	1.05	1.17	1.00	1.04	1.06	1.08	1.07	1.04	1.02	1.09	1.01	1.08	1.06	1.07	1.18	1.10	1.01	1.00	4.7
				GC-MS/MS	1.03	1.12	1.03	1.03	1.07	1.01	1.08	1.06	1.01	1.08	1.10	1.09	1.09	1.09	1.18	1.12	1.05	1.00	4.2
			Oolong tea	GC-MS	1.24	1.18	1.15	1.30	1.17	1.12	1.14	1.30	1.24	1.16	1.18	1.16	1.17	1.22	1.25	1.17	1.22	1.13	4.6
				GC-MS/MS	1.23	1.18	1.13	1.28	1.19	1.15	1.24	1.29	1.24	1.09	1.15	1.14	1.13	1.33	1.23	1.10	1.24	1.11	5.9

(Continued)

Appendix Table 5.3 Youden Pair Ratios of Each Average Content From 201 Pesticides and RSD for 19 Circulative Determinations (cont.)

No.	Pesticides	SPE	Sample	Method	Youden pair ratio																			RSD%
					Nov 9, 2009 (n=5)	Nov 14, 2009 (n=3)	Nov 19, 2009 (n=3)	Nov 24, 2009 (n=3)	Nov 29, 2009 (n=3)	Dec 4, 2009 (n=3)	Dec 9, 2009 (n=3)	Dec 14, 2009 (n=3)	Dec 19, 2009 (n=3)	Dec 24, 2009 (n=3)	Dec 14, 2009 (n=3)	Jan 3, 2010 (n=3)	Jan 8, 2010 (n=3)	Jan 13, 2010 (n=3)	Jan 18, 2010 (n=3)	Jan 23, 2010 (n=3)	Jan 28, 2010 (n=3)	Feb 2, 2010 (n=3)	Feb 7, 2010 (n=3)	
145	Dibutyl	TPT	Green tea	GC-MS	1.33	1.21	1.17	1.17	1.12	1.23	1.26	1.20	1.19	1.17	1.30	1.26	1.24	1.1	1.24	1.18	1.34	1.17	1.26	5.0
				GC-MS/MS	1.10	1.17	1.17	1.19	1.10	1.25	1.27	1.17	1.22	1.17	1.38	1.30	1.17	1.18	1.22	1.18	1.22	1.19	1.15	5.5
			Oolong tea	GC-MS	1.33	1.18	1.15	1.14	1.28	1.21	1.27	1.20	1.26	1.29	1.23	1.15	1.20	1.24	1.31	1.25	1.27	1.33	1.24	4.8
				GC-MS/MS	1.34	1.20	1.11	1.18	1.23	1.23	1.18	1.23	1.26	1.27	1.11	1.13	1.17	1.20	1.42	1.27	1.17	1.25	1.15	6.4
		ENVI-Carb + PSA	Green tea	GC-MS	1.08	1.01	1.12	1.10	1.06	1.11	1.12	1.08	1.00	1.09	1.07	1.15	1.10	1.00	1.10	1.05	1.00	1.03	1.13	4.4
				GC-MS/MS	1.10	1.26	1.05	1.11	1.00	1.08	1.02	1.11	1.12	1.05	1.12	1.12	1.16	1.21	1.12	1.16	1.12	1.11	1.07	5.5
			Oolong tea	GC-MS	1.08	1.03	1.10	1.08	1.11	1.10	1.06	1.09	1.04	1.08	1.12	1.08	1.08	1.06	1.10	1.04	1.19	1.03	1.14	3.8
				GC-MS/MS	1.13	1.18	1.02	1.10	1.03	1.10	0.99	1.05	1.10	1.07	1.07	1.26	1.09	1.09	1.13	1.19	1.11	1.17	1.04	5.9
146	Diethofen-carb	TPT	Green tea	GC-MS	0.92	1.02	1.07	1.08	1.06	0.99	1.05	1.08	0.98	1.13	1.11	1.07	0.97	0.99	1.06	1.01	0.96	1.01	1.13	5.8
				GC-MS/MS	1.01	1.05	1.10	1.07	1.10	1.08	1.10	1.10	0.96	1.06	1.13	1.20	1.10	0.96	1.11	1.04	0.96	1.00	1.11	6.0
			Oolong tea	GC-MS	1.11	1.13	1.07	1.13	1.12	1.22	1.26	1.00	0.95	1.56	1.84	1.54	1.41	1.0	1.14	1.18	1.33	1.02	1.11	18.7
				GC-MS/MS	1.19	1.08	1.13	1.17	1.08	1.21	1.22	1.10	1.34	1.16	1.42	1.33	1.17	1.12	1.16	1.19	1.26	1.12	1.10	7.8
		ENVI-Carb + PSA	Green tea	GC-MS	0.86	1.13	1.03	1.16	1.01	1.13	0.96	1.15	0.96	1.01	1.24	1.38	1.22	1.00	1.11	1.19	1.15	0.82	0.94	13.0
				GC-MS/MS	0.89	1.00	1.06	1.15	1.04	1.07	0.96	1.12	1.07	1.01	1.18	1.07	1.16	1.11	1.12	1.19	1.15	0.98	0.95	7.9
			Oolong tea	GC-MS	1.11	1.05	1.06	1.07	1.34	1.21	1.14	1.37	1.47	1.60	1.55	1.48	1.44	1.45	1.16	1.31	1.16	1.22	1.12	13.9
				GC-MS/MS	1.14	1.13	1.04	1.11	1.28	1.22	1.12	1.34	1.27	1.27	1.11	1.11	1.12	1.11	1.31	1.25	1.05	1.23	1.10	7.9
147	Diflufenican	TPT	Green tea	GC-MS	1.01	1.05	1.11	1.07	1.09	1.08	1.03	1.04	0.93	1.05	1.09	1.13	1.04	0.99	1.03	0.99	0.92	0.96	1.15	6.0
				GC-MS/MS	1.05	1.05	1.09	1.08	1.09	1.11	1.07	1.05	0.92	1.00	1.08	1.22	1.05	1.06	1.09	1.04	0.96	1.01	1.10	5.8
			Oolong tea	GC-MS	1.12	1.11	1.04	1.20	1.03	1.10	1.05	1.01	1.22	1.09	1.40	1.23	1.09	1.1	1.17	1.16	1.19	1.10	1.08	8.2
				GC-MS/MS	1.23	1.10	1.13	1.21	1.06	1.20	1.10	1.06	1.29	1.09	1.37	1.26	1.11	1.13	1.16	1.19	1.22	1.19	1.11	7.1
		ENVI-Carb + PSA	Green tea	GC-MS	0.94	1.04	1.05	1.20	1.03	1.06	1.09	1.13	1.01	1.01	1.15	1.13	1.19	0.99	1.10	1.19	1.13	0.88	0.91	8.8
				GC-MS/MS	0.91	0.96	1.06	1.19	0.99	1.06	1.02	1.15	1.09	1.04	1.23	1.04	1.20	1.10	1.10	1.21	1.12	1.00	0.95	8.7
			Oolong tea	GC-MS	1.12	1.05	1.06	1.06	1.30	1.15	1.20	1.24	1.24	1.21	1.28	1.01	1.17	1.17	1.21	1.27	1.20	1.11	1.04	7.4
				GC-MS/MS	1.08	1.13	1.01	1.11	1.27	1.19	1.17	1.24	1.22	1.17	1.11	1.09	1.11	1.14	1.31	1.31	1.12	1.15	1.11	7.0
148	Dimepiperate	TPT	Green tea	GC-MS	0.97	1.04	1.13	1.10	1.11	1.04	1.04	1.09	1.00	1.08	1.12	1.06	1.08	0.96	1.07	1.00	0.95	1.04	1.14	5.4
				GC-MS/MS	1.06	1.03	1.10	1.06	1.08	1.08	1.09	1.08	0.97	1.10	1.06	1.17	1.06	1.03	1.12	1.04	0.99	1.02	1.14	4.6
			Oolong tea	GC-MS	1.16	1.13	1.12	1.09	1.08	1.21	1.41	1.11	1.24	1.17	1.29	1.21	1.19	1.1	1.15	1.13	1.17	1.07	1.09	7.2
				GC-MS/MS	1.21	1.16	1.17	1.19	1.07	1.23	1.33	1.20	1.24	1.13	1.39	1.27	1.26	1.13	1.18	1.17	1.24	1.12	1.06	6.9
		ENVI-Carb + PSA	Green tea	GC-MS	0.89	1.21	1.02	1.09	1.05	1.18	1.03	1.09	1.04	1.00	1.04	1.45	1.08	1.05	1.06	1.22	1.10	1.10	1.00	10.7
				GC-MS/MS	0.99	1.15	1.03	1.10	1.03	1.09	0.92	1.07	1.07	0.99	1.07	1.05	1.16	1.07	1.12	1.20	1.12	1.07	0.98	6.4
			Oolong tea	GC-MS	1.16	1.11	1.23	1.13	1.31	1.19	1.16	1.23	1.20	1.24	1.20	1.01	1.12	1.16	1.22	1.19	1.17	1.17	1.11	5.4
				GC-MS/MS	1.23	1.17	1.13	1.18	1.28	1.24	1.15	1.21	1.31	1.21	1.08	1.11	1.18	1.15	1.33	1.24	1.10	1.22	1.07	6.3

149	Dimethametryn	TPT	Green tea	GC-MS	1.03	1.00	1.12	1.12	1.13	0.99	1.09	1.08	1.00	1.03	1.07	1.07	1.07	1.00	1.13	1.01	1.00	1.02	1.11	4.8
				GC-MS/MS	1.08	1.00	1.16	1.21	1.12	1.08	1.12	1.08	1.01	1.00	1.12	1.01	1.02	0.99	1.11	1.04	0.99	0.98	1.18	6.6
			Oolong tea	GC-MS	1.13	1.09	1.12	1.09	1.04	1.15	1.21	1.14	1.35	1.27	1.27	1.22	1.22	1.2	1.16	1.23	1.27	1.07	1.12	6.9
				GC-MS/MS	1.08	1.08	1.22	1.19	1.05	1.21	1.23	1.14	1.20	1.18	1.39	1.28	1.26	1.16	1.19	1.15	1.24	1.09	1.17	6.8
		ENVI-Carb + PSA	Green tea	GC-MS	1.02	1.18	1.04	1.12	1.03	1.19	0.98	1.07	0.98	0.99	1.06	1.53	1.10	1.10	1.05	1.24	1.15	1.17	1.01	11.7
				GC-MS/MS	1.07	1.24	1.05	1.14	1.04	1.12	1.01	1.11	1.06	1.00	1.12	1.06	1.09	1.16	1.12	1.21	1.19	1.15	1.00	6.3
			Oolong tea	GC-MS	1.13	1.10	1.11	1.13	1.33	1.18	1.17	1.19	1.23	1.25	1.16	1.03	1.18	1.25	1.28	1.21	1.18	1.27	1.12	6.2
				GC-MS/MS	1.13	1.13	1.07	1.15	1.24	1.22	1.15	1.19	1.33	1.23	1.06	1.16	0.97	1.15	1.41	1.22	1.11	1.17	1.17	8.2
150	Dimethomorph	TPT	Green tea	GC-MS	0.77	1.04	1.06	1.08	1.14	1.09	1.05	1.11	1.10	1.15	1.23	1.03	1.01	0.94	1.16	1.00	1.07	1.10	1.14	9.2
				GC-MS/MS	0.97	0.98	1.09	1.13	1.15	1.08	1.04	1.12	1.06	1.12	1.19	1.11	0.96	0.87	1.13	1.01	1.04	1.08	1.13	7.5
			Oolong tea	GC-MS	1.09	1.08	1.11	1.16	1.10	1.15	1.10	1.06	1.27	1.17	1.24	1.23	1.15	1.1	1.17	1.28	1.20	1.12	1.21	5.6
				GC-MS/MS	1.22	1.08	1.12	1.14	1.01	1.15	1.08	1.07	1.31	1.16	1.33	1.28	1.08	1.15	1.15	1.23	1.22	1.12	1.10	7.5
		ENVI-Carb + PSA	Green tea	GC-MS	0.91	1.06	0.89	1.09	0.99	1.15	0.73	1.10	1.11	1.04	1.01	1.31	1.07	1.18	1.07	1.18	1.20	1.22	1.01	12.4
				GC-MS/MS	0.87	1.09	1.03	1.14	0.93	1.09	0.69	1.15	1.18	1.02	1.02	1.08	1.12	1.18	1.08	1.17	1.19	1.14	1.03	11.7
			Oolong tea	GC-MS	1.09	1.08	1.02	1.07	1.24	1.18	1.28	1.17	1.14	1.49	1.23	1.17	1.15	1.10	1.34	1.18	1.13	1.28	1.10	9.4
				GC-MS/MS	1.15	1.13	0.94	1.11	1.14	1.23	1.18	1.13	1.14	1.47	1.03	1.21	1.08	1.09	1.50	1.18	1.07	1.26	1.10	11.5
151	Dimethylphthalate	TPT	Green tea	GC-MS	1.05	1.59	1.07	1.07	1.05	1.08	1.07	1.05	1.01	1.05	1.08	1.07	1.05	1.05	1.05	1.00	1.01	1.02	1.11	11.6
				GC-MS/MS	1.06	1.03	1.09	1.06	1.06	1.11	1.11	1.09	0.98	1.03	1.06	1.15	1.08	0.95	1.07	1.02	0.98	1.02	1.12	4.9
			Oolong tea	GC-MS	1.37	1.23	1.20	1.21	1.14	1.25	1.21	1.18	1.21	1.16	1.29	1.23	1.21	1.2	1.22	1.18	1.19	1.13	1.19	4.5
				GC-MS/MS	1.14	1.20	1.20	1.23	1.11	1.27	1.27	1.18	1.21	1.16	1.35	1.28	1.18	1.17	1.20	1.18	1.18	1.12	1.11	5.2
		ENVI-Carb + PSA	Green tea	GC-MS	1.10	1.18	1.04	1.08	1.00	1.06	0.99	1.08	1.09	1.06	1.16	1.05	1.10	1.08	1.13	1.17	1.12	1.12	1.01	4.9
				GC-MS/MS	1.05	1.23	1.04	1.10	0.98	1.06	1.02	1.12	1.11	1.04	1.15	1.05	1.09	1.21	1.12	1.17	1.10	1.08	1.04	5.9
			Oolong tea	GC-MS	1.37	1.18	1.14	1.16	1.28	1.19	1.21	1.19	1.27	1.25	1.17	1.16	1.16	1.19	1.26	1.23	1.19	1.23	1.14	4.7
				GC-MS/MS	1.36	1.21	1.12	1.18	1.24	1.23	1.16	1.23	1.28	1.26	1.08	1.12	1.14	1.16	1.34	1.24	1.10	1.20	1.11	6.6
152	Diniconazole	TPT	Green tea	GC-MS	1.04	1.05	1.11	1.11	1.15	0.96	1.08	1.08	1.01	1.06	1.08	0.97	1.05	1.03	1.09	0.96	1.01	1.05	1.17	5.5
				GC-MS/MS	1.06	1.03	1.15	1.03	1.17	1.09	1.08	1.10	1.03	1.02	1.12	1.11	1.07	0.96	1.11	1.03	0.99	1.02	1.15	5.5
			Oolong tea	GC-MS	1.12	0.96	1.00	1.07	1.03	1.16	1.10	1.00	1.50	1.28	1.10	1.24	1.17	1.1	1.24	1.26	1.32	1.07	1.13	11.6
				GC-MS/MS	1.17	1.10	1.13	1.22	1.05	1.20	1.15	1.07	1.30	1.15	1.35	1.30	1.16	1.14	1.19	1.21	1.23	1.12	1.12	6.7
		ENVI-Carb + PSA	Green tea	GC-MS	0.91	1.10	1.05	1.18	1.06	1.22	0.92	1.07	1.06	0.95	1.05	1.50	1.08	1.09	1.12	1.20	1.07	1.06	1.01	11.7
				GC-MS/MS	0.98	1.15	1.05	1.17	1.05	1.10	0.95	1.11	1.09	1.04	1.11	1.07	1.15	1.17	1.09	1.24	1.14	1.10	1.00	6.6
			Oolong tea	GC-MS	1.12	1.04	1.06	1.04	1.30	1.15	1.20	1.21	1.14	1.27	1.30	1.09	1.17	1.22	1.31	1.25	1.20	1.22	1.02	7.8
				GC-MS/MS	1.10	1.12	1.01	1.11	1.30	1.20	1.16	1.20	1.25	1.24	1.07	1.14	1.27	1.13	1.35	1.27	1.10	1.21	1.12	7.5

(Continued)

Appendix Table 5.3 Youden Pair Ratios of Each Average Content From 201 Pesticides and RSD for 19 Circulative Determinations (cont.)

Youden pair ratio

No.	Pesticides	SPE	Sample	Method	Nov 9, 2009 (n=5)	Nov 14, 2009 (n=3)	Nov 19, 2009 (n=3)	Nov 24, 2009 (n=3)	Nov 29, 2009 (n=3)	Dec 4, 2009 (n=3)	Dec 9, 2009 (n=3)	Dec 14, 2009 (n=3)	Dec 19, 2009 (n=3)	Dec 24, 2009 (n=3)	Dec 29, 2009 (n=3)	Jan 3, 2010 (n=3)	Jan 8, 2010 (n=3)	Jan 13, 2010 (n=3)	Jan 18, 2010 (n=3)	Jan 23, 2010 (n=3)	Jan 28, 2010 (n=3)	Feb 2, 2010 (n=3)	Feb 7, 2010 (n=3)	RSD%
153	Diphenamid	TPT	Green tea	GC–MS	1.04	1.04	1.10	1.10	0.90	0.96	1.00	0.85	0.93	1.06	1.11	1.06	1.09	0.98	1.17	0.99	1.01	1.04	1.13	7.9
				GC–MS/MS	1.04	1.00	1.16	1.14	1.09	1.10	1.11	1.13	1.02	1.03	1.14	1.09	0.97	0.92	1.10	1.04	0.98	1.03	1.13	6.2
			Oolong tea	GC–MS	1.17	1.17	1.23	1.16	1.14	1.13	1.21	1.05	1.22	1.15	1.33	1.29	1.20	1.1	1.20	1.20	1.19	1.05	1.11	5.9
				GC–MS/MS	1.07	1.15	1.20	1.21	1.09	1.25	1.30	1.17	1.19	1.18	1.43	1.28	1.29	1.16	1.16	1.19	1.26	1.09	1.12	7.4
		ENVI-Carb + PSA	Green tea	GC–MS	1.05	1.16	1.03	1.05	1.02	1.19	0.41	1.20	1.03	1.02	1.13	1.28	1.10	1.14	1.11	1.25	1.14	1.15	1.00	16.9
				GC–MS/MS	1.09	1.20	1.04	1.10	1.02	1.10	0.93	1.09	1.05	1.00	1.12	1.13	1.10	1.09	1.12	1.22	1.14	1.13	0.99	6.6
			Oolong tea	GC–MS	1.17	1.14	1.11	1.11	1.25	1.19	0.63	1.35	1.06	1.27	1.15	1.15	1.13	1.19	1.25	1.20	1.14	1.23	1.08	12.6
				GC–MS/MS	1.17	1.18	1.08	1.16	1.22	1.21	1.12	1.21	1.26	1.29	1.04	1.19	1.11	1.18	1.34	1.18	1.05	1.24	1.12	6.7
154	Dipropetryn	TPT	Green tea	GC–MS	1.03	1.04	1.12	1.15	1.15	1.00	1.09	1.10	0.99	1.14	1.13	1.03	1.05	1.02	1.10	1.01	1.03	1.03	1.14	5.2
				GC–MS/MS	1.07	0.99	1.18	1.05	1.12	1.09	1.12	1.08	1.01	1.06	1.14	1.05	1.07	0.96	1.08	1.02	0.97	1.00	1.12	5.6
			Oolong tea	GC–MS	1.07	1.54	1.18	1.08	1.02	1.22	1.20	1.15	1.20	1.33	1.22	1.18	1.25	1.1	1.18	1.21	1.25	1.05	1.12	9.7
				GC–MS/MS	1.07	1.24	1.18	1.19	1.06	1.22	1.30	1.13	1.17	1.17	1.30	1.34	1.23	1.11	1.12	1.21	1.35	1.07	1.09	7.7
		ENVI-Carb + PSA	Green tea	GC–MS	0.98	1.20	1.02	1.08	1.02	1.21	1.01	1.07	1.01	1.02	0.97	1.45	1.08	1.09	1.09	1.26	1.13	1.16	0.99	10.7
				GC–MS/MS	1.07	1.23	1.07	1.10	1.05	1.11	1.01	1.11	1.06	0.99	1.10	0.84	1.10	1.20	1.04	1.21	1.17	1.15	0.99	8.4
			Oolong tea	GC–MS	1.07	1.05	1.05	1.13	1.17	1.16	1.18	1.16	1.18	1.46	1.18	1.21	1.22	1.21	1.31	1.26	1.21	1.35	0.99	9.2
				GC–MS/MS	1.15	1.11	1.04	1.34	1.18	1.26	1.14	1.20	1.25	1.35	1.10	1.19	1.30	1.45	1.76	1.26	1.09	1.19	1.09	13.4
155	Ethalfluralin	TPT	Green tea	GC–MS	1.04	1.06	1.11	1.11	0.96	1.06	1.05	1.03	0.83	1.05	1.07	1.10	1.08	0.99	1.04	1.03	0.98	0.99	1.10	6.3
				GC–MS/MS	1.02	1.05	1.12	0.78	1.03	1.14	1.08	1.05	0.85	1.04	1.07	1.21	1.08	0.91	1.12	1.02	0.96	1.00	1.19	10.3
			Oolong tea	GC–MS	1.25	1.13	1.14	1.21	1.19	1.45	1.31	1.13	1.28	1.17	1.34	1.31	1.19	1.0	1.22	1.05	1.20	1.11	1.09	8.6
				GC–MS/MS	1.13	1.15	1.19	1.19	1.13	1.29	1.31	1.17	1.27	1.19	1.48	1.39	1.22	1.13	1.21	1.15	1.21	1.16	1.16	7.7
		ENVI-Carb + PSA	Green tea	GC–MS	1.11	1.00	1.05	1.13	1.08	1.26	1.35	1.03	1.08	1.07	1.14	1.37	1.06	1.00	1.09	1.20	1.17	1.05	1.03	9.6
				GC–MS/MS	1.01	1.03	1.08	1.11	1.05	1.05	1.84	1.08	1.06	1.04	1.10	1.08	0.97	1.19	1.02	1.28	1.14	0.96	1.04	17.3
			Oolong tea	GC–MS	1.25	1.16	1.25	1.10	1.49	1.18	1.15	1.23	1.31	1.21	1.23	1.21	1.23	0.94	1.00	1.56	1.13	1.17	1.17	12.0
				GC–MS/MS	1.30	1.22	1.19	1.16	1.28	1.25	1.12	1.27	1.34	1.21	1.16	1.11	1.17	1.21	1.29	1.35	1.14	1.09	1.16	6.4
156	Etofenprox	TPT	Green tea	GC–MS	1.03	1.05	1.02	1.00	1.13	1.03	1.06	0.98	0.97	0.95	1.09	1.01	1.00	0.99	1.06	1.03	0.99	0.98	1.04	4.4
				GC–MS/MS	1.05	1.01	1.12	1.06	1.11	1.06	1.12	0.96	0.98	1.01	1.07	1.01	1.06	0.98	1.07	0.98	0.95	0.98	1.06	5.2
			Oolong tea	GC–MS	1.17	1.13	1.08	1.22	1.06	1.17	1.08	1.10	1.18	1.13	1.23	1.24	1.13	1.0	1.12	1.19	1.18	1.15	1.14	5.6
				GC–MS/MS	1.11	1.09	1.14	1.19	1.03	1.20	1.09	1.08	1.22	1.14	1.29	1.24	1.15	1.13	1.16	1.19	1.24	1.14	1.13	5.6
		ENVI-Carb + PSA	Green tea	GC–MS	1.04	1.18	0.98	1.03	0.95	1.04	0.88	0.97	1.07	0.99	0.98	0.94	1.03	1.02	1.03	1.09	1.03	1.11	1.01	6.5
				GC–MS/MS	0.99	1.18	0.91	1.03	0.89	1.01	0.96	0.95	1.09	0.96	1.03	1.06	1.35	0.85	0.94	1.14	0.98	1.02	0.99	11.2
			Oolong tea	GC–MS	1.17	1.14	1.06	1.07	1.23	1.17	1.29	1.12	1.26	1.25	1.20	1.12	1.14	1.12	1.20	1.17	1.22	1.20	1.13	5.4
				GC–MS/MS	1.06	1.04	0.99	1.04	0.99	0.92	0.89	1.17	1.24	1.22	1.09	1.15	1.05	1.21	1.38	0.88	1.18	1.05	0.99	12.1

157	Etridiazol	TPT	Green tea	GC–MS	1.06	1.14	1.10	1.05	0.99	1.14	1.11	1.02	0.96	1.10	1.05	1.11	1.09	0.99	1.05	1.03	0.99	0.95	1.02	5.6
				GC–MS/MS	0.99	1.15	0.94	1.25	1.00	1.09	1.11	1.06	0.88	1.16	1.11	1.19	0.99	0.90	1.08	1.01	0.92	0.96	1.11	10.0
			Oolong tea	GC–MS	1.30	1.22	1.12	1.30	1.18	1.28	1.25	1.17	1.27	1.12	1.40	1.35	1.19	1.1	1.23	1.04	1.22	1.26	1.17	7.2
				GC–MS/MS	1.21	1.23	1.16	1.36	1.28	1.32	1.19	1.21	1.31	1.08	1.46	1.32	1.15	1.18	1.23	1.20	1.27	1.27	1.10	7.5
		ENVI-Carb + PSA	Green tea	GC–MS	1.00	1.03	1.01	1.17	1.03	0.80	1.14	1.13	1.14	1.03	1.30	0.90	1.16	0.95	1.13	1.10	1.13	1.03	0.94	11.0
				GC–MS/MS	0.74	1.11	1.04	1.18	1.04	1.00	1.35	1.04	1.05	1.01	1.29	1.04	1.13	1.04	1.20	1.13	1.17	1.00	0.99	12.0
			Oolong tea	GC–MS	1.30	1.25	1.05	1.11	1.35	1.18	0.99	1.27	1.52	1.23	1.12	1.14	1.21	1.19	1.22	1.45	1.12	1.16	1.19	10.6
				GC–MS/MS	1.76	1.27	1.06	1.14	1.38	1.25	1.07	1.30	1.51	1.18	1.06	1.13	1.34	1.10	1.28	1.53	1.09	1.31	1.10	15.1
158	Fenazaquin	TPT	Green tea	GC–MS	1.05	1.04	1.12	1.10	1.17	1.08	1.10	1.09	1.11	0.94	1.10	1.11	1.04	1.06	1.10	1.05	1.02	0.99	1.14	5.1
				GC–MS/MS	1.07	1.01	1.15	1.21	1.12	1.11	1.11	1.09	1.02	1.04	1.10	1.03	1.06	0.93	1.11	1.04	0.99	1.09	1.14	6.0
			Oolong tea	GC–MS	1.17	1.14	1.13	1.18	1.07	1.18	1.07	1.09	1.31	1.17	1.26	1.19	1.19	1.1	1.25	1.13	1.27	1.14	1.14	5.7
				GC–MS/MS	1.10	1.11	1.16	1.22	1.02	1.21	1.16	1.08	1.20	1.15	1.33	1.28	1.23	1.15	1.18	1.21	1.23	1.15	1.16	6.1
		ENVI-Carb + PSA	Green tea	GC–MS	1.07	1.15	1.07	1.09	1.05	1.21	0.99	1.06	1.06	1.01	0.97	1.56	1.10	1.12	1.09	1.19	1.10	1.19	1.01	11.5
				GC–MS/MS	1.10	1.21	1.04	1.15	1.02	1.09	1.05	1.13	1.10	1.02	1.09	1.04	1.19	1.24	1.05	1.22	1.15	1.13	1.02	6.3
			Oolong tea	GC–MS	1.17	1.12	1.08	1.12	1.27	1.23	1.23	1.13	1.31	1.27	1.27	1.03	1.15	1.21	1.20	1.26	1.25	1.19	1.10	6.5
				GC–MS/MS	1.13	1.14	1.01	1.14	1.24	1.22	1.20	1.20	1.26	1.22	1.11	1.14	0.94	1.18	1.39	1.25	1.14	1.11	1.11	8.3
159	Fenchlorphos	TPT	Green tea	GC–MS	1.01	1.02	1.09	1.09	1.02	1.03	1.10	1.07	0.95	1.04	1.10	1.08	1.06	1.07	1.05	1.05	1.03	0.99	1.07	3.8
				GC–MS/MS	0.99	1.15	0.98	0.95	0.95	1.10	1.14	1.07	0.95	1.02	1.18	1.15	1.06	0.99	1.08	1.08	0.97	0.99	1.06	7.2
			Oolong tea	GC–MS	1.21	1.16	1.19	1.15	1.08	1.18	1.34	1.19	1.15	1.15	1.20	1.18	1.23	1.1	1.15	1.12	1.19	1.10	1.13	4.9
				GC–MS/MS	1.15	1.17	1.19	1.22	1.14	1.24	1.26	1.15	1.18	1.08	1.25	1.18	1.24	1.15	1.14	1.17	1.19	1.13	1.08	4.3
		ENVI-Carb + PSA	Green tea	GC–MS	1.09	1.16	1.02	1.04	1.04	0.91	1.12	1.13	1.07	1.00	1.07	1.18	1.09	1.07	1.00	1.15	1.10	0.99	1.01	6.3
				GC–MS/MS	0.82	1.15	1.05	1.05	1.01	1.00	1.35	1.04	1.07	1.02	1.06	1.04	1.04	1.28	1.02	1.21	1.15	0.97	0.98	11.1
			Oolong tea	GC–MS	1.21	1.16	1.12	1.10	1.25	1.13	0.98	1.17	1.25	1.23	1.12	1.07	1.09	1.13	1.22	1.18	1.14	1.16	1.09	6.0
				GC–MS/MS	1.52	1.16	1.05	1.16	1.21	1.16	1.11	1.21	1.26	1.17	1.08	1.12	1.17	1.07	1.30	1.24	1.08	1.23	1.08	9.2
160	Fenoxanil [6,0]	TPT	Green tea	GC–MS	1.01	1.02	1.11	1.09	1.15	1.08	1.02	1.13	1.05	1.08	1.16	1.03	1.00	1.03	1.10	1.01	1.03	1.06	1.17	5.1
				GC–MS/MS	1.00	0.88	1.25	1.00	1.18	1.09	1.14	1.04	1.06	1.17	1.14	1.14	1.11	1.20	1.14	1.04	1.00	1.04	1.16	8.1
			Oolong tea	GC–MS	1.21	1.21	1.19	1.08	1.08	1.18	1.10	1.11	1.18	1.21	1.21	1.20	1.22	1.2	1.15	1.25	1.24	0.96	1.13	6.2
				GC–MS/MS	1.22	1.23	1.19	1.15	1.01	1.21	1.40	1.08	1.19	1.20	1.35	1.27	1.22	1.22	1.16	1.25	1.18	1.09	1.11	7.5
		ENVI-Carb + PSA	Green tea	GC–MS	1.02	1.23	1.04	1.12	0.99	1.27	0.87	1.17	1.09	1.11	1.05	1.27	1.15	1.10	1.11	1.22	1.15	1.16	0.95	9.5
				GC–MS/MS	1.08	1.21	1.04	1.15	1.14	1.27	0.79	1.18	0.96	0.99	1.12	1.14	1.23	1.03	1.08	1.20	1.12	1.10	1.01	10.3
			Oolong tea	GC–MS	1.21	1.10	1.00	1.16	1.24	1.19	1.20	1.17	1.13	1.35	1.19	1.12	1.25	1.28	1.33	1.15	1.13	1.28	1.10	7.2
				GC–MS/MS	1.07	1.13	1.04	1.35	1.10	1.29	1.16	1.18	1.13	1.58	1.05	1.18	1.12	1.21	1.57	1.12	1.06	1.18	1.11	13.2

(Continued)

Appendix Table 5.3 Youden Pair Ratios of Each Average Content From 201 Pesticides and RSD for 19 Circulative Determinations (cont.)

No.	Pesticides	SPE	Sample	Method	Nov 9, 2009 (n=5)	Nov 14, 2009 (n=3)	Nov 19, 2009 (n=3)	Nov 24, 2009 (n=3)	Nov 29, 2009 (n=3)	Dec 4, 2009 (n=3)	Dec 9, 2009 (n=3)	Dec 14, 2009 (n=3)	Dec 19, 2009 (n=3)	Dec 24, 2009 (n=3)	Jan 3, 2010 (n=3)	Jan 8, 2010 (n=3)	Jan 13, 2010 (n=3)	Jan 18, 2010 (n=3)	Jan 23, 2010 (n=3)	Jan 28, 2010 (n=3)	Feb 2, 2010 (n=3)	Feb 7, 2010 (n=3)	RSD%
161	Fenpropidin	TPT	Green tea	GC–MS	1.05	1.01	1.12	1.09	1.18	1.03	1.09	1.21	1.12	1.16	1.02	1.05	0.96	1.21	1.06	1.08	1.07	1.13	6.3
				GC–MS/MS	1.11	0.99	1.11	1.01	1.23	1.11	1.10	1.12	1.07	1.16	1.00	1.08	0.92	1.18	1.05	1.05	1.07	1.12	6.7
			Oolong tea	GC–MS	1.32	1.26	1.55	1.53	1.68	1.64	1.69	1.77	2.34	2.98	2.29	2.21	2.2	2.34	2.29	2.41	2.01	2.33	31.3
				GC–MS/MS	1.02	1.29	1.52	1.50	1.67	1.90	2.04	1.95	2.22	2.20	2.35	2.68	2.51	2.66	2.76	2.97	2.56	2.88	26.6
		ENVI-Carb +PSA	Green tea	GC–MS	1.05	1.08	1.07	1.15	0.96	1.33	1.59	1.03	1.04	1.05	1.34	1.14	1.23	1.11	1.29	1.18	1.20	1.00	13.3
				GC–MS/MS	1.10	1.11	1.04	1.16	0.95	1.10	0.79	1.14	1.08	1.01	1.05	1.18	1.46	1.12	1.26	1.17	1.12	1.01	13.1
			Oolong tea	GC–MS	1.32	1.27	1.58	1.66	2.26	1.75	2.08	1.68	2.19	3.59	4.24	2.13	2.43	2.57	2.44	2.31	2.65	2.40	34.7
				GC–MS/MS	1.31	1.35	1.47	1.82	2.21	2.15	2.03	2.27	2.19	2.82	2.40	3.11	2.89	3.57	2.89	2.81	3.21	2.88	26.9
162	Fenson	TPT	Green tea	GC–MS	1.00	1.08	1.08	1.01	0.98	1.05	1.03	1.09	0.89	0.99	1.14	1.03	1.10	0.86	1.01	0.98	1.03	1.11	6.9
				GC–MS/MS	1.05	1.09	1.05	1.01	1.00	1.13	1.08	1.05	0.83	0.99	1.29	1.02	0.92	1.02	1.03	0.96	0.98	1.13	9.0
			Oolong tea	GC–MS	1.19	1.12	1.11	1.20	1.16	1.20	1.41	1.00	1.24	1.05	1.22	1.09	1.1	1.22	1.18	1.22	1.13	1.10	8.4
				GC–MS/MS	1.31	1.15	1.15	1.21	1.14	1.33	1.23	1.16	1.27	1.03	1.33	1.17	1.14	1.23	1.15	1.24	1.19	1.12	7.6
		ENVI-Carb +PSA	Green tea	GC–MS	1.06	1.00	1.06	1.14	1.06	1.08	1.43	1.07	1.06	1.04	1.07	1.09	1.01	0.97	1.19	1.11	0.84	0.96	11.4
				GC–MS/MS	0.96	0.90	1.11	1.15	1.05	0.98	1.28	1.12	1.11	1.07	1.08	1.04	1.07	1.01	1.26	1.13	0.91	0.98	10.6
			Oolong tea	GC–MS	1.19	1.13	1.12	1.04	1.26	1.08	1.18	1.19	1.18	1.11	1.12	1.18	1.02	1.12	1.42	1.18	1.14	1.17	7.4
				GC–MS/MS	1.14	1.16	1.09	1.11	1.29	1.15	1.17	1.26	1.29	1.12	1.10	1.24	1.05	1.22	1.46	1.12	1.14	1.12	8.3
163	Flutenacet	TPT	Green tea	GC–MS	1.01	1.07	1.10	1.14	0.94	1.04	1.08	1.03	0.95	0.98	1.19	1.12	1.05	1.01	0.95	1.01	0.96	1.01	6.8
				GC–MS/MS	0.92	1.29	0.93	1.07	0.90	1.11	1.24	1.05	0.89	0.93	1.17	0.98	0.82	1.10	1.08	0.93	0.96	1.12	12.5
			Oolong tea	GC–MS	1.20	1.18	1.17	1.28	1.13	1.17	1.45	1.04	0.90	1.05	1.29	1.09	1.1	1.22	0.99	1.12	1.17	1.07	11.3
				GC–MS/MS	1.04	1.30	1.30	1.44	1.25	1.32	1.14	1.20	1.36	1.07	1.21	1.38	1.15	1.17	1.18	1.06	1.24	1.05	9.8
		ENVI-Carb +PSA	Green tea	GC–MS	1.00	0.97	1.19	1.08	1.07	0.80	1.04	1.14	1.13	1.05	1.15	1.18	0.99	1.09	1.36	1.14	0.86	0.99	12.2
				GC–MS/MS	0.94	1.19	1.08	1.09	1.15	1.00	1.28	1.04	0.98	1.00	1.09	1.13	1.32	1.07	1.24	1.34	0.73	0.79	14.8
			Oolong tea	GC–MS	1.20	1.23	1.05	1.03	1.29	1.11	1.05	1.28	1.53	1.45	1.04	1.08	1.06	1.16	1.16	1.16	1.24	1.11	12.0
				GC–MS/MS	1.38	1.16	0.99	1.15	1.44	1.15	1.03	1.27	1.48	1.18	1.13	1.22	0.97	1.37	1.38	1.30	1.34	1.05	13.0
164	Furalaxyl	TPT	Green tea	GC–MS	1.01	1.09	1.12	1.10	1.10	1.06	1.07	1.12	1.04	1.08	1.05	1.05	0.98	1.14	0.99	1.01	1.04	1.14	4.8
				GC–MS/MS	1.05	1.01	1.13	1.03	1.10	1.11	1.12	1.12	1.01	1.02	1.16	1.03	0.92	1.12	1.04	0.98	1.03	1.13	6.0
			Oolong tea	GC–MS	1.17	1.16	1.17	1.15	1.10	1.18	1.27	1.17	1.21	1.20	1.28	1.27	1.1	1.18	1.20	1.26	1.05	1.11	5.4
				GC–MS/MS	1.13	1.12	1.18	1.13	1.06	1.25	1.28	1.17	1.17	1.16	1.31	1.33	1.15	1.19	1.22	1.25	1.07	1.07	7.8
		ENVI-Carb +PSA	Green tea	GC–MS	1.06	1.14	1.04	1.03	1.03	1.08	0.89	1.07	1.06	1.00	1.19	1.10	1.11	1.08	1.25	1.14	1.14	0.99	7.2
				GC–MS/MS	1.03	1.18	1.05	1.10	1.02	1.09	0.90	1.09	1.06	0.99	1.08	1.13	1.20	1.09	1.24	1.14	1.12	1.01	7.3
			Oolong tea	GC–MS	1.17	1.15	1.10	1.11	1.26	1.17	1.12	1.16	1.27	1.33	1.14	1.17	1.17	1.29	1.17	1.13	1.27	1.11	5.8
				GC–MS/MS	1.18	1.15	1.06	1.14	1.23	1.26	1.11	1.19	1.33	1.32	1.16	1.09	1.11	1.40	1.19	1.07	1.27	1.07	8.6

No.	Compound	Cleanup	Matrix	Instrument																				(%)
165	Heptachlor	TPT	Green tea	GC-MS	1.09	1.04	1.15	1.04	1.05	1.07	1.08	1.03	0.92	1.06	1.10	1.09	1.07	1.05	1.07	1.05	0.97	0.98	1.07	4.9
				GC-MS/MS	1.03	1.09	1.05	1.04	1.20	1.09	1.09	1.06	0.94	1.09	1.12	1.15	1.09	1.00	1.08	1.04	0.99	0.97	1.11	6.0
			Oolong tea	GC-MS	1.26	1.20	1.19	1.10	1.17	1.21	1.35	1.18	1.19	1.13	1.33	1.30	1.20	1.1	1.19	1.09	1.22	1.11	1.12	6.5
				GC-MS/MS	1.13	1.17	1.18	1.13	1.16	1.23	1.27	1.12	1.26	1.16	1.40	1.30	1.24	1.14	1.19	1.16	1.18	1.16	1.10	6.2
		ENVI-Carb + PSA	Green tea	GC-MS	1.06	1.09	1.01	0.99	1.12	0.94	1.10	1.10	1.05	1.01	1.13	0.92	1.07	0.99	1.04	1.17	1.07	0.97	0.99	6.4
				GC-MS/MS	0.79	1.08	1.03	1.02	1.13	1.00	1.42	1.03	1.00	1.00	1.11	1.01	1.08	1.18	1.07	1.14	1.13	1.00	0.98	11.3
			Oolong tea	GC-MS	1.26	1.23	1.17	1.37	1.15	1.17	0.92	1.23	1.33	1.22	1.15	1.12	1.14	1.17	1.14	1.31	1.16	1.18	1.13	8.2
				GC-MS/MS	1.52	1.23	1.13	1.33	1.16	1.20	1.12	1.27	1.33	1.20	1.09	1.14	1.28	1.12	1.29	1.29	1.10	1.32	1.10	9.1
166	Iprobenfos	TPT	Green tea	GC-MS	1.05	1.02	1.05	1.18	1.05	0.98	1.06	1.08	0.94	1.08	1.10	1.02	1.09	0.91	1.02	1.03	0.95	1.03	1.11	6.3
				GC-MS/MS	1.01	1.03	1.10	1.12	1.08	1.07	1.11	1.11	0.99	1.05	1.13	1.18	1.09	0.97	1.12	1.04	0.98	1.03	1.12	5.5
			Oolong tea	GC-MS	1.20	1.04	1.08	1.06	1.07	1.19	1.48	1.11	1.37	1.36	1.29	1.19	1.23	1.1	1.20	1.15	1.29	1.08	1.12	10.3
				GC-MS/MS	1.17	1.13	1.19	1.14	1.15	1.23	1.47	1.14	1.29	1.21	1.42	1.37	1.28	1.17	1.18	1.19	1.28	1.11	1.16	8.3
		ENVI-Carb + PSA	Green tea	GC-MS	1.02	0.92	0.98	1.09	1.21	1.25	1.01	1.15	1.07	1.00	1.10	1.31	1.09	1.07	1.05	1.21	1.08	1.02	1.02	9.1
				GC-MS/MS	0.83	1.09	1.05	1.05	1.11	1.10	1.02	1.11	1.07	0.99	1.07	1.07	1.14	1.30	1.09	1.23	1.15	1.05	0.99	8.9
			Oolong tea	GC-MS	1.20	1.10	1.28	1.45	1.07	1.21	1.03	1.24	1.19	1.30	1.23	1.09	1.16	1.20	1.22	1.22	1.18	1.26	1.19	7.8
				GC-MS/MS	1.44	1.21	1.19	1.38	1.19	1.29	1.12	1.29	1.31	1.26	1.11	1.16	1.21	1.21	1.36	1.24	1.11	1.33	1.16	7.7
167	Isazofos	TPT	Green tea	GC-MS	1.01	1.06	1.07	1.07	1.13	1.00	1.09	1.08	0.97	1.06	1.08	1.01	1.04	1.09	1.07	1.03	1.01	1.02	1.11	3.9
				GC-MS/MS	1.01	1.01	1.09	1.07	1.16	1.06	1.12	1.11	1.00	1.04	1.14	1.03	1.12	1.01	1.08	1.02	0.99	0.99	1.15	5.3
			Oolong tea	GC-MS	1.24	1.12	1.15	1.09	1.15	1.20	1.30	1.11	1.15	1.05	1.32	1.16	1.15	1.1	1.19	1.11	1.22	1.10	1.16	6.2
				GC-MS/MS	1.14	1.17	1.17	1.12	1.17	1.28	1.33	1.15	1.21	1.16	1.38	1.27	1.24	1.19	1.15	1.10	1.26	1.15	1.08	6.7
		ENVI-Carb + PSA	Green tea	GC-MS	1.07	0.89	1.00	0.99	1.11	1.19	1.05	1.11	1.02	1.02	1.06	1.14	1.01	1.14	1.08	1.17	1.06	1.12	1.01	6.9
				GC-MS/MS	0.91	1.12	1.07	1.01	1.10	1.08	1.12	1.07	1.09	1.03	1.12	1.16	1.05	1.16	1.04	1.20	1.07	1.00	1.01	7.4
			Oolong tea	GC-MS	1.24	1.15	1.15	1.35	1.12	1.18	1.57	1.07	1.08	1.17	1.06	1.00	1.07	1.10	1.27	1.18	1.16	1.27	1.16	10.7
				GC-MS/MS	1.37	1.21	1.18	1.27	1.17	1.26	1.13	1.27	1.27	1.20	1.08	1.17	1.27	1.16	1.35	1.19	1.11	1.26	1.08	6.9
168	Isoprothiolane	TPT	Green tea	GC-MS	1.02	1.04	1.13	1.10	1.11	1.08	1.06	1.10	1.03	1.04	1.11	1.06	1.07	1.04	1.13	1.00	1.01	1.05	1.15	4.0
				GC-MS/MS	1.05	1.01	1.19	1.12	1.06	1.11	1.11	1.10	1.01	1.02	1.13	1.00	1.09	0.97	1.13	1.05	1.00	1.03	1.14	5.5
			Oolong tea	GC-MS	1.19	1.13	1.15	1.10	1.15	1.20	1.24	1.14	1.21	1.19	1.25	1.26	1.41	1.0	1.21	1.20	1.24	1.08	1.11	7.1
				GC-MS/MS	1.12	1.11	1.18	1.07	1.20	1.21	1.27	1.17	1.20	1.17	1.37	1.32	1.23	1.16	1.16	1.19	1.22	1.11	1.11	6.3
		ENVI-Carb + PSA	Green tea	GC-MS	1.07	1.17	1.04	1.03	1.09	1.12	0.96	1.06	1.05	1.00	1.05	1.32	1.07	1.14	1.09	1.25	1.13	1.13	1.01	8.0
				GC-MS/MS	1.08	1.18	1.06	1.04	1.10	1.10	1.00	1.09	1.07	0.99	1.10	1.15	1.12	1.23	1.08	1.24	1.13	1.08	1.02	6.2
			Oolong tea	GC-MS	1.19	1.10	1.12	1.28	1.11	1.20	1.17	1.15	1.25	1.24	1.20	1.08	1.17	1.13	1.27	1.22	1.18	1.25	1.12	5.0
				GC-MS/MS	1.12	1.15	1.09	1.25	1.14	1.21	1.13	1.20	1.28	1.23	1.09	1.17	1.01	1.15	1.37	1.22	1.07	1.23	1.11	7.3

(Continued)

Appendix Table 5.3 Youden Pair Ratios of Each Average Content From 201 Pesticides and RSD for 19 Circulative Determinations (cont.)

No.	Pesticides	SPE	Sample	Method	Youden pair ratio																			RSD%
					Nov 9, 2009 (n=5)	Nov 14, 2009 (n=3)	Nov 19, 2009 (n=3)	Nov 24, 2009 (n=3)	Nov 29, 2009 (n=3)	Dec 4, 2009 (n=3)	Dec 9, 2009 (n=3)	Dec 14, 2009 (n=3)	Dec 19, 2009 (n=3)	Dec 24, 2009 (n=3)	Dec 14, 2009 (n=3)	Jan 3, 2010 (n=3)	Jan 8, 2010 (n=3)	Jan 13, 2010 (n=3)	Jan 18, 2010 (n=3)	Jan 23, 2010 (n=3)	Jan 28, 2010 (n=3)	Feb 2, 2010 (n=3)	Feb 7, 2010 (n=3)	
169	Kresoximmethyl	TPT	Green tea	GC-MS	0.90	1.05	1.12	1.12	1.17	0.97	1.06	1.09	1.08	1.17	1.06	1.03	0.99	1.03	0.90	1.06	0.98	1.00	0.96	7.6
				GC-MS/MS	1.04	1.03	1.08	1.17	1.08	1.06	1.22	1.15	0.96	1.02	1.11	1.15	1.01	0.95	1.05	1.05	1.02	1.10	1.22	7.2
			Oolong tea	GC-MS	1.16	1.12	1.18	1.03	0.89	0.90	1.31	1.17	1.16	1.24	1.01	1.25	1.13	1.1	1.19	1.16	1.25	1.07	1.13	9.8
				GC-MS/MS	1.18	1.11	1.50	1.16	0.94	1.21	1.15	1.04	1.10	1.12	1.26	1.24	1.21	1.11	1.12	1.18	1.14	1.15	1.14	9.4
		ENVI-Carb + PSA	Green tea	GC-MS	1.31	1.00	1.16	1.02	0.83	1.38	0.77	1.28	1.08	0.83	0.87	1.47	1.06	1.05	1.07	1.25	1.08	0.78	1.06	18.9
				GC-MS/MS	1.17	0.85	1.16	0.85	0.88	1.17	0.94	1.24	1.21	0.89	1.10	1.06	0.92	1.08	0.94	1.25	1.03	0.90	1.02	13.3
			Oolong tea	GC-MS	1.16	1.13	1.09	1.55	1.04	1.13	1.00	1.26	1.21	1.17	1.32	1.07	1.17	1.09	1.12	1.34	1.13	1.11	1.23	10.7
				GC-MS/MS	1.11	1.15	0.95	1.15	1.02	1.18	1.22	1.14	1.27	1.11	1.07	1.03	0.96	1.07	1.20	1.39	1.12	1.03	1.14	9.5
170	Mefenacet	TPT	Green tea	GC-MS	0.88	1.05	1.08	0.97	1.03	1.05	1.03	1.07	0.89	0.90	1.10	1.11	0.97	0.91	1.00	0.92	1.14	1.19	1.24	10.0
				GC-MS/MS	0.91	1.09	0.82	1.14	0.97	1.05	1.19	1.09	0.88	0.95	1.05	1.35	1.00	1.02	1.08	0.99	0.96	0.94	1.08	11.7
			Oolong tea	GC-MS	1.00	1.08	1.06	1.23	1.07	1.11	1.13	1.05	1.09	1.05	1.31	1.38	1.01	1.0	1.20	1.11	1.29	1.19	1.26	9.7
				GC-MS/MS	1.51	1.10	1.03	1.19	1.11	1.22	1.11	1.03	1.55	1.06	1.36	1.21	1.17	1.09	1.18	1.24	1.27	1.24	1.23	11.8
		ENVI-Carb + PSA	Green tea	GC-MS	0.83	0.85	1.02	1.15	1.05	0.82	0.92	1.15	1.28	0.97	1.34	0.95	1.08	0.87	0.95	1.28	1.03	1.11		15.4
				GC-MS/MS	1.05	1.39	1.05	1.21	0.87	0.93	1.29	1.15	1.12	1.07	1.25	1.13	1.16	1.00	0.99	1.12	0.99	0.97	0.96	12.3
			Oolong tea	GC-MS	1.00	1.03	1.00	0.96	1.38	1.14	0.91	0.97	1.23	1.14	1.07	0.96	1.12	0.92	0.97	1.59	1.22	1.08	1.14	15.6
				GC-MS/MS	1.18	1.03	0.93	1.07	1.39	1.13	1.23	1.32	1.24	1.20	1.13	1.06	1.26	1.02	1.18	1.46	1.17	1.41	1.23	11.7
171	Mepronil	TPT	Green tea	GC-MS	0.98	1.05	1.08	1.07	1.15	1.08	1.07	1.13	1.12	1.01	1.13	1.09	1.03	1.03	1.00	1.03	1.04	1.05	1.16	5.0
				GC-MS/MS	1.05	1.00	1.12	1.05	1.12	1.10	1.10	1.14	1.09	1.05	1.06	1.17	1.03	0.99	1.10	1.06	1.01	1.06	1.14	4.6
			Oolong tea	GC-MS	1.21	1.15	1.12	1.13	1.03	1.11	1.07	1.10	1.16	1.10	1.31	1.20	1.10	1.1	1.17	1.20	1.20	1.11	1.15	5.7
				GC-MS/MS	1.06	1.09	1.14	1.13	0.98	1.17	1.10	1.05	1.21	1.11	1.33	1.23	1.22	1.14	1.16	1.22	1.23	1.13	1.03	7.4
		ENVI-Carb + PSA	Green tea	GC-MS	1.04	0.81	1.02	1.11	0.94	1.12	0.93	1.12	1.17	0.99	1.11	1.33	1.13	1.02	1.21	1.20	1.05	1.17	0.99	11.2
				GC-MS/MS	1.08	1.25	1.02	1.10	0.98	1.14	0.90	1.15	1.09	1.05	1.12	1.05	1.24	1.15	1.07	1.18	1.08	1.09	1.02	7.8
			Oolong tea	GC-MS	1.21	1.00	0.92	1.04	1.16	1.18	1.20	1.06	1.22	1.25	1.16	1.06	1.14	1.15	1.25	1.21	1.17	1.19	1.11	7.8
				GC-MS/MS	1.18	1.09	0.95	1.13	1.14	1.22	1.20	1.10	1.14	1.23	1.04	1.09	1.04	1.13	1.38	1.21	1.09	1.18	1.07	8.1
172	Metribuzin	TPT	Green tea	GC-MS	1.01	1.03	1.13	1.13	0.97	1.02	1.02	1.08	0.86	1.02	1.12	1.05	1.07	1.12	0.94	1.03	1.01	0.99	1.15	7.1
				GC-MS/MS	1.05	1.07	1.06	0.98	0.97	1.14	1.14	1.11	0.88	0.98	1.18	1.28	1.08	0.91	1.05	1.06	0.99	1.00	1.14	9.1
			Oolong tea	GC-MS	1.09	1.11	1.13	1.21	1.19	1.32	1.30	1.14	1.42	1.29	1.37	1.29	1.28	1.2	1.31	1.26	1.35	1.22	1.16	7.6
				GC-MS/MS	1.23	1.20	1.32	1.27	1.20	1.37	1.28	1.23	1.38	1.20	1.46	1.41	1.35	1.27	1.25	1.30	1.31	1.26	1.21	5.8
		ENVI-Carb + PSA	Green tea	GC-MS	0.96	0.84	1.13	1.13	1.07	1.52	1.15	1.06	1.00	1.02	1.13	1.41	1.17	1.00	1.09	1.25	1.13	0.83	0.96	15.5
				GC-MS/MS	1.03	0.93	1.12	1.08	1.07	1.00	1.32	1.14	1.08	1.04	1.05	1.07	1.03	1.34	1.04	1.34	1.18	0.93	0.95	11.5
			Oolong tea	GC-MS	1.09	1.11	1.17	1.06	1.38	1.18	1.20	1.32	1.27	1.32	1.27	1.14	1.30	1.09	1.24	1.49	1.20	1.30	1.22	9.0
				GC-MS/MS	1.10	1.20	1.09	1.19	1.41	1.24	1.20	1.38	1.41	1.28	1.10	1.20	1.30	1.15	1.46	1.44	1.09	1.30	1.21	9.9

173	Molinate	TPT	Green tea	GC-MS	1.03	1.05	1.10	1.07	1.09	1.07	1.07	1.10	1.04	1.06	1.09	1.07	1.10	1.05	1.18	1.03	1.08	1.04	1.10	3.3
				GC–MS/MS	1.08	1.01	1.12	1.09	1.08	1.11	1.11	1.06	0.99	1.11	1.08	1.01	1.13	0.95	1.10	1.04	0.99	0.98	1.13	5.5
			Oolong tea	GC-MS	1.32	1.30	1.25	1.23	1.14	1.26	1.24	1.23	1.23	1.22	1.28	1.20	1.20	1.1	1.21	1.17	1.17	1.06	1.14	5.4
				GC–MS/MS	1.10	1.18	1.20	1.22	1.11	1.27	1.30	1.16	1.24	1.19	1.39	1.33	1.17	1.16	1.22	1.20	1.21	1.12	1.11	6.5
		ENVI-Carb + PSA	Green tea	GC-MS	1.12	1.28	1.02	1.10	1.02	1.10	0.99	1.03	1.04	1.05	1.09	1.15	1.08	1.08	1.16	1.16	1.11	1.13	1.04	6.2
				GC–MS/MS	1.08	1.27	1.05	1.10	0.99	1.07	1.03	1.10	1.12	1.03	1.15	1.08	1.12	1.00	1.14	1.16	1.09	1.13	1.05	5.9
			Oolong tea	GC-MS	1.32	1.21	1.10	1.16	1.35	1.24	1.28	1.23	1.30	1.28	1.18	1.12	1.14	1.19	1.30	1.19	1.19	1.21	1.08	6.3
				GC–MS/MS	1.37	1.23	1.09	1.19	1.24	1.27	1.15	1.23	1.29	1.27	1.08	1.11	1.10	1.20	1.40	1.27	1.15	1.20	1.11	7.6
174	Napropamide	TPT	Green tea	GC-MS	1.02	1.04	1.13	1.10	1.14	1.06	1.09	1.12	1.06	1.07	1.14	1.03	1.14	1.03	1.16	1.02	1.04	1.04	1.19	4.9
				GC–MS/MS	1.06	0.99	1.21	1.07	1.14	1.13	1.13	1.11	1.03	1.03	1.12	1.14	1.05	0.94	1.11	1.05	0.98	1.02	1.15	6.4
			Oolong tea	GC-MS	1.19	1.16	1.18	1.17	1.11	1.12	1.15	1.18	1.13	1.15	1.28	1.28	1.32	1.2	1.21	1.20	1.25	1.12	1.12	5.1
				GC–MS/MS	1.09	1.09	1.19	1.19	1.08	1.25	1.29	1.17	1.18	1.17	1.38	1.33	1.30	1.15	1.16	1.19	1.26	1.08	1.08	7.5
		ENVI-Carb + PSA	Green tea	GC-MS	1.10	1.17	1.04	1.09	1.05	1.13	0.90	1.10	1.03	1.02	1.06	1.37	1.02	1.13	1.07	1.25	1.12	1.17	1.10	8.8
				GC–MS/MS	1.09	1.21	1.06	1.12	1.03	1.12	0.94	1.11	1.06	1.01	1.11	1.04	1.16	1.18	1.10	1.25	1.14	1.11	1.11	6.7
			Oolong tea	GC-MS	1.19	1.12	1.09	1.14	1.33	1.16	1.18	1.19	1.29	1.25	1.17	1.09	1.16	1.17	1.29	1.19	1.20	1.31	1.17	5.7
				GC–MS/MS	1.14	1.16	1.10	1.17	1.25	1.25	1.12	1.20	1.31	1.25	1.08	1.17	1.05	1.08	1.36	1.20	1.08	1.21	1.08	7.1
175	Nuarimol	TPT	Green tea	GC-MS	1.03	1.05	1.11	1.11	1.10	1.02	1.11	1.04	1.08	1.08	1.10	1.01	1.04	1.17	1.08	1.02	1.05	1.06	1.19	4.5
				GC–MS/MS	1.05	1.01	1.20	1.19	1.10	1.14	1.09	1.10	1.03	1.02	1.09	1.14	1.04	0.89	1.11	1.03	1.00	1.03	1.15	6.9
			Oolong tea	GC-MS	1.20	1.18	1.14	1.13	1.03	1.20	1.14	1.10	1.17	1.13	1.25	1.14	1.05	1.1	1.23	1.23	1.24	1.16	1.19	5.4
				GC–MS/MS	1.14	1.13	1.15	1.22	1.04	1.20	1.17	1.08	1.19	1.12	1.34	1.28	1.21	1.14	1.21	1.25	1.23	1.12	1.11	6.3
		ENVI-Carb + PSA	Green tea	GC-MS	1.06	0.72	1.14	1.10	0.96	1.08	0.82	1.09	1.22	1.00	1.02	1.51	1.05	1.17	1.09	1.21	1.16	1.18	1.09	15.1
				GC–MS/MS	1.01	1.20	1.06	1.13	1.02	1.09	0.90	1.12	1.10	1.05	1.09	1.05	1.11	1.33	1.08	1.22	1.13	1.11	1.11	8.2
			Oolong tea	GC-MS	1.20	1.06	0.99	1.06	1.14	1.22	1.26	1.22	1.21	1.36	1.13	1.08	1.14	1.17	1.33	1.18	1.17	1.21	1.14	7.9
				GC–MS/MS	1.11	1.15	0.97	1.10	1.21	1.16	1.16	1.16	1.23	1.27	1.05	1.18	0.89	1.09	1.41	1.22	1.06	1.23	1.11	10.0
176	Permethrin	TPT	Green tea	GC-MS	1.05	1.06	1.13	1.10	1.09	1.10	1.09	1.07	0.99	1.01	1.12	1.00	1.06	1.03	1.08	1.02	1.05	1.03	1.15	4.2
				GC–MS/MS	1.07	1.00	1.15	1.15	1.09	1.10	1.10	1.06	0.94	1.06	0.99	1.12	1.07	1.06	1.07	1.03	0.99	0.99	1.16	5.8
			Oolong tea	GC-MS	1.15	1.12	1.09	1.18	1.08	1.18	1.07	1.05	1.21	1.12	1.27	1.20	1.10	1.1	1.18	1.19	1.21	1.15	1.16	5.0
				GC–MS/MS	1.15	1.09	1.14	1.22	1.04	1.21	1.10	1.06	1.25	1.11	1.31	1.19	1.18	1.10	1.17	1.17	1.22	1.16	1.11	5.9
		ENVI-Carb + PSA	Green tea	GC-MS	0.96	0.91	1.05	1.06	1.03	1.06	1.07	1.05	1.15	1.02	1.13	1.20	1.16	1.10	1.04	1.25	1.14	1.13	1.01	7.6
				GC–MS/MS	1.09	1.15	1.03	1.14	1.02	1.07	1.14	1.13	1.13	1.05	1.08	1.06	1.11	0.99	1.02	1.23	1.15	1.08	1.02	5.7
			Oolong tea	GC-MS	1.15	1.09	1.04	1.08	1.27	1.17	1.27	1.13	1.23	1.17	1.19	1.10	1.17	1.19	1.25	1.26	1.22	1.16	1.14	5.7
				GC–MS/MS	1.06	1.12	0.99	1.14	1.21	1.21	1.21	1.16	1.23	1.21	1.10	1.12	1.00	1.11	1.37	1.28	1.20	1.15	1.11	7.9

(Continued)

Appendix Table 5.3 Youden Pair Ratios of Each Average Content From 201 Pesticides and RSD for 19 Circulative Determinations (cont.)

No.	Pesticides	SPE	Sample	Method	Youden pair ratio																			RSD%
					Nov 9, 2009 (n=5)	Nov 14, 2009 (n=3)	Nov 19, 2009 (n=3)	Nov 24, 2009 (n=3)	Nov 29, 2009 (n=3)	Dec 4, 2009 (n=3)	Dec 9, 2009 (n=3)	Dec 14, 2009 (n=3)	Dec 19, 2009 (n=3)	Dec 24, 2009 (n=3)	Dec 14, 2009 (n=3)	Jan 3, 2010 (n=3)	Jan 8, 2010 (n=3)	Jan 13, 2010 (n=3)	Jan 18, 2010 (n=3)	Jan 23, 2010 (n=3)	Jan 28, 2010 (n=3)	Feb 2, 2010 (n=3)	Feb 7, 2010 (n=3)	
177	Phenothrin	TPT	Green tea	GC-MS	1.11	1.18	1.09	1.09	1.13	1.08	1.08	1.08	1.03	1.00	1.14	1.00	1.07	1.02	1.14	1.03	1.08	1.04	1.13	4.7
				GC-MS/MS	1.12	1.10	1.14	1.22	1.07	1.14	1.08	1.11	1.20	1.12	1.26	1.20	1.21	1.2	1.25	1.15	1.21	1.11	1.19	4.8
			Oolong tea	GC-MS	1.03	1.18	1.15	1.17	1.02	1.19	1.11	0.95	1.10	0.81	1.13	1.09	1.41	1.17	1.25	1.06	0.81	0.92	1.12	13.3
				GC-MS/MS	1.06	1.06	1.04	1.13	1.10	1.18	1.00	1.07	1.07	1.04	1.03	1.16	1.09	1.16	1.08	1.26	1.12	1.14	0.98	6.1
		ENVI-Carb + PSA	Green tea	GC-MS	1.12	1.10	1.05	1.09	1.23	1.19	1.26	1.15	1.19	1.20	1.17	1.13	1.16	1.17	1.26	1.24	1.23	1.16	1.12	5.1
				GC-MS/MS	1.08	0.96	0.99	1.13	1.39	1.00	1.25	1.19	1.10	1.09	0.76	0.94	1.09	1.13	0.77	0.66	0.91	0.67	0.59	21.9
			Oolong tea	GC-MS	1.22	0.97	1.10	1.05	1.11	1.09	1.09	1.06	1.03	1.04	1.00	1.13	0.87	0.87	0.79	0.98	1.09	1.07	1.17	10.3
				GC-MS/MS	1.14	1.23	1.10	1.09	1.22	1.12	1.04	1.10	1.10	1.03	1.03	1.05	1.12	1.05	1.06	1.34	1.13	1.09	1.03	7.2
178	Piperonyl	TPT	Green tea	GC-MS	1.11	1.00	1.04	1.09	1.02	1.15	1.15	1.12	1.34	1.15	1.31	1.16	1.17	1.1	1.19	1.20	1.24	1.09	1.14	7.7
				GC-MS/MS	1.11	1.10	1.13	1.19	1.09	1.21	1.20	1.09	1.21	1.15	1.35	1.28	1.21	1.14	1.18	1.18	1.22	1.12	1.12	5.7
			Oolong tea	GC-MS	1.11	1.06	1.08	1.09	1.34	1.19	1.22	1.17	1.23	1.21	1.34	1.02	1.15	1.22	1.33	1.18	1.26	1.20	1.15	7.7
				GC-MS/MS	1.11	1.14	1.02	1.15	1.26	1.21	1.17	1.20	1.24	1.23	1.10	1.15	0.94	1.19	1.37	1.23	1.13	1.19	1.12	7.9
		ENVI-Carb + PSA	Green tea	GC-MS	1.08	1.01	1.17	1.27	1.13	1.11	1.10	1.07	1.01	1.06	1.11	1.06	1.06	1.00	1.10	1.04	0.99	1.05	1.15	6.3
				GC-MS/MS	1.10	1.22	1.04	1.12	1.04	1.12	0.99	1.11	1.10	1.00	1.10	1.07	1.21	1.05	1.12	1.22	1.14	1.13	1.04	6.1
			Oolong tea	GC-MS	1.06	0.98	1.12	1.10	1.14	1.02	1.08	1.06	0.94	1.05	1.03	1.08	1.08	1.03	1.12	1.01	1.02	1.06	1.16	5.2
				GC-MS/MS	1.13	1.11	1.03	1.13	1.02	1.21	0.97	1.13	1.11	1.01	1.15	1.67	1.11	1.14	1.11	1.25	1.13	1.18	1.02	13.0
179	Pretilachlor	TPT	Green tea	GC-MS	1.03	1.04	1.09	1.17	1.07	1.02	1.10	1.10	0.98	1.05	1.14	1.11	1.06	1.09	1.12	1.06	1.00	1.01	1.05	4.5
				GC-MS/MS	1.01	1.22	1.01	1.73	1.00	1.13	1.17	1.11	0.98	1.00	1.21	1.13	1.03	0.93	1.13	1.12	1.00	1.00	1.12	15.4
			Oolong tea	GC-MS	1.26	1.23	1.23	1.18	1.08	1.14	1.45	1.11	1.12	1.10	1.21	1.27	1.27	1.1	1.13	1.12	1.20	1.07	1.17	7.8
				GC-MS/MS	1.10	1.14	1.25	1.25	1.12	1.26	1.32	1.18	1.24	1.11	1.31	1.19	1.41	1.18	1.13	1.22	1.24	1.10	1.05	7.6
		ENVI-Carb + PSA	Green tea	GC-MS	1.13	1.24	1.00	1.02	1.02	0.85	1.04	1.15	1.07	0.96	1.05	1.09	1.09	1.12	0.98	1.20	1.07	1.03	1.03	8.3
				GC-MS/MS	0.75	1.21	1.07	1.03	1.00	0.95	1.39	1.02	1.03	0.97	1.06	1.06	1.14	1.28	1.05	1.25	1.20	0.99	1.03	13.3
			Oolong tea	GC-MS	1.26	1.19	1.07	1.11	1.25	1.15	0.88	1.23	1.34	1.28	1.10	1.20	1.10	1.19	1.20	1.17	1.12	1.19	1.04	8.7
				GC-MS/MS	1.60	1.16	1.00	1.20	1.25	1.21	1.10	1.23	1.36	1.23	1.12	1.21	1.03	1.10	1.37	1.25	1.08	1.31	1.05	11.9
180	Prometon	TPT	Green tea	GC-MS	1.01	1.05	1.12	1.10	1.13	1.01	1.09	1.11	1.03	1.08	1.12	1.05	1.04	1.02	1.15	1.01	1.03	1.03	1.14	4.6
				GC-MS/MS	1.06	0.99	1.18	1.10	1.13	1.10	1.13	1.09	1.02	1.03	1.13	1.10	1.07	0.96	1.09	1.05	0.98	1.02	1.13	5.5
			Oolong tea	GC-MS	1.31	1.08	1.12	1.11	1.08	1.20	1.23	1.22	1.35	1.24	1.30	1.32	1.27	1.2	1.23	1.22	1.24	1.07	1.09	7.4
				GC-MS/MS	1.09	1.10	1.20	1.19	1.07	1.23	1.33	1.18	1.14	1.16	1.40	1.32	1.33	1.16	1.16	1.19	1.25	1.08	1.11	7.9
		ENVI-Carb + PSA	Green tea	GC-MS	1.11	1.20	1.03	1.08	1.04	1.17	0.96	1.06	0.98	1.01	1.07	1.83	1.09	1.14	1.10	1.27	1.12	1.14	1.00	16.6
				GC-MS/MS	1.09	1.23	1.06	1.12	1.03	1.10	0.98	1.12	1.07	1.01	1.11	1.10	1.12	1.25	1.11	1.25	1.14	1.11	1.02	6.8
			Oolong tea	GC-MS	1.31	1.07	1.11	1.18	1.38	1.17	1.17	1.19	1.25	1.27	1.19	1.02	1.14	1.23	1.26	1.22	1.20	1.28	1.12	7.3
				GC-MS/MS	1.21	1.17	1.13	1.19	1.28	1.26	1.12	1.21	1.33	1.23	1.06	1.14	1.20	1.22	1.38	1.22	1.12	1.24	1.11	6.5

No.	Compound	Sorbent	Tea	Method	C1	C2	C3	C4	C5	C6	C7	C8	C9	C10	C11	C12	C13	C14	C15	C16	C17	C18	C19	RSD
181	Pronamide	TPT	Green tea	GC-MS	1.02	1.02	1.24	1.02	1.09	1.03	1.07	1.13	0.99	1.05	1.16	1.06	1.06	1.00	1.09	1.00	1.03	1.03	1.13	6.1
				GC-MS/MS	1.05	0.99	1.15	1.24	1.08	1.09	1.10	1.12	0.99	1.04	1.13	1.12	1.06	0.99	1.09	1.04	0.98	1.02	1.14	6.2
			Oolong tea	GC-MS	1.29	1.09	1.18	1.17	1.02	1.20	1.17	1.14	1.21	1.07	1.23	1.25	1.14	1.1	1.14	1.14	1.18	1.16	1.06	5.8
				GC-MS/MS	1.06	1.10	1.12	1.15	1.03	1.21	1.17	1.12	1.15	1.03	1.28	1.28	1.18	1.15	1.14	1.11	1.15	1.07	1.12	6.1
		ENVI-Carb + PSA	Green tea	GC-MS	1.04	0.95	1.05	1.11	0.97	1.27	0.96	1.06	1.06	0.99	1.07	1.38	1.10	1.11	1.08	1.22	1.13	1.09	0.95	10.3
				GC-MS/MS	1.05	1.17	1.04	1.11	1.05	1.08	0.99	1.09	1.05	1.00	1.09	1.05	1.10	1.25	1.09	1.25	1.14	1.06	1.01	6.6
			Oolong tea	GC-MS	1.29	1.15	1.01	1.06	1.24	1.09	1.06	1.13	1.44	1.13	1.10	1.09	1.09	1.09	1.18	1.16	1.04	1.13	1.13	8.7
				GC-MS/MS	1.24	1.20	1.03	1.10	1.18	1.11	1.03	1.17	1.35	1.10	1.02	1.10	0.90	1.09	1.28	1.17	1.01	1.12	1.12	9.3
182	Propetam-phos	TPT	Green tea	GC-MS	1.05	1.04	1.15	1.05	0.98	0.96	1.07	1.17	0.80	1.13	0.99	1.04	1.10	0.93	0.95	1.02	1.01	0.99	1.13	8.7
				GC-MS/MS	1.04	1.03	1.12	1.08	1.01	1.09	1.13	1.07	0.86	1.00	1.06	1.18	1.12	0.91	1.05	1.05	0.95	0.96	1.11	7.7
			Oolong tea	GC-MS	1.20	1.04	1.11	1.15	1.09	1.23	1.26	1.35	1.70	1.89	1.13	0.99	1.21	1.1	1.23	1.18	1.24	1.11	1.12	17.9
				GC-MS/MS	1.17	1.12	1.19	1.20	1.16	1.30	1.25	1.22	1.34	1.14	1.43	1.31	1.31	1.17	1.20	1.16	1.25	1.16	1.13	6.8
		ENVI-Carb + PSA	Green tea	GC-MS	1.07	1.04	1.10	1.17	1.06	1.18	1.37	1.11	1.01	1.05	1.25	1.40	1.10	1.01	1.07	1.24	1.12	0.82	0.99	12.2
				GC-MS/MS	1.00	0.97	1.10	1.12	1.09	0.99	1.60	1.10	1.08	1.08	1.21	1.10	0.96	1.34	1.02	1.27	1.20	0.88	0.99	14.7
			Oolong tea	GC-MS	1.20	1.06	1.24	1.08	1.37	1.11	1.23	1.83	1.63	1.79	1.06	1.22	1.20	1.12	1.16	1.37	1.16	1.17	1.16	18.2
				GC-MS/MS	1.26	1.21	1.17	1.16	1.37	1.19	1.16	1.30	1.33	1.19	1.11	1.06	1.12	1.13	1.31	1.36	1.08	1.21	1.13	7.8
183	Propoxur-1	TPT	Green tea	GC-MS	0.99	1.03	1.13	1.05	1.13	1.09	1.07	1.15	1.08	1.08	1.14	1.04	1.02	1.02	1.13	1.02	1.03	1.07	1.14	4.7
				GC-MS/MS	1.05	0.97	1.13	1.08	1.08	1.10	1.10	1.15	1.02	1.00	1.10	1.07	1.01	0.95	1.10	1.03	1.01	1.05	1.14	5.5
			Oolong tea	GC-MS	1.28	1.21	1.23	1.16	1.11	1.19	1.27	1.22	1.20	1.20	1.33	1.23	1.26	1.2	1.20	1.22	1.20	1.07	1.15	4.9
				GC-MS/MS	1.09	1.14	1.21	1.17	1.06	1.24	1.35	1.18	0.90	1.19	0.88	1.29	1.20	1.19	1.20	1.20	1.23	1.09	1.09	10.1
		ENVI-Carb + PSA	Green tea	GC-MS	1.08	1.17	1.01	1.08	0.99	1.07	0.84	1.05	1.06	1.03	1.09	1.02	1.08	1.11	1.08	1.19	1.11	1.16	1.01	7.3
				GC-MS/MS	1.10	1.24	1.03	1.08	0.96	1.08	0.87	1.09	1.08	1.01	1.10	1.08	1.12	1.33	1.09	1.18	1.10	1.11	1.02	9.0
			Oolong tea	GC-MS	1.28	1.21	1.17	1.16	1.26	1.23	1.18	1.13	1.22	1.33	1.14	1.21	1.13	1.20	1.32	1.17	1.16	1.24	1.11	5.3
				GC-MS/MS	1.21	1.21	1.11	1.18	1.21	1.24	1.18	1.21	1.27	1.33	1.08	1.12	1.11	1.17	1.41	1.17	1.08	1.23	1.09	7.2
184	Propoxur-2	TPT	Green tea	GC-MS	1.03	1.09	1.06	1.15	0.80	1.10	1.08	1.03	1.06	0.96	1.01	0.98	0.96	1.08	1.15	0.96	1.01	1.08	1.09	7.9
				GC-MS/MS	2.95	1.20	0.96	1.08	0.85	1.13	1.33	0.92	0.59	0.87	1.11	1.17	1.09	0.77	0.92	1.05	0.98	0.90	1.14	17.2
			Oolong tea	GC-MS	1.23	1.17	1.12	1.30	0.89	1.00	0.94	1.07	1.05	1.04	1.31	1.37	1.02	1.0	0.99	0.99	0.99	0.92	1.08	13.0
				GC-MS/MS	1.06	1.21	1.27	1.56	1.49	1.40	1.12	1.14	1.06	1.02	1.38	1.27	1.33	1.11	1.19	1.23	1.15	1.32	1.12	12.1
		ENVI-Carb + PSA	Green tea	GC-MS	1.28	0.82	1.12	0.90	1.12	1.13	0.79	1.16	0.96	1.17	1.30	0.92	1.01	0.99	0.99	1.02	0.99	0.93	0.95	13.6
				GC-MS/MS	1.01	0.66	1.11	1.12	1.27	0.93	4.52	1.09	0.87	0.99	1.41	1.30	1.03	1.26	0.93	1.54	1.45	0.51	0.70	67.2
			Oolong tea	GC-MS	1.23	1.20	1.05	1.12	1.09	1.01	1.27	1.36	1.58	1.42	1.10	1.02	1.26	0.98	1.07	1.46	1.02	1.21	1.13	14.3
				GC-MS/MS	1.37	1.25	1.09	1.03	1.37	1.29	1.06	1.36	1.44	1.24	1.05	1.16	1.07	0.86	1.26	1.48	0.98	1.03	1.12	14.5

(Continued)

Appendix Table 5.3 Youden Pair Ratios of Each Average Content From 201 Pesticides and RSD for 19 Circulative Determinations (cont.)

No.	Pesticides	SPE	Sample	Method	Youden pair ratio																		RSD%
					Nov 9, 2009 (n=5)	Nov 14, 2009 (n=3)	Nov 19, 2009 (n=3)	Nov 24, 2009 (n=3)	Nov 29, 2009 (n=3)	Dec 4, 2009 (n=3)	Dec 9, 2009 (n=3)	Dec 14, 2009 (n=3)	Dec 19, 2009 (n=3)	Dec 24, 2009 (n=3)	Jan 3, 2010 (n=3)	Jan 8, 2010 (n=3)	Jan 13, 2010 (n=3)	Jan 18, 2010 (n=3)	Jan 23, 2010 (n=3)	Jan 28, 2010 (n=3)	Feb 2, 2010 (n=3)	Feb 7, 2010 (n=3)	
185	Prothiophos	TPT	Green tea	GC-MS	1.02	1.03	1.11	1.09	1.06	1.01	1.07	1.08	0.98	1.09	1.06	1.10	1.05	1.06	1.02	1.01	1.02	1.14	4.0
				GC-MS/MS	1.04	1.03	1.11	1.05	1.02	1.07	1.15	1.06	0.98	1.07	1.12	1.05	1.00	1.11	1.05	1.02	1.05	1.16	4.7
			Oolong tea	GC-MS	1.18	1.12	1.15	1.14	1.08	1.20	1.32	1.12	1.28	1.23	1.24	1.24	1.1	1.18	1.17	1.23	1.09	1.11	5.5
				GC-MS/MS	1.16	1.19	1.19	1.19	1.08	1.22	1.28	1.12	1.19	1.18	1.28	1.25	1.20	1.21	1.13	1.26	1.09	1.19	5.5
		ENVI-Carb + PSA	Green tea	GC-MS	1.04	1.13	1.03	1.04	1.05	1.11	1.18	1.08	1.12	1.00	1.52	1.08	1.12	1.04	1.22	1.12	1.02	1.01	10.5
				GC-MS/MS	0.97	1.17	1.07	1.07	1.05	1.05	1.24	1.08	1.13	1.03	1.12	1.04	1.22	1.04	1.28	1.18	1.00	1.02	7.8
			Oolong tea	GC-MS	1.18	1.13	1.13	1.13	1.29	1.19	1.14	1.20	1.22	1.23	1.07	1.15	1.18	1.24	1.22	1.18	1.19	1.14	4.4
				GC-MS/MS	1.26	1.17	1.10	1.16	1.28	1.18	1.17	1.20	1.30	1.17	1.16	1.28	1.18	1.31	1.27	1.09	1.19	1.19	5.9
186	Ptrimethanil	TPT	Green tea	GC-MS	1.02	1.00	1.03	0.97	0.95	1.05	1.14	0.95	0.91	0.87	0.95	0.99	0.99	1.08	0.98	0.89	0.92	1.10	7.1
				GC-MS/MS	1.05	1.35	0.94	0.97	0.91	0.97	1.19	1.16	1.07	0.94	1.05	0.90	0.94	1.01	1.12	0.96	0.98	0.97	11.1
			Oolong tea	GC-MS	1.42	1.22	0.95	1.00	0.99	1.11	1.04	1.01	0.84	0.91	1.20	1.08	1.03	0.90	1.02	0.88	1.01	1.09	13.0
				GC-MS/MS	1.01	1.04	1.03	1.27	1.13	1.26	0.97	1.09	1.23	1.00	1.29	1.03	0.9	1.26	1.14	1.25	1.18	1.08	10.9
		ENVI-Carb + PSA	Green tea	GC-MS	0.89	1.00	1.01	1.16	0.90	1.04	0.89	1.06	1.12	1.02	1.02	1.00	0.97	0.92	1.01	1.01	0.99	0.86	8.2
				GC-MS/MS	1.85	0.57	1.18	0.99	1.12	1.01	1.37	1.12	1.21	1.16	1.18	1.08	0.96	1.08	1.07	1.10	0.86	0.94	22.3
			Oolong tea	GC-MS	1.01	1.00	1.02	0.95	1.39	1.08	1.16	1.57	1.31	1.04	0.97	1.27	0.92	0.97	1.60	1.19	1.04	1.25	17.5
				GC-MS/MS	1.81	1.04	1.00	1.10	1.06	0.95	0.99	1.18	1.31	1.08	0.94	0.78	1.08	1.15	1.05	1.00	0.92	0.86	20.7
187	Pyridaben	TPT	Green tea	GC-MS	1.00	1.26	0.94	1.25	1.00	1.04	1.07	1.14	0.93	0.95	1.32	0.97	0.89	1.07	1.08	0.96	1.01	1.09	11.5
				GC-MS/MS	0.70	0.94	1.07	1.12	0.97	0.95	1.20	1.15	1.13	1.02	0.95	1.20	1.30	1.03	1.24	1.31	0.92	0.93	14.5
			Oolong tea	GC-MS	0.99	1.02	1.07	1.07	1.05	0.98	1.05	1.09	0.88	1.04	1.14	1.07	0.99	1.06	1.01	1.05	1.01	1.06	5.7
				GC-MS/MS	1.03	1.07	0.99	1.16	1.01	1.12	1.26	0.99	1.25	1.21	1.28	1.12	1.1	1.19	1.16	1.20	1.16	1.24	8.3
		ENVI-Carb + PSA	Green tea	GC-MS	1.30	1.12	1.09	1.21	1.10	1.21	1.15	1.08	1.52	1.07	1.22	1.17	1.11	1.17	1.28	1.30	1.20	1.20	9.0
				GC-MS/MS	0.96	0.96	1.03	1.14	1.02	0.85	0.98	1.22	1.13	1.00	1.43	1.14	1.04	1.02	1.09	1.10	0.96	0.98	11.9
			Oolong tea	GC-MS	1.03	1.03	1.02	1.05	1.22	1.14	0.92	1.30	1.24	1.33	0.96	1.19	1.00	1.17	1.31	1.16	1.04	1.06	10.9
				GC-MS/MS	1.40	1.07	0.95	1.11	1.34	1.21	1.13	1.32	1.30	1.25	1.21	1.52	1.06	1.29	1.37	1.14	1.42	1.20	11.7
188	Pyridaphenthion	TPT	Green tea	GC-MS	1.06	0.98	1.18	1.11	1.09	1.10	1.12	1.06	0.99	1.06	1.12	1.10	0.99	1.09	1.05	1.01	1.02	1.15	5.2
				GC-MS/MS	1.09	1.22	1.06	1.10	1.03	1.10	0.99	1.10	1.06	1.00	1.05	1.07	1.20	1.09	1.21	1.10	1.09	1.03	5.8
			Oolong tea	GC-MS	1.03	1.04	1.14	1.11	1.12	1.01	1.08	1.12	1.01	1.05	1.04	1.07	1.04	1.10	1.01	1.04	1.06	1.13	4.2
				GC-MS/MS	1.22	1.15	1.18	1.13	1.10	1.20	1.28	1.21	1.20	1.18	1.30	1.20	1.1	1.21	1.19	1.22	1.08	1.13	4.8
		ENVI-Carb + PSA	Green tea	GC-MS	1.08	1.13	1.19	1.19	1.11	1.23	1.31	1.16	1.14	1.15	1.30	1.24	1.09	1.21	1.13	1.20	1.15	1.13	6.8
				GC-MS/MS	1.11	1.15	1.05	1.08	1.02	1.17	0.94	1.06	1.05	1.02	1.65	1.10	1.09	1.08	1.21	1.10	1.16	1.01	13.0
			Oolong tea	GC-MS	1.22	1.15	1.21	1.13	1.31	1.19	1.16	1.14	1.28	1.25	1.05	1.10	1.26	1.30	1.26	1.16	1.26	1.13	6.0
				GC-MS/MS	1.20	1.20	1.13	1.19	1.25	1.24	1.16	1.24	1.28	1.20	1.11	1.28	1.20	1.33	1.22	1.12	1.30	1.05	6.2

189	Pyriproxyfen	TPT	Green tea	GC-MS	1.02	1.04	1.12	1.11	1.13	1.08	1.06	1.06	1.05	1.03	1.11	1.07	1.06	1.05	1.10	0.99	1.05	1.08	1.14	3.7
				GC-MS/MS	1.02	0.97	1.04	0.96	0.96	1.05	1.12	1.06	0.92	0.88	0.97	1.07	0.96	1.00	1.09	0.98	1.00	0.92	1.10	6.7
			Oolong tea	GC-MS	1.14	1.12	1.12	1.18	1.06	1.03	1.13	1.10	1.21	1.14	1.23	1.22	1.14	1.1	1.24	1.22	1.23	1.17	1.18	5.1
				GC-MS/MS	1.10	1.10	1.14	1.20	1.04	1.19	1.11	1.07	1.23	1.13	1.31	1.22	1.13	1.14	1.16	1.19	1.24	1.16	1.12	5.6
		ENVI-Carb + PSA	Green tea	GC-MS	1.10	1.28	1.05	1.16	1.02	1.16	0.96	1.08	1.09	1.02	1.04	1.25	1.04	1.11	1.03	1.24	1.16	1.17	1.17	8.1
				GC-MS/MS	1.04	1.34	0.94	0.97	0.92	0.98	0.99	0.96	1.07	0.94	0.98	1.06	0.91	0.93	1.01	1.11	0.96	0.98	0.98	9.7
			Oolong tea	GC-MS	1.14	1.08	1.05	1.07	1.26	1.21	1.26	1.12	1.23	1.20	1.24	1.15	1.14	1.22	1.29	1.20	1.24	1.16	1.16	5.8
				GC-MS/MS	1.06	1.13	0.98	1.16	1.20	1.23	1.21	1.13	1.22	1.23	1.09	1.13	1.08	1.18	1.38	1.25	1.15	1.17	1.12	7.4
190	Quinalphos	TPT	Green tea	GC-MS	0.98	1.01	1.09	1.07	1.03	1.03	1.02	1.07	0.94	1.06	1.07	1.05	1.01	1.00	0.98	1.07	1.00	1.00	1.08	4.0
				GC-MS/MS	1.01	1.11	1.01	1.19	1.00	1.03	1.10	1.10	0.91	1.00	1.17	1.16	1.04	0.98	1.12	1.09	0.98	1.04	1.12	7.2
			Oolong tea	GC-MS	1.17	1.15	1.06	1.19	1.07	1.12	1.07	1.07	1.19	1.16	1.09	1.30	1.19	1.0	1.15	1.07	1.10	1.05	1.10	6.2
				GC-MS/MS	1.23	1.14	1.19	1.21	1.13	1.28	1.17	1.17	1.30	1.13	1.25	1.29	1.33	1.15	1.16	1.21	1.09	1.15	1.15	5.6
		ENVI-Carb + PSA	Green tea	GC-MS	0.97	1.14	1.04	1.11	1.06	0.90	1.26	1.11	1.03	0.98	1.07	1.73	1.14	1.04	1.13	1.00	1.04	1.14	1.03	15.8
				GC-MS/MS	0.87	1.08	1.07	1.06	1.03	1.00	1.23	1.09	1.05	1.01	1.20	1.08	1.07	1.10	1.03	1.22	1.10	1.07	1.00	8.5
			Oolong tea	GC-MS	1.17	1.13	1.11	1.10	1.32	1.14	1.04	1.27	1.29	1.22	1.42	1.04	1.13	1.14	1.24	1.21	1.14	1.19	1.08	8.3
				GC-MS/MS	1.37	1.16	1.09	1.11	1.31	1.18	1.13	1.28	1.30	1.18	1.12	1.07	1.20	1.19	1.36	1.24	1.04	1.28	1.15	8.1
191	Quinoxyphen	TPT	Green tea	GC-MS	1.02	1.04	1.13	1.07	1.10	1.09	1.09	1.06	1.03	1.06	1.07	1.04	1.06	1.04	1.09	1.02	1.01	1.04	1.15	3.6
				GC-MS/MS	1.07	1.01	1.15	1.13	1.09	1.12	1.10	1.08	1.01	1.03	1.13	1.18	1.04	0.98	1.10	1.05	1.00	1.03	1.15	5.3
			Oolong tea	GC-MS	1.17	1.16	1.16	1.19	1.06	1.13	1.24	1.14	1.21	1.12	1.30	1.22	1.21	1.1	1.26	1.22	1.24	1.10	1.13	5.2
				GC-MS/MS	1.08	1.11	1.16	1.20	1.05	1.21	1.19	1.12	1.20	1.14	1.34	1.29	1.17	1.15	1.17	1.19	1.19	1.12	1.13	5.8
		ENVI-Carb + PSA	Green tea	GC-MS	1.05	1.02	1.04	1.09	1.01	1.11	0.99	1.06	1.06	0.98	1.10	1.13	1.05	1.11	1.10	1.21	1.13	1.14	0.98	5.8
				GC-MS/MS	1.07	1.21	1.04	1.12	1.02	1.11	0.98	1.10	1.08	1.02	1.10	1.03	1.15	1.12	1.11	1.23	1.12	1.12	1.02	6.0
			Oolong tea	GC-MS	1.17	1.16	1.04	1.12	1.29	1.17	1.21	1.16	1.24	1.24	1.20	1.14	1.16	1.22	1.27	1.23	1.21	1.22	1.14	4.9
				GC-MS/MS	1.09	1.14	1.04	1.14	1.23	1.22	1.17	1.18	1.28	1.23	1.09	1.16	1.27	1.18	1.33	1.25	1.12	1.21	1.12	6.2
192	Telodrin	TPT	Green tea	GC-MS	1.06	1.03	1.11	1.04	0.99	1.13	1.10	1.05	0.86	1.02	1.08	1.12	1.08	0.97	0.99	1.03	0.99	0.98	1.11	6.6
				GC-MS/MS	1.05	1.03	1.10	1.03	0.94	1.09	1.13	1.05	0.85	1.02	1.08	1.19	1.13	0.96	1.08	1.02	0.97	0.95	1.13	7.8
			Oolong tea	GC-MS	1.25	1.18	1.16	1.17	1.13	1.27	1.30	1.23	1.15	1.10	1.34	1.23	1.21	1.1	1.21	1.13	1.22	1.07	1.10	6.4
				GC-MS/MS	1.17	1.21	1.21	1.24	1.13	1.32	1.26	1.21	1.20	1.12	1.39	1.28	1.26	1.16	1.16	1.17	1.21	1.14	1.16	5.7
		ENVI-Carb + PSA	Green tea	GC-MS	1.08	0.96	1.07	1.05	1.07	1.03	1.51	1.08	1.10	1.08	1.21	1.09	1.06	1.09	0.95	1.28	1.12	0.93	1.06	11.7
				GC-MS/MS	1.03	1.02	1.07	1.08	1.08	0.97	1.82	1.11	1.10	1.10	1.14	1.10	0.94	1.34	1.01	1.37	1.12	0.89	1.05	18.3
			Oolong tea	GC-MS	1.25	1.27	1.21	1.16	1.30	1.14	1.12	1.17	1.36	1.17	1.16	1.13	1.21	1.11	1.17	1.25	1.15	1.19	1.19	5.4
				GC-MS/MS	1.26	1.21	1.14	1.19	1.29	1.14	1.21	1.22	1.36	1.17	1.06	1.12	1.35	1.09	1.31	1.35	1.10	1.18	1.16	7.6

(Continued)

Appendix Table 5.3 Youden Pair Ratios of Each Average Content From 201 Pesticides and RSD for 19 Circulative Determinations (cont.)

No.	Pesticides	SPE	Sample	Method	Youden pair ratio																			RSD%
					Nov 9, 2009 (n = 5)	Nov 14, 2009 (n = 3)	Nov 19, 2009 (n = 3)	Nov 24, 2009 (n = 3)	Nov 29, 2009 (n = 3)	Dec 4, 2009 (n = 3)	Dec 9, 2009 (n = 3)	Dec 14, 2009 (n = 3)	Dec 19, 2009 (n = 3)	Dec 24, 2009 (n = 3)	Dec 14, 2009 (n = 3)	Jan 3, 2010 (n = 3)	Jan 8, 2010 (n = 3)	Jan 13, 2010 (n = 3)	Jan 18, 2010 (n = 3)	Jan 23, 2010 (n = 3)	Jan 28, 2010 (n = 3)	Feb 2, 2010 (n = 3)	Feb 7, 2010 (n = 3)	
193	Tetrasul	TPT	Green tea	GC-MS	0.98	1.05	1.11	1.09	1.10	1.07	1.07	1.06	1.04	1.05	1.12	1.08	1.11	1.05	1.13	1.02	1.02	1.04	1.17	4.2
				GC-MS/MS	1.05	0.98	1.17	0.96	1.08	1.11	1.11	1.07	1.01	1.09	1.10	1.11	1.06	1.07	1.10	1.07	0.94	1.03	1.14	5.6
			Oolong tea	GC-MS	1.22	1.19	1.19	1.18	1.09	1.17	1.21	1.17	1.20	1.14	1.28	1.25	1.21	1.1	1.04	1.17	1.20	1.07	1.09	5.4
				GC-MS/MS	1.08	1.17	1.18	1.18	1.08	1.22	1.25	1.14	1.18	1.14	1.33	1.27	1.21	1.10	1.17	1.15	1.23	1.08	1.03	6.4
		ENVI-Carb + PSA	Green tea	GC-MS	1.08	1.19	1.00	1.09	1.03	1.11	0.99	1.04	1.06	1.01	1.04	1.04	1.07	1.16	1.12	1.30	1.12	1.12	1.01	7.0
				GC-MS/MS	1.09	1.27	1.03	1.10	1.04	1.12	1.05	1.08	1.04	0.99	1.08	1.06	1.17	1.08	1.10	1.23	1.12	1.09	1.01	6.4
			Oolong tea	GC-MS	1.22	1.18	1.05	1.14	1.23	1.14	1.17	1.11	1.26	1.24	1.18	1.16	1.15	1.22	1.31	1.20	1.16	1.18	1.10	5.0
				GC-MS/MS	1.14	1.16	1.06	1.14	1.25	1.20	1.14	1.21	1.27	1.22	1.05	1.15	1.10	1.17	1.36	1.23	1.08	1.20	1.03	7.0
194	Thiazopyr	TPT	Green tea	GC-MS	1.14	1.01	1.12	1.10	1.08	1.07	1.09	1.11	1.02	1.05	1.15	1.07	1.06	1.02	1.12	1.02	1.02	1.03	1.12	4.2
				GC-MS/MS	1.11	1.02	1.17	0.76	1.15	1.11	1.14	1.10	1.01	1.07	1.13	1.19	1.03	0.95	1.17	1.04	0.98	1.08	1.16	9.5
			Oolong tea	GC-MS	1.23	1.18	1.16	1.19	1.12	1.20	1.36	1.19	1.23	1.23	1.36	1.27	1.28	1.1	1.21	1.20	1.25	1.12	1.14	5.6
				GC-MS/MS	1.08	1.08	1.22	1.17	1.09	1.28	1.34	1.18	1.23	1.18	1.37	1.31	1.36	1.22	1.17	1.13	1.17	1.18	1.11	7.6
		ENVI-Carb + PSA	Green tea	GC-MS	1.07	1.22	1.02	1.05	1.04	1.11	0.98	1.07	1.04	1.02	1.07	1.14	1.07	1.08	1.09	1.22	1.11	1.10	0.97	6.2
				GC-MS/MS	1.04	1.21	1.07	1.04	1.04	1.09	0.99	1.05	1.06	1.01	1.06	1.09	1.08	1.11	1.05	1.27	1.16	1.11	1.09	6.2
			Oolong tea	GC-MS	1.23	1.18	1.21	1.10	1.30	1.20	1.19	1.21	1.34	1.30	1.21	1.20	1.15	1.21	1.25	1.22	1.18	1.24	1.11	4.8
				GC-MS/MS	1.21	1.18	1.10	1.23	1.28	1.21	1.17	1.22	1.35	1.29	1.15	1.16	0.92	1.09	1.27	1.21	1.09	1.13	1.11	8.2
195	Tolclofos-methyl	TPT	Green tea	GC-MS	1.01	1.03	1.10	1.09	1.07	1.03	1.09	1.09	0.99	1.05	1.11	1.06	1.02	1.04	1.07	1.04	1.02	1.02	1.10	3.4
				GC-MS/MS	0.99	1.09	1.02	1.05	1.01	1.11	1.14	1.09	0.98	1.15	1.14	1.13	1.07	1.01	1.12	1.05	0.98	0.98	1.12	5.8
			Oolong tea	GC-MS	1.24	1.16	1.19	1.14	1.10	1.20	1.33	1.23	1.23	1.17	1.21	1.20	1.32	1.1	1.15	1.17	1.22	1.10	1.15	5.2
				GC-MS/MS	1.18	1.14	1.25	1.23	1.11	1.27	1.33	1.18	1.16	1.11	1.30	1.31	1.36	1.18	1.17	1.15	1.18	1.14	1.10	6.6
		ENVI-Carb + PSA	Green tea	GC-MS	1.05	1.17	1.01	1.04	1.04	0.99	1.07	1.11	1.06	1.04	1.07	1.31	1.04	1.10	1.03	1.12	1.10	1.06	1.01	6.7
				GC-MS/MS	0.84	1.21	1.04	1.06	1.03	1.06	1.17	1.07	1.05	1.04	1.08	1.11	1.05	1.23	1.07	1.17	1.16	1.10	0.99	8.1
			Oolong tea	GC-MS	1.24	1.17	1.17	1.12	1.30	1.17	1.05	1.20	1.24	1.24	1.17	1.08	1.13	1.15	1.24	1.18	1.18	1.21	1.12	5.1
				GC-MS/MS	1.47	1.20	1.14	1.19	1.27	1.22	1.14	1.21	1.28	1.20	1.02	1.19	1.28	1.17	1.36	1.17	1.12	1.24	1.10	8.1

196	Transfluthrin	TPT	Green tea	GC–MS	1.06	1.04	1.05	1.09	1.07	1.07	1.06	1.08	1.01	1.05	1.11	1.07	1.07	0.90	1.06	1.03	1.20	1.05	1.15	5.5
				GC–MS/MS	1.06	0.98	1.15	1.11	1.08	1.14	1.12	1.08	0.98	1.06	1.08	1.10	1.09	1.03	1.04	1.09	1.00	1.02	1.14	4.8
			Oolong tea	GC–MS	1.19	1.17	1.14	1.15	1.11	1.20	1.27	1.23	1.15	1.15	1.26	1.29	1.21	1.1	1.22	1.21	1.22	1.10	1.11	4.8
				GC–MS/MS	1.09	1.12	1.20	1.21	1.07	1.24	1.32	1.18	1.18	1.18	1.38	1.30	1.32	1.19	1.21	1.19	1.24	1.11	1.09	7.2
		ENVI-Carb + PSA	Green tea	GC–MS	1.16	0.94	1.08	0.96	1.03	1.19	1.02	1.04	1.07	1.01	1.09	1.25	1.10	1.15	1.06	1.21	1.12	1.12	1.02	7.6
				GC–MS/MS	1.13	1.19	1.04	1.08	1.03	1.07	1.08	1.09	1.10	1.03	1.08	1.07	1.08	1.29	1.06	1.18	1.12	1.08	1.03	6.0
			Oolong tea	GC–MS	1.19	1.19	1.16	1.09	1.27	1.21	1.18	1.18	1.27	1.25	1.17	1.12	1.15	1.22	1.24	1.20	1.18	1.22	1.13	4.1
				GC–MS/MS	1.23	1.19	1.14	1.16	1.25	1.24	1.15	1.21	1.29	1.22	1.09	1.12	1.16	1.16	1.35	1.25	1.09	1.18	1.09	5.9
197	Triadimefon	TPT	Green tea	GC–MS	0.99	1.06	1.12	1.11	1.04	1.00	1.06	1.11	1.04	1.07	1.07	1.05	1.09	0.97	1.01	1.01	1.00	1.02	1.11	4.4
				GC–MS/MS	1.17	1.10	1.15	1.15	1.08	1.21	1.24	1.15	1.23	1.30	0.87	1.26	1.11	1.0	1.19	1.14	1.11	1.07	1.11	8.5
			Oolong tea	GC–MS	1.13	1.14	1.21	1.17	1.08	1.26	1.34	1.16	1.17	1.13	1.48	1.32	1.26	1.18	1.26	1.22	1.22	1.17	1.05	8.2
				GC–MS/MS	1.02	1.07	1.03	1.11	1.02	1.21	0.96	1.03	0.99	0.98	1.04	1.30	1.11	1.15	1.13	1.24	1.10	1.05	1.00	8.5
		ENVI-Carb + PSA	Green tea	GC–MS	1.05	1.18	1.04	1.11	1.06	1.11	0.93	1.05	1.03	0.99	1.13	1.08	1.11	1.12	1.08	1.26	1.19	1.01	0.97	7.4
				GC–MS/MS	1.17	1.05	1.19	1.09	1.35	1.19	1.13	1.28	1.22	1.29	1.02	0.97	1.18	1.04	1.22	1.18	1.15	1.31	1.10	8.8
			Oolong tea	GC–MS	1.12	1.16	1.10	1.13	1.22	1.24	1.13	1.21	1.30	1.24	1.01	1.15	1.26	1.18	1.38	1.23	1.09	1.21	1.05	7.5
				GC–MS/MS	1.02	0.98	1.16	1.09	1.10	1.07	1.10	1.11	1.03	1.03	1.11	1.12	1.05	0.96	1.11	1.04	1.03	1.03	1.17	5.3
198	Triadimenol	TPT	Green tea	GC–MS	1.04	1.06	1.11	1.10	1.09	0.95	1.06	1.10	0.96	0.96	1.08	0.99	1.05	0.95	1.10	1.01	1.03	1.07	1.08	5.5
				GC–MS/MS	1.09	0.94	0.97	0.99	0.95	1.25	1.20	0.99	1.26	1.27	1.18	1.24	1.20	1.1	1.16	1.31	1.28	1.08	0.98	10.6
			Oolong tea	GC–MS	1.12	1.10	1.16	1.16	1.05	1.20	1.23	1.11	1.20	1.13	1.35	1.25	1.27	1.14	1.17	1.21	1.25	1.11	1.08	6.4
				GC–MS/MS	0.89	0.96	1.01	1.16	1.00	1.25	0.87	1.10	1.02	1.06	0.93	1.66	1.10	1.12	1.09	1.25	1.14	1.13	0.98	16.0
		ENVI-Carb + PSA	Green tea	GC–MS	1.00	1.15	1.05	1.15	1.04	1.10	0.82	1.12	1.08	1.02	1.12	1.10	1.17	1.18	1.09	1.23	1.15	1.12	0.99	8.2
				GC–MS/MS	1.09	1.01	0.98	0.96	0.99	1.18	1.14	1.36	1.07	1.39	1.24	0.97	1.12	1.17	1.36	1.19	1.15	1.27	1.12	11.8
			Oolong tea	GC–MS	1.11	1.15	0.98	1.10	1.25	1.23	1.14	1.16	1.22	1.31	1.04	1.14	1.29	1.11	1.39	1.18	1.08	1.22	0.99	8.6
				GC–MS/MS	1.02	0.99	1.17	1.05	1.17	1.13	1.09	1.10	1.04	0.97	1.12	1.04	1.02	0.95	1.08	1.04	1.01	1.05	1.15	6.0

(Continued)

Appendix Table 5.3 Youden Pair Ratios of Each Average Content From 201 Pesticides and RSD for 19 Circulative Determinations (cont.)

Youden pair ratio

No.	Pesticides	SPE	Sample	Method	Nov 9, 2009 (n=5)	Nov 14, 2009 (n=3)	Nov 19, 2009 (n=3)	Nov 24, 2009 (n=3)	Nov 29, 2009 (n=3)	Dec 4, 2009 (n=3)	Dec 9, 2009 (n=3)	Dec 14, 2009 (n=3)	Dec 19, 2009 (n=3)	Dec 24, 2009 (n=3)	Dec 29, 2009 (n=3)	Jan 3, 2010 (n=3)	Jan 8, 2010 (n=3)	Jan 13, 2010 (n=3)	Jan 18, 2010 (n=3)	Jan 23, 2010 (n=3)	Jan 28, 2010 (n=3)	Feb 2, 2010 (n=3)	Feb 7, 2010 (n=3)	RSD%
199	Triallate	TPT	Green tea	GC-MS	1.04	1.04	1.11	1.06	1.02	1.08	1.09	1.05	0.93	1.09	1.07	1.07	1.06	1.02	1.02	1.02	1.02	0.99	1.14	4.4
				GC-MS/MS	1.29	1.18	1.17	1.18	1.12	1.24	1.25	1.20	1.22	1.13	1.27	1.30	1.25	1.1	1.19	1.17	1.17	1.11	1.14	4.9
			Oolong tea	GC-MS	1.11	1.16	1.19	1.21	1.10	1.27	1.25	1.19	1.16	1.12	1.37	1.28	1.22	1.13	1.22	1.17	1.17	1.14	1.14	5.7
				GC-MS/MS	1.15	1.09	1.06	1.07	1.04	1.06	1.30	1.06	1.09	1.07	1.17	1.40	1.03	1.10	1.04	1.25	1.09	1.03	0.99	9.3
		ENVI-Carb + PSA	Green tea	GC-MS	1.09	1.12	1.06	1.09	1.05	1.02	1.47	1.11	1.12	1.06	1.12	1.13	0.99	1.27	1.03	1.29	1.14	0.99	1.06	10.5
				GC-MS/MS	1.29	1.19	1.18	1.13	1.30	1.16	1.20	1.17	1.28	1.19	1.17	1.10	1.18	1.13	1.23	1.25	1.19	1.13	1.15	4.6
			Oolong tea	GC-MS	1.28	1.21	1.12	1.17	1.24	1.22	1.16	1.20	1.31	1.19	1.07	1.12	1.18	1.16	1.34	1.24	1.11	1.14	1.14	6.0
				GC-MS/MS	1.04	1.01	1.15	0.99	1.01	1.10	1.13	1.05	0.90	1.11	1.04	1.28	1.06	0.97	1.08	1.05	0.99	1.04	1.14	7.7
200	Tribenuron-methyl	TPT	Green tea	GC-MS	1.07	1.01	1.12	1.02	1.01	1.06	1.02	1.05	0.85	1.04	1.08	1.06	1.01	0.77	1.06	0.97	0.97	1.05	1.17	8.7
				GC-MS/MS	1.04	1.04	1.09	1.03	1.04	1.07	0.98	1.13	0.81	1.07	1.05	1.22	1.06	0.89	0.99	0.99	0.95	0.97	1.16	8.9
			Oolong tea	GC-MS	1.06	0.93	0.84	1.30	1.21	1.29	1.06	1.04	1.18	1.00	1.29	1.26	1.03	1.0	1.18	1.06	1.17	1.15	0.99	12.1
				GC-MS/MS	1.14	0.97	0.90	1.24	1.15	1.28	1.03	1.06	1.19	0.99	1.39	1.34	1.05	1.00	1.16	1.11	1.26	1.23	1.19	11.5
		ENVI-Carb + PSA	Green tea	GC-MS	1.21	0.80	1.20	1.22	1.04	1.09	2.03	1.10	1.08	1.20	1.63	1.32	1.04	1.04	1.24	1.22	1.07	1.03	0.97	22.3
				GC-MS/MS	1.02	0.77	1.13	1.29	1.02	0.97	0.88	1.28	1.14	1.15	1.48		1.00	1.21	1.26	1.40	1.19	0.93	0.98	16.6
			Oolong tea	GC-MS	1.06	1.09	1.05	0.93	1.39	1.07	1.34	1.19	1.30	1.22	1.05	1.01	1.09	0.95	1.05	1.45	1.04	1.16	1.13	12.9
				GC-MS/MS	1.07	1.15	0.93	0.97	1.40	1.04	1.19	1.27	1.26	1.28	1.03	1.06	1.07	0.96	1.17	1.41	1.00	1.16	1.19	12.5
201	Vinclozolin	TPT	Green tea	GC-MS	1.00	1.02	1.10	1.08	1.05	1.07	1.06	1.07	0.94	0.98	1.08	1.09	1.06	0.99	1.10	1.02	1.09	0.99	1.16	5.0
				GC-MS/MS	1.05	1.00	1.15	1.06	1.02	1.11	1.12	1.10	0.90	1.02	1.11	1.20	1.05	0.99	1.07	1.05	0.98	0.98	1.13	6.9
			Oolong tea	GC-MS	1.20	1.17	1.19	1.15	1.15	1.25	1.21	1.21	1.19	1.15	1.32	1.21	1.22	1.1	1.23	1.16	1.18	1.07	1.12	4.5
				GC-MS/MS	1.15	1.12	1.21	1.24	1.09	1.24	1.26	1.14	1.15	1.17	1.44	1.28	1.21	1.09	1.14	1.21	1.29	1.11	1.11	7.3
		ENVI-Carb + PSA	Green tea	GC-MS	1.15	1.11	1.10	1.06	1.07	1.04	1.15	1.08	1.08	1.07	1.16	1.14	1.06	1.18	1.03	1.22	1.12	1.05	1.00	5.3
				GC-MS/MS	1.03	1.04	1.09	1.09	1.08	1.03	1.40	1.08	1.12	1.07	1.14	1.14	1.01	1.27	0.97	1.24	1.06	1.00	0.96	10.1
			Oolong tea	GC-MS	1.20	1.18	1.17	1.08	1.28	1.16	1.19	1.20	1.28	1.19	1.14	1.14	1.15	1.18	1.20	1.26	1.16	1.23	1.15	4.3
				GC-MS/MS	1.14	1.19	1.14	1.14	1.22	1.18	1.17	1.20	1.35	1.17	1.07	1.11	1.30	1.17	1.37	1.29	1.05	1.16	1.11	7.5

REFERENCES

[1] Significant dates in food and drug law history (1956) Public Health Rep.71(6),558-559.

[2] Pure Food and Drug Act (1906). United States Statutes at Large (59th Cong., Sess. I, Chp. 3915, pp. 768–772; cited as 34 U.S. Stats. 768).

[3] WHO Global Strategy for Food Safety (2002), ISBN 9245545741 (NLM Classification: WA 695).

[4] Food Safety, Fifty-Third World Health Assembly, WHA53.15 (2007) Geneva, Switzerland.

[5] Available from http://www.epa.gov/history/topics/ddt/01.htm. (Accessed December 4, 2007).

[6] Regulation (EC) No 396/2005 of the European Parliament and of the Council, I–IV.

[7] Food and Agriculture Organization, World Health Organization. Report of the Forty First Session of the Codex Committee on Pesticide Residues. Beijing, China, 2009.

[8] US Environmental Protection Agency Office of Pesticide Programs, Index to Common Names and Part 180 Tolerance Information of Pesticide Chemicals in Food and Feed Commodities (2009).

[9] Song, W.C., Shan, W.L., Ye, J.M., Li, Z.A., Zhou, Z.Q., Pan, C.P., 2009. Chin. J. Pesticide Sci. 11 (4), 414–420.

[10] Standardization Administration of PR China. GB2763-2005 Maximum limits for pesticides in food (2005).

[11] Chen, Z.M., 2002. Chin. Tea 24 (1), 8–11.

[12] Available from http://m5.ws001.squarestart.ne.jp/foundation/foodlist.php. (Accessed October 10, 2012).

[13] Available from http://ec.europa.eu/sanco_pesticides/public/index.cfm?event=commodity.resultat. (Accessed October 10, 2012).

[14] Lehotay, S.J., de Kok, A., Hiemstra, M., Van Bodegraven, P., 2005. J. AOAC Int. 88 (2), 595–614.

[15] Wong, J.W., Webster, M.G., Halverson, C.A., Hengel, M.J., Ngim, K.K., Ebeler, S.E., 2003. J. Agr. Food Chem. 51 (5), 1148–1161.

[16] Kitagawa, Y., Okihashi, M., Takatori, S., Okamoto, Y., Fukui, N., Murata, H., Sumimoto, T., Obana, H., 2009. Shok. Eiseigaku Zasshi 50 (5), 198–207.

[17] Okihashi, M., Takatori, S., Kitagawa, Y., Tanaka, Y., 2007. J. AOAC Int. 90 (4), 1165–1179.

[18] Huang, Z., Zhang, Y., Wang, L., Ding, L., Wang, M., Yan, H., Li, Y., Zhu, S., 2009. J. Sep. Sci. 32 (9), 1294–1301.

[19] Fillion, J., Sauvé, F., Selwyn, J., 2000. J. AOAC Int. 83 (3), 698–713.

[20] Walorczyk, S., 2008. J. Chromatogr. A 1208, 202–214.

[21] Nguyen, T.D., Han, E.M., Seo, M.S., Kim, S.R., Yun, M.Y., Lee, D.M., Lee, G.H., 2008. Anal. Chim. Acta 619 (1), 67–74.

[22] Przybylski, C., Segard, C., 2009. J. Sep. Sci. 32 (11), 1858–1867.

[23] Koesukwiwat, U., Lehotay, S.J., Miao, S., Leepipatpiboon, N., 2010. High throughput analysis of 150 pesticides in fruits and vegetables using QuEChERS and low-pressure gas chromatography–time-of-flight mass spectrometry. J. Chromatogr. A 1217 (43), 6692–6703.

[24] Schenck, F., Wong, J., Lu, C., Li, J., Holcomb, J.R., Mitchell, L.M., 2009. J. AOAC Int. 92 (2), 561–573.

[25] Nguyen, T.D., Yu, J.E., Lee, D.M., Lee, G.H., 2008. Food Chem. 110 (1), 207–213.

[26] Saito, Y., Kodama, S., Matsunaga, A., Yamamoto, A., 2004. J. AOAC Int. 87 (6), 1356–1367.

[27] Hirahara, Y., Kimura, M., Inoue, T., Uchikawa, S., Otani, S., Hirose, H., Suzuki, S., Uchida, Y., 2006. Screening method for the determination of 199 pesticides in agricultural products by gas chromatography/ion trap mass spectrometry (GC/MS/MS). Shok. Eiseig. Zasshi 47 (5), 213–221.

[28] Pizzutti, I.R., de Kok, A., Hiemstra, M., Wickert, C., Prestes, O.D., 2009. J. Chromatogr. A 1216 (21), 4539–4552.

[29] Nguyen, T.D., Lee, K.J., Lee, M.H., Lee, G.H., 2010. Microchem. J. 95 (1), 43–49.

[30] Hu, X., Jianxin, Y., Zhigang, Y., Lansun, N., Yanfei, L., Peng, W., Jing, L., Xin, H., Xiaogang, C., Yibin, Z., 2004. J. AOAC Int. 87 (4), 972–985.

[31] Chu, X.G., Hu, X.Z., Yao, H.Y., 2005. J. Chromatogr. A 1063 (1–2), 201–210.

[32] Nguyen, T.D., Yun, M.Y., Lee, G.H., 2009. J. Agr. Food Chem. 57 (21), 10095–10101.

[33] Matsumoto, N., Yoshikawa, M., Eda, K., Kobayashi, A., Yokoshima, M., Murakami, M., Kanekita, H., 2008. Shok. Eiseig. Zasshi 49 (3), 211–222.

[34] Standardization Administration of PR China, 2008. GB/T 19426-2006, Method for the Determination of 497 Pesticides and Related Chemicals Residues in Honey, Fruit Juice and Wine by Gas Chromatography–Mass Spectrometry Method. Standards Press, Beijing,China.

[35] Standardization Administration of PR China, 2008. GB/T 19648-2006, Method for Determination of 500 Pesticides and Related Chemicals Residues in Fruits and Vegetables by Gas Chromatography–Mass Spectrometry Method. Standards Press, Beijing, China.

[36] Standardization Administration of PR China, 2008. GB/T 19649-2006, Method for Determination of 475 Pesticides and Related Chemicals Residues in Grains by Gas Chromatography–Mass Spectrometry Method. Standards Press, Beijing, China.

[37] Standardization Administration of PR China, 2006. GB/T 19650-2006, Method for Determination of 478 Pesticides and Related Chemicals Residues in Animal Muscles""GC–MS Method. Standards Press, Beijing, China.

[38] Standardization Administration of PR China, 2008. GB/T 23204-2008, Determination of 519 Pesticides and Related Chemicals Residues in Tea""GC–MS Method. Standards Press, Beijing, China.

[39] Standardization Administration of PR China, 2008. GB/T 23216-2008, Determination of 503 Pesticides and Related Chemicals Residues in Mushrooms""GC–MS Method. Standards Press, Beijing, China.

[40] Standardization Administration of PR China, 2008. GB/T 23207-2008, Determination of 485 Pesticides and Related Chemicals Residues in Fugu Eel and Prawn""GC–MS Method. Standards Press, Beijing, China.

[41] Standardization Administration of PR China, 2008. GB/T 23200-2008, Determination of 488 Pesticides and Related Chemicals Residues in Mulberry Twig, Honeysuckle Barbary Wolfberry Fruit and Lotus Leaf""GC–MS Method. Standards Press, Beijing, China.

[42] Standardization Administration of PR China, 2008. GB/T 23210-2008, Determination of 511 Pesticides and Related Chemicals Residues in Milk and Milk Powder""GC–MS Method. Standards Press, Beijing, China.

6 Study on the Degradation of Pesticide Residues in Tea

Chapter 6.1

Study on the Degradation of 271 Pesticide Residues in Aged Oolong Tea by Gas Chromatography-Tandem Mass Spectrometry and Its Application in Predicting the Residue Concentrations of Target Pesticides

6.1.1 INTRODUCTION

As one of the world's three major health drinks, tea makes up the majority of traditional export commodities of China. The growing tea plant is prone to be attacked by various pests and diseases because it is mostly planted in warm temperate zones and subtropical areas. Different kinds of pesticides have been widely used on tea to control pests and plant diseases to increase harvest productivity. However, the pesticide residue on tea may harm human health [1–4]. With the strengthening interest in food safety in strategic places across different countries and regions of the world, pesticide residue in tea has awakened great public concern [5–8].

The research of degradation regularity of pesticides together with the model to simulate the dynamics in tea sample could be applied to analyze and predict the pesticide residues in tea. It's significant in guiding the farmer to spray pesticides on tea plants reasonably and helpful for predicting the risk of pesticide residues in the tea trade. The degradation of pesticides is a complex process affected by many factors, such as temperature, humidity, sunshine, and metals [9,10]. Many

Analysis of Pesticide in Tea. http://dx.doi.org/10.1016/B978-0-12-812727-8.00012-2

studies have been carried out on the degradation of pesticides in agriculture products with different kinetic models [11–13]. Ozbey et al. [14] investigated the behavior of some organophosphorus pesticide residues in peppermint tea during the infusion process. Manikandan et al. [15] studied the leaching of certain pesticides, such as ethion, endosulfan, dicofol, chlorpyrifos, deltamethrin, hexaconazole, fenpropathrin, propargite, quinalphos, and *lambda*-cyhalothrin from powdered black tea into the brew. Lin et al. [16] studied the natural degradation dynamics of bifenthrin, fenpropathrin, cypermethrin, and buprofezinon on oolong tea plant new shoots by gas chromatography. Chen et al. [17] developed a gas chromatography-mass spectrometry (GC-MS) method for the analysis of bifenthrin, cyhalothrin, teflubenzuron, flufenoxuron, and chlorfluazuron in dried oolong tea leaf samples, and then the natural degradation of these pesticides in the leaves of olong tea trees and the effect of processing steps on the residue of them were researched in detail.

Less attention has been given to the degradation of pesticide in aged tea. In order to research the degradation regularity of pesticides in aged tea samples on the basis of our previous studies [18–20], the developed GC-MS/MS method was used to determine the multiresidue of 271 pesticides, including organonitrogen, organophosphorus, organochlorine, organosulfur, carbamates, and pyrethroids, in 3–4 month aged oolong tea. In addition, the regularity of 271 pesticides in aged oolong tea in at days and 120 days was studied from different aspects according to fitting curves. Subsequently, 20 representative pesticides from different classes were optimized for further study. At higher spray concentrations, the residues of the selected 20 pesticides in aged oolong tea were studied again at 90 days to investigate the degradation regularity at different concentations. The degradation values of target pesticides on a specific day could be predicted by a logarithmic function obtained from taking the determination time (day) as the *x*-axis and the difference between each determined value and the first time-determined value of the target pesticide as the *y*-axis, based on the degradation results of 20 pesticides at higher concentration for 90 days.

Finlly, the proposed procedure was validated by predicting the pesticide residue at one of the Youden pair concentrations according to the logarithmic function from another concentration. The predicted values were compared with the measured results and evaluated by the deviation ratios.

6.1.2 REAGENTS AND MATERIAL

6.1.2.1 Reagents

1. Solvents. Acetonitrile, dichloromethane, isooctane, and methanol (HPLC grade), purchased from Dikma Co. (Beijing, China).
2. Anhydrous sodium sulfate. Analytically pure. Baked at 650°C for 4 h and stored in a desiccator.
3. Pesticides standard and internal standard (heptachlor-epoxide). Purity, ≥95% (LGC Promochem, Wesel, Germany).
4. Stock standard solutions. Accurately weigh 5–10 mg individual pesticide and chemical pollutant standards (accurate to 0.1 mg) into a 10 mL volumetric flask. Dissolve and dilute to volume with methanol, toluene, acetone, acetonitrile, isooctane, and so on, depending on each individual compound's solubility. All standard stock solutions are stored in the dark at 0–4°C.
5. Mixed standard solution. Depending on properties and retention timc of cach pesticide, all the 271 pesticides for GC-MS/MS analysis are divided into three groups. The concentration of mixed standard solutions was dependent on the sensitivity of each compound for the instrument used for analysis. Mixed standard solutions should be stored in the dark below 4°C.

6.1.2.2 Material

1. SPE cartridge. Cleanert-TPT (10 mL, 2000 mg, Agela, Tianjin, China).
2. Homogenizer. Rotational speed higher than 13 500 r/min (report also in g-force units; T-25B (Janke & Kunkel, Staufen, Germany), or equivalent.
3. Rotary evaporator. Buchi EL131 (Flawil, Switzerland) or equivalent.
4. Centrifuge. Centrifugal force higher than 2879 $\times g$ (Z320; B. HermLe AG, Gosheim, Germany), or equivalent.
5. Nitrogen evaporator. EVAP 112 (Organomation Associates, Inc., New Berlin, MA), or equivalent.

6.1.3 APPARATUS AND CONDITIONS

GC-MS/MS system. Model 7890A gas chromatograph connected to a Model 7000B triple quadrupole mass spectrometer with electron ionization (EI) source and equipped with a Model 7693 autosampler equipped with Mass Hunter data processing software system (Agilent Technologies, Wilmington, DE). Gas chromatographic separation was achieved on a DB-1701 capillary column (30 m × 0.25 mm × 0.25 μm, Agilent J&W Scientific, Folsom, CA).

The oven temperature was programmed as follows: 40°C hold for1 min, to 130°C at 30°C/min, to 250°C at 5°C/min, to 300°C at 10°C/min, hold for 5 min; carrier gas, helium, purity ≥99.999%; flow rate, 1.2 mL/min; injection port temperature, 290°C; injection volume, 1 μL; injection mode, splitless, purge on after 1.5 min; ionization voltage, 70 eV; ion source temperature, 230°C; GC/MS interface temperature, 280°C; the ion monitoring mode is multi reaction monitor (MRM) mode. Each compound is monitored by one quantifying precursor/product ion transition, and one qualifying precursor/product ion transition.

6.1.4   SAMPLE PRETREATMENT

6.1.4.1   The Preparation Procedures for Aged Tea Samples

Take oolong tea that is found to be free from a target pesticide after testing and pass them respectively through 10 mesh and 16 mesh sieves after initial blending via blender. Take 10–16 mesh 500 g oolong tea, and spread them uniformly over the bottom of a stainless steel vessel with 40 cm diameter and wait for spraying. Accurately transfer a certain amount of pesticide mixed standard solutions into the full-glass sprayer and spray the tea leaves. Spray while stirring the tea leaves with a glass rod so as to make a uniform spray. After completion of spraying, continue to stir the tea leaves for half an hour to wait for the volatile solvents in the tea leaves, after which, place the sprayed tea leaves into a 4 L brown bottle, and avoid exposure to light for storage at room temperature and keep oscillation blending for 12 h.

Spread the aged tea samples onto the bottom of a flat-bottomed vessel, draw a cross and weigh a total of five portions of aged tea samples located the symmetrical four points of the cross and the central part. Submit them for GC-MS/MS determination, and calculate the average value of the pesticide content of the aged samples and RSD. When the RSD < 4% for GC-MS/MS, it could be judged that tea samples have been sprayed and mixed homogeneously.

6.1.4.2   Extraction

Weigh 5 g dry tea powder (accurate to 0.01 g) into 80 mL centrifuge tube, add 15 mL acetonitrile, homogenize at 13,500 rpm for 1 min, centrifuge 2879 $\times g$ for 5 min and transfer the supernatants into a pear-shaped flask. Reextract the residue with 15 mL acetonitrile, centrifuge, combine the two extracts, and rotary evaporate in water bath at 40°C to about 1 mL for cleanup.

6.1.4.3   Cleanup

Place a pear-shaped flask under the 5-port flask vacuum manifold, and mount a Cleanert-TPT cartridge onto the manifold. Add about 2 cm anhydrous sodium sulfate onto the Cleanert-TPT cartridge packing material, prewash with 10 mL acetonitrile-toluene (3 + 1, v/v) and discard the effluents to activate the cartridge. Stop the flow through the cartridge when the liquid level in the cartridge barrel has just reached the top of the sodium sulfate packing. Discard the waste solution collected in the pear-shaped flask and replace with a clean pear-shaped flask.

Transfer the aforementioned sample concentrates into an SPE cartridge, rinse the sample solution bottle with 2 mL acetonitrile-toluene (3 + 1, v/v), repeat this step three times, transfer the rinsing liquids into the cartridge, attach a 50 mL storage device onto the cartridge, then elute with 25 mL acetonitrile–toluene (3 + 1, v/v). Collect the aforementioned effluents into a pear-shaped flask by gravity feeding and rotary evaporate in water bath at 40°C to about 0.5 mL. Add 40 μL heptachlor epoxide (internal standard; ISTD) working standard solution to the sample. Evaporate to dryness under a stream of nitrogen in a 35°C water bath. Dissolve the dried residue in 1.5 mL hexane, ultrasonicate the samples to mix, and filter through a 0.2 μm membrane filter. The sample is ready for GC-MS/MS analysis.

6.1.5   THE DEGRADATION OF 271 PESTICIDES IN AGED OOLONG TEA

The residues of 271 pesticides in aged oolong tea were determined 25 times by GC-MS/MS in 120 days (every 5 days) to monitor the degradation behavior of them. In order to study the degradation regularity of the 271 pesticides in aged oolong tea, the determination days were taken as the horizontal ordinate and concentrations of pesticide residue as the vertical ordinate to draw scatter diagrams at a and b spray concentration (conc.) in 40 and 120 days. The degradation equations are summarized in Supplemental Table 1 (at a conc.) and Supplemental Table 2 (at b conc.).

By comparing the degradation equations of pesticides at a and b conc. in 40 days and 120 days, the degradation trend is found to be different. They include the following aspects: (A) the residues of pesticide dropped exponentially in both 40 days and 120 days. (B) The residues of pesticide dropped exponentially in 40 days and logarithmically in 120 days.

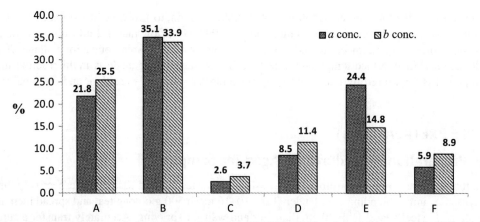

FIGURE 6.1.1   The ratio of pesticides accord with A–F degradation trend in 271 pesticides at *a* and *b* concentration (%). % of y-axis: the number of pesticides accord with type A or others × 100/total number of pesticides (271).

(C) The residues of pesticide dropped logarithmically in 40 days and logarithmically/polynomially in 120 days. (D) No trend in 40 days ($R^2 < 0.4$) and showed as the scatter points, while presented some dropping trends in 120 days. (E) Presented some dropping trends in 40 days, while no trend in 120 days and showed as the scatter points ($R^2 < 0.4$). (F) No trend in both 40 days and 120 days, which showed as the scatter points ($R^2 < 0.4$). In addition, five pesticides were not detected at both *a* and *b* conc.

According to the A–F types, the degradation regularity of 271 pesticides was studied. The ratios of pesticides accorded with degradation trend A–F in 271 pesticides are shown in Fig. 6.1.1. It was observed that most of the pesticides accorded with the A, B, and E degradation trend. At *a* conc., the ratio for A, B, and E was 21.8%, 35.1%, and 24.4%, respectively, and the total number of them was 220, accounting for 81.2% of 271 pesticides. At *b* conc., the ratio for A, B, and E was 25.5%, 33.9%, and 14.8%, respectively, and the total number of them was 201, accounting for 74.2% of 271 pesticides. These results demonstrated that A, B, and E degradation trends could represent the main aspects of the 271 pesticides. That means most of the pesticides dropped exponentially in 40 days, and they presented dropping trends exponentially or logarithmically in 120 days. Although the ratio of pesticides with other degradation trends was small, they did represent certain of degradation regularity. Therefore, all of the A–F types will be discussed.

6.1.5.1   Degradation Trend A

Taking propachlor as an example, the degradation trend of type A in 40 days and 120 days is shown in Fig. 6.1.2(i) and (ii), respectively. It is clearly observed that the concentrations of propachlor in aged oolong tea exponentially decrease with the increase of intervals. The degradation kinetics of group A is a first-order reaction as shown by Eq. (6.1.1):

$$C = C_0 e^{-kt} \tag{6.1.1}$$

where C is the concentration of each pesticide in aged oolong tea, C_0 is the initial concentration of each pesticide in aged oolong tea, k is the degradation rate constant, and t is the determination time (day).

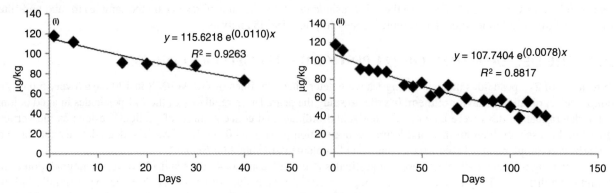

FIGURE 6.1.2   Degradation profiles of degradation trend A (e.g., propachlor) in 40 days (i) and 120 days (ii).

The conclusion is that without the influence of other factors the degradation rate of pesticides in group A is in direct ratio to the initial concentration of pesticides in aged oolong tea. Based on the first-order reaction model, half-life of pesticides in type A are calculated by Eq. (6.1.2):

$$t_{1/2} = \ln 2 \, / \, k \tag{6.1.2}$$

where $t_{1/2}$ is the half-life of the determined pesticide.

According to the degradation equations of pesticides in type A listed in Supplemental Table 1 (at a conc.) and Supplemental Table 2 (at b conc.), the half-life is calculated. The degradation rate constant and half-life are also shown in Supplemental Tables 1 and 2. The resulting k values of a and b conc., by comparison, in both 40 and 120 days, showed that most of the pesticides in type A had higher k values in 40 days than in 120 days, except 12 and 19 pesticides were opposite at a and b conc., respectively. This indicates that the concentration of most of the pesticides in type A dropped fast in the first 40 days, while decreasing slowly in the remaining days. However, chlorfenapyr, bupirimate, fonons, and furalaxyl had similar degradation equations in both 40 days and 120 days. For these four pesticides, therefore, the degradation trend could be expressed by the degradation equations in 40 days. At a conc., the half-life of pesticides in degradation trend A was 24.4–223.6 and 44.4–203.9 days, according to the degradation equations in 40 and 120 days, respectively. Except for the previously mentioned four pesticides, the half-life of the other pesticides was varied in 40 days and 120 days; the biggest difference was 156.9 days. At b conc., the half-life of pesticides in degradation trend A was 40.8–315.1 and 46.8–330.1 days, according to the degradation equations in 40 and 120 days, respectively. On the whole, whether at a or b conc., there were great differences at the half-life in 40 and 120 days. Based on the actual condition, the half-life calculated by degradation equations in 120 days was considered to be reasonable. In addition, there were 28 pesticides in accordance with degradation trend A at both a and b conc. By comparison of k values from degradation equation in 120 days, 10 pesticides had the higher k values at b conc. than a conc., while the remaining 18 pesticides were the opposite.

6.1.5.2 Degradation Trend B

For degradation trend B, the concentration of pesticides in aged oolong tea dropped faster in 40 days than in the remaining days. From the data in Supplemental Tables 1 and 2, it was clearly summarized that the exponential equation was suitable for the first 40 days, while logarithmical equation for 120 days. Taking chlorfenvinphos, for example, the degradation profiles of type B are shown in Fig. 6.1.3(i), (ii) for 40 and 120 days, respectively.

There were 95 and 92 pesticides in accordance with the degradation trend B at a and b conc., respectively, and accounting for 35.1% and 33.9%, respectively. Among them, 50 pesticides accorded with the degradation trend B at both a and b conc. Comparing the k values from the degradation equations in 40 days showed that most of the 36 pesticides had higher k values at a conc. than b conc.

6.1.5.3 Degradation Trend C

Degradation trend C was similar to degradation trend B, with the difference that the concentration of pesticides decreased logarithmically in 40 days, as shown in Fig. 6.1.4(i), (ii). There are 7 and 10 pesticides in accordance with a degradation trend C at a and b conc., respectively. It can be found from the raw data that the concentration of pesticides at the 5th or 10th day had a bigger difference than the first day. This can be considered as the explanation of degradation trend C.

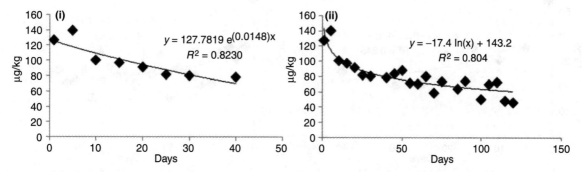

FIGURE 6.1.3 Degradation profiles of degradation trend B (e.g., chlorfenvinphos) in 40 days (i) and 120 days (ii).

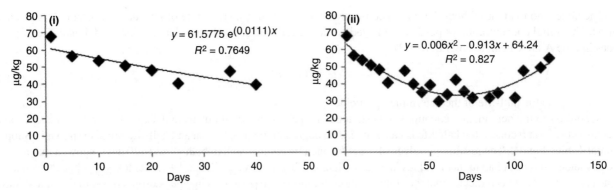

FIGURE 6.1.4 Degradation profiles of degradation trend C (e.g., dibutylsuccinate) in 40 days (i) and 120 days (ii).

6.1.5.4 Degradation Trend D

For degradation trend D, the concentration of pesticides presented as scatter points with a fitting coefficient (R^2) less than 0.4 in 40 days, and most of the dropped trend could be fitted by exponential and logarithmical curves and a few by polynomial curves in 120 days, which are shown in Fig. 6.1.5(i), (ii). At a conc., there were 23 pesticides in accordance with the degradation trend D and among them the 15 pesticides decreased exponentially and the other 8 pesticides decreased logarithmically or polynomially in 120 days. At b conc., there were 31 pesticides in accordance with the degradation trend D, and there were 23 pesticides decreased exponentially in 120 days. By comparison, it was also found that 10 pesticides accorded with degradation trend D at a and b conc.

6.1.5.5 Degradation Trend E

The ratios of pesticides accorded with degradation trend E in 271 pesticides were third-ranked among A–F aspects. The degradation profiles of 4,4-dibromobenzophenonein 40 days and 120 days were taken as examples of degradation trend E (in Fig. 6.1.6(i), (ii)). For degradation of trend E, the concentrations of pesticides decreased exponentially or logarithmically in 40 days, while no suitable equations could be used to fit the scatter points in 120 days. Supplemental Tables 1 and 2 show that there were 66 pesticides in accordance with degradation trend E at a conc., and among them there were 9 and 21 pesticides in accordance with degradation trend A and B at b conc., respectively, accounting for 45.5% of the 66 pesticides. In addition, there were 40 pesticides in accordance with degradation trend E at b conc., but only 3 and 5 pesticides corresponded with degradation trend A and B at a conc., respectively. At the same time, there were 24 pesticides in accordance with degradation trend E at both a and b conc., accounting for 36.4% and 60.0% of the pesticides, which accorded with the degradation trend E at a and b conc., respectively.

6.1.5.6 Degradation Trend F

Taking kresoxim-methyl as an example, the profiles of degradation trend F are shown in Fig. 6.1.7(i), (ii). The concentrations of the degraded pesticides presented as scatter points, and no degradation trend, can be found by any of the

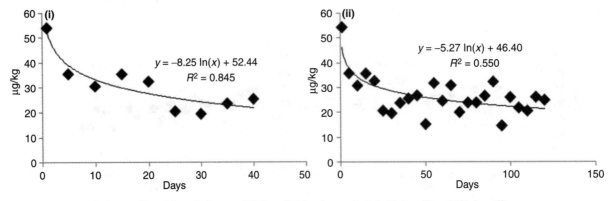

FIGURE 6.1.5 Degradation profiles of degradation trend D (e.g., dichlorofop-methyl) in 40 days (i) and 120 days (ii).

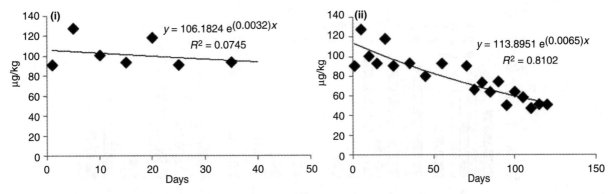

FIGURE 6.1.6   Degradation profiles of degradation trend E (e.g., cycluron) in 40 days (i) and 120 days (ii).

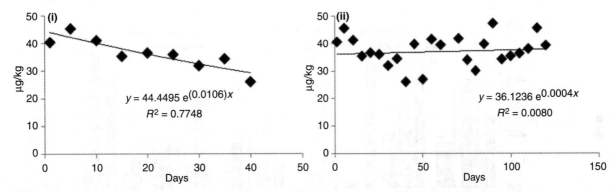

FIGURE 6.1.7   Degradation profiles of degradation trend F (e.g., 4,4-dibromobenzophenone) in 40 days (i) and 120 days (ii).

aforementioned fitting curves with $R^2 \geq 0.4$. The unsteadiness properties of these pesticides in aged oolong tea during storage might be the reason. There were 16 and 24 pesticides in accordance with degradation trend F at a and b conc., respectively, accounting for 5.9% and 8.9% of 271 pesticides, respectively. In addition, of them 9 pesticides accorded with degradation trend F at both concentrations.

The earlier discussion indicated that the 271 pesticides studied here had relatively complex degradation trends in aged oolong tea with various fitting curves at different concentrations. Among the degradation trend of type A–F, degradation trends A, B, and F had higher ratios than the others. All of the pesticides accorded with degradation trends A, B, and E decreased exponentially in 40 days and mainly exponentially or logarithmically in 120 days. In addition, although there was no degradation trend for the pesticides of degradation trend D in 40 days, they decreased mainly exponentially in 120 days. The conclusion derived is that pesticides in tea will degrade slowly and the concentration of pesticide will decrease within 4 months, wherein at concentration a the deviation for each pesticide from day 1 to day 120 falls within the range of 0.2%–85.6% with a mean value of 36.4%, while at concentration b the deviation for each pesticide from day 1 to day 120 falls within the range of 4.0%–92.7%, with the mean value 50.8%.

6.1.6   PESTICIDES IN DIFFERENT CLASSES

For further investigating the degradation trend, pesticides of A–F aspects were divided into different classes according to organonitrogen, organophosphorus, organochlorine, organosulfur, carbamates, pyrethroids, and others. The distributions of them can be found in Figs. 6.1.8 and 6.1.9.

Most of the 271 pesticides were organonitrogen, organophosphorus, and organochlorine, so the pesticides are discussed according to these three classes. For degradation trend A, most of the pesticides were organonitrogen and organophosphorus. For degradation trend B, the number of organophosphorus and organochlorine pesticides was equal, but far below the number of organonitrogen pesticides. The number of organochlorine pesticides for degradation trend E was far more than organonitrogen and organophosphorus pesticides, and the number of organophosphorus pesticides was 5 and 3 at a and b conc., respectively. The organophosphorus and organochlorine pesticides are discussed in detail for degradation trend A + B and E. At a conc., 37.9% and 71.4% pesticides were organochlorine and organophosphorus for degradation trend

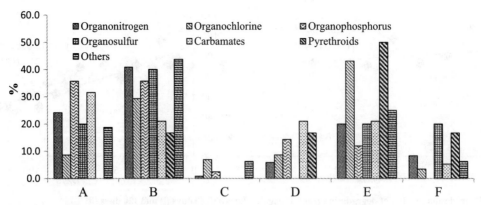

FIGURE 6.1.8 Distributions of pesticides according to different classes in A–F aspects at *a* concentration.

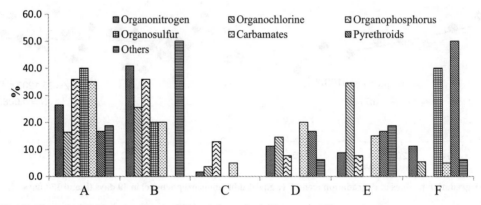

FIGURE 6.1.9 Distributions of pesticides according to different classes in A–F aspects at *b* concentration.

A + B, respectively. At the same time, 43.1% and 11.9% pesticides were organochlorine and organophosphorus for degradation trend E, respectively. At *b* conc., the percentage of organochlorine and organophosphorus pesticides for degradation trend A + B was 41.8% and 71.8%, respectively. For degradation trend E, the percentage for organochlorine and organophosphorus pesticides was 34.5% and 7.7%, respectively. Most of the organophosphorus pesticides degraded in accordance with degradation trend A and B in aged oolong tea. Conversely, most of the organochlorine pesticides decreased according to degradation trend E in aged oolong tea.

6.1.7 THE PRACTICAL APPLICATION OF DEGRADATION REGULARITY

6.1.7.1 The Degradation Regularity of 20 Representative Pesticides

The single lab validation results of AOAC priority research project "high-throughput analytical techniques for the determination and confirmation of residues of 653 multiclass pesticides and chemical pollutants in tea by GC-MS, GC-MS/MS, and LC-MS/MS" showed that the method could be used for determination of as many as 653 target pesticides. In order to reduce the workload and guarantee the smooth AOAC collaborative study, an alternative "shrunken" protocol has been proposed by AOAC International. In the "shrunken" protocol, the pesticides are commonly used in the process of tea growth as well as being needed to determine the tea international trade, which was selected from 271 pesticides determined by GC-MS/MS. In addition, these pesticides after being sprayed onto tea are all of relatively good stability and their polarities are also widely representative. On the basis of the degradation equations in Supplemental Tables 1 and 2, 20 representative pesticides were optimized (Table 6.1.1).

In addition, the degradation regularity of the 20 representative pesticides in aged oolong tea was studied at *c* and *d* conc. (*d* > *c* > *b* > *a*) in another 90 days by GC-MS/MS. Similarly, the degradation equations of them were obtained by taking the determination time (every 5 days) as *x*-axis and the concentration as *y*-axis (see Table 6.1.2). Table 6.1.2 shows that the degradation of the 20 representative pesticides agrees with the 90-day logarithmic equation mentioned earlier. This means they degraded slowly in 3 months in aged oolong tea.

TABLE 6.1.1 Retention Time and Monitored Ion Transitions for the 20 Pesticides by gas chromatography-mass spectrometry (GC-MS)/MS

No.	Pesticides	Retention time (min)	Quantifying precursor/product ion transition	Qualifying precursor/ product ion transition
ISTD	Heptachlor-epoxide	22.15	353/263	353/282
1.	Trifluralin	15.41	306/264	306/206
2.	Tefluthrin	17.4	177/127	177/101
3.	Pyrimethanil	17.42	200/199	183/102
4.	Propyzamide	18.91	173/145	173/109
5.	Pirimicarb	19.02	238/166	238/96
6.	Dimethenamid	19.73	230/154	230/111
7.	Fenchlorphos	19.83	287/272	287/242
8.	Tolclofos-methyl	19.87	267/252	267/93
9.	Pirimiphos-methyl	20.36	290/233	290/125
10.	2,4'-DDE	22.79	318/248	318/246
11.	Bromophos-ethyl	23.16	359/303	359/331
12.	4,4'-DDE	23.9	318/248	318/246
13.	Procymidone	24.7	283/96	283/255
14.	Picoxystrobin	24.75	335/173	335/303
15.	Quinoxyfen	27.18	237/208	237/182
16.	Chlorfenapyr	27.37	408/59	408/363
17.	Benalaxyl	27.66	148/105	148/79
18.	Bifenthrin	28.63	181/166	181/165
19.	Diflufenican	28.73	266/218	266/246
20.	Bromopropylate	29.46	341/185	341/183

In Supplemental Tables 1 and 2, except for pirimicarb, fenchlorphos, and 4,4'-DDE, the other 17 optimized pesticides dropped exponentially or logarithmically in 120 days at *a* and *b* conc. On the other hand, at higher concentrations, *c* and *d*, all of the 20 optimized pesticides dropped logarithmically in 90 days. Therefore, it is concluded that the degradation regularity of them are in accordance with logarithmical equation with the spray concentration increasing.

6.1.7.2   The Prediction of Pesticide Residues in Aged Oolong Tea

The aforementioned discussion indicates that the degradation behavior of pesticide in aged oolong tea has certain regularity. However, it should be noted that this process is time-consuming for multiple determinations. Therefore, it is very necessary to propose a method to predict the residue of pesticide in aged oolong tea. In this study, the prediction method is developed and validated by taking the raw degradation data of *c* and *d* conc. for a particular instance.

Based on the results of pesticides in aged oolong tea determined by GC-MS/MS in 90 days (from the raw data of *d* conc.), trend charts (e.g., dimethenamid, see Fig. 6.1.10) were plotted, in which the determination time (day) was *x*-axis and the difference between each measured value and the first time-measured value (degradation value) of target pesticides was *y*-axis. The logarithmic equations were obtained by fitting the 90 days determination results. From these equations, the degradation value of any of the 20 target pesticides at any specific day could be calculated and applied to the raw data generated for that pesticide in a particular laboratory.

The logarithmic functions of the 20 pesticides at *d* conc. are listed in Table 6.1.3, and they were applied to predict the residue concentrations of pesticides in aged oolong tea at *c* conc. The predicted residue of each pesticide on one day could be obtained by subtracting the degradation value of this day from the concentration of the first day.

Accordingly, the residue concentrations of 20 pesticides in aged oolong tea at different degradation intervals of 5th, 10th, 15th, 20th, 25th, 30th, 35th, 40th, 45th, 50th, 55th, 60th, 65th, 70th, 75th, 80th, 85th, and 90th day after spraying at *c* conc.

TABLE 6.1.2 Degradation Equations of 20 Representative Pesticides in Aged Oolong Tea at *C* and *D* Concentrations in 90 Days by GC-MS/MS

No.	Pesticides	c concentration			d concentration		
		Initial value	Equations in 90 days	R^2	Initial value	Equations in 90 days	R^2
1.	Trifluralin	214.4	$y = -6.93 \ln(x) + 222.0$	0.548	248.6	$y = -14.1 \ln(x) + 249.5$	0.787
2.	Tefluthrin	106.1	$y = -4.52 \ln(x) + 105.4$	0.742	122.6	$y = -7.94 \ln(x) + 117.8$	0.832
3.	Pyrimethanil	116.9	$y = -5.96 \ln(x) + 120.1$	0.693	124.7	$y = -8.06 \ln(x) + 126.1$	0.77
4.	Propyzamide	119.6	$y = -4.06 \ln(x) + 125.0$	0.512	126.6	$y = -5.73 \ln(x) + 130.7$	0.598
5.	Pirimicarb	114.7	$y = -6.65 \ln(x) + 119.5$	0.822	126.2	$y = -9.11 \ln(x) + 128.6$	0.873
6.	Dimethenamid	46.2	$y = -2.23 \ln(x) + 47.66$	0.702	51.9	$y = -3.52 \ln(x) + 52.19$	0.86
7.	Fenchlorphos	240.3	$y = -10.1 \ln(x) + 245.1$	0.533	267.4	$y = -15.5 \ln(x) + 263.7$	0.785
8.	Tolclofos-methyl	116.7	$y = -4.00 \ln(x) + 118.1$	0.543	130.5	$y = -7.08 \ln(x) + 129.6$	0.78
9.	Pirimiphos-methyl	113.0	$y = -5.43 \ln(x) + 117.0$	0.74	126.3	$y = -8.31 \ln(x) + 127.9$	0.867
10.	2,4'-DDE	451.7	$y = -21.7 \ln(x) + 463.0$	0.794	466.9	$y = -26.4 \ln(x) + 476.8$	0.824
11.	Bromophos-ethyl	116.0	$y = -5.74 \ln(x) + 119.4$	0.686	124.2	$y = -7.74 \ln(x) + 126.7$	0.784
12.	4,4'-DDE	451.7	$y = -20.2 \ln(x) + 460$	0.768	466.9	$y = -24.6 \ln(x) + 473.3$	0.805
13.	Procymidone	117.9	$y = -4.22 \ln(x) + 120.5$	0.587	120.3	$y = -5.34 \ln(x) + 123.3$	0.733
14.	Picoxystrobin	234.0	$y = -10.7 \ln(x) + 240.3$	0.717	236.0	$y = -12.3 \ln(x) + 244.3$	0.774
15.	Quinoxyfen	115.9	$y = -6.64 \ln(x) + 122.1$	0.667	116.4	$y = -6.57 \ln(x) + 119.3$	0.653
16.	Chlorfenapyr	965.0	$y = -36.0 \ln(x) + 984.2$	0.579	966.0	$y = -42.4 \ln(x) + 995.3$	0.65
17.	Benalaxyl	116.8	$y = -6.24 \ln(x) + 121.7$	0.763	118.9	$y = -7.50 \ln(x) + 124.2$	0.746
18.	Bifenthrin	112.9	$y = -5.87 \ln(x) + 116.2$	0.653	114.8	$y = -7.09 \ln(x) + 119.1$	0.728
19.	Diflufenican	118.6	$y = -5.93 \ln(x) + 122.9$	0.571	119.4	$y = -5.86 \ln(x) + 122.3$	0.576
20.	Bromopropylate	237.4	$y = -11.3 \ln(x) + 249.2$	0.524	240.0	$y = -12.1 \ln(x) + 248.9$	0.647

Note: The degradation equations of them were obtained by taking the determination time (every 5 days) as *x*-axis and the concentration as *y*-axis.

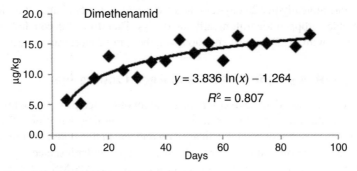

FIGURE 6.1.10   Logarithmic chart of dimethenamid in aged Oolong tea.

were predicted, and they were compared with the measured results determined by GC-MS/MS (see Tables 6.1.4 and 6.1.5). Tables 6.1.4 and Table 6.1.5 show that the deviation ratios of trifluralin, tefluthrin, and dimethenamid were higher than others at different intervals, the range of deviation ratio for them was −23.1% to 21.5%, 12.2%–26.0%, and 6.6%–24.0%, respectively. They were followed by pirimiphos-methyl, tolclofos-methyl, and fenchlorphos with the range of deviation ratio −2.8% to 21.8%, 5.8%–23.1%, and −2.0%–24.4%, respectively. The remaining 14 pesticides had relatively lower deviation ratios, except for part of the intervals. It can be also found that the lowest deviation ratios of the 20 pesticides were different. The numbers of pesticides, which had the lowest deviation ratio at intervals of 20th, 30th, and 45th day, were 4, 9, and 3, respectively. In addition, the highest deviation ratio of 2,4'-DDE, 4,4'-DDE and bromopropylate was found at intervals of 80th, 85th, and

TABLE 6.1.3 Logarithmic Functions of 20 Pesticides in Aged Oolong Tea at *d* Concentration

No.	Pesticides	Logarithmic functions	R^2
1.	Trifluralin	$y = 12.21 \ln(x) + 3.401$	0.705
2.	Tefluthrin	$y = 6.336 \ln(x) + 10.94$	0.641
3.	Pyrimethanil	$y = 8.822 \ln(x) - 5.055$	0.758
4.	Propyzamide	$y = 6.399 \ln(x) - 4.981$	0.620
5.	Pirimicarb	$y = 9.929 \ln(x) - 5.560$	0.800
6.	Dimethenamid	$y = 3.836 \ln(x) - 1.264$	0.807
7.	Fenchlorphos	$y = 18.07 \ln(x) - 9.564$	0.768
8.	Tolclofos-methyl	$y = 7.461 \ln(x) + 0.113$	0.698
9.	Pirimiphos-methyl	$y = 8.866 \ln(x) - 3.763$	0.781
10.	2,4′-DDE	$y = 29.77 \ln(x) - 22.61$	0.748
11.	Bromophos-ethyl	$y = 9.26 \ln(x) - 7.603$	0.757
12.	4,4′-DDE	$y = 29.07 \ln(x) - 23.00$	0.705
13.	Procymidone	$y = 6.414 \ln(x) - 7.533$	0.600
14.	Picoxystrobin	$y = 15.28 \ln(x) - 19.24$	0.740
15.	Quinoxyfen	$y = 6.511 \ln(x) - 3.341$	0.744
16.	Chlorfenapyr	$y = 45.47 \ln(x) - 38.61$	0.630
17.	Benalaxyl	$y = 9.157 \ln(x) - 12.91$	0.722
18.	Bifenthrin	$y = 8.334 \ln(x) - 9.895$	0.761
19.	Diflufenican	$y = 7.221 \ln(x) - 8.177$	0.730
20.	Bromopropylate	$y = 15.97 \ln(x) - 23.75$	0.596

Note: The determination time (day) was x-axis and the difference between each measured value and the first time-measured value (degradation value) of target pesticides was y-axis, and then the Logarithmic functions can be fitted.

35th day (they were close with the deviation ratios at 80th day), respectively, with the 80th day for all of the other pesticides. It could be due to the results at the interval of the 80th day were abnormal. In order to evaluate the deviation ratio accurately, the ranges of the deviation ratio for the pesticides were re-statistics without the data of interval of 80th day (Table 6.1.5).

Among the aforementioned remaining 14 pesticides, the deviation ratios were less than 15%, and most of them less than 10%, except pirimicarb and bromophos-ethyl, which had the highest deviation ratios of 17.1% and 16.8%, respectively, and quinoxyfen and benalaxyl had the lowest deviation ratios of −24.3% and −16.0%, respectively. It's evident that the proposed method for predicting the residue concentrations of pesticides in aged oolong tea in this study was accurate.

In addition, the initial values of pesticides at *c* and *d* conc. together with the deviation ratio of them were also listed in Table 6.1.5. The six pesticides with relatively higher deviation ratios of predicted and measured results also have relatively higher deviation ratios of their initial values, 10.1%–13.7%. However, the deviation ratios of initial values of the other 14 pesticides were less than 10% with the range of 0.1%–9.1%. With the decrease of the deviation ratio of initial values, the deviation ratio of predicted and measured results gets smaller. It can be concluded that the closer the initial values of them, the better results will be obtained.

6.1.8   CONCLUSIONS

In summary, the degradation regularity of 271 pesticides in aged oolong tea in 120 days was studied by the developed GC-MS/MS method. The results indicate that more than 70% of the 271 pesticides decreased exponentially or logarithmically in aged oolong tea. The conclusion is that pesticides in aged oolong tea will degrade slowly and the concentration of pesticide will decrease with at 4 month intervals. Further discussion in different classes suggests that most of the organophosphorus pesticides in aged oolong tea degraded in accordance with degradation trend A and B and most of the organochlorine pesticides decreased according to degradation trend E.

TABLE 6.1.4 Deviation Ratios of Predicted and Determined Results of c Concentration at Different Intervals (%)

No.	Pesticides	5 days	10 days	15 days	20 days	25 days	30 days	35 days	40 days	45 days	50 days	55 days	60 days	65 days	70 days	75 days	80 days	85 days	90 days
1.	Trifluralin	13.3	14.1	11.6	−22.9	21.0	15.4	14.1	18.0	11.6	16.0	−23.1	17.8	10.0	19.2	11.4	21.5	20.3	20.2
2.	Tefluthrin	15.2	17.4	16.0	14.1	15.1	25.5	16.1	19.7	12.2	23.5	20.3	20.7	17.8	20.6	21.2	26.0	25.3	21.4
3.	Pyrimethanil	5.4	11.2	9.7	3.6	3.5	7.7	6.2	9.0	0.3	11.1	6.6	8.8	6.3	11.7	10.4	20.0	13.7	14.1
4.	Propyzamide	6.9	11.4	9.9	3.6	4.9	7.6	9.2	7.1	0.0	11.4	3.4	8.6	8.9	8.6	7.8	15.4	12.4	10.4
5.	Pirimicarb	7.4	11.9	10.7	5.7	9.5	13.7	11.6	11.2	8.1	15.7	8.5	12.6	11.5	13.2	13.4	20.6	17.1	14.5
6.	Dimethenamid	9.4	13.0	11.4	6.6	15.3	18.0	17.6	16.3	7.9	17.5	12.3	17.2	13.7	19.9	17.0	24.0	21.4	18.9
7.	Fenchlorphos	6.5	12.8	11.3	−2.0	9.2	12.8	4.1	12.0	8.0	13.9	3.8	15.6	16.0	17.7	12.7	24.4	19.3	18.5
8.	Tolclofos-methyl	9.5	13.7	11.8	5.8	11.9	9.4	8.2	13.6	9.2	16.1	9.7	17.0	15.8	17.4	14.5	23.1	19.9	17.6
9.	Pirimiphos-methyl	9.2	13.9	13.0	6.1	11.2	10.1	9.2	13.1	−2.8	13.1	11.3	13.5	12.9	16.0	14.9	21.8	20.6	17.6
10.	2,4′-DDE	0.8	5.6	7.0	4.5	4.1	−0.1	5.3	2.6	0.2	7.3	5.7	4.1	0.4	4.3	6.7	10.8	11.3	4.6
11.	Bromophos-ethyl	3.8	10.5	10.8	2.4	8.0	1.2	5.6	6.0	5.7	10.5	3.4	11.6	9.2	13.2	10.7	17.8	16.8	12.4
12.	4,4′-DDE	0.4	5.2	6.4	3.9	3.5	−0.9	4.5	1.8	−0.7	6.4	5.4	4.5	0.4	5.2	7.9	10.8	11.4	5.6
13.	Procymidone	−2.2	4.0	5.7	4.4	3.7	−3.9	6.3	−0.5	−2.3	4.8	2.0	3.9	−2.6	1.9	6.2	9.8	8.6	2.9
14.	Picoxystrobin	−3.6	2.7	5.5	4.0	2.5	−3.2	8.4	−2.9	−0.7	3.2	0.4	3.2	−4.9	2.1	6.2	8.9	7.1	0.1
15.	Quinoxyfen	2.3	11.2	10.1	4.1	1.2	−24.3	8.1	−6.0	−8.9	0.7	−1.2	2.1	−2.4	4.5	5.6	14.7	6.1	3.2
16.	Chlorfenapyr	−1.9	4.6	6.2	2.0	2.9	−5.6	6.7	−4.7	−1.8	3.1	−2.7	3.7	−1.0	2.3	4.0	8.4	6.0	−1.6
17.	Benalaxyl	−4.2	4.2	4.6	4.0	3.0	−16.0	6.8	−4.2	−9.7	4.6	1.7	4.3	−3.4	3.0	7.4	8.7	6.5	−0.2
18.	Bifenthrin	−4.6	4.3	5.4	7.8	3.1	−0.1	10.0	−4.2	−9.5	3.5	4.2	4.7	−3.0	3.7	8.9	11.1	7.8	−0.3
19.	Diflufenican	−2.2	9.1	7.3	2.0	1.8	−11.8	9.5	−5.6	−7.9	0.5	−2.6	2.3	−3.1	4.7	6.6	13.5	8.2	2.5
20.	Bromopropylate	−0.8	9.1	7.3	−0.5	6.0	−8.6	10.5	−6.0	−2.3	2.1	−1.7	4.5	−3.8	10.0	6.9	10.2	7.4	0.3

Deviation ratio, % = (determined result−predicted value) × 100/determined result.

TABLE 6.1.5 Initial Values of *c* and *d* Concentration and the Range of Deviation Ratio of Predicted and Determined Results

No.	Pesticides	Initial value		Deviation ratio (%)	Range of deviation ratio[a] (%)	Revised range of deviation ratio[b] (%)
		c concentration (µg/kg)	*d* concentration (µg/kg)			
1.	Trifluralin	214.4	248.6	13.7	−23.1 to 21.5	−23.1 to 21.0
2.	Tefluthrin	106.1	122.6	13.5	12.2 to 26.0	12.2 to 25.5
3.	Pyrimethanil	116.9	124.7	6.3	0.3 to 20.0	0.3 to 14.1
4.	Propyzamide	119.6	126.6	5.5	0 to 15.4	0 to 12.4
5.	Pirimicarb	114.7	126.2	9.1	5.7 to 20.6	5.7 to 17.1
6.	Dimethenamid	46.2	51.9	11.1	6.6 to 24.0	6.6 to 21.4
7.	Fenchlorphos	240.3	267.4	10.1	−2.0 to 24.4	−2.0 to 19.3
8.	Tolclofos-methyl	116.7	130.5	10.5	5.8 to 23.1	5.8 to 19.9
9.	Pirimiphos-methyl	113.0	126.3	10.6	−2.8 to 21.8	−2.8 to 20.6
10.	2,4′-DDE	451.7	466.9	3.3	−0.1 to 11.3	−0.1 to 11.3
11.	Bromophos-ethyl	116.0	124.2	6.6	1.2 to 17.8	1.2 to 16.8
12.	4,4′-DDE	451.7	466.9	3.3	−0.9 to 11.4	−0.9 to 11.4
13.	Procymidone	117.9	120.3	2.0	−3.9 to 9.8	−3.9 to 8.6
14.	Picoxystrobin	234.0	236.0	0.9	−4.9 to 8.9	−4.9 to 8.4
15.	Quinoxufen	115.9	116.4	0.4	−24.3 to 14.7	−24.3 to 11.2
16.	Chlorfenapyr	965.0	966.0	0.1	−5.6 to 8.4	−5.6 to 6.7
17.	Benalaxyl	116.8	118.9	1.7	−16.0 to 8.7	−16.0 to 7.4
18.	Bifenthrin	112.9	114.8	1.7	−9.5 to 11.1	−9.5 to 10.0
19.	Diflufenican	118.6	119.4	0.7	−11.8 to 13.5	−11.8 to 9.5
20.	Bromopropylate	237.4	240.0	1.1	−8.6 to 10.5	−8.6 to 10.5

Deviation ratio, % = (*d* concentration − *c* concentration) × 100/*d* concentration.
[a]Between predicted and determined results.
[b]Between predicted and determined results without interval of 80th days.

In addition, the pesticide residues in aged oolong tea were predicted accurately by subtracting the degradation value of target pesticides on a specific day from the logarithmical curves, which took determination time (days) and the difference between each measured value and the first time-measured value of target pesticides as *x*-axis and *y*-axis, respectively. The predicted results of 14 pesticides were satisfactory by comparing with measured results at one of the concentrations in Youden pair.

Furthermore, we hope that the obtained degradation regularity of 271 pesticides in aged oolong tea will be helpful for studying the stability of the standard material of multipesticide residues in tea. What's more, we also propose that the prediction procedure of pesticide in aged oolong tea in this study may offer new insight insight into the error analysis of multiresidue pesticides in other complex matrices of international collaborative study.

REFERENCES

[1] Ballesteros, E., Parrado, M.J., 2004. Continuous solid-phase extraction and gaschromatographic determination of organophosphorus pesticides in natural and drinking waters. J. Chromatogr. A 1029, 267–273.

[2] Kamel, F., Hoppin, J.A., 2004. Association of pesticide exposure with neurologic dysfunction and disease. Environ. Health Persp. 112 (9), 950–958.

[3] Gros, M., Petrovic´, M., Barceló, D., 2006. Development of a multi-residue analytical methodology based on liquid chromatography–tandem mass spectrometry (LC–MS/MS) for screening and trace level determination of pharmaceuticals in surface and wastewaters. Talanta 70, 678–690.

[4] Park, J.Y., Choi, J.H., Abd El-Aty, A.M., Kim, B.M., Oh, J.H., Do, J.A., Kwon, K.S., Shim, K.H., Choi, O.J., Shin, S.C., Shim, J.H., 2011. Simultaneous multiresidue analysis of 41 pesticide residues in cooked foodstuff using QuEChERS: comparison with classical method. Food Chem. 128, 241–253.

[5] Lu, C.H., Liu, X.G., Dong, F.S., 2010. Simultaneous determination of pyrethrins residues in teas by ultra-performance liquid chromatography/tandem mass spectrometry. Anal. Chim. Acta 678, 56–62.

[6] Chen, G.Q., Cao, P.Y., Liu, R.J., 2011. A multi-residue method for fast determination of pesticides in tea by ultra performance liquid chromatography-electrospray tandem mass spectrometry combined with modified QuEChERS sample preparation procedure. Food Chem. 125, 1406–1411.

[7] European Union, Informal coordination of MRLs established in Directives 76/895/EEC, 86/362/EEC, 86/363/EEC, and 90/642/EEC. Available from: http://europa.eu.int/comm/food/plant/protection/pesticides/index.en.htm

[8] Simo, O.P., Qi, Z., 2002. The degradation of organophosphorus pesticides in natural waters: a critical review. Environ. Sci. Technol. 32, 17–72.

[9] Liu, T.F., Sun, C., Ta, N., Hong, J., Yang, S.G., Chen, C.X., 2007. Effect of copper on the degradation of pesticides cypermethrin and cyhalothrin. J. Environ. Sci. 19, 1235–1238.

[10] Juraske, R., Antón, A., Castells, F., 2008. Estimating half-lives of pesticides in/on vegetation for use in multimedia fate and exposure models. Chemosphere 70, 1748–1755.

[11] Sniegowski, K., Mertens, J., Diels, J., Smolders, E., Springael, D., 2009. Inverse modeling of pesticide degradation and pesticide-degrading population size dynamics in a bioremediation system: parameterizing the Monod model. Chemosphere 75, 726–731.

[12] Uygun, U., Senoz, B., Öztürk, S., Koksel, H., 2009. Degradation of organophosphorus pesticides in wheat during cookie processing. Food Chem. 117, 261–264.

[13] Ozbey, A., Uygun, U., 2007. Behaviour of some organophosphorus pesticide residues in peppermint tea during the infusion process. Food Chem. 104, 237–241.

[14] Manikandan, N., Seenivasan, S., Ganapathy, M.N.K., Muraleedharan, N.N., Selvasundaram, R., 2009. Leaching of residues of certain pesticides from black tea to brew. Food Chem. 113, 522–525.

[15] Lin, J.K., Li, X.F., Lin, X.D., Tu, L.J., 2008. The natural degradation dynamics of 4 pesticides on cultivars suitable to oolong tea. Chin. Agric. Sci. Bull. 24 (1), 104–111.

[16] Chen, L., ShangGuan, L.M., Wu, Y.N., Xu, L.J., Fu, F.F., 2012. Study on the residue and degradation of fluorine-containing pesticides in oolong tea by using gas chromatography-mass spectrometry. Food Control 25, 433–440.

[17] Pang, G.F., Fan, C.L., Zhang, F., Li, Y., Chang, Q.Y., Cao, Y.Z., Liu, Y.M., Li, Z.Y., Wang, Q.J., Hu, X.Y., Liang, P., 2011. High-throughput GC/MS and HPLC/MS/MS techniques for the multiclass, multiresidue determination of 653 pesticides and chemical pollutants in tea. J. AOAC Int. 94 (4), 1253–1296.

[18] Fan, C.L., Chang, Q.Y., Pang, G.F., Li, Z.Y., Kang, J., Pan, G.Q., Zheng, S.Z., Wang, W.W., Yao, C.C., Ji, X.X., 2013. High-throughput analytical techniques for determination of residues of 653 multiclass pesticides and chemical pollutants in tea part II: comparative study of extraction efficiencies of three sample preparation techniques. J. AOAC Int. 96 (2), 432–440.

[19] Pang, G.F., Fan, C.L., Chang, Q.Y., Li, Y., Kang, J., Wang, W.W., Cao, J., Zhao, Y.B., Li, N., Li, Z.Y., 2013. High-throughput analytical techniques for determination of residues of 653 multiclass pesticides and chemical pollutants in tea, part III: the evaluation of the cleanup efficiency of SPE cartridge newly developed for multiresidues in tea. J. AOAC Int. 96 (4), 887–896.

Chapter 6.2

A GC-MS, GC-MS/MS and LC-MS/MS Study of the Degradation Profiles of Pesticide Residues in Green Tea

Chapter Outline

6.2.1 INTRODUCTION

Tea originated in China and is one of the most popular drinks in the world. China, India, Kenya, Sri Lanka, and Turkey are the world's biggest producers, with a volume contribution of 76% of the total world export; of this volume, Chinese tea ranks first across the globe (http://faostat.fao.org). Tea is an agricultural product that is also susceptible to contamination by pesticides, because of the need to control pest-produced diseases. In addition, pesticide residue analysis in tea is difficult because it is complex and contains several organic and inorganic compounds, such as proteins, amino acids, alkaloids, phenols, sugars, organic acids, fats, pigments, aromatic substances, vitamins, and enzymes, and inorganic substances, such as calcium, iron, zinc [1]. Because tea is consumed extensively, significant research has been directed at developing multi-residue methods for the analysis of tea.

Since 2009, our research team has conducted [2–7]:

1. A 3-month stability study on 460 pesticides in six groups of mixed standard solutions;
2. A 3-month stability study on 345 pesticides in tea;
3. A 3-month study on the detection of deviation ratios of 275 pesticides in Youden paired tea samples;
4. A 3-month degradation kinetic study on 271 pesticides in aged tea samples as well as a study of applied kinetic equation on the deviation ratios of predicted values and actual values for pesticide residues;
5. A 3-month ruggedness study on the method using eight different analytical procedures;
6. An evaluation of the efficiencies of different SPE cleanup columns for pesticide residues in matrices and development of Cleanert TPT tea column with proprietary intellectual right, which is a unique cleanup effect for more than 600 pesticide residues in tea;
7. A 3-month verification study on the applicability of EU standards (EU Document No. SANCO/10684/2009) in an AOAC collaborative study;

8. A matrix effect study on the determination of 200 pesticide residues in tea;
9. A systematic study by GC-MS/MS on the effect on the analysis of pesticide residues in tea with and without hydration.

On the basis of this extensive research, a collaborative study on the method for the determination of 653 pesticide residues in tea was organized in 2014, with 30 laboratories from 11 countries or regions across America, Europe, and Asia using GC-MS, GC-MS/MS, and/or LC-MS/MS; 550 samples were analyzed by the 30 laboratories. This collaborative study was adopted in 2015 as the AOAC's first official action method for the determination of 653 pesticide residues in tea.

Several research studies have been conducted to elucidate the degradation characteristics of individual pesticides in tea. For instance, Lin et al. [8] used GC-MS to study the natural degradation of bifenthrin, fenpropathrin, cypermethrin, and buprofezin residues on new branches of two tea tree cultivars (Tieguanyin and Wuyirougui) from which suitable oolong tea is derived. Chen et al. [9] developed a GC-MS method for five pesticides in oolong tea and studied the natural degradation of five pesticides in tea tree and the effect of different processing modes on the distribution profile of the five pesticides. Satheshkumar et al. [10] studied the degradation of bifenazate in black tea at three sampling points and derived kinetic degradation equations and sampling intervals.

The rate of degradation of pesticide residues on tea trees in a tea plantation can be relatively complex and susceptible to the influence of physical and chemical factors, such as light, heat, acidity, and humidity [11,12]. Few published studies exist on the subject. In the literature currently available, Tewary et al. [13] studied the degradation of bifenthrin in new leaves of tea trees, mature tea, and tea soup in the dry and wet seasons in 2005 and recommended sampling intervals, which were based on their analytical results.

In this chapter, the degradation of pesticide residues in incurred green tea samples were investigated for different field trials, one in spring and the other in autumn, and under storage conditions at room temperatures. The new leaves from the tea trees were collected from different sampling intervals after pesticide application, and then processed into finished green tea. The authors used GC-MS, GC-MS/MS, and LC-MS/MS for the study:

1. To derive mathematical equations for the degradation of target pesticide residues from tea in the field trials;
2. Use the derived mathematical equations to predict the days needed for the concentrations of the target pesticides to degrade to the level of ML;
3. Investigate the concentration change of target pesticides in incurred green tea samples within a period of 12 weeks in storage at room temperature to provide a scientific basis for the preparation and stability of the incurred tea samples for the AOAC collaborative study.

6.2.2 REAGENTS AND MATERIALS

1. Solvents—acetonitrile, toluene, hexane, and methanol are all chromatic pure and purchased from Dikma Co. (Beijing, China).
2. Nineteen pesticide standards (including internal standards)—purity and purchased from LGC Promochem (Wesel, Germany).
3. Standard reserve solutions—accurately weigh 5–10 mg (accurate to 0.1 mg) standard solutions and place it in a 10 mL volumetric bottle, respectively, and select appropriate solvents per solubility of standard matters and dilute to graduation. Keep the standard reserve solutions at 4°C away from exposure to light.

6.2.3 DESIGN OF FIELD TRIALS

6.2.3.1 Selection of Pesticide Varieties

The pesticides should be in conformity with three conditions: relatively stable pesticides in the early stage; pesticides with *Maximum* Residue Limit (*MRL*) requirements in tea; pesticides readily available on the market. Based on these conditions, 19 pesticides were selected for the first trial and 11 pesticides for the second trial. The pesticide application amount is listed in Table 6.2.1. Each *mu* is provided with 60 kg and in six groups, with a preparation using a two-step method (prepare mother solutions before further dilution). (See Table 6.2.1.)

6.2.3.2 Selection and Planning of Tea Plantations

A 2 *mu* experimental zone was selected in Youshan Tea Plantations, Fuzhou City, Fujian Province, China, with a low elevation (average altitude of eastern longitude 115°50′~120°44′ and northern latitude 23°31′-28°19′), which was guarded from polluting the tea trees close by (Fig. 6.2.1). Field trials were carried out according to Good Agricultural Practices, GAP.

TABLE 6.2.1 Pesticides for Filed Trails

No.	Pesticides	Manufacturer	Packing specification	Stage I	Stage II
First group (tree spray)					
1	Trifloxystrobin (50%)	Bayer (Germany)	5 g/bag	15 g/*mu*	7.5 g/*mu*
2	Cyprodini (25%)	Syngenta (Switzerland)	15 g/bag	50 g/*mu*	
3	Boscalid (50%)	BASF SE (Germany)	12 g/bag	30 g/*mu*	6 g/*mu*
4	Azoxystrobin (50%)	BASF SE (Germany)	5 g/bag	2 g/*mu*	
5	Pyrimethanil (40%)	Zhongbao High-Tech. Co., Ltd. (Beijing, China)	15 mL/bag	60 mL/*mu*	30 mL/*mu*
Second group (tree spray)					
6	Dimethomorph (50%)	BASFSE (Germany)	100 g/bag	40 g/*mu*	
7	Metalaxyl (35%)	Jiangsu Baoling Chemical Co., Ltd (China)	40 g/bag	40 g/*mu*	30 g/*mu*
8	Pyridaben	Huizhou Zhongxun Chemical Co., Ltd (China)	10 g/bag	5 g/*mu*	
Third group (tree spray)					
9	Bifenthrin	Hunan Dafang Agricultural Chemicals Co., Ltd	200 mL/bag	40 ml/*mu*	20 ml/*mu*
10	Endosulfan (35%)	Jingbo Agrochem (China)	200 g/bottle	40 g/*mu*	
Fourth group (spray on tree)					
11	Triadimefon (25%)	Beijing Zhongbao High-tech (China)	50 g/bag	20 g/*mu*	20 g/*mu*
12	Diniconazole (5%)	Beijing Zhongbao High-tech (China)	180 mL/bottle	40 mL/*mu*	20 mL/*mu*
13	Epoxiconazole	BASF SE (Germany)	10 mL/bag	30 mL/*mu*	30 mL/*mu*
14	Propiconazole (25%)	Syngenta (Switzerland)	100 mL/bottle	40 mL/*mu*	
Fifth group (tree spray)					
15	Ametryn	Changxing Chemical(China)	120 g/bag	130 g/*mu*	65 g/*mu*
16	Pendimethalin	BASF SE (Germany)	200 mL/bottle	110 mL/*mu*	
Sixth group (soil spray)					
17	Napropamide	Jiangsu Kuaida Agrochemical (China)	100 g /bag	100 g/*mu*	100 g/*mu*
18	Butralin (48%)	Longjiang (China)	200 mL/bottle	200 mL/*mu*	
19	Acetochlor	Shandong Qiaochang Chemical Co., Ltd (China)	15 g/bag	100 mL/*mu*	100 mL/*mu*

FIGURE 6.2.1 Experimental zone of Youshan tea plantations, Fuzhou, China.

For the field trials, a 0.2 *mu* blank zone (i.e., a zone to which no pesticide were applied) was set up, which were three lines of trees away from the pesticide applied zone. The zone to which pesticides were applied and divided into 20 portions for sample collection on days 1, 2, 3, 5, 7, 10, 15, 20, 25, 30 after pesticide application. Samples were collected from 2 of the 20 zones on each assigned sample collection day.

6.2.3.3 Pesticide Application and Incurred Tea Sample Preparations

Pesticide application: pesticides were divided into six groups per their properties: one group containing three herbicides was applied in the soils and the other five were sprayed onto the trees (Table 6.2.1). Pesticide applications were carried out strictly in accordance with operational procedures.

Sampling: a total of 10 duplicate sample collections were made over the 30-day period following the pesticide application. These were sealed and packed independently and delivered to laboratories after drying, grinding, homogenizing, and sieving. The samples were analyzed once a week of a period of 12 weeks.

Weather conditions: the first field trial spraying date was September 21, 2011, and the weather was fine within 15 days after the pesticide application; the second field trial spraying date was May 24, 2012, and it had a rainy drizzle on days 1 and 2, with moderate rain on days 15–18 after the pesticide application.

6.2.4 ANALYTICAL METHOD

6.2.4.1 Extraction and Cleanup

Weigh a 5 g test sample (accurate to 0.01 g) into 80 mL centrifugal tube, add 15 mL acetonitrile; make a homogeneous extraction at 13,500 r/min for 1 min; centrifuge at 4200 r/min for 5 min and take supernatants and transfer into an 80 mL pear-shaped flask. Extract the dregs with 15 mL one more time, centrifuge, and merge the two extractions and place in a water bath at 40°C and rotary evaporate to 1 mL and await cleanup.

SPE column conditioning: add 2 cm high anhydroussodiumsulfate into Cleanert-TPT column and place it on a fixture attached with a pear-shaped flask underneath. Rinse the Cleanert-TPT column with 10 mL acetonitrile–toulene and discard the effluents. Clean up test sample extraction fluids: when the rinsing liquid surface reaches the high anhydroussodiumsulfate top, transfer the aforementioned test sample concentrations into Cleanert-TPT and collect the elutes with a new pear-shaped flask; wash the test concentration bottle with 3 × 2 mL acetonititrile–toulene and transfer the rinsing liquid into the column when the test sample concentration surface reaches the top of the high anhydroussodiumsulfate. Connect the top of SPE column with a 30 mL reservoir, rinse the column with 25 mL acetonitrile–toulene, and concentrate the eluates in the pear-shaped flask in a water bath at 40°C with rotary evaporation.

GC-MS (GC-MS/MS): in the aforementioned concentration and cleanup solutions, add 40 heptachlor epoxide internal standard working solutions, evaporate to dryness in a water bath at 35°C, dissolve the pesticide with 1.5 mL hexane, and homogenize with ultrasound and filter with 0.2 membrane before being submitted for determination.

LC-MS/MS: in the aforementioned concentrated cleanup solutions, add 40 chlorpyrifos-methyl internal standard working solutions, blow with nitrogen gas to dryness in a water bath at 35, use 1.5 mL acetonitrile–water to dissolve the dregs, homogenize with ultrasound and filter with 0.2 μm membrane before being submitted for determination.

6.2.4.2 Apparatus Conditions

6.2.4.2.1 GC-MS System

Connect a Model 7890A gas chromatograph to a Model 5975C mass selective detector with electron ionization (EI) source and equip with a Model 7683 autosampler and Chemstation data processing software system (Agilent Technologies), or equivalent. The column used is a DB-1701 capillary column (30 m × 0.25 mm × 0.25 μm; J&W Scientific). Temperature raising procedure: keep 1 min at 40°C, raise it to 130°C at 30°C/min, then raise to 250°C at 5°C/min, raise it to 300°C at 10°C/min and keep 5 min. Carrier gas: helium gas, purity ≥99.999%, with a flow rate of 1.2 mL/min. The injection port temperature is 290°C and 1 μL samples are injected splitless with the purge on after 1.5 min electron impact source: 70 eV, ion source temperature: 230°C, interface temperature: 280°C, ion monitoring mode: selected ion monitoring (SIM).

6.2.4.2.2 GC-MS/MS System

Connect Model 7890 gas chromatograph to a Model 7000A triple quadrupole mass spectrometer and equip with a Model 7693 autosampler (Agilent Technologies, Wilmington, DE). The column used is a DB-1701 capillary column (30 m × 0.25 mm × 0.25 μm; J&W Scientific). The ion-monitoring mode is multireaction monitoring mode (MRM). Other analytical conditions are identical with those of GC-MS.

6.2.4.2.3 LC-MS/MS System

Couple an Agilent Series 1200 HPLC system directly to a 6430 triple quadrupole mass spectrometer equipped with an electrospray ionization source and Model G1367D autosampler with a Mass Hunter data processing software system (Agilent Technologies). Column: Zorbax SB-C18, 2.1 × 100 mm × 3.5 μm, or equivalent. Column temperature: 40°C. Injection volume: 10 μL. Ionization mode: ESI. Ion source polarity: positive ion. Nebulizer gas: nitrogen gas. Nebulizer gas pressure: 0.28 MPa. Ion spray voltage: 4000 V. Dry gas temperature: 350°C. Dry gas flow rate: 10 L/min. mobile phase: A: 0.1%

formic acid in water, B: acetonitrile; gradient elution: 0~3 min, 1%~30% B; 3~6 min, 30%~40% B; 6~9 min, 40% B; 9~15 min, 40%~60% B; 15~19 min, 60%~99% B; 19~23 min, 99% B; 23~23.01 min, 99%~1% B.

6.2.5 THE PESTICIDE DEGRADATION PROFILES WITHIN 30 DAYS FOR THE FIELD TRIALS

Regarding the incurred tea samples collected over 10 days from days 1, 2, 3, 5, 7, 10, 15, 20, 25, 30, the content of the pesticide tested once every week was taken as Y coordinate while the collection dates were adopted as X coordinate. A pesticide degradation curve diagram was plotted, with 12 (tested over 12 weeks) degradation curves derived for each pesticide and a chart of pesticide degradation trends in the two stages (Figs. 6.2.2 and 6.2.3).

Fig. 6.2.2 shows that the pesticide degradation profiles for incurred tea samples over the test of 12 weeks were basically consistent, and the pesticides were capable of degrading to the level of MRL below within 30 days and tended to go stable. In Fig. 6.2.3, a turning point in the degradation profile curve occurred on day 2 as it continued to rain on days 2 and 3 after the pesticide application. The pesticide had not yet been totally absorbed by the tea leaves but had been washed away by rain.

6.2.6 STUDY ON PESTICIDE DEGRADATION BY THREE TECHNIQUES

Pesticide degradation characteristics were studied by GC-MS, GC-MS/MS, and LC-MS/MS for two filed trials, which can be fitted by power function. See Table 6.2.2.

For Stage I, the correlation coefficient (R^2) of power function regression equation of GC-MS exceeding 0.90 made up 50%, that exceeding 0.80 made up 78%; the correlation coefficient (R^2) of power function regression equation of GC-MS/MS exceeding 0.90 made up 44%, that exceeding 0.80 made up 78%; the correlation coefficient (R^2) of power function regression equation of LC-MS/MS exceeding 0.90 made up 60%, that exceeding 0.80 made up 66%; the regression analysis proved that the degradation curve fitting is relatively good by the three techniques. For Stage II, the correlation coefficient (R^2) of power

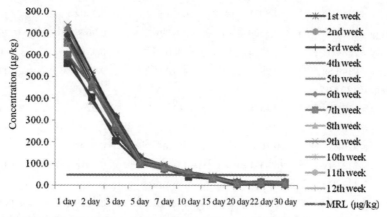

FIGURE 6.2.2 Degradation profiles of epoxiconazole in the first field trial by LC-MS/MS.

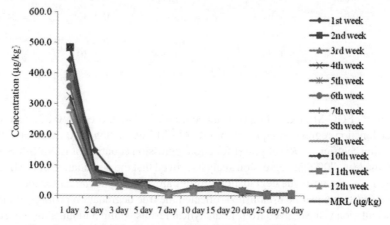

FIGURE 6.2.3 Degradation profiles of epoxiconazole in the second field trial by LC-MS/MS.

TABLE 6.2.2 Kinetic Equation of Pesticides Degradation for the Field Trails in the Tea Plantations

No	Pesticides	GC-MS		GC-MS/MS		LC-MS/MS	
		Power function	Correlation coefficient	Power function	Correlation coefficient	Power function	Correlation coefficient
Stage I (Dec. 27, 2011–Feb. 20, 2012)							
1	Acetochlor	$y = 222.14x^{-1.17}$	0.7940	$y = 2281.8x^{-1.471}$	0.8800	$y = 1301.8x^{-2.712}$	0.9271
2	Ametryn	$y = 909.78x^{-1.001}$	0.7323	$y = 951.49x^{-0.984}$	0.7416	$y = 1016.9x^{-1.002}$	0.8178
3	Azoxystrobin					$y = 3745.7x^{-2.326}$	0.9433
4	Bifenthrin	$y = 10489x^{-1.623}$	0.8068	$y = 11655x^{-1.552}$	0.8385		
5	Boscalid	$y = 67902x^{-1.381}$	0.8136	$y = 61722x^{-1.339}$	0.7746	$y = 19098x^{-0.862}$	0.7251
6	Butralin	$y = 3076.9x^{-1.694}$	0.7969	$y = 3790.3x^{-1.724}$	0.8014	$y = 621.38x^{-1.061}$	0.6687
7	Cyprodinil	$y = 39183x^{-1.244}$	0.9854	$y = 63744x^{-1.437}$	0.9841	$y = 5698.9x^{-0.803}$	0.9859
8	Dimethomorph	$y = 160851x^{-1.982}$	0.9146	$y = 170270x^{-2.076}$	0.8327	$y = 20485x^{-1.424}$	0.8378
9	Diniconazole	$y = 2359.7x^{-1.384}$	0.9618	$y = 2032.3x^{-1.145}$	0.9349	$y = 8335.5x^{-2.8}$	0.9558
10	Endosulfan	$y = 307501x^{-2.674}$	0.9440	$y = 347971x^{-2.731}$	0.9146		
11	Epoxiconazole	$y = 1053.6x^{-1.334}$	0.9845	$y = 755.05x^{-1.205}$	0.9856	$y = 953.2x^{-1.309}$	0.9715
12	Metalaxyl	$y = 5195.3x^{-1.429}$	0.9300	$y = 4955.6x^{-1.517}$	0.8757	$y = 6413.8x^{-1.579}$	0.9447
13	Napropamide	$y = 797.33x^{-1.421}$	0.8132	$y = 900.73x^{-1.421}$	0.8328	$y = 1656.5x^{-2.055}$	0.7214
14	Pendimethalin	$y = 3471.6x^{-1.219}$	0.6309	$y = 4392.7x^{-1.286}$	0.6359		
15	Propiconazole	$y = 6055.2x^{-1.496}$	0.9901	$y = 6314.8x^{-1.504}$	0.9830	$y = 4142.6x^{-1.397}$	0.9795
16	Pyridaben	$y = 6410.6x^{-1.305}$	0.8031	$y = 6955.5x^{-1.333}$	0.7802		
17	Pyrimethanil	$y = 14294x^{-1.297}$	0.9629	$y = 20479x^{-1.445}$	0.9606	$y = 9768.6x^{-1.118}$	0.9624
18	Triadimefon	$y = 2784x^{-1.581}$	0.8443	$y = 3480.8x^{-1.749}$	0.9761	$y = 3113.9x^{-1.894}$	0.9558
19	Trifloxystrobin	$y = 9349.8x^{-2.023}$	0.9913	$y = 9218x^{-1.953}$	0.9863	$y = 10338x^{-2.679}$	0.8584
Stage II (May 23, 2012–Sep. 13, 2012)							
1	Acetochlor	$y = 118.81x^{-0.964}$	0.7576	$y = 481.6x^{-1.093}$	0.9252	$y = 143.15x^{-1.287}$	0.8384
2	Ametryn	$y = 4.4829x^{-0.181}$	0.6542	$y = 4.592x^{-0.192}$	0.6309	$y = 3.6009x^{-0.394}$	0.6352
3	Bifenthrin	$y = 426.8x^{-0.697}$	0.7068	$y = 401.87x^{-0.683}$	0.6804		
4	Boscalid	$y = 2321.1x^{-1.028}$	0.9311	$y = 2562.4x^{-1.038}$	0.9016	$y = 1975.5x^{-0.836}$	0.9418
5	Diniconazole	$y = 157.73x^{-0.575}$	0.7022	$y = 171.68x^{-0.495}$	0.6126	$y = 184.43x^{-1.346}$	0.7969
6	Epoxiconazole	$y = 127.42x^{-0.75}$	0.5644	$y = 134.95x^{-0.735}$	0.5319	$y = 134.71x^{-0.768}$	0.4970
7	Metalaxyl	$y = 1274.4x^{-1.085}$	0.9276	$y = 1187.8x^{-1.098}$	0.9286	$y = 1635.4x^{-1.357}$	0.8558
8	Napropamide	$y = 1862x^{-1.698}$	0.9528	$y = 2017.8x^{-1.744}$	0.9547	$y = 1005.7x^{-1.544}$	0.9486
9	Pyrimethanil	$y = 3485.2x^{-0.965}$	0.9414	$y = 4260.3x^{-1.055}$	0.9191	$y = 2458.5x^{-0.862}$	0.8603
10	Triadimefon	$y = 273.31x^{-1.049}$	0.6859	$y = 233.36x^{-1.115}$	0.7013	$y = 630.59x^{-2.733}$	0.9138
11	Trifloxystrobin	$y = 1175.2x^{-1.582}$	0.9641	$y = 1009.9x^{-1.408}$	0.9130	$y = 1079.6x^{-1.794}$	0.9667

function regression equation of GC-MS exceeding 0.90 made up 45%, that exceeding 0.80 made up 45%; the correlation coefficient (R^2) of power function regression equation of GC-MS/MS exceeding 0.90 made up 54%, that more than 0.80 made up 54%; the correlation coefficient (R^2) of power function regression equation of LC-MS/MS exceeding 0.90 made up 40%, that more than 0.80 made up 70%; the regression analysis proved that the degradation curve fitting is relatively good by the three techniques. In general, the pesticide degradation profiles are basically identical for the two field trials by the three techniques, and the power function equation established reflected the true degradation profiles of these pesticides.

Take the average concentration value derived from 12 analytical results within 3 months for the tea sample collected at the same time as Y-coordinate and collection time as X-coordinate and plot a degradation trend line. See Fig. 6.2.4 for an example.

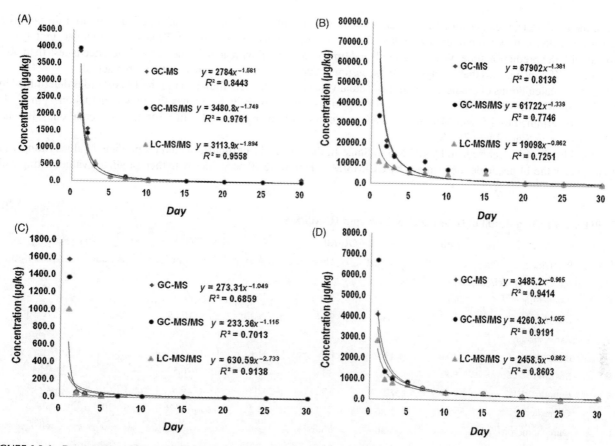

FIGURE 6.2.4 Degradation profiles of pesticides by GC-MS, GC-MS/MS and LC-MS/MS. (A) Triadimefon in stage I, (B) Boscalid in stage I, (C) Triadimefon in stage II, and (D) Pyrimethanil in stage II.

For Stage I, the fitting pesticide degradation trend lines of the analytical values for some pesticides (e.g., trifloxystrobin) by GC-MS, GC-MS/MS, and LC-MS/MS were basically identical, for instance, Fig. 6.2.4A; for other pesticides (e.g., boscalid), the concentrations from days 1–7 by LC-MS/MS were very low, causing the degradation trend lines to be below GC-MS and GC-MS/MS. The author considers that due to the lack of experience, the pesticide spraying quantity for the first field trials was relatively great, leading to the very high content of residual pesticides in the tea samples collected within 7 days after the pesticide application, and certain pesticide contents exceeding the linear range of instruments; besides, the differences of working principles and structures with GC-MS (GC-MS/MS) and LC-MS/MS resulted in the marked differences of their linear range and matrix effect, causing the content of specific pesticides to exceed the linear range of the instrument and the analytical results from GC-MS (GC-MS/MS) and LC-MS/MS to have differences. For instance, Fig. 6.2.4B.

For Stage II, based on the findings from the experiment in Stage I, the pesticide spraying quantity was reduced for certain pesticides with their content exceeding the linear range of the instrument, but it rained continuously after the pesticide application and some pesticides were washed away before they were totally absorbed by the teas, leading to pesticide loss as well. Therefore, pesticide residual content in tea decreased markedly compared with that in the first time. But from the degradation trend line, the pesticide degradation profiles in Stage II were basically identical with those in Stage I, and see Figs. 6.2.4C and 6.2.4D.

6.2.7 COMPARISON OF THE PESTICIDES DEGRADATION TO MRL ACTUAL VALUES IN FIELD TRIALS WITH POWER FUNCTION EQUATIONS PREDICTED VALUES DETECTED BY THREE TECHNIQUES

A weather pesticide degradation power function equation is capable of accurately predicting the degradation process of pesticide content in tea is an important criterion in evaluating if pesticide degradation equation is scientifically rational. In this chapter, a comparative evaluation is adopted for the time needed for pesticide content to degrade to MRL value using the prediction of pesticide degradation power function equation and actual values.

For the field trials, the collection time of tea samples was on days 1, 2, 3, 5, 7, 10, 15, 20, 25, and 30 after the pesticide application, with the maximum interval of 5 days. Therefore, the actual detection time in Table 6.2.3 cannot accurately correspond with the prediction time. Where the predicted time is compared with actual time, if the difference is less than 30% or the time difference is less than 3 days, it can be assumed that the predicted is consistent with the actual.

Statistical calculations of predicted values and actual values in Table 6.2.3 found that in the 18 pesticides determined by GC-MS and GC-MS/MS in Stage I, there were 15 pesticides with their predicted values fitting with actual ones; in the 15 pesticides determined by LC-MS/MS, there were 9 pesticides with their predicted values fitting with actual ones; in total, the fittings rate in Stage I was 76.5%.

In the 11 pesticides determined by GC-MS in Stage II, there were 11 pesticides with their predicted values fitting with actual ones; in the 11 pesticides determined by GC-MS/MS, there were 10 pesticides with their predicted values fitting with

TABLE 6.2.3 Testing Result of GC-MS, GC-MS/MS and LC-MS/MS

No.	Pesticides	MRL (µg/kg)	GC-MS Actual	GC-MS Theoretic	GC-MS/MS Actual	GC-MS/MS Theoretic	LC-MS/MS Actual	LC-MS/MS Theoretic
Stage I (Dec. 27, 2011–Feb. 20, 2012)								
1	Acetochlor	50	3	2.0	15	8.4	5	2.6
2	Ametryn		>30	90.6	>30	102.5	>30	100.8
3	Azoxystrobin						7	4.7
4	Bifenthrin	300	15	8.9	15	10.6		
5	Boscalid	500	30	35.0	30	36.5	30	68.4
6	Butralin		20	19.5	20	20.9	20	25.5
7	Cyprodinil		>30	212.1	>30	144.9	>30	364.3
8	Dimethomorph		>30	58.8	>30	50.3	>30	68.3
9	Diniconazole	50	20	16.2	20	25.4	10	6.2
10	Endosulfan		5	2.4	5	2.5		
11	Epoxiconazole	50	15	9.8	15	9.5	15	9.5
12	Metalaxyl	100	15	15.9	15	13.1	10	13.9
13	Napropamide	50	20	5.8	15	7.6	15	5.5
14	Pendimethalin		20	18.4	20	18.9		
15	Propiconazole		20	15.5	20	15.7	20	14.4
16	Pyridaben	50	>30	41.2	>30	40.5		
17	Pyrimethanil		>30	45.9	>30	39.8	>30	60.2
18	Triadimefon	200	5	5.3	5	5.1	5	4.3
19	Trifloxystrobin	50	15	13.3	15	14.5	15	7.3
Stage II (May 23, 2012–Sep. 13, 2012)								
1	Acetochlor	50	2	1.3	5	4.7	2	1.4
2	Ametryn	-	1	0.1	1	0.0	1	0.1
3	Bifenthrin	300	2	1.7	3	1.9		
4	Boscalid	500	5	4.5	5	5.6	7	4.9
5	Diniconazole	50	7	6.5	7	11.9	3	2.2
6	Epoxiconazole	50	3	3.5	3	4.9	3	3.7
7	Metalaxyl	100	15	10.6	10	10.4	15	7.5
8	Napropamide	50	7	8.5	7	9.1	7	7.0
9	Pyrimethanil	50	>30	41.9	>30	40.5	>30	44.1
10	Triadimefon	200	2	1.4	2	1.4	2	1.5
11	Trifloxystrobin	50	7	7.3	10	8.6	7	5.5

actual ones; in the 10 pesticides determined by LC-MS/MS, there were 9 pesticides with their predicted values fitting with actual ones; in total, the fitting rate in Stage II was 93.8%; the comparison of the power function predicted values and actual ones of GC-MS, GC-MS/MS, and LC-MS/MS from the pesticide field trials in these two stages found that the predicted values of the majority pesticides were fitted with actual ones.

6.2.8   STABILITY STUDY ON THE PESTICIDES IN INCURRED TEA SAMPLES AT ROOM TEMPERATURE STORAGE CONDITIONS (18–25°C)

The stability of pesticides in tea at room temperature storage conditions plays a decisive role in whether this can be regarded as the standard of "intercollaborative study blind samples." For this purpose, the three techniques of GC-MS, GC-MS/MS, and LC-MS/MS were adopted in this chapter to make a statistical computation of pesticide concentrations and RSD values of the identical incurred tea samples determined 12 times over 3 months (Table 6.2.4).

TABLE 6.2.4 Stability of Pesticides in Incurred Tea Samples by GC-MS, GC-MS/MS and LC-MS/MS

No.	Pesticides	Concentration (µg/kg)	RSDs (%) (n = 12)		
			GC-MS	GC-MS/MS	LC-MS/MS
Stage I (Dec. 27, 2011–Feb. 20, 2012)					
1	Acetochlor	18.8–334.6	15.2–22.2	5.8–19.5	16.0
2	Ametryn	27.6–253.5	9.3–27.1	7.4–16.3	15.7–23.1
3	Azoxystrobin	29.3–447.2			14.9–21.6
4	Bifenthrin	45.1–608.7	10.4–19.5	11.3–19.8	
5	Boscalid	241.6–978.0	9.8–15.6	18.6–20.1	14.9–20.5
6	Butralin	7.0–71.5	22.9–37.4	14.8–23.0	7.1–24.7
7	Dimethomorph	70.3–303.9	13.1–35.3	20.0–21.3	23.2–26.0
8	Diniconazole	89.0–1180.7	11.6–19.4	23.1–28.7	12.0–18.0
9	Endosulfan	899.0–2178.7	27.7	9.2–10.9	
10	Epoxiconazole	12.6–253.6	14.4–26.8	18.7–23.4	7.8–26.3
11	Metalaxyl	45.5–875.2	10.6–24.1	14.2–27.6	12.3–15.1
12	Napropamide	9.6–277.9	21.6–28.7	10.5–28.6	11.0–26.6
13	Pendimethalin	31.2–152.2	17.6–22.6	18.3–22.7	
14	Propiconazole	6.7–422.7	10.6–25.7	6.9–21.2	7.5–21.3
15	Pyridaben	29.0–136.8	6.4–20.3	19.2–21.4	
16	Pyrimethanil	158.5–1405.1	8.5–12.7	14.2–22.3	10.1–22.2
17	Triadimefon	46.6–561.7	10.0–22.0	12.4–20.5	13.3–16.2
18	Trifloxystrobin	22.5–360.7	11.5–15.0	11.0–22.7	6.9–15.4
Stage II (May 23, 2012–Sep. 13, 2012)					
1	Acetochlor	13.9–126.5	14.4–17.5	9.4–22.0	18.5–20.1
2	Ametryn	6.7–7.5	21.5	24.5	21.8
3	Bifenthrin	54.4–720	11.4–25.4	7.7–28.8	
4	Boscalid	66.4–850	7.2–21.4	11.6–27.0	13.8–26.6
5	Diniconazole	74–544.9	14.3–21.3	19.5–28.1	17.8–17.8
6	Epoxiconazole	8.8–62.9	7.4–20.7	13.1–23.7	16.5–24.7
7	Metalaxyl	54.3–349.5	9.0–19.0	9.2–38.9	12.9–24.4
8	Napropamide	9.6–194.2	7.0–19.0	12.7–18.0	10.7–30.5
9	Pyrimethanil	76.8–1377.7	7.6–16.0	9.9–25.6	13.6–27.5
10	Triadimefon	43.0–58.4	14.6–19.0	24.0–24.8	12.0
11	Trifloxystrobin	17.4–240.6	7.8–19.4	16.3–28.8	22.2–25.9

It can be viewed from Table 6.2.4 for Stage I: GC-MS: RSD < 25%, accounting for 80.95%; (2)RSD > 30%, accounting for 3.57%. RSD maximum is 37.4%. GC-MS/MS: RSD < 25%, accounting for 90.22%; RSD maximum is 28.7%; LC-MS/MS: RSD < 25%, accounting for 93.12%; RSD maximum is 26.6%.

For Stage II: GC-MS: RSD < 25%, accounting for 98.36%, RSD maximum is 25.4%. GC-MS/MS: RSD < 25%, accounting for 81.16%; RSD > 30%, accounting for 1.45%. RSD maximum is 38.9%: LC-MS/MS: RSD < 25%, accounting for 91.11%; RSD > 30%, accounting for 2.22%. RSD maximum is 30.5%.

Table 6.2.4 shows pesticides with RSD < 25% in Stage I made up 87.61%, RSD > 30% made up 1.28%; pesticides with RSD < 25% in Stage II made up 89.71%, RSD > 30% made up 1.14%. It can be seen that pesticide residues in incurred tea samples at room temperature storage for 3 months did not change markedly, and the stability of the majority pesticides was capable of fulfilling the requirements of AOAC collaborative study.

6.2.9 CONCLUSIONS

The degradation profiles and speed of the 19 pesticides (first time) and the 11 pesticides (second time) and the stability of incurred samples over 3 months were studied using GC-MS, GC-MS/MS, and LC-MS/MS in two field trials in 2 years. The degradation profiles study found that the majority of pesticide residues could degrade to below MRL 30 days after pesticide application; the stability study of pesticides in the incurred samples found that 10 samples from different days with RSD less than 25% from successive 12 determinations once a week for 3 months accounted for 88%. Therefore, the results of such a study not only prove that the stability of incurred tea samples prepared from pesticide field trials in tea plantations satisfy the requirements of the AOAC collaborative study, but also provide the proficiency testing of international laboratories with a scientific method of preparing incurred tea standard reference samples.

6.2.10 ACKNOWLEDGMENTS

Acknowledgments are made to Youshan Tea Plantations, Fujian for providing the tea plantation for study purposes free of charge as well as to the financial support from General Administration of Quality Supervision, Inspection and Quarantine of the People's Republic of China (AQSIQ) special public welfare project of China (Project No. 201010080).

REFERENCES

[1] Zhang, J.Y., Jiang, H.Y., Cui, H.C., Jiang, Y.W., Wang, B., Huang, Y.D., 2011. Hunan Agric. Sci. 3, 104–108.

[2] Pang, G.F., Fan, C.L., Zhang, F., Li, Y., Chang, Q.Y., Cao, Y.Z., Liu, Y.M., Li, Z.Y., Wang, Q.J., Hu, X.Y., Liang, P., 2011. J. AOAC Int. 94 (4), 1253–1296.

[3] Fan, C.L., Chang, Q.Y., Pang, G.F., Li, Z.Y., Kang, J., Pan, G.Q., Zheng, S.Z., Wang, W.W., Yao, C.C., Ji, X.X., 2013. J. AOAC Int. 96 (2), 432–440.

[4] Pang, G.F., Fan, C.L., Chang, Q.Y., Li, Y., Kang, J., Wang, W.W., Cao, J., Zhao, Y.B., Li, N., Li, Z.Y., 2013. J. AOAC Int. 96 (4), 887–896.

[5] Chang, Q.Y., Pang, G.F., Fan, C.L., Chen, H., Wang, Z.B., 2016. J. AOAC Int.

[6] Li, Y., Chen, X., Fan, C.L., Pang, G.F., 2012. J. Chromatogr. A 1266, 131–142.

[7] Chen, X., Li, Y., Chang, Q.Y., Hu, X.Y., Pang, G.F., Fan, C.L., 2015. J. AOAC Int. 98 (1), 149–159.

[8] Lin, J.K., Li, X.F., Lin, X.D., Tu, L.J., 2008. Chin. Agric. Sci. Bull. 24 (1), 104–111.

[9] Chen, L., Shang Guan, L.M., Wu, Y.N., Xu, L.J., Fu, F.F., 2012. Food Control 25, 433–440.

[10] Satheshkumar, A., Senthurpandian, V.K., Shanmugaselvan, V.A., 2014. Food Chem. 145, 1092–1096.

[11] Chen, Z.M., Wan, H.B., Wang, Y., Xue, Y., Xia, H., 1987. Chin. Acad. Agric. Sci. 4–9, 146–149.

[12] Agnihothrudu, V., Muraleedharan, N., 1990. Planters Chron. 85, 125–127.

[13] Tewary, D.K., Kumar, V., Ravindranath, S.D., Shanker, A., 2005. Food Control 16, 231–237.

7 High-Throughput Analytical Techniques for Determination of Residues of 653 Multiclass Pesticides and Chemical Pollutants in Tea by GC-MS, GC-MS/MS and LC-MS/MS: Collaborative Study

Chapter 7.1

The Pre-Collaborative Study of AOAC Method Efficiency Evaluation

Chapter Outline

7.1.1 INTRODUCTION

Various tea beverages are popular drinks besides the bottle water; they are one of the most consumed healthy drinks everywhere in the world. Because there are plenty of functional ingredients in teas, people all over the world like to drink teas. The consumption of teas has increased year by year. At the same time, as more and more countries and organizations are concerned about the food safety issues, the regulations related to pesticide residues in teas become more stringent. Pesticide residues limits become the threshold of the entry in international trade. Until March in 2011, the regulations on MRLs (Maximum Residue Limits) in various countries and organizations were 443 in EU, 268 in Japan, 530 in Germany, 11 in CAC, 32 in Korea respectively [1–5]. Until now, there are 17 countries and organizations in the world that have set up more than 800 MRLs for pesticides in teas [6]. Due to such large number of testing items, it is needed and imperative for the research and development of high-throughput pesticide residue analysis technology. Meanwhile, the efficiency evaluation of analytical methods is an important tool to analytical technique standardization. A lot of international organizations, such as Association of Official Analytical Chemists (AOAC), American Society for Testing and Material (ASTM), United States Food and Drug Administration (FDA), Food and Agricultural Organization (FAO), International Organization for Standardization (ISO), the International Union of Pure and Applied Chemistry (IUPAC) [7–12], have established operation guidelines for method efficiency evaluation. Researchers choose these evaluation guidelines to evaluate the effectiveness of analytical methods. In 1996, Hsu et al. organized nine labs to conduct an interlaboratory assessment of the determination of Cephalexin preparations. Dixon was used for testing; RSDr and RSD_R were 0.32% and 1.94%, respectively [13]. In 1997, Nilsson et al. organized 20 laboratories to evaluate solid phase micro-extraction (SPME) quantitative methods for the determination of volatile organic solvent in water samples. Accuracy and precision were evaluated according to ISO standards, analyzed by Analysis of Variance (ANOVA), Cochran and Grubbs conduct inspection [14]. Hoogerbrugge et al. organized 24 laboratories in 8 EU countries to conduct interlaboratory assessment of chlorophenol in water determination methods in 1999. Mandel's plots, Cochran and Grubbs were applied for data analysis and outlier elimination. Statistic results show the repeatability, reproducibility, and proportion of outliers meet the requirements of ISO 5725 [15]; in the same year, Ferrari et al. organized 10 laboratories to do interlaboratories evaluation of triazines and their metabolites in water at ng/L level by SPME, the average of linear correlation coefficient R^2 was above 0.99. In accordance with ISO 5725, it was found that there was no outlier in Gochran test, two outliers in Grubbs test, RSD_r between 6 and 14%, and RSD_R between 10 and 17% [16]. In 2002, Lehotay organized an interlaboratory assessment of supercritical fluid extraction methods for pesticide residues in fruits and vegetables from 17 laboratories in 7 countries. Five pesticides were tested in apple, green bean, and carrot. The recoveries of the five pesticides ranged from 58 to 105%, RSD_r was 7 to 15%, and RSD_R was 7 to 27%. Except polar pesticide Trifluralin, p, p'-DDE and Bifenthrin, the average Horwitz ratios of other pesticides is less than 1.0 [17]. In 2004, Kakimoto et al. organized eight laboratories to do interlaboratory evaluations on the determination methods of 139 pesticides in six fruits and vegetables. The statistical results showed that 111 pesticides can be accurately determined by their method, and 118 pesticides can be analyzed by scanning methods, in the linear range of 0.5–5 g/L, correlation coefficient between 0.983–1.000 [18]. In 2006, Ratola et al. conducted an interlaboratory evaluation of the determination of Ochratoxin A in wine. The participating laboratories consisted of 24 laboratories in 17 countries on 5 continents. Data were analyzed by Cochran and Grubbs, and exceptions in the laboratory were eliminated, studies results showed that the determination of Ochratoxin A in wine has good reproducibility [19]. In the same year, Simon et al. conducted an interlaboratory study on the HPLC method for the determination of 15 polycyclic aromatic hydrocarbons (PAHs) in cigarette condensates. Twenty-five laboratories from the European Union participated in the study, and determined by liquid chromatography fluorescence and GC-MS [20]. In 2007, Lehotay conducted an interlaboratory research of the QuEChERS method in 13 laboratories from 7 countries. The research conducted pretreatment of 20 representative pesticides in fruits and vegetables using GC/MS and LC/MS/MS methods. Three types of substrates (grapes, lettuces, and oranges) were selected to the study at three unknown levels and eight contaminated samples were tested for collaborative study. Cochran, single Grubbs and double Grubbs were used to test the results. The results showed that the QuEChERS method is suitable for the monitoring of pesticide residues in fruits and vegetables [21]. In 2010, Wong et al. conducted an interlaboratory evaluation of the QuEChERS method for the determination of 191 pesticides in fruits and vegetables. The U.S. FDA, National Research Centre for Grapes (NRCG), India, and Ontario Ministry of the Environment (MOE) laboratories conducted the collaborative study. The recovery of more than 79% of the pesticides was between 80 and 120%, and the principal component analysis (PCA) was applied for statistical analysis, experimental results proved the similarity of the results and the applicability of the method [22]. To evaluate the efficiency of pesticide multiresidues high-throughput analytical method for 653 pesticides and chemical contaminants in teas developed in our laboratory, as general trend of current analytical methods evaluation, according to the internationally acceptable method evaluation rules, our research group organized 16 domestic laboratories with experience in the detection of pesticide residues in tea to conduct a collaborative study from Dec. 2010 to Apr. 2011. Experiment protocol designed that each of the participating laboratories was required to test six collaborative study samples, including green tea and oolong tea. Each type of tea contained an aged sample and blank sample. Samples were quantified using a 5-point matrix-matched calibration curve. In the 16 laboratories, all data from two laboratories are excluded due to deviation operation. The statistical analysis result is from the remaining 14 laboratories. The AOAC method efficiency evaluation shows that the method has good repeatability and reproducibility, and is fully applicable to the qualitative and quantitative analysis of 653 pesticide residues in tea.

7.1.2 DESIGN OF COLLABORATIVE STUDY

7.1.2.1 Selection of Representative Pesticides

If hundreds of pesticides were used for OAC collaborative study, it will be unimaginably difficult to the participants in terms of resources, time, and personnel. Therefore, as determined by the International AOAC Method Committee, 20 representative pesticides were selected from 653 pesticides for collaborative study, which are shown in Table 7.1.1. These 20 pesticides can be classified by different methods. By chemical composition, they include organophosphorus, organonitrogen, organohalogen, carbamates, pyrethroid, and others; by functions they include herbicide, insecticide, bactericide, acaricide, and so on. These selections are pesticides, which are commonly used in the growth of tea, and they are also often required to be tested in the international trade of tea. In addition, these pesticides have good stability after being sprayed onto tea leaves, and their polarity is also extensive representativeness. A few pesticides were detected by both GC-MS (GC-MS/MS) and LC-MS/MS.

TABLE 7.1.1 Pesticides to Be Testing in Collaborative Study

No.	GC-MS and GC-MS/MS		LC-MS/MS	
	Pesticides	*MRL/(µg/kg)*	*Pesticides*	*MRL/(µg/kg)*
1	2,4'-DDE	200	Acetochlor	10
2	Benalaxyl	100	Benalaxyl	100
3	Bifenthrin	5000	Bensulide	30
4	Bromophos-ethyl	100	Butralin	20
5	Bromopropylate	100	Chlorpyrifos	100
6	Chlorfenapyr	50,000	Clomazone	20
7	Diflufenican	50	Diazinon	20
8	Dimethenamid	20	Ethoprophos	20
9	Fenchlorphos	100	Fenazaquin	10
10	Picoxystrobin	100	Flutolanil	50
11	Pirimicarb	50	Imidacloprid	50
12	Pirimiphos-methyl	50	Indoxacarb	50
13	Propyzamide	50	Kresoxim-methyl	100
14	Pyrimethanil	100	Phenothrin	50
15	Quinoxyfen	50	Picoxystrobin	100
16	Tebufenpyrad	100	Pirimiphos methyl	50
17	Tefluthrin	50	Propoxur	100
18	Tolclofos-methyl	100	Tebufenpyrad	100
19	Trifluralin	100	Triadimefon	200
20	Vinclozolin	100	Trifloxystrobin	50

7.1.2.2 Preparation and Homogeneity Evaluation of Aged Samples

See Section 6.1.4.1.

7.1.2.3 Types and Quantity of Collaborative Study Samples

Each of the participating collaborative study laboratories obtained the collaborative study samples shown as Fig. 7.1.1. Six collaborative study samples, including green tea and oolong tea, each of which includes aged samples and blank samples. The aging samples were prepared via the Youden Pair principle. The 20 target pesticide concentrations ranged from 20 µg/kg to 1.7 mg/kg. The Youden Pair samples had similar pesticide concentrations, with the difference of no more than 5%.

7.1.2.4 Organizing and Implementing of Collaborative Study

The collaborative laboratory through open recruitment on the Internet, after the verification of the qualifications of collaborative laboratories, finally selected 16 laboratories with experience in the determination of pesticide residues in tea (including 6 government owned laboratories; 5 foreign owned laboratories; 2 enterprises laboratories; 2 research institute laboratories; 1 university laboratory).

Collaborative study organizer provided to each participating laboratory: AOAC collaborative study protocol (OMA-2010-Jan-001); 1 mL Ampoule with quantified mixed standard solution; two of 1 mL Ampoule with internal standard solution (one for GC-MS and GC-MS/MS, the other for LC-MS/MS); blank tea sample; ampoules of a standard mixture solution for calibration curve; and one aged sample.

Participate laboratory should submit to the organizer: Analytical instrument parameter settings; Aged sample experimental data report; Original spectrum and text records.

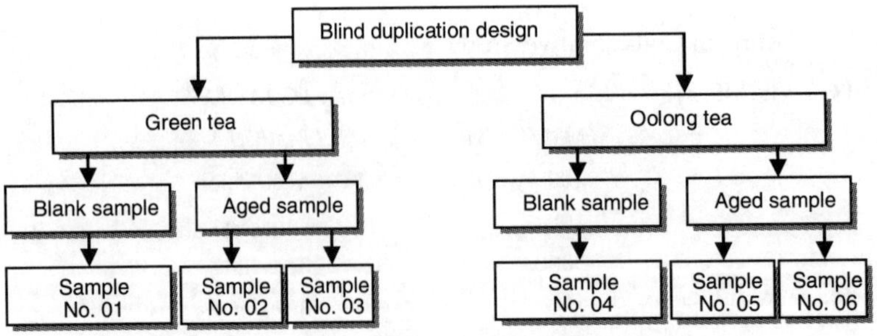

FIGURE 7.1.1 Number of blank, fortified, and pesticide aged samples for the collaborative study.

Submission deadline: 2 months from the date of shipment of the sample.

7.1.2.5 System Suitability Check

Instrument sensitivity testing: according to the internal standard signal to noise ratio (S/N) GC-MS> 1000, GC-MS/MS> 2000, LC-MS/MS> 500.

Instrument stability testing (calculated according to internal standard): retention time deviation < 3% (two successive injections), and peak area deviation < 6% (two successive injections).

GC-MS, GC-MS/MS, and LC-MS/MS are all equipped with data processing software system.

7.1.2.6 Requirement for Participating in Collaborative Laboratories

Skilled pesticide residue analysis sample preparation technique; expert GC-MS (GC-MS/MS) and LC-MS/MS detection technology and experience (operating, error diagnostic, and maintenance experience); professional mass spectrometry analysis technology.

The laboratory that intends to participate in collaborative study must pass the precollaborative study test and conduct self-ability testing through blind samples before the official testing of collaborative study samples. Only when the precollaborative study sample acceptance criteria are achieved can the official collaborative study samples be detected.

7.1.3 EXPERIMENT

7.1.3.1 Instruments and Reagents

Gas chromatography-mass spectrometer (GC-MS) is equipped with an electron ionization source (EI); gas chromatography-tandem mass spectrometer (GC-MS/MS) is equipped with an electron ionization source (EI); liquid chromatography-tandem mass spectrometer is equipped with electrospray ionization source (LC-MS/MS); rotary evaporator; homogenizer; vortex mixer; high purity water generator; centrifuge; ultrasonic cleaner; electronic balance; nitrogen concentrator; and solid phase extraction unit.

Acetonitrile, n-hexane, toluene (HPLC grade); formic acid chromatographic purity; anhydrous sodium sulfate for analytical purity, anhydrous sodium sulfate burning at 650 °C 4 h, stored in a desiccator; water is deionized water; SPE cartridge: Cleanert-TPT (or equivalent).

7.1.3.2 Preparation of Standard Solutions

Erect any of the three reserve solutions ampoule bottles till the solutions flow down the inner walls of the bottle from the top thoroughly before opening the bottle to carefully transfer the solutions to a 10 mL volumetric bottle. To ensure a full transfer, wash and cleanse the ampoule bottle with 3 × 1 mL toluene and transfer the cleansing liquid into the 10 mL volumetric bottle. Then use toluene to dilute the solutions to the graduations and mix homogeneously to form a corresponding standard working solution. The concentrations of quality control standard working solutions are shown in Table 7.1.2, while the concentrations of pesticide working solutions are listed in Table 7.1.3. The concentrations of internal standard heptachlor-epoxide standard working solutions are 35 µg/mL, while the concentrations of internal standard chlorpyrifos-methyl standard working solutions are 40 µg/mL, with both standard working solutions kept from exposure to light in a 4 °C fridge.

TABLE 7.1.2 The Concentration of Quality Control Standard Working Solutions

	GC-MS(GC-MS/MS)			LC-MS/MS	
No.	Pesticides	Concentration/µg/mL	No.	Pesticides	Concentration/µg/mL
ISTD	Heptachlor-epoxide	28.0	ISTD	chlorpyrifos-methyl	32.0
1	Bifenthrin	10.0	1	Butralin	2.0
2	Propyzamide	10.0	2	Ethoprophos	2.0
3	Pyrimethanil	10.0	3	Picoxystrobin	2.0
4	Tebufenpyrad	10.0	4	Pirimiphos-methyl	2.0
5	Trifluralin	20.0	5	Triadimefon	2.0

TABLE 7.1.3 The Concentration of Pesticide Mixed Standard Working Solutions

	GC-MS(GC-MS/MS)			LC-MS/MS	
No.	Pesticides	Concentration/µg/mL	No.	Pesticides	Concentration/µg/mL
1	2,4'-DDE	40.0	1	Acetochlor	4.0
2	Benalaxyl	10.0	2	Benalaxy	2.0
3	Bifenthrin	10.0	3	Bensulide	6.0
4	Bromophos-ethyl	10.0	4	Butralin	2.0
5	Bromopropylate	20.0	5	Clomazone	2.0
6	Chlorfenapyr	80.0	6	Diazinon	2.0
7	Diflufenican	10.0	7	Chlorpyrifos	20
8	Dimethenamid	4.0	8	Ethoprophos	2.0
9	Fenchlorphos	20.0	9	fenazaquin	2.0
10	Picoxystrobin	20.0	10	Flutolanil	2.0
11	Pirimicarb	10.0	11	Indoxacarb	2.0
12	Pirimiphos-methyl	10.0	12	Imidacloprid	4.5
13	Propyzamide	10.0	13	Kresoxim-methyl	20.0
14	Pyrimethanil	10.0	14	phenothrin	10.0
15	Quinoxyfen	10.0	15	Picoxystrobin	2.0
16	tebufenpyrad	10.0	16	Pirimiphos methyl	2.0
17	Tefluthrin	10.0	17	Propoxur	5.0
18	Tolclofos-methyl	10.0	18	Tebufenpyrad	2.0
19	Trifluralin	20.0	19	Triadimefon	2.0
20	vinclozolin	20.0	20	Trifloxystrobin	2.0

7.1.3.3 Sample Preparation

Weigh 5 g sample, put into an 80 mL centrifuge tube; add 15 mL acetonitrile, homogenize at 13,500 rpm for 1 min, centrifuge at 4200 rpm for 5 min, transfer the supernatant into a pear-shaped flask; re-extract the sample with 15 mL acetonitrile, centrifuge, and combine the supernatants from the two extractions; concentrate the extract to approximately 1 mL in a rotary evaporator (or TurboVap) in a 40 °C water bath; add anhydrous sodium sulfate (approximately 2 cm) onto the Cleanert-TPT packing material, add 10 mL acetonitrile–toluene (3:1, v/v) to activate the cartridge; discard the waste solution.

Sample extract cleaning: stop the flow through the cartridge when the liquid level in the cartridge barrel has just reached the top of the sodium sulfate packing; load the above concentrated extract (about 1 mL) into the preconditioned Cleanert-TPT cartridge, collect the elute in another clean pear-shaped flask; rinse the flask that contained the concentrated extract with 3 × 2 mL acetonitrile–toluene (3:1, v/v); load the rinse into the cartridge; connect a 30 mL reservoir onto the upper part of the cartridge using an adapter; elute the cartridge with 25 mL acetonitrile–toluene (3:1, v/v); evaporate the eluate to approximately 0.5 mL using a rotary evaporator in a 40 °C water bath.

Add 40 μL heptachlor-epoxide (ISTD) working standard solution to the sample prepared for GC-MS and GC-MS/MS analysis; add 40 μL chlorpyrifos-methyl (ISTD) working standard solution to the sample prepared for LC-MS/MS analysis; evaporate to dryness under a stream of nitrogen in a 35 °C water bath; GC-MS and GC-MS/MS solution residue redissolve in 1.5 mL n-hexane; LC-MS/MS solution residue re-dissolve in 1.5 mL of acetonitrile–water (3:2, v/v), ultrasonicate the samples to mix, and filter through a 0.2 μm membrane filter. The sample is then ready for analysis.

7.1.3.4 Test Condition

7.1.3.4.1 GC-MS (GC-MS/MS) Test Condition

Column: DB-1701 capillary column (30 m × 0.25 mm × 0.25 μm), or equivalent; column temperature program: 40 °C hold 1 min, at 30 °C/min–130 °C, at 5 °C/min–250 °C, at 10 °C/min–280 °C, hold 5 min; carrier gas: helium, purity ≥ 99.999%, flow rate: 1.2 mL/min; injection port temperature: 290 °C; injection volume: 1 μL; injection mode: splitless, purge on after 1.5 min; ionization mode: EI; ion source polarity: positive ion; ionization voltage: 70 eV; ion source temperature: 230 °C; GC-MS interface temperature: 280 °C. The retention times, quantifying ions, qualifying ions, ion abundances, LODs and LOQs of 20 pesticides are shown in Tables 7.1.4 and 7.1.5.

TABLE 7.1.4 Retention Times, Quantifying Ions, Qualifying Ions, Ion Abundances, LODs and LOQs for 20 Pesticides by GC-MS

No.	Pesticides	Retention Time, Min	Quantifying Ion	Qualifying Ion 1	Qualifying Ion 2	LOQ, μg/kg	LOD, μg/kg
ISTD	*Heptachlor-Epoxide*	*22.15*	*353(100)*	*355(79)*	*351(52)*	Δ	Δ
1	Trifluralin	15.35	306(100)	264(72)	335(7)	20.0	10.0
2	Tefluthrin	17.42	177(100)	197(26)	161(5)	10.0	5.0
3	Pyrimethanil	17.45	198(100)	199(45)	200(5)	10.0	5.0
4	Propyzamide	18.73	173(100)	255(23)	240(9)	10.0	5.0
5	Pirimicarb	18.83	166(100)	238(23)	138(8)	20.0	10.0
6	Dimethenamid	19.57	154(100)	230(43)	203(21)	10.0	5.0
7	Tolclofos-methyl	19.77	265(100)	267(36)	250(10)	10.0	5.0
8	Fenchlorphos	19.80	285(100)	287(69)	270(6)	40.0	20.0
9	Vinclozolin	20.27	285(100)	212(109)	198(96)	10.0	5.0
10	Pirimiphos-methyl	20.28	290(100)	276(86)	305(74)	20.0	10.0
11	2,4′-DDE	22.84	246(100)	318(34)	176(26)	25.0	12.5
12	Bromophos-ethyl	23.12	359(100)	303(77)	357(74)	10.0	5.0
13	Picoxystrobin	24.53	335(100)	303(43)	367(9)	20.0	10.0
14	Chlorfenapyr	27.08	247(100)	328(54)	408(51)	200.0	100.0
15	Quinoxyphen	27.13	237(100)	272(37)	307(29)	10.0	5.0
16	Benalaxyl	27.48	148(100)	206(32)	325(8)	10.0	5.0
17	Bifenthrin	28.61	181(100)	166(32)	165(35)	10.0	5.0
18	Diflufenican	28.73	266(100)	394(25)	267(14)	10.0	5.0
19	Tebufenpyrad	29.19	318(100)	333(78)	276(44)	10.0	5.0
20	Bromopropylate	29.39	341(100)	183(54)	339(51)	20.0	10.0

TABLE 7.1.5 Retention Times, Monitored Ion Transitions, Collision Energies, LODs and LOQs for 20 Pesticides by GC-MS/MS

No.	Pesticides	Retention Time, Min	Quantifying Precursor/ Production Transition	Qualifying Precursor/ Production Transition	Collision Energy, V	LOQ, µg/kg	LOD, µg/kg
ISTD	Heptachlor-Epoxide	22.15	353/282	353/263	17;17	Δ	Δ
1	Trifluralin	15.41	306/264	306/206	12;15	4.8	2.4
2	Tefluthrin	17.40	177/127	177/101	13;25	0.8	0.4
3	Pyrimethanil	17.42	200/199	183/102	10;30	6.0	3.0
4	Propyzamide	18.91	173/145	173/109	15;25	1.0	0.5
5	Pirimicarb	18.98	238/166	238/96	15;25	4.0	2.0
6	Dimethenamid	19.73	230/154	230/111	8;25	2.0	1.0
7	Fenchlorphos	19.83	287/272	287/242	15;25	16.0	8.0
8	Tolclofos-methyl	19.87	267/252	267/93	15;25	10.0	5.0
9	Pirimiphos-methyl	20.36	290/233	290/125	5;15	10.0	5.0
10	Vinclozolin	20.43	285/212	285/178	10;10	12.0	6.0
11	2,4'-DDE	22.79	318/248	318/246	15;15	6.0	3.0
12	Bromophos-ethyl	23.16	359/331	359/303	10;10	10.0	5.0
13	Picoxystrobin	24.75	335/303	335/173	10;10	10.0	5.0
14	Quinoxyfen	27.18	237/208	237/182	25;25	80.0	40.0
15	Chlorfenapyr	27.37	408/363	408/59	5;15	140.0	70.0
16	Benalaxyl	27.66	148/105	148/79	15;25	2.0	1.0
17	Bifenthrin	28.63	181/166	181/165	10;5	10.0	5.0
18	Diflufenican	28.73	266/246	266/218	10;25	20.0	10.0
19	Tebufenpyrad	29.26	333/276	333/171	5;15	8.0	4.0
20	Bromopropylate	29.46	341/185	341/183	15;15	8.0	4.0

7.1.3.4.2   LC-MS Testing Condition

Column: ZORBAX SB-C_{18}, 3.5 µm, 100 mm × 2.1 mm or equivalent; mobile phase program and the flow rate: 0.4 mL/min, gradient condition: 0.00 min-99% A(A: 0.1% formic acid, B: acetonitrile); 3.00 min-70% A; 6.00 min-60%A; 9.00 min-60% A; 15.00 min-40% A; 19.00 min-1% A; 23.00 min-1% A; 23.01 min-99% A; column temperature: 40 °C; injection volume: 10 µL; ionization mode: ESI; ion source polarity: positive ion; nebulizer gas: nitrogen gas; nebulizer gas pressure: 0.28 MPa; ion spray voltage: 4000 V; dry gas temperature: 350 °C; dry gas flow rate: 10 L/min. the retention times, monitored ion transitions, collision energies, fragmentations, LODs and LOQs are presented in Table 7.1.6.

7.1.4   METHOD EFFICIENCY ACCEPTANCE CRITERIA

7.1.4.1   General Principle

Interlaboratories collaborative study is an international standard operation process in the world. Each laboratory cannot deviate from a unified operation protocol or procedure, otherwise it is dealt as a laboratory that deviates from the operation protocol and will be excluded. In addition, there are four following detail technical indicators acceptance criteria.

7.1.4.2 Standard Curve Linear Correlation Coefficient Acceptance Criteria

At least five points of the matrix-matched the internal standard calibration curve, the linear correlation coefficient $R^2 \geq 0.995$.

7.1.4.3 Target Pesticide Ion Abundance Ratio Criteria

The target pesticide ion abundance ratio must meet the EU standard [23], see Table 7.1.7.

TABLE 7.1.6 Retention Times, Monitored Ion Transitions, Collision Energies, Fragmentations, LODs and LOQs for 20 Pesticides by LC-MS/MS

No.	Pesticides	Retention Time, Min	Quantifying Precursor/ Production Transition	Qualifying Precursor/ Production Transition	Collision Energy, V	Fragmentation/V	LOQ, µg /kg	LOD, µg/kg
ISTD	Chlorpyrifos-methyl	16.9	322.0/125.0Δ	322.0/290.0Δ	15;15Δ	80	Δ	Δ
1	Imidacloprid	4.6	256.1/209.1	256.1 /175.1	10;10	80	22.0	11.0
2	Propoxur	6.8	210.1/111.0	210.1/168.1	10;5	80	24.4	12.2
3	Clomazone	9.3	240.1/125.0	240.1/89.1	20;50	100	0.4	0.2
4	Ethoprophos	12.4	243.1/173.0	243.1/215.0	10;10	120	2.8	1.4
5	Triadimefon	12.7	294.2/69.0	294.2/197.1	20;15	100	7.9	3.9
6	Acetochlor	13.9	270.2/224.0	270.2/148.2	5;20	80	10.0	5.0
7	Flutolanil	14.2	324.2/262.1	324.2/282.1	20;10	120	1.1	0.6
8	Benalaxyl	15.3	326.2/148.1	326.2/294.0	15;5	120	1.2	0.6
9	Kresoxim-methyl	15.5	314.1/267	314.1/206.0	5;5	80	100.6	50.3
10	Picoxystrobin	15.9	368.1/145.0	368.1/205.0	20;5	80	8.4	4.2
11	Pirimiphos methyl	16.0	306.2/164.0	306.2/108.1	20;30	120	0.2	0.1
12	Diazinon	16.1	305.0/169.1	305.0/153.2	20;20	160	0.7	0.4
13	Bensulide	16.4	398.0/158.1	398.0/314.0	20;5	80	34.2	17.1
14	Tebufenpyrad	17.6	334.3/147.0	334.3/117.1	25;40	160	0.3	0.1
15	Indoxacarb	17.6	528.0/150.0	528.0/218.0	20;20	120	7.5	3.8
16	Trifloxystrobin	17.6	409.3/186.1	409.3/206.2	15;10	120	2.0	1.0
17	Chlorpyrifos	18.4	350.0/198.0	350.0/79.0	20;35	100	53.8	26.9
18	Butralin	18.8	296.1/240.1	296.1/222.1	10;20	100	1.9	1.0
19	Fenazaquin	19.0	307.2/57.1	307.2/161.2	20;15	120	0.3	0.2
20	Phenothrin	19.7	351.1/183.2	351.1/237.0	15;5	100	50.0	25.0

TABLE 7.1.7 Recommended Maximum Permitted Tolerances for Relative Ion Intensities Using a Range of Spectrometric Techniques From EU

Relative Intensity (% of base peak)	GC-MS-SIM (relative)	GC-MS-MS, LC-MS/MS (relative)
>50%	±10%	±20%
>20%–50%	±15%	±25%
>10%–20%	±20%	±30%
≤10%	±50%	±50%

TABLE 7.1.8 The Judging Criteria of Recoveries, RSD$_r$ and RSD$_R$

Concentration	Recovery Limits	Repeatability (RSD$_r$)	Reproducibility (RSD$_R$)
10 µg/g (ppm)	80%–115%	6%	11%
1 µg/g	75%–120%	8%	16%
10 µg/kg (ppb)	70%–125%	15%	32%
Concentration	Recovery limits	Repeatability (RSD$_r$)	Reproducibility (RSD$_R$)

7.1.4.4 Recovery, RSDr, and RSDR Acceptance Criteria

According to AOAC Guidelines for single laboratory validation of chemical methods for dietary supplements and botanicals [24], acceptable recovery limits, repeatability (RSDr), and reproducibility (RSDR) are in Table 7.1.8.

7.1.4.5 Outliers Elimination Via Grubbs and Dixon Double Checking

When each of the above four indicators fails to reach 70% of the corresponding items, it means that this laboratory is incapable of participating in the evaluation of the efficiency of this method, and the laboratory's data will be excluded in the final statistics.

7.1.5 RESULT AND DISCUSSION

7.1.5.1 The Labs That Deviated from the Operational Method Have Been Eliminated

First, the raw data of these 16 labs were investigated one by one, with some questions verified through email and telephone calls with collaborators. The results found that the following three labs deviated from the protocol on some major operational procedures.

Lab 7: (1) The pesticide of GC-MS-SIM matrix internal standard calibration curve correlation coefficient $R^2 \geq 0.995$ established in this lab only accounts for 25% of the total number, failing to meet the stipulations in the protocol. (2) The pretreatment of samples for GC-MS-SIM was executed by two operators one after another. The first operator added internal standards before extraction while the second one added them once more before the nitrogen blow to dryness and redissolving, which caused confusion. (3) When preparing aged samples for LC-MS/MS determination, internal standards were added before extraction, which also deviated from the method.

Lab 3: To reduce the workload, the operator only established a green tea matrix internal standard calibration curve, by which the oolong tea samples were quantified. This also violated the stipulations of the protocol.

Lab 9: (1) This lab established only the green tea matrix matching calibration curve like Lab 3, by which oolong tea samples were quantified. (2) The mixed standard reserve solutions in an ampoule bottle provided by the study director are to be used only after thorough transfer and dilution, but this lab transferred only a portion, which violated operational stipulations. (3) For GC-MS-SIM determination, this lab adopted an external calibration curve for quantification and deviated from the protocol. The earlier described causes led to the fact that 87% of the data from the test results by this lab were the maximum values or next-to-maximum values of the collaborators' data of the same group.

Based on these reasons, we eliminated 360 data of Lab 7 and Lab 9 as well as the test data of oolong tea of Lab 3 from statistical calculation.

7.1.5.2 Eliminations of Outliers Via Double-Checking of Grubbs and Dixon

Regarding the 1960 data of the remaining 14 labs, outliers were eliminated per Grubbs and Dixon inspection, and statistical calculation was conducted per AOAC Youden pair statistical method (Table 7.1.9). A sole GC-MS/MS statistical calculation was not carried out and the data of Lab 5 and Lab 11 were merged into GC-MS-SIM calculation because labs that applied for GC-MS/MS did not total 8.

7.1.5.3 Method Efficiency

The parameters of method efficacy RSD$_r$, RSD$_R$, and HorRat values in Table 7.1.9 are summed up per the range in Table 7.1.10.

TABLE 7.1.9 Method Efficiency for Determination of 40 Pesticides in Tea by Both GC-MS-SIM (GC-MS/MS) and LC-MS/MS

No.	Analyte	GC-MS-SIM (GC-MS/MS) Average (C, μg/kg)	S_r (μg/kg)	RSD_r (%)	S_R (μg/kg)	RSD_R (%)	HorRat	No.	Analyte	LC-MS/MS Average (C, μg/kg)	S_r (μg/kg)	RSD_r (%)	S_R (μg/kg)	RSD_R (%)	HorRat
									Green Tea						
1	Trifluralin	348.1	13.3	3.8	69.3	19.9	1.06	1	Imidacloprid	74.5	10.0	13.4	19.6	26.3	1.11
2	Tefluthrin	169.7	8.1	4.8	36.1	21.3	1.02	2	Propoxur	64.7	5.3	8.2	10.8	16.7	0.69
3	Pyrimethanil	186.5	6.4	3.4	35.8	19.2	0.93	3	Clomazone	24.6	2.3	9.5	4.4	17.8	0.64
4	Pirimicarb	178.0	8.1	4.5	24.3	13.6	0.66	4	Ethoprophos	22.7	2.2	9.6	4.2	18.5	0.65
5	Propyzamide	203.9	7.8	3.8	25.3	12.4	0.61	5	Triadimefon	26.0	1.9	7.2	4.3	16.4	0.59
6	Dimethenamid	72.5	4.5	6.3	20.0	27.6	1.16	6	Acetochlor	50.3	3.4	6.8	11.2	22.2	0.88
7	Fenchlorphos	361.7	16.0	4.4	69.3	19.2	1.03	7	Flutolanil	31.9	2.9	9.0	4.6	14.6	0.54
8	Tolclofos-methyl	179.5	8.8	4.9	47.3	26.3	1.27	8	Benalaxyl	27.8	2.0	7.1	3.8	13.6	0.50
9	Pirimiphos-methyl	179.7	8.2	4.6	34.6	19.3	0.93	9	Kresoxim-methyl	276.1	10.0	3.6	23.0	8.3	0.43
10	Vinclozolin	397.7	15.5	3.9	66.1	16.6	0.90	10	Pirimiphos-methyl	24.0	1.8	7.5	3.1	12.8	0.46
11	2,4-DDE	722.3	32.3	4.5	151.6	21.0	1.25	11	Picoxys-trobin	28.2	2.0	6.9	2.6	9.2	0.34
12	Bromophos-ethyl	177.2	9.5	5.3	43.8	24.7	1.19	12	Diazinon	21.7	1.0	4.6	2.0	9.0	0.32
13	Picoxys-trobin	399.1	16.5	4.1	73.2	18.3	1.00	13	Bensulide	91.4	4.4	4.8	9.1	9.9	0.43
14	Quinoxyphen	190.7	7.0	3.7	26.4	13.9	0.68	14	Indoxacarb	31.1	4.3	13.9	5.7	18.4	0.68
15	Chlorfenapyr	1616.0	86.1	5.3	216.9	13.4	0.90	15	Trifloxystrobin	28.4	3.9	13.7	4.9	17.1	0.63
16	Benalaxyl	190.4	7.9	4.1	26.3	13.8	0.67	16	Tebufenpyrad	28.1	2.7	9.7	4.9	17.4	0.63
17	Bifenthrin	220.4	16.4	7.4	96.4	43.7	2.18	17	Chlorpyrifos	260.2	17.9	6.9	50.8	19.5	1.00

No.	Compound							No.	Compound						
18	Diflufenican	194.9	6.2	3.2	23.9	12.2	0.60	18	Butralin	24.9	2.4	9.5	5.5	22.0	0.79
19	Tebufenpyrad	194.2	10.3	5.3	18.1	9.3	0.45	19	Fenazaquin	24.9	3.1	12.5	2.9	11.7	0.42
20	Bromopropylate	386.0	17.7	4.6	70.2	18.2	0.98	20	Phenothrin	156.7	21.6	13.8	33.0	21.1	1.00
Oolong Tea															
1	Trifluralin	176.6	8.4	4.7	30.0	17.0	0.82	1	Imidacloprid	108.1	13.3	12.3	26.5	24.5	1.10
2	Tefluthrin	81.9	3.5	4.3	15.3	18.7	0.80	2	Propoxur	108.5	9.4	8.7	22.2	20.5	0.92
3	Pyrimethanil	96.0	3.7	3.9	12.1	12.6	0.55	3	Clomazone	39.8	3.1	7.8	8.5	21.4	0.82
4	Pirimicarb	92.0	4.4	4.7	15.8	17.1	0.75	4	Ethoprophos	37.7	2.0	5.4	8.5	22.7	0.87
5	Propyzamide	109.3	3.5	3.2	12.7	11.7	0.52	5	Triadimefon	44.2	2.9	6.7	7.7	17.4	0.68
6	Dimethenamid	38.0	1.6	4.1	7.0	18.5	0.71	6	Acetochlor	80.1	4.4	5.5	14.3	17.9	0.76
7	Fenchlorphos	180.9	5.1	2.8	13.2	7.3	0.35	7	Flutolanil	50.1	3.3	6.7	8.0	16.0	0.64
8	Tolclofos-methyl	94.2	2.5	2.6	12.1	12.8	0.56	8	Benalaxyl	45.0	3.3	7.3	7.9	17.6	0.69
9	Pirimiphos-methyl	90.5	2.2	2.4	13.4	14.9	0.65	9	Kresoxim-methyl	469.6	31.4	6.7	50.3	10.7	0.60
10	Vinclozolin	210.2	3.3	1.6	8.7	4.2	0.21	10	Pirimiphos-methyl	40.5	3.1	7.7	8.3	20.7	0.79
11	2,4'-DDE	361.0	8.5	2.4	18.7	5.2	0.28	11	Picoxystrobin	45.9	3.2	6.9	7.5	16.3	0.64
12	Bromophos-ethyl	89.6	3.4	3.8	13.9	15.5	0.67	12	Diazinon	37.6	2.2	5.9	7.8	20.7	0.79
13	Picoxystrobin	205.3	5.4	2.6	25.0	12.2	0.60	13	Bensulide	150.5	7.9	5.2	12.8	8.5	0.40
14	Quinoxyphen	100.5	3.4	3.4	12.9	12.8	0.57	14	Indoxacarb	50.4	3.3	6.6	7.1	14.2	0.56
15	Chlorfenapyr	841.9	22.5	2.7	71.9	8.5	0.52	15	Trifloxystrobin	47.3	3.3	7.1	6.8	14.3	0.57
16	Benalaxyl	100.5	3.4	3.4	13.8	13.8	0.61	16	Tebufenpyrad	45.2	3.6	7.9	6.4	14.3	0.56

(Continued)

TABLE 7.1.9 Method Efficiency for Determination of 40 Pesticides in Tea by Both GC-MS-SIM (GC-MS/MS) and LC-MS/MS (Cont.)

No.	Analyte	GC-MS-SIM (GC-MS/MS) Average (C, µg/kg)	S_r (µg/kg)	RSD_r (%)	S_R (µg/kg)	RSD_R (%)	HorRat	No.	Analyte	LC-MS/MS Average (C, µg/kg)	S_r (µg/kg)	RSD_r (%)	S_R (µg/kg)	RSD_R (%)	HorRat
									Green Tea						
17	Bifenthrin	100.0	3.9	3.9	11.4	11.4	0.50	17	Chlorpyrifos	431.4	26.1	6.0	71.7	16.6	0.92
18	Diflufenican	104.4	2.8	2.6	12.4	11.9	0.53	18	Butralin	41.0	1.8	4.3	8.5	20.7	0.80
19	Tebufenpyrad	104.6	4.8	4.6	5.3	5.1	0.23	19	Fenazaquin	44.9	2.2	4.8	4.6	10.3	0.41
20	Bromopropylate	214.4	7.9	3.7	11.7	5.4	0.27	20	Phenothrin	286.7	68.0	23.7	160.1	55.9	2.89

TABLE 7.1.10 Distribution Range of the Parameters of Method Efficacy RSD$_r$ RSD$_R$ and HorRat values

Parameters of Method Efficacy	Range	GC-MS-SIM (GC-MS/MS)				LC-MS/MS					
		Green Tea	Percentage (%)	Oolong Tea	Percentage (%)	Green Tea	Percentage (%)	Oolong Tea	Percentage (%)	Total	Percentage (%)
RSD$_r$ (%)	<8	20	100.0	20	100.0	9	45.0	17	85.0	66	82.5
	8~15	0	0.0	0	0.0	11	55.0	2	10.0	13	16.3
	>15	0	0.0	0	0.0	0	0.0	1	5.0	1	1.3
RSD$_R$ (%)	<16	7	35.0	16	80.0	8	40.0	6	30.0	37	46.3
	16~25	10	50.0	4	20.0	11	55.0	13	65.0	38	47.5
	>25	3	15.0	0	0.0	1	5.0	1	5.0	5	6.3
HorRat	<0.50	1	5.0	5	25.0	6	30.0	2	10.0	14	17.5
	0.50~1.00	11	55.0	15	75.0	13	65.0	16	80.0	55	68.8
	1.01~2.00	7	35.0	0	0.0	1	5.0	1	5.0	9	11.3
	>2.00	1	5.0	0	0.0	0	0.0	1	5.0	2	2.5

In terms of GC-MS-SIM (GC-MS/MS), Table 7.1.10 shows (1) the within-lab repeatability: RSD$_r$ < 8% accounts for 100%, which explains that this method is of very good repeatability in every lab. (2) Between-lab reproducibility: RSD$_R$ < 16% accounts for 57.5%; RSD$_R$ < 25% accounts for 92.5%, which demonstrates a good between-lab reproducibility. (3) HorRat value: HorRat values that are less than 1.0 account for 80.0%; HorRat values that are less than 2.0 account for 97.5%. This demonstrates that the method has had a good reproducibility under corresponding analytical concentrations. There is only one compound with HorRat values greater than 2, which is Bifenthrin with RSD$_R$ 43.7%. An examination found that the green tea blank samples used for preparing a matrix-matching calibration curve contained a small amount of such pesticide residues that led to the problems with the calibration curve and relatively big differences among the between-lab quantification results as well as the HorRat values of Bifenthrin exceeding 2.

In terms of LC-MS/MS, (1) within-lab repeatability: RSDr < 8% accounts for 65.0%, RSD$_r$ < 15% for 97.5%, which shows that this method is of good repeatability in every lab. (2) Between-lab reproducibility: RSD$_R$ < 16% accounts 35.0%; RSD$_R$ < 25% makes up 95.0%, demonstrating good between-lab reproducibility. (3) HorRat values: HorRat values that are less than 1 account for 92.5%; HorRat values that are less than 2 account for 97.5%, which explains that this method has good reproducibility under the corresponding analytical concentrations. There was only one compound, Phenothrin, with HorRat values greater than 2.89, RSD$_r$ 23.7% and RSD$_R$ 55.9%. These errors were traced to the change of concentration in pesticide-mixed standard solutions in the ampoule because blue flames occurred when sealing the ampoule bottle containing the pesticide standard solutions using high-temperature flames. To investigate the source of the blue flames, the concentrations of pesticide standard solutions in 10 ampoules were parallel-tested, and it was discovered that all the phenotherin concentrations were affected by high flames, with relatively big changes incurring. Based on this result, this pesticide will be eliminated when conducting an official collaborative study.

To sum up the aforementioned analysis, this simulated collaborative study result indicates that the efficiency of the proposed method is acceptable.

7.1.5.4 Qualification and Quantification

The protocol points out clearly that the deviations of ion abundance of target pesticide-qualifying ions shall be in conformity with the EU standard. For this simulated collaborative study, a total of 3600 target ion abundance data were obtained using eight green tea and oolong tea samples (excluding blank samples) with three analytical methods of GC-MS-SIM, GC-MS/MS, and LC-MS/MS, and the conformity ratios of ion abundance with the EU standard were checked one by one. See Table 7.1.11 for results.

7.1.5.4.1 The qualification Analysis of Target Pesticide

Concerning GC-MS-SIM, Table 7.1.11 shows that the total number of target pesticide ions detected by 13 labs is 2080, of which 1955 conform to the EU standard in terms of ion abundance ratios, making up 94.0%. Regarding LC-MS/MS, the

TABLE 7.1.11 Conformity Ratios of Ion Abundance of Target Pesticide With EU Standard

| | GC-MS-SIM | | | | GC-MS/MS | | | | LC-MS/MS | | | |
| | Green Tea | | Oolong Tea | | Green Tea | Oolong Tea | | | Green Tea | | Oolong Tea | |
Lab	No. 2 Number[a] (percentage)[b]	No. 3 Number[a] (percentage)[b]	No. 5 Number[a] (percentage)[b]	No. 6 Number[a] (percentage)[b]	No. 2 Number[a] (percentage)[b]	No. 3 Number[a] (percentage)[b]	No. 5 Number[a] (percentage)[b]	No. 6 Number[a] (percentage)[b]	No. 2 Number[a] (percentage)[b]	No. 3 Number[a] (percentage)[b]	No. 5 Number[a] (percentage)[b]	No. 6 Number[a] (percentage)[b]
1	38 (95.0%)	39 (97.5%)	39 (97.5%)	38 (95.0%)	–	–	–	–	20 (100.0%)	20 (100.0%)	20 (100.0%)	20 (100.0%)
2	40 (100%)	39 (97.5%)	39 (97.5%)	40 (100.0%)	–	–	–	–	–	–	–	–
3	40 (100%)	40 (100.0%)	40 (100.0%)	40 (100.0%)	–	–	–	–	–	–	–	–
4	40 (100%)	39 (97.5%)	38 (95.0%)	38 (95.0%)	–	–	–	–	14 (70.0%)	17 (85.0%)	20 (100.0%)	20 (100.0%)
5	–	–	–	–	20 (100.0%)	20 (100.0%)	20 (100.0%)	20 (100.0%)	20 (100.0%)	20 (100.0%)	20 (100.0%)	20 (100.0%)
6	38 (95.0%)	38 (95.0%)	39 (97.5%)	38 (95.0%)	20 (100.0%)	20 (100.0%)	20 (100.0%)	20 (100.0%)	20 (100.0%)	20 (100.0%)	20 (100.0%)	20 (100.0%)
7	35 (87.5%)	35 (87.5%)	35 (87.5%)	35 (87.5%)	–	–	–	–	19 (95.0%)	19 (95.0%)	20 (100.0%)	20 (100.0%)
8	40 (100.0%)	40 (100.0%)	40 (100.0%)	40 (100.0%)	–	–	–	–	20 (100.0%)	20 (100.0%)	20 (100.0%)	20 (100.0%)
9	40 (100.0%)	40 (100.0%)	40 (100.0%)	40 (100.0%)	20 (100.0%)	20 (100.0%)	20 (100.0%)	20 (100.0%)	18 (90.0%)	18 (90.0%)	19 (95.0%)	19 (95.0%)
10	28 (70.0%)	29 (72.5%)	25 (62.5%)	30 (75.0%)	20 (100.0%)	20 (100.0%)	20 (100.0%)	20 (100.0%)	20 (100.0%)	20 (100.0%)	20 (100.0%)	20 (100.0%)
11	–	–	–	–	20 (100.0%)	20 (100.0%)	20 (100.0%)	20 (100.0%)	19 (95.0%)	19 (95.0%)	18 (90.0%)	19 (95.0%)
12	37 (92.5%)	37 (92.5%)	37 (92.5%)	37 (92.5%)	–	–	–	–	20 (100.0%)	20 (100.0%)	20 (100.0%)	20 (100.0%)
13	37 (92.5%)	38 (95.0%)	35 (87.5%)	35 (87.5%)	–	–	–	–	20 (100.0%)	20 (100.0%)	19 (95.0%)	19 (95.0%)
14	–	–	–	–	–	–	–	–	19 (95.0%)	19 (95.0%)	20 (100.0%)	20 (100.0%)
15	40 (100.0%)	40 (100.0%)	40 (100.0%)	39 (97.5%)	–	–	–	–	19 (95.0%)	19 (95.0%)	18 (90.0%)	18 (90.0%)
16	37 (92.5%)	37 (92.5%)	39 (97.5%)	38 (95.0%)	20 (100.0%)	20 (100.0%)	20 (100.0%)	20 (100.0%)	20 (100.0%)	20 (100.0%)	20 (100.0%)	20 (100.0%)
Number of conformity ion	1955				400				1088			

Total number of Ion	2080	400	1120
Total conformity ratios%	94.0	100.0	97.1
number of labs	13	5	14

aIon number refers to the number of 20 pesticide qualifying ions monitored conforming to EU standard.
bPercentage refers to the ratio of ion number conforming to EU standard over the total number of ions monitored.

total number of target pesticide ions detected by 14 labs is 1120, of which 1088 conforms to the EU standard, making up 97.1%. For GC-MS/MS, the total number of target pesticide ions detected by 5 labs is 400, all of which conform to the EU standard. The earlier stated analysis states that the 16 labs have detected 3600 target pesticide ions, of which 3443 conform to the EU standard in abundant ratios, accounting for 95.6%.

7.1.5.4.2 The Quantification Analysis of Target Pesticide

This study adopted a matrix-matching internal standard calibration curve for quantification, with 1220 target pesticide calibration curves established by 16 labs. The conformance ratios of their correlation coefficients $R^2 \geq 0.995$ are shown in Table 7.1.12.

Table 7.1.12 indicates that for GC-MS-SIM, 13 labs established 480 matrix-matching standard curves for 20 pesticides in green tea and oolong tea, respectively, of which there were 428 pesticides with $R^2 \geq 0.995$, making up 89.2%, 49 with R^2 0.990–0.995, making up 10.2%, 3 with R^2 under 0.990, making up 0.6%.

In terms of LC-MS/MS, 14 labs established 540 matrix-matching internal standard calibration curves for 20 pesticides in green tea and oolong tea, respectively, of which there were 513 with $R^2 \geq 0.995$, making up 95.0%, 24 with R^2 0.990–0.995, making up 4.4%, 3 with R^2 under 0.990, making up 0.6%.

TABLE 7.1.12 Conformity Rate of Correlation Coefficient of R^2 with $R^2 \geq 0.995$ Per Calibration Curve Data

Lab	GC-MS-SIM Green Tea Number (percentage)[a]	GC-MS-SIM Oolong Tea Number (percentage)[a]	GC-MS/MS Green Tea Number (percentage)[a]	GC-MS/MS Oolong Tea Number (percentage)[a]	LC-MS/MS Green Tea Number (percentage)[a]	LC-MS/MS Oolong Tea Number (percentage)[a]
1	20 (100.0%)	20 (100.0%)	–	–	17 (85.0%)	18 (90.0%)
2	20 (100.0%)	20 (100.0%)	–	–	–	–
3	20 (100.0%)	–	–	–	–	–
4	20 (100.0%)	20 (100.0%)	–	–	19 (95.0%)	20 (100.0%)
5	–	–	20 (100.0%)	20 (100.0%)	20 (100.0%)	19 (95.0%)
6	20 (100.0%)	20 (100.0%)	20 (100.0%)	20 (100.0%)	19 (95.0%)	20 (100.0%)
7	8 (40.0%)	2 (10.0%)	–	–	20 (100.0%)	20 (100.0%)
8	20 (100.0%)	20 (100.0%)	–	–	20 (100.0%)	19 (95.0%)
9	19 (95.0%)	–	–	–	20 (100.0%)	–
10	14 (70.0%)	17 (85.0%)	19 (95.0%)	20 (100.0%)	20 (100.0%)	19 (95.0%)
11	–	–	19 (95.0%)	15 (75.0%)	20 (100.0%)	20 (100.0%)
12	20 (100.0%)	20 (100.0%)	–	–	20 (100.0%)	20 (100.0%)
13	20 (100.0%)	20 (100.0%)	–	–	19 (95.0%)	17 (85.0%)
14	–	–	–	–	17 (85.0%)	20 (100.0%)
15	8 (40.0%)	20 (100.0%)	–	–	16 (80.0%)	16 (80.0%)
16	20 (100.0%)	20 (100.0%)	20 (100.0%)	18 (90.0%)	19 (95.0%)	19 (95.0%)
Conformity quantity of correlation coefficients	428		191		513	
Total number of correlation coefficients	480		200		540	
Total conformity rate %	89.2		95.5		95.0	
Number of labs	13		5		14	

aPercentage refers to the ratio of the number of linear correlation coefficient $R^2 \geq 0.995$ over the total number of correlation coefficients

In terms of GC-MS/MS, five labs have established 200 matrix-matching internal standard calibration curves for 20 pesticides in green tea and oolong tea, of which there were 191 with $R^2 \geq 0.995$, making up 95.5%; 8 with R^2 0.990–0.995, making up 4.0%, and 1 with R^2 under 0.990, making up 0.5%.

Summing up the earlier information, there were 16 labs that established 1132 standard curves, of which there were 1128 with R^2 values satisfying the requirements of ≥ 0.995, making up 92.8%.

7.1.5.5 Error Analysis and Tracing

7.1.5.5.1 Error Analysis

For this simulated study, 16 labs provided us with 2560 analytical data (excluding blank samples), eliminating 360 data (excluding 240 data determined by GC-MS/MS) from three labs that deviated from the method, and inspecting the remaining 1960 analytical data with Grubbs and Dixon for outliers, with 138 outliers detected, making up 7.0%. Distribution and percentage of outliers in different pesticide varieties and different samples are shown in Table 7.1.13; the distribution and percentage of outliers in different labs and different samples are shown in Table 7.1.14.

Table 7.1.14 shows that 138 outliers come from 10 labs, of which Lab 11 and Lab 3 outliers make up the relatively large proportion of the total experimental data, with some cause for error.

There are 71 outlier values among the 160 data provided by Lab 11, making up 44.4% as well as 51.4% of the total number of outliers. The raw data from this lab found for GC-MS/MS, in establishing a green tea matrix-matching internal standard curve, the internal standard at the third point was added repeatedly and this point was abandoned with only four points for matrix-matching standard curve. In establishing the calibration curve, the concentrations were entered incorrectly for primiphos-methy by GC-MS/MS and diazinon, chlorpyrifos, and imidacloprid by LC-MS/MS. Therefore, large quantities of outlier values occurred with the analytical data of this lab.

There are 10 outlier values among the 40 data provided by Lab 3, making up 25% as well as 7.2% of the total number of outlier data. An inquiry with the operator found that this lab used Envi-Carb/PSA for cleanup of tea samples and Envi-Carb cartridge for the cleanup of matrix-matching blank samples because of the lack of sufficient SPE cartridges, which led to differences of matrix effects. Besides, this lab adopted isocarbophos instead of heptachlor-eporxide as an internal standard during actual determinations. Large errors occurred with the final results as isocarbophos had low sensitivity and interferences during determinations.

7.1.5.5.2 Error Tracing

We conducted error tracing to 138 outliers from 10 labs, with five error sources.

Most errors were caused by matrix-matching internal standard calibration curves: (1) the matrix-matching internal standard calibration curves established using green tea for quantification target pesticides in oolong tea, such as Lab 3 and Lab 9; (2) pesticide concentrations were entered incorrectly when establishing matrix-matching internal standard curves, for example, Lab 11; (3) a lab had more than 30% pesticides whose R^2 values failed to meet the requirements of ≥ 0.995, for example, Lab 7; and (4) an external standard calibration curve was adopted quantification, such as Lab 9.

The improper use of internal standards caused many errors as well. (1) To amortize the loss of pesticides during the process of cleanup extraction, internal standards were added before the sample extraction, such as Lab 7; (2) when adding internal standard solutions, the improper operation caused the inconsistency of adding quantities of internal standards, such as Lab 8; (3) choosing other compounds to replace the internal standards specified in the method. The compound chosen had a low sensitivity during determinations and was interfered with, leading to errors in the final test results, for example, Lab 3.

Errors were also caused by cleanup conditions. For blank samples and aged samples, different SPE cartridges were used, leading to errors, for example, Lab 3.

In addition, errors were caused by testing in violation of operational procedures: (1) the mixed standard reserve solutions in the ampoule bottle provided by the study director were required to be thoroughly transferred and diluted before use, but certain laboratories transferred only a portion, which breached the operational stipulations and led to systematic bias, for example, Lab 9; (2) reparations of lab samples were made by two operators one after another, which caused confusion, for example, Lab 7.

There were also errors caused by other factors. Sample extracting solutions for LC-MS/MS determinations were not blown to dryness at a nitrogen blow; there were separating layers when using acetonitrile:water (3:2) to dilute, which caused the deviations of analytical results, for example, Lab 14.

TABLE 7.1.13 Distribution of outlier values in different pesticides

No.	Pesticides	GC-MS-SIM (GC-MS/MS)				Pesticides	LC-MS/MS			
		Green Tea		Oolong Tea			Green Tea		Oolong Tea	
		No. 2	No. 3	No. 5	No. 6		No. 2	No. 3	No. 5	No. 6
1	Trifluralin		11-D, 11-G			Imidacloprid	11-G		11-D,11-G	
2	Tefluthrin		11-D, 11-G			Propoxur		8-D, 8-G	11-D,11-G	
3	Pyrimethanil	11-D	11-D, 11-G		11-D, 11-G	Clomazone	15-D		11-D,11-G	
4	Pirimicarb	3-D, 3-G, 11-D, 11-G, 15-D	11-D			Ethoprophos		8-D, 8-G	11-D,11-G	
5	Propyzamide	3-D, 11-D			11-G	Triadimefon		8-D, 8-G	11-D,11-G	
6	Dimethenamid					Acetochlor			11-D,11-G	
7	Fenchlorphos	11-D	11-D, 11-G	4-D, 11-D, 16-D	11-D	Flutolanil			11-D,11-G	
8	Tolclofos-methyl				11-D	Benalaxyl		8-D, 8-G	11-D,11-G	
9	Pirimiphos-methyl	11-D, 11-G	11-D,11-G	11-D, 11-G	11-D, 11-G	Kresoxim-methyl	8-D, 14-D, 15-D	8-D, 8-G, 15-D, 15-G	11-D,11-G	14-G
10	Vinclozolin	11-D, 11-G	11-D, 11-G	11-D, 11-G, 16-D, 2-D, 4-D	11-D, 11-G	Pirimiphos-methyl	8-G, 15-G			
11	2,4'-DDE		11-D	4-D, 8-D, 11-D, 16-D	11-D	Picoxystrobin	8-D, 14-D, 15-D, 15-G	8-D, 8-G	11-D,11-G	
12	Bromophos-ethyl					Diazinon	8-D, 11-D, 11-G, 14-D, 15-D	8-D, 11-D, 11-G, 15-D	11-D,11-G	11-D,11-G
13	Picoxystrobin	11-D	11-G	11-G	11-D, 11-G	Bensulide	8-D, 8-G	8-D, 14-D	4-D,6-D,11-D	
14	Quinoxyphen	3-D, 3-G, 11-D, 11-G	3-D, 11-D		11-D, 11-G	Indoxacarb			11-D,11-G	
15	Chlorfenapyr	3-G, 11-D, 11-G	3-D, 11-D	11-D, 11-G	4-D, 4-G, 11-D, 11-G	Trifloxystrobin				14-G
16	Benalaxyl	3-D, 11-D			11-D	Tebufenpyrad				
17	Bifenthrin	3-D, 11-D	11-D, 11-G		11-D, 11-G	Chlorpyrifos	11-D, 11-G	11-D, 14-D, 14-G	11-D,11-G	11-D,11-G
18	Diflufenican	11-D, 12-D, 3-D	11-D, 12-D	11-D, 11-G, 12-D	11-D, 12-D	Butralin	12-D, 12-G	12-D, 12-G	11-G	

19 Tebufenpyrad	3-D, 3-G, 15-D, 16-D		11-D, 11-G	4-D, 4-G, 11-D, 11-G, 16-D
20 Bromopropylate	3-G			4-G, 11-D, 11-G, 16-D, 16-G
Number of Labs	13	13	12	12
Total number of data	260	260	240	240
Number of outliers	23	16	16	21
Outlier values over the total number data/ ratio, percentage	8.8	6.2	6.7	8.8

Fenazaquin	11-D, 12-D, 12-G, 8-D, 15-DDDD	8-D, 12-D, 12-G, 15-D	11-D, 11-G	4-D
Phenothrin	6-D, 11-D			
Number of Labs	12	12	12	12
Total number of data	240	240	240	240
Number of outliers	23	17	17	5
outlier values over the total number data/ ratio, percentage	9.6	7.1	7.1	2.1

TABLE 7.1.14 Distribution of Outlier Values in Different Labs

	GC-MS-SIM (GC-MS/MS)				LC-MS/MS				Total Number of Experimental Data	Number of Outlier Values	Outlier Values Over Individual Lab Data's Ratio (%)	Outlier Values Over the Total Number of Outlier Values' Ratio
	Green Tea		Oolong Tea		Green Tea		Oolong Tea					
Lab	No. 2	No. 3	No. 5	No. 6	No. 2	No. 3	No. 5	No. 6				
1	0	0	0	0	0	0	0	0	160	0	0.0	0
2	0	0	1	0	–	–	–	–	80	1	1.3	0.7
3	8	2	–	–	–	–	–	–	40	10	25.0	7.2
4	0	0	3	3	0	0	1	1	160	8	5.0	5.8
5	0	0	0	0	0	0	0	0	160	0	0.0	0
6	0	0	0	0	1	0	1	0	160	2	1.3	1.4
7	–	–	–	–	–	–	–	–	–	–	–	–
8	0	0	0	0	6	8	0	0	160	14	8.8	10.1
9	–	–	–	–	–	–	–	–	–	–	–	–
10	0	0	0	0	0	0	0	0	160	0	0	0
11	11	13	8	15	5	2	15	2	160	71	44.4	51.4
12	1	1	1	1	0	2	0	0	160	8	5.0	5.8
13	0	0	0	0	0	0	0	0	160	0	0	0
14	–	–	–	–	3	2	0	2	80	7	8.8	5.1
15	2	0	0	0	6	3	0	0	160	11	6.9	8.0
16	1	0	3	2	0	0	0	0	160	6	3.8	4.3
Total	23	16	16	21	21	17	17	5	1960	138	7.0	100

7.1.6   CONCLUSIONS

This simulated collaborative study not only achieved successful experiences, but also found ways to avoid outliers. The satisfaction of the analysis results exceeded expectations. This fully shows that this method is applicable to GC-MS, GC-MS/MS and LC-MS/MS analysis of pesticide residues in green tea and oolong tea, meets AOAC or EU standards on technical indicators, and has achieved good method efficiency. The experiment results also confirm that the multi-residue analysis method is accurate and reliable, and has good efficiency.

REFERENCES

[1]   http://ec.europa.eu/sanco_pesticides/public/index.cfm.

[2]   http://www.ffcr.or.jp/zaidan/FFCRHOME.nsf/pages/MRLs-n.

[3]   http://www.kennzeichnungsrecht.de/english/mrlsearch.htm.

[4]   http://eng.kfda.go.kr/index.php.

[5]   http://eng.kfda.go.kr/index.php.

[6]   X.Y. Wu, Studies on maximum residue limits for pesticides in tea and relative risk assessment, thesis for the doctoral degree of Anhui Agricultural University, China, 2007.

[7]   AOAC Validation Guidelines Appendix D: Guidelines for Collaborative Study Procedures to Validate Characteristics of a Method of Analysis, 2005.

[8]   ASTM E1301-95, Standard Guide for Proficiency Testing by Interlaboratory Comparisons, 2003.

[9]   Food and Drug Administration of the United States, Guidance for Industry-Bioanalytical Method Validation, 2001.

[10]   CX/MAS 02/12, Codex Alimentarius Commission, 2002. Codex Committee on Methods of Analysis and Sampling (FAO/WHO), Validation of methods through the use of results from proficiency testing schemes. Agenda Item 8c on the 24th Session, Budapest, Hungary, 18-22 November 2002.

[11] ISO 5725-1994, Accuracy (Trueness and Precision) of Measurement Methods and Results, International Standardization Organization, Geneva, 1994.

[12] IUPAC Nomenclature of interlaboratory analytical studies.(IUPAC Recommendations 1994).

[13] Hsu, M.C., Chung, H.C., Lin, Y.S., 1996. J. Chromatogr. A 727 (2), 239–244.

[14] Nilsson, T., Ferrari, R., Facchetti, S., 1997. Anal. Chim. Acta. 356 (2–3), 113–123.

[15] Hoogerbrugge, R., Gort, S.M., van der Velde, E.G., Piet van, Z., 1999. Anal. Chim. Acta. 388 (1–2), 119–135.

[16] Ferrari, R., Nilsson, T., Arena, R., Arlati, P., Bartolucci, G., Basla, R., Cioni, F., Del, C.G., Dellavedova, P., Fattore, E., Fungi, M., Grote, C., Guidotti, M., Morgillo, S., Muller, L., Volante, M., 1998. J. Chromatogr. A 795 (2), 371–376.

[17] Lehotay, S.J., 2002. J. AOAC Int. 85 (5), 1148–1166.

[18] Kakimoto, Y., Naetoko, Y., Hara, H., Miyatake, M., Sato, A., Tatsuguchi, H., Takahata, R., Yamamoto, R., Joh, T., 2004. Shokuhin Eiseigaku Zasshi 45 (3), 165–174.

[19] Ratola, N., Barros, P., Simoes, T., Cerdeira, A., Venancio, A., Alves, A., 2006. Talanta 70 (4), 720–731.

[20] Simon, R., Palme, S., Anklam, E., 2007. Food Chem. 104 (2), 876–887.

[21] Lehotay, S.J., 2007. J. AOAC Int. 90 (2), 485–520.

[22] Wong, J., Hao, C., Zhang, K., Yang, P., Banerjee, K., Hayward, D., Iftakhar, I., Schreiber, A., Tech, K., Sack, C., Smoker, M., Chen, X., Utture, S.C., Oulkar, D.P., 2010. J. Agric. Food Chem. 58 (10), 5897–5903.

[23] Document No. SANCO/10684/2009 Method Validation and Quality Central Procedures for Pesticide Residues Analysis in Food and Feed.

[24] AOAC Guidelines for single laboratory validation of chemical methods for dietary supplements and botanicals.).

Chapter 7.2

Collaborative Study

Chapter Outline

7.2.1 INTRODUCTION

Tea is considered to be one of the three most consumed beverages in the world and is enjoyed by over 2 billion people from more than 160 countries and regions drink tea, which is referred to as one of the three most popular drinks in the world [1]. At present, 60-plus countries around the globe grow tea, with the area taken up by tea plantations around 250 ha and annual output of 3.5 million tons. India, China, Sri Lanka, Kenya, and Indonesia are the world's five largest tea producers, and their tea output makes up about 80% of total production [2]. Tea mostly grows in warm temperate zones and subtropical regions and is subject to threats from diseases and pest infestations. Pesticides are widely used, hence, pesticide residue contamination is incurred. At present, 16 countries and international organizations (CAC, EU, Germany, Holland, Switzerland, Hungary, Israel, Italy, China, Japan, Korea, USA, Australia, India, Kenya, and South Africa, as of 2006), stipulate maximum residue limit (MRL) for more than 800 pesticide residues [3]. With the quickening of the current global economy

Analysis of Pesticide in Tea. http://dx.doi.org/10.1016/B978-0-12-812727-8.00015-8

integration and strengthening of food safety strategies in different countries, as well as the building of higher thresholds for the limits of detection for pesticide residues in international trade, all these call for a deeper study of high-throughput pesticide residual analytical techniques for tea.

At the beginning of this century, the author's team focused on the study of high-throughput analytical techniques for 400–500 pesticide residues in different kinds of agricultural products from the world's commonly used 1000-plus pesticide and chemical contaminants and established analytical methods of GC–MS and LC–MS/MS determination for 400–500 pesticide residues in 20 items. These included fruits and vegetables [4–5], grains [6–7], teas [8–9], Chinese medicinal herbs [10–11], edible fungi (mushrooms) [12–13], animal tissues [14–15], aquatic products [16–17], raw milk and milk powders [18–19], honey, fruit juices, and fruit wines [20–22], and drinkable water [23] and their study resulted in a series of papers published [24–29].

Coming from such a basis, a profound study has been conducted on the methodology of analytical techniques for 653 pesticide residues in tea in eight aspects, including a 3-month stability study on 460 pesticides in six groups of mixed standard solutions, a 3-month stability study on 345 pesticides in tea, and a 3-month detection study on deviation ratios of 275 pesticides in tea Youden pair samples. Others include a 3-month study on the ruggedness of the method, a 3-month study on degradation kinetics of 227 pesticides in aged tea samples as well as the deviation ratios of predicted values and actual values from applying kinetic equations for pesticides, a 3-month verification study on the applicability of the EU standard (EU Document No. SANCO/10684/2009) [30] in the AOAC collaborative study, and a 3-month study on degradation patterns of pesticides in tea in a field experiment and the stability of tea-incurred samples. Sixteen laboratories were organized to participate in the simulated AOAC collaborative study. In addition, a comparative study with three different sample preparation methods was conducted on the applicability of tea hydration for the efficiency of pesticide multiresidue method, and so on. This research went through eight stages lasting 4 years, with more than 500,000 test data obtained. The research results in different stages were introduced and discussed at four AOAC annual meetings in 2009–12, and a series of related papers were published subsequently [31–34].

This AOAC collaborative study was taken up by 30 laboratories from 11 countries and regions. A total of 560 samples were analyzed, including three categories of samples, such as fortification samples from green tea and oolong tea, aged samples from oolong tea, and incurred samples from green tea, using three methods of GC–MS, GC–MS/MS, and LC–MS/MS. Review of 6868 test data and related information from 30 laboratories has found that one laboratory deviated from the collaborative study operational procedures, entering directly into the official collaborative study without undertaking the prestudy, resulting in a great deal of outliers from its sample inspection; this laboratory has been eliminated. A total of 6638 data from the remaining 29 laboratories were effective results and inspected with Grubbs and Dixon to eliminate 187 outliers (2.8% of the total effective data). There were 1977 test data from GC–MS determination, with 65 outliers detected, making up 3.3%; 1704 test data from GC–MS/MS determination, with 65 outliers detected, accounting for 3.8%; 2957 test data from LC–MS/MS determination, with 57 outliers detected, making up 1.9%. The efficiency of three methods was derived from the statistical analysis.

GC–MS: For the fortified green tea samples: Avg. C.: 37.5–759.6 μg/kg, Avg. Rec% of blind duplication samples: 87.7%–96%, at 2 MRL level for each pesticide; RSD_r: 2.1%–4.9%; RSD_R: 6.5%–9.9%; and HorRat: 0.3–0.5. For the fortified oolong tea samples: Avg. C.: 17.3–335.7 μg/kg; Avg. Rec% of blind duplication samples: 81%–91.1%; RSD_r: 2.8%–7.8%; RSD_R: 12.5%–25%; and HorRat: 0.5–1.3. For the aged oolong tea samples: Avg. C.: 77.6–1642.6 μg/kg; RSD_r: 2%–5.8% and RSD_R: 8.9%–16.9%, HorRat: 0.4–0.9. For the incurred green tea samples: Avg. C.: 77.8 μg/kg, 613.3 μg/kg; RSD_r: 3.3%, 4%; RSD_R: 12.2%, 23%, HorRat: 0.7, 1.

GC–MS/MS: For the fortified green tea samples: Avg. C.: 37.5–749.1 μg/kg; Avg. Rec% of blind duplication samples: 87%–97.1%; RSD_r: 3.1%–6%; RSD_R: 6.6%–14.8%; and HorRat: 0.3–0.7. For the fortified Oolong tea samples: Avg. C.: 17.5–335.8 μg/kg; Avg. Rec% of blind duplication samples: 77.1%–90.8%, at MRL level for each pesticide; RSD_r: 1.4%–5.4%; RSD_R: 7%–32.7%; and HorRat: 0.4–1.3. For the aged oolong tea samples: Avg. C.: 72.4–1511.9 μg/kg; RSD_r: 4.6%–9.6%, RSD_R: 21.7%–34.7%, HorRat: 1.1–1.8. For the incurred green tea samples: Avg. C.: 78.6 μg/kg, 575.4 μg/kg; RSD_r: 5.6%, 5.6%, RSD_R: 14.1%, 22.2%; HorRat: 0.8, 0.9.

LC–MS/MS: For the fortified green tea samples: Avg. C.: 18.2–191.8 μg/kg; Avg. Rec% of blind duplication samples: 91.7%–97.7%; RSD_r: 4.9%–9.4%; RSD_R: 8.4%–17.1%; and HorRat: 0.3–0.7. For the fortified oolong tea samples: Avg. C.: 8.5–84.3 μg/kg; Avg. Rec% of blind duplication samples: 82.5%–93.7%; RSD_r: 3.6%–10.2%; RSD_R: 13.6%–29.7%; and HorRat: 0.4–1.3. For the aged oolong tea samples: Avg. C.: 34.2–441.6 μg/kg; RSD_r: 5%–9.1%; RSD_R: 16.8%–34.6%, HorRat: 0.7–1.6. For the incurred green tea samples: Avg. C.: 14.1–90.7 μg/kg; RSD_r: 8.4%–10.6%, RSD_R: 21.3%–23.6%, HorRat: 0.8–0.9.

To sum up the aforementioned analysis, Avg. C., Avg. Rec., RSD_r, RSD_R, and HorRat values all met AOAC technical requirements except for specific data. AOAC recommends this method as an official first action.

7.2.2 COLLABORATIVE STUDY PROTOCOL

7.2.2.1 Need/Purpose

With the integration of the global economy stepping things up, the strategic positions of the world food safety and pesticide residue limit threshold for international trade are increasing, which calls for high throughput residue analytical techniques. The purpose of this collaborative study is to evaluate whether the sensitivity, repeatability, and reproducibility of the new analytical method of GC–MS-SIM, GC–MS/MS, and LC–MS/MS for determination of 653 multigroup pesticide residues in tea will be able to satisfy the criteria designed to be the AOAC official method.

7.2.2.2 Scope/Applicability

This method is applicable for the qualitative and quantitative determination of 653 residue pesticides and chemical pollutants in tea (green tea, black tea, oolong tea, and pu'er tea). The limits of quantification for 490 pesticides determined by GC–MS-SIM were 1–500 µg/kg, and for 448 pesticides determined by LC–MS/MS were 0.03–4820 µg/kg. This method will satisfy the absolute majority of MRL testing requirements of the EU, USA, Japan, and China.

7.2.2.3 Materials/Matrices

It will be unimaginably difficult in terms of resources, time, and personnel for each laboratory to participate in a collaborative study on hundreds of pesticides, so AOAC International has proposed an alternative "shrunken" protocol, which is to select two teas and three methods (GC–MS-SIM, GC–MS/MS, and LC–MS/MS) for a respective determination of 20 pesticides, with a total of 40 pesticides to be involved in the collaborative study.

7.2.2.4 Concentrations/Ranges of Analytes

The 40 pesticides selected out of the 653 pesticides are listed in Table 7.2.1.

The concentration range of 40 target pesticides is 10 µg/kg–2 mg/kg. (Note: the concentration ranges of target pesticide related to the study were decided at two exclusive meetings organized by the AOAC Pesticide and Chemical Contaminants Community, one of which was at the Florida Pesticide Residue Workshop held July 18–21, 2011, and the other at 125th AOAC Annual Meeting.)

7.2.2.5 Spiked Samples of Blind Duplication

Every participant will make a determination of a total of 10 collaborative study samples (four fortified samples, two incurred samples, two aged samples, and two blank samples). The compositions of these 10 collaborative study samples are shown in Fig. 7.2.1.

7.2.2.6 Naturally Incurred Residues Matrices

During 2011–12 a tea plantation named "Youshan Garden" was chosen in our first tea-producing province, Fujian, where growing teas are sprayed with pesticides as per the relevant procedure and the teas sprayed and picked per different phases are then processed into incurred samples.

7.2.2.7 Sample Preparation and Handing Homogeneity

The dried raw teas are pulverized in a mixer and then passed through 10-mesh and 16-mesh sieves, respectively, before the tea powders after between 10-mesh and 16-mesh sieves will be fully uniformly mixed and adopted as analytical samples for laboratories (incurred, aged samples, and blank samples).

7.2.2.7.1 Preparation of Collaborative Fortified Samples for Analysis (Samples Nos. 01 and 02 for Green Tea; Nos. 06 and 07 for Oolong Tea)

1. Label two 80 mL centrifuge tubes Nos. 01 and 02 and add the corresponding blank green tea samples supplied by the study director as shown in Table 7.2.2.

TABLE 7.2.1 Pesticides to Be Testing in Collaborative Study

No.	GC–MS and GC–MS/MS			LC–MS/MS		
	Pesticides	MRL (µg/kg)	Source	Pesticides	MRL (µg/kg)	Source
1	2, 4′-DDE	200	EU	Acetochlor	10	EU
2	4, 4′-DDE	200	EU	Benalaxyl	100	EU
3	Benalaxyl	100	EU	Bensulide	30	Japan
4	Bifenthrin	5000	EU	Butralin	20	EU
5	Bromophos-ethyl	100	EU	Chlorpyrifos	100	EU
6	Bromopropylate	100	EU	Clomazone	20	EU
7	Chlorfenapyr	50,000	EU	Diazinon	20	EU
8	Diflufenican	50	EU	Ethoprophos	20	EU
9	Dimethenamid	20	EU	Flutolanil	50	EU
10	Fenchlorphos	100	EU	Imidacloprid	50	EU
11	Picoxystrobin	100	EU	Indoxacarb	50	EU
12	Pirimicarb	50	EU	Kresoxim-methyl	100	EU
13	Pirimiphos-methyl	50	EU	Monolinuron	50	EU
14	Procymidone	100	EU	Picoxystrobin	100	EU
15	Propyzamide	50	EU	Pirimiphos methyl	50	EU
16	Pyrimethanil	100	EU	Propoxur	100	EU
17	Quinoxyfen	50	EU	Quinoxyfen	50	EU
18	Tefluthrin	50	EU	Tebufenpyrad	100	EU
19	Tolclofos-methyl	100	EU	Triadimefon	200	EU
20	Trifluralin	100	EU	Trifloxystrobin	50	EU

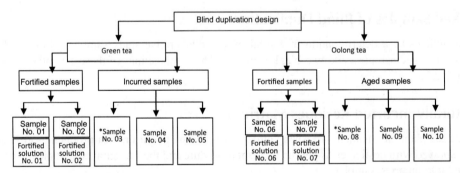

FIGURE 7.2.1   Number of blank, fortified, incurred and aged samples for the collaborative study. (Fortified samples are designed per principle of Blinds Duplication).

2. Label two 80 mL centrifuge tubes Nos. 06 and 07 and add the corresponding blank oolong tea samples supplied by the study director as shown in Table 7.2.2.

3. Stand the ampoules containing fortification solutions in front of the corresponding centrifuge tubes until the solutions drain down the inner walls of the ampoule from the top down.

4. Carefully open the ampoule and quantitatively transfer the fortification solutions into the corresponding centrifuge tubes with a pipette or syringe. Remarkably, please do not dump directly.

5. Rinse each ampoule three times with 0.5 mL acetonitrile and add the rinses to the centrifuge tubes.

6. Let the fortified tea samples mix and sit for 30 min.

TABLE 7.2.2 The Corresponding Serial Numbers of Centrifuge Tubes, Blank Samples and Fortification Solutions in Ampoules for Preparing "Collaborative Fortified Samples"

	Green tea		Oolong tea	
Serial numbers	No. 01	No. 02	No. 06	No. 07
Centrifugal tubes	No. 01	No. 02	No. 06	No. 07
Fortified solutions	No. 01	No. 02	No. 06	No. 07

7. Subject the fortified samples to sample preparation procedures as described under sample extraction and cleanup (see Section 7.2.5.5 of the AOAC collaborative study method).

Because the concentrations of fortified solutions in each ampoule will be unknown to each participating laboratory, it is recommended that good laboratory practice (GLP) techniques are used in preparing the samples. The volume of fortification solution provided is sufficient for only one analysis. It means that there is one collaborative sample for only one opportunity, just one shot.

7.2.2.7.2 Preparation of Collaborative Incurred or Aged Samples for Analysis (Samples Nos. 03–05 for Green Tea; Nos. 08–10 for Oolong Tea)

1. Subject the incurred green tea samples (Nos. 03–05) and aged oolong tea (Nos. 08–10) samples to sample preparation procedures as described under sample extraction and cleanup (see Section 7.2.5.5 of the AOAC collaborative study method).
2. The amount of the tea sample in each bag provided by the study director is exactly 5 g, so just pour the tea into the centrifuge tube and carry out the analysis.

7.2.2.7.3 Shipping and Storage

The shipping of samples will comply with the Dangerous Good Regulations from the International Air Transport Association (IATA).

The testing materials shipped by SD shall include standard reserve solutions, fortified solutions, blank samples, and incurred/aged samples, which will be placed and kept at 4°C in a fridge upon reception.

7.2.2.7.4 Aged Sample Stability

Spray the blank tea with the 40 pesticides to be studied and prepare into multiresidue aged samples. Make a respective determination with GC–MS, GC–MS/MS, and LC–MS/MS, run three parallel samples each time, test once every 5 days for a continuous period of 3 months, and obtain 6840 raw data. Statistical analysis of precision is made with regard to the 19 test results from two kinds of tea, three methods, 40 pesticides, and a period of 3 months. Results show that RSD values fall within the range of 4.3%–31.1%. That is also to say, be it green tea or oolong tea, these 40 pesticides degrade very slowly within 3 months; deviation rate is about 20%. This is good enough to demonstrate that the stability of incurred samples is well guaranteed in the predicted 3 months of collaborative study sample testing (Fig. 7.2.2).

7.2.2.8 Quality Assurance

7.2.2.8.1 Precollaborative Study

Prior to the determination of collaborative samples, collaborators should practice testing the samples in advance in their own laboratories (i.e., precollaborative study) and shall not undertake official collaborative samples until acceptance criteria is reached for the test results.

7.2.2.8.1.1 Precollaborative Study Content

Collaborators will conduct fortified recovery tests with blank tea samples (green tea or oolong tea) and the mixed pesticide working standard solutions supplied by SD. Add 50 µL mixed pesticide working standard solutions into each portion of 5 g blank tea samples and at least add five portions to carry out parallel tests. For analytical results, it is required that each pesticide recovery, RSD ($n = 5$), ion abundance, and linear correlation coefficient (R^2) values be calculated, which will then be transmitted to SD.

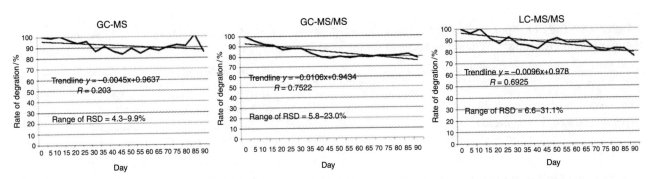

FIGURE 7.2.2 Degradation of 40 pesticides in aged sample.

7.2.2.8.1.2 Precollaborative Study Acceptance Criteria

The results of the practice must reach the acceptance criteria for precollaborative study as:

1. R^2 of 20 pesticides at least 5-point matrix-matched calibration standard curves shall be greater than 0.995 for GC–MS-SIM, GC–MS/MS, or LC–MS/MS.
2. Recoveries fall within 70%–120% with RSD < 15% ($n = 5$).
3. Ion abundance of the targeted pesticides should be in accordance with the recommended maximum permitted tolerances of the AOAC regulation or the EU regulation "EU Document No. SANCO/10684/2009 method validation and quality central procedures for pesticide residues analysis in food and feed."

No matter whether it is recovery or RSD, or ion abundance, or R^2 values, if 30% data from any of these technical indexes or the totaling of these four indexes fails to reach the acceptance criteria, this laboratory proves to be lacking in the mastery of the key technique of this method, which will be rendered unable to undertake official collaborative study samples unless test results finally reach the acceptance criteria.

For those laboratories that fail to meet the acceptance criteria in practice, they will, in principle, forfeit their right to continue with the official collaborative samples and will not be considered as collaborative laboratories accordingly.

7.2.2.8.2 Calibration Check

Prepare 5-point matrix-matching internal standard calibration mixed solutions and set up a 5-point matrix-matching internal standard calibration curve. $R^2 \geq 0.995$ will satisfy the requirements.

7.2.2.8.3 System Suitability Check

Before making determinations of each consignment sample, use quality control standard working solutions supplied by the study director to check the sensitivity and stability of the instrument.

1. Instrumental sensibility (calculated per internal standard reference noise ratios (S/N)): GC–MS-SIM >500, GC–MS/MS >1000, LC–MS/MS >500.
2. Instrumental stability (calculated per internal standard peaks): retention time deviation <3% (two successive injections); peak area deviation <10% (two successive injections).
3. GC–MS-SIM, GC–MS/MS, and LC–MS/MS are all equipped with data processing software system.

7.2.2.8.4 Blank Test

1. Confirm there is no existence of any interference from reagents by running a blank test of the whole process.
2. Run the blank the same as described in the Collaborative Study Method (Section 7.2.5.5) without adding the sample.

7.2.3 METHOD PERFORMANCE PARAMETERS FOR SINGLE LABORATORY

7.2.3.1 Limit of Detection and Limit of Quantification

LOD and LOQ of the GC–MS-SIM methods were obtained with blank samples fortified with 490 pesticides for GC–MS-SIM and 448 pesticides for LC–MS/MS at different concentrations. The fortified concentration gave a signal to noise ratio (S/N) ≥ 5 for each pesticide was defined as the LOD of the method while the fortified concentration gave an S/N ≥ 10 was defined as the LOQ of the method.

7.2.3.2 Accuracy and Precision for a Single Laboratory

The accuracy and precision of the method were evaluated through recovery experiments by spiking pesticides to black tea, green tea, puer tea, and oolong tea that contained no pesticides or related chemicals. Based on the low fortification level that is taken for an instant, it can be seen that at low fortification levels there are 424 pesticides with recoveries 60%–120% among the 451 pesticides tested by GC–MS-SIM, accounting for 94% of the pesticides and related chemicals, of which pesticides with RSD <20% account for 77% of the pesticides and the related chemicals tested. There are 91% of 439 pesticides and related chemicals tested by LC–MS/MS with average recoveries 60%–120%, 76% pesticides and related chemicals have RSD <20%. The 270 pesticides that can be determined by both GC–MS-SIM and LC–MS/MS with average recoveries 60%–120% account for 96% and 94%, respectively; pesticides with RSD <20% account for 75% and 79%, respectively, which means that LC–MS/MS and GC–MS-SIM are of very good reproducibility and repeatability for determination of tea matrix pesticide residues.

7.2.3.3 Linearity of 653 Pesticides for Single Laboratory

A good linear relationship was shown in the linear regression analysis. The linear correlation coefficient for 96% pesticides determined by GC–MS-SIM method is $R^2 \geq 0.980$; the linear correlation coefficient for 90% pesticides determined by LC–MS/MS method is $R^2 \geq 0.980$.

7.2.3.4 Ruggedness of Multiresidue Method

November 2010–March 2011: to investigate the scientificity and maneuverability of the protocol, the study director organized a simulated AOAC collaborative study in China. Sixteen laboratories (including five overseas wholly foreign-owned labs stationed in China) participated in this simulated AOAC collaborative study. The study director provided each lab with the protocol (OMA-2011-Jan-001) and shipped pesticide age samples and standard solutions for quantification. The 16 labs finished this study on time and provided SD with the feedback of analytical parameters of instruments and raw data, including 1617 data of monitoring ions, 2560 raw data of target pesticides, 3660 data of calibration curve, and 12,100 data of ion abundance (including 8500 standard curve). The statistical analytical data showed that the simulated AOAC collaborative study had achieved very good results. See Table 7.2.3.

November 2009–February 2010: for the purpose of evaluating the efficiency of the method, the study director chose 227 pesticides, two teas (green tea and oolong tea), two kinds of instruments (GC–MS and GC–MS/MS), two kinds of cartridges Cartridge (Envi-Carb + PSA and Cleanert triple phase of tea (TPT)), ran two parallel samples each time, and conducted 19 circulative determinations over 3 months to obtain 41,768 test data. Test results were obtained from less than a total of 16 different conditions by evaluating different instruments, different cartridges, different teas, and different concentrations. The comprehensive statistical analytical results demonstrate that the ruggedness of the method is very good. See Table 7.2.4.

7.2.4 COLLABORATORS

Thirty laboratories, including 8 brand laboratories, 9 government-owned labs, 5 independent laboratories, 3 research institutes, 2 colleges and universities, and 3 enterprises, from 11 countries and regions from Europe, Asia, and the Americas participated in the collaborative study of multiresidue pesticides in tea using GC–MS, GC–MS/MS, and LC–MS/MS.

7.2.4.1 Instrument and Materials Used by Collaborators

The instrument and material used in the collaborative study will be introduced according to GC–MS, GC–MS/MS, and LC–MS/MS, respectively.

7.2.4.1.1 GC–MS Used by Collaborators

For GC–MS (16 labs), these are the details.

Four types of Agilent instrument were used by all of the labs: (1) Agilent 7890A-Agilent 5975C (Lab 1, Lab 3, Lab 4, Lab 6, Lab 9, Lab 10, Lab 11, Lab 12, Lab 14, and Lab 15); (2) Agilent 6890-Agilent 5975C (Lab 2); (3) Agilent 6890-Agilent 5973 (Lab 8, Lab 22, Lab 25); and (4) Agilent 6890N-Agilent 5975B (Lab 28). Lab 19, using Varian (CP-3800)-Varian Ion Trap (Saturn 2200) was excluded.

The specification of chromatagraphic column for 16 labs is 30 m × 0.25 mm × 0.25 μm. Five brands of chromatographic column were used: (1) DB-1701 (Lab 1, Lab 3, Lab 4, Lab 8, Lab 11, Lab 12, Lab 15, Lab 22, Lab 25, and Lab 28); (2) Restek Rtx 1701 (Lab 2); (3) ZB-MR2 (Lab 6); (4) HP-5 ms (Lab 10 and Lab 14); (5) HP-5MSUI (Lab 9 and Lab 19).

TABLE 7.2.3 Method Efficacy RSD_r, RSD_R and HorRat Values for the Simulated AOAC Collaborative Study (2010.11–2011.3)

Parameter	Range	GC–MS-SIM (GC–MS/MS)					LC–MS/MS				
		Green tea	Percentage (%)	Oolong tea	Percentage (%)	Total (%)	Green tea	Percentage (%)	Oolong tea	Percentage (%)	Total (%)
RSD_r (%)	<8	20	100	20	100	100	9	45	17	85	65
	8 ~ 15	0	0	0	0	0	11	55	2	10	32.5
	>15	0	0	0	0	0	0	0	1	5	2.5
RSD_R (%)	<16	7	35	16	80	57	8	40	6	30	35
	16 ~ 25	10	50	4	20	35	11	55	13	65	60
	>25	3	15	0	0	7	1	5	1	5	5
HorRat	<0.50	1	5	5	25	15	6	30	2	10	20
	0.50 ~ 10	11	55	15	75	65	13	65	16	80	72.5
	1.01 ~ 2	7	35	0	0	17	1	5	1	5	5
	>2	1	5	0	0	2	0	0	1	5	2.5

TABLE 7.2.4 RSD Values From 19 Determinations Over 3 months With Eight Different Analytical Conditions

RSD%	Cartridges		Samples		Instruments	
	Cleanert TPT	Envi-Carb + PSA	Green tea	Oolong tea	GC–MS	GC–MS/MS
≤5	20.3 (163)[a]	6.3 (51)	17.5 (141)	9.1 (73)	19 (153)	7.6 (61)
5 ~ 10	66.8 (537)	62.1 (499)	59.3 (477)	69.5 (559)	59.6 (479)	69.3 (557)
10 ~ 15	10.2 (82)	20.9 (168)	15.8 (127)	15.3 (123)	14.9 (120)	16.2 (130)
≤15	97.3 (782)	89.3 (718)	92.6 (745)	93.9 (755)	93.5 (752)	93.1 (748)
15 ~ 20	1.9 (15)	7.3 (59)	4.9 (39)	4.4 (35)	4.7 (38)	4.5 (36)
>20	0.9 (7)	3.4 (27)	2.5 (20)	1.7 (14)	1.7 (14)	2.5 (20)
Total	100 (804)	100 (804)	100 (804)	100 (804)	100 (804)	100 (804)

[a]Percentage (quantity of pesticide).

There were five temperature programs for GC: (1) 40°C hold 1 min, at 30°C/min to 130°C, at 5°C/min to 250°C, at 10°C/min to 300°C, hold 5 min (Lab 1, Lab 4, Lab 8, Lab 9, Lab 10, Lab 11, Lab 12, Lab 14, Lab 15, Lab 19, and Lab 28); (2) 40°C hold 1 min, 30°C/min to 130°C, 5°C/min to 250°C, 10°C/min to 290°C, hold 5 min (Lab 2); (3) 40°C (1 min), 30°C/min to 130°C, 5°C/min to 250 (4), 10°C/min to 280°C, hold 2 min (Lab 3 and Lab 25); (4) 70°C hold 1 min, at 30°C/min to 130°C, at 5°C/min to 250°C, at 10°C/min to 300°C, hold 5 min (Lab 22); and (5) 80°C, at 20°C/min to 140°C, at 5°C/min to 210°C, at 1°C/min to 220°C, at 5°C/min to 240°C, at 30°C/min to 300°C, hold 5 min (Lab 6).

Three solid phase extraction (SPE) cartridges were used in the sample cleanup: (1) Cleanert TPT (Lab 1, Lab 3, Lab 4, Lab 6, Lab 8, Lab 9, Lab 10, Lab 11, Lab 12, Lab 15, Lab 19, Lab 22, and Lab 26); (2) UCT (ECPSACB506) (Lab 2); and (3) Supelco, Envi-Carb/PSA (Lab 14 and Lab 28).

7.2.4.1.2 GC–MS/MS Used by Collaborators

For GC–MS/MS (14 labs), here is the breakdown.

Five types instruments were used by 14 labs: (1) Agilent 7890A-Agilent 7000 (Lab 1, Lab 5, Lab 13, Lab 16, Lab 18, Lab 21, Lab 27, Lab 28 and Lab 30); (2) Agilent 7890A-Waters Quattro micro (Lab 10); (3) Thermo Fisher Trace GC Ultra-Thermo Fisher TSQ Quantum (Lab23); (4) Bruker450-GC, 300-MS (Lab 26); (5) Agilent 6890-Waters Quattro micro (Lab 29).

There were eight brands of chromatographic column: (1) DB-1701 (Lab 1, Lab 5, Lab 13, and Lab 28); (2) HP-5 ms (Lab 10, Lab 21, and Lab 29); (3) HP-5MSUI (Lab 16 and Lab 27); (4) HP-5MSI (Lab 18); (5) DB-5MSUI (Lab 23); (6) DB5-MS (Lab 24); (7) BR-5ms (Lab 26); (8) Rxi5 Sil (Lab 30).

Three specifications of chromatographic column were used: (1) 30 m × 0.25 mm × 0.25 μm (Lab 1, Lab 5, Lab 10, Lab 13, Lab 16, Lab 21, Lab 26, Lab 27, Lab 28, Lab 29, and Lab 30); (2) 15 m × 0.25 mm × 0.25 μm (Lab 18); (3) length 20 m/diam. 0.180 mm/film 0.18 μm (Lab 23).

There were nine temperature programs for GC: (1) 40°C hold 1 min, at 30°C/min to 130°C, at 5°C/min to 250°C, at 10°C/min to 300°C, hold 5 min (Lab 1, Lab 5, Lab 10, Lab 13, Lab 24, and Lab 28); (2) 70°C hold 1 min, at 50°C/min to 150°C, at 6°C/min to 200°C, at 16°C/min to 280°C, hold 5.5 min (Lab 16); (3) 40°C (1 min), 130°C (60°C/min), 250°C (10°C/min), 300°C (20°C/min, 2.5 min) (Lab 18); (4) 70°C hold 2 min, at 25°C/min to 150°C, at 3°C/min to 200°C, at 8°C/min to 280°C, hold 10 min (Lab 21); (5) 60°C, hold 1.5 min, at 30°C/min to 150°C, at 10°C/min to 260°C, at 30°C/min to 290°C, hold 5.5 min (Lab 23); (6) 120°C, hold 1 min, at 15°C/min to 210°C, at 10°C/min to 300°C, hold 2 min (Lab 26); (7) 70°C, hold 2 min, at 25°C/min to 150°C, at 3°C/min to 200°C, at 8°C/min to 280°C, hold 15 min (Lab 27); (8) 70°C, hold 2 min, at 25°C/min to 150°C, at 3°C/min to 200°C, at 8°C/min to 280°C, hold 13 min (Lab 30); (9) 90°C, hold 2 min, at 7°C/min to 244°C, hold 1 min, at 25°C/min to 300°C, hold 2 min (Lab 29).

Four SPE cartridges were used in the sample cleanup: (1) Cleanert TPT (Lab 1, Lab 10, Lab 13, Lab 18, Lab 21, Lab 26, and Lab 27); (2) GL Sciences (InertSep GC/PSA) (Lab 5); (3) Agilent Technologies BE Carbon 500 mg/PSA (Lab 16); (4) Supelco, Envi-Carb/PSA (Lab 23, Lab 24, Lab 28, Lab 29, and Lab 30).

7.2.4.1.3 LC–MS/MS Used by Collaborators

For LC–MS/MS (24 labs), note the following instruments and materials.

Twelve types instruments were used by 24 labs: (1) Agilent 1200-Agilent 6430 (Lab 1 and Lab 13); (2) Agilent 1200-Agilent 6410 (Lab 3, Lab 4, Lab 10, Lab 12, Lab 16 and Lab 30); (3) SHIMADZU LC20AB-SHIMADZU 8030 (Lab 5); (4) Agilent 1200-Agilent 6460 (Lab 6 and Lab 11); (5) ACQUITY UPLC system-Quattro Premier XE (Lab 7, Lab 17 and Lab 22); (6) Agilent1100-3200Q TRAP (Lab 8); (7) Waters ACQUTY UPLC-Waters Xevo TQS (Lab 9, Lab 28 and Lab 29); (8) Agilent1200-Agilent (6490) (Lab 18); (9) Thermo Scientific, Accela 1250-Thermo Scientific, TSQ Quantum Access (Lab 21 and Lab 23); (10) Dionex ultimate 3000-4000Qtrap (Lab 25); (11) Agilent1290-5500 API (Lab 24); (12) Agilent 1200-AB Sciex 5000 (Lab 27).

Fourteen brands of chromatographic column were used by 24 labs: (1) ZORBAX SB-C18 (Lab 1, Lab 3, Lab 4, Lab 5, Lab 10, Lab 12, Lab 13, Lab 16, Lab 22 and Lab 25); (2) phenomenex Luna C8 (Lab 6); (3) Acquity HSS T3 (Lab 7); (4) Waters Atlantis C18 (Lab 8); (5) Waters ACQUITY UPLC BEH C18 (Lab 9); (6) Agilent Eclipse plus (Lab 11); (7) Acquity BEH C18 (Lab 17); (8) Agilent Zorbax C8 (Lab 18); (9) Thermo Scientific Hypersil Gold C8 (Lab 21 and Lab 23); (10) x-Terra C8 (Lab 24); phenomenex Luna C18 (Lab 27); ACQUITY UPLC HSS T3 (Lab 28); (11) Acquity UPLCTM BEH C18 (Lab 29); (12) phenomenex Gemini C18 (Lab 30).

Fifteen specifications of chromatographic column were used: (1) 3.5 μm × 100 mm × 2.1 mm (Lab 1, Lab 5, Lab 10, Lab 11, Lab 12, Lab 13 and Lab 22); (2) 3.5 μm × 150 mm × 2.1 mm (Lab 3, Lab 4, Lab 8 and Lab 25); (3) 150 mm × 2 mm × 3 μm (Lab 6); (4) 100 mm × 2.1 mm × 1.8 μm (Lab 7); (5) 3 mm × 100 mm × 1.7 μm (Lab 9); (6) 2.1 mm × 100 mm × 1.8 μm (Lab 16); (7) 1.7 μm × 2.1 mm × 100 mm (Lab 17); (8) 5 μm, 4.6 mm × 150 mm (Lab 18); (9) 150 mm × 2.1, Part Sz. 5 μm (Lab 21); (10) Hypersil GOLD 50 mm × 2.1 mm Particle Sz 1.9 μm (Lab 23); (11) 3 μm × 100 mm × 2.1 mm (Lab 24); (12) 150 mm × 2 mm × 5 μm (Lab 27); (13) 1.8 μm × 100 mm × 2.1 mm (Lab 28); (14) 1.7 μm × 100 mm × 2.1 mm (Lab 29); (15) 3 μm × 150 mm × 3 mm (Lab 30).

Eleven temperature programs were used for LC: (1) 0.00 min-99% A; 3.00 min-70% A; 6.00 min-60% A; 9.00 min-60% A; 15.00 min-40% A; 19.00 min-1% A; 23.00 min-1% A; 23.01 min-99% A (Lab 1, Lab 4, Lab 5, Lab 6, Lab 10, Lab 11, Lab 12, Lab 13, Lab 16, Lab 17, Lab 18, Lab 22, and Lab 25); (2) 0.01 min-99% A, 0.50 min-99% A, 2.00 min-60% A, 15.00 min-20% A, 18.00 min-100% A, 20.00 min-100% A, 20.10 min-99% A, 25.00 min-99% A (Lab 3); (3) 0.00 min-10% A, 1.50 min-70% A, 5.00 min-85% A, 7.50 min-99% A, 9.00 min-100% A, 9.80 min-100% A, 10.30 min-10% A, hold 2 min (Lab 7); (4) 0.00 min-50% A, 5.00 min-50% A, 10.00 min-10% A, 15.00 min-10% A, 15.01 min-50% A, 25.00 min-50% A (Lab 8); (5) 0.00 min-95% A, 1.00 min-95% A, 8.00 min-2% A, 12.50 min-2% A, 12.80 min-95% A, 15.00 min-95% A (Lab 9); (6) 0.00 min-90% A, 1.00 min-90% A, 10.00 min-20% A, 20.00 min-20% A, 21.00 min-90% A, 25.00 min-90% A (Lab 21); (7) 0.00 min-60A%, 2.00 min-50A%, 7.00 min-15A%, 10.00 min-15A%, 12.00 min-5% A, 14.00 min-5% A, 15.00 min-60% A, 17.00 min-60% A (Lab 23); (8) 0.00 min-99% A; 2.00 min-99% A; 3.75 min-70% A; 5.75 min-60% A; 8.75 min-60% A; 14.75 min-40% A; 18.75 min-1% A; 22.75 min-1% A; 23.00 min-99% A; 28.00 min-99% A (Lab 24); (9) 0.00 min-99% A, 1.00 min-70% A, 2.00 min-60% A, 8.00 min-45% A, 9.00 min-1% A, 11.00 min-1% A, 13.00 min-99% A (Lab 28); (10) 0.00 min-99% A; 1.00 min-50% A; 6.00 min-30% A; 8.00 min-10% A; 9.00 min-1% A; 12.00 min-1% A; 13.00 min-99% A (Lab 29); (11) 0.00 min-85% A; 1.00 min-15% A; 17.00 min-5% A; 24.00 min-5% A; posttime 7 min (Lab 30).

For the aforementioned 24 labs, five SPE cartridges were used in the sample cleanup: (1) Cleanert TPT (Lab 1, Lab 3, Lab 4, Lab 6, Lab 7, Lab 8, Lab 9, Lab 10, Lab 11, Lab 12, Lab 13, Lab 17, Lab 18, Lab 21, Lab 22, Lab 25 and Lab 27); (2) GL Sciences (InertSep GC/PSA) (Lab 5); (3) Agilent Technologies BE Carbon 500 mg/PSA (Lab 16); (4) Supelco, Envi-Carb (Lab 24); (5) Supelco, Envi-Carb/PSA (Lab 23, Lab 28, Lab 29 and Lab 30).

7.2.5 AOAC OFFICIAL METHOD: HIGH-THROUGHPUT ANALYTICAL TECHNIQUES FOR DETERMINATION OF RESIDUES OF 653 MULTICLASS PESTICIDES AND CHEMICAL POLLUTANTS IN TEA BY GC–MS, GC–MS/MS, AND LC–MS/MS

The method is applicable for determining 40 representative pesticides in green tea and oolong tea: 2,4'-DDE; 4,4'-DDE; benalaxyl; bifenthrin; bromophos-ethyl; bromopropylate; chlorfenapyr; diflufenican; dimethenamid; fenchlorphos; picoxystrobin; pirimicarb; pirimiphos-methyl; procymidone; propyzamide; pyrimethanil; quinoxyfen; tefluthrin; tolclofos-methyl; trifluralin by GC–MS and GC–MS/MS, and acetochlor; benalaxyl; bensulide; butralin; chlorpyrifos; clomazone; diazinon; ethoprophos; flutolanil; imidacloprid; indoxacarb; kresoxim-methyl; monolinuron; picoxystrobin; pirimiphos-methyl; propoxur; quinoxyfen; tebufenpyrad; triadimefon; and trifloxystrobin by LC–MS/MS.

According to the results of a single laboratory, the method was popularized and widely used in the analysis of 653 pesticide residues in green tea, oolong tea, black tea, and pu'er tea, with the LOQ: 0.03 µg/kg~1.21 mg/kg.

LOD of the GC–MS-SIM method is 1–500 µg/kg, LOQ 2–1000 µg/kg; LOD of the LC–MS/MS method 0.03–4820 µg/kg and LOQ 0.06–9640 µg/kg. For the LOD comparison data of both GC–MS-SIM and LC–MS/MS, there are 482 pesticides with LOD ≤ 100 µg/kg for the GC–MS-SIM method, accounting for 98% of the pesticides tested; there are 417 pesticides for LC–MS/MS, accounting for 93% of the pesticides analyzed; there are 264 pesticides with LOD ≤ 10 µg/kg for the GC–MS-SIM method, accounting for 54% of the pesticides tested; 325 pesticides for LC–MS/MS method, making up 73% of the pesticides analyzed. There are 270 pesticides that can be analyzed by both GC–MS-SIM and LC–MS/MS. There are 264 pesticides with LOD ≤ 100 µg/kg for the GC–MS-SIM method, accounting for 98% of the pesticides tested; 247 for LC–MS/MS, making up 91% of the pesticides analyzed. There are, however, 133 pesticides with LOD ≤ 10 µg/kg for the GC–MS-SIM method, accounting for 49% of the pesticides tested; 200 pesticides for LC–MS/MS method, making up 74% of the pesticides analyzed.

7.2.5.1 Principle

Test samples are extracted with acetonitrile using a homogenizer and the extracts are cleaned up with Cleanert TPT cartridge, Envi-Carb/PSA or ECPSACB506 (UCT) or BE Carbon 500g/PSA (Agilent) or InertSep GC/PSA (GL Sciences) or equivalent cartridge. The pesticides are eluted with acetonitrile–toluene (3:1, v/v), after concentration, dryness, and dissolution, the samples are analyzed by GC–MS, GC–MS/MS, or LC–MS/MS, and quantified with a matrix-matched standard calibration curve.

7.2.5.2 Apparatus and Conditions

7.2.5.2.1 GC–MS Analytical

GC–MS system description: Model 7890A gas chromatograph connected to a Model 5975C mass selective detector with electron ionization (EI) source and equipped with a Model 7683 autosampler, a Chemstation data processing software system (Agilent Technologies, Wilmington, DE, USA), or equivalent. Column: DB-1701 capillary column (30 m × 0.25 mm × 0.25 µm), or equivalent; column temperature: 40°C hold 1 min, at 30°C/min–130°C, at 5°C/min–250°C, at 10°C/min–300°C, hold 5 min; carrier gas: helium, purity ≥ 99.999%, flow rate: 1.2 mL/min; injection port temperature: 290°C; injection volume: 1 µL; injection mode: splitless, purge on after 1.5 min; ionization mode: EI; ion source polarity: positive ion; ionization voltage: 70 eV; ion source temperature: 230°C; GC–MS interface temperature: 280°C; Solvent delay: 14 min.

Ion monitoring mode: selected ion monitoring (SIM), one quantifying ion, and two qualifying ions are selected for each compound. The retention times, quantifying ions, qualifying ions, and the expected ion abundances for each of the 20 pesticides included in the study and heptachlor epoxide are listed in Table 7.2.5. SIM acquisition parameters for ions monitored by GC–MS are shown in Table 7.2.6.

TABLE 7.2.5 GC–MS Retention Times, Quantifying Ions, Qualifying Ions, Ion Abundances, LODs and LOQs for the 20 Pesticides

No.	Pesticides	Retention time (min)	Quantifying ion	Qualifying ion 1	Qualifying ion 2	LOQ (µg/kg)	LOD (µg/kg)
ISTD	Heptachlor-epoxide	22.15	353 (100)	355 (79)	351 (52)		
1	Trifluralin	15.43	306 (100)	264 (72)	335 (7)	20	10
2	Tefluthrin	17.35	177 (100)	197 (26)	161 (5)	10	5
3	Pyrimethanil	17.43	198 (100)	199 (45)	200 (5)	10	5
4	Propyzamide	18.94	173 (100)	255 (23)	240 (9)	10	5
5	Pirimicarb	19	166 (100)	238 (23)	138 (8)	20	10
6	Dimethenamid	19.77	154 (100)	230 (43)	203 (21)	10	5
7	Tolclofos-methyl	19.83	265 (100)	267 (36)	250 (10)	10	5
8	Fenchlorphos	19.90	285 (100)	287 (69)	270 (6)	40	20
9	Pirimiphos-methyl	20.37	290 (100)	276 (86)	305 (74)	20	10
10	2,4'-DDE	22.75	246 (100)	318 (34)	176 (26)	25	12.5
11	Bromophos-ethyl	23.12	359 (100)	303 (77)	357 (74)	10	5
12	4,4'-DDE	23.95	318 (100)	316 (80)	246 (139)	10	5
13	Procymidone	24.57	283 (100)	285 (70)	255 (15)	10	5
14	Picoxystrobin	24.79	335 (100)	303 (43)	367 (9)	20	10
15	Chlorfenapyr	27.40	247 (100)	328 (54)	408 (51)	200	100
16	Quinoxyfen	27.15	237 (100)	272 (37)	307 (29)	10	5
17	Benalaxyl	27.68	148 (100)	206 (32)	325 (8)	10	5
18	Bifenthrin	28.62	181 (100)	166 (32)	165 (35)	10	5
19	Diflufenican	28.73	266 (100)	394 (25)	267 (14)	10	5
20	Bromopropylate	29.46	341 (100)	183 (54)	339 (51)	20	10

7.2.5.2.2 GC–MS/MS Analytical

GC–MS/MS system description: Model 7890A gas chromatograph connected to a Model 7000B triple quadrupole mass spectrometer with electron ionization (EI) source and equipped with a Model 7693 autosampler equipped with Mass Hunter data processing software system (Agilent Technologies, Wilmington, DE, USA), or equivalent.

Operating conditions are the same as for GC–MS with the exception that the ion monitoring mode is by selected reaction monitoring (SRM) and monitoring one precursor ion and two product ion transitions.

The monitored ion transitions and the collision energies for the 20 pesticides of interest in the study and heptachlor epoxide are shown in Table 7.2.7. The SRM acquisition parameters for the precursor and product ion transitions monitored by GC–MS/MS are shown in Table 7.2.8.

7.2.5.2.3 LC–MS/MS Analytical

LC–MS/MS system description: an Agilent Series 1200 HPLC system directly coupled to a 6430 triple quadrupole mass spectrometer equipped with electrospray ionization (ESI) source, a Model G1367D autosampler with a Mass Hunter data processing software system (Agilent Technologies, Santa Clara, CA, USA). Gas chromatographic separation was achieved on a Zorbax SB-C18, 2.1×100 mm, 3.5 µm (Agilent Technologies, Santa Clara, CA, USA), or equivalent. Column: ZOR-BAX SB-C18, 3.5 µm, 100 mm×2.1 mm or equivalent; mobile phase program and the flow rate: refer to Table 7.2.9; column temperature: 40°C; injection volume: 10 µL; ionization mode: ESI; ion source polarity: positive ion; nebulizer gas: nitrogen gas; nebulizer gas pressure: 0.28 MPa; ion spray voltage: 4000 V; dry gas temperature: 350°C; dry gas flow rate: 10 L/min.

TABLE 7.2.6 SIM Acquisition Parameters by GC–MS for the 20 Pesticides of Interest in This Study

Group	Start time (min)	Monitored Ions (m/z)	Dwell time (ms)
1	14.85	306, 264, 335	80
2	16.85	177, 197, 161, 198, 199, 200	80
3	17.97	173, 255, 240, 166, 238, 138	80
4	19.43	154, 230, 203, 285, 287, 270, 265, 267, 250	40
5	20	290, 276, 305	80
6	21.77	246, 318, 176, 353, 355, 351	80
7	22.93	359, 303, 357, 318, 316, 246	80
8	24.20	335, 303, 367, 283, 285, 255	80
9	25.87	237, 272, 307, 247, 328, 408, 148, 206, 325	40
10	28.49	181, 166, 165, 266, 394, 267, 341, 183, 339	40

TABLE 7.2.7 GC–MS/MS Retention Times, Monitored Ion Transitions, Collision Energies, LODs and LOQs for the 20 Pesticides

No.	Pesticides	Retention time (min)	Quantifying precursor/production transition	Qualifying precursor/product ion transition	Collision energy (V)	LOQ (µg/kg)	LOD (µg/kg)
ISTD	Heptachlor-epoxide	22.15	353/263	353/282	17; 17		
1	Trifluralin	15.41	306/264	306/206	12; 15	4.8	2.4
2	Tefluthrin	17.40	177/127	177/101	13; 25	0.8	0.4
3	Pyrimethanil	17.42	200/199	183/102	10; 30	6	3
4	Propyzamide	18.91	173/145	173/109	15; 25	1	0.5
5	Pirimicarb	19.02	238/166	238/96	15; 25	4	2
6	Dimethenamid	19.73	230/154	230/111	8; 25	2	1
7	Fenchlorphos	19.83	287/272	287/242	15; 25	16	8
8	Tolclofos-methyl	19.87	267/252	267/93	15; 25	10	5
9	Pirimiphos-methyl	20.36	290/233	290/125	5; 15	10	5
10	2,4'-DDE	22.79	318/248	318/246	15; 15	6	3
11	Bromophos-ethyl	23.16	359/303	359/331	10; 10	10	5
12	4,4'-DDE	23.90	318/248	318/246	25; 25	4	2
13	Procymidone	24.70	283/96	283/255	10:10	2	1
14	Picoxystrobin	24.75	335/173	335/303	10; 10	10	5
15	Quinoxyfen	27.18	237/208	237/182	25; 25	80	40
16	Chlorfenapyr	27.37	408/59	408/363	15; 5	140	70
17	Benalaxyl	27.66	148/105	148/79	15; 25	2	1
18	Bifenthrin	28.63	181/166	181/165	10; 5	10	5
19	Diflufenican	28.73	266/218	266/246	25; 10	20	10
20	Bromopropylate	29.46	341/185	341/183	15; 15	8	4

TABLE 7.2.8 SRM Acquisition Parameters by GC–MS/MS Analysis for the 20 Pesticides of Interest

Group	Start time (min)	Monitored ion transitions (m/z)	Dwell time (ms)
1	14.76	306/264, 306/206	50
2	15.87	177/127, 177/101, 200/199, 183/102	50
3	18.06	173/145, 173/109, 238/166, 238/96	50
4	19.26	230/154, 230/111, 287/272, 287/242, 267/252, 267/93	25
5	20.07	290/233, 290/125	50
6	21.87	353/282, 353/263	50
7	22.60	359/331, 359/303, 318/248, 318/246	50
8	23.59	335/303, 335/173, 318/248, 318/246, 283/96, 283/255	50
9	26.71	148/105, 148/79, 408/363, 408/59, 237/208, 237/182	25
10	27.88	266/246, 266/218, 181/166, 181/165	50
11	28.96	341/185, 341/183	50

TABLE 7.2.9 Gradient Conditions for LC–MS/MS Analysis

Step	Time (min)	Flow rate/ (μL/min)	Mobile phase A (0.1% Formic acid water) (%)	Mobile phase B (Acetonitrile) (%)
0	0	400	99	1
1	3	400	70	30
2	6	400	60	40
3	9	400	60	40
4	15	400	40	60
5	19	400	1	99
6	23	400	1	99
7	23.01	400	99	1

Monitored ion transitions, collision energies, and fragmentation energies for the 20 pesticides and chlorpyrifos methyl are shown in Table 7.2.10. SRM acquisition parameters by LC–MS/MS for the precursor and product ion transitions monitored are shown in Table 7.2.11.

Homogenizer: rotational speed higher than 3500 r/min (report also in g-force units), T-25B (Janke & Kunkel, Staufen, Germany) or equivalent. Rotary evaporator. Buchi EL131 (Flawil, Switzerland) or equivalent. Centrifuge: centrifugal force higher than 2879 ×g, Z320 (B. HermLe AG, Gosheim, Germany) or equivalent. Nitrogen evaporator. EVAP 112 (Organomation Associates, Inc., New Berlin, MA, USA) or equivalent. TurboVap. LV Evaporation System (Caliper Life Sciences, Hopkinton, MA, USA) or equivalent. Visiprep 5-port flask vacuum manifold. RS-SUPELCO 57101-U (Sigma Aldrich Trading Co., Ltd). See Fig. 7.2.3 or equivalent. Variable volume pipette: 10 μL, 200 μL, and 1 mL. Balance: capable of accurately measuring weights from 0.05 to 100 g within ±0.01 g.

7.2.5.3 Reagents and Materials

Solvents: acetonitrile, toluene, and *n*-hexane: HPLC grade. Acetonitrile–toluene (3:1, v/v). Ultrapure water is obtained in a Milli-RO plus system together with a Milli-Q system from Millipore (Bedford, MA, USA). Anhydrous sodium sulfate: analytically pure. Baked at 650°C for 4 h and stored in a desiccator.

SPE cartridges: Cleanert–TPT (2000 mg, 12 mL, Agela, China) or Envi-Carb/PSA (500 mg/500 mg, 6 mL, Supelco, USA) or equivalent. SPE Tube Adapter: for 12 mL SPE Tubes (57267), for 6 mL SPE Tubes (57020-U) (Sigma Aldrich Trading Co., Ltd) or equivalent. Disposable flow control valve liners for visiprep TM-DL (57059) (Sigma Aldrich Trading Co., Ltd). Pear-shaped flask: 80 mL (Z680346-1EA, Sigma Aldrich Trading Co., Ltd), see Fig. 7.2.3 or equivalent. Reservoir: 30 mL (A82030, Agela, China). See Fig. 7.2.3 or equivalent. Centrifuge tube: 80 mL. Millipore filter membrane (nylon): 13 mm × 0.2 μm.

TABLE 7.2.10 LC–MS/MS Retention Times, Ion Transitions, Collision Energies, LODs and LOQs for the 20 Pesticides of Interest in This Study

No.	Pesticides	Retention time (min)	Quantifying precursor/ product ion transition	Qualifying precursor/ product ion transition	Collision energy (V)	Fragmentation (V)	LOQ (µg/kg)	LOD (µg/kg)
ISTD	Chlorpyrifos-methyl	16.01	322/125	322/290	15; 15	80		
1	Imidacloprid	03.81	256.1/209.1	256.1/175.1	10; 10	80	22	11.0
2	Propoxur	05.89	210.1/111.0	210.1/168.1	10; 5	80	24.4	12.2
3	Monolinuron	6.83	215.1/126.0	215.1/148.1	15; 10	100	3.6	1.8
4	Clomazone	8.30	240.1/125.0	240.1/89.1	20; 50	100	0.4	0.2
5	Ethoprophos	11.37	243.1/173.0	243.1/215.0	10; 10	120	2.8	1.4
6	Triadimefon	11.64	294.2/69.0	294.2/197.1	20; 15	100	7.9	3.9
7	Acetochlor	12.94	270.2/224.0	270.2/148.2	5; 20	80	47.4	23.7
8	Flutolanil	13.25	324.2/262.1	324.2/282.1	20; 10	120	1.1	0.6
9	Benalaxyl	14.40	326.2/148.1	326.2/294	15; 5	120	1.2	0.6
10	Kresoxim-methyl	14.58	314.1/267	314.1/206	5; 5	80	100.6	50.3
11	Picoxystrobin	14.99	368.1/145	368.1/205	20; 5	80	8.4	4.2
12	Pirimiphos methyl	15.05	306.2/164	306.2/108.1	20; 30	120	0.2	0.1
13	Diazinon	15.20	305/169.1	305/153.2	20; 20	160	0.7	0.4
14	Bensulide	15.45	398/158.1	398/314	20; 5	80	34.2	17.1
15	Quinoxyfen	16.60	308/197	308/272	35; 35	180	153.4	76.7
16	Tebufenpyrad	16.82	334.3/147	334.3/117.1	25; 40	160	0.3	0.1
17	Indoxacarb	16.76	528/150	528/218	20; 20	120	7.5	3.8
18	Trifloxys-trobin	16.82	409.3/186.1	409.3/206.2	15; 10	120	2	1
19	Chlorpyrifos	17.65	350/198	350/97	20; 35	100	53.8	26.9
20	Butralin	17.98	296.1/240.1	296.1/222.1	10; 20	100	1.9	1

TABLE 7.2.11 SRM Acquisition Parameters by LC–MS/MS Analysis for the 20 Pesticides

Group	Start time (min)	Monitored ion transitions (m/z)	Dwell time (ms)
1	0	256.1/209.1, 256.1/175.1, 210.1/111, 210.1/168.1, 240.1/125, 240.1/89.1, 243.1/173, 243.1/215, 294.2/69, 294.2/197.1, 215.1/126; 215.1/148.1	30
2	12	270.2/224, 270.2/148.2, 306.2/164, 306.2/108.1, 324.2/262.1, 324.2/282.1, 326.2/148.1, 326.2/294, 305/169.1, 305/153.2, 314.1/267, 314.1/206, 322/125, 322/290, 368.1/145, 368.1/205, 398/158.1, 398/314	20
3	16.4	334.3/147, 334.3/117.1, 528/150, 528/218, 409.3/186.1, 409.3/206.2, 296.1/240.1, 296.1/222.1, 350/198, 350/97, 308/197; 308/272	25

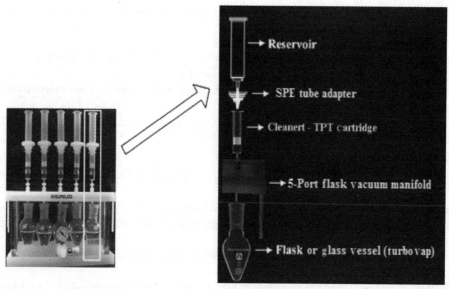

FIGURE 7.2.3 Solid phase extraction equipment.

7.2.5.4 Preparation of Standard Solutions

7.2.5.4.1 Preparation of Stock Solutions

Accurately weigh 5–10 mg of individual pesticide and chemical pollutants standards (accurate to 0.1 mg) into a 10 mL volumetric flask. Dissolve and dilute to volume with methanol, toluene, acetone, acetonitrile, isooctane, and so on, depending on each individual compound's solubility. All standard stock solutions are stored in the dark at 0–4°C and can be used for 1 year.

7.2.5.4.2 Preparation of Mixed Standard Solution

Depending on properties and retention times of compounds, all compounds are divided into a series of groups. The concentration of each compound is determined by its sensitivity on the instrument for analysis. Mixed standard solutions are stored in the dark below 4°C.

7.2.5.4.3 Working Standard Mixed Solution in a Matrix

Working standard mixture solution in a matrix of pesticide and chemical pollutants is prepared by diluting an appropriate amount of mixed standard solution with blank extract, which has been taken through the method with the rest of the samples. Mix thoroughly. They are used for plotting the standard curve.

Working standard mixture solution in a matrix must be prepared fresh.

7.2.5.5 Extraction and Cleanup Procedure

7.2.5.5.1 Sample Extraction

1. Weigh 5 g dry tea powder (accurate to 0.01 g) into an 80 mL centrifuge tube.
2. Add 15 mL acetonitrile.
3. Homogenize at 13,500 r/min for 1 min.
4. Centrifuge at 2879×g for 5 min at room temperature.
5. Transfer the supernatant into a pear-shaped flask.
6. Re-extract the sample with 15 mL acetonitrile, homogenize, centrifuge, and combine the supernatants from the two extractions.

7. Concentrate the extract to approximately 1 mL in a rotary evaporator (or TurboVap) in a 40°C water bath.
8. Place a pear-shaped flask in the vacuum manifold.
9. Mount a Cleanert-TPT cartridge onto the manifold.
10. Add anhydrous sodium sulfate (approximately 2 cm) onto the Cleanert-TPT packing material.
11. Add 10 mL acetonitrile–toluene (3:1, v/v) to activate the cartridge.
12. Stop the flow through the cartridge when the liquid level in the cartridge barrel has just reached the top of the sodium sulfate packing.
13. Discard the waste solution collected in the pear-shaped flask and replace with a clean pear-shaped flask.

7.2.5.5.2 SPE Cleanup

1. Load the concentrated extract from Step 7.2.3.5.1(7) into the conditioned Cleanert-TPT cartridge collecting the eluate into the clean pear-shaped flask.
2. Rinse the pear-shaped flask that contained the concentrated extract with 3×2 mL acetonitrile–toluene (3:1, v/v).
3. Load the rinse into the cartridge when the level of the loading solution in the cartridge reaches the top of the anhydrous sodium sulfate packing.
4. Connect a 30 mL reservoir onto the upper part of the cartridge using an adapter (Fig. 7.2.3).
5. Elute the cartridge with 25 mL acetonitrile–toluene (3:1, v/v).
6. Evaporate the eluate to approx. 0.5 mL using a rotary evaporator (or TurboVap) in a 40°C water bath.
   Follow steps 7.2.3.5.2(7) to 7.2.3.5.2(9) if the samples are being prepared for GC–MS and/or GC–MS/MS analysis.
7. Add 40 μL heptachlor epoxide (internal standard (ISTD)) working standard solution to the sample in step 7.2.3.5.2(6).(f).
8. Evaporate to dryness under a stream of nitrogen in a 35°C water bath (or Turbo Vap).
9. Dissolve the dried residue in 1.5 mL of hexane, ultrasonicate the samples to mix, filter through a 0.2 μm membrane filter. The sample is ready for GC–MS or GC–MS/MS analysis.
   Follow steps 7.2.3.5.2(10) to 7.2.3.5.2(12) if the sample is being prepared for LC–MS/MS analysis.
10. Add 40 μL chlorpyrifos methyl (ISTD) working standard solution to the sample prepared in step 7.2.3.5.2(6).
11. Evaporate to dryness under a stream of nitrogen in a 35°C water bath (or Turbo Vap).
12. Dissolve the dried residue in 1.5 mL of acetonitrile–water (3:2, v/v), ultrasonicate the samples to mix, and filter through a 0.2 μm membrane filter. The sample is then ready for LC–MS/MS analysis.

7.2.5.6 Qualitative and Quantitative

7.2.5.6.1 Criteria for Qualitative Identification and Confirmation

Measure the retention time of the monitored peaks and match them with the same peaks on the pesticide standard chromatograms. Measure the ion abundances for the qualifier ions for the detected pesticides and verify that they are within the expected limits; if the peaks match, the presence of the pesticide is confirmed. See Table 7.2.12.

TABLE 7.2.12 Recommended Maximum Permitted Tolerances for Relative Ion Intensities Using A Range of Spectrometric Techniques

Relative intensity (percentage of base peak)	GC–MS (relative, %)	GC–MS–MS, LC–MS/MS (relative, %)
>50	±10	±20
>20–50	±15	±25
>10–20	±20	±30
≤10	±50	±50

7.2.5.6.2 *Quantitative Calculations*

Use instrument data processing software for GC–MS (SIM), GC–MS/MS, and/or LC–MS/MS to calculate a response ratio (measured adundance of pesticide/measured abundance of heptachlor epoxide for GC and chlorpyrifos methyl for LC) and construct a 5-point matrix-matched calibration curve of concentration of pesticide in standard solution versus response ratio. Using the regression data from the appropriate matrix-matched calibration curve. Calculate the concentration of each pesticide found in the samples.

If a validated computer system is not being used for calculations, follow the next steps:

1. Measure the peak area of each respective standard level for each pesticide and the peak area of corresponding internal standard.
2. Calculate the ratio of the analyte response to that of the internal standard.
3. Run a linear regression analysis using the ratio of each pesticide at five different levels with no weighting or $1/x$ weighting, where x = concentration.
4. Measure the peak area of each pesticide found in the sample and the peak area of corresponding internal standard.
5. Calculate the amount of each pesticide in the solution injected from the standard curve.
6. Calculate the amount of each pesticide present in the sample.

Test results should be retained two decimal places or four significant digits.

7.2.6 RESULTS AND DISCUSSION

7.2.6.1 Evaluation of Collaborative Study Results

In 2010, the multiresidue determination method for 653 pesticides in tea was selected to be one of the priority study projects of AOAC. After 3 years preparation, 30 laboratories from 11 countries and regions participated in this collaborative study from March 1 through June 30, 2013. A total of 560 samples have been analyzed, including three categories of samples, such as fortification samples from green tea and oolong tea, aged samples from oolong tea, and incurred samples from green tea, using the three methods of GC–MS, GC–MS/MS, and LC–MS/MS. Thirty laboratories have submitted to SD clusters of data totaling 82,459 (Table 7.2.13). Table 7.2.13 includes 6868 test results of first-class target pesticides, which are used directly for analysis of method efficiency; 41,238 second-class monitoring ion data and 23,205 third-class ion abundance ratio data, two classes of which are used for qualification of target pesticides to judge if target pesticides are identified correctly; the fourth-class 2233 calibration curve data are used for quantification of target pesticides to judge if target pesticides are quantified correctly; the fifth-class 8915 prestudy results are used to judge if collaborators have mastered well the collaborative study method, procedures, and key control points before they undertake the official collaborative samples. In a word, the method efficiency can be evaluated comprehensively with these five classes of data, and errors can be analyzed and traced. In addition, it includes a 5000-page report of raw data and chromatographs.

First of all, the raw data and the information from 30 laboratories submitted was checked item by item, and discussion and verification was conducted for certain problems through emails and phone calls with collaborators. Then Grubbs and Dixon were adopted to test the outlier data from different laboratories, with the most outlier data from Lab 20 reaching 44, accounting for 19.1% of total effective data (230).

In terms of GC–MS/MS, there are 16 out of 44 outlier data from the method, with No. 1 green tea fortification samples making up 13, and the other 3 are respectively distributed in Nos. 4 and 5 green tea-incurred samples and No. 7 oolong tea fortification samples. Review of the data from this laboratory found in terms of GC MS/MS outlier data (1) Nos. 1 and 2 are parallel samples fortified with 20 pesticides, but the average values for No. 1 test results fail to reach 70% of the average values for No. 2, which led to the occurrence of 13 outlier data in the 20 pesticide test results for No. 1. (2) Nos. 9 and 10 are oolong tea-aged parallel samples containing 20 pesticides, but 20 pesticides in No. 9 samples have all been detected, while none of the 20 pesticides in No. 10 samples has been detected. Regarding GC–MS/MS calibration curve, review of the green tea 20 pesticides matrix-matching calibration curve found that the calibration curve R^2 for 12 pesticides of them is less than 0.995, accounting for 60%. The R^2s for the three pesticides op-DDE, pp-DDE, and bifenthrin are all less than 0.990. Examination of the raw data reports found that there are tremendous differences at the minimum concentration points Exp. Conc. and Calc. Conc., for instance, op-DDE's Exp. Conc. is 160, while its Calc. Conc. is only 84.8, with a deviation rate of 47%; pp-DDE's Exp. Conc. is 160, while its Calc. Conc. is only 89.9, with a deviation rate of 43.8%; biferthrin's Exp. Conc. is 40, while its Calc. Conc. is only 25.5, with a deviation rate of 36.3%. Review of oolong tea matrix-matching calibration curve has found that there are also relatively big differences at the minimum low concen-

TABLE 7.2.13 The Number of Test Data of AOAC Collaborative Study Provided by 30 Labs From 11 Countries and Regions

Lab	Results of target pesticides			Monitoring ions			Ion abundance			Calibration curve			Results of pre-collaborative study		
	GC-MS	GC-MS/MS	LC-MS/MS	GC-MS	GC-MS/MS	LC-MS/MS	GC-MS	GC-MS/MS	LC-MS/MS	GC-MS	GC-MS/MS	LC-MS/MS	GC-MS	GC-MS/MS	LC-MS/MS
1	124	124	126	972	648	652	648	324	326	40	40	40	200	200	200
2	124	–	–	972	–	–	648	–	–	40	–	–	200	–	–
3	124	–	126	972	–	652	648	–	326	40	–	40	200	–	200
4	124	–	126	972	–	652	648	–	326	40	–	40	100	–	100
5	–	124	126	–	648	652	–	324	326	–	40	40	–	200	200
6	124	–	126	972	–	652	648	–	326	40	–	40	200	–	100
7	–	–	126	–	–	652	–	–	326	–	–	40	–	–	200
8	124	–	126	972	–	652	648	–	326	40	–	40	200	–	100
9	124	–	126	972	–	652	648	–	326	40	–	40	200	–	200
10	124	124	126	972	648	652	648	324	326	40	40	40	200	200	200
11	124	–	126	972	–	652	648	–	326	40	–	40	100	–	–
12	124	–	126	972	–	652	648	–	326	40	–	40	100	–	100
13	–	124	126	–	648	652	–	324	326	–	40	40	40	200	200
14	124	–	–	972	–	–	648	–	–	40	–	–	–	–	–
15	124	–	–	972	–	–	648	–	–	40	–	–	200	–	–
16	–	124	126	–	648	652	–	324	326	–	40	40	–	200	200
17	–	–	126	–	–	652	–	–	326	–	–	40	–	–	100
18	–	124	126	–	648	652	–	324	326	–	40	40	200	–	200
19	124	–	–	972	–	–	648	–	–	40	–	–	200	–	–
20	–	104	126	–	608	652	–	304	326	–	40	40	–	–	–
21	–	104	67	–	608	534	–	304	267	–	40	40	–	140	140
22	124	–	126	972	–	652	648	–	326	40	–	40	200	–	200
23	–	124	126	–	648	652	–	324	326	–	40	40	–	200	200
24	–	124	124	–	648	648	–	324	324	–	40	40	–	200	200
25	123	–	126	969	–	652	646	–	326	40	–	40	–	–	100
26	–	118	–	–	616	–	–	308	–	–	38	–	–	200	–
27	–	124	126	–	648	652	–	324	326	–	40	40	–	100	100
28	118	118	120	939	616	620	626	308	310	39	38	38	200	200	95
29	–	124	126	–	648	652	–	324	326	–	40	40	–	200	200
30	–	124	126	–	648	652	–	324	326	–	40	40	–	200	200
Sum	1977	1808	3083	15,516	9576	16,146	10,344	4788	8073	639	596	998	2740	2440	3735
Total	6868	41,238	23,205				2233	8915							

tration point Exp. Conc. and Conc. for 2,4'-DDE, 4,4'-DDE, and bifenthrin; 2.4'-DDE's Exp. Conc. is 160, while its Calc. Conc. is only 97.9, with a deviation rate of 38.8%; 4, 4'-DDE's Exp. Conc. is 160, while its Calc. Conc. is only 105.2, with a deviation rate of 34.3%; bifenthrin's Exp. Conc. is 40, while its Calc. Conc. is only 22.2, with a deviation rate of 44.5%. The aforementioned analysis shows that the laboratory has relatively big deviations for GC–MS/MS determination of parallel samples, leading to outlier data and invalid data. Its matrix calibration curve linear correlation is poor, with the poor precision at the minimum concentration point also the cause of such deviations.

In terms of LC–MS/MS, there were 28 out of 44 outlier data for LC–MS/MS determination concerning Lab 20, of which green tea fortification samples Nos. 1 and 2 made up 7, oolong tea fortification samples Nos. 6 and 7 made up 15, and oolong tea-aged samples Nos. 9 and 10 made up 6. It can be seen that the outlier data by LC–MS/MS determination is ubiquitous in green tea fortified samples, oolong tea fortified samples, and aged samples. Review of the LC–MS/MS analytical results and raw data from the laboratory found: (1) the average values of analytical results for Nos. 1 and 2 green tea fortification samples were about 30% higher than those from other laboratories; the average values from Nos. 6 and 7 oolong tea fortification samples were about 30% lower than those from other laboratories; and the average values from Nos. 9 and 10 oolong tea-aged samples were about 45% lower than those from other laboratories. These observations reflect relatively big within-lab standard deviations with said laboratory. (2) There are also relatively big differences between Calc. Conc. and Exp. Conc. at the minimum concentration point for oolong tea matrix-matching calibration curves for the said laboratory. For instance, regarding ethoprophos, acetolachlor, and picoxystrobin, Calc. Conc. even fails to reach 50% of Exp. Conc. at the minimum concentration point, while propoxur is higher by about 50% with Exp. Conc. at the minimum concentration point Calc. Conc. Statistical analysis found that the deviation rate at the minimum concentration point exceeds 20% (Table 7.2.14).

As for the causes of the aforementioned deviations with Lab 20, the study director is of the opinion that the laboratory did not carry out the precollaborative study in accordance with the collaborative study method, not having submitted any prestudy results to the study director so far. Therefore, the procedure of the collaborative study method was not well understood and the calibration curve linear correlation established was poor, causing the laboratory to have relatively big within-lab errors and a large quantity of data to be outliers. Therefore, according to the stipulations of the collaborative study method: "For those laboratories failing to meet the acceptance criteria in practice, they, in principle, will forfeit their right to continue with the official collaborative samples, which will not be considered as collaborative laboratories accordingly, " the data from the said laboratory will be excluded from participation in the data statistical analysis.

Inspection of 6638 effective data from the remaining 29 laboratories with Grubbs and Dixon for elimination of outliers found: 1977 test data from GC–MS determination, with 65 outliers detected, making up 3.3%; 1704 test data from GC–MS/MS determination, with 65 outliers detected, accounting for 3.8%; 2957 test data from LC–MS/MS determination, with 57 outliers detected, making up 1.9%. A total of 187 outliers were derived by the three methods, GC–MS, GC–MS/MS, and LC–MS/MS, making up 2.8%. Statistical calculation was made with the software provide by AOAC, with evaluation results of the method efficiency in Tables 7.2.15–7.2.19.

7.2.6.2 The Method Efficiency of the Fortification Samples by GC–MS, GC–MS/MS, and LC–MS/MS

The method efficiency parameters, such as Rec., RSD_R, RSD_R, and HorRat values in Tables 7.2.15–7.2.17 are summarized in Table 7.2.20, per sectors.

Table 7.2.20 shows that no matter whether it is green tea or oolong tea, the average recoveries for the fortification samples by the three methods of GC–MS, GC–MS/MS, and LC–MS/MS fall within the range of 75%–100%, $RSD_r < 15\%$ accounting for 100%, $RSD_R < 25\%$ over 94%. For oolong tea samples only, RSD_R from GC–MS/MS determination of two pesticides and LC–MS/MS determination of five pesticides exceeds 25%, totaling seven, accounting for 6%. There are two points when it comes to analyzing its causes: the first one is that the fortification concentrations of oolong tea samples are equivalent to the concentration levels of MRL, and for instruments manufactured at an earlier date, its concentration level is close to the limit of detection of the instrument, which is susceptible to causing relatively big deviations. The second one is that our review of the raw data has found that owing to relatively low fortification concentrations of oolong tea, which is close to the minimum concentration point of the standard curve, its linear precision is relatively poor. For instance, in Lab 25 the minimum concentration point Calc. Conc. of matrix standard curve was obviously lower than Exp. Conc., causing the quantification results of low concentration samples to be rather small and the within-lab standard deviations to increase. The reasons for this follow in the next few paragraphs.

TABLE 7.2.14 Comparison of the Deviation Rate of Calc. Conc. and Exp. Conc. at the Minimum Concentration Point for LC-MS/MS Matrix Matching Calibration Curve for Lab 20

| No. | Pesticides | Oolong tea calibration curve | | | |
		Exp. Conc. (µg/kg)	Calc. Conc. (µg/kg)	Difference	Deviation ratio (%)
1	Imidacloprid	18	14.2	3.8	21.1
2	Propoxur	20	29.3	−9.3	−46.3
3	Monolinuron	8	8.9	−0.9	−10.9
4	Clomazon	8	5.5	2.5	31.7
5	Ethoprophos	8	3.6	4.4	55
6	Triadimefon	8	4.8	3.2	39.8
7	Acetolachlor	16	7.2	8.8	55
8	Flutolanil	8	5.9	2.1	25.9
9	Benalaxyl	8	5.7	2.3	28.6
10	Kresoxim-methyl	80	69.5	10.5	13.1
11	Picoxystrobin	8	3.3	4.7	58.8
12	Pirimiphos-methyl	8	6.4	1.6	19.8
13	Diazinon	8	5.9	2.1	25.9
14	Bensulide	24	16.3	7.7	32.1
15	Quinoxyfen	40	49	−9	−22.4
16	Tebufenpyrad	8	6.9	1.1	13.7
17	Indoxacarb	8	9	−1	−13
18	Trifloxystrobin	8	7.3	0.7	8.3
19	Chlorpyrifos	80	110	−30	−37.5
20	Butralin	8	10.2	−2.2	−26.9

For the purpose of reflecting the high sensitivity and precision of the said method, the study director designed a pair of parallel fortification samples of relatively low concentrations (sample Nos. 6 and 7) for LC–MS/MS determination of oolong tea samples, even as low as 10 µg/kg and also lower than MRL values for fortification concentrations of majority target pesticides. In the single-laboratory validation of the study director, very good recoveries and precision were obtained for fortification samples of this concentration. The minimum point concentrations of calibration curves used for quantification of fortification samples are 80% of the fortification concentrations of samples (Table 7.2.21).

Unexpectedly, however, certain laboratories did not report the test results when the pesticide content in the samples was found to be lower than the minimum concentration point of calibration curves, which made the test results incomplete, as in for instance, Lab 21. The cause for such errors was that when designing the minimum concentration points of calibration curves, we should have designed a much lower minimum concentration point (e.g., being 50% of fortification concentrations) and tried to ensure that pesticide concentrations derived after samples were extracted were still above the minimum concentration points of calibration curves. In addition, as a test method, whether test results were reported or not, the LOD of the method should be adopted to judge it, and the test results should be reported if concentrations of the test results are higher than LOD. In Table 7.2.10 of the collaborative study method, LODs of 20 pesticides for LC–MS/MS determinations are listed, and it should be considered more scientific and reasonable if collaborators report their test results according to these LOD.

Concerning the aforementioned issues, we will pay more attention in our future organization or participation in collaborative studies, and we also hope that readers of this chapter give it more concern in their future organizations or when participating in collaborative studies.

TABLE 7.2.15 Method Efficiency for Determination of 20 Pesticides in Fortified Samples by GC–MS (16 Labs)

No.	Pesticides	Green tea												Oolong tea											
		No. of labs	Average C (µg/kg) No.1	No.2	Average, C (µg/kg)	Rec (%) No.1	No.2	Average, Rec (%)	S_r (µg/kg)	RSD$_r$ (%)	S_R (µg/kg)	RSD$_R$ (%)	Hor-Rat	No. of labs	Average, C (µg/kg) No.6	No.7	Average, C (µg/kg)	Rec (%) No.6	No.7	Average, Rec (%)	S_r (µg/kg)	RSD$_r$ (%)	S_R (µg/kg)	RSD$_R$ (%)	HorRat
1	Trifluralin	15	186.7	180.1	184.6	93.4	90	91.7	8.3	4.5	15.4	8.4	0.4	16	85.6	85.8	85.7	85.6	85.8	85.7	2.4	2.8	19	22.2	1
2	Tefluthrin	16	93.4	92	92.7	93.4	92	92.7	3.5	3.8	9.1	9.8	0.4	16	42.9	42.8	42.8	85.7	85.6	85.6	1.7	4	7.3	17	0.7
3	Pyrimethanil	16	90.8	89.9	90.3	90.8	89.9	90.3	2.5	2.7	7.9	8.8	0.4	16	41.7	42.3	42	83.4	84.5	83.9	1.7	3.9	8.7	20.7	0.8
4	Propyzamide	16	95.4	95	95.2	95.4	95	95.2	4	4.2	8.1	8.5	0.4	14	46	45.1	45.1	92	90.2	91.1	1.4	3.2	6.9	15.3	0.6
5	Pirimicarb	15	92.4	92.3	92.4	92.4	92.3	92.4	2.6	2.8	9.1	9.9	0.4	13	44.6	43.2	44.6	89.3	86.4	87.8	2.1	4.6	8.3	18.6	0.7
6	Dimethenamid	16	37.5	37.5	37.5	93.8	93.6	93.7	1.2	3.1	3	7.9	0.3	16	17.6	17	17.3	88.1	85	86.6	1.4	7.8	3.6	20.9	0.7
7	Fenchlorphos	16	175.7	175	175.3	87.8	87.5	87.7	5.7	3.2	13.4	7.6	0.4	16	82.1	79.8	81	82.1	79.8	81	5	6.2	14.4	17.8	0.8
8	Tolclofos-methyl	14	92.5	92.6	92.7	92.5	92.6	92.5	2.5	2.7	6.1	6.5	0.3	14	44.7	42.7	44.1	89.5	85.4	87.4	2.8	6.5	5.5	12.5	0.5
9	Pirimiphos-methyl	16	94	93.1	93.5	94	93.1	93.5	2.9	3.1	6.4	6.9	0.3	16	42.8	41.7	42.2	85.5	83.4	84.5	2.8	6.7	7.6	18	0.7
10	2,4-DDE	16	372.1	371.2	371.7	93	92.8	92.9	10.4	2.8	34.3	9.2	0.5	16	174	171.9	172.9	87	85.9	86.5	5.2	3	28.9	16.7	0.8
11	Bromophos-ethyl	16	92.8	92.3	92.6	92.8	92.3	92.6	2.3	2.5	7.2	7.8	0.3	16	43.7	43.7	43.7	87.4	87.4	87.4	1.7	3.9	9.3	21.3	0.8
12	4,4-DDE	16	373	373	373	93.3	93.3	93.3	10.9	2.9	31.8	8.5	0.5	16	171.7	170.8	171.3	85.9	85.4	85.6	5.8	3.4	34.1	19.9	1
13	Procymidone	15	94.3	94.3	94.1	94.3	94.3	94.3	1.9	2.1	8	8.5	0.4	16	45	43.9	44.5	90	87.8	88.9	3	6.7	8.3	18.6	0.7
14	Picoxystrobin	16	191.7	192.2	191.9	95.8	96.1	96	6.4	3.4	15.3	7.9	0.4	15	90.4	88.4	88.9	90.4	88.4	89.4	3.4	3.8	12.3	13.8	0.6
15	Quinoxyfen	15	90.8	91.8	91.3	90.8	91.8	91.3	4.4	4.9	8.4	9.2	0.4	16	43.2	43.4	43.3	86.4	86.8	86.6	2.7	6.2	10.7	24.8	1
16	Chlorfenapyr	16	764.2	754.9	759.6	95.5	94.4	94.9	21.9	2.9	61.1	8	0.5	16	333.2	338.2	335.7	83.3	84.6	83.9	22.5	6.7	84	25	1.3
17	Benalaxyl	15	93.7	95.4	94.1	93.7	95.4	94.6	4.5	4.8	8	8.5	0.4	13	42.9	44.8	43.6	85.9	83.7	87.8	2.9	6.6	5.8	13.4	0.5
18	Bifenthrin	15	94.7	95.2	95.1	94.7	95.2	94.9	2.8	2.9	8.7	9.2	0.4	15	45.7	45.3	45.5	91.3	93.6	91	3.3	7.2	8.5	18.6	0.7
19	Diflufenican	15	94.4	94.3	94.5	94.4	94.3	94.4	3.4	3.6	8.1	8.6	0.4	15	43.4	42.6	43.4	86.7	85.3	86	2.4	5.4	6.4	14.8	0.6
20	Bromopropylate	15	189.5	190.1	190.2	94.8	95.1	94.9	5.4	2.9	15.1	7.9	0.4	16	88	87.1	87.6	88	87.1	87.6	5.9	6.7	12.3	14.1	0.6

TABLE 7.2.16 Method Efficiency for Determination of 20 Pesticides in Fortified Samples by GC–MS//MS (14 Labs)

		Green tea												Oolong tea											
No.	Pesticides	No. of labs	Average, C (µg/kg) No. 1	No. 2	Average, C (µg/kg)	Rec (%) No. 1	No. 2	Averag. Rec (%)	S_r (µg/kg)	RSD$_r$ (%)	S_R (µg/kg)	RSD$_R$ (%)	HorRat	No. of labs	Average, C (µg/kg) No. 6	No. 7	Average, C (µg/kg)	Rec (%) No. 6	No. 7	Average, Rec (%)	S_r (µg/kg)	RSD$_r$ (%)	S_R (µg/kg)	RSD$_R$ (%)	HorRat
1	Trifluralin	11	183.7	182.5	183.1	91.9	91.2	91.5	7.3	4	22.4	12.2	0.6	13	87.3	87.1	87.2	87.3	87.1	87.2	1.2	1.4	14.2	16.2	0.7
2	Tefluthrin	13	92.1	92.6	92.2	92.1	92.6	92.3	3.6	3.9	10.7	11.6	0.5	13	43.8	43.8	43.8	87.5	87.6	87.6	1	2.3	7.4	16.8	0.7
3	Pyrimethanil	13	89.3	90.1	90.1	89.3	90.1	89.7	3.6	4	10.6	11.8	0.5	12	41	41.3	43.4	81.9	82.6	82.3	1	2.4	5.8	13.3	0.5
4	Propyzamide	13	93.9	95.2	94.6	93.9	95.2	94.6	3.3	3.5	9.5	10.1	0.4	14	41.9	41.6	41.8	83.8	83.3	83.5	1.1	2.7	9.9	23.8	0.9
5	Pirimicarb	12	91.7	93	92.5	91.7	93	92.4	2.9	3.2	8.6	9.3	0.4	12	42.9	43.1	43	85.8	86.2	86	1.1	2.7	4	9.4	0.4
6	Dimethenamid	13	37.4	37.6	37.5	93.4	94.1	93.7	1.4	3.7	4.5	12	0.5	12	17	17.5	17.5	84.8	87.7	86.3	0.6	3.5	2.3	13.4	0.5
7	Fenchlorphos	12	175.4	172.8	172.6	87.7	86.4	87	7.2	4.2	15.4	8.9	0.4	14	77	77.1	77.1	77	77.1	77.1	1.6	2.1	18.2	23.6	1
8	Tolclofos-methyl	13	92	93.3	92.6	92	93.3	92.6	3.2	3.4	8.1	8.8	0.4	14	40.4	40.4	40.4	80.7	80.8	80.8	1	2.6	9.5	23.6	0.9
9	Pirimiphos-methyl	12	91.9	93	92.5	91.9	93	92.4	3.3	3.6	6.4	6.9	0.3	12	41.8	43.6	43.2	83.6	87.1	85.4	1	2.4	6	14	0.5
10	2,4'-DDE	12	358	364.6	361	89.5	91.2	90.3	13.6	3.8	36.2	10	0.5	11	171.3	172.1	171.7	85.7	86.1	85.9	2.9	1.7	19.6	11.4	0.5
11	Bromophos-ethyl	12	92.7	92.3	92	92.7	92.3	92.5	2.8	3.1	7.9	8.6	0.4	12	44.3	43.8	44.6	88.7	87.5	88.1	0.9	2	6.5	14.6	0.6
12	4,4'-DDE	11	357.7	364.6	360.6	89.4	91.2	90.3	15	4.2	23.7	6.6	0.4	12	174.9	176.8	175.9	87.5	88.4	87.9	3.5	2	20	11.4	0.5
13	Procymidone	13	91.8	93.1	92.5	91.8	93.1	92.4	3.5	3.8	12.1	13.1	0.6	14	39.3	40.2	39.7	78.6	80.4	79.5	1.4	3.6	13	32.7	1.3
14	Picoxystrobin	13	183.9	187.5	185.7	91.9	93.7	92.8	7.9	4.3	20.9	11.3	0.5	12	86.7	90.6	90	86.7	90.6	88.6	2.1	2.3	13	14.5	0.6
15	Quinoxyfen	13	90.3	92.9	91.6	90.3	92.9	91.6	4.3	4.7	11.9	13	0.6	12	40.6	42.3	42.1	81.2	84.5	82.9	1.2	2.8	6.2	14.7	0.6
16	Chlorfenapyr	12	744.9	755.1	749.1	93.1	94.4	93.8	36.4	4.9	67.4	9	0.5	10	334.1	337.4	335.8	83.5	84.3	83.9	10.3	3.1	23.5	7	0.4
17	Benalaxyl	13	90.8	92.7	91.7	90.8	92.7	91.8	4.7	5.1	12.2	13.3	0.6	12	44.5	43.8	44.8	89.1	87.6	88.3	2.4	5.4	6.5	14.6	0.6
18	Bifenthrin	13	96.9	97.3	97.2	96.9	97.3	97.1	5.5	5.6	14.4	14.8	0.7	11	45	45.8	45.4	90	91.6	90.8	2.2	4.9	5.1	11.1	0.4
19	Diflufenican	12	93.5	97	94.5	93.5	97	95.3	5	5.3	10.5	11.2	0.5	14	41.4	41.6	41.5	82.8	83.3	83	1.4	3.4	12	28.8	1.1
20	Bromopropylate	12	188.5	191.7	188.9	94.3	95.8	95.1	11.3	6	17.3	9.1	0.4	12	83.5	87.2	86.6	83.5	87.2	85.3	2.9	3.3	14.3	16.5	0.7

TABLE 7.2.17 Method Efficiency for Determination of 20 Pesticides in Fortified Samples by LC–MS/MS (24 Labs)

No.	Pesticides	Green tea												Oolong tea											
		No. of labs	Average, C (µg/kg) No. 1	No. 2	Average, C (µg/kg)	Rec (%) No. 1	No. 2	Average, Rec (%)	S_r (µg/kg)	RSD_r (%)	S_R (µg/kg)	RSD_R (%)	HorRat	No. of labs	Average, C (µg/kg) No. 6	No. 7	Average, C (µg/kg)	Rec (%) No. 6	No. 7	Average, Rec (%)	S_r (µg/kg)	RSD_r (%)	S_R (µg/kg)	RSD_R (%)	HorRat
1	Acetochlor	20	37.9	37.6	37.9	94.8	94	94.4	2.2	5.7	3.2	8.4	0.3	23	18.1	18.3	18.2	90.6	91.6	91.1	1.1	6.2	4	21.7	0.7
2	Benalaxyl	22	19.3	19.7	19.5	96.6	98.7	97.7	1.3	6.7	2.3	12	0.4	23	9	9.1	9	90.2	90.8	90.5	0.5	5.8	1.5	17	0.5
3	Bensulide	22	55	57.8	56.4	91.6	96.3	94	4.1	7.3	7.8	13.9	0.6	23	25.8	25.7	25.8	86	85.7	85.8	2.3	8.9	6.2	24.2	0.9
4	Butralin	21	18.3	18.2	18.2	91.3	91.2	91.3	1.4	7.4	2.3	12.8	0.4	22	8.5	8.5	8.5	84.9	84.6	84.7	0.5	6.2	2.2	26.2	0.8
5	Chlorpyrifos	19	185.2	183.7	185.3	92.6	91.8	92.2	9.6	5.2	18.7	10.1	0.5	22	83.5	86.9	84.3	83.5	86.9	85.2	5.7	6.8	25	29.7	1.3
6	Clomazone	23	18.9	18.7	18.9	94.7	93.3	94	1.3	6.7	2.1	11.2	0.4	22	8.9	9	8.9	88.6	90.4	89.5	0.3	3.8	2	22.7	0.7
7	Diazinon	22	19.4	19	19.2	96.9	95	96	0.9	4.9	2.6	13.4	0.5	20	9.2	9.2	9.2	92.3	92.2	92.2	0.3	3.6	1.3	13.6	0.4
8	Ethoprophos	23	19	18.7	18.9	95.1	93.6	94.3	1.2	6.5	2.5	13.2	0.5	22	8.7	8.8	8.7	86.8	87.8	87.3	0.5	5.5	1.9	21.5	0.7
9	Flutolanil	23	19.1	19.1	19.2	95.3	95.7	95.5	1.4	7.5	2.5	13	0.4	23	8.7	8.9	8.8	87.3	88.8	88	0.4	5.1	1.9	21.3	0.7
10	Imidacloprid	23	42.7	42.1	42.6	94.8	93.5	94.2	3.3	7.7	5.3	12.4	0.5	23	19.5	19.9	19.7	86.9	88.3	87.6	1.2	6.2	5.4	27.4	0.9
11	Indoxacarb	23	19	18.7	18.9	95	93.6	94.3	1.3	6.7	2.4	12.7	0.4	22	9.1	9.6	9.3	91.4	96	93.7	0.8	8.2	2.1	22.7	0.7
12	Kresoximmethyl	20	200.5	189.6	191.8	100.2	94.8	97.5	15.9	8.3	25.1	13.1	0.6	22	81.6	88	83.7	81.6	88	84.8	8.5	10.2	18.9	22.5	1
13	Monolinuron	23	19	18.6	18.9	95	93.1	94	1.5	7.9	2.5	13.4	0.5	22	9	9.2	9.1	90	92.4	91.2	0.6	6.2	1.8	19.2	0.6
14	Picoxystrobin	21	19.4	18.8	19	97.2	94	95.6	1.3	6.8	2.3	12.1	0.4	23	8.9	9	8.9	88.5	89.6	89.1	0.8	8.9	2.1	24.1	0.7
15	Pirimiphos-methyl	23	19.3	19.2	19.3	96.6	96	96.3	1	5.3	2.6	13.5	0.5	22	9.1	8.9	9.1	91.2	89.3	90.2	0.4	4.6	1.7	19.1	0.6
16	Propoxur	23	47.7	47.4	47.9	95.3	94.9	95.1	4.5	9.4	8.2	17.1	0.7	23	21.9	22.4	22.2	87.7	89.8	88.7	1.4	6.1	6.1	27.4	1
17	Quinoxyfen	22	92.9	92.1	92.7	92.9	92.1	92.5	4.7	5	12.8	13.8	0.6	23	41.1	41.4	41.2	82.2	82.8	82.5	2.5	6	9.5	23.1	0.9
18	Tebufenpyrad	23	18.7	18.8	18.8	93.6	94.1	93.8	1.1	6	2.5	13.5	0.5	23	8.9	8.9	8.9	89.2	89	89.1	0.6	6.4	2.4	26.8	0.8
19	Triadimefon	22	18.9	19.1	19.1	94.4	95.7	95.1	1.2	6.4	2.7	13.9	0.5	22	9.5	9.3	9.4	94.8	92.5	93.7	0.5	4.8	1.9	20.1	0.6
20	Trifloxystrobin	21	19.3	19.1	19.3	96.4	95.4	95.9	1.2	6.1	2	10.5	0.4	23	8.9	8.9	8.9	89.2	89.4	89.3	0.4	4.9	2.1	23.9	0.7

TABLE 7.2.18 Method Efficiency for Determination of 20 Pesticides in Aged Oolong tea by GC–MS, GC–MS/MS and LC–MS/MS

GC–MS (16 labs)

No.	Pesticides	No. of labs	Average, C (µg/kg)	S_r (µg/kg)	RSD$_r$ (%)	S_R (µg/kg)	RSD$_R$ (%)	HorRat
1	Trifluralin	16	359.6	14.2	3.9	60.1	16.7	0.9
2	Tefluthrin	14	169.3	5.1	3	15.1	8.9	0.4
3	Pyrimethanil	15	173.2	4.2	2.4	25.2	14.6	0.7
4	Propyzamide	15	209.9	9.1	4.3	31.1	14.8	0.7
5	Pirimicarb	14	189	6	3.2	20.6	10.9	0.5
6	Dimethenamid	15	77.6	2.2	2.8	11.5	14.8	0.6
7	Fenchlorphos	16	320.6	11	3.4	54	16.9	0.9
8	Tolclofos-methyl	13	178.8	3.6	2	18.2	10.2	0.5
9	Pirimiphos-methyl	15	170.1	4.2	2.5	24.9	14.7	0.7
10	2,4'-DDE	14	700	25.2	3.6	70.7	10.1	0.6
11	Bromophos-ethyl	16	175.1	5	2.9	22.3	12.8	0.6
12	4,4'-DDE	15	692.4	24	3.5	85.5	12.4	0.7
13	Procymidone	15	204.1	6.1	3	24.6	12	0.6
14	Picoxystrobin	15	416.2	18.3	4.4	42.1	10.1	0.6
15	Quinoxyfen	15	193.6	10.1	5.2	22.4	11.6	0.6
16	Chlorfenapyr	15	1642.6	58.5	3.6	199.4	12.1	0.8
17	Benalaxyl	15	208	11.8	5.6	30.9	14.9	0.7
18	Bifenthrin	15	199.2	11.6	5.8	24.9	12.5	0.6
19	Diflufenican	15	209.4	9.5	4.5	28.1	13.4	0.7
20	Bromopropylate	15	409.7	21	5.1	59.9	14.6	0.8

GC–MS/MS (14 labs)

No.	Pesticides	No. of labs	Average, C (µg/kg)	S_r (µg/kg)	RSD$_r$ (%)	S_R (µg/kg)	RSD$_R$ (%)	HorRat
1	Trifluralin	14	325.8	19.1	5.9	113	34.7	1.8
2	Tefluthrin	14	154.5	8.9	5.8	48.7	31.5	1.5
3	Pyrimethanil	14	167	9.2	5.5	49.2	29.4	1.4
4	Propyzamide	13	192	10.9	5.7	42.5	22.1	1.1
5	Pirimicarb	13	179.9	9.3	5.2	46.4	25.8	1.2
6	Dimethenamid	14	72.4	3.9	5.4	20.4	28.1	1.2
7	Fenchlorphos	14	298.8	19	6.4	101.9	34.1	1.8
8	Tolclofos-methyl	13	169.7	8.6	5.1	45.5	26.8	1.3
9	Pirimiphos-methyl	14	163.1	15.6	9.6	53.4	32.7	1.6
10	2,4'-DDE	13	631.5	33.3	5.3	196	31	1.8
11	Bromophos-ethyl	14	155.6	9	5.8	47	30.2	1.4
12	4,4'-DDE	14	630.1	30.8	4.9	183.4	29.1	1.7
13	Procymidone	14	186.1	9.8	5.3	54.2	29.2	1.4
14	Picoxystrobin	14	373.6	19.4	5.2	85.3	22.8	1.2
15	Quinoxyfen	14	182.4	9.2	5	54.1	29.6	1.4
16	Chlorfenapyr	13	1511.9	84.7	5.6	332.7	22	1.5
17	Benalaxyl	14	194.1	8.9	4.6	48.1	24.8	1.2
18	Bifenthrin	14	183.8	9.3	5	51	27.7	1.3
19	Diflufenican	13	191.3	11.6	6.1	41.6	21.7	1.1
20	Bromopropylate	14	387.2	22.1	5.7	84.9	21.9	1.2

LC–MS/MS (24 labs)

No.	Pesticides	No. of labs	Average, C (µg/kg)	S_r (µg/kg)	RSD$_r$ (%)	S_R (µg/kg)	RSD$_R$ (%)	HorRat
1	Acetochlor	23	77.5	3.9	5	19.9	25.6	1.1
2	Benalaxyl	22	44.5	2.3	5.1	7.5	16.8	0.7
3	Bensulide	24	131.7	10.4	7.9	36.4	27.6	1.3
4	Butralin	22	34.2	2.8	8.2	9.3	27.3	1
5	Chlorpyrifos	24	341.9	24	7	103.3	30.2	1.6
6	Clomazone	22	40.2	3.3	8.1	9.1	22.7	0.9
7	Diazinon	23	34.6	1.8	5.3	9.8	28.3	1.1
8	Ethoprophos	24	37.9	3.1	8.1	13.1	34.6	1.3
9	Flutolanil	23	44.5	2.7	6	9.4	21.1	0.8
10	Imidacloprid	24	94.6	6	6.3	30.6	32.3	1.4
11	Indoxacarb	23	44	3.5	8	8.2	18.6	0.7
12	Kresoxim-methyl	22	441.6	22.7	5.1	84.4	19.1	1.1
13	Monolinuron	24	40.7	3.7	9.1	12.6	31	1.2
14	Picoxystrobin	22	44.3	3.1	7	9	20.2	0.8
15	Pirimiphos-methyl	23	36.2	2.5	6.8	8.9	24.5	0.9
16	Propoxur	24	101.2	6.7	6.6	31.5	31.2	1.4
17	Quinoxyfen	24	192.9	13	6.7	54.8	28.4	1.4
18	Tebufenpyrad	24	41	2.6	6.3	11.4	27.7	1.1
19	Triadimefon	23	42.1	2.5	5.8	9.5	22.6	0.9
20	Trifloxystrobin	24	40.9	2.8	6.9	11.5	28	1.1

TABLE 7.2.19 Method Efficiency for Determination of 20 Pesticides in Incurred Green Tea by GC–MS, GC–MS/MS and LC–MS/MS

No.	Pesticides	No. of labs	Average C (µg/kg)	S_r (µg/kg)	RSD_r (%)	S_R (µg/kg)	RSD_R (%)	HorRat
GC–MS (16 Labs)								
1	Pyrimethanil	13	613.3	20.2	3.3	74.5	12.2	0.7
2	Bifenthrin	15	77.8	3.1	4	17.9	23	1
GC–MS/MS (14 labs)								
1	Pyrimethanil	12	575.4	32.2	5.6	80.9	14.1	0.8
2	Bifenthrin	14	78.6	4.4	5.6	17.5	22.2	0.9
LC–MS/MS (24 labs)								
1	Acetochlor	22	14.1	1.4	9.8	3.3	23.6	0.8
2	Triadimefon	24	41.3	4.4	10.6	8.9	21.6	0.8
3	Trifloxystrobin	24	90.7	7.6	8.4	19.4	21.3	0.9

TABLE 7.2.20 Distribution Range of Rec, RSD_r, RSD_R and HorRat Values for Fortified Samples

Parameters of method efficacy	Rec (%)			RSD_r (%)			RSD_R (%)			HorRat			
Range	<75	75 ~ 100	>100	<8	8 ~ 15	>15	<16	16 ~ 25	>25	<0.50	0.5 ~ 1	1.01 ~ 2	>2
GC–MS (16 labs)													
Green tea	0	40 (100)	0	20 (100)	0	0	20 (100)	0	0	17 (85)	3 (15)	0	0
Oolong tea	0	40 (100)	0	20 (100)	0	0	6 (30)	14 (70)	0	0	19 (95)	1 (5)	0
GC–MS/MS (14 labs)													
Green tea	0	40 (100)	0	20 (100)	0	0	20 (100)	0	0	8 (40)	12 (60)	0	0
Oolong tea	0	40 (100)	0	20 (100)	0	0	12 (60)	6 (30)	2 (10)	3 (15)	15 (15)	2 (10)	0
LC–MS/MS (24 Labs)													
Green tea	0	40 (100)	0	18 (90)	2 (10)	0	19 (95)	1 (5)	0	8 (40)	12 (60)	0	0
Oolong tea	0	40 (100)	0	16 (80)	4 (20)	0	1 (5)	14 (70)	5 (25)	1 (5)	18 (90)	1 (5)	0
Total	0	240 (100)	0	114 (95)	6 (5)	0	78 (65)	35 (29)	7 (6)	37 (31)	79 (66)	4 (3)	0

7.2.6.3 The Method Efficiency of the Aged Samples by GC–MS, GC–MS/MS, and LC–MS/MS

The method efficiency parameters, such as Rec., RSD_R, RSD_R, and HorRat values in Table 7.18 are summarized in Table 7.2.22, per sectors.

For 16 collaborators using GC–MS, the statistical results of oolong tea-aged samples in Table 7.2.22 show the following: (1) Within-lab repeatability: RSD_r < 8% accounting for 100%. (2) Between-lab reproducibility: RSD_R < 16% accounting for 90%, RSD_R < 25% accounting for 100%. (3) HorRat values: less than 1 account for 100%.

Concerning GC–MS/MS used by 14 collaborators: (1) Within-lab repeatability: RSD_r < 8% accounting for 95%, RSD_r < 15% accounting for 100%. (2) Between-lab reproducibility: RSD_R < 25% reach 6, making up 30%, RSD_R > 25% reach 14, making up 70%, which shows that there are relatively larger deviations for the analytical results from relatively more pesticides. (3) HorRat values: all the pesticides are less than 2.

TABLE 7.2.21 The MRL, Fortified Concentration, LOD and the Minimum Point Concentrations of Calibration Curves for Fortified Samples by LC/MS-MS

No.	Pesticides	MRL (µg/kg)	Fortified concentration (µg/kg)		LOD/µg/kg	The minimum point concentrations of calibration curves/µg/kg
			Oolong tea	Green tea		
1	Acetochlor	10	20	40	16	23.7
2	Benalaxyl	100	10	20	8	0.6
3	Bensulide	30	30	60	24	17.1
4	Butralin	20	10	20	8	1
5	Chlorpyrifos	100	100	200	80	26.9
6	Clomazone	20	10	20	8	0.2
7	Diazinon	20	10	20	8	0.4
8	Ethoprophos	20	10	20	8	1.4
9	Flutolanil	50	10	20	8	0.6
10	Imidacloprid	50	22.5	45	18	11
11	Indoxacarb	50	10	20	8	3.8
12	Kresoxim-methyl	100	100	200	80	50.3
13	Monolinuron	50	10	20	8	1.8
14	Picoxystrobin	100	10	20	8	4.2
15	Pirimiphos-methyl	50	10	20	8	0.1
16	Propoxur	100	25	50	20	12.2
17	Quinoxyfen	50	50	100	40	76.7
18	Tebufenpyrad	100	10	20	8	0.1
19	Triadimefon	200	10	20	8	3.9
20	Trifloxystrobin	50	10	20	8	1

TABLE 7.2.22 Distribution Range of RSD_r, RSD_R and HorRat Values for Aged Samples

Parameters of method efficacy	RSD_r (%)			RSD_R (%)			HorRat			
Range	<8	8 ~ 15	>15	<16	16 ~ 25	>25	<0.50	0.50 ~ 1	1.01~2	>2
GC–MS (16 labs)										
Oolong tea	20 (100)	0	0	18 (90)	2 (10)	0	1 (5)	19 (95)	0	0
GC–MS/MS (14 labs)										
Oolong tea	19 (95)	1 (5)	0	0	6 (30)	14 (70)	0	0	20 (100)	0
LC–MS/MS (24 labs)										
Oolong tea	15 (75)	5 (25)	0	0	8 (40)	12 (60)	0	8 (40)	12 (60)	0
Total	54 (90)	6 (10)	0	17 (28)	17 (28)	26 (43)	1 (2)	27 (45)	32 (53)	0

Concerning 24 collaborators using LC–MS/MS: (1) Within-lab repeatability: RSD_r < 8% accounting for 75%, RSD_r < 15% accounting for 100%. (2) Between-lab reproducibility: RSD_R < 5% making up 40%. There are 12 pesticides with its between-lab reproducibility RSD_R > 25%, making up 60%. (3) HorRat values < 1 account for 40%, HorRat values < 2 account for 100%.

The cause of oolong tea-aged samples GC–MS/MS and LC–MS/MS RSD_R > 25% exceeding the good scopes is that statistical analysis has found that there is relatively more GC–MS/MS and LC–MS/MS RSD_R > 25% for aged samples. There are 14 with GC–MS/MS $RSD_R > 25\%$, accounting for 70%, 12 with LC–MS/MS $RSD_R > 25\%$, making up 60% (Table 7.2.22). The reasons are: aged samples are prepared by spraying pesticides onto dry tea powders in advance, which is then mixed uniformly. In a certain period after sample preparation, pesticides in tea are always in the process of slow degradation during storage and transit, so in our single-laboratory validation at an earlier stage, a two-time study was conducted on the pesticide degradation kinetics in dry tea ahead of time. The conclusion derived is: within these 3 months pesticides in tea will degrade slowly and pesticide content will decrease, leading to the increase of RSD turbulence in the range of approximately 4.3%–31.1%. According to this conclusion, taking GC–MS/MS as an example for this collaborative study, the determination time for collaborators' laboratories is taken as a horizontal ordinate, while GC–MS/MS test results from different collaborators' laboratories are taken as vertical ordinates to establish regression equations for change of 20 pesticides test results with time, with a tendency chart drawn (Fig. 7.2.4).

On the basis of the stability results of pesticides in aged oolong tea determined by GC–MS/MS in continuous 3 months (see Section 7.2.2.7.4) and the aforementioned tendency chart, the determination time (day) was taken as x-axis and the difference between each determined value and first time-determined value of target pesticides was taken as y-axis. The logarithmic function (Fig. 7.2.5) was obtained by fitting the 3 months determination results, and then the degradation value of the target pesticide at a specific day could be calculated (Fig. 7.2.6).

In this collaborative study, the determination time is different from lab to lab. From the start of determination by the first collaborator's laboratory (March 28) until the ending of the last collaborator (June 27), it lasts 120 days, which is a large time span. Therefore, the degradation value of each pesticide from various labs could be calculated according to the determination time by the aforementioned logarithmic function. In order to eliminate the influence from the natural degradation of pesticides, the degradation value of each pesticide from various labs could be added to the corresponding raw data and the raw data of different labs could be corrected. The raw data and corrected data of aged oolong tea by GC–MS/MS are listed in Table 7.2.23. Table 7.2.23 shows that among the parameters of method performance, such as Sr, RSDr, SR, RSD_R, and HorRat, the values of RSD_R have relatively great variation. So, only the raw data and corrected data of RSD_R are discussed and shown in Table 7.2.24. For the raw data, the range of RSD_R was 21.7%–34.7% with the average of 27.8%, and the RSD_R of six pesticides falls within the range of 16%–25%; the other 14 pesticides have the RSD_R more than 25%. For the corrected data, the range of RSD_R was 18.5%–29% with the average of 22.8%, and the RSD_R of 14 pesticides falls within the range of 16%–25%; the other 6 pesticides have the RSD_R more than 25%. Before and after correction, the dif-

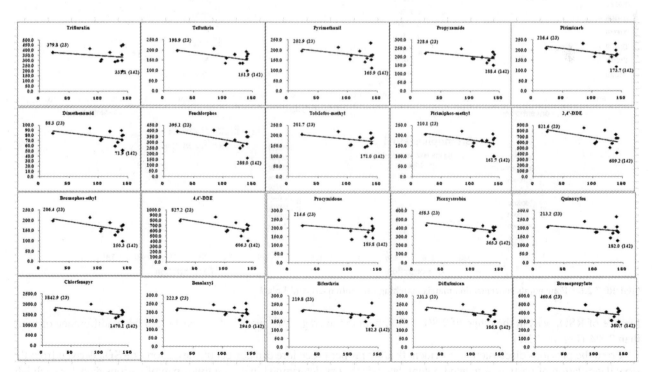

FIGURE 7.2.4 **Degradation trends of 20 pesticides in aged oolong tea from 23 to 142 days by GC–MS/MS.**

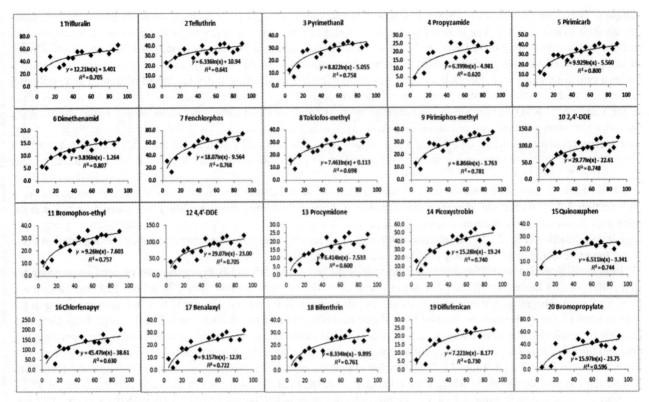

FIGURE 7.2.5 Logarithmic function of 20 pesticides in aged oolong tea of single laboratory by GC–MS/MS.

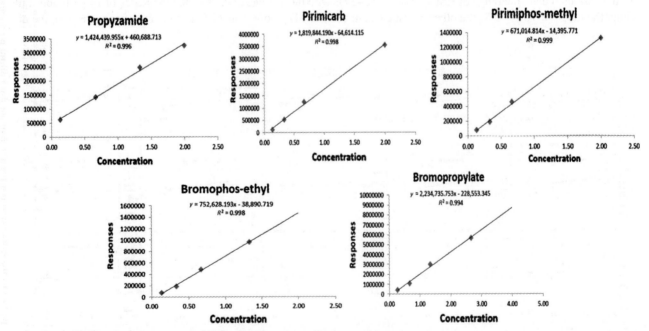

FIGURE 7.2.6 Four points concentration matrix matching calibration curve of Lab. 25.

ference of RSD_R was in the range of 2.8%–7.9%, with the average 5%, and all of the pesticides have a difference of more than 2.8% (Figure 7.2.7).

From the previous discussion, it is evident that the proper procedure can correct the deviation of RSD_R derived from the slow degradation of pesticides in aged oolong tea, which was determined in a long time span of various days for each lab. In short, this procedure could reflect the value of RSD_R objectively.

TABLE 7.2.23 Comparison of Method Efficiency for Determination of 20 Pesticides in Aged Oolong Tea by GC–MS/MS With and Without Correction

No.	Pesticides	No. of labs	Average C_r (µg/kg)		S_r (µg/kg)		RSD_r (%)		S_R (µg/kg)		RSD_R (%)		HorRat	
			Without correction	With correction	Without correction	With correction	Without correction	With correction	Without correction	With correction	Without correction	With correction	Without correction	With correction
1	Trifluralin	14	325.8	386.3	19.1	19.1	5.9	4.9	113	112	34.7	29	1.8	1.6
2	Tefluthrin	14	154.5	195.1	8.9	8.9	5.8	4.6	48.7	47.4	31.5	24.3	1.5	1.2
3	Pyrimethanil	14	167	203.2	9.2	9.2	5.5	4.5	49.2	48.1	29.4	23.6	1.4	1.2
4	Propyzamide	13	192	216.9	10.9	10.9	5.7	5	42.5	41.3	22.1	19.1	1.1	0.9
5	Pirimicarb	13	179.9	220.9	9.3	9.3	5.2	4.2	46.4	44.5	25.8	20.2	1.2	1
6	Dimethenamid	14	72.4	89.1	3.9	3.9	5.4	4.4	20.4	19.9	28.1	22.3	1.2	1
7	Fenchlorphos	14	298.8	373.8	19	19	6.4	5.1	101.9	97.8	34.1	26.2	1.8	1.4
8	Tolclofos-methyl	13	169.7	204.6	8.6	8.6	5.1	4.2	45.5	44.4	26.8	21.7	1.3	1.1
9	Pirimiphos-methyl	14	163.1	200.8	15.6	15.6	9.6	7.8	53.4	51.8	32.7	25.8	1.6	1.3
10	2,4'-DDE	13	631.5	747.5	33.3	33.3	5.3	4.5	196	189.2	31	25.3	1.8	1.5
11	Bromophos-ethyl	14	155.6	191.3	9	9	5.8	4.7	47	44.9	30.2	23.4	1.4	1.1
12	4,4'-DDE	14	630.1	743.1	30.8	30.8	4.9	4.1	183.4	176.3	29.1	23.7	1.7	1.4
13	Procymidone	14	186.1	208.5	9.8	9.8	5.3	4.7	54.2	53.6	29.2	25.7	1.4	1.3
14	Picoxystrobin	14	373.6	425.9	19.4	19.4	5.2	4.6	85.3	82.4	22.8	19.3	1.2	1.1
15	Quinoxyfen	14	182.4	209.6	9.2	9.2	5	4.4	54.1	53.5	29.6	25.5	1.4	1.3
16	Chlorfenapyr	13	1511.9	1685.2	84.7	84.7	5.6	5	332.7	323.4	22	19.2	1.5	1.3
17	Benalaxyl	14	194.1	224	8.9	8.9	4.6	4	48.1	47.4	24.8	21.2	1.2	1.1
18	Bifenthrin	14	183.8	212.9	9.3	9.3	5	4.4	51	49.9	27.7	23.5	1.3	1.2
19	Diflufenican	13	191.3	216.7	11.6	11.6	6.1	5.4	41.6	40.1	21.7	18.5	1.1	0.9
20	Bromopropylate	14	387.2	438.2	22.1	22.1	5.7	5	84.9	82.4	21.9	18.8	1.2	1

TABLE 7.2.24 Raw Data and Corrected Data of RSD$_R$ for Aged Oolong Tea by GC–MS/MS

No.	Pesticides	Before correction (%)	After correction (%)	Difference (%)	Deviation rate (%)
1	Trifluralin	34.7	29	5.7	16.4
2	Tefluthrin	31.5	24.3	7.2	22.8
3	Pyrimethanil	29.4	23.6	5.8	19.8
4	Propyzamide	22.1	19.1	3	13.6
5	Pirimicarb	25.8	20.2	5.6	21.7
6	Dimethenamid	28.1	22.3	5.8	20.7
7	Fenchlorphos	34.1	26.2	7.9	23.2
8	Tolclofos-methyl	26.8	21.7	5.1	19.1
9	Pirimiphos-methyl	32.7	25.8	6.9	21.2
10	2,4'-DDE	31	25.3	5.7	18.5
11	Bromophos-ethyl	30.2	23.4	6.8	22.5
12	4,4'-DDE	29.1	23.7	5.4	18.6
13	Procymidone	29.2	25.7	3.5	11.9
14	Picoxystrobin	22.8	19.3	3.5	15.5
15	Quinoxyfen	29.6	25.5	4.1	14
16	Chlorfenapyr	22	19.2	2.8	12.7
17	Benalaxyl	24.8	21.2	3.6	14.5
18	Bifenthrin	27.7	23.5	4.2	15.3
19	Diflufenican	21.7	18.5	3.2	14.9
20	Bromopropylate	21.9	18.8	3.1	14.3

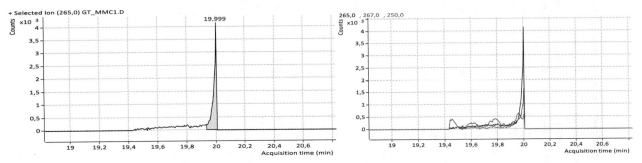

FIGURE 7.2.7 Deviations occurred in the setting of MRM-acquisition window of toclophos-mehtyl.

7.2.6.4 The Method Efficiency of the Incurred Samples by GC–MS, GC–MS/MS, and LC–MS/MS

The method efficiency parameters, such as Rec., RSD$_R$, RSD$_R$, and HorRat values in Table 7.2.19 are summarized in Table 7.2.25 per sectors.

The statistical results of green tea-incurred samples in Table 7.2.25 show the following in regard to GC–MS. (1) Within-lab repeatability: RSD$_r$ <8% accounting for 100%, which illustrates that this method's repeatability is very good in different laboratories. (2) Between-lab reproducibility: RSD$_R$ < 25% making up 100%, which demonstrates that the between-lab reproducibility is very good. (3) HorRat values < 1 accounting for 100%, which shows that the method has achieved very good reproducibility under the corresponding analytical conditions. Concerning GC–MS/MS, note the following. (1) Within-lab repeatability: RSD$_R$ < 8% accounting for 100%, which shows that the method's repeatability is very good in different laboratories. (2) Between-lab reproducibility: RSD$_R$ < 25% making up 100%, which proves that the between-lab reproducibility is very good. (3) HorRat values <1 accounting for 100%, which indicates that the method has achieved

TABLE 7.2.25 Distribution Range of RSD$_r$ RSD$_R$ and HorRat Values for Incurred Samples

Parameters of method efficacy	RSD$_r$ (%)			RSD$_R$ (%)			HorRat			
Range	<8	8 ~ 15	>15	<16	16 ~ 25	>25	<0.50	0.50 ~ 1	1.01 ~ 2	>2
GC–MS (16 labs)										
Green tea	2 (100)	0	0	1 (50)	1 (50)	0	0	2 (100)	0	0
GC–MS/MS (14 Labs)										
Green tea	2 (100)	0	0	0	2 (100)	0	0	2 (100)	0	0
LC–MS/MS (24 Labs)										
Green tea	0	3 (100)	0	0	3 (100)	0	0	3 (100)	0 (0)	0
Total	4 (57)	3 (43)	0	2 (29)	5 (71)	0	0	7 (100)	0 (0)	0

very good reproducibility under the corresponding analytical concentrations. Statistical results for LC–MS/MS include the following. (1) Within-lab repeatability: RSD$_r$ < 15% accounting for 100%, which shows that the method's repeatability is very good in different laboratories. (2) Between-lab reproducibility: RSD$_R$ < 25% making up 100%.

To summarize the aforementioned analysis, this collaborative study result shows that the method efficiency is acceptable.

7.2.7   QUALIFICATION AND QUANTIFICATION

The collaborative study protocol stipulates that the deviations of qualifying ion abundance of the target pesticides shall have to comply with EU standards. For this collaborative study, three methods of GC–MS, GC–MS/MS, and LC–MS/MS were adopted by 29 labs for determination for green tea and oolong tea pesticide samples (excluding blank samples), with 10,231 target ion abundance ratio data obtained, and ion abundance ratios are checked item by item against EU standard conformity rates, with results tabulated in Tables 7.2.26–7.2.28.

7.2.7.1   Qualification of Target Pesticides

Tables 7.2.26–7.2.28 show that concerning GC–MS, 16 labs participated in the study with a total of 4414 (2324 + 2090) target pesticide ions obtained from determination of green tea and oolong tea samples, among which the ion abundance ratios that comply with EU standards reached 4154 (2229 + 1925), accounting for 94%. For GC–MS/MS, 14 laboratories participated in the study, with a total of 2268 (1041 + 1227) target pesticide ions of green tea and oolong tea, among which the ion abundance ratios that comply with EU standards reached 2247 (1032 + 1215), accounting for 99%. For LC–MS/MS, 24 laboratories participated in the study, with a total of 3549 (1819 + 1730) target pesticide ions of green tea and oolong tea samples, among which the ion abundance ratios that comply with EU standards reached 3494 (1794 + 1700), making up 98%.

The aforementioned analysis reveals that a total of 10,231 of target pesticide ions were obtained by the 30 participating laboratories, among which the ion abundance ratios that comply with EU standards reach 9895 accounting for 97%, which proves that the said method enables an accurate qualification inspection for the absolute majority of pesticides varied different instruments by different laboratories.

7.2.7.2   Quantification of Target Pesticides

For this collaborative study, a matrix-matching internal standard calibration curve quantification method is adopted, with 2153 target pesticide calibration curves established by 29 laboratories. The conformity rates of their correlation coefficients $R^2 \geq 0.995$ are tabulated in Table 7.2.29.

Table 7.2.29 shows that concerning GC–MS: 16 laboratories established 639 matrix-matching internal standard calibration curves respectively for 20 pesticides in green tea and oolong tea, among which those with $R^2 \geq 0.995$ reach 637, accounting for 99.7%; 1 with R^2 0.990–0.995, 1 with R^2 less than 0.990. Regarding GC–MS/MS: 14 laboratories estab-

TABLE 7.2.26 Conformity Ratios of Ion Abundance of Target Pesticide With EU Standard by GC–MS-SIM for Precollaborative Study

Green tea

Lab.	No. 1 Ion 1	No. 1 Ion 2	No. 2 Ion 1	No. 2 Ion 2	No. 3 Ion 1	No. 3 Ion 2	No. 4 Ion 1	No. 4 Ion 2	No. 5 Ion 1	No. 5 Ion 2	Total
1	20 (100%)	20 (100%)	20 (100%)	20 (100%)	20 (100%)	20 (100%)	20 (100%)	20 (100%)	20 (100%)	20 (100%)	200 (100%)
2	19 (95%)	19 (95%)	19 (95%)	19 (95%)	19 (95%)	19 (95%)	19 (95%)	20 (100%)	19 (95%)	20 (100%)	192 (96%)
3	20 (100%)	20 (100%)	20 (100%)	20 (100%)	20 (100%)	20 (100%)	20 (100%)	20 (100%)	20 (100%)	20 (100%)	200 (100%)
4	18 (90%)	18 (95%)	18 (90%)	18 (95%)	18 (90%)	18 (95%)	18 (90%)	18 (95%)	18 (90%)	18 (95%)	180 (92%)
6	19 (95%)	15 (75%)	19 (95%)	15 (75%)	19 (95%)	18 (90%)	19 (95%)	17 (85%)	19 (95%)	18 (90%)	178 (89%)
8	19 (95%)	18 (90%)	19 (95%)	17 (85%)	20 (100%)	19 (95%)	18 (90%)	19 (95%)	19 (95%)	19 (95%)	187 (94%)
9	19 (100%)	19 (100%)	19 (100%)	19 (100%)	19 (100%)	19 (100%)	19 (100%)	19 (100%)	19 (100%)	19 (100%)	190 (100%)
10	19 (100%)	19 (100%)	19 (100%)	19 (100%)	19 (100%)	19 (100%)	19 (100%)	19 (100%)	19 (100%)	19 (100%)	190 (100%)
12	20 (100%)	20 (100%)	19 (95%)	20 (100%)	20 (100%)	17 (85%)	18 (90%)	19 (95%)	20 (100%)	18 (90%)	189 (95%)
14	19 (100%)	20 (100%)	18 (95%)	18 (95%)	–	–	–	–	–	–	74 (97%)
15	19 (95%)	20 (100%)	19 (95%)	20 (100%)	–	–	2 (100%)	2 (100%)	2 (100%)	2 (100%)	86 (98%)
18	–	–	–	–	–	–	–	–	–	–	–
19	–	–	–	–	–	–	–	–	–	–	–
22	20 (100%)	18 (90%)	19 (95%)	17 (85%)	18 (90%)	15 (75%)	17 (85%)	19 (95%)	17 (85%)	15 (75%)	175 (88%)
28	19 (100%)	19 (100%)	19 (100%)	19 (100%)	18 (95%)	19 (100%)	19 (100%)	18 (95%)	19 (100%)	19 (100%)	188 (99%)
Number of conformity ion	250	244	247	239	210	203	208	210	211	207	2229
Total number of Ion	256	254	256	254	217	215	219	217	219	217	2324
Total conformity ratios%	98	96	97	94.	97	94	95	97	96	95	96

Oolong tea

Lab.	No. 6 Ion 1	No. 6 Ion 2	No. 7 Ion 1	No. 7 Ion 2	No. 8 Ion 1	No. 8 Ion 2	No. 9 Ion 1	No. 9 Ion 2	No. 10 Ion 1	No. 10 Ion 2	Total
1	20 (100%)	20 (100%)	20 (100%)	20 (100%)	20 (100%)	20 (100%)	20 (100%)	20 (100%)	20 (100%)	20 (100%)	200 (100%)
2	17 (85%)	18 (90%)	19 (95%)	19 (95%)	17 (85%)	19 (95%)	16 (80%)	20 (100%)	17 (85%)	19 (95%)	182 (91%)
3	19 (95%)	20 (100%)	20 (100%)	20 (100%)	20 (100%)	20 (100%)	20 (100%)	20 (100%)	20 (100%)	20 (100%)	199 (99.5%)

Lab.	No. 6		No. 7		No. 8		No. 9		No. 10		Total
	Ion 1	Ion 2	Ion 1	Ion 2	Ion 1	Ion 2	Ion 1	Ion 2	Ion 1	Ion 2	
6	15 (75%)	14 (70%)	16 (80%)	14 (70%)	-	-	19 (95%)	19 (95%)	19 (95%)	19 (95%)	135 (84%)
8	19 (95%)	18 (90%)	19 (95%)	17 (85%)	20 (100%)	19 (95%)	18 (90%)	19 (95%)	19 (95%)	19 (95%)	187 (93%)
9	19 (100%)	19 (100%)	19 (100%)	19 (100%)	19 (100%)	19 (100%)	19 (100%)	19 (100%)	19 (100%)	19 (100%)	190 (100%)
10	18 (95%)	18 (95%)	18 (95%)	18 (95%)	18 (95%)	18 (95%)	18 (95%)	18 (95%)	18 (95%)	18 (95%)	180 (95%)
11	19 (95%)	17 (85%)	19 (95%)	18 (90%)	20 (100%)	17 (85%)	18 (90%)	19 (95%)	20 (100%)	18 (90%)	185 (93%)
15	15 (75%)	15 (75%)	16 (80%)	14 (70%)	-	-	19 (95%)	19 (95%)	19 (95%)	19 (95%)	136 (85%)
18	-	-	-	-	-	-	-	-	-	-	-
19	-	-	-	-	-	-	-	-	-	-	-
22	16 (80%)	14 (70%)	15 (75%)	14 (70%)	16 (80%)	13 (65%)	14 (70%)	16 (80%)	14 (70%)	14 (70%)	146 (73%)
28	18 (95%)	19 (100%)	18 (95%)	18 (95%)	17 (89%)	19 (100%)	19 (100%)	19 (100%)	19 (100%)	19 (100%)	185 (97%)
Number of conformity ion	195	192	199	192	167	164	200	208	204	204	1925
Total number of Ion	211	211	211	177	177	177	211	211	211	211	2090
Total conformity ratios percentage	92	91	94	91	94	93	95	99	97	97	92

-, This method was not used.

TABLE 7.2.27 Conformity Ratios of Ion Abundance of Target Pesticide With EU Standard by GC–MS/MS for Pre-Collaborative Study

Lab	Green tea						Oolong tea					
	No. 1	No. 2	No. 3	No. 4	No. 5	Total	No. 6	No. 7	No. 8	No. 9	No. 10	Total
1	20 (100%)	20 (100%)	20 (100%)	20 (100%)	20 (100%)	100 (100%)	20 (100%)	20 (100%)	20 (100%)	20 (100%)	20 (100%)	100 (100%)
5	20 (100%)	20 (100%)	20 (100%)	20 (100%)	20 (100%)	100 (100%)	20 (100%)	20 (100%)	20 (100%)	20 (100%)	20 (100%)	100 (100%)
10	20 (100%)	20 (100%)	20 (100%)	20 (100%)	20 (100%)	100 (100%)	20 (100%)	20 (100%)	20 (100%)	20 (100%)	20 (100%)	100 (100%)
13	20 (100%)	20 (100%)	20 (100%)	20 (100%)	20 (100%)	100 (100%)	20 (100%)	20 (100%)	20 (100%)	20 (100%)	20 (100%)	100 (100%)
16	20 (100%)	20 (100%)	-	20 (100%)	20 (100%)	80 (100%)	18 (90%)	18 (90%)	-	18 (90%)	18 (90%)	72 (90%)
21	20 (100%)	-	-	2 (100%)	2 (100%)	24 (100%)	20 (100%)	20 (100%)	-	20 (100%)	20 (100%)	80 (100%)
23	20 (100%)	20 (100%)	20 (100%)	20 (100%)	20 (100%)	100 (100%)	19 (95%)	20 (100%)	20 (100%)	20 (100%)	19 (95%)	98 (98%)
24	19 (95%)	19 (95%)	19 (95%)	19 (95%)	19 (95%)	95 (95%)	20 (100%)	20 (100%)	20 (100%)	19 (95%)	20 (100%)	99 (99%)
26	18 (95%)	18 (95%)	-	2 (100%)	2 (100%)	40 (95%)	17 (94%)	18 (100%)	-	18 (100%)	18 (100%)	71 (99%)
27	-	-	-	-	-	-	20 (100%)	20 (100%)	20 (100%)	20 (100%)	20 (100%)	100 (100%)
28	19 (100%)	19 (100%)	19 (100%)	19 (100%)	19 (100%)	95 (100%)	19 (100%)	19 (100%)	19 (100%)	19 (100%)	19 (100%)	95 (100%)
29	20 (100%)	20 (100%)	20 (100%)	19 (95%)	19 (95%)	98 (98%)	20 (100%)	20 (100%)	20 (100%)	20 (100%)	20 (100%)	100 (100%)
30	20 (100%)	20 (100%)	20 (100%)	20 (100%)	20 (100%)	100 (100%)	20 (100%)	20 (100%)	20 (100%)	20 (100%)	20 (100%)	100 (100%)
Number of conformity ion	236	216	178	201	201	1032	253	255	199	254	254	1215
Total number of Ion	238	218	179	203	203	1041	257	257	199	257	257	1227
Total conformity ratios percentage	99.2	99.1	99.4	99	99	99.1	98.4	99.2	100	98.8	98.8	99

-, This method was not used.

TABLE 7.2.28 Conformity Ratios of Ion Abundance of Target Pesticide With EU Standard by LC–MS/MS for Pre-Collaborative Study

Lab	Green tea						Oolong tea					
	No. 1	No. 2	No. 3	No. 4	No. 5	Total	No. 6	No. 7	No. 8	No. 9	No. 10	Total
1	20 (100%)	20 (100%)	20 (100%)	20 (100%)	20 (100%)	100 (100%)	20 (100%)	20 (100%)	20 (100%)	20 (100%)	20 (100%)	100 (100%)
3	18 (90%)	18 (90%)	20 (100%)	20 (100%)	20 (100%)	96 (96%)	20 (100%)	20 (100%)	19 (95%)	20 (100%)	20 (100%)	99 (99%)
4	–	–	–	–	–	–	19 (95%)	19 (95%)	19 (95%)	17 (85%)	19 (95%)	93 (93%)
5	20 (100%)	20 (100%)	20 (100%)	20 (100%)	20 (100%)	100 (100%)	20 (100%)	20 (100%)	19 (95%)	18 (90%)	19 (95%)	96 (96%)
6	20 (100%)	20 (100%)	19 (95%)	19 (95%)	19 (95%)	97 (97%)	–	–	–	–	–	–
7	20 (100%)	20 (100%)	20 (100%)	20 (100%)	20 (100%)	100 (100%)	20 (100%)	20 (100%)	20 (100%)	20 (100%)	20 (100%)	100 (100%)
8	20 (100%)	20 (100%)	20 (100%)	20 (100%)	20 (100%)	100 (100%)	–	–	–	–	–	–
9	20 (100%)	20 (100%)	20 (100%)	20 (100%)	20 (100%)	100 (100%)	20 (100%)	20 (100%)	20 (100%)	20 (100%)	20 (100%)	100 (100%)
10	20 (100%)	20 (100%)	20 (100%)	20 (100%)	20 (100%)	100 (100%)	19 (100%)	19 (100%)	19 (100%)	19 (100%)	18 (94.7%)	94 (98.9%)
12	20 (100%)	20 (100%)	20 (100%)	20 (100%)	19 (95%)	99 (99%)	–	–	–	–	–	–
13	20 (100%)	20 (100%)	20 (100%)	20 (100%)	20 (100%)	100 (100%)	20 (100%)	19 (95%)	20 (100%)	19 (95%)	19 (95%)	97 (97%)
16	20 (100%)	20 (100%)	20 (100%)	20 (100%)	20 (100%)	100 (100%)	20 (100%)	20 (100%)	20 (100%)	20 (100%)	20 (100%)	100 (100%)
17	–	–	–	–	–	–	18 (90%)	18 (90%)	18 (90%)	19 (95%)	19 (95%)	92 (92%)
18	20 (100%)	20 (100%)	20 (100%)	20 (100%)	20 (100%)	100 (100%)	20 (100%)	20 (100%)	20 (100%)	20 (100%)	20 (100%)	100 (100%)
21	20 (100%)	–	–	2 (100%)	2 (100%)	24 (100%)	–	20 (100%)	20 (100%)	20 (100%)	20 (100%)	40 (100%)
22	20 (100%)	20 (100%)	20 (100%)	20 (100%)	20 (100%)	100 (100%)	20 (100%)	20 (100%)	20 (100%)	20 (100%)	20 (100%)	100 (100%)
23	20 (100%)	20 (100%)	20 (100%)	20 (100%)	20 (100%)	100 (100%)	20 (100%)	20 (100%)	20 (100%)	20 (100%)	20 (100%)	100 (100%)
24	20 (100%)	20 (100%)	20 (100%)	20 (100%)	20 (100%)	100 (100%)	20 (100%)	20 (100%)	20 (100%)	20 (100%)	20 (100%)	100 (100%)
25	17 (89.5%)	17 (89.5%)	16 (84.2%)	16 (84.2%)	16 (84.2%)	82 (86.3%)	–	–	–	–	–	–
27	–	–	–	–	–	–	–	–	–	–	–	–
28	–	–	–	–	–	–	–	–	–	–	–	–
29	19 (95%)	19 (95%)	19 (95%)	20 (100%)	19 (95%)	96 (96%)	19 (95%)	18 (90%)	19 (95%)	19 (95%)	19 (100%)	95 (100%)
30	20 (100%)	20 (100%)	20 (100%)	20 (100%)	20 (100%)	100 (100%)	20 (100%)	19 (95%)	20 (100%)	20 (100%)	20 (100%)	95 (95%)
Number of conformity ion	374	354	354	357	355	1794	334	334	332	350	353	1700
Total number of Ion	379	359	359	361	361	1819	338	338	338	358	358	1730
Total conformity ratios%	98.7	98.6	98.6	98.9	98.3	98.6	99.8	98.8	98.2	97.8	98.6	98.3

–, This method was not used.

TABLE 7.2.29 Conformity Rate of Correlation Coefficient of R^2 with $R^2 \geq 0.995$ Per Calibration Curve Data

Lab	GC–MS (16 Labs)		GC–MS/MS (14 Labs)		LC–MS/MS (24 Labs)	
	Green tea	Oolong tea	Green tea	Oolong tea	Green tea	Oolong tea
	Number (%)	Number (%)	Number (%)	Number (%)	Number (%)	Number (%)
1	20 (100%)	20 (100%)	20 (100%)	20 (100%)	19 (95%)	20 (100%)
2	20 (100%)	20 (100%)	–	–	–	–
3	20 (100%)	20 (100%)	–	–	20 (100%)	20 (100%)
4	20 (100%)	20 (100%)	–	–	20 (100%)	18 (90%)
5	–	–	20 (100%)	20 (100%)	20 (100%)	20 (100%)
6	20 (100%)	20 (100%)	–	–	20 (100%)	20 (100%)
7	–	–	–	–	20 (100%)	20 (100%)
8	20 (100%)	20 (100%)	–	–	20 (100%)	20 (100%)
9	20 (100%)	20 (100%)	–	–	20 (100%)	20 (100%)
10	20 (100%)	20 (100%)	20 (100%)	19 (95%)	19 (95%)	20 (100%)
11	20 (100%)	20 (100%)	–	–	20 (100%)	20 (100%)
12	20 (100%)	20 (100%)	–	–	20 (100%)	19 (95%)
13	–	–	19 (95%)	20 (100%)	16 (80%)	16 (80%)
14	20 (100%)	20 (100%)	–	–	–	–
15	20 (100%)	20 (100%)	–	–	–	–
16	–	–	20 (100%)	20 (100%)	20 (100%)	20 (100%)
17	–	–	–	–	14 (70%)	16 (80%)
18	–	–	20 (100%)	20 (100%)	19 (95%)	20 (100%)
19	20 (100%)	20 (100%)	–	–	–	–
20	–	–	–	–	–	–
21	–	–	20 (100%)	20 (100%)	10 (50%)	14 (70%)
22	20 (100%)	19 (95%)	–	–	10 (50%)	10 (50%)
23	–	–	20 (100%)	20 (100%)	20 (100%)	20 (100%)
24	–	–	20 (100%)	20 (100%)	16 (80%)	19 (95%)
25	20 (100%)	19 (95)	–	–	20 (100%)	20 (100%)
26	–	–	19 (100%)	18 (94.7%)	–	–
27	–	–	16 (80%)	20 (100%)	20 (100%)	20 (100%)
28	19 (100%)	20 (100%)	19 (100%)	19 (100%)	19 (100%)	19 (100%)
29	–	–	20 (100%)	20 (100%)	20 (100%)	20 (100%)
30	–	–	20 (100%)	20 (100%)	20 (100%)	20 (100%)
Conformity quantity of correlation coefficients	637		549		893	
Total number of correlation coefficients	639		556		958	
Total conformity rate (%)	99.7		98.7		93.2	

–, This method was not used.

lished 556 matrix-matching internal standard calibration curves respectively for 20 pesticides in green tea and oolong tea, among which those with $R^2 \geq 0.995$ reach 549, making up 98.7%; 7 with R^2 0.990–0.995, accounting for 1.3%, and with no R^2 less than 0.990. Concerning LC–MS/MS: 24 laboratories established 958 matrix-matching internal standard calibration curves respectively for 20 pesticides in green tea and oolong tea samples, among which those with $R^2 \geq 0.995$ reach 893, accounting for 93.2%; 34 with R^2 0.990–0.995, making up 3.7%, and 11 with $R^2 < 0.990$, making up 1.2%. The afore-mentioned analysis shows that there are 2079 with $R^2 \geq 0.995$ out of the 2153 matrix-matching internal standard curves established by 29 laboratories, accounting for 96.6%. It demonstrates that the matrix-matching internal standard calibration curves adopted for the method are capable of realizing accurate quantification for the absolute majority pesticides using three different types of instruments by different laboratories.

7.2.8   ERROR ANALYSIS AND TRACEABILITY

The 6638 effective data derived in this study are inspected with Grubbs and Dixon for outliers, with 187 outliers obtained, accounting for 2.8%. The distribution of outliers derived from these three methods GC–MS, GC–MS/MS, and LC–MS/MS for different teas, different pesticide varieties, and different samples are tabulated in Tables 7.2.30–7.2.32. The outliers from 10 samples of three categories determined by three different methods of GC–MS, GC–MS/MS, and LC–MS/MS for two teas in Tables 7.2.30–7.2.32 are summarized and tabulated in Table 7.2.33. Distribution of outliers for different laboratories and different samples is listed in Table 7.2.34.

7.2.8.1   GC–MS Data Error Analysis and Traceability

Tables 7.2.30, 7.2.33 and 7.2.34 show: each of 16 laboratories analyzed 20 pesticide residues in 8 samples (excluding two blank samples) and obtained 1977 effective data; the inspection of Grubbs and Dixon was adopted and 65 outliers were discovered, making up 3.3%: they came from 11 laboratories, with 32 from Lab 19, accounting for 49.2% of the total outliers; outliers from the other 10 laboratories are 33, totally make up 50.8%, and moreover outliers from each of these laboratories are less than 8, belonging to accidental deviations.

Thirty-two outliers from Lab 19 came from Nos. 4 and 5 green tea-incurred samples and Nos. 9 and 10 oolong tea-aged samples. Test results from these four samples are 20%–50% higher than those from other laboratories. Review of the experimental raw data record has found that this laboratory established the matrix calibration curves on June 8, and then the fortified sample of green tea (Nos. 1 and 2) and oolong tea (Nos. 6 and 7) were determined, while Nos. 4 and 5 green tea-incurred samples and Nos. 9 and 10 aged samples were tested on June 12, with an interval of 4 days. Precise reasons that caused such deviations failed to be found in the raw data record, and it is assumed that during these 4 days the interval instrument condition had changed possible, which may be the cause of such systematic deviations. For this, the collaborator considered that their instrument was stable because comparison of the quality control sample had been conducted, but no reasonable explanation has been yet offered for the cause of such errors.

All in all, Lab 19 did not carry out a continuous inspection of the samples after establishing calibration curves, with a time interval for 4 days before continuing inspection, which led to the results from their laboratories 20%–50% higher than those from other laboratories. Such a practice is not recommended by the study director, who points out in the collaborative study method: the matrix-matched calibration standards should be prepared and used *only* for the quantitative analysis of the samples prepared at the same time under the same conditions. As far as this point is concerned, Lab 19 has deviated from the collaborative study operational procedures unconsciously, which inevitably led to relatively large deviations in test results with those from other laboratories.

7.2.8.2   GC–MS/MS Data Error Analysis and Traceability

Tables 7.2.31, 7.2.33 and 7.2.34 show: 1704 effective data were obtained from determination of 20 pesticides in 8 samples (excluding 2 blank samples) by 14 laboratories; Grubbs and Dixon inspection was adopted and 65 outliers were found (making up 3.8%). Sixty-five outliers came from 11 laboratories, of which 25 were from Lab 21, accounting for 38.5% of total outliers; 19 from Lab 18, making up 29.2% of total outliers; 7 from Lab 27, making up 10.8% of total outliers. Outliers from these three laboratories total 51, accounting for 78.5% of the total outliers; outliers from the other 8 laboratories total 14, only making up 21.5%, and outliers from each laboratory are less than 3, belonging to accidental errors.

Regarding Lab 21, 23 of 25 outliers from Nos. 6 and 7 oolong tea fortification samples, and the analytical results from these two samples, were 40% lower than those from other laboratories. In addition, there were very big differences in test

TABLE 7.2.30 Distribution of Outlier Data in Different Pesticides by GC–MS (16 Labs)

No.	Pesticides	Green tea							Oolong tea			
		Fortified samples		Incurred samples		Sum	Fortified samples		Aged samples		Sum	Total
		No. 1	No. 2	No. 4	No. 5		No. 6	No. 7	No. 9	No. 10		
1	Trifluralin	28-G, 28-D				1						1
2	Tefluthrin								19-G, 19-D, 25-G, 25-D	25-D	3	3
3	Pyrimethanil			19-G, 19-D	2-D, 6-D, 19-G, 19-D	4			19-G, 19-D	19-G, 19-D	2	6
4	Propyzamide							8-G, 25-G, 25-D	19-G, 19-D	19-G, 19-D	4	4
5	Pirimicarb	2-G, 2-D	2-G, 2-D			2	2-D, 14-D		2-G, 2-D, 19-G, 19-D	2-G, 2-D, 19-G, 19-D	6	8
6	Dimethenamid								19-G, 19-D	19-G, 19-D	2	2
7	Fenchlorphos											
8	Tolclofos-methyl		25-G, 25-D			1	25-G		19-G, 19-D	19-G, 19-D, 25-D	4	5
9	Pirimiphos-methyl								19-G	19-G, 19-D	2	2
10	2,4'-DDE								19-G, 19-D, 25-G, 25-D	19-G, 19-D, 25-G	4	4
11	Bromophos-ethyl											
12	4,4'-DDE	6-D				1						1
13	Procymidone								19-G, 19-D	19-G, 19-D		2
14	Picoxystrobin							11-G, 11-D	19-G, 19-D	19-G, 19-D		3
15	Quinoxyfen	11-D	11-G, 11-D			2			19-G, 19-D	19-G, 19-D		4
16	Chlorfenapyr								19-G, 19-D	19-G, 19-D		2
17	Benalaxyl	11-G, 11-D				1	2-D, 14-D, 22-G, 22-D	22-G, 22-D		19-G, 19-D		6
18	Bifenthrin		3-G		19-G, 19-D	3	12-G, 12-D	12-G, 12-D	19-G, 19-D	19-G, 19-D	4	7
19	Diflufenican		3-G, 3-D			1		22-G, 22-D	19-D	19-D	2	3
20	Bromopropylate		3-G			1			19-D		1	2
	Total number of data	319	32	32	702	318	319	319	319	1275	1977	
	Number of outliers (%)	5 (1.6%)	2 (6.3%)	4 (12.5%)	17 (2.4%)	7 (2.2%)	6 (1.9%)	18 (5.6%)	17 (5.3%)	48 (3.8%)	65 (3.3%)	

G, Grubbs outlier; D, Dixon outlier.

TABLE 7.2.31 Distribution of Outlier Data in Different Pesticides by GC–MS/MS (14 Labs)

| No. | Pesticides | Green tea | | | | | Oolong tea | | | | | Total |
| | | Fortified samples | | Incurred samples | | Sum | Fortified samples | | Aged samples | | Sum | |
		No. 1	No. 2	No. 4	No. 5		No. 6	No. 7	No. 9	No. 10		
1	Trifluralin	27-D, 28-D				2	10-G, 10-D	10-G, 10-D			2	4
2	Tefluthrin						18-G, 18-D	18-D			2	2
3	Pyrimethanil				5-D, 26-D	2						2
4	Propyzamide								26-D		1	1
5	Pirimicarb		10-D			1	18-G, 18-D, 21-G, 21-D	18-G, 18-D, 21-G, 21-D	21-G, 21-D	21-G, 21-D	6	7
6	Dimethenamid						21-D	18-D, 21-D			3	3
7	Fenchlorphos		27-G			1						1
8	Tolclofos-methyl								30-G		1	1
9	Pirimiphos-methyl	27-G, 27-D	27-G, 27-D			2	18-D	18-D, 21-D			3	5
10	2,4'-DDE						18-G, 18-D, 21-G, 21-D	18-G, 18-D, 21-G, 21-D			4	4
11	Bromophos-ethyl		27-G			1	18-D, 21-D	21-G, 21-D			3	4
12	4,4'-DDE		13-D, 30-D			2	18-G, 21-G, 21-D	18-G, 21-G, 21-D			4	6
13	Procymidone											0
14	Picoxystrobin						21-D	18-D, 21-D			3	3
15	Quinoxyfen						21-D	18-D, 21-D			3	3
16	Chlorfenapyr						18-D, 21-D, 24-D	21-G, 21-D			4	4
17	Benalaxyl						18-G, 18-D, 21-G, 21-D	21-D			3	3
18	Bifenthrin						18-D, 21-D, 29-D	18-D, 21-D, 29-D			6	6
19	Diflufenican	27-G, 27-D				1			26-D		1	2
20	Bromopropylate		27-G, 27-D			1	21-D	18-D, 21-D			3	4
	Total number of data	258	28	28	592	278	278	278	278	1112	1704	278
	Number of outliers (%)	7 (2.7%)	0 (0%)	2 (7.1%)	13 (2.2%)	23 (8.3%)	24 (8.6%)	4 (1.4%)	1 (0.4%)	52 (4.7%)	65 (3.8%)	4 (1.4%)

TABLE 7.2.32 Distribution of Outlier Data in Different Pesticides by LC–MS/MS (24 Labs)

No.	Pesticides	Green tea — Fortified samples No. 1	No. 2	Green tea — Incurred samples No. 4	No. 5	Sum	Oolong tea — Fortified samples No. 6	No. 7	Oolong tea — Aged samples No. 9	No. 10	Sum	Total
1	Acetochlor	25-D, 30-D	25-G, 28-D, 30-D		23-G, 23-D	6				24-G	1	7
2	Benalaxyl	30-G, 30-D				1			24-G	21-G, 21-D, 24-G, 24-D	3	4
3	Bensulide	30-G				1					0	1
4	Butralin		18-G, 18-D			1				9-D	1	2
5	Chlorpyrifos	22-G, 22-D, 25-G, 25-D, 28-G, 28-D	17-G, 21-G, 21-D, 28-G, 28-D			6	28-G, 28-D				1	7
6	Clomazone					0	28-G, 28-D			21-G, 24-G	3	3
7	Diazinon		10-G			1	8-G, 8-D, 24-D, 25-G, 25-D	25-G, 25-D		24-G	5	6
8	Ethoprophos	21-G				1	5-G, 5-D	5-G, 5-D		24-D	2	3
9	Flutolanil					0					0	0
10	Imidacloprid					0				25-G	1	1
11	Indoxacarb					0	28-G, 28-D		24-G		2	2
12	Kresoxim-methyl		22-G, 28-G, 28-D, 30-G, 30-D			3	28-G, 28-D			21-G, 24-G, 24-D	4	7
13	Monolinuron		28-G, 30-G			2	28-G	28-G, 28-D			2	4
14	Picoxystrobin					0				21-D, 24-D	2	2
15	Pirimiphos-methyl					0	25-G			24-G	2	2
16	Propoxur					0					0	0
17	Quinoxyfen	10-G, 10-D				1					0	1
18	Tebufenpyrad					0					0	0
19	Triadimefon		17-G, 17-D			1	25-G			24-G, 24-D	2	3
20	Trifloxystrobin	25-G	17-G, 17-D			2					0	2
	Total number of data	479	460	71	71	1081	459	459	479	479	1876	2957
	Number of outliers (%)	10 (2.1%)	15 (3.3%)	0 (0%)	1 (1.4%)	26 (2.4%)	11 (2.4%)	3 (0.7%)	2 (0.4%)	15 (3.1%)	31 (1.7%)	57 (1.9%)

TABLE 7.2.33 The Statistical Results of Outliers for Fortified Samples, Aged Sample and Incurred Samples in Green Tea and Oolong Tea by GC–MS, GC–MS/MS and LC–MS/MS

| | Green tea | | | | | Oolong tea | | | | | |
| Item | Fortified samples | | Incurred samples | | | Fortified samples | | Aged samples | | | |
	No. 1	No. 2	No. 4	No. 5	Sum	No. 6	No. 7	No. 9	No. 10	Sum	Total
GC–MS											
Total number of data	319	319	32	32	702	318	319	319	319	1275	1977
Number of outliers (%)	5 (1.6%)	6 (1.9%)	2 (6.3%)	4 (12.5%)	17 (2.4%)	7 (2.2%)	6 (1.9%)	18 (5.6%)	17 (5.3%)	48 (3.8%)	65 (3.3%)
GC–MS/MS											
Total number of data	278	258	28	28	592	278	278	278	278	1112	1704
Number of outliers (%)	4 (1.4%)	7 (2.7%)	0 (0%)	2 (7.1%)	13 (2.2%)	23 (8.3%)	24 (8.6%)	4 (1.4%)	1 (0.4%)	52 (4.7%)	65 (3.8%)
LC–MS/MS											
Total number of data	479	460	71	71	1081	459	459	479	479	1876	2957
Number of outliers (%)	10 (2.1%)	15 (3.3%)	0 (0%)	1 (1.4%)	26 (2.4%	11 (2.4%)	3 (0.7%)	2 (0.4%)	15 (3.1%)	31 (1.7%)	57 (1.9%)

results for two parallel samples of Nos. 1 and 2 green tea by GC–MS/MS determination, with normal recoveries for 20 pesticides of No. 1 samples but with no detection of any test results for 20 pesticides of No. 2 sample. Likewise, regarding LC–MS/MS, there are 4 samples fortified with 20 pesticides, of which Nos. 1 and 2 are green tea parallel fortification samples and Nos. 6 and 7 are oolong tea parallel samples. Recoveries for 20 pesticides of the No. 1 sample are normal, but none of the 20 pesticides has been detected with the No. 2 sample. The same is true of Nos. 6 and 7 fortification samples, with no detection of any pesticides either, belonging to overall deviations.

Nineteen outliers from Lab 18 all came from Nos. 6 and 7 oolong tea fortification samples, and the test results from these two samples are 37% on average compared with those from other laboratories. As for this issue, the collaborator has failed to find out the cause for it.

Seven outliers from Lab 27 came from the 5 pesticides in Nos. 1 and 2 green tea fortification samples, and it is judged tentatively that these pesticides were interfered with, causing accidental errors.

To summarize the aforementioned analysis, in the GC–MS/MS test results, overall deviations occurred with several specific samples for Lab 21 and Lab 18, resulting in two-thirds of the outliers concentrated in these two laboratories. Errors from other laboratories were accidental deviations distributed in multiple kinds of pesticides.

7.2.8.3 LC–MS/MS Data Error Analysis and Traceability

Tables 7.2.32–7.2.34 show: 2957 effective data were obtained from the determination of 20 pesticide residues in 8 samples (excluding 2 blank samples) by 24 laboratories; Grubbs and Dixon inspection was adopted and 57 outliers were discovered (making up 1.9%). These 57 outliers came from 13 laboratories. There are 12 outliers from Lab 24, accounting for 21.1%; 11 from Lab 28, accounting for 19.3%; 9 from Lab 25, making up 15.8%, 6 from Lab 30, accounting for 10.5%. Outliers from the other 9 laboratories total 19, accounting for 33.3%. Outliers from each of these 9 laboratories are less than 6, belonging to accidental errors.

Eleven of 12 outliers with Lab 24 from Nos. 9 and 10 oolong tea aged samples and the analytical results from these two samples were 40% lower than those from other laboratories. There were 11 outliers with Lab 28. These outliers came from seven pesticides in four fortification samples of Nos. 1, 2, 6, and 7 of two kinds of tea. It is mainly traced to accidental

TABLE 7.2.34 Distribution of Outlier Data in Different Labs

Lab	GC-MS No.1	No.2	No.4	No.5	No.6	No.7	No.9	No.10	GC-MS/MS No.1	No.2	No.4	No.5	No.6	No.7	No.9	No.10	LC-MS/MS No.1	No.2	No.4	No.5	No.6	No.7	No.9	No.10	Number of outlier data	Total number of experimental data	Outlier data over individual lab data's ratio percentage	Outlier data over the total number of outlier data's ratio percentage
1	0	0	0	0	0	0	0	0	0	0	0	0	0	0	0	0	0	0	0	0	0	0	0	0	0	374	0	0
2	1	1	0	1	2	0	1	1	–	–	–	–	–	–	–	–	–	–	–	–	–	–	–	–	7	124	5.6	3.7
3	0	3	0	0	0	0	0	0	–	–	–	–	–	–	–	–	0	0	0	0	0	0	0	0	3	250	1.2	1.6
4	0	0	0	0	0	0	0	0	–	–	–	–	–	–	–	–	0	0	0	0	0	0	0	0	0	250	0	0
5	1	1	1	1	–	1	0	1	0	0	0	1	0	1	0	0	–	0	0	0	1	1	0	0	3	250	1.2	1.6
6	1	0	0	1	0	0	0	1	–	–	–	–	–	–	–	–	0	0	0	0	0	0	0	0	2	250	0.8	1.1
7	1	1	–	–	–	0	1	0	–	–	–	–	–	–	–	–	0	0	0	0	0	0	0	0	0	126	0	0
8	0	0	0	0	0	1	0	1	–	–	–	–	–	–	–	–	0	0	0	0	0	0	0	0	2	250	0.8	1.1
9	0	0	0	0	0	0	0	0	0	0	1	0	1	0	0	0	0	0	0	0	0	0	0	1	1	250	0.4	0.5
10	0	0	0	0	0	1	1	0	–	–	–	–	–	1	0	0	0	0	0	0	0	0	0	1	5	374	1.3	2.7
11	2	1	0	0	0	1	0	0	–	–	–	–	–	–	–	–	0	0	0	0	0	0	0	0	4	250	1.6	2.1
12	0	0	0	0	1	1	0	1	–	–	–	–	0	0	0	0	0	0	0	0	0	0	0	0	2	250	0.8	1.1
13	–	–	–	–	–	–	–	–	–	1	–	–	1	–	0	–	0	0	0	0	0	0	0	1	1	250	0.4	0.5
14	0	0	0	0	2	0	0	0	0	0	0	0	0	0	0	0	0	0	1	0	0	0	0	1	2	124	1.6	1.1
15	0	0	0	0	0	0	0	0	0	0	0	0	0	0	0	0	1	1	1	0	1	1	1	1	0	124	0	0
16	–	–	–	–	–	–	–	–	0	0	0	0	0	0	0	0	0	0	0	0	0	0	0	0	0	250	0	0
17	–	–	–	–	–	–	–	–	0	0	0	0	0	0	0	1	0	3	–	0	0	0	0	0	3	126	2.4	1.6
18	–	–	–	–	–	–	–	–	0	0	0	0	9	10	0	0	0	1	–	0	0	0	0	0	20	250	8	10.7
19	0	0	2	2	–	0	15	13	–	–	–	–	–	–	–	–	0	–	1	1	1	1	1	0	32	124	25.8	17.1
20	–	–	–	–	–	–	–	–	–	–	–	–	–	–	–	–	–	–	–	–	–	–	–	–	–	–	–	–
21	–	–	–	–	–	–	–	–	0	0	0	0	11	12	1	1	1	1	1	0	0	0	0	4	31	171	18.1	16.6
22	–	–	–	–	–	–	–	–	0	0	0	0	0	0	0	0	1	1	0	1	0	0	0	0	5	250	2	2.7
23	0	0	0	0	1	2	0	0	0	0	0	0	1	0	0	0	0	0	0	0	0	0	2	9	1	250	0.4	0.5
24	0	1	0	0	1	1	1	3	0	0	0	0	1	1	2	0	0	0	0	0	1	0	0	1	13	248	5.2	7
25	0	1	0	0	1	1	2	3	1	1	0	0	0	1	0	1	3	1	0	0	3	1	0	1	17	249	6.8	9.1
26	–	–	–	–	–	–	1	1	3	0	0	0	0	0	2	0	0	0	0	0	0	0	0	0	3	118	2.5	1.6
27	1	1	1	0	0	0	0	0	1	0	0	0	0	0	0	0	0	0	0	0	5	1	0	0	7	250	2.8	3.7
28	1	0	0	0	0	0	0	0	0	0	0	0	0	0	0	0	1	4	0	0	5	0	0	0	13	356	3.7	7
29	–	–	–	–	0	0	0	0	0	1	0	0	0	1	0	0	0	3	0	0	0	0	0	0	2	250	0.8	1.1
30	–	–	–	–	–	–	1	0	0	1	0	0	0	0	1	1	3	3	0	1	0	0	0	0	8	250	3.2	4.3
Total	5	6	2	4	7	6	18	17	4	7	0	2	23	24	4	1	10	15	0	1	11	3	2	15	187	6638	2.8	100

errors with a determination of certain pesticides. There were 9 outliers with Lab 25. They came from seven pesticides in five samples of Nos. 1, 2, 6, 7, and 10, belonging to accidental errors with certain pesticides. Six outliers with Lab 30 came from green tea fortification samples of Nos. 1 and 2, and the test results of acetochlor, benalaxyl, bensulide, kresoxim-methyl, and picoxystrobin in these two samples are about 40% greater than those from other laboratories, which also belong to accidental errors with certain pesticides.

To summarize the aforementioned descriptions, there are 57 outliers from the 2957 effective data by LC–MS/MS, accounting for 1.9%. Fifty-seven outliers came from 13 laboratories, proving that they are relatively dispersed. No big systematic deviations happened with test results.

7.2.9 COLLABORATORS' COMMENTS ON METHOD

In the sheet of collaborative results, a column of collaborators' comments on the method was created, under which 12 collaborators filled out what they found in the process of the study. In the next sections, replies are made to the main questions item by item.

7.2.9.1 Collaborators' Comments on GC–MS

Lab 19 comments: N.D. = "Below Limit of Quantification" as described in the supplied collaborative study method.

SD reply: In the table of collaborative sample results filled out by the laboratory, pesticides marked with "BLQ" as test results refer to the samples free from these pesticides. Such marking is correct, and these samples are indeed free from pesticides.

Lab 25 comments: For oolong Tea, one point from the calibration curve was taken out for propyzamide, pirimiphos methyl, pirimicarb, bromophos-ethyl and bromopropylate to make the R^2 value 0.995 or greater.

SD reply: If, at the time of establishing the 5-point concentration matrix-matching calibration curve, numerical values at 1 concentration point deviate obviously, causing R^2 value of calibration curve to fail 0.995, this concentration point can be eliminated and the 4-point matrix-matching calibration curve is adopted for quantification. In the matrix calibration curve established by the laboratory, propyzamide eliminated the second concentration point, primicarb the fourth point, prirmiphos-methyl the fourth point, bromophos-ethyl the fifth point, and bromopropylate the sixth point. Details are as follows:

SD considers that such handling is scientific and reasonable.

Lab 28 comments: No results for tolclofos-methyl due to falsely determined MRM-acquisition window.

SD reply: At the time of GC–MS determination by this laboratory, deviations occurred in the setting of MRM-acquisition window of tolclofos-methyl, so only part of the peaks were integrated, leading to an absence of any test results, such as the later diagram. SD considers that when this problem was discovered by Lab 28, MRM-acquisition window should be reset according to the retention time of tolclofos-methyl in order to be able to collect a complete set of peaks and to obtain accurate results. This problem was easy to solve, but Lab 28 did not use any remedial measures, which led to the absence of any test results of tolclofor-methyl. This loss should not have been committed, and such problems will not occur if measures are adopted as those that are described earlier.

7.2.9.2 Collaborators' Comments on GC–MS/MS

Lab 18 comments: Bifenthrin has been found in *all samples*. In samples 3 and 8 the area of bifenthrin is *very low* and *similar to the blank* employed for the calibration curve. For this reason the quantification has been made by a standard addition method.

SD reply: Bifenthrin is a commonly used pesticide in tea plantations in China. If using an instrument of relatively high sensitivity for determination, it is understandable for Lab 18 to say: Bifenthrin has been found in *all samples*. Lab 18 pointed out: "In samples 3 and 8 the area of bifenthrin is VERY LOW and SIMILAR TO THE BLANK employed for the calibration curve." Because Nos. 3 and 8 are blank samples and appear as blind samples and are unknown to any of collaborators, Lab 18's finding is very correct. Moreover, Lab 18 adopted a standard fortification method for quantification, and this thinking is very meticulous and scientific, which is worth learning from.

Lab 26 comments: We can not develop the instrumental condition for chlorfenapyr because *m/z* 408 in chlorfenapyr has low sensitivities.

SD reply: *m/z* 408 in chlorfenapyr is the ion recommended in the collaborative method. No similar problems have been found when determining this ion by other laboratories. Our review of Appendix B provided by this laboratory has found that the precursor ion monitored by Lab 26 was 480, so it is judged that the cause of this problem might be that this laboratory entered incorrectly 480 for 408 when the method was established.

Lab 28 comments: No results for 2,4'-DDE due to falsely determined MRM-acquisition window.

SD reply: When using GC–MS determination of tolclofos-methyl by the aforementioned Lab 28, only part of the peaks were integrated for the target pesticide owing to errors with setting of the MRM-acquisition window, leading to an absence of test results. At the time of GC–MS/MS determination by the laboratory, they made the mistake again when setting 2,4′-DDE MRM. They established the collection method according to the retention time of 4,4′-DDE, resulting in no detection of 2,4′-DDE, and we regret that they made this mistake.

7.2.9.3 Collaborators' Comments on LC–MS/MS

Lab 17 comments: For the chromatography, our experience is that the maximum injection volume on an acquity column is 5 μL. The 10 μL injection to follow in the method could explain some bad chromatographic peak shape. It was difficult to obtain $R^2 > 0,995$. For several pesticides one calibration has to be removed (clearly outlier) but even with 4 points of calibration, the 0.995 criteria was not always satisfied.

SD considers that their experience is maximum injection being 5 μL for chromatograph, while 10 μL injection was adopted in the collaborative study method, causing certain pesticide chromatograph peak shapes to worsen. Lab 17 also considers that they eliminated the point of obvious deviation for calibration curves of some compounds in the specific determinations so as to ensure R^2 to reach 0.995. However, although the 4-point calibration curve was adopted, the criteria of $R^2 > 0.995$ could not be met.

SD considers that injection quantity mainly depends on the sensitivity of the instrument, and the instrument used by Lab 18 has very high sensitivity, and the sample was diluted 5 times before injection for determination. It is also acceptable that 5 μL injection adopted by Lab 17 can satisfy the technical indexes of the method. Regarding the criteria of $R^2 \geq 0.995$ being hard to meet, there are 2113 with R^2 values submitted by 30 collaborators, with those in conformity with the criteria of $R^2 > 0.995$ reach 2059, accounting for 97%. As for this requirement, it should be strengthened all the more in future collaborative studies, and in so doing the accuracy of the quantification of target pesticides will be greatly increased while measuring errors will be reduced, so to speak, minimizing outliers.

Lab 18 comments: Before injection, the samples were diluted 5 times with acetonitrile: water (3:2, v/v) because the system is very sensitive and we have some pesticides that were saturated.

SD reply: Agilent 6490 adopted by this laboratory is the latest model of Agilent Tri-Quadruple MS. This instrument adopts i-funnel technique, about 10 times higher than the sensitivity of 6460. Therefore, it is understandable that dilution is needed before injection. At the same time, it was also found that Lab 22 used an old instrument from the 1990s, and its sensitivity is relatively low. To ensure that the calibration curve reaches above $R^2 > 0.995$, it adopted the method of increasing the injection to solve the problems encountered in the experiment. Lab 18 used the dilution approach to solve the problems encountered, while Lab 22 used the method of increasing injection to tackle the problems that arose. In a nutshell, Lab 18 and Lab 22 investigated the ruggedness of the method in two aspects. Based on the conformance with big principle, they used some skills to tackle the practical problems encountered in the experiment, which is advisable, very good, and praiseworthy.

Lab 23 comments: Retention times slightly change from the green tea to the oolong tea matrix, a column has been added to show the exact retention time for both the matrix.

SD reply: When using LC–MS/MS for respective determination of green tea and oolong tea samples, the laboratory discovered that there existed a system difference of a maximum of 3/100 s for retention time of pesticides in the two matrices. Lab 23's observation for the experiment was so sharp that we should learn from their meticulous scientific research attitude in the first place. Because green tea and oolong tea are both teas all right but their processing technologies are different; they belong to two teas of totally different matrices. Owing to the different matrix, the system deviation of a maximum of 3/100 s for pesticide retention time is normal, which will not affect the test results. Here we mainly focus on learning from Lab 23's pursuit of meticulous scientific experiment.

Lab 24 comments: In the case of the contents of 10 μg/kg peak areas below the areas of the lowest calibration point; oolong tea: after dissolution of the residue's orange and oily drops, the solution is light yellow and clear; after the membrane filtration solution clears.

SD reply: Our review of the collaborative result table provided by this lab has found that the detection concentrations for the two compounds of triadimefon and trifloxystrobin in green tea-incurred samples from this lab turned out to be 40%–50% lower than those from other laboratories, from which we judge that the determination results of acetochlor are lower than 10 μg/kg, so no specific detection results are reported. For this problem, we also found that similar things also happened with Lab 21. Based on the aforementioned descriptions, Lab 24 also discovered that there are orange and oily drops with oolong tea pesticide extraction solutions, which became clear after a filtration membrane. We also encountered such a phenomenon, which is normal and will not affect test results.

Lab 25 comments: Oolong had to increase the amount of extraction solution to get a better extraction. For oolong tea, the high standard (5) had to be removed due to it having a good calibration fit (picoxystrobin, indoxacarb, chlorpyrifos, and butralin). For oolong tea, they used linear $1/x$ to give more weight to a lower standard.

SD reply: Lab 25 proposed increasing the amount of extraction solution to get a better extraction for pesticides in oolong tea, and SD has not discovered similar problems from other laboratories. SD considers that pesticide-extracting efficiencies have a lot to do with the choice of extraction solutions and amounts as well as extraction utensils were used. As for the homogenizer recommended in this collaborative study method, we compared the efficiencies of 3-time extraction and 2-time extraction. Our findings are only that 2-time extraction will suffice for the requirements for recovery indexes. If Lab 25 found in their own laboratory that a 2-time extraction cannot suffice recoveries, an additional extraction may also be conducted. For calibration curves of oolong tea, the high concentration points with deviations for picoxystrobin, in dopoxacarb, chlorpyrifos, and butralin are eliminated from their own calibration curves so as to obtain better calibration curves, which is feasible. It is a good way "For Oolong tea, using liner 1/X to gave more weight to lower stand."

Lab 28 comments: No results for butralin due to constant signals in the given MRM-transitions.

SD reply: Regarding no detection of butralin, our fault is the major cause for the problem. Because instruments from Waters are unavailable in our laboratory, the precursor ion 337 of butralin given by a local employee from Waters was not verified, so using 337 is a mistake, which was not found timely. In this collaborative study, Lab 9, Lab 29, and Lab 25 also used instruments from Waters, they found 337 is a mistake through scanning, and the correct one should be 296. Therefore, an accurate determination was conducted on butralin. In terms of learning this lesson, the precise ion information can be obtained and such mistakes can be discovered by scanning the target pesticides when encountering similar problems in future experiments. If it was handled like this, this problem could have been solved earlier. Having said so, SD is not shirking responsibility but discussing the problem encountered in order to find a scientific thinking and solve the problem. In a word, as SD said before, it is our fault and we caused the trouble to the lab involved.

7.2.10 CONCLUSIONS

This collaborative study was one of the priority research items of AOAC in 2010 and a gigantic systematic project with three features as listed in the next paragraphs.

First, a review of 6868 test data and related information from 30 laboratories found that one laboratory deviated from the collaborative study operational procedures, entering directly into an official collaborative study with no undertaking of the prestudy, with a great deal of outliers from its sample inspection; this laboratory was eliminated. A total of 6638 data from the remaining 29 laboratories were effective results and inspected with Grubbs and Dixon to eliminate 187 outliers, making up 2.8% of the total effective data. There were 1977 test data from GC–MS determination, with 65 outliers detected, making up 3.3%; 1704 test data from GC–MS/MS determination, with 65 outliers detected, accounting for 3.8%; 2957 test data from LC–MS/MS determination, with 57 outliers detected, making up 1.9%. Statistical analysis with AOAC software also proved that method efficiencies for GC–MS, GC–MS/MS, and LC–MS/MS were acceptable (1) for green tea 40 pesticides fortification samples by the three methods: Avg. C.: 18.2–759.6 µg/kg, Avg. Rec 87%–97.7%, RSD_r: 2.1%–9.4%, RSD_R: 6.5%–17.1%, HorRat: 0.3–0.7; (2) for oolong tea 40 pesticides fortification samples by the three methods: Avg. C.: 8.5–335.8 µg/kg, Avg. Rec 77.1%–93.7%, RSD_r: 1.4%–10.2%, RSD_R: 7%–32.7%, HorRat: 0.4–1.3; (3) for oolong tea 20 pesticides aged samples by the three methods: Avg. C.: 34.2–1642.6 µg/kg, RSD_r: 2%–9.6%, RSD_R: 8.9%–34.7%, HorRat: 0.4–1.8; (4) for green tea 5 pesticides incurred samples by the three methods: Avg. C.: 14.1–613.3 µg/kg, RSD_r: 3.3%–10.6%, RSD_R: 12.2%–23.6%, HorRat: 0.7–1. To sum up the aforementioned analysis, Avg. C., Avg. Rec, RSD_R, RSD_R, and HorRat values all met AOAC technical requirements except for specific data. SD recommends this method as an official first action.

Second, one important experience achieved in this collaborative study is that a prestudy stage must be added in designing the collaborative study protocol for a complex topic. Tea matrices are relatively complicated, with a certain difficulty for using three techniques for determination of hundreds of pesticide residues. SD very much appreciated the suggestion of Dr. Jo Marie Cook regarding adding a prestudy stage in the collaborative study protocol first proposed in 2010. The protocol stipulated that the four indexes of recoveries, RSD, R^2, and ion abundance with target pesticides in the prestudy would have to meet the acceptance criteria, otherwise collaborators would, in principle, forfeit their right to continue determination of official collaborative study samples. An expert once wrote to SD during the collaborative study period: "If we understood well the final amount is five grams. It means only one testing opportunity, just one shot. We consider important to get some more material (e.g., 10 g) for duplicate analysis. Is it possible?" SD replied: "What you comprehend is absolutely correct. For the official collaborative study, there is only one sample for one preparation, which is what you called 'just one shot.' However, in the AOAC collaborative study method, we have provided you with sufficient practice samples and request the collaborators to practice strictly in the first place. Only after they meet the acceptance criteria can they start the official collaborative study. If your practice results have met with the criteria, I believe that you will be able to achieve very good collaborative study results with only one shot." Now we have finally achieved the triumphant results of "one sample good only for one shot." The reason for it is that the prestudy played a vital role, which is an important ingredient for success.

Third, several laboratories used Waters' LC–MS/MS in this collaborative study and the transitions of the target pesticide in Appendix F of the collaborative study method were recommended by other colleagues because in SD's laboratory such an instrument was unavailable. We did not verify the transitions, with the result that mistakes and losses were incurred by the collaborators, owing to our supplying the incorrect ion information for several laboratories. This reflects SD's negligence, which provides an important lesson, and here SD also extends a deep apology to the collaborators involved.

All in all, this is a very complicated and important systematic project, which has been unforgettable by any of the experts involved with the collaborative study, especially those who directly participated in the study. SD considers that the AOAC intercollaborative study is a great undertaking and hopes that more and more analytical chemists will be involved in it, making greater contributions to the development of an AOAC official method.

REFERENCES

[1] Chen, D.L., Su, Z., 2009. Marketing strategy of Chinese tea market under the background of upgraded consumption. Tea Science and Technology Innovation and Industrial Development Symposiumpp. 639–647.

[2] Liu, Y., 2008. Analysis on International Competitiveness of Chinese Tea Industry, Master's thesis. Ocean University of China.

[3] Wu, X.Y., Studies on Maximum Residue Limits for Pesticides in Tea and Relative Risk Assessment, 2007, Doctoral Dissertation of Anhui Agricultural University of China.

[4] Standardization Administration of PR China, 2006. GB/T 19648-2006, Method for Determination of 500 Pesticides and Related Chemicals Residues in Fruits and Vegetables by Gas Chromatography-Mass Spectrometry Method. Standards Press, Beijing, China.

[5] Standardization Administration of PR China, 2008. GB/T 20769-2008, Determination of 450 Pesticides and Related Chemicals Residues in Fruits and Vegetables: LC–MS-MS Method. Standards Press, Beijing, China.

[6] Standardization Administration of PR China, 2006. GB/T 19649-2006, Method for Determination of 475 Pesticides and Related Chemicals Residues in Grains by Gas Chromatography-Mass Spectrometry Method. Standards Press, Beijing, China.

[7] Standardization Administration of PR China, 2008. GB/T 20770-2008, Determination of 486 Pesticides and Related Chemicals Residues in Grains: LC–MS-MS Method. Standards Press, Beijing, China.

[8] Standardization Administration of PR China, 2008. GB/T 23204-2008, Determination of 519 Pesticides and Related Chemicals Residues in Teas—GC–MS Method. Standards Press, Beijing, China.

[9] Standardization Administration of PR China, 2008. GB/T 23205-2008, Determination of 448 Pesticides and Related Chemicals Residues in Tea: LC–MS-MS Method. Standards Press, Beijing, China.

[10] Standardization Administration of PR China, 2008. GB/T 23200-2008, Determination of 488 Pesticides and Related Chemicals Residues in Mulberry Twig, Honeysuckle, Barbary Wolfberry Fruit and Lotus Leaf: GC–MS Method. Standards Press, Beijing, China.

[11] Standardization Administration of PR China, 2008. GB/T 23201-2008, Determination of 413 Pesticides and Related Chemicals Residues in Mulberry Twig, Honeysuckle, Barbary Wolfberry Fruit and Lotus Leaf: LC–MS-MS Method. Standards Press, Beijing, China.

[12] Standardization Administration of PR China, 2008. GB/T 23216-2008, Determination of 503 Pesticides and Related Chemicals Residues in Mushrooms: GC–MS Method. Standards Press, Beijing, China.

[13] Standardization Administration of PR China, 2008. GB/T 23202-2008, Determination of 440 Pesticides and Related Chemicals Residues in Mushrooms—LC–MS-MS Method. Standards Press, Beijing, China.

[14] Standardization Administration of PR China, 2008. GB/T 19650-2006, Method for Determination of 478 Pesticides and Related Chemicals Residues in Animal Muscles by Gas Chromatography-Mass Spectrometry Method. Standards Press, Beijing, China.

[15] Standardization Administration of PR China, 2008. GB/T 20772-2008, Determination of 461 Pesticides and Related Chemicals Residues in Animal Muscles: LC–MS-MS Method. Standards Press, Beijing, China.

[16] Standardization Administration of PR China, 2008. GB/T 23207-2008, Determination of 485 Pesticides and Related Chemicals Residues in Fugu, Eel and Prawn: GC–MS Method. Standards Press, Beijing, China.

[17] Standardization Administration of PR China, 2008. GB/T 23208-2008, Determination of 450 Pesticides and Related Chemicals Residues in Fugu, Eel and Prawn: LC–MS-MS Method. Standards Press, Beijing, China.

[18] Standardization Administration of PR China, 2008. GB/T 23210-2008, Determination of 511 Pesticides and Related Chemicals Residues in Milk and Milk Powder: GC–MS Method. Standards Press, Beijing, China.

[19] Standardization Administration of PR China, 2008. GB/T 23211-2008, Determination of 493 Pesticides and Related Chemicals Residues in Milk and Milk Powder: LC–MS-MS Method. Standards Press, Beijing, China.

[20] Standardization Administration of PR China, 2008. GB/T 19426-2006, Method for the Determination of 497 Pesticides and Related Chemicals Residues in Honey, Fruit Juice and Wine by Gas Chromatography-Mass Spectrometry Method. Standards Press, Beijing, China.

[21] Standardization Administration of PR China, 2006. GB/T 20771-2008, Determination of 486 Pesticides and Related Chemicals Residues in Honey: LC–MS-MS Method. Standards Press, Beijing, China.

[22] Standardization Administration of PR China, 2008. GB/T 23206-2008, Determination of 512 Pesticides and Related Chemicals Residues in Fruit Juice, Vegetable Juice and Fruit Wine: LC–MS-MS Method. Standards Press, Beijing, China.

[23] Standardization Administration of PR China, 2008. GB/T 23214-2008, Determination of 450 Pesticides and Related Chemicals Residues in Drinking Water: LC–MS-MS Method. Standards Press, Beijing, China.

[24] Pang, G.F., Cao, Y.Z., Fan, C.L., Jia, G.Q., Zhang, J.J., Li, X.M., Liu, Y.M., Shi, Y.Q., Li, Z.Y., Zheng, F., Lian, Y.J., 2009. Analysis method study on 839 pesticide and chemical contaminant multiresidues in animal muscles by gel permeation chromatography cleanup GC/MS, and LC/MS/MS. J. AOAC Int. 92 (3), 1–72.

[25] Pang, G.F., Cao, Y.Z., Zhang, J.J., Fan, C.L., Liu, Y.M., Li, X.M., Jia, G.Q., Li, Z.Y., Shi, Y.Q., Wu, Y.P., Guo, T.T., 2006. Validation study on 660 pesticide residues in animal tissues by gel permeation chromatography cleanup/gas chromatography–mass spectrometry and liquid chromatography–tandem mass spectrometry. J. Chromatogr. A 1125, 1–30.

[26] Pang, G.F., Fan, C.L., Liu, Y.M., Cao, Y.Z., Zhang, J.J., Fu, B.L., Li, X.M., Li, Z.Y., Wu, Y.P., 2006. Multiresidue method for determination of 450 pesticide residues in honey, fruit juice and wine by double-cartridge solid-phase extraction/gas chromatography-mass spectrometry and liquid chromatography-tandem mass spectrometry. Food Addit. Contam. 23 (8), 777–810.

[27] Pang, G.F., Fan, C.L., Liu, Y.M., Cao, Y.Z., Zhang, J.J., Li, X.M., Li, Z.Y., Wu, Y.P., Guo, T.T., 2006. Simultaneous determination of 446 pesticide residues in fruits and vegetables by solid phase extraction-gas chromatography mass spectrometry/liquid chromatography-tandem mass spectrometry. J. AOAC Int. 89 (3), 740–771.

[28] Pang, G.F., Liu, Y.M., Fan, C.L., Zhang, J.J., Cao, Y.Z., Li, X.M., Li, Z.Y., Wu, Y.P., Guo, T.T., 2006. Simultaneous determination of 405 pesticide residues in grains by accelerated solvent extractor-gas chromatography mass spectrometry/liquid chromatography-tandem mass spectrometry. Anal. Bioanal. Chem. 384, 1366–1408.

[29] Lian, Y.J., Pang, G.F., Shu, H.R., Fan, C.L., Liu, Y.M., Feng, J., Wu, Y.P., Chang, Q.Y., 2010. Simultaneous determination of 346 multiresidue pesticides in grapes by PSA-MSPD and GC–MS-SIM. J. Agric. Food Chem. 58 (17), 9428–9453.

[30] Document No. SANCO/10684/2009, Method validation and quality control procedures for pesticide residues analysis in food and feed. Available from: http://www.eurl-pesticides.eu/docs/public/tmplt_article.asp?CntID=727& LabID=100&Lang=EN.

[31] Fan, C.L., Chang, Q.Y., Pang, G.F., Li, Z.Y., Kang, J., Pan, G.Q., Zheng, S.Z., Wang, W.W., Yao, C.C., Ji, X.X., 2013. High-throughput analytical techniques for determination of residues of 653 multiclass pesticides and chemical pollutants in tea part II: comparative study of extraction efficiencies of three sample preparation techniques. J. AOAC Int. 96 (2), 432–440.

[32] Pang, G.F., Fan, C.L., Chang, Q.Y., Li, Y., Kang, J., Wang, W.W., Cao, J., Zhao, Y.B., Li, N., Li, Z.Y., 2013. High-throughput analytical techniques for determination of residues of 653 multiclass pesticides and chemical pollutants in tea part III: the evaluation of the cleanup efficiency of SPE cartridge newly developed for multiresidues in tea. J. AOAC Int. 96 (4), 887–896.

[33] Li, Y., Chen, X., Fan, C., Pang, G., 2012. Compensation for matrix effects in the gas chromatography–mass spectrometry analysis of 186 pesticides in tea matrices using analyte protectants. J. Chromatogr. A 1266, 131–142.

[34] Pang, G.F., Fan, C.L., Zhang, F., Li, Y., Chang, Q.Y., Cao, Y.Z., Liu, Y.M., Li, Z.Y., Wang, Q.J., Hu, X.Y., Liang, P., 2011. High-throughput GC/MS and HPLC/MS/MS techniques for the multiclass, multiresidue determination of 653 pesticides and chemical pollutants in tea. J. AOAC Int. 94 (4), 1253–1296.

Index